Anatomy & Physiology

AN INTEGRATIVE APPROACH

3e

Michael P. McKinley
Glendale Community College

Valerie Dean O'Loughlin
Indiana University

Theresa Stouter Bidle
Hagerstown Community College

Digital Author
Justin York
Glendale Community College

Mc Graw Hill Education

ANATOMY & PHYSIOLOGY: AN INTEGRATIVE APPROACH, THIRD EDITION

Published by McGraw-Hill Education, 2 Penn Plaza, New York, NY 10121.

This book is printed on acid-free paper.

2 3 4 5 6 7 8 9 LWI 21 20 19 18

ISBN 978-1-259-39862-9
MHID 1-259-39862-5

Managing Director: *Thomas Timp*
Portfolio Manager: *Amy Reed*
Product Developer: *Donna Nemmers*
Director of Digital Content: *Michael G. Koot, Ph.D*
Marketing Manager: *James Connely*
Content Project Managers: *Jessica Portz, Brent dela Cruz, and Sandra Schnee*
Buyer: *Sandy Ludovissy*
Design: *David W. Hash*
Content Licensing Specialist: *Lori Hancock*
Cover Image: *©shironosov/iStock/Getty Images Plus*
Compositor: *MPS Limited*

Library of Congress Cataloging-in-Publication Data

McKinley, Michael P., author. | O'Loughlin, Valerie Dean, author. |
Bidle, Theresa Stouter, author.
Anatomy & physiology : an integrative approach / Michael P. McKinley,
Glendale Community College, Valerie Dean O'Loughlin, Indiana University,
Theresa Stouter Bidle, Hagerstown Community College.
Anatomy and physiology
3e. | New York, NY : MHE, [2019] | Includes index.
LCCN 2017040498 | ISBN 9781259398629 (alk. paper)
LCSH: Human anatomy. | Human physiology.
LCC QM25 .M32 2019 | DDC 612–dc23 LC record available
at https://lccn.loc.gov/2017040498

mheducation.com/highered

about the authors

MICHAEL MCKINLEY received his undergraduate degree from the University of California at Berkeley, and both his M.S. and Ph.D. degrees from Arizona State University. He did his postdoctoral fellowship at the University of California Medical School–San Francisco (UCSF) in the laboratory of Dr. Stanley Prusiner, where he worked for 12 years investigating prions and prion diseases. During this time, he was also a member of the UCSF Medical School anatomy faculty and taught medical histology for 10 years. In 1991, Michael became a member of the biology faculty at Glendale Community College (GCC) in Glendale, Arizona. He taught undergraduate anatomy and physiology, general biology, and genetics at the GCC Main Campus. In 2009, he moved to the GCC North Campus, where he taught anatomy and physiology courses exclusively until he retired in 2012. Between 1991 and 2000, Michael also participated in Alzheimer disease research and served as director of the Brain Donation Program at the Sun Health Research Institute. During this time he also taught developmental biology and genetics at Arizona State University West Campus. He has been an author and co-author of more than 80 scientific papers. Mike's vast experience in histology, neuroanatomy, and cell biology greatly shaped the related content in the market-leading textbook McKinley/O'Loughlin/Pennefather-O'Brien, *Human Anatomy,* 5th edition. Mike is an active member of the Human Anatomy and Physiology Society (HAPS). He resides in Tempe, Arizona with his wife Jan.

VALERIE DEAN O'LOUGHLIN received her undergraduate degree from the College of William and Mary, and her M.A. and Ph.D. degrees in biological anthropology from Indiana University. She is a professor of anatomy at Indiana University, where she teaches human gross anatomy to medical students, basic human anatomy to undergraduates, and human anatomy for medical imaging evaluation to undergraduate and graduate students. She also teaches a pedagogical methods course and mentors Ph.D. students pursuing anatomy education research. She is active in the American Association of Anatomists (AAA) and the Society for Ultrasound in Medical Education (SUSME). She is a President Emeritus of the Human Anatomy and Physiology Society (HAPS) and currently serves on the Steering Committee. She received the AAA Basmajian Award for excellence in teaching gross anatomy and outstanding accomplishments in scholarship in education, and recently was selected for the AAA Henry Gray Distinguished Educator award. Valerie is co-author of the market-leading textbook McKinley/O'Loughlin/Pennefather-O'Brien, *Human Anatomy,* 5th edition.

THERESA STOUTER BIDLE received her undergraduate degree from Rutgers University, her M.S. degree in biomedical science from Hood College in Maryland, and has completed additional graduate coursework in genetics at the National Institutes of Health and in science education at the University of Maryland. She is a professor at Hagerstown Community College, where she teaches anatomy and physiology and nutrition to pre–allied health students. She also mentors new full-time and adjunct faculty who teach anatomy and physiology. Before joining the faculty in 1990, she was the coordinator of the Science Learning Center, where she developed study materials and a tutoring program for students enrolled in science classes. Terri has been a developmental reviewer, has written supplemental materials for both textbooks and lab manuals, and is co-author for Eckel/Ross/Bidle, *Anatomy and Physiology Laboratory Manual,* 2nd edition.

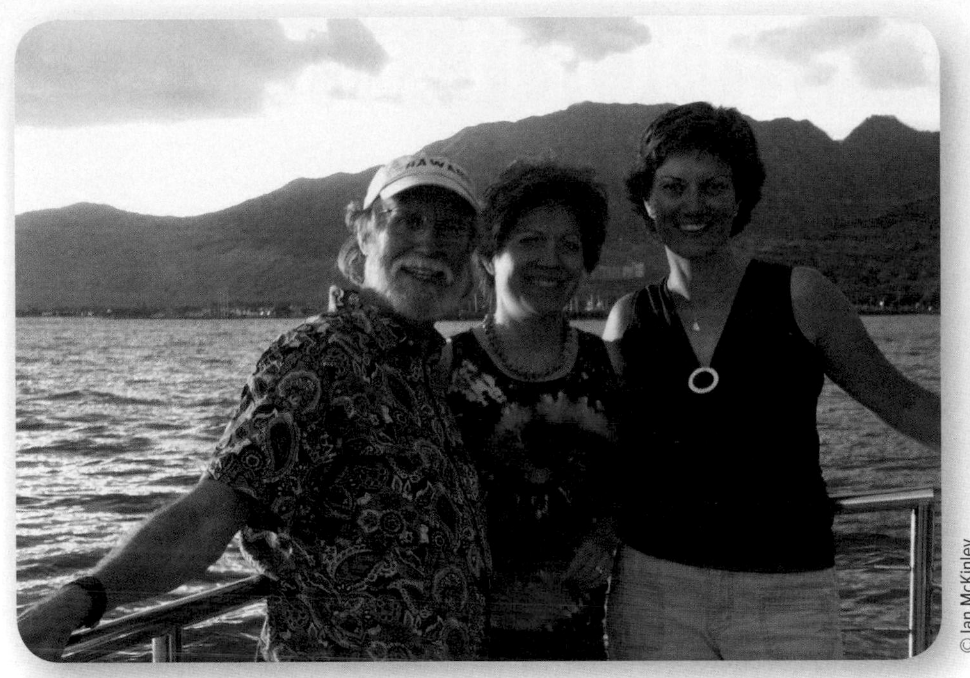

©Jan McKinley

Author team: Michael McKinley, Valerie Dean O'Loughlin, and Theresa Bidle

Dedications

I am indebted to Jan (my wife); Renee, Ryan, and Shaun (my children); and Connor, Eric, Patrick, Keighan, Aydan, and Abbygail (my grandchildren). They are the love of my life and my inspiration always.

—Michael P. McKinley

To my husband Bob and my daughter Erin: Thank you for always being there for me.

—Valerie Dean O'Loughlin

With love and thanks to my husband Jay and my daughter Stephanie for the many ways that they have supported me during this project.

—Terri Stouter Bidle

brief contents

PART I: ORGANIZATION OF THE HUMAN BODY

PART II: SUPPORT AND BODY MOVEMENT

PART III: COMMUNICATION AND CONTROL

PART IV: MAINTENANCE AND REGULATION

PART V: REPRODUCTION

contents

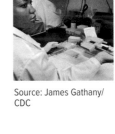

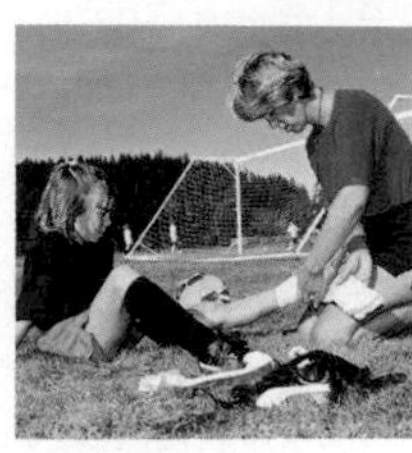
©Russell Illig/Getty Images RF

©Tony McConnell/Science Source

©Fotosearch/agefotostock

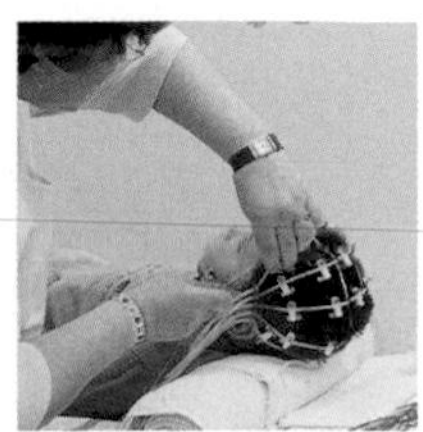
©BSIP/Science Source

©Corbis RF

©Lea Paterson/Science Source

MAINTENANCE AND REGULATION

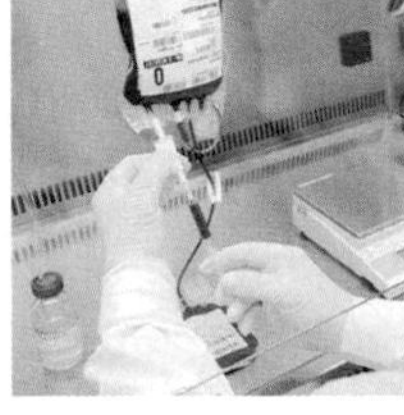

©Arno Massee/Science Source

©Robin Nelson/PhotoEdit

©Mark Harmel/Getty Images

©Marty Heitner/The Image Works

©Phanie/Science Source

CHAPTER 23
Respiratory System 900

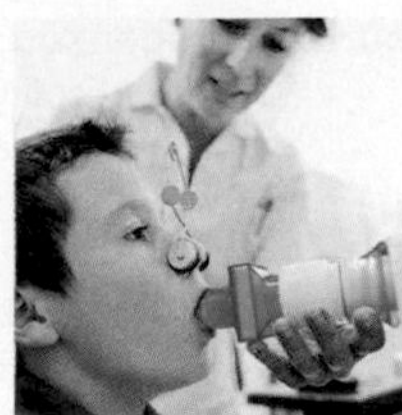

©Science Photo Library/ Alamy RF

CHAPTER 24
Urinary System 954

©Javier Larrea/agefotostock

CHAPTER 25
Fluid and Electrolytes 1000

©CC Studio/Science Source

CHAPTER 26
Digestive System 1034

©By Ian Miles-Flashpoint Pictures/Alamy

©Burger/Phanie/Science Source

REPRODUCTION

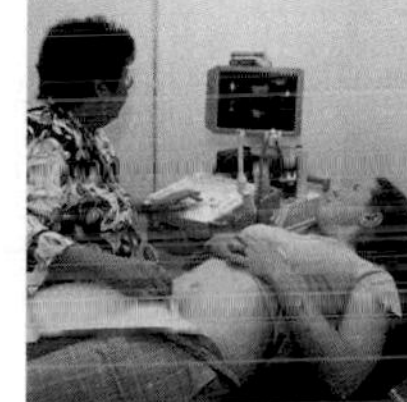

©Jason Edwards/National Geographic Creative

©BSIP/Science Source

CLINICAL VIEW BOXES

preface

Human anatomy and physiology is a fascinating subject. However, students can be overwhelmed by the complexity, the interrelatedness of concepts from different chapters, and the massive amount of material in the course. Our goal was to create a textbook to guide students on a clearly written and expertly illustrated beginner's path through the human body.

An Integrative Approach

One of the most daunting challenges that students face in mastering concepts in an anatomy and physiology course is integrating related content from numerous chapters. Understanding a topic like blood pressure, for example, requires knowledge from the chapters on the heart, blood vessels, kidneys, and how these structures are regulated by the nervous and endocrine systems. The usefulness of a human anatomy and physiology text is dependent in part on how successfully it helps students integrate these related concepts. Without this, students are only acquiring what seems like unrelated facts without seeing how they fit into the whole.

To adequately explain such complex concepts to beginning students in our own classrooms, we as teachers present multiple topics over the course of many class periods, all the while balancing these detailed explanations with refreshers of content previously covered and intermittent glimpses of the big picture. Doing so ensures that students learn not only the individual pieces, but also how the pieces ultimately fit together. This book represents our best effort to replicate this teaching process. In fact, it is the effective integration of concepts throughout the text that makes this book truly unique from other undergraduate anatomy and physiology texts.

Our goal of emphasizing the interrelatedness of body systems and the connections between form and function necessitates a well-thought-out pedagogical platform to deliver the content. First and foremost, we have written a very user-friendly text with concise, accurate descriptions that are thorough, but don't overwhelm readers with nonessential details. The text narrative is deeply integrated with corresponding illustrations drawn specifically to match the textual explanations. In addition, we have included a set of "Integrate" features that support our theme and work together to give the student a well-rounded introduction to anatomy and physiology. **Integrate: Concept Overview** figures are one- or two-page visual summaries that aggregate related concepts in a big-picture view. These comprehensive figures link multiple sections of a chapter together in a cohesive snapshot ideal for study and review. **Integrate: Concept Connections** boxes provide glimpses of how concepts at hand will play out in upcoming chapters, and also pull vital information from earlier chapters back into the discussion at crucial points when relevant to a new topic. **Integrate: Clinical View** discussions apply concepts from the surrounding narrative to practical or clinical contexts, providing examples of what can go wrong in the human body to help crystallize understanding of the "norm." **Integrate: Learning Strategy** boxes infuse each chapter with practical study tips to understand and remember information. Learning strategies include mnemonics, analogies, and kinesthetic activities that students can perform to relate the anatomy and physiology to their own bodies. Finally, the digital assets that accompany our book are tied to each section's learning objectives and previewed in the **Integrate: Online Study Tools** boxes at the end of each chapter.

Chapter Organization

In order to successfully execute an integrative approach, foundational topics must be presented at the point when it matters most for understanding. This provides students with a baseline of knowledge about a given concept before it comes time to apply that information in a more complex situation. Topics are thus subdivided and covered in this sequence:

- **Chapter 2: Atoms, Ions, and Molecules** Most students taking an A&P course have limited or no chemistry background, which requires a textbook to provide a detailed, organized treatment of atomic and molecular structure, bonding, water, and biological macromolecules as a basis to understanding physiological processes.
- **Chapter 3: Energy, Chemical Reactions, and Cellular Respiration** ATP is essential to all life processes. A solid understanding of ATP furthers student comprehension of movement of materials across a membrane, muscle contractions, production of needed replacement molecules and structures in cells, action potentials in nerves, pumping of the heart, and removal of waste materials in the kidneys. This textbook elevates the importance of the key concept of ATP by teaching it early. We then utilize this knowledge in later chapters as needed, expanding on what has already been introduced rather than re-teaching it entirely.
- **Chapter 13: Nervous System: Brain and Cranial Nerves and Chapter 14: Nervous System: Spinal Cord and Spinal Nerves** Instead of subdividing the nervous system discussion into separate central nervous system (CNS) and peripheral nervous system (PNS) chapters, nervous system structures are grouped by region. Thus, students can integrate the cranial nerves with their respective nuclei in the brain, and they can integrate the spinal cord regions with the specific spinal nerves that originate from these regions.
- **Chapter 17: Endocrine System** We have organized both the endocrine system chapter and the specific coverage of the many hormones released from endocrine glands to most effectively and efficiently guide students in understanding how this system of control functions in maintaining homeostasis. Within the chapter on the endocrine system, we provide an introduction and general discussion of the endocrine system's central concepts and describe selected representative hormones that maintain body homeostasis. Details of the actions of most other hormones—which require an understanding of specific anatomic structures covered in other chapters—are described in those chapters; for example, sex hormones are discussed in Chapter 28: Reproductive System. Learning the various hormones is facilitated by the inclusion of a "template" figure for each major hormone; each visual template includes the same components (stimulus,

receptor, control center, and effectors) organized in a similar layout. In addition, information on each major hormone described in this text can be quickly accessed in the summary tables following chapter 17.

- **Chapter 21: Lymphatic System and Chapter 22: Immune System and the Body's Defense** A single chapter that discusses both the lymphatic system and immune system is overwhelming for most students. Thus, we separated the discussion into two separate chapters. The lymphatic system chapter focuses on the anatomic structures that compose the system, and provides a brief functional overview of each structure. This allows us to provide a thorough discussion and overview of the immune system in a separate chapter, where we frequently reference and integrate material from the earlier chapter.
- **Chapter 29: Development, Pregnancy, and Heredity** Coverage of heredity is included in the chapter on pregnancy and human development as a natural extension of Chapter 28: Reproductive System. This introduction will serve well as a precursor for students who follow their A&P course with a genetics course.

Changes to the Third Edition

Real student data points derived from thousands of SmartBook users have guided the revision process for this edition. In addition, this revision has been informed by dozens of chapter reviews by A&P instructors. The following global changes have been implemented throughout all chapters:

- Additional references were added to concepts previously covered, as well as to related material in upcoming sections and chapters, to further connect concepts.
- Terminology has been updated and definitions are added throughout.
- New "What Do You Think?" and "What Did You Learn?" questions were added throughout the text.

Chapter 1

- New section added about how best to study A&P
- Revised: figure 1.8, figure 1.9, figure 1.13
- Clinical View 1.1: Etiology (Causes) and Pathogenesis (Development) of Disease updated to include more detail on sonography and imaging; added new photos and labeling to images

Chapter 2

- Moved description of inorganic and organic molecules from section 2.7: Biological Molecules to beginning of section 2.4: Molecular Structure of Water and the Properties of Water
- Revised: figure 2.1b, figure 2.9, figure 2.10, figure 2.13, figure 2.17, figure 2.19, figure 2.20, figure 2.21, figure 2.22, figure 2.24, figures within table 2.4

Chapter 3

- Revised: figure 3.1, figure 3.3, figure 3.6, figure 3.7, figure 3.10, figure 3.13, figure 3.14, figure 3.16, figure 3.18, figure 3.19
- Revised coverage of ATP cycling in section 3.2b: Classification of Chemical Reactions

Chapter 4

- Revised: figure 4.4, figure 4.5, figure 4.6, figure 4.7, figure 4.9, figure 4.10, figure 4.11, figure 4.14, figure 4.15, figure 4.17, figure 4.18, figure 4.19, figure 4.20, figure 4.21, figure 4.23, figure 4.27, figure 4.28, figure 4.30, figure 4.32, figure 4.39, figure 4.42 (several new photos)
- New figure 4.38 on genetic code
- Modified section 4.1a: How Cells Are Studied
- Updated section 4.2b: Membrane Proteins
- Revised section 4.3: Membrane Transport
- Revised and reorganized section 4.6b: Non-Membrane-Bound Organelles

Chapter 5

- Updated art in tables 5.2, 5.3
- Revised: figure 5.4, figure 5.8, figure 5.10, figure 5.11, figure 5.12
- Modified section 5.2b: Functions of Connective Tissue
- Updated text in tables 5.6, 5.10
- Removed coverage of perichondrium (covered in Chapter 7)
- Modified section 5.5b: Body Membranes
- Updated Clinical View 5.4: Stem Cells
- Revised Clinical View 5.5: Gangrene
- Updated Chapter Summary to include lymph in discussion of fluid connective tissue

Chapter 6

- Updated terminology to use "keratinocyte" instead of "cell" where appropriate
- Updated section on stratum corneum to include coverage of dermidicin
- Removed reference to human pheromones

Chapter 7

- In section 7.2, added information on the appearance of living bone
- Revised: figure 7.4, figure 7.11, figure 7.15
- Revised discussion of exercise in space
- Clinical View 7.7: Osteoporosis—added information on DEXA scans

Chapter 8

- Table 8.1 revised to include parietal foramina
- Revised: figure 8.5, figure 8.7a, figure 8.11, figure 8.12a, figure 8.14, figure 8.22a, figure 8.31, figure 8.34b
- In section 8.2b: Views of the Skull and Landmark Features, revised sella turcica information to include hypophyseal fossa
- Table 8.2 and table 8.3 figures were revised to improve clarity and accuracy
- Revised Clinical View 8.2: Craniosynostosis and Plagiocephaly
- In section 8.3: Bones Associated with the Skull, revised information on functions of the hyoid bone's cornua and body
- Revised figure in Clinical View 8.4: Herniated Discs
- Table 8.6 text and figures updated regarding width of pelvis and ilia

Chapter 9

- Reversed order of sections 9.5 and 9.6, so Movements of Synovial Joints are discussed prior to Synovial Joints and Levers
- Table 9.1 simplified
- Revised: figure 9.2, figure 9.3, figure 9.6, figure 9.14
- Modified description of synovial membrane

Chapter 10

- Updated section 10.1a: Functions of Skeletal Muscle and section 10.1b: Characteristics of Skeletal Muscle Cells
- Revised section 10.2a: Gross Anatomy of Skeletal Muscle
- Revised section 10.2b: Microscopic Anatomy of Skeletal Muscle to add details of triad
- Revised section 10.4: Skeletal Muscle Metabolism
- Revised section 10.8: Effects of Exercise and Aging on Skeletal Muscle
- New Concept Connection on Excitability
- New Learning Strategy on prefixes *myo-*, *mys-*, *sarco-*
- New Concept Connection on connective tissue coverings
- New Clinical View 10.1: Muscular Dystrophy
- New Concept Connection on nervous system diseases that influence muscle function
- New Clinical View 10.9: Unbalanced Skeletal Muscle Development
- New Concept Connection on Myogenic Response
- Revised: figure 10.1, figure 10.3, figure 10.6, figure 10.7, figure 10.9, figure 10.10, figure 10.11, figure 10.12, figure 10.14, figure 10.15, figure 10.16, figure 10.17, figure 10.20, figure 10.21, figure 10.22, figure 10.30
- Revised figure in Learning Strategy on sarcomere shortening
- New: figure 10.17 on ATP for muscle metabolism; figure 10.21 on recruitment; figure 10.24 comparing isometric and isotonic contractions; figure 10.26 on maximizing force of contractions

Chapter 11

- Replaced the terms *origin* and *insertion* with *superior attachment* and *inferior attachment* where appropriate
- Revised Clinical View 11.2: Idiopathic Facial Nerve Paralysis (Bell Palsy) to include information on Lyme disease effects
- New photo in Clinical View 11.2: Idiopathic Facial Nerve Paralysis (Bell Palsy)
- Revised: table 11.3, table 11.8, table 11.18
- Revised: figure 11.17, figure 11.19, figure 11.21b, figure 11.22
- New photo in Clinical View 11.7: Lateral Epicondylitis ("Tennis Elbow")
- Revised figure in Clinical View 11.8: Carpal Tunnel Syndrome

Chapter 12

- Revised section 12.2b: Neuron Structure
- Added section 12.1c: Nerves and Ganglia
- New Concept Connection on bundling by connective tissue
- New Concept Connection on excitability
- New Learning Strategy on glial cells
- Clinical View 12.3: Nervous System Disorders Affecting Myelin now includes Zika virus as it relates to Guillain-Barré syndrome
- New Learning Strategy on states of voltage-gated Na^+ channels
- Revised Clinical View 12.6: Altered Acetylcholine Function and Changes in Breathing
- Section 12.1c: Nerves and Ganglia was previously section 12.2e (moved forward in the chapter)
- Revised: figure 12.3, figure 12.5, figure 12.6, figure 12.7, figure 12.9b, figure 12.10, figure 12.11, figure 12.14, figure 12.16, figure 12.17, figure 12.18, figure 12.21, figure 12.22, figure 12.23, figure 12.24, figure 12.25
- New Clinical View 12.1: Pathogenic Agents and Fast Axonal Transport figure of axonal transport for how pathogenic organisms move to cell body
- New illustration of battery for text discussion on Ohm's Law (pg. 457)
- New figure 12.13: Neuron's and Ohm's Law
- New photo in figure 12.17

Chapter 13

- Revised section 13.1: Brain Organization and Development
- Revised Clinical View 13.1: Traumatic Brain Injuries: Concussion and Contusion
- Revised section 13.1c: Gray Matter and White Matter Distribution
- Revised: figure 13.1, figure 13.4, figure 13.5, figure 13.9, figure 13.12, figure 13.14, figure 13.21, figure 13.23b, figure 13.24a, figure 13.25, figure 13.30, figure 13.31
- Revised table 13.2, table 13.4, table 13.5
- Updated Clinical View 13.3: Meningitis and Encephalitis
- Revised Clinical View 13.6: Mapping Functional Brain Regions
- New Learning Strategy on functions of the hypothalamus
- Revised coverage of medulla autonomic centers

Chapter 14

- Revised section 14.1: Overview of the Spinal Cord and Spinal Nerves reorganized to provide an overview of the chapter content; section 14.1c: Spinal Nerve Identification and Gross Anatomy now contains content on spinal roots that was moved from section 14.5a
- Section 14.2: Protection and Support of the Spinal Column revised to include more explanation on vertebral column
- Section 14.3: Sectional Anatomy of the Spinal Cord and Spinal Roots provides more information on functional relationship between gray matter and spinal nerve roots, sensory receptors, and effectors.
- Section 14.5a: General Distribution of Spinal Nerves
- Revised: figure 14.1, figure 14.2, figure 14.4, figure 14.5, figure 14.6, figure 14.10, figure 14.14, figure 14.18c

Chapter 15

- Section 15.3b: Pelvic Splanchnic Nerves rearranged to discuss parasympathetic cranial nerve physiology first, followed by anatomy
- Section 15.5: Autonomic Plexuses and the Enteric Nervous System moved forward to appear after discussion of sympathetic and parasympathetic systems
- New section 15.5b: Enteric Nervous System
- New Learning Strategy on parasympathetic division

- Updated Clinical View 15.3: Raynaud Syndrome, with photo added
- Revised: figure 15.4, figure 15.6, figure 15.7b, figure 15.10, figure 15.11

Chapter 16

- New Clinical View 16.2: Eye Infections
- Revised: figure 16.2, figure 16.3, figure 16.5, figure 16.7, figure 16.8, figure 16.10, figure 16.15, figure 16.18, figure 16.22, figure 16.26, figure 16.27, figure 16.28

Chapter 17

- Section 17.1a: Overview of Endocrine System rewritten and expanded
- Revised: Section 17.1b: Comparison of the Two Control Systems
- Section 17.1c: General Functions of the Endocrine System contains added examples for each function of the endocrine system
- Section 17.3b: Local Hormones rewritten, now includes new Concept Connection on local hormones that act as vasoactive substances; new Clinical View 17.1: Synthesis of Eicosanoids
- Section 17.7b: Interactions Between the Hypothalamus and the Posterior Pituitary Gland contains expanded description of ADH and oxytocin
- Section 17.7c: Interactions Between the Hypothalamus and the Anterior Pituitary Gland reorganized and updated to add explanation for figure 17.12
- Revised: figure 17.1, figure 17.5, figure 17.7, figure 17.8, figure 17.10, figure 17.11, figure 17.12, figure 17.14, figure 17.16, figure 17. 17, figure 17.18, figure 17.19, figure 17.20, figure 17.22, figure 17.23

Chapter 18

- Clinical View 18.3: Transfusions updated to include information on donor wait times between transfusions
- New Clinical View 18.4: Whole Blood Versus Plasma Donations: What's the Difference?
- New Clinical View 18.5: Fetal Hemoglobin and Physiologic Jaundice
- New Learning Strategy on Blood Types
- Revised: figure 18.5a, 18.7, figure 18.8, figure 18.11 (colorized micrograph of blood clotting), figure 18.12

Chapter 19

- Section 19.1a: General Function contains expanded content of the function of the cardiovascular system in transporting blood, including a new Concept Connection that provides examples of body systems dependent on blood transport
- Section 19.1b: Overview of Components revised to further describe circulation routes through the right and left sides of the heart
- New Clinical View 19.1: Congestive Heart Failure
- Section 19.2b: The Pericardium contains added description of pericardial sac
- Section 19.3a: Superficial Features of the Heart contains added explanation to align with figure 19.7
- Revised Clinical View 19.5: Coronary Heart Disease, Angina Pectoris, and Myocardial Infarction to add content on coronary heart disease
- Added Concept Connection to describe a sinus
- Added Concept Connection on resting membrane potential
- Added Concept Connection on nodal cells compared to neurons
- Added Concept Connection on conductivity
- Section 19.7b: Electrical and Mechanical Events of Cardiac Muscle Cells has been revised to add details on triad
- Revised: figure 19.2, figure 19.3, figure 19.7, figure 19.11, figure 19.13, figure 19.14, figure 19.16, figure 19.18, figure 19.19, figure 19.20, figure 19.21, figure 19.22

Chapter 20

- Revised: figure 20.1, figure 20.3, figure 20.4, figure 20.5, figure 20.7, figure 20.14, figure 20.18, figure 20.19, figure 20.20, figure 20.22, figure 20.26, figure 20.27

Chapter 21

- Revised Learning Strategy on lymphocytes to describe origin of name for B-lymphocytes
- Revised: figure 21.1, figure 21.6, figure 21.9

Chapter 22

- New Table 22.5: Actions of Antibodies Following Antigen Binding
- Section 22.1: Overview of Diseases Caused by Infectious Agents revised to update definition of bacteria
- New Learning Strategy on the study of the immune system
- Revised Section 22.3a: Preventing Entry
- New Learning Strategy on CD4/CD8 with MHC Class I/II
- Section 22.5b: Selection and Differentiation of T-Lymphocytes rearranged for better alignment of illustration and text
- Section 22.8b: Action of Antibodies reorganized into table format for clarity
- Revised: figure 22.1, figure 22.2, figure 22.3, figure 22.4, figure 22.7, figure 22.8, figure 22.10, figure 22.12, figure 22.13, figure 22.14, figure 22.15, figure 22.16, figure 22.17, figure 22.19, figure 22.20

Chapter 23

- New Learning Strategy on structural and functional organization of the respiratory system
- Revised Section 23.3b: Trachea
- New Clinical View 23.4: Tracheotomy
- Clinical View 23.5: Bronchitis now includes coverage on exercise-induced asthma
- Revised Section 23.3c: Bronchial Tree
- Revised Section 23.3d: Respiratory Zone: Respiratory Bronchioles, Alveolar Ducts, and Alveoli
- New Clinical View 23.7: Pneumonia
- New photos of lungs in Clinical View 23.8: Smoking
- Revised: figure 23.3, figure 23.5, figure 23.6, figure 23.8, figure 23.9, figure 23.10, figure 23.11, figure 23.14, figure 23.19, figure 23.14, figure 23.16, figure 23.19, figure 23.21, figure 23.23, figure 23.25, figure 23.26, figure 23.31
- New figure 23.27: Changes in Respiratory Gas Partial Pressures Within the Blood

Chapter 24

- New figure in Clinical View 24.1: Renal Ptosis and Hydronephrosis
- Section 24.3c: Juxtaglomerular Apparatus revised to expand coverage of mesangial cells
- Concept Connection on blood pressure expanded
- Revised: figure 24.2, figure 24.3, figure 24.4, figure 24.5, figure 24.6, figure 24.7, figure 24,9, figure 24.11, figure 24.13, figure 24.14, figure 24.15, figure 24.16, figure 24.18, figure 24.19, figure 24.22, figure 24.23, figure 24.24, figure 24.26, figure 24.27, figure 24.28

Chapter 25

- Section 25.1: Body Fluids introduction revised to emphasize the main points of the chapter
- Figure 25.2: Percentages of Solutes in Body Fluids is new and contains the information from previous edition's Table 25.1
- New Clinical View 25.2: Hemorrhaging
- New Clinical View 25.4: Angiotensin-Converting Enzyme (ACE) Inhibitors
- Section 25.3b: Major Electrolytes: Location, Functions, and Regulation revised to integrate concepts of acid-base balance and hyperkalemia and hypokalemia
- Section 25.5: Acid-Base Balance revised for increased readability
- Section 25.5c: Respiration and Regulation of Volatile Acid has tighter integration with concepts in Chapter 23: Respiratory System
- New Learning Strategy on Type A and Type B Intercalated Cells
- Section 25.6: Disturbances to Acid-Base Balance is revised for increased readability and tighter integration with Clinical View 25.8: Arterial Blood Gas (ABG) and Diagnosing Different Types of Acid-Base Disturbances
- Section 25.6b: Respiratory-Induced Acid-Base Disturbances includes normal values for arterial blood gas
- New: figure 25.2, figure 25.12, figure 25.16
- Revised: figure 25.1, figure 25.4, figure 25.5, figure 25.6, figure 25.7, figure 25.8, figure 25.9, figure 25.10, figure 25.11, figure 25.12, figure 25.13, figure 25.14, figure 25.15

Chapter 26

- Sections 26.1b, 26.1c, 26.1d, and 26.1e revised for readability
- New Learning Strategy for layers of muscularis in GI tract wall
- New Clinical View 26.3: Achalasia
- New Learning Strategy on gastric gland secretions for parietal cells and chief cells
- Sections 26.3b: Small Intestine and 26.3c: Accessory Digestive Organs and Ducts revised
- New Clinical View 26.10: Pancreatic Cancer
- New Clinical View 26.13: Fecal Transplant
- New Learning Strategy on lipid digestion and absorption
- New Clinical View 26.17: Cystic Fibrosis and the Pancreas
- New Section 26.4e: Water, Electrolyte, and Vitamin Absorption
- Revised: figure 26.2c, figure 26.6, figure 26.8, figure 26.13, figure 26.14, figure 26.15, figure 26.16, figure 26.18, figure 26.20, figure 26.22, figure 26.29

Chapter 27

- New Clinical View 27.3: Obesity
- New Clinical View 27.5: Heat Related Illnesses
- Expanded Clinical View 27.6: Hypothermia, Frostbite, and Dry Gangrene

Chapter 28

- Sections 28.3a: Ovaries and 28.3b: Oogenesis and the Ovarian Cycle discussion on primary and secondary follicles revised, and discussion on antral follicles expanded
- Revised information on ovarian ligaments
- Revised Clinical View 28.2: Ovarian Cancer with current statistics
- Revised Clinical View 28.5: Cervical Cancer to include information on recommendations for vaccination
- Updated Clinical View 28.6: Breast Cancer
- Revised: figure 28.4, figure 28.5, figure 28.6, figure 28.7, figure 28.9, figure 28.11, figure 28.12, figure 28.15, figure 28.16, figure 28.18, figure 28.19d

Chapter 29

- Revised Clinical View 29.2: Infertility and Infertility Treatments
- Revised Clinical View 29.3: Gestational Diabetes
- Revised: figure 29.1, figure 29.3, figure 29.4, figure 29.7, figure 29.16

We Welcome Your Input!

We hope you enjoy reading this textbook, and that it becomes central to mastering the concepts in your anatomy and physiology course. This text is a product that represents over 75 years of combined teaching experience in anatomy and physiology. We are active classroom instructors, and are well aware of the challenges that current students face in mastering these subjects. We have taken what we have learned in the classroom and have created a textbook truly written for students.

Please let us know what you think about this text. We welcome your thoughts and suggestions for improvement, and look forward to your feedback!

Michael P. McKinley

Glendale Community College, retired

mpmckinley@hotmail.com

Valerie Dean O'Loughlin

Medical Sciences

Indiana University

vdean@indiana.edu

Terri Stouter Bidle

Science Division

Hagerstown Community College

tsbidle@hagerstowncc.edu

ACKNOWLEDGMENTS

Many people have worked with us over the last several years to produce this text. We would like to thank the many individuals at McGraw-Hill who worked with us to create this textbook. We are especially grateful to Donna Nemmers, our Product Developer, Amy Reed, our Portfolio Manager, and Jayne Klein and Jessica Portz, our Content Project Managers, for expertly guiding the project through its production phases; David Hash, Designer, for his beautiful interior and cover designs; and Jim Connely, Marketing Manager, for his marketing expertise. We would also like to thank our copyeditor, Deb DeBord, and our proofreaders, Wendy Nelson and Lauren Timmer. We are very grateful for the enthusiasm and expertise of Dr. Justin York, digital author and collaborator on multiple portions of the supporting online assessment and instructor tools that accompany this textbook. Justin's eye to detail also helped us improve the accuracy of several sections of the text.

Finally, we could not have performed this effort were it not for the love and support of our families. Jan, Renee, Ryan, and Shaun McKinley; Bob and Erin O'Loughlin; and Jay and Stephanie Bidle—thank you and we love you! We are blessed to have you all.

Many instructors and students across the country have positively affected this text through their careful reviews of manuscript drafts, art proofs, and page proofs, as well as through class tests and through their attendance at focus groups and symposia. We gratefully acknowledge their contributions to this text.

Reviewers

Tim Ballard
University of North Carolina—Wilmington

Charles Benton
Madison College

Lindsay Biga
Oregon State University

Jeff Bolles
University of North Carolina—Pembroke

Chester Brown
University of Illinois—Urbana-Champaign

Susan C. Burgoon
Amarillo College

Dennis Burke
Quincy College—Plymouth

Charles Cahill
West Kentucky Community and Technical College

Melody Candler-Catt
Vincennes University

Ronald A. Canterbury
University of Cincinnati

Rajeev Chandra
Norfolk State University

Cathleen Ciesielski
Gardner-Webb University

Jordan Clark
Sam Houston State University

Beth Collins
Iowa Central Community College

Andrew Corless
Vincennes University

Smruti Desai
Lone Star College—CyFair

Cynthia Doffitt
Northwestern State University

Bruce Evans
Huntington University

David Evans
Penn College

Colin Everhart
St. Petersburg College

John Fishback
Ozarks Technical Community College

Margaret Flemming
Austin Community College

Caroline Garrison
Carroll Community College

Nola Kelly-Gondek
Hudson Valley Community College

Elizabeth Granier
St. Louis Community College

Siabhon M. Harris
Tidewater Community College

Cynthia A. Herbrandson
Kellogg Community College

Lisa Hight
Baptist College of Health Sciences

Kendricks D. Hooker
Baptist College of Health Sciences

Shahdi Jalilvand
Tarrant County College—Southeast

Kurt E. Kwast
University of Illinois—Urbana-Champaign

Marta Klesath
North Carolina State University

Kristine N. Kraft
University of Akron

Paul Luyster
Tarrant County College—South

Benjamin Navia
Andrews University

Raffaella Pernice
Hudson County Community College

Julie L. Posey
Columbus State Community College

Laura H. Ritt
Rowan College—Burlington County

Corinna Ross
Texas A&M San Antonio

Merideth Sellars
Columbus State Community College

Michael A. Silva
El Paso Community College

William G. Sproat, Jr.
Walters State Community College

Michael Thompson
Jefferson Community and Technical College

Emmanuel Vrotsos
Broward College

Chad Wayne
University of Houston

Robinlyn Wright
Houston Community College

guided tour

Fully Integrated Content and Pedagogy

Anatomy and Physiology: An Integrative Approach is structured around a tightly integrated learning system that combines illustrations and photos with textual descriptions; focused discussions with big-picture summaries; previously learned material with new content; factual explanations with practical and clinical examples; and bite-sized topical sections with multi-tiered assessment.

Unparalleled Art Program

In a visually oriented subject like A&P, quality illustrations are crucial to understanding and retention. The brilliant illustrations in *Anatomy and Physiology: An Integrative Approach* have been carefully rendered to convey realistic, three-dimensional detail while incorporating pedagogical conventions that help deliver a clear message. Each figure has been meticulously reviewed for accuracy and consistency, and precisely labeled to coordinate with the text discussions.

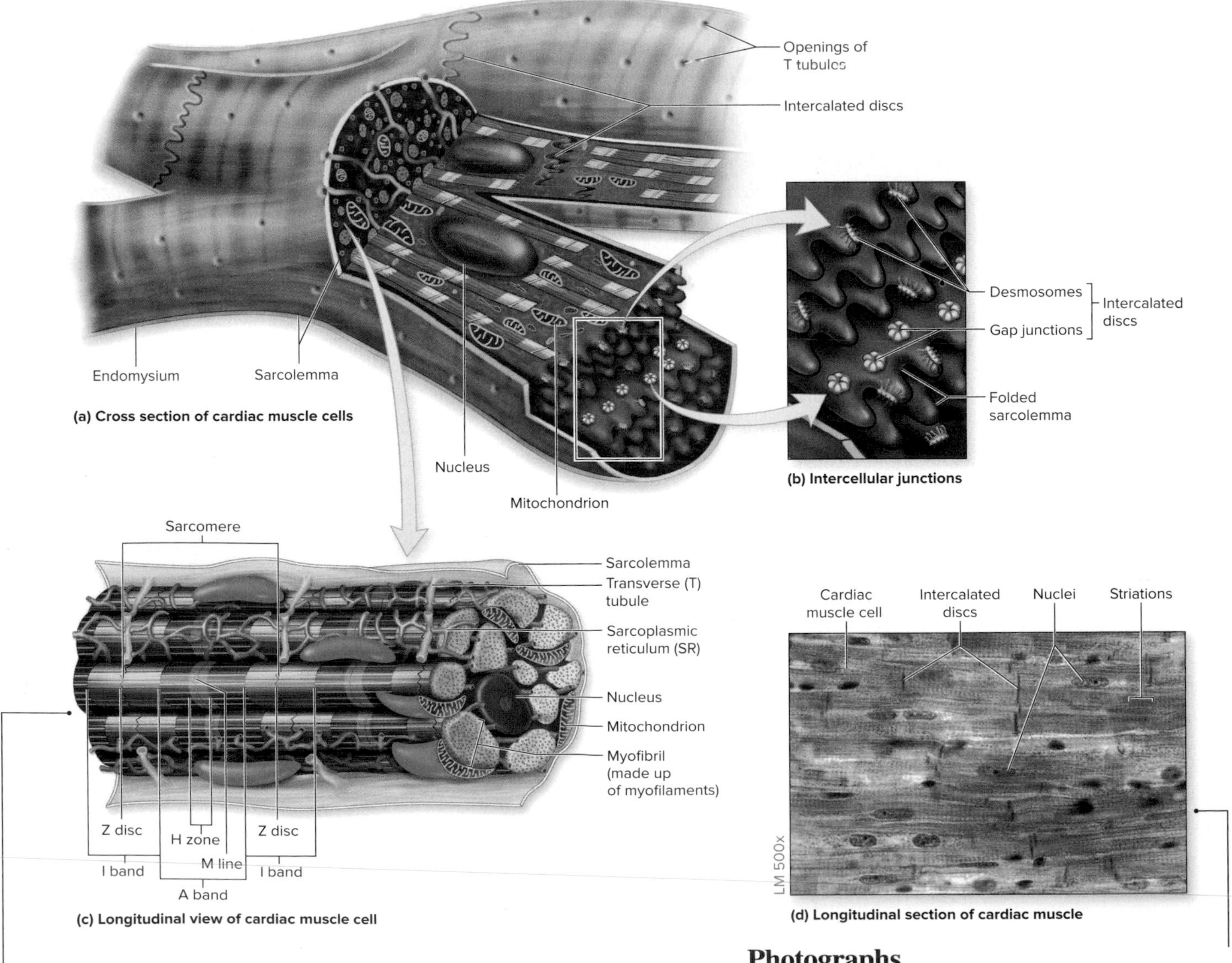

(a) Cross section of cardiac muscle cells

(b) Intercellular junctions

(c) Longitudinal view of cardiac muscle cell

(d) Longitudinal section of cardiac muscle

Rich Detail
Vibrant colors and three-dimensional shading make it easy to envision body structures and processes.

Photographs
Atlas-quality micrographs and cadaver images are frequently paired with illustrations to expose students to the appearance of real anatomic structures.

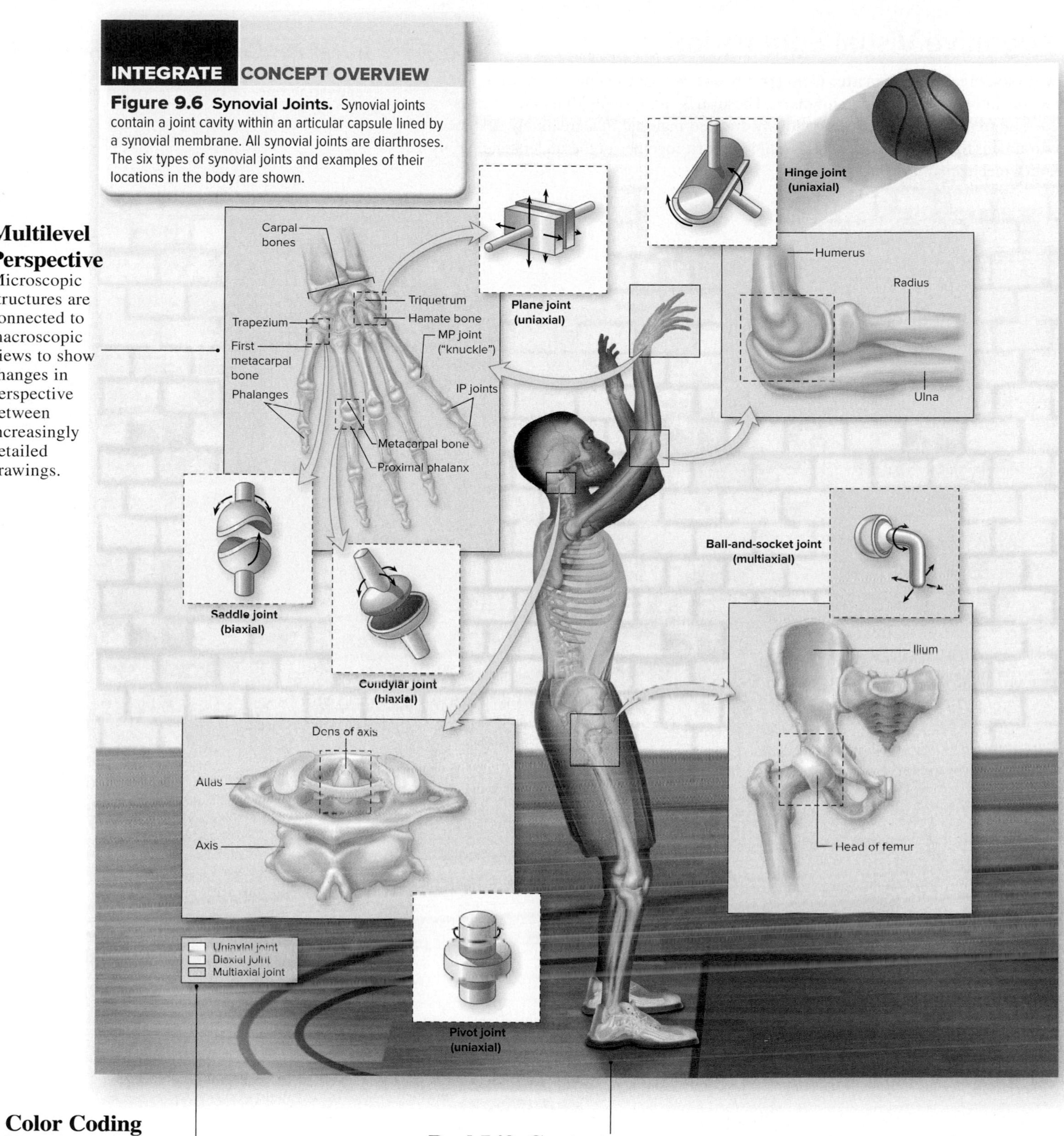

Figure 9.6 Synovial Joints. Synovial joints contain a joint cavity within an articular capsule lined by a synovial membrane. All synovial joints are diarthroses. The six types of synovial joints and examples of their locations in the body are shown.

Multilevel Perspective

Microscopic structures are connected to macroscopic views to show changes in perspective between increasingly detailed drawings.

Color Coding

Many figures use color coding to organize information and clarify concepts for visual learners.

Real-Life Context

Illustrations include depictions of realistic people and situations to make figures more relevant and memorable.

Integrative Visual Summaries

The groundbreaking **Integrate: Concept Overview** figures combine multiple concepts into one big-picture summary. These striking, visually dynamic presentations offer a review of previously covered material in a creatively designed environment to emphasize how individual parts fit together in the understanding of a larger mechanism or concept.

Integrate: Concept Overview Figures
Multifaceted concepts are brought together in captivating one- or two-page visual presentations.

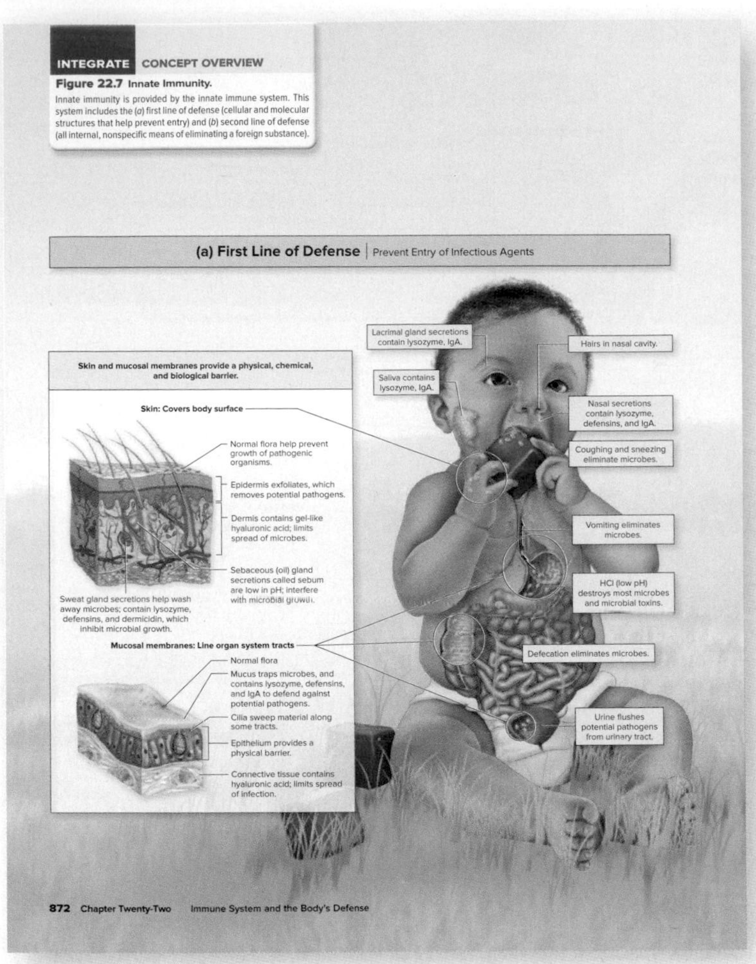
INTEGRATE CONCEPT OVERVIEW

Figure 22.7 Innate Immunity.
Innate immunity is provided by the innate immune system. This system includes the (a) first line of defense (cellular and molecular structures that help prevent entry) and (b) second line of defense (all internal, nonspecific means of eliminating a foreign substance).

(a) First Line of Defense | Prevent Entry of Infectious Agents

Skin and mucosal membranes provide a physical, chemical, and biological barrier.
Skin: Covers body surface
Normal flora help prevent growth of pathogenic organisms.
Epidermis exfoliates, which removes potential pathogens.
Dermis contains gel-like hyaluronic acid; limits spread of microbes.
Sebaceous (oil) gland secretions called sebum are low in pH; interfere with microbial growth.
Sweat gland secretions help wash away microbes; contain lysozyme, defensins, and dermicidin, which inhibit microbial growth.
Mucosal membranes: Line organ system tracts
Normal flora
Mucus traps microbes, and contains lysozyme, defensins, and IgA to defend against potential pathogens.
Cilia sweep material along some tracts.
Epithelium provides a physical barrier.
Connective tissue contains hyaluronic acid; limits spread of infection.
Lacrimal gland secretions contain lysozyme, IgA.
Hairs in nasal cavity.
Saliva contains lysozyme, IgA.
Nasal secretions contain lysozyme, defensins, and IgA.
Coughing and sneezing eliminate microbes.
Vomiting eliminates microbes.
HCl (low pH) destroys most microbes and microbial toxins.
Defecation eliminates microbes.
Urine flushes potential pathogens from urinary tract.

872 Chapter Twenty-Two Immune System and the Body's Defense

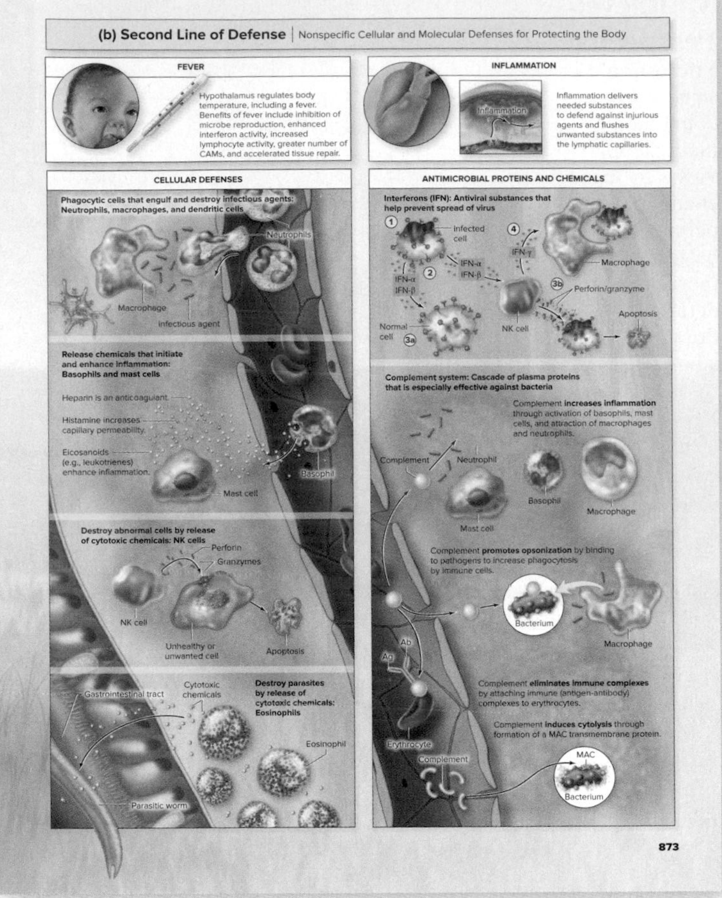
(b) Second Line of Defense | Nonspecific Cellular and Molecular Defenses for Protecting the Body

FEVER
Hypothalamus regulates body temperature, including a fever. Benefits of fever include inhibition of microbe reproduction, enhanced interferon activity, increased lymphocyte activity, greater number of CAMs, and accelerated tissue repair.
INFLAMMATION
Inflammation
Inflammation delivers needed substances to defend against injurious agents and flushes unwanted substances into the lymphatic capillaries.
CELLULAR DEFENSES
Phagocytic cells that engulf and destroy infectious agents: Neutrophils, macrophages, and dendritic cells
Neutrophils
Macrophage
Infectious agent
Release chemicals that initiate and enhance inflammation: Basophils and mast cells
Heparin is an anticoagulant.
Histamine increases capillary permeability.
Eicosanoids (e.g., leukotrienes) enhance inflammation.
Mast cell
Basophil
Destroy abnormal cells by release of cytotoxic chemicals: NK cells
Perforin
Granzymes
NK cell
Unhealthy or unwanted cell
Apoptosis
Gastrointestinal tract
Cytotoxic chemicals
Destroy parasites by release of cytotoxic chemicals: Eosinophils
Eosinophil
Parasitic worm
ANTIMICROBIAL PROTEINS AND CHEMICALS
Interferons (IFN): Antiviral substances that help prevent spread of virus
Infected cell
IFN-α
IFN-β
IFN-γ
Macrophage
Perforin/granzyme
Normal cell
NK cell
Apoptosis
Complement system: Cascade of plasma proteins that is especially effective against bacteria
Complement increases inflammation through activation of basophils, mast cells, and attraction of macrophages and neutrophils.
Complement
Neutrophil
Basophil
Mast cell
Macrophage
Complement promotes opsonization by binding to pathogens to increase phagocytosis by immune cells.
Bacterium
Macrophage
Ab
Ag
Complement eliminates immune complexes by attaching immune (antigen-antibody) complexes to erythrocytes.
Complement induces cytolysis through formation of a MAC transmembrane protein.
Erythrocyte
Complement
MAC
Bacterium

873

Practical and Clinical Applications

Integrating familiar contexts into the study of A&P makes seemingly abstract concepts more relevant and memorable. **Integrate: Learning Strategy** boxes provide simple, practical advice for learning the material. **Integrate: Clinical View** readings offer insight on how complex physiologic processes or anatomic relationships affect body functioning.

Learning Strategies
Classroom tried-and-tested learning strategies offer everyday analogies, mnemonics, and useful tips to aid understanding and memory.

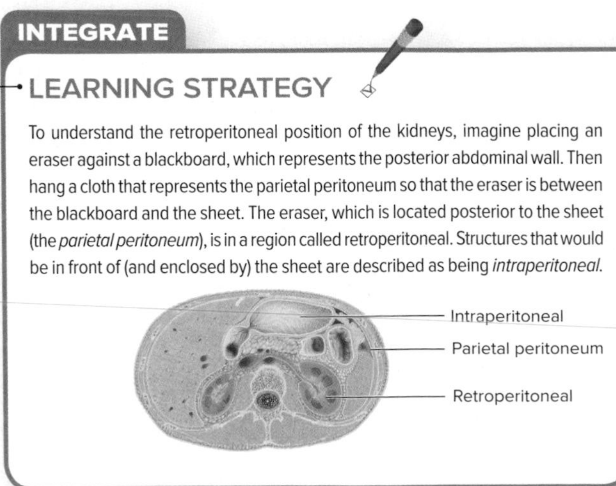
INTEGRATE

LEARNING STRATEGY

To understand the retroperitoneal position of the kidneys, imagine placing an eraser against a blackboard, which represents the posterior abdominal wall. Then hang a cloth that represents the parietal peritoneum so that the eraser is between the blackboard and the sheet. The eraser, which is located posterior to the sheet (the *parietal peritoneum*), is in a region called retroperitoneal. Structures that would be in front of (and enclosed by) the sheet are described as being *intraperitoneal.*

Intraperitoneal
Parietal peritoneum
Retroperitoneal

INTEGRATE

CLINICAL VIEW 12.2

Tumors of the Central Nervous System

Neoplasms resulting from unregulated cell growth, commonly known as **tumors,** sometimes occur within the central nervous system (CNS). A tumor that originates within the organ where it is found is called a **primary tumor.** Because most mature neurons cannot undergo mitosis, primary CNS tumors typically originate in supporting tissues within the brain or spinal cord that have retained the capacity to undergo mitosis: the meninges (protective membranes of the CNS) or the glial cells. Glial cell tumors, termed **gliomas,** may be either relatively benign and slow-growing or malignant (capable of metastasizing or spreading to distant sites).

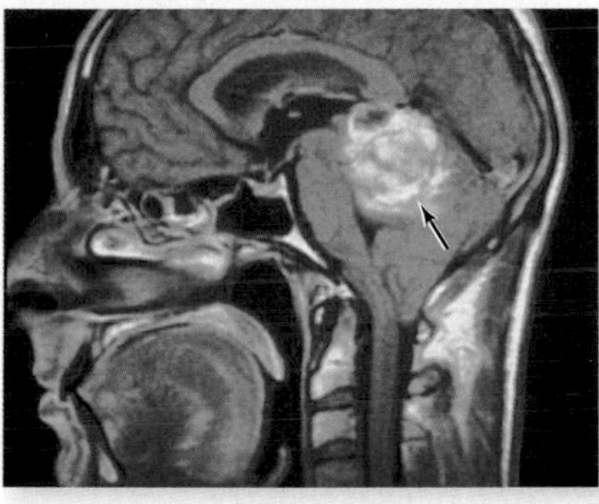

An MRI shows a glioma (arrow).
©Simon Fraser/Science Source

Clinical View

Interesting clinical sidebars reinforce or expand upon the facts discussed within the narrative. The clinical views are adjacent to the facts in the narrative (rather than placed at the end of the chapter) so students may immediately make connections between the narrative and real-life applications.

Concept Integration

Both backward and forward references are supplied throughout the text to remind the reader of the significance of previously covered material, and to foreshadow how knowledge of a topic at hand will come into play in a later discussion. Simple references appear in the flow of the text, while more detailed refreshers are presented in **Integrate: Concept Connection** boxes.

©Digital Vision RF

11.9c Leg Muscles That Move the Ankle, Foot, and Toes

LEARNING OBJECTIVES

35. Compare and contrast the muscles of the three compartments of the leg and their actions.
36. Distinguish between the muscles of the superficial layer and deep layer of the leg's posterior compartment.

The muscles that move the ankle, foot, and toes are housed within the leg and are called the **crural muscles.** Some of these muscles also help flex the leg. The deep fascia partitions the leg musculature into three compartments (anterior, lateral, and posterior), each with its own blood supply and innervation, and muscles in the same compartment tend to share common functions (see figure 11.23).

INTEGRATE

CONCEPT CONNECTION

You will learn in section 20.5a that venous circulation of the lower limbs is reliant upon the muscular system. Specifically, the regular contraction and relaxation of the leg muscles works as a skeletal muscle "pump" to propel venous blood from the lower limb back to the torso. When the lower limbs are immobile for long periods of time (e.g., during long plane rides or when a person is bedridden), the skeletal muscle pump is inactive, and the risk of developing a blood clot in the lower limb veins increases (see Clinical View 20.6: "Deep Vein Thrombosis").

INTEGRATE

CLINICAL VIEW 20.6

Deep Vein Thrombosis

Deep vein thrombosis (throm-bō´sis; a clotting) **(DVT)** refers to a **thrombus** (blood clot) in a vein. The most common site for the thrombus is a vein in the sural region (calf). DVT typically occurs in individuals with heart disease or those who are inactive or immobile for a long period of time, such as bedridden patients. Even healthy individuals who have been on a long airline trip may develop DVT.

Initial signs of DVT include fever, tenderness and redness in the affected area, severe pain and swelling in the areas drained by the affected vein, and rapid heartbeat. The most serious complication of DVT is a **pulmonary embolus** (em´bō-lŭs; a plug), in which a blood clot breaks free and is transported to the lung, eventually blocking a branch of the pulmonary artery and potentially causing respiratory failure and death. If a DVT is diagnosed, the patient is given anticoagulation medication, such as *low-molecular-weight heparin,* to help prevent further clotting and break up the existing clot.

Integrated Assessments

Throughout each chapter, sections begin with learning objectives and end with questions intended to assess whether those objectives have been met. Critical-thinking questions within the narrative prompt students to apply the material as they read. A set of tiered questions at the end of the chapter, as well as additional online problems, further challenge students to master the material.

What Do You Think?

These critical-thinking questions engage students in application or analysis and encourage them to think more globally about the content.

WHAT DO YOU THINK?

3 What type of connective tissue have you damaged when you sprain your ankle?

WHAT DID YOU LEARN?

12 Compare loose connective tissue to dense connective tissue with respect to fiber density, fiber distribution, and the amount of ground substance.

13 Describe the composition and location of fibrocartilage.

14 Why is blood considered a connective tissue?

What Did You Learn?

These mini self-tests at the end of each section help students determine whether they have a sufficient grasp of the information before moving on to the next section.

CHALLENGE YOURSELF

Do You Know the Basics?

1. Which tissue contains a calcified ground substance and is specialized for structural support?
   a. muscle tissue
   b. dense regular connective tissue
   c. areolar connective tissue
   d. bone connective tissue

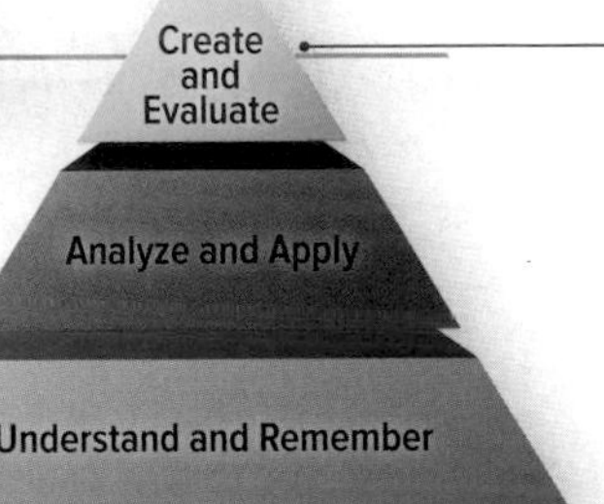

Challenge Yourself

Assessments at the end of each chapter are correlated with Bloom's Taxonomy and progress through knowledge-, application-, and synthesis-level questions. The "Can You Apply ..." and "Can You Synthesize ..." question sets are clinically oriented to encourage concept application, and expose students who may be pursuing health-related careers to problem solving in clinical contexts.

Can You Apply What You've Learned?

1. John is a 53-year-old construction worker who has come into your office complaining of a sore knee joint. You see a buildup of fluid close to the patella (kneecap) but deep to the skin and suspect the soreness is due to bursitis, an inflammation of membranes that surround some joints. Which type of body membrane is inflamed?
   a. cutaneous membrane
   b. serous membrane
   c. synovial membrane
   d. mucous membrane

Can You Synthesize What You've Learned?

1. During a microscopy exercise in the anatomy laboratory, a student makes the following observations about a tissue section: (a) The section contains some different types of scattered protein fibers—that is, they exhibit different widths, some are branched, and some are long and unbranched. (b) The observed section has some "open spaces"—that is, places between both cells and the fibers that appear clear with no recognizable features. (c) Several connective tissue cell types are scattered throughout the section, but these cells are not grouped tightly

Integrated Digital and Textbook

Each chapter ends with a listing of online tools that may be used to study and master the concepts presented.

INTEGRATE

ONLINE STUDY TOOLS

connect | SMARTBOOK® | AP|R

The following study aids may be accessed through Connect.

Concept Overview Interactive: Figure 10.16: Skeletal Muscle Contraction

Clinical Case Study: Progressive Weakness in a Young Woman

Interactive Questions: This chapter's content is served up in a number of multimedia question formats for student study

SmartBook: Topics and terminology include introduction to skeletal muscle; anatomy and physiology of skeletal muscle; skeletal muscle metabolism; skeletal muscle fiber types; measurement of skeletal muscle tension; factors affecting skeletal muscle tension; effects of exercise and aging on skeletal muscle; cardiac muscle tissue; smooth muscle tissue

Anatomy & Physiology Revealed: Topics include skeletal muscle; skeletal muscle striations; sarcomere; sliding filament; neuromuscular junction; excitation-contraction coupling; crossbridge cycle; cardiac muscle; smooth muscle

Animations: Topics include skeletal muscle; function of the neuromuscular junction; action potentials and muscle contraction; sarcomere shortening; breakdown of ATP and crossbridge movement during muscle contraction; mechanics of single fiber contraction; activation of contraction in smooth vs. skeletal muscle

Concept Overviews into Digital Learning

Selected **Concept Overview Figures** from the textbook have been transformed into interactive study modules. This digital transformation process was guided by anatomy and physiology professors who reviewed the modules throughout the development process. Interactive Concept Overview Figures also have assessable, autograded learning activities in Connect®, and are also provided separately to instructors as classroom presentation tools.

Concept Overview Interactives are available for the following topics:

Membrane Transport
Muscle Contraction
Neuron Physiology
Endocrine System (New)
Cardiac Cycle
Blood Pressure (New)
Innate Immunity (New)
Adaptive Immunity (New)
Respiration
Glomerular Filtration
Tubular Resorption/Secretion

Concept Overview from Textbook

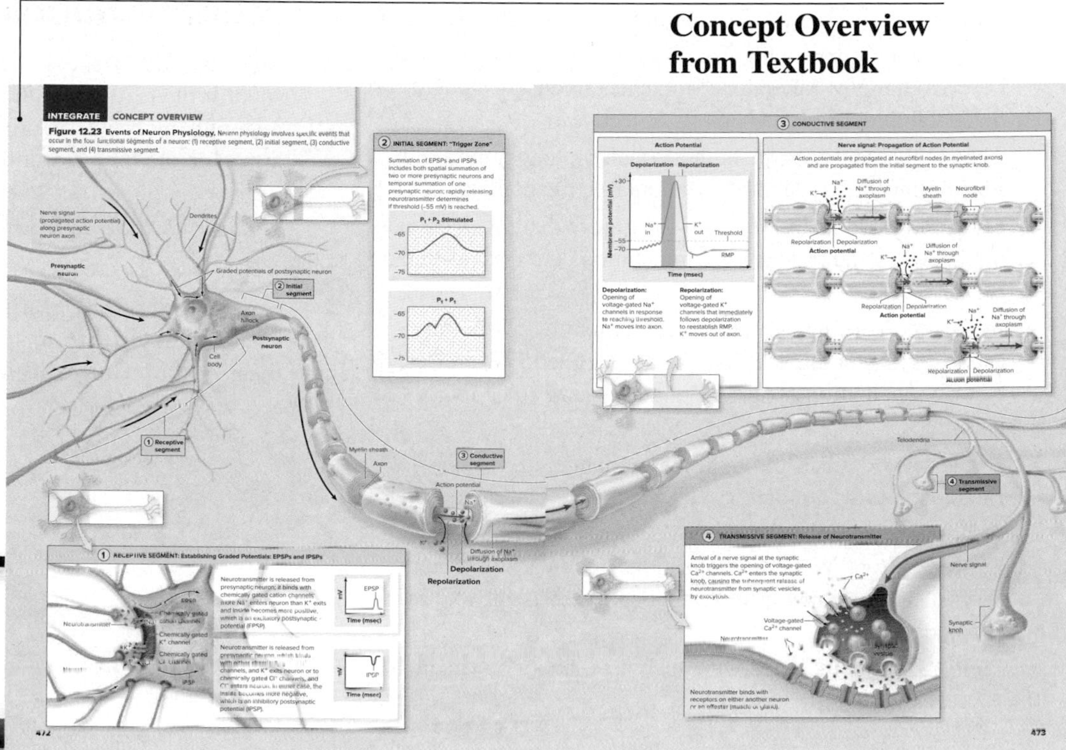

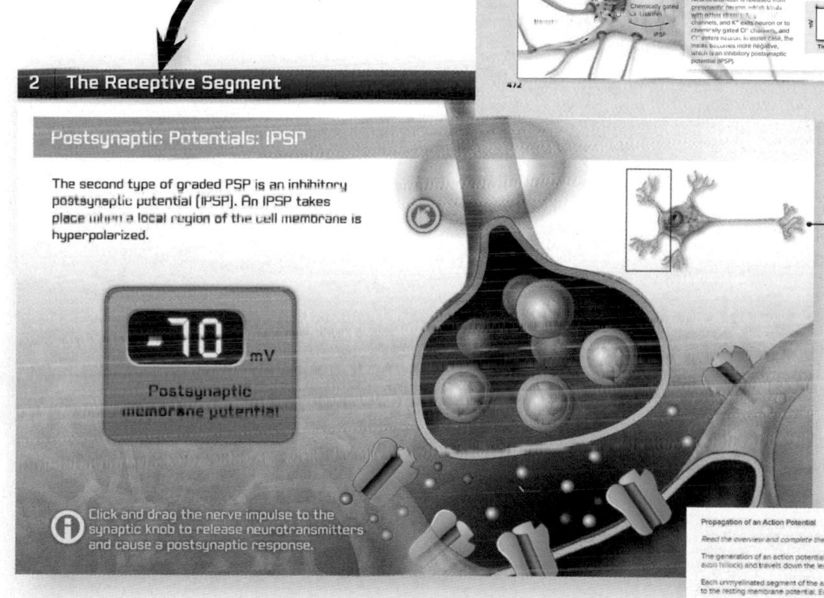

Interactive Presentation Study Tool

Assessable Autograded Activity in Connect

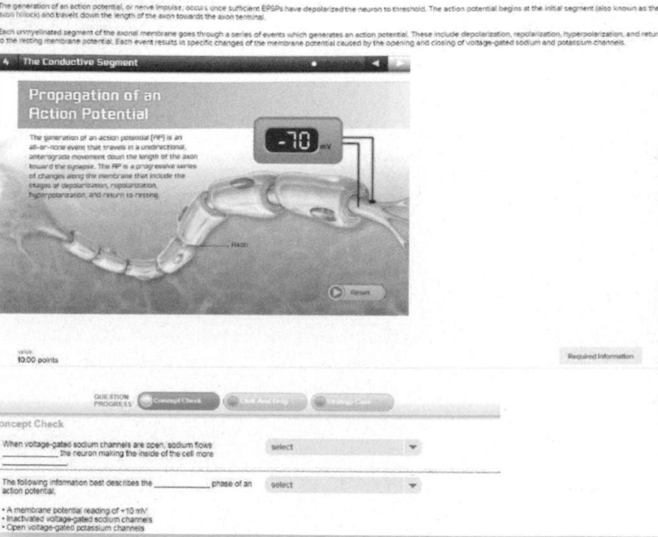

Lab Manual Options to Fit Your Course

Anatomy & Physiology Laboratory Manual by Christine Eckel, Kyla Ross, and Theresa Bidle is a laboratory manual specifically developed for the McKinley/O'Loughlin/Bidle *Anatomy and Physiology: An Integrative Approach* text:

- Three versions are available including main, cat, and fetal pig.
- Each chapter opens with a set of learning objectives that are keyed to the post-laboratory worksheet to ensure student understanding of each chapter's objectives.
- The manual includes the highest-quality photographs and illustrations of any laboratory manual in the market.
- Laboratory exercises are "how-to" guides that involve touch, dissection, observation, experimentation, and critical-thinking exercises.
- In-chapter learning activities offer a mixture of labeling exercises, sketching activities, table completion exercises, data recoring, palpation of surface anatomy, and other sources of learning.
- Numerous exercises throughout the manual utilize Physiology Interactive Lab Simulations (Ph.I.L.S.) 4.0 Online to provide additional student understanding of physiology.
- Pre-Laboratory Worksheet questions and Post-Laboratory Worksheet questions from each chapter are assignable in Connect.
- Ph.I.L.S. 4.0 is included with each new laboratory manual.

Laboratory Manual for Human Anatomy & Physiology by Terry Martin is written to coincide with any A&P textbook:

- Three versions available, including main, cat, and fetal pig
- Includes Ph.I.L.S. 4.0 Online
- Outcomes and assessments format
- Clear, concise writing style

Student Supplements

McGraw-Hill offers various tools and technology products to support the textbook. Students can order supplemental study materials by contacting their campus bookstore or online at https://create.mheducation.com/shop/

Instructor Supplements

Instructors can obtain teaching aids by calling the McGraw-Hill Customer Service Department at 1-800-338-3987, vising our online catalog at https://create.mheducation.com/shop/, or by contacting their local McGraw-Hill sales representative.

Top 10 Tips

to Thrive in Your Anatomy & Physiology Course

© Blend Images/Image Source

1 Preview

Preview assigned material before lecture. Lecture will make much more sense if you've previewed what will be discussed.

2 Look

Look at the images to be covered in lecture. Anatomy and Physiology is a very visual course.

3 Review

Review the prior lecture's material by re-writing your notes or making summary tables/flow charts.

4 Visuals and Notes

Always study your notes along with related visuals. You need to be able to combine visual images with black and white text.

5 Avoid Cramming

Study anatomy and physiology every day or at least every other day. More frequent studying is preferable to studying only two or three days per week. Set a schedule where you spend some time every day either previewing or reviewing anatomy and physiology information.

6 Organize

Organize the course material in a manner that makes the most sense to you. You can create notecards that summarize similar information, design a flow chart, draw simple line diagrams, create mnemonics, or create a table or chart of information.

7 Quiz Yourself

Make your own exam questions. This is a great technique to utilize once you have done a fair amount of studying. Ensure you don't shy away from quizzing yourself on topics you're not confident about.

8 Explain

You master a concept best when you are able to explain it. Practice explaining what you've learned–a process or concept–to someone who knows nothing about anatomy and physiology, or to a fellow classmate.

9 Study Group

Meet weekly or before every exam with several other students to learn the material. Assign each member different challenging topics and have that person teach it to the others in the group. You could also create a few questions on certain topics and then meet and share them with your group. Through the process of creating the questions you will become an "expert" in those topics and could better explain/clarify this information to each other.

10 Office Hours

Make appointments to meet with your instructor to clarify information.

McGraw-Hill Connect® is a highly reliable, easy-to-use homework and learning management solution that utilizes learning science and award-winning adaptive tools to improve student results.

Homework and Adaptive Learning

- Connect's assignments help students contextualize what they've learned through application, so they can better understand the material and think critically.
- Connect will create a personalized study path customized to individual student needs through SmartBook®.
- SmartBook helps students study more efficiently by delivering an interactive reading experience through adaptive highlighting and review.

Over **7 billion questions** have been answered, making McGraw-Hill Education products more intelligent, reliable, and precise.

Connect's Impact on Retention Rates, Pass Rates, and Average Exam Scores

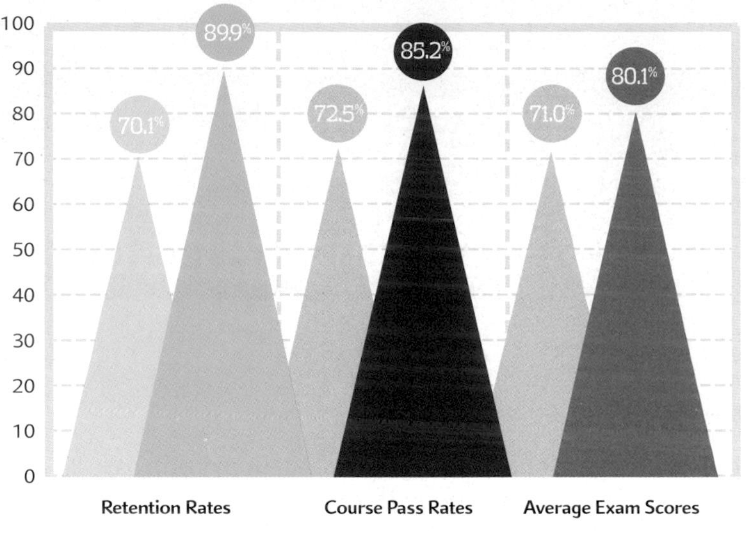

Using **Connect** improves retention rates by **19.8%**, passing rates by **12.7%**, and exam scores by **9.1%**.

73% of instructors who use **Connect** require it; instructor satisfaction **increases** by 28% when **Connect** is required.

Quality Content and Learning Resources

- Connect content is authored by the world's best subject matter experts, and is available to your class through a simple and intuitive interface.
- The Connect eBook makes it easy for students to access their reading material on smartphones and tablets. They can study on the go and don't need internet access to use the eBook as a reference, with full functionality.
- Multimedia content such as videos, simulations, and games drive student engagement and critical thinking skills.

Robust Analytics and Reporting

©Hero Images/Getty Images

- Connect Insight® generates easy-to-read reports on individual students, the class as a whole, and on specific assignments.
- The Connect Insight dashboard delivers data on performance, study behavior, and effort. Instructors can quickly identify students who struggle and focus on material that the class has yet to master.
- Connect automatically grades assignments and quizzes, providing easy-to-read reports on individual and class performance.

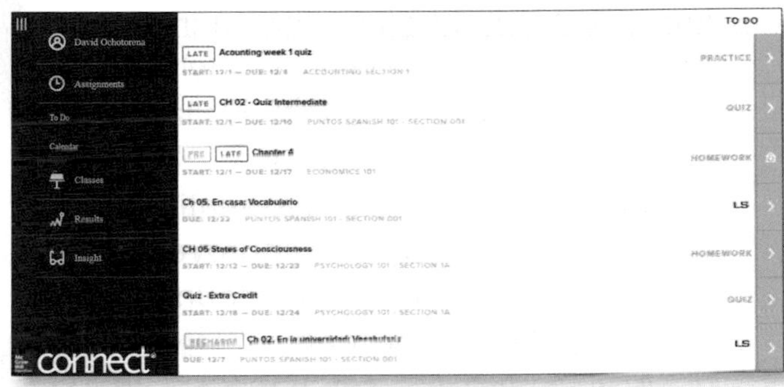

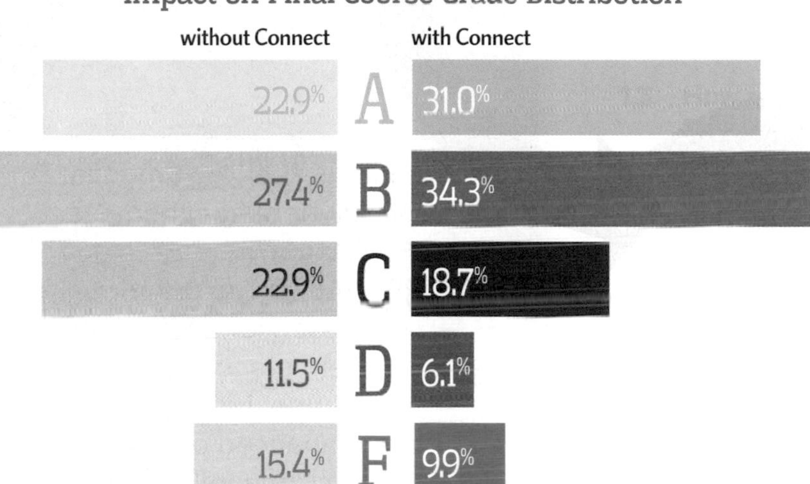

More students earn **As** and **Bs** when they use **Connect.**

Trusted Service and Support

- Connect integrates with your LMS to provide single sign-on and automatic syncing of grades. Integration with Blackboard®, D2L®, and Canvas also provides automatic syncing of the course calendar and assignment-level linking.
- Connect offers comprehensive service, support, and training throughout every phase of your implementation.
- If you're looking for some guidance on how to use Connect, or want to learn tips and tricks from super users, you can find tutorials as you work. Our Digital Faculty Consultants and Student Ambassadors offer insight into how to achieve the results you want with Connect.

1.1 Anatomy and Physiology Compared

In this section, we compare anatomy and physiology and present the general subdivisions of these sciences.

Anatomy is the study of structure and form. The word *anatomy* is derived from the Greek word *anatome,* which means to cut apart or dissect. Anatomists are scientists who study the form and structure of organisms. Specifically, they examine the relationships among parts of the body as well as the structure of individual organs. **Physiology** is the study of function of the body parts. Physiologists are scientists who examine how organs and body systems function under normal circumstances, as well as how the functioning of these organs may be altered via medication or disease. For example, when studying blood capillaries (the smallest of blood vessels), an anatomist may describe the composition of the thin wall. In contrast, a physiologist will explain how the thin wall promotes gas and nutrient exchange between the blood within the capillary and the tissue cells external to the capillary.

Anatomists and physiologists are professionals who use the scientific method to explain and understand the workings of the body. The **scientific method** is a systematic and rigorous process by which scientists:

- Examine natural events (or phenomena) through observation
- Develop a hypothesis (possible explanation) for explaining these phenomena
- Experiment and test the hypothesis through the collection of data
- Determine if the data support the hypothesis, or if the hypothesis needs to be rejected or modified

For example, early anatomists and physiologists used the scientific method to explain how blood circulates through the body. Today, we continue to use the scientific method for a variety of topics, such as to understand how the brain stores memories or explain how cancer may spread throughout the body.

Throughout this text, we have attempted to integrate the study of both anatomy and physiology, showing how form and function are interrelated.

1.1a Anatomy: Details of Structure and Form

LEARNING OBJECTIVES

1. Describe the science of anatomy.
2. List the subdivisions in both microscopic and gross anatomy.

The discipline of anatomy is extremely broad and can be divided into several more specific fields. **Microscopic anatomy** examines structures that cannot be seen by the unaided eye. For most of these studies, scientists prepare individual cells or thin slices of body structures and examine these specimens under the microscope. Microscopic anatomy has several subdivisions with two main divisions:

- **Cytology** (sī-tol′ō-jē; *kytos* = a hollow [cell], *logos* = study), or *cellular anatomy,* is the study of body cells and their internal structure.
- **Histology** (his-tol′ō-jē; *histos* = web, tissue) is the study of body tissues.

Gross anatomy, also called *macroscopic anatomy,* investigates the structure and relationships of body parts that are visible to the unaided eye, such as the intestines, stomach, brain, heart, and kidneys. In these macroscopic investigations, specimens or their parts are often dissected (cut open) for examination. Gross anatomy may be approached in several ways:

- **Systemic anatomy** studies the anatomy of each functional body system. For example, studying the urinary system would involve examining the kidneys (where urine is formed) and the organs of urine transport (ureters and urethra) and storage (urinary bladder). Most undergraduate anatomy and physiology classes use this systemic approach.
- **Regional anatomy** examines all of the structures in a particular region of the body as a complete unit. For example, one may study the axillary (armpit) region of the body, and in so doing examine the blood vessels (axillary artery and vein), nerves (branches of the brachial plexus), lymph nodes (axillary lymph nodes), musculature, connective tissue, and skin. Most medical school gross anatomy courses are taught using a regional anatomy approach.
- **Surface anatomy** focuses on both superficial anatomic markings and the internal body structures that relate to the skin covering them. Health-care providers use surface features to identify and locate important landmarks, such as pulse locations or the proper body region on which to perform cardiopulmonary resuscitation (CPR). Most anatomy and physiology classes also instruct students on important surface anatomy locations.
- **Comparative anatomy** examines the similarities and differences in the anatomy of different species. For example, a comparative anatomy class may examine limb structure in humans, chimps, dogs, and cats.
- **Embryology** (em′brē-ol′o-jē; *embryon* = young one) is the discipline concerned with developmental changes occurring from conception to birth.

Several specialized branches of anatomy focus on the diagnosis of medical conditions or the advancement of basic scientific research. **Pathologic** (path-ō-loj′ik; *pathos* = disease) **anatomy** examines all anatomic changes resulting from disease. Both gross anatomic changes and microscopic structures are examined. **Radiographic anatomy** investigates the relationships among internal structures that may be visualized by specific scanning procedures, such as sonography, magnetic resonance imaging (MRI), or x-ray. (See Clinical View 1.4: "Medical Imaging.")

It may seem as though nothing new can be learned about anatomy—after all, the body has been much the same for thousands of years. Yet in fact, new information is being learned from ongoing anatomic studies, some of which displace the traditional thinking about the workings of various organs. Never forget that anatomy is a dynamic, changing science, not a static, unchanging one.

WHAT DID YOU LEARN?

1. How might knowledge of surface anatomy be important for a health-care worker during a CPR emergency?

1.1b Physiology: Details of Function

LEARNING OBJECTIVES

3. Describe the science of physiology.
4. List the subdivisions in physiology.

Physiologists examine the function of various organ systems, and they typically focus on the molecular or cellular level. Thus, a basic knowledge of both chemistry and cells is essential in understanding physiology, and that's why we've included several early chapters on these topics. Mastery of these early chapters on chemistry and cells is critical to understanding the physiologic concepts that are covered throughout the text.

The discipline of physiology parallels anatomy because it also is very broad and may be subdivided into smaller groups. Many specific physiology subdisciplines focus their studies on a particular body system. For example, **cardiovascular physiology** examines the functioning of the heart, blood vessels, and blood. Cardiovascular physiologists examine how the heart pumps the blood, what are the parameters for healthy blood pressure, and details of the cellular exchange mechanisms by which respiratory gases, nutrients, and wastes move between blood and body structures. Other examples include **neurophysiology** (which examines how nerve impulses are propagated throughout the nervous system), respiratory physiology (which studies how respiratory gases are transferred by gas exchange between the lungs and the blood vessels), and **reproductive physiology** (which explores how the regulation of reproductive hormones can drive the reproductive cycle and influence sex cell production and maturation).

Pathophysiology investigates the relationship between the functioning of an organ system and disease or injury to that organ system. For example, a pathophysiologist would examine how blood pressure, contractile force of the heart, and both gas and nutrient exchange may be affected in an individual afflicted with heart disease.

WHAT DID YOU LEARN?

2 What is the relationship between anatomy and physiology?

3 __________ physiology examines how the heart, blood vessels, and blood function.

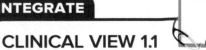

INTEGRATE

CLINICAL VIEW 1.1

Etiology (Causes) and Pathogenesis (Development) of Disease

All health-care professionals must understand both how body structures function normally and how disease or injury can affect them. Throughout the chapters in this book, Clinical View boxes (which are always enclosed in the color *blue*) provide you with selected pathologies and how these pathologies affect the anatomy and physiology of those structures.

1.2 Anatomy and Physiology Integrated

LEARNING OBJECTIVE

5. Explain how the studies of form and function are interrelated.

The sciences of anatomy and physiology are intertwined; one must have some understanding of anatomic form to study physiologic function of a structure. Likewise, one cannot adequately describe and understand the anatomic form of an organ without learning that organ's function. This interdependence of the study of anatomy and physiology reflects the inherent and important interrelationship of how the structure and form of a component of the body determine how it functions. This concept is central to mastering the study of anatomy and physiology.

Integrating the disciplines of anatomy and physiology, rather than trying to separate discussion of form and function, is the most effective way to learn about both fields. Anatomists and physiologists may be describing the organs slightly differently, but both disciplines must use information from the other field for a full understanding of the organ system. You cannot fully understand *how* the small intestine propels food and digests or absorbs nutrients unless you also know about the *structure* of the small intestine wall. **Figure 1.1** visually compares how anatomists and physiologists examine the human body, using the small intestine as an example. Note that anatomists (left side of the figure) tend to focus on the form and structure, whereas physiologists (right side of figure) focus on the mechanisms and functions of these structures. However, both anatomists and physiologists understand that the form and function of structures are interrelated. Throughout this text, we integrate these disciplines so you can more easily see that anatomic form and physiologic function are inseparable.

Note that figure 1.1 is an example of a central feature of this text called **Concept Overview (COV) figures.** These specialized illustrations are included in each chapter (e.g., figures 4.19 and 23.31) and are designed to help you to visually connect and integrate content that has been previously discussed within the chapter.

WHAT DID YOU LEARN?

4 Compare and contrast how anatomists and physiologists specifically describe the small intestine.

1.3 How to Study Anatomy and Physiology Effectively

LEARNING OBJECTIVE

6. Describe best practices for studying anatomy and physiology effectively.

Anatomy and Physiology (A&P) is a content dense course that sometimes may overwhelm learners new to the subject. Success in the course requires careful time management and appropriate study skills for comprehending the material. When we teach our courses, we often encounter students who simply need to adopt more effective study strategies to perform well. In this section, we discuss some of these strategies.

INTEGRATE CONCEPT OVERVIEW

Figure 1.1 **Comparing How Anatomists and Physiologists Examine the Human Body.**

(*a*) Anatomists typically focus on the *form* and *structure* of an organ, such as the small intestine. (*b*) Physiologists tend to focus on the *function* of an organ or a system. However, both anatomists and physiologists recognize that form and function are interrelated.

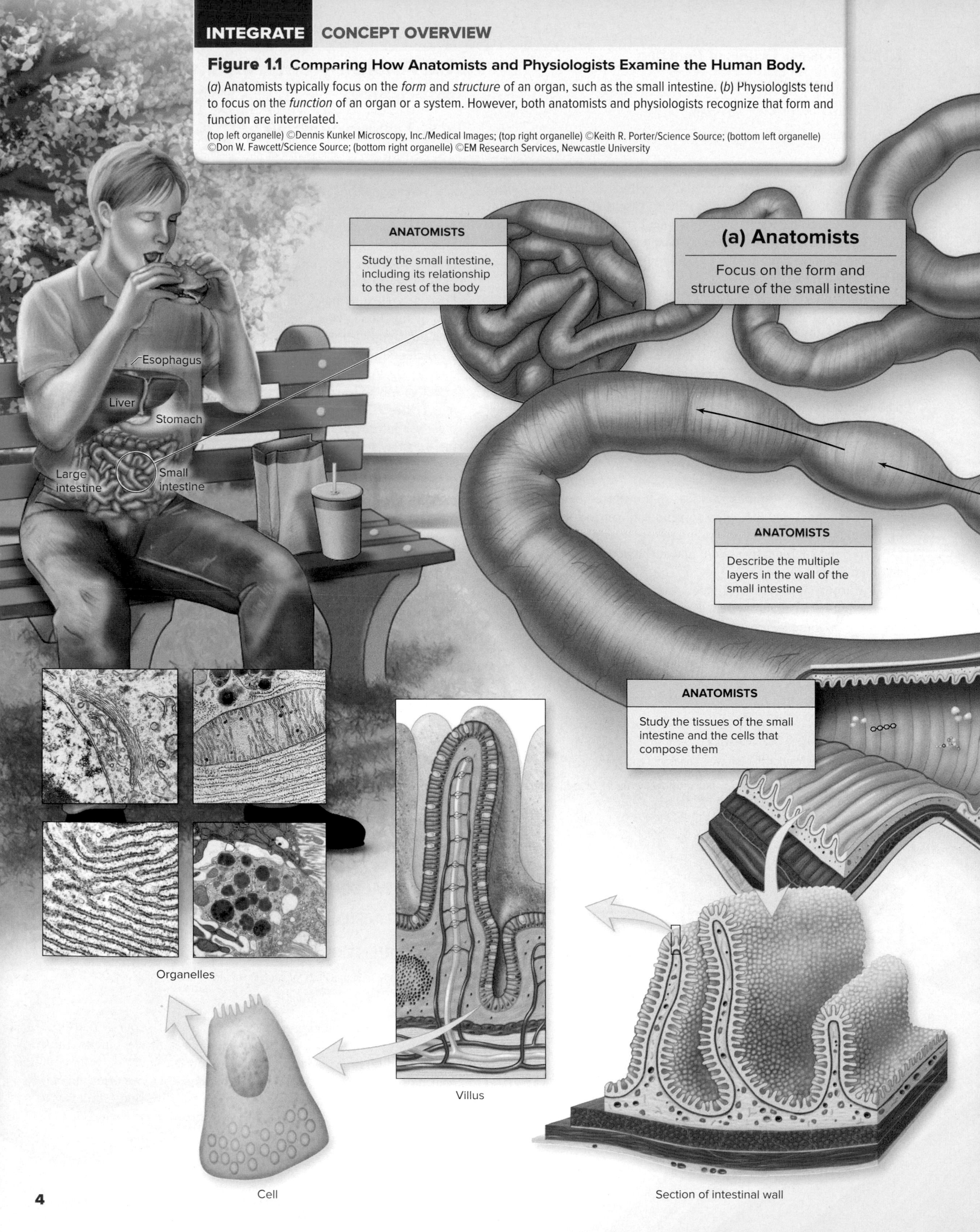

(a) Anatomists
Focus on the form and structure of the small intestine
ANATOMISTS
Study the small intestine, including its relationship to the rest of the body
Esophagus
Liver
Stomach
Large intestine
Small intestine
ANATOMISTS
Describe the multiple layers in the wall of the small intestine
ANATOMISTS
Study the tissues of the small intestine and the cells that compose them
Organelles
Villus
Cell
Section of intestinal wall

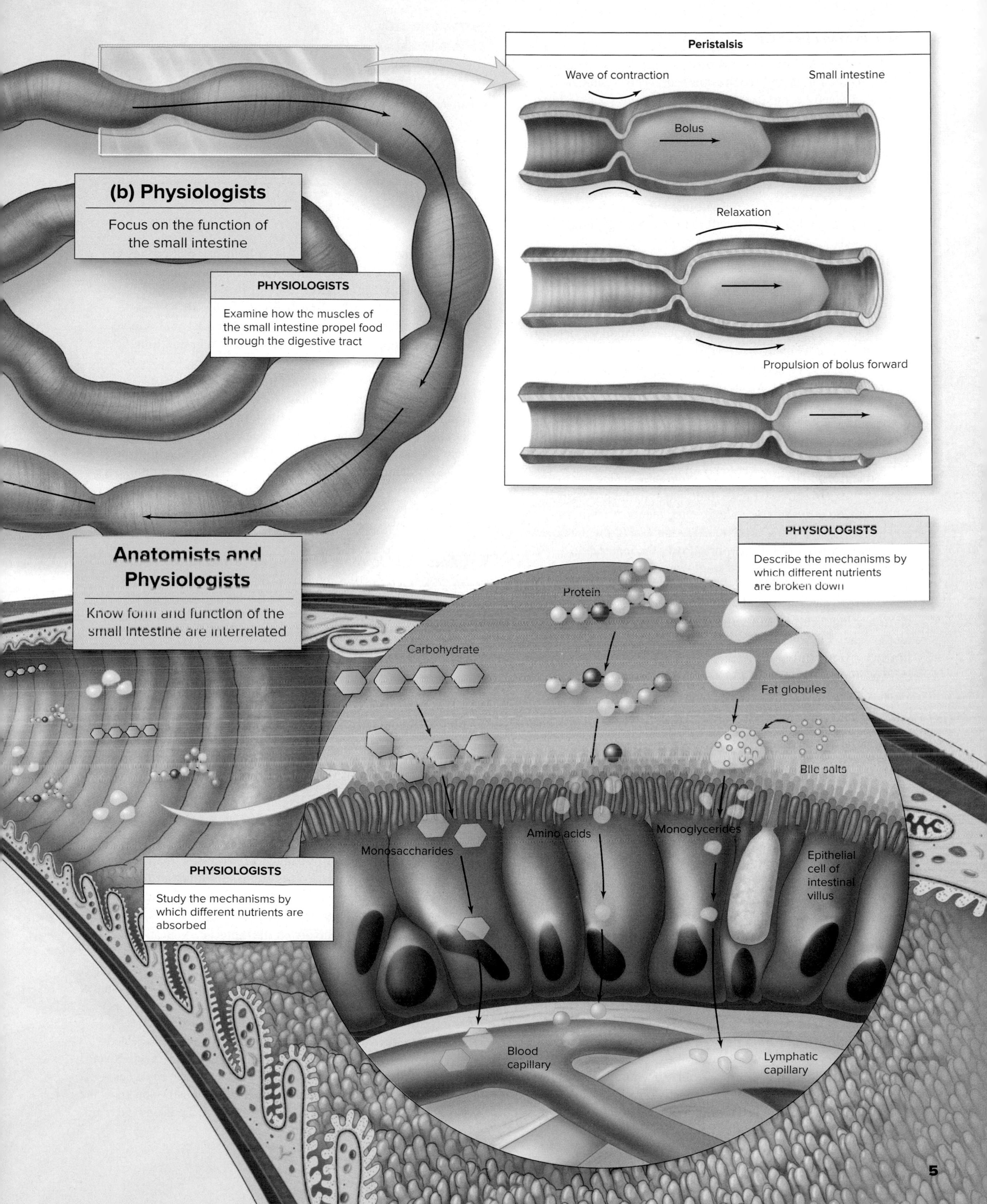

Peristalsis
Wave of contraction
Small intestine
Bolus
Relaxation
Propulsion of bolus forward
(b) Physiologists
Focus on the function of the small intestine
PHYSIOLOGISTS
Examine how the muscles of the small intestine propel food through the digestive tract
Anatomists and Physiologists
Know form and function of the small intestine are interrelated
PHYSIOLOGISTS
Describe the mechanisms by which different nutrients are broken down
Protein
Carbohydrate
Fat globules
Bile salts
Amino acids
Monoglycerides
Monosaccharides
Epithelial cell of intestinal villus
PHYSIOLOGISTS
Study the mechanisms by which different nutrients are absorbed
Blood capillary
Lymphatic capillary

How *NOT* to Study for A&P

1. **Wait until the last minute to study.** As previously mentioned, A&P is content rich and requires the learner to be able to understand many complex processes. Beginning your studying a few days before an exam is simply *not* enough time for you to understand the material and truly learn it.
2. **Study for long periods of time without breaks.** Your brain works best if you study for shorter periods of time (½ hour or less) and then take a short break before studying again. A 4-hour marathon study session will just leave you feeling overwhelmed, and you likely will not remember anything you studied.
3. **Study with multiple distractions.** Do you try to study with the TV on, your phone available to answer texts, and your computer open to your social media account? If so, the time you think you are spending studying is not effective. For each time you take a break to answer a text, check email, or listen to TV, you are not focusing on the material. Multitasking is a myth—in reality, you are quickly switching from one task to another without staying focused on any one thing. This type of study method is disjointed and will prevent you from engaging in the material.
4. **Simply passively read over your notes.** Don't simply read over your notes multiple times as a form of studying. This study method is referred to as *passive learning*—it is called *passive* because the person doesn't have to do much in the process! Although you may *think* you are learning the material, in fact, you are only acquiring a superficial recognition of the material. Unless you practice quizzing yourself over the material to repeatedly *retrieve* the material from your memory and do other active learning methods, your brain will not be able to quickly access what you've learned for an exam. Students who rely solely on re-reading their notes often will say, "I recognized the material on the exam, but I wasn't sure of what answer to choose."
5. **Study by yourself only.** When you study by yourself only, you can't accurately gauge if you know the material well and can explain it to others. You also are more likely to reinforce a misconception if you don't have a study partner who can help you work through some of the more difficult concepts.

So now that we've discussed some of the big mistakes in studying A&P, what are more effective ways of studying? The following is a list of best practices for studying.

Best Practices for Studying A&P

1. **Schedule regular daily study sessions well before the upcoming exam.** Your studying should begin the first week of class and should be a part of a daily or every-other-day schedule. Do not wait until the week prior to an exam to first become acquainted with the material! The night after a lecture or lab, review the material you've learned with some of the methods outlined in this list. Connect the material you are learning with A&P material previously covered. If you follow this plan, then you may spend the week prior to the exam reviewing material you've already studied, rather than starting your study process.
2. **Study for multiple, *short* periods of time.** During these daily (or every-other-day) study periods, set a timer for ½ hour or a little less and promise yourself you will focus just on the A&P material at hand. Select a study topic that you can review effectively in that ½ hour. For example, you could compare and contrast the epidermis and the dermis of the skin during that time. After ½ hour has passed, reward yourself with a short (~5-minute) break, and then reset the timer to study again. After three of these short periods, reward yourself with a longer break. You will be able to review more material, and *remember* the material you've reviewed, better than if you tried to study in one long 4-hour block.
3. **Minimize your distractions.** Put away the phone, turn off the TV, and shut down your email. Research has shown that people don't multitask—rather, the brain jumps from one task to another quickly, so the activity for each task is disjointed and may not be well organized. You will be amazed at how much more efficient your studying becomes when you minimize the distractions and focus on the material. If you use the timer technique mentioned previously (study for ½ hour with no distractions), you can reward yourself during those short breaks by looking at your texts or social media.
4. **Utilize *active learning* methods when you study.** *Active learning* is defined as a process by which you are engaged in the material, problem solving, and applying what you've learned to previous knowledge. It is the opposite of passive learning. Examples of active learning include
   a. **Make your own tables to organize material.** Take your lecture notes and reorganize them into tabular form. For example, you can group muscles of similar functions. The act of writing out the muscles and reorganizing the information in tabular form will help you remember the material better than if you just read over your notes.
   b. **Draw and label anatomic structures.** Make your own sketches of organs and tissues, and label the key features. When you draw, you are integrating multiple pieces of information into one diagram. You don't have to be an artist and the drawing doesn't have to be pretty—rather, it simply has to make sense to you.
   c. **Make flowcharts of physiological processes.** Map out the pathway that filtrate becomes urine in the kidney. Create a flowchart to illustrate how blood travels from the heart to the lungs, and back to the heart.
   d. **Quiz yourself repeatedly on the material.** Educational research has shown that long-term learning is most likely to occur when an individual practices and retrieves that material on multiple occasions. Your textbook has multiple opportunities to quiz yourself—you can use the end-of-chapter questions, LearnSmart modules associated with the e-text, and the quizzing feature in the Anatomy and Physiology | Revealed program associated with the McGraw-Hill Connect site. If you are studying with a partner, take turns quizzing each other. When you can retrieve the information accurately, you know the material. You do not want to wait until you are taking the exam to determine if you can do this.

INTEGRATE

LEARNING STRATEGY

Learning Strategy boxes like this one (which are always enclosed in the color *green*) provide you with helpful analogies, memory aids, and other study tips to help you better understand and learn the material. Look for these boxes throughout each chapter.

INTEGRATE

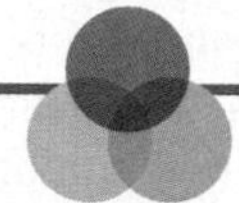

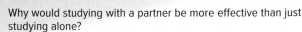

CONCEPT CONNECTION

Throughout future chapters, **Concept Connection** boxes like this one (which are always enclosed in the color *orange*) will highlight how various organ systems do not work in isolation, but rather are interconnected to carry out overlapping functions. For example, the cardiovascular system and respiratory system work together in the transport of respiratory gases (oxygen and carbon dioxide) by the blood throughout the body.

e. **Explain/teach a concept to a partner.** There is a saying that when one person teaches another, both learn. Your teachers have reinforced their A&P knowledge by teaching students year after year. As you are learning new concepts, meet with a study partner and explain that concept to him or her in your own words. The act of explaining the concept and answering your partner's questions will help *you* solidify your knowledge. Utilize the Concept Overview (COV) figures (e.g., figure 1.1) in the textbook to explain concepts to someone else.

5. **Study with a partner or group.** A lot of the active study methods mentioned work best when you are studying with a partner. It is difficult to quiz yourself and know for sure if you truly understand a concept. You and your study partner can each help determine where gaps in knowledge are, keep study sessions focused and on track, and serve as a "sounding board" when trying to explain a concept.

6. **Utilize *all* of the resources your textbook has to offer.** Your textbook and its accompanying digital platform contain numerous resources to help you learn anatomy and physiology more efficiently. So don't just read the text—use the following aids provided in each chapter of the text:

   a. **Integrate: Learning Strategy boxes.** These boxes provide analogies, mnemonics, and study tips to help you learn the material.

   b. **Integrate: Concept Connection boxes.** These boxes provide summaries of topics that may be discussed and presented across multiple chapters, such as acid-base balance or hormonal regulation of growth. Read these boxes to help you connect material among different chapters.

   c. **Integrate: Concept Overview figures.** Each chapter has one or more of these figures, designed to provide a big-picture summary of a major concept in that chapter. For example, Figure 1.1 provides a comparison of how anatomists and physiologists study the body. Let these figures guide your explanation of a concept to a study partner.

   d. **Integrated, multiple assessments in each chapter.** As you read, write out your answers for the What Did You Learn? questions at the end of each section of text. When you are done reading a chapter, use the end-of-chapter questions to test your knowledge.

   e. **LearnSmart.** Each chapter is associated with an interactive e-module that allows you to test yourself on concepts you have read. The program will highlight topics you have not yet mastered and create a study plan for you about these topics.

   f. **Anatomy and Physiology | REVEALED (APR).** APR is an interactive cadaver dissection tool that allows you to highlight anatomic features and review lab and lecture concepts. You can view both gross anatomy and histology concepts, watch animations about particular physiologic processes, and test yourself with the lab quiz tool.

This list of best practices is not exhaustive; you may have some additional study strategies that are equally effective. Although we cannot guarantee you will earn an A, we *can* be reasonably certain that your understanding of anatomy and physiology will greatly increase if you adopt the best practices outlined here. We encourage you to use these best practices for your other courses as well.

WHAT DID YOU LEARN?

5 Why would studying with a partner be more effective than just studying alone?

1.4 The Body's Levels of Organization

Scientists group the body's components into an organizational hierarchy of form and function. In thinking about these levels, it is helpful to know the characteristics common to living things and how each organizational level supports these characteristics. For example, the organ system concept allows functions to be considered as an interaction between many organs.

1.4a Characteristics That Describe Living Things

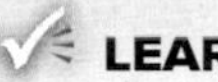

LEARNING OBJECTIVE

7. List the characteristics common to all living things.

Several distinctive properties are common to all organisms, including humans:

- **Organization.** All organisms exhibit a complex structure and order. In section 1.4b, we describe the increasingly complex levels of organization of the human body.
- **Metabolism.** All organisms engage in **metabolism** (mĕ-tab′ō-lizm; *metabole* = change), which is defined as the sum of all of the chemical reactions that occur within the body. Metabolism consists of both **anabolism** (ă-nab′ō-lizm; *anabole* = a raising up), in which small molecules are joined to form larger molecules, and **catabolism** (kă-tab′ō-lizm; *katabole* = a casting down), in which large molecules are broken down into smaller molecules. An example of a metabolic reaction is the use of cellular energy (called ATP; see section 2.7d) for muscle contraction (see section 10.3). The concepts of chemical reactions and metabolism are discussed in sections 3.2a and 3.2b, respectively.

WHAT DO YOU THINK?

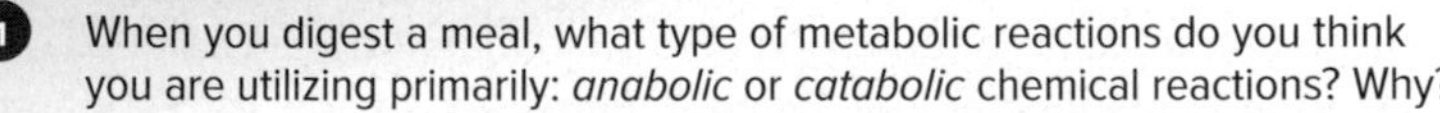

1 When you digest a meal, what type of metabolic reactions do you think you are utilizing primarily: *anabolic* or *catabolic* chemical reactions? Why?

- **Growth and development** During their lifetime, organisms assimilate materials from their environment and often exhibit increased size (growth) and increased specialization as related to form and function (development). As the human body grows and develops, structures such as the brain become more complex and elaborately integrated.

- **Responsiveness.** All organisms exhibit **responsiveness,** which is the ability to detect and react to **stimuli** (changes in the external or internal environment). A stimulus to the skin of the hands, such as an extremely hot temperature, causes the human to withdraw the hand from the stimulus so as to prevent injury or damage. Responsiveness occurs at almost all levels of organization.
- **Regulation.** An organism must be able to adjust internal bodily function in response to environmental changes. When body temperature rises, more blood is circulated near the body's surface to facilitate heat loss, and thus return body temperature to within the normal range. (The process of maintaining body structures and function is called homeostasis, which is discussed in greater depth in section 1.6.)
- **Reproduction.** All organisms produce new cells for growth, maintenance, and repair. The somatic (body) cells divide by a process called mitosis (see section 4.9), whereas sex cells (called gametes) are produced by another type of cell division called meiosis (see section 28.2). The sex cells, under the right conditions, have the ability to develop into a new living organism.

WHAT DID YOU LEARN?

6 What does it mean if an organism is "responsive," and how does this characteristic relate to the survival of this organism?

1.4b The View from Simplest to Most Complex

LEARNING OBJECTIVE

8. Describe the levels of organization in the human body.

Anatomists and physiologists recognize several levels of increasingly complex organization in humans, as illustrated in **figure 1.2**. These levels, from simplest to most complex, are the chemical level, cellular level, tissue level, organ level, organ system level, and organismal level.

The **chemical level** is the simplest level, and it involves atoms and molecules. **Atoms** are the smallest units of matter that exhibit the characteristics of an element, such as carbon and hydrogen. When two or more atoms combine, they form a **molecule.** Examples of molecules include a sugar, a water molecule, or a vitamin. More complex molecules are called **macromolecules** and include some proteins and the deoxyribonucleic acid (DNA) molecules. Macromolecules form specialized microscopic subunits in cells, called **organelles.** Chemical structures are described in chapter 2.

The **cellular level** consists of **cells,** which are the smallest living structures and serve as the basic units of structure and function in organisms. Cells and their components are formed from the atoms and molecules from the chemical level. The structures of cells vary widely, reflecting the specializations needed for their different functions. For example, a skeletal muscle cell may be very long and contain numerous organized protein filaments that aid in muscle contraction, whereas a red blood cell is small and has a flattened disc shape that facilitates the quick and effective ex-change of respiratory gases. Cells and cellular organelles are discussed in chapter 4.

The **tissue level** consists of **tissues,** which are groups of similar cells that perform common functions. There are four major types of tissues. Epithelial tissue covers exposed surfaces and lines body cavities. Connective tissue protects, supports, and binds structures and organs. Muscle tissue produces movement. Finally, nervous tissue conducts nerve impulses for communication.

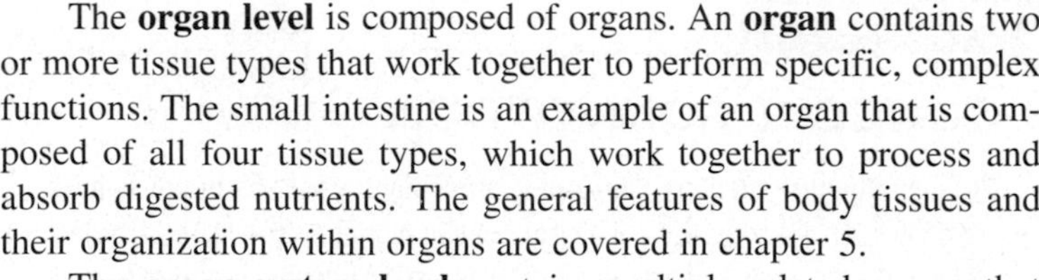

The **organ level** is composed of organs. An **organ** contains two or more tissue types that work together to perform specific, complex functions. The small intestine is an example of an organ that is composed of all four tissue types, which work together to process and absorb digested nutrients. The general features of body tissues and their organization within organs are covered in chapter 5.

The **organ system level** contains multiple related organs that work together to coordinate activities and achieve a common function. For example, the organs of the digestive system (e.g., oral cavity, stomach, small and large intestine, and liver) work together to digest food particles, absorb nutrients, and expel the waste products. The 11 organ systems are introduced in section 1.4c.

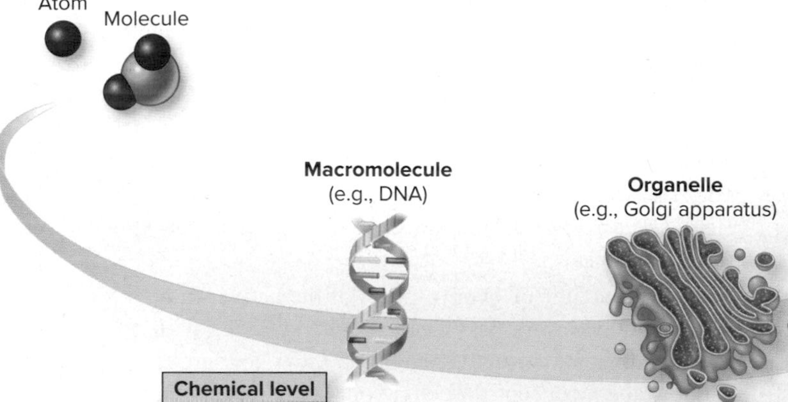

The highest level of structural organization in the body is the **organismal level.** All body systems function interdependently in an **organism,** which is the living person.

WHAT DID YOU LEARN?

7 Does a higher level of organization contain all the levels beneath it? Explain.

1.4c Introduction to Organ Systems

LEARNING OBJECTIVE

9. Compare the organ systems of the human body.

All organisms must exchange nutrients, wastes, and gases with their environment to remain alive and healthy. Simple organisms (e.g., bacteria) may exchange these substances directly across their surface cell boundaries. In contrast, complex, multicellular organisms require sophisticated organ systems with specialized structures and functions to perform the many activities required for the routine events of life. In humans, 11 **organ systems** are commonly denoted, each composed of interrelated organs that work in concert to perform specific functions **(figure 1.3)**. A person maintains a healthy body through the intricate interworkings of all of its organ systems. Subsequent chapters examine each of these organ systems in detail.

WHAT DID YOU LEARN?

8 Which organ system is responsible for filtering the blood and removing the waste products of the blood in the form of urine?

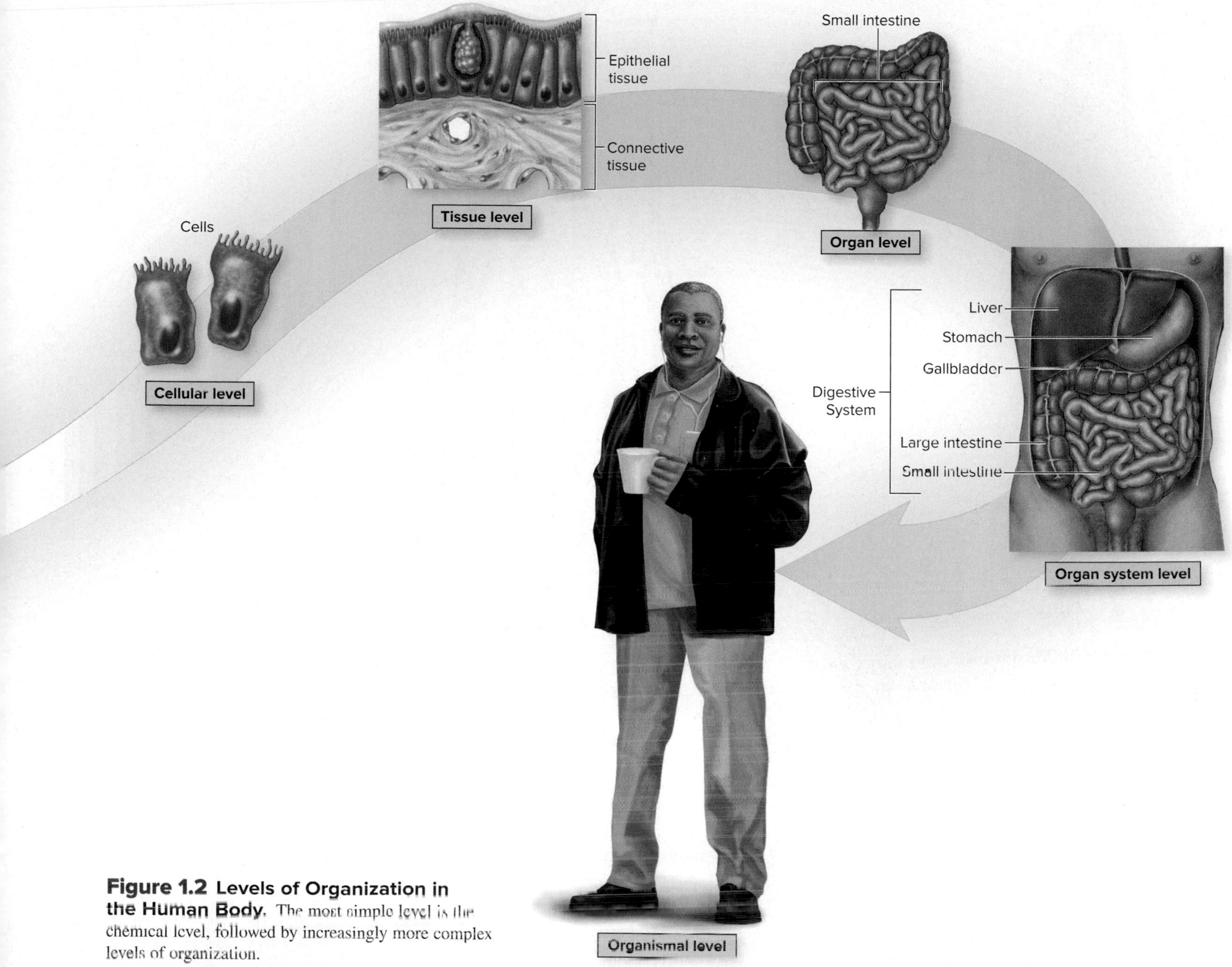

Figure 1.2 Levels of Organization in the Human Body. The most simple level is the chemical level, followed by increasingly more complex levels of organization.

1.5 The Precise Language of Anatomy and Physiology

Clinicians and researchers in anatomy and physiology require a precise language to ensure that they are all discussing the same features and functions. A technical terminology has been developed that describes body position, direction, regions, and body cavities. These technical terms are different from those used in everyday conversation, because the more conversational terms often do not accurately describe location and position or identify structures. For example, the term *arm* in everyday conversation refers to the entire upper limb, but in anatomy the specific portions of the upper limb are named, and the term *arm* or *brachium* refers only to that part of the upper limb between the shoulder and the elbow.

Most anatomic and physiologic terms are derived from Greek or Latin, and we frequently provide word origins, pronunciations, and definitions of terms where appropriate throughout this text. We've used *Stedman's Medical Dictionary* (which defines all medical terms) and *Terminologia Anatomica* (which lists and categorizes the modern, proper anatomic terms) as references. If you actively practice the vocabulary and descriptive terminology presented here, your understanding and appreciation of body structure and function will be enhanced significantly.

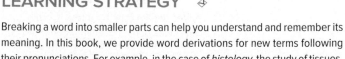

INTEGRATE

LEARNING STRATEGY

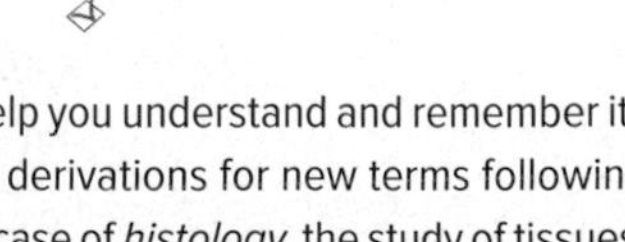

Breaking a word into smaller parts can help you understand and remember its meaning. In this book, we provide word derivations for new terms following their pronunciations. For example, in the case of *histology,* the study of tissues, we provide the following: (*histos* = web, tissue, *logos* = study).

Many biological terms share some of the same prefixes, suffixes, and word roots, so learning the meanings of these common terms can help you figure out the meanings of unfamiliar terms.

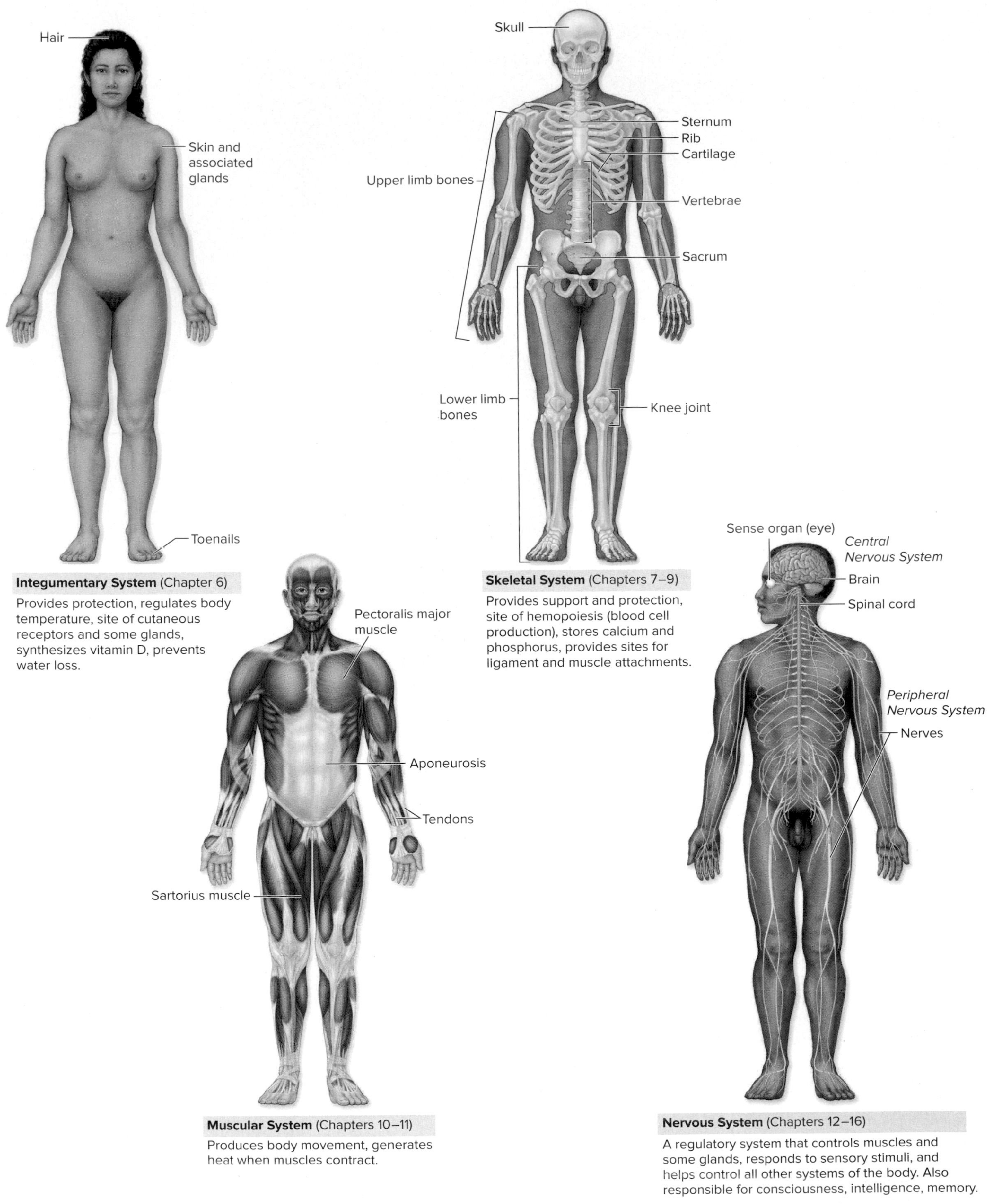
Hair
Skin and associated glands
Toenails
Integumentary System (Chapter 6)
Provides protection, regulates body temperature, site of cutaneous receptors and some glands, synthesizes vitamin D, prevents water loss.
Skull
Sternum
Rib
Cartilage
Upper limb bones
Vertebrae
Sacrum
Lower limb bones
Knee joint
Skeletal System (Chapters 7–9)
Provides support and protection, site of hemopoiesis (blood cell production), stores calcium and phosphorus, provides sites for ligament and muscle attachments.
Pectoralis major muscle
Aponeurosis
Tendons
Sartorius muscle
Muscular System (Chapters 10–11)
Produces body movement, generates heat when muscles contract.
Sense organ (eye)
Central Nervous System
Brain
Spinal cord
Peripheral Nervous System
Nerves
Nervous System (Chapters 12–16)
A regulatory system that controls muscles and some glands, responds to sensory stimuli, and helps control all other systems of the body. Also responsible for consciousness, intelligence, memory.

Figure 1.3 Organ Systems. Major components and characteristics of the 11 organ systems of the human body are presented. AP|R

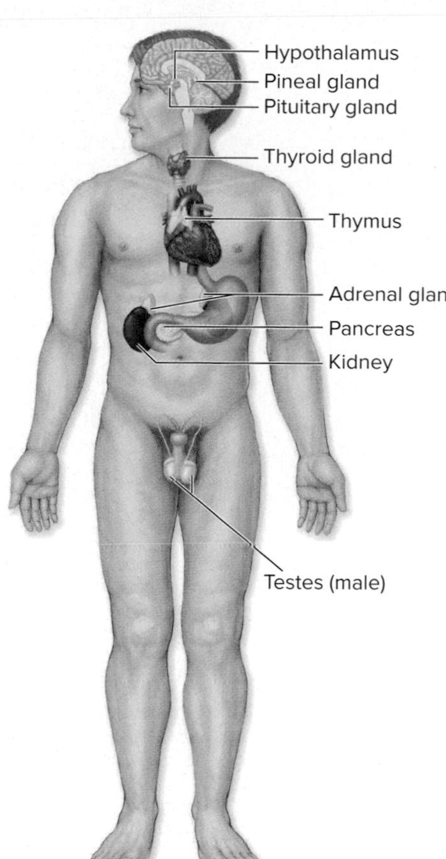

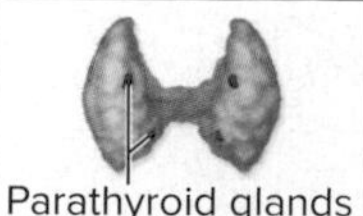

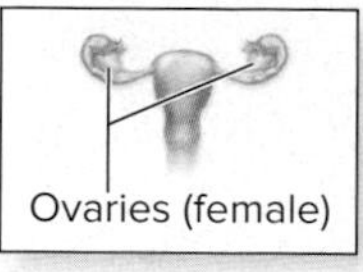

Endocrine System (Chapter 17)

Consists of glands and cell clusters that secrete hormones, (some of which regulate development, growth and metabolism); maintain homeostasis of blood composition and volume, control digestive processes, and control reproductive functions.

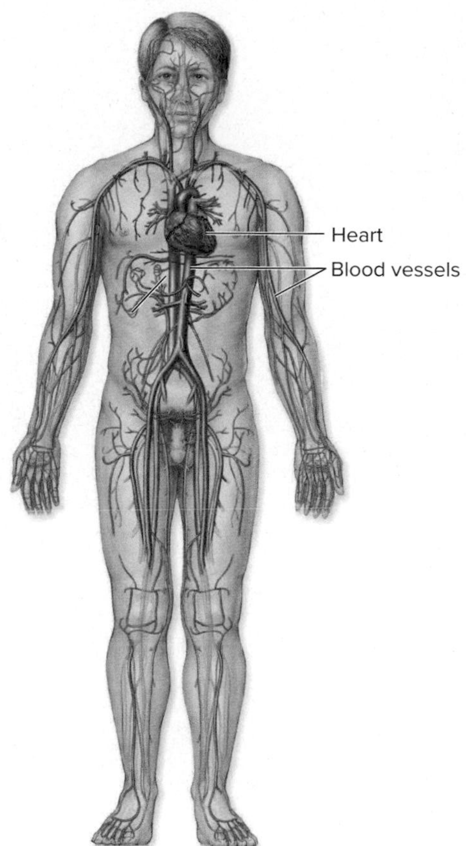

Cardiovascular System (Chapters 18–20)

Consists of the heart (a pump) and blood vessels; the heart moves blood through blood vessels in order to distribute hormones, nutrients, gases, and pick up waste products.

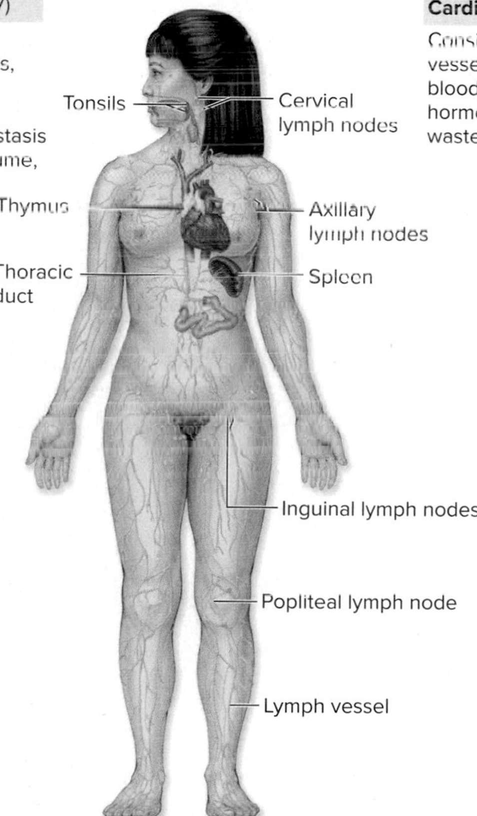

Lymphatic System (Chapters 21–22)

Transports and filters lymph (interstitial fluid that is collected in and transported through lymph vessels) and may participate in an immune response.

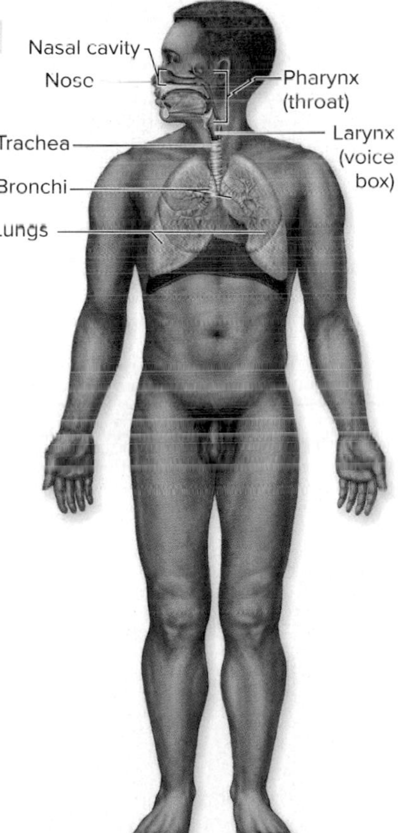

Respiratory System (Chapter 23)

Responsible for exchange of gases (oxygen and carbon dioxide) between blood and the air in the lungs.

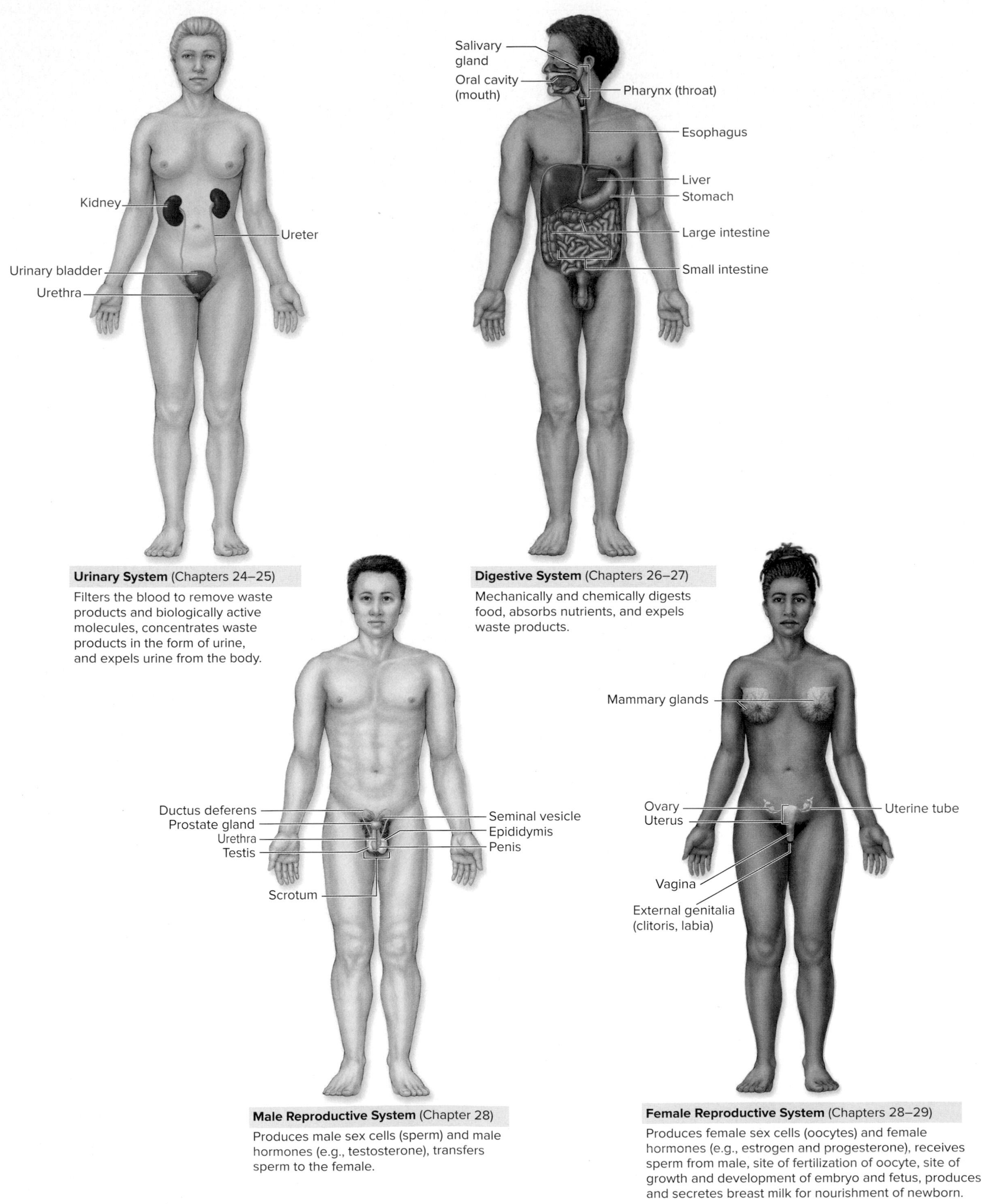
Kidney
Ureter
Urinary bladder
Urethra
Urinary System (Chapters 24–25)
Filters the blood to remove waste products and biologically active molecules, concentrates waste products in the form of urine, and expels urine from the body.
Salivary gland
Oral cavity (mouth)
Pharynx (throat)
Esophagus
Liver
Stomach
Large intestine
Small intestine
Digestive System (Chapters 26–27)
Mechanically and chemically digests food, absorbs nutrients, and expels waste products.
Ductus deferens
Prostate gland
Urethra
Testis
Scrotum
Seminal vesicle
Epididymis
Penis
Male Reproductive System (Chapter 28)
Produces male sex cells (sperm) and male hormones (e.g., testosterone), transfers sperm to the female.
Mammary glands
Ovary
Uterus
Uterine tube
Vagina
External genitalia (clitoris, labia)
Female Reproductive System (Chapters 28–29)
Produces female sex cells (oocytes) and female hormones (e.g., estrogen and progesterone), receives sperm from male, site of fertilization of oocyte, site of growth and development of embryo and fetus, produces and secretes breast milk for nourishment of newborn.

Figure 1.3 Organ Systems. ***(continued)*** AP|R

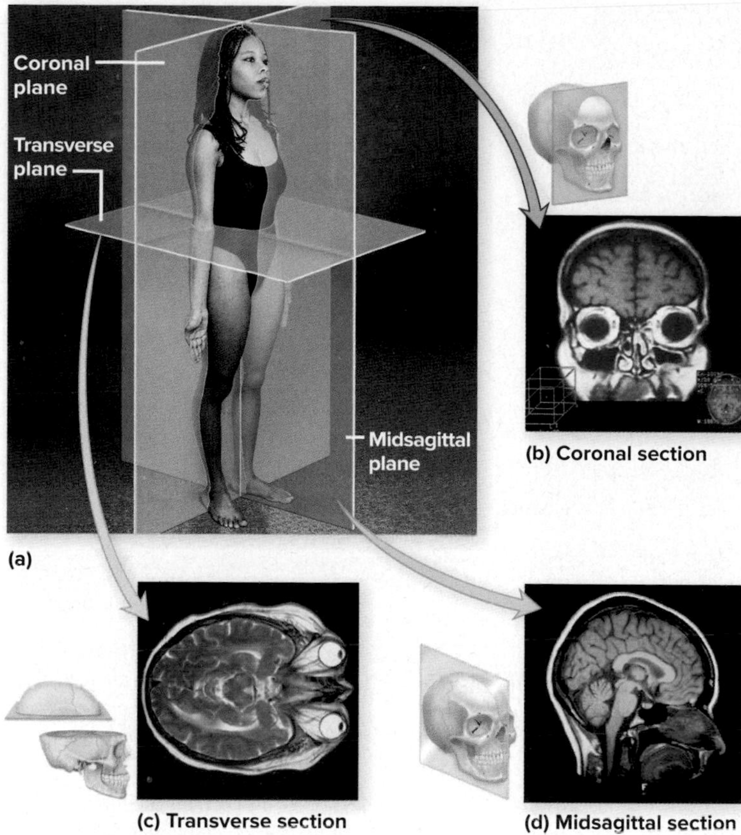

Figure 1.4 Anatomic Position and Body Planes. (*a*) In the anatomic position, the body is upright, and the forearms are positioned so the palms are facing anteriorly. A plane is an imaginary surface that slices the body into specific sections. Sections are shown from each of the three major anatomic planes of reference: (*b*) coronal, (*c*) transverse, and (*d*) midsagittal planes. AP|R

(a) ©McGraw-Hill Education/Joe DeGrandis; (b) ©James Cavallini/Science Source; (c) ©Trevor Lush/Getty Images; (d) ©Stevie Grand/Science Source

1.5a Anatomic Position

LEARNING OBJECTIVE

10. Describe the anatomic position and its importance in the study of anatomy.

Descriptions of any body region or part require a common initial point of reference. Note that terms such as *superior* and *inferior* can be relative terms. For example, when a person is standing it would be accurate to say "the heart is superior to the stomach," yet if that person were in a **supine** (lying down, face upward) position, this statement would seem not to be true. For accuracy and clarity, anatomists and physiologists describe these parts based on the premise that the body is in what is termed the anatomic position, which is then the point of common reference. An individual in the **anatomic position** stands upright with the feet parallel and flat on the floor, the upper limbs are at the sides of the body, and the palms face anteriorly (toward the front); the head is level, and the eyes look forward toward the observer **(figure 1.4)**. All of the anatomic and directional terms used in this book refer to the body in anatomic position.

1.5b Sections and Planes

LEARNING OBJECTIVE

11. Describe the anatomic sections and planes through the body.

Anatomists and physiologists refer to real or imaginary "slices" of the body, called sections or planes, to examine the internal anatomy and describe the position of one body part relative to another. The term **section** implies an actual cut or slice to expose the internal anatomy, whereas the word **plane** implies an imaginary flat surface passing through the body. The three major anatomic planes are the coronal, transverse, and midsagittal planes (figure 1.4).

A **coronal** (kōr′o-năl; *korone* = crown) **plane,** also called a *frontal plane,* is a vertical plane that divides the body or organ into *anterior* (front) and *posterior* (back) parts. When a coronal plane is taken through the trunk, the anterior portion contains the chest and the posterior portion contains the back and buttocks.

A **transverse plane,** also called a *horizontal plane* or *cross-sectional plane,* divides the body or organ into *superior* (top) and *inferior* (bottom) parts. If a transverse plane is taken through the middle of the trunk, the superior portion contains the chest and the inferior portion contains the abdomen.

A **midsagittal** (mid-saj′ĭ-tăl; *sagitta* = arrow) **plane,** or *median plane,* is a vertical plane and divides the body or organ into equal *left* and *right halves.* A midsagittal plane through the head will split it into a left half and a right half (each containing one eye, one ear, and half of the nose and mouth). A plane that is parallel to the midsagittal plane, but either to the left or right of the midsagittal plane, is termed a **sagittal plane.** A sagittal plane divides a structure into left and right portions that are not equal. Although there is only one midsagittal plane, an infinite number of sagittal planes are possible.

In addition to these major planes, there are numerous minor planes called **oblique** (ob-lēk′) **planes** that pass through a structure at an angle **(figure 1.5)**.

Interpreting body sections has become increasingly important for health-care professionals. Technical advances in medical imaging have produced sectional images of internal body structures (figures 1.4*b*-*d*). To determine the shape of any object within a section, we must be able to reconstruct its three-dimensional shape by observing many serial sections.

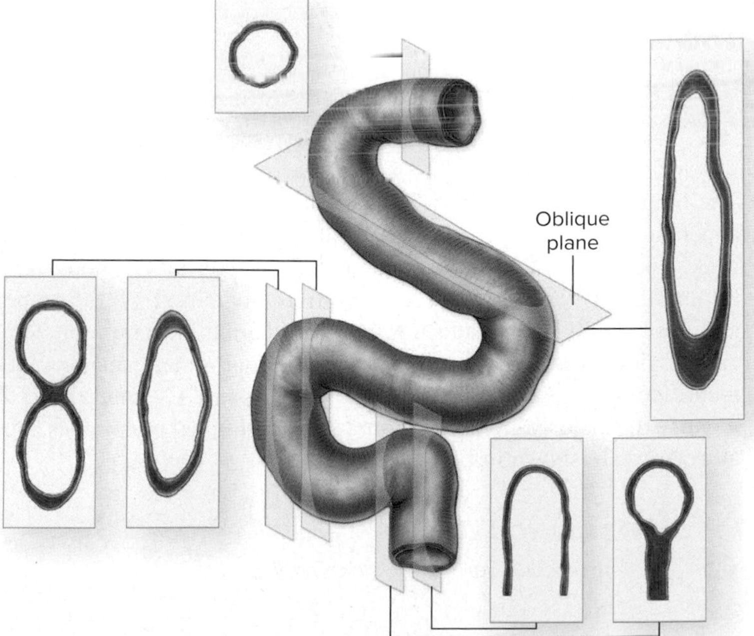

Figure 1.5 Sections from a Three-Dimensional Structure. Serial sections through an object are used to reconstruct its three-dimensional structure, as in these sections of the small intestine. Often a single section, such as the plane at the lower part of this figure, misrepresents the complete structure of the object. An oblique plane is labeled for reference.

Sectioning the body or an organ along different planes often results in very different views of that organ or region. For example, different sections through the abdominal cavity exhibit multiple profiles of the long, twisted tube that is the small intestine. These sections may appear as circles, ovals, a figure eight, or maybe a long tube with parallel sides, depending on where the section was taken (figure 1.5). Being able to convert and interpret two-dimensional images into three-dimensional structures is especially important when comparing and understanding histologic and gross anatomic views of the same organ.

WHAT DID YOU LEARN?

 What type of plane would separate the nose and mouth into superior and inferior structures?

1.5c Anatomic Directions

LEARNING OBJECTIVE

12. Define the different anatomic directional terms.

When the body is in the anatomic position, we can precisely describe the relative positions of structures by using specific directional terms. These directional terms are precise and usually presented in opposing pairs. Examples include **anterior** (in front of) and **posterior** (in back of), and **proximal** (nearer to the trunk) and **distal** (farther from the trunk). **Table 1.1** and **figure 1.6** describe some commonly used directional terms. Studying the table and figure together, and referring back to them as needed, will maximize your understanding of anatomic directions and aid your study of anatomy throughout the rest of this book.

WHAT DID YOU LEARN?

 Which directional term would be most appropriate in the sentence "The elbow is ___________ to the wrist"?

1.5d Regional Anatomy

LEARNING OBJECTIVE

13. Identify the major regions of the body, using proper anatomic terminology.

The human body is partitioned into two main regions, the axial and appendicular regions. The **axial** (ak′sē-ăl) **region** includes the head, neck, and trunk; it forms the main vertical axis of the body. The **appendicular** (ap′en-dik′ū-lăr) **region** is composed of the upper and lower limbs, which attach to the axial region. Several more specific regions are located within these two main ones, and they are identified by proper anatomic terminology. **Figure 1.7** and **table 1.2** identify the major regional terms and some additional minor ones. Not all regions are shown in figure 1.7.

WHAT DID YOU LEARN?

 The term *antebrachial* refers to which body region?

1.5e Body Cavities and Membranes

LEARNING OBJECTIVES

14. Describe the body cavities and their subdivisions.

15. Explain the role of serous membranes in the ventral cavities.

Table 1.1 Anatomic Directional Terms

Direction	Term	Meaning	Example
Relative to front (belly side) or back of the body	Anterior	In front of; toward the front surface	The stomach is *anterior* to the spinal cord.
	Posterior	In back of; toward the back surface	The heart is *posterior* to the sternum.
	Dorsal	Toward the back side of the human body	The spinal cord is on the *dorsal* side of the body.
	Ventral	Toward the belly side of the human body	The umbilicus (navel, belly button) is on the *ventral* side of the body.
Relative to the head or bottom of the body	Superior	Closer to the head	The chest is *superior* to the pelvis.
	Inferior	Closer to the feet	The stomach is *inferior* to the heart.
	Cranial (cephalic)	Toward the head end	The shoulders are *cranial* to the feet.
	Caudal	Toward the rear or tail end	The buttocks are *caudal* to the head.
	Rostral	Toward the nose or mouth	The frontal lobe of the brain is *rostral* to the back of the head.
Relative to the midline or center of the body	Medial	Toward the midline of the body	The lungs are *medial* to the shoulders.
	Lateral	Away from the midline of the body	The arms are *lateral* to the heart.
	Ipsilateral	On the same side	The right arm is *ipsilateral* to the right leg.
	Contralateral	On the opposite side	The right arm is *contralateral* to the left leg.
	Deep	Closer to the inside, internal to another structure	The heart is *deep* to the rib cage.
	Superficial	Closer to the outside, external to another structure	The skin is *superficial* to the biceps brachii muscle.
Relative to point of attachment of appendage	Proximal	Closer to point of attachment to trunk	The elbow is *proximal* to the hand.
	Distal	Farther away from point of attachment to trunk	The wrist is *distal* to the elbow.

Internal organs and organ systems are housed within enclosed spaces, or cavities. These body cavities are named according to either the bones that surround them or the organs they contain. For purposes of discussion, these body cavities are grouped into a posterior aspect and a ventral cavity.

Posterior Aspect

The **posterior aspect** of the body is different from the ventral cavity, in that the posterior aspect contains cavities that are completely encased in bone and are physically and developmentally different

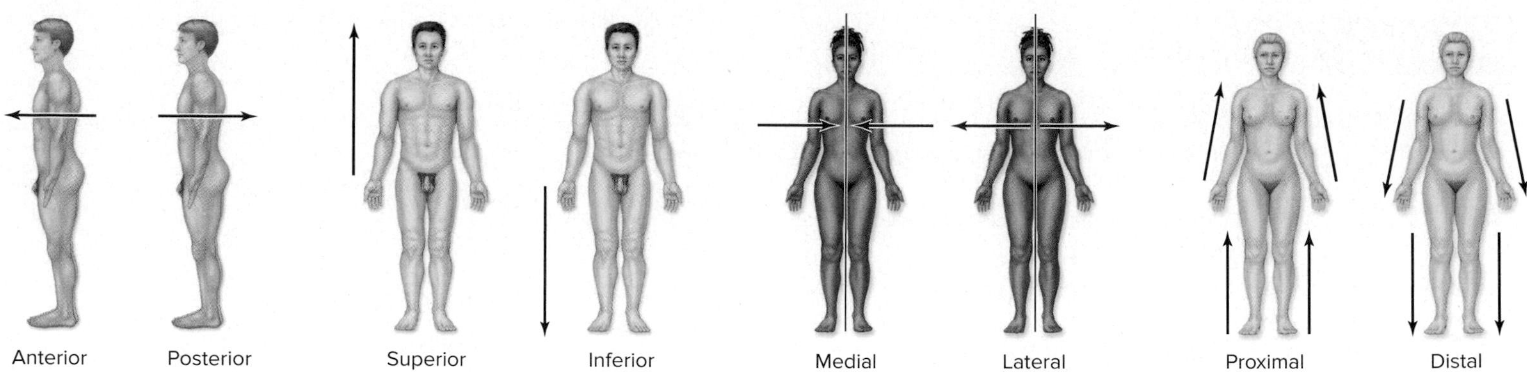

Figure 1.6 Directional Terms in Anatomy. Directional terms precisely describe the location and relative relationships of body parts. (See also table 1.1.) AP|R

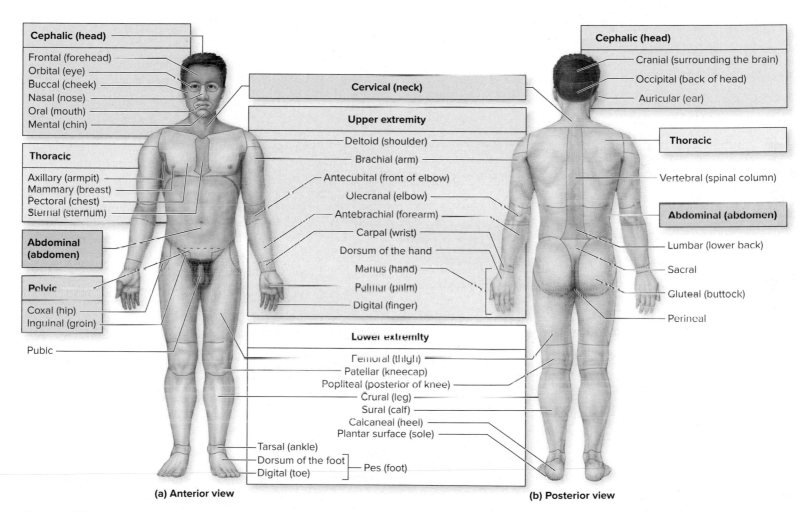

Cephalic (head)
Frontal (forehead)
Orbital (eye)
Buccal (cheek)
Nasal (nose)
Oral (mouth)
Mental (chin)
Thoracic
Axillary (armpit)
Mammary (breast)
Pectoral (chest)
Sternal (sternum)
Abdominal (abdomen)
Pelvic
Coxal (hip)
Inguinal (groin)
Pubic
Cervical (neck)
Upper extremity
Deltoid (shoulder)
Brachial (arm)
Antecubital (front of elbow)
Olecranal (elbow)
Antebrachial (forearm)
Carpal (wrist)
Dorsum of the hand
Manus (hand)
Palmar (palm)
Digital (finger)
Lower extremity
Femoral (thigh)
Patellar (kneecap)
Popliteal (posterior of knee)
Crural (leg)
Sural (calf)
Calcaneal (heel)
Plantar surface (sole)
Tarsal (ankle)
Dorsum of the foot
Digital (toe)
Pes (foot)
Cephalic (head)
Cranial (surrounding the brain)
Occipital (back of head)
Auricular (ear)
Thoracic
Vertebral (spinal column)
Abdominal (abdomen)
Lumbar (lower back)
Sacral
Gluteal (buttock)
Perineal
(a) Anterior view
(b) Posterior view

Figure 1.7 Regional Terms. (*a*) Anterior and (*b*) posterior views show key regions of the body. Their common names appear in parentheses. AP|R

from the ventral cavity. The term *dorsal body cavity* has been used by others to describe this posterior aspect but is not used here because of these differences between the ventral cavity and posterior aspect.

The posterior aspect is subdivided into two enclosed cavities **(figure 1.8*a*)**. A **cranial cavity** is formed by the bones of the cranium, and so it also goes by the name *endocranium.* The cranial cavity houses the brain. The second cavity is the **vertebral** (ver′te-brăl) **canal,** which is formed by the bones of the vertebral column. The vertebral canal houses the spinal cord.

Ventral Cavity

The **ventral cavity** is the larger, anteriorly placed cavity in the body (figure 1.8). Unlike the posterior aspect, the ventral cavity and its subdivisions do not completely encase their organs in bone. The ventral cavity is partitioned by the **thoracic diaphragm** into a superior **thoracic** (thō-ras′ik) **cavity** and an inferior **abdominopelvic** (ab-dom′i-nō-pel′vik) **cavity.**

Another significant difference between the posterior aspect and the ventral cavity is that the subdivisions of the ventral cavity are

Table 1.2	Human Body Regions[1]
Region Name	**Description**
Abdominal	Region inferior to the thorax (chest) and superior to the pelvic brim of the hip bones
Antebrachial	Forearm (the portion of the upper limb between the elbow and the wrist)
Antecubital	Region anterior to the elbow; also known as the cubital region
Auricular	Visible surface structures of the ear
Axillary	Armpit
Brachial	Arm (the portion of the upper limb between the shoulder and the elbow)
Buccal	Cheek
Calcaneal	Heel of the foot
Carpal	Wrist
Cephalic	Head
Cervical	Neck
Coxal	Hip
Cranial	Skull
Crural	Leg (the portion of the lower limb between the knee and the ankle)
Deltoid	Shoulder
Digital	Fingers or toes (also called phalangeal)
Dorsal/ Dorsum	Back
Facial	Face
Femoral	Thigh
Fibular	Lateral aspect of the leg
Frontal	Forehead
Gluteal	Buttock
Hallux	Great toe
Inguinal	Groin (sometimes used to indicate the crease or junction of the thigh with the trunk)
Lumbar	The "small of the back": the inferior part of the back between the ribs and the pelvis
Mammary	Breast
Manus	Hand
Mental	Chin
Nasal	Nose
Occipital	Posterior aspect of the head
Olecranal	Posterior aspect of the elbow
Oral	Mouth
Orbital	Eye
Palmar	Palm (anterior surface) of the hand
Patellar	Kneecap
Pectoral	Chest, includes mammary region
Pelvic	Pelvis; region inferior to the pelvic brim of the hip bones
Perineal	Diamond-shaped region between the thighs that contains the anus and external reproductive organs
Pes	Foot
Plantar	Sole of the foot
Pollex	Thumb
Popliteal	Area posterior to the knee
Pubic	Anterior region of the pelvis
Radial	Lateral aspect (thumb side) of forearm
Sacral	Posterior region between the hip bones
Scapular	Shoulder blade
Sternal	Anterior middle region of the thorax
Sural	Calf (posterior part of the leg)
Tarsal	Proximal part of the foot and ankle
Thoracic	Part of torso superior to thoracic diaphragm; contains the pectoral, axillary, and sternal regions
Tibial	Medial aspect of leg
Ulnar	Medial aspect (pinky side) of the forearm
Umbilical	Navel
Vertebral	Spinal column

1. The word *region* should follow each region name listed in the table (e.g., femoral region).

lined with thin **serous membranes.** (Posterior aspect cavities have no serous membranes.) In this usage, a *membrane* is a continuous layer of cells, as compared to the plasma membrane that surrounds a single cell (see section 4.1c). Serous membranes form two layers: (1) a **parietal** (pă-rī′ĕ-tăl) **layer** that typically lines the internal surface of the body wall and (2) a **visceral** (vis′er-ăl) **layer** that covers the external surface of the organs **(viscera)** within that cavity. Between the parietal and visceral serous membrane layers is a potential space called the **serous cavity.** (Note: A potential space is capable of becoming a larger opening under certain physiological or pathological conditions.) Serous membranes secrete a liquid called **serous fluid** within a serous cavity. Serous fluid has the consistency of oil and serves as a lubricant. In a living person, organs (e.g., heart, lungs, intestines) move and rub against each other and the body wall. Friction caused by this movement is reduced by the serous

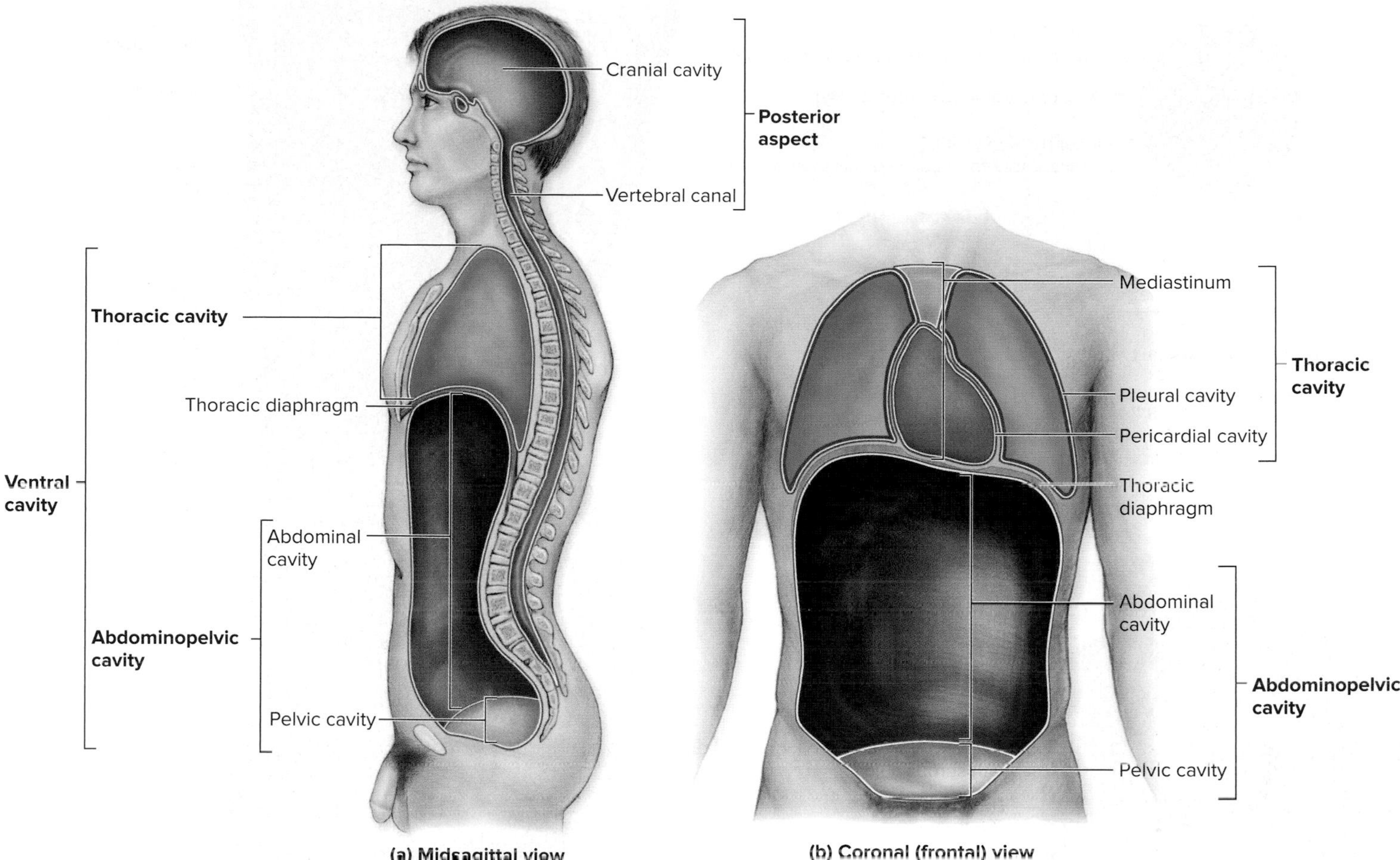

Figure 1.8 Body Cavities. The body is composed of two main spaces: the posterior aspect and the ventral cavity. Many vital organs are housed within these spaces. (*a*) A midsagittal view shows both the posterior aspect and the ventral cavity. (*b*) A coronal view shows the relationship between the thoracic and abdominopelvic cavities within the ventral cavity. AP|R

fluid so the organs move more smoothly against one another and the body walls. Serous membranes will be discussed again in section 5.5b.

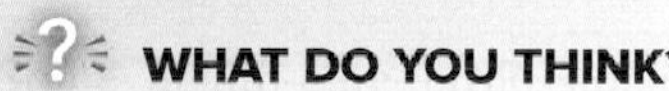

WHAT DO YOU THINK?

What do you think would happen to your body organs if there were no serous fluid between the parietal and visceral layers?

Figure 1.9*a* provides a helpful analogy for visualizing the serous membrane layers. The closed fist is comparable to an organ, and the balloon is comparable to a serous membrane. When a fist is pushed against the wall of the balloon, the inner balloon wall that surrounds the fist is comparable to the visceral layer of the serous membrane. The outer balloon wall is comparable to the parietal layer of the serous membrane. The thin, air-filled space within the balloon, between the two "walls," is comparable to the serous cavity. Note that the organ is not *inside* the serous cavity; it is actually *outside* the cavity and merely covered by the serous membrane.

Thoracic Cavity Within the thoracic cavity, the median space between the lungs is called the **mediastinum** (mē-dē-as-tī′nŭm; *medius* = middle) (figure 1.8*b*). It contains the heart, thymus, esophagus, trachea, and major blood vessels that connect to the heart.

Within the mediastinum, the heart is enclosed by a two-layered serous membrane called the serous **pericardium** (per-ĭ-kar′dē-ŭm; *peri* = around, *kardia* = heart). The **parietal pericardium** is the outermost layer of the serous membrane and forms the sac around the heart, whereas the **visceral pericardium** forms the heart's external surface (figure 1.9*b*). The **pericardial cavity** is the serous cavity between the parietal and visceral layers of the pericardium, and it contains serous fluid. (see section 19.2b).

The right and left sides of the thoracic cavity house the lungs, which are associated with a two-layered serous membrane called the **pleura** (plūr′ă; a rib) (figure 1.9*c*).

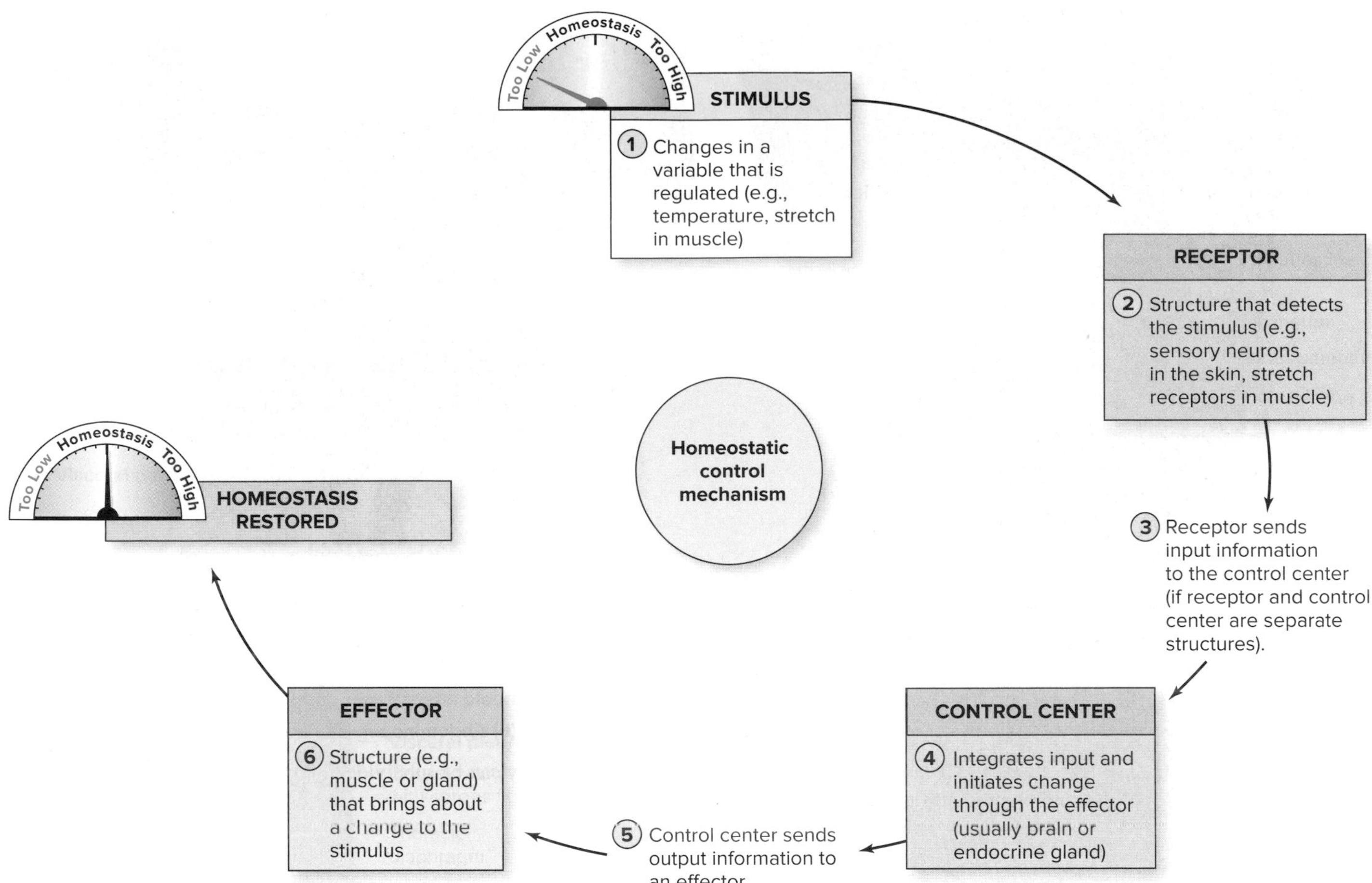

Figure 1.11 Components of a Homeostatic Control Mechanism. A homeostatic control mechanism consists of a receptor (detects a stimulus), a control center (integrates input and initiates change through the effector), and an effector (brings about a change in response to the stimulus).

typically consists of sensory neurons (nerve cells). These neurons may be in the skin, internal organs of the body, or specialized organs such as the eye, ear, tongue, or nose. A **stimulus** is a change in the variable (a physical or chemical factor), such as a change in light, temperature, chemicals (e.g., glucose or oxygen levels), or stretch in muscle. Thus, a receptor is the structure that detects a stimulus. For example, the retina of the eye (receptor) detects a change in light (stimulus) entering the eye.

Control Center

The **control center** is the structure that both interprets input from the receptor and initiates changes through the effector. You can think of it as the "go between" for the other two components of a homeostatic system. The control center is generally a portion of the nervous system (brain or spinal cord) or an organ of the endocrine system (e.g., the thyroid gland). A homeostatic system involving the nervous system provides a relatively quick means of responding to change. An example is regulating blood pressure when you rise from bed in the morning. In contrast, the endocrine system usually provides a means of a more sustained response over several hours or days through the release of hormones. An example is when the parathyroid hormone continuously regulates blood calcium levels, a process that is essential for the normal function of both muscles and nerves (see section 17.10b). Note that the control center is sometimes the same structure as the receptor because it both detects the stimulus and causes a response to regulate it. For example, the pancreas acts as a receptor because it detects an increase in blood glucose and acts as a control center because it releases the hormone insulin in response (see section 17.9b).

Effector

The **effector** is the structure that brings about the change to alter the stimulus (i.e., the effector causes an "effect"). Most body structures can serve as effectors, although muscles and exocrine glands (see section 5.1d) are often the effectors. For example, smooth muscle in the walls of air passageways (bronchioles) regulates airflow into and out of the lungs. Salivary glands increase their release of saliva to moisten the mouth.

As you view figure 1.11, notice that the response of a homeostatic system occurs through a feedback loop that includes the following:

- A *stimulus,* which is the change in the variable
- A receptor that detects the stimulus
- The *control center,* which both integrates input information from the receptor and initiates output to the effectors
- The *effectors* that cause the change (or effect)
- *Homeostasis restored* as a result of the changes from the effectors

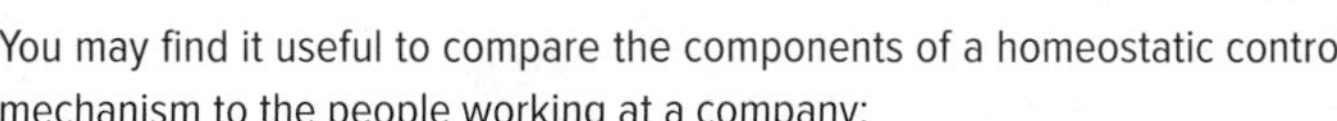

LEARNING STRATEGY

You may find it useful to compare the components of a homeostatic control mechanism to the people working at a company:

- The *receptor* is the worker who first detects a change or problem in workflow and reports to the boss of the company.
- The *control center* is the boss of the company. After receiving information from the receptor, the boss will decide what action needs to be implemented.
- The *effectors* are the workers who receive the boss's plan of action and implement the plan to cause the effect or change.

Homeostatic control systems are separated into two broad categories based on whether the system maintains the variable within a normal range by moving the stimulus in the opposite direction, or amplifies the stimulus in the same direction. These two types of feedback control are called negative feedback and positive feedback, respectively.

WHAT DID YOU LEARN?

 List and describe the three components of a homeostatic system, and give examples of each in the human body.

1.6b Homeostatic Systems Regulated by Negative Feedback

LEARNING OBJECTIVES

19. Define negative feedback.
20. Explain how homeostatic mechanisms regulated by negative feedback detect and respond to environmental changes.

Most processes in the body are controlled by negative feedback. If a homeostatic system is controlled by **negative feedback,** the resulting action will always be in the *opposite* direction of the stimulus. In this way, the variable is maintained within a normal level, or what is called its **set point.**

How a variable that is regulated by negative feedback fluctuates over time can be viewed in **figure 1.12**. Notice that the variable does not remain constant over time but instead fluctuates, and its fluctuation occurs around the set point. If the stimulus increases, the homeostatic system is activated to cause a decrease in the stimulus until it returns to the set point. In contrast, if the stimulus decreases, the homeostatic system causes an increase in the stimulus until it returns to normal. This idea is generally better understood by describing a specific example, such as temperature regulation.

Temperature Regulation

We begin by first explaining how a negative feedback mechanism works to maintain the temperature of your home at a set point of 70°F. On a very cold day, the indoor temperature drops. This drop in temperature is detected by the thermostat. The drop in temperature is relayed through the electrical wiring of your home to the heat pump, which then turns on. The heat pump continues to heat your home until the thermostat reaches 70°F. An electrical signal is then sent from the thermostat to shut off the heat pump.

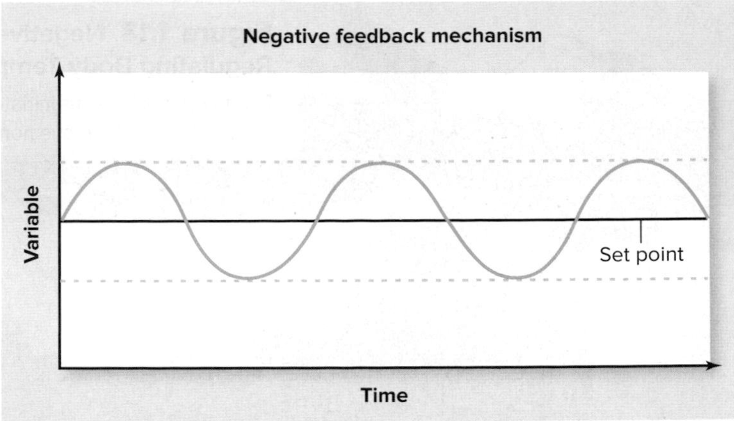

Figure 1.12 Negative Feedback. Note that when a variable is regulated by negative feedback, the variable fluctuates around a set point (rather than being a constant).

Body temperature is regulated in an analogous way to how the temperature of your home is regulated (**figure 1.13*a***). If you venture outside on a cold day, body temperature may begin to drop. This decrease in body temperature is detected by the sensory receptors of the skin, which send nerve impulses to the hypothalamus (a component of the brain; see section 13.4c). (The hypothalamus can also directly detect changes in body temperature by monitoring blood temperature as it passes through this region of the brain.) The hypothalamus compares sensory input to body temperature set point (e.g., 37°C or 98.6°F), and initiates motor output to blood vessels in the skin to decrease the diameter of the inside opening (lumen) of the vessels, thus decreasing the amount of blood circulating to the surface of the body. As a result, less heat is released through the skin. Nerve impulses are also sent to skeletal muscles, which cause shivering, and perhaps to smooth muscle associated with hair follicles of the skin, causing "goose bumps."

In contrast, on a very hot day (figure 1.13*b*), or when you are engaging in strenuous exercise, an increase in body temperature is detected by the sensory receptors of the skin or hypothalamus. The hypothalamus detects the difference between the increased body temperature and the original temperature set point, and initiates motor output to the blood vessels of the skin. This change increases the lumen diameters of blood vessels so that additional blood is brought near the surface of the body for the release of heat through the skin. Nerve impulses are also sent from the hypothalamus to the sweat glands to initiate sweating. Both responses help cool the body by the loss of heat from its surface. In these examples, regulation occurs through the nervous system.

Other examples of homeostatic regulation through the nervous system include the withdrawal reflex in response to injury from stepping on glass or burning your hand (see section 14.6), regulating heart rate and blood pressure when you exercise (see section 20.6a), or changing breathing rate in response to an increase in carbon dioxide levels (see section 23.5).

Recall that the control center may also be an organ of the endocrine system. Examples of homeostatic systems that regulate through the endocrine system include the parathyroid gland release of parathyroid hormone in response to a decrease in blood calcium (see section 7.6b) or pancreas release of insulin in response to an increase in blood glucose (see section 17.9b).

WHAT DID YOU LEARN?

 On a cold day, what are some of the strategies the body uses to conserve heat?

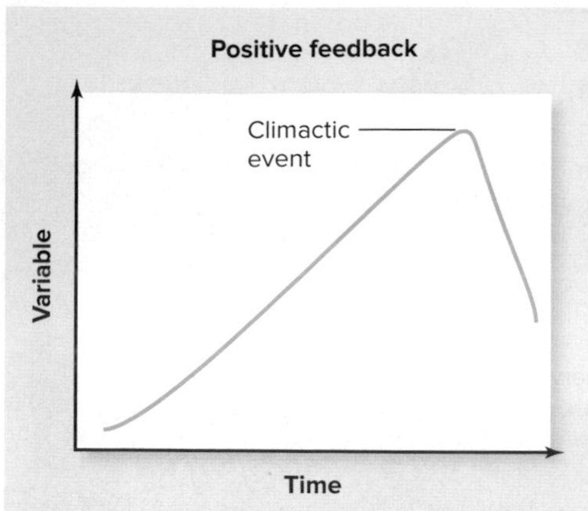

Figure 1.14 Positive Feedback. Positive feedback results in the stimulus being reinforced until a climactic event occurs, and then the body returns to homeostasis.

1.6c Homeostatic Systems Regulated by Positive Feedback

LEARNING OBJECTIVES

21. Define positive feedback.

22. Describe the actions of a positive feedback loop.

A homeostatic system may also be controlled by **positive feedback.** The stimulus here is reinforced to continue in the *same* direction until a climactic event occurs **(figure 1.14)**. Following the climactic event, the body again returns to homeostasis. Because their end result is to increase the activity (instead of initially returning the body to homeostasis), positive feedback mechanisms occur much less frequently than negative feedback mechanisms.

Figure 1.15 illustrates one example of a positive feedback mechanism in the human body, when a mother breastfeeds her baby. The baby suckling at the breast is the initial stimulus detected by sensory receptors in the skin of the nipple region. The receptors transmit this input to the control center, which is the hypothalamus of the brain. The hypothalamus signals the posterior pituitary (an endocrine gland) to release the hormone oxytocin into the blood. Oxytocin is the "output" that is sent to the effector, which is the glandular tissue of the breast. Oxytocin stimulates the mammary gland to eject the breast milk. The baby feeds and the cycle repeats as long as the baby suckles. Once the baby stops suckling (and thus the initial stimulus is removed), then the cycle will stop.

Other examples of positive feedback mechanisms include the blood clotting cascade (see section 18.4c) and uterine contractions involved in labor and childbirth (see section 29.6c).

WHAT DID YOU LEARN?

16 What is the main difference between a homeostatic system regulated by negative feedback and one regulated by positive feedback?

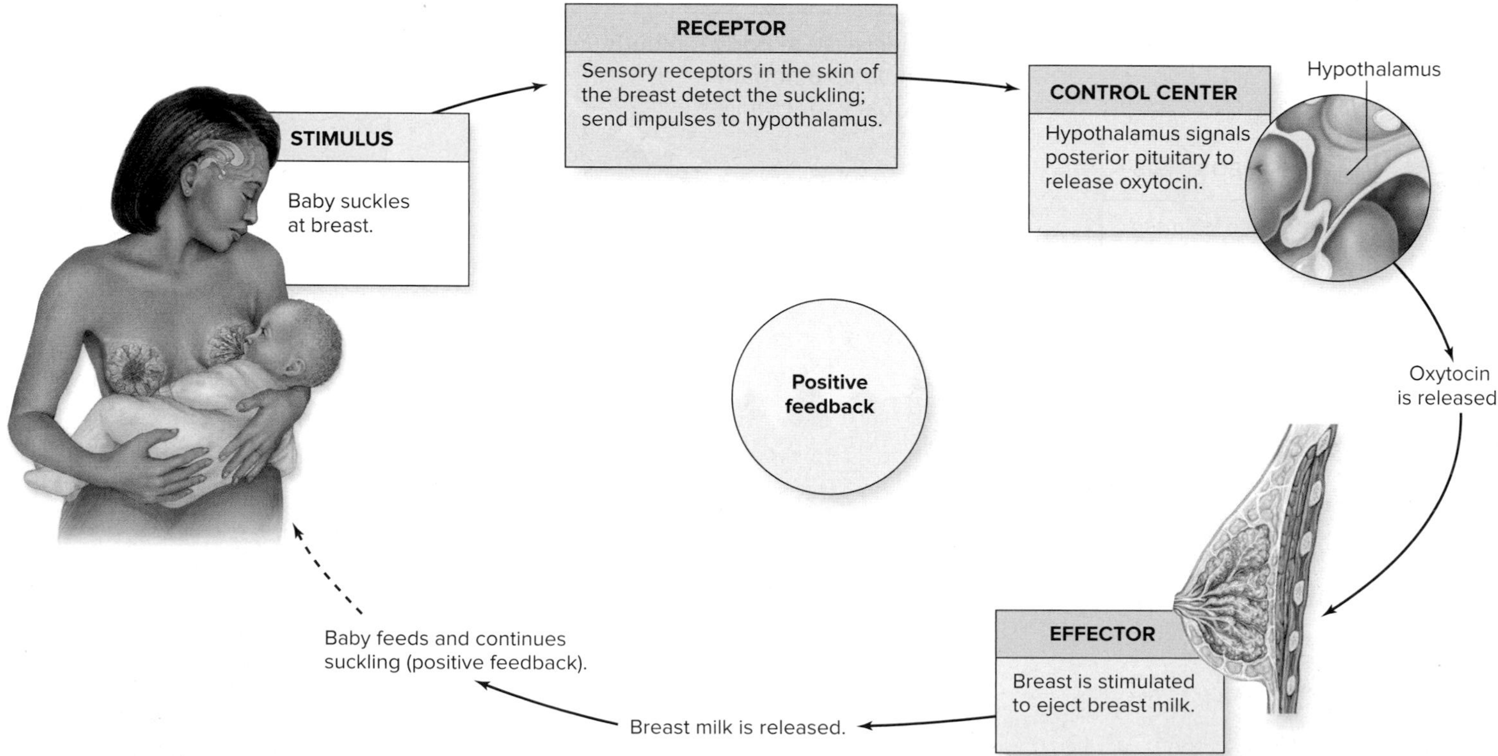

Figure 1.15 Positive Feedback. Positive feedback mechanisms often work in loops, where the initial step in the pathway is the stimulus, and the end product of the pathway is to stimulate (not turn off) the pathway activity. In this example of a mother breastfeeding her child, the stimulus of the baby suckling initiates nerve impulses to the brain to cause release of hormones that stimulate the breast to secrete more breast milk.

1.7 Homeostasis, Health, and Disease

LEARNING OBJECTIVE

23. Explain the general relationship of maintaining homeostasis to health and disease.

In summary, *homeostasis* is a term that describes the many physiologic processes to maintain the health of the body. These characteristics are noted about homeostatic systems:

- They are dynamic.
- The control center is generally the nervous system or the endocrine system.
- There are three components: receptor, control center, and effector.
- They are typically regulated through negative feedback to maintain a normal value or set point.
- It is when these systems fail that a homeostatic imbalance or disease potentially results and ultimately may threaten an individual's survival.

Diabetes is an example of a homeostatic imbalance. Diabetes occurs when the homeostatic mechanisms for regulating blood glucose are not functioning normally, and blood glucose fluctuates out of the normal range, sometimes resulting in extremely high blood glucose readings. High blood glucose results in damage to anatomic structures throughout the body. Patients with diabetes must rely on other methods, such as diet restriction, exercise, and perhaps a medication, to lower blood glucose.

Sometimes a homeostatic imbalance results when critical changes from aging or disease cause a variable that is normally controlled by negative feedback to be abnormally controlled by positive feedback. An example is when there is extensive damage to the heart, perhaps from a heart attack. This heart is less able to pump blood to the structures of the body, including the heart itself. Consequently, the heart receives reduced amounts of nutrients and oxygen. The heart becomes progressively weaker, and even less able to pump blood to the body's structure. Ultimately, the heart becomes so weak that the heart stops beating.

Treating patients generally involves determining a **diagnosis,** or a specific cause of the homeostatic imbalance. Once diagnosed, the patient is treated through the administration of medications or through other therapeutic avenues to facilitate the body in maintaining homeostasis.

Health-care practitioners also need to understand how the drugs patients are taking may affect the normal homeostatic control mechanisms. For example, one type of medication for the treatment of depression is an SSRI, which stands for selective serotonin reuptake inhibitor. Paroxetine (Paxil), fluoxetine (Prozac), and sertraline (Zoloft) are examples of SSRIs. Serotonin is a type of neurotransmitter. Normally, a neurotransmitter is released from one nerve cell in response to a stimulation (nerve impulse). The neurotransmitter accomplishes its communication task, and then is taken up again by the nerve cell for future use. Some depressed individuals may have lower levels of serotonin, so an SSRI blocks the reuptake of serotonin into the nerve cell. Therefore, serotonin stays outside the nerve cell for a longer period of time and its effects are prolonged, which may elevate the mood of the patient taking the SSRI.

However, like all drugs, SSRIs come with some undesirable side effects. Some SSRI side effects include digestive system distress, such as nausea, upset stomach, diarrhea, or combinations of all three. As it turns out, serotonin is also used in the nerve cells of the digestive system. By tinkering with the serotonin reuptake in the brain, the drug also affects serotonin reuptake in the digestive system. Essentially, the digestive system becomes more excitable due to the intake of the SSRI drug, with the symptoms just described.

Virtually all medications have some benefits and some side effects, many of which can be explained by examining the homeostatic control mechanisms with which they interact. Thus, an understanding of these mechanisms is a must for anatomists, physiologists, and health-care practitioners.

WHAT DID YOU LEARN?

17 What is an example of a disease process by which homeostasis is disrupted?

INTEGRATE

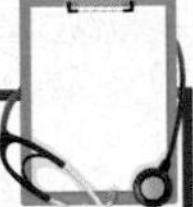

CLINICAL VIEW 1.2

Establishing Normal Ranges for Clinical Practice

What is clinically accepted as the "normal range" for a variable, such as body temperature of 98.6°F, blood glucose of 80–110 milligrams/deciliter (mg/dL), or blood pressure of 90–120/60–80 mm Hg is determined by sampling healthy individuals in a population. A normal range for a variable is determined by the value for 95% of the individuals sampled. Health-care practitioners should be aware that this means that 5% of the population, although healthy, will have values for a given variable considered outside of the normal range.

INTEGRATE

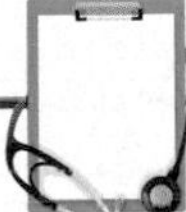

CLINICAL VIEW 1.3

Clinicians' Use of Scientific Method

Clinicians regularly apply the principles of the scientific method when interacting with patients. Consider what typically occurs when a patient with a health problem or complaint goes in for a doctor's appointment. First, information is gathered. The nurse obtains the patient's weight, blood pressure, and other vital signs. The physician solicits the patient's medical history, asks about his or her specific complaint(s), and completes a physical examination. Based on the information gathered, the clinician forms a hypothesis—a tentative explanation of any specific symptoms the patient may be experiencing. As a follow-up to the initial hypothesis, the clinician orders tests and evaluates the test results. After all information is gathered, the clinician draws a conclusion to make a diagnosis. (Sometimes additional tests may be ordered if the test results are inconclusive.) Following a definitive diagnosis, the clinician treats the patient, and additional information is gathered as the patient's response to the treatment is monitored.

INTEGRATE

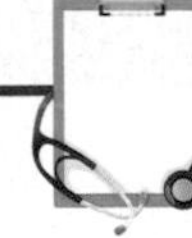

CLINICAL VIEW 1.4

Medical Imaging

Health-care professionals have taken advantage of sophisticated medical imaging techniques to extend their ability to visualize internal body structures noninvasively (i.e., without inserting an instrument into the body). Some of the most common techniques are radiography, sonography, computed tomography, digital subtraction angiography, dynamic spatial reconstruction, magnetic resonance imaging, and positron emission tomography.

Radiography

Radiography (rā-dē-og′ra-fē; *radius* = ray, *grapho* = to write) is the primary method of obtaining an image of a body part for diagnostic purposes. A beam of **x-rays,** which are a form of high-energy radiation, penetrates solid structures within the body. X-rays can pass through soft tissues but they are absorbed by dense tissues, including bone, teeth, and tumors. Film images produced by x-rays passing through soft tissues leave the film lighter in the areas where x-rays are absorbed. Hollow organs can be visualized if they are filled with a radiopaque (rā-dē-ō-pāk; *opacus* = shady) substance that absorbs x-rays.

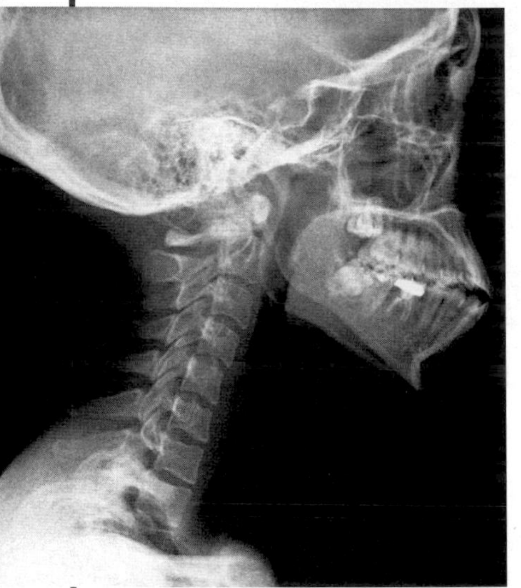

Radiograph (x-ray) of the head and neck, viewed from the right lateral surface.

©Medical Body Scans/Science Source

The term *x-ray* also applies to the photograph (radiograph) made by this technique. Originally, x-rays got their name because they were an unknown type of radiation, but they are also called roentgen rays in honor of Wilhelm Roentgen, the German physicist who accidentally discovered them. Radiography is commonly used in dentistry, mammography, diagnosis of fractures, and chest examination. Disadvantages of x-rays are that they are difficult to interpret when organs overlap in the images, and they are unable to reveal slight differences in tissue density. In addition, the radiation of an x-ray is not without risk.

Sonography

The second most widely used imaging method is **sonography** (sō-nog′rā-fē; *sonus* = sound), also known as *ultrasound.* A technician slowly moves a small, handheld device across the body surface. This device produces high-frequency ultrasound waves and then receives signals that are reflected from internal organs. The image produced is called a **sonogram.** Sonography is the method of choice in obstetrics, where a sonogram can visualize the placenta, examine the fetus, and evaluate fetal age, position, and development. Sonography avoids the harmful effects of x-rays, and the equipment is relatively inexpensive and portable. Once limited to the fields of obstetrics and cardiology, sonography is now widely used in most fields of medicine. For example, orthopedists may quickly examine a shoulder with ultrasound for a rotator cuff tear, and emergency medicine doctors use ultrasound imaging to quickly assess the status of a patient's internal organs.

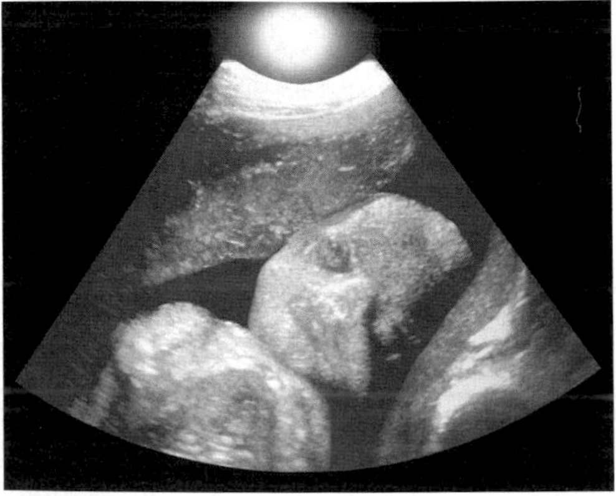

Sonogram of a fetus.

©ATL/Science Source

Improvements in sonography include three-dimensional and four-dimensional ultrasound. In three-dimensional ultrasound, sound waves are emitted in various angles and processed in a computer. This creates a three-dimensional view. A two-dimensional ultrasound is a flat image and a three-dimensional ultrasound shows depth, contour, and detail. Four-dimensional ultrasound shows movement using a compilation of three-dimensional images. Movements like heart motion and yawning can be seen in real time. When radiography or sonography fail to produce the desired images, other more detailed but much more expensive imaging techniques are available.

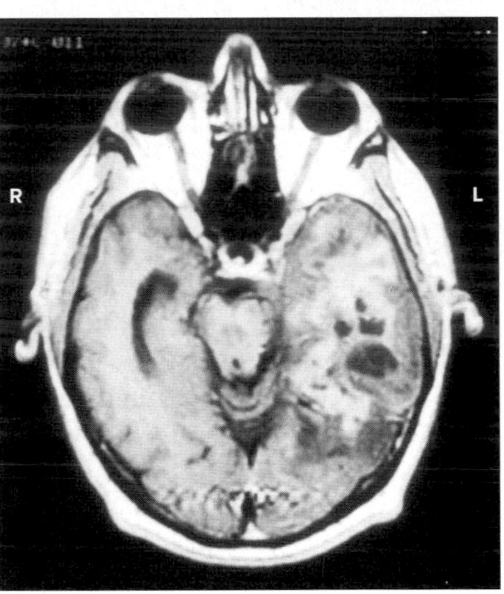

Computed tomography (CT) scan of the head at the level of the eyes, viewed in transverse section.

©SIU BioMed Com/Custom Medical Stock Photo

Computed Tomography (CT)[1]

A **computed tomography (CT)** (tō′mō-graf-ē; *tomos* = a section) scan, previously termed a computerized axial tomography (CAT) scan, is a more sophisticated application of x-rays. A patient is slowly moved through a cylindrical, doughnut-shaped machine while low-intensity x-rays are emitted on one side of the cylinder, passed through the body, collected by detectors, and then processed and analyzed by a computer. These signals produce an image of the body that is about the thickness of a dime. Continuous thin "slices" can be used to reconstruct a three-dimensional image of the body. Little overlap of organs occurs in these thin sections, and the image is much sharper than one obtained by a conventional x-ray. CT scanning is useful for identifying tumors, aneurysms, kidney stones, cerebral hemorrhages, and other abnormalities. A drawback to CTs is that they expose the patient to higher doses of radiation than a traditional x-ray.

Digital Subtraction Angiography (DSA)

Digital subtraction angiography (DSA) is a modified three-dimensional x-ray technique used primarily to view blood vessels. It involves taking radiographs both prior to and after injecting an opaque medium into a blood vessel. The computer compares the before and after images, and removes or subtracts the data from the before image from the data generated by the after image, thus leaving an image that may indicate evidence of vessel blockages. DSA is useful in the procedure in which a physician directs a catheter through a blood vessel and puts a stent in the area where a blood vessel is blocked. The image produced by the DSA allows the physician to accurately guide the catheter to the blockage.

1. CT and MRI films taken in the transverse plane are usually, but not always, read from an inferior view. So the right side of the body is on the left side of the image, and the left side of the body is on the right side of the image. Thus, when reading a CT or MRI scan in transverse section, check the orientation of the image. There should be L and R letters to let you know which side of the film corresponds to the left or right side of the patient.

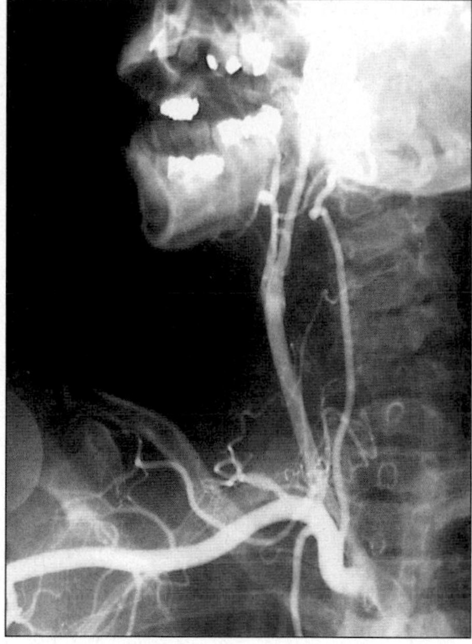

Digital subtraction angiography (DSA) shows three-dimensional images of blood vessels and normal changes in these vessels.
©ISM/Athenais/Medical Images

Dynamic Spatial Reconstruction (DSR)

Using modified CT scanners, a special technique called **dynamic spatial reconstruction (DSR)** provides two important pieces of medical information: (1) three-dimensional images of body organs, and (2) information about the normal organ movement as well as changes in its internal volume. Unlike traditional static CT scans, DSR allows the physician to see the movement of an organ. This type of observation, at slow speed or halted in time completely, has been invaluable in observations of the heart and the flow of blood through blood vessels.

Magnetic Resonance Imaging (MRI)[1]

Magnetic resonance imaging (MRI), previously called *nuclear magnetic resonance (NMR) imaging,* was developed as a noninvasive technique to visualize soft tissues. The patient is placed in a supine position within a cylindrical chamber that is surrounded by a large electromagnet. The magnet generates a strong magnetic field that causes protons in the nuclei of hydrogen atoms in the tissues to align. Thereafter, upon exposure to radio waves, the protons absorb additional energy and align in a different direction. The hydrogen atoms then abruptly realign themselves to the magnetic field immediately after the radio waves are turned off. This results in the release of the atoms' excess energy at different rates, depending on the type of tissue. A computer analyzes the emitted energy to produce an image of the body. MRI is better than CT for distinguishing between soft tissues, such as the white and gray matter of the nervous system. However, dense structures (e.g., bone) do not show up well in MRI. Formerly, another disadvantage of MRI was that patients felt claustrophobic while isolated in the closed cylinder. However, newer MRI technology has improved the hardware and lessened this effect.

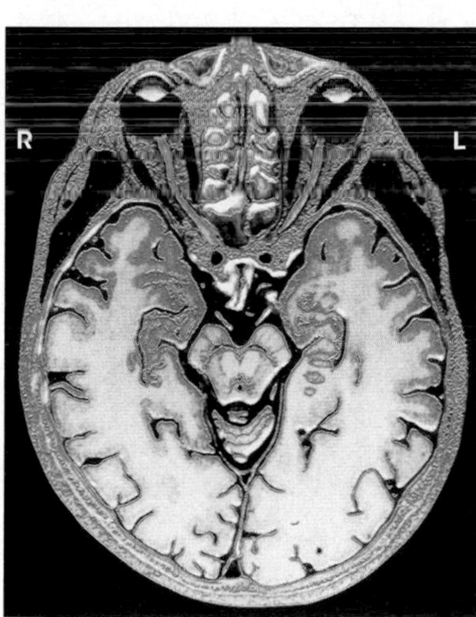

Magnetic resonance imaging (MRI) scan of the head at the level of the eyes, viewed in transverse section.
©Alfred Pasieka/Science Source

A specific type of MRI, called **functional MRI (fMRI),** maps brain function based on local oxygen concentration differences in blood flow. Increased blood flow relates to increased brain activity and is detected by a decrease in deoxyhemoglobin (the form of hemoglobin lacking oxygen) in the blood.

Positron Emission Tomography (PET)

The **positron emission tomography (PET) scan** is used both to analyze the metabolic state of a tissue at a given moment in time and to determine which tissues are most active. The procedure begins with an injection of radioactively labeled glucose (sugar), which emits particles called positrons (like electrons, but with a positive charge). Collisions between positrons and electrons cause the release of gamma rays that can be detected by sensors and analyzed by computer. The result is a brilliant color image that shows which tissues were using the most glucose at that moment. In cardiology, the image can reveal the extent of damaged heart tissue—because damaged heart tissue consumes little or no glucose, the damaged tissue will appear dark. PET scans have been used to illustrate activity levels in the brain and, in so doing, have been useful in examining the effects of neurologic ailments (e.g., schizophrenia, Alzheimer disease). They also may detect whether certain cancers have metastasized throughout the body, because cancerous cells will take up more glucose and show up as a *hot spot* on the scan. The PET scan is an example of nuclear medicine, which uses radioisotopes (see section 2.1b) to form anatomic images of the body.

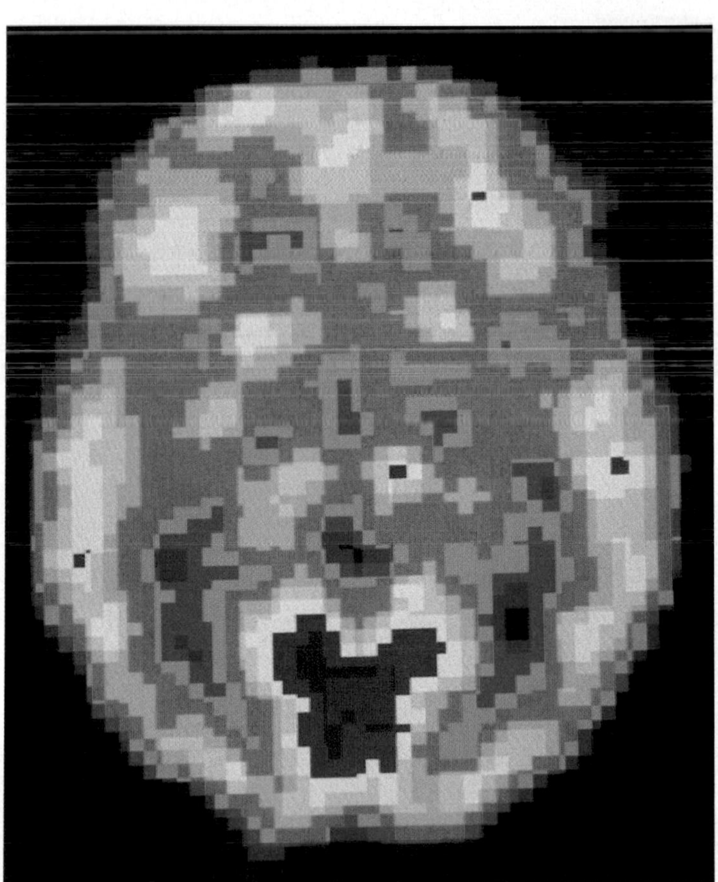

Positron emission tomography (PET) scan of the brain of an unmedicated schizophrenic patient. Red areas indicate high glucose use (metabolic activity). The visual center at the posterior region of the brain was especially active when the scan was made.
©Hank Morgan/Science Source

CHAPTER SUMMARY

1.1 Anatomy and Physiology Compared	• Anatomy is the study of structure and form of the human body, whereas physiology is the study of function of these parts. **1.1a Anatomy: Details of Structure and Form** • Anatomy may be subdivided into microscopic anatomy (anatomic study of materials using the microscope) and gross anatomy (the study of structures visible to the unaided eye). **1.1b Physiology: Details of Function** • Physiologists may examine the function of specific body systems (e.g., cardiovascular physiology) and may focus on problems or pathologies of such systems (pathophysiology).
1.2 Anatomy and Physiology Integrated	• Form and function are interrelated. Anatomists cannot gain a full appreciation of anatomic form without first understanding the structure's function. Likewise, physiologists cannot fully appreciate body functions without learning about the structure's form. • It is effective to learn anatomy and physiology by integrating the two disciplines, rather than trying to separate the study of form from the study of function.
1.3 How to Study Anatomy and Physiology Effectively	• Utilize active learning techniques, draw out structures, make tables, and work with a partner to learn the material. If you can accurately explain a concept to your partner, then you truly understand the concept.
1.4 The Body's Levels of Organization	• Scientists group the body's components into an organizational hierarchy of form and function. **1.4a Characteristics That Describe Living Things** • All living organisms exhibit several common properties: organization, metabolism, growth and development, responsiveness, adaptation, and reproduction. **1.4b The View from Simplest to Most Complex** • Anatomic structure is organized in an increasingly complex series of levels: the chemical level, cellular level, tissue level, organ level, organ system level, and organismal level. **1.4c Introduction to Organ Systems** • The human body contains 11 organ systems: integumentary, skeletal, muscular, nervous, endocrine, cardiovascular, lymphatic, respiratory, urinary, digestive, and reproductive.
1.5 The Precise Language of Anatomy and Physiology`	• Clear, exact terminology accurately describes body structures and helps us identify and locate them. **1.5a Anatomic Position** • The anatomic position is used as a standard reference point for the human body. **1.5b Sections and Planes** • Three planes section the body and help describe relationships among the parts of the three-dimensional human body: the coronal plane, the transverse plane, and the midsagittal plane. **1.5c Anatomic Directions** • Specific directional terms indicate body structure locations. **1.5d Regional Anatomy** • Specific anatomic terms identify body regions. **1.5e Body Cavities and Membranes** • Body cavities are spaces that enclose organs and organ systems. • The posterior aspect of the body contains the cranial cavity and the vertebral canal. • The ventral cavity is subdivided into a thoracic cavity and an abdominopelvic cavity (which is partitioned into an abdominal cavity and a pelvic cavity). • The ventral cavity is lined by thin serous membranes. A parietal layer lines the internal body wall surface, and a visceral layer covers the organs. **1.5f Abdominopelvic Regions and Quadrants** • Regions and quadrants are two aids for describing locations of the abdominopelvic viscera. • There are nine abdominopelvic regions and four abdominopelvic quadrants.
1.6 Homeostasis: Keeping Internal Conditions Stable	• *Homeostasis* refers to the body's ability to maintain a relatively stable internal environment in response to changes in either internal or external environmental factors. **1.6a Components of Homeostatic Systems** • The three components are the receptor (detects a stimulus), control center (interprets input from the receptor and initiates changes through the effector), and effector (the structure that brings about a change to the stimulus). **1.6b Homeostatic Systems Regulated by Negative Feedback** • Negative feedback mechanisms or loops are initiated by either an increase or a decrease in the stimulus, and the end result is to return the stimulus to within its normal range or set point. Most feedback mechanisms in the human body work by negative feedback. **1.6c Homeostatic Systems Regulated by Positive Feedback** • Positive feedback mechanisms are initiated by a stimulus, and they maintain or increase the activity of the original stimulus until a climatic event.
1.7 Homeostasis, Health, and Disease	• An understanding of the concept of homeostasis is essential when understanding the structure and function of a normal, healthy body, the mechanisms of disease, and how the body reacts to pharmaceutical agents.

CHALLENGE YOURSELF

Create and Evaluate

Analyze and Apply

Understand and Remember

Do You Know the Basics?

1. Examining the superficial anatomic markings and internal body structures as they relate to the covering skin is called
   a. regional anatomy.
   b. surface anatomy.
   c. pathologic anatomy.
   d. comparative anatomy.
2. The ___________ level of organization is composed of two or more tissue types that work together to perform a common function.
   a. cellular
   b. molecular
   c. organ
   d. organismal
3. The term ___________ refers to the sum of all chemical reactions in the body.
   a. metabolism
   b. responsiveness
   c. stimulus
   d. reproduction
4. A midsagittal plane separates the body into
   a. anterior and posterior portions.
   b. superior and inferior portions.
   c. right and left halves.
   d. unequal right and left portions.
5. The term used to describe an appendage structure that is closest to its point of attachment to the trunk is
   a. distal.
   b. lateral.
   c. superior.
   d. proximal.
6. The ___________ region is the anterior part of the knee.
   a. patellar
   b. popliteal
   c. pes
   d. inguinal
7. Which body cavity is located inferior to the thoracic diaphragm and superior to the pelvic brim of the hip bones?
   a. abdominal cavity
   b. pelvic cavity
   c. pleural cavity
   d. pericardial cavity
8. The ___________ is the serous membrane layer that covers the surface of the lungs.
   a. parietal pleura
   b. visceral pericardium
   c. visceral peritoneum
   d. visceral pleura
9. The state of maintaining a constant internal environment within an organism is called
   a. reproduction.
   b. homeostasis.
   c. imbalance.
   d. life.
10. In a negative feedback mechanism, which of the following events does *not* occur?
    a. A stimulus (i.e., a change in some variable) occurs.
    b. A receptor perceives a stimulus.
    c. The control center sends output to an effector.
    d. The effector stimulates or increases the stimulus, so the cycle continues.
11. What are the similarities and differences between anatomy and physiology?
12. List the levels of organization in a human, starting at the simplest level and proceeding to the most complex. Give an example of a body structure in each level.
13. What properties are common to all living things?
14. Name the eleven organ systems in the human body.
15. Describe the body in the anatomic position. Why is the anatomic position used?
16. List the anatomic terms that describe the following regions: forearm, wrist, chest, armpit, thigh, and foot.
17. What are the two body cavities within the posterior aspect, and what does each cavity contain?
18. Describe the structure and function of serous membranes in the body.
19. What are the main components in a homeostatic control system?
20. Compare and contrast negative and positive feedback mechanisms.

Can You Apply What You've Learned?

Use the following paragraph to answer questions 1–3.

Your friend Eric complains of some pain in his "belly area." You ask him to point to the precise location of the pain. He points to a region that is below his umbilicus, on the lower right side of his abdomen, and just medial to his hip bones.

1. Which abdominal quadrant contains Eric's pain?
   a. right upper quadrant
   b. right lower quadrant
   c. left upper quadrant
   d. left lower quadrant
2. You also could describe the pain as being in the ___________ abdominopelvic region.
   a. right lumbar
   b. right hypochondriac
   c. right umbilical
   d. right iliac

3. Eric goes to the doctor to determine the cause and source of the pain. The physician orders a CT scan, which shows that Eric has an enlarged and inflamed appendix (an organ associated with the digestive system). Eric asks the physician why she didn't just take an x-ray of his belly region. She explains an x-ray would not be the best diagnostic imaging tool in this case because
   a. x-ray images are more expensive to produce than CT scans.
   b. soft-tissue structures don't show up well on basic x-rays.
   c. the x-rays could inflame the appendix further and cause it to burst.
   d. x-rays now are used primarily for bone injuries only.
4. When you are outside on a hot, humid day, what body changes occur to help your body temperature return to normal?
   a. The blood vessels in your skin constrict.
   b. The sweat glands release sweat.
   c. Nerve impulses are sent to muscles to cause shivering.
   d. The smooth muscle associated with hair follicles contracts, causing goose bumps.
5. A friend just started taking Zoloft (an SSRI) and is experiencing an upset stomach and diarrhea. Your friend asks if the drug is causing her symptoms and you respond:
   a. Yes, because the drug is irritating her stomach lining and that explains her symptoms.
   b. Yes, because serotonin is located in both the brain and the digestive tract, so the drug is altering digestive system functioning.
   c. No, because the drug is supposed to elevate mood and affect brain function, and it shouldn't have any effect on the digestive system.
   d. No, because the drug is quickly absorbed from the digestive tract and does not remain in the digestive system long enough to have any effect there.

Can You Synthesize What You've Learned?

1. Lynn was knocked off her bicycle during a race. She broke some bones in her right antebrachial region, suffered an abrasion on her mental region, and had severe bruising on her right gluteal and femoral regions. Explain where each of these injuries is located.
2. Carly was stung by a bee and was taken to the emergency room because she was undergoing anaphylactic shock (e.g., her breathing became more rapid and more difficult, her heartbeat increased). She was given a shot of epinephrine, which reduced her allergic reactions and brought her breathing and heartbeat back to normal. Did the dose of epinephrine result in a negative feedback mechanism occurring, or a positive feedback mechanism occurring? Explain your answer.
3. Your grandmother is being seen by a radiologist to diagnose a possible tumor in her small intestine. Explain to your grandmother what imaging techniques would best determine whether a tumor exists, and which techniques would be inadequate for determining the placement of the tumor.

INTEGRATE

ONLINE STUDY TOOLS

connect | SMARTBOOK® | AP|R

The following study aids may be accessed through Connect.

Clinical Case Study: A Clinical Case of Hyperthermia in an Athlete

Interactive Questions: This chapter's content is served up in a number of multimedia question formats for student study

SmartBook: Topics and terminology include anatomy and physiology and details of form and function; the body's levels of organization; terminology for anatomic position, sections, and planes; homeostasis, health, and disease

Anatomy & Physiology Revealed: Topics include organ systems, planes of section, directional terms, body regions, body cavities, pleura and pericardium; abdominal quadrants and regions

Animations: Topics include positive and negative feedback

chapter

2

Atoms, Ions, and Molecules

Source: Brenda J. Jones/CDC

CAREER PATH
Clinical Chemist

A clinical chemist utilizes laboratory tests to evaluate blood samples, study DNA, and examine both tissues and cells. Samples obtained from patients are sent to the laboratory and evaluated by the clinical chemist. The results of these tests provide more precise information to the physician for both determining a diagnosis and implementing an appropriate treatment.

Module 2: Cells and Chemistry

In later chapters we describe many fascinating physiologic processes that occur within the human body, including: the transmission of an impulse along a nerve cell, the transport of oxygen within the blood, and nutrient digestion in the gastrointestinal tract. By reading and comprehending the material in this chapter on atoms, ions, and molecules—and information on energy, enzymes, and metabolism in chapter 3—you will develop a working knowledge of the basic chemical concepts needed to understand these concepts and other important physiologic processes. We have also included some of the common medical diagnostic tests (e.g., complete blood cell counts and urinalysis tests) in later chapters. Interpreting these tests requires an understanding of these same chemical concepts. We will refer back to the content within this chapter throughout the rest of the text, so that you will more easily see the connections between chemistry and concepts presented in later chapters.

Our coverage of chemistry is not comprehensive. Rather, our discussion is tailored to focus primarily on chemical structures relevant to the study of anatomy and physiology, and we have included examples of their application in the human body. The general characteristics of atoms, ions, and molecules are presented in the earlier sections of this chapter. Then we describe specific molecules, including water and water mixtures, and the four major biological macromolecules: lipids, carbohydrates, nucleic acids, and proteins. The chapter concludes with a section devoted to proteins, the most versatile biological macromolecules of living systems. Now, let's embark upon an adventure to explore the body's simplest constituents: atoms and molecules.

2.1 Atomic Structure

The human body at its simplest level of organization is composed of chemical structures that include atoms, ions, and molecules. We begin our discussion on the human body's chemical composition by describing matter, atoms, elements, and the position of each element in the periodic table. This information allows us to make certain predictions regarding the chemical properties of each element, including whether and how the element forms ions and molecules.

2.1a Matter, Atoms, Elements, and the Periodic Table

LEARNING OBJECTIVES

1. Define matter, and list its three forms.
2. Describe and differentiate among the subatomic particles that compose atoms.
3. Explain the arrangement of elements in the periodic table based on atomic number.
4. Diagram the structure of an atom.

The human body is composed of **matter,** which is generally defined as a substance that has mass and occupies space. Matter is present in the body in three forms: solid, liquid, and gas. For example, bone is a solid, blood is a liquid, and oxygen (O_2) and carbon dioxide (CO_2) are gases.

All matter is composed of atoms. An **atom** is the smallest particle that exhibits the chemical properties of an **element.** There are 92 naturally occurring elements. Hydrogen is the smallest and lightest element; uranium is the largest and heaviest element. Technical advances in both chemistry and physics have resulted in the ability to produce "ultraheavy" elements that are larger than uranium. All elements, including the scientifically manufactured elements, are organized into a chart called the **periodic table of elements (figure 2.1*a*).**

Elements are grouped into major, minor, and trace elements based on the percentage each composes by weight in the human body. *Major elements* make up almost 99% of our body weight and *minor elements* less than 1% (figure 2.1*b*). In comparison, *trace elements* appear in the body in only limited amounts (less than 0.01%). Only 12 elements occur in living organisms in greater than trace amounts. Six are major elements, which include oxygen, carbon, hydrogen, nitrogen, calcium, phosphorus and six are minor elements, which include sulfur, potassium, sodium, chlorine, magnesium, and iron.

Note in figure 2.1*a* that the 12 major and minor elements are elevated above the other elements. Each element is color-coded, and these color-codes are used throughout the text.

The Components of an Atom

Atoms are composed of three **subatomic particles:** protons, neutrons, and electrons **(figure 2.2).** Two major criteria differentiate subatomic particles—namely, mass and charge. The mass of an atom is expressed as the **atomic mass unit (amu),** or *dalton.* (An atomic mass unit is very small: consider that 1 amu is equal to 1.66×10^{-27} kilograms.)

Protons and neutrons each have a mass of 1 amu. A **proton** has a positive charge of one (+1), whereas a **neutron** is uncharged (meaning it is neutral). Protons and neutrons compose almost the entire mass of an atom and are located at the center, or core, of the atom, called the **nucleus.** The nucleus contains mass because it has both protons and neutrons, and it is positively charged because of the protons.

The **electron** is the third component of an atom. An electron has a very small mass—only about 1/1800th of the mass of a proton or neutron—and makes a negligible (very small) contribution to the total mass of the atom. Each electron has a negative charge of one (−1). Electrons are located at varying distances from the nucleus in regions called *orbitals,* often depicted either as an electron cloud (in a *cloud model*) or as discrete energy shells (in a *shell model*). Both the cloud model and shell model indicate where the electrons are most likely found, as depicted in figure 2.2. For simplicity, we diagram atoms with the shell model in this text.

Elements and the Periodic Table

Elements differ in the number of subatomic particles. The periodic table may be used to obtain the number of subatomic particles in an atom of a specific element. Several important features for each element appear in the periodic table, including the element's chemical symbol, atomic number, and average atomic mass.

A unique **chemical symbol** has been assigned to each element. An element is usually identified either by its first letter or by its first letter plus an additional letter of its English name. For example, H is for hydrogen, C is for carbon, and O is for oxygen. A second letter, which is always lowercase, distinguishes elements that begin with the same letter (e.g., Ca for calcium and Cl for chlorine). The symbol of some elements is derived from their Latin name. For instance, the symbol for sodium is Na from the Latin *natrium* and potassium is K from the Latin *kalium.*

The **atomic number** of an element indicates the number of protons in an atom of that element and is located above its symbol in the periodic table. Note that elements are arranged by the atomic number in consecutive order within rows (figure 2.1*a*). The atomic number is designated as a subscript at the left of the chemical symbol when it is written. For example, ${}_{1}H$ shows that the nucleus of a hydrogen atom has one proton, and ${}_{6}C$ indicates that the carbon nucleus has six protons.

The **atomic mass** indicates the mass of both protons and neutrons in the atomic nucleus, and it reflects the "heaviness" of an element's atoms relative to other atoms. (The electrons are not included because of their relatively small mass.) It is shown below the element's symbol on the periodic table. The average atomic mass (described in section 2.1b) is rounded to the nearest whole number, and it is designated by a superscript to the left of the chemical symbol when it is written. For example, a sodium atom, with an atomic number of 11 and an average atomic mass of 22.99, is designated as ${}^{23}_{11}Na$.

 **WHAT DO YOU THINK?**

 How would the chemical shorthand for oxygen be written?

Determining the Number of Subatomic Particles

The number of each type of subatomic particle is determined as follows:

- The **number of protons** is the atomic number—thus, carbon has six protons and oxygen has eight protons.
- The **number of neutrons** can be determined by subtracting the atomic number (number of protons) from the atomic mass

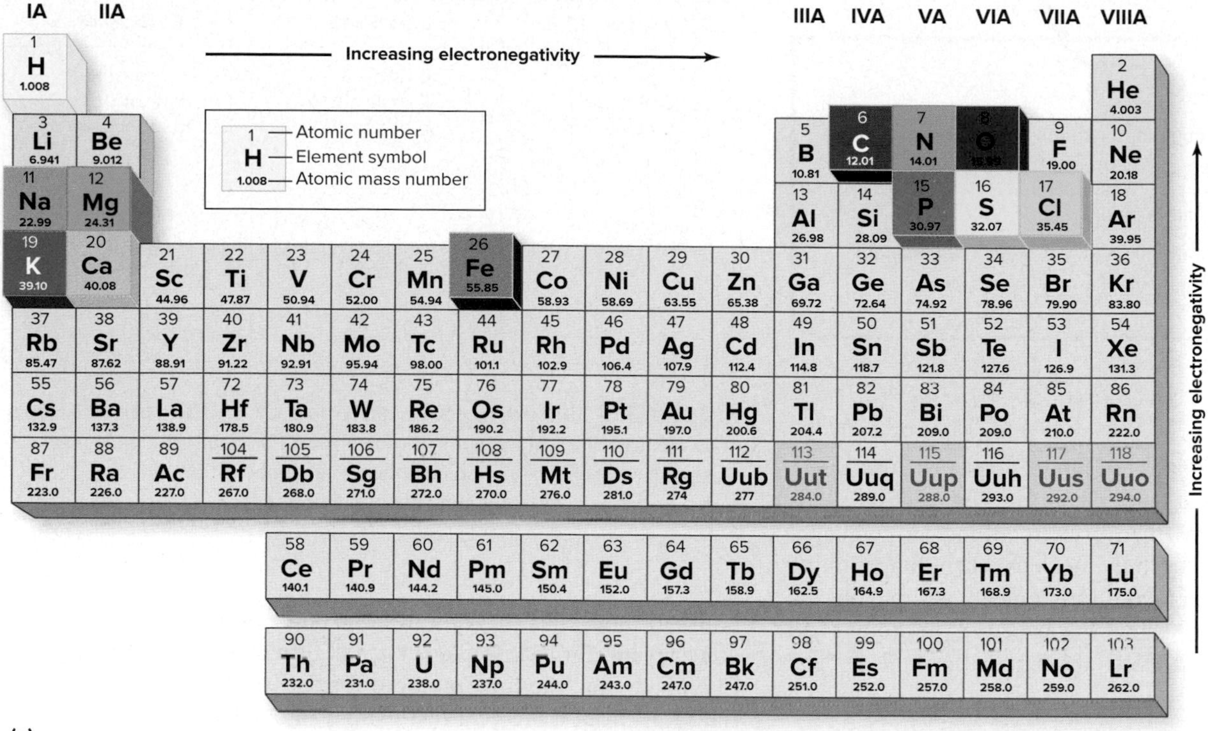

(a)

Most Common Elements of the Human Body

Major elements (collectively compose almost 99% of body weight)			Minor elements (collectively compose less than 1% of body weight)		
Symbol		% Body weight	Symbol		% Body weight
O	Oxygen	65.0	S	Sulfur	0.25
C	Carbon	18.5	K	Potassium	0.20
H	Hydrogen	9.5	Na	Sodium	0.15
N	Nitrogen	3.0	Cl	Chlorine	0.15
Ca	Calcium	1.5	Mg	Magnesium	0.05
P	Phosphorus	1.0	Fe	Iron	0.006

(b)

Figure 2.1 Periodic Table of Elements. (*a*) The periodic table includes all of the elements arranged in chart form. The grayed out boxes at the bottom of the table are for proposed elements that as of yet do not have sufficient evidence to support their official inclusion as elements. (*b*) Twelve of the elements that compose the human body are present in greater than trace amounts. These 12 elements are color-coded in both (*a*) and (*b*), and this color scheme is used throughout the book to represent these elements.

(protons and neutrons). For example, to calculate the number of neutrons in sodium ($^{23}_{11}Na$), you would take the total of 23 protons and neutrons, and subtract the number of protons = 11; thus, sodium has 12 neutrons (23 – 11 = 12).

- The **number of electrons** in an atom is also determined by the atomic number. This is possible because all atoms are *neutral.* The number of negatively charged electrons must equal the number of positively charged protons for an atom to be neutral. Each electron has a charge of –1 and counters the +1 charge of a proton in the nucleus—therefore, the atom has zero charge, meaning it has no net charge.

Diagramming Atomic Structures

The shell model represents the atom as having shells of electrons that surround the atomic nucleus. The electrons within a shell have a given energy level. Each shell can hold only a limited number of electrons. The innermost shell, or *first shell,* may hold up to two

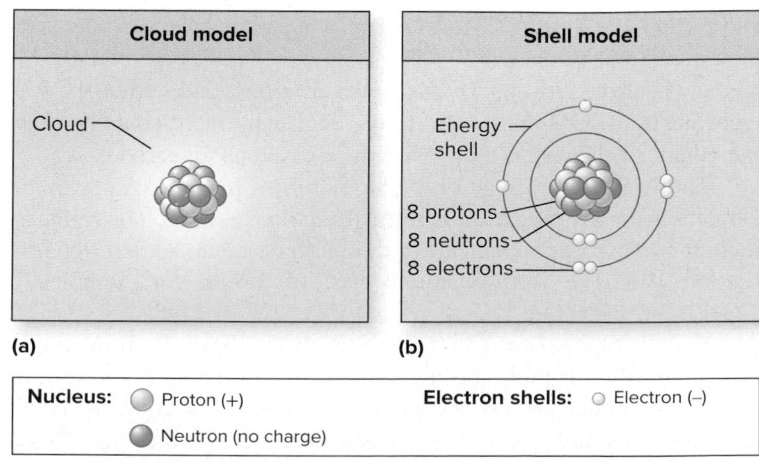

Figure 2.2 General Atomic Structure. Oxygen atoms are depicted with protons and neutrons in the nucleus. Electrons are shown in both a cloud model (*a*) and an energy shell model (*b*). AP|R

INTEGRATE

LEARNING STRATEGY

To diagram an atom, the number of protons, neutrons, and electrons must be known. How these are determined can be summarized as follows:

- **Proton number (p)** = Atomic number
- **Neutron number (n)** = Atomic mass (p + n) − atomic number (p)
- **Electron number (e)** = Proton number (p)

electrons and the *second shell* up to eight electrons. All subsequent shells have a capacity of at least eight electrons. The electron shells closest to the nucleus must be filled prior to filling any potential shells at some greater distance from the nucleus when diagramming an atom. (Note: This is a general rule for elements with an atomic number of 18 or less.) Figure 2.2 shows the placement of protons and neutrons within the nucleus and electrons in specific shells for the atomic structure of oxygen.

WHAT DID YOU LEARN?

1. What subatomic particles determine the mass of an atom? What subatomic particles determine the charge of an atom?
2. Diagram the atomic structure of chlorine— the atomic number is 17 and the mass number is 35.

2.1b Isotopes

LEARNING OBJECTIVES

5. Describe an isotope.
6. Explain how radioisotopes differ from other types of isotopes.

Elements in nature usually occur as a mixture of isotopes. **Isotopes** (ī′sō-tōp; *iso* = same) are atoms of the same element that have the same number of protons and electrons but differ in the number of neutrons. Isotopes of an element exhibit essentially identical chemical characteristics but have different atomic masses.

For example, carbon exists in three isotopes: carbon-12, carbon-13, and carbon-14 (**figure 2.3**). All isotopes of carbon have six protons in their nuclei and six electrons in their atomic shells. However, carbon-12 has six neutrons, carbon-13 has seven neutrons, and carbon-14 has eight neutrons. Generally, one isotope is usually more common than the others. Carbon-12 is the most common isotope for carbon.

The weighted average of the atomic mass for all isotopes of an element is the **average atomic mass** (or *atomic weight*). This value is included below each element on the periodic table, as described in section 2.1a. The average atomic mass of carbon, for example, is 12.01 amu.

Some isotopes are referred to as **radioisotopes.** Radioisotopes are generally unstable because their nuclei contain an excess number of neutrons. For example, carbon-14 has eight neutrons in its nucleus and is unstable and radioactive, whereas carbon-12 has six neutrons in its nucleus and is stable and not radioactive. Radioisotopes usually lose nuclear components in the form of high-energy radiation that includes alpha particles, beta particles, or gamma rays as they decay or break down into a more stable isotope. The time it takes for 50% of the radioisotope to become stable is its **physical half-life.** This time may vary from a few

Carbon-12	Carbon-13	Carbon-14
6 protons 6 neutrons 6 electrons	6 protons 7 neutrons 6 electrons	6 protons 8 neutrons 6 electrons

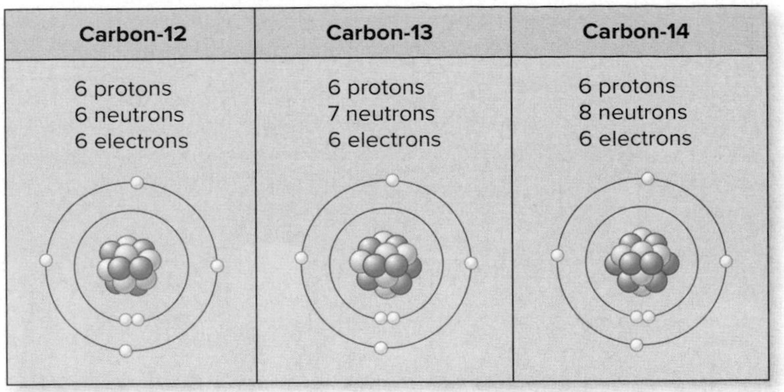

Figure 2.3 Differences in Isotopes of an Element. Isotopes of an element differ in the number of neutrons. For example, notice that these three carbon isotopes contain the same number of protons and electrons but have different numbers of neutrons.

hours to thousands of years. Radioisotopes produced in a nuclear power plant, for example, have a half-life of at least 10,000 years. In comparison, the time required for half of the radioactive material (e.g., from a medical test using radioactive contrast material; see Clinical View 2.1: "Medical Imaging of the Thyroid Gland Using Iodine Radioisotopes") to be eliminated from the body is the **biological half-life.** Biological half-life is also applied to substances that are nonradioactive, such as hormones or drugs (see section 17.4b).

WHAT DID YOU LEARN?

3. Do isotopes represent the same element? Do they have the same number of protons, neutrons, or electrons? Describe a radioisotope.

2.1c Chemical Stability and the Octet Rule

LEARNING OBJECTIVES

7. Describe how elements are organized in the periodic table based on the valence electron number.
8. State the octet rule.

Recall that the periodic table is organized into rows based on atomic number. It is also organized into columns based on the number of electrons in the outer shell, or what is referred to as the **valence shell.** This organization is presented in **figure 2.4,** showing the atomic structure of elements 1 to 20 arranged as they appear in a periodic table. Column IA shows hydrogen, lithium, sodium, and potassium: All the atomic structures of these elements contain one electron in their outer shell. Each consecutive column thereafter (columns IIA–VIIIA) has one additional electron in its outer shell for each element in that column. This organization allows us to make predictions about the chemical characteristics of a given element simply by its location in the periodic table.

Notice that the elements in column VIIIA each have a valence shell that is full or complete. These elements (only helium, neon, and argon are shown in figure 2.4) have an outermost shell housing eight electrons, except for helium, which has only two electrons. A complete outer shell results in chemical stability. These stable atoms are relatively chemically inert and exhibit low reactivity; thus, these elements do not usually combine with other elements. The atoms in column VIIIA are referred to as **noble gases** because they do not typically react with the "common" elements in other columns of the periodic table.

INTEGRATE

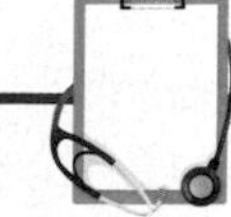

CLINICAL VIEW 2.1

Medical Imaging of the Thyroid Gland Using Iodine Radioisotopes

Radioisotopes including ^{14}C, ^{32}P, and ^{123}I have been extensively used in basic biology research efforts, medical investigations, and diagnoses. Radioisotopes introduced into the body during medical procedures are used by cells in a manner similar to a substance that is a nonradioisotope (e.g., ^{12}C, ^{31}P, and ^{127}I). Thus, these elements may be traced and quantified by the high-energy radiation the radioisotopes emit following their uptake into the body. **Scintigraphy** refers to the nuclear medicine imaging procedure whereby a body structure's metabolic activity may be assessed with these radioisotopes. Scintigraphy of the **thryoid gland** (shown) illustrates the gland's uptake of radioactive ^{123}I. Most of the thyroid appears whitish because the cells within these areas are metabolically active with greater amounts of radioactive ^{123}I uptake. The darker-colored area is a benign nodule that is less metabolically active than the surrounding thyroid tissue with noticeably less radioactive ^{123}I uptake.

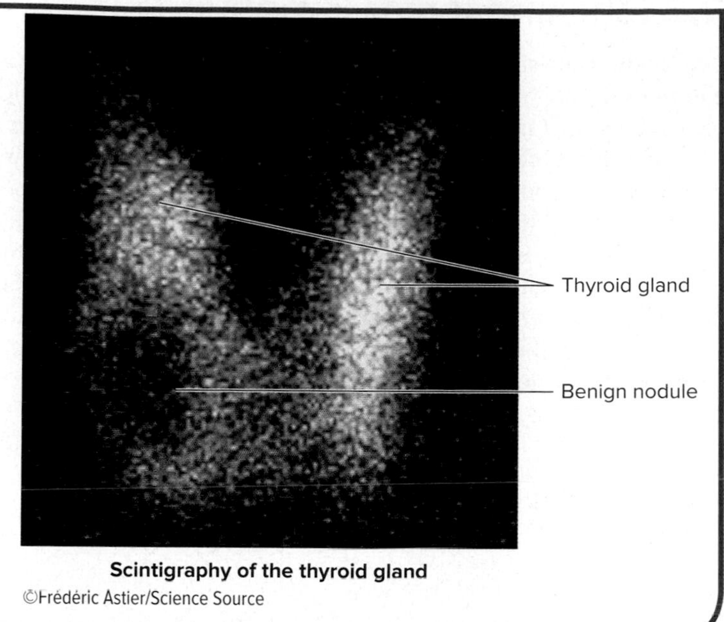

Scintigraphy of the thyroid gland

Examine the atomic structure of the other elements in figure 2.4 (i.e., those that are not noble gases) and note that they lack a full outer shell with eight electrons. As we will see, these elements tend to lose, gain, or share electrons to obtain a complete outer shell. Chemists call this tendency the **octet rule:** Atoms obtain an outer shell with eight electrons and gain chemical stability through the loss, gain, or sharing of electrons. Not all elements follow the octet rule, but it is accurate for our purposes here. An ion is formed from an atom with either the loss or the gain of electrons, whereas a covalent bond is formed by the sharing of electrons between two atoms. We describe the formation of ions and covalent bonds in the next two sections.

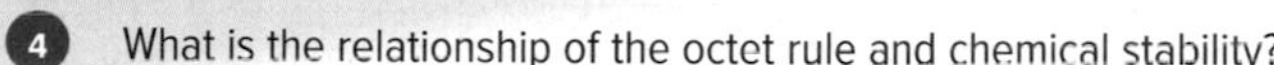

WHAT DID YOU LEARN?

4 What is the relationship of the octet rule and chemical stability?

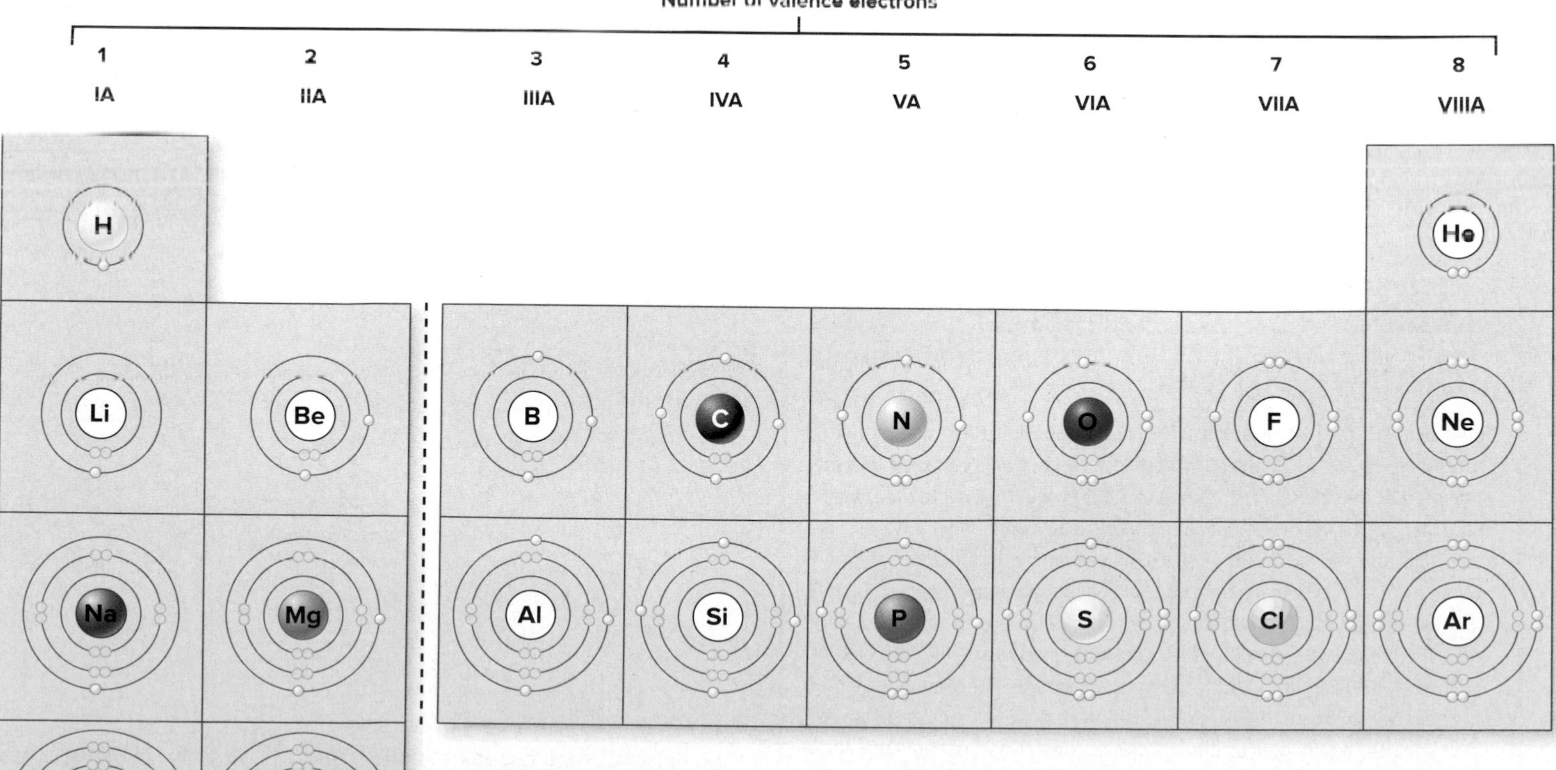

Figure 2.4 Organization of the Periodic Table Based on Valence Shell Electrons. The atomic structures of elements 1 to 20 of the periodic table are depicted, showing the nucleus and electrons in energy shells. The valence shell electrons are shown in yellow. Valence shell number increases by one in each column as you move from left to right from columns IA through VIIIA.

2.2 Ions and Ionic Compounds

The body consists mostly of chemical compounds. **Chemical compounds** are stable associations between two or more elements combined in a fixed ratio. These associations are classified as either ionic compounds or molecular compounds. Here we describe ionic compounds, which are structures composed of ions that are held together in a lattice by electrostatic interactions called *ionic bonds*. We first define ions and list some common ions of the human body, then explain how ions are formed and their charge determined, and finally we describe the electrostatic interaction between ions within an ionic compound. Molecular compounds are described in section 2.3.

2.2a Ions

LEARNING OBJECTIVES

9. Define an ion.
10. List some common ions in the body.
11. Differentiate between cations and anions.
12. Describe how charges are assigned to ions.

Ions are either individual atoms or groups of atoms that have a positive or negative charge. An ion can have either a positive charge from the loss of one or more electrons or a negative charge from the gain of one or more electrons. Examples of ions include sodium ion (Na^+), chloride ion (Cl^-), and bicarbonate ion (HCO_3^-). **Table 2.1** lists the most common ions in the human body along with their significant physiologic functions—notice the diversity of body structures (e.g., nerves, muscles, liver, stomach) that require specific ions to function normally.

Losing Electrons and the Formation of Cations

A sodium atom, found in column IA of the periodic table (figure 2.4), is a good example of an element that can reach stability by losing an electron. The atomic structure of sodium has one electron in its outer shell. By giving up or donating that electron, sodium now satisfies the octet rule and becomes stable **(figure 2.5*a*)**. But is the structure still neutral? Recall that *neutral* means that the number of positively charged protons is equal to the number of negatively charged electrons. Because an electron has been donated by the sodium atom, the newly formed sodium ion has 11 protons but only 10 electrons, and

Table 2.1 Common Ions in the Human Body and Their Physiologic Significance

COMMON CATIONS (POSITIVELY CHARGED IONS)		
Cation	**Structure**	**Physiologic Significance**
Sodium ion	Na^+	• Most common extracellular cation • Participant in conducting electrical signals in nerves and muscle • Most important in osmotic movement of water • Sodium gradient involved in cotransport of other substances across a plasma membrane
Potassium ion	K^+	• Most common intracellular cation • Participant in conducting electrical signals in nerves and muscle • Role in glycogen storage in liver and muscle • Function in pH balance
Calcium ion	Ca^{2+}	• Hardness of bone and teeth • Muscle contraction • Exocytosis (including release of neurotransmitter) • Blood clotting • Second messenger in hormonal stimulation of cells
Magnesium ion	Mg^{2+}	• Required for ATP production
Hydrogen ion	H^+	• Concentration determines pH of blood and other fluids of the body
COMMON ANIONS (NEGATIVELY CHARGED IONS)		
Anion	**Structure**	**Physiologic Significance**
Chloride ion	Cl^-	• Alters nerve cell responsiveness to stimulation • Component of stomach acid (HCl) • Chloride shift in erythrocytes
Bicarbonate ion	HCO_3^- O ‖ HO—C—O⁻	• Conversion of CO_2 gas to HCO_3^-, which is transported in the blood • Buffering of pH in blood
Phosphate ion	PO_4^{3-} O ‖ ⁻O—P—O⁻ | O⁻	• As $Ca_3(PO_4)_2$, it hardens bone and teeth • Component of phospholipids (membranes) • Component of nucleotides, including ATP and nucleic acids (DNA and RNA) • Most common intracellular anion • Intracellular buffer

INTEGRATE

CONCEPT CONNECTION

Ions, including the common ions in the human body, function as **electrolytes** (substances that when dissolved in water can conduct an electric current; see sections 2.4c and 25.3a). Optimal health requires **electrolyte balance,** which is the state of maintaining homeostatic blood levels of the different ions (see table 18.3 and section 25.3b). For example, drinking orange juice (a good source of K^+) helps to maintain blood K^+ concentration by replacing K^+ lost from the body through sweat and urine. An **electrolyte imbalance** occurs when the blood concentration of an electrolyte becomes either too high or too low. These changes, in some cases, may be fatal. For example, a *lethal injection* (i.e., an injection with a significantly large dose of K^+) causes a severe K^+ imbalance, which induces a fatal cardiac arrest. Good patient care often includes knowing the causes, effects, and symptoms associated with electrolyte imbalances of the common ions (see table 25.1), and understanding how to prevent and treat them.

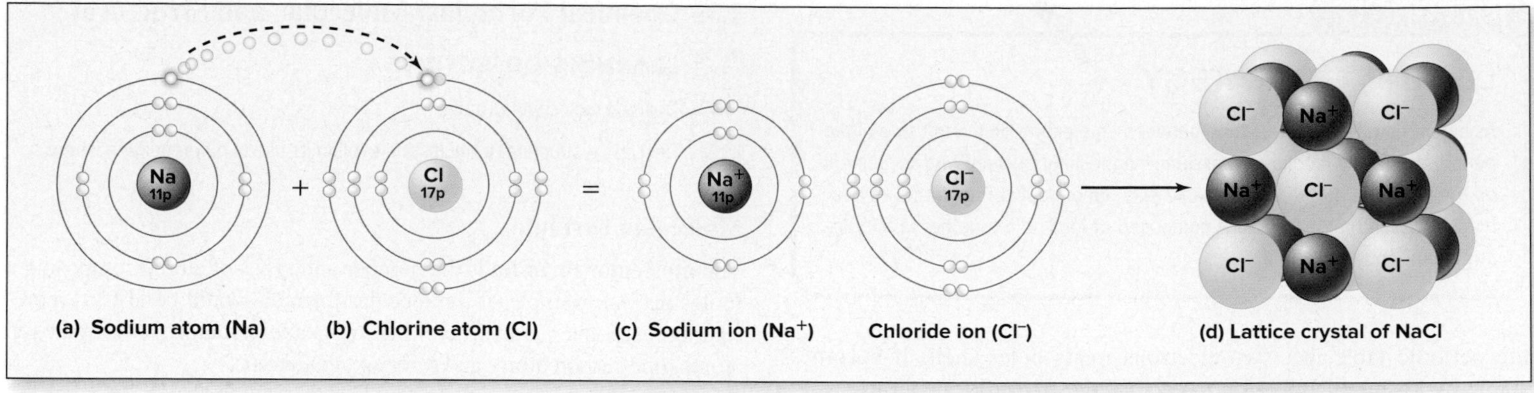

Figure 2.5 Formation of an Ionic Bond Involving Sodium and Chloride. (*a*) A sodium atom donates an electron to a chlorine atom (*b*). (*c*) The loss of an electron from a sodium atom results in the formation of a positively charged sodium ion (Na^+), and the gain of an electron by a chlorine atom results in the formation of a negatively charged chloride ion (Cl^-). (*d*) The NaCl ionic compound (a salt) is composed of a lattice crystal formed by the ionic bonds between Na^+ and Cl^-. AP|R

the charge is calculated as follows: 11(+) and 10 (−) = +1. Ions with a positive charge are called **cations.** Thus, sodium ion is a cation with a +1 charge and is designated as Na^+. Other examples of common cations, as listed in table 2.1, include K^+, Ca^{2+}, Mg^{2+}, and H^+.

Gaining Electrons and the Formation of Anions

A chlorine atom, found in column VIIA of the periodic table (figure 2.4), is a good example of an element that can reach stability by gaining an electron. The atomic structure of chlorine contains seven electrons in its outer shell, and by gaining one electron, stability is reached (figure 2.5*b*). The structure formed is referred to as a chloride ion. Because an electron has been gained, the chloride ion has 17 protons and now has 18 electrons, and the charge is calculated as follows: 17(+) and 18(−) = −1, written as Cl^-. Negatively charged ions are called **anions.** Common anions are listed in table 2.1.

Two of the common anions included in table 2.1, bicarbonate ion (HCO_3^-) and phosphate ion (PO_4^{3-}), are composed of more than one atom and are referred to as **polyatomic** (pol′ē-ă-tom′ik; *poly* = many) **ions.** Bicarbonate ion contains one oxygen atom that has gained an electron, whereas phosphate ion contains three oxygen atoms that have each gained an electron.

General Rules for Assigning Charges

A general rule helps you determine which atoms gain or lose electrons and the charge that they develop (figure 2.4). Atoms with one, two, or three electrons in the outer shell generally *donate* electrons and become positively charged cations. The amount of charge depends upon the number of electrons donated—namely, one, two, or three. For example, because calcium has two electrons in its outer shell, it reaches stability by losing two negatively charged electrons, develops a charge of +2, and in ionic form is written as Ca^{2+}.

In comparison, atoms with five, six, or seven electrons in the outer shell tend to *gain* electrons and become negatively charged anions. The amount of charge depends upon the number of electrons gained to meet the octet rule criterion—namely, three, two, or one. An atom with seven electrons in the outer shell, such as chlorine, reaches stability and becomes chloride (Cl^-) when it gains one electron and develops a −1 charge.

WHAT DO YOU THINK?

2 Can you predict the specific charge for magnesium ion (atomic number 12) based on its position in the periodic table?

Using the Periodic Table to Assign Charges

The periodic table (see figure 2.1*a*) may be used to quickly assess whether an atom will become a cation or anion and the amount of its specific charge. Usually, elements on the left side of the periodic table, and in column IIIA, tend to lose electrons. The specific positive charge is dependent upon the position of the element in the periodic table: Group IA = +1, group IIA = +2, and group IIIA = +3. In contrast, elements on the right side of the periodic table (columns VA–VIIA) tend to gain electrons. The specific amount of negative charge is as follows: Group VA = −3, group VIA = −2, and group VIIA = −1. Sometimes there are anomalies, such as iron (Fe), which is the metal in the hemoglobin molecule within red blood cells (erythrocytes). It can form more than one type of ion, either a ferrous (Fe^{2+}) or a ferric (Fe^{3+}) ion.

WHAT DID YOU LEARN?

5 List the common cations and anions of the human body, including their name and symbol.

6 On the periodic table (see figure 2.1), highlight the elements that form the common ions of the human body (do not include the polyatomic ions).

7 Explain how and why ions form based on the octet rule.

2.2b Ionic Bonds

LEARNING OBJECTIVES

13. Define an ionic bond.
14. Describe an ionic compound of NaCl.
15. List other examples of ionic compounds.

Positively charged cations and negatively charged anions may bind together by electrostatic interactions called **ionic bonds.** The structure formed is typically a **salt.** A classic example involves the formation of common table salt from atoms of sodium and atoms of chlorine. Each sodium atom loses one outer shell electron to a chlorine atom. The sodium atom then becomes a sodium ion (Na^+), and the chlorine atom becomes a chloride ion (Cl^-). The oppositely charged Na^+ and Cl^- ions are held together by ionic bonds in a precise, lattice crystal structure composing an **ionic compound** (figure 2.5*d*). NaCl is the smallest repeating structure of this ionic compound and represents its chemical formula (the chemical constituents in a compound and their ratios).

Consider also the ionic compound magnesium chloride ($MgCl_2$). Notice that the chemical formula contains one magnesium and two chloride ions. This is because magnesium, which is in column IIA on

INTEGRATE

LEARNING STRATEGY

As a general rule, the interaction between an element on the left side of the periodic table (e.g., sodium, potassium, magnesium, calcium) with an element on the right side of the periodic table (e.g., chlorine, fluorine) forms an ionic bond. The salt that is formed is composed of ions held together in a lattice arrangement by ionic bonds.

the periodic table, has two electrons in its outer shell. It becomes stable by losing one electron to each of the two chlorine atoms.

Other examples of ionic compounds include those involving polyatomic anions, sodium bicarbonate ($NaHCO_3$), and the most common ionic compound in the body, calcium phosphate, $Ca_3(PO_4)_2$, which helps harden bones and teeth. Notice that each of these examples is a combination of a common cation and common anion (table 2.1).

WHAT DID YOU LEARN?

8 Could an ionic bond form between two cations or between two anions? Explain.

2.3 Covalent Bonding, Molecules, and Molecular Compounds

Instead of losing or gaining electrons to form ionic compounds that are arranged in a lattice, atoms also can reach chemical stability by sharing electrons. The sharing of electrons between two atoms results in a *covalent bond.* The resulting structure of the covalently bonded atoms is a **molecule.** Molecules composed of two or more different elements are more specifically called **molecular compounds.** Thus, molecules of carbon dioxide (CO_2) and water (H_2O) are molecular compounds. However, molecules composed of only one element are not, such as molecular oxygen (O_2) and molecular hydrogen (H_2).

Here we describe covalent bonds, the different types of covalently bonded molecules, and interactions between covalently bonded molecules, after first discussing how molecules are represented with molecular and structural formulas.

2.3a Chemical Formulas: Molecular and Structural

LEARNING OBJECTIVES

16. Define a molecular formula.

17. Describe a structural formula, and explain its use in differentiating isomers.

Molecular Formula

The **molecular formula** is the number and types of atoms composing a molecule. An example of a molecular formula would be H_2CO_3 representing carbonic acid, where the molecule contains two hydrogen atoms, one carbon atom, and three oxygen atoms.

Structural Formula

The **structural formula** of a molecule is complementary to its molecular formula and exhibits not only the numbers and types of atoms but also their spacial arrangements within the molecule. Atoms are always arranged in a defined manner within each molecule. The molecular formula for carbon dioxide (CO_2) thus is complemented by its structural formula (O=C=O).

Structural formulas provide a means for differentiating **isomers,** which are molecules composed of the same number and types of elements but arranged differently in space. Two important isomers in humans are glucose and galactose (**figure 2.6**). These sugar molecules both have a molecular formula of $C_6H_{12}O_6$, indicating that each molecule has 6 carbon, 12 hydrogen, and 6 oxygen atoms. However, the atoms are arranged differently in space, as shown in their structural formulas. Notice the different arrangement of atoms on carbon four in the six-sided ring structure of glucose and galactose. (These sugars in a ring-structure have carbon atoms that are numbered beginning with the carbon to the right of the oxygen as carbon one, and then are numbered consecutively clockwise around the ring.) Although fructose is a five-sided ring, it is also an isomer of glucose and galactose because it has the same molecular formula ($C_6H_{12}O_6$), but it has a different structural arrangement.

Isomers may have very different properties from one another—therefore, structural formulas are an important piece of chemical information.

WHAT DID YOU LEARN?

9 What information about a molecule is gained by a structural formula? How does a structural formula differ from a molecular formula?

10 What is an isomer?

Molecular formula

Structural formula

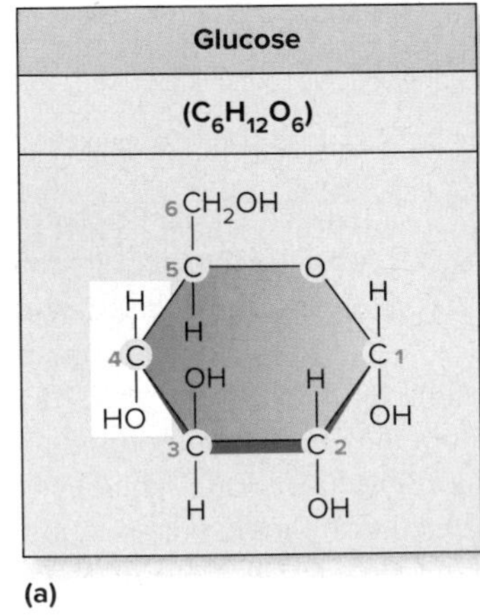

(a)

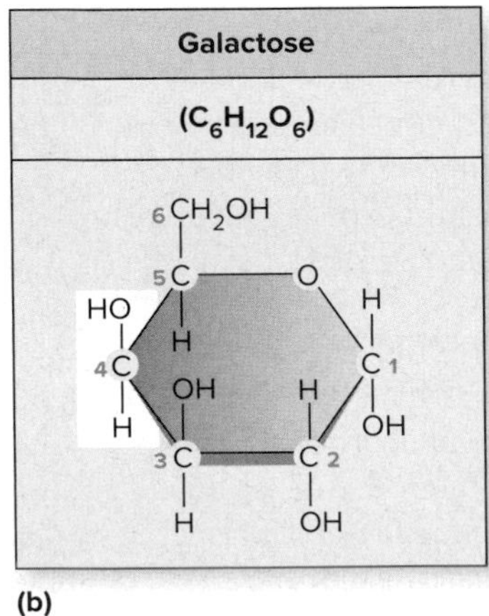

(b)

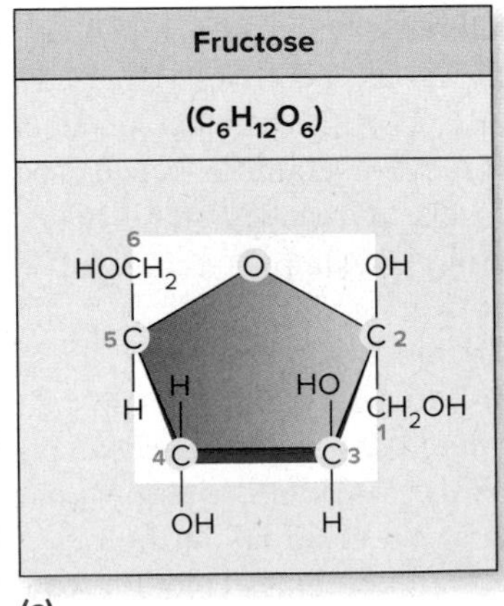

(c)

Figure 2.6 Isomers. Isomers contain the same number and types of elements, but are arranged differently in space. (*a*) Glucose, (*b*) galactose, and (*c*) fructose are isomers that share the same molecular formula ($C_6H_{12}O_6$) but have different structural formulas (highlighted in yellow).
Note: The carbon atoms are numbered consecutively in glucose and galactose, beginning with the carbon to the right of the oxygen in the ring. In fructose, carbon one is the carbon attached to the ring.

2.3b Covalent Bonds

LEARNING OBJECTIVES

18. Describe a covalent bond, and explain its formation based on the octet rule.
19. List the four most common elements in the human body.
20. Distinguish between single, double, and triple covalent bonds.
21. Explain polar and nonpolar covalent bonds.

The bond that is formed when atoms *share* electrons is a **covalent bond.** A covalent bond forms when both atoms require electrons to become stable. This takes place when the participating atoms that form the chemical bond have four, five, six, or seven electrons in the outer shell. View figure 2.4 and note that this applies to the elements on the right side of the periodic table. (The exception is hydrogen because only two electrons are needed to complete its outermost electron shell.)

INTEGRATE

LEARNING STRATEGY

The interaction between elements on the right side of the periodic table (e.g., carbon, oxygen, nitrogen), or between these elements and a hydrogen atom, typically forms a covalent bond. The resulting structure, composed of elements that are covalently bonded, is a molecule.

The four most common elements of the human body that form covalent bonds are hydrogen (H), oxygen (O), nitrogen (N), and carbon (C). These four elements account for over 96% of the body's weight. The simplest example of covalent bond formation occurs when hydrogen gas is formed from two hydrogen atoms. The covalent bond formation between two hydrogen atoms fills the outer shell of each atom because each hydrogen atom shares its single electron (and only two electrons are required for the first shell to be complete).

The Number of Bonds an Atom Can Form

Although hydrogen can share only one pair of electrons and thus produce one covalent bond to become stable, some elements are able to share more than one pair. The number of covalent bonds formed by an atom may be determined by examining the number of electrons needed to complete the outer shell (see figure 2.4).The atomic structure of the four most common elements that compose molecules shows that hydrogen needs one electron, oxygen needs two electrons, nitrogen needs three electrons, and carbon needs four electrons. Thus, hydrogen can form one covalent bond, oxygen two, nitrogen three, and carbon four.

INTEGRATE

LEARNING STRATEGY

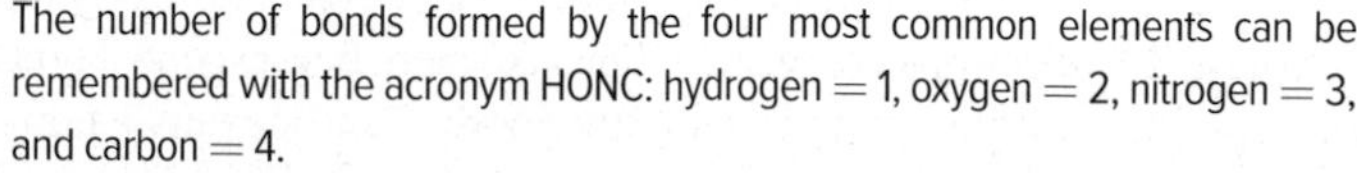

The number of bonds formed by the four most common elements can be remembered with the acronym HONC: hydrogen = 1, oxygen = 2, nitrogen = 3, and carbon = 4.

Single, Double, and Triple Covalent Bonds

Atoms of elements that can form more than one covalent bond may do so through combinations of single, double, or triple covalent bonds **(figure 2.7)**. A **single covalent bond** is one pair of electrons shared between two atoms. The bond just described between two

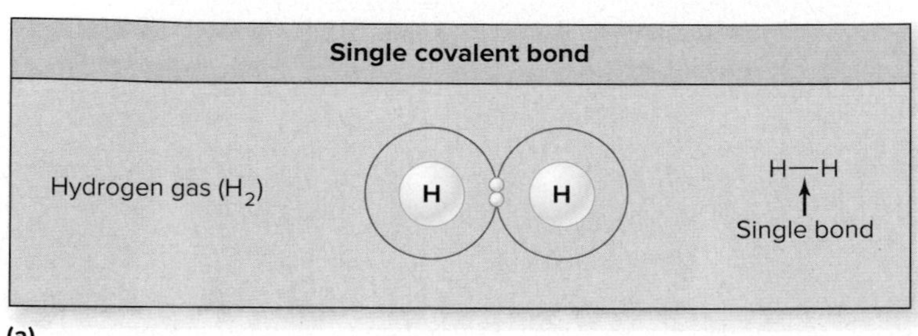

(a)

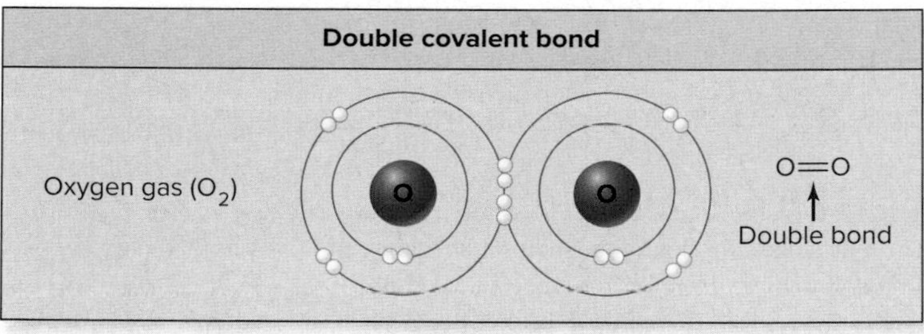

(b)

Triple covalent bond

Nitrogen gas (N_2) N N N≡N Triple bond

(c)

Figure 2.7 Single, Double, and Triple Covalent Bonds. Covalent bonds involve the sharing of electrons. (*a*) One pair of electrons is shared in a single covalent bond, (*b*) two pairs of electrons are shared in a double covalent bond, and (*c*) three pairs of electrons are shared in a triple covalent bond.

hydrogen atoms is a single covalent bond. A **double covalent bond** involves the sharing of two pairs of electrons between two atoms. An example is the double covalent bond between two oxygen atoms. The sharing of two pairs of electrons between oxygen atoms is necessary to achieve stability because each oxygen atom has only six electrons in its outer shell but needs eight electrons to satisfy the octet rule. Three pairs of electrons are shared between atoms in some molecules, forming a **triple covalent bond.** A prime example is the triple covalent bond between two nitrogen atoms. The order of stability (i.e., the more energy required to break the bonds) is from least to greatest: single covalent bond, double covalent bond, triple covalent bond.

An atom of a specific element might share electrons in a variety of ways to satisfy the octet rule. For example, a carbon atom contains four electrons in the outer electron shell and thus needs four electrons to satisfy the octet rule. These four electrons can be obtained in a number of ways to form different types of molecules **(figure 2.8)**.

Carbon Skeleton Formation

In later sections of this chapter and in subsequent chapters we will see examples of molecules that have numerous carbon atoms bonded together within their chemical structure. The arrangement of these carbon atoms is referred to as the molecule's **carbon skeleton.** The carbon skeleton can be thought of as the molecule's "backbone." Three common carbon skeleton arrangements are: a straight chain, a branched chain, and a ring **(figure 2.9)**. Notice that in a structural formula of a carbon chain or ring skeleton, the letter C is often not included with the understanding that a carbon atom lies where lines meet at an angle. In addition, if a carbon atom does not have its

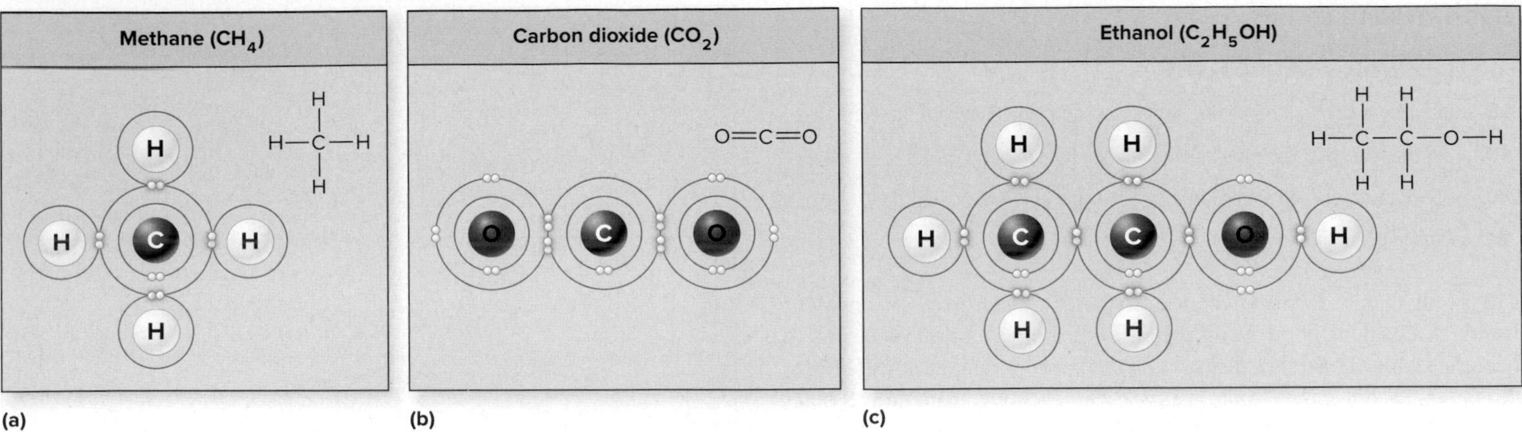

Figure 2.8 Carbon-Containing Molecules. Carbon can form four covalent bonds and may do so in many ways. For example, carbon can form (*a*) four single covalent bonds with four hydrogen atoms in methane; (*b*) two double bonds with oxygen atoms in carbon dioxide; and (*c*) four single covalent bonds in a more complex structure like ethanol.

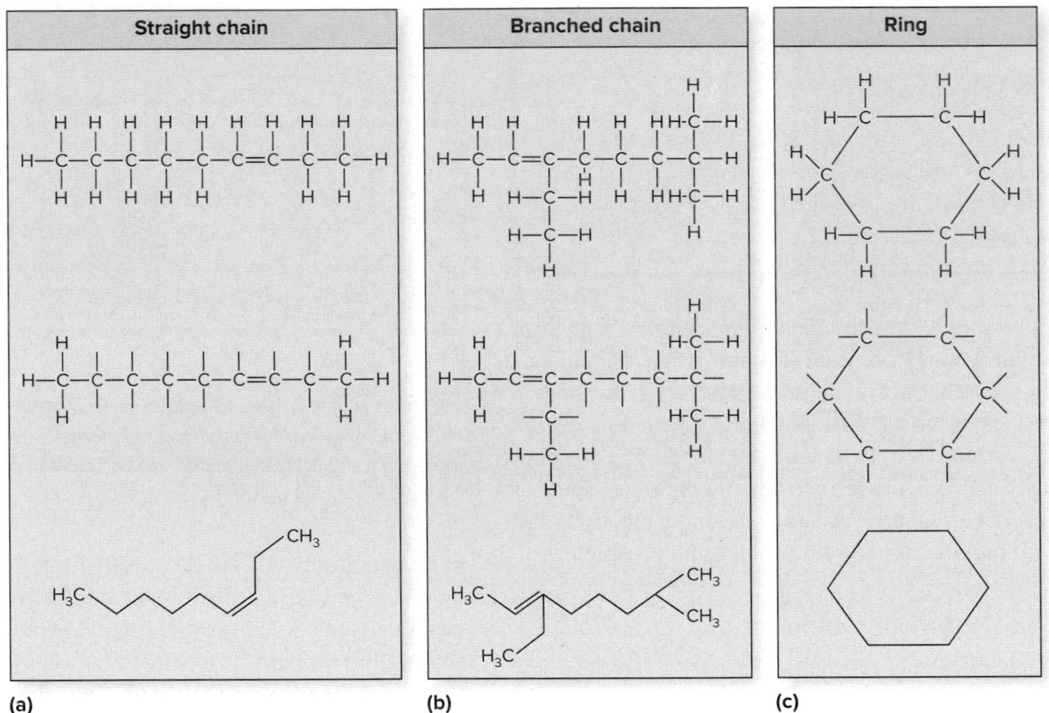

Figure 2.9 Carbon Skeleton. A carbon skeleton has three major types of arrangements: (*a*) a straight chain, (*b*) a branched chain, and (*c*) a ring. Remember the shorthand convention that a carbon atom is present where the lines meet at an angle. In addition, each carbon atom must have four bonds; the bonds that are not shown for each carbon atom are bonds between the carbon atom and a hydrogen atom.

required four bonds shown in the structural formula, each bond that is not shown is a bond between the carbon and a hydrogen atom.

Nonpolar and Polar Covalent Bonds

Atoms share electrons in a covalent bond either equally or unequally between the atoms. How they share is determined by the relative attraction each atom has for electrons, a concept referred to as **electronegativity.** Because two atoms of the same element, such as two hydrogen atoms, two oxygen atoms, or two carbon atoms, have equal attraction for electrons, they share the electrons equally. The resultant bond is a **nonpolar covalent bond.**

Different types of atoms have varying degrees of electronegativity, or attraction for electrons, and thus may share the electrons unequally. The resultant bond is a **polar covalent bond** (an important exception will be described at the end of this section). As a general rule, electronegativity increases both from left to right across a row of the periodic table and from bottom to top in a column (see figure 2.1*a*). These trends reflect that, for elements across a specific row, a greater number of protons in the nucleus of the atom are "pulling" on the electrons. For elements in a column, the elements closer to the top of the periodic table have electrons in shells closer to the nucleus (see figure 2.4). Thus, electronegativity is determined both by the number of protons in the nucleus and by the proximity of the valence electron shells to the nucleus.

Consequently, for the four most common elements composing living organisms, the order of electronegativity from least to greatest is as follows: hydrogen $<$ carbon $<$ nitrogen $<$ oxygen. Therefore, of these four elements, hydrogen has the least attraction for electrons and oxygen has the greatest. The electrons being shared in a covalent bond spend a greater amount of time orbiting the nucleus of the more electronegative atom.

Because electrons have a negative charge, the more electronegative atom develops a partial negative charge, and the less electronegative atom develops a partial positive charge. Partial charges are written using the Greek letter delta (δ) followed by a superscript + or – to designate the relative charge. For example, a polar covalent bond formed between oxygen and hydrogen would be assigned and written as $^{\delta-}O—H^{\delta+}$. Note that the term **polar** (or *dipole*) refers to the "poles" of partial electrical charges, which are analogous to poles of a magnet.

Polar bonds have varying degrees of unequal sharing of electrons. Thus, polar bonds are along a continuum with ionic bonds (which involve a complete donation of electrons) on one end and nonpolar bonds (with equal sharing of electrons) on the other.

One important exception exists to the rule that a polar bond generally forms between two different atoms. The exception is carbon bonding with hydrogen. Because the electronegativity difference between carbon and hydrogen is relatively small, a covalent bond produced between carbon and hydrogen (C—H) has approximately equal sharing of electrons, and an essentially *nonpolar* covalent bond forms between them.

WHAT DID YOU LEARN?

11 Explain covalent bond formation in terms of chemical stability.

12 Assign the partial charges between nitrogen and hydrogen (N—H) in a polar covalent bond.

13 Why are some covalent bonds nonpolar and others polar? Identify the exception to the rule that polar covalent bonds are formed between two different types of atoms.

2.3c Nonpolar, Polar, and Amphipathic Molecules

LEARNING OBJECTIVES

22. Describe the difference between a nonpolar molecule and a polar molecule.
23. Define an amphipathic molecule.

Covalent bonds may be nonpolar or polar. To determine whether an entire molecule is nonpolar or polar, the molecule as a whole is evaluated to determine the prevalence (and relative strength) of its nonpolar and polar bonds. The most important concept to remember is that **nonpolar molecules** contain primarily nonpolar covalent bonds between the atoms within the molecule. As previously described, nonpolar covalent bonds are bonds formed between the same elements (e.g., C—C, O—O), by C—H bonds, or both. Oxygen (O_2) and triglyceride (fat) molecules are examples of nonpolar molecules (**figure 2.10*a***).

In comparison, **polar molecules** contain relatively more polar covalent bonds between the atoms within the molecule. Recall from

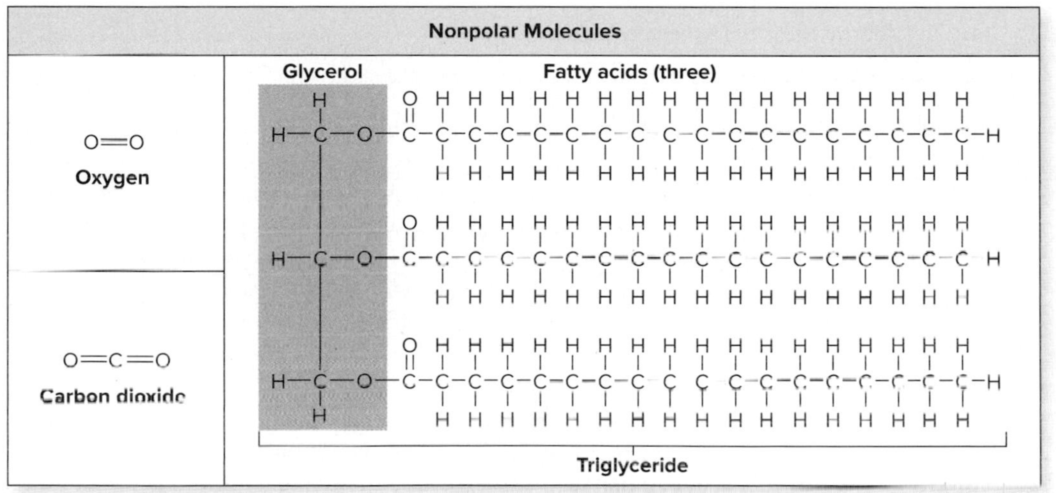

(a)

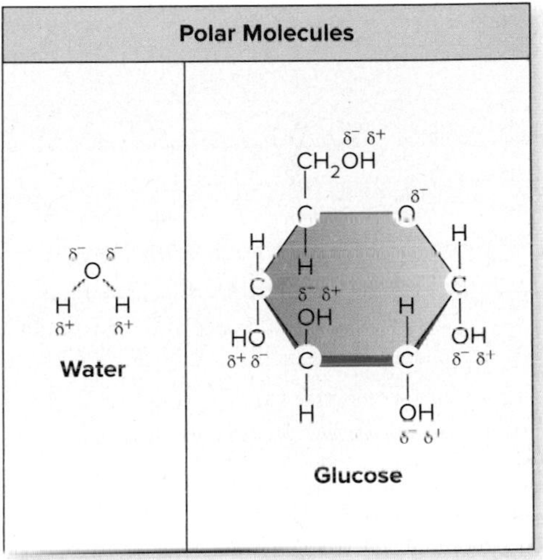

(b)

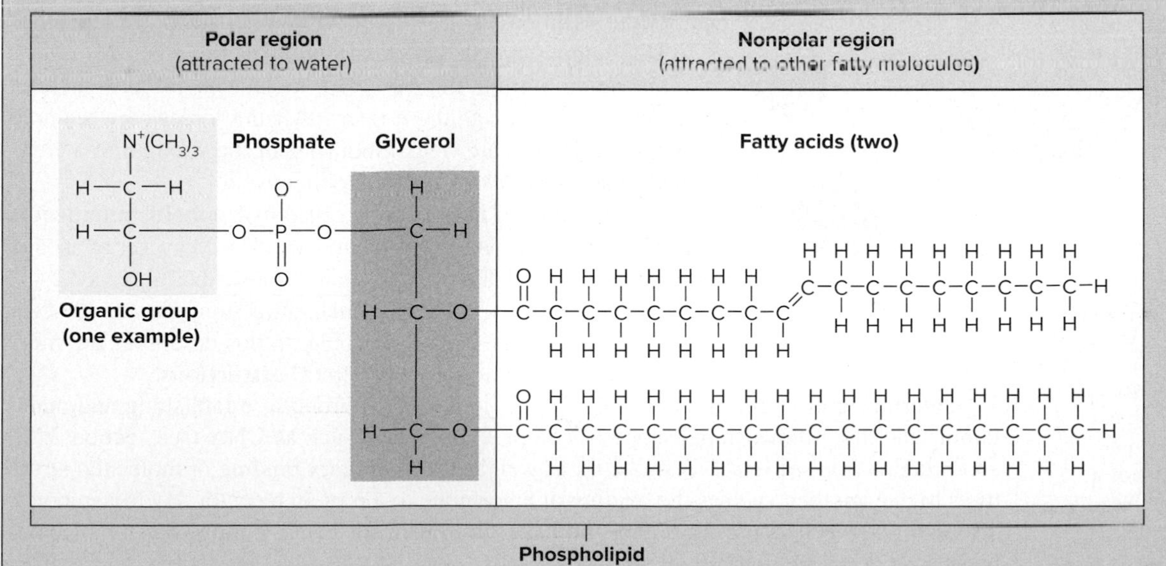

(c)

Figure 2.10 Nonpolar, Polar, and Amphipathic Molecules. (*a*) Oxygen, carbon dioxide, and a triglyceride are examples of nonpolar molecules. (*b*) Water and glucose are examples of polar molecules. (*c*) A phospholipid is an example of an amphipathic molecule.

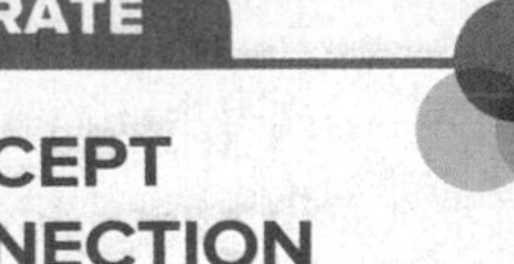

INTEGRATE

CONCEPT CONNECTION

The two respiratory gases, O_2 and CO_2, are both nonpolar molecules. This chemical characteristic—and the fact that they are small in size—permits both to easily cross cell membranes. The diffusion of respiratory gases is discussed in other chapters of the text, including section 4.3a on membrane transport and section 23.6 on the exchange of O_2 and CO_2 between the blood and alveoli (air sacs) and between the blood and systemic cells.

the previous section that polar covalent bonds are between different elements, such as O—H, C—O, N—H, and N—O. Water (H_2O) and glucose ($C_6H_{12}O_6$) are both polar molecules (figure 2.10*b*). Notice that oxygen is bonded to two hydrogen atoms in a water molecule, and several C—O and O—H bonds are within a glucose molecule. One exception to this general pattern exists: A molecule containing polar covalent bonds that extend in opposite directions can be nonpolar because the partial charges cancel each other. Carbon dioxide ($^{\delta-}O{=}C{=}O^{\delta-}$) is an example.

Sometimes a molecule is large enough that it can have one major part that is nonpolar and another part that is polar. Molecules that contain both nonpolar and polar components are called **amphipathic** (*amphi* = both, *patheia* = feeling) **molecules.** A phospholipid molecule is an example of an amphipathic molecule (figure 2.10*c*).

WHAT DO YOU THINK?

3 Is the fatty acid portion of a triglyceride (figure 2.10*a*) a nonpolar or polar molecule? Explain your answer. Would you predict that it will or will not dissolve in water?

WHAT DID YOU LEARN?

14 Are O_2 and CO_2 nonpolar or polar molecules?

2.3d Intermolecular Attractions

LEARNING OBJECTIVES

24. Describe hydrogen bonding between polar molecules.

25. List and define the intermolecular attractions between nonpolar molecules.

Molecules sometimes have weak chemical attractions to other molecules called **intermolecular** (*inter* = between) **attractions.** One important intermolecular attraction is called a hydrogen bond. A **hydrogen bond** forms between polar molecules. It is a weak attraction between a partially positive (δ^+) hydrogen atom within a polar molecule and a partially negative (δ^-) atom within a polar molecule. The partially negative atom is usually oxygen, but sometimes nitrogen. A hydrogen bond is designated in this text with a dotted or broken line. Several hydrogen bonds are shown in **figure 2.11.** Although an individual hydrogen bond is a weak bond or attraction (approximately 5–10% the strength of a covalent bond), hydrogen bonds are strong collectively.

Another type of intermolecular attraction involves nonpolar molecules. These occur when electrons orbiting the nucleus of an atom of a nonpolar molecule are distributed unequally for a brief instant. One portion of the atom is slightly negative and another portion is slightly positive. This momentary unequal distribution of charge in the atom induces an unequal distribution of electrons in an adjacent atom of another nonpolar molecule. This temporary unequal distribution of the electrons allows the atoms within the nonpolar molecules to form a brief attraction. Although these intermolecular forces between nonpolar molecules are weak (about 1% of the strength of a covalent bond), they also are strong collectively.

A third type of interaction is called **hydrophobic interaction,** which results when nonpolar molecules are placed in water or another polar substance. (Hydrophobic interactions are described in section 2.4c.)

Note that all of these intermolecular attractions may also occur between different portions of a large molecule. In this case, they are more appropriately called **intramolecular** (*intra* = within) **attractions.**

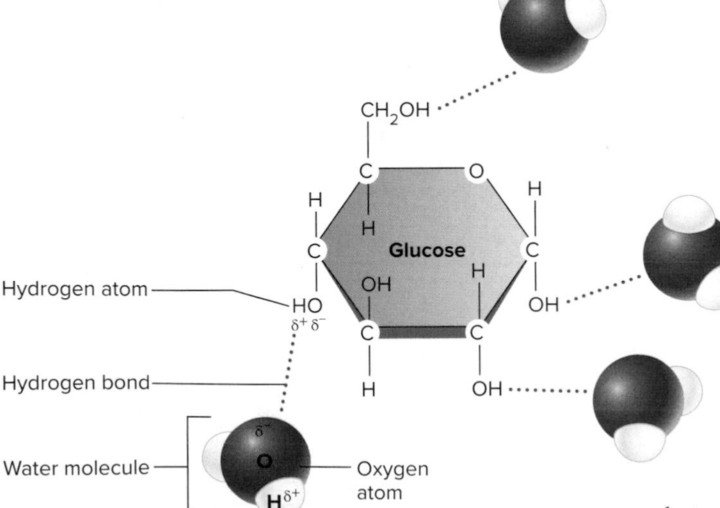

Figure 2.11 Hydrogen Bonding. Several hydrogen bonds between glucose and water are shown. Each is formed between a partially positive charged hydrogen atom of a glucose molecule and the partially negative charged oxygen atom of a water molecule.

Intermolecular (and intramolecular) attractions are important in establishing and maintaining the three-dimensional shape of complex molecules such as DNA (see section 2.7d) and proteins (see sections 2.7e and 2.8), as well as the temporary binding of molecular structures to one another, such as the binding of a hormone to a protein receptor. Hydrogen bonds also form between water molecules and have significant influence in how water molecules behave, as described in section 2.4.

WHAT DID YOU LEARN?

15 What is the name of the intermolecular attraction between a partially charged hydrogen of one polar molecule and a partially negative atom of another polar molecule?

2.4 Molecular Structure and Properties of Water

Chemists classify molecules into two broad categories—organic molecules and inorganic molecules. **Organic molecules** are defined as molecules that contain carbon, which are (or have been) components of living organisms (e.g., glucose, protein, triglycerides). All other types of molecules are **inorganic molecules.** Examples of inorganic molecules include water, salts (e.g., sodium chloride), acids (e.g., hydrochloric acid), and bases (e.g., sodium hydroxide). The remaining portions of this chapter discuss in detail specific types of molecules—including water, acids, bases, and the four primary organic biological macromolecules (lipids, carbohydrates, nucleic acids, and proteins).

Water is the first molecule that we examine in detail. It is appropriate to begin with water because it is the substance that composes approximately two-thirds of the human body by weight. We first look closely at the molecular structure of water molecules and then describe several important water properties. The relevance to normal body activities is included for each property.

2.4a Molecular Structure of Water

LEARNING OBJECTIVE

26. Describe the molecular structure of water and how each water molecule can form four hydrogen bonds.

Water is a polar molecule composed of one oxygen atom bonded to two hydrogen atoms. It exhibits polarity because there is an unequal sharing of electrons between the oxygen atom and each of the two hydrogen atoms **(figure 2.12*a*)**. The oxygen atom is more electronegative, and it has two partial negative charges. In contrast, each hydrogen atom exhibits a single partial positive charge.

Every water molecule has the ability to form four hydrogen bonds with adjacent water molecules. This is because each of the two hydrogen atoms forms one hydrogen bond, and each oxygen atom forms two hydrogen bonds (figure 2.12*b*). Recall from our earlier discussion of hydrogen bonds that these intermolecular attractions are individually weak but collectively very strong. Thus, hydrogen bonding between water molecules is central to the properties exhibited by water.

WHAT DID YOU LEARN?

16 What is the intermolecular bond that is significant in determining the properties of water?

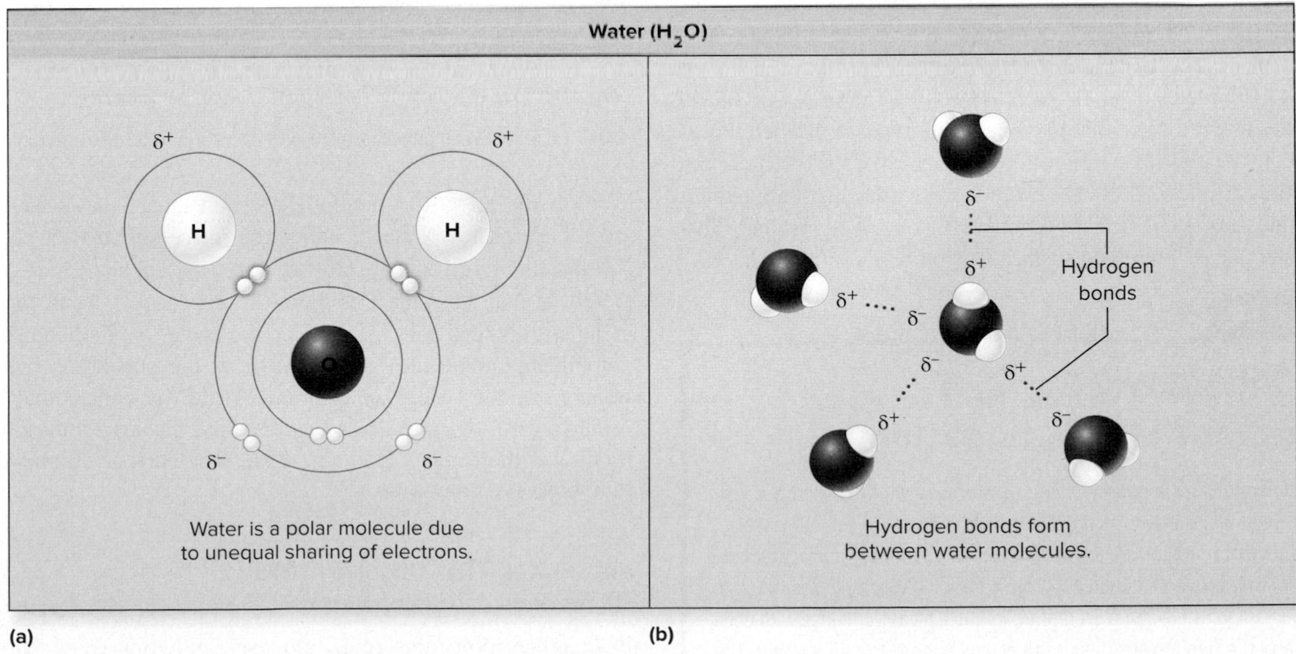

Figure 2.12 Water Molecule. (*a*) A water molecule is a polar molecule due to the unequal sharing of electrons between the oxygen atom and each of the two hydrogen atoms. Partial charges are shown. (*b*) Hydrogen bonds form between the partial positive (δ^+) hydrogen atom of one water molecule and the partial negative (δ^-) oxygen atom of a different water molecule.

2.4b Properties of Water

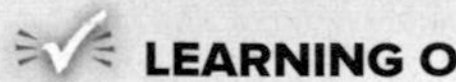

27. List the different properties of water, and provide an example of the importance of each property within the body.

Phases of Water

Water is present in three phases, depending upon the temperature: a *gas* (water vapor), a *liquid* (water), and a *solid* (ice). Substances that have a low molecular mass (described in section 2.6b) such as water generally are present in the gaseous phase at room temperature. However, water is liquid at room temperature because hydrogen bonds hold the water molecules in the liquid phase and limit their escape into the gaseous phase. Almost all water within the human body occurs within the liquid phase, although small amounts of water are present as water vapor within the air passageways. As a liquid, water serves the following functions:

- **Transports.** Substances are dissolved in water and moved throughout the body in water-based fluids (e.g., blood [see section 18.1b] and lymph [see section 21.1a]).
- **Lubricates.** Water-based fluids located between body structures decrease friction (e.g., serous fluid between the heart and its sac [see sections 1.5e and 19.2b], synovial fluid within joints [see section 9.4a]).
- **Cushions.** The force of sudden body movements is absorbed by water-based fluids (e.g., cerebrospinal fluid surrounding the brain and spinal cord [see section 13.2c]).
- **Excretes wastes.** Unwanted substances are eliminated in the body dissolved in water (e.g., urine [see section 24.8a]).

Cohesion, Surface Tension, and Adhesion

Cohesion is the attraction between water molecules. They are inclined to "stick together" because hydrogen bonds form between these molecules. **Surface tension** is the inward pulling of cohesive forces at the surface of water. This inward attraction occurs because water molecules at the surface are pulled by hydrogen bonds in only three directions, whereas water molecules that are internal in the liquid are pulled by hydrogen bonds in four directions. **Adhesion** is the attraction between water molecules and a substance other than water. This occurs when hydrogen bonds form between water molecules and the molecules that compose those other substances.

Surface tension can be demonstrated readily. Try this experiment: First, stack two flat plates of glass, such as two clean microscope slides, and then lift them apart. Note that they are easily separated. Then repeat this experiment, but first place one or two drops of water between the slides before they are stacked. Note it is much more difficult, if not impossible, to separate the plates of glass without first wedging them apart. This is because water causes increased surface tension between the surfaces.

INTEGRATE

CLINICAL VIEW 2.2

Surface Tension and Surfactant

Surface tension can be exhibited within the walls of the alveoli (air sacs) because opposing alveolar walls are moist and if collapsed will stick together. However, we produce a mixture of lipids and proteins called *pulmonary surfactant* (see section 23.3d) that prevents the alveoli from collapsing when we breathe out. Without surfactant (a risk of some premature infants), alveoli collapse with each breath out, and the two moist sides of the alveoli adhere to one another. The next breath in requires breaking of the surface tension within alveoli and their reinflation— a situation that requires much greater effort.

High Specific Heat and High Heat of Vaporization

Temperature is a measure of the kinetic energy, or random movement, of atoms or molecules within a substance. The relationship between temperature and kinetic energy is direct—the temperature is higher when there is a greater amount of kinetic energy in an object. Two properties of water influence water temperature: its specific heat and the heat of vaporization.

Specific heat is the amount of energy (measured in calories) required to increase the temperature of 1 gram of a substance by 1 degree Celsius (C). The specific heat of water has one of the highest values of any substance (1 calorie/gram/°C). This is because most of the energy imparted into water during heating is first used to break hydrogen bonds. Thereafter, the energy from heating increases the kinetic energy of water molecules. As we change between cool and warm environments, or generate large amounts of heat during physical exertion, much of this heat (energy) is used to break hydrogen bonds and not increase the random movement of molecules (kinetic energy) within the body. Thus, body temperature remains relatively constant.

Heat of vaporization is the energy required for the release of molecules from a liquid phase into the gaseous phase for 1 gram of a substance. Water has a high heat of vaporization because the hydrogen bonds between individual water molecules must first be broken before these molecules can be released from the liquid phase into the gaseous phase. This is the reason sweating is an effective measure in helping to cool the body. As water molecules evaporate from the surface of the skin, excess heat is dissipated from the body as liquid water is changed to a gas.

WHAT DID YOU LEARN?

17 Which property of water contributes to the need to produce surfactant and prevent collapse of the alveoli? Which property contributes to body temperature regulation through sweating?
Why is sweating less effective in cooling the body on a humid day?

2.4c Water as the Universal Solvent

LEARNING OBJECTIVES

28. Compare substances that dissolve in water with those that both dissolve and dissociate in water. Distinguish between electrolytes and nonelectrolytes.

29. Describe the chemical interactions of nonpolar substances and water.

30. Explain how amphipathic molecules interact in water to form chemical barriers.

Water is the **solvent** of the body, and substances that dissolve in water are called **solutes.** Water is called the **universal solvent** because most substances dissolve in it. However, not all substances dissolve completely or at all in water. The chemical properties of a substance (whether it is a charged ion, a nonpolar molecule, a polar molecule, or an amphipathic molecule) determine how the substance interacts with water molecules. Here we examine (1) the substances that dissolve in water (polar molecules and ions), (2) those that do not dissolve in water (nonpolar molecules), and (3) those that partially dissolve in water (amphipathic molecules).

Substances That Dissolve in Water (Polar Molecules and Ions)

Both polar molecules (e.g., glucose) and ions (e.g., Na^+, HCO_3^-) favorably interact with water molecules to disperse or **dissolve** within water. For this reason they are appropriately called **hydrophilic** (meaning *water-loving*).

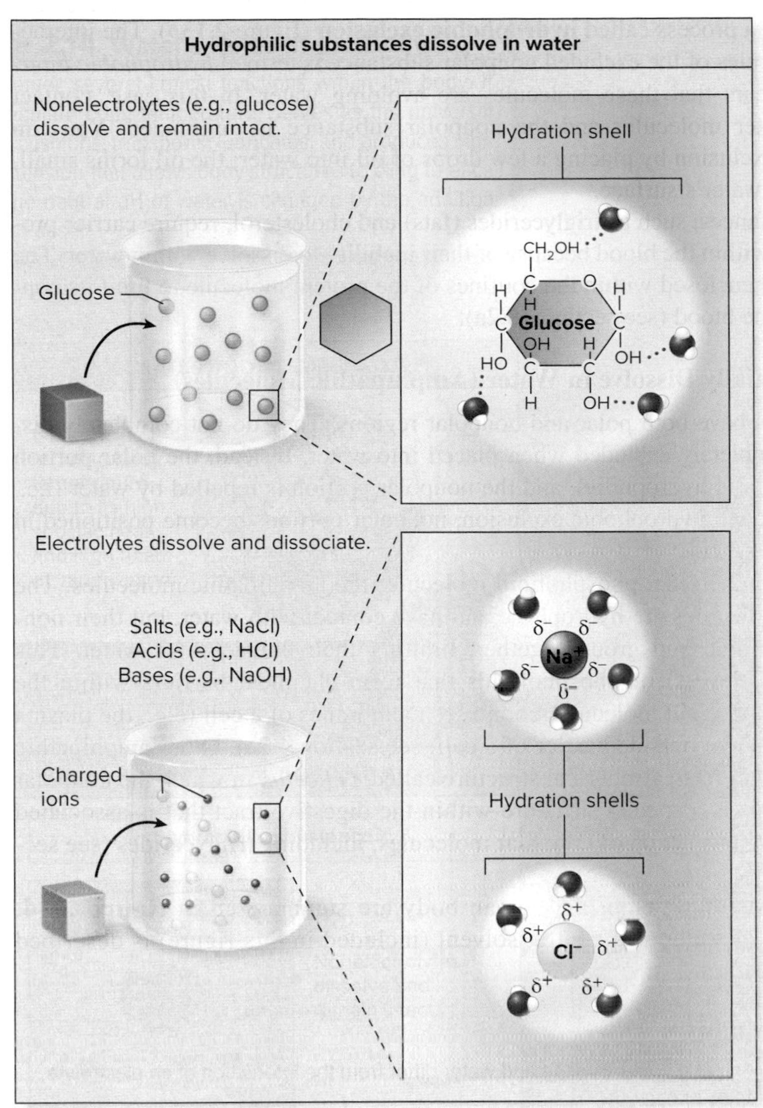

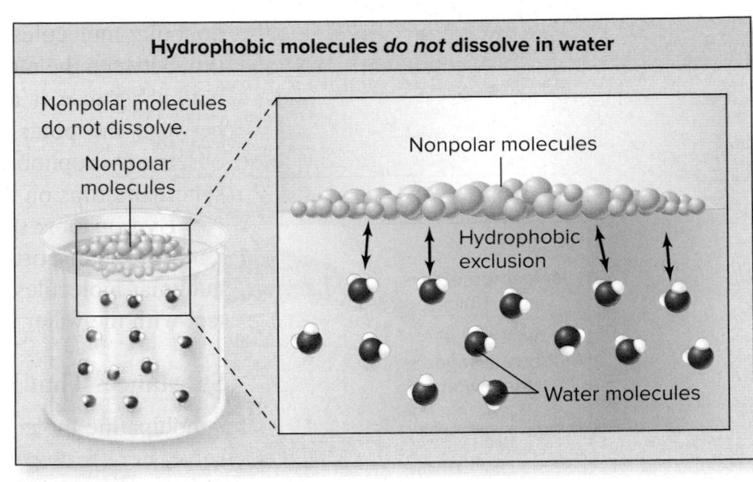

Figure 2.13 Substance Interaction with Water. Chemical properties of the substance combined with water will determine their interaction. (*a*) Hydrophilic substances dissolve in water. Nonelectrolytes, which include polar molecules, dissolve and remain intact within water as water molecules form a hydration shell around each molecule. Electrolytes, which include salts, acids, and bases, dissolve and dissociate in water to a certain extent as water molecules form hydration shells around each ion. (*b*) Hydrophobic molecules are nonpolar and do not dissolve in water; rather, the nonpolar molecules are "pushed" out of the water by hydrophobic exclusion. (*c*) Amphipathic molecules are unique in that the polar portion dissolves in water and the nonpolar portion does not. Membranes (e.g., plasma membrane, nuclear envelope) formed from phospholid molecules and micelles formed from bile salts are two common arrangements formed within the body by amphipathic molecules.

Polar molecules such as glucose dissolve in water as a result of the hydrogen bonds that form between those molecules and water molecules. Observe in **figure 2.13*a*** how water molecules surround each polar molecule and form a **hydration shell** around it. Also notice how the individual polar molecules remain intact as they are surrounded by water molecules.

Ionic compounds such as NaCl salt also dissolve in water. Here the water molecules' partial negative charges ($O^{\delta-}$) interact with the cations (Na^+), and the water molecule's partial positive charges ($H^{\delta+}$) interact with the anions (Cl^-) of the salt. A hydration shell forms around each ion. Notice, however, that the interaction of the water molecules and ions of the salt have resulted in the separation of Na^+ and Cl^-. Thus, these substances do not remain intact, but come apart, or **dissociate.** An important characteristic of ionic compounds is that they both dissolve *and* dissociate (separate) when placed in water.

Acids and bases also dissolve and dissociate in water. Hydrochloric acid (HCl) dissociates to form both H^+ and Cl^-, whereas sodium bicarbonate dissociates to form both Na^+ and HCO_3^- ions. Acids and bases are described in detail in section 2.5.

Substances that both dissolve and dissociate in water, such as salts, acids, and bases, can readily conduct an electric current. For this reason, they are called **electrolytes.** In contrast, substances that remain intact when introduced into water, such as glucose, do not conduct an electric current and are called **nonelectrolytes.** Maintaining normal levels of electrolytes including salts, acids, and bases is discussed in section 25.3a.

Substances That Do Not Dissolve in Water (Nonpolar Molecules)

Nonpolar molecules do not dissolve in water, and so they are called **hydrophobic** (meaning *water-fearing*). The hydrogen bonds between water molecules cause the water molecules to be cohesive and attract each other; at the same time, they exclude, or "force out," the

Equal numbers of positively charged hydrogen ions (H^+) and negatively charged hydroxide ions (OH^-) are produced from the dissociation of water. Consequently, water has no net charge and is neutral.

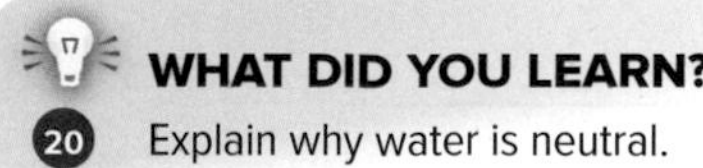

WHAT DID YOU LEARN?

20 Explain why water is neutral.

2.5b Acids and Bases

LEARNING OBJECTIVE

32. Explain the difference between an acid and a base.

An **acid** is a substance that dissociates in water to produce both an H^+ and an anion. An acid increases the concentration of H^+ (written as $[H^+]$) that is free in solution **(figure 2.15)**. Because H^+ is a proton, an acid is also called a **proton donor.** The equation is as follows:

$$\text{Substance A (an acid in water)} \longrightarrow H^+ + \text{Anion}$$

Strong acids dissociate to a greater extent and produce more H^+. Hydrochloric acid (HCl), secreted by cells lining the stomach, is a good example of a strong acid. Weak acids, such as carbonic acid (H_2CO_3) within the blood, dissociate to a lesser extent, producing fewer H^+ ions.

In contrast, a **base** accepts H^+ when added to a solution. Therefore, a base is also called a **proton acceptor.** A base decreases the concentration of H^+ free in solution. Thus,

$$\text{Substance B (a base in water)} + H^+ \longrightarrow \text{B—H}$$

Stronger bases dissociate to a greater extent and bind more H^+ than do weak bases, leaving less H^+ in solution. Sodium hydroxide (NaOH) is an example of a strong base. Weak bases bind less H^+, leaving more H^+ in solution. Bicarbonate (HCO_3^-) is one of the most important weak bases in the body; it is both transported in the blood and released into secretions from the liver and pancreas that enter the small intestine.

WHAT DID YOU LEARN?

21 Which type of substance releases H^+ when added to water?

2.5c pH, Neutralization, and the Action of Buffers

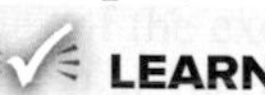

LEARNING OBJECTIVES

33. Define pH, and explain the relative pH values of both acids and bases.

34. Explain neutralization, and describe how the neutralization of both an acid and a base occurs.

35. Describe the action of a buffer.

The **pH** of a solution is a measure of the relative amounts of H^+ it contains; it is expressed as a number between 0 and 14. (The term *pH* is an abbreviation of the phrase *potential of hydrogen.* The unit of measurement for the pH scale is moles/liter.) pH is the inverse relationship of the logarithmic values for a given *hydrogen ion concentration* $[H^+]$:

$$\text{Negative log of } [H^+] = \frac{1}{\text{log of } [H^+]}$$

Recall that water readily dissociates to produce 10^{-7} H^+ and OH^- ions (or 1/10,000,000 ions) per liter. If this concentration of H^+ is placed into the formula just given for pH, then the value for the pH of water is calculated to be equal to 7. In other words:

$$[H^+] = 1 \times 10^{-7}$$

$$[H^+] = 0.0000001$$

$$\text{pH} = 7$$

Notice that as $[H^+]$ changes, pH changes. For example, if the $[H^+]$ increases so that $[H^+] = 1 \times 10^{-4}$, $[H^+] = 0.0001$, then the pH decreases to 4. Conversely, if the $[H^+]$ decreases so that $[H^+] = 1 \times 10^{-9}$, $[H^+] = 0.000000001$, then the pH increases to 9. Thus, $[H^+]$ and pH value are *inversely related.* The inverse relationship between $[H^+]$ and pH is an extremely important concept. Remember, as $[H^+]$ increases, pH decreases, and as $[H^+]$ decreases, pH increases.

Interpreting the pH Scale

Pure water and other solutions that have equal concentrations of H^+ and OH^- are neutral and have a pH of 7. Solutions with a pH below 7 are acidic, and solutions with a pH above 7 are basic, or alkaline.

Moving from one increment to another (e.g., 7 to 6) represents a 10-fold change in $[H^+]$. Therefore, a pH 6 solution has a $[H^+]$ that is

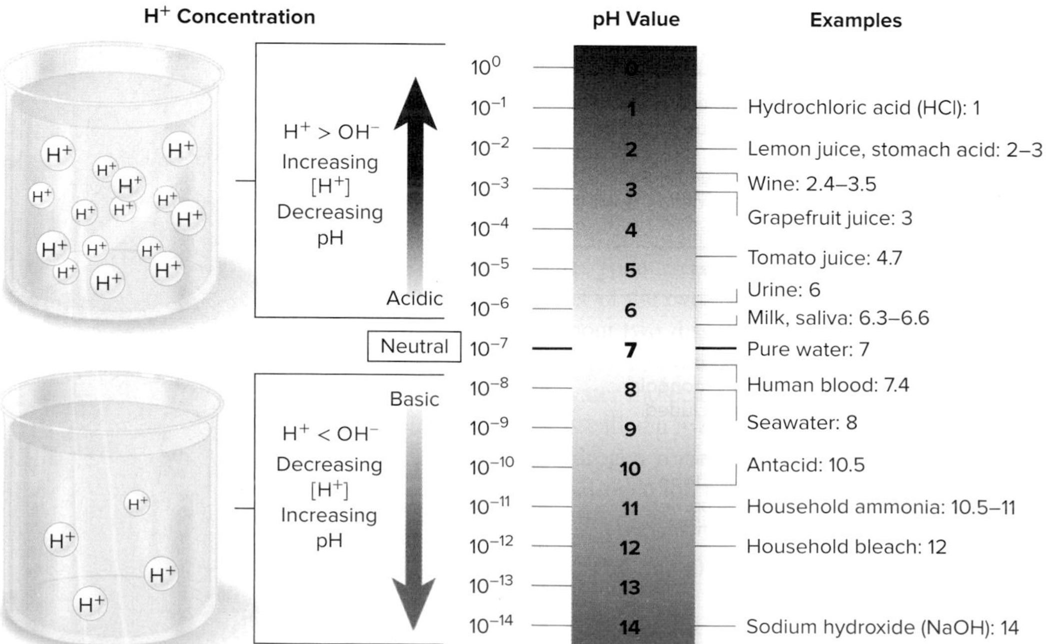

Figure 2.15 pH. pH scale is a measure of the relative concentrations of H^+ and OH^-. A neutral solution has equal amounts of H^+ and OH^-, acidic solutions contain greater amounts of H^+ than OH^-, and basic solutions contain lesser amounts of H^+ than OH^-. Examples of common solutions that exhibit a specific pH are shown.

10 times greater than pure water. Changing by two units (e.g., 7 to 5) is a 100-fold increase in [H^+]. This relationship is accurate for changes above 7 as well, except that each increment is a 10-fold decrease in [H^+].

WHAT DO YOU THINK?

 If stomach acid has a pH of 2, how much more acidic is stomach acid than water (pH 7)? Predict what would happen to the stomach if it were not protected from the effects of hydrochloric acid.

Neutralization

Neutralization occurs when a solution that is either acidic or basic becomes neutral (i.e., has a pH of 7). The neutralization of an acidic solution is accomplished by adding a base, whereas a basic solution is neutralized by adding an acid. Thus, over-the-counter antacids must contain a base to neutralize stomach acid.

Buffers

A **buffer** is either a single type of molecule or two or more different types of molecules that helps prevent pH changes if either acid or base is added. A buffer acts either to accept H^+ from added acid or to donate H^+ to neutralize added base. Both bicarbonate (HCO_3^-) and carbonic acid (H_2CO_3), for example, are present within the blood and serve as buffers. Bicarbonate (HCO_3^-) accepts H^+ as acid is added to the blood and carbonic acid (H_2CO_3) releases H^+ as base is added to the blood to maintain the pH of the blood within the normal range of 7.35 to 7.45 (see section 18.1b). It is critical to maintain acid-base balance in the body because even small changes in pH (called an acid-base disturbance or imbalance) can be fatal. Acid-base balance and disturbances to acid-base balance are described in detail in sections 25.5 and 25.6, respectively.

WHAT DID YOU LEARN?

 What is the general relationship of [H^+] and pH?

 Why are buffers important, and how do they function to help maintain pH?

INTEGRATE

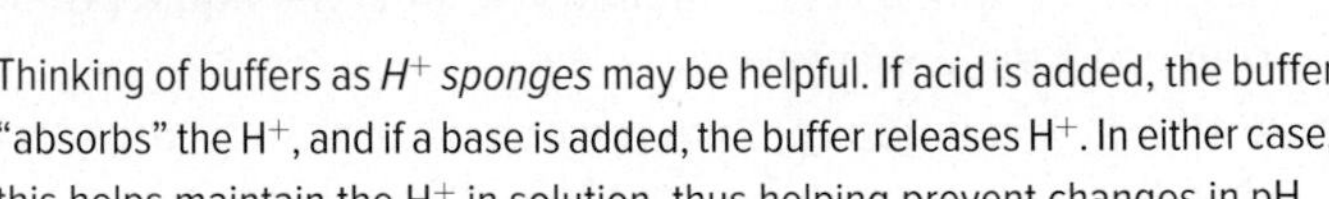

LEARNING STRATEGY

Thinking of buffers as *H^+ sponges* may be helpful. If acid is added, the buffer "absorbs" the H^+, and if a base is added, the buffer releases H^+. In either case, this helps maintain the H^+ in solution, thus helping prevent changes in pH.

2.6 Water Mixtures

Mixtures are formed from the combining or mixing of two or more substances. Several types of mixtures have already been described, including sugar water, salt water, acidic solutions, and basic solutions. Two defining features of mixtures are (1) the substances that are mixed are not chemically changed and (2) the substances in the mixture can be separated by physical means, such as by evaporation or by filtering. Our emphasis here is on water mixtures. We describe how water mixtures are categorized into three types and explain how solution concentration is expressed.

2.6a Categories of Water Mixtures

LEARNING OBJECTIVES

36. Compare and contrast the three different types of water mixtures.

37. Explain how an emulsion differs from other types of mixtures.

Water mixtures are placed into three categories based on the relative size of the substance mixed with water and include suspensions, colloids, and solutions (**figure 2.16**).

- **Suspension.** A suspension is a mixture composed of particles that are relatively large (e.g., greater in size than 1 millimeter). These types of mixtures do not remain mixed together unless the mixture is in motion. The large solutes or cells settle out

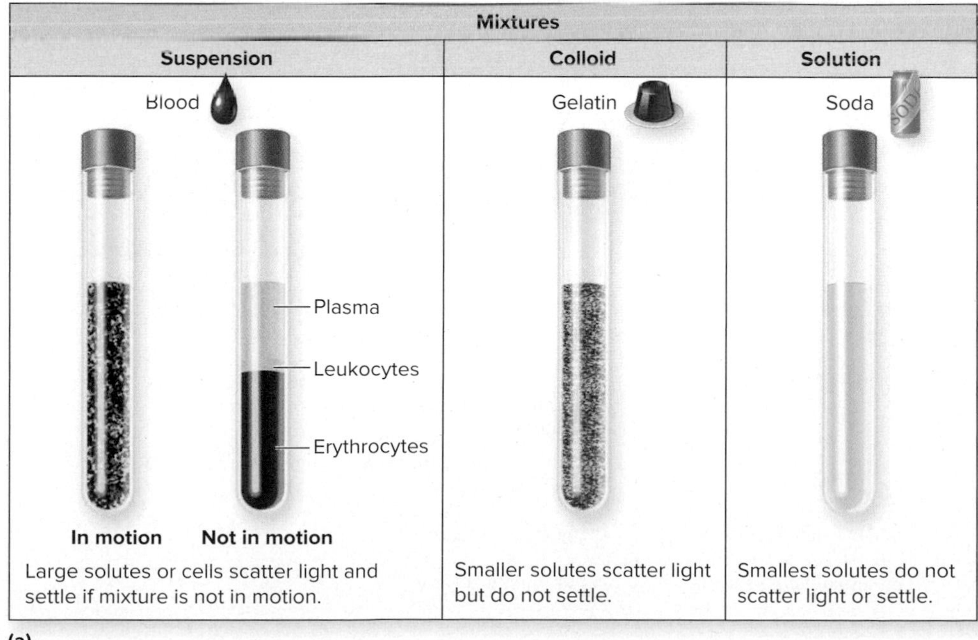

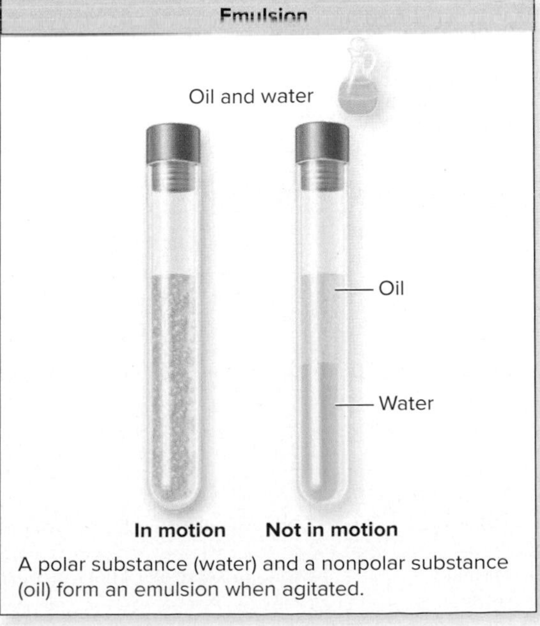

Figure 2.16 Mixtures and an Emulsion. (*a*) Suspensions, colloids, and solutions are types of mixtures. (*b*) An emulsion is more specifically a type of colloid composed of water (or other polar substance) and a nonpolar substance, such as vegetable oil, that mix only when agitated.

of a suspension when it is not in motion. A suspension appears cloudy or opaque and scatters light until the particles drop out of the liquid. Sand in water is an example of a suspension. Blood cells within the plasma (the liquid portion) of blood form a suspension.

- **Colloid.** A colloid is a mixture composed of smaller particles than those in a suspension (but larger than those in a solution). A colloid may appear either opaque or milky and scatters light like a suspension—but, unlike a suspension, remains mixed when not in motion. Some colloids, including gelatin and agar media (used in the microbiology laboratory), display an interesting feature—namely, they contain protein and become liquid when heated but change to a gel-like state when left standing and cooled. Colloids within the body include fluid within a cell (cytosol) and fluid within the blood (plasma).
- **Solution.** A solution is a homogeneous mixture in which the dissolved substance is very small (less than 1 nanometer in diameter). The water is the solvent, and the dissolved substance is the solute. Both salt water and sugar water are examples of solutions. The small size of the solutes in a solution results in a mixture with these characteristics: The solutes are not visible, do not scatter light, and do not settle if the solution is not in motion. Blood plasma is an example of a body solution, with salts, glucose, HCO_3^-, and other dissolved nonprotein substances.

INTEGRATE

CONCEPT CONNECTION

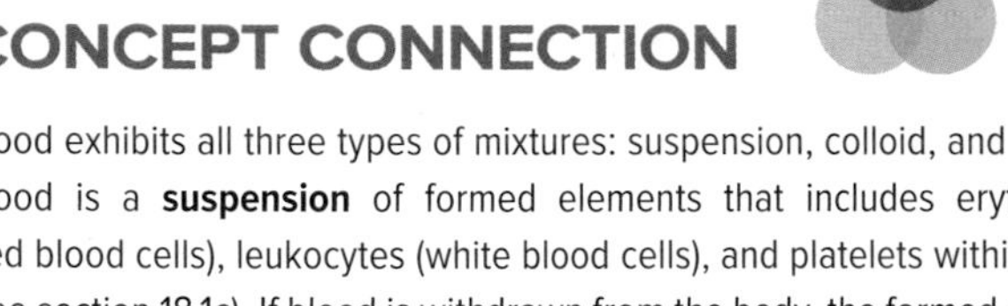

Blood exhibits all three types of mixtures: suspension, colloid, and solution. Blood is a **suspension** of formed elements that includes erythrocytes (red blood cells), leukocytes (white blood cells), and platelets within plasma (see section 18.1c). If blood is withdrawn from the body, the formed elements will settle out from the plasma. Blood is also a **colloid** of proteins (e.g., albumin) dissolved in plasma, and it is a **solution** of ions (e.g., Na^+, K^+, Ca^{2+}) and molecules (e.g., glucose, amino acids) dissolved in plasma (see section 18.2).

An **emulsion** is composed specifically of water and a nonpolar (hydrophobic) liquid substance such as vegetable oil, which does not "mix" unless shaken or agitated (e.g., oil and vinegar salad dressing). Breast milk is an example of an emulsion in the body. Emulsions are classified as a specific type of colloid.

WHAT DID YOU LEARN?

24 When left standing, erythrocytes (red blood cells) settle to the bottom of a tube of blood. Based on this characteristic alone, blood would be characterized as a (a) suspension, (b) colloid, or (c) solution?

25 Why is blood also considered the other two types of water mixtures?

2.6b Expressions of Solution Concentration

LEARNING OBJECTIVE

38. Explain the different ways to express the concentration of solute in a solution.

The amount of solute dissolved in a solution determines the concentration of a solution. Concentration of a solution may be expressed in several ways, including mass/volume, mass/volume percent, molarity, and molality. The units for expressing concentration are summarized in **table 2.2**, which includes a description, units of measurement, and an example for each.

Mass/volume is mass of solute per volume of solution. Results from a blood test are often expressed in mass/volume. **Mass/volume percent** is grams of solute per 100 milliliters of solution. For example, mass/volume percent is the unit of measurement for intravenous (IV) solutions.

Molarity is a measure of number of moles per *liter of solution.* A molar solution of glucose is made by placing 180.10 grams (its molecular mass) of glucose into a container and adding enough water until it measures 1 liter. (Both moles and molecular mass are described shortly.) **Molality** is the moles *per kilogram of solvent.* A solution of one molality is made by placing 180.10 grams of glucose into a container and adding 1 kilogram of water. Molarity and molality may be used interchangeably, subject to this caveat: The two values are the closest when the measurements are taken at 4°C. At this temperature, 1 liter of water is at its most dense, and its mass is exactly equal to 1 kilogram of water.

Table 2.2 Expressing Solution Concentrations

Solution Concentration	Expressed As	Unit of Measurement	Examples
Mass/volume	Mass of solute per volume of solution	μg solute/dL solution	Normal blood concentration of iron is within the range of 40 to 150 μg/dL.
		mg solute/dL solution	Normal blood concentration of glucose is between 70 and 110 mg/dL.
Mass/volume percent	Mass of solute per 100 milliliters (mL) of solution	grams/100 mL	5% dextrose intravenous (IV) solution (D5W) has a concentration of 5 grams of dextrose (glucose) per 100 mL of solution.
			Physiologic saline (0.9% NaCl) has 0.9 gram of NaCl per 100 mL of solution.
Molarity	Moles of solute per liter of solution	moles solute/L solution	0.164 mol/L solution
Molality	Moles of solute per kilogram of solvent	moles solute/kg solvent	0.164 mol/kg solvent

kg = kilogram = 1000 grams
mg = milligrams = 1/1000 of a gram
μg = microgram = 1/1,000,000 of a gram
dL = deciliter = 1/10th of a liter

Molarity alters with changes in temperature (and to a limited degree with changes in pressure). When water temperature increases above 4°C, the solution expands, and molarity (moles per liter of solution) decreases slightly because it is based on volume (liter) of solution. In contrast, molality (moles per kilogram of solvent) does not alter with changes in temperature because it is based on mass (kilograms) of solvent. Consequently, molality is the more accurate of the two values. However, molality is more difficult to measure in the human body; thus, molarity—the slightly less accurate unit of concentration—is more often used.

Osmoles, Osmolarity, and Osmolality

Another means of expressing concentration is with **osmoles (osm),** which reflect whether a substance either dissolves, or dissolves and dissociates, when placed into a solution (i.e., whether it is a nonelectrolyte or an electrolyte). It is the unit of measurement for the number of particles in solution. The term *osmole* generally is used to reflect the extent a solution is able to alter water movement through osmosis (a concept described in section 4.3b). If a solute dissolves but does not dissociate, as occurs when glucose, amino acids, or proteins are placed in water, then the osmolarity = 1 osmole (osm). However, if the solute both dissolves and dissociates, as occurs with both NaCl and $CaCl_2$, then there is a change in the number of particles after the solute is in solution. A 1-molar (1 M) solution of NaCl dissociates in solution to form both 1 M Na^+ and 1 M Cl^-. The osmoles of 1 M of a NaCl solution = 2 osm. What would you predict for the osmoles of $CaCl_2$? Each molecule of $CaCl_2$ dissolves into three particles: 1 Ca^{2+} and 2 Cl^-; thus, 1 M $CaCl_2$ = 3 osm.

Osmoles can be expressed as either osmolarity or osmolality. **Osmolarity** is the number of particles in a 1-liter solution, whereas **osmolality** is the number of particles in 1 kilogram of water. Osmolality more accurately reflects the osmotic movement of water (see section 4.3b) in the body, but it is difficult to measure. Osmolarity is a close approximation to osmolality. Although osmolarity is less accurate, it is often the preferred unit of measurement for the same reason that molarity is used over molality.

The range of values associated with the body is generally much smaller than osmoles and given as milliosmoles (mOsm). Note that 1 osm = 1000 mOsm. Thus, normal blood serum may be expressed as either 275 to 295 mOsm per liter or 275 to 295 mOsm per kilogram.

Moles and Molecular Mass

Notice in table 2.2 that molarity and molality are based on the number of particles in units called a **mole.** The specific value of 1 mole is 6.022×10^{23} atoms, ions, or molecules. The number of particles may seem huge, but the particles themselves are minuscule.

A mole is the mass in grams that is equal to either the average atomic mass of an element or the molecular mass of a compound. For example, a mole of carbon is equal to 12.01 grams. **Molecular mass** is determined using the molecular formula and the average atomic mass for each element. To find a compound's molecular mass, multiply the number of units of each element by its average atomic mass and add together the totals. The molecular mass of glucose ($C_6H_{12}O_6$), for example, is determined as follows:

6 carbon atoms × 12.01 amu =	72.06 amu
12 hydrogen atoms × 1.008 amu =	12.10 amu
6 oxygen atoms × 15.99 amu =	95.94 amu
Molecular mass =	180.10 amu

Thus, 1 mole of glucose ($C_6H_{12}O_6$) would be equal to 180.10 grams (with some variation due to isotopes).

WHAT DID YOU LEARN?

26 What are four ways solution concentration may be expressed?

2.7 Biological Macromolecules

The remaining sections of this chapter describe *biological macromolecules,* which are organized into four primary classes that include: lipids, carbohydrates, nucleic acids, and proteins. We begin this section by discussing the similarities among the four classes, then describe each class in detail.

INTEGRATE

LEARNING STRATEGY

You can use the acronym **CLaPiN** to remember the four major classes of macromolecules—**C**arbohydrates, **L**ipids, **P**roteins, and **N**ucleic acids.

2.7a General Characteristics

LEARNING OBJECTIVES

39. Identify the six chemical elements that generally compose biological macromolecules.
40. Describe a hydrocarbon and its chemical properties, and explain and give examples of functional groups and their chemical properties.
41. Define the terms *monomer, dimer,* and *polymer.*
42. Describe the role of water in both dehydration and hydrolysis reactions in altering biological macromolecules.

Biological macromolecules are large organic molecules (see section 2.4) that are synthesized by the human body. These molecules always contain the elements carbon, hydrogen, and oxygen. Some biological macromolecules may also have one or more of the following: nitrogen (N), phosphorus (P), or sulfur (S). Notice that all of these elements (except hydrogen) are clustered on the right side of the periodic table (see figure 2.1*a*). (You can remember the six elements that compose biological macromolecules with the acronym CHON P.S.)

The carbon component of biological macromolecules may simply be an individual carbon atom or numerous carbon atoms arranged in a carbon skeleton as a chain, branch, or ring (see section 2.3b).

A single carbon atom or a carbon skeleton can have only hydrogen atoms attached. These molecules are more specifically called **hydrocarbons.** They are nonpolar molecules because they contain only C—C and C—H bonds (see section 2.3c). Consequently, hydrocarbons are hydrophobic and not soluble in water (see section 2.4c). Methane gas (CH_4) is an example of a hydrocarbon (see figure 2.8*a*).

Biological macromolecules are not simply hydrocarbons because these molecules also contain what are called functional groups. A **functional group** is two or more atoms that when present together on a molecule always exhibit the same specific chemical characteristics. Functional groups include hydroxyl (—OH), carboxyl or carboxylic acid (—COOH), amine ($-NH_2$), and phosphate (PO_4^{3-}). Each of these functional groups is polar and able to form hydrogen bonds (see section 2.3d), increasing the molecule's solubility in water (see section 2.4c). In addition, some functional groups may act as an acid and release H^+, such as a carboxyl group, whereas others may act as a base

INTEGRATE

LEARNING STRATEGY

A polymer is like a necklace. This necklace might be a pearl necklace, where every subunit is identical, or it might be a charm necklace, where every unit is slightly different. However, like the individual pieces in a necklace, the chemical subunits combine to produce the polymer that is the finished product.

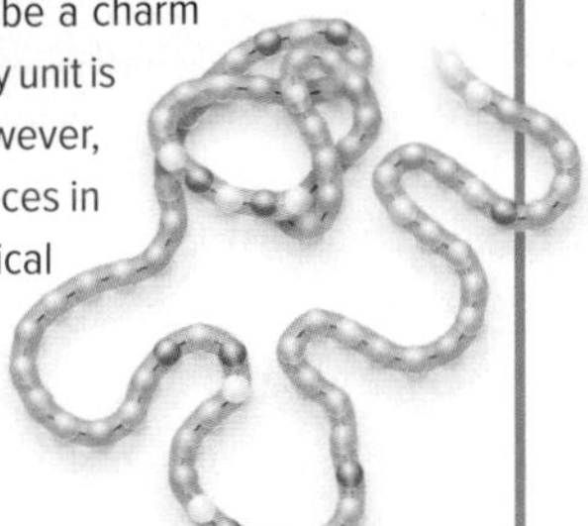

by binding H^+, such as an amine group. Biological macromolecules typically have more than one functional group within them. These four important functional groups, their chemical properties, and the biological macromolecules that contain them are presented in **table 2.3**.

Polymers

Many important biological macromolecules are polymers. **Polymers** are molecules that are made up of repeating subunits called **monomers,** and each monomer is either identical or similar in its chemical structure. Some important carbohydrates (e.g., glycogen, starch), nucleic acids, and proteins are polymers, whereas lipids are not. Carbohydrate polymers contain sugar monomers, nucleic acids have nucleotide monomers, and proteins are composed of amino acid monomers. Note that two monomers bonded together are called a **dimer.**

Process of Dehydration Synthesis and Hydrolysis

Two processes are associated with both the synthesis and the breakdown of complex biological macromolecules: dehydration synthesis and hydrolysis, respectively. During the synthesis of complex molecules from simpler subunits, one specific subunit loses an —H, and the other subunit loses an —OH, to form a water molecule (which is released) as a new covalent bond is

Table 2.3 Functional Groups and Their Chemical Properties

Functional Group	Structural Formula	Properties	Representative Molecules	Structural Diagram of Example Molecule
Hydroxyl	—OH	• Polar • Forms hydrogen bonds • Increases molecules' solubility in water	• Carbohydrates • Proteins • Nucleic acids • Lipids	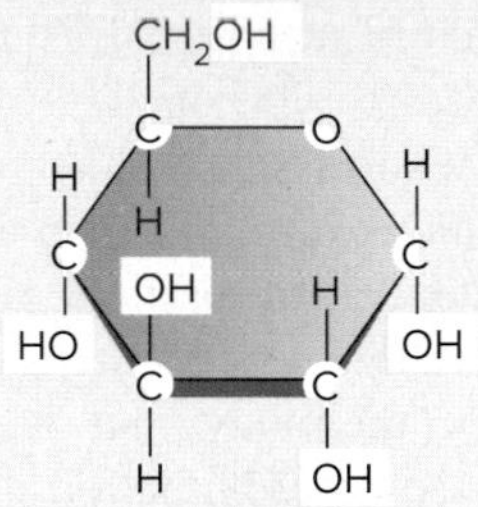 Glucose
Carboxyl (carboxylic acid)	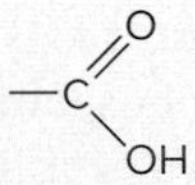 —C(=O)OH	• Polar • Forms hydrogen bonds • Increases molecules' solubility in water • Acts as an acid	• Proteins • Lipids	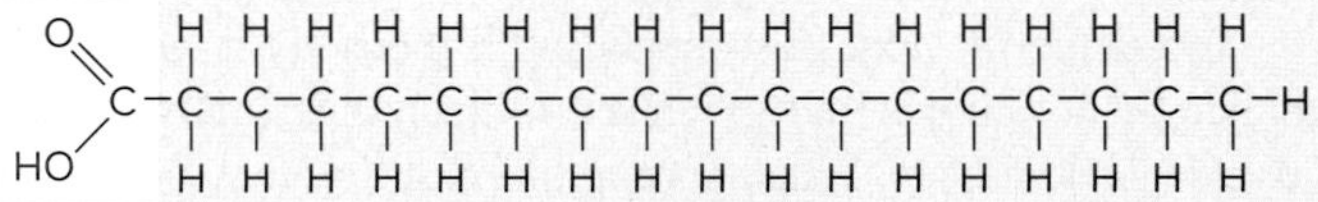 Fatty acid
Amine	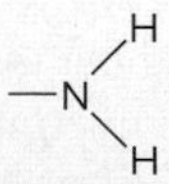 —N(H)H	• Polar • Forms hydrogen bonds • Increases molecules' solubility in water • Acts as a base	• Proteins • Nucleic acids	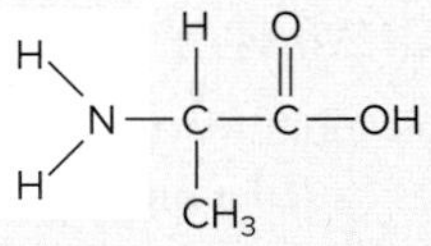 Alanine (an amino acid)
Phosphate	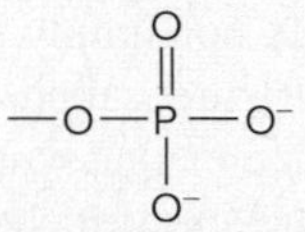 —O—P(=O)(O⁻)—O⁻	• Polar • Forms hydrogen bonds • Increases molecules' solubility in water • Forms phosphodiester bonds • Acts as an acid (shown here with hydrogens released)	• Nucleic acids • Phospholipids • ATP	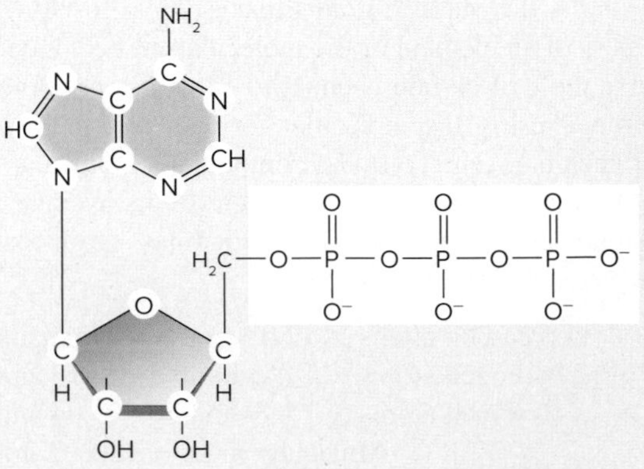 Adenosine triphosphate (ATP)

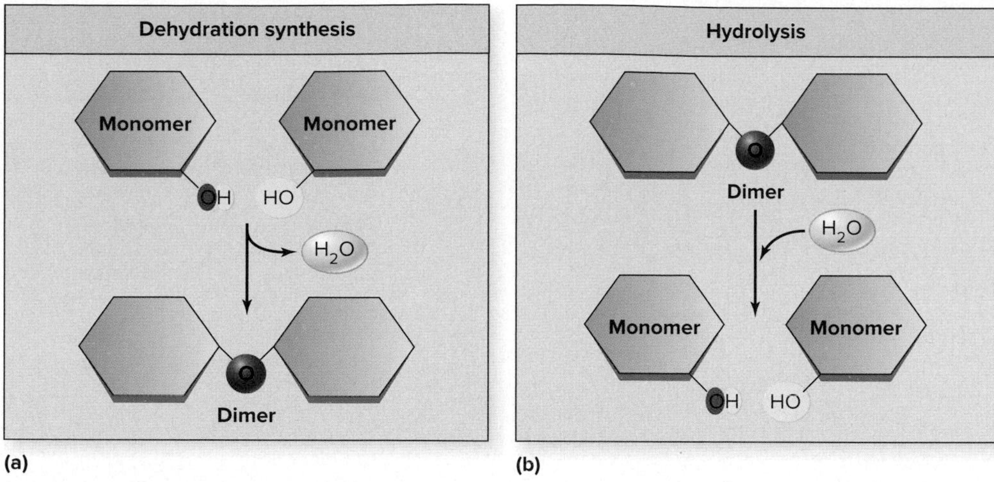

Figure 2.17 Dehydration Synthesis and Hydrolysis. (*a*) Dehydration synthesis involves the loss of a water molecule from simpler components as they are formed into a complex molecule. (*b*) Hydrolysis occurs with the addition of a water molecule to a complex molecule as it is digested into simpler components.

produced. This type of reaction is called **dehydration** (*de* = away, *hydro* = water) **synthesis** or *condensation* because the equivalent of a water molecule is "lost" from the original structures **(figure 2.17)**.

During the breakdown of complex molecules, H_2O molecules are split. An —H is added to one subunit, and an —OH is added to another subunit in the complex molecule, and the chemical bond is broken between them. Because the equivalent of water is added to break a bond within the molecule, this process is referred to as **hydrolysis** (hī-drol′i-sis; *lysis* = destruction) or a *hydrolysis reaction.* We will see examples of both of these types of reactions in the following sections on the specific classes of biological macromolecules.

WHAT DID YOU LEARN?

27 Using a different color than you did for highlighting the common ions on the periodic table, highlight the six common elements that form biological macromolecules. Which element both (a) forms a common ion and (b) is a common element in biological macromolecules?

28 What functional groups may act as an acid?

29 What defines a polymer? List the three biological macromolecules that are polymers and the monomers that compose them.

2.7b Lipids

LEARNING OBJECTIVES

43. Describe the general characteristics of a lipid.
44. Identify the four types of lipids and their physiologic roles.

Lipids are a very diverse group of fatty, water-insoluble (hydrophobic) molecules that function as stored energy, components of cellular membranes, and hormones. Triglycerides (neutral fats), phospholipids, steroids, and eicosanoids are the four primary classes of lipids. They are the only category of biological macromolecules that are not polymers because they are not formed from repeating monomers. Lipids are summarized in **table 2.4**.

INTEGRATE

LEARNING STRATEGY

The acronym **PETS** may help you remember these four classes of lipids:

Phospholipids
Eicosanoids
Triglycerides
Steroids

Triglycerides: Energy Storage

Triglycerides (trī-glis′ĕr-idz; *tri* = three), or *triacylglycerols*, are the most common form of lipids in living things. They are used for long-term energy storage in adipose connective tissue (see section 5.2) and for structural support, cushioning, and insulation of the body. Adipose connective tissue deep to the skin in the abdomen, for example, serves as long-term energy storage and insulates the abdomen against heat loss. Adipose connective tissue posterior to the eye cushions the eye within the eye's bony orbit (see section 16.4a).

Table 2.4 Major Classes of Lipids[1]

Class	Structure	Description	Function
Triglycerides	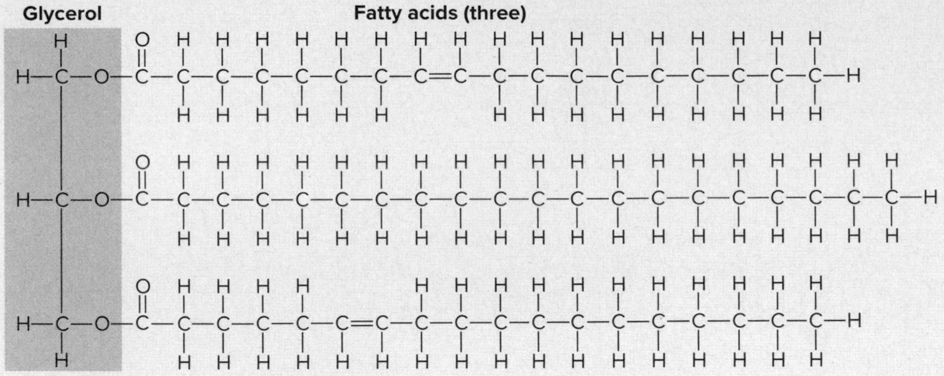 Glycerol; Fatty acids (three)	Composed of glycerol and three fatty acids Fatty acids may be saturated or unsaturated	Long-term energy storage in adipose connective tissue Structural support, cushioning, and insulation of the body
Phospholipids	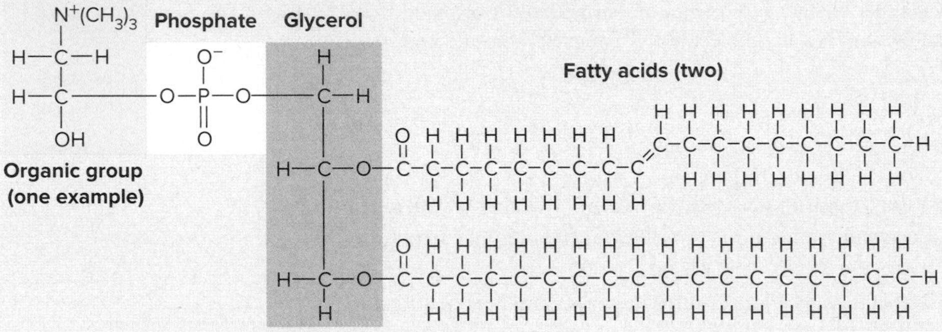 $N^+(CH_3)_3$; Phosphate; Glycerol; Fatty acids (two); Organic group (one example)	Composed of glycerol, two fatty acids, a phosphate, and various organic groups Glycerol, phosphate, and organic groups form a polar head, and fatty acids form two nonpolar tails	Major component of membranes, including the plasma membrane, which forms the chemical barrier between the inside and outside of a cell
Steroids (include cholesterol, steroid hormones, and bile salts)	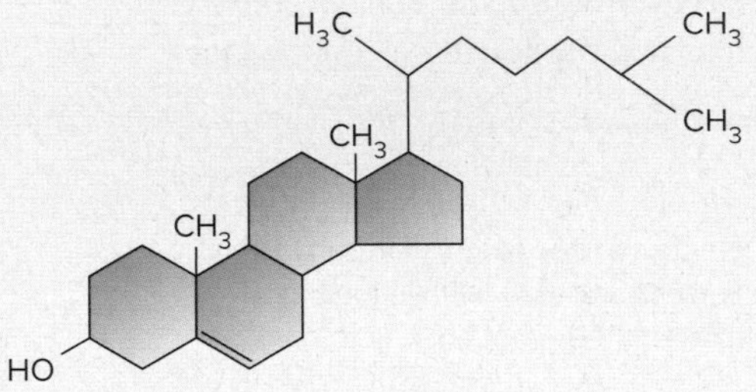 H_3C; CH_3; CH_3; CH_3; CH_3; HO	Four rings composed predominantly of hydrocarbons that differ in the side chains extending from the rings	Cholesterol is a component of plasma membranes and is the precursor molecule for synthesis of other steroids Steroid hormones are regulatory molecules released by certain endocrine glands Bile salts facilitate micelle formation in the digestive tract
Eicosanoids (include prostaglandins, prostacyclins, thromboxanes, and leukotrienes)	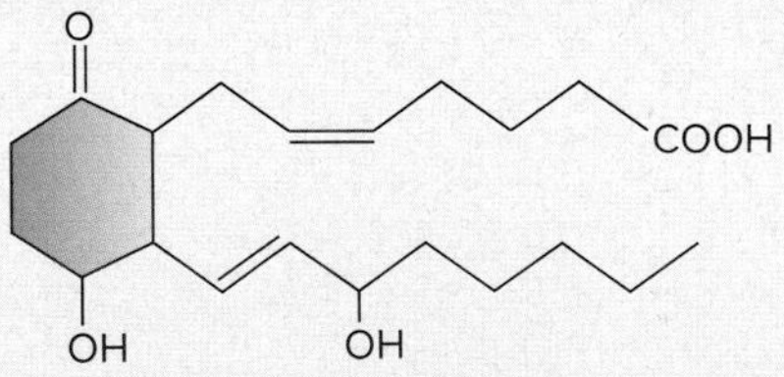 O; COOH; OH; OH	Modified 20-carbon fatty acids	Locally acting signaling molecules associated with all body systems; have primary functions in both the inflammatory response of the immune system and communication within the nervous system

1. Glycolipids and fat-soluble vitamins are also types of lipids (not shown here).

Triglycerides are formed from a glycerol molecule and three fatty acids. Glycerol is a three-carbon molecule with a hydroxyl functional group attached to each carbon. A fatty acid is composed of a long chain of hydrocarbons with a carboxylic acid functional group on one end. Triglyceride molecules form by the process of dehydration synthesis, during which the equivalent of a water molecule is lost for each fatty acid added to the glycerol: an —H from the glycerol and an —OH from the fatty acid (**figure 2.18**).

Fatty acids may vary in length—commonly ranging in even numbers from 14 to 20 carbons—and may differ in the number and position of double bonds between the carbons in the chain. The fatty acid is **saturated** if it lacks double bonds—that is, every carbon has the maximum number of hydrogen atoms bound to it. An **unsaturated** fatty acid has one double bond, and a **polyunsaturated** fatty acid has two or more double bonds.

Adipose connective tissue is used to store triglycerides. When conditions of excess nutrients exist, adipose connective tissue binds fatty acids to glycerol to form triglycerides in a dehydration synthesis process called **lipogenesis** (līp′o-jĕn′e-sis; *lipos* = fat, *genesis* = production). Adipose connective tissue breaks down triglycerides and releases the products into the blood when nutrients are needed. This process is a hydrolysis reaction called **lipolysis** (li-pol′i-sis).

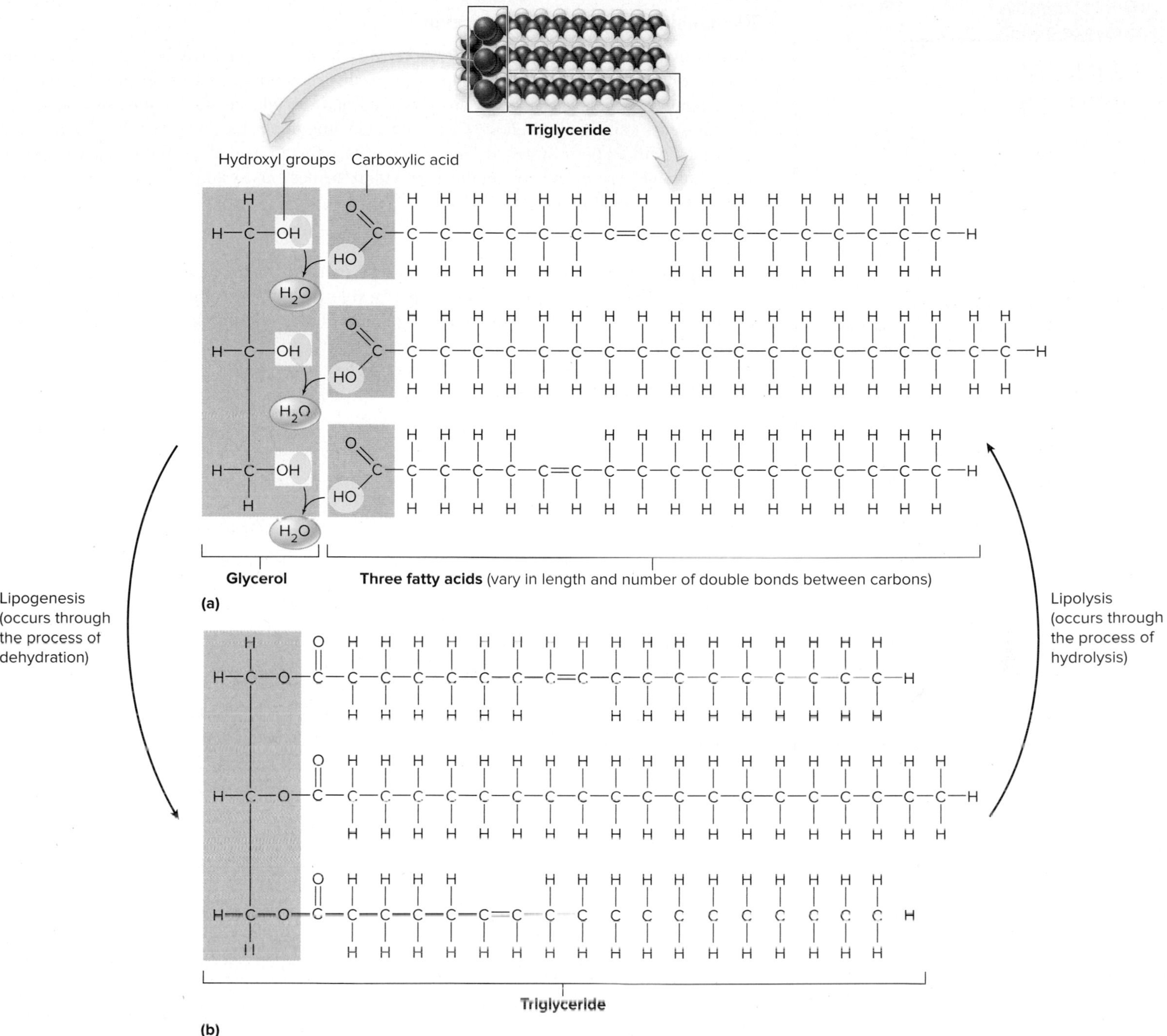

Figure 2.18 Triglycerides. (*a*) A glycerol and three fatty acid molecules. The middle fatty acid is a saturated fatty acid, whereas the other two are unsaturated fatty acids. (*b*) A triglyceride molecule. Lipogenesis occurs through a dehydration synthesis reaction that involves the removal of a water molecule between each fatty acid and a glycerol. Lipolysis is a hydrolysis reaction that splits the glycerol and three fatty acids by the addition of water at each of the fatty acids.

Phospholipids: Form Cell Membranes

Phospholipids were previously described as amphipathic molecules that form chemical barriers of cell membranes, including plasma membranes that form the outer barrier of a cell (see figure 2.13*c*). The chemical structure of a phospholipid is similar to a triglyceride, except that one end of the glycerol has a polar phosphate group with various organic groups (choline, ethanolamine, or the amino acid serine) attached to it instead of a fatty acid (table 2.4). The glycerol, phosphate, and organic groups are polar and form the water-soluble hydrophilic part of the molecule referred to as the **hydrophilic (polar) head.** The two fatty acid molecules attached to the glycerol form a water-insoluble hydrophobic end called the **hydrophobic (nonpolar) tails.** Phospholipid molecules within a plasma membrane are conventionally depicted with an icon of what appears like a balloon with two tails (see figure 2.13*c*).

Steroids: Ringed Structures, Including Some Hormones

Steroids are composed predominantly of hydrocarbons arranged in a distinct multiringed structure. A steroid has four attached carbon rings; three rings have six carbon atoms and one ring has five carbon atoms. Steroids differ in the side chains extending from their rings. Steroids include cholesterol, steroid hormones (e.g., testosterone, estrogen), and bile salts. Cholesterol is a component of animal plasma membranes as well as the precursor used to synthesize other steroids. Cholesterol is synthesized in the liver from fatty acids and may be obtained from eating animal products such as meat, eggs, and milk.

INTEGRATE

CLINICAL VIEW 2.3

Fatty Acids: Saturated, Unsaturated, and Trans Fats

Most naturally occurring animal fats are **saturated,** whereas most naturally occurring vegetable fats (except palm and coconut tropical oils) are **unsaturated.** You can usually distinguish between a saturated and unsaturated fat by whether it is solid (most saturated fats) or liquid (most unsaturated fats) at room temperature. Unsaturated fats are generally healthier than saturated fats—that is why we are encouraged to use olive oil instead of butter for our cooking.

Unsaturated fats are converted to saturated fats through the process of **hydrogenation**, which increases the amount of hydrogen within the molecule. For example, vegetable oil may be hydrogenated to form margarine. Numerous studies have shown that partial hydrogenation of fatty acids may lead to the production of **trans fats** (which have their hydrogen atoms on the opposite sides of a double bond). These fats are solid at a lower temperature than their counterparts, *cis fats* (which have their hydrogen atoms on the same side of a double bond). Studies have shown that trans fats in the diet raise the level of low-density lipoprotein (LDL) within the blood, which increases the risk of developing lipid buildup on the inside of blood vessels (atherosclerosis), a condition that increases the risk of heart attack or stroke. The Food and Drug Administration (FDA) has taken measures to have the food industry remove trans fat from our foods.

Eicosanoids: Locally Acting Hormones

Eicosanoids (ī′kō-să-noydz; *eicosa* = 20, *eido* = form) are modified 20-carbon fatty acids that are obtained from the phospholipids of plasma membranes, which form the barrier of cells. Four classes of eicosanoids are produced and include prostaglandins, prostacyclins, thromboxanes, and leukotrienes. These are signaling molecules that act locally and are associated with all body systems: They have primary functions in the inflammatory response of the immune system and communication within the nervous system. The synthesis of eicosanoids is described in section 17.3b.

Other Lipids

Glycolipids are lipid molecules with an attached carbohydrate. These molecules are associated with plasma membranes and serve several roles, including cellular recognition to form tissues (see section 4.5a). Fat-soluble vitamins such as vitamins A, E, and K (see section 27.3a) are also lipids.

WHAT DID YOU LEARN?

30 Do lipid molecules typically dissolve in water? Explain.

31 Which class of lipids forms cell membranes? What characteristic allows it to perform this function?

2.7c Carbohydrates

LEARNING OBJECTIVES

45. Describe the distinguishing characteristics of carbohydrates.

46. Explain the relationship between glucose and glycogen.

47. Name some other carbohydrates found in living systems.

The term **carbohydrate** means *hydrated carbon.* Nearly every carbon is hydrated with the equivalent of a water molecule—that is, both an —H and an —OH are usually attached to every carbon. The general chemical formula for carbohydrates is $(CH_2O)n$, where n indicates the number of carbon atoms that are in a molecule. The number of carbon atoms typically ranges from three to seven. The least complex carbohydrates are simple sugar monomers called **monosaccharides.** All monosaccharides have between three and seven carbon atoms. Carbohydrates that are dimers formed from two monosaccharides are **disaccharides,** and those with many monosaccharides are **polysaccharides.**

Glucose and Glycogen

Glucose is a six-carbon (hexose) carbohydrate that is the most common monosaccharide. It is shown here (and throughout the text) in the ring form (**figure 2.19**). Glucose is crucial to life processes because it is the primary nutrient supplying energy to cells. In fact, the brain and other nervous tissue use glucose almost exclusively as their nutrient fuel molecule. The concentration of blood glucose must be carefully maintained by homeostasis (see section 1.6) to ensure a continual, adequate energy supply for cellular activities. One way the body ensures its supply is to store excess glucose immediately following a meal. Liver and skeletal muscle tissue absorb the excess glucose—and then bind the glucose monomers together to form a polysaccharide called **glycogen** by a process called **glycogenesis** (glī′kō-jen′ĕ-sis). The glycogen polymer is shown in figure 2.19*b*—although only several glucose molecules are shown, glycogen may contain thousands of glucose monomers.

When blood glucose levels drop between meals, the liver hydrolyzes some of the glycogen into glucose and releases it into the blood. This process is called **glycogenolysis** (glī′kō-jĕ-nol′i-sis). Thus, the liver serves the role of a "glucose bank" by storing glycogen, then breaking down glycogen as needed to release glucose (see section 17.10b). Nutrient storage and release are important in the endocrine regulation of nutrient blood levels of both triglycerides and glucose. Note: The liver can also form glucose from noncarbohydrate sources (e.g., fats, proteins) through a process called **gluconeogenesis.** See Clinical View 17.7: "The Stress Response (General Adaptation Syndrome)."

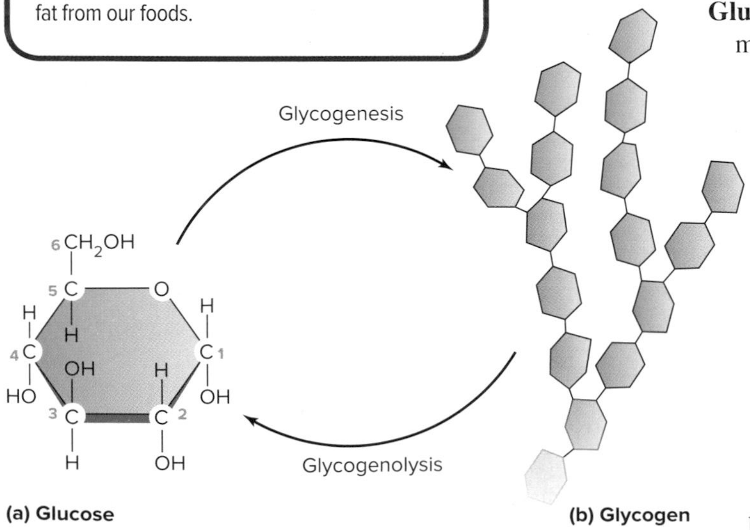

Glycogenesis

Glycogenolysis

(a) Glucose

(b) Glycogen

Figure 2.19 Glucose and Glycogen. (*a*) The monomer glucose is typically shown in ring form. (*b*) Glycogen is a polysaccharide composed of multiple glucose monomers. Glycogen is formed from glucose molecules through glycogenesis, whereas glycogen is digested into glucose through glycogenolysis.

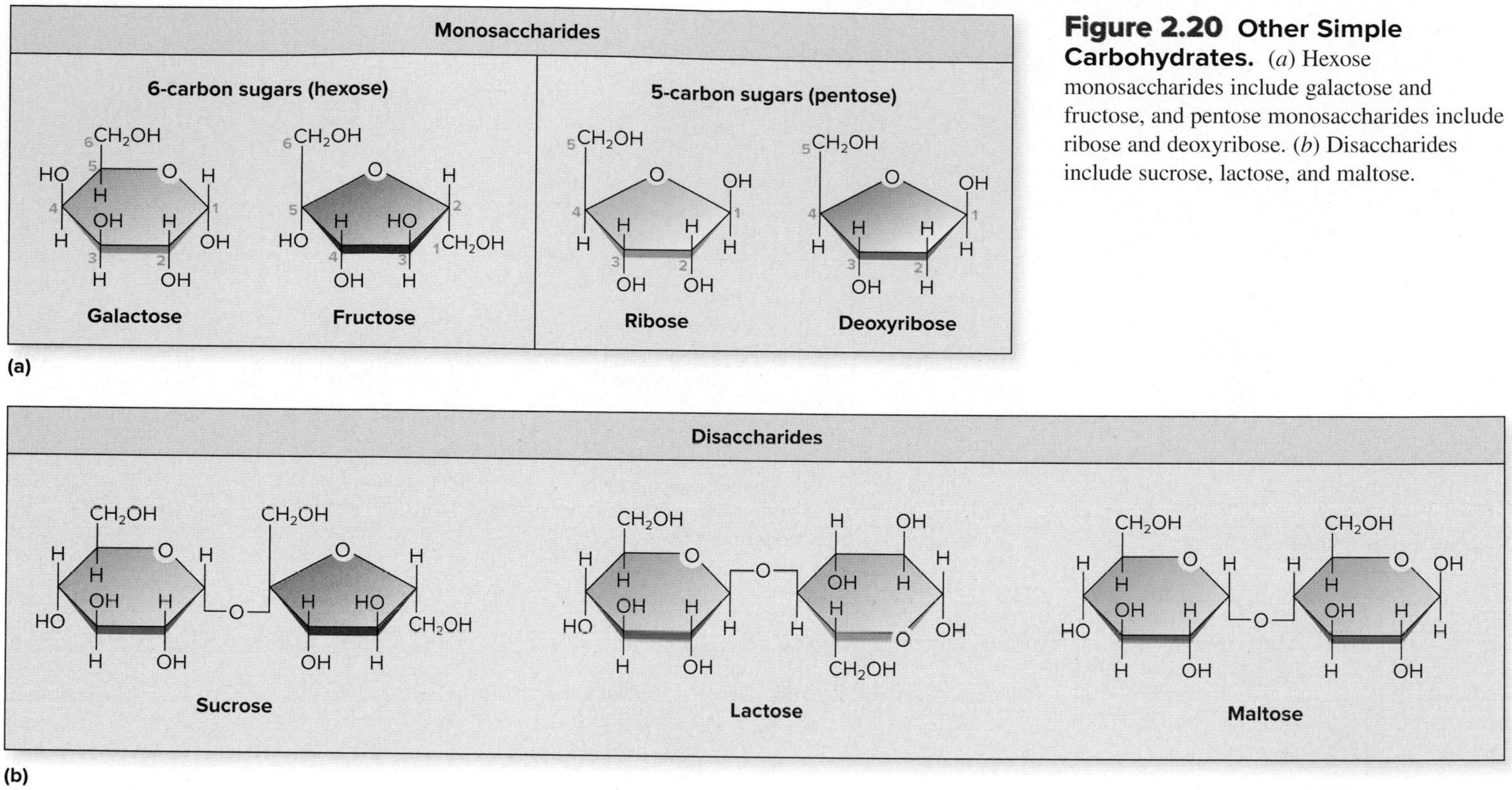

Figure 2.20 Other Simple Carbohydrates. (*a*) Hexose monosaccharides include galactose and fructose, and pentose monosaccharides include ribose and deoxyribose. (*b*) Disaccharides include sucrose, lactose, and maltose.

Other Types of Carbohydrates

Other hexose monosaccharides include galactose and fructose, which are glucose isomers, as described in section 2.3a (**figure 2.20*a***). Some monosaccharides, such as ribose and deoxyribose, are composed of five carbons. These five-carbon monosaccharides are called *pentose sugars*. The pentose sugars ribose and deoxyribose are structural components of nucleic acids, which are discussed in the section 2.7d. The only structural difference between these pentose sugars is the lack of an oxygen atom on carbon two of the deoxyribose sugar.

Disaccharides (which you will recall are dimers composed of two monosaccharides bonded together) include: sucrose (table sugar), lactose (milk sugar), and maltose (malt sugar, which is found in sprouting grains). All three disaccharides contain a glucose monosaccharide bonded to a second hexose monosaccharide (figure 2.20*b*).

Polysaccharides (which you will recall are composed of three or more monosaccharides) include glycogen (figure 2.19*b*), which is the storage form of glucose within liver and muscle cells of animals. Polysaccharides in plants include starch and cellulose, which are also composed of repeating glucose monomers. Plant starch is a major nutritional source of glucose for humans. It is found in potatoes, grains, and many other plant foods. Glucose released from the breakdown of starch within the digestive tract is absorbed into the blood. Cellulose, a structural polysaccharide of plant cell walls, however, is a source of fiber (nondigestible substances). Humans cannot digest cellulose because of the unique chemical bonds between the glucose molecules. The breakdown of both disaccharides and polysaccharides is described in section 26.4a.

WHAT DID YOU LEARN?

32 What is the repeating monomer of glycogen? Where is glycogen stored in the body?

33 For each of the following, indicate if it is a monosaccharide, disaccharide, or polysaccharide: fructose, galactose, glucose, glycogen, lactose, maltose, starch, and sucrose.

2.7d Nucleic Acids

LEARNING OBJECTIVES

48. Describe the general structure of a nucleic acid.

49. Describe the structure of a nucleotide monomer.

50. Distinguish between DNA and RNA.

51. Name other important nucleotides.

Nucleic acids (**figure 2.21*c, d***) are biological macromolecules within cells that store and transfer genetic, or hereditary, information. Originally discovered within the cell nucleus

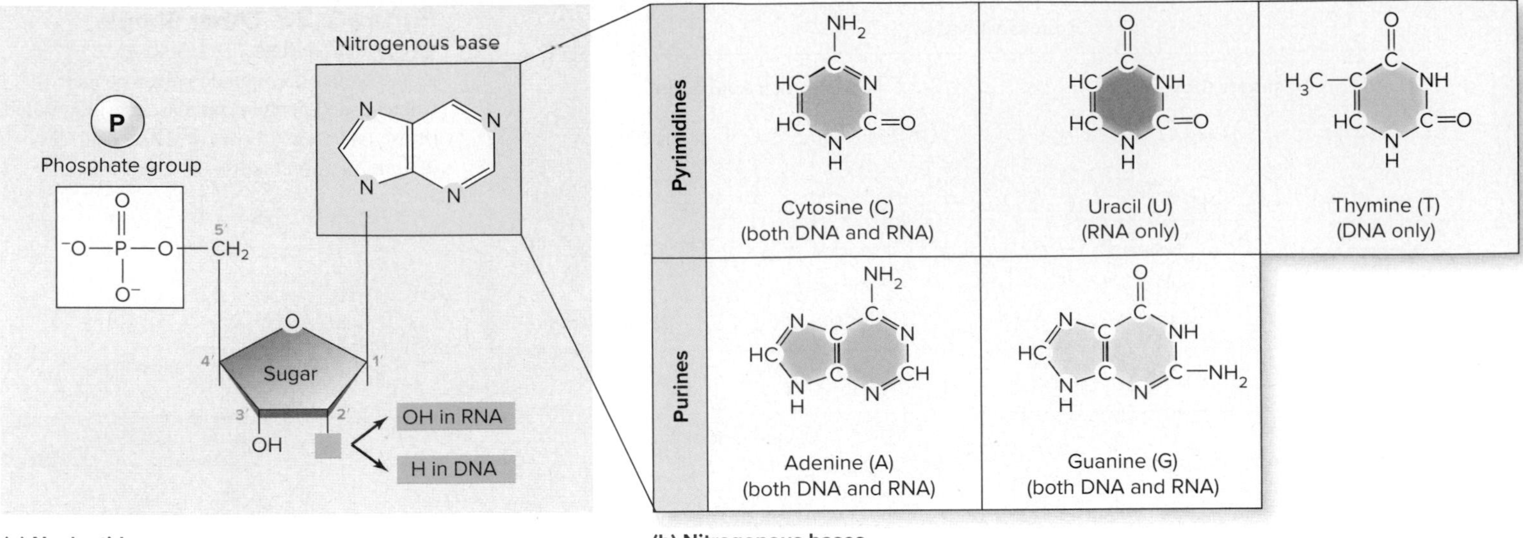

(a) Nucleotide monomer

(b) Nitrogenous bases

Unique to RNA

5′

Phosphate group

Nitrogenous base

Nucleotide

Ribose sugar

Phosphodiester bonds

OH

3′

Sugar-phosphate "backbone"

Nitrogenous base

Deoxyribose sugar

Nucleotide

Phosphate group

5′

Unique to DNA

Hydrogen bonds between nitrogenous bases

3′

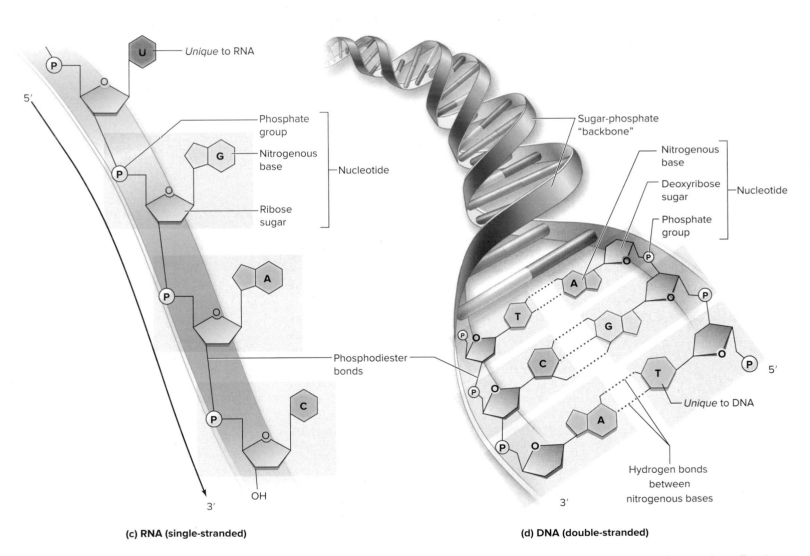

(c) RNA (single-stranded)

(d) DNA (double-stranded)

Figure 2.21 Nucleic Acids. (*a*) A general representation of a nucleotide monomer, which is composed of a pentose sugar (ribose or deoxyribose), a phosphate functional group, and a nitrogenous base. Nucleotides that contain ribose are ribonucleotides, and nucleotides that contain deoxyribose are deoxyribonucleotides. Note the numbering of the carbon atoms in the ribose/deoxyribose of a nucleotide monomer. (*b*) The five nitrogenous bases. (*c*) RNA is a single-stranded nucleic acid formed from repeating units of ribonucleotides linked together by phosphodiester bonds. Each ribonucleotide contains one of the nitrogenous bases uracil, guanine, adenine, or cytosine. (*d*) DNA is a double-stranded nucleic acid with each strand composed of repeating units of deoxyribonucleotides linked together by phosphodiester bonds. Each deoxyribonucleotide contains one of the nitrogenous bases thymine, guanine, adenine, or cytosine. Hydrogen bonds between complementary bases (T:A and C⋮G) hold the two strands together. Both RNA and DNA participate in protein formation.

(see section 4.7), nucleic acids ultimately determine the types of proteins synthesized within cells—a process described in detail in section 4.8.

The two classes of nucleic acid are deoxyribonucleic acid (DNA) and ribonucleic acid (RNA). Both DNA and RNA are polymers composed of **nucleotide** monomers. These monomers are linked together through covalent bonds between one nucleotide monomer and an adjacent nucleotide monomer. This covalent bond between the phosphate of one nucleotide and the sugar of the adjacent nucleotide is called a **phosphodiester bond.**

Nucleotide

A nucleotide has three components: a sugar, a phosphate functional group, and a nitrogenous base (figure 2.21*a*). The sugar is a five-carbon pentose sugar (deoxyribose for DNA and ribose for RNA). A phosphate functional group is attached at carbon number five; a nitrogenous base is attached to the same sugar but at carbon number one. A **nitrogenous base** has either a single-ring or a double-ring structure that contains both carbon and nitrogen within the ring.

Five different nitrogenous bases commonly occur in nucleic acids (figure 2.21*b*). Single-ring nitrogenous bases are called **pyrimidines,** and they include cytosine (C), uracil (U), and thymine (T). (You can remember these bases with the acronym CUT.) Double-ring nitrogenous bases are called **purines,** which include adenine (A) and guanine (G). The nitrogenous bases within either group—pyrimidines or purines—differ in the functional groups attached to the ring.

Deoxyribonucleic Acid (DNA)

Deoxyribonucleic acid (DNA) is a double-stranded nucleic acid (figure 2.21*d*); it can be found as a component of chromosomes within the nucleus (see section 4.7). A small, circular strand of DNA is also within mitochondria (see sections 3.4c and 4.6a). The nucleotides that form DNA have a deoxyribose sugar, a phosphate, and one of four nitrogenous bases: adenine, guanine, cytosine, or thymine. Notice that DNA does not contain uracil. The double strands of nucleic acid are held together by hydrogen bonds formed between complementary nitrogenous bases: thymine with adenine and guanine with cytosine.

Ribonucleic Acid (RNA)

Ribonucleic acid (RNA) is a single-stranded nucleic acid located both within the cell nucleus and within the cytoplasm (see section 4.1c) of the cell (figure 2.21*c*). The nucleotides that are part of RNA molecules are composed of the sugar ribose, a phosphate, and one of four nitrogenous bases: adenine, guanine, cytosine, or uracil. RNA does not contain thymine. **Table 2.5** compares the chemical structural differences between RNA and DNA.

Table 2.5 Differences Between RNA and DNA

Characteristic	RNA	DNA
Number of strands	1	2
Sugar	Ribose	Deoxyribose
Nitrogenous base	Uracil (unique to RNA)	Thymine (unique to DNA)

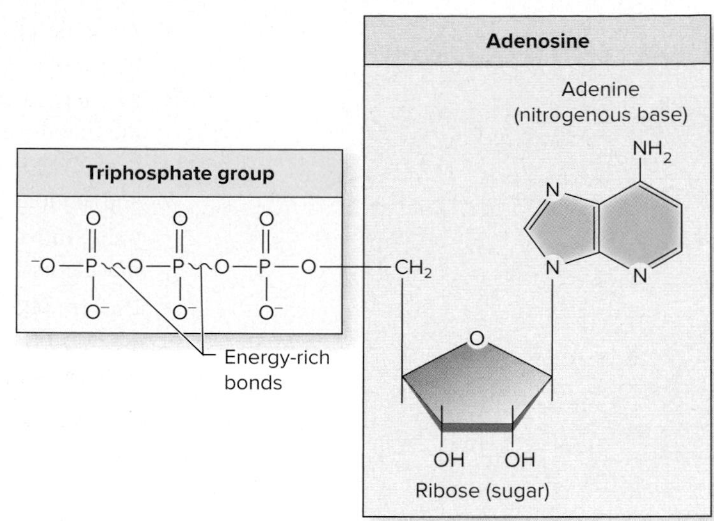

Figure 2.22 ATP. Adenosine triphosphate (ATP) is composed of the sugar ribose and the nitrogenous base adenine, which together are referred to as adenosine. Three phosphates attached to adenosine form adenosine triphosphate.

Other Important Nucleotides and Nucleotide-Containing Molecules

An important nucleotide is **adenosine triphosphate** (a-den′ō-sēn trī-fos′fat), or **ATP.** It is composed of the nitrogenous base adenine, a ribose sugar, and three phosphate groups covalently linked (**figure 2.22**). ATP is the central molecule in the transfer of chemical energy within cells. Biologists often refer to this molecule as the "energy currency" of a cell. The covalent phosphate bond linkages between the last two phosphate groups are unique, energy-rich bonds. ATP is produced continuously and only stored in limited amounts within cells (e.g., muscle cell ATP stores lasts for approximately 4–6 seconds during high intensity exercise). When ATP molecules are split into adenosine diphosphate (ADP) and phosphate, energy is released. This energy is then used by the cell (e.g., muscle contraction).

Two important nucleotide-containing molecules are *nicotinamide adenine dinucleotide* (NAD^+) and *flavin adenine dinucleotide* (*FAD*). Both molecules participate in the production of ATP that occurs in cellular mitochondria. (ATP, NAD^+, and FAD are discussed in more detail in chapter 3.)

WHAT DID YOU LEARN?

34 What is the general function of nucleic acids?

35 What are the structural differences between RNA and DNA?

2.7e Proteins

LEARNING OBJECTIVES

52. List the general functions of proteins.

53. Describe the general structure of amino acids and proteins.

Scientists estimate that approximately 50,000 different proteins are synthesized (produced) by cells, and that proteins account for about one-fifth of the human body by weight. Following their synthesis, proteins may function within a cell, a plasma membrane, or blood plasma and other body fluids. Proteins serve a vast array of functions. For example, proteins

- Serve as **catalysts** (enzymes) in most metabolic reactions of the body (see section 3.3)

- Act in **defense,** which occurs, for example, when immunoglobulins (antibodies) attach to foreign substances for their elimination (see section 22.8)
- Aid in **transport,** as when hemoglobin molecules transport respiratory gases within the blood (see section 18.3b)
- Contribute to structural **support,** such as collagen, a major component of ligaments and tendons (see section 5.2d)
- Cause **movement,** when myosin and actin proteins interact during contraction of muscle tissue (see section 10.2b)
- Perform **regulation,** as occurs when insulin helps control blood glucose levels (see section 17.10.b)
- Provide **storage,** such as ferritin, which stores iron in liver cells (see section 18.3b)

Table 2.6 organizes the functions of proteins into several categories providing the general function, class of protein, and examples for each.

General Protein Structure

Proteins are polymers composed of one or more linear strands of **amino acid** monomers that may number in the thousands **(figure 2.23)**. Twenty standard amino acids are normally found in the proteins of living organisms. Each amino acid has both an amine (—NH_2)

Table 2.6 Protein Functions

Function	Class of Protein	Examples of Proteins and Their Functions
Catalysts	Enzymes	Hydrolytic enzymes: digest polysaccharide macromolecules
		Isomerases: convert a molecule to its isomer—a different chemical structure (arrangement of atoms)
		DNA polymerase: synthesizes DNA
		Kinases: transfer phosphate functional groups
Defense	Immunoglobulins	Antibodies: "tag" foreign antigen for elimination
	Cell surface antigens	Major histocompatibility complex (MHC) proteins: serve as "self" recognition molecules
Transport	Circulating transporters	Hemoglobin: transports O_2 and CO_2 in blood
		Transferrin: transports iron in the blood
	Membrane transporters	Cytochromes: participate in electron transport
		Sodium-potassium pump: participates in establishing a resting membrane potential
		Glucose transporter: transports glucose across plasma membranes
Support	Supporting proteins	Collagen: forms ligaments, tendons
		Keratin: forms hair, nails
		Fibrin: forms blood clots
Movement	Contractile proteins	Actin: participates in contraction of muscle fibers
		Myosin: participates in contraction of muscle fibers
Regulation	Osmotic proteins	Albumin: maintains osmotic concentration of blood
	Hormones	Insulin: controls blood glucose levels
		Antidiuretic hormone (ADH): increases water retention by kidneys
		Oxytocin: stimulates uterine contractions and milk ejection
	Molecular chaperones	Protein disulfide isomerase: participates in folding of proteins that involve intramolecular interactions
Storage	Metal binding proteins	Ferritin: stores iron in liver cells Lactoferrin: binds iron in breast milk
	Ion binding proteins	Calmodulin: binds calcium ions in sarcoplasmic reticulum (SR) of muscle cells

(a) Amino acid

Amine — Carboxylic acid — R group (1 of 20 different structures)

(b) Dipeptide

Peptide bond — H_2O

(c) Polymer protein

Amine — N-terminal — Carboxylic acid — C-terminal

Figure 2.23 Proteins. (*a*) Amino acids are the monomers of proteins. (*b*) Dehydration synthesis reaction occurs as a hydroxyl functional group is removed from the carboxylic acid of one amino acid and a hydrogen atom is removed from the amine group of another amino acid to form a peptide bond between two amino acids. A molecule of water is released during the process. (*c*) The polymer protein is composed of repeating amino acid monomers that are held together by peptide bonds.

functional group and a carboxylic acid (—COOH) functional group (see table 2.3). Both functional groups are covalently linked to the same carbon atom, which accounts for the general name "amino acid" for these monomers. The other two covalent bonds to this carbon are a hydrogen (—H) and different side-chain structures that are simply referred to as the **R (Remainder) group.** The R groups distinguish different amino acids from one another. Properties of the R groups form the basis for classifying amino acids (described in section 2.8a).

Amino acids are linked covalently by **peptide bonds** that form during dehydration synthesis reactions between the amine functional group of one amino acid and the carboxylic acid functional group of a second amino acid. An —H is lost from the amine group and an OH from the carboxylic acid of another amino acid. The ends of a protein are distinguished as the **N-terminal** end, which has a free amine group, and the **C-terminal** end, which has a free carboxyl group.

A strand of amino acids that includes between 3 and 20 amino acids is termed an **oligopeptide,** whereas a **polypeptide** has a strand composed of between 21 and 199 amino acids. If more than 200 amino acids are linked, the structure is called a **protein.** (Note that the specific numbers of amino acids for each category of protein are debated by protein chemist experts.) In this text, we use the general term *protein* to apply to all three types of these structures.

Proteins with carbohydrate attached are called **glycoproteins.** ABO blood groups are based on structural differences in glycoproteins within the plasma membrane of erythrocytes (see section 18.3b). The glycocalyx of cells (see section 4.2b), which is composed of both glycoproteins and glycolipids, is used to identify the cell as self (see section 22.4a).

Lipids, carbohydrates, nucleic acids, and proteins are four major classes of biological macromolecules that compose the human body. They are summarized in **figure 2.24**.

INTEGRATE

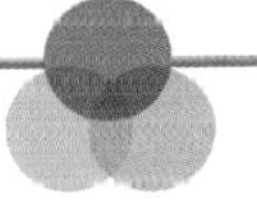

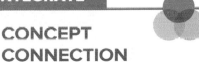

CONCEPT CONNECTION

The amine group of an amino acid can be removed from the amino acid by a process known as **deamination.** Nitrogen in the amine group is converted into **urea,** one form of nitrogenous waste, through a metabolic process called the *urea cycle*. Other types of nitrogenous wastes include uric acid, a waste product of the breakdown of nucleic acid, and creatinine, a waste product of muscle tissue breakdown. Nitrogenous waste is eliminated from the body by the urinary system (see section 24.6e).

WHAT DID YOU LEARN?

36 What are the monomers of proteins and the name of the bond between them?

37 What are the names of structures that contain 2 amino acids, 3 to 20 amino acids, 21 to 199 amino acids, and 200 or more amino acids? What general term is used to refer to any of these structures, except a structure composed of 2 amino acids?

2.8 Protein Structure

Here we describe protein structure in more depth, including categories of amino acids, amino acid sequence, and three-dimensional protein shape. We also discuss how a protein can lose its three-dimensional shape and the consequences of this occurrence.

2.8a Categories of Amino Acids

LEARNING OBJECTIVES

54. Name the categories of amino acids.

55. Distinguish between nonpolar, polar, and charged amino acids.

56. Give examples of amino acids with special characteristics.

Amino acids are organized into groups based on the chemical characteristics of their R group (see section 2.7e). Each amino acid is categorized as nonpolar, polar, charged, and those having special functions **(figure 2.25)**.

- **Nonpolar amino acids** contain R groups with either hydrogen (glycine) or hydrocarbons (alanine, valine, isoleucine, leucine, phenylalanine, and tryptophan). They tend to group with other nonpolar amino acids by hydrophobic interactions within the body's aqueous environment (see section 2.3c).
- **Polar amino acids** contain R groups with elements in addition to carbon and hydrogen (i.e., O, N, S) (serine, threonine, asparagine, glutamine, and tyrosine). They form interactions with other polar amino acids and with water molecules.
- **Charged amino acids** can have either a negative charge or a positive charge. Those with a negatively charged R group include glutamate and aspartate, and those with a positively charged R group include histidine, lysine, and arginine. An ionic bond (electrostatic interaction) can form between an R group with a negative charge and an R group with a positive charge. Groups of amino acids that are either polar or charged are hydrophilic, and their presence increases the solubility of the protein in water.
- **Amino acids with special functions**—three amino acids have unique characteristics. The R group in proline attaches to the amine group, forming a ring. Proline amino acids cause a bend in the protein chain. The sulfhydryl (—S—H) functional groups of two cysteine amino acids form **disulfide bonds.** These covalent bonds are significant in stabilizing the folding of a protein (described in section 2.8b). Methionine is always the first amino acid positioned when a protein is synthesized (see section 4.8b), and it may or may not be removed later.

WHAT DID YOU LEARN?

38 Why is the amino acid leucine (figure 2.25) classified as a nonpolar amino acid?

INTEGRATE

LEARNING STRATEGY

The formation of proteins from amino acids is similar to the building of words from the letters of an alphabet. Just as English, Spanish, and many other languages are built upon words formed by stringing letters together, the "language of proteins" is built upon the stringing of amino acids together.

Nonpolar						
	Glycine (Gly)	Valine (Val)	Isoleucine (Ile)	Leucine (Leu)	Phenylalanine (Phe)	Tryptophan (Trp)
	Alanine (Ala)					

Polar					
	Serine (Ser)	Threonine (Thr)	Asparagine (Asn)	Glutamine (Gln)	Tyrosine (Tyr)

Charged	(–) Charge		(+) Charge		
	Glutamate (Glu)	Aspartate (Asp)	Histidine (His)	Lysine (Lys)	Arginine (Arg)

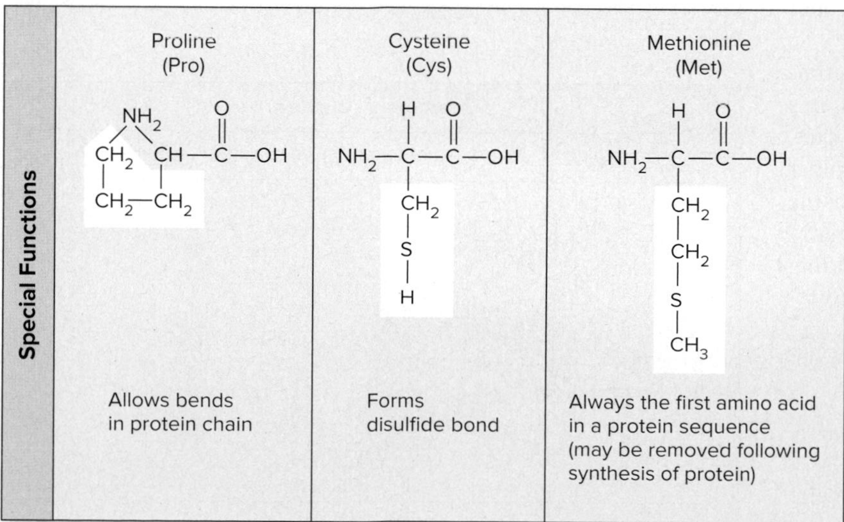

Figure 2.25 Amino Acids. Amino acids are categorized based on the chemical properties of the R groups. The categories include nonpolar, polar, charged, and those with special functions.

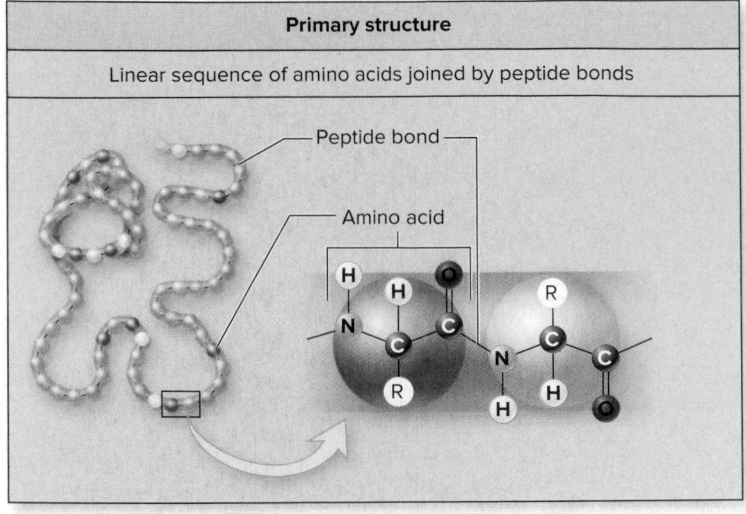

(a)

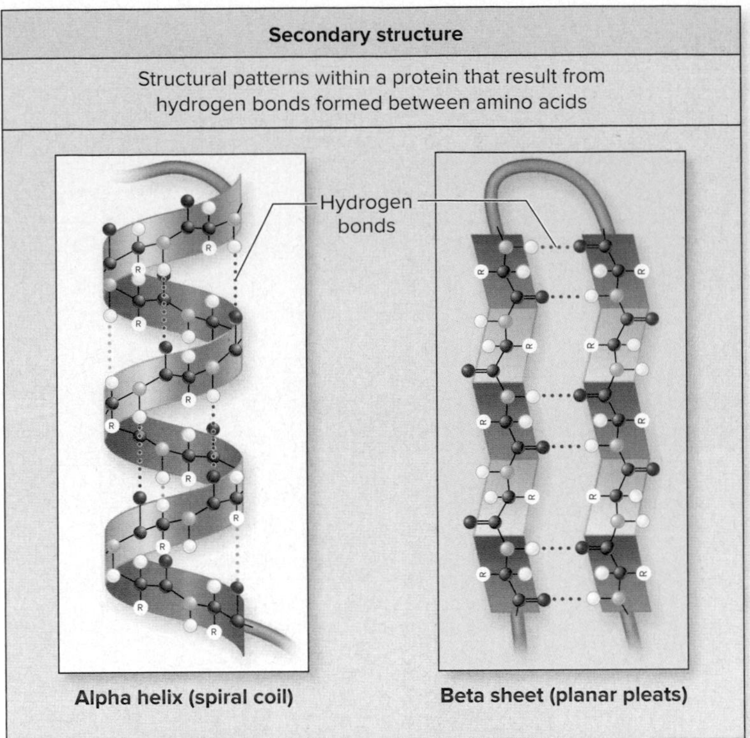

(b)

Figure 2.26 Levels of Protein Structure. Amino acids bonded together form protein polymers. There are four increasingly complex levels of organization. (*a*) The primary structure is the linear sequence of amino acids in the protein. (*b*) The secondary structures of a protein may include alpha helixes and beta sheets. (*c*) The tertiary structure is the completed 3-dimensional shape or conformation of the protein, which may be a globular or fibrous protein. (*d*) A quaternary structure is formed in some complex proteins when two or more protein molecules associate to form the final protein.

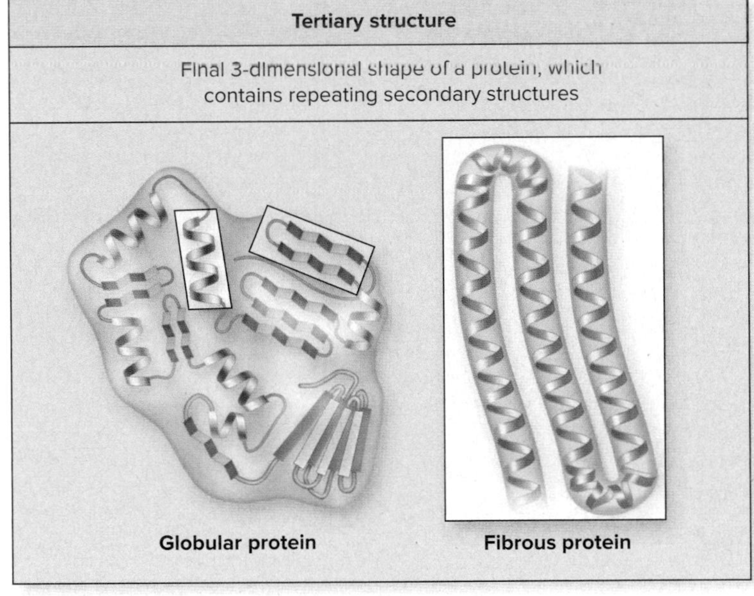

(c)

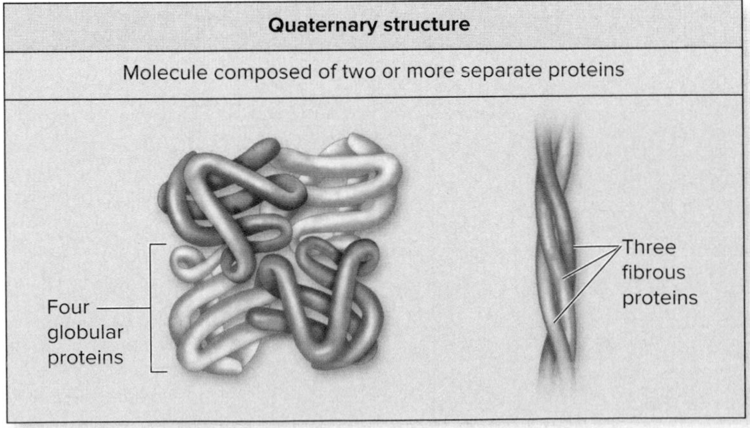

(d)

2.8b Amino Acid Sequence and Protein Conformation

LEARNING OBJECTIVES

57. Describe the different types of intramolecular (or intermolecular) attractions that participate in both folding a protein and maintaining its three-dimensional shape.

58. Distinguish between the four structural hierarchy levels of proteins.

59. Explain what is meant by denaturation, and list factors that can cause it.

We describe a protein as being composed of a linear sequence of amino acids that are bonded together through covalent peptide bonds. This sequence is called its **primary structure** (**figure 2.26*a***). The protein then folds to form into its three-dimensional shape, or **conformation.** Protein conformation is crucial for its proper function.

The conformational structure of proteins has increasingly complex hierarchies, or levels, of organization beyond a protein's primary structure, including secondary, tertiary, and perhaps quaternary structural levels. These more complex structural organizations are dependent upon intramolecular attractions between the amino acids in the linear sequence for the proper folding and maintaining of a protein's conformation. The process of protein folding is assisted by specialized proteins called *chaperones* that "direct" the folding process.

The intramolecular interactions that contribute to the final conformation of a protein are as follows:

- The **primary structure** of a protein is forced into its initial shape as hydrophobic exclusion "tucks" amino acids with nonpolar R groups into a more central location, limiting their contact with water.

Figure 2.27 Denaturation. Proteins denature in response to (*a*) increased H^+ (decrease in pH) as H^+ is added to the negatively charged R group or (*b*) decreased H^+ (increase in pH) as H^+ is removed from the positively charged R group.

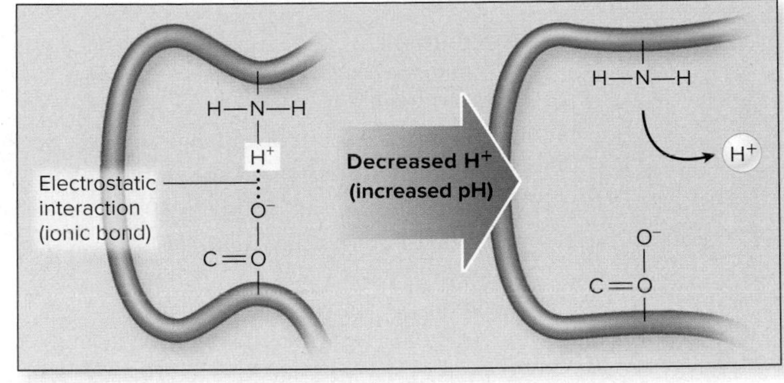

(a) (b)

- Hydrogen bonds form between polar R groups, and between the amine and carboxylic acid functional groups of closely positioned amino acids.
- Electrostatic interactions (ionic bonds) form between negatively charged and positively charged R groups.
- Disulfide bonds form between the sulfhydryl (—S—H) groups of two cysteine amino acids.

Higher levels of protein organization result from folding. The **secondary structure** is a pattern within a protein that may repeat several times. The two distinctive secondary structures are a spiral coil, called an **alpha helix,** and a planar pleat arrangement, called a **beta sheet.** These secondary structures confer unique characteristics on the regions of the protein where they occur. Alpha helix gives some elasticity to fibrous proteins that are located, for example, in skin or hair. In contrast, beta sheets give some degree of flexibility to many globular proteins (e.g., enzymes).

Tertiary structure is the final three-dimensional shape exhibited by one completed protein chain. Two categories of proteins—either globular or fibrous—are distinguished by their molecular shape. **Globular proteins** fold into a compact, often nearly spherical shape such as enzymes (see figure 3.9), antibodies (see figure 22.18), and some hormones. In comparison, **fibrous proteins** are extended linear molecules that are constituents of ligaments and tendons (see figure in table 5.6) and contractile proteins within muscle cells (see figure in table 5.10).

The **quaternary structure** of a protein is present only in those proteins with *two or more* protein strands. The protein hemoglobin is an example because it is composed of four protein chains (see section 18.3b). Each of these chains has its own primary, secondary, and tertiary structures. Only when the four separate strands associate through intermolecular attractions to form the quaternary structure does the biological molecule of hemoglobin become active. Therefore, hemoglobin is functional only when all four polypeptide chains are present in the correct association.

The normal function of a protein may also require a **prosthetic group.** This is a nonprotein structure covalently bonded to the protein. The lipid heme group in the hemoglobin protein is an example of a prosthetic group (see figure 18.6).

WHAT DO YOU THINK?

5 Why might proteins be rendered nonfunctional by exposure to high temperatures?

The biological activity of a protein is usually disturbed or terminated when its conformation (or structure) is changed. This change in protein conformation is called **denaturation** (dē-na′tyū-ra′shŭn). The tertiary structure of the protein is disrupted if the protein is heated or chemically altered. Usually, the loss of biological activity cannot be reversed. Denaturation may occur as a consequence of increased temperature because this increase weakens the intramolecular interactions that hold the protein in its three-dimensional shape. The risks of an elevated body temperature and denaturation are discussed in the section on the risks of a high fever (see section 22.3e).

Alternatively, denaturation may take place in response to changes in pH. Changes in pH can denature proteins because the alteration of hydrogen ion concentration $[H^+]$ interferes with electrostatic interactions (and other intramolecular bonds) that hold the protein in its three-dimensional shape. If pH decreases as a consequence of an increase in $[H^+]$, the excess H^+ binds to negatively charged R groups, as shown in **figure 2.27*a*.** The electrostatic interaction is disrupted because an H^+ attaches to the negatively charged R group, with an accompanying loss of the negative charge on the R group that previously participated in the electrostatic interaction. If pH increases, reflecting a decrease in $[H^+]$, the electrostatic interaction is disrupted because the H^+ that previously participated in the electrostatic interaction has been removed, as shown in figure 2.27*b*. Changes in blood pH out of the normal homeostatic range can be lethal because of the potential disruption of protein structure and function (acid-base balance is described in section 25.5). Proteins are essential in the normal processes of the body, and the structure and function of many different proteins will be discussed throughout this text.

WHAT DID YOU LEARN?

39 What distinguishes the tertiary and quaternary levels of organization of a protein?

40 What happens to a protein when it denatures? How does a protein denature when it is exposed to higher than normal H^+ concentration?

CHAPTER SUMMARY

- Atoms, ions, and molecules form the human body at its simplest level.

2.1 Atomic Structure

- Diagramming atomic structure from the periodic table provides the foundation to understand isotopes, ions, and molecules.

2.1a Matter, Atoms, Elements, and the Periodic Table

- An atom is the smallest particle that exhibits the chemical properties of an element.
- Atomic structure—composed of protons, neutrons, and electrons—can be drawn from the information obtained from the periodic table.

2.1b Isotopes

- Atoms having the same number of protons and electrons—but differing in the number of neutrons and thus atomic mass—are called isotopes.
- Unstable isotopes resulting from an excess number of neutrons are called radioisotopes.

2.1c Chemical Stability and the Octet Rule

- Atoms with a complete outer shell of eight electrons are chemically stable. Atoms bond to reach chemical stability.

2.2 Ions and Ionic Compounds

- Chemical compounds, which include ionic compounds and molecular compounds, are stable associations between two or more elements combined in a fixed ratio.

2.2a Ions

- An ion is a charged atom that is either positively charged or negatively charged as a result of having lost or gained electrons, respectively.
- Common ions of the human body include sodium (Na^+), potassium (K^+), calcium (Ca^{2+}), magnesium (Mg^{2+}), hydrogen (H^+), chloride (Cl^-), bicarbonate (HCO_3^-), and phosphate (PO_4^{3-}).
- Cations are positively charged ions that are formed by the loss of electron(s) from atoms with one, two, or three electrons in their outer shell, or elements in columns IA, IIA, or IIIA in the periodic table, respectively.
- Anions are negatively charged ions that are formed by the gain of electron(s) by atoms typically with five, six, or seven electrons in the outer shell, or elements in columns VA, VIA, or VIIA in the periodic table, respectively.

2.2b Ionic Bonds

- Ionic bonds are electrostatic attractions between positively charged cations and negatively charged anions that hold the ions in a lattice crystal (or salt).

2.3 Covalent Bonding, Molecules, and Molecular Compounds

- Molecules formed with two or more atoms of the same or different elements are covalently bonded together.
- Molecular compounds are formed by two different elements covalently bonded together.

2.3a Chemical Formulas: Molecular and Structural

- The molecular formula is the number and kinds of atoms within a molecule.
- The structural formula demonstrates both the number and kind of atoms within a molecule and their arrangement within the molecule—and can be used to distinguish isomers.
- Isomers have the same number and kind of atoms, or molecular formula, but differ in their arrangement in space.

2.3b Covalent Bonds

- Covalent bonds are formed between two atoms in which both atoms have four, five, six, or seven electrons in their outer shell (as well as hydrogen atoms, which require only two electrons to have a complete valence shell).
- A single covalent bond is the typical covalent bond and involves the sharing of a pair of electrons. A double or triple covalent bond can be formed if both atoms, respectively, need at least two or three electrons to become stable.
- Nonpolar covalent bonds occur when the two atoms share electrons equally, as occurs between atoms of the same element, or almost equally, as occurs when hydrogen bonds to carbon. Polar covalent bonds occur between atoms of different elements that share electrons unequally.

2.3c Nonpolar, Polar, and Amphipathic Molecules

- Nonpolar molecules generally include those molecules composed predominantly of nonpolar bonds between the atoms.
- Polar molecules generally include those molecules composed of relatively more polar bonds between the atoms.
- Amphipathic molecules are large molecules that have both nonpolar and polar regions.

2.3d Intermolecular Attractions

- Intermolecular attractions occur between molecules, and intramolecular attractions occur between regions within large molecules.
- Hydrogen bonds occur between a partially positively charged hydrogen atom and a partially negatively charged atom of polar molecules.
- Intermolecular attractions between nonpolar molecules occur with a temporary unequal distribution of electrons. Hydrophobic interactions occur when nonpolar molecules are placed in water.

2.4 Molecular Structure and Properties of Water

- Water makes up on average about two-thirds of the human body by weight.

2.4a Molecular Structure of Water

- Water is a polar molecule that can form four hydrogen bonds with other water molecules.

2.4b Properties of Water

- Water is present in three phases (water vapor, liquid, and ice), depending upon the temperature.
- Cohesion is attraction between water molecules, surface tension is the inward pull of water molecules at the surface, and adhesion is the attraction of water molecules for substances other than water.
- Water's high specific heat and high heat of vaporization help maintain a normal body temperature.

2.4 Molecular Structure and Properties of Water (continued)	**2.4c Water as the Universal Solvent** • Polar organic molecules, such as glucose, dissolve and remain intact within water and are nonelectrolytes. • Salts, acids, and bases dissolve and dissociate in water and are electrolytes. • Nonpolar substances do not form hydrogen bonds with water and are separated from water molecules by hydrophobic exclusion. • The polar portion of an amphipathic molecule dissolves in water, and the nonpolar portion does not. Amphipathic molecules form chemical barriers within the body, including membranes and micelles.
2.5 Acidic and Basic Solutions, pH, and Buffers	• Water has a neutral pH, and its pH is changed by the addition of either an acid or a base.
	2.5a Water: A Neutral Solvent • Positively charged hydrogen ions are equal to negatively charged hydroxide ions in pure water; thus, water is neutral.
	2.5b Acids and Bases • Acids increase and bases decrease the hydrogen ion concentration in solution.
	2.5c pH, Neutralization, and the Action of Buffers • pH is a measure of hydrogen ion concentration in solution. pH values are inversely related to hydrogen ion concentration. • Neutralization is the changing of an acidic or basic solution to neutral. • A buffer helps prevent pH changes by "absorbing" and "releasing" hydrogen ions.
2.6 Water Mixtures	• Mixtures are formed from the mixing of two or more substances. • Mixtures are not chemically changed, and the components can be separated by physical means.
	2.6a Categories of Water Mixtures • Types of water mixtures include suspensions, colloids, and solutions. • Emulsions are a type of colloid formed by water and a liquid nonpolar substance when it is agitated.
	2.6b Expressions of Solution Concentration • The concentration of solutions can be expressed in several ways, including mass/volume, mass/volume percent, molarity, and molality. • Osmoles, osmolarity, and osmolality reflect number of particles in solution and are relevant in the movement of water by osmosis.
2.7 Biological Macromolecules	• The four major classes of organic biological macromolecules are lipids, carbohydrates, nucleic acids, and proteins.
	2.7a General Characteristics • Biological macromolecules have diverse carbon skeletons with varying amounts of hydrogen and functional groups attached to them. • Three of the four major classes of biological macromolecules—including complex carbohydrates, nucleic acids, and proteins—exist as polymers, chemical structures that are composed of repeating identical or similar monomers. Lipids are not polymers. • All of the classes of biological macromolecules are formed by dehydration synthesis reactions and digested through hydrolysis.
	2.7b Lipids • Lipids are a very diverse group of fattylike, water-insoluble molecules that include triglycerides, phospholipids, steroids, and eicosanoids. • Triglycerides are composed of glycerol and three fatty acids and are generally used for long-term energy storage. • Phospholipids are made up of glycerol, two fatty acids, and a phosphate functional group with various organic groups attached to it. Phospholipids are amphipathic molecules containing a polar head and two nonpolar tails that form membranes. • Steroids are distinct multiringed structures formed predominantly of hydrocarbons and include cholesterol, steroid hormones, and bile salts. • Eicosanoids are modified 20-carbon fatty acids that are synthesized as needed from arachidonic acid, a common component of plasma membranes. Eicosanoids act locally. • Other lipids include glycolipids and fat-soluble vitamins.
	2.7c Carbohydrates • Carbohydrates are molecules with the following chemical formula: $(CH_2O)_n$. Carbohydrates exist in increasing levels of complexity that include monosaccharides, disaccharides, and polysaccharides. • Glucose is the most common monosaccharide in the human body and is used for energy. When in excess, glucose is stored as the polysaccharide called glycogen in liver and skeletal muscle tissue. • Other types of carbohydrates include the monosaccharides galactose, fructose, ribose, and deoxyribose; the disaccharides sucrose, lactose, and maltose; and the polysaccharides starch and cellulose.
	2.7d Nucleic Acids • Nucleic acids include deoxyribonucleic acid (DNA) and ribonucleic acid (RNA), polymers formed from nucleotide monomers. These molecules ultimately determine the type of proteins synthesized by cells. • Adenosine triphosphate (ATP) is a nucleotide that serves as the energy currency molecule of a cell.
	2.7e Proteins • Proteins serve many different functions in the body. • Proteins are polymers that differ in the number and sequence of 20 different amino acid monomers arranged in a unique linear sequence.

(continued on next page)

CHAPTER SUMMARY (continued)

2.8 Protein Structure

- The three-dimensional structure of proteins is dependent upon the linear sequence of its amino acids.

2.8a Categories of Amino Acids

- The 20 amino acids can be categorized based on the chemical composition of their R group as nonpolar, polar, charged, and those with special functions.

2.8b Amino Acid Sequence and Protein Conformation

- Protein organization includes the primary, secondary, and tertiary structures—as well as a quaternary structure if there are two or more protein chains. These levels of organization ultimately determine the structure and function of a protein.
- Denaturation usually results in the loss of biological activity of a protein in response to the change in its three-dimensional shape that may have occurred through an increase in temperature or a change in pH.

CHALLENGE YOURSELF

Create and Evaluate

Analyze and Apply

Understand and Remember

Do You Know the Basics?

1. Atoms composed of the same numbers of protons and electrons, but different numbers of neutrons, are called
   a. isomers. c. isotopes.
   b. ions. d. organic atoms.

2. Substances that dissolve in water include all of the following *except*
   a. lipids. c. proteins.
   b. glucose. d. salts.

3. Temperature stabilization is dependent upon which properties of water?
   a. cohesion and adhesion
   b. capillary action and adhesion
   c. specific heat and heat of vaporization
   d. cohesion and specific heat

4. All of the following are accurate about H^+ concentration and pH *except*
   a. acids contain more H^+ than water.
   b. H^+ concentration and pH are inversely related.
   c. neutralizing an acidic solution requires that base is added.
   d. a pH of 6 is basic, or alkaline.

5. Blood is a mixture that is more specifically described as a
   a. suspension. c. solution.
   b. colloid. d. All of these are correct.

6. Which biological macromolecule is *not* a polymer?
   a. triglyceride c. glycogen
   b. protein d. DNA

7. Glucose is stored as which molecule within the liver and skeletal muscle tissue?
   a. starch c. glycogen
   b. phospholipid d. glucagon

8. All of the following are common ions of the human body *except*
   a. Na^+. c. Ca^{2+}.
   b. P^+. d. Cl^-.

9. Intermolecular attractions between polar molecules are
   a. hydrophobic interactions.
   b. hydrogen bonds.
   c. van der Waals forces.
   d. ionic bonds.

10. When a protein permanently unfolds, it has been
    a. polymerized. c. converted to nucleic acids.
    b. denatured. d. made more efficient.

11. List the common ions of the human body by name, symbol, and charge.

12. Describe a polar bond and a polar molecule.

13. Diagram two water molecules, and label the polar covalent bonds and a hydrogen bond.

14. Compare and contrast what occurs when a substance dissolves in water with a substance that dissolves and dissociates in water. Include examples of specific substances.

15. Define the terms *acid, base, pH,* and *buffers.*

16. Explain the units for expressing a concentration that are included in this chapter.

17. List the four biological macromolecules and the building blocks that compose them.

18. Which two biological macromolecules contain nitrogen atoms and contribute to the formation of nitrogenous waste that must be eliminated by the urinary system?

19. Describe how phospholipid molecules form the plasma membrane of a cell.

20. Explain protein denaturation, including how it occurs and the consequences in response to an increase in temperature and a change in pH.

Can You Apply What You've Learned?

1. Which property of water is significant in children born prematurely because it causes the air sacs to collapse in the lungs, making breathing difficult?
   a. specific heat
   b. water reactivity
   c. surface tension
   d. capillary action

2. A young boy playing outside on a very hot day has become dehydrated. When he enters the house, he appears lethargic. The mother is a nurse and becomes concerned that he may be experiencing a fluid and electrolyte imbalance. Electrolytes include all of the following *except*
   a. sodium ion.
   b. glucose.
   c. potassium ion.
   d. chloride ion.

3. A young woman has noticed that her thyroid gland appears enlarged. One of the diagnostic procedures used to produce an image of her thyroid gland requires this type of substance, which emits high-energy radiation.
   a. ions
   b. radioisotopes
   c. radioisomers
   d. isomers

4. The condition of rickets involves bones that have insufficient amounts of this common ion, resulting in the bones bending under a child's weight.
   a. Na^+
   b. K^+
   c. Cl^-
   d. Ca^{2+}

5. The hormone insulin is a ______________ composed of repeating units of amino acids and cannot be administered orally because the enzymes within the gastrointestinal tract will break the peptide bond through the process of hydrolysis, releasing individual amino acids.
   a. nucleic acid
   b. glycogen
   c. protein
   d. steroid

Can You Synthesize What You've Learned?

1. An individual is exposed to high-energy radiation. Which biological macromolecule that regulates the process of protein synthesis may have been mutated?

2. The lab results from a diabetic patient show a lower than normal pH (a condition referred to as acidosis). Explain the change in H^+ concentration in the blood, and describe how this change may affect the folding of proteins in the blood plasma (and elsewhere).

3. A patient is given a new drug that decreases blood sugar levels. This drug is regulating which specific molecule?

INTEGRATE

ONLINE STUDY TOOLS

connect | SMARTBOOK® | 

The following study aids may be accessed through Connect.

Clinical Case Study: A True Story of Killing Patients with Potassium Chloride

Interactive Questions: This chapter's content is served up in a number of multimedia question formats for student study

SmartBook: Topics and terminology include atomic structure; ions and ionic compounds; covalent bonding, molecules, and molecular compounds; molecular structure of water and the properties of water; acidic and basic solutions, pH, and buffers; water mixtures; biological macromolecules; and protein structure

Anatomy & Physiology Revealed: Topics include atomic structure, bonds

Animations: Topics include ionic solutions, ionic bond, solutions; making solutions; buffers

chapter

3 Energy, Chemical Reactions, and Cellular Respiration

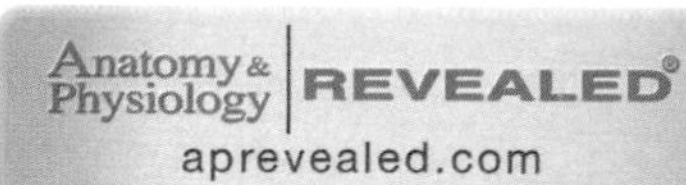

Module 2: Cells and Chemistry

All living organisms require energy. In humans, energy is needed to power muscle contractions when walking, pump blood through the body, absorb nutrients from the gastrointestinal tract, and exchange respiratory gases. Energy is also required both to synthesize (produce) new molecules for maintenance, growth, and repair and to establish ion concentrations in cells. Here we discuss the chemical principles of energy, chemical reactions, enzymes, and metabolic pathways. Many of these concepts are then integrated to describe the metabolic pathway that breaks down glucose molecules. It is during the breakdown of glucose (and other fuel molecules) that energy is released to synthesize adenosine triphosphate (ATP). ATP is the molecule that serves as the "energy currency" of a cell for all energy-requiring cellular processes.

INTEGRATE

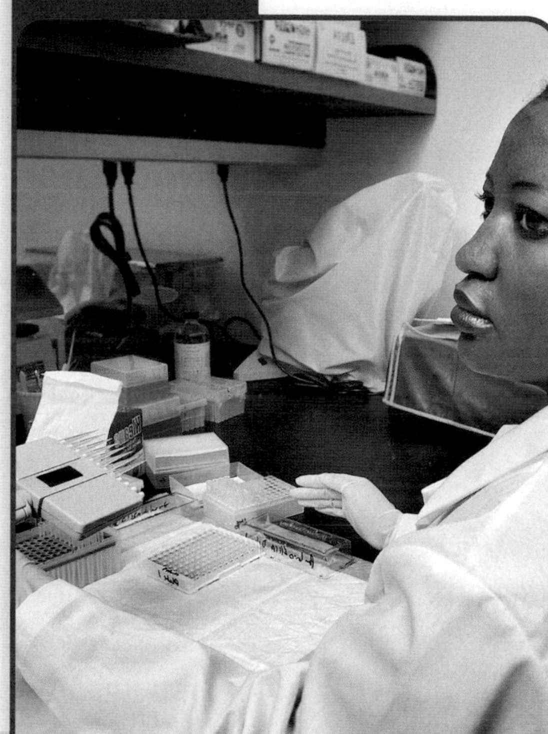

Source: James Gathany/CDC

CAREER PATH
Biochemists

Biochemists study the chemical composition and physical principles of living cells and organisms, including those associated with cell metabolism, development, growth, reproduction, and heredity. Their work may involve determining the effects of substances (e.g., foods, vitamins, enzymes, hormones, allergens) on living organisms or studying the effects of mutations that lead to cancer.

3.1 Energy

Energy is defined as the capacity to do work. Energy differs from matter in that it has no mass and does not take up space. It is invisible, but its effects on matter may be observed. This section describes the two major classes of energy, the various forms of energy, and physical laws that govern energy.

3.1a Classes of Energy

LEARNING OBJECTIVE

1. Describe the two classes of energy.

Two classes of energy exist: potential energy and kinetic energy. **Potential energy** is the energy of *position* or *stored* energy. **Kinetic energy** is the energy of *motion.* Potential energy can be converted, or changed, to kinetic energy, and vice versa. For example, water behind a dam has potential energy because of its position; when the water falls over the dam it now has kinetic energy because of its motion. The kinetic energy of falling water can be harnessed to do work if it drives a water wheel at the bottom of the dam.

A bow and arrow also provides an example of an energy conversion from potential energy to kinetic energy. When the arrow is pulled back in the bow, it has potential energy because of the tension in the bowstring. This potential energy is converted to kinetic energy as the string is released and the arrow flies. The kinetic energy of the flying arrow can do work when it knocks an apple from a tree.

Potential energy is exhibited in cells of living organisms when a concentration gradient (i.e., a difference in the amount of a substance in two areas) exists across the plasma membrane, which is the boundary between the inside and outside of a cell (**figure 3.1*a***). Sodium ion (Na^+) concentration typically is greater outside the cell than inside. This difference in Na^+ concentration across the membrane is analogous to the water at the top of a dam because it represents potential energy. The movement of Na^+ from a high concentration outside the cell to a low concentration inside the cell is an example of kinetic energy. Like water falling over a dam, the kinetic energy of Na^+ movement may be harnessed to do work. You will see applications of this principle both in this chapter and in later chapters (see section 24.6c).

Potential energy is also exhibited by the position of electrons in electron shells relative to an atom's nucleus (figure 3.1*b*). Electrons can move from a higher-energy shell to a lower-energy shell. Note that when electrons move during a chemical reaction, they may do so either within the same chemical structure or from one chemical structure to another (e.g., in the electron transport chain of a mitochondrion [see figure 3.19], a cell organelle that synthesizes ATP). The kinetic energy of electron movement can be harnessed to do work—movement of electrons is critical to the formation of ATP molecules.

Note that potential energy has the ability, or *potential,* to do work because of its position. Potential energy must be converted to kinetic energy to be actively engaged in doing work.

WHAT DID YOU LEARN?

1. Both the movement of Na^+ from an area of higher concentration to an area of lower concentration and the movement of an electron from a higher energy to a lower energy state are examples of (a) potential energy or (b) kinetic energy?

3.1b Forms of Energy

LEARNING OBJECTIVES

2. Describe chemical energy (one form of potential energy) and the various forms of kinetic energy.
3. List the three important molecules within the body that function primarily in chemical energy.

Potential and kinetic energy exist in several forms. We first describe chemical energy, which is one form of potential energy, and then describe several forms of kinetic energy.

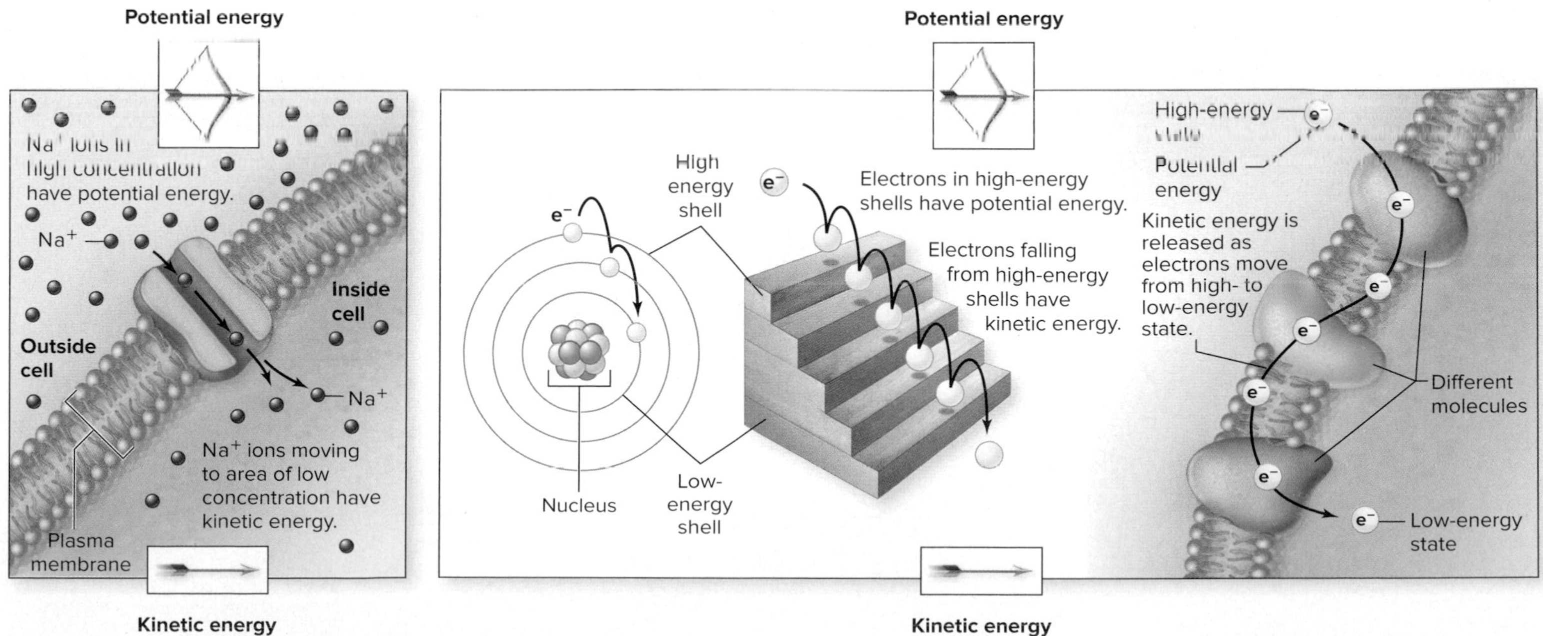

Figure 3.1 Conversion of Potential Energy to Kinetic Energy. Potential energy that is converted to kinetic energy may be harnessed to do work. (*a*) The Na^+ gradient across the plasma membrane has potential energy that changes to kinetic energy when Na^+ ions move from where they are in high concentration outside of the cell to where they are in low concentration inside the cell. (*b*) A high-energy electron has potential energy and may be converted to kinetic energy in the same atom or be transferred to different molecules as the electron "falls" from high-energy shells to low-energy shells.

Chemical Energy (a Form of Potential Energy)

Chemical energy is one form of potential energy. **Chemical energy** is the energy stored in a molecule's chemical bonds, and is the most important form of energy in the human body. It specifically is used for the energy-requiring processes of movement, synthesis of molecules, and establishment of concentration gradients. The chemical bonds of all molecules have chemical energy. This energy is released when bonds are broken during chemical reactions.

Three important molecules in the human body function primarily in chemical energy storage: triglycerides, glucose, and adenosine triphosphate (ATP). These molecules differ in their chemical structure, where they are stored, and the length of time each generally stores energy. Recall the following from chapter 2:

- Triglycerides are involved in long-term energy storage in adipose connective tissue (see section 2.7b).
- Glucose is stored in the liver and muscle tissue in the form of the polymer glycogen (see section 2.7c).
- ATP is stored in all cells in limited amounts and is produced continuously and used immediately for cells' energy-requiring processes (see section 2.7d).

Note that protein also stores chemical energy and can be used as a fuel molecule. However, as described in section 2.7e, proteins are primarily important as the structural and functional components of the body.

Kinetic Energy Forms

Other forms of energy—electrical, mechanical, sound, radiant, and heat—exist as kinetic energy.

Electrical energy is the movement of charged particles. Examples of electrical energy include electricity, which is the movement of electrons along a wire, and the propagation of a nerve signal (an impulse) due to the movement of ions across the plasma membrane of a neuron (nerve cell; see section 12.2a).

Mechanical energy is exhibited by an object in motion due to an applied force. Examples of mechanical energy include muscle contraction for walking and the pumping action of the heart to circulate blood.

Sound energy occurs when the compression of molecules that move in a solid, liquid, or gas is caused by a vibrating object, such as the head of a drum or the vibration of the vocal cords. The sense of hearing is initiated when sound waves cause vibration of the tympanic membrane (eardrum) in the ear (see section 16.5a).

Radiant energy is the energy of electromagnetic waves traveling in the universe. Radiant energy consists of a spectrum of different energy forms that vary in wavelength and frequency (**figure 3.2**). The higher the frequency in the spectrum, the greater the

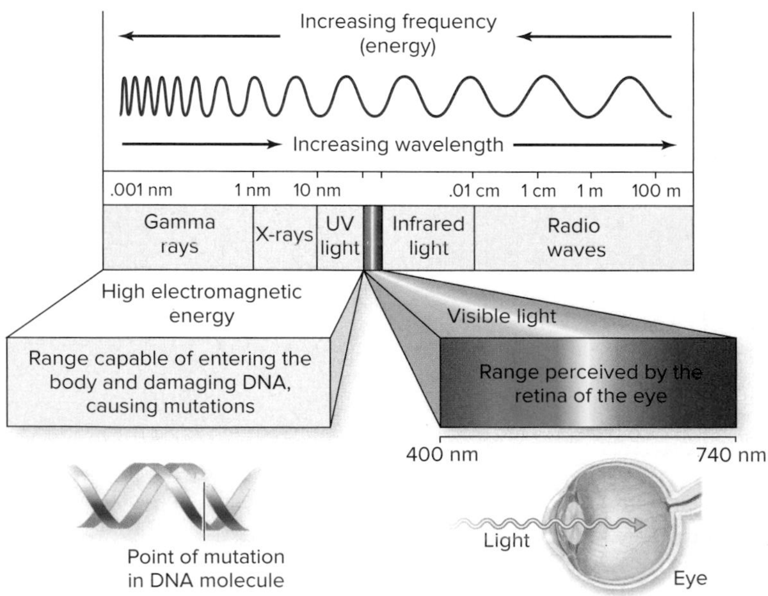

Figure 3.2 The Electromagnetic Spectrum. Different forms of radiant energy are represented on an electromagnetic spectrum. The forms of energy are arranged from shortest wavelength to longest wavelength. Notice that this arrangement is also from highest frequency (energy) to lowest frequency (energy).

amount of radiant energy associated with it. Gamma rays have the highest amount of radiant energy, whereas radio waves have the lowest. All forms of radiant energy with a frequency higher than visible light (gamma rays, x-rays, and ultraviolet [UV] light) have sufficient energy to penetrate the body and mutate (change) the DNA of living organisms. Cells of the skin normally protect themselves from everyday UV light exposure by producing the pigment melanin (see section 6.1a). This process commonly darkens skin color and is referred to as tanning. Visible light is a lower-frequency radiant energy that is detected by retinal cells of the eye. This visual input is then relayed along the optic nerve to the brain for interpretation.

Heat is the kinetic energy associated with random motion of atoms, ions, or molecules. It is usually considered an unusable form of energy, or a *waste product* that accompanies all changes in energy form because heat is the only type of energy that is not available to do work. Heat is measured as the **temperature** of a substance.

WHAT DID YOU LEARN?

2 Muscle contraction is an example of what form of energy?

3.1c Laws of Thermodynamics

LEARNING OBJECTIVES

4. State the first law and second law of thermodynamics.
5. Explain why energy conversion is always less than 100%.

Energy can be converted from one form to another. For example:

- When a candle is burned, chemical energy in the burning wax is converted to both light and heat.
- Light energy from the sun is converted by the retinal cells into electrical energy associated with nerve signals involved in sight (see section 16.4d).
- Chemical energy in the food we eat is first converted into another chemical form (ATP), which is then converted in our cells into mechanical energy used to power muscle contraction (see section 10.4a)

In these examples, energy is simply changing from one form to another. The study of energy transformations is called **thermodynamics** (ther′mō-dī-nam′iks; *thermo* = heat, *dynamis* = force).

Two laws of thermodynamics describe energy transformations. The **first law of thermodynamics** states that energy can neither be created nor destroyed—it can only be transformed, or converted, from one form to another.

The **second law of thermodynamics** states that every time energy is transformed from one form to another, some of that energy is converted to heat. That means there is never 100% conversion of one form of usable energy to another. Energy conversions have a price, and that price always appears as heat. Because heat is not available to do work, the usable amount of energy is decreased each time an energy conversion occurs. For example, the conversion of the chemical energy in gasoline to the mechanical energy (movement) of a car is approximately 25%. This means that approximately 75% of gasoline's chemical energy is converted to sound and heat.

Heat is produced when the chemical energy stored in the foods we eat is used to power our muscle contractions. One of the functions of muscle tissue is to produce heat that keeps the body warm (see section 10.1a). When the environmental temperature drops, and we begin to move around in the hope of generating enough heat to keep ourselves warm, we are applying the second law of thermodynamics.

Figure 3.3 summarizes the two major classes of energy, the various forms of energy as they relate to the body, and the laws of energy.

WHAT DID YOU LEARN?

3 Energy can neither be created nor destroyed. However, according to the first and second laws of thermodynamics, what can happen to it, and what is always generated?

Enzymes in the oxidoreductase class, for example, participate in oxidation-reduction reactions. **Dehydrogenase** (dē-hī′drō-jen-ās; *de* = from) enzymes are a subcategory of enzymes within the oxidoreductase class. These enzymes participate in oxidation-reduction reactions by moving hydrogen from one molecule to a different molecule.

Another example of a class of enzymes is the transferase class. All enzymes in this class transfer atoms or molecules between chemical structures. **Kinase** (kī′nās) enzymes belong to this class because they specifically transfer a phosphate functional group, usually from ATP to another molecule. You will learn about specific examples of both dehydrogenase and kinase enzymes in section 3.4.

WHAT DO YOU THINK?

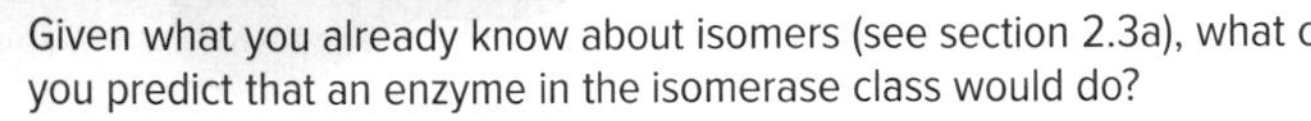

2 Given what you already know about isomers (see section 2.3a), what can you predict that an enzyme in the isomerase class would do?

The name of a given enzyme is generally based upon the name of the substrate or product involved in the chemical reaction, sometimes the name of the subclass, and the suffix *-ase*, added to the final word of the name. Here are three examples:

- **Pyruvate dehydrogenase** is an enzyme that transfers a hydrogen, specifically from a pyruvate molecule (see section 3.4c).
- **DNA polymerase** is central to the formation of the polymer DNA from deoxyribonucleotides (see section 4.9b).
- **Lactase** digests the disaccharide lactose. (see section 26.4a).

Although the name of an enzyme generally contains an *-ase* suffix and the rest of the name reflects its function, there are exceptions. For example, pepsin, trypsin, and chymotrypsin (see section 26.4b) are all protein-digesting enzymes with names that do not provide clear clues to these molecules' enzymatic nature or specific activity.

WHAT DID YOU LEARN?

12 Explain how enzymes are generally named.

3.3e Enzymes and Reaction Rates

LEARNING OBJECTIVES

21. Define how enzyme and substrate concentration affect reaction rates.
22. Explain the effect of temperature on enzymes.
23. Describe how pH changes affect enzymes.

Several conditions influence the reaction rates catalyzed by enzymes. The most significant factors are enzyme and substrate concentration, temperature, and pH.

Effect of Enzyme and Substrate Concentration

The rate of a chemical reaction may be accelerated by either an increase in enzyme concentration or an increase in substrate concentration. An increase in substrate concentration, however, increases the rate of reaction only up to the point of saturation of the enzyme (**figure 3.11*a***). **Saturation** occurs when so much substrate is present that all enzyme molecules are actively engaged in the chemical reaction, resulting in no further (notable) increase in reaction rate.

Effect of Temperature

Enzymes typically are proteins, and their three-dimensional shape is dependent upon environmental variables, including temperature and pH, as described in section 2.8b. Each enzyme has a specific environment in which it can most effectively participate in a chemical reaction. Human enzymes function efficiently at normal body temperature, which, for most individuals is around 37°C (98.6°F) (figure 3.11*b*). The activity of human enzymes increases with a rise in body temperature (within several degrees) and continues to increase until the **optimal temperature** is reached. This is generally around 40°C (104°F) for enzymes within the human body. These increases in body

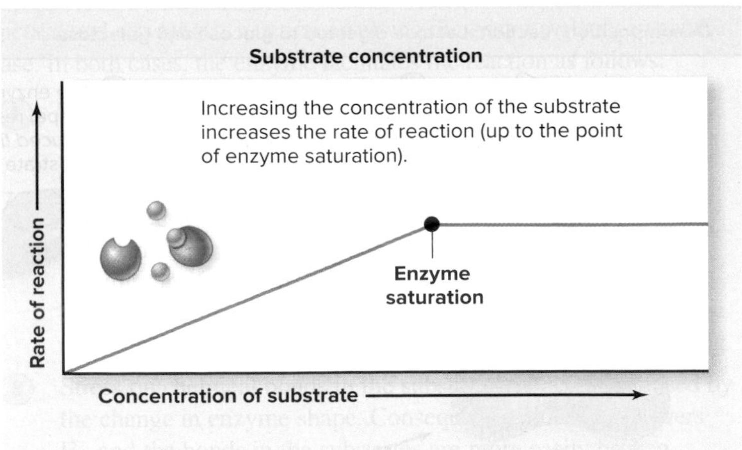

(a)

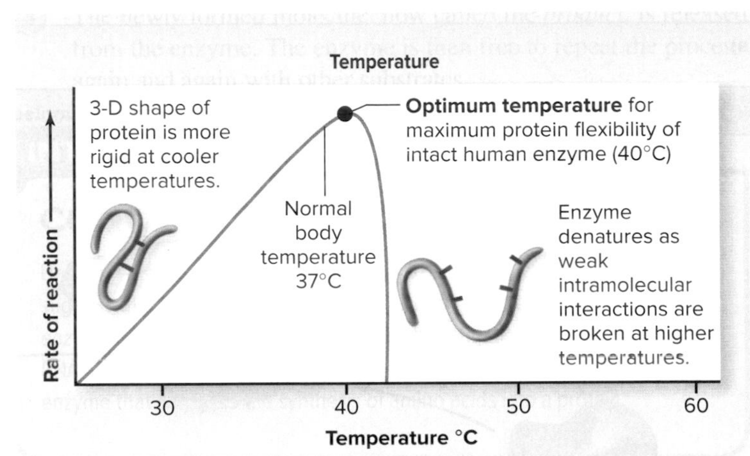

(b)

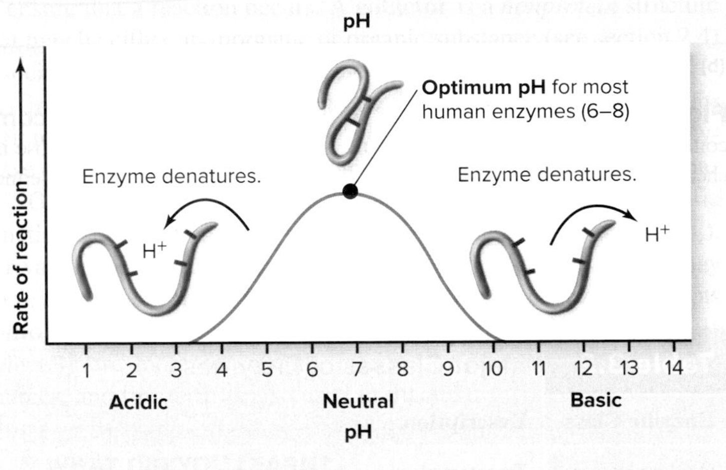

(c)

Figure 3.11 Environmental Conditions That Influence Reaction Rates of Enzymes. Reaction rates are influenced by changes in (*a*) the concentration of substrates, (*b*) temperature, and (*c*) pH. AP|R

temperature, as observed with a low fever, facilitate the eliminating of infectious agents (e.g., bacteria, viruses) (see section 22.3e).

However, more severe increases in temperature—meaning temperatures greater than 40°C (104°F) in humans—weaken the intramolecular bonds that hold an enzyme's protein structure in its three-dimensional shape. The enzyme subsequently denatures, permanently losing function. The greater the increase in temperature, the more likely this is to occur. Observe in figure 3.11*b* the severe decrease in the rate of reaction with temperatures greater than 40°C (104°F).

Effect of pH

Enzymes function most efficiently at their **optimal pH.** Optimal pH for most human enzymes is between pH 6 and 8, and changes in pH can affect the enzyme (figure 3.11*c*). An increase in H^+ (which causes a decrease in pH) results in additional H^+ binding to the enzyme. In contrast, a decrease in H^+ (which causes an increase in pH) results in the release of H^+ from an enzyme. In either case, the change in amount of H^+ attached to the enzyme disrupts the electrostatic interactions that hold the enzyme protein in its shape. A significant disruption results in denaturation of the enzyme (see section 2.8b).

Not all enzymes have an optimal pH between 6 and 8. The pH in the stomach, for example, is between 2 and 4; thus, the optimal pH for the stomach enzyme pepsin corresponds to this pH range. In comparison, the pH in the small intestine is between 6 and 9. Thus, when stomach contents are moved into the small intestine, the stomach enzyme pepsin is inactivated (see section 26.4b).

 **WHAT DID YOU LEARN?**

 13 How do changes in substrate concentration, temperature, and pH affect the reaction rate of enzyme-catalyzed chemical reactions?

3.3f Controlling Enzymes

 **LEARNING OBJECTIVE**

24. Describe how competitive and noncompetitive inhibitors control enzyme action.

An enzyme continues to facilitate the conversion of its substrate(s) to product as long as ample substrate is present and environmental conditions are close to normal. However, uncontrolled enzymes would result in depleted substrate levels and concentration of products that exceeds what is needed. Thus, enzymes must be temporarily "turned off" to prevent overproduction of the product. Control of enzymes occurs through **inhibitors,** which are substances that bind to an enzyme and turn it off, thus preventing it from catalyzing the reaction (**figure 3.12**). Later, the release of the inhibitor from the enzyme allows it to function and continue catalyzing the reaction. This switching occurs in different ways, depending upon whether the inhibitor is competitive or noncompetitive.

A **competitive inhibitor** resembles the substrate and binds to the active site of the enzyme. Consequently, the substrate and the regulatory compound compete with each other for occupation of the enzyme's active site (figure 3.12*b*). The amount of substrate relative to the amount of competitive inhibitor determines the degree of inhibition. The greater the concentration of the substrate, the less likely the competitive inhibitor will occupy the enzyme's active site. In contrast, if substrate concentration decreases, the competitive inhibitor is more likely to occupy the enzyme's active site, and lower amounts of product are formed.

Noncompetitive inhibitors do not resemble the substrate, but rather function to inhibit an enzyme by binding to a site on the enzyme other than the active site. This type of inhibition is not influenced by the concentration of substrate. Rather, it is simply dependent upon the amount of inhbitor. The greater the amount of inhibitor, the greater the degree of inhibition. In contrast, the lesser the amount of inhibitor, the lesser the degree of inhibition. Most noncompetitive enzymes bind to a specific part of an enzyme termed the **allosteric** (al′ō-ster′ik; *allo* = other, *stereo* = three-dimensionality) **site.** Binding of a noncompetitive inhibitor to the allosteric site induces a conformational change in the enzyme with an accompanying change in the shape of the enzyme's active site (figure 3.12*c*). This type of noncompetitive inhibitor is more specifically called an **allosteric inhibitor.**

 **WHAT DID YOU LEARN?**

 14 How are enzymes regulated through competitive and noncompetitive inhibitors?

3.3g Metabolic Pathways and Multienzyme Complexes

LEARNING OBJECTIVES

25. Distinguish between a metabolic pathway and a multienzyme complex.
26. Explain the role of negative feedback in enzyme regulation.
27. Identify and explain the processes involving phosphate that commonly are used to regulate enzymes.

Usually, multiple enzymes are required to convert an initial substrate to a final product. Depending upon both the substrate and sequence of conversion, these multiple enzymes are arranged either in a metabolic pathway or as a multienzyme complex.

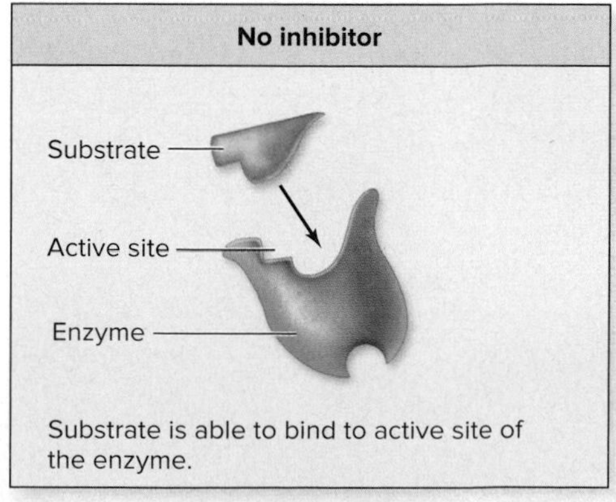

(a)

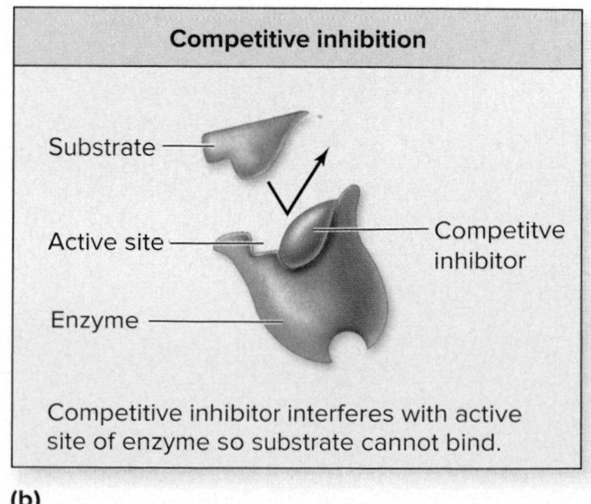

(b)

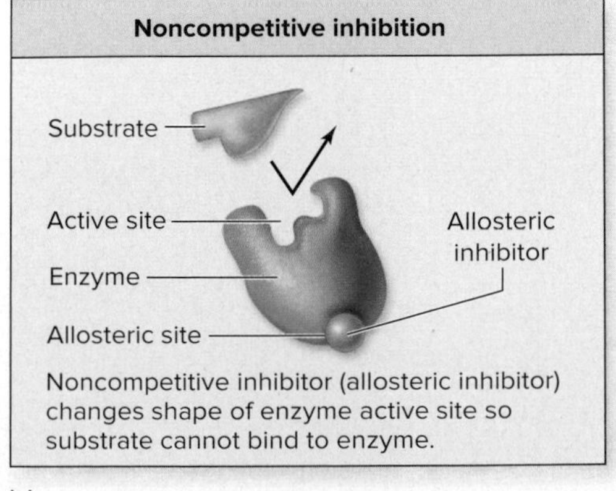

(c)

Figure 3.12 Enzyme Inhibition. (*a*) The substrate binds to the active site of the enzyme with no inhibitor present. A substrate can be prevented from binding to an active site by (*b*) a competitive inhibitor that enters the active site or (*c*) a noncompetitive inhibitor (functioning as an allosteric inhibitor) binds to an allosteric site to induce a conformational change in the enzyme with an accompanying change in its active site.

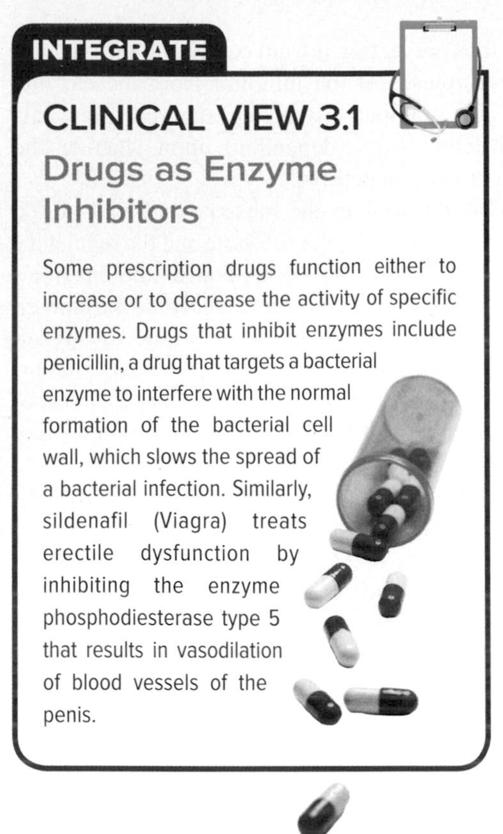

INTEGRATE
CLINICAL VIEW 3.1
Drugs as Enzyme Inhibitors
Some prescription drugs function either to increase or to decrease the activity of specific enzymes. Drugs that inhibit enzymes include penicillin, a drug that targets a bacterial enzyme to interfere with the normal formation of the bacterial cell wall, which slows the spread of a bacterial infection. Similarly, sildenafil (Viagra) treats erectile dysfunction by inhibiting the enzyme phosphodiesterase type 5 that results in vasodilation of blood vessels of the penis.

©Creatas/PunchStock RF

A **metabolic pathway** is formed by numerous enzymes **(figure 3.13)**. Each enzyme catalyzes one progressive change to its specific substrate molecule and then releases the product. In turn, the product of one enzyme becomes the substrate of the next enzyme. For example, there are numerous enzymes involved in the chemical breakdown of glucose to produce carbon dioxide and water during the production of ATP (see figures 3.16 and 3.18).

A **multienzyme complex** is a group of enzymes that are physically attached to each other through noncovalent bonds to form the complex. These attached enzymes work in a sequence of reactions. Pyruvate dehydrogenase, the multienzyme complex involved in breakdown of glucose, is an example, as shown in figure 3.17.

A multienzyme complex has two major advantages. First, the product from one chemical reaction is immediately bound to the next enzyme in the multienzyme complex. This makes it more likely that the needed product is formed and less likely that the substance will diffuse away and come into contact with an enzyme from a different biochemical pathway. Second, the enzymatic pathway can be regulated by controlling the single complex rather than multiple individual enzymes.

Metabolic pathways and multienzyme complexes must be regulated to prevent overproduction of an unneeded product and exhaustion of substrates that could be used elsewhere. This regulation occurs through the process of negative feedback. Here, the product from a metabolic pathway acts as an allosteric inhibitor to turn off an enzyme early in the metabolic pathway. As the product accumulates, it is more likely to become bound to the enzyme and inhibit the metabolic pathway, with progressively less and less product being formed. Over time, as the amount of product decreases, the amount of the allosteric inhibitor bound to the enzyme decreases, and activity of that enzymatic pathway increases once again. In this way, a steady state of product is produced.

One specific mechanism for regulating enzymes is by either phosphorylation or dephosphorylation of the enzyme. **Phosphorylation** (fos′fōr-i-lā′shŭn) is the addition of a phosphate group, whereas **dephosphorylation** is the removal of a phosphate group. Note that phosphorylation may turn on some enzymes but turn off other enzymes. Equally, dephosphorylation may cause opposite effects in activity by different types of enzymes. The enzymes that add phosphate are generally called either **phosphorlyases** or **kinases,**

Figure 3.13 Metabolic Pathway. A metabolic pathway is composed of numerous enzymes to convert a specific substrate to the final product. The product of one enzyme is the substrate for the next enzyme in the pathway. Metabolic pathways can be regulated by negative feedback that involves a product that serves as an allosteric inhibitor binding to an enzyme early in the pathway.

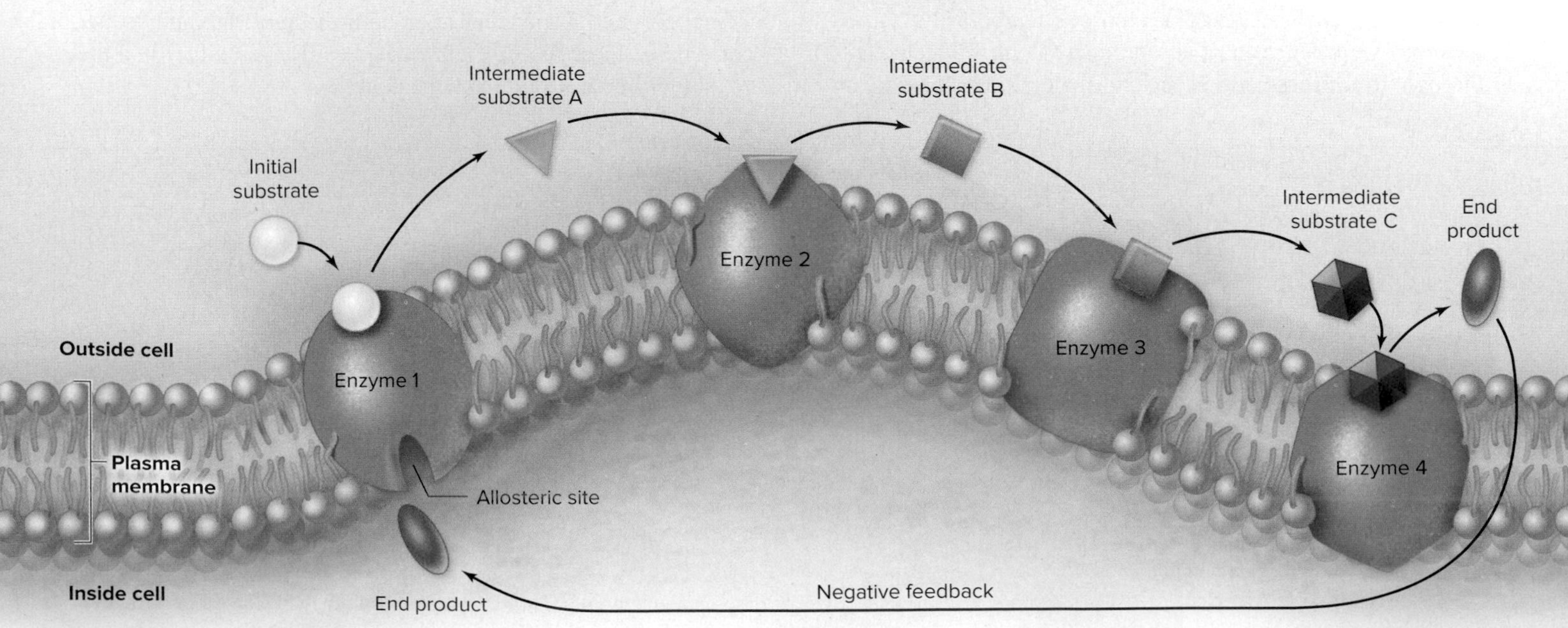

whereas enzymes that remove the phosphates are called **phosphatases** (see figure 4.21 in section 4.5b). This concept is described in more detail in section 17.5b in discussion of the endocrine system.

Figure 3.14 summarizes some important concepts and features of enzymes, including their function, structure and location, mechanism of action, and other features.

WHAT DID YOU LEARN?

15 What is a metabolic pathway? Explain the role of negative feedback in enzyme regulation.

16 What two processes involve phosphate and are commonly used to regulate enzymes in a metabolic pathway or a multienzyme complex?

INTEGRATE

CONCEPT CONNECTION

Cellular respiration typically requires an uninterrupted supply of O_2 and the continuous removal of CO_2. In section 23.5, we describe the systemic processes of **respiration,** which involve both the respiratory system and the cardiovascular system. These systems function in both the delivery of O_2 from the atmosphere to the body's cells and the movement of CO_2 from the body's cells to the atmosphere. Individuals with either impaired respiratory function (e.g., emphysema) or cardiovascular disease (e.g., congestive heart failure) may experience difficulty in the delivery of O_2 to the body's cells. These individuals often experience "energy problems" and feel lethargic and tired. This decrease in energy results from the dependence of cellular respiration on O_2 delivery for maximum production of ATP as described in this chapter.

3.4 Cellular Respiration

Cellular respiration is a multistep metabolic pathway whereby organic molecules (e.g., glucose, fatty acids, amino acids) are disassembled (broken down) in a controlled manner by a series of enzymes. During this disassembly, potential energy stored in the molecule's chemical bonds is released; the energy is then used to make new bonds between ADP and P_i (free phosphate) to form ATP (see figure 3.7). It is important to note the following about the processes of cellular respiration:

- These processes are exergonic, or energy-releasing.
- The organic molecule that has given up its energy has done so by releasing high-energy electrons; thus, the molecule is said to be oxidized.
- The energy released is used to synthesize ATP, which is an endergonic, or energy-requiring, process.
- Oxygen is required for maximum ATP production.

Although different types of organic molecules may be chemically digested during the collective processes of cellular respiration, our discussion here focuses on the oxidation of glucose.

3.4a Overview of Glucose Oxidation

LEARNING OBJECTIVES

28. Write the overall formula for glucose oxidation.

29. Name the two pathways that generate ATP.

30. List the four stages of glucose oxidation and where each stage occurs within a cell.

Glucose oxidation occurs within cells and is a step-by-step enzymatic breakdown of glucose with the accompanying release of energy to synthesize ATP. If oxygen is available, glucose is completely broken down and carbon dioxide and water are formed. Here we describe several significant features of glucose oxidation.

Overall Chemical Reaction Glucose has the molecular formula $C_6H_{12}O_6$. It is an energy-rich molecule because of its many C—C, C—H, and C—O chemical bonds as can be viewed in its structural formula (see figure 2.19*a*). When enzymes completely disassemble glucose, the net chemical reaction for the process is

$$C_6H_{12}O_6 + 6\ O_2 \rightarrow 6\ CO_2 + 6\ H_2O$$

Pathways for ATP Production Glucose oxidation is an exergonic reaction (see section 3.2b). During the many enzymatic reaction steps that accomplish the breakdown of glucose, some of the energy of the broken bonds is captured to attach P_i to ADP to synthesize ATP: The energy transfer from bonds in the glucose molecule can be used either directly (least common way) or indirectly (most common way) to form ATP. The direct method of synthesizing ATP is called **substrate-level phosphorylation.** The indirect method—in which the energy is first released to coenzymes (i.e., NAD^+, FAD), which then transfer the energy to form ATP—is called **oxidative phosphorylation.**

INTEGRATE CONCEPT OVERVIEW

Figure 3.14 How Enzymes Work. Characteristics of enzymes include (*a*) enzyme function; (*b*) structure and location; (*c*) naming of enzymes; (*d*) mechanism of action; (*e*) reaction rate; (*f*) metabolic pathway, multienzyme complex, and their regulation; and (*g*) controlling enzymes.

(a) Enzyme function

Enzymes decrease the activation energy (E_a) so reaction rate increases and more product is formed in a given time period.

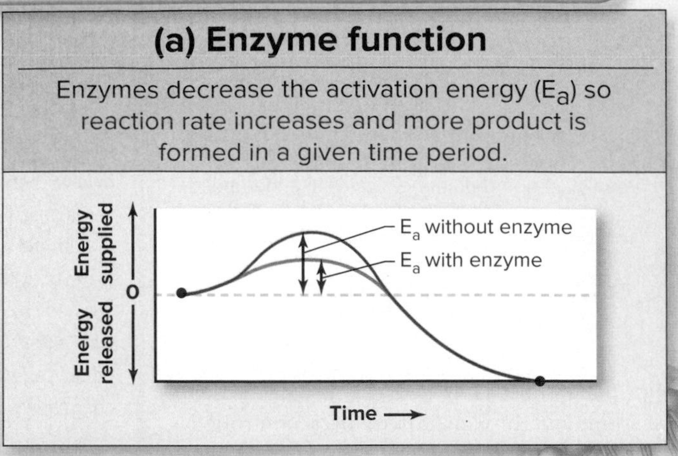

(b) Structure and location

Most enzymes are globular proteins with a unique active site.

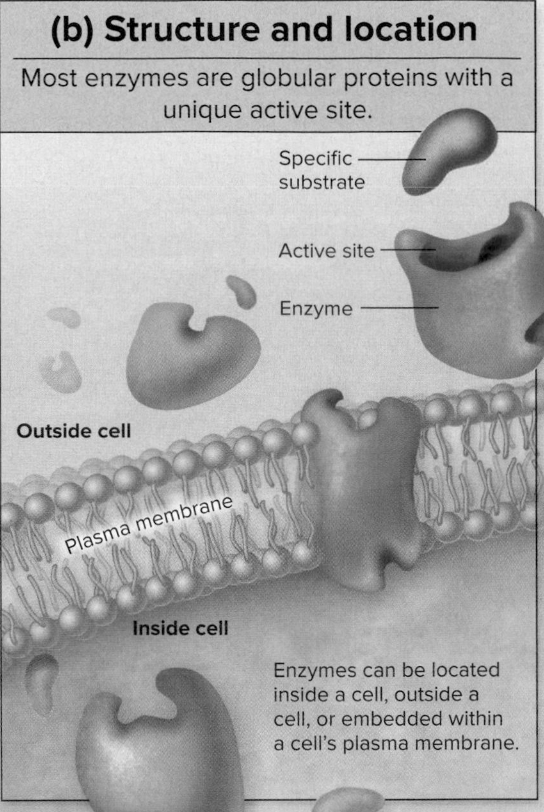

Enzymes can be located inside a cell, outside a cell, or embedded within a cell's plasma membrane.

(c) Naming of enzymes

The enzyme name typically includes the name of the substrate or product, or sometimes the name of the class or subclass involved in the chemical reaction with an *-ase* ending.

Example:

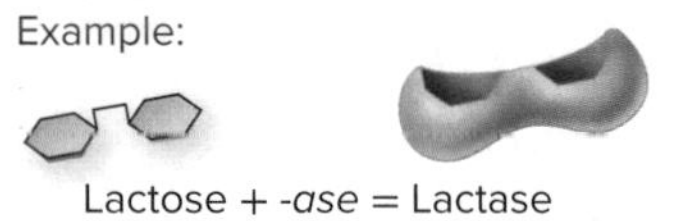

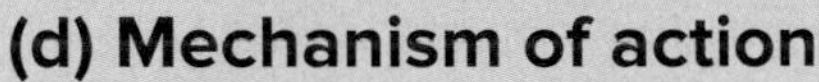

(d) Mechanism of action

Enzymes participate in either decomposition or synthesis reactions.

Decomposition Reaction

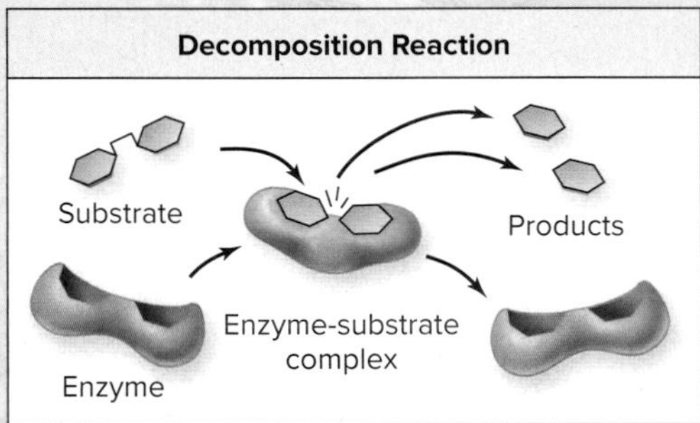

Synthesis Reaction

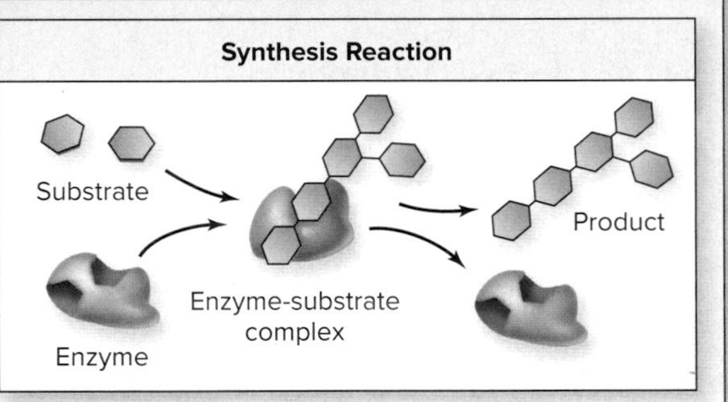

(e) Reaction rate (speed of a chemical reaction)

Reaction rate is influenced by substrate or enzyme concentration, temperature, and pH.

Substrate / Enzyme Concentration

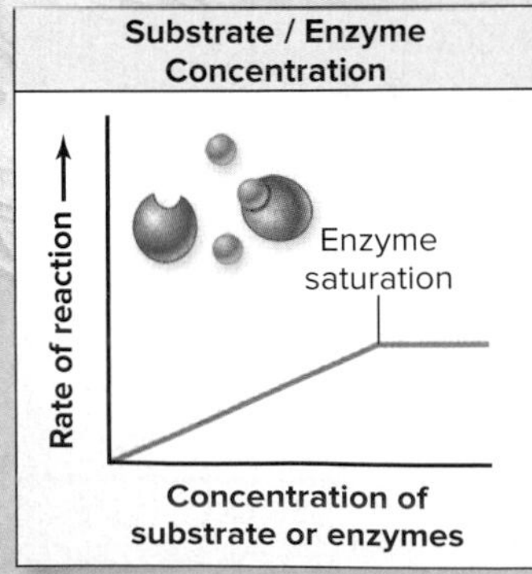

Increasing the substrate or enzyme concentration (to the point of enzyme saturation) increases the rate of reaction.

Temperature

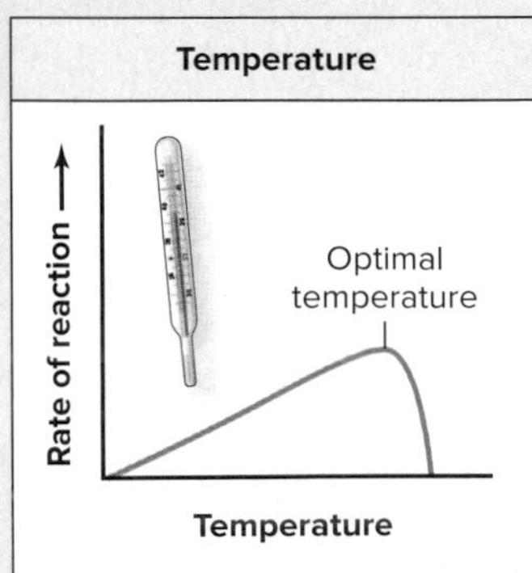

Increasing the temperature increases the rate of reaction up to the point of denaturation of the enzyme.

pH

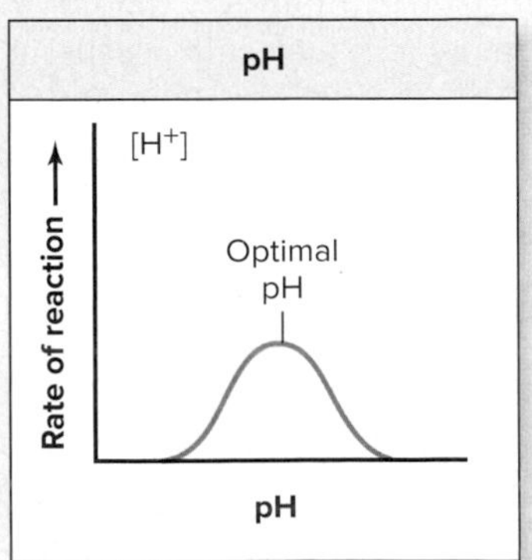

Enzymes are most effective at their optimal pH. Either an increase or a decrease from the optimal pH decreases the rate of the reaction.

(f) Metabolic pathway and multienzyme complex

Metabolic pathway: Series of enzymes

Many metabolic pathways are regulated by negative feedback by a product.

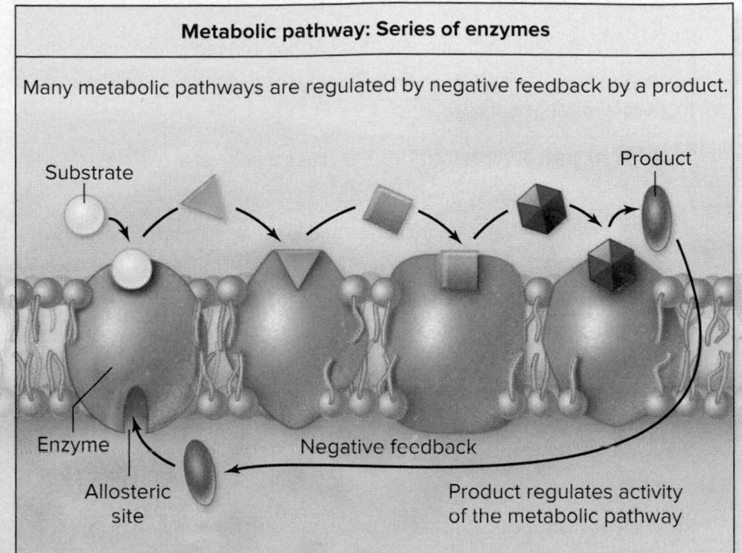

Multienzyme complex: Physically linked enzymes

The product of one enzyme becomes the substrate of a different enzyme in the complex.

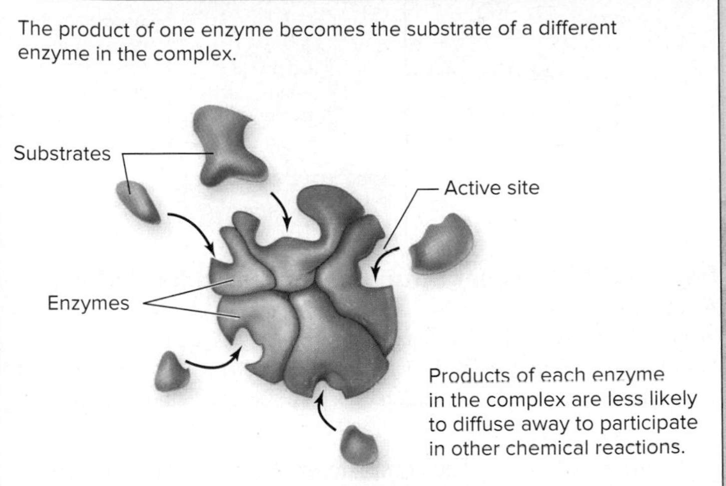

(g) Controlling enzymes

Competitive inhibitor

Competitive inhibitors interfere with active site directly.

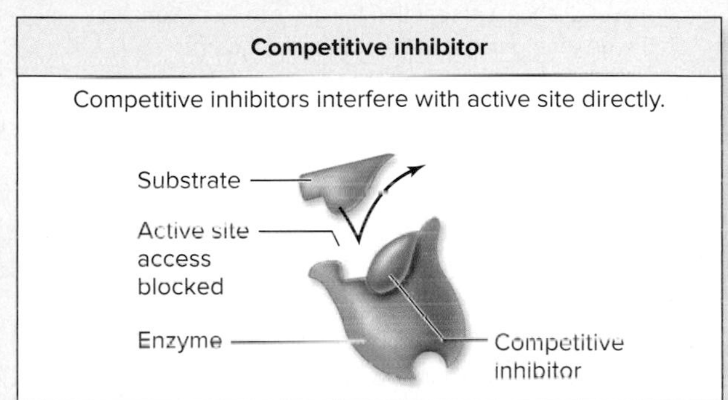

Noncompetitive inhibitor

Noncompetitive inhibitors are allosteric inhibitors that change the shape of an enzyme so the substrate cannot bind to the active site.

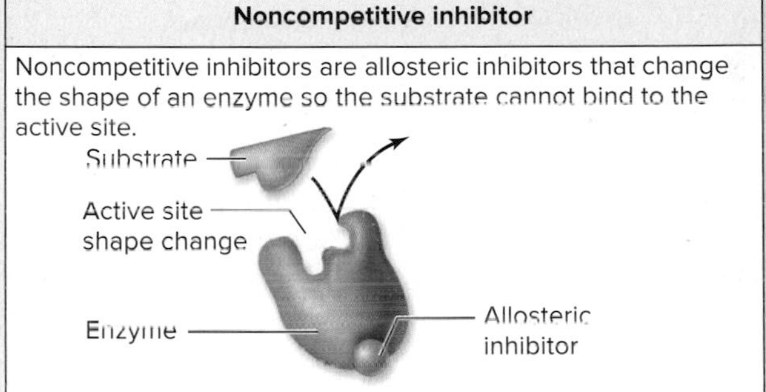

INTEGRATE

CLINICAL VIEW 3.2

Lactose Intolerance

Lactase is an enzyme required to break the bond in the disaccharide lactose (milk sugar) into glucose and galactose for its absorption from the digestive tract into the blood. **Lactose intolerance** (the inability to break down lactose) is caused by either a deficiency in the enzyme lactase or an abnormal (low-functioning) lactase enzyme. Differences in lactase enzyme function are due to genetic variation within the population. For example, individuals of northern European descent have a low incidence of lactose intolerance. Lactose intolerance is also more common in older adults because they produce less lactase over time. Abdominal upset including nausea, diarrhea, bloating, and gas are the most common symptoms. Avoidance of foods containing milk, drinking milk with lactose removed (lactose-free milk), or the oral administration of products containing lactase enzymes are recommended to avoid symptoms of lactose intolerance.

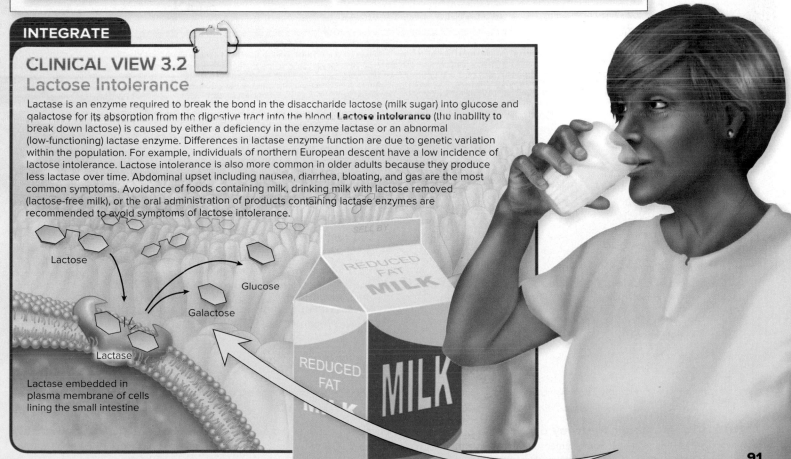

Lactose

Glucose

Galactose

Lactase

Lactase embedded in plasma membrane of cells lining the small intestine

REDUCED FAT MILK

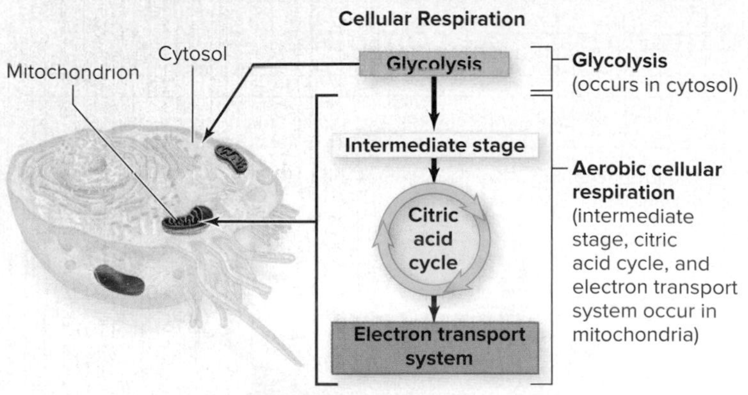

Figure 3.15 Cellular Structures Required for Cellular Respiration. Components of the cell associated with glucose oxidation include the cytosol, where the enzymes for glycolysis are located, and mitochondria, where the enzymes for aerobic cellular respiration (intermediate stage, citric acid cycle, and electron transport system) are housed. AP|R

INTEGRATE

LEARNING STRATEGY

Keep in mind the following questions as you read through each of the first three stages of cellular respiration (the stages required for completely oxidizing [disassembling] glucose):

1. Does it occur in the cytosol or mitochondria of a cell?
2. Does it require oxygen (i.e., is it aerobic)?
3. What are the initial substrate and the final product?
4. Is energy released to produce ATP directly (substrate-level phosphorylation), transferred to a coenzyme that serves as a "temporary holder" that will participate in oxidative phosphorylation, or both?

Cellular Location The complete oxidation of glucose requires at least 20 different enzymes that are located in both the cell's cytosol and its mitochondria (**figure 3.15**). **Cytosol** is the semifluid contents of the cell. **Mitochondria** (sing., mitochondrion) are small organelles within the cell, described in section 3.4c.

Four Stages of Cellular Respiration We separate the processes of glucose oxidation into four stages: glycolysis, intermediate stage, citric acid cycle, and electron transport system. Glycolysis occurs in the cytosol and does not require oxygen; thus, glycolysis can occur either in the presence of oxygen or in the absence of oxygen. The other three stages occur in the mitochondria and require oxygen to proceed.

WHAT DID YOU LEARN?

17 Write the overall chemical reaction for glucose oxidation, and explain the general process of what is occurring.

18 What are the four stages of cellular respiration for glucose oxidation, and where does each occur in the cell?

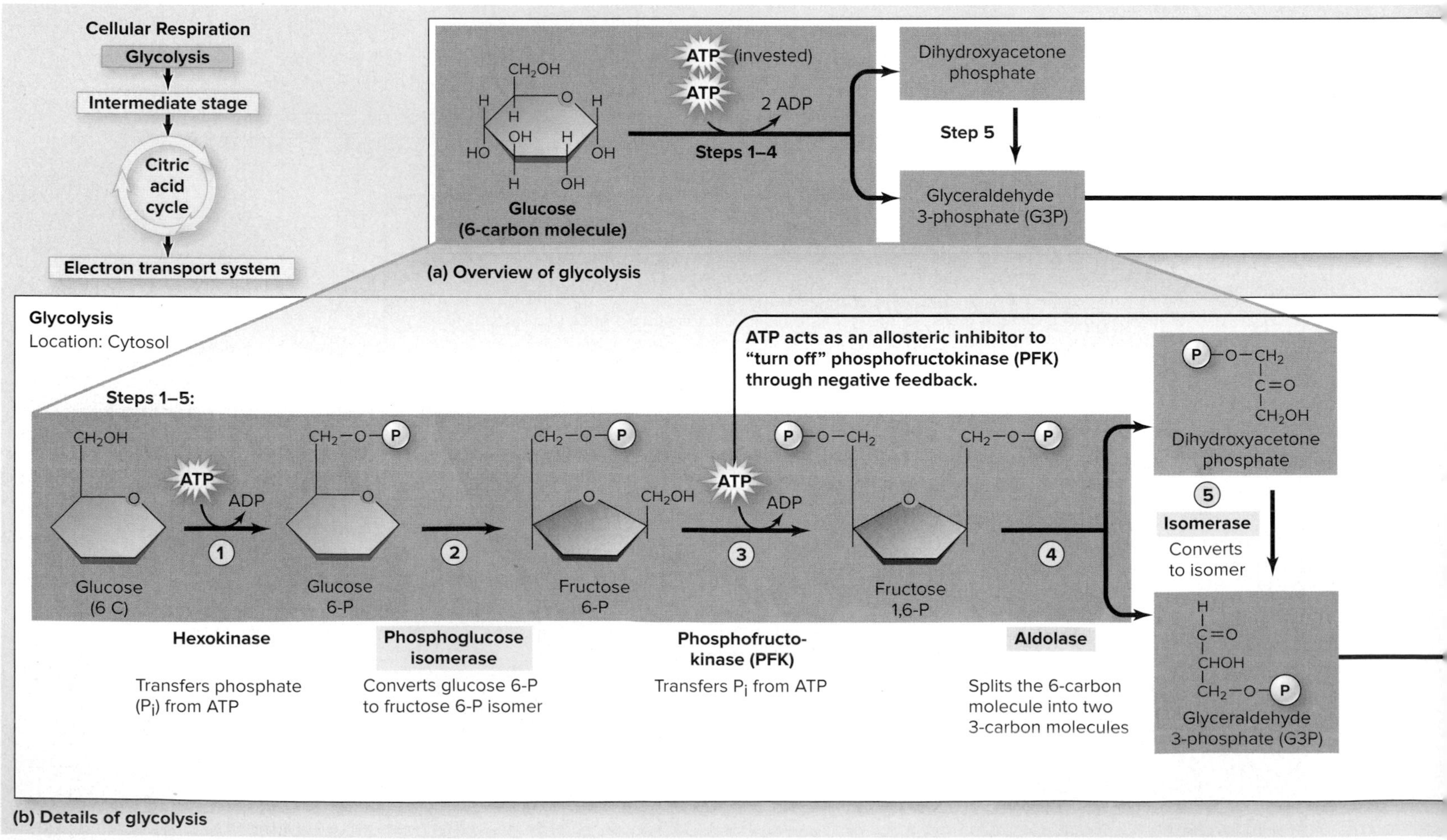

Cellular Respiration
Glycolysis
Intermediate stage
Citric acid cycle
Electron transport system

Glucose (6-carbon molecule)
ATP (invested)
ATP
2 ADP
Steps 1–4
Dihydroxyacetone phosphate
Step 5
Glyceraldehyde 3-phosphate (G3P)

(a) Overview of glycolysis

Glycolysis
Location: Cytosol

ATP acts as an allosteric inhibitor to "turn off" phosphofructokinase (PFK) through negative feedback.

Steps 1–5:

Glucose (6 C) → (1) ATP, ADP → Glucose 6-P → (2) → Fructose 6-P → (3) ATP, ADP → Fructose 1,6-P → (4) → Dihydroxyacetone phosphate / Glyceraldehyde 3-phosphate (G3P)

(1) Hexokinase — Transfers phosphate (P_i) from ATP

(2) Phosphoglucose isomerase — Converts glucose 6-P to fructose 6-P isomer

(3) Phosphofructokinase (PFK) — Transfers P_i from ATP

(4) Aldolase — Splits the 6-carbon molecule into two 3-carbon molecules

(5) Isomerase — Converts to isomer

(b) Details of glycolysis

Figure 3.16 Metabolic Pathway of Glycolysis. Glycolysis is a metabolic pathway that occurs within the cytosol. The pathway requires ten enzymes involved in the conversion of glucose to pyruvate with a net production of 2 ATP molecules and 2 NADH molecules. The fate of pyruvate is dependent upon the availability of oxygen to the cell. (*a*) An overview of glycolysis. (*b*) The detailed pathway of glycolysis. AP|R

3.4b Glycolysis

LEARNING OBJECTIVE

31. Summarize the metabolic pathway of glycolysis, including (a) where it occurs in a cell, (b) if it requires oxygen, (c) the initial substrate and final product, and (d) the molecules formed during energy transfer.

Glycolysis (glī-kol′i-sis; *glykys* = sweet, *lysis* = loosening) does not require oxygen. Ten enzymes within the cytosol of a cell participate in the metabolic pathway of glycolysis. Glucose is broken down in this pathway into two pyruvate molecules with an accompanying energy transfer to form a net production of 2 ATP molecules and 2 NADH molecules.

Steps of Glycolysis

The 10 enzymatically regulated chemical reactions of glycolysis are shown in **figure 3.16**; both an overview figure (figure a) and the detailed steps (figure b) are included. Note that steps 1 through 5 occur once per glucose, and steps 6 through 10 occur *twice* per glucose because glucose is split into *two* three-carbon molecules.

(1–5) Steps 1 through 5 of glycolysis involve splitting glucose into two molecules of glyceraldehyde 3-phosphate (G3P) through the action of the first five enzymes. ATP is "invested" when kinase enzymes transfer P_i from ATP to glucose and the breakdown products of glucose (steps 1 and 3). Thus, an investment of 2 ATP molecules occurs at these two steps.

(6–7) Steps 6 and 7 of glycolysis occur twice in oxidation of a glucose molecule. Step 6 involves transferring an unattached P_i to the substrate (so this molecule now has two phosphates), and two hydrogen atoms are released to NAD^+ to form an NADH (and H^+). This transfer of hydrogen is catalyzed by a hydrogenase enzyme. In step 7, the original P_i is transferred to ADP to form ATP through substrate-level phosphorylation by a kinase enzyme.

(8–10) Steps 8 through 10 of glycolysis also occur twice in oxidation of a glucose molecule. These steps involve converting the molecule produced in step 7 to an isomer (step 8) and then the loss of a water molecule (step 9). The remaining P_i is transferred to ADP to form ATP through substrate-level phosphorylation by a kinase enzyme (step 10), forming the final product of pyruvate.

WHAT DO YOU THINK?

3 What is the net energy transfer during glycolysis?

Summary of Glycolysis

Glycolysis is a metabolic process that occurs in the cytosol without the requirement of oxygen. Glucose is the initial substrate and pyruvate is the final product. The net transfer of energy is used in the formation of 2 ATP and 2 NADH molecules.

- **Formation of ATP.** Two ATP molecules are invested early in glycolysis (steps 1 and 3). Four ATP molecules are formed during glycolysis (steps 7 and 10, which occur twice per the original glucose molecule). Thus, there is a net of 2 ATP molecules formed during glycolysis (2 ATP invested, 4 ATP formed).
- **Formation of NADH.** Two NADH molecules are formed from glucose breakdown during glycolysis (step 6, which occurs twice per the original glucose molecule).

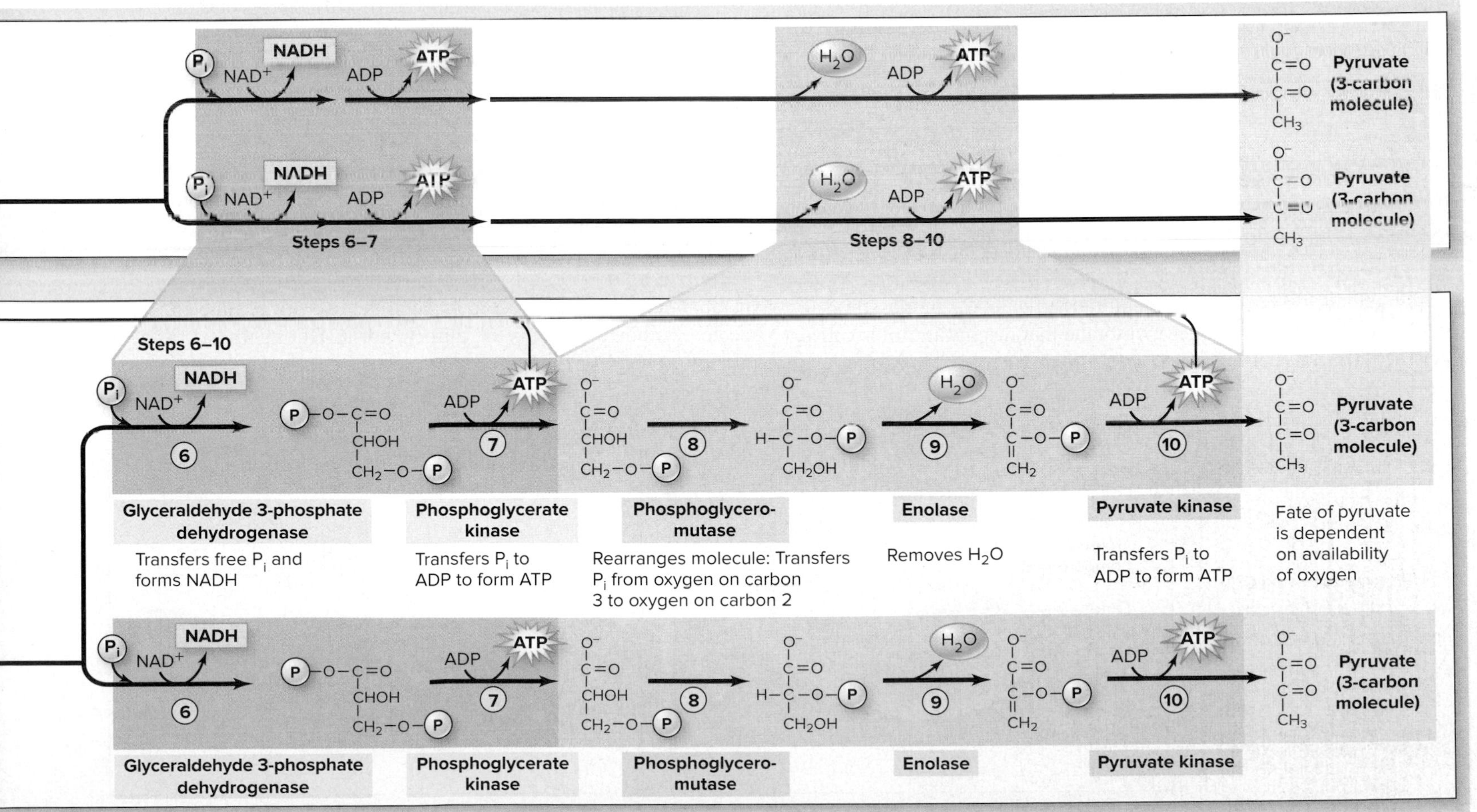

Regulation of Glycolysis

Glycolysis is regulated through negative feedback, like many metabolic pathways. ATP acts as an allosteric inhibitor to "turn off" phosphofructokinase (PFK) (step 3).

As ATP levels increase in the cell cytosol, ATP binding inhibits PFK, and the glycolytic pathway is progressively shut down. In contrast, as ATP decreases, glycolysis increases. Phosphofructokinase is also regulated in a similar way by other substances that indicate the "energy status" of the cell, including NADH, citrate (an intermediate in the citric acid cycle), fatty acids, and other fuel molecules. Increased levels of these substances result in a decrease in the processes of glycolysis.

The Fate of Pyruvate

Pyruvate is the final product of glycolysis. What chemical changes are then made to pyruvate depend upon the availability of oxygen. If sufficient oxygen is available, pyruvate enters a mitochondrion to complete its aerobic breakdown yielding carbon dioxide and water (described next). In contrast, if sufficient oxygen is not available, pyruvate is converted to lactate (described in section 3.4g).

WHAT DID YOU LEARN?

19 Describe glycolysis—where it occurs, if the process requires oxygen (i.e., is aerobic), the net chemical reaction, and the net energy transfer.

20 What are the two general fates of pyruvate, and what is the criterion that determines its fate?

3.4c Intermediate Stage

LEARNING OBJECTIVES

32. Explain the enzymatic reaction of the intermediate stage, including (a) where it occurs in a cell, (b) if it requires oxygen, (c) the initial substrate and final product, and (d) the molecules formed during energy transfer.

33. Define decarboxylation.

The remaining stages of cellular respiration, including the intermediate stage, citric acid cycle, and electron transport system, are aerobic processes that occur within mitochondria.

Mitochondrion Structure

A mitochondrion is a double-membrane organelle, composed of an outer membrane and an inner membrane that has inward folds called **cristae** (**figure 3.17*a***). The fluid-filled space between the two membranes is the **outer compartment.** The innermost space in a mitochondrion is called the **matrix.** Both the multienzyme complex of the intermediate stage and the enzymes of the citric acid cycle metabolic pathway reside in the matrix. The significant molecules that participate in the electron transport system are embedded in the cristae.

Intermediate Stage and Pyruvate Dehydrogenase

The **intermediate stage** (figure 3.17*b*) is the "link" between the multistep metabolic processes of glycolysis (first stage) and the multistep metabolic processes that occur in the citric acid cycle (third stage). The intermediate stage is catalyzed by a multienzyme complex called **pyruvate dehydrogenase.**

During the intermediate stage, pyruvate dehydrogenase brings together pyruvate and a molecule of coenzyme A (CoA) that is already present within the matrix to form acetyl CoA (a two-carbon molecule with CoA attached). Concurrently, a carboxyl group, consisting of one carbon atom and two oxygen atoms, is released from the pyruvate as CO_2. This process is termed **decarboxylation** (dē′kar-bok′si-lā′shŭn; *de* = away). Energy is released during decarboxylation as two hydrogen atoms (two electrons plus two hydrogen ions) are transferred to the coenzyme NAD^+ to form NADH (and H^+) during this process. The acetyl CoA then enters the third stage of glucose oxidation, termed the citric acid cycle.

Note that the intermediate stage must occur twice for the complete digestion of the original glucose molecule because two pyruvate molecules were produced from each glucose that went through glycolysis. Thus, two NADH are produced from the original glucose molecule.

WHAT DID YOU LEARN?

21 Explain the enzymatic reaction involving pyruvate dehydrogenase in the intermediate stage—where it occurs, if the process requires oxygen, the net chemical reaction, and the net energy transfer.

3.4d Citric Acid Cycle

LEARNING OBJECTIVE

34. Summarize the metabolic pathway of the citric acid cycle, including (a) where it occurs in a cell, (b) if it requires oxygen, (c) the initial substrate and final product, and (d) the molecules formed during energy transfer.

The **citric acid cycle** (also known as the *Krebs cycle*) is a cyclic metabolic pathway that occurs through the activity of nine enzymes located within the matrix of mitochondria. During the citric acid cycle, the acetyl CoA produced in the intermediate stage is converted

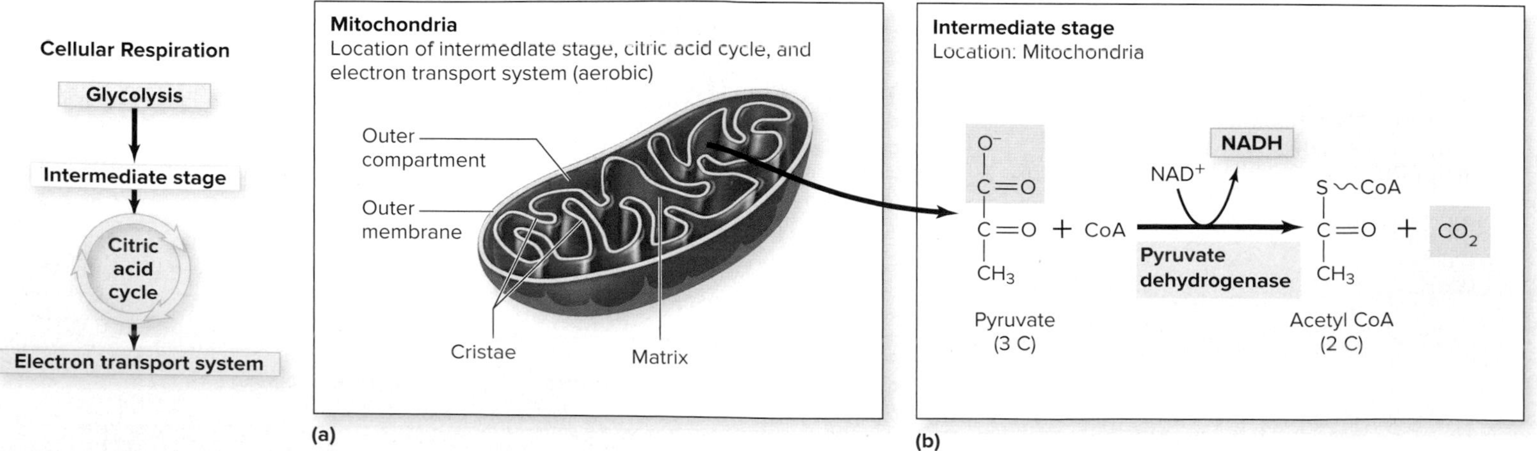

Figure 3.17 Intermediate Stage. (*a*) Mitochondria are the cellular organelles where aerobic cellular respiration occurs. (*b*) The intermediate stage involves a multienzyme complex called pyruvate dehydrogenase, which converts pyruvate to acetyl CoA and carbon dioxide with formation of NAD^+ to NADH.

to two CO_2 molecules and a CoA molecule is released. Energy is transferred to form 1 ATP molecule, 3 NADH molecules, and 1 $FADH_2$ molecule during one "turn" of the citric acid cycle.

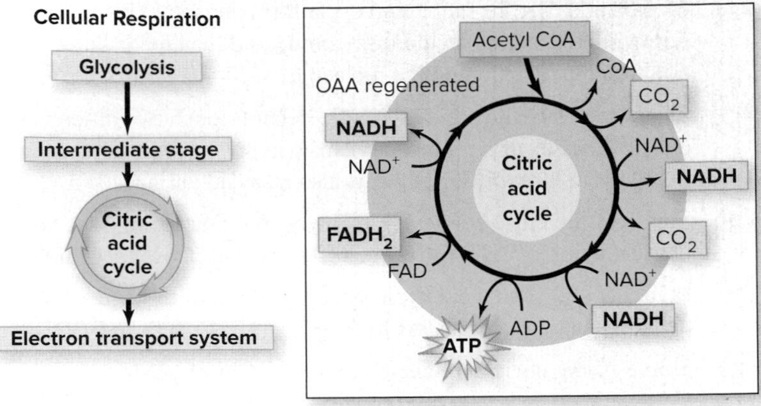

(a) Net chemical reaction of citric acid cycle

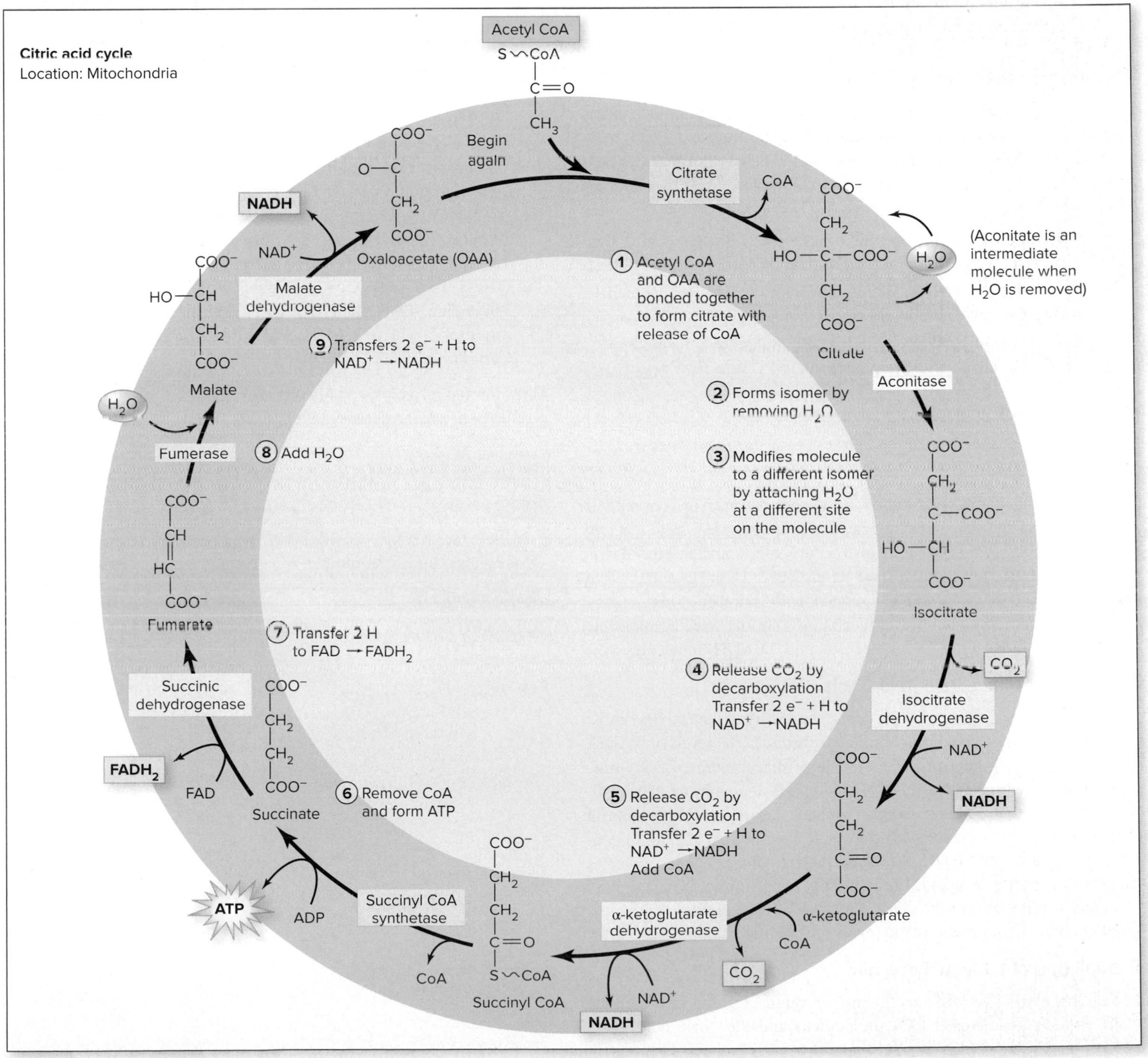

Steps of the Citric Acid Cycle

The nine enzymatic steps of the citric acid cycle are shown in **figure 3.18**; both an overview figure and the detailed steps are included.

1. Step 1 of the citric acid cycle uses the first enzyme to combine an acetyl CoA molecule produced in the intermediate stage with a molecule of **oxaloacetate** (ok′să-lō-ă-sē′tāt) to form **citrate.** (Note that the addition of a hydrogen ion to citrate forms citric acid.) The first step of this cycle gives this enzymatic pathway its name.

(b) Details of citric acid cycle

Figure 3.18 Citric Acid Cycle. The citric acid cycle is the metabolic pathway that occurs within the matrix of mitochondria for the chemical breakdown of acetyl CoA with the net transfer of energy to form 1 ATP, 3 NADH, and 1 $FADH_2$ molecules. (*a*) An overview of the citric acid cycle. (*b*) The detailed pathway of the citric acid cycle. AP|R

2–3 Steps 2 and 3 of the citric acid cycle form an isomer by removing a water molecule from citrate and then reattaching it to a different location on the molecule.

4–5 Steps 4 and 5 of the citric acid cycle occur through two different dehydrogenase enzymes that participate in the transfer of hydrogen to NAD^+ to form NADH. CoA is also attached during step 5.

6 Step 6 of the citric acid cycle involves the removal of CoA and the formation of ATP through substrate-level phosphorylation.

7 Step 7 of the citric acid cycle occurs through the action of a dehydrogenase that transfers hydrogens to FAD to form $FADH_2$.

 Step 8 of the citric acid cycle is the removal of water.

 Step 9 of the citric acid cycle is catalyzed by a dehydrogenase that transfers hydrogen to NAD^+ to form NADH. Oxaloacetate (OAA) is regenerated in this final step.

Summary of the Citric Acid Cycle

The citric acid cycle is a metabolic process that occurs in mitochondria and requires oxygen. Acetyl CoA is the initial substrate and 2 CO_2 molecules and 1 CoA molecule are the products. The net transfer of energy produced in this cycle is used to form 1 ATP molecule, 3 NADH molecules, and 1 $FADH_2$ molecule.

- **Formation of ATP.** One ATP molecule is formed during the citric acid cycle (step 6) through substrate-level phosphorylation.
- **Formation of NADH.** Three NADH molecules are formed during the citric acid cycle (steps 4, 5, and 9).
- **Formation of $FADH_2$.** One $FADH_2$ molecule is formed during the citric acid cycle (step 7).

WHAT DO YOU THINK?

4 Why is the enzymatic pathway of the citric acid cycle considered to be a "cycle"?

This enzymatic pathway is called a "cycle" because oxaloacetate is involved in the first step and is regenerated in the last step. Note that two "turns" of the citric acid cycle must also occur for the complete breakdown. of the original glucose molecule (one per each acetyl CoA generated from the original glucose molecule). Consequently, the number of high-energy molecules produced within the citric acid cycle from 1 glucose molecule is 2 ATP molecules, 6 NADH molecules, and 2 $FADH_2$ molecules.

Regulation of the Citric Acid Cycle

Regulation of the citric acid cycle occurs primarily at the enzyme in the first step of the citric acid cycle (citrate synthetase). The levels of NADH, ATP, and pathway intermediates are the primary regulators of citrate synthetase activity. A low level of NADH, ATP, and pathway intermediates indicates that cellular energy demands are high. This results in increased activity of citrate synthetase and the citric acid cycle. In contrast, a high level of NADH, ATP, and pathway intermediates indicates that cellular energy demands are low. This results in decreased activity of citrate synthetase and the citric acid cycle. These physiologic adjustments help maintain homeostatic levels of ATP molecules.

Completion of Glucose Digestion

Following glycolysis and two "turns" through both the intermediate stage (which generates 2 CO_2 molecules) and the citric acid cycle (which produces 4 CO_2 molecules), glucose has been completely digested and the 6 carbon atoms of glucose ($C_6H_{12}O_6$) have been released as 6 CO_2 molecules. Notice that the carbon atoms in the glucose (and other fuel molecules) that you eat are converted to carbon dioxide within the mitochondria of your cells!

Summary of the Chemical Breakdown of Glucose

Table 3.3 provides a summary of each of the first three stages of glucose oxidation that result in the chemical breakdown of glucose into carbon dioxide. The critical aspects of this process include:

- **Glycolysis.** Glycolysis occurs in the cytosol and does not require oxygen. Energy is transferred to form 2 ATP molecules (net) and 2 NADH molecules. If sufficient oxygen is available, the pyruvate formed enters a mitochondrion and is further metabolized in the intermediate stage and the citric acid cycle.
- **The intermediate stage.** The intermediate stage occurs in a mitochondrion and requires oxygen. It involves a multienzyme complex that converts pyruvate to acetyl CoA and 1 CO_2 molecule. Energy is transferred to form 1 NADH molecule. Remember, 1 NADH molecule is formed per pyruvate entering the intermediate stage. Recall that 2 pyruvates are produced from 1 glucose molecule. Thus, the intermediate stage must occur twice—so a total of 2 NADH molecules are formed from the original glucose molecule.
- **The citric acid cycle.** The citric acid cycle also occurs in a mitochondrion, requires oxygen, and completes the breakdown of glucose. Two CO_2 molecules are produced per turn of the cycle. Energy is transferred during this process to form 1 ATP, 3 NADH, and 1 $FADH_2$. Remember, this reflects the energy transferred per each acetyl CoA entering the citric acid cycle. Two acetyl CoA molecules are produced from 1 glucose molecule. Thus, the citric acid cycle must occur twice—so a total of 2 ATP, 6 NADH, and 2 $FADH_2$ are formed from the original glucose molecule.

Thus, through glycolysis and two turns of both the intermediate stage and the citric acid cycle, the 6 carbons in the original glucose have

Table 3.3 Comparison of the First Three Stages of Glucose Breakdown

Characteristics	Glycolysis	Intermediate Stage	Citric Acid Cycle
Where it occurs	Cytosol	Mitochondria	Mitochondria
Requires oxygen (is aerobic)?	No	Yes (aerobic)	Yes (aerobic)
Substrate	Glucose	Pyruvate (2 pyruvates from each glucose)	Acetyl CoA (2 acetyl CoA from each glucose)
Product	2 pyruvate molecules	Acetyl CoA and 1 CO_2 per pyruvate	2 CO_2 per acetyl CoA
Pathway or complex	Metabolic pathway	Multienzyme complex	Metabolic pathway
Net energy molecules produced	2 ATP (net) and 2 NADH	1 NADH per pyruvate	1 ATP per acetyl CoA 3 NADH per acetyl CoA 1 $FADH_2$ per acetyl CoA
How lack of oxygen affects the stage	Lactate produced (to regenerate NAD^+ so glycolysis can continue)	Pathway inhibited by lack of oxygen	Pathway inhibited by lack of oxygen

been released as 6 CO_2 molecules and the energy has been transferred to form

Glycolysis	→ 2 ATP	2 NADH	
Intermediate stage	→	2 NADH	
Citric acid cycle	→ 2 ATP	6 NADH	2 $FADH_2$

Cellular Respiration

Glycolysis
↓
Intermediate stage
↓
Citric acid cycle
↓
Electron transport system

WHAT DID YOU LEARN?

22 Summarize the metabolic pathway of the citric acid cycle—where it occurs, if the process requires oxygen, the net chemical reaction, and the net energy transfer.

23 What energy molecules are produced from the chemical breakdown of glucose during each of the three steps of cellular respiration?

3.4e The Electron Transport System

LEARNING OBJECTIVES

35. Describe the importance of NADH and $FADH_2$ in energy transfer.

36. Explain the actions that take place in the electron transport system.

Given that the breakdown of glucose is completed by the end of the citric acid cycle, what is the function of the electron transport system, the final stage of cellular respiration? The **electron transport system** involves the transfer of electrons (energy) from the coenzymes NADH and $FADH_2$ that are produced during the first three stages of cellular respiration. The energy released from these coenzymes is used to form ATP. This is a critical stage in cellular respiration because most of the energy captured in glucose oxidation is initially transferred to form NADH from NAD^+, as well as the smaller amounts of energy to form $FADH_2$ from FAD. Thus, it is through the electron transport system that the majority of chemical energy that was originally in the glucose molecule and transferred to the coenzymes (in the previous three stages of cellular respiration) is now released from the coenzymes and transferred to form the high energy bond between ADP and P_i as ATP is synthesized. The electron transport system involves structures located in the inner folded membrane (or cristae) of mitochondria.

Structures of the Electron Transport System

Several significant types of molecules are embedded within the cristae of the mitochondria: H^+ pumps, electron carriers, and ATP synthase enzymes (**figure 3.19**). H^+ pumps are proteins that transport H^+ from the matrix to the outer membrane compartment. This maintains a H^+ gradient between the outer compartment and the matrix within a mitochondrion, with more H^+ in the outer compartment than in the matrix. H^+ pumps also bind and release electrons (e^-).

Electron carriers ubiquinone (Q) and cytochrome *c* (C) are located between proteins serving as H^+ pumps. Electron carriers transport electrons (e^-) between the H^+ pumps. You may be familiar with these electron carriers because they are purchased as antioxidant supplements by cyclists, bodybuilders, and other athletes (e.g., CoQ10 as a ubiquinone supplement).

This series of H^+ pumps and electron carriers is collectively called the **electron transport chain.** ATP synthase allows for the passage of H^+ from the outer compartment back into the matrix. During this process, the kinetic energy from the flow of H^+ down its concentration gradient is harnessed to bond P_i to ADP to form ATP.

Steps of the Electron Transport System

The processes of the electron transport system are organized into three major steps (figure 3.19):

(1) **Electrons are transferred from coenzymes to O_2.** The coenzyme, either NADH or $FADH_2$, releases hydrogen (e^- and H^+) and it is oxidized. The released electrons are

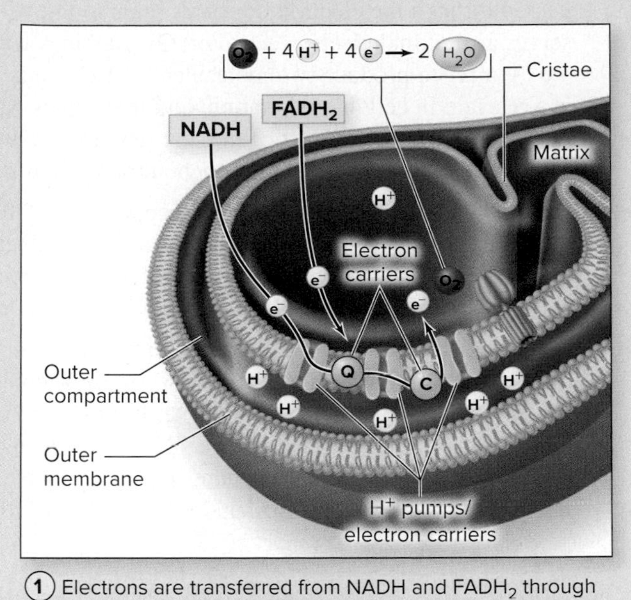

(1) Electrons are transferred from NADH and $FADH_2$ through a series of electron carriers within the cristae. O_2 is the final electron acceptor. (O_2, e^-, and H^+ join to form H_2O.)

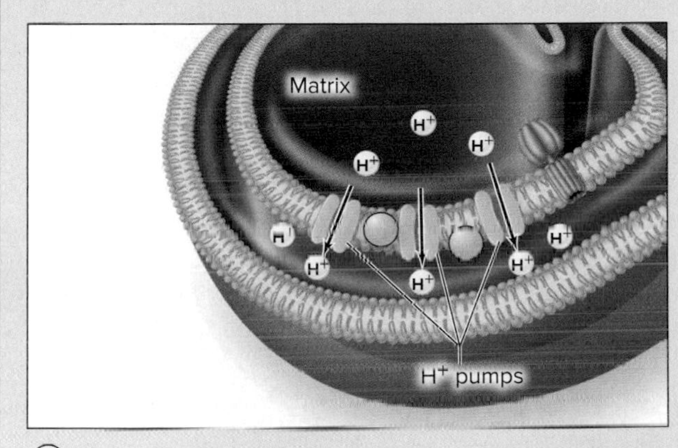

(2) Energy of electrons "falling" (in step 1) is used by H^+ pumps to move H^+ up its concentration gradient from the matrix to the outer compartment.

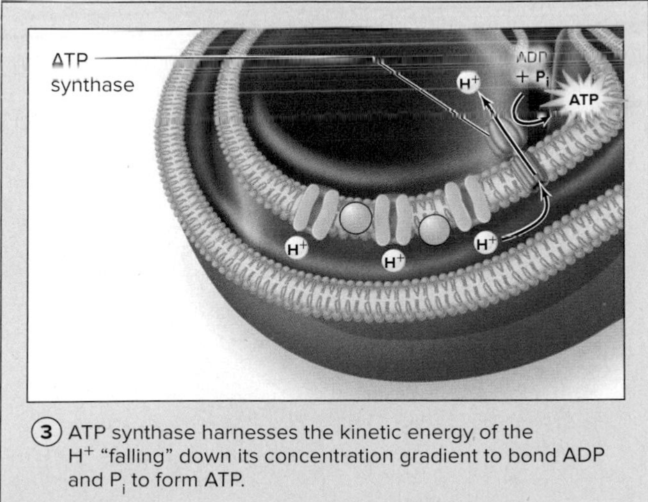

(3) ATP synthase harnesses the kinetic energy of the H^+ "falling" down its concentration gradient to bond ADP and P_i to form ATP.

Figure 3.19 The Electron Transport System. The processes of the electron transport system can be organized into three steps in which energy temporarily held by the coenzymes NADH and $FADH_2$ is transferred to form a bond between ADP and P_i, yielding ATP. AP|R

passed through the electron transport chain to O_2, which serves as the final electron acceptor. O_2 combines with $4\ e^- + 4\ H^+$ to produce two molecules of H_2O. Thus, oxygen is a reactant in cellular respiration, and it becomes part of the water that is produced. Notice that the oxygen you breathe is converted to water within the mitochondria of your cells!

2 **Proton gradient is established.** As electrons are "falling" and passed through the electron transport chain, their kinetic energy is harnessed by H^+ pumps to move H^+ from the mitochondrial matrix into the outer compartment, maintaining a proton gradient.

3 **Proton gradient is harnessed to form ATP.** H^+ moves down its concentration gradient as it is transported across the inner membrane by ATP synthase. It moves from the outer compartment into the matrix. (Note that H^+ moves back into the area of the mitochondrion from which it was just pumped.) This process is analogous to water falling over a dam and turning a water wheel. The kinetic energy of the falling H^+ is harnessed by ATP synthase to form a new bond between ADP and P_i, producing an ATP.

The process of forming ATP is referred to as **oxidative phosphorylation** because it involves oxygen as the final electron acceptor, and ATP is formed from the phosphorylation of ADP. This process is distinguished from substrate-level phosphorylation, which forms ATP from energy directly released from a substrate, as occurs in specific steps of glycolysis (see figure 3.16, steps 7 and 10) and the citric acid cycle (see figure 3.18, step 6). The steps of cellular respiration are summarized in **figure 3.20**.

WHAT DID YOU LEARN?

24 What is the importance of NADH and $FADH_2$ in energy transfer?

25 What are the three primary steps that take place in the electron transport system?

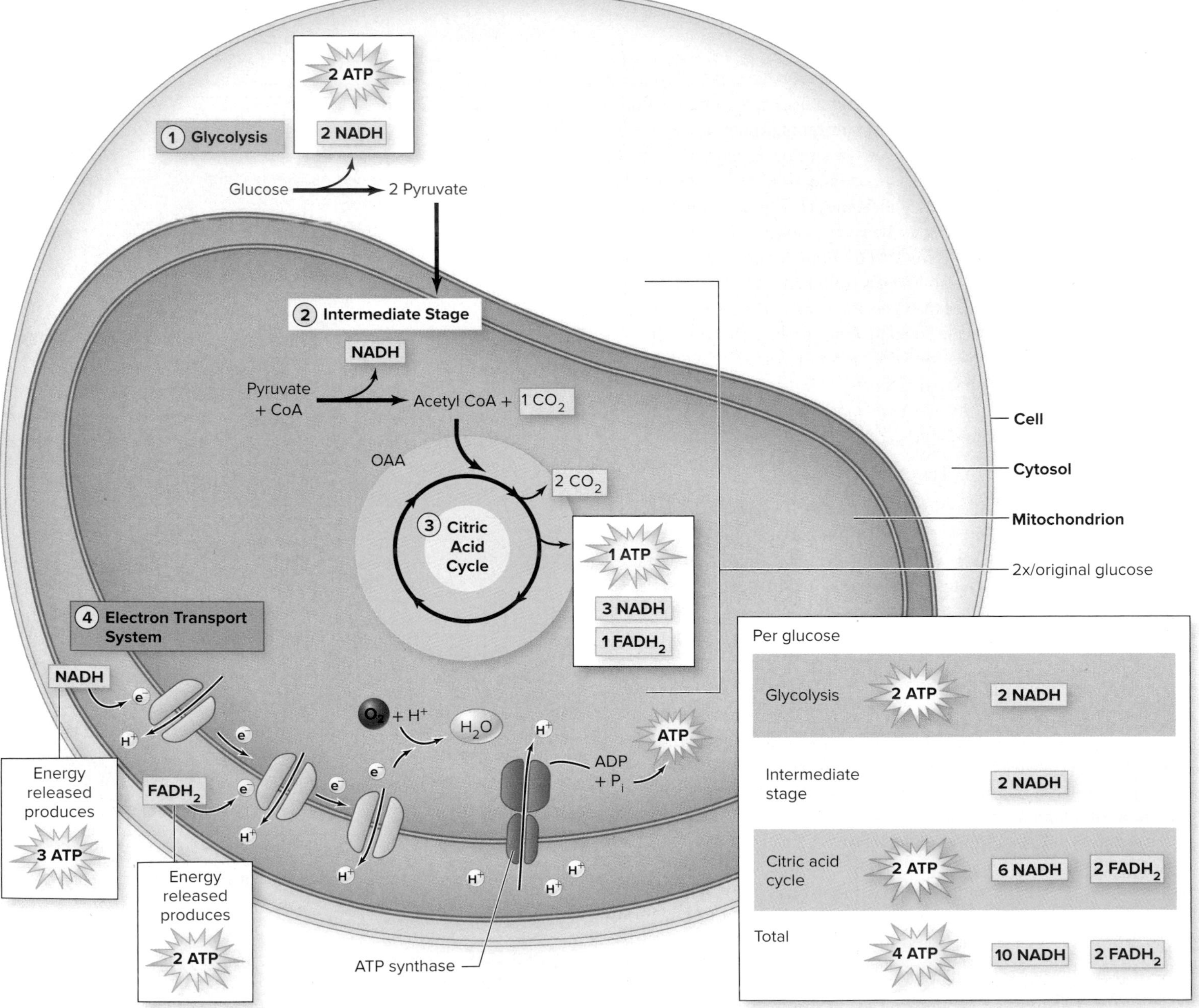

Per glucose			
Glycolysis	2 ATP	2 NADH	
Intermediate stage		2 NADH	
Citric acid cycle	2 ATP	6 NADH	2 $FADH_2$
Total	4 ATP	10 NADH	2 $FADH_2$

Figure 3.20 Summary of Stages of Cellular Respiration. (*1*) Glycolysis (an enzymatic pathway within the cytosol that initiates the breakdown of glucose to two pyruvate molecules). (*2*) Intermediate stage (a multienzyme complex within mitochondria that converts pyruvate to acetyl CoA). (*3*) Citric acid cycle (an enzymatic pathway within mitochondria that completes the breakdown of glucose to CO_2). (*4*) Electron transport system (specialized structures in cristae of mitochondria that provide the means to transfer energy from NADH and $FADH_2$ coenzymes to bond P_i to ADP to form ATP molecules—a process in which oxygen accepts electrons and combines with H^+ to form H_2O.

INTEGRATE

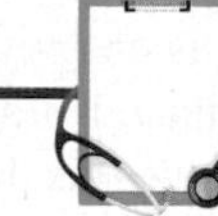

CLINICAL VIEW 3.3

Cyanide Poisoning

A **cyanide** is formed whenever a nitrogen atom is triple bonded with a carbon atom. Cyanides are highly toxic if they bind with a specific cristae electron carrier of the electron transport chain (called *cytochrome c oxidase*). This binding inhibits the processes of the electron transport chain and the subsequent production of ATP. Although oxygen is available for "catching electrons" within the electron transport chain, the inhibition of one of the electron carriers within the electron transport chain prevents the electrons from reaching oxygen. Thus, the aerobic components of cellular respiration progressively decrease as the cells cannot utilize the oxygen. Cyanide poisoning is most harmful to the organs that require the most oxygen (e.g., heart and lungs). Treatment for nonlethal cyanide doses (doses as low as 1.5 mg/kg body weight can be fatal) involves the administration of substances that bind cyanide (e.g., nitrites), which are then eliminated in the urine.

3.4f ATP Production

LEARNING OBJECTIVE

37. Calculate the number of ATP molecules produced in cellular respiration if oxygen is not available and if oxygen is available.

The number of ATP molecules generated when electrons are released from coenzymes is dependent upon the entry point of the electrons into the electron transport chain (figure 3.19). The electrons from NADH enter at the top of the electron transport chain and are passed through three H^+ pumps, which results in enough energy being released to generate 3 ATP molecules. In contrast, the electrons from $FADH_2$ enter at the second H^+ pump; this results in the generation of 2 ATP molecules. Consequently, each NADH generates 3 ATP and each $FADH_2$ generates 2 ATP.

WHAT DO YOU THINK?

5 Given that energy from each NADH produces 3 ATP molecules and each $FADH_2$ produces 2 ATP molecules, calculate the number of ATP molecules generated from glucose during cellular respiration.

We are able to calculate the specific number of ATP molecules produced in breakdown of a glucose molecule by knowing the following: (1) the specific number of energy molecules (i.e., ATP, NADH, and $FADH_2$) that are generated from glucose breakdown in each of the first three stages of cellular respiration and (2) the specific number of ATP generated by the oxidation of each type of coenzyme in the electron transport system (NADH = 3 ATP molecules and FADH = 2 ATP molecules).

The following is a summary of the number of ATP molecules produced by substrate-level phosphorylation and the number of ATP produced by oxidative phosphorylation from glucose oxidation.

Stage/Total	Substrate-Level Phosphorylation	Oxidative Phosphorylation
Glycolysis	2 ATP	2 NADH → 6 ATP
Intermediate stage	—	2 NADH → 6 ATP
Citric acid cycle	2 ATP	6 NADH → 18 ATP 2 $FADH_2$ → 4 ATP
Total Number of ATP/ Based on Method of Formation	**4 ATP**	**34 ATP**

In theory, the total ATP produced from 1 glucose molecule is 38 ATP. However—and this is very important for energy totals—the actual net yield is lower. There are several energy-requiring steps for transporting molecules during cellular respiration that decrease the actual net ATP produced. These energy-requiring processes include moving (1) pyruvate from the cytosol into mitochondria, (2) phosphate and ADP into mitochondria for its use in ATP synthesis, and (3) NADH produced during glycolysis into mitochondria for oxidative phosphorylation in the electron transport system. Note: The *net* ATP produced from a glucose molecule is 30 ATP.

WHAT DID YOU LEARN?

26 How many net ATP molecules are generated through glycolysis (i.e., without participation of mitochondria)? How many net ATP molecules are formed through the combined processes that occur in the cytosol and those that occur within mitochondria?

3.4g The Fate of Pyruvate with Insufficient Oxygen

LEARNING OBJECTIVES

38. Explain the fate of pyruvate when oxygen is in short supply.

39. Describe the impact on ATP production if there is insufficient oxygen.

We followed pyruvate in the previous discussion assuming that sufficient oxygen was present for oxidative phosphorylation to occur within mitochondria. If sufficient oxygen is not available, we must consider the following:

1. Cellular respiration processes requiring oxygen (i.e., aerobic cellular respiration) decrease, including the activity of the electron transport chain. Electrons remain with the NADH and $FADH_2$ molecules and NADH and $FADH_2$ accumulate. This is accompanied by decreased levels of NAD^+ and FAD.
2. The cell becomes increasingly dependent upon glycolysis, a metabolic pathway that requires NAD^+ to continue.
3. Extended low-oxygen conditions would ultimately result in the complete shutdown of glycolysis within the cell because of the lack of NAD^+.
4. NAD^+ must be regenerated if glycolysis is to continue.

Regenerating NAD^+ involves the transfer of hydrogen from NADH. Two electrons and hydrogen are transferred from NADH to pyruvate, which is converted to lactate. (Note that the addition of a hydrogen ion to lactate forms lactic acid.) This enzymatic reaction is catalyzed by lactate dehydrogenase (**figure 3.21**).

Although this process is an effective means to permit glycolysis to continue (because it makes NAD^+ available), we must keep in mind that without the availability and use of mitochondria, only 2 ATP are produced per glucose. Compared to the net 30 ATP produced with sufficient oxygen, there is significantly less ATP generated (2 ATP

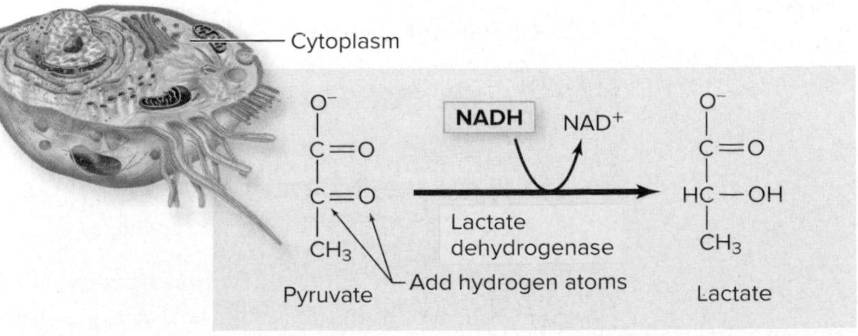

Figure 3.21 Conversion of Pyruvate to Lactate. Lactate dehydrogenase converts pyruvate to lactate to regenerate NAD^+ molecules.

versus 30 ATP). This is a 15-fold difference! Remember this important association: Low O_2 equals low energy. So, in clinical practice when you work with individuals with decreased ability to deliver oxygen to cells (e.g., those with impaired respiratory or cardiovascular function), keep in mind that they will have less available ATP to meet the body's energy needs.

WHAT DID YOU LEARN?

27 Pyruvate is converted to what molecule if there is insufficient oxygen? Explain why this occurs.

3.4h Other Fuel Molecules That Are Oxidized in Cellular Respiration

LEARNING OBJECTIVE

40. Describe the entry point in the metabolic pathway of cellular respiration for both fatty acids and amino acids.

There are other fuel molecules, such as fatty acids and amino acids that may be oxidized to generate ATP (see figure 27.6).

Fatty acids are enzymatically changed two carbon units at a time to form acetyl CoA. This process is called **beta-oxidation.** Acetyl CoA then enters the cell respiration metabolic pathway at the citric acid cycle. Because fatty acids enter this metabolic pathway in the mitochondria, they can only be oxidized aerobically. (Note that a by-product of fatty acid metabolism is the production of ketoacids, with significant amounts produced in individuals with uncontrolled diabetes mellitus; see Clinical View 25.7: "Lactic Acidosis and Ketoacidosis").

A different pathway is employed if protein is used for fuel. The point of entry of deaminated amino acids (amino acids with the amine group [$-NH_2$] removed) is dependent upon the specific type of amino acids. Different amino acids enter the metabolic pathway at glycolysis, the intermediate stage, or the citric acid cycle. The amine group is a waste product that is converted to urea and excreted by the kidneys.

WHAT DID YOU LEARN?

28 Why is oxygen required to burn fatty acids?

INTEGRATE

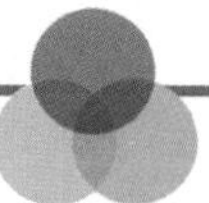

CONCEPT CONNECTION

What happens to the increased levels of lactate produced by skeletal muscle tissue? Lactate is either absorbed by surrounding muscle tissue or transported by the blood to the liver. Muscle cells may use lactate immediately for ATP synthesis by either converting it back to pyruvate or converting it to glucose and storing it as glycogen. Liver cells convert lactate into glucose. Glucose formed in the liver is either stored as glycogen in the liver or released back into the blood (for muscles or other cells to take up). This cycling of lactate from muscles to the liver, the conversion of lactate to glucose, and the subsequent transport of glucose from the liver to muscle is called the **Cori cycle.**

CHAPTER SUMMARY

	• The concepts discussed are energy, chemical reactions, enzymes, metabolic pathways, and the production of ATP through cellular respiration.
3.1 Energy	• Energy is the capacity to do work.
	3.1a Classes of Energy • Energy exists in two classes: energy based upon position, called potential energy, and energy of motion, called kinetic energy. • Energy can be converted from potential energy to kinetic energy (or vice versa). Examples are the movement of a substance down its concentration gradient and the movement of electrons from a higher-energy state to a lower-energy state.
	3.1b Forms of Energy • Energy exists in different forms, including chemical, electrical, mechanical, sound, radiant, and heat.
	3.1c Laws of Thermodynamics • The first law of thermodynamics states that energy cannot be created or destroyed, only converted from one form to another. • The second law of thermodynamics states that some energy is lost as heat with every energy conversion.
3.2 Chemical Reactions	• Chemical reactions are expressed in chemical equations and are classified using various criteria.
	3.2a Chemical Equations • *Metabolism* is the collective term for all biochemical reactions that occur within the body. • Reactants become products in a chemical reaction, and an arrow indicates the direction of change.
	3.2b Classification of Chemical Reactions • Chemical reactions can be classified using different criteria: change in chemical structure, change in chemical energy, and if the reaction is irreversible or reversible.
	3.2c Reaction Rates and Activation Energy • Reaction rate is the measure of how quickly a chemical reaction takes place; this determines the amount of product formed per unit of time. • Activation energy is the energy required to break existing chemical bonds for the chemical reaction to proceed.

3.3 Enzymes

- Enzymes facilitate chemical reactions.

3.3a Function of Enzymes

- Enzymes are biologically active catalysts that increase reaction rates by lowering activation energy (E_a).

3.3b Enzyme Structure and Location

- Generally, enzymes are globular proteins with an active site that bind a specific substrate. Enzymes can be located within cells, embedded in the plasma membrane, or present in fluid outside of cells.

3.3c Mechanism of Enzyme Action

- Enzymes are catalysts that participate in both decomposition and synthesis reactions.

3.3d Classification and Naming of Enzymes

- Enzymes are usually named based upon the name of the substrate, product, or type of chemical reaction (as indicated by the enzyme class or subclass) and contain the *-ase* suffix.

3.3e Enzymes and Reaction Rates

- The reaction rate is dependent upon the concentrations of both the enzyme and substrate, temperature, and pH.

3.3f Controlling Enzymes

- Enzymes can be controlled by either competitive inhibitors or noncompetitive inhibitors.

3.3g Metabolic Pathways and Multienzyme Complexes

- A metabolic pathway involves numerous enzymes that subsequently convert a substrate to a final product. Metabolic pathways are regulated by negative feedback to maintain the needed amount of the final product.
- A multienzyme complex is a structure composed of enzymes physically linked to convert a substrate to a final product.
- Phosphorylation is the addition of a phosphate group, and dephosphorylation is the removal of a phosphate group. This is a common means of regulating enzymes.

3.4 Cellular Respiration

- Cellular respiration is the metabolic process for oxidizing organic molecules (e.g., glucose) to release energy to synthesize ATP.

3.4a Overview of Glucose Oxidation

- The net chemical reaction for glucose oxidation is $C_6H_{12}O_6 + 6\ O_2 \rightarrow 6\ CO_2 + 6\ H_2O$.
- ATP is produced directly through substrate-level phosphorylation and indirectly by oxidative phosphorylation.
- Glucose oxidation occurs within a cell: glycolysis within the cytosol and aerobic cellular respiration (the intermediate stage, the citric acid cycle, and the electron transport system) within mitochondria.

3.4b Glycolysis

- Glycolysis, which occurs within the cytosol, is a metabolic pathway that uses 10 enzymes and does not require oxygen. Glucose is converted to 2 pyruvate molecules, producing 2 net ATP and 2 NADH per glucose.
- The fate of pyruvate is dependent upon the availability of oxygen.

3.4c Intermediate Stage

- In the intermediate stage, which occurs within mitochondria, the multienzyme complex pyruvate dehydrogenase converts pyruvate to acetyl CoA and releases 1 CO_2 (by decarboxylation). One NADH is produced per pyruvate.
- The intermediate stage occurs twice per the original glucose.

3.4d Citric Acid Cycle

- The citric acid cycle, which also occurs within mitochondria, is a metabolic pathway that breaks down acetyl CoA with 2 CO_2 and CoA released. One ATP molecule, 3 NADH molecules, and 1 $FADH_2$ molecule are produced per acetyl CoA.
- The citric acid cycle occurs twice per the original glucose and completes the digestion of glucose.

3.4e The Electron Transport System

- The electron transport system involves several significant structures that are embedded in the cristae membrane of a mitochondrion, including H^+ pumps, electron carriers, and the enzyme ATP synthase.
- Electrons from NADH and $FADH_2$ are transferred to electron carriers in the electron transport chain of a mitochondrion, and ultimately to O_2. The electrons, oxygen, and hydrogen ions form H_2O.
- A H^+ ion gradient is formed; H^+ ions then flow down this gradient and the energy is harnessed by ATP synthase to produce ATP through oxidative phosphorylation.

3.4f ATP Production

- Glucose oxidation produces a net of 2 ATP in glycolysis and 30 ATP in the complete breakdown of glucose.

3.4g The Fate of Pyruvate with Insufficient Oxygen

- If insufficient oxygen is available, pyruvate will be converted by lactate dehydrogenase to lactate to regenerate NAD^+ so glycolysis can continue.

3.4h Other Fuel Molecules That Are Oxidized in Cellular Respiration

- The digestion of other fuel molecules, such as fatty acids and amino acids, can also be oxidized to generate ATP.

CHALLENGE YOURSELF

Create and Evaluate

Analyze and Apply

Understand and Remember

Do You Know the Basics?

1. Energy in ATP is used to power skeletal muscle contraction. This is an example of what type of energy conversion?
   a. chemical energy to mechanical energy
   b. light energy to mechanical energy
   c. chemical energy to light energy
   d. electrical energy to chemical energy
2. Oxidation-reduction can be best classified as a(n) _______________ reaction.
   a. exchange
   b. endergonic
   c. synthesis
   d. reversible
3. All of the following increase enzymatic activity *except*
   a. an increase in temperature.
   b. an increase in pH.
   c. an increase in concentration of the substrate.
   d. an increase in concentration of the enzyme that catalyzes the reaction.
4. ATP inhibits phosphofructokinase by binding to an allosteric site in glycolysis. ATP is functioning as a
   a. competitive inhibitor.
   b. competitive activator.
   c. noncompetitive inhibitor.
   d. noncompetitive activator.
5. All of the following are accurate about enzymes *except*
   a. enzymes are typically globular proteins with an active site.
   b. enzymes decrease activation energy.
   c. enzymes can be used over and over to catalyze a substrate to a product.
   d. enzymes are versatile and can catalyze different types of chemical reactions.
6. Glucose is converted to pyruvate in which stage of cellular respiration?
   a. glycolysis
   b. intermediate stage
   c. citric acid cycle
   d. electron transport system
7. NAD^+ and FAD are examples of
   a. enzymes.
   b. high-energy organic molecules that are digested in cellular respiration.
   c. allosteric inhibitors, which participate in regulating enzymes.
   d. coenzymes that are required by some enzymes to function.
8. All stages of cellular respiration are decreased in conditions of insufficient oxygen *except*
   a. glycolysis.
   b. the intermediate stage.
   c. the citric acid cycle.
   d. the electron transport system.
9. In glycolysis, _______ ATP are formed, and if sufficient oxygen is present, _______ ATP are formed.
   a. 2, 2
   b. 36, 38
   c. 2, 30
   d. 10, 30
10. Oxidative phosphorylation involves
    a. electrons transported in the electron transport chain and accepted by O_2.
    b. ATP synthase harnessing the energy in a H^+ gradient.
    c. coenzymes NADH and $FADH_2$ giving up their electrons.
    d. all of these processes.
11. List and define the different forms of energy, and give one example each of their application in the human body.
12. Describe the different ways of classifying chemical reactions, and explain the category to which oxidation-reduction belongs.
13. Explain ATP cycling.
14. Describe the structure and mechanism of enzymes.
15. Describe a metabolic pathway, and explain how it is controlled by negative feedback.
16. Summarize glycolysis, including where it occurs in a cell, if it requires oxygen, the substrate and final product, and the formation of energy-containing molecules (i.e., ATP, NADH, $FADH_2$).
17. In general terms, explain the fate of pyruvate if there is (a) sufficient oxygen and (b) insufficient oxygen.
18. Describe how oxygen becomes part of water during cellular respiration.
19. Identify the source of carbon in carbon dioxide.
20. Based on what you know about glycolysis and aerobic cellular respiration, explain the advantage in terms of ATP production of a healthy respiratory and cardiovascular system.

Can You Apply What You've Learned?

1. Albinism (achromia) is a genetic condition in which an individual cannot synthesize melanin from tyrosine (an amino acid), a brown pigment of the hair, skin, and eyes. These individuals lack
   a. specific fatty acids.
   b. a protein that contains tyrosine.
   c. an enzyme that converts tyrosine to melanin.
   d. cofactors that convert tyrosine to melanin.

2. If an individual has impaired respiratory function, as occurs with emphysema, you would expect all of the following *except*
   a. production of additional lactate.
   b. an impaired ability to make ATP.
   c. low energy levels and complaints of being tired.
   d. increased aerobic cellular respiration.
3. Another challenge to a patient with impaired respiratory function is the buildup of CO_2 in the blood. What would you predict, given the following reversible enzymatic reaction that occurs in the blood?

   $H_2O + CO_2 \leftrightarrows H_2CO_3 \leftrightarrows H^+ + HCO_3^-$

   a. increased production of H_2O
   b. increased production of H^+ (with an accompanying decrease in pH)
   c. decreased production of H^+ (with an accompanying increase in pH)
   d. all of the above
4. You would expect decreased production of ATP in all of the following individuals *except*
   a. an individual with impaired ability to transport oxygen in the blood, such as a person with anemia.
   b. an individual with severe asthma.
   c. an individual in congestive heart failure.
   d. an athlete.
5. Brown adipose tissue contains cells that allow H^+ to fall down the concentration gradient in the electron transport chain without producing ATP. Instead, all of the kinetic energy is converted to heat. If scientists could increase the amount of brown adipose tissue in our bodies, then
   a. our body temperature would be cooler than 98.6°F.
   b. brown adipose tissue cells would be more efficient at producing ATP.
   c. we could eat more and not gain weight.
   d. we would be able to run faster.

Can You Synthesize What You've Learned?

1. Tiffany had returned to her college dorm and was having difficulty breathing. She knew she was having an asthma attack. What changes in her energy level are predicted?
2. Provide a general explanation to a patient on the advantages of aerobic fitness in terms of ATP production.
3. What occurs to the amount of product formed in a metabolic pathway if inhibition does not occur?

INTEGRATE

ONLINE STUDY TOOLS

connect | SMARTBOOK | 

The following study aids may be accessed through Connect.

Clinical Case Study: Mitochondrial Dysfunction: A Special Form of Inheritance

Interactive Questions: This chapter's content is served up in a number of multimedia question formats for student study

SmartBook: Topics and terminology include energy; chemical reactions, enzymes, cellular respiration

Anatomy & Physiology Revealed: Topics include NADH oxidative-reduction reactions; enzymes; glycolysis; citric acid cycle (Kreb's cycle); electron transport and ATP synthesis

Animations: Topics include enzyme structure and action; coenzymes; metabolic pathways; glycolysis

chapter

4 Biology of the Cell

Module 2:
Cells and Chemistry

INTEGRATE

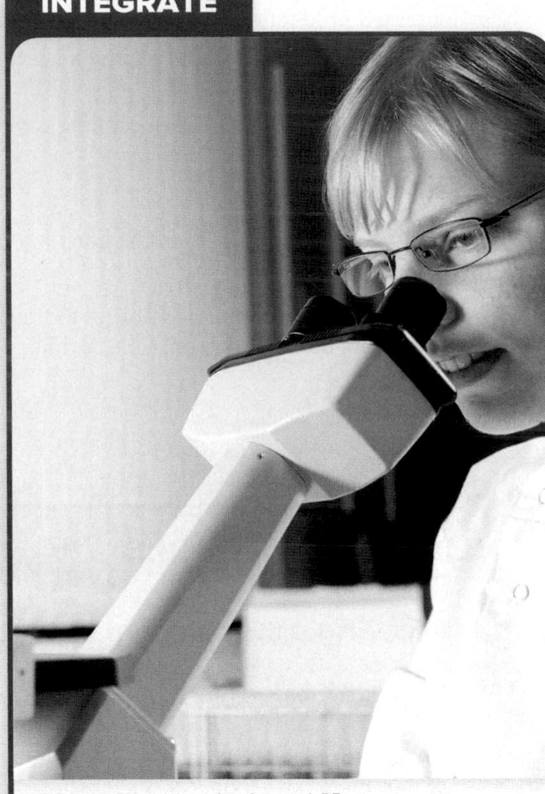

©Science Photo Library/agefotostock RF

CAREER PATH
Cytologist

A cytologist examines cells under a microscope to detect abnormalities that may indicate cancer or other diseases. Cytologists use specialized equipment to prepare cell samples, and then they apply stains or other techniques to enhance the specimens for detailed observation. Drawing upon an extensive knowledge of cell structure and function, they then analyze cell samples and submit their findings to a pathologist, who ultimately makes a diagnosis.

Heart cells contract to pump blood out of the heart chambers; retinal cells of the eye detect light; phagocytic white blood cells engulf foreign substances (e.g., bacteria, viruses); and pancreatic cells synthesize and secrete the hormone insulin. All human body processes are ultimately dependent upon cells and their activities. For this reason, the cell is often referred to as the "functional unit of the body." Knowledge of cellular structure and function is critical for understanding the concepts of all later chapters.

Throughout this chapter, we present a broad discussion of a cell by describing how cells are studied and the general structural components and functions of cells. Subsequent chapters examine specialized cells and provide details of their unique functions.

4.1 Introduction to Cells

Our examination of cells begins with a description of how we study them. We then describe how the cells that compose the human body vary in both size and shape, and that these differences reflect their specific function. This section concludes with a discussion of the common structural features all cells possess and the general functions that all cells must perform.

4.1a How Cells Are Studied

LEARNING OBJECTIVE

1. Distinguish among light microscopy, scanning electron microscopy, and transmission electron microscopy.

The study of cells is called **cytology** (sī-tol′ō-jē; *kytos* = a hollow [cell]). The small size of cells is the greatest obstacle to determining their nature. Cells were discovered after microscopes were invented because high-magnification microscopes are required to see the smallest human body cells.

Microscopy is the use of a microscope to view small-scale structures, and it is an invaluable asset in anatomic investigations. The most commonly used instruments are the light microscope, the scanning electron microscope, and the transmission electron microscope.

One of the challenges with viewing anatomic specimens with a microscope is that microscopy samples prepared from body tissues have no inherent contrast (difference between specimen and background). To provide contrast so microscopic structures can be seen more clearly, colored dye stains are used with light microscopes, and heavy-metal stains are used with both scanning electron and transmission electron microscopes. **Figure 4.1** allows us to compare the images produced when each of these types of microscopes is used to examine the same specimen. Notice the remarkable difference between these images of the same body structure—in this case, hairlike cilia (see section 4.6c) on the inner lining of the respiratory tract.

The **light microscope (LM)** produces a two-dimensional image by passing visible light through the specimen stained with colored dyes. Glass lenses focus and magnify the image as it is projected toward the eye. A stained image of the magnified specimen is produced with an LM (figure 4.1*a*). A typical LM used in a college laboratory might magnify a specimen 40×, 100×, or 1000× (depending upon the objective lens that is used). The specimen in figure 4.1*a* is magnified 720× (as indicated on the side of the image).

The **electron microscope (EM)** uses a beam of electrons to "illuminate" the specimen stained with heavy metal. Electron microscopes easily exceed the magnification obtained by light microscopy—but more importantly, they improve the resolution (ability to see details) by more than a thousand-fold over the light microscope. A **scanning electron microscope (SEM)** directs electrons across the surface of a specimen to produce a three-dimensional image that is captured on a screen. An SEM allows us to visualize the specimen's surface features (figure 4.1*b*). The specimen in figure 4.1*b* is magnified 3000×.

A **transmission electron microscope (TEM)** directs an electron beam through a thin-cut section of the specimen. A two-dimensional image of the specimen is focused either onto a screen or onto photographic film. A TEM allows us to visualize the details of the specimen's internal structures (figure 4.1*c*). The specimen in figure 4.1*c* is magnified 50,000×!

WHAT DID YOU LEARN?

1. What is the advantage of using a TEM instead of an LM to study intracellular structure?

4.1b Cell Size and Shape

LEARNING OBJECTIVES

2. Describe the range in size of human cells.
3. Identify some of the shapes cells may exhibit.

Cells are typically depicted as being of one size and either spherical or cubelike in shape, when in reality the structure of the approximately 75 trillion cells of the adult human shows great variety. Most cells are microscopic in size, but some are large enough to be seen with the naked eye (**figure 4.2**). For example, red blood cells (erythrocytes) are relatively small with a diameter of about 7–8 μm, whereas a human oocyte has a diameter of about 120 μm. To help you to relate to how small some cells are, consider that about 5 million erythrocytes would fit on the head of a pin. Cells also vary greatly in shape (**figure 4.3**). Although some cells are spherical or

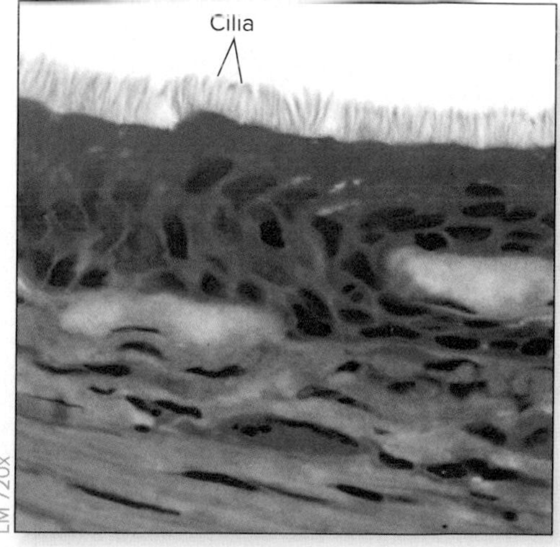

(a) Light microscopy (LM)

(b) Scanning electron microscopy (SEM)

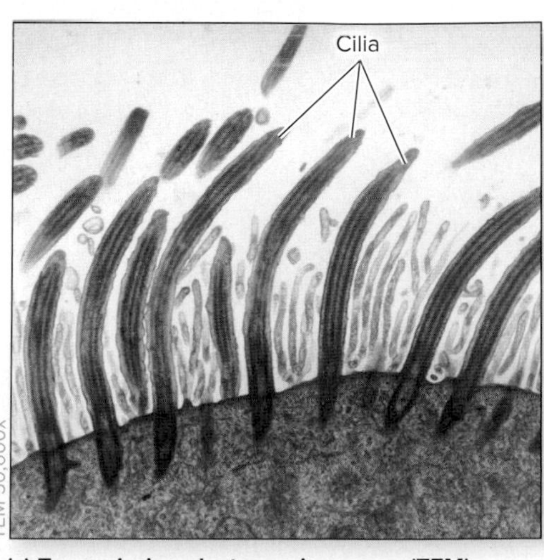

(c) Transmission electron microscopy (TEM)

Figure 4.1 Microscopic Techniques for Cellular Studies. Different microscopic techniques are used to investigate cellular anatomy. (*a*) A light microscope (LM) shows hairlike structures, termed cilia, that project from cells lining the respiratory tract. (*b*) A scanning electron microscope (SEM) shows the three-dimensional image of the cilia of the same type of cells. (*c*) A transmission electron microscope (TEM) reveals the ultrastructure. Note: The dimensional unit often used to measure the specimen when viewing with an LM is the micrometer (μm). Ten thousand (10^4) μm are equal to 1 centimeter. In comparison, when viewing with an electron microscope, the unit often used is the nanometer (nm). Ten million (10^7) nm are equal to 1 centimeter.

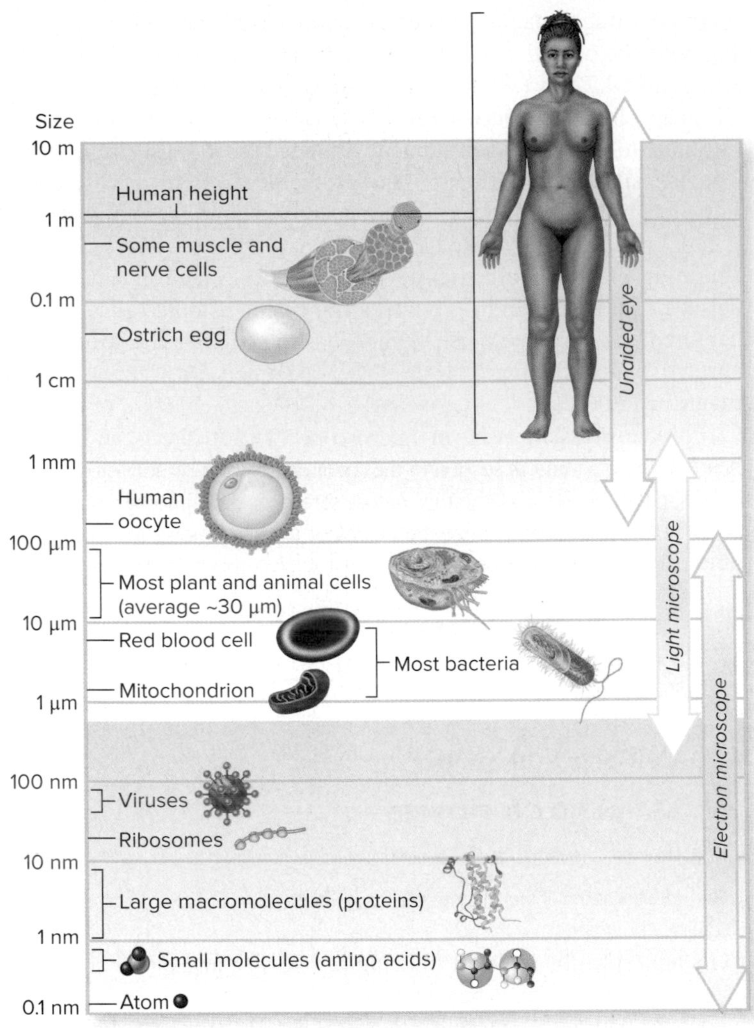

Figure 4.2 The Range of Cell Sizes. Most cells in the human body are between 1 micrometer (µm) and 100 µm in diameter.

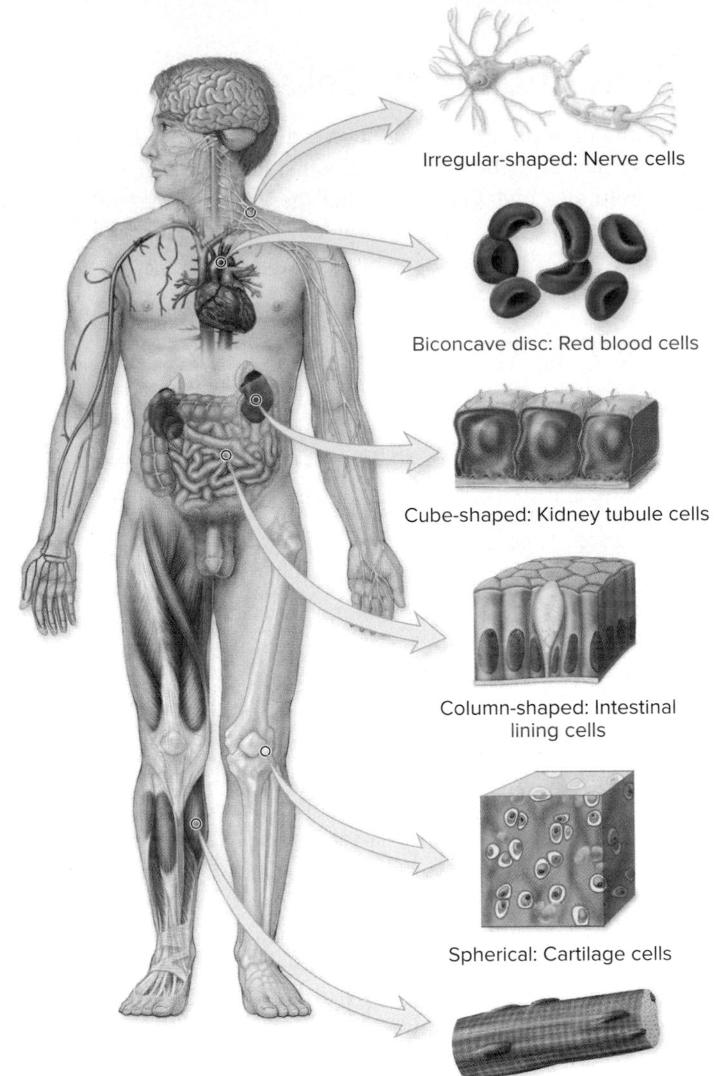

Figure 4.3 The Variety of Cell Shapes. Cells throughout the body exhibit different shapes that support various functions.

cubelike, others are columnlike, cylindrical, disc-shaped, or irregular-shaped. Note that a relationship exists between the size and shape of a cell and its function in the body.

WHAT DID YOU LEARN?

2 Which cell is larger, an erythrocyte or a human oocyte? What are their respective sizes?

4.1c Common Features and General Functions

LEARNING OBJECTIVES

4. Describe the three main structural features of a cell.
5. Identify the membrane-bound and non-membrane-bound organelles.
6. Distinguish between organelles and cell inclusions.
7. Explain the general functions that cells must perform.

Most cells are composed of characteristic parts that work together to allow each cell type in the body to perform certain common functions.

Overview of Cellular Components

The generalized cell shown in **figure 4.4** is not an actual body cell, but rather a representation of a cell that combines features of different types of body cells. The common features include the following:

- **Plasma membrane.** The plasma (plaz′mă; *plasso* = to form) membrane is the cell membrane that forms the outer, limiting barrier (of approximately 5–10 nm) that separates the internal contents of the cell from the **interstitial fluid** (fluid that surrounds the cell). Modified extensions of the plasma membrane include cilia, a flagellum, and microvilli.

- **Nucleus.** The nucleus (nū′klē-ūs; *nux* = the kernel) is the largest structure within the cell (approximately 5–10 μm in diameter) and is enclosed by a nuclear envelope. Much of the internal content of the nucleus is the genetic material, deoxyribonucleic acid (DNA). The fluid within the nucleus is called the nucleoplasm. Within the nucleus is a dark-staining body called the nucleolus.
- **Cytoplasm.** Cytoplasm (sī′tō-plazm; *plasma* = a thing formed) is a general term for all cellular contents located between the plasma membrane and the nucleus. The three primary components of the cytoplasm are the cytosol, organelles, and inclusions.

Cytoplasmic Components

The **cytosol** (sī′tō-sol; *sol* = soluble), also called the *intracellular fluid (ICF)* or *cytoplasmic matrix,* is the viscous, syruplike fluid of the cytoplasm. It has a high water content and contains many dissolved macromolecules that include carbohydrates, lipids, proteins, and small molecules such as glucose and amino acids. Cytosol also contains various types of ions, such as potassium ion (K^+) and phosphate ion (PO_4^{3-}).

Organelles (or′gă-nel; *organon* = organ, *elle* = the diminutive suffix), meaning *little organs,* are complex, organized structures within cells that have unique characteristic shapes and functions. Two categories of organelles are recognized: membrane-bound organelles and non-membrane-bound organelles. **Membrane-bound organelles,** or *membranous organelles,* are enclosed by a membrane similar to the plasma membrane. The membrane separates the organelle's contents from the cytosol so that the specific activities of the organelle can proceed without disruption from other cellular activities. Membrane-bound organelles include the endoplasmic

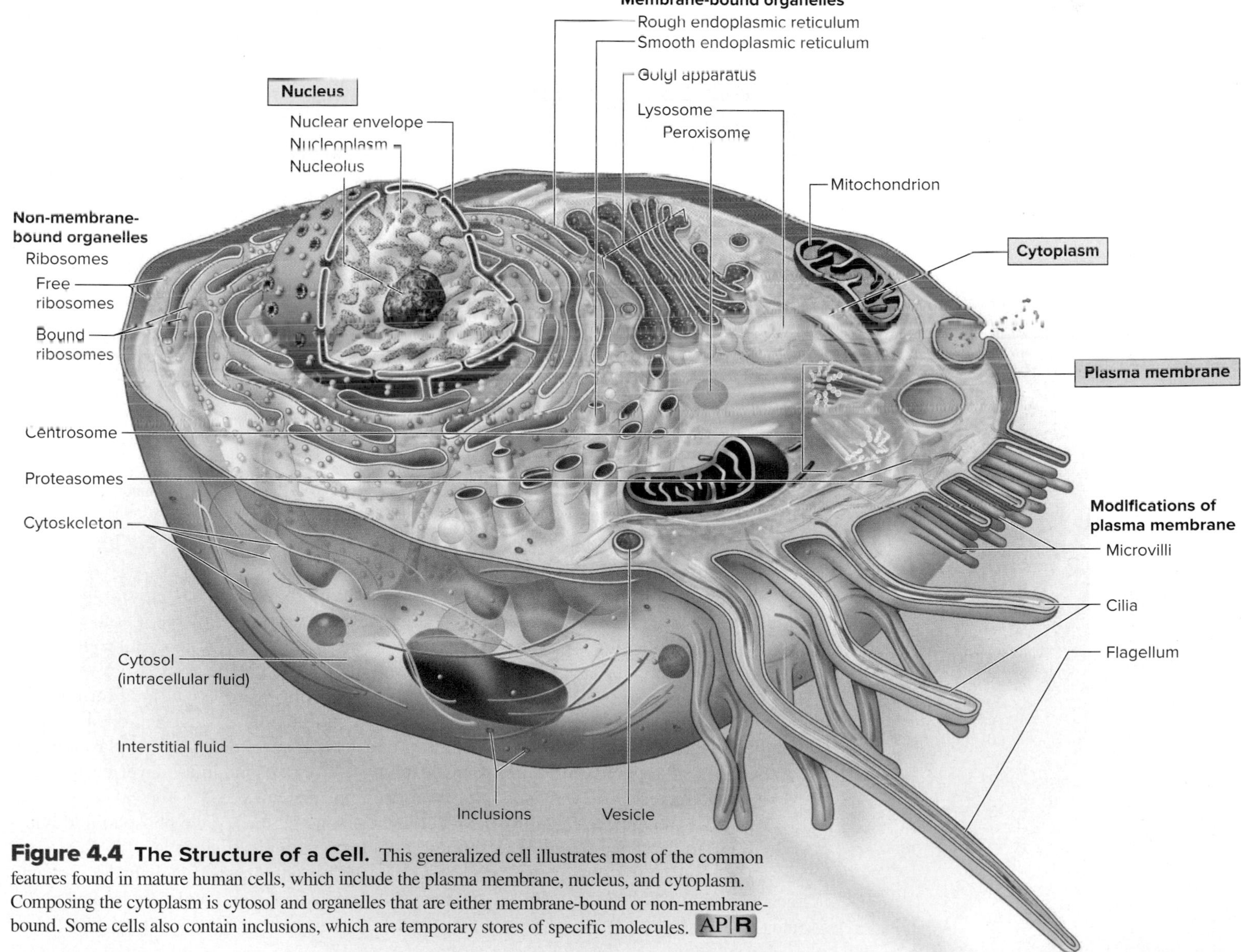

Figure 4.4 The Structure of a Cell. This generalized cell illustrates most of the common features found in mature human cells, which include the plasma membrane, nucleus, and cytoplasm. Composing the cytoplasm is cytosol and organelles that are either membrane-bound or non-membrane-bound. Some cells also contain inclusions, which are temporary stores of specific molecules. AP|R

reticulum (rough and smooth), Golgi apparatus, lysosomes, peroxisomes, and mitochondria (figure 4.4). Vesicles are temporary membrane-bound structures formed from the endoplasmic reticulum, Golgi apparatus, and plasma membrane. The **non-membrane-bound organelles,** or *nonmembranous organelles,* are not enclosed within a membrane. These structures are generally composed of protein and include ribosomes (either attached [bound] to the external surface of the endoplasmic reticulum or free within the cytosol), the centrosome, proteasomes, and the cytoskeleton. Each of these organelles is discussed in detail in section 4.6.

The cytosol of some cells temporarily stores **inclusions.** Cell inclusions are not considered organelles, but rather are aggregates (clusters) of a single type of molecule. Molecules are continuously being added to and removed from inclusions. Pigments (e.g., melanin, a stored pigment in some skin, hair, and eye cells) and nutrient stores (e.g., glycogen in liver cells and triglycerides in adipose connective tissue) are examples of cell inclusions.

General Cell Functions

Cells must perform general functions, including:

- **Maintain integrity and shape of a cell.** The integrity and shape of a cell are dependent upon both the plasma membrane, which forms the external boundary of the cell, and the internal contents, which function to support the cell.
- **Obtain nutrients and form chemical building blocks.** Each cell must get nutrients and other needed substances from its surrounding fluid. Cells form new chemical structures and harvest the energy necessary for survival through diverse metabolic processes.
- **Dispose of wastes.** Cells must dispose of the waste products they produce so they do not accumulate and disrupt normal cellular activities.

In addition, some cells are capable of undergoing cell division to make more cells of the same type, as described in section 4.9. These new cells help to maintain the tissue or organ to which they belong by providing cells for new growth and replacing cells that die. However, during development, some cells do not retain this ability (e.g., most nerve cells typically do not; see section 12.2a).

WHAT DID YOU LEARN?

What are the three main structural features of a cell?

What cellular structure is responsible for forming the boundary of a cell and maintaining its integrity?

4.2 Chemical Structure of the Plasma Membrane

The plasma membrane is not a rigid boundary, but rather is a fluid matrix composed of approximately an equal mixture, by weight, of lipids and proteins. It regulates the movement of most substances both into and out of a cell.

4.2a Lipid Components

LEARNING OBJECTIVE

8. List the lipid components of the plasma membrane, and explain the actions of each component.

The plasma membrane contains several different types of lipids (see section 2.7b), including phospholipids, cholesterol, and glycolipids **(figure 4.5)**.

Most of the plasma membrane lipids are **phospholipids** (see section 2.3c). Often these molecules are artistically portrayed in the membrane as an icon that looks similar to a balloon with two tails (see table 2.4). The balloonlike "head" is polar and hydrophilic. In contrast, the two "tails" are nonpolar and hydrophobic. Phospholipid molecules readily associate to form two parallel sheets of molecules lying tail-to-tail, with the hydrophobic tails forming the internal environment of the membrane and their hydrophilic polar heads positioned adjacent to either the cell's cytoplasm or the interstitial fluid. This basic structure of the plasma membrane framework is called the **phospholipid bilayer.** The phospholipid bilayer ensures that cytosol remains inside the cell, and interstitial fluid remains outside.

Cholesterol is scattered within the inner hydrophobic regions of the phospholipid bilayer. It strengthens the membrane and stabilizes it at temperature extremes.

Glycolipids (glī′kō-lip′id; *glykys* = sweet) are lipids with attached carbohydrate groups. Each carbohydrate group of a glycolipid is attached to a phospholipid molecule located on the outer phospholipid layer of the plasma membrane. These carbohydrates extend like "sugar antennae" from the cell's external phospholipid surface, where they are exposed to the interstitial fluid. These molecules contribute to the glycocalyx, which is described at the end of this section.

The lipid portion of the plasma membrane is insoluble in water, which ensures that the plasma membrane will not simply dissolve when it comes into contact with water. Rather, this boundary is an effective nonpolar physical barrier to most substances. Only small and nonpolar substances can readily penetrate (move through) this barrier without assistance (see section 4.3a).

WHAT DID YOU LEARN?

5. How do lipids maintain the basic physical barrier of the plasma membrane?

4.2b Membrane Proteins

LEARNING OBJECTIVES

9. Differentiate between the two types of membrane proteins based upon their relative position in the plasma membrane.
10. Name the six major roles played by membrane proteins.

Although lipids form the main component of the plasma membrane, the proteins dispersed within the lipids make up about half of the plasma membrane by weight. Proteins can "float" and move about the phospholipid bilayer, much like a beach ball floating on the water surface in a swimming pool. Most of the membrane's specific functions are determined by its resident proteins.

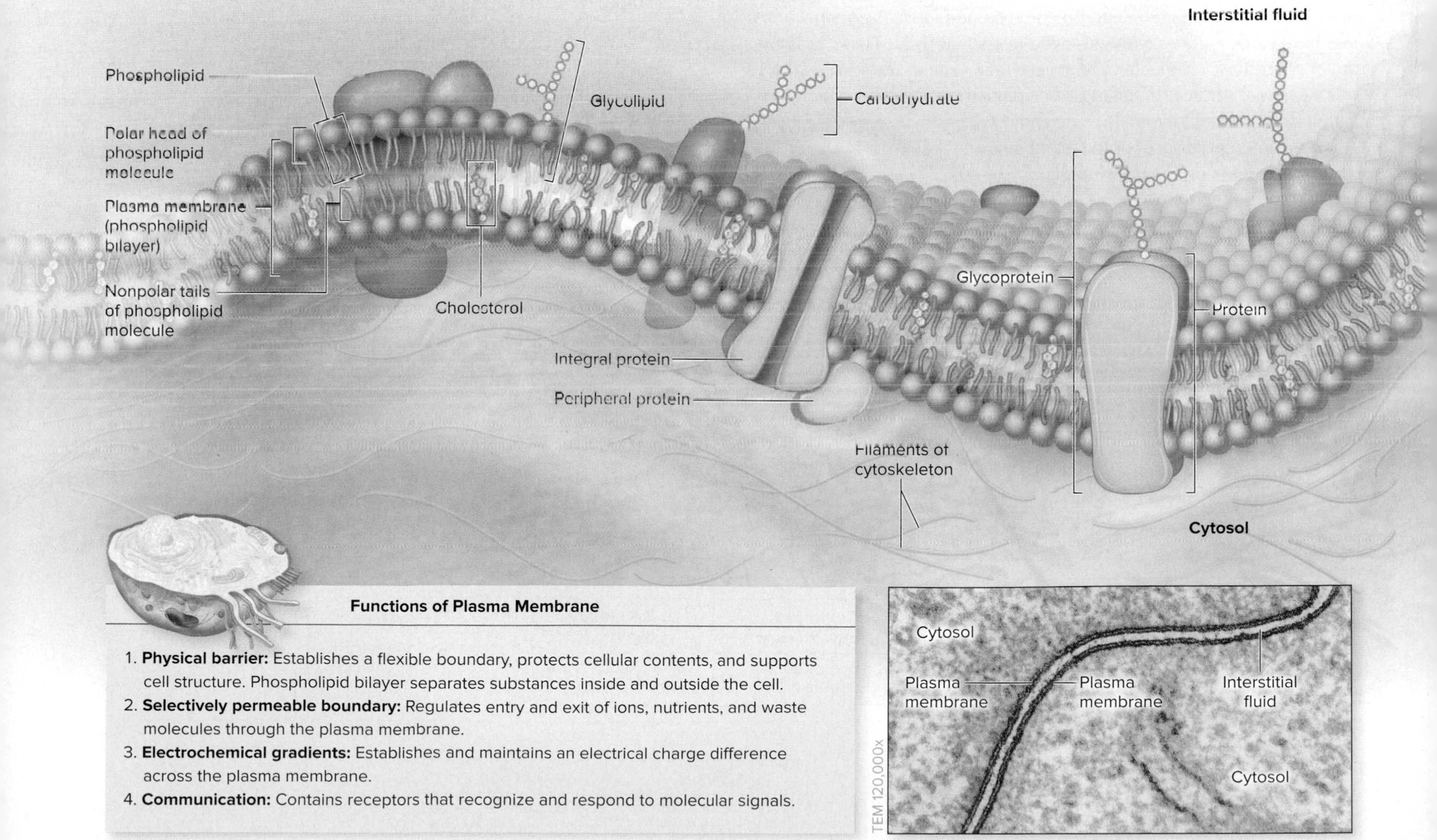

INTEGRATE

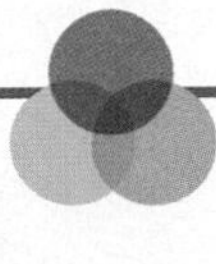

CONCEPT CONNECTION

Cholesterol is a component of plasma membranes found only in animal cells, not in plant cells. Consequently, any animal-derived food, such as eggs, milk, and meat, contains cholesterol. Foods obtained from plants, including carrots, corn, and potato chips cooked in vegetable oil, do not contain cholesterol.

Functions of Plasma Membrane

1. **Physical barrier:** Establishes a flexible boundary, protects cellular contents, and supports cell structure. Phospholipid bilayer separates substances inside and outside the cell.
2. **Selectively permeable boundary:** Regulates entry and exit of ions, nutrients, and waste molecules through the plasma membrane.
3. **Electrochemical gradients:** Establishes and maintains an electrical charge difference across the plasma membrane.
4. **Communication:** Contains receptors that recognize and respond to molecular signals.

(a) Plasma membrane

(b) Adjacent plasma membranes

Figure 4.5 Structure and Functions of the Plasma Membrane. (*a*) The plasma membrane is a phospholipid bilayer, with cholesterol and proteins scattered throughout. (*b*) The plasma membranes (phospholipid bilayers) of two adjacent cells are visible in a TEM. Each plasma membrane has a width of approximately 5–10 nm. AP|R

(*b*) ©Don W. Fawcett/Science Source

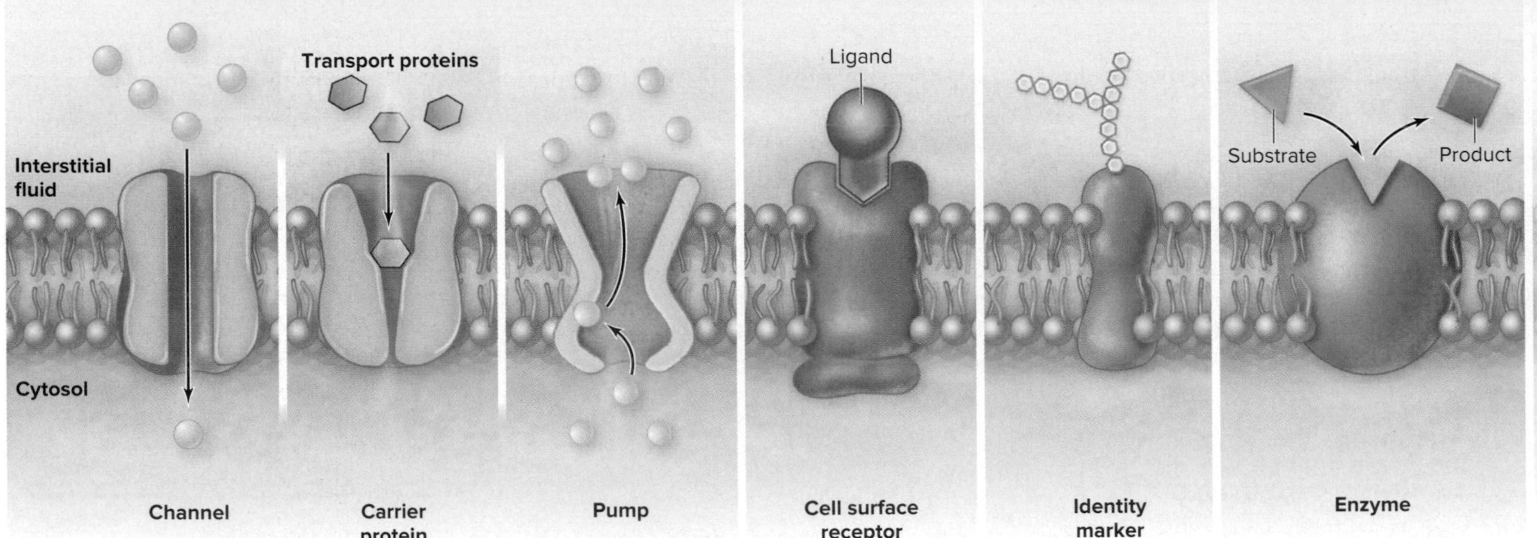

Figure 4.6 Plasma Membrane Proteins. The major functional categories of plasma membrane proteins include the several types of transport proteins (e.g., channels, carrier proteins, pumps), cell surface receptors, identity markers, enzymes, anchoring sites for the cytoskeleton, and cell-adhesion proteins.

Membrane proteins are classified as one of two structural types: integral or peripheral. **Integral proteins** are embedded within, and extend completely across, the phospholipid bilayer (figure 4.5). Hydrophobic regions within the integral proteins interact with the hydrophobic interior of the membrane. In contrast, the hydrophilic regions of the integral proteins are exposed to the aqueous environments on either side of the membrane. Many integral membrane proteins are **glycoproteins** that have carbohydrates exposed to the interstitial fluid. These carbohydrates (like those of glycolipids) extend like "sugar antennae" from a cell's external surface. In contrast, **peripheral proteins** are not embedded within the lipid bilayer. They are attached loosely to either the external or the internal surfaces of the membrane and are often "anchored" to the exposed parts of an integral protein.

Membrane proteins are also categorized functionally based upon the specific role they serve **(figure 4.6)**.

- **Transport proteins** provide a means of regulating the movement of substances across the plasma membrane. Different types of transport proteins include **channels, carriers, pumps, symporters,** and **antiporters** (see section 4.3).
- **Cell surface receptors** bind specific molecules called ligands. **Ligands** are molecules that bind to macromolecules (e.g., binding to a receptor). An example of a ligand is a neurotransmitter released from a nerve cell that binds to the cell surface receptor of a muscle cell to initiate contraction (see section 10.3a).

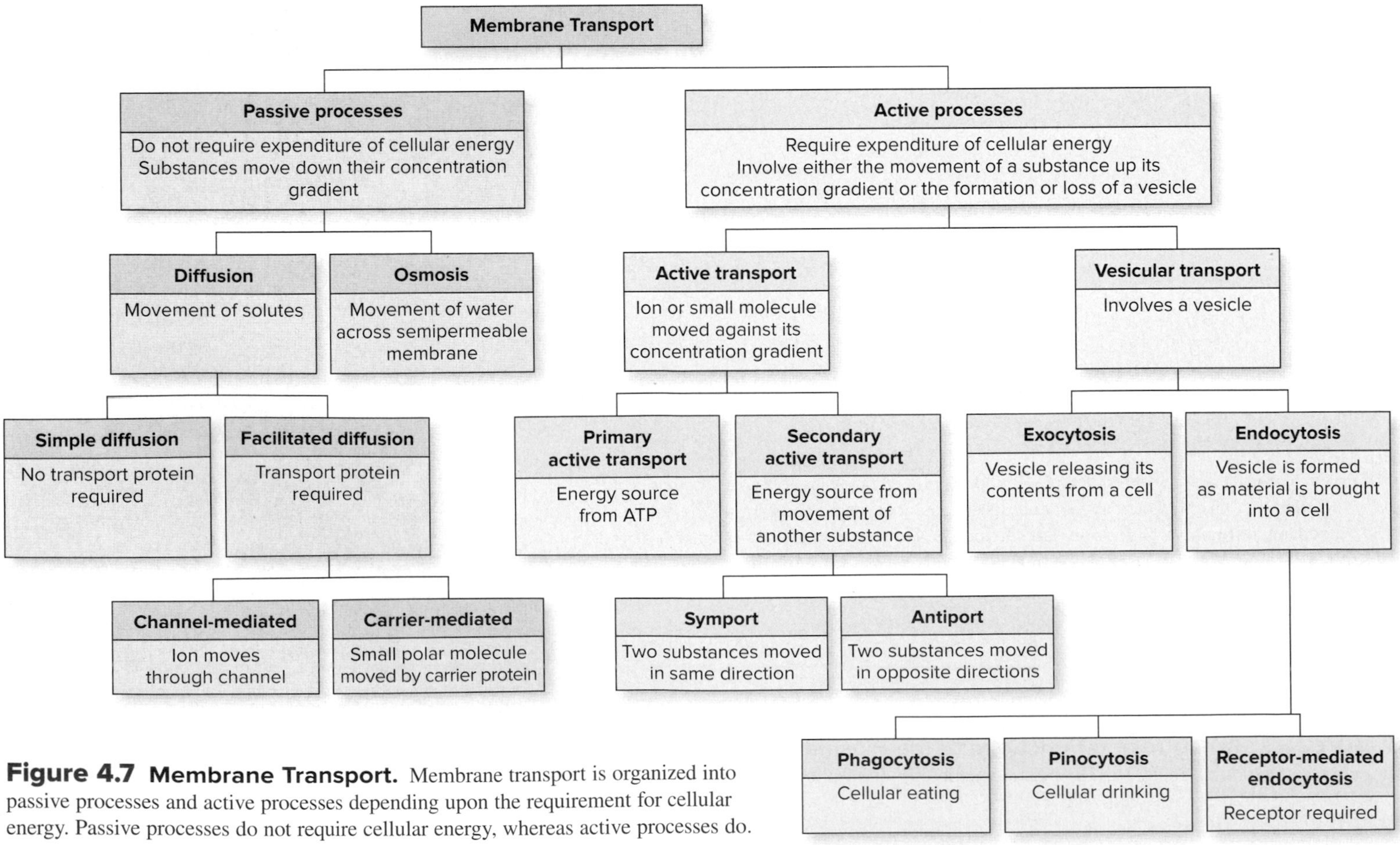

Membrane Transport

- **Passive processes** — Do not require expenditure of cellular energy. Substances move down their concentration gradient
  - **Diffusion** — Movement of solutes
    - **Simple diffusion** — No transport protein required
    - **Facilitated diffusion** — Transport protein required
      - **Channel-mediated** — Ion moves through channel
      - **Carrier-mediated** — Small polar molecule moved by carrier protein
  - **Osmosis** — Movement of water across semipermeable membrane
- **Active processes** — Require expenditure of cellular energy. Involve either the movement of a substance up its concentration gradient or the formation or loss of a vesicle
  - **Active transport** — Ion or small molecule moved against its concentration gradient
    - **Primary active transport** — Energy source from ATP
    - **Secondary active transport** — Energy source from movement of another substance
      - **Symport** — Two substances moved in same direction
      - **Antiport** — Two substances moved in opposite directions
  - **Vesicular transport** — Involves a vesicle
    - **Exocytosis** — Vesicle releasing its contents from a cell
    - **Endocytosis** — Vesicle is formed as material is brought into a cell
      - **Phagocytosis** — Cellular eating
      - **Pinocytosis** — Cellular drinking
      - **Receptor-mediated endocytosis** — Receptor required

Figure 4.7 Membrane Transport. Membrane transport is organized into passive processes and active processes depending upon the requirement for cellular energy. Passive processes do not require cellular energy, whereas active processes do.

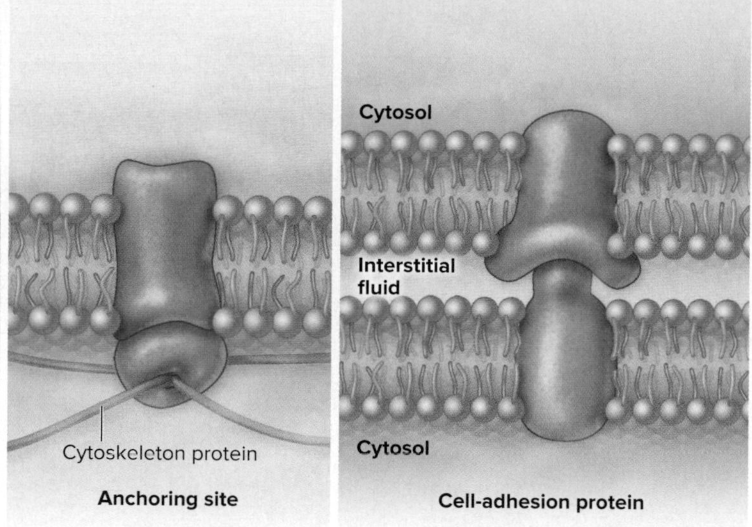

- **Identity markers** communicate to other cells that they belong to the body. Cells of the immune system use identity markers to distinguish normal, healthy cells from foreign, damaged, or infected cells that are to be destroyed (see section 22.4c).
- **Enzymes** may be attached to either the internal or the external surface of a cell for catalyzing chemical reactions (see section 3.3b).
- **Anchoring sites** secure the cytoskeleton (the internal, protein support of a cell) to the plasma membrane (see section 4.6b).
- **Cell-adhesion proteins** are for cell-to-cell attachments. Proteins that form membrane junctions perform a number of functions, including binding cells to one another (see section 22.3d).

One additional cellular feature of a cell's plasma membrane is a "coating of sugar" at the cell's external surface called the **glycocalyx** (glī′kō kā′liks; *kalyx* = husk). The carbohydrates of both glycolipids and glycoproteins that extend outward from the plasma membrane compose the glycocalyx (figure 4.5). The glycocalyx is unique to each cellular type and is important in cell-to-cell recognition (see sections 4.5a, 22.4c, and 29.2a).

WHAT DID YOU LEARN?

6 What type of plasma membrane protein provides the means for moving materials across the plasma membrane? What are three subtypes?

4.3 Membrane Transport

The plasma membrane has four primary functions (figure 4.5). It serves as the *physical barrier* between a cell and the fluid that surrounds it, called the interstitial fluid, as described in section 4.2. In addition, the plasma membrane is a *selectively permeable boundary* that regulates movement of materials into and out of a cell through membrane transport, *establishes and maintains electrochemical gradients* across the plasma membrane, and functions in *cell communication.* In this section we discuss its role in membrane transport. Its functions in establishing electrochemical gradients and in cell communication are described in sections 4.4 and 4.5, respectively. Refer to **figure 4.7**—an organizational flowchart of membrane transport processes—as you read through this section.

Substances (e.g., ions, molecules, water) move or are moved across the plasma membrane through several different processes collectively called **membrane transport.** These processes are organized into two major categories based upon the requirement for expending cellular energy. **Passive processes** do not require expenditure of cellular energy. Instead, these processes simply depend upon the kinetic energy (or random movement) of ions and molecules as each moves down its concentration gradient (i.e., from where there is more of it to where there is less). Diffusion (which involves the movement of *solutes* down their concentration gradient) and osmosis (which involves the movement of *water* across a semipermeable membrane down its concentration gradient) are the two major types of passive processes.

Active processes differ from passive processes in that they require expenditure of cellular energy. The cellular energy could be used for active transport, where a solute (ion or molecule) is moved up its concentration gradient (i.e., from where there is less of it to where there is more). The cellular energy could also be used for vesicular transport, which involves either a vesicle releasing its contents from a cell or a vesicle being formed as material is moved into a cell.

4.3a Passive Processes: Diffusion

LEARNING OBJECTIVES

11. Summarize the general concept of diffusion.
12. Distinguish between the cellular processes of simple diffusion and facilitated diffusion.

Molecules and ions have kinetic energy (see section 3.1a) due to their random, constant motion. They continuously move about, and when they strike obstacles such as other molecules and ions they bounce off, moving in a different direction. If a concentration gradient exists for a given type of ion or molecule, it becomes more evenly distributed over time. This net movement of a substance from where it is more concentrated to where it is less concentrated is called **diffusion** (di-fū′zhun; *diffundo* = to pour in different directions). Diffusion, if unopposed, occurs until the substance reaches **equilibrium** (i.e., the molecules or ions become evenly distributed throughout a given area or areas (**figure 4.8**).

The rate at which substances diffuse is not constant, but is dependent upon environmental conditions, including

- **The "steepness" of its concentration gradient.** Steepness of a concentration gradient is a measure of the difference in concentration of a substance between two areas. A steeper concentration gradient causes a faster rate of diffusion.
- **Temperature.** Temperature reflects the kinetic energy (or random movement) of a substance. When the temperature is higher, there is a greater random movement of the molecules and ions composing a substance, resulting in a faster rate of diffusion.

Cellular Diffusion

Several concepts are important in understanding diffusion involving a cell. (1) The distribution of most molecules and ions associated with

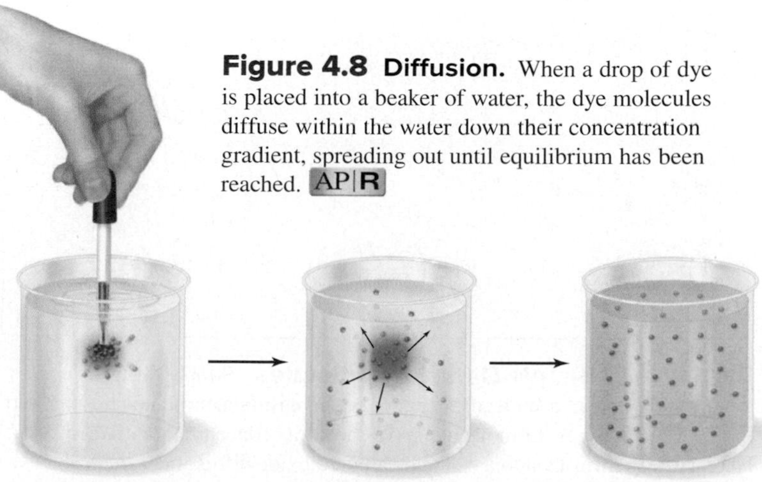

Figure 4.8 Diffusion. When a drop of dye is placed into a beaker of water, the dye molecules diffuse within the water down their concentration gradient, spreading out until equilibrium has been reached. AP|R

cells is not equal between the inside and the outside of a cell. Some substances generally have a greater concentration inside the cell than outside the cell (e.g., CO_2, K^+), whereas others have a greater concentration outside the cell than inside (e.g., O_2, Na^+). (2) The concentration gradient between the inside and the outside of the cell determines whether the solute diffuses into the cell or out of the cell. As noted, solutes always diffuse from an area where they are more concentrated to an area where they are less concentrated. (3) The chemical characteristics of the diffusing solute dictate whether it moves across the plasma membrane unassisted by the process of simple diffusion or it is assisted with a protein by the process of facilitated diffusion.

Simple Diffusion Molecules that are small and nonpolar move into or out of a cell down their concentration gradient by **simple diffusion.** These molecules *move unassisted* across the plasma membrane (i.e., they do not require a transport protein). Their chemical characteristics (i.e., small and nonpolar) allow these molecules to simply pass between the phospholipid molecules of the plasma membrane (**figure 4.9**). Molecules that move via simple diffusion include respiratory gases (O_2 and CO_2), small fatty acids, ethanol, and urea (a nitrogenous waste produced from amino acids).

The plasma membrane cannot regulate simple diffusion—rather, the movement of these molecules is dependent only upon the concentration gradient. Each type of molecule continues to diffuse across the plasma membrane as long as its concentration gradient exists. Impaired respiratory and cardiovascular function can alter the concentration gradients of oxygen and carbon dioxide, resulting in decreased diffusion of these gases (see sections 23.6b and c).

Facilitated Diffusion Small solutes that are charged ions or polar molecules are effectively blocked from passing through the plasma membrane by the nonpolar phospholipid bilayer. Their transport either into or out of the cell, down their concentration gradient, must be assisted by plasma membrane proteins in a process called **facilitated** (fa-sil′i-tā-ted) **diffusion.** Two types of facilitated diffusion—channel-mediated diffusion and carrier-mediated diffusion—are distinguished by the type of transport protein used to move the substance across the plasma membrane (see figure 4.7).

Channel-mediated diffusion is the movement of small ions across the plasma membrane through water-filled protein channels (**figure 4.10*a***). Each channel is typically specific for one type of ion. The channel is either a **leak channel,** which (as a general rule) is continuously open, or a **gated channel.** A gated channel is usually closed, opens only in response to a stimulus (e.g., chemical, light, voltage change), and then stays open for just a fraction of a second before it closes. For example, Na^+ leak channels allow Na^+ to pass through continuously. In contrast, chemically gated Na^+ channels open to allow Na^+ to move through the channel only when it temporarily opens in response to the presence of a particular chemical (e.g., neurotransmitter). Channel-mediated diffusion of ions is important in the normal function of both muscle cells and nerve cells.

Carrier-mediated diffusion is the movement of polar molecules (e.g., glucose or amino acids) across the plasma membrane. The relatively larger size of polar molecules (in comparison to ions) requires that their movement across the plasma membrane be assisted by a **carrier protein.** Three primary events take place in carrier-mediated diffusion: (1) The carrier protein within the plasma membrane binds the polar molecule, which (2) induces the carrier protein to change shape and move or carry the polar molecule to the other side of the plasma membrane, where (3) it is released. Like channels, a carrier moves a substance *down* its gradient; however, note that carrier-mediated diffusion involves a conformational change in the carrier protein for the transport of the

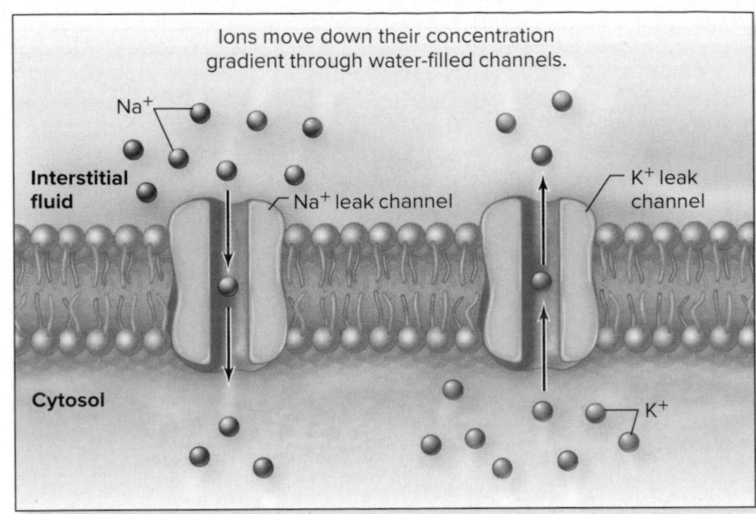

(a) Channel-mediated diffusion

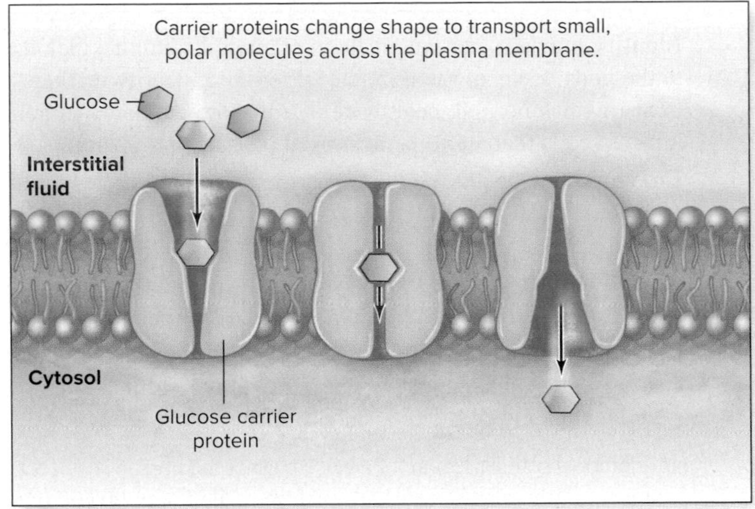

(b) Carrier-mediated diffusion

Figure 4.10 Facilitated Diffusion of Solutes. Facilitated diffusion occurs when ions or small, polar molecules are transported down their concentration gradient either through or by plasma membrane proteins. (*a*) Channel-mediated diffusion: Ions (e.g., Na^+, K^+) move through specific water-filled protein channels. (*b*) Carrier-mediated diffusion: Small, polar molecules (e.g., glucose) are transported by protein carriers. AP|R

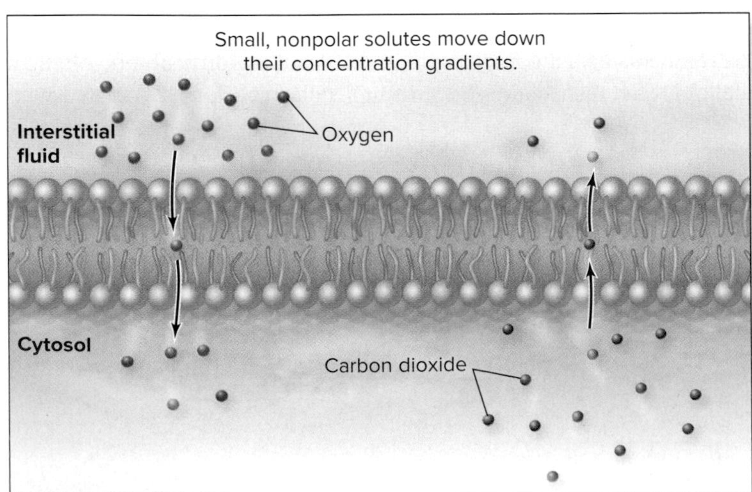

Figure 4.9 Simple Diffusion of Solutes. Simple diffusion occurs when small, nonpolar molecules pass between plasma membrane phospholipid molecules. Each type of molecule moves down its concentration gradient. Here, oxygen diffuses into a cell and carbon dioxide diffuses out of a cell.

molecule across the plasma membrane. Figure 4.10*b* depicts how a carrier protein binds the substance, changes shape, and then releases the substance on the other side of the membrane. A carrier that transports only one substance is called a **uniporter.** Glucose carriers are uniporters that normally prevent the loss of glucose in the urine (see section 24.6c).

The number of channels and carriers in a plasma membrane determines the maximum rate at which a substance can be transported. Thus, a cell can alter the transport rate of a given substance down its concentration gradient by changing the number of channel or carrier proteins in the plasma membrane. A greater rate occurs with increased numbers of these transport proteins, and a lesser rate with decreased numbers (see section 24.6b).

WHAT DID YOU LEARN?

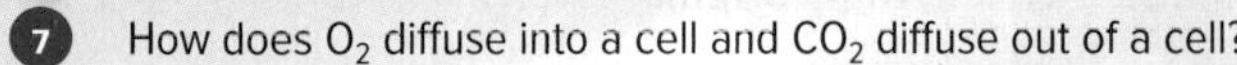

7 How does O_2 diffuse into a cell and CO_2 diffuse out of a cell?

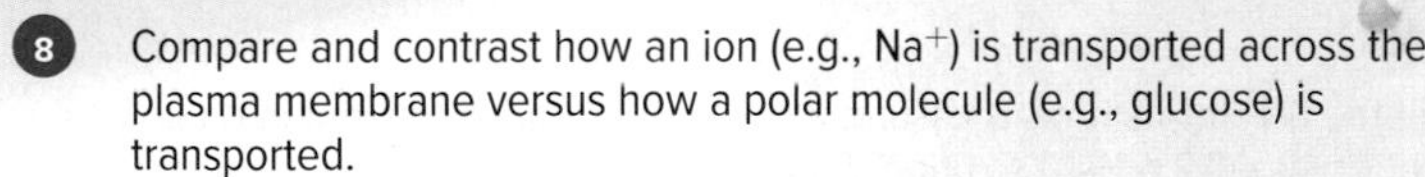

8 Compare and contrast how an ion (e.g., Na^+) is transported across the plasma membrane versus how a polar molecule (e.g., glucose) is transported.

4.3b Passive Processes: Osmosis

LEARNING OBJECTIVES

13. Define osmosis.

14. Define osmotic pressure.

15. Describe the relationship of osmosis and tonicity.

Osmosis is unlike the other types of passive membrane transport, because it involves water movement and does not involve the movement of solutes (see figure 4.7). **Osmosis** (os-mō′sis; *osmos* = a thrusting) is the passive movement of water through a semipermeable (or *selectively permeable*) membrane. This movement occurs in response to a difference in relative concentration of water on either side of a membrane. Please refer to **figure 4.11** as you read through this section.

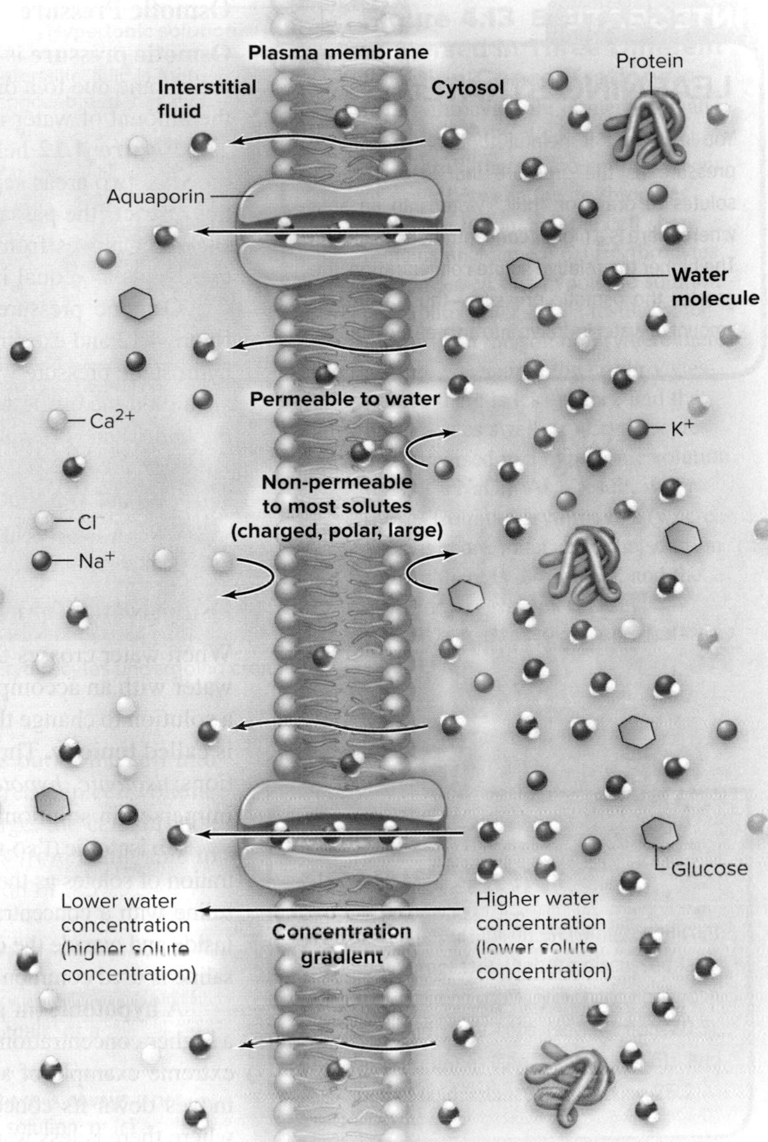

Figure 4.11 Osmosis in Cells. Osmosis occurs in cells across the plasma membrane, which is permeable to water but non-permeable to most solutes. Water always moves across the plasma membrane from an area of high water concentration to an area of low water concentration until equilibrium is reached. AP|R

Plasma Membrane: A Selectively Permeable Membrane

The plasma membrane is a **semipermeable membrane** that allows the passage of water, but its phospholipid bilayer prevents the movement of most solutes. A plasma membrane is also more specifically a **selectively permeable membrane** because the movement of most solutes is *regulated* (selectively) by this barrier.

Water molecules cross the plasma membrane in one of two ways: Either they "slip between" the molecules of the phospholipid bilayer (limited amounts) or they move through integral protein water channels called **aquaporins** (ak-kwă-pōr′in; *aqua* = water, *porus* = channel). Thus, cells can alter the amount of water that crosses the plasma membrane by changing the number of aquaporins.

The phospholipid bilayer of the plasma membrane is non-permeable to most solutes. In the context of osmosis, solutes are classified into two categories based upon whether their passage across the plasma membrane is prevented by the phospholipid bilayer. **Permeable** solutes (e.g., small and nonpolar solutes such as oxygen, carbon dioxide, and urea) pass through the bilayer, and **non-permeable** solutes (e.g., charged, polar, or large solutes such as ions, glucose, and proteins) are prevented from crossing the bilayer. (The term *solutes* in this discussion on osmosis will refer to non-permeable solutes.)

Concentration Gradients Across the Plasma Membrane

A difference in solute concentration can exist between the cytosol and the interstitial fluid because solutes are prevented from moving across the phospholipid bilayer of the plasma membrane. Note that when a solute concentration exists, a water concentration also exists. A solution with a greater concentration of solutes contains a lower concentration of water. For example, a solution containing 3% solutes has a lower water concentration (97% water) than a solution with 1% solutes (99% water). Note that solute percentage reflects the collective percentage of all of the solutes (e.g., glucose, proteins, Na^+).

Movement of Water into or Out of a Cell by Osmosis

The net movement of water by osmosis is dependent upon the concentration gradient between the cytosol and the solution in which the cell is immersed. For example, water moves down its concentration gradient from the solution containing 1% solutes (and 99% water) into the solution containing 3% solutes (and 97% water). Water continues to move until equilibrium is reached (the concentration of water in the cell equals the concentration of the surrounding fluid). Note that water moves toward the solution with the lower water concentration (stated another way, water moves toward the solution with the greater solute concentration).

Figure 4.11 shows water moving across the plasma membrane by osmosis from an area of high water concentration to an area of low water concentration.

with what was previously part of the plasma membrane. This newly formed vesicle typically fuses with a lysosome (a cellular organelle containing digestive enzymes, described in section 4.6a). The molecules composing the ingested material are broken down or digested by the enzymes within the lysosome. Only a few types of cells are able to perform phagocytosis. For example, it occurs regularly when a white blood cell engulfs and digests a microbe (e.g., bacterium; see section 22.3b).

Pinocytosis (pin′ō-sī-tō′sis or pī′nō-; *pineo* = to drink) is known as *cellular drinking*. This process occurs when multiple, small regions of the plasma membrane invaginate and multiple, small vesicles are formed as the cell internalizes interstitial fluid that contains dissolved solutes (figure 4.18*b*). This process is considered nonspecific because all solutes dissolved within the interstitial fluid are taken into the cell. Most cells perform this type of membrane transport.

Receptor-mediated endocytosis uses receptors on the plasma membrane to bind molecules within the interstitial fluid and bring the molecules into the cell. This enables the cell to obtain bulk quantities of certain substances, even though those substances may not be very concentrated in the interstitial fluid.

Receptor-mediated endocytosis begins when specific molecules in the interstitial fluid attach to their distinct integral membrane protein receptors in the plasma membrane to form a ligand-receptor complex (figure 4.18*c*). Following the binding of the ligand, the ligand-receptor complexes move laterally in the plane of the plasma membrane and accumulate at special membrane regions that contain **clathrin** protein on the internal surface of the membrane. The clathrin-coated regions of the plasma membrane housing the ligand-receptor complex folds inward to form an invagination called a *clathrin-coated pit*. This invagination deepens and pinches off, and the lipid bilayer of the plasma membrane fuses to form a *clathrin-coated vesicle,* which then moves into the cytosol. Following the formation of the clathrin-coated vesicles, the clathrin coat must be enzymatically removed before the vesicle may proceed to its intracellular destination. Again, it is the fusion of these lipid bilayers that requires the cell to expend energy. Following entry, receptors and ligands are uncoupled. Ligands may be stored, modified, or destroyed, and receptors (unless damaged) are returned to the plasma membrane.

The transport of cholesterol from the blood to a cell is an example of receptor-mediated endocytosis. When cholesterol is transported in the blood, it is bound to protein molecules in structures called low-density lipoproteins (or LDLs). LDLs move from the blood into the interstitial fluid and then bind to LDL receptors in the cell's plasma membrane. LDLs are then internalized by the process of receptor-mediated endocytosis just described (see section 27.6b).

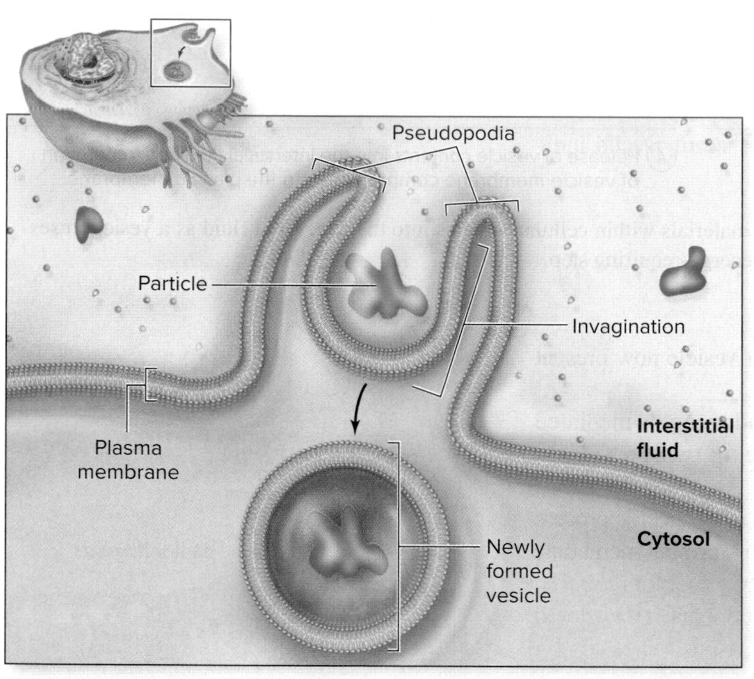

(a) Phagocytosis (cellular eating)

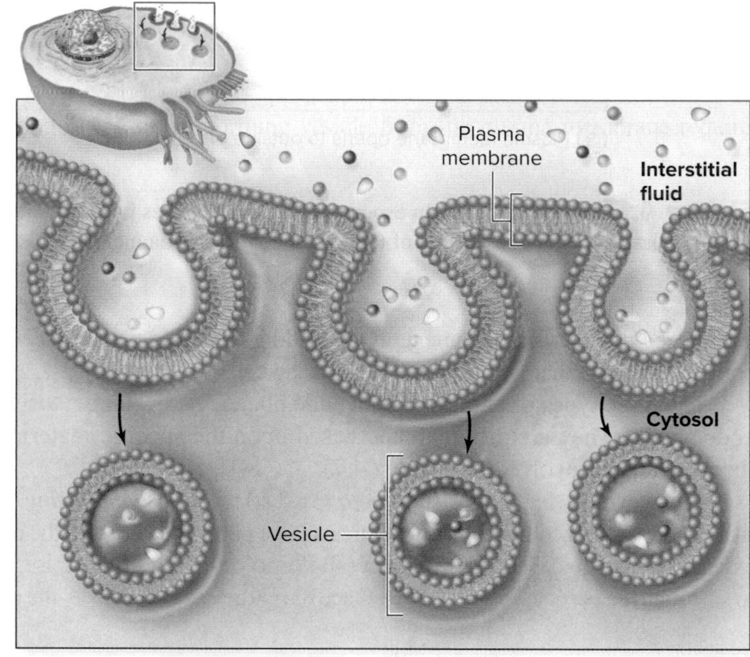

(b) Pinocytosis (cellular drinking)

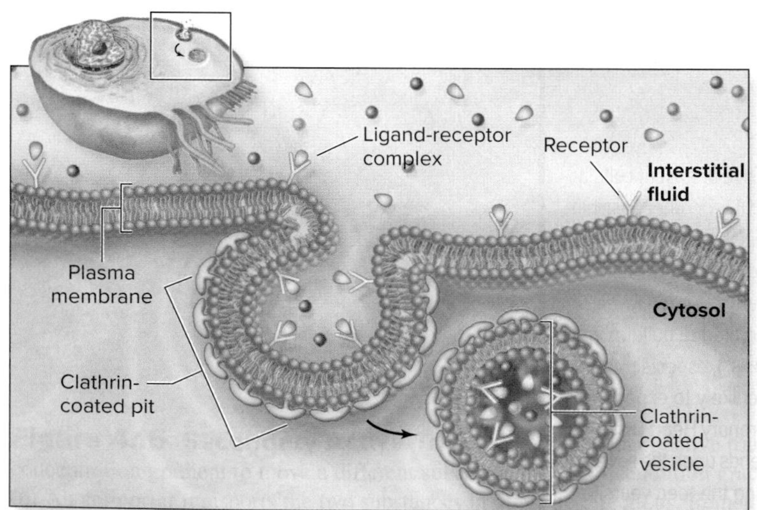

(c) Receptor-mediated endocytosis

Figure 4.18 Three Forms of Endocytosis. Endocytosis is the process whereby a vesicle is formed as the cell acquires materials from the interstitial fluid. (*a*) Phagocytosis occurs when plasma membrane extensions called pseudopodia engulf a relatively large particle and internalize it into a vesicle. (*b*) Pinocytosis is the incorporation of numerous droplets of interstitial fluid into the cell in small vesicles as many regions of the plasma membrane invaginate. (*c*) Receptor-mediated endocytosis occurs when specific molecules bind to receptors in the plasma membrane. These receptors with bound molecules then aggregate within the membrane, and are internalized when the membrane invaginates, forming a vesicle. The formation of the vesicle is the energy-requiring step.

The various types of membrane transport mechanisms are integrated in **figure 4.19**. Passive processes are depicted on the left and active processes on the right in this two-page summary figure. Note as you observe this figure that the passive processes of diffusion and osmosis (which allow molecules and ions to move down their concentration gradient) *facilitate* the reaching of equilibrium (equal distribution of ions and molecules) across the plasma membrane. In comparison, active transport (which moves molecules and ions up their concentration gradient) *opposes* the reaching of equilibrium across the plasma membrane. It is the active transport processes that maintains cellular concentration gradients at the plasma membrane, which are necessary for normal cellular function.

WHAT DID YOU LEARN?

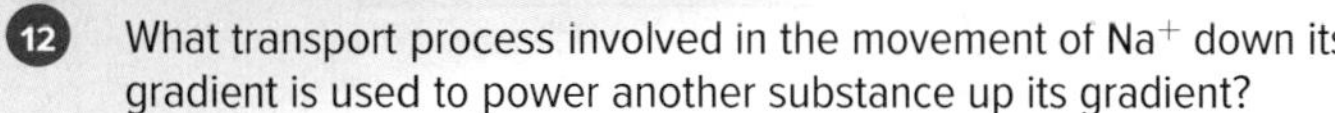

12 What transport process involved in the movement of Na^+ down its gradient is used to power another substance up its gradient?

13 Engulfing of a bacterium by a white blood cell occurs by what type of cellular transport?

4.4 Resting Membrane Potential

The plasma membrane also functions in establishing and maintaining an electrochemical gradient at the plasma membrane called the *resting membrane potential (RMP)*, which is essential in the normal function of both muscle cells (see section 10.2d) and nerve cells (see section 12.7b). We first define an RMP and then discuss how resting membrane potentials are established and maintained. Refer to **figure 4.20** as you read through this section.

4.4a Introduction

LEARNING OBJECTIVES

19. Define a resting membrane potential (RMP).

20. Describe the cellular conditions that are significant for establishing and maintaining a resting membrane potential.

Cells have an electrical charge difference at the plasma membrane. This electrical charge difference represents potential energy (see section 3.1a), and thus is appropriately called the **membrane potential.** The membrane potential when a cell is at rest is more specifically called the **resting membrane potential (RMP).** Two cellular conditions are significant in establishing and maintaining an RMP.

First, a cell has an unequal distribution of ions and charged molecules across the plasma membrane. The cytosol close to the plasma membrane contains relatively more K^+ than does the surrounding interstitial fluid that is close to the plasma membrane. In comparison, the interstitial fluid close to the plasma membrane contains relatively more Na^+ than the cytosol close to the plasma membrane. These relative distributions of K^+ (more inside) and Na^+ (more outside) are the result of the activity of Na^+/K^+ pumps (described in section 4.3c). In addition, the cytosol has negatively charged protein molecules, which are formed by protein synthesis (see section 4.8). Note that these negatively charged proteins are too large to pass through the plasma membrane.

Second, the relative amounts of positive and negative charges are not equally distributed at the plasma membrane. There is relatively more positive charge on the outside of a cell than on the inside of a cell. Thus, the inside of the cell is *relatively negative* compared to the outside of the cell near the plasma membrane. This difference in charge can be measured by electrodes, which are positioned with one just inside the cell and the other outside the cell. Cell types vary in the specific value of their resting membrane potential, which typically ranges between –50 millivolts (mV) and –100 mV. Nerve cells (i.e., neurons), for example, have a resting membrane potential of –70 mV (see section 12.7b).

INTEGRATE

LEARNING STRATEGY

To understand the concept of "relatively negative," consider the following simple arrangement. There are 100 positive charges (+) outside a membrane and 30 positive charges (+) inside a membrane—thus, the inside is less positive, or it is *relatively* negative to the outside.

WHAT DID YOU LEARN?

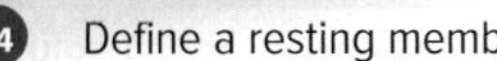

14 Define a resting membrane potential.

4.4b Establishing and Maintaining an RMP

LEARNING OBJECTIVES

21. Explain the role of both K^+ and Na^+ in establishing an RMP.

22. Discuss how Na^+/K^+ pumps are necessary in maintaining an RMP.

A resting membrane potential (RMP) is primarily a consequence of the relative movement of ions across the plasma membrane. The two most significant ions are K^+ and Na^+. The net movement of each of these is dependent upon both the number of its leak channels and the **electrochemical gradient,** which is the combination of the electrical gradient at the plasma membrane and the chemical concentration gradient of the specific ion. Here we first discuss the role of K^+ and the role of Na^+ in establishing an RMP. We then describe the role of Na^+/K^+ pumps.

Establishing an RMP

The Role of K^+ Potassium diffusion is the most important factor in establishing the specific value of the RMP. Movement of K^+ is dependent upon its electrochemical gradient. Potassium ions exit the cell through K^+ leak channels moving down their relatively steep chemical concentration gradient from the cytosol into the interstitial fluid. This loss of K^+ leaves relatively more negatively charged structures (e.g., proteins) inside the cell. They remain within the cell because they are too large to cross the plasma membrane. The movement of K^+ to the outside of a cell is, however, opposed by the electrical gradient. The positive charge on the outside of the cell repels the movement of K^+, and the negative charge on the inside of the cell attracts K^+. Thus, K^+ movement out of the cell is facilitated by its chemical concentration gradient but opposed by the electrical gradient. As additional K^+ diffuses out of the cell into the interstitial fluid, the inside becomes more negative. Consequently, the pull to keep K^+ within the cell is greater. At some point, the electrical gradient that opposes the movement of K^+ out of the cell becomes equal to the force of the chemical concentration gradient allowing K^+ out of a cell. Thus, K^+ movement has reached equilibrium. In a nerve cell, for example, if only K^+ leak channels were present, the loss of the K^+ from the nerve cell would result in an RMP with a specific value of –90 mV.

The Role of Na^+ Sodium diffusion into cells occurs simultaneously to the loss of K^+ from the cell, and it is dependent upon its electrochemical gradient. Sodium ions enter the cell through Na^+ leak channels moving down their chemical concentration gradient from the interstitial fluid into the cytosol. Sodium ions are also "pulled" into the cell by the electrical gradient. Both of these forces (the chemical gradient and the electrical gradient) facilitate the

Microfilaments (mī-krō-fil′ă-ment; *micros* = small) are the smallest components of the cytoskeleton, with a diameter of about 7 nanometers. They are composed of globular *actin* protein monomers that are organized into two thin, intertwined protein filaments (actin filaments) similar to two twisted pearl strands. Individual globular actin proteins are added to one end of the microfilament for growth in a particular direction and removed from the other end for shortening. They form an interlacing web or network of protein on the cytoplasmic side of the plasma membrane. This network of protein provides internal structural support of the plasma membrane, including the plasma membrane extensions called microvilli (described in section 4.6c). A contractile ring of these proteins separates the two cells formed during cytokinesis (a process of cell division described in section 4.9b). Microfilaments are also located throughout the cell to facilitate movement of organelles, vesicles, and molecules within the cell (through cytoplasmic streaming) and participate in muscle contraction (see section 10.3c).

Microtubules (mī-krō-tū′būl; *tubus* = tube) are the largest components of the cytoskeleton, with a diameter of about 25 nanometers. They are composed of globular *tubulin* protein monomers that are organized into hollow cylinders. Microtubules, like microfilaments, may be elongated or shortened as needed, by the addition or removal of tubulin monomers. Microtubules are arranged like railway tracks for directing the movement of organelles and vesicles within a cell (e.g., fast axonal transport in neurons; see section 12.2c). They also extend into the core of both cilia (see figure 4.1*b*) and flagella, where they slide past one another for the movement of these hairlike cellular projections (see section 4.6c). New microtubules are formed during cellular division (as spindle fibers) to separate chromosomes during mitosis (see section 4.9b).

Intermediate filaments are intermediate in size relative to the microfilaments and microtubules, with a diameter between 8 and 12 nanometers. These less flexible proteins extend across the inside of the cell and function as rigid rods to both support the cell and stabilize junctions between them (see desmosomes in section 4.6d). Their protein composition differs, depending upon the cells in which they occur. Keratin, a protein of the skin, hair, and nails, is an example of one type of intermediate filament (see section 6.1a); another type forms neurofilaments of nerve cells (see section 12.2b).

WHAT DID YOU LEARN?

20 Which non-membrane-bound organelle functions to (a) digest unwanted proteins, (b) form the structural support of cell, (c) synthesize proteins, and (d) participate in cell division?

4.6c Structures of the Cell's External Surface

LEARNING OBJECTIVES

29. Distinguish between cilia and flagella in terms of both structure and function.
30. Describe the structure and function of microvilli.

The structures that extend from the surface of some cells include cilia, flagella, and microvilli. Cilia and flagella are extensions of the plasma membrane involved in movement, whereas microvilli are structures that increase the surface area of the plasma membrane.

Cilia and Flagella

Cilia (sil′ē-ă; sing., sil′ē-ŭm; *cilium* = an eyelash) are small (5 μm to 10 μm in length and 0.2 μm in width), hairlike projections extending from the exposed surfaces of some cells. They contain supportive microtubule proteins (see section 4.6b) and are enclosed by the plasma membrane. Cilia are usually found in large numbers on the exposed surfaces of specific cells such as those that line portions of the respiratory passageways (see figure 4.1). The beating of these cilia moves mucus and any adherent substances along the cell surface toward the throat, where it may then be expelled from the respiratory system (see section 23.1c).

Flagella (flă-jel′ă; sing., *flagellum,* flă-jel′ŭm; a whip) are similar to cilia in basic structure—however, they are longer and wider (50 μm in length and about 0.5 μm in width), and when present, there is usually only one. The function of a flagellum is to help propel an entire cell. In humans, the only example of a cell with a flagellum is sperm (see figure 28.18), which moves through the female reproductive tract to reach the oocyte (developing egg). The movement of both cilia and flagella occurs through the microtubules within their core, a process that requires energy provided by the splitting of ATP molecules.

Microvilli

Microvilli are thin, microscopic membrane extensions of the plasma membrane. Microvilli are shorter and narrower than cilia (average about 1 μm high and 0.08 μm wide), are more densely packed together, and lack powered movement (**figure 4.31**). Each microvillus is supported by microfilaments (actin proteins are cross-linked into a dense bundle that serves as its structural core). Microvilli provide a more extensive plasma membrane surface area for more efficient membrane transport (see section 4.3). Just as not all cells have cilia, not all cells have microvilli. Cells with microvilli occur, for example, throughout the small intestine, where increased surface area is needed to absorb digested nutrients (see section 26.3b).

WHAT DID YOU LEARN?

21 Which cellular surface structure functions in (a) increasing the cell's surface area and (b) moving material past the cell?

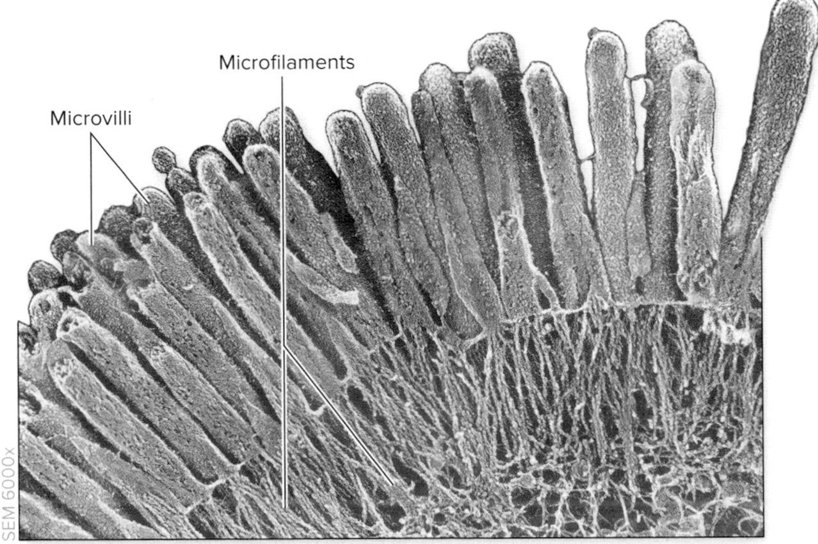

Figure 4.31 Microvilli. Microvilli are thin, microscopic projections. They extend from the surface of the plasma membrane and are supported by microfilaments. Microvilli function to increase the surface area of the plasma membrane for more efficient membrane transport. AP|R

©Don W. Fawcett/Science Source

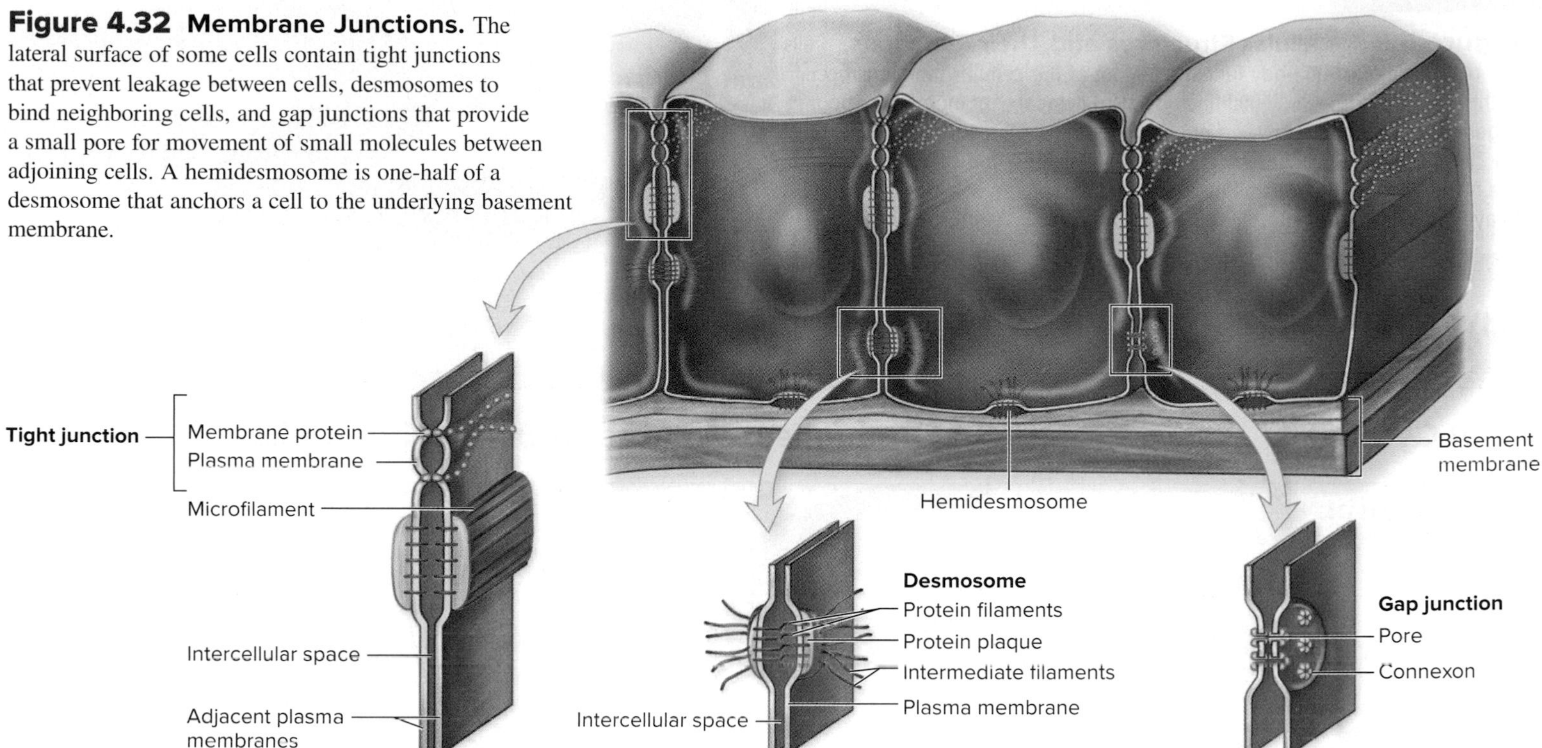

Figure 4.32 Membrane Junctions. The lateral surface of some cells contain tight junctions that prevent leakage between cells, desmosomes to bind neighboring cells, and gap junctions that provide a small pore for movement of small molecules between adjoining cells. A hemidesmosome is one-half of a desmosome that anchors a cell to the underlying basement membrane.

4.6d Membrane Junctions

LEARNING OBJECTIVE

31. Compare and contrast the structure and function of the three major types of membrane junctions.

Membrane junctions are composed of both integral and peripheral membrane proteins (see section 4.2b), which function to connect and support cells. Most of our cells are arranged into structural units called *tissues* that act together in a common function (see section 1.4b). To provide an orderly arrangement between some cells and coordinate their interactions, membrane junctions are located between adjacent cells. There are three major types of membrane junctions: tight junctions, desmosomes, and gap junctions (**figure 4.32**).

Tight Junctions

A **tight junction** (or *zonula occludens*) is composed of plasma membrane proteins that form strands or rows of proteins (e.g., claudins and occludins). These cell membrane junctions are positioned at the apical surfaces around the circumference of each of the adjacent cells. Tight junctions function as spot welds to seal off the intercellular space and prevent substances from passing between the cells; this requires all materials to move through, rather than between, the cells. For example, tight junctions associated with epithelial cells that form the lining of the small intestine prevent corrosive digestive enzymes that are within the lumen of the intestine from moving between cells and damaging other body structures (see section 26.3b). These junctions also prevent leakage of urine through the urinary bladder wall (see section 24.8b).

Desmosomes

A **desmosome** (dez′mō-sōm; *desmos* = a band), also called *macula adherens* ("adhering spot"), is composed of several different proteins that bind neighboring cells. A thickened structure called a protein plaque is located on the internal surface of the plasma membrane of adjoining cells. Numerous protein filaments (e.g., desmogleins) extend from the plaque through the plasma membrane of both neighboring cells. These filamentous proteins extend across the small space between the two cells, where they are anchored to one another. Also anchored to each protein plaque are intermediate filaments of the cytoskeleton that extend from the protein plaque throughout the cell to provide support and strength. Observe in figure 4.32 the pattern of these proteins forming a desmosome between two adjoining cells: intermediate filaments, protein plaque, adjoining protein filaments, protein plaque, and intermediate filaments. This anatomic arrangement provides structural integrity to cells that are exposed to stress, such as the external layer of the skin (see section 6.1a) and cardiac muscle (see section 19.3f). **Hemidesmosomes** (i.e., half of a desmosome) anchor the basal layer (attached surface) of cells of the epidermis to the underlying basement membrane (see section 6.1a).

Gap Junctions

A **gap junction** is composed of six integral plasma membrane proteins, called **connexons** (kon-neks′onz), that form a tiny, fluid-filled tunnel or pore that extends across a small gap (about 2 nanometers) between adjacent cells. Gap junctions provide a direct passageway for substances to move between neighboring cells. Ions, glucose, amino acids, and other small solutes can pass directly from the cytosol of one cell into the neighboring cell through these pores. The flow of ions between cells allows spread of electrical activity in cardiac muscle of the heart (see section 19.3f).

Cellular structures and their associated functions are integrated in **figure 4.33**. This figure includes the three primary components of a cell: the cytoplasm, the nucleus, and the plasma membrane.

WHAT DID YOU LEARN?

22 Which cellular junction (a) provides resistance to mechanical stress, (b) allows the passage of ions between cells, and (c) prevents leakage between cells?

4.7 Structure of the Nucleus

The nucleus is the largest structure in the cell, typically averaging about 5 μm to 7 μm in diameter. It is often called the cell's control center (**figure 4.34**). A cell typically has one nucleus. However, erythrocytes contain no nucleus, and skeletal muscle cells have many nuclei. The shape of the nucleus generally mirrors the shape of a cell. For example, a cuboidal cell has a spherical nucleus in the center of the cell, whereas a thin, flattened cell has a nucleus elongated in the same direction as

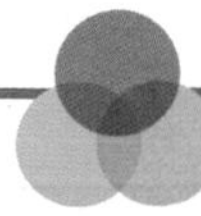

INTEGRATE

CONCEPT CONNECTION

During the initial stages of development, all cells have the ability to reproduce. However, during development as cells change into their specific types (a process called **differentiation**), some cells retain the ability to replicate, whereas others decrease or lose the ability. Damaged or diseased cells are not replaced, and this results in a lack of function. For example, epithelial skin cells are replaced frequently. If you cut your finger, the cells go through cell division, replacing the damaged cells of the skin (see section 6.4). However, other cells, such as cardiac muscle cells, undergo cell division infrequently or not at all. Thus, if an individual has a heart attack, and cardiac muscle cells die from lack of oxygen, these cells are not replaced or are replaced to a limited extent (see section 19.4a).

During the **S phase** (also called *synthesis*), the 46 double helix strands of DNA are replicated. Forming DNA requires large numbers of deoxyribonucleotides and the enzyme **DNA polymerase.** All of these components are located in the nucleoplasm within the nucleus.

The following steps in DNA replication (**figure 4.41**) involve unwinding, breaking, assembly, and restoration:

1. **Unwinding of DNA molecule.** The spiral, complementary DNA strands are unwound from each other by specific enzymes.
2. **Breaking the parent strands apart.** The hydrogen bonds holding the complementary bases together in the DNA strands are broken. Once the portions of strands are separated, binding proteins (not shown) ensure that the strands remain separated.
3. **Assembly of new DNA strands.** Both strands of DNA are read as templates by DNA polymerase enzymes that move along both parental strands (one called the *leading strand* and the other the *lagging strand* because of how transcription takes place on each). DNA polymerase assembles new strands of DNA as complementary deoxyribonucleotides are paired. For example, if the base sequence of a small portion of a DNA strand is TTAGCTAGC, then the base sequence of the newly formed complementary DNA strand assembled by DNA polymerase would be AATCGATCG. Complementary base pairs are held together by hydrogen bonds. The bond between nucleotides in the DNA polymer is a phosphodiester bond.
4. **Restoration of DNA double helix.** The DNA double strands are returned to their coiled, helix structure.

The process continues until the entire length of both strands of DNA are replicated. The replicated DNA strands, now called **sister chromatids,** remain attached at a region called a **centromere** (sen′trō-mēr; *kentron* = center, *meros* = part). The joined sister chromatids form a chromosome. The sister chromatids are separated at the centromeres during mitosis; following their separation, each is called a chromosome.

WHAT DO YOU THINK?

4 Describe the difference in DNA replication and transcription in terms of (a) the type of nucleic acid formed and (b) the amount of DNA copied.

The last part of interphase, called the **G_2 phase** (or the *second gap phase*), is brief (figure 4.40). During this phase, centriole replication is completed (having produced two centriole pairs present within the cell) and enzymes and other structures needed for cell division are synthesized.

INTEGRATE

LEARNING STRATEGY

To prevent confusion between the process of DNA replication and transcription (formation of RNA from DNA), remember that transcription of mRNA is analogous to copying *one* recipe from a cookbook. The recipe is written as RNA. In contrast, DNA replication is analogous to printing a replica of an *entire* cookbook. The cookbook is printed as DNA.

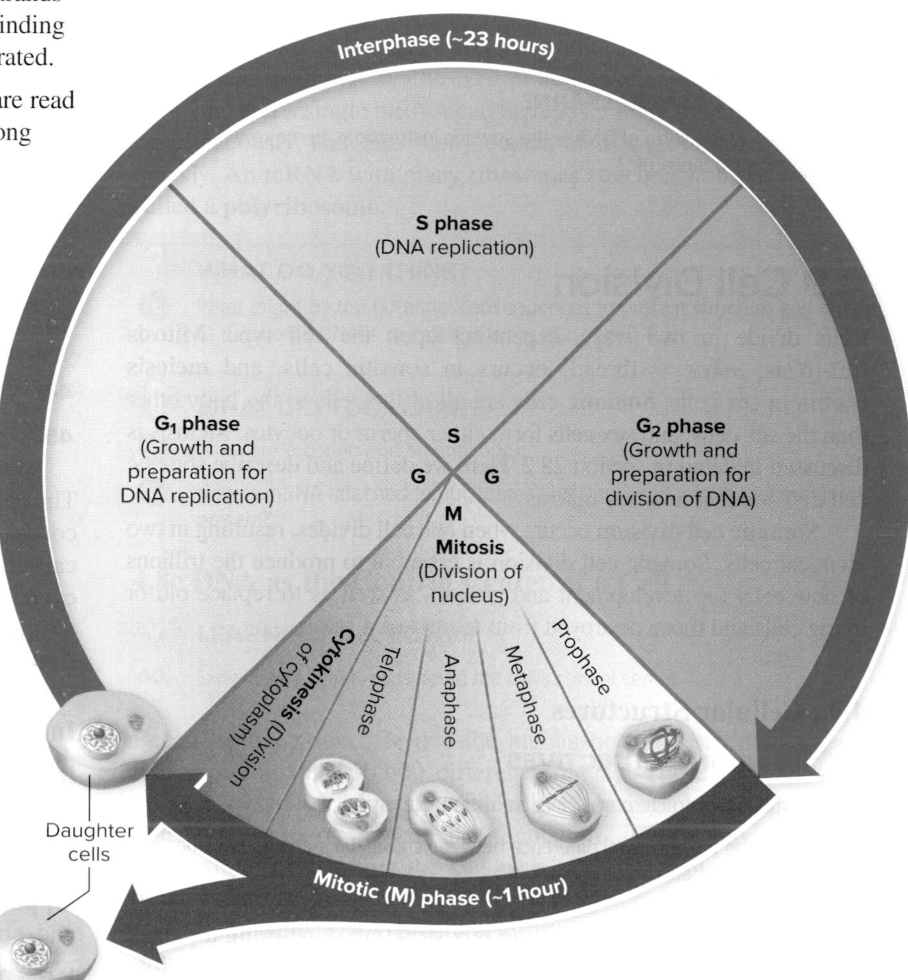

Figure 4.40 The Cell Cycle. The cell cycle includes two basic phases: interphase and the mitotic (M) phase. Interphase is a growth period that is subdivided into G_1, S, and G_2. Synthesis of DNA occurs during the S phase of interphase. The mitotic phase includes mitosis, the process of nuclear division, and cytokinesis, the division of the cytoplasm. Note that this figure does not accurately reflect the relative time a cell would spend in each phase. AP|R

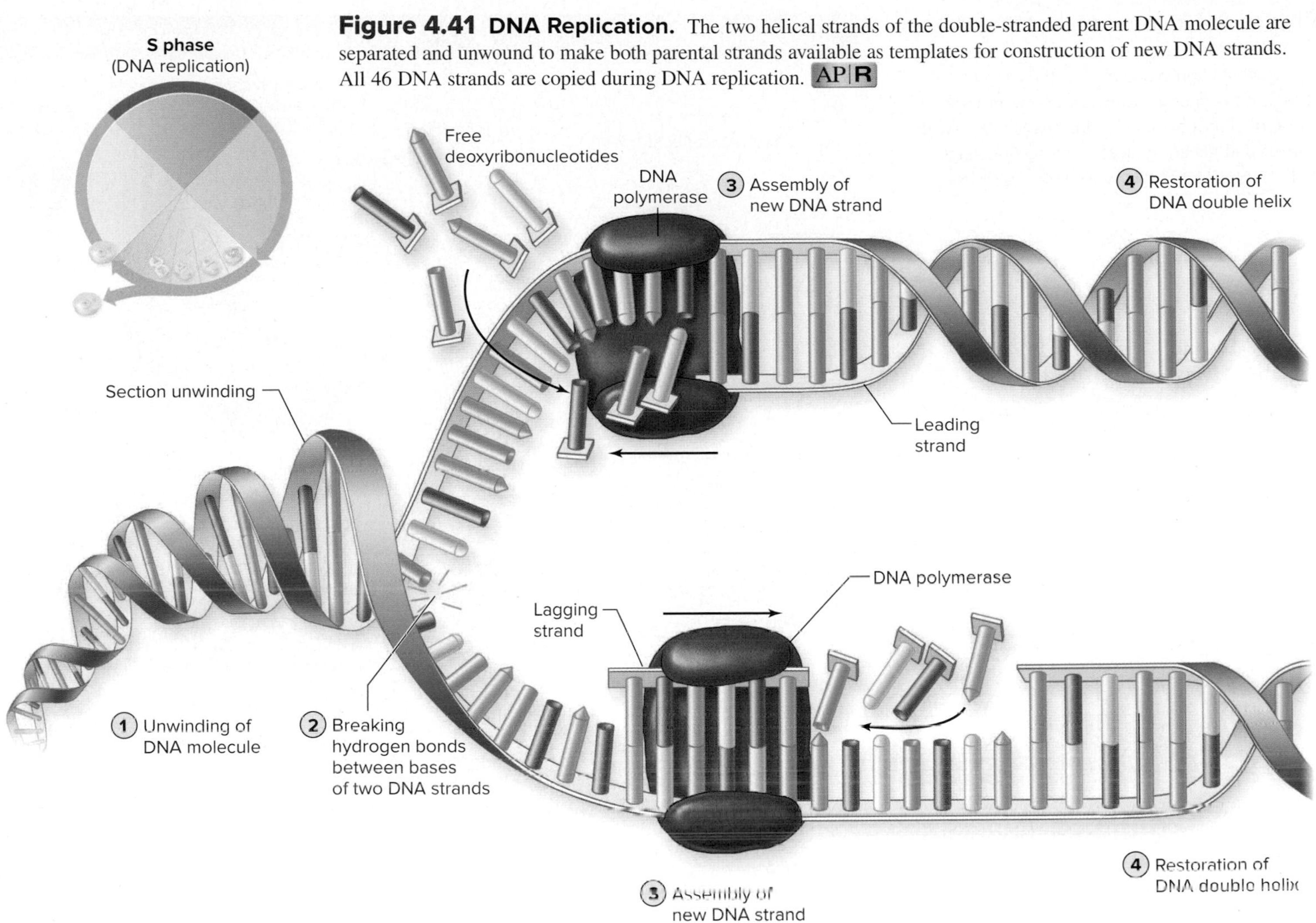

Figure 4.41 DNA Replication. The two helical strands of the double-stranded parent DNA molecule are separated and unwound to make both parental strands available as templates for construction of new DNA strands. All 46 DNA strands are copied during DNA replication. AP|R

M Phase (Mitotic Phase)

Following interphase, cells enter the M (mitotic) phase. Two distinct events occur in this phase to produce two new cells: mitosis, which is division of the nucleus, and cytokinesis, which is the division of the cytoplasm. **Mitosis** begins first with cytokinesis starting and overlapping with later stages of mitosis.

Four consecutive phases take place during mitosis: prophase, metaphase, anaphase, and telophase, which can be remembered with the acronym P-MAT. Each phase merges smoothly into the next in a nonstop process. The events of each stage are summarized in **figure 4.42**.

Prophase is the first stage of mitosis (step b). Chromatin becomes supercoiled into the chromosomes that are more maneuverable and are less likely to become tangled during cell division. The DNA and protein within the chromatin coil, wrap, and twist, forming the chromosomes. Chromosomes are composed of the two sister chromatids, which resemble relatively short, thick rods, and chromosomes become noticeable with a light microscope during the prophase stage as dark-staining structures within the nucleus.

In addition, the nucleolus breaks down and disappears. Elongated microtubules (see figure 4.30) called *spindle fibers* begin to grow from the centrioles. The two centriole pairs are pushed apart by the elongating microtubules composing the spindle fibers; eventually, the centriole pairs come to lie at opposite poles (ends) of the cell. The end of prophase is marked by the dissolution (disassembly) of the nuclear envelope. This permits the chromosomes to be moved by spindle fibers through the cytoplasm during the next stages of mitosis.

Metaphase is the second stage of mitosis (step c), during which the chromosomes are aligned along an imaginary line in the middle of the cell (region called the equatorial plate). This alignment occurs through the growth of spindle fibers from each centriole toward the chromosomes. Some fibers attach to the centromere of each chromosome, directing their movement to the equatorial plate.

Anaphase, which is the third stage of mitosis, initiates as the spindle fibers cause the sister chromatids to be moved apart toward the cell's poles; the centromere leads the way, and its "arms" trail behind (step d). Each chromatid is now a chromosome composed of one DNA double helix with its own centromere.

Telophase begins with the arrival of a group of chromosomes at each cell pole. Essentially, the processes of prophase are reversed in telophase. The chromosomes begin to uncoil and return to the form of dispersed threads of chromatin, each new nucleus forms a nucleolus, the mitotic spindle breaks up and disappears, and a new nuclear envelope forms around each set of chromosomes. Telophase signals the end of nuclear division.

Cytokinesis **Cytokinesis** (sī′tō-ki-nē′sis; *kinesis* = movement) is the other major event in the mitotic phase, and it is the division of the cytoplasm between the two newly forming cells. This phase usually overlaps with anaphase and telophase of mitosis. A ring of microfilament proteins (see figure 4.30) on the inner surface of the cell's

Figure 4.42 Interphase, Mitosis, and Cytokinesis. Drawings and micrographs depict what happens inside a cell during the stages of (*a*) interphase and (*b–e*) the mitotic phase of cell division. Cytokinesis overlaps with mitosis and generally begins during anaphase.
(*a,e*) ©Michael Abbey/Science Source; (*b,c,d*) ©Ed Reschke/Getty Images

Interphase (G_1, S, G_2) | **Mitotic Phase (Mitosis and Cytokinesis)**

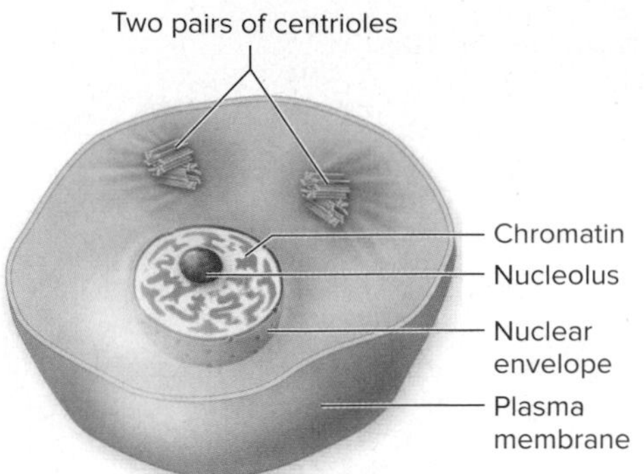

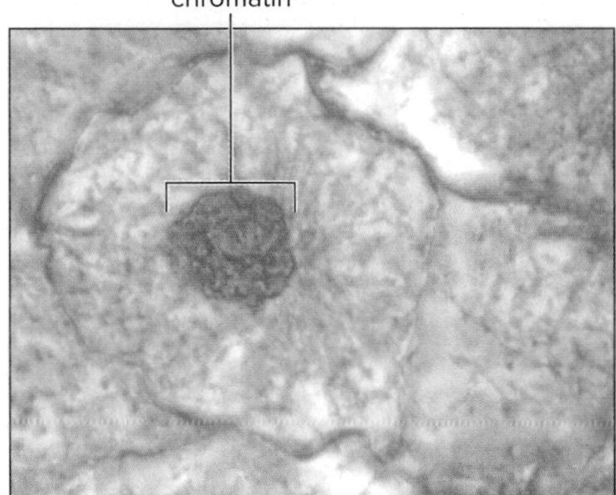

(a) Interphase

Synthesis of cellular components needed for cell division, including synthesis of DNA, which occurs during the S phase, and duplication of centrioles, which occurs throughout interphase

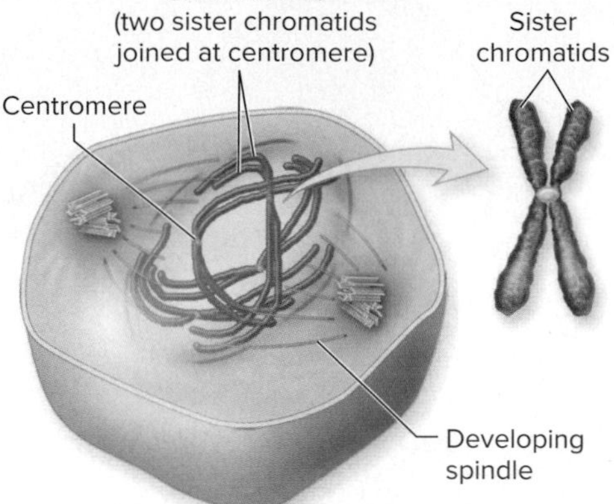

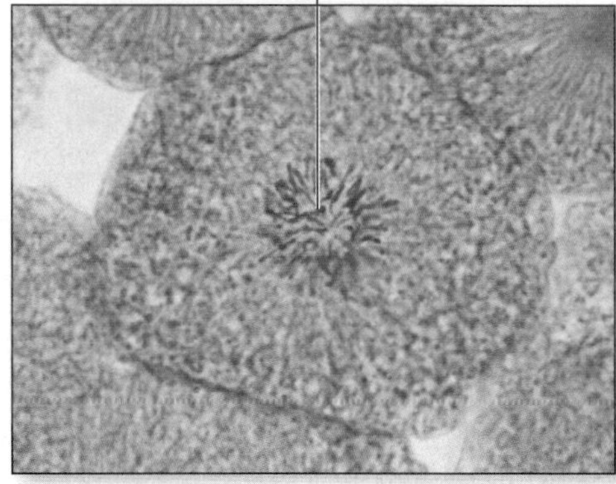

(b) Prophase

Chromosomes appear due to coiling of chromatin.
Nucleolus breaks down.
Spindle fibers begin to form from centrioles.
Centrioles move toward opposing cell poles.
Nuclear envelope breaks down at the end of this stage.

plasma membrane contracts at the cell's equator. It pinches the mother cell into two separate cells in a manner analogous to the tightening of a belt. The resulting **cleavage furrow** that appears indicates where the cytoplasm is dividing (figure 4.42*e*). Two new daughter cells are formed and cell division is complete.

INTEGRATE

LEARNING STRATEGY

These tips should help you remember the major events of each phase of mitosis:

- The **p** in **p**rophase stands for the **puffy** ball of chromosomes that forms within the nucleus.
- The **m** in **m**etaphase stands for **middle:** During this phase, the chromosomes align along the **middle** of the cell.
- The **a** in **a**naphase stands for **apart:** During this phase, the sister chromatids are pulled **apart.**
- The **t** in **t**elophase stands for **two:** During this phase, **two** new cells become noticeable as a cleavage furrow deepens to divide the cytoplasm.

WHAT DID YOU LEARN?

31 Describe the process of DNA replication that occurs during the S phase of interphase.

32 What are the events that occur during the mitotic phase (mitosis and cytokinesis)? Explain each.

4.10 Cell Aging and Death

LEARNING OBJECTIVES

46. Define apoptosis.

47. List the actions that occur in a cell during apoptosis.

Aging is a normal, continuous process that often exhibits obvious body signs. In contrast, changes within cells at the molecular level due to aging are neither obvious nor well understood. The reduced metabolic functions of normal cells often have wide-ranging effects throughout the body, including a reduced ability of cells to maintain homeostasis. These signs of aging reflect a lower number of normally functioning body cells, and may even suggest abnormal functions of some remaining cells.

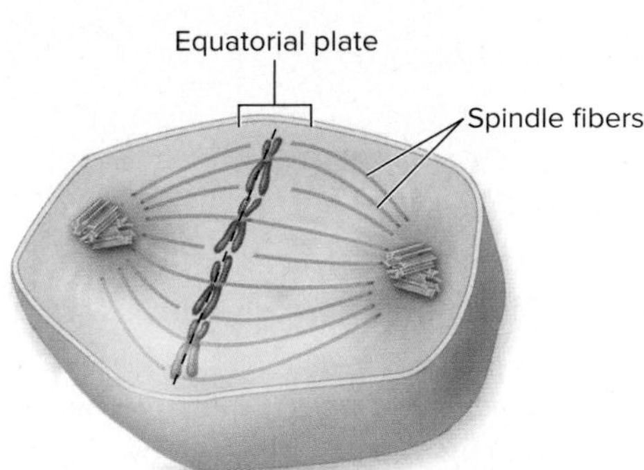

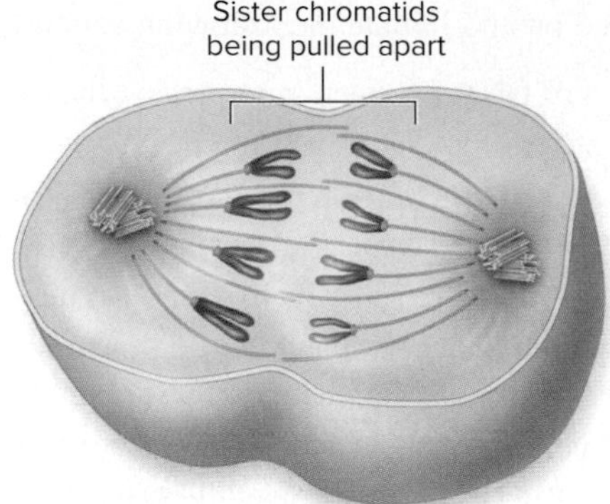

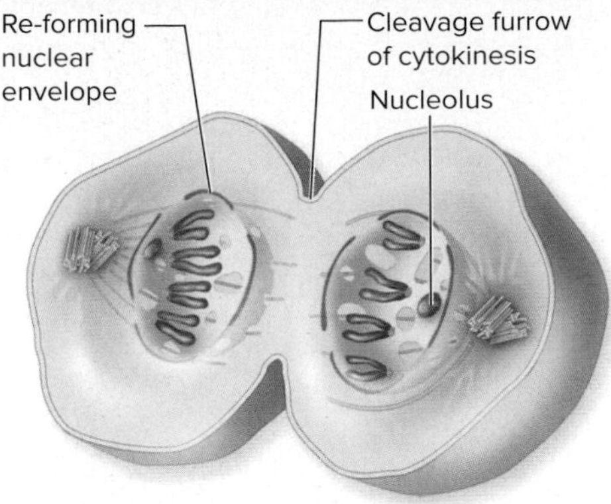

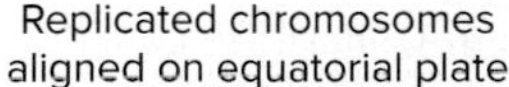

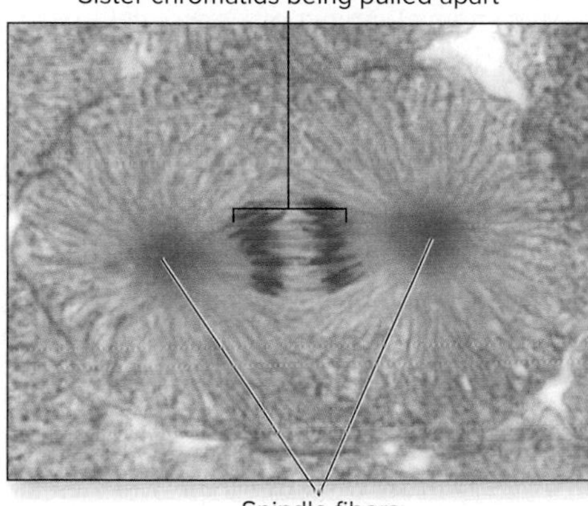

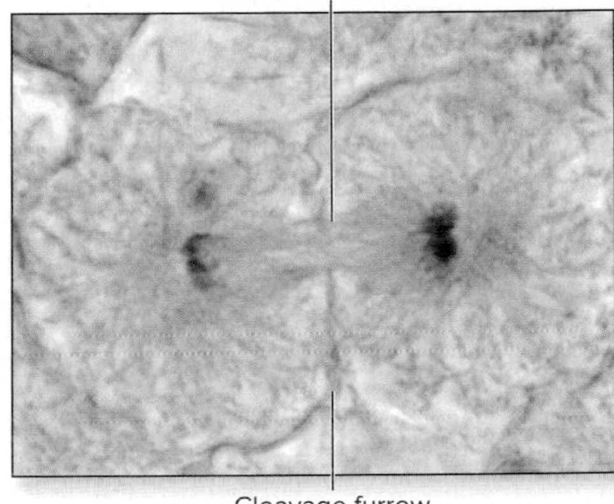

(c) Metaphase

Spindle fibers attach to the centromeres of the chromosomes extending from the centrioles.

Chromosomes are aligned at the equatorial plate of the cell by spindle fibers.

(d) Anaphase

Sister chromatids are separated by spindle fibers and moved toward opposite ends of the cell.

During this process centromeres that held sister chromatids together separate; each sister chromatid is now a chromosome with its own centromere.

Cytokinesis begins.

(e) Telophase

Chromosomes uncoil to form chromatin.

A nucleolus re-forms within each nucleus.

Spindle fibers break up and disappear.

New nuclear envelope forms around each set of chromosomes.

Cytokinesis continues as cleavage furrow deepens.

INTEGRATE

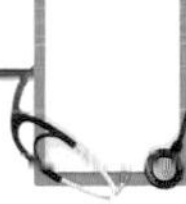

CLINICAL VIEW 4.3

Tumors

Normally, many regulatory mechanisms signal a cell when it should divide and when to stop dividing. **Tumors** arise when cells either proceed through the cell cycle without a start signal or fail to respond to signals that normally stop cell division. The tumor may, due to its size, interfere with the function of the normal surrounding cells. A cancerous tumor is invasive, and cells may enter the blood or lymph and metastasize to other areas of the body and establish secondary tumors.

Various types of cancer are discussed in later chapters of this text (e.g., Table 6.8: Skin Cancer; Clinical View 12.2: "Tumors of the Central Nervous System"; Clinical View 18.5: "Leukemia"; Clinical View 21.1: "Metastasis"; Clinical View 21.3: "Lymphoma"; Clinical View 23.9: "Lung Cancer"; Clinical View 28.2: "Ovarian Cancer"; Clinical View 28.5: "Cervical Cancer"; Clinical View 28.6: "Breast Cancer"; Clinical View 28.10: "Benign Prostate Hyperplasia and Prostate Cancer").

Cells affected by aging may exhibit alteration in either the structure or number of specific organelles. For example, if mitochondrial function begins to fail, the cell's ability to synthesize ATP diminishes. Additionally, changes in the distribution and structure of the chromatin and chromosomes within the nucleus may occur. Often, both chromatin and chromosomes clump, shrink, or fragment as a result of repeated divisions.

Essentially, cells die by one of two mechanisms: (1) they are killed by harmful agents or mechanical damage or (2) they are induced to commit suicide, a process of programmed cell death called **apoptosis** (ap′op-tō′sis; *apo* = off, *ptosis* = a falling).

Apoptosis occurs in orderly, well-defined, continuous, degradative steps to destroy and remove cellular components and eventually cell remnants. This biochemical mechanism is initiated by ligand-receptor signaling. Upon binding of ligand to a receptor, inactive, self-destructive enzymes within the cytosol are turned on and initiate the following actions:

- Destruction of DNA polymerase to prevent the synthesis of new DNA
- Digestion of the DNA into small fragments
- Digestion of the cytoskeleton, thus destroying structural support for organelles and the nucleus—the cell appears to shrink and become rounded and the nuclear shape changes
- Formation of small, irregular blebs (bubbles) on the plasma membrane surface
- Condensation of the cytosol and destruction of organelles—specifically, the mitochondria to deprive the cell of ATP needed for energy-requiring processes
- Release of proteins within mitochondria, activating specific digestive enzymes (caspase proteaseses) in cytosol; these enzymes digest cellular structures and signal for engulfment of the cell by phagocytosis

Programmed cell death occurs both to promote proper development and to remove harmful cells. For example, the proper development of fingers and toes begins with the formation of a paddlelike structure at the distal end of the developing limb. Programmed cell death removes the cells and tissues between the fingers and toes developing within this paddle structure.

Programmed cell death sometimes destroys harmful cells, reducing potential health threats. Cells of our immune system promote programmed cell death in some virus-infected cells to reduce the further spread of infection (see section 22.3c). Cells that have damage to their DNA often appear to promote events leading to apoptosis, presumably to prevent these cells from causing developmental defects or becoming cancerous. Some cancer therapy treatments lead to apoptosis in certain types of cancer cells.

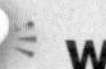

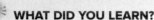

WHAT DID YOU LEARN?

33 What are the specific changes that occur to DNA during apoptosis?

CHAPTER SUMMARY

	• Cells are the structural and functional units of the body.
4.1 Introduction to Cells	• Cells vary in size and shape, but they have certain common features and functions.
	4.1a How Cells Are Studied • Cells are microscopic and can be studied using a light microscope (LM), scanning electron microscope (SEM), and transmission electron microscope (TEM).
	4.1b Cell Size and Shape • Although some cells are round or cubelike, other cells are flat, cylindrical, oval, or quite irregular in shape.
	4.1c Common Features and General Functions • The three major structural components of a cell include the nucleus, plasma membrane, and cytoplasm (composed of cytosol, organelles, and perhaps cell inclusions). • All cells must maintain their integrity and shape, obtain nutrients and form chemical building blocks, dispose of wastes, and if possible, replace cells.
4.2 Chemical Structure of the Plasma Membrane	• The plasma membrane is a fluid matrix that has approximately an equal mixture of lipids and proteins by weight.
	4.2a Lipid Components • The plasma membrane is composed of a bilayer of phospholipids with embedded cholesterol molecules. Glycolipids are lipids with carbohydrates extending from the outer surface of the cell.
	4.2b Membrane Proteins • Plasma membrane proteins are integral proteins that extend through the plasma membrane, whereas peripheral proteins reside on either the internal or external surface of the plasma membrane. • Functionally, the plasma membrane proteins include several types of transport proteins, receptors, identity markers, enzymes, attachment sites for the cytoskeleton, and cell-adhesion proteins.
4.3 Membrane Transport	• Substances are moved into and out of cells by processes of membrane transport, and are organized into passive processes and active processes. Passive processes do not require the expenditure of cellular energy and active processes do.
	4.3a Passive Processes: Diffusion • Diffusion is the movement of a solute from an area where it is more concentrated to an area where it is less concentrated. • Simple diffusion is the unassisted movement of small, nonpolar molecules through the phospholipid bilayer. • Channel-mediated facilitated diffusion is the transport of ions through channels that either are always open (leak channels) or open and close as a result of a stimulus (gated channels). • Carrier-mediated facilitated diffusion is the transport of polar molecules through a carrier that is induced to change shape to move the molecules across the plasma membrane.
	4.3b Passive Processes: Osmosis • Osmosis is the passive movement of water across a semipermeable membrane down a water concentration gradient. • Osmotic pressure is the pressure exerted by the movement of water across a semipermeable membrane due to a difference in solution concentration; the greater the difference, the greater the osmotic pressure. • The terms *isotonic*, *hypotonic*, and *hypertonic* describe the relative concentration of solutions.
	4.3c Active Processes • Active processes require the expenditure of cellular energy and include both active transport and vesicular transport. • The two types of active transport are primary active transport, which obtains its energy directly from ATP, and secondary active transport, which is "powered" by the movement of a second substance (usually sodium ion) down its concentration gradient. • Vesicular transport occurs through energy-requiring processes that involve a vesicle for transporting large materials or relatively large amounts of a substance either out of a cell or into a cell. • Exocytosis moves material out of a cell, and endocytosis moves substances into a cell. • The three types of endocytosis are phagocytosis, pinocytosis, and receptor-mediated endocytosis.
4.4 Resting Membrane Potential	• The plasma membrane functions in establishing and maintaining the resting membrane potential (RMP).
	4.4a Introduction • The RMP is the electrical charge difference at the plasma membrane when a cell is at rest; it typically ranges between –50 millivolts (mV) and –100 mV.
	4.4b Establishing and Maintaining an RMP • K^+ leaks out of a cell through K^+ leak channels, and Na^+ leaks into a cell through Na^+ leak channels. This movement of ions is primarily responsible for establishing an RMP. The Na^+/K^+ pumps maintain ion gradients following movement of these ions.

(continued on next page)

2. A young man in his 20s has a heart attack and is rushed to the hospital. Blood is drawn, and his cholesterol level is tested and found to be very high. The doctor tells him that he has a genetic condition in which he is unable to effectively remove LDL particles containing cholesterol from his blood and into his cells. Which cellular process is not functioning normally?
    a. channel-mediated facilitated diffusion
    b. receptor-mediated endocytosis
    c. exocytosis
    d. simple diffusion
3. Tumors involve a malfunction in this cellular process.
    a. transcription
    b. translation
    c. phagocytosis
    d. mitosis
4. A rare genetic disease involves the inability of the cells to respond to testosterone. The cells are lacking
    a. enzymes for testosterone.
    b. protein carriers for testosterone.
    c. receptors for testosterone.
    d. all plasma membrane proteins.
5. The hormone insulin is a protein composed of repeating units of amino acids. It is produced through the process(es) of
    a. transcription and translation.
    b. DNA replication.
    c. mitosis.
    d. differentiation.

Can You Synthesize What You've Learned?

1. The liver produces a protein called albumin. The major function of albumin is to exert osmotic pressure to pull fluid back into the blood. Predict what happens to blood osmotic pressure in a patient who has cirrhosis of the liver and is not producing adequate levels of albumin.
2. In a patient with pneumonia (a respiratory condition that results in lower levels of oxygen in the blood), will diffusion of oxygen increase, decrease, or stay the same in comparison to normal? Explain.
3. Explain to a young man, with reduced numbers of receptors for LDL, why his cholesterol level is elevated.

INTEGRATE

ONLINE STUDY TOOLS

connect | SMARTBOOK® | APR

The following study aids may be accessed through Connect.

Concept Overview Interactive: Figure 4.19 Passive and Active Processes of Membrane Transport

Clinical Case Study: A Young Man with Scaly Skin

Interactive Questions: This chapter's content is served up in a number of multimedia question formats for student study

SmartBook: Topics and terminology include chemical structure of plasma membrane; membrane transport; cell communication; cellular structures; structure and function of the nucleus and ribosomes; cell division; cell aging and death

Anatomy & Physiology Revealed: Topics include structure of plasma membrane; diffusion; osmosis and tonicity; sodium-potassium pump; smooth and rough endoplasmic reticulum; Golgi apparatus; lysosome; intermediate filament; microvilli; nucleus; protein synthesis; cell cycle and mitosis; DNA replication

Animations: Topics include osmosis; primary active transport; sodium-potassium exchange pump; endocytosis and exocytosis; lysosomes; transcription; how translation works; how the cell cycle works

Design Elements: **Learning Objective** icon: ©McGraw-Hill Education; **What Do You Think?** icon: ©McGraw-Hill Education; **What Did You Learn?** icon: ©McGraw-Hill Education; **Learning Strategy** icon: ©Slavoljub Pantelic/Shutterstock RF; **Clinical View** icon: ©Laia Design Studio/Shutterstock RF; **Concept Connection** icon: ©McGraw-Hill Education; **Challenge Yourself** icon: ©McGraw-Hill Education; **Do You Know the Basics?** icon: ©McGraw-Hill Education; **Can You Apply What You've Learned?** icon: ©McGraw-Hill Education; **Can You Synthesize What You've Learned?** icon: ©McGraw-Hill Education; **Online Study Tools** icon: ©ayzek/Shutterstock RF

Tissue Organization

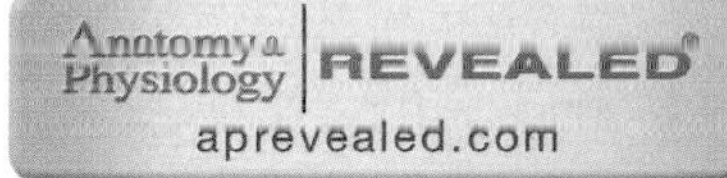

Module 3: Tissues

INTEGRATE

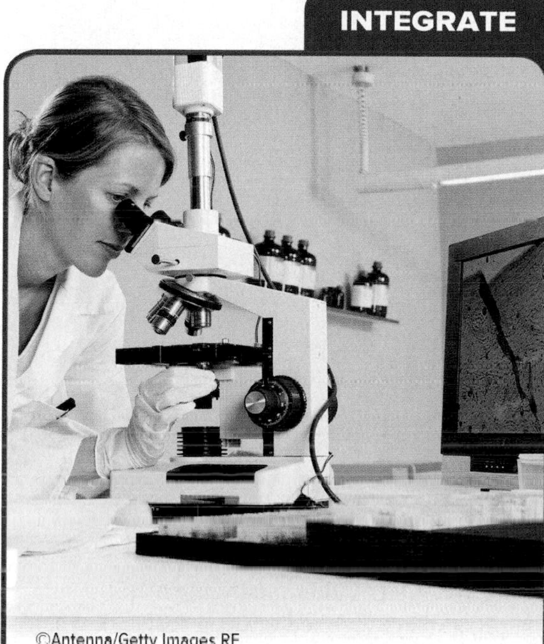

©Antenna/Getty Images RF

CAREER PATH
Histologist

The microscopic anatomy of cells and tissues is studied by a histologist. This professional uses various techniques such as light microscopy (LM) and electron microscopy (EM). In the hospital setting, a histologist may prepare frozen tissue sections taken from patient biopsies for rapid analysis by a pathologist. A thorough understanding of the characteristics of the four basic tissue types is essential for this health-care professional. Having an in-depth knowledge and understanding of normal tissue structure assists the histologist in recognizing potential tissue abnormalities, and thus possible evidence of disease or infection.

The trillions of cells in the human body are organized into more complex units called tissues. **Tissues** typically are groups of similar cells and extracellular material that perform a common function, such as providing protection or facilitating body movement. The study of tissues is called **histology** (his-tol′ō-jē; *histos* = web). Histological images are viewed using various types of microscopes, which provide different levels of detail (see section 4.1a).

Tissues in the body are classified into four major types: epithelial tissue, connective tissue, muscle tissue, and nervous tissue **(table 5.1)**. You can remember the four major tissue types with the acronym Con-MEN. These four tissue types vary in the structure of their cells, the functions of these cells, and the composition of an **extracellular matrix** (mā′triks; *matrix* = womb). The extracellular matrix is composed of varying amounts of protein fibers, water, and dissolved molecules (e.g., glucose, oxygen). Its consistency ranges from fluid to semisolid to solid.

Table 5.1 Overview of Tissues

Tissue Type	Epithelial Tissue	Connective Tissue	Muscle Tissue	Nervous Tissue
Composition	Tightly packed cells with minimal extracellular matrix	Contains cells, protein fibers, and ground substance	Cells that may be cylindrical, branching, or spindle-shaped; contain contractile proteins (myofilaments)	Contains neurons and glial cells
Functions	Covers body and organ surfaces, lines body cavities and organ cavities, forms glands	Binds, supports, and protects other tissues and organs	Moves the skeleton, organ walls, or body structures	Transmits nerve impulses and processes information
Subtypes	Simple Epithelium Simple squamous Simple cuboidal Simple columnar Pseudostratified columnar Stratified Epithelium Stratified squamous Stratified cuboidal Stratified columnar Transitional	Connective Tissue Proper Loose (areolar, adipose, reticular) Dense (regular, irregular, elastic) Supporting Connective Tissue Cartilage (hyaline, elastic, fibrocartilage) Bone Fluid Connective Tissue Blood Lymph	Skeletal Cardiac Smooth	*(None)*

5.1 Epithelial Tissue: Surfaces, Linings, and Secretory Functions

An **epithelium** (ep-i-thē′lē-um; *epi* = upon, *thele* = nipple; pl., epithelia), also referred to as **epithelial tissue,** is composed of one or more layers of closely packed cells, and it contains little to no extracellular matrix between these cells. Epithelial tissue covers the body surfaces, lines the body cavities and organ cavities, and forms glands.

5.1a Characteristics of Epithelial Tissue

LEARNING OBJECTIVE

1. Describe the common features of epithelial tissue.

All epithelia exhibit the following common characteristics, some of which are shown in **figure 5.1**:

- **Cellularity.** Epithelial tissue is composed almost entirely of tightly packed cells. There is a minimal amount of extracellular matrix between the cells.
- **Polarity.** An epithelium has an **apical** (ap′i-kăl) **surface** (*free,* or *superficial*), which is exposed either to the external environment or to some internal body space. The apical surface may have either microvilli or cilia. Recall from section 4.6c that cilia are numerous, slightly longer, membranous projections that move fluid, mucus, and materials past the cell surface, whereas microvilli are small, membranous projections on the apical surface of the cell that increase its surface area for secretion and absorption. The lateral surfaces may contain membrane (intercellular) junctions (see section 4.6d). Additionally, each epithelium has a **basal** (bā′săl) **surface** (a fixed or deep surface), where the epithelium is attached to a basement membrane with underlying connective tissue.
- **Attachment to a basement membrane.** The epithelial layer is bound at its basal surface to a thin **basement membrane.** It may be seen as a single noncellular (or molecular) layer using

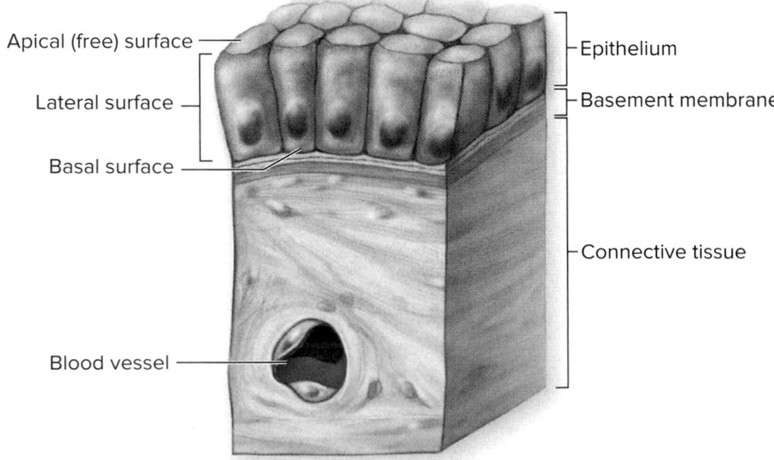

Figure 5.1 Characteristics of Epithelia. An epithelium is composed mostly of cells, it exhibits its polarity, and the lateral surfaces of cells are connected by membrane junctions (see figure 4.32). An epithelium attaches to underlying tissue via a basement membrane. AP|R

the light microscope—however, in reality it consists of three molecular layers that can be viewed using an electron microscope: the *lamina lucida,* the *lamina densa,* and the *reticular lamina.* These molecular layers are formed by secretions of both the epithelium and the underlying connective tissue, and are composed of collagen, glycoproteins (e.g., laminin, fibronectin), and proteoglycans. Together, these basement membrane components strengthen the attachment and form a selective molecular barrier between the epithelium and the underlying connective tissue.

- **Avascularity.** All epithelial tissues lack blood vessels. Nutrients for epithelial cells are obtained either directly across the apical surface or by diffusion across the basal surface from blood vessels within the underlying connective tissue.
- **Extensive innervation.** Epithelia are richly innervated (supplied with nerves) to detect changes in the environment at that body or organ region.
- **High regeneration capacity.** Epithelial cells undergo cell division frequently (see section 4.9b). This characteristic allows this tissue to regenerate itself at a high rate, a necessary condition for a tissue that is often exposed to the environment and lost by abrasion and damage. The continual replacement occurs through cell division of the deepest epithelial cells (called *stem cells*), which are adjacent to the basement membrane.

WHAT DID YOU LEARN?

Why does an epithelium need to be highly regenerative?

5.1b Functions of Epithelial Tissue

LEARNING OBJECTIVE

2. Explain the four functions of epithelial tissues.

Epithelia have several functions, although no one epithelium performs all of them.

- **Physical protection.** Epithelial tissues protect both external and internal surfaces from dehydration, abrasion, and destruction by physical, chemical, or biological agents.
- **Selective permeability.** An epithelium typically exhibits a range of permeability; it may be relatively non-permeable to some substances, while promoting and assisting the passage of other ions and molecules. All substances that enter or leave the body must pass through an epithelium, and thus epithelial cells act as "gatekeepers."
- **Secretions.** Some epithelial cells are specialized to produce and release secretions. These cells form *glands.* Glands may be individual cells scattered among other cell types in an epithelium (e.g., goblet cells) or arranged in small, organized clusters within a multicellular gland (see section 5.1d).
- **Sensations.** Epithelial tissues are innervated by sensory nerve endings to detect or respond to a stimulus (see section 16.1a). These nerve endings—and those in the underlying connective tissue—continuously relay sensory input to the central nervous system (i.e., brain and spinal cord) concerning touch, pressure, temperature, and pain. Additionally, several organs contain a specialized epithelium, called a *neuroepithelium,* that houses specific cells responsible for the senses of sight, taste, smell, hearing, and equilibrium (as described in chapter 16).

WHAT DO YOU THINK?

1 Why do you think epithelial tissue does not contain any blood vessels? Can you think of any epithelial function that could be compromised if blood vessels were running through the tissue?

WHAT DID YOU LEARN?

Why is an epithelium considered selectively permeable?

5.1c Classification of Epithelial Tissue

LEARNING OBJECTIVES

3. Name the classes of epithelia based on cell layers and cell shapes.
4. Give examples of each type of epithelium.

The body contains many different types of epithelia, and the classification of each type is indicated by a two-part name. The first part of the name refers to the *number* of epithelial cell layers, and the second part describes the *shape* of cells at the apical (superficial) surface of the epithelium.

Classification by Number of Cell Layers

Epithelia are classified as either simple or stratified (**figure 5.2*a***). A **simple epithelium** is one layer of epithelial cells, and all of the epithelial cells are in direct contact with the basement membrane. A simple epithelium is found in areas where stress is minimal and filtration, absorption, or secretion is the primary function.

A **stratified epithelium** contains two or more layers of epithelial cells. Only the cells in the deepest (basal) layer are in direct contact with the basement membrane. A stratified epithelium resembles a brick wall, where the bricks in contact with the ground represent the basal layer and the bricks at the top of the wall represent the apical (superficial) layer. This tissue provides either more structural support or better protection for underlying tissue. A stratified epithelium is found in areas likely to be subjected to abrasive activities or mechanical stresses, as multiple layers of cells are better able to resist the wear and tear. Cells in the basal layer continuously regenerate as the cells in the apical layer are lost due to abrasion or stress.

A **pseudostratified** (sū′dō-strat′i-fīd; *pseudo* = false, *stratum* = layer) **epithelium** appears layered (stratified) because the cells' nuclei are distributed at different levels between the apical and basal surfaces. Although all of these epithelial cells are attached to the basement membrane, some of them do not reach its apical surface. For our purposes, we have classified pseudostratified epithelium as a type of simple epithelium, because all of the cells are attached to the basement membrane.

Classification by Cell Shape

Epithelia are also classified by the shape of the cell at the apical surface. In a simple epithelium, all of the cells display the same shape, whereas in a stratified epithelium, a difference in shape can be seen between cells within the basal layer and those within the apical layer. Figure 5.2*b* shows the three common cell shapes seen in epithelia: squamous, cuboidal, and columnar. (Note that the cells in this figure all appear hexagonal when viewed from their apical surface. Thus, these terms describe the cells' shapes when viewed laterally, or from the side.)

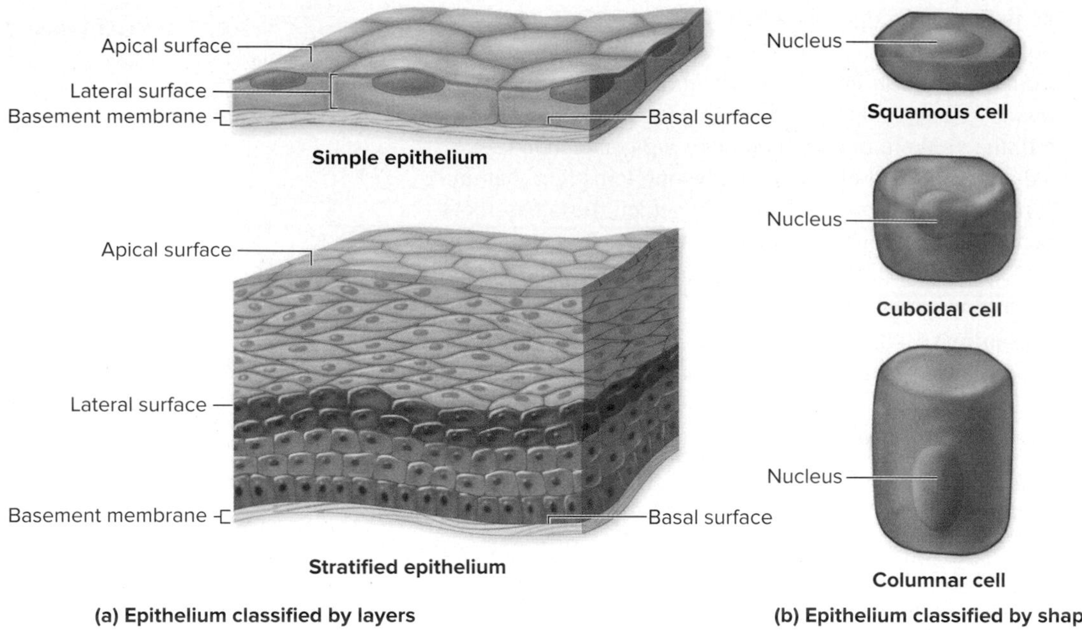

Figure 5.2 Classification of Epithelia. Two criteria are used to classify epithelia: the number of cell layers and the shape of the cell at the apical surface. (*a*) An epithelium is simple if it is one cell layer thick, and stratified if it has two or more layers of cells. (*b*) Epithelial cell shapes include squamous (thin, flattened cells), cuboidal (cells about as tall as they are wide), and columnar (cells taller than they are wide).

Squamous (skwā′mŭs; *squamosus* = scaly) **cells** are flat, wide, and somewhat irregular in shape. The cells are arranged like floor tiles, and the nucleus is somewhat flattened. **Cuboidal** (kū-boy′dăl) **cells** are about as tall as they are wide. The cells do not resemble perfect cubes because their edges are somewhat rounded. The cell nucleus is spherical and located within the center of the cell. **Columnar** (kol-ŭm′năr) **cells** are slender and taller than they are wide. The cell nucleus is oval and usually oriented lengthwise and in the basal region of the cell. Another shape classification that occurs in epithelial cells is called **transitional.** These cells can readily change their shape from polyhedral to more flattened, depending upon the degree to which the epithelium is stretched. The shape change occurs when the epithelium cycles between distended and relaxed states, such as in the lining of the bladder, which fills with urine and is later emptied.

Using the classification system just described, epithelium can be broken down into the primary types shown in **figure 5.3**.

Simple Squamous Epithelium

A **simple squamous epithelium** consists of a single layer of flattened cells **(table 5.2*a*)**. When viewed *en face* (looking onto the surface), the irregularly shaped cells display a spherical to oval nucleus, and the cells are tightly bound together. Each squamous cell resembles a fried egg, with the slightly bulging nucleus of the cell representing the yolk. This epithelium is extremely delicate and represents the thinnest possible barrier to allow rapid movement of molecules and ions across the epithelium by membrane transport processes (see section 4.3). Simple squamous epithelium forms the lining of the air sacs (alveoli) of the lung (see section 23.3d), where this thin epithelium is well suited for the exchange of oxygen and carbon dioxide between the blood and the inhaled air. Simple squamous epithelium also is found lining the lumen (inside space) of blood vessel walls (see section 20.1a), where it allows for rapid exchange of nutrients and waste between the blood and the interstitial fluid surrounding the blood vessels. Serous membranes, which cover body organs and secrete serous fluid, are also formed by a simple squamous epithelium.

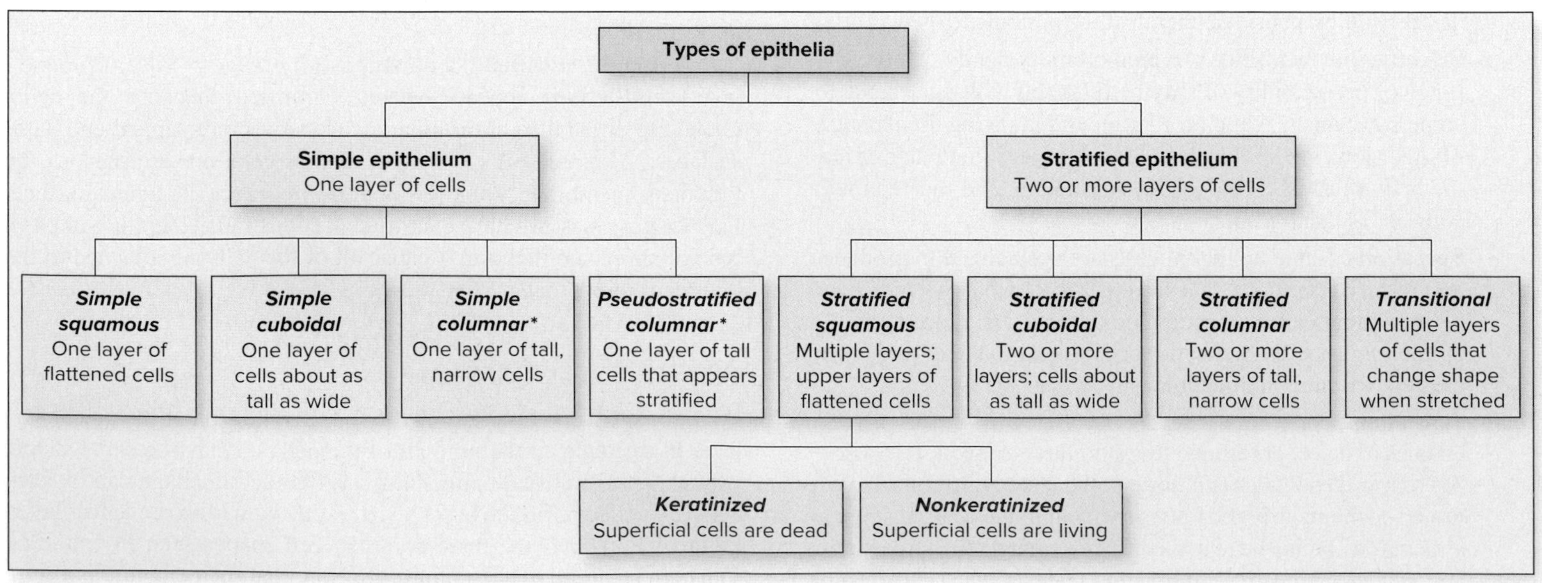

Figure 5.3 Organization and Relationship of Epithelia Types. Epithelia are classified by (1) the number of cell layers and (2) the cell shape at the surface.

Table 5.2 Simple Epithelia

(a) SIMPLE SQUAMOUS EPITHELIUM AP|R

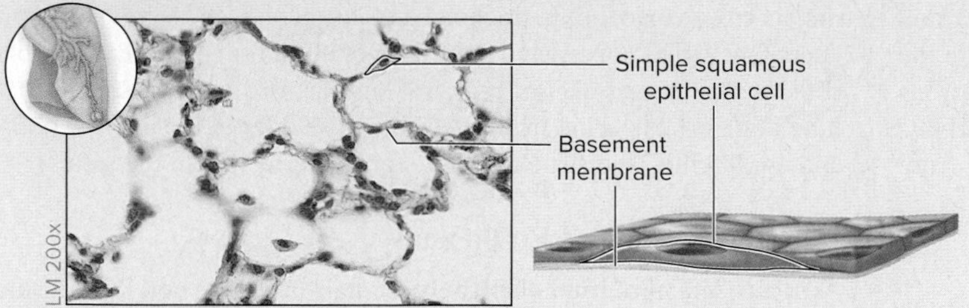

Structure
Single layer of thin, flat cells resembling irregular floor tiles; the single nucleus of each cell bulges at its center

Function
Thinnest possible barrier to allow for rapid diffusion and filtration; secretion in serous membranes

Location
Air sacs in lungs (alveoli); lining of lumen of blood vessels and lymph vessels (endothelium); serous membranes of body cavities (mesothelium)

(b) SIMPLE CUBOIDAL EPITHELIUM AP|R

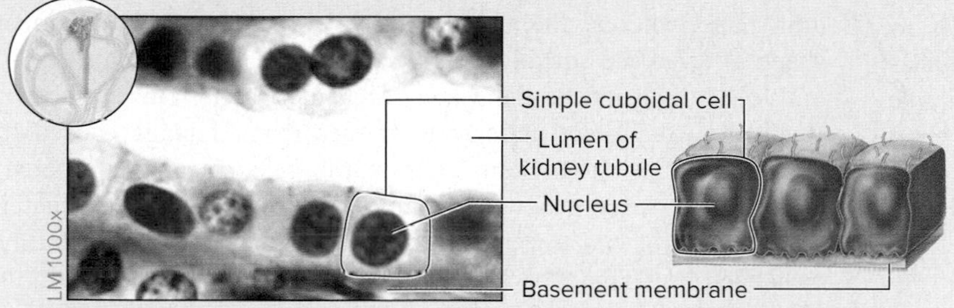

Structure
Single layer of cells about as tall as they are wide; spherical and centrally located nucleus

Function
Absorption and secretion; forms secretory tissue of most glands and small ducts

Location
Kidney tubules, thyroid gland follicles; surface of ovary; secretory regions and ducts of most glands

(c) NONCILIATED SIMPLE COLUMNAR EPITHELIUM AP|R

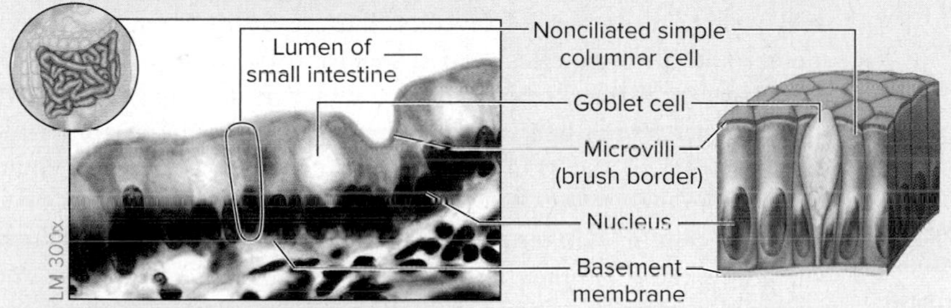

Structure
Single layer of cells taller than they are wide; oval-shaped nucleus oriented lengthwise in basal region of cell; apical regions of cell may have microvilli; may contain goblet cells that secrete mucin

Function
Absorption and secretion; secretion of mucin

Location
Inner lining of most of digestive tract (stomach, small intestine, and large intestine)

(d) CILIATED SIMPLE COLUMNAR EPITHELIUM AP|R

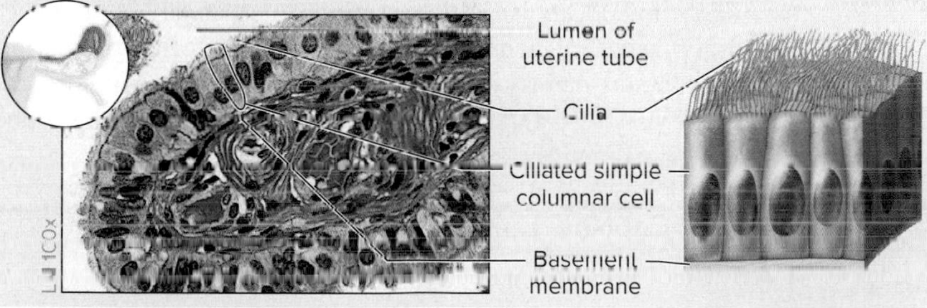

Structure
Single layer of ciliated cells taller than they are wide; oval-shaped nucleus oriented lengthwise in basal region of cell; may contain goblet cells

Function
Secretion of mucin and movement of mucus along apical surface of epithelium by cilia; oocyte movement through uterine tube

Location
Lining of the larger bronchioles (air passageways) of the lung and the uterine tubes

(e) CILIATED PSEUDOSTRATIFIED COLUMNAR EPITHELIUM AP|R

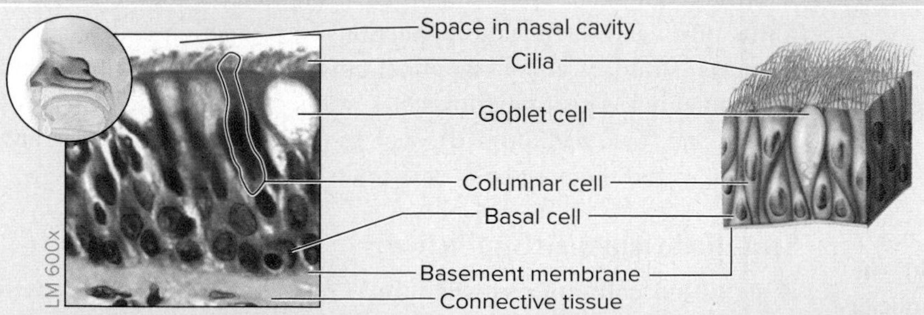

Structure
Single layer of cells with varying heights; all cells connect to the basement membrane, but not all cells reach the apical surface; has goblet cells and cilia

Function
Protection; secretion of mucin and movement of mucus along apical surface of epithelium by cilia

Location
Lining of the larger airways of respiratory tract, including nasal cavity, part of pharynx, parts of larynx, trachea, and bronchi

(f) NONCILIATED PSEUDOSTRATIFIED COLUMNAR EPITHELIUM AP|R

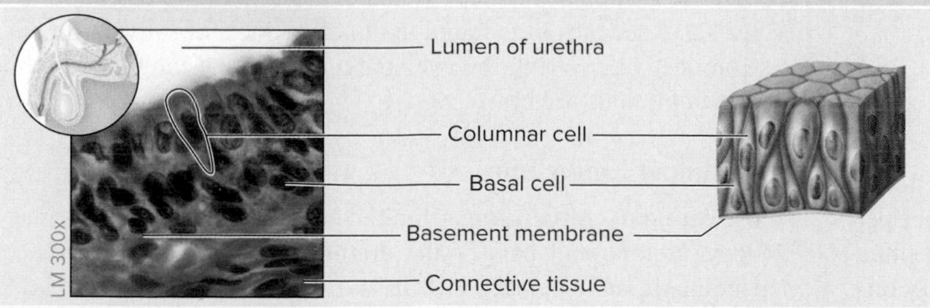

Structure
Single layer of cells with varying heights; all cells connect to the basement membrane, but not all cells reach the apical surface; lacks goblet cells and cilia

Function
Protection

Location
Rare—lining of part of the male urethra and epididymis

Specific names are used to refer to the simple squamous epithelia in certain locations within the body. **Endothelium** (en-dō-thē′lē-ŭm; *endon* = within) is the name of the simple squamous epithelium that lines both blood vessels and lymph vessels (see sections 20.1a and 21.1a), and **mesothelium** (mez-ō-thē′lē-ŭm; *mesos* = middle) is the name given to the simple squamous epithelium that forms the serous membranes of body cavities (see section 1.4e). Mesothelium gets its name from the embryonic primary germ layer called mesoderm, from which it is derived (see section 5.6a).

Simple Cuboidal Epithelium

A **simple cuboidal epithelium** contains one layer of uniformly shaped cells that are about as tall as they are wide with a centrally located, spherical nucleus (table 5.2*b*). This epithelium allows for both absorption and secretion. Its cells' uniformity in shape makes them ideal to form the structural components of glands. For example, a simple cuboidal epithelium forms the follicles (spherical structures) of the thyroid gland and covers each ovary. Simple cuboidal epithelium also composes the walls of small ducts (or tubules), including those of kidney tubules.

Simple Columnar Epithelium

A **simple columnar epithelium** is composed of a single layer of cells that are taller than they are wide. The nucleus is oval, oriented lengthwise, and located in the basal region of the cell. This type of epithelium is ideal for both absorptive and secretory functions. Simple columnar epithelium has two forms: One type has no cilia, whereas the apical surface of the other type is covered with cilia.

Nonciliated simple columnar epithelium often contains **microvilli** (see section 4.6c) and a scattering of unicellular glands called **goblet cells** (table 5.2*c*). Individual microvilli cannot be distinguished under the microscope; rather, the microvilli collectively appear as a bright, fuzzy structure known as a **brush border.** Goblet cells secrete **mucin** (mū′sin), which is a glycoprotein that when hydrated (mixed with water) forms **mucus.** Nonciliated simple columnar epithelium lines most of the digestive tract from the stomach to the anal canal.

Ciliated simple columnar epithelium has cilia that project from the apical surfaces of the cells (table 5.2*d*). Mucus covers the apical surface and is moved along by the beating of the cilia. Goblet cells typically are interspersed throughout this epithelium. Ciliated columnar epithelium lines the larger bronchioles (air passageways) in the lung. It also lines the luminal (internal) surface of the uterine tubes, where it helps move an oocyte (egg) from the ovary to the uterus.

Pseudostratified Columnar Epithelium

A **pseudostratified columnar epithelium** (sū′dō-strat′i-fīd; *pseudes* = false, *stratum* = layer) is so named because upon first glance, it *appears* to consist of multiple layers of cells. However, this epithelium is not really stratified because all of its cells are in direct contact with the basement membrane. Although it may look stratified because the nuclei are scattered at different distances from the basal surface, not all of the cells reach the apical surface in this epithelium. Its columnar cells always reach the apical surface, and the shorter cells are stem cells (see Clinical View 5.4: "Stem Cells") that give rise to the columnar cells.

Pseudostratified columnar epithelium consists of two forms: **pseudostratified ciliated columnar epithelium,** which contains cilia on its apical surface (table 5.2*e*), and **pseudostratified nonciliated columnar epithelium,** which lacks cilia (table 5.2*f*). Both types perform protective functions. Pseudostratified ciliated columnar epithelium houses goblet cells that secrete mucin, which hydrates to become the mucus that traps foreign particles and is moved by the beating cilia. This type is found in the larger air passageways of the respiratory system (e.g., the nasal cavity, part of the pharynx [throat], part of the larynx [voice box], trachea, and bronchi [see section 23.1b]). Pseudostratified nonciliated columnar epithelium is rare, lacks goblet cells and cilia, and occurs primarily in part of the male urethra (the tube that conveys urine from the urinary bladder to outside the body) and epididymis (the structure that stores sperm within the testes).

Stratified Squamous Epithelium

A **stratified squamous epithelium** has multiple cell layers, and only the deepest layer of cells is in direct contact with the basement membrane. The cells in the basal layers have a cuboidal or polyhedral shape, whereas the apical cells display a flattened, squamous shape. A stratified squamous epithelium is so named because of its multiple cell layers and the shape of its apical cells. This epithelium is adapted to protect underlying tissues from damage caused by abrasion and friction. Stem cells in the basal layer continuously divide, to produce a new stem cell and a committed cell that is gradually displaced toward the surface to replace those cells that have been lost. This type of epithelium exists in two forms: keratinized and nonkeratinized.

In **keratinized stratified squamous epithelium,** the superficial layers are composed of cells that are dead. These cells lack nuclei and all organelles, and instead are filled with the protein **keratin** (ker′ă-tin; *keras* = horn), which is a tough, protective protein that strengthens the tissue (table 5.3*a*). The epidermis (outer layer) of the skin consists of keratinized stratified squamous epithelium (see section 6.1a).

The cells in **nonkeratinized stratified squamous epithelium,** including those at the tissue's apical surface, lack keratin and remain alive. Because all of the cells are alive, the nuclei characteristic of squamous cells are visible throughout the tissue (table 5.3*b*). This tissue is kept moist with secretions such as saliva or mucus. Nonkeratinized stratified squamous epithelium forms the surface tissue of mucous membranes that line the oral cavity (mouth), part of the pharynx (throat), part of the larynx (voicebox), the esophagus, the vagina, and the anus (see figure 5.12).

Stratified Cuboidal Epithelium

A **stratified cuboidal epithelium** contains two or more layers of cells, and the superficial cells tend to be cuboidal in shape (table 5.3*c*). Stratified cuboidal epithelium, like simple cuboidal epithelium, forms tubes and ducts, and it functions in protection and secretion. However, stratified cuboidal epithelium is thicker than simple cuboidal epithelium. This tissue forms the walls of the ducts of most exocrine glands (see section 5.1d), such as the ducts of the sweat glands in the skin and the periphery of ovarian follicles.

Stratified Columnar Epithelium

A **stratified columnar epithelium** is relatively rare in the body. It consists of two or more layers of cells, but only the cells at the apical surface are columnar in shape (table 5.3*d*). This type of epithelium protects and secretes. It is found in the large ducts of salivary glands, the conjunctiva covering the eye, and a segment of the male urethra (i.e., membranous urethra).

Transitional Epithelium

A **transitional epithelium** is limited to the urinary tract (urinary bladder, ureters, and part of the urethra). It varies in appearance, depending upon whether it is in a relaxed state or a distended

Table 5.3 Stratified Epithelia

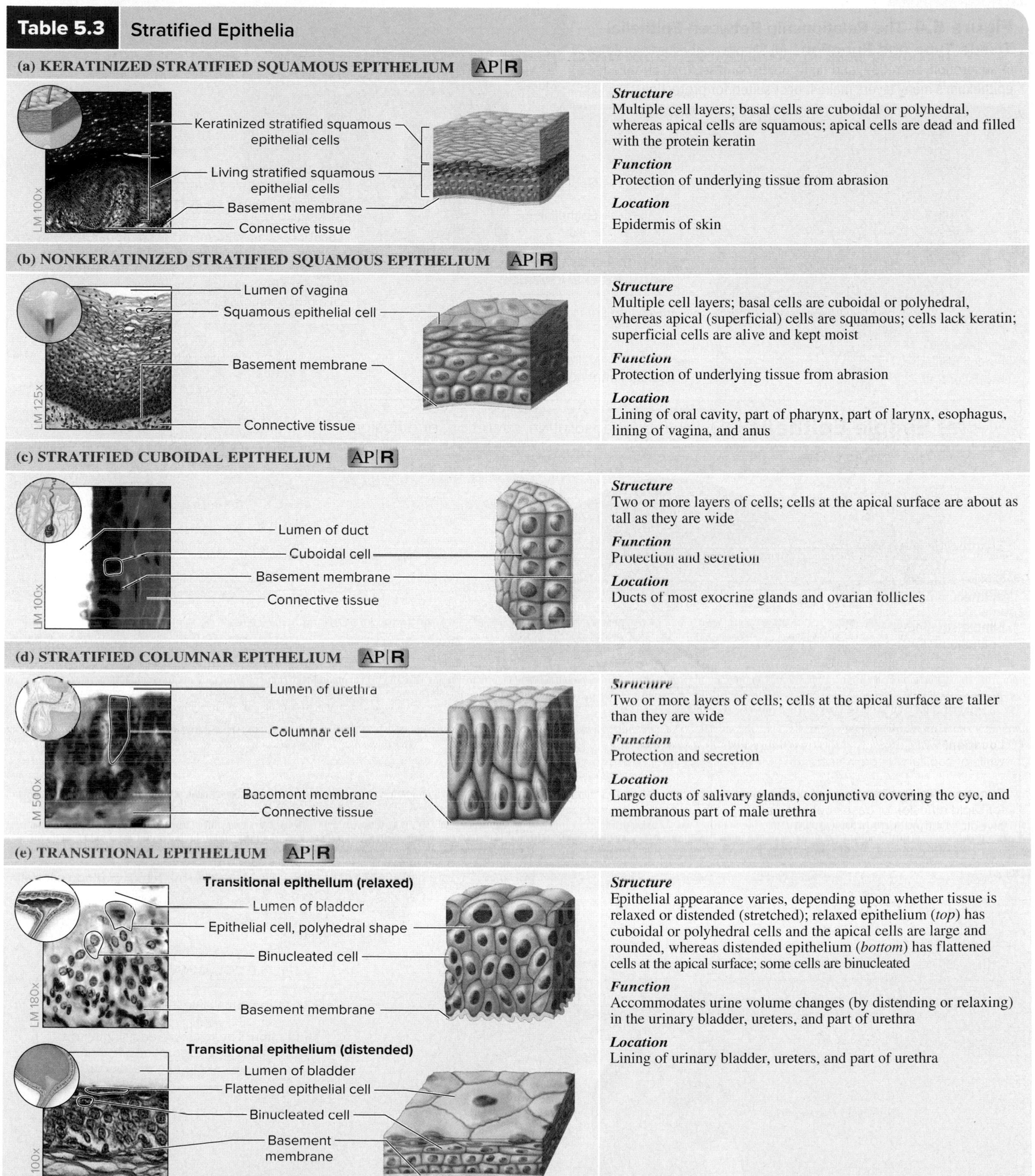

(a) KERATINIZED STRATIFIED SQUAMOUS EPITHELIUM AP|R

Structure
Multiple cell layers; basal cells are cuboidal or polyhedral, whereas apical cells are squamous; apical cells are dead and filled with the protein keratin

Function
Protection of underlying tissue from abrasion

Location
Epidermis of skin

(b) NONKERATINIZED STRATIFIED SQUAMOUS EPITHELIUM AP|R

Structure
Multiple cell layers; basal cells are cuboidal or polyhedral, whereas apical (superficial) cells are squamous; cells lack keratin; superficial cells are alive and kept moist

Function
Protection of underlying tissue from abrasion

Location
Lining of oral cavity, part of pharynx, part of larynx, esophagus, lining of vagina, and anus

(c) STRATIFIED CUBOIDAL EPITHELIUM AP|R

Structure
Two or more layers of cells; cells at the apical surface are about as tall as they are wide

Function
Protection and secretion

Location
Ducts of most exocrine glands and ovarian follicles

(d) STRATIFIED COLUMNAR EPITHELIUM AP|R

Structure
Two or more layers of cells; cells at the apical surface are taller than they are wide

Function
Protection and secretion

Location
Large ducts of salivary glands, conjunctiva covering the eye, and membranous part of male urethra

(e) TRANSITIONAL EPITHELIUM AP|R

Structure
Epithelial appearance varies, depending upon whether tissue is relaxed or distended (stretched); relaxed epithelium (*top*) has cuboidal or polyhedral cells and the apical cells are large and rounded, whereas distended epithelium (*bottom*) has flattened cells at the apical surface; some cells are binucleated

Function
Accommodates urine volume changes (by distending or relaxing) in the urinary bladder, ureters, and part of urethra

Location
Lining of urinary bladder, ureters, and part of urethra

(photos): (*a, b, c, d*) ©McGraw-Hill Education/Al Telser; (*e*) ©McGraw-Hill Education/Al Telser; ©Victor P. Eroschenko

(stretched) state (table 5.3*e*). In a relaxed state, the basal cells appear cuboidal or polyhedral, and the apical cells are large and rounded. When transitional epithelium stretches, it thins and the apical cells flatten and become almost squamous in shape. One distinguishing feature of transitional epithelium is the presence of some **binucleated** (containing two nuclei) cells. By being able to stretch as the bladder fills, this tissue ensures that urine does not seep into the underlying tissues of these organs.

WHAT DO YOU THINK?

2 Which types of epithelia are well suited for protection and why?

INTEGRATE

LEARNING STRATEGY

Integrate lab and lecture material: Ask yourself the following questions to help distinguish the type of epithelium under the microscope:

1. *Is the epithelium one layer or many layers thick?* If it is one layer thick, then you are looking at some type of simple epithelium. If it appears many layers thick, you are looking at either some type of stratified epithelium or pseudostratified epithelium.
2. *What is the shape of the cells at the apical surface of the epithelium?* If the cells are flattened, it is a squamous epithelium; if they are about as tall as they are wide, the cells are cuboidal; and if the cells are tall and narrow, they are columnar.

Your answer for question 1 gives you the first part of the epithelium name (e.g., simple). Your answer for question 2 gives you the second part of the epithelium name (e.g., squamous). Put these answers together, and you get the name of the tissue (e.g., simple squamous).

Now that you have examined all the different types of epithelia, refer to **figure 5.4** to reexamine the relationship between epithelial type and function. Note in this figure that the simple epithelia function in diffusion, absorption, and secretion functions, because these epithelia are thinner than stratified epithelium. Stratified epithelia are best suited for protective functions. Thus, when you examine organs with different epithelia, you will have a clue about the organ's function based on its epithelium.

WHAT DID YOU LEARN?

3 How does simple epithelium differ from stratified epithelium?

4 What type of epithelial tissue lines the air sacs of the lungs?

5 What epithelial tissue contains multiple layers of cells, and the most superficial cells are squamous, dead, and filled with the protein keratin?

5.1d Glands

LEARNING OBJECTIVES

5. Define glands.
6. Distinguish between endocrine and exocrine glands.
7. List exocrine gland types based on both anatomic form and physiologic method of secretion.

Glands are either individual cells or multicellular organs composed predominantly of epithelial tissue. They secrete substances either for use elsewhere in the body or for elimination from the body. Glandular secretions may include mucin, ions, hormones, enzymes, or urea (a nitrogenous waste produced by the body).

Endocrine and Exocrine Glands

Endocrine (en′dō-krin; *krino* = to separate) **glands** lack ducts and secrete their products, called *hormones,* into the blood to be transported throughout the body. Hormones act as chemical messengers (or ligands) to influence cell communication (see section 4.5b). Endocrine glands, such as the thyroid and adrenal glands, are discussed in depth in chapter 17.

Exocrine (ek′sō-krin) **glands** typically originate from an invagination of epithelium that burrows into the underlying connective tissue. These glands usually maintain their connection with the epithelial surface by means of a **duct,** an epithelium-lined tube through which the gland secretions are discharged onto the epithelial surface. Examples of exocrine glands include sweat glands, mammary glands, and salivary glands.

Exocrine glands may be unicellular (one-celled) or multicellular. **Unicellular exocrine glands** typically do not contain a duct, and they are located close to the surface of the epithelium in which they reside. The most common type of unicellular exocrine gland is the goblet cell, which is usually found in both simple columnar epithelium and pseudostratified ciliated columnar epithelium (refer to table 5.2*c* and *e* for examples). In contrast, **multicellular exocrine glands** contain numerous cells that work together to produce a secretion (**figure 5.5**).The gland often consists of **acini** (as′i-nī; *acinus* = grape), which are the clusters of cells that produce the secretion, and one or more smaller **ducts,** which merge to form a larger duct that transports the secretion to the epithelial surface. Multicellular exocrine glands typically are surrounded by a fibrous capsule, and extensions of the capsule called *septa* partition the gland into **lobes.**

Classification of Exocrine Glands

Multicellular exocrine glands may be classified either by anatomic form or by method of secretion, which may be thought of as a physiologic classification.

Classification by Anatomic Form Exocrine glands may be classified anatomically based on the structure and complexity of their ducts. **Simple glands** have a single, unbranched duct; **compound glands** have branched ducts. In addition, glands may be classified according to the shape of their secretory portions. The gland is called

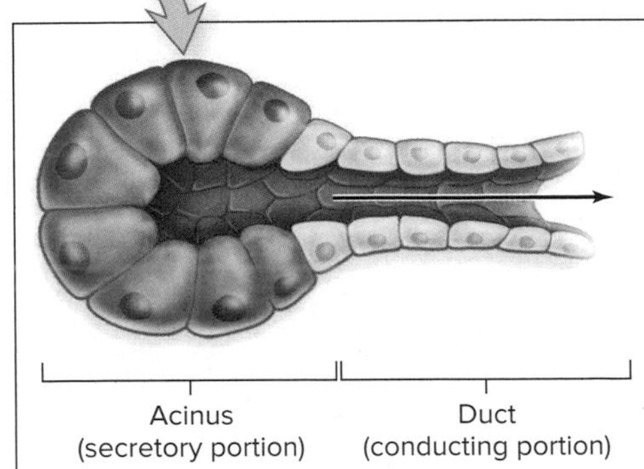

Figure 5.5 General Structure of Multicellular Exocrine Glands. Exocrine glands may contain secretory portions called acini, and conducting portions composed of many ducts that merge to form a larger duct that transports the secretion to the epithelial surface.

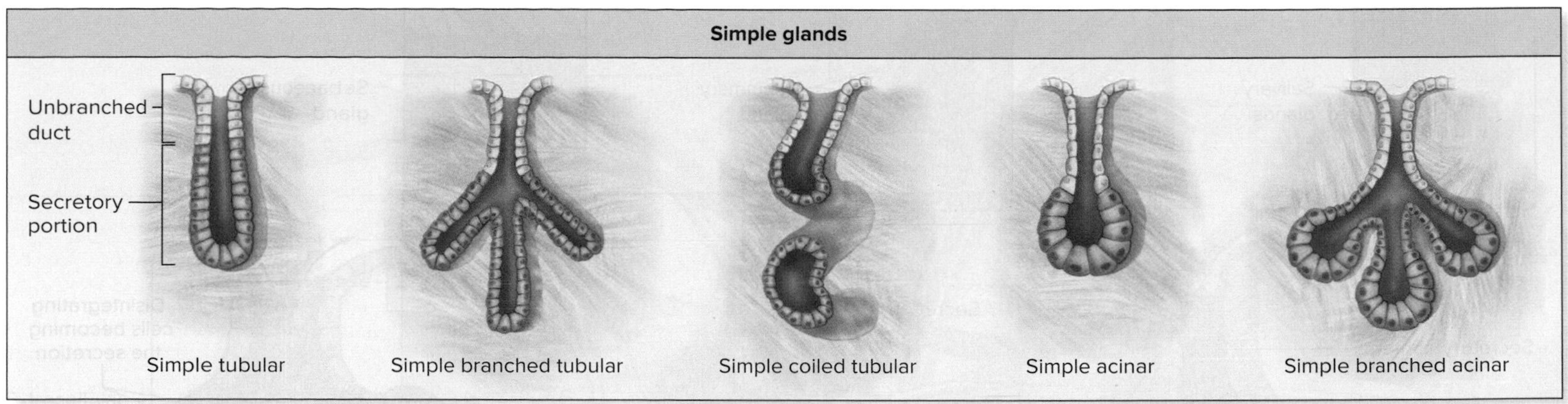

(a)

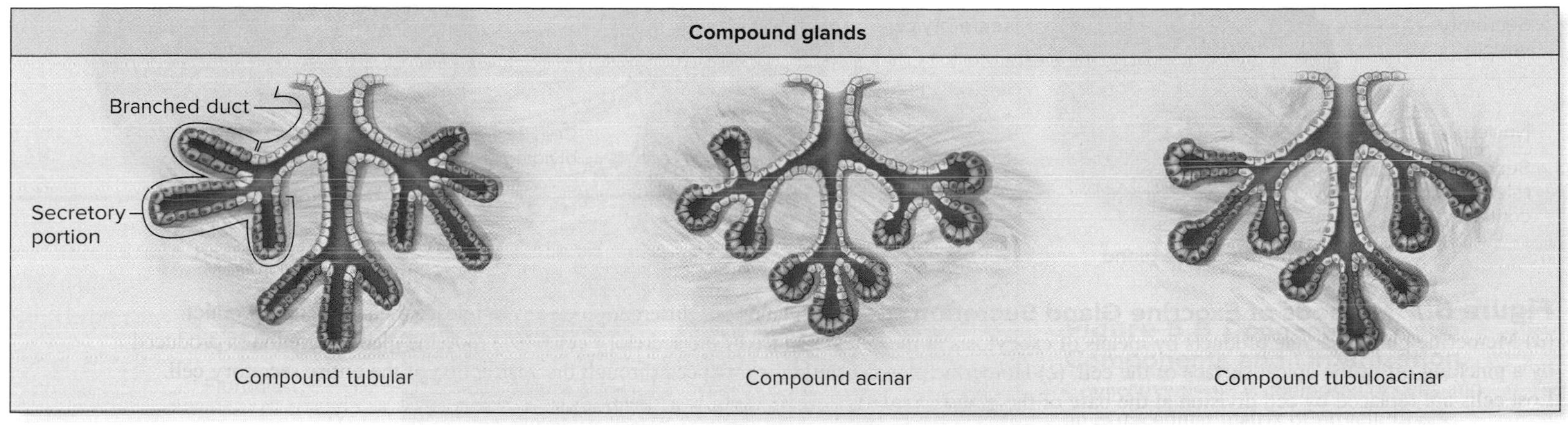

(b)

Figure 5.6 Structural Classification of Multicellular Exocrine Glands. Simple glands (*a*) have unbranched ducts, whereas compound glands (*b*) have ducts that branch. These glands also exhibit different forms: Tubular glands have secretory cells in a space with a uniform diameter, acinar glands have secretory cells arranged in saclike acini, and tubuloacinar glands have secretory cells in both the tubular and acinar regions.

tubular if the secretory portion and the duct have the same diameter. If the secretory portion forms an expanded sac, the gland is called **acinar.** Finally, a gland with both tubules and acini is called a **tubuloacinar gland. Figure 5.6** shows several types of exocrine glands based on their anatomic form.

Classification by Method of Secretion Exocrine glands may be classified physiologically by their method of secretion. The three basic types of exocrine glands in this classification are merocrine glands, apocrine glands, and holocrine glands **(figure 5.7)**.

Merocrine (mer′ō-krin; *meros* = share) **glands** package their secretions into secretory vesicles and release the secretions by exocytosis (see section 4.3c). The glandular cells remain intact and are not damaged in any way by producing the secretion. Examples of merocrine glands include lacrimal (tear) glands; salivary glands; some sweat glands, also known as *eccrine glands*; the exocrine glands of the pancreas (see section 26.3c); and the gastric glands of the stomach (see section 26.2d).

Apocrine (ap′ō-krin; *apo* = away from, off) **glands** produce their secretory material in the following way: Secretion occurs when the cell's apical portion pinches off, releasing cytoplasmic content. Thereafter, the cell repairs itself in order to repeat its secretory activity. Examples include the mammary glands (see section 28.3f) and ceruminous glands of the ear (see section 16.5a).

Holocrine (hōl′ō-krin; *holos* = whole) **glands** are formed from cells that accumulate a product; the entire cell then disintegrates. Thus, a holocrine secretion is a viscous mixture of both cell fragments and the product the cell produced prior to its disintegration. The ruptured, dead cells are continuously replaced by other epithelial cells undergoing cellular division. The oil-producing glands (sebaceous glands) in the skin are examples of holocrine glands (see section 6.2c).

INTEGRATE

LEARNING STRATEGY

The *apo* part of *apocrine* sounds like "a part." Apocrine gland secretions are produced when "a part" of the cell is pinched off and becomes the secretion. The *hol* part of *holocrine* sounds like the word *whole*. Holocrine gland secretions are produced when the *whole* cell ruptures, dies, and becomes the secretion. Finally, merocrine glands "merrily" form vesicles to release secretions without damage to the gland's cells.

WHAT DID YOU LEARN?

6 What are the two basic parts of a multicellular exocrine gland?

7 What are the differences between holocrine and merocrine glands?

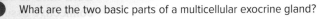

6.1 Composition and Functions of the Integument

The integument is the body's largest organ and is composed of all the tissue types that function in concert to protect internal body structures. Its surface is an epithelium that protects underlying body layers. The connective tissue that is deep to the epithelium provides strength and resilience to the skin. This connective tissue also contains smooth muscle associated with hair follicles (arrector pili) that alters hair position. Finally, nervous tissue detects and monitors sensory stimuli in the skin, thus providing information about touch, pressure, temperature, and pain.

The integument accounts for 7% to 8% of the body weight and covers the entire body surface with an area that varies between about 1.5 and 2.0 square meters (m^2). Its thickness ranges between 1.5 millimeters (mm) and 4 mm or more, depending on body location. (For comparison, a sheet of copier paper is about 0.1 mm thick, so the thickness of the skin would range between 15 and 40 sheets of paper.) The integument consists of two distinct layers: a layer of stratified squamous epithelium called the *epidermis*, and a deeper layer of both areolar and dense irregular connective tissue called the *dermis* (**figure 6.1**). Deep to the integument is a layer of areolar and adipose connective tissue called the *subcutaneous layer*, or *hypodermis*. The subcutaneous layer is not part of the integumentary system; however, it is described in this chapter because it is closely involved with both the structure and function of the skin.

6.1a Epidermis

LEARNING OBJECTIVES

1. Describe the five layers (strata) of the epidermis.
2. Differentiate between thick skin and thin skin.
3. Explain what causes differences in skin color.

The epithelium of the integument is called the **epidermis** (ep-i-derm′is; *epi* = on, *derma* = skin). It is a keratinized, stratified squamous epithelium (see section 5.1c).

Careful examination of the epidermis, from the basement membrane to its surface, reveals several specific layers, or strata. From deep to superficial, these layers are the stratum basale, stratum spinosum, stratum granulosum, stratum lucidum (found in thick skin only), and stratum corneum (**figure 6.2**). The first three strata listed are composed of living keratinocytes, whereas the most superficial two strata contain dead keratinocytes.

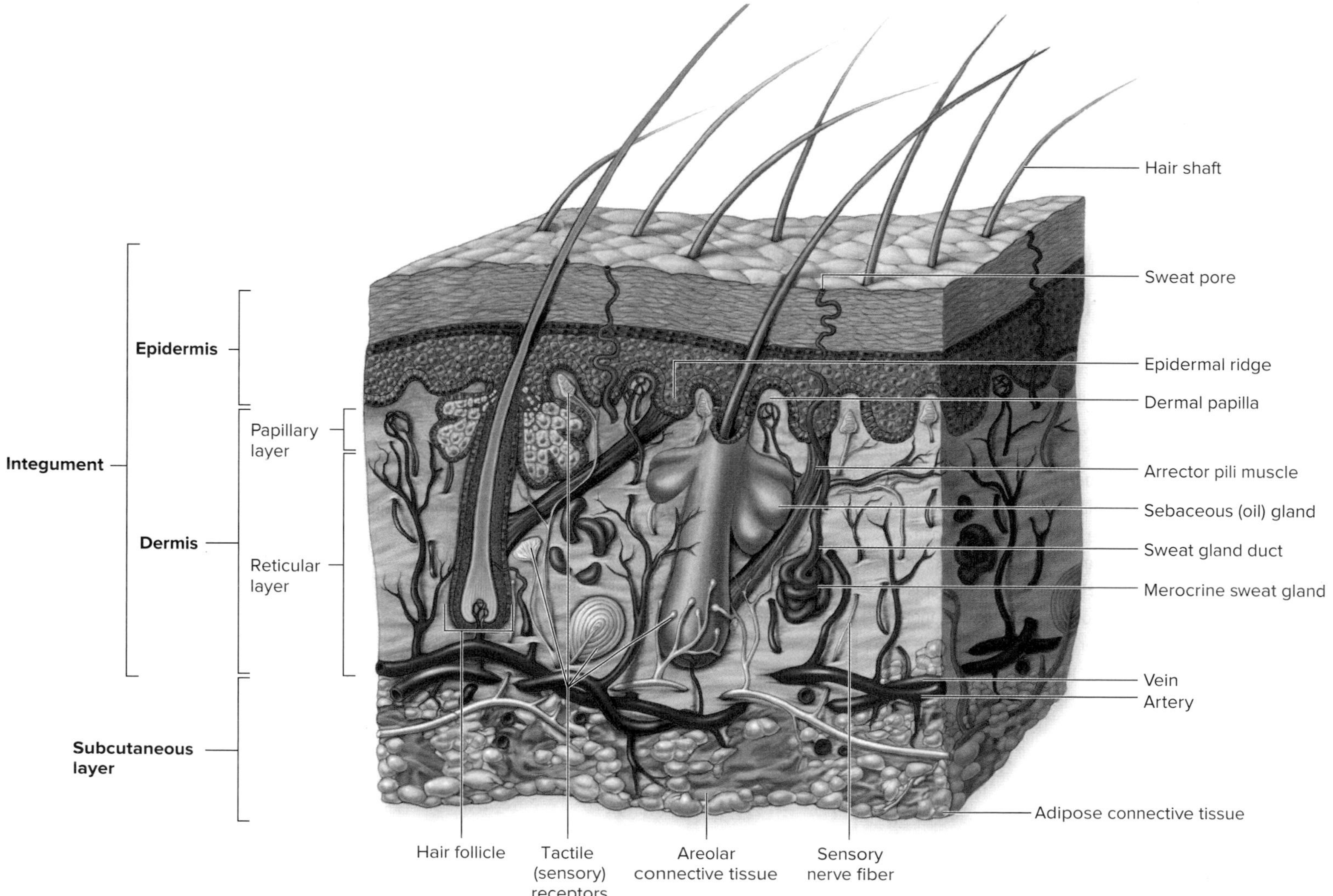

Figure 6.1 Layers of the Integument. A diagrammatic sectional view through the integument shows its relationship to the underlying subcutaneous layer. AP|R

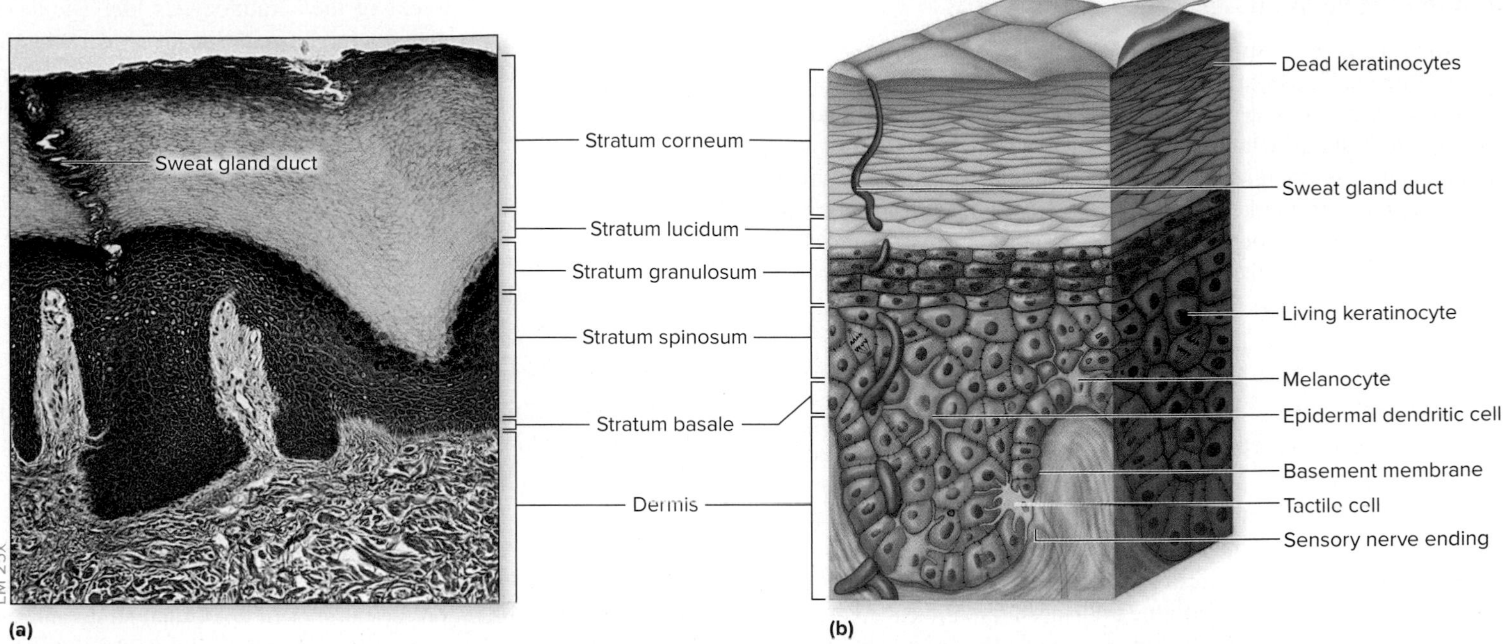

Figure 6.2 Epidermal Strata. (*a*) Photomicrograph and (*b*) diagram compare the order and relationship of the epidermal strata in thick skin. AP|R

(*a*) ©Ed Reschke/Getty Images

Stratum Basale

The deepest epidermal layer is the **stratum basale** (strat′ŭm bah-sa′lē), also known as the *stratum germinativum,* or *basal layer.* This single layer of cuboidal to low columnar cells is tightly attached by hemidesmosomes (see section 4.6d) to an underlying basement membrane that separates the epidermis from the connective tissue of the dermis. Three types of cells occupy the stratum basale (figure 6.2*b*):

1. **Keratinocytes** (ke-ra′ti-nō-sīt; *keras* = horn) are the most abundant cell type in the epidermis and are found throughout all epidermal strata. The stratum basale is dominated by large keratinocyte stem cells, which divide to generate new keratinocytes that replace dead keratinocytes shed from the surface. Their name is derived from their synthesis of **keratin,** a protein that strengthens the epidermis considerably. Keratin is one of a family of fibrous structural proteins that are both tough and insoluble. Fibrous keratin molecules can twist and intertwine around each other to form helical intermediate filaments of the cytoskeleton (see section 4.6b). The keratin proteins found in keratinocytes are called *cytokeratins.* Their structure in these keratinocytes gives skin its strength and makes the epidermis water resistant.
2. **Melanocytes** (mel′ă-nō-sīt; *melano* = black) have long, branching processes and are scattered among the keratinocytes of the stratum basale. They produce and store the pigment **melanin** (mel′ă-nin) in response to ultraviolet light exposure. Their cytoplasmic processes transfer melanin pigment within granules called **melanosomes** (mel′ă-nō-sōmes) to the keratinocytes within the basal layer and sometimes in more superficial layers. This pigment (which includes the colors black, brown, tan, and yellow-brown) accumulates around the nucleus of the keratinocyte and shields the nuclear DNA from ultraviolet radiation. The darker tones of the skin result from melanin produced by the melanocytes. Thus, "tanning" is the result of the melanocytes producing melanin to block UV light from causing mutations in the DNA of your keratinocytes (in the epidermis) and fibroblasts (in the dermis).
3. **Tactile cells,** also called *Merkel cells,* are few in number and found scattered among the cells within the stratum basale (see section 16.2a). Tactile cells are sensitive to touch and, when compressed, they release chemicals that stimulate sensory nerve endings, providing information about objects touching the skin.

Stratum Spinosum

Several layers of polygonal keratinocytes form the **stratum spinosum** (spī-nō′sŭm), or *spiny layer.* Each time a keratinocyte stem cell in the stratum basale divides, a daughter cell is pushed toward the external surface from the stratum basale, while the other cell remains as a stem cell in the stratum basale. Once this new cell enters the stratum spinosum, it begins to differentiate into a nondividing, highly specialized keratinocyte. The keratinocytes in the stratum spinosum attach to their neighbors by many membrane junctions called desmosomes (described in section 4.6d).

The process of preparing epidermal tissue for observation on a microscope slide shrinks the cytoplasm of the cells in the stratum spinosum. Because the cytoskeletal elements and desmosomes remain intact, the shrunken keratinocytes in the stratum spinosum resemble miniature porcupines that are attached to their neighbors. This spiny appearance accounts for the name of this layer.

In addition to the keratinocytes, the stratum spinosum also contains the fourth epidermal cell type, called **epidermal dendritic** (*Langerhans*) **cells** (figure 6.2*b*). Epidermal dendritic cells are immune cells that help fight infection in the epidermis. These immune cells are often present in the stratum spinosum and stratum granulosum, but they are not identifiable in standard histologic preparations. Their phagocytic activity initiates an immune response to protect the body against pathogens that have penetrated the superficial epidermal layers as well as epidermal cancer cells (see section 22.3b).

Stratum Granulosum

The **stratum granulosum** (gran-ū-lō′sum), or *granular layer,* consists of three to five layers of keratinocytes superficial to the stratum spinosum. Within this stratum begins a process called **keratinization** (ker′ă-tin-i-zā′shŭn), where the keratinocytes fill up with the protein keratin, and in so doing, cause both the cell's nucleus and organelles to disintegrate and the keratinocyte dies. Keratinization is not complete until the keratinocytes reach the more superficial epidermal layers. A fully keratinized cell is dead (because it has neither a nucleus nor organelles), but it is structurally strong because of the keratin it contains.

Stratum Lucidum

The **stratum lucidum** (lū′sĭ-dum), or *clear layer,* is a thin, translucent region of about two to three keratinocyte layers that is superficial to the stratum granulosum. This stratum is found only in the thick skin within the palms of the hands and the soles of the feet. Keratinocytes occupying this layer are flattened, pale cells with indistinct boundaries. They are filled with the translucent protein called **eleidin** (ē-lē′i-din), which is an intermediate product in the process of keratin maturation. This layer helps protect the skin from ultraviolet light.

Stratum Corneum

The **stratum corneum** (kōr′nē-ŭm; *corneus* = horny), or *hornlike layer,* is the most superficial layer of the epidermis. It is the stratum you see when you look at your skin. The stratum corneum consists of about 20 to 30 layers of dead, scaly, interlocking, keratinized cells. The dead keratinocytes are **anucleate** (lacking a nucleus) and are tightly packed together.

A keratinized, or *cornified*, epithelium contains large amounts of keratin. After keratinocytes are formed from stem cells within the stratum basale, they change in structure and in their relationship to their neighbors as they progress through the different strata until they eventually reach the stratum corneum and are sloughed off its external surface. Major changes during keratinocyte migration include synthesis of keratin and loss of the nucleus and organelles as described. What remains of these keratinocytes in the stratum corneum is essentially keratin protein enclosed in a thickened plasma membrane. Migration of the keratinocyte to the stratum corneum occurs during the first 2 weeks of the keratinocyte's life. The dead, keratinized cells usually remain for an additional 2 weeks in the exposed stratum corneum layer, providing a barrier before they are shed. Overall, individual keratinocytes are present in the integument for about 1 month following their formation.

The stratum corneum presents a thickened surface unsuitable for the growth of many microorganisms. Additionally, some exocrine gland secretions (e.g., sweat, which contains *demicidin*, an antimicrobial peptide) help prevent the growth of microorganisms on the epidermis, thus supporting its barrier function (see section 22.3a).

Variations in the Epidermis

The epidermis exhibits variations between different body regions within one individual as well as differences between individuals. The epidermis varies in its thickness, coloration, and skin markings.

Thick Skin Versus Thin Skin Over most of the body, the skin ranges in thickness between 1 mm and 2 mm. Skin is classified as either thick or thin based on the number of epidermal strata and the relative thickness of the epidermis, rather than the thickness of the entire integument (**figure 6.3**).

Thick skin is found on the palms of the hands and the soles of the feet. All five epidermal strata occur in the thick skin. The epidermis of thick skin ranges between 0.4 mm and 0.6 mm thick. It houses sweat glands but has no hair follicles or sebaceous (oil) glands.

Thin skin covers most of the body. It lacks a stratum lucidum, so it has only four specific layers in the epidermis. Thin skin contains the following structures: hair follicles, sebaceous glands, and sweat glands. The epidermis of thin skin ranges from 0.075 mm to 0.150 mm thick.

WHAT DO YOU THINK?

1. Why do you think thick skin lacks hair follicles and sebaceous glands? Think about where thick skin is found and how that may interfere with the function of skin in that area.

Skin Color Normal skin color results from a combination of the colors of hemoglobin, melanin, and carotene. **Hemoglobin** (hē-mō-glō′bin; *haima* = blood) is an oxygen-binding protein present in red blood cells (see section 18.3b). It exhibits a bright red color upon

INTEGRATE

LEARNING STRATEGY

Integrate lab and lecture material: Follow these steps to help you identify the epidermal strata under the microscope:

1. **Determine if the layer is closer to the free surface or is deeper.** Remember the stratum corneum forms the free surface, whereas the stratum basale forms the deepest epidermal layer.
2. **Examine the shape of the keratinocytes.** The stratum basale contains cuboidal to low columnar keratinocytes, the stratum spinosum contains polygonal keratinocytes, and the stratum lucidum and corneum contain squamous keratinocytes.
3. **See if the keratinocytes have a nucleus or are anucleate.** When they are still alive (as in the strata basale, spinosum, and granulosum), you are able to see nuclei. The stratum lucidum and corneum layers contain dead, anucleate keratinocytes.
4. **Count the layers of keratinocytes in the stratum.** The stratum basale has only one layer of keratinocytes, and the stratum corneum contains 20 to 30 layers of keratinocytes. The other layers contain about two to five layers of keratinocytes.
5. **Determine if the cytoplasm of each keratinocyte contains visible granules.** If the keratinocytes contain visible granules, you likely are looking at the stratum granulosum.

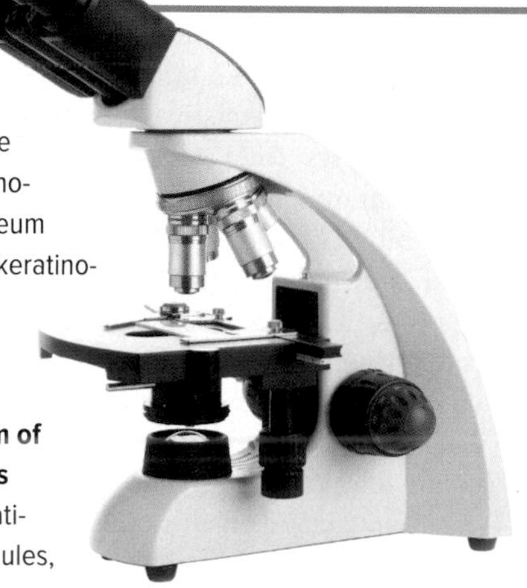

©Science Photo Library/Getty Images RF

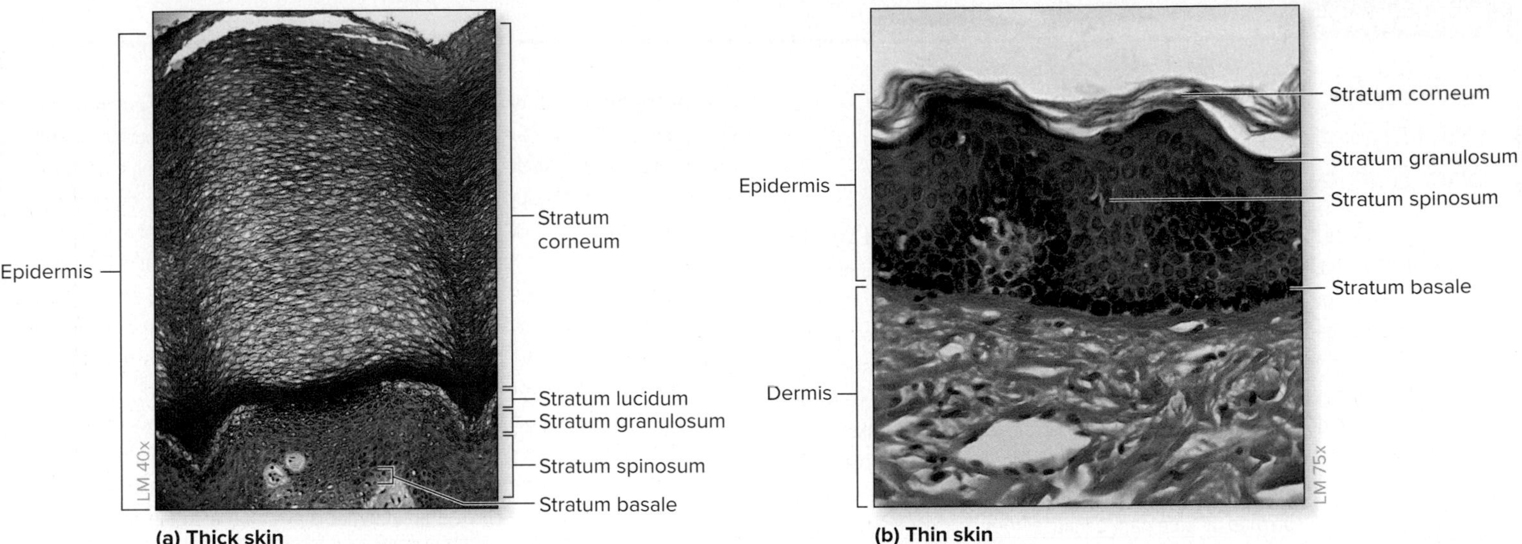

Figure 6.3 Thick Skin and Thin Skin. The stratified squamous epithelium of the epidermis varies in thickness, depending upon the region of the body in which it is located. (*a*) Thick skin contains all five epidermal strata and covers the palms of the hands and the soles of the feet. (*b*) Thin skin covers most body surfaces; it contains only four epidermal strata and lacks a stratum lucidum. AP|R

(*a, b*) ©Carolina Biological Supply Company/Medical Images

binding oxygen, thus giving blood vessels in the dermis a reddish tint that is seen most easily in lightly pigmented individuals. If the blood vessels in the superficial layers vasodilate (i.e., the blood vessel diameter increases), such as during physical exertion, then the red tones are much more visible.

Melanin is a pigment produced and stored in melanocytes (described earlier in this section), and it occurs in a variety of black, brown, tan, and yellow-brown shades. Recall that melanin is transferred in melanosomes from melanocytes to keratinocytes in the stratum basale. Because keratinocytes are displaced toward the stratum corneum, melanocyte activity affects the color of the entire epidermis (**figure 6.4**).

The amount of melanin in the skin is determined by both heredity and light exposure. All people have about the same *number* of melanocytes. However, melanocyte *activity* and the *color* of the melanin produced by these cells vary among individuals and races, resulting in different skin color tones. Darker-skinned individuals have melanocytes that produce relatively more and darker melanin than lighter-skinned individuals. Further, these more active melanocytes tend to package melanin into cells in the more superficial epidermal layers, such as the stratum granulosum. Recall that exposure to ultraviolet light stimulates melanocytes to make more melanin.

Carotene (kar′ō-tēn) is a yellow-orange pigment that is acquired from various yellow-orange vegetables, such as carrots, corn, and squashes. Normally, it accumulates inside keratinocytes of the stratum corneum and in the subcutaneous fat. Within the body, carotene is converted into vitamin A, which plays an important role in normal vision (see section 16.4d). Vitamin A has also been thought to reduce potentially dangerous molecules called *free radicals* that form during normal metabolic activity in the body. Additionally, carotene may improve immune cell number and activity.

Albinism (al′bi-nizm) is an inherited recessive condition where the enzyme needed to produce melanin is nonfunctional. As a result, melanocytes are unable to produce melanin. Individuals who have albinism typically have white hair, pale skin, and pink irises.

Skin Markings A **nevus** (nē-vŭs; pl., *nevi*), commonly called a *mole,* is a harmless, localized overgrowth of melanocytes. On rare

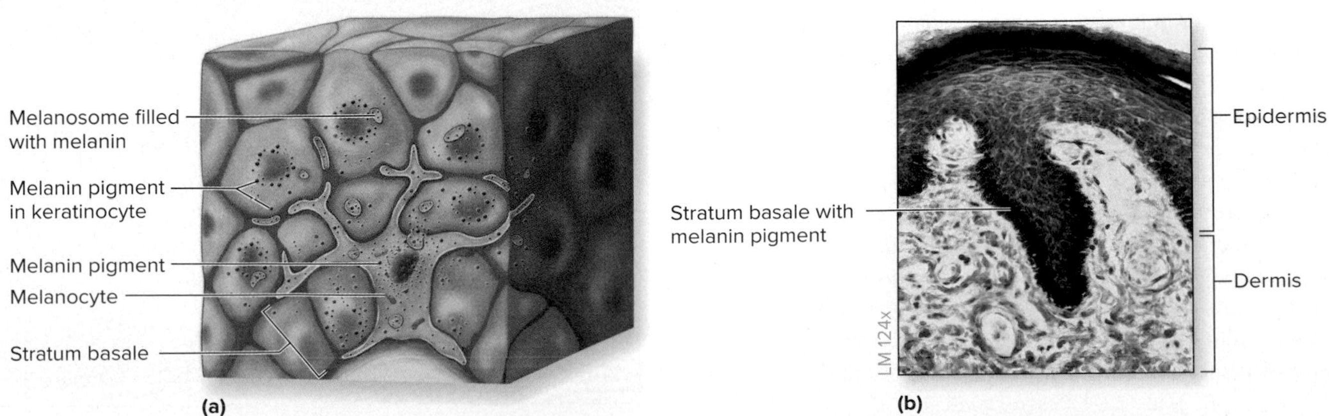

Figure 6.4 Production of Melanin by Melanocytes. Melanin gives a yellow-brown to tan to brown or black color to the skin. (*a*) Melanosomes in melanocytes transport the melanin pigment to the keratinocytes, where the pigment surrounds the nucleus. (*b*) Melanin is incorporated into the keratinocytes primarily of the stratum basale.

(*b*) ©John Burbidge/Science Source

INTEGRATE

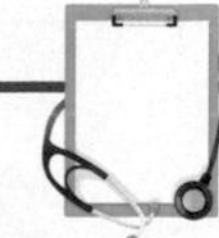

CLINICAL VIEW 6.1

UV Radiation, Sunscreens, and Sunless Tanners

The sun generates three forms of ultraviolet radiation: **UVA (ultraviolet A), UVB (ultraviolet B),** and **UVC (ultraviolet C).** The wavelength of UVA ranges between 320 and 400 nanometers (nm), that of UVB ranges between 290 and 320 nm, and UVC ranges between 100 and 280 nm (see figure 3.2). In contrast, visible light ranges begin at about 400 nm (the deepest violet). UVC rays are absorbed by the upper atmosphere and do not reach the earth's surface; UVA and UVB rays do reach the surface and can affect individuals' skin color. UVA light commonly is termed *tanning rays,* whereas UVB often is called *burning rays.* Many tanning salons claim to provide a safe tan because they use only UVA rays. However, UVA rays can cause burning as well as tanning, and they inhibit the immune system. Both UVA and UVB rays are known to cause skin cancer. Thus, there is no such thing as a healthy suntan.

Sunscreens are lotions that contain materials that absorb or block UVA and UVB rays. Sunscreens can help protect the skin, not only for light-skinned people but also for those with darker skin colors—but *only* if they are used correctly. First, sunscreen must be applied liberally over all exposed body surfaces, and reapplied after being in the water or perspiring. Second, a sunscreen must have a high enough **SPF (sun protection factor).** SPF is a number determined experimentally by exposing subjects to a light spectrum. The amount of light that induces redness in sunscreen-protected skin, divided by the amount of light that induces redness in unprotected skin, equals the SPF. For example, a sunscreen with an SPF of 15 will delay the onset of sunburn in a person who would otherwise burn in 10 minutes to 150 minutes. However, it is never safe to assume that a sunscreen will protect you completely from the sun's harmful rays.

Sunless tanners produce a tanned, bronzed skin without UV light exposure. The most effective ones contain **dihydroxyacetone (DHA)** as their active ingredient. Sunless tanners do not affect melanin production. Instead, when applied to the epidermis, DHA interacts with the amino acids in the cells to produce a darkened, brownish color. Because only the most superficial epidermal cells are affected, the color change is temporary and lasts about 5 to 7 days. There are some sunless tanners on the market that contain other chemicals, but they do not appear to be as effective.

Most sunless tanners contain no sunscreen and offer no protection against UV rays because protective melanin has not been produced. Thus, individuals who use sunless tanners should also apply sunscreen to protect their skin.

occasions, a nevus may become malignant, typically as a consequence of excessive UV light exposure. Thus, nevi should be monitored for changes that may suggest malignancy (see table 6.2). **Freckles** are yellowish or brown spots that represent localized areas of increased melanocyte activity, not an increase in melanocyte numbers. A freckle's degree of pigmentation varies and is dependent upon both sun exposure and heredity.

A **hemangioma** (he-man′jē-ō′mă) is an anomaly that results in skin discoloration due to blood vessels that proliferate to form a benign tumor. **Capillary hemangiomas,** or *strawberry-colored birthmarks,* appear in the skin as bright red to deep purple nodules that are usually present at birth and disappear in childhood. However, their development may occur in adults. **Cavernous hemangiomas,** also known as *port-wine stains,* involve larger dermal blood vessels and may last a lifetime.

Friction ridges are another type of skin marking. These ridge patterns follow the contours of the skin, varying from small, conical pegs (in thin skin) to the complex arches and whorls. Friction ridges are found on the fingers (**fingerprints**), palms, soles, and toes (**figure 6.5**). These ridges are formed from large folds and valleys of both dermis and epidermis. They help increase friction on contact, so that our hands can firmly grasp items and our feet do not slip when we walk barefoot. Some researchers have suggested friction ridges also provide flexibility to the skin and allow it to deform without being damaged. When sweat glands and oil glands release their secretions, noticeable fingerprints may be left on touched surfaces. Each individual has a unique pattern of friction ridges, allowing matching of prints and identification of individuals.

WHAT DID YOU LEARN?

1. As you trim your roses, a thorn penetrates your palm through all epidermal strata. What are the layers of the epidermis penetrated, starting from the surface of the skin?
2. Briefly describe the process of keratinization. Where does it occur? Why is it important?
3. How does hemoglobin contribute to skin color?
4. What is the function of friction ridges?

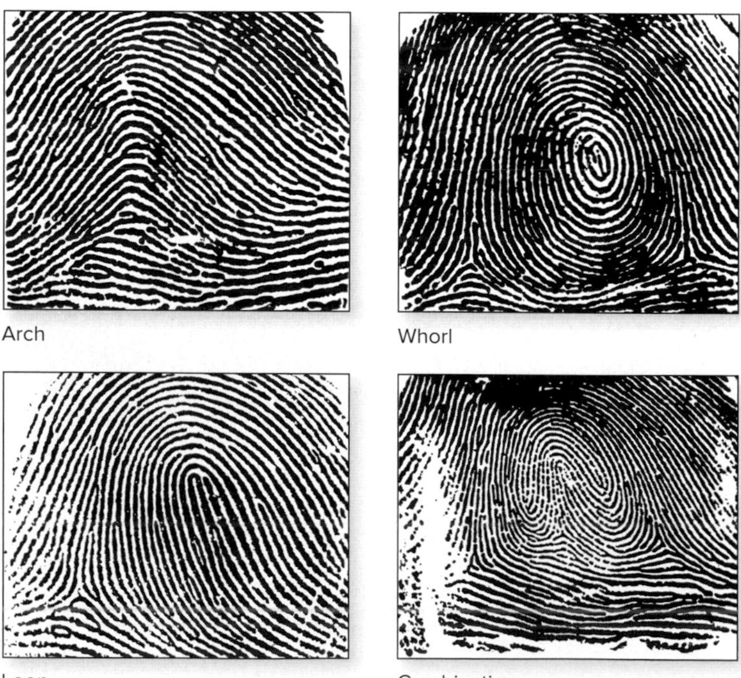

Figure 6.5 Friction Ridges of Thick Skin. Friction ridges form fingerprints, palm prints, and toe prints. Shown here are four basic fingerprint patterns.

6.1b Dermis

LEARNING OBJECTIVES

4. Characterize the two layers of the dermis.
5. Explain the significance of cleavage lines.
6. Describe how dermal blood vessels function in temperature regulation.

The **dermis** (der′mis) is deep to the epidermis and ranges in thickness from 0.5 mm to 3.0 mm. This layer of the integument is composed of connective tissue proper (see section 5.2d) and contains primarily collagen fibers, although both elastic and reticular fibers also are found within the dermis. Additionally, researchers recently have discovered motile cells in the dermis called *dendritic cells*. These cells are similar to the epidermal dendritic cells in that they serve an immune function, except they are located in the dermis (see section 22.3b). Other structures within the dermis are blood vessels, sweat glands, sebaceous glands, hair follicles, nail roots, sensory nerve endings, and smooth muscle tissue associated with hair follicles (arrector pili). Two major regions of the dermis can be distinguished: a superficial papillary layer and a deeper reticular layer (**figure 6.6**).

Papillary Layer of the Dermis

The **papillary** (pap′i-lār-ē) **layer** is the superficial region of the dermis that is deep to the epidermis. It is composed of areolar connective tissue, and it derives its name from the projections of the dermis called **dermal papillae** (der′măl pă-pil′ē; *papilla* = a nipple). The dermal papillae interdigitate with deep projections of the epidermis called **epidermal ridges,** much like two sets of egg crate foam stacked on top of one another. Together, the epidermal ridges and dermal papillae increase the area of contact between the two layers and interlock them. Each dermal papilla contains the capillaries that supply nutrients to the cells of the epidermis. Additionally, dermal papillae contain sensory nerve endings that serve as tactile receptors (see figure 6.1); these receptors continuously monitor touch on the surface of the epidermis. Tactile receptors are discussed in detail in section 16.2a.

Reticular Layer of the Dermis

The **reticular layer** forms the deeper, major portion of the dermis that extends from the papillary layer to the underlying subcutaneous layer. The reticular layer consists primarily of dense irregular connective tissue through which large bundles of collagen fibers extend in all directions. These fibers are interwoven into a meshwork that surrounds structures in the dermis, such as the hair follicles, sebaceous glands and sweat glands, nerves, and blood vessels. The word *reticular* means "network" and refers to this meshwork of collagen fibers. Note that the reticular layer is different from reticular connective tissue, described in section 5.2d.

Lines of Cleavage and Stretch Marks

The majority of the collagen and elastic fibers in the skin are oriented in parallel bundles at specific body locations. The alignment of fiber bundles within the dermis is a result of the direction of applied stress during routine movement; therefore, the function of the bundles is to resist stress. **Lines of cleavage** (*tension lines*) in the skin identify the predominant orientation of collagen fiber bundles (**figure 6.7**). These are clinically and surgically significant because any procedure resulting in a cut perpendicular to a cleavage line usually is pulled

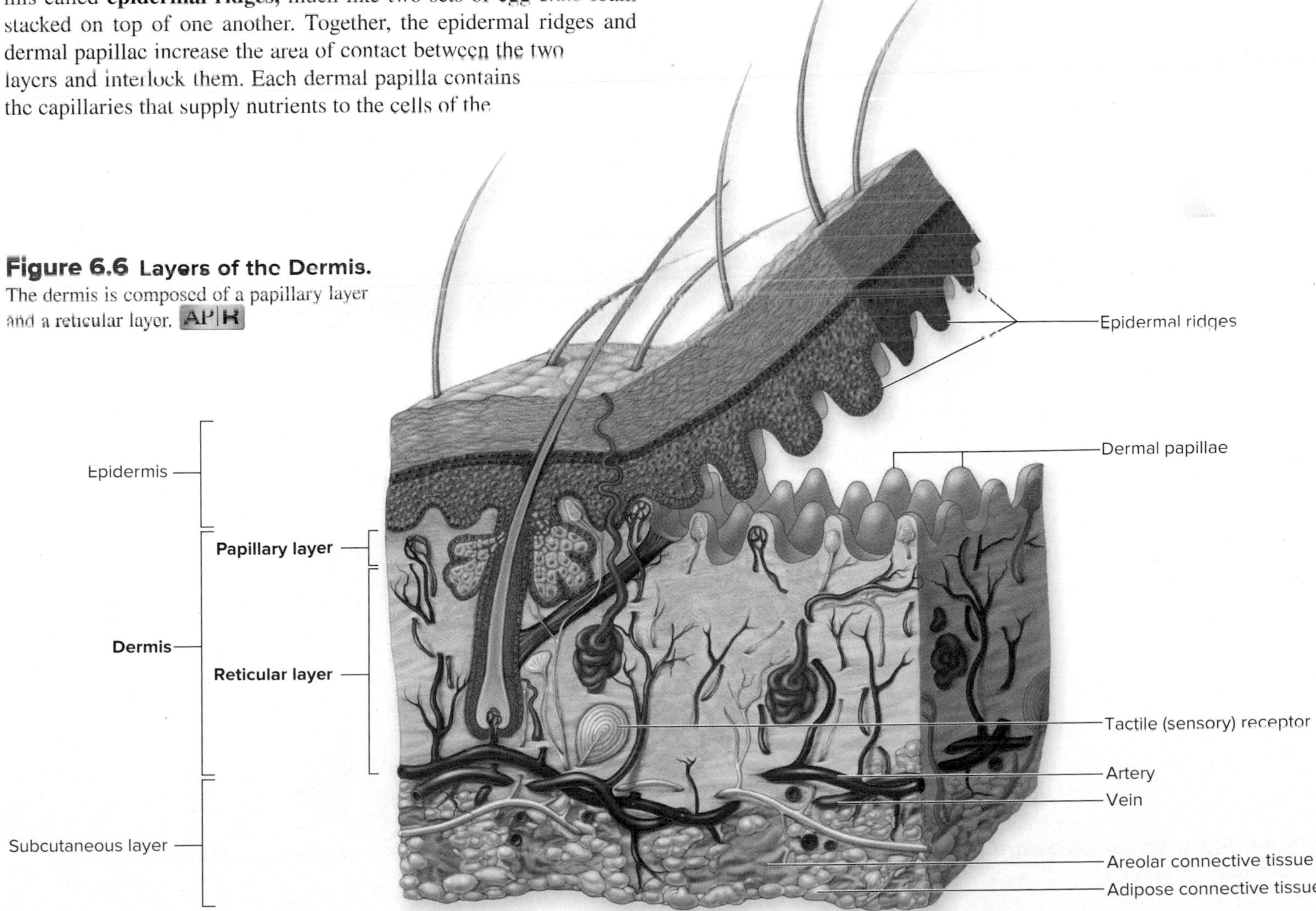

Figure 6.6 Layers of the Dermis. The dermis is composed of a papillary layer and a reticular layer. AP|R

open as a result of the recoil resulting from cut elastic fibers. This often results in slow healing and increased scarring. In contrast, an incision made parallel to a cleavage line usually will remain closed. Therefore, surgical procedures should be planned to consider lines of cleavage, thus ensuring rapid healing and prevention of scarring.

Collagen fibers and elastic fibers together contribute to the visible physical characteristics of the skin. Whereas the collagen fibers impart tensile strength, elastic fibers allow some stretch and recoil in the dermis during normal movement activities. Stretching of the skin that may occur as a result of excessive weight gain or pregnancy often exceeds the elastic capabilities of the skin. When the skin is stretched beyond its capacity, some collagen fibers are torn and result in stretch marks, called **striae** (strī′ē; *stria* = furrow). Both the flexibility and thickness of the dermis are diminished by effects of exposure to ultraviolet light and aging, causing either sagging or wrinkled skin.

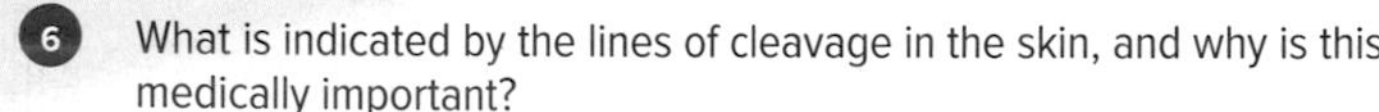

WHAT DID YOU LEARN?

5 Compare and contrast the papillary versus reticular layer of the dermis, with respect to their tissue type and the structures they contain.

6 What is indicated by the lines of cleavage in the skin, and why is this medically important?

INTEGRATE

CLINICAL VIEW 6.2

Tattoos

©Ingram Publishing RF

Tattoos are permanent images produced on the integument through the process of injecting a dye into the dermis. The dermis doesn't have a rapid cell turnover, so the injected dye remains in this area for a long time. Scar tissue surrounds the dye granules, which are too large for dendritic cells to ingest, and they become a permanent part of the dermis layer.

It is usually impossible to completely remove a tattoo, and some scarring may occur. Older methods include excision (cutting out the tattoo), **dermabrasion** (sanding down the tattooed skin), and **cryosurgery** (freezing the area of tattooed skin prior to its removal). Lasers are now used in some cases to break down the tattoo pigments. Newer tattoo inks have been introduced that allow for easier tattoo removal.

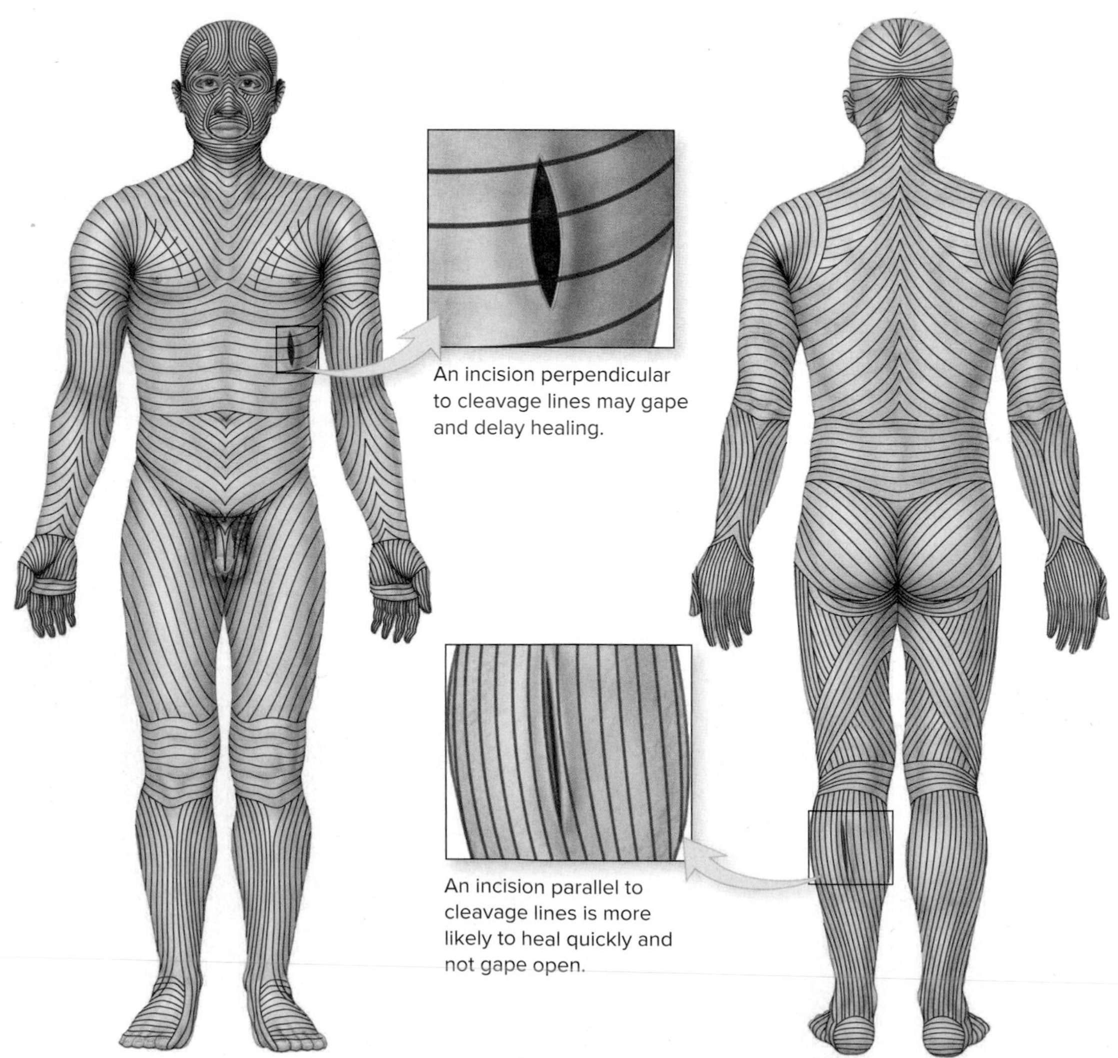

An incision perpendicular to cleavage lines may gape and delay healing.
An incision parallel to cleavage lines is more likely to heal quickly and not gape open.

Figure 6.7 Lines of Cleavage. Lines of cleavage partition the skin and indicate the predominant direction of underlying collagen fibers in the reticular layer of the dermis.

6.1c Subcutaneous Layer

LEARNING OBJECTIVE

7. List the functions of the subcutaneous layer.

Deep to the integument is the **subcutaneous** (sŭb-kū-tā′nē-ŭs; *sub* = beneath, *cutis* = skin) **layer,** also called the *hypodermis,* or *superficial fascia.* It is not considered a part of the integument. This layer consists of both areolar connective tissue and adipose connective tissue (see figure 6.1). In some locations of the body, adipose connective tissue predominates; thus, the subcutaneous layer is called **subcutaneous fat.** The connective tissue fibers of the reticular layer of the dermis are extensively interwoven with those of the subcutaneous layer to stabilize the position of the skin and bind it to the underlying structures. The subcutaneous layer pads and protects the body, acts as an energy reservoir, and provides thermal insulation. Drugs often are injected into the subcutaneous layer because its extensive vascular network promotes rapid absorption of the drugs.

Normally, the subcutaneous layer is thicker in women than in men, and its regional distribution differs between the sexes. Adult males tend to accumulate subcutaneous fat primarily at the neck, upper arms, abdomen, lower back, and buttocks, whereas adult females accumulate adipose connective tissue primarily in the breasts, buttocks, hips, and thighs.

Table 6.1 reviews the layers of the integument and the subcutaneous layer.

WHAT DID YOU LEARN?

What types of tissue form the subcutaneous layer?

Table 6.1 Integument Layers and the Subcutaneous Layer APR

Layer	Specific Layer	Description
INTEGUMENT: EPIDERMIS		
Stratum corneum; Stratum lucidum; Stratum granulosum; Stratum spinosum; Stratum basale	Stratum corneum	Most superficial layer of epidermis; 20–30 layers of dead, flattened, anucleate, keratin-filled keratinocytes
	Stratum lucidum	2–3 layers of anucleate, dead keratinocytes; seen only in thick skin (i.e., palms of hands, soles of feet)
	Stratum granulosum	3–5 layers of keratinocytes with distinct granules in cytoplasm; keratinization begins in this layer
	Stratum spinosum	Several layers of keratinocytes attached to neighbors by desmosomes; epidermal dendritic cells present
	Stratum basale	Deepest, single layer of cuboidal to low columnar keratinocytes in contact with basement membrane; cell division occurs here; also contains melanocytes and tactile cells
INTEGUMENT: DERMIS		
Papillary layer; Reticular layer	Papillary layer	Superficial layer of dermis; composed of areolar connective tissue; forms dermal papillae; houses capillaries and tactile receptors
	Reticular layer	Deeper layer of dermis; composed of dense irregular connective tissue surrounding and supporting hair follicles, sebaceous glands and sweat glands, nerves, and blood vessels
SUBCUTANEOUS LAYER		
	No specific layers	Not considered part of the integument; deep to dermis; composed of areolar and adipose connective tissue

INTEGRATE

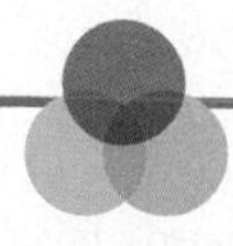

CONCEPT CONNECTION

The integumentary system protects almost all body systems:

- The integument provides the first line of defense against pathogens and toxins trying to enter the body (see section 22.3a).
- The integument helps in preventing fluid loss, and thus assists the cardiovascular system in maintaining blood volume (see section 25.2a).
- When muscular exertion generates heat (see section 10.1a), the integumentary system helps release excess heat via sweating and vasodilation.
- The integument provides abundant sensory information to the nervous system (see section 16.2a).
- The integument can synthesize vitamin D, which is required for calcium homeostasis (see section 7.6a).
- Dendritic cells are components of the immune system that initiate an immune response (see section 22.2a).
- Hairs in the nasal cavity help the respiratory system filter inspired air (see section 23.2a).
- The integumentary system and the urinary system both excrete nitrogenous waste products (see section 24.1).

6.1d Functions of the Integument

LEARNING OBJECTIVES

8. Name ways in which the integument protects the body and prevents water loss.
9. Describe the integument's involvement in calcium and phosphorus utilization.
10. Describe the integument's role in secretion and absorption.
11. Identify the immune cells that reside in the integument, and describe their actions.
12. Explain how the skin helps cool the body or retain warmth.
13. List the sensations detected by the skin's sensory receptors.

The epidermis serves a protective function, helps prevent water loss, and is involved with metabolic regulation. The epidermis and the dermis together secrete and absorb materials, and both play a role in immunity. We explore these functions in detail.

Protection

The epidermis acts as a physical barrier that protects the entire body from injury and trauma. It offers protection against harmful chemicals, toxins, microbes, and excessive heat or cold. The skin also protects deeper tissues from solar radiation, especially UV rays. When exposed to the sun, the melanocytes become more active and produce more melanin, thus giving the skin a tanned look. Even when you get a sunburn, the deeper tissues (muscles and internal organs) remain unaffected.

Prevention of Water Loss and Water Gain

The epidermis is water resistant, but not entirely waterproof. Some water is always lost through the skin when you sweat. More water is typically lost through transpiration, a process in which fluids slowly penetrate through the epidermis and then evaporate into the surrounding air (see section 25.2a). We realize how important the skin is in preventing water loss when treating individuals with severe burns. One of the main dangers is dehydration, because without the protective skin barrier, much larger amounts of water can escape from body tissues.

The skin also helps prevent water gain. If the skin were *not* water resistant, then each time you took a bath you would swell up like a sponge as the skin absorbed water!

Metabolic Regulation

Vitamin D_3, also called **cholecalciferol** (kō′lē-kal-sif′er-ol), is synthesized from a steroid precursor by the keratinocytes when they are exposed to ultraviolet radiation. Vitamin D_3 is then released into the blood and transported to the liver, where it is converted to another intermediate molecule (calcidiol), and then transported to the kidney, where it is converted to **calcitriol** (kal-si-trī′ol). Calcitriol is the active form of vitamin D and is considered a hormone (see section 7.6a). It increases absorption of calcium and phosphate from the small intestine into the blood, which results in a greater amount of the calcium being absorbed from the foods we eat. Thus, the synthesis of vitamin D_3 is important in regulating the levels of calcium and phosphate in the blood. As little as 10 to 15 minutes of direct sunlight a day provides your body with its daily vitamin D through this process.

The skin also is involved in other forms of metabolic regulation. It is able to convert some compounds to slightly different forms that may be used by the skin. For example, when topical corticosteroids (e.g., hydrocortisone) are applied to the skin, the corticosteroid medication enters the keratinocytes, where the cells convert and use the medication to stop inflammation and itching. (For further information about corticosteroids, see section 17.9a.)

WHAT DO YOU THINK?

During the Industrial Revolution, many children in cities spent little time outdoors and most of their time working in factories, leading to an increase in a disorder called rickets. Rickets is a bone disorder caused by inadequate vitamin D. Based on your knowledge of skin function, why do you think these children developed rickets?

Secretion and Absorption

Skin exhibits a secretory function when it discharges substances from the body during sweating. Sweating occurs to release excess heat from the body. Notice that sweat sometimes feels "gritty" because of the waste products being secreted onto the skin surface. The substances secreted in sweat include water, salts, and *urea*, a nitrogenous waste product of amino acid breakdown (see section 24.6e). The amount of water, salts, and urea secreted can be adjusted by the skin, and in so doing, the skin alters electrolyte levels in the body (see section 25.3). Thus, the skin also plays a role in electrolyte homeostasis. Sebum, excreted by sebaceous glands, lubricates the epidermis and hair, and helps make the integument water resistant.

The skin can absorb certain chemicals and drugs, such as estrogen from a birth control patch or nicotine from a nicotine patch. We refer to the skin as being **selectively permeable** because some materials are able to pass through it, whereas others are effectively blocked. In a process called **transdermal administration,** drugs that are soluble either in oils or in lipid-soluble carriers may be administered transdermally by an adhesive patch that keeps the drug in contact with the skin surface. Drugs administered in this way slowly penetrate the epidermis and can be absorbed into the blood vessels of the dermis. Transdermal patches are especially useful because they release a continual, slow absorption of the drug into the blood over a relatively long period of time. The barrier to drug diffusion through the epidermis requires that the concentration of the drug in the patch be relatively high.

Immune Function

In section 6.1a, we described a small population of immune cells within the stratum spinosum called epidermal dendritic cells. These cells play an important role in initiating an immune response against pathogens that have penetrated the skin (see section 22.2a). These cells, along with dendritic cells of the dermis, also mount an attack against epidermal cancer cells.

Temperature Regulation

We previously discussed temperature regulation in section 1.6b. Body temperature can be influenced by the vast capillary networks and sweat glands in the dermis. Dermal blood vessels have an important role in body temperature and blood pressure regulation. **Vasoconstriction** (vā′sō; *vas* = a vessel) means that the diameters of the vessels narrow, so relatively less blood is transported through them. Because less blood can flow through these dermal blood vessels, relatively more blood is transported through blood vessels deeper under the skin. The net effect is a shunting of blood *away* from the periphery of the body and toward deeper structures. Vasoconstriction of dermal blood vessels occurs to conserve heat. This is why we appear paler when we are exposed to cold temperatures. In addition, arrector pili (small bands of smooth muscle associated with hair follicles) are stimulated to contract to produce heat.

Conversely, **vasodilation** of the dermal blood vessels means that the diameter of the vessels increases, so relatively more blood is transported through them. The dermal blood vessels vasodilate so that more blood is transported close to the body surface. Here the net effect is a shunting of blood *to* the periphery of the body and *away* from deeper structures. Vasodilation of dermal blood vessels occurs to release excess heat. This additional blood flow through the dermis results in a more reddish/pinkish hue to the skin. This additional blood flow accounts for your face becoming flushed when you exercise. Consider how on a cold day your face may be paler than normal. When you come in from the cold, your face may become flushed as your body adjusts to the warmer temperature.

Sensory Reception

The dermis of the skin has an extensive **innervation,** which refers to its distribution of nerve fibers. Sensory nerve fibers in the skin (called *sensory receptors*) monitor stimuli in both the dermis and epidermis. For example, **tactile cells** (or *Merkel cells*) are large, specialized epithelial cells that stimulate specific sensory nerve endings when they are distorted by fine touch or pressure. In fact, seven major types of sensory receptors are housed within the skin to detect, distinguish, and interpret the many stimuli of our external environment that interact with our skin (see section 16.2a). In addition, motor nerve fibers extend through the skin to control glandular secretions and blood flow in blood vessels.

Figure 6.8 illustrates how the anatomy of the integument supports these various functions.

WHAT DID YOU LEARN?

8. How does the skin produce vitamin D?
9. Is the skin entirely waterproof? Explain.
10. What are some ways the skin can dissipate excess heat?

(a) Epidermis Functions

PROTECTION

Epidermal strata provide layers of protection against harmful chemicals, toxins, microbes, and excessive heat or cold. Skin also protects deeper tissues from UV radiation as melanocytes are stimulated to produce more melanin.

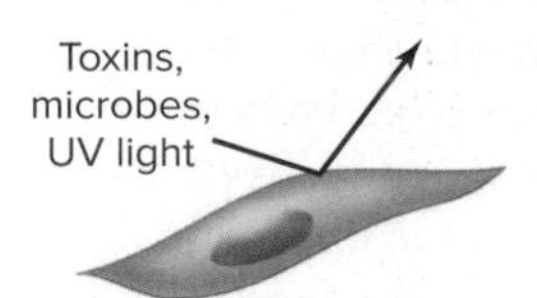

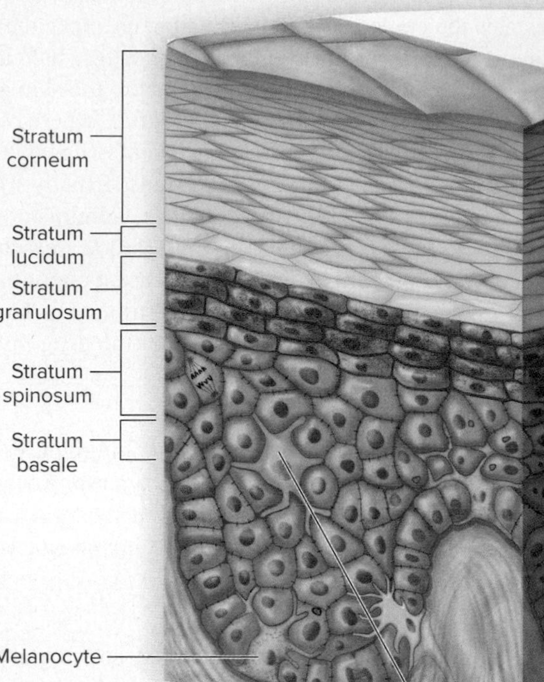

PREVENTION OF WATER LOSS AND WATER GAIN

The epidermis is water resistant and keeps water from either exiting or entering the skin easily.

METABOLIC REGULATION

Sunlight

Upon exposure to UV rays, keratinocytes produce vitamin D_3 and melanocytes are stimulated to produce more melanin, giving the skin a more tanned look.

SECRETION AND ABSORPTION

Materials (e.g., sebum, sodium, water, urea) secreted by dermal structures are released onto the epidermal surface. The skin is selectively permeable because some materials (e.g., certain drugs, like nicotine and estrogen within transdermal patches) may be absorbed while others are blocked.

Transdermal nicotine patch

IMMUNE FUNCTION

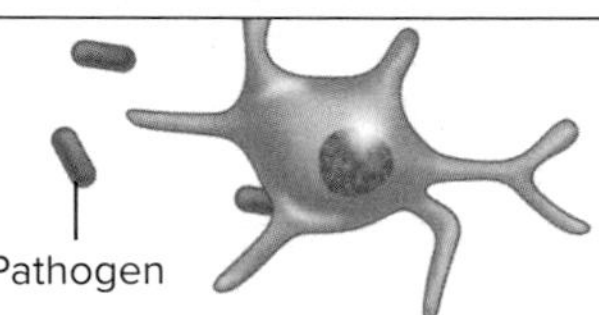

Epidermal dendritic cells engulf and destroy pathogens, alert the immune system to the presence of pathogens, and initiate an immune response. (Note: The dermis also contains its own dendritic cells.)

Most of the nail body appears
ing in the underlying capillaries;
white because there are no underl
and the proximal end of the nail bo
the **nail matrix,** which is the activ
lunula (lū′nū-lă; *luna* = moon) is
cular) area of the proximal end o
appearance because a thickened st
lying blood vessels.

Along the lateral and proximal
called **nail folds** overlap the nail.
onyx = nail), also known as the **cu**
epidermis extending from the mar
body. The **hyponychium** (hi-pō-n
epithelium underlying the free edg

WHAT DID YOU LEARN?

11 What is the difference between th
a fingernail?

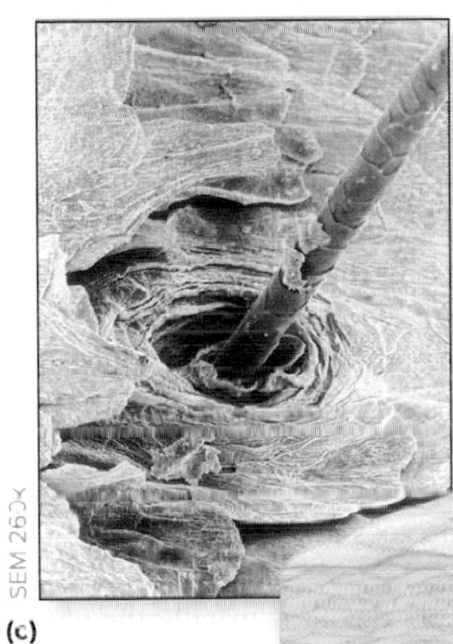

(c)

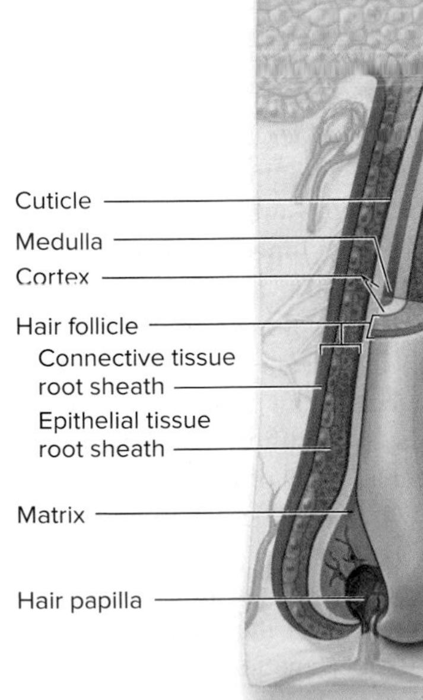

(a)

Figure 6.10 Hair. Hair is a de
(*b*) Photomicrographs of a hair folli
(*top b, bottom b*) ©McGraw-Hill Education/Al T

INTEGRATE **CONCEPT OVERVIEW**

Figure 6.8 How Integument Form Influences Its Functions.
(*a*) The epidermis is composed of multiple layers of keratinized epithelial cells that make it well suited for protection and prevention of water loss. It also participates in secretion and absorption, metabolic regulation, and immunity. (*b*) The dermis is composed of well-vascularized connective tissue that participates in secretion and absorption, temperature regulation, and sensory reception.

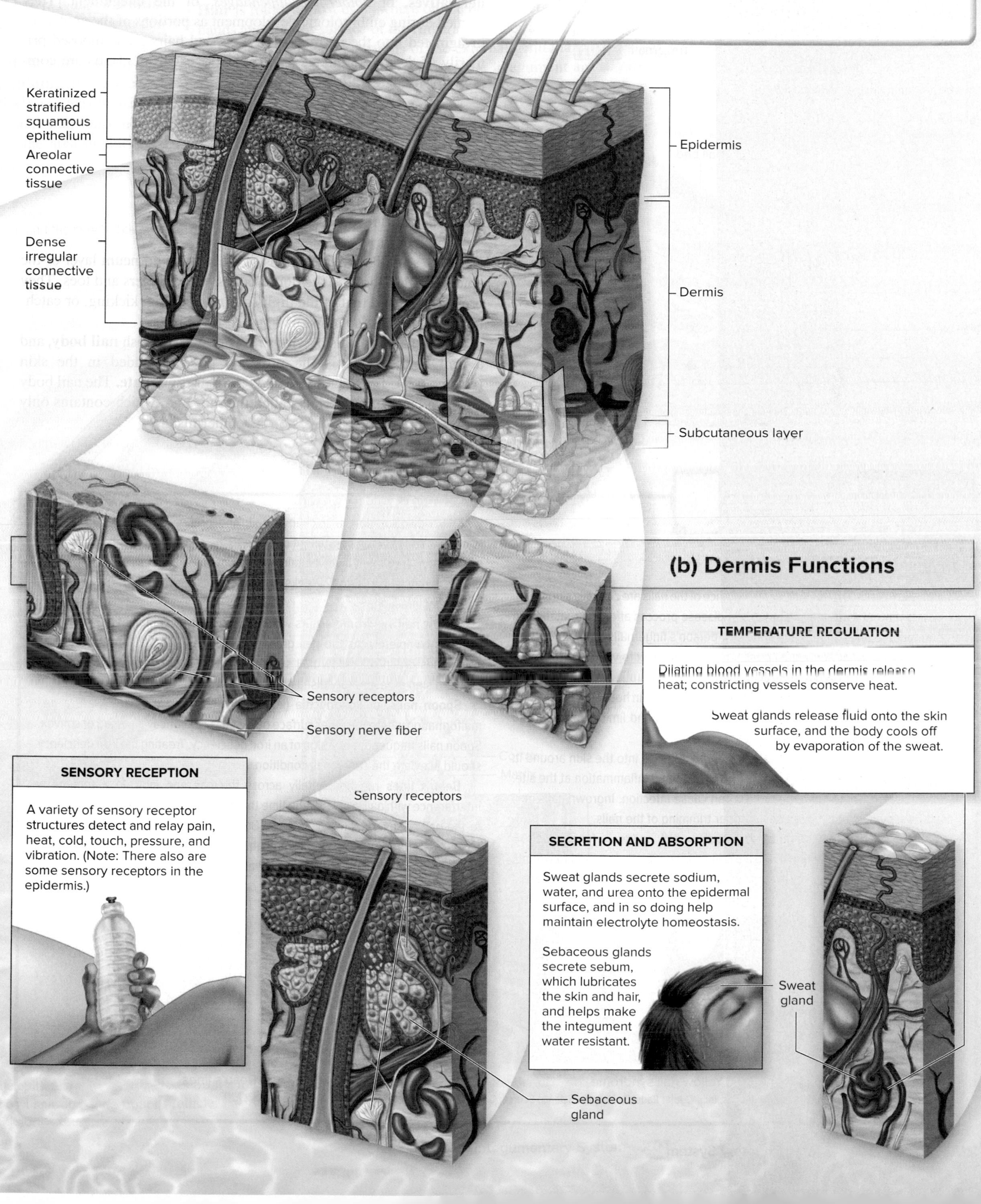

Keratinized stratified squamous epithelium
Areolar connective tissue
Dense irregular connective tissue
Epidermis
Dermis
Subcutaneous layer

(b) Dermis Functions

Sensory receptors
Sensory nerve fiber

SENSORY RECEPTION

A variety of sensory receptor structures detect and relay pain, heat, cold, touch, pressure, and vibration. (Note: There also are some sensory receptors in the epidermis.)

TEMPERATURE REGULATION

Dilating blood vessels in the dermis release heat; constricting vessels conserve heat.

Sweat glands release fluid onto the skin surface, and the body cools off by evaporation of the sweat.

Sensory receptors

SECRETION AND ABSORPTION

Sweat glands secrete sodium, water, and urea onto the epidermal surface, and in so doing help maintain electrolyte homeostasis.

Sebaceous glands secrete sebum, which lubricates the skin and hair, and helps make the integument water resistant.

Sebaceous gland
Sweat gland

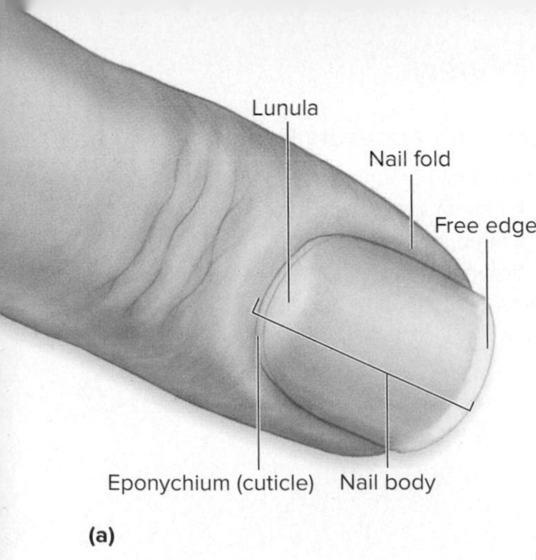

(a)

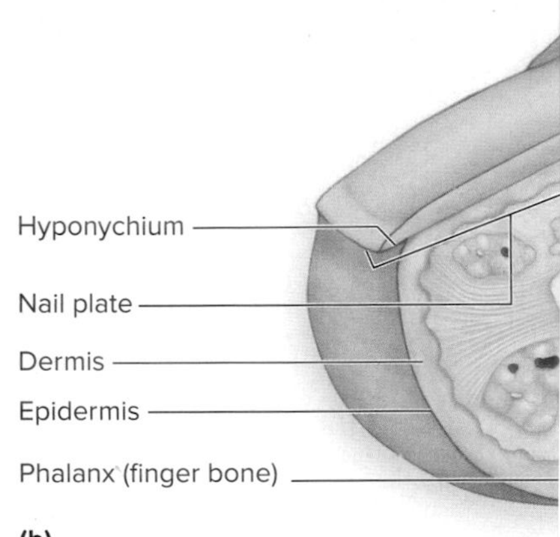

(b)

INTEGRATE

CLINICAL VIEW 6.3

Nail Disorders

Changes in the shape, structure, or appeara
A change may indicate the existence of a
throughout the body. In fact, the state of a
be indicative of his or her overall health. N

Brittle nails are prone to vertical spli
layers at the free edge. Overexposure to w
can cause brittle nails. Keeping the nails
water and chemicals can alleviate brittle r

An **ingrown nail** occurs when the edg
This painful condition is first characterized
If left untreated, some ingrown toenails c
result from overly tight shoes and improp

Onychomycosis (on′i-kō-mī-kō′sis; *o*
condition) is a fungal infection that occurs
and moisture. The fungus starts to grow

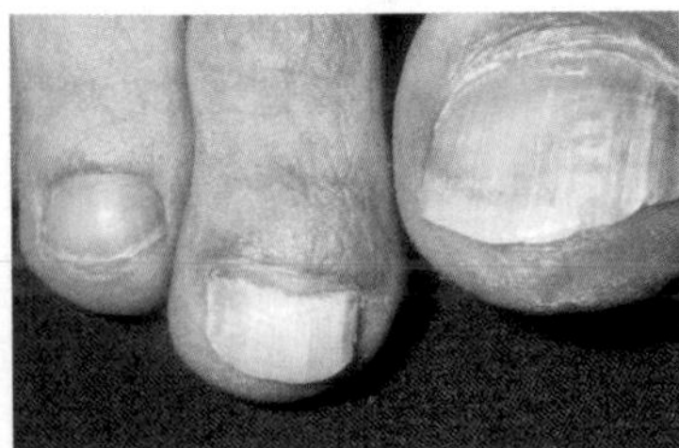

Onychomycosis
©Dr. P. Marazzi/Science Source; ©Logical Image

Merocrine Sweat Glands **Merocrine** (*eccrine*) **sweat glands** (figure 6.11*b*) are the most numerous and widely distributed sweat glands. The adult integument contains between 3 and 4 million merocrine sweat glands. They are simple, coiled, tubular glands that discharge their secretions directly onto the surface of the skin. The clear secretion they release by exocytosis (see section 5.1d) is called **sweat;** it consists of approximately 99% water and 1% other chemicals that include electrolytes (primarily sodium and chloride), metabolites (e.g., lactate), and waste products (urea and ammonia).

The major function of merocrine sweat glands is **thermoregulation,** which is the regulation of body temperature by evaporation of fluid from the skin (see sections 1.5b, 25.2a). Merocrine sweat gland secretions provide a means for the loss of both water and electrolytes. The secretions also may help eliminate a number of ingested drugs. Finally, merocrine sweat gland secretions provide some protection from environmental hazards both by diluting harmful chemicals and by preventing the growth of microorganisms (antibacterial and antifungal activity; see section 22.3a).

Apocrine Sweat Glands **Apocrine sweat glands** (figure 6.11*c*) are coiled, tubular glands that release their secretion into hair follicles in the axillae, around the nipples, in the pubic region, and in the anal region. Originally, these glands were called *apocrine* because their cells were thought to secrete their product by an apocrine mechanism (meaning that the apical portion of the cell's cytoplasm pinches off and, along with cellular components of the apical region, becomes the secretory product—see section 5.1d). Now, researchers have shown that both apocrine and merocrine sweat glands produce their secretion by exocytosis. However, the secretory portion of an apocrine gland has a much larger lumen than that of a merocrine gland, and the secretions they produce are different, so these glands continue to be called apocrine glands. The secretion they produce is viscous and cloudy, and it contains both proteins and lipids that are acted upon by bacteria to produce a distinct, noticeable odor. (Underarm deodorant is designed to mask the odor, whereas a deodorant with antiperspirant also helps prevent the formation of sweat.) These sweat glands become active and produce secretions beginning around puberty.

Sebaceous Glands

Sebaceous glands are holocrine glands (see section 5.1d) that produce an oily, waxy secretion called **sebum** (sē′bŭm; *sebum* = tallow) that is usually discharged into a hair follicle and onto the hair itself. Sebum acts as a lubricant to keep the skin and hair from becoming dry, brittle, and cracked. Sebum also has some bactericidal (bacteria-killing) properties. Several sebaceous glands may open onto a single follicle.

The secretion of sebum in both sexes is stimulated by hormones, especially androgens (male sex hormones). Sebaceous glands are relatively inactive during childhood; however, they are activated during puberty in both sexes, when the production of sex hormones increases (see section 28.1b).

Other Integumentary Glands

Some specialized glands of the integument are restricted to specific locations. Two important examples are the ceruminous glands and the mammary glands.

INTEGRATE

CLINICAL VIEW 6.4

Acne and Acne Treatments

Acne (ak′nē) is the term used to describe plugged sebaceous ducts. Acne typically becomes pronounced beginning at puberty, because increases in hormone levels stimulate sebaceous gland secretion, making the pores more prone to blockage. Acne is prevalent during the teenage years, although any age group may experience acne.

The types of acne lesions include

- **Comedo** (kom′ē-dō). A sebaceous gland plugged with sebum. An open comedo is called a **blackhead,** because the plugged material has a dark, blackish appearance. A closed comedo is called a **whitehead,** because the top surface is whitish in color.
- **Papule** (pap′ūl) and **pustule** (pŭs′tyūl). Both are dome-shaped lesions. Papules typically are fluid-filled, form red elevations on the skin, and do not contain pus. Papules may become pustules, which are filled with a mixture of white blood cells, dead skin cells, and bacteria (called pus).
- **Nodule** (nod′ūl). Similar to a pustule, but extending into the deeper skin layers and usually rupturing the hair follicle wall. Nodules can be prone to scarring.
- **Cyst.** A large, fluid-filled nodule that can become severely inflamed and painful and can lead to scarring of the skin.

Many medicinal treatments are available for acne, depending upon the type and severity. The effectiveness of the following medications varies from individual to individual. These treatments include benzoyl peroxide, salicylic (sal-i-sil′ik) acid, topical and oral antibiotics, topical vitamin A–like compounds (e.g., retinoids such as tretinoin [Retin-A]), and systemic retinoids such as isotretinoin (e.g., Accutane).

Other treatments include light chemical skin peels and comedo extraction by a dermatologist. Untreated acne, if severe, can lead to scarring, as can picking at the acne lesions.

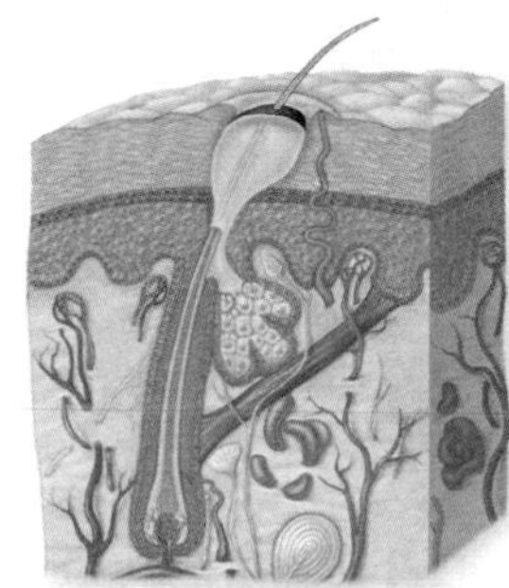

Blackhead (open comedo)

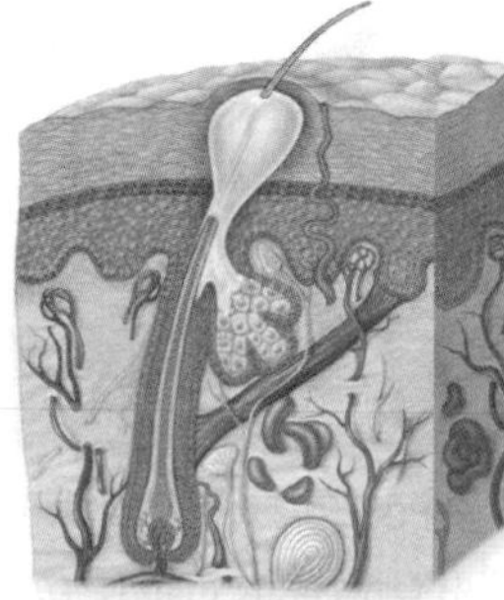

Whitehead (closed comedo)

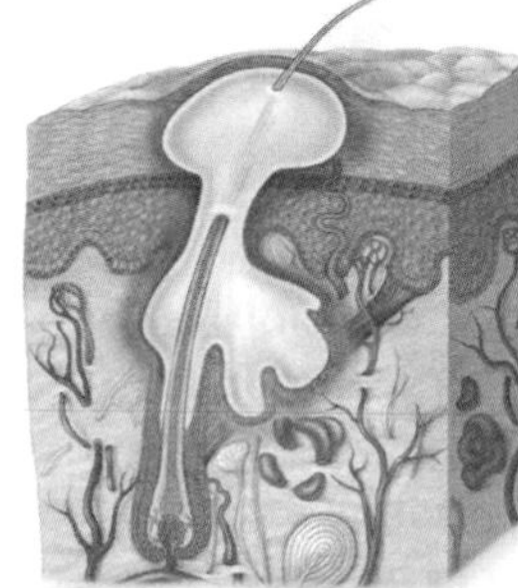

Pustule

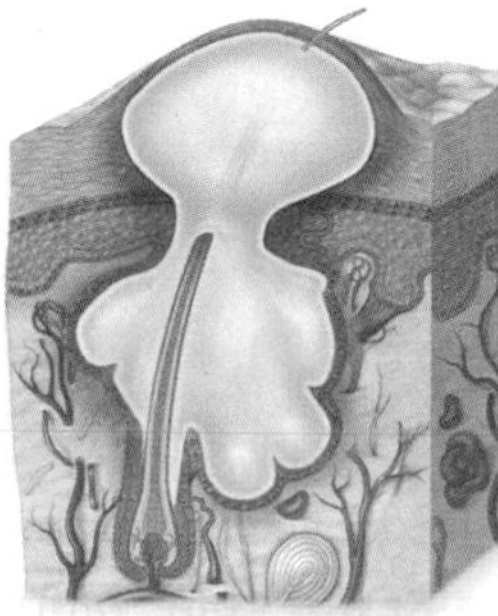

Nodule

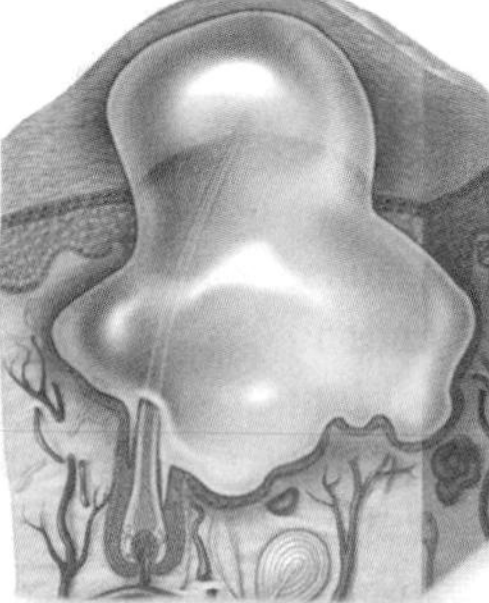

Cyst

Ceruminous (sĕ-rū′mi-nŭs) **glands** are modified apocrine sweat glands located only in the external acoustic meatus (ear canal), where their secretion forms a waterproof earwax called **cerumen** (sĕ-rū′men). Both cerumen and the tiny hairs in the meatus help trap foreign particles or small insects and keeps them from reaching the eardrum. Cerumen also helps lubricate the external acoustic meatus and eardrum (see section 16.5a).

The **mammary glands** of the breasts are modified apocrine sweat glands. Both males and females have mammary glands, but these glands become functional only in pregnant and lactating females, when they produce milk, a secretion that nourishes offspring. The development of the glands and its secretions are controlled by a complex interaction between gonadal and pituitary hormones, discussed in section 28.3f.

WHAT DID YOU LEARN?

14 How do apocrine sweat glands differ from merocrine sweat glands in terms of their location, secretions, and function?

15 What do sebaceous glands secrete, and where is this material secreted?

6.3 Repair and Regeneration of the Integumentary System

LEARNING OBJECTIVES

21. Distinguish between regeneration and fibrosis.
22. Describe the process of wound healing.

The components of the integumentary system exhibit a tremendous ability to respond to stressors, trauma, and damage. Repetitive mechanical stresses applied to the integument stimulate cell division in the stem cells of the stratum basale, resulting in a thickening of the epidermis and an improved ability to withstand stress. For example, walking about without shoes causes the soles of the feet to thicken, thus providing more protection for the underlying tissues.

Damaged tissues are normally repaired in one of two ways. The replacement of damaged or dead cells with the same cell type is called **regeneration.** This restores organ function. When regeneration is not possible because part of the organ is too severely damaged or its cells lack the capacity to divide, the body fills in the gap with scar (fibrous) tissue. This process of scar tissue deposition in connective tissue during healing is referred to as **fibrosis,** and it binds the damaged parts together. The replacement scar tissue is produced by fibroblasts and is composed primarily of collagen fibers. Some structural restoration occurs; however, functional activities are not restored.

Both regeneration and fibrosis may occur in the healing of damage to the skin. **Figure 6.12** illustrates stages in wound healing of the skin:

1. Cut blood vessels initiate bleeding into the wound. The blood brings clotting proteins, numerous leukocytes (white blood cells; see section 18.3c), and antibodies (see section 22.8).

INTEGRATE

CONCEPT CONNECTION

The process of wound repair requires stimulation and activation of the immune system. The immune system is described in detail in chapter 22.

INTEGRATE

CLINICAL VIEW 6.5

Psoriasis

Psoriasis (sō-rī′-ă-sis) is a chronic autoimmune disease (see Clinical View 22.4: "Autoimmune Disorders") of the skin that has periods of flare-ups and remissions throughout a person's lifetime. A type of leukocyte called a T-lymphocyte mistakenly attacks the keratinocytes, causing rapid overgrowth and overproduction of new skin cells. The normal sloughing-off cycle of keratinocytes is thrown out of balance, and the proliferation of cells develops into patches of typically whitish, scaly skin (called *plaques*) that form on the epidermal surface. The plaques are not contagious, but for the affected individual they may cause severe itching, pain, and skin cracking and bleeding. Any part of the skin may be affected, but common areas of involvement include the scalp, the limbs, and the buttocks. Treatments for psoriasis include topical corticosteroids, ultraviolet light therapy, and certain oral medications (e.g., methotrexate) that may interfere with the production of skin cells.

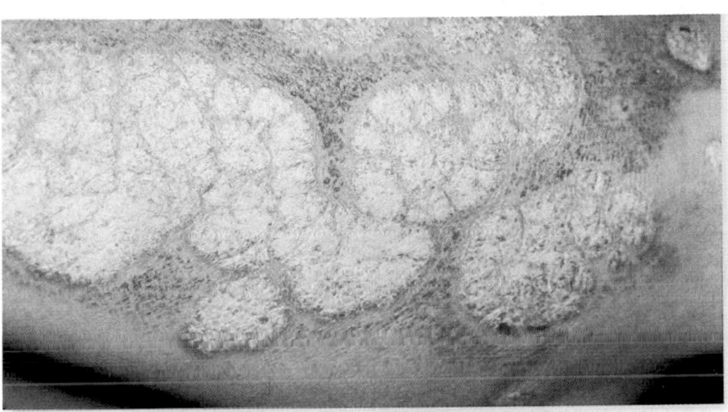

Multiple whitish plaques on the forearm of a person with psoriasis.

2. A blood clot forms, temporarily patching the edges of the wound together and acting as a barrier to prevent the entry of pathogens into the body. Internal to the clot, macrophages and neutrophils (two types of leukocytes; see section 18.3c) clean the wound of cellular debris. (For more information about blood clotting, see section 18.4.)
3. The cut blood vessels regenerate and grow in the wound. A soft mass deep in the wound becomes **granulation** (gran′ū-lā-shŭn) **tissue,** which is a vascular connective tissue that initially forms in a healing wound. Macrophages within the wound begin to remove the clotted blood. Fibroblasts produce new collagen fibers in the region.
4. Epithelial regeneration of the epidermis occurs due to division of epithelial cells at the edge of the wound. These new epithelial cells migrate over the wound, moving internally to the now superficial remains of the clot (the scab). The connective tissue is replaced by fibrosis.

The skin repair and regeneration process is dependent on the extent of the injury. The wider and deeper the surface affected, the longer it takes for skin to be repaired. Additionally, the area under repair usually is more susceptible to complications due to fluid loss and infection. As the severity of the damage increases, the repair and regeneration ability of the integument is strained and its return to its original condition becomes much less likely. Some integumentary

INTEGRATE

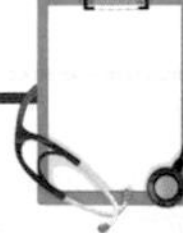

CLINICAL VIEW 6.7

Botox and Wrinkles

Many individuals have explored ways to diminish the appearance of age-related wrinkles. One popular treatment for wrinkles caused by repeated facial muscle expression is **botulinum toxin type A** (Botox). This medicine is derived from the toxin produced by the bacterium *Clostridium botulinum*. Although large doses of this toxin may be deadly, small therapeutic doses temporarily block nerve impulses to the facial expression muscles, thereby decreasing or eliminating the wrinkles they produce.

The procedure is done in a doctor's office, where the doctor injects Botox in the specific facial muscles responsible for the wrinkles, such as frown lines and crows' feet. The effect is temporary, and an individual must repeat the procedure after about 4 months, as the muscles regain their function.

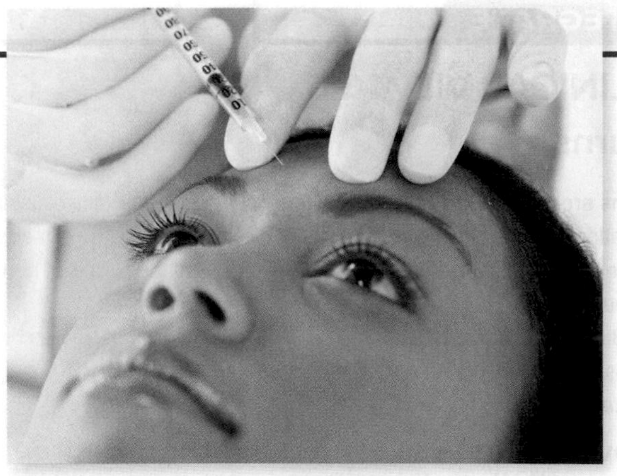

Botox is relatively safe, but some individuals may experience adverse effects, and overuse of Botox can produce a face that appears frozen and devoid of facial expression.

6.4a Development of the Integument and Its Derivatives

 **LEARNING OBJECTIVES**

23. Describe how integument develops from two germ layers.

24. Explain the developmental origins of nails, hair, and glands.

By the end of week 7 of development, the ectoderm forms a layer of squamous epithelium that flattens and then becomes both a covering layer, called the **periderm,** and an underlying basal layer. The basal layer will form the stratum basale and all other epidermal layers. By week 21, the stratum corneum and friction ridges form. During the fetal period (weeks 9-38), the periderm is eventually sloughed off. The sloughed-off cells mix with sebum secreted by the sebaceous glands, producing a waterproof protective coating called the **vernix caseosa** that coats the skin of the fetus.

The dermis is derived from mesoderm. During weeks 3-8, this mesoderm becomes **mesenchyme.** The mesenchymal cells begin to form the components of the dermis at about 11 weeks.

Fingernails and toenails start to form in the tenth week of development. The fingernails reach the tips of the fingers by 32 weeks, whereas the toenails become fully formed by about 36 weeks.

Hair follicles begin to appear between 9 and 12 weeks of development as pockets of cells, called **hair buds,** invade the dermis from the overlying stratum basale of the epidermis. These hairs do not become easily recognizable in the fetus until about 20 weeks. Finally, sweat and sebaceous glands develop from the stratum basale of the epidermis and first appear at about 20 weeks on the palms and soles and later in other regions.

 **WHAT DID YOU LEARN?**

 17 What two primary germ layers form the integument?

6.4b Aging of the Integument

 **LEARNING OBJECTIVES**

25. Explain changes to the skin with age.

26. List factors that contribute to skin aging.

Although some adolescents develop acne when they enter puberty, most skin changes do not become obvious until an individual reaches middle age. Then, the skin repair processes take longer to complete because of a reduced number and activity of stem cells. Skin repair and regeneration activities that took 3 weeks in a healthy young person often take twice that time for a person in the seventh decade of life. Additionally, the reduced stem cell activity in the epidermis results in a thinner skin that is less likely to protect against abrasive, mechanical trauma.

Collagen fibers in the dermis decrease in number and organization, and elastic fibers lose elasticity. Years of particular facial expressions (squinting, smiling) produce crease lines in the integument. As a result, the skin forms wrinkles and becomes less resilient. In addition, the skin's immune responsiveness is diminished by a decrease in the number and efficiency of epidermal dendritic cells. Also, hair follicles either produce thinner hairs or stop production entirely.

Chronic overexposure to UV rays can damage the DNA in epidermal cells and accelerate aging, and it is the predominant factor in the development of nearly all skin cancers. Skin cancer is the most common type of cancer. It occurs most frequently on the head and neck regions, followed by other regions commonly exposed to the sun. Fair-skinned individuals, especially those who had severe sunburns as children, are most at risk for skin cancer.

Skin cancer can arise in anyone at any age. Individuals should use sunscreen regularly and avoid prolonged exposure to the sun. An individual should regularly and thoroughly inspect his or her skin for any changes, such as an increase in the number or size of moles or the appearance of new skin lesions. In addition, a person should be examined routinely by a dermatologist. **Table 6.2** compares and describes the three main types of skin cancer.

WHAT DID YOU LEARN?

18 How do UV rays contribute to skin aging?

Table 6.2 Skin Cancer

BASAL CELL CARCINOMA

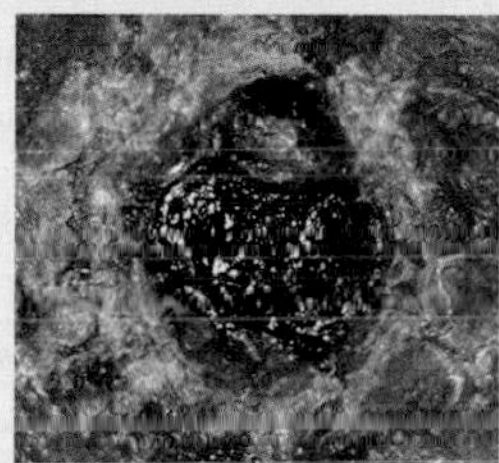

- Most common type of skin cancer
- Least dangerous type, as it seldom metastasizes (i.e., spreads to other locations within the body)
- Originates in stratum basale
- First appears as small, shiny elevation that enlarges and develops central depression with pearly edge
- Usually occurs on face
- Treated by surgical removal of lesion

SQUAMOUS CELL CARCINOMA

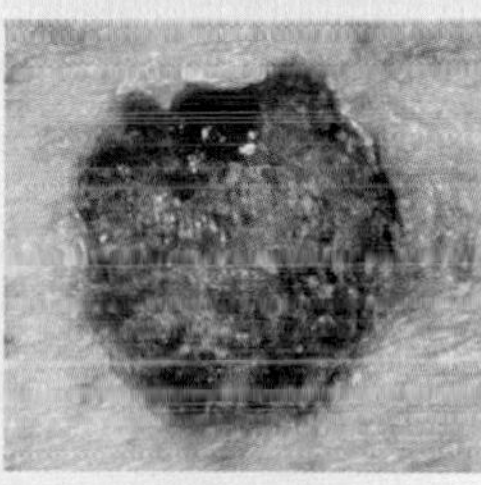

- Arises from keratinocytes of stratum spinosum
- Lesions usually appear on scalp, ears, lower lip, or dorsum of hand.
- Early lesions are raised, reddened, scaly; later lesions form concave ulcers with elevated edges.
- Treated by early detection and surgical removal of lesion
- May metastasize to other parts of the body

MALIGNANT MELANOMA

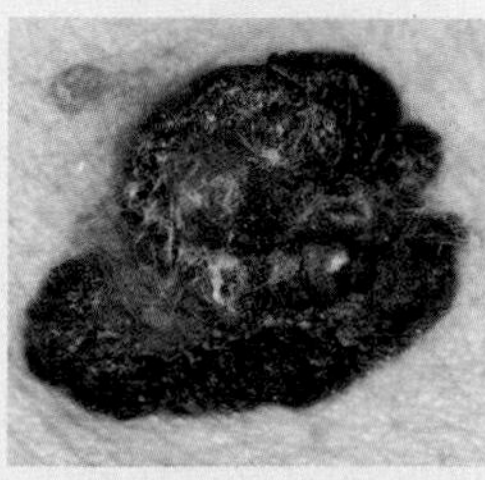

- Most deadly type of skin cancer due to aggressive growth and metastasis
- Arises from melanocytes, usually in a preexisting mole
- Individuals at increased risk include those who have had severe sunburns, especially as children.
- Characterized by change in mole diameter, color, shape of border, and symmetry
- Survival rate improved by early detection and surgical removal of lesion
- Advanced cases (metastasis of disease) are difficult to cure and are treated with chemotherapy, interferon therapy, and radiation therapy.

The usual signs of melanoma may be easily remembered using the **ABCDE rule.** Report any of the following changes in a birthmark or mole to your physician:

A = **A**symmetry: One-half of a mole or birthmark does not match the other.

B = **B**order: Edges are notched, irregular, blurred, or ragged.

C = **C**olor: Color is not uniform; differing shades (usually brown or black and sometimes patches of white, blue, or red) may be seen.

D = **D**iameter: Affected area is larger than 6 mm (about 1/4 inch) or is growing larger.

E = **E**volving: Change in the size, shape, or color of a mole or a change in symptoms, such as how a mole feels (how itchy or tender it feels) or what happens on the surface of a mole (especially bleeding)

2. During anatomy lab, Susan scratched her arm and noticed that some skin cells had sloughed off. She prepared a slide of these cells and examined them under a microscope. What characteristics do you expect the cells to have?
   a. polygonal cells with prominent nuclei
   b. cuboidal cells, some undergoing mitosis
   c. flattened anucleate cells
   d. oval cells surrounded by abundant collagen
3. While running to class, Jennifer slipped and skinned her knee. The wound appeared superficial, yet there was extensive bleeding. Based on this information, and her knowledge of integument composition, she determined that the wound penetrated
   a. the stratum corneum layer of the epidermis only.
   b. all layers of the epidermis but not the dermis.
   c. all layers of the epidermis and part of the dermis.
   d. all layers of the epidermis and dermis, as well as the subcutaneous layer.

Can You Synthesize What You've Learned?

1. When you are outside on a cold day, your skin is much paler than normal. Later, when you enter a warm room, your face becomes flushed. What are the reasons for the change in color of your face?
2. Teri was involved in a chemical accident where she suffered third-degree burns on 30% of her body. What potential complications could Teri develop as a result of these burns? As Teri's physician, how would you help minimize these complications?
3. At the age of 50, John noticed that one of the moles on his face looked different than normal. The mole appeared larger than normal, darkened, and asymmetrical. John's dermatologists suspected skin cancer. Based on this description, what type of skin cancer is most likely, and should John be concerned?

INTEGRATE

ONLINE STUDY TOOLS

connect | SMARTBOOK® | APR

The following study aids may be accessed through Connect.

Clinical Case Study: A Case of Malignant Melanoma in a Young Woman

Interactive Questions: This chapter's content is served up in a number of multimedia question formats for student study

SmartBook: Topics and terminology include composition of the integument; integumentary structures derived from the epidermis; functions of the epidermis; repair and regeneration of the integumentary system; development and aging of the integumentary system

Anatomy & Physiology Revealed: Topics include thick skin and subcutaneous tissue; thick and thin skin magnifications of dermis and epidermis; fingernail structure; hair follicle structure; sebaceous glands

Design Elements: **Learning Objective** icon: ©McGraw-Hill Education; **What Do You Think?** icon: ©McGraw-Hill Education; **What Did You Learn?** icon: ©McGraw-Hill Education; **Learning Strategy** icon: ©Slavoljub Pantelic/Shutterstock RF; **Clinical View** icon: ©Laia Design Studio/Shutterstock RF; **Concept Connection** icon: ©McGraw-Hill Education; **Challenge Yourself** icon: ©McGraw-Hill Education; **Do You Know the Basics?** icon: ©McGraw-Hill Education; **Can You Apply What You've Learned?** icon: ©McGraw-Hill Education; **Can You Synthesize What You've Learned?** icon: ©McGraw-Hill Education; **Online Study Tools** icon: ©ayzek/Shutterstock RF

Bone resorption is a proces
by substances released from oste
adjacent to the bone. Proteolytic
within the osteoclasts chemically
lagen fibers and proteoglycans)
acid (HCl) dissolves the mineral p
bone matrix. The liberated calc

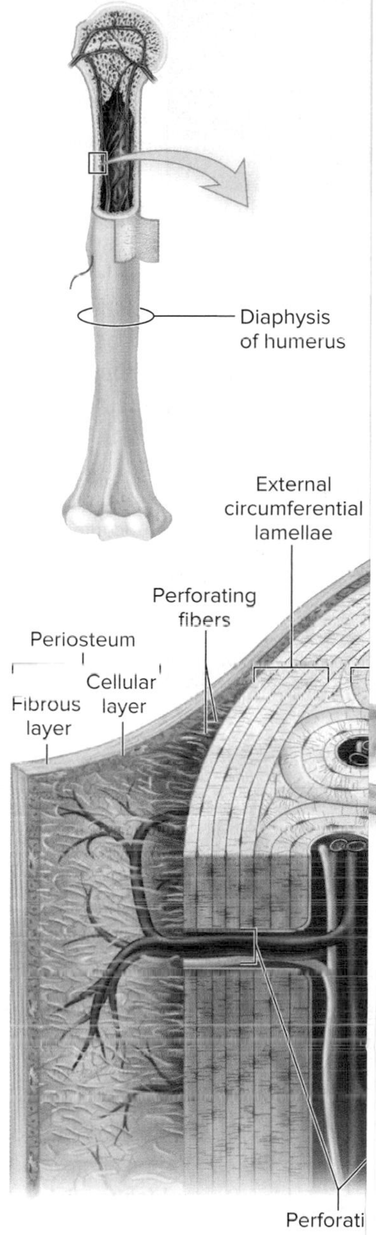

(a) Section of humerus

Figure 7.7 Components of arrangement of osteons within (*b*) th spongy bone. AP|R

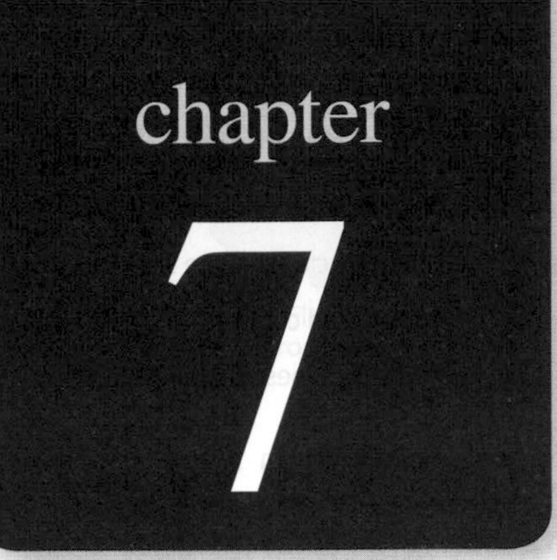

Skeletal System: Bone Structure and Function

Module 5: Skeletal System

INTEGRATE

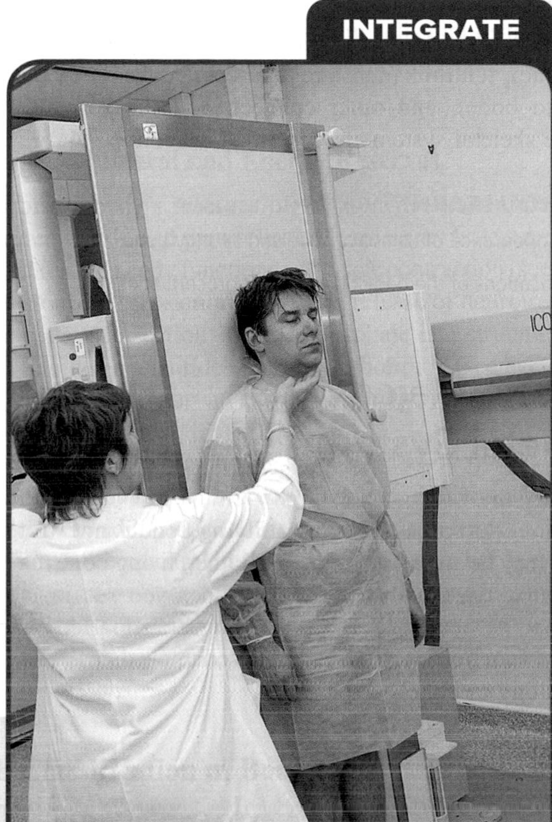

©AJPhoto/Hôpital Américain/Science Source

CAREER PATH
Radiologist

Radiologists and radiology technicians use an array of medical imaging techniques (such as radiography, ultrasound, CT, and MRI) to produce images that view details of internal body structures. These imaging techniques tend to be noninvasive and help in the diagnosis of internal pathology and trauma. All radiologists must be able to accurately identify skeletal structures and distinguish them from associated muscles and soft tissues. In addition, the radiologist must be able to tell the difference between a broken bone and a growing bone, because both may appear similar in radiographs to the layperson.

Mention of the skeletal system conjures up images of dry, supposedly lifeless bones in various sizes and shapes. But the **skeleton** (skel′ĕ-ton; *skeletos* = dried) is much more than a supporting framework for the soft tissues of the body. The skeletal system is composed of dynamic, living tissues; it interacts with all of the other organ systems and continually rebuilds and remodels itself.

We begin this chapter with a brief description of the skeletal system and a detailed discussion of bone anatomy, the primary organ of this system. Then we examine several important concepts of bone physiology, including cartilage growth, bone formation, bone growth and bone remodeling, the regulation of blood calcium, and the effects of aging on the skeletal system. Our chapter concludes with a discussion of bone fracture and repair.

INTEGRATE

CLINICAL VIEW 7.2
Osteitis Deformans

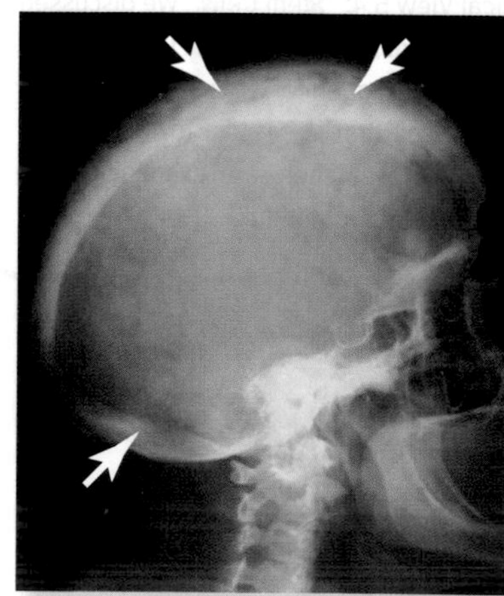

Lateral x-ray of a skull with osteitis deformans. White arrows indicate areas of excessive bone deposition.

normal osteoclasts). As a result, these lar higher rate. In response to this excessive deposit additional bone, but this new bon making it more susceptible to deformatio commonly affected include the pelvis, sku and tibia (leg bone). Initial symptoms in Eventually, the lower limb bones may bow, a and enlarged.

reflecting the activity of these cells. C tant function of synthesizing and s organic form of bone matrix called resemblance). Osteoid later calcifie deposition. As a consequence of this oid, osteoblasts become entrapped w and secrete, and thereafter they diffe

Osteocytes (*kytos* = a hollow [derived from osteoblasts that have lo when enveloped by calcified osteoid of the original neighboring osteobl become osteocytes. Osteocytes ma detect mechanical stress on a bone. blasts are signaled, and it may result i matrix at the surface.

Osteoclasts (os′tē-ō-klast; *klastc* nuclear, phagocytic cells. They are d row cells similar to those that produ section 18.3c). These cells exhibit a contact the bone, which increases th the bone. An osteoclast is often loc depression or pit on the bone surface (*Howship's lacuna*). Osteoclasts are bone in an important process called shortly).

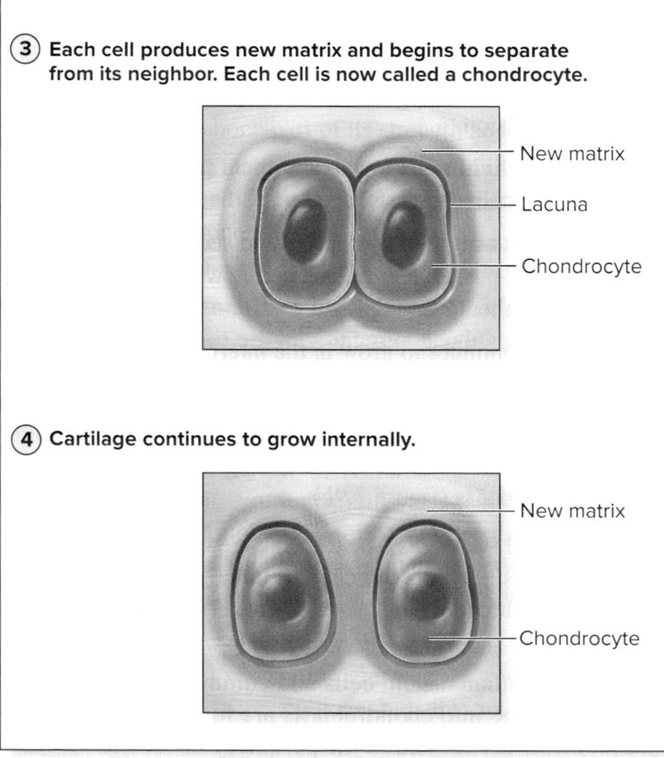

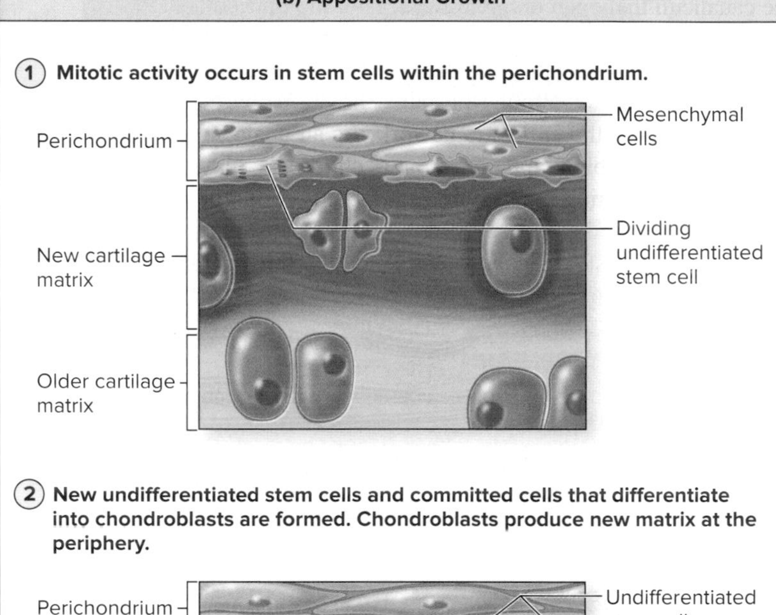

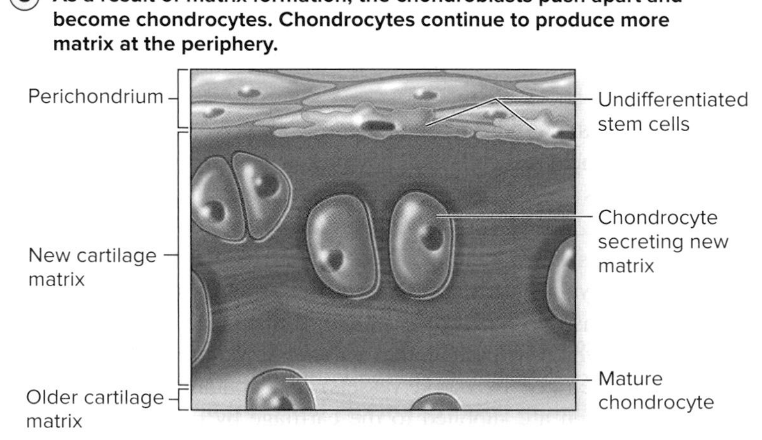

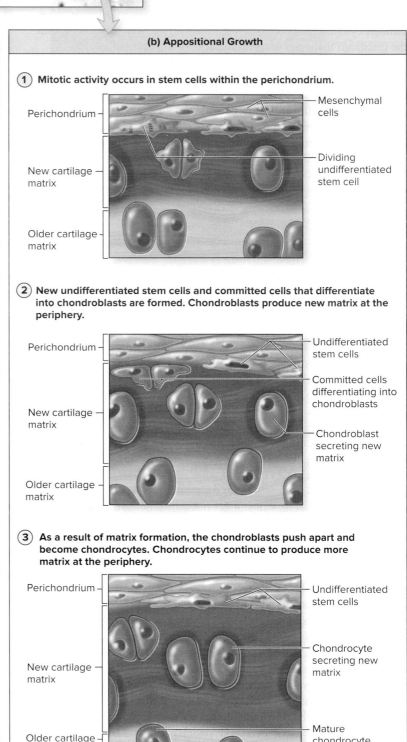

Figure 7.9 Formation and Growth of Cartilage. Cartilage grows either (*a*) from within by interstitial growth or (*b*) at its periphery (edge) by appositional growth. AP|R

During early embryonic development, both interstitial and appositional cartilage growth occur simultaneously. Note that interstitial growth declines rapidly as the cartilage matures because the cartilage becomes semirigid, and it is no longer able to expand. Further growth can occur only at the periphery of the tissue, so later growth is primarily appositional. Once the cartilage is fully mature, new cartilage growth typically stops. Thereafter, cartilage growth usually occurs only after injury to the cartilage, yet this growth is limited due to the lack of blood vessels in the tissue.

WHAT DID YOU LEARN?

12 Where do interstitial and appositional growth of cartilage occur?

7.4 Bone Formation

Ossification (os′i-fi-kā′shŭn; *facio* = to make), or **osteogenesis** (os′tē-ō-jen′ĕ-sis; *genesis* = beginning), refers to the formation and development of bone connective tissue. Ossification begins in the embryo and continues as the skeleton grows during childhood and adolescence. By the eighth through twelfth weeks of embryonic development, the skeleton begins forming from either thickened condensations of mesenchyme (intramembranous ossification) or a hyaline cartilage model of bone (endochondral ossification).

7.4a Intramembranous Ossification

LEARNING OBJECTIVES

16. Identify bones that are produced by intramembranous ossification.
17. Explain the four main steps in intramembranous ossification.

Intramembranous (in′tră-mem′brā-nŭs) **ossification** literally means "bone growth within a membrane." It is so named because the thin layer of mesenchyme in these areas is sometimes referred to as a membrane. Intramembranous ossification also is called *dermal ossification* because the mesenchyme that is the source of these bones is in the area of the future dermis. Recall from sections 5.2a and 5.2c that mesenchyme is an embryonic connective tissue that has mesenchymal cells and abundant ground substance.

Intramembranous ossification produces the flat bones of the skull (e.g., frontal bone), some of the facial bones (e.g., zygomatic bone, maxilla), the mandible (lower jaw), and the central part of the clavicle (collarbone). It begins when mesenchyme becomes thickened and condensed with a dense supply of blood capillaries, and it continues in the following steps **(figure 7.10)**:

1. **Ossification centers form within thickened regions of mesenchyme beginning at the eighth week of development.** Some cells in the thickened, condensed mesenchyme divide, and the committed cells that are formed then differentiate into osteoprogenitor cells. Some osteoprogenitor cells become osteoblasts and begin to secrete osteoid. Multiple ossification centers develop within the thickened mesenchyme as the number of osteoblasts increases.
2. **Osteoid undergoes calcification.** Osteoid formation is quickly followed by calcification, as calcium salts are deposited onto the osteoid and then they crystallize (solidify). When calcification entraps osteoblasts within lacunae in the matrix, the entrapped cells become osteocytes.
3. **Woven bone and its surrounding periosteum form.** Initially, the newly formed bone connective tissue is immature and not well organized, a type called **woven**

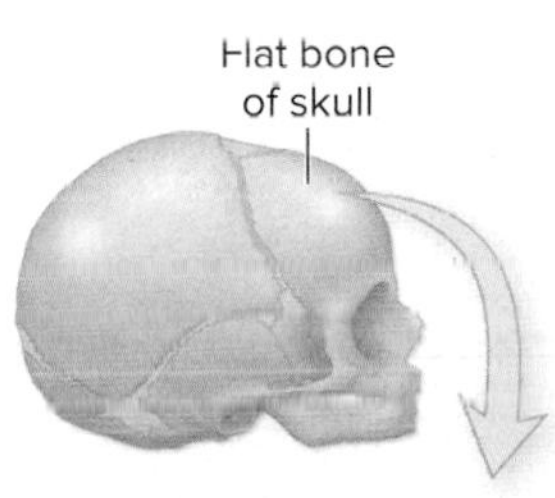

Figure 7.10 Intramembranous Ossification. A flat bone in the skull forms from mesenchymal cells in a series of continuous steps.

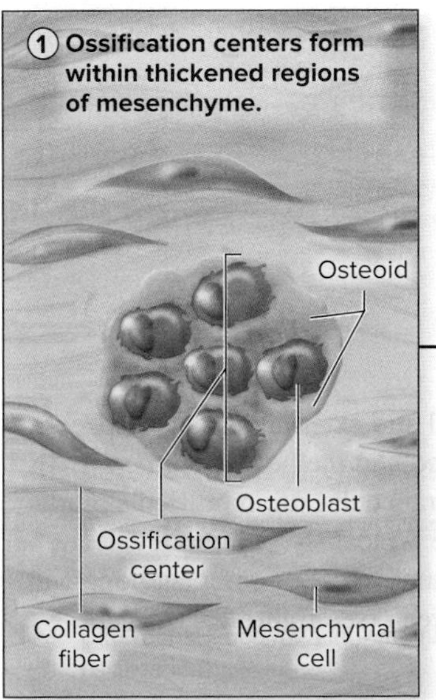

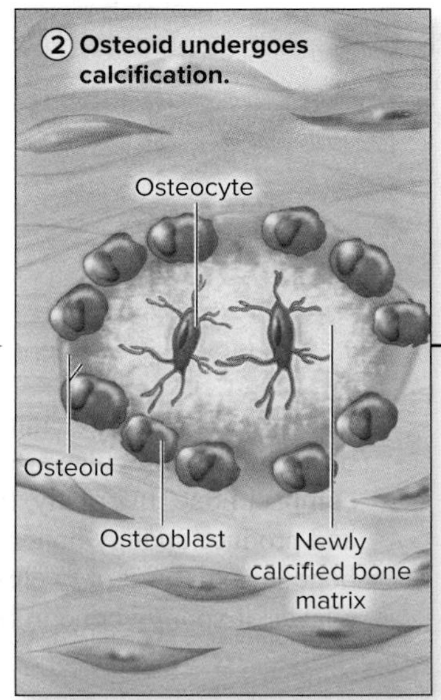

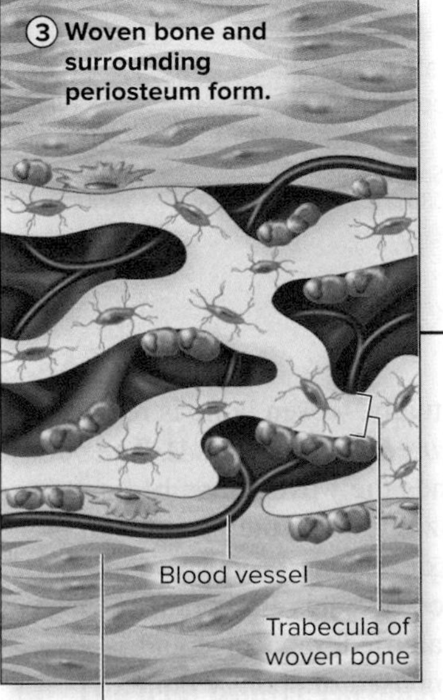

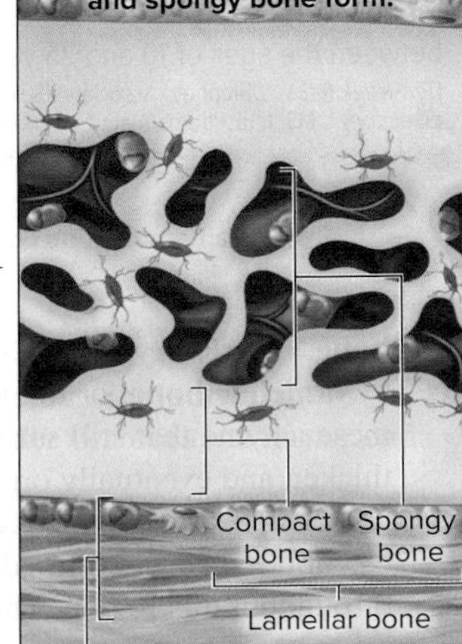

INTEGRATE

LEARNING STRATEGY

Endochondral bone growth is a complex process. Before trying to remember every detail, first learn the following basics:

1. A hyaline cartilage model of bone forms.
2. Bone first replaces hyaline cartilage in the diaphysis.
3. Next, bone replaces hyaline cartilage in the epiphyses.
4. Eventually, bone replaces hyaline cartilage everywhere, except the epiphyseal plates and articular cartilage.
5. By a person's late 20s, all epiphyseal plates typically have ossified, and lengthwise bone growth is complete.

forms. Chondrocytes are trapped within lacunae, and a perichondrium surrounds the cartilage.

(2) **Cartilage calcifies, and a periosteal bone collar forms.** Within the center of the cartilage model (future diaphysis), chondrocytes start to hypertrophy (enlarge) and resorb (eat away) some of the surrounding cartilage matrix, producing larger holes in the matrix. As these chondrocytes enlarge, the cartilage matrix begins to calcify. Chondrocytes in this region die and disintegrate because nutrients cannot diffuse to them through this calcified matrix. The result is a calcified cartilage shaft with large holes where living chondrocytes had been.

As the cartilage in the shaft is calcifying, blood vessels grow toward the cartilage and start to penetrate the perichondrium around the shaft. Stem cells within the perichondrium divide to form osteoblasts. The osteoblasts develop as this supporting connective tissue becomes highly vascularized, and the perichondrium becomes a periosteum. The osteoblasts within the internal layer of the periosteum start secreting a layer of osteoid around the calcified cartilage shaft. The osteoid hardens and forms a periosteal bone collar around this shaft.

(3) **The primary ossification center forms in the diaphysis.** A growth of capillaries and osteoblasts, called a **periosteal bud,** extends from the periosteum into the core of the cartilage shaft, invading the spaces where the living chondrocytes had been. The remains of the calcified cartilage serve as a template on which osteoblasts begin to produce osteoid. This region is called the **primary ossification center** because it is the first major center of bone formation. Bone development extends in both directions toward the epiphyses from the primary ossification center. Healthy bone connective tissue quickly displaces the calcified, degenerating cartilage in the shaft. Most, but not all, primary ossification centers have formed by the twelfth week of development.

(4) **Secondary ossification centers form in the epiphyses.** The same basic process that formed the primary ossification center occurs later in the epiphyses. Beginning around the time of birth, the hyaline cartilage in the center of each epiphysis calcifies and begins to degenerate. Epiphyseal blood vessels and osteoprogenitor cells enter each epiphysis. **Secondary ossification centers** form as bone displaces calcified cartilage. Note that not all secondary ossification centers form at birth; some form later in childhood. As the secondary ossification centers form, osteoclasts resorb some bone matrix within the diaphysis, creating a hollow medullary cavity.

(5) **Bone replaces almost all cartilage, except the articular cartilage and epiphyseal cartilage.** By late bone development, almost all of the hyaline cartilage has been

INTEGRATE

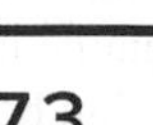

CLINICAL VIEW 7.3

Forensic Anthropology: Determining Age at Death

During endochondral ossification, the epiphyseal plates ossify and fuse to the rest of the bone in an orderly manner. The timing of such fusion is well known. If an epiphyseal plate has not yet ossified, the diaphysis and epiphysis are still two separate pieces of bone. Thus, a skeleton that displays separate epiphyses and diaphyses (as opposed to whole fused bones) is that of a juvenile rather than an adult. Forensic anthropologists use this anatomic information to help determine the age of skeletal remains.

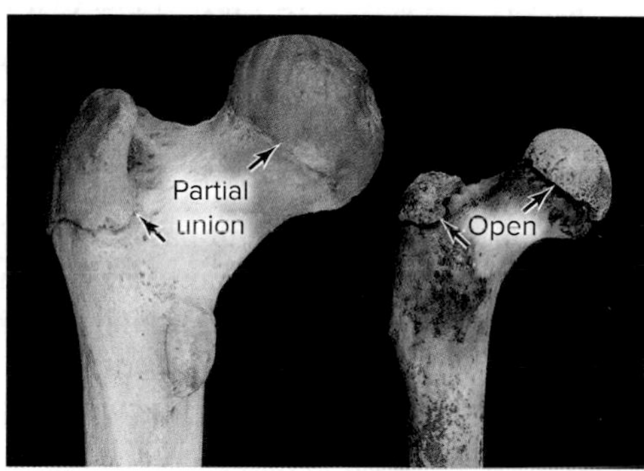

(Left) *Partial union—a femur with partially fused epiphyses.*
(Right) *Open—no fusion between the epiphyses and the diaphysis.*
©David Hunt/Smithsonian Institution

Fusion of an epiphyseal plate is progressive and is usually scored as follows:

- **Open** (no bony fusion or union between the epiphysis and the other bone end)
- **Partial union** (some fusion between the epiphysis and the rest of the bone, but a distinct line of separation may be seen)
- **Complete union** (all visible aspects of the epiphysis are united to the rest of the bone)

When determining the age at death from skeletal remains, the skeleton will be older than the oldest complete union and younger than the youngest open center. For example, if one epiphyseal plate that typically fuses at age 17 is completely united, but another plate that typically fuses at age 19 is open, the skeleton is that of a person between the ages of 17 and 19.

Current standards for estimating age based upon epiphyseal plate fusion have primarily used male skeletal remains. Female epiphyseal plates tend to fuse approximately 1 to 2 years earlier than those of males, so this fact needs to be considered when estimating the age of a female skeleton. Further, population differences may exist with some epiphyseal plate unions. With these caveats in mind, the following table lists standards for selected epiphyseal plate unions.

Bone	Male Age at Complete Epiphyseal Union
Humerus, lateral epicondyle	11–16 (female: 9–13)
Humerus, medial epicondyle	11–16 (female: 10–15)
Humerus, head	14.5–23.5
Proximal radius	14–19
Distal radius	17–22
Distal fibula and tibia	14.5–19.5
Proximal tibia	15–22
Femur, head	14.5–21.5
Distal femur	14.5–21.5
Clavicle	19–30

displaced by bone. Hyaline cartilage remains as articular cartilage only on the articular surface of each epiphysis and at the epiphyseal plates.

⑥ **Lengthwise growth continues until the epiphyseal plates ossify and form epiphyseal lines.** Lengthwise bone growth continues into puberty until the epiphyseal plate is converted to the epiphyseal line, indicating that the bone has reached its adult length. Depending upon the bone, most epiphyseal plates ossify to become epiphyseal lines between the ages of 10 and 25. (The last epiphyseal plates to ossify are those of the clavicle in the late 20s.)

WHAT DO YOU THINK?

1 Why does endochondral bone formation involve so many complex steps? Instead of having the hyaline cartilage model followed by the separate formation of the diaphysis and epiphyses, why can't bone simply be completely formed in the fetus?

WHAT DID YOU LEARN?

14 Briefly describe the process by which a long bone forms by endochondral ossification.

INTEGRATE

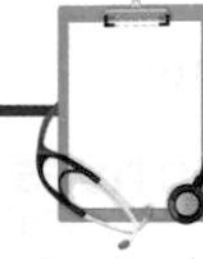

CLINICAL VIEW 7.5
Why Are Males Typically Taller Than Females?

Two processes determine why males typically are taller than females. First, the growth spurt in females is triggered primarily by estrogen, and in males it is caused primarily by testosterone. Ironically, estrogen is a more powerful growth stimulant than testosterone. However, the growth plate closes more quickly in response to estrogen, which consequently provides a smaller window of time for accelerated growth. Second, females enter puberty about 2 to 3 years earlier than males, so females have fewer years of childhood growth prior to puberty than do males.

In contrast, removal or significant decrease of mechanical stress weakens bone through both reduction of collagen formation and demineralization. When a person has a fractured bone and wears a cast or is bedridden, the strength of the unstressed bone decreases in the immobilized limb. Thus, while in space, astronauts must exercise to reduce the effects of loss of bone mass due to lack of gravity.

 **WHAT DID YOU LEARN?**

 16 What is bone remodeling, where does it occur, and when does it occur?

7.5c Hormones That Influence Bone Growth and Bone Remodeling

 **LEARNING OBJECTIVE**

24. Identify the hormones that influence bone growth and bone remodeling, and describe their effects.

Hormones are molecules that are released from one cell into the blood and are transported throughout the body to affect other cells (see section 17.1a). Certain hormones influence bone composition and growth patterns by altering the rates of chondrocyte, osteoblast, and osteoclast activity **(table 7.2)**.

Table 7.2 Effects of Hormones on Bone Maintenance and Growth

Hormone	Effect on Bone
Growth hormone	Stimulates liver to produce the hormone IGF, which causes cartilage proliferation at epiphyseal plate and resulting bone elongation
Thyroid hormone	Stimulates bone growth by stimulating metabolic rate of osteoblasts
Calcitonin	Promotes calcium deposition in bone and inhibits osteoclast activity
Calcitriol	Stimulates absorption of calcium ions from the small intestine into the blood
Parathyroid hormone	Increases blood calcium levels by encouraging bone resorption by osteoclasts
Sex hormones (estrogen and testosterone)	Stimulate osteoblasts; promote epiphyseal plate growth and closure
Glucocorticoids	Increase bone loss and, in children, impair bone growth when there are chronically high levels of glucocorticoids
Serotonin	Inhibits osteoprogenitor cells from differentiating into osteoblasts when there are chronically high levels of serotonin

Growth hormone, also called *somatotropin* (sō′mă-tō-trō-pin), is produced by the anterior pituitary gland (see section 17.7d). It affects bone growth by stimulating the liver to form another hormone called **insulin-like growth factor (IGF)** (also called *somatomedin;* sō′mă-tō-mē′din). Both growth hormone and IGF directly stimulate growth of cartilage in the epiphyseal plate.

Thyroid hormone is secreted by the thyroid gland and stimulates bone growth by influencing the basal metabolic rate of bone cells (see section 17.8b). If maintained in proper balance, growth hormone and thyroid hormone regulate and maintain normal activity at the epiphyseal plates until puberty.

Sex hormones (**estrogen** and **testosterone;** see tables R.9 and R.10), which begin to be secreted in relatively large amounts at puberty (see section 28.1b), dramatically accelerate bone growth. Sex hormones increase the rate of both cartilage growth and bone formation within the epiphyseal plate. Ironically, the appearance of high levels of sex hormones at puberty also signals the beginning of the end for growth at the epiphyseal plate. This happens because bone formation occurs at a faster rate than cartilage growth. Bone growth eventually overcomes the region of cartilage, replacing all cartilage with bone at the epiphyseal plates.

INTEGRATE

CONCEPT CONNECTION

Many hormones secreted by the endocrine system (see section 17.1) are responsible for normal growth and homeostasis of bone tissue. Growth hormone, thyroid hormone, calcitonin, and sex hormones help promote bone growth, whereas parathyroid hormone, calcitriol, glucocorticoids, and serotonin may either inhibit bone growth or increase bone resorption. Disorders of the endocrine system, as a result, often are manifested in part by skeletal system disorders.

 **WHAT DO YOU THINK?**

 3 Given what you know about the effects of testosterone, explain why there is a risk of stunted growth to a young boy (pre-puberty) taking anabolic steroids (substances that have effects similar to those of testosterone).

Glucocorticoids are a group of steroid hormones that are released from the adrenal cortex and regulate blood glucose levels (see section 17.9b). High amounts increase bone loss and, in children, impair growth at the epiphyseal plate. It is because of this relationship that a child's growth is monitored if receiving high doses of glucocorticoids as an anti-inflammatory, such as a treatment for severe asthma.

Serotonin (ser-ō-tō′nin) was previously discussed in section 1.7. Researchers have discovered that most bone cells have serotonin receptors and specifically that, when levels of circulating serotonin are too high, osteoprogenitor cells are prevented from differentiating into osteoblasts. Thus, serotonin appears to play a role in the rate and regulation of normal bone remodeling because it affects osteoblast differentiation. Further research is ongoing to see if abnormally high levels of serotonin are linked to low bone density disorders.

Three additional hormones—parathyroid hormone, calcitriol, and calcitonin—participate in both regulating bone remodeling and regulating blood calcium levels. These hormones are discussed in detail in section 7.6.

 **WHAT DID YOU LEARN?**

 17 What are the effects of growth hormone and thyroid hormone on bone growth and bone mass?

7.6 Regulating Blood Calcium Levels

Regulating calcium concentration in blood (between 8.9 and 10.1 milligrams per deciliter [mg/dL]) is essential because calcium is required for numerous physiologic processes such as initiation of muscle contraction (see section 10.3a); exocytosis of molecules from cells (see section 4.3c), including nerve cells (neurons) (see section 12.8d); stimulation of the heart by pacemaker cells (see section 19.6a); and blood clotting (see section 18.4c). The two primary hormones that regulate blood calcium are calcitriol (an active form of vitamin D) and parathyroid hormone. We describe blood calcium regulation here because of the role of the skeleton in storage of calcium. (Also see table R.2 for information about the hormones involved in regulating blood calcium levels.)

7.6a Activation of Vitamin D to Calcitriol

LEARNING OBJECTIVE

25. Explain the activation of vitamin D to calcitriol.

To effectively describe the actions of calcitriol and parathyroid hormone we first describe the enzymatic pathway of activating vitamin D to calcitriol. The three steps are as follows **(figure 7.14)**:

1. Ultraviolet light converts the precursor molecule in keratinocytes of the skin (*7-dehydrocholesterol,* a modified cholesterol molecule) to **vitamin D_3 (cholecalciferol),** which is released into the blood. (Vitamin D_3 also is absorbed from the small intestine into the blood from the diet.)
2. Vitamin D_3 circulates throughout the blood. As it passes through the blood vessels of the liver, it is converted by liver enzymes to **calcidiol** by the addition of a hydroxyl group (—OH). Both steps 1 and 2 occur continuously with limited regulation.
3. Calcidiol circulates in the blood: As it passes through blood vessels of the kidney, it is converted to **calcitriol** by kidney enzymes (when another —OH group is added). Calcitriol is the active form of vitamin D_3. The presence of parathyroid hormone increases the rate of this final enzymatic step in the kidney. Thus, greater amounts of calcitriol are formed when parathyroid hormone is present.

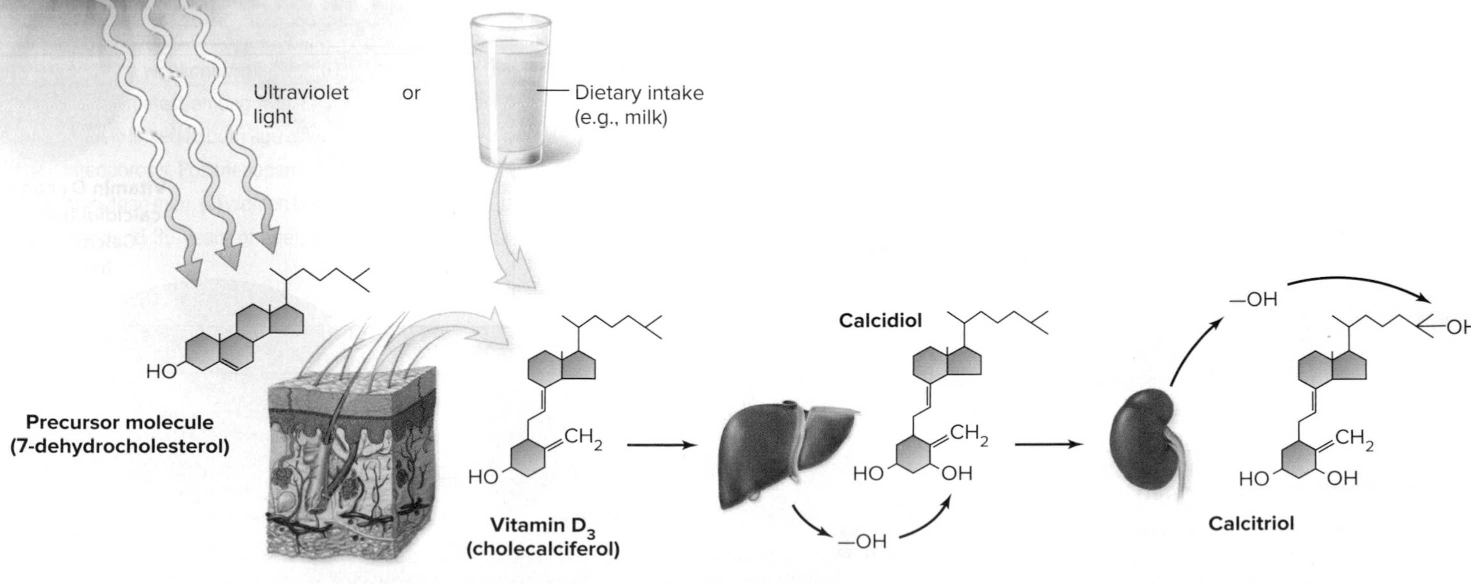
Ultraviolet light
or
Dietary intake (e.g., milk)
HO
Precursor molecule (7-dehydrocholesterol)
CH2
HO
Vitamin D3 (cholecalciferol)
Calcidiol
CH2
HO
OH
—OH
—OH
—OH
CH2
HO
OH
Calcitriol
1 The precursor molecule is converted to vitamin D3 (cholecalciferol).
2 Vitamin D3 is converted to calcidiol in the liver (when an —OH group is added).
3 Calcidiol is converted to calcitriol in the kidney (when another —OH group is added).

Figure 7.14 Calcitriol Production. Calcitriol is produced as follows: When keratinocytes are exposed to UV rays, a precursor molecule (7-dehydrocholesterol) in keratinocytes is transformed to vitamin D_3 (cholecalciferol). Humans also may obtain vitamin D_3 from dietary sources, such as milk. The liver then synthesizes calcidiol from the vitamin D_3. Finally, the kidneys will convert calcidiol to calcitriol.

Second, bone loses calcium and other minerals (demineralization). The bones of the skeleton become thinner and weaker, resulting in insufficient ossification, a condition called **osteopenia** (os'tē-ō-pen'ē-ă; *penia* = poverty). Aging causes all people to become slightly osteopenic. This reduction in bone mass may begin as early as 35–40 years of age, when osteoblast activity declines, while osteoclast activity continues at previous levels. Different parts of the skeleton are affected unequally. Vertebrae, jaw bones, and epiphyses lose large amounts of mass, resulting in reduced height, loss of teeth, and fragile limbs.

Every decade, women lose roughly more of their skeletal mass than do men. A significant percentage of older women and a smaller proportion of older men suffer from **osteoporosis** (os'tē-ō-pō-rō'sis; *poros* = pore, *osis* = condition), a condition characterized by reduction in bone mass sufficient to compromise normal function (see Clinical View 7.7: "Osteoporosis").

In addition, vitamin D and numerous hormones, including growth hormone, estrogen, and testosterone, decrease with age. This decrease in hormone levels contributes to reduction in bone mass.

WHAT DID YOU LEARN?

21 Explain why women are more likely than men to develop osteoporosis.

7.8 Bone Fracture and Repair

LEARNING OBJECTIVE

30. Explain the four steps by which fractures heal.

Bone has great mineral strength, but it may break as a result of unusual stress or a sudden impact. Breaks in bones are called **fractures** and are classified in several ways. A **stress fracture** is a thin break caused by increased physical activity in which the bone experiences repetitive loads (e.g., as seen in some runners). A **pathologic fracture** usually occurs in bone that has been weakened by disease. In a **simple fracture,** the broken bone does not penetrate the skin, whereas in a **compound fracture,** one or both ends of the broken bone pierce the overlying skin. **Figure 7.16** shows the classifications of fractures.

The healing of a simple fracture takes about 2 to 3 months, whereas a compound fracture takes longer to heal. Fractures heal much more quickly in young children (average healing time, 3 weeks) and become slower to heal as we age. In the elderly, the normal thinning and weakening of bone increase the incidence of fractures, and

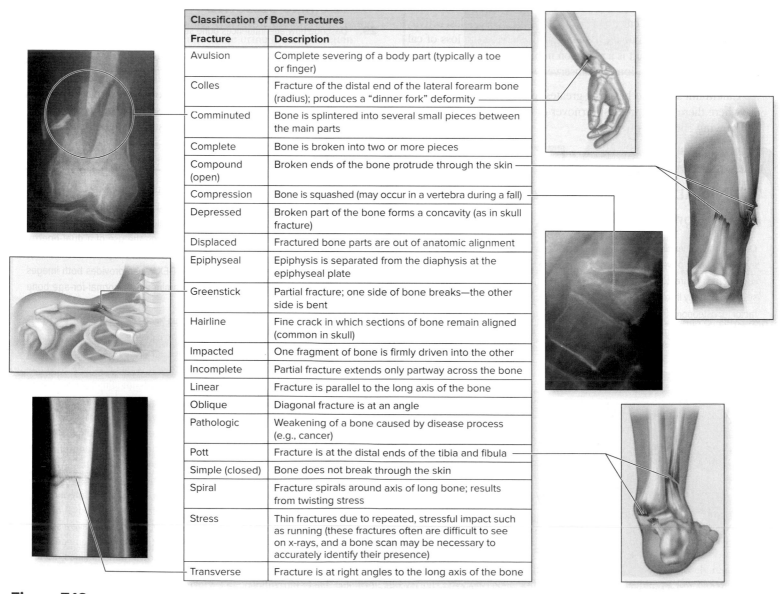

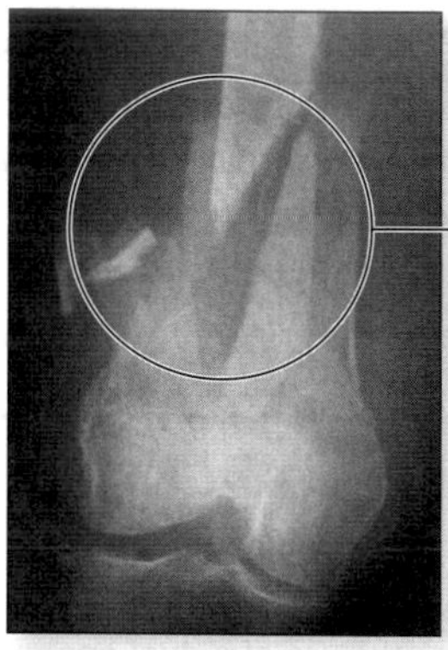

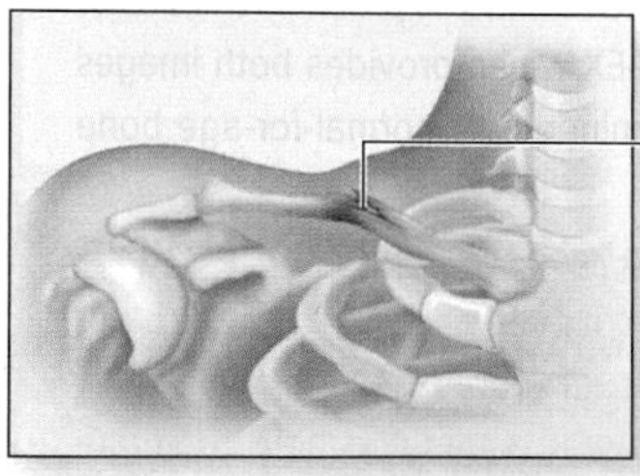

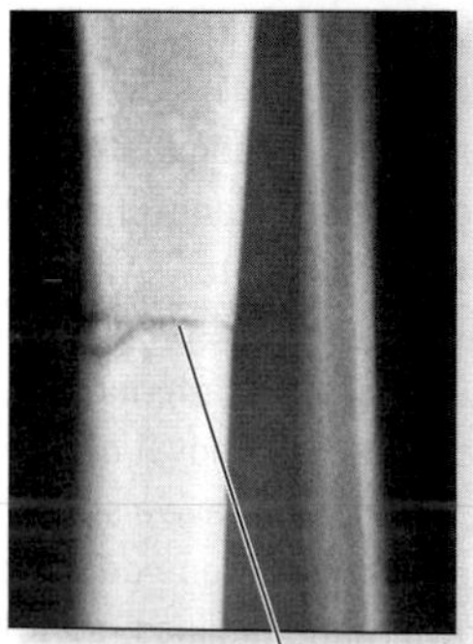

Classification of Bone Fractures

Fracture	Description
Avulsion	Complete severing of a body part (typically a toe or finger)
Colles	Fracture of the distal end of the lateral forearm bone (radius); produces a "dinner fork" deformity
Comminuted	Bone is splintered into several small pieces between the main parts
Complete	Bone is broken into two or more pieces
Compound (open)	Broken ends of the bone protrude through the skin
Compression	Bone is squashed (may occur in a vertebra during a fall)
Depressed	Broken part of the bone forms a concavity (as in skull fracture)
Displaced	Fractured bone parts are out of anatomic alignment
Epiphyseal	Epiphysis is separated from the diaphysis at the epiphyseal plate
Greenstick	Partial fracture; one side of bone breaks—the other side is bent
Hairline	Fine crack in which sections of bone remain aligned (common in skull)
Impacted	One fragment of bone is firmly driven into the other
Incomplete	Partial fracture extends only partway across the bone
Linear	Fracture is parallel to the long axis of the bone
Oblique	Diagonal fracture is at an angle
Pathologic	Weakening of a bone caused by disease process (e.g., cancer)
Pott	Fracture is at the distal ends of the tibia and fibula
Simple (closed)	Bone does not break through the skin
Spiral	Fracture spirals around axis of long bone; results from twisting stress
Stress	Thin fractures due to repeated, stressful impact such as running (these fractures often are difficult to see on x-rays, and a bone scan may be necessary to accurately identify their presence)
Transverse	Fracture is at right angles to the long axis of the bone

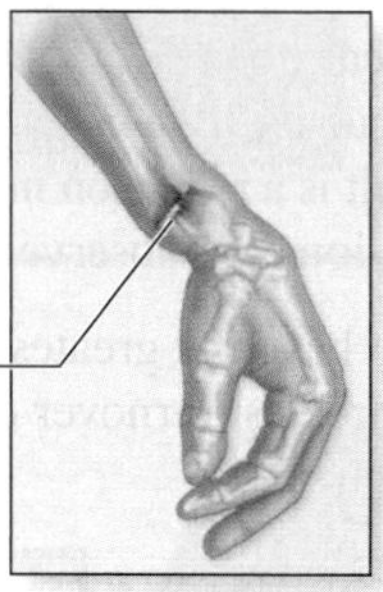

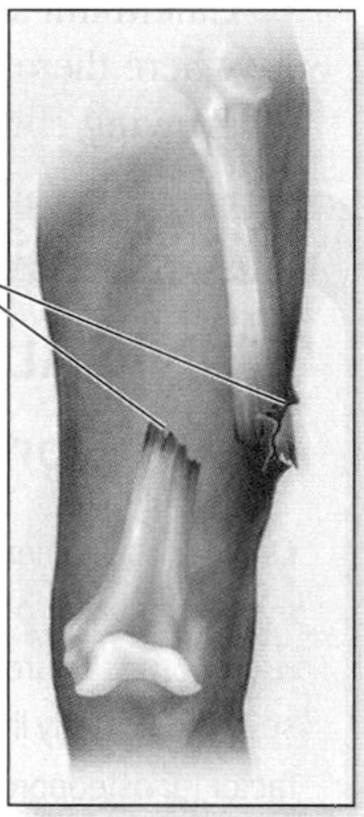

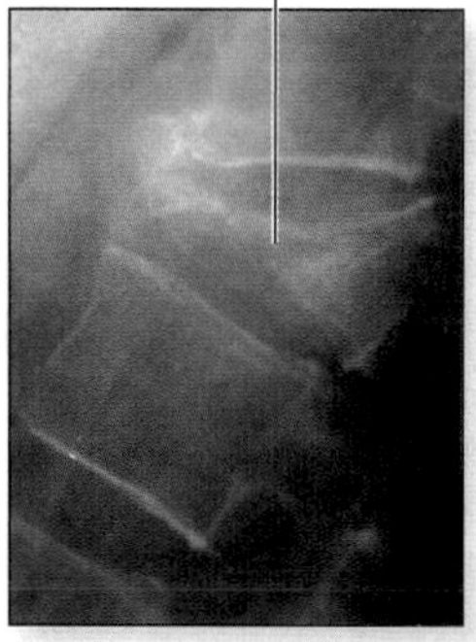

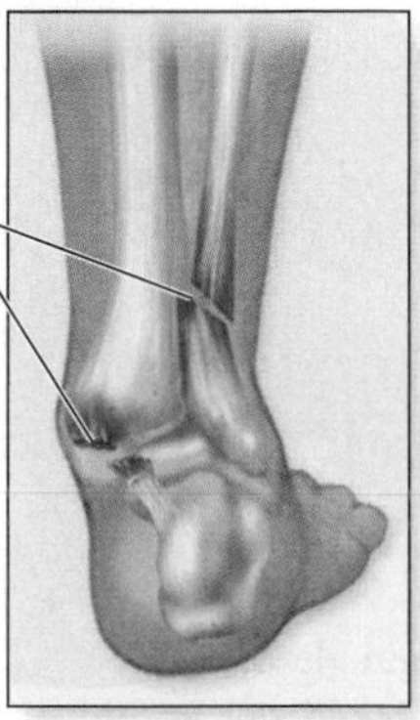

Figure 7.16 Classification of Bone Fractures.
(*Comminuted*) ©Medical-on-Line/Alamy; (*Compression*) ©ISM/Medical Images; (*Transverse*) ©Wellcome Photo Library, Wellcome Images

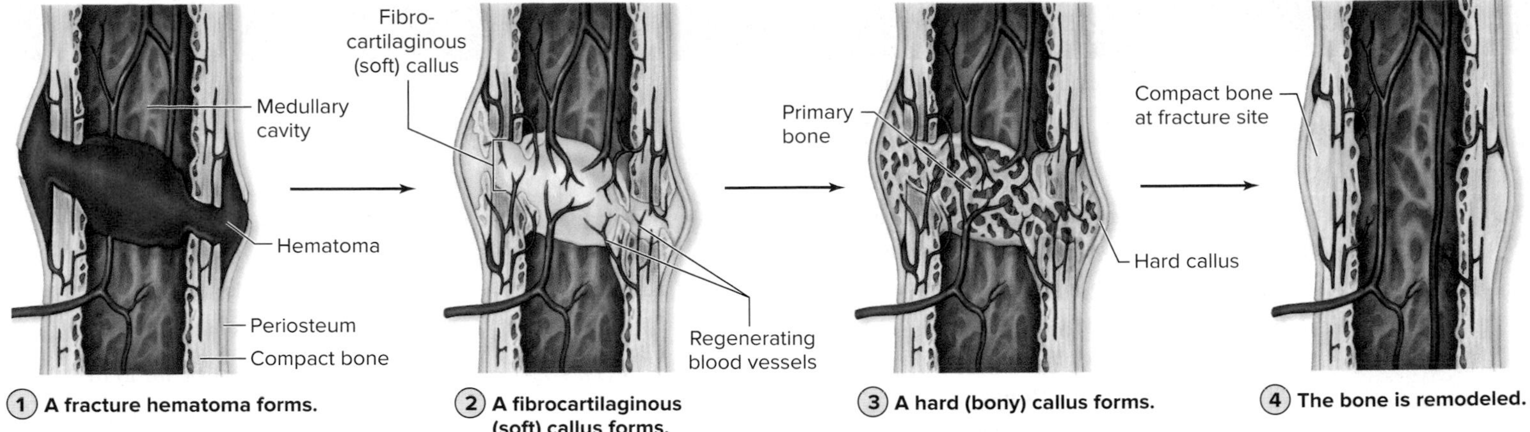

Figure 7.17 Fracture Repair. The repair of a bone fracture occurs in a series of steps.

some complicated fractures require surgical intervention to heal properly. Bone fracture repair can be described as a series of four steps **(figure 7.17)**:

1. **A fracture hematoma forms.** A bone fracture tears blood vessels inside the bone and within the periosteum, causing bleeding. This bleeding results in a **fracture hematoma** that forms from the clotted blood.
2. **A fibrocartilaginous (soft) callus forms.** Regenerated blood capillaries infiltrate the fracture hematoma. First, the fracture hematoma is reorganized into an actively growing connective tissue called a *procallus*. Fibroblasts within the procallus produce collagen fibers that help connect the broken ends of the bones. Chondroblasts in the newly growing connective tissue form a dense regular connective tissue associated with the cartilage. Eventually, the procallus becomes a **fibrocartilaginous (soft) callus** (kal′ŭs; hard skin). The fibrocartilaginous callus stage lasts at least 3 weeks.
3. **A hard (bony) callus forms.** Within a week after the injury, osteoprogenitor cells in areas adjacent to the fibrocartilaginous callus become osteoblasts and produce trabeculae of primary bone. The fibrocartilaginous callus is then replaced by this bone, which forms a **hard (bony) callus.** The trabeculae of the hard callus continue to grow and thicken for several months.
4. **The bone is remodeled.** Remodeling is the final phase of fracture repair. The hard callus persists for at least 3 to 4 months as osteoclasts remove excess bony material from both exterior and interior surfaces. Compact bone replaces primary bone. The fracture usually leaves a slight thickening of the bone (as detected by x-ray); however, in some instances, healing occurs with no persistent obvious thickening.

WHAT DID YOU LEARN?

22 What are the four basic steps in fracture repair?

23 Explain the risk of bearing weight on a bone when the fibrocartilage callus is forming.

INTEGRATE

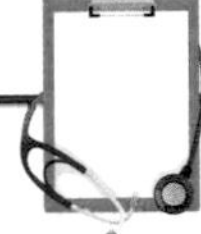

CLINICAL VIEW 7.8

Bone Scans

Bone scans are tests that can detect bone pathologies. The patient often is injected intravenously with a small amount of a radioactive tracer compound that is absorbed by bone. A scanning camera then detects and measures the radiation emitted from the bone. The images produced show normal bone tissue with an even distribution of gray coloring (although the axial skeleton typically appears uniformly darker than the appendicular skeleton). An abnormal bone scan may have focal darker areas called *hot spots*, or lighter areas called *cold spots*. Hot spots typically indicate increased metabolism or greater turnover of the bone tissue, as may be experienced by a stress fracture or cancer metastasis to bone. Cold spots indicate less metabolic activity of bone tissue, as may occur with avascular necrosis (death due to lack of blood supply) of a bone. AP|R

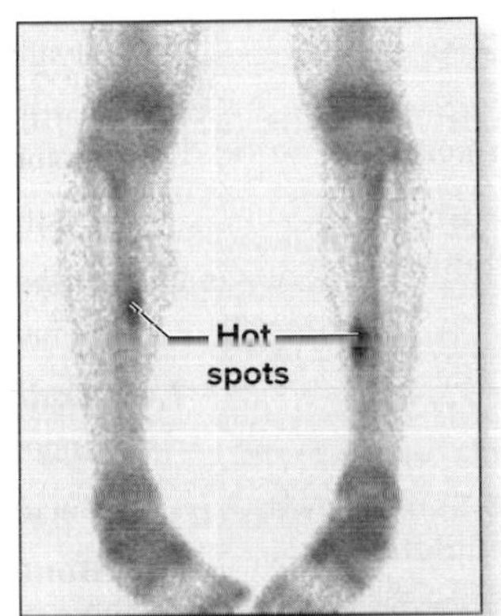

Hot spots in both tibiae indicative of stress fractures.

Source: John Pape/MedPlx™. Website: rad.usuhs.edu/medpix

Lateral View

A lateral view of the skull (**figure 8.6**) shows one parietal bone, **temporal bone,** and **zygomatic** (zī′gō-mat′ik; *zygoma* = a joining, a yoke) **bone.** This view also shows part of the maxilla, mandible, frontal bone, and occipital bone. The **superior** and **inferior temporal**

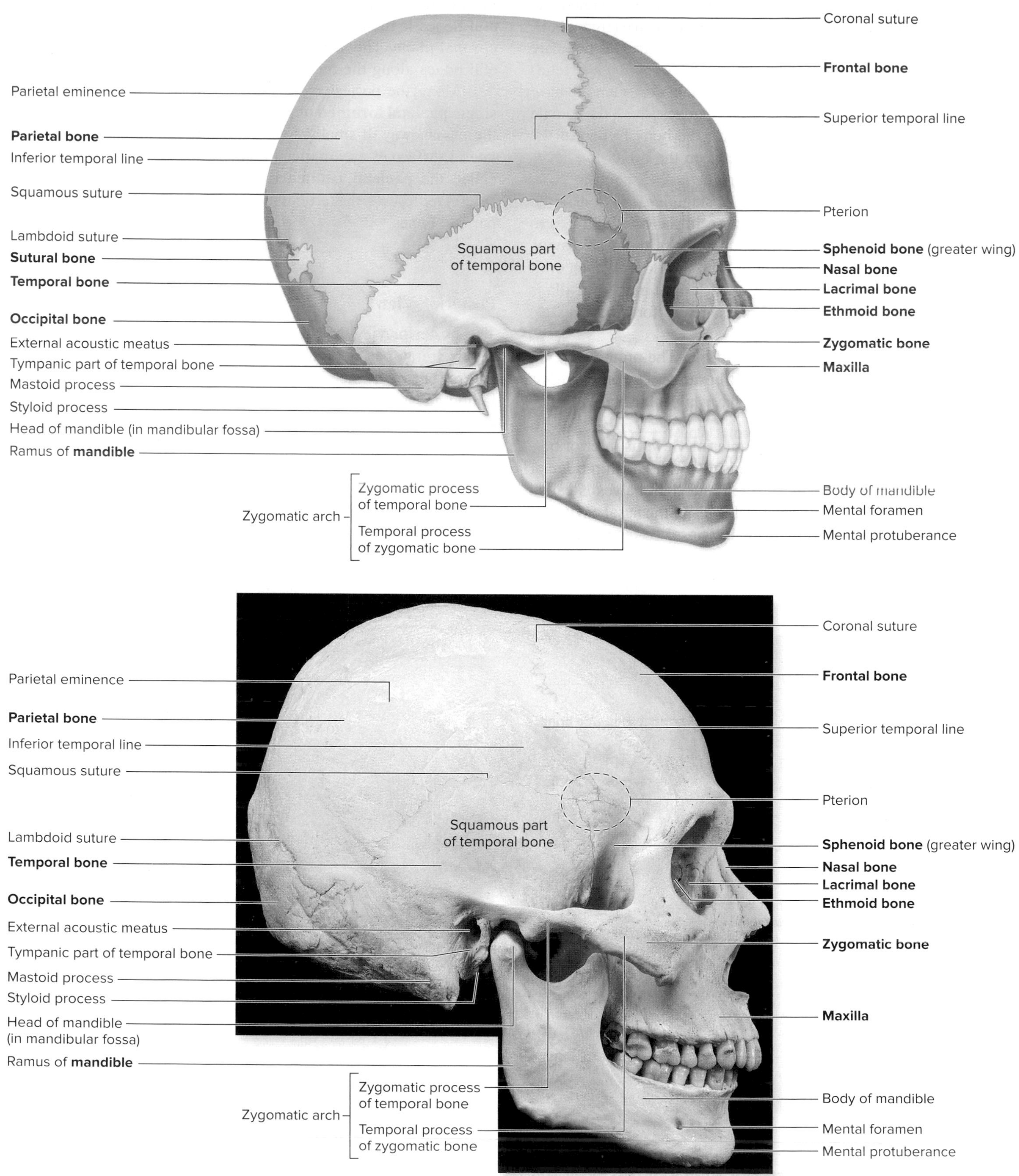

Figure 8.6 Lateral View of the Skull. The parietal, temporal, zygomatic, frontal, and occipital bones, as well as the maxilla and mandible, are prominent in this view. AP|R

lines arc across the surface of the parietal and frontal bones and mark the attachment site of the temporalis muscle (see section 11.3c). The small **lacrimal** (lak′ri-măl; *lacrima* = a tear) **bone** articulates with the maxilla anteriorly and with the ethmoid bone posteriorly. A portion of the **sphenoid** (sfē′noyd; wedge-shaped) **bone** articulates with the frontal, parietal, and temporal bones. This region is called the **pterion** (tĕ′rē-on; *ptéron* = wing) and is circled on figure 8.6. Pterion includes the H-shaped set of sutures of these four articulating bones.

The **temporal process** of the **zygomatic bone** and the **zygomatic process** of the **temporal bone** fuse to form the **zygomatic arch.** Put your fingers along the bony prominences ("apples") of your cheeks and move your fingers posteriorly toward your ears; you are feeling the zygomatic arch. The zygomatic arch terminates superior to the point where the mandible articulates with the **mandibular** (man-dib′ū-lăr) **fossa** of the temporal bone. This articulation is called the **temporomandibular joint (TMJ)** and is described further in section 9.7a. By putting your finger anterior to your external ear opening and then opening and closing your jaw, you can feel that joint moving.

The **squamous part** of the temporal bone lies directly inferior to the squamous suture. Immediately posterolateral to the mandibular fossa is the **tympanic** (tim-pan′ik; *tympanon* = drum) **part** of the temporal bone. This is a small, bony ring surrounding the external ear opening called the **external acoustic meatus** (mē-ā′tŭs; a passage), or *external auditory canal* (see section 16.5a). Inferior and posterior to this meatus is the **mastoid** (mas′toyd; *masto* = breast, *eidos* = resemblance) **process,** the bump you feel behind your external ear opening. The **styloid** (stī′loyd; *stylos* = pillar, post) **process** is a thin, pointed projection of bone located anteromedial to the mastoid process. It serves as an attachment site for several hyoid and tongue muscles (see section 11.3c).

Sagittal Sectional View

Cutting the skull along a sagittal sectional plane reveals bones that form the endocranium and the nasal cavity (**figure 8.7*a***). The cranial cavity is formed from a complex articulation of the frontal, parietal, temporal, occipital, **ethmoid** (eth′moyd; *ethmos* = sieve), and sphenoid bones.

Vessel impressions may be visible on the internal surface of the skull. The **frontal sinus** (a space within the frontal bone) and the **sphenoidal sinus** (open space within the sphenoid bone) are visible in a sagittal view.

A sagittal sectional view also shows the bones that form the nasal septum more clearly. The **perpendicular plate** of the ethmoid forms the posterosuperior portion of the nasal septum, whereas the **vomer** (vō′mer; plowshare) forms the posteroinferior portion. (The anterior part of the nasal septum is cartilaginous.) The ethmoid bone serves as the division between the anterior floor of the cranial cavity and the roof of the nasal cavity. The **palatine process** of the maxillae and the **palatine** (pal′a-tīn) **bones** form the hard palate (figure 8.7*b*), which acts as both the floor of the nasal cavity and a portion of the roof of the mouth. Move your tongue along the roof of your mouth; you are palpating the maxillae anteriorly and palatine bones posteriorly.

Inferior (Basal) View

In an inferior (basal) view, the most anterior structure is the hard palate (figure 8.7*b*). On the posterior aspect of either side of the palate are the **medial** and **lateral pterygoid** (ter′i-goyd; *pteryx* = winglike) **plates** of the sphenoid bone. Together, both plates form a **pterygoid process.** Medially adjacent to these structures are the internal openings of the nasal cavity, called the **choanae** (kō′an-ē; sing., *choana,* kō′an-ă; funnel).

Between the mandibular fossa and the pterygoid processes are several paired foramina and canals. Typically, these openings provide passage for specific blood vessels and nerves. For example, the **jugular** (jŭg′ū-lar; *jugulum* = throat) **foramen** is an opening between the temporal and occipital bones that provides a passageway for the internal jugular vein and several nerves. The **foramen lacerum** (anteromedial to the carotid canal) extends between the occipital and temporal bones. This opening is covered by cartilage in a living individual. The entrance to the **carotid** (ka-rot′id; *karoo* = to put to sleep) **canal** is anteromedial to the jugular foramen; the internal carotid artery passes through this canal.

The **stylomastoid foramen** lies between the mastoid process and the styloid process. The facial nerve (CN VII) extends through the stylomastoid foramen to innervate the facial muscles (see sections 11.3a and 13.9).

The largest foramen of all is the **foramen magnum,** literally meaning *big hole*. Through this opening, the spinal cord enters the cranial cavity and is continuous superiorly with the brainstem. On either side of the foramen magnum are the rounded **occipital condyles,** which articulate with the first cervical vertebra of the vertebral column. At the anteromedial edge of each condyle is a **hypoglossal canal** through which the hypoglossal nerve (CN XII) extends to innervate tongue muscles (see sections 11.3c and 13.9).

the pelvic girdle (i.e., both ossa coxae) articulates with the lower limb. The proximal part of each limb has one large bone: the humerus in the upper limb and the femur in the lower limb. The distal part of each limb contains two bones; these bones are able to pivot slightly about one another. Both the wrist and the proximal foot contain multiple bones (carpal and tarsal bones, respectively) that allow for a range of movement. Finally, the feet and hands are very similar in that both contain either 5 metacarpals (palm of hand) or 5 metatarsals (arch of foot), and each contains a total of 14 phalanges (bones of the fingers and toes, respectively).

The structural differences between the upper and lower limb skeletons arise from the functional differences. Understanding these general differences between upper and lower limbs will make the study of their individual bones easier. Because the lower limb is weight bearing and is used for locomotion, some mobility at specific joints has been lost for greater stability. The upper limb is not weight bearing, so both arm and forearm bones are relatively smaller and lighter than the similar respective lower limb bones. Additionally, the upper limb joints are relatively more mobile than the respective lower limb joints, so we may utilize the upper limbs for a wide range of activities. Unfortunately, more mobile joints are less stable, and that is why some of the upper limb joints (such as the shoulder joints) are the most frequently injured.

WHAT DID YOU LEARN?

18 What are some of the functional differences between the upper and lower limbs?

8.8 The Pectoral Girdle and Its Functions

The **pectoral** (pek′tŏ-răl; *pectus* = breastbone) **girdle** articulates with the trunk and supports the upper limbs. A pectoral girdle consists of the clavicles and the scapulae.

8.8a Clavicle

LEARNING OBJECTIVE

26. Identify and locate the clavicle and its landmarks.

The **clavicle** (klav′i-kĕl; *clavis* = key), commonly known as the collarbone, is an elongated, S-shaped bone that extends between the manubrium of the sternum and the acromion of the scapula (**figure 8.23**). Its **sternal end** (medial end) is roughly pyramidal in shape and articulates with the manubrium of the sternum, forming the sternoclavicular joint (see section 9.7b). The **acromial end** (lateral end) of the clavicle is broad and flattened. The acromial end articulates with the acromion of the scapula, forming the acromioclavicular joint. You can palpate your own clavicle by first locating the superior aspect of your sternum and then moving your hand laterally. The curved bone you feel under your skin, and close to the neck opening of your shirt, is your clavicle.

The superior surface of the clavicle is relatively smooth and the inferior surface is roughened (figure 8.23). On the inferior surface, near the acromial end, is a rough tuberosity called the **conoid** (kō′noyd; *konoeides* = cone-shaped) **tubercle** for the conoid ligament (part of the coracoclavicular ligament of the shoulder joint; see section 9.7b). The inferiorly located prominence at the sternal end of the clavicle is the **costal tuberosity,** for the attachment of the shoulder's costoclavicular ligament.

WHAT DID YOU LEARN?

19 How do the sternal end and acromial end of the clavicle differ?

8.8b Scapula

LEARNING OBJECTIVE

27. Describe the landmarks and features of the scapula.

The **scapula** (skap′yū-lă) is a broad, flat, triangular bone that forms the *shoulder blade* (**figure 8.24**). You can palpate your scapula by putting your hand on your superolateral back region and moving

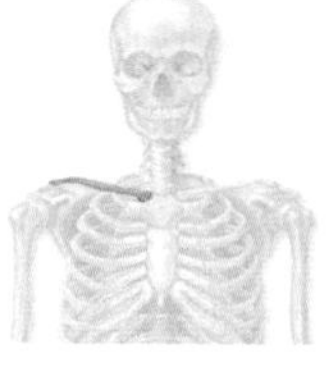

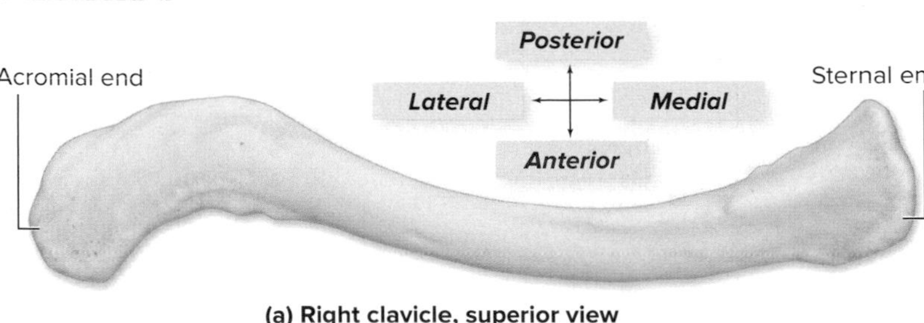

(a) Right clavicle, superior view

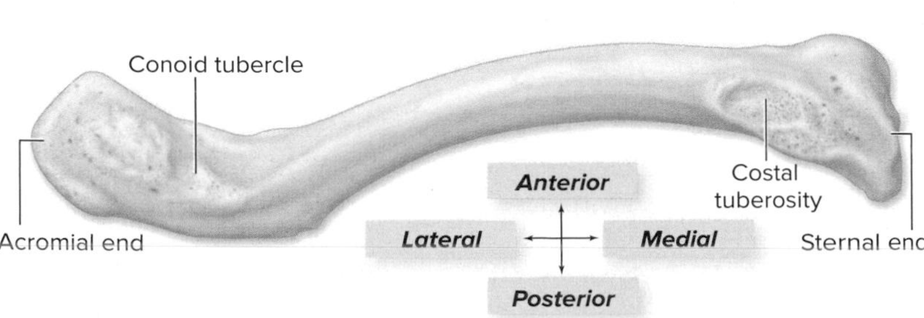

(b) Right clavicle, inferior view

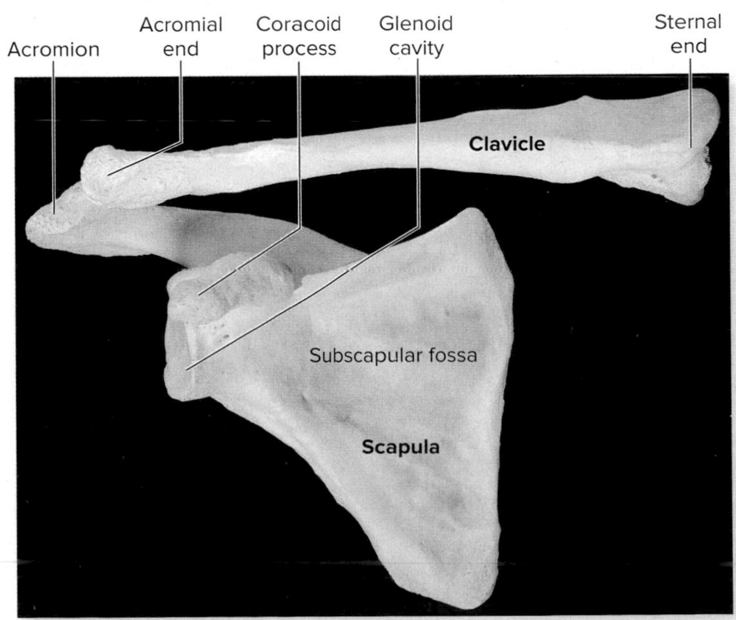

(c) Right scapula and clavicle articulation, anterior view

Figure 8.23 Clavicle. The S-shaped clavicle represents the only direct connection between the pectoral girdle and the axial skeleton. (*a*) Superior and (*b*) inferior views of the right clavicle. (*c*) Anterior view of an articulated right clavicle and scapula. AP|R

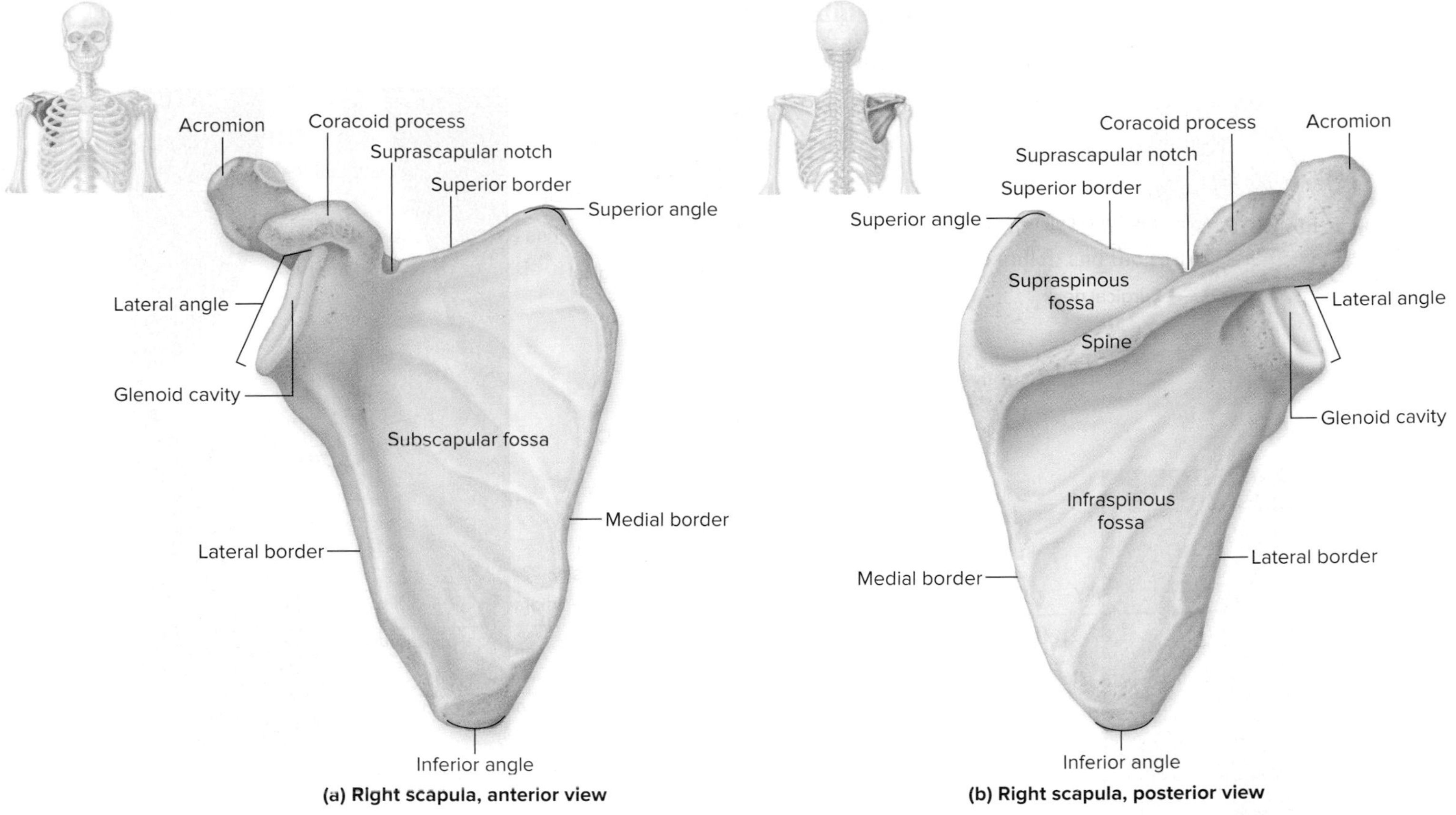

(a) Right scapula, anterior view

(b) Right scapula, posterior view

Figure 8.24 Scapula. The right scapula is known as the "shoulder blade" and is shown in (*a*) anterior and (*b*) posterior views. AP|R

your upper limb; the bone you feel moving is the scapula. The **spine** of the scapula is a ridge of bone on the posterior aspect of the scapula. It is easily palpated under the skin. The spine is continuous with a larger, posterior process called the **acromion** (a-krō′mē-on; *akron* = tip, *omos* = shoulder), which forms the bony tip of the shoulder. Palpate your upper shoulder; the prominent bump you feel is the acromion. The **coracoid** (kōr′ă-koyd) **process** is the smaller, more anterior, hook-shaped projection that is a site for muscle attachment.

The triangular shape of the scapula forms three sides, or borders. The **superior border** is the horizontal edge of the scapula superior to the spine of the scapula; the **medial border** (also called the *vertebral border*) is the edge of the scapula closest to the vertebrae; and the **lateral border** (also called the *axillary border*) is closest to the axilla. A **suprascapular notch** (which in some individuals is a **suprascapular foramen**) in the superior border provides passage for the suprascapular nerve and blood vessels.

Between these borders are the superior, inferior, and lateral angles. The **superior angle** is located between the superior and medial, while the **inferior angle** is positioned between the medial and lateral borders. The **lateral angle** is primarily made up of the cup-shaped, shallow **glenoid** (glē′noyd; glen′oyd; resembling a socket) **cavity,** or *glenoid fossa*, which articulates with the humerus, the bone of the arm.

The broad, relatively smooth anterior surface of the scapula is called the **subscapular** (sŭb-skap′yū-lăr; *sub* = under) **fossa.** A large muscle called the subscapularis overlies this fossa. The spine subdivides the posterior surface of the scapula into two shallow fossae. The depression superior to the spine is the **supraspinous** (sū-pră-spī′nŭs; *supra* = above) **fossa;** inferior to the spine is a broad, extensive surface called the **infraspinous fossa.** The supraspinatus and infraspinatus muscles, respectively, occupy these fossae (see section 11.8b).

WHAT DID YOU LEARN?

20 What fossae are located on the scapula, and what is located in each fossa?

8.9 Bones of the Upper Limb

The upper limb consists of the brachium (arm), antebrachium (forearm), and hand. The complex structure of the hand in particular gives humans capabilities beyond those of most other vertebrates.

Each upper limb contains a total of 30 bones:

- 1 humerus, located in the brachium region
- 1 radius and 1 ulna, located in the antebrachium region
- 8 carpal bones, which form the wrist
- 5 metacarpal bones, which form the palm of the hand
- 14 phalanges, which form the fingers

8.9a Humerus

LEARNING OBJECTIVES

28. Describe the articulations of the humerus.

29. List landmarks and features of the humerus.

The **humerus** (hyū′mĕr-ŭs) is the longest and largest upper limb bone (**figure 8.25**). Its proximal end has a hemispherical **head** that

a wheel, ray) is located more laterally. The proximal end of the radius has a distinctive disc-shaped **head** that articulates with the capitulum of the humerus. A narrow **neck** extends from the radial head to the **radial tuberosity** (or *bicipital tuberosity*). The radial tuberosity is an attachment site for the biceps brachii muscle.

The **shaft** of the radius curves slightly and leads to a wider distal end, where there is a laterally placed **styloid** (stī′loyd) **process.** This bony projection can be palpated on the lateral side of the wrist, just proximal to the thumb. On the distal medial surface of the radius is an **ulnar notch,** which articulates with the medial surface of the distal end of the ulna at the distal radioulnar joint.

The **ulna** (ŭl′nă) is the longer, medially placed bone of the forearm. At the proximal end of the ulna, a C-shaped **trochlear notch** interlocks with the trochlea of the humerus. The posterosuperior aspect of the trochlear notch has a prominent projection called the **olecranon.** The olecranon articulates with the olecranon fossa of the humerus and forms the posterior "bump" of the elbow. The inferior lip of the trochlear notch, called the **coronoid process,** articulates with the humerus at the coronoid fossa. Lateral to the coronoid process, a smooth, curved **radial notch** accommodates the head of the radius and helps form the proximal radioulnar joint. Also at the proximal end of this bone is the **tuberosity of ulna.** At the distal end of the ulna, the shaft narrows and terminates in a knoblike **head** that has a posteromedial **styloid process.** The styloid process of the ulna may be palpated on the medial (*little finger*) side of the wrist.

Both the radius and the ulna exhibit **interosseous borders,** which face each other; the ulna's interosseous border faces laterally, whereas the interosseous border on the radius faces medially. These interosseous borders are connected by an **interosseous membrane** (*interosseous ligament*) composed of dense regular connective tissue. This membrane helps keep the radius and ulna a fixed distance apart from one another and provides a pivot of rotation for the forearm. The bony joints that move during this rotation are the proximal and distal radioulnar joints.

In anatomic position, the palm of the hand is facing anteriorly, and the bones of the forearm are said to be in **supination** (sū′pi-nā′shŭn) (figure 8.26*c*). Note that the radius and the ulna are parallel with one another. If you view your own supinated forearm, the radius is on the lateral (thumb) side of the forearm, and the ulna is on the medial (little finger) side.

Pronation (prō-nā′shŭn) of the forearm requires that the radius cross over the ulna and that both bones pivot along the interosseous membrane (figure 8.26*d*). When the forearm is pronated, the palm of the hand is facing posteriorly and the head of the radius is still along the lateral side of the elbow, but the distal end of the radius has crossed over and become a more medial structure.

When an individual is in anatomic position (i.e., has the upper limbs extended and forearms supinated), note that the bones of the forearm may angle laterally from the elbow joint. This positioning is referred to as the **carrying angle** of the elbow, and this angle positions the bones of the forearms such that the forearms will clear the hips during walking as the forearms swing. Females have wider carrying angles than males, presumably because they have wider hips than males.

INTEGRATE

LEARNING STRATEGY

No matter what the position of the forearm (pronated or supinated), the distal end of the radius is always near the thumb, and the distal end of the ulna is near the little finger.

WHAT DID YOU LEARN?

23 What are some bony features that the radius and ulna share?

24 Describe how the radius and ulna are positioned when the forearm is pronated.

8.9c Carpals, Metacarpals, and Phalanges

LEARNING OBJECTIVES

33. Locate and identify the carpals and metacarpals.

34. Describe the phalanges and their relative locations.

The bones that form the wrist and hand are the carpals, metacarpals, and phalanges (**figure 8.27**). The **carpals** (kar′păl; *karpus* = wrist) are small, short bones that form the wrist. They are arranged in two rows (a proximal row and a distal row) of four bones each and allow for the multiple movements possible at the wrist.

The *proximal row* of carpal bones, listed from lateral to medial, are the **scaphoid** (skaf′oyd; *skaphe* = boat), **lunate** (lū′nāt; *luna* = moon), **triquetrum** (trī-kwē′trŭm; *triquetrus* = three-cornered), and **pisiform** (pis′i-fōrm; *pisum* = pea, *forma* = appearance).

The *distal row* of the carpal bones, listed from lateral to medial, are the **trapezium** (tra-pē′zē-ŭm; *trapeza* = table), **trapezoid** (trap′ĕ-zoyd), **capitate** (kap′i-tāt), and **hamate** (ha′māt; *hamus* = hook).

Bones in the palm of the hand are called **metacarpals** (met′ă-kar′păl; *meta* = beyond). Five metacarpal bones articulate with the distal carpal bones and support the palm. Roman numerals I–V denote the metacarpal bones, with metacarpal I located at the base of the thumb, and metacarpal V at the base of the little finger.

A total of 14 bones are present in the digits; these are called **phalanges** (fă-lan′jēz; sing., *phalanx,* fā′langks; line of soldiers). Three

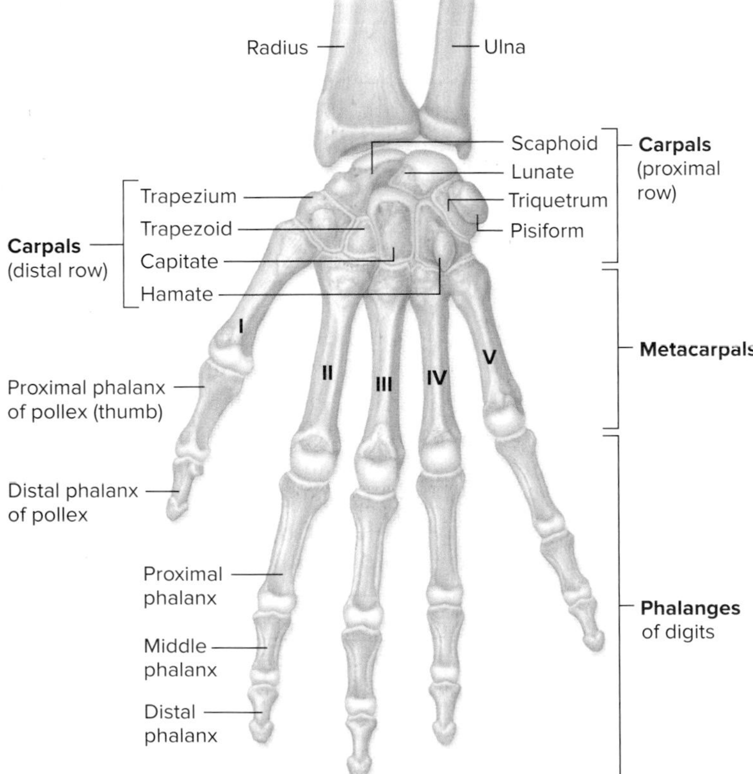

Figure 8.27 Bones of the Carpals, Metacarpals, and Phalanges. The carpal bones form the wrist, and the metacarpals and phalanges form the hand. Anterior (palmar) view of the right wrist and hand. AP|R

phalanges are found in each of the second through fifth fingers, but only two phalanges are present within the thumb, also known as the **pollex** (pol′eks; thumb). The **proximal phalanx** articulates with the head of a metacarpal, whereas the **distal phalanx** is the bone in the very tip of the finger. The **middle phalanx** of each finger lies between the proximal and distal phalanges; however, a middle phalanx is not present in the pollex.

WHAT DID YOU LEARN?

25 List the eight carpal bones. Which of these bones may be prone to developing avascular necrosis if fractured?

INTEGRATE

CLINICAL VIEW 8.8

Scaphoid Fractures

The scaphoid bone is one of the more commonly fractured carpal bones. A fall on the outstretched hand may cause the scaphoid to fracture into two separate pieces. Usually, blood vessels are torn on the proximal part of the scaphoid, resulting in **avascular necrosis,** which is death of the bone tissue due to inadequate blood supply. Scaphoid fractures take a very long time to heal properly due to this complication.

8.10 The Pelvic Girdle and Its Functions

The adult **pelvis** (pel′vis; pl., *pelves,* pel′vēz; basin) is composed of four bones: the sacrum, the coccyx, and the right and left **ossa coxae** (os′ă kok′să; sing., *os coxae;* hip bone) (**figure 8.28**). The pelvis protects and supports the viscera in the inferior part of the ventral body cavity (see section 1.5e).

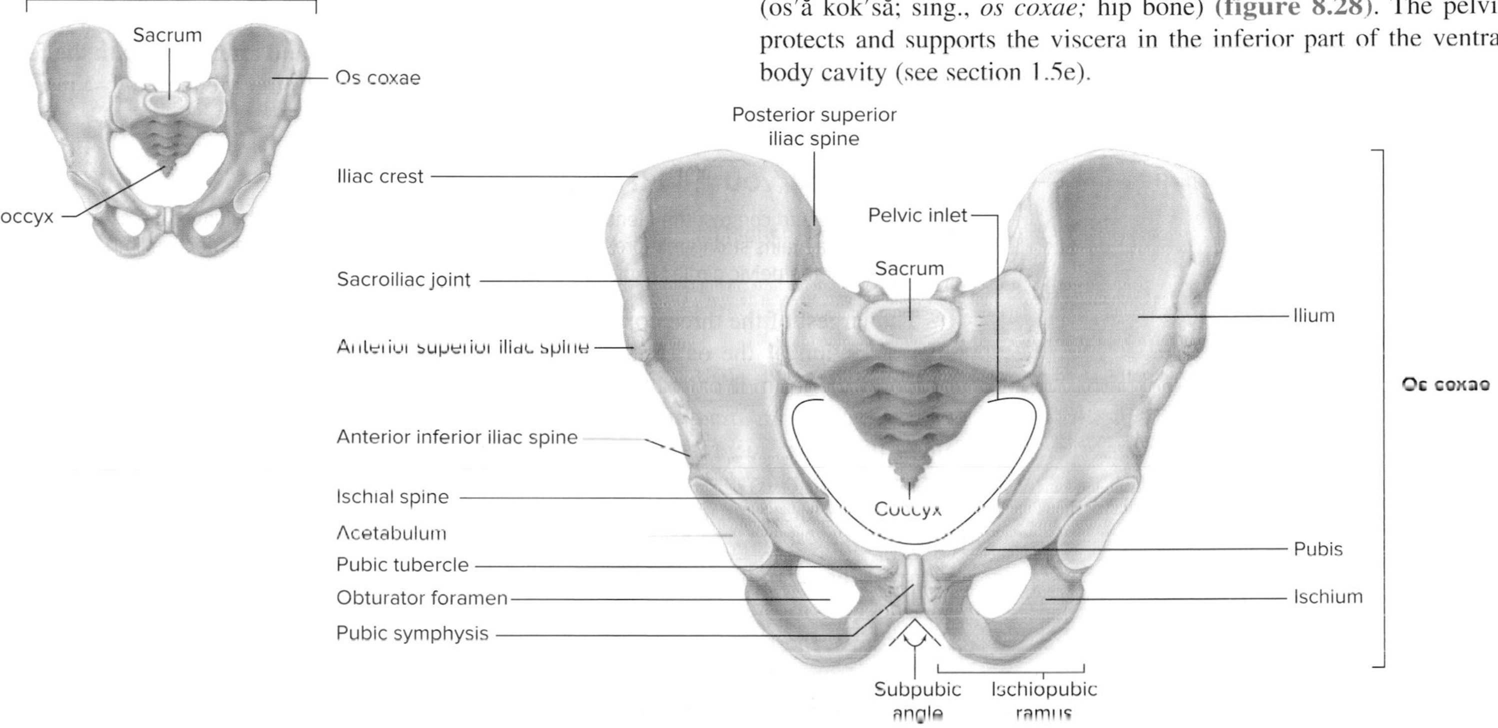

(a) Anterior view

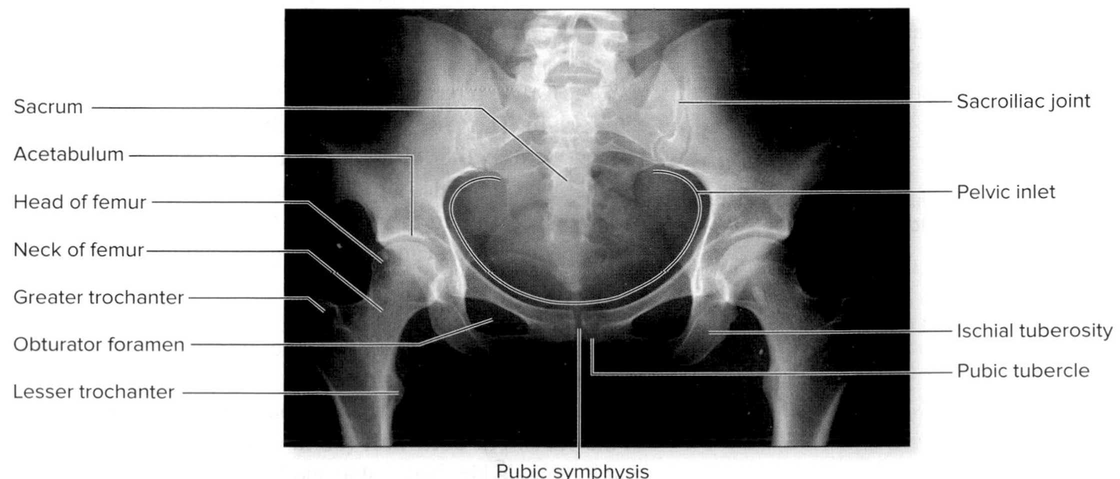

(b) Pelvis radiograph, anterior view

Figure 8.28 Pelvis. The complete pelvis consists of (*a*) the two ossa coxae, the sacrum, and the coccyx. (*b*) A radiograph shows an anterior view of the articulation between the pelvis and the femur. AP|R

(*b*) ©Image Source/Getty Images RF

CHAPTER SUMMARY (continued)

8.10 The Pelvic Girdle and Its Functions	• The pelvic girdle consists of two ossa coxae, whereas the pelvis is composed of the ossa coxae, sacrum, and coccyx.
	8.10a Os Coxae • Each os coxae forms through the fusion of an ilium, an ischium, and a pubis. The acetabulum of the os coxae articulates with the head of the femur.
	8.10b True and False Pelves • The pelvic brim is an oval ridge of bone that divides the pelvis into a true pelvis and a false pelvis.
	8.10c Sex Differences in the Pelvis • The ossa coxae are the most sexually dimorphic bones of the body. • The female pelvis is wider and has a broader greater sciatic notch and a more rectangular-shaped pubis than the male pelvis.
	8.10d Age Differences in the Ossa Coxae • As a person ages, the symphysial surface transforms from a billowed surface to more flattened.
8.11 Bones of the Lower Limb	• Each lower limb is composed of the femur, the patella, the tibia, the fibula, 7 tarsals, 5 metatarsals, and 14 phalanges.
	8.11a Femur and Patella • The femur has a rounded head and an elongated neck. • The medial and lateral condyles of the femur articulate with the condyles on the tibia. • The patella is the kneecap and is located within the quadriceps femoris tendon.
	8.11b Tibia and Fibula • The tibia is the medially located, thick, and strong bone of the leg, and its medial malleolus forms the medial bump of the ankle. • The fibula is the laterally located, slender bone, and its lateral malleolus forms the lateral bump of the ankle.
	8.11c Tarsals, Metatarsals, and Phalanges • The 7 tarsal bones form the proximal foot, the 5 metatarsals form the bones of the arch of the foot, and the 14 phalanges form the bones of the toes.
	8.11d Arches of the Foot • The three arches of the foot support the body's weight and ensure plantar structures do not get compressed when we are standing.
8.12 Development of the Skeleton	• The appendicular skeleton forms from limb buds beginning in the fourth week. Development of the limbs is mostly complete by week 8.

CHALLENGE YOURSELF

Create and Evaluate

Analyze and Apply

Understand and Remember

Do You Know the Basics?

1. The bony portion of the nasal septum is formed by the
   a. perpendicular plate of the ethmoid bone and vomer.
   b. perpendicular plate of the ethmoid bone only.
   c. nasal bones and perpendicular plate of the ethmoid bone.
   d. vomer and sphenoid bones.
2. Which bone marking is matched with its correct description?
   a. foramen; bony prominence
   b. facet; opening in the bone
   c. tubercle; small, bony projection
   d. alveolus; narrow groove
3. The frontal and parietal bones articulate at the _____ suture.
   a. coronal
   b. sagittal
   c. lambdoid
   d. squamosal
4. The compression of an infant's skull bones at birth is facilitated by spaces between unfused cranial bones called
   a. ossification centers.
   b. fontanelles.
   c. foramina.
   d. fossae.
5. Most _____ vertebrae have a heart-shaped body and a long spinous process that is angled inferiorly.
   a. cervical
   b. thoracic
   c. lumbar
   d. sacral
6. The female pelvis typically has which of the following characteristics?
   a. narrow, U-shaped greater sciatic notch
   b. wide subpubic angle, greater than 100 degrees
   c. short, triangular pubic body
   d. smaller, heart-shaped pelvic inlet

7. When the forearm is supinated,
   a. the pollex is laterally placed.
   b. the radius and ulna are crossed.
   c. the little finger is laterally placed.
   d. the pisiform is facing posteriorly.
8. The spine of the scapula separates which two fossae?
   a. supraspinous, subscapular
   b. subscapular, infraspinous
   c. infraspinous, supraspinous
   d. supraspinous, glenoid
9. The femur articulates with the tibia at the
   a. linea aspera.
   b. medial and lateral condyles.
   c. head of the femur.
   d. greater trochanter of the femur.
10. When sitting upright, you are resting on your
   a. pubic bones.
   b. ischial tuberosities.
   c. sacroiliac joints.
   d. iliac crest.
11. What are sutures, and how do they affect skull shape and growth?
12. What cranial and facial bones may be seen easily on the inferior surface of the skull?
13. What are the functions of the paranasal sinuses?
14. What are the spinal curvatures, when do they form, and what are their functions?
15. Describe similarities and differences among true, false, and floating ribs.
16. Compare and contrast the anatomic and functional features of the pectoral and pelvic girdles.
17. What are the primary similarities and differences between the upper and lower limbs?
18. Distinguish between the true and false pelves. What bony landmark separates the two?
19. What are the functions of the arches of the foot?
20. Discuss the development of the limbs. What primary germ layers form the limb bud? List the major events during each week of limb development.

Can You Apply What You've Learned?

Use the following paragraph to answer questions 1–5.

You are called to a crime scene reported in the woods. A hiker found a skeleton underneath some leaves—and as the osteologist of the team, it is up to you to identify the bones, as well as determine the age and sex of the skeleton. You begin examining the skeleton.

1. The first bone you identify is long and rather large. It has a rounded head, an elongated neck, and smooth condyles on its distal surface. There also are large, bony projections near the neck of this bone. Based on these features, you determine that the bone in question is a
   a. humerus.
   b. radius.
   c. femur.
   d. tibia.
2. As you examine the rest of the skeleton, you notice an S-shaped bone that appears to have been fractured prior to death. This bone likely was fractured due to a fall on the outstretched hand. What bone is it?
   a. clavicle
   b. metacarpal
   c. rib
   d. phalanx
3. You pick up the skull and begin examining it. The mastoid process is rather small, and the external occipital protuberance is not well defined. The supraorbital margins are rather sharp, and the mental protuberance is pointed (instead of squared-off). All of these features lead you to the following conclusion:
   a. The skeleton has been buried for a long time.
   b. The skull is from a female.
   c. The skull is from a male.
   d. The skull is from a young child.
4. Based on the answer you gave to question 3, what feature(s) would you expect to see in the pelvis?
   a. narrow pelvic inlet
   b. elongated, rectangular pubic bones
   c. narrow, U-shaped sciatic notch
   d. 90-degree subpubic arch
5. The police officers also want to know if you can determine the age at death of the skeleton. You determine that all long bone epiphyses have fused to their diaphyses, and all permanent teeth are erupted. The cranial sutures are still open, and the symphysial surface is flattened, but there is no complete rim around the symphysis. Based on these features, a likely age range for the skeleton would be
   a. younger than 10 years.
   b. 10–20 years.
   c. 20–35 years.
   d. 35–50 years.

Can You Synthesize What You've Learned?

1. Paul viewed his newborn daughter through the nursery window at the hospital and was distressed because the infant's skull was badly misshapen. A nurse told him not to worry—the shape of the infant's head would return to normal in a few days. What caused the misshapen skull, and what anatomic feature of the neonatal skull allows it to return to a more rounded shape?

The spheno-occipital synchondrosis is found between the body of the sphenoid and the basilar part of the occipital bone. This synchondrosis typically fuses between 18 and 25 years of age, making it a useful tool for assessing the age of the skull (see section 8.4b).

Other examples of synchondroses are formed from costal cartilage. The **costochondral** (kos-tō-kon′drăl; *costa* = rib) **joint,** the joint between each bony rib and its respective costal cartilage, is a synchondrosis. Finally, the attachment of the first rib to the sternum by costal cartilage (called the *first sternocostal joint*) is another synchondrosis. Here, the first rib and its costal cartilage are united firmly to the manubrium of the sternum to provide stability to the rib cage. (Note that the sternocostal joints between the sternum and the costal cartilage of ribs 2–7 are synovial joints and not synchondroses.)

WHAT DID YOU LEARN?

6 What is the composition of a synchondrosis, and where in the body is it found?

9.3b Symphyses

LEARNING OBJECTIVE

8. Name the locations of symphyses and their functions in these locations.

A **symphysis** (sim′fi-sis; pl., *symphyses,* -sēz; growing together) has a pad of fibrocartilage between the articulating bones (figure 9.3*b*). The fibrocartilage resists both compression and tension stresses and acts as a resilient shock absorber. All symphyses are amphiarthroses—thus, they allow slight mobility.

One example of a symphysis is the pubic symphysis, which is located between the right and left pubic bones. In pregnant females, the pubic symphysis becomes more mobile to allow the pelvis to change shape slightly as the fetus passes through the birth canal.

Other examples of symphyses are the intervertebral joints, where the bodies of adjacent vertebrae are both separated and united by intervertebral discs. Individual intervertebral discs allow only slight movements between the adjacent vertebrae; however, the collective movements of all the intervertebral discs afford the spine considerable flexibility.

WHAT DID YOU LEARN?

7 Into what functional category is a symphysis placed? Why is it in this category?

9.4 Synovial Joints

Synovial joints are freely mobile articulations (i.e., diarthroses). Most of the commonly known joints in the body are synovial joints, including the glenohumeral (shoulder) joint, the temporomandibular joint, the elbow joint, and the knee joint.

9.4a Distinguishing Features and Anatomy of Synovial Joints

LEARNING OBJECTIVES

9. Identify the characteristics common to all synovial joints.
10. List the basic features of a synovial joint.
11. Describe the composition and function of synovial fluid in a typical synovial joint.

Unlike the joints previously discussed, the bones in a synovial joint are separated by a space called a *joint cavity*. Functionally, all synovial joints are classified as diarthroses, because all are freely mobile. Often, the terms *synovial joint* and *diarthrosis* are equated. All synovial joints include several basic features: an articular capsule, a joint cavity, synovial fluid, articular cartilage, ligaments, nerves, and blood vessels (**figure 9.4**).

Each synovial joint is composed of a double-layered capsule called the **articular** (ar-tik′yū-lăr) **capsule,** or *joint capsule.* Its outer layer is called the **fibrous layer,** and the inner layer is a **synovial membrane** (or *synovium*) (see section 5.5b). The fibrous layer is formed from dense connective tissue. It strengthens the joint to prevent the bones from being pulled apart. The synovial membrane is a specialized type of connective tissue, the cells of which help produce and secrete synovial fluid (described shortly). This membrane covers all the internal joint surfaces not covered by cartilage and lines the articular capsule.

All articulating bone surfaces in a synovial joint are covered by a thin layer of hyaline cartilage called **articular cartilage.** This cartilage has numerous functions: It reduces friction in the joint during movement, acts as a spongy cushion to absorb compression placed on the joint, and prevents damage to the articulating ends of the bones. This special hyaline cartilage lacks a perichondrium (see section 7.2c). Mature cartilage is avascular, so it does not have blood vessels to bring nutrients to and remove waste products from the cartilage. The repetitious compression and expansion that occurs during exercise is vital to maintaining healthy articular cartilage because this action enhances its obtaining nutrition and its waste removal.

Only synovial joints house a **joint cavity** (or *articular cavity*), a space that permits separation of the articulating bones. The articular cartilage and synovial fluid (described next) within the joint cavity together reduce friction as bones move at a synovial joint.

Synovial fluid is a viscous, oily substance located within a synovial joint. It is a product of both the synovial membrane cells

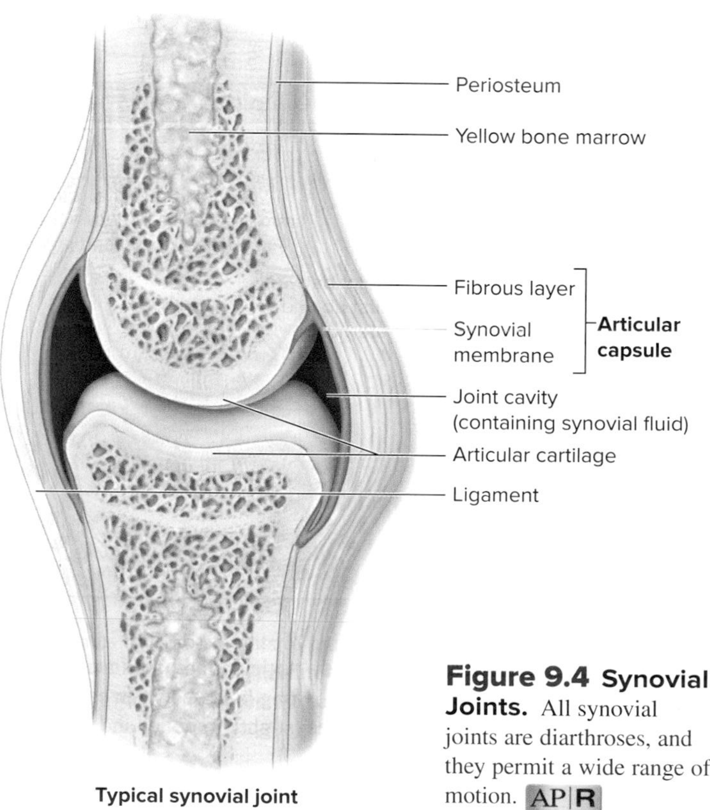

Figure 9.4 Synovial Joints. All synovial joints are diarthroses, and they permit a wide range of motion. **AP|R**

Skeletal System: Articulations

Module 5: Skeletal System

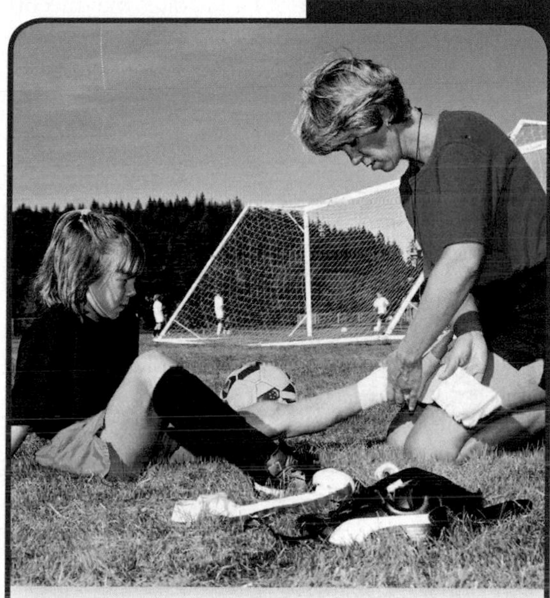

CAREER PATH
Athletic Trainer

An athletic trainer promotes both fitness and wellness, encourages prevention of illness and injury, and can diagnose musculoskeletal injuries, especially those involving the articulations (joints) we discuss in this chapter. This health-care professional must be aware of the normal range of movement at each joint, the muscular and ligamentous structures that support the joint, and how injury may impact mobility and ultimate healing of the joint. Here, we see an athletic trainer examining and taping a soccer player's ankle after an injury on the field.

Our skeleton protects vital organs and supports soft tissues. Its marrow cavity is the source of new blood cells. When it interacts with the muscular system, the skeleton helps the body move. Bones are too rigid to bend—but they meet at joints that anatomists call articulations. In this chapter, we examine how bones articulate (connect) and permit varying degrees of freedom of movement, depending upon both the shapes and the supporting structures of the different joints.

9.1 Classification of Joints

LEARNING OBJECTIVES

1. Define a joint.
2. Compare the structural and the functional classification of joints.
3. Explain the inverse relationship between mobility and stability within a joint.

A **joint,** or an **articulation** (ar-tik′yū-lā′shŭn), is the place of contact between bones, between bone and cartilage, or between bones and teeth. Bones are said to **articulate** with each other at a joint. The scientific study of joints is called **arthrology** (ar-throl′ō-jē; *arthron* = joint, *logos* = study).

Joints are classified by both their structural characteristics and their functional characteristics, which are the movements they allow (**table 9.1**). Joints are categorized structurally on the basis of whether a space occurs between the articulating bones and the type of connective tissue that binds the articulating surfaces of the bones:

- A **fibrous** (fī′brŭs) **joint** has no joint cavity and occurs where bones are held together by dense regular (fibrous) connective tissue.
- A **cartilaginous** (kar-ti-laj′i-nŭs) **joint** has no joint cavity and occurs where bones are joined by cartilage.
- A **synovial** (si-nō′vē-ăl) **joint** has a fluid-filled joint cavity that separates the articulating surfaces of the bones. The articulating surfaces are enclosed within a connective tissue capsule, and the bones are attached to each other by various ligaments.

Joints are classified functionally based on the extent of movement they permit.

- A **synarthrosis** (sin′ar-thrō′sis; pl., *synarthroses,* -sēz; *syn* = joined, together; *osis* = condition) is an immobile joint. Two types of fibrous joints and one type of cartilaginous joint are synarthroses.
- An **amphiarthrosis** (am′fē-ar-thrō′sis; pl., *amphiarthroses,* -sēz; *amphi* = around) is a slightly mobile joint. One type of fibrous joint and one type of cartilaginous joint are amphiarthroses.
- A **diarthrosis** (dī-ar-thrō′sis; pl., *diarthroses,* -sēz; *di* = two) is a freely mobile joint. All synovial joints are diarthroses.

The motion permitted at a joint ranges from no movement, such as where some skull bones interlock at a suture, to extensive movement, such as that seen at the shoulder, where the humerus articulates with the scapula. The structure of each joint determines both its mobility and its stability. There is an inverse relationship between mobility and stability in articulations. When the mobility of a joint increases, its stability decreases. In contrast, if a joint is immobile, it has maximum stability. **Figure 9.1** illustrates the tradeoff between mobility and stability for various joints. It allows you to view and compare the structural versus functional classification of some common joints.

INTEGRATE

LEARNING STRATEGY

You can logically figure out the names of most joints by putting together the names of the bones that form them. For example, the *glenohumeral* joint is where the glenoid cavity of the scapula meets the head of the humerus, and the *sternoclavicular* joint is where the manubrium of the sternum articulates with the sternal end of the clavicle.

The following detailed discussion of articulations is based upon their structural classification, with functional categories included as appropriate.

WHAT DID YOU LEARN?

1. What is the relationship between mobility and stability in a joint?

2. Are all fibrous joints also synarthroses? Explain why or why not.

Table 9.1 Joint Classifications

Structural Classification	Structural Characteristics	Structural Categories	Functional Classification
Fibrous (see figure 9.2)	Dense regular connective tissue holds together the ends of bones and bone parts; no joint cavity	Gomphosis Suture Syndesmosis	Synarthrosis (immobile) or amphiarthrosis (slightly mobile)
Cartilaginous (see figure 9.3)	Pad of cartilage is wedged between the ends of bones; no joint cavity	Synchondrosis Symphysis	Synarthrosis (immobile) or amphiarthrosis (slightly mobile)
Synovial (see figure 9.6)	Ends of bones covered with articular cartilage; joint cavity separates the articulating bones; joint enclosed by an articular capsule, lined by a synovial membrane; contains synovial fluid	Plane Hinge Pivot Condylar Saddle Ball-and-socket	Diarthrosis (freely mobile)

INTEGRATE CONCEPT OVERVIEW

Figure 9.1 The Relationship Between Mobility and Stability in Joints. The more mobile a joint is, the less stable it is, and vice versa.

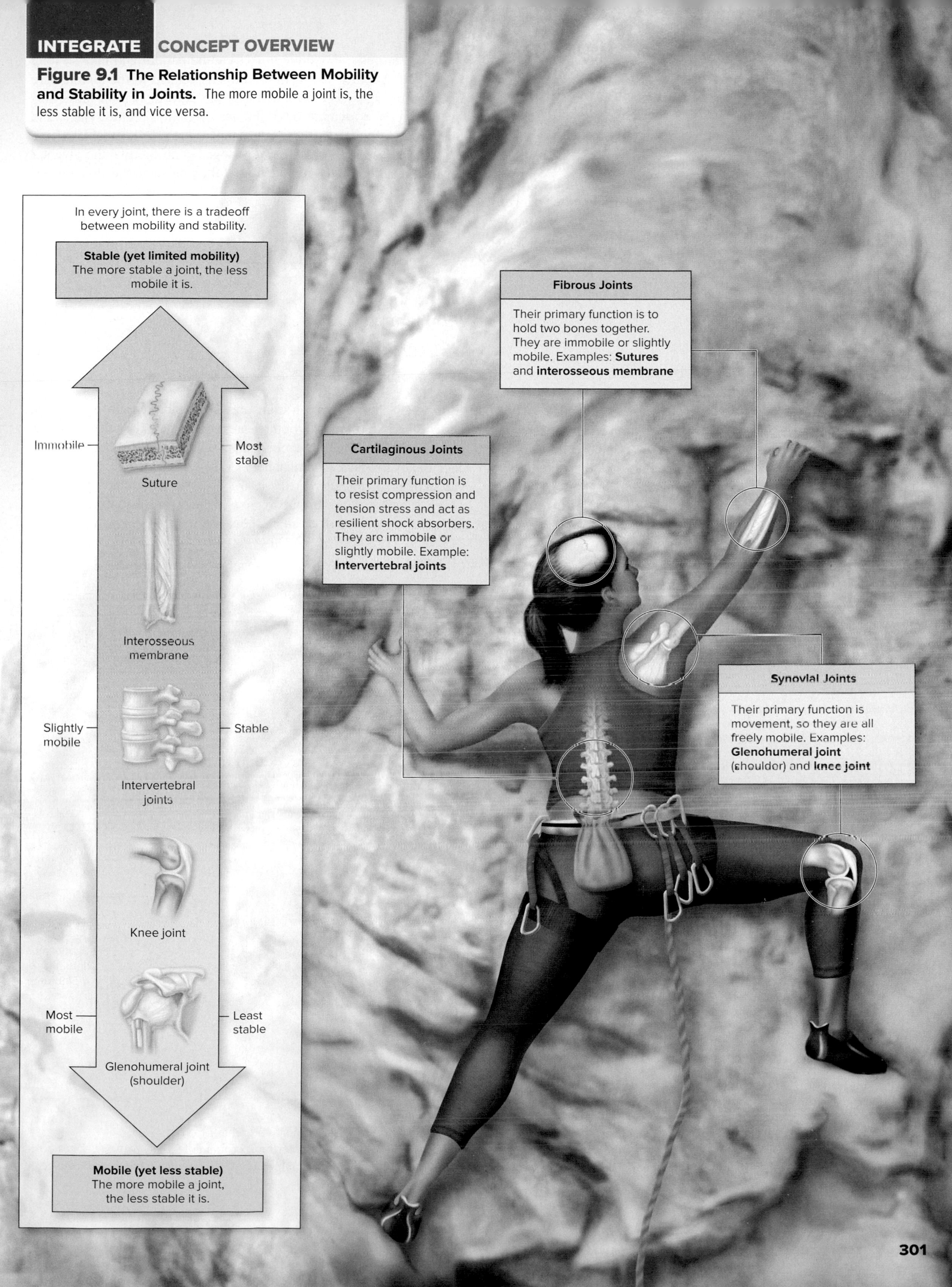

In every joint, there is a tradeoff between mobility and stability.
Stable (yet limited mobility)
The more stable a joint, the less mobile it is.
Immobile
Most stable
Suture
Interosseous membrane
Slightly mobile
Stable
Intervertebral joints
Knee joint
Most mobile
Least stable
Glenohumeral joint (shoulder)
Mobile (yet less stable)
The more mobile a joint, the less stable it is.
Fibrous Joints
Their primary function is to hold two bones together. They are immobile or slightly mobile. Examples: Sutures and interosseous membrane
Cartilaginous Joints
Their primary function is to resist compression and tension stress and act as resilient shock absorbers. They are immobile or slightly mobile. Example: Intervertebral joints
Synovial Joints
Their primary function is movement, so they are all freely mobile. Examples: Glenohumeral joint (shoulder) and knee joint

9.2 Fibrous Joints

Articulating bones in fibrous joints are connected by dense regular connective tissue. Fibrous joints have no joint cavity; thus, they lack a space between the articulating bones. Most fibrous joints are immobile or at most only slightly mobile; their primary function is to hold together two bones. Examples include the articulations of the teeth in their sockets, sutures between skull bones, and articulations between either the radius and ulna or the tibia and fibula. In this section, we look at the three types of fibrous joints: gomphoses, sutures, and syndesmoses (**figure 9.2**).

9.2a Gomphoses

LEARNING OBJECTIVE

4. Explain the location and characteristics of gomphoses.

A **gomphosis** (gom-fō′sis; pl., *gomphoses*, -sēz; *gomphos* = bolt) resembles a "peg in a socket." The only gomphoses in the human body are the articulations of the roots of individual teeth with the alveolar processes (sockets) of the mandible and the maxillae. A tooth is held firmly in place by fibrous **periodontal** (per′ē-ō-don′tăl; *peri* = around, *odous* = tooth) **membranes.** This joint is immobile and thus is functionally classified as a synarthrosis.

The reasons orthodontic braces can be painful and take so long to correctly position the teeth are directly related to the architecture of the gomphosis. The orthodontist's job is to reposition these normally immobile joints through the use of clamps, bands, rings, and braces. In response to these mechanical stressors, osteoblasts and osteoclasts (see section 7.2e) work together to modify the alveolar process, resulting in the remodeling of the joints and the slow repositioning of the teeth.

WHAT DID YOU LEARN?

3. Where are gomphoses located, and what type of movement do they allow?

9.2b Sutures

LEARNING OBJECTIVE

5. Describe the location and functions of sutures.

Sutures (sū′chūr; *sutura* = a seam) are fibrous joints found only between certain bones of the skull. Sutures are functionally classified as synarthroses, since they are immobile joints. Sutures have distinct, interlocking, usually irregular edges that both increase their strength and decrease the number of fractures at these articulations. In addition to joining bones, sutures permit the skull to grow as the brain increases in size during childhood. In an older adult, the dense regular connective tissue in

Fibrous Joints: Dense regular connective tissue holds bones together; no joint cavity

(a) Gomphosis: Periodontal membranes hold a tooth to bony jaw (*synarthrosis*)

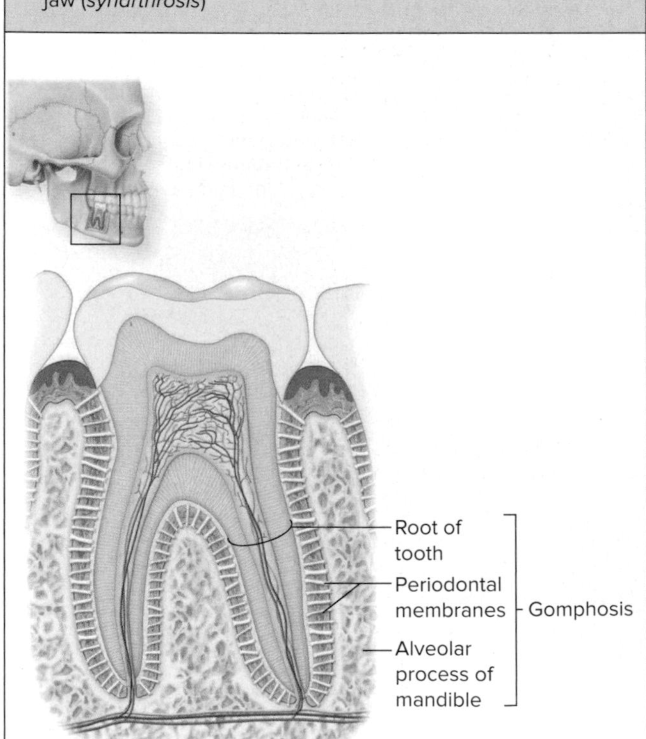

(b) Suture: Connects skull bones (*synarthrosis*)

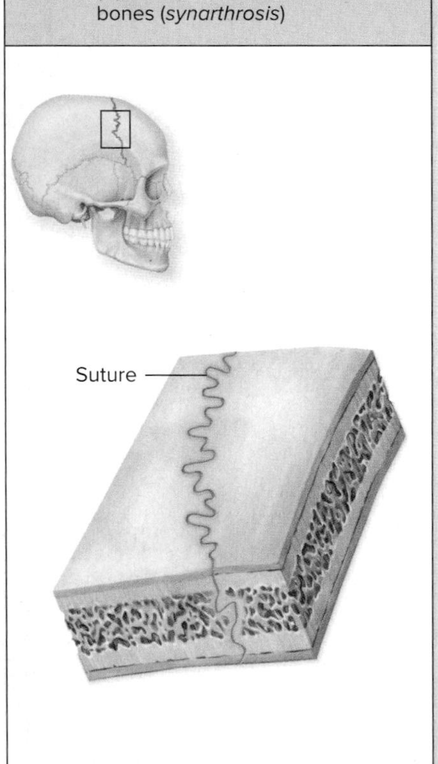

(c) Syndesmosis: Interosseous membranes (dense regular CT) between bones (*amphiarthrosis*)

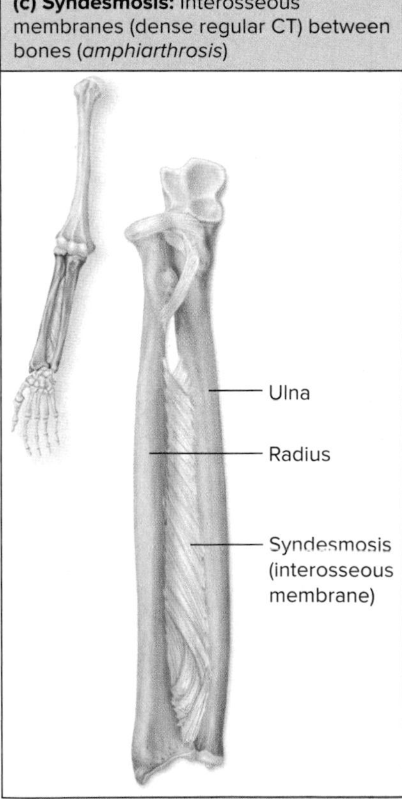

Figure 9.2 Fibrous Joints. Dense regular connective tissue binds the articulating bones in fibrous joints to prevent or restrict movement. (*a*) A gomphosis is a synarthrosis (immobile joint) between a tooth and the jaw. (*b*) A suture is a synarthrosis (immobile joint) between bones of the skull. (*c*) A syndesmosis is classified functionally as an amphiarthrosis, as it permits slight mobility between the radius and the ulna (shown) and between the tibia and fibula (not shown). AP|R

the suture becomes ossified, fusing the skull bones together. When the bones have completely fused across the suture line, these obliterated sutures are now called **synostoses** (sin-os-tō′sēz; sing., -sis; *osteon* = bone). (See Clinical View 8.2: "Craniosynostosis and Plagiocephaly.")

WHAT DID YOU LEARN?

4 What is the composition of a suture, and where in the body is it found?

9.2c Syndesmoses

LEARNING OBJECTIVE

6. List the locations of syndesmoses, and describe their function.

Syndesmoses (sin′dez-mō′sēz; sing., *syndesmosis*; *syndesmos* = a fastening) are fibrous joints in which articulating bones are joined by long strands of dense regular connective tissue only. Because syndesmoses allow for slight mobility, they are classified functionally as amphiarthroses. Syndesmoses are found both between the radius and ulna and between the tibia and fibula. The shafts of the two articulating bones are bound by a broad, ligamentous sheet called an **interosseous membrane** (or *interosseous ligament*). The interosseous membrane provides a pivot where the radius and ulna (or the tibia and fibula) can move relative to one another.

WHAT DID YOU LEARN?

5 What type of movement is allowed at a syndesmosis?

9.3 Cartilaginous Joints

Cartilaginous joints have cartilage between the articulating bones. Like fibrous joints, cartilaginous joints also lack a joint cavity. They may be either immobile or slightly mobile. The cartilage found between the articulating bones is either hyaline cartilage or fibrocartilage. The two types of cartilaginous joints are synchondroses and symphyses (**figure 9.3**).

INTEGRATE

CLINICAL VIEW 9.1

Costochondritis

Costochondritis is the inflammation and irritation of the costochondral joints, resulting in localized chest pain. The cause of costochondritis is usually unknown, but some documented cases include repeated minor trauma to the chest wall (e.g., from forceful, repeated coughing or from overexertion during exercise) or either bacterial or viral infection of the joints themselves. The localized chest pain may be mistaken for pain resulting from a myocardial infarction (heart attack; see Clinical View 19.4: "Angina Pectoris and Myocardial Infarction"). Costochondritis may be treated with NSAIDs (nonsteroidal anti-inflammatory drugs, such as aspirin). With proper rest and treatment, symptoms typically disappear after several weeks.

9.3a Synchondroses

LEARNING OBJECTIVE

7. Describe the locations and functions of synchondroses.

An articulation in which bones are joined by hyaline cartilage is called a **synchondrosis** (sin′kon-drō′sis; pl., *synchondroses*, -sēz; *chondros* = cartilage). Functionally, all synchondroses are immobile and thus are classified functionally as synarthroses.

WHAT DO YOU THINK?

1 Why is a synchondrosis also a synarthrosis? Why would you want a synchondrosis to be immobile?

The hyaline cartilage of epiphyseal plates in children forms synchondroses that bind the epiphyses and diaphysis of long bones (figure 9.3*a*). When the hyaline cartilage stops growing, bone replaces the cartilage and a synchondrosis no longer exists (see section 7.4b).

Cartilaginous Joints: Cartilage between ends of articulating bones; no joint cavity

(a) Synchondroses: Contain hyaline cartilage (*synarthroses*)

Epiphyseal plate
Costochondral joints (immobile joints between the rib and its costal cartilage)
Joint between first rib and sternum

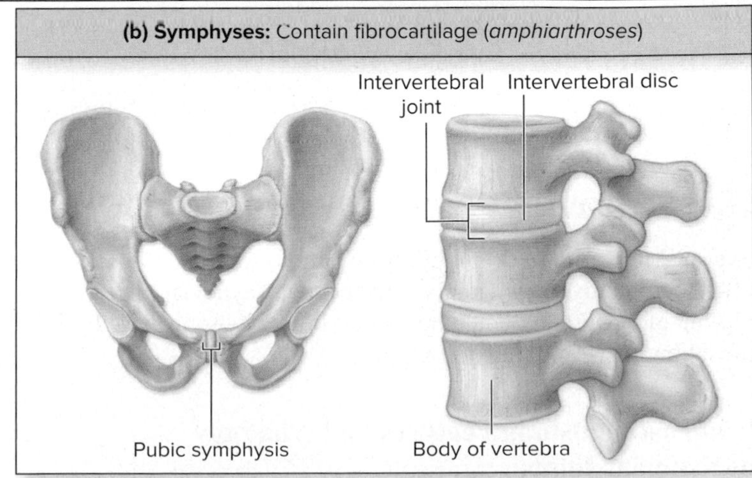

Figure 9.3 Cartilaginous Joints. Articulating bones are joined by cartilage in cartilaginous joints. (*a*) Synchondroses are synarthroses (immobile joints) that occur in an epiphyseal plate in a long bone, between each rib and its respective costal cartilage, and in the joint between the first rib and the sternum. (*b*) Symphyses are amphiarthroses (slightly mobile) and occur in the pubic symphysis and the intervertebral joints. AP|R

The spheno-occipital synchondrosis is found between the body of the sphenoid and the basilar part of the occipital bone. This synchondrosis typically fuses between 18 and 25 years of age, making it a useful tool for assessing the age of the skull (see section 8.4b).

Other examples of synchondroses are formed from costal cartilage. The **costochondral** (kos-tō-kon′drăl; *costa* = rib) **joint,** the joint between each bony rib and its respective costal cartilage, is a synchondrosis. Finally, the attachment of the first rib to the sternum by costal cartilage (called the *first sternocostal joint*) is another synchondrosis. Here, the first rib and its costal cartilage are united firmly to the manubrium of the sternum to provide stability to the rib cage. (Note that the sternocostal joints between the sternum and the costal cartilage of ribs 2–7 are synovial joints and not synchondroses.)

WHAT DID YOU LEARN?

6 What is the composition of a synchondrosis, and where in the body is it found?

9.3b Symphyses

LEARNING OBJECTIVE

8. Name the locations of symphyses and their functions in these locations.

A **symphysis** (sim′fi-sis; pl., *symphyses,* -sēz; growing together) has a pad of fibrocartilage between the articulating bones (figure 9.3*b*). The fibrocartilage resists both compression and tension stresses and acts as a resilient shock absorber. All symphyses are amphiarthroses—thus, they allow slight mobility.

One example of a symphysis is the pubic symphysis, which is located between the right and left pubic bones. In pregnant females, the pubic symphysis becomes more mobile to allow the pelvis to change shape slightly as the fetus passes through the birth canal.

Other examples of symphyses are the intervertebral joints, where the bodies of adjacent vertebrae are both separated and united by intervertebral discs. Individual intervertebral discs allow only slight movements between the adjacent vertebrae; however, the collective movements of all the intervertebral discs afford the spine considerable flexibility.

WHAT DID YOU LEARN?

7 Into what functional category is a symphysis placed? Why is it in this category?

9.4 Synovial Joints

Synovial joints are freely mobile articulations (i.e., diarthroses). Most of the commonly known joints in the body are synovial joints, including the glenohumeral (shoulder) joint, the temporomandibular joint, the elbow joint, and the knee joint.

9.4a Distinguishing Features and Anatomy of Synovial Joints

LEARNING OBJECTIVES

9. Identify the characteristics common to all synovial joints.
10. List the basic features of a synovial joint.
11. Describe the composition and function of synovial fluid in a typical synovial joint.

Unlike the joints previously discussed, the bones in a synovial joint are separated by a space called a *joint cavity.* Functionally, all synovial joints are classified as diarthroses, because all are freely mobile. Often, the terms *synovial joint* and *diarthrosis* are equated. All synovial joints include several basic features: an articular capsule, a joint cavity, synovial fluid, articular cartilage, ligaments, nerves, and blood vessels (**figure 9.4**).

Each synovial joint is composed of a double-layered capsule called the **articular** (ar-tik′yū-lăr) **capsule,** or *joint capsule.* Its outer layer is called the **fibrous layer,** and the inner layer is a **synovial membrane** (or *synovium*) (see section 5.5b). The fibrous layer is formed from dense connective tissue. It strengthens the joint to prevent the bones from being pulled apart. The synovial membrane is a specialized type of connective tissue, the cells of which help produce and secrete synovial fluid (described shortly). This membrane covers all the internal joint surfaces not covered by cartilage and lines the articular capsule.

All articulating bone surfaces in a synovial joint are covered by a thin layer of hyaline cartilage called **articular cartilage.** This cartilage has numerous functions: It reduces friction in the joint during movement, acts as a spongy cushion to absorb compression placed on the joint, and prevents damage to the articulating ends of the bones. This special hyaline cartilage lacks a perichondrium (see section 7.2c). Mature cartilage is avascular, so it does not have blood vessels to bring nutrients to and remove waste products from the cartilage. The repetitious compression and expansion that occurs during exercise is vital to maintaining healthy articular cartilage because this action enhances its obtaining nutrition and its waste removal.

Only synovial joints house a **joint cavity** (or *articular cavity*), a space that permits separation of the articulating bones. The articular cartilage and synovial fluid (described next) within the joint cavity together reduce friction as bones move at a synovial joint.

Synovial fluid is a viscous, oily substance located within a synovial joint. It is a product of both the synovial membrane cells

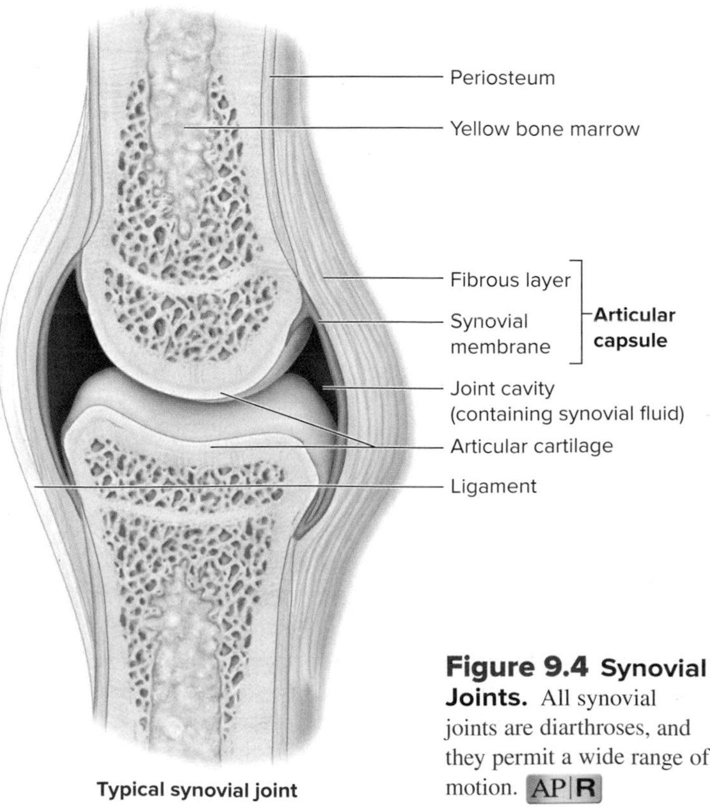

Figure 9.4 Synovial Joints. All synovial joints are diarthroses, and they permit a wide range of motion. AP|R

and the filtrate formed from blood plasma. Synovial fluid has three functions:

1. Synovial fluid lubricates the articular cartilage on the surface of articulating bones (in the same way that oil in a car engine lubricates the moving engine parts).
2. Synovial fluid nourishes the articular cartilage's chondrocytes. The relatively small volume of synovial fluid must be circulated continually to provide nutrients to and remove wastes from these cells. Whenever movement occurs at a synovial joint, the combined compression and re-expansion of the articular cartilage circulates the synovial fluid into and out of the cartilage matrix.
3. Synovial fluid acts as a shock absorber, distributing stresses and force evenly across the articular surfaces when the pressure in the joint suddenly increases.

Ligaments (lig′ă-ment; *ligamentum* = a band) are composed of dense regular connective tissue, and they connect one bone to another bone. Ligaments function to stabilize, strengthen, and reinforce most synovial joints. **Intrinsic ligaments** represent thickenings of the articular capsule itself. Intrinsic ligaments include *extracapsular ligaments* outside the joint capsule and *intracapsular ligaments* within the joint capsule. **Extrinsic ligaments** are outside of, and physically separate from, the joint capsule. The specific intrinsic and extrinsic ligaments are specific to each type of joint (e.g., knee, shoulder).

All synovial joints have numerous **sensory nerves** that innervate the articular capsule and associated ligaments. The sensory nerves detect painful stimuli in the joint and report on the amount of movement and stretch within the joint. By monitoring stretching at a joint, the nervous system can detect changes in our posture and adjust body movements. Synovial joints also contain **blood vessels** that provide oxygen and nutrients and remove wastes.

Tendons (ten′dŏn; *tendo* = extend) are like ligaments and are composed of dense regular connective tissue, but they are not part of the synovial joint itself. Whereas a ligament binds bone to bone, a tendon attaches a muscle to a bone. When a muscle contracts, the tendon from that muscle moves the bone to which it is attached, thus causing movement at the joint. Tendons help stabilize joints because they pass across or around a joint to provide mechanical support, and sometimes they limit the range or amount of movement permitted at a joint.

Synovial joints usually have bursae and fat pads as accessory structures in addition to the main components just described. A **bursa** (ber′să; pl., *bursae,* ber′sē; a purse) is a fibrous, saclike structure that contains synovial fluid and is lined internally by a synovial membrane (**figure 9.5*a***). Bursae are associated with most

INTEGRATE

CLINICAL VIEW 9.2

"Cracking Knuckles"

Stretching or pulling on a synovial joint (e.g., *cracking your knuckles*) causes the joint volume to immediately expand and the pressure on the fluid within the joint to decrease, so that a partial vacuum exists within the joint. As a result, the gases dissolved in the fluid become less soluble, and they form bubbles, a process called *cavitation.* When the joint is stretched to a certain point, the pressure in the joint drops even lower, so the bubbles in the fluid burst, resulting in a popping or cracking sound. It typically takes about 20 to 30 minutes for the gases to dissolve back into the synovial fluid. You cannot crack your knuckles again until these gases dissolve. Contrary to popular belief, cracking your knuckles does *not* cause arthritis.

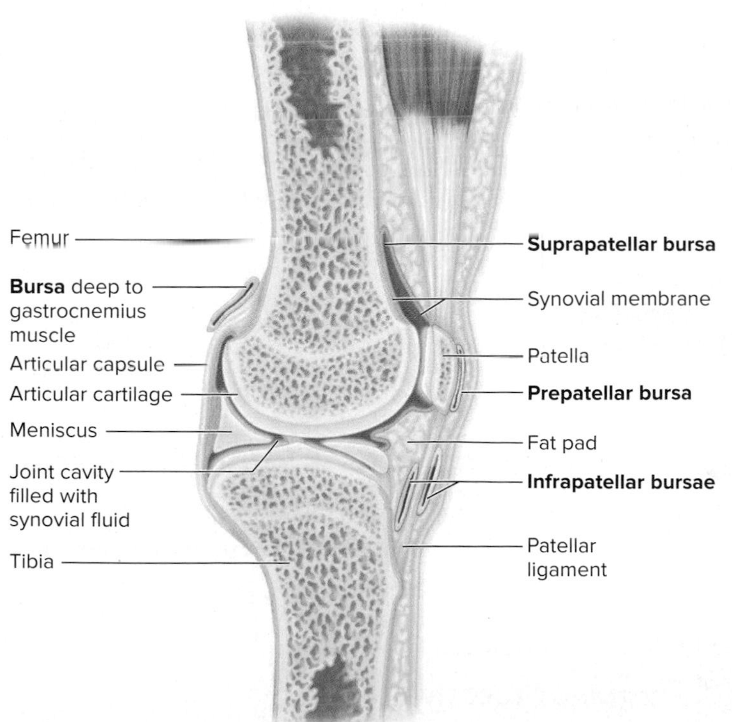

(a) Bursae of the knee joint, sagittal section

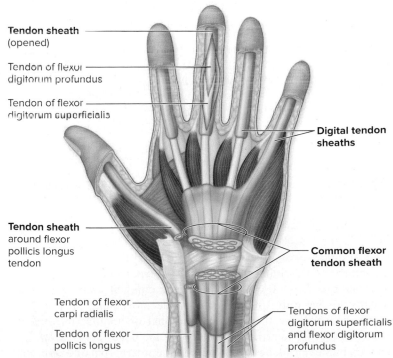

Tendon sheath (opened)
Tendon of flexor digitorum profundus
Tendon of flexor digitorum superficialis
Digital tendon sheaths
Tendon sheath around flexor pollicis longus tendon
Common flexor tendon sheath
Tendon of flexor carpi radialis
Tendon of flexor pollicis longus
Tendons of flexor digitorum superficialis and flexor digitorum profundus

(b) Tendon sheaths of wrist and hand, anterior view

Figure 9.5 Bursae and Tendon Sheaths. Synovial fluid–filled structures called bursae and tendon sheaths reduce friction where ligaments, muscles, tendons, and bones rub together. (*a*) The knee joint contains a number of bursae (blue and purple). (*b*) The wrist and hand contain numerous tendon sheaths (blue).

synovial joints and are where bones, ligaments, muscles, skin, or tendons overlie each other and rub together. Bursae may be either connected to the joint cavity or completely separate from it. They alleviate the friction resulting from the various body movements, such as where a tendon or ligament rubs against bone. An elongated bursa called a **tendon sheath** wraps around a tendon where there may be excessive friction. Tendon sheaths are especially common in the confined spaces of the wrist and ankle (figure 9.5*b*).

Fat pads are often distributed along the periphery of a synovial joint. They act as packing material and provide some protection for the joint. Often, fat pads fill the spaces that form when bones move and the joint cavity changes shape (figure 9.5*a*).

WHAT DID YOU LEARN?

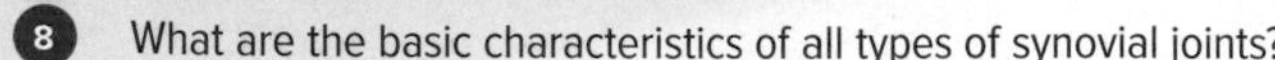

8 What are the basic characteristics of all types of synovial joints?

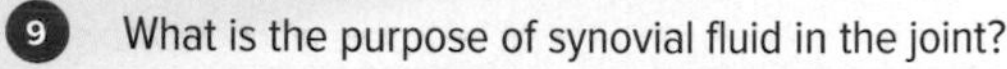

9 What is the purpose of synovial fluid in the joint?

9.4b Classification of Synovial Joints

LEARNING OBJECTIVES

12. Explain the movement of a joint with respect to the three perpendicular axes of space.

13. Compare and contrast the six types of synovial joints.

Synovial joints are classified by the shapes of their articulating surfaces and the types of movement they allow. Movement of a bone at a synovial joint is best described with respect to three intersecting perpendicular planes or axes:

- A joint is said to be **uniaxial** (yū-nē-ak′sē-ăl; *unus* = one) if the bone moves in just one plane or axis.
- A joint is **biaxial** (bī-ak′sē-ăl; *bi* = double) if the bone moves in two planes or axes.
- A joint is **multiaxial** or **triaxial** (trī-ak′sē-ăl; *tri* = three) if the bone moves in multiple planes or axes.

All synovial joints are diarthroses, as mentioned, but some are more mobile than others. From least mobile to most freely mobile, the six specific types of synovial joints are plane joints, hinge joints, pivot joints, condylar joints, saddle joints, and ball-and-socket joints. These joints and examples of where they are found in the body are shown in **figure 9.6**.

WHAT DO YOU THINK?

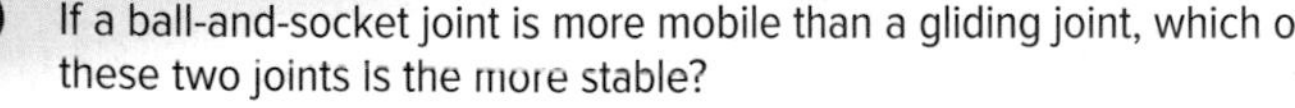

2 If a ball-and-socket joint is more mobile than a gliding joint, which of these two joints is the more stable?

A **plane** (*planus* = flat) **joint,** also called a *planar* or *gliding joint,* is the simplest synovial articulation and the least mobile type of diarthrosis. Anatomists have debated how to describe the movement of this joint with respect to perpendicular planes. We describe this type of synovial joint as a uniaxial joint because it usually allows only limited side-to-side movements in a single plane, and because there is no rotational or angular movement with this joint. The articular surfaces of the bones are flat, or planar. Examples of plane joints include the intercarpal and intertarsal joints (the joints between the carpal bones and tarsal bones, respectively).

A **hinge joint** is formed by the convex surface of one articulating bone fitting into a concave depression on the other bone in the joint. Movement is confined to a single axis, like the movement seen at the hinge of a door, so a hinge joint is considered a uniaxial joint. An example is the elbow joint. The trochlear notch of the ulna fits directly into the trochlea of the humerus, so the forearm can be moved only anteriorly toward the arm or posteriorly away from the arm. Other hinge joints occur in the knee and the finger (interphalangeal [IP]) joints.

A **pivot joint** is a uniaxial joint in which one articulating bone with a rounded surface fits into a ring formed by a ligament and another bone. The first bone rotates on its longitudinal axis relative to the second bone. An example is the proximal radio-ulnar joint, where the rounded head of the radius pivots along the ulna and permits the radius to rotate. Another example is the atlantoaxial joint between the first two cervical vertebrae. The rounded dens of the axis fits snugly against an articular facet on the anterior arch of the atlas. This joint pivots when you shake your head "no."

Condylar (kon′di-lar) **joints,** also called *condyloid* or *ellipsoid joints,* are biaxial joints with an oval, convex surface on one bone that articulates with a concave articular surface on the second bone of the joint. Biaxial joints can move in two axes, such as back-and-forth and side-to-side. Examples of condylar joints are the metacarpophalangeal (met′ă-kar′pō-fă-lan′jē-ăl) (MP) joints of fingers 2 through 5. The MP joints are commonly referred to as *knuckles*. Examine your hand and look at the movements along the MP joints; you can flex and extend the fingers at this joint, which is one axis of movement. You also can move your fingers apart from one another and move them closer together, which is the second axis of movement.

A **saddle joint** is so named because the articular surfaces of the bones have convex and concave regions that resemble the shape of a saddle. This biaxial joint allows a greater range of movement than either a condylar or hinge joint. The carpometacarpal joint of the thumb (between the trapezium, which is a carpal bone, and the first metacarpal) is an example of a saddle joint. This joint permits the thumb to move toward the other fingers so that we can grasp objects.

Ball-and-socket joints are multiaxial joints in which the spherical articulating head of one bone fits into the rounded, cuplike socket of a second bone. Examples of these joints are the coxal (hip) and glenohumeral (shoulder) joints. The multiaxial nature of these joints permits movement in three planes. Move your arm at your shoulder, and note the wide range of movements that can be produced. The ball-and-socket joint is considered the most freely mobile type of synovial joint.

WHAT DID YOU LEARN?

10 What types of movements do each of the six kinds of joints allow?

9.5 The Movements of Synovial Joints

Four types of motion occur at synovial joints: gliding motion, angular motion, rotational motion, and special movements (motions that occur only at specific joints) **(table 9.2)**.

9.5a Gliding Motion

LEARNING OBJECTIVE

14. Describe gliding motion, and name joints in which it occurs.

Gliding is a simple movement in which two opposing surfaces slide slightly back-and-forth or side-to-side with respect to one another. In a

INTEGRATE CONCEPT OVERVIEW

Figure 9.6 Synovial Joints. Synovial joints contain a joint cavity within an articular capsule lined by a synovial membrane. All synovial joints are diarthroses. The six types of synovial joints and examples of their locations in the body are shown.

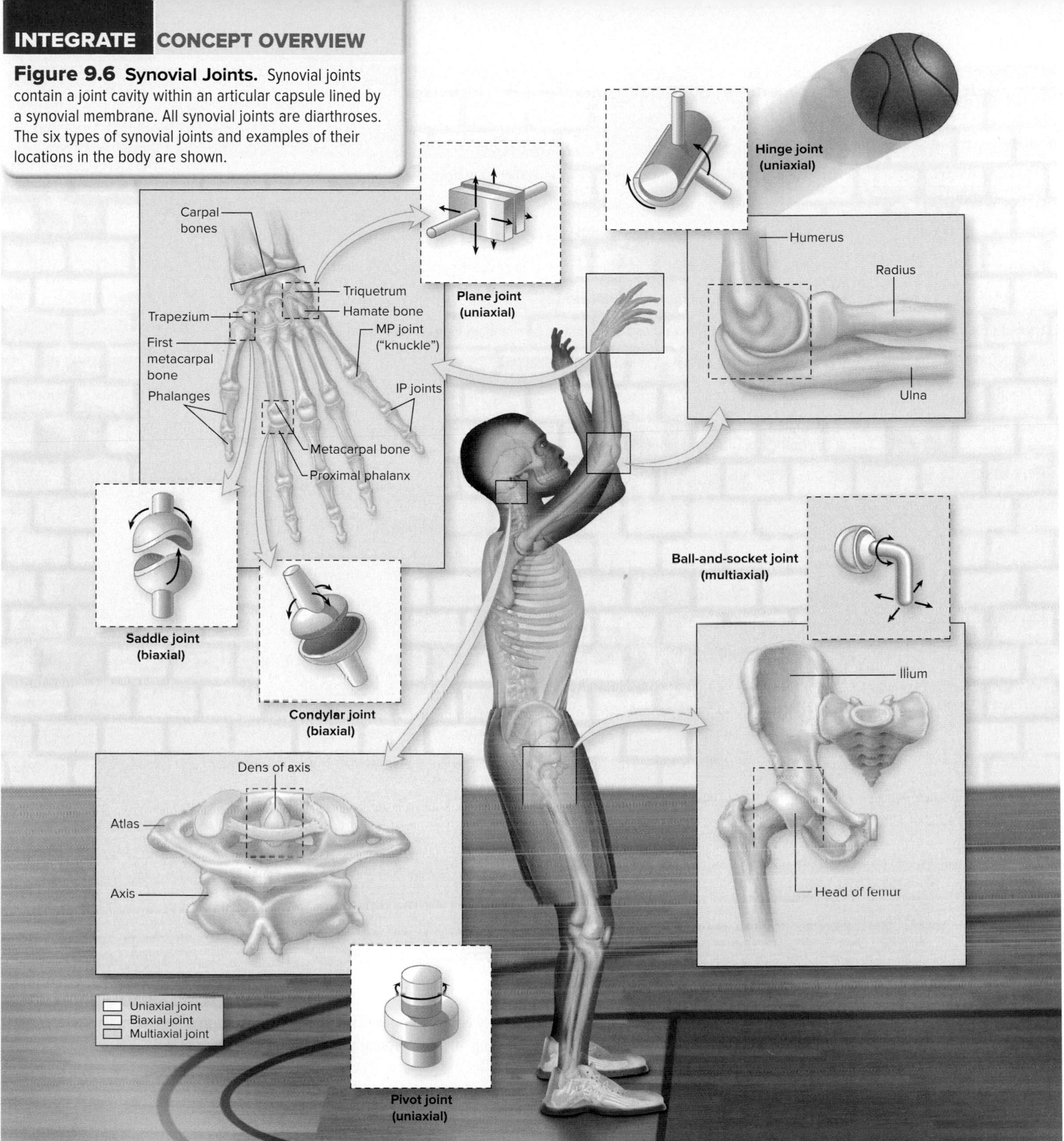

Synovial Joints

Structural Categories	Example(s)	Functional Classification
Uniaxial **Plane joint:** Flattened or slightly curved faces slide across one another. **Hinge joint:** Convex feature of one bone fits into concave depression of another bone. **Pivot joint:** Bone with a rounded surface fits into a ring formed by a ligament and another bone.	**Plane joint:** Intercarpal joints, intertarsal joints **Hinge joint:** Elbow joint, knee joint, IP (interphalangeal) joints **Pivot joint:** Atlantoaxial joint	Diarthrosis (freely mobile)
Biaxial **Condylar joint:** Oval articular surface on one bone closely interfaces with a depressed oval surface on another bone. **Saddle joint:** Saddle-shaped articular surface on one bone closely interfaces with a saddle-shaped surface on another bone.	**Condylar joint:** MP (metacarpophalangeal or metatarsophalangeal) joints **Saddle joint:** Articulation between a carpal bone (the trapezium) and first metacarpal	Diarthrosis (freely mobile)
Multiaxial (Triaxial) **Ball-and-socket joint:** Round head of one bone rests within cup-shaped depression in another bone.	**Ball-and-socket joint:** Glenohumeral (shoulder) joint, hip joint	Diarthrosis (freely mobile)

Table 9.2	Movements at Synovial Joints APR	
Movement	**Description**	**Opposing Movement[1]**
GLIDING MOTION	Two opposing articular surfaces slide past each other in almost any direction; the amount of movement is slight	
ANGULAR MOTION	The angle between articulating bones increases or decreases	
Flexion	The angle between articulating bones decreases in an anterior-posterior (AP) plane	Extension
Extension	The angle between articulating bones increases in an anterior-posterior (AP) plane	Flexion
Hyperextension	Extension movement continues past 180 degrees	Flexion
Lateral flexion	The vertebral column moves (bends) in a lateral direction along a coronal plane	None
Abduction	Lateral movement of a body part away from the midline	Adduction
Adduction	Lateral movement of a body part toward the midline	Abduction
Circumduction	A continuous movement that combines flexion, abduction, extension, and adduction in succession; the distal end of the limb or digit moves in a circle	None
ROTATIONAL MOTION	A bone pivots around its own longitudinal axis	
Pronation	Rotation of the forearm where the palm is turned posteriorly or inferiorly	Supination
Supination	Rotation of the forearm in which the palm is turned anteriorly or superiorly	Pronation
SPECIAL MOVEMENTS	Types of movement that do not fit into the previous categories	
Depression	Movement of a body part inferiorly	Elevation
Elevation	Movement of a body part superiorly	Depression
Dorsiflexion	Ankle joint movement where the dorsum (superior surface) of the foot is brought toward the anterior surface of the leg	Plantar flexion
Plantar flexion	Ankle joint movement where the sole of the foot is brought toward the posterior surface of the leg	Dorsiflexion
Eversion	Twisting motion of the foot that turns the sole laterally or outward	Inversion
Inversion	Twisting motion of the foot that turns the sole medially or inward	Eversion
Protraction	Anterior movement of a body part from anatomic position	Retraction
Retraction	Posterior movement of a body part from anatomic position	Protraction
Opposition	Special movement of the thumb across the palm toward the fingers to permit grasping and holding of an object	Reposition

1. Some movements (e.g., circumduction) do not have an opposing movement.

gliding motion, the angle between the bones does not change, and only limited movement is possible in any direction. Gliding motion typically occurs along plane joints, such as between the carpals or the tarsals.

WHAT DID YOU LEARN?

11 What joints typically use gliding motion?

9.5b Angular Motion

LEARNING OBJECTIVES

15. Describe angular motion.
16. Name the specific types of angular motion.
17. Give examples of joints that exhibit angular motion.

Angular motion either decreases or increases the angle between two bones. These movements may occur at many of the synovial joints. They include the following specific types:

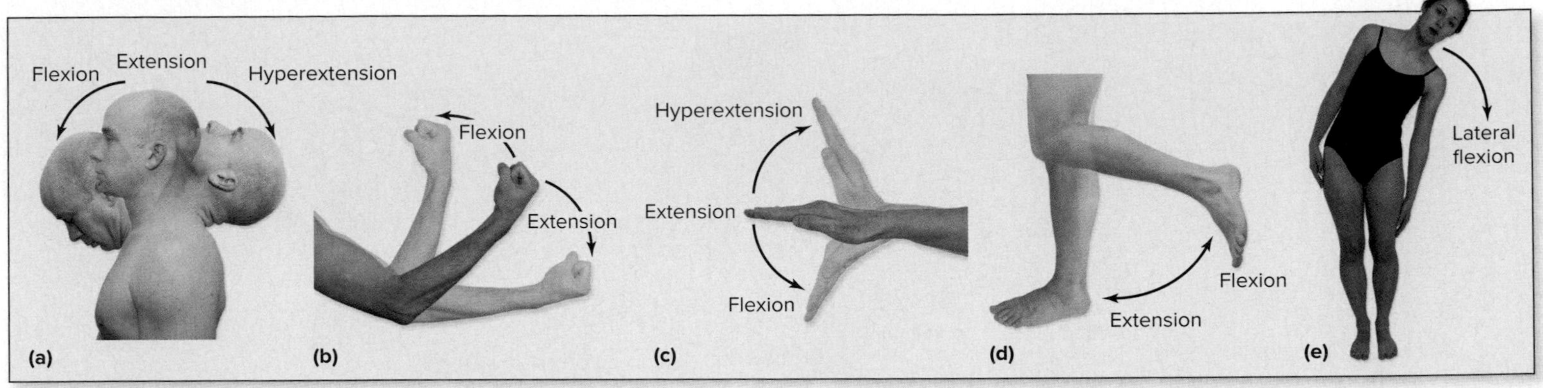

Figure 9.7 Flexion, Extension, Hyperextension, and Lateral Flexion. Flexion decreases the joint angle in an anterior-posterior (AP) plane, whereas extension increases the joint angle in the AP plane. Hyperextension is extension of a joint beyond 180 degrees. Lateral flexion decreases a joint angle, but in a coronal plane. Examples of joints that allow some of these movements are (*a*) the atlanto-occipital joint, (*b*) the elbow joint, (*c*) the radiocarpal joint, (*d*) the knee joint, and (*e*) the intervertebral joints. AP|R

flexion and extension, hyperextension, lateral flexion, abduction and adduction, and circumduction.

Flexion (flek′shŭn; *flecto* = to bend) is movement in an anterior-posterior (AP) plane of the body that *decreases* the angle between the bones. Bones are brought closer together as the angle between them decreases. Examples are the bending of the fingers toward the palm to make a fist, the bending of the forearm toward the arm at the elbow, flexion at the shoulder when the arm is raised anteriorly, and flexion of the neck when the head is bent anteriorly and you look down at your feet.

The opposite of flexion is **extension** (eks-ten′shŭn; *extensio* = a stretching out), which is movement in an anterior-posterior (AP) plane that *increases* the angle between the articulating bones. Extension is a straightening action that occurs in an AP plane. Straightening the arm and forearm until the upper limb projects directly away from the anterior side of your body, and straightening the fingers after making a clenched fist, are examples of extension.

When a joint is extended more than 180 degrees, the movement is termed **hyperextension** (hī′per-eks-ten′shŭn; *hyper* = above normal). For example, if you extend your arm and hand with the palm facing inferiorly, and then raise the back of your hand as if admiring a new ring on your finger, the wrist is hyperextended. If you glance up at the ceiling while standing, your neck is hyperextended. Flexion, extension, and hyperextension of various body parts are illustrated in **figure 9.7*a–d***.

Lateral flexion occurs when the trunk of the body moves in a coronal plane laterally away from the body. This type of movement occurs primarily between the vertebrae in the cervical and lumbar regions of the vertebral column (figure 9.7*e*).

WHAT DO YOU THINK?

3 When sitting upright in a chair, are your hip and knee joints flexed or extended?

Abduction (ab-dŭk′shŭn; *duco* = to draw) means to *move away*, and it is a lateral movement of a body part *away from* the body midline. Abduction occurs when either the arm or the thigh is moved laterally away from the body midline. Abduction of either the fingers or the toes means that you spread them apart, away from the longest digit that acts as the midline. Abducting the wrist (also known as *radial* deviation) involves pointing the hand and fingers laterally, away from the body. The opposite of abduction is **adduction** (ad-dŭk′shŭn), meaning to *move toward*. This is the medial movement of a body part *toward* the body midline. Adduction occurs when the raised arm or thigh is brought back toward the body midline, or in the case of the digits, toward the midline of the hand. Adducting the wrist (also known as *ulnar* deviation) involves pointing the hand and fingers medially, toward the body. Abduction and adduction of various body parts are shown in **figure 9.8**.

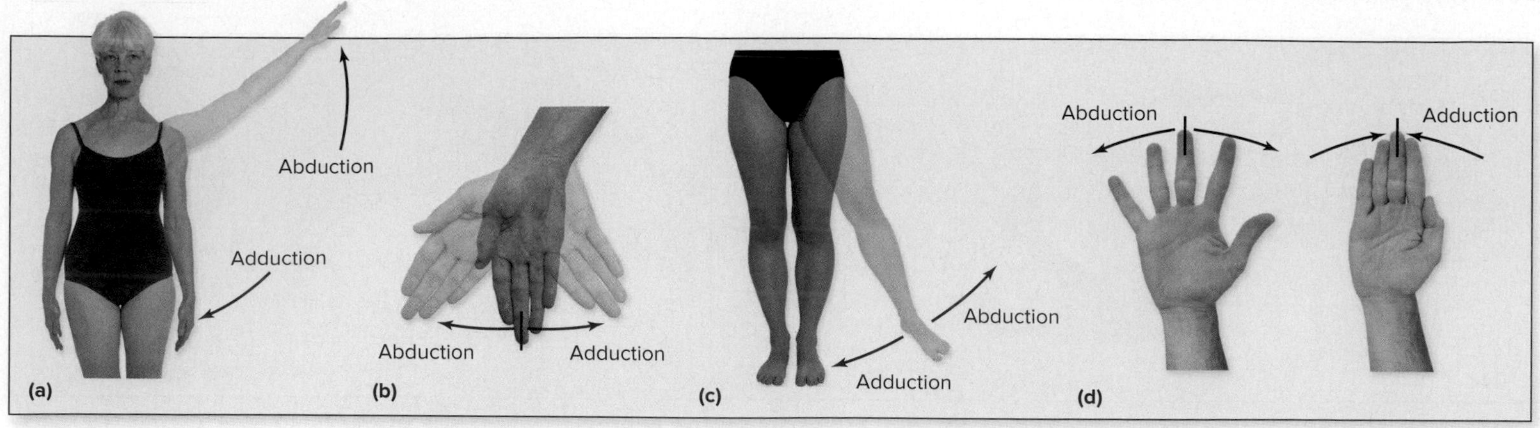

Figure 9.8 Abduction and Adduction. Abduction moves a body part away from the trunk in a lateral direction, whereas adduction moves the body part toward the trunk in a medial direction. Some examples occur at (*a*) the glenohumeral joint, (*b*) the radiocarpal joint, (*c*) the hip joint, and (*d*) the metacarpophalangeal (MP) joints. AP|R

Circumduction (ser-kŭm-dŭk′shŭn; *circum* = around) is a sequence of movements in which the proximal end of an appendage remains relatively stationary while the distal end makes a circular motion **(figure 9.9)**. The resulting movement makes an imaginary cone shape. This is demonstrated when you draw a circle on the blackboard. The shoulder remains stationary while your hand moves. The tip of the imaginary cone is the stationary shoulder, while the rounded "base" of the cone is the circle made by the hand. Circumduction is a complex movement that occurs as a result of a continuous sequence of flexion, abduction, extension, and adduction.

WHAT DID YOU LEARN?

12 How do flexion and extension differ? What movements are involved in circumduction?

9.5c Rotational Motion

LEARNING OBJECTIVE

18. Explain rotational motion, and name joints in which it occurs.

Rotation is a pivoting motion in which a bone turns on its own longitudinal axis **(figure 9.10)**. Rotational movement occurs at the atlantoaxial joint, which pivots when you rotate your head to gesture "no." Some limb rotations are described as either away from the median plane or toward it. For example, **lateral rotation** (or *external* rotation) turns the anterior surface of the femur or humerus laterally, whereas **medial rotation** (or *internal* rotation) turns the anterior surface of the femur or humerus medially.

Pronation (prō-nā′shŭn) is the medial rotation of the forearm so that the palm of the hand is directed posteriorly or inferiorly. The radius and ulna are crossed to form an X (see section 8.9b).

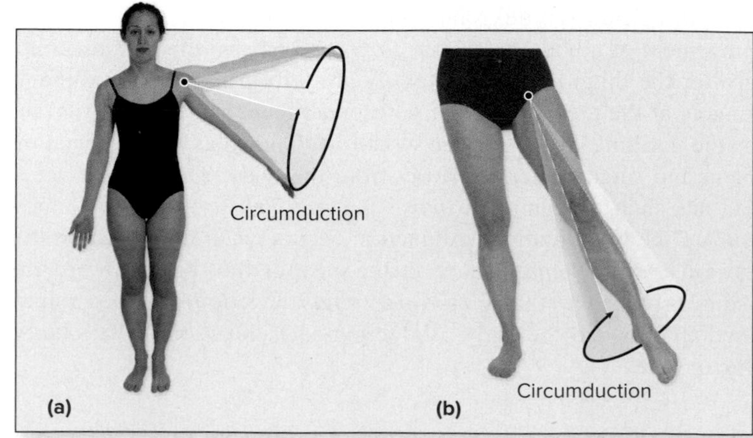

Figure 9.9 Circumduction. Circumduction is a complex movement that involves flexion, abduction, extension, and adduction in succession. Examples of joints that allow this movement are (*a*) the glenohumeral joint and (*b*) the hip joint.

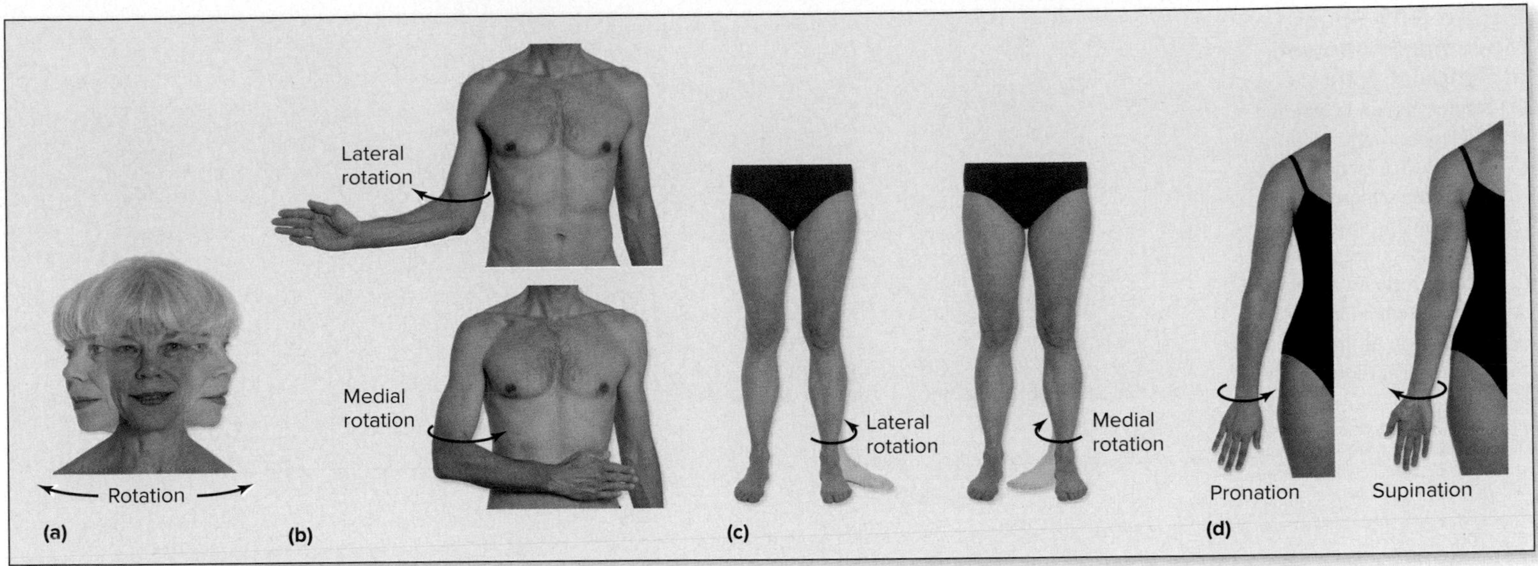

Figure 9.10 Rotational Movements. Rotation allows a bone to pivot on its longitudinal axis. Examples of joints that allow this movement are (*a*) the atlantoaxial joint, (*b*) the glenohumeral joint, and (*c*) the hip joint. (*d*) Pronation and supination occur at the forearm. AP|R

Supination (sū′pi-nā′shŭn) occurs when the forearm rotates laterally so that the palm faces anteriorly or superiorly. In the anatomic position, the forearm is supinated. Figure 9.10*d* illustrates pronation and supination.

WHAT DID YOU LEARN?

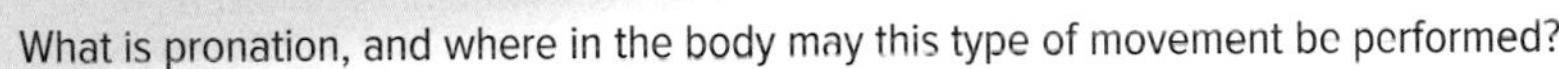

13 What is pronation, and where in the body may this type of movement be performed?

9.5d Special Movements

LEARNING OBJECTIVE

19. Explain what is meant by special movements, and give examples of joints at which they occur.

Some movements occur only at specific joints and do not readily fit into any of the functional categories previously discussed. These special movements include depression and elevation, dorsiflexion and plantar flexion, eversion and inversion, protraction and retraction, and opposition.

Depression (dē-presh′ŭn, *de* = away, down, *presso* = to press) is the inferior movement of a part of the body. Examples of depression include opening your mouth (by depressing your mandible) to chew food and the movement of your shoulders in an inferior direction. **Elevation** (el-ĕ-vā′shŭn) is the superior movement of a body part. Examples of elevation include the superior movement of the mandible while closing the mouth and the movement of the shoulders in a superior direction (shrugging your shoulders). **Figure 9.11*a*** illustrates depression and elevation at the glenohumeral joints.

Dorsiflexion and plantar flexion are limited to the ankle joint (figure 9.11*b*). **Dorsiflexion** (dōr-si-flek′shŭn) occurs when the talocrural (ankle) joint is bent such that the dorsum (superior surface) of the foot and the toes moves toward the leg. This movement occurs when you dig in your heels, and it prevents your toes from scraping the ground when you take a step. **Plantar flexion** (plan′tăr; *planta* = sole of foot) is a movement of the foot at the talocrural joint so that the toes point inferiorly. When a ballerina is standing on her tiptoes, her ankle joint is in full plantar flexion.

Eversion and inversion are movements that occur at the intertarsal joints of the foot only (figure 9.11*c*). During **eversion** (ē-ver′zhŭn), the sole of the foot turns to face laterally or outward, whereas the sole of the foot turns medially or inward during **inversion** (in-ver′zhŭn). (*Note:* Some orthopedists and runners use the terms *pronation* and *supination* when describing foot movements as well, instead of using *eversion* and *inversion*. Simply put, eversion is foot pronation, whereas inversion is foot supination.)

Figure 9.11 Special Movements Allowed at Synovial Joints.
(*a*) Depression and elevation at the glenohumeral joint. (*b*) Dorsiflexion and plantar flexion at the talocrural joint. (*c*) Eversion and inversion at the intertarsal joints. (*d*) Protraction and retraction at the temporomandibular joint. (*e*) Opposition at the carpometacarpal joints. AP|R

(*a*) ©McGraw-Hill Education/Tamara Klein; (*b, c*) ©McGraw-Hill Education/Jw Ramsey; (*d,e*) ©McGraw-Hill Education/ Tamara Klein

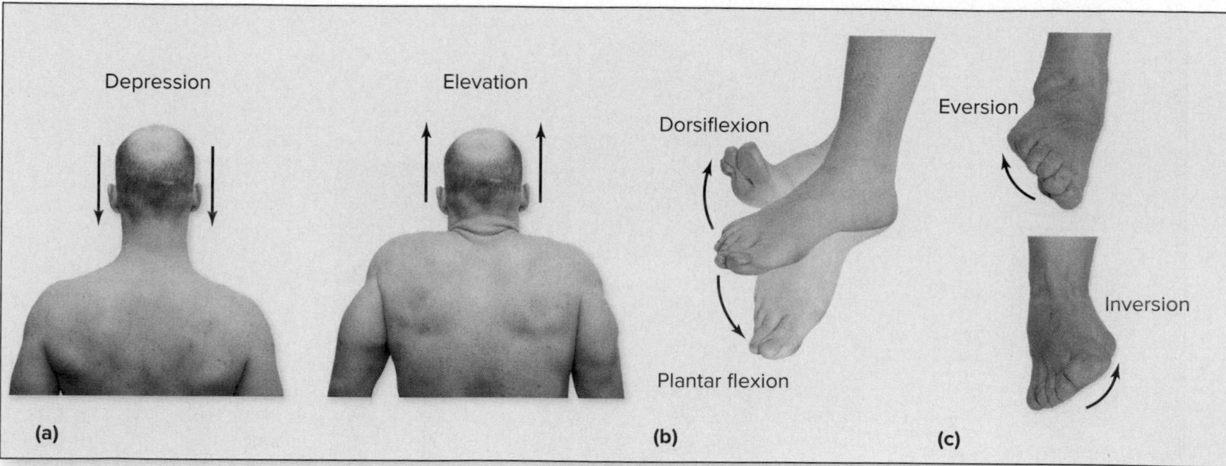

Protraction (prō-trak′shŭn) is the anterior movement of a body part from anatomic position, as when jutting your jaw anteriorly at the temporomandibular joint or hunching the shoulders anteriorly by crossing the arms. In the latter case, the clavicles move anteriorly due to movement at both the acromioclavicular and sternoclavicular joints. **Retraction** (rē-trak′shŭn) is the posteriorly directed movement of a body part from the anatomic position. Figure 9.11*d* illustrates protraction and retraction at the temporomandibular joint.

At the carpometacarpal joint, the thumb moves toward the palmar tips of the fingers as it crosses the palm of the hand. This movement is called **opposition** (op′ō-si′shŭn) (figure 9.11*e*). It enables the hand to grasp objects and is the most distinctive digital movement in humans. The opposite movement is called **reposition.**

WHAT DID YOU LEARN?

14 What is the difference between inversion and eversion, and which joints allow these movements?

9.6 Synovial Joints and Levers

When analyzing synovial joint movement and muscle contraction, anatomists often compare the movement to the mechanics of a lever; this practice of applying mechanical principles to biology is known as **biomechanics.**

9.6a Terminology of Levers

LEARNING OBJECTIVES

20. Define a lever.

21. Discriminate between the effort arm and the resistance arm in a lever.

A **lever** (lev′er, lē′ver; to lift) is an elongated, rigid object that rotates around a fixed point called the **fulcrum** (ful′krum). A seesaw is a familiar example of a lever. Levers have the ability to alter the speed and distance of movement produced by a force, the direction of an applied force, and the force strength.

Movement occurs when an **effort** applied to one point on the lever exceeds a **resistance** located at some other point. The part of a lever from the fulcrum to the point of effort is called the **effort arm,** and the lever part from the fulcrum to the point of resistance is the **resistance arm.** In the body, a long bone acts as a lever, a joint serves as the fulcrum, and the effort is generated by a muscle attached to the bone.

INTEGRATE

LEARNING STRATEGY

Use the acronym **FRE** to help you remember which part of the lever system is in between the other two parts:

- In a first-class lever, the fulcrum (and the first letter of this acronym) is in between the resistance and the effort.
- In a second-class lever, the resistance (and the second letter of the acronym) is in between the fulcrum and the effort.
- In a third-class lever, the effort (and the third letter of this acronym) is in between the fulcrum and the resistance.

WHAT DID YOU LEARN?

15 What is the difference between the effort arm and the resistance arm in a lever?

9.6b Types of Levers

LEARNING OBJECTIVE

22. Compare and contrast the three types of levers in the human body.

Three classes of levers are found in the human body: first-class, second-class, and third-class (**figure 9.12**).

First-Class Levers

A **first-class lever** has a fulcrum in the middle, between the effort (force) and the resistance. An example of a first-class lever is a pair of scissors. The effort is applied to the handle of the scissors while the resistance is at the cutting end of the scissors. The fulcrum (pivot for movement) is along the middle of the scissors, between the handle and the cutting ends. In the body, an example of a first-class lever is the atlanto-occipital joint of the neck, where the muscles on the posterior side of the neck (effort) pull inferiorly on the nuchal lines of the skull and oppose the tendency of the head (resistance) to tip anteriorly.

Second-Class Levers

The resistance in a **second-class lever** is between the fulcrum and the applied effort. A common example of this type of lever is lifting the handles of a wheelbarrow, allowing it to pivot on its wheel at the opposite end and lift a load in the middle. The load weight is the resistance, and the upward lift on the handle is the effort. A small force

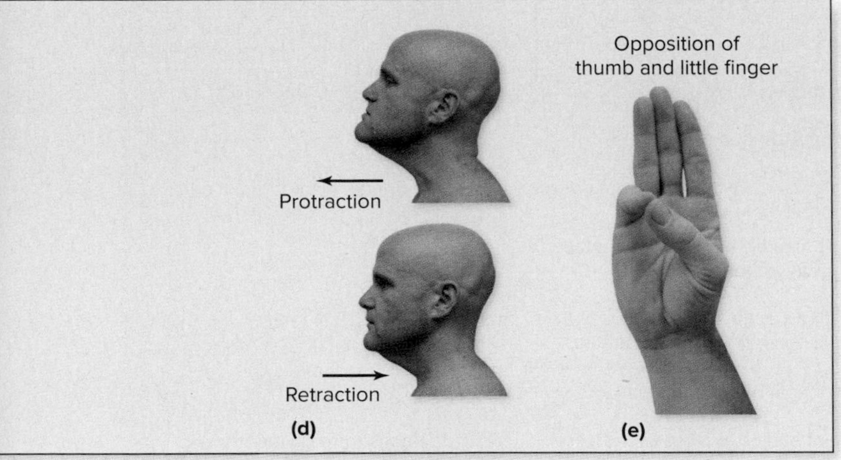

can balance a larger weight in this type of lever, because the effort is always farther from the fulcrum than the resistance. Second-class levers are rare in the body, but one example occurs when the foot is depressed (plantar flexed) so that a person can stand on tiptoe. The contraction of the calf muscle causes a pull superiorly by the calcaneal tendon attached to the heel (calcaneus).

Third-Class Levers

A **third-class lever** is noted when the effort is applied between the resistance and the fulcrum, as when picking up a small object with a pair of forceps. Third-class levers are the most common levers in the body. A third-class lever is found at the elbow where the fulcrum is the joint between the humerus and ulna, the effort is applied by the biceps brachii muscle, and the resistance is provided by any weight in the hand or by the weight of the forearm itself. The mandible acts as a third-class lever when you bite with your incisors on a piece of food. The temporomandibular joint is the fulcrum, and the temporalis muscle exerts the effort, whereas the resistance is the item of food being bitten.

WHAT DID YOU LEARN?

16 How does the position of the fulcrum, resistance, and effort vary in first-class, second-class, and third-class levers?

9.7 Features and Anatomy of Selected Joints

Both the axial skeleton and appendicular skeleton exhibit many more joints than are individually discussed here. **Table 9.3** summarizes the main features of major joints of the axial skeleton.

The structure and function of the more commonly known articulations of the axial and appendicular skeletons are examined in this section. These are the temporomandibular joint of the skull; the shoulder joint and elbow joint; and the hip joint, knee joint, and talocrural (ankle) joint.

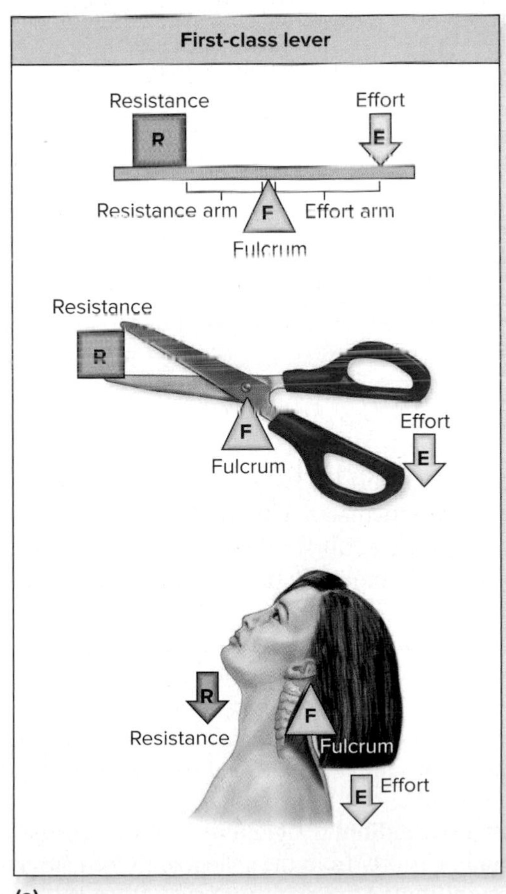

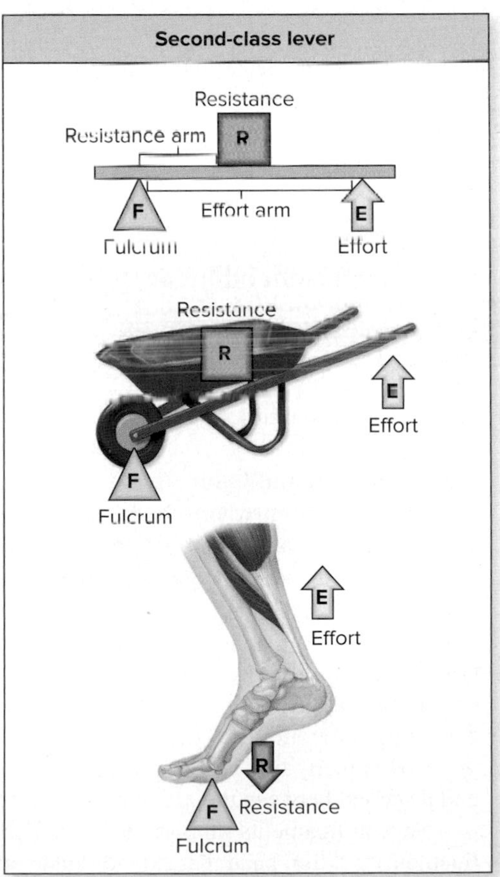

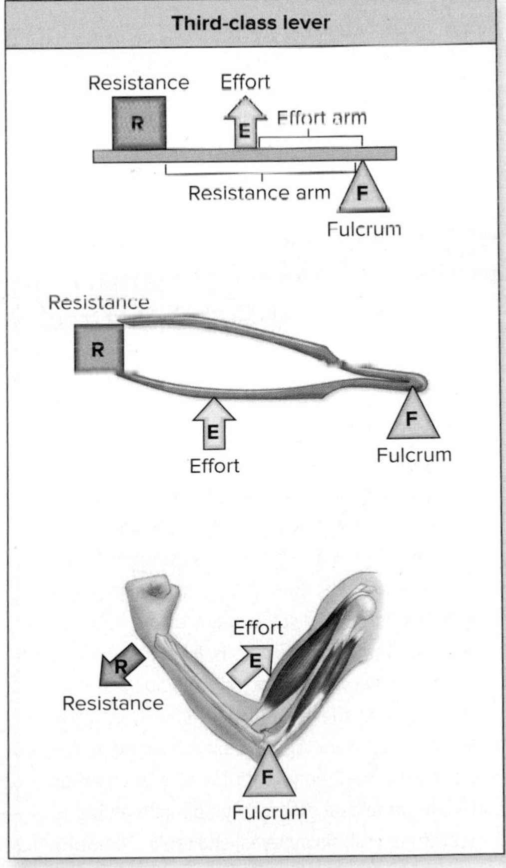

Figure 9.12 Classes of Levers. (*a*) In a first-class lever, the fulcrum is located between the resistance and effort, such as with a pair of scissors or the trapezius muscle (in the neck). (*b*) In a second-class lever, the resistance is between the fulcrum and effort, such as with a wheelbarrow or the calf muscles. (*c*) The most common type of lever is the third-class lever, where effort is applied between the resistance and the fulcrum, such as with forceps (tweezers) or the arm muscles.

Table 9.3 Axial Skeleton Joints

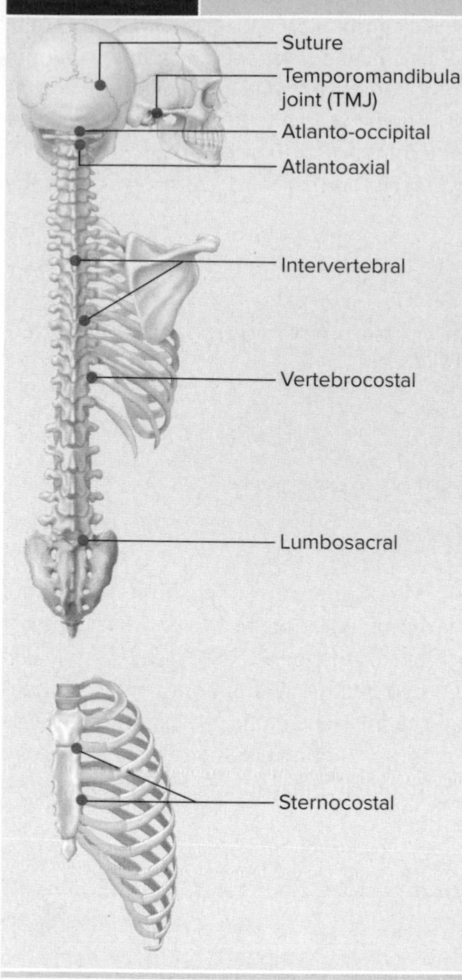

Joint	Articulation Components	Structural Classification
Suture	Adjacent skull bones	Fibrous joint
Temporomandibular	Head of mandible and mandibular fossa of temporal bone Head of mandible and articular tubercle of temporal bone	Synovial (hinge, plane) joints
Atlanto-occipital	Superior articular facets of atlas and occipital condyles of occipital bone	Synovial (condylar) joint
Atlantoaxial	Anterior arch of atlas and dens of axis	Synovial (pivot) joint
Intervertebral	Vertebral bodies of adjacent vertebrae Superior and inferior articular processes of adjacent vertebrae	Cartilaginous joint (symphysis) between vertebral bodies; synovial (plane) joint between articular processes
Vertebrocostal	Facets of heads of ribs and bodies of adjacent thoracic vertebrae and intervertebral discs between adjacent vertebrae Articular part of tubercles of ribs and facets of transverse processes of thoracic vertebra	Synovial (plane) joint
Lumbosacral	Body of the fifth lumbar vertebra and the base of the sacrum Inferior articular facets of fifth lumbar vertebra and superior articular facets of first sacral vertebra	Cartilaginous joint (symphysis) between lumbar body and base of sacrum; synovial (plane) joint between articular processes
Sternocostal	Sternum and first seven pairs of ribs	Cartilaginous joint (synchondrosis) between sternum and first ribs; synovial (plane) joint between sternum and ribs 2–7

INTEGRATE

CLINICAL VIEW 9.3

TMJ Disorders

The temporomandibular joint (TMJ) is subject to various disorders. TMJ disorders are often seen in people who habitually chew gum or grind or clench their teeth. The most common TMJ disorder occurs as a result of alterations in the ligaments that secure the joint, causing progressive internal displacement of the articular disc. As the articular disc is forced out of its normal position, a clicking or popping noise may be heard as the person opens or closes the mouth. Pain from the TMJ disorder may be felt not only within the joint but also in such areas as the paranasal sinuses, tympanic membrane (eardrum), oral cavity, eyes, and teeth. The widespread distribution of pain occurs because all of these structures, including the chewing muscles, mandible, and maxilla, are innervated by numerous sensory branches of the trigeminal nerve (see section 13.9).

9.7a Temporomandibular Joint

LEARNING OBJECTIVES

23. Describe the features of the temporomandibular joint (TMJ).

24. List the movements of the TMJ.

The **temporomandibular** (tem′pŏ-rō-man-dib′yū-lăr) **joint (TMJ)** is the articulation formed at the point where the head of the mandible articulates with the temporal bone—specifically, the articular tubercle of the temporal bone anteriorly and the mandibular fossa posteriorly. This small, complex articulation is the only mobile joint between bones in the skull (**figure 9.13** and table 9.3).

The temporomandibular joint has several unique anatomic features. A loose **articular capsule** surrounds the joint and promotes an extensive range of motion. It contains an **articular disc,** which is a thick pad of fibrocartilage separating the articulating bones and extending horizontally to divide the synovial cavity into two separate chambers. As a result, the TMJ is really two synovial joints—one between the temporal bone and the articular disc, and a second between the articular disc and the mandible.

Several ligaments support the TMJ. The **sphenomandibular ligament** (an extracapsular ligament) is a thin band that extends anteriorly and inferiorly from the sphenoid to the medial surface of the mandibular ramus. The **temporomandibular ligament** (or *lateral ligament*) is composed of two short bands that extend inferiorly and posteriorly from the articular tubercle of the temporal bone to the mandible.

The temporomandibular joint functions as a hinge during jaw depression and elevation while chewing. It also glides slightly forward during protraction of the jaw for biting, and glides slightly from side to side to grind food between the teeth during chewing.

Functional Classification	Description of Movement
Synarthrosis	None allowed
Diarthrosis	Depression, elevation, lateral displacement, protraction, retraction, slight rotation
Diarthrosis	Extension and flexion of the head; slight lateral flexion of head to sides
Diarthrosis	Head rotation
Amphiarthrosis between vertebral bodies; diarthrosis between articular processes	Extension, flexion, lateral flexion of vertebral column
Diarthrosis	Some slight gliding
Amphiarthrosis between lumbar body and base of sacrum; diarthrosis between articular processes	Extension, flexion, lateral flexion of vertebral column
Synarthrosis between sternum and first ribs; diarthrosis between sternum and ribs 2–7	No movement between sternum and first ribs; some gliding movement permitted between sternum and ribs 2–7

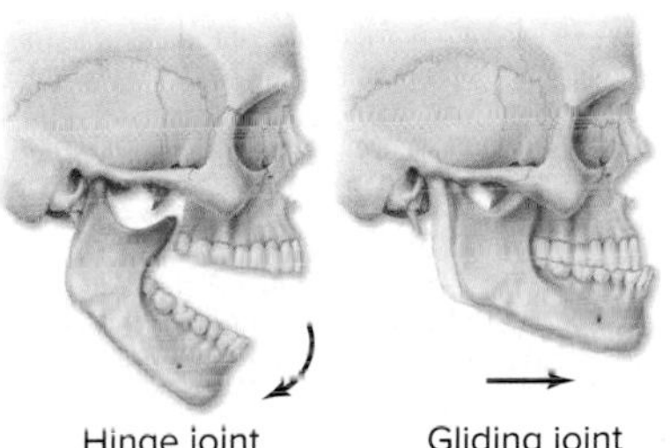

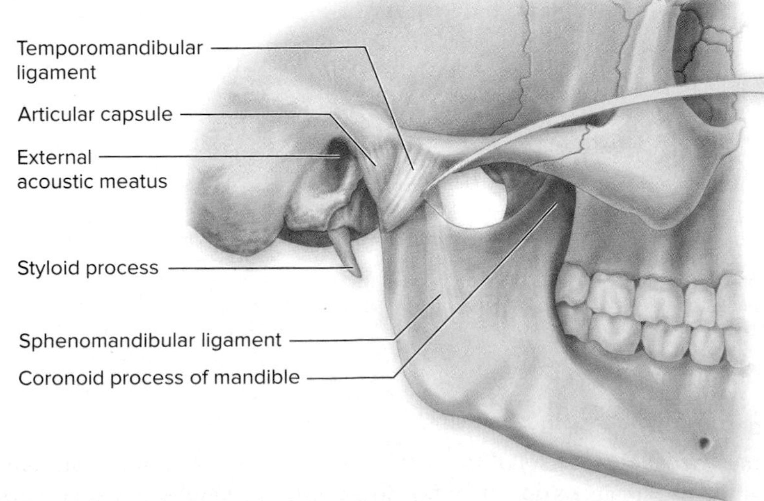

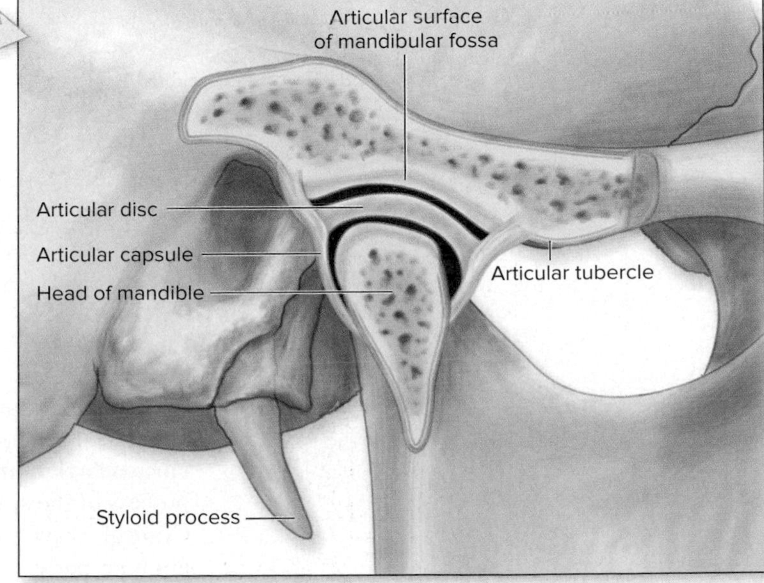

Figure 9.13 Temporomandibular Joint. The articulation between the head of the mandible and the mandibular fossa of the temporal bone exhibits a wide range of movements. AP|R

WHAT DID YOU LEARN?

17 What movements are allowed at the temporomandibular joint?

9.7b Shoulder Joint

LEARNING OBJECTIVES

25. Describe the three individual joints that make up the shoulder articulation.

26. Explain why the glenohumeral joint is relatively unstable.

The joints associated with movement at the shoulder include the sternoclavicular joint, the acromioclavicular joint, and the glenohumeral joint. (**Table 9.4** lists the features of the major joints of the pectoral girdle and upper limbs.)

Sternoclavicular Joint

The **sternoclavicular** (ster′nō-kla-vik′yū-lăr) **joint** is a saddle joint formed by the articulation between the manubrium of the sternum and the sternal end of the clavicle (**figure 9.14**). An **articular disc** partitions the sternoclavicular joint into two parts and forms two separate synovial cavities. As a result, a wide range of movement is possible, including depression, elevation, and circumduction.

Support and stability are provided to this articulation by the fibers of the articular capsule and by multiple extracapsular ligaments, such as the sternoclavicular and costoclavicular ligaments. This anatomic arrangement makes the sternoclavicular joint very stable. If you fall on an outstretched hand so that force is applied to the joint, the clavicle will fracture before this joint dislocates.

Acromioclavicular Joint

The **acromioclavicular** (ă-krō′mē-ō-kla-vik′yū-lăr) **joint** is a plane joint between the acromion of the scapula and the lateral end of the clavicle (**figure 9.15**). A fibrocartilaginous articular disc lies within the joint cavity between these two bones. This joint works with both the sternoclavicular joint and the glenohumeral joint to give the upper limb a full range of movement.

Table 9.4 Pectoral Girdle and Upper Limb Joints

Joint	Articulation Components	Structural Classification	Functional Classification	Description of Movement
Sternoclavicular	Manubrium of sternum and sternal end of clavicle	Synovial (saddle)	Diarthrosis	Depression, elevation, and circumduction
Acromioclavicular	Acromion of scapula and acromial end of clavicle	Synovial (plane)	Diarthrosis	Gliding of scapula on clavicle
Glenohumeral	Glenoid cavity of scapula and head of humerus	Synovial (ball-and-socket)	Diarthrosis	Abduction, adduction, circumduction, flexion, extension, lateral rotation, and medial rotation of arm
Elbow	**Humeroulnar joint:** Trochlea of humerus and trochlear notch of ulna **Humeroradial joint:** Capitulum of humerus and head of radius	Synovial (hinge)	Diarthrosis	Flexion and extension of forearm
Radioulnar	**Proximal joint:** Head of radius and radial notch of ulna **Distal joint:** Distal end of ulna and ulnar notch of radius	Synovial (pivot)	Diarthrosis	Rotation of radius with respect to ulna
Radiocarpal	Distal end of radius; scaphoid, lunate, and triquetrum	Synovial (condylar)	Diarthrosis	Abduction, adduction, circumduction, flexion and extension of wrist
Intercarpal	Adjacent bones in proximal row of carpal bones Adjacent bones in distal row of carpal bones Adjacent bones between proximal and distal rows (midcarpal joints)	Synovial (plane)	Diarthrosis	Gliding
Carpometacarpal	**Thumb:** Trapezium and first metacarpal **Other digits**: Carpals and metacarpals II–V	Synovial (saddle) at thumb; synovial (plane) at other digits	Diarthrosis	Flexion, extension, abduction, adduction, circumduction, and opposition at thumb; gliding at other digits
Metacarpo-phalangeal (MP joints, "knuckles")	Heads of metacarpals and bases of proximal phalanges	Synovial (condylar)	Diarthrosis	Flexion, extension, abduction, adduction, and circumduction of phalanges
Interphalangeal (IP joints)	Heads of proximal and middle phalanges with bases of middle and distal phalanges, respectively	Synovial (hinge)	Diarthrosis	Flexion and extension of phalanges

Several ligaments provide great stability to this joint. The fibrous joint capsule is strengthened superiorly by an **acromioclavicular ligament.** In addition, a very strong **coracoclavicular** (kōr′ă-kō-kla-vik′yū-lăr) **ligament** binds the clavicle to the coracoid process of the scapula. If this ligament is torn (as occurs in severe shoulder separations; see Clinical View 9.4: "Shoulder Joint Dislocations"), the acromion and clavicle no longer align properly.

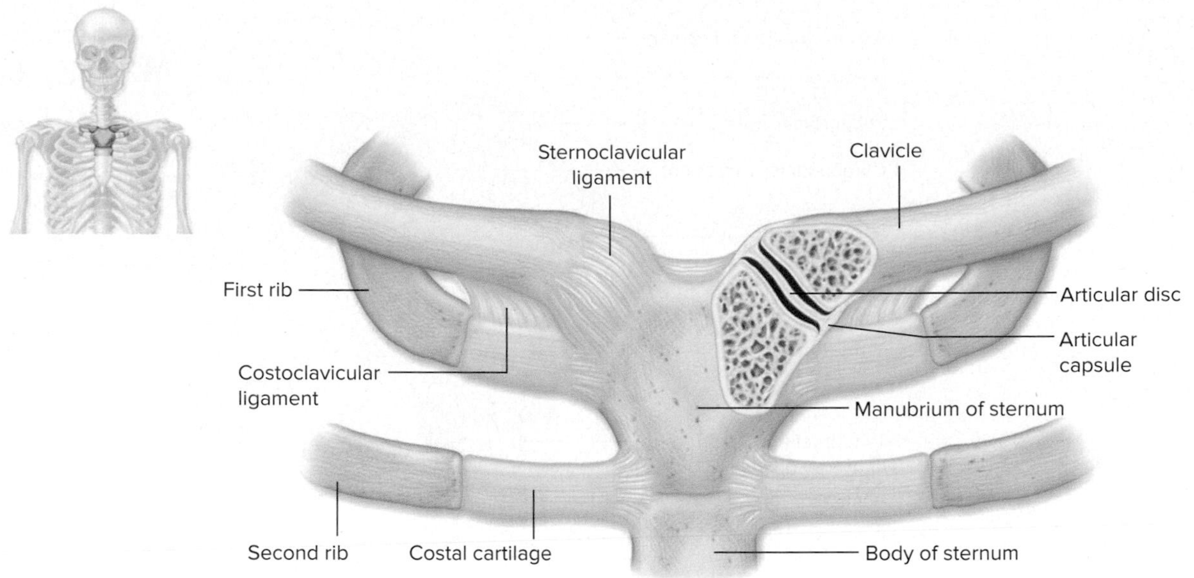

Figure 9.14 Sternoclavicular Joint. The sternoclavicular joint helps stabilize movements of the entire shoulder.

INTEGRATE

CLINICAL VIEW 9.4

Shoulder Joint Dislocations

Dislocation (dis′lō-kā′shŭn; *dis* = apart, *locatio* = placing), a joint injury in which the articulating bones have separated, is common in the shoulder. Although this injury can occur at any of the three shoulder joints, it is more common at either the acromioclavicular joint or the glenohumeral joint than at the sternoclavicular joint.

The term **shoulder separation** refers to a dislocation of the acromioclavicular joint. This injury often results from a hard blow to the joint, as when a hockey player is "slammed into the boards" or one falls onto the shoulder. Symptoms include tenderness and edema (swelling) in the area of the joint and pain when the arm is abducted more than 90 degrees, because in this position significant movement occurs between the separated bone surfaces. Additionally, the acromion will appear very prominent and pointed. Treatment can range from rest to surgery, depending upon the severity of the dislocation.

A **glenohumeral joint dislocation** is a very common injury because this joint is very mobile and unstable. This dislocation usually occurs when a fully abducted humerus is struck hard—for example, when a quarterback is hit as he is about to release a football, or when a person falls on an outstretched hand. The initial blow pushes the humerus into the inferior part of the articular capsule and tears the capsule as the humerus dislocates. (The inferior part of the capsule is relatively weak and not protected by muscle tendons as are the other surfaces of the capsule.) Once the humeral head is no longer held in place by the capsule, the anterior thorax (chest) muscles pull superiorly and medially on the humeral head, causing it to lie just inferior to the coracoid process. The result is that the shoulder appears flattened and "squared-off," because the humeral head is dislocated anteriorly and inferiorly to the glenohumeral joint capsule. Some glenohumeral dislocations can be repaired by "popping" the humerus back into the glenoid cavity. More severe dislocations may need surgical repair.

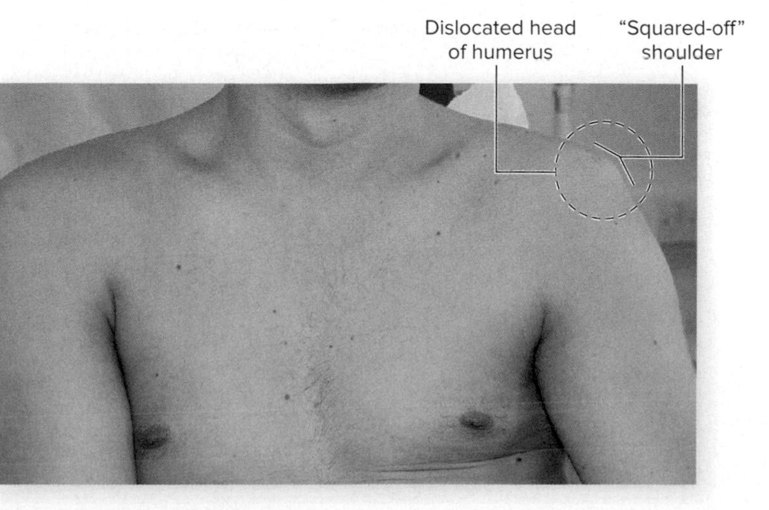

Glenohumeral joint dislocation.
Courtesy of Dr. Mike Langran/www.ski-injury.com

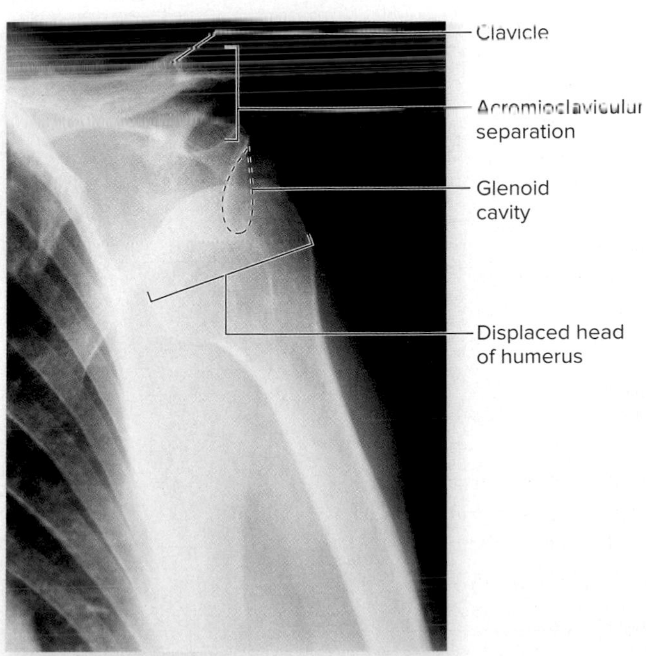

Radiograph of acromioclavicular and glenohumeral dislocations.
©Giancarlo Zuccotto/Medical Images

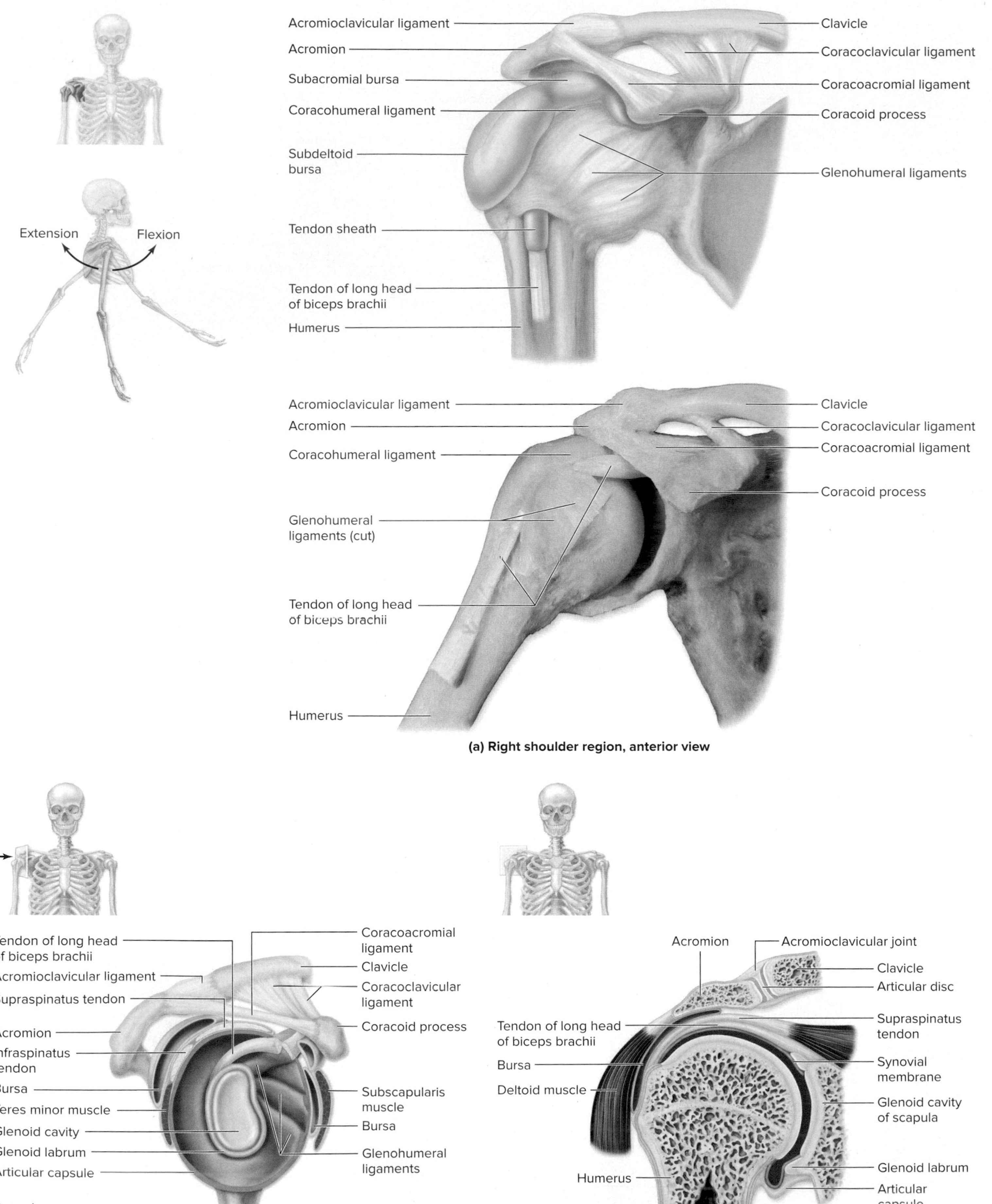

(a) Right shoulder region, anterior view

(b) Right shoulder, lateral view

(c) Right shoulder, coronal section

Figure 9.15 Acromioclavicular and Glenohumeral Joints. (*a*) Anterior diagrammatic view and cadaver photo of both joints on the right side of the body. (*b*) Right lateral view and (*c*) right coronal section show the articulating bones and supporting structures at the shoulder. AP|R

Glenohumeral (Shoulder) Joint

The **glenohumeral** (glē′nō-hyū′mer-ăl) **joint** is commonly referred to as the shoulder joint. It is a ball-and-socket joint formed by the articulation of the head of the humerus and the glenoid cavity of the scapula (figure 9.15). It permits the greatest range of motion of any joint in the body, and so it is both the most unstable joint in the body and the one most frequently dislocated.

The fibrocartilaginous **glenoid labrum** encircles and covers the surface of the glenoid cavity. A relatively loose articular capsule attaches to the surgical neck of the humerus. The glenohumeral joint has several major ligaments. The **coracoacromial** (kōr′ă-kō-ă-krō′mē-ăl) **ligament** extends across the space between the coracoid process and the acromion. The large **coracohumeral** (kōr′ă-kō-hyū′mer-ăl) **ligament** is a thickening of the superior part of the joint capsule. It extends from the coracoid process to the humeral head. The **glenohumeral ligaments** are three thickenings of the anterior portion of the articular capsule. These ligaments are often indistinct or absent and provide only minimal support. In addition, the tendon of the long head of biceps brachii is within the articular capsule and helps stabilize the humeral head in the joint.

Ligaments of the glenohumeral joint strengthen the joint only minimally. Most of the joint's strength is due to the **rotator cuff** muscles surrounding it (see section 11.8b). The rotator cuff muscles (i.e., subscapularis, supraspinatus, infraspinatus, and teres minor) work as a group to hold the head of the humerus in the glenoid cavity. The tendons of these muscles encircle the joint (except for its inferior portion) and fuse with the articular capsule. Because the inferior portion of the joint lacks support from rotator cuff muscles, this area is weak and is the most likely site of injury.

Bursae help decrease friction at the specific places on the shoulder where both tendons and large muscles extend across the joint capsule. The shoulder has a relatively large number of bursae.

WHAT DID YOU LEARN?

18 Why is the shoulder joint considered the most mobile and at the same time the most unstable joint in the human body?

9.7c Elbow Joint

LEARNING OBJECTIVES

27. Describe the elbow joint and its motion.

28. Explain why the elbow joint is relatively stable.

The **elbow joint** is a hinge joint composed of two articulations: (1) the humeroulnar joint, where the trochlea of the humerus articulates with the trochlear notch of the ulna, and (2) the humeroradial joint, where the capitulum of the humerus articulates with the head of the radius. Both joints are enclosed within a single articular capsule (**figure 9.16**; table 9.4).

The elbow is an extremely stable joint for several reasons. First, the articular capsule is fairly thick, and thus effectively protects the articulations. Second, the bony surfaces of the humerus and ulna interlock very well, and thus provide a solid bony support. Finally, multiple strong supporting ligaments help reinforce the articular capsule. Because of the tradeoff between stability and mobility, the elbow joint is very stable but is not as mobile as some other joints, such as the glenohumeral joint.

The elbow joint has two main supporting ligaments. The **radial collateral ligament** (or *lateral collateral ligament*) is responsible for stabilizing the joint at its lateral surface; it extends around the head of the radius between the anular ligament and the lateral epicondyle of the humerus. The **ulnar collateral ligament** (or *medial collateral ligament*) stabilizes the medial side of the joint and extends from the medial epicondyle of the humerus to both the coronoid process and the olecranon of the ulna. In addition, an **anular** (an′ū-lăr; *anulus* = ring) **ligament** surrounds the neck of the radius and binds the proximal head of the radius to the ulna. The anular ligament helps hold the head of the radius in place.

INTEGRATE

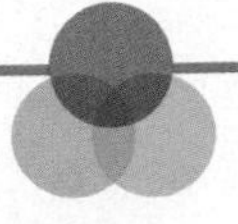

CONCEPT CONNECTION

The shoulder joint illustrates the interrelatedness of the skeletal system (see chapter 8) and muscular system (see chapter 11). The stability of the glenohumeral joint primarily comes from the musculature, as opposed to the skeletal elements. Injury to this musculature will affect the movement of this joint.

INTEGRATE

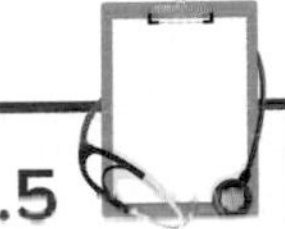

CLINICAL VIEW 9.5

Subluxation of the Head of the Radius

The term **subluxation** refers to an incomplete dislocation, in which the contact between the bony joint surfaces is altered but they are still in partial contact. In **subluxation of the head of the radius,** the head is pulled out of the anular ligament. The layman's terms for this injury include *pulled elbow, nursemaid's elbow,* and *slipped elbow.* This injury occurs commonly and almost exclusively in children (typically those younger than age 5), because a child's anular ligament is thin and the head of the radius is not fully formed. After age 5, both the ligament and the radial head are more fully formed, so the risk of this type of injury lessens dramatically. However, it may occur if an individual pulls suddenly on a child's pronated forearm. Luckily, treatment is simple: The pediatrician applies posteriorly placed pressure to the head of the radius while slowly supinating and extending the child's forearm. This movement literally screws the radial head back into the anular ligament.

©Floresco Productions/age fotostock RF

Figure 9.16 Elbow Joint. The elbow joint is a hinge joint. The right elbow is shown here in (*a*) anterior view, (*b*) lateral view, (*c*) medial view, and (*d*) sagittal section. **AP|R**

Despite the support from the capsule and ligaments, the elbow joint is subject to damage from severe impacts or unusual stresses. For example, if you fall on an outstretched hand and the elbow joint is partially flexed, the posterior stress on the ulna combined with contractions of muscles that extend the elbow may break the ulna at the center of the trochlear notch. Sometimes dislocations result from stresses to the elbow. This is particularly true when growth is still occurring at the epiphyseal plate, so children and teenagers may be prone to humeral epicondyle dislocations or fractures.

WHAT DID YOU LEARN?

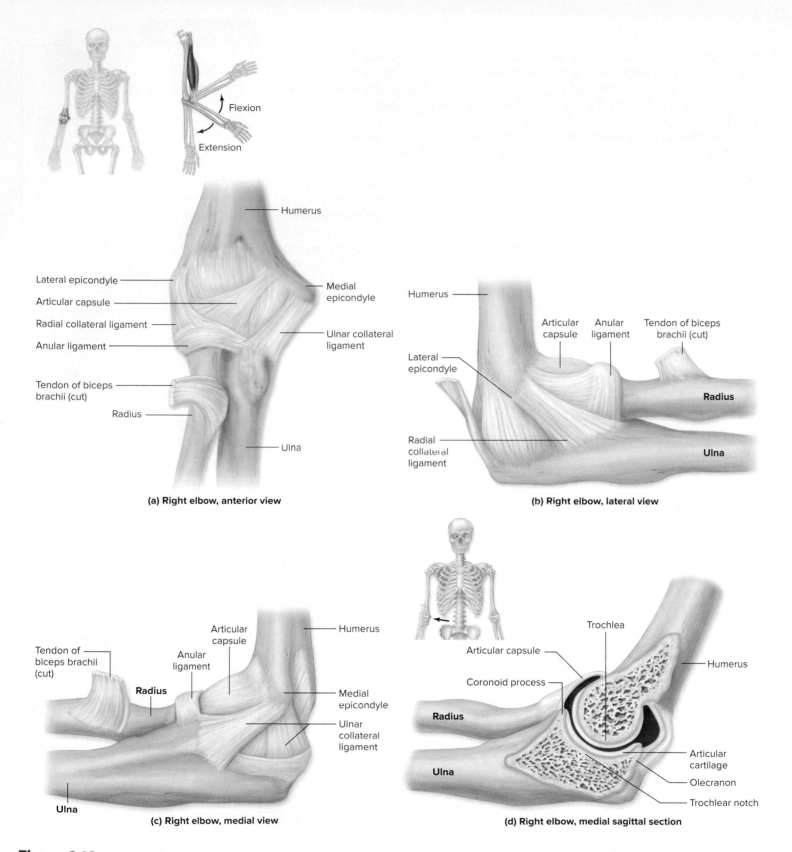

19 What is the function of the anular ligament in the elbow joint, and what injury may occur to this ligament and joint in young children?

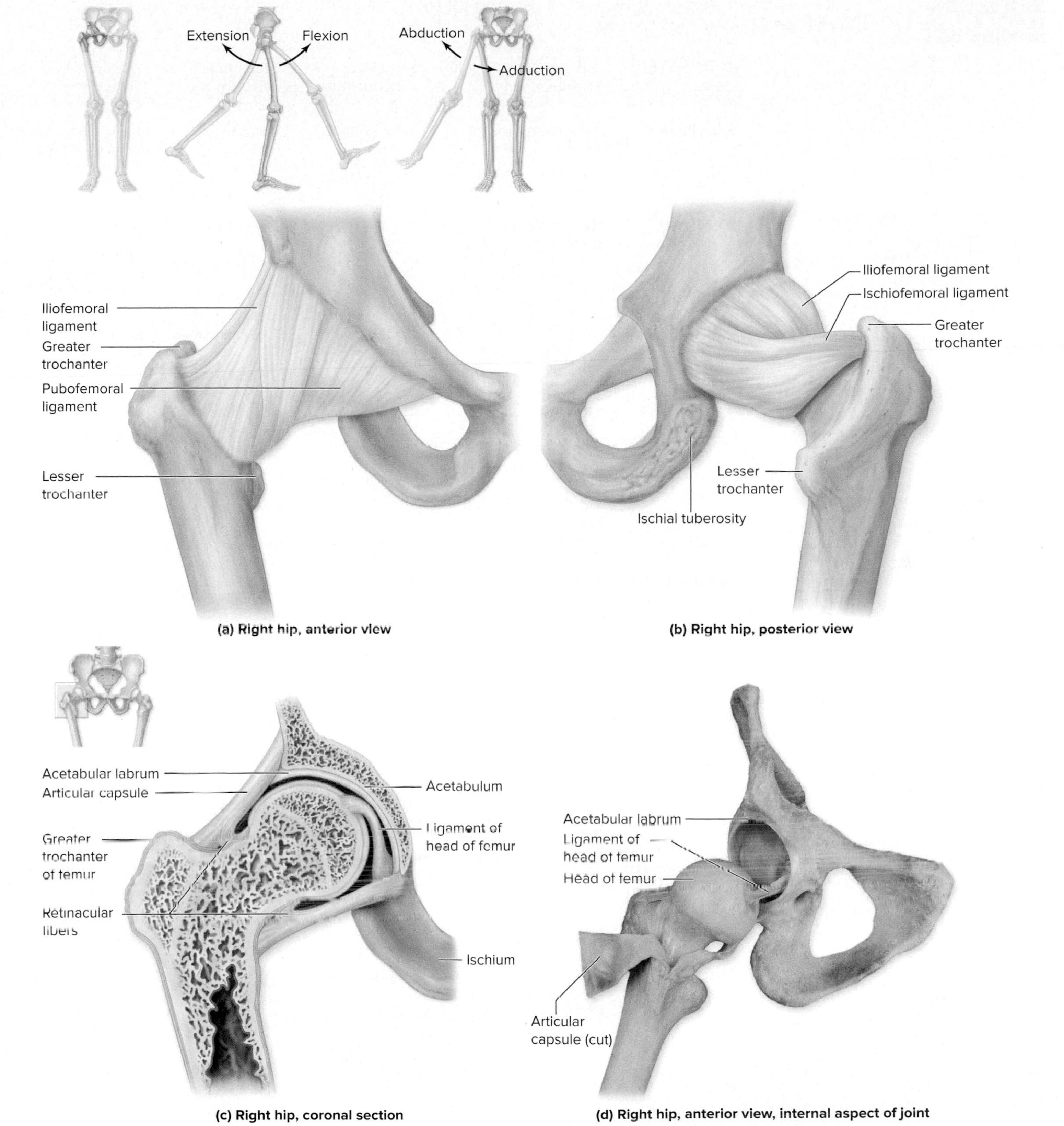

Extension
Flexion
Abduction
Adduction

Iliofemoral ligament
Greater trochanter
Pubofemoral ligament
Lesser trochanter

(a) Right hip, anterior view

Iliofemoral ligament
Ischiofemoral ligament
Greater trochanter
Lesser trochanter
Ischial tuberosity

(b) Right hip, posterior view

Acetabular labrum
Articular capsule
Greater trochanter of femur
Retinacular fibers
Acetabulum
Ligament of head of femur
Ischium

(c) Right hip, coronal section

Acetabular labrum
Ligament of head of femur
Head of femur
Articular capsule (cut)

(d) Right hip, anterior view, internal aspect of joint

Figure 9.17 Hip Joint. The hip joint is formed by the head of the femur and the acetabulum of the os coxae. The right hip joint is shown in (*a*) anterior view, (*b*) posterior view, and (*c*) coronal section. (*d*) Cadaver photo of the coxal joint, with the articular capsule cut to show internal structures. 

9.7d Hip Joint

LEARNING OBJECTIVES

29. Describe the hip joint and its motions.

30. Explain why the hip joint is more stable than the glenohumeral joint.

The **hip joint,** also known as the *coxal joint,* is the articulation between the head of the femur and the relatively deep, concave acetabulum of the os coxae (**figure 9.17**; **table 9.5**). A fibrocartilaginous **acetabular labrum** further deepens this socket. The hip joint's more extensive bony architecture is therefore much stronger and more

Table 9.5 Pelvic Girdle and Lower Limb Joints

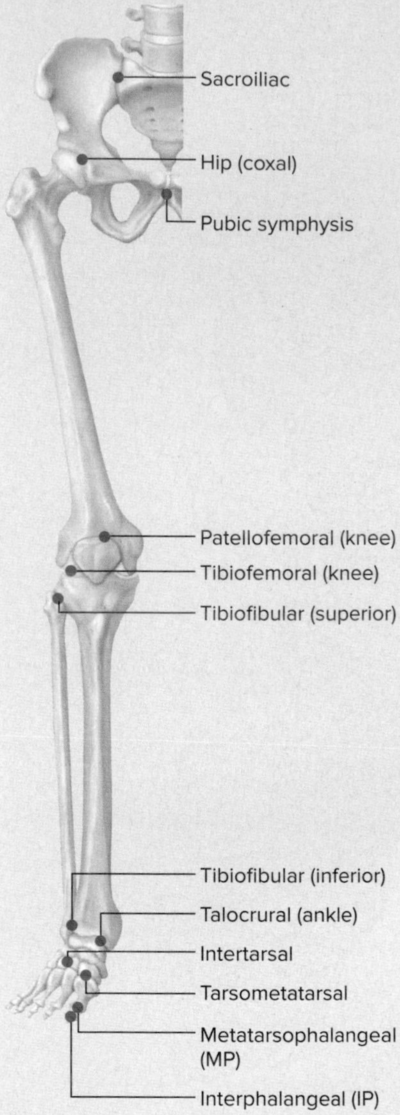

Sacroiliac
Hip (coxal)
Pubic symphysis
Patellofemoral (knee)
Tibiofemoral (knee)
Tibiofibular (superior)
Tibiofibular (inferior)
Talocrural (ankle)
Intertarsal
Tarsometatarsal
Metatarsophalangeal (MP)
Interphalangeal (IP)

Joint	Articulation Components	Structural Classification	Functional Classification	Description of Movement
Sacroiliac	Auricular surfaces of sacrum and ilium	Synovial (plane)	Diarthrosis	Slight gliding; more movement during pregnancy and childbirth
Hip (coxal)	Head of femur and acetabulum of os coxae	Synovial (ball-and-socket)	Diarthrosis	Flexion, extension, abduction, adduction, circumduction, medial and lateral rotation of thigh
Pubic symphysis	Symphyseal surface of both pubic bones	Cartilaginous (symphysis)	Amphiarthrosis	Very slight gliding; more movement during childbirth
Knee	**Tibiofemoral joint:** Medial condyle of femur, medial meniscus, and medial condyle of tibia **Patellofemoral joint:** Patella and patellar surface of femur	Synovial (hinge) at tibiofemoral joint,[1] both synovial (hinge) and synovial (plane) at patellofemoral joint	Diarthrosis	Flexion, extension, lateral rotation of leg in flexed position, slight medial rotation
Tibiofibular	**Superior joint:** Head of fibula and lateral condyle of tibia **Inferior joint:** Distal end of fibula and fibular notch of tibia	**Superior joint:** Synovial (plane) **Inferior joint:** Fibrous (syndesmosis)	Amphiarthrosis	Slight rotation of fibula during dorsiflexion of foot
Talocrural (ankle)	Distal end of tibia and medial malleolus of tibia with talus Lateral malleolus of fibula and talus	Synovial (hinge)	Diarthrosis	Dorsiflexion and plantar flexion
Intertarsal	Between the tarsal bones	Synovial (plane)	Diarthrosis	Eversion and inversion of foot
Tarsometatarsal	Three cuneiforms and cuboid (tarsals) and bases of five metatarsals	Synovial (plane)	Diarthrosis	Slight gliding
Metatarso-phalangeal (MP joints)	Heads of metatarsals and bases of proximal phalanges	Synovial (condylar)	Diarthrosis	Flexion, extension, abduction, adduction, and circumduction of phalanges
Interphalangeal (IP joints)	Heads of proximal and middle phalanges with bases of middle and distal phalanges, respectively	Synovial (hinge)	Diarthrosis	Flexion and extension of phalanges

1. Although anatomists classify the tibiofemoral joint as a hinge joint, some kinesiologists and exercise scientists prefer to classify the tibiofemoral joint as a modified condylar joint.

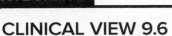

CLINICAL VIEW 9.6

Fracture of the Femoral Neck

Fracture of the femoral neck is often incorrectly referred to as a "fractured hip," because the break is in the femoral neck, not the os coxae. With this injury, the pull of the lower limb muscles causes the leg to rotate laterally and appear shorter by several inches. Fractures of the femoral neck are of two types: intertrochanteric and subcapital.

Intertrochanteric fractures of the femoral neck occur distally to or outside the hip joint capsule—in other words, these fractures are *extracapsular*. The fracture line runs between the greater and lesser trochanters. This type of injury typically occurs in younger and middle-aged individuals. It usually occurs in response to trauma.

Subcapital (*or intracapsular*) **fractures** of the femoral neck occur within the hip articular capsule, very close to the head of the femur itself. This type of fracture usually occurs in elderly people whose bones have been weakened by osteoporosis and thus are more susceptible to fracture.

Subcapital fractures result in a tearing of the retinacular arteries that supply the head and neck of the femur. The ligament of the head of the femur may be torn as well. As a result, the head and neck of the femur lose their blood supply and may develop **avascular necrosis,** which is death of the bone due to lack of blood. Frequently, hip replacement surgery is needed, whereby a metal femoral head and neck replace the dying bone. This surgery is not without risk, however, and many elderly individuals do not survive because of surgical complications.

stable than that of the glenohumeral joint. Conversely, the hip joint's increased stability means that it is less mobile than the glenohumeral joint. The hip joint must be more stable (and thus less mobile) because it supports the body weight.

The hip joint is secured by a strong articular capsule, several ligaments, and a number of powerful muscles. The articular capsule extends from the acetabulum to the trochanters of the femur, enclosing both the femoral head and neck. This arrangement prevents the head from moving away from the acetabulum. The ligamentous fibers of the articular capsule reflect around the neck of the femur. These reflected fibers, called **retinacular** (ret-i-nak′yū-lăr) **fibers,** provide additional stability to the capsule. Located within the retinacular fibers are retinacular arteries (branches of the deep femoral artery), which supply almost all of the blood to the head and neck of the femur.

The articular capsule is reinforced by three spiraling intracapsular ligaments: The **iliofemoral** (il′ē-ō-fem′ŏ-răl) **ligament** is a Y-shaped ligament that provides strong reinforcement for the anterior region of the articular capsule. The **ischiofemoral** (is-kē-ō-fem′ŏ-răl) **ligament** is a spiral-shaped, and posteriorly located. The **pubofemoral** (pyū′bō-fem′ŏ-răl) **ligament** is a triangular thickening of the capsule's inferior region. All of these spiraling ligaments become taut when the hip joint is extended, so the hip joint is most stable in the extended position. Try this experiment: Flex your hip joint, and try to move the femur; you may notice a great deal of mobility. Now extend your hip joint (stand up), and try to move the femur. Because these ligaments are taut, you don't have as much mobility in the joint as you did when the hip joint was flexed.

Another tiny ligament, the **ligament of head of femur,** also called the *ligamentum teres,* originates along the acetabulum. Its attachment point is the fovea of the head of the femur. This ligament does not provide strength to the joint; rather, it typically contains a small artery that supplies the head of the femur.

The combination of a deep bony socket, a strong articular capsule, supporting ligaments, and muscular padding gives the hip joint its stability. Movements possible at the hip joint include flexion, extension, abduction, adduction, circumduction, and medial and lateral rotation.

WHAT DID YOU LEARN?

20 How do the glenohumeral and hip joints compare with respect to their mobility and stability?

9.7e Knee Joint

LEARNING OBJECTIVES

31. Describe the knee joint and its motion.

32. Name the ligaments that support the knee joint.

The **knee joint** is the largest and most complex diarthrosis of the body (**figure 9.18**; table 9.5; see also figure 9.5*a*). It is primarily a hinge joint, but when the knee is flexed, it is

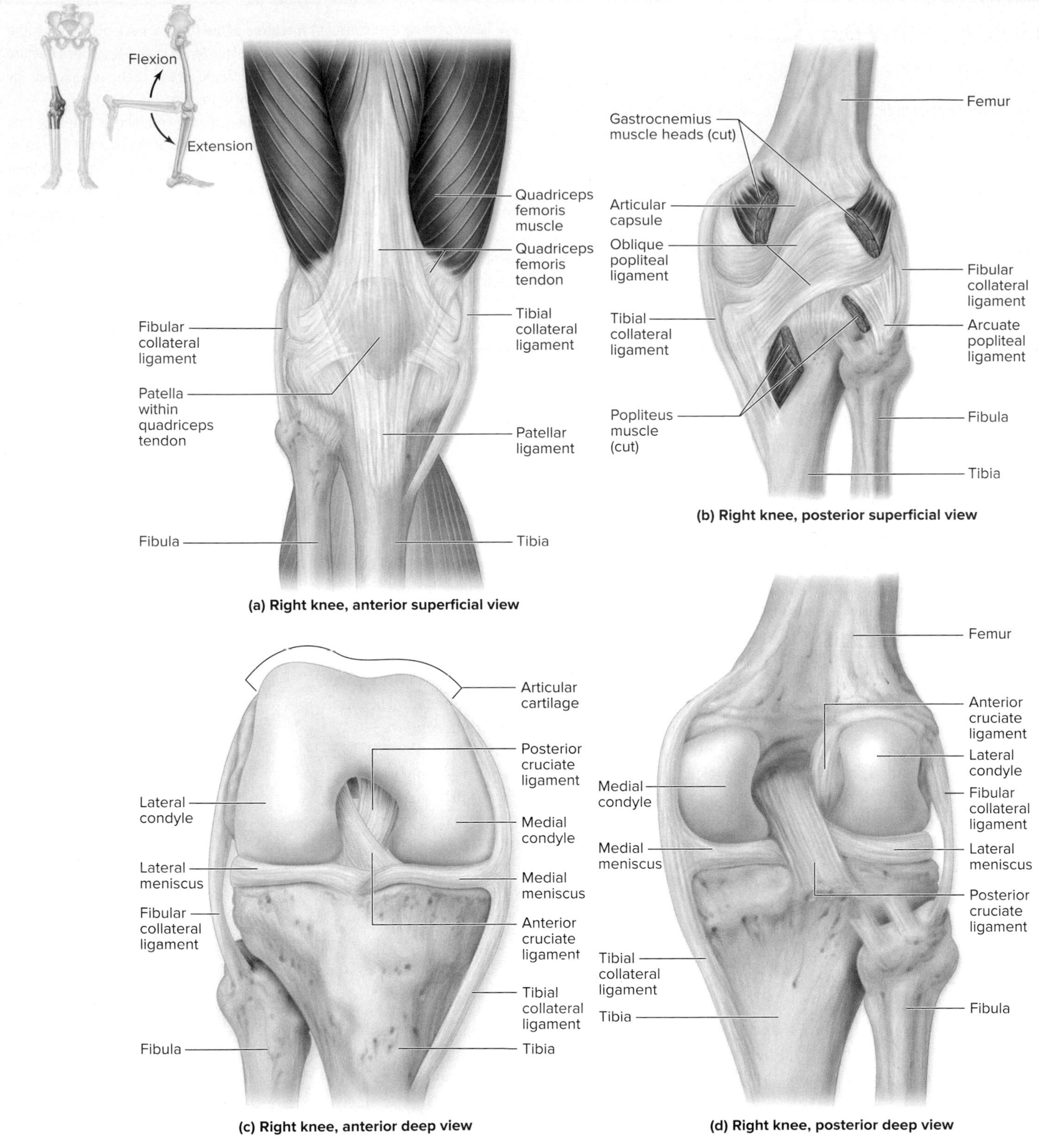

Figure 9.18 Knee Joint. This joint is the most complex diarthrosis of the body. (*a*) Anterior superficial, (*b*) posterior superficial, (*c*) anterior deep, and (*d*) posterior deep views reveal the complex interrelationships of the parts of the right knee. AP|R

also capable of slight rotation and lateral gliding. Structurally, the knee is composed of two separate articulations: (1) The **tibiofemoral** (tib-ē-ō-fem′ŏ-răl) **joint** is between the condyles of the femur and the condyles of the tibia, and (2) the **patellofemoral joint** is between the patella and the patellar surface of the femur.

The knee joint has an articular capsule that encloses only the medial, lateral, and posterior regions of the knee joint. The articular capsule does not cover the anterior surface of the knee joint; rather, the quadriceps femoris muscle tendon passes over the knee joint's anterior surface. The patella is embedded within this tendon, and the **patellar ligament** extends beyond the patella and continues to where it attaches on the tibial tuberosity of the tibia. Thus, there is no single unified capsule in the knee, nor is there a common joint cavity. Posteriorly, the capsule is strengthened by several popliteal ligaments.

INTEGRATE

CLINICAL VIEW 9.7

Knee Ligament and Cartilage Injuries

Although the knee is capable of bearing much weight and has numerous strong supporting ligaments, it is highly vulnerable to injury, especially among athletes. Because the knee is reinforced by tendons and ligaments only, ligamentous injuries to the knee are very common.

The tibial collateral ligament is frequently injured when the leg is forcibly abducted at the knee, such as when a person's knee is hit on the lateral side. Because the tibial collateral ligament is attached to the medial meniscus, the medial meniscus may be injured as well.

Injury to the fibular collateral ligament can occur if the medial side of the knee is struck, resulting in hyperadduction of the leg at the knee. This type of injury is fairly rare, in part because the fibular collateral ligament is very strong and because medial blows to the knee are not common.

The anterior cruciate ligament (ACL) can be injured when the leg is hyperextended—for example, if a runner's foot hits a hole. Because the ACL is rather weak compared to the other knee ligaments, it is especially prone to injury. To test for ACL injury, a physician gently tugs anteriorly on the tibia. In this so-called **anterior drawer test,** too much forward movement indicates an ACL tear.

Posterior cruciate ligament (PCL) injury may occur if the leg is hyperflexed or if the tibia is driven posteriorly on the femur. PCL injury occurs rarely, because this ligament is rather strong. To test for PCL injury, a physician gently pushes posteriorly on the tibia. In this **posterior drawer test,** too much posterior movement indicates a PCL tear.

The menisci also may be prone to injury. Tears in the menisci may occur due to blows to the knee or due to general overuse of the joint. Because the menisci are composed of fibrocartilage, they cannot regenerate and often must be surgically treated.

The **unhappy triad** of injuries refers to a triple injury of the tibial collateral ligament, medial meniscus, and anterior cruciate ligament. This is the most common type of football injury. It occurs when a player is illegally "clipped" by a lateral blow to the knee, and the leg is forcibly abducted and laterally rotated. If the blow is severe enough, the tibial collateral ligament tears, followed by tearing of the medial meniscus, as these two structures are connected. The force that tears the tibial collateral ligament and the medial meniscus is thus transferred to the ACL. Because the ACL is relatively weak, it tears as well.

The treatment of ligamentous knee injuries depends upon the severity and type of injury. Conservative treatment involves immobilizing the knee for a period of time to rest the joint. Surgical treatment can include repairing the torn ligaments or replacing the ligaments with a graft from another tendon or ligament (such as the quadriceps tendon). Many knee surgeries may be performed with arthroscopy. **Arthroscopy** is a type of conservative surgical treatment where a small incision is made in the knee and then an **arthroscope** (an instrument with a camera and light source) is inserted into the knee, allowing the surgeon to clearly see the surgical area without having to make large incisions.

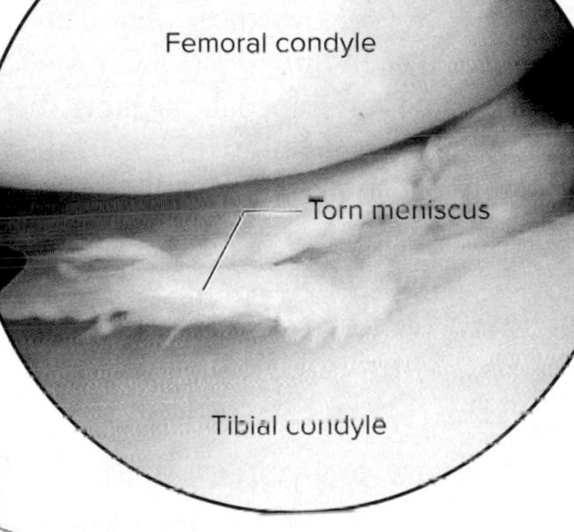

Arthroscopic view of knee joint, showing torn meniscus

©CNRI/Science Source

Lateral blow to knee

Torn anterior cruciate ligament

Torn tibial collateral ligament

Torn medial meniscus

"Unhappy triad" of injuries to the right knee.

On either side of the knee joint are two collateral ligaments that become taut on extension and provide additional stability to the joint. The **fibular collateral ligament** (*lateral collateral ligament*) reinforces the lateral surface of the joint. This ligament extends from the femur to the fibula and prevents hyperadduction of the leg at the knee. (In other words, it prevents the leg from moving too far medially relative to the thigh). The **tibial collateral ligament** (*medial collateral ligament*) reinforces the medial surface of the knee joint. This ligament runs from the femur to the tibia and prevents hyperabduction of the leg at the knee. (In other words, it prevents the leg from moving too far laterally relative to the thigh.) This ligament is attached to the medial meniscus of the knee joint as well, so an injury to the tibial collateral ligament usually affects the medial meniscus.

Deep to the articular capsule and within the knee joint itself are a pair of C-shaped fibrocartilage pads positioned on the condyles of the tibia. These pads are called the **medial meniscus** and the **lateral meniscus.** They partially stabilize the joint medially and laterally, act as cushions between articular surfaces, and continuously change shape to conform to the articulating surfaces as the femur moves.

Two **cruciate** (krū′shē-āt) **ligaments** are deep to the articular capsule of the knee joint. They limit the anterior and posterior movement of the femur on the tibia. These ligaments cross each other in the form of an X, hence the name *cruciate* (which means *cross*). The **anterior cruciate ligament (ACL)** extends from the posterior femur to the anterior side of the tibia. When the knee is extended, the ACL is pulled tight and prevents hyperextension. The ACL prevents the tibia from moving too far anteriorly relative to the femur. The **posterior cruciate ligament (PCL)** attaches from the anteroinferior femur to the posterior side of the tibia. The PCL becomes taut on flexion, and so it prevents hyperflexion of the knee joint. The PCL also prevents posterior displacement of the tibia relative to the femur.

Humans are bipedal, meaning that we walk on two feet. An important aspect of bipedal locomotion is the ability to "lock" the knees in the extended position and stand erect without tiring the leg muscles. At full extension, the tibia rotates laterally so as to tighten the anterior cruciate ligament and squeeze the menisci between the tibia and femur. Muscular contraction by the popliteus muscle (see section 11.9c) unlocks and flexes the knee joint. (This ability should be distinguished from being unable to bend or straighten the knee because of injury or disease, which is sometimes also called locking.)

WHAT DID YOU LEARN?

21 What are the functions of each of the intracapsular ligaments of the knee joint?

9.7f Talocrural (Ankle) Joint

LEARNING OBJECTIVE

33. Describe the talocrural joint and its motion.

The **talocrural (ankle) joint** is a highly modified hinge joint that permits both dorsiflexion and plantar flexion. It includes two articulations within one joint capsule. One articulation is between the distal end of the tibia and the talus; the other is between the distal end of the fibula and the lateral aspect of the talus (**figure 9.19**; table 9.5). The medial and lateral malleoli of the tibia and fibula, respectively, form extensive medial and lateral margins and prevent the talus from sliding side-to-side.

The talocrural joint includes several distinctive anatomic features. Its articular capsule covers the distal surfaces of the tibia, the medial malleolus, the lateral malleolus, and the talus. A multipart **deltoid ligament** (or *medial ligament*) binds the tibia to the foot on the medial side. This ligament prevents overeversion of the foot. It is incredibly strong and rarely tears; in fact, it typically will pull off the medial malleolus before it ever tears (see Clinical View 9.8: "Ankle Sprains and Pott Fractures"). A much thinner, multipart **lateral ligament** binds the fibula to the foot on the lateral side. This ligament prevents overinversion of the foot. It is not as strong as the deltoid ligament and is prone to sprains and tears. Two **tibiofibular** (tib′ē-ō-fib′yū-lăr) **ligaments** (**anterior** and **posterior**) bind the tibia to the fibula.

WHAT DID YOU LEARN?

22 What bones articulate at the talocrural joint, and what movements are permitted at this joint?

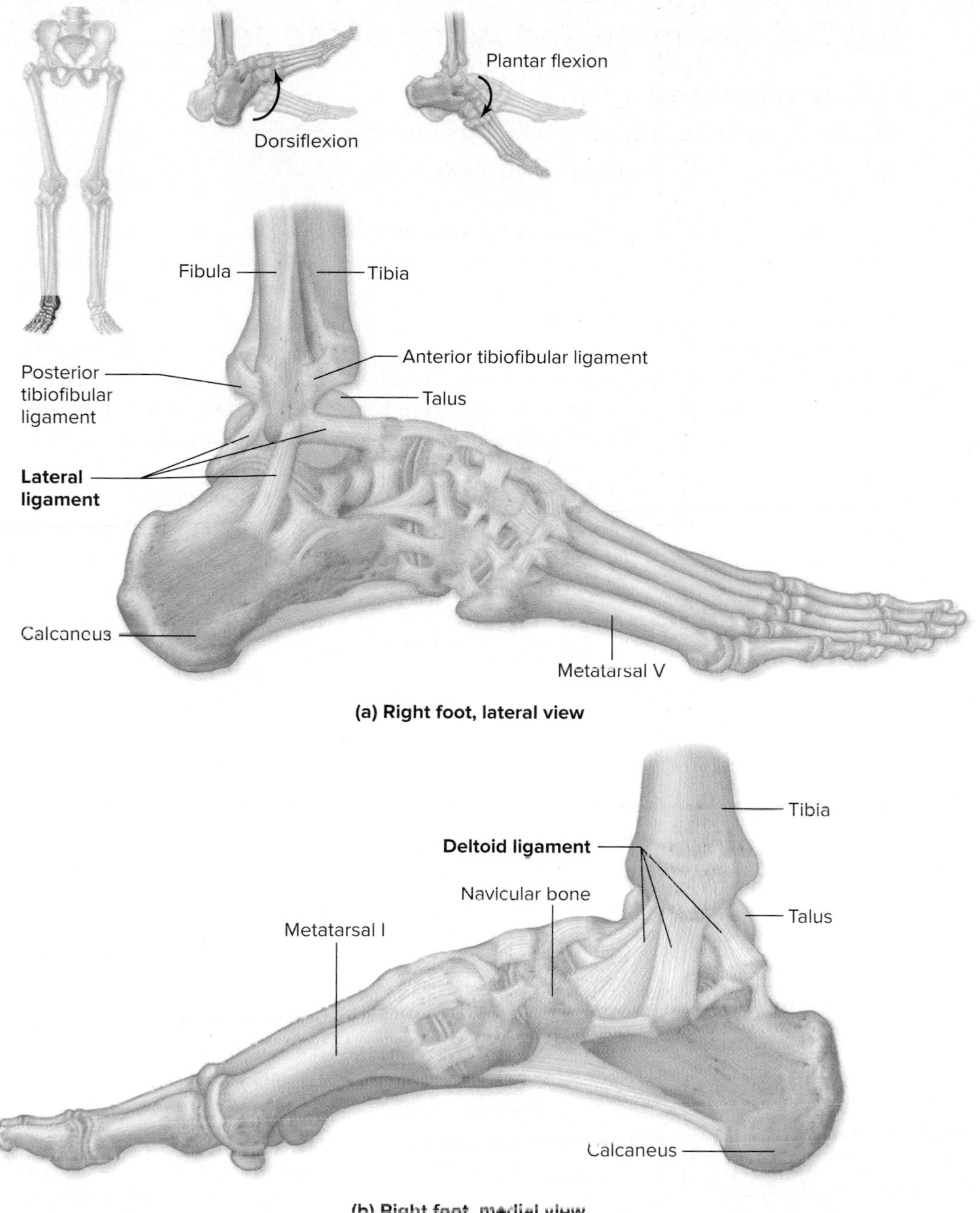

Figure 9.19 Talocrural Joint. (*a*) Lateral and (*b*) medial views of the right foot show that the talocrural joint contains articulations among the tibia, fibula, and talus. This joint permits dorsiflexion and plantar flexion only. AP|R

INTEGRATE

CLINICAL VIEW 9.8

Ankle Sprains and Pott Fractures

A **sprain** is a stretching or tearing of ligaments, without fracture or dislocation of the joint. An ankle sprain results from twisting of the foot, almost always due to *overinversion.* Fibers of the lateral ligament are either stretched (in mild sprains) or torn (in more severe sprains), producing localized swelling and tenderness anteroinferior to the lateral malleolus. *Overeversion* sprains rarely occur due to the strength of the deltoid (medial) ligament. Recall that ligaments are composed of dense regular connective tissue, which is poorly vascularized (see section 5.2d). Poorly vascularized tissue takes a long time to heal, and that is the case with ankle sprains. These structures are also prone to reinjury.

If overeversion *does* occur, the injury that usually results is called a **Pott fracture** (see section 7.8). If the foot is overeverted, it pulls on the deltoid ligament, which is very strong and doesn't tear. Instead, the pull can avulse (pull off) the medial malleolus of the tibia. The force from the injury then continues to move the talus laterally, as the medial malleolus can no longer restrict side-to-side movements of the ankle. As the talus moves laterally and puts force on the fibula, the fibula fractures as well (usually at its distal end or by the lateral malleolus). Thus, both the tibia and the fibula fracture in this injury, yet the deltoid ligament remains intact.

9.8 Development and Aging of the Joints

LEARNING OBJECTIVES

34. Explain how the three major types of joints form in the embryo and fetus.
35. Describe some of the common age-related changes in joints.

Joints start to form by the sixth week of development and progressively differentiate during the fetal period. In the area of future fibrous joints, the mesenchyme around the developing bone differentiates into dense regular connective tissue, whereas in cartilaginous joints it differentiates either into fibrocartilage or hyaline cartilage.

The development of the synovial joints is more complex. The most *laterally* placed mesenchyme forms the articular capsule and supporting ligaments of the joint. Just medial to this region, the mesenchyme forms the synovial membrane, which then starts secreting synovial fluid into the joint cavity. The *centrally* located mesenchyme may be reabsorbed or can form menisci or articular discs, depending upon the type of synovial joint.

Prior to the closure of the epiphyseal plates, some injuries to a young person may result in subluxation or fracture of an epiphysis, with potential adverse effects on the future development and health of the joint; the bone may not reach its potential full length, or the individual may develop arthritic-like changes in the joint.

Arthritis is a rheumatic (i.e., referring to the joints or muscles) disease that involves damage to articular cartilage (see Clinical View 9.9: "Arthritis"). The primary problem that develops in

INTEGRATE

CLINICAL VIEW 9.9

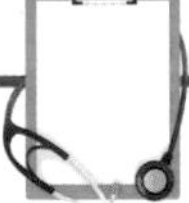

Arthritis

Arthritis (ar-thrī′tis) is a group of inflammatory or degenerative diseases of joints that occur in various forms. Each form presents the same symptoms: swelling of the joint, pain, and stiffness. It is the most prevalent crippling disease in the United States. Some common forms of arthritis are gouty arthritis, osteoarthritis, and rheumatoid arthritis.

Gouty arthritis is typically seen in middle-aged and older individuals, and it is more common in males. Often called *gout,* this disease occurs as a result of an increased level of *uric acid* (a normal cellular waste product formed from breakdown of nitrogenous bases; see section 2.1d) in the blood. This abnormal level causes urate crystals to accumulate in the blood, synovial fluid, and synovial membranes. The body's inflammatory response to the urate crystals results in joint pain. Gout usually begins with an attack on a single joint (often in the great toe) and later progresses to other joints. Eventually, gouty arthritis may immobilize joints by causing fusion between the articular surfaces of the bones.

Osteoarthritis (OA), also known as *degenerative arthritis,* is the most common type of arthritis. This chronic degenerative joint condition also is termed *wear-and-tear arthritis* because repeated use of a joint gradually wears down the articular cartilage, much as the repeated use of a pencil eraser wears down the eraser. If the cartilage is worn down enough, osteoarthritis results. Eventually, bone rubs against bone, causing abrasions on the bony surfaces. Without the protective articular cartilage, movements at the joints become stiff and painful. The joints most affected by osteoarthritis are those of the fingers, knuckles, hips, knees, and shoulders. Osteoarthritis is typically seen in older individuals, although more and more athletes are experiencing arthritis at an earlier age due to the repetitive stresses placed on their joints.

Rheumatoid (rū′mă-toyd) **arthritis (RA)** is typically seen in younger and middle-aged adults, and it is much more prevalent in women. The common age at onset is approximately 40–50 (although individuals as young as their teens have been diagnosed with the disorder). Symptoms include pain and joint swelling, muscle weakness, osteoporosis, and assorted problems with both the heart and the blood vessels. Rheumatoid arthritis is an *autoimmune disorder* in which the body's immune system targets its own tissues for attack (see Clinical View 22.4: "Autoimmune Disorders"). Rheumatoid arthritis starts with synovial membrane inflammation. Fluid and white blood cells leak from small blood vessels into the joint cavity, causing an increase in synovial fluid volume. As a consequence, the joint swells and the inflamed synovial membrane thickens; eventually, the articular cartilage and, often, the underlying bone become eroded. Scar tissue later forms and ossifies, and bone ends fuse together—a process called **ankylosis** (ang′ki-lō′sis)—immobilizing the joint. Two types of medications often are prescribed to treat RA. The faster-acting, first-line medications include NSAIDs and corticosteroids and are used to relieve joint pain. The slower, but longer-acting, second-line medications, such as methotrexate and hydroxychloroquine, help put the disease into remission and slow down the destruction of the joints.

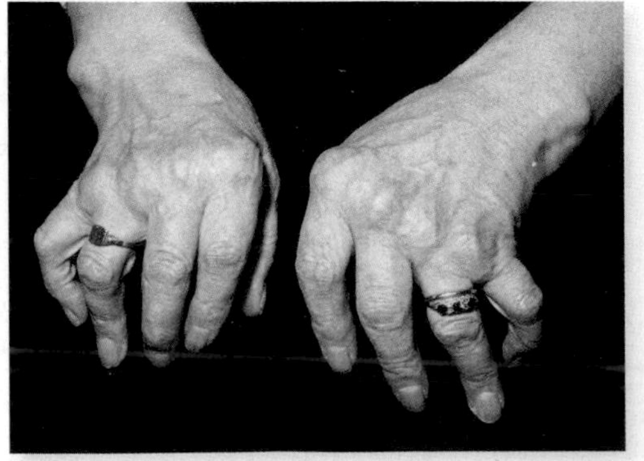

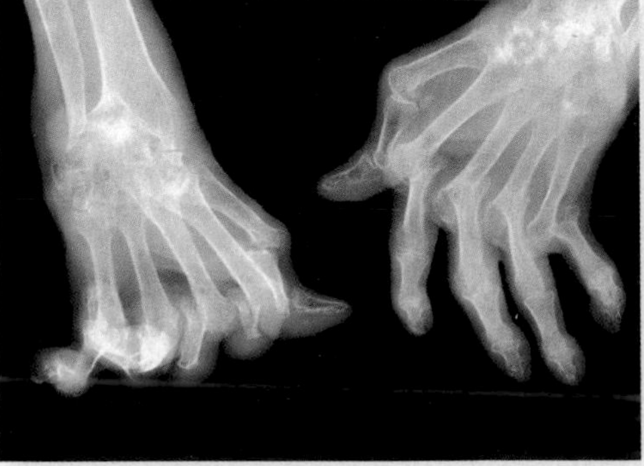

A photograph and colorized radiograph of hands with rheumatoid arthritis.
(*a*) ©John Watney/Science Source; (*b*) ©CNRI/Science Source

an aging joint is osteoarthritis, also known as degenerative arthritis. The damage can have various causes, but it usually results from cumulative wear and tear at the joint surface.

Just as the strength of a bone is maintained by continual application of stress, the health of joints is directly related to moderate exercise. Exercise compresses the articular cartilages, causing synovial fluid to be squeezed out of the cartilage and then pulled back inside the cartilage matrix. This flow of fluid gives the chondrocytes within the cartilage the nourishment required to maintain their health. Exercise also strengthens the muscles that support and stabilize the joint. However, extreme exercise should be avoided, because it aggravates potential joint problems and may worsen osteoarthritis.

WHAT DID YOU LEARN?

23 What are some ways that joints change as a person ages?

CHAPTER SUMMARY

	• Articulations are the joints where bones are in contact. Joints differ in structure, function, and the amount of movement they allow.
9.1 Classification of Joints	• The three structural categories of joints are fibrous, cartilaginous, and synovial. • The three functional categories of joints are synarthroses (immobile joints), amphiarthroses (slightly mobile joints), and diarthroses (freely mobile joints).
9.2 Fibrous Joints	• Fibrous joints lack a joint cavity, and they interconnect articulating bones by dense regular connective tissue.
	9.2a Gomphoses • A gomphosis is a synarthrosis between the tooth and either the mandible or the maxillae.
	9.2b Sutures • A suture is a synarthrosis that tightly binds the bones of the skull. Fused sutures are called synostoses.
	9.2c Syndesmoses • A syndesmosis is an amphiarthrosis, and the bones are connected by interosseous membranes.
9.3 Cartilaginous Joints	• Cartilaginous joints lack a joint cavity; the cartilage between the articulating bones may be either hyaline cartilage or fibrocartilage.
	9.3a Synchondroses • A synchondrosis is a synarthrosis where hyaline cartilage is wedged between the articulating bones.
	9.3b Symphyses • A symphysis is an amphiarthrosis, where a disc of fibrocartilage is wedged between the articulating bones.
9.4 Synovial Joints	• All synovial joints are diarthroses.
	9.4a Distinguishing Features and Anatomy of Synovial Joints • Synovial joints contain an articular capsule, a joint cavity, synovial fluid, articular cartilage, ligaments, nerves, and blood vessels.
	9.4b Classification of Synovial Joints • The six types of synovial joints are plane, hinge, pivot, condylar, saddle, and ball-and-socket.
9.5 The Movements of Synovial Joints	• Motions that occur at synovial joints include gliding, angular, rotational, and special.
	9.5a Gliding Motion • Gliding is a simple movement where two opposing surfaces slide back-and-forth or side-to-side against one another.
	9.5b Angular Motion • Angular movements involve decreasing or increasing the angle of a joint. Angular movements include flexion, extension, hyperextension, lateral flexion, abduction, adduction, and circumduction.
	9.5c Rotational Motion • Rotational movements involve a pivoting motion. Examples of rotational movements are lateral rotation, medial rotation, pronation, and supination.
	9.5d Special Movements • Special movements include depression and elevation, dorsiflexion and plantar flexion, eversion and inversion, protraction and retraction, and opposition.

(continued on next page)

CHAPTER SUMMARY (continued)

9.6 Synovial Joints and Levers	• Biomechanics is the practice of applying mechanical principles to biology.
	9.6a Terminology of Levers • Synovial joints may be compared to levers, which have a fixed point (fulcrum) around which movement occurs when effort applied to one point exceeds resistance at another point.
	9.6b Types of Levers • A first-class lever has a fulcrum between the effort and the resistance. • A second-class lever has the resistance placed between the fulcrum and the effort. • A third-class lever, the most common type of lever in the human body, has the effort applied between the resistance and the fulcrum.
9.7 Features and Anatomy of Selected Joints	• In each articulation, unique features of the articulating bones support the intended movements.
	9.7a Temporomandibular Joint • The temporomandibular joint is an articulation between the head of the mandible and the mandibular fossa of the temporal bone.
	9.7b Shoulder Joint • The sternoclavicular joint and acromioclavicular joint support movement of the shoulder. • The glenohumeral joint is a ball-and-socket joint between the glenoid cavity of the scapula and the head of the humerus.
	9.7c Elbow Joint • The elbow is a hinge joint among the humerus, radius, and ulna.
	9.7d Hip Joint • The hip joint is a ball-and-socket joint between the head of the femur and the acetabulum of the os coxae.
	9.7e Knee Joint • The knee joint is primarily a hinge joint but is capable of slight rotation and gliding.
	9.7f Talocrural (Ankle) Joint • The talocrural joint is a hinge joint that permits dorsiflexion and plantar flexion of the ankle.
9.8 Development and Aging of the Joints	• Joints begin to form during week 6 of development. • Osteoarthritis is a common joint problem that occurs with aging.

CHALLENGE YOURSELF

Create and Evaluate

Analyze and Apply

Understand and Remember

Do You Know the Basics?

1. The greatest range of mobility of any joint in the body is found in the
   a. knee joint.
   b. hip joint.
   c. glenohumeral joint.
   d. elbow joint.
2. A movement of the foot that turns the sole outward or laterally is called
   a. dorsiflexion.
   b. inversion.
   c. eversion.
   d. plantar flexion.
3. A ______ is formed when two bones previously connected by a suture fuse.
   a. gomphosis
   b. synostosis
   c. symphysis
   d. syndesmosis
4. The ligament that helps to maintain the alignment of the condyles between the femur and tibia and to limit the anterior movement of the tibia on the femur is the
   a. tibial collateral ligament.
   b. posterior cruciate ligament.
   c. anterior cruciate ligament.
   d. fibular collateral ligament.
5. Which joint is a diarthrosis?
   a. symphysis
   b. synchondrosis
   c. syndesmosis
   d. saddle
6. In this type of lever, the effort is located between the resistance and the fulcrum. An example would be your knee joint.
   a. first-class
   b. second-class
   c. third-class
   d. Two of the above are correct.

7. A metacarpophalangeal (MP) joint, which has oval articulating surfaces and permits movement in two planes, is what type of synovial joint?
   a. condylar
   b. plane
   c. hinge
   d. saddle
8. All of the following ligaments provide stability to the hip joint *except* the
   a. ischiofemoral ligament.
   b. pubofemoral ligament.
   c. iliofemoral ligament.
   d. ligament of the head of the femur.
9. Which of the following is a function of synovial fluid?
   a. lubricates the joint
   b. provides nutrients for the articular cartilage
   c. absorbs shock within the joint
   d. All of these are correct.
10. Plantar flexion and dorsiflexion are movements permitted at the __________ joint.
    a. hip
    b. knee
    c. sternoclavicular
    d. talocrural
11. Discuss the factors that influence both the stability and mobility of a joint. What is the relationship between a joint's mobility and its stability?
12. Both fibrous joints and synovial joints have dense regular connective tissue holding the bones together. So how are these two joints different, both structurally and functionally?
13. List and describe all joints that are functionally classified as synarthroses.
14. How do a hinge joint and a pivot joint compare with respect to structure, function, and location within the body?
15. Compare and contrast first-, second-, and third-class levers.
16. Describe and compare the movements of abduction, adduction, pronation, and supination.
17. Most ankle sprains are overinversion injuries. What are the anatomic reasons that overeversion ankle sprains are relatively uncommon? Are there any overeversion injuries that occur to the ankle?
18. What are the main supporting ligaments of the elbow joint?
19. Compare the functions of the tibial and the fibular collateral ligaments in the knee joint. Which of the two is injured more frequently and why?
20. What are the similarities and differences between osteoarthritis and rheumatoid arthritis?

Can You Apply What You've Learned?

Use the following paragraph to answer questions 1–3.

A mother and her 4-year-old son were visiting a toy store, and the child did not want to leave. As the child threw a temper tantrum and resisted his mother, the mother pulled on the boy's arm to lead him out of the store. Immediately after the pull, the boy cried out in pain and displayed a prominent bump on the lateral side of the elbow. The mother drove the boy to the doctor in a panic. The doctor examined the boy's elbow and determined the boy had a subluxated head of the radius.

1. Which ligament failed to keep the head of the radius in place when the mother pulled on the boy's elbow?
   a. anular ligament
   b. ulnar collateral ligament
   c. radial collateral ligament
   d. coronoid ligament
2. The doctor mentions that this type of injury is common in children younger than 5 years of age. What is one reason for this?
   a. The olecranon of the ulna does not fit in properly with the olecranon fossa of the radius.
   b. The head of the radius is not fully formed.
   c. The medial and lateral epicondyle epiphyseal plates have not yet fused to the rest of the humerus.
   d. The articular capsule of the elbow joint is weak in its anterior surface.
3. What bony feature caused the prominent bump on the lateral side of the elbow?
   a. lateral epicondyle of the humerus
   b. coronoid process of the ulna
   c. head of the radius
   d. radial collateral ligament
4. While Robert was running, he stepped into a pothole and twisted his right ankle. A swelling appeared along the lateral side of this ankle. Which ligament was injured, and what movement resulted in the injury?
   a. deltoid ligament, caused by overeversion of the foot
   b. lateral ligament, caused by overeversion of the foot
   c. deltoid ligament, caused by overinversion of the foot
   d. lateral ligament, caused by overinversion of the foot
5. Most knee ligaments become taut upon extension of the joint, except for one. Which knee ligament becomes taut upon *flexion* of the joint and prevents hyperflexion of the joint?
   a. anterior cruciate ligament
   b. posterior cruciate ligament
   c. patellar ligament
   d. tibial collateral ligament

Can You Synthesize What You've Learned?

1. During soccer practice, Erin tripped over the outstretched leg of a teammate and fell directly onto her shoulder. She was taken to the hospital in excruciating pain. Examination revealed that the head of the humerus had moved inferiorly and anteriorly into the axilla. What happened to Erin in this injury?
2. While Lucas and Omar were watching a football game, a player was penalized for "clipping," meaning that he had hit an opposing player on the lateral knee, causing hyperabduction at the knee joint. Lucas asked Omar what the big deal was about clipping. What joint is most at risk,

and what kinds of injuries can occur if a player gets clipped?

3. Jackie visits her physician because she is experiencing pain by her right ear. The doctor checks her ears and sees no sign of infection. She asks Jackie to open and close her mouth while she palpates the portions of her face adjacent to her ears. Why is the doctor having Jackie move her mouth, when she is experiencing ear pain? How may the two be related? What do you think the doctor will discover when Jackie opens and closes her mouth?

INTEGRATE

ONLINE STUDY TOOLS

connect | SMARTBOOK® | APR

The following study aids may be accessed through Connect.

Clinical Case Study: A Historical Case of Spinal Involvement with Tuberculosis

Interactive Questions: This chapter's content is served up in a number of multimedia question formats for student study

SmartBook: Topics and terminology include classification of joints; fibrous joints; cartilaginous joints; synovial joints; synovial joints and levers; movements of synovial joints; features and anatomy of selected joints; development and aging of joints

Anatomy & Physiology Revealed: Topics include skull, synovial joint; temporomandibular joint; glenohumeral joint; elbow joint; hand; hip joint; knee joint; tibiofibular joint

Animation: Synovial joints

Muscle Tissue

INTEGRATE

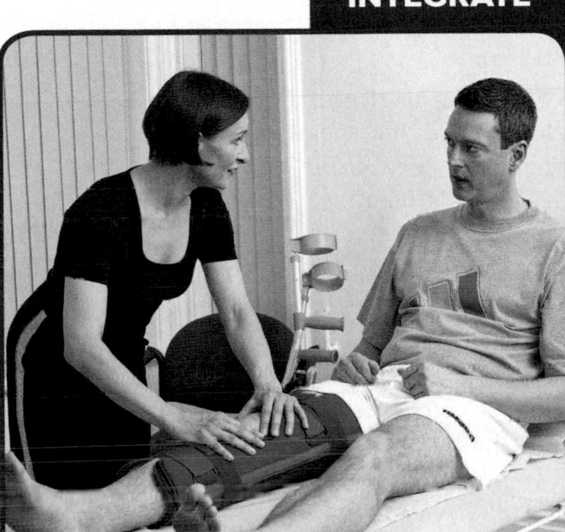

CAREER PATH

Sports Medicine Physician

Sports medicine physicians are highly educated and trained medical doctors who work with athletes and physically active individuals to improve their performance. Their education following medical school includes specializing in pediatrics, internal medicine, emergency medicine, neuromusculoskeletal systems, or rehabilitation with an additional two years of training in an accredited fellowship program in sports medicine. Sports medicine physicians earn the Certificate of Added Qualifications in Sports Medicine after passing the national Sports Medicine certification examination.

Those in this medical care specialty focus on the prevention, diagnosis, and treatment of acute and overuse chronic injuries, and develop and administer workout programs for maximizing athletic conditioning. They are also interested in enhancing their patients' overall health, which includes providing nutritional guidance for maintaining health and maximizing athletic performance. Here we see a sports medicine physician providing care for the treatment of a knee injury.

Anatomy & Physiology REVEALED®
aprevealed.com

Module 6: Muscular System

When we hear the word *muscle,* most of us think of the muscles that move the skeleton. Over 700 skeletal muscles have been named, and together they form the **muscular system.** Skeletal muscles, however, are not the only places where muscle tissue is found. Muscle tissue is distributed almost everywhere in the body and is responsible for the movement of materials within and throughout the body. This vital tissue propels the food we eat through the gastrointestinal tract, expels the waste products we produce, adjusts the diameter of blood vessels to regulate blood pressure, and pumps blood to body tissues.

The three types of muscle tissue—skeletal muscle, cardiac muscle, and smooth muscle—were first introduced and compared in section 5.3. Here we describe the details of skeletal muscle anatomy and physiology. The chapter finishes with a brief description of cardiac muscle (which is described in detail in section 19.3f) and a general discussion about smooth muscle. The structure and actions of individual skeletal muscles of the muscular system are explained in chapter 11.

10.1 Introduction to Skeletal Muscle

Skeletal muscle typically composes 40–50% of the weight of a healthy adult. It is primarily attached to the skeleton but is also found, for example, at the openings of the gastrointestinal and urinary tracts. We begin our discussion on skeletal muscle by describing both its general functions and the characteristics of the skeletal muscle cells that primarily compose it.

10.1a Functions of Skeletal Muscle

LEARNING OBJECTIVE

1. Explain the five general functions of skeletal muscle.

The hundreds of skeletal muscles within your body perform a wide range of functions. These include

- **Body movement.** Contraction of your skeletal muscles generates large body movements, such as those of walking, and the smaller, more precise body movements such as picking up an object. It is also responsible for the highly developed movements involved in communicating that occur when speaking, writing, and changing facial expressions; the movements associated with breathing (see section 23.5b); and those involved in the voluntary phase of swallowing (see section 26.2c).
- **Maintenance of posture.** Contraction of specific skeletal muscles stabilizes your trunk, pelvis, legs, neck, and head to keep you erect. These postural muscles contract continuously when you are awake to keep you from collapsing.
- **Protection and support.** Skeletal muscle is arranged in layers within the walls of the abdominal cavity (see figure 11.16) and the floor of the pelvic cavity (see figure 11.17). These layers of muscle protect the internal organs and support their normal position within the abdominopelvic cavity.
- **Regulating elimination of materials.** Circular muscle bands, called **sphincters** (sfingk′ter; *sphincter* = a band) contract and relax to regulate passage of material. These skeletal muscle sphincters at the **orifices** (or′i-fis; *orificium* = opening) of the gastrointestinal and urinary tracts allow you to voluntarily control the expulsion of feces and urine, respectively (see figures 26.23*b* and 24.28).
- **Heat production.** Energy is required for muscle tissue contraction, and heat is always produced by this energy use (the second law of thermodynamics; see section 3.1c). Thus, muscles are like small furnaces that continuously generate heat and function to help maintain your normal body temperature. You shiver when you are cold because involuntary skeletal muscle contraction gives off heat. Likewise, you sweat during exercise to release the additional heat produced by your working muscles (see sections 1.6b and 6.1d).

WHAT DID YOU LEARN?

1. What are the five major functions of skeletal muscle?

10.1b Characteristics of Skeletal Muscle

LEARNING OBJECTIVE

2. Describe the five characteristics of skeletal muscle.

Skeletal muscle is composed primarily of muscle cells that exhibit these characteristics—excitability, conductivity, contractility, extensibility, and elasticity:

- **Excitability** is the ability of a cell to respond to a stimulus (e.g., chemical, stretch). The stimulus causes a local change in the resting membrane potential (see section 4.4) by triggering the movement of ions across the plasma membrane of the excitable cell. A skeletal muscle cell responds when its receptors bind neurotransmitter (acetylcholine), which is released from a motor neuron (see section 10.3a).
- **Conductivity** involves an electrical signal that is propagated along the plasma membrane as voltage-gated channels open sequentially during an *action potential.* These electrical signals functionally connect the plasma membrane of the muscle cell (where stimulation occurs) to the interior of the muscle cell (where contraction occurs; see section 10.3b).
- **Contractility** is exhibited when contractile proteins within skeletal muscle cells slide past one another. Contractility is what enables muscle cells to cause body movement and to perform the other functions of muscles (see section 10.3c). The excitability, conductivity, and contractility of muscle cells may be viewed collectively in figure 10.9 at step 1, step 2, and step 3, respectively.
- **Extensibility** is the lengthening of a muscle cell (see figure 10.25). This lengthening is possible because the contractile proteins slide past one another to decrease their degree of overlap. Muscle's extensibility is exhibited when we stretch our muscles, such as before exercising.
- **Elasticity** is the ability of a muscle cell to return to its original length following either shortening or lengthening of the muscle. Elasticity of muscle cells is dependent upon the release of tension in the springlike connectin protein associated with contractile proteins (see figure 10.5*b*).

WHAT DID YOU LEARN?

2. Explain the skeletal muscle characteristics of contractility, extensibility, and elasticity. How do these differ?

INTEGRATE

CONCEPT CONNECTION

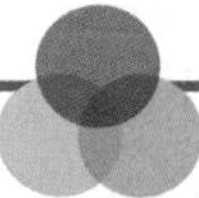

The characteristic of *excitability* is exhibited in other body cells—such as nerve cells, called neurons, which may respond to a neurotransmitter (see section 12.8a) and sensory receptors that respond to a specific type of sensory stimulus (see section 16.1a).

10.2 Anatomy of Skeletal Muscle

A single muscle, such as the gracilis muscle on the medial thigh (see figure 11.1), may be composed of thousands of muscle cells that are typically as long as the entire muscle. Because of their potentially extraordinary length, skeletal muscle cells are often referred to as **muscle fibers** (or *myofibers*). Here we describe the gross anatomy of skeletal muscle, the microscopic anatomy of individual skeletal muscle fibers, and innervation of skeletal muscle fibers.

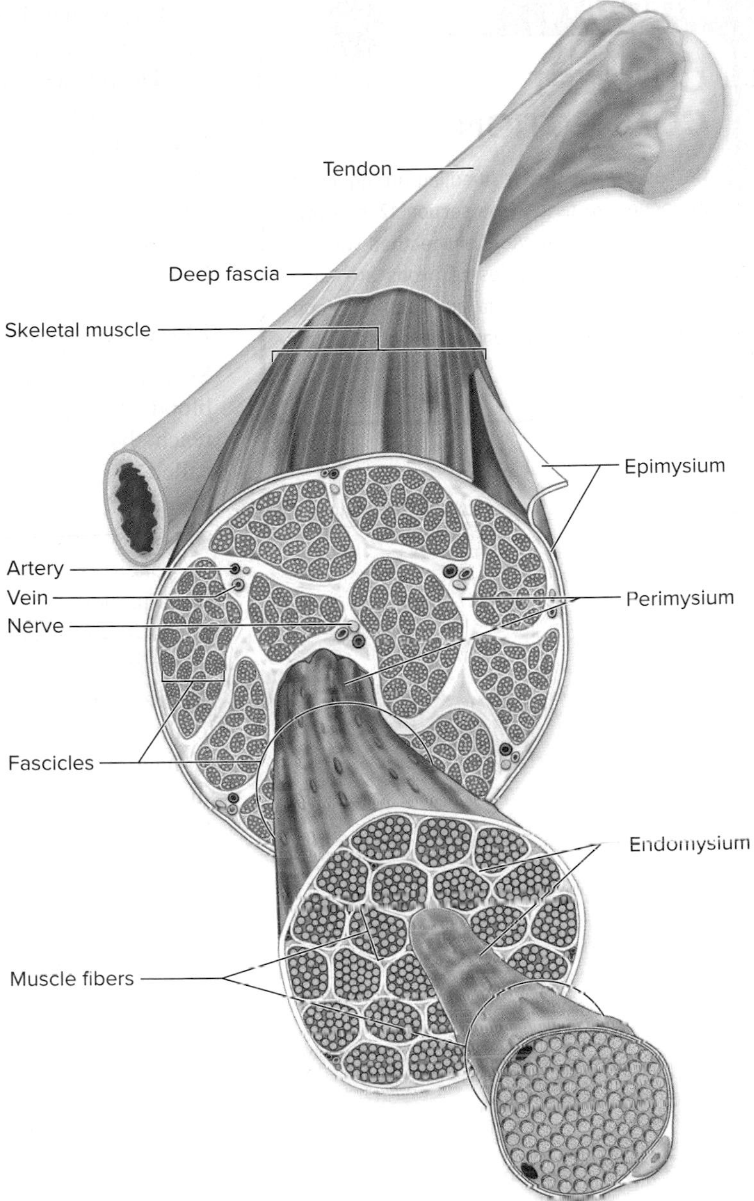

Figure 10.1 Structural Organization of Skeletal Muscle. A skeletal muscle consists of multiple fascicles, which are composed of bundles of muscle fibers. The entire muscle is ensheathed in a tough outer connective tissue called the epimysium. Each fascicle is wrapped in a connective tissue layer called the perimysium. Within fascicles, each muscle fiber is surrounded by a delicate connective tissue layer termed the endomysium. AP|R

10.2a Gross Anatomy of Skeletal Muscle

 **LEARNING OBJECTIVES**

3. Identify and describe the three connective tissue layers associated with a skeletal muscle.
4. Describe the structure and function of a tendon and an aponeurosis.
5. Explain the function of blood vessels and nerves serving a muscle.

A skeletal muscle is an **organ,** which, recall from section 1.4b, is two or more types of tissue that work together to perform a specific function. Each skeletal muscle is composed of skeletal muscle fibers, connective tissue layers, blood vessels, and nerves. The organization of a muscle is shown in **figure 10.1**. Observe the specific anatomic arrangement of muscle fibers within a muscle. Notice that many muscle fibers are bundled within a **fascicle** (fas′i-kl; *fascis* = bundle); and many fascicles are bundled within the whole skeletal muscle.

 **WHAT DO YOU THINK?**

1. List the structures of skeletal muscle from largest to smallest: muscle fiber, muscle, and fascicle.

Connective Tissue Components

Three layers of connective tissue are within muscles: the epimysium, the perimysium, and the endomysium. These layers provide protection and support, a means of attachment of the muscle to the skeleton or other structures within the body, and sites for distribution of blood vessels and nerves:

- The **epimysium** (ep-i-mis′ē-ŭm; *epi* = upon, *mys* = muscle) is a layer of dense irregular connective tissue (see section 5.2d) that surrounds the whole skeletal muscle. This fibrous tissue ensheathes the entire skeletal muscle to protect and support it like a tough leather sleeve.
- The **perimysium** (per-i-mis′ē-ŭm; *peri* = around) is a layer of dense irregular connective around each fascicle. These tough, fibrous connective tissue sleeves also provide protection and support, but to each bundle of muscle fibers.
- The **endomysium** (en′dō-mis′ē-ŭm; *endon* = within) is composed of areolar connective tissue that surrounds each muscle fiber. These more delicate coverings function to electrically insulate the muscle fibers.

The epimysium, perimysium, and endomysium collectively extend past the muscle fibers to form either a tendon or an aponeurosis. A **tendon** is a thick, cordlike structure composed of dense regular connective tissue, whereas an **aponeurosis** (ap′ō-nū-rō′sis; *apo* = from, *neuron* = sinew) is a thin, flattened sheet of dense regular connective tissue (see figures 11.5 and 11.16). Both tendons and aponeuroses attach a muscle either to a skeletal component (bone or ligament) or to fascia (described next). Imagine the typical scenario

INTEGRATE

LEARNING STRATEGY

The prefixes *myo, mys,* and *sarco* are used often in this chapter, and each refers to some feature of muscle (e.g., myofibril, epimysium, sarcolemma).

stimulated to differentiate and then fuse with a damaged skeletal muscle fiber to assist to a limited extent in its repair and regeneration.

Sarcolemma and T-tubules

The plasma membrane of a skeletal muscle fiber is called the **sarcolemma** (sar′kō-lem′ă; *lemma* = husk) (figure 10.3*a*). Deep invaginations of the sarcolemma, called **T-tubules** or **transverse** (trans-vers′; *trans* = across, *versus* = to turn) **tubules,** extend into the skeletal muscle fiber as a network of narrow, membranous tubules to the sarcoplasmic reticulum, which is the endoplasmic reticulum (ER) of the muscle (described shortly). Located within the membrane of both the sarcolemma (along its length) and the T-tubules are voltage-gated channels (figure 10.3*c*). These channels include both voltage-gated Na^+ channels and voltage-gated K^+ channels, which participate in conducting an electrical signal (an action potential) as described in section 10.3b. (Note: Channels within the plasma membrane are first introduced in section 4.3a, and voltage-gated and chemically gated channels are discussed in detail in section 12.6a.)

Myofibrils

Approximately 80% of the volume of a skeletal muscle fiber is composed of long, cylindrical structures termed **myofibrils** (mī′ō-fī′bril) (figure 10.3*a*). A skeletal muscle fiber contains hundreds to thousands of myofibrils. Each myofibril extends the entire length of the skeletal muscle fiber (and is about 1 to 2 micrometers in diameter). Note that each myofibril is composed of bundles of contractile proteins called *myofilaments* and is enclosed in portions of the sarcoplasmic reticulum (figure 10.3*b*).

Sarcoplasmic Reticulum

The **sarcoplasmic reticulum** (sar-kō′-plaz′mik re-tik′ū-lŭm; *rete* = a net) is an internal membrane complex that is similar to the smooth endoplasmic reticulum of other cells (see section 4.6a). Segments of the sarcoplasmic reticulum (SR) fit around the myofibril like a sleeve of membrane netting. At either end of individual sections of the sarcoplasmic reticulum are blind sacs called **terminal cisternae** (sis-ter′nē; sing., sis-ter′nă; *cista* = a box), which are much like the hem of a sleeve. Terminal cisternae serve as the reservoirs for calcium ions (Ca^{2+}) and are immediately adjacent to each T-tubule (figure 10.3*d*). Together, two terminal cisternae and a centrally located T-tubule form a structure called a **triad.** Within the triad, the T-tubule membrane contains **voltage-sensitive Ca^{2+} channels** (dihydropyridine receptors), which are responsive to electrical signals (i.e., action potentials). The terminal cisternae membrane of the sarcoplasmic reticulum contain **Ca^{2+} release channels** (ryanodine receptors). It is here in the triad that the connection occurs between the electrical signals (action potentials) propagated along the sarcolemma and T-tubule and the release of calcium from the sarcoplasmic reticulum (see section 10.3b). This release of Ca^{2+} initiates muscle contraction, as described in section 10.3c.

Also embedded within the membrane of the sarcoplasmic reticulum are Ca^{2+} pumps, which move Ca^{2+} from the cytosol into the sarcoplasmic reticulum, where it is stored bound to specialized proteins called **calmodulin** (kal-mod′ū-lin) and **calsequestrin** (kal′sē-kwes′trin). Calcium pumps function through primary active transport (see section 4.3c) to maintain low cytosol levels of calcium. These pumps return Ca^{2+} to the terminal cisternae of the sarcoplasmic reticulum following its release to initiate muscle contraction.

Myofilaments

Myofilaments (mī′ō-fil′ă-ment; *filum* = thread) are contractile proteins that are bundled within myofibrils (figure 10.3*b*). A myofilament is not as long as a myofibril; rather, it takes *many* successive units of myofilaments to extend the entire length of the myofibril. Myofibril bundles contain two types of myofilaments: thick filaments and thin filaments **(figure 10.4)**.

Thick Filaments **Thick filaments** (or *thick myofilaments*) are assembled from bundles of 200 to 500 **myosin** protein molecules (figure 10.4*a*). Each myosin protein consists of two strands; each strand has a globular head and an elongated tail. The myosin head contains a binding site for actin of the thin filaments. The head also has a catalytic ATPase site where adenosine triphosphate (ATP) attaches and is split into

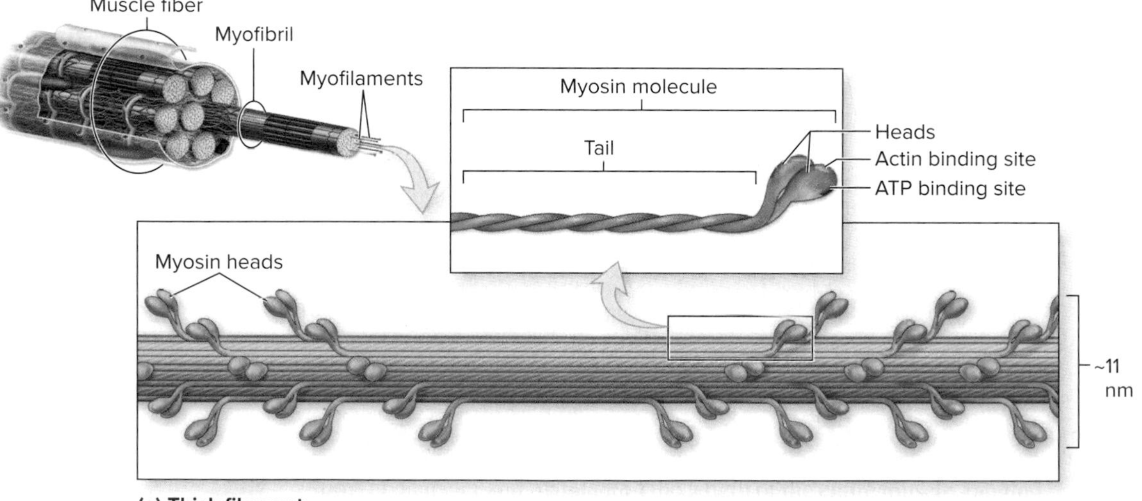

(a) Thick filament

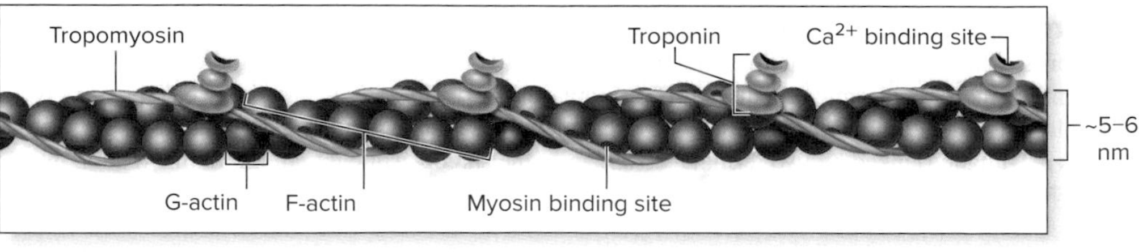

(b) Thin filament

Figure 10.4 Molecular Structure of Thick and Thin Filaments. Myofilaments, which include thick filaments and thin filaments, are the contractile proteins bundled within myofibrils. (*a*) A thick filament consists of 200 to 500 myosin protein molecules. (*b*) A thin filament is composed of actin, tropomyosin, and troponin proteins. **AP|R**

adenosine diphosphate (ADP) and phosphate (P_i). (It is because the head of myosin functions as an ATPase enzyme that myosin is often referred to more specifically as **myosin ATPase.**) The tails of two strands of a myosin molecule are intertwined. Each myosin molecule composing a thick filament is oriented so that its tails point toward the center of the thick filaments and its heads point toward the ends of the thick filaments. You may find it helpful to think of the myosin protein molecules as two intertwined golf clubs, where many are grouped together with the golf club shafts in the center and the club heads on each end.

Thin Filaments **Thin filaments** (or *thin myofilaments*) are approximately half of the diameter of thick filaments (about 5 to 6 nanometers). Thin filaments are primarily composed of two strands of **actin** protein twisted around each other to form a helical shape (figure 10.4*b*). In each strand of actin, many (about 300 to 400) small, spherical molecules (G, or globular, actin) are connected to form a fibrous strand (F, or filamentous, actin). F-actin resembles two beaded necklaces that are twisted and intertwined together, with G-actin as the individual beads. Each G-actin molecule has a significant feature called a **myosin binding site.** The myosin head attaches to the myosin binding site of actin during muscle contraction.

Tropomyosin and troponin are regulatory proteins associated with thin filaments. Together they form the **troponin-tropomyosin complex. Tropomyosin** (trō-pō-mī′ō-sin) is a short, thin, twisted filament that is a "stringlike" protein. Consecutive tropomyosin molecules cover small regions of the actin strands, including the myosin binding sites in a noncontracting muscle. **Troponin** (trō′pō-nin) is a globular, or "ball-like," protein attached to tropomyosin. Troponin contains the binding site for Ca^{2+}.

Organization of a Sarcomere

Myofilaments within myofibrils are arranged in repeating, microscopic, cylindrical units (2 micrometers in length) called **sarcomeres** (sar′kō-mēr; *meros* = part). **Figure 10.5*a*** shows several repeating sarcomeres in a section of a myofibril within a skeletal muscle fiber. The number of sarcomeres varies with the length of the myofibril within the skeletal muscle fiber. Each sarcomere is composed of overlapping thick filaments and thin filaments.

A two-dimensional, longitudinal view of a cylindrical sarcomere is shown in figure 10.5*b*. Here we see that each sarcomere is delineated at both ends by Z discs. **Z discs** (also called *Z lines*) are composed of specialized proteins that are positioned perpendicular to the myofilaments and serve as anchors for the thin filaments. Although the Z disc appears as a flat disc when the myofibril is viewed from its end, only the edge of the disc is visible in a side view, and it sometimes looks like a zigzagged line.

The thick filaments and thin filaments overlap within a sarcomere, forming the following regions:

- **I bands** extend from both directions of a Z disc and are bisected by the Z disc. These end regions contain only thin filaments; this region appears light when viewed with a microscope. At

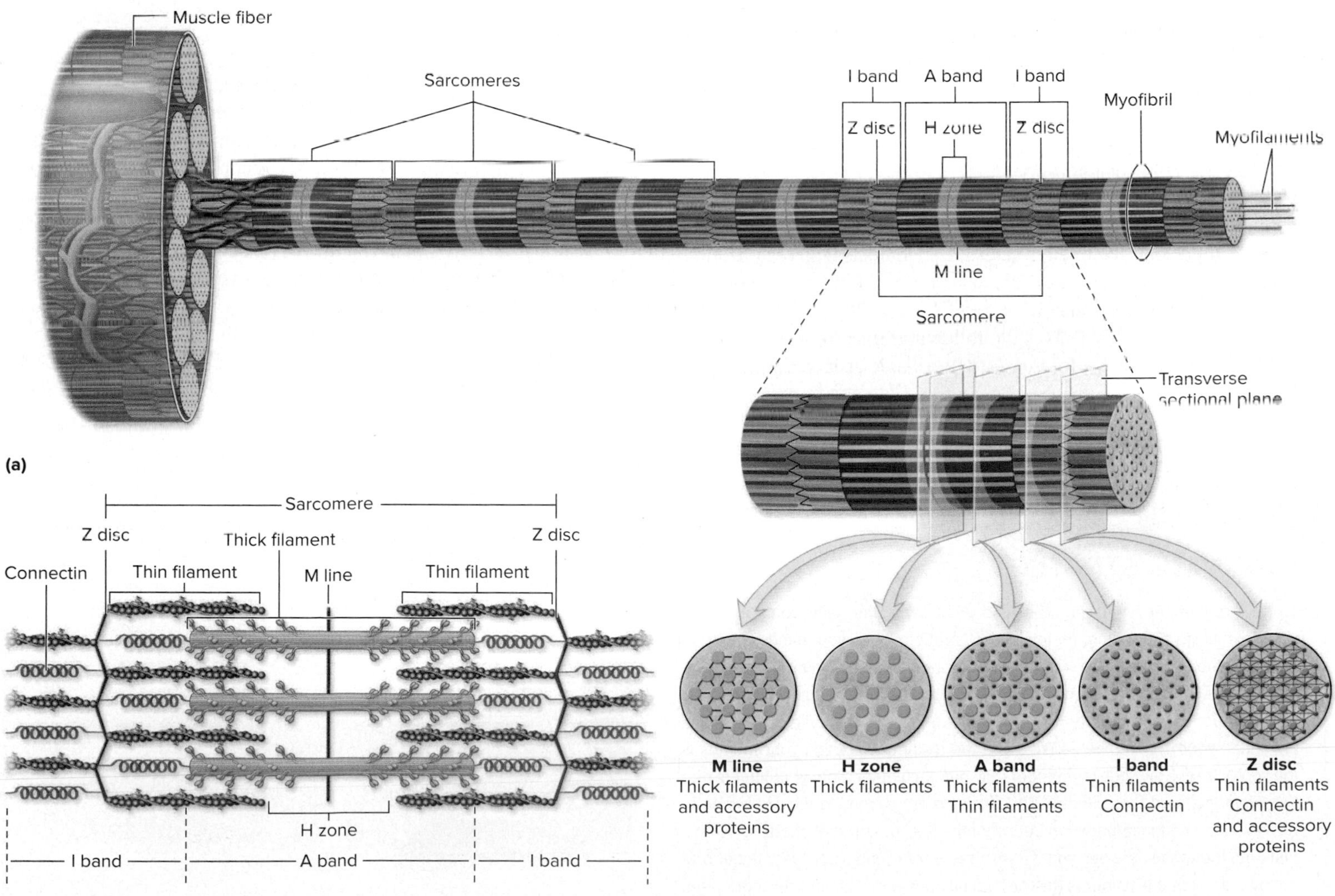

Figure 10.5 Structure of a Sarcomere. (*a*) Numerous sarcomeres extend the length of a myofibril. (*b*) Longitudinal section of a sarcomere. (*c*) Cross sections of different regions of a sarcomere. AP|R

INTEGRATE

LEARNING STRATEGY

You can remember A bands and I bands in the following manner: The word *dark* contains the letter "A," just as A bands are *dark* colored. In contrast, the word *light* contains the letter "I," reminding you that I bands are *light* colored.

maximal muscle shortening, the thin filaments are pulled parallel along the thick filaments, causing the I bands to disappear.

- The **A band** is the central region of a sarcomere that contains the entire thick filament. Thin filaments partially overlap the thick filament on each end of an A band. The A band appears dark when viewed with a microscope. The A band does not change in length during muscle contraction.
- The **H zone** (also called the *H band*) is the most central portion of the A band in a resting sarcomere. This region does not have thin filament overlap; only thick filaments are present. During maximal muscle shortening, this zone disappears when the thin filaments are pulled past thick filaments.
- The **M line** is a thin transverse protein meshwork structure in the center of the H zone. It serves as an attachment site for the thick filaments and keeps the thick filaments aligned during contraction and relaxation events.

The repeating light and dark bands of the overlapping myofilaments form unique striped patterns within a skeletal muscle fiber called **striations.** Striations are visible when viewing a longitudinal section of skeletal muscle tissue in an image produced by either a light microscope (see figure 10.6*b*) or an electron microscope (see figure 10.15). This striated appearance is due to both the size and density differences between thin filaments and thick filaments.

Figure 10.5*c* shows cross sections through various regions in a sarcomere. It presents the relative sizes, arrangements, and organization of thick and thin filaments at different locations within the sarcomere. Notice that in a cross section of an A band the arrangement of thick filaments relative to thin filaments is the following: Each thin filament has three thick filaments around it that form a triangle at its periphery, and each thick filament is sandwiched by six thin filaments.

WHAT DO YOU THINK?

2 If a muscle is contracted and shortening, what happens to the following: (a) width of the A band, (b) width of the H zone, (c) relationship of the Z discs, and (d) width of the I band?

Other Structural and Functional Proteins Other proteins have structural and functional roles within muscle fibers. These include connectin and dystrophin (only connectin is shown in figure 10.5).

Connectin (kon-nek′tin), also called *titin,* is a "cablelike" protein that extends from the Z discs to the M line through the core of each thick filament (figure 10.5*b*). It stabilizes the position of the thick filament and maintains thick filament alignment within a sarcomere. Additionally, portions of the connectin molecules are coiled and "springlike" so that during sarcomere shortening they are compressed to produce passive tension. This passive tension is then released to return the sarcomere to its normal resting length. Thus, connectin contributes to skeletal muscle fiber elasticity (see section 10.1b).

Dystrophin (dis-trō′fin) is part of a protein complex that anchors myofibrils that are adjacent to the sarcolemma to proteins within the sarcolemma. These proteins of the sarcolemma also extend to the connective tissue of the endomysium that encloses the muscle fiber. Thus, dystrophin links internal myofilament proteins of a muscle fiber to external proteins. The genetic disorder of *muscular dystrophy* is caused by abnormal structure, or amounts, of dystrophin protein (see Clinical View 10.1: "Muscular Dystrophy").

Mitochondria and Other Structures Associated with Energy Production

Skeletal muscle fibers have a great demand for energy and contain several components that facilitate the production of ATP (see section 10.4a). Skeletal muscle fibers have abundant mitochondria for aerobic cellular respiration (see section 3.4); a typical skeletal muscle fiber contains approximately 300 mitochondria. The fibers also contain glycogen stores (granules called glycosomes) for use as an immediate fuel molecule. **Myoglobin** (mī-ō-glō′bin) is a molecule unique to muscle tissue. Myoglobin is a reddish, globular protein that is somewhat similar to hemoglobin. It binds oxygen when the muscle is at rest and releases it for use during muscular contraction. This additional source of oxygen provides the means to enhance aerobic cellular respiration and the production of ATP. Skeletal muscle fibers also contain another type of molecule called *creatine phosphate* (see figure 3.4*c*). Creatine phosphate provides muscle fibers with a very rapid means of supplying ATP.

INTEGRATE

LEARNING STRATEGY

To help remember what shortens (and what stays the same length) during skeletal muscle shortening, try this demonstration. Place your hands with your palms facing you and your thumbs pointed upward, as shown in the figure. Now imagine there are cylinders placed between and partially overlapping your fingers. Your fingers represent the thin filaments, the cylinders the thick filaments, your thumbs the *Z discs,* and the distance between your thumbs is the *sarcomere*. Now slide your fingers toward each other. Both your fingers (the *thin filaments*) and the cylinders (the *thick filaments*) remain the same length. However, the distance between your thumbs (the sarcomere) shortens. Notice that both the space between your fingers (the *H zone*) and where your fingers did not overlap with the cylinders (the *I band*) disappear during muscle shortening. Thus, note that the length of each individual structure stays the same but the relative relationship between structures changes.

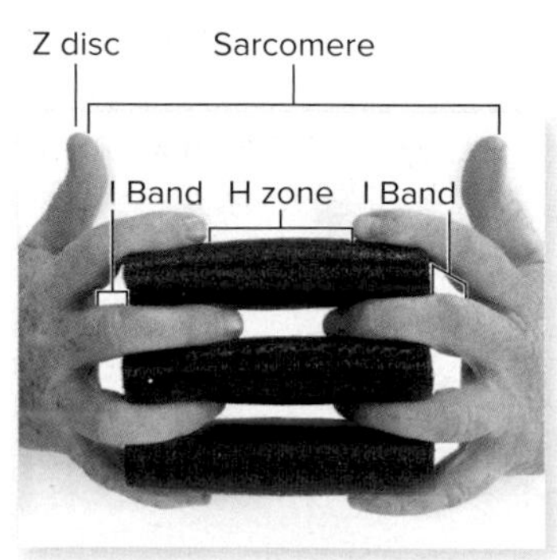

(a) Relaxed sarcomere

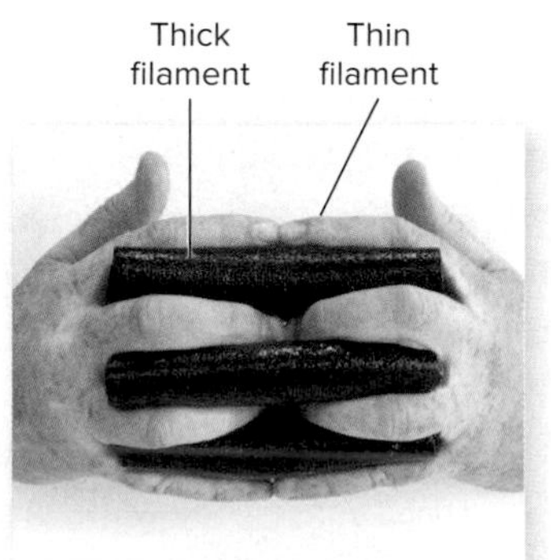

(b) Contracted sarcomere

(*a, b*) ©McGraw-Hill Education/Mark Dierker

INTEGRATE

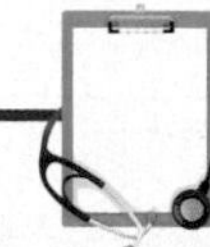

CLINICAL VIEW 10.1

Muscular Dystrophy

Muscular dystrophy (dis′tro-fe) is a collective term for several hereditary diseases in which the skeletal muscles degenerate, lose strength, and are gradually replaced by adipose and fibrous connective tissue. In a viscous cycle, the new connective tissues impede blood circulation, which further accelerates muscle degeneration.

Duchenne muscular dystrophy (DMD) is the most common form of the illness. It is almost exclusively a disease of males and occurs in about 1 in 3500 live births. DMD results from the expression of a sex-linked recessive allele (see section 29.9c), which is the gene that directs the synthesis of dystrophin. In DMD, dystrophin either has an abnormal structure or is produced in insufficient amounts. The defective, reduced, or absent dystrophin results in an unstable sarcolemma, which is susceptible to damage by forces generated during muscle contraction. Excess calcium ions then enter the muscle fibers, damaging the contractile proteins with an accompanying loss of muscle fibers.

For these individuals, muscular difficulties become apparent in early childhood. Walking is a problem; the child falls frequently and has difficulty standing up again. The hips are affected first, followed by the lower limbs, and eventually the abdominal and vertebral muscles. Muscular atrophy causes shortening of the muscles, which results in postural abnormalities such as scoliosis (lateral curvature of the spine; see Clinical View 8.3: "Spinal Curvature Abnormalities"). DMD is an incurable disease, with patients confined to a wheelchair by adolescence. An individual with DMD rarely lives beyond the age of 30, and death typically results from respiratory or heart complications.

The details of how ATP is provided to meet the high energy needs of skeletal muscle fibers are described in section 10.4a.

WHAT DID YOU LEARN?

4 Draw and label a diagram of a sarcomere.

5 Place the following gross anatomic and microscopic anatomic structures in order from largest to smallest: fascicle, myofibril, myofilament, muscle, muscle fiber, and sarcomere. Describe their anatomic relationship.

10.2c Innervation of Skeletal Muscle Fibers

LEARNING OBJECTIVES

11. Define a motor unit, and describe its distribution in a muscle and why it varies in size.
12. Describe the three components of a neuromuscular junction.

The anatomic relationship of skeletal muscle fibers and somatic motor neurons that control them is described in this section.

Motor Unit

Somatic motor neurons are nerve cells that transmit electrical signals (nerve signals) from the brain or spinal cord to control skeletal muscle activity (see section 12.1b). The axon of each motor neuron divides into many individual branches to innervate numerous skeletal muscle fibers. A single motor neuron and the skeletal muscle fibers it controls is called a **motor unit (figure 10.6)**.

The number of skeletal muscle fibers a single motor neuron innervates—and thus the size of the motor unit—varies and can range from small motor units that have less than five muscle fibers to large motor units that have several thousand muscle fibers. The size of the motor unit determines the degree of control. There is an inverse relationship between the size of a motor unit and the degree of control. For example, motor neurons innervating extrinsic eye muscles (see section 11.3b) are small because greater control is essential in the muscles that move the eye. In contrast, a single motor neuron controls several thousand individual skeletal muscle fibers in the power-generating muscles in our lower limbs, where less precise control is required.

The skeletal muscle fibers of a motor unit are not clustered within one area of a muscle, but rather are dispersed throughout most of a muscle. Normally, the stimulation of a motor unit does not produce a strong contraction in a localized area within the muscle, but a weak contraction over a wide area.

Neuromuscular Junctions

Each skeletal muscle fiber is typically described as having one neuromuscular junction. A **neuromuscular** (nūr-ō-mŭs′kū-lăr) **junction** is the specific location, usually in the mid-region of the skeletal muscle fiber where it is innervated by a motor neuron **(figure 10.7*a*)**.

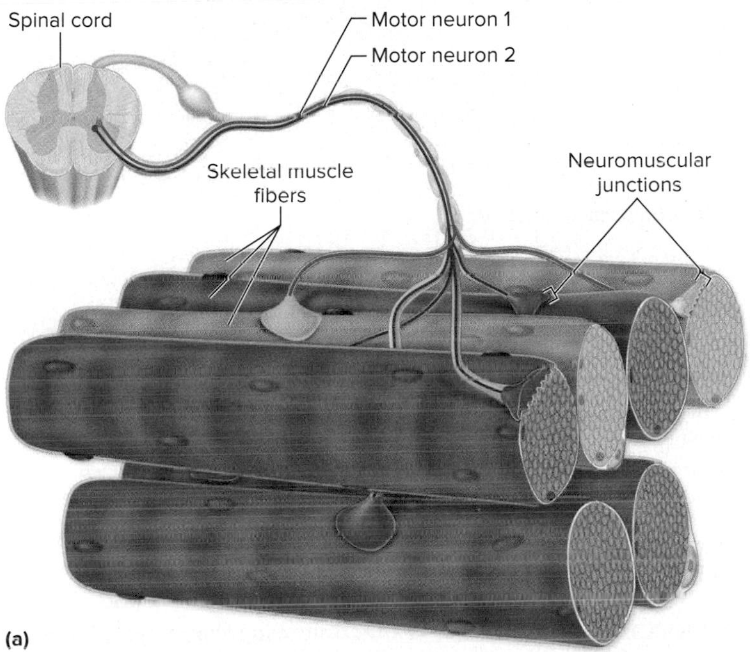

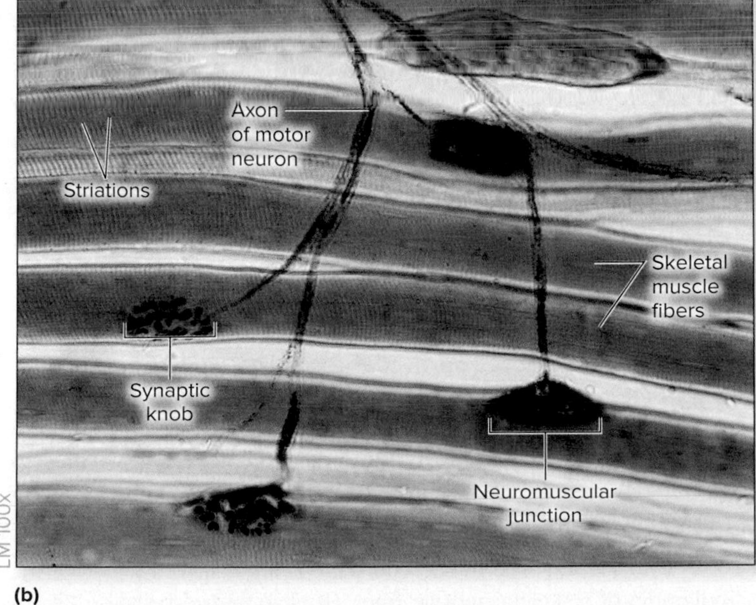

Figure 10.6 A Motor Unit. (*a*) A motor unit is a motor neuron and all of the skeletal muscle fibers it innervates. Different colors distinguish the two motor units in this figure. (*b*) Light micrograph of the terminal ends of a motor neuron axon and skeletal muscle fibers in a motor unit. AP|R

(*b*) ©Dr. Thomas Caceci, Virginia-Maryland Regional College of Veterinary Medicine

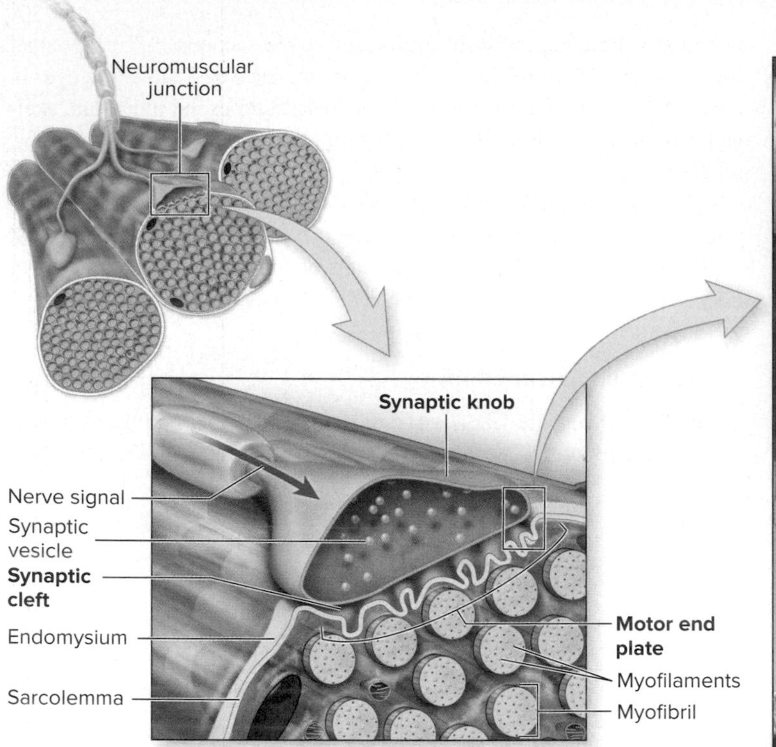

(a) Neuromuscular junction

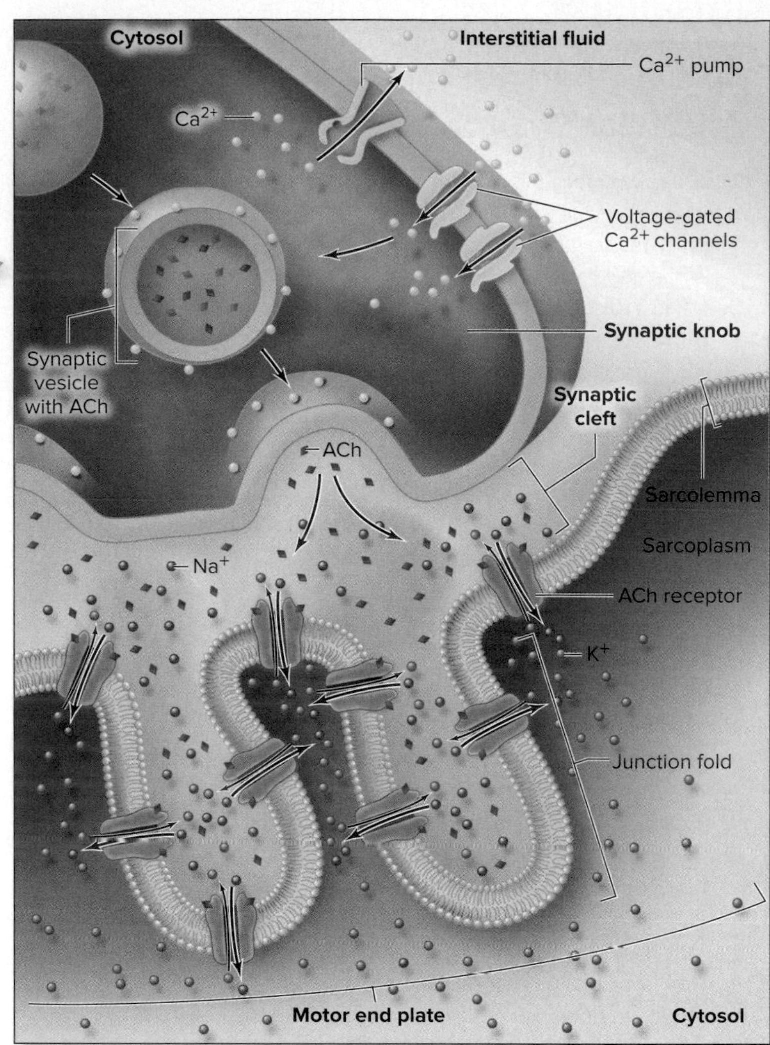

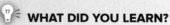

(b) Close-up of neuromuscular junction

Figure 10.7 Structure and Organization of a Neuromuscular Junction. The synaptic knob of an axon meets a skeletal muscle fiber to form a neuromuscular junction. (*a*) Three primary components are within a neuromuscular junction and include a synaptic knob, motor end plate, and synaptic cleft. (*b*) Synaptic knobs house synaptic vesicles containing the neurotransmitter acetylcholine (ACh). Embedded in the plasma membrane of the synaptic knob are Ca^{2+} pumps and voltage-gated Ca^{2+} channels. The motor end plate contains ACh receptors, which are chemically gated ion channels. AP|R

The neuromuscular junction has the following parts: synaptic knob, motor end plate, and synaptic cleft.

Synaptic Knob The **synaptic** (si-nap′tik) **knob** of a motor neuron is an expanded tip of an axon. Where the axon nears the sarcolemma of a muscle fiber, the synaptic knob enlarges and flattens to cover a relatively large surface area of the sarcolemma. The synaptic knob cytosol houses numerous **synaptic vesicles** (small membrane sacs) filled with molecules of the neurotransmitter **acetylcholine** (a-sē′til-kō′lēn) **(ACh).**

Several points can be made about synaptic knobs (figure 10.7*b*). First, Ca^{2+} pumps are embedded within the plasma membrane of the synaptic knob. Prior to the arrival of the electrical signal (nerve signal) at the synaptic knob, Ca^{2+} pumps within its plasma membrane have established a Ca^{2+} concentration gradient, with more Ca^{2+} outside the synaptic knob than inside it. Second, voltage-gated Ca^{2+} channels are also embedded in the membrane of the synaptic knob. Opening of these channels allows Ca^{2+} to flow down its concentration gradient from the interstitial fluid into the synaptic knob, which will trigger exocytosis of acetylcholine from the vesicles. Third, vesicles are normally repelled from the synaptic knob plasma membrane.

Motor End Plate The **motor end plate** is a specialized region of the sarcolemma of a skeletal muscle fiber. (It is so named because "motor end" reflects that it is located at the end of a motor neuron and "plate" describes its large, saucerlike appearance.) It has numerous folds and indentations (junction folds) to increase the membrane surface area covered by the synaptic knob. The motor end plate has vast numbers of **ACh receptors.** These plasma membrane protein channels are *chemically gated ion channels* (see section 4.3a). Binding of ACh opens these channels, allowing Na^+ entry into the muscle fiber and K^+ to exit. ACh receptors are like doors; ACh is the only "key" to open these receptor doors.

Synaptic Cleft The **synaptic cleft** is an extremely narrow (30 nanometers), fluid-filled space separating the synaptic knob and the motor end plate. The enzyme **acetylcholinesterase** (a-sē′til-kō′lēn-es′ter-ās) **(AChE)** resides within the synaptic cleft (not shown in figure 10.7) and quickly breaks down ACh molecules following their release into the synaptic cleft. (See the electron micrograph of neuromuscular junction in figure 12.22. A detailed view of the breakdown of acetylcholine by acetylcholinesterase is shown in figure 12.25.)

WHAT DID YOU LEARN?

6. What is a motor unit, and why does it vary in size?
7. Diagram and label the anatomic structures of a neuromuscular junction.

10.2d Skeletal Muscle Fibers at Rest

LEARNING OBJECTIVE

13. Describe a skeletal muscle fiber at rest.

Skeletal muscle fibers exhibit several significant features when the muscle is at rest. See **figure 10.8** as your read through this section.

One essential feature of skeletal muscle fibers is the electrical charge difference across the sarcolemma; the cytosol right inside the

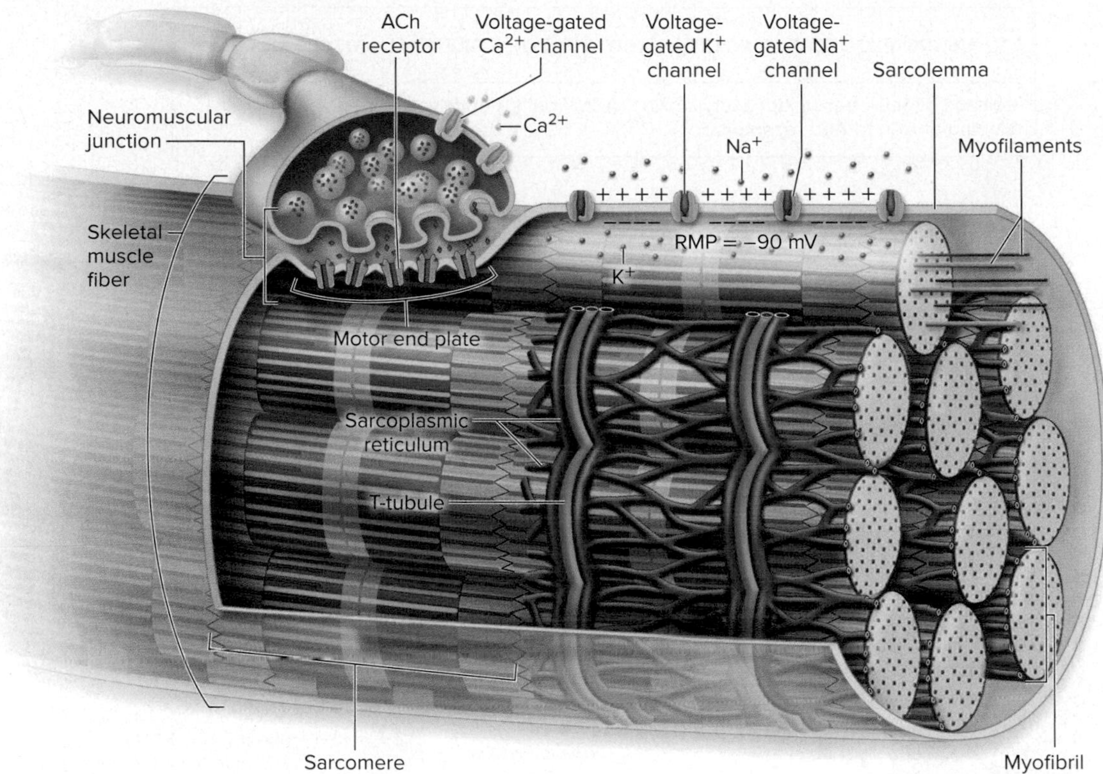

Figure 10.8 Skeletal Muscle Fiber at Rest. At rest, a skeletal muscle fiber has a negative resting membrane potential (RMP) of –90 mV, with more Na^+ outside the cell and more K^+ inside the cell. All gated channels are closed, Ca^{2+} is stored within the sarcoplasmic reticulum, and the contractile proteins (myofilaments) are in their relaxed position.

plasma membrane is relatively negative in comparison to the interstitial fluid outside of the cell. This electrical charge difference when the cell is at rest is called the **resting membrane potential (RMP)** (see section 4.4). Skeletal muscle fibers have an RMP of about –90 millivolts (mV). An RMP is established and maintained by both leak channels and Na^+/K^+ pumps (not shown in figure 10.8). The primary function of the Na^+/K^+ pumps is to maintain the concentration gradients for Na^+ (with more Na^+ outside the cell) and K^+ (with more K^+ inside the cell).

The acetylcholine receptors (chemically gated ion channels) within the motor end plate and the voltage-gated Na^+ channels and voltage-gated K^+ channels in the sarcolemma and T-tubules are closed, Ca^{2+} ions are stored within the terminal cisternae of the sarcoplasmic reticulum, and the contractile proteins (myofilaments) within the sarcomeres are in their relaxed position.

WHAT DID YOU LEARN?

8 Describe the distribution or Na^+ and K^+ at the sarcolemma.

INTEGRATE

CONCEPT CONNECTION

Either an impaired ability or an inability of the nervous system to stimulate skeletal muscle fibers can result in decreased or absent muscle fiber contraction. Causes include damage to any the following: (1) components of the *brain,* which initiate nerve impulses for muscle contraction (see Clinical View 13.9: "Cerebrovascular Accident"); (2) the *spinal cord,* which relays nerve impulses from the brain to many of the skeletal muscles (see Clinical View 14.3: "Treating Spinal Cord Injuries"); and (3) *somatic motor neurons,* which stimulate the skeletal muscle fibers (see Clinical View 12.5: "Neurotoxicity" and Clinical View 14.2: "Poliomyelitis"). Toxins can also interfere with skeletal muscle contraction (see Clinical View 10.3: "Muscular Paralysis and Neurotoxicity").

10.3 Physiology of Skeletal Muscle Contraction

A motor neuron stimulates skeletal muscle fibers. This stimulation ultimately results in the interaction between myofilaments within the skeletal muscle fibers to produce tension. The resulting tension is exerted on the portions of the skeleton (or other body structures) where the muscle is attached to cause movement in the body.

The anatomic structures and associated physiologic processes of skeletal muscle contraction include the events that occur at the (1) neuromuscular junction; (2) sarcolemma, T-tubules, and sarcoplasmic reticulum; and (3) sarcomeres. An overview of these processes is included in **figure 10.9**.

INTEGRATE

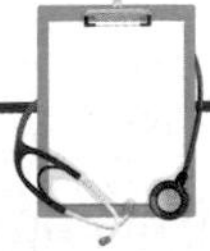

CLINICAL VIEW 10.2

Myasthenia Gravis (MG)

Myasthenia (mī′as-thē′nē-ă; *asthenia* = weakness) **gravis (MG)** is an autoimmune disease that occurs in about 1 in 10,000 people, primarily women between 20 and 40 years of age. A person's own antibodies attack the neuromuscular junctions, binding ACh receptors into clusters. The abnormally clustered ACh receptors are removed from the muscle fiber sarcolemma by endocytosis, thus significantly diminishing the number of receptors within the sarcolemma. The resulting decreased muscle stimulation causes rapid fatigue and muscle weakness. Eye and facial muscles are often attacked first, producing double vision and drooping eyelids. These symptoms are usually followed by swallowing problems, limb weakness, and overall low physical stamina. Some patients with MG have a normal life span, whereas others die within a short time from paralysis of the respiratory muscles.

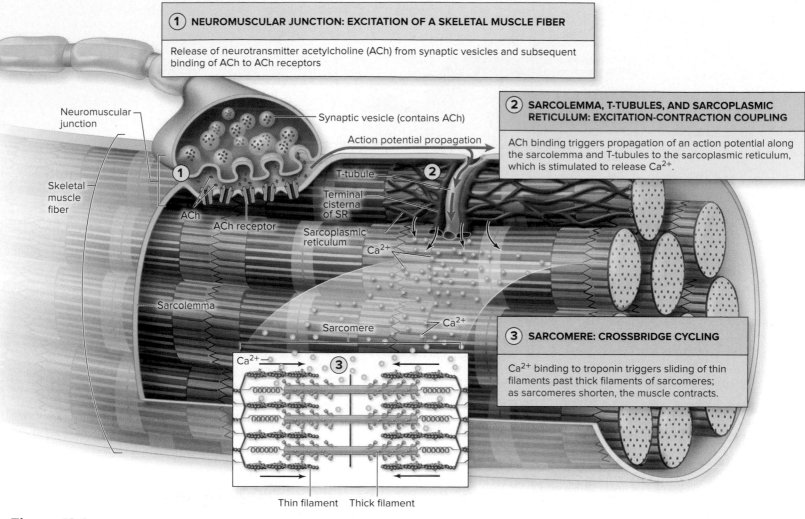

① NEUROMUSCULAR JUNCTION: EXCITATION OF A SKELETAL MUSCLE FIBER

Release of neurotransmitter acetylcholine (ACh) from synaptic vesicles and subsequent binding of ACh to ACh receptors

② SARCOLEMMA, T-TUBULES, AND SARCOPLASMIC RETICULUM: EXCITATION-CONTRACTION COUPLING

ACh binding triggers propagation of an action potential along the sarcolemma and T-tubules to the sarcoplasmic reticulum, which is stimulated to release Ca^{2+}.

③ SARCOMERE: CROSSBRIDGE CYCLING

Ca^{2+} binding to troponin triggers sliding of thin filaments past thick filaments of sarcomeres; as sarcomeres shorten, the muscle contracts.

Neuromuscular junction
Synaptic vesicle (contains ACh)
Action potential propagation
Skeletal muscle fiber
T-tubule
Terminal cisterna of SR
ACh
ACh receptor
Sarcoplasmic reticulum
Ca^{2+}
Sarcolemma
Sarcomere
Thin filament
Thick filament

Figure 10.9 Overview of Events in Skeletal Muscle Contraction. Skeletal muscle contraction involves the physiologic events that occur (1) at the neuromuscular junction, (2) along the sarcolemma and T-tubules to the sarcoplasmic reticulum of a skeletal muscle fiber, and (3) within a sarcomere. AP|R

10.3a Neuromuscular Junction: Excitation of a Skeletal Muscle Fiber

LEARNING OBJECTIVE

14. Explain the events that lead to release of the neurotransmitter ACh from a motor neuron.

The first physiologic event of skeletal muscle contraction is muscle fiber *excitation* by a somatic motor neuron—an event that occurs at the neuromuscular junction and results in release of ACh and its subsequent binding to ACh receptors. These events are summarized in **figure 10.10**.

Calcium Entry at Synaptic Knob

A **nerve signal** (or *nerve impulse*), which is the electrical signal that is propagated down an axon, is sent along a motor neuron of the somatic nervous system. (Nerve signals are discussed in detail in section 12.8c.) The nerve signal triggers the opening of voltage-gated Ca^{2+} channels within the synaptic knob plasma membrane, and calcium moves down its concentration gradient from the interstitial fluid through the open channels into the synaptic knob. Calcium binds with membrane proteins (synaptotagmin [not shown]) exposed on the external surface of synaptic vesicles (step 1a).

Release of ACh from Synaptic Knob

The binding of Ca^{2+} to synaptic vesicles triggers the merging of synaptic vesicles with the synaptic knob plasma membrane, resulting in exocytosis of ACh into the synaptic cleft. Acetylcholine is released from approximately 300 vesicles per nerve signal with each vesicle releasing thousands of molecules of ACh (step 1b).

Binding of ACh at Motor End Plate

ACh diffuses across the fluid-filled synaptic cleft to bind with ACh receptors within the motor end plate. This causes excitation of a skeletal muscle fiber (step 1c).

Note that nerve signals are repeatedly propagated along the motor axon (at about 10 to 40 times per second). Thus, these events (steps 1a–1c) will continue until stimulation of the skeletal muscle fiber by the neuron ceases (stops) and acetylcholinesterase catalyzes breakdown of ACh that is within the synaptic cleft (see Clinical View 12.6: "Altered Acetylcholine Function and Changes in Breathing").

WHAT DID YOU LEARN?

9 What triggers the binding of synaptic vesicles to the synaptic knob membrane to cause exocytosis of ACh?

10.3b Sarcolemma, T-tubules, and Sarcoplasmic Reticulum: Excitation-Contraction Coupling

LEARNING OBJECTIVE

15. Describe the steps in excitation-contraction coupling.

The second physiologic event of muscle contraction is *excitation-contraction coupling*—an event that involves the sarcolemma, T-tubules,

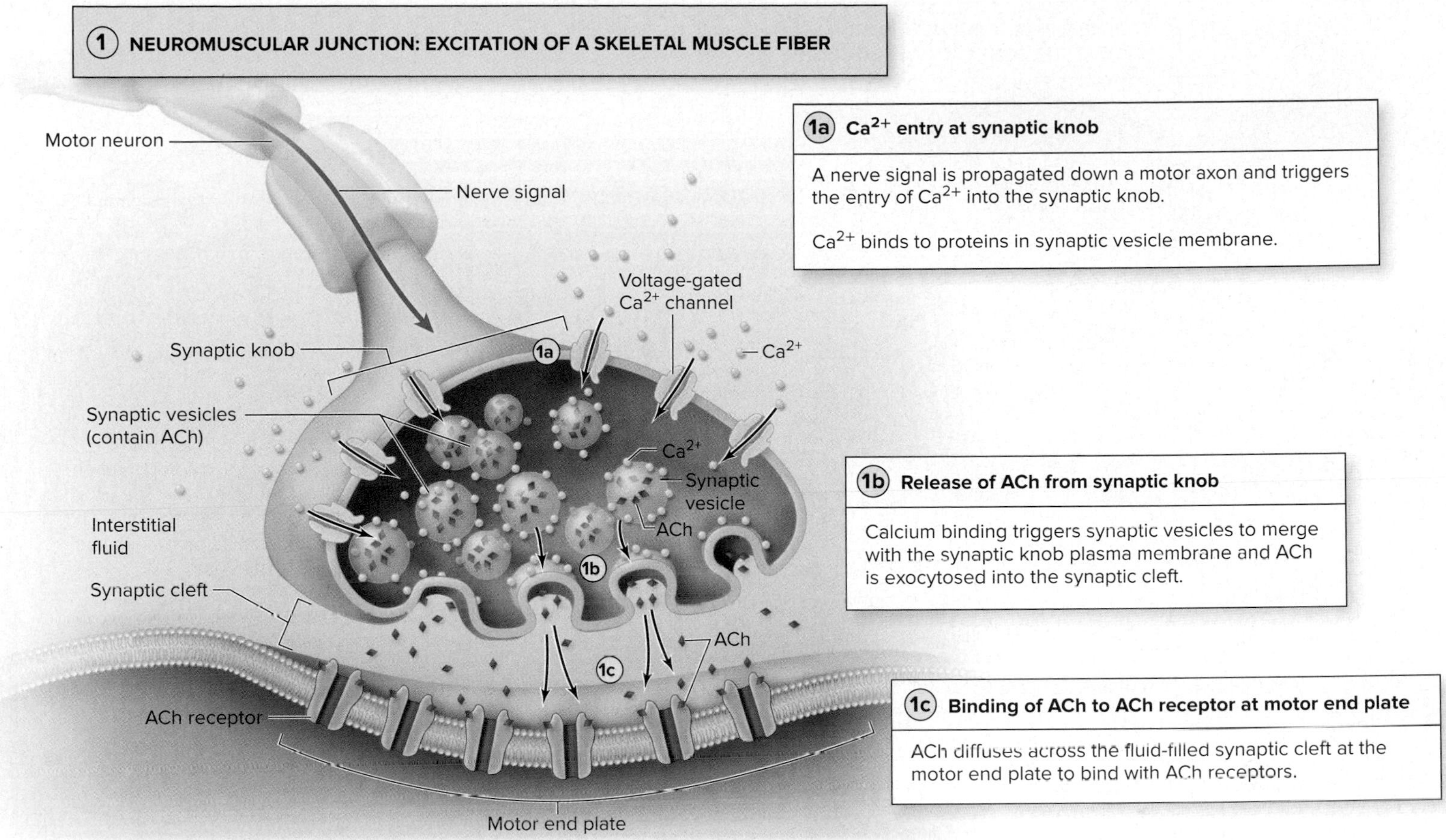

(1) NEUROMUSCULAR JUNCTION: EXCITATION OF A SKELETAL MUSCLE FIBER

(1a) Ca^{2+} entry at synaptic knob

A nerve signal is propagated down a motor axon and triggers the entry of Ca^{2+} into the synaptic knob.

Ca^{2+} binds to proteins in synaptic vesicle membrane.

(1b) Release of ACh from synaptic knob

Calcium binding triggers synaptic vesicles to merge with the synaptic knob plasma membrane and ACh is exocytosed into the synaptic cleft.

(1c) Binding of ACh to ACh receptor at motor end plate

ACh diffuses across the fluid-filled synaptic cleft at the motor end plate to bind with ACh receptors.

Figure 10.10 Neuromuscular Junction: Excitation of a Skeletal Muscle Fiber. A skeletal muscle fiber is excited by the release of the neurotransmitter acetylcholine (ACh) from a motor neuron synaptic knob. (Proteins associated with synaptic vesicles are not shown.) **AP|R**

and sarcoplasmic reticulum. This event "couples," or links, the events of skeletal muscle stimulation at the neuromuscular junction (first step) to the events of contraction caused by sliding myofilaments within the sarcomeres of skeletal muscle fiber (third step). Three events occur during excitation-contraction coupling: development of an end-plate potential at the motor end plate, initiation and propagation of an action potential along the sarcolemma and T-tubules, and release of Ca^{2+} from the sarcoplasmic reticulum. These events are summarized in **figure 10.11**.

Development of an End-Plate Potential at the Motor End Plate

The ACh receptors, which are *chemically gated* ion channels, are stimulated to open temporarily when ACh binds to them (step 2a). The opening of these channels allows relatively small amounts of both Na^+ to rapidly diffuse into the skeletal muscle fiber and K^+ to slowly diffuse out of the skeletal muscle fiber. More Na^+ diffuses in than K^+ diffuses out, and there is a net gain of positive charge on the inside of the skeletal muscle fiber. The flow of both Na^+ and K^+ ions quickly slows and then ceases as the ions meet with resistance. Thus, these changes in membrane potential in the motor end plate are both transient (short-lived) and local. However, if there is sufficient gain of positive charge to change the RMP of about –90 mV to –65 mV, an end-plate potential is produced. An **end-plate potential (EPP)** is the minimum voltage change (or threshold) in the motor end plate that can trigger opening of voltage-gated channels in the sarcolemma to initiate an action potential.

Initiation and Propagation of Action Potential Along the Sarcolemma and T-tubules

The EPP triggers an action potential that is propagated along the sarcolemma and T-tubules of the skeletal muscle fiber (step 2b). An **action potential** involves two events: *depolarization,* which causes the inside of the sarcolemma of the skeletal muscle fiber to become positive due to the influx of Na^+, and *repolarization,* which is the returning of the inside of the sarcolemma to its relatively negative resting membrane potential due to the outward flow of K^+.

The electrical change of the EPP in the motor end plate stimulates the opening of *voltage-gated* Na^+ channels in the adjacent area of the sarcolemma. The opening of voltage-gated Na^+ channels allows Na^+ to move rapidly across the sarcolemma down its concentration gradient into the skeletal muscle fiber. Sufficient Na^+ enters to cause a reversal of the membrane potential of the sarcolemma. The inside, which was relatively negative, becomes relatively positive with a change in the membrane potential from the threshold value of –65 mV to +30 mV. This reversal in polarity at the sarcolemma is referred to as **depolarization.**

The propagation of depolarization along the length of the sarcolemma and T-tubules involves the sequential opening of voltage-gated Na^+ channels. The inflow of Na^+ at the initial portion of the sarcolemma causes adjacent regions of the sarcolemma to experience electrical changes that initiate voltage-gated Na^+ channels in these areas to open. Sodium flows in to cause depolarization in this region of the sarcolemma. Adjacent depolarization is repeated rapidly down the sarcolemma and T-tubules. The propagation of an action potential along the sarcolemma and T-tubule is similar to the falling of a series of stacked dominoes—once started, it does not stop until it reaches the end.

Voltage-gated K^+ channels located along the sarcolemma and T-tubules open immediately following the opening of the voltage-gated Na^+ channels. The opening of voltage-gated K^+ channels allows K^+ to move across the sarcolemma down its concentration gradient and out of the skeletal muscle fiber. Sufficient K^+ exits so that the membrane potential at the sarcolemma and T-tubules reverses and the negative resting membrane potential (–90 mV) is reestablished. This process, which changes the membrane potential from +30 mV to reestablish the RMP of –90 mV, is referred to as **repolarization.** The

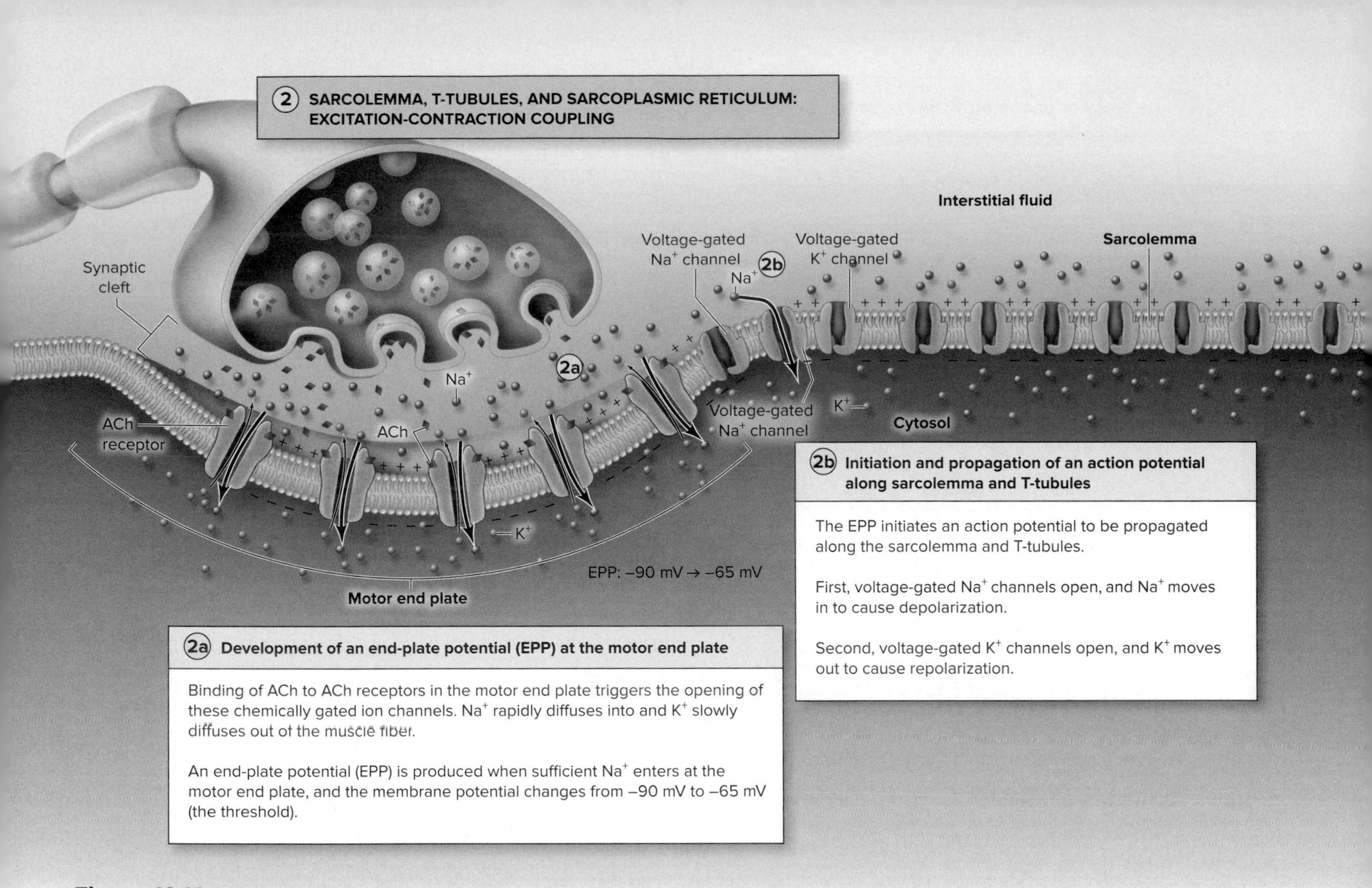

(2) SARCOLEMMA, T-TUBULES, AND SARCOPLASMIC RETICULUM: EXCITATION-CONTRACTION COUPLING

Interstitial fluid

Synaptic cleft

Voltage-gated Na^+ channel

(2b)

Voltage-gated K^+ channel

Sarcolemma

Na^+

(2a)

Na^+

ACh receptor

ACh

Voltage-gated Na^+ channel

K^+

Cytosol

K^+

EPP: −90 mV → −65 mV

Motor end plate

(2a) Development of an end-plate potential (EPP) at the motor end plate

Binding of ACh to ACh receptors in the motor end plate triggers the opening of these chemically gated ion channels. Na^+ rapidly diffuses into and K^+ slowly diffuses out of the muscle fiber.

An end-plate potential (EPP) is produced when sufficient Na^+ enters at the motor end plate, and the membrane potential changes from −90 mV to −65 mV (the threshold).

(2b) Initiation and propagation of an action potential along sarcolemma and T-tubules

The EPP initiates an action potential to be propagated along the sarcolemma and T-tubules.

First, voltage-gated Na^+ channels open, and Na^+ moves in to cause depolarization.

Second, voltage-gated K^+ channels open, and K^+ moves out to cause repolarization.

Figure 10.11 Skeletal Muscle Fiber: Excitation-Contraction Coupling. Skeletal muscle fiber excitation by a motor neuron is coupled to the contraction of myofilaments within the muscle fiber. (Note: Normally, there is no space between the T-tubule and terminal cisternae. Space is for illustration purposes only.)

opening of voltage-gated K^+ channels also occurs sequentially, and repolarization is propagated along the sarcolemma and T-tubules. Repolarization allows the skeletal muscle fiber to propagate a new action potential when stimulated again by a motor neuron.

Note that an action potential is a self-sustaining electrical change in the membrane potential that is propagated along the sarcolemma and is caused by the sequential opening of *voltage*-gated channels. Action potential propagation at the sarcolemma is similar to action potential propagation that occurs in neurons (see section 12.8c).

Figure 10.12 is a graph of the electrical changes at the sarcolemma. These electrical changes include reaching the threshold, depolarization, and repolarization. The period of time that includes depolarization and repolarization is called the **refractory period.** The refractory period is significant because during this brief period of time the muscle cannot be restimulated. A new action potential can occur only when the resting membrane potential at the sarcolemma has been reestablished.

Release of Calcium from the Sarcoplasmic Reticulum

When the action potential reaches the sarcoplasmic reticulum, it (1) stimulates a conformational change to voltage-sensitive Ca^{2+} channels (dihydropyridine receptors) within the T-tubule membrane, which (2) causes a conformational change in Ca^{2+} release channels (ryanodine receptors) located in the terminal cisternae of the sarcoplasmic reticulum, causing them to open (figure 10.11, step 2c). This allows Ca^{2+} to diffuse out of the cisternae of the sarcoplasmic reticulum into the cytosol. Calcium now "mingles" with the thick filaments and thin filaments within myofibrils.

WHAT DID YOU LEARN?

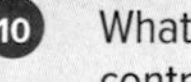

10 What two events are linked in the physiologic process called excitation-contraction coupling?

11 Provide a description of the events of excitation-contraction coupling.

10.3c Sarcomere: Crossbridge Cycling

LEARNING OBJECTIVE

16. Summarize the changes that occur within a sarcomere during contraction.

The third physiologic event in skeletal muscle contraction involves binding of Ca^{2+} and crossbridge cycling. These events are summarized in **figure 10.13**.

Calcium Binding

Calcium released from the sarcoplasmic reticulum binds to a subunit of globular troponin, a component of thin filaments. This induces a conformational change in troponin. Recall that troponin is attached to tropomyosin, forming the troponin-tropomyosin complex. When troponin changes shape, the entire troponin-tropomyosin complex is moved and the myosin binding sites of actin are exposed. Crossbridge cycling is initiated (step 3a).

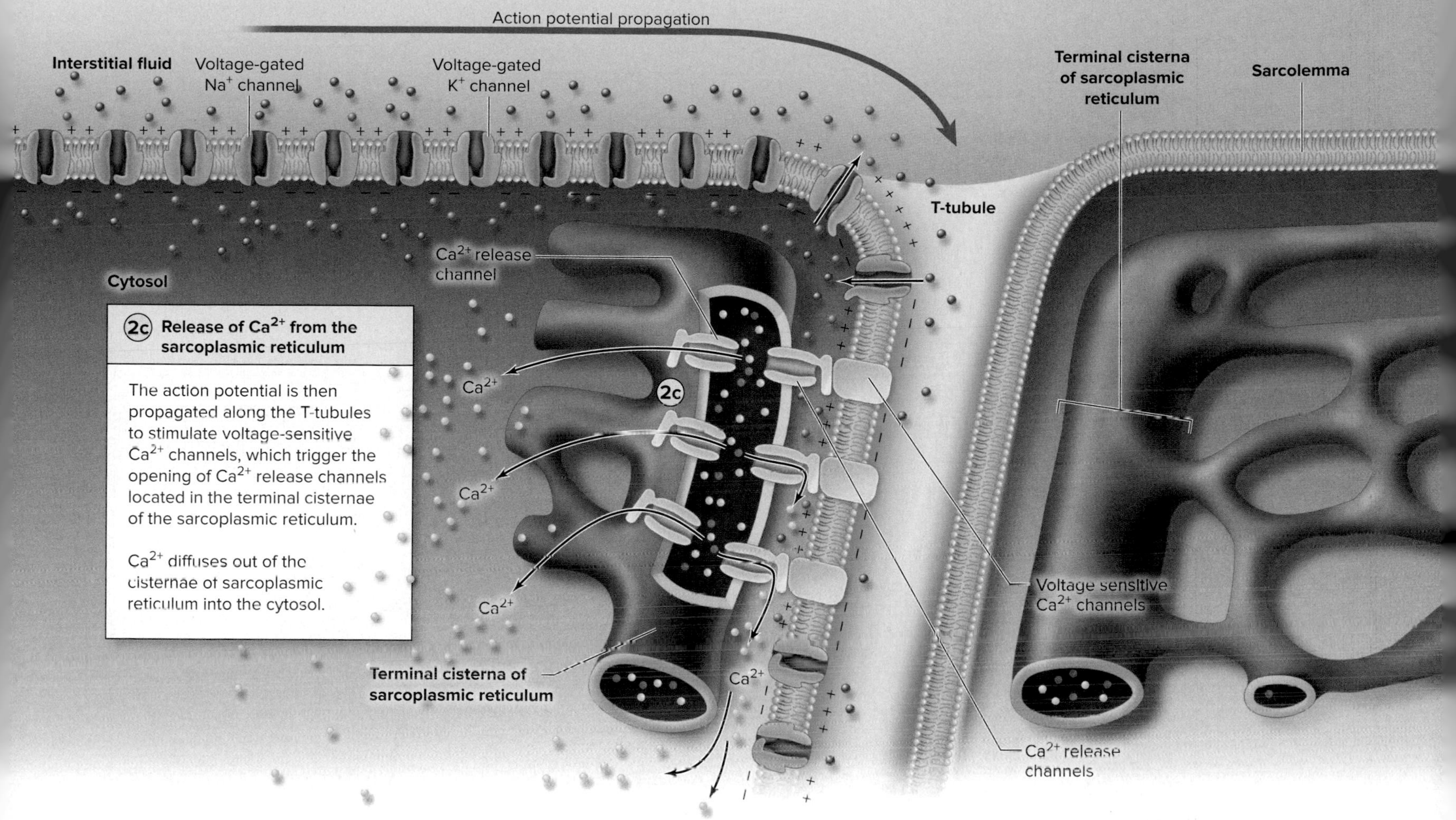

Action potential propagation

Interstitial fluid

Voltage-gated Na^+ channel

Voltage-gated K^+ channel

Terminal cisterna of sarcoplasmic reticulum

Sarcolemma

T-tubule

Cytosol

Ca^{2+} release channel

2c Release of Ca^{2+} from the sarcoplasmic reticulum

The action potential is then propagated along the T-tubules to stimulate voltage-sensitive Ca^{2+} channels, which trigger the opening of Ca^{2+} release channels located in the terminal cisternae of the sarcoplasmic reticulum.

Ca^{2+} diffuses out of the cisternae of sarcoplasmic reticulum into the cytosol.

Ca^{2+}

2c

Voltage sensitive Ca^{2+} channels

Terminal cisterna of sarcoplasmic reticulum

Ca^{2+} release channels

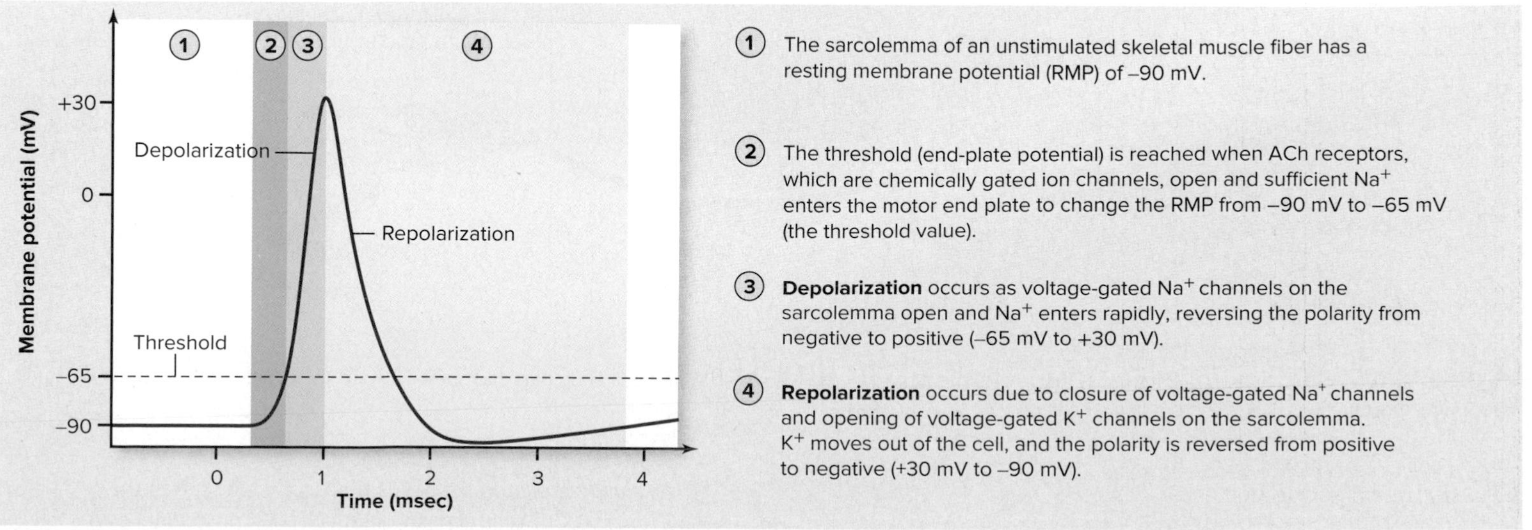

1
2
3
4
+30
0
-65
-90
Membrane potential (mV)
Depolarization
Repolarization
Threshold
0
1
2
3
4
Time (msec)
1 The sarcolemma of an unstimulated skeletal muscle fiber has a resting membrane potential (RMP) of –90 mV.
2 The threshold (end-plate potential) is reached when ACh receptors, which are chemically gated ion channels, open and sufficient Na^+ enters the motor end plate to change the RMP from –90 mV to –65 mV (the threshold value).
3 **Depolarization** occurs as voltage-gated Na^+ channels on the sarcolemma open and Na^+ enters rapidly, reversing the polarity from negative to positive (–65 mV to +30 mV).
4 **Repolarization** occurs due to closure of voltage-gated Na^+ channels and opening of voltage-gated K^+ channels on the sarcolemma. K^+ moves out of the cell, and the polarity is reversed from positive to negative (+30 mV to –90 mV).

Figure 10.12 Events of an Action Potential at the Sarcolemma. A tracing of the membrane voltage (mV) changes associated with an action potential initiated at the neuromuscular junction. Changes occur in just a few milliseconds and result from the opening and closing of voltage-gated Na^+ channels and voltage-gated K^+ channels in the sarcolemma.

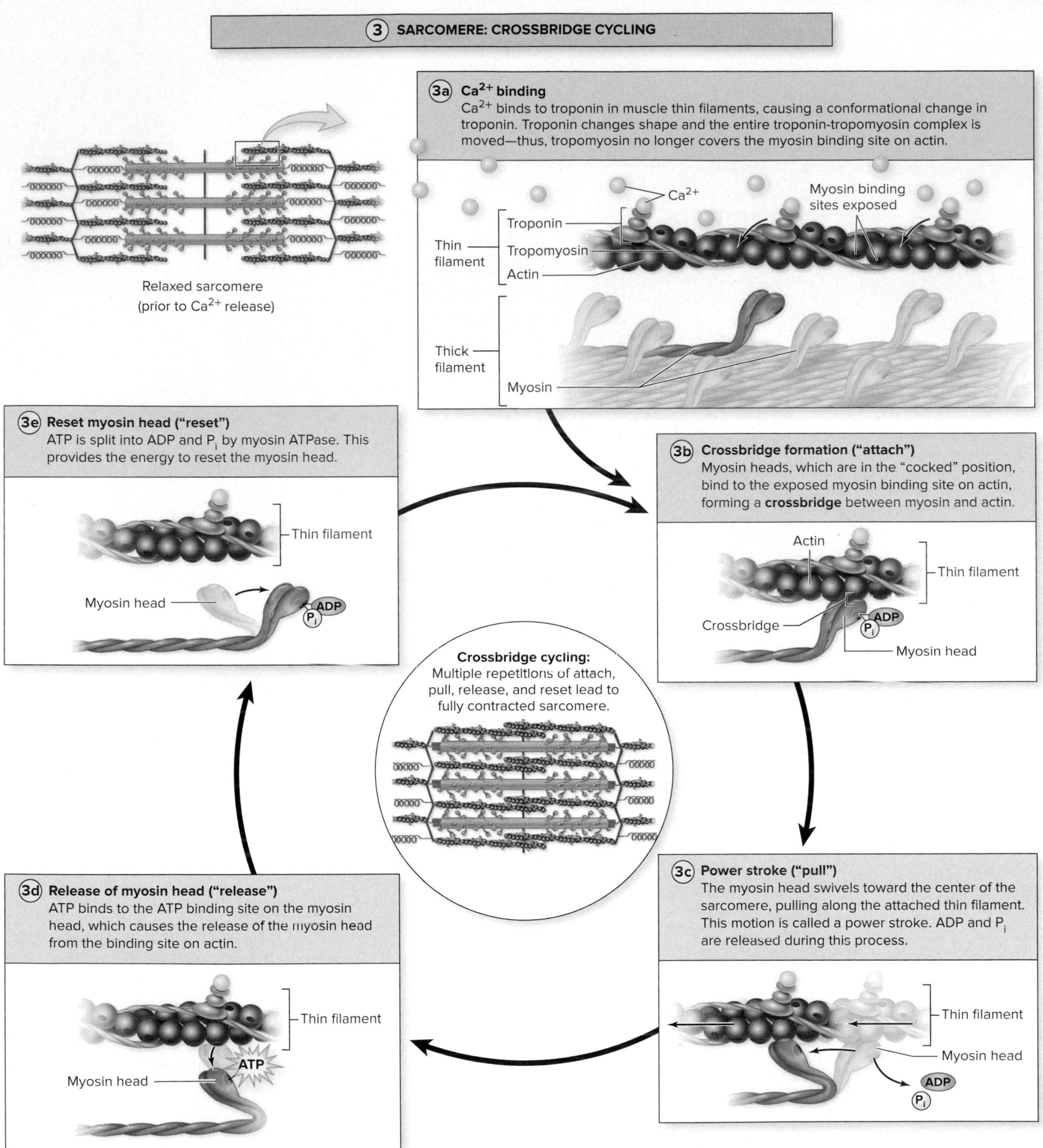

Figure 10.13 Sarcomere: Skeletal Muscle Contraction. Contractile proteins slide past each other toward the center of the sarcomere, and the sarcomere shortens. **AP|R**

Crossbridge Cycling

Crossbridge cycling refers to a four step process that is repeated (step 3b–e): (1) crossbridge formation (attaching of myosin head to actin), (2) power stroke (pulling thin filament by movement of myosin head), (3) release of myosin head from actin, and (4) resetting of myosin head:

Crossbridge formation Myosin heads, which are in the "cocked," or ready, position attach to exposed myosin binding sites of actin. Binding of each myosin head results in formation of a **crossbridge** between the thick and thin filament (step 3b).

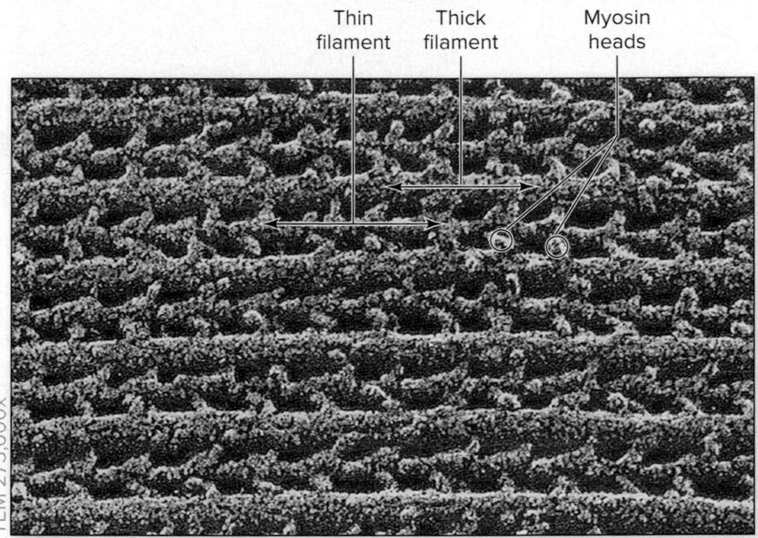

Figure 10.14 Portion of a Sarcomere. Electron micrograph of a sarcomere shows interaction between the myosin heads of the thick filament with the thin filament.
©John Heuser, Washington University School of Medicine, St. Louis, MO

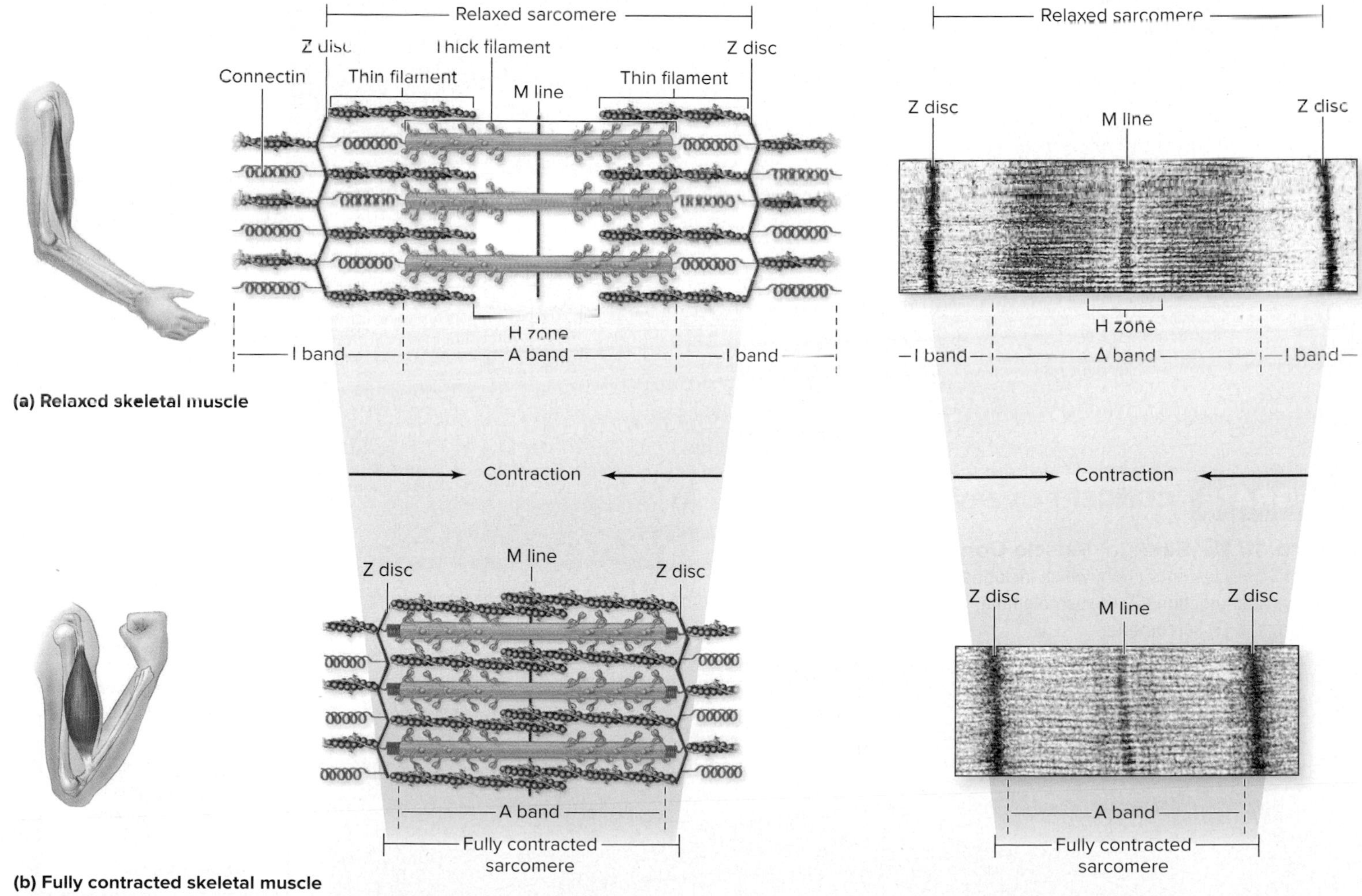

Power stroke After forming a crossbridge, the myosin head swivels (or ratchets) in what is called a **power stroke.** The swiveling of a myosin head pulls the thin filament a small distance past the thick filament toward the center of the sarcomere. ADP and P_i are released during this process, and the ATP binding site becomes available again (step 3c).

Release of myosin head ATP then binds to the ATP binding site of a myosin head, which causes the release of the myosin head from the binding site on actin (step 3d).

Resetting of myosin head Myosin ATPase splits ATP into ADP and P_i, providing the energy to reset the myosin head in the cocked position (step 3e).

If Ca^{2+} is still present, and the myosin binding sites are still exposed, then these four steps involving the myosin heads continue: attach, pull, release, and reset. It is the repetitive action of these steps that results in sarcomere shortening, and a sarcomere moves from its relaxed state into a contracted state. Note that calcium levels remain elevated because the skeletal muscle fiber is repeatedly stimulated by the motor neuron at a very rapid rate. **Figure 10.14** is an electron micrograph of crossbridges between myosin heads and myosin binding sites in actin.

A sarcomere in both a relaxed and a contracted skeletal muscle fiber that has shortened is shown in **figure 10.15**. The following changes to the sarcomere occur in the contracted muscle: The H zone disappears, the I band narrows in width and may disappear, and the Z discs in each sarcomere move closer together. However, the thin and thick filaments do not shorten. A description of the repetitive movement of thin filaments sliding past thick filaments is called the **sliding filament theory.** The three major events of skeletal muscle contraction are integrated in **figure 10.16**.

Figure 10.15 Sarcomere Shortening. Illustrations and corresponding electron micrographs demonstrate sarcomere shortening during skeletal muscle contraction. (*a*) In a relaxed skeletal muscle, the A band, I band, and H zone are all visible. (*b*) In a fully contracted muscle, the sarcomere shortens, the Z discs move closer together, the I band narrows and may disappear, and the H zone disappears.
(*a, b*) ©Dr. H. E. Huxley

INTEGRATE

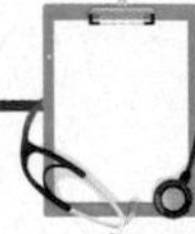

CLINICAL VIEW 10.3

Muscular Paralysis and Neurotoxins

Muscular paralysis (inability of skeletal muscles to contract) may occur if either nervous system function at the neuromuscular junction or excitation-contraction coupling is impaired. This damage may be the result of **neurotoxins,** which are toxins that damage nervous system components. Two paralysis conditions caused by toxins are tetanus and botulism.

Tetanus is a form of spastic paralysis caused by a toxin produced by the bacterium *Clostridium tetani*. The toxin blocks the release of glycine (an inhibitory neurotransmitter in the spinal cord), resulting in overstimulation by motor neurons of the muscles and excessive muscle contractions. Penetrating wounds contaminated with soil and vegetable matter are especially prone to developing *C. tetani* infection. This condition is potentially life-threatening, and so we routinely are vaccinated against it.

Botulism (bot′ū-lizm), a potentially fatal muscular paralysis, is caused by a toxin produced by the bacterium, *Clostridium botulinum.* The toxin prevents the release of acetylcholine (ACh) at synaptic knobs and leads to muscular paralysis. Like *C. tetani, C. botulinum* is common in the environment and produces its toxin under anaerobic conditions. Most cases of botulism poisoning result from ingesting the toxin in canned foods that were not processed at temperatures high enough to kill the botulism spores. Similarly, ingestion of unpasteurized honey by infants in the first year of life can introduce *C. botulinum* spores into their immature gastrointestinal tracts.

The Food and Drug Administration (FDA) approved the use of botulinum toxin type A (Botox) for temporary diminishing of wrinkles (see Clinical View 6.7: "Botox and Wrinkles"). Botox is also used clinically to help reduce overcontraction of muscle (or spasticity) associated with certain disorders or conditions (e.g., cerebral palsy, multiple sclerosis, torticollis, changes following a stroke or spinal cord injury). Botox injections have become one of the most important treatments for spasticity and are most effective 1 to 2 weeks after the injections, with spasticity reduced for up to 3 to 6 months. Treatments may be repeated as often as every 3 months.

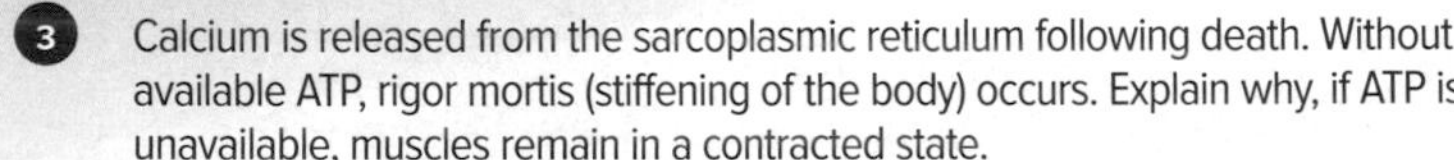

WHAT DO YOU THINK?

3. Calcium is released from the sarcoplasmic reticulum following death. Without available ATP, rigor mortis (stiffening of the body) occurs. Explain why, if ATP is unavailable, muscles remain in a contracted state.

WHAT DID YOU LEARN?

12. What is the function of Ca^{2+} in skeletal muscle contraction?
13. Describe the four processes that repeat in crossbridge cycling to cause sarcomere shortening.
14. What causes the release of the myosin head from actin? What resets the myosin head?

10.3d Skeletal Muscle Relaxation

LEARNING OBJECTIVES

17. Discuss what happens to each of the following to allow for skeletal muscle relaxation: ACh, action potential, Ca^{2+} concentration in sarcoplasm, and troponin-tropomyosin complex.
18. Explain the relationship of skeletal muscle elasticity and muscle relaxation.

The first step in skeletal muscle relaxation is the termination of the rapid nerve signals propagated along the motor neuron. When the nerve signals stop, there is no additional release of acetylcholine, and the acetylcholine remaining in the synaptic cleft is hydrolyzed by acetylcholinesterase (see figure 12.25). The ACh receptors close, and both the end-plate potentials at the motor end plate and the action potentials along the sarcolemma and T-tubules cease.

Calcium channels in the triads (both those in T-tubules and those in terminal cisternae of the sarcoplasmic reticulum) return to their normal (unstimulated) position with no further release of Ca^{2+}. The Ca^{2+} already released from the sarcoplasmic reticulum is continuously returned into the terminal cisternae by Ca^{2+} pumps. After cessation of skeletal muscle fiber stimulation, the remaining Ca^{2+} in the sarcoplasm is transported back into storage within the sarcoplasmic reticulum, where it is bound by both calmodulin and calsequestrin proteins.

Troponin returns to its original shape when Ca^{2+} is removed, and simultaneously the tropomyosin moves over the myosin binding sites on actin. This prevents myosin-actin crossbridge formation. Through the natural elasticity of the skeletal muscle fiber, the muscle may return to its original relaxed position, a process facilitated by the

INTEGRATE

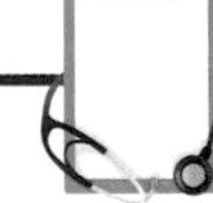

CLINICAL VIEW 10.4

Rigor Mortis

Within a few hours after the heart stops beating, ATP levels in skeletal muscle fibers have been completely exhausted. The sarcoplasmic reticulum loses its ability to return Ca^{2+} from the sarcoplasm and move it back into the sarcoplasmic reticulum by the Ca^{2+} pumps, which require ATP to function. Remember that ATP is also needed to detach the myosin head of the thick filament from the myosin binding site of actin on the thin filaments. Because ATP is no longer available, the crossbridges between thick and thin filaments cannot detach. As a result, the Ca^{2+} already present in the sarcoplasm, as well as the Ca^{2+} that continues to leak out of the sarcoplasmic reticulum, triggers a sustained contraction in the skeletal muscle fibers. All skeletal muscles lock into a contracted position and the body of the deceased individual becomes rigid. This physiologic state, termed **rigor mortis** (rig′er mōr′tis), continues for about 15 to 24 hours. Rigor mortis gradually disappears because lysosomal enzymes are released within the muscle fibers, causing autolysis (self-destruction and breakdown) of the myofibrils.

Forensic pathologists often use the development and resolution of rigor mortis to establish an approximate time of death. Because a number of factors affect the rate of development and resolution of rigor mortis, environmental conditions need to be taken into consideration. For example, a warmer body will develop and resolve rigor mortis much more quickly than a body of normal temperature. The following chart provides rough guidelines for estimating the death interval, assuming that body temperature and ambient (surrounding environment) temperature are within normal range.

Death Interval	Body Temperature	Stiffness
Dead less than 3 hours	Warm	No stiffness
Dead 3–8 hours	Warm, but cooling	Developing stiffness
Dead 8–24 hours	Ambient temperature	Stiff, but resolving
Dead 24–36 hours	Ambient temperature	No stiffness

release of passive tension that developed in connectin proteins that were compressed during shortening.

It is interesting to note that a significant amount of ATP is used by the Ca^{2+} pumps of the sarcoplasmic reticulum. Calcium levels within the cytosol of muscle fibers must be kept low to prevent Ca^{2+} from binding with phosphate ions (which are released from ATP) to form hydroxyapatite (see section 7.2e), which would calcify and harden muscles in a process similar to that in bone tissue. Thus, ATP is required for both contraction (by myosin ATPase) and relaxation. In fact, if sufficient ATP is not available (as occurs following death), muscle relaxation cannot occur and the muscle remains in a contracted state (see Clinical View 10.4: "Rigor Mortis").

WHAT DID YOU LEARN?

15 How do acetylcholinesterase and Ca^{2+} pumps function in the relaxation of a muscle?

10.4 Skeletal Muscle Metabolism

We discussed how cells form ATP through the process of cellular respiration in section 3.4. Here those concepts are integrated to describe specifically how a muscle fiber meets its very-high-energy needs—and how its various means of supplying ATP are used to classify skeletal muscle fibers into three primary types.

10.4a Supplying Energy for Skeletal Muscle Metabolism

LEARNING OBJECTIVES

19. Describe how ATP is made available within skeletal muscle through myosin kinase, creatine kinase, glycolysis, and aerobic cellular respiration.
20. Explain how the means of supplying ATP is related to intensity and duration of exercise.

Most of the ATP required by skeletal muscle fibers is used to reset the mysosin heads of the thick filaments during muscle contraction (see section 10.3c), which demands very large amounts of ATP (approximately 2500 ATP molecules per thick filament per second). ATP is also required by the calcium pump within the sarcoplasmic reticulum membrane to return Ca^{2+} to the terminal cisternae for storage, as described in section 10.3d.

A very limited amount of ATP is already present within skeletal muscle fibers, and additional small amounts can be rapidly produced as phosphate (P_i) is transferred from one ADP to another ADP, yielding ATP and adenosine monophosphate (AMP), an enzymatic reaction catalyzed by **myokinase** (**figure 10.17*a***). This usually provides only enough energy for about 5 to 6 seconds of maximal exertion. Thus, meeting these high energy demands requires forming ATP from other sources. These include ATP formed from creatine phosphate, by glycolysis, and through aerobic cellular respiration.

Creatine Phosphate

Creatine phosphate (see figure 3.4*c*) is a molecule with a high-energy chemical bond between creatine and P_i and is present in tissues with both large and fluctuating energy needs (e.g., muscle, brain). When skeletal muscle is actively contracting, the P_i in creatine phosphate is readily transferred to ADP to form additional ATP (and creatine), an enzymatic reaction catalyzed by **creatine kinase** (figure 10.17*a*). This provides an additional 10 to 15 seconds of energy during maximum exertion.

INTEGRATE

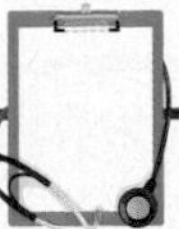

CLINICAL VIEW 10.5

Creatine Kinase Blood Levels as a Diagnostic Tool

Creatine kinase, also called *creatine phosphokinase,* is the enzyme that helps transfer a phosphate between creatine and ATP. Different forms of creatine kinase are present within cardiac muscle and skeletal muscle. The heart muscle form of creatine kinase is found in elevated blood levels in patients suffering from a myocardial infarction (heart attack; see Clinical View 19.5: "Coronary Heart Disease, Angina Pectoris, and Myocardial Infarction"). This provides a specific diagnostic tool for identifying damage to the heart. In contrast, elevated levels of the skeletal muscle form of creatine kinase are used to diagnose degenerative skeletal muscle disease, such as muscular dystrophy (see Clinical View 10.1: "Muscular Dystrophy"). However, note that elevated levels of the skeletal muscle form of creatine kinase may also occur after intense exercise and therefore is not always a sign of disease.

Later during times of rest, the limited stores of ATP and CP within skeletal muscle are replenished. ATP is formed through cellular respiration, and some of those ATP molecules are used to regenerate creatine phosphate. The process that happens at rest is the reverse of the process that happens during exercise. The P_i in ATP is transferred to creatine to form additional creatine phosphate and ADP, an enzymatic reaction also catalyzed by creatine kinase. These enzymatic reactions involving P_i transfer to ADP to form ATP are not dependent upon the presence of oxygen.

WHAT DO YOU THINK?

4 When skeletal muscle tissue is damaged, creatine kinase is released. What general conclusions can be drawn with increasing blood levels of creatine kinase?

Glycolysis

Recall from section 3.4b that **glycolysis** is a metabolic pathway, which involves the breakdown of glucose into two pyruvate molecules, producing a net of 2 ATP molecules (figure 10.17*b*). It occurs within the cytosol, and although it can function in the presence of oxygen, oxygen is not required. Glucose is made available either directly from glycogen stores within the muscle fiber (through glycogenolysis; see figure 2.19) or delivered by the blood.

One of the main advantages of producing ATP through glycolysis is that it does not require oxygen (i.e., it is nonoxidative). The other is its *rapid rate* of ATP production (i.e., the amount of ATP produced per time—at almost twice the rate of aerobic cellular respiration). Although lower total amounts of ATP are produced (compared to aerobic cellular respiration), ATP is produced more quickly, a necessary requirement for short bursts of maximum exercise (e.g., running a 100-meter dash).

What happens to the pyruvate molecules that are produced during glycolysis? Recall from section 3.4g that the fate of pyruvate molecules is dependent upon oxygen availability. Pyruvate molecules either (1) enter a mitochondrion to be broken down through aerobic cellular respiration (if sufficient oxygen is available) or (2) are converted to lactate molecules (under conditions of low oxygen availability).

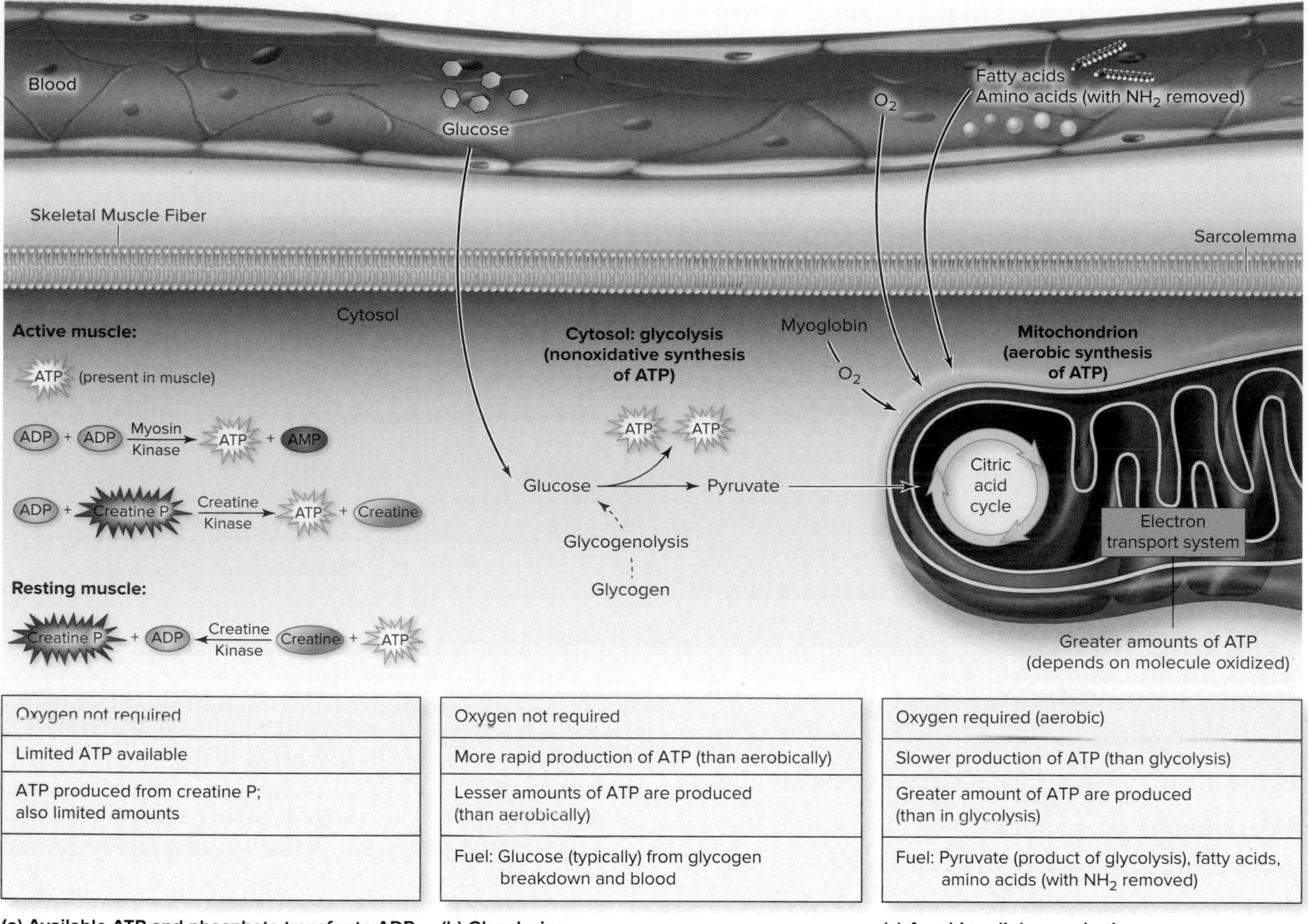

Figure 10.17 Metabolic Processes for Generating ATP. (*a*) Limited amount of ATP molecules is available in muscle. Additional ATP is generated through phosphate transfer to ADP from either another ADP or creatine phosphate. (*b*) Glycolysis, which occurs within the cytosol, is the means for rapid production of limited amounts of ATP production. (*c*) Aerobic cellular respiration that occurs within mitochondria is the means for slower production of much greater amounts of ATP production.

Aerobic Cellular Respiration

Aerobic cellular respiration occurs within mitochondria and requires oxygen, which is made available from the blood or released from myoglobin (figure 10.17*c*). It involves three stages, including the intermediate step, the citric acid cycle, and the electron transport system (see sections 3.4c–f). One of the primary advantages of producing ATP through aerobic cellular respiration is the variety of nutrients that can be oxidized, which include pyruvate (made available through glycolysis), fatty acids, and amino acids (which are deaminated (NH_2 is removed); see section 3.4h). The other advantage is that, although the rate of ATP formation is slower (than in glycolysis), *greater amounts* of ATP are produced. The specific amounts are dependent upon the nutrient that is oxidized (e.g., pyruvate generates 17 ATP molecules; the fatty acid palmitate generates 129 ATP molecules). These higher amounts are a necessary requirement for longer, more moderate levels of activity (e.g., jogging several miles).

Lactate Formation and Its Fate

Lactate formation from pyruvate occurs under conditions of low oxygen availability. This occurs, for example, during intense exercise when skeletal muscle's oxygen demands for aerobic cellular respiration cannot be met. The pyruvate molecules are instead converted to lactate molecules, an enzymatic reaction catalyzed by **lactate dehydrogenase** (see section 3.4g).

What happens to lactate following its formation? Lactate can either enter a mitochondrion within a skeletal muscle fiber, where it is converted *back* to pyruvate and oxidized to carbon dioxide through aerobic cellular respiration, or leave the skeletal muscle fiber and enter the blood. Lactate that enters the blood can then either be (1) taken up by cardiac muscle of the heart, where (as in skeletal muscle) it is converted back to pyruvate and oxidized through aerobic cellular respiration (see section 19.3f) or (2) taken up by the liver to be converted to glucose through gluconeogenesis (see section 27.6c). Glucose molecules are then released by the liver back into the blood, where they may be taken up into a skeletal muscle fiber. This cycling of lactate to the liver, where it is converted to glucose, and the subsequent transport of glucose from the liver to the muscle is called the **lactic acid cycle** (or *Cori cycle*). Observe that lactate does *not* accumulate in skeletal muscles and thus is *not* the cause of muscle soreness (see Clinical View 10.7: "Muscle Pain Associated with Exercise").

Energy Supply and Varying Intensity of Exercise

The use of creatine phosphate, glycolysis, and aerobic cellular respiration as the primary means for supplying ATP during physical

Figure 10.18 Utilization of Energy Sources. The intensity and duration of an activity are important factors in energy utilization. For short sprints, the available ATP and the ATP made available through phosphate transfer are primarily used, whereas for longer runs, glycolysis is used initially but will be replaced by aerobic cellular respiration.

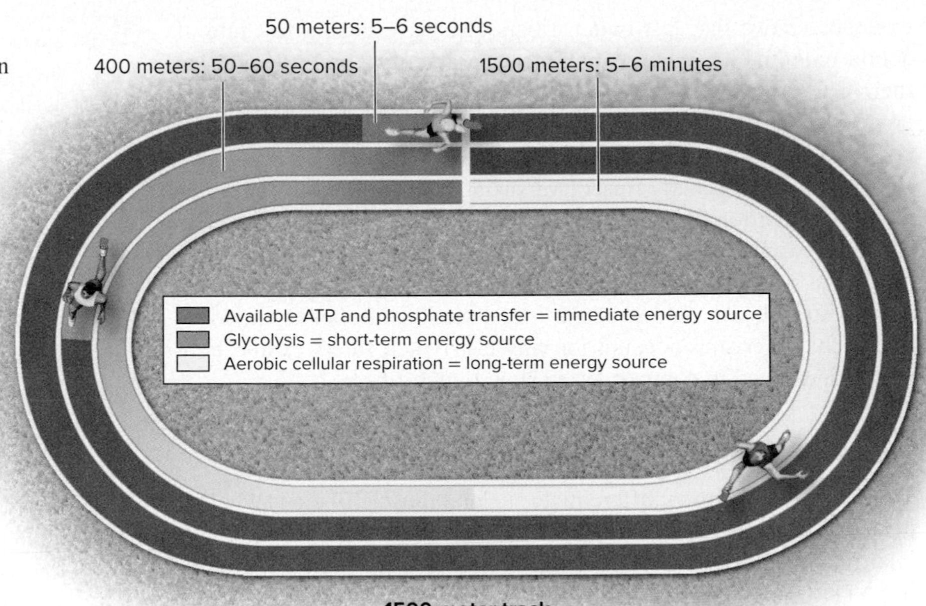

activity is dependent upon both the intensity and the duration of the activity. At rest, skeletal muscle obtains the needed ATP almost exclusively through aerobic cellular respiration involving oxidation of fatty acids. To illustrate the use of energy during exercise, we describe the primary means of supplying ATP for runners at a track meet in which individuals run different distances (**figure 10.18**).

When an individual participates in a 50-meter sprint, an event that may take 5 to 6 seconds, ATP is supplied primarily by available ATP and P_i transfer between two ADP molecules and between creatine phosphate and ADP. In a longer sprint of 400 meters, an event that may take 50 to 60 seconds, ATP is supplied initially by ATP and P_i transfer and then primarily by glycolysis. Finally, in a 1500-meter run, an event that may take 5 to 6 minutes, ATP is supplied by all three means, but primarily by aerobic processes after about the first minute. Keep in mind, however, that there is overlap between the three different energy sources.

Intense exercise that is sustained longer than approximately 1 minute is dependent upon the body's ability to deliver sufficient oxygen through the cardiovascular and respiratory systems. One consequence of participating in regular aerobic exercise (defined as a sustained exercise of moderate intensity that involves raising the heart rate above the baseline) is that it produces changes within both the respiratory system (see section 23.8b) and the heart and blood vessels of the cardiovascular system (see section 20.7) that enhance oxygen delivery. This allows an individual to more effectively provide ATP through aerobic cellular respiration and, thus, to be able to exercise both at greater levels of intensity and for longer periods of time (see section 10.8a).

 **WHAT DID YOU LEARN?**

 Additional ATP is made immediately available in skeletal muscle through which phosphate-containing molecules?

 What are the various means for making ATP available in a 1500-meter race?

10.4b Oxygen Debt

 **LEARNING OBJECTIVE**

21. Define oxygen debt, and explain why it occurs.

There are limitations to how much oxygen can be supplied to skeletal muscle in a given time period. When an individual participates in exercise during which the demand for oxygen exceeds the availability of oxygen, an oxygen debt is incurred. **Oxygen debt** is the amount of additional oxygen that is consumed following exercise to restore pre-exercise conditions. This additional oxygen is required primarily by

- Skeletal muscle fibers to replace oxygen on myoglobin molecules, replenish ATP and creatine phosphate, and replace glycogen stores
- Liver cells to convert lactate back to glucose through gluconeogenesis

Additional oxygen is also required by the respiratory muscles that are engaging in forced breathing (see section 23.5b), the heart as it pumps more blood through the body, and the overall higher metabolic rate. The next time you see individuals breathing hard following exercise, you will realize that they are "paying off" their oxygen debt, helping to return the body to its state prior to exercise.

 **WHAT DID YOU LEARN?**

 What is oxygen debt, and how is the additional oxygen used following intense exercise?

10.5 Skeletal Muscle Fiber Types

Skeletal muscle fiber types are organized into three primary categories. Here we discuss the criteria used to classify them, then we describe the three primary skeletal muscle fiber types.

10.5a Criteria for Classification of Muscle Fiber Types

 **LEARNING OBJECTIVE**

22. Explain the two primary criteria used to classify skeletal muscle fiber types.

Skeletal muscle fibers that compose a muscle are differentiated into three categories based on two criteria: (1) the type of contraction generated and (2) the primary means used for supplying ATP.

Type of Contraction Generated

Skeletal muscle fibers differ in the power, speed, and duration of the muscle contraction generated. **Power** is related to the diameter of a muscle fiber; large muscle fibers have a larger number of myofibrils in parallel, allowing them to produce a more powerful contraction.

Speed has traditionally been described based on whether the skeletal muscle fiber expresses the relatively slow or fast genetic variant of myosin ATPase, the enzyme that splits ATP (see section 10.3c). Those with a fast variant are called **fast-twitch fibers,** and those with the slow variant are called **slow-twitch fibers.** However, recent evidence shows that fast-twitch fibers also have both a fast rate of action potential propagation along the sarcolemma and are quick in their Ca^{2+} release and reuptake by the sarcoplasmic reticulum in comparison to slow-twitch fibers. Thus, fast-twitch fibers initiate a contraction more quickly following stimulation than a slow-twitch fiber (0.01 milliseconds [msec] versus at least 0.02 msec) and produce a contraction of shorter **duration** (7.5 msec versus 100 msec).

Fast-twitch fibers typically have all three characteristics: They produce a strong contraction, initiate a contraction more quickly following stimulation, and produce a contraction of shorter duration. These characteristics account for why fast-twitch fibers exhibit both power and speed in comparison to slow-twitch fibers.

Means for Supplying ATP

The second criterion to differentiate skeletal muscle fibers is whether the primary means the fiber uses to supply ATP is either aerobic cellular respiration or glycolysis. **Oxidative fibers** specialize in providing ATP through aerobic cellular respiration and have several features that support these processes, including an extensive capillary network, large numbers of mitochondria, and a large supply of the red pigment myoglobin. (The presence of both myoglobin and mitochondria gives these skeletal muscle fibers a red appearance, and they are sometimes called **red fibers.**) The higher levels of ATP generated provide energy for oxidative fibers to continue contracting for extended periods of time without tiring, or fatiguing—thus, these fibers are also called **fatigue-resistant.**

In contrast, **glycolytic fibers** specialize in providing ATP more rapidly through glycolysis. Generally, they have fewer structures needed for aerobic cellular respiration—thus, they have less extensive capillary networks, fewer mitochondria, and smaller amounts of myoglobin. (The relatively small amount of myoglobin and mitochondria is why these skeletal muscle fibers have a white appearance and are sometimes called **white fibers.**) However, they do have large glycogen reserves for supplying glucose for glycolysis, which is useful when oxygen stores are low. Glycolytic fibers generally tire easily after a short time of sustained muscular activity—thus, these fibers are also called **fatigable.**

WHAT DID YOU LEARN?

19 Explain how a fast-twitch fiber differs from a slow-twitch fiber and how an oxidative fiber differs from a glycolytic fiber.

10.5b Classification of Muscle Fiber Types

LEARNING OBJECTIVE

23. Compare and contrast the three skeletal muscle fiber types.

Physiologists use both the type of contraction generated and the primary means to supply ATP to differentiate skeletal muscle fibers into three subtypes (**table 10.1**):

- **Slow oxidative (SO) fibers,** also called *type I fibers,* typically have half the diameter of other skeletal muscle fibers and contain slow myosin ATPase. These fibers produce contractions that are slower and less powerful. However, they can contract over long periods of time without fatigue because ATP is supplied primarily through aerobic cellular respiration. These fibers appear dark red because of the presence of large amounts of both myoglobin molecules and mitochondria.
- **Fast oxidative (FO) fibers,** also called *intermediate fibers* or *type IIa,* are the least numerous of the skeletal muscle fiber types. They are intermediate in size and contain fast myosin ATPase. They produce a fast, powerful contraction with ATP provided primarily through aerobic cellular respiration. However, the vascular supply to fast oxidative fibers is less extensive than the network of capillaries serving SO fibers—thus, the delivery rate of nutrients and oxygen is lower. These fibers also contain myoglobin, but less than the amount found in SO fibers. Consequently, these fibers can be distinguished from SO fibers on a microscopic image because they appear a lighter red than SO fibers.
- **Fast glycolytic (FG) fibers,** also called *fast anaerobic fibers* or *type IIb,* are the most prevalent skeletal muscle fiber type. They are largest in diameter, contain fast myosin ATPase, and

Table 10.1 Structural and Functional Characteristics of Different Types of Skeletal Muscle Fibers

Skeletal Muscle Fiber Characteristics	Slow Oxidative (SO) Fibers (Type I Fibers)	Fast Oxidative (FO) Fibers (Type IIa Fibers)	Fast Glycolytic (FG) Fibers (Type IIb Fibers)
ATP Use	Slow	Fast	Fast
Capacity to Make ATP	High	Moderate	Limited
Concentration of Capillaries	Extensive	Moderately extensive	Sparse
Color of Fibers	Red	Lighter red	White (pale)
Contraction Velocity	Slow	Fast	Fast
Resistance to Fatigue	Highest	High	Low
Fiber Diameter	Smallest	Intermediate	Largest
Number of Mitochondria	Many	Many	Few
Amount of Myoglobin	Large	Medium	Small
Primary Fiber Function	Endurance (e.g., maintaining posture, marathon running)	Medium duration, moderate movement (e.g., walking, biking)	Short duration, intense movement (e.g., sprinting, lifting weights)
Muscles with a Large Abundance of Fiber Type	Trunk and lower limb muscles	Lower limb muscles	Upper limb muscles

provide both power and speed. However, they can contract for only short bursts because ATP is provided primarily through glycolysis. These fibers appear white because of the relative lack of myoglobin and mitochondria.

WHAT DID YOU LEARN?

20 Which skeletal muscle fiber type is slow and fatigue-resistant? What is the advantage of this skeletal muscle fiber type?

10.5c Distribution of Muscle Fiber Types

LEARNING OBJECTIVE

24. Describe the distribution of skeletal muscle fiber types in a muscle and how this distribution relates to the muscle's function.

A mixture of muscle fiber types in a typical skeletal muscle is shown in **figure 10.19**. Although most muscles contain a mixture of all three fiber types, the relative percentage of the muscle fiber types varies among different skeletal muscles of the body and reflects the function of the muscle. For example, the extrinsic muscles of the eye and hand require swift but brief contractions and so they contain a high percentage of FG fibers. In contrast, SO fibers dominate many postural back and calf muscles, which contract almost continually to help us maintain an upright posture.

Variations are also present between individuals, and this is seen most dramatically in high-caliber athletes. Elite distance runners have higher proportions of SO fibers in their lower limb muscles, and top athletes who participate in brief periods of intense activity, such as sprinting or weight lifting, have a higher percentage of FG fibers. These variations in the proportion of the skeletal muscle fiber types are determined primarily by a person's genes and less so by the type of training. A proportion of FG fibers may develop the appearance and functional capabilities of FO fibers with physical conditioning if the muscle is used repeatedly for endurance events. Whether this shift actually represents a change of skeletal muscle fiber type—or is simply a temporary alteration to the muscle, which reverts back when the training ceases—remains controversial.

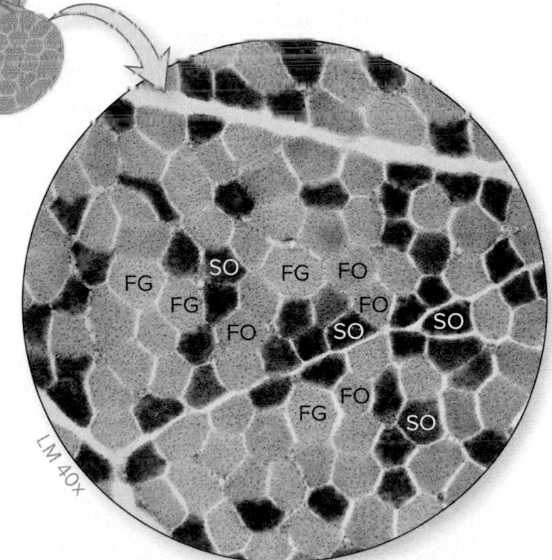

Figure 10.19 Comparison of Fiber Types in Skeletal Muscle. A cross section of a skeletal muscle using a specific staining technique demonstrates the types of fibers in the muscle. The fibers are distinguished by their shade of color. Slow oxidative (SO) fibers are the darkest; fast oxidative (FO) fibers are less dark than the SO fibers, and fast glycolytic (FG) fibers are the lightest.

WHAT DID YOU LEARN?

21 Muscles that maintain posture are composed primarily of what type of skeletal muscle fibers?

10.6 Muscle Tension in Skeletal Muscle

Muscle tension is the force generated when a skeletal muscle is stimulated to contract. The term *tension* is used to describe the force that a muscle exerts because a muscle can only pull on a structure. The tension generated by the contractile proteins within a muscle is transferred to its connective tissue coverings, which move a body part (see section 10.2a).

Muscle tension produced in a contracting muscle is measured in several classic laboratory experiments. One variation of these experiments uses the specimen of a gastrocnemius (calf) muscle (see section 11.9c) with an attached sciatic nerve that is excised from a frog. The gastrocnemius muscle is then anchored to an apparatus that produces a **myogram,** a graphic recording of changes in muscle tension when it is stimulated. Here we describe the generation and graphic recording of (1) a muscle twitch; (2) motor unit recruitment; and (3) wave summation, incomplete tetany, and tetany.

10.6a Muscle Twitch

LEARNING OBJECTIVE

25. Describe what occurs in a skeletal muscle during a single twitch, and relate each event to a graph of a twitch.

Electrodes that are in direct contact with either the skeletal muscle or the sciatic nerve are used to apply single, brief episodes of stimulation to the muscle, and the resulting muscle contraction is recorded using a myograph (**figure 10.20**). The voltage is increased in increments until the muscle responds, producing a twitch. A **twitch** is defined as a single, brief contraction period and then relaxation period of a skeletal muscle in response to a single stimulation. The minimum voltage needed to stimulate the skeletal muscle to generate a twitch is the **threshold.** The voltage below the threshold is called a *subthreshold stimulus.*

There is a delay called a **latent period** (*lag period*) that occurs after the stimulus is applied and before the contraction of the skeletal muscle fiber begins. There is no change in fiber length during the latent period. This delay can be accounted for by the time necessary for all of the events in excitation-contraction coupling, Ca^{2+} release from the sarcoplasmic reticulum into the cytosol, and the beginning of tension generation within the skeletal muscle fiber. The **contraction period** begins as repetitive power strokes pull the thin filaments past the thick filaments, shortening the sarcomeres; muscle tension increases during muscle contraction. The **relaxation period** begins with release of crossbridges as Ca^{2+} is returned to its storage within the sarcoplasmic reticulum; muscle tension decreases during muscle relaxation. Relaxation depends upon the elasticity of connectin (see figure 10.5*b*) within muscle tissue to return to its original length following shortening of the muscle.

WHAT DO YOU THINK?

5 Based on skeletal muscle fiber type distribution, would you predict that the duration of a muscle twitch in an extrinsic eye muscle is shorter or longer than it is in the gastrocnemius muscle? Explain.

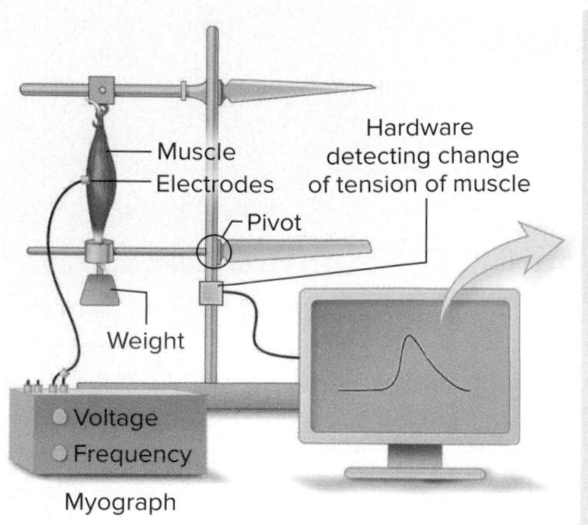

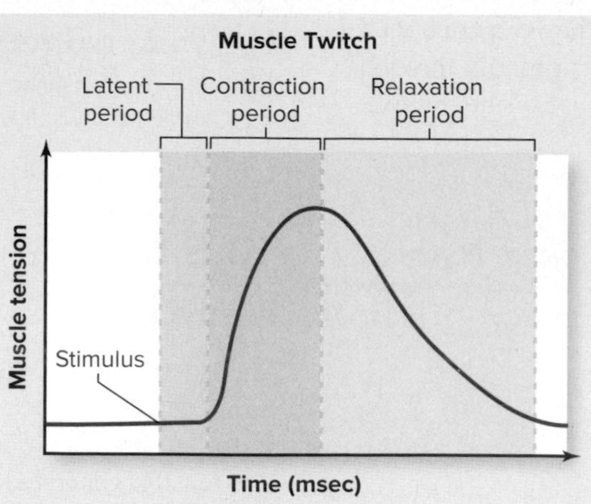

Figure 10.20 Muscle Twitch. A myogram of a single brief stimulation of a skeletal muscle results in a single contraction event called a muscle twitch, which is recorded using a myograph. The latent period is the elapsed time between stimulation of the muscle fiber and the generation of contractile force. The contraction period is the time during which there is an increase in muscle tension. The relaxation period is the time when there is a decrease in muscle tension.

The time required for a twitch varies based on the predominant type of skeletal muscle fibers composing the muscle (see section 10.5b). The extrinsic eye muscles are predominantly fast-twitch fibers producing a twitch that is as rapid as 7.5 milliseconds, whereas the soleus (deep calf muscle; see figure 11.35) is predominantly slow-twitch fibers producing a twitch that lasts about 100 milliseconds.

WHAT DID YOU LEARN?

22 What events are occurring in a muscle that produce the different components of a muscle twitch (latent period, contraction, and relaxation)?

10.6b Changes in Stimulus Intensity: Motor Unit Recruitment

LEARNING OBJECTIVE

26. Explain the events that occur in motor unit recruitment as the intensity of stimulation is increased.

The gastrocnemius muscle is stimulated repeatedly in a set of experiments to demonstrate motor unit recruitment, and each stimulation event is at a greater voltage. The frequency of stimulation remains the same, and the time between stimulation events is sufficient for the muscle to contract and relax before it is stimulated again. Because motor units vary in their sensitivity to stimulation, each increase in voltage causes a greater number of motor units to contract (**figure 10.21**). Consequently, the tension generated with each muscle contraction increases until the point of maximum contraction is reached when all motor units have been stimulated. This increase in muscle tension that occurs with an increase in stimulus intensity is called **recruitment,** or **multiple motor unit summation.**

Recruitment helps to account for how our muscles can both exhibit the all-or-none law and exert varying degrees of force. The **all-or-none law** states that if a skeletal muscle fiber contracts in response to stimulation, it will contract completely (all)—and if the stimulus is not sufficient, it will not contract (none). This means that the skeletal muscle fiber contracts maximally or not at all.

The difference in the force and precision of skeletal muscle movement is varied primarily by changing the number of motor units that are activated. If a reduced number of motor units are activated, then fewer skeletal muscle fibers contract and less force is exerted. In contrast, if a greater number of motor units are activated, more skeletal muscle fibers contract and a greater force is exerted. The recruitment of motor units is not random, however. Rather, it is based on size of motor units within a muscle (figure 10.21*c*). Smallest motor units (which are the most sensitive) are recruited first, with subsequent stimulation of progressively less sensitive larger motor units. This allows for fine motor control when less force is

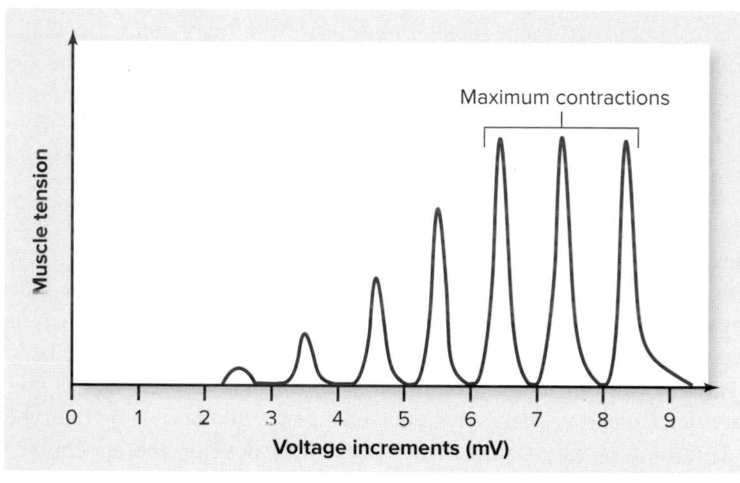

(a) Muscle tension

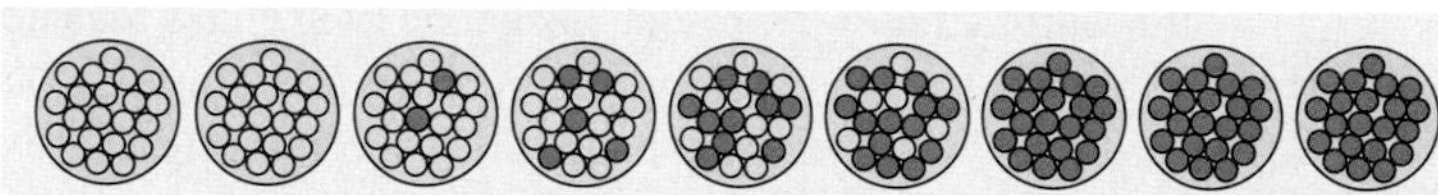

(b) Relative proportion of excited motor units

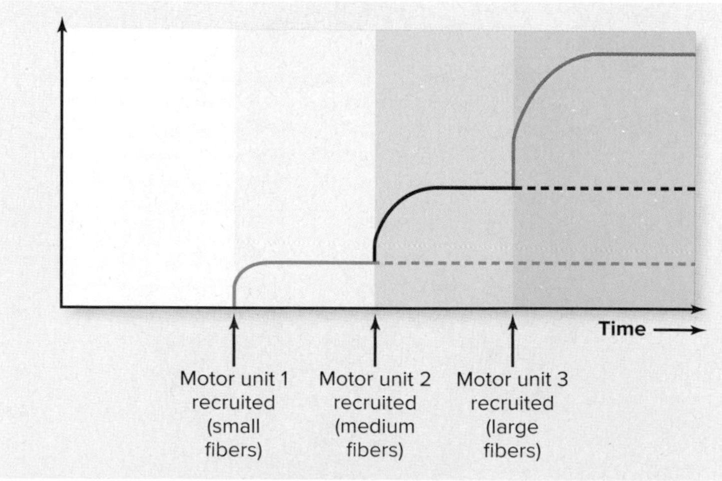

(c) Relative size of excited motor units

Figure 10.21 Skeletal Muscle Response to Change in Stimulus Intensity. Increasing the intensity of stimulation (*a*) increases muscle tension, (*b*) causes a progressive increase in the number of motor units contracting, and (*c*) activates progressively larger motor units. This phenomenon is referred to as either recruitment or multiple motor unit summation.

required (e.g., picking up a pencil) and the most power when more force is required (e.g., lifting a suitcase).

WHAT DID YOU LEARN?

23 What is recruitment? Explain its importance in the body.

10.6c Changes in Stimulus Frequency: Wave Summation, Incomplete Tetany, and Tetany

LEARNING OBJECTIVE

27. Distinguish between wave summation, incomplete tetany, and tetany that occur with an increase in frequency of stimulation.

A different set of experiments subjects the skeletal muscle to increasing frequency of stimulation while the voltage remains the same. Note that each of the two graphs in **figure 10.22** has a different frequency of stimulation.

The first graph in the figure illustrates the stimulation frequency of a skeletal muscle that occurs at a relatively slow rate (less than 10 stimuli per second) (figure 10.22*a*). At this rate, each muscle twitch shows that the muscle is contracting and completely relaxing before the next stimulation event. The muscle tension produced in each muscle twitch is the same.

Stimulation can occur so rapidly (e.g., 20 to 50 stimuli per second) that complete relaxation of the skeletal muscle does not occur before the next stimulation event (figure 10.22*b*). The rapidly restimulated muscle displays a summation of contractile forces as the effect of each new wave is added to the previous wave. This effect is often called either **wave summation** because contraction waves are added together, or "summed," or **temporal summation** because it depends upon increasing frequency (tempo, or timing) of stimulation.

Further increases in stimulation frequency allow less time for relaxation between contraction cycles, and now **incomplete tetany** (the tension tracing continues to increase and the distance between waves decreases) is noted. Stimulation frequency is further increased (e.g., 40 to 50 stimuli per second) until ultimately the contractions of the skeletal muscle fiber "fuse" and form a continuous contraction that lacks any relaxation. This continuous contraction is called **tetany** (the tension tracing is a smooth line). If stimulation continues, the muscle reaches **fatigue**—a decrease in muscle tension occurs from repetitive stimulation (muscle fatigue is described in section 10.7d). Changes in muscle stimulation frequency by the nervous system primarily allow skeletal muscle contraction to exert a coordinated action that gradually increases in force.

Nervous stimulation of skeletal muscles in the human body usually does not exceed 25 stimuli per second. Therefore, muscle tetany is seen only in laboratory experiments. Sustained contractions in the body occur when you are holding something you do not want to drop because the nervous system stimulates different motor units within the same muscle in an overlapping pattern, so the muscle tension can be maintained for a longer period.

WHAT DID YOU LEARN?

24 What happens to skeletal muscle during wave summation? Explain its importance in the body.

10.7 Factors Affecting Skeletal Muscle Tension Within the Body

The discussion of skeletal muscle tension continues with a description of several factors that influence the action of muscles within the human body, including muscle tone, the length-tension relationship, and whether the muscle tension is generated during an isometric or an isotonic contraction. How muscle fatigue influences our ability to generate muscle tension is described at the end of this section.

10.7a Muscle Tone

LEARNING OBJECTIVE

28. Describe muscle tone, and explain its significance.

Skeletal muscles do not completely relax, even when at rest. **Muscle tone** is the resting tension in a skeletal muscle generated by involuntary somatic nervous stimulation of the muscle. Limited numbers of motor units within a muscle are usually stimulated randomly at any given time to maintain a constant tension; the specific motor units being stimulated during rest change continuously so motor units do not become fatigued.

This random contraction of small numbers of motor units causes the skeletal muscle to develop tension, called **resting muscle tone.** These random contractions do not generate enough tension to cause

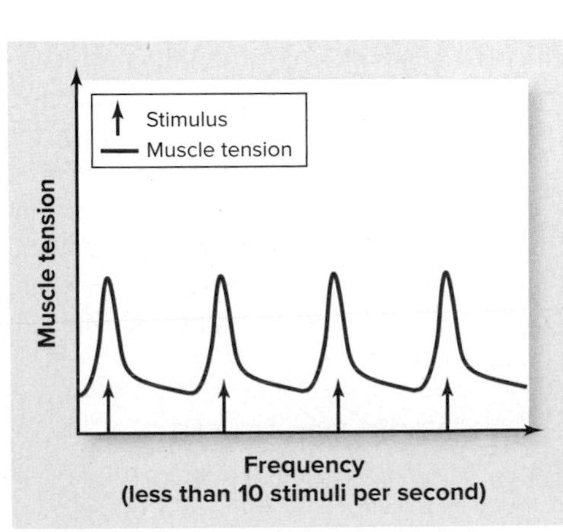

(a) Twitch

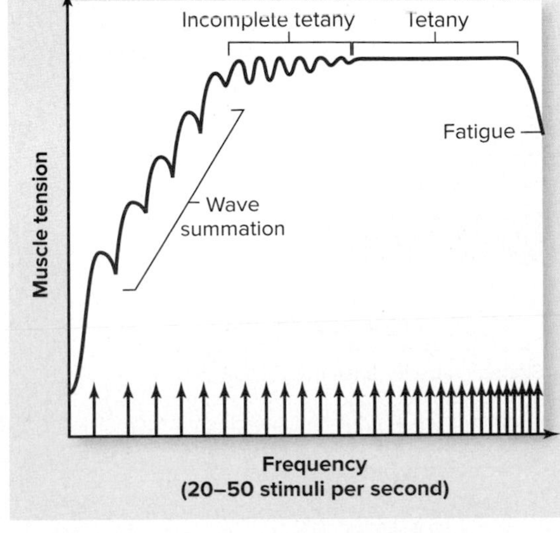

(b) Wave summation, incomplete tetany, and tetany

Figure 10.22 Skeletal Muscle Response to Change in Stimulus Frequency. (*a*) A twitch always produces the same amount of muscle tension. (*b*) Wave summation, incomplete tetany, and tetany are seen when the muscle is stimulated at varying frequencies that allow different degrees of relaxation.

INTEGRATE

CLINICAL VIEW 10.6

Isometric Contraction and Increase in Blood Pressure

©Ed Scott/agefotostock RF

An increase in blood pressure is generally associated with sustained isometric contractions of skeletal muscles. Thus, individuals who already have high blood pressure may experience an additional spike in pressure during intensive exercise, placing them at an increased risk of having a heart attack (see Clinical View 19.4: "Coronary Heart Disease, Angina Pectoris, and Myocardial Infarction"). Thus those at risk of having a heart attack should be extra careful while shoveling snow, because both the sustained isometric contractions and the general peripheral vasoconstriction (that results from being in the cold) can raise blood pressure to unhealthy levels. (For further information about blood pressure, see section 20.5a.)

movement. The resting muscle tone establishes constant tension on the muscle's tendon, thus stabilizing the position of the bones and joints. Another function of muscle tone is to "prime" a muscle for contraction, so that it can respond more readily to stimulation requiring muscle movement. Note that muscle tone decreases during deep sleep (sleep associated with rapid eye movement, or REM, sleep [see section 13.8c]). Consider the difference in muscle tone when carrying an awake child (who has muscle tone) compared to carrying a sleeping child (who temporarily lacks muscle tone). You may have noticed that it is more difficult to carry a sleeping child because the child's body is less rigid.

WHAT DID YOU LEARN?

25 What is the function of skeletal muscle tone?

10.7b Isometric Contractions and Isotonic Contractions

LEARNING OBJECTIVE

29. Distinguish between isometric and isotonic contractions, and give examples of both.

Two primary factors must be considered when describing the consequences of consciously initiating skeletal muscle contraction: (1) force generated by the muscle and (2) resistance (load) that must be overcome. When skeletal muscle tension is insufficient to overcome the resistance (i.e., force generated is less than the load), there is no movement of the muscle. This type of muscle contraction is called an **isometric** (ī-sō-met′rik; *iso* = same, *metron* = measure) **contraction.** Thus, the skeletal muscle contracts and muscle tension increases, but muscle length stays the same. Some examples of isometric contractions include pushing on a wall (a posture for stretching one's leg muscles), holding a very heavy weight in the gym while your arm does not move, attempting to move a shovel load of snow that is too heavy, and holding a baby in one position (**figure 10.23*a***).

When skeletal muscle tension results in movement of the muscle, this type of muscle contraction is called an **isotonic** (ī-sō-ton′ik; *tonos* = tension) **contraction.** The tone in the skeletal muscle remains the same as the length of muscle changes. Examples of an isotonic contraction include walking, lifting a baby, and swinging a tennis racket. Isotonic contractions are differentiated into two subclasses based on whether the muscle is shortening or lengthening as it contracts (figure 10.23*b*). The shortening of muscle length is a **concentric contraction.** It occurs because the muscle tension is greater than the resistance. It may occur in the biceps brachii (muscle of the anterior arm) when lifting a baby. In contrast, lengthening of muscle is an **eccentric contraction.** During an eccentric contraction, the muscle exerts *less* force than that needed to move the load, and the muscle lengthens. If you are holding a 10-pound weight in your hand, and the muscles of your anterior upper arm (e.g., biceps brachii) exert 5 pounds of force, then the biceps muscle lengthens in an eccentric contraction (as your arm extends). This also occurs, for example, in the biceps brachii when placing a baby into a crib. See **figure 10.24**, showing three graphs that visually represent the relationship of muscle tension and muscle length in the three types of muscle contractions.

WHAT DID YOU LEARN?

26 When you flex your biceps brachii while doing "biceps curls," what is the type of movement?

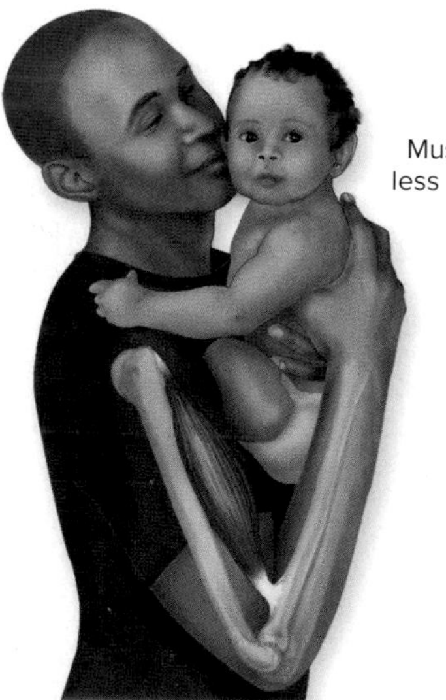
Muscle tension is
less than resistance.

Isometric contraction
Skeletal muscle tension is less than the resistance. Although tension is generated, the muscle does not shorten, and no movement occurs.

(a)

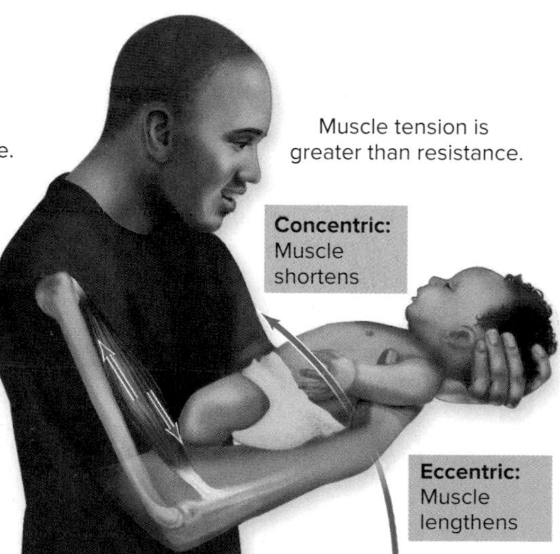
Muscle tension is
greater than resistance.
Concentric:
Muscle
shortens
Eccentric:
Muscle
lengthens

Isotonic contraction
Skeletal muscle tension is greater than the resistance. The muscle shortens (concentric) or lengthens (eccentric), and movement occurs.

(b)

Figure 10.23 Isometric Versus Isotonic Contraction.

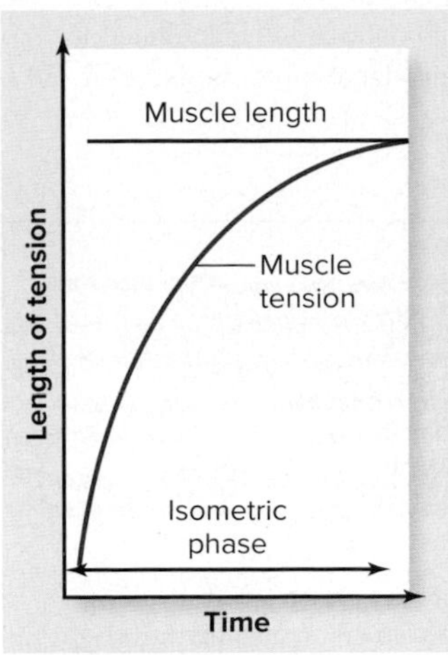

(a) Isometric contraction

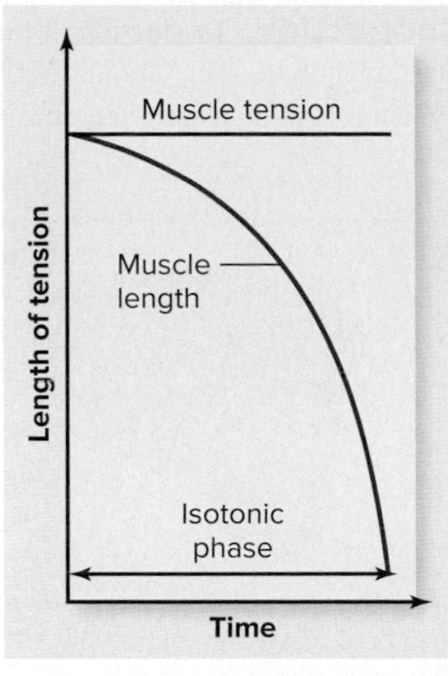

(b) Isotonic contraction: Concentric

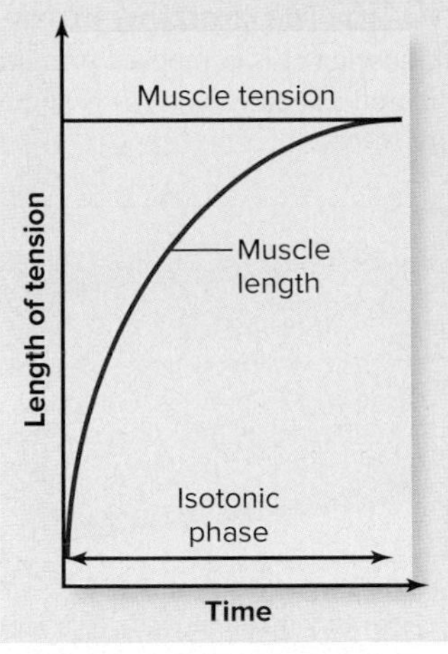

(c) Isotonic contraction: Eccentric

Figure 10.24 Muscle Length and Tension Relationship During Muscle Contraction. (*a*) During an isometric contraction, muscle length remains the same as muscle tension increases. (*b*) Muscle tension remains the same and muscle length *decreases* during a concentric isotonic contraction. (*c*) Muscle tension remains the same and muscle length *increases* during an eccentric isotonic contraction.

10.7c Length-Tension Relationship

LEARNING OBJECTIVE

30. Explain the length-tension relationship in skeletal muscle contraction.

The amount of tension a skeletal muscle can generate when stimulated is influenced significantly by the amount of overlap of thick and thin filaments within its muscle fibers when the muscle begins its contraction. This principle is termed the **length-tension relationship.** A skeletal muscle generates different amounts of tension, depending upon its length at the time of stimulation. The graphical presentation of the length-tension relationship is called the **length-tension curve** (figure 10.25).

A skeletal muscle fiber stimulated when it is at a normal resting length generates a maximum contractile force because there is optimal overlap of thick and thin filaments, allowing for the largest number of crossbridges to form. In contrast, a muscle that is either already contracted or overly stretched produces a weaker contraction when stimulated. Weaker contractions in muscles that are already contracted occur because the thick filaments are close to the Z discs, and sliding filaments are limited in their movement. Weaker contractions occur in muscles that are overly stretched because there is minimal thick and thin filament overlap for crossbridge formation. So,

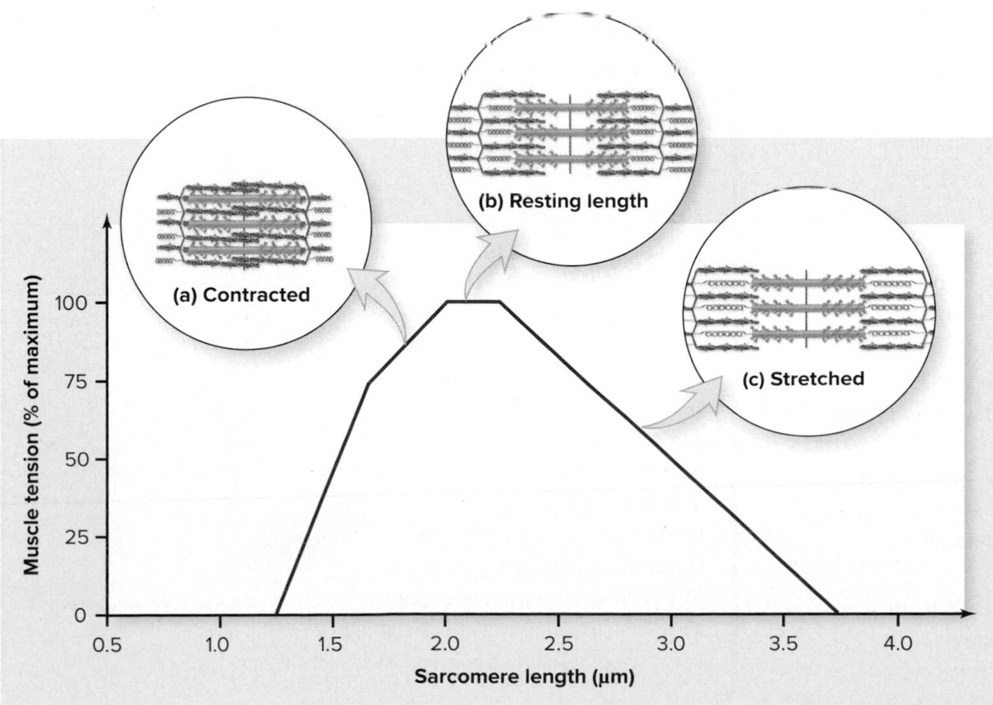

Figure 10.25 Length-Tension Curve. The tension generated by a skeletal muscle fiber within a muscle is graphically related to its precontraction resting length. (*a*) If the skeletal muscle fiber is already contracted at the time of stimulation, it does not have the ability to shorten much more, and it exhibits a weak contraction. (*b*) A skeletal muscle fiber at its normal resting length is generally capable of exhibiting the strongest contraction because of optimal overlap of myofilaments. (*c*) If the skeletal muscle fiber is very stretched when stimulated, relatively little contraction may occur because myofilaments have minimal overlap.

Figure 10.26 Maximizing Force of Contraction. Production of the most forceful contraction is facilitated if the skeletal muscle (*a*) is composed primarily of fast, glycolytic skeletal muscle fibers, (*b*) contains large motor units, (*c*) is stimulated more frequently, and (*d*) is stimulated when it is at its resting length.
(*a*) ©ISM/Medical Images

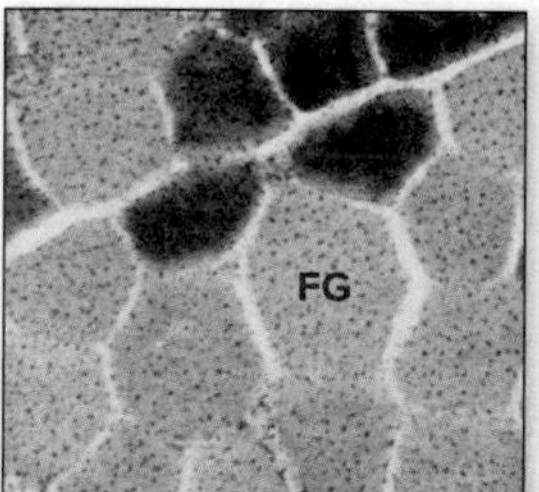

(a) Fast, glycolytic fibers: Larger in diameter (and provide both power and speed)

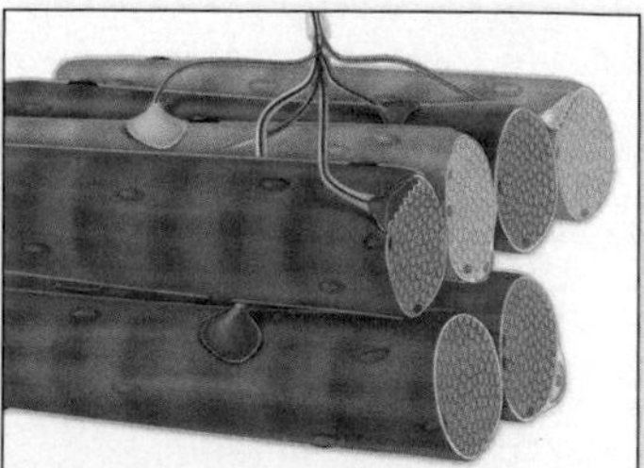

(b) Large motor units: For any given muscle, recruitment stimulates progressively larger motor units

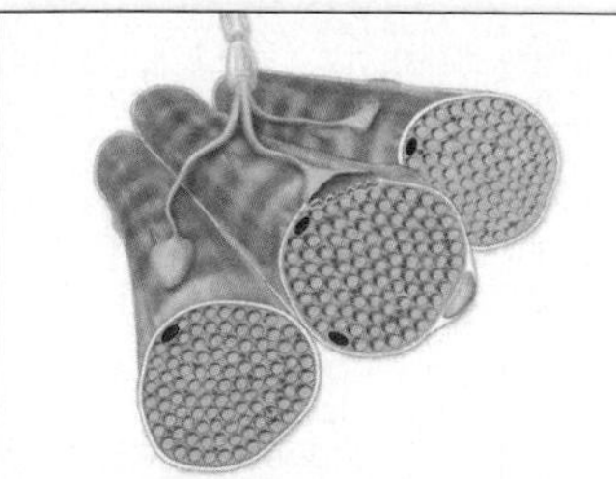

(c) Greater stimulus frequency: Due to wave summation because muscle fibers do not relax completely between contractions.

(d) Muscle at resting length: Due to maximum myofilament overlap

for example, you may be able to lift a heavier dumbbell when your elbow is partially flexed than when your elbow is fully extended, because there is minimal overlap of thick and thin filaments during the full extension.

Numerous factors have been discussed that influence the amount of muscle tension that is generated by skeletal muscle fibers within the body. The primary four factors are visually summarized in **figure 10.26**.

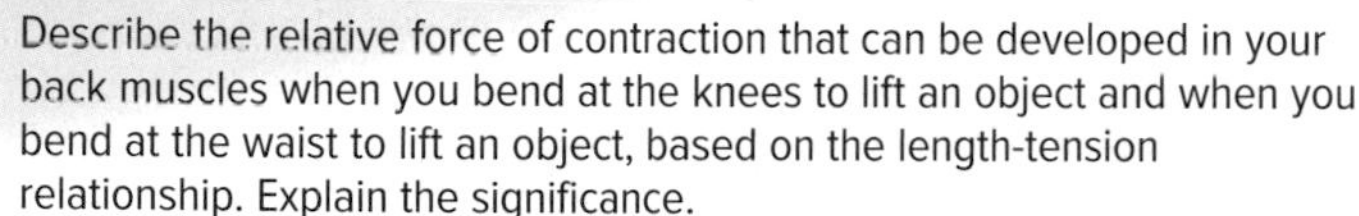

WHAT DID YOU LEARN?

27 Describe the relative force of contraction that can be developed in your back muscles when you bend at the knees to lift an object and when you bend at the waist to lift an object, based on the length-tension relationship. Explain the significance.

10.7d Muscle Fatigue

LEARNING OBJECTIVE

31. Define muscle fatigue, and explain some of its causes.

Muscle fatigue is the reduced ability or the inability of the skeletal muscle to produce muscle tension. The primary cause of muscle fatigue during excessive or sustained exercise (e.g., running a marathon) is a decrease in glycogen stores. However, there are many other causes of muscle fatigue, which are still being debated. Here they are organized by the specific physiologic event of muscle contraction that is affected.

- **Excitation at the neuromuscular junction.** Muscle fatigue may be caused either by insufficient free Ca^{2+} at the neuromuscular junction to enter the synaptic knob or by a decreased number of synaptic vesicles to release neurotransmitter (see section 10.3a). Both limit the ability of somatic motor neurons to stimulate a skeletal muscle.
- **Excitation-contraction coupling.** Muscle fatigue may be due to a change in ion concentration (e.g., Na^+, K^+) that interferes with the ability of the muscle fiber to conduct an action potential along the sarcolemma (see section 10.3b). This interferes with stimulating release of Ca^{2+} from the sarcoplasmic reticulum.
- **Crossbridge cycling.** Muscle fatigue may result from increased phosphate ion (P_i) concentration. Elevated P_i concentration in the muscle sarcoplasm interferes with P_i release from the myosin head during crossbridge cycling, and this slows the rate of cycling. Muscle fatigue also may occur when lower amounts of Ca^{2+} are available for release from the sarcoplasmic reticulum (part of which is due to its binding with the excess P_i). Lower Ca^{2+} levels result in less Ca^{2+} binding to troponin, reducing crossbridge formation, which results in a weaker muscle contraction (see section 10.3c). Thus, both an increase in P_i concentration and lower Ca^{2+} levels result in a weaker force generated during muscle contraction.

Lack of ATP is not currently thought to be a primary cause of muscle fatigue. This is because ATP levels are generally maintained through aerobic cellular respiration in mitochondria during sustained exercise. It remains to be determined if ATP may still be a factor because of its location in the cell—that is, within the mitochondria and not in proximity to myofilaments.

WHAT DID YOU LEARN?

28 How can muscle fatigue result from changes in each of the three primary events of skeletal muscle contraction?

INTEGRATE

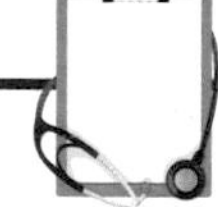

CLINICAL VIEW 10.7

Muscle Pain Associated with Exercise

©George Doyle/Getty Images RF

Accumulation of lactate produced during conditions of low oxygen in skeletal muscle (see section 3.4b) is commonly (but incorrectly) thought to cause muscle pain as a result of exercise. Studies have shown that lactate does not accumulate in skeletal muscle fibers (see section 10.4a) and muscle pain is due in part to some minor tearing of the skeletal muscle fibers, which results in a buildup of fluid and inflammation.

10.8 Effects of Exercise and Aging on Skeletal Muscle

Skeletal muscle is affected by the process of exercise and aging. Here we describe the effects on skeletal muscle of both a sustained exercise program and a lack of exercise, along with the changes that occur as we age.

10.8a Effects of Exercise

 **LEARNING OBJECTIVE**

32. Compare and contrast the changes in skeletal muscle that occur as a result of the two primary types of exercise programs or from the lack of exercise.

Changes in Muscle from a Sustained Exercise Program

The outcome of exercise that involves repetitive stimulation of skeletal muscle fibers depends upon the type of exercise—either endurance exercise or resistance exercise. **Endurance** (or *aerobic*) **exercise** involves sustained, moderate activity that increases heart rate (e.g., running several miles). This type of exercise causes changes that primarily alter how skeletal muscle fibers are supplied with energy (see section 10.4a). The changes that specifically occur to *skeletal muscle fibers* include (1) an increase in the number of mitochondria and the enzymes within mitochondria, which enhances ATP production through aerobic cellular respiration; (2) an increase in enzymes for using fatty acids in aerobic cellular respiration; and (3) an increase in the amounts of lactate dehydrogenase enzyme (for converting lactate back to pyruvate). The greater availability of fatty acids and pyruvate delays glycogen depletion within skeletal muscle fiber and, thus, fatigue (see section 10.7d).

Endurance exercise also induces changes to the *cardiovascular system.* The heart wall thickens, which increases the amount of blood that can be pumped by the heart (see section 19.9a), and additional blood vessels form within skeletal muscle through angiogenesis (see section 20.4a). Both of these changes provide more efficient delivery of blood, and additional oxygen is supplied to skeletal muscle. These changes also enhance ATP production through aerobic cellular respiration.

In comparison, **resistance exercise,** which involves producing forceful muscle contractions (e.g., weight lifting or power lifting), primarily results in stronger skeletal muscles. Resistance exercise stimulates skeletal muscle fibers to increase contractile proteins (myosin, actin), especially in fast glycolytic muscle fibers. These changes primarily result in an increase in muscle size, which is called **hypertrophy** (hī-pĕr′trō-fē; *hyper* = above or over, *trophe* = nourishment). Hypertrophy also results from an increase in the number of mitochondria, greater amounts of myoglobin, and larger glycogen reserves. A slight increase in both ATP and creatine phosphate stores also occurs. Recent evidence suggests that some (limited) increase in the *number* of muscle fibers also may occur, a process called **hyperplasia.** An athlete who competes as a bodybuilder or weight lifter might consider using anabolic steroids to stimulate greater muscle mass than would normally occur. However, numerous and significant side effects are associated with their use (see Clinical View 10.9: "Anabolic Steroids as Performance-Enhancing Compounds").

Changes in Muscle from Lack of Exercise

Lack of exercise (and, thus, lack of muscle use) results primarily in decreasing the skeletal muscle fiber size, a process called **atrophy** (at′rō-fē; *a* = without). This causes a decrease in muscle fiber size, tone, and power, and the muscle becomes flaccid. Even a temporary reduction in muscle use can lead to muscular atrophy. Comparing limb muscles before and after a cast that has been worn for a fracture reveals the loss of muscle tone and size for the casted limb. Individuals who suffer damage to the nervous system or are paralyzed by spinal injuries gradually lose skeletal muscle tone and size in the areas affected. Although skeletal muscle atrophy is initially reversible, dead or dying skeletal muscle fibers are not replaced. When extreme atrophy occurs, the loss of gross skeletal muscle function is permanent because muscle is replaced with connective tissue, including adipose connective tissue. For these reasons, physical therapy is required for patients who suffer temporary loss of mobility.

INTEGRATE

CLINICAL VIEW 10.8

Unbalanced Skeletal Muscle Development

Young athletes today are more likely than in the past to participate in one sport, and to play that sport for more sport seasons; sometimes they play the same sport all year long. Consequently, physical therapists are treating more sports-related injuries. They note that the increase in injuries is often due to unbalanced skeletal muscle development. For example, soccer players are more likely to have overdeveloped hamstring muscles (see figure 11.1*b*) and underdeveloped quadriceps muscles (see figure 11.1*a*), putting them at higher risk of injury.

 **WHAT DID YOU LEARN?**

 What anatomic changes occur in a skeletal muscle fiber when it undergoes hypertrophy?

10.8b Effects of Aging

 **LEARNING OBJECTIVE**

33. Summarize the effects of aging on skeletal muscle.

A slow, progressive loss of skeletal muscle mass typically begins in a person's mid-30s and becomes more noticeable after the age of 50. There is some loss in skeletal muscle fiber diameter, which is due to a reduction in both the size and the number of myofibrils and the number of myofilaments. However, most loss of muscle mass associated with aging is caused by a decrease in skeletal muscle fiber number. This loss is partly due to decreased physical activity. There is also a progressive loss of motor neurons that stimulate skeletal muscle fibers, which results in skeletal muscle fiber atrophy or loss. Other motor neurons can innervate the skeletal muscle fibers; however, this results in a decrease in fine motor control (because of the increase in motor unit size). Thus, elderly individuals progressively have more difficulty with physical tasks and balance. Strength training can slow the progress by increasing muscle size.

The ability to produce ATP also decreases with aging due to a variety of factors, including a reduction in myoglobin (less oxygen storage capacity) and decreased glycogen storage. Overall, muscle strength and endurance are impaired, and the individual has a tendency to fatigue more quickly. Decreased cardiovascular performance often accompanies aging; thus, the blood supplied to active skeletal muscles is much less in elderly people when they exercise.

Skeletal muscle tissue has a reduced capacity to recover from disease or injury as a person grows older. The number of satellite cells

INTEGRATE

CLINICAL VIEW 10.9

Anabolic Steroids as Performance-Enhancing Compounds

©Ingram Publishing RF

Anabolic (an-ă-bol′ik) **steroids** are synthetic substances that mimic the actions of natural testosterone. Recall from section 3.2b that the term *anabolism* refers to the synthesis of complex molecules (e.g., protein) from simple molecules (e.g., amino acids). To date, over 100 compounds have been developed with anabolic properties, but they all require a prescription for legal use in the United States. Anabolic steroids have only a few accepted medical uses—among them, the treatment of delayed puberty, certain types of impotence, and the wasting condition associated with HIV infection and other diseases. Because anabolic steroids stimulate the manufacture of muscle proteins, these compounds have been used by some athletes as performance enhancers.

Relatively large doses of anabolic steroids are needed to stimulate excessive muscle development that results in extra strength and speed. But this enhanced muscle power and speed comes at a price. Medical professionals have reported many devastating side effects associated with extended anabolic steroid use, including increased risk of heart disease and stroke, kidney damage, liver tumors, testicular atrophy and a reduced sperm count, gynecomastia (enlargement of breast glandular tissue in males), acne, high blood pressure, aggressive behavior, and personality aberrations. Because anabolic steroids mimic the effects of testosterone, female athletes who use them often experience menstrual irregularities, growth of facial hair, and in extreme circumstances atrophy of the uterus and mammary glands; sterility has even been reported. Adding to these problems is the route of administration. Because many of these steroid preparations must be injected, the improper use or sharing of needles raises the possibility of transferring pathogens that cause disease (e.g., AIDS or hepatitis). For all of these reasons, the use of anabolic steroids as performance-enhancing compounds has been widely banned.

in skeletal muscle (see section 10.2b) steadily decreases, and scar tissue often forms due to diminished repair capabilities. Finally, the elasticity of skeletal muscle also decreases as we age. Muscle mass is often replaced by adipose connective tissue and dense regular (fibrous) connective tissue, a process called **fibrosis** (fī-brō′sis). The increasing amounts of this connective tissue decrease the flexibility of muscle; an increase in collagen fibers can restrict movement and circulation.

All of us will eventually experience a decline in muscular performance, regardless of our lifestyle or exercise patterns. But we can improve our chances of being in good shape later in life by striving for physical fitness throughout life.

WHAT DID YOU LEARN?

30 What changes in skeletal muscle occur as a result of aging?

10.9 Cardiac Muscle Tissue

LEARNING OBJECTIVE

34. List and describe the similarities and differences between skeletal muscle and cardiac muscle.

In addition to skeletal muscle, two other types of muscle occur in the body: cardiac muscle and smooth muscle (see section 5.3). Here we briefly review the characteristics of cardiac muscle (for a detailed description of cardiac muscle, see section 19.3f).

Cardiac muscle cells are individual muscle cells arranged in thick bundles within the heart wall (**figure 10.27**). Cardiac muscle cells branch and are both shorter and thicker than skeletal muscle fibers. (They typically have a diameter of about 15 μm and range in length from 50 to 100 μm.) Individual cells are joined to adjacent muscle cells at junctions termed intercalated discs. **Intercalated** (in-ter′kă-lā′ted) **discs** are unique to cardiac muscle; they are composed of desmosomes and gap junctions (see section 4.6d). These cells have only one or two nuclei. Cardiac muscle cells are striated because, like skeletal muscle fibers, cardiac muscle cells also contain sarcomeres. Cardiac muscle cells contain a large number of mitochondria, and they use aerobic cellular respiration almost exclusively to generate the ATP required for their unceasing work. In addition, the sarcoplasmic reticulum is not as well developed as in skeletal muscle, so most Ca^{2+} enters the cardiac muscle fibers from interstitial fluid instead.

Cardiac muscle cells are stimulated by a specialized **autorhythmic** pacemaker. This feature is responsible for the repetitious, rhythmic heartbeat. The autonomic nervous system (a division of the nervous system that controls cardiac muscle contraction, smooth muscle contraction, and gland secretion, which is discussed in chapter 15) controls both the rate and the force of contraction of cardiac muscle.

WHAT DID YOU LEARN?

31 What are three anatomic or physiologic differences between skeletal muscle and cardiac muscle?

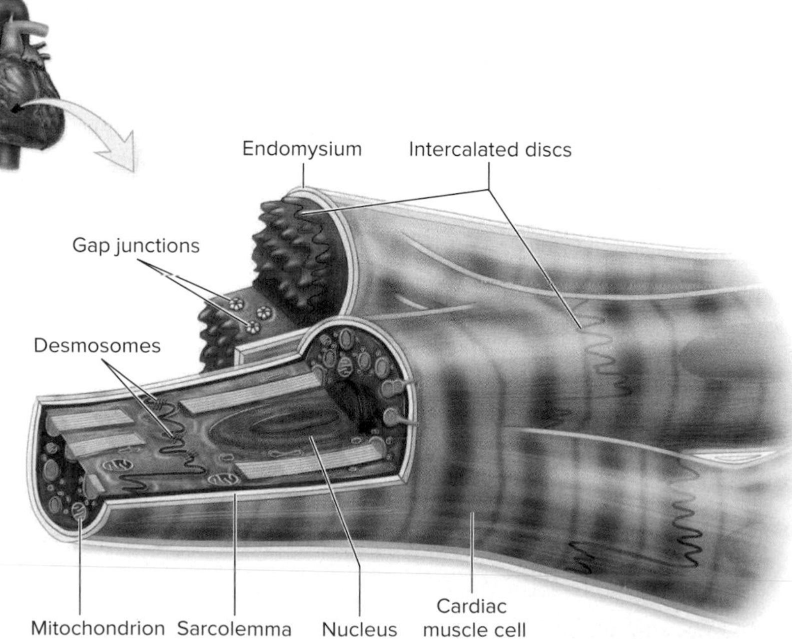

Figure 10.27 Cardiac Muscle. Cardiac muscle is found only in the heart walls. Cardiac muscle cells branch and are connected by intercalated discs. AP|R

10.10 Smooth Muscle Tissue

Smooth muscle tissue is located throughout the body, typically composing approximately 2% of the weight of an adult. Here we describe the general features of smooth muscle, including locations, microscopic anatomy, mechanism of contraction, how it is controlled, and its functional categories.

10.10a Location of Smooth Muscle

LEARNING OBJECTIVE

35. Identify organs of various body systems where smooth muscle is located.

Smooth muscle tissue is found in the walls of organs of many different body systems. Smooth muscle function is determined by its location. The following are examples (**figure 10.28**):

- Cardiovascular system: Blood vessels alter blood pressure and the distribution of blood flow.
- Respiratory system: Bronchioles (air passageways) control the resistance to airflow that enters and exits the air sacs (alveoli) of the lungs.
- Digestive system: The stomach, small intestine, and large intestine mix and propel ingested material as it is moved along the gastrointestinal tract.
- Urinary system: Ureters propel urine from the kidney to the urinary bladder, which eliminates urine from the body.
- Female reproductive system: The uterus helps expel the baby during delivery.

These are but a few examples of smooth muscle distribution in our bodies. Smooth muscle also composes other specialized structures, including the iris of the eye to control the amount of light entering the eye, the ciliary body of the eye for focusing on an object (see section 16.4b), and the arrector pili muscles that produce *goose bumps* (see section 6.2b).

Smooth muscle is similar to both skeletal muscle and cardiac muscle because it has cells that can increase in size (hypertrophy). Both skeletal muscle and cardiac muscle, however, have limited ability to increase in cell number through mitosis (hyperplasia), whereas smooth muscle retains this ability. The smooth muscle in the wall of the uterus exemplifies this characteristic as it increases in thickness during pregnancy from a combination of hypertrophy and hyperplasia (see section 29.5c). There is a significant advantage for any tissue that retains mitotic ability because following an injury the damaged tissue is replaced with the original tissue (not scar tissue) (see section 4.9). Thus, as you consider the various locations of smooth muscle in the body, keep in mind that if smooth muscle is damaged, it is typically replaced with new smooth muscle tissue and has the potential to continue functioning as it did before.

WHAT DID YOU LEARN?

Where is smooth muscle located in the human body?

10.10b Microscopic Anatomy of Smooth Muscle

LEARNING OBJECTIVE

36. Compare the microscopic anatomy of smooth muscle to skeletal muscle.

Smooth muscle cells are small and fusiform shaped (widest in the middle and tapered on the ends) with a centrally located nucleus (**figure 10.29**). They typically have a diameter of 5 to 10 micrometers and a length of 50 to 200 micrometers. Thus, their diameter is up to 10 times smaller and their length thousands of times shorter than

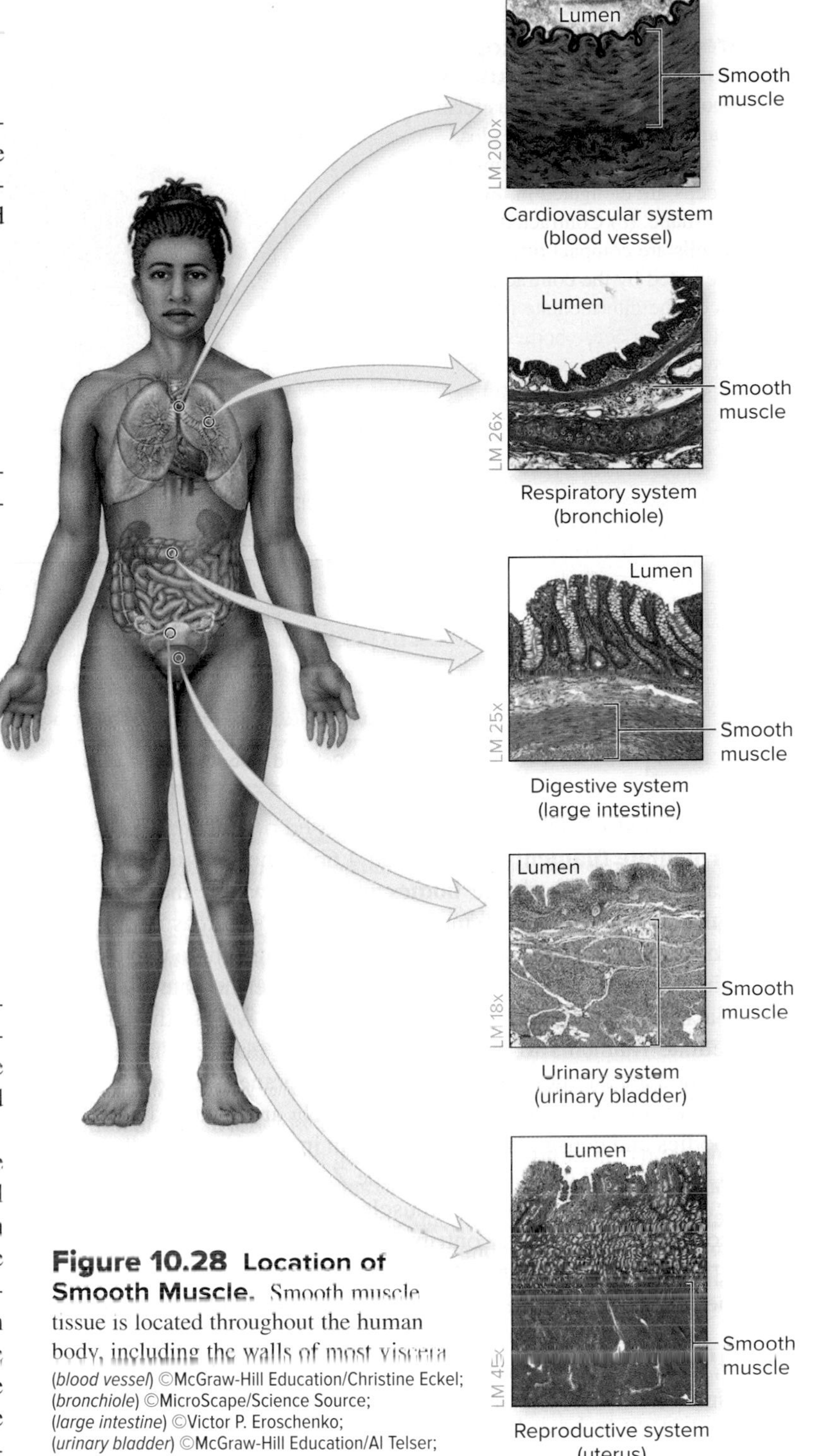

Figure 10.28 Location of Smooth Muscle. Smooth muscle tissue is located throughout the human body, including the walls of most viscera.
(blood vessel) ©McGraw-Hill Education/Christine Eckel; *(bronchiole)* ©MicroScape/Science Source; *(large intestine)* ©Victor P. Eroschenko; *(urinary bladder)* ©McGraw-Hill Education/Al Telser; *(uterus)* ©MicroScape/Science Source

a skeletal muscle fiber. An endomysium wraps around each smooth muscle cell. The small, tapered ends of the cells overlap the larger, middle area of adjacent cells to provide for close packing of cells.

The sarcolemma contains various types of Ca^{2+} channels (e.g., voltage-gated, chemically gated, modality gated) that allow these cells to respond to different types of stimuli. Transverse tubules are absent in smooth muscle cells. Instead, the sarcolemmal surface area is increased by invaginations called **caveolae** (kav′ē-ŏ-lē; pl. *caveola;* small pocket). The sarcoplasmic reticulum is sparse and located close to the sarcolemma, with some caveolae in contact with it. The source of Ca^{2+} comes both from the interstitial fluid outside the cell and the sarcoplasmic reticulum.

Arrangement of Anchoring Proteins and Contractile Proteins of Smooth Muscle

Smooth muscle contains a unique arrangement of anchoring protein structures that includes the cytoskeleton, dense bodies, and dense

CHALLENGE YOURSELF

Do You Know the Basics?

1. The unit of skeletal muscle structure that is composed of bundles of myofibrils, enclosed within a sarcolemma, and surrounded by a connective tissue covering called endomysium is a
   a. myofibril.
   b. fascicle.
   c. myofilament.
   d. skeletal muscle fiber.
2. The physiologic event that takes place at the plasma membrane of a skeletal muscle fiber is
   a. release of calcium.
   b. propagation of an action potential.
   c. binding of calcium by troponin.
   d. crossbridge cycling.
3. In a skeletal muscle fiber, Ca^{2+} is released from
   a. ACh receptors.
   b. the motor end plate.
   c. the sarcoplasmic reticulum.
   d. the sarcolemma and T-tubules.
4. The bundle of dense regular connective tissue that attaches a skeletal muscle to bone is called a(n)
   a. tendon.
   b. ligament.
   c. endomysium.
   d. fascicle.
5. In excitation-contraction coupling, the T-tubules function to
   a. conduct an action potential into the sarcoplasmic reticulum to cause release of calcium.
   b. uptake and store excess Na^+ and K^+ from the sarcoplasm.
   c. keep the thin and thick myofilaments separated.
   d. provide structural support for sarcomeres.
6. During muscle contraction, the I band
   a. hides the H zone.
   b. shortens or narrows.
   c. overlaps the Z line.
   d. always remains the same length.
7. During a concentric contraction of a skeletal muscle fiber, myofibrils
   a. lengthen.
   b. remain the same length.
   c. increase in diameter.
   d. shorten.
8. What event causes a troponin-tropomyosin complex to regain its original shape in muscle relaxation?
   a. stimulation of ACh receptors
   b. diffusion of Na^+ back into transverse tubules
   c. return of Ca^{2+} into the sarcoplasmic reticulum
   d. breaking of the bond with tropomyosin
9. In sustained, moderate exercise, skeletal muscle is predominantly using ______________ as its energy source.
   a. amino acids
   b. glucagon
   c. fatty acids
   d. creatine phosphate
10. Skeletal muscle and cardiac muscle are similar in that both types of muscle
    a. have cells that branch.
    b. contain intercalated discs.
    c. are under involuntary control.
    d. are striated.
11. Explain the structural relationship between a sarcomere, a myofibril, a myofilament, and a skeletal muscle fiber.
12. Diagram and label a sarcomere, including a thick filament, thin filament, A band, H zone, I band, Z disc, and M line.
13. Explain why the ratio of motor neurons to skeletal muscle fibers is greater in muscles that control eye movement than in postural muscles of the leg.
14. Put the following skeletal muscle contraction events in the order that they occur:
    a. The myosin head swivels toward the center of the sarcomere.
    b. Calcium ions are released from the sarcoplasmic reticulum and bind to troponin.
    c. An action potential is propagated along the sarcolemma and transverse tubules.
    d. Myosin binds to actin, forming crossbridges.
    e. Myosin heads bind ATP molecules and release from actin.
    f. Tropomyosin molecules are moved off active sites on actin.
    g. ATPase splits ATP, providing the energy to reset the myosin head.
15. Explain the various means of providing ATP for skeletal muscle contraction.
16. Explain why athletes who excel at short sprints probably have fewer slow-twitch fibers in their lower limb muscles.
17. Explain why skeletal muscle generates the most force when it is at its resting length at the time of stimulation based on the length-tension relationship.
18. Describe the characteristics of smooth muscle that allow it to contract for extended periods of time and not fatigue.
19. Describe the response of smooth muscle to sustained stretch.
20. Identify the location of both multiunit smooth muscle and single-unit smooth muscle, and explain the difference in how each is regulated.

Can You Apply What You've Learned?

1. A bacterial toxin is known to block the release of ACh at the motor end plate of skeletal muscle. Consequently,
   a. the skeletal muscle contracts with increasing force.
   b. the skeletal muscle contracts with increasing frequency.
   c. the ability to stimulate the muscle is impaired.
   d. other neurotransmitters would stimulate the muscle.
2. One of the primary reasons that one individual is faster in a 50-meter sprint than another is
   a. a greater number of muscle fibers of smaller diameter.
   b. more oxidative fibers in the lower limb muscles.
   c. an enhanced ability to deliver oxygen to the muscles.
   d. a greater percentage of fast-twitch fibers in the lower limb muscles.
3. Which electrolyte imbalance is least likely to impair muscle contraction because it is not required in muscle contraction?
   a. F^-
   b. Na^+
   c. K^+
   d. Ca^{2+}
4. Rigor mortis occurs following death because
   a. tropomyosin remains over the myosin binding sites of actin.
   b. myosin heads attach to actin and are not released due to lack of ATP.
   c. the myosin becomes misshapen.
   d. all of the Ca^{2+} remains within the sarcoplasmic reticulum.
5. An athlete participates in aerobic exercise three times a week. One of the changes is an increased ability to deliver oxygen to her skeletal muscles. Over time she notices that she can continue the exercise with greater intensity and duration. The reason for this change is that there is a(n)
   a. greater response from phosphate transfer.
   b. greater production of ATP from glycolysis and less from aerobic cellular respiration.
   c. greater production of ATP from aerobic cellular respiration.
   d. increased production of lactate.

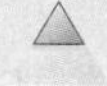

Can You Synthesize What You've Learned?

1. Your anatomy and physiology course is required for a career in forensics, and one of the short essays is an explanation of why the body becomes stiff after death. Provide an answer for an individual who has some understanding of skeletal muscle physiology.
2. Describe the effect of the botulinum toxin, which inhibits the release of acetylcholine at the neuromuscular junction. Would the poison curare, which competes for acetylcholine receptors (by attaching to the acetylcholine receptors and preventing acetylcholine from binding) have a similar effect? Explain.
3. Smooth muscle is within the urinary bladder wall. Explain why, if you initially have the sensation of having to urinate, the sensation sometimes passes. Base your answer on the stress-relaxation response.

INTEGRATE

ONLINE STUDY TOOLS

connect | SMARTBOOK® | APR

The following study aids may be accessed through Connect.

Concept Overview Interactive: Figure 10.16: Skeletal Muscle Contraction

Clinical Case Study: Progressive Weakness in a Young Woman

Interactive Questions: This chapter's content is served up in a number of multimedia question formats for student study

SmartBook: Topics and terminology include introduction to skeletal muscle; anatomy and physiology of skeletal muscle; skeletal muscle metabolism; skeletal muscle fiber types; measurement of skeletal muscle tension; factors affecting skeletal muscle tension; effects of exercise and aging on skeletal muscle; cardiac muscle tissue; smooth muscle tissue

Anatomy & Physiology Revealed: Topics include skeletal muscle; skeletal muscle striations; sarcomere; sliding filament; neuromuscular junction; excitation-contraction coupling; crossbridge cycle; cardiac muscle; smooth muscle

Animations: Topics include skeletal muscle; function of the neuromuscular junction; action potentials and muscle contraction; sarcomere shortening; breakdown of ATP and crossbridge movement during muscle contraction; mechanics of single fiber contraction; activation of contraction in smooth vs. skeletal muscle

Design Elements: **Learning Objective** icon: ©McGraw-Hill Education; **What Do You Think?** icon: ©McGraw-Hill Education; **What Did You Learn?** icon: ©McGraw-Hill Education; **Learning Strategy** icon: ©Slavoljub Pantelic/Shutterstock RF; **Clinical View** icon: ©Laia Design Studio/Shutterstock RF; **Concept Connection** icon: ©McGraw-Hill Education; **Challenge Yourself** icon: ©McGraw-Hill Education; **Do You Know the Basics?** icon: ©McGraw-Hill Education; **Can You Apply What You've Learned?** icon: ©McGraw-Hill Education; **Can You Synthesize What You've Learned?** icon: ©McGraw-Hill Education; **Online Study Tools** icon: ©ayzek/Shutterstock RF

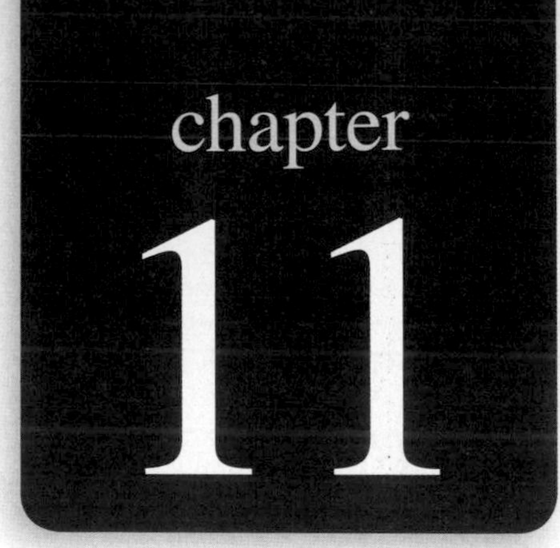

Muscular System: Axial and Appendicular Muscles

The partitioning of the skeletal system into axial and appendicular divisions provides a useful guideline for subdividing the muscular system. **Axial muscles** have both their attachments on parts of the axial skeleton. Axial muscles support and move the head and vertebral column, function in nonverbal communication by affecting facial features, move the mandible during chewing, assist in food processing and swallowing, aid in breathing, and both support and protect the abdominal and pelvic organs. The **appendicular muscles** control the movements of the upper and lower limbs, and stabilize and control the movements of the pectoral and pelvic girdles. These muscles are organized into groups based upon their location in the body or the part of the skeleton they move. Some muscles of both divisions are shown in **figure 11.1**.

The muscles in this chapter have been organized into groups according to their location in the body. For each group, tables provide descriptions of the muscles as well as information about their action, attachments, and innervation. (Note: The word *innervation* refers to the nerve(s) that control(s) a muscle and stimulate(s) it to contract. For further information about the nerves listed in the tables, see sections 13.9 and 14.5.)

INTEGRATE

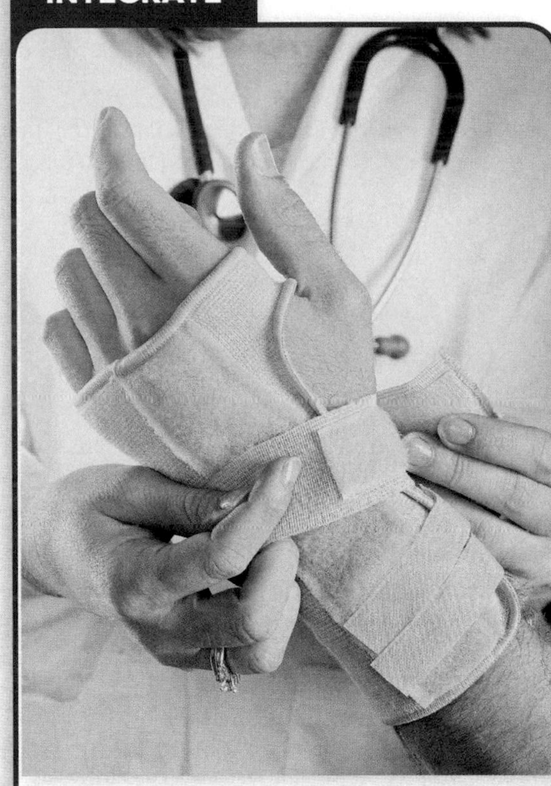

CAREER PATH
Physical Therapist

Physical therapists help individuals achieve greater mobility and improvement of their quality of life. Orthopedic physical therapists (a specialty of physical therapy) treat musculoskeletal disorders and help patients regain mobility after orthopedic surgery. An understanding of skeletal muscle function and which muscles work synergistically and antagonistically is essential for orthopedic physical therapists. These health-care professionals utilize this knowledge to develop treatment plans for the patient, such as the patient shown here (who is recovering from a wrist injury).

Module 6: Muscular System

Figure 11.1 Body Musculature. (*a*) Anterior view shows superficial muscles on the right side of the body and some deeper muscles on the left side. (*b*) Posterior view shows superficial muscles on the left side of the body and some deeper muscles on the right side. Labels for the axial muscles are in bold; not all muscles shown in the figure are identified.

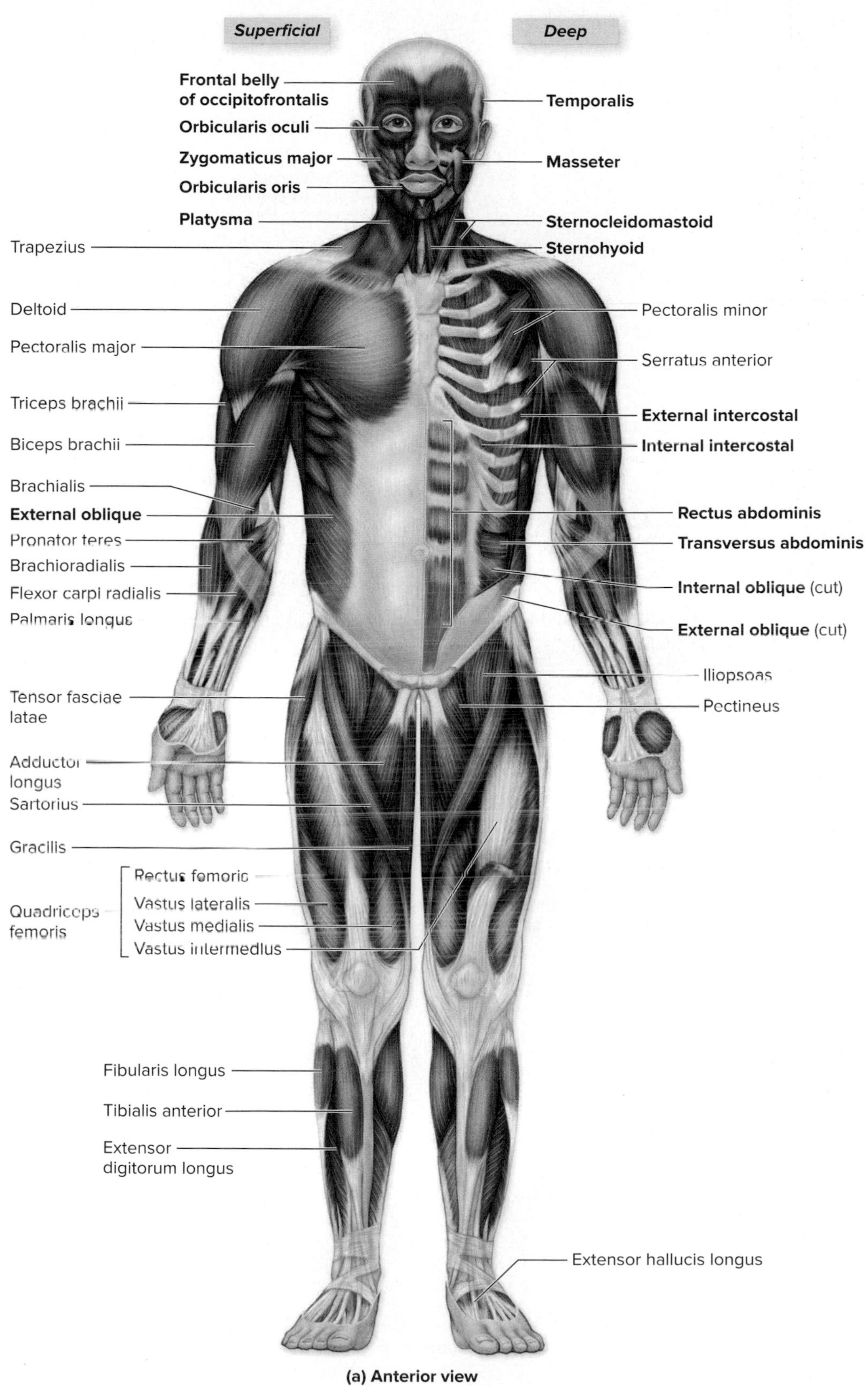

(a) Anterior view

(continued on next page)

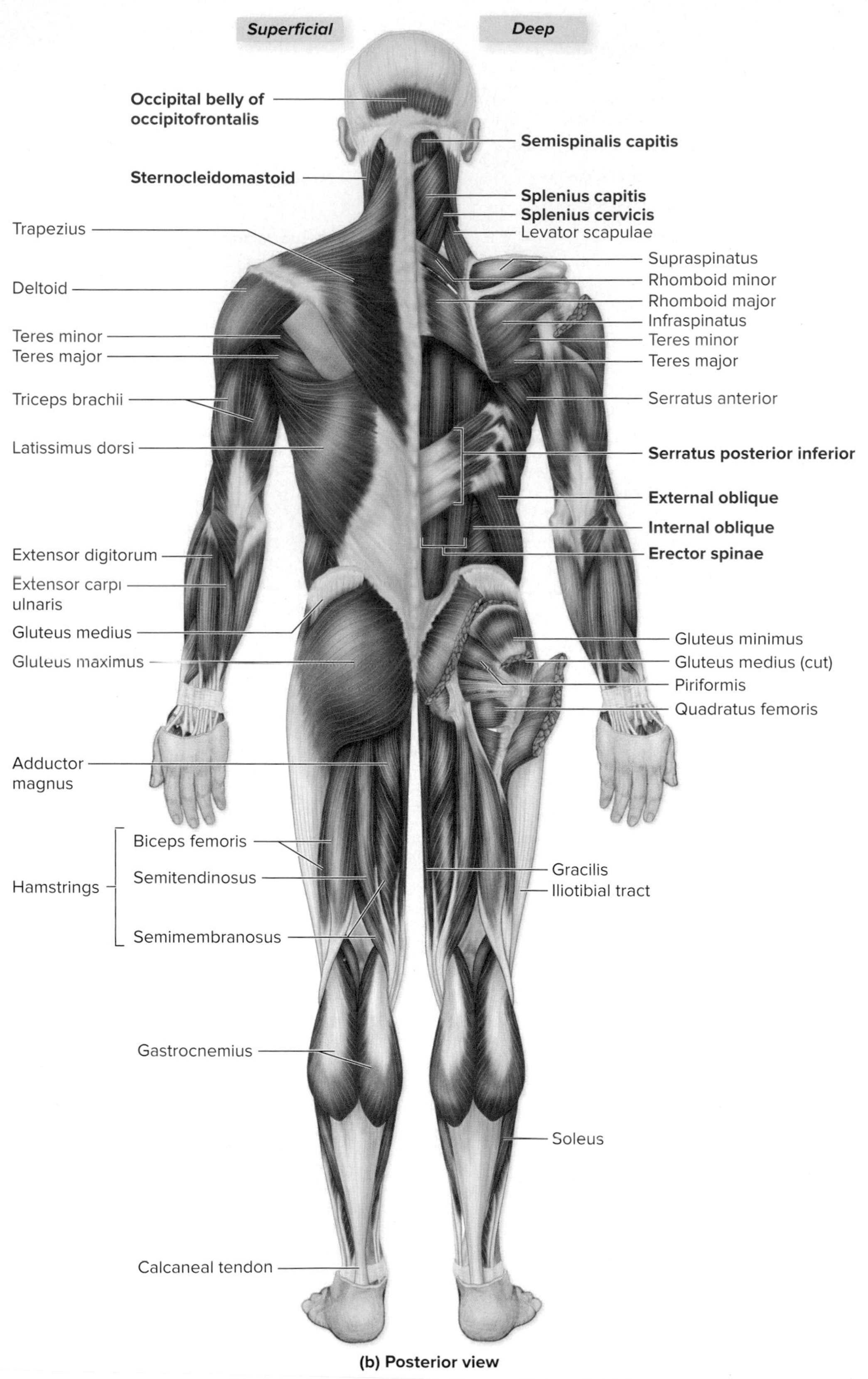

Superficial
Deep
Occipital belly of occipitofrontalis
Semispinalis capitis
Sternocleidomastoid
Splenius capitis
Splenius cervicis
Levator scapulae
Trapezius
Supraspinatus
Rhomboid minor
Deltoid
Rhomboid major
Infraspinatus
Teres minor
Teres minor
Teres major
Teres major
Triceps brachii
Serratus anterior
Latissimus dorsi
Serratus posterior inferior
External oblique
Internal oblique
Extensor digitorum
Erector spinae
Extensor carpi ulnaris
Gluteus medius
Gluteus minimus
Gluteus maximus
Gluteus medius (cut)
Piriformis
Quadratus femoris
Adductor magnus
Biceps femoris
Gracilis
Hamstrings
Semitendinosus
Iliotibial tract
Semimembranosus
Gastrocnemius
Soleus
Calcaneal tendon

(b) Posterior view

11.1 Skeletal Muscle Composition and Actions

We examined the macroscopic and microscopic anatomy of skeletal muscle in section 10.2. Here we discuss the attachments of a skeletal muscle, the organizational patterns exhibited by skeletal muscle fibers, and the general actions of skeletal muscles.

11.1a Skeletal Muscle Attachments

LEARNING OBJECTIVE

1. Compare and contrast the superior (or proximal) and inferior (or distal) attachments of a skeletal muscle.

At the ends of a muscle, the three connective tissue layers (epimysium, perimysium, and endomysium) merge to form a fibrous **tendon,** which attaches the muscle to bone, skin, or another muscle. Tendons usually have a thick, cordlike structure. Sometimes, the tendon forms a thin, flattened sheet, termed an **aponeurosis** (ap′ō-nū-rō′sis; *apo* = from, *neuron* = sinew).

Most skeletal muscles extend between bones and cross at least one mobile joint. Upon contraction, one of the bones moves while the other bone usually remains fixed. Previously, the less movable attachment was called the *origin* and the more movable attachment was called the *insertion*. However, the origin and insertion of muscles are not always easily determined by either movement or position, and some anatomists and clinicians are no longer using this terminology. Thus, we typically will use the terms **superior attachment** and **inferior attachment** when discussing most axial muscles and **proximal attachment** and **distal attachment** when discussing the muscles moving the appendicular skeleton. For muscles moving the axial skeleton, the superior attachment often (but not always) is more movable than the inferior attachment; thus, the superior portion of the body is pulled toward the inferior portion when the muscles contract, as in flexing your neck or doing an abdominal crunch. In muscles moving the appendicular skeleton, the distal attachment typically is more movable than the proximal attachment, and when muscles contract, the distal attachment moves toward the proximal attachment, as shown in **figure 11.2**. The biceps brachii muscle's proximal attachment is the scapula, and the distal attachment is the radius. As the biceps brachii contracts, the radius is pulled toward the scapula, causing the arm to flex at the elbow.

There are some exceptions where there is no clearly defined superior or inferior attachment of an axial muscle. With those exceptions, slightly alternative terminology will be used and noted in the appropriate tables.

WHAT DID YOU LEARN?

1. What is the difference between the proximal and distal attachment of a skeletal muscle?

11.1b Organizational Patterns of Skeletal Muscle Fibers

LEARNING OBJECTIVE

2. Describe and differentiate between the organizational patterns in muscle fascicles.

As mentioned in section 10.2a, bundles of muscle fibers, termed *fascicles,* lie parallel to each other within each muscle. However, the organization of fascicles in different muscles often varies. There are four different patterns of fascicle arrangement: circular, parallel, convergent, and pennate (**figure 11.3**).

Circular Muscles

A **circular muscle** has concentrically arranged muscle fascicles around an opening or recess. A circular muscle is also called a *sphincter,* and

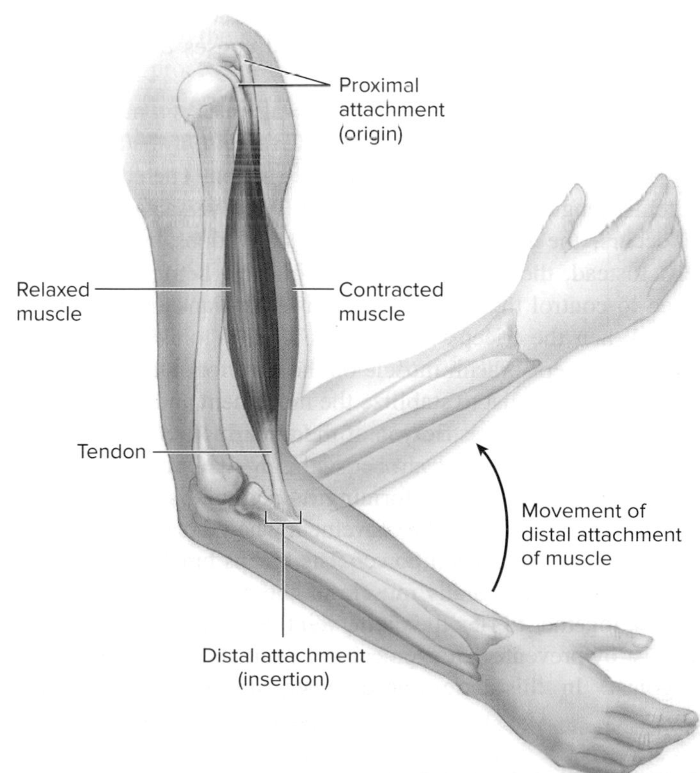

Figure 11.2 Muscle Attachments. For appendicular muscles, the proximal attachment is less movable than the distal attachment, as shown in this view of the biceps brachii muscle.

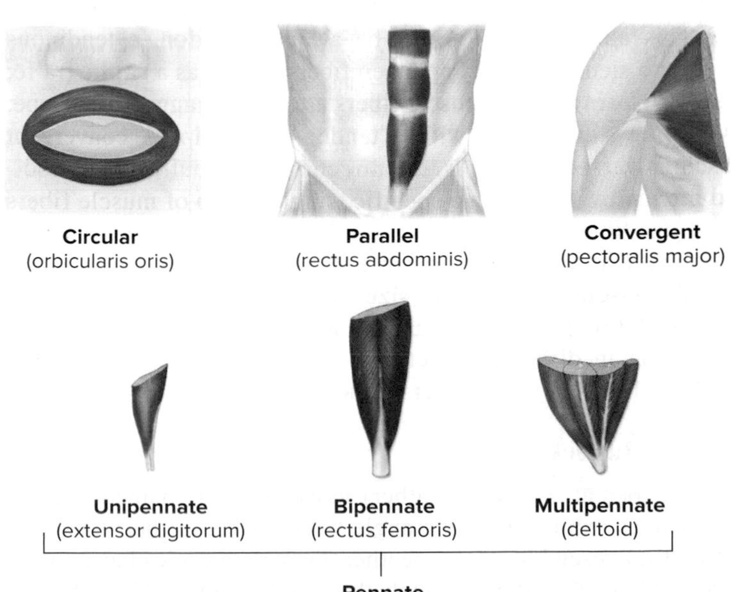

Figure 11.3 Organization of Muscle Fibers. Muscle fascicles may be organized into one of four basic patterns: circular, parallel, convergent, or pennate.

INTEGRATE

LEARNING STRATEGY

The following suggestions may help you learn the muscles:

- Muscles may be organized into groups. Learning the muscles in groups is easiest.
- When studying a particular muscle, try to palpate it on yourself, and look in a mirror so you can better visualize the muscle's location. Contract the muscle to sense its action.
- Repeat the name of a muscle aloud to become familiar with it.
- Associate visual images from models, cadavers, a photographic atlas, or dissected animals with muscle names.
- Locate the attachments of muscles on an articulated skeleton to understand how they produce their actions.
- Learn the derivation of each muscle name.

11.3 Muscles of the Head and Neck

The muscles of the head and neck are separated into several specific groups. Almost all of these muscles (except for a few muscles of the anterior neck) attach to either the bones of the skull or the hyoid bone.

11.3a Muscles of Facial Expression

LEARNING OBJECTIVES

6. Name the muscles that move the forehead, the skin around the eyes, and the nose, and describe their actions.
7. List the muscles that move the mouth and cheeks and their actions.

The muscles of facial expression attach from the superficial fascia or skull bones to the superficial fascia of the skin (**figure 11.5**). When these muscles contract, they pull on the skin, causing it to move. Most of these muscles are innervated by the seventh cranial nerve (CN VII), termed the *facial nerve* (see section 13.9).

The **epicranius** is composed of the **occipitofrontalis muscle** and a broad **epicranial aponeurosis,** also called the *galea aponeurotica.* The **frontal belly** of the occipitofrontalis is superficial to the frontal bone on the forehead. When this muscle contracts, it raises the eyebrows and wrinkles the skin of the forehead. The **occipital belly** of the occipitofrontalis covers the posterior aspect of the skull. When this muscle contracts, it retracts the scalp slightly.

Deep to the frontal belly of the occipitofrontalis is the **corrugator supercilii.** This muscle draws the eyebrows together and forms vertical wrinkle lines around the nose. The **orbicularis oculi** consists of circular muscle fibers that surround the eye's orbit. When this muscle contracts, the eyelid closes, as when you wink, blink, or squint. The **levator palpebrae superioris** elevates the upper eyelid when you open your eyes.

Several muscles of facial expression are associated with the nose. The **nasalis** elevates the corners of the nostrils. When you "flare" your nostrils, you are using the nasalis muscles. If you wrinkle your nose in distaste after smelling a foul odor, you have used your **procerus** muscle. This muscle is continuous with the frontal belly of the occipitofrontalis muscle and runs over the bridge of the nose, where it produces transverse wrinkles when it contracts.

The mouth is the most expressive part of the face. The **orbicularis oris** consists of muscle fibers that encircle the opening of the mouth. When this muscle contracts, you close your mouth. In addition, when you "pucker up" for a kiss, you are using this muscle. The **depressor labii inferioris** does what its name suggests—it pulls the lower lip inferiorly. The **depressor anguli oris** is considered the "frown" muscle, because it pulls the corners of the mouth inferiorly. (Note, however, that it takes more muscles than just this one to produce a frown.)

In contrast, some muscles of the mouth elevate part or all of the upper lips. The **levator labii superioris** pulls the upper lip superiorly, as if a person were sneering or snarling. The **levator anguli oris** pulls the corners of the mouth superiorly and laterally. The **zygomaticus major** and **zygomaticus minor** work with the levator anguli oris

INTEGRATE

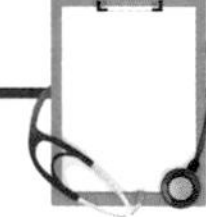

CLINICAL VIEW 11.2

Idiopathic Facial Nerve Paralysis (Bell Palsy)

Unilateral paralysis of the muscles of facial expression is termed **facial nerve (CN VII) paralysis.** If the cause of the condition is unknown, doctors refer to it as **idiopathic** (id′ē-ō-path′ik; *idios* = one's own, *pathos* = suffering) **facial nerve paralysis,** or *Bell palsy.* Facial nerve paralysis has sometimes been associated with herpes simplex 1 viral infection. Facial nerve paralysis is also associated with exposure to cold temperatures and is commonly seen in individuals who sleep with one side of their head facing an open window. Whatever the underlying cause, the nerve becomes inflamed and compressed within the narrow stylomastoid foramen (see figure 8.7), through which the facial nerve extends. The muscles on the same side of the face are paralyzed as a result. Individuals with later stages of Lyme disease may develop bilateral (instead of unilateral) facial nerve paralysis.

Treatment of facial nerve paralysis usually means alleviating the symptoms. Doctors often use prednisone (a type of steroid) to reduce the inflammation and swelling of the nerve. If herpes simplex infection is suspected, an antiviral medication known as acyclovir (Zovirax) is also given. Facial nerve paralysis caused by Lyme disease (a bacterial infection) is treated with antibiotics. Like its underlying cause, recovery from idiopathic facial nerve paralysis is equally mysterious. Over 50% of all patients experience a complete, spontaneous recovery within 30 days of their first symptoms. Recovery may take longer for some, whereas other patients may never recover.

Facial nerve (CN VII) paralysis on the left side of the face. Note the drooping left side of the mouth (black arrow) and the lack of elevation of the eyebrow by the left frontal belly muscle (white arrow) when the woman tries to smile.

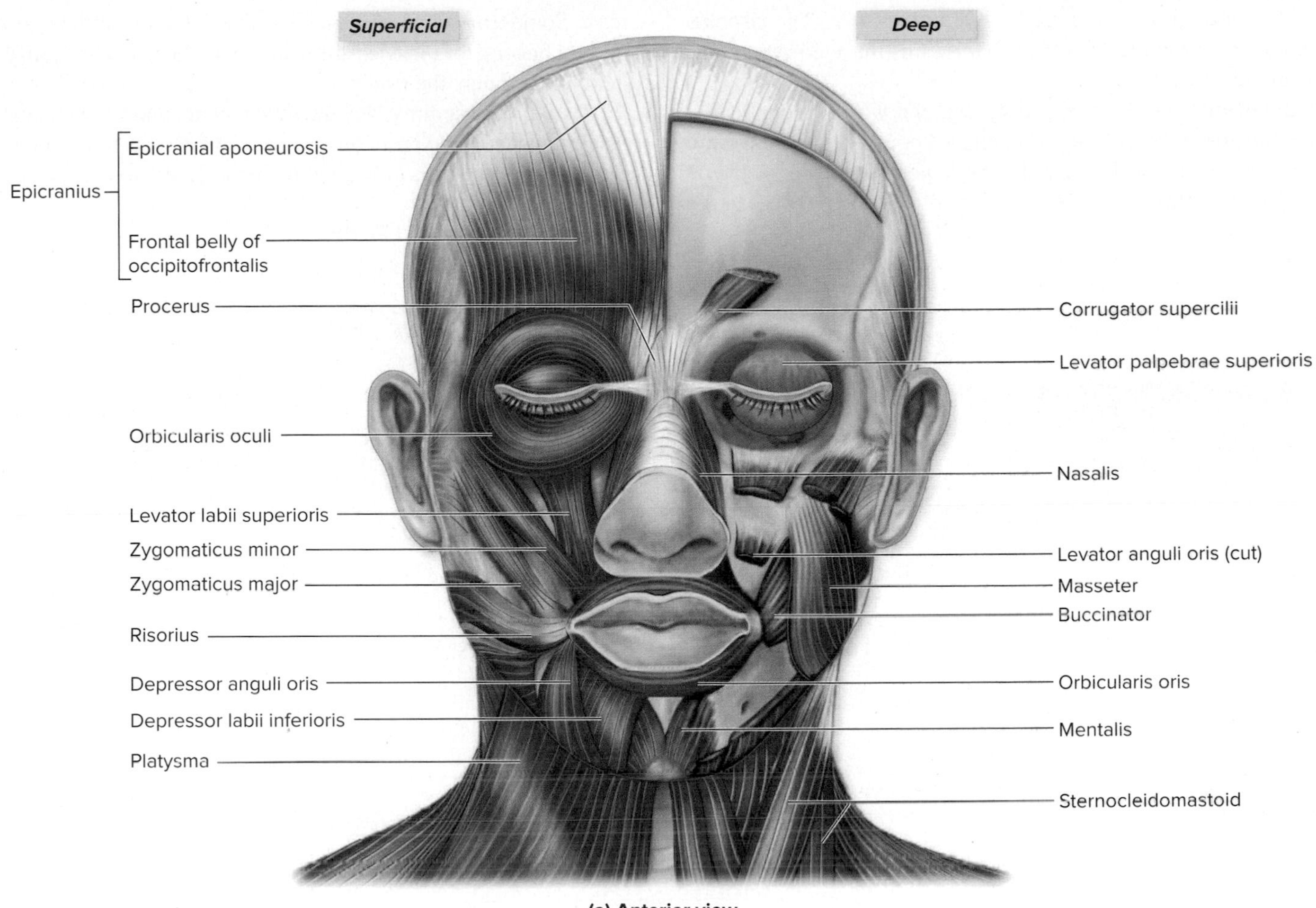

(a) Anterior view

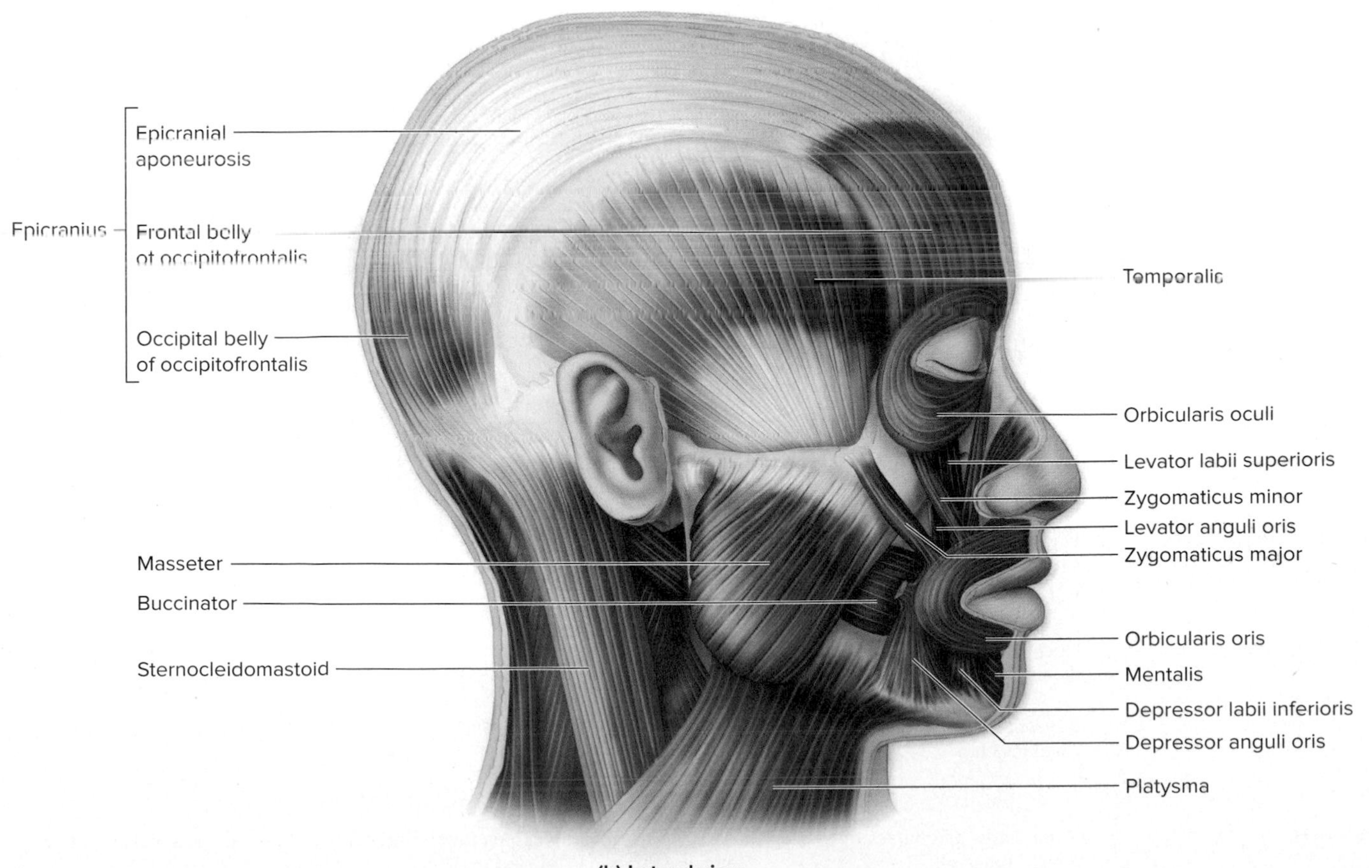

(b) Lateral view

Figure 11.5 **Muscles of Facial Expression.** APIR

muscles. You use all of these muscles when you smile. The **risorius** pulls the corner of the lips laterally; you use this muscle if you make a closed-mouth smile.

The **mentalis** attaches to the lower lip, and when it contracts, it protrudes the lower lip (as when a person "pouts"). The **platysma** tenses the skin of the neck and pulls the lower lip inferiorly. If you stand in front of a mirror and tense the skin of your neck, you can see these thin muscles bulging out.

The **buccinator** compresses the cheek against the teeth when we chew (and is the reason our cheeks don't bulge out like a squirrel's when we eat). Infants use the buccinator when they suckle at the breast. Some trumpet players (such as Dizzy Gillespie) have stretched out their buccinator muscles, allowing their cheeks to be "puffy" with air when they play the trumpet.

Table 11.1 summarizes the attachments and movements of the muscles of facial expression. **Figure 11.6** illustrates how these muscles produce some of the more characteristic expressions.

WHAT DID YOU LEARN?

6 What muscles of facial expression must contract for you to smile?

7 The corners of the mouth are pulled inferiorly into a frown position by the contraction of what muscle?

Table 11.1 Muscles of Facial Expression

Region/Muscle	Action(s)	Bony Attachment(s) (B)/Soft Tissue Attachment(s) (S)[1]	Innervation (see section 13.9)
SCALP			
Epicranius (ep′ĭ-krā′nē-us; *epi* = over, *cran* = skull): Composed of an epicranial aponeurosis and the two bellies of the occipitofrontalis muscle			
Frontal belly of occipitofrontalis (ok-sip′i-tō-fron-tā′lis) *front* = forehead	Moves scalp, eyebrows; wrinkles skin of forehead	B: Epicranial aponeurosis S: Skin and subcutaneous layer of forehead and eyebrows	CN VII (facial nerve)
Occipital belly of occipitofrontalis *occipito* = base of skull	Retracts scalp	B: Superior nuchal line S: Epicranial aponeurosis	CN VII (facial nerve)
NOSE			
Nasalis (nā′ză-lis)	Compresses bridge and depresses tip of nose; elevates corners of nostrils	B: Maxillae and alar cartilage of nose S: Dorsum of nose	CN VII (facial nerve)
Procerus (prō-sē′rŭs) *procerus* = long	Moves and wrinkles nose	B: Nasal bone and lateral nasal cartilage S: Aponeurosis at bridge of nose and skin of forehead	CN VII (facial nerve)
MOUTH			
Buccinator (buk′sĭ-nā′tōr) *bucco* = cheek	Compresses cheek, holds food between teeth during chewing	B: Alveolar processes of mandible and maxillae S: Orbicularis oris muscle	CN VII (facial nerve)
Depressor anguli oris (dĕ-pres′ŏr ang′gyū-lī ōr-is) *depressor* = depresses *angul* = angle *oris* = mouth	Draws corners of mouth inferiorly and laterally ("frown" muscle)	B: Body of mandible S: Skin at inferior corner (angle) of mouth	CN VII (facial nerve)
Depressor labii inferioris (lā′bē-ī in-fēr′ē-ōr-is) *labi* = lip *infer* = below	Draws lower lip inferiorly	B: Body of mandible lateral to midline S: Skin at inferior lip	CN VII (facial nerve)
Levator anguli oris (lē-vā′tor, le-vā′ter) *leva* = raise	Draws corners of mouth superiorly and laterally ("smile" muscle)	B: Lateral maxilla S: Skin at superior corner (angle) of mouth	CN VII (facial nerve)
Levator labii superioris (sū-pēr′ē-ōr-is)	Opens lips; raises and furrows the upper lip ("Elvis" lip snarl)	B: Zygomatic bone; maxilla S: Skin and muscle of superior lip	CN VII (facial nerve)
Mentalis (men-tă′lis) *ment* = chin	Protrudes lower lip ("pout"); wrinkles chin	B: Central mandible S: Skin of chin	CN VII (facial nerve)
Orbicularis oris (ōr-bik′yū-lā′ris) *orb* = circular	Compresses and purses lips ("kiss" muscle)	B: Maxilla and mandible; blend with fibers from other facial muscles S: Encircling mouth; skin and muscles at angles to mouth	CN VII (facial nerve)

1. The muscles of facial expression arise on bone, and their more movable attachment typically is on muscle, tendon, subcutaneous tissue, or integument. Thus, it is more appropriate to refer to *bony attachment* and *soft tissue attachment* for these muscles.

Table 11.1	Muscles of Facial Expression (continued)		
Region/Muscle	**Action(s)**	**Bony Attachment(s) (B)/Soft Tissue Attachment(s) (S)[1]**	**Innervation (see section 13.9)**
Risorius (ri-sōr′ē-ūs) *risor* = laughter	Draws corner of lip laterally; tenses lips	B: Deep fascia associated with masseter muscle S: Skin at angle of mouth	CN VII (facial nerve)
Zygomaticus major (zī′gō-mat′i-kus) *zygomatic* = cheekbone *major* = greater	Elevates corner of mouth ("smile" muscle)	B: Zygomatic bone S: Skin at superolateral edge of mouth	CN VII (facial nerve)
Zygomaticus minor *minor* = lesser	Elevates corner of mouth ("smile" muscle)	B: Zygomatic bone S: Skin of superior lip	CN VII (facial nerve)
		EYE	
Corrugator supercilii (kōr′ū-gā-ter sū′per′sil′ē-ī) *corrugo* = to wrinkle *cilium* = eyelid	Pulls eyebrows inferiorly and medially; forms vertical wrinkles above nose	B: Medial end of superciliary arch S: Skin superior to supraorbital margin and superciliary arch	CN VII (facial nerve)
Levator palpebrae superioris (pal-pē′brā) *palpebra* = eyelid	Elevates superior eyelid	B: Lesser wing of sphenoid bone S: Superior tarsal plate and skin of superior eyelid	CN III (oculomotor nerve)
Orbicularis oculi (ok′yū-lī) *ocul* = eye	Closes eye; produces winking, blinking, squinting ("blink" muscle)	B: Medial wall or margin of the orbit S: Skin surrounding eyelids	CN VII (facial nerve)
		NECK	
Platysma[2] (plă-tiz′mă) *platy* = flat	Pulls lower lip inferiorly; tenses skin of neck	B: Fascia of deltoid and pectoralis major muscles and acromion of scapula[2] S: Skin of cheek and mandible	CN VII (facial nerve)

2. Platysma is an exception to the "bony attachment" rule. Its traditional "origin" arises primarily from muscle.

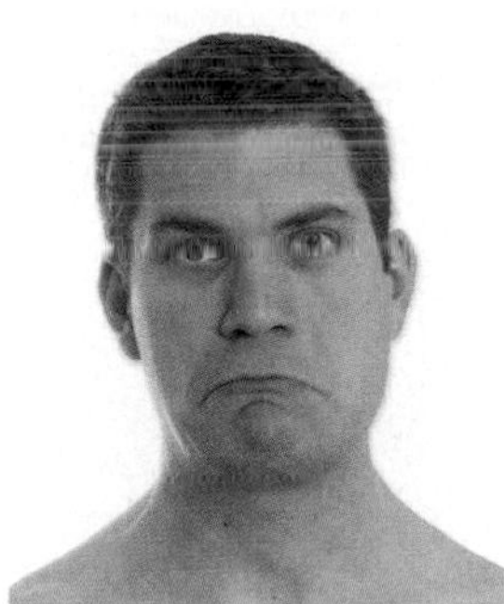
Depressor anguli oris (frown)

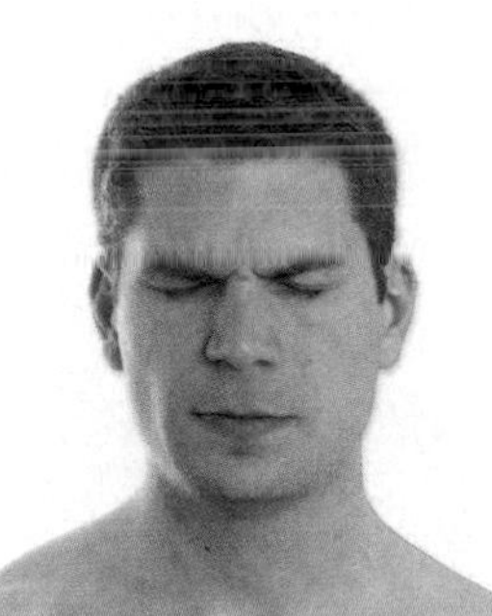
Orbicularis oculi (blink/close eyes)

Zygomaticus major (smile)

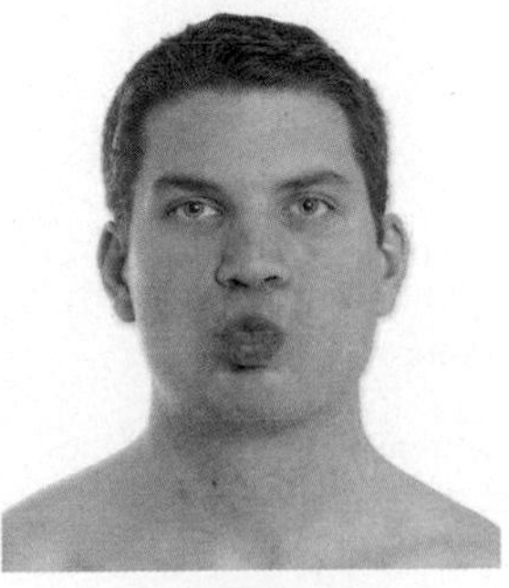
Orbicularis oris (close mouth/kiss)

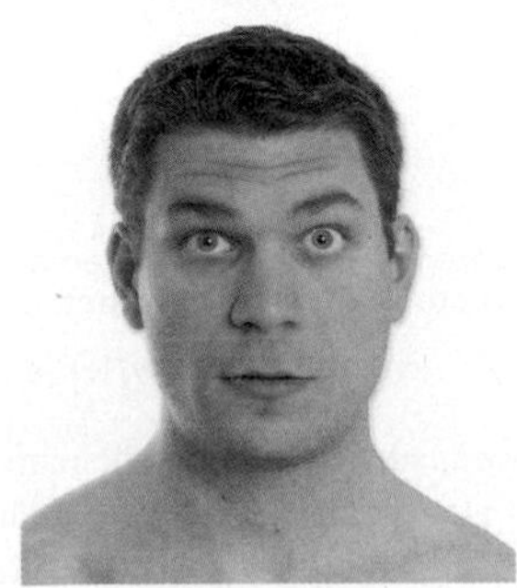
Frontal belly of occipitofrontalis (wrinkle forehead, raise eyebrows)

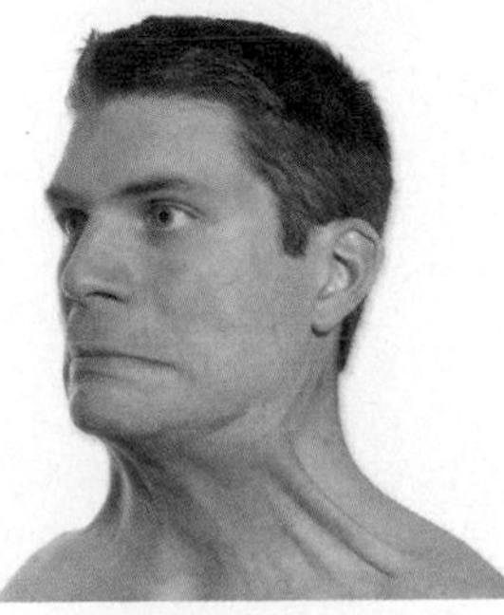
Platysma (tense skin of neck)

Figure 11.6 Surface Anatomy of Some Muscles of Facial Expression. These muscles permit complex expressions that often are used as a means of communication.

11.3b Extrinsic Eye Muscles

LEARNING OBJECTIVES

8. Become familiar with the six extrinsic muscles of the eye, and describe how each affects eye movement.
9. Name the three cranial nerves that innervate the extrinsic eye muscles, and identify which muscles they act upon.

The **extrinsic eye muscles,** often called *extraocular muscles,* move the eyes. They are termed extrinsic because they attach to the outer white surface of the eye, called the *sclera* (see section 16.4b). There are six extrinsic eye muscles: the four rectus muscles (medial, lateral, inferior, and superior) and the two oblique muscles (inferior and superior) (**figure 11.7**).

The four rectus eye muscles arise from a **common tendinous ring** in the orbit. These muscles attach to the *anterior* part of the eye and are named according to which part of the eye they are located (medial, lateral, inferior, or superior).

The **medial rectus** attaches to the anteromedial surface of the eye and pulls the eye medially (adducts the eye). It is innervated by CN III (oculomotor nerve). The **lateral rectus** attaches to the

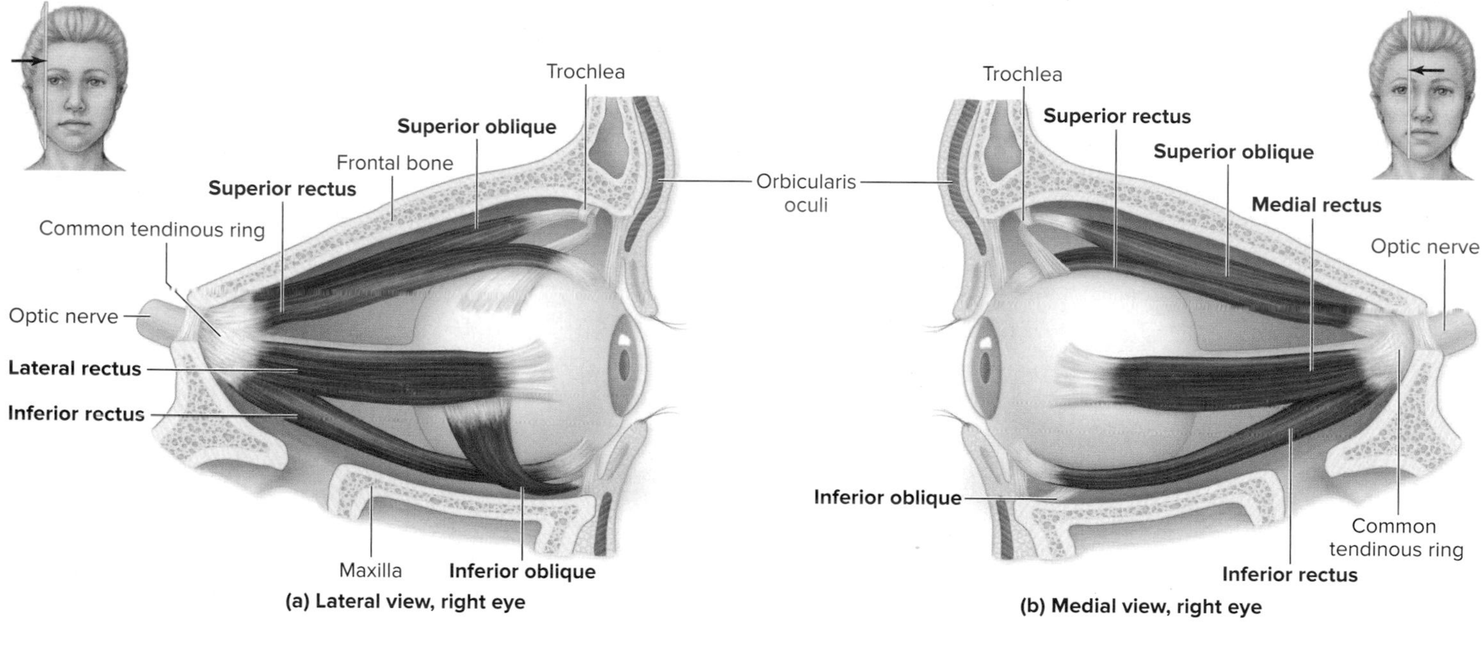

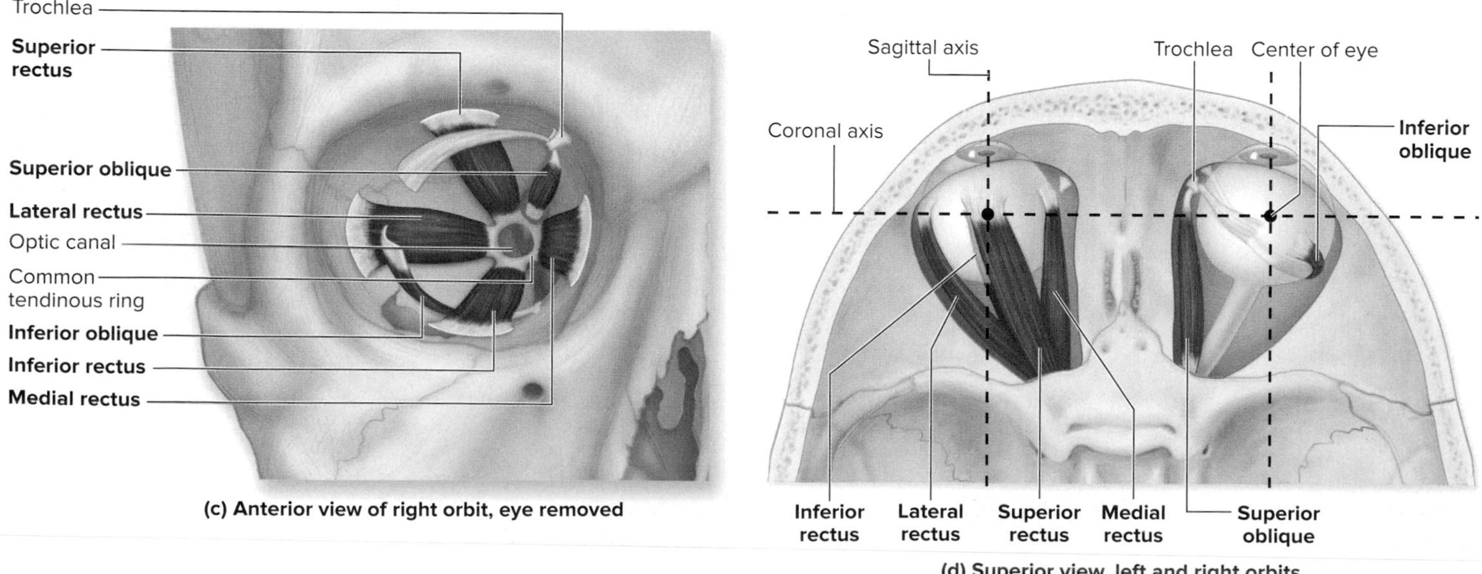

Figure 11.7 Extrinsic Muscles of the Eye. The extrinsic eye muscles control movements of the eye. The names of these muscles are in bold in the figure. (*a*) The attachments for most of the extrinsic eye muscles are noticeable in a lateral view of the right eye. (*b*) The medial rectus muscle appears prominently in a medial view of the right eye. (*c*) The four rectus muscles arise from a common tendinous ring, shown here in an anterior view of the right orbit. (*d*) A superior view of the left and right orbits illustrates the eye attachment differences between the rectus and oblique muscles and how these eye attachment differences affect their movement of the eye. AP|R

anterolateral surface of the eye and pulls the eye laterally (abducts the eye). This muscle is innervated by CN VI (abducens). (Notice that this nerve's name tells you the action of the muscle it innervates—the muscle abducts the eye.) The **inferior rectus** attaches to the anteroinferior part of the eye. The inferior rectus pulls the eye inferiorly (as when you look down) and medially (as when you look at your nose). The **superior rectus** is located superiorly and attaches to the anterosuperior part of the sclera. The superior rectus pulls the eye superiorly (as when you look up) and medially (as when you look at your nose). The inferior and superior rectus muscles are innervated by CN III. Figure 11.7*d* illustrates that the superior and inferior rectus muscles do not pull directly parallel to the long axis of the eye; that is why both muscles also move the eye slightly in the medial direction.

The oblique eye muscles arise from within the orbit and attach to the *posterolateral* part of the sclera of the eye. The **inferior oblique** elevates the eye and turns the eye laterally. Since this muscle attaches to the inferior *posterior* part of the eye, contracting this muscle pulls the posterior part of the eye inferiorly (but *elevates* the *anterior* part of the eye). This muscle is innervated by CN III. The **superior oblique** depresses the eye and turns the eye laterally. This muscle passes through a pulleylike loop, called the **trochlea,** in the anteromedial orbit. This muscle attaches to the superior *posterior* part of the eye, so contracting this muscle pulls the posterior part of the eye superiorly (but *depresses* the *anterior* surface of the eye). This muscle is innervated by CN IV (trochlear). (Notice that this nerve's name is derived from the trochlea that holds the superior oblique in place.)

Table 11.2 compares the extrinsic muscles of the eye. Refer to section 13.9 to review the cranial nerves mentioned in this section.

WHAT DID YOU LEARN?

8 Which extrinsic eye muscles abduct the eye (move the eye laterally)?

INTEGRATE

CONCEPT CONNECTION

Vision problems may be due to musculoskeletal issues, nervous system issues, or both. For example, if a person cannot abduct his right eye, that indicates that CN VI (abducens nerve) may be injured (see section 13.9). In addition, some vision problems may be due to a single weak extrinsic muscle, and correcting this muscle imbalance (with either exercises or wearing a patch over the eye with the stronger extrinsic muscles) helps alleviate the vision problem. Thus, a physician must integrate both the innervation and muscle information to properly diagnose a patient's visual disturbance.

INTEGRATE

LEARNING STRATEGY

The following "chemical formula" can help you remember the eye muscle innervation:

$[(SO_4)(LR_6)]_3$

In words, the superior oblique **(SO)** is innervated by cranial nerve IV **(4),** the lateral rectus **(LR)** is innervated by cranial nerve VI **(6),** and the rest of the eye muscles are innervated by cranial nerve III **(3).**

INTEGRATE

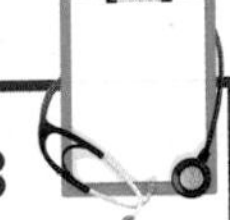

CLINICAL VIEW 11.3

Strabismus and Diplopia

The extrinsic eye muscles move the left and right eyes in unison, so both eyes focus on the same image. **Strabismus** (strā-biz′mŭs; *strabismos* = a squinting) is a condition where both eyes cannot focus on the same image, and the gaze of one eye is displaced. Strabismus may be caused by cranial nerve injury (see section 13.9) or weak eye muscles of one eye, or it may occur when the brain favors vision in the stronger eye (which is the cause of *lazy eye,* where the weaker eye muscle doesn't track properly and may favor one position). Individuals with strabismus may experience **diplopia** (dip′lō′pē′ă; diploos = double, *ops* = eye), or *double vision.*

Table 11.2 Extrinsic Eye Muscles

Group/Muscle	Action(s)	Orbit Attachment (O)/Eyeball Attachment (E)[1]	Innervation (see section 13.9)
RECTUS MUSCLES			
Medial rectus (mē′dē-ăl rek′tus) *rectus = straight*	Moves eye medially (adducts eye)	O: Common tendinous ring E: Anteromedial surface of eye	CN III (oculomotor nerve)
Lateral rectus (lat′er-ăl)	Moves eye laterally (abducts eye)	O: Common tendinous ring E: Anterolateral surface of eye	CN VI (abducens nerve)
Inferior rectus	Moves eye inferiorly (depresses eye) and medially (adducts eye)	O: Common tendinous ring E: Anteroinferior surface of eye	CN III (oculomotor nerve)
Superior rectus	Moves eye superiorly (elevates eye) and medially (adducts eye)	O: Common tendinous ring E: Anterosuperior surface of eye	CN III (oculomotor nerve)
OBLIQUE MUSCLES			
Inferior oblique *obliquus = slanting*	Moves eye superiorly (elevates eye) and laterally (abducts eye)	O: Anterior orbital surface of maxilla E: Posteroinferior, lateral surface of eye	CN III (oculomotor nerve)
Superior oblique	Moves eye inferiorly (depresses eye) and laterally (abducts eye)	O: Sphenoid bone E: Posterosuperior, lateral surface of eye	CN IV (trochlear nerve)

1. The less movable attachment for an eye muscle is on the bony orbit or tendinous ring, whereas the more movable attachment is on the eyeball.

Figure 11.8 Muscles of Mastication. (*a*) Superficial and (*b*) deep lateral views of the muscles of mastication (shown in bold), which move the mandible. AP|R

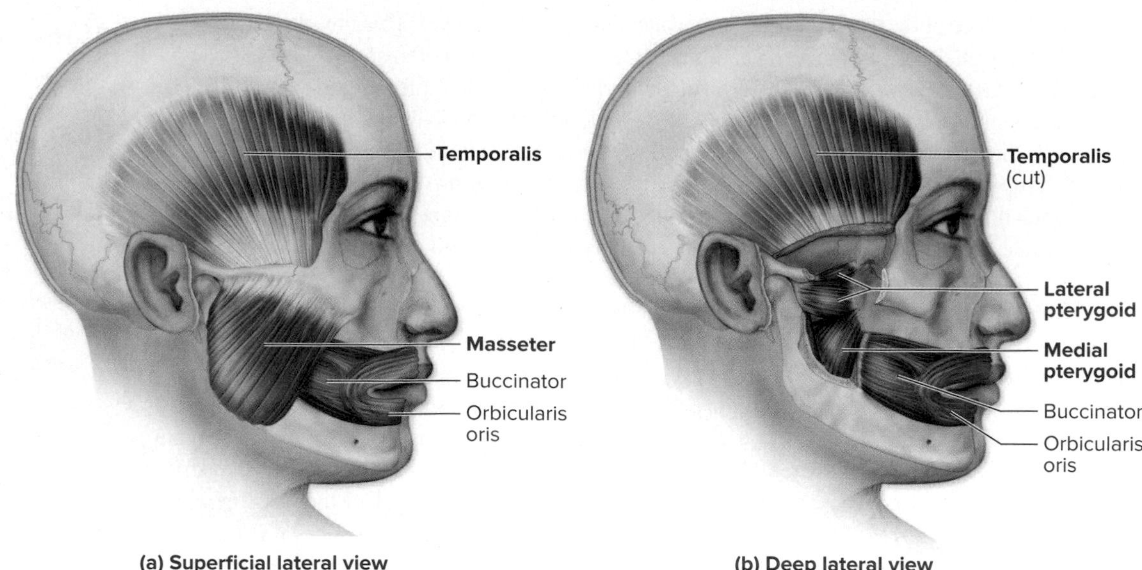

11.3c Muscles of the Oral Cavity and Pharynx

LEARNING OBJECTIVES

10. Describe how each of the four muscles of mastication affects mandibular movement.
11. Describe the actions of the intrinsic muscles and the four paired extrinsic muscles of the tongue.
12. Explain what function the three primary muscles of the pharynx accomplish.

The muscles of the oral cavity and pharynx include muscles that help us chew, move the tongue, and swallow.

Muscles of Mastication

The term **mastication** (mas-ti-kā′shŭn) refers to the process of chewing. These muscles move the mandible at the temporomandibular joint (TMJ). There are four paired muscles of mastication: the temporalis, the masseter, and the lateral and medial pterygoids **(figure 11.8)**. The muscles of mastication are innervated by the mandibular division of CN V (trigeminal [see section 13.9]).

The **temporalis** (or *temporal* muscle) is a broad, fan-shaped muscle that extends from the temporal lines of the skull and attaches to the coronoid process of the mandible. It elevates and retracts (pulls posteriorly) the mandible. You can palpate the temporalis by placing your fingers along your temple (lateral skull at same level of orbits) as you open and close your jaw. The muscle you feel contracting is the temporalis.

The **masseter** elevates and protracts (pulls anteriorly) the mandible. It is the most powerful and important of the masticatory muscles. This short, thick muscle is superficial to the temporalis. You can feel the movement of the masseter by palpating near the angle of the mandible as you open and close your mouth.

The **lateral** and **medial pterygoid** muscles arise from the pterygoid processes of the sphenoid bone and attach to the mandible. Both pterygoids protract the mandible and move it from side to side during chewing. These movements maximize the effectiveness of the teeth while chewing or grinding foods of various consistencies. The medial pterygoid also elevates the mandible.

Table 11.3 summarizes the characteristics of the muscles of mastication.

Table 11.3 Muscles of Mastication

Muscle	Action(s)	Superior Attachment(s) (S)/Inferior Attachment(s) (I)[1]	Innervation (see section 13.9)
Temporalis (tem-pŏ-rā′lis) *tempora* = pertaining to temporal bone	Elevates and retracts mandible	S: Superior and inferior temporal lines I: Coronoid process of mandible	CN V_3 (trigeminal nerve, mandibular division)
Masseter (mas′ĕ-tĕr) *maseter* = chewer	Elevates and protracts mandible; agonist (prime mover) of mandible elevation	S: Zygomatic arch I: Ramus (lateral surface) and angle of mandible	CN V_3 (trigeminal nerve, mandibular division)
Medial pterygoid (ter′i-goyd) *pterygoid* = winglike	Elevates and protracts mandible; produces side-to-side movement of mandible	S: Maxilla, palatine, and medial surface of lateral pterygoid plate I: Medial surface of mandibular ramus	CN V_3 (trigeminal nerve, mandibular division)
Lateral pterygoid	Protracts mandible; produces side-to-side movement of mandible	S: Greater wing of sphenoid and lateral surface of lateral pterygoid plate I: Condylar process of mandible	CN V_3 (trigeminal nerve, mandibular division)

1. The superior attachments of the mastication muscles are less movable than the inferior attachments.

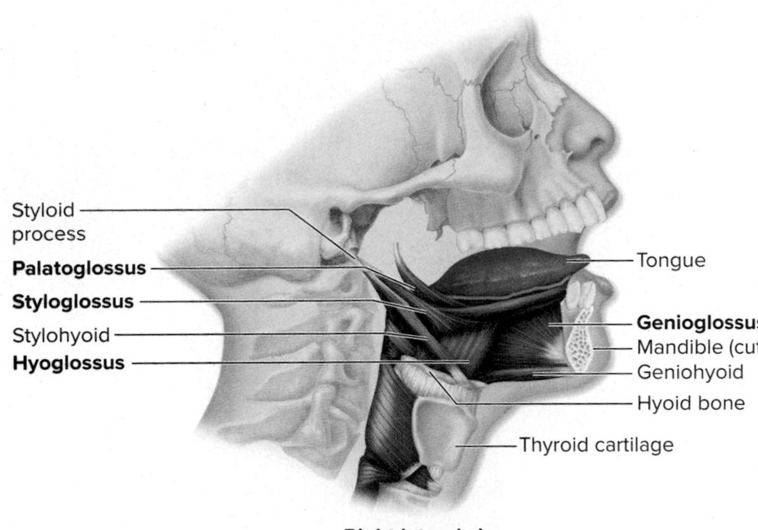

Figure 11.9 Muscles That Move the Tongue. Extrinsic tongue muscles (shown in bold) arise from structures other than the tongue and attach to the tongue to allow gross tongue movement.

Muscles That Move the Tongue

The tongue is an agile, highly mobile organ. It is composed of **intrinsic muscles** that compose the tongue, which curl, squeeze, and fold the tongue during chewing and speaking.

The **extrinsic muscles** of the tongue arise from other head and neck structures and attach to the tongue. The extrinsic muscles end in the suffix *-glossus,* meaning *tongue* (**figure 11.9**). These extrinsic tongue muscles are used in various combinations to accomplish the precise, complex, and delicate tongue movements required for proper speech and food manipulation within the mouth. Most of these muscles are innervated by CN XII, the hypoglossal nerve (see section 13.9).

The left and right **genioglossus** muscles arise from the mandible and protract the tongue. You use these muscles when you stick out your tongue. The left and right **styloglossus** muscles arise from the styloid processes of the temporal bone. These muscles elevate and retract the tongue (pull the tongue posteriorly, back into the mouth). The left and right **hyoglossus** muscles arise from the hyoid bone and depress and retract the tongue. The left and right **palatoglossus** muscles arise from the soft palate and elevate the posterior portion of the tongue.

Table 11.4 summarizes the characteristics of the muscles that move the tongue.

Pharynx Muscles

The **pharynx** (far'ingks), commonly known as the *throat,* is a funnel-shaped tube that lies posterior to both the oral and nasal cavities (see figure 26.4). Several muscles help form or attach to this tube and aid in swallowing (**figure 11.10**). Most pharyngeal muscles are innervated by CN X (vagus nerve).

The primary pharynx muscles are the **pharyngeal constrictors** (**superior, middle,** and **inferior**). When food is swallowed and enters the pharynx, these muscles contract sequentially to initiate swallowing and force the bolus inferiorly into the esophagus (see section 26.2c). Other pharyngeal muscles help elevate or tense the palate when swallowing. These muscles are summarized in **table 11.5**.

WHAT DID YOU LEARN?

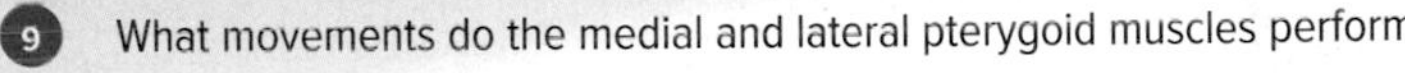

9. What movements do the medial and lateral pterygoid muscles perform?

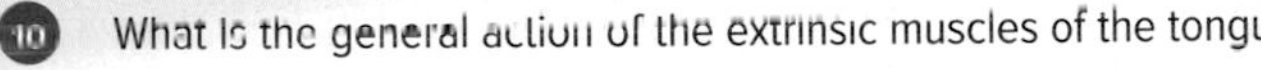

10. What is the general action of the extrinsic muscles of the tongue?

11.3d Muscles of the Anterior Neck: The Hyoid Muscles

LEARNING OBJECTIVE

13. Contrast the actions of the four suprahyoid muscles and the four infrahyoid muscles.

Table 11.4 Muscles That Move the Tongue

Muscle	Action(s)	Head or Neck Attachment(s) (H)/Tongue Attachment(s) (T)[1]	Innervation (see section 13.9)
Genioglossus (jē'nē-ō-glos'ŭs) *geni* = chin *glossus* = tongue	Protracts tongue	H: Mental spines of mandible T: Inferior region of tongue; hyoid bone	CN XII (hypoglossal nerve)
Styloglossus (stī'lō-glos'ŭs) *stylo* = pertaining to styloid process of temporal bone	Elevates and retracts tongue	H: Styloid process of temporal bone T: Side and inferior aspect of tongue	CN XII (hypoglossal nerve)
Hyoglossus (hī'ō-glos'ŭs) *hyo* = pertaining to hyoid bone	Depresses and retracts tongue	H: Hyoid bone T: Inferolateral side of tongue	CN XII (hypoglossal nerve)
Palatoglossus (pal-ă-tō-glos'ŭs) *palato* = palate	Elevates posterior part of tongue	H: Anterior surface of soft palate T: Side and posterior aspect of tongue	CN X (vagus nerve) via pharyngeal plexus of nerves

1. The tongue attachments are the more movable parts of these muscles.

Figure 11.10 Pharyngeal Constrictors, Palate Muscles, and Laryngeal Elevators. A right lateral view reveals some of the muscles that constrict the pharynx when swallowing, move the palate, and elevate the larynx (palatopharyngeus and salpingopharyngeus not shown). AP|R

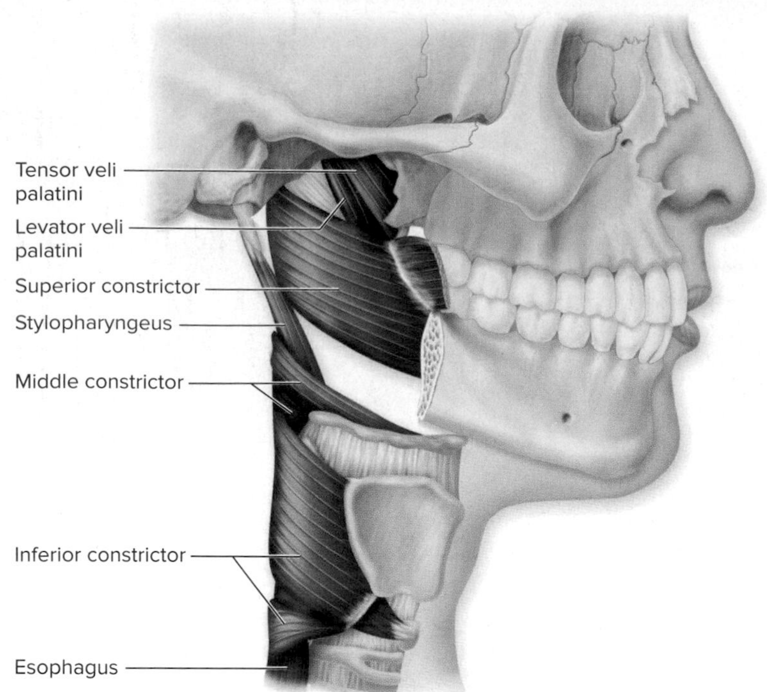

Table 11.5	Muscles of the Pharynx[1]		
Region/Muscle	**Action(s)**	**Origin/Insertion[2]**	**Innervation (see section 13.9)**
	PALATE MUSCLES		
Levator veli palatini (vel′ī pal′ă-tē′nī) *velum* = veil	Elevates soft palate when swallowing	O: Petrous part of temporal bone I: Soft palate	CN X (vagus nerve)
Tensor veli palatini (ten′sōr) *tensus* = to stretch	Tenses soft palate and opens auditory tube when swallowing or yawning	O: Sphenoid bone; region around auditory tube I: Soft palate	CN V_3 (trigeminal nerve, mandibular division)
	PHARYNGEAL CONSTRICTORS		
Superior constrictor (kon-strik′ter, -tōr) *constringo* = to draw together	Constricts pharynx in sequence to force bolus into esophagus; superior is innermost	O: Pterygoid process of sphenoid bone; medial surface of mandible I: Posterior median raphe (muscle fiber union from both sides)	CN X (vagus nerve) via branches of pharyngeal plexus
Middle constrictor	Constricts pharynx in sequence	O: Hyoid bone I: Posterior median raphe	CN X (vagus nerve) via branches of pharyngeal plexus
Inferior constrictor	Constricts pharynx in sequence; inferior is outermost	O: Thyroid and cricoid cartilage I: Posterior median raphe	CN X (vagus nerve) via branches of pharyngeal plexus
	LARYNGEAL (VOICE BOX) ELEVATORS		
Palatopharyngeus (pal′ă-tō-fă-rin′jē′ŭs) *pharynx* = pharynx	Elevates pharynx and larynx	O: Soft palate I: Side of pharynx and thyroid cartilage of larynx	CN X (vagus nerve) via branches of pharyngeal plexus
Salpingopharyngeus (sal-ping′gō-fă-rin′jē-ŭs) *salpinx* = trumpet	Elevates pharynx and larynx	O: Auditory tube I: Blends with palatopharyngeus on lateral wall of pharynx	CN X (vagus nerve) via branches of pharyngeal plexus
Stylopharyngeus (stī′lō-fă-rin′jē-ŭs)	Elevates pharynx and larynx	O: Styloid process of temporal bone I: Side of pharynx and thyroid cartilage of larynx	CN IX (glossopharyngeal nerve) via branches of pharyngeal plexus

1. Only the pharyngeal constrictors are discussed in the text.
2. Here, we use the terms *origin* and *insertion* because the terms *superior attachment* and *inferior attachment* are not applicable for all of the muscles in this table.

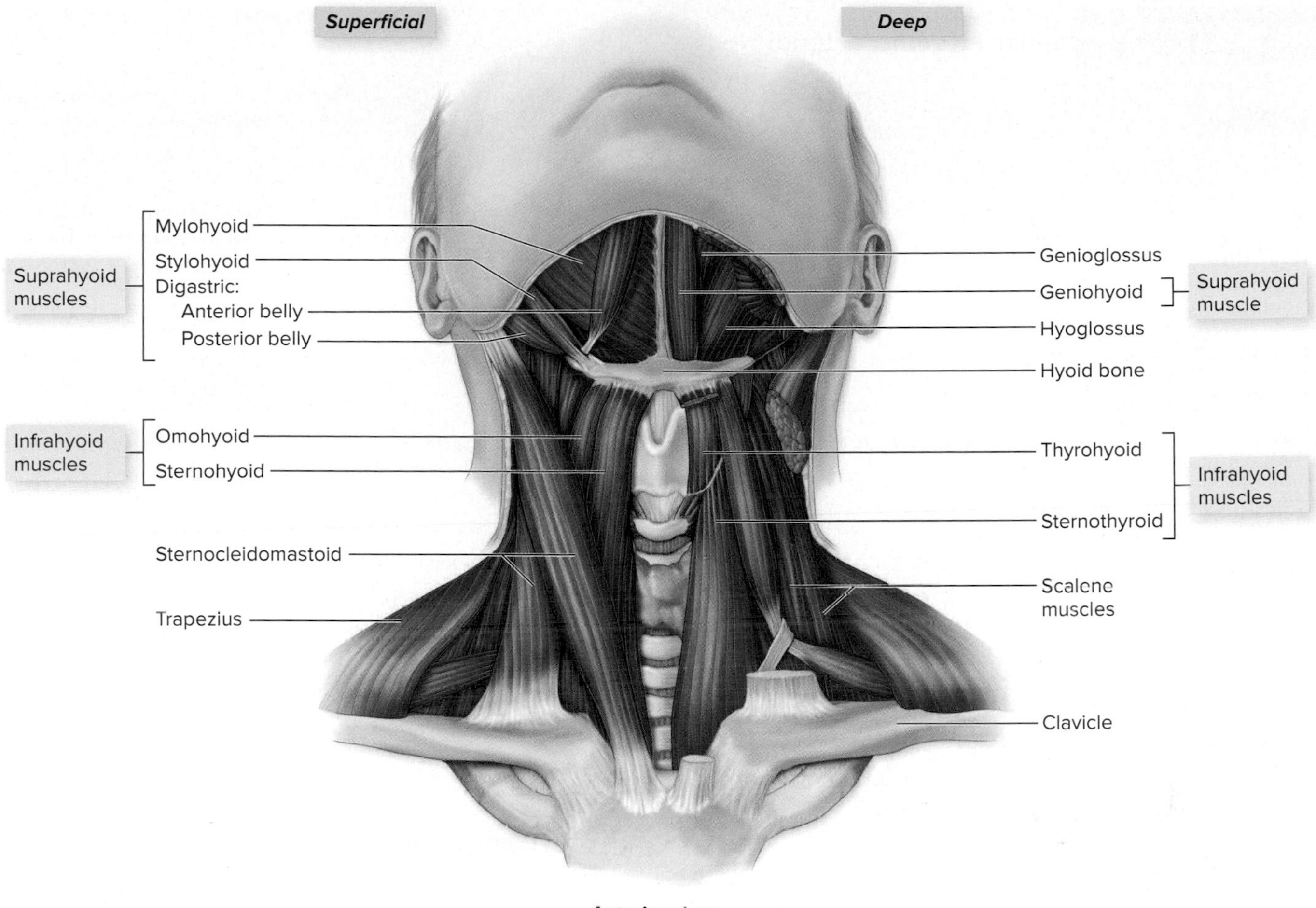

Figure 11.11 Muscles of the Anterior Neck. The anterior neck muscles move the hyoid bone and thyroid cartilage (of the larynx; see section 23.3a). Superficial muscles are shown on the right side of the body, and deeper muscles are shown on the left side of the body. AP|R

The muscles of the anterior neck are divided into the **suprahyoid muscles,** which are superior to the hyoid bone, and the **infrahyoid muscles,** which are inferior to the hyoid bone **(figure 11.11)**.

The suprahyoid muscles are associated with the floor of the mouth. In general, these muscles act as a group to elevate the hyoid bone during swallowing or speaking. Some of these muscles perform additional functions. The **digastric** has two bellies, anterior and posterior. The anterior belly extends from the mental protuberance of the mandible to the hyoid bone, and the posterior belly continues from the hyoid bone to the mastoid part of the temporal bone. The two bellies are united by an intermediate tendon that is held in position by a fibrous loop. In addition to elevating the hyoid bone, the digastric muscle can also depress the mandible. The **geniohyoid** attaches to the mental spines of the mandible and the hyoid bone. This muscle elevates the hyoid bone. The broad, flat **mylohyoid** attaches to the mylohyoid line of the mandible and the hyoid bone, and it forms the muscular floor to the mouth. When this muscle contracts, it both elevates the hyoid bone and raises the floor of the mouth. The muscle fibers of the left and right mylohyoid are aligned in a V shape. The **stylohyoid** connects the styloid process of the skull and the hyoid bone. Upon contraction, it elevates the hyoid bone, causing the floor of the oral cavity to elongate during swallowing.

As swallowing is completed, the infrahyoid muscles contract to influence the position of the hyoid bone and the larynx (see section 23.3a). In general, these muscles either depress the hyoid bone or depress the thyroid cartilage of the larynx. The **omohyoid** contains two thin muscle bellies anchored in place by a fascia "sling." This muscle is lateral to the sternohyoid and extends from the superior border of the scapula to the hyoid, where it depresses the hyoid bone. The **sternohyoid** extends from the sternum to the hyoid bone, where it depresses the hyoid bone. The **sternothyroid** is deep to the sternohyoid. It extends from the sternum to the thyroid cartilage of the larynx. It depresses the thyroid cartilage to return it to its original position after swallowing. The **thyrohyoid** extends from

Table 11.6 Muscles of the Anterior Neck

Region/Muscle	Action(s)	Superior Attachment(s) (S)/Inferior Attachment(s) (I)[1]	Innervation (see sections 13.9 and 14.5d)
SUPRAHYOID MUSCLES			
Digastric (dī-gas′trik) *di* = two *gaster* = belly	Depresses mandible; elevates hyoid bone	S: Anterior belly: Mandible near mental protuberance; posterior belly, mastoid process I: Hyoid bone via fascia sling	Anterior belly: CN V_3 (trigeminal nerve, mandibular division) Posterior belly: CN VII (facial nerve)
Geniohyoid (jĕ-nē-ō-hī′oyd)	Elevates hyoid bone	S: Mental spines of mandible I: Hyoid bone	First cervical spinal nerve (C1) via CN XII (hypoglossal nerve)
Mylohyoid (mī′lō-hī′oyd) *myle* = molar	Elevates hyoid bone; elevates floor of mouth	S: Mylohyoid line of mandible I: Hyoid bone	CN V_3 (trigeminal nerve, mandibular division)
Stylohyoid (stī-lō-hī′oyd)	Elevates hyoid bone	S: Styloid process of temporal bone I: Hyoid bone	CN VII (facial nerve)
INFRAHYOID MUSCLES			
Omohyoid (ō-mō-hī′oyd) *omo* = shoulder	Depresses hyoid bone; fixes hyoid during opening of mouth (depression of mandible)	S: Hyoid bone I: Superior border of scapula	Cervical spinal nerves C1–C3 through ansa cervicalis (from cervical plexus)
Sternohyoid (ster′nō-hī′oyd) *sterno* = sternum	Depresses hyoid bone	S: Hyoid bone I: Manubrium of sternum and medial end of clavicle	Cervical spinal nerves C1–C3 through ansa cervicalis (from cervical plexus)
Sternothyroid (ster′nō-thī′royd) *thyro* = thyroid cartilage	Depresses thyroid cartilage of larynx; fixes hyoid	S: Thyroid cartilage of larynx I: Posterior surface of manubrium of sternum	Cervical spinal nerves C1–C3 through ansa cervicalis (from cervical plexus)
Thyrohyoid (thī-rō-hī′oyd)	Depresses hyoid bone and elevates thyroid cartilage of larynx; fixes hyoid	S: Hyoid bone I: Thyroid cartilage of larynx	First cervical spinal nerve C1 via CN XII (hypoglossal nerve)

1. Note that with most of the infrahyoid muscles, the superior attachment is the more movable part of each muscle, whereas the inferior attachment tends to be more movable for the suprahyoid muscles.

the thyroid cartilage of the larynx to the hyoid bone. It depresses the hyoid bone and elevates the thyroid cartilage to close off the larynx during swallowing. In addition, the omohyoid, sternohyoid, and thyrohyoid help anchor the hyoid so the digastric can depress the mandible.

Table 11.6 summarizes the characteristics of these muscles.

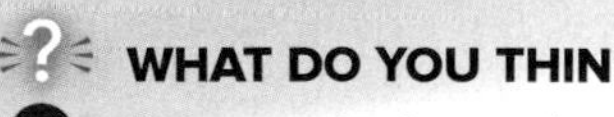

WHAT DO YOU THINK?

1 Since muscles frequently are named for their attachment sites, what do you think the prefix *-omo* in *omohyoid* means?

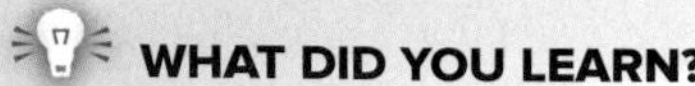

WHAT DID YOU LEARN?

11 List the four suprahyoid muscles, and describe a common action for them.

11.3e Muscles That Move the Head and Neck

LEARNING OBJECTIVE

14. Compare and contrast the actions of the anterolateral neck muscles and the posterior neck muscles.

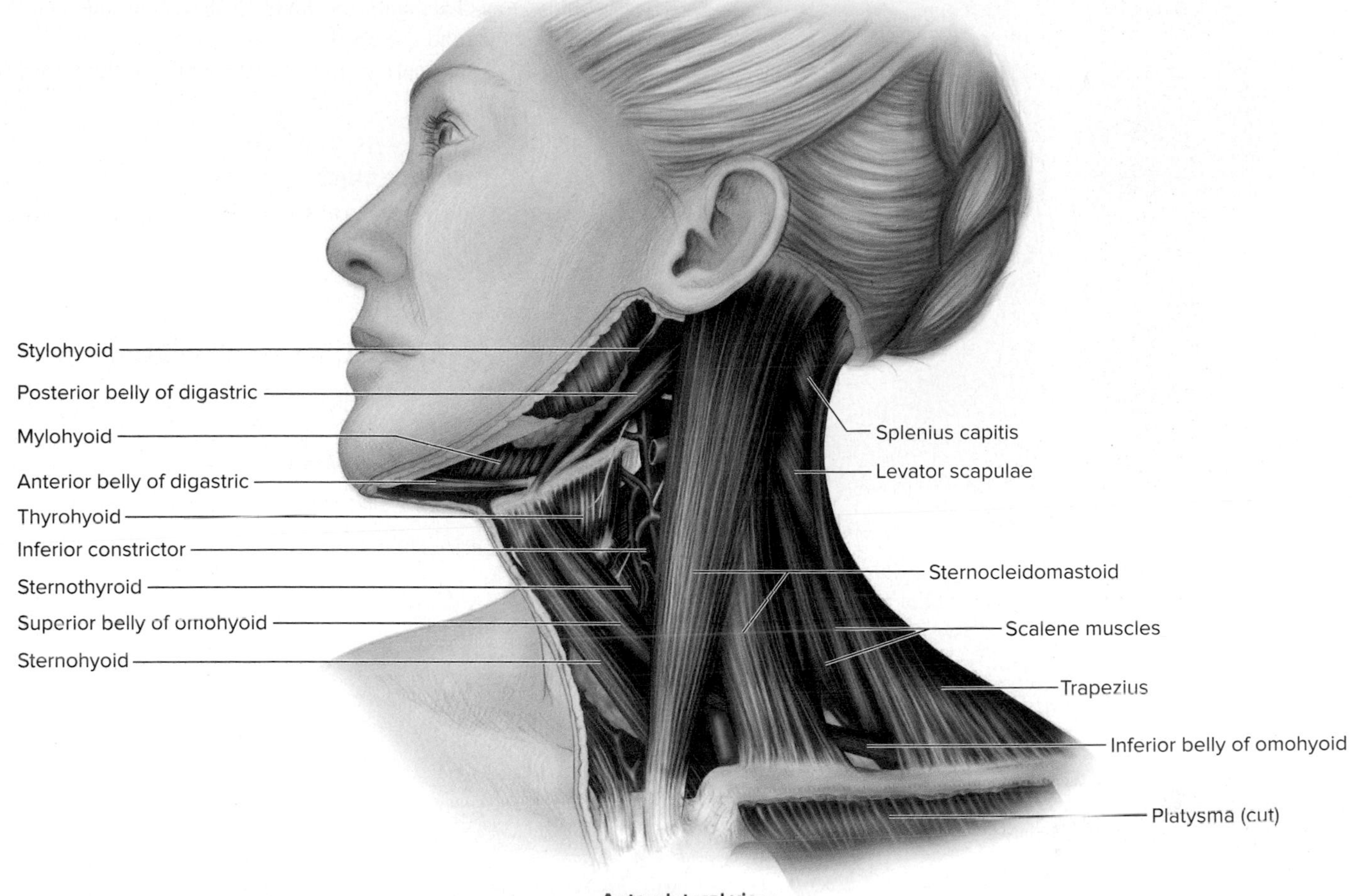

Figure 11.12 Muscles That Move the Head and Neck. Anterolateral muscles collectively flex the neck, whereas posterior neck muscles extend the neck. APR

Muscles that move the head and neck arise from the vertebral column, the thoracic cage, and the pectoral girdle, and attach to bones of the cranium (figures 11.11 and **11.12**).

Anterolateral Neck Muscles

The anterolateral neck muscles all flex the neck. The main muscles in this group are the sternocleidomastoid and the three scalenes.

The **sternocleidomastoid** is a thick, cordlike muscle that extends from the sternum and clavicle to the mastoid process posterior to the ear. Contraction of both sternocleidomastoid muscles (called **bilateral contraction**) flexes the neck. Contraction of just one sternocleidomastoid muscle (termed **unilateral contraction**) results in lateral flexion of the neck to its own side (as occurs when you tilt your head as if to touch your ear to your shoulder), as well as rotation of the head to the opposite side. Thus, if the left sternocleidomastoid muscle contracts, it rotates the head to the right side of the body. The three **scalene muscles (anterior, middle,** and **posterior)** work with the sternocleidomastoid to flex the neck. In addition, the scalene muscles elevate the first and second ribs during forced inspiration (see section 23.5b).

Posterior Neck Muscles

Many of the posterior neck muscles work together to extend the neck (**figure 11.13**). The trapezius attaches to the skull and helps extend the neck, but its primary function is to help move the pectoral girdle.

When the left and right **splenius capitis, splenius cervicis, semispinalis capitis,** and **longissimus capitis** muscles bilaterally contract, they extend the neck. Unilateral contraction rotates (turns) the head and neck to the same side.

A group of muscles called the *suboccipital muscles* includes the obliquus capitis superior, obliquus capitis inferior, rectus capitis posterior major, and rectus capitis posterior minor.

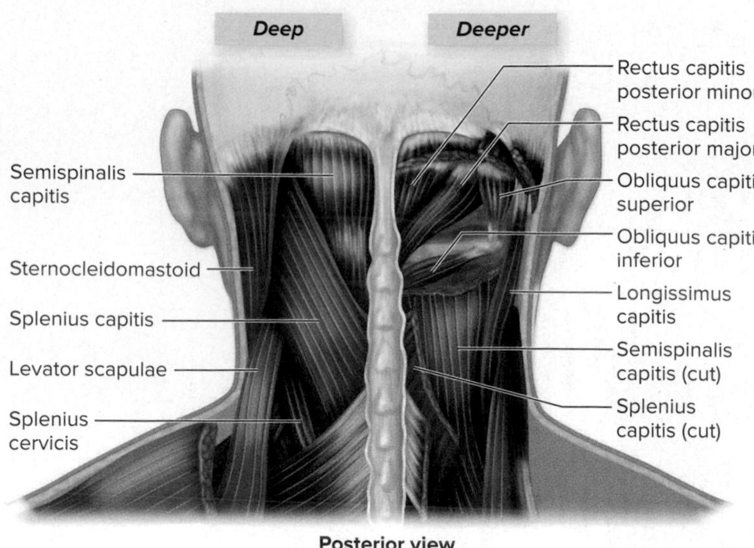

Figure 11.13 Posterior Neck Muscles. An illustration shows the deep and deeper muscles that extend and rotate the head and neck. AP|R

The oblique muscles rotate the head to the same side, whereas the rectus muscles extend the head and neck.

Table 11.7 summarizes the characteristics of the muscles of the head and neck.

WHAT DID YOU LEARN?

12 Which neck muscles extend the neck? Which neck muscles flex the neck?

11.4 Muscles of the Vertebral Column

LEARNING OBJECTIVES

15. Name and describe the three groups of erector spinae muscles.
16. Describe the actions of the transversospinalis and quadratus lumborum muscles.

Table 11.7 Muscles That Move the Head and Neck

Muscle	Action(s)	Superior Attachment(s) (S)/Inferior Attachment(s) (I)[1]	Innervation (see sections 13.9 and 14.5c, d)
Sternocleidomastoid (ster′nō-klī′dō-măs-toyd) *sterno* = sternum *cleido* = clavicle *masto* = mastoid process	Bilateral action[2]: Flexes neck Unilateral action[3]: Lateral flexion of neck, rotation of head to opposite side	S: Mastoid process I: Manubrium of the sternum and sternal end of clavicle	CN XI (accessory nerve)
Scalene muscles (anterior, middle, posterior) (see also table 11.9) (skā′lēnz) *scalene* = uneven	Flex neck (when 1st rib is fixed); elevate 1st and 2nd ribs during forced inspiration when neck is fixed	S: Transverse processes of cervical vertebrae I: Superior surface of 1st and 2nd ribs	Cervical spinal nerves
Splenius capitis and cervicis (splē′nē-ŭs ka-pi-tis) (ser′vi′sis) *splenion* = bandage	Bilateral action: Extends neck Unilateral action: Rotates head to same side	S: Occipital bone and mastoid process of temporal bone I: Ligamentum nuchae	Cervical spinal nerves
Longissimus capitis (lon-gis′i-mŭs) *longissimus* = longest *caput* = head	Bilateral action: Extends neck Unilateral action: Rotates head to same side	S: Mastoid process I: Transverse process of T_1–T_4 and articular processes of C_4–C_7 vertebrae	Cervical and thoracic spinal nerves
Obliquus capitis superior (ob-lī′kwŭs) *obliquus* = slanting	Rotates head to same side	S: Inferior nuchal line of occipital bone I: Transverse process of atlas	Suboccipital nerve (posterior ramus of C1 spinal nerve)
Obliquus capitis inferior	Rotates head to same side	S: Transverse process of atlas I: Spinous process of axis	Suboccipital nerve (posterior ramus of C1 spinal nerve)
Rectus capitis posterior major	Extends neck	S: Inferior nuchal line of occipital bone I: Spinous process of axis	Suboccipital nerve (posterior ramus of C1 spinal nerve)
Rectus capitis posterior minor	Extends neck	S: Inferior nuchal line of occipital bone I: Posterior tubercle of atlas	Suboccipital nerve (posterior ramus of C1 spinal nerve)

1. Here, the superior attachments are more movable than the inferior attachments.
2. *Bilateral action* means both the left and right muscles are contracting together.
3. *Unilateral action* means only one muscle (either the left or right muscle) is contracting.

INTEGRATE

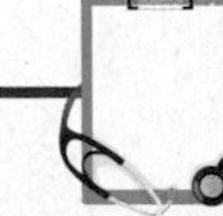

CLINICAL VIEW 11.4

Congenital Muscular Torticollis

Congenital muscular torticollis (CMT), often known as *wryneck,* is a condition where a newborn presents with a shortened and tightened sternocleidomastoid muscle that may persist into childhood. It is thought to be a result of trauma resulting from either a difficult birth or prenatal position of the fetus. The trauma causes a hematoma and fibrosing of the muscle tissue. Pediatricians also have seen an increase in acquired muscular torticollis among newborns who are kept in infant seats for extended periods of time outside of the car. Children with CMT often tilt their heads to the affected side and their chins to the unaffected side. Because the child typically favors a particular head position when sleeping, **plagiocephaly** (flattening of the head; see Clinical View 8.2: "Craniosynostosis and Plagiocephaly") often accompanies CMT.

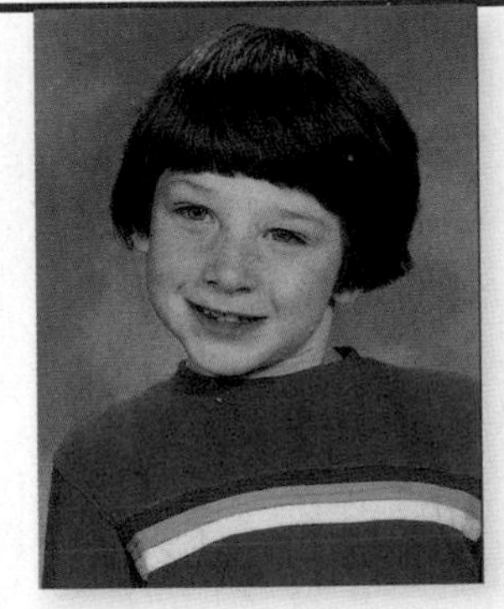

Photo of 7-year-old boy with CMT.
©SC Photo

CMT treatment typically involves stretching the affected muscle several times a day, changing sleeping positions, and making the child use the affected side more frequently. A newer approach to treatment of CMT is the use of botulinum toxin type A (Botox), which impairs contraction of the affected muscle), combined with stretching. In cases that do not respond to the treatments mentioned, sternocleidomastoid release surgery is recommended. The sternocleidomastoid is cut from at least one attachment point, repositioned, and reattached, and Botox may be injected into the muscle.

The muscles of the vertebral column are very complex; they have multiple attachments, and they exhibit extensive overlap (**figure 11.14** and **table 11.8**). All of these muscles are covered by the most superficial back muscles, which move the upper limb.

Note that the neck is the cervical portion of the vertebral column. Thus, the posterior muscles discussed previously in connection with neck extension (splenius cervicis, splenius capitis, longissimus capitis, semispinalis capitis) extend the *cervical region* of the vertebral column.

The **erector spinae** are used to maintain posture and to help an individual stand erect. When the left and right erector spinae muscles contract together, they extend the vertebral column. If the erector spinae muscles on only one side contract, the vertebral column flexes laterally toward that side.

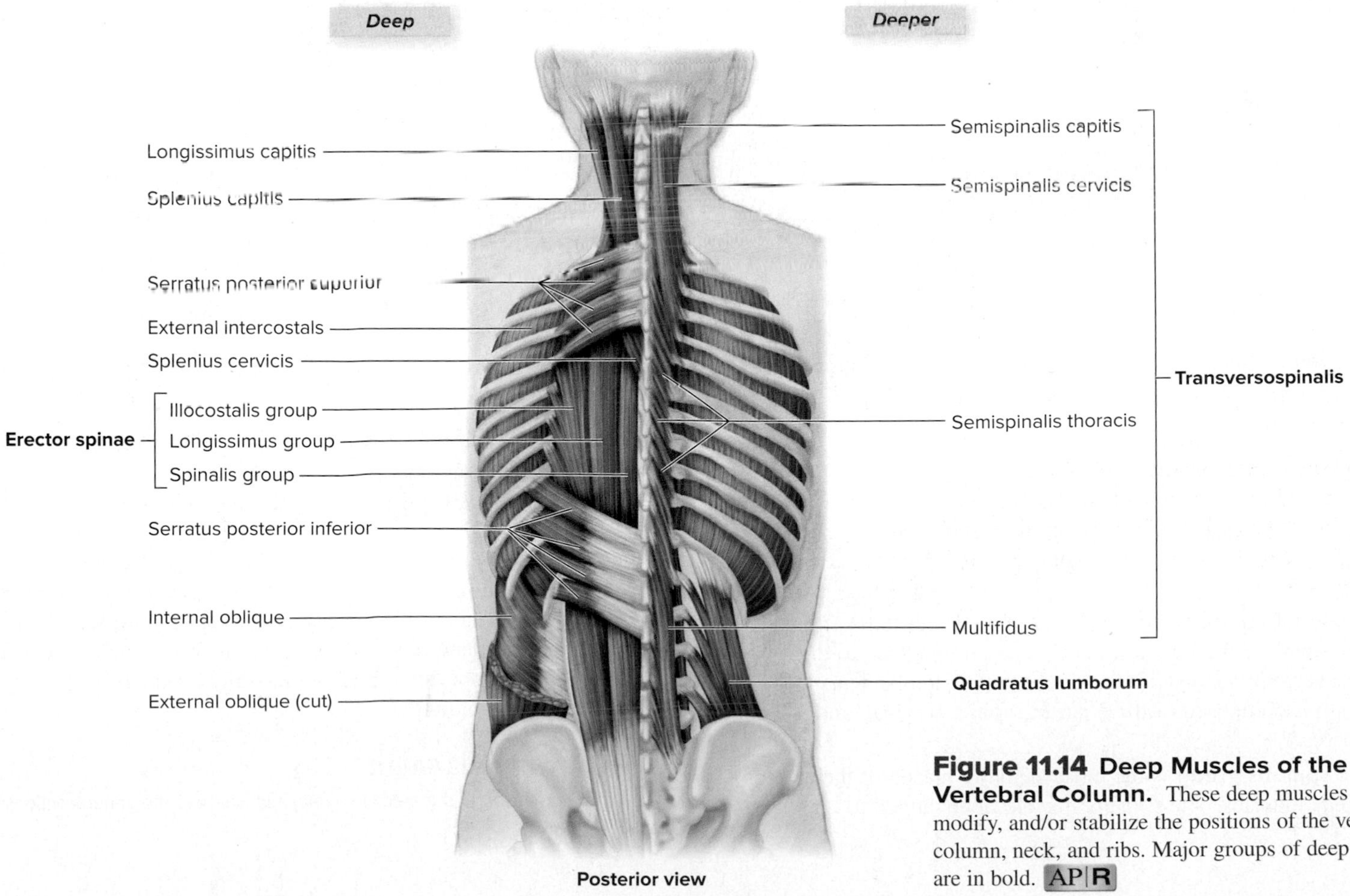

Figure 11.14 Deep Muscles of the Vertebral Column. These deep muscles extend, modify, and/or stabilize the positions of the vertebral column, neck, and ribs. Major groups of deep muscles are in bold. AP|R

Table 11.8	Muscles of the Vertebral Column		
Group/Muscle	**Action(s)**	**Superior Attachment(s) (S)/Inferior Attachment(s) (I)**[1]	**Innervation (see sections 14.5c, d, f)**
		ERECTOR SPINAE	
Iliocostalis group (il-ē-ō-kos-tă′lis) *ilio* = ilium *cost* = rib	Bilateral action[2]: Extends vertebral column; maintains posture Unilateral action[3]: Laterally flexes vertebral column	S: Angles of ribs; transverse processes of cervical vertebrae I: Tendon from posterior part of iliac crest, posterior sacrum, and lumbar spinous processes	Cervical, thoracic, and lumbar spinal nerves
Longissimus group (lon-gis′i-mŭs)	Bilateral action: Extends vertebral column; maintains posture Unilateral action: Rotates head and laterally flexes vertebral column	S: Mastoid process of temporal bone and transverse processes of cervical and thoracic vertebrae I: Tendon from posterior part of iliac crest, posterior sacrum, and lumbar spinous processes	Cervical and thoracic spinal nerves
Spinalis group (spī-nā′lis) *spin* = spine	Bilateral action: Extends vertebral column; maintains posture Unilateral action: Laterally flexes vertebral column	S: Spinous process of axis and thoracic vertebrae I: Lumbar spinous processes (thoracic part) and C_7 spinous process (cervical part)	Cervical and thoracic spinal nerves
		TRANSVERSOSPINALIS GROUP	
Multifidus (mul-tif′i-dŭs) *multus* = much *findo* = to cleave	Bilateral action: Extends vertebral column Unilateral action: Rotates vertebral column toward opposite side	S: Spinous processes of vertebrae located 2–4 segments superior to inferior attachment I: Sacrum and transverse processes of each vertebra	Cervical, thoracic, and lumbar spinal nerves
Rotatores (rō-tā′tōrz)	Bilateral action: Extends vertebral column Unilateral action: Rotates vertebral column toward opposite side	S: Spinous process of immediately superior vertebra I: Transverse processes of immediately inferior vertebra	Cervical, thoracic, and lumbar spinal nerves
Semispinalis group (sem′ē-spī-nā′lis)	Bilateral action: Extends vertebral column and neck Unilateral action: Laterally flexes vertebral column and neck	S: Occipital bone and spinous processes of cervical and thoracic vertebrae I: Transverse processes of C_4–T_{12} vertebrae	Cervical and thoracic spinal nerves
		SPINAL EXTENSORS AND LATERAL FLEXORS	
Quadratus lumborum (kwah-drā′tūs lŭm-bōr′ŭm) *quad* = four-sided	Bilateral action: Extends vertebral column Unilateral action: Laterally flexes vertebral column	S: 12th rib; transverse processes of lumbar vertebrae I: Iliac crest	Thoracic and lumbar spinal nerves

1. The superior attachments for vertebral column muscles are more movable typically than the inferior attachments.
2. *Bilateral action* means both the left and right muscles are contracting together.
3. *Unilateral action* means only one muscle (either the left or right muscle) is contracting.

The erector spinae muscles are organized into three groups. The muscles are named based upon the body region with which they are associated.

- The **iliocostalis group** is the most laterally placed of the three erector spinae components. It is composed of three parts: cervical, thoracic, and lumbar, which attach either to the angles of the ribs or to the transverse processes of cervical vertebrae.
- The **longissimus group** is medial to the iliocostalis group. The fibers of the longissimus muscle group attach to the transverse processes of the vertebrae. The longissimus group is composed of three parts: capitis, cervical, and thoracic.
- The **spinalis group** is the most medially placed of the erector spinae muscles. The spinalis muscle fibers attach to the spinous processes of the vertebrae. The spinalis group is composed of cervical and thoracic parts.

Deep to the erector spinae, a group of muscles collectively called the **transversospinalis muscles** connect and stabilize the vertebrae. This group includes several specific muscles, which are listed in table 11.8.

A final pair of muscles help move the vertebral column. The **quadratus lumborum muscles** are located primarily in the lumbar region. When the left and right quadratus lumborum muscles bilaterally contract, they extend the vertebral column. When either the left or right quadratus lumborum muscle unilaterally contracts, it laterally flexes the vertebral column.

WHAT DID YOU LEARN?

13 Which muscles form the erector spinae, and what are the general actions of the erector spinae?

11.5 Muscles of Respiration

LEARNING OBJECTIVES

17. List the posterior and anterior thoracic muscle groups involved in respiration, and describe their actions.

18. Describe the role of the diaphragm in breathing and in raising intra-abdominal pressure.

The process of respiration involves inspiration and expiration. During **inspiration,** several muscles contract to increase the dimensions of the thoracic cavity to allow the lungs to fill with air. During **expiration,** some respiratory muscles contract and others relax, collectively decreasing the dimensions of the thoracic cavity and forcing air out of the lungs (see section 23.5b).

The muscles of respiration are on the posterior and anterior surfaces of the thorax. These muscles are covered by more superficial muscles (such as the pectoral muscles, trapezius, and latissimus dorsi) that move the upper limb.

Two posterior thorax muscles assist with respiration. The **serratus posterior superior** attaches to ribs 2–5 (figure 11.14) and elevates these ribs during forced inspiration, thereby increasing the lateral dimensions of the thoracic cavity. The **serratus posterior inferior** attaches to ribs 8–12 and depresses those ribs during forced expiration. (Normal, quiet expiration takes no active muscular effort.)

Several groups of anterior thorax muscles change the dimensions of the thorax during respiration **(figure 11.15)**. The scalene muscles (discussed previously with other neck muscles) help elevate the first and second ribs during forced inspiration, thereby increasing the dimensions of the thoracic cavity.

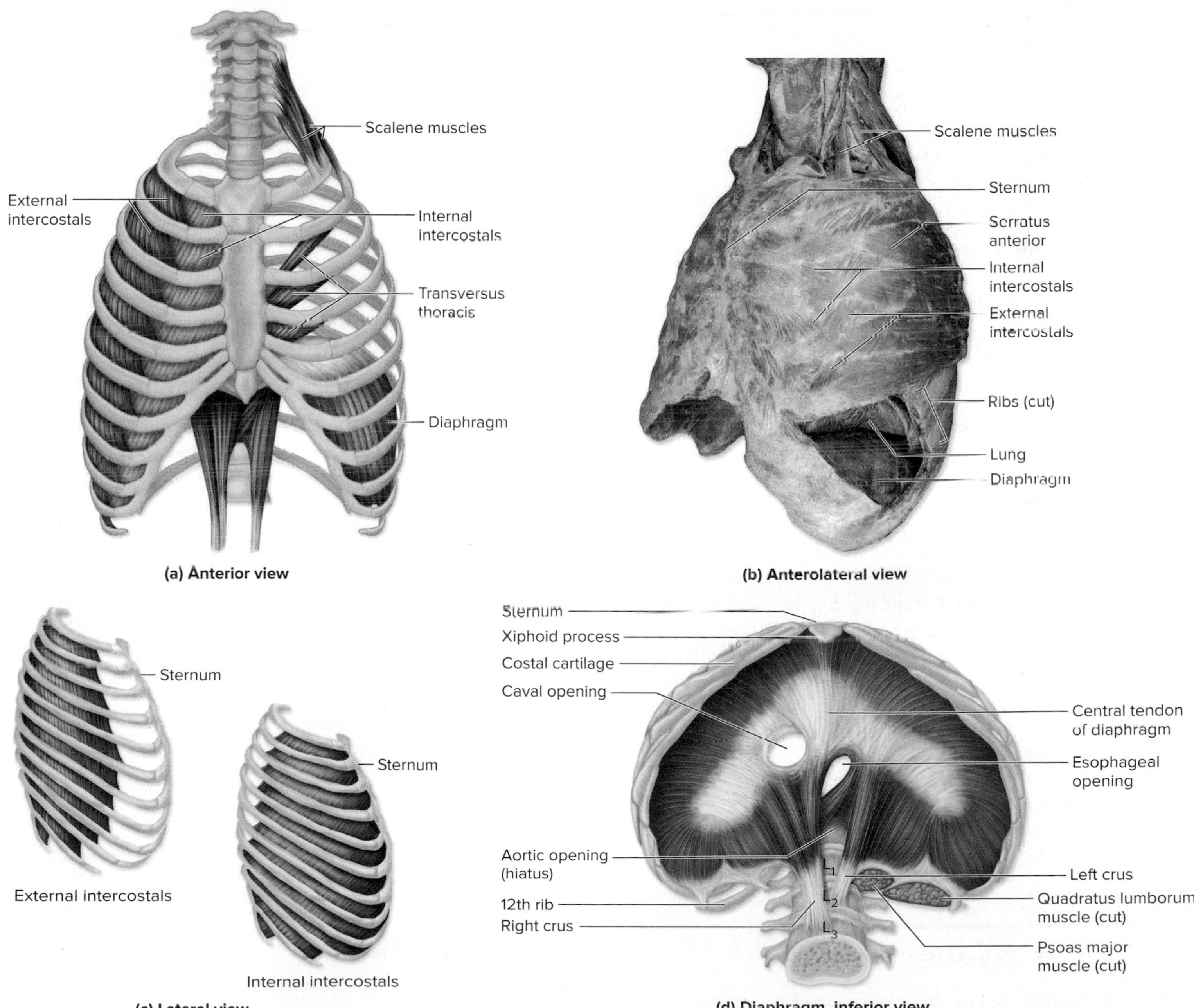

Figure 11.15 Muscles of Respiration. These skeletal muscles contract rhythmically to alter the size of the thoracic cavity and facilitate respiration. (*a*) Anterior view. (*b*) A cadaver photo provides an anterolateral view, with the inferior ribs cut to expose the thoracic cavity and the superior surface of the diaphragm. (*c*) Lateral views demonstrate fiber directions of the external and internal intercostals. (*d*) Inferior view of the diaphragm. AP|R

Table 11.9	Muscles of Respiration		
Muscle	**Action(s)**	**Superior Attachment(s) (S)/Inferior Attachment(s) (I)[1]**	**Innervation (see sections 14.5c, d)**
Serratus posterior superior (sĕr-ā′tŭs) *serratus* = a saw	Elevates ribs during forced inspiration	S: Spinous processes of C_7–T_3 vertebrae I: Lateral borders of ribs 2–5	Thoracic spinal nerves
Serratus posterior inferior	Depresses ribs during forced expiration	S: Inferior borders of ribs 8–12 or 9–12 I: Spinous processes of T_{11}–L_3 vertebrae	Thoracic spinal nerves
Scalene muscles (anterior, middle posterior) (described in table 11.7)			
External intercostals (in′ter-kos′talz) *inter* = between *cost* = rib	Elevate ribs during quiet and forced inspiration	S: Inferior border of superior rib I: Superior border of inferior rib	Thoracic spinal nerves
Internal intercostals	Depresses ribs during forced expiration	S: Superior border of inferior rib I: Inferior border to superior rib	Thoracic spinal nerves
Transversus thoracis (trans-ver′sŭs thō-ra′sis)	Depresses ribs during forced expiration	S: Costal cartilages 2–6 I: Posterior surface of xiphoid process and inferior region of sternum	Thoracic spinal nerves
Diaphragm (dī′ă-fram) *dia* = across *phragm* = partition	Contraction causes flattening of diaphragm (moves inferiorly) during inspiration and thus expands thoracic cavity; increases pressure in abdominopelvic cavity	S: Central tendon I: Inferior internal surface of ribs 7–12; xiphoid process of sternum and costal cartilages of inferior 6 ribs; lumbar vertebrae	Phrenic nerves (C_3–C_5)

1. The superior attachment of these muscles is more movable, except for serratus posterior superior and internal intercostals.

The **external intercostals** extend inferomedially from the superior rib to the adjacent inferior rib. The external intercostals assist in expanding the thoracic cavity by elevating the ribs during inspiration. This movement is like lifting a bucket handle—that is, as the bucket handle (rib) is elevated, its distance from the center of the bucket (thorax) increases. Thus, contraction of the external intercostals increases the transverse dimensions of the thoracic cavity. The **internal intercostals** lie deep to the external intercostals, and their muscle fibers are at right angles to the external intercostals. The internal intercostals depress the ribs only during forced expiration. A small **transversus thoracis** extends across the inner surface of the thoracic cage and attaches to ribs 2–6. It helps depress the ribs during forced expiration.

Finally, the **diaphragm** is an internally placed, dome-shaped muscle that forms a partition between the thoracic and abdominal cavities. (The term *diaphragm* refers to a muscle or group of muscles that covers or partitions an opening.) The diaphragm is the most important muscle associated with breathing. The muscle fibers of the diaphragm converge from its margins toward a fibrous **central tendon.** During inspiration, the diaphragm contracts and the central tendon is pulled inferiorly toward the abdominal cavity, thereby increasing the vertical dimensions of the thoracic cavity.

Table 11.9 summarizes the characteristics of the muscles of respiration. Further details about muscles of respiration are found in section 23.5b.

WHAT DO YOU THINK?

2 After you've eaten a very large meal, it is sometimes difficult to take a big, deep breath. Why is it more difficult to breathe deeply with a full GI tract?

WHAT DID YOU LEARN?

14 Compare the actions of the external intercostals and the internal intercostals.

15 How is the diaphragm involved in respiration?

INTEGRATE

CONCEPT CONNECTION

When the diaphragm contracts, it compresses the abdominal cavity, thus increasing intra-abdominal pressure. Increased intra-abdominal pressure is necessary for urination (see section 24.8c), defecation (see section 26.3d), and childbirth (see section 29.6). Beyond respiration, diaphragm movements are also important in helping return venous blood to the heart from the inferior half of the body (see section 20.5a).

11.6 Muscles of the Abdominal Wall

LEARNING OBJECTIVES

19. List the four pairs of abdominal muscles.
20. Compare the actions of the rectus abdominis muscle with the oblique muscles and transversus abdominis.

The anterolateral wall of the abdomen is reinforced by four pairs of muscles that collectively compress and hold the abdominal organs in place: the external oblique, internal oblique, transversus abdominis, and rectus abdominis (**figure 11.16**). These muscles also work

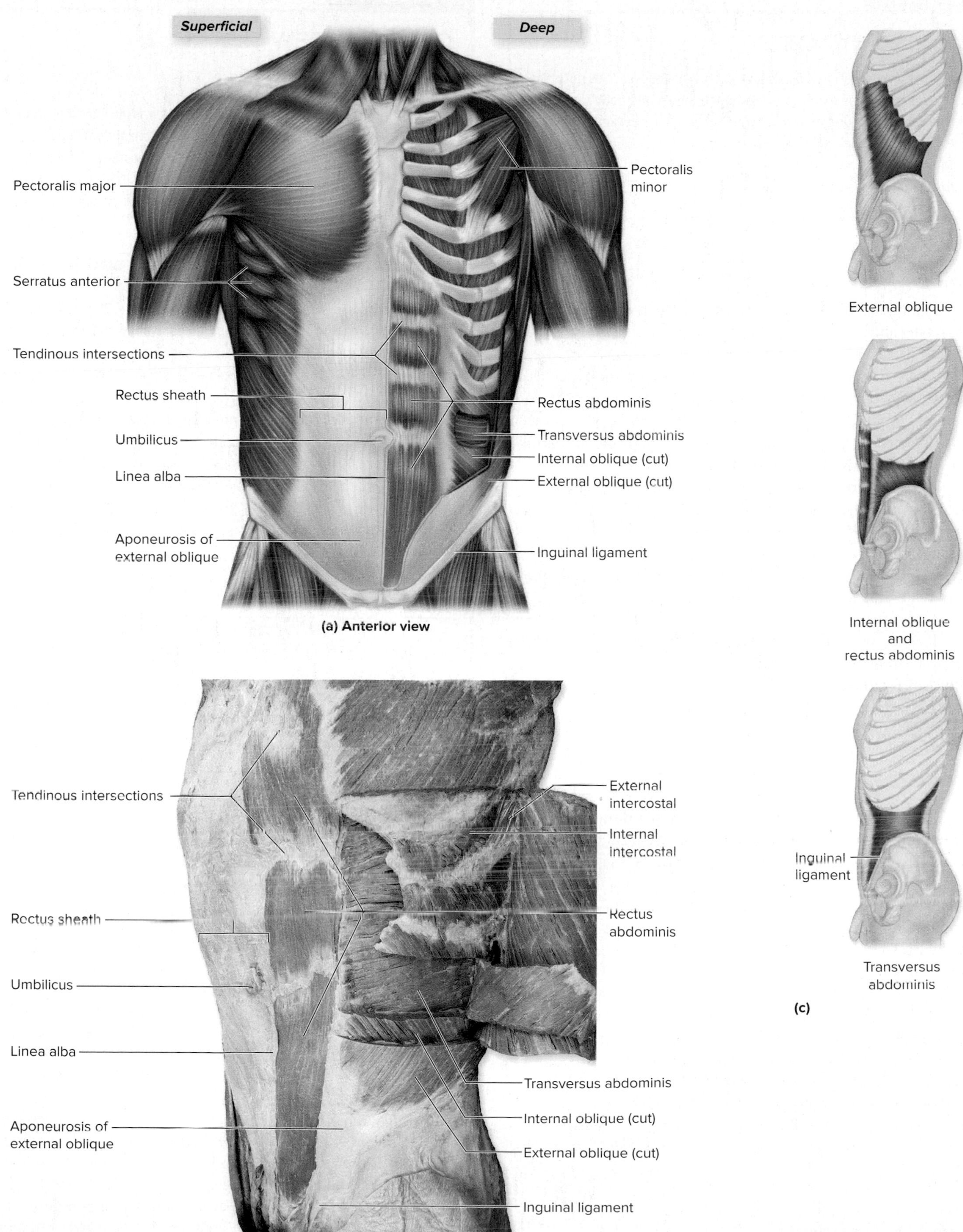

Figure 11.16 Muscles of the Abdominal Wall. The abdominal muscles compress abdominal contents and flex the vertebral column. (*a*) An illustration depicts the anterior view of some superficial and deep muscles. (*b*) A cadaver photo provides an anterolateral view of the muscles of the abdominal wall. (*c*) Diagrams show some individual abdominal muscles, ranging from superficial to deep. AP|R

INTEGRATE

LEARNING STRATEGY

The fibers of the external intercostals and external oblique muscles run in the same direction—inferomedially (*downward* and toward the body's midline). This is the same direction that you put your hands in your pockets.

The fibers of the internal intercostals and internal oblique muscles run perpendicular (in the opposite direction) to the external muscles—superomedially (*upward* and toward the body's midline).

together to flex and stabilize the vertebral column. When the oblique and transversus abdominis muscles unilaterally contract, they laterally flex the vertebral column.

The muscle fibers of the superficial **external oblique** are directed inferomedially. The external oblique is composed of muscle along the lateral abdominal wall and forms an aponeurosis as it projects anteriorly. Inferiorly, the aponeurosis of the external oblique forms a strong, cordlike **inguinal ligament** that extends from the anterior superior iliac spine to the pubic tubercle. Immediately deep to the external oblique is the **internal oblique.** Its muscle fibers project superomedially, which is at right angles to the external oblique. Like the external oblique, this muscle forms an aponeurosis as it projects anteriorly. Unilaterally, the external oblique on one side of the body and the internal oblique on the opposite side of the body work together to rotate the vertebral column.

The deepest muscle is the **transversus abdominis,** whose fibers project transversely across the abdomen and which has an aponeurosis as it projects anteriorly. When the transversus abdominis unilaterally contracts, it laterally flexes the vertebral column.

The **rectus abdominis** is a long, straplike muscle that extends vertically the entire length of the anteromedial abdominal wall between the sternum and the pubic symphysis. It is partitioned into four segments by three fibrous **tendinous intersections,** which form the traditional "six-pack" of a muscular, toned abdominal wall. The rectus abdominis is enclosed within a fibrous sleeve called the **rectus sheath,** which is formed from the aponeuroses of the external oblique, internal oblique, and transversus abdominis muscles. The left and right rectus sheaths are connected by a vertical fibrous strip termed the **linea alba.**

Table 11.10 summarizes the characteristics of the muscles of the abdominal wall.

You have probably noticed that multiple muscles may work together to perform a common function. For example, several neck muscles and back muscles work together to extend the vertebral column. Learning muscles in groups according to common function helps most students assimilate the anatomy information. **Table 11.11** summarizes the actions of various axial muscles and groups them according to common function. Note that a muscle that has multiple functions is listed in more than one group.

WHAT DID YOU LEARN?

16 What are the main actions of the abdominal muscles?

INTEGRATE

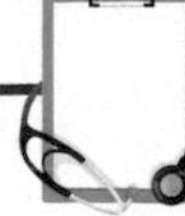

CLINICAL VIEW 11.5

Hernias

The condition in which a portion of the viscera, such as the intestine, protrudes through a weakened point of the muscular wall of the abdominopelvic cavity is called a **hernia** (her′nē-ă; rupture). A significant medical problem may develop if the herniated portion of the intestine swells, becoming trapped. Blood flow to the trapped segment may diminish, causing that portion of the intestine to die.

How Do Inguinal Hernias Form, and Why Are Males More Prone to Get Them?

An **inguinal hernia** is the most common type of hernia. The inguinal region is one of the weakest areas of the abdominal wall. Within this region is a canal (inguinal canal) that allows the passage of the spermatic cord in males (see figure 28.15 in section 28.4a), and a smaller structure called the *round ligament* in females (see figure 28.9 in section 28.3c). The inguinal canal, or the **superficial inguinal ring** associated with it, is often the site of a rupture or separation of the abdominal wall. Males are more likely to develop inguinal hernias than females because their inguinal canals and superficial inguinal rings are larger to allow room for the spermatic cord. Rising pressure in the abdominal cavity, as might develop while straining to lift a heavy object, provides the force to push a segment of the small intestine into the canal.

How Do Physicians Test for an Inguinal Hernia?

The physician inserts a finger in the depression formed by the superficial inguinal ring and asks the patient to turn his or her head and cough, because the act of coughing increases intra-abdominal pressure and would potentially induce a portion of intestine to poke through the ring if there was a problem. While the patient coughs, the physician palpates the superficial inguinal ring to make sure no intestine is protruding through the ring.

11.7 Muscles of the Pelvic Floor

LEARNING OBJECTIVES

21. Describe the functions of the pelvic floor muscles.
22. Identify the boundaries of the perineum.

The floor of the pelvic cavity is formed by three layers of muscles and associated fasciae, collectively known as the **pelvic diaphragm.** The pelvic diaphragm extends from the ischium and pubis of the ossa coxae across the pelvic outlet to the sacrum and coccyx. These muscles collectively form the pelvic floor and support the pelvic viscera (**figure 11.17**).

The diamond-shaped region between the lower appendages is called the **perineum** (per′i-nē′ŭm). The perineum has four significant bony landmarks: the pubic symphysis anteriorly, the coccyx posteriorly, and both ischial tuberosities laterally. A transverse line drawn between the ischial tuberosities partitions the perineum into an anterior **urogenital triangle,** which contains the external genitalia and urethra, and a posterior **anal triangle,** which contains the anus (figure 11.17*b, c*).

Table 11.10 Muscles of the Abdominal Wall

Muscle	Action(s)	Superior Attachment(s) (S)/Inferior Attachment(s) (I)	Innervation (see sections 14.5c, f)
External oblique	Bilateral action[1]: Flexes vertebral column and compresses abdominal wall Unilateral action[2]: Laterally flexes vertebral column; rotates vertebral column to same side	S: External and inferior borders of the inferior 8 ribs (ribs 5–12) I: Linea alba by a broad aponeurosis; some to iliac crest	Spinal nerves T8–T12, L1
Internal oblique	Bilateral action: Flexes vertebral column and compresses abdominal wall Unilateral action: Laterally flexes vertebral column; rotates vertebral column to opposite side	S: Lumbar fascia, inguinal ligament, and iliac crest I: Linea alba, pubic crest, inferior rib surfaces of the inferior 4 ribs (9–12); costal cartilages of ribs 8–10	Spinal nerves T8–T12, L1
Transversus abdominis (ab-dom′i-nis) *transverse* = across	Bilateral action: Flexes vertebral column and compresses abdominal wall Unilateral action: Laterally flexes vertebral column	S: Iliac crest, cartilages of inferior 6 ribs (ribs 7–12); lumbar fascia; inguinal ligament I: Linea alba and pubic crest	Spinal nerves T8–T12, L1
Rectus abdominis	Flexes vertebral column and compresses abdominal wall	S: Xiphoid process of sternum; inferior surfaces of ribs 5–7 I: Superior surface of pubis near pubic symphysis	Spinal nerves T7–T12

1. *Bilateral action* means both the left and right muscles are contracting together.
2. *Unilateral action* means only one muscle (either the left or right muscle) is contracting.

Table 11.11 Muscle Actions on the Axial Skeleton

Extend the Neck and/or Vertebral Column	Flex the Neck and/or Vertebral Column	Laterally Flex the Vertebral Column	Rotate the Head to One Side	Elevate the Ribs	Depress the Ribs
Splenius muscles[1]	Sternocleidomastoid[1]	Quadratus lumborum[2]	Sternocleidomastoid[2]	Serratus posterior superior	Serratus posterior inferior
Erector spinae[1] (iliocostalis, longissimus, spinalis)	Scalene muscles[1]	External oblique[2]	Splenius muscles[2]	External intercostals	Internal intercostals
Quadratus lumborum[1]	External oblique[1]	Internal oblique[2]	Longissimus capitis[2]	Scalene muscles (1st and 2nd ribs only)	Transversus thoracis
Transversospinalis group[1]	Internal oblique[1]	Transversus abdominis[2]	Obliquus capitis inferior[2]		
Rectus capitis posterior major and minor[1]	Transversus abdominis[1] Rectus abdominis[1]		Obliquus capitis superior[2]		

1. Bilateral action
2. Unilateral action of muscles

Figure 11.17 Muscles of the Pelvic Floor. The pelvic cavity floor is composed of muscle layers that form the urogenital and anal triangles, extend across the pelvic outlet, and support the organs in the pelvic cavity. (The puborectalis muscle is not shown.) (*a*) Superior view of the female pelvic floor. Inferior views show (*b*) both the superficial and deep muscles of the male and (*c*) female perineal regions. Muscles of the pelvic floor are in bold.

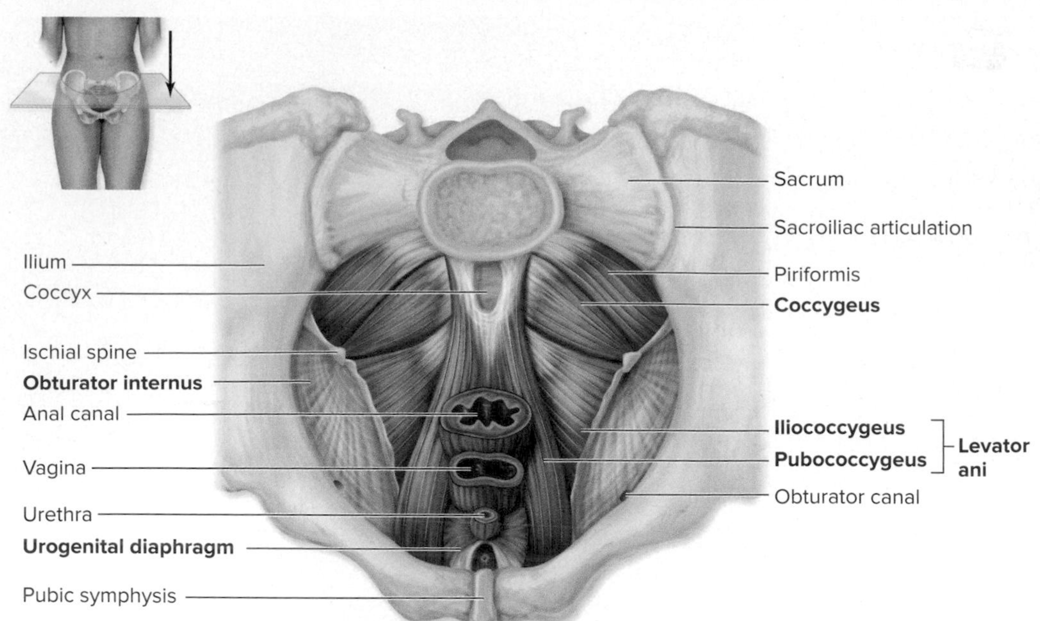

(a) Female, superior view

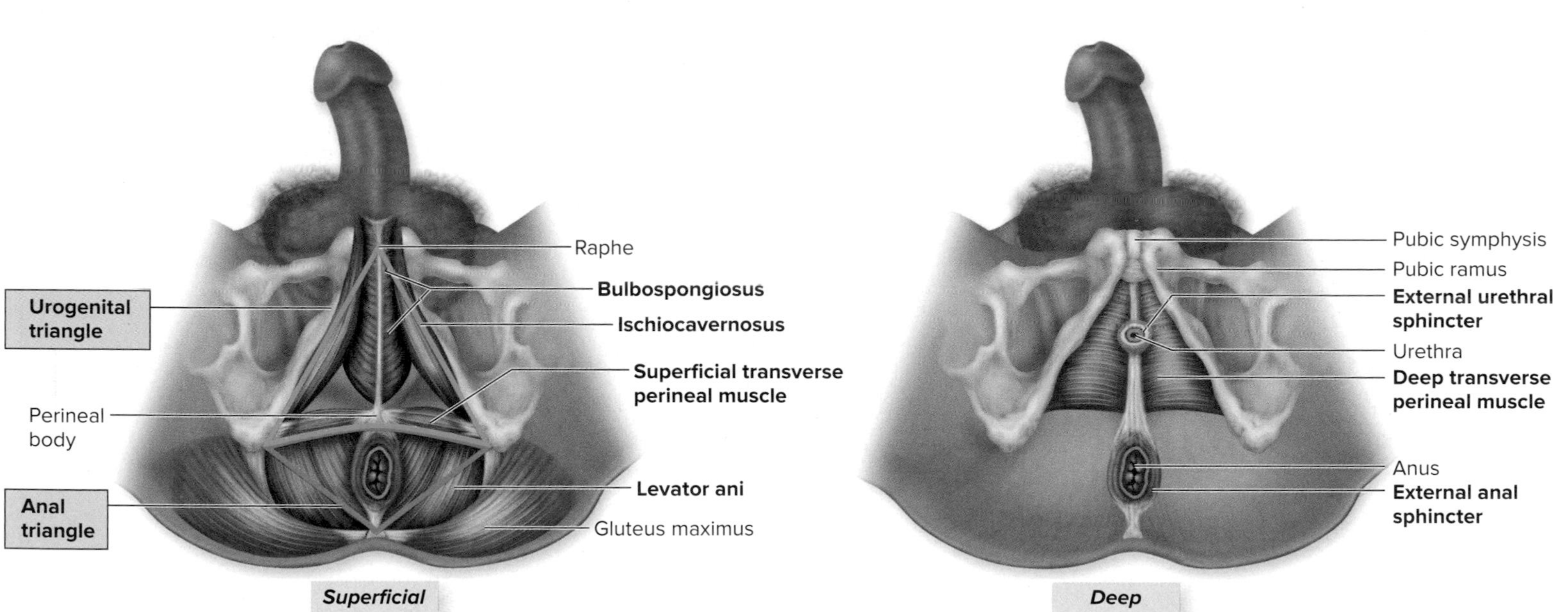

(b) Male, inferior view

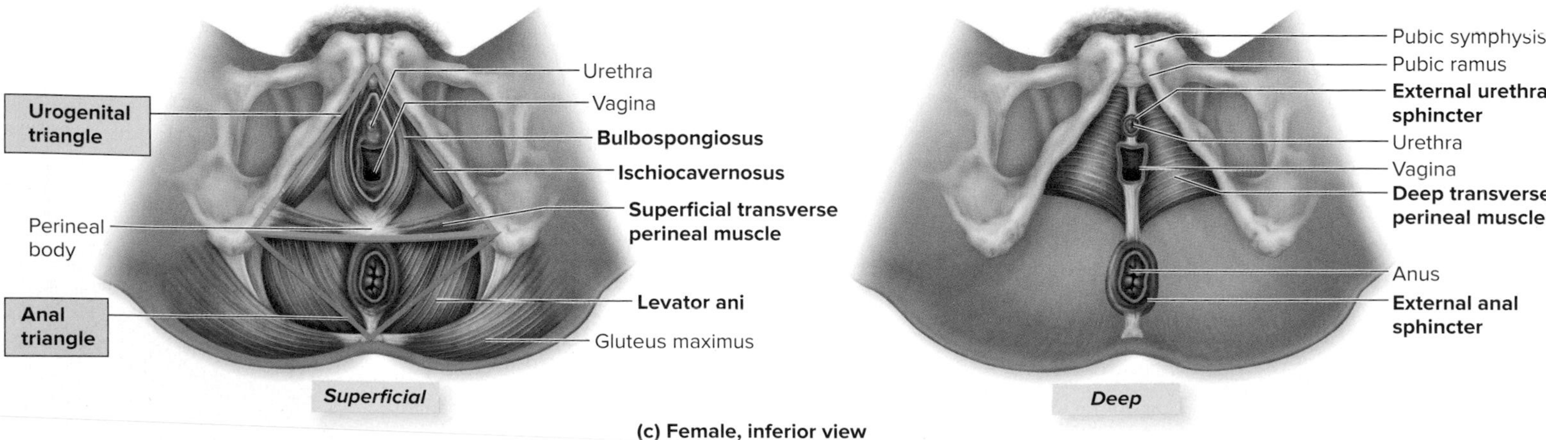

(c) Female, inferior view

Table 11.12	Muscles of the Pelvic Floor		
Group/Muscle	**Action(s)**	**Origin/Insertion**[1]	**Innervation (see section 14.5g)**
ANAL TRIANGLE			
Coccygeus (kok-sij′ē-ŭs) *coccy* = coccyx	Forms pelvic floor and supports pelvic viscera	O: Ischial spine I: Lateral and inferior borders of sacrum and coccyx	Spinal nerves S4–S5
External anal sphincter (ā′năl sfingk′tĕr) *anal* = referring to anus *sphin* = squeeze	Constricts anal opening; must voluntarily relax to defecate	O: Perineal body I: Encircles anal opening	Pudendal nerve (S2–S4)
Levator ani (lē-vā′tor, lē-vā-ter ā′nī; *levator* = raises; *ani* = anus): Group of muscles that form the anterior and lateral parts of the pelvic diaphragm			
Iliococcygeus (il′ē-ō-kok-si′jē-ŭs)	Forms pelvic floor and supports pelvic viscera	O: Pubis and ischial spine I: Coccyx and median raphe	Pudendal nerve (S2–S4)
Pubococcygeus (pyū′bō-kok-si′jē-ŭs)	Forms pelvic floor and supports pelvic viscera	O: Pubis and ischial spine I: Coccyx and median raphe	Pudendal nerve (S2–S4)
Puborectalis (pyū′bō-rek′tăl-is) *rectal* = rectum	Supports anorectal junction; must relax to defecate	O: Pubis and ischial spine I: Coccyx and median raphe	Pudendal nerve (S2–S4)
UROGENITAL TRIANGLE			
Superficial layer			
Bulbospongiosus (female) (bul′bō-spŭn′jē-ō′sŭs) *bulbon* = bulb *spongio* = sponge	Narrows vaginal opening; compresses and stiffens clitoris	O: Sheath of collagen fibers at base of clitoris I: Perineal body	Pudendal nerve (S2–S4)
Bulbospongiosus (male)	Ejects urine or semen; compresses base of penis; stiffens penis	O: Sheath of collagen fibers at base of penis I: Median raphe and perineal body	Pudendal nerve (S2–S4)
Ischiocavernosus (ish′ē-ō-kav′er-nō-sŭs) *caverna* = hollow chamber	Assists with erection of penis or clitoris	O: Ischial tuberosities and ischial ramus I: Pubic symphysis	Pudendal nerve (S2–S4)
Superficial transverse perineal muscle (per′i-nē′ăl)	Supports pelvic organs	O: Ramus of ischium I: Perineal body	Pudendal nerve (S2–S4)
Deep layer (urogenital diaphragm)			
Deep transverse perineal muscle	Supports pelvic organs	O: Ischial ramus I: Median raphe of urogenital diaphragm	Pudendal nerve (S2–S4)
External urethral sphincter (yū-rē′thrăl)	Constricts urethra; must voluntarily relax in order to urinate	O: Rami of ischium and pubis I: Median raphe of urogenital diaphragm	Pudendal nerve (S2–S4)

1. Here, we use the terms *origin* and *insertion* because the terms *superior attachment* and *inferior attachment* are not applicable for all of the muscles in this table.

Table 11.12 describes the specific muscles of the pelvic floor and perineum, and summarizes their characteristics.

WHAT DID YOU LEARN?

17 What are the functions of the pelvic floor muscles?

INTEGRATE

CONCEPT CONNECTION

The muscles of the pelvic floor may become stretched or torn after childbirth (see section 29.6), so women may not have adequate support for the pelvic organs. These issues may become more problematic as a woman ages. As a result, incontinence (leakage of urine) is a common complaint among women who have given birth. Strengthening the pelvic floor muscles (with exercises such as Pilates or Kegel exercises) often alleviates these problems.

Table 11.14 Muscles That Move the Glenohumeral Joint/Arm (continued)

Group/Muscle	Action(s)	Proximal Attachment(s) (P)/Distal Attachment(s) (D)	Innervation (see section 14.5e)
Teres major (ter′ēz) *teres* = round	Extends, adducts, and medially rotates arm	P: Inferior lateral border and inferior angle of scapula D: Lesser tubercle and intertubercular sulcus of humerus	Lower subscapular nerve (C5–C6)
Triceps brachii (long head) (trī′seps brā′kē-ī) *triceps* = three heads	Extends and adducts arm	P: Infraglenoid tubercle of scapula D: Olecranon of ulna	Radial nerve (C5–C7 nerve fibers)
Biceps brachii (long head) (bī′seps) *biceps* = two heads	Flexes arm	P: Supraglenoid tubercle of scapula D: Radial tuberosity and bicipital aponeurosis	Musculocutaneous nerve (C5–C6 nerve fibers)
Rotator cuff muscles (rō-tā′ter, tōr kŭf; *rotatio* = to revolve): Collectively, these four muscles stabilize the glenohumeral joint			
Subscapularis (sŭb-skap′yū-lăr′ris)	Medially rotates arm	P: Subscapular fossa of scapula D: Lesser tubercle of humerus	Upper and lower subscapular nerves (C5–C6)
Supraspinatus (sū-pră-spī-nā′tŭs) *supra* = above, over	Abducts arm	P: Supraspinous fossa of scapula D: Greater tubercle of humerus	Suprascapular nerve (C5–C6)
Infraspinatus (in-fră-spī-nā′tŭs) *infra* = below	Adducts and laterally rotates arm	P: Infraspinous fossa or scapula D: Greater tubercle of humerus	Suprascapular nerve (C5–C6)
Teres minor	Adducts and laterally rotates arm	P: Upper dorsal lateral border of scapula (superior to teres major attachment) D: Greater tubercle of humerus	Axillary nerve (C5–C6)

The muscles that move the arm at the glenohumeral joint are grouped in **table 11.15** according to different types of actions. Note that a muscle that has multiple functions is listed in more than one group. For example, the deltoid may abduct, extend, or flex the humerus, depending upon whether its middle, posterior, or anterior fibers are contracting. Pectoralis major and coracobrachialis both adduct and flex the humerus, so they are listed in both the adduction and flexion columns. We recommend that you copy the columns multiple times and then test your knowledge by trying to write out all of the muscles in a group without looking at your notes.

WHAT DID YOU LEARN?

19 Which muscles extend the arm at the glenohumeral (shoulder) joint?

20 Identify the rotator cuff muscles, and describe their actions.

Table 11.15 Summary of Muscle Actions at the Glenohumeral Joint/Arm

Abduction	Adduction	Extension	Flexion	Lateral Rotation	Medial Rotation
Deltoid (middle fibers)[1]	**Latissimus dorsi**	**Latissimus dorsi**	**Pectoralis major**	**Infraspinatus**	**Subscapularis**
Supraspinatus	**Pectoralis major**	**Deltoid (posterior fibers)**	**Deltoid (anterior fibers)**	**Teres minor**	Deltoid (anterior fibers)
	Coracobrachialis	Teres major	Coracobrachialis	Deltoid (posterior fibers)	Latissimus dorsi
	Teres major	Long head of triceps brachii	(Long head of biceps brachii)		Pectoralis major
	Teres minor				Teres major
	Infraspinatus				
	(Long head of triceps brachii)				

1. Boldface indicates an agonist; others are synergists. Parentheses around an entire muscle name, such as the long head of the biceps brachii under "Flexion," indicate only a slight effect.

11.8c Arm and Forearm Muscles That Move the Elbow Joint/Forearm

LEARNING OBJECTIVES

26. Name the muscles in the arm's anterior and posterior compartments, and contrast their common functions.

27. Describe the muscles that pronate and supinate the forearm.

When you move the elbow joint, you move the bones of the forearm. Thus, the phrase "flexing the elbow joint" is synonymous with "flexing the forearm." Keep this point in mind as we discuss the muscles that move the elbow joint and forearm.

The muscles in the limbs are organized into **compartments,** which are surrounded by deep fascia. Each compartment houses functionally related skeletal muscles, as well as their associated nerves and blood vessels. In general, muscles in the same compartment tend to perform similar functions. **Figure 11.23** provides a visual overview of how the muscles are organized into compartments. Note how muscles in opposite compartments tend to be antagonists. For example, the anterior forearm muscles are primarily flexors and pronators, whereas the posterior forearm muscles are primarily extensors and supinators. Likewise, in the lower limb, knee extensors are in the anterior compartment of the thigh, whereas knee flexors are in the posterior compartment of the thigh. Hip adductors are in the medial compartment of the thigh, whereas a hip abductor is in the lateral compartment of the thigh. These compartments may help you to learn the muscles in common functional groups. If you know what compartment a muscle is in, you likely can figure out what action that muscle performs, and vice versa.

WHAT DO YOU THINK?

3 The brachialis is on the anterior surface of the arm. Without looking at the muscle tables, determine whether this muscle flexes or extends the elbow joint. How did you reach your conclusion?

The muscles of the arm may be subdivided into an anterior compartment and a posterior compartment. The **anterior compartment** primarily contains elbow flexors, so it is also known as the *flexor compartment.* Muscles in this compartment are supplied by the deep brachial artery and are innervated by the musculocutaneous nerve. Muscles in this compartment include the coracobrachialis (note that this muscle is an arm flexor and not an elbow flexor), biceps brachii, brachialis, and brachioradialis muscles. Put your hand on your anterior arm, and then flex your elbow. Observe how these muscles bulge as they contract, affirming that these anterior compartment arm muscles flex the elbow.

The **posterior compartment** contains elbow extensors, so this compartment is also called the *extensor compartment*. These muscles receive their blood supply from the deep brachial artery and are innervated by the radial nerve. The primary muscle in this compartment is the triceps brachii. Put your hand on your posterior arm, and then extend your elbow. You can feel your muscles contracting as you perform this action, affirming that posterior arm muscles extend the elbow. Perform these activities each time we discuss muscles in a particular limb compartment so you can see and feel for yourself how these muscles move.

Muscles of the Arm's Anterior Compartment

On the anterior side of the humerus are the principal flexors of the forearm: the biceps brachii and brachialis **(figure 11.24)**. The **biceps brachii** is a large, two-headed muscle on the anterior surface of the humerus. This muscle flexes the elbow joint, and it is a powerful supinator of the forearm when the elbow is flexed. (An example of this supination movement occurs when you tighten a screw with your right hand.) The tendon of the long head of the biceps brachii crosses the shoulder joint, and so this muscle helps flex the humerus as well (albeit weakly).

The **brachialis** is deep to the biceps brachii on the anterior surface of the humerus. It is the most powerful flexor of the forearm at the elbow. The **brachioradialis** is another prominent muscle on the anterolateral surface of the forearm. It is a synergist in elbow flexion, effective primarily when the prime movers of forearm flexion have already partially flexed the elbow.

Muscles of the Arm's Posterior Compartment

The posterior compartment of the arm contains two muscles that extend the forearm at the elbow: the triceps brachii and the anconeus **(figure 11.25)**. The **triceps brachii** is the large, three-headed muscle on the posterior surface of the arm. The long head of the triceps brachii also crosses the glenohumeral joint, where it helps extend the humerus. All three parts of this muscle merge to form a common distal attachment on the olecranon of the ulna. A weak elbow extensor is the small **anconeus** that crosses the posterolateral region of the elbow.

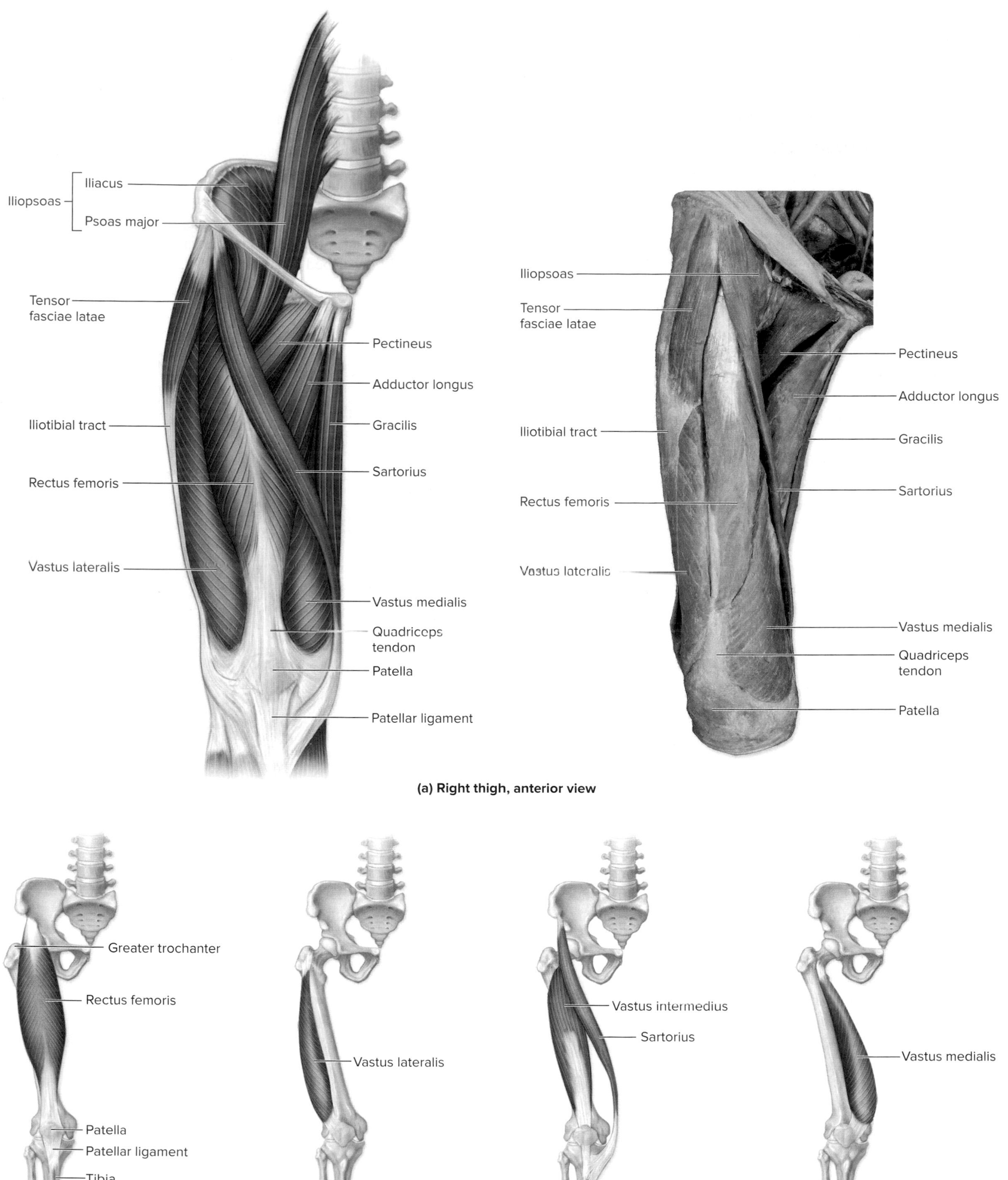

Figure 11.32 Muscles of the Anterior Thigh. Muscles of the anterior thigh flex the thigh and extend the leg. (*a*) Illustration and cadaver photo show an anterior view of the right thigh. (*b*) Individual muscles of the right anterior thigh. AP|R

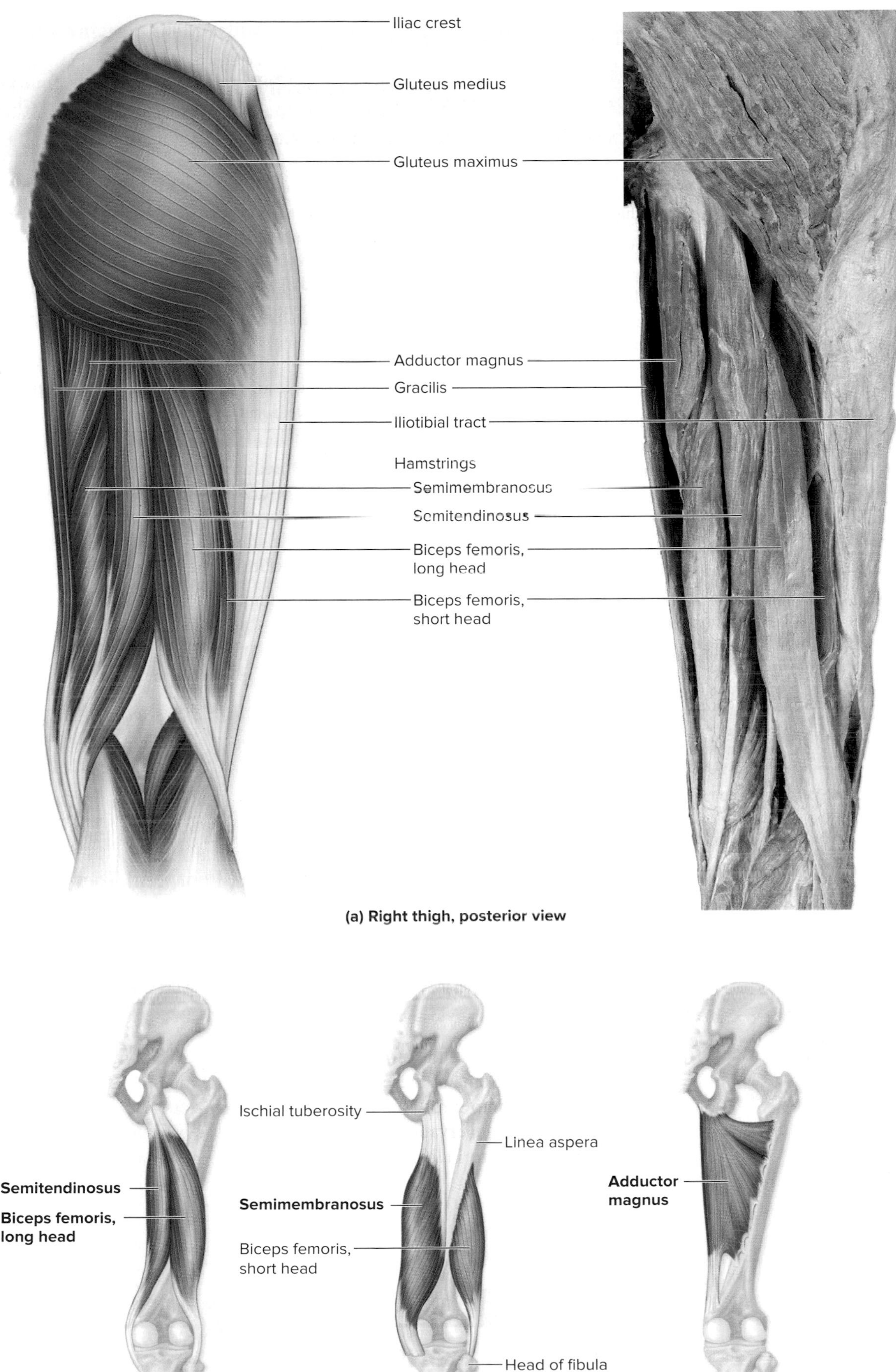

Figure 11.33 Muscles of the Gluteal Region and Posterior Thigh. Muscles of the posterior thigh extend the thigh and flex the leg. (*a*) Illustration and cadaver photo show the gluteal and posterior muscles of the right thigh. (*b*) Individual muscles that extend the thigh are shown in bold (note the short head of biceps femoris does not participate in thigh extension). APR

(a) ©McGraw-Hill Education/Christine Eckel

CHAPTER SUMMARY (continued)

	11.8e Intrinsic Muscles of the Hand • The intrinsic muscles may be divided into three groups: (1) the thenar group (moves the thumb), (2) hypothenar group (moves the little finger), and (3) the midpalmar group (moves fingers 2–5).
11.9 Muscles of the Pelvic Girdle and Lower Limb	• Four groups of muscles are associated with the pelvis and lower limb: (1) muscles that move the hip joint/thigh, (2) thigh muscles that move the knee joint/leg, (3) leg muscles, and (4) intrinsic muscles of the foot.
	11.9a Muscles That Move the Hip Joint/Thigh • Anterior thigh muscles flex the thigh. • Gluteus maximus and the posterior thigh muscles (hamstrings) extend the thigh. • Gluteus medius, gluteus minimus, and tensor fasciae latae abduct the thigh. • Most medial thigh muscles adduct and flex the thigh.
	11.9b Thigh Muscles That Move the Knee Joint/Leg • The quadriceps femoris extends the leg. • The gracilis, sartorius, and posterior thigh muscles (hamstrings) flex the leg.
	11.9c Leg Muscles That Move the Ankle, Foot, and Toes • Anterior leg muscles dorsiflex the foot, extend the toes, or both. One muscle (tibialis anterior) also inverts the foot. • Lateral leg muscles evert the foot and plantar flex the foot. • Posterior leg muscles plantar flex the foot, flex the toes, or both. One muscle (tibialis posterior) also inverts the foot.
	11.9d Intrinsic Muscles of the Foot • Dorsal muscles extend the toes. • The four layers of plantar muscles can potentially flex, extend, abduct, or adduct the toes.

CHALLENGE YOURSELF

Create and Evaluate
Analyze and Apply
Understand and Remember

Do You Know the Basics?

1. Which statement is accurate about an agonist?
   a. It opposes the function of the prime mover.
   b. It is the primary muscle that produces a movement.
   c. It functions primarily to stabilize a joint.
   d. Its muscle fibers are always circular in structure.
2. When the left and right ________ contract, they flex the head and neck.
   a. sternocleidomastoid
   b. longissimus group
   c. splenius
   d. rectus abdominis
3. When this large muscle contracts, the vertical dimensions of the thoracic cavity increase.
   a. external intercostal
   b. internal intercostal
   c. diaphragm
   d. transversus thoracis
4. This muscle depresses and adducts the eye.
   a. inferior rectus
   b. inferior oblique
   c. lateral rectus
   d. superior oblique
5. Which of these muscles extends the vertebral column?
   a. external oblique
   b. transversus abdominis
   c. spinalis
   d. internal oblique
6. The dorsal interossei muscles in the hand
   a. adduct fingers 2–5.
   b. abduct fingers 2–5.
   c. flex the PIP and DIP joints.
   d. extend the MP joints.
7. Muscles in the anterior compartment of the leg
   a. evert the foot.
   b. dorsiflex the foot and extend the toes.
   c. plantar flex the foot.
   d. flex the toes.
8. All of the following muscles flex the forearm *except* the
   a. brachialis.
   b. biceps brachii.
   c. brachioradialis.
   d. anconeus.
9. Which muscles arise from the ischial tuberosity and extend the thigh plus flex the leg?
   a. adductor
   b. fibularis
   c. hamstring
   d. quadriceps
10. The _____ plantar flexes the foot.
    a. iliopsoas
    b. gastrocnemius
    c. fibularis tertius
    d. vastus intermedius
11. What are some of the ways that muscles are named?
12. Which muscles of facial expression do you use to (a) smile, (b) close your eyes, and (c) close your mouth?
13. Distinguish between suprahyoid and infrahyoid muscles, and describe the functions of each group.
14. What is the effect of contracting the abdominal oblique muscles?
15. What movements are possible at the glenohumeral joint, and which muscles are the prime movers for each of these movements?

16. Identify the compartments of the arm (brachium), the muscles in each compartment, and their function.
17. Compare and contrast the flexor digitorum superficialis and the flexor digitorum profundus; where does each attach distally, how are their tendons interrelated, and what muscle actions do they perform?
18. What muscles are responsible for thigh extension? Which of these is the prime mover of thigh extension?
19. What leg muscles allow a ballet dancer to rise up and balance on her toes?
20. Which muscles are responsible for foot inversion?

Can You Apply What You've Learned?

1. A 50-year-old woman was concerned about the appearance of "crow's-feet" by her eyes. She was told by her physician that these wrinkles were caused by years of squinting and blinking using which muscle?
   a. frontal belly of occipitofrontalis
   b. orbicularis oculi
   c. risorius
   d. orbicularis oris
2. Eliza complained of double vision and went to see her optometrist. The optometrist tested the function of various eye muscles and discovered that Eliza could not move her right eye medially. Which muscle may be injured?
   a. superior oblique
   b. lateral rectus
   c. medial rectus
   d. inferior oblique
3. After an intensive workout, George felt especially sore in his posterior arm regions. What repetitive workout activity most likely caused the soreness?
   a. flexing the humerus
   b. flexing the forearm
   c. pronating the forearm
   d. extending the forearm
4. While Carly was playing soccer, she was kicked in the anterior thigh by an opposing teammate. Due to this injury, what muscle function may have been difficult to perform?
   a. extending the knee
   b. flexing the knee
   c. extending the thigh
   d. dorsiflexing the foot
5. Joshua broke his fibula and had to wear a cast for 6 weeks. The muscles attaching to this bone atrophied during this time. What muscle function was Joshua unable to perform as a result?
   a. plantar flexing the foot
   b. flexing the knee
   c. everting the foot
   d. dorsiflexing the foot

Can You Synthesize What You've Learned?

1. Albon is a 45-year-old male who characterizes himself as a "couch potato." He exercises infrequently and has a rounded abdomen ("beer belly"). While helping a friend move some heavy furniture, he felt a sharp pain deep within his abdominopelvic cavity. An emergency room resident told Albon that he had suffered an inguinal hernia. What is this injury, how did it occur, and how might Albon's poorly developed abdominal musculature have contributed to it?
2. While training on the balance beam, Pat slipped during her landing from a back flip and fell, straddling the beam. Although only slightly sore from the fall, she became concerned when she suddenly lost the ability to completely control her urination. What might have happened to Pat's pelvic floor structures during the fall?
3. After falling while skateboarding, Karen had surgery on her elbow. During her recovery, she must visit the physical therapist to improve muscle function around the elbow. Develop a series of exercises that may improve all of Karen's elbow movements, and determine which muscles are being helped by each exercise.
4. Why is it more difficult for Eric to lift a heavy weight when his forearm is pronated than when his forearm is in the supine position?

INTEGRATE

ONLINE STUDY TOOLS

connect | SMARTBOOK® | AP|R

The following study aids may be accessed through Connect.

Clinical Case Study: A Young Boy Who Is Losing His Ability to Walk

Interactive Questions: This chapter's content is served up in a number of multimedia question formats for student study

SmartBook: Topics and terminology include skeletal muscle composition and actions; skeletal muscle naming; muscles of the head and neck; muscles of the vertebral column; muscles of respiration; muscles of the abdominal wall; muscles of the pelvic floor; muscles of the pectoral girdle and upper and lower limbs

Anatomy & Physiology Revealed: Topics include dissection of the head and neck, orbit, back, thorax, abdomen, shoulder, forearm, arm, wrist, hand, hip, thigh, and foot; muscle actions of pectoralis major, latissimus dorsi, flexor digitorum superficialis, and gastrocnemius

which is the signal receiver, or target (binds neurotransmitter). The synapse may be between the axon of the presynaptic neuron and any portion of the surface of a postsynaptic neuron (dendrite, cell body, or axon), except those regions that are covered by a myelin sheath (described in section 12.4c). Most commonly, a synapse is with a dendrite of the postsynaptic neuron. The synaptic knob of the presynaptic neuron does not quite make contact with the postsynaptic neuron (figure 12.3*c*). The two neurons are separated by an extremely narrow, fluid-filled gap (of about 30 nanometers) called the **synaptic cleft.**

Transmission between a presynaptic and postsynaptic neuron occurs when *neurotransmitter* molecules stored in synaptic vesicles are released from the synaptic knob of a presynaptic neuron into the synaptic cleft. Some of the neurotransmitter diffuses across the synaptic cleft to bind to receptors within the plasma membrane of the postsynaptic neuron to initiate a graded potential (see section 12.8a). There is a **synaptic delay** associated with neurotransmitter release at chemical synapses. The delay is the time between the neurotransmitter release from the presynaptic cell, its diffusion across the synaptic cleft, and neurotransmitter binding to receptors in the postsynaptic neuron plasma membrane. This delay is usually between 0.3 and 0.5 millisecond. Note that one postsynaptic neuron may, and often does, receive signals from more than one presynaptic neuron simultaneously.

A second type of synapse (which is much less common in humans) is an electrical synapse. An **electrical synapse** is composed of a presynaptic neuron and a postsynaptic neuron physically bound together. Gap junctions (see section 4.6d) are present in the plasma membranes of both neurons and facilitate the flow of ions between the cells. The cells act as though they shared a plasma membrane. Thus, the electrical signal passes between the cells with essentially no synaptic delay. Electrical synapses are located within limited regions of the brain and the eyes.

WHAT DID YOU LEARN?

9 What is a chemical synapse within the nervous system, and how does it function?

12.4 Nervous Tissue: Glial Cells

Glial cells are the other distinct cell type within nervous tissue. These nonexcitable cells primarily support and protect the neurons.

12.4a General Characteristics of Glial Cells

LEARNING OBJECTIVE

14. List the distinguishing features of glial cells.

Glial (glī′ăl; *glia* = glue) **cells** are sometimes referred to as *neuroglia* (nū-rog′lē-a). They are found within both the CNS and the PNS. Glial cells are both smaller than neurons and capable of producing new glial cells through cell division (i.e., glial cells have mitotic ability; see section 4.9b). Glial cells do not transmit electrical signals, but they do assist neurons with their functions. The glial cells cooperate to physically protect and help nourish neurons as well as provide an organized, supporting scaffolding for all the nervous tissue. During development, glial cells form the framework that guides young, migrating neurons to their final destinations. Recent evidence has

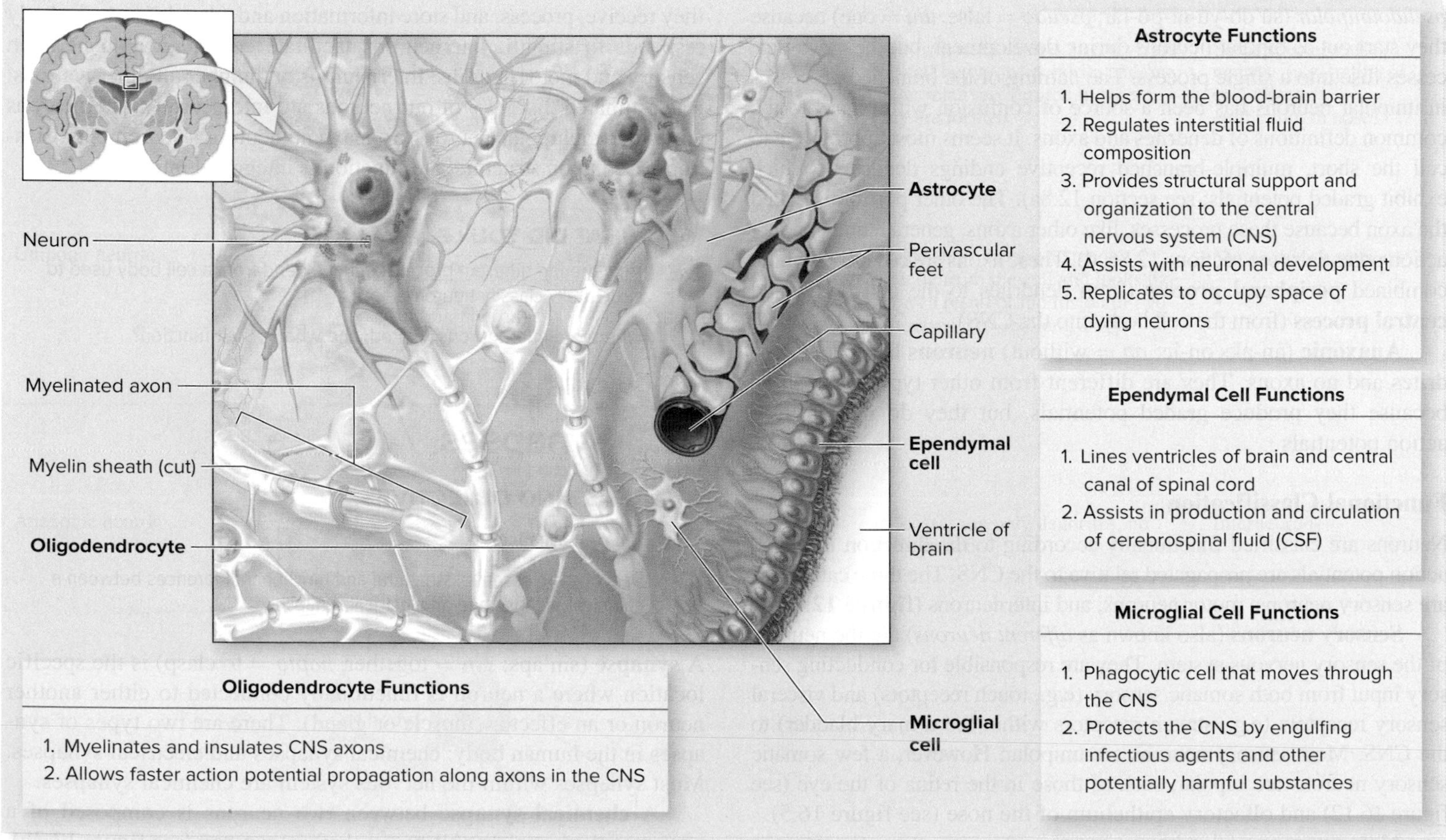

Neuron
Myelinated axon
Myelin sheath (cut)
Oligodendrocyte
Astrocyte
Perivascular feet
Capillary
Ependymal cell
Ventricle of brain
Microglial cell
Astrocyte Functions
1. Helps form the blood-brain barrier
2. Regulates interstitial fluid composition
3. Provides structural support and organization to the central nervous system (CNS)
4. Assists with neuronal development
5. Replicates to occupy space of dying neurons
Ependymal Cell Functions
1. Lines ventricles of brain and central canal of spinal cord
2. Assists in production and circulation of cerebrospinal fluid (CSF)
Microglial Cell Functions
1. Phagocytic cell that moves through the CNS
2. Protects the CNS by engulfing infectious agents and other potentially harmful substances
Oligodendrocyte Functions
1. Myelinates and insulates CNS axons
2. Allows faster action potential propagation along axons in the CNS

Figure 12.5 Glial Cells of the Central Nervous System (CNS). Four types of glial cells are located within the CNS: astrocytes, ependymal cells, microglia, and oligodendrocytes. These cells differ in both structure and function.

shown that glial cells are critical for the normal function at neural synapses, both maintaining the anatomic structure of synapses and modifying transmission that occurs there.

Glial cells far outnumber neurons. The nervous tissue of a young adult may contain 35 to 100 billion neurons and 100 billion to 1 trillion glial cells. Collectively, glial cells account for roughly half the volume of the nervous system.

WHAT DID YOU LEARN?

10 If a person has a brain tumor, is it more likely to have developed from neurons or from glial cells? Why?

12.4b Types of Glial Cells

LEARNING OBJECTIVE

15. Describe the structure and function of the four types of glial cells within the CNS and the two types of glial cells within the PNS.

INTEGRATE

CLINICAL VIEW 12.2

Tumors of the Central Nervous System

Neoplasms resulting from unregulated cell growth, commonly known as **tumors,** sometimes occur within the central nervous system (CNS). A tumor that originates within the organ where it is found is called a **primary tumor.** Because most mature neurons cannot undergo mitosis, primary CNS tumors typically originate in supporting tissues within the brain or spinal cord that have retained the capacity to undergo mitosis: the meninges (protective membranes of the CNS) or the glial cells. Glial cell tumors, termed **gliomas,** may be either relatively benign and slow-growing or malignant (capable of metastasizing or spreading to distant sites).

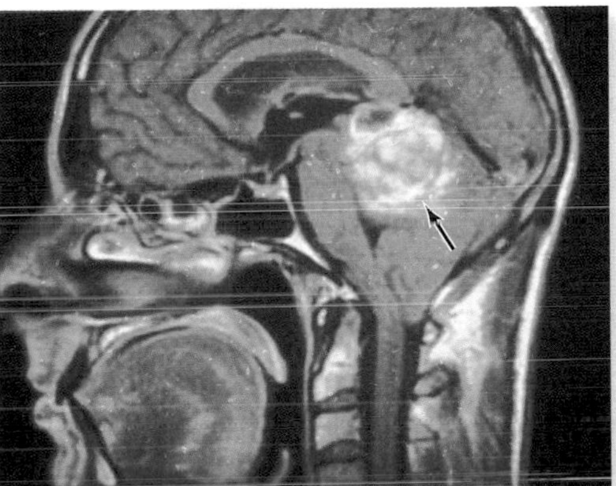

An MRI shows a glioma (arrow).

Glial Cells of the CNS

Four types of glial cells are found in the central nervous system (CNS). These different cells are astrocytes, ependymal cells, microglia, and oligodendrocytes (**figure 12.5**). They can be distinguished based upon size, intracellular organization, and the presence of specific cytoplasmic processes.

Astrocytes (as′trō-sīt; *astron* = star) exhibit a starlike shape due to projections from their surface. These numerous cell processes have contact with both neurons and blood capillaries (smallest blood vessels) (see section 20.1c). Astrocytes are the most abundant glial cells in the CNS and constitute over 90% of the nervous tissue in some areas of the brain. Astrocytes nurture, protect, support, and guide neurons, as follows:

- **Help form the blood-brain barrier.** The ends of astrocyte processes are called **perivascular feet:** They both cover and wrap around capillaries in the brain. The perivascular feet and the brain capillaries together contribute to a **blood-brain barrier (BBB).** The BBB strictly controls movement of substances from exiting the blood and entering the nervous tissue in the brain. The BBB protects the delicate neurons of the brain from toxins, but at the same time allows needed nutrients to pass through (see section 13.2d).
- **Regulate interstitial fluid composition.** Astrocytes help maintain an optimal chemical composition of the interstitial fluid (fluid around cells) within the brain. For example, astrocytes regulate potassium ion concentration by absorbing these ions to maintain a constant potassium ion concentration that is critical to electrical activity of neurons.
- **Form structural support.** The cytoskeleton in astrocytes strengthens these cells to provide a structural framework to support and organize neurons within the CNS.
- **Assist neuronal development.** Astrocytes help direct the development of neurons in the fetal brain by secreting chemicals that regulate the formation of connections between neurons.
- **Occupy the space of dying neurons.** When neurons are damaged and die, the space they formerly occupied is often filled by astrocytes that replicate through cell division.

Ependymal (e-pen′di-măl) **cells** are ciliated simple cuboidal or simple columnar epithelial cells (see table 5.2) that line the internal cavities (ventricles) of the brain (see figure 13.7) and the central canal of the spinal cord (see figure 14.3). These cells have slender processes that branch extensively to make contact with other glial cells in the surrounding nervous tissue.

Ependymal cells and nearby blood capillaries together form a network called the **choroid** (ko′royd) **plexus** (see figure 13.8). The choroid plexus helps produce cerebrospinal fluid (CSF), a clear liquid that bathes the external surfaces of the CNS and fills its internal cavities. The cilia of ependymal cells help circulate the CSF (see section 13.2c).

Microglia (mī-krog′lē-ă; *micros* = small) are typically small cells that have slender branches extending from the main portion of the cell. They represent the smallest percentage of CNS glial cells, with some estimates of their prevalence as low as 5%. They are classified as phagocytic cells (macrophages) of the immune system (see section 22.2a). Microglial cells wander through the CNS and replicate in response to an infection. They protect the CNS against microorganisms (e.g., bacteria) and other potentially harmful substances by engulfing and destroying them through phagocytosis (see figure 22.3*a*). Microglia also function in removing debris from dead or damaged nervous tissue that results from infections, inflammation, trauma, and brain tumors.

Oligodendrocytes (ol′i-gō-den′drō-sīt; *oligos* = few) are large cells with a bulbous (round) body and slender cytoplasmic extensions or processes. The extensions of oligodendrocytes wrap around and insulate axons within the CNS to form a myelin sheath through a process called *myelination* (see section 12.4c). This insulation allows for faster propagation of action potentials along the axon.

Glial Cells of the PNS

Two types of glial cells are found in the peripheral nervous system (PNS). These specialized glial cells function in insulating neurons and include satellite cells and neurolemmocytes (**figure 12.6**).

INTEGRATE CONCEPT OVERVIEW

Figure 12.23 Events of Neuron Physiology. Neuron physiology involves specific events that occur in the four functional segments of a neuron: (1) receptive segment, (2) initial segment, (3) conductive segment, and (4) transmissive segment.

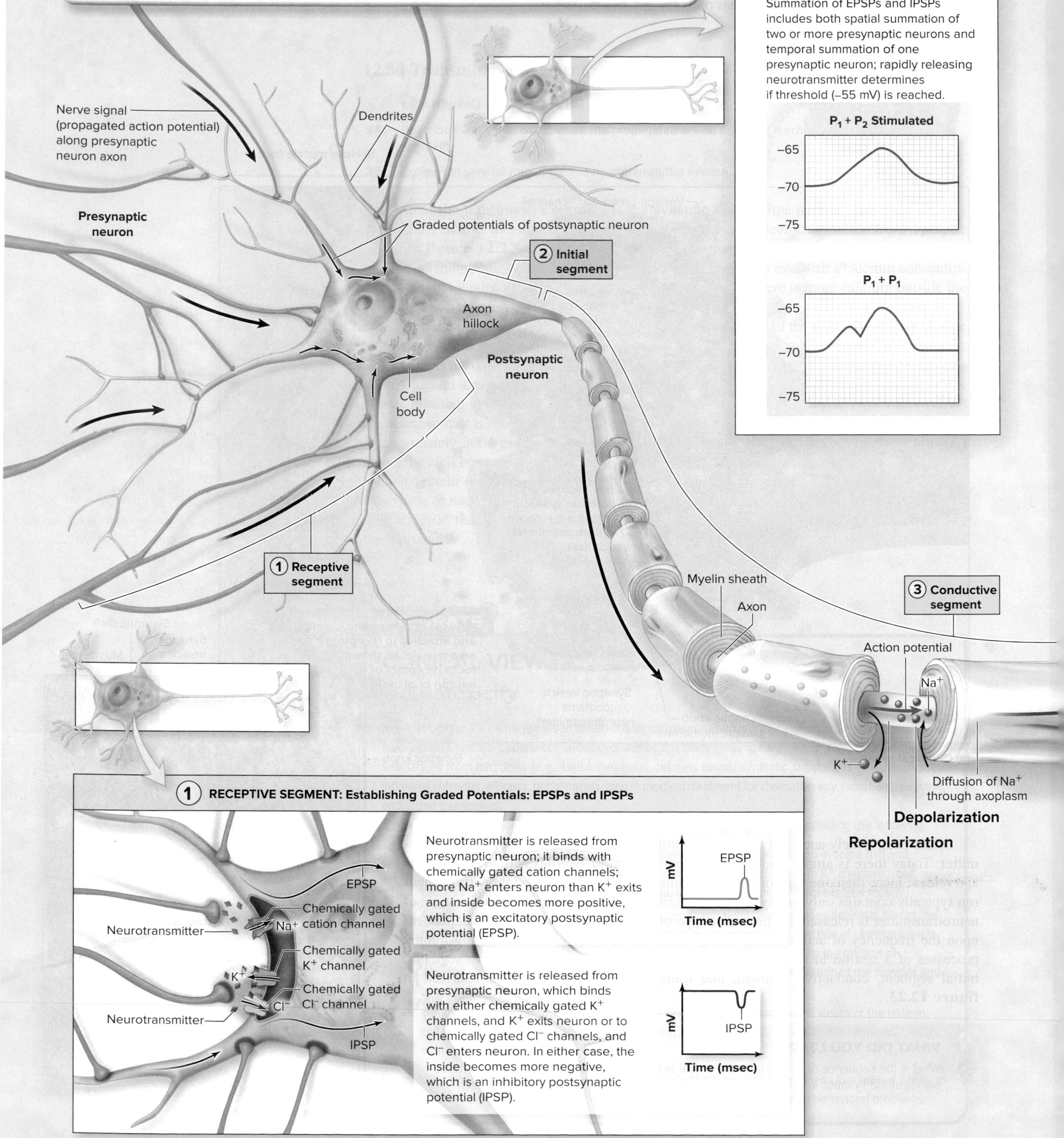

Nerve signal (propagated action potential) along presynaptic neuron axon

Dendrites

Presynaptic neuron

Graded potentials of postsynaptic neuron

2 Initial segment

Axon hillock

Postsynaptic neuron

Cell body

1 Receptive segment

Myelin sheath

Axon

3 Conductive segment

Action potential

Na^+

K^+

Diffusion of Na^+ through axoplasm

Depolarization

Repolarization

2 INITIAL SEGMENT: "Trigger Zone"

Summation of EPSPs and IPSPs includes both spatial summation of two or more presynaptic neurons and temporal summation of one presynaptic neuron; rapidly releasing neurotransmitter determines if threshold (–55 mV) is reached.

$P_1 + P_2$ Stimulated

–65
–70
–75

$P_1 + P_1$

–65
–70
–75

1 RECEPTIVE SEGMENT: Establishing Graded Potentials: EPSPs and IPSPs

EPSP

Chemically gated cation channel

Neurotransmitter

Na^+

Chemically gated K^+ channel

K^+

Chemically gated Cl^- channel

Cl^-

Neurotransmitter

IPSP

Neurotransmitter is released from presynaptic neuron; it binds with chemically gated cation channels; more Na^+ enters neuron than K^+ exits and inside becomes more positive, which is an excitatory postsynaptic potential (EPSP).

mV
EPSP
Time (msec)

Neurotransmitter is released from presynaptic neuron, which binds with either chemically gated K^+ channels, and K^+ exits neuron or to chemically gated Cl^- channels, and Cl^- enters neuron. In either case, the inside becomes more negative, which is an inhibitory postsynaptic potential (IPSP).

mV
IPSP
Time (msec)

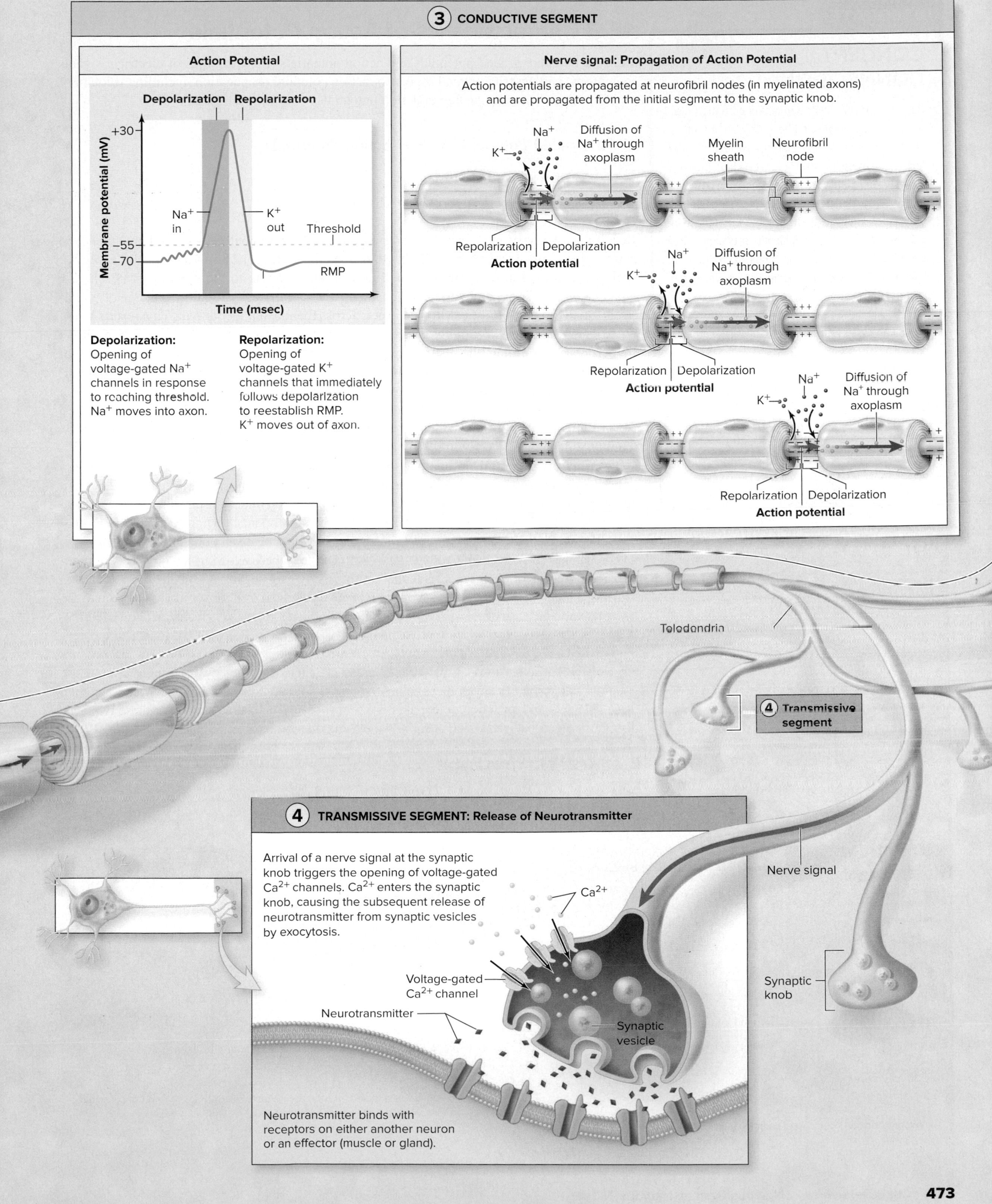

3 CONDUCTIVE SEGMENT
Action Potential
Depolarization
Repolarization
+30
Membrane potential (mV)
Na+ in
K+ out
Threshold
−55
−70
RMP
Time (msec)
Depolarization:
Opening of voltage-gated Na+ channels in response to reaching threshold. Na+ moves into axon.
Repolarization:
Opening of voltage-gated K+ channels that immediately follows depolarization to reestablish RMP. K+ moves out of axon.
Nerve signal: Propagation of Action Potential
Action potentials are propagated at neurofibril nodes (in myelinated axons) and are propagated from the initial segment to the synaptic knob.
Na+
K+
Diffusion of Na+ through axoplasm
Myelin sheath
Neurofibril node
Repolarization
Depolarization
Action potential
Repolarization
Depolarization
Action potential
Repolarization
Depolarization
Action potential
Telodendria
4 Transmissive segment
4 TRANSMISSIVE SEGMENT: Release of Neurotransmitter
Arrival of a nerve signal at the synaptic knob triggers the opening of voltage-gated Ca2+ channels. Ca2+ enters the synaptic knob, causing the subsequent release of neurotransmitter from synaptic vesicles by exocytosis.
Ca2+
Voltage-gated Ca2+ channel
Neurotransmitter
Synaptic vesicle
Neurotransmitter binds with receptors on either another neuron or an effector (muscle or gland).
Nerve signal
Synaptic knob

INTEGRATE

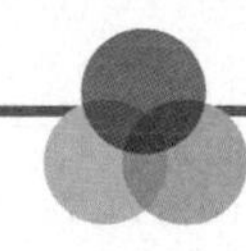

CONCEPT CONNECTION

Both graded potentials and action potentials also occur in skeletal muscle fibers. Graded potentials occur in the motor end plate of skeletal muscle fibers, and when the threshold is reached, they are called an end-plate potential. An end-plate potential initiates an action potential along the sarcolemma of a skeletal muscle fiber to trigger the release of calcium from the sarcoplasmic reticulum (see section 10.3b). Calcium ion binds to troponin to initiate the sliding of contractile proteins within the skeletal muscle fiber (see section 10.3c).

12.9 Characteristics of Action Potentials

Here we compare graded potentials and action potentials (the two types of electrical signals in neurons) and describe several aspects of action potential propagation, including velocity of action potentials and frequency of action potentials.

12.9a Graded Potentials Versus Action Potentials

 **LEARNING OBJECTIVE**

36. Compare graded potentials and action potentials.

Recall that two types of electrical signals are associated with neurons—graded potentials and action potentials. **Graded potentials** (described in section 12.8a) occur in the receptive segment of a neuron (dendrites and cell bodies) and are due to the opening of *chemically gated* channels. The chemically gated channels open temporarily to allow passage of a relatively small amount of a specific type of ion across the plasma membrane. This results in the membrane potential becoming either more positive (depolarization) or more negative (hyperpolarization) than the resting membrane potential. The degree of change is dependent upon the magnitude of the stimulus; thus, it is graded. A larger stimulus opens more chemically gated channels, and more ions flow across the plasma membrane than occurs during a weaker stimulus. The established **local current** of ions exhibits decreased intensity (decreased flow of ions) as the ions move along the plasma membrane; thus, graded potentials are short-lived (1 millisecond to a few milliseconds) and travel relatively short distances.

An **action potential,** in comparison, is generated within the initial segment (see section 12.8b) and propagated along the conductive segment of the neuron (see section 12.8c). An action potential is initiated when *voltage-gated* channels open in response to a minimum voltage change (threshold value). First, voltage-gated Na^+ channels open, allowing Na^+ into a neuron to cause depolarization (reversal of membrane potential from negative to positive). These voltage-gated channels then close and voltage-gated K^+ channels open, allowing K^+ out to cause repolarization (return of membrane potential from positive to negative). An action potential is self-propagating and maintains its intensity (charge difference) as it moves along the axon to the synaptic knob because of the successive opening of voltage-gated Na^+ channels followed immediately by the successive opening of voltage-gated K^+ channels. Propagation of an action potential is called a *nerve signal* (or *nerve impulse*). Action potentials obey the all-or-none law because any voltage sufficient to open the voltage-gated channels (threshold value) initiates an action potential (all), whereas any voltage below the threshold (subthreshold) is not sufficient to open these channels and an action potential is not sent (none). The characteristics of these two very distinctive electrical events that occur at the plasma membrane are summarized in **table 12.2**.

 **WHAT DID YOU LEARN?**

26 Explain how action potentials differ from graded potentials.

Table 12.2 Graded Potential Versus Action Potential

Characteristics	Graded Potential	Action Potential
Neuron segment[1]	Dendrites and cell body	Axon
Channels	Chemically gated channels	Voltage-gated channels
Direction of voltage change	Positive or negative	Positive then negative
Amount of voltage change	Relatively small change	Relatively large change that causes temporary reversal of polarity
Degree of voltage change	Dependent upon magnitude of the stimulus	Generally does not vary
Duration	1 msec to a few msec	Self-propagating along axon (time varies by length of axon)
Distance traveled	Relatively short distance	Length of axon
Change in intensity	Decreases with distance	Same intensity (because voltage-gated channels continue to open in sequence)

1. The regions listed reflect the most common location for that type of potential.

12.9b Velocity of Action Potential Propagation

LEARNING OBJECTIVES

37. Describe the two primary factors that influence the velocity of action potential propagation.

38. Identify the criteria used to distinguish the groups of nerve fibers.

Propagation of an action potential (the nerve signal) along an axolemma (axon plasma membrane) varies in its velocity and is influenced primarily by two factors—the diameter of an axon and the myelination of an axon:

- **Diameter of the axon.** Nerve signal velocity is generally faster in axons with a larger diameter. This is because there is less resistance to the movement of ions within the larger axon, allowing these axons to reach threshold more rapidly than smaller axons.
- **Myelination of the axon.** Myelination of the axon was described in section 12.4c and is the more important factor influencing nerve signal velocity. Nerve signal velocity occurs more rapidly in myelinated axons than in unmyelinated axons (see figure 12.21).

A **nerve fiber** is an axon and its myelin sheath. Nerve fibers are classified into three major groups called A, B, and C, based upon their nerve signal velocity. *Group A* nerve fibers each have a nerve signal velocity that may be as fast as 150 meters per second; these fibers have a large diameter and are myelinated. Most somatic sensory neurons that extend from sensory receptors to the CNS (e.g., those relaying visual input), and all somatic motor neurons that extend from the CNS to skeletal muscles, are included in this group. *Group B* nerve fibers each propagate nerve signals at approximately 15 meters per second, and *group C* nerve fibers each propagate nerve signals at 1 meter per second; group B and group C nerve fibers are generally small in diameter, unmyelinated, or both. Visceral sensory and autonomic motor neurons, as well as small somatic sensory neurons that extend from the receptors of the skin to the CNS, are included in groups B and C.

WHAT DID YOU LEARN?

27 What are the general characteristics of group A nerve fibers, and what functions do they normally serve?

12.9c Frequency of Action Potentials

LEARNING OBJECTIVE

39. Describe how action potentials vary in frequency.

Action potentials are always propagated along an axon (as nerve signals) at the same *amplitude* (change in voltage). However, action potential *frequency* can vary and is dependent upon the stimulus strength. As the stimulus strength increases, the frequency of action potentials increases (up to the point of maximum frequency). Consider the following: A brighter light initiates more nerve signals to be relayed from the eye along the optic nerve to the brain, and a loud sound initiates more nerve signals to be relayed along the vestibulocochlear nerve from the inner ear to the brain. The brain then interprets the increased frequency of nerve signals as a more intense stimulus (see section 16.1c). In addition, frequency of action potentials relayed along somatic motor neurons to skeletal muscle increases muscle tension, as described in section 10.6c and shown in figure 10.22.

Recall from section 12.8d that varying frequency of action potentials can also influence the type of neurotransmitter released from the synaptic knobs for those neurons that store and release more than one type of neurotransmitter.

WHAT DID YOU LEARN?

28 Explain how frequency of action potentials differs from the velocity of an action potential propagation.

12.10 Neurotransmitters and Neuromodulation

Neurotransmitters are released into the synaptic cleft and their action is modified by neuromodulation. Here we describe the different means of classifying neurotransmitters based upon chemical structure and function, prior to reviewing critical features of acetylcholine and other specific types of neurotransmitters. We then discuss how the action of neurotransmitters can be altered through the process of neuromodulation.

12.10a Classification of Neurotransmitters

 **LEARNING OBJECTIVES**

40. Identify the four classes of neurotransmitters based upon chemical structure.

41. Describe how neurotransmitters are classified based upon function.

Conventionally, *neurotransmitters* have been defined as small, organic molecules that (1) are synthesized by neurons and stored within vesicles in synaptic knobs; (2) are released from the vesicles when an action potential triggers calcium entry into the synaptic knob; (3) bind to a specific receptor in a target cell (neuron, muscle, or gland); and (4) trigger a physiologic response in the target cell. There are estimated to be approximately 100 known neurotransmitters. However, it should be noted that some molecules that are called neurotransmitters (e.g., nitric oxide) do not meet all of these criteria.

Neurotransmitters are classified based upon their chemical structure and function (**figure 12.24**). There are four categories of neurotransmitters based upon the chemical structure (figure 12.24*a*):

- **Acetylcholine (ACh).** The structure of ACh is significantly different from the other neurotransmitters and for this reason is placed in its own category.
- **Biogenic amines** (also called *monoamines*). They are derived from certain amino acids (see figure 2.25) by the removal of a carboxyl group (—COOH) and the addition of another functional group (e.g., an hydroxyl group) by enzymatic pathways within the cytosol. The functional group added determines whether the molecule belongs to either **catecholamines** (dopamine, norepinephrine, and epinephrine) that are synthesized from the amino acid tyrosine or **indolamines,** which include histamine (synthesized from histidine) and serotonin (synthesized from tryptophan).
- **Amino acids.** These include glutamate, aspartate, serine, glycine (see figure 2.25), and gamma aminobutyric acid (GABA, a modified amino acid). Some controversy remains about how chemical structures that are so plentiful in the cell for protein synthesis also can serve the neurotransmitter communication function.
- **Neuropeptides** (or *peptides*). These are chains of amino acids that range in length from 2 to 40 amino acids. Examples of neuropeptides include the natural opiates (e.g., enkephalins, beta-endorphins), and substance P.

Neurotransmitter classification based upon function (figure 12.24*b*) reflects the specific effect that a neurotransmitter has on the membrane potential of a target cell. Neurotransmitters are considered **excitatory** if they induce an EPSP, whereas they are **inhibitory** if they induce an IPSP (see section 12.8a). (Note that some neurotransmitters may be either excitatory or inhibitory depending upon the specific response they cause in their target organs.)

Another neurotransmitter classification based upon function reflects whether the target cell response is either **direct** (i.e., the neurotransmitter directly binds to the receptor of the target cell to cause opening of an ion channel) or **indirect** (i.e., the neurotransmitter binds to a receptor that activates the second messenger pathway involving G protein; see G proteins in section 4.5b). The second messenger ultimately can trigger much more diverse effects, including the opening of ion channels, the activation of an existing enzymatic pathway, or transcription of genes for the synthesis of new proteins.

 **WHAT DID YOU LEARN?**

 Describe how neurotransmitters are classified based upon structure and function.

12.10b Features of Neurotransmitters

LEARNING OBJECTIVE

42. Describe how acetylcholine functions as a neurotransmitter.

43. Discuss the different mechanisms for removing neurotransmitter from the synaptic cleft.

Here we discuss the synthesis and function of acetylcholine (ACh) in detail because it exhibits all of the classical features of neurotransmitters (as described in section 12.10a) and is the most understood. Several attributes about acetylcholine are considered, including its (1) synthesis, (2) removal from the synaptic cleft, and (3) interaction

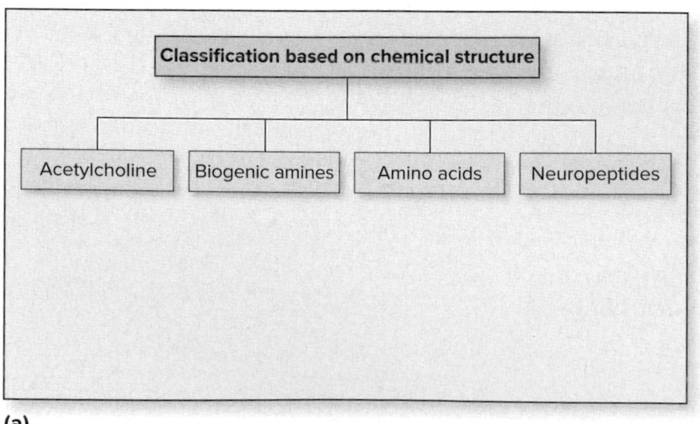

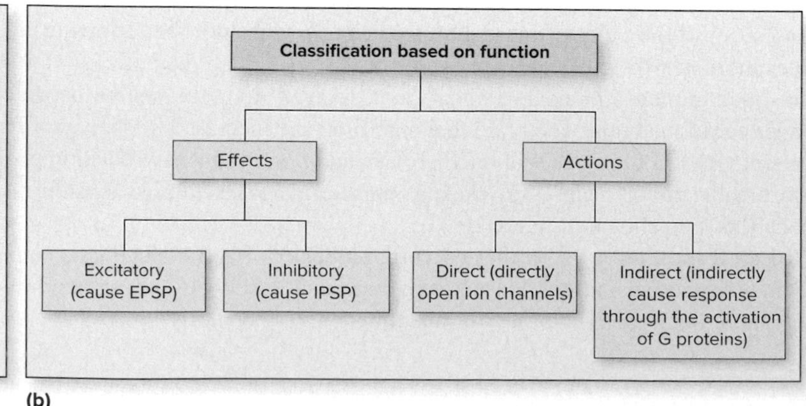

Figure 12.24 Classification of Neurotransmitters. Neurotransmitters are classified based upon both their chemical structure and their function. (*a*) Three of the chemical structures (except acetylcholine) are amino acids or synthesized from amino acids (biogenic amines and peptides). (*b*) Neurotransmitters also may be classified by their function—either by their effects (produce an EPSP or an IPSP) or their action on the target organ (directly opens ion channels or indirectly by acting through G proteins to cause more diverse effects).

with target cells. Please view **figure 12.25** as you read through this section. Specific diseases, drugs, and poisons that alter the normal function of ACh are also included in this figure (see Clinical View 12.6: "Altered Acetylcholine Function and Changes in Breathing").

The neurotransmitter acetylcholine is released from neurons located throughout the body. These include the somatic motor neurons at the neuromuscular junction (as described in section 10.3a) and many neurons of the autonomic nervous system (see chapter 15). Acetylcholine also acts as a neuromodulator (see section 12.10c) within the central nervous system, where it acts to increase attention and arousal. The process of acetylocholine synthesis, release and removal is discussed below and shown in figure 12.25.

Synthesis and release (step a). Acetylcholine is synthesized from acetate and choline and then stored inside synaptic vesicles within the synaptic knobs of a neuron. Thousands of molecules of ACh are released by exocytosis into the synaptic cleft in response to the arrival of a nerve signal in the presynaptic neuron. The more frequent the nerve signals, the greater the amount of acetylcholine released.

Removal from synaptic cleft (step b). Some ACh molecules will be immediately digested by **acetylcholinesterase,** an enzyme that resides in the synaptic cleft. Acetylcholine is digested into acetate and choline, and then the choline is taken up into the neuron that released the ACh. Some of the ACh molecules cross the synaptic cleft and bind to target cell receptors. These ACh molecules will quickly dissociate from the receptors usually within 1 millisecond to then be digested by acetylcholinesterase.

WHAT DO YOU THINK?

2 Predict the general effect of a drug that crosses the blood-brain barrier and inhibits the action of acetylcholinesterase.

There are other means of removing different types of neurotransmitter from the synaptic cleft. These include (1) the reuptake of the neurotransmitter into the synaptic knob and subsequent enzymatic digestion (e.g., biogenic amines and amino acids are digested by monoamine oxidase [MAO]) and (2) diffusion from the synaptic cleft and its uptake by surrounding glial cells.

Certain prescription medications were developed based upon their ability to influence the amount of neurotransmitter in a synaptic cleft (e.g., selective serotonin reuptake inhibitors [SSRIs] block the reuptake of serotonin and are used in the treatment of depression [see section 1.7]). Other antidepressants function by inhibiting MAO enzymes. The result is that neurotransmitter reuptake or breakdown is slowed and the neurotransmitter remains active in the synaptic cleft for a longer period of time.

Interaction with Target Cells (step c). The effect ACh has on a target cell depends upon the specific type of receptor embedded within the plasma membrane of the target cell. The receptors that bind acetylcholine are either nicotinic receptors or one of several subtypes of muscarinic receptors (see section 15.5b). ACh interacts directly with **nicotinic receptors,** causing the opening of ion channels and the production of an EPSP (see section 12.8a). In comparison, the interaction of ACh with **muscarinic receptors** causes ion channels to open indirectly through the second messenger pathway that involves G protein (see section 4.5b). Interestingly, the result may be the formation of an EPSP or an IPSP depending upon the specific subtype of muscarinic receptor to which ACh binds.

WHAT DID YOU LEARN?

30 How is it possible for acetylcholine to generate either an EPSP or an IPSP?

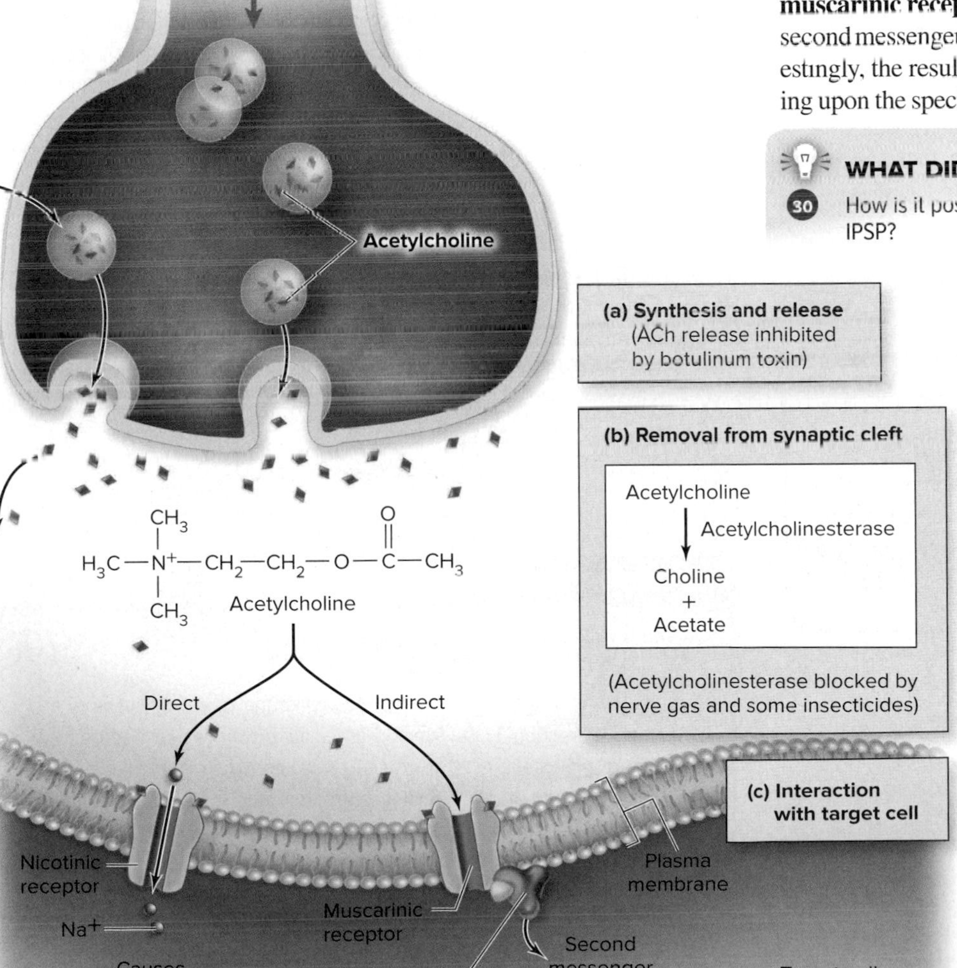

Figure 12.25 Acetylcholine Release, Removal from Synaptic Cleft, and Action. Acetylcholine (*a*) is synthesized and released from synaptic knobs into a synaptic cleft, (*b*) is removed from the synaptic cleft through enzymatic breakdown by acetylcholinesterase, and (*c*) interacts with target cells either directly or indirectly. Each of these processes is altered by specific types of toxins, chemical poisons, or drugs.

INTEGRATE

CLINICAL VIEW 12.6

Altered Acetylcholine Function and Changes in Breathing

Some substances or conditions may alter the response of acetylcholine at the neuromuscular junction of skeletal muscles. The response is increased (or excited) in some cases, whereas in others it may be decreased (or inhibited). This altered response differentially affects the skeletal muscle of breathing at the neuromuscular junction (NMJ).

Substances or Conditions that *Decrease* Response at the NMJ:

Some substances that *decrease* the response at the NMJ include (a) botulinum toxin (figure 12.25, step a), which decreases the release of ACh (see Clinical View 10.3: "Muscular Paralysis and Neurotoxins"), (b) cobratoxin and curare (figure 12.25, step c), which act as competitive inhibitors of ACh receptors (preventing binding of ACh), and (c) loss of acetylcholine receptors in myasthenia gravis (see Clinical View 10.2: "Myasthenia Gravis"). The outcome of this decreased response at the NMJ is ultimately expressed as a weakness in skeletal muscles, including muscles of breathing (e.g., diaphragm; see figure 23.19). Impaired breathing muscle contractions may be fatal if sufficiently severe.

Substances or Conditions that *Increase* Response at the NMJ:

Substances or conditions that increase the response at the NMJ include **acetylcholinesterase inhibitors,** which interfere with the breakdown of acetylcholine within the neuromuscular junction. Organophosphates (e.g., chemicals found in some insecticides) are acetylcholinesterase inhibitors. The outcome of this increased response at the NMJ is ultimately expressed as overstimulation of skeletal muscles, including muscles of breathing. This prevents breathing muscles from relaxing and may be fatal. Unfortunately, overexposure to common insecticides often is the cause of this poisoning.

INTEGRATE

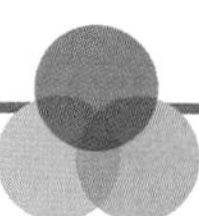

CONCEPT CONNECTION

Other neurotransmitters include **adenosine triphosphate** (**ATP;** a molecule described in section 2.7d) and adenosine (which is merely adenosine without the three phosphates). **Adenosine,** for example, when bound to adenosine receptors in the brain, has an inhibitory effect. Caffeine acts as a stimulant by blocking adenosine receptors and acting as a competitive inhibitor (see section 3.3f).

12.10c Neuromodulation

LEARNING OBJECTIVES

44. Define neuromodulation, including its function in facilitation and inhibition.

45. Describe how nitric oxide and endocannabinoids function as neuromodulators.

Neuromodulation (nūr′ō-mod-yū-lā′shŭn) is the release of chemicals (other than classical neurotransmitters) from cells that locally regulate or alter the response of neurons to neurotransmitters. The substances released are called **neuromodulators,** and they become participants in the "decision making" that occurs during the transmission of information by the nervous system. Neuromodulation generally results in either facilitation or inhibition. **Facilitation** (or *presynaptic facilitation*) occurs when there is greater response from a postsynaptic neuron because of the release of neuromodulators. Facilitation may result from either an increased amount of neurotransmitter in the synaptic cleft (greater release, slower breakdown, or slower reuptake) or an increased number of receptors on postsynaptic neurons. **Inhibition** (or *presynaptic inhibition*) occurs when there is less response from a postsynaptic neuron because of the release of neuromodulators. This results from either a decreased amount of neurotransmitter in the synaptic cleft (decreased release, faster breakdown, or faster reuptake) or decreased number of receptors on postsynaptic neurons. Enkephalins, for example, act as neuromodulators to decrease the perception of pain by blocking transmission of the sensory input at the level of the spinal cord.

Nitric oxide is an unusual neurotransmitter, classified by some experts as a neuromodulator. Nitric oxide is different from the classical neurotransmitter for several reasons. It is a short-lived gas that is not stored in a vesicle but synthesized from the amino acid arginine on an "as needed" basis. This small, nonpolar molecule is produced and released from *postganglionic* neurons. It diffuses (to more than 100 μm) and enters cells in all directions. Its entry into presynaptic neurons provides a retrograde means of communication (i.e., a means of communication in the reverse direction that typically occurs between a presynaptic and postsynaptic neuron). Nitric oxide functions within the brain in developing memories by stimulating presynaptic neurons to increase the release of neurotransmitter. Nitric oxide is also released from motor neurons that innervate blood vessels, causing relaxation of smooth muscle within the blood vessel wall. This results in vasodilation (increasing blood vessel diameter) and increased blood flow. (Endothelial cells that line blood vessels also release nitric oxide to cause vasodilation; see section 20.4c.)

Endocannabinoids (en′dō-ka′nab′i-noydz) are molecules that bind with the same receptors as the active ingredient tetrahydrocannabinol (THC) within cannabis (i.e., marijuana). Endocannabinoids are similar to nitric oxide in that they are small, nonpolar molecules, which are produced and released on demand from postsynaptic neurons. Binding of endocannabinoid decreases neurotransmitter release from presynaptic neurons, altering learning and memory, affecting appetite, and suppressing nausea. Neurotransmitters and neuromodulators are summarized in **table 12.3**.

WHAT DID YOU LEARN?

How does nitric oxide act as a neuromodulator?

INTEGRATE

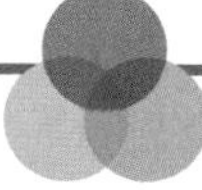

CONCEPT CONNECTION

A region of the brain called the *hippocampus* and its role in memory are described in section 13.8e. Numerous cellular changes to neurons are associated with memory. These occur in response to increased activity between a presynaptic and postsynaptic neuron. Two of these changes include (1) an increase in Ca^{2+} concentration in the synaptic knobs of presynaptic neurons, which causes release of additional neurotransmitter, and (2) an increase in both the number and sensitivity of receptors that bind neurotransmitter in the postsynaptic neurons. These modifications enhance the magnitude of graded potentials established in the postsynaptic neurons—making it either more likely (if EPSPs are generated) or less likely (if IPSPs are generated) that the postsynaptic neuron will reach the threshold and initiate an action potential along the axon.

Table 12.3 Neurotransmitters

Neurotransmitter	Description/Action
ACETYLCHOLINE (ACh)	
$H_3C-N^+(CH_3)_2-CH_2-CH_2-O-C(=O)-CH_3$	The primary neurotransmitter used at the neuromuscular junction (see section 10.3a) and by most of the autonomic nervous system (see section 15.6).
BIOGENIC AMINES	
$NH_2-CH_2-CH(OH)-$ (Aromatic ring) $-OH$, $-OH$	Molecules synthesized from an amino acid by removal of the carboxyl group and retaining the single amine group; also called monoamines
Catecholamines	A distinct group of biogenic amines that contain a catechol chemical group; originally described as hormones (chemicals produced by a gland in one part of the body that affect cells in other parts of the body)
Dopamine	Produces inhibitory activity in the brain; important roles in cognition (learning, memory), motivation, behavior, and mood; decreased levels in Parkinson disease; amphetamines increase release; cocaine decreases removal from synaptic cleft; ecstasy increases release
Norepinephrine (noradrenaline)	Neurotransmitter of peripheral autonomic nervous system (sympathetic division) and various regions of the CNS; amphetamines increase release; cocaine decreases removal from synaptic cleft; ecstasy increases release
Epinephrine (adrenaline)	Has various effects in the thalamus, hypothalamus, and spinal cord
Indolamines	A distinct group of biogenic amines that contain an indole chemical group.
Histamine	Neurotransmitter of the CNS; plays a role in sleep and memory
Serotonin	Has various functions in the brain related to sleep, appetite, cognition (learning, memory), and mood; fluoxetine (Prozac) decreases reuptake; ecstasy increases the release; LSD binds to most serotonin receptors
AMINO ACIDS	
$NH_2-CH_2(R)-C(=O)OH$	Molecules with both carboxyl (—COOH) and amine (—NH_2) groups and various R groups; building blocks of proteins; act as signaling molecules in the nervous system
Glutamate	Excites activity in nervous system to promote cognitive function in the brain (learning and memory); most common neurotransmitter in the brain; stroke causes excessive release, resulting in neuron death
Aspartate	Excites activity primarily in descending motor pathways through the spinal cord to skeletal muscle
Serine	Activates diverse areas in the brain
Gamma-aminobutyric acid (GABA)	Modified amino acid that is synthesized from glutamate; primary inhibitory neurotransmitter in the brain; also influences muscle tone; alcohol, diazepam (Valium), and barbiturates increase inhibitory effects of GABA
Glycine	Inhibits activity between neurons in the brain, spinal cord, and eye; strychnine blocks receptors that bind glycine
NEUROPEPTIDES	
Tyr–Gly–Gly–Phe–Met	Small molecules made of chains of amino acids; generally act through G proteins to cause more diverse effects; opioids include enkephalins, endorphins, endomorphins, dynorphins, and nociceptin; methadone binds at opiate receptors
Enkephalin (an opioid)	Helps regulate response to something that is perceived to be noxious or potentially painful
Beta-endorphin (an opioid)	Prevents release of pain signals from neurons and fosters a feeling of well-being; morphine mimics endorphins; heroin is converted to morphine in the body
Neuropeptide Y	Involved in memory regulation and energy balance (increased food intake and decreased physical activity)
Somatostatin	Inhibits activities of neurons in specific brain areas
Substance P	Assists with pain information transmission into the brain
Cholecystokinin	Stimulates neurons in the brain to help mediate satiation (fullness) and repress hunger
Neurotensin	Helps control and moderate the effects of dopamine
OTHERS	
Adenosine	Part of a nucleotide (a building block of nucleic acid); has an inhibitory effect on neurons in the brain and spinal cord
Nitric oxide	Involved in learning and memory; relaxation of muscle in the digestive tract; relaxation of smooth muscle in blood vessels
Endocannabinoids	Most prevalent receptors in the brain

12.11 Neural Integration and Neuronal Pools of the CNS

LEARNING OBJECTIVE

46. Identify the four different types of neuronal pools, and explain how they function.

The nervous system coordinates and integrates neuronal activity in part because billions of interneurons within the nervous system are grouped in complex patterns called **neuronal pools,** or *neuronal circuits* or *pathways*. Neuronal pools are identified based upon function into four types of circuits: converging, diverging, reverberating, and parallel-after-discharge (**figure 12.26**). A pool may be localized, with its neurons confined to one specific location, or its neurons may be distributed in several different regions of the CNS. However, all neuronal pools are restricted in their number of input sources and output destinations.

The **converging circuit** involves inputs that come together (converge) at a single postsynaptic neuron (figure 12.26*a*). This neuron receives input from several presynaptic neurons. For example, multiple sensory neurons synapse on the neurons in the salivary nucleus in the brainstem, causing the salivary nucleus to alter activity of salivary glands to produce saliva at mealtime. The various inputs originate from more than one stimulus: smelling food, seeing dinnertime on the clock, hearing food preparation activities, or seeing pictures of food in a magazine. These multiple inputs lead to a single output: the production of saliva.

A **diverging circuit** spreads information from one presynaptic neuron to several postsynaptic neurons, or from one pool to multiple pools (figure 12.26*b*). The neurons in the brain control the movements of skeletal muscles in the legs during walking and stimulate the muscles in the back to maintain posture and balance while walking. In this case, a single or a few inputs lead to multiple outputs.

Reverberating circuits utilize feedback to produce a repeated, cyclical stimulation of the circuit: This cyclical feedback is termed *reverberation* (figure 12.26*c*). Once activated, a reverberating circuit may continue to function until the cycle is broken by either inhibitory stimuli or synaptic fatigue. The repetitious nature of a reverberating circuit ensures that we continue breathing while we are asleep.

In a **parallel-after-discharge circuit,** input is transmitted simultaneously along several neuron pathways to a common postsynaptic cell (figure 12.26*d*). Note that neuron pathways in a parallel-after-discharge circuit vary in the number of neurons within the pathway and thus the number of synapses within the pathway. Recall from section 12.3 that neuron-to-neuron communication at a synapse involves a synaptic delay (the time delay for the events at a synapse). Consequently, the greater the number of neurons in the pathway, the greater the number of synapses and the greater the amount of time required to transmit the information. This results in the information arriving from the point of stimulus input to the common postsynaptic cell at varying times. You might find it helpful to think of the arrival of information from each group of neurons to the common postsynaptic cell as an "echo" of the original stimulus input. This type of circuit is believed to be involved in higher-order thinking; for example, it reinforces the repetitive neural activity needed for performing precise mathematical calculations.

WHAT DID YOU LEARN?

32. How are neurons arranged in a converging circuit?
33. What are the differences between a reverberating circuit and a parallel-after-discharge circuit?

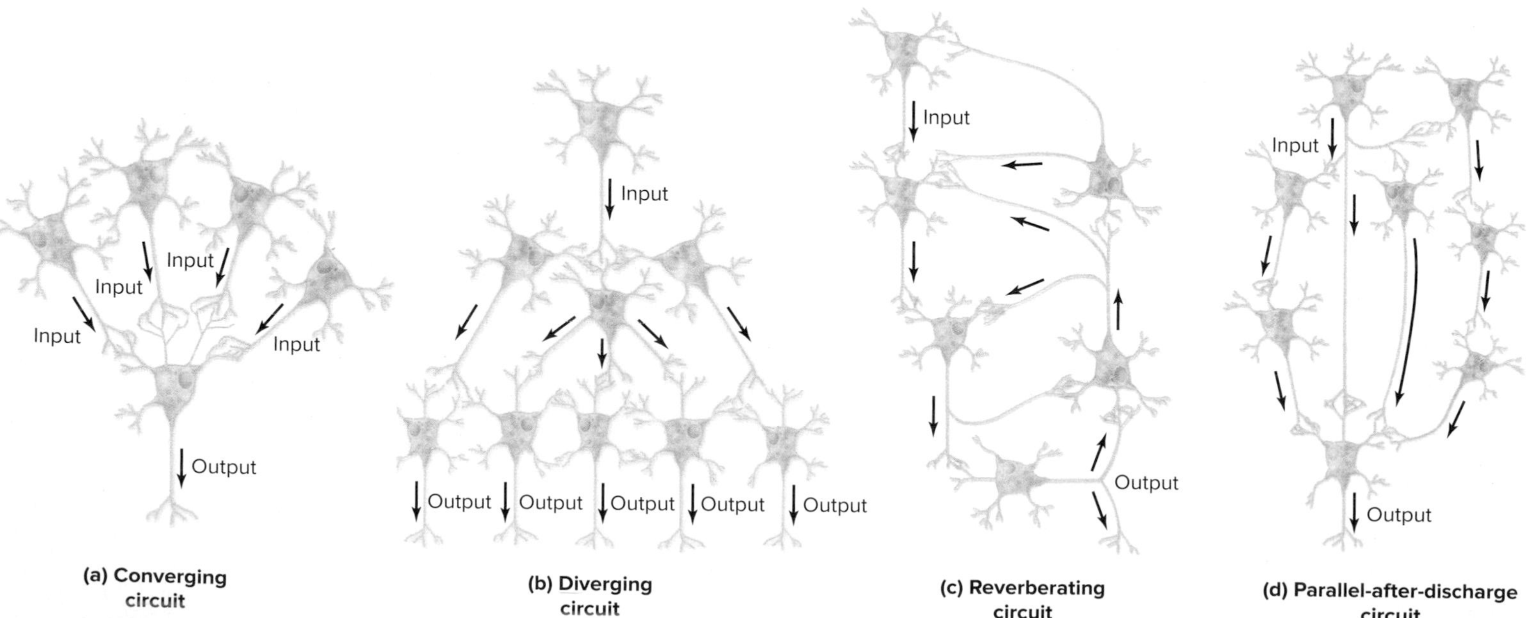

Figure 12.26 Neuronal Pools. Neuronal pools are groups of neurons arranged in specific patterns (circuits) through which input is conducted and distributed. Four types of neuronal pools are recognized.

CHAPTER SUMMARY

	• The nervous system interprets and controls all sensory input from receptors and motor output to effectors.
12.1 Introduction to the Nervous System	• The nervous system is composed of the brain, spinal cord, nerves, and ganglia.
	12.1a General Functions of the Nervous System • The nervous system collects information through receptors and sensory input, processes and evaluates information, and responds through motor output to effectors (muscles or glands).
	12.1b Organization of the Nervous System • The nervous system is organized structurally into the central nervous system (CNS) and the peripheral nervous system (PNS). • The nervous system is organized functionally into a sensory component and a motor component.
	12.1c Nerves and Ganglia • A nerve is a collection of axons that are wrapped in connective tissue. • The entire nerve is enclosed with an epineurium. Fascicles of axons are ensheathed with a perineurium, and each axon (and its surrounding neurolemmocyte) is wrapped with an endoneurium. • A ganglion is a cluster of neuron cell bodies located along a nerve.
12.2 Nervous Tissue: Neurons	• Nervous tissue is composed of excitable neurons that initiate and transmit graded potentials and action potentials, and glial cells that support and protect them.
	12.2a General Characteristics of Neurons • General characteristics of neurons include excitability, conductivity, secretion, and longevity; in addition, they are typically amitotic.
	12.2b Neuron Structure • A generalized neuron has a cell body. Processes typically extending from the cell body are numerous short, tapering dendrites and a single and often long axon.
	12.2c Neuron Transport • Neurons transport substances between the cell body and synaptic knobs by fast axonal transport and slow axonal transport.
	12.2d Classification of Neurons • Neurons are classified structurally as multipolar, bipolar, unipolar, and anaxonic by the number of processes attached to the cell body. • The three functional categories of neurons are sensory neurons, motor neurons, and interneurons.
12.3 Synapses	• A synapse is the functional junction of a neuron with either another neuron or an effector. • Synapses are either chemical synapses or electrical synapses.
12.4 Nervous Tissue: Glial Cells	• Glial cells are the other distinct cell type of nervous tissue.
	12.4a General Characteristics of Glial Cells • Glial cells are nonexcitable cells that primarily support and protect the neurons.
	12.4b Types of Glial Cells • Four types of glial cells within the central nervous system are astrocytes, ependymal cells, microglia, and oligodendrocytes. • Two types of glial cells within the peripheral nervous system are satellite cells and neurolemmocytes.
	12.4c Myelination • Myelination is the process by which part of an axon is wrapped and insulated with myelin. • Neurolemmocytes myelinate axons in the PNS, and oligodendrocytes myelinate axons in the CNS.
12.5 Axon Regeneration	• Regeneration of damaged neurons is limited to PNS axons. • A PNS axon can regrow to reestablish innervation if the cell body is intact and a critical amount of neurilemma remains.
12.6 Plasma Membrane of Neurons	• Establishing and changing a resting membrane potential is dependent upon various types of pumps and channels within a neuron's plasma membrane.
	12.6a Types of Pumps and Channels • Pumps and channels are membrane proteins that facilitate movement of ions across the neuron plasma membrane.
	12.6b Distribution of Pumps and Channels • Some membrane transport proteins are located along the entire neuron, and some are primarily in specific functional neuron segments.
12.7 Introduction to Neuron Physiology	• Neuron physiology involves the initiation and transmission of electric currents.
	12.7a Neurons and Ohm's Law • Ohm's law (current = voltage/resistance) has application in the principles of neuron physiology.

(continued on next page)

CHAPTER SUMMARY (continued)

	12.7b Neurons at Rest • Neurons at rest have a negative resting membrane potential (RMP), which on average is –70 mV. • Other characteristics of neurons at rest include closed gated channels; Na^+, K^+, and Cl^- concentration gradients along the length of the axon; and a Ca^{2+} concentration gradient at the synaptic knob.
12.8 Physiologic Events in the Neuron Segments	• Physiologic events that occur in a neuron's functional segments begin with stimulation of the receptive segment until the release of neurotransmitter from the transmissive segment.
	12.8a Receptive Segment • The receptive segment includes the dendrites and cell body. This segment involves the formation and propagation of graded potentials: both excitatory postsynaptic potentials (EPSPs) and inhibitory postsynaptic potentials (IPSPs).
	12.8b Initial Segment • The summation (or adding together) of EPSPs and IPSPs that reach the initial segment determines whether the threshold value (–55 mV) is reached and an action potential is initiated.
	12.8c Conductive Segment • The conductive segment is involved in the propagation of an action potential (nerve signal), a process that involves depolarization and repolarization. • The brief period of time that an axon is either incapable of generating an action potential or a greater than normal amount of stimulation is required to generate another action potential is called the refractory period. • Saltatory conduction occurs if the axon is myelinated.
	12.8d Transmissive Segment • The transmissive segment involves exocytosis of neurotransmitter from synaptic vesicles, which are located within synaptic knobs.
12.9 Characteristics of Action Potentials	• Action potentials differ from graded potentials and can vary in both velocity and frequency.
	12.9a Graded Potentials Versus Action Potentials • Graded potentials are short-lived electrical signals that occur in the dendrites and cell body due to opening of chemically gated channels, whereas action potentials are self-propagating electrical signals that are initiated in the initial segment, propagated along an axon, and result from the sequential opening of voltage-gated channels.
	12.9b Velocity of Action Potential Propagation • The velocity of action potentials (or nerve signals) is greater in larger and myelinated axons. • Nerve fibers, which are axons and their myelin sheath, are classified into three groups based upon the velocity of the nerve signal propagation.
	12.9c Frequency of Action Potentials • Frequency of action potentials occurs with increased stimulation of the neuron.
12.10 Neurotransmitters and Neuro-modulation	• Neurotransmitters are released at synaptic clefts and their action is modified by neuromodulation.
	12.10a Classification of Neurotransmitters • Neurotransmitters are conventionally described as molecules synthesized by neurons, which are then stored within vesicles in synaptic knobs and, when released, bind to specific receptors in a target cell to trigger a physiologic response. • Major classes of neurotransmitters include acetylcholine, biogenic amines, amino acids, and neuropeptides.
	12.10b Features of Neurotransmitters • Acetylcholine (ACh) is discussed in detail because it exhibits all of the classical features of neurotransmitters and is the most understood.
	12.10c Neuromodulation • Neuromodulation is the release of chemicals other than neurotransmitters that either increase the responsiveness to a neurotransmitter (facilitation) or decrease the responsiveness to a neurotransmitter (inhibition).
12.11 Neural Integration and Neuronal Pools of the CNS	• Interneurons are organized into neuronal pools, which are groups of interconnected neurons with specific functions and are classified as converging, diverging, reverberating, and parallel-after-discharge.

CHALLENGE YOURSELF

Create and Evaluate

Analyze and Apply

Understand and Remember

Do You Know the Basics?

1. The cell body of a neuron does all of the following *except*
   a. release neurotransmitter into the synaptic cleft.
   b. produce synaptic vesicles containing neurotransmitter that are subsequently transported to the synaptic knob.
   c. conduct graded potentials to the initial segment.
   d. receive graded potentials from dendrites.

2. Neurons that have only two processes attached to the cell body are called
   a. unipolar.
   b. bipolar.
   c. multipolar.
   d. efferent.

3. Which neurons are located only within the CNS?
   a. afferent neurons
   b. glial cells
   c. sensory neurons
   d. interneurons

4. EPSPs are caused by the movement of
   a. Na^+ out of the cell.
   b. Na^+ into the cell.
   c. K^+ into the cell.
   d. both Na^+ and K^+ into the cell.

5. The glial cells that help produce and circulate cerebrospinal fluid in the CNS are
   a. satellite cells.
   b. microglia.
   c. ependymal cells.
   d. astrocytes.

6. Which of the following is a part of the PNS?
   a. microglia
   b. spinal cord
   c. brain
   d. neurolemmocyte

7. An action potential is generated when threshold is reached, at which time
   a. voltage-gated K^+ channels close.
   b. voltage-gated Na^+ channels open.
   c. chemically gated Na^+ channels open.
   d. Ca^{2+} enters the cell.

8. Which type of neuronal pool utilizes feedback to repeatedly stimulate the circuit?
   a. converging circuit
   b. diverging circuit
   c. reverberating circuit
   d. parallel-after-discharge circuit

9. At an electrical synapse, presynaptic and postsynaptic membranes interface through
   a. neurofibril nodes.
   b. gap junctions.
   c. telodendria.
   d. neurotransmitters.

10. The two primary factors that influence the speed of an action potential propagation are axon diameter and
    a. myelination.
    b. the type of associated glial cell(s).
    c. concentration of K^+ in the cell.
    d. the length of the axon.

11. What are the four structural types of neurons? How do they compare to the three functional types of neurons?

12. Identify the principal glial cell types, and briefly discuss the function of each type.

13. How does myelination differ between the CNS and the PNS?

14. Describe the procedure by which a PNS axon may repair itself (axon regeneration).

15. Describe how the resting membrane potential is established and maintained in a neuron.

16. Compare and contrast graded potentials and action potentials.

17. Explain summation of EPSPs and IPSPs and the relationship to the initiation of an action potential.

18. Graph and explain the events associated with an action potential.

19. Explain the mechanism for the release of neurotransmitter from the synaptic knob.

20. List and briefly describe the major types of neurotransmitters.

Can You Apply What You've Learned?

1. Andrew was taken to the doctor's office after he was bitten by a stray dog. The concern was that the dog might be infected with the rabies virus. The rabies virus infects neurons by using which method that normally transports materials from the synaptic knob to the cell body?
   a. anterograde transport
   b. fast axonal transport
   c. slow axonal transport
   d. All of these are correct.

2. An elderly neighbor was diagnosed with an astrocytoma tumor in the brain. This cancer affects what types of cell?
   a. ependymal cells
   b. microglia
   c. astrocytes
   d. satellite cells

3. Cynthia has received her lab results and is told that her blood calcium levels are abnormal. The event in neuron transmission most likely to be affected is
   a. summation of graded potentials in the initial segment.
   b. production of graded potentials in the dendrites and cell body.
   c. release of neurotransmitter from the synaptic knob.
   d. propagation of an action potential in the axon.
4. Heidi's physician prescribed a medication that is known to block the reuptake of serotonin neurotransmitter from the synaptic cleft. This medication affects what segment in neuron transmission that is responsible for releasing the neurotransmitter?
   a. initial segment
   b. conductive segment
   c. transmissive segment
   d. receptive segment
5. Sarah wants to call her new friend, Julie, and needs to write down her phone number but cannot find a pen. She continues to repeat the number over and over. This is most likely occurring in what type of neuronal pool?
   a. reverberating circuit
   b. divergent circuit
   c. convergent circuit
   d. parallel-after-discharge circuit

Can You Synthesize What You've Learned?

1. Over a period of 6 to 9 months, Marianne began to experience vision problems as well as weakness and loss of fine control of the skeletal muscles in her leg. Blood tests revealed the presence of antibodies (immune system proteins) that attack myelin. Beyond the presence of the antibodies, what was the cause of Marianne's vision and muscular difficulties?
2. Surgeons were able to reattach Irving's amputated limb, sewing both the nerves and the blood vessels back together. After the surgery, which proceeded very well, the limb regained its blood supply almost immediately, but the limb remained motionless and Irving had no feeling in it for several months. Why did it take longer to reestablish innervation than circulation?
3. Certain types of neurotoxins prevent depolarization of the axon. What specific type of channel is impaired?

INTEGRATE

ONLINE STUDY TOOLS

connect | SMARTBOOK® | AP|R

The following study aids may be accessed through Connect.

Concept Overview Interactive Figure 12.23: Events of Neuron Physiology

Clinical Case Study: A Young Man Who Gets Weak When Overheated

Interactive Questions: This chapter's content is served up in a number of multimedia question formats for student study

SmartBook: Topics and terminology include neurons; synapses; glial cells; axon regeneration; ultrastructure of neurons; neuron physiology; physiologic events in the neuron segments; velocity of a nerve signal; neurotransmitters and neuromodulation; neural integration and neuronal pools of the CNS

Anatomy & Physiology Revealed: Topics include multipolar neuron; Schwann cell (neurolemmocyte); unmyelinated axon and myelin sheath; action potential generation and propagation; chemical synapse

Animations: Topics include PSPs (postsynaptic potentials); action potential propagation; action potential generation; action potential propagation in myelinated neurons

Design Elements: **Learning Objective** icon: ©McGraw-Hill Education; **What Do You Think?** icon: ©McGraw-Hill Education; **What Did You Learn?** icon: ©McGraw-Hill Education; **Learning Strategy** icon: ©Slavoljub Pantelic/Shutterstock RF; **Clinical View** icon: ©Laia Design Studio/Shutterstock RF; **Concept Connection** icon: ©McGraw-Hill Education; **Challenge Yourself** icon: ©McGraw-Hill Education; **Do You Know the Basics?** icon: ©McGraw-Hill Education; **Can You Apply What You've Learned?** icon: ©McGraw-Hill Education; **Can You Synthesize What You've Learned?** icon: ©McGraw-Hill Education; **Online Study Tools** icon: ©ayzek/Shutterstock RF

chapter

13

Nervous System: Brain and Cranial Nerves

Anatomy & Physiology REVEALED
aprevealed.com

Module 7: Nervous System

INTEGRATE

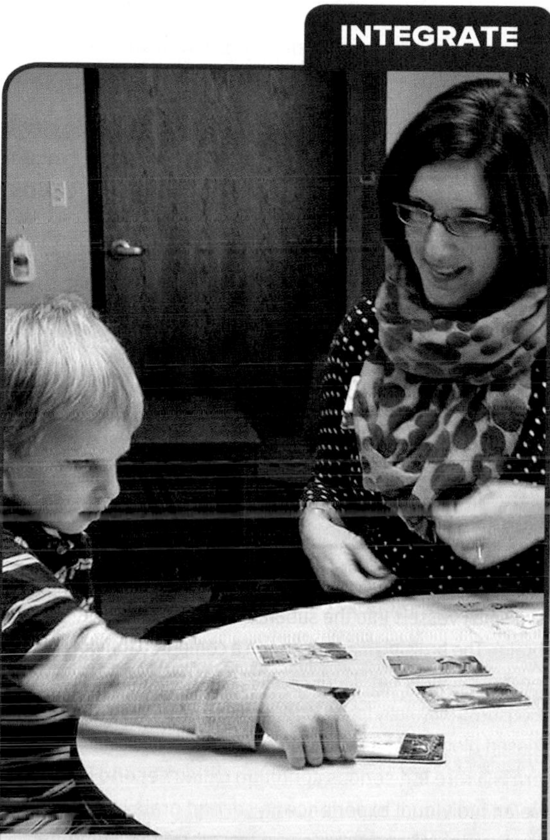

Courtesy Kristine Queck

CAREER PATH

Speech-Language Pathologist

A speech-language pathologist (SLP) assesses and diagnoses a range of communication and swallowing disorders. This health professional evaluates and develops a plan to improve a patient's speech and language skills. Patients range from the very young to the elderly, and the communication difficulties vary from stuttering to abnormal intonations in speech. SLPs work in a variety of health-care settings, such as hospitals, nursing homes, day-care centers, and schools. An SLP must understand relevant brain anatomy related to speech and communication, as well as be aware of the anatomic structures and neurologic pathways involved in the production of speech and language.

The human brain, while weighing on average about 3 pounds, is able to simultaneously process billions of pieces of information. It is continuously receiving sensory input from sensory receptors and initiating motor output to control effectors. This organ allows us to understand complex information, write poetry, and compute mathematical problems. The brain, which is part of the central nervous system (CNS), is associated with the 12 pairs of cranial nerves, which are considered part of the peripheral nervous system (PNS). The brain and cranial nerves are described in this chapter.

13.1 Brain Org and Developr

We begin our study of tl and surface landmark s embryonic brain develo the adult brain are name tion of gray and white r

13.1a Overview of I

LEARNING OBJ

1. Describe the general

The brain is composed cephalon, brainstem, ai parts of the adult brain (figure 13.1*a*), inferio (figure 13.1*c*). The *ce* left and right cerebral l subdivided into five fu limits the size of the b

INTEGRATE

CLINICAL V

Traumatic B and Contusi

Traumatic brain injury of an accident or trau characterized by tempo or the sudden stop of a confusion, and amnes cumulative effect, cau ability with each episoc to long-term personalit are prone to concussi for these detrimental c the NFL (National Foo enough in protecting p coaches and athletic t an athlete play if a co

INTEGRATE

LEARNING STRATEGY

Use the following to help you to remember the specific cerebral lobe that receives and interprets each type of sensory input:

Parietal = touch and proprioception (Imagine someone patting a child on the top of the head.)

Temporal lobe = hearing and smell (This lobe is essentially between our ears and nose.)

Occipital lobe = vision (We have "eyes" in the back of our head.)

Insula = taste (Imagine a tongue suppressor is needed to pull down the temporal lobe to reach the insula.)

texture of objects being touched and sensory input regarding body position from proprioceptors within our joints and muscles.

The **temporal lobe** lies internal to the temporal bone and inferior to the lateral sulcus. The cerebral cortex of this lobe is involved with hearing and smell.

The **occipital lobe** lies internal to the occipital bone and forms the posterior region of each hemisphere. The cerebral cortex of the occipital lobe is responsible for processing incoming visual information and storing visual memories.

The **insula** (in′sū-lă; island) is a small lobe deep to the lateral sulcus. It can be observed by laterally reflecting (pulling aside) the temporal lobe. The insula's lack of accessibility has prevented aggressive studies of its function, but the cerebral cortex of the insula is apparently involved in memory and the interpretation of taste.

WHAT DID YOU LEARN?

13 List the five cerebral lobes and the main functions of each.

13.3c Functional Areas of the Cerebrum

LEARNING OBJECTIVES

16. Locate and list the functions of the motor cortical regions and their association areas.
17. Differentiate among the sensory cortical regions and their association areas.
18. Explain the functions of the prefrontal cortex, and hypothesize why this brain region may function differently in adults and teenagers.
19. Describe the main actions of the Wernicke area.

Research has shown that specific structural areas of the cerebral cortex have distinct motor and sensory functions (as introduced in section 13.3b). In contrast, some higher mental functions, such as language and memory, are dispersed over large areas. We have organized the functions of the cerebral cortex into motor functions and their association areas and sensory functions and their association areas. Note that the central sulcus serves as an anatomic landmark that separates motor functions that are controlled by the frontal lobe, and sensory input that is relayed to the parietal lobe, occipital lobe, and temporal lobe, as well as the internal insula.

Motor Areas

The cortical areas that control motor functions are housed within the frontal lobes, as just described. The **primary motor cortex,** also called the *somatic motor area,* is specifically located within the precentral gyrus of the frontal lobe (figure 13.12). Neurons in this area control voluntary skeletal muscle activity. The axons of these neurons project contralaterally (to the opposite side) within either the brainstem or the spinal cord. Thus, the left primary motor cortex controls the skeletal muscles on the right side of the body and the right primary motor cortex controls the skeletal muscles on the left side of the body.

The distribution of the primary motor cortex innervation to various body parts can be diagrammed as a **motor homunculus** (hō-mŭngk′ū-lŭs; little man) on the precentral gyrus (**figure 13.13*a***).

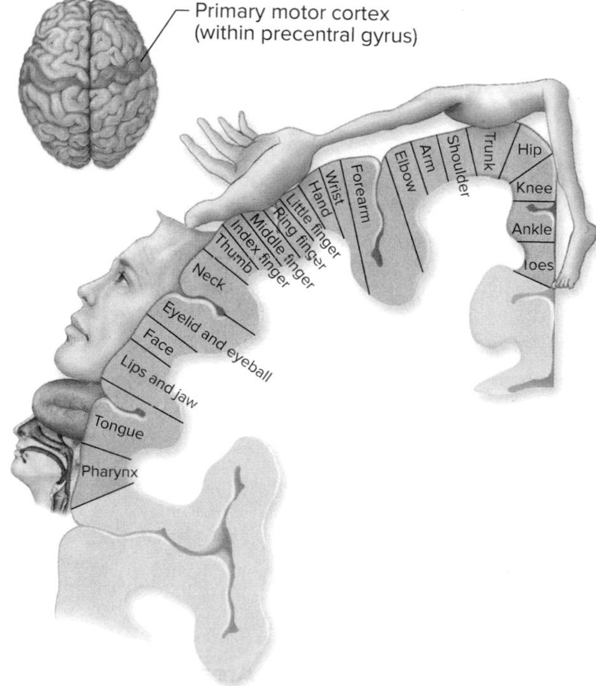

(a) Primary motor cortex

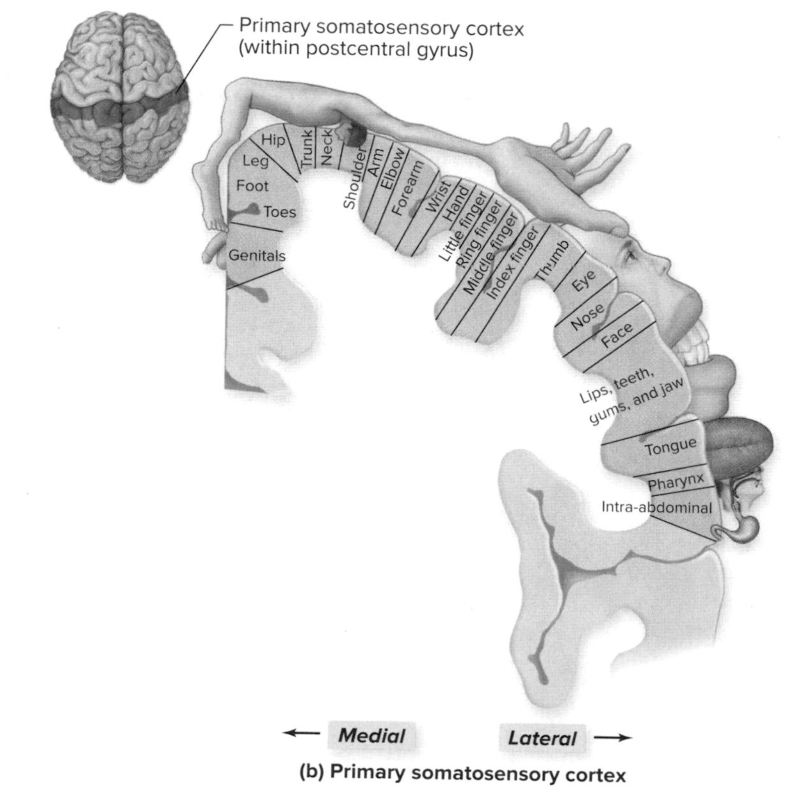

(b) Primary somatosensory cortex

Figure 13.13 Primary Motor and Somatosensory Cortices. Body maps called the motor homunculus and the sensory homunculus illustrate the topography of (*a*) the primary motor cortex and (*b*) the primary somatosensory cortex in coronal section. The figure of the body (homunculus) depicts the nerve distributions; the size and location of each body region indicate relative innervation. AP|R

INTEGRATE

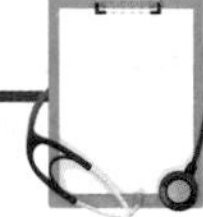

CLINICAL VIEW 13.6

Mapping Functional Brain Regions

Scientists and physicians have long been interested in developing functional brain maps that correlate specific parts of the cerebral cortex to specific functions. Among the first to develop such a map was Korbinian Brodmann. He studied the comparative anatomy of the mammalian brain cortex in the early 1900s and produced a map that shows the specific areas of the cerebral cortex where certain functions occur. Brodmann developed a numbering system that correlates with his map and shows that similar cognitive functions are usually sequential. For example, areas 3, 1, and 2 represent the primary somatosensory cortex; area 17 overlaps the primary visual cortex; and areas 44 and 45 form the motor speech area. In 1907 Brodmann published his Brodmann areas map of 52 specialized brain regions (figure *a*).

With the advent of new and more sophisticated medical imaging techniques, researchers have been able to update and revise the functional brain areas map. The Human Connectome Project (HCP; https://www.humanconnectome.org/), a consortium of researchers from three universities in the United States and United Kingdom, examined the brains of 1200 volunteers using multimodal MRI techniques. The data collected by the HCP were analyzed and used to develop a modern map of the human brain, the first version of which was published in 2016. This map includes 97 additional functional brain areas (figure *b*) and better delineates some of the brain regions first mapped by Brodmann. The HCP continues to revise and update its functional brain map.

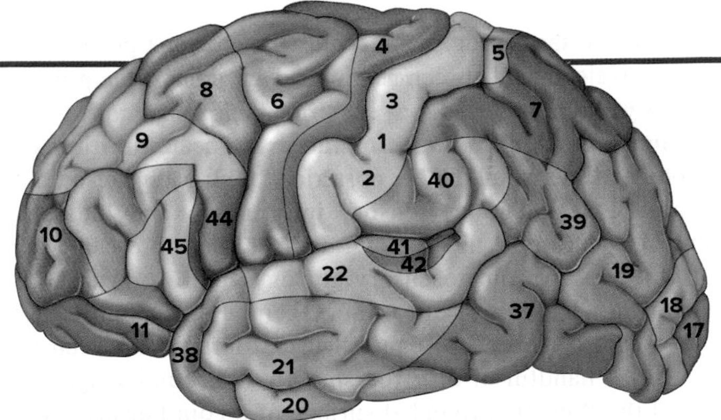

(a) Brodmann areas map of brain

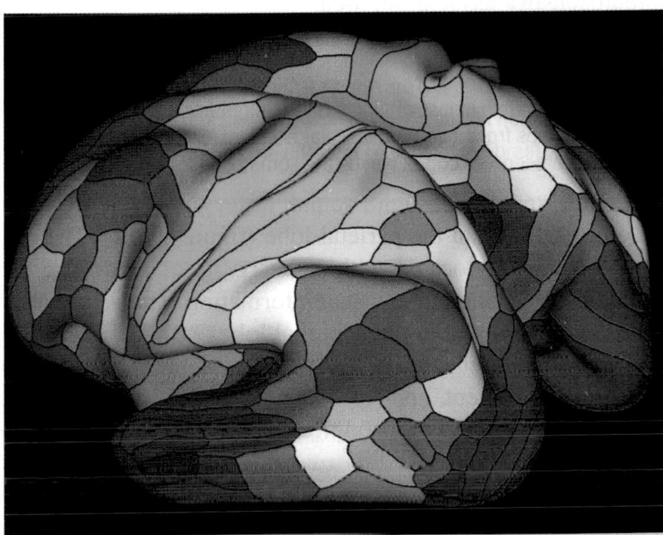

(b) Multimodal MRI of functional brain areas

A comparison of (a) Brodmann's map of the brain, showing selected Brodmann areas, and (b) a multimodal MRI from the Human Connectome Project, highlighting both previously known and newly discovered functional brain regions.

The bizarre, distorted proportions of the homunculus body reflect the amount of cortex dedicated to the motor activity of each body part. For example, the hands are represented by a much larger area of cortex than the trunk, because the hand muscles perform much more detailed, precise movements than do the trunk muscles. From a functional perspective, more motor activity is devoted to the human hand than in other animals because our hands are adapted for the precise, fine motor movements needed to manipulate the environment, and many motor units are devoted to muscles that move the hand and fingers.

Certain motor functions have been mapped to specific areas of the frontal lobe, including the motor speech area and the frontal eye field. The **motor speech area** (also known as the *Broca area*) is located in most individuals within the inferolateral portion of the left frontal lobe (figure 13.12). This region is responsible for regulating the breathing and controlling the muscular movements necessary for vocalization. The **frontal eye field** is within the frontal lobe immediately superior to the motor speech area. This cortical area controls and regulates the eye movements needed for reading and coordinating binocular vision. Some investigators include the frontal eye fields within the premotor area.

The primary motor cortical regions are connected to adjacent **association areas** that coordinate discrete skeletal muscle movement (figure 13.12). The **premotor cortex** also is called the *somatic motor association area,* and it is located within the frontal lobe immediately anterior to the precentral gyrus. It is primarily responsible for coordinating learned, skilled motor activities, such as moving the eyes in a coordinated fashion when reading a book or playing the guitar. An individual who has sustained trauma to this area would still be able to understand written letters and words but would have difficulty reading because his or her eyes couldn't follow the lines on a printed page.

Sensory Areas

The cortical areas within the parietal, temporal, and occipital lobes and the insula are involved with conscious awareness of sensation, as described in section 13.3b. Each of the major senses has a distinct cortical area. Keep in mind as you read through this section that each primary sensory cortical region is the specific area of the cortex that receives sensory input from a specific type of receptor. In addition, each primary cortical region typically has an association area. The general function of association areas is to receive input from the primary region and integrate the current sensory input with previous experiences and memories.

The **primary somatosensory cortex** is housed within the postcentral gyrus of the parietal lobes. Neurons within this cortex receive general somatic sensory information from receptors of the skin regarding touch, pressure, pain, and temperature, as well as sensory input from proprioceptors from the joints and muscles regarding the conscious interpretation of body position. We typically are conscious of the sensations received by this cortex. A **sensory homunculus** may be traced on the postcentral gyrus surface, similar to a motor homunculus (figure 13.13*b*). The surface area of somatosensory cortex devoted to a body region indicates the amount of sensory information collected within that region. Thus, the lips, fingers, and genital region occupy larger portions of the homunculus, whereas the trunk of the body has proportionately fewer receptors, so its associated homunculus region is

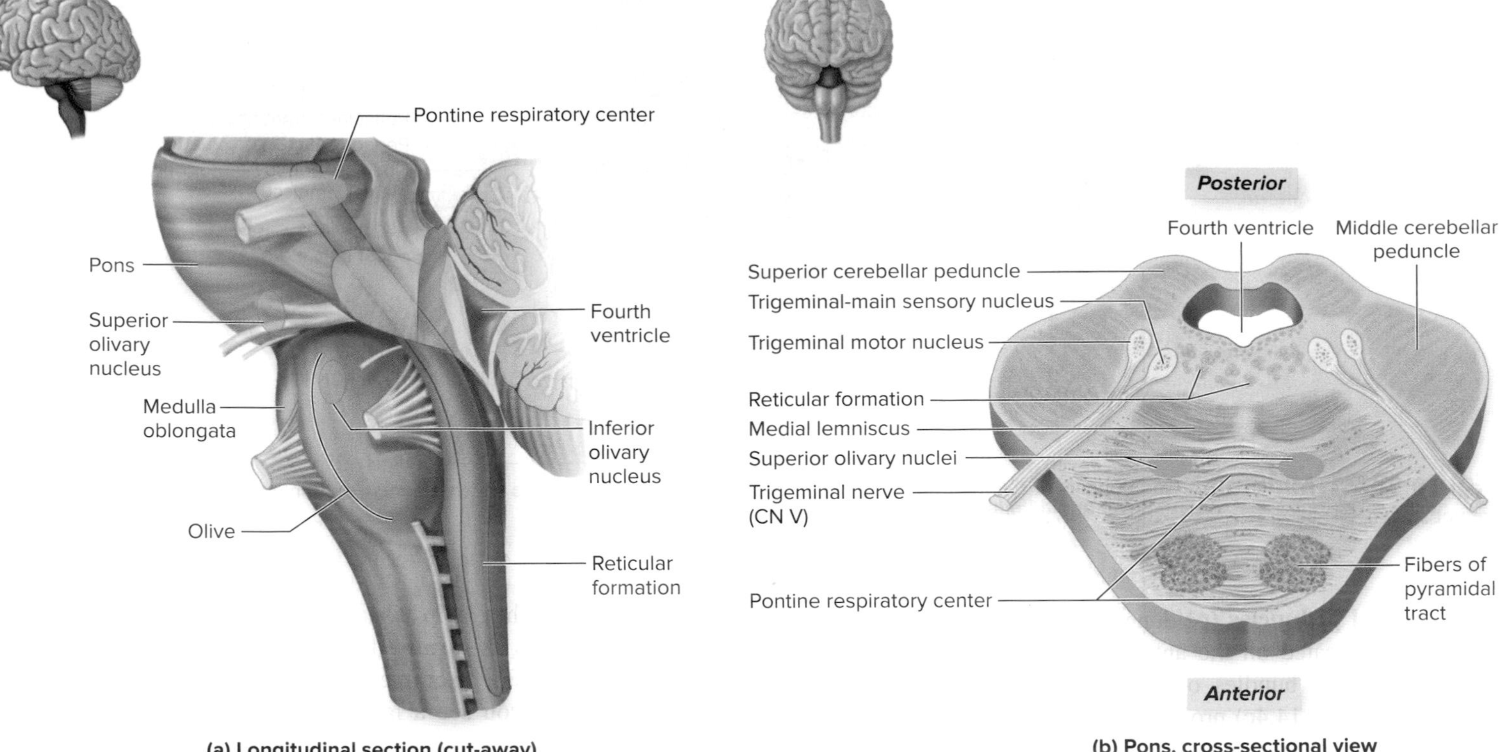

Figure 13.22 Pons. The pons forms a bulge on the ventral side of the brainstem that contains tracts and nuclei. The reticular formation, which extends through the brainstem, is also visible. (*a*) A partially cut-away longitudinal section identifies the pontine respiratory center and the superior olivary nucleus. (*b*) A cross section through the pons shows its associated tracts (pyramidal tracts, medial lemniscus, and middle cerebellar peduncles), the pontine respiratory center, the superior olivary nuclei, and some cranial nerve nuclei.

WHAT DID YOU LEARN?

26 What is the general function of the pontine respiratory center in the pons?

13.5c Medulla Oblongata

LEARNING OBJECTIVES

33. Describe the main features of the medulla oblongata.

34. List the autonomic centers of the medulla and the function of each.

The **medulla oblongata** (me-dūl′ā ob-long-gah′tă; *medulla* = marrow or middle, *oblongus* = rather long) is often simply called the **medulla.** It is the most inferior part of the brainstem and is continuous with the spinal cord inferiorly. The most inferior portion of the medulla has a flattened, rounded shape and narrow central canal. As this tubelike opening extends (superiorly and anteriorly) toward the pons, the central canal enlarges and becomes the inferior portion of the fourth ventricle. All communication between the brain and spinal cord involves tracts that ascend or descend through the medulla oblongata (**figure 13.23**; see also figures 13.20 and 13.22).

Several external landmarks are readily visible on the medulla oblongata. The anterior surface exhibits two longitudinal ridges called the **pyramids** (pir′ă-mid), which house the motor projection tracts called the corticospinal (pyramidal) tracts that extend through the medulla oblongata (see section 14.4c). In the anterior region of the medulla, most of the axons of the pyramidal tracts cross to the opposite side of the brain at a point called the **decussation** (dē-kŭ-sā′shŭn; *decussate* = to cross in the form of an X) **of the pyramids.** As a result of the crossover, each cerebral hemisphere controls the voluntary movements of the opposite side of the body. Immediately lateral to each pyramid is a distinct bulge, called the **olive,** which contains a large fold of gray matter called the **inferior olivary nucleus.** The inferior olivary nuclei relay ascending sensory nerve signals, especially proprioceptive information, to the cerebellum. Additionally, paired **inferior cerebellar peduncles** (see figure 13.20) are tracts that connect the medulla oblongata to the cerebellum.

The medulla oblongata contains several autonomic nuclei, which group together to form centers that regulate functions vital for life. The most important autonomic centers in the medulla oblongata and their functions include

- The **cardiovascular center,** composed of both the **cardiac center,** which regulates both the heart's rate and its force of contraction to alter cardiac output (see section 19.5b), and the **vasomotor center,** which controls the contraction and relaxation of smooth muscle within the walls of the smallest arteries (the *arterioles*) to alter these vessels' diameter. Both cardiac output and blood vessel diameter influence blood pressure (see section 20.6).
- The **medullary respiratory center,** which regulates the respiratory rate. It is composed of a ventral respiratory group and a dorsal respiratory group. These groups are influenced by the pontine respiratory center (see section 23.5c). The primary function of the medullary respiratory center is to rhythmically initiate nerve signals that cause contraction of breathing muscles (see section 23.5c).
- Other nuclei in the medulla, which are involved in coughing, sneezing, salivation, swallowing, gagging, and vomiting reflexes.

WHAT DO YOU THINK?

5 Based on your understanding of the medulla oblongata's functions, would you expect severe injury to the medulla oblongata to cause death or merely be disabling? Why?

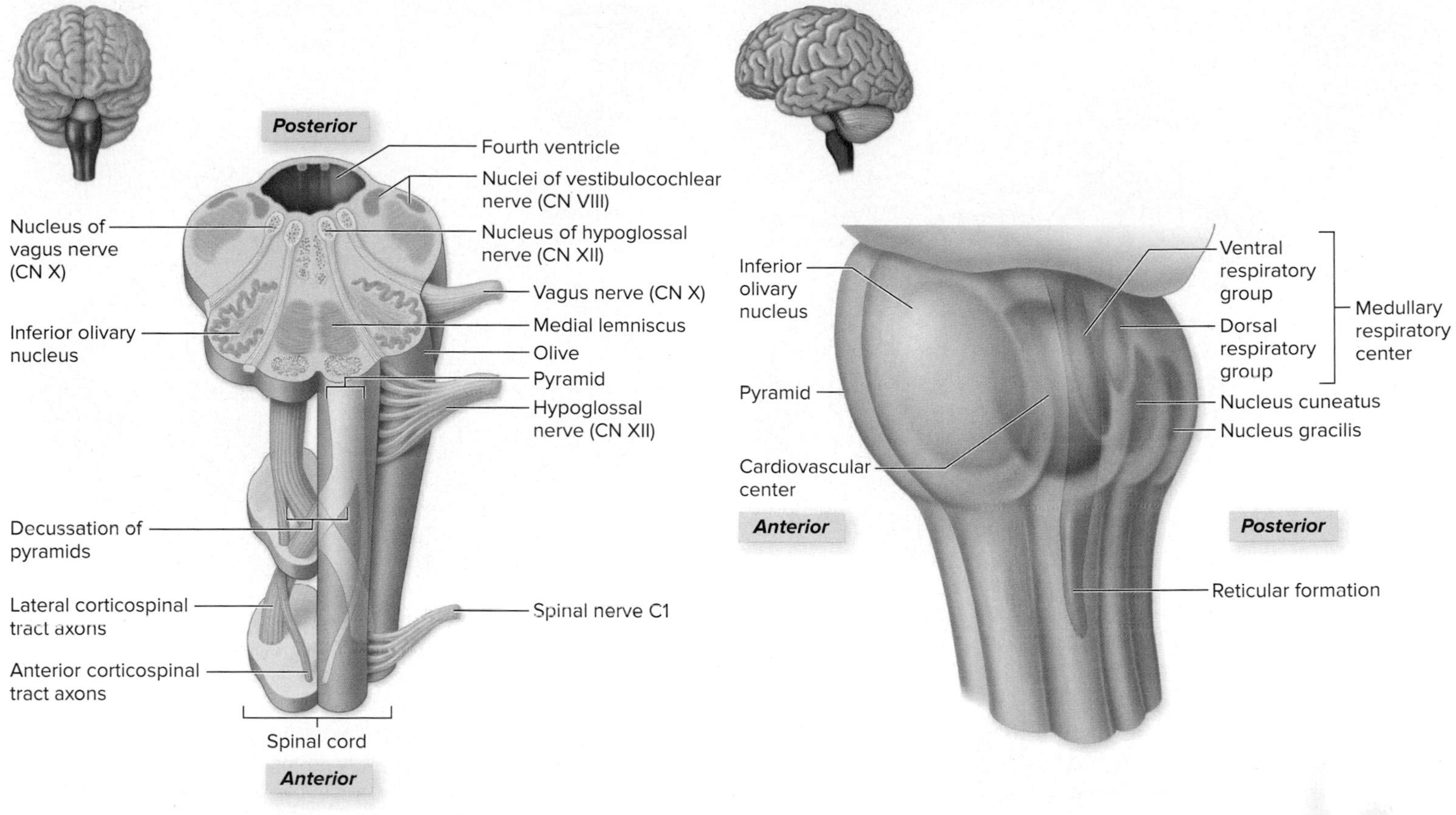

Figure 13.23 Medulla Oblongata. The medulla oblongata connects the brain to the spinal cord. (*a*) A cross section illustrates decussations of the pyramids and some cranial nerve nuclei (of CN VIII, CN X, CN XII). (*b*) The medulla contains several autonomic nuclei (cardiovascular center, medullary respiratory center), nuclei associated with sensory input from proprioceptors relayed to the cerebellum (inferior olivary nucleus), and nuclei associated with somatic sensory input (nucleus cuneatus, nucleus gracilis). AP|R

Finally, the medulla oblongata contains cranial nerve nuclei that are associated with the vestibulocochlear (CN VIII), glossopharyngeal (CN IX), vagus (CN X), accessory (CN XI), and hypoglossal (CN XII) cranial nerves. The medulla oblongata also contains the **nucleus cuneatus** (kū-nē-ā′tŭs; wedge) and the **nucleus gracilis** (gras-i′lis; slender), which relay somatic sensory information to the thalamus. The medial lemniscus exits from these nuclei and projects through the brainstem to the ventral posterior nucleus of the thalamus.

WHAT DID YOU LEARN?

 Where are the pyramids located, and what is their function?

 What are the main autonomic centers located in the medulla?

13.6 Cerebellum

The **cerebellum** (ser-e-bel′ŭm; little brain) is the second largest part of the brain. It coordinates fine control over skeletal muscle actions and stores memories of movement patterns, such as the playing of scales on a piano.

13.6a Structural Components of the Cerebellum

LEARNING OBJECTIVES

35. Name the parts and landmarks of the cerebellum.
36. Identify the three tracts through which the brainstem is linked to the cerebellum.

The cerebellum is composed of left and right **cerebellar hemispheres** (**figure 13.24**). Each hemisphere consists of two lobes, the **anterior lobe** and the **posterior lobe,** which are separated by the **primary fissure.** A narrow band of nervous tissue known as the **vermis** (ver′mis; worm) lies along the midline between the left and right cerebellar lobes. The cerebellar hemispheres and vermis have surface folds called **folia** (fō′lē-ă; *folium* = leaf). (These folds are similar to the gyri of the cerebrum.)

The cerebellum is partitioned internally into three regions: an outer gray matter called the **cerebellar cortex,** an internal region of white matter, and the deepest gray matter layer that is composed of cerebellar nuclei. The internal region of white matter is called the **arbor vitae** (ar′bōr vī′tē; *arbor* = tree, *vita* = life) because its distribution pattern resembles the branches of a tree (see section 13.1c).

Three thick nerve tracts, called peduncles, connect the cerebellum with the brainstem (see figure 13.20*b*). The superior cerebellar peduncles connect the cerebellum to the midbrain (see section 13.5a). The middle cerebellar peduncles connect the cerebellum to the pons (see section 13.5b). The inferior cerebellar peduncles connect the cerebellum to the medulla oblongata (see section 13.5c).

WHAT DID YOU LEARN?

 What are the primary anatomic components of the cerebellum?

 The middle cerebellar peduncles connect the cerebellum to which part of the brainstem?

13.6b Functions of the Cerebellum

LEARNING OBJECTIVE

37. Explain the functions of the cerebellum.

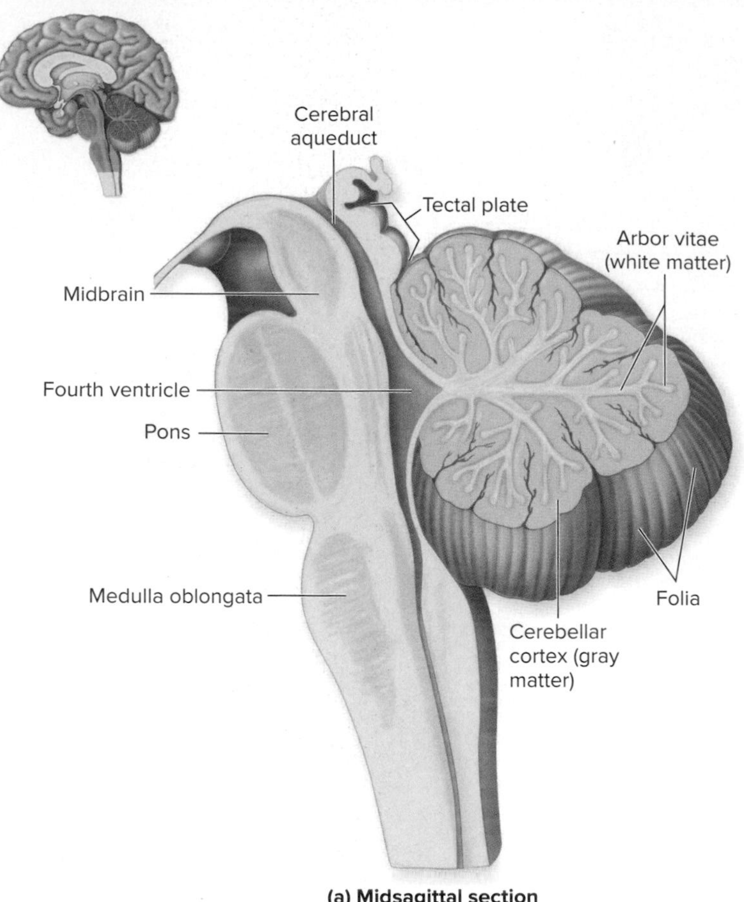

(a) Midsagittal section

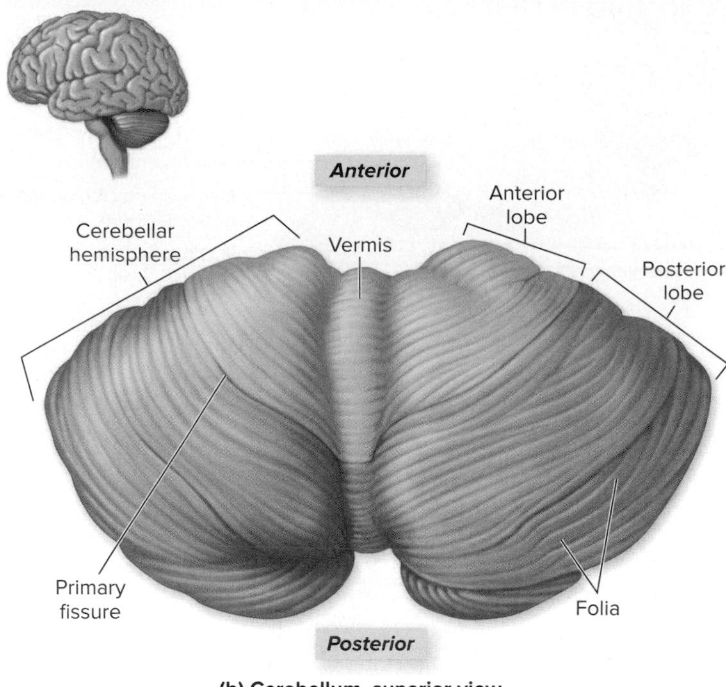

(b) Cerebellum, superior view

Figure 13.24 Cerebellum. The cerebellum lies posterior to the pons and medulla oblongata of the brainstem. (*a*) A midsagittal section shows the relationship of the cerebellum to the brainstem. (*b*) A superior view compares the anterior and posterior lobes of the cerebellum. (Note: The cerebrum and diencephalon have been removed.) AP|R

The cerebellum does not *initiate* skeletal muscle movement. Rather, it coordinates and fine-tunes skeletal muscle movements that were initiated by the cerebrum, and it ensures that skeletal muscle contraction follows the correct pattern, leading to smooth, coordinated movements. The cerebellum stores memories of previously learned movement patterns. This function is performed indirectly, by regulating activity along both the voluntary and involuntary motor pathways at the cerebral cortex, cerebral nuclei, and motor centers in the brainstem. The cerebrum initiates a movement and sends a "rough draft" of the movement to the cerebellum, which then coordinates and adjusts it. For example, the controlled, precise movements a classical guitarist makes when playing a concerto result from fine-tuning by the cerebellum. Without the cerebellum, the guitarist's movements would be choppy and sloppy, without precise coordination between the two hands.

INTEGRATE

CONCEPT CONNECTION

The precise movements of our fingers rely on multiple systems. The skeletal system provides structural support for the muscles and tissues within the fingers, the muscular system is responsible for the movement of the fingers, and the nervous system sends the nerve signals to the muscular system to control the patterned movements.

The cerebellum has several additional functions. Skeletal muscle activity is adjusted to maintain equilibrium and posture. It also receives proprioceptive (sensory) information from the muscles and joints and uses this information to regulate the body's position. For example, you are able to balance on one foot because the cerebellum takes the proprioceptive information from the body joints and maps out a muscle tone plan to keep the body upright. Finally, the proprioceptive information from the body's muscles and joints is sent to the cerebellum and then to the cerebrum. Thus, the cerebrum is made aware of the position of each body joint and its muscle tone, even if the person is not looking at the joint. For example, if you close your eyes, you are still aware of which body joints are flexed and which are extended.

INTEGRATE

CLINICAL VIEW 13.11

Effects of Alcohol and Drugs on the Cerebellum

A variety of drugs, and alcohol in particular, can temporarily or permanently impair cerebellar function. Alcohol intoxication leads to the following symptoms of impaired cerebellar function, which are used in the classic sobriety tests performed by police officers:

- **Disturbance of gait.** A person under the influence of alcohol or drugs rarely walks in a straight line but appears to sway and stagger. In addition, falling and bumping into objects are likely, due to the temporary cerebellar disturbance.
- **Loss of balance and posture.** When attempting to stand on one foot, a person who is intoxicated usually tips and falls over.
- **Inability to detect proprioceptive information.** When asked to close the eyes and touch the nose, a person under the influence of drugs or alcohol frequently misses the mark. This impairment is due to reduced ability to sense proprioceptive information, compounded by uncoordination of skeletal muscles.

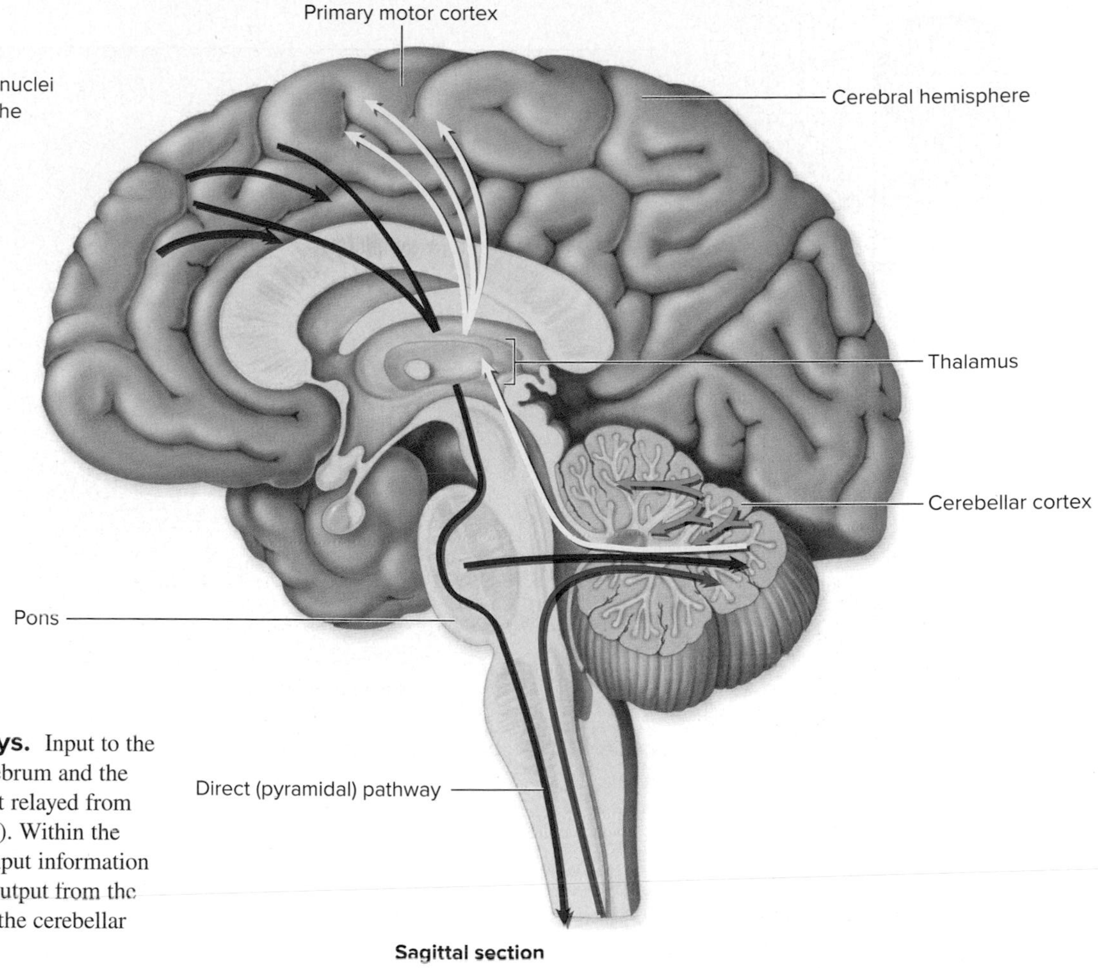

Voluntary movements
The primary motor cortex and the basal nuclei in the forebrain send impulses through the nuclei of the pons to the cerebellum.

Assessment of voluntary movements
Proprioceptors in skeletal muscles and joints report degree of movement to the cerebellum.

Integration and analysis
The cerebellum compares the planned movements (motor signals) against the results of the actual movements (sensory signals).

Corrective feedback
The cerebellum sends impulses through the thalamus to the primary motor cortex and to motor nuclei in the brainstem.

Primary motor cortex
Cerebral hemisphere
Thalamus
Cerebellar cortex
Pons
Direct (pyramidal) pathway
Sagittal section

Figure 13.25 Cerebellar Pathways. Input to the cerebellum from the motor cortex of the cerebrum and the pons (dark red arrows), and the sensory input relayed from proprioceptors to the cerebellum (blue arrow). Within the cerebellum, the integration and analysis of input information occurs (green arrows). Corrective feedback output from the cerebellum (yellow arrows) extends through the cerebellar peduncles (not shown).

The cerebellum continuously receives convergent input from both the various sensory pathways and the motor pathways in the brain (**figure 13.25**). In this way, the cerebellum unconsciously perceives the position of the body, receives the plan for movement, and then follows the activity to see if it was carried out correctly. When the cerebellum detects a disparity between the intended and actual movement, it may generate error-correcting nerve signals. These nerve signals are transmitted to both the premotor and primary motor cortices via the brainstem and the thalamus. Descending pathways then transmit these error-correcting signals to the motor neurons. Thus, the cerebellum influences and controls movement by indirectly affecting the excitability of motor neurons.

WHAT DID YOU LEARN?

31 What are the functions of the cerebellum?

13.7 Functional Brain Systems

The brain has two important functional systems that work together for a common purpose. These are considered functional brain systems because their structures are not confined to one major region of the brain but are located throughout two or more regions of the brain. These systems are the limbic system and the reticular formation.

13.7a Limbic System

LEARNING OBJECTIVES

38. Describe the main functions of the limbic system.

39. List the seven structures that compose the limbic system, and summarize their actions.

The **limbic** (lim′bik; *limbus* = edge) **system** is composed of multiple cerebral and diencephalic structures that collectively process and experience emotions. Thus, the limbic system is sometimes referred to as the *emotional brain*. The structures of the limbic system form a ring or border around the diencephalon. Although neuroanatomists continue to debate the components of the limbic system, the brain structures commonly recognized are shown in **figure 13.26** and listed here:

1. The **cingulate** (sin′gū-lāt; *cingulum* = girdle, to surround) **gyrus** is an internal mass of cerebral cortex located within the longitudinal fissure and superior to the corpus callosum. This cortical mass may be seen only in sagittal section, and it surrounds the diencephalon. It receives input from the other components of the limbic system.
2. The **parahippocampal gyrus** is a mass of cerebral cortical tissue in the temporal lobe. Its function is associated with the hippocampus.
3. The **hippocampus** (hip-ō-kam′pŭs; seahorse) is a component of the cerebrum located superior to the parahippocampal gyrus. It connects to the diencephalon via the fornix. As its name implies, this nucleus is shaped like a seahorse. Both the hippocampus and the parahippocampal gyrus are essential in storing memories and forming long-term memory.
4. The **amygdaloid body** (of the cerebral nuclei) connects to the hippocampus. The amygdaloid body is involved in several aspects of emotion, especially fear. It can also help store and code memories based on how a person emotionally perceives them—for example, as related to fear, extreme happiness, or sadness.

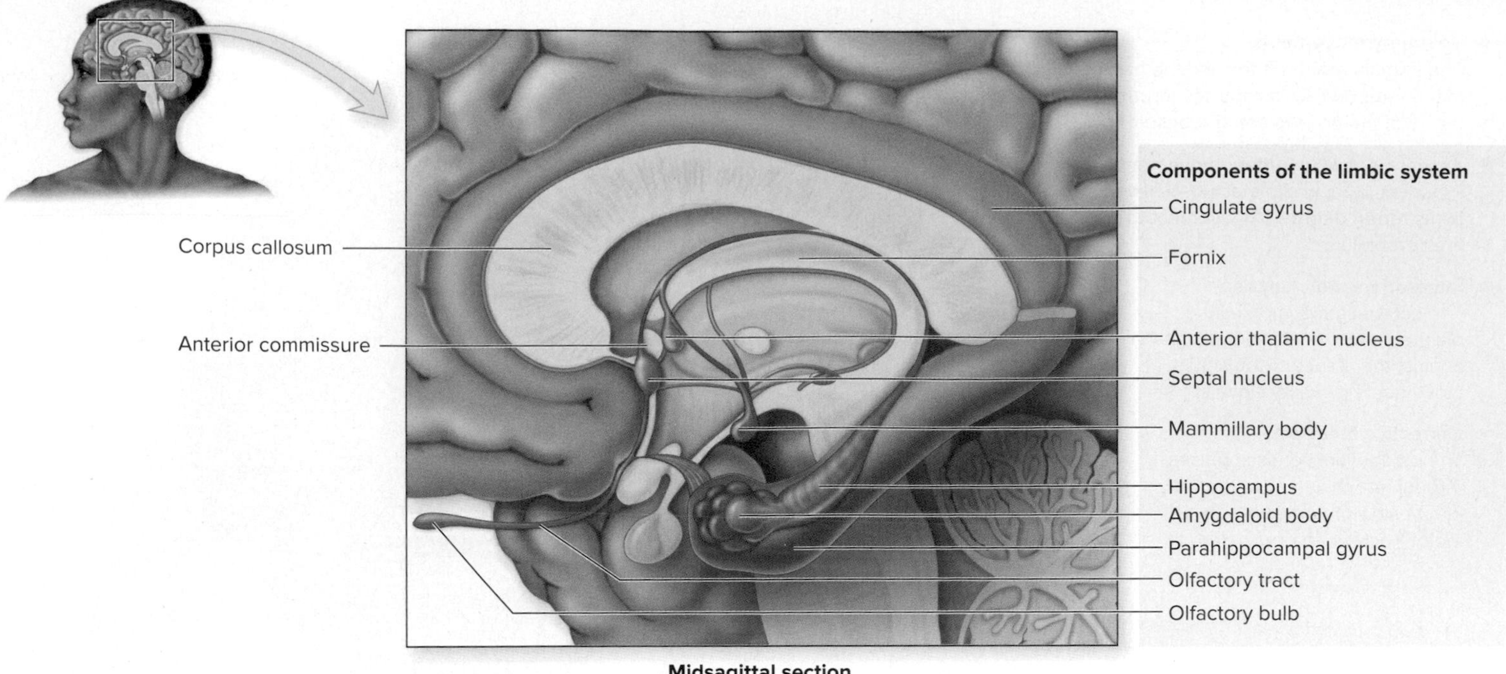

Figure 13.26 Limbic System. The components of the limbic system affect behavior and emotions. The olfactory cortex of the temporal lobe and the habenular nuclei of the epithalamus are not shown. APR

5. The **olfactory bulbs, olfactory tracts,** and **olfactory cortex** are part of the limbic system as well. You have probably experienced how particular odors can provoke certain emotions or be associated with certain memories (see section 16.3a).
6. The **fornix** (fōr′niks; arch) is a thin tract of white matter that connects the hippocampus with limbic system structures of the diencephalon.
7. Various nuclei in the diencephalon, such as the **anterior thalamic nuclei,** the **habenular nuclei** of the epithalamus, the **septal nuclei,** and the **mammillary** (mam′i-lār-ē; *mammilla* = nipple) **bodies** of the hypothalamus, interconnect other parts of the limbic system and contribute to its overall function.

WHAT DID YOU LEARN?

32. What are the components of the limbic system?
33. What are the main functions of the limbic system?

13.7b Reticular Formation

LEARNING OBJECTIVES

40. Describe the components and function of the reticular formation.

41. Explain the anatomy and function of the reticular activating system (RAS).

Projecting vertically through the core of the midbrain, pons, and medulla is a loosely organized mass of gray matter called the **reticular formation** (**figure 13.27**). The reticular formation extends slightly into the diencephalon and the spinal cord as well. This functional brain system has both motor and sensory components.

The motor component of the reticular formation communicates with the spinal cord and is responsible for regulating muscle tone (especially when the muscles are at rest). This motor component also assists in autonomic motor functions, such as respiration, blood pressure, and heart rate, by working with the autonomic centers in the medulla and pons.

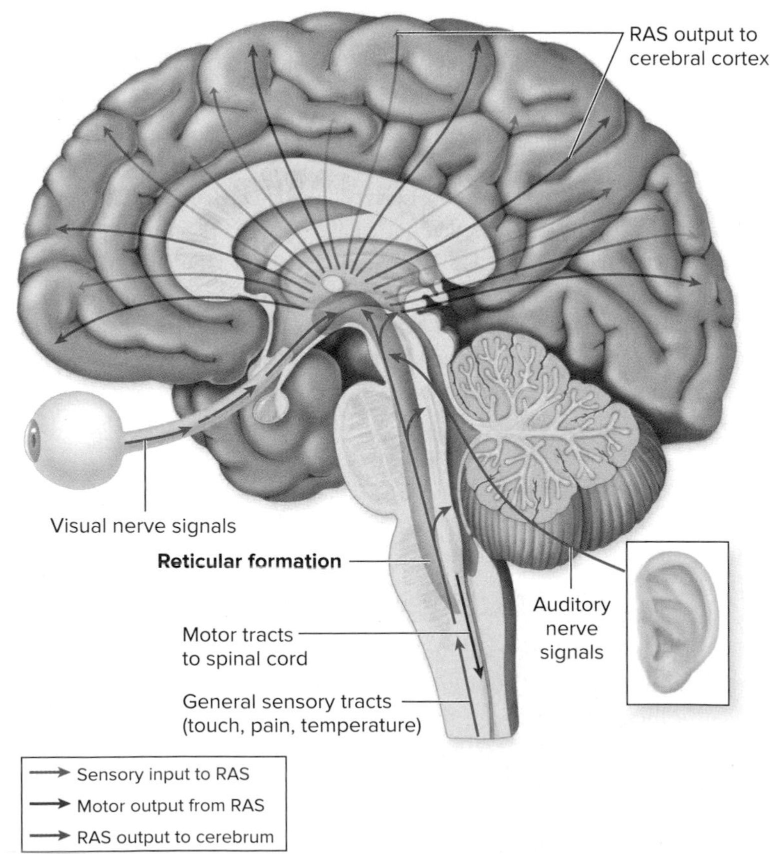

Figure 13.27 The Reticular Formation. The reticular formation receives and processes various types of input from sensory receptors (blue arrows). It participates in cyclic activities such as arousing the cortex to consciousness (purple arrows) and controlling the sleep-wake cycle. Some motor output from the reticular formation influences muscle activity (red arrow).

The sensory component of the reticular formation is responsible for alerting the cerebrum to incoming sensory information. This sensory component is called the **reticular activating system (RAS),** and it contains sensory axons that project to the cerebral cortex. The RAS processes visual, auditory, and touch stimuli and uses this information to keep us in a state of mental alertness. Additionally, the RAS arouses us from sleep. The sound of an alarm clock can awaken us because the RAS receives this sensory stimulus and sends it to the cerebrum. Conversely, under conditions of little or no stimuli, such as when you are in bed with the lights out and no sounds are disturbing you, the RAS is not stimulated and you find it easier to sleep.

Consciousness includes an awareness of sensation, voluntary control of motor activities, and the activities necessary for higher mental processing. It involves the simultaneous activity of large areas of the cerebral cortex. Levels of consciousness exist on a continuum. The highest state of consciousness and cortical activity is **alertness,** in which the individual is responsive, aware of self, and well oriented to person, place, and time.

 **WHAT DID YOU LEARN?**

 34 How is the reticular activating system related to the reticular formation?

13.8 Integrative Functions and Higher-Order Brain Functions

Higher-order brain functions include learning, memory, and reasoning. These functions occur within the cerebral cortex and involve multiple brain regions connected by complicated networks and arrays of axons. Both conscious and unconscious processing of information are involved in higher-order brain functions, and this processing may be continually adjusted or modified.

13.8a Development of Higher-Order Brain Functions

 **LEARNING OBJECTIVE**

42. Describe the relationship between age and higher-order brain functioning.

From infancy on, our motor control and processing capabilities become increasingly complex as we grow and mature. During the first year of life, the number of cortical neurons continues to increase. The myelination of many CNS axons continues throughout the first 2 years. (As a result, pediatricians recommend that infants and toddlers drink whole milk instead of skim milk, so their bodies will have adequate fat intake to support the development of myelin and the brain in general.) The brain grows rapidly in size and complexity so that by the age of 5, brain growth is 95% complete. (The rest of the body doesn't reach its adult size until puberty.)

As the CNS continues to develop, some neurons expand their number of connections, providing the increased number of synaptic junctions required for increasingly complex reflex activities and processing. During this same period, the brain will "prune" various synaptic connections, so only the most commonly used connections will remain. Some CNS axons remain unmyelinated until the teenage years (e.g., some of the axons in the prefrontal cortex). In general, the axons of PNS neurons continue to myelinate past puberty. A person's ability to carry out higher-order brain functions is a direct result of the level of nervous system maturation.

 **WHAT DID YOU LEARN?**

 35 What are some implications of the brain's anatomic development not being complete until the mid-teens?

13.8b Electroencephalogram

 **LEARNING OBJECTIVE**

43. Describe how an electroencephalogram examines brain activity.

An **electroencephalogram (EEG)** is a diagnostic test where electrodes are attached to the head to record the electrical activity of the brain (**figure 13.28**). This procedure is performed to investigate sleep disorders and lesions, and to determine if an individual is in a coma or a persistent vegetative state (see Clinical View 13.12: "Pathologic States of Consciousness"). EEGs also may evaluate a **seizure,** which is an event of abnormal electrical activity in the brain. There are different types of seizures, some of which may result in a brief blackout and others that may result in shaking and muscle spasms. **Epilepsy** is the condition where a person experiences repeated seizures over time (see Clinical View 13.8: "Epilepsy and Cerebral Lateralization").

An EEG measures and plots four types of brain waves (i.e., alpha, beta, theta, and delta). The distribution and frequency of these waves vary, depending upon whether the person is a child or an adult and if the individual is in a deep sleep, having a seizure, or experiencing a pathologic state of consciousness. For example, alpha and beta waves are typically seen in an awake or alert state, whereas theta and delta waves are more common during sleep. The presence of theta and delta waves in an awake adult is suggestive of a brain abnormality. Each electrode attached to a person's head will register a brain wave over that region of the head, so a patient's EEG printout will show multiple brain waves over a period of time.

INTEGRATE

CLINICAL VIEW 13.12

Pathologic States of Unconsciousness

When a person is asleep, he or she is technically unconscious, but not pathologically so. However, other unconscious conditions are pathologic. These pathologic states of unconsciousness may occur due to traumatic brain injury (TBI), CVA, low blood sugar, or diseases of the liver or kidney.

A brief loss of consciousness, termed **fainting** or **syncope** (sin′kō-pē; cutting short), often signals inadequate cerebral blood flow due to low blood pressure, as might follow hemorrhage or sudden emotional stress. **Stupor** (stū′per; *stupeo* = to be stunned) is a moderately deep level of unconsciousness from which the person can be aroused only by extreme repeated or painful stimuli.

A **coma** is a deep and profound state of unconsciousness from which the person cannot be aroused, even by repeated or painful stimuli. A person in a coma is alive but unable to respond to the environment and is not aware. A coma represents the deepest level of unconsciousness.

A **persistent vegetative state** is a long-term condition where the person is may be unconscious or awake, but has no self-awareness or awareness of one's surroundings. Some people in this state exhibit reflexive or spontaneous movements, such as moving their eyes, grimacing, crying, and smiling. However, there is no purposeful movement.

 **WHAT DID YOU LEARN?**

 36 An EEG may be used to evaluate what clinical conditions?

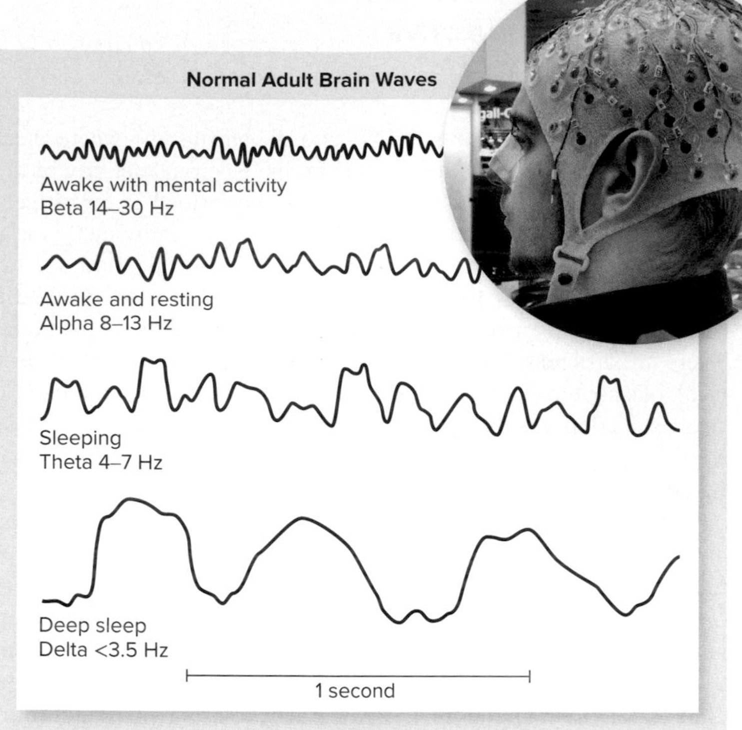

Figure 13.28 Electroencephalograms (EEGs). An individual wears an EEG cap that contains multiple electrodes to record brain wave activity, and a sample EEG recording is shown.

13.8c Sleep

LEARNING OBJECTIVES

44. Describe the main characteristics of sleep.

45. Compare and contrast non-REM and REM sleep.

People normally alternate between periods of alertness and **sleep,** which is the natural, temporary absence of consciousness from which a person can be aroused by normal stimulation. Cortical activity is depressed during sleep, but functions continue in the vital centers within the brainstem. Sleep is a natural, repeated condition of rest for both the body and brain. During periods of sleep, the decreased body activities are accompanied by conservation of energy, increasing growth, and renewal of strength in an individual.

Sleep may be subdivided into two main types, **non-REM** (nonrapid eye movement) and **REM** (rapid eye movement) **sleep.** Both types are distinguished by their EEG patterns and the absence or presence of rapid eye movements, respectively. In addition, it is during REM sleep that we have our most memorable dreams (although we may not remember all of our dreams when we awake). We spend about 75% of our total sleep time in non-REM sleep, and the remaining 25% in REM sleep. Some sleep scientists believe that non-REM sleep is meant for body repair, whereas REM sleep is consolidating and organizing memories, because the brain is very active during this period and it uses as much oxygen as when an individual is awake.

Non-REM sleep may be further subdivided into four stages. The EEG has helped scientists detect these four stages. We cycle through these non-REM stages and REM sleep multiple times throughout a normal-length sleep cycle, as shown in **figure 13.29**. The different stages of non-REM sleep differ in the types of brain waves present (e.g., alpha, beta, theta, and delta) and the ease at which one may be awakened. After about 90 minutes of non-REM sleep, the first incidence of REM sleep occurs and typically lasts about 10 minutes. The body then cycles back into non-REM sleep and then a longer period of REM sleep.

The amount of sleep a person needs varies based on age and health. Infants typically need up to 17 to 18 hours of sleep a day, and this number drops as we get older. Teens typically need between 8.5 and 9.5 hours of sleep a night, whereas the average adult needs about 7 to 8 hours of sleep. Lack of sleep has been associated with depression, impaired memory, and decreased immune function.

The term **insomnia** refers to the difficulty in falling asleep and staying asleep. Insomnia becomes more prevalent as we age, and certain medications may interfere with sleep (including how frequently we experience REM sleep) as well. **Sleep apnea** is where a person experiences repeated breathing interruptions during sleep. These breathing interruptions cause the individual to wake repeatedly throughout the night, so the individual is at risk to a variety of ailments associated with lack of sleep. Sleep apnea may be treated with a **CPAP (continuous positive airway pressure) machine,** where air is pumped through a mask that the patient wears during sleep so as to keep the airways open and allow the individual to sleep uninterrupted (see Clinical View 23.12: "Apnea").

WHAT DID YOU LEARN?

37 What are the main differences between non-REM and REM sleep? During what percentage of our sleep cycle are we in each type of sleep?

13.8d Cognition

LEARNING OBJECTIVES

46. Identify the brain areas in which cognition occurs.

47. Explain how lesions to different regions of the cortex affect cognition.

Mental processes such as awareness, knowledge, memory, perception, and thinking are collectively called **cognition** (kog-ni′shŭn). The association areas of the cerebrum (see section 13.3c), which form about 70% of the nervous tissue in the brain, are responsible for both cognition and the processing and integration of information between sensory input and motor output areas.

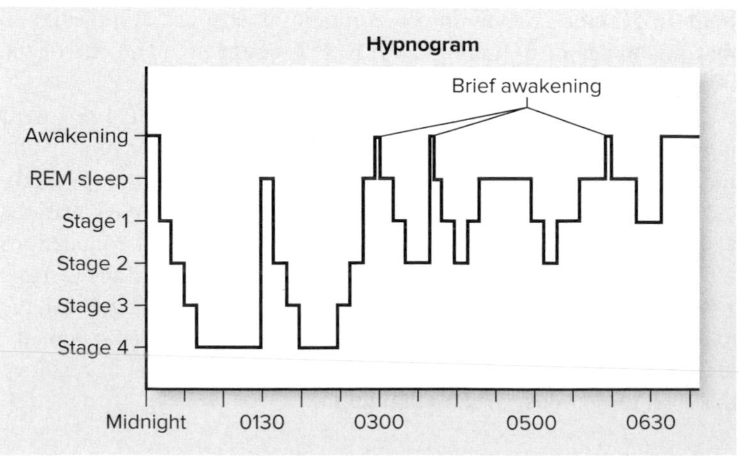

Figure 13.29 Hypnogram. A hypnogram documents the length of time a person is in each sleep stage.

Various studies of individuals suffering from brain lesions (caused by cancer, infection, stroke, and trauma) have provided insight into the functions of these areas of the brain. For example, the prefrontal cortex (frontal association area) integrates information from the sensory, motor, and association areas to enable the individual to think, plan, and execute appropriate behavior. Thus, an individual with a frontal lobe lesion exhibits personality abnormalities.

If an individual loses the ability to detect and identify stimuli (termed loss of awareness) on one side of the body, or on the limbs on that side, the primary somatosensory area in the hemisphere opposite the affected side of the body has been damaged.

An individual who has **agnosia** (ag-nō′zē-ă; *a* = without, *gnosis* = knowledge) displays an inability either to recognize or to understand the meaning of various stimuli. For example, a lesion in the temporal lobe may result in an inability to recognize or understand the meaning of sounds or words. Specific symptoms of agnosia vary, depending upon the location of the lesion within the cerebrum.

WHAT DID YOU LEARN?

 What is the definition of cognition?

13.8e Memory

LEARNING OBJECTIVES

48. Compare and contrast short term and long-term memory, and describe the parts of the brain involved with each.

49. Name the two regions of the limbic system involved in conversion of short-term memory to long-term memory.

Memory is a versatile element of human cognition involving different lengths of time and different storage capacities. Storing and retrieving information requires higher-order brain functions and depends upon complex interactions among different brain regions. On a broader scale, in addition to memory, information management by the brain entails both learning (the acquisition of new information) and forgetting (the elimination of trivial or nonuseful information).

Neuroscientists classify memory in various ways. For example, **sensory memory** occurs when we form important associations based on sensory input from the environment, such as the sounds coming from a crowded cafeteria, the smell from the food line, and the bright lights from the room. Sensory memory typically lasts for milliseconds to 1 second at most.

Short-term memory (STM) is generally characterized by limited capacity (approximately seven small segments of information) and brief duration (typically lasting less than 1 minute unless the information is rehearsed). Suppose that, in a Friday morning anatomy and physiology lecture, your instructor lists the general functions of the cerebral lobes on the board. Unless you study this information over the weekend, you will probably not recall it by Monday's lecture.

Some short-term memory, if adequately repeated and assessed, may be converted to long-term memory. Once information is placed into **long-term memory (LTM),** it may exist for limitless periods of time. So, for example, if over the weekend you practice retrieving the information from lecture (by rewriting your notes from memory or quizzing yourself) and/or work with a study partner to explain a lecture concept, you likely will store the information as LTM within the cerebral lobes. Not only will these practices help you to be well prepared for your next examination, but you may even remember this information for years to come. (However, information in LTM needs to be retrieved occasionally or it can be "lost," and our ability to store and retrieve information declines with aging.)

It appears that our brain must organize complex information in short-term memory prior to storing it in long-term memory (**figure 13.30**). Conversion from STM to LTM is called **encoding,** or *memory consolidation.* Encoding requires the proper functioning of two components of the limbic system: the hippocampus and the amygdaloid body (see section 13.7a). The hippocampus is required for the formation of STM, whereas LTM is stored primarily in the corresponding association areas of the cerebral cortex. For example, voluntary motor activity memory is stored in the premotor cortex, whereas memory of sounds is stored in the auditory association area.

Because STM and LTM involve different anatomic structures, loss of the ability to form STM does not affect the maintenance or accessibility of LTM.

WHAT DID YOU LEARN?

 What types of study habits best convert short-term memories into long-term memories?

13.8f Emotion

LEARNING OBJECTIVE

50. Explain the interactions of the prefrontal cortex and the limbic system in expression of emotions.

Emotional expression varies widely. For example, an automobile accident may cause those involved and some observers to cry, scream, or totally lose "emotional control," whereas the responding emergency personnel generally appear stoic, wearing masked expressions as they go about their professional duties.

Expression of our emotions is interpreted by our limbic system but ultimately is controlled by the prefrontal cortex. Irrespective of how we feel, this cortical region decides the appropriate way to show our feelings. Researchers have identified the emotional control centers of the brain by using traditional techniques as well as by examining the behavior of experimental animals and individuals with brain lesions. Although interpreting the results is often difficult because of the complexities of both the brain and our behavior, researchers have learned that many important aspects of emotion also depend upon an intact, functional amygdaloid body and hippocampus (components of the limbic system). If specific regions of either of these structures are damaged or artificially stimulated, we exhibit either deadened or

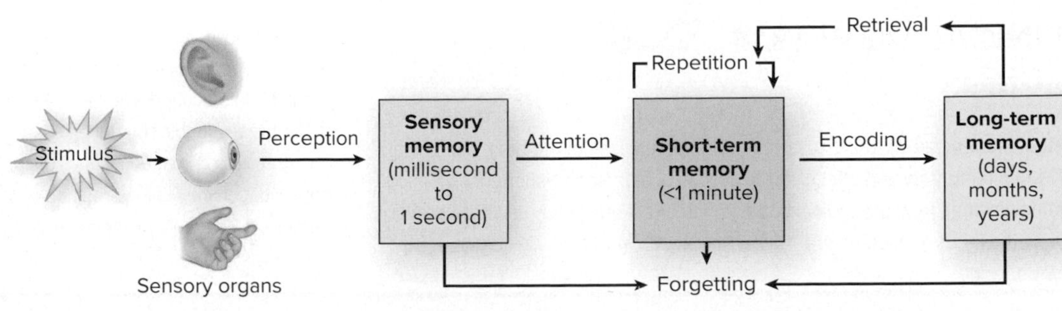

Figure 13.30 Model of Information Processing. Cognitive psychologists have proposed a model to show the relationships between sensory memory, short-term memory, and long-term memory, which develops later.

INTEGRATE

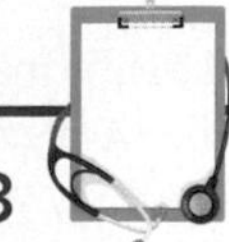

CLINICAL VIEW 13.13

Alzheimer Disease: The "Long Goodbye"

Alzheimer disease (AD) has become the leading cause of dementia in the developed world. (**Dementia** refers to a general loss of cognitive abilities, including memory, language, and decision-making skills.)

What are the classic symptoms of AD?

AD typically becomes clinically apparent in one's 70s or later; **early onset Alzheimer disease** is the diagnosis when an individual develops symptoms before the age of 65. An AD diagnosis is often delayed because of confusion with other forms of cognitive impairment. Symptoms include slow, progressive loss of higher intellectual functions and changes in mood and behavior. AD gradually causes language deterioration, impaired visuospatial skills, indifferent attitude, and poor judgment, while leaving motor function intact. Patients become confused and restless, often asking the same question repeatedly. AD progresses relentlessly over months and years, and thus has come to be known as "the long goodbye." Eventually, it robs its victims of their memory, their former personality, and even the capacity to speak.

What causes AD?

The underlying cause of AD remains a mystery, although both genetics and environment seem to play a role. Postmortem examinations of the brains of AD patients show marked and generalized cerebral atrophy. Microscopic examinations of brain tissue reveal a profound decrease in the number of cerebral cortical neurons, and a proliferation of two abnormal types of structures: **amyloid** (am′i-loyd) **plaques** and **neurofibrillary tangles.** Amyloid plaques are insoluble deposits of a protein termed beta amyloid as well as portions of neurons and microglial cells (see section 12.4b). The neurofibrillary tangles are formed from a protein called **tau** that is hyperphosphorylated (contains excess amounts of phosphate). Researchers are not sure if the plaques and tangles are the cause of AD, or merely a by-product of the disease's manifestation. Biochemical alterations also occur, most significantly a decreased level of the neurotransmitter acetylcholine in the cerebrum.

Is there a cure or test for AD?

There is no cure for AD, although some medications help alleviate the symptoms and seem to slow the progress of the disease. In the meantime, researchers are trying to develop diagnostic tests that can better predict who may be at risk for AD. Until recently, the only way to definitively diagnose AD was at autopsy, when the brain could be macroscopically and microscopically examined. Now, positron emission tomography (PET) scans appear to be able to identify the early brain changes seen with AD.

Recent research has suggested that difficulty or loss in identifying common smells (e.g., lemon, cinnamon) is linked with an increased risk in developing AD. In fact, this loss of smell may be one of the first signs of developing the disease, presumably because the brain regions involved with smell are among the first regions to develop the amyloid plaques and neurofibrillary tangles of AD.

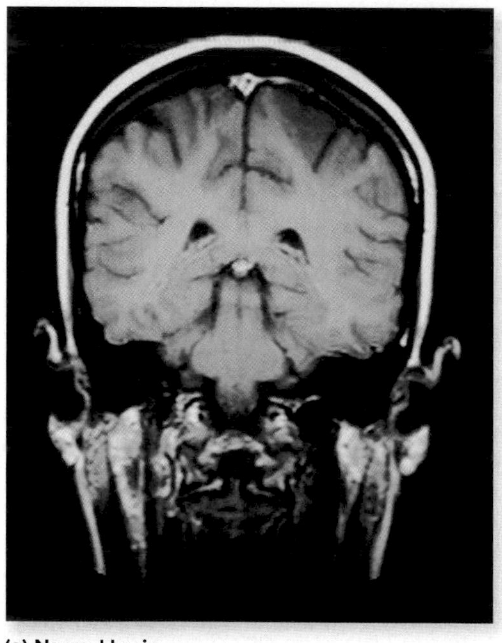

(a) Normal brain

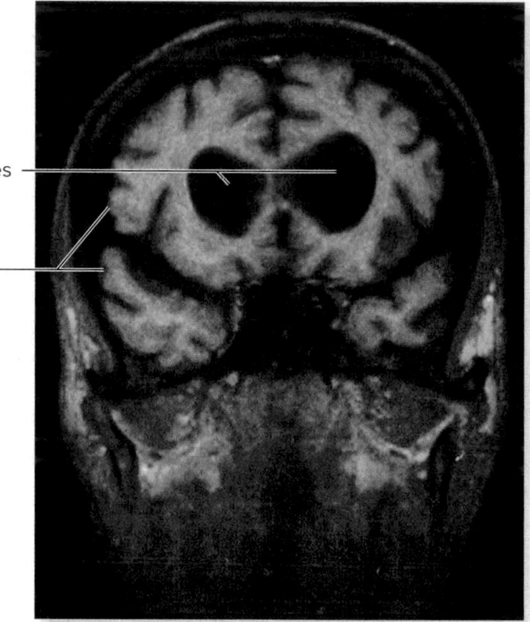

(b) Alzheimer brain

Magnetic resonance imaging (MRI) scans show coronal sections of (a) a normal brain and (b) an Alzheimer brain. (Note the large ventricles and wide spaces between gyri in the Alzheimer brain.)

INTEGRATE

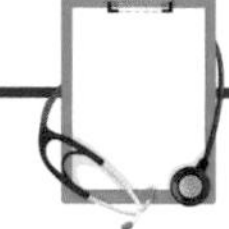

CLINICAL VIEW 13.14

Amnesia

Amnesia (am-nē′zē-ă) refers to complete or partial loss of memory. Most often, amnesia is temporary and affects only a portion of a person's experiences. Causes of amnesia range from psychological trauma to direct brain injury, such as a severe blow to the head or even a cerebrovascular accident (CVA).

Because memory processing and storage involve numerous regions of the brain, the type of memory loss that occurs in an episode of amnesia depends upon the area of the brain damaged. The most serious forms of amnesia result from damage to the thalamus and limbic structures, especially the hippocampus. If one or more of these structures are damaged, serious disruption or complete lack of memory storage and consolidation may follow. The nature of the underlying problem determines whether amnesia is complete or partial, and to what degree recovery, if any, is possible.

exaggerated expressions of aggression, affection, anger, fear, love, pain, pleasure, or sexuality, as well as anomalies in learning and memory.

WHAT DID YOU LEARN?

40 What portions of the brain and limbic system are involved with modulation of emotion?

INTEGRATE

CLINICAL VIEW 13.15

Dyslexia

Dyslexia (dis-lek'sē-ă; *dys* = bad, *lexis* = word) is an inherited learning disability characterized by problems with single-word decoding. It often runs in families. Affected individuals not only have trouble reading but may also have problems writing and spelling accurately. These individuals may be able to recognize letters normally, but they demonstrate a level of reading competence far below that expected for their level of intelligence. Their writing may be disorganized and uneven, with the letters of words in incorrect order or even completely reversed. Some individuals appear to outgrow this condition, or at least develop improved reading ability over time. This improvement may reflect neural maturation or retraining of parts of the brain to better decode words and symbols. Some researchers have postulated that dyslexia is a form of **disconnect syndrome,** in which transfer of information between the cerebral hemispheres through the corpus callosum is impaired.

13.8g Language

LEARNING OBJECTIVE

51. List the cerebral centers involved in written and spoken language, and describe how these centers work together.

The higher-order processes involved in language include reading, writing, speaking, and understanding words. Recall that two important cortical areas involved in speech integration are the Wernicke area and the motor speech area (Broca area) (see section 13.3c). The Wernicke area is involved in interpreting what we read or hear, whereas the motor speech area receives nerve signals originating from the Wernicke area and then helps regulate the motor activities needed for us to speak. Thus, the Wernicke area is central to our ability to recognize written and spoken language. Immediately posterior to the Wernicke area is the **angular gyrus,** a region that processes the words we read into a form that we can speak (**figure 13.31**). First, the Wernicke area sends a speech plan to the motor speech area, which initiates a specific patterned motor program that is transmitted to the primary motor cortex. Next, the primary motor cortex signals other motor neurons, which then stimulate the muscles of the cheeks, larynx, lips, and tongue to produce speech.

The Wernicke area is in the categorical hemisphere in most people (see section 13.3e). In the representational hemisphere, a cortical region opposite the Wernicke area recognizes the emotional content of speech. A lesion in this area of the cerebrum can make a person unable to understand emotional nuances, such as bitterness or happiness, in spoken words. A lesion in the cortical region of the representational hemisphere opposite the motor speech area results in **aprosodia,** which causes dull, emotionless speech.

Several speech disorders affect the interpretation, processing, and execution of language (including sign language). For example, **apraxia** (a-prak'sē-ă; *pratto* = to do) **of speech** is a motor function disorder. Individuals are consciously aware of what they want to say but are unable to coordinate and execute the motor commands needed to produce the speech. As a result, these individuals may have difficulty in producing recognizable sounds and sequencing them properly for normal speech. In contrast, an individual with **aphasia** (a-fā'zē-ă; speechlessness) has difficulty understanding speech or writing, or is unable to produce comprehensible speech. Aphasic individuals may have consistent difficulties

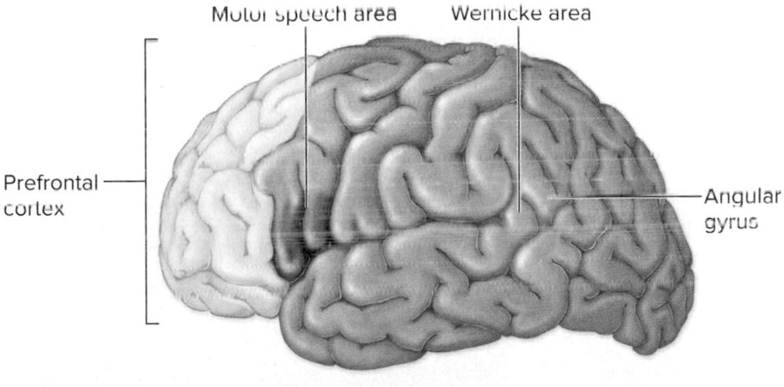

(a) Lateral view

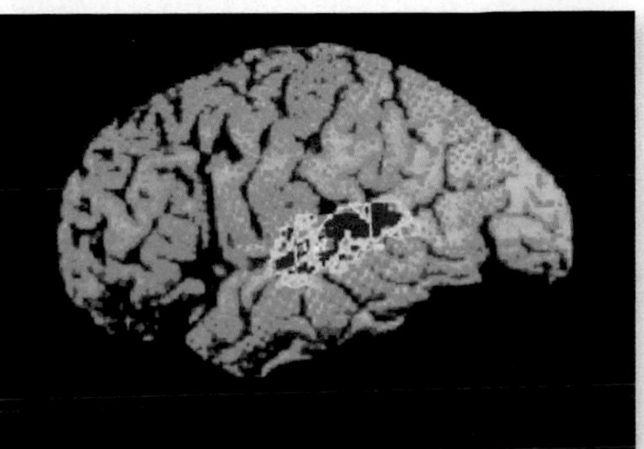

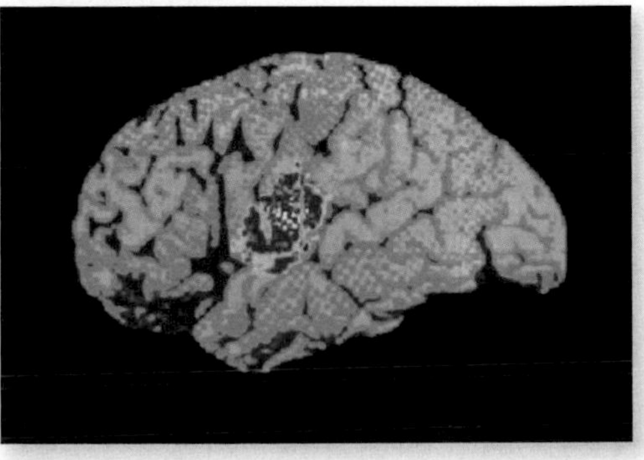

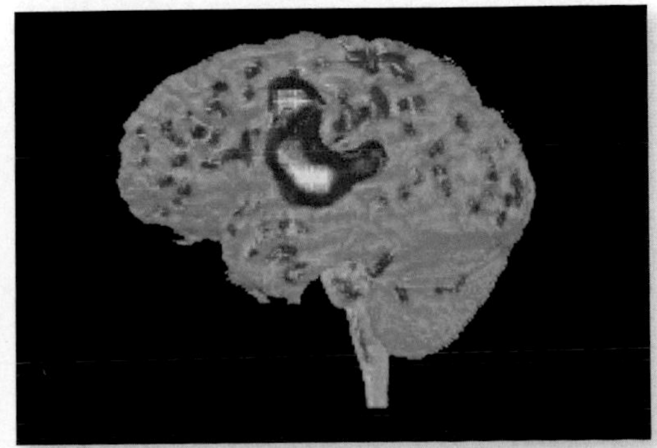

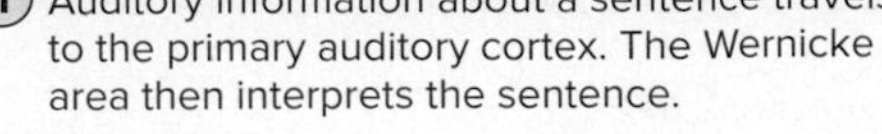

(1) Auditory information about a sentence travels to the primary auditory cortex. The Wernicke area then interprets the sentence.

(2) Information from the Wernicke area travels to the motor speech area.

(3) Information travels from the motor speech area to the primary motor cortex, where motor commands involving muscles used for speech are given.

(b) PET scans

Figure 13.31 Functional Areas Within the Cerebral Cortex. (*a*) The left cerebral hemisphere in most people houses the Wernicke area, the motor speech area, and the prefrontal cortex. (*b*) A PET scan shows the areas of the brain that are most active during speech.
(*b1, 2*) ©WDCN/Univ. College London/Science Source; (*b3*) ©National Cancer Institute/Science Source

finding the correct words and may use nonsense words to describe certain objects. Often, they may not realize that others can't understand what they are saying. Many cases of aphasia are due to head injury or strokes that damage brain centers responsible for interpreting language. Speech-language pathologists (SLPs) work extensively with those suffering from either apraxia of speech or aphasia to help them improve language production and understanding.

WHAT DID YOU LEARN?

41 How is the Wernicke area involved in language processing?

13.9 Cranial Nerves

LEARNING OBJECTIVES

52. List the names and locations of the 12 pairs of cranial nerves.

53. Compare the functions of each of the cranial nerves.

There are 12 pairs of cranial nerves, which are designated with both a number and a name. They are numbered with Roman numerals according to their positions originating on the brain, beginning with the most anteriorly placed nerve. Note that the number is sometimes preceded by the prefix *CN* (**figure 13.32**).

The name of each nerve generally has some relation to its function. The 12 pairs of cranial nerves are the olfactory (CN I), optic (CN II), oculomotor (CN III), trochlear (CN IV), trigeminal (CN V), abducens (CN VI), facial (CN VII), vestibulocochlear (CN VIII), glossopharyngeal (CN IX), vagus (CN X), accessory (CN XI), and hypoglossal (CN XII).

Table 13.4 summarizes the main motor and sensory functions of each cranial nerve. For easier reference, each main function of a nerve is color-coded. Blue represents a sensory function,

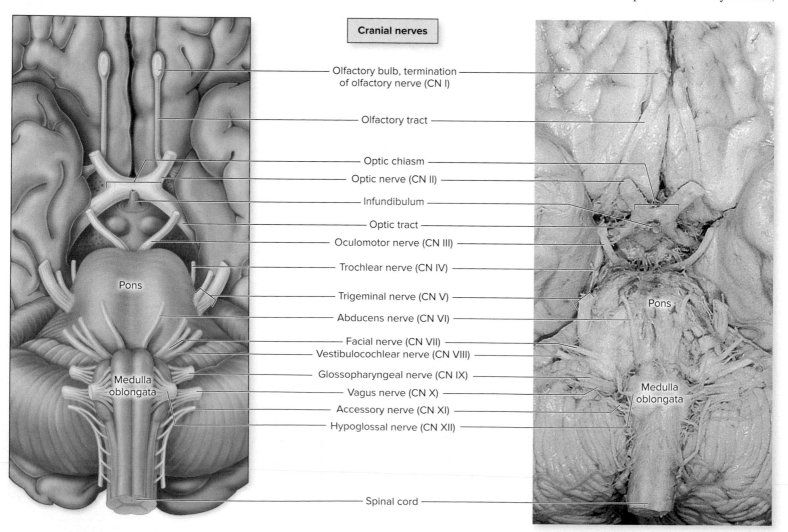

Figure 13.32 Cranial Nerves. A view of the inferior surface of the brain shows the 12 pairs of cranial nerves. AP|R

Table 13.4 Primary Functions of Cranial Nerves

Cranial Nerve	Sensory Function	Somatic Motor Function	Parasympathetic Motor (Autonomic) Function[1]
I (olfactory)	Olfaction (smell)	*None*	*None*
II (optic)	Vision	*None*	*None*
III (oculomotor)	*None*[2]	Four extrinsic eye muscles (medial rectus, superior rectus, inferior rectus, inferior oblique); levator palpebrae superioris muscle (elevates eyelid)	Innervates sphincter pupillae muscle in eye to make pupil constrict; contracts ciliary muscles to make lens of eye more rounded (as needed for near vision)
IV (trochlear)	*None*[2]	Superior oblique extrinsic eye muscle	*None*
V (trigeminal)	General sensory from anterior scalp, nasal cavity, nasopharynx, entire face, most of oral cavity, teeth, anterior two-thirds of tongue; part of auricle of ear; meninges	Muscles of mastication, mylohyoid, digastric (anterior belly), tensor tympani, tensor veli palatini	*None*
VI (abducens)	*None*[2]	Lateral rectus extrinsic eye muscle	*None*
VII (facial)	Taste from anterior two-thirds of tongue	Muscles of facial expression, digastric (posterior belly), stylohyoid, stapedius	Increases secretion from lacrimal gland of eye, submandibular and sublingual salivary glands
VIII (vestibulocochlear)	Hearing (cochlear branch); equilibrium (vestibular branch)	*None*[3]	*None*
IX (glossopharyngeal)	General sensory and taste from posterior one-third of tongue, general sensory from part of pharynx, visceral sensory from carotid bodies	One pharyngeal muscle (stylopharyngeus)	Increases secretion from parotid salivary gland
X (vagus)	Visceral sensory information from heart, lungs, most abdominal organs General sensory information from external acoustic meatus, tympanic membrane, part of pharynx, laryngopharynx, and larynx	Most pharyngeal muscles; all laryngeal muscles	Innervates smooth muscle and glands of heart, lungs, larynx, trachea, most abdominal organs
XI (accessory)	*None*[2]	Trapezius muscle, sternocleidomastoid muscle	*None*
XII (hypoglossal)	*None*[2]	Intrinsic and extrinsic tongue muscles	*None*

1. The autonomic nervous system contains a parasympathetic division and sympathetic division. Some cranial nerves contain parasympathetic axons and are listed in this table. Detailed information about these divisions is found in section 15.2.

2. These nerves do contain some tiny proprioceptive sensory axons from the muscles, but in general, these nerves tend to be described as motor only.

3. A few motor axons travel with this nerve to the inner ear, but they are not considered a significant component of the nerve.

INTEGRATE

LEARNING STRATEGY

Developing a code or phrase called a **mnemonic** (nē-mon′ik) may help you remember the cranial nerves. Here is a sample mnemonic for the cranial nerves:

Oh (*olfactory*)
once (*optic*)
one (*oculomotor*)
takes (*trochlear*)
the (trigeminal)
anatomy (abducens)
final (*facial*)
very (*vestibulocochlear*)
good (*glossopharyngeal*)
vacations (*vagus*)
are (*accessory*)
heavenly! (*hypoglossal*))

pink stands for a somatic motor function (see section 12.1b), and orange denotes a parasympathetic motor function. **Table 13.5** lists the individual cranial nerves and discusses their functions, origins, and pathways. The color-coding in table 13.4 carries over to table 13.5, so you can easily determine whether a cranial nerve has sensory function, motor function, or both.

WHAT DID YOU LEARN?

42 Which cranial nerves have sensory functions only?

Table 13.5 Cranial Nerves

CN I OLFACTORY NERVE (ol-fak′tō-rē; *olfacio* = to smell)

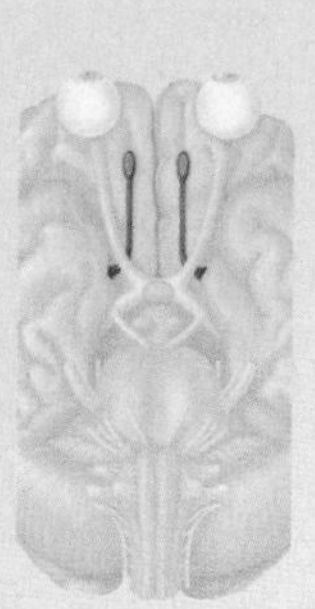

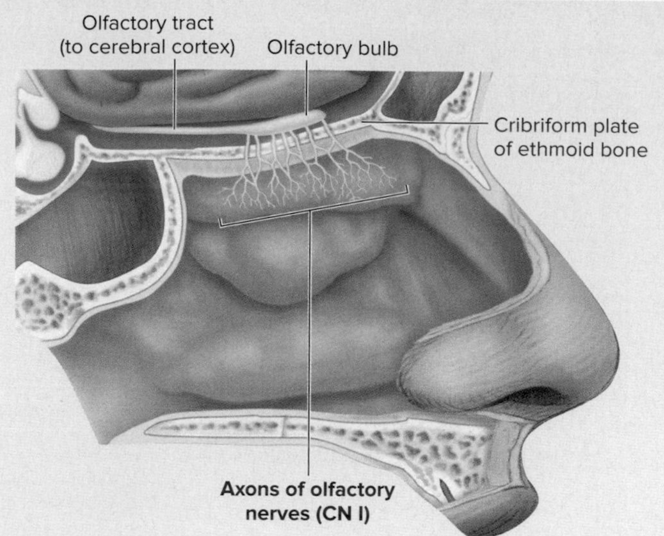

Description	Special sensory nerve that conducts olfactory (smell) sensation from the nose to the brain
Sensory function	Olfaction (smell)
Origin	Receptors (bipolar neurons) in olfactory epithelium of nasal cavity
Pathway	Travels through the cribriform foramina of ethmoid bone and synapses in the olfactory bulbs, which extend to various locations within the brain, including the primary olfactory cortex of the temporal lobe
Conditions caused by nerve damage	Anosmia (partial or total loss of smell)
How to test for nerve damage	Test smell (have patient close eyes, close one nostril, and inhale an odor with the other nostril).

CN II OPTIC NERVE (op′tik; *ops* = eye)

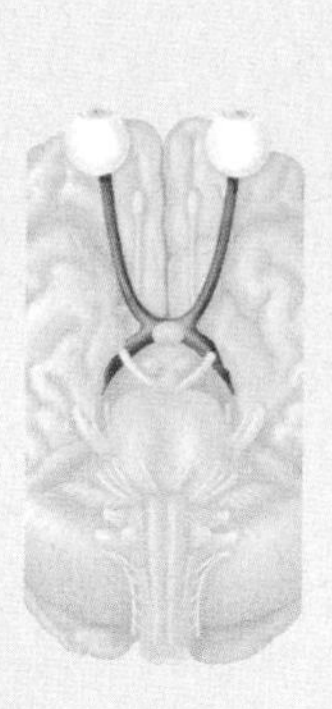

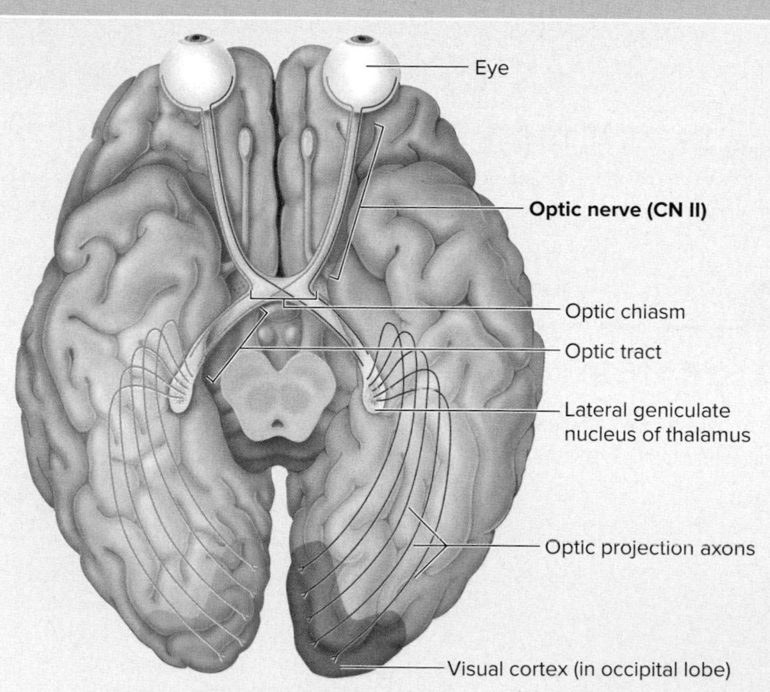

Description	Special sensory nerve that conducts visual information from the retina of the eye to the brain
Sensory function	Vision
Origin	Retina of the eye
Pathway	Enters cranium via optic canal of sphenoid bone; left and right optic nerves unite at optic chiasm; optic tract travels to lateral geniculate nucleus of thalamus; nerve fibers project to the primary visual cortex of the occipital lobe
Conditions caused by nerve damage	Anopsia (visual defects)
How to test for nerve damage	Test vision (cover one eye and have patient view a visual acuity chart with the other eye).

Table 13.5 Cranial Nerves (continued)

CN III OCULOMOTOR NERVE (ok′ū-lō-mō′tŏr; *oculus* = eye, *motorius* = moving)

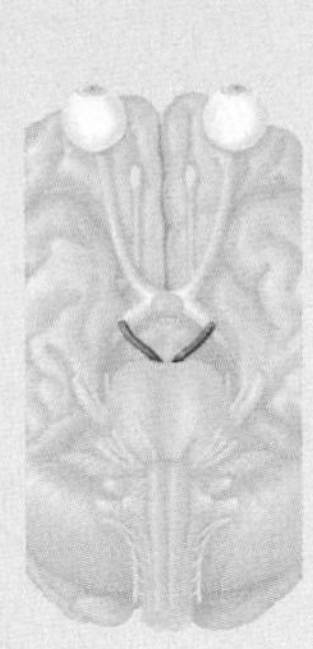

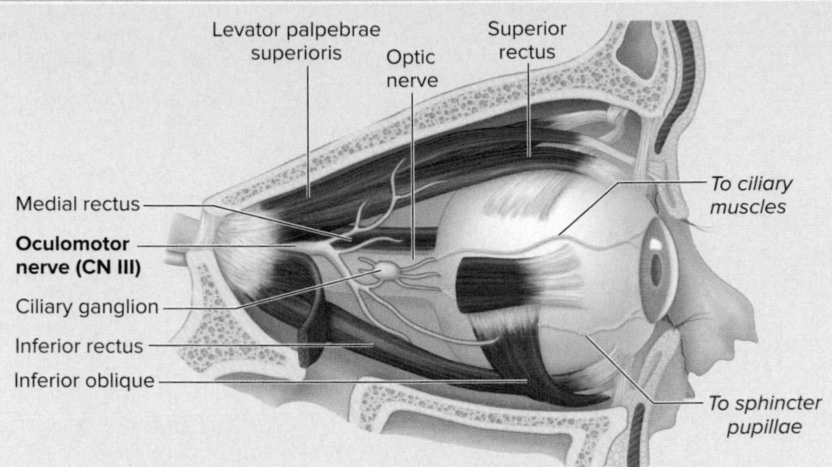

Description	Motor nerve that innervates four of the six extrinsic eye muscles, an upper eyelid muscle, and intrinsic eye muscles (smooth muscle within the eye)
Somatic motor function	Contracts four extrinsic eye muscles (superior rectus, inferior rectus, medial rectus, inferior oblique) to move eye Contracts levator palpebrae superioris muscle to elevate eyelid
Parasympathetic motor function	Contracts sphincter pupillae muscle of iris to make pupil constrict Contracts smooth muscle of ciliary body to make lens of eye more spherical (as needed for near vision)
Origin	Oculomotor and Edinger Westphal nuclei within the midbrain
Pathway	Leaves cranium via superior orbital fissure and travels to eye and eyelid (parasympathetic axons travel to ciliary ganglion, and postganglionic parasympathetic axons then travel to iris and ciliary muscles)
Conditions caused by nerve damage	Ptosis (upper eyelid droop); paralysis of most eye muscles, leading to strabismus (eyes not in parallel/deviated improperly), diplopia (double vision), focusing difficulty, dilated pupil (mydriasis)
How to test for nerve damage	Determine if the upper eyelid droops, examine if pupil constricts in response to light, examine eye movement (have patient follow a moving object with eyes).

CN IV TROCHLEAR NERVE (trōk′lē-ar; *trochlea* = a pulley)

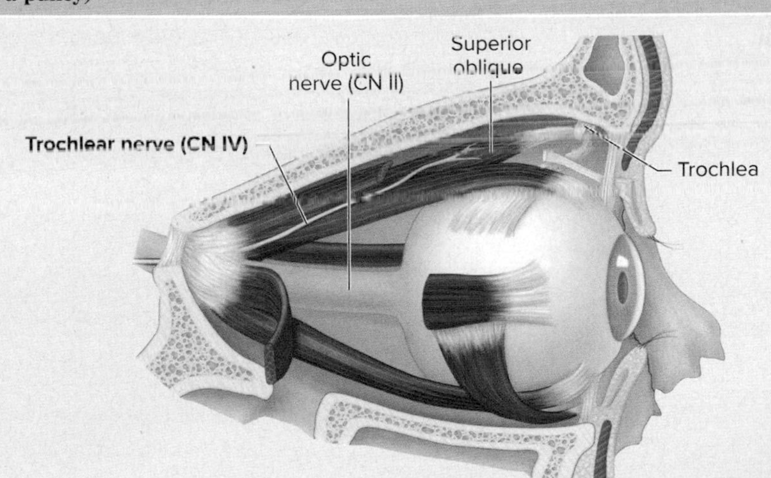

Description	Motor nerve that innervates one extrinsic eye muscle (superior oblique) that loops through a pulley-shaped ligament called a trochlea
Somatic motor function	Contracts one extrinsic eye muscle (superior oblique) to move eye inferiorly and laterally
Origin	Trochlear nucleus within the midbrain
Pathway	Leaves cranium via superior orbital fissure and travels to superior oblique muscle
Conditions caused by nerve damage	Paralysis of superior oblique, leading to strabismus (eyes not in parallel/deviated improperly), diplopia (double vision)
How to test for nerve damage	Examine eye movement (have patient follow a moving object with eyes).

(continued on next page)

Table 13.5 Cranial Nerves (continued)

CN V TRIGEMINAL NERVE (trī-jem′i-năl; *trigeminus* = threefold)

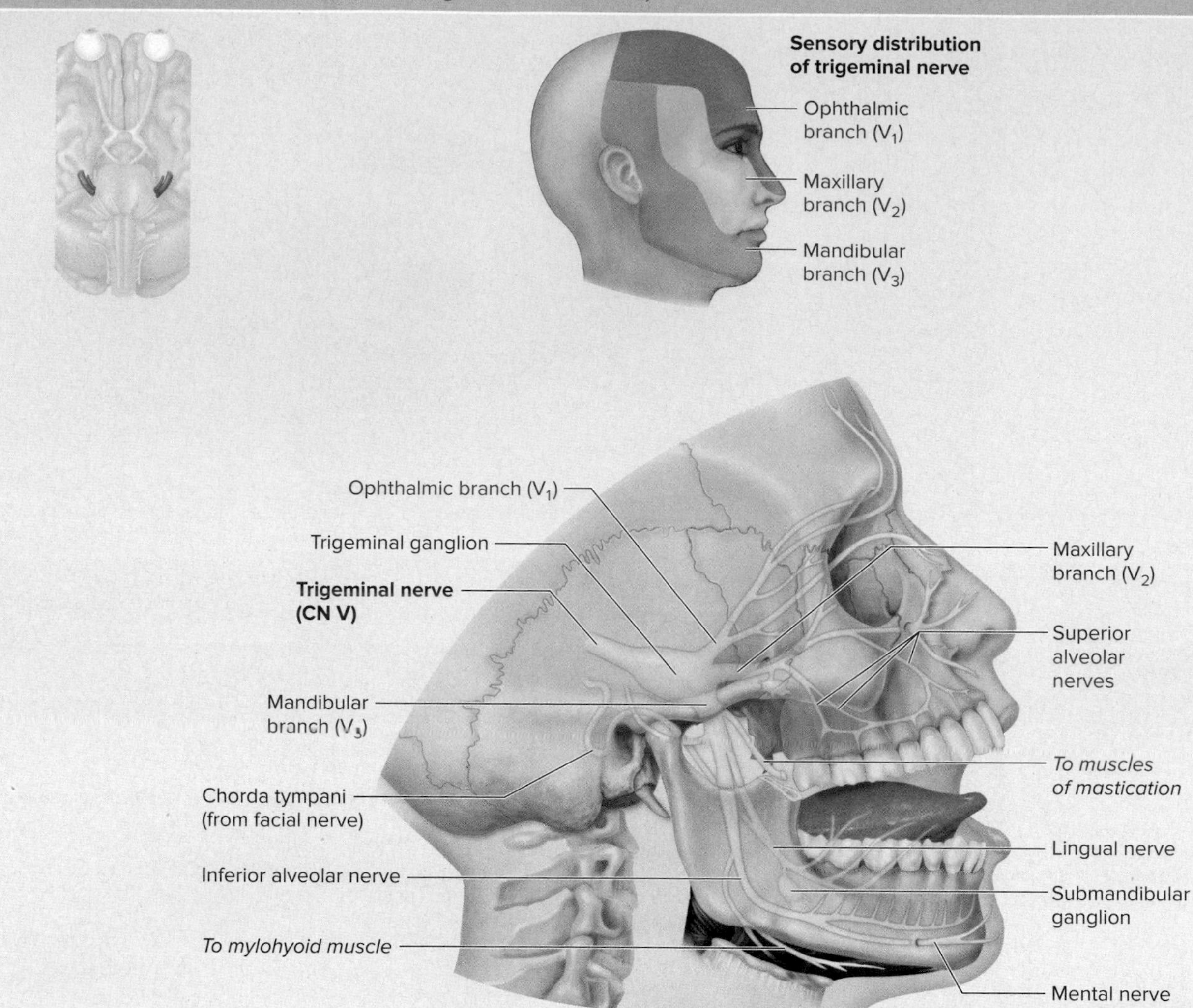

Description	Mixed nerve that consists of three divisions: ophthalmic (V_1), maxillary (V_2), and mandibular (V_3); receives sensory nerve signals from face, oral cavity, nasal cavity, meninges, and anterior scalp and innervates muscles of mastication
Sensory function	Sensory stimuli for this nerve are touch, temperature, and pain. V_1: Conducts sensory nerve signals from cornea, nose, forehead, anterior scalp, meninges V_2: Conducts sensory nerve signals from nasal mucosa, palate, gums, cheek, meninges V_3: Conducts sensory nerve signals from anterior two-thirds of tongue, meninges, skin of chin, lower jaw, lower teeth; one-third from sensory axons of auricle of ear
Somatic motor function	Innervates muscles of mastication (temporalis, masseter, lateral and medial pterygoids), mylohyoid, anterior belly of digastric, tensor tympani muscle, and tensor veli palatini
Origin	Nuclei in the pons
Pathway	V_1: Sensory axons enter cranium via superior orbital fissure and travel to trigeminal ganglion before entering pons. V_2: Sensory axons enter cranium via foramen rotundum and travel to trigeminal ganglion before entering pons. V_3: Sensory axons travel through foramen ovale to trigeminal ganglion before entering pons. Motor axons leave pons and exit cranium via foramen ovale to supply muscles.
Conditions caused by nerve damage	Trigeminal neuralgia (tic douloureux) is caused by inflammation of the sensory components of the trigeminal nerve and results in intense, pulsating pain lasting from minutes to several hours.
How to test for nerve damage	Have patient close mouth against resistance; also have patient close eyes and then determine if an object (such as a feather) moved along the face can be felt.

Table 13.5 Cranial Nerves (continued)

CN VI ABDUCENS NERVE (ab-dū′senz; to move away from)

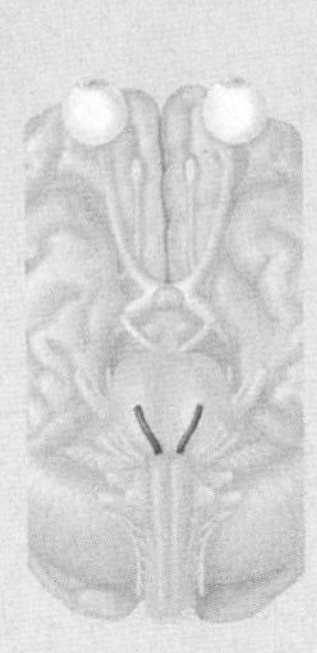

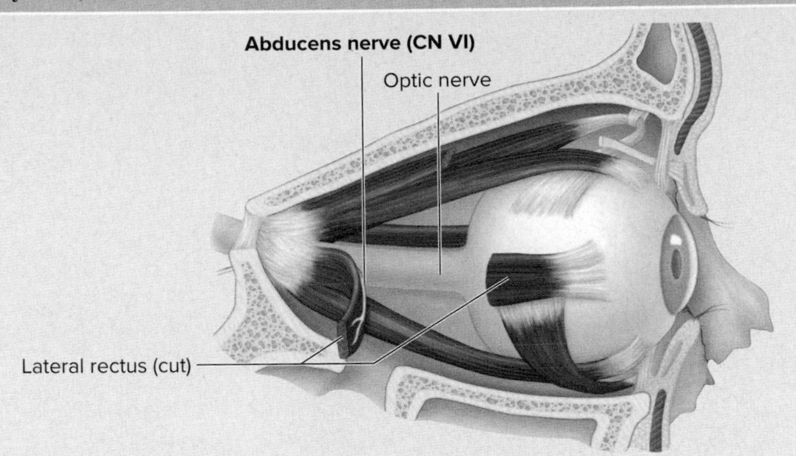

Description	Motor nerve that innervates one extrinsic eye muscle (lateral rectus) to move the eye
Somatic motor function	Contracts lateral rectus for eye abduction (moving eye laterally)
Origin	Pontine (abducens) nucleus in the pons
Pathway	Leaves cranium through superior orbital fissure and travels to lateral rectus muscle
Conditions caused by nerve damage	Paralysis of lateral rectus limits lateral movement of eye, diplopia (double vision)
How to test for nerve damage	Examine eye movement (have patient follow a moving object with eyes) and determine if the eye is able to be abducted.

CN VII FACIAL NERVE (fā′shăl)

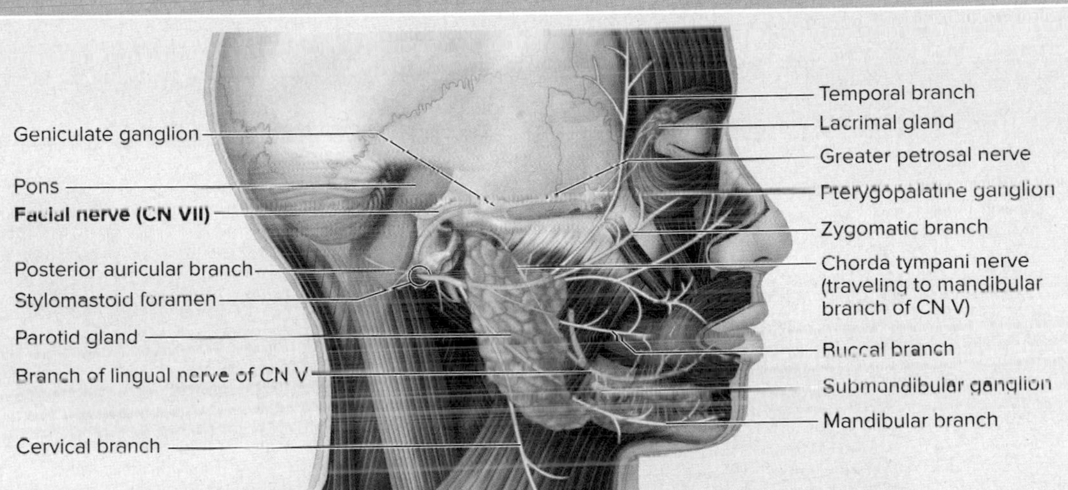

Description	Mixed nerve that conducts taste sensations from anterior two-thirds of tongue; relays motor output to muscles of facial expression, lacrimal (tear) gland, and salivary glands inferior to the tongue (submandibular salivary gland and sublingual salivary gland)
Sensory function	Conducts taste sensations from anterior two-thirds of tongue
Somatic motor function	The five major motor branches (temporal, zygomatic, buccal, mandibular, and cervical) innervate the muscles of facial expression, the posterior belly of the digastric muscle, and the stylohyoid and stapedius muscles.
Parasympathetic motor function	Increases secretions of the lacrimal gland of the eye; increases secretions of the submandibular and sublingual salivary glands
Origin	Nuclei in the pons
Pathway	Sensory axons travel from the tongue via the chorda tympani branch of the facial nerve through a tiny foramen to enter the skull, and axons synapse at the geniculate ganglion of the facial nerve. Somatic motor axons leave the pons and enter the temporal bone through the internal acoustic meatus, project through temporal bone, and emerge through the stylomastoid foramen to innervate the muscles of facial expression. Parasympathetic motor axons leave the pons, enter the internal acoustic meatus, leave with either the greater petrosal nerve or chorda tympani nerve, and travel to an autonomic ganglion before innervating their respective glands.
Conditions caused by nerve damage	Decreased tearing (dry eye) and decreased salivation (dry mouth); loss of taste sensation to anterior two-thirds of tongue; nerve paralysis (also known as Bell palsy; see Clinical View 11.2: "Idiopathic Facial Nerve Paralysis") characterized by paralyzed facial muscles, lack of orbicularis oculi contraction, sagging at corner of mouth
How to test for nerve damage	Have patient smile, blink, and squint—inability to move facial muscles on one side indicates nerve damage.

(continued on next page)

Table 13.5	**Cranial Nerves (continued)**

CN VIII VESTIBULOCOCHLEAR NERVE (ves-tib′ū-lō-kok′lē-ăr; relating to vestibule and cochlea of the ear)

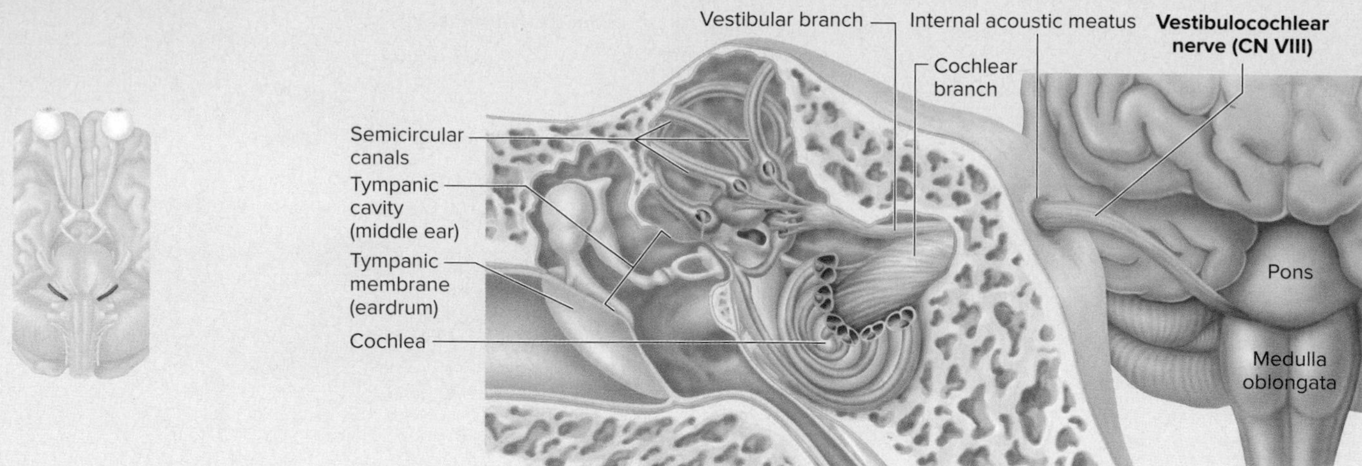

Description	Sensory nerve with two branches that conducts equilibrium and auditory (hearing) sensations from inner ear to brain
Sensory function	Vestibular branch conducts nerve signals for equilibrium, while cochlear branch conducts nerve signals for hearing
Origin	Vestibular branch: Hair cells in the vestibule of the inner ear Cochlear branch: Cochlea of the inner ear
Pathway	Sensory cell bodies of the vestibular branch are located in the vestibular ganglion, whereas sensory cell bodies of the cochlear branch are located in the spiral ganglion near the cochlea. The vestibular and cochlear branches merge, and together enter cranial cavity through internal acoustic meatus and travel to junction of the pons and the medulla oblongata.
Conditions caused by nerve damage	Lesions in vestibular branch produce loss of balance, nausea, vomiting, and dizziness; lesions in cochlear branch result in deafness (loss of hearing).
How to test for nerve damage	Test hearing.

CN IX GLOSSOPHARYNGEAL NERVE (glos′ō-fă-rin′jē-ăl; *glossa* = tongue, *pharynx* = throat)

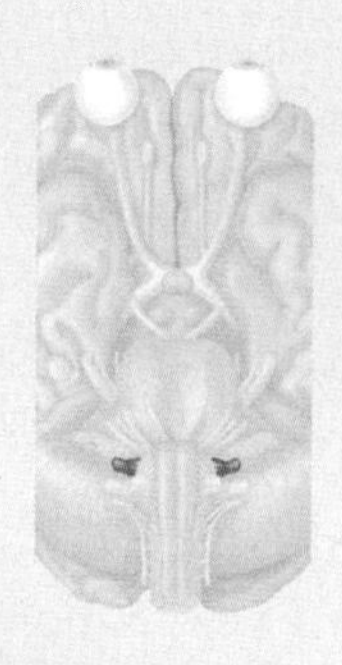

Description	Mixed nerve that receives taste and touch sensations from posterior one-third of the tongue; innervates one pharynx muscle and the parotid salivary gland
Sensory function	General sensation and taste from the posterior one-third of tongue; general sensation from most of pharynx; relays sensory information from the carotid arteries—from both chemoreceptors (which monitor blood levels of CO_2, H^+, and O_2) and baroreceptors (which monitor blood pressure)
Somatic motor function	Contracts a pharynx muscle (stylopharyngeus) for swallowing
Parasympathetic motor function	Increases secretion of parotid salivary gland
Origin	Sensory axons originate on taste buds and mucosa of posterior one-third of tongue, as well as the carotid bodies. Motor axons originate in nuclei in the medulla oblongata.
Pathway	Sensory axons travel from posterior one-third of tongue and carotid bodies along nerve through the inferior or superior ganglion into the jugular foramen, and travel to pons. Somatic motor axons leave cranium via jugular foramen and travel to stylopharyngeus. Parasympathetic motor axons travel to otic ganglion and then to parotid gland.
Conditions caused by nerve damage	Reduced salivary secretion (dry mouth); loss of taste sensations to posterior one-third of tongue
How to test for nerve damage	Have patient open mouth and say "ahhh"—the soft palate should elevate and the uvula should remain in the midline under normal conditions.

Table 13.5	Cranial Nerves (continued)

CN X VAGUS NERVE (vā′gŭs; wandering)

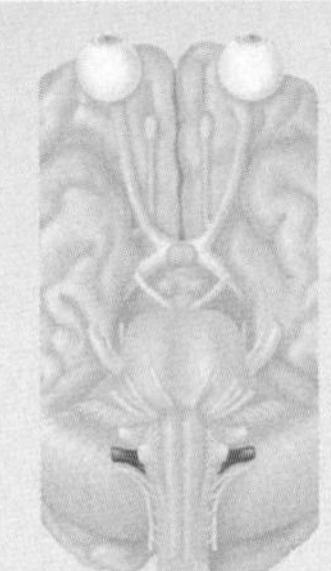

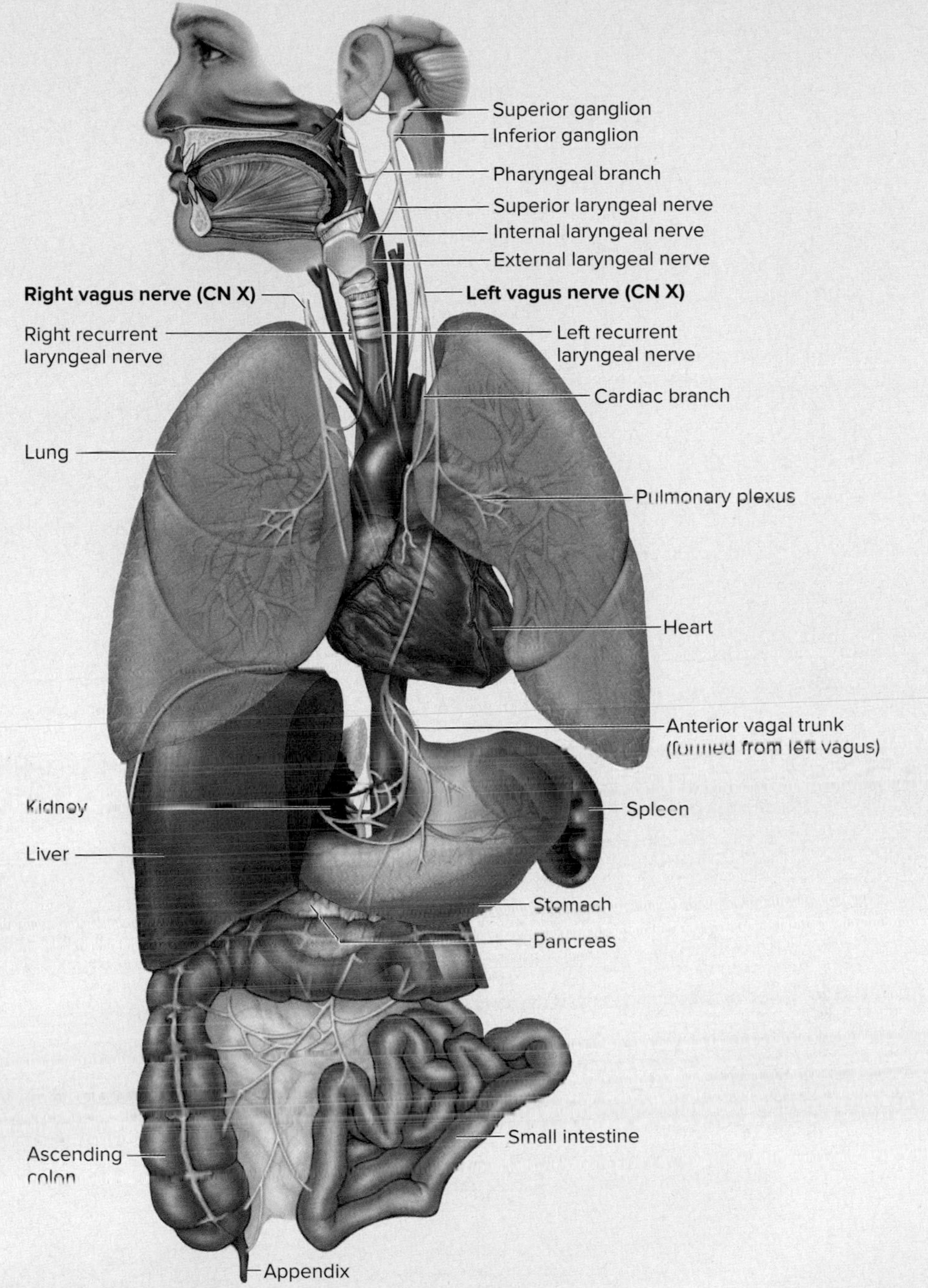

Description	Mixed nerve that innervates structures in the head and neck and in the thoracic and abdominal cavities
Sensory function	Visceral sensory information from heart, lungs, and most abdominal organs; general sensory information from external acoustic meatus, tympanic membrane (eardrum), laryngopharynx (inferior part of throat), and larynx (voice box)
Somatic motor function	Controls most pharynx muscles (for swallowing) and all larynx muscles (for production of speech)
Parasympathetic motor function	Innervates smooth muscle and glands of thoracic and most abdominal organs and cardiac muscle of the heart
Origin	Motor nuclei in medulla oblongata
Pathway	Leaves cranium via jugular foramen before traveling and branching extensively in neck, thorax, and abdomen; sensory neuron cell bodies are located in the superior and inferior ganglia associated with the nerve
Conditions caused by nerve damage	Paralysis leads to a variety of larynx problems, including hoarseness, monotone voice, or complete loss of voice. Other lesions may cause difficulty in swallowing or impaired gastrointestinal tract motility.
How to test for nerve damage	Ask patient if he has difficulties in swallowing. Determine if voice is hoarse or monotone. Have patient open mouth and say "ahhh"— the soft palate should elevate and the uvula should remain in the midline under normal conditions.

(continued on next page)

Table 13.5 Cranial Nerves (continued)

CN XI ACCESSORY NERVE (ak-ses′ō-rē)

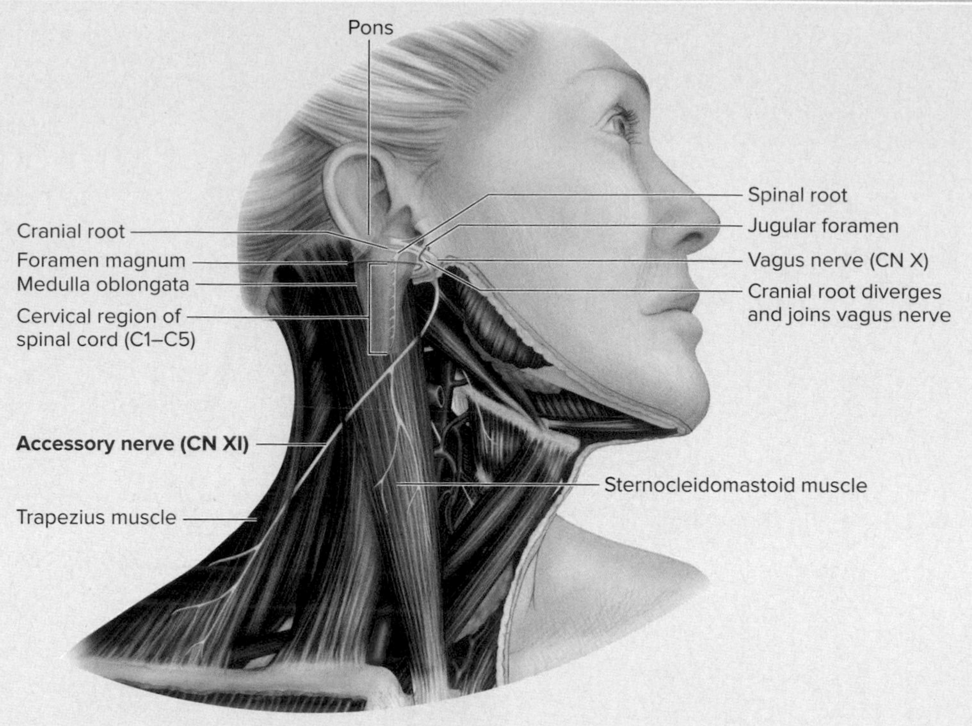

Description	Motor nerve that innervates trapezius and sternocleidomastoid muscles, also assists CN X to innervate pharynx muscles; formerly called the *spinal accessory nerve*
Somatic motor function	Cranial root: Travels with CN X to assist in innervating the pharynx muscles Spinal root: Contracts trapezius and sternocleidomastoid muscles
Origin	Cranial root: Nucleus in medulla oblongata Spinal root: Nucleus in spinal cord
Pathway	Spinal root travels superiorly to enter skull through foramen magnum; there, cranial and spinal roots merge and leave the skull via jugular foramen. Once outside the skull, cranial root splits to travel with CN X (vagus) to innervate pharynx muscles, and spinal root travels to sternocleidomastoid and trapezius.
Conditions caused by nerve damage	Paralysis of trapezius and sternocleidomastoid
How to test for nerve damage	Have patient elevate or shrug shoulders (tests trapezius function) or have patient turn head to opposite side (tests sternocleidomastoid function).

CN XII HYPOGLOSSAL NERVE (hī-pō-glos′ăl; *hypo* = below, *glossus* = tongue)

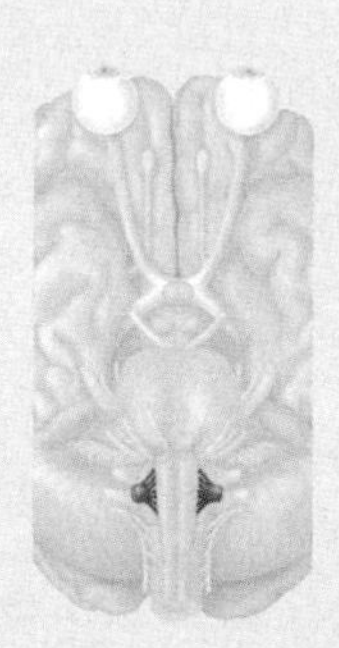

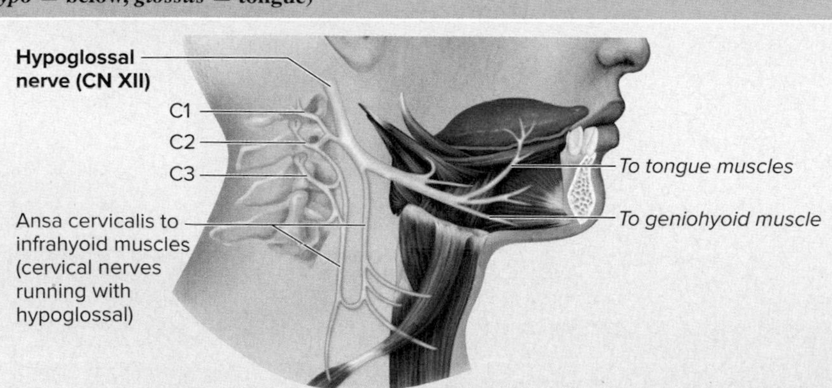

Description	Motor nerve that innervates both intrinsic and extrinsic tongue muscles
Somatic motor function	Contracts intrinsic and extrinsic tongue muscles to move tongue
Origin	Hypoglossal nucleus in medulla oblongata
Pathway	Leaves cranium via hypoglossal canal; travels inferior to mandible and to inferior surface of tongue
Conditions caused by nerve damage	Swallowing and speech difficulties due to impaired tongue movement
How to test for nerve damage	Have patient protrude (stick out) tongue: If a single hypoglossal nerve (either left or right) is paralyzed, a protruded (stuck-out) tongue deviates to the side of the damaged nerve.

CHAPTER SUMMARY

	• The brain is composed of the cerebrum, diencephalon, brainstem, and cerebellum. Twelve cranial nerves are associated with the brain.
13.1 Brain Organization and Development	**13.1a Overview of Brain Anatomy** • The brain has two cerebral hemispheres that exhibit folds called gyri with shallow sulci and deep fissures in between.
	13.1b Development of Brain Divisions • Three primary vesicles (prosencephalon, mesencephalon, and rhombencephalon) form from the neural tube by the late fourth week of development. • Five secondary vesicles (telencephalon, diencephalon, mesencephalon, metencephalon, and myelencephalon) form from the primary vesicles by the fifth week of development.
	13.1c Gray Matter and White Matter Distribution • Gray matter is composed primarily of dendrites and cell bodies of neurons and functions in processing and integrating information, whereas white matter is composed primarily of bundles of myelinated axons and functions in relaying nerve signals to and from gray matter.
13.2 Protection and Support of the Brain	• The brain is protected and isolated by the cranium, cranial meninges, cerebrospinal fluid, and a blood-brain barrier.
	13.2a Cranial Meninges • The cranial meninges are the pia mater, arachnoid mater, and dura mater. • The cranial dural septa are folds of dura mater that project between the major parts of the brain and stabilize the brain's position.
	13.2b Brain Ventricles • Fluid-filled spaces in the brain are the paired lateral ventricles, the third ventricle, the cerebral aqueduct, and the fourth ventricle.
	13.2c Cerebrospinal Fluid • Cerebrospinal fluid (CSF) is a clear fluid that provides buoyancy, protection, and a stable environment for the brain and spinal cord. • The choroid plexus (formed from ependymal cells and capillaries) produces CSF in the ventricles. • CSF leaves the ventricles and enters the subarachnoid space, where it circulates around the brain and spinal cord. Excess CSF returns to the venous circulation through the arachnoid villi.
	13.2d Blood-Brain Barrier • The blood-brain barrier regulates movement of materials between the blood and the interstitial fluid of the brain.
13.3 Cerebrum	• The cerebrum is the center of our sensory perception, thought, memory, judgment, and voluntary motor actions.
	13.3a Cerebral Hemispheres • The left and right cerebral hemispheres are separated by a longitudinal fissure.
	13.3b Lobes of the Cerebrum • Each hemisphere contains five lobes: four superficial lobes (frontal, parietal, temporal, occipital lobes) and the insula, which is not visible from the surface.
	13.3c Functional Areas of the Cerebrum • The primary motor cortex in the frontal lobe directs voluntary movements. • The primary somatosensory cortex in the parietal lobe collects somatic sensory information from skin and proprioceptors. • Other primary cortices and association areas are housed in each of the five lobes.
	13.3d Central White Matter • The central white matter contains three major groups of axons: association tracts, commissural tracts, and projection tracts.
	13.3e Cerebral Lateralization • The left hemisphere is the categorical hemisphere in most individuals, and the right is the representational hemisphere.
	13.3f Cerebral Nuclei • The cerebral nuclei are masses of gray matter located within the cerebrum.
13.4 Diencephalon	• The diencephalon is composed of the epithalamus, thalamus, and hypothalamus.
	13.4a Epithalamus • The epithalamus forms part of the posterior roof of the diencephalon; it contains the pineal gland (which secretes melatonin) and habenular nuclei (help relay signals from the limbic system to the midbrain).
	13.4b Thalamus • The thalamus is the main relay point for integrating, assimilating, and amplifying sensory signals sent to the cerebrum.
	13.4c Hypothalamus • The hypothalamus oversees the endocrine and autonomic nervous systems and houses many control and integrative centers.

(continued on next page)

CHAPTER SUMMARY (continued)

13.5 Brainstem	• The brainstem is composed of the midbrain, pons, and medulla oblongata.
	13.5a Midbrain • The midbrain contains cerebral peduncles, medial lemniscus, substantia nigra, tegmentum, tectal plate, and nuclei for two cranial nerves.
	13.5b Pons • The pons contains axon tracts, the pontine respiratory center, and nuclei of four cranial nerves.
	13.5c Medulla Oblongata • The medulla oblongata connects the brain to the spinal cord. It contains a cardiovascular center, a medullary respiratory center, sensory processing centers, and nuclei for four cranial nerves.
13.6 Cerebellum	**13.6a Structural Components of the Cerebellum** • The cerebellum is composed of left and right cerebellar hemispheres, with the vermis in between. • Cerebellar peduncles are thick axon tracts that connect the cerebellum to different parts of the brainstem.
	13.6b Functions of the Cerebellum • The cerebellum helps maintain posture and balance and fine-tunes skeletal muscle contractions, leading to smooth, coordinated movement.
13.7 Functional Brain Systems	• A functional brain system consists of separate components of the brain that work together toward a common function. Examples include the limbic system and reticular formation.
	13.7a Limbic System • The limbic system (i.e., the emotional brain) includes a group of structures that surround the corpus callosum and thalamus. The limbic system functions in memory and emotional behavior.
	13.7b Reticular Formation • The reticular formation is gray matter that extends through the brainstem. It participates in cyclic activities such as arousing the cortex to consciousness and controlling the sleep-wake cycle.
13.8 Integrative Functions and Higher-Order Brain Functions	**13.8a Development of Higher-Order Brain Functions** • Higher-order functions mature and increase in complexity as development proceeds.
	13.8b Electroencephalogram • An electroencephalogram monitors brain activity by measuring brain waves through the use of electrodes.
	13.8c Sleep • Sleep is a period of rest for the brain and involves cycles of REM (rapid eye movement) and non-REM (non-rapid eye movement) activity.
	13.8d Cognition • Mental processes such as awareness, knowledge, memory, perception, and thinking are collectively called cognition.
	13.8e Memory • Memory is a higher-order brain function involving the storage and retrieval of information gathered through previous activities.
	13.8f Emotion • Emotion is controlled by the limbic system and is regulated by the prefrontal cortex.
	13.8g Language • The motor speech area initiates a specific motor program for the movements involved in speech, whereas the Wernicke area is responsible for recognition of spoken and written language.
13.9 Cranial Nerves	• Twelve pairs of nerves, called cranial nerves, project from the brain. Each nerve has a specific name and function and is designated by a Roman numeral.

CHALLENGE YOURSELF

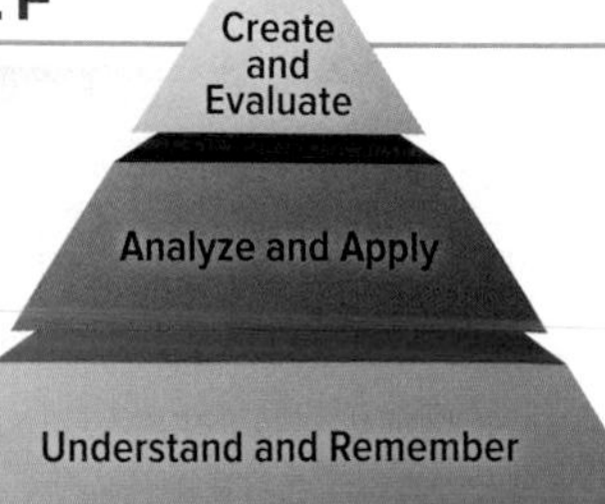

Do You Know the Basics?

1. Which cranial nerve is responsible for innervating the intrinsic and extrinsic tongue muscles?
   a. accessory (CN XI)
   b. glossopharyngeal (CN IX)
   c. trigeminal (CN V)
   d. hypoglossal (CN XII)

2. The subdivision of the brain that does not initiate somatic motor movements, but rather coordinates and fine-tunes those movements, is the
   a. medulla oblongata.
   b. cerebrum.
   c. cerebellum.
   d. diencephalon.

3. Which of these is the least likely to affect information transfer from STM (short-term memory) to LTM (long-term memory)?
   a. emotional state
   b. repetition or rehearsal
   c. auditory association cortex
   d. cerebral nuclei
4. All of the following are functions of the hypothalamus *except*
   a. controls endocrine system.
   b. regulates sleep-wake cycle.
   c. controls autonomic nervous system.
   d. initiates voluntary skeletal muscle movement.
5. All of the following statements are accurate about the choroid plexus *except*
   a. it is located within the ventricles of the brain.
   b. it is composed of ependymal cells and blood capillaries.
   c. it receives and filters all sensory information.
   d. it produces and circulates cerebrospinal fluid.
6. The ________ are descending motor tracts on the anterolateral surface of the midbrain.
   a. cerebral peduncles
   b. inferior colliculi
   c. pyramids
   d. tegmenta
7. Which cerebral lobe is located immediately posterior to the central sulcus and superior to the lateral sulcus?
   a. frontal lobe
   b. parietal lobe
   c. temporal lobe
   d. occipital lobe
8. The primary motor cortex is located in which cerebral structure?
   a. precentral gyrus
   b. postcentral gyrus
   c. motor speech area
   d. prefrontal cortex
9. The ________ are the isolated, innermost gray matter areas near the base of the cerebrum, inferior to the lateral ventricles.
   a. auditory association areas
   b. cerebral nuclei
   c. substantia nigra
   d. corpus callosum axons
10. Which structure contains autonomic nuclei involved in regulating respiration?
    a. pons
    b. superior colliculi
    c. cerebellum
    d. thalamus
11. Describe (a) how and where the cerebrospinal fluid is formed, (b) its subsequent circulation, and (c) how and where it is reabsorbed into the vascular system.
12. Which specific area of the brain may be impaired if you cannot tell the difference between a smooth and a rough surface using your hands only?
13. What activities occur in the visual association area?
14. Describe the relationship between the cerebral nuclei and the cerebellum in motor activities.
15. List the functions of the hypothalamus.
16. Describe the pathway by which the pressure applied to the right hand during a handshake is transmitted and perceived in the left primary somatosensory cortex.
17. Identify the components of the limbic system.
18. During surgery to remove a tumor from the occipital lobe of the left cerebrum, a surgeon must cut into the brain to reach the tumor. List in order (starting with the covering skin) all the layers that must be cut through to reach the tumor.
19. What is the difference between apraxia of speech and aphasia?
20. Which cranial nerves are associated with some aspect of eye movements or vision?

Can You Apply What You've Learned?

Use the following paragraph to answer questions 1 and 2.

Alex went to the dentist to get a cavity filled. The dentist used an anesthetic to numb the teeth prior to drilling.

1. Which nerve likely was anesthetized?
   a. CN IV (trochlear)
   b. CN V (trigeminal)
   c. CN VII (facial)
   d. CN IX (glossopharyngeal)
2. After the filling was inserted, Alex's gums on the same side of the tooth remained numb for a short while. What other problems may Alex temporarily experience until the anesthetic wears off?
   a. inability to close the mouth
   b. inability to protrude the tongue
   c. dry mouth
   d. numbness of the lips

Use the following paragraph to answer questions 3 and 4.

Shannon was the pitcher on her softball team. During one game, a batter hit the ball and it ricocheted off the left side of Shannon's head, knocking her temporarily unconscious. She eventually regained consciousness but experienced severe pain near her left temple. Within a few hours, Shannon had trouble moving her right upper limb and became lethargic. Her team captain took her to the emergency room, where the physician diagnosed her with an epidural hematoma.

3. An epidural hematoma causes an accumulation of blood to develop between what two structures?
   a. the skull and the periosteal layer of the dura mater
   b. the periosteal and meningeal layers of the dura mater
   c. the meningeal layer of the dura mater and the arachnoid mater
   d. the arachnoid mater and the pia mater
4. Shannon experienced problems with moving her right upper limb because the hematoma likely was impinging on what brain structure?
   a. left precentral gyrus
   b. left postcentral gyrus
   c. left cerebral nuclei
   d. left cerebellum

5. A 25-year-old male named Carlos went to the optometrist with the complaint of double vision (diplopia). The optometrist performed various eye tests on Carlos. Carlos was able to read an eye chart with each eye but experienced the double vision when he tried to use both eyes to focus. The optometrist noticed that when he had Carlos look laterally with each eye, Carlos's right eye did not move as far laterally as his left could. Based on these tests, the optometrist suspected that the muscle innervated by _________ was not working properly.
   a. CN II (optic)
   b. CN III (oculomotor)
   c. CN IV (trochlear)
   d. CV VI (abducens)

Can You Synthesize What You've Learned?

1. Peyton felt strange when she awoke one morning. She could not hold a pen in her right hand when trying to write an entry in her diary, and her muscles were noticeably weaker on the right side of her body. Additionally, her husband noticed that she was slurring her speech, so he took her to the emergency room. What does the ER physician suspect has occurred? Where in the brain might the physician suspect that abnormal activity or perhaps a lesion is located, and why?
2. Parkinson disease is the result of decreased levels of the neurotransmitter dopamine in the brain. However, these patients cannot take dopamine in drug form because the drug cannot reach the brain. What anatomic structure prevents the drug from reaching the brain? How could this anatomic structure be beneficial to an individual under normal circumstances?
3. During a robbery at his convenience store, Dustin was shot in the right cerebral hemisphere. He survived, although some specific functions were impaired. Would Dustin have been more likely or less likely to have survived if he had been shot in the medulla oblongata? Why?

INTEGRATE

ONLINE STUDY TOOLS

connect | SMARTBOOK® |

The following study aids may be accessed through Connect.

Clinical Case Study: A Case of the Shaking Palsy in an Elderly Woman

Interactive Questions: This chapter's content is served up in a number of multimedia question formats for student study

SmartBook: Topics and terminology include brain development and organization; protection and support of the brain; cerebrum; diencephalon; brainstem; cerebellum; functional brain systems; integrative functions and higher-order brain functions; cranial nerves

Anatomy & Physiology Revealed: Topics include brain; divisions of the brain; meninges; CSF flow; MRI; thalamus; hypothalamus; brainstem; cerebral peduncle; cerebellum; hippocampus; inferior brain

Nervous System: Spinal Cord and Spinal Nerves

Module 7: Nervous System

INTEGRATE

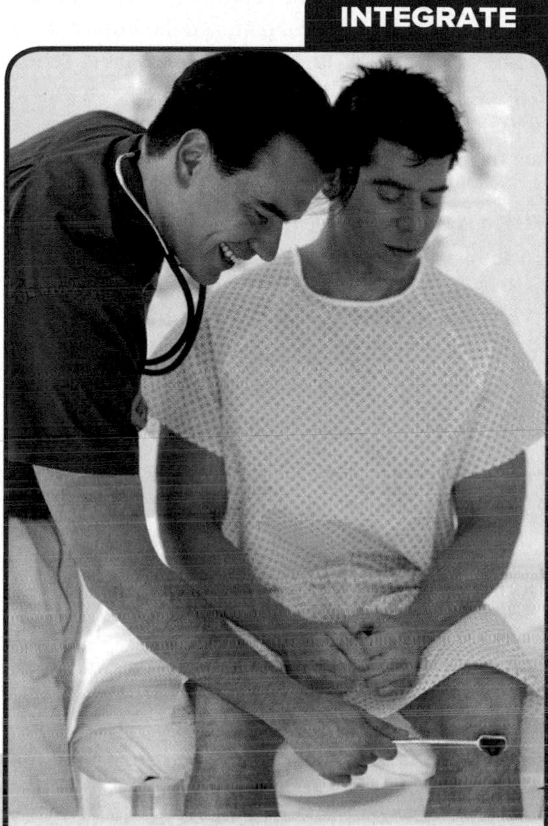

©Creatas Images/PunchStock RF

CAREER PATH
Neurologist

A neurologist is a medical doctor who specializes in the diagnosis and treatment of nervous system disorders. A typical neurologic exam involves testing a patient's reflexes and detecting the patient's perception of various sensations. Hyperactive or hypoactive reflexes, or loss of sensation in specific dermatomes, may help pinpoint disorders in individual nerves, specific segments of the spinal cord, or regions of the brain. The neurologist has a thorough knowledge of the central nervous system, as well as motor and sensory components of the body's nerves, and uses this information to aid in an accurate diagnosis of the patient.

The spinal cord is about 18 inches long and is about as wide as a piece of rope. It is continuously relaying (1) sensory input from the body to the brain as well as (2) motor output from the brain to the body. Thus, it serves as the means of communication between the body and the brain. The spinal cord, which is part of the central nervous system (CNS), is associated with 31 pairs of spinal nerves. We explore the structure and function of the spinal cord and spinal nerves in this chapter.

14.1 Overview of the Spinal Cord and Spinal Nerves

Our study of the spinal cord and spinal nerves begins by providing an overview of their two primary functions. We then describe the general structure of both the spinal cord and the spinal nerves, which will help you to begin to integrate how these nervous system components operate together in performing these functions.

14.1a General Functions

LEARNING OBJECTIVE

1. Describe the two primary functions of the spinal cord and spinal nerves.

The spinal cord and its attached spinal nerves serve two important functions. Their first function is to provide an essential structural and functional link between the brain and the torso and limbs of the body. Sensory input is relayed from the torso and limbs to the brain, and motor output is relayed from the brain to the torso and limbs. These vital inputs and outputs are relayed along neuron *pathways* that are within the spinal cord and spinal nerves. For example, when you pick up an object, sensory information about that object (shape, weight, temperature) is relayed along one or more spinal nerves and then through the spinal cord to reach the brain, where the information is interpreted. Additionally, the movement of your limbs (e.g., while you are walking) is controlled by nerve signals initiated in the brain; these nerve signals are then relayed through the spinal cord and then the spinal nerves to the skeletal muscles of your arms and legs. Notice that *both* sensory input and motor output are relayed along the pathways within the spinal cord and spinal nerves.

The second important function of the spinal cord and spinal nerves is their role in *spinal reflexes.* These involve nervous system responses that do not require the involvement of the brain, but instead have the spinal cord as the integration center. Spinal reflexes initiate our quickest reactions to a stimulus. It is through spinal reflexes that the spinal cord exhibits some functional independence from the brain. Consider that when you pull your hand back in response to a hot stimulus (e.g., touching a hot stove), this involves a spinal reflex. It is only when this sensory input reaches the brain that you consciously perceive that it is painful. The functional pathways and the spinal reflexes are discussed in detail in this chapter.

WHAT DID YOU LEARN?

What are the two primary functions of the spinal cord and spinal nerves?

14.1b Spinal Cord Gross Anatomy

LEARNING OBJECTIVE

2. Describe the general structure of the spinal cord and its four anatomic subdivisions.

The **spinal cord** is a roughly cylindrical nervous system structure that extends through the vertebral column to the inferior border of the L1 vertebra (**figure 14.1**). The superior end of the spinal cord is continuous with the medulla oblongata of the brain, and its inferior end tapers (narrows) to form the **conus medullaris** (kō′nŭs med′ū-lār′is; *kōnos* = cone, *medulla* = middle). A typical adult spinal cord is approximately ¾ inch in diameter and ranges between 16 and 18 inches (42 and 45 centimeters) in length. (These dimensions are similar to a width of your finger that would have the length from your fingertip to your elbow.) Observe in figure 14.1 that the spinal cord does *not* extend the entire length of the vertebral column, but ends at approximately the superior border of the small of your back. This is because growth of the individual vertebrae continues longer than the growth of the spinal cord; thus, an adult spinal cord is shorter than the vertebral column.

Two longitudinal depressions extend the full length of the spinal cord. The **posterior median sulcus** is a narrow groove on the posterior surface, whereas the **anterior median fissure** is a slightly wider groove on the anterior surface (not shown in figure 14.1).

There are four continuous subdivisions (parts) of the spinal cord (from superior to inferior): (1) the **cervical part,** which is continuous with the medulla oblongata, (2) the **thoracic part,** (3) the **lumbar part,** and (4) the **sacral part.** (Some references further divide the sacral part into a sacral part and a coccygeal part.) Notice in figure 14.1 that two areas of the spinal cord are wider than other cord areas: The **cervical enlargement** is the wider area in the cervical part, and the **lumbosacral enlargement** is the wider area in the lumbar and sacral parts. These regions of the spinal cord are enlarged due to the presence of a greater number of neurons within the spinal nerves extending from these spinal cord parts to innervate the upper and lower limbs, respectively.

WHAT DID YOU LEARN?

What are the general shape, diameter, and length of the spinal cord, and where is it housed?

14.1c Spinal Nerve Identification and Gross Anatomy

LEARNING OBJECTIVES

3. Discuss the naming of the 31 pairs of spinal nerves, and provide a general description of a spinal nerve and its composition.
4. Explain how the cauda equina arises during development.

General Description of a Nerve

Nerves were described in detail in section 12.1c. Recall that a **nerve** is an organ composed of a cablelike bundle of axons, which are enclosed within successive connective tissue wrappings. These include the **epineurium**, which ensheathes the entire nerve; the **perineurium,** which encloses each fascicle (bundle) of axons; and the **endoneurium,** which electrically insulates each axon (see figure 12.2*a, b*).

Naming Spinal Nerves

The spinal cord is associated with *31 pairs* of **spinal nerves** (figure 14.1*a*). Each spinal nerve is typically identified by the first letter of the spinal cord part to which it attaches, followed by a number. Thus, each side of the spinal cord contains 8 cervical nerves (called C1–C8), 12 thoracic nerves (T1–T12), 5 lumbar nerves (L1–L5), 5 sacral nerves (S1–S5), and 1 coccygeal nerve (Co1). Spinal nerve names are readily distinguished from cranial nerve names (discussed in section 13.9) because cranial nerves are designated by either CN followed by a roman numeral (e.g., CN I, CN II) or a specific name (e.g., olfactory nerve, optic nerve). The nerve plexuses (which are labeled in figure 14.1*a,* such as the brachial plexus) are extensions of spinal nerves and are discussed in section 14.5.

Gross Anatomy of a Spinal Nerve

Each spinal nerve anchors to the spinal cord by two roots, a posterior root and an anterior root, and each of these roots is composed of multiple **rootlets** (**figure 14.2**). The **posterior root** houses sensory neurons (see section 12.2d) that extend from sensory receptors. These sensory neurons relay nerve signals from the sensory receptors *to* the spinal cord. Observe that the sensory neurons that compose the spinal nerves are *unipolar neurons* (see section 12.2d). The dendrites of

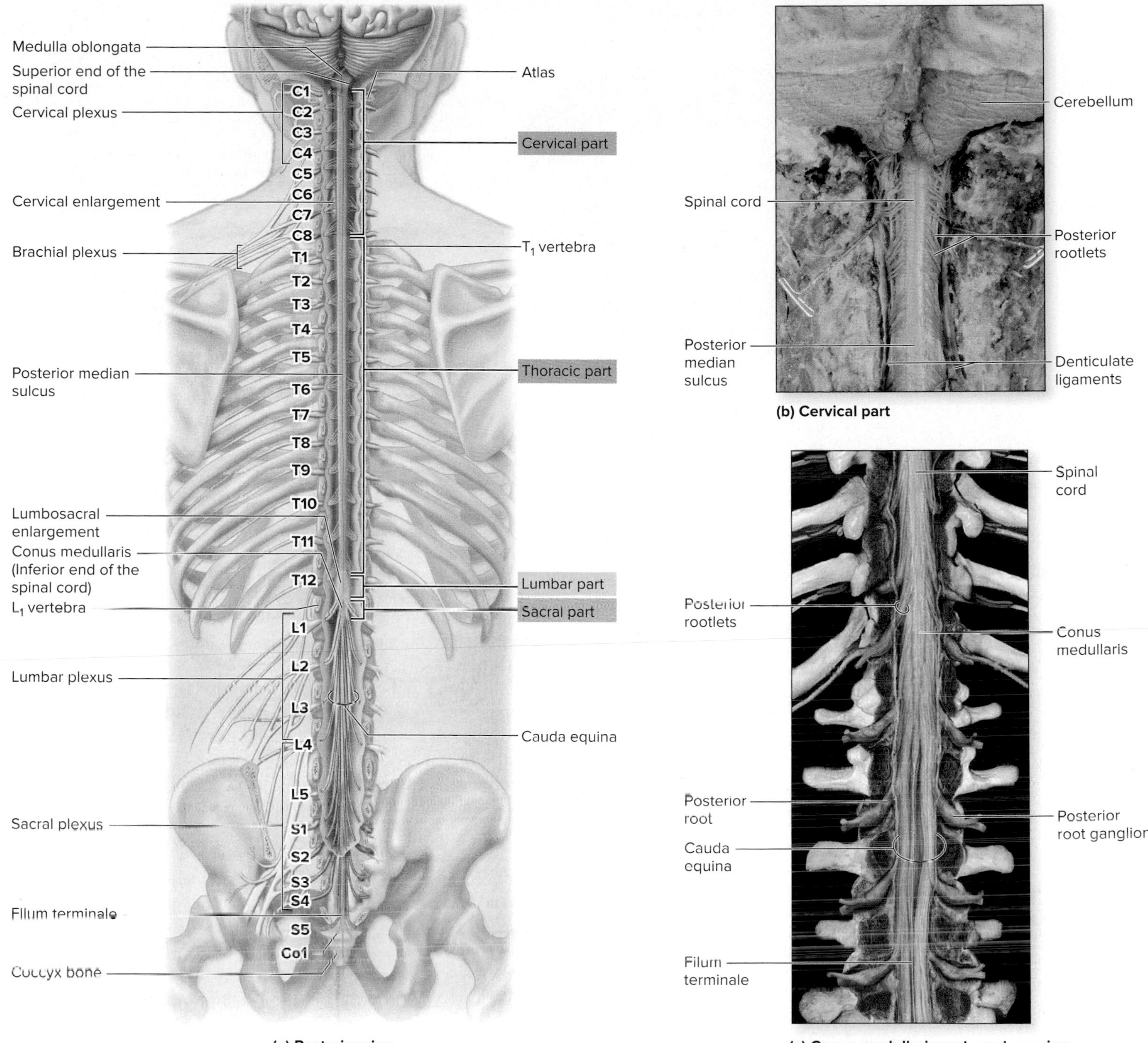

(a) Posterior view

(b) Cervical part

(c) Conus medullaris and cauda equina

Figure 14.1 Gross Anatomy of the Spinal Cord and Spinal Nerves. The spinal cord extends inferiorly from the medulla oblongata through the vertebral canal. (*a*) The vertebral arches have been removed to reveal the anatomy of the adult spinal cord and its spinal nerves. (*b*) Cadaver photo of the cervical part of the spinal cord. (*c*) Cadaver photo of the conus medullaris and the cauda equina. AP|R

these sensory neurons form the sensory receptors (see section 16.1b), and their axons extend from the dendrites to the spinal cord. It is critical to note that the cell bodies of these sensory neurons (which are positioned along the length of the axon) are located external to the spinal cord and form the **posterior root ganglion** (see figure 12.2*c*).

The **anterior root** contains motor neurons that extend to effectors (muscle or glands; figure 14.2). These motor neurons relay nerve signals *from* the spinal cord and control muscles and glands. Observe that the motor neurons that compose the spinal nerves are *multipolar neurons* (see section 12.2d). Both the dendrites and the cell bodies of motor neurons, unlike those of sensory neurons, are housed within the spinal cord; thus, the anterior root does *not* contain a ganglion along its length. Motor neuron axons exit from the spinal cord within the anterior root and extend through the spinal nerve to their terminal ends, which innervate an effector. Thus, a significant difference between anterior and posterior roots is that each *anterior root* lacks a ganglion along its length. This is because, as mentioned, the dendrites and cell bodies of motor neurons are within the spinal cord and the anterior root contains only the axons of these neurons.

Each *spinal nerve* forms where the posterior root (containing sensory neurons) and the anterior root (containing motor neurons) join. Thus, both sensory and motor neurons compose each spinal nerve, and it is classified as a **mixed nerve** (see section 12.1c). If you compare a spinal nerve to a cable composed of multiple wires, the "wires" within a spinal nerve are the sensory and motor axons, and each "wire" transmits signals in one direction only.

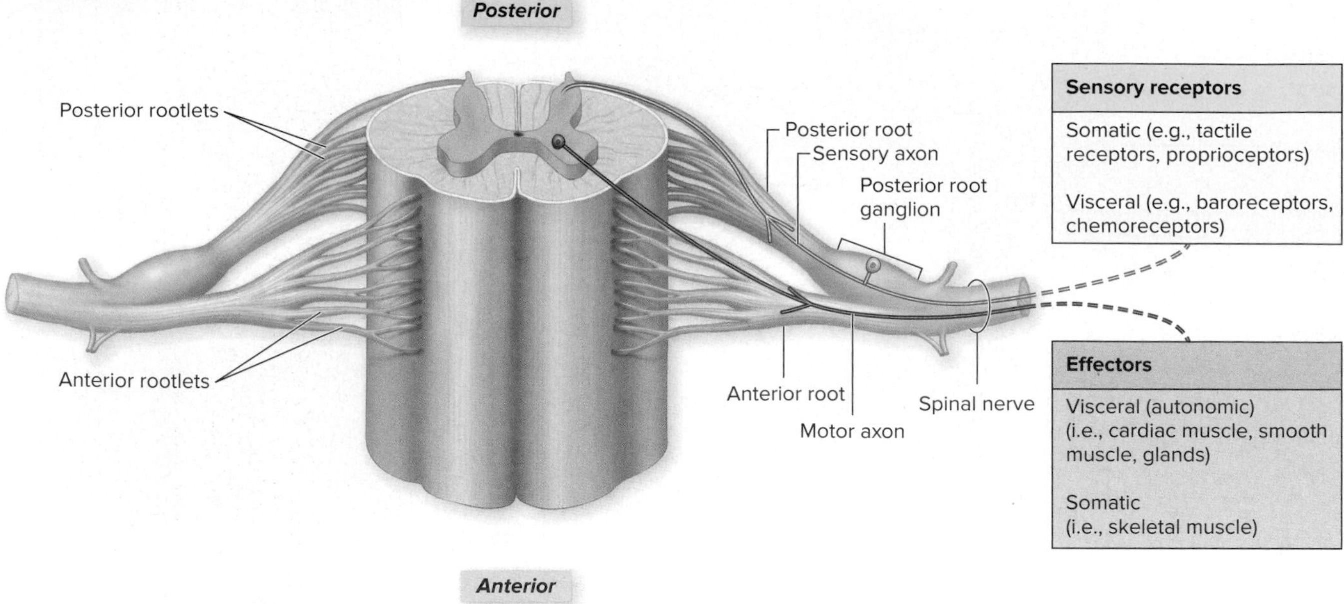

Figure 14.2 **Spinal Roots and Spinal Nerve.**

Cauda Equina

The spinal cord does not extend the entire length of the vertebral column, but typically ends at the inferior border of the L_1 vertebra, as described (figure 14.1). Consequently, roots of the lumbar, sacral and coccygeal spinal nerves do not extend horizontally from the spinal cord (as illustrated in figure 14.2 for the more superior spinal nerves). Instead, these spinal nerve roots extend inferiorly from the conus medullaris until the specific location where they exit the vertebral column (figure 14.1*a, c*). Collectively, these spinal nerve roots form a structure called the **cauda equina** (kaw′dă ē-kwī′nă). They are so named because they resemble a horse's tail (*cauda* = tail, *equus* = horse). Note that the conus medullaris (at the L_1 vertebra) represents (1) the inferior end of the spinal cord and (2) the most superior portion of the spinal roots forming the cauda equina.

WHAT DID YOU LEARN?

3. What is the total number of spinal nerves, and how are they specifically identified?
4. How is the cauda equina formed? What composes the cauda equina?

14.2 Protection and Support of the Spinal Cord

LEARNING OBJECTIVES

5. Discuss the relationship of the spinal cord and spinal nerves to the vertebral column.
6. Describe the locations and functions of the spinal cord meninges, and compare and contrast the three spaces associated with the spinal cord meninges.

The spinal cord is collectively protected by a bony structure, meninges, and cerebrospinal fluid, just like the brain.The bony framework that houses the spinal cord is the **vertebral column** (see figure 8.1 in section 8.5a). It is formed by 26 stacked vertebrae and the intervertebral discs between vertebrae. The vertebral column physically protects the spinal cord and is flexible enough to allow for movement of the torso. All of the stacked vertebral foramina collectively form the **vertebral canal,** which houses both the spinal cord and the cauda equina. Note that the different parts of the spinal cord described in section 14.1b do not match up exactly with the vertebrae of the same name (figure 14.1*a*). For example, the lumbar part of the spinal cord is actually closer to the inferior thoracic vertebrae than to the lumbar vertebrae. This apparent discrepancy is due to the continued growth of individual vertebrae after spinal cord growth is complete.

Each spinal nerve exits the vertebral column through an **intervertebral foramen** (which is a lateral opening between two adjacent stacked vertebrae; **figure 14.3*a***). Note that each of the more superior spinal nerves (i.e., cervical and thoracic) extends horizontally through its associated intervertebral foramen at the same level (as shown in figure 14.3*a*). The more inferior spinal nerves (i.e., lumbar, sacral and coccygeal) have roots that extend inferiorly as part of the cauda equina, and each of these spinal nerves then extends through its intervertebral foramen, which is inferior to where the roots are anchored to the spinal cord (see figure 14.1*a, c*).

You might wonder why there are eight cervical nerves and only seven cervical vertebrae (see section 8.5c). The first seven cervical spinal nerves (C1–C7) exit the vertebral canal and extend through an intervertebral foramen that is *superior* to the vertebra of the same number. For example, the C2 spinal nerve exits the vertebral canal

INTEGRATE

LEARNING STRATEGY 

With one exception, the number of spinal nerves matches the number of vertebrae in a particular body region. For example, the 12 pairs of thoracic spinal nerves correspond to the 12 thoracic vertebrae. The sacrum forms from 5 sacral vertebrae, and there are 5 pairs of sacral spinal nerves. The coccygeal vertebrae tend to fuse into one structure, and there is 1 pair of coccygeal nerves. The exception to this rule is that there are 8 pairs of cervical spinal nerves, but only 7 cervical vertebrae. This is because the first cervical pair emerges inferior to the occipital bone and superior to the atlas (the first cervical vertebra), and the eighth cervical nerve arises inferior to the seventh cervical vertebra.

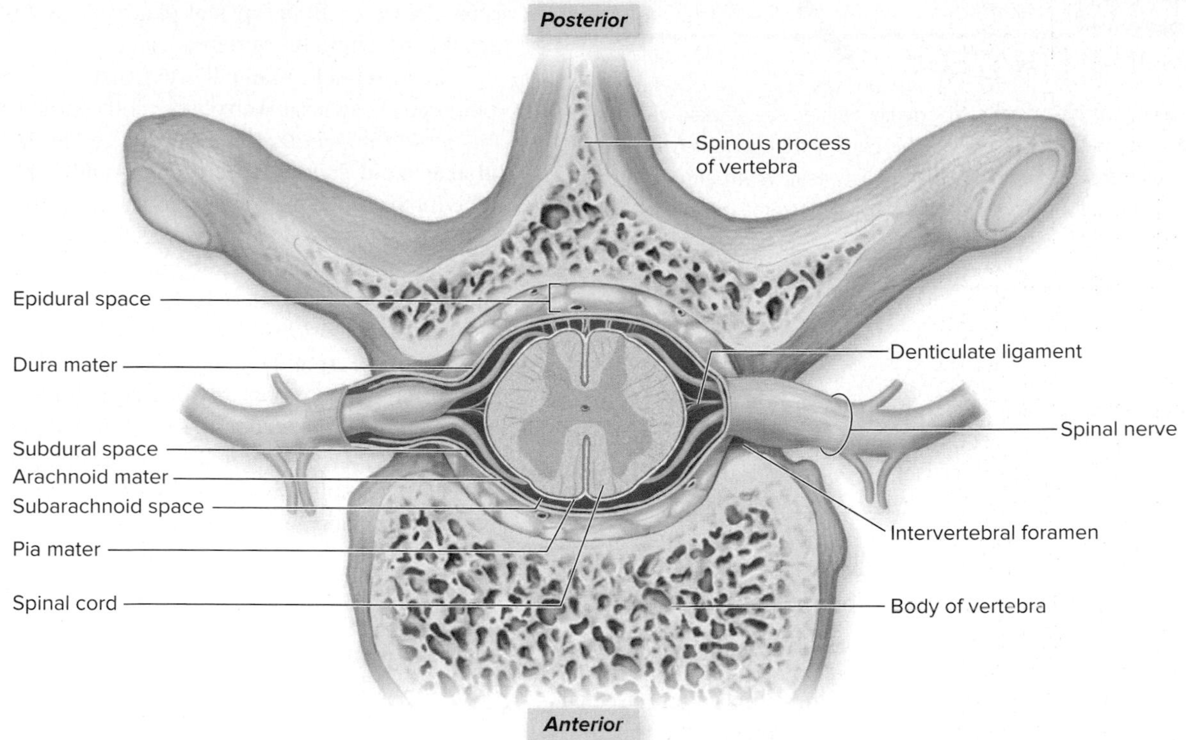

(a) Cross section of vertebra and spinal cord

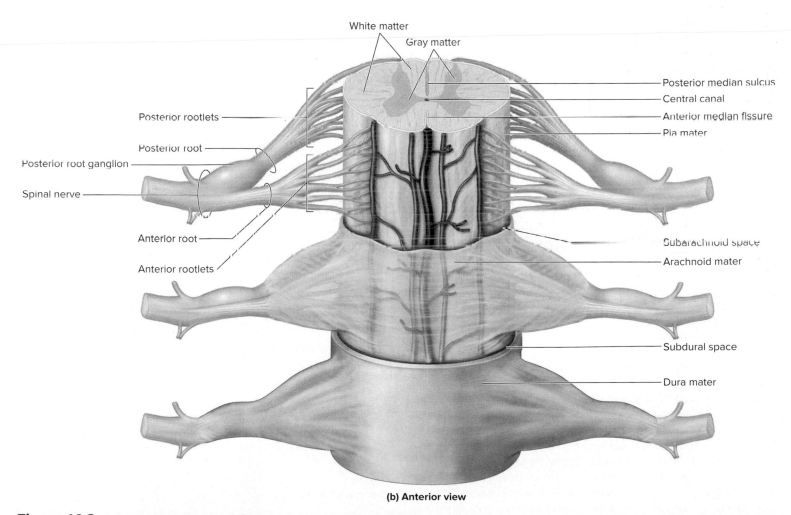

White matter
Gray matter
Posterior median sulcus
Central canal
Posterior rootlets
Anterior median fissure
Pia mater
Posterior root
Posterior root ganglion
Spinal nerve
Anterior root
Subarachnoid space
Arachnoid mater
Anterior rootlets
Subdural space
Dura mater

(b) Anterior view

Figure 14.3 Spinal Meninges and Structure of the Spinal Cord. (*a*) A cross section of the spinal cord shows the relationship between the meningeal layers and the superficial landmarks of the spinal cord and vertebral column. (*b*) Anterior view shows the spinal cord and meninges. AP|R

INTEGRATE

LEARNING STRATEGY

Keep in mind that the spinal cord extends through the stacked **vertebral foramina,** which medially form the vertebral canal, and the spinal nerves extend through **intervertebral foramina** (*inter* = between), which are the lateral openings formed between two adjacent stacked vertebrae on both the right side and left side of the vertebral column.

through the intervertebral foramen between the C_1 and C_2 vertebrae. The eighth cervical spinal nerve, in comparison, exits the intervertebral foramen inferior to the C_7 vertebra. All remaining spinal nerves inferior to the C8 nerve exit the vertebral canal and extend through an intervertebral foramen *inferior* to the vertebra of the same number. Thus, for example, the T2 spinal nerve exits the vertebral canal through the intervertebral foramen inferior to the T_2 vertebra.

The **spinal cord meninges** (mĕ-nin′jes, mē′nin-jēz; sing., *meninx,* men′ingks; membrane) are connective tissue membranes that protect and encapsulate the spinal cord within the vertebral canal. (They are continuous with the cranial meninges, described in section 13.2a.) The spinal cord meninges, layered from innermost to outermost, are: pia mater, arachnoid mater, and dura mater.

The **pia mater** directly adheres to the external surface of the spinal cord. It is the delicate, innermost meningeal layer, which is a meshlike membrane composed of both elastic and collagen fibers. Pia mater extensions form two structures: denticulate ligaments and the filum terminale. **Denticulate** (den-tik′ū-lāt; *dentatus* = toothed) **ligaments** are the numerous paired, triangular extensions present along the spinal cord. These pia mater extensions suspend and anchor the spinal cord laterally to the arachnoid and dura mater (see figures 14.1*b* and 14.3*a*). The **filum terminale** (fī′lŭm ter′mi-năl; *terminus* = end) is a thin strand of pia mater that anchors the conus medullaris to the coccyx bone. It extends within the cauda equina and can be viewed in figure 14.1*a, c*). Both types of pia mater extensions help stabilize the spinal cord within the vertebral canal.

The **arachnoid mater** lies external to the pia mater. It is partially composed of a delicate web of both collagen and elastic fibers termed the *arachnoid trabeculae.* Immediately deep to the arachnoid mater is the **subarachnoid space.** Cerebrospinal fluid (CSF) circulates within this space (both around the spinal cord and around the brain). Cerebrospinal fluid can be analyzed (e.g., for infectious agents) following its removal from the subarachnoid space by the clinical procedure called a lumbar puncture (see Clinical View 14.1: "Lumbar Puncture").

The outermost layer of meninges is the **dura mater.** It is composed of dense irregular connective tissue. The dura mater associated with the spinal cord has only one layer (unlike the dura mater covering the brain, which is composed of both a periosteal layer and meningeal layer; see section 13.2a). Extensions of the dura mater ensheathe the spinal nerve roots and merge with the connective tissue layer that surrounds the spinal nerves (i.e., the epineurium; see section 12.1c). Two spaces are associated with the dura mater: the subdural space and the epidural space. The **subdural space** is a potential space internal to the dura mater (between the arachnoid mater and dura mater). The **epidural space** is a space external to the dura mater. The epidural space is a clinically significant area that houses adipose and areolar connective tissue, as well as blood vessels. Epidural anesthetics, such as may be used to lessen pain during childbirth, are introduced into this space. (See Clinical View: 29.7: "Anesthetic Procedures to Facilitate True Labor.")

WHAT DO YOU THINK?

1 What are the similarities and differences among the meningeal layers and spaces that a hypodermic needle must pass through (or enter) for a *lumbar puncture* to remove cerebrospinal fluid from the subarachnoid space and for an *epidural* to deliver a drug to the epidural space?

WHAT DID YOU LEARN?

5 Where are the epidural, subdural, and subarachnoid spaces located? Which space contains CSF?

INTEGRATE

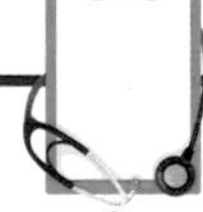

CLINICAL VIEW 14.1
Lumbar Puncture

It is sometimes necessary to analyze the cerebrospinal fluid (CSF) to determine whether an infection or a disorder of the central nervous system is present. The clinical procedure for obtaining CSF is known as a **lumbar puncture** (commonly referred to as a *spinal tap*). The needle must be inserted through the skin, back muscles, and ligamentum flavum (between vertebrae). Then, the needle must pass through the epidural space, dura mater, and arachnoid mater and enter the subarachnoid space to obtain approximately 3 to 9 milliliters of CSF.

Since the adult spinal cord typically ends at the level of the L_1 vertebra, a lumbar puncture must be performed inferior to this level to ensure that the needle does not pierce the spinal cord. A lumbar puncture typically is made at the level of either the L_3 and L_4 vertebrae or the L_4 and L_5 vertebrae. To locate this level, the physician palpates the highest points of the iliac crests, which are at the same horizontal level as the spinous process of the L_4 vertebra. The physician can then insert the lumbar puncture needle either directly above or directly below the spinous process of L_4 when the vertebral column is flexed.

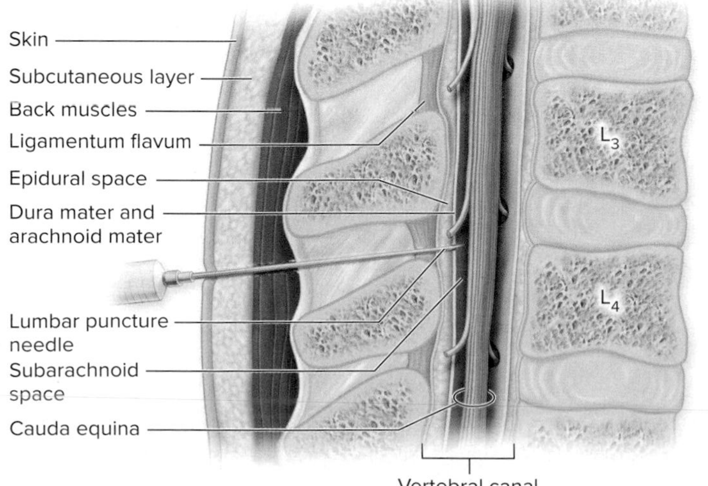

Site of needle insertion for a lumbar puncture.

14.3 Sectional Anatomy of the Spinal Cord and Spinal Roots

We now examine the spinal cord and a pair of associated spinal roots in cross section to explore their structural and functional relationship. The spinal cord as shown in **figure 14.4** appears in cross section as a roughly cylindrical structure that is slightly flattened both posteriorly and anteriorly. It has a relatively narrow posterior median sulcus and a slightly wider anterior median fissure that are readily visible. Observe that the spinal cord is partitioned into two areas: an inner gray matter region and an outer white matter region. Here we discuss the general composition and function of these regions.

14.3a Distribution of Gray Matter

LEARNING OBJECTIVES

7. Identify the four anatomic locations of gray matter on either side of the spinal cord.
8. Describe the structures that form each gray matter region.
9. Trace sensory input to the spinal cord and motor output from the spinal cord.

Gray matter was first discussed in section 13.1c, where it was described as (1) being primarily composed of the dendrites and cell bodies of neurons and (2) functioning as a processing center. The gray matter within the spinal cord is centrally located, and its shape resembles a letter H or a butterfly. The gray matter is subdivided into the following components on each side of the spinal cord: a posterior horn, a lateral horn, an anterior horn, and a bar of gray matter that connects the left and right sides called the gray commissure. The gray matter of each horn is discussed first. You will find it helpful to refer to figure 14.4*b* as you read through this section.

Posterior horns are both the left and right posterior masses of gray matter. The gray matter forming the posterior horns is due to the presence of the dendrites and cell bodies of *interneurons* (the neurons that are located completely within the CNS; see section 12.2d). Sensory neurons within the spinal nerves extend through the posterior root and synapse with the dendrites and cell bodies of the interneurons within the posterior horn. The posterior horn gray matter on each side of the spinal cord is subdivided regionally into both somatic sensory nuclei and visceral sensory nuclei based upon the specific type of sensory neurons that synapse there.

The **somatic sensory nuclei** (light blue–shaded region) is the site for synapses between *somatic sensory neurons* (light blue line) that extend from somatic sensory receptors (e.g., tactile receptors within the skin; see section 16.2a) and the interneurons within the posterior horns of the spinal cord. (Recall that *somatic* refers to "body" in general).

The **visceral sensory nuclei** (dark blue–shaded region) is the location for synapses between *visceral sensory neurons* (dark blue line) that extend from visceral sensory receptors (e.g., baroreceptors

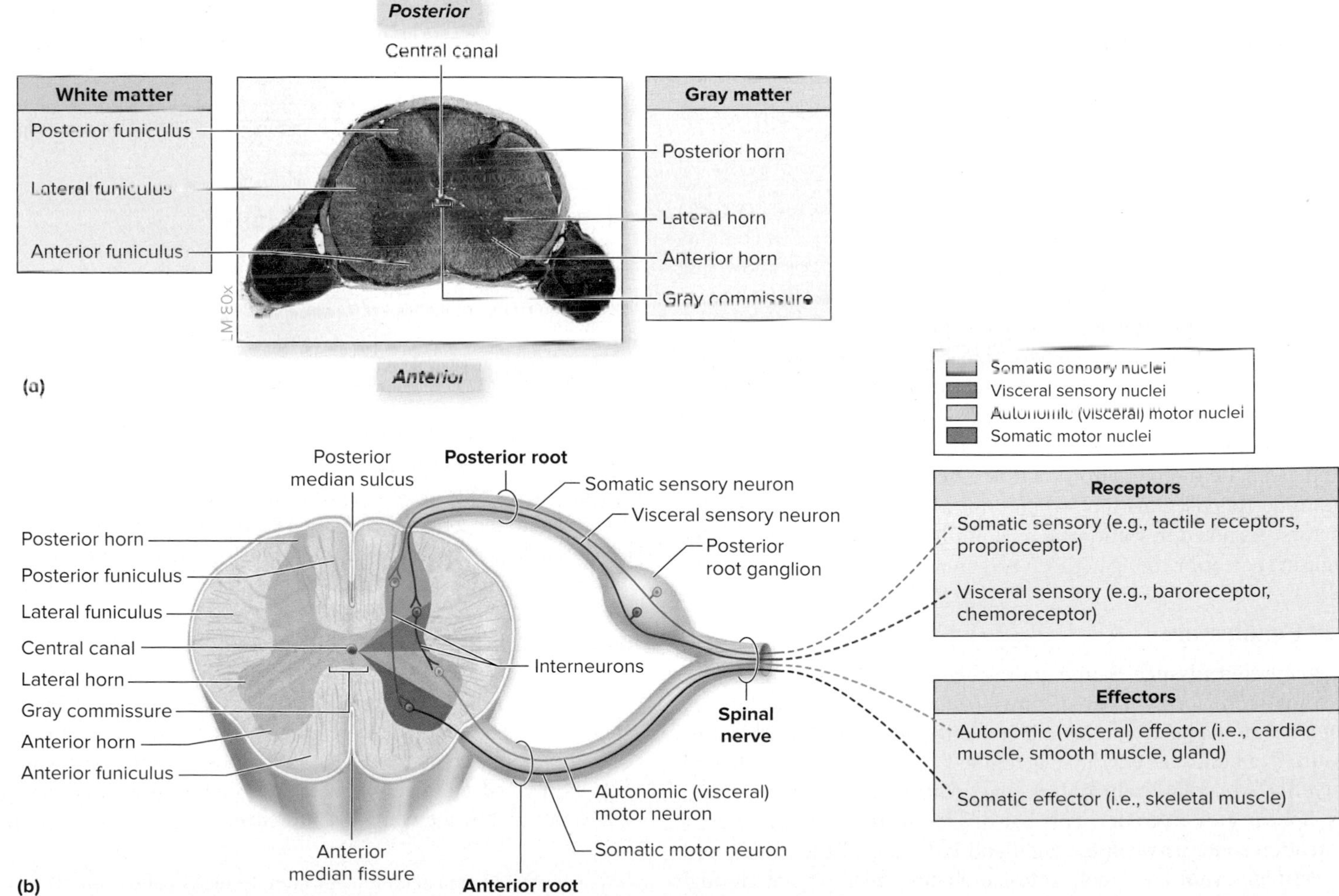

Figure 14.4 Cross Section of Spinal Cord and Spinal Roots. (*a*) Histology of a cross section of the spinal cord. (*b*) Illustration of a cross section of the spinal cord and spinal roots. The collections of neuron cell bodies within the CNS form specific nuclei. Neurons and their respective nuclei are color-coded on the drawing.

of the urinary wall) to the interneurons within the posterior horns of the spinal cord. (Recall that *visceral* refers to an internal organ.)

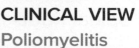

Anterior horns are both the left and right anterior masses of gray matter. The gray matter of the anterior horns is due to the presence of the dendrites and cell bodies of *somatic motor neurons*. Collectively, they form the **somatic motor nuclei** (red shaded area), which forms the entire anterior horn on each side of the spinal cord. The axons of *somatic motor neurons* (red line) extend to and innervate a somatic effector. The *somatic effector* includes only the muscle that can be controlled consciously or voluntarily (i.e., skeletal muscle). Note: A type of virus that specifically targets somatic motor neurons within the spinal cord and potentially result in muscle paralysis is the poliovirus (see Clinical View 14.2: "Poliomyelitis").

Lateral horns are both the left and right lateral masses of gray matter. They are located only within the T1–L2 parts of the spinal cord, not the entire length of the spinal cord. The gray matter of the lateral horns is due to the presence of the dendrites and cell bodies of *autonomic motor neurons*. Collectively, they form the **autonomic motor nuclei** (orange shaded region), which makes up the entire lateral horn on each side of the spinal cord. The axons of *autonomic motor neurons* (orange line) extend to and innervate autonomic (or visceral) effectors. *Autonomic effectors* include those body structures that are not controlled consciously or voluntarily (i.e., cardiac muscle, smooth muscle, and glands). (Examples of autonomic effectors include cardiac muscle of the heart, smooth muscle of the stomach wall, and exocrine glands of the pancreas. Autonomic effectors are components of the autonomic nervous system and are discussed in detail in chapter 15.)

WHAT DO YOU THINK?

2 How do the posterior horns differ in their composition and function from both the anterior horns and lateral horns?

Posterior Root and Anterior Root Composition

The composition of both the posterior root and the anterior root was first described in section 14.1c. Recall that the *posterior root* contains sensory neurons that extend from sensory receptors. We know that these sensory neurons include somatic sensory neurons, which relay nerve signals from somatic sensory receptors, and visceral sensory neurons, which relay nerve signals from visceral sensory receptors. Both types of sensory neurons synapse with interneurons within the gray matter of the posterior horns. In comparison, the *anterior root* contains motor neurons that extend to effectors. These motor neurons include somatic motor neurons, which extend to skeletal muscle, and autonomic (or visceral) motor neurons, which extend to autonomic effectors. As noted in section 14.1c, a posterior root has an associated ganglion (posterior root ganglion) and an anterior root does not.

Gray Commissure

The **gray commissure** (kom′i-shūr; *commissura* = a seam) within the spinal cord forms a bar of gray matter connecting the left and right sides of the posterior, lateral, and anterior horns. The gray commissure is an unusual gray matter region because it primarily houses *un*myelinated axons (which lack the whitish-colored myelin and, thus, appear gray in color). This bar of gray matter serves as a communication route between the right and left sides of the spinal cord. The **central canal** is a small, internal channel that extends through the center of the gray commissure along the entire length of the spinal cord. As with the ventricles of the brain, the central canal is formed during embryonic development from the neural canal within the neural tube (see section 14.7). The central canal contains cerebrospinal fluid (CSF), which enters this space from the fourth ventricle of the brain (see figure 13.7).

INTEGRATE

CLINICAL VIEW 14.2

Poliomyelitis

Poliomyelitis (pō′lē-ō-mī′e-lī′tis; *polio* = gray, *myelos* = marrow, *itis* = infection) is an infection caused by one of the three strains of poliovirus. Infection is by oral-fecal or oral-oral route, and common routes of transmission are through contaminated food or water supplies. Most cases of polio are mild and may result in digestive or flulike symptoms. However, in about 1% of the cases, the virus spreads to the nervous system and attacks somatic motor neurons in the anterior horn of the spinal cord. (The disease got its name by describing the inflammation of the gray matter of the spinal cord.) These cases are referred to as **paralytic polio.** Here, the motor neurons are damaged or destroyed, resulting in paralysis of the muscles innervated by those segments of the spinal cord. Paralysis may be temporary or permanent, depending upon the extent of somatic motor neuron damage. Polio is rare in the Western world due to an active vaccination program, but it is still endemic in Pakistan, Afghanistan, Nigeria, Syria, and Chad (where vaccination programs have been disrupted or incomplete).

WHAT DID YOU LEARN?

6 Place the following structures in order for relaying sensory input to the spinal cord: posterior horn, sensory receptor (e.g., tactile receptors of the skin), spinal nerve, and posterior root.

7 Place the following structures in order for relaying motor output from the spinal cord to skeletal muscle: anterior horn, effector (skeletal muscle), spinal nerve, and anterior root.

14.3b Distribution of White Matter

LEARNING OBJECTIVES

10. Identify the locations of white matter within the spinal cord.

11. List the three anatomic divisions of the white matter, and explain their general composition.

White matter was first discussed in section 13.1c, where it was described as being (1) primarily composed of myelinated axons and (2) functioning to relay nerve signals. The white matter of the spinal cord is external to the gray matter and on each side of the cord is partitioned into three distinct anatomic structural regions based upon their location within the spinal cord. Each of these regions is called a **funiculus** (fū-nik′ū-lŭs; pl. *funiculi*,[1] fū-nik′ū-lī; *funis* = cord) (figure 14.4). Each **posterior funiculus** is white matter that lies between the posterior gray horns on the posterior side of the cord and the posterior median sulcus. The **lateral funiculus** is the white matter on each lateral side of the spinal cord. The **anterior funiculus** is composed of white matter that occupies the space on each anterior side of the cord between anterior gray horns and the anterior median fissure; the anterior funiculi are interconnected by the **white commissure.**

[1] *Note:* Anterior and lateral funiculi were formerly called *columns*. The Federative Committee on Anatomical Terminology (FCAT) now states that the term *column* refers to structures within the gray matter of the spinal cord, whereas *funiculus* refers to the white matter regions.

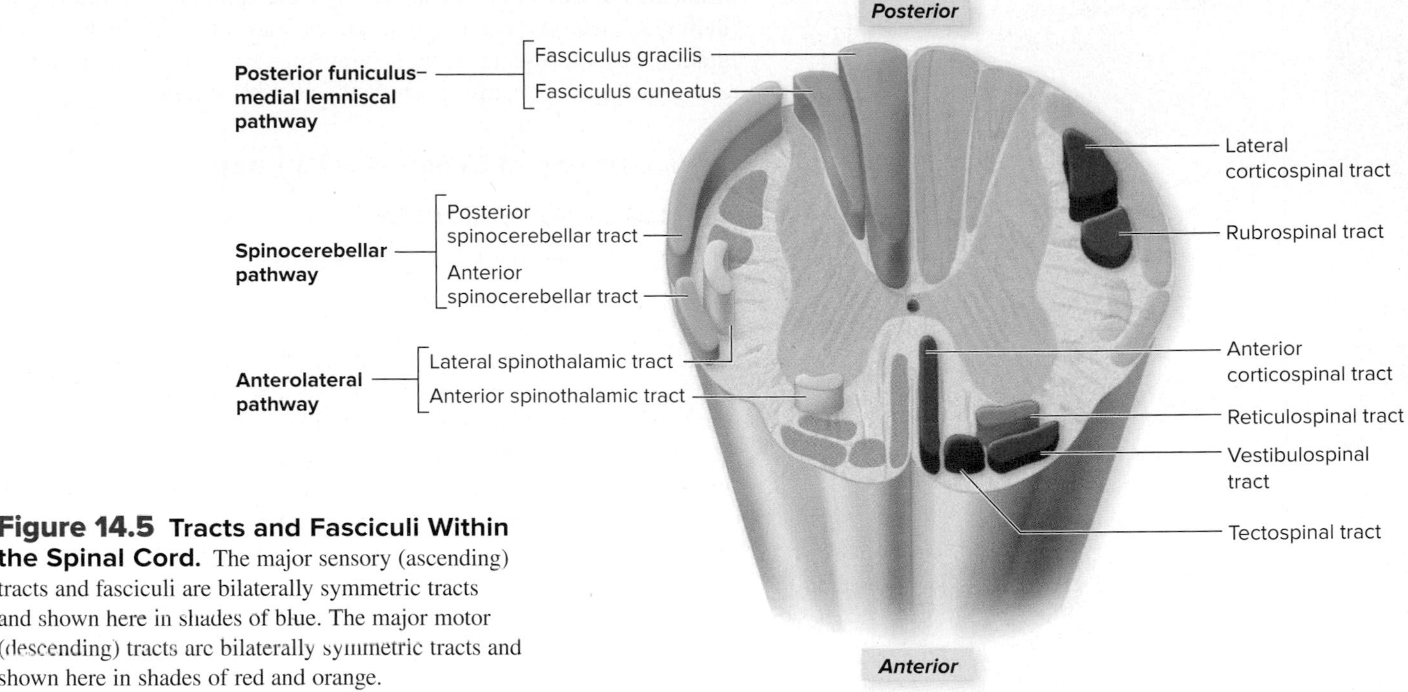

Figure 14.5 Tracts and Fasciculi Within the Spinal Cord. The major sensory (ascending) tracts and fasciculi are bilaterally symmetric tracts and shown here in shades of blue. The major motor (descending) tracts are bilaterally symmetric tracts and shown here in shades of red and orange.

The axons within each funiculus are organized into smaller structural units (or bundles of myelinated axons) called **fasciculi** (fă-sik′ū-lī; *fascis* = bundle) **(figure 14.5)**. White matter on each side of the cord can also be referred to as tracts, which have common *functions.* Individual **tracts** are either (1) **sensory** (or *ascending*) **tracts,** which conduct nerve signals *from* the spinal cord to the brain, or (2) **motor** (or *descending*) **tracts,** which conduct nerve signals *from* the brain to the spinal cord. Figure 14.5 shows sensory tracts in blue shading on one side of the spinal cord and motor tracts in red shading on the other side of the spinal cord. However, keep in mind that the spinal cord is symmetric; both sensory tracts and motor tracts are on both sides of the cord.

Observe in figure 14.5 that each funiculus region (posterior, lateral, and anterior funiculus) contains sensory tracts (blue shading), and the lateral and anterior funiculi contain motor tracts (orange and red shading). Thus, sensory input is relayed to the brain within each funiculus, whereas motor output from the brain is relayed only within the lateral and anterior funiculi (not within the posterior funiculus).

Tracts are myelinated axons that have a common origin, a common destination, and a similar function. The name of each tract reflects its origin and destination. For example, sensory tracts usually begin with the prefix *spino-,* indicating that they originate in the spinal cord. The second part of the name provides its destination. An example is the sensory *spinothalamic tract,* which extends from the spinal cord to the thalamus. Another example is the *spinocerebellar tract,* which extends from the spinal cord to the cerebellum. (The fasciculus gracilis and fasciculus cuneatus are exceptions to how sensory tracts are named.) Motor pathways begin either with *cortico-,* indicating an origin in the cerebral cortex, or with the name of a brainstem nucleus (such as *rubro-,* indicating an origin within the red nucleus of the midbrain; see section 13.5a). Thus, the *corticospinal tract* extends from the cerebral cortex to the spinal cord. How tracts within the spinal cord function in sensory and motor pathways is discussed in section 14.4.

Figure 14.6 presents representative cross sections through each spinal cord part. This illustration helps us to realize that both the size and shape of the spinal cord in each section along its length vary in each spinal cord part. The difference in the relative amounts of gray matter and white matter reflects the function of that part of the spinal cord. For example, the lumbar part of the spinal cord has a greater amount of gray matter because more neuron cell bodies are located there that have axons extending from there to innervate the lower limbs.

WHAT DID YOU LEARN?

8 What are the three types of funiculi? List the specific tracts found in each.

INTEGRATE

LEARNING STRATEGY

The white matter within the spinal cord is identified by the use of four terms based upon either structure or function.

Terms to identify structure:

Funiculus: The specific *location* of the white matter (e.g., posterior funiculus is the white matter forming the posterior aspect of the spinal cord)

Fasciculus: A structural subdivision of a funiculus that shares common features (e.g., fasciculus gracilis, which is within a posterior funiculus)

Terms to identify function:

Tract: Myelinated axons that relay nerve signals from a common origin to a common destination (e.g., spinothalamic tract, which relays nerve signals from the spinal cord to the thalamus)

Pathway: A more general term that includes all of the neurons (and associated structures) that relay nerve signals between the brain and the body. Spinal pathways include components within the brain (e.g., cerebrum), components of the spinal cord (identified as either fasciculi or tracts), and spinal nerves.

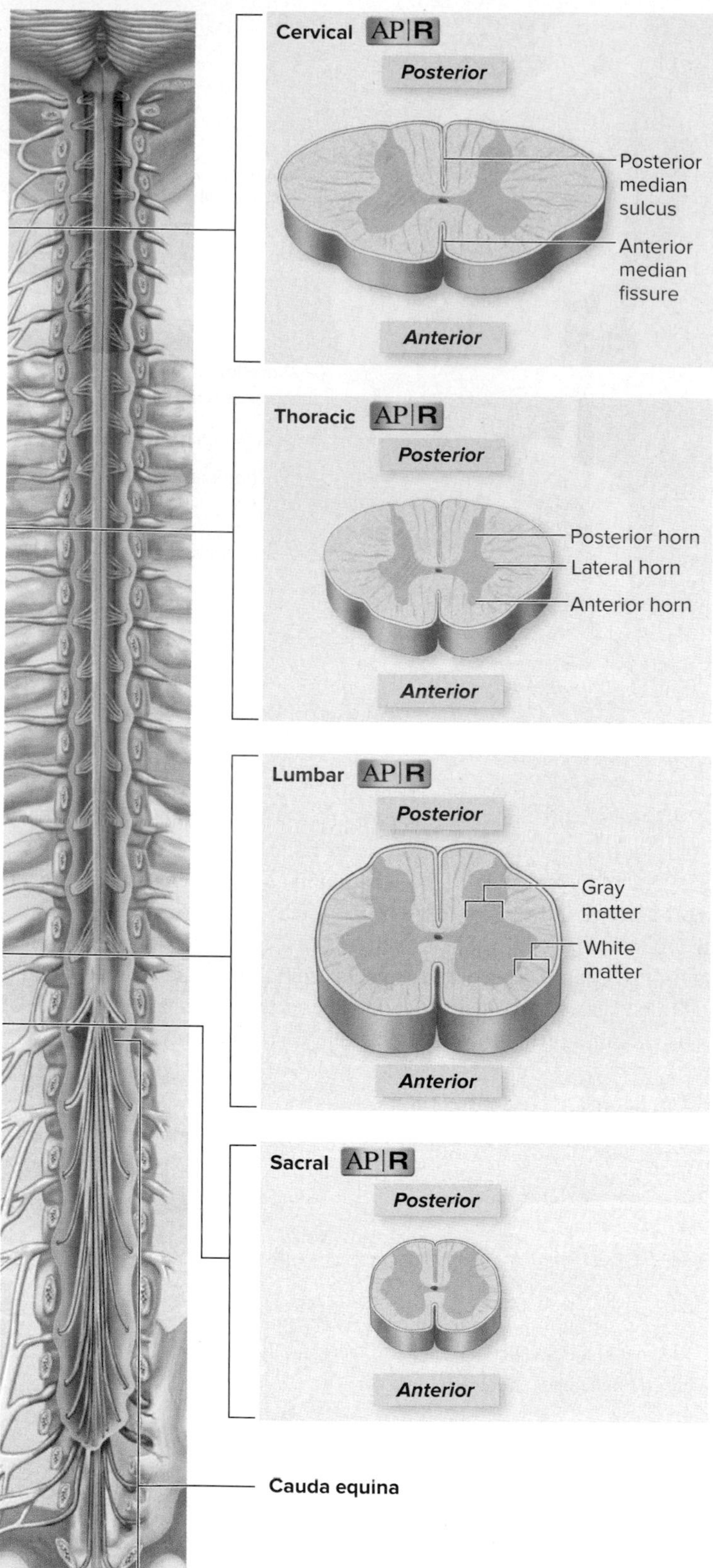

Figure 14.6 Representative Cross Sections of the Spinal Cord. The cross sections of the cervical, thoracic, lumbar, and sacral parts of the spinal cord vary in shape and size. APIR

14.4 Sensory and Motor Pathways

Conduction pathway is an inclusive term that refers to all of the series of neurons (and their associated structures) that relay signals between the brain and the body. Pathways that extend between the brain and the torso and limbs include components that are (1) within the brain (e.g., cerebrum), (2) within the spinal cord (either a fasciculus or tract that extends through the spinal cord), and (3) the individual neurons within spinal nerves. Pathways also include integration and processing centers (gray matter) at different locations along each type of pathway where the neurons synapse.

14.4a Overview of Conduction Pathways

LEARNING OBJECTIVES

12. Define a sensory pathway and a motor pathway.
13. List the features common to all pathways.

Conduction pathways are identified as either sensory or motor pathways, depending upon the direction nerve signals are relayed relative to the brain. **Sensory pathways** include the sensory neurons that relay sensory input *to* the brain. Sensory pathways are also called *ascending pathways* because the nerve signals are relayed from the sensory receptors superiorly to the brain. **Motor pathways** include the series of motor neurons that relay motor output *from* the brain. Motor pathways are also called *descending pathways* because the nerve signals are relayed from the brain inferiorly to the body's muscles and glands. Most conduction pathways—whether sensory or motor—share several general characteristics:

- **Paired tracts.** All pathways are composed of paired tracts. Thus, a pathway on one side of the CNS has a matching tract on the other side of the CNS.
- **Composed of two or more neurons.** Most pathways are composed of a series of two or three neurons that form the pathway.
- **Common location of neuron cell bodies.** Neuron cell bodies are located in one of three general places: the posterior root ganglion, the gray horns within the spinal cord, or nuclei within the brain along the pathway.
- **Common location of axons.** The axons of the different neurons extend through spinal nerves, the spinal cord (as named tracts or fasciculi), and the brain.
- **Decussation.** Most pathways include neurons that cross over, or decussate (dē-kŭ-sāt′; *decusseo* = to make in the form of an X), from one side of the body to the other side at some point along the pathway—within either the spinal cord or the brain. This means that the left side of the brain receives sensory input from or initiates motor output to the right side of the body, whereas the right side of the brain receives sensory input from or initiates motor output to the left side of the body. The term **contralateral** (kon-tră-lat′er-ăl; *contra* = opposite, *latus* = side) is used to indicate the relationship to the opposite side. Over 90% of all neurons within pathways decussate.
- **Limited ipsilateral pathway.** Pathways may have some neurons (about 10%) that remain on the same side of the body. The term **ipsilateral** (ip-si-lat′er-ăl; *ipse* = same) is used to indicate the relationship to the same side.

WHAT DID YOU LEARN?

9. What characteristics are common to most conduction pathways?

14.4b Sensory Pathways

LEARNING OBJECTIVES

14. Describe the general characteristics of the general sense receptors.
15. List the neurons in the sensory pathway chain.
16. Describe the three major somatosensory pathways.

Sensory pathways have been described as ascending pathways that relay sensory information from sensory receptors to the brain. Sensory input transmitted through the spinal cord (discussed in this section) is detected by *general sense receptors*. Understanding these pathways requires some preliminary discussion regarding general sense receptors (which are described in detail in sections 16.1 and 16.2).

Overview of Sensory Receptors

General sense receptors are sensory receptors located throughout the body (and are distinguished from the *special sense receptors,* which are limited to the head—the eyes, ears, nose, and tongue). General sense receptors are subdivided into two categories: somatic sensory (or somatosensory) receptors and visceral sensory receptors. **Somatic sensory** (or **somatosensory**) **receptors** are tactile receptors or proprioceptors. **Tactile receptors** are housed within both the skin and mucous membranes that line body cavities. These sensory receptors monitor characteristics of an object (e.g., texture). **Proprioceptors** are located within joints, muscles, and tendons to detect stretch and pressure relative to position and movement of the skeleton and skeletal muscles. **Visceral sensory receptors** are located in the walls of the viscera (internal organs) and blood vessels. They detect changes to an organ or a blood vessel (e.g., stretch).

Categorization of Sensory Pathways

Consequently, *sensory pathways* are organized into two categories depending upon the type of general sensory receptor involved. **Somatosensory pathways** process stimuli received from somatosensory receptors (e.g., tactile receptors within the skin, proprioceptors), whereas **viscerosensory pathways** process stimuli received from visceral sensory receptors (i.e., from the receptors of internal organs). We limit our discussion to somatosensory pathways.

Sensory pathways use a series of two or three neurons to transmit nerve signals from the sensory receptors to the brain, which are the primary neuron, secondary neuron, and tertiary neuron.

- The **primary neuron** (or *first-order neuron*) is the first neuron in the chain of neurons. The primary neuron extends from the sensory receptor to the CNS (brain or spinal cord), where it synapses with a secondary neuron.
- The **secondary neuron** (or *second-order neuron*) is an interneuron that extends from the primary neuron to either the tertiary neuron or the cerebellum.
- The **tertiary neuron** (or *third-order neuron*) is also an interneuron. It extends from the secondary neuron to the cerebrum (specifically, the primary somatosensory cortex of the parietal lobe; see section 13.3c). Pathways that lead to the cerebellum do not have a tertiary neuron.

There are three major types of somatosensory pathways: the posterior funiculus–medial lemniscal pathway, the anterolateral pathway, and the spinocerebellar pathway. As you read the discussion of these sensory pathways, consider the following for each: (1) what type of sensory receptors are involved and what type of sensory information they are providing to the brain, (2) the location of each sensory neuron within the chain of two or three neurons that compose the pathway, and (3) what region of the brain receives and processes this sensory information.

Posterior Funiculus–Medial Lemniscal Pathway

The **posterior funiculus–medial lemniscal pathway** uses a chain of three sensory neurons to communicate with the brain about a specific stimulus (**figure 14.7**). This pathway originates at either of the two types of somatosensory receptors: (1) *tactile receptors* housed within both the skin and mucous membranes or (2) *proprioceptors* within joints, muscles, and tendons. This sensory input is providing information to the brain (specifically, the cerebral cortex) about discriminative touch, precise pressure, and vibration sensations from the tactile receptors of the skin and with conscious perception of the skeleton and skeletal muscles from proprioceptors. For example, this pathway provides the information to your brain to identify an object in your hand (even if your eyes are closed) and where your arms are positioned (even if your eyes are closed).

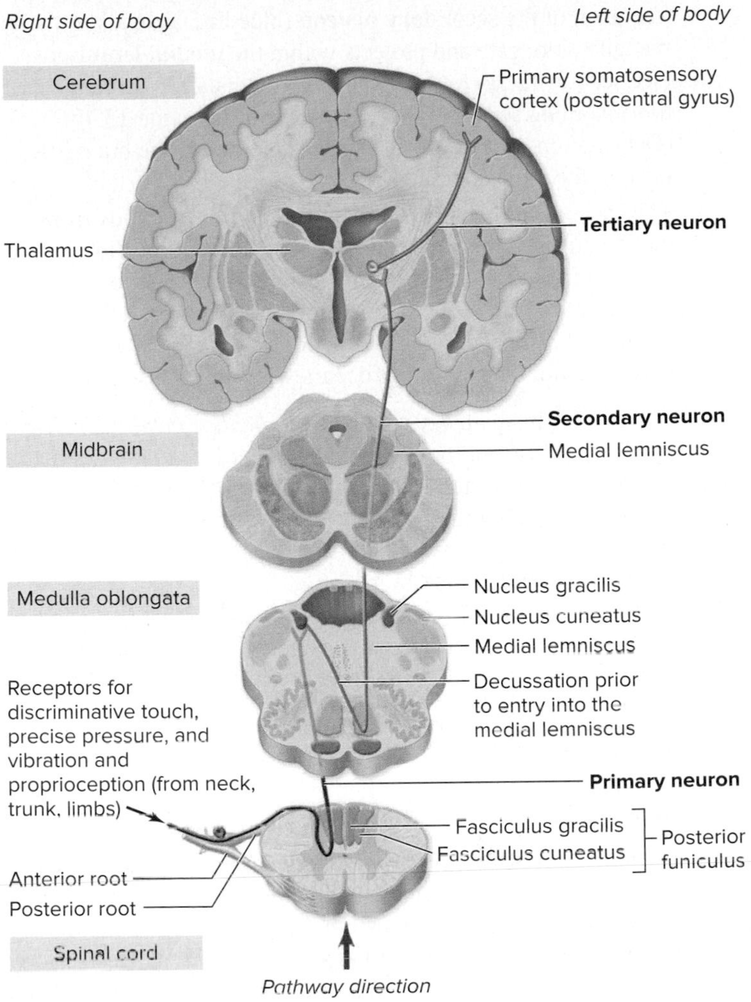

Figure 14.7 Posterior Funiculus–Medial Lemniscal Pathway. This pathway transmits sensory information about discriminative touch, precise pressure, and vibration and limb position (proprioception). This pathway is bilaterally symmetric—but to avoid confusion, only sensory input from the right side of the body is shown here. Decussation of axons occurs just prior to the medial-lemniscus in the medulla oblongata. The primary neuron is purple, the secondary neuron is blue, and the tertiary neuron is green.

Three sensory neurons compose the chain of neurons within this pathway:

- The axon of the **primary neuron** (purple line) extends from the somatosensory receptor into the spinal cord (via the posterior root) and ascends within the *posterior funiculus* within the spinal cord. (The specific fasciculus is either the **fasciculus cuneatus** or the **fasciculus gracilis.**) The primary neuron synapses with the secondary neuron within the gray matter of the medulla oblongata of the brain (specifically, the **nucleus cuneatus** and **nucleus gracilis,** respectively).

- The axon of the **secondary neuron** (blue line) extends from the medulla oblongata and projects within the **medial lemniscus** (see section 13.5a) to the thalamus. The thalamus "filters" this incoming sensory input as described in section 13.4b. (Decussation occurs to the opposite side within the brain just prior to the medial lemniscus.)
- The axon of the **tertiary neuron** (green line) extends from the thalamus to the cerebrum (specifically to a location within the primary somatosensory cortex housed within the postcentral gyrus of the parietal lobe; see figure 13.13 in section 13.3c). Conscious perception of the tactile or proprioceptor sensory input occurs within the parietal lobe.

The name of this pathway (*posterior funiculus–medial lemniscal pathway*) is derived from the two components of white matter that it extends through: the posterior funiculus within the spinal cord and medial lemniscus within the brain.

WHAT DO YOU THINK?

3 With respect to the posterior funiculus–medial lemniscal pathway, (1) what type of sensory receptors are involved, and what type of sensory information are they providing to the brain, (2) what is the location of each of the sensory neurons within the chain of three neurons that compose this pathway, and (3) what region of the brain receives the sensory information?

Anterolateral Pathway

The **anterolateral pathway** (or *spinothalamic pathway*) uses a chain of three neurons to communicate with the brain about a specific stimulus (**figure 14.8**). This pathway originates at tactile somatosensory receptors within both the skin and mucous membranes. This sensory input is providing information to the brain (specifically, the cerebral cortex) about crude touch and pressure as well as pain and temperature. Typically, sensations that require us to act in response to the stimulus (such as either an itch that makes us want to scratch or tickling that makes us jerk away) are relayed through the anterolateral pathway.

Three sensory neurons compose the chain of neurons within this pathway:

- The axon of the **primary neuron** (purple line) extends from the somatosensory receptor into the spinal cord (via the posterior root). The primary neuron synapses with the secondary neuron within the posterior horn of the spinal cord, as described in section 14.3a. (Note: The synapsing of the primary neuron at the level of the spinal cord is the most significant structural difference of the anterolateral pathway when compared to the posterior funiculus–medial lemniscal pathway.)
- The axon of the **secondary neuron** (blue line) extends from the spinal cord to the thalamus. The axons project within the spinothalamic tract—either within the anterior funiculus (via the anterior spinothalamic tract) or the lateral funiculus (via the lateral spinothalamic tract) to the thalamus. The thalamus "filters" the incoming sensory input as described in section 13.4b. (Decussation occurs to the opposite side within the spinal cord as the axons extend into the spinothalamic tract.)
- The axon of the **tertiary neuron** (green line) extends from the thalamus to the cerebrum (specifically, to a location within the primary somatosensory cortex housed within the postcentral gyrus of the parietal lobe (see figure 13.13 in section 13.3c). Conscious perception of the tactile or proprioceptor sensory input occurs within the parietal lobe.

The principal name of this pathway (*anterolateral pathway*) is derived from the location of the two funiculi through which it ascends (anterior funiculus and lateral funiculus). Its secondary name (spinothalamic pathway) is derived from the tracts that relay the nerve signals within the spinal cord to the thalamus.

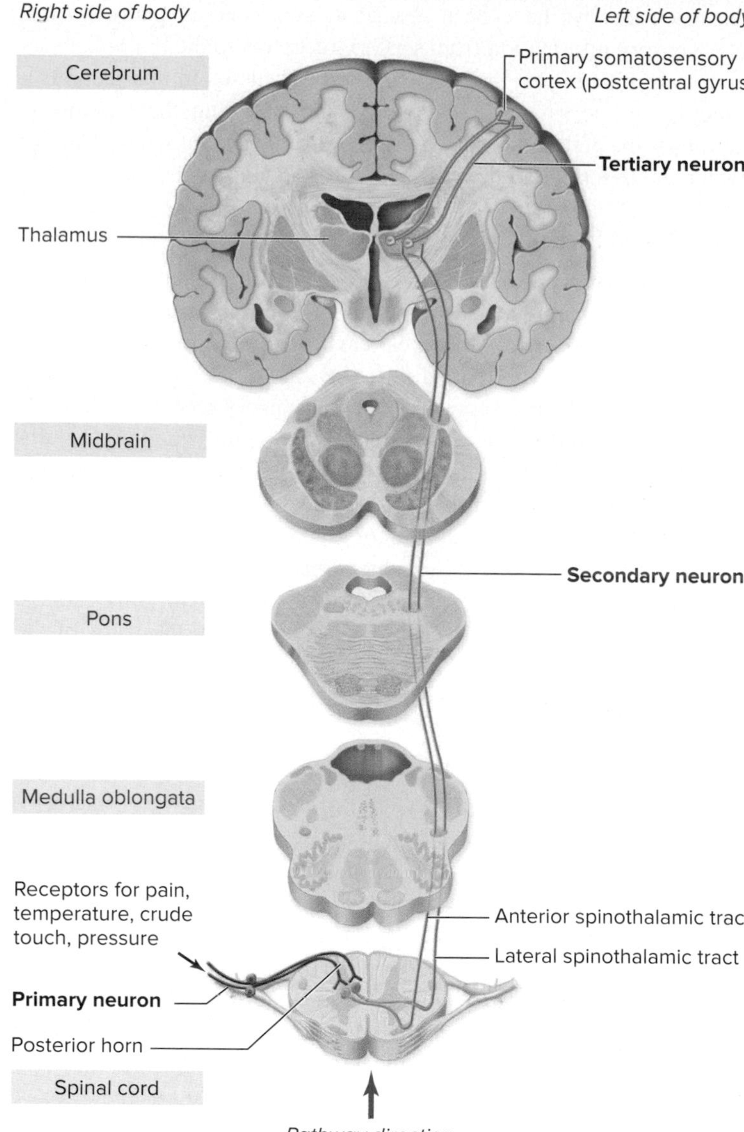

Figure 14.8 Anterolateral Pathway. This pathway conducts crude touch, pressure, pain, and temperature sensations toward the brain. Decussation of axons occurs at the level where the primary neuron axon enters the spinal cord. The primary neuron is purple, the secondary neuron is blue, and the tertiary neuron is green.

WHAT DO YOU THINK?

4 Regarding the anterolateral pathway, (1) what type of sensory receptor is involved, and what type of sensory information is being provided to the brain, (2) what is the location of each of the sensory neurons within the chain of three neurons that compose this pathway, and (3) what specific region of the brain receives the sensory information?

Spinocerebellar Pathway

The **spinocerebellar pathway** uses a chain of only two neurons to communicate with the brain about a specific stimulus (**figure 14.9**). This pathway originates at proprioceptors within joints, muscles, and tendons at different locations in the body. This sensory input is providing information to the brain (specifically, the cerebellum) related to subconscious postural input, which helps in maintaining balance and posture (see section 13.6b).

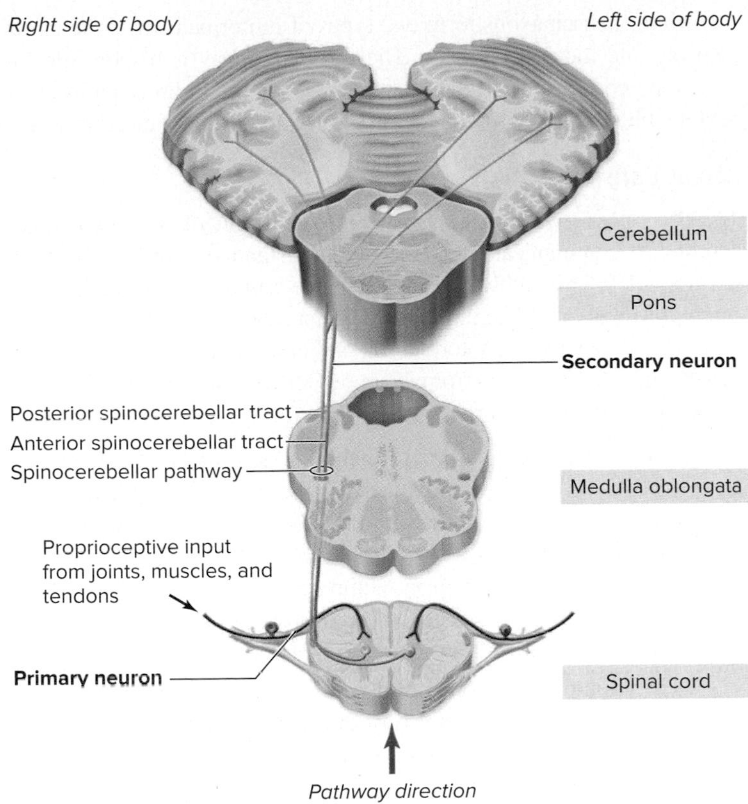

Figure 14.9 Spinocerebellar Pathway. This pathway conducts proprioceptive information to the cerebellum through both the anterior and posterior spinocerebellar tracts. Only some axons decussate, and these do so at the level where the primary neuron axon enters the spinal cord. Only primary (purple) and secondary (blue) neurons are found in this type of pathway.

Two sensory neurons compose the chain of neurons within this pathway. (There is no tertiary neuron.)

- The axon of the **primary neuron** (purple line) extends from a proprioceptor into the spinal cord (via the posterior root). The primary neuron synapses with the secondary neuron within the posterior horn of the spinal cord. (This is similar to the synapse between primary and secondary neurons within the anterolateral pathway.)
- The axon of the **secondary neuron** (blue line) extends from the spinal cord within the spinocerebellar tract—either within the anterior or within the posterior portion of the lateral funiculus to the cerebellum.

The name of this pathway (*spinocerebellar pathway*) is derived from the origin of its tracts that ascend from the spinal cord to the cerebellum.

WHAT DO YOU THINK?

5 Regarding the spinocerebellar pathway, (1) what type of sensory receptor is involved, and what type of sensory information are they providing to the brain, (2) what is the location of each of the sensory neurons within the chain of two neurons that compose this pathway, and (3) which region of the brain receives the sensory information?

Table 14.1 summarizes the characteristics of the three major types of sensory pathways.

WHAT DID YOU LEARN?

10 What are the general locations and functions of primary, secondary, and tertiary neurons in sensory pathways?

11 What types of information does the posterior funiculus–medial lemniscal pathway transmit?

Table 14.1 Functions and Neuron Locations of Principal Sensory Spinal Cord Pathways

Pathway	Posterior Funiculus–Medial Lemniscal		Anterolateral		Spinocerebellar	
Components of Pathway	**Fasciculus Cuneatus**	**Fasciculus Gracilis**	**Anterior Spinothalamic Tract**	**Lateral Spinothalamic Tract**	**Anterior Spinocerebellar Tract**	**Posterior Spinocerebellar Tract**
Function	Sensory input for limb position and discriminative touch, precise pressure, and vibration sensation		Sensory input for crude touch, pressure, pain, and temperature		Sensory input sent from proprioceptors to cerebellum for subconscious interpretation	
	Relays input from upper limb, superior trunk, neck, posterior head	Relays input from lower limb, inferior trunk	Relays input for crude touch and pressure	Relays input for pain and temperature	Relays input from inferior regions of trunk and lower limbs	Relays input from lower limbs, regions of trunk and upper limbs
Primary neuron	Extends from receptor to medulla oblongata Cell bodies within posterior root ganglion		Extends from receptor to spinal cord Cell bodies within posterior root ganglion		Extends from receptor to spinal cord Cell bodies within posterior root ganglion	
Secondary neuron	Extends from medulla oblongata to thalamus Cell bodies within medulla oblongata (nucleus cuneatus or nucleus gracilis)		Extends from spinal cord to thalamus Cell bodies within spinal cord (posterior horn)		Extends from spinal cord to cerebellum Cell bodies within spinal cord (posterior horn)	
Tertiary neuron	Extends from thalamus to cerebral cortex Cell bodies within thalamus		Extends from thalamus to cerebral cortex Cell bodies within thalamus		N/A	
Structures involved in decussation	Axons of secondary neurons decussate just prior to medial lemniscus		Axons of secondary neurons decussate within spinal cord at level of entry		Some axons decussate in spinal cord and pons, whereas other axons do not decussate	Axons do not decussate

14.4c Motor Pathways

LEARNING OBJECTIVES

17. Define a motor pathway, and describe its actions.
18. Distinguish between an upper motor neuron and a lower motor neuron, based upon function and cell body location.
19. Compare and contrast the direct and indirect motor pathways.

Motor pathways are the descending pathways that originate within the brain and act to control effectors. Here we discuss the motor pathways that specifically control skeletal muscle of the torso and limbs. These motor pathways originate from the cerebral cortex, the cerebral nuclei, or the brainstem **(figure 14.10)**.

At least two motor neurons are present within the motor pathway to transmit signals from the brain to the body: an upper motor neuron and a lower motor neuron.

- An **upper motor neuron** is the first neuron in a chain of neurons. The cell body of the upper motor neuron is housed within the cerebral cortex, cerebral nuclei, or a specific nucleus within the brainstem. Axons of the upper motor neuron synapse either directly upon lower motor neurons (in *direct pathways*) or upon interneurons that ultimately synapse upon lower motor neurons (in *indirect pathways*). The upper motor neurons either excite or inhibit the activity of lower motor neurons.
- The **lower motor neuron** is the last neuron in the chain of neurons. The cell body of a lower motor neuron is housed within the anterior horn of the spinal cord (as described in section 14.3a). Axons of the lower motor neurons exit the spinal cord through the anterior root and project to and innervate a specific skeletal muscle. The lower motor neuron always excites the skeletal muscle fibers to contract.

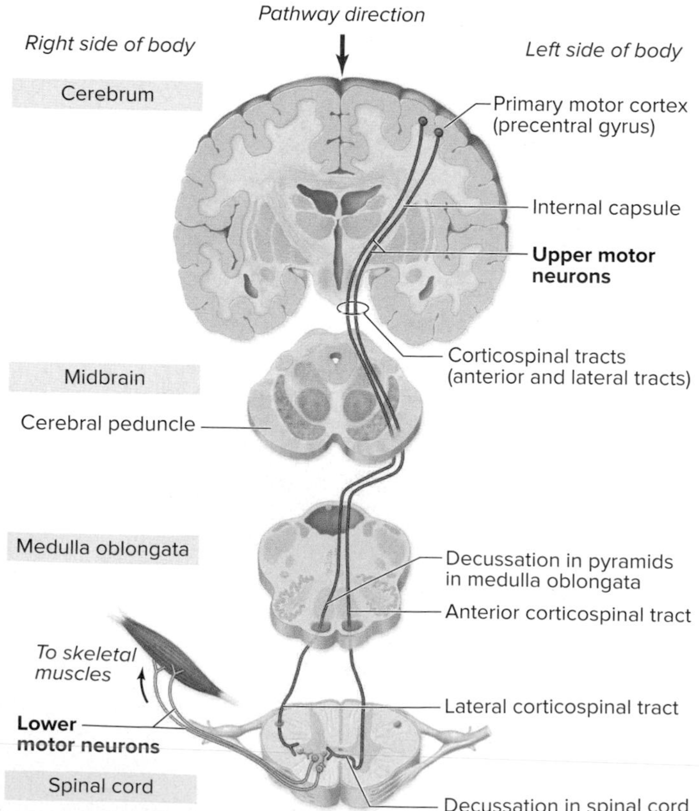

Figure 14.10 Corticospinal Tracts. Corticospinal tracts originate within the cerebrum, decussate in the pyramids or the spinal cord, and synapse on lower motor neurons within the anterior horns of the spinal cord. The upper motor neurons are red, and the lower motor neurons are orange.

Motor neuron axons form two types of motor pathways: the direct pathway and indirect pathway. The direct pathway is responsible for conscious control of skeletal muscle activity; the indirect pathway is responsible for subconscious (or reflexive) control of skeletal muscle.

Direct Pathway

The **direct** (or *pyramidal*) **pathway** uses a chain of only two motor neurons to communicate between the brain and the skeletal muscles (figure 14.10). This pathway originates in the primary motor cortex of the cerebral frontal lobe. The direct pathway name is derived from the presence of only one upper motor neuron and one lower motor neuron. The name *pyramidal* is derived from the pyramid-like shape of the cell bodies of the upper motor neurons within gray matter of the cerebral cortex.

- The axon of the upper motor neuron extends from the frontal lobe of the cerebral cortex through the internal capsule and cerebral peduncles of the brain (see section 13.3d) and through a corticospinal tract within the spinal cord. This axon synapses on the lower motor neuron within the anterior horn of the spinal cord. The dendrites and cell bodies of the lower motor neurons form the gray matter of the anterior horn.
- The axon of the lower motor neuron extends from the spinal cord through the anterior root into the spinal nerve to innervate the target skeletal muscle.

The direct pathways are housed within one of two pathways within the spinal cord: the lateral corticospinal tract and anterior corticospinal tract. These two pathways differ in several significant ways, including the specific muscles they innervate and control:

- The **lateral corticospinal tracts** (which composes 85% of the direct pathway) innervates skeletal muscles that control skilled movements in the *limbs*, such as playing a guitar, dribbling a soccer ball, or typing on your computer keyboard.
- The **anterior corticospinal tracts** (which composes the other 15% of the direct pathway) innervate *axial* skeletal muscle.

Decussation of the *lateral* corticospinal tracts occurs to the opposite side within the brain at the medulla oblongata (specifically, at pyramids of the medulla oblongata; see section 13.5c), whereas the *anterior* corticospinal tracts decussate through the anterior gray commissure at the level of a spinal cord segment.

INTEGRATE

CONCEPT CONNECTION

The **corticobulbar** (kōr′ti-kō-bŭl′bar) **pathways** are another type of direct pathway, but they originate from the facial region of the motor homunculus within the primary motor cortex to help regulate activity of muscles of the face and neck. Note that these tracts differ from the others because (1) they do not pass through the spinal cord and (2) they involve cranial nerves (instead of spinal nerves). Axons of these upper motor neurons extend to the brainstem, where they synapse with lower motor neuron cell bodies that are housed within brainstem *cranial nerve nuclei.* Axons of these lower motor neurons help form some of the *cranial nerves* (see section 13.9).

Indirect Pathway

Several nuclei within the brainstem initiate motor commands for skeletal muscle activities that occur at a subconscious, or reflexive, level. The **indirect pathway** is so named because upper motor neurons originate within brainstem nuclei and take a complex, circuitous route through the

brain to the spinal cord that involves more than one upper motor neuron. The indirect pathway modifies or helps control the pattern of somatic motor activity by exciting or inhibiting the lower motor neurons that innervate the muscles.

The different tracts of the indirect pathway are grouped according to their primary functions as either a lateral pathway or a medial pathway. The **lateral pathway** regulates and controls precise, discrete movements and tone in flexor muscles of the limbs—for example, the type of movement required to gently lay a baby in a crib. This pathway consists of the **rubrospinal** (rū′brō-spī′năl; *rubro* = red) **tracts** that originate in the red nucleus of the midbrain (see section 13.5a).

The **medial pathway** regulates reflexive muscle tone and gross movements of the muscles of the head, neck, proximal parts of the limbs, and trunk. Within the medial pathway, three groups of tracts originate in the midbrain, pons, or medulla oblongata.

- The **reticulospinal** (re-tik-ū-lō-spī′năl) **tracts** originate from the reticular formation in the midbrain (see section 13.5a). They help control reflexive movements related to posture and maintaining balance.
- The **tectospinal** (tek-tō-spī′năl) **tracts** extend from the superior and inferior colliculi in the tectum of the midbrain to help regulate reflexive positional changes of the upper limbs, eyes, head, and neck as a consequence of visual and auditory stimuli.
- The **vestibulospinal** (ves-tib′ū-lō-spī′năl) **tracts** originate within vestibular nuclei of the brainstem. Nerve signals conducted within these tracts regulate reflexive muscular activity that helps maintain balance during sitting, standing, and walking.

Table 14.2 summarizes the characteristics of the principal types of motor pathways. **Figure 14.11** summarizes the main differences between the sensory and motor pathways.

WHAT DID YOU LEARN?

12. What are the locations and functions of upper and lower motor neurons in the motor pathways?
13. What are the differences between direct and indirect motor pathways?

INTEGRATE

CLINICAL VIEW 14.3

Treating Spinal Cord Injuries

Spinal cord injuries frequently leave individuals paralyzed and unable to perceive sensations to varying degrees, depending upon the location and extent of the injury. In recent years, advances have been made in the treatment of spinal cord injuries (although some of the findings are still preliminary). Prompt use of steroids immediately after the injury appears to preserve some muscular function that might otherwise be lost. Early use of antibiotics has substantially reduced the number of deaths caused by pulmonary and urinary tract infections that accompany spinal cord injuries. Recent research with rats has achieved reconnection and partial restoration of function of severed spinal cords. In addition, other research indicates that neural stem cells may be able to regenerate spinal cord axons.

Table 14.2 Principal Motor Spinal Cord Pathways

Tract	Manner of Decussation	Destination of Upper Motor Neurons	Termination Site	Function
DIRECT PATHWAY				
Corticobulbar tracts	All cranial nerve motor nuclei receive bilateral (both ipsilateral and contralateral) input except CN VI, VII to the lower face, and XII (these nerves receive only contralateral input)	Brainstem only	Cranial nerve nuclei; reticular formation	Voluntary movement of cranial and facial muscles
Lateral corticospinal tracts	All decussate at the pyramids	Lateral funiculus	Gray matter region between posterior and anterior horns; anterior horn; all levels of spinal cord	Voluntary movement of appendicular muscles
Anterior corticospinal tracts	Decussation occurs in spinal cord at level of lower motor neuron cell body	Anterior funiculus	Gray matter region between posterior and anterior horns; anterior horn; cervical part of spinal cord	Voluntary movement of axial muscles
INDIRECT PATHWAY				
Lateral Pathway				
Rubrospinal tract	Decussate at ventral tegmentum of midbrain	Lateral funiculus	Lateral region between posterior and anterior horns; anterior horn; cervical part of spinal cord	Regulates and controls precise discrete movements and tone in flexor muscles of the limbs
Medial Pathway				
Reticulospinal tract	No decussation (ipsilateral)	Anterior funiculus	Medial region between posterior and anterior horns; anterior horn; all parts of spinal cord	Controls reflexive movements related to posture and maintaining balance
Tectospinal tract	Decussate at dorsal tegmentum of midbrain	Anterior funiculus	Medial region between posterior and anterior horns; anterior horn; cervical part of spinal cord	Regulates reflexive positional changes of the upper limbs, eyes, head, and neck due to visual and auditory stimuli
Vestibulospinal tract	Some decussate (contralateral) and some do not (ipsilateral)	Anterior funiculus	Medial region between posterior and anterior horns; anterior horn; medial tracts to cervical and superior thoracic parts of spinal cord; lateral tracts to all parts of spinal cord	Regulates reflexive muscular activity that helps maintain balance during sitting, standing, and walking

Figure 14.13 Dermatome Maps.
A dermatome is an area of skin supplied by a single spinal nerve. These diagrams only approximate the dermatomal distribution. AP|R

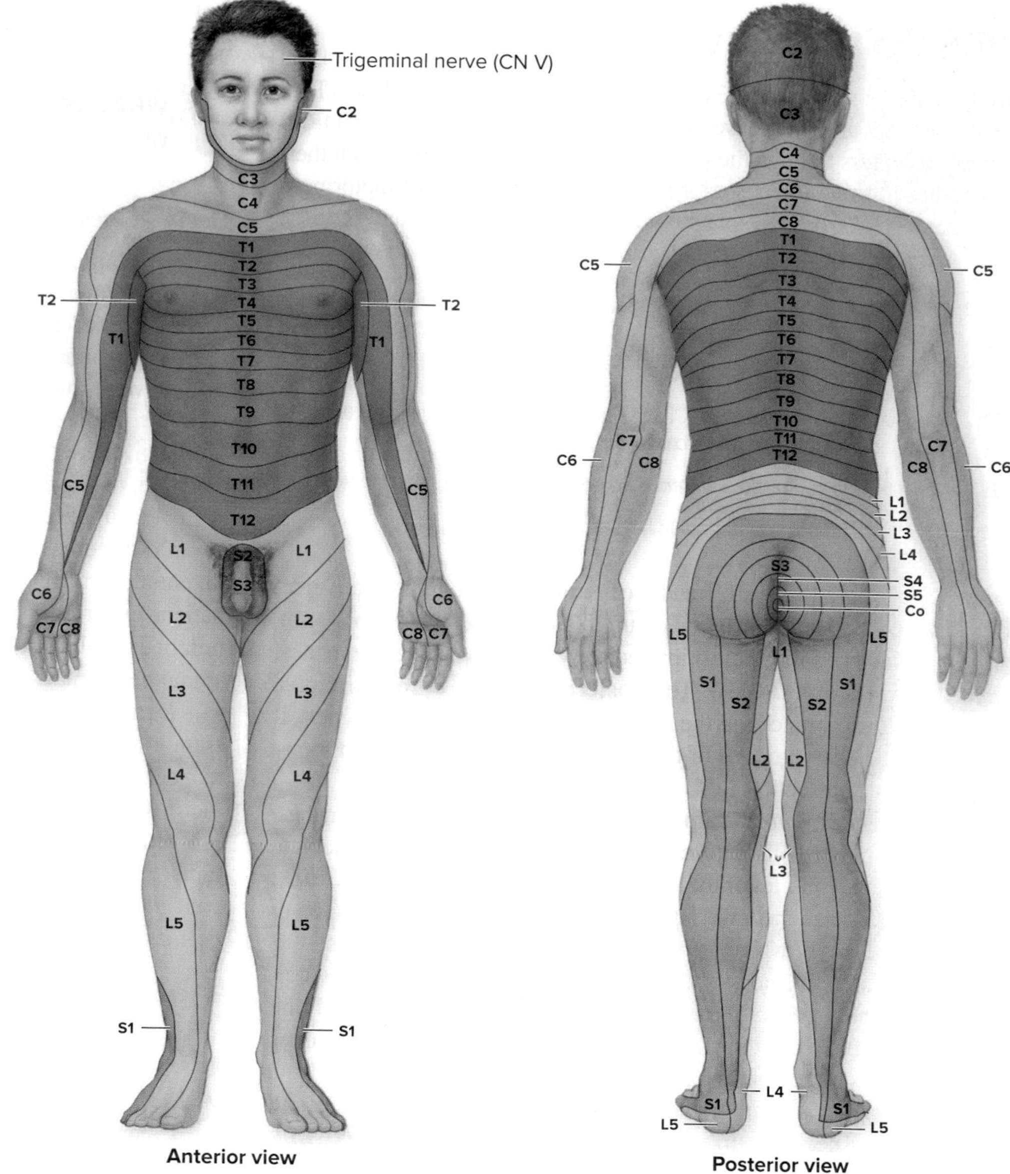

INTEGRATE

CLINICAL VIEW 14.4

Shingles

Some individuals (usually adults over 50) experience a reactivation of their childhood chicken pox infection, a condition termed **shingles** (shing′glz). Psychological stress, other infections (such as a cold or the flu), and even a sunburn can trigger the development of shingles.

During the initial infection, the chicken pox virus (varicella-zoster) sometimes leaves the skin and invades the posterior root ganglia. There, the virus remains latent until adulthood, when it becomes reactivated and proliferates, traveling through the sensory axons to the dermatome. (The word *shingles* is derived from the Latin word *cingulum,* meaning "girdle," reflecting the dermatomal pattern of its spread.) The virus gives rise to a rash and blisters along the dermatome, which are often accompanied by intense burning or tingling pain.

Antiviral medication (e.g., acyclovir) may reduce the severity and duration of symptoms of shingles. Additionally, older adults may receive a vaccine for shingles, which may help prevent or reduce the severity of the disease.

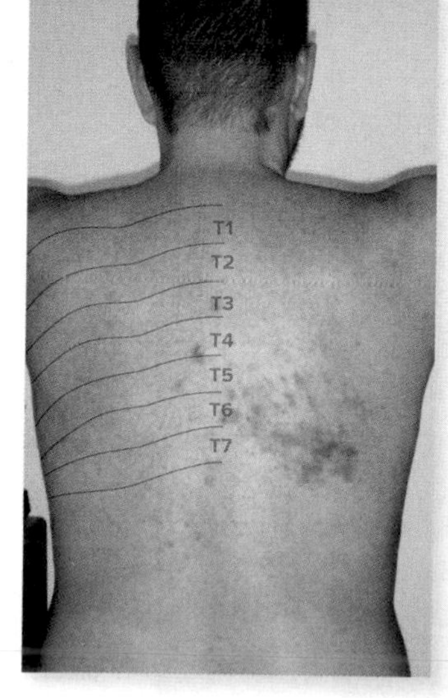

Typical dermatomal spread of a shingles rash in a 49-year-old man.

14.5b Nerve Plexuses

LEARNING OBJECTIVE

23. Define a nerve plexus.

A **nerve plexus** (plek′sŭs; a braid) is a network of interweaving anterior rami of spinal nerves. The anterior rami of most spinal nerves form nerve plexuses on both the right and left sides of the body. These nerve plexuses then split into multiple "named" nerves that innervate various body structures. The main plexuses are the cervical plexuses, brachial plexuses, lumbar plexuses, and sacral plexuses (see figure 14.1).

WHAT DO YOU THINK?

7 What is the benefit of having an intricate nerve plexus, rather than a single spinal nerve that innervates a structure?

Nerve plexuses are organized such that axons from each anterior ramus extend to body structures through several different branches. In addition, each terminal branch of the plexus houses axons from several different spinal nerves. Thus, damage to a single segment of the spinal cord or damage to a single spinal nerve generally does not result in complete loss of innervation to a particular muscle or region of skin.

Most of the thoracic spinal nerves, as well as nerves S5–Co1, do not form plexuses. We discuss the anterior rami of thoracic spinal nerves (called intercostal nerves) first, followed by the individual nerve plexuses.

WHAT DID YOU LEARN?

16 What is the composition of a typical nerve plexus?

14.5c Intercostal Nerves

LEARNING OBJECTIVE

24. Identify the distribution of the intercostal nerves.

The anterior rami of spinal nerves T1–T11 are called **intercostal nerves** because they are located within the intercostal space sandwiched between two adjacent ribs (**figure 14.14**). (T12 is

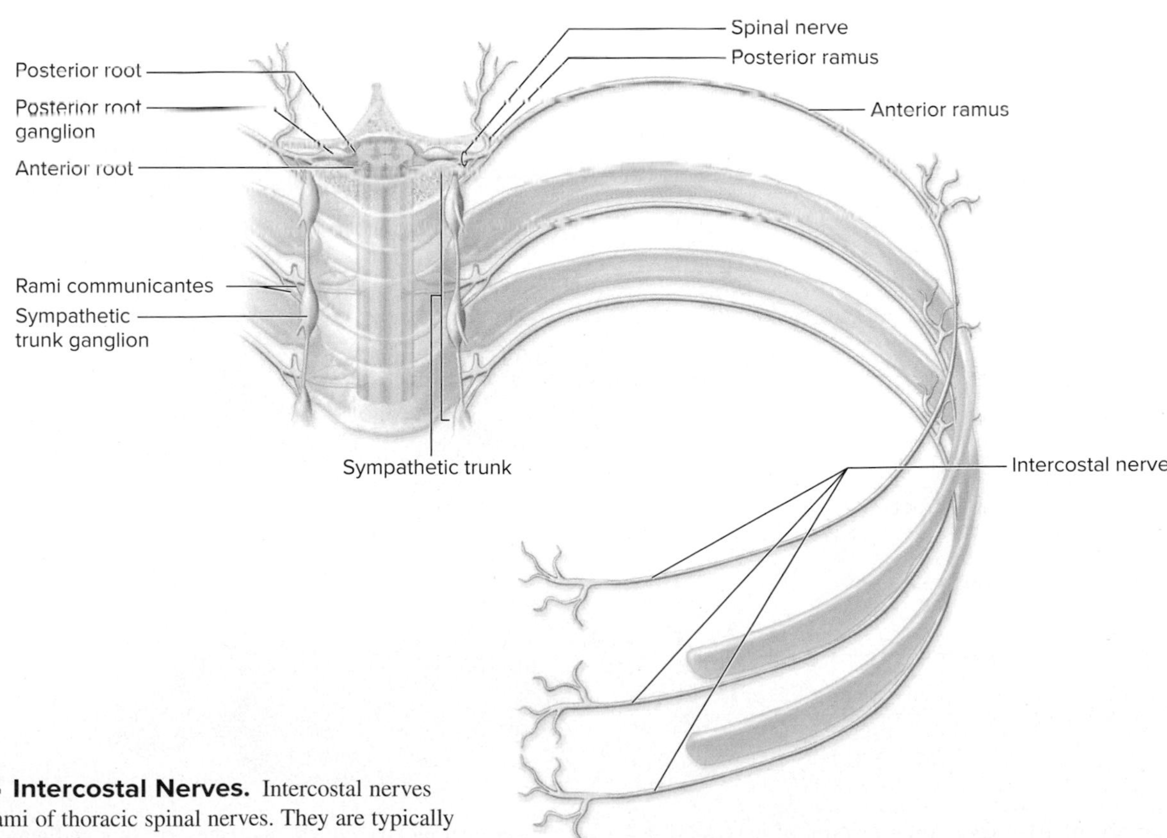

Figure 14.14 Intercostal Nerves. Intercostal nerves are the anterior rami of thoracic spinal nerves. They are typically distributed as shown here.

called a **subcostal nerve,** because it arises inferior to the ribs, not between two ribs.) With the exception of T1, the intercostal nerves do not form plexuses. The intercostal nerves innervate much of the torso wall and portions of the upper limb (see the dermatomal map in figure 14.13). The specific innervation pattern of the T1–T12 nerves is as follows:

- A portion of the anterior ramus of T1 helps form the brachial plexus, but a branch of it is housed within the first intercostal space.
- The anterior ramus of nerve T2 emerges from its intervertebral foramen and innervates the intercostal muscles of the second intercostal space. Additionally, a branch of T2 transmits sensory information from the skin covering the axilla and the medial surface of the arm.
- Anterior rami of nerves T3–T6 follow the costal grooves of the ribs to innervate the intercostal muscles and receive sensations from the anterior and lateral chest wall.
- Anterior rami of nerves T7–T12 innervate not only the inferior intercostal spaces but also the abdominal muscles and their overlying skin.

WHAT DID YOU LEARN?

17 In general, what do the intercostal nerves innervate?

14.5d Cervical Plexuses

LEARNING OBJECTIVES

25. List the nerves of the cervical plexuses.
26. Explain the action of the phrenic nerve.

The left and right **cervical plexuses** are located deep on each side of the neck, immediately lateral to cervical vertebrae C_1–C_4 **(figure 14.15)**. They are formed primarily by the anterior rami of spinal nerves C1–C4. The fifth cervical spinal nerve is not considered part of the cervical plexus, although it contributes some axons to one of the plexus branches. Branches of the

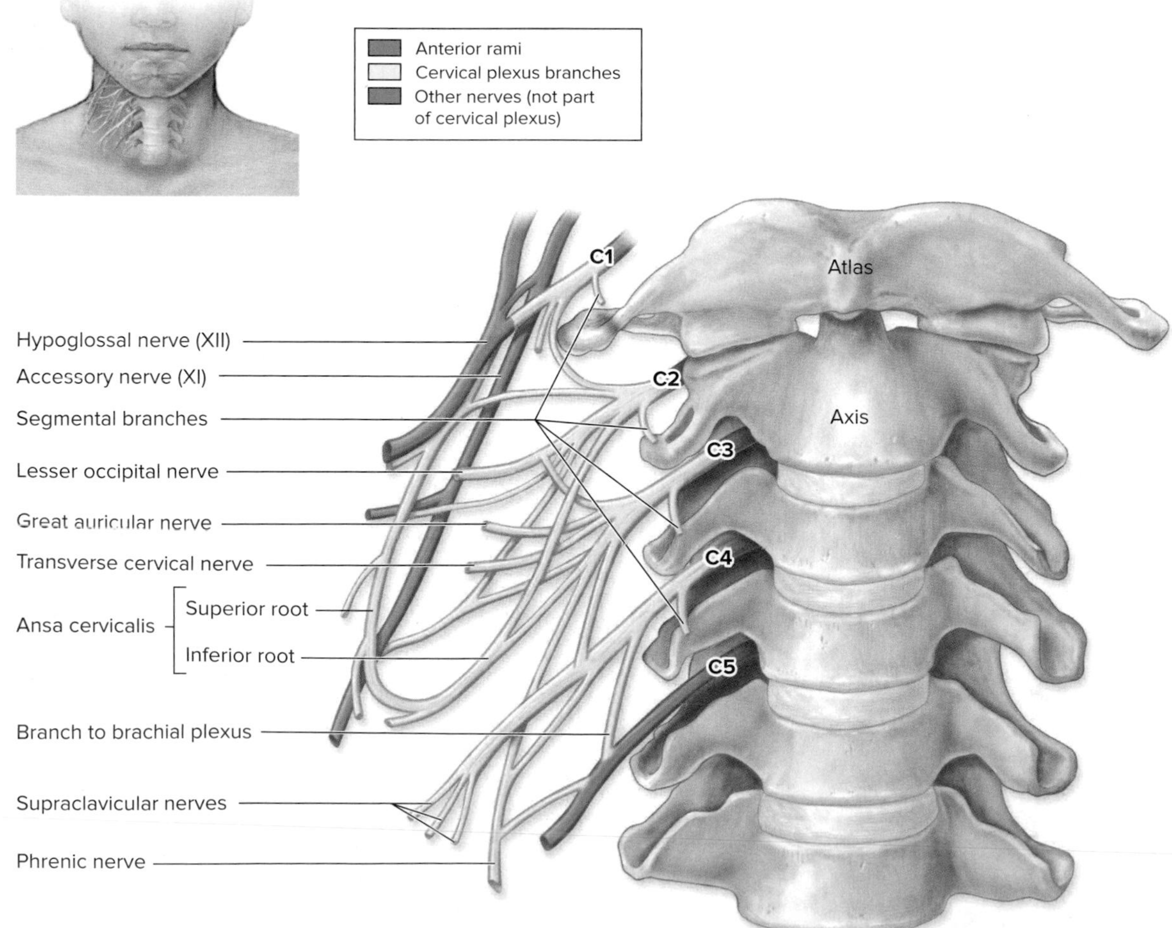

Figure 14.15 Cervical Plexus. Anterior rami of nerves C1–C4 form the cervical plexus, which innervates the skin and many muscles of the neck. AP|R

Table 14.3	Branches of the Cervical Plexuses	
Nerves	**Anterior Rami**	**Innervation**
	MOTOR BRANCHES	
Ansa cervicalis		Geniohyoid; infrahyoid muscles (omohyoid, sternohyoid, sternothyroid, and thyrohyoid)
Superior root	C1, C2	
Inferior root	C2, C3	
Segmental branches	C1–C4	Anterior and middle scalenes
	CUTANEOUS BRANCHES	
Greater auricular	C2, C3	Skin on ear; connective tissue capsule covering parotid gland
Lesser occipital	C2	Skin of scalp superior and posterior to ear
Supraclavicular	C3, C4	Skin on superior part of chest and shoulder
Transverse cervical	C2, C3	Skin on anterior part of neck

Note: Although CN XII (hypoglossal) and the phrenic nerve (C3, C4, C5) travel with the nerves of the cervical plexus, hypoglossal and phrenic are not considered part of this plexus.

INTEGRATE

CONCEPT CONNECTION

Recall from chapter 11 that the named skeletal muscles are innervated by specific spinal nerves. We are now learning the composition of these nerves in this chapter and which muscles they innervate. Should one or more of these spinal nerves become damaged, there will be a loss in sensory and motor function. Thus, if you recall the movements of a skeletal muscle and now you learn what nerve innervates that muscle, you may deduce the clinical symptoms a patient would have if one of these nerves were damaged.

INTEGRATE

LEARNING STRATEGY

This mnemonic will help you remember the nerves that innervate the diaphragm: **C three, four, and five keep the diaphragm alive.**

cervical plexuses innervate anterior neck muscles (see section 11.3d) as well as the skin of the neck and portions of the head and shoulders. The branches of the cervical plexuses are described in detail in **table 14.3**. Note that in the tables, *Motor branches* indicates the portion that relays motor output to skeletal muscles and *Cutaneous branches* indicates the portion that relays sensory input from the skin.

One important branch of the cervical plexus is the **phrenic** (fren′ik; *phren* = diaphragm) **nerve,** which is formed primarily from the C4 nerve and some contributing axons from C3 and C5. The phrenic nerve extends through the thoracic cavity to innervate the thoracic diaphragm, which is the primary skeletal muscle of breathing (see section 11.5).

WHAT DO YOU THINK?

8 Spinal cord injuries often alter an individual's ability to breathe. At what level of spinal cord injury would an individual no longer be able to breathe: C1, T1, L1, or S1? Explain.

WHAT DID YOU LEARN?

18 What is the action of the phrenic nerve?

14.5e Brachial Plexuses

LEARNING OBJECTIVES

27. Explain the structure of the brachial plexus, including the three trunks, two divisions, and three cords.
28. Describe the distribution of the five major nerve branches that arise from the three cords.

The left and right **brachial plexuses** are networks of nerves that supply the upper limb. Each brachial plexus is formed by the anterior rami of spinal nerves C5–T1 **(figure 14.16)**. The components of the brachial plexus extend laterally from the neck, pass superior to the first rib, and then continue into the axilla. Each brachial plexus innervates the pectoral girdle and the entire upper limb of one side.

Structure of the Brachial Plexus

Structurally, each brachial plexus is more complex than a cervical plexus and is composed of anterior rami, trunks, divisions, and cords when examined from a medial to lateral perspective. The **anterior rami** (sometimes called *roots*) of the brachial plexus are simply the continuations of the anterior rami of spinal nerves C5–T1. These rami emerge through the intervertebral foramina and extend to the neck. The five rami unite in the posterior triangle of the neck to form the **superior, middle,** and **inferior trunks.** Nerves C5 and C6 unite to form the superior trunk; nerve C7 remains as the middle trunk, and nerves C8 and T1 unite to form the inferior trunk.

Portions of each trunk divide deep to the clavicle into an **anterior division** and a **posterior division** (shown in green and purple, respectively, in figure 14.16). These contain axons that primarily innervate the anterior and posterior parts of the upper limb, respectively.

At the axilla, these anterior and posterior divisions converge to form three cords. They are named with respect to their position near the axillary artery:

- The **posterior cord** is posterior to the axillary artery and is formed by the posterior divisions of the superior, middle, and inferior trunks; therefore, it contains portions of C5–T1 nerves.
- The **medial cord** is medial to the axillary artery and is formed by the anterior division of the inferior trunk; it contains portions of nerves C8–T1.

INTEGRATE

LEARNING STRATEGY

In general, nerves from the *anterior division* of the brachial plexus tend to innervate muscles that *flex* the parts of the upper limb. Nerves from the *posterior division* of the brachial plexus tend to innervate muscles that *extend* the parts of the upper limb.

Table 14.4 Branches of the Brachial Plexus (continued)

Smaller Branches of the Brachial Plexus	Anterior Rami	Motor Innervation	Cutaneous Innervation
Dorsal scapular	C5	Rhomboids, levator scapulae	
Long thoracic	C5–C7	Serratus anterior	
Lateral pectoral	C5–C7	Pectoralis major	
Medial pectoral	C8–T1	Pectoralis major Pectoralis minor	
Medial cutaneous nerve of arm	C8–T1		Medial side of arm
Medial cutaneous nerve of forearm	C8–T1		Medial side of forearm
Nerve to subclavius	C5–C6	Subclavius	
Suprascapular	C5–C6	Supraspinatus, infraspinatus	
Subscapular nerves	C5–C6	Subscapularis, teres major	
Thoracodorsal (nerve to latissimus dorsi)	C6–C8	Latissimus dorsi	

INTEGRATE

CLINICAL VIEW 14.5

Brachial Plexus Injuries

Injuries to parts of the brachial plexus are fairly common, especially in individuals aged 15–25. Minor plexus injuries may be treated by simply resting the limb. More severe brachial plexus injuries may require nerve grafts or nerve transfers, and for very severe injuries, no effective treatment exists.

Axillary Nerve Injury

The axillary nerve can be compressed within the axilla, or it can be damaged if the surgical neck of the humerus is broken (recall that the axillary nerve extends posterior to the surgical neck of the humerus). A patient whose axillary nerve is damaged has great difficulty abducting the arm due to paralysis of the deltoid muscle, as well as anesthesia (loss of sensation) along the superolateral skin of the arm.

Radial Nerve Injury

The radial nerve is especially subject to injury during humeral shaft fractures or in injuries to the lateral elbow. Nerve damage results in paralysis of the extensor muscles of the forearm, wrist, and fingers. A common clinical sign of radial nerve injury is *wrist drop,* where the patient is unable to extend his or her wrist. The patient also experiences anesthesia along the posterior arm, the forearm, and the part of the hand normally supplied by this nerve.

Posterior Cord Injury

The posterior cord of the brachial plexus (which includes the axillary and radial nerves) may be injured by improper use of crutches, a condition called **crutch palsy.** Similarly, the posterior cord can also be compressed if a person drapes the upper limb over the back of a chair for an extended period of time. Because this can happen if someone passes out in a drunken stupor, this condition is also referred to as **drunkard's paralysis.**

Median Nerve Injury

The median nerve may be impinged or compressed as a result of carpal tunnel syndrome (see Clinical View: 11.8 "Carpal Tunnel Syndrome") or by any deep laceration of the wrist. Median nerve injury often results in paralysis of the thenar group of muscles. The classic sign of median nerve injury is the **ape hand deformity,** which develops over time as the thenar eminence wastes away until the hand eventually resembles that of an ape (apes lack well-developed thumb muscles). The lateral two lumbricals are also paralyzed, and sensation is lost in the part of the hand supplied by the median nerve.

Ulnar Nerve Injury

The ulnar nerve may be injured by fractures or dislocations of the elbow because of this nerve's close proximity to the medial epicondyle of the humerus. When you "hit your funny bone," you actually have hit your ulnar nerve. Most of the intrinsic hand muscles are paralyzed, so the person is unable to adduct or abduct the fingers. In addition, the person experiences sensory loss along the medial side of the hand. A clinician can test for ulnar nerve injury by having a patient hold a piece of paper tightly between the fingers as the doctor tries to pull it away. If the person has weak or paralyzed interossei muscles, the paper can be easily extracted.

Superior Trunk Injury

The superior trunk of the brachial plexus can be injured by excessive separation of the neck and shoulder, as when a person riding a motorcycle is flipped from the bike and lands on the side of the head. A superior trunk injury affects the C5 and C6 anterior rami, so any brachial plexus branch that has these nerves is also affected to some degree.

Inferior Trunk Injury

The inferior trunk of the brachial plexus can be injured if the arm is excessively abducted, as when a neonate's arm is pulled too hard during delivery. In children and adults, inferior trunk injuries happen when grasping something above the head in order to break a fall—for example, grabbing a branch to keep from falling out of a tree. An inferior trunk injury involves the C8 and T1 anterior rami, so any brachial plexus branch that is formed from these nerves (such as the ulnar nerve) also is affected to some degree.

14.5f Lumbar Plexuses

LEARNING OBJECTIVES

29. Identify the spinal nerves that make up the lumbar plexus.
30. Compare and contrast the femoral and obturator nerve composition and distribution.

The left and right **lumbar plexuses** are formed from the anterior rami of spinal nerves L1–L4 located lateral to the L_1–L_4 vertebrae and along the psoas major muscle in the posterior abdominal wall (**figure 14.17**). This plexus innervates the inferior abdominal wall, anterior thigh, medial thigh, and skin of the medial leg. The lumbar plexus is structurally less complex than the brachial plexus. However, like the brachial plexus, the lumbar plexus is subdivided into an anterior division and a posterior division. The primary nerves of the lumbar plexus are listed in **table 14.5**.

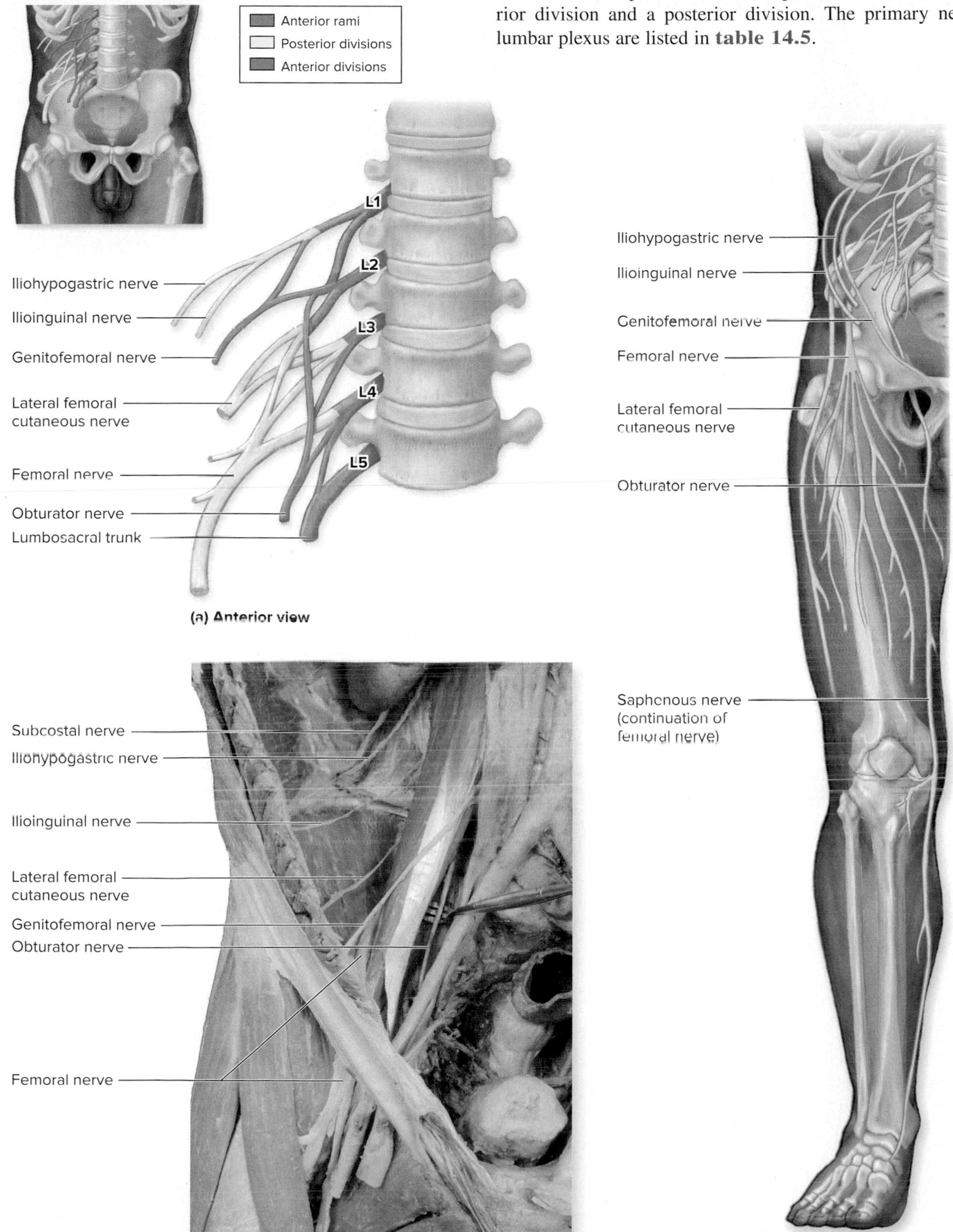

Figure 14.17 Lumbar Plexus. (*a*) Anterior rami of nerves L1–L4 form the lumbar plexus. (*b*) Cadaver photo shows the components of the lumbar plexus. (*c*) Pathways of lumbar plexus nerves. AP|R

Table 14.5	Branches of the Lumbar Plexus		
Main Branch	**Anterior Rami**	**Motor Innervation**	**Cutaneous Innervation**
Femoral Nerve 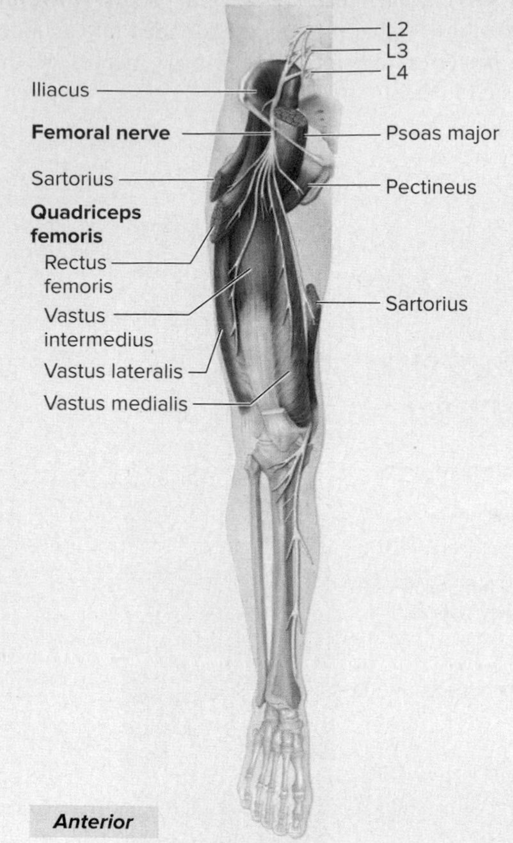 	L2–L4	**Anterior thigh muscles** Quadriceps femoris (*extends knee*) Rectus femoris (*flexes hip*) Vastus lateralis Vastus intermedius Vastus medialis Iliopsoas (*flexes hip*) Sartorius (*flexes hip and knee*) Pectineus[1] (*flexes hip*)	Anterior thigh Inferomedial thigh Medial side of leg Most medial aspect of foot 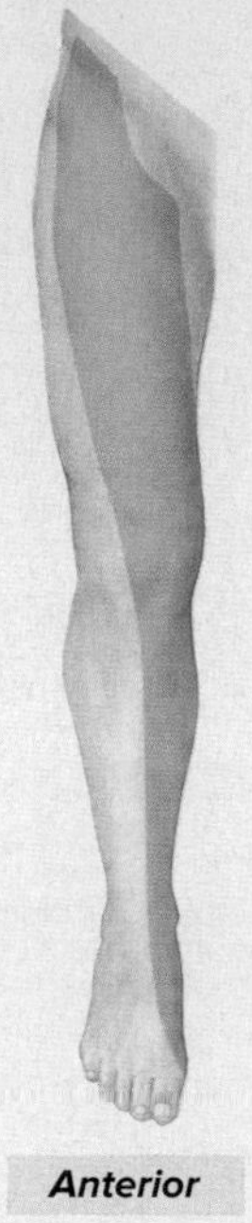
Obturator Nerve 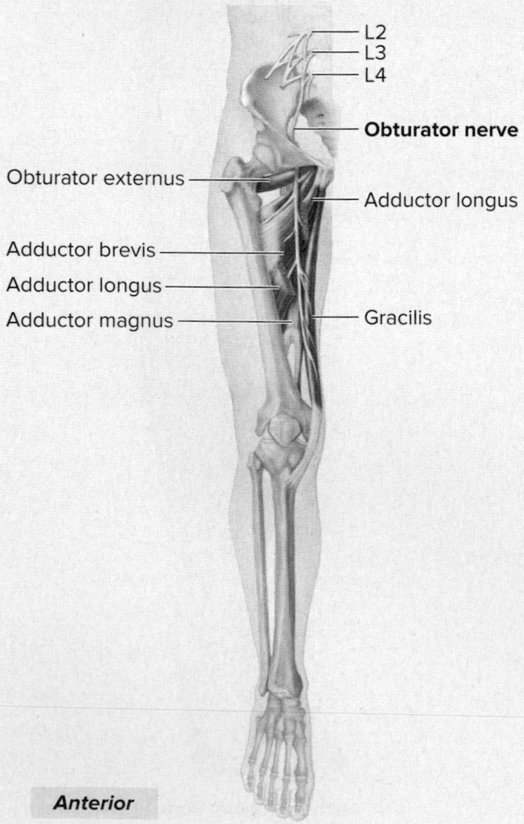 	L2–L4	**Medial thigh muscles** (*adduct and flex hip*) Adductors Gracilis Pectineus[1] Obturator externus (*laterally rotates thigh*)	Superomedial thigh 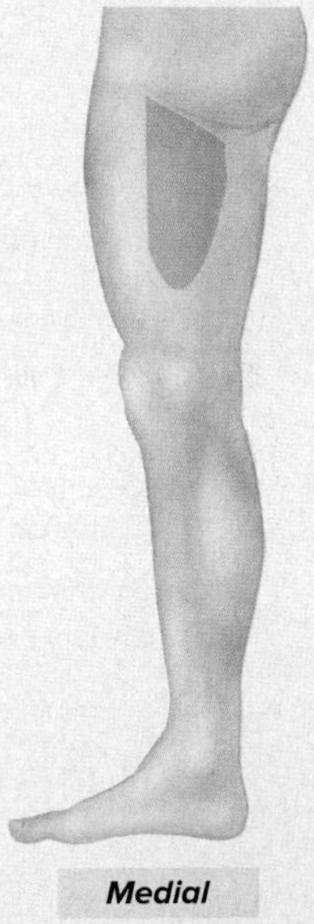

1. Pectineus may be innervated by the femoral nerve, obturator nerve, or branches from both nerves.

Table 14.5	Branches of the Lumbar Plexus (continued)		
Smaller Branches of the Lumbar Plexus	**Anterior Rami**	**Motor Innervation**	**Cutaneous Innervation**
Iliohypogastric	L1	Partial innervation to abdominal muscles (*flex vertebral column*)	Superior lateral gluteal region Inferior abdominal wall
Ilioinguinal	L1	Partial innervation to abdominal muscles (*flex vertebral column*)	Inferior abdominal wall Scrotum (males) or labia majora (females)
Genitofemoral	L1, L2		Small area in anterior superior thigh Scrotum (males) or labia majora (females)
Lateral femoral cutaneous	L2, L3		Anterolateral thigh

The main nerve of the posterior division of the lumbar plexus is the **femoral nerve.** This nerve innervates the anterior thigh muscles, such as the quadriceps femoris (knee extensor) and the sartorius, psoas, and iliacus (hip flexors, see sections 11.9a and b). It also receives sensory input from the skin of the anterior and inferomedial thigh as well as the medial aspect of the leg.

The main nerve of the anterior division is the **obturator nerve,** which extends through the obturator foramen of the os coxae to the medial thigh. There, the nerve innervates the medial thigh muscles (which adduct the thigh, see section 11.9a) and conducts sensory input from the superomedial skin of the thigh. Smaller branches of each lumbar plexus innervate the abdominal wall, portions of the external genitalia, and the inferior portions of the abdominal muscles (table 14.5; see also section 11.6).

WHAT DID YOU LEARN?

22 Which nerve of the lumbar plexus might you have damaged if you have difficulty extending your knee?

14.5g Sacral Plexuses

LEARNING OBJECTIVES

31. List the spinal nerves that form the sacral plexus.
32. Describe the composition of the sciatic nerve, and compare its branches.

The left and right **sacral plexuses** are formed from the anterior rami of spinal nerves L4–S4 and are located immediately inferior to the lumbar plexuses **(figure 14.18)**. The lumbar and sacral plexuses are sometimes considered together as the *lumbosacral plexus*. The nerves emerging from a sacral plexus innervate the gluteal region, pelvis, perineum, posterior thigh, and almost all of the leg and foot.

The anterior rami of the sacral plexus are organized into an anterior division and a posterior division. The nerves that are formed from the anterior division tend to innervate muscles that flex (or plantar flex) parts of the lower limb, whereas the posterior division nerves tend to innervate muscles that extend (or dorsiflex) part of the lower limb. **Table 14.6** lists the main and smaller nerves of the sacral plexus.

The **sciatic** (sī-at′ik) **nerve,** also known as the *ischiadic* (is-kē-at′ik; hip joint) *nerve,* is the largest and longest nerve in the body. It is formed from portions of both the anterior and posterior divisions of the sacral plexus. This nerve projects from the pelvis through the greater sciatic notch of the os coxae and extends into the posterior region of the thigh. The sciatic nerve is actually composed of two divisions—the *tibial division* and the *common fibular division*—wrapped in a common sheath.

Just superior to the popliteal fossa, the two divisions of the sciatic nerve split into two nerves. The **tibial nerve** is formed from the anterior divisions of the sciatic nerve. In the posterior thigh, the tibial division of the sciatic nerve innervates the hamstrings (except for the short head of the biceps femoris) and the hamstring part of the adductor magnus. It extends within the posterior compartment of the leg, where it innervates the plantar flexors of the foot and the toe flexors (see sections 11.9c and d). In the foot, the tibial nerve splits into the lateral and medial plantar nerves, which innervate the plantar muscles of the foot and receive sensory input from the skin covering the sole of the foot.

The **common fibular** (*common peroneal*) **nerve** is formed from the posterior division of the sciatic nerve. As the common fibular division of the sciatic nerve, it innervates the short head of the biceps femoris muscle (see section 11.9b). Along the lateral knee, as it wraps around the neck of the fibula, this nerve splits into two main branches: the deep fibular nerve and the superficial fibular nerve.

INTEGRATE

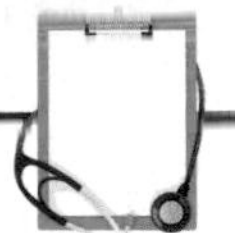

CLINICAL VIEW 14.6
Sacral Plexus Injuries

Some branches of the sacral plexus are readily subject to injury. For example, a poorly placed gluteal intramuscular injection can injure the superior or inferior gluteal nerves, and in some cases even the sciatic nerve. Additionally, a herniated intervertebral disc may impinge on the nerve branches that form the sciatic nerve. Injury to the sciatic nerve produces a condition known as **sciatica** (sī-at′i-kă), which is characterized by pain down the posterior of the thigh and leg.

The common fibular nerve is especially prone to injury due to fracture of the neck of the fibula or compression from a leg cast that is too tight. The anterior and lateral leg muscles may be paralyzed and leave the person unable to dorsiflex and evert the foot. One classic sign of fibular nerve injury is **foot drop.** Because the person cannot dorsiflex the foot to walk normally, he or she compensates by flexing the hip to lift the affected area and keep from tripping or stubbing the toes.

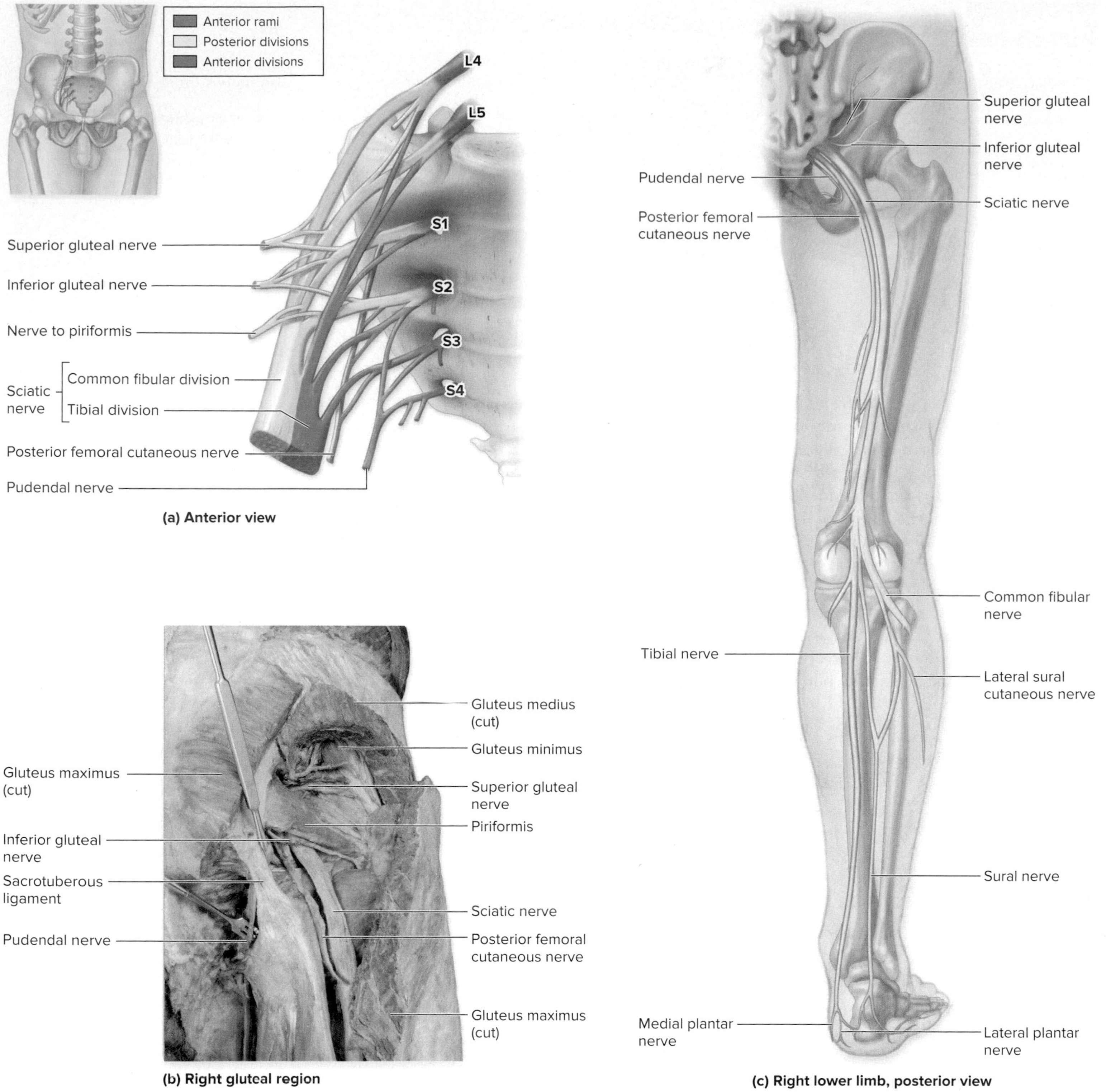

Anterior rami
Posterior divisions
Anterior divisions
L4
L5
S1
S2
S3
S4
Superior gluteal nerve
Inferior gluteal nerve
Nerve to piriformis
Sciatic nerve
Common fibular division
Tibial division
Posterior femoral cutaneous nerve
Pudendal nerve

(a) Anterior view

Gluteus maximus (cut)
Inferior gluteal nerve
Sacrotuberous ligament
Pudendal nerve
Gluteus medius (cut)
Gluteus minimus
Superior gluteal nerve
Piriformis
Sciatic nerve
Posterior femoral cutaneous nerve
Gluteus maximus (cut)

(b) Right gluteal region

Pudendal nerve
Posterior femoral cutaneous nerve
Superior gluteal nerve
Inferior gluteal nerve
Sciatic nerve
Common fibular nerve
Tibial nerve
Lateral sural cutaneous nerve
Sural nerve
Medial plantar nerve
Lateral plantar nerve

(c) Right lower limb, posterior view

Figure 14.18 Sacral Plexus. Anterior rami of nerves L4, L5, and S1–S4 form the sacral plexus. (*a*) The sacral plexus has six roots and both anterior and posterior divisions. (*b*) Cadaver photo reveals the major sacral plexus nerves of the right gluteal region. (*c*) A posterior view shows the distribution of the major nerves of the sacral plexus.

The **deep fibular** (*deep peroneal*) **nerve** extends through the anterior compartment of the leg and terminates between the first and second toes. It innervates the anterior leg muscles (which dorsiflex the foot and extend the toes) and the muscles on the dorsum of the foot (which extend the toes, see sections 11.9c and d). In addition, this nerve receives sensory input from the skin between the first and second toes on the dorsum of the foot.

The **superficial fibular** (*superficial peroneal*) **nerve** extends through the lateral compartment of the leg. Just proximal to the ankle, this nerve becomes superficial along the anterior part of the ankle and dorsum of the foot. The superficial fibular nerve innervates the lateral compartment muscles of the leg (foot evertors and weak plantar flexors, see sections 11.9c and d). It also receives sensory input from most of the dorsal surface of the foot and the anteroinferior part of the leg.

WHAT DID YOU LEARN?

23 What anterior rami form the sacral nerve plexus, and what general areas of the body does it innervate?

Table 14.6 Branches of the Sacral Plexus

Main Branch	Anterior Rami	Motor Innervation	Cutaneous Innervation
Sciatic Nerve (Composed of tibial and common fibular divisions wrapped in a common sheath)	L4–S3	(See tibial and common fibular nerves)	(See tibial and common fibular nerves)
Tibial Nerve 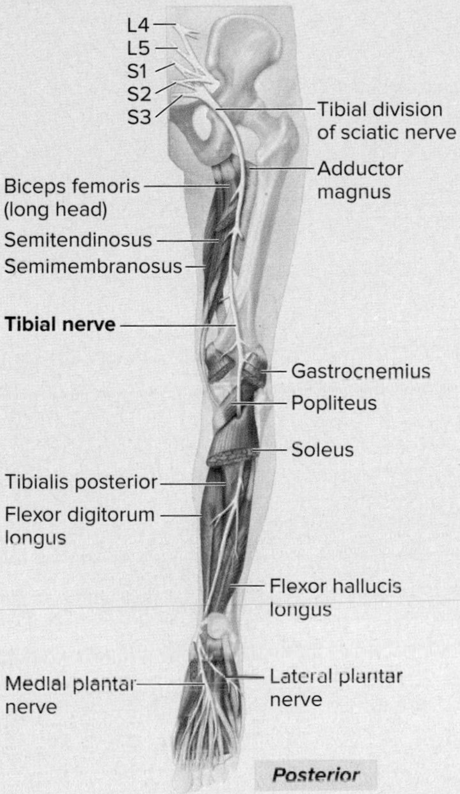	L4–S3	**Posterior thigh muscles** (*extend hip and flex knee*) Long head of biceps femoris Semimembranosus Semitendinosus Part of adductor magnus **Posterior leg muscles** (*flex knee and plantar flex foot*) Flexor digitorum longus Flexor hallucis longus Gastrocnemius Soleus Popliteus Tibialis posterior (*inverts foot*) **Plantar foot muscles** (*via medial and lateral plantar nerve branches*)	Branches to the heel, and via its medial and lateral plantar nerve branches (which supply the sole of the foot) 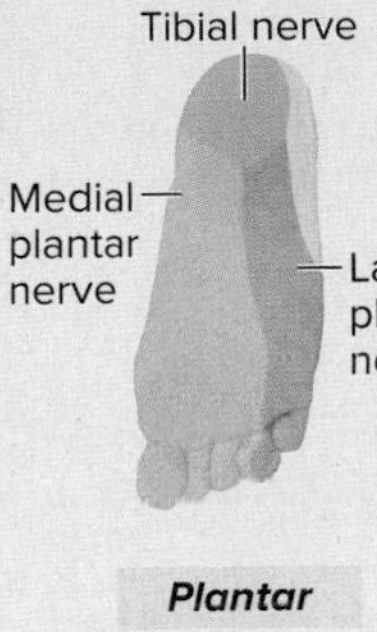
Common Fibular Nerve (Divides into deep fibular and superficial fibular branches) 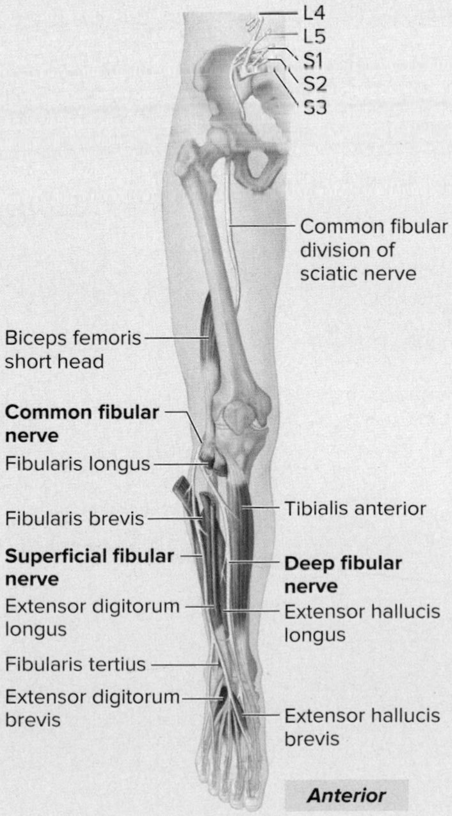	L4–S2	**Short head of biceps femoris** (*flexes knee*); see also deep fibular and superficial fibular nerves	(See deep fibular and superficial fibular nerves)

(continued on next page)

Table 14.6 Branches of the Sacral Plexus (continued)

Main Branch	Anterior Rami	Motor Innervation	Cutaneous Innervation
Deep Fibular Nerve 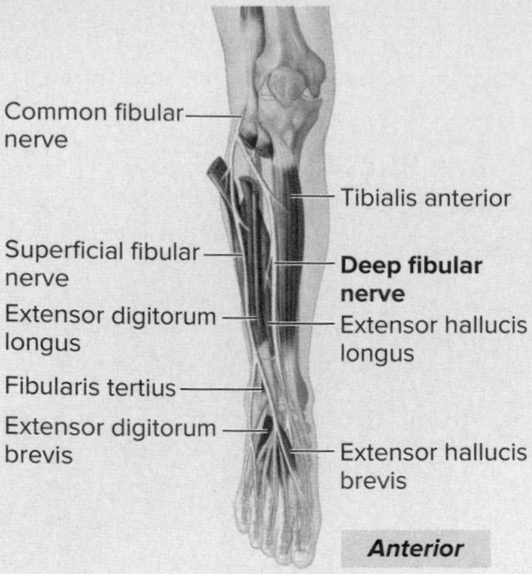 Common fibular nerve; Tibialis anterior; Superficial fibular nerve; **Deep fibular nerve**; Extensor digitorum longus; Extensor hallucis longus; Fibularis tertius; Extensor digitorum brevis; Extensor hallucis brevis; *Anterior*	L4–S1	**Anterior leg muscles** (*dorsiflex foot, extend toes*) Tibialis anterior (*inverts foot*) Extensor hallucis longus Extensor digitorum longus Fibularis tertius **Dorsum foot muscles** (*extend toes*) Extensor hallucis brevis Extensor digitorum brevis	Dorsal interspace between first and second toes 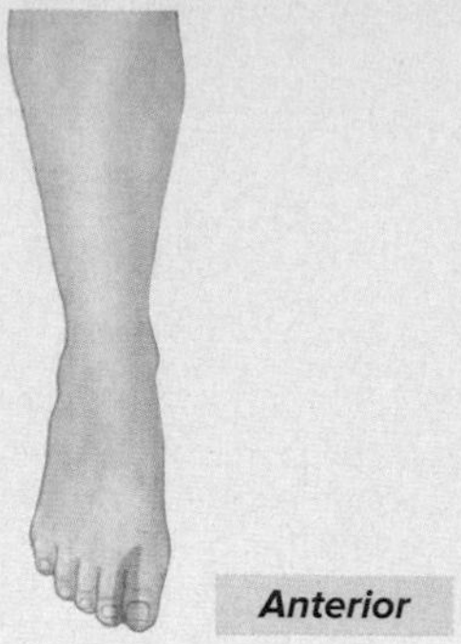 *Anterior*
Superficial Fibular Nerve 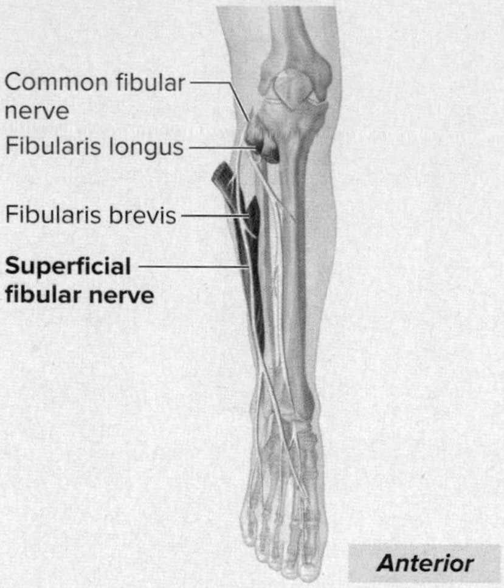 Common fibular nerve; Fibularis longus; Fibularis brevis; **Superficial fibular nerve**; *Anterior*	L5–S2	**Lateral leg muscles** (*evert foot and weakly plantar flex foot*) Fibularis longus Fibularis brevis	Anteroinferior part of leg; most of dorsum of foot 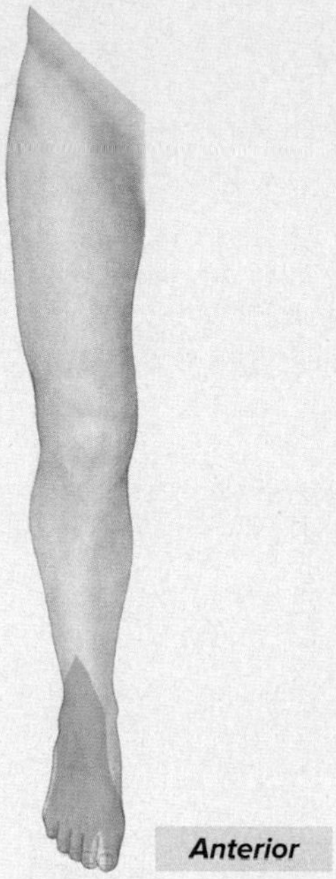 *Anterior*

Smaller Branches of the Sacral Plexus	Anterior Rami	Motor Innervation	Cutaneous Innervation
Inferior gluteal nerve	L5–S2	Gluteus maximus (*extends thigh*)	
Superior gluteal nerve	L4–S1	Gluteus medius, gluteus minimus, and tensor fasciae latae (*abduct thigh*)	
Posterior femoral cutaneous nerve	S1–S3		Skin on posterior thigh
Pudendal nerve	S2–S4	Muscles of perineum, external anal sphincter, external urethral sphincter	Skin on external genitalia

14.6 Reflexes

Here we consider the characteristics of a reflex, the components of a reflex arc, the classification of reflexes, the different spinal reflexes, and how reflexes are tested in a clinical setting.

14.6a Characteristics of Reflexes

LEARNING OBJECTIVES

33. Describe the properties of a reflex.

34. Explain the general function of a reflex.

Reflexes are rapid, preprogrammed, involuntary responses of muscles or glands to a stimulus. An example of a reflex occurs when you accidentally touch a hot burner on a stove. Instantly and automatically, you remove your hand from the stimulus (the hot burner), even before you are completely aware that your hand was touching something extremely hot.

All reflexes have similar properties:

- A *stimulus* is required to initiate a reflex.
- A *rapid response* requires that few neurons are involved and synaptic delay (see section 12.3) is minimal.
- A *preprogrammed response* occurs the same way every time.
- An *involuntary response* requires no conscious intent or preawareness of the reflex activity. Thus, reflexes are usually not suppressed.

A reflex is a survival mechanism; it allows us to quickly respond to a stimulus that may be detrimental to our well-being without having to wait for the brain to process the information. Awareness of the stimulus occurs after the reflex action has been completed, in time to correct or avoid a potentially dangerous situation. (This is possible because sensory input has reached the cerebral cortex.)

WHAT DID YOU LEARN?

24 What are the four main properties of a reflex?

14.6b Components of a Reflex Arc

LEARNING OBJECTIVE

35. List the structures involved in a reflex arc and the steps in its action.

A **reflex arc** includes a sensory receptor, an effector, and the neural wiring between the two. It always begins at a receptor in the PNS, communicates with the CNS, and ends at a peripheral effector, either a muscle or a gland. The number of intermediate steps varies, depending upon the complexity of the reflex. Generally, five steps are involved in a reflex, as illustrated in **figure 14.19** and described here:

1. **A stimulus activates a sensory receptor.** A sensory receptor (dendritic endings of a sensory neuron or specialized receptor cells) responds to external and internal stimuli, such as temperature, pressure, or tactile changes. Proprioceptors are sensory receptors found in muscles and tendons, and a stimulus to a proprioceptor (such as the tapping of tendon) may initiate a reflex as well.
2. **The sensory neuron transmits a nerve signal to the CNS.** A sensory neuron transmits a nerve signal from the receptor to the spinal cord (or brain).
3. **Information from the nerve signal is processed in the integration center by interneurons.** More complex reflexes may use a number of interneurons within the CNS to integrate and process incoming sensory nerve signals and transmit information to the motor neuron. The simplest reflexes do not involve interneurons; rather, the sensory neuron synapses directly on a motor neuron in the CNS.
4. **The motor neuron transmits a nerve signal from the CNS to an effector.** A motor neuron transmits a nerve signal from the CNS to a peripheral effector organ—a gland or a muscle.
5. **The effector responds to the nerve signal from the motor neuron.** An effector is a muscle or a gland that responds to the nerve signal from the motor neuron. This response is intended to counteract or remove the original stimulus.

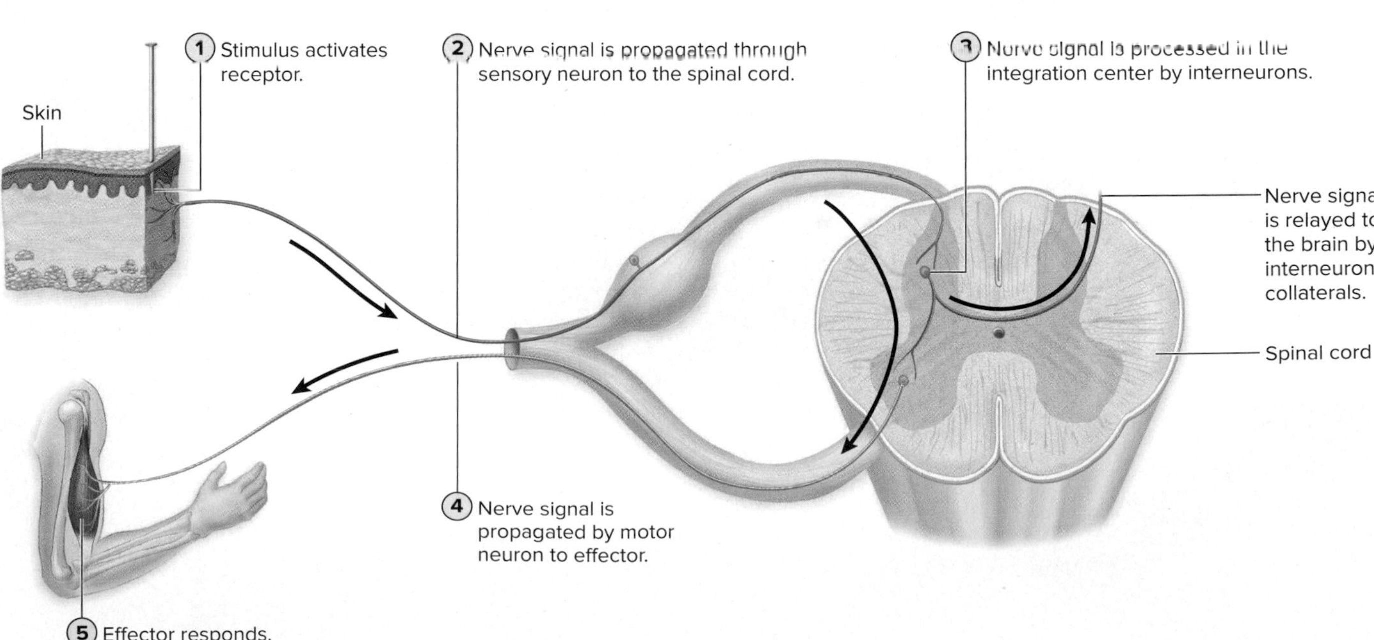

1 Stimulus activates receptor.
2 Nerve signal is propagated through sensory neuron to the spinal cord.
3 Nerve signal is processed in the integration center by interneurons.
Skin
Nerve signal is relayed to the brain by interneuron collaterals.
Spinal cord
4 Nerve signal is propagated by motor neuron to effector.
5 Effector responds.

Figure 14.19 Simple Reflex Arcs. A reflex arc is a neural pathway composed of neurons that control rapid, subconscious, preprogrammed responses to a stimulus. APR

WHAT DID YOU LEARN?

 Identify (in order) the five steps of a reflex arc.

14.6c Classifying Spinal Reflexes

LEARNING OBJECTIVE

36. Explain the five ways a reflex may be classified.

The specific components (or attributes) of a reflex can vary. Some reflexes involve the spinal cord, whereas others involve the brain. Some involve skeletal muscle and some involve other muscle types or glands. Some have only two neurons and some have more. We describe five different ways of classifying a reflex:

- **Spinal reflex or cranial reflex.** A reflex may be identified by the specific area of the central nervous system (integration center) that serves as the processing site. *Spinal reflexes* involve the spinal cord, whereas *cranial reflexes* involve the brain.
- **Somatic reflex or visceral reflex.** This classification criterion is determined by the type of effector that is stimulated by the motor neurons involved in the reflex. *Somatic reflexes* involve skeletal muscle as the effector. *Visceral* (or *autonomic*) *reflexes* involve cardiac muscle, smooth muscle, or a gland as the effector.
- **Monosynaptic reflex or polysynaptic reflex.** A reflex may also be classified by the number of neurons participating in the reflex. A *monosynaptic* (mon′ō-si-nap′tik; *monos* = single) *reflex* has only a sensory neuron and a motor neuron (**figure 14.20**). The axon of the sensory neuron synapses directly on the motor neuron, whose axon projects to the effector. Thus, there is only one synapse between neurons. Monosynaptic reflexes are the simplest, and they are the most rapid. With only one synaptic delay, the response is very prompt. A *polysynaptic* (pol′ē-si-nap′tik; *polys* = many) *reflex* has one or more interneurons positioned between the sensory and the motor neuron. These reflex arcs are more complicated and not as rapid.
- **Ipsilateral reflex or contralateral reflex.** The reflex may also be classified based upon whether it involves only one side of the body. An *ipsilateral reflex* is a reflex in which both the receptor and effector organs are on the same side of the spinal cord. A *contralateral reflex* is a reflex that involves an effector on the opposite side of the body from the receptor that detected the stimulus. Note that this terminology is only applicable to reflexes that involve the limbs. For example, an ipsilateral effect occurs when the muscles in your left arm contract to pull your left hand away from a hot object. In comparison, a contralateral effect occurs when you step on a sharp object with your left foot and then contract the muscles in your right leg to maintain balance as you withdraw your left leg from the damaging object.
- **Innate reflex or acquired reflex.** The reflex may be classified based upon whether you are born with it. An *innate reflex* is a reflex that you are born with, whereas an *acquired reflex* is one that is developed after birth.

As you read about the reflexes described in this section, see if you can determine how each would be classified based upon these five criteria. Keep in mind that there may be instances where you do not have enough information to classify the reflex in all five ways. All of the reflexes discussed in section 14.6d are spinal reflexes with skeletal muscle as the effector. Thus, these reflexes are classified as both spinal reflexes and somatic reflexes.

WHAT DID YOU LEARN?

 How are somatic reflexes and visceral reflexes distinguished?

 What is the major difference between monosynaptic and polysynaptic reflexes?

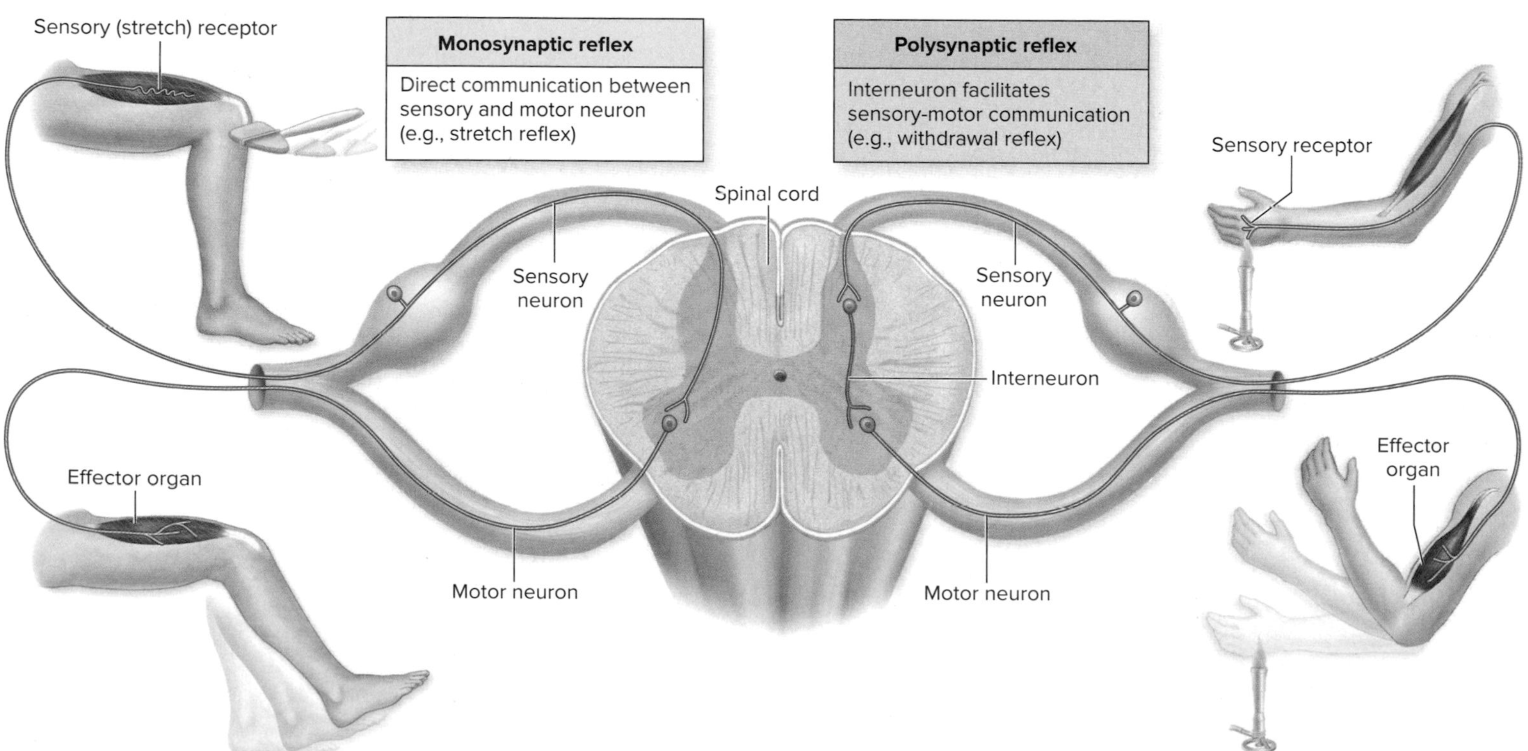

Figure 14.20 Monosynaptic and Polysynaptic Reflexes. The minimal number of neurons and the pathways of a monosynaptic reflex (*left*) are compared to those of a polysynaptic reflex (*right*).

14.6d Spinal Reflexes

LEARNING OBJECTIVE

37. Name and describe four common spinal reflexes.

Four common spinal reflexes, which involve skeletal muscle as the effector, include the stretch reflex, the Golgi tendon reflex, the withdrawal (flexor) reflex, and the crossed-extensor reflex. These reflexes can be initiated by proprioceptors or pain receptors (nociceptors).

Those reflexes involving proprioceptors include stretch reflexes and Golgi tendon reflexes. Recall from section 14.4a that a **proprioceptor** resides in a joint, muscle, or tendon. These sensory receptors specifically detect any change to that structure, such as change in stretch or tension associated with muscle contraction and muscle stretching.

A **muscle spindle** is a proprioceptor that detects changes in stretch within a muscle; for this reason, a muscle spindle is also known as a *stretch receptor* **(figure 14.21)**. A muscle spindle is composed of **intrafusal muscle fibers** surrounded by a connective tissue capsule. These intrafusal muscle fibers lack myofilaments in their central regions and are contractile only at their distal regions. (Actin and myosin are found only at the ends of these fibers.) These muscle fibers are innervated by both sensory neurons (which relay nerve signals to the spinal cord) and **gamma (γ) motor neurons,** so named because *gamma* refers to motor neurons with small-diameter axons. Gamma motor neurons stimulate the contractile fibers at the distal ends of the intrafusal muscle fibers to contract, which elongates the inner portion of the muscle spindle fiber, causing the muscle spindle to be more sensitive to any additional stretch. (Thus, the gamma motor neurons function prior to the reflex to increase the sensitivity of a muscle to stretch.)

Around the muscle spindle are **extrafusal muscle fibers,** which are innervated by **alpha (α) motor neurons,** so named because these motor neurons have the largest diameter axons. (One study tip to remember the difference between the two neuron types is that **g**amma **g**oes within the muscle, and **a**lpha wraps **a**round the muscle.) A muscle spindle is associated with a type of reflex called a stretch reflex. Alpha motor neurons stimulate the extrafusal muscle fiber of a skeletal muscle to contract.

Stretch Reflex

The **stretch reflex** is a reflex that is initiated by a muscle spindle proprioceptor and involves a muscle reflexively contracting in response to stretching of a muscle (figure 14.21). When the muscle spindle is stretched, the sensation is detected by sensory neurons that are wrapped around the intrafusal muscle fibers of the muscle spindle. The sensory neurons transmit nerve signals to the spinal cord (CNS), where they synapse with the alpha motor neurons associated with that muscle. The alpha motor neurons then transmit nerve signals to the extrafusal muscle fibers, which causes the muscle to contract and thus resist the stretch.

The triceps reflex is an example of a stretch reflex. The triceps reflex may be classified as follows: It is a spinal reflex (it involves the spinal cord as the integration center), a somatic reflex (skeletal muscle is the effector), a monosynaptic reflex (involves no interneurons), and

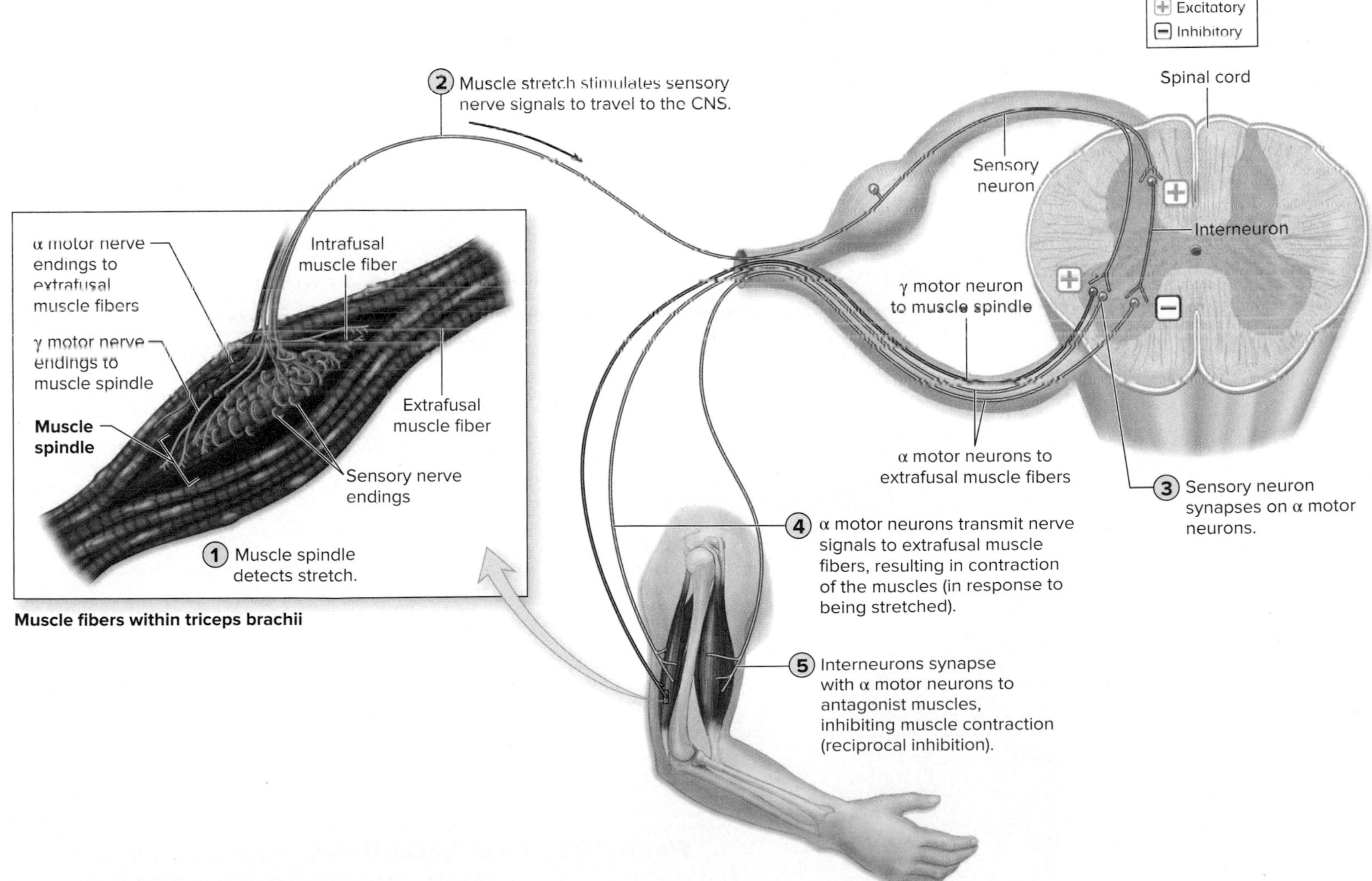

Figure 14.21 Stretch Reflex. A stretch reflex is a simple monosynaptic reflex. A stretching force detected by a muscle spindle results in the contraction of that muscle. Conversely, antagonistic muscle contraction is dampened, in a process called reciprocal inhibition.

an ipsilateral reflex (the receptor and effector are located on the same side of the spinal cord) and is innate (we are born with this reflex). The stimulus (e.g., the reflex hammer tap on the triceps brachii tendon) stretches the muscle spindle in the triceps brachii muscle. Sensory neurons transmit nerve signals to the spinal cord, where they synapse with the alpha motor neurons. The alpha motor neurons transmit nerve signals to the extrafusal muscle fibers in triceps brachii, thereby initiating contraction of the muscle and extending the elbow joint.

Note in figure 14.21 that a stretch reflex also is indirectly involved in a process called **reciprocal inhibition.** When the sensory nerve signals reach the spinal cord, some of the sensory axons synapse with interneurons. These interneurons synapse with alpha motor neurons that inhibit antagonistic muscle contraction. In the case of the triceps reflex, the biceps brachii muscle is the inhibited antagonistic muscle. Thus, as the triceps brachii is stimulated, reciprocal inhibition results in the biceps brachii contraction being dampened, so the triceps movement will not be opposed by the biceps brachii.

The stretch reflex is a monosynaptic reflex, but the corresponding reciprocal inhibition is polysynaptic, because it uses an interneuron within the circuit.

Golgi Tendon Reflex

A **Golgi tendon reflex** is a reflex that is initiated by a Golgi tendon organ proprioceptor. **A Golgi tendon organ** is composed of sensory nerve endings within a tendon or near a muscle-tendon junction and detects change in tension (stretch) in a muscle tendon when a muscle contracts. Whereas the stretch reflex prevents muscles from stretching excessively, the Golgi tendon reflex prevents muscles from doing the opposite: tensing or contracting excessively. The Golgi tendon reflex is a polysynaptic reflex that results in muscle relaxation in response to increased tension at a Golgi tendon organ (**figure 14.22**).

As a muscle contracts, its associated tendon stretches, resulting in increased tension in the tendon and activation of the Golgi tendon organ. Sensory neurons in the Golgi tendon organ transmit nerve signals to interneurons in the spinal cord, which in turn inhibit the alpha motor neurons in the same muscle. When the motor neurons are inhibited, the associated muscle is allowed to relax, thus protecting the muscle and tendon from excessive tension damage.

Note that the sensory neurons also communicate with other interneurons in the spinal cord that stimulate alpha motor neurons for the *antagonistic* muscles. This process is called **reciprocal activation.**

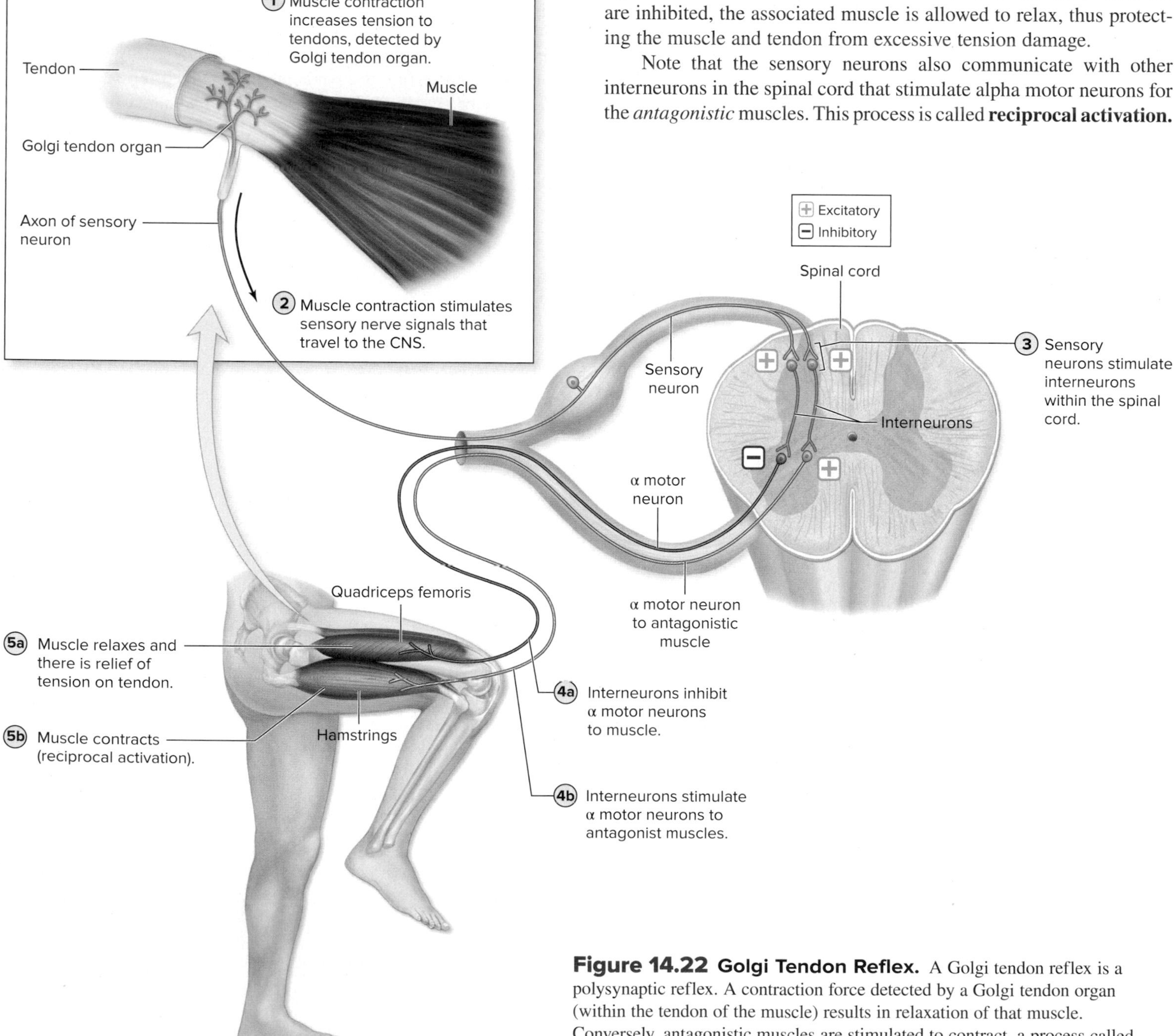

Figure 14.22 Golgi Tendon Reflex. A Golgi tendon reflex is a polysynaptic reflex. A contraction force detected by a Golgi tendon organ (within the tendon of the muscle) results in relaxation of that muscle. Conversely, antagonistic muscles are stimulated to contract, a process called reciprocal activation.

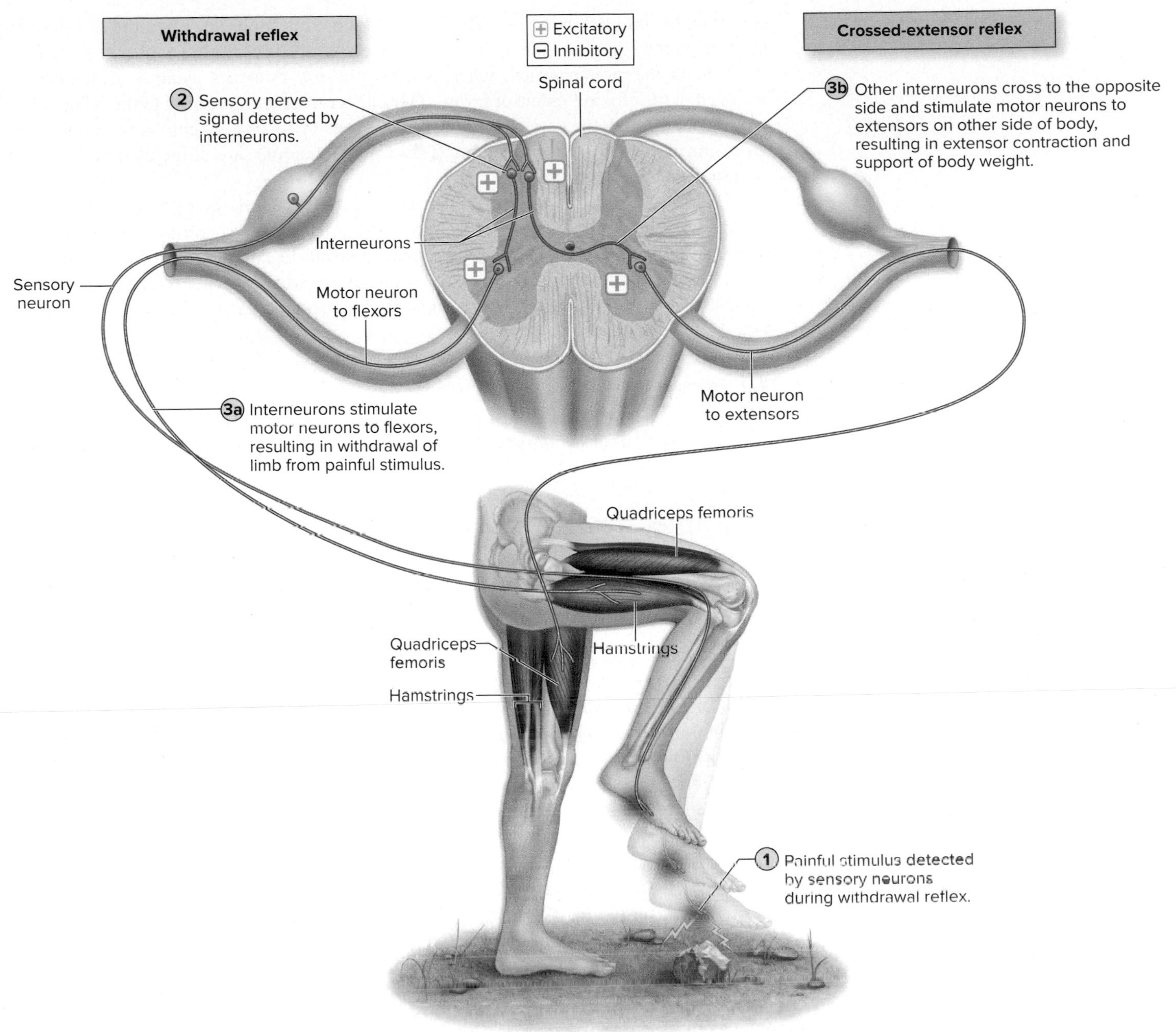

Figure 14.23 Withdrawal and Crossed-Extensor Reflexes. A withdrawal reflex is a polysynaptic reflex that is initiated by a painful stimulus. The crossed-extensor reflex occurs in response to the withdrawal reflex, by stimulating the extensor muscles in the opposite limb and thereby ensuring the opposite limb supports the body's weight.

So, for example, if a Golgi tendon organ in the quadriceps femoris muscle detects excessive tension, then the Golgi tendon reflex ultimately relaxes the quadriceps femoris muscle, and reciprocal activation results in the hamstrings being stimulated to contract.

Golgi tendon reflexes help protect a muscle or tendon from injury due to excessive tension and help ensure that the muscle contraction process occurs smoothly and efficiently. In cases where the muscle is under extreme tension (such as when one lifts a very heavy weight), a Golgi tendon reflex may nullify a stretch reflex (and thus the person drops the weight due to the excessive tension on the muscle). Thus, weightlifters are encouraged to have "spotters" who can help prevent the heavy weight from dropping unexpectedly.

Withdrawal Reflex

A **withdrawal** (*flexor*) **reflex** involves muscles contracting to withdraw the body part away from a painful stimulus. This reflex involves pain receptors, termed nociceptors (see section 16.1d). It is initiated by a painful stimulus, such as touching something very hot or painful (**figure 14.23**). This stimulation initiates a nerve signal that is transmitted by a sensory neuron to the spinal cord. Interneurons receive the sensory nerve signal and stimulate motor neurons to the flexor muscles. These flexor muscles contract in response.

For example, when you step on a sharp object, sensory neurons detect the sensation and transmit nerve signals to the spinal cord. They synapse with interneurons, which stimulate motor neurons to contract the flexor muscles (in this case, the hamstrings of the lower limb). You lift the lower limb up and away from the painful stimulus. In addition, reciprocal inhibition occurs with the extensor (quadriceps) muscles of the same leg, so the hamstrings can contract unimpeded. (Reciprocal inhibition is not shown in figure 14.23.)

Crossed-Extensor Reflex

The **crossed-extensor reflex** often occurs in conjunction with the withdrawal reflex, usually in the lower (weight-bearing) limbs (figure 14.23). In essence, when the withdrawal reflex is occurring in one limb, the crossed-extensor reflex occurs in the other limb. So, when the sensory neurons transmit nerve signals to the spinal cord, some sensory branches synapse with interneurons involved in the stretch reflex, whereas other sensory branches synapse with interneurons involved in the crossed-extensor reflex. These latter interneurons cross to the other side of the spinal cord through the gray commissure and synapse with motor neurons that control antagonistic muscles in the opposite limb. These motor neurons are stimulated and cause these antagonistic muscles to contract.

Figure 14.23 illustrates how the withdrawal reflex causes the right lower limb to flex at the knee, due to contraction of the right hamstring muscles. In contrast, the crossed-extensor reflex stimulates the left quadriceps femoris muscle to contract, so the left limb remains extended and supports the body weight. Thus, the crossed-extensor reflex helps us maintain balance and shift body weight accordingly in these situations.

WHAT DID YOU LEARN?

28 What are the four common spinal reflexes?

29 Identify the Golgi tendon reflex (which is an innate reflex) as (a) spinal or cranial, (b) somatic or visceral, (c) monosynaptic or polysynaptic, and (d) ipsilateral or contralateral.

14.6e Reflex Testing in a Clinical Setting

LEARNING OBJECTIVE

38. Explain the indications of a hypoactive reflex versus those of a hyperactive reflex.

Reflexes can be an important diagnostic tool. Clinicians use them to test specific muscle groups and specific spinal nerves or segments of the spinal cord (**table 14.7**). Although some variation in reflexes is normal, a consistently abnormal reflex response may indicate damage to the nervous system or muscles.

A reflex response may be normal, hypoactive, or hyperactive. A **hypoactive reflex** refers to a reflex response that is diminished or absent. A hypoactive spinal reflex may indicate damage to a segment of the spinal cord, or it may suggest muscle disease or damage to the neuromuscular junction.

A **hyperactive reflex** refers to an abnormally strong response. A hyperactive spinal reflex may indicate damage somewhere in either the brain or spinal cord, especially if it is accompanied by **clonus** (klō′nŭs; *klonus* = tumult), rhythmic oscillations between flexion and extension, when the muscle reflex is tested.

WHAT DID YOU LEARN?

30 What problems could a hypoactive reflex suggest?

Table 14.7 Some Clinically Important Reflexes

Reflex	Spinal Nerve Segments Tested	Normal Action of Effector
Biceps reflex	C5, C6	Flexes elbow when biceps brachii tendon is tapped
Triceps reflex	C6, C7	Extends elbow when triceps brachii tendon is tapped
Abdominal reflexes	T8–T12	Contract abdominal muscles when one side of the abdominal wall is briskly stroked
Cremasteric reflex	L1, L2	Elevates testis (due to contraction of cremaster muscle in scrotum) when medial side of thigh is briskly stroked
Patellar (knee-jerk) reflex	L2–L4	Extends knee when patellar ligament is tapped
Ankle (Achilles) reflex	S1	Plantar flexes ankle when calcaneal tendon is tapped
Plantar reflex	L5, S1	Flexes toes when plantar side of foot is briskly stroked[1]

1. This is the normal reflex response in adults; in adults with spinal cord damage and in normal infants, the **Babinski sign** occurs, which is extension of the great toe and fanning of the other toes.

14.7 Development of the Spinal Cord

LEARNING OBJECTIVE

39. Describe how the neural tube forms the gray matter structures in the spinal cord.

Recall from section 13.1b that the caudal (inferior) part of the neural tube forms the spinal cord. As the caudal part of the neural tube differentiates and specializes, the spinal cord begins to develop (**figure 14.24**). However, this developmental process is much less complex than that of the brain. A hollow **neural canal** in the neural tube develops into the central canal of the spinal cord. Note that the neural canal doesn't shrink in size; rather, the neural tube around it grows at a rapid rate. Thus, as the neural tube walls grow and expand, the neural canal in the newborn appears as a tiny channel called the central canal.

During the fourth and fifth weeks of embryonic development, the neural tube starts to grow rapidly and unevenly. Part of the neural tube forms the outer white matter of the spinal cord, whereas other components form the inner gray matter. By the sixth week of development, a horizontal groove called the **sulcus limitans** (lim′i-tanz; *limes* = boundary) forms in the lateral walls of the central canal (figure 14.24). The sulcus limitans also represents a dividing point in the neural tube as two specific regions become evident: the basal plates and alar plates.

The **basal plates** lie anterior to the sulcus limitans. The basal plates develop into the anterior and lateral horns, motor structures of the gray matter. They also form the anterior part of the gray commissure.

The **alar** (ā′lăr; *ala* = wing) **plates** lie posterior to the sulcus limitans. By about the ninth week of development, the alar plates develop into posterior horns, sensory structures of the gray matter. They also form the posterior part of the gray commissure.

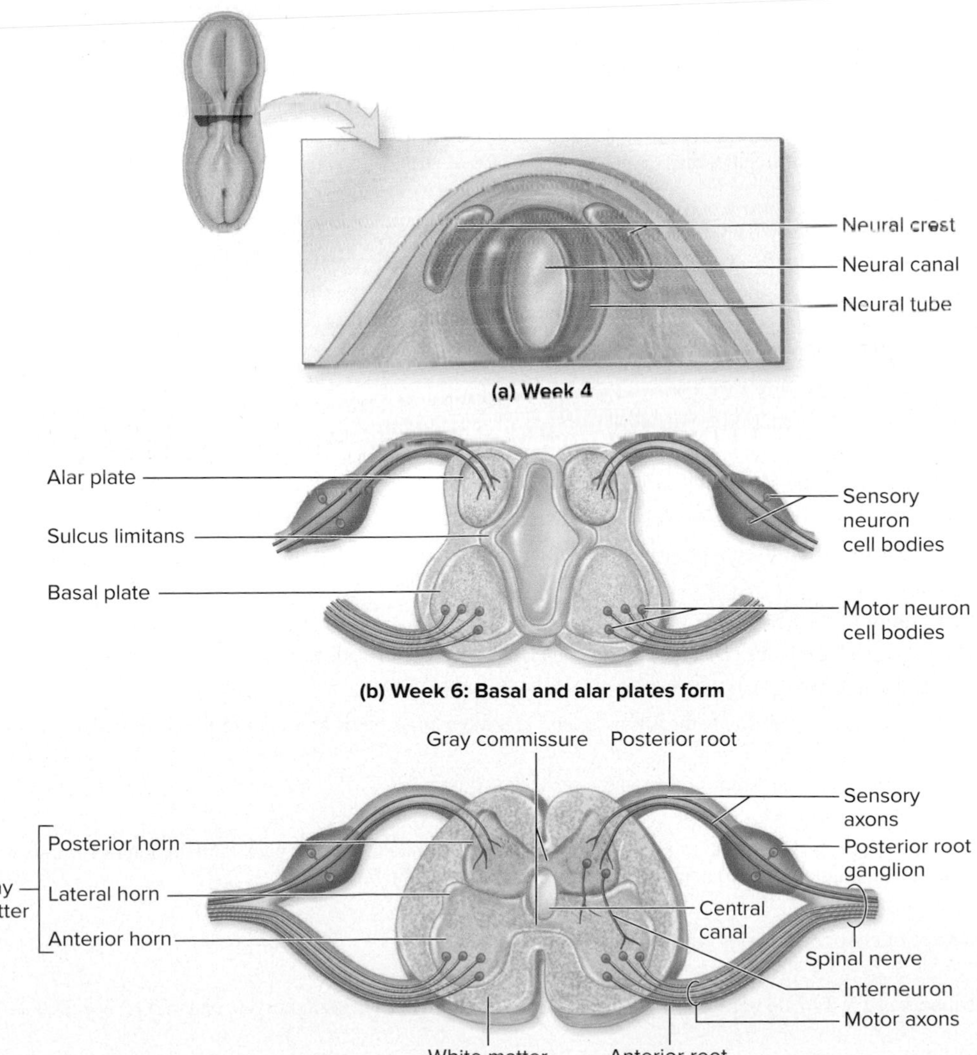

Figure 14.24 Spinal Cord Development. The spinal cord begins development as a tubular extension of the brain. (*a*) A cross section shows the structures of the neural tube of an embryo in week 4 of development. Transverse sections show (*b*) the formation of the basal and alar plates at week 6 and (*c*) the developing spinal cord at week 9.

During the embryonic period, the spinal cord extends the length of the vertebral canal. However, during the fetal period, growth of the vertebral column (and its vertebral canal) outpaces that of the spinal cord. By the sixth fetal month, the spinal cord is at the level of the S_1 vertebra; in contrast, a newborn's spinal cord ends at about the L_3 vertebra. By adulthood, the spinal cord length extends only to the level of the L_1 vertebra. This disproportionate growth explains why the lumbar, sacral, and coccygeal parts of the spinal cord and its associated nerve roots do not lie next to their respective vertebrae (as described in section 14.1).

WHAT DID YOU LEARN?

31 What structures develop from the alar plates, and what structures develop from the basal plates?

CHAPTER SUMMARY

	• The spinal cord is part of the central nervous system (CNS) and has 31 pairs of spinal nerves that extend from it.
14.1 Overview of the Spinal Cord and Spinal Nerves	**14.1a General Functions** • The spinal cord and spinal nerves serve as a pathway for sensory and motor nerve signals and are responsible for reflexes.
	14.1b Spinal Cord Gross Anatomy • The adult spinal cord traverses the vertebral canal and typically ends at the level of the L_1 vertebra.
	14.1c Spinal Nerve Identification and Gross Anatomy • There are 31 pairs of spinal nerves: 8 pairs of cervical nerves, 12 pairs of thoracic nerves, 5 pairs of lumbar nerves, 5 pairs of sacral nerves, and 1 pair of coccygeal nerves.
14.2 Protection and Support of the Spinal Cord	• Protection and support of the spinal cord includes the bony vertebral column, meninges, and cerebrospinal fluid (CSF). • The meninges include the pia mater, arachnoid mater, and dura mater. • A potential subdural space lies between the dura mater and arachnoid mater, the epidural space is external to the dura mater and contains adipose connective tissue and blood vessels, and the subarachnoid space lies between the arachnoid mater and pia mater and contains cerebrospinal fluid (CSF).
14.3 Sectional Anatomy of the Spinal Cord and Spinal Roots	• Gray matter is centrally located and composed of neuron cell bodies, dendrites, unmyelinated axons, and glial cells. • White matter is peripherally located and composed of myelinated axons.
	14.3a Distribution of Gray Matter • Gray matter is composed of three horns: posterior (contains sensory axons and interneurons), anterior (contains cell bodies of somatic motor neurons), and lateral (contains cell bodies of autonomic motor neurons).
	14.3b Distribution of White Matter • White matter is organized into three pairs of funiculi, most of which contain sensory (ascending) tracts and motor (descending) tracts.
14.4 Sensory and Motor Pathways	• The central nervous system (CNS) communicates with the body through conduction pathways that extend through the spinal cord.
	14.4a Overview of Conduction Pathways • Sensory pathways transmit ascending information from sensory receptors to the CNS, whereas motor pathways transmit descending information from the brain to muscles and glands.
	14.4b Sensory Pathways • Sensory pathways use primary, secondary, and tertiary neurons. • The posterior funiculus–medial lemniscal pathway transmits stimuli of fine proprioception, discriminitive touch, precise pressure, and proprioception pressure to the cerebrum (parietal lobe). • The anterolateral pathway transmits stimuli related to crude touch, pressure, pain, and temperature to the cerebrum (parietal lobe). • The spinocerebellar pathway transmits stimuli from proprioceptors to the cerebellum.
	14.4c Motor Pathways • Motor pathways use upper and lower motor neurons. • Somatic motor commands extend through either the direct system (for conscious control) or the indirect system (for unconscious control). • The direct system that extends through the spinal cord consists of the corticospinal tracts. • The indirect system consists of the lateral pathway (rubrospinal tracts) and the medial pathway (reticulospinal, tectospinal, and vestibulospinal tracts).
14.5 Spinal Nerves	• A spinal nerve is formed from the union of an anterior root and a posterior root.
	14.5a General Distribution of Spinal Nerves • Spinal nerves have two branches: A posterior ramus innervates the skin and deep muscles of the back, and an anterior ramus innervates the skin and muscle of the anterior and lateral portions of the trunk and the limbs.
	14.5b Nerve Plexuses • A nerve plexus is a network of interwoven anterior rami. Nerve plexuses occur in pairs.
	14.5c Intercostal Nerves • The anterior rami of spinal nerves T1–T11 do not form a plexus, but rather form the intercostal nerves. Nerve T12 is called a subcostal nerve.
	14.5d Cervical Plexuses • Each cervical plexus is formed from the anterior rami of C1–C4 spinal nerves. It innervates the anterior neck muscles and the skin along the neck and shoulders.

CHAPTER SUMMARY (continued)

14.5 Spinal Nerves (continued)	**14.5e Brachial Plexuses** • Each brachial plexus is formed from the anterior rami of spinal nerves C5–T1. It innervates an upper limb.
	14.5f Lumbar Plexuses • Each lumbar plexus is formed from the anterior rami of spinal nerves L1–L4. It innervates the anterior and medial thigh, the lower abdominal wall, and the skin of the medial leg.
	14.5g Sacral Plexuses • Each sacral plexus is formed from the anterior rami of spinal nerves L4–S4. It innervates most of the lower limb as well as the perineum.
14.6 Reflexes	• A reflex is a rapid, preprogrammed response of muscles or glands to a stimulus.
	14.6a Characteristics of Reflexes • A reflex is rapid, preprogrammed, involuntary response to a stimulus.
	14.6b Components of a Reflex Arc • The five steps of a reflex are (1) activation of a receptor by a stimulus, (2) nerve signal propagation along a sensory neuron to the CNS, (3) integration and processing of information by interneurons, (4) nerve signal propagation along a motor neuron, and (5) effector response.
	14.6c Classifying Spinal Reflexes • Reflexes may be classified as to whether they are (a) spinal or cranial, (b) somatic or visceral, (c) monosynaptic or polysynaptic, (d) ipsilateral or contralateral, and (e) innate or acquired.
	14.6d Spinal Reflexes • A stretch reflex is monosynaptic and contracts the muscle in response to increased stretch in a muscle spindle. • A Golgi tendon reflex is polysynaptic and prevents muscles from tensing excessively. • A withdrawal reflex is polysynaptic and activates flexor muscles to immediately remove a body part from a painful stimulus. • A crossed-extensor reflex stimulates the extensor muscles in the opposite limb, in response to the withdrawal reflex.
	14.6e Reflex Testing in a Clinical Setting • Reflex testing can help diagnose nervous system or muscular disorders. • Hypoactive reflexes, in which the response is diminished, may indicate spinal cord damage or muscle pathology. • Hyperactive reflexes, in which the response is abnormally strong, may indicate damage to the brain or spinal cord.
14.7 Development of the Spinal Cord	• The neural tube forms basal plates and alar plates. • Basal plates form the anterior horns, lateral horns, and anterior half of the gray commissure. • Alar plates form the posterior horns and the posterior half of the gray commissure.

CHALLENGE YOURSELF

Create and Evaluate

Analyze and Apply

Understand and Remember

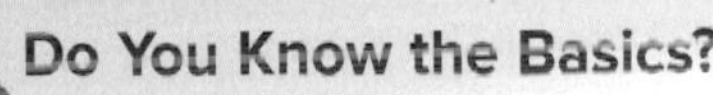

Do You Know the Basics?

1. Identify the meningeal layer immediately deep to the subdural space.
   a. arachnoid mater
   b. pia mater
   c. dura mater
   d. epidural space

2. The anterior root of a spinal nerve contains
   a. axons of both motor and sensory neurons.
   b. axons of sensory neurons only.
   c. interneurons.
   d. axons of motor neurons only.

3. Where are tertiary neurons found?
   a. extending between the posterior horn and anterior horn
   b. extending between the posterior horn and the brainstem
   c. extending between the thalamus and the primary somatosensory cortex
   d. extending between the primary motor cortex and brainstem

4. Which of the following is an example of a sensory pathway?
   a. reticulospinal tract
   b. spinocerebellar tract
   c. corticobulbar tract
   d. tectospinal tract

5. The radial nerve originates from the ____________ plexus.
   a. cervical
   b. brachial
   c. lumbar
   d. sacral

6. Which structure sends motor nerve signals to the deep back muscles and receives sensory nerve signals from the skin of the back?
   a. posterior root
   b. posterior ramus
   c. anterior root
   d. anterior ramus

7. Which statement is accurate about intercostal nerves?
   a. They are formed from the posterior rami of spinal nerves.
   b. They form a thoracic plexus of nerves.
   c. They originate from the thoracic part of the spinal cord.
   d. They innervate the lower limb.

8. The ____________ nerve innervates the anterior thigh muscles and the skin on the anterior thigh.
   a. femoral
   b. obturator
   c. sciatic
   d. tibial

9. A ____________ reflex is monosynaptic and responds to stretching in a muscle spindle.
   a. withdrawal
   b. crossed-extensor
   c. Golgi tendon
   d. stretch

10. Which statement is correct about reflexes?
   a. The patellar reflex tests the S1–S3 spinal nerve segments.
   b. A hypoactive reflex may indicate damage to the neuromuscular junction.
   c. The more hyperactive the reflex, the healthier the individual.
   d. A normal biceps reflex response is extension of the elbow when the biceps tendon is tapped.

11. Identify the spinal cord parts, which spinal nerves are associated with them, and their relationship to the corresponding vertebrae.

12. List the three gray matter horns on each side of the spinal cord, and discuss the neuronal composition of each. In addition, list which types of nuclei (motor or sensory) are located in each horn.

13. Compare the main differences between the posterior funiculus–medial lemniscal pathway and the anterolateral pathway.

14. Describe the location and function of upper and lower motor neurons in the motor pathways.

15. What are the main terminal branches of the brachial plexus, and what muscles do these terminal branches innervate?

16. What anterior rami form the lumbar plexus, and what in general does this plexus innervate?

17. What muscles do the tibial and common fibular nerves innervate?

18. What are the five basic steps involved in a reflex arc?

19. What are the differences between a stretch and Golgi tendon reflex?

20. Where are the basal and alar plates of the neural tube located, and what does each form?

Can You Apply What You've Learned?

Use the following paragraph to answer questions 1–3.

Madeline is an active 18-year-old who fell off her bike, fracturing the medial epicondyle of her humerus. In addition to experiencing severe pain in her elbow, she had numbness along the medial side of her hand. She went to the emergency room and the physician examined her.

1. After x-raying her elbow, the physician performed further tests on Madeline to determine what nerve damage she may have experienced. What nerve likely was injured when she fractured her medial epicondyle?
   a. radial
   b. ulnar
   c. musculocutaneous
   d. median

2. What diagnostic test would best help determine the potential damage to this nerve?
   a. Have Madeline extend her elbow against resistance.
   b. Have Madeline flex her wrist and fingers against resistance.
   c. Have Madeline hold a piece of paper between her fingers while the physician tries to pull the paper away.
   d. Have Madeline flex her elbow while the physician applies gentle pressure to the anterior forearm muscles.

3. What other muscle function could be impaired in this type of injury?
   a. adduction of the thumb
   b. flexion of the thumb
   c. extension of the thumb
   d. abduction of the thumb

4. George tried to lift a heavy box and in the process herniated one of his lumbar intervertebral discs. The herniated disc pinched on nerve roots and most likely caused which of the following ailments?
   a. pain down the back of the leg
   b. inability to flex the thigh
   c. inability to adduct the thigh
   d. pain down the medial side of the leg

5. Carlos experienced some muscle weakness in his right lower limb and went to see his physician. The physician noticed Carlos could not evert his foot and had anesthesia along most of the dorsum of his right foot. Carlos was still able to dorsiflex his foot and invert his foot. Based on these symptoms, what specific nerve likely was injured?
   a. common fibular
   b. tibial
   c. deep fibular
   d. superficial fibular

Can You Synthesize What You've Learned?

1. Arthur dove off a small cliff into water that was shallower than he expected and he hit his head. He is now a quadriplegic, which means that both his upper and lower limbs are paralyzed. Approximately where is the location of his injury? What is the likelihood that Arthur will recover from this injury?

2. Jessica fractured her fibula and had to wear a leg cast for several weeks. When the cast was removed, Jessica had trouble walking normally and experienced "foot drop." What structure likely was pinched by the leg cast, resulting in the foot drop?

3. Juanita was walking barefoot on the sidewalk when she stepped on a piece of glass with her right foot. Her right leg reflexively lifted up, away from the shard of glass. What is this type of reflex called? In addition, Juanita did not fall down when she lifted her leg. What action in her left lower limb helped stabilize her?

INTEGRATE

ONLINE STUDY TOOLS

connect | SMARTBOOK® | AP|R

The following study aids may be accessed through Connect.

Clinical Case Study: A Young Man with Rapidly Developing Muscle Weakness

Interactive Questions: This chapter's content is served up in a number of multimedia question formats for student study

SmartBook: Topics and terminology include spinal cord gross anatomy; protection and support of the spinal cord; sectional anatomy of the spinal cord; spinal cord conduction pathways; spinal nerves; reflexes; development of the spinal cord

Anatomy & Physiology Revealed: Topics include cervical, thoracic, lumbar, and sacral regions of spinal cord; spinal nerves; brachial plexus; cervical plexus, lumbosacral plexus; reflex arc

Design Elements: **Learning Objective** icon: ©McGraw-Hill Education; **What Do You Think?** icon: ©McGraw-Hill Education; **What Did You Learn?** icon: ©McGraw-Hill Education; **Learning Strategy** icon: ©Slavoljub Pantelic/Shutterstock RF; **Clinical View** icon: ©Laia Design Studio/Shutterstock RF; **Concept Connection** icon: ©McGraw-Hill Education; **Challenge Yourself** icon: ©McGraw-Hill Education; **Do You Know the Basics?** icon: ©McGraw-Hill Education; **Can You Apply What You've Learned?** icon: ©McGraw-Hill Education; **Can You Synthesize What You've Learned?** icon: ©McGraw-Hill Education; **Online Study Tools** icon: ©ayzek/Shutterstock RF

chapter

15 Nervous System: Autonomic Nervous System

Module 7: Nervous System

INTEGRATE

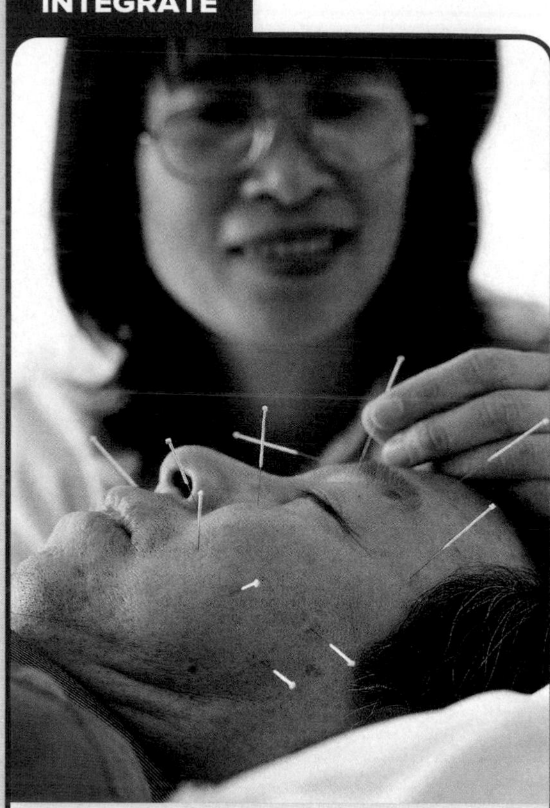

©Tek Image/Science Source

CAREER PATH
Acupuncturist

Acupuncture is one of the mainstays of traditional Chinese medicine (TCM) as well as complementary and alternative medicine (CAM). TCM is based on the premise that disease and pathology are due to a blockage of *qi* (chē; vital energy) at some location in the body. An acupuncturist inserts long, thin, metallic needles into the skin at specific acupuncture sites to promote *qi* circulation. Acupuncture is effective at relieving some types of pain, and many studies suggest it can relieve nausea associated with chemotherapy or other medical treatments. Acupuncture has also been used to treat depression, anxiety, infertility and some neurologic disorders. Some researchers hypothesize that acupuncture manipulates our autonomic nervous system (ANS) in ways that are not yet understood. Because the ANS controls our internal environment and innervates our viscera, practices such as acupuncture may help the workings of these structures.

On a twisting downhill slope, an Olympic skier is concentrating on controlling her body to negotiate the course faster than anyone else in the world. Compared to the spectators in the viewing areas, her pupils are dilated, and her heart is beating faster and pumping more blood to her skeletal muscles. At the same time, all organ system functions not needed in the race are practically shut down. Digestion, urination, and defecation can wait until the race is over. The skier exhibits a state of heightened readiness, called the fight-or-flight response, because the sympathetic division of the autonomic nervous system (ANS) is dominant. When the race is finished and she stops to rest and eat, the parasympathetic division of the ANS will dominate to meet the demands of digesting the meal. Thus, both divisions of the ANS—the sympathetic division and the parasympathetic division—function to regulate the body's response to changing demands—whether they are the demands of exercise or the demands of supplying nutrients to the body.

15.1 Comparison of the Somatic and Autonomic Nervous Systems

We introduce this chapter on the autonomic nervous system by first describing how the nervous system is organized into a somatic nervous system and an autonomic nervous system. Here we compare the somatic and autonomic nervous systems' functional organization, their lower motor neurons, and the CNS regions that control the autonomic nervous system.

15.1a Functional Organization

LEARNING OBJECTIVE

1. List the similarities and differences between the SNS and the ANS.

Recall from section 12.1 that anatomists and physiologists have devised various ways to organize the nervous system both structurally and functionally. Here, we introduce a slightly different way of functionally organizing the nervous system into two systems based upon whether we are conscious of the process. The two components are the somatic nervous system and the autonomic nervous system **(figure 15.1)**.

The **somatic nervous system (SNS)** includes processes that are perceived or controlled consciously (figure 15.1*a*). The **somatic sensory** portion includes detection of stimuli and transmission of nerve signals from the *special senses* (i.e., vision, hearing, equilibrium, smell, and taste), *skin*, and *proprioceptors* (receptors in joints and muscles that detect body position) to the CNS. The sensory portion of the somatic nervous system, for example, allows us to see a beautiful mountain, smell a baby's skin, or taste a delicious dinner. The **somatic motor** portion involves initiation and transmission of nerve signals from the CNS to control *skeletal muscles*. This is exemplified by voluntary activities such as getting out of a chair, picking up a book, and throwing a ball for a dog to chase. Both the sensory input we consciously perceive and the motor output we voluntarily initiate to skeletal muscle involve the cerebrum (see section 13.3c). Reflexive skeletal muscle activity is controlled by the brainstem and spinal cord (see section 14.6), whereas coordination of skeletal muscle movements is the function of the cerebellum (see section 13.6b).

The **autonomic** (aw-tō-nom′ik; *auto* = self, *nomos* = law) **nervous system (ANS),** also called the **autonomic motor** or *visceral motor system,* includes processes regulated below the conscious level (figure 15.1*b*). This system is a motor system only (see section 12.1b). These autonomic motor components initiate and transmit nerve signals from the CNS to cardiac muscle, smooth muscle, and glands. The autonomic nervous system often responds to input from **visceral sensory** components, such as receptors that detect stimuli associated with blood vessels and internal organs (viscera). Some of these sensory neurons, for example, monitor carbon dioxide concentration in the blood, whereas others detect pressure by measuring stretch in smooth muscles of visceral walls. Note that the visceral sensory structures are not part of the ANS per se, but instead simply transmit input from sensory receptors that may result in motor output by the ANS.

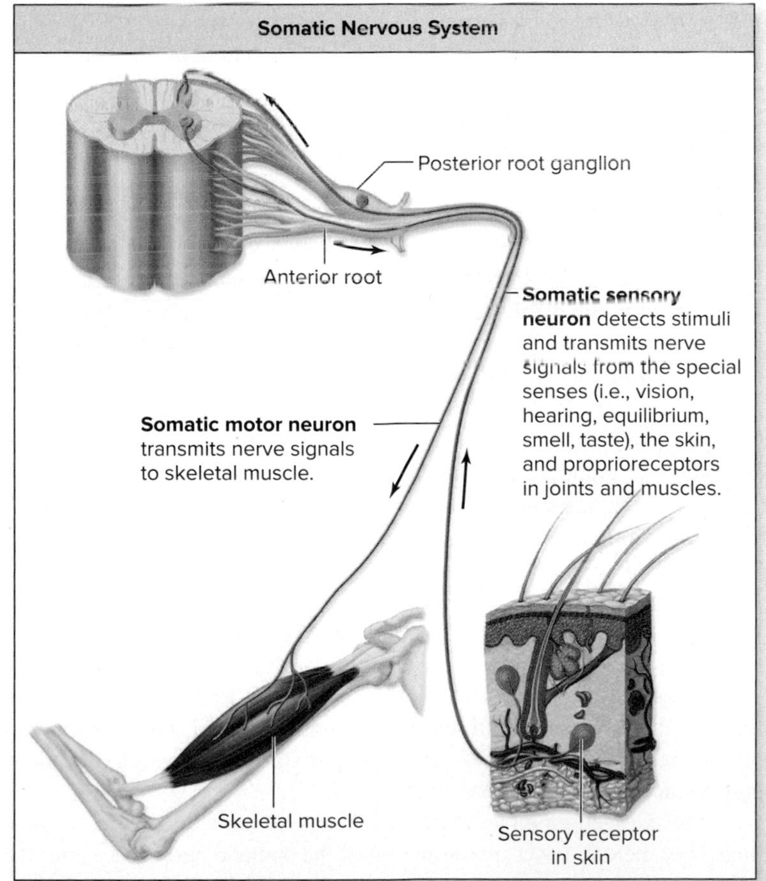

Somatic Nervous System
Posterior root ganglion
Anterior root
Somatic sensory neuron detects stimuli and transmits nerve signals from the special senses (i.e., vision, hearing, equilibrium, smell, taste), the skin, and proprioreceptors in joints and muscles.
Somatic motor neuron transmits nerve signals to skeletal muscle.
Skeletal muscle
Sensory receptor in skin
(a)

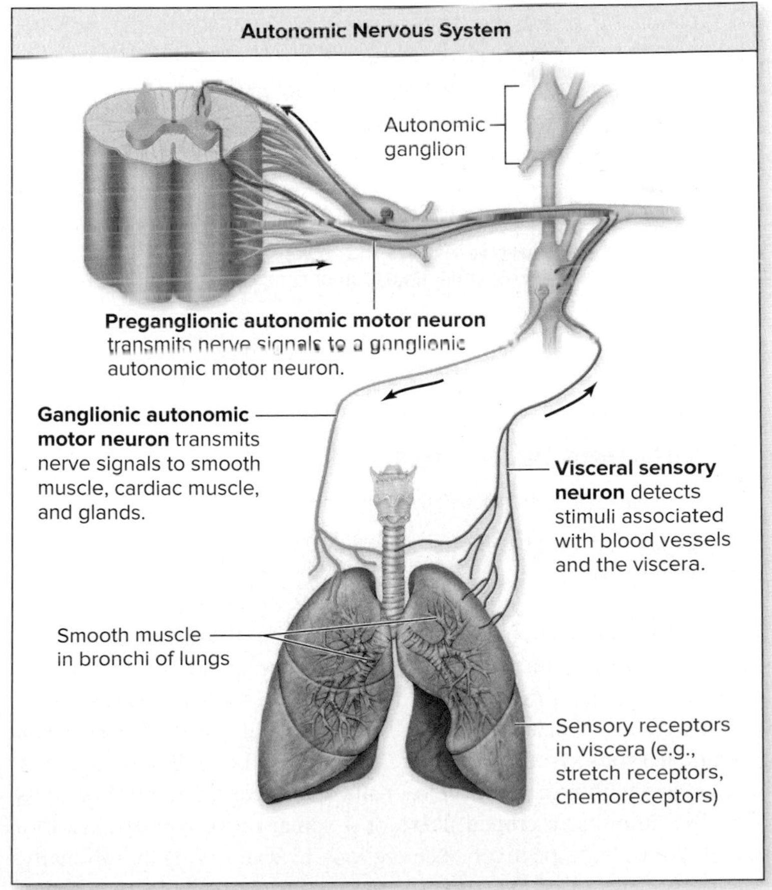

Autonomic Nervous System
Autonomic ganglion
Preganglionic autonomic motor neuron transmits nerve signals to a ganglionic autonomic motor neuron.
Ganglionic autonomic motor neuron transmits nerve signals to smooth muscle, cardiac muscle, and glands.
Visceral sensory neuron detects stimuli associated with blood vessels and the viscera.
Smooth muscle in bronchi of lungs
Sensory receptors in viscera (e.g., stretch receptors, chemoreceptors)
(b)

Figure 15.1 Comparison of Somatic and Autonomic Nervous Systems. The nervous system is functionally organized into the (*a*) somatic nervous system, which involves processes that we consciously perceive and control, and the (*b*) autonomic nervous system, which involves processes that are regulated below the conscious level. (Note that the visceral sensory structures are not part of the ANS per se.)

INTEGRATE

LEARNING STRATEGY

A good way to understand the two-neuron ANS chain is to compare it to the U.S. airline system, which uses connecting flights and airport hubs to transport the maximum number of people in the most cost-effective way.

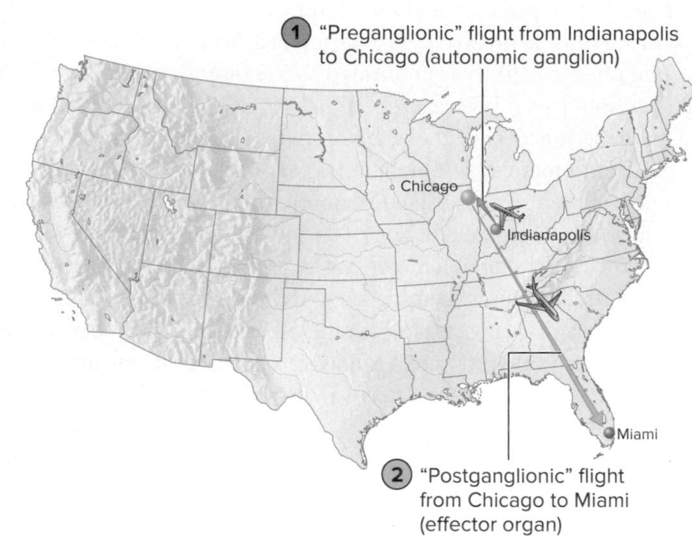

Imagine that you are flying from Indianapolis to Miami for spring break: Your first flight from Indianapolis to Chicago is the **preganglionic axon.** Although flying north to Chicago is out of your way, the airline sends you to an airport hub because it is more efficient to send all Indianapolis passengers to this main location before they take different flights throughout the United States.

The airport hub in Chicago is the **autonomic ganglion,** the point where preganglionic and postganglionic flights meet up. Other preganglionic flights are arriving at the airport hub, and here all these passengers will connect with other flights.

Your connecting flight from Chicago to Miami is the **ganglionic neuron.** This flight will take you to your final destination, just as a postganglionic axon sends a nerve signal to an effector organ. On the plane with you are people from other preganglionic flights who all want to go to Miami as well.

Is using two different flights the most *direct* way for you to get from Indianapolis to Florida? Of course not. But it is the most *cost-efficient* (or *energy-efficient*) way for the airlines to transport and disperse many passengers with a limited number of planes.

The connecting-flight arrangement also allows for the extensive convergence of passengers to an airport hub, and for the extensive divergence of passengers from the hubs to the final destinations. Thus, the two motor neurons used in the ANS allow for neuronal convergence and divergence.

The autonomic nervous system is similar to connecting airline flights and airport hubs in that both try to group and disperse many different items (nerve signals or passengers) with a limited number of neurons or flights.

The function of the ANS is to maintain **homeostasis,** or a constant internal environment (see section 1.6). Thus, the ANS regulates all physiologic processes that must be maintained by the nervous system to keep the body alive, including the regulation of heart rate, blood pressure, body temperature, sweating, and digestion. The ANS keeps these processes within optimal ranges and adjusts the variables to meet changing body needs.

WHAT DID YOU LEARN?

1. What criterion is used to organize the nervous system into the SNS and the ANS? What sensory and motor components are associated with each of the two systems?

15.1b Lower Motor Neurons of the Somatic Versus Autonomic Nervous System

LEARNING OBJECTIVES

2. Compare and contrast lower motor neurons in the SNS and ANS.
3. Describe how the two-neuron chain in the ANS facilitates communication and control.

One significant anatomic difference between the somatic nervous system and autonomic nervous system is the number of lower motor neurons that extend from the CNS (see section 14.4c). A single lower motor neuron extends from the CNS to skeletal muscle fibers in the somatic nervous system (figure 15.1*a*). The cell body of a lower motor neuron lies within the brainstem or the spinal cord, and its axon exits the CNS in either a cranial nerve or a spinal nerve and extends to a skeletal muscle, respectively (see sections 13.9 and 14.5). Additionally, the motor neurons of the somatic nervous system (1) are composed of myelinated axons with a large diameter, allowing for fast propagation of a nerve signal (see section 12.9b), and (2) always release acetylcholine (ACh) neurotransmitter from the motor neuron's synaptic knobs to stimulate or excite the skeletal muscle fiber (see section 10.3a).

In comparison, a chain of two lower motor neurons extends from the CNS to innervate cardiac muscle, smooth muscle, and glands in the ANS (figure 15.1*b* and **figure 15.2**). The first of the two ANS motor neurons is the **preganglionic** (prē′gang-lē-on′ik) **neuron.** Its cell body lies within the brainstem or the spinal cord. A **preganglionic axon** extends from this cell body and exits the CNS in either a cranial nerve or a spinal nerve. This axon projects to an autonomic ganglion in the peripheral nervous system. Preganglionic neurons have myelinated axons that typically are small in diameter, and nerve signals always result in the release of acetylcholine to excite the second neuron.

The second neuron in this pathway is called a **ganglionic neuron,** sometimes referred to as a *postganglionic neuron.* (Note that the term *postganglionic* is not accurate, as the cell body of the neuron resides *within* the ganglion, not *after* the ganglion.) Its cell body resides within an autonomic ganglion. A **postganglionic axon** extends from the cell body to an effector (cardiac muscle, smooth muscle, or a gland). (It is appropriate to call the axon *postganglionic* because it extends from the autonomic ganglion.) Ganglionic neurons have unmyelinated axons that are even smaller in diameter than preganglionic axons. The neurotransmitter released from the ganglionic neuron in response to a nerve signal is either ACh or norepinephrine (NE). Both neurotransmitters can either excite or inhibit an effector depending upon the type of receptors present within the effector (a concept described in section 15.6 in detail). Because motor neurons of the ANS are small and mostly unmyelinated, propagation of nerve signals is relatively slow in comparison to nerve signal propagation along somatic motor axons.

The two-neuron motor pathway in the ANS has a distinctive advantage over the one lower motor neuron of the somatic nervous system: It allows for increasing communication and control. This occurs because of neuronal convergence and neuronal divergence (see section 12.11). **Neuronal convergence** (kon-ver′jens; *convergo* = to incline together) occurs because axons from numerous preganglionic neurons synapse with and influence a single ganglionic neuron. **Neuronal divergence** (di-ver′jens;

Table 15.1	Comparison of Somatic and Autonomic Motor Nervous Systems	
Feature	**Somatic Nervous System**	**Autonomic Nervous Systems**
Functional Organization		
Sensory input	Special senses, skin, proprioceptors	Visceral senses (and some somatic senses)
Effectors	Skeletal muscle fibers	Cardiac muscle cells, smooth muscle cells, glands
CNS regions of control	Cerebrum, thalamus, cerebellum, brainstem, spinal cord	Hypothalamus, brainstem, spinal cord Cerebrum, thalamus, limbic system (regulated by hypothalamus, brainstem, spinal cord)
Motor Neurons		
Number of neurons in pathway	One neuron from CNS: Somatic motor neuron axon extends from CNS to effector	Two neurons from CNS: Preganglionic neuron has preganglionic axon that projects to ganglionic neuron; ganglionic neuron has postganglionic axon that projects to effector
Axon properties	Myelinated and thicker in diameter; fast nerve signal propagation	Preganglionic axons are myelinated and small in diameter Postganglionic axons are unmyelinated and smaller in diameter; both have relatively slow nerve signal propagation
Neurotransmitter released	Acetylcholine (ACh)	Preganglionic axons release ACh Postganglionic axons release either ACh or norepinephrine (NE)
Response of effector	Excitation only	Either excitation or inhibition
Ganglia associated with motor neurons	None	Parasympathetic division: terminal ganglia, intramural ganglia Sympathetic division: sympathetic trunk ganglia, prevertebral ganglia

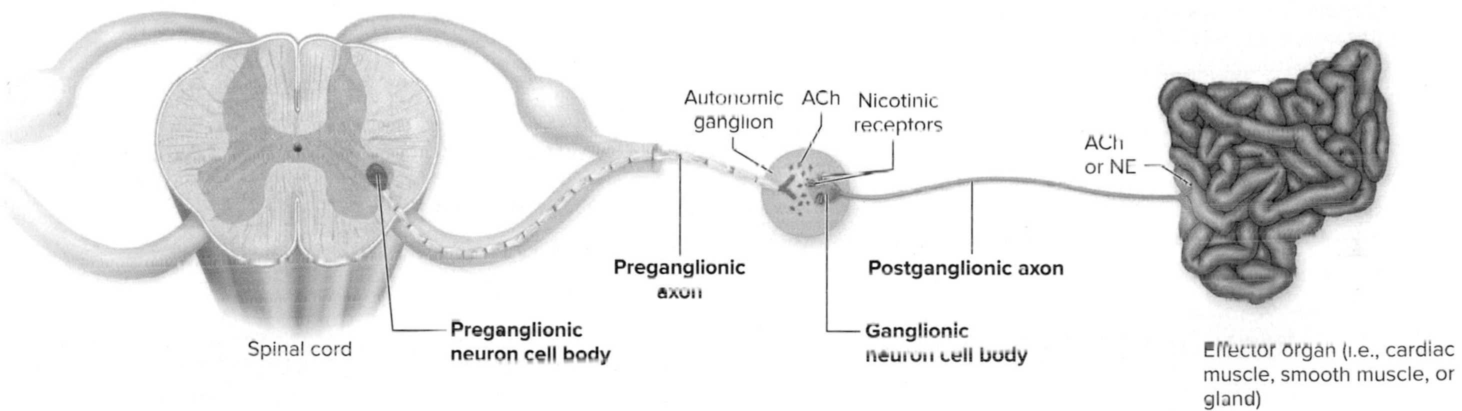

Figure 15.2 Lower Motor Neurons of the Autonomic Nervous System. The autonomic nervous system employs a chain of two lower motor neurons, a preganglionic neuron and a ganglionic neuron. The dendrites and cell body of a preganglionic neuron are housed within the CNS (brain or spinal cord). The preganglionic axon synapses with a ganglionic neuron within an autonomic ganglion. The postganglionic axon extends to an effector organ, which includes cardiac muscle, smooth muscle, and glands.

di = apart) occurs because axons from one preganglionic cell synapse with and influence numerous ganglionic neurons.

Table 15.1 compares the characteristics of the somatic and autonomic nervous systems.

WHAT DID YOU LEARN?

2 What are the anatomic features that distinguish the motor neurons in the SNS and ANS?

15.1c CNS Control of the Autonomic Nervous System

LEARNING OBJECTIVE

4. Describe the CNS hierarchy that controls the autonomic nervous system.

Several levels of CNS complexity are required to coordinate and regulate ANS function. Thus, despite the name *autonomic,* the ANS is a regulated nervous system, not an independent one. Autonomic function is regulated by three CNS regions: the hypothalamus, brainstem, and spinal cord **(figure 15.3)**. These CNS regions may be influenced by the cerebrum, thalamus, and limbic system.

The hypothalamus is the integration and command center for autonomic functions (see section 13.4c). It contains nuclei that control visceral functions in both divisions of the ANS, and it communicates with other CNS regions, including the brainstem and spinal cord. The hypothalamus is the central brain structure involved in emotions and physiologic processes, which are regulated through the ANS. For example, the sympathetic division's fight-or-flight response originates in the sympathetic nucleus in this brain region.

The brainstem nuclei mediate visceral reflexes (see section 13.5). These reflex centers control changes in blood pressure, blood vessel diameter, digestive activities, heart rate, pupil size, and eye lens shape for focusing on close-up objects.

Some autonomic responses, notably the parasympathetic activities associated with defecation and urination (in children), are processed and controlled at the level of the spinal cord without the involvement of the brain. However, the higher centers in the brain may consciously prevent defecation and urination by controlling the external sphincters.

ANS activities are affected by conscious activities in the cerebral cortex and subconscious communications between association areas in the cortex and the centers of parasympathetic and sympathetic control in the hypothalamus. Additionally, sensory processing in the thalamus (see section 13.4b) and emotional states controlled in the limbic system (see section 13.7a) directly affect the hypothalamus.

WHAT DID YOU LEARN?

3 What CNS structure is the integration and command center for autonomic function?

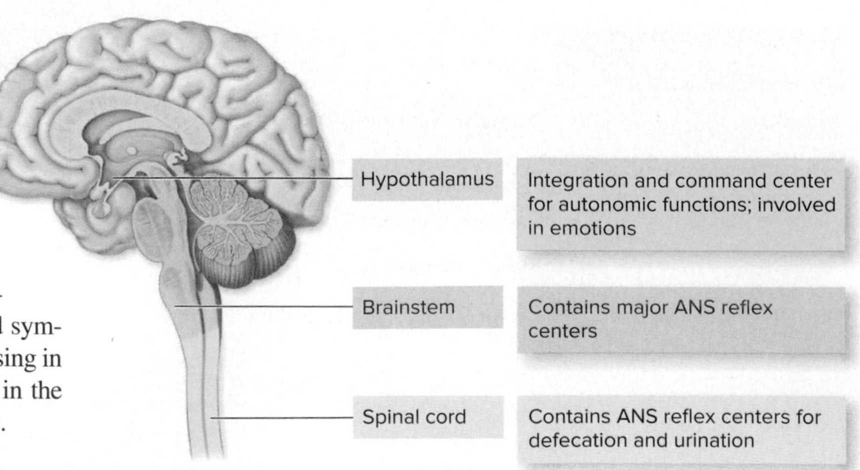
Hypothalamus
Integration and command center for autonomic functions; involved in emotions
Brainstem
Contains major ANS reflex centers
Spinal cord
Contains ANS reflex centers for defecation and urination

Figure 15.3 Control of Autonomic Functions by Higher Brain Centers. ANS functions are regulated by the hypothalamus, which in turn controls ANS regions in the brainstem and spinal cord.

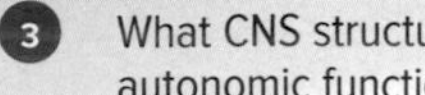

INTEGRATE

LEARNING STRATEGY

The analogy of a corporation can help you understand the hierarchy of control of the ANS:

- The **hypothalamus** is the president of the Autonomic Nervous System corporation. It oversees all activity in this system.
- The **autonomic reflex centers** in the brainstem and spinal cord are the vice presidents of the corporation. They have a lot of control and power, but ultimately they must answer to the president (hypothalamus).
- **Preganglionic and ganglionic neurons** are the workers in the corporation. They are ultimately under the control of both the president and the vice presidents.

15.2 Divisions of the Autonomic Nervous System

The motor component of the ANS is further subdivided into the parasympathetic division and the sympathetic division. Here we discuss the general functional and anatomic differences between the two divisions, as well as the degree of response (local or mass activation) that is possible when each division is activated.

15.2a Functional Differences

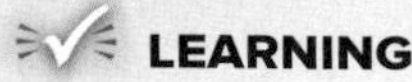

LEARNING OBJECTIVE

5. Describe the general functions of the parasympathetic and sympathetic divisions of the autonomic nervous system.

The divisions perform dramatically different functions—but instead of being thought of as antagonistic, they should be considered complementary. The **parasympathetic** (par-ă-sim-pa-thet′ik; *para* = alongside, *sympatheo* = to feel with) **division** functions to maintain homeostasis when we are at rest. This division is primarily concerned with conserving energy and replenishing nutrient stores. Because it is most active when the body is at rest or digesting a meal, the parasympathetic division has been nicknamed the *rest-and-digest* division.

The **sympathetic** (sim-pă-thet′ik) **division** functions to maintain homeostasis during exercise or times of stress or emergency, which includes the release of nutrients from stores (e.g., glucose released from the liver). Because of its function in regulating the more active states, the sympathetic division has been nicknamed the *fight-or-flight* division. (To help you identify when the sympathetic division is active, remember the "three *E*s": exercise, excitement, or emergency.)

WHAT DID YOU LEARN?

4 The parasympathetic division is responsible for what main functions?

15.2b Anatomic Differences in Lower Motor Neurons

LEARNING OBJECTIVE

6. Compare and contrast the anatomic differences in the lower motor neurons and associated ganglia of the parasympathetic and sympathetic divisions.

Anatomically, these two divisions are similar in that they typically both use a preganglionic neuron and a ganglionic neuron to innervate cardiac muscle, smooth muscle, or glands. Additionally, both divisions have autonomic ganglia that house the ganglionic neuron cell bodies. One of the major differences is where the preganglionic neuron cell bodies are housed in the CNS (**figure 15.4**). Parasympathetic preganglionic cell bodies are located in either the brainstem or the lateral gray matter of the S2–S4 spinal cord segments, and for this reason this division is also termed the *craniosacral* (krā′nē-ō-sā′krăl) *division.* In comparison, sympathetic preganglionic neuron cell bodies are located in the lateral horns of the T1–L2 spinal cord segments, and so this division also goes by the phrase the *thoracolumbar* (thō′ră-kō-lŭm′bar) *division.*

Other anatomic differences between the parasympathetic and sympathetic nervous system include:

- **Length of preganglionic and postganglionic axons.** Parasympathetic preganglionic axons are longer, and postganglionic axons are shorter, when compared to their counterparts in the sympathetic division. In the sympathetic division, preganglionic axons are shorter and postganglionic axons are longer.
- **Number of preganglionic axon branches.** Parasympathetic preganglionic axons tend to have few (less than 4) branches,

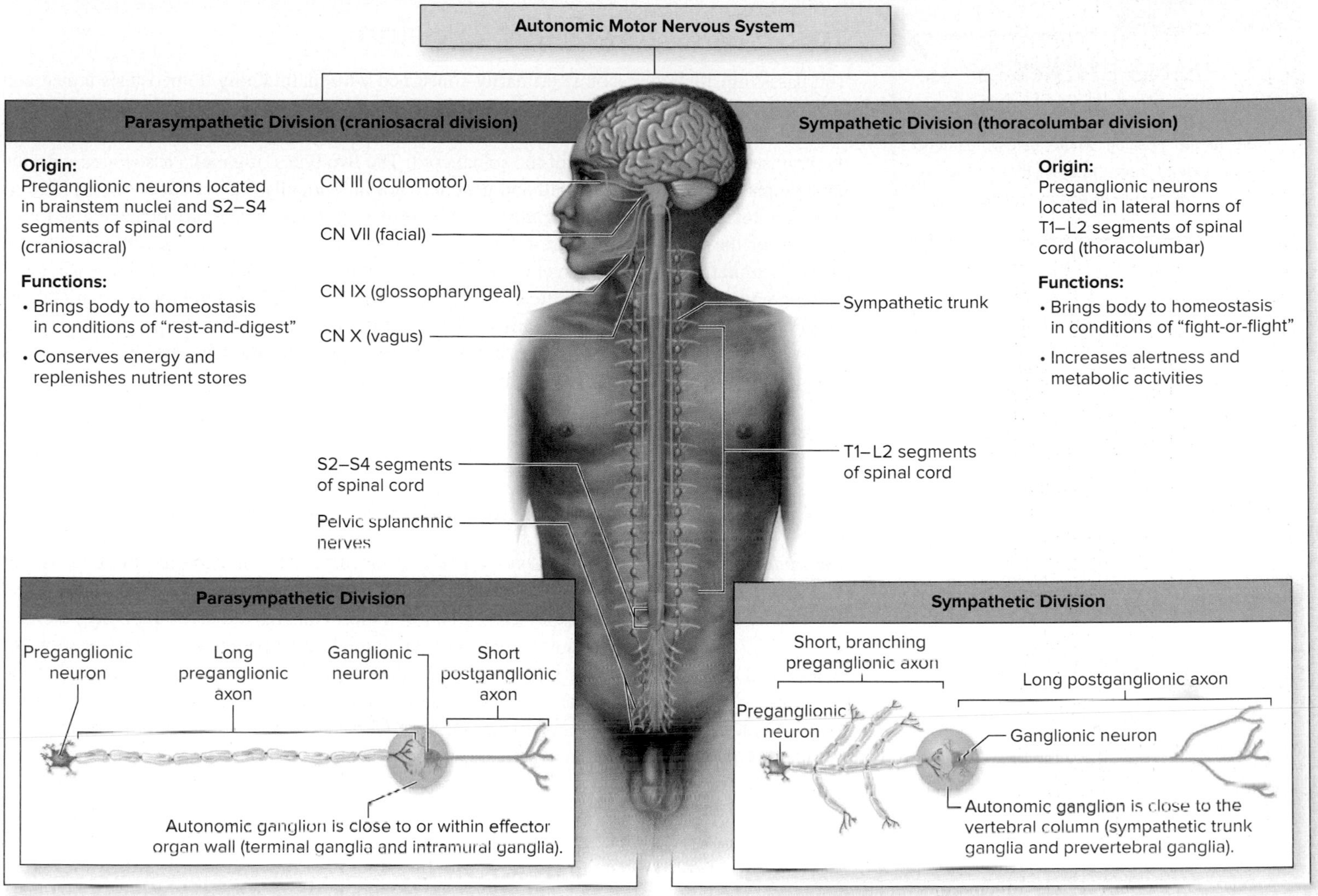

Autonomic Motor Nervous System

Parasympathetic Division (craniosacral division)

Origin:
Preganglionic neurons located in brainstem nuclei and S2–S4 segments of spinal cord (craniosacral)

Functions:
- Brings body to homeostasis in conditions of "rest-and-digest"
- Conserves energy and replenishes nutrient stores

CN III (oculomotor)
CN VII (facial)
CN IX (glossopharyngeal)
CN X (vagus)
S2–S4 segments of spinal cord
Pelvic splanchnic nerves

Sympathetic Division (thoracolumbar division)

Origin:
Preganglionic neurons located in lateral horns of T1–L2 segments of spinal cord (thoracolumbar)

Functions:
- Brings body to homeostasis in conditions of "fight-or-flight"
- Increases alertness and metabolic activities

Sympathetic trunk
T1–L2 segments of spinal cord

Parasympathetic Division

Preganglionic neuron
Long preganglionic axon
Ganglionic neuron
Short postganglionic axon
Autonomic ganglion is close to or within effector organ wall (terminal ganglia and intramural ganglia).

Sympathetic Division

Short, branching preganglionic axon
Long postganglionic axon
Preganglionic neuron
Ganglionic neuron
Autonomic ganglion is close to the vertebral column (sympathetic trunk ganglia and prevertebral ganglia).

Figure 15.4 Comparison of Parasympathetic and Sympathetic Divisions. The neurons of the parasympathetic division extend from the brainstem and sacral region, whereas the neurons of the sympathetic division extend from the thoracic and lumbar regions of the cord. The parasympathetic division axons exhibit very little branching, and autonomic ganglia lie close to or within the effector organ. The sympathetic division axons of both neurons show much branching, and autonomic ganglia lie close to the vertebral column. AP|R

whereas sympathetic preganglionic axons tend to have many branches (more than 20).

- **Location of ganglia.** Parasympathetic autonomic ganglia are either close to or within the effector (terminal ganglia and intramural ganglia, respectively). In comparison, sympathetic autonomic ganglia are relatively close to the spinal cord, and are on either side of the spinal cord or anterior to the spinal cord (sympathetic trunk ganglia and prevertebral ganglia, respectively). (These ganglia are described in sections 15.3a and 15.4a in detail.)

WHAT DID YOU LEARN?

5 Describe the general anatomic differences in the parasympathetic and sympathetic divisions.

15.2c Degree of Response

LEARNING OBJECTIVE

7. Explain why parasympathetic activation is local and discrete, and sympathetic activation can result in mass activation.

It is the combination of long preganglionic axons with limited branches that results in a local response when the parasympathetic division is activated. Parasympathetic activity regulates either one or only a few structures at the same time without having to "turn on" or "turn off" all the other organs.

In comparison, the combination of short preganglionic axons with more extensive branching within the sympathetic division allows for significant neuronal divergence and facilitates the activation of many structures simultaneously, a process called **mass activation.** This process is facilitated when the adrenal medulla is stimulated by the sympathetic division, which causes this gland to release norepinephrine and epinephrine into the blood (see sections 17.2b and 17.9a). Mass activation is especially important in response to stress, when it is necessary to coordinate rapid changes in activity with numerous structures at once. Think of all the bodily changes that are initiated when you are exercising or scared: changes that include an increase in heart rate and blood pressure, increases in the amount of air that enters the lungs, dilation of the pupils, and mobilization of energy reserves from the liver. Keep in mind, however, that there are times when the sympathetic division may activate a single effector. For example, only a single effector is involved when the sympathetic division stimulates smooth muscle to increase the diameter of the pupil of the eye during low-light conditions (see section 16.4b).

WHAT DID YOU LEARN?

6 What is the difference between the degrees of response for the parasympathetic division and the sympathetic division?

INTEGRATE

LEARNING STRATEGY

The name *parasympathetic* (*para* = alongside) refers to the position where parasympathetic neurons extend from the brainstem and sacral region relative to those of the sympathetic division.

15.3 Parasympathetic Division

The parasympathetic division is primarily concerned with maintaining homeostasis at rest and is functionally considered the *rest-and-digest* division. The parasympathetic division is also called the *craniosacral division* because of the anatomic origin of its preganglionic neuron from the brainstem and sacral region of the spinal cord. The two types of ganglia associated with the parasympathetic division are small and generally not individually named but are referred to as either the **terminal** (ter′mi-năl; *terminus* = a boundary) **ganglia,** which are located close to the effector, or the **intramural** (in′tră-mū′răl; *intra* = within, *murus* = wall) **ganglia,** which are located within the wall of the target organ. The exceptions are the four relatively large parasympathetic ganglia associated with the head and neck, which are individually named (e.g., ciliary ganglion). Here we discuss the details of the structure and function of the cranial and sacral components of the parasympathetic division. **Table 15.2** summarizes the different nerves associated with the parasympathetic division.

15.3a Cranial Components

LEARNING OBJECTIVE

8. Name the four cranial nerves associated with the parasympathetic division, and describe their actions.

The cranial nerves containing neurons from the parasympathetic division are the oculomotor (CN III), facial (CN VII), glossopharyngeal (CN IX), and vagus (CN X) nerves. (Recall that these cranial nerves are paired; they are found on the left and right sides of the body.) Review section 13.9 for summaries and illustrations of the cranial nerve pathways and the locations of their associated parasympathetic ganglia. The first three of these nerves transmit parasympathetic innervation to the head, whereas the vagus nerve is the source of parasympathetic innervation for the thoracic and most abdominal organs (**figure 15.5**). Here we discuss the function and then the anatomic pathway for the parasympathetic component of each of these four cranial nerves.

Oculomotor Nerve (CN III)

The **oculomotor nerve (CN III)** innervates both (1) the ciliary muscle (within the eye) to adjust the shape of the lens to see close-up objects and (2) the sphincter pupillae muscle of the iris (which constricts the pupil) to allow less light into the eye, such as when we first walk outside on a bright, sunny day (see section 16.4b).

The axons of the preganglionic neurons extend from cell bodies housed in nuclei within the midbrain to the **ciliary** (sil′ē-ar-ē; *ciliaris* = eyelash) **ganglion** within the orbit. Postganglionic axons project from this ganglion to the effectors (ciliary muscle and iris).

Facial Nerve (VII)

The **facial nerve (CN VII)** innervates the submandibular and sublingual salivary glands in the floor of the mouth (see section 26.2b), lacrimal glands in the superior portion of each orbit (see section 16.4a), and small glands of the nasal cavity, oral cavity, and palate. Stimulation by the parasympathetic division increases release of secretion by these glands. Your mouth waters when you smell an aromatic meal due in part to these parasympathetic neurons within the facial nerve.

Table 15.2 Parasympathetic Division Outflow

Nerve(s)	Origin of Preganglionic Neurons	Autonomic Ganglia	Effectors Innervated
CN III (oculomotor)	Midbrain	Ciliary ganglion	Eye: ciliary muscles to alter the shape of the lens for close vision; iris (sphincter pupillae muscle) to constrict pupil
CN VII (facial)	Pons	Pterygopalatine ganglion	Lacrimal glands; glands of nasal cavity, palate, oral cavity
		Submandibular ganglion	Submandibular and sublingual salivary glands
CN IX (glossopharyngeal)	Medulla oblongata	Otic ganglion	Parotid salivary glands
CN X (vagus)	Medulla oblongata	Terminal and intramural ganglia	Thoracic viscera and most abdominal viscera
Pelvic splanchnic nerves	S2–S4 segments of spinal cord	Terminal and intramural ganglia	Some abdominal viscera and most pelvic viscera

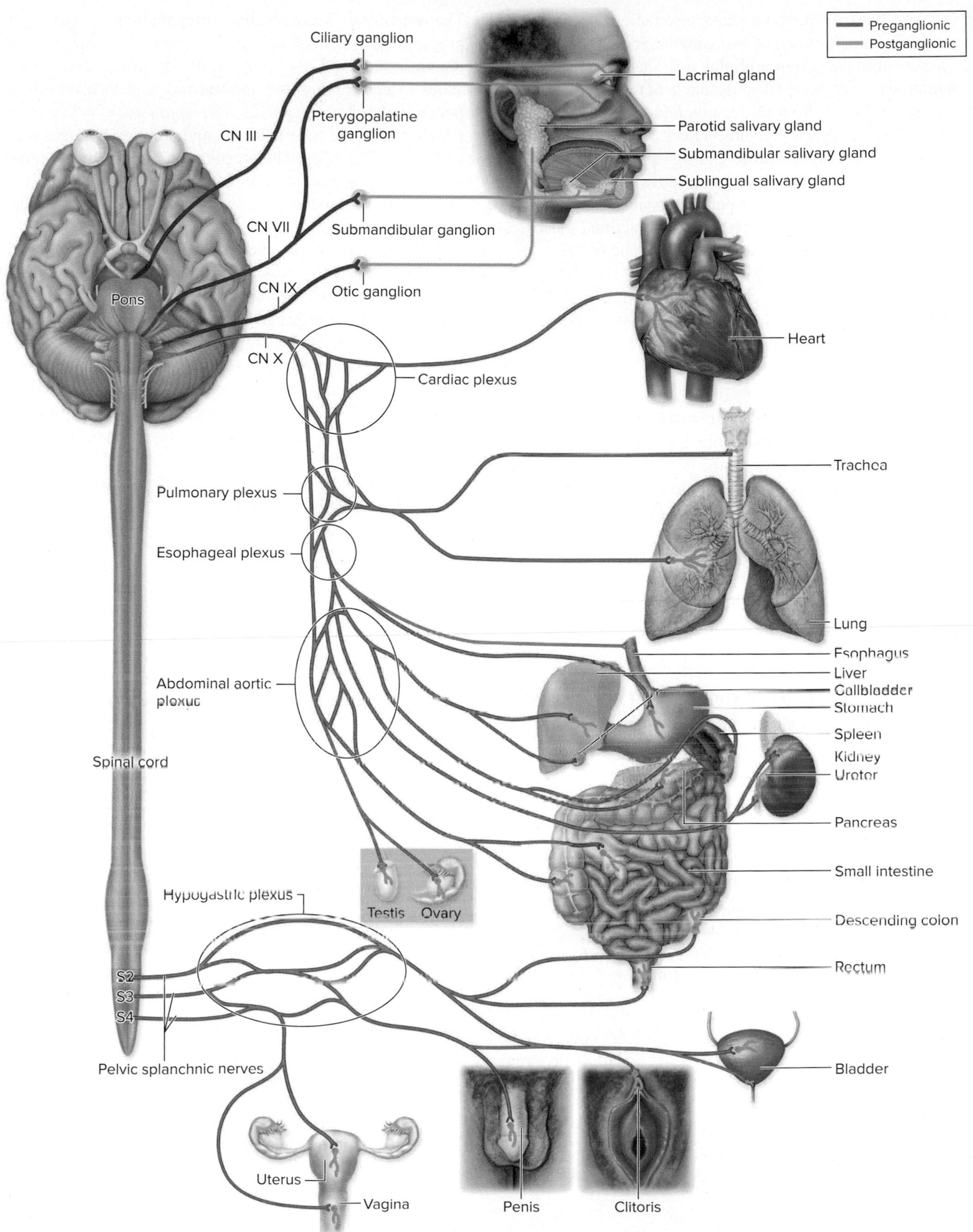

Figure 15.5 Overview of Parasympathetic Pathways. Preganglionic neurons of the parasympathetic division extend from the brain and sacral region of the spinal cord. Ganglionic neurons are located in terminal and intramural ganglia, and extend from these ganglia to innervate the viscera. AP|R

The axons of the preganglionic parasympathetic neurons extend from the pons to terminate at one of two ganglia. Some preganglionic axons terminate within the **submandibular** (sŭb-man-dib′ū-lăr; *sub* = under) **ganglion,** which is located near the angle of the mandible. Postganglionic axons projecting from this ganglion innervate the submandibular and sublingual salivary glands. Other preganglionic axons terminate within the **pterygopalatine** (ter′i-gō-pal′ă-tīn; *pterygo* = wing-shaped, *palatine* = of the palate) **ganglion,** which is positioned near the junction of the maxilla and palatine bones. Postganglionic axons projecting from this ganglion innervate the lacrimal glands, as well as small glands of the nasal cavity, oral cavity, and palate.

 **WHAT DO YOU THINK?**

 The pterygopalatine ganglion is sometimes nicknamed the *hay fever ganglion*. Why is this nickname appropriate?

Glossopharyngeal Nerve (CN IX)

The **glossopharyngeal nerve (CN IX)** innervates the parotid salivary glands and stimulates these glands to increase the release of their secretions. The axons of the preganglionic neurons extend from cell bodies within the brainstem to the **otic** (ō′tik; *ous* = ear) **ganglion,** which is anterior to the ear. Postganglionic axons project from this ganglion to innervate the parotid salivary glands.

Vagus Nerve (CN X)

The **vagus nerve (CN X)** innervates the thoracic organs and most of the abdominal organs, as well as the gonads (ovaries and testes).[1] The term *vagus* means "wanderer," which describes the wandering pathway the vagus nerve makes as it projects inferiorly through the neck and extends throughout the trunk. The influence of the vagus nerve is extensive, given the significant number of autonomic effectors this cranial nerve innervates. The vagus affects the following structures:

- **Heart:** decreases heart rate (see section 19.9b)
- **Bronchi/bronchioles:** constricts to decrease airflow into the air sacs (alveoli) of the lungs (see section 23.5d)
- **Gastrointestinal (GI) tract:** stimulates secretions released from the GI tract wall, increases motility (or movement) of the contents through the GI tract, and relaxes sphincters to allow the passage of contents within the GI tract (see section 26.1c)
- **Liver:** stimulates glycogenesis, which is glycogen formation from glucose (see section 27.6c)

The axons of the preganglionic neurons of the vagus nerve extend through one or more plexuses (see section 15.5a) and continue to either a terminal or an intramural ganglion. Postganglionic axons project from the ganglion to the effector organ.

 **WHAT DID YOU LEARN?**

 Which four cranial nerves have a parasympathetic component? What organs are innervated and what physiologic response is caused by each?

15.3b Pelvic Splanchnic Nerves

 **LEARNING OBJECTIVE**

9. Explain the actions of the pelvic splanchnic nerves.

The remaining parasympathetic preganglionic axons originate from preganglionic neuron cell bodies housed within the lateral gray regions of the S2–S4 spinal cord segments. These preganglionic axons extend through the anterior root and then branch to form the **pelvic splanchnic** (splangk′nik; *splanchnic* = visceral) **nerves,** which contribute to a superior and inferior hypogastric plexus on each side of the body. The preganglionic axons that continue through each plexus project to the ganglionic neurons within either the terminal or intramural ganglia. The postganglionic axons extend to the effector.

The target organs innervated include the distal portion of the large intestine, the rectum, the urinary bladder, the distal part of the ureter, and most of the reproductive organs. The parasympathetic regulation of these target organs causes increased smooth muscle motility (muscle contraction) and secretory activity in these portions of the digestive tract (see chapter 26), contraction of smooth muscle in the urinary bladder wall and relaxation of the smooth muscle of the internal urethral sphincter (which facilitates urination; see section 24.8c), and erection of the female clitoris and the male penis (see sections 28.3g and 28.4f).

 **WHAT DID YOU LEARN?**

 What organs are innervated by the pelvic splanchnic nerves?

15.4 Sympathetic Division

The sympathetic division is primarily concerned with preparing the body for exercise and emergencies, and is functionally considered the fight-or-flight division. Recall from section 15.2b that the lower motor neurons of the sympathetic division extend only from the thoracic and lumbar regions of the spinal cord (T1–L2), which is why this division is also called the *thoracolumbar division.* The two types of ganglia associated with the sympathetic division are the *sympathetic trunk ganglia* and the *prevertebral ganglia.*

15.4a Organization and Anatomy of the Sympathetic Division

LEARNING OBJECTIVES

10. Give the location of the sympathetic preganglionic neuron cell bodies.
11. Describe the left and right sympathetic trunks and ganglia.
12. Compare and contrast white and gray rami regarding their location and composition.
13. Explain the differences between the sympathetic trunk ganglia and the prevertebral ganglia.

The sympathetic division is much more anatomically complex than the parasympathetic division, so we first describe its anatomic components and then its pathways **(figure 15.6)**. The sympathetic preganglionic neuron cell bodies are housed in the **lateral horn** of the T1–L2 regions of the spinal cord. (See figure 14.4*b* in section 14.3a, which compares the location of autonomic motor and the somatic motor cell bodies in the spinal cord.) From there, the preganglionic sympathetic axons travel with somatic motor axons to exit the spinal cord through first the anterior roots and then the T1–L2 spinal nerves. However, these preganglionic sympathetic axons remain with the spinal nerve for only a short distance before they branch from the spinal nerve.

[1]It is unclear what function, if any, these parasympathetic axons have on the gonads.

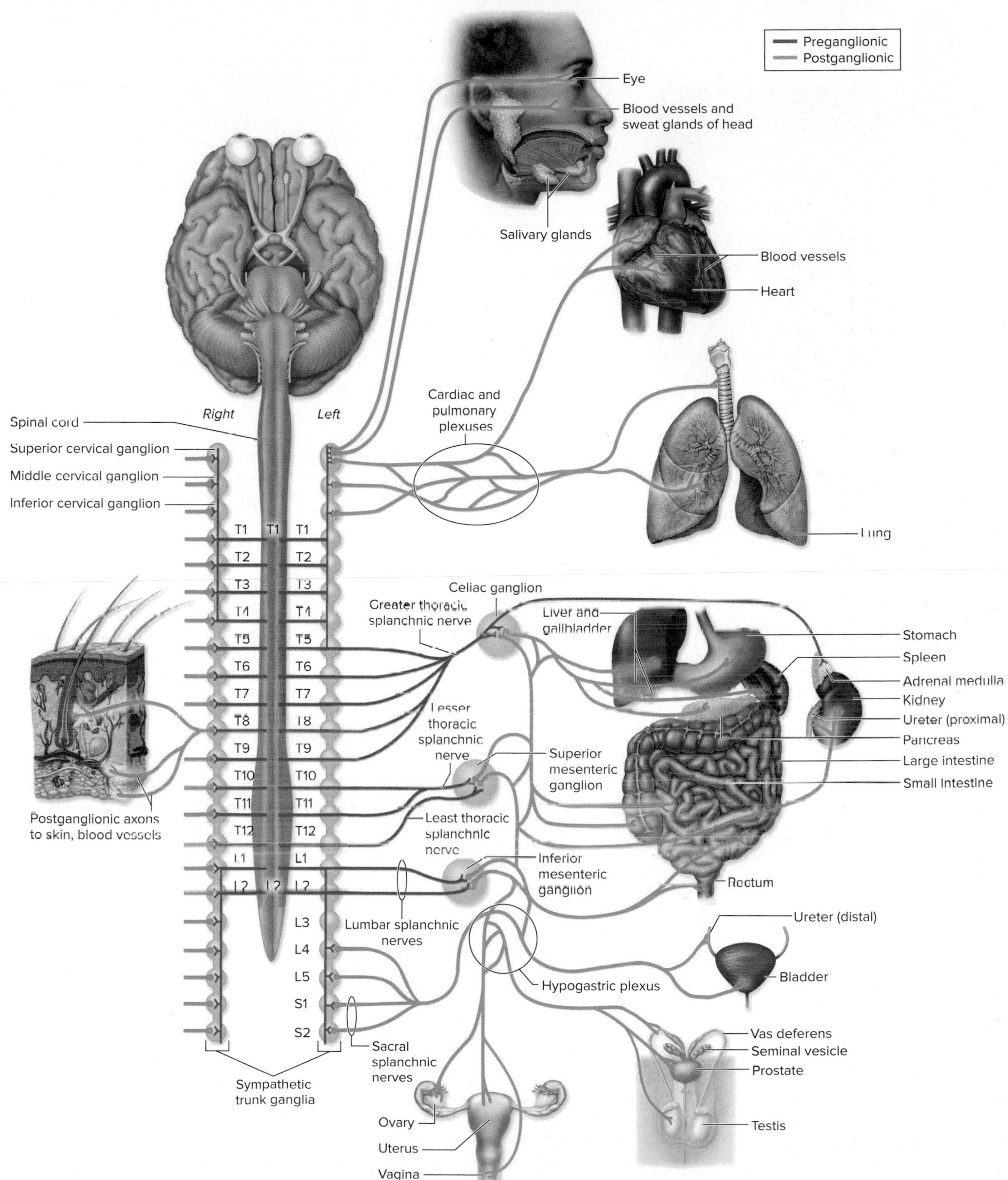

Figure 15.6 Overview of Sympathetic Pathways. Preganglionic axons of the sympathetic division extend from the T1–L2 regions of the spinal cord. Ganglionic axons are located in the sympathetic trunk and prevertebral ganglia, and extend to the target organs. *(Left)* The outflow of preganglionic axons and the distribution of postganglionic axons innervating the skin. *(Right)* Sympathetic postganglionic axon pathways to internal organs. AP|R

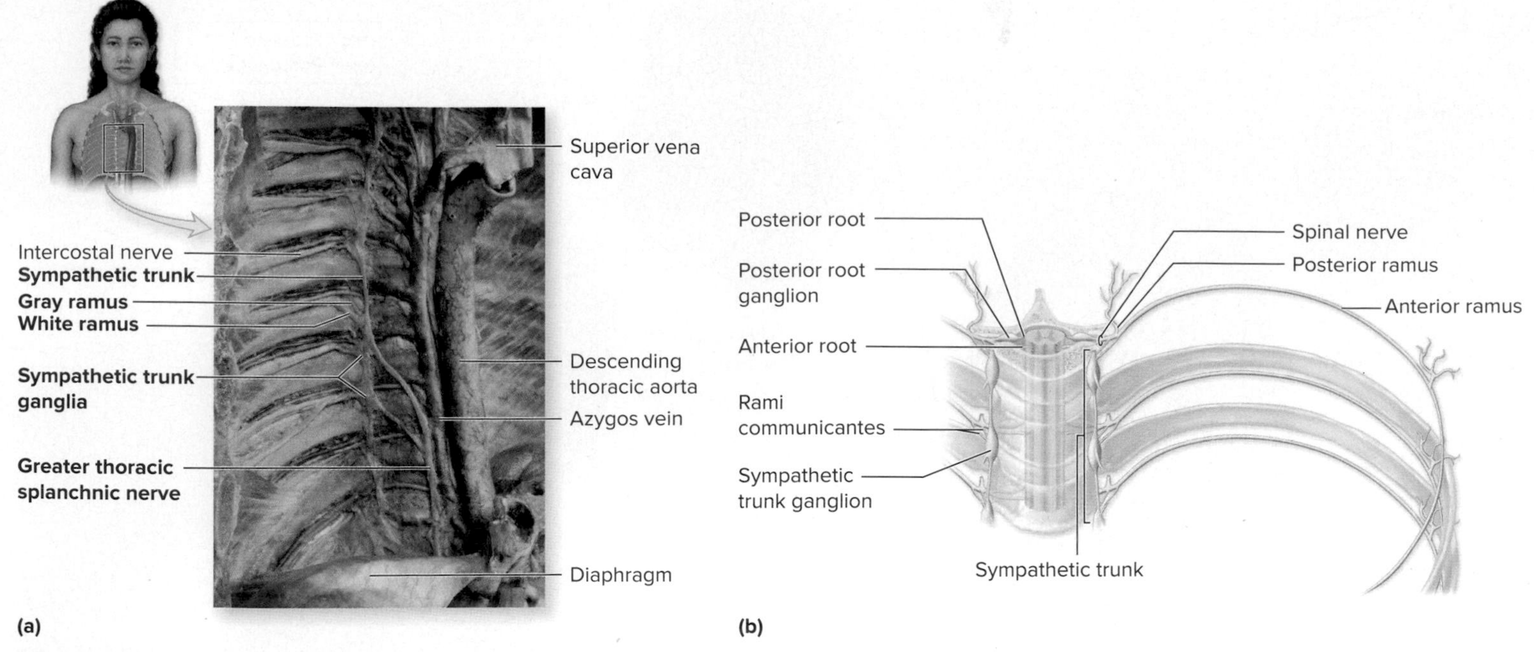

Figure 15.7 Sympathetic Trunk. (*a*) An anterolateral photo of the right side of the thoracic cavity shows the selected sympathetic division structures in bold. Note how the gray and white rami attach between the sympathetic trunk and each spinal nerve. (*b*) An illustration of the anatomic relationship of the spinal cord, which is housed within the vertebral column, and the sympathetic trunks, which are not. AP|R

(*a*) ©McGraw-Hill Education/Christine Eckel

INTEGRATE

LEARNING STRATEGY

The name *sympathetic* for this division of the ANS refers to how its sympathetic trunks "follow the spinal cord in sympathy" because these trunks run parallel to the spinal cord.

Sympathetic Trunks and Sympathetic Trunk Ganglia

Immediately lateral to the vertebral column and anterior to the paired spinal nerves are the left and right **sympathetic trunks** (**figure 15.7**). A sympathetic trunk looks much like a pearl necklace. The "string" of the necklace is composed of bundles of axons, whereas the "pearls" are the **sympathetic trunk ganglia** (also known as *paravertebral* or *sympathetic chain ganglia*), which house sympathetic ganglionic neuron cell bodies.

One sympathetic trunk ganglion typically is associated with each spinal nerve. However, the cervical portion of each sympathetic trunk is partitioned into only three sympathetic trunk ganglia—the superior, middle, and inferior cervical ganglia—as opposed to the eight cervical spinal nerves (figure 15.6). The **superior cervical ganglion** contains postganglionic sympathetic neuron cell bodies whose axons are distributed primarily to structures within the head and neck and to some thoracic viscera. These postganglionic axons innervate the sweat glands and smooth muscle in blood vessels of the head and neck, the dilator pupillae muscle of the eye, and the superior tarsal muscle of the eye (which elevates the eyelid). The **middle** and **inferior cervical ganglia** also house neuron cell bodies that extend postganglionic axons to the thoracic viscera (e.g., heart, bronchi and bronchioles of the lungs).

White and Gray Rami

Connecting the spinal nerves to each sympathetic trunk are *rami communicantes* (rā′mī kŏ-mū-ni-kan′tēz; sing., *ramus communicans; rami* = branches, *communico* = to share with someone) (figure 15.7; see also figure 14.13). **White rami communicantes** (or simply **white rami**) are composed of *pre*ganglionic sympathetic axons from the T1–L2 spinal nerves to the sympathetic trunk. Thus, white rami are associated only with the T1–L2 spinal nerves. Recall that preganglionic axons are myelinated, giving these rami a whitish appearance. White rami are similar to entrance ramps onto a highway.

Gray rami communicantes (or simply **gray rami**) are composed of *post*ganglionic sympathetic axons that extend from the sympathetic trunk to the spinal nerve. Postganglionic axons are unmyelinated, so these rami have a grayish appearance. Gray rami connect to all spinal nerves, including the cervical, sacral, and coccygeal spinal nerves. By these routes, the sympathetic information that started out in the thoracolumbar region can be dispersed to all parts of the body. Gray rami are similar to exit ramps from a highway.

Sympathetic Splanchnic Nerves

Sympathetic splanchnic nerves are composed of preganglionic sympathetic axons that did not synapse in a sympathetic trunk ganglion (figure 15.6). They run anteriorly from the

INTEGRATE

CLINICAL VIEW 15.1

Horner Syndrome

Horner syndrome is a condition caused by impingement of or injury to the cervical sympathetic trunk or the T1 sympathetic trunk ganglion, where postganglionic sympathetic axons extending to the head originate. Symptoms typically are limited to the same side of the head where the original injury occurred. The patient presents with **ptosis** (tō′sis; a falling), in which the superior eyelid droops because the superior tarsal muscle is paralyzed. Paralysis of the pupil dilator muscle of the eye results in **miosis** (mī-ō′sis; *muein* = to close the eyes), which is a constricted pupil. **Anhydrosis** (an-hī-drō′sis; *an* = without, *hidros* = sweat) occurs because the sweat glands no longer receive sympathetic stimulation. A fourth symptom is distinct facial flushing due to lack of sympathetic stimulation to blood vessel walls, resulting in vasodilation.

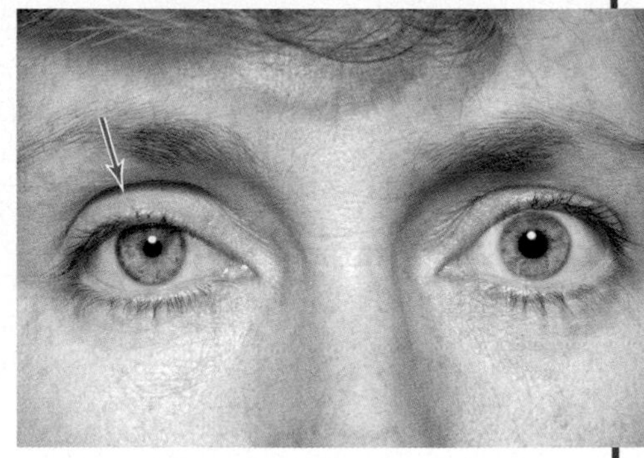

Horner syndrome in right eye. Note the ptosis of the superior eyelid (blue arrow) and constriction of the pupil.

sympathetic trunk to most of the abdominal and pelvic viscera. These splanchnic nerves should not be confused with the pelvic splanchnic nerves associated with the parasympathetic division, described in section 15.3. Some of the larger splanchnic nerves have specific names, such as thoracic splanchnic nerves, lumbar splanchnic nerves, or sacral splanchnic nerves.

Prevertebral Ganglia

Splanchnic nerves typically terminate in **prevertebral** (or *collateral*) ganglia differ from the sympathetic trunk ganglia in that (1) they are anterior to the vertebral column (prevertebral) on the anterolateral surface of the aorta and (2) they are located only in the abdominopelvic cavity. Prevertebral ganglia include the celiac, superior mesenteric, and inferior mesenteric ganglia.

The **celiac ganglia** are adjacent to the origin of the celiac artery. The greater thoracic splanchnic nerves (composed of axons from T5–T9 segment of the spinal cord) synapse on ganglionic neurons within each celiac ganglion. Postganglionic axons from the celiac ganglia innervate the stomach, spleen, liver, gallbladder, and proximal part of the duodenum (first part of small intestine) and part of the pancreas.

The **superior mesenteric** (mez-en-ter′ik; *mesos* = middle, *enteron* = intestine) **ganglia** are adjacent to the origin of the superior mesenteric artery. The lesser and least thoracic splanchnic nerves project to and terminate within the superior mesenteric ganglia. Thus, these ganglia receive preganglionic sympathetic neurons from the T10–T12 segments of the spinal cord. Postganglionic axons extending from the superior mesenteric ganglia innervate the distal half of the duodenum, the remainder of the small intestine, the proximal part of the large intestine, part of the pancreas, the kidneys, and the proximal parts of the ureters.

The **inferior mesenteric ganglia** are adjacent to the origin of the inferior mesenteric artery. They receive sympathetic preganglionic axons via the lumbar splanchnic nerves, which originate in the L1–L2 segments of the spinal cord. The postganglionic axons project to and innervate the distal part of the large intestine, the rectum, the urinary bladder, distal parts of the ureters, and most of the reproductive organs.

WHAT DID YOU LEARN?

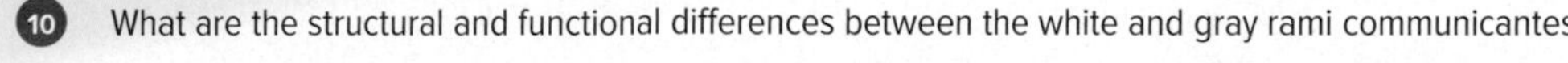

9. What is the difference between sympathetic trunk ganglia and prevertebral ganglia?
10. What are the structural and functional differences between the white and gray rami communicantes?

15.4b Sympathetic Pathways

LEARNING OBJECTIVES

14. Describe the four pathways of sympathetic neurons.
15. Compare and contrast which general effector organs are innervated by each pathway.

All sympathetic preganglionic neurons originate in the lateral gray horns of the T1–L2 regions of the spinal cord. The axons of the preganglionic sympathetic neurons travel with somatic motor axons to exit the spinal cord within the anterior roots and then through the spinal nerves. However, these preganglionic sympathetic axons remain with the spinal nerve for only a short distance before they leave the spinal nerve within the white ramus. It is at this point where the major pathways of the sympathetic division differ. Each type of pathway is dependent upon the location of the effector organ being innervated. Axons exit the sympathetic trunk by one of four pathways.

Spinal Nerve Pathway

The **spinal nerve pathway** extends from the spinal cord to effectors of the skin of the neck, torso, and limbs. Skin effectors include sweat glands, smooth muscle forming arrector pili muscles (which produce "goose bumps"), and smooth muscle cells within the walls of blood vessels (see section 6.2a). In this pathway, a preganglionic neuron synapses with a ganglionic neuron in a sympathetic trunk ganglion at either the same or different level (**figure 15.8*a***). The postganglionic axon extends through a gray ramus that is at the same "level" as the ganglionic neuron. For example, if the preganglionic and ganglionic neurons synapse in the L4 ganglion, the postganglionic axon extends through the gray ramus at the level of the L4 spinal nerve. After the postganglionic axon extends through the gray ramus, it enters the spinal nerve and extends to its target organ.

Postganglionic Sympathetic Nerve Pathway

The **postganglionic sympathetic nerve pathway** extends from the spinal cord to the internal organs of the thoracic cavity (including the esophagus, heart, lungs, and thoracic blood vessels), the effectors of the skin of the head (sweat glands, arrector pili, and blood vessels of the skin), the neck viscera, and the superior tarsal and dilator pupillae muscles in the eye (for increasing the amount of light entering the eye) (see section 16.4b). In this pathway, the preganglionic neuron synapses with a ganglionic neuron in a sympathetic trunk ganglion, but the postganglionic axon does *not* leave the trunk via a gray ramus (figure 15.8*b*). Instead, the postganglionic axon extends away from the sympathetic trunk ganglion and projects directly to the effector organ.

Splanchnic Nerve Pathway

The **splanchnic nerve pathway** extends from the spinal cord to the abdominal and pelvic organs (e.g., stomach, small intestines, kidney). In this pathway, a preganglionic neuron does not synapse with a ganglionic neuron in a sympathetic trunk ganglion. Rather, the preganglionic axons pass through the sympathetic trunk ganglia without synapsing and extend to the prevertebral ganglia (figure 15.8*c*). There, the preganglionic axon synapses with a ganglionic neuron. The postganglionic axon then projects to the effector organs.

Adrenal Medulla Pathway

The final pathway is the **adrenal medulla pathway** (figure 15.8*d*). In this pathway, the internal region of the adrenal gland, called the **adrenal** (ă-drē′năl; *ad* = to, *ren* = kidney) **medulla,** is directly innervated by preganglionic sympathetic axons. (There is no ganglionic neuron.) The axons of the preganglionic neuron extend through both the sympathetic trunk and the prevertebral ganglia and then synapse on neurosecretory cells within the adrenal medulla. Stimulation of these cells causes the release of **epinephrine** (ep′i-nef′rin) and

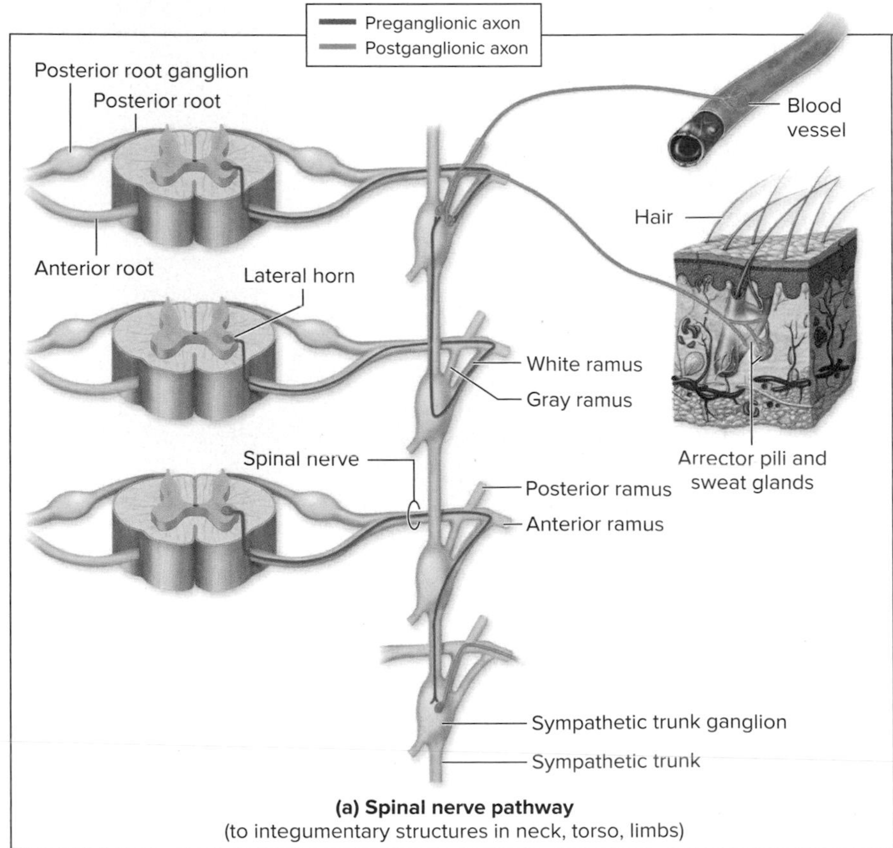

(a) Spinal nerve pathway
(to integumentary structures in neck, torso, limbs)

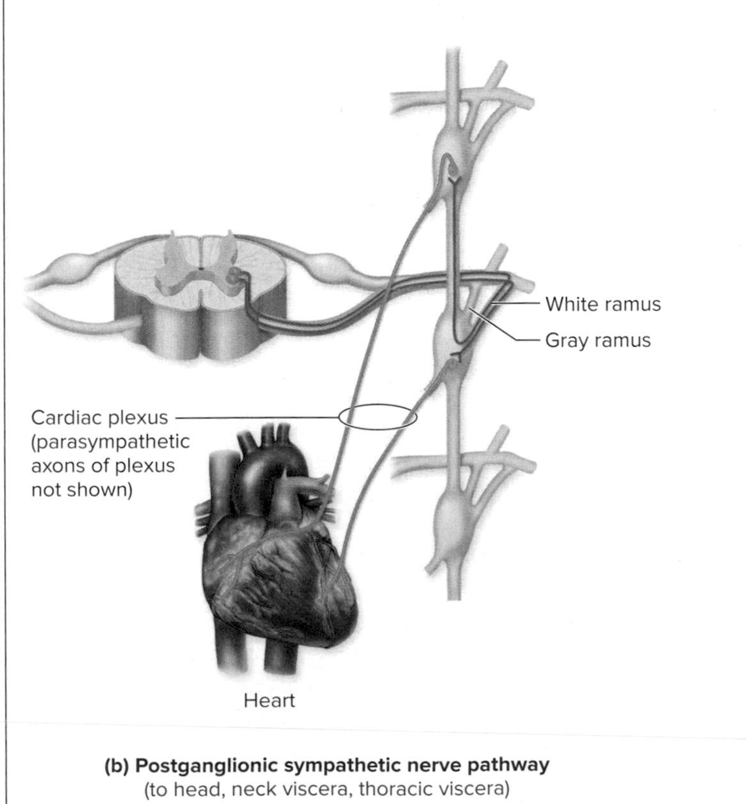

(b) Postganglionic sympathetic nerve pathway
(to head, neck viscera, thoracic viscera)

Figure 15.8 Types of Sympathetic Pathways. Pathways of (*a*) a spinal nerve, (*b*) a postganglionic sympathetic nerve, (*c*) a splanchnic nerve, and (*d*) the adrenal medulla.

Table 15.3 Sympathetic Division Pathways

Pathway	Destination	Spinal Segment Origin	Postganglionic Axon Pathway from Sympathetic Trunk	Effectors Innervated[1]
Spinal nerve	Integumentary structures	T1–L2	Via cervical gray rami to all spinal nerves	Sweat glands, arrector pili muscles, blood vessels in skin of neck, torso, and limbs
Postganglionic sympathetic nerve	Head and neck viscera	T1–T2 (primarily from T1)	Via superior cervical ganglion and travel with blood vessels to the head and neck viscera	Sweat glands, arrector pili muscles, and blood vessels in skin of head; dilator pupillae muscle of eye; superior tarsal muscle of eye; neck viscera
	Thoracic organs	T1–T5	Via cervical and thoracic ganglia to autonomic nerve plexuses near organs	Esophagus, heart, lungs, blood vessels within thoracic cavity
Splanchnic nerve	Most abdominal organs	T5–T12	Via thoracic splanchnic nerves to prevertebral ganglia (e.g., celiac, superior mesenteric, and inferior mesenteric ganglia)	Abdominal portion of esophagus, stomach, liver, gallbladder, spleen, pancreas, small intestine, most of large intestine, kidneys, ureters, adrenal glands, blood vessels within abdominopelvic cavity
	Pelvic organs	T10–L2	Via lumbar and sacral splanchnic nerves to autonomic nerve plexuses that travel to effectors	Distal part of large intestine, anal canal, and rectum; distal part of ureters; urinary bladder; reproductive organs
Adrenal medulla	Adrenal gland	T8–T12	No ganglionic neuron; axons of preganglionic neurons extend via thoracic splanchnic nerves directly to adrenal medulla	Neurosecretory cells of adrenal medulla

1. Sympathetic axons innervate smooth muscle, cardiac muscle, and glands associated with the organs listed.

norepinephrine (nōr′ēp-i-nef′rin) into the blood (see section 17.9a). (The relative amounts of these two hormones released are not equal. Typically, epinephrine accounts for approximately 80% and norepinephrine accounts for 20% of the hormone molecules that are released.) Both of these hormones then circulate within the blood and bind to many of the same receptors as norepinephrine, and in so doing help potentiate (prolong) the effects of the sympathetic stimulation. For example, if you narrowly miss getting into a car accident, your heart continues to beat quickly, you breathe rapidly, and you feel tense and alert well after the event because of prolonged effects of the sympathetic stimulation of the adrenal medulla. Details for epinephrine and norepinephrine are listed at the end of chapter 17 in **Table R.5**: "Regulating the Stress Response with Catecholamines and Glucocorticoids." **Table 15.3** summarizes the sympathetic division pathways.

WHAT DO YOU THINK?

When a person is very stressed and tense, his or her blood pressure typically rises. What aspect of the sympathetic division causes this rise in blood pressure?

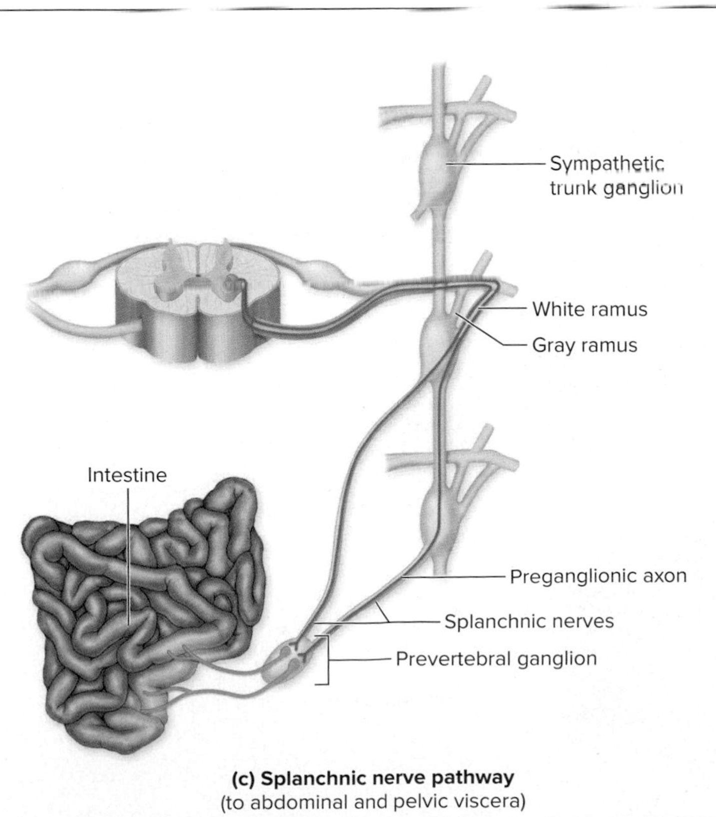

(c) Splanchnic nerve pathway (to abdominal and pelvic viscera)

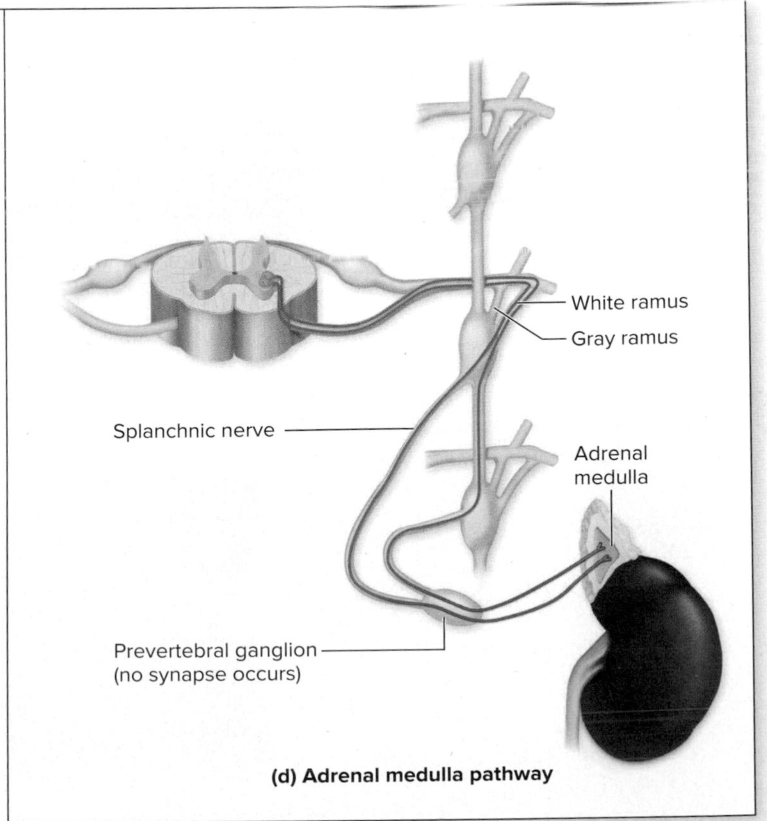

(d) Adrenal medulla pathway

WHAT DID YOU LEARN?

11 How do the spinal nerve and splanchnic nerve sympathetic pathways differ, regarding both the pathway and the organs innervated?

12 In what ways does the adrenal medulla pathway help prolong the effects of the sympathetic stimulation?

15.5 Autonomic Plexuses and the Enteric Nervous System

Both divisions of the autonomic nervous system innervate organs through specific axon bundles called *autonomic plexuses.* Communication between neurons and effectors is through specific neurotransmitters (described in detail in section 15.6). The enteric nervous system is an autonomic array of neurons that may act independently or be affected by ANS innervation. We discuss autonomic plexuses first.

15.5a Autonomic Plexuses

LEARNING OBJECTIVE

16. Describe the structure and location of the five autonomic plexuses.

Autonomic plexuses are collections of sympathetic postganglionic axons and parasympathetic preganglionic axons, as well as some visceral sensory axons. These sympathetic and parasympathetic axons are close to one another, but they do not interact or synapse with each another. Although these plexuses look like disorganized masses of axons, they provide a complex innervation pattern to their target organs (**figure 15.9**).

In the mediastinum of the thoracic cavity, the **cardiac plexus** consists of sympathetic postganglionic axons that originate in the cervical and thoracic sympathetic trunk ganglia, as well as parasympathetic preganglionic axons from the vagus nerve. Increased sympathetic activity increases heart rate and blood pressure, whereas increased parasympathetic activity decreases heart rate (see section 19.5b).

The **pulmonary plexus** consists of sympathetic postganglionic axons from the cervical and thoracic sympathetic trunk ganglia and parasympathetic preganglionic axons from the vagus nerve. The axons project to the bronchi and bronchioles of the lungs. Sympathetic innervation causes bronchodilation (increase in the diameter of the bronchi and bronchioles of the lung), whereas stimulation of this parasympathetic pathway causes bronchoconstriction (reduction in the diameter of the bronchi and bronchioles; see section 23.3c).

The **esophageal plexus** consists of sympathetic postganglionic axons from the cervical and sympathetic trunk ganglia and parasympathetic preganglionic axons from the vagus nerve. Sympathetic innervation will inhibit muscle motility. Smooth muscle activity in the inferior esophageal wall is coordinated by parasympathetic axons that control the swallowing reflex in the inferior region of the esophagus by innervating the *cardiac sphincter,* a valve through which swallowed food and drink must pass (see section 26.2c).

The **abdominal aortic plexus** consists of the **celiac plexus, superior mesenteric plexus,** and **inferior mesenteric plexus.** It

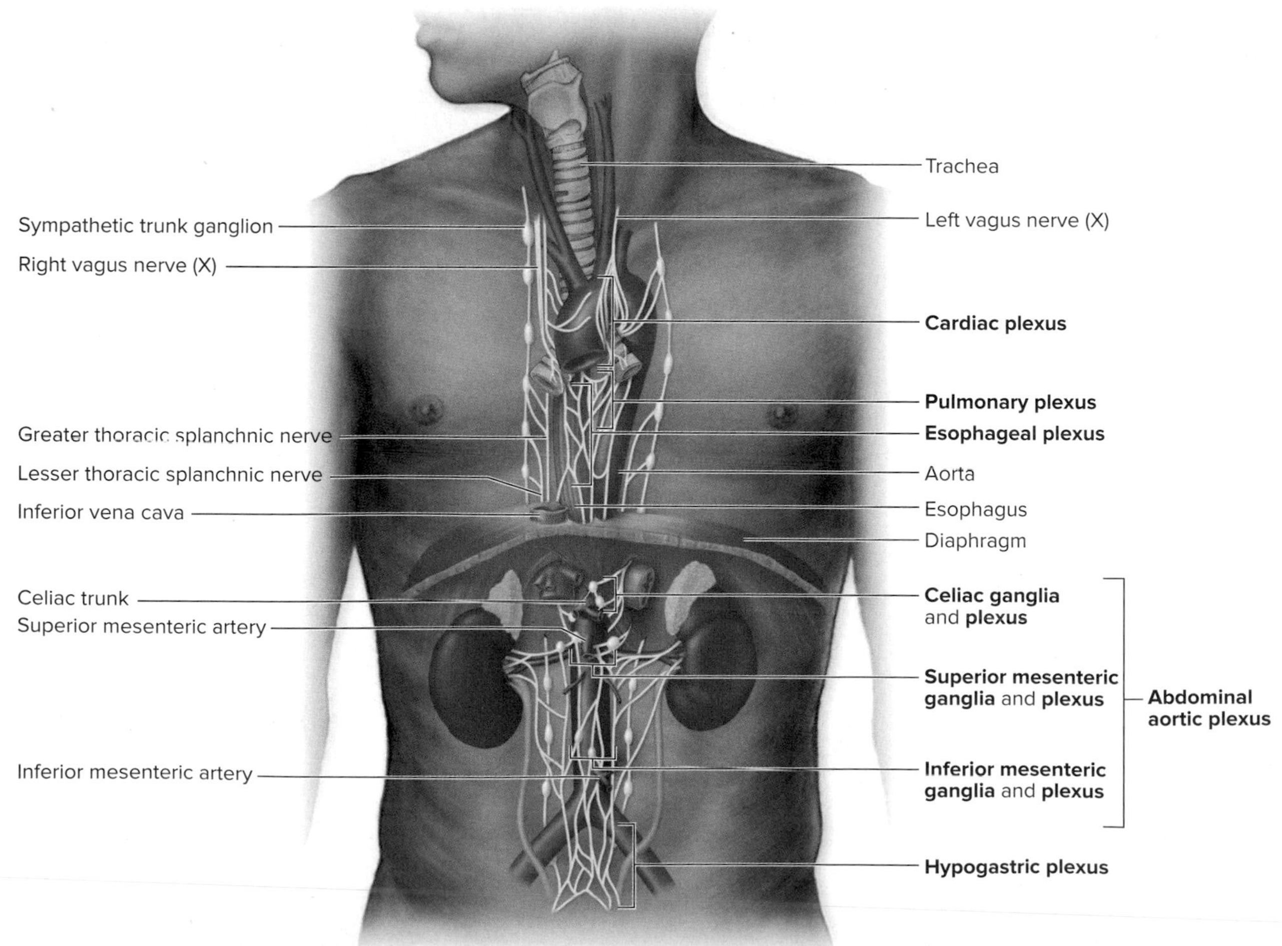

Figure 15.9 Autonomic Plexuses. Autonomic plexuses are located in both the thoracic and abdominopelvic cavities. This anterior view shows the cardiac, pulmonary, and esophageal plexuses in the thoracic cavity and the abdominal aortic plexus (celiac, superior mesenteric, inferior mesenteric plexuses) in the abdominopelvic cavity. AP|R

innervates all abdominal and some pelvic organs. The abdominal aortic plexus is composed of sympathetic postganglionic axons projecting from the prevertebral ganglia and parasympathetic preganglionic axons from either the vagus nerve or the pelvic splanchnic nerves. Note that the celiac plexus is also known as the *solar plexus,* and it is partly responsible for the sensation of "getting the wind knocked out of you" when you are hit hard in the epigastric region.

The **hypogastric plexus** consists of a complex meshwork of sympathetic postganglionic axons (from the aortic plexus and the lumbar region of the sympathetic trunk) and preganglionic parasympathetic axons from the pelvic splanchnic nerves. Its axons innervate viscera within the pelvic region.

15.5b Enteric Nervous System

LEARNING OBJECTIVE

17. Explain the function and location of the enteric nervous system (ENS).

The **enteric nervous system (ENS)** is an array of neurons (both autonomic motor and visceral sensory) that are arranged throughout the wall of the gastrointestinal (GI) tract, from the esophagus to the anus. The ENS not only innervates the smooth muscle and glands of the GI tract but also mediates the complex coordinated reflexes for *peristalsis,* or movement of materials through the GI tract (see section 26.1c). ENS neurons are located both within numerous small ganglia throughout the GI tract wall and within two large plexuses: (1) the **submucosal plexus** (also called the *Meissner plexus*) and (2) the **myenteric plexus** (also called the *Auerbach plexus*). Although the enteric nervous system can function independently of the rest of the ANS, the ANS division may adjust its activity. In general, the parasympathetic division increases ENS activity, and the sympathetic division decreases ENS activity.

WHAT DID YOU LEARN?

13 What basic structures form an autonomic plexus?

14 Where is the enteric nervous system located, and what is its function?

15.6 Comparison of Neurotransmitters and Receptors of the Two Divisions

Transmission of a nerve signal to synaptic knobs causes the release of neurotransmitters into the synaptic cleft. The ANS utilizes several types of neurotransmitters, which are discussed here.

15.6a Overview of ANS Neurotransmitters

LEARNING OBJECTIVE

18. Identify the targets of the cholinergic and adrenergic neurotransmitters of the ANS.

Acetylcholine (ACh) and **norepinephrine (NE)** are the main neurotransmitters used in the ANS **(figure 15.10)**. These neurotransmitters will bind to specific receptors on the postsynaptic cell. Depending upon the receptor type, the neurotransmitter may cause either stimulation or inhibition.

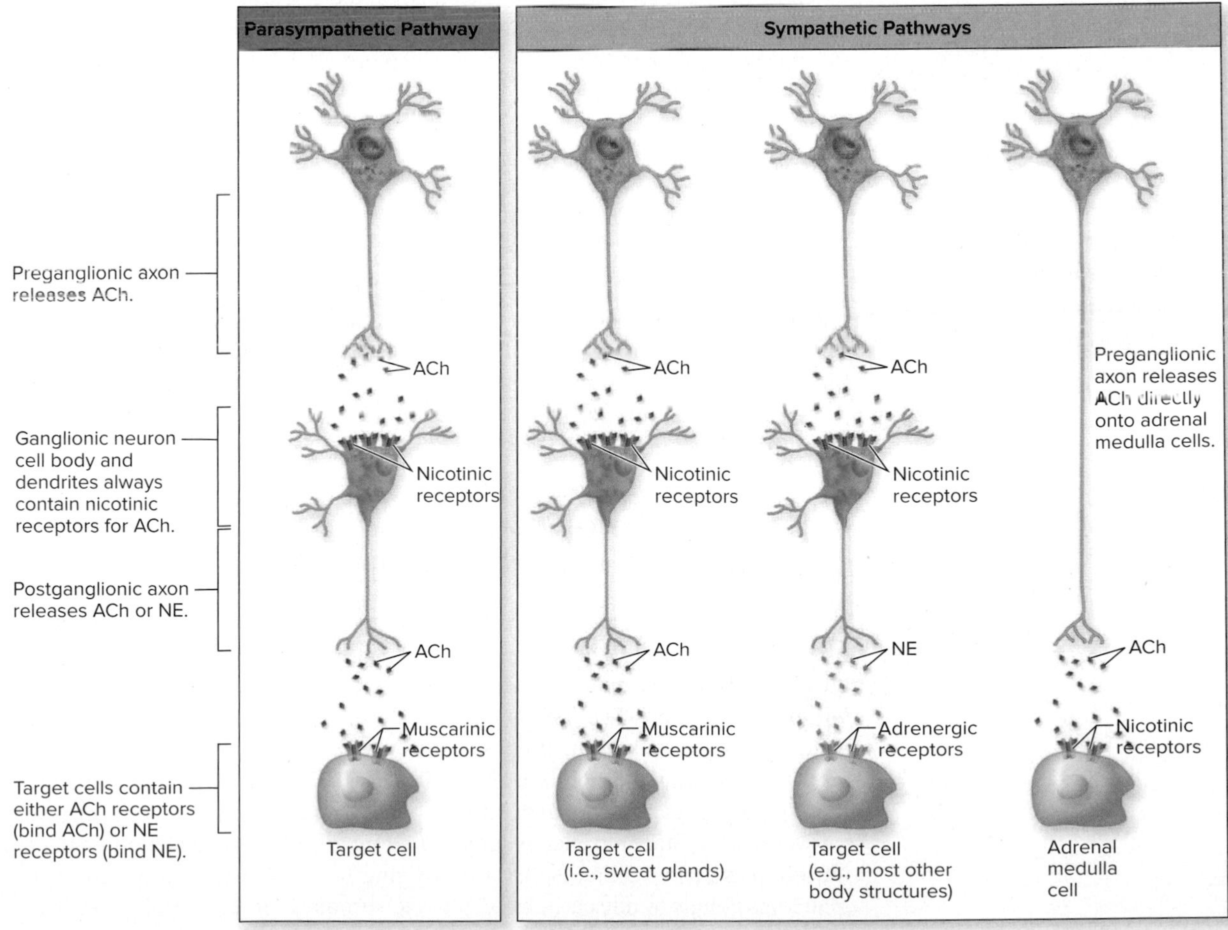

Figure 15.10 Comparison of Neurotransmitters in the Autonomic Nervous System. In the parasympathetic pathway, both the preganglionic and postganglionic axons release acetylcholine (ACh). In the sympathetic pathways, all preganglionic axons and a few specific postganglionic axons release ACh. Most postganglionic sympathetic axons release norepinephrine (NE).

Neurons that synthesize and release acetylcholine (ACh) are called **cholinergic** (kol-in-er′jik; *ergon* = work) **neurons.** Receptors that bind ACh are called **cholinergic receptors.** We will see in section 15.6b that there are two types of cholinergic receptors (nicotinic and muscarinic), and each responds differently to binding of ACh. Cholinergic neurons include the following:

- All sympathetic and parasympathetic preganglionic neurons
- All parasympathetic ganglionic neurons
- The specific sympathetic ganglionic neurons that innervate sweat glands of the skin *(Note that there are nicotinic receptors at the neuromuscular junctions of skeletal muscle, but the neurons involved are somatic, not autonomic.)

Neurons that synthesize and secrete norepinephrine (NE) are called **adrenergic** (ad-re-ner′jik) **neurons.** Most other sympathetic ganglionic neurons are adrenergic. Receptors that bind NE (or a related molecule, like epinephrine) are called **adrenergic receptors.** These receptors are subdivided into alpha (α) and beta (β) types and are discussed in further detail in section 15.6c.

WHAT DID YOU LEARN?

15 Which ANS neurons are cholinergic? Which are adrenergic?

15.6b Cholinergic Receptors

LEARNING OBJECTIVE

19. Describe the two types of cholinergic receptors and the action of each when the neurotransmitter acetylcholine binds to them.

Two categories of cholinergic receptors, nicotinic and muscarinic, have been identified in the CNS and PNS. They were differentiated and named (nicotinic or muscarinic) because molecules that are similar to ACh bind to them and cause their stimulation:

Nicotinic receptors were so named because **nicotine** (i.e., the chemical compound found in tobacco) selectively binds to these receptors. These receptors are found on the cell bodies and dendrites of all ganglionic neurons (figure 15.10), as well as on adrenal medulla cells. When nicotinic receptors bind ACh, they open ion channels to allow greater movement of sodium ions (Na^+) into the cell than potassium ions (K^+) out of the cell. Thus, the membrane depolarizes and an excitatory postsynaptic potential (EPSP) is produced (see section 12.8a). In other words, ACh binding to nicotinic receptors always produces a stimulatory or excitatory response.

INTEGRATE

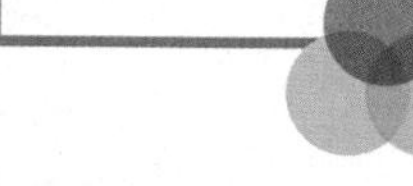

CONCEPT CONNECTION

ACh binding to nicotinic receptors always produces an stimulatory or excitatory response, whether in the autonomic nervous system (ANS) or somatic nervous system. The nicotonic receptors of the ANS are stimulated by ACh. Conversely, the ACh released by somatic nervous system axons excites skeletal muscle fibers (see section 10.3b).

Nicotinic receptors have various subtypes. These subtypes account for the difference in response that a given neurotransmitter may have at different locations. For example, the nicotinic receptor at the neuromuscular junction is blocked by the toxin curare, but it is not blocked by the cholinergic drugs hexamethonium and mecamylamine. The reverse conditions exist for nicotinic receptors on postganglionic neurons. Here, hexamethonium and mecamylamine bind, but curare does not.

WHAT DO YOU THINK?

3 What effect, if any, would smoking have on the nicotinic receptors of the ANS?

Muscarinic receptors were so named because they respond to **muscarine,** a mushroom toxin. They are found in all target organs stimulated by the parasympathetic division and in the few selected sympathetic target cells (e.g., sweat glands in the skin and blood vessels within skeletal muscle). There are different subtypes of muscarinic receptors, which have different effects on various body systems. These different subclasses of muscarinic receptors are either stimulated or inhibited by binding ACh. For example, the binding of ACh to smooth muscle in the gastrointestinal (GI) tract will result in contraction and increased motility (movement) of the muscle, whereas binding of ACh to muscarinic receptors on cardiac muscle pacemaker cells results in decreasing heart rate. The drug pilocarpine (used to treat glaucoma) binds to muscarinic receptors to stimulate ciliary muscles to contract, which facilitates drainage of aqueous humor from the anterior chamber of the eye (see figure 16.15). All muscarinic receptors use second messenger systems (see section 17.5b).

Nicotinic and muscarinic receptors are compared in **table 15.4**.

Table 15.4	Cholinergic Receptors	
Characteristics	**Nicotinic**	**Muscarinic**
Primary locations	All postganglionic neurons in the ANS (both parasympathetic division and sympathetic division) Adrenal medulla (gland innervated by preganglionic neuron of sympathetic division) Neuromuscular junction of skeletal muscle Some neurons of central nervous system (neurons involved in learning and memory)	All effectors of parasympathetic target organs Limited sympathetic target organs that have muscarinic receptors (i.e., sweat glands)
Excitatory or inhibitory	Always excitatory	Generally excitatory (except for selected structures, such as cardiac muscle pacemaker cells)
Examples of drugs that interact with receptor	*Nicotine* binds with all nicotinic receptors. *Curare* binds to specific subtype of nicotinic receptors at the neuromuscular junctions. *Hexamethonium* or *mecamylamine* binds to specific subtype of nicotinic receptors of postganglionic neurons in the ANS.	*Muscarine* binds with all muscarinic receptors. *Pilocarpine* binds to specific subtype of muscarinic receptors of ciliary muscles within the eye.

WHAT DID YOU LEARN?

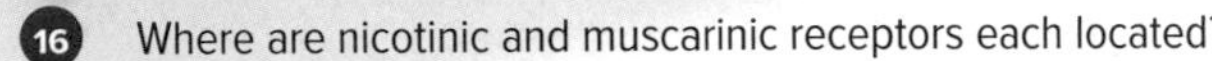

16 Where are nicotinic and muscarinic receptors each located?

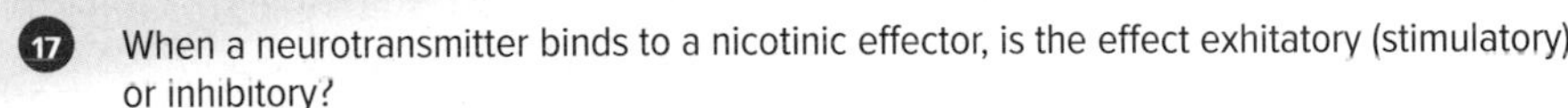

17 When a neurotransmitter binds to a nicotinic effector, is the effect exhitatory (stimulatory) or inhibitory?

15.6c Adrenergic Receptors

LEARNING OBJECTIVES

20. List the neurotransmitters categorized as catecholamines.

21. Name the four adrenergic receptors, and give the locations of each.

Recall from section 4.5b that signaling molecules (e.g., neurotransmitters, hormones) are called **ligands** when they specifically bind to receptors in the plasma membrane. The class of ligands that bind to adrenergic receptors in neurons is called **biogenic amines,** or *monoamines.* One category of biogenic amines is called **catecholamines** because of the presence of a catechol ring structure in the molecule. Catecholamines include dopamine, norepinephrine, and epinephrine (see section 17.3a).

As mentioned in section 15.6a, the two types of adrenergic receptors are **alpha (α)** and **beta (β) receptors,** which may be further divided into subclasses such as α_1, α_2, β_1, and β_2. Target cells with α receptors typically are stimulated, whereas those cells with β receptors may be either stimulated or inhibited in response to binding neurotransmitter (or hormone). (Note that there are exceptions to this rule.) These types of receptors may be further subdivided, as follows:

- **α_1 receptors** are the most common adrenergic receptors of the sympathetic division. These receptors are located within plasma membranes of most smooth muscle cells and stimulate smooth muscle contraction. The specific organs with α_1 receptors include most blood vessels (including those going to the skin, GI tract, and kidneys), arrector pili muscles, uterus, ureters, internal urethral sphincter, and dilator pupillae muscle of the eye. These receptors are involved with vasoconstriction of the above blood vessels, contraction of arrector pili muscles, contraction of the uterine wall, contraction of the ureters, closing of the internal urethral sphincter, and dilation of the pupil, respectively.
- **α_2 receptors** are located throughout the CNS (e.g., brainstem) and decrease norepinephrine release, thereby inhibiting sympathetic activity. Stimulation of these CNS α_2 receptors also causes sedation and analgesia (i.e., the inability to sense pain). The receptors also are in the pancreas to inhibit insulin secretion. In addition, these receptors are located in GI sphincters, and when they are stimulated they constrict the sphincter.

Table 15.5 Adrenergic Receptors

Characteristics	Alpha (α) Receptors		Beta (β) Receptors		
	α_1	α_2	β_1	β_2	β_3
Primary locations and specific actions	Almost all effectors of sympathetic division (exceptions include cardiac muscle and bronchioles) Causes contraction of most smooth muscle, including blood vessels of the skin, blood vessels of GI tract, blood vessels of kidneys, arrector pili, uterus, ureters, internal urethral sphincter, dilator pupillae muscle of eye	Pancreas (inhibits insulin release) CNS (decreases norepinephrine release and thus inhibits sympathetic activity; causes sedation and analgesia) GI sphincters (cause contraction)	Heart (both sinoatrial node and cardiac muscle; increases heart rate and force of contraction) Kidney (stimulates release of renin to increase blood pressure)	Causes smooth muscle relaxation in blood vessels of the heart wall (coronary arteries), liver, and skeletal muscle; results in vasodilation Causes smooth muscle relaxation within bronchioles, uterus, GI tract	Adipose connective tissue (stimulates lipolysis) Urinary bladder (relaxes detrusor muscle)
General effect	Excitatory	Inhibitory or excitatory	Excitatory	Primarily inhibitory	Excitatory or inhibitory
Examples of drugs that interact with receptor	*Phenylephrine* causes vasoconstriction of nasal blood vessels, decreasing nasal secretions.	*Clonidine* is used to treat high blood pressure by stimulating α_2 receptors in the vasomotor center of the brainstem.	*Propranolol* is a nonselective beta blocker used to treat high blood pressure.	*Albuterol* dilates bronchioles; used to treat asthma. *Terbutaline* relaxes uterine wall to delay preterm labor.	*Mirabegron* relaxes urinary bladder wall; used to treat overactive bladder.

INTEGRATE

CLINICAL VIEW 15.2

Epinephrine for Treatment of Asthma

Asthma is a condition in which airflow into the lung is decreased due to the narrowing of the air passages called bronchioles (see Clinical View 23.6: "Asthma"). Bronchioles contain β_2 receptors, and epinephrine (hormone) binds more effectively to β_2 receptors and causes greater relaxation of the smooth muscles of bronchioles than does norepinephrine. The greater the degree of bronchodilation, the greater the rate of airflow into and out of the lungs. Consequently, epinephrine (not norepinephrine) is the active ingredient in medicines for treating asthma (e.g., the medicine in an *EpiPen*®).

- **β_1 receptors** primarily have excitatory (stimulatory) effects. They are found within the heart, where they increase heart rate and force of contraction (see section 19.5b), and in the kidney, where they stimulate secretion of renin (see sections 20.6b and 25.4a) to ultimately cause an increase in blood pressure.
- **β_2 receptors** primarily have inhibitory effects. They are present in smooth muscle of blood vessels supplying the heart wall (coronary arteries) and liver. Unlike activation of α_1 receptors, activation of β_2 receptors causes smooth muscle relaxation, resulting in vasodilation of these specific blood vessels. Their stimulation also causes bronchodilation within the lung, as well as relaxation of uterine and GI tract smooth muscle.
- **β_3 receptors** may have excitatory or inhibitory effects. They are found mainly in adipose connective tissue, where they stimulate lipolysis (breakdown of triglycerides into fuel molecules). They also cause smooth muscle relaxation in the urinary bladder.

Alpha and beta receptors are compared in **table 15.5**.

WHAT DID YOU LEARN?

18 What are the different types of catecholamines?

19 How is it possible for the stimulation of adrenergic receptors to result in either vasoconstriction or vasodilation of selected blood vessels?

15.7 Interactions Between the Parasympathetic and Sympathetic Divisions

Most organs are innervated by both divisions of the autonomic nervous system, and these control the targets continuously to varying degrees. The parasympathetic and sympathetic division effects are compared in **table 15.6** and **figure 15.11**.

15.7a Autonomic Tone

LEARNING OBJECTIVE

22. Discuss the nature of autonomic tone and its effects.

The parasympathetic and sympathetic divisions both continuously release neurotransmitter to regulate specific target organs for either sustained stimulation or inhibition, a process referred to as the **autonomic tone.** Activity of an organ may be controlled merely by the change in tone within a single ANS division.

Table 15.6	Main Effects of the Parasympathetic and Sympathetic Divisions	
Effector	**Parasympathetic Innervation Effects**	**Sympathetic Innervation Effects**
	CARDIOVASCULAR SYSTEM	
Heart	Decreases heart rate	Increases heart rate and force of contraction
Blood vessels		
Coronary	None	Vasodilation (β receptors) or vasoconstriction (α receptors)
To skeletal muscle	None	Vasodilation (β receptors)
Integumentary and most other blood vessels	None	Vasoconstriction (α receptors)
	DIGESTIVE SYSTEM	
Salivary glands	Stimulates watery secretion	Stimulates more viscous secretion; ultimately reduces saliva secretion
GI tract gland secretion	Stimulates	Inhibits
Smooth muscle in GI tract	Stimulates peristalsis (motility)	Inhibits peristalsis (motility)
Sphincters	Relax (open, to allow passage of materials)	Contract (close to prevent passage of materials)
GI tract blood vessels	Vasodilation	Vasoconstriction
Liver	Stimulates glycogenesis (formation of glycogen from glucose)	Stimulates glycogenolysis (breakdown of glycogen into glucose)
	RESPIRATORY SYSTEM	
Bronchi/bronchioles of lungs	Bronchoconstriction (airway narrows)	Bronchodilation (airway widens)
	URINARY SYSTEM	
Kidneys	None	Stimulates release of renin (to raise blood pressure)
Bladder (muscle wall)	Contraction (to facilitate urination)	Relaxation
Internal urethral sphincter	Relaxation (opens, to facilitate urination)	Contraction (closes, to inhibit urination)
	REPRODUCTIVE SYSTEM	
Penis	Stimulates erection	Stimulates ejaculation
Clitoris	Stimulates erection	None
Uterine muscle	None	Contraction
Gland secretion	Stimulates	Inhibits
Vaginal muscular wall	None	Contraction
	INTEGUMENTARY SYSTEM	
Arrector pili	None	Contraction to cause hair elevation (i.e., "goose bumps")
Sweat glands	None	Secretion
	EYE	
Pupil size	Constriction (allows less light into the eye)	Dilation (allows more light into the eye)
Ciliary muscle	Contraction for near vision	None
Lacrimal gland	Stimulates secretion	None
	ADRENAL MEDULLA	
	None	Stimulates release of epinephrine and norepinephrine
	ADIPOSE CONNECTIVE TISSUE	
	None	Stimulates lipolysis (β receptors)

INTEGRATE CONCEPT OVERVIEW

Figure 15.11 Comparison of the Parasympathetic and Sympathetic Divisions of the ANS. (*a*) The parasympathetic division, also known as the *rest-and-digest* division, has its preganglionic neurons located in the cranial region of the brainstem and sacral region of the spinal cord. (*b*) The sympathetic division, also known as the *fight-or-flight* division, contains preganglionic neurons located in the T1–L2 regions of the spinal cord.

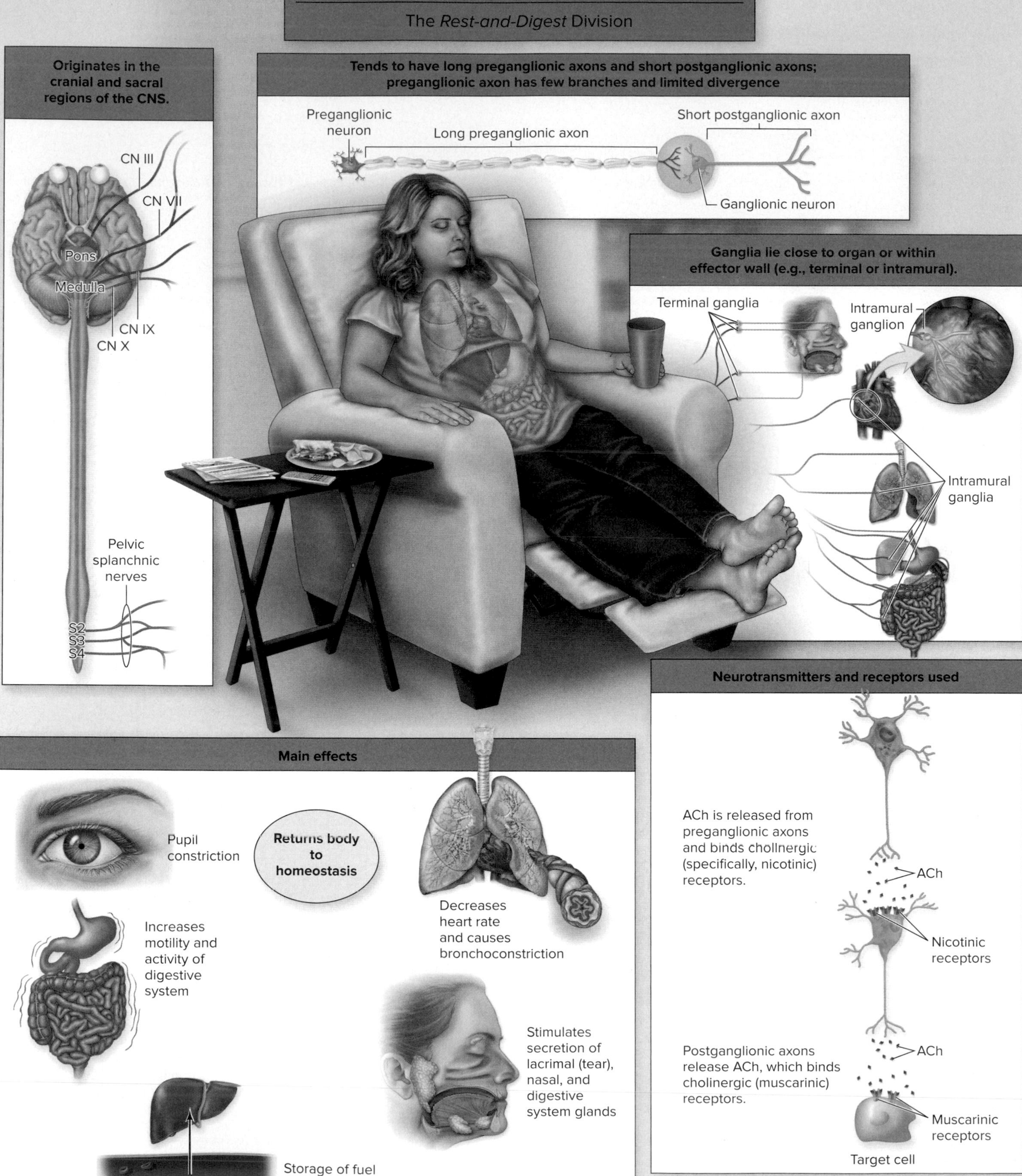

(a) Parasympathetic Division
The Rest-and-Digest Division
Originates in the cranial and sacral regions of the CNS.
CN III
CN VII
Pons
Medulla
CN IX
CN X
Pelvic splanchnic nerves
S2
S3
S4
Tends to have long preganglionic axons and short postganglionic axons; preganglionic axon has few branches and limited divergence
Preganglionic neuron
Long preganglionic axon
Short postganglionic axon
Ganglionic neuron
Ganglia lie close to organ or within effector wall (e.g., terminal or intramural).
Terminal ganglia
Intramural ganglion
Intramural ganglia
Neurotransmitters and receptors used
ACh is released from preganglionic axons and binds cholinergic (specifically, nicotinic) receptors.
ACh
Nicotinic receptors
Postganglionic axons release ACh, which binds cholinergic (muscarinic) receptors.
ACh
Muscarinic receptors
Target cell
Main effects
Pupil constriction
Returns body to homeostasis
Decreases heart rate and causes bronchoconstriction
Increases motility and activity of digestive system
Stimulates secretion of lacrimal (tear), nasal, and digestive system glands
Storage of fuel molecules in liver

(b) Sympathetic Division

The *Fight-or-Flight* Division

Tends to have shorter preganglionic axons and longer postganglionic axons; preganglionic axon has many branches and extensive divergence

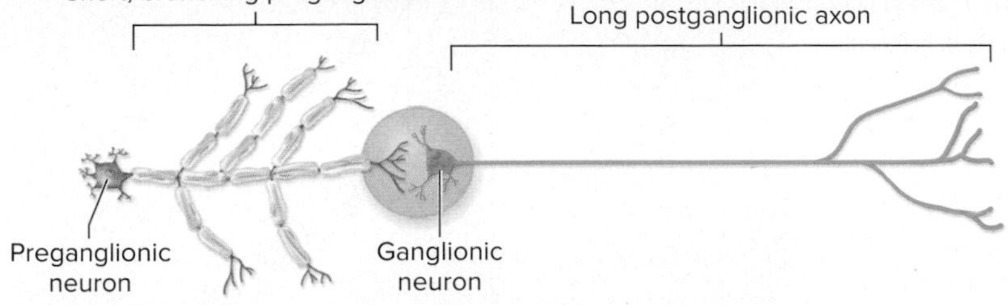

Short, branching preganglionic axon

Long postganglionic axon

Preganglionic neuron

Ganglionic neuron

Originates in the T1–L2 segments of the spinal cord

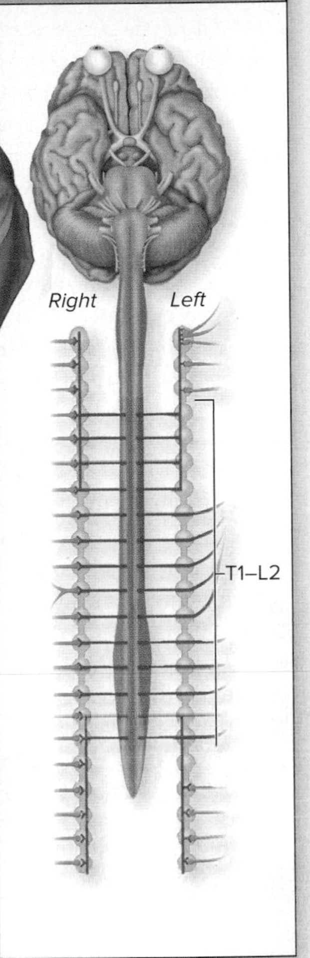

Right

Left

T1–L2

Ganglia (sympathetic trunk or prevertebral) lie close to the vertebral column.

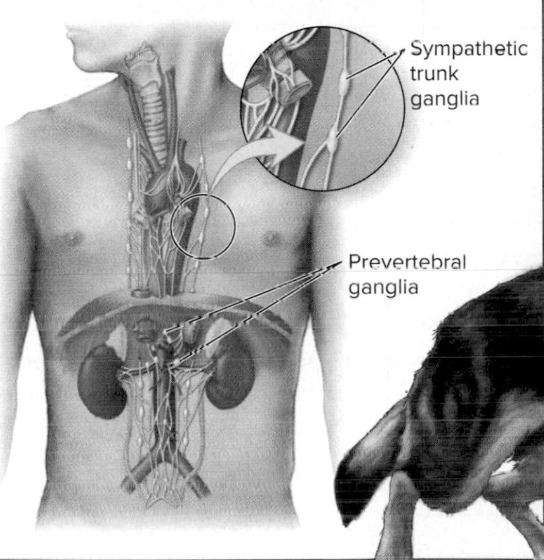

Sympathetic trunk ganglia

Prevertebral ganglia

Neurotransmitters and receptors used

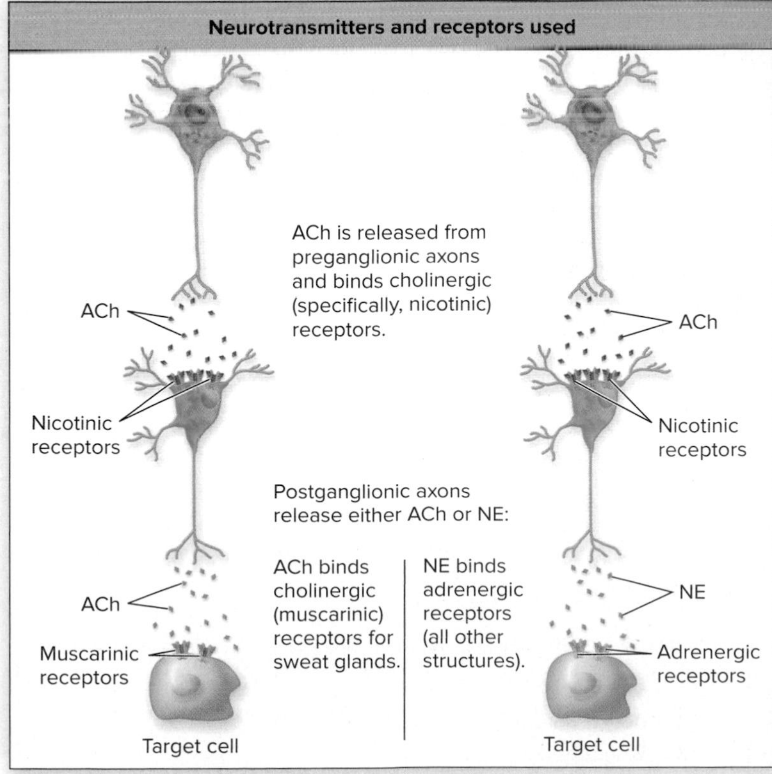

ACh is released from preganglionic axons and binds cholinergic (specifically, nicotinic) receptors.
ACh
ACh
Nicotinic receptors
Nicotinic receptors
Postganglionic axons release either ACh or NE:
ACh
ACh binds cholinergic (muscarinic) receptors for sweat glands.
NE binds adrenergic receptors (all other structures).
NE
Muscarinic receptors
Adrenergic receptors
Target cell
Target cell

Main effects

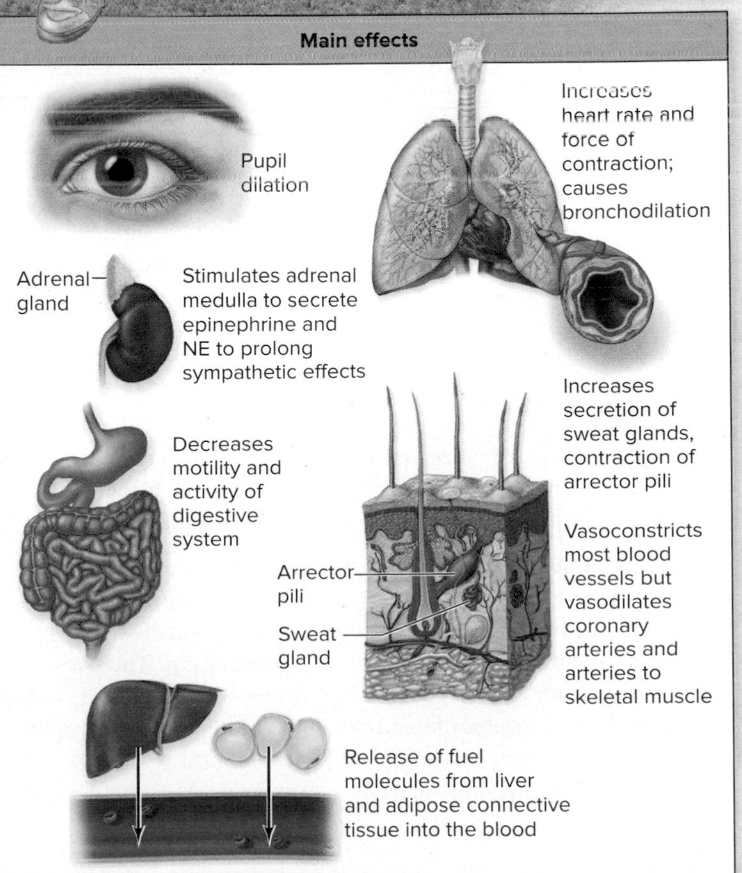

Pupil dilation
Increases heart rate and force of contraction; causes bronchodilation
Adrenal gland
Stimulates adrenal medulla to secrete epinephrine and NE to prolong sympathetic effects
Decreases motility and activity of digestive system
Increases secretion of sweat glands, contraction of arrector pili
Arrector pili
Sweat gland
Vasoconstricts most blood vessels but vasodilates coronary arteries and arteries to skeletal muscle
Release of fuel molecules from liver and adipose connective tissue into the blood

INTEGRATE

CLINICAL VIEW 15.3

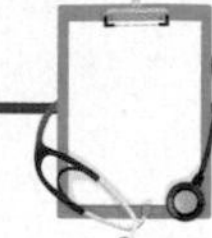

Raynaud Syndrome

Raynaud syndrome, or *primary Raynaud phenomenon,* is a rapid constriction of the small arteries of the digits and occasionally the lips, nose, and ears. The immediate decrease in blood flow results in blanching (loss of the normal hue) of the skin distal to the area of vascular constriction. This blanching is followed by a cyanotic period (the affected body parts turn blue) and then a quick blood reflow period, in which the affected body parts turn red. The initial constriction is accompanied by pain, which may even continue for a while after the vessels have dilated and restored the local blood flow. Episodes are typically triggered by exposure to cold, although emotional stress has been known to precipitate a Raynaud attack. This condition is more common in women than in men. It is believed to result from an exaggerated local sympathetic response: specifically, overactivity of α_2 receptors in the smooth muscle of these blood vessels, resulting in the sudden vasoconstriction. The severity of this medical condition determines the frequency and length of time of each occurrence. Most people affected with Raynaud syndrome must minimize exposure to the cold and other triggering circumstances.

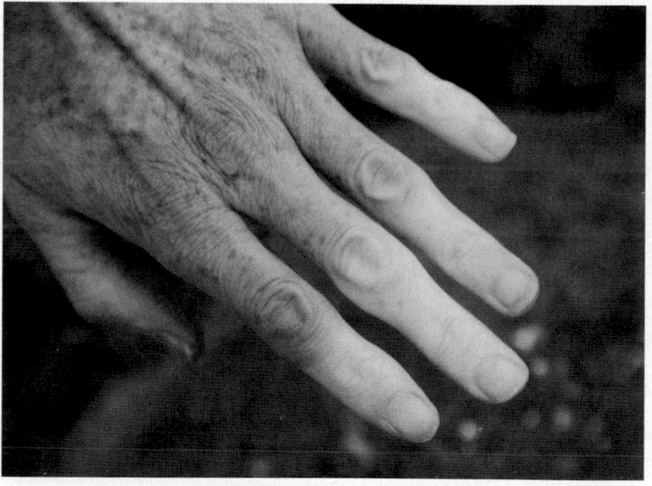

Blanching of distal tips of fingers in primary Raynaud phenomenon.

For example, the diameter of most blood vessels is maintained in a partially constricted state by the effects of sympathetic **vasomotor tone.** A decrease in stimulation below the sympathetic tone causes vessel dilation, whereas an increase above sympathetic tone brings about vessel constriction (see section 20.1b). If the initial level of sympathetic tone were not present, then only vasoconstriction could occur as a result of sympathetic division activity. Another example of autonomic tone results from continued stimulation of the heart pacemaker (sinoatrial node) by the parasympathetic division through **vagal tone,** which decreases the heart rate (see section 19.6b).

WHAT DID YOU LEARN?

20 How does autonomic tone permit the control of blood vessel diameter by sympathetic innervation?

15.7b Dual Innervation

LEARNING OBJECTIVES

23. Explain dual innervation.
24. Describe the antagonistic and cooperative effects of dual innervation.

Many effectors of the ANS have **dual innervation,** meaning that they are innervated by postganglionic axons from both parasympathetic and sympathetic divisions. The actions caused by activities of both the divisions on the same organ usually result in effects that are antagonistic or cooperative.

Antagonistic Effects

Generally, the effects of parasympathetic and sympathetic innervation to the same organ are antagonistic—that is, they oppose each other. Some variations occur in how antagonistic effects are expressed, as follows:

- **Control of heart rate.** Parasympathetic stimulation slows down the heart rate, whereas sympathetic stimulation speeds up the heart rate (see section 19.9b). The same heart muscle effector cells receive this opposing stimulation. The two divisions are able to cause different responses because cardiac muscle cells contain more than one type of cellular receptor (e.g., muscarinic receptors and β_1 receptors).
- **Control of muscular activity in the gastrointestinal tract.** Parasympathetic stimulation of smooth muscle cells in the gastrointestinal tract wall increases their force of contraction and thus increases gastrointestinal tract motility. Conversely, sympathetic stimulation decreases the force of contraction and thus decreases motility. Again, both ANS divisions innervate the same effector cells, but they house two different types of receptors (see section 26.1e).
- **Control of pupil diameter in the iris of the eye.** Here, different effectors are innervated by the ANS divisions. Parasympathetic stimulation of the circular muscle layer in the iris causes pupil constriction, whereas sympathetic stimulation of the radial muscle layer in the iris causes pupil dilation (see section 16.4b).

Cooperative Effects

Cooperative effects are seen when both parasympathetic and sympathetic stimulation cause different responses that together produce a single, distinct result. The best example of cooperative effects occurs in the sexual function of the male reproductive system. The male penis becomes erect as a result of parasympathetic stimulation, and ejaculation of semen from the penis is facilitated by stimulation from the sympathetic division (see section 28.4f). This synergistic effort of ANS divisions facilitates reproduction.

WHAT DID YOU LEARN?

21 What are the antagonistic effects of the sympathetic and parasympathetic divisions on the heart?

22 How do the sympathetic and parasympathetic divisions exhibit cooperative effects on the male reproductive system?

INTEGRATE

LEARNING STRATEGY

The mnemonic **ABE** may help you remember which structures receive only sympathetic innervation:

A = **A**drenal medulla
B = **B**lood vessels
E = **E**ffectors of the skin (arrector pili muscles and sweat glands)

15.7c Systems Controlled Only by the Sympathetic Division

LEARNING OBJECTIVE

25. Describe the systems innervated only by the sympathetic division and how they function.

In some ANS effectors, opposing effects are achieved without dual innervation. For example, the neurosecretory cells of the adrenal medulla also are innervated only by the sympathetic division. These cells have nicotinic receptors, so upon release of ACh by the postganglionic axon, the adrenal medulla cells are stimulated to release epinephrine and norepinephrine into the blood. There, these molecules act as hormones and prolong the fight-or-flight effects of the sympathetic division.

In addition, many blood vessels are innervated by sympathetic axons only. Increased sympathetic stimulation increases smooth muscle contraction, resulting in increased blood pressure. An analogy is pressing down on a gas pedal to cause a car to accelerate. Decreasing the sympathetic stimulation below the autonomic tone will result in vasodilation, just as lifting your foot off the gas pedal may slow a car down (because you are not supplying the gas). Thus, opposing effects are achieved merely by increasing or decreasing the autonomic tone in the sympathetic division.

Other examples of innervation by only the sympathetic division are seen in sweat glands in the trunk (stimulates sweating) and innervation of arrector pili muscles in the skin to cause "goose bumps" (see section 6.2b).

WHAT DID YOU LEARN?

23 What are the body structures innervated by the sympathetic division only?

INTEGRATE

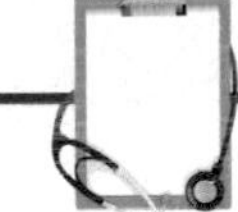

CLINICAL VIEW 15.4

Autonomic Dysreflexia

Autonomic dysreflexia is a potentially dangerous vascular condition that causes blood pressure to rise profoundly, sometimes so high that blood vessels rupture. Specifically, it stimulates a sympathetic division reflex that causes systemic vasoconstriction and a marked increase in blood pressure. Autonomic dysreflexia is caused by hyperactivity of the autonomic nervous system in the weeks and months after a spinal cord injury. Often, the initial reaction to spinal cord injury is spinal shock, which is characterized by the loss of autonomic reflexes. However, paradoxically, this decrease in reflex activities may cause certain viscera to respond abnormally to the lack of innervation, a phenomenon called **denervation hypersensitivity.** For example, when a person loses the ability to voluntarily evacuate the bladder, the bladder may continue to fill with urine to the point of overdistension. This induces a spinal cord reflex that causes involuntary relaxation of the internal urethral sphincter, allowing the bladder to empty.

15.8 Autonomic Reflexes

LEARNING OBJECTIVES

26. Discuss how autonomic reflexes help maintain homeostasis.

27. Describe some major examples of autonomic reflexes.

All responses of the autonomic nervous system are regulated through reflexes (below the conscious level of control). Recall from section 14.6b that a reflex is a rapid, preprogrammed response of a muscle or gland to a stimulus, which includes five components arranged in a reflex arc. These components include (1) a receptor (that detects the stimulus); (2) a sensory neuron that relays nerve signals from the receptor to the CNS; (3) an integration center (brain or spinal cord) that integrates the sensory input and initiates motor output; (4) motor neurons that relay nerve signals from the CNS to the effectors; and (5) effectors that bring about a response. Also note that whereas somatic reflexes involve skeletal muscles, autonomic reflexes involve cardiac muscle, smooth muscle, or glands. The autonomic nervous system helps maintain homeostasis through the involuntary activity of **autonomic reflexes,** also termed *visceral reflexes*. These reflexes enable the ANS to control visceral function. Autonomic reflexes consist of smooth muscle contractions (or relaxation), cardiac muscle contractions, or stimulation or inhibition of secretion by glands, which are mediated by autonomic reflex arcs in response to a specific stimulus. Some autonomic reflexes are described here. See if you are able to determine the five components of the reflex arc for each of the examples:

- **Cardiovascular reflex.** A classic autonomic reflex involves the reduction of blood pressure. When blood pressure elevates, stretch receptors in the walls of large blood vessels (e.g., aorta) are stimulated and nerve signals are propagated along visceral sensory neurons to the cardiac center in the medulla oblongata. These nerve signals inhibit sympathetic output and activate parasympathetic output to the heart to slow heart rate and decrease the volume of blood ejected, resulting in a decrease in blood pressure (see section 19.5b).
- **Gastrointestinal reflex.** Autonomic reflexes control defecation. Fecal matter entering the rectum causes stretch of the rectal wall. Sensory neurons relay increased nerve signals to the spinal cord, which initiates a change in nerve signals along motor neurons to the rectum and anal sphincter. The rectum contracts and the internal anal sphincter relaxes (see section 26.3d).
- **Micturition reflex.** The mechanism leading to bladder emptying is similar to that for fecal emptying of the colon. In a young child who is not yet toilet trained, stretch receptors send nerve signals to the sacral spinal cord when urine fills the bladder **(figure 15.12)**. The reflex results in contraction of the smooth muscle in the bladder wall and relaxation of the urinary sphincters. (In a toilet-trained individual, sensory nerve signals end at the pons instead of the sacral region of the spinal cord, and urination occurs following voluntary relaxation of the external urethral sphincter. We discuss micturition in greater detail in section 24.8c).

Other autonomic reflexes include changing the size of bronchioles to regulate the amount of air entering the lungs, regulating digestive system activities, and changing pupil diameter.

WHAT DID YOU LEARN?

24 How does the cardiovascular reflex affect blood pressure?

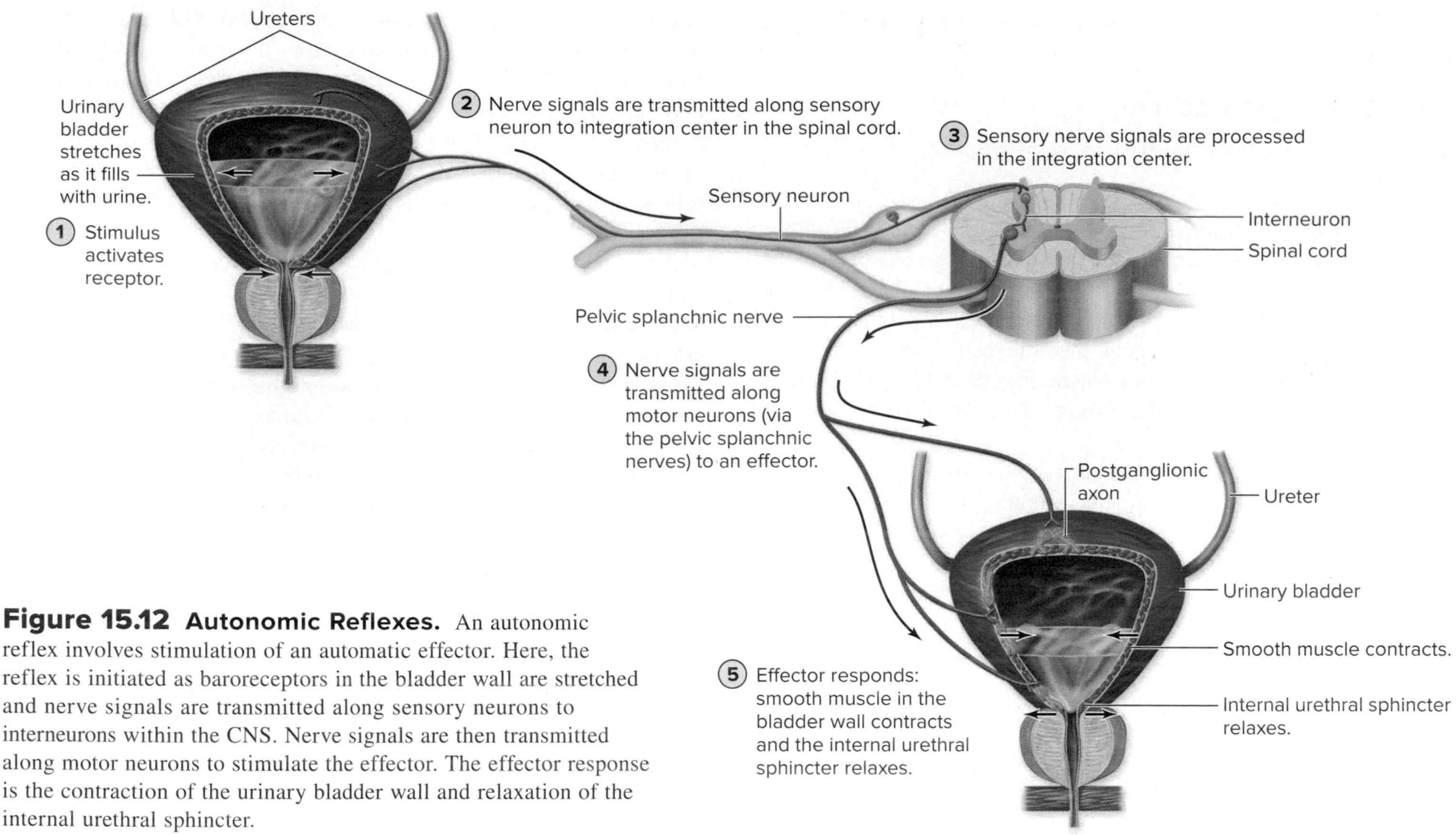
Ureters
Urinary bladder stretches as it fills with urine.
1 Stimulus activates receptor.
2 Nerve signals are transmitted along sensory neuron to integration center in the spinal cord.
3 Sensory nerve signals are processed in the integration center.
Sensory neuron
Interneuron
Spinal cord
Pelvic splanchnic nerve
4 Nerve signals are transmitted along motor neurons (via the pelvic splanchnic nerves) to an effector.
Postganglionic axon
Ureter
Urinary bladder
Smooth muscle contracts.
5 Effector responds: smooth muscle in the bladder wall contracts and the internal urethral sphincter relaxes.
Internal urethral sphincter relaxes.

Figure 15.12 Autonomic Reflexes. An autonomic reflex involves stimulation of an automatic effector. Here, the reflex is initiated as baroreceptors in the bladder wall are stretched and nerve signals are transmitted along sensory neurons to interneurons within the CNS. Nerve signals are then transmitted along motor neurons to stimulate the effector. The effector response is the contraction of the urinary bladder wall and relaxation of the internal urethral sphincter.

CHAPTER SUMMARY

	• The autonomic nervous system controls the internal environment and helps maintain homeostasis.
15.1 Comparison of the Somatic and Autonomic Nervous Systems	• The nervous system can be functionally organized into the somatic nervous system (SNS) and autonomic nervous system (ANS) based upon whether the sensory input and motor output can be consciously regulated.
	15.1a Functional Organization • The somatic nervous system includes sensory input from the special senses, skin, muscles and joints, and motor output to control skeletal muscle. • The autonomic nervous system includes involuntary motor output to cardiac muscle, smooth muscle, and glands and responds to sensory input from visceral sensory components.
	15.1b Lower Motor Neurons of the Somatic Versus Autonomic Nervous System • A single motor neuron innervates skeletal muscle fibers in the SNS, whereas the ANS has a two-neuron pathway consisting of preganglionic neurons in the central nervous system (CNS) and ganglionic neurons in the peripheral nervous system (PNS).
	15.1c CNS Control of the Autonomic Nervous System • Autonomic function is regulated by three CNS regions: hypothalamus, brainstem, and spinal cord.
15.2 Divisions of the Autonomic Nervous System	**15.2a Functional Differences** • The parasympathetic division is primarily concerned with maintaining homeostasis when the body is at rest, which includes conserving energy and replenishing nutrient stores. • The sympathetic division is primarily concerned with maintaining homeostasis in conditions of fight-or-flight.
	15.2b Anatomic Differences in Lower Motor Neurons • Parasympathetic preganglionic neurons reside in the brainstem and sacral regions of the spinal cord, whereas sympathetic preganglionic axons reside in the thoracic and lumbar regions of the spinal cord.
	15.2c Degree of Response • The parasympathetic response tends to be discrete and localized, whereas the sympathetic response has the potential to produce a mass activation effect.

15.3 Parasympathetic Division	• The parasympathetic division is also known as the craniosacral system, because of the location of its preganglionic neurons. **15.3a Cranial Components** • Parasympathetic preganglionic axons extend through the oculomotor, facial, glossopharyngeal, and vagus cranial nerves. **15.3b Pelvic Splanchnic Nerves** • The remaining preganglionic parasympathetic cell bodies are housed within the S2–S4 segments of the spinal cord and form pelvic splanchnic nerves.
15.4 Sympathetic Division	• The sympathetic division is also known as the thoracolumbar division, because its preganglionic neurons reside in the T1–L2 segments of the spinal cord. • A single effector may control one tissue, but many effectors often respond together, a phenomenon called mass activation. **15.4a Organization and Anatomy of the Sympathetic Division** • Preganglionic neuronal cell bodies are housed within the lateral gray horn of the spinal gray matter and their axons extend through white rami communicantes to the sympathetic trunk. • Gray rami communicantes are composed of postganglionic sympathetic axons from the sympathetic trunk to the spinal nerve. • Some preganglionic axons pass through the sympathetic trunk without synapsing and form splanchnic nerves that project to the prevertebral ganglia. Postganglionic axons extend from the prevertebral ganglia to the target organ. **15.4b Sympathetic Pathways** • In the spinal nerve pathway, the postganglionic axon enters the spinal nerve through the gray ramus and extends to target organs (blood vessels and glands of the skin of the neck, torso, and limbs). • In the postganglionic sympathetic nerve pathway, the postganglionic axon extends from the sympathetic trunk and projects directly to the target organs (e.g., head, neck viscera, and thoracic viscera). • In the splanchnic nerve pathway, the preganglionic axon passes through the sympathetic trunk and travels to the prevertebral ganglia, where it synapses with a ganglionic neuron, which extends to the target organs (most abdominal and pelvic viscera). • In the adrenal medulla pathway, the preganglionic axon extends through both a sympathetic trunk ganglion and a prevertebral ganglion without synapsing. It synapses on secretory cells in the adrenal medulla that release epinephrine and norepinephrine.
15.5 Autonomic Plexuses and the Enteric Nervous System	**15.5a Autonomic Plexuses** • Autonomic plexuses are meshworks of sympathetic postganglionic axons and parasympathetic preganglionic axons, as well as some visceral sensory axons. **15.5b Enteric Nervous System** • The gastrointestinal (GI) tract has an enteric nervous system to regulate digestive functions. • The enteric nervous system may operate independently, although ANS divisions may stimulate or inhibit this system.
15.6 Comparison of Neuro-transmitters and Receptors of the Two Divisions	• The two main types of receptors are cholinergic and adrenergic receptors. **15.6a Overview of ANS Neurotransmitters** • Acetylcholine (ACh) is the neurotransmitter released by cholinergic neurons, which include all preganglionic neurons as well as all ganglionic parasympathetic neurons. ACh also is used by some ganglionic sympathetic neurons to sweat glands. • Norepinephrine (NE) is the neurotransmitter released from adrenergic neurons, which include all other sympathetic postganglionic neurons. **15.6b Cholinergic Receptors** • Nicotinic receptors are located on all ganglionic neurons and cells in the adrenal medulla, and they are always excitatory (stimulatory). • Muscarinic receptors are located on all parasympathetic target cells, on sweat glands in the skin, and on blood vessels in skeletal muscle. Their effect may be excitatory or inhibitory depending upon the muscarinic receptor subtype. **15.6c Adrenergic Receptors** • Adrenergic receptors include α and β receptors. There are several subsets for both α and β types.
15.7 Interactions Between the Parasympathetic and Sympathetic Divisions	• Most organs are innervated by both divisions of the ANS. **15.7a Autonomic Tone** • Both ANS divisions maintain some continual activity, which is called the autonomic tone for that division. **15.7b Dual Innervation** • Many visceral effectors have dual innervation, meaning they are innervated by both ANS divisions. The actions of the divisions mostly are antagonistic (having opposite effects on a target organ), but a few are cooperative (working together to cause a single result). **15.7c Systems Controlled Only by the Sympathetic Division** • The adrenal medulla, most blood vessels, and sweat glands of the skin are innervated by sympathetic axons only.
15.8 Autonomic Reflexes	• The autonomic nervous system helps maintain homeostasis through autonomic reflexes, which are also called visceral reflexes.

CHALLENGE YOURSELF

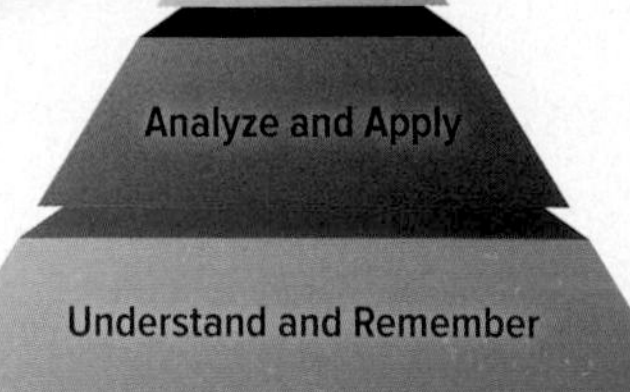

Do You Know the Basics?

1. A splanchnic nerve in the sympathetic division of the ANS
   a. connects neighboring sympathetic trunk ganglia.
   b. controls parasympathetic functions in the thoracic cavity.
   c. is formed by preganglionic axons that extend to prevertebral ganglia.
   d. travels through parasympathetic pathways in the head.
2. Some parasympathetic preganglionic neuron cell bodies are housed within the
   a. hypothalamus.
   b. sacral region of the spinal cord.
   c. cerebral cortex.
   d. thoracolumbar region of the spinal cord.
3. Which of the following is a function of the para sympathetic division of the ANS?
   a. increases heart rate and breathing rate
   b. prepares for emergency
   c. increases digestive system motility and activity
   d. dilates pupils
4. Maintaining a resting level of ANS activity in a cell is called
   a. autonomic tone.
   b. cooperative effect.
   c. dual innervation.
   d. antagonistic effect.
5. Sympathetic division preganglionic axons travel to the ________ ganglia via the ________ rami.
   a. terminal, white
   b. sympathetic trunk, gray
   c. prevertebral, gray
   d. sympathetic trunk, white
6. All parasympathetic division synapses use ________ as a neurotransmitter.
   a. dopamine
   b. acetylcholine
   c. norepinephrine
   d. epinephrine
7. Which autonomic nerve plexus innervates the pelvic organs?
   a. cardiac plexus
   b. esophageal plexus
   c. hypogastric plexus
   d. inferior mesenteric plexus
8. A sympathetic postganglionic axon is
   a. long and unmyelinated.
   b. short and myelinated.
   c. short and unmyelinated.
   d. long and myelinated.
9. Nicotinic receptors are located on which of the following?
   a. plasma membranes of ganglionic neurons
   b. target cells that receive parasympathetic innervation
   c. blood vessels in skeletal muscles
   d. sweat glands
10. Which of the following is accurate about a beta receptor?
    a. It binds acetylcholine.
    b. Its effects are excitatory (stimulatory) only.
    c. It causes general vasoconstriction.
    d. It increases heart rate.
11. What are the three CNS regions that regulate autonomic function?
12. For the following ganglia, identify the location and the division of the ANS each is a part of: sympathetic trunk ganglia, prevertebral ganglia, and terminal ganglia.
13. Compare and contrast the postganglionic axons of the parasympathetic and sympathetic divisions. Examine the axon length, myelination (or lack thereof), and the neurotransmitter used.
14. Compare and contrast sympathetic and parasympathetic innervation effects on digestive system structures.
15. Explain responses of nicotinic receptors and muscarinic receptors to simulation by ACh.
16. Describe the differences between cooperative effects and antagonistic effects in dual innervation of target organs.
17. Describe how the general functions of the sympathetic and parasympathetic divisions of the ANS differ.
18. What may occur with the mass activation of the sympathetic division of the ANS?
19. Describe the process of the micturition reflex.
20. How does sympathetic innervation regulate vasoconstriction or vasodilation in the same blood vessels?

Can You Apply What You've Learned?

Use the following paragraph to answer questions 1 and 2.

Arlene was crossing the street when a car ran a red light and nearly hit her. Arlene was not hurt, but she was very frightened and was in a heightened state of alertness well after the incident.

1. Arlene likely experienced all of the following physiologic effects *except*
   a. increased heart rate.
   b. pupil constriction.
   c. goose bumps.
   d. sweaty palms.

2. Arlene was in a heightened state of alertness well after the incident because
   a. the adrenal medulla secreted epinephrine and norepinephrine.
   b. the parasympathetic division stimulated regions of the brain.
   c. the sympathetic division decreased overall autonomic tone of blood vessels.
   d. All of these are correct.
3. George has hypertension (high blood pressure). His physician prescribed the drug propranolol, which is described as a beta-blocker, to reduce his blood pressure What could be a side effect of propranolol?
   a. reduced heart rate
   b. increased blood clotting
   c. vasoconstriction of blood vessels to the skin
   d. bronchodilation
4. Albuterol is a drug designed to counteract the effects of asthma—namely, the medication, which may be used in an inhaler, facilitates bronchodilation. What receptors would you expect this drug to bind?
   a. α_1 receptors
   b. α_2 receptors
   c. β_1 receptors
   d. β_2 receptors
5. One surgical treatment for gastric ulcers is a selective vagotomy, where branches of the vagus nerve to the upper GI tract are cut. How would you suppose a vagotomy would help the treatment of a gastric ulcer?
   a. It would stimulate vasodilation of the blood vessels serving the stomach.
   b. It would reduce gastric gland secretion.
   c. It would promote faster movement of materials through the stomach.
   d. All of these are correct.

Can You Synthesize What You've Learned?

1. Our body expends a lot of energy activating and propagating the mass activation response of the sympathetic nervous system. Why is it necessary for us to have such an "expensive" mechanism at our disposal?
2. When you were younger, your parents may have told you to wait until 1 hour after you've eaten before you go swimming. Based on what you've learned about the ANS, can you hypothesize why swimming right after a meal may be problematic?
3. Some faculty dislike teaching lecture classes after lunch, complaining that the students do not pay attention at this time. From a physiologic viewpoint, what is happening to these students?

INTEGRATE

ONLINE STUDY TOOLS

connect | SMARTBOOK® | AP|R

The following study aids may be accessed through Connect.

Clinical Case Study: A Secret Life Leading to an Unexpected Death

Interactive Questions: This chapter's content is served up in a number of multimedia question formats for student study

SmartBook: Topics and terminology include comparison of the somatic and autonomic nervous systems; divisions of the autonomic nervous system; parasympathetic division; sympathetic division; comparison of neurotransmitters and receptors of the two divisions; interactions between the parasympathetic and sympathetic divisions; control and integration of autonomic system function

Anatomy & Physiology Revealed: Topics include sympathetic and parasympathetic overviews; thoracic region

Design Elements: **Learning Objective** icon: ©McGraw-Hill Education; **What Do You Think?** icon: ©McGraw-Hill Education; **What Did You Learn?** icon: ©McGraw-Hill Education; **Learning Strategy** icon: ©Slavoljub Pantelic/Shutterstock RF; **Clinical View** icon: ©Laia Design Studio/Shutterstock RF; **Concept Connection** icon: ©McGraw-Hill Education; **Challenge Yourself** icon: ©McGraw-Hill Education; **Do You Know the Basics?** icon: ©McGraw-Hill Education; **Can You Apply What You've Learned?** icon: ©McGraw-Hill Education; **Can You Synthesize What You've Learned?** icon: ©McGraw-Hill Education; **Online Study Tools** icon: ©ayzek/Shutterstock RF

chapter

16 Nervous System: Senses

Module 7: Nervous System

INTEGRATE

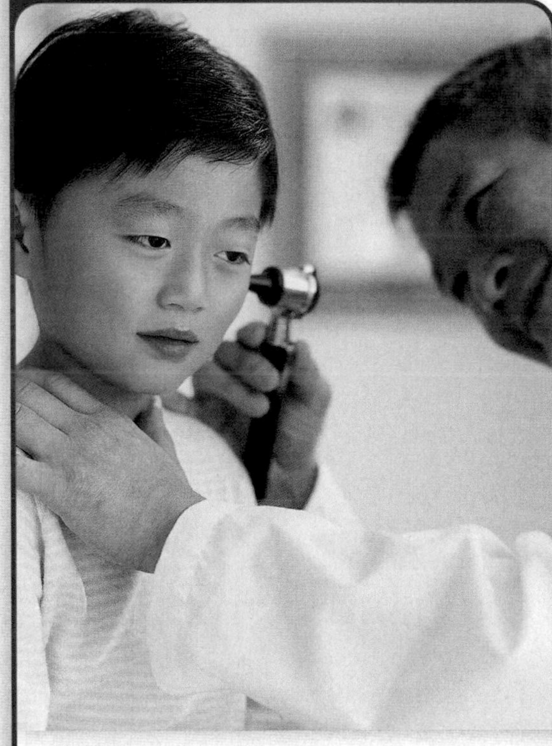

©Corbis RF

CAREER PATH
Pediatrician

A pediatrician focuses on the health and well-being of children, from birth to early adulthood. One of the more common ailments of children is an earache, which typically occurs as a result of a respiratory infection that has spread to the middle ear. Thus, knowledge of ear anatomy and physiology is essential for this health-care professional. The pediatrician uses an instrument called an otoscope to examine the ear canal, eardrum, and middle ear for possible infection. This chapter provides examples of how the otoscope can be used to visualize healthy versus infected middle ears, and why middle ear infections are more prevalent in children under 5.

We are barraged continuously with sensory information about the environment—both outside and inside our bodies. This information is detected by various sensory receptors and then transmitted to the brain and spinal cord so that it can be interpreted and the appropriate responses initiated. The sensory information comes in many different forms. Touch receptors react to physical contact, pressure receptors within blood vessels respond to stretch, taste receptors detect chemicals in the food we eat, visual receptors respond to light, and hearing receptors detect and react to sound waves. In this chapter on the senses, we first provide an introduction to sensory receptors and describe the structure and function of the general senses. We then examine in detail the special senses (smell, taste, vision, hearing, and equilibrium).

16.1 Introduction to Sensory Receptors

Sensory receptors (*recipio* = to receive) are components of the nervous system that provide us with information about our external and internal environments. Here we describe the general function and structure of sensory receptors, the type of information they provide, and how the various types of sensory receptors are classified.

16.1a General Function of Sensory Receptors

LEARNING OBJECTIVE

1. Describe the general function of sensory receptors as transducers.

The general function of all sensory receptors is to respond to a **stimulus** (stim′ū-lus; pl., *stimuli;* a goad) and initiate sensory input to the central nervous system (CNS). This involves converting stimulus energy into an electrical signal. The original energy form detected is specific to the type of sensory receptor (e.g., light energy is detected by the eye, sound energy by the ear, and mechanical energy by blood vessels). However, the form the energy is transduced, or changed, to is always electrical energy, and it is sent along a sensory neuron. This sensory information is propagated as nerve signals; see section 12.8c) to the CNS for interpretation. It is because sensory receptors transduce stimulus energy to electrical energy that sensory receptors are referred to as **transducers** (tranz-dū′sĕr; *trans* = across, *duco* = to lead).

Two features are critical to allow sensory receptors to function as transducers: (1) Sensory receptors, like neurons and muscle cells, establish and maintain a resting membrane potential (RMP) across their plasma membrane (see section 4.4). (2) Sensory receptors contain modality gated channels within their plasma membranes. A **modality gated channel** opens in response to a stimulus other than a neurotransmitter or a voltage change at the plasma membrane. (Recall that chemically gated channels open in response to a neurotransmitter and voltage-gated channels open in response to a voltage change; see section 12.6a). The specifics for the various types of sensory receptors and the opening of their specialized modality gated channels are explained in detail throughout the later sections of this chapter.

WHAT DID YOU LEARN?

1 How does a sensory receptor function as a transducer?

INTEGRATE

CONCEPT CONNECTION

Our sensory receptors initiate communication among most body systems. For example, cutaneous receptors (i.e. sensory receptors in our integument) signal our central nervous system to initiate cooling or heating mechanisms if the environmental temperature is either too hot or too cold (see sections 1.6b and 27.8b). Chemoreceptors in blood vessels monitor carbon dioxide and hydrogen ion levels within the blood, and when their balance is changed both the cardiovascular system and respiratory system are stimulated to bring these levels back to homeostasis (see sections 20.6a and 23.5c). Proprioceptors (i.e., sensory receptors associated with our musculoskeletal system) alert the nervous system about the body's position and balance (see section 13.6b). Finally, sensory receptors in the digestive tract allow us to perceive tastes in our food, detect textures and consistencies of foods, and let us know when our stomach has been extensively stretched when we have eaten too much (see sections 16.1d, 26.2d, and 26.3b).

16.1b General Structure of Sensory Receptors

LEARNING OBJECTIVE

2. Describe the general structure of a sensory receptor, and explain the significance of a receptive field.

Sensory receptors range in complexity from the relatively simple, bare dendritic endings of a single sensory neuron (e.g., some touch receptors; see figure 16.2) to specialized, complex structures called sense organs (e.g., the eye; see figure 16.10). Regardless of their anatomic complexity, however, all are functionally connected to the CNS by sensory neurons. This provides the means of relaying sensory information from the sensory receptors to the brain and spinal cord.

A **receptive field** is the area within which the dendritic endings of a single sensory neuron are distributed. The concept of a receptive field and its significance is most clearly shown with a comparison of receptive fields within the skin (**figure 16.1**). Note the relative amount of area that sensory neurons of the skin are distributed in two different regions of the body—the skin of the fingertips and the skin of the upper back. The size of the receptive field will determine the ability of the CNS to identify the exact location of a stimulus. A small receptive field provides us with the ability to identify the stimulus location more specifically. In contrast, a large receptive field allows us to determine only the general region of the stimulus.

Although it might seem advantageous for all sensory neurons to have small receptive fields (because it would provide us with

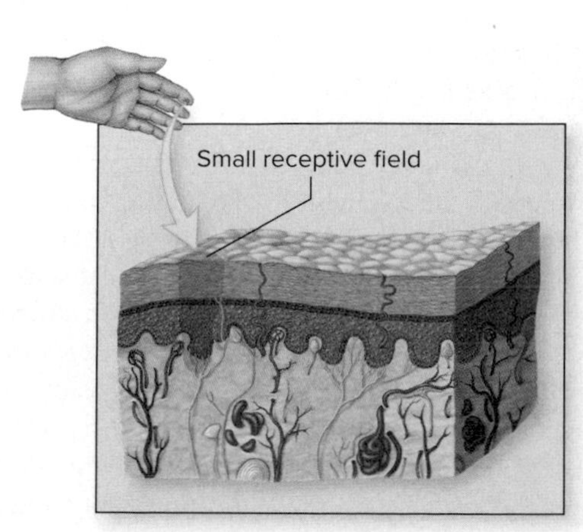

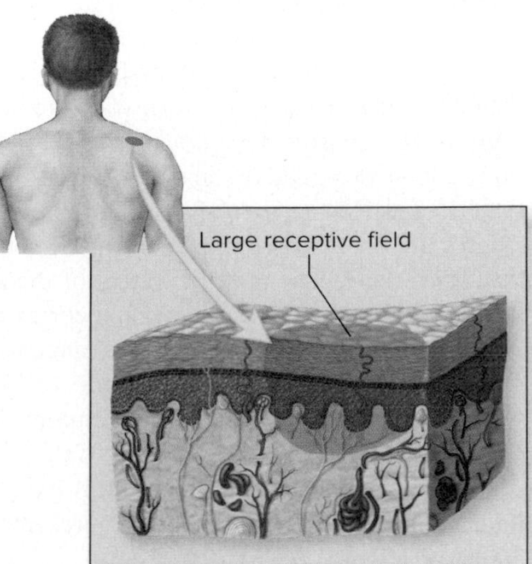

Figure 16.1 Receptive Fields. The specific area monitored by each sensory neuron is called its receptive field. A smaller receptive field (*left*) offers greater specificity of localization than does a larger receptive field (*right*).

enhanced perceptive abilities), the number of sensory neurons in the body would have to markedly increase for us to gain this advantage. This greater number of sensory neurons would require a significant increase in body size, and the energy costs to maintain their activity would be enormous.

WHAT DID YOU LEARN?

Describe the general range in structural complexity of sensory receptors, and identify what is associated with all sensory receptors.

16.1c Sensory Information Provided by Sensory Receptors

LEARNING OBJECTIVES

3. Define a sensation.
4. Explain the various characteristics of a stimulus that sensory receptors provide to the CNS.

Sensory input is relayed from sensory receptors to the CNS for interpretation. Whether we consciously perceive a stimulus is dependent upon which specific region of the CNS receives that sensory information. Only nerve signals that reach the *cerebral cortex* of the brain result in our conscious awareness. A stimulus that we are consciously aware of is called a **sensation.** A sensation occurs when we recognize a child's face or realize that the room is too warm. Although your body is constantly bombarded by numerous sensory stimuli, you are consciously aware of only a fraction of them. Much of the sensory input is relayed to other areas of the CNS (e.g., the hypothalamus, brainstem, spinal cord), where a response is initiated without your awareness. Sensory stimuli regarding your blood pressure, blood carbon dioxide levels, and chemical composition of material within the small intestine are examples of sensory information that is detected and responded to on a subconscious level.

A sensory receptor must be able to provide the CNS with several characteristics regarding a stimulus (whether it is consciously perceived or not). These characteristics include its modality, location, intensity, and duration. The **modality** (mō-dal′i-tē; *modus* = a mode) or form of a stimulus is provided by a given type of sensory receptor relaying sensory input along designated sensory neurons to specific regions of the CNS. For example, the sensory receptors of the eye (the retina) initiate nerve signals along the optic nerve to the occipital lobe (visual cortex), and sensory receptors of the ear (spiral organ) initiate nerve signals along the cochlear branch of the vestibulocochlear nerve to the temporal lobe (auditory cortex). In comparison, baroreceptors within the aorta send nerve signals along the vagus nerve to the cardiovascular center within the brainstem as part of blood pressure regulation. The brain is like a switchboard, and it interprets the source based upon which "line" the signal arrived.

The specific *location* of a stimulus is able to be determined by the CNS because sensory information is relayed either from different regions of a sensory receptor or from different locations within the body along designated sensory neurons within a given nerve that reaches specific regions of the CNS. For example, the optic nerve (see figure 16.10) is composed of many sensory neurons that extend from different portions of the retina of the eye to communicate with designated regions within the visual cortex of the occipital lobe. The inner ear has a similar anatomic arrangement between the different regions of the spiral organ (see figure 16.27) and designated regions within the auditory cortex of the temporal lobe. Recall that the specific location of sensory receptors of the skin relay sensory input to designated regions of the postcentral gyrus of the parietal lobe for interpretation, a concept visually represented in the sensory homunculus (see figure 13.13).

The CNS is able to interpret the relative *intensity* of the stimulus because of the change in number of nerve signals that are arriving along a designated nerve. A greater or more intense stimulus results in both the most sensitive sensory receptors initiating nerve signals more frequently and the less sensitive sensory receptors (which are not typically active) initiating nerve signals. A lesser stimulus would result in fewer nerve signals being relayed by its associated sensory neurons. For example, a bright light results in a greater frequency of nerve signals relayed along the optic nerve to the visual cortex, whereas a softer sound results in a lesser frequency of nerve signals relayed along the vestibulocochlear nerve to the auditory cortex.

The CNS is able to determine the *duration* of stimulus because all sensory receptors become less sensitive to a constant stimulus and initiate a progressive decrease in nerve signals. This decrease in sensitivity to a continuous stimulus is called **adaptation.** However, the rate of decrease is different for the various types of sensory receptors. This difference in adaption is used to categorize sensory receptors as either tonic receptors or phasic receptors. **Tonic receptors** demonstrate limited adaptation. In response to a constant stimulus, tonic receptors continuously generate nerve signals and only *slowly* decrease the number relayed to the CNS. Examples of tonic receptors include sensory receptors within the inner ear that determine head position and proprioceptors in the joints and muscles that provide information of where your body is in space. In addition, all pain receptors are tonic receptors so as to provide the motivation to address the cause of the pain and hopefully eliminate it so that the pain will stop. In comparison, **phasic receptors** exhibit rapid adaptation to a constant stimulus. Phasic receptors generate nerve signals only in response to a new (or changing) stimulus and *quickly* decrease the number of nerve signals relayed to the CNS. Examples include the deep pressure receptors that sense the increased pressure when we first sit down in a chair. We are immediately aware of the pressure increase wherever our body contacts the chair. But soon, we do not notice this pressure because adaptation has occurred in these receptors. You may have experienced adaptation after placing your glasses on the top of your head and then forgetting that they were there. It is advantageous for us to not be continuously bombarded by this type of sensory information.

WHAT DID YOU LEARN?

Explain how sensory receptors provide input regarding the modality, location, intensity, and duration of a stimulus.

16.1d Sensory Receptor Classification

LEARNING OBJECTIVES

5. Identify and describe the three criteria used to classify receptors.
6. Classify the various types of sensory receptors based upon each of the three criteria.

Three criteria are used to categorize sensory receptors—receptor distribution, stimulus origin, and modality of stimulus. These classification criteria are summarized in **table 16.1**.

Sensory Receptor Distribution

Sensory receptors may be classified based upon their distribution in the body. The two distribution types are the general senses and the special senses.

General sense receptors are distributed throughout the body and are simple in structure. The receptors for general senses are subdivided into two categories based upon their location in the body and include somatic sensory receptors and visceral sensory receptors. **Somatic sensory** (or *somatosensory*) **receptors** are tactile receptors housed within both the skin and mucous membranes, which line the nasal cavity, oral cavity, vagina, and anal canal. These sensory receptors monitor

Table 16.1 Criteria for Classifying Sensory Receptors

Classification	Description	Examples
SENSORY RECEPTOR DISTRIBUTION (LOCATION OF RECEPTOR)		
General senses	Distributed throughout the body; structurally simple	
Somatic sensory receptors	Located in skin and mucous membranes Located in joints, muscles, and tendons	Tactile (touch) receptors Joint receptors, muscle spindles, Golgi tendon organs
Visceral sensory receptors	Located within walls of viscera and blood vessels	Stretch receptors in stomach wall, chemoreceptors in blood vessels
Special senses	Located only in the head; structurally complex sense organs	Sensory receptors for smell, taste, vision, hearing, and equilibrium
STIMULUS ORIGIN (LOCATION OF STIMULUS)		
Exteroceptors	Detect stimuli in the external environment	Sensory receptors within skin or mucous membranes Sensory receptors for smell, taste, vision, hearing, and equilibrium
Interoceptors	Detect stimuli within the body	Sensory receptors within walls of viscera and blood vessels
Proprioceptors	Detect stimuli within joints, skeletal muscles, and tendons that sense body or limb movement	Joint receptors, muscle spindles, Golgi tendon organs
MODALITY OF STIMULUS (STIMULATING AGENT)		
Chemoreceptors	Detect chemicals (molecules or ions) dissolved in fluid	Taste receptors, receptors in blood vessels that monitor carbon dioxide levels
Thermoreceptors	Detect changes in temperature	Sensory receptors within skin and hypothalamus
Photoreceptors	Detect changes in light intensity, color, and movement	Retina of the eye
Mechanoreceptors	Detect physical deformation of the plasma membrane due to touch, pressure, vibration, and stretch; subtypes include baroreceptors, proprioceptors, tactile receptors, and other specialized cells such as the hair cells in the cochlea of the ear	Tactile receptors in the skin
Nociceptors	Detect painful stimuli	Pain receptors present in almost all organs

several types of stimuli, including texture of an object, pressure, stretch, vibration, temperature, and pain. Somatic sensory receptors are also within joints, muscles, and tendons and include joint receptors, muscle spindles, and Golgi tendon organs, respectively. These detect stretch and pressure relative to position and movement of the skeleton and skeletal muscles. A Golgi tendon organ (see figure 14.22), for example, monitors stretch of a tendon when a muscle contracts. **Visceral sensory receptors** are located in the walls of the viscera (internal organs) and blood vessels. These sensory receptors detect stretch in the smooth muscle within the walls of internal organs (e.g., stretch of the stomach wall), chemical changes in the contents within their lumen (e.g., a change in carbon dioxide levels in the blood), temperature, and pain.

In comparison, receptors of the **special senses** are located only within the head and are specialized, complex sense organs. The five special senses are olfaction (smell), gustation (taste), vision (sight), hearing (audition), and equilibrium (head position and acceleration).

Stimulus Origin

Sensory receptors can also be classified based upon where the stimulus originates. These classifications include exteroceptors, interoceptors, and proprioceptors.

Exteroceptors (eks′ter-ō-sep′ter, -tōr; *exterus* = external) detect stimuli from the external environment. Exteroceptors include the somatic sensory receptors of the skin and mucous membranes, as well as the receptors of the special senses. All of these types of sensory receptors respond to a stimulus that is outside of the body.

Interoceptors (in′ter-ō-sep′ter; *inter* = between) detect stimuli from within our internal environment. Interoceptors include the visceral sensory receptors within the wall of internal organs and blood vessels. Interoceptors keep our CNS informed about the changes that are occurring within our bodies.

Proprioceptors (prō′prē-ō-sep′ter; *proprius* = one's own) detect body and limb movements and include only the somatic sensory receptors within joints, muscles, and tendons.

Modality of Stimulus

Sensory receptors may also be classified according to the stimulus they respond to, which is called the **modality of stimulus,** or the *stimulating agent.* Thus, some sensory receptors respond only to temperature changes, whereas others respond to chemical changes. There are five groups of sensory receptors, based upon their modality of stimulus: chemoreceptors, thermoreceptors, photoreceptors, mechanoreceptors, and nociceptors.

- **Chemoreceptors** (kē′mō-rē-sep′tōr, -ter) detect chemicals, either molecules or ions dissolved in fluid. These chemicals include the food and drink we have ingested, the composition of our body fluids, and the relative components of our inhaled air. The sensory receptors in the taste buds on our tongue are chemoreceptors, because they respond to the specific molecules and ions in our ingested food to provide information to us about what is present in what we are eating. Likewise, chemoreceptors in some of our blood vessels monitor the

INTEGRATE

LEARNING STRATEGY

When you eat a very spicy meal, your mouth stings because the nociceptors in your mouth are somatic nociceptors. When you swallow and the food moves through your gastrointestinal (GI) tract, you may not experience a stinging, burning sensation because visceral nociceptors respond only to abnormal muscle stretch, oxygen deprivation, or chemical imbalance in the tissue. When the waste products from that spicy meal are expelled from the body, the anus may sting because the nociceptors around the anus and inferior anal canal are somatic nociceptors.

©Getty Images/Stockbyte RF

concentration of carbon dioxide in our blood, to help influence our respiratory rate (see section 23.5c).

- **Thermoreceptors** (ther′mō-rē-sep′tōr; *therme* = heat) respond to changes in temperature. Thermoreceptors are present in both the skin and the hypothalamus. These sensory receptors are components of reflexes that regulate and maintain body temperature.
- **Photoreceptors** (fō′tō-rē-sep′tōr; *phot* = light) are located in the eye, where they detect changes in light intensity, color, and movement.
- **Mechanoreceptors** (me-kăn′ō-rē-sep′tōr; *mechane* = machine) respond to distortion of the plasma membrane that occurs due to touch, pressure, vibration, and stretch. The various types of mechanoreceptors include baroreceptors, proprioceptors, tactile receptors, and other specialized cells such as the hair cells in the cochlea of the ear that move in response to sound waves. For example, **baroreceptors** (bar′ō-rē-sep′tōr; *baros* = weight) are a type of mechanoreceptor, which are stimulated by changes in stretch or distension within the wall of body structures. Baroreceptors located within blood vessel walls monitor stretch of these structures, as part of blood pressure regulation (see section 20.6a).
- **Nociceptors** (nō′si-sep′tōr; *noci* = pain) respond to painful stimuli. The purpose of nociceptors is to inform the brain of injury or other damage so that an appropriate response may be made. **Somatic nociceptors** detect chemical, heat, or mechanical damage to the body surface, joints, or skeletal muscles. For example, touching a hot pan or suffering a sprained ankle stimulates somatic nociceptors. **Visceral nociceptors** detect internal body damage within the viscera. When we feel discomfort in our internal organs, it often occurs because (1) a tissue has been deprived of oxygen—for example, as a result of a heart attack; (2) the smooth muscle in the wall of the organ has been stretched so much that we are uncomfortable—for example, as a result of eating too much; or (3) the tissue has suffered trauma, and damaged cells have released chemicals that stimulate specific nociceptors.

Classifying a Sensory Receptor

A given sensory receptor is described based upon each classification criterion: receptor distribution, stimulus origin, and modality of stimulus. The eyes, for example, are special senses because they are located in the head (sensory receptor distribution); exteroceptors because they detect stimuli outside the body (stimulus origin); and photoreceptors because they detect light (modality of stimulus). In comparison, sensory receptors that detect stretch of blood vessels are classified as general senses because they are distributed throughout the body, interoceptors because they detect stimuli within the body, and mechanoreceptors (specifically, baroreceptors) because they detect changes in distension of the organ wall.

We use sensory receptor distribution (general senses versus special senses) as the criterion for organizing the remaining sections in this chapter on the senses. General senses are first described in section 16.2, followed by the details of the special senses in sections 16.3 to 16.5.

 **WHAT DID YOU LEARN?**

 Describe the following sensory receptors based upon the three classification criteria: the ear and the sense of hearing, the tongue and the sense of taste, and stretch receptors in the urinary bladder wall.

16.2 The General Senses

Receptors for general senses, as described in section 16.1, are organized into somatic sensory receptors (tactile receptors of the skin and mucous membranes and proprioceptors) and visceral sensory receptors. Here we discuss somatic sensory receptors (tactile receptors only) and referred pain. (Proprioceptors are discussed in sections 14.4b and 14.6d, and visceral sensory receptors are included throughout the text when discussing the physiology of the various body systems.)

16.2a Tactile Receptors

LEARNING OBJECTIVE

7. Compare and contrast unencapsulated and encapsulated tactile receptors.

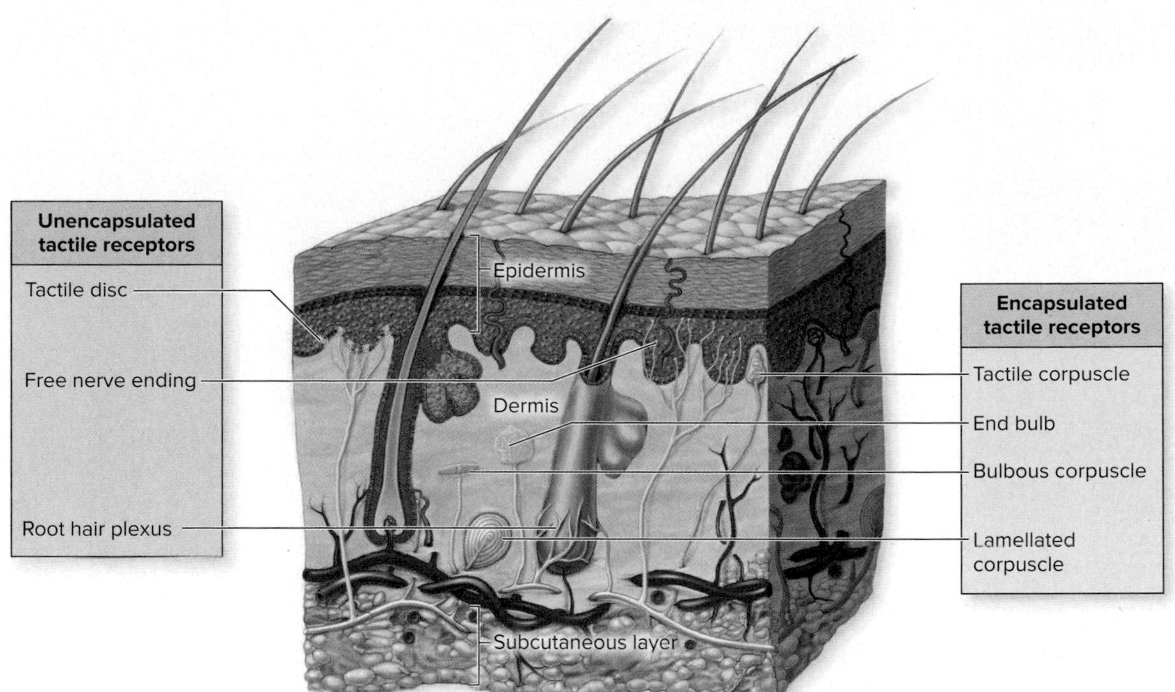

Figure 16.2 Tactile Receptors. Various types of tactile receptors in the skin receive information about touch, pressure, vibration, or temperature in our immediate environment. Unencapsulated tactile receptors are simply bare dendritic endings of sensory neurons, whereas encapsulated tactile receptors are dendritic endings of sensory neurons that are enclosed by either connective tissue or connective tissue and glial cells.

Tactile (tak′til; *tango* = to touch) **receptors** are the most numerous type of sensory receptor (**figure 16.2**). They are mechanoreceptors located in the skin and mucous membranes. The dendritic endings that compose these sensory receptors are either unencapsulated or encapsulated (**table 16.2**).

Unencapsulated Tactile Receptors

Unencapsulated tactile receptors are simply dendritic endings of sensory neurons with no protective covering. The three types of unencapsulated receptors are free nerve endings, root hair plexuses, and tactile discs.

Free nerve endings are the least complex of the tactile receptors and reside closest to the surface of the skin, usually in the papillary layer (superficial layer) of the dermis (see section 6.1b). Often, some branches extend into the deepest epidermal strata and terminate between the epithelial cells. Free nerve endings are also located in mucous membranes. These tactile receptors primarily detect temperature and pain stimuli, but some also detect light touch and pressure. Free nerve endings can be either tonic receptors (adapt slowly) or phasic receptors (adapt quickly).

Root hair plexuses are specialized dendritic endings of sensory neurons that form a weblike sheath around hair follicles in the reticular layer (deeper layer) of the dermis. Any movement or displacement of the hair changes the arrangement of these dendritic endings, initiating nerve signals. These phasic receptors quickly adapt; thus, although we feel the initial contact of a long-sleeved shirt on our arm hairs when we put on the garment, our conscious awareness subsides immediately until we move and the root hair plexuses are restimulated.

Tactile discs, previously called *Merkel discs,* are flattened dendritic endings of sensory neurons that extend to **tactile cells** (*Merkel cells*), which are specialized epithelial cells located in the stratum basale (deepest layer) of the epidermis. These discs function as tonic receptors for light touch. (Note that tactile cells are the only specialized tactile receptor cells; the remaining tactile receptors are simply the dendritic endings of sensory neurons.)

Encapsulated Tactile Receptors

Encapsulated tactile receptors are dendritic endings of sensory neurons that are wrapped either by connective tissue or by connective tissue and specialized glial cells called neurolemmocytes (previously called Schwann cells; see section 12.4b). Encapsulated tactile receptors include end bulbs, lamellated corpuscles, bulbous corpuscles, and tactile corpuscles.

End bulbs, or *Krause bulbs,* are dendritic endings of sensory neurons ensheathed in connective tissue. They are located both in the dermis of the skin and in the mucous membranes of the

Table 16.2 Types of Tactile Receptors

	UNENCAPSULATED TACTILE RECEPTORS		
	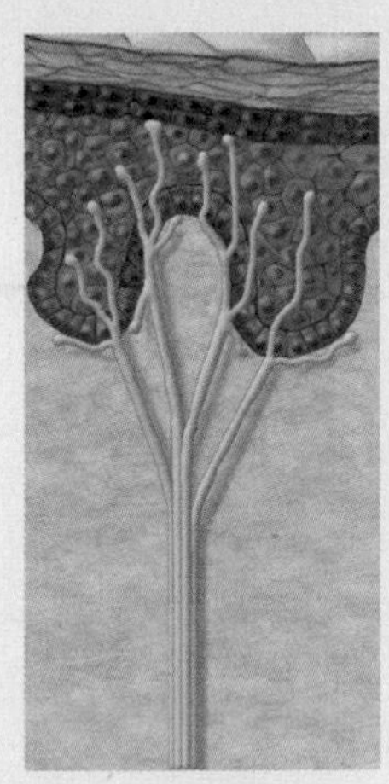	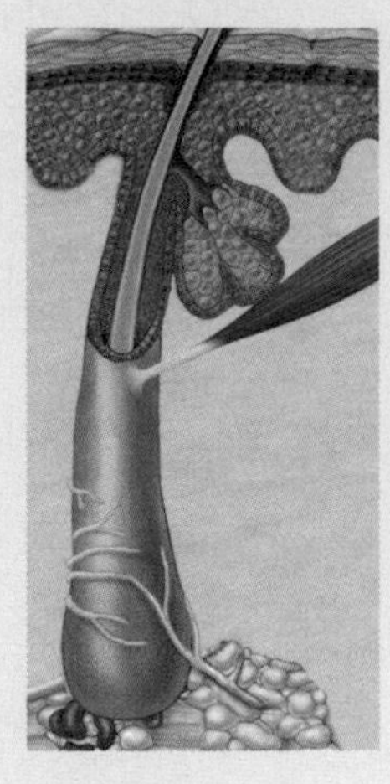	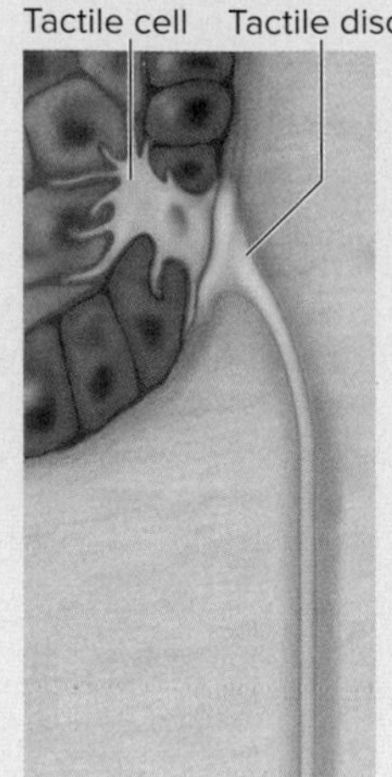
Receptor Type	**Free Nerve Ending**	**Root Hair Plexus**	**Tactile Disc**
Structure	Dendritic endings of sensory neurons	Dendritic endings of sensory neurons that surround hair follicles	Flattened dendritic endings of sensory neurons that end adjacent to specialized tactile cells
Location	Closest to skin surface (papillary layer of the dermis and some dendritic endings extend into the deepest layers of the epidermal strata); mucous membranes	Reticular layer of the dermis	Stratum basale of epidermis
Function	Detects temperature, pain, light touch and pressure	Detects movement of the hair	Detects light touch
Rate of Adaptation	Phasic or tonic	Phasic	Tonic

	ENCAPSULATED TACTILE RECEPTORS			
	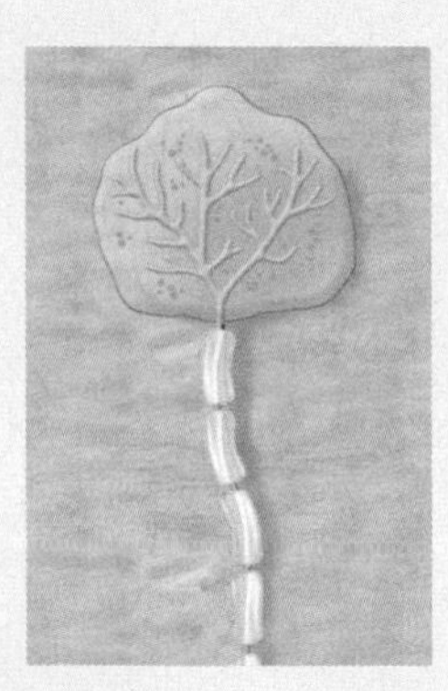		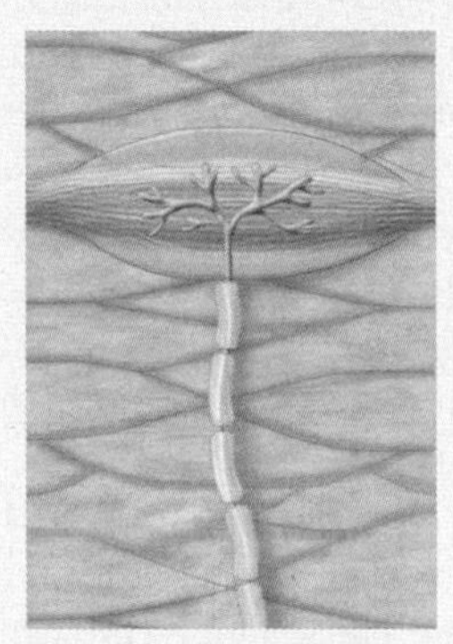	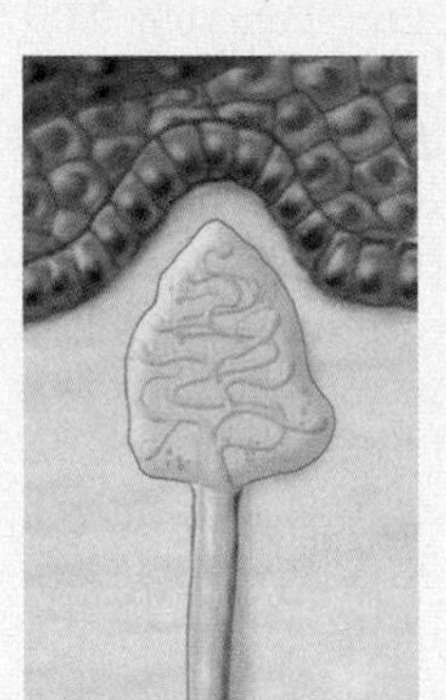
Receptor Type	**End Bulb**	**Lamellated Corpuscle**	**Bulbous Corpuscle**	**Tactile Corpuscle**
Structure	Dendritic endings of sensory neurons ensheathed in connective tissue	Dendritic endings of sensory neurons ensheathed with an inner core of neurolemmocytes and outer concentric layers of connective tissue	Dendritic endings of sensory neurons within connective tissue	Highly intertwined dendritic endings of sensory neurons enclosed by modified neurolemmocytes and dense irregular connective tissue
Location	Dermis, mucous membranes of oral cavity, nasal cavity, vagina, and anal canal	Reticular layer of the dermis; hypodermis of the palms of the hands, soles of the feet, breasts, and external genitalia; and walls of some organs	Dermis and subcutaneous layer	Dermal papillae, especially in lips, palms, eyelids, nipples, and genitals
Function	Detects light pressure and low-frequency vibration	Functions in coarse touch; detects continuous deep pressure and high-frequency vibration	Detects continuous deep pressure and skin distortion	Discriminative touch for distinguishing texture and shape of an object; light touch
Rate of Adaptation	Tonic	Phasic	Tonic	Phasic

oral cavity, nasal cavity, vagina, and anal canal. End bulbs are tonic receptors that detect light pressure stimuli and low-frequency vibration.

Lamellated (lam′e-lāt-ed; *lamina* = leaf) **corpuscles** (*corpus* = body), previously called *Pacinian corpuscles,* are large, leaf-shaped tactile receptors composed of several dendritic endings ensheathed with an inner core of neurolemmocytes and outer concentric layers of connective tissue. They are phasic receptors found deep within the reticular layer of the dermis of the skin; in the hypodermis of the palms of the hands, soles of the feet, breasts, and external genitalia; and in the walls of some organs. The structure and location of lamellated corpuscles allow them to function in coarse touch, and sensing continuous deep-pressure and high-frequency vibration stimuli.

Bulbous corpuscles, or *Ruffini corpuscles,* are dendritic endings of sensory neurons ensheathed within connective tissue that are housed within the dermis and subcutaneous layer. They are tonic receptors that detect both continuous deep pressure and distortion in the skin.

Tactile corpuscles, previously called *Meissner corpuscles,* are receptors formed from highly intertwined dendritic endings of sensory neurons enclosed by modified neurolemmocytes, which are then covered with dense irregular connective tissue. They are housed within the dermal papillae (which are projections of the dermis; see section 6.1b), especially in the lips, palms, eyelids, nipples, and genitals. Tactile corpuscles are phasic receptors for discriminative touch to distinguish texture and shape of an object and for detecting light touch.

WHAT DID YOU LEARN?

5 What are the three types of unencapsulated tactile receptors, and where are they located within the integument?

16.2b Referred Pain

LEARNING OBJECTIVE

8. Define referred pain, and explain its significance in diagnosis.

Referred pain occurs when sensory nerve signals from certain viscera are perceived as originating not from the organ, but from somatic sensory receptors within the skin and skeletal muscle (see section 14.5a). Numerous somatic sensory neurons and visceral sensory neurons conduct nerve signals on the same ascending tracts within the spinal cord (**figure 16.3**). As a result, the somatosensory cortex in the brain (see section 13.3c) is unable to accurately determine the actual source of the stimulus, and thus the stimulus may be localized incorrectly.

Clinically, some common sites of referred pain are useful in medical diagnosis (**figure 16.4**). For example, cardiac problems are often a source of referred pain because the heart receives its sympathetic innervation from the T1–T5 segments of the spinal cord (see section 15.4). Pain associated with a myocardial infarction (heart attack) may be referred to the skin innervated by the T1–T5 spinal nerves, which lie along the pectoral region and the medial side of the arm. Thus, some individuals who are experiencing heart problems may perceive pain along the medial side of the left upper limb where the T1 spinal nerve innervates (see Clinical View 19.5: "Coronary Heart Disease, Angina Pectoris, and Myocardial Infarction"). By the same token, kidney and ureter pain may be referred along the T10–L2 spinal nerves, which typically overlie the inferior abdominal wall in the groin and loin regions (see Clinical View 24.7: "Renal Calculi").

Visceral pain is usually referred along the sympathetic nerve pathways, but sometimes it can follow parasympathetic pathways as well. Referred pain from the urinary bladder often can follow the parasympathetic pathways (via the pelvic splanchnic nerves; see section 15.3b). Because the pelvic splanchnic nerves lie in the S2–S4 region of the spinal cord, pain may be referred to the S2–S4 spinal nerves, which overlie the medial buttocks regions (see Clinical View 24.8: "Urinary Tract Infections").

WHAT DID YOU LEARN?

6 Explain the clinical significance of referred pain.

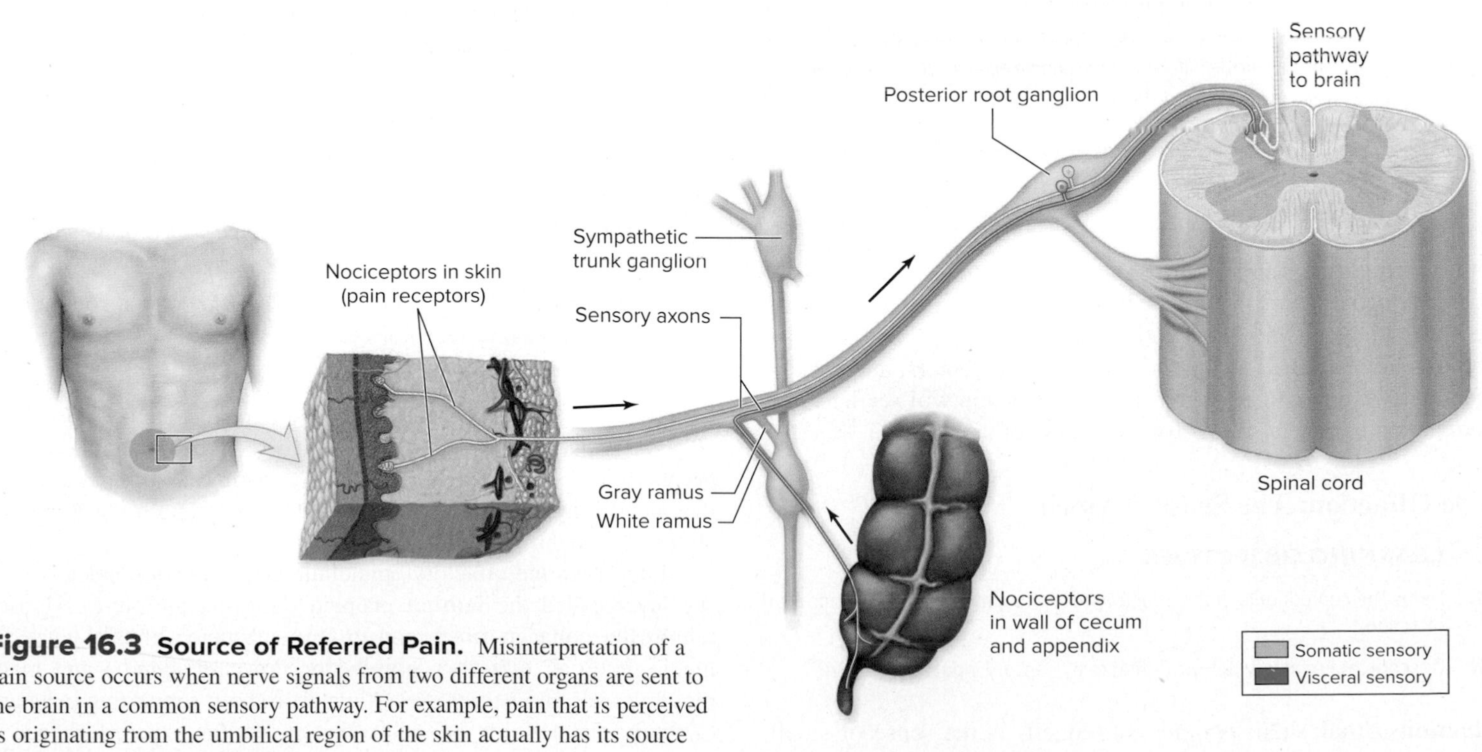

Figure 16.3 Source of Referred Pain. Misinterpretation of a pain source occurs when nerve signals from two different organs are sent to the brain in a common sensory pathway. For example, pain that is perceived as originating from the umbilical region of the skin actually has its source within the cecum and appendix.

Figure 16.4 Common Sites of Referred Pain Note: Appendicitis typically refers pain to the umbilical region. It is only in the later stages of appendicitis, when the pain becomes localized (due to the parietal peritoneum becoming inflamed as well), that the pain may be felt in the lower right abdominal quadrant.

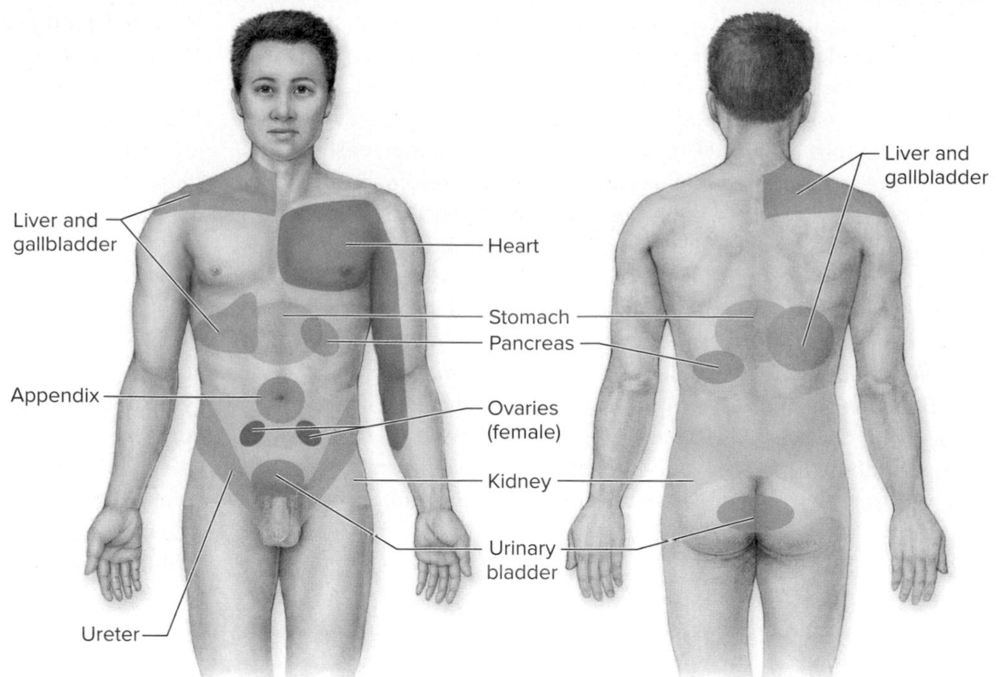

INTEGRATE

CLINICAL VIEW 16.1

Phantom Pain

Phantom (fan′tŏm; a ghost) **pain** is a sensation associated with a body part that has been removed. Following the amputation of an appendage, a patient often continues to experience pain from what is perceived as the removed part. For example, a patient may feel pain that seems localized in a foot that is no longer there. The stimulation of a sensory neuron pathway from the removed limb anywhere on the remaining intact portion of the pathway initiates nerve signals to the CNS, where they are interpreted as originating in the amputated limb. In other words, the cell bodies of the sensory neurons that had previously innervated the limb remain alive because they were not part of the limb but, rather, reside in posterior root ganglia by the spinal cord (see figure 14.12). This so-called **phantom limb syndrome** can be quite debilitating. Some people experience extreme pain, whereas others have an insatiable desire to scratch a nonexistent itch.

16.3 Olfaction and Gustation

Both olfaction (the sense of smell) and gustation (the sense of taste) occur through sensory receptors in the head that detect dissolved chemicals in the air and in our food, respectively. Thus, both of these receptors are classified as special senses, exteroceptors, and chemoreceptors. We cover these together because as you will see later, the appreciation of taste also involves our sense of smell.

16.3a Olfaction: The Sense of Smell

 **LEARNING OBJECTIVES**

9. Name the components of the olfactory receptors, and discuss their mode of action.
10. Describe the olfactory pathways that relay sensory input to the brain.

Olfaction (ol-fak′shŭn; *olfacio* = to smell) is the sense of smell, whereby volatile molecules (called **odorants**) must be dissolved in the mucus in our nasal cavity to be detected by chemoreceptors. We use this sense to sample our environment for information about the food we will eat, the presence of other individuals in the room, or potential danger (e.g., spoiled food, smoke from a fire).

Compared to many other animals, our olfactory ability is much less sensitive and not as highly developed. Consequently, we do not rely as greatly on olfactory information to find food or communicate with others. Yet we do have the ability to distinguish one odor among thousands of different ones, a capability that we may appreciate as we walk through a garden of flowers.

Olfactory Epithelium

The sensory receptor organ for smell is the **olfactory epithelium.** This epithelium lines the superior region of the nasal cavity, covering both the inferior surface of the cribriform plate and superior nasal conchae of the ethmoid bone (see figure 8.12*a*). The olfactory epithelium is composed of three distinct cell types **(figure 16.5)**:

- **Olfactory receptor cells** (also called *olfactory neurons*), which detect odors
- **Supporting cells** (also called *sustentacular cells*), which sustain the olfactory receptor cells
- **Basal cells**, which function as neural stem cells to continually replace olfactory receptor cells

Olfactory receptor cells are one of the few neuronal types that are replaced. Olfactory receptor neurons are regenerated every 40 to 60 days by basal cells within the olfactory epithelium. This process decreases with age, and the olfactory receptor cells that remain lose their sensitivity to odors. Thus, an elderly individual has a decreased ability to recognize odor molecules.

Internal to the olfactory epithelium is an areolar connective tissue layer called the **lamina propria** (lam′i-nă prō′prē-ă). Housed within the collagen fibers and ground substance of this layer are mucin-secreting **olfactory glands** (or *Bowman glands*) and many blood vessels and nerves. Secretions from both supporting cells and olfactory glands form the mucus that covers the exposed surface of the olfactory epithelium.

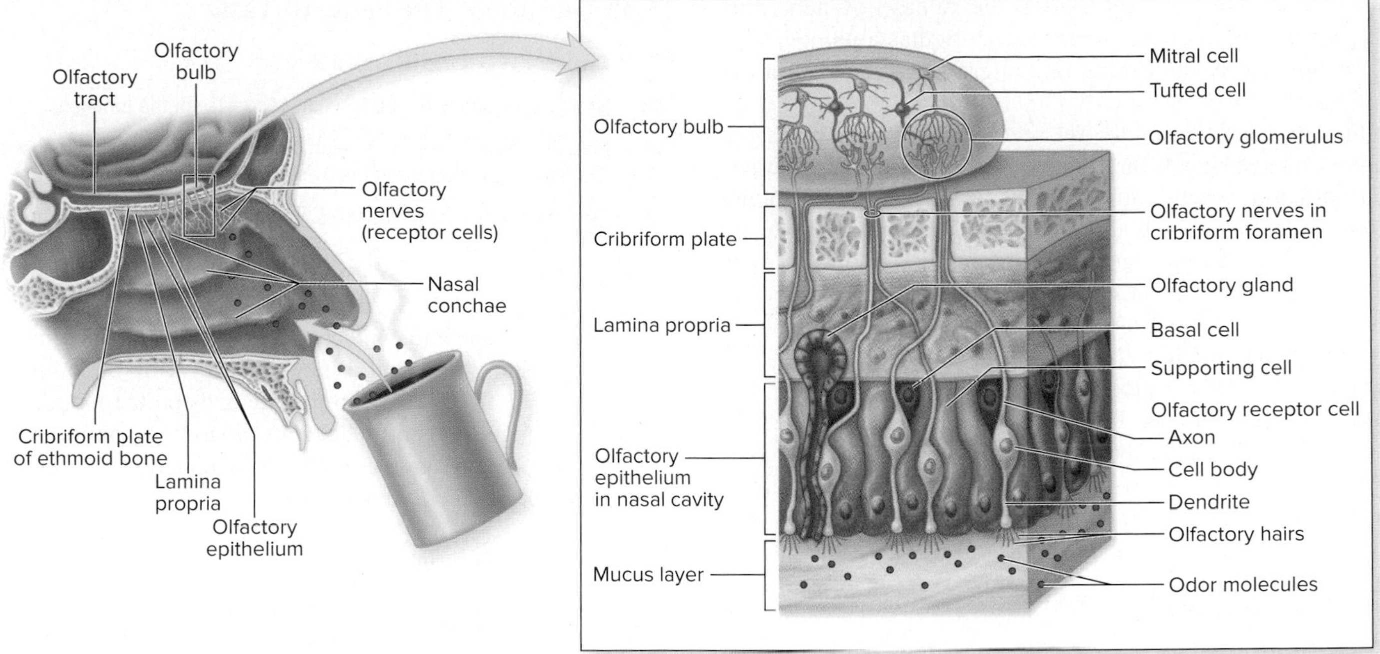

Figure 16.5 Olfactory Epithelium. The olfactory epithelium detects chemical stimuli in the air we breathe. When receptor cells within the olfactory epithelium are stimulated, they initiate nerve signals in axons that pass through the cribriform plate and synapse on neurons within the olfactory bulb. AP|R

Olfactory Receptor Cells

Olfactory receptor cells are bipolar neurons that have undergone extensive differentiation and modification, and serve as the primary neuron in the sensory pathway for smell. Olfactory receptor cells have both a single dendrite and an unmyelinated axon. Projecting from the dendrites are numerous thin, nonmotile cilia called **olfactory hairs,** which extend into the layer of mucus. Olfactory hairs contain chemoreceptors within their plasma membrane that detect one specific odorant molecule. There are approximately 400 types of scent receptors. Depending upon which olfactory receptor cells are stimulated, different smells will be perceived. Axons of olfactory receptor cells form bundles (fascicles) of the **olfactory nerves (CN I)** (see section 13.9). These fascicles project through foramina (holes) in the cribriform plate of the ethmoid bone to enter an olfactory bulb.

Olfactory Nerve Structures and Pathways

The **olfactory bulbs** are the terminal ends of olfactory tracts located inferior to the frontal lobes of the brain (see section 13.9). Axons of olfactory nerves synapse with both *mitral cells* and *tufted cells* (which are secondary neurons) within the olfactory bulbs. The resulting spherical structures are called **olfactory glomeruli** (glō-mar′yū-lī; sing., glomerulus; small ball). We have about 2000 glomeruli in our olfactory bulbs, and numerous olfactory receptor cells converge on each olfactory glomerulus. The convergence of signals within the glomerulus facilitates our ability to detect faint odors.

Axon bundles of the mitral and tufted cells form the paired **olfactory tracts** that project posteriorly along the inferior frontal lobe surface directly to the primary olfactory cortex in the temporal lobe (see table 13.5) and selected different regions of the brain, including the hypothalamus and amygdala. Unlike other sensory information, olfactory pathways do not project to the thalamus and therefore do not undergo any thalamic processing prior to reaching the cerebrum.

Detecting Smells

During normal, relaxed breathing, most inhaled air does not pass across the olfactory epithelium. To be sure to detect different smells, we must sniff repeatedly or breathe deeply, which causes the inhaled air to mix and swirl in the superior region of the nasal cavity, so that odor molecules diffuse into the mucus layer covering the olfactory receptor cells. Within the mucus, soluble proteins called **odorant-binding proteins** display an affinity for a variety of odorants.

Olfactory receptor cells are stimulated by contact of odorant-binding proteins with olfactory cell receptors. The olfactory pathway is so sensitive that only a few stimulating odorant-binding proteins with bound odorants are needed to bind to receptors and initiate olfactory sensation. Stimulation of olfactory receptor cells activates G proteins within these cells (see section 17.5b). In turn, activated G proteins stimulate adenylate cyclase enzymes, which convert ATP to cAMP (an intracellular second messenger). cAMP stimulates the opening of cation channels that allow the inflow of both Na^+ and Ca^{2+}. This net inflow of positive ions results in generation of local receptor potentials (a type of graded potential, which is described in section 12.8a) within the olfactory hairs of the olfactory receptor cells. These local potentials will initiate an action potential that is propagated along the axon of olfactory receptor cells, causing the release of neurotransmitter from the terminal ends of the axon. This results in stimulation of different patterns of the approximately 2000 glomeruli. Consider, for

INTEGRATE

CONCEPT CONNECTION

Stimulation of the hippocampus and amygdala, which are part of the limbic system involved with memory (see section 13.8e) and emotion (see section 13.8f), respectively, links emotional reactions with smells like specific foods or perfumes encountered at the same time. This explains why scents are so strongly associated with either nostalgic memories or bad experiences.

example, that cooking a meal causes the release of numerous odorants, but the smell of the meal concoction has a unique "signature" recognized by the excitation pattern observed within the glomeruli.

Binding of neurotransmitter by secondary neurons results in propagation of nerve signals through the various olfactory pathways. Sensory information reaches different regions of the brain, including the (1) cerebral cortex, which allows us to consciously perceive and identify the smell; (2) hypothalamus, which controls visceral reactions to smell, such as salivation, sneezing, or gagging; and (3) amygdala, which is a center for recognition of odors and often associating those odors to a particular emotion.

Note that once the receptors are stimulated, changes to the ion channels alter the flow of ions. This results in interfering with the subsequent generation of local receptor potentials within olfactory receptor cells, and adaptation of smell occurs rapidly. Thus, an initially strong smell (such as rotting food in a trash can) may seem to dissipate as your olfactory receptor cells quickly adapt to the foul odor.

WHAT DID YOU LEARN?

7. What is the role of the mucus in detection of smells?
8. Why do some smells stimulate an emotional reaction?

16.3b Gustation: The Sense of Taste

LEARNING OBJECTIVES

11. Describe the structure and function of papillae of the tongue.
12. Discuss the structure and location of gustatory receptors, and describe the gustatory pathways that relay sensory input to the brain.
13. Describe the five types of tastes, and explain the association of smell with taste.

Our sense of taste, called **gustation** (gŭs-tā′shŭn; *gusto* = to taste), occurs when we come in contact with the taste-producing molecules and ions of what we eat and drink (called **tastants**). Gustatory cells are chemoreceptors located within taste buds on the tongue and soft palate. The tongue and soft palate also house mechanoreceptors and thermoreceptors to provide us with information about the texture and temperature of our food, respectively.

Papillae of the Tongue

On the dorsal surface of the tongue are epithelial and connective tissue elevations called **papillae** (pă-pil′ē; *papula* = a small nipple), which are of four types: filiform, fungiform, foliate, and vallate (**figure 16.6*a*, *b***):

Filiform (fil′i-fōrm; *filum* = thread) **papillae** are short and spiked; they are distributed on the anterior two-thirds of the tongue surface. These papillae do not house taste buds and

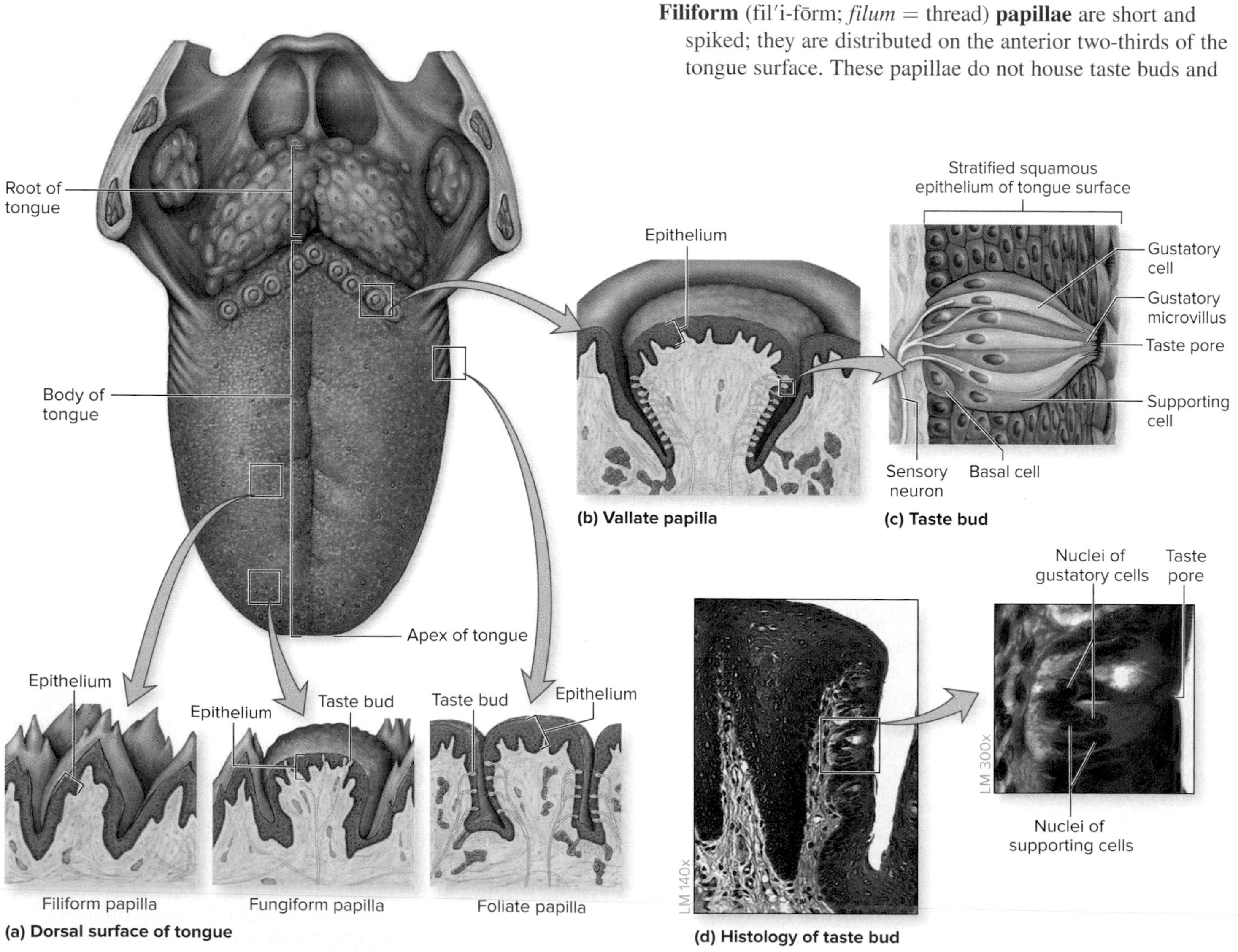

Figure 16.6 Tongue Papillae and Taste Buds. (*a*) Papillae are small elevations on the tongue surface that exist in four types: filiform, fungiform, foliate, and vallate. (*b*) A vallate papilla exhibits many taste buds, one of which is shown in detail (*c*). (*d*) Photomicrographs show the histologic structure of a taste bud on a vallate papilla. AP|R

(*d*) ©McGraw-Hill Education/Al Telser

thus, have no role in gustation. Instead, their bristlelike structure serves a mechanical function; they assist in detecting texture and manipulating food.

Fungiform (fŭn′ji-fōrm; mushroom-shaped) **papillae** are blocklike projections primarily located on the tip and sides of the tongue. Each contains only a few taste buds.

Foliate (fō′lē-āt; leaflike) **papillae** are not well developed on the human tongue. They extend as ridges on the posterior lateral sides of the tongue and house only a few taste buds during infancy and early childhood.

Vallate (val′āt; *vallo* = to surround) **papillae,** or *circumvallate papillae,* are the least numerous (about 10 to 12) yet are the largest papillae on the tongue. They are arranged in an inverted V shape on the posterior dorsal surface of the tongue. Each papilla is surrounded by a deep, narrow depression. Most of our taste buds are housed within the walls of these papillae along the side facing the depression.

Taste Buds

Taste buds are cylindrical sensory receptor organs containing cells that have the appearance of an onion (figure 16.6*c, d*). Each taste bud is composed of three distinct cell types:

- **Gustatory cells** (also called gustatory receptors), which detect *tastants* (taste-producing molecules and ions) in our food
- **Supporting cells** that sustain the gustatory cells
- **Basal cells,** which function as neural stem cells to continually replace the relatively short-lived gustatory cells

Gustatory cells are regenerated every 7 to 9 days by basal cells within the taste bud. This process decreases with age, and our sensitivity to taste also decreases. Beginning at about age 50, our ability to distinguish between different tastes declines.

Gustatory Cells

Gustatory cells within the taste buds are specialized neuroepithelial cells. The dendritic ending of each gustatory cell is formed by a slender **gustatory microvillus,** sometimes called a *taste hair.* Many gustatory microvilli extend through an opening in the taste bud, called the **taste pore,** to the surface of the tongue. This is the receptive portion of the cell. Within the oral cavity, saliva keeps the environment moist; the tastants in our food dissolve in the saliva, and then they are able to contact and stimulate gustatory cells.

Gustatory Pathways

Dendritic endings of primary neurons (see section 14.4b) are associated with gustatory cells, with each neuron contacting several gustatory cells. These sensory neurons are primarily components of the facial nerve (CN VII), which innervates taste buds from the anterior two-thirds of the tongue and the glossopharyngeal nerve (CN IX), which innervates taste buds from the posterior one-third of the tongue (**figure 16.7**). Axons within these nerves project to the medulla oblongata (specifically, the nucleus solitarius) to synapse with secondary neurons. These secondary neurons project to the thalamus, and axons of tertiary neurons project to the primary gustatory cortex in the insula of the cerebrum.

Gustatory Discrimination and Physiology of Taste

In contrast to the large number of olfactory receptors we have in our nose, our tongue detects just five basic **taste sensations:** sweet, salty, sour, bitter, and umami:

- **Sweet** tastes are produced by organic compounds such as sugar or other molecules (e.g., artificial sweeteners).
- **Salt** tastes are produced by metal ions, such as sodium (Na^+) and potassium (K^+).
- **Sour** tastes are associated with acids in the ingested material, such as hydrogen ions (H^+) in vinegar.
- **Bitter** tastes are produced primarily by alkaloids such as quinine, unsweetened chocolate, nicotine, and caffeine.
- **Umami** stimuli: *Umami* (u′ma mē) is a Japanese word meaning "delicious flavor." It is a taste related to amino acids, such as glutamate and aspartate, to produce a meaty flavor.

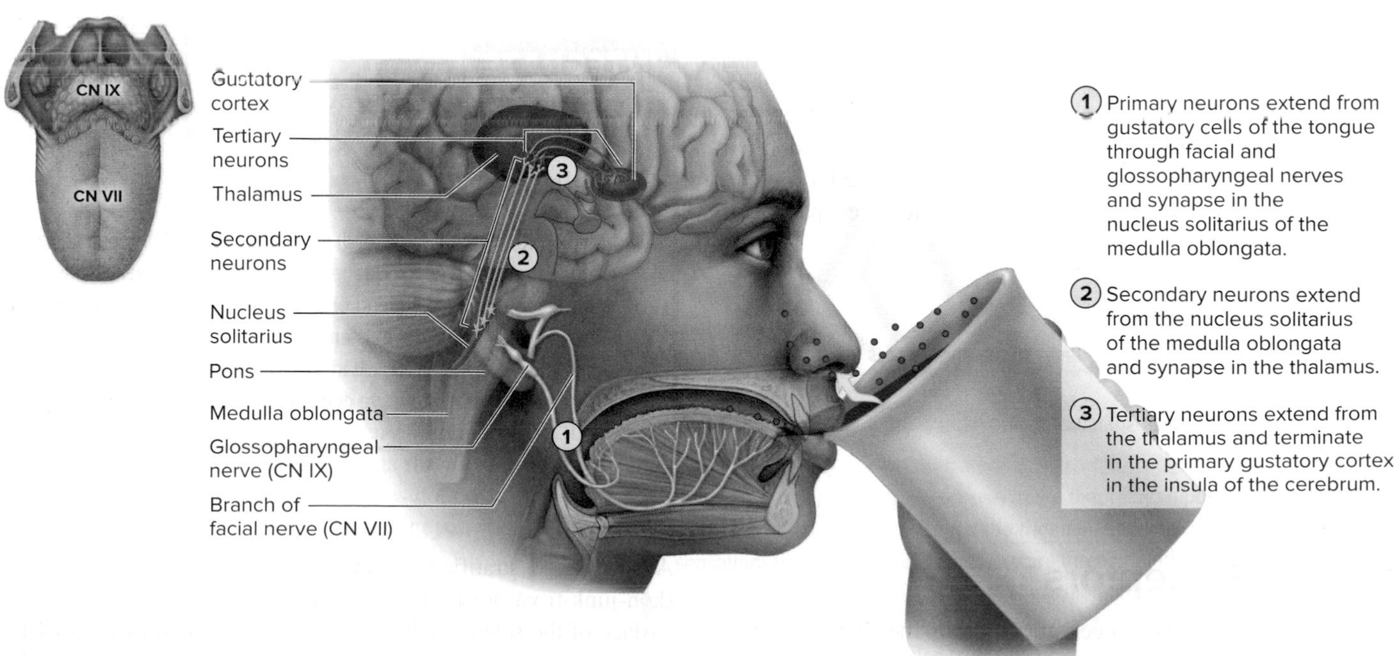

CN IX
CN VII
Gustatory cortex
Tertiary neurons
Thalamus
Secondary neurons
Nucleus solitarius
Pons
Medulla oblongata
Glossopharyngeal nerve (CN IX)
Branch of facial nerve (CN VII)
1 Primary neurons extend from gustatory cells of the tongue through facial and glossopharyngeal nerves and synapse in the nucleus solitarius of the medulla oblongata.
2 Secondary neurons extend from the nucleus solitarius of the medulla oblongata and synapse in the thalamus.
3 Tertiary neurons extend from the thalamus and terminate in the primary gustatory cortex in the insula of the cerebrum.

Figure 16.7 Gustatory Pathway. Taste sensations are transmitted by the facial nerves (CN VII) from the anterior two-thirds of the tongue and by the glossopharyngeal nerves (CN IX) from the posterior one-third of the tongue. These taste sensations are transmitted to the nucleus solitarius of the medulla oblongata before being transmitted to the thalamus and finally entering the gustatory cortex of the cerebrum.

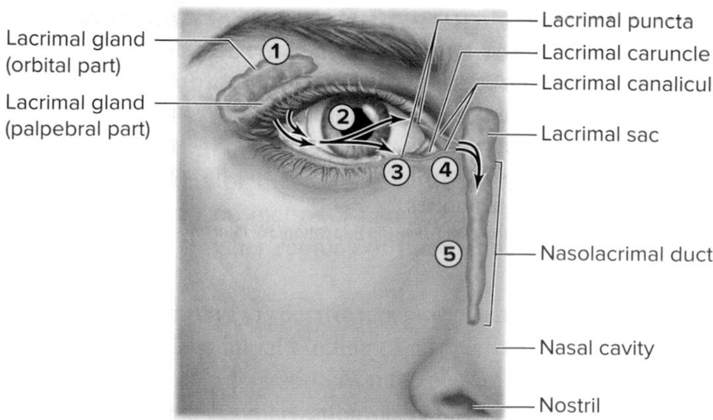

Figure 16.9 Lacrimal Apparatus. The lacrimal apparatus continually produces lacrimal fluid (tears) that cleanses and maintains a moist condition on the anterior surface of the eye. Lacrimal fluid production and drainage occur in a series of steps, which are described in the text.

fluid lubricates the anterior surface of the eye to reduce friction from eyelid movement; continuously cleanses and moistens the eye surface; helps prevent bacterial infection; and provides oxygen and nutrients to the corneal epithelium (described in section 16.4b).

The production and movement of lacrimal fluid occurs as follows (figure 16.9):

1. A **lacrimal gland,** which is about the size and shape of an almond and located within the superolateral depression of each orbit, continuously produces lacrimal fluid that drains through short ducts to the eye surface.
2. Blinking, which occurs about 15 to 20 times per minute, but less frequently when we are focusing on something such as reading a book, "washes" the lacrimal fluid over the eyes. Gradually, the lacrimal fluid is transferred to the lacrimal caruncle at the medial surface of the eye.
3. The lacrimal fluid drains into the **lacrimal puncta** (pungk′tă; sing., *punctum;* to prick), which are two small openings on the superior and inferior side of the lacrimal caruncle. (If you examine the lacrimal caruncle within your own eye, each punctum appears as a "hole.")
4. Each lacrimal punctum has a **lacrimal canaliculus** (kan-ă-lik′yū-lŭs; small canal) that drains lacrimal fluid into a rounded **lacrimal sac.**
5. The fluid drains from the lacrimal sac into a **nasolacrimal duct.** This duct drains lacrimal fluid into the nasal cavity, where it mixes with mucus and then moves into the pharynx (throat) and is swallowed. Excess production of lacrimal fluid produces tears.

WHAT DID YOU LEARN?

11 Where is the conjunctiva located, and what are its functions?

12 How is lacrimal fluid spread across the eye surface and removed from the orbital region?

16.4b Eye Structure

LEARNING OBJECTIVE

15. Describe the structures of the eye.

The eye is an almost spherical organ that measures about 2.5 centimeters (1 inch) in diameter. Most of the eye is receded into the orbit of the skull **(figure 16.10),** a space also occupied by the lacrimal gland, extrinsic eye muscles, numerous blood vessels, and the cranial nerves that innervate the eye and other structures in the orbit. **Orbital fat** cushions the posterior and lateral sides of the eye (see figure 16.8*b*), providing support and protection and facilitating oxygen and nutrient delivery by the blood through its associated blood vessels.

The interior of the eye consists of two fluid-filled cavities. These two cavities are separated by the *lens,* which is a transparent, biconvex structure enclosed in a fibrous capsule (figure 16.10). The **posterior cavity,** which lies behind the lens, contains a permanent fluid called *vitreous humor*. The **anterior cavity,** which is in front of the lens, contains a circulating fluid called *aqueous humor*. The anterior cavity is subdivided into two chambers (or rooms) by the *iris* (the colored part of the eye): the **anterior chamber,** which is located between the iris and cornea, and the **posterior chamber,** which is located between the iris and the lens. The lens, iris, and both fluids (vitreous humor and aqueous humor) are described in detail later in this section.

The wall of the eye is formed by three principal tunics (or layers): the fibrous tunic (external layer), the vascular tunic (middle layer), and the retina (inner layer).

Fibrous Tunic

The external layer of the eye wall is called the **fibrous tunic,** or *external tunic*. It is composed of the posterior sclera and the anterior cornea.

Most of the fibrous tunic (the posterior five-sixths) is the tough **sclera** (sklĕr′ă; *skleros* = hard), a part of the outer layer that is called the "white" of the eye. It is composed of dense irregular connective tissue containing numerous blood vessels and nerves. The sclera provides for eye shape, protects the eye's delicate internal components, and serves as an attachment site for extrinsic eye muscles. Posteriorly, the sclera is continuous with the **optic nerve sheath,** which is an extension of dura mater that surrounds the optic nerve. ("Bloodshot" eyes occur with vasodilation of the scleral blood vessels, which become visible through the transparent conjunctiva.)

The **cornea** (kōr′nē-ă) is a convex, transparent structure that forms the anterior one-sixth of the fibrous tunic; its convex shape refracts (bends) light rays coming into the eye. The cornea is composed of an inner simple squamous epithelium, a middle layer of collagen fibers, and an outer stratified squamous epithelium, called the **corneal epithelium.** (Think of the cornea as a collagen protein sandwich with epithelial layers as the bread.) The cornea contains no blood vessels. Nutrients and oxygen are supplied to the internal epithelium of the cornea by aqueous humor within the anterior cavity of the eye, whereas the surface corneal epithelium receives its oxygen and nutrients from lacrimal fluid.

The cornea merges with the sclera at its outer edge; this region is called the **limbus** (lim′bŭs), or the *corneal scleral junction*. The corneal epithelium forming the external portion of the cornea is continuous with the ocular conjunctiva that covers the sclera. Thus, the entire eye is covered with an epithelium.

Vascular Tunic

The middle layer of the eye wall is the **vascular tunic,** also called the *uvea* (ū′vē-ă; *uva* = grape). The vascular tunic houses an extensive array of blood vessels, lymph vessels, and the intrinsic muscles of the eye. It is composed of three distinct regions; from posterior to anterior, they are the choroid, the ciliary body, and the iris.

The **choroid** (kōr′oyd) is the most extensive and posterior region of the vascular tunic, and it is composed of areolar connective tissue that houses both an extensive network of capillaries and melanocytes (cells that produce melanin pigment). Two primary functions are associated with the choroid. Its vast network of blood vessels supplies oxygen and nutrients to the retina (inner adjacent layer of the eye, which contains photoreceptors), and melanin produced by its

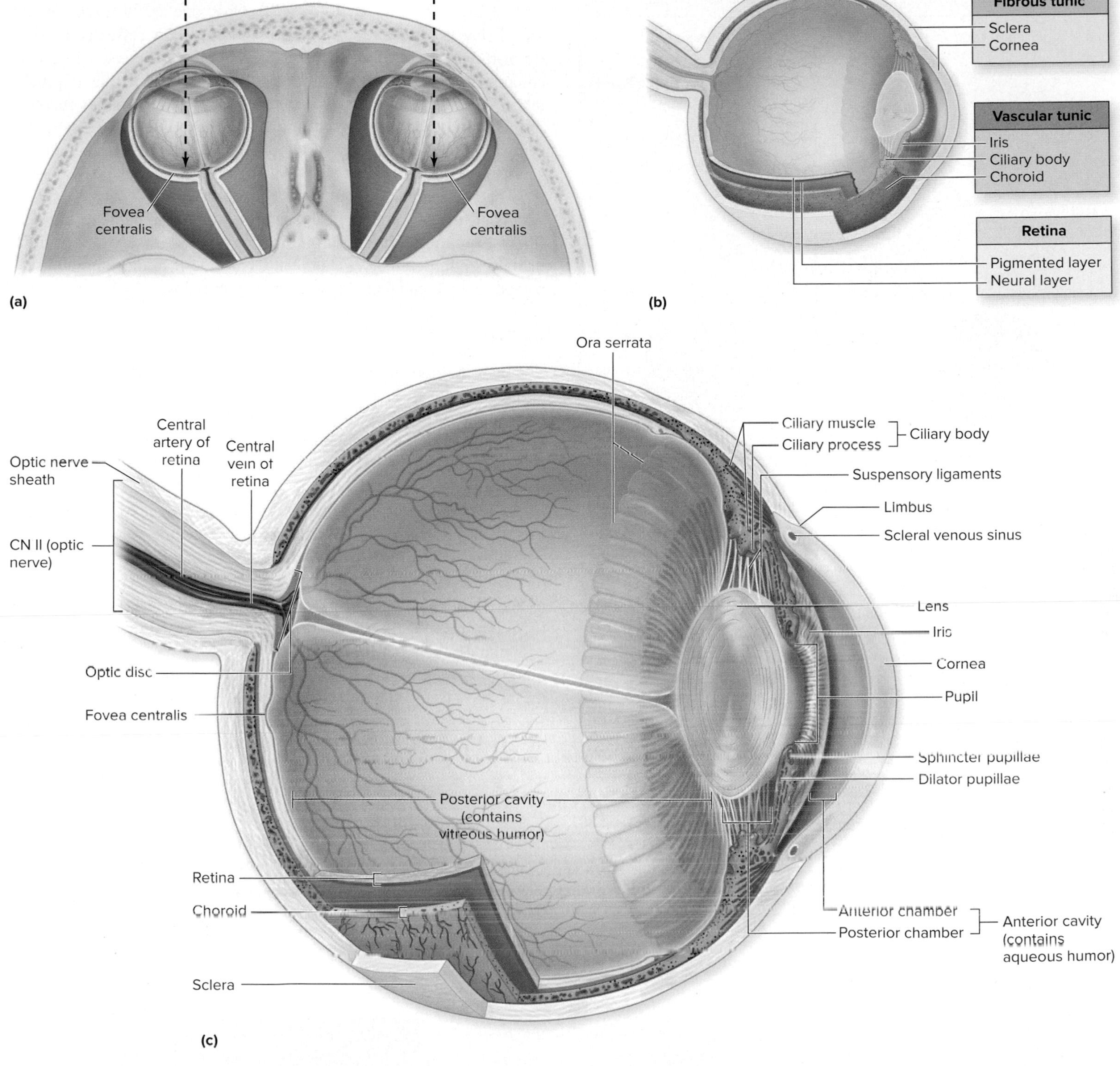

Figure 16.10 Anatomy of the Internal Eye. (*a*) A superior view of the skull and orbits shows the anatomic position of the eyes within the skull. Sagittal views depict, (*b*) the three tunics of the eye, and (*c*) internal eye structures. AP|R

melanocytes absorbs extraneous light to prevent it from scattering within the eye.

The **ciliary** (sil′ē-ar-ē; *cilium* = eyelid) **body** is located immediately anterior to the choroid and is composed of both a ciliary muscle and ciliary processes. The **ciliary muscle** is a ring of smooth muscle. Extending from the ciliary muscle to the lens are *suspensory ligaments,* which anchor the lens. Relaxation and contraction of the ciliary muscles change the tension on the suspensory ligaments, thereby altering the shape of the lens. The **ciliary processes** contain capillaries that secrete aqueous humor (both functions of the ciliary body are discussed in detail in section 16.4b).

The most anterior region of the vascular tunic is the **iris** (ī′ris; rainbow), which is the colored portion of the eye. The iris is composed of two layers of smooth muscle fibers, melanocytes, and an array of vascular and nervous structures. In the center of the iris is an opening called the **pupil** (pū′pil), which allows light to enter the eye to reach the retina. The iris controls pupil size, or diameter—and thus the amount of light entering the eye—using its two smooth muscle layers (**figure 16.11**). The **sphincter pupillae** (pyū-pil′ē) **muscle** (or *pupillary constrictor*) is arranged in a pattern that resembles concentric circles around the pupil. This muscle contracts (and the pupil becomes *smaller*) when stimulated by visceral motor

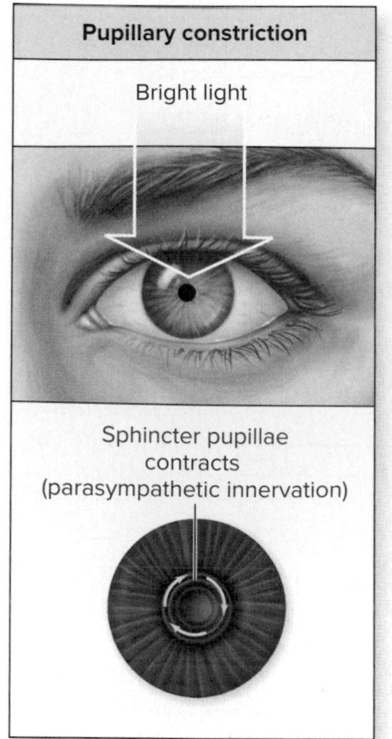

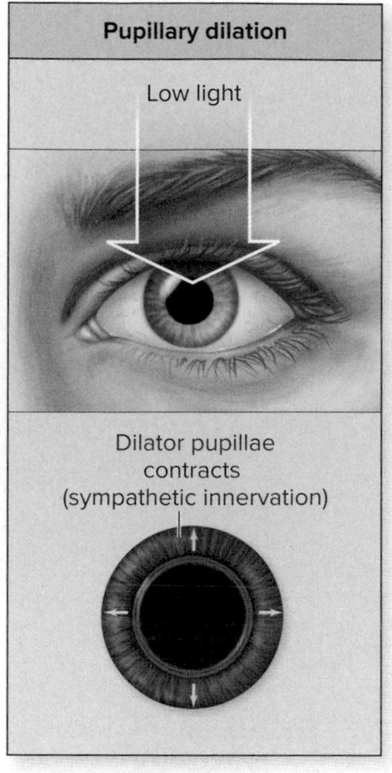

Figure 16.11 Iris Control of Pupil Diameter. Pupillary constriction is caused by contraction of the sphincter pupillae muscle of the iris, which is controlled by the parasympathetic division of the ANS. This narrows the diameter of the pupil to decrease the amount of light entering the eye. Pupillary dilation is caused by contraction of the dilator pupillae muscle of the iris, which is controlled by the sympathetic division of the ANS. This widens the pupil diameter to increase light entering the eye.

neurons of the parasympathetic division of the ANS that are within the oculomotor nerve (CN III). In comparison, the **dilator pupillae muscle** (or *pupillary dilator*) is organized in a radial pattern extending peripherally through the iris. This muscle contracts (and the pupil becomes *larger*) when stimulated by neurons of the sympathetic division of the ANS (see section 15.7b). Only one set of these smooth muscle layers can contract at a time. The **pupillary reflex** is the ability of the iris to change the size of the pupil in response to varying amounts of light. When stimulated by bright light, the pupillary reflex involves the relaying of sensory input from the photoreceptors of the eye to the brain, which initiates nerve signals along the parasympathetic division fibers of the oculomotor nerve to stimulate the sphincter pupillae muscle to contract, which decreases pupil diameter. When stimulated by low light levels, the pupillary reflex ultimately involves initiating nerve signals along sympathetic division fibers to stimulate the dilator pupillae muscle to contract, which increases pupil diameter. This reflex is tested if brain trauma is suspected (e.g., from a car accident or drug overdose).

Retina

The internal layer of the eye wall, called the **retina** (ret′i-nă; *rete* = a net) also is known as the *internal tunic* or *neural tunic.* It is composed of two layers: an outer pigmented layer and an inner neural layer (**figure 16.12**). The **pigmented layer** is immediately internal to the choroid and attached to it. Two primary functions are associated with the pigmented layer. It provides vitamin A for the photoreceptor (light-detecting) cells of the neural layer and absorbs extraneous light to prevent it from scattering within the eye (a function it shares with the choroid). The inner **neural layer** (or *neural retina*) houses all of the photoreceptor cells and their associated neurons. This layer of the retina is responsible for vision by absorbing light rays and converting them into nerve signals that are transmitted to the brain.

The **ora serrata** (ōră sē-ră′tă; *serratus* = sawtooth) is a jagged margin between the photosensitive posterior region of the retina and the nonphotosensitive anterior region of the retina. This nonphotosensitive portion continues anteriorly to cover the ciliary body and the posterior aspect of the iris (see figure 16.10*c*).

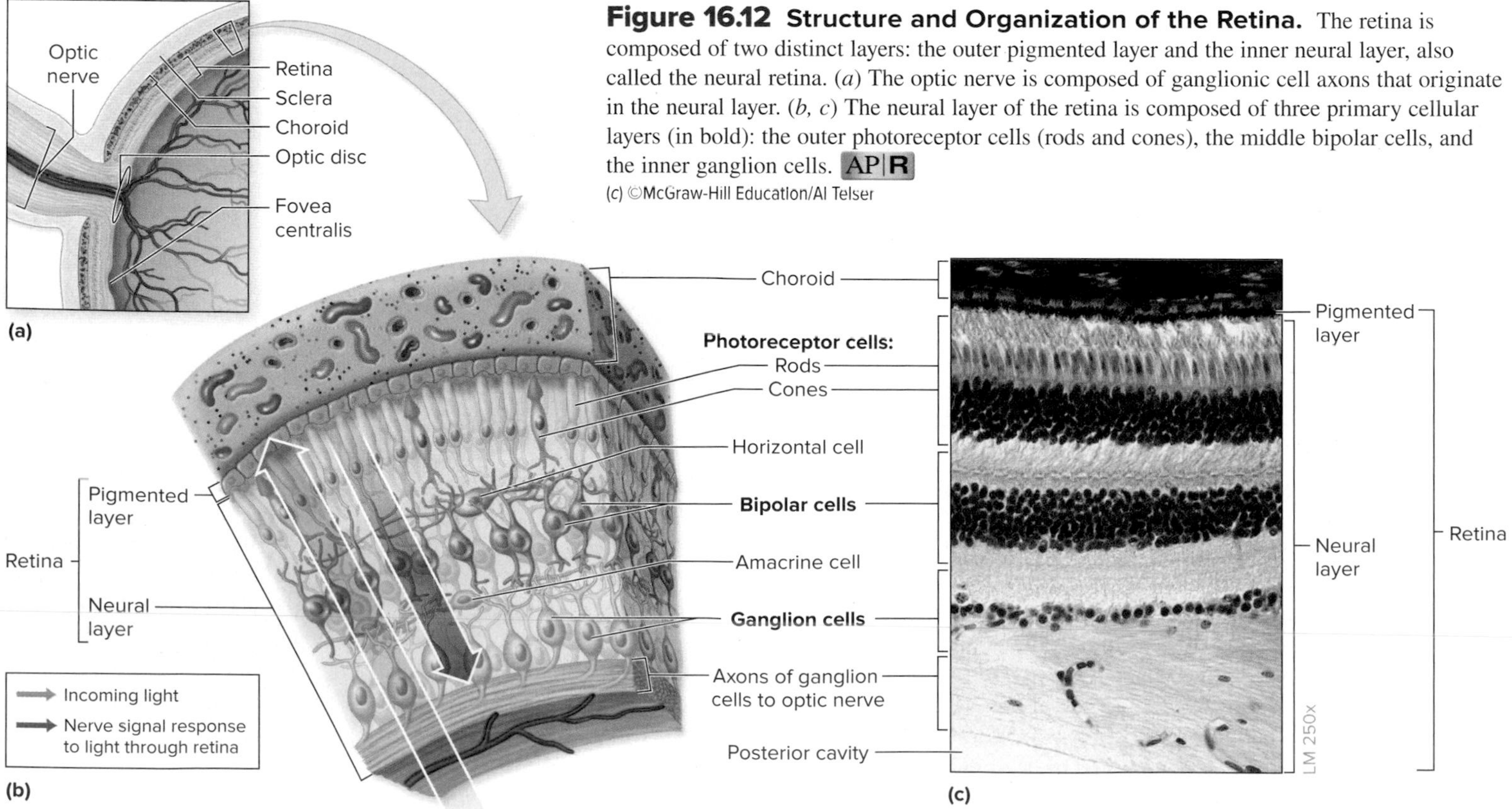

Figure 16.12 Structure and Organization of the Retina. The retina is composed of two distinct layers: the outer pigmented layer and the inner neural layer, also called the neural retina. (*a*) The optic nerve is composed of ganglionic cell axons that originate in the neural layer. (*b, c*) The neural layer of the retina is composed of three primary cellular layers (in bold): the outer photoreceptor cells (rods and cones), the middle bipolar cells, and the inner ganglion cells. AP|R

Cells of the Neural Layer Three distinct layers of neurons form the neural layer: photoreceptor cells, bipolar cells, and ganglion cells. The outermost layer of cells in the neural layer is composed of **photoreceptor** (*phot* = light) **cells,** which contain pigment molecules that react to light energy. The two types of photoreceptor cells are *rods,* which have a rod-shaped outer portion and function in dim light, and *cones,* which have a cone-shaped outer portion and function in high-intensity light and in color vision. These cells are described in more detail in section 16.4d.

Immediately internal to the photoreceptor cells is a layer of **bipolar cells.** Bipolar cells are located between photoreceptor cells and ganglion cells. The dendrites of the bipolar cells synapse with rods and cones. The axons of bipolar cells synapse with dendrites of ganglion cells.

Ganglion cells form the innermost layer in the neural layer and are adjacent to the posterior cavity. Axons of the ganglionic cells extend into and through the optic disc to form the optic nerve. Note that incoming light must pass through the ganglion cells and bipolar cells before reaching the photoreceptor cells. Electrical signals (which are initiated by transduction of light by photoreceptor cells) are then relayed to bipolar cells and then to ganglion cells.

Other cells that function in transmission of light stimuli include horizontal cells and amacrine cells. **Horizontal cells** are sandwiched between the photoreceptor and bipolar cells in a thin web. These horizontal cells regulate and integrate the electrical signals sent from the photoreceptor cells to the other cell layers. **Amacrine** (am′ă-krin) **cells** are positioned between the bipolar and ganglion cells and help process and integrate electrical signals between bipolar and ganglion cells. The electrical signals are either action potentials or graded potentials (see section 12.9a). Only the amacrine and ganglion cells in the retina produce action potentials; the other cells generate graded potentials.

Components of the Retina The distribution of rods and cones, the two types of photoreceptor cells, is not uniform throughout the retina. Three specific regions are identified: the optic disc, macula lutea, and peripheral retina **(figure 16.13)**. The **optic disc** contains no photoreceptors. This is where axons of the ganglion cells extend from the back of the eye as the optic nerve (figure 16.12*a*). It is commonly called the **blind spot** because it lacks photoreceptor cells, and no image forms there. The **macula lutea** (mak′ū-lă lū′tē-a; *macula* = small spot, *lutea* = saffron-yellow) is a rounded, yellowish region just lateral to the optic disc (figure 16.13). Within the macula lutea is

INTEGRATE

CLINICAL VIEW 16.3

Detached Retina

A **detached retina** occurs when the outer pigmented and inner neural layers of the retina separate. Detachment may result from head trauma (soccer players and high divers are especially susceptible), or it may have no overt cause. Individuals who are nearsighted, due to a more elliptical eyeball, are at increased risk for detachment because their retina is typically thinned or stretched more than that of a normal eye. There is also increased risk for retinal detachment in diabetics and older individuals. A detached retina results in deprivation of nutrients for cells in the inner neural layer because it is pulled away from the vascularized choroid layer. Degeneration and death of the neural layer of the retina occur if the blood supply is not restored.

Symptoms of a detached retina include a large number of *floaters* (small, particle-like objects) in the vision; the appearance of a "curtain" in the affected eye; flashes of light; and decreased, watery, or wavy vision.

Pneumatic retinopexy is a treatment for upper retinal detachment. The physician inserts a needle into the anesthetized eye and injects a gas bubble into the vitreous humor. The gas bubble rises and pushes the neural layer back into its normal position. The gas bubble is absorbed and disappears over 1 to 2 weeks, and then a laser may be used to tack the two layers of the retina together. The **scleral buckle** is another treatment, which uses a silicone band to press inward on the sclera to hold the retina in place. A laser is then used to reattach the retina.

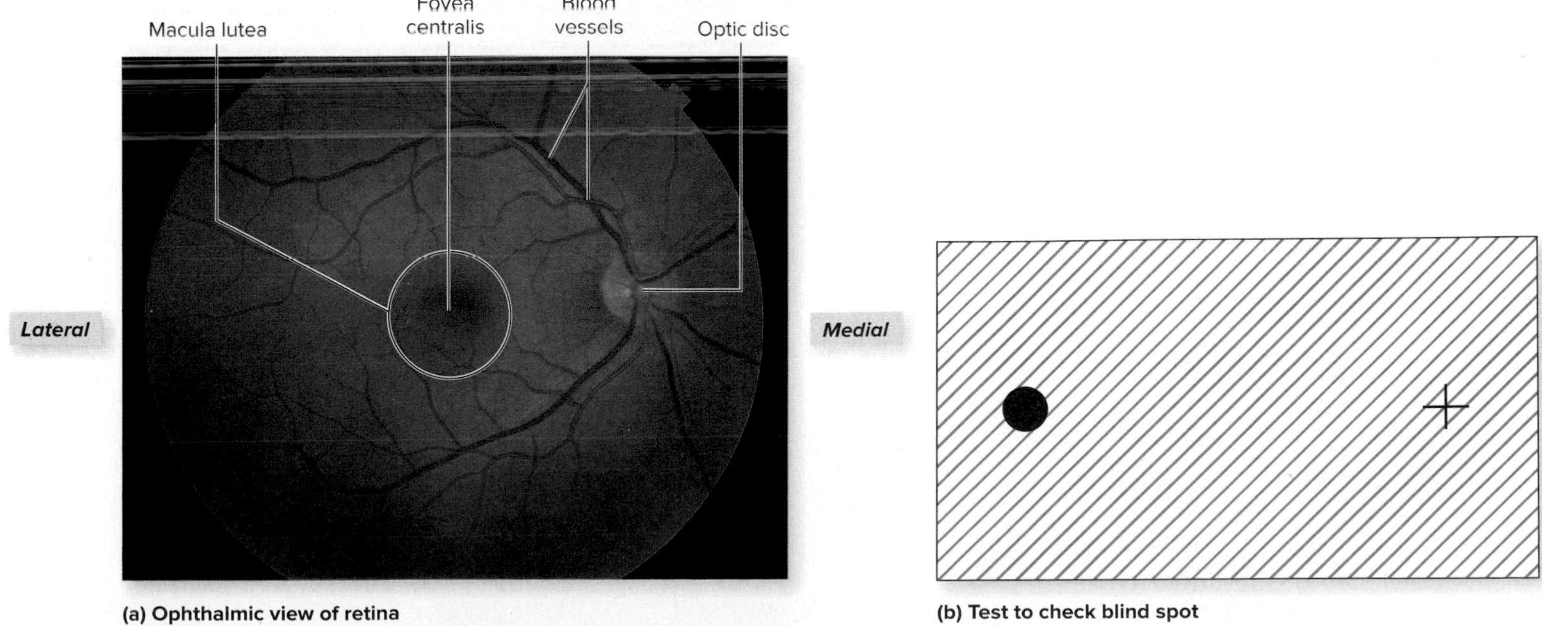

Figure 16.13 Internal View of the Retina Showing the Optic Disc (Blind Spot). (*a*) An ophthalmoscope is used to view the retina through the pupil. Blood vessels accompany the optic nerve as it enters the eye at the optic disc. (*b*) Check your blind spot! Close your left eye. Hold this figure in front of your right eye, and stare at the black spot. Move the figure toward your open eye. At approximately 6 inches from your eye, the image of the plus sign is over the optic disc and the plus sign seems to disappear.

(*a*) ©Paul Whitten/Science Source

a depressed pit called the **fovea centralis** (fō′vē-ă sen′tră′lis; *fovea* = pit, *centralis* = central), which contains the highest proportion of cones and almost no rods. This pit is the area of sharpest vision; when you read the words in your text, they are precisely focused here. Although the other regions of the retina also receive and interpret light rays, no other region can focus as precisely as can the fovea centralis because of its high concentration of cones. The remaining most extensive region of the retina is called the **peripheral retina,** which contains primarily rods and functions most effectively in low light.

Lens

The **lens** is a strong yet deformable, transparent structure. It is composed of precisely arranged layers of cells that have lost their organelles and are filled completely by a protein called *crystallin,* which are enclosed by a dense, fibrous, elastic capsule. The lens focuses incoming light onto the retina, and its shape determines the degree of light refraction.

The **suspensory** (sŭs-pen′sŏ-rē; *suspendo* = to hang up) **ligaments** attach to the lens capsule at its periphery, where they transmit tension that enables the lens to change shape. The relative tension in the suspensory ligaments is altered by relaxation and contraction of the ciliary muscles in the ciliary body. When we view objects greater than 20 feet away, the ciliary muscles relax, the ciliary body moves away from the lens, and so the tension on the suspensory ligaments increases. This constant tension causes the lens to flatten (**figure 16.14*a***). This flattened shape of the lens is the "resting" position of the lens.

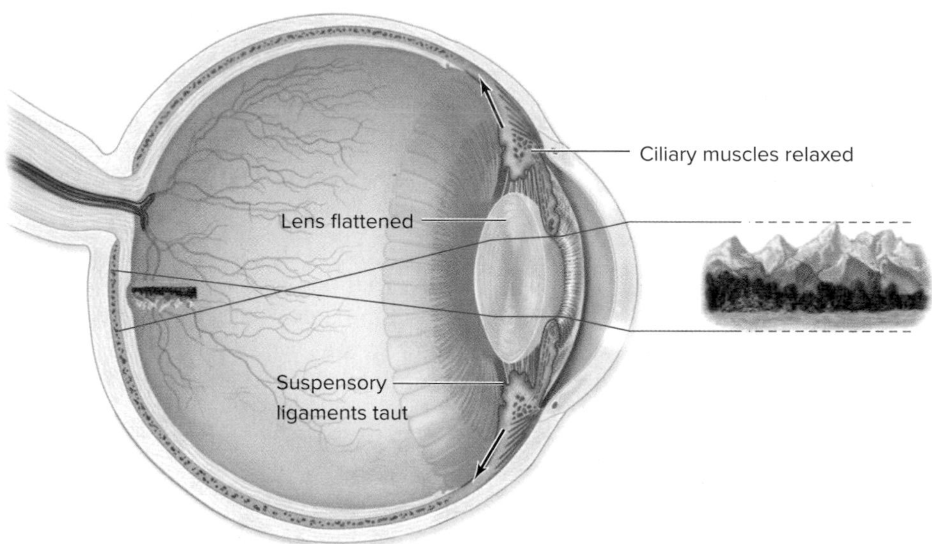

(a) Lens shape for distant vision

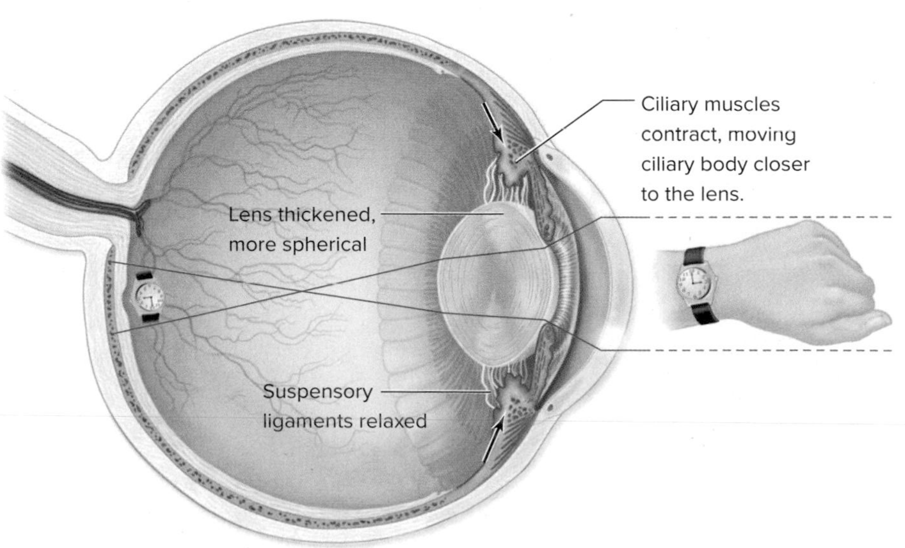

(b) Lens shape for near vision (accommodation)

Figure 16.14 Lens Shape in Far Vision and Near Vision. (*a*) To focus a distant object on the retina, the ciliary muscles within the ciliary body relax, which tenses the suspensory ligaments and flattens the lens. (*b*) To focus a near object on the retina, the ciliary muscles contract, causing release of tension on the suspensory ligaments and the lens to thicken (become more spherical or "puffy"). This process is called accommodation. Note: All focused images are inverted on the retina.

INTEGRATE

CLINICAL VIEW 16.4

Macular Degeneration

Macular degeneration, which is the physical deterioration of the macula lutea, is the leading cause of blindness in developed countries. Although the majority of cases are reported in people over 55, the condition may occur in younger people as well. Most non-age-related cases are associated with conditions such as diabetes, an ocular infection, hypertension, or trauma to the eye.

Macular degeneration is typically associated with the loss of visual acuity in the center of the visual field, diminished color perception, and "floaters."

At present, there is no cure for macular degeneration. However, its progression may be slowed. Early detection has become an important element in treatment. To track the progression of the disease, doctors rely heavily on self-monitoring, in which the patient regularly performs a simple visual test using the **Amsler grid.** While staring at the dot, the patient looks for wavy lines, blurring, or missing parts of the grid, which would indicate degenerative changes in the macula.

Normal vision.

The same scene as viewed by a person with macular degeneration.

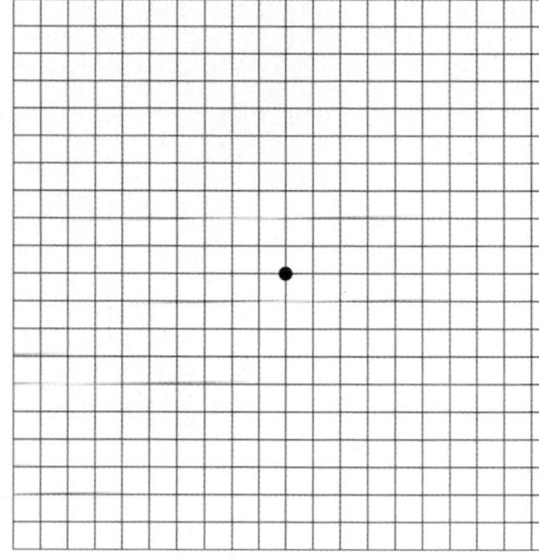

Amsler grid, seen with normal vision.

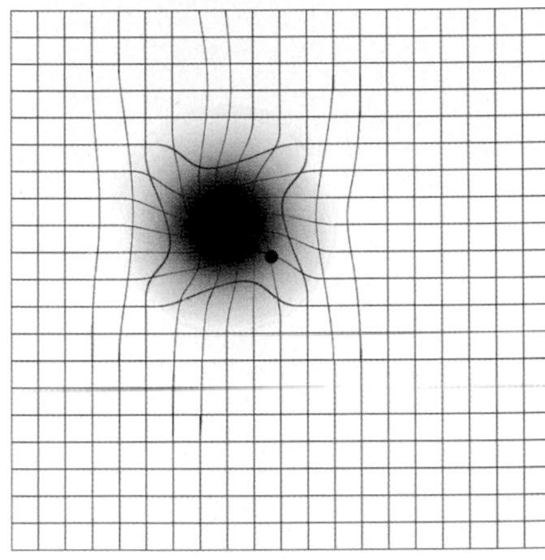

Amsler grid, as viewed by a person with macular degeneration.

In contrast, when we wish to view objects closer than 20 feet, the ciliary muscles contract, the ciliary body moves closer to the lens, and the tension on the suspensory ligaments decreases. This releases some of their pull on the lens so the lens can become more spherical, or curved. The process of making the lens more spherical to view close-up objects is called **accommodation** (ă-kom′ŏ-dā′shŭn; *accommodo* = to adapt) (figure 16.14*b*). Accommodation is controlled by autonomic motor neurons of the parasympathetic division that extend within the oculomotor nerve (CN III; see section 13.9).

Vitreous Humor and Aqueous Humor

Vitreous humor (vit′rē-ŭs; *vitrum* = glassy), or *vitreous body,* is the transparent, gelatinous fluid that completely fills the posterior cavity. This permanent fluid is produced during embryonic development and helps to both maintain eye shape and support the retina to keep it flush against the back of the eye (see Clinical View 16.3: "Detached Retina").

Aqueous humor (ak′qwē-ŭs hū′mer; watery fluid) is a transparent, watery fluid that circulates within the anterior cavity **(figure 16.15)**. It is continuously produced by the ciliary processes. The circulation of aqueous humor provides nutrients and oxygen to both the avascular cornea (specifically, its inner epithelium) and the lens.

Blood plasma (see section 18.2) is filtered across the walls of capillaries of ciliary processes and enters the posterior chamber to form aqueous humor. (This process is similar to the formation of cerebrospinal fluid by the choroid plexuses within the ventricles of the brain; see section 13.2c.) The aqueous humor circulates from the posterior chamber through the pupil and into the anterior chamber. Aqueous fluid drains from the anterior chamber into a circular canal at the limbus called the **scleral venous sinus** (previously called the *canal of Schlemm*). This fluid then drains into nearby veins. Thus, as with cerebrospinal fluid, aqueous humor is produced from capillaries, circulates, and then enters the venous circulation. Normally, the rate of formation by the ciliary processes is equal to the drainage into the scleral venous sinus; thus, a normal intraocular pressure is maintained. *Glaucoma* results from the blockage of aqueous humor drainage (see Clinical View 16.6: "Glaucoma").

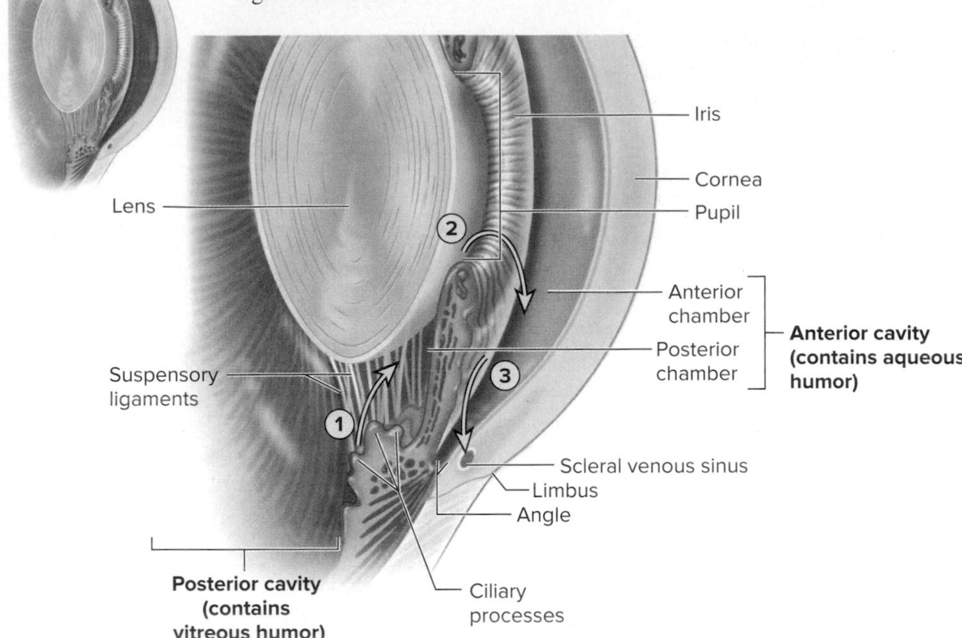

Figure 16.15 Aqueous Humor: Secretion and Resorption. Aqueous humor is a watery secretion that is continuously produced and circulated through the anterior cavity of the eye, to provide oxygen and nutrients to the inner portion of the cornea and the adjacent region of the lens.

INTEGRATE

CLINICAL VIEW 16.5

Cataracts

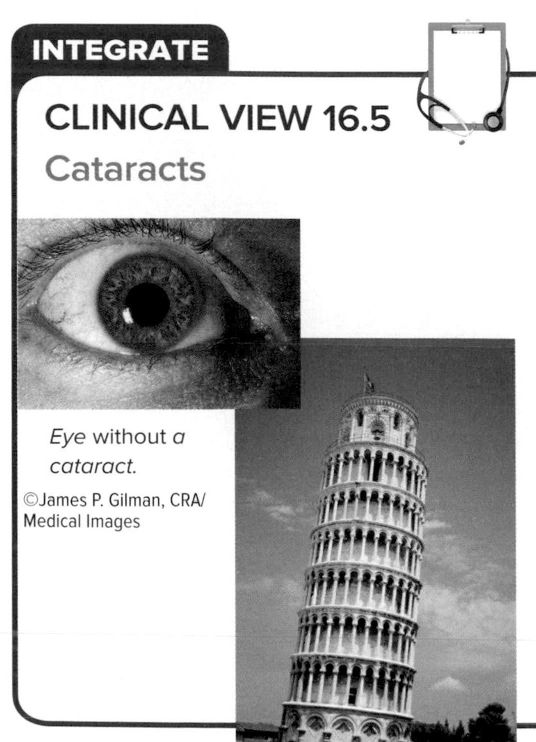

Eye without a cataract.

©James P. Gilman, CRA/ Medical Images

Normal vision.

©David Buffington/Getty Images RF

Cataracts (kat′ă-rakt) are small opacities within the lens that, over time, may coalesce to completely obscure the lens. Most cases occur as a result of aging, although other factors include diabetes, intraocular infections, excessive ultraviolet light exposure, and glaucoma. The resulting vision problems include difficulty focusing on close objects, reduced visual clarity due to clouding of the lens, "milky" vision, and reduced intensity of colors.

A cataract needs to be removed only when it interferes with normal daily activities. Newer surgical techniques include **phacoemulsification** (fāk′ō-i-mul′se-fi-kā′shen), a process by which the opacified center of the lens is fragmented using ultrasonic sound waves, thus making it easier to remove. The destroyed lens is then replaced with an artificial intraocular lens, which becomes a permanent part of the eye.

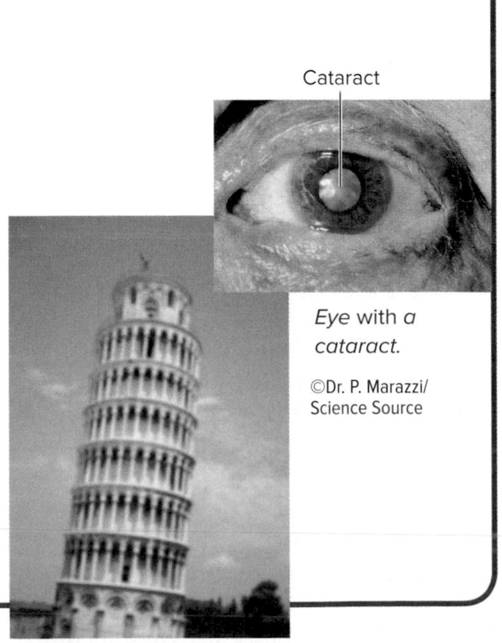

Eye with a cataract.

©Dr. P. Marazzi/ Science Source

Image seen through cataract.

©David Buffington/Getty Images RF

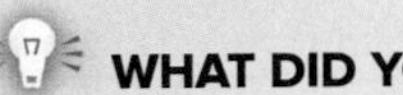

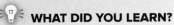

WHAT DID YOU LEARN?

13 What are the three eye tunics, and what is the primary function of each tunic?

14 Compare the anatomic structure of the cornea and the lens. Explain how oxygen and nutrients are supplied to these two structures.

15 What are the functions of the vitreous humor and the aqueous humor?

16.4c Physiology of Vision: Refraction and Focusing of Light

LEARNING OBJECTIVES

16. Describe refraction of light.

17. Discuss how light is focused on the retina.

The physiology of vision requires that light enters the eye and is transduced into electrical signals, which are then sent to the brain for integration and interpretation. The primary processes of vision include the refraction and focusing of light on the retina (which is discussed in this section) and phototransduction and relaying of sensory input along visual pathways (which are discussed in sections 16.4d and 16.4e).

Refraction of Light

Light rays are straight as they first enter the eye. However, the ability to see clearly requires refraction (or bending) of the light rays so that they hit on the retina—specifically, at the fovea centralis, the portion of the retina that is predominantly composed of cones and provides the sharpest vision (figure 16.14). Light rays are refracted when (1) they pass between two media of different densities and (2) these media meet at a *curved* surface. Each medium—such as air, water, and other clear fluids, and even clear solids such as glass—is assigned a *refractive index,* a number that represents its comparative density. The refraction of light rays is greater when there is a larger difference in the refractive index between adjacent media, such as between air and water **(figure 16.16)** and with increasing curvature of the media surface.

Before light can reach the photoreceptor cells, it must pass from the air through the cornea, aqueous humor, lens, and vitreous humor, as well as through the cells forming the inner layers of the retina. Both the cornea and lens play a significant role in refraction of light for vision—the cornea because of the relatively large refraction of light that occurs as light passes

Figure 16.16 Refraction. Light waves change in speed as they pass between media of different densities. This results in the distortion of the image.

©Charles D. Winters/Science Source

INTEGRATE

CLINICAL VIEW 16.6

Glaucoma

Glaucoma (glaw-kō′mă) is a disease that exists in three forms, all characterized by increased intraocular pressure: angle-closure glaucoma, open-angle glaucoma, and congenital or juvenile glaucoma. Angle-closure and open-angle glaucoma both involve the angle formed in the anterior chamber of the eye by the union of the choroid and the sclera (see figure 16.10c), or corneal-scleral junction (i.e., limbus). This angle is the important passageway for draining the aqueous humor. If it narrows, fluid and pressure build up within the anterior chamber. About one-third of all cases of glaucoma develop as a direct consequence of the narrowing of this angle, a condition called *angle-closure glaucoma. Open-angle glaucoma* accounts for about two-thirds of glaucoma cases. In this instance, although the drain angles are adequate, fluid transport out of the anterior chamber is impaired. *Congenital glaucoma* (or *childhood glaucoma*) occurs only rarely and is due to hereditary factors or intrauterine infection.

Regardless of the cause, fluid buildup in the anterior cavity causes a posterior dislocation of the lens and a substantial increase in pressure in the posterior cavity. Compression of the choroid layer may occur, constricting the blood vessels that nourish the retina. Retinal cell death and increased pressure may distort the axons of ganglionic cells that form the optic nerve, leading to impaired vision. Eventually, the patient may experience such symptoms as reduced field of vision, dim vision, and halos around lights. These symptoms are often unrecognized until it is too late and the damage is irreversible, so it is essential individuals get regular eye screenings where early stages may be detected by an optometrist or ophthalmologist.

TE

CAL VIEW 16.7

tional Visual Impairments

ropia (em-ē-trō′pē-ă; *emmetros* = according to measure, *ops* = eye) is ndition of normal vision, in which parallel rays of light are focused exactly e retina. Any variation in the curvature of either the cornea or the lens, or e overall shape of the eye, causes entering light rays to form an abnormal al point. Conditions that can result include hyperopia, myopia, and stigmatism.

People with **hyperopia** (hī-per-ō′pē-ă) have trouble seeing close-up objects and so are called *farsighted.* In this optical condition, only convergent rays (those that come together from distant points) can be brought to focus on the retina. The cause of hyperopia is a short eyeball; parallel light rays from objects close to the eye focus posterior to the retina. By contrast, people with **myopia** (mī-ō′pē-ă; *myo* = to shut) have trouble seeing faraway objects and so are called *nearsighted.* In myopia, only rays relatively close to the eye focus on the retina. The cause of this condition is a long eyeball; parallel light rays from objects at some distance from the eye focus anterior to the retina within the vitreous body. Another variation is **astigmatism** (ă-stig′mă-tizm), which causes unequal focusing and blurred

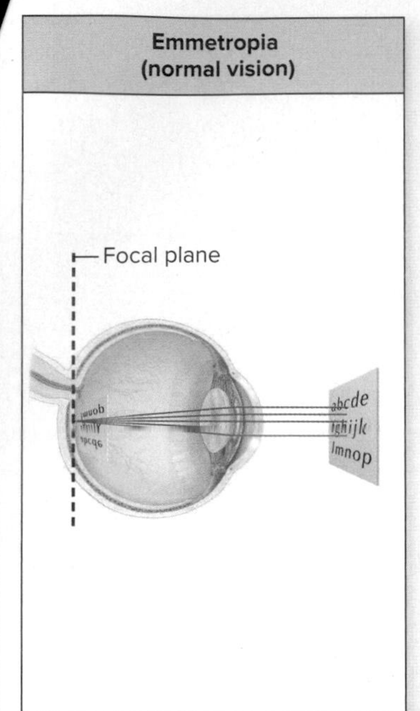

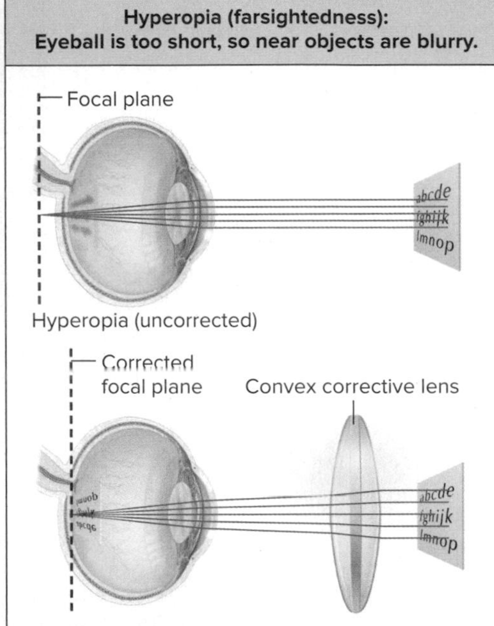

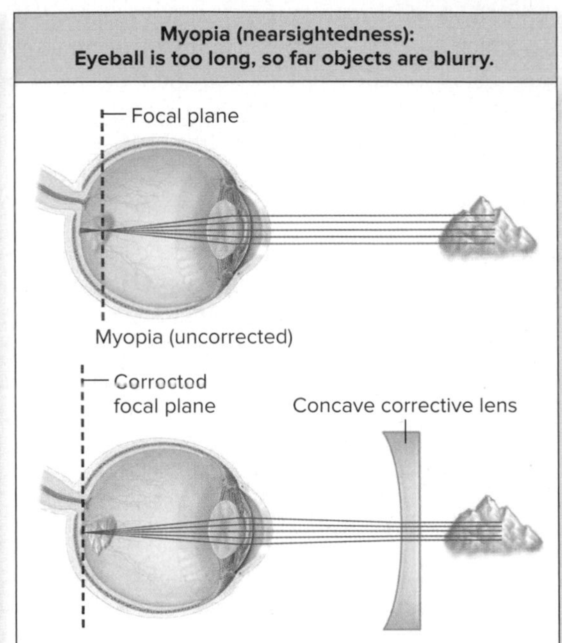

Vision correction using convex and concave lenses.

from air into the cornea and the lens because of its ability to change shape. Recall that the curvature of the lens (and thus the refraction of light) can be altered through accommodation by relaxation and contraction of the ciliary muscles.

Focusing of Light

The processes that occur in the eye for us to see clearly are dependent upon how far away the object is that we are viewing. Let us first consider the three significant changes that occur when we view objects that are closer than 20 feet. These include convergence of the eyes, accommodation of the lens, and constriction of the pupil.

Convergence of the Eyes **Convergence** of the eyes is the voluntary contraction of the extrinsic eye muscles to move the eyes medially. (The extreme case of this is when your eyes are *cross-eyed* such as occurs when you try to focus on your finger a few inches from your eyes.) This positions the eyes so that the image of the object being viewed is directed onto the fovea centralis. Individuals with extrinsic eye muscles that are weaker in one eye than in the other may be unable to converge the eyes and as a result will have **diplopia** (di-plō′pē-ă) or *double vision.*

Accommodation of the Lens Recall from section 16.4b that accommodation involves stimulation of the ciliary muscles by the parasympathetic division of the ANS when viewing objects closer than 20 feet. In response, the ciliary muscles contract to reduce tension in the suspensory ligaments, so the lens becomes more spherical, or curved. Consequently, the light is refracted to a greater extent. Note that this change in

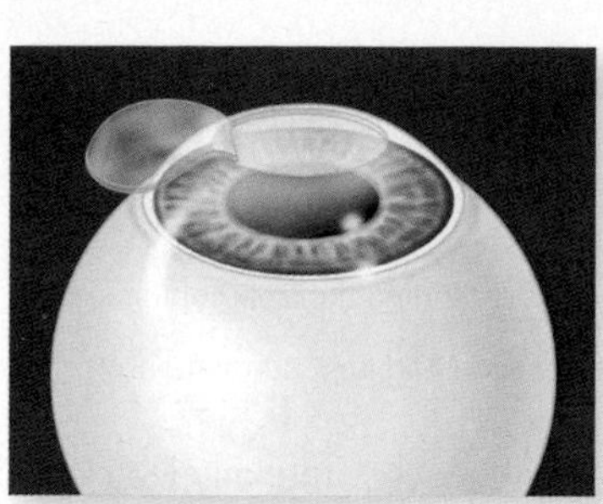

(1) Cornea is sliced with a sharp knife. Flap of cornea is reflected, and deeper corneal layers are exposed.

(2) A laser removes microscopic portions of the deeper corneal layers, thereby changing the shape of the cornea.

(3) Corneal flap is put back in place, and the edges of the flap start to fuse within 72 hours.

LASIK laser vision correction procedure.

images due to unequal curvatures in one or more of the refractive surfaces (cornea, anterior surface, or posterior surface of the lens).

With increasing age, the lens becomes less resilient and less able to become spherical. Thus, even if the suspensory ligaments relax, the lens may not be able to spring out of the flattened position into the more spherical shape needed for near vision, and reading close-up words becomes difficult. This age-related change is called **presbyopia** (prez-bē-ō′pē-ă; *presbys* = old man, *ops* = eye).

The typical treatment for vision disturbances is eyeglasses. A concave eyeglass lens is used to treat myopia, because the concave lens bends the light rays to make the focused image appear directly on the retina, instead of too far in front of it. A convex lens is used to treat both hyperopia and presbyopia.

Corneal incision surgical techniques may help treat hyperopia, myopia, and astigmatism. These techniques involve cutting the cornea to change its shape, and thereby changing its ability to refract light. One type of procedure, called a **radial keratotomy** (ker′ă-tot′ō-mē; *tome* = incision), or **RK,** treats nearsightedness. The ophthalmologist makes radial-oriented cuts in the cornea, which flatten the cornea and allow it to refract the light rays so that they focus on the retina.

Laser vision correction uses a laser to change the shape of the cornea. Two of the more popular types of laser vision correction are photorefractive keratectomy and laser-assisted in situ keratomileusis.

Photorefractive keratectomy (PRK) is called a *photoablation procedure* because the laser removes (ablates) tissue directly from the surface of the cornea. The removal of corneal tissue results in a newly shaped cornea that can focus better. This procedure is becoming less popular, because its regression rate is high—that is, the epithelial tissue that covers the surface of the cornea can regrow and regenerate, leading to partial return to uncorrected vision.

Laser-assisted in situ keratomileusis (ker′ă-tō-mī-lū′sis) (or **LASIK**) is rapidly becoming the most popular laser vision correction procedure. It can treat nearsightedness, farsightedness, and astigmatism. LASIK removes tissue from the inner, deeper layer of the cornea, which is less likely to regrow than surface tissue, so less vision regression occurs.

light refraction is necessary because as objects become increasingly closer, the light rays reflecting from these objects must be bent to a greater degree so that the light rays hit on the retina.

Constriction of the Pupil The parasympathetic division also stimulates the sphincter pupillae muscle to contract to decrease the light rays passing through the edges of the lens (see section 15.7b). This is required when looking at objects closer than 20 feet because the lens must become more curved, but the edges of the lens are unable to curve to the extent that occurs at the center of the lens. Thus, light rays are not refracted at the edges of the lens to the same extent as the center of the lens. Consequently, light passing through the edges of the lens are not focused on the retina, and this portion of the object appears blurry. By constricting the pupil and allowing less light into the eye, light is passing only through the center of the lens. Collectively, these changes associated with focusing on objects that are closer than 20 feet is called the **near response.**

In comparison, when viewing objects at a distance greater than 20 feet away, the near response is not occurring. Instead, the following is noted:

- The eyes are facing forward and are not converging.
- The ciliary muscles are relaxed and the lens is flatter, so the light is refracted to a lesser extent (i.e., there is no accommodation).
- The pupil is relatively dilated to allow a greater amount of light into the eye for maximizing visual input regarding the environment. Note that when viewing objects at any distance, it is an inverted image that hits the retina (see figure 16.14).

 **WHAT DID YOU LEARN?**

16 Describe how light is focused on the retina when viewing an object that is closer than 20 feet.

16.4d Physiology of Vision: Phototransduction

LEARNING OBJECTIVES

18. Define phototransduction.

19. Compare and contrast the two general types of photoreceptors, including their photopigments.

20. Explain the bleaching reaction and how it relates to dark adaptation and light adaptation.

Phototransduction is the converting (or transducing) of light energy into an electrical signal. *Photoreceptor cells* (rods and cones) are the specific cells within the neural layer of the retina that engage in phototransduction. We first describe the anatomic details of photoreceptor cells (and other cells of the neural layer of the retina) and then discuss the process of phototransduction and the initiating of nerve signals that are sent to the brain.

Photoreceptors and Other Cells of the Neural Layer of the Retina

Both types of photoreceptor cells are composed of an outer segment, an inner segment, a cell body, and synaptic terminals. The **outer segment** extends into the pigmented layer of the retina (rod-shaped in rods and conical-shaped in cones) **(figure 16.17)**. The outer segment is composed of hundreds of discs that are flattened, membranous sacs, which are constantly being replaced. New discs are added at the base of the outer segment and begin to move externally toward the tip, where old, worn-out discs are removed by phagocytic cells within the pigmented layer. Usually, it takes about 10 days for a disc to traverse this distance. The outer segment is connected to the **inner segment,** which contains the organelles for the cell, such as mitochondria. The inner segment connects to the **cell body,** which contain the nucleus. **Synaptic terminals** on the other side of the cell body house synaptic vesicles with glutamate neurotransmitter.

Rods and Cones **Rods** are longer and narrower than cones. Each eye contains more than 100 million rods, and they are primarily located in the peripheral retina. Rods are activated by dim (low-intensity) light, such as when you are in an unlit room at night, and provide no color recognition. We describe how rods produce limited sharpness of vision in section 16.4d.

Cones occur at a density of less than 10 million per eye and are concentrated in the fovea centralis (the place of our most acute vision). Cones are activated by high-intensity light and provide color recognition and precise visual sharpness. Thus, when you notice the fine details in a colorful picture, the cones of your retina are responsible.

Why are there two types of photoreceptor cells (rods and cones)? One explanation is based upon both the arrangement of the three primary layers of neurons within the different regions of the retina and the varying degrees of sensitivity of rods and cones to light stimuli. In the peripheral retina, many rods converge on fewer bipolar cells, which synapse on one ganglion cell. In this arrangement, one ganglion cell is receiving input from many rods. In addition, recall that rods are more sensitive to light than are cones and can be stimulated by low light. There is both an advantage and a disadvantage to this anatomic arrangement and characteristic of rods. The advantage to this arrangement can be seen in conditions of low light. Rods are stimulated and spatial summation (see section 12.8b) occurs at both the bipolar cells and ganglion cells. The additive effect can cause sufficient neurotransmitter to be released from bipolar cells to stimulate ganglion cells to initiate an action potential that is propagated to the brain. However, because one ganglion is stimulated by numerous rods that cover a relatively wide area (1 mm^2), the brain's perception of this sensory input is of a slightly blurry image.

In comparison, in the fovea centralis there is a one-to-one relationship between each cone and a bipolar cell, and between a bipolar cell and each ganglion cell. However, recall that cones are less sensitive to light than are rods and require bright light to be stimulated. There is also both an advantage and a disadvantage to this anatomic arrangement and characteristic of cones. The advantage is having visual input from a very small area of the retina (about 1 μm^2) that allows the brain to perceive a sharp image. However, cones are stimulated only in

INTEGRATE

CLINICAL VIEW 16.8

Color Blindness

Color blindness is an inherited genetic trait (see section 29.9c) that occurs when an individual has an absence or a deficit in one type of cone cell. The most common form of color blindness is the X-linked recessive trait, which involves the red and green cone cells, resulting in red-green color blindness. For these individuals, red and green colors appear similar and are difficult to distinguish. For example, in the adjacent image, a person with color blindness cannot distinguish the green number 74 from the rest of the speckled red background. The image instead would appear as a bunch of different-sized dots without great differences in color. Color blindness is much more common in males (seen in about 8% of the male population) because it is an X-linked recessive trait.

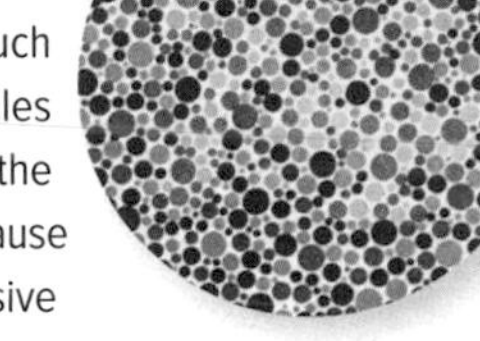

©Steve Allen/Getty Images RF

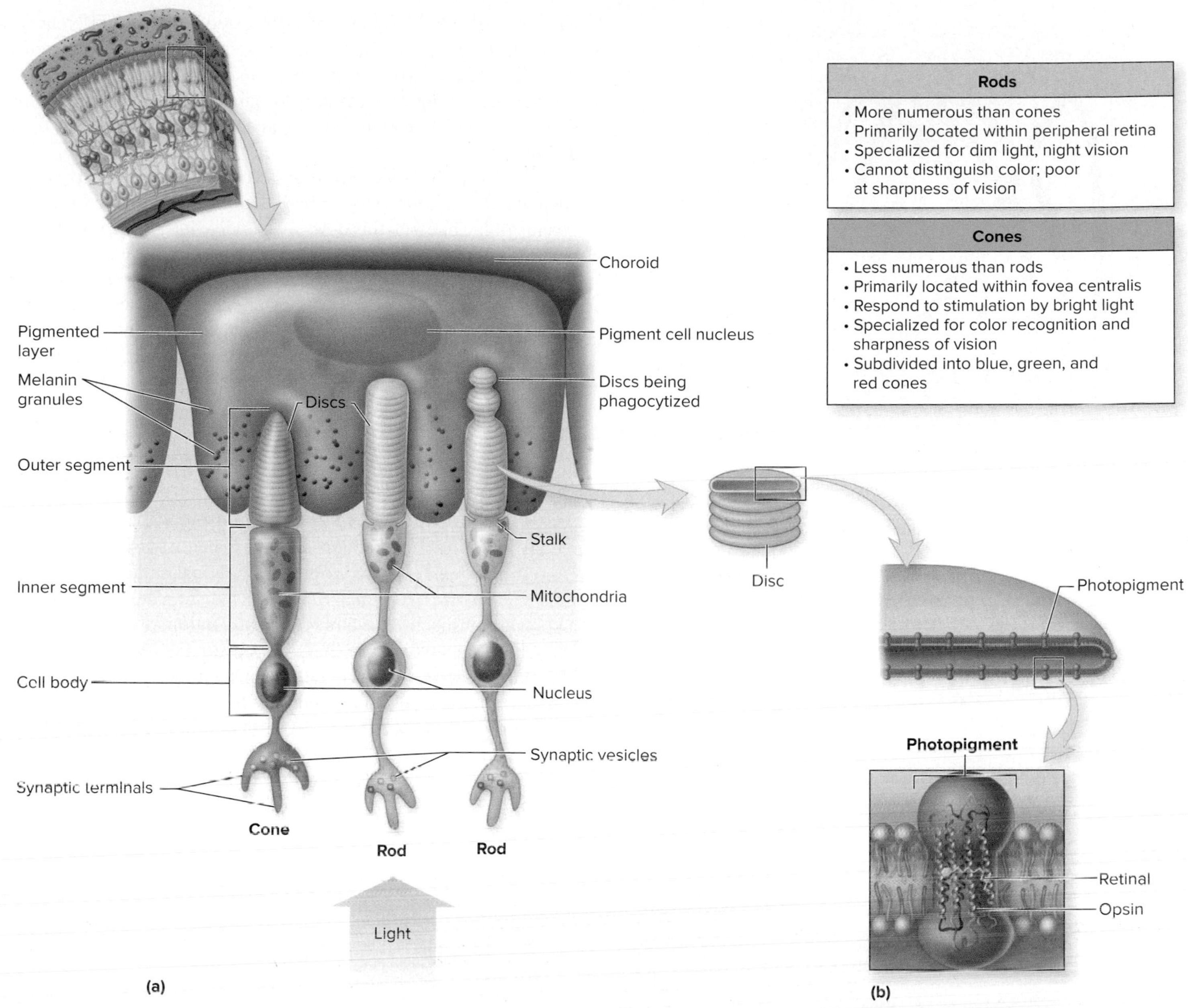
Rods
• More numerous than cones
• Primarily located within peripheral retina
• Specialized for dim light, night vision
• Cannot distinguish color; poor at sharpness of vision
Cones
• Less numerous than rods
• Primarily located within fovea centralis
• Respond to stimulation by bright light
• Specialized for color recognition and sharpness of vision
• Subdivided into blue, green, and red cones
Choroid
Pigmented layer
Pigment cell nucleus
Melanin granules
Discs
Discs being phagocytized
Outer segment
Stalk
Inner segment
Mitochondria
Cell body
Nucleus
Synaptic vesicles
Synaptic terminals
Cone
Rod
Rod
Light
(a)
Disc
Photopigment
Photopigment
Retinal
Opsin
(b)

Figure 16.17 Photoreceptors. (*a*) The outer segments of both rods and cones consist of stacks of discs embedded in the pigmented layer. (*b*) Membranes of the discs contain photopigments. Each photopigment is composed of an opsin and a retinal.

bright light, and sufficient light must be present for our cones to function. Thus, rods allow us to see in dim light but produce a blurry image, whereas cones produce a sharp image but require bright light.

Photopigments Photopigments are the specific molecules that absorb light and that are embedded within the plasma membrane of the outer segment of both rods and cones (figure 16.17*b*). A photopigment is composed of a protein called an **opsin** (op′sin; *opsis* = vision) and a light-absorbing molecule called a **retinal** (or a *retineme*), which is formed from vitamin A. There are several different types of photopigments that contain different opsins, and each type transduces different wavelengths of light. Thus, some photopigments may transduce light of longer wavelengths like reds, whereas other photopigments may transduce light of shorter wavelengths like blues. However, each photoreceptor cell expresses only one photopigment type.

The photopigment in rods is called **rhodopsin** (rō-dop′sin; *rhodon* = a rose). Rhodopsin is involved in the transduction of dim light and is most sensitive to light at a 500-nm wavelength **(figure 16.18)**. It is less sensitive to other light wavelengths.

The photopigment in cones is called **photopsin** (fō-top′sin; *phos* = light). There are three different photopsin proteins, and each type of photopsin protein maximally absorbs different

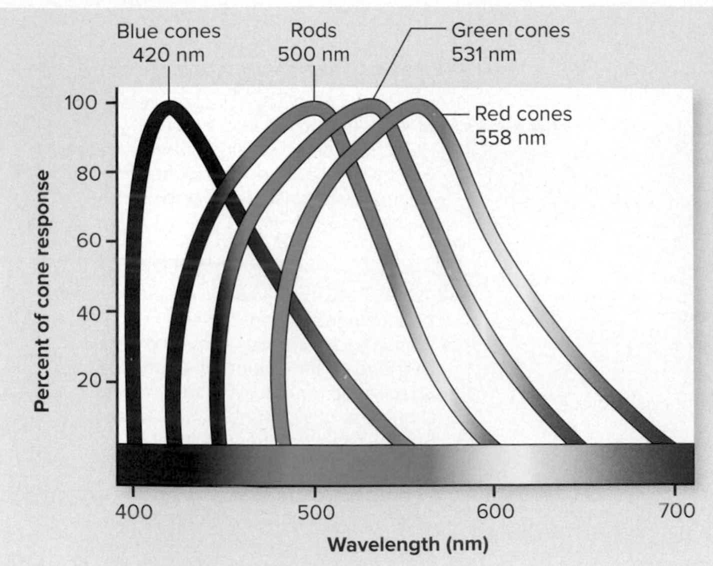

Figure 16.18 Absorption Wavelengths. Each photoreceptor detects a range of wavelengths, but not all wavelengths are detected to the same degree. For example, blue cones reach their peak stimulation at wavelengths of 420 nm—yet at 450 nm these blue cones respond at about 70% of their maximum. Mixtures of colors (such as a blue-green, at 470 nm) stimulate green and blue cones at about 50% of their maximum.

wavelengths of light. Cones are categorized into three different types on the basis of the specific type of photopsin protein they contain and the wavelength to which it is most sensitive. **Blue cones** best detect wavelengths of light at about 420 nanometers (nm); **green cones** maximally absorb light at 531 nm; and **red cones** best detect light at 558 nm.

Thus, certain colors are best perceived by specific cones. But what about conditions where there is a mixture of colors (such as a greenish-blue hue)? In this case, we perceive intricate colors based on patterns of nerve signals arriving from a combination of these cones. Note in figure 16.18 that at about 470 nm, only the blue and green cones are stimulated; however, they are not stimulated at their peak. Instead, both types of cones are stimulated at about 50% capacity. When the blue and green cones initiate nerve signals, the brain interprets the color as a blue-green.

Phototransduction

Phototransduction occurs as light enters the eye and is transduced to an electrical signal by photoreceptor cells. We first discuss what happens with rods **(figure 16.19)**. Prior to being activated by light, the retinal portion of the rhodopsin is in a bent, twisted shape called ***cis*-retinal.** Upon exposure to light, the retinal straightens out into a form called ***trans*-retinal.** *Trans*-retinal dissociates from the opsin and phototransduction occurs. This dissociation of rhodopsin into its two components (*trans*-retinal and opsin) is a process termed a **bleaching reaction** because the rhodopsin goes from a bluish-purple color to colorless. Bleaching reduces rhodopsin amounts in rods and temporarily affects our ability to see in dim light conditions.

Rhodopsin must be regenerated for the rod cell to continue to function. Its regeneration occurs as follows: The dissociated *trans*-retinal is transported from the rod within the neural layer to the

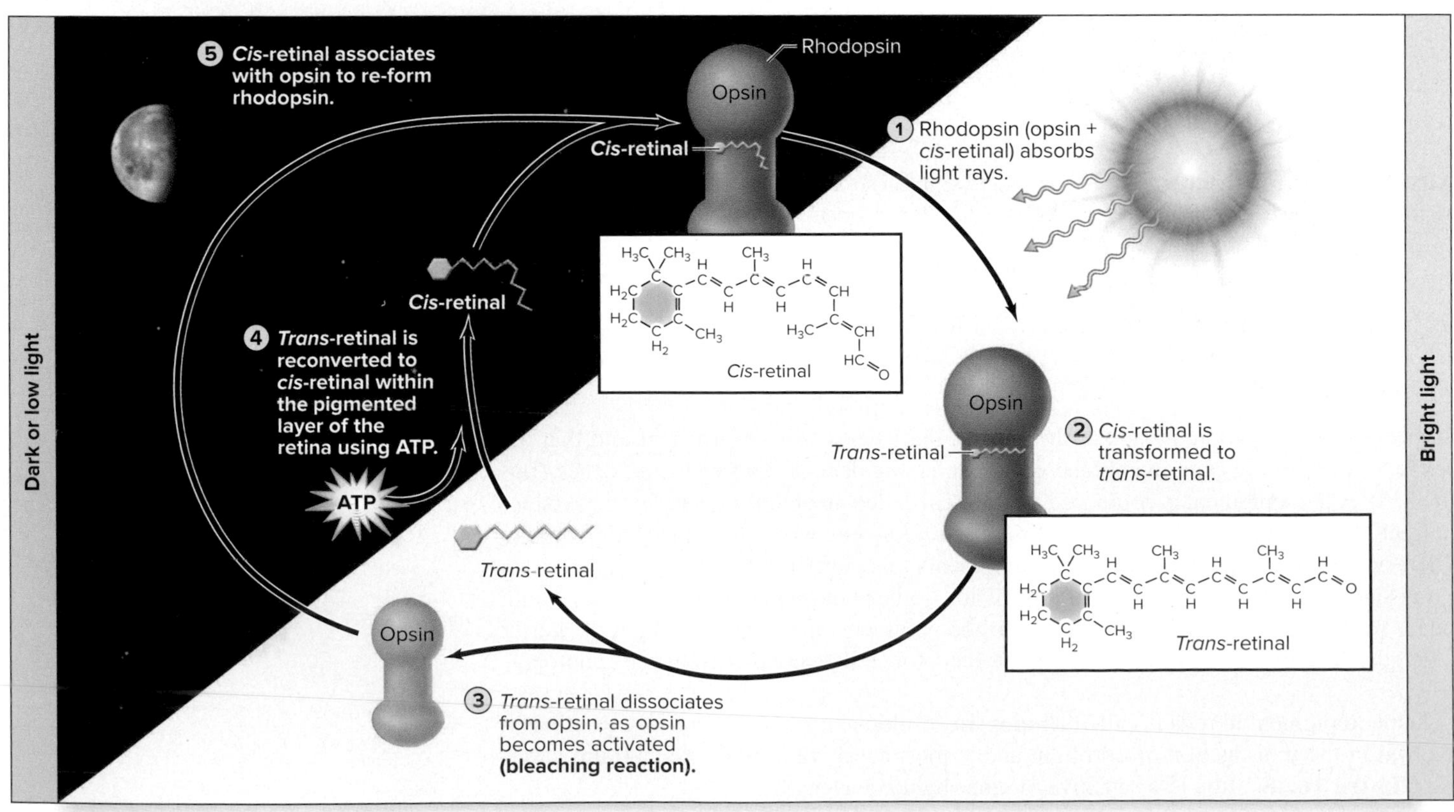

Figure 16.19 Bleaching Reaction and Regeneration of Rhodopsin. When light waves reach the rod, *cis*-retinal is transformed to *trans*-retinal, which then disassociates from opsin in a process called the *bleaching reaction*. Rhodopsin is re-formed when *trans*-retinal is converted back to *cis*-retinal, which then reunites with opsin.

pigmented layer, where it is converted back to its bent *cis*-retinal form—a process that requires ATP. The *cis*-retinal is then transported back to the rod, where it associates with the opsin and re-forms the rhodopsin. This process is relatively slow; typically, only half of the bleached rhodopsin is regenerated after about 5 minutes. Light will interfere with this process, and rhodopsin will bleach as fast as it is re-formed when we are in high-intensity light. For this reason, rods are essentially nonfunctional in bright light situations; the bleaching reaction in the neural retina occurs more quickly than rhodopsin re-formation in the pigmented layer.

A similar process occurs for the photopsin of cone cells. *Cis*-retinal transforms to *trans*-retinal, and a bleaching reaction occurs, but a more intense light is required. However, the regeneration of photopsin occurs much more quickly than the regeneration of rhodopsin; thus, cone cells are not as negatively affected by bright light as rods.

Recall a time when you went from bright light conditions outside on a sunny day into a darkened movie theater. You probably remember the slow return of your sensitivity to low light levels. This phenomenon, called **dark adaptation,** occurs because initially our cones become nonfunctional in the low light (because they require a more intense stimulus), but our rods are still bleached from the bright light conditions from outside. It may take 20 to 30 minutes for your rhodopsin to be regenerated sufficiently that you can see well in low-light conditions.

In comparison, **light adaptation** is the process by which your eyes adjust from low light to bright light conditions, such as when you wake up at night and turn on the bright light in the bathroom. Even though your pupils constrict to reduce the amount of light entering your eyes, you are temporarily blinded as the rods become inactive and the cones, which initially were overstimulated, gradually adjust to the brighter light. In about 5 to 10 minutes, the cones can produce sufficient visual acuity and color vision.

Initiating Nerve Signals

How does the splitting of photopigments within photoreceptor cells in the retina initiate nerve signals that are relayed to the brain? We first describe what is occurring in the rods, bipolar cells, and ganglion cells within the retina in the absence of light (**figure 16.20*a***). In the dark, the outer segment of rods continuously produces cyclic GMP (cGMP) from guanosine triphosphate (GTP). This reaction is catalyzed by the enzyme guanylate cyclase. cGMP binds to cation channels in the plasma membrane of the outer segment, allowing an influx of both Na^+ and Ca^{2+}. This influx of cations (called the dark current) depolarizes the photoreceptor cells to about −40 mV. These local currents of ions diffuse from the outer segment, reaching voltage-gated Ca^{2+} channels at the synaptic terminals of the rods; this change in voltage triggers these channels to open. Calcium ion enters the synaptic terminals, triggering the continuous (tonic) release of glutamate neurotransmitter. Glutamate binds with receptors of the bipolar cells to cause hyperpolarization of the bipolar cells. This inhibits the bipolar cells and prevents them from releasing glutamate neurotransmitter from their synaptic terminals. Thus, no nerve signals are generated by ganglion cells.

When exposed to light (figure 16.20*b*), rhodopsin is split, and through a G protein second messenger pathway (see figure 4.21), the enzyme (phosphodiesterase) that breaks down cGMP is activated. Lower levels of cGMP result in the closing of the cation channels, preventing the influx of both Na^+ and Ca^{2+}. Thus, the dark current ceases, and the photoreceptor cells hyperpolarize. Voltage-gated Ca^{2+} channels at the synaptic terminals of the photoreceptor cells now close, and release of glutamate neurotransmitter ceases. Bipolar cells are no longer inhibited and now release glutamate neurotransmitter that binds to receptors of ganglion cells. If sufficient neurotransmitter is released from the bipolar cells and the threshold is reached in a ganglion cell, a nerve signal is propagated along the axon of the ganglion cell into the brain.

WHAT DID YOU LEARN?

17 What are the differences between rods and cones with respect to their anatomy, their photopigments, and the light they process?

18 How does dark adaptation differ from light adaptation?

19 What occurs during phototransduction of light?

16.4e Visual Pathways

LEARNING OBJECTIVES

21. Describe the visual pathway from the photoreceptors to the brain.
22. Explain how stereoscopic vision provides depth perception.

Figure 16.21 depicts the visual pathways. The visual pathways begin at the retina, where the photoreceptor cells transduce light stimuli to an electrical signal. The bipolar cells are the primary neurons, and the ganglion cells are the secondary neurons with the visual pathways (see sensory pathways from section 14.4b). The axons of the ganglion cells form the **optic nerve,** which exits the back of the eye at the optic disc (see figure 16.10). Optic nerves project from each eye and converge at the **optic chiasm** (kī′azm) (immediately anterior to the pituitary gland; see figure 13.1*b*). The optic chiasm is a flattened structure anterior to the infundibulum where many of the optic nerve axons decussate (cross) to the other side. The ganglionic axons originating from the medial region of each retina cross to the opposite side of the brain at the optic chiasm, whereas ganglionic axons originating from the lateral region of each retina remain on the same side of the brain and do not cross. **Optic tracts** extend laterally from the optic chiasm as a composite of ganglionic axons originating from the retina of each eye.

The majority of the optic tract axons extend to the thalamus, specifically to the **lateral geniculate** (je-nik′ū-lāt; *genu* = knee, referring to its appearance) **nucleus,** where visual information is processed within each thalamic body (see section 13.4b). Tertiary neurons project axons (projection fibers) from the thalamus to the visual cortex of the occipital lobe for conscious interpretation of incoming visual stimuli.

Note that the left and right eyes have somewhat overlapping visual fields. For the brain to interpret these two distinct visual images, it must process or unite them into one. These overlapping images then provide us with **stereoscopic vision,** or **depth perception,** which is the ability to determine how close or far away an object is. Animals that don't have overlapping visual fields (such as horses and deer) may have a greater visual range, but they cannot perceive visual depths.

In addition to the neural pathway that relays input to the visual cortex, a limited number of axons within each optic tract project to the midbrain as part of reflexes. Those that extend to the **superior colliculi** (ko-lik′yū-lī) of the midbrain coordinate the reflexive movements of the extrinsic eye muscles (see section 13.5a), whereas those that extend to the **pretectal nuclei** of the midbrain function as the control centers in both the pupillary reflex to regulate the amount of the light that enters the eye and the accommodation reflex for focusing the lens. (Input sent to the pretectal nucleus is initiated by specialized ganglion cells that respond directly to light through the visual pigment melanopsin, a light-sensitive retinal protein.)

In the dark

Rod

Cation channels are kept open by high levels of continuously produced cGMP.

cGMP

Ca^{2+}

Na^+

Cation channel (open)

Dark current is caused by Na^+ and Ca^{2+} entry.

1. Dark current (caused by entry of Na^+ and Ca^{2+}) causes the photoreceptor cell to be depolarized at –40 mV. A graded potential (which is a local current of ions) is transmitted along the inside of the cell.

Graded potential

OFF

Ca^{2+}

Voltage-gated Ca^{2+} channel (open)

Glutamate neurotransmitter is released continuously in the dark.

Glutamate neurotransmitter

2. Voltage-gated Ca^{2+} channels open, and the neurotransmitter glutamate is released from photoreceptor.

3. Binding of glutamate hyperpolarizes the bipolar cell, causing inhibition.

Bipolar cell

Ca^{2+} channel (closed)

4. There is no release of glutamate neurotransmitter from the bipolar cells.

Ganglion cell

5. No nerve signal is generated by the ganglion cells.

(a)

In the light

Rod

Rhodopsin bleaches, leading to a drop in cGMP. Lack of cGMP binding closes cation channels.

Cation channel (closed)

Dark current stops because of decrease in Na^+ and Ca^{2+} entry.

1. Stimulation by light causes the photoreceptor cell to be hyperpolarized at –70 mV because of decreased entry of Na^+ and Ca^{2+}.

ON

Voltage-gated Ca^{2+} channel (closed)

No glutamate neurotransmitter is released in the light.

2. Voltage-gated Ca^{2+} channels close, and no glutamate neurotransmitter is released.

3. The bipolar cell is no longer inhibited and thus it depolarizes.

Bipolar cell

4. Bipolar cell releases neurotransmitter (glutamate).

Ca^{2+} channel (open)

Neurotransmitter

Ganglion cell

Nerve signal

5. The glutamate neurotransmitter binds to receptors in the ganglion cell, and a nerve signal is initiated to the brain.

(b)

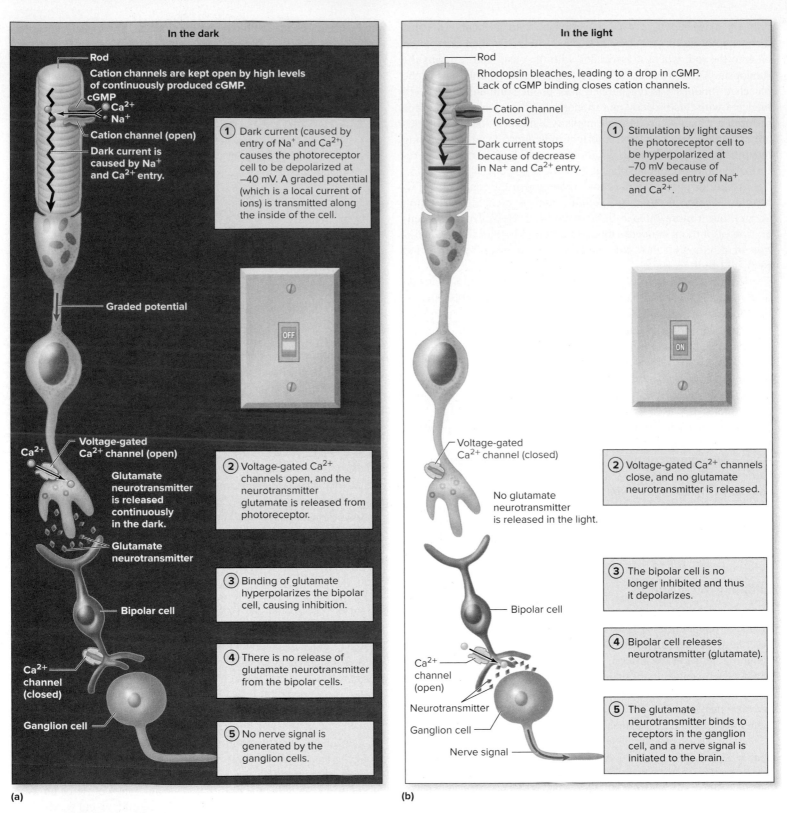

Figure 16.20 Phototransduction in Rod Photoreceptors. (*a*) Events associated with dim light that prevent initiation of nerve signals to the brain. (*b*) Events associated with bright light that initiate nerve signals to the brain.

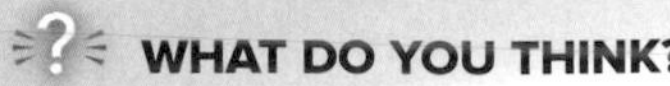

WHAT DO YOU THINK?

2 Why would it be an advantage for a deer to have a wide visual field, as compared to overlapping, stereoscopic vision?

The physiology of vision, which involves the eyes, visual pathways, and brain is integrated in **figure 16.22**. Vision loss may result from either disease or disorders in any structures involved with vision, including the eye, optic nerve, optic tract, or brain.

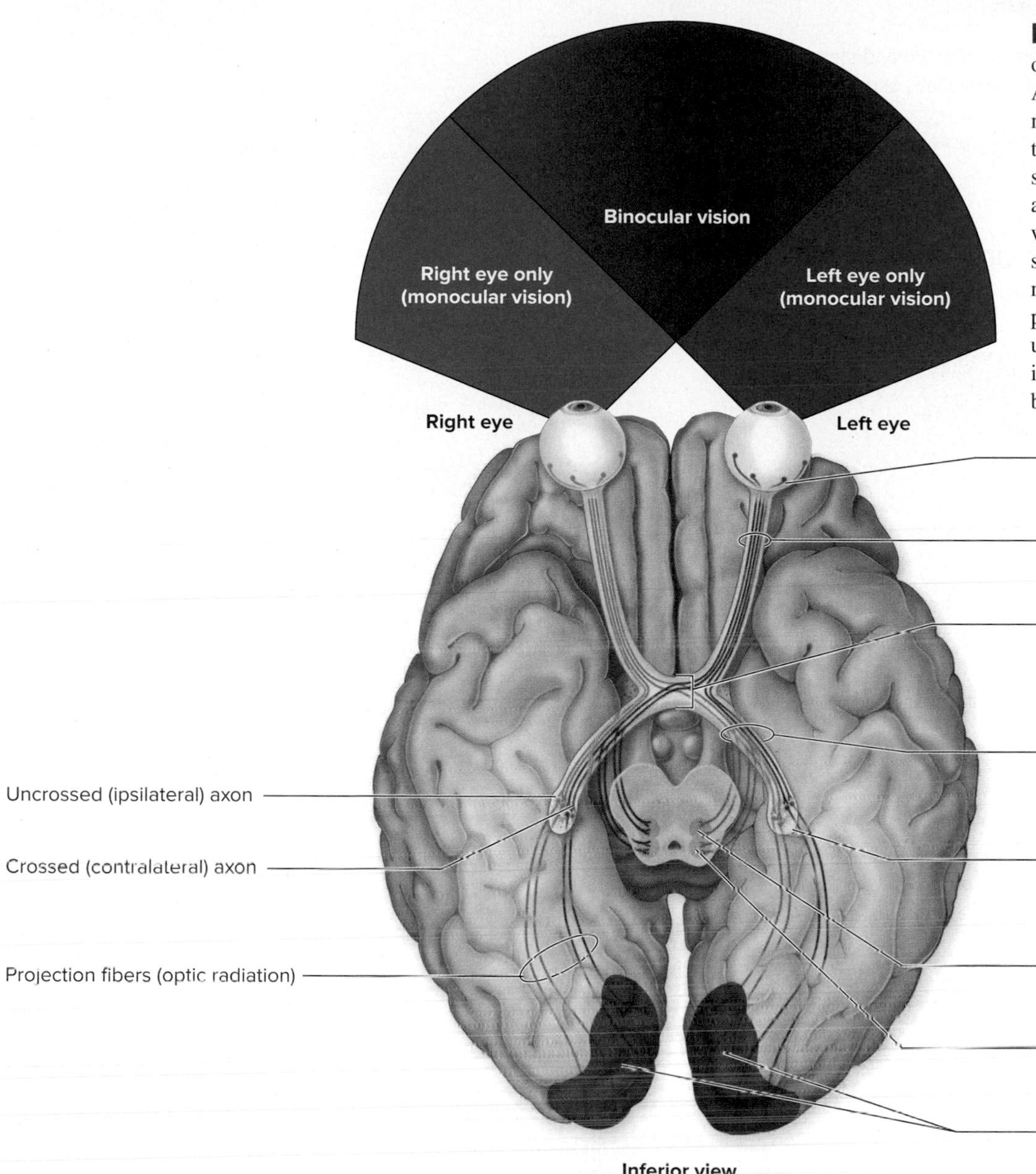

Binocular vision
Right eye only
(monocular vision)
Left eye only
(monocular vision)
Right eye
Left eye
Retina
Photoreceptors and neurons in the retina process the stimulus from incoming light.
Optic nerve
Axons of retinal ganglion cells form optic nerves and exit the eye.
Optic chiasm
Optic nerve axons from the medial region of the retina cross at the optic chiasm; the axons from the lateral region of the retina remain uncrossed.
Optic tract
The optic tract contains axons from both eyes, and these axons will project to either the thalamus or the midbrain.
Lateral geniculate nucleus of thalamus
The majority of the optic tract axons project to the lateral geniculate nucleus in the thalamus.
Pretectal nucleus of the midbrain
Limited number of optic tract axons project to the pretectal nucleus of the midbrain.
Superior colliculus of midbrain
Some optic tract axons project to the superior colliculus in the midbrain.
Primary visual cortex of occipital lobe
Receives processed information from the thalamus
Uncrossed (ipsilateral) axon
Crossed (contralateral) axon
Projection fibers (optic radiation)
Inferior view

Figure 16.21 Visual Pathways. Each optic nerve conducts visual stimuli information. At the optic chiasm, some axons from the optic nerve decussate. The optic tract on each side then contains axons from both eyes. Visual stimuli information is processed by the thalamus and then interpreted by visual association areas within the occipital lobe of the cerebrum. Visual sensory input involved in reflexes is relayed to nuclei within the midbrain (superior colliculi and pretectal nuclei). Note: In this image we have used the color red (previously used in other images for motor pathways) to distinguish between the two eye sensory pathways. AP|R

WHAT DID YOU LEARN?

20 What areas of the brain consciously perceive visual stimuli, and which areas respond reflexively (below the conscious level)?

21 What is the significance of some ganglionic axons crossing to the opposite side of the brain?

16.5 Hearing and Equilibrium Receptors

The **ear** is the organ that detects both sound and movements of the head. These stimuli are transduced into nerve signals that are transmitted by the vestibulocochlear nerve (CN VIII) (see section 13.9), resulting in the sensations of hearing and equilibrium.

16.5a Ear Structure

LEARNING OBJECTIVES

23. Describe the structures of the outer, middle, and inner ear.
24. Name the auditory ossicles, and explain how they function in hearing.
25. Compare and contrast the bony labyrinth and the membranous labyrinth.

The ear is partitioned into three distinct anatomic regions: external, middle, and inner (**figure 16.23**). The **external ear** is located mostly on the outside of the body, and both the **middle ear** and **inner ear** are housed within the petrous part of the temporal bone (see section 8.2b).

INTEGRATE CONCEPT OVERVIEW

Figure 16.22 How We See. (*a*) Light is refracted and then focused on the retina. (*b*) Light rays are transduced to nerve signals, and (*c*) these nerve signals are transmitted to the brain.

(a) Refraction and Focusing of Light

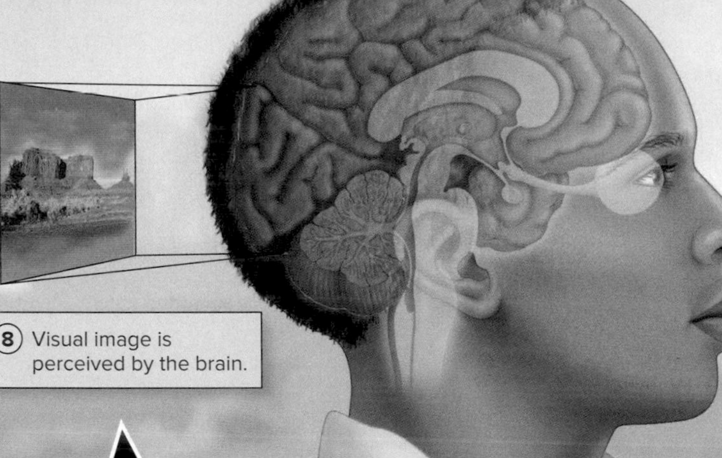

Flattened lens
Far/distant vision
Puffy lens
Near vision
③ Light passes through the pupil and is refracted by the lens. The lens flattens for distant vision, whereas the lens becomes more curved as the eye accommodates for near vision.
① Light rays are refracted by the cornea as they enter the eye.
② Pupils dilate for low light or constrict for bright light.
⑧ Visual image is perceived by the brain.

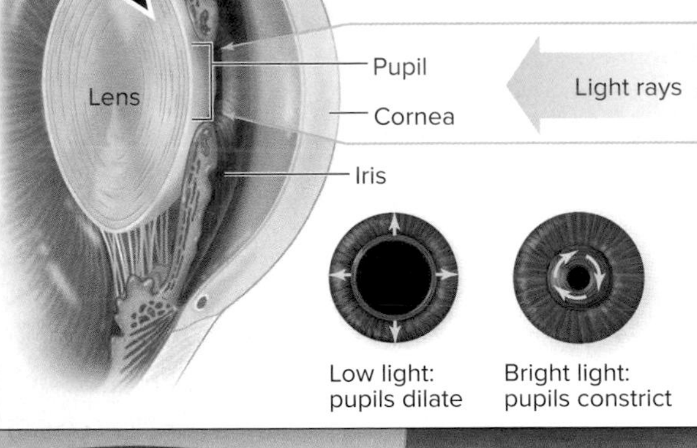

Lens
Pupil
Cornea
Iris
Light rays
Low light: pupils dilate
Bright light: pupils constrict

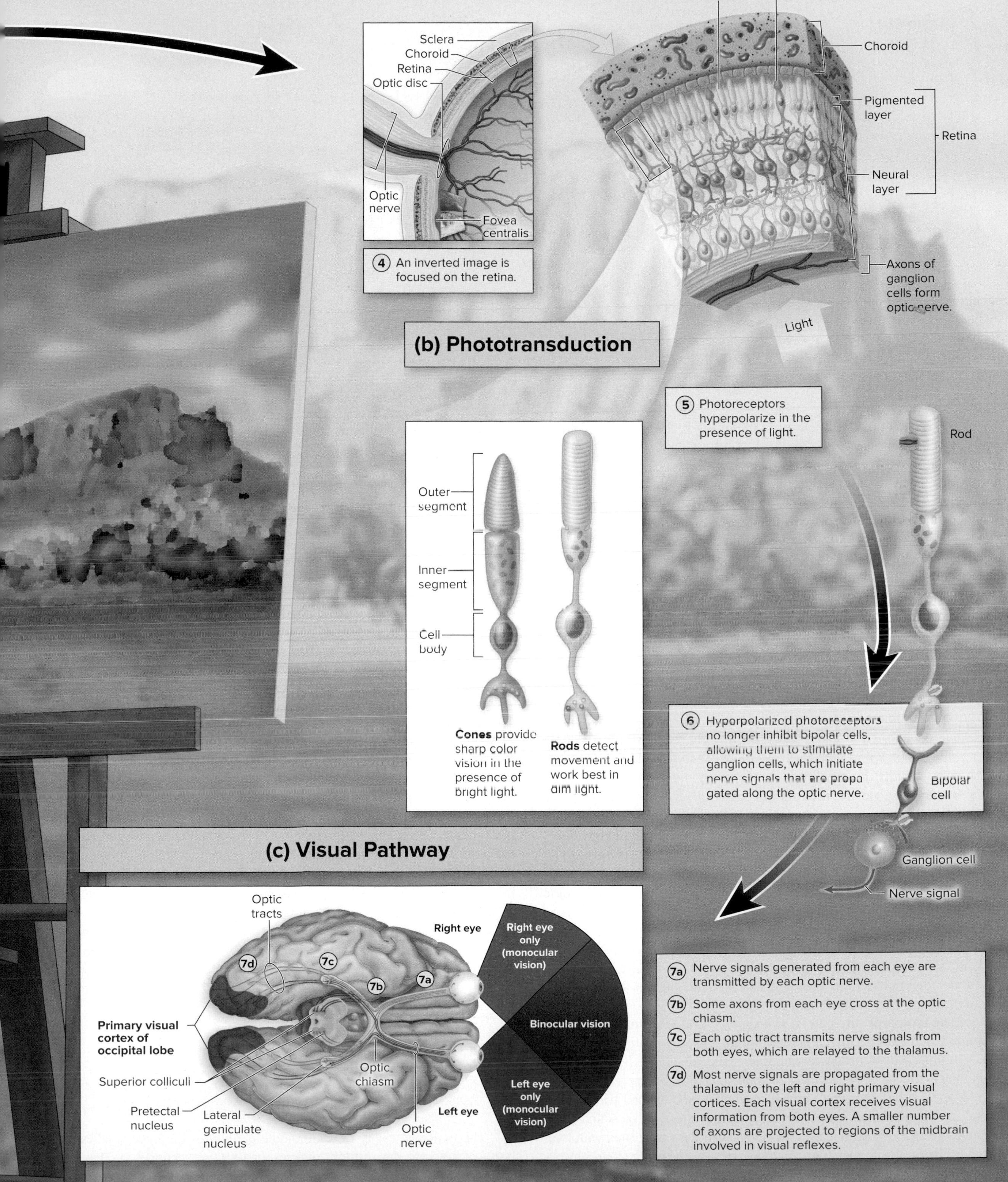
Sclera
Choroid
Retina
Optic disc
Optic nerve
Fovea centralis
4 An inverted image is focused on the retina.
Rod
Cone
Choroid
Pigmented layer
Retina
Neural layer
Axons of ganglion cells form optic nerve.
Light
(b) Phototransduction
5 Photoreceptors hyperpolarize in the presence of light.
Outer segment
Inner segment
Cell body
Cones provide sharp color vision in the presence of bright light.
Rods detect movement and work best in dim light.
Rod
6 Hyperpolarized photoreceptors no longer inhibit bipolar cells, allowing them to stimulate ganglion cells, which initiate nerve signals that are propagated along the optic nerve.
Bipolar cell
Ganglion cell
Nerve signal
(c) Visual Pathway
Optic tracts
7d
7c
7b
7a
Right eye
Right eye only (monocular vision)
Binocular vision
Left eye only (monocular vision)
Left eye
Primary visual cortex of occipital lobe
Superior colliculi
Pretectal nucleus
Lateral geniculate nucleus
Optic chiasm
Optic nerve
7a Nerve signals generated from each eye are transmitted by each optic nerve.
7b Some axons from each eye cross at the optic chiasm.
7c Each optic tract transmits nerve signals from both eyes, which are relayed to the thalamus.
7d Most nerve signals are propagated from the thalamus to the left and right primary visual cortices. Each visual cortex receives visual information from both eyes. A smaller number of axons are projected to regions of the midbrain involved in visual reflexes.

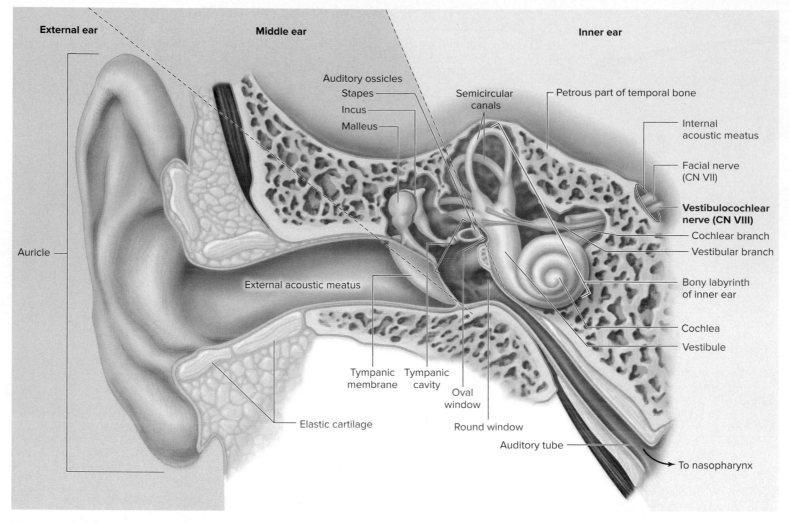

Figure 16.23 Anatomic Regions of the Right Ear. The ear is divided into external, middle, and inner regions. **AP|R**

External Ear

The most visible portion of the external ear is a skin-covered, elastic cartilage–supported structure called the **auricle** (aw′ri-kl; *auris* = ear), or *pinna* (pin′ă; wing). The auricle is funnel-shaped, and it both protects the entry into the ear and directs sound waves into the bony tube called the **external acoustic meatus** (or *external auditory meatus;* see section 8.2b). This canal, which is about 2.5 centimeters (1 inch) in length and 7.5 millimeters (0.3 inch) in diameter, extends slightly superiorly from the lateral surface of the head to the tympanic membrane.

The narrow external opening in the external acoustic meatus prevents large objects from entering and damaging the tympanic membrane. Near its entrance, fine hairs help guard the opening. Deep within the canal, **ceruminous glands** produce a waxlike secretion called **cerumen** (sĕ-rū′men; *cera* = wax), which combines with dead, sloughed skin cells to form **earwax.** This material may help reduce infection within the external acoustic meatus by impeding microorganism (e.g., bacteria) growth.

The **tympanic** (tim-pan′ik; *tympanon* = drum) **membrane,** or *eardrum*, is a funnel-shaped membrane (approximately 1 centimeter in diameter) composed of fibrous connective tissue sandwiched between two epithelial sheets. It serves as the boundary between the external and middle ear. The tympanic membrane vibrates when sound waves hit it, and its vibrations provide the means for transmission of sound wave energy from the external ear to the middle ear. Pain associated with trauma to the tympanic membrane is transmitted to the brain along sensory neurons within both the vagus and trigeminal nerves (see section 13.9).

Middle Ear

The middle ear contains an air-filled **tympanic cavity (figure 16.24).** Medially, a bony wall separates the middle ear from the inner ear. Two membrane-covered openings are located within this wall: the oval window and round window (discussed in detail in section 16.5b). Inferiorly, the **auditory tube** (also called the *pharyngotympanic tube* or *Eustachian tube*), which is approximately 3.5 centimeters (1.5 inches) in length, serves as a passageway that extends from the middle ear into the nasopharynx (portion of upper throat posterior to the nasal cavity; see figure 23.5). This tube is normally closed. Air movement through this tube occurs as a result of chewing, yawning, and swallowing, which equalize pressure on either side of the tympanic membrane—allowing the tympanic membrane to vibrate freely. The tympanic cavity, auditory tube, and nasopharynx are lined with a continuous mucous membrane. Middle ear infections result when infectious agents (e.g., a cold virus) move from the nasopharynx through the auditory tube into the middle ear (see Clinical View 16.9: "Otitis Media").

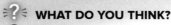

WHAT DO YOU THINK?

3 When an airplane descends to a lower altitude, you may feel greater pressure in your ears, followed by a popping sensation—then more normal pressure resumes. What do you think has happened?

The tympanic cavity of each middle ear houses the three smallest bones of the body, the **auditory ossicles** (os′i-kl) (see section 8.3). These three bones, which are positioned between the tympanic membrane and

INTEGRATE

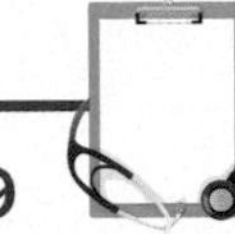

CLINICAL VIEW 16.9

Otitis Media

Otitis (ō-tī′tis; *itis* = inflammation) **media** (mē′dē-ă; *medius* = middle) is an infection of the middle ear. It is most often experienced by young children, whose auditory tubes are horizontal, relatively short, and underdeveloped. If a young child has a respiratory infection, the causative agent may spread from the pharynx (throat) through the auditory tube. Fluid then accumulates in the middle ear cavity, resulting in pressure, pain, and sometimes impaired hearing.

Classic symptoms of otitis media include fever (sometimes over 104°F), pulling on or holding the affected ear, and general irritability. An **otoscope** (ō′tō-skōp; *oto* = ear, *skopeo* = to view) is an instrument used to examine the tympanic membrane, which normally appears white and pearly but in cases of severe otitis media is red (due to inflammation and sometimes bleeding) and may even bulge due to fluid pressure in the middle ear.

Repeated ear infections, or a chronic ear infection that does not respond to antibiotic treatment, usually requires a surgical procedure called a **myringotomy** (mir-ing-got′ō-mē; *myringa* = membrane), whereby a ventilation tube is inserted into the tympanic membrane. This procedure allows the infection to heal and the pus and mucus to drain from the middle ear into the external acoustic meatus and offers immediate relief from the pressure. Eventually, the inserted tube is sloughed, and the tympanic membrane heals.

When a child is about 5 years old, the auditory tube has become larger, more vertically angled, and better able to drain fluid and prevent infection from reaching the middle ear. Thus, the occurrence rate for ear infections drops dramatically at this time.

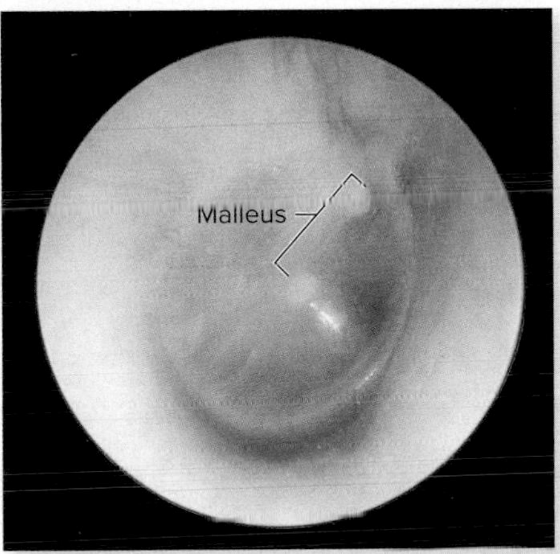

Normal tympanic membrane

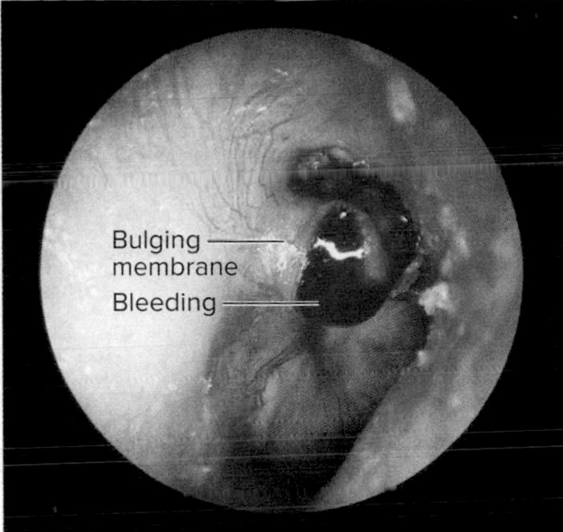

Otitis media

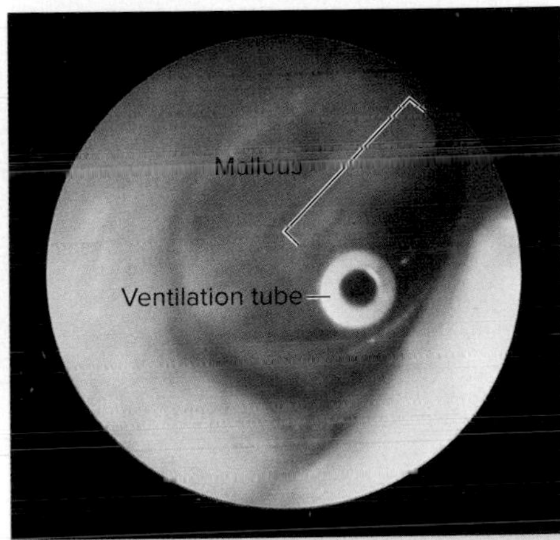

Myringotomy

Middle ear views as seen with an otoscope.

the oval window, are, from lateral to medial, the malleus (hammer), the incus (anvil), and the stapes (stirrup). (You can remember the order with the acronym MIS.) The **malleus** (mal′ē-ŭs), with its long handle and expanded end, resembles a large hammer in shape; the long handle has contact with the tympanic membrane. The **incus** (ing′kŭs), which is approximately triangular in shape and resembles an anvil, is the middle auditory ossicle. The **stapes** (stā′pēz), composed of an arch and a plate of bone, resembles a stirrup on a saddle. Its cylindrical, disclike footplate fits into the oval window. The ossicles are anchored by various small ligaments to the surrounding structures.

The auditory ossicles are responsible for amplifying sound waves from the tympanic membrane to the oval window. When sound waves strike the tympanic membrane, the three middle ear ossicles vibrate along with the tympanic membrane, causing the footplate of the stapes to move in and out of the oval window. The movement of this ossicle initiates pressure waves in the fluid within the closed compartment of the inner ear. Two tiny skeletal muscles, the **tensor tympani** (attached to the malleus) and the **stapedius** (attached to the stapes), are located within the middle ear. These muscles reflexively restrict ossicle movement when loud sounds occur (including when we are speaking) and thus protect the sensitive sensory receptors within the inner ear. This reflexive response that involves contraction of the tensor tympani and stapedius to loud sounds takes approximately 40 milliseconds and thus is not able to protect the

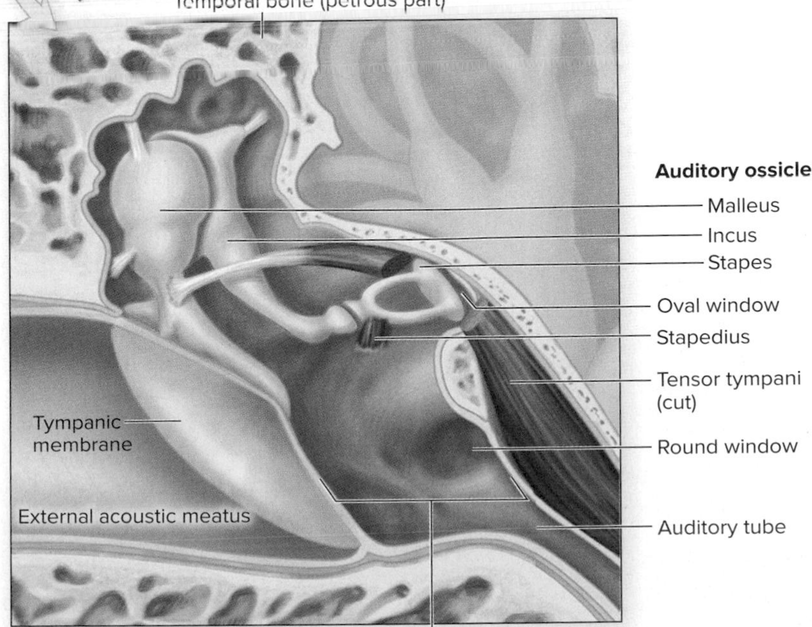

Figure 16.24 Middle Ear. The middle ear contains the auditory ossicles and associated structures within the tympanic cavity of the temporal bone.

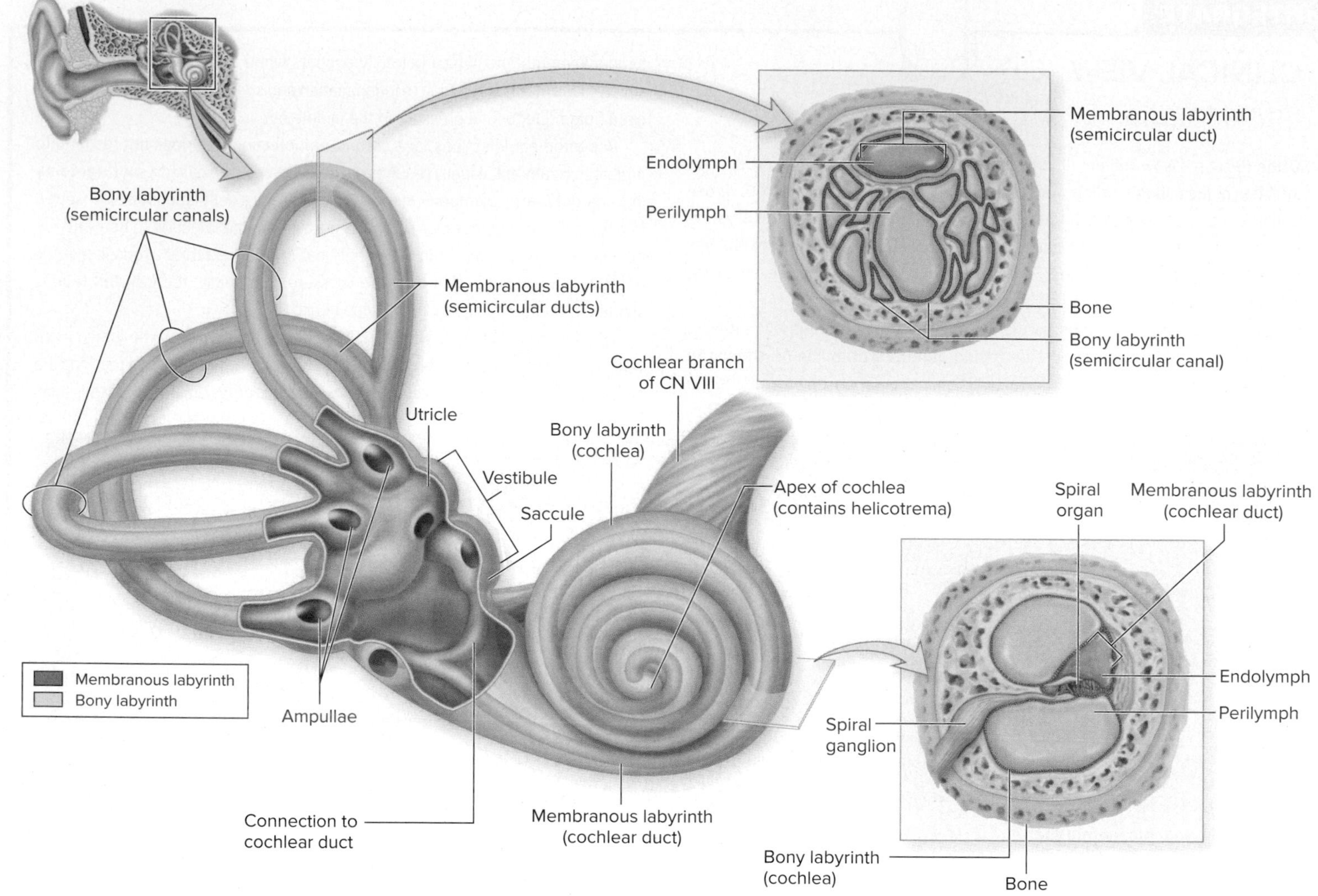

Figure 16.25 **Inner Ear.** The inner ear is composed of a bony labyrinth cavity that houses a fluid-filled membranous labyrinth. Within the bony labyrinth are the portions of the membranous labyrinth for hearing (the cochlear duct) and equilibrium and balance (saccule, utricle, and semicircular ducts). AP|R

sensory receptors from blasts of sounds, as occurs with a gunshot. For this reason, loud blasts of sound are especially damaging to the sensory receptors within the inner ear.

Inner Ear

The inner ear is located within the petrous part of the temporal bone, where there are spaces, or cavities, called the **bony labyrinth** (lab′i-rinth; an intricate, mazelike passageway) **(figure 16.25)**. Within the bony labyrinth are membranous, fluid-filled tubes and sacs called the **membranous labyrinth.** Receptors for both hearing and equilibrium are within the membranous labyrinth.

The space between the outer walls of the bony labyrinth and the membranous labyrinth is filled with a fluid called **perilymph** (per′i-limf) that is similar in composition to interstitial fluid. The perilymph suspends, supports, and protects the membranous labyrinth from the wall of the bony labyrinth. The space within the membranous labyrinth contains a fluid called **endolymph** (en′dō-limf), which is an unusual extracellular fluid because it is similar in composition to intracellular fluid with relatively high levels of K^+ (see section 4.4a).

The bony labyrinth is structurally and functionally partitioned into three distinct regions, including the cochlea, vestibule, and semicircular canals:

- The **cochlea** (kok′lē-ă; snail shell) houses a membranous labyrinth structure called the **cochlear duct** (or *scala media*).
- The **vestibule** (ves′ti-b ūl; *vestibulum* = entrance court) contains two saclike, membranous labyrinth structures—the **utricle** (ū′tri-kl; *uter* = leather bag) and the **saccule** (sak′ūl; *saccus* = sack).
- The **semicircular canals** each contain a membranous labyrinth structure called the **semicircular duct.**

The structures of the membranous labyrinth are continuous. The cochlear duct is continuous with the saccule, which is connected through a narrow passageway to the utricle, and the utricle is continuous with the semicircular ducts.

WHAT DID YOU LEARN?

22 What is the function of the external acoustic meatus?

23 Where are the auditory ossicles located, and what is their function?

24 What are the membranous labyrinth structures, and what is the specific bony labyrinth structure in which each resides?

16.5b Hearing

LEARNING OBJECTIVES

26. Explain the components of the cochlea and how they function in the sense of hearing.
27. Trace the path of a sound wave from outside the ear to stimulation of the vestibulocochlear nerve (CN VIII).
28. Distinguish between frequency and intensity of sound.

Hearing is the ability to detect and perceive sound. Here we follow the progression of how sound from the environment enters the ear and is transduced in the inner ear into electrical signals that are relayed to the brain. First, we consider the structures in the inner ear that are related to sound detection.

Structures for Hearing

Hearing organs are housed within the cochlea in both inner ears. View both **figure 16.26**, which is a cross section of the cochlea, and **figure 16.27**, which is a longitudinal section of a cochlea (partially "unrolled" so the structure can be viewed more easily), as you read through this section.

Cochlea The **cochlea** is a snail-shaped, spiral chamber within the bone of the inner ear. Figure 16.26*a* depicts how this chamber "wraps" approximately 2.5 times around a spongy bone axis called the **modiolus** (mō-dī′ō-lŭs; hub of a wheel), giving the cochlea its snail-shaped appearance. The membranous labyrinth called the cochlear duct is housed within the cochlea. The roof of the cochlear duct is formed by the **vestibular membrane,** and the floor is formed by the **basilar membrane** (figures 16.26*b* and 16.27). These membranes partition the bony labyrinth of the cochlea into two smaller chambers on either side of the cochlear duct; both are filled with perilymph. The chamber adjacent to the vestibular membrane is the **scala vestibuli** (*vestibular duct*), and the chamber adjacent to the basilar membrane is the **scala tympani** (*tympanic duct*). The scala vestibuli and scala tympani merge at the **helicotrema** (hel′i-kō-trē′mă; *helix* = spiral, *trema* = hole) at the apex of the cochlea (see figure 16.27).

Spiral Organ Protected within the membranous cochlear duct is the **spiral organ** (formerly called the *organ of Corti*), which is the sensory structure for hearing (figures 16.26*c, d* and 16.27). The cochlear duct contains endolymph. The spiral organ is a thick sensory epithelium, consisting of both **hair cells** and supporting cells, that rests on the basilar membrane. Two categories of hair cells rest on the basilar membrane, including a single row of **inner hair cells** (which function as the sensory receptors for hearing) and three rows of **outer hair cells** (which alter the response of the spiral organ to sound). The apical surface of each hair cell has a covering of numerous (more than 50) long, stiff microvilli that are called **stereocilia** (ster′ē-ō-sil′ē-ă; *stereos* = solid) and one long cilium, called a **kinocilium** (kī-nō-sil′ē-ŭm; *kino* = movement). The stereocilia and kinocilium are embedded in an overlying gelatinous structure called the **tectorial** (tek-tōr′ē-ăl; *tectus* = to cover) **membrane.** Extending from the base of these hair cells are primary sensory neurons (about 90% from the inner hair cells and about 10% from the outer hair cells). The cell bodies of these sensory neurons are housed within the **spiral ganglia** in the modiolus, which is located medial to the cochlear duct (figure 16.26*a, b*).

Pathway from Sound Wave to Nerve Signal

How we detect sound waves, transduce sound energy to an electrical signal, and then transmit a nerve signal along the vestibulocochlear nerve to the brain is described here. See figure 16.27 as you read through this section.

Sound waves are collected and funneled by the auricle of the external ear to enter the external acoustic meatus, which make the tympanic membrane vibrate. The vibration of the tympanic membrane causes movement of the auditory ossicles (malleus, incus, and stapes), which makes

INTEGRATE

LEARNING STRATEGY

You may find it helpful to imagine elongated balloons of different shapes are located within the elaborate space of the inner ear within the petrous part of the temporal bone. These balloons contain the specialized sensory receptors that are bathed in endolymph and are surrounded by perilymph, which prevents their contact with the temporal bone.

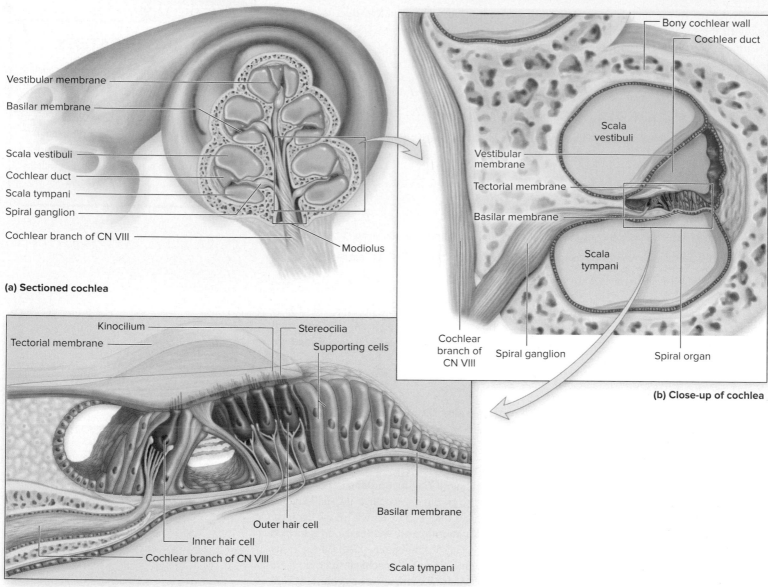

(a) Sectioned cochlea

(b) Close-up of cochlea

(c) Spiral organ

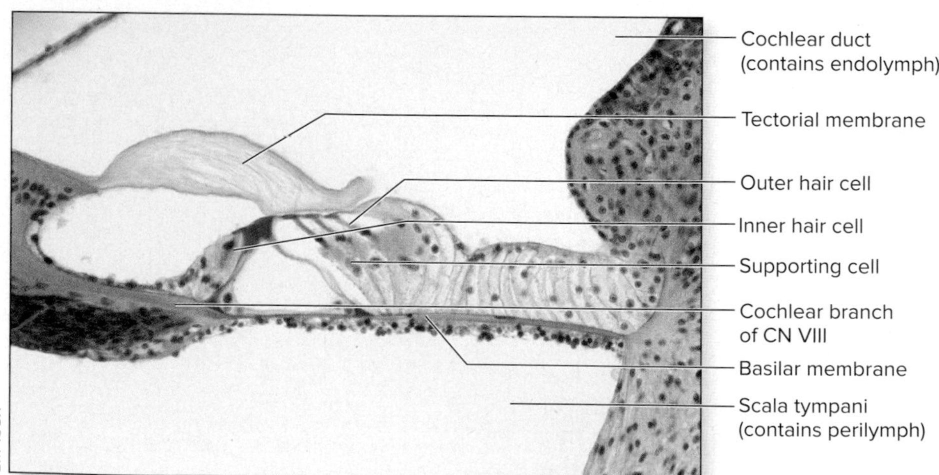

(d) Spiral organ

Figure 16.26 Structure of the Cochlea and Spiral Organ. The cochlea exhibits a snail-like, spiral shape and is composed of three fluid-filled ducts. (*a*) A section through the cochlea details the relationship among the three ducts: the cochlear duct, scala vestibuli, and scala tympani. (*b*) A magnified view of the cochlea. (*c*) Hair cells rest on the basilar membrane of the spiral organ within the cochlear duct. (*d*) Light micrograph of the spiral organ. AP|R

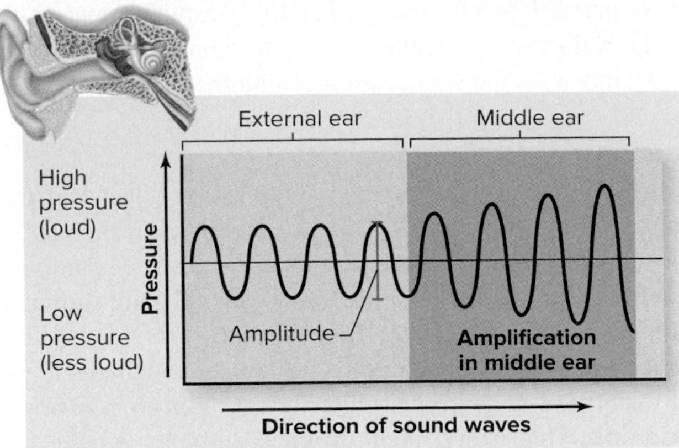

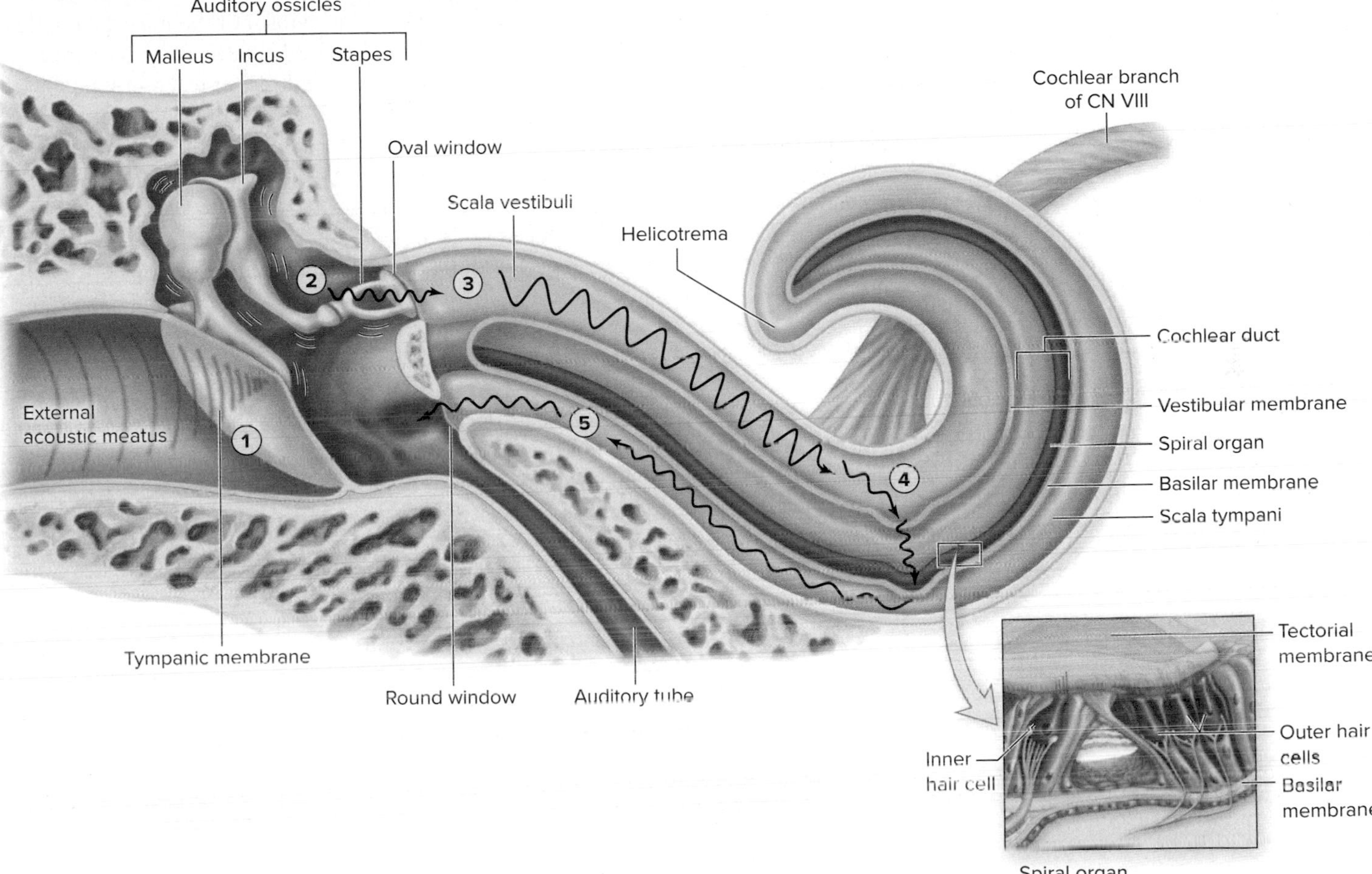

(1) Sound waves are directed by the auricle into the external acoustic meatus, causing the tympanic membrane to vibrate.

(2) Tympanic membrane vibration moves auditory ossicles (malleus, incus, and stapes); sound waves are amplified.

(3) The stapes at the oval window generates pressure waves in the perilymph within the scala vestibuli.

(4) Pressure waves cause the vestibular membrane to move, resulting in pressure wave formation in the endolymph within the cochlear duct and displacement of a specific region of the basilar membrane. Hair cells in the spiral organ are distorted, initiating nerve signals in the cochlear branch of the vestibulocochlear nerve (CN VIII).

(5) Remaining pressure waves are transferred to the perilymph within the scala tympani and are absorbed as the round window bulges slightly.

Figure 16.27 Sound Wave Pathways Through the Ear. Sound waves enter the external ear and vibrate the tympanic membrane to move the ossicles of the middle ear, which ultimately causes movement of a specific region of the spiral organ within the inner ear. AP|R

the oval window vibrate. Because the tympanic membrane is 20 times greater in diameter than the oval window, sounds transmitted across the middle ear are amplified 20-fold. This is a required step as the energy is transferred from air in the middle ear to fluid within the inner ear.

Vibration of the oval window causes pressure waves in the perilymph within the scala vestibuli. Pressure waves in the scala vestibuli deform the vestibular membrane, resulting in pressure waves in the endolymph within the cochlear duct. Depending upon the specific frequency of the sound waves, localized regions of the basilar membrane move. Hair cells in the spiral organ of this region are distorted, initiating nerve signals in the cochlear branch of the vestibulocochlear nerve (CN VIII, see section 13.9). Simultaneously, the pressure wave vibrations within the cochlear duct are transmitted to the perilymph within the scala tympani, and are absorbed at the round window, as the window bulges slightly.

Cochlear Hair Cell Stimulation

Recall that inner hair cells, which function in converting sound energy into an electrical signal, are components of the spiral organ within the cochlea. These cells reside on the basilar membrane, and the tips of their stereocilia and kinocilium are embedded within the gelatinous tectorial membrane (figures 16.26 and 16.27). The details of the inner hair cells are shown in **figure 16.28**. Note that stereocilia of the inner hair cells are aligned, with each stereocilium progressively taller. Also notice that the tips of each stereocilium contain an ion channel, and that each ion channel is connected to the tip of the adjacent taller stereocilium by a filamentous protein appropriately called a **tip link.** This arrangement of inner hair cells is bathed in endolymph. The endolymph is a fluid high in potassium ion (K^+), contributing to the endolymph's electrical potential of $+80$ mV. In comparison, the cytosol of the inner hair cell has an electrical potential of -40 mV. This electrical difference between the endolymph and cytosol of the inner hair cells that exists when the inner hair cells are at rest represents potential energy (similar to the potential energy in resting neurons; see section 12.7b).

How does this arrangement result in the production of an electrical signal? Motion of the endolymph repeatedly moves the basilar membrane upward and downward. When the basilar membrane moves upward, the inner hair cells are pushed into the tectorial membrane, causing each stereocilium to tilt toward its taller neighboring stereocilium. This pulls on the tip link, and the movement of each tip link opens an ion channel on its shorter neighboring stereocilium. Potassium ion (primarily) diffuses from the endolymph through the open ion channels into the inner hair cell. This movement of K^+ causes depolarization (becoming more positive) of the inner hair cell, which triggers the release of neurotransmitter from the base of the inner hair cell. Neurotransmitter binds to dendrites of the primary neuron, causing graded potentials to be established, which move toward the axon. When the threshold is reached, an action potential is initiated along the axon of the primary neuron to the brain. In comparison, the downward movement of the basilar membrane pulls the inner hair cells away from the tectorial membrane, causing the inner cells to straighten. The pull by the tip links decreases, and the ion channels close. The inner hair cells temporarily hyperpolarize (become more negative), and neurotransmitter is no longer released. Amazingly, this upward and downward movement of the basilar membrane can occur at a rate of up to 20,000 times per second.

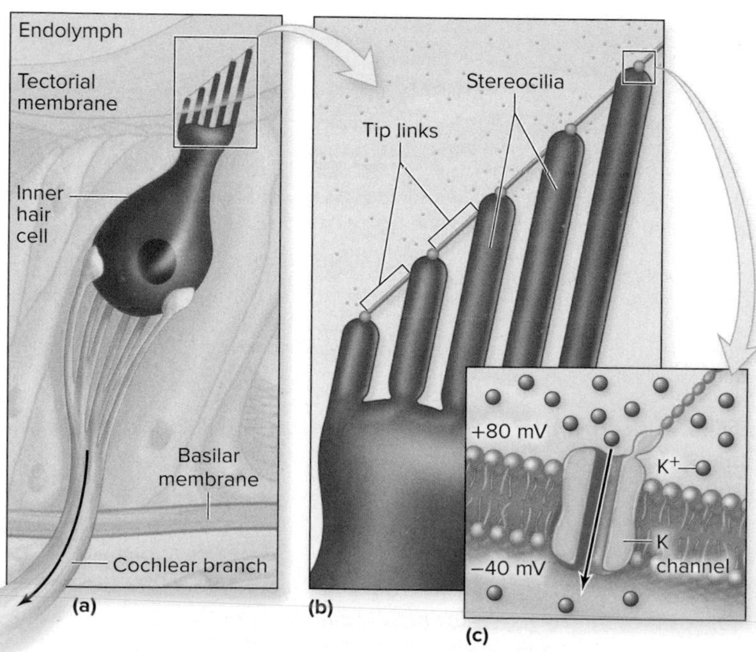

Figure 16.28 Inner Hair Cells. (*a*) The inner hair cells contain ion channels at the tips of their stereocilia. (*b*) The ion channel is attached to its taller neighboring stereocilia by a tip link protein. (*c*) Open ion channels allow K^+ to move into the inner hair cell.

Perception of Sound

Sound is the perception of pressure waves that are established by a vibrating object (e.g., a drum, guitar string, vocal cord). These waves of pressure move through any medium, including air, liquid, or a solid. The vibrating object pushes molecules within the medium. Those molecules push adjacent molecules, which in turn push adjacent molecules. However, no one molecule moves very far—it simply transfers the energy of its movement to other molecules. (This would be similar to a line of individuals, with each person pushing the next.)

Two properties of sound that we perceive are pitch and loudness. **Pitch** refers to the perception of a sound as high or low. Pitch is ultimately dependent upon the frequency of the vibrating object. **Frequency** is the rate of back-and-forth motion of the vibrating object, and is measured in cycles per second and expressed in *Hertz* (*Hz*). The spiral organ of the human ear can perceive sounds with frequencies that range from about 20 Hz to 20,000 Hz, but we are most sensitive to sounds with frequencies between 1500 and 4000 Hz. We are able to perceive variations in pitch because there is a continuous difference in relative "stiffness" of the basilar membrane, as it extends from being thick and short at the oval window to thin and long at the cochlear apex. Different regions of the basilar membrane move in response to the frequency of the sound waves. High-frequency sounds cause movement of the basilar membrane near the oval window (where the basilar membrane is relatively stiff), whereas low-frequency sounds move the basilar membrane far away from the oval window (where the basilar membrane is relatively flexible) (**figure 16.29**). (This is analogous to the difference in pitch of piano strings from one side of the piano to the other.) We cannot perceive frequencies either lower or higher than this because the basilar membrane does not move in response to these rates of vibration.

Loudness (or *intensity* of sound) is dependent upon the amount of back-and-forth motion of the vibrating object that establishes the degree of compression of the molecules (or the amplitude of the sound waves). Soft sounds cause relatively small movements of the basilar membrane in a relatively smaller area of the spiral organ, whereas loud sounds cause relatively larger movement of the basilar membrane in a wider area of the spiral organ. The greater movement of the basilar membrane associated with louder sounds increases both the rate of nerve signals that are initiated by the inner hair cells and the number of hair cells that are stimulated. The auditory cortex within the temporal lobe of the brain interprets this change in sensory input as a louder sound. The loudness of a sound is measured in decibels (dB). Zero decibels is the minimum sound (or threshold) that is heard by humans. Consider that the energy associated with sound increases ten times with every 10-decibel increase. This would mean that a sound of 20 dB has 10 times the energy of a sound with 10 dB, and a sound of 30 dB has 100 times the energy of a sound of 10 dB. A normal conversation usually occurs at about 60 dB, and prolonged

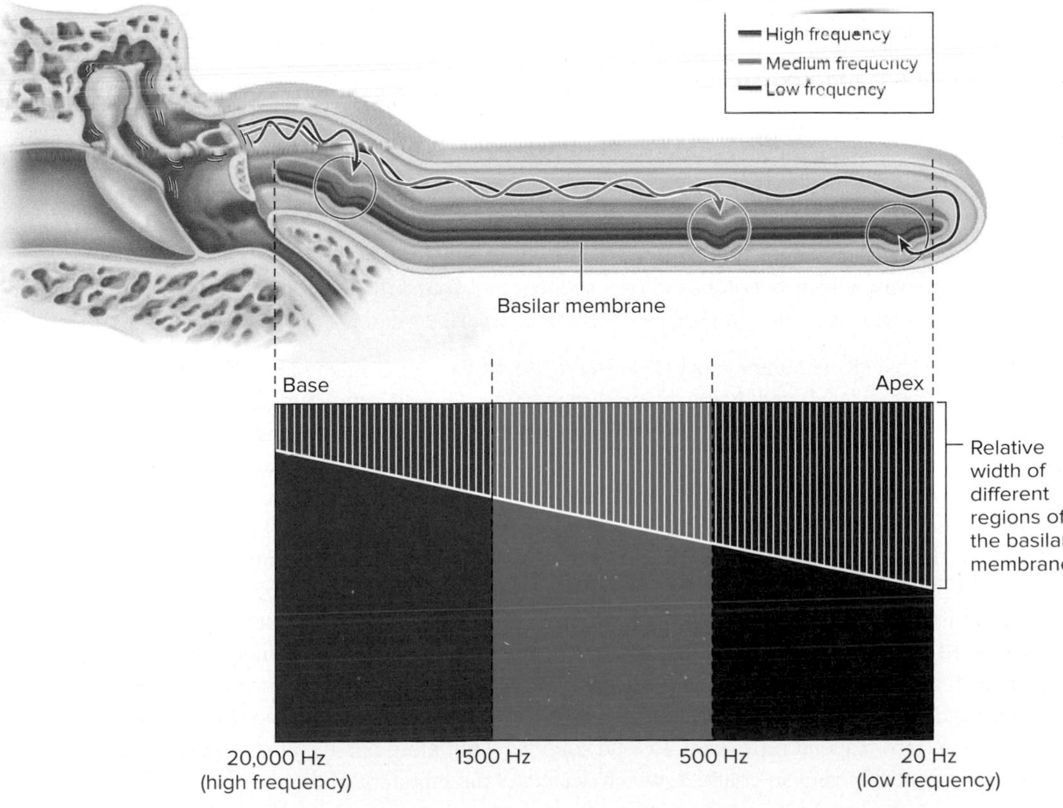

Figure 16.29 Sound Wave Interpretation at the Basilar Membrane. Sound waves are interpreted at specific sites along the basilar membrane of the spiral organ (which is shown here straightened from its normal spiral shape). High-frequency sounds (red arrow) generate pressure waves that cause the basilar membrane to displace close to the base of the cochlea. Medium-frequency sounds (green arrow) generate pressure waves that cause the basilar membrane to displace near the center of the cochlea. Low-frequency sounds (blue arrow) generate pressure waves that cause the basilar membrane to displace near the helicotrema.

INTEGRATE

CLINICAL VIEW 16.10

Cochlear Implants

A **cochlear implant** is an electronic device that assists some hearing-impaired people. It is not a hearing aid and does not restore normal hearing; rather, it compensates for damaged or nonfunctioning parts of the inner ear. The components of a cochlear implant include (1) an external microphone to detect sound (typically worn behind one ear), (2) a speech processor to arrange the sounds from the microphone, and (3) a transmitter connected to a receiver/stimulator (placed within the cochlea) to convert the processed sound into electrical impulses.

A cochlear implant parallels normal hearing by selecting sounds, processing them into electrical signals, and then sending sound information to the brain for interpretation. However, this type of hearing is considerably different from normal hearing. Patients with implants report that voices sound squeaky and high-pitched, but this limitation does not prevent people with implants from having oral communication capabilities.

Although tens of thousands of people have received cochlear implants, some controversy accompanies their increasing use. First, any surgical installation procedure is always accompanied by a small risk for infection or complication. Second, cochlear implants do not always improve hearing quality. Cochlear implant opponents also have expressed concern that the implant represents a threat to the unique, sign-based culture of deaf people. Others question the use of implants in children, who may not have made the decision for themselves.

1 Antenna, transmitter, and receiver are inserted in skin posterior and superior to the auricle. The transmitter lead is inserted into the inner ear.

Transmitter lead
Spiral organ
Cochlea cutaway
Transmitter
Receiver
Antenna
Transmitter lead

2 Sound waves are detected by the receiver and turned into electrical signals, which travel through the transmitter.

3 Electrical signals from the transmitter stimulate the cochlear nerve, which then transmits nerve signals to the brain.

Cochlear implant.

exposure to sounds of 90 dB or greater has sufficient energy to damage the sensory receptors that detect sound within the inner ear.

WHAT DID YOU LEARN?

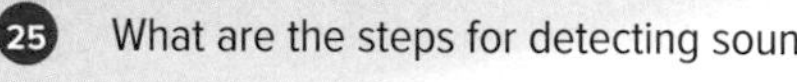

25 What are the steps for detecting sounds?

26 Compare the difference in how we perceive pitch versus how we perceive loudness.

16.5c Auditory Pathways

LEARNING OBJECTIVE

29. Describe the auditory pathway from stimulation of the vestibulocochlear nerve (CN VIII) to the brain.

We discussed how sound waves travel through the inner ear and how an electrical signal is initiated by the inner hair cells within the cochlear duct in the previous section. But what is the nerve pathway on which the initiated nerve signals are transmitted to the brain? This **auditory pathway,** which is composed of a sequence of four sensory neurons instead of the normal two or three, is shown in **figure 16.30** and described here:

1. When the basilar membrane moves, the stereocilia of the spiral organ hair cells distort because they are anchored by the tectorial membrane. This distortion initiates nerve signals that are transmitted through the cochlear branch of the vestibulocochlear nerve (the primary neurons in this sensory pathway) to the **cochlear nucleus** within the medulla oblongata.
2. After integration and processing of the incoming information within the cochlear nucleus, nerve signals are transmitted along secondary neurons to the inferior colliculus in the midbrain (see section 13.5a). These nerve signals are relayed either (a) directly to the inferior colliculus or (b) first to the superior olivary nucleus in the pons before transmission to the inferior colliculus. Nerve signals arriving at the **inferior colliculus** are involved in a reflex to loud sounds; here nerve signals are relayed along motor neurons to skeletal muscles of the body that cause us to jump and turn our head in response to loud sounds. Nerve signals arriving at the **superior olivary nucleus** function to (a) localize the sound and (b) respond reflexively to loud sounds by initiating nerve signals to the tensor tympani and stapedius to contract, which decreases the vibration of the ossicles.

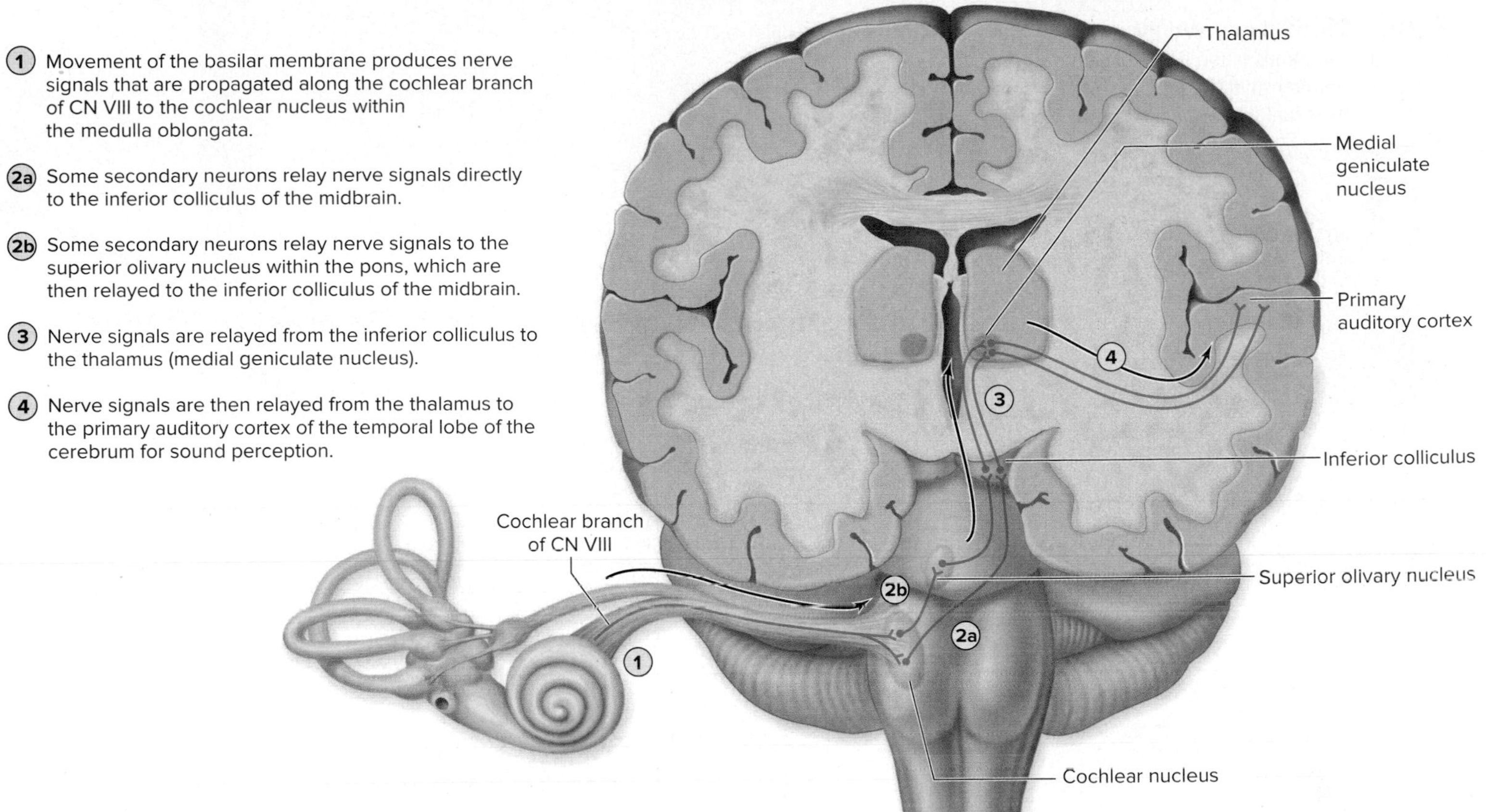

(1) Movement of the basilar membrane produces nerve signals that are propagated along the cochlear branch of CN VIII to the cochlear nucleus within the medulla oblongata.

(2a) Some secondary neurons relay nerve signals directly to the inferior colliculus of the midbrain.

(2b) Some secondary neurons relay nerve signals to the superior olivary nucleus within the pons, which are then relayed to the inferior colliculus of the midbrain.

(3) Nerve signals are relayed from the inferior colliculus to the thalamus (medial geniculate nucleus).

(4) Nerve signals are then relayed from the thalamus to the primary auditory cortex of the temporal lobe of the cerebrum for sound perception.

Thalamus
Medial geniculate nucleus
Primary auditory cortex
Inferior colliculus
Cochlear branch of CN VIII
Superior olivary nucleus
Cochlear nucleus

Figure 16.30 Central Nervous System Pathways for Hearing. Nerve signals are propagated along the cochlear branch of CN VIII to the brainstem, then relayed to the thalamus before being transmitted to the primary auditory cortex. Note: Although not shown in the figure, some of the axons decussate to the other side of the brain.

INTEGRATE

CLINICAL VIEW 16.11

Deafness

Deafness is defined as any hearing loss. Hearing loss is categorized as either conductive deafness or sensorineural deafness. **Conductive deafness** involves any interference with the transmission of sound waves from the auricle through the external acoustic meatus, tympanic membrane, and ossicles. Causes of conductive deafness include rupture of the tympanic membrane, fusion of the ossicles, and inflammation of the middle ear (otitis media). **Sensorineural deafness** involves malformation of or damage to either the structures of the inner ear or the cochlear nerve. Examples include damage to cilia of the inner ear from loud sounds (e.g., rock concerts, firing a weapon) and trauma to the cochlear nerve from a blow to the head.

3. Nerve signals are transmitted from the inferior colliculus to the **medial geniculate nucleus** of the thalamus (see section 13.4b) for initial processing and filtering of auditory sensory information.
4. Nerve signals are then relayed from the thalamus to the **primary auditory cortex** within the temporal lobe of the cerebrum, where nerve signals are consciously perceived as sounds.

To integrate what you've learned about how sounds are processed and interpreted by the brain, refer to **figure 16.31** to review the anatomy and physiology of hearing.

WHAT DID YOU LEARN?

27 What are the major brain structures involved in the auditory pathway, and what is the function of each?

INTEGRATE CONCEPT OVERVIEW

Figure 16.31 How We Hear. Sound waves enter the external ear and then (*a*) are transmitted to the middle ear. (*b*) Sound is amplified within the middle ear and transmitted to the inner ear, where (*c*) sound energy is transduced to nerve signals that (*d*) are transmitted along the auditory pathway to the brain.

Primary auditory cortex

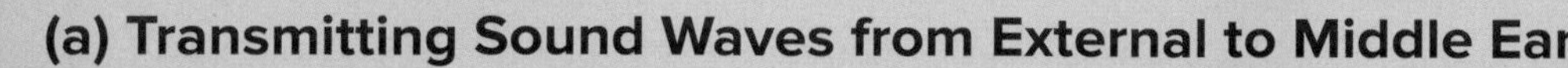

(a) Transmitting Sound Waves from External to Middle Ear

(1) Sound waves are directed to the external ear.

(2) Sound waves are funneled into the external acoustic meatus and vibrate the tympanic membrane.

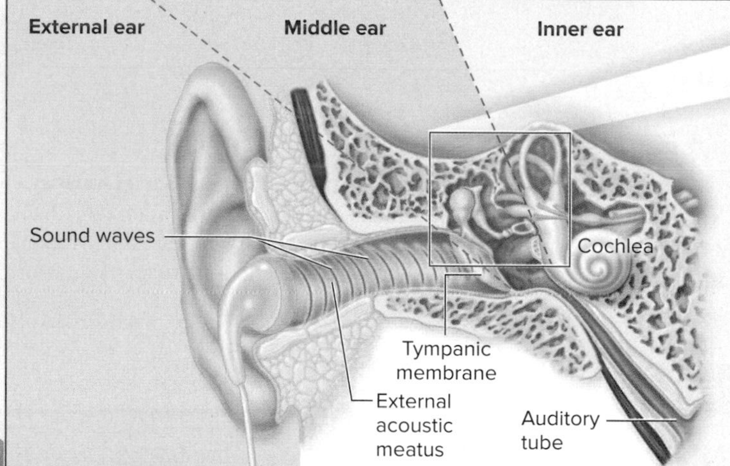

(d) Auditory Pathway

(8) Nerve signals are transmitted along the cochlear branch of CN VIII to the brainstem where nuclei that control reflexive responses to sound are located. Nerve signals are then relayed through the medial geniculate nucleus of the thalamus, and ultimately go to the primary auditory cortex of the temporal lobe, where they are perceived as sounds.

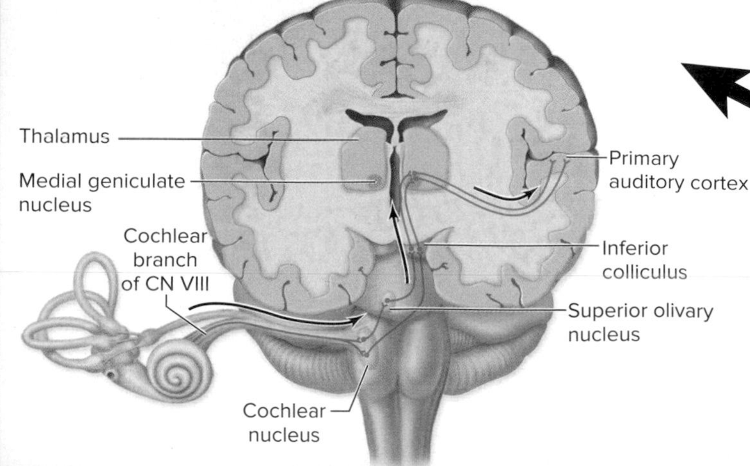

(b) Amplification and Transmission of Sound from Middle to Inner Ear

3 The tympanic membrane vibration causes the auditory ossicles to vibrate, and they amplify the sound.

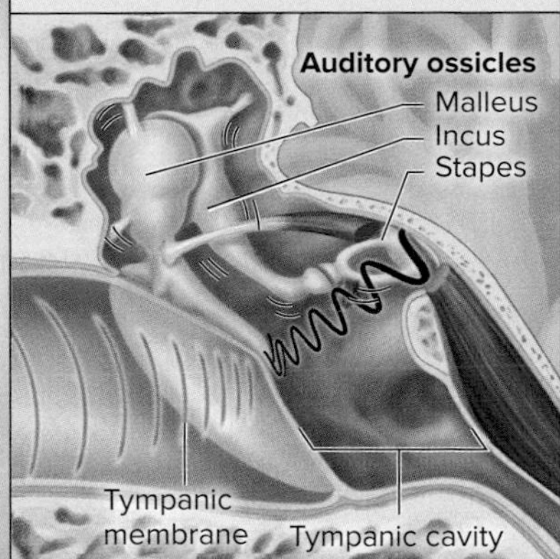

4 The stapes moves the oval window, generating pressure waves in the perilymph of the scala vestibuli.

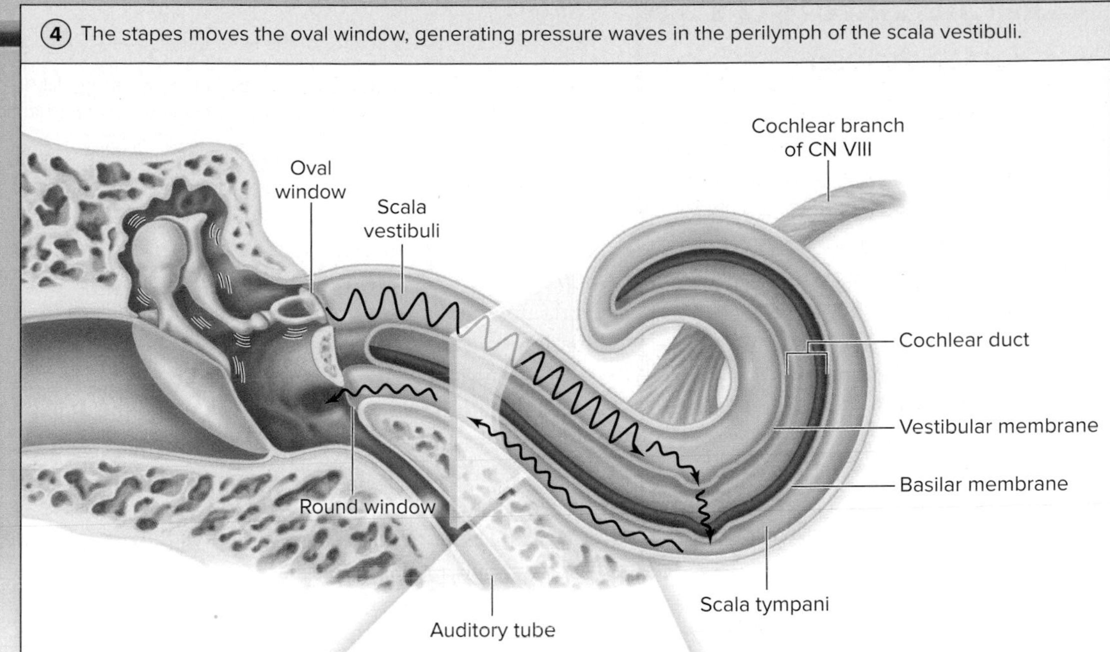

(c) Transduction of Sound Energy to Nerve Signals

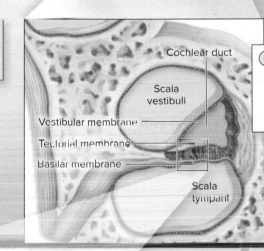

Cochlear duct
Scala vestibuli
Vestibular membrane
Tectorial membrane
Basilar membrane
Scala tympani

5 Pressure waves in the perilymph within the scala vestibuli cause vibration of the vestibular membrane and movement of endolymph within the cochlear duct.

Tectorial membrane
Stereocilia
Inner hair cell
Cochlear branch of CN VIII
Basilar membrane
Nerve signal
Outer hair cell

6 The pressure waves displace specific regions of the basilar membrane, depending upon the frequency of the sounds.

High frequency
Medium frequency
Low frequency
Basilar membrane

7 The displacement of the basilar membrane causes the stereocilia on the hair cells to bend, and nerve signals are propagated along the cochlear branch of CN VIII.

INTEGRATE

CLINICAL VIEW 16.12

Motion Sickness

Motion sickness is a sense of nausea, mild disorientation, and dizziness that some of us have felt while flying in an airplane or riding in an automobile. It develops when a person is subjected to acceleration and directional changes, but there is limited or discrepant visual contact with the outside horizon. In this situation, the vestibular complex of the inner ear is sending nerve signals to the brain that conflict with the visual reference. The eyes inform the brain we are standing still in an airplane or a ship's cabin, but the inner ear is relaying something completely different.

Motion sickness may be alleviated by seeking a place of less movement and reestablishing the visual reference. Some people drink a carbonated beverage or eat soda crackers, although the reason this lessens the symptoms is not clear. Antihistamines are effective in reducing symptoms, and a number of nonprescription oral or transdermal preparations are available as well, including dimenhydrinate (Dramamine), meclizine (Bonine), and cyclizine (Marezine).

16.5d Equilibrium and Head Movement

LEARNING OBJECTIVES

30. Describe the structures of the inner ear involved in equilibrium.

31. Explain how the utricle and saccule detect static equilibrium and linear movements of the head, and explain how the semicircular ducts function to detect rotational movements of the head.

32. Summarize the nerve pathways involved in equilibrium.

The term **equilibrium** refers to our awareness and monitoring of head position. Sensory receptors in the utricle, saccule, and semicircular ducts, collectively called the **vestibular apparatus,** help monitor and adjust our equilibrium. Our brain receives this sensory input and, along with visual sensory input and input from our proprioceptors, integrates this information so we can keep our balance and make positional adjustments as necessary.

The utricle and saccule detect head position during **static equilibrium—**that is, when the head is stationary. For example, when you are standing in the anatomic position, the utricle and saccule inform your brain that your head is upright. The utricle and saccule also detect **linear acceleration** changes of the head. This occurs, for example, when you tilt your head downward to look at your shoes.

The semicircular ducts, in contrast, are responsible for detecting **angular acceleration,** or rotational movements of the head. The sensory receptors in the semicircular ducts are stimulated when you shake your head "no," or when a figure skater does a spin on the ice.

Static Equilibrium and Linear Acceleration

Both static equilibrium and linear acceleration are detected by the sensory receptors housed within the vestibule of the inner ear. The sensory receptor, called a **macula,** is located along the internal wall of both the membranous utricle and saccule. Each macula is composed of a mixed layer of hair cells and supporting cells **(figure 16.32)**. These hair cells not only have stereocilia, but each hair cell also has one long kinocilium on its apical surface. Stereocilia and kinocilia projecting from the hair cells embed within a gelatinous mass that completely covers the apical surface of the epithelium. This gelatinous layer is covered by a mass of small calcium carbonate crystals called **otoliths** (ō′tō-lith; *lithos* = stone), or *statoconia.* Together,

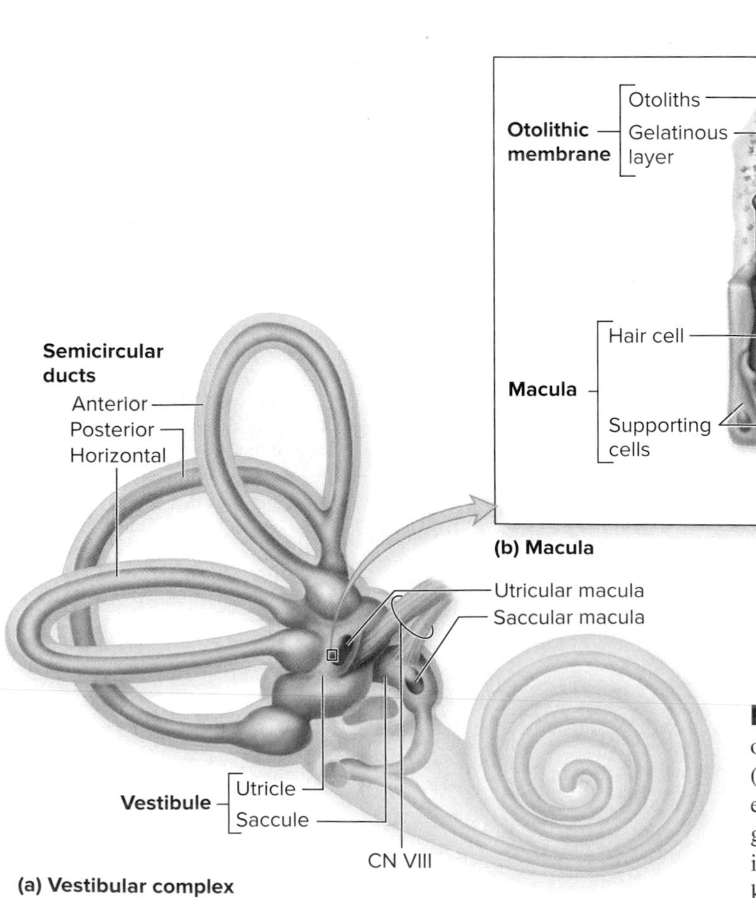

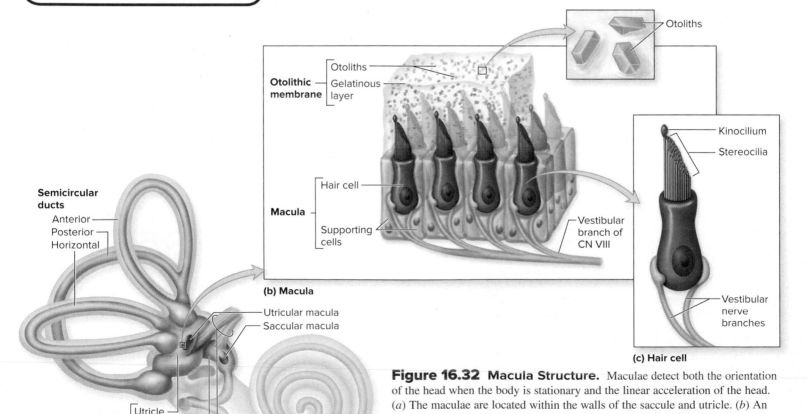

Figure 16.32 Macula Structure. Maculae detect both the orientation of the head when the body is stationary and the linear acceleration of the head. (*a*) The maculae are located within the walls of the saccule and utricle. (*b*) An enlarged view of a macula shows the apical surface of the hair cells covered by a gelatinous layer overlaid with otoliths, called the otolithic membrane. (*c*) An individual hair cell has numerous microvilli called stereocilia and a single long kinocilium.

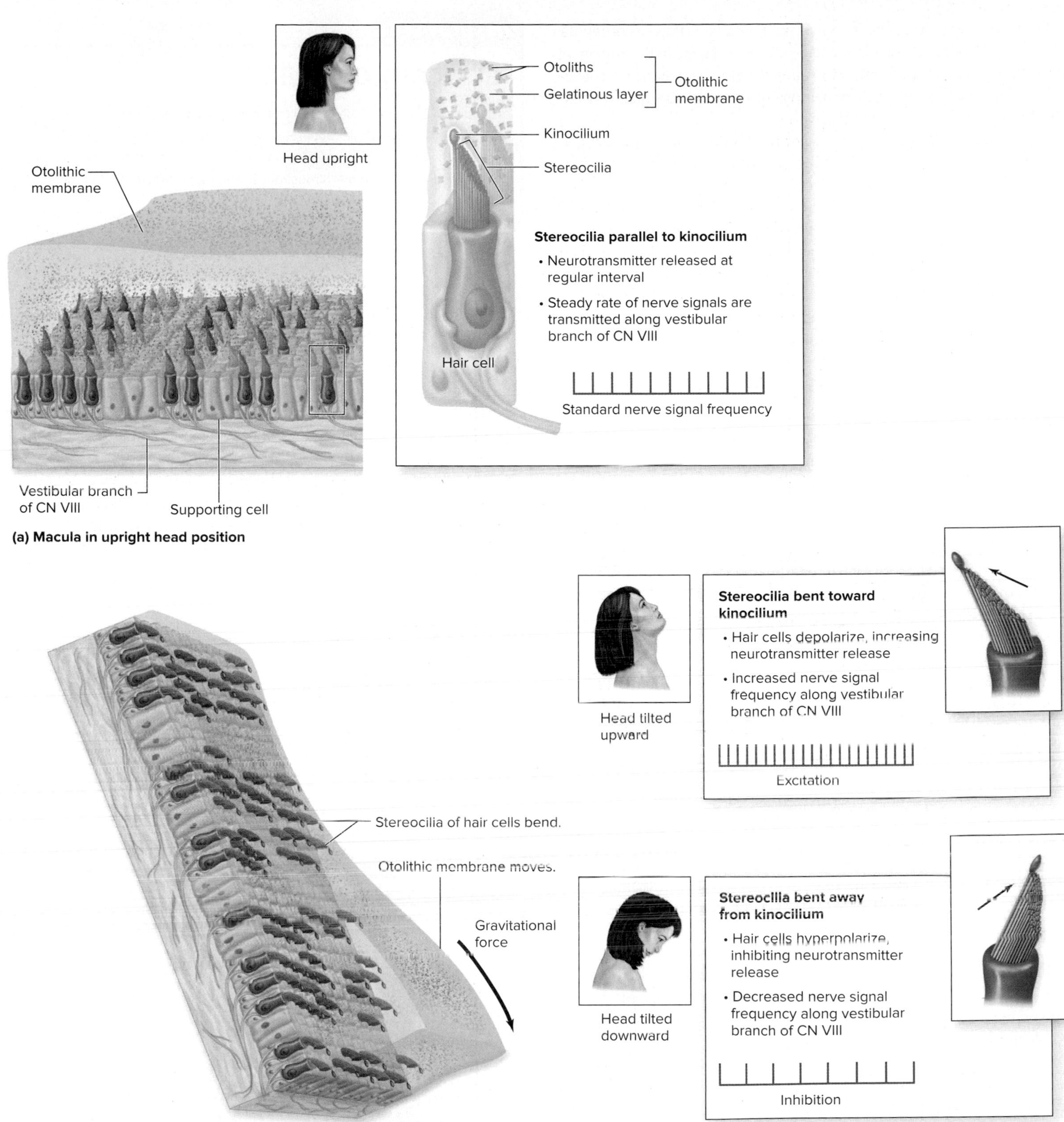

Figure 16.33 How the Maculae Detect Head Position and Linear Acceleration of the Head. (*a*) When the head is in an upright position, hair cells and stereocilia are in parallel and supported by the otolithic membrane. (*b*) Tilting the head causes the otolithic membrane to move slightly, causing the stereocilia to bend and alter the frequency of propagated nerve signals, which reports the change in head position. When the stereocilia are bent toward the kinocilium, the hair cell depolarizes and thus nerve signal propagation increases in frequency. Conversely, when stereocilia are bent away from the kinocilium, the hair cell hyperpolarizes and thus nerve signal propagation decreases in frequency.

the gelatinous layer and the crystals form the **otolithic membrane** (or *statoconic membrane*). The otoliths push on the underlying gelatinous layer, thereby increasing the weight of the otolithic membrane covering the hair cells.

The position of the head influences the position of the otolithic membrane (**figure 16.33**). When the head is held erect, the otolithic membrane applies pressure directly onto the hair cells, and minimal stimulation of the hair cells occurs. However, tilting the head causes the otolithic membrane to shift its position on the macula surface, thus distorting the stereocilia. Bending of the stereocilia results in a change in the amount of neurotransmitter released from the hair cells and a simultaneous change in the stimulation of the sensory neurons of the vestibular branch of the vestibulocochlear nerve (CN VIII).

Bending of the stereocilia toward the kinocilium causes the hair cells to depolarize and increase their rate of neurotransmitter release (figure 16.33*b*). A greater frequency (rate) of nerve signals is generated in the vestibular branches of vestibulocochlear nerves (CN VIII) as a result. In contrast, bending of the stereocilia away from the kinocilium causes the hair cells to hyperpolarize and thus decrease their rate of neurotransmitter release. Ultimately, fewer and less frequent nerve signals are generated in the axons of the vestibular branch as time elapses. The brain interprets the change in nerve signals to determine the direction the head has tilted.

Angular Acceleration

Angular acceleration is detected by the sensory receptors housed within the semicircular ducts of the semicircular canals. At the base of each semicircular canal is an enlarged region called the **ampulla** (am-pul′lă; pl., *ampullae,* am-pul′lē; bottle-shaped) **(figure 16.34)**. The ampulla contains an elevated region, called the **crista ampullaris** (or *ampullary crest*), which is covered by an epithelium of hair cells and supporting cells. Both the stereocilia and kinocilia are embedded into an overlying gelatinous dome called the **cupula** (kū′pū-lă).

The crista ampullaris in each semicircular duct detects rotational movements of the head **(figure 16.35)**. When the head first rotates, inertia causes endolymph to lag behind. This endolymph pushes against the cupula, causing bending of the stereocilia. Stereocilia bending results in altered neurotransmitter release from the hair cells and simultaneous stimulation of the sensory neurons. As previously described for sensory receptors within the vestibule, bending of the stereocilia in the direction of the kinocilium results in depolarization of the hair cells and increased frequency of nerve signals, whereas bending in the opposite direction results in hyperpolarization of the hair cells and decreased frequency of nerve signals (see figure 16.33*b*). Interestingly, these sensory receptors respond primarily to changes in velocity—meaning they respond to rotational acceleration or deceleration. If the head is rotating at a constant speed, eventually the movement of the endolymph catches up with the ampulla, so the stereocilia on the hair cells are no longer bent and the hair cells are no longer stimulated.

You may have done the following as a child: You stood upright, closed your eyes, and then spun in a clockwise (or counterclockwise) direction for about 1 minute. Initially, you felt your head spinning, but after about 30 seconds that spinning sensation may have disappeared. The reason is that the endolymph movement caught up with the ampulla's rotational movement. When you stopped spinning, your head may still have felt like it was moving even though you were standing still. When you came to a stop (and decelerated), the endolymph's momentum resulted in endolymph movement after the body stopped.

INTEGRATE

CONCEPT CONNECTION

Multiple systems are involved in maintaining balance and equilibrium of the entire body. The vestibular apparatus detects motion of the head, our eyes provide the brain with visual information about body position (see section 16.4e), and proprioceptors throughout the musculoskeletal system detect muscle and tendon tension (see section 14.6d).

Equilibrium Pathways

The nerve pathway that relays sensory input from the vestibular apparatus is shown in **figure 16.36** and described here:

1. When the stereocilia of either the maculae (within the utricle and saccule housed within the vestibule) or the crista ampullaris (within the ampulla of a semicircular duct) distort, nerve signals are initiated through the vestibular branch of the vestibulocochlear nerve (CN VIII).
2. The sensory axons of the vestibular branch (the primary neurons in this sensory pathway) terminate at either the medulla oblongata (vestibular nuclei) or cerebellum. (a) The paired **vestibular nuclei**

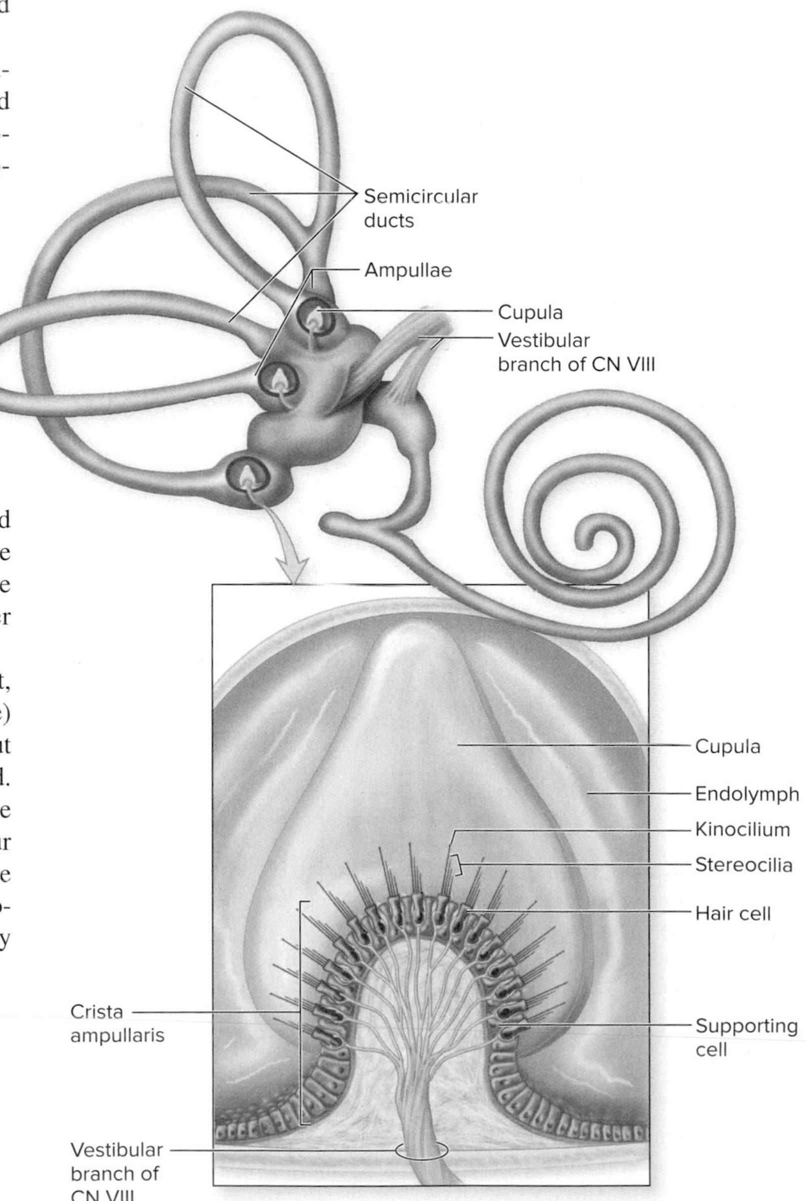

Figure 16.34 Ampulla. A diagrammatic section through an ampulla in a semicircular duct details the relationships among the hair cells and supporting cells, the cupula, and the endolymph in detecting rotational movement of the head.

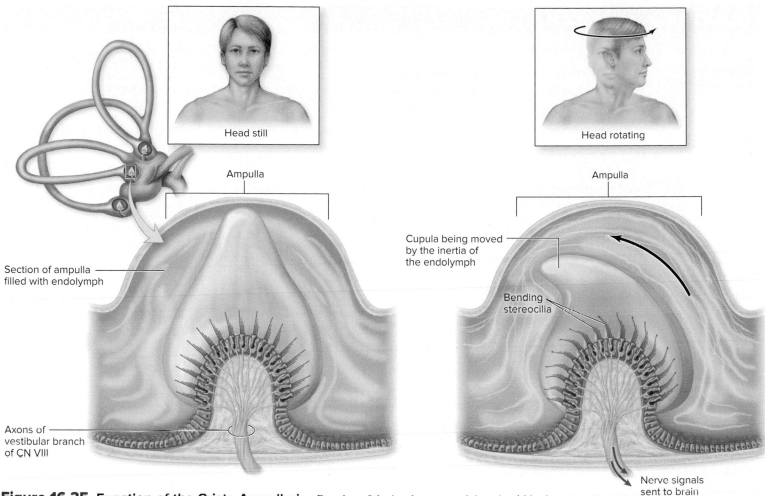

Figure 16.35 Function of the Crista Ampullaris. Rotation of the head causes endolymph within the semicircular duct to push against the cupula covering the hair cells, resulting in bending of their stereocilia and an alteration in the frequency of nerve signal propagation.

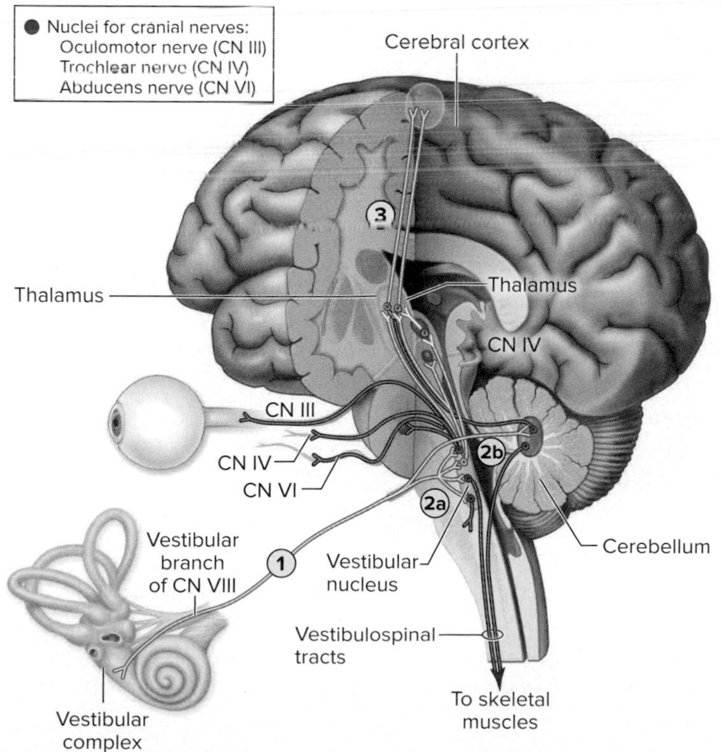

Figure 16.36 Equilibrium Pathways. Information from the vestibular complex is sent to multiple parts of the brain, so posture and body movement may be adjusted accordingly. Sensory information pathways are shown in blue, and motor pathways are shown in red.

within the medulla oblongata integrate the stimuli from the vestibular apparatus (as well as the eyes and proprioceptors) to reflexively control movements of the eye (via the oculomotor [CN III], trochlear [CN IV], and abducens [CN VI] cranial nerves) and to control skeletal muscle contraction for maintaining balance (via the descending vestibulospinal motor pathway; see section 14.4c). (b) The cerebellum integrates the sensory input from the vestibular apparatus (as well as input from the eyes, proprioceptors, and cerebrum) to coordinate skeletal muscle to maintain balance and muscle tone (via the descending vestibulospinal motor pathways).

3. Nerve signals are also sent from the vestibular nuclei and cerebellum to the thalamus and eventually the cerebral cortex for further processing and for our awareness of body position.

WHAT DID YOU LEARN?

28 What type of movement do the maculae detect, and how do they detect this movement?

29 What type of movement do the ampullae detect, and how do they detect this movement?

CHAPTER SUMMARY

	• Sensory receptors provide us with information about our external and internal environments.
16.1 Introduction to Sensory Receptors	**16.1a General Function of Sensory Receptors** • Sensory receptors function as transducers to change stimulus energy to an electrical signal that is transmitted to the CNS.
	16.1b General Structure of Sensory Receptors • Sensory receptors range in complexity from the bare dendritic ending of a single sensory neuron to complex sense organs.
	16.1c Sensory Information Provided by Sensory Receptors • A sensation is a stimulus that is consciously perceived by the cerebral cortex. Most stimuli are not consciously perceived. • A sensory receptor provides the CNS with several characteristics regarding a stimulus, including its modality, location, intensity, and duration.
	16.1d Sensory Receptor Classification • Sensory receptors are defined and characterized by receptor distribution (general senses and special senses); stimulus origin (exteroceptors, interoceptors, and proprioceptors); and modality of stimulus (chemoreceptors, thermoreceptors, photoreceptors, mechanoreceptors, and nociceptors).
16.2 The General Senses	**16.2a Tactile Receptors** • Tactile receptors are the most numerous type of sensory receptors and are organized into unencapsulated and encapsulated receptors. • Unencapsulated receptors include free nerve endings, root hair plexuses, and tactile discs. • Encapsulated receptors include end bulbs, lamellated corpuscles, bulbous corpuscles, and tactile corpuscles.
	16.2b Referred Pain • Referred pain is pain that is perceived as if it originates in the skin and skeletal muscle but actually is caused by sensory nerve signals originating from an internal organ.
16.3 Olfaction and Gustation	• Both olfactory and gustatory receptors are classified as special senses, exteroceptors, and chemoreceptors.
	16.3a Olfaction: The Sense of Smell • The olfactory epithelium is the sensory receptor for the sense of smell and is located in the nasal cavity; it is composed of olfactory receptor cells, supporting cells, and basal cells. • The sense of smell is transmitted by olfactory nerves to olfactory tracts, which project to the hypothalamus, amygdala, and directly to the olfactory cortex without first synapsing in the thalamus.
	16.3b Gustation: The Sense of Taste • Taste buds house the sensory receptors for the sense of taste, which are located primarily on the tongue; they are composed of gustatory cells, supporting cells, and basal cells. • The five basic taste sensations are sweet, salt, sour, bitter, and umami. • The sense of taste is transmitted by both the facial nerves (CN VII) and glossopharyngeal nerve (CN IX) through the gustatory pathway.
16.4 Visual Receptors	• The eyes house photoreceptors that detect light, color, and movement.
	16.4a Accessory Structures of the Eye • Accessory structures include the extrinsic eye muscles, eyebrows, eyelashes, eyelids, conjunctiva, and lacrimal glands. • The conjunctiva is a membrane that covers the anterior surface of the sclera of the eye (the ocular conjunctiva) and the internal surface of the eyelid (the palpebral conjunctiva); it does not cover the cornea. • Each lacrimal apparatus includes a lacrimal gland that distributes lacrimal fluid to the eye surface. Lacrimal fluid is collected into the lacrimal puncta, which drain into the nasolacrimal duct and then the nasal cavity.
	16.4b Eye Structure • The eye is divided into two cavities by the lens: the posterior cavity, which contains the permanent, gelatinous vitreous humor, and the anterior cavity, which contains the circulating, watery aqueous humor. • The fibrous tunic is composed of the sclera, which helps protect and maintain the shape of the eye, and the cornea, which helps focus light on the retina. • The vascular tunic of the eye wall has three regions: the choroid, which contains blood vessels to nourish the retina; the ciliary body, which assists lens shape changes with the suspensory ligaments and secretes aqueous humor; and the iris, which controls pupil diameter. • The neural tunic, or retina, is composed of an outer pigmented layer (that absorbs stray light rays) and an inner neural layer that houses all of the photoreceptors and their associated neurons. • The retina has a posterior, yellowish region called the macula lutea. Vision is sharpest at a depression within the center of the macula lutea called the fovea centralis.

16.4c Physiology of Vision: Refraction and Focusing of Light

- Refraction is the bending of light rays—a process that occurs primarily at the cornea and lens so that light hits the retina.
- Focusing of light when viewing objects closer than 20 feet involves convergence of the eyes, accommodation of the lens, and constriction of the pupil.

16.4d Physiology of Vision: Phototransduction

- Phototransduction involves photoreceptor cells (rods and cones), which are located in the neural layer of the retina, converting light energy into an electrical signal.
- Rods, which are primarily located in the peripheral retina, are activated by dim light and provide no color recognition. The image produced has limited visual sharpness.
- Cones, which are primarily located within the fovea centralis, are activated by bright light and provide color recognition. The image produced has visual sharpness. There are three types of cones, and each type maximally absorbs at different wavelengths of light.

16.4e Visual Pathways

- The optic nerves are formed by ganglionic axons that project from each eye, converge at the optic chiasm, and extend into the brain as optic tracts.
- The optic tracts transmit nerve signals to the thalamus and superior colliculi. The thalamus then transmits visual information to the occipital lobe for conscious interpretation.

16.5 Hearing and Equilibrium Receptors

- Receptors in the ear provide for the senses of hearing and equilibrium.

16.5a Ear Structure

- The external acoustic meatus funnels sound waves to the tympanic membrane, which then directs sound waves to the middle ear.
- The middle ear is an air-filled space occupied by three auditory ossicles. The auditory ossicles (malleus, incus, and stapes) transmit and amplify the sound waves from the tympanic membrane to the oval window.
- The inner ear houses the structures for hearing and equilibrium. Specialized receptors are housed in the membranous labyrinth, which lies within a cavernous space of the temporal bone called the bony labyrinth.

16.5b Hearing

- The spiral organ (organ of hearing) is housed within the cochlea. Hair cells in the spiral organ of the cochlea rest on the basilar membrane and are anchored into the tectorial membrane.
- Sound is perceived when sound waves vibrate the tympanic membrane, resulting in auditory ossicle vibrations that lead to vibrations of the basilar membrane, which bend cilia to initiate nerve signals that are conducted along the cochlear branch of CN VIII.

16.5c Auditory Pathways

- Secondary axons from the cochlear nuclei project directly to the inferior colliculus, or to the superior olivary nuclei and then the inferior colliculus. Axons then project to the thalamus, where thalamic axons project to the primary auditory cortex so sound is consciously perceived.

16.5d Equilibrium and Head Movement

- The maculae within the saccule and utricle detect both the stationary position of the head and linear acceleration of the head.
- The semicircular canals are parts of the bony labyrinth that house the membranous semicircular ducts. Each semicircular duct has an expanded region called an ampulla that houses the crista ampullaris for detecting rotational movements of the head.
- Nerve signals from the vestibular apparatus (sensory receptors within the vestibule and semicircular canals) are transmitted via the vestibular branch of CN VIII either to the vestibular nuclei (in the medulla oblongata) or directly to the cerebellum. Nerve signals are then sent from the vestibular nuclei and cerebellum to the thalamus and eventually to the cerebral cortex for our conscious awareness of head position and movement.

CHALLENGE YOURSELF

Create and Evaluate

Analyze and Apply

Understand and Remember

Do You Know the Basics?

1. Unencapsulated dendritic endings of sensory neurons are called
   a. lamellated corpuscles.
   b. free nerve endings.
   c. bulbous corpuscles.
   d. end bulbs.

2. Each of these sensory receptors is accurately matched with its function *except*
   a. chemoreceptor; detects specific molecules and ions dissolved in fluid
   b. nociceptor; detects pain
   c. proprioceptor; detects change in muscle tension
   d. thermoreceptor; detects change in pressure

3. All of the following are accurate about the conjunctiva *except*
   a. it contains blood vessels that nourish the sclera.
   b. it covers the cornea.
   c. it contains sensory neurons that detect foreign objects on the eye.
   d. it is composed of stratified columnar epithelium.

4. Lacrimal fluid performs all of the following functions *except*
   a. cleansing the eye surface.
   b. preventing bacterial infection.
   c. humidifying the orbit.
   d. moistening the eye surface.

5. The arrangement of tunics in the eye, from the innermost to outermost aspect of the eye, is
   a. retina, vascular, fibrous.
   b. vascular, retina, fibrous.
   c. vascular, fibrous, retina.
   d. retina, fibrous, vascular.

6. When viewing an object closer than 20 feet, all of the following processes occur *except*
   a. the eyes accommodate to focus on the close-up object.
   b. the ciliary body contracts, so the suspensory ligaments become less taut.
   c. the lens becomes more rounded to better refract light rays.
   d. the pupils dilate to let in more light for the retina.

7. Which ear structure is correctly matched with its function?
   a. round window; transmits sound waves into the inner ear
   b. external acoustic meatus; directs sound waves to the tympanic membrane
   c. auditory ossicles; dampen sound waves before they reach the inner ear
   d. vestibular membrane; bends the stereocilia on hair cells to produce a nerve signal

8. Which statement is accurate about the cochlear duct?
   a. It detects linear acceleration of the head when the otolithic membrane bends the hair cells.
   b. It is filled with perilymph.
   c. It contains hair cells that convert sound waves into nerve signals.
   d. It contains a spiral organ that rests on a vestibular membrane.

9. Which of the following is accurate about the maculae of the vestibular apparatus?
   a. They detect rotational movements of the head.
   b. They are located in the semicircular canal.
   c. Nerve signals are generated when the otolithic membrane bends the stereocilia of the hair cells.
   d. They are the organs of hearing.

10. The only sensation to reach the cerebral cortex without first processing through the synapses in the thalamus is
    a. olfaction.
    b. vision.
    c. proprioception.
    d. touch.

11. What are the five classifications of receptors according to modality of stimulus? Give an anatomic example of each.

12. How are visceral nociceptors different from somatic nociceptors, and how do they relate to the phenomenon known as referred pain?

13. What is the pathway by which taste sensations reach the brain?

14. Describe the pathway by which olfactory stimuli are transmitted from the nasal cavity to the brain.

15. What structure in the wall of the eye helps control the amount of light entering the eye?

16. How is the lens able to focus images from a book that you are reading, and then immediately also focus the image of children playing in your backyard?

17. Compare lacrimal fluid and aqueous humor in terms of their formation, circulation, and function.

18. Briefly describe the structural relationship between the membranous labyrinth and the bony labyrinth.

19. Describe the pathway by which sound waves enter the ear and how that sound is converted into a nerve signal.

20. Explain how the vestibule and semicircular canals detect equilibrium.

Can You Apply What You've Learned?

1. You are babysitting a 5-year-old child who does not want to eat his broccoli. He states that the broccoli tastes yucky. You tell him to hold his nose closed while he eats a piece of broccoli and see how it tastes. He says it doesn't taste as yucky when he holds his nose (although he still doesn't want to eat the broccoli). Why did the broccoli taste "less yucky" when the child held his nose?
   a. Cranial nerve IX was pinched when the child held his nose, so tastes could not be perceived as clearly.
   b. The activity of holding his nose took the child's mind off of eating the broccoli.
   c. Olfaction is responsible for a large component of taste, so when the child cannot smell the broccoli, its taste is diminished.
   d. Holding the nose closed facilitated odiferous molecules to enter the mouth, thereby increasing the number of molecules that could be tasted by the taste buds.

2. Horner syndrome is a condition where sympathetic innervation to one side of the head and neck is damaged. What visual disturbances would you expect a person to have if she had Horner syndrome?
   a. inability to accommodate the eyes for near vision
   b. constricted pupil
   c. permanently flattened lens
   d. abduction of the affected eye

3. You may be familiar with cell phone "mosquito tones"—ringtones of a relatively high pitch that most children and teenagers can hear but most adults cannot. Why do you think adults cannot hear these tones as well?
   a. Hair cells close to the oval window of the inner ear become damaged as we age.

b. The tympanic membrane loses its flexibility as we age, so it is not as able to transmit sound waves to the middle ear.
c. Adults likely have damaged the spiral organ near the helicotrema, resulting in their inability to hear the tones.
d. As we age, there is a gradual accumulation of cerumen that blocks the external auditory meatus and makes hearing sounds more difficult.

4. Individuals with macular degeneration may experience loss of vision in their central vision area, but their peripheral vision either is not affected or is not as greatly affected. Why is this the case?
   a. Macular degeneration damages the optic disc, making central vision impaired but leaving peripheral vision intact.
   b. Only a single optic tract typically is damaged, so using your peripheral vision allows you to use the normal optic tract.
   c. Only rods are damaged in macular degeneration, so peripheral vision (where there are fewer rods) is not as affected.
   d. The central vision loss is due to damage to the fovea centralis; the peripheral retina is not greatly affected.

5. An elderly gentleman is taken to the emergency room. He is complaining of pain radiating down his left arm and has shortness of breath. The ER physician suspects the gentleman is suffering a heart attack. Why is the man experiencing pain down his left arm if the problem is with his heart?
   a. Chemoreceptors in the blood vessels of the left arm have detected a change in oxygen levels in the blood and initiated the pain sensations.
   b. The cerebral cortex has mistakenly located the source of the stimulus as being from skin and muscle in the arm instead of the heart.
   c. The patient is experiencing phantom pain in the upper limb.
   d. A heart attack stimulates baroceptors in the blood vessels of the upper limb.

Can You Synthesize What You've Learned?

1. Savannah is an active 3-year-old who began to cough and sniffle. Then she experienced an earache and a marked decrease in hearing acuity. During a physical examination, her pediatrician noted the following signs: elevated temperature, a reddened tympanic membrane with a slight bulge, and some inflammation in the pharynx. What might the pediatrician call this condition, and how would it be treated? How does Savannah's age relate to her illness?
2. After Alejandro quit smoking, he discovered that foods seemed much more flavorful than when he had smoked. How are smoking and taste perceptions linked?
3. George is unclear why nearsighted persons may need bifocal eyeglasses (which have two lenses—one for reading and one for nearsightedness) when they get older. He does not understand why the same set of glasses wouldn't work for both reading and the nearsighted correction. Can you explain to George the difference between the two conditions, and why the treatment for both may appear to be similar?

INTEGRATE

ONLINE STUDY TOOLS

connect | SMARTBOOK® | AP|R

The following study aids may be accessed through Connect.

Clinical Case Study: An Elderly Man Gets His Vision Back

Interactive Questions: This chapter's content is served up in a number of multimedia question formats for student study

SmartBook: Topics and terminology include introduction to sensory receptors; the general senses; olfaction and gustation; visual receptors; hearing and equilibrium receptors

Anatomy & Physiology Revealed: Topics include nasal cavity; tongue; eye; retina; vision; ear; cochlea; hearing

Design Elements: **Learning Objective** icon: ©McGraw-Hill Education; **What Do You Think?** icon: ©McGraw-Hill Education; **What Did You Learn?** icon: ©McGraw-Hill Education; **Learning Strategy** icon: ©Slavoljub Pantelic/Shutterstock RF; **Clinical View** icon: ©Laia Design Studio/Shutterstock RF; **Concept Connection** icon: ©McGraw-Hill Education; **Challenge Yourself** icon: ©McGraw-Hill Education; **Do You Know the Basics?** icon: ©McGraw-Hill Education; **Can You Apply What You've Learned?** icon: ©McGraw-Hill Education; **Can You Synthesize What You've Learned?** icon: ©McGraw-Hill Education; **Online Study Tools** icon: ©ayzek/Shutterstock RF

17.1b Comparison of the Two Control Systems

 **LEARNING OBJECTIVE**

2. Compare and contrast the actions of the endocrine system and the nervous system to control body function.

The nervous system and endocrine system serve as the two complementary systems of control. The methods and effects of the two control systems differ, however. The nervous system exercises control between two specific locations in the body by way of neurons (figure 17.1*d*). Nerve signals trigger the release of neurotransmitter, which crosses the synaptic cleft and binds to another neuron, a muscle cell, or a gland cell to initiate a localized response of the target cell, such as muscle contraction or gland secretion. The neurotransmitter is quickly degraded or taken back into the neuron. The response induced by the nervous system is generally both rapid and short-lived. Notice, however, the two central features in common: (1) In response to stimuli, specialized cells of both systems release chemical substances called **ligands** (hormones or neurotransmitters) to communicate with particular target cells (see section 4.5b) and (2) the ligand binds to a cellular receptor in target cells to initiate a cellular change.

The general characteristics of these two complementary control systems are compared in **table 17.1**.

 **WHAT DID YOU LEARN?**

2 How does the endocrine system differ from the nervous system with respect to their target cells?

17.1c General Functions of the Endocrine System

 **LEARNING OBJECTIVE**

3. Describe the general functions controlled by the endocrine system.

The endocrine system (through the release of hormones) can communicate with any body cell that has a receptor for that hormone. Consequently, its functions are very diverse. We have organized the functions of the endocrine system into the following four broad categories:

- **Regulating development, growth, and metabolism.** Hormones have regulatory roles in both cell division and cell differentiation, which occur during development and growth of the body. They also control our metabolic activities—both anabolic (synthesis) and catabolic (degradation) processes. For example, growth hormone controls growth and the thyroid hormone regulates cellular metabolism.
- **Maintaining homeostasis of blood composition and volume.** Hormones regulate the amount of specific substances dissolved within blood plasma, such as glucose, amino acids, and ions (e.g., Na^+, Ca^{2+}). Additionally, hormones also regulate other characteristics of blood, including its volume, its cellular concentration (erythrocytes and leukocytes), and number of platelets. *Insulin,* for example, regulates blood glucose levels (see section 17.10b).
- **Controlling digestive processes.** Several hormones influence both the secretory processes and the movement of materials through the gastrointestinal tract in our digestive system. An example is *gastrin,* which is released from the stomach and controls digestive processes by increasing stomach contractions and secretions (see section 26.2d).
- **Controlling reproductive activities.** Hormones affect both development and function of the reproductive system as well as expression of sexual behaviors. One example is *prolactin*. Prolactin, a hormone that influences reproductive activities, stimulates formation of breast milk by the mammary glands (see section 28.3f).

 **WHAT DID YOU LEARN?**

 Diabetes mellitus is noted by sustained high blood glucose levels. Which of the four functions listed is the most directly affected?

Table 17.1 Comparing the Endocrine System and the Nervous System

Features	Endocrine System	Nervous System
Communication Method	Secretes hormones; hormones are transported within the blood and distributed to target cells throughout body	A nerve signal causes neurotransmitter release from a neuron into a synaptic cleft
Target of Stimulation	Any cell in the body with a receptor for the hormone	Other neurons, muscle cells, and gland cells
Response Time	Relatively slow reaction time: Seconds to minutes to hours	Rapid reaction time: Typically milliseconds or seconds
Range of Effect	Typically has widespread effects throughout the body	Typically has localized, specific effects in the body
Duration of Response	Long-lasting: Minutes to days to weeks; may continue after stimulus is removed	Short-term: Milliseconds; terminates with removal of stimulus

17.2 Endocrine Glands

This section and the following three sections provide a general overview of four central features of the endocrine system. These include a discussion of (1) the location of the major endocrine glands and how they are stimulated to release their hormones (see section 17.2); (2) the general categories of hormones and their chemical structures (see section 17.3); (3) how hormone molecules are transported within the blood (see section 17.4); and (4) how hormones interact with their target cells (see section 17.5). Here we describe the major endocrine glands.

17.2a Location of the Major Endocrine Glands

 **LEARNING OBJECTIVE**

4. Identify the major endocrine glands and their location within the body.

Endocrine glands are typically composed of a connective tissue framework, which houses and supports epithelial tissue that produces and releases hormones from their secretory cells. The secretory cells are organized in two general ways: either as a single organ with only an endocrine function, or as cells housed in small clusters within organs or tissues that have some other primary function **(figure 17.2)**.

An **endocrine organ** is a single organ that is entirely endocrine in function. Endocrine organs include the pituitary gland, pineal gland, thyroid gland, parathyroid glands, and adrenal glands.

Some endocrine cells are housed in tissue clusters within specific organs or tissues. These cells secrete one or more hormones, but

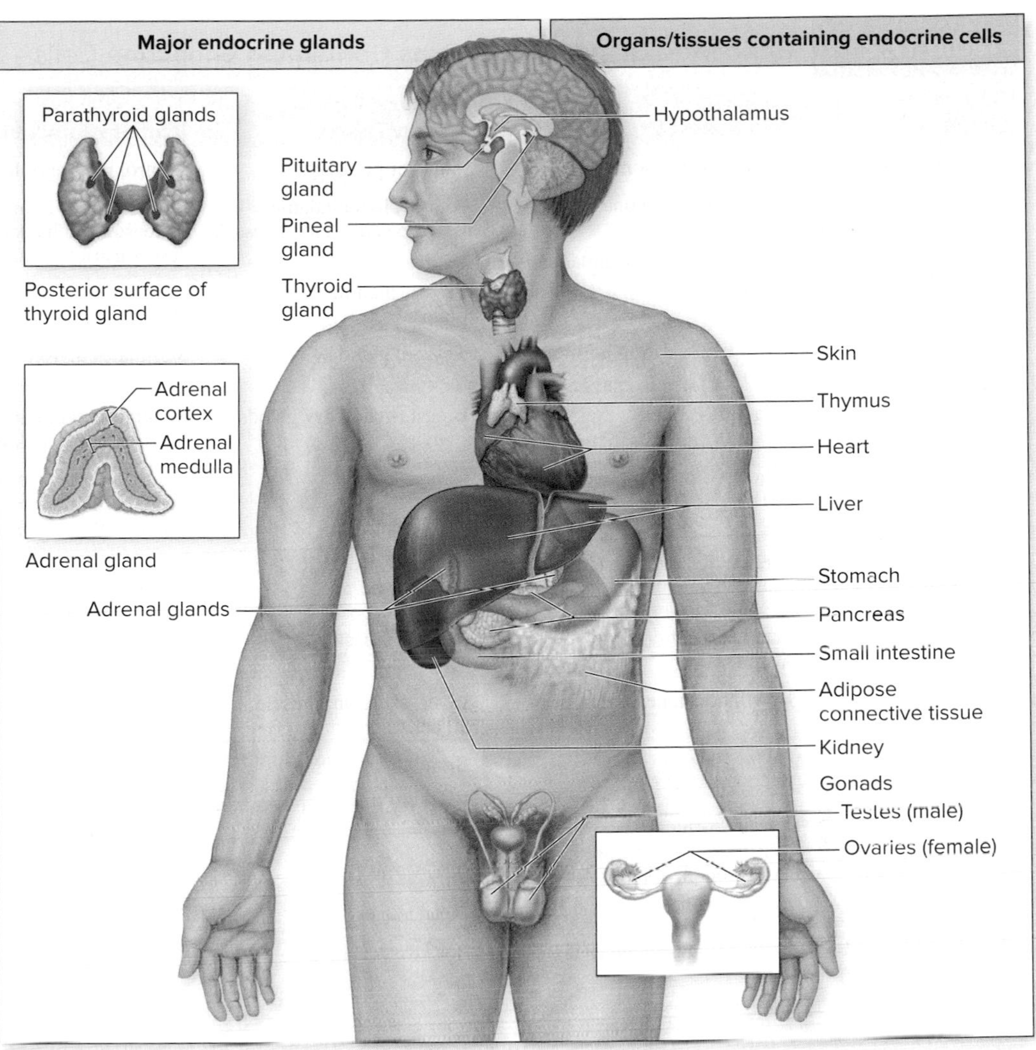

Figure 17.2 Location of the Major Endocrine Glands, Organs, and Tissues Containing Endocrine Cells. AP|R (Note: The placenta is not shown in this figure; see figure 29.7.)

the organs have some other primary function as well. These organs include the hypothalamus, skin, thymus, heart, liver, stomach, pancreas, small intestine, adipose connective tissue, kidneys, and gonads (testes and ovaries).

Note that the term *endocrine gland* will be used throughout the rest of this chapter when referring either to an endocrine organ or an organ or tissue containing endocrine cells. The major endocrine glands of the body are listed in **table 17.2**. This table includes the hormones produced by each gland, the primary target organs or tissue, and the primary function(s) of each hormone.

WHAT DID YOU LEARN?

4 What are the major endocrine organs in the human body? What are the organs (or tissues) that have another primary function and contain endocrine cells?

17.2b Stimulation of Hormone Synthesis and Release

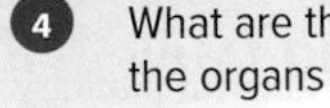

LEARNING OBJECTIVE

5. Explain the three reflex mechanisms for regulating secretion of hormones.

The regulated secretion of a hormone from an endocrine gland is controlled through a reflex (see section 14.6a). Reflexes occur in both the nervous system and the endocrine system. Endocrine reflexes are initiated by one of three types of stimulation: hormonal stimulation, humoral stimulation, or nervous system stimulation (**figure 17.3**).

- **Hormonal stimulation.** The stimulus for the release of many hormones from an endocrine gland is the binding of another hormone. An example occurs when thyroid-stimulating hormone (which is released from the anterior pituitary) binds to the thyroid gland to cause release of thyroid hormone (figure 17.3*a*).
- **Humoral stimulation.** Some endocrine glands are stimulated to release their hormones in response to a changing level of nutrient molecules (e.g., glucose) or ions (e.g., Ca^{2+}) within the blood. (The term *humoral* is a historical term related to blood as one of the four "humors," or fluids, of the body.) When either nutrient or ion levels decrease or increase within the blood, an endocrine gland is stimulated to release its hormone molecules. An example of humoral stimulation occurs when blood glucose increases and the pancreas releases insulin (figure 17.3*b*).
- **Nervous system stimulation.** A few endocrine glands are stimulated to release hormone(s) by direct stimulation from the nervous system. The classic example is the release of epinephrine and norepinephrine by the adrenal medulla in response to stimulation by the sympathetic division of the autonomic nervous system (figure 17.3*c*).

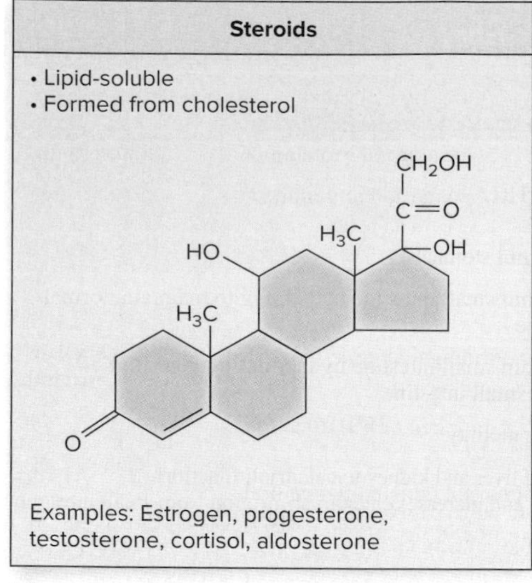

(a)

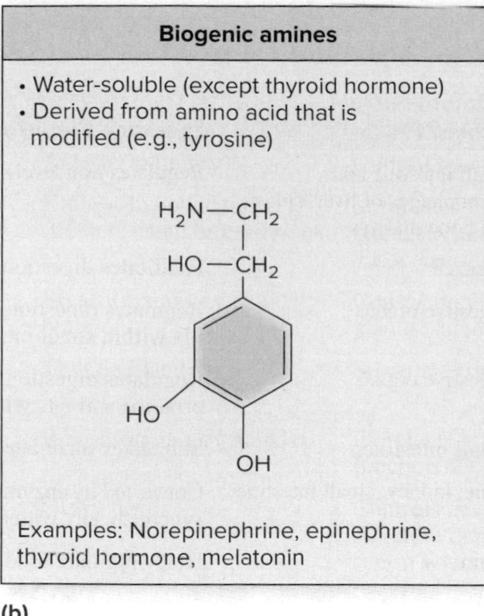

(b)

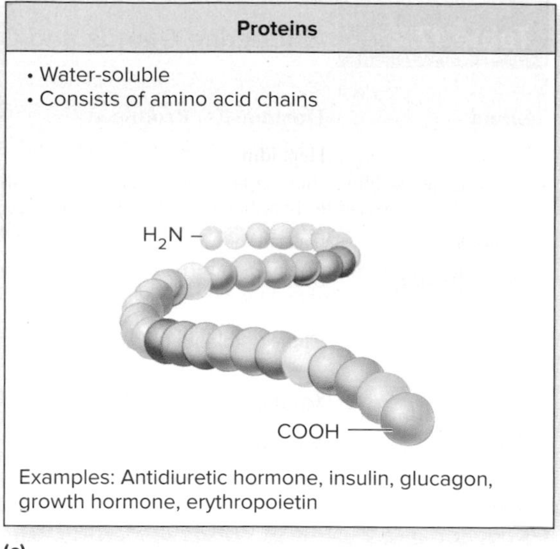

(c)

Figure 17.4 Hormone Types. (*a*) Steroids (which include hormones released from both the gonads and adrenal cortex) are lipid-soluble and formed from cholesterol. In comparison, (*b*) biogenic amines (which include hormones released from the adrenal medulla, thyroid hormone, and melatonin) and (*c*) proteins (which include most hormones) typically are water-soluble and formed from amino acids.

which are the monomers that compose proteins (see section 2.7e). Cholesterol molecules are modified in the synthesis of *steroid hormones.* Amino acids are the building blocks for both *biogenic amines* and *protein hormones.* An example of each category is shown in **figure 17.4**. As you read about the chemical structure of these hormones, please note whether the molecules in each category are lipid-soluble or water-soluble. This difference in solubility influences both the transport of the hormone in the blood and how it interacts with its target cells.

Steroids

Steroids are lipid-soluble molecules synthesized from cholesterol (figure 17.4*a*). This category includes both the steroids produced within the gonads (e.g., estrogen and progesterone in the ovaries and testosterone in the testes) and the hormones synthesized by the adrenal cortex (e.g., corticosteroids such as cortisol, and mineralocorticoids such as aldosterone).

Calcitriol is the hormone produced from vitamin D (see section 7.6a), and it is sometimes classified as a steroid hormone. However, it is more accurately classified as a *sterol* hormone, a slightly different molecule but one that is also lipid-soluble.

Biogenic Amines

Biogenic amines are also called *monoamines.* They are modified amino acids (modification involves removal of a carboxyl functional group from an amino acid) (figure 17.4*b*). Biogenic amines include the catecholamines (e.g., epinephrine and norepinephrine) released from the adrenal medulla, thyroid hormone released from the thyroid gland, and melatonin from the pineal gland. Biogenic amines are water-soluble except for thyroid hormone. Thyroid hormone is lipid-soluble because it is produced from two tyrosine amino acids (see figure 17.16), which are amino acids that contain a nonpolar ring.

Proteins

Most hormones are **proteins.** These are molecules composed of small chains of amino acids and include small peptides, large polypeptides, and glycoproteins (figure 17.4*c*). All hormones in this category are water-soluble.

WHAT DID YOU LEARN?

6. Identify which of the following hormone categories are lipid-soluble: (a) reproductive hormones produced in the gonads, (b) adrenal cortex hormones, and (c) thyroid hormone.
7. What two events or processes associated with a hormone are influenced by whether a hormone is lipid-soluble or water-soluble?

17.3b Local Hormones

LEARNING OBJECTIVES

8. Describe the general structure, formation, and function of local hormones.
9. Compare autocrine and paracrine signaling that occurs through local hormones.

Local hormones are a large group of short-lived signaling molecules that do not circulate within the blood. Instead, cells synthesize and release these molecules, which then bind with either the same cell that produced them (**autocrine stimulation**) or neighboring cells (**paracrine stimulation**). These signaling molecules have properties similar to hormones because the released ligands (signaling molecules) initiate and regulate cellular changes. They just do so "in the tissue neighborhood."

Eicosanoids (ī'kō-să-noydz; *eicosa* = twenty, *eidos* = formed), which were first introduced in section 2.7b (see table 2.4), are the primary type of local hormones. Eicosanoids include prostaglandins, thromboxanes, and leukotrienes **(figure 17.5)**. These signaling

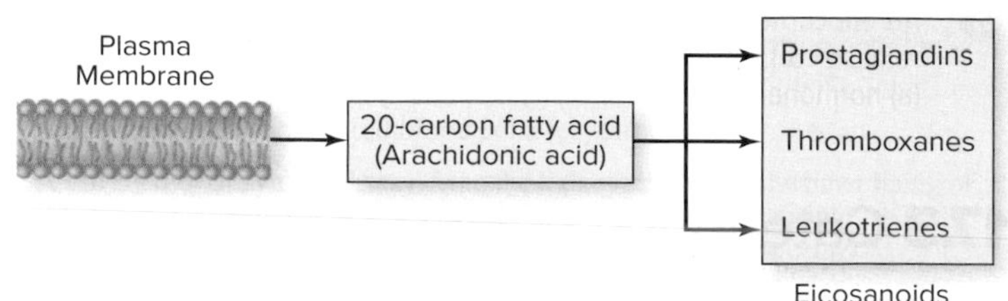

Figure 17.5 Eicosanoid Formation. Eicosanoids include prostaglandins, thromboxane, and leukotrienes. These local hormones are synthesized by an enzymatic cascade that begins with phospholipids from within a plasma membrane.

molecules are formed from *fatty acids,* which are cleaved from phospholipid molecules of a cell's plasma membrane. Their production is not limited to one endocrine gland (as circulating hormones typically are) but instead are synthesized by cells composing many tissues located throughout the body. Thus, eicosanoid synthesis and release provides the means for all tissues to locally regulate cellular responses.

Prostaglandins are the most diverse category of eicosanoids (e.g., prostaglandin D, E, and F with various subtypes such as prostaglandin E_2 [PGE_2]), and their effects are wide-ranging, depending upon both the specific type of prostaglandin molecule and the specific type of cellular receptor to which it binds. Examples of prostaglandin activity include (1) stimulating the hypothalamus to raise body temperature to induce a fever, (2) inhibiting stomach acid secretion, (3) acting on mast cells to release molecules that increase inflammation, and (4) stimulating pain receptors.

WHAT DO YOU THINK?

1 Given the examples of prostaglandin activity just described, provide examples of why an individual might take aspirin, which blocks the synthesis of prostaglandins (see Clinical View 17.1: "Synthesis of Eicosanoids").

INTEGRATE

CONCEPT CONNECTION

Prostaglandins, thromboxanes, and leukotrienes are local hormones that can act as local vasoactive substances (substances that change the size of blood vessels by narrowing blood vessels through vasoconstriction or widening blood vessels through vasodilation). The role of prostaglandins and thromboxanes as a vasoconstrictor to prevent blood loss is described in section 20.4c. In comparison, the role of prostaglandins and leukotrienes as vasodilators (as part of the inflammatory response) is discussed in section 22.3d.

INTEGRATE

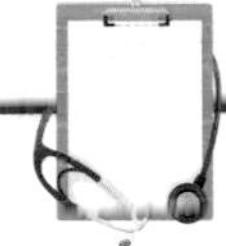

CLINICAL VIEW 17.1

Synthesis of Eicosanoids

Eicosanoids are formed from phospholipid molecules within the plasma membrane of a cell (figure 17.5). The enzyme **phospholipase A_2** releases a 20-carbon fatty acid called arachidonic acid. (This enzyme is inhibited by *steroid drugs* [e.g., hydrocortisone] to prevent formation of all eicosanoids.) Arachidonic acid is then acted upon by either (1) **cyclooxygenase,** which converts arachidonic acid to prostaglandins and thromboxanes (this enzyme is inhibited by *aspirin* to prevent formation of both of these eicosanoids) or (2) **lipooxygenase,** which converts arachidonic acid to leukotrienes (this enzyme is inhibited by *St. John's wort* to prevent formation of leukotrienes).

WHAT DID YOU LEARN?

8 Leukotrienes from damaged tissue cause smooth muscle in local blood vessels to vasodilate (increase the diameter of the vessel lumen). Is this an example of (a) autocrine stimulation or (b) paracrine stimulation? Explain.

17.4 Hormone Transport

Hormone molecules are transported within the blood to target cells following their release from the endocrine gland that synthesized them (see figure 17.1). Here we consider how both lipid-soluble and water-soluble hormones are transported and the factors that influence the level of circulating hormone.

17.4a Transport in the Blood

LEARNING OBJECTIVE

10. Compare the transport of lipid-soluble hormones with that of water-soluble hormones.

Lipid-soluble hormones (e.g., steroids, calcitriol, thyroid hormone) do not dissolve readily within the aqueous environment of the blood (blood plasma), and so they require carrier molecules. These molecules are water-soluble proteins synthesized by the liver. Think of these proteins as boats that "ferry" the hormone molecules within the blood. Some hormone carrier proteins may be very selective, binding and transporting only one specific lipid-soluble molecule (e.g., thyroxine-binding globulin), whereas other carrier proteins are nonselective (e.g., albumin), meaning they bind and transport numerous lipid-soluble molecules. A transport carrier protein has the added benefit of protecting the hormone molecule and helping to prevent its early destruction (or, for small hormones, their loss in the urine). Thus, the association of a hormone with a carrier protein often acts as a safeguard to help prolong the life of the hormone.

The binding between a lipid-soluble hormone and a carrier protein is temporary. Hormone molecules bind to the carrier molecule, detach from the carrier and float free within the blood, and then may later reattach to a different carrier molecule. Any hormone that is attached to a carrier is a **bound hormone,** whereas an unattached hormone is an **unbound (free) hormone.** Only unbound hormone, which represents a very small fraction of the hormone transported within the blood (0.1% to 10%), is generally able to exit the blood and bind to cellular receptors of target organs. It is the maintaining of a given homeostatic level (or steady state) of the physiologically active, unbound hormone that is required. The bound hormone simply serves as a readily available source within the blood.

Water-soluble hormones (i.e., most biogenic amines and proteins), in comparison, readily dissolve within the aqueous environment of the blood, and so these hormones do not generally require carrier proteins. Thus, water-soluble hormones are generally released directly into the blood and transported to target cells. Note that some water-soluble hormones (e.g., insulin-like growth factor) are transported by carrier proteins, which function to prolong the life of these hormones.

WHAT DID YOU LEARN?

9 Why are carrier proteins necessary for lipid-soluble hormones?

10 What is the added benefit of a carrier protein?

17.4b Levels of Circulating Hormone

LEARNING OBJECTIVES

11. Describe the two primary factors that affect the concentration level of a circulating hormone.

12. Explain what is meant by the half-life of a hormone.

Hormones exert their physiologic effects primarily as a result of their blood concentration. Consequently, the amount of each hormone

Figure 17.12 Anterior Pituitary Hormones. The hypothalamus releases regulatory hormones, to control the release of hormones from the anterior pituitary. Each regulatory hormone released from the hypothalamus is the same highlighted color as the specific hormone(s) from the anterior pituitary that it controls (Note: The anterior pituitary also secretes melanocyte-stimulating hormone (MSH), but because it normally has little effect in humans and secretion ceases prior to adulthood, it is not included in this figure.) AP|R

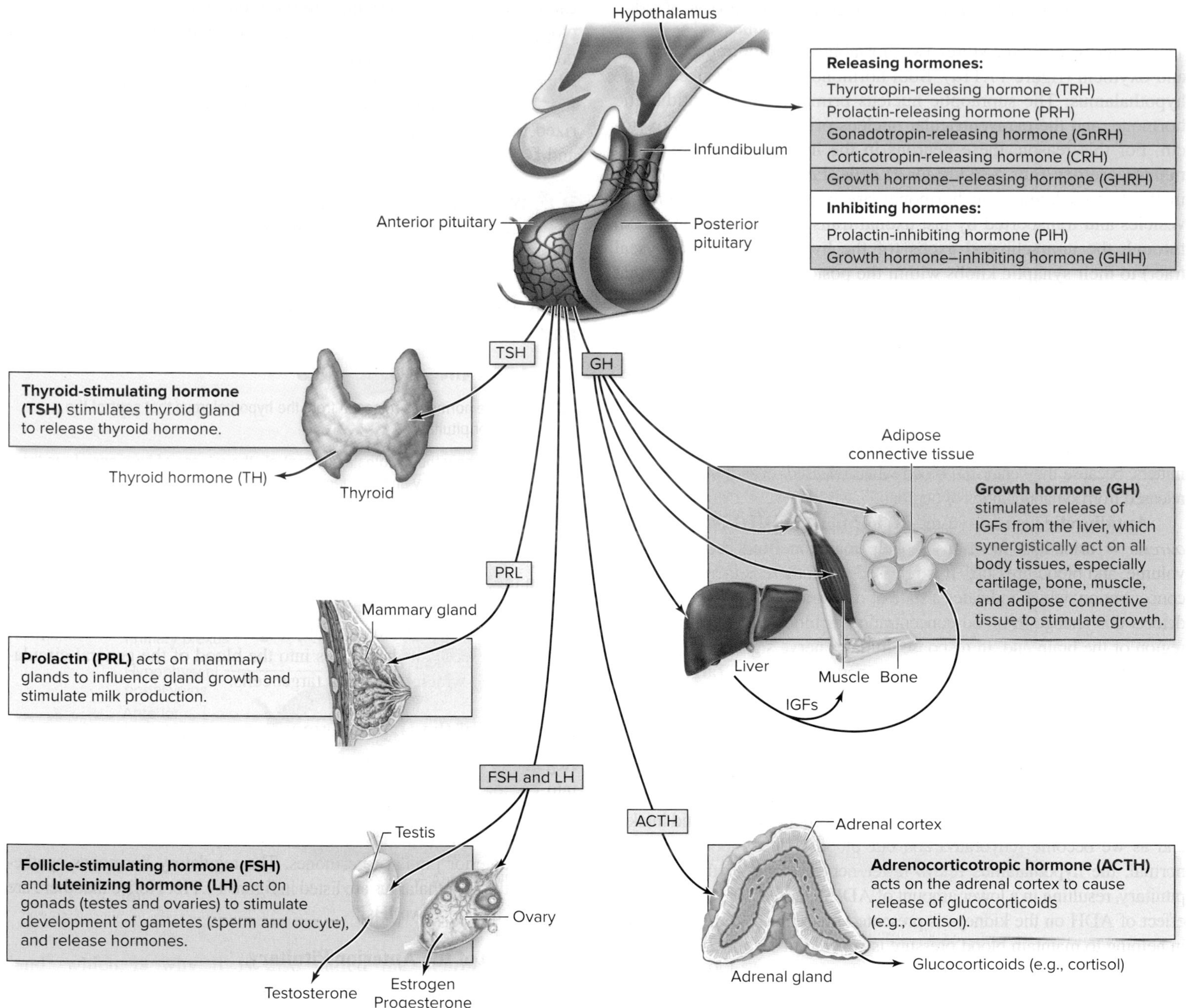

Hypothalamus
Releasing hormones:
Thyrotropin-releasing hormone (TRH)
Prolactin-releasing hormone (PRH)
Gonadotropin-releasing hormone (GnRH)
Corticotropin-releasing hormone (CRH)
Growth hormone–releasing hormone (GHRH)
Inhibiting hormones:
Prolactin-inhibiting hormone (PIH)
Growth hormone–inhibiting hormone (GHIH)
Infundibulum
Anterior pituitary
Posterior pituitary
TSH
GH
Thyroid-stimulating hormone (TSH) stimulates thyroid gland to release thyroid hormone.
Thyroid hormone (TH)
Thyroid
Adipose connective tissue
Growth hormone (GH) stimulates release of IGFs from the liver, which synergistically act on all body tissues, especially cartilage, bone, muscle, and adipose connective tissue to stimulate growth.
PRL
Mammary gland
Prolactin (PRL) acts on mammary glands to influence gland growth and stimulate milk production.
Liver
Muscle
Bone
IGFs
FSH and LH
ACTH
Testis
Follicle-stimulating hormone (FSH) and luteinizing hormone (LH) act on gonads (testes and ovaries) to stimulate development of gametes (sperm and oocyte), and release hormones.
Ovary
Testosterone
Estrogen
Progesterone
Adrenal cortex
Adrenocorticotropic hormone (ACTH) acts on the adrenal cortex to cause release of glucocorticoids (e.g., cortisol).
Adrenal gland
Glucocorticoids (e.g., cortisol)

- **Thyrotropin-releasing hormone (TRH)** stimulates the anterior pituitary to release **thyroid-stimulating hormone (TSH)** (also called *thyrotropin*). TSH stimulates the thyroid gland to release **thyroid hormone (TH),** which regulates the metabolic rate. Thus, the general sequence of hormone release is TRH → TSH → TH (see section 17.9b).
- **Prolactin-releasing hormone (PRH)** stimulates the anterior pituitary to release **prolactin** (prō-lak′tin; *lac* = milk) **(PRL),** which acts on mammary glands to influence gland growth and stimulate milk production. Prolactin release is inhibited by **prolactin-inhibiting hormone** (see section 29.8c).
- **Gonadotropin-releasing hormone (GnRH)** stimulates the anterior pituitary to release both **follicle-stimulating hormone (FSH)** and **luteinizing** (lū′tēin-i-zing) **hormone (LH),** which are collectively called **gonadotropins.** These hormones act on the gonads in both females and males to stimulate development of gametes (oocyte and sperm). The ovaries release estrogen and progesterone; the testes release testosterone (see sections 28.3b and 28.4b).
- **Corticotropin-releasing hormone (CRH)** stimulates release of **adrenocorticotropic** (ă-drē′nō-kōr′ti-kō-trō′pik) **hormone (ACTH)** (also called *corticotropin*). ACTH stimulates the adrenal cortex to produce and secrete glucocorticoids (e.g., cortisol), which raise blood levels of nutrient molecules (e.g., glucose). Thus, the general sequence of hormone release is CRH → ACTH → glucocorticoids (see section 17.10b).
- **Growth hormone–releasing hormone (GHRH)** stimulates the anterior pituitary to release **growth hormone (GH)** (or *somatotropin*). GH stimulates the liver to release both **insulin-like growth factor 1** and **2 (IGF-1** and **IGF-2)** (also called *somatomedins*) (sō′mă-tō-mē′din). GH and IGFs function synergistically to cause growth. Thus, the general sequence of hormone release is GHRH → GH → IGFs. Release of GH is inhibited by **growth hormone–inhibiting hormone (GHIH)**

INTEGRATE

CLINICAL VIEW 17.3

Hypophysectomy

The surgical removal of the pituitary gland (the hypophysis) is called a **hypophysectomy** (hī′pof-i-sek′tō-mē), which is performed for tumors in the pituitary gland. The preferred route of entry in a hypophysectomy is through the nasal cavity and sphenoidal sinus, directly into the sella turcica (see figure 8.12). This approach requires very small instruments and allows complete removal of the pituitary with minimal trauma. Continuous postoperative care is required following a hypophysectomy, including replacement of various hormones. Individuals are monitored by an endocrinologist to maintain the proper levels of each hormone normally released from the pituitary gland.

Both are secreted by cells called *gonadotropes.* These hormones act on the gonads in both females and males. In the female, FSH and LH act on the ovaries to control (1) development of the oocyte and the follicle (the spherical structure that encloses the oocyte), (2) ovulation, which is the release of oocyte from the follicle, and (3) the release of estrogen and progesterone (see section 28.3b). In the male, FSH and LH act on the testes to regulate the development of sperm and the release of testosterone (see section 28.4b).

- **Adrenocorticotropic** (ă-drē′nō-kōr′ti-kō-trō′pik) **hormone (ACTH)** (also called *corticotropin*) is secreted by cells called *corticotropes.* ACTH stimulates the adrenal cortex to produce and secrete glucocorticoids (e.g., cortisol), which increase blood levels of nutrient molecules, including glucose, glycerol, fatty acids, and amino acids (see section 17.9b).
- **Growth hormone (GH)** (or *somatotropin*) is secreted by cells called *somatotropes.* GH stimulates the liver to release both **insulin-like growth factor 1** and **2 (IGF-1** and **IGF-2,** also called *somatomedins*) (sō′mă-to-mē′dĭn). Both GH and IGFs function synergistically to stimulate cell growth and cell division, particularly within the skeletal and muscular systems (see section 17.7d).

All of these hormones (except prolactin and growth hormone[1]) are called **tropic** (trōp′ik) **hormones,** which are hormones that target another endocrine gland to secrete its hormone(s). Note: The anterior pituitary also secretes **melanocyte-stimulating hormone (MSH),** but because it normally has little effect in humans and secretion ceases prior to adulthood, it will not be discussed further in this text.

Regulation of the Anterior Pituitary by the Hypothalamus

The hormones released from the hypothalamus into the hypothalamo-hypophyseal portal system, which control specific cells of the anterior pituitary to release their hormones into the general circulation, are shown in **figure 17.12**. Take a moment to read the descriptions at the bottom of figure 17.12, which explain the functional relationship between the hypothalamus and each of the specific hormones released from the anterior pituitary. Each hormone released from the hypothalamus is colorcoded with the specific hormone(s) that it triggers to release from the anterior pituitary.

1. Some experts consider growth hormone to be a tropic hormone because it stimulates the liver to release IGFs (in addition to GH stimulating bone and cartilage growth).

It is interesting to note that the pituitary gland is often dubbed the *master gland* because of controlling many other endocrine organs, including the thyroid, gonads (testes, ovaries), and adrenal cortex. As we have observed, however, the release of hormones from both the posterior pituitary and the anterior pituitary is regulated by the hypothalamus. Thus, the "master gland" does not function independently.

WHAT DID YOU LEARN?

18 What are the six primary hormones released from the anterior pituitary? How is the release of each of these hormones regulated by the hypothalamus?

INTEGRATE

LEARNING STRATEGY

Remember the primary anterior pituitary hormones with the **TP-FLAG** mnemonic:

T = Thyroid-stimulating hormone	**F** = Follicle-stimulating hormone
	L = Luteinizing hormone
P = Prolactin	**A** = Adrenocorticotropic hormone
	G = Growth hormone

17.7d Growth Hormone: Its Regulation and Effects

LEARNING OBJECTIVES

23. Describe how the release of growth hormone is regulated.
24. Describe the effects of growth hormone on its primary target organs.

Regulation of Growth Hormone Release

This section provides a detailed examination of the homeostatic system involving growth hormone to integrate many of the concepts that have been discussed in this chapter and in previous chapters. We describe its regulated release by the hypothalamus and its effects (as a water-soluble hormone) on its target cells. An overview of the homeostatic system involving growth hormone is presented in **figure 17.13**.

The release of GH from the anterior pituitary is controlled through hormonal stimulation (see section 17.2b) by the release of growth hormone–releasing hormone (GHRH) from the hypothalamus. GHRH enters the hypothalamo-hypophyseal portal system and is transported by the blood directly to the anterior pituitary (see section 17.7c). GHRH binds to receptors in specific cells of the anterior pituitary (somatotropic cells) and stimulates the release of GH into the general circulation to be transported throughout the body (see figure 17.1). The release of both GHRH from the hypothalamus and GH from the anterior pituitary (steps 3 and 4 in figure 17.13) is regulated by negative feedback (step 8 in figure 7.13). In response to increased levels of either GH or IGFs, the hypothalamus is stimulated to release **growth hormone-inhibiting hormone (GHIH),** which inhibits the release of GH from the anterior pituitary. GH also directly inhibits its own release from the anterior pituitary.

The amount of GHRH released from the hypothalamus (step 3 in figure 17.13) occurs in response to a number of other factors (as measured by growth hormone levels in the blood) (**figure 17.14**):

- **A person's age.** Growth hormone levels fluctuate with age. Notice in figure 17.14*a* that children and adolescents experience the highest amounts of GH, with progressive

INTEGRATE

CLINICAL VIEW 17.5

Disorders of Thyroid Hormone Secretion

Thyroid hormone (TH) release is very tightly controlled in the healthy state, but should the amount vary by even a little, a person could become either hyperactive and heat-intolerant or sluggish and overweight. Disorders of thyroid activity are among the most common metabolic problems clinicians see.

Hyperthyroidism results from excessive production of TH and is characterized by increased metabolic rate, weight loss, hyperactivity, and heat intolerance. Although there are a number of causes of hyperthyroidism, the more common ones are (1) ingestion of T_4 (weight control clinics sometimes use TH to increase metabolic activity); (2) excessive stimulation of the thyroid by the pituitary gland; and (3) loss of feedback control by the thyroid. This last condition, called **Graves disease,** is an autoimmune disorder involving the formation of autoantibodies that mimic TSH hormone. The autoantibodies bind to TSH receptors on the follicular cells of the thyroid, causing an abnormally high level of TH release. Graves disease includes all the symptoms of hyperthyroidism plus a peculiar change in the eyes known as **exophthalmos** (protruding and bulging eyeballs). Hyperthyroidism is treated by removing the thyroid gland, either by surgery or by intravenous injections of radioactive iodine (I-131). In the procedure for treating hyperthyroidism, the thyroid literally "cooks itself" as it sequesters the I-131, but other organs are not damaged because they do not store iodine as does the thyroid. Patients whose thyroid glands have been removed or destroyed must take daily hormone supplements.

Hypothyroidism results from decreased production of TH. It is characterized by low metabolic rate, lethargy, a feeling of being cold, weight gain (in some patients), and photophobia (the disdain and avoidance of light). Hypothyroidism may be caused by decreased iodine intake, loss of pituitary stimulation of the thyroid, post-therapeutic hypothyroidism (resulting from either surgical removal or radioactive iodine treatments), or destruction of the thyroid by the person's own immune system (**Hashimoto thyroiditis**). Oral replacement of TH may be needed for most of these cases of hypothyroidism.

A **goiter** is the enlargement of the thyroid, typically due to an insufficient amount of dietary iodine. Although the pituitary releases more TSH in an effort to stimulate the thyroid, the lack of dietary iodine prevents the thyroid from producing the needed TH. The long-term consequence of the excessive TSH stimulation is overgrowth of the thyroid follicles and the thyroid itself. Goiter was a relatively common deformity in the United States until food processors began adding iodine to table salt. It still occurs in parts of the world where iodine is lacking in the diet and, as such, is referred to as endemic goiter. Unfortunately, goiters do not readily regress once iodine is restored to the diet, and surgical removal of the thyroid gland is often required.

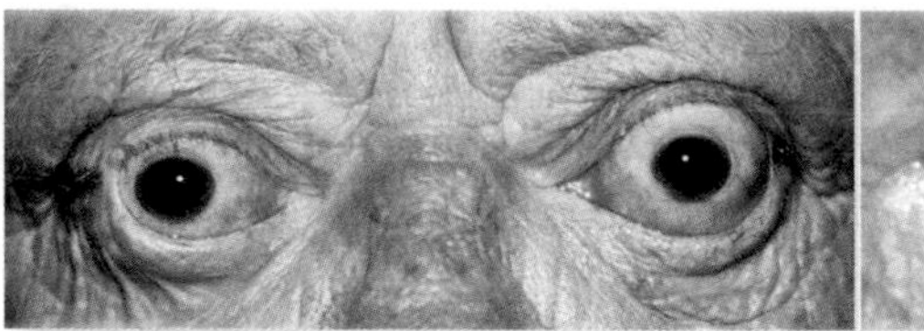

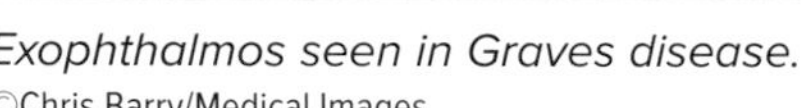
Exophthalmos seen in Graves disease.
©Chris Barry/Medical Images

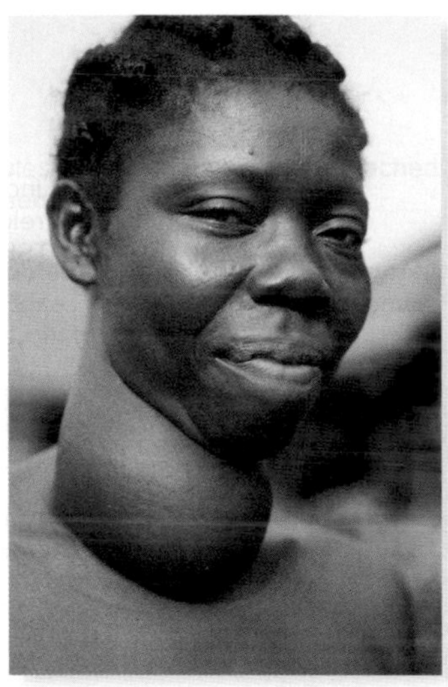
Goiter.
©Scott Camazine/Science Source

INTEGRATE

CONCEPT CONNECTION

The second law of thermodynamics (see section 3.1c) helps explain the relationship between increased TH and an increase in body temperature. TH stimulates the synthesis of sodium-potassium (Na^+/K^+) pumps to move these ions across the membrane. Chemical energy in the form of ATP is converted to mechanical energy to power the pumps. The second law of thermodynamics states that heat is produced during any energy conversion. With greater numbers of Na^+/K^+ pumps, more chemical energy is converted to mechanical energy, additional heat is produced, and body temperature rises.

genesis = production) **effect.** Increased amino acid uptake by cells provides the structural building blocks for these processes. TH also stimulates all cells to increase their glucose uptake. Concomitant with this increase is buildup in the number of cellular respiration enzymes within mitochondria because additional ATP is needed to support the higher metabolic rate.

TH stimulates other target cells to help meet the additional requirements for ATP. Hepatocytes are stimulated to increase both *glycogenolysis* and *gluconeogenesis* (see section 2.7c), while glycogenesis is inhibited, thus additional glucose is released into the blood. Adipose connective tissue cells respond to TH by both stimulating *lipolysis* and inhibiting lipogenesis (see section 2.7b). As a result, both glycerol and fatty acids are released into blood as alternative nutrient molecules for cellular respiration. This saves blood glucose for the brain and is called the **glucose-sparing effect.**

Respiration rate increases in response to TH to meet the additional demands for oxygen. Additionally, both heart rate and force of heart contraction increase blood flow to the tissues to deliver more nutrients and oxygen. This is a result of the cardiac muscle cells increasing their number of cellular receptors for the hormones epinephrine and norepinephrine, ensuring that the heart continues to be more responsive to these hormones (see section 19.9b). The homeostatic system involving thyroid hormone and its primary effects are summarized in figure 17.17. Additional responses to the release of thyroid hormone are listed in the summary table on thyroid hormone in the reference section, which directly follows this chapter (see **table R.4**).

WHAT DO YOU THINK?

7. Predict the signs/symptoms that a person with hyperthyroidism would exhibit in each of the following circumstances: (a) high temperature or low temperature, (b) elevated pulse or decreased pulse, (c) elevated breathing or decreased breathing, and (d) plump or thin body.

WHAT DID YOU LEARN?

22. What is the relationship of TRH, TSH, and TH in regulating metabolism?

23. What are the primary target organs/tissues of TH? Describe the effect on each.

17.8c Calcitonin: Its Regulation and Effects

LEARNING OBJECTIVE

28. Explain the role of calcitonin in regulating blood calcium.

Calcitonin is synthesized and released from the less numerous cells of the thyroid gland call the *parafollicular cells*. These cells are located around the follicular cells that produce thyroid hormone (see figure 17.15*b*). The stimulus for calcitonin release from parafollicular cells is a high blood calcium level; it is also secreted in response to stress from exercise. Calcitonin primarily inhibits osteoclast activity within bone tissue (which decreases the breakdown of bone tissue) and stimulates the kidneys to increase the loss of calcium in the urine. The net effect of calcitonin is a reduction in blood calcium levels. The relationship of calcitonin to parathyroid hormone and calcitriol in regulating blood calcium is discussed in section 7.6c.

WHAT DID YOU LEARN?

24 Does calcitonin decrease or increase blood calcium? Explain.

17.9 Adrenal Glands

The adrenal glands, like the hypothalamus and pituitary gland, are composed of both nervous tissue and hormone-producing cells. The inner portion of each gland, called the adrenal medulla, is composed of nervous tissue. The outer portion, called the adrenal cortex, is composed of hormone-producing cells. Numerous types of hormones are released from this gland. We first describe adrenal gland anatomy and its associated hormones.

17.9a Anatomy of the Adrenal Glands

LEARNING OBJECTIVES

29. Describe the structure and location of the adrenal glands.
30. Name the three zones of the adrenal cortex and the hormones produced in each zone.

The **adrenal** (ă-drē′năl; *ad* = to, *ren* = kidney) **glands,** or *suprarenal glands,* are paired, pyramid-shaped endocrine glands anchored on the superior surface of each kidney (**figure 17.18**). These glands (like the kidney that each is superior to) are retroperitoneal, which is posterior to the parietal peritoneum (see section 24.2a). Each adrenal gland is embedded within fat and fascia to minimize their movement. Two regions constitute an adrenal gland: the adrenal medulla and the adrenal cortex (figure 17.18*b*).

Adrenal Medulla

The **adrenal medulla** forms the inner core of each adrenal gland. It has a pronounced red-brown color due to its extensive vascularization. It releases the catecholamines epinephrine and norepinephrine (which are biogenic amines; see section 17.3a). The stimulus for their release is activation by the sympathetic division (see figure 17.3*c*). Approximately 80% of the hormone released is epinephrine and about 20% is norepinephrine. Both hormones circulate within the blood and help prolong the fight-or-flight response, which is caused by the activation of the sympathetic division (see section 15.4).

Adrenal Cortex

The **adrenal cortex** exhibits a distinctive yellow color as a consequence of the stored lipids within its cells. These cells synthesize more than 25 different lipid-soluble corticosteroids (see section 17.3a). The adrenal cortex is partitioned into three separate regions: the outer zona glomerulosa, the middle zona fasciculata, and the inner zona reticularis (figure 17.18*c*). Different functional categories of steroids are synthesized and secreted in the separate zones.

Hormones of the Adrenal Cortex

The **zona glomerulosa** (zō′nă glō-mĕr-ū-lōs′ă; *glomerulus* = ball of yarn) is the thin, outer cortical layer composed of dense, spherical clusters of cells. These cells synthesize **mineralocorticoids** (min′er-al-ō-kōr′ti-koyd), a group of hormones that help regulate the

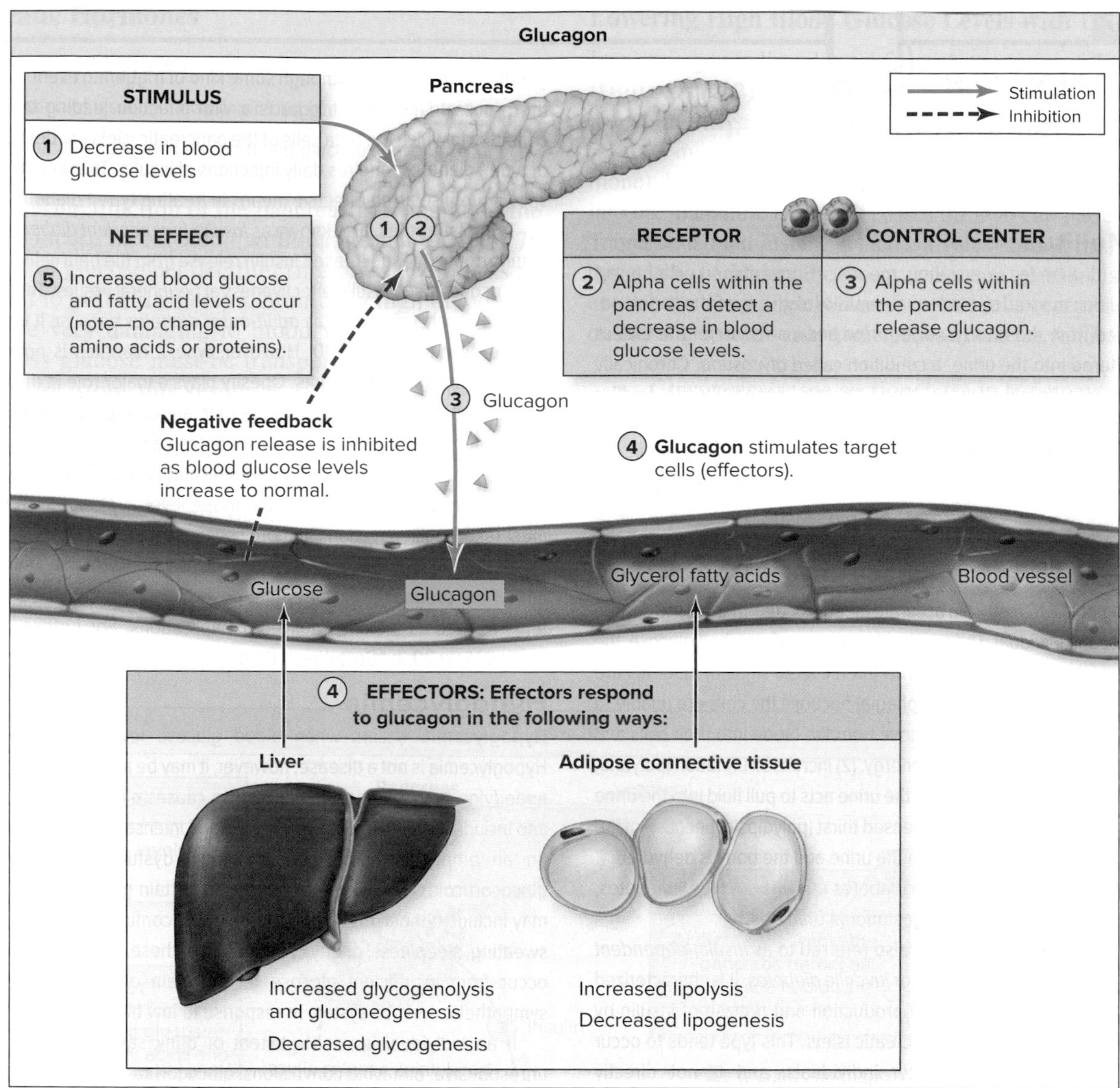

Figure 17.23 Regulation and Action of Glucagon. Glucagon is released from alpha cells within pancreatic islets in response to low blood glucose levels. Glucagon binds with target cells that increase glucose, glycerol, and fatty acid levels within the blood.

- *Lipolysis* (see section 2.7b) in adipose connective tissue cells is stimulated, and lipogenesis is inhibited. Fatty acids and glycerol are released from fat storage and are increased within the blood.

In summary, the release of glucagon results in an increase in glucose, glycerol, and fatty acids in the blood and in a decrease in the storage form of these molecules within body tissues. An increase in blood glucose levels results in a decrease in glucagon release by negative feedback.

Note that glucagon has no effect on the structural and functional protein components of the body. The physiologic significance of this lack of effect is that the ongoing, and regular, release of this hormone (e.g., during periods between meals) does not tear down the muscles and other protein components of the body to maintain blood glucose levels in "nonemergency" situations.

Interestingly, paramedics may administer glucagon subcutaneously under certain conditions when low blood glucose is detected. This may be done if the individual is unconscious and is unable to be given sugar orally to directly raise blood glucose.

WHAT DID YOU LEARN?

29 Is the stimulus for insulin and glucagon release from the pancreas hormonal, humoral, or nervous?

30 What are the stimulus, receptor, control center, and effector response to the release of insulin? Indicate what happens to nutrient levels in the blood.

31 Which of these hormones causes release of glucose into the blood: growth hormone, thyroid hormone, cortisol, insulin, or glucagon?

17.11 Other Endocrine Glands

Here we describe other endocrine glands and the functions of the hormones that are released by each.

17.11a Pineal Gland

LEARNING OBJECTIVE

36. Describe the general structure, location, and function of the pineal gland.

The **pineal** (pin′ē-ăl) **gland** (or *pineal body*) is a small, cone-shaped structure forming the posterior region of the epithalamus within the diencephalon (see figure 17.2 and section 13.4a). The pineal gland secretes **melatonin** (mel-ă-tōn′in; *melas* = dark hue, *ton as* = contraction), which makes us drowsy. Melatonin production tends to be cyclic; it increases at night, decreases during the day, and has the lowest levels around lunchtime. Melatonin helps regulate the circadian rhythm (24-hour body clock). Studies have linked low melatonin levels with mood (affective) disorders, such as *seasonal affective disorder (SAD)*, a condition that may be treated with light therapy.

Melatonin also appears to affect the synthesis of gonadotropin-releasing hormone (GnRH) from the hypothalamus. This hormone is responsible for regulating the synthesis of follicle-stimulating hormone (FSH) and luteinizing hormone (LH) from the anterior pituitary, which in turn regulate the reproductive system. The role of melatonin in sexual maturation is not well understood. However, excessive melatonin secretion is known to delay puberty in humans.

WHAT DID YOU LEARN?

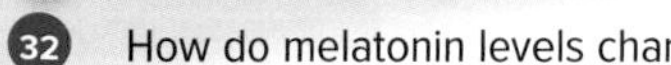

32 How do melatonin levels change throughout the day?

17.11b Parathyroid Glands

LEARNING OBJECTIVE

37. Describe the general structure, location, and function of the parathyroid glands.

The small, brownish-red **parathyroid** (par-ā-thī′royd) **glands** are located on the posterior surface of the thyroid gland (see figure 17.2). These glands are usually four small nodules, but some individuals may have as few as two or as many as six of these glands. There are two different types of cells in the parathyroid glands: chief cells and oxyphil cells.

The more common **chief cells,** or *principal cells*, are the source of parathyroid hormone (PTH), which is released from the parathyroid gland in response to a decrease in blood calcium levels. Parathyroid hormone functions to increase blood calcium levels. It stimulates release of calcium from bone tissue, decreases loss of calcium in urine, and causes the kidney to release an enzyme to convert the inactive calcidiol hormone to the active calcitriol hormone (see section 17.11c). The functional details of PTH and calcitriol are described in section 7.6b and summarized in the reference tables following this chapter (see **table R.2**). (See Clinical View 7.6: "Rickets" and Clinical View 7.7: "Osteoporosis.")

The role of oxyphil cells is not known, although these cells are associated with a rare form of cancer called *oxyphil cell adenoma*.

WHAT DID YOU LEARN?

33 What is the primary hormone released from the parathyroid gland? What is its general function?

17.11c Structures with an Endocrine Function

LEARNING OBJECTIVE

38. Identify and provide a description of the general function of the hormone(s) released from each of the organs discussed in this section.

Thymus

The **thymus** (thī′mŭs) is a bilobed organ that is located anterior to the heart on its superior aspect (see figure 17.2). The thymus is relatively large in infants, continues to grow until puberty, and then begins to regress (decrease in size) after puberty. A connective tissue framework houses both epithelial cells and maturing T-lymphocytes (a specific type of white blood cell). The T-lymphocytes migrate to the thymus following their formation in the bone marrow, and epithelial cells there secrete **thymic hormones** (i.e., thymosin, thymulin, and thymopoietin), which participate in the maturation of T-lymphocytes (see sections 21.3b and 22.5).

Heart

Endocrine tissue within the atria of the heart synthesizes and releases the hormone **atrial natriuretic** (nā′trē-yū-ret′ik; *natrium* = to carry, *ouron* = urine) **peptide (ANP)** in response to increased stretch of the atrial wall (which indicates an increase in blood volume and blood pressure). This peptide hormone stimulates both the kidneys to increase urine output (which decreases blood volume) and the blood vessels to dilate. Both of these actions facilitate blood pressure to decrease and return it to within normal homeostatic limits. Thus, the primary function of ANP is to lower blood pressure. The functional details of ANP are described in sections 24.5e, 24.6d, and 25.4d and summarized in the reference tables following this chapter (see **table R.7**).

Kidneys

Endocrine tissue within the kidneys release **erythropoietin (EPO)** (ĕ-rith′rō-poy′ĕ-tin) when specialized receptors (chemoreceptors) within the kidney detect low blood oxygen levels. EPO stimulates red bone marrow to increase the production rate of red blood cells (erythrocytes), which are the oxygen-carrying cells. The functional details of erythropoietin are described in section 18.3b and summarized in the reference tables following this chapter (see **table R.6**).

Liver

Recall that the liver releases insulin-like growth factors (IGFs) in response to growth hormone, as described in section 17.7d. The liver also releases **angiotensinogen** (an′jē-ō-ten-sin′ō-jen), an inactive hormone. The activation of angiotensinogen to **angiotensin II** (the active form of the hormone) requires both an enzyme released from the kidney (renin) and an enzyme anchored within the inner lining of blood vessels (angiotensin-converting enzyme, or ACE) (see sections 20.6b and 25.4a and Clinical View 25.4: "ACE Inhibitors and Lowering Blood Pressure"). The primary function of angiotensin II is to increase blood pressure. Angiotensin II has several effects. It (1) is a powerful blood vessel constrictor, (2) stimulates the kidney to decrease urine output, and (3) stimulates the thirst center. All of these processes function to keep blood pressure within normal homeostatic limits. (Note: The liver also secretes EPO; however, the primary producers of EPO are the kidneys.) The functional details of angiotensin II are described in sections 20.6b, 24.5e, and 25.4a and summarized in the reference tables following this chapter (see **table R.7**).

Stomach and Small Intestine

The stomach and small intestine are regions of the gastrointestinal (GI) tract. The stomach both synthesizes and releases gastrin. A primary function of **gastrin** (gas′trin; *gaster* = stomach) is to increase stomach activity (both its motility and its release of secretions) to facilitate digestion within the stomach (see section 26.2d). The small intestine is a long tube that is inferior to the stomach and located medially within the abdominal cavity. The small intestine releases both secretin and cholecystokinin, both of which function to facilitate digestion within the small intestine. A primary function of **secretin** (se-krē′tin) is to stimulate release of secretions from both the liver (bile) and pancreas (pancreatic juice) into the small intestine. A primary function of **cholecystokinin (CCK)** (ko′lē-sis-to-kī′nin; *chole* = bile, *cyst* = sac, *kinin* = to move), and what gives the hormone its name, is to stimulate release of bile from the gallbladder (a muscular sac on the inferior surface of the liver). The functional details of gastrin, secretin, and cholecystokinin are described in sections 26.2d and 26.3c and summarized in the reference tables following this chapter (see **table R.8**).

18.1 Functions and General Composition of Blood

Blood is the specialized fluid that is transported through the **cardiovascular** (kar′dē-ō-vas′kū-lăr; *cardio* = heart, *vascular* = vessels) **system,** which is composed of the heart and blood vessels. Blood vessels form a circuit away from the heart and back to the heart that includes the arteries, capillaries, and veins. **Arteries** transport blood away from the heart, whereas **veins** transport blood toward the heart. **Capillaries** (kap′i-lăr-ēz; *capillaries* = relating to hair) are permeable, microscopic vessels between arteries and veins. Capillaries serve as the sites of exchange between the blood and body tissues; it is at our capillaries that oxygen and nutrients exit the blood, and carbon dioxide and cellular wastes enter the blood.

Blood is composed of **formed elements** (erythrocytes, leukocytes, and platelets) and plasma. **Erythrocytes** (ĕ-rith′rō-sīt; *erthyros* = red, *kytes* = cell), also known as *red blood cells,* function to transport respiratory gases in the blood. **Leukocytes** (lu′kō-sit; *leuko* = white), also known as *white blood cells,* contribute to defending the body against pathogens, and **platelets** help clot the blood and prevent blood loss from damaged vessels. **Plasma** is the fluid portion of blood containing plasma proteins and dissolved solutes. We begin by describing the general functions of blood, its physical characteristics, and its general components.

18.1a Functions of Blood

 **LEARNING OBJECTIVE**

1. Describe the general functions of blood.

Blood carries out a variety of important functions as it circulates throughout the body; these functions can be grouped as transportation, regulation, and protection.

Transportation

Blood transports formed elements and dissolved molecules and ions throughout the body. Consider that as blood is transported through the blood vessels it transports oxygen from and carbon dioxide to the lungs for gas exchange (see section 23.1a), nutrients absorbed from the gastrointestinal (GI) tract (see section 26.1a), hormones released by endocrine glands (see section 17.1a), and heat and waste products from the systemic cells. Even when you take a medication, it is the blood that delivers it to the cells of your body. Thus, the blood serves as the "delivery system" for the body.

INTEGRATE

CONCEPT CONNECTION

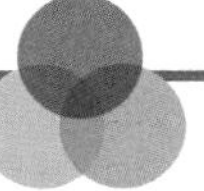

Recall from sections 1.6b and 6.1d that the amount of heat absorbed and released through the skin is regulated by the hypothalamus. This thermoregulation is accomplished by both (1) stimulating muscle tissue contraction to increase the amount of heat generated (e.g., shivering and formation of goose bumps) and (2) redistributing blood flow to the dermis through vasoconstriction (to retain heat) and vasodilation (to release heat).

Regulation

Blood participates in the regulation of body temperature, body pH, and fluid balance:

- **Body temperature.** Blood helps regulate body temperature. This is possible because blood absorbs heat from body cells, especially skeletal muscle, as it passes through blood vessels of body tissues. Heat is then released from blood at the body surface as blood is transported through blood vessels of the skin (see section 1.6b).
- **Body pH.** Blood, because it absorbs acid and base from body cells, helps maintain the pH of cells. Blood contains chemical buffers (e.g., proteins, bicarbonate) that bind and release hydrogen ions (H^+) to maintain blood pH until the excess is eliminated from the body (see section 25.6a).
- **Fluid balance.** Water is added to the blood from the GI tract and lost in numerous ways (including in urine, sweat, and respired air). In addition, there is a constant exchange of fluid between the blood plasma in the capillaries and the interstitial fluid surrounding the cells of the body's tissues. Blood contains proteins and ions that exert osmotic pressure to pull fluid back into the capillaries to help maintain normal fluid balance (see section 20.3b).

Protection

Blood contains leukocytes, plasma proteins, and various molecules that help protect the body against potentially harmful substances. These substances are part of the immune system, which is described in detail in chapter 22. Components of blood, including platelets and plasma proteins, also protect the body against blood loss, as described in section 18.4.

 **WHAT DID YOU LEARN?**

 What are some of the materials that blood transports?

 How does blood help regulate body temperature and fluid levels in the body?

18.1b Physical Characteristics of Blood

 **LEARNING OBJECTIVE**

2. Name six characteristics that describe blood, and explain the significance of each to health and homeostasis.

Blood is a type of connective tissue (see section 5.2d) that can be described based on its physical characteristics, including color, volume, viscosity, plasma concentration, temperature, and pH:

- **Color.** The color of blood depends upon whether it is oxygen-rich or oxygen-poor. Oxygen-rich blood is bright red or almost scarlet. Contrary to popular belief, oxygen-poor blood is not blue; rather, oxygen-poor blood is dark red. The bluish appearance of our veins can be attributed to both (1) the fact that we can see the blood moving through the superficial veins in the skin and (2) how light is reflected back to the eye from different colors. Lower-energy light wavelengths, like red, are absorbed by the skin and *not* reflected back to the eye, but higher-energy wavelengths, like blue, *are* reflected back to the eye, so the eyes can perceive only the blue coloration from the veins.

- **Volume.** The average volume of blood in an adult is 5 liters (L). Males tend to have, on average, 5 to 6 L, whereas females have, on average, 4 to 5 L. The greater amount of blood in males is due to their larger average size. Sustaining a normal blood volume is essential in maintaining blood pressure.
- **Viscosity.** Blood is about four to five times more *viscous* than water, meaning that it is thicker. Viscosity of blood depends upon the amount of dissolved substances in the blood relative to the amount of fluid; that is, viscosity is increased if the amount of substances—primarily erythrocytes—increases, the amount of fluid decreases, or both.
- **Plasma concentration.** Although viscosity is a characteristic of whole blood, another important physical characteristic is the plasma concentration, which is the relative concentration of solutes (e.g., proteins, ions) in plasma. This is normally a 0.09% concentration, and it determines whether fluids move into or out of the plasma by osmosis as blood is transported through capillaries. For example, when an individual is dehydrated, the plasma becomes hypertonic (see section 4.3b), and fluid moves into the plasma from the surrounding tissues. Additionally, the plasma concentration is used to determine intravenous (IV) solution concentrations, which are usually isotonic to plasma.
- **Temperature.** The temperature of blood is almost 2°C higher than measured body temperature; thus, if your body temperature is 37°C (98.6°F), your blood temperature is about 38°C (100.4°F). Therefore, blood warms areas through which it travels.
- **Blood pH.** Blood plasma is slightly alkaline, with a pH between 7.35 and 7.45. Plasma proteins, like all proteins of the body, have a three-dimensional shape that is dependent upon H^+ concentration. If the pH is altered from the normal range, plasma proteins become denatured and are unable to carry out their functions (see section 2.8b).

The physical characteristics of blood are summarized in **table 18.1**.

WHAT DO YOU THINK?

If a woman has 5 L of blood and she donates 1 pint (about 0.5 L), what approximate percentage is she donating: 1%, 5%, 10%, or 15%? Why do you think individuals below a certain weight (i.e., less than 110 pounds) are not allowed to give blood?

Table 18.1 Physical Characteristics of Blood

Characteristics	Normal Values
Color	Scarlet (oxygen-rich) to dark red (oxygen-poor)
Volume	4–5 L (females) 5–6 L (males)
Viscosity (relative to water)	4.5–5.5× (whole blood)
Plasma concentration	0.09%
Temperature	38°C (100.4°F)
pH	7.35–7.45

WHAT DID YOU LEARN?

Will blood be able to properly carry out its functions if blood pH is significantly altered? Why or why not?

18.1c Components of Blood

LEARNING OBJECTIVES

3. List the three components of a centrifuged blood sample.
4. Define hematocrit, and explain how the medical definition differs from the clinical usage.
5. Name the three formed elements of the blood, and compare their relative abundance.

Centrifuged Blood

Whole blood, which is both plasma and formed elements, can be separated into its liquid and cellular components by using a **centrifuge,** a device that spins the sample of blood in a tube so that heavier components collect at the bottom. **Figure 18.1** shows the resulting three components separated in the test tube. From bottom to top, these components are as follows:

- Erythrocytes form the lower layer of the centrifuged blood. They typically make up about 44% of a blood sample.
- A thin **buffy coat** makes up the middle layer. This slightly gray-white layer is composed of both leukocytes and platelets. The buffy coat forms less than 1% of a blood sample.
- Plasma is a pale yellowish liquid that rises to the top in the test tube; it generally makes up about 55% of blood.

INTEGRATE

CONCEPT CONNECTION

Recall from section 2.5b that an **acid** increases the concentration of H^+ by releasing it into the solution. Examples include hydrochloric acid (HCl) and carbonic acid (H_2CO_3). In contrast, a **base** decreases the concentration of H^+ in the solution. Examples include bicarbonate ions (HCO_3^-) and hydroxide ions (OH^-). The **pH** is a measure of the relative amounts of H^+ in solution. A **buffer** helps prevent pH changes by binding or releasing excess H^+ to maintain the normal H^+ concentration in a solution.

The percentage of the volume of all formed elements (erythrocytes, leukocytes, and platelets) in the blood is called the **hematocrit** (hē′mă-tō-krit, hem′ă-; *hemato* = blood, *krino* = to separate). This medical dictionary definition of the *true hematocrit* differs slightly from the *clinical hematocrit* definition, which equates the hematocrit to the percentage of only erythrocytes. (In practice, the true hematocrit and the clinical hematocrit are considered the same.)

Hematocrit values vary somewhat and are dependent upon the age and sex of the individual. A very young child's hematocrit may vary from 30% to 60%, and that range will narrow to 35% to 50% as

Erythropoiesis

Erythrocytes make up more than 99% of formed elements, with a concentration between 4.2 and 6.2 million per cubic millimeter. The process of erythrocyte production is called **erythropoiesis** (ĕ-rith′rō-poy-ē′sis). Normally, erythrocytes are produced at the rate of about 3 million per second. The hormone erythropoietin (EPO) controls this rate by increasing the rate of erythrocyte formation (described in section 18.3b). Dietary requirements for normal erythropoiesis include iron, B vitamins (e.g., folic acid, riboflavin), and amino acids (to build proteins).

The process of erythropoiesis begins with a **myeloid stem cell,** which under the influence of multi-CSF forms a progenitor cell. The progenitor cell forms a **proerythroblast,** which is a large, nucleated cell. It then becomes an **erythroblast,** which is a slightly smaller cell that is producing hemoglobin in its cytosol. The next stage, called a **normoblast,** is a still smaller cell with more hemoglobin in the cytosol; its nucleus has been ejected. A cell called a **reticulocyte** (re-tik′ū-lō-sīt) eventually is formed. The reticulocyte has lost all organelles except some ribosomes, so it can continue to produce hemoglobin (through protein synthesis). The transformation from myeloid stem cell to reticulocyte takes about 5 days.

Some reticulocytes finish maturation while circulating in blood vessels (and in normal circumstances, make up 0.5–2.0% of the circulating blood). One to two days after entering the circulation, the ribosomes in the reticulocyte degenerate, and the reticulocyte becomes a mature erythrocyte. Without a nucleus and cellular organelles, the mature erythrocyte is essentially a plasma membrane "bag" containing hemoglobin.

Leukopoiesis

Leukocytes make up less than 0.01% of formed elements, with a concentration between 4500 and 11,000 per cubic millimeter. The production of leukocytes is called **leukopoiesis** (lū′kō-poy-ē′sis). Leukopoiesis involves three different types of maturation processes: granulocyte maturation, monocyte maturation, and lymphocyte maturation.

All three types of granulocytes (neutrophils, basophils, and eosinophils) are derived from a myeloid stem cell. This stem cell is stimulated by multi-CSF and GM-CSF to form a progenitor cell. The granulocyte line develops when the progenitor cell forms a **myeloblast** (mī′ĕ-lō-blast) under the influence of G-CSF. The myeloblast ultimately differentiates into one of the three types of granulocytes.

Like granulocytes, monocytes are also derived from a myeloid stem cell. The myeloid stem cell differentiates into a progenitor cell, and under the influence of M-CSF this cell forms a **monoblast.** This is the monocyte line. Eventually, the monoblast forms a promonocyte that differentiates and matures into a **monocyte.**

Lymphocytes are derived from a **lymphoid stem cell** through the lymphoid line. The lymphoid stem cell differentiates into **B-lymphoblasts** and **T-lymphoblasts.** B-lymphoblasts mature into B-lymphocytes, whereas T-lymphoblasts mature into T-lymphocytes. Some lymphoid stem cells differentiate directly into NK (natural killer) cells. (Lymphocyte development and maturation are further described in section 22.5.)

Thrombopoiesis

Platelets (or *thrombocytes*) make up less than 1% of formed elements, with a concentration between 150,000 and 400,000 per cubic millimeter. The production of platelets is called **thrombopoiesis** (throm′bō-poy-ē′sis; *thrombos* = clot). From the myeloid stem cell, a committed cell called a **megakaryoblast** (meg-ă-kar′ē-ō-blast; *mega* = big) is produced. It matures under the influence of thrombopoietin to form a **megakaryocyte** (meg-ă-kar′ē-ō-sīt) (**figure 18.4**). Megakaryocytes are easily distinguished both by their large size (about 100 micrometers [μm] in diameter) and by their dense, multilobed nucleus. Each megakaryocyte then produces thousands of platelets.

Megakaryocytes produce platelets by forming long extensions from themselves called *proplatelets.* While still attached to the megakaryocyte, these proplatelets extend through the blood vessel wall (between the endothelial cells) in the red bone marrow. The force from the blood flow "slices" these proplatelets into the fragments we know as platelets.

Researchers previously thought that most, if not all, megakaryocytes were located within red bone marrow. However, research on mice published in 2017 demonstrated that megakaryocytes circulate through the lung vasculature and that while in the lung, these cells release platelets. This mouse model indicates that the lungs may be the site of up to 50% of platelet production. Research is ongoing to determine if this finding exists in humans.

WHAT DID YOU LEARN?

9 Describe the process of erythropoiesis, beginning with the stem cell and then placing the precursor cells in order until a mature erythrocyte is produced.

10 What are the two main types of precursor cells for formed element development, and what mature formed elements are derived from each?

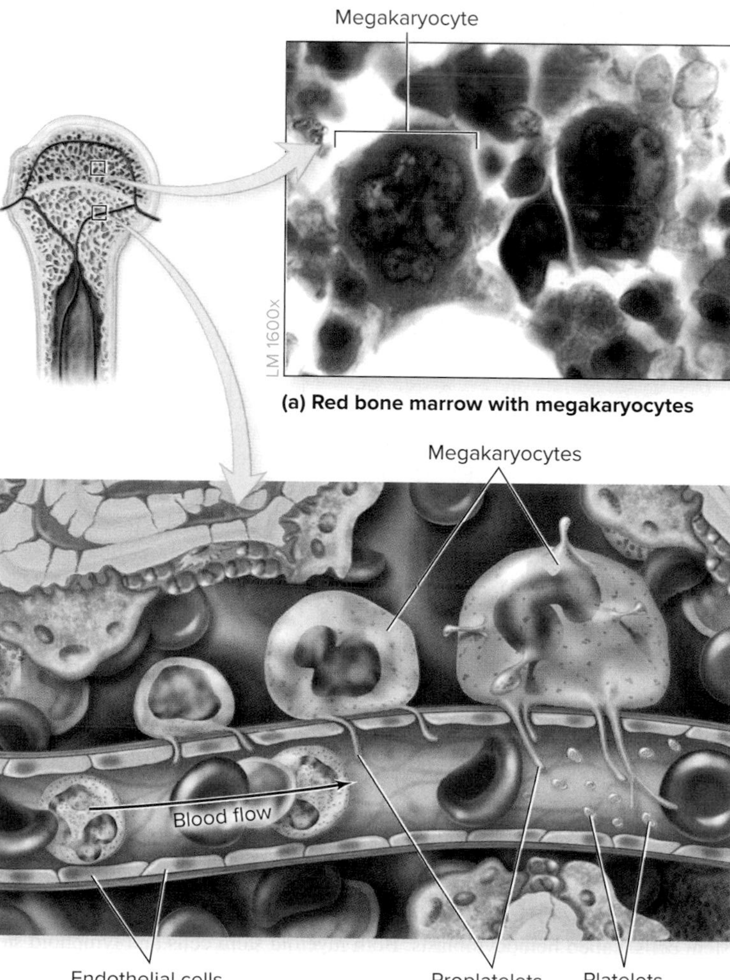

Figure 18.4 Platelet Formation. Platelets are derived from megakaryocytes. (*a*) Photomicrograph of megakaryocytes in red bone marrow. (*b*) Megakaryocytes extend long processes (called proplatelets) through the blood vessel wall. These proplatelets are spliced (by the force of the blood flow) into platelets, which then circulate in the blood. AP|R

(*a*) ©McGraw-Hill Education/Al Telser

18.3b Erythrocytes

LEARNING OBJECTIVES

13. Describe the structure of erythrocytes.
14. List the events by which erythrocyte production is stimulated.
15. Explain the process by which erythrocyte components are recycled.
16. Compare and contrast the different blood types and their importance when transfusing blood.

Erythrocytes are very small, flexible cells, with a diameter of approximately 7.5 μm (**figure 18.5**). Although erythrocytes are commonly referred to as red blood cells, or RBCs, the term *red blood cell* is a misnomer because a mature erythrocyte lacks a nucleus and cellular organelles. As a result, it is more appropriate to refer to an erythrocyte as a *formed element.* An erythrocyte has a unique biconcave disc structure (at its narrowest point about 0.75 μm and at its widest point about 2.6 μm). It is composed of a plasma membrane within which are housed about 280 million hemoglobin molecules.

Erythrocytes transport oxygen and carbon dioxide between the tissues and the lungs. The fact that erythrocytes lack a nucleus and organelles enables them to carry respiratory gases more efficiently. The biconcave shape and flexibility of erythrocytes allow them to stack and line up in single file. This single file of erythrocytes is termed a **rouleau** (rū-lō′; pl., *rouleaux*; cylinder), as they pass through capillaries. A latticework of spectrin protein supports the plasma membrane of the erythrocyte on its internal surface and provides flexibility to the erythrocyte as it moves through the capillaries.

WHAT DO YOU THINK?

2 What does an erythrocyte gain by the loss of its nucleus and organelles? What three cellular processes can it no longer engage in due to the loss of its (a) nucleus and (b) mitochondria?

Hemoglobin

Hemoglobin (hē′mō-glō-bin) is a red-pigmented protein that transports oxygen and carbon dioxide. When blood is maximally loaded with oxygen, it is termed **oxygenated** (and appears bright red). Conversely, when some oxygen is lost and carbon dioxide is gained during systemic cellular gas exchange, blood is called **deoxygenated** (and appears dark red).

Each hemoglobin molecule consists of four protein molecules called **globins.** Two of these globins are called **alpha (α) chains,** and the other two, which are slightly different, are called **beta (β) chains (figure 18.6).** All globin chains contain a **heme** group, which is composed of a porphyrin (organic compound) ring, with an iron ion (Fe^{2+}) in its center. Oxygen binds to the Fe^{2+} in heme groups for transport in the blood.

Sectional view

Superior view

~280 million hemoglobin molecules

~0.75 μm

~2.6 μm

~7.5 μm

(a) Erythrocyte structure

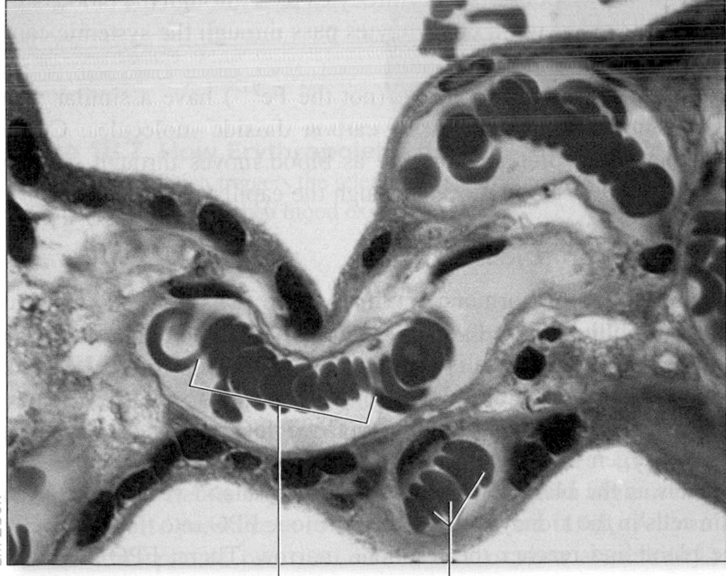

(b) Erythrocyte rouleau in a blood vessel

Figure 18.5 Erythrocyte Structure. (*a*) Erythrocytes have the gross structure of a biconcave disc, as shown here in sectional and superior views. (*b*) Micrograph of erythrocytes shows their three-dimensional structure and a rouleau. **AP|R**

(*b*) ©Ed Reschke/Getty Images

INTEGRATE

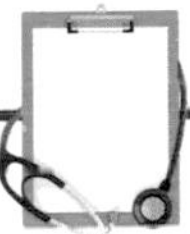

CLINICAL VIEW 18.1

Blood Doping

To enhance their performance in endurance events, some athletes may try to boost their bodies' ability to deliver oxygen to the muscles by increasing the number of erythrocytes in their blood. One way the number of erythrocytes can be increased naturally is by living and training at high altitude. The body compensates for the decreased oxygen concentration in the atmosphere by increasing the rate of erythrocyte production, thus increasing the number of erythrocytes per unit volume of blood.

An illegal procedure used by some athletes is called **blood doping.** There are two different methods for blood doping. In the first (and older) method, the athlete essentially donates erythrocytes to himself or herself. Prior to competition, the athlete has a unit of blood removed and stored. As the kidneys detect the decreased blood oxygen, the hormone erythropoietin (EPO) is released and the bone marrow responds by increasing production of erythrocytes. This causes the body to increase erythrocyte production to make up for the ones just removed. A few days before the competition, the erythrocytes from the donated unit are transfused back into the athlete's body. The increased number of erythrocytes increases the amount of oxygen transported in the blood and is thought to favorably affect muscle performance, thereby improving athletic performance. The second method of blood doping has occurred with the development of pharmaceutical EPO, which is used to treat anemia. In this method of blood doping, an athlete is injected with pharmaceutical EPO to further increase erythrocyte levels.

Potentially deadly dangers are inherent in blood doping. By increasing the number of erythrocytes per measured volume of blood, blood doping also increases the viscosity of the blood. Thus, the heart must work harder to pump this more viscous blood. Eventually, temporary athletic success may be overshadowed by permanent cardiovascular damage that can even lead to death, as has occurred with some athletes. Thus, blood doping has now been banned by athletic competition governing bodies.

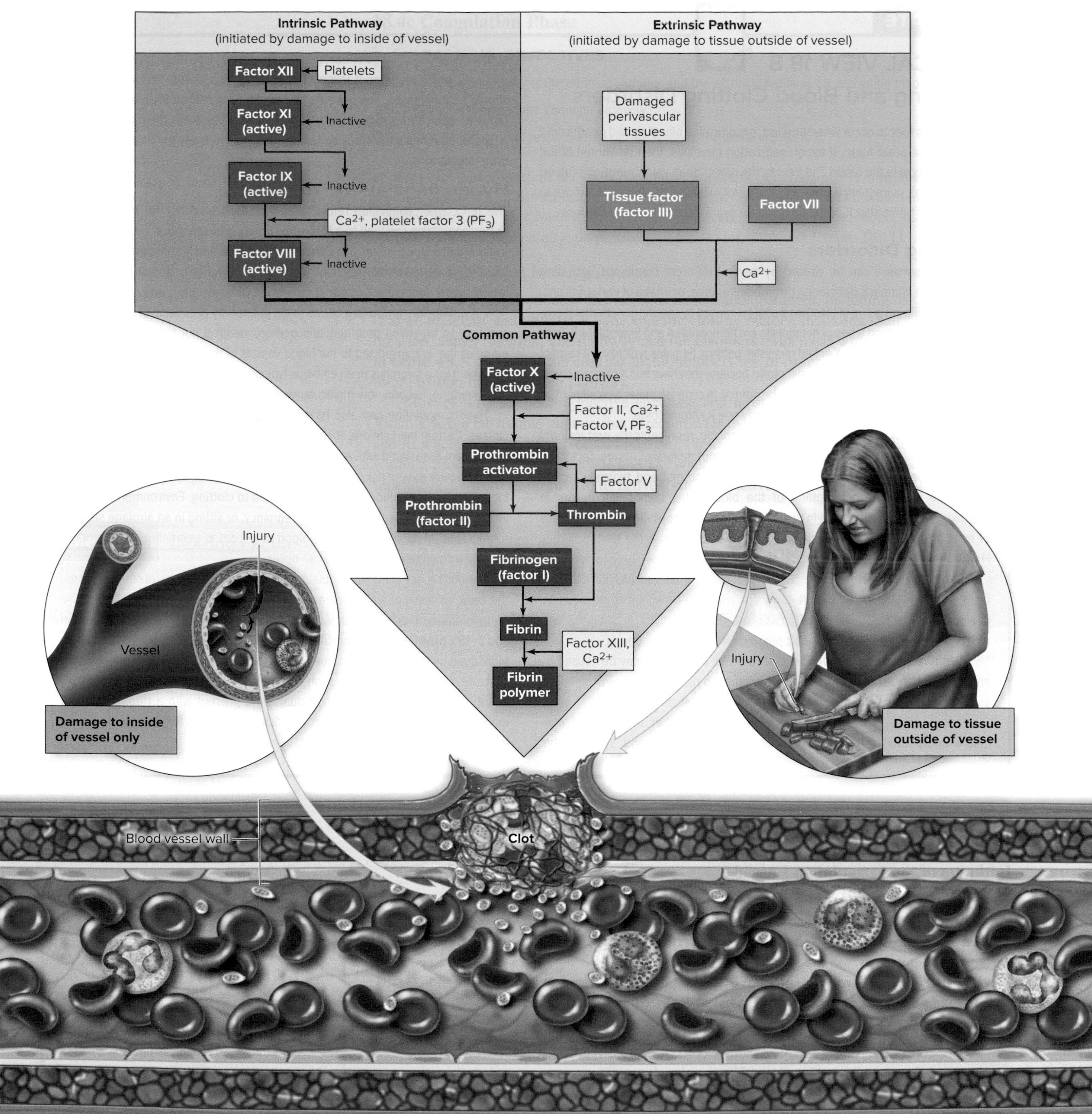

Figure 18.13 Coagulation Pathways. The process of coagulation typically involves both an intrinsic and an extrinsic pathway. Both pathways "merge" into a common pathway that ultimately converts fibrinogen (factor I) into fibrin.

Factor X, activated by either the intrinsic or extrinsic pathway, is the first step in the **common pathway:**

1. Active factor X combines with factors II and V, Ca^{2+}, and platelet factor 3 (PF_3) to form prothrombin activator.
2. Prothrombin activator activates prothrombin to thrombin.
3. Thrombin converts soluble fibrinogen into insoluble fibrin.
4. In the presence of Ca^{2+}, factor XIII is activated. Factor XIII cross-links and stabilizes the fibrin monomers into a fibrin polymer that serves as the framework of the clot.

Other components of blood become trapped in this spiderweb-like protein mesh. Like platelet plug formation, the clotting cascade is regulated by positive feedback (see section 1.6c). Once initiated by the intrinsic or extrinsic pathway, the events of the clotting cascade continue until a clot is formed (the climactic event). The size of the clot is limited because thrombin is either trapped within the clot or quickly degraded by enzymes within the blood.

The Sympathetic Response to Blood Loss

In severe cases, when over 10% of the blood volume has been lost, a survival response is initiated.

As blood volume decreases, blood pressure decreases. If greater than 10% of the blood volume is lost from the blood vessels, the sympathetic division of the autonomic nervous system (ANS) is activated (see section 15.4), bringing about increased vasoconstriction of blood vessels, increased heart rate, and increased force of heart contraction in an attempt to maintain blood pressure. Blood flow is also redistributed to the heart and brain to keep these vital structures functioning. These processes are effective in maintaining blood pressure until approximately 40% of the blood is lost. Blood loss greater than 40% results in insufficient blood volume within the blood vessels, and blood pressure decreases to levels unable to support life.

WHAT DID YOU LEARN?

20 In what ways do the intrinsic and extrinsic pathways of the clotting cascade differ?

21 At what point in blood loss is the sympathetic division of the ANS typically activated, and what physiologic changes occur?

18.4d Elimination of the Clot

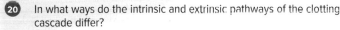

LEARNING OBJECTIVE

27. Explain the processes of clot retraction and fibrinolysis.

A blood clot is a temporary measure to stop blood loss through a damaged vessel wall. To return to normal, the blood vessel wall must be repaired and the clot eliminated. Elimination of the blood clot includes both clot retraction and fibrinolysis.

Clot retraction occurs as the clot is forming when actinomyosin, a contractile protein within platelets, contracts and squeezes the serum out of the developing clot. This makes the clot smaller as the sides of the vessel wall are pulled closer together.

To destroy the fibrin framework of the clot, **plasmin** (plaz′min) (or *fibrinolysin*) degrades the fibrin strands through **fibrinolysis** (fī′bri-nol′i-sis; *lysis* = dissolution). This is a process that begins within 2 days of the clot formation and occurs slowly over a number of days.

A "balancing act" is occurring continually within your blood between clot formation processes and those processes that prevent clot formation. The balance can be tipped so that blood clotting is initiated. A damaged blood vessel, impaired blood flow, atherosclerosis, or inflammation of the blood vessels can initiate blood clotting. Additionally, certain nutrients and vitamins must be present and available for blood clotting to occur normally. For example, calcium is used during the clotting process, and vitamin K is required for the synthesis of certain plasma proteins by the liver. Problems with the balance can lead to either bleeding or blood clotting disorders.

WHAT DO YOU THINK?

5 Predict why an individual confined to a wheelchair is more likely to develop unwanted clots.

WHAT DID YOU LEARN?

22 What is fibrinolysis, and what is its purpose?

18.5 Development and Aging of Blood

LEARNING OBJECTIVES

28. Describe when and how blood is formed in the embryo, fetus, childhood, and adulthood.
29. List some conditions that occur with the bone marrow and blood in the elderly.

The first primitive hemopoietic stem cells develop in the yolk sac wall of the embryo by the third week of development. The primitive hemopoietic stem cells go on to colonize other organs, such as the liver, spleen, and thymus. In these organs, these very primitive stem cells develop into the hemocytoblasts that produce all of the formed elements. Later in fetal development (perhaps beginning at 10 weeks), the hemocytoblasts begin to colonize red bone marrow, although the liver doesn't completely cease its blood cell production until close to birth.

Recall from section 7.2d that hemopoiesis occurs in most bones in young children, but as an individual reaches adulthood, hemopoiesis is restricted to selected bones in the axial skeleton. More red bone marrow is replaced with fat as individuals continue to age. Thus, older individuals have relatively less red bone marrow and may be more prone to developing anemia, which is a decrease in the ability to deliver oxygen to body cells. This decreased ability may be due to either a decrease in the numbers of circulating erythrocytes or lower than normal amounts of hemoglobin. In addition, older red bone marrow may be less able to meet any demands for an increased number of leukocytes. The leukocytes in the elderly may be less efficient and active than those in younger individuals, and the elderly also may have decreased numbers of leukocytes. Certain types of leukemias also are more prevalent among the elderly, probably due to the immune system (and its leukocytes) being less efficient.

WHAT DID YOU LEARN?

23 Where does hemopoiesis occur in (a) a fetus, (b) a child, and (c) an adult?

24 What are some ways that red bone marrow changes in the elderly?

19.1 Introduction to the Cardiovascular System

The **cardiovascular** (kar′dē-ō-vas′kū-lăr; *cardio* = heart, *vascular* = vessels) **system** is composed of both the heart and the blood vessels (**figure 19.1**). Here we describe the general function of the cardiovascular system, provide an overview of its components, and then examine how these components are organized into two circuits: pulmonary circulation and systemic circulation.

19.1a General Function

LEARNING OBJECTIVE

1. Describe the general function of the cardiovascular system.

The general function of the cardiovascular system is to circulate blood throughout the body to meet the changing needs of body cells. To remain healthy, all cells require (1) a continuous delivery of oxygen and nutrients and (2) the removal of carbon dioxide and other waste products. Additionally, cells of each body system have their own specific needs that must be met. Examples of the different requirements of specific body systems are presented in the Concept Connection on this page.

INTEGRATE

CONCEPT CONNECTION

All body systems have specific needs that are dependent upon the transport of blood throughout the body by the cardiovascular system. Examples include the *nervous system* for forming cerebrospinal fluid (see section 13.2c), the *endocrine system* for transporting hormones from endocrine glands to target organs (see section 17.4a), the *immune system* for delivering immune cells to the site of infection (see section 22.3d), the *respiratory system* for oxygen delivery and carbon dioxide removal (see section 23.6b), the *urinary system* for the removal of nitrogenous wastes from the body (see section 24.6e), and the *digestive system* for nutrient absorption (see section 26.4a).

Providing Adequate Perfusion

The cardiovascular system is effective in fulfilling its function if it provides *adequate perfusion* to maintain the health of all body cells. Perfusion (per-fyū′zhŭn; *perfusio* = pouring) is the delivery of blood per unit time per gram of tissue. It is typically expressed in milliliters per minute per gram (mL/min/g). Maintaining adequate perfusion requires both the continual pumping action of the heart

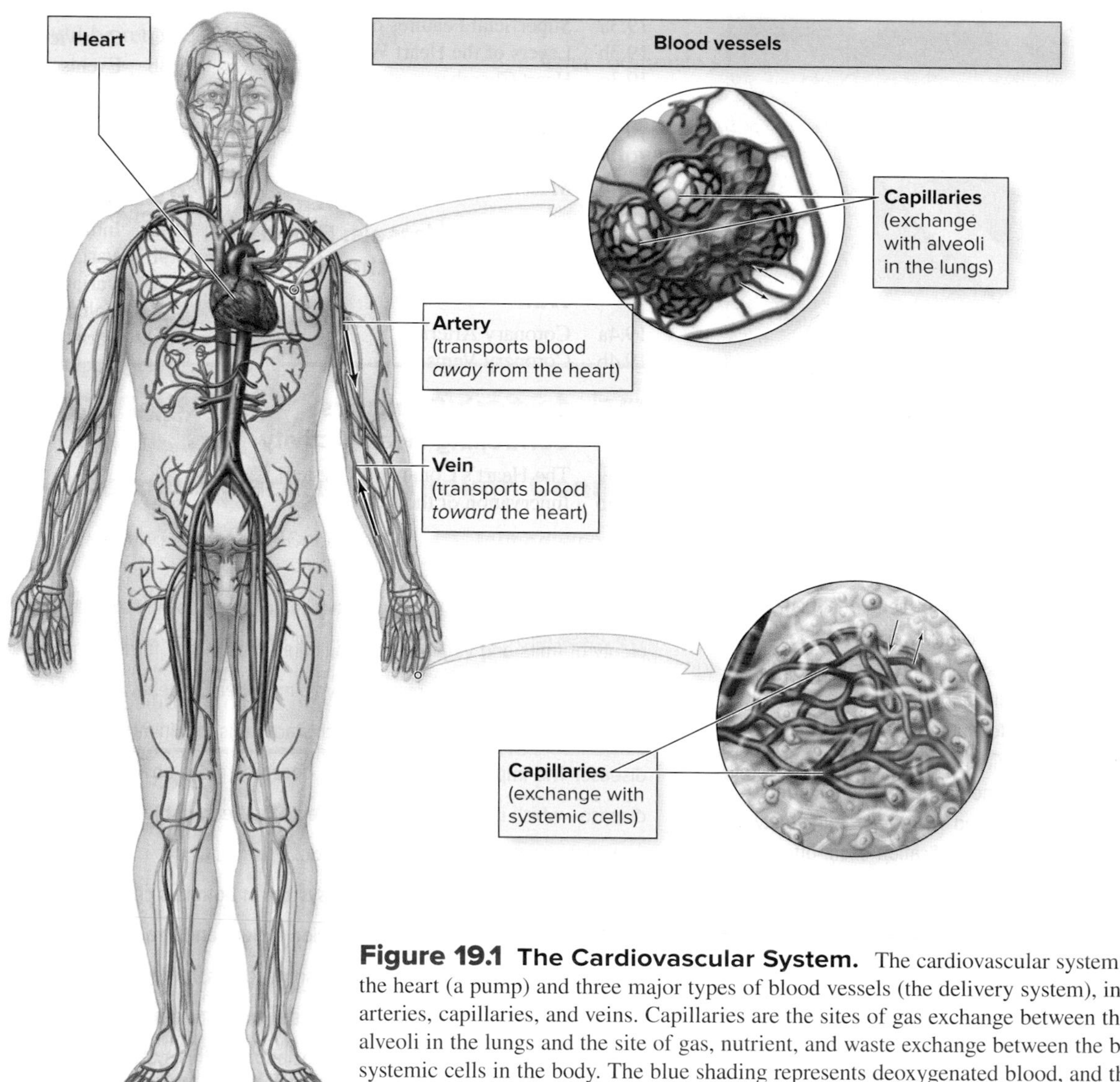

Figure 19.1 The Cardiovascular System. The cardiovascular system consists of the heart (a pump) and three major types of blood vessels (the delivery system), including arteries, capillaries, and veins. Capillaries are the sites of gas exchange between the blood and alveoli in the lungs and the site of gas, nutrient, and waste exchange between the blood and systemic cells in the body. The blue shading represents deoxygenated blood, and the red shading represents oxygenated blood. The violet shading represents where gas exchange occurs. AP|R

and healthy, patent (open and unblocked) blood vessels. If the heart fails to pump sufficient volumes of blood, or the vessels become hardened or occluded (blocked), then an adequate amount of blood may not reach the body's cells. Thus, tissues will be deprived of needed oxygen and nutrients, waste products accumulate, and cell death may occur.

WHAT DID YOU LEARN?

1. Define perfusion. Why would it be significant if the cardiovascular system failed to maintain adequate perfusion?

19.1b Overview of Components

LEARNING OBJECTIVES

2. Differentiate among the three primary types of blood vessels.
3. Describe the general structure and function of the heart.

Here we provide an overview of the components of the cardiovascular system: (1) the primary types of blood vessels, through which blood is transported, and (2) the heart, which pumps the blood.

Blood Vessels

Blood vessels are the conduits, or "soft pipes," of the cardiovascular system that transport blood throughout the body (figure 19.1). They are categorized into three primary types: **Arteries** (ar′ter-ēz) transport blood away from the heart; **veins** (vānz) transport blood toward

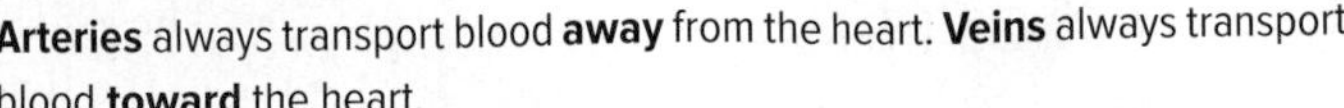

INTEGRATE

LEARNING STRATEGY

Arteries always transport blood **away** from the heart. **Veins** always transport blood **toward** the heart.

the heart; and **capillaries** (kap′i-lār-ēz; relating to hair) serve as the sites of exchange, either between the blood and the alveoli (air sacs) of the lungs or between the blood and the systemic cells. (The details of blood vessel anatomy are described in section 20.1.)

A common error is to describe arteries as the vessels that always transport **oxygenated** (ok′si-je-nāt-ed) blood (blood high in O_2 and low in CO_2) and veins as the vessels that always transport **deoxygenated** blood (blood low in O_2 and high in CO_2). As you will see, this is *not* an accurate generalization; in some parts of the cardiovascular system, the reverse is true. The defining factor is whether the blood is moving away from the heart (as it does in arteries) or toward the heart (as it does in veins).

The Heart

The **heart** is the center of the cardiovascular system. It is a hollow, four-chambered organ that pumps blood throughout the body. Three anatomic features are significant in the normal function of the heart: (1) the two sides of the heart, (2) the great vessels attached to the heart, and (3) the two sets of valves located within the heart **(figure 19.2)**.

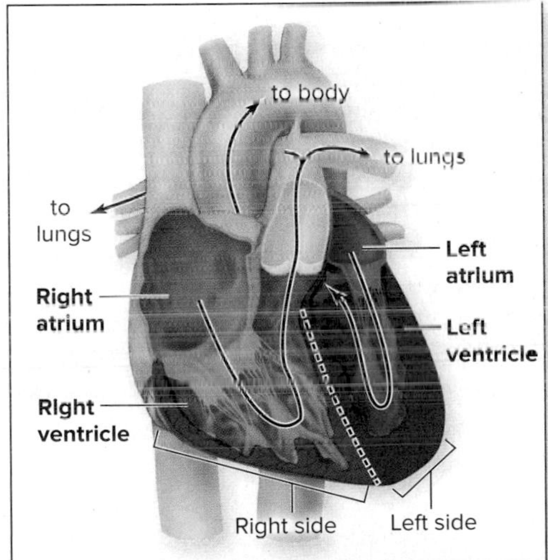

Two sides

Each side has a receiving chamber (atrium) and a pumping chamber (ventricle).

Right side

- **Right atrium:** receives deoxygenated blood from the body
- **Right ventricle:** pumps deoxygenated blood to the lungs

Left side

- **Left atrium:** receives oxygenated blood from the lungs
- **Left ventricle:** pumps oxygenated blood to the body

(a)

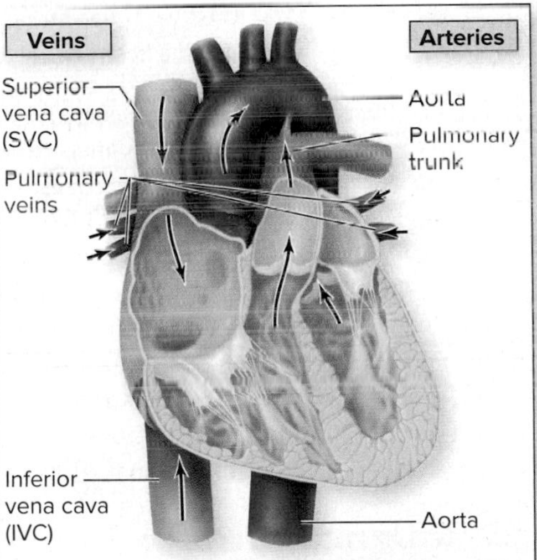

Great vessels

The great vessels are the large arteries and veins that are directly attached to the heart.

Arteries (arterial trunks) transport blood away from the heart.

- Pulmonary trunk transports from right ventricle.
- Aorta transports from left ventricle.

Veins transport blood toward the heart.

- Venae cavae (SVC and IVC) drain into right atrium.
- Pulmonary veins drain into left atrium.

(b)

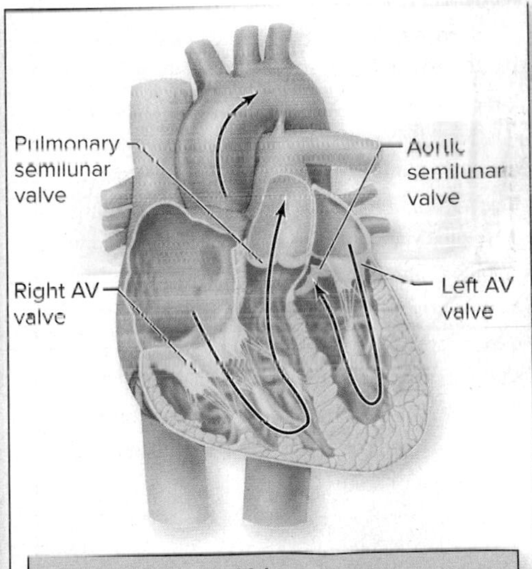

Valves

Heart valves prevent backflow to ensure one-way blood flow.

- **Atrioventricular (AV) valves** (i.e., right AV valve and left AV valve) are between an atrium and a ventricle.
- **Semilunar valves** (i.e., pulmonary semilunar valve and aortic semilunar valve) are between a ventricle and an arterial trunk.

(c)

Figure 19.2 Significant Anatomic Features of the Heart. The heart is composed of the following features seen in an anterior view: (*a*) a right side and a left side, (*b*) great vessels that are directly attached to it, and (*c*) two sets of valves that permit one-way flow of blood. AP|R

INTEGRATE

CLINICAL VIEW 19.1

Congestive Heart Failure

Congestive heart failure (or *heart failure*) is the impaired ability of the heart to pump blood. It may result from a decreased ability of the right ventricle, left ventricle, or both ventricles to pump blood effectively. One of the specific signs of congestive heart failure is **edema** (e-dē′mă), which is the accumulation of fluid in the interstitial space surrounding the cells. Edema can be specific to either the body or the lungs if only one side of the heart is involved.

If the *right* ventricle is impaired, it cannot keep up with the pumping action of the left ventricle. Consequently, blood returning from the systemic circulation into the right side of the heart accumulates within the systemic circulation. Additional fluid leaves the systemic capillaries to enter the interstitial space surrounding the body's cells, resulting in fluid accumulation within the body's tissue—a condition called **systemic edema.** Clinical signs of systemic edema include swelling of body tissues, especially in the legs, ankles, and feet (see section 25.2b).

If the *left* ventricle is impaired, it cannot keep up with the pumping action of the right ventricle, and blood returning from the pulmonary circulation into the left side of the heart accumulates within the pulmonary circulation. Additional amounts of fluid leave the pulmonary circulation to enter the interstitial space around the alveoli, resulting in swelling and fluid accumulation within the lungs—a condition called **pulmonary edema.** Clinical signs of pulmonary edema include pale skin, coughing up blood, and breathing difficulties. Pulmonary edema may lead to impaired gas exchange and may cause respiratory failure and death if not treated.

Notice that fluid accumulates in the area that blood is returning *from*. This is because that side of the heart is unable to pump the blood, and the "inflow" backs up. Consider the basic flow pattern: body ⟶ right side of heart ⟶ lungs ⟶ left side of heart

If the right side of the heart is impaired, fluid accumulates at the systemic cells in the body, whereas if the left ventricle is impaired, fluid accumulates within the lungs.

First, the heart is composed of two sides: the right side and the left side (figure 19.2*a*). The two sides of the heart allow separation of circulating deoxygenated and oxygenated blood:

- The *right* side of the heart receives deoxygenated blood from the body and pumps it to the lungs. (The blue shading in figure 19.2 [and throughout this chapter] represents deoxygenated blood.)
- The *left* side of the heart receives oxygenated blood from the lungs and pumps it to the body. (The red shading in figure 19.2 [and throughout this chapter] represents oxygenated blood.)

Notice that each side of the heart must both receive blood and pump blood. These two tasks are the functions of different types of chambers present on each side of the heart. A smaller chamber called an **atrium** receives blood, and a larger chamber called a **ventricle** pumps blood. Thus, four chambers in the heart are identified. On the right side of the heart are the **right atrium** and **right ventricle,** and on the left side of the heart are the **left atrium** and **left ventricle.**

Second, blood is transported directly to and from the chambers of the heart by the **great vessels** that directly connect with specific chambers of the heart (figure 19.2*b*). These vessels include both arteries and veins. There are two large arteries (or *arterial trunks*) attached to the superior border of the ventricles: They transport blood from a ventricle away from the heart. The **pulmonary trunk** transports blood from the right ventricle, whereas the **aorta** (ā-ōr′tă) transports blood from the left ventricle. Large veins deliver blood to the heart into an atrium. These include the two venae cavae: the **superior vena cava (SVC),** which drains blood from the head, neck, upper limbs, and superior region of the trunk (essentially from areas above the heart), and the **inferior vena cava (IVC),** which drains blood from the lower limbs and inferior regions of the trunk (essentially from areas below the heart). Both venae cavae drain blood into the right atrium. In addition, the **pulmonary veins** drain blood from the lungs into the left atrium. Remember, ventricles pump blood into the arterial trunks, and atria receive blood from the veins.

WHAT DO YOU THINK?

1. What vessels attached to the heart contain oxygenated blood? Are they both arteries? Explain.

Third, two sets of valves are located within the heart (figure 19.2*c*). The **atrioventricular** (ā′trē-ō-ven-trik′ū-lăr) **(AV) valves** are between the atrium and ventricle of each side of the heart. The **right AV valve,** also called the *tricuspid valve,* is located between the right atrium and the right ventricle. (Remember that the **TRI**cuspid is on the **RI**ght side.) The **left AV valve,** also called the *bicuspid valve,* or *mitral valve,* is located between the left atrium and left ventricle.

The other set of valves is the **semilunar valves**, which mark the boundary between a ventricle and its associated arterial trunk. (Note that each is named for its associated arterial trunk.) The **pulmonary semilunar valve** is between the right ventricle and the pulmonary trunk, and the **aortic semilunar valve** is between the left ventricle and the aorta. The valves open to allow blood to flow through the heart and then close to prevent its backflow. This ensures one-way, or unidirectional, flow of blood through the heart.

WHAT DID YOU LEARN?

2. What generalization can be made about all arteries? What generalization can be made about all veins?
3. Where does the right ventricle pump blood? Where does the left ventricle pump blood?
4. What great veins deliver blood to the right atrium? What great veins deliver blood to the left atrium?
5. Where are AV valves located within the heart? Where are semilunar valves located within the heart?

19.1c Pulmonary and Systemic Circulation

LEARNING OBJECTIVE

4. Compare and contrast pulmonary circulation and systemic circulation of the cardiovascular system. Trace blood flow through both circulations.

The two sides of the heart and the blood vessels are arranged in two circuits: the pulmonary circulation and systemic circulation. View **figure 19.3** as you read through this section.

The **pulmonary** (pŭl′mō-nār-ē; *pulmo* = lung) **circulation** (or *pulmonary circuit*) includes the movement of blood to and from the *lungs* for gas exchange. Deoxygenated blood is transported from (1) the right side of the heart through blood vessels to (2) the lungs. Exchange of respiratory gases occurs within the lungs as oxygen moves from the alveoli into the blood and

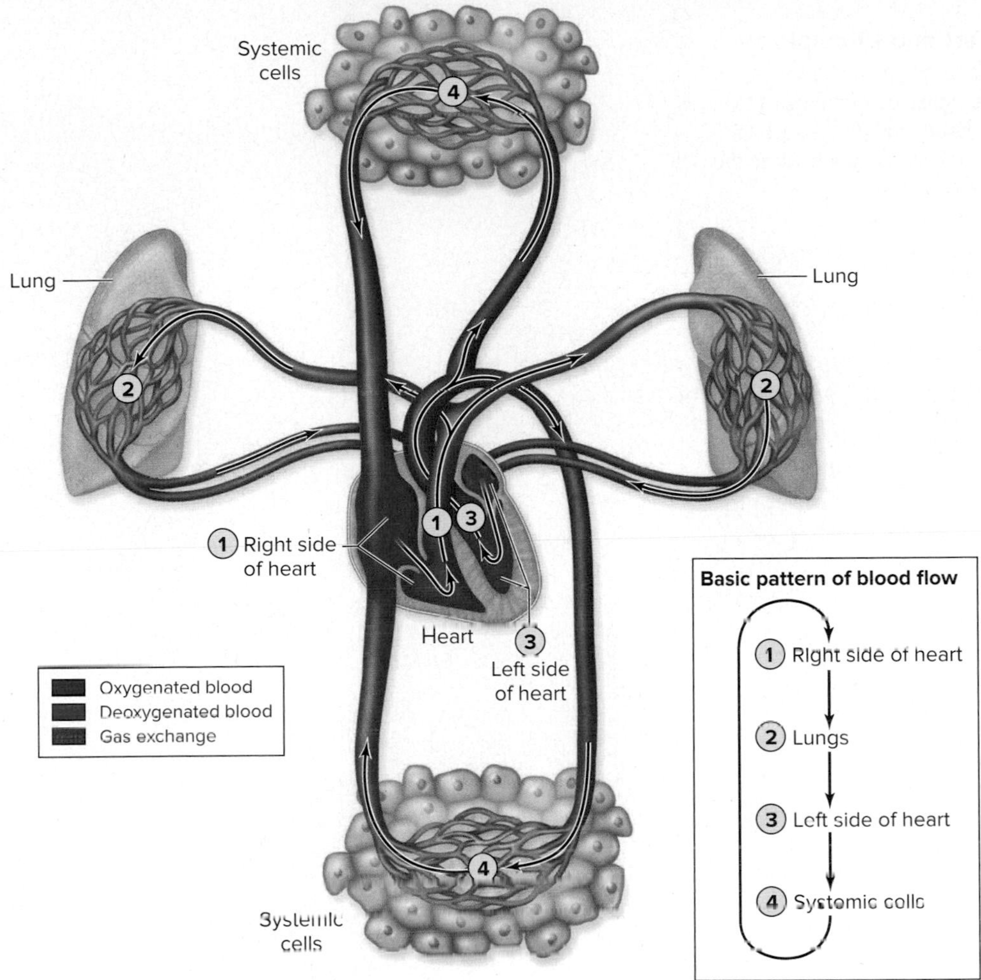

Figure 19.3 Cardiovascular System Circulations. The cardiovascular system is composed of the pulmonary circulation (steps 1, 2 and back to step 3) and the systemic circulation (steps 3, 4 and back to step 1). AP|R

carbon dioxide moves from the blood into the alveoli. Oxygenated blood is then transported through blood vessels to (3) the left side of the heart. Thus, the pulmonary circulation is movement of blood from the right side of the heart, to the lungs, and back to the left side of the heart.

The **systemic circulation** (or *systemic circuit*) includes the movement of blood to and from the *systemic cells*. Oxygenated blood is transported from (3) the left side of the heart through blood vessels to (4) the systemic cells such as those of the liver, skin, muscle, and brain. Exchange of respiratory gases occurs at systemic cells as oxygen moves from the blood into systemic cells and carbon dioxide moves from systemic cells into the blood. Deoxygenated blood is then transported through blood vessels that return blood to the (1) right side of the heart. Thus, the systemic circulation is movement of blood from the left side of the heart, to the systemic cells of the body, and back to the right side of the heart. The overall pattern of blood flow as shown in the inset in figure 19.3 is (1) the right side of the heart ⟶ (2) the lungs ⟶ (3) the left side of the heart ⟶ (4) the systemic cells of the body ⟶ back to the right side. Notice that the right side of the heart pumps blood to the lungs and the left side of the heart pumps blood to the systemic cells. This has important clinical implications if one side of the heart fails (see Clinical View 19.1: "Congestive Heart Failure").

The details of the blood flow pattern are summarized in **figure 19.4.** Consider spending some time reviewing this figure and becoming familiar with the details of both the pulmonary circulation and the systemic circulation. This will provide you with the "big picture" of the cardiovascular system to keep in mind as the details of the heart are covered in the remaining sections of this chapter and the details of blood vessels are covered in chapter 20.

WHAT DID YOU LEARN?

6 What path does blood follow through the heart and through the two circulations? Identify all structures that it passes through, including each chamber, valve, and great vessel. Begin at the right atrium (figure 19.4).

7 Which of the great vessels is an artery and transports deoxygenated blood? Which of the great vessels is a vein and transports oxygenated blood?

INTEGRATE CONCEPT OVERVIEW

Figure 19.4 Blood Flow Through the Heart and Circulatory Routes. The two circulatory routes for blood flow are (*a*) the pulmonary circulation, which includes the pumping of blood by the right side of the heart to the lungs and the return of blood to the left side of the heart, and (*b*) the systemic circulation, which includes the pumping of blood by the left side of the heart to the systemic cells and the return of blood to the right side of the heart.

(a) Pulmonary Circulation

Transports blood from the right side of the heart to the alveoli of the lungs for gas exchange and back to the left side of the heart

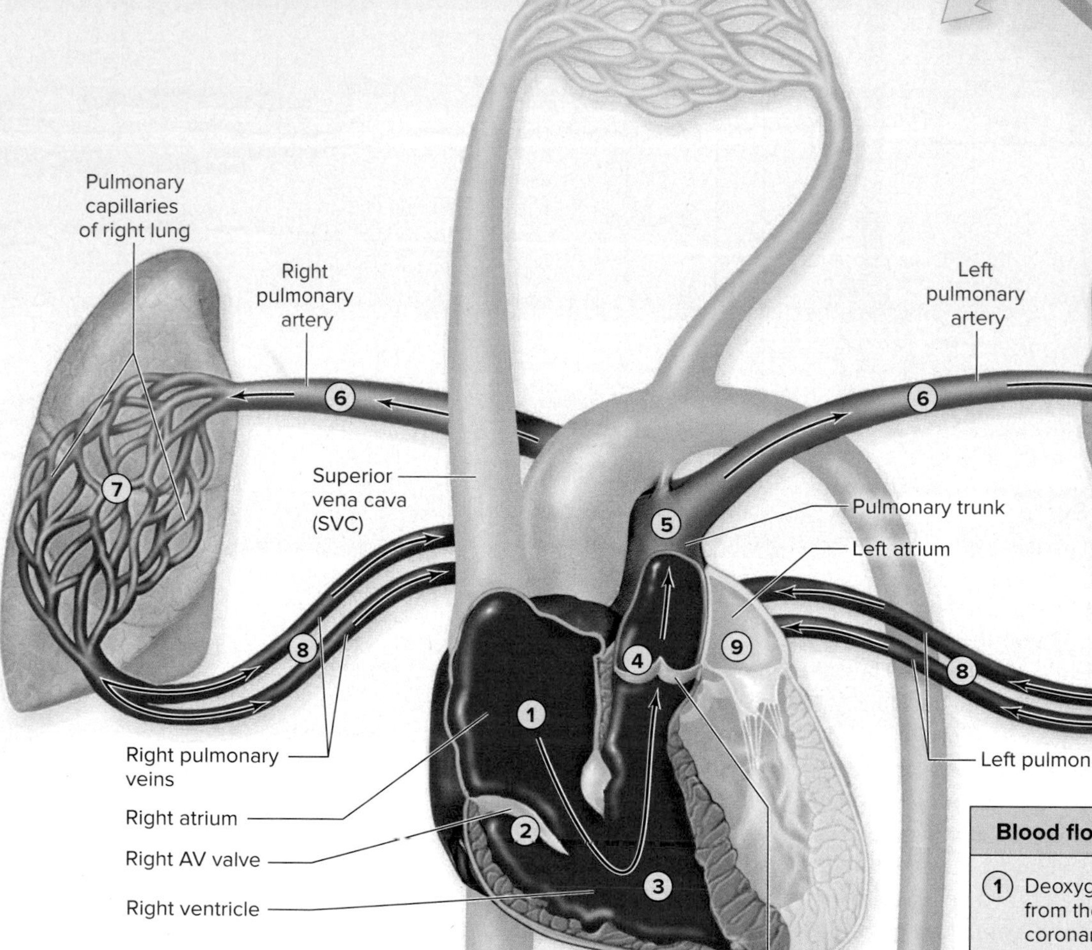

Blood flow through pulmonary circulation

1. Deoxygenated blood enters the **right atrium** from the venae cavae (SVC and IVC) and coronary sinus (not shown). This blood then
2. passes through the **right AV valve** (tricuspid valve),
3. enters the **right ventricle,**
4. passes through the **pulmonary semilunar valve,** and
5. enters the **pulmonary trunk.**
6. This blood continues through the **right and left pulmonary arteries** to both lungs, and
7. enters **pulmonary capillaries** of both lungs for gas exchange.
8. This blood, which is now oxygenated, enters **right and left pulmonary veins,** and is returned to
9. the **left atrium** of the heart.

(b) Systemic Circulation

Transports blood from the left side of the heart to the systemic cells of the body for nutrient and gas exchange, and back to the right side of the heart

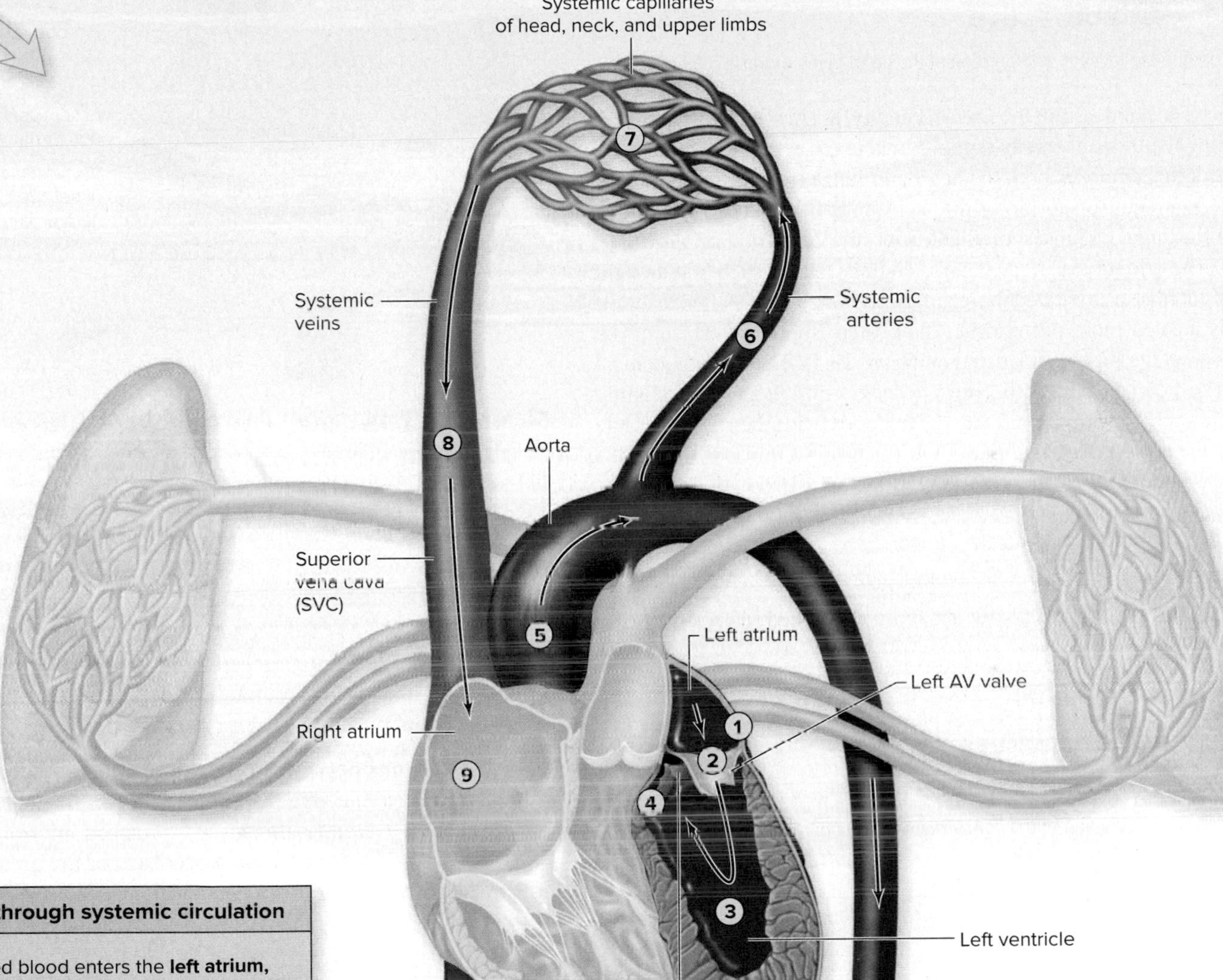

Blood flow through systemic circulation

1. Oxygenated blood enters the **left atrium,**
2. passes through the **left AV valve** (bicuspid or mitral valve),
3. enters the **left ventricle,**
4. passes through **aortic semilunar valve,** and
5. enters the **aorta.**
6. This blood is distributed by the **systemic arteries,** and
7. enters **systemic capillaries** for nutrient and gas exchange.
8. This blood, which is now deoxygenated, ultimately drains into the **SVC, IVC,** and **coronary sinus** (not shown), and
9. enters the **right atrium.**

19.2 The Heart Within the Thoracic Cavity

The heart is the center of the cardiovascular system, and the remaining sections of this chapter discuss both the anatomic and physiologic details of the heart. The location and position of the heart, as well as the pericardium that encloses it, are described in this section. Here our emphasis is on the specific orientation of the heart and the anatomic features that protect and support it.

19.2a Location and Position of the Heart

LEARNING OBJECTIVE

5. Describe the location and position of the heart in the thoracic cavity.

The heart is located within the thoracic cavity, as shown in **figure 19.5**. Supporting the thoracic cavity is the **thoracic cage,** the protective bony framework that encloses both the heart and lungs (see section 8.6). The heart is positioned posterior to the sternum left of the body midline between the lungs within the **mediastinum** (me′dē-as-tī′nŭm; *medius* = middle). The orientation of the heart is slightly rotated such that its right side or right border is located more anteriorly, whereas its left side or left border is located more posteriorly. The postero-superior surface of the heart is called the **base** (not labeled on figure 19.5). The inferior, conical end of the heart is called the **apex** (ā′peks; tip). It projects slightly anteroinferiorly toward the left side of the body with the right ventricle lying on the diaphragm. You may find it helpful to think of the heart's position as an upside down pyramid with the apex below the base.

WHAT DID YOU LEARN?

8. What is the bony structure that protects both the heart and the lungs?
9. Where is the heart positioned, and how is it oriented within the thoracic cavity?

19.2b The Pericardium

LEARNING OBJECTIVES

6. List the structural components of the pericardium.
7. Describe the function of the pericardium and the purpose of the serous fluid within the pericardial cavity.

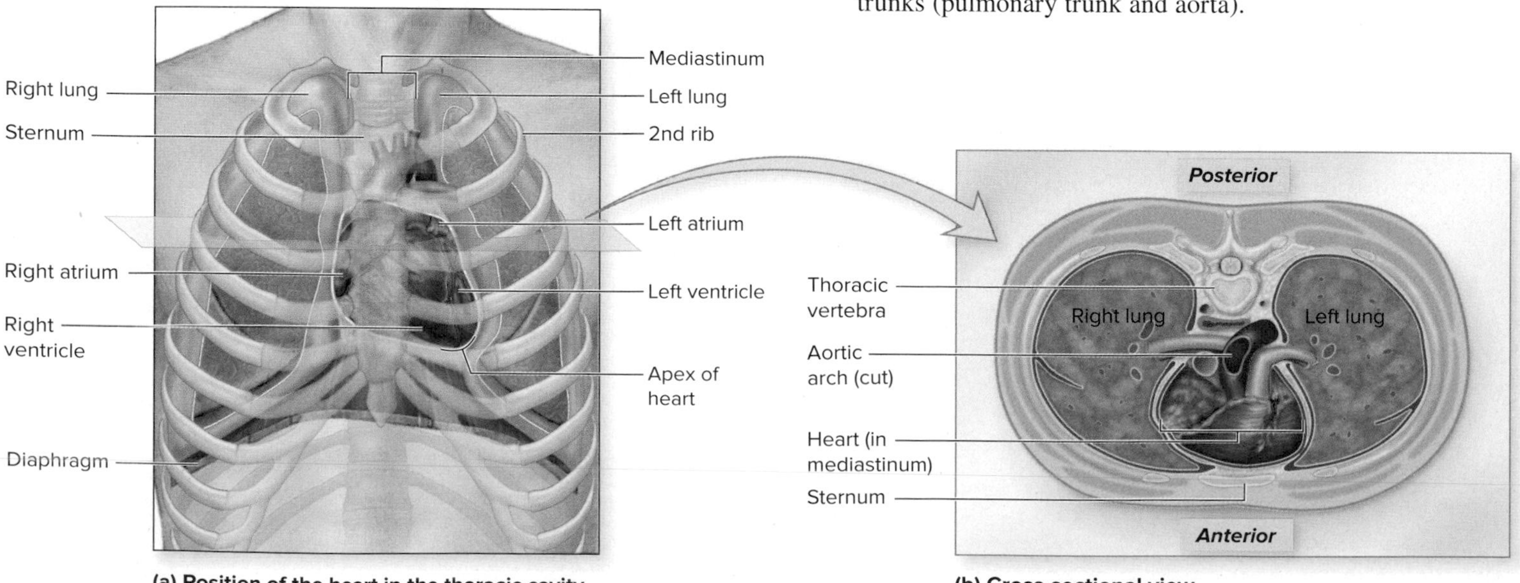

Figure 19.5 Heart Position Within the Thoracic Cavity. (*a*) The heart is located within the mediastinum of the thoracic cavity between the lungs. (*b*) A cross-sectional view depicts the heart's relationship to the other organs in the thoracic cavity. APIR

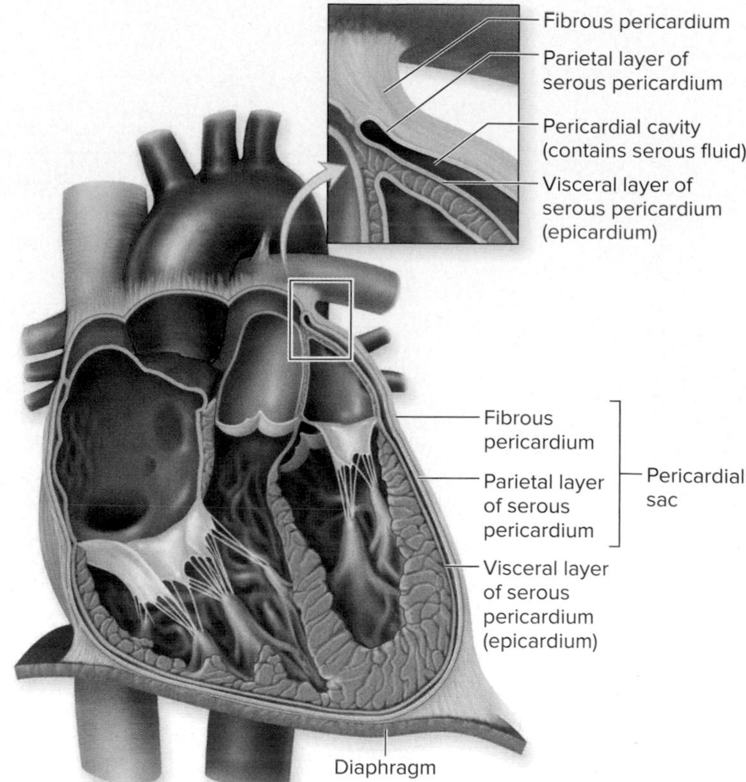

Figure 19.6 Pericardium. The protective layers of the heart include the pericardial sac, composed of an outer fibrous pericardium and an inner serous membrane called the parietal layer of serous pericardium. Tightly adhered to the heart is a serous membrane called the visceral layer of serous pericardium. The space between the parietal and visceral layers is called the pericardial cavity, which contains serous fluid produced by both serous membranes. APIR

The heart is enclosed in three layers, collectively called the **pericardium** (per-i-kar′dē-ŭm; *peri* = around, *kardia* = heart) **(figure 19.6)**. These layers are (from outside to inside) as follows:

- The **fibrous pericardium**, which is composed of dense irregular connective tissue that encloses the heart but does not attach to it. Rather, this layer is attached inferiorly to the diaphragm and superiorly to the base of the great arterial trunks (pulmonary trunk and aorta).

- The **parietal layer of the serous pericardium,** which is composed of simple squamous epithelium and an underlying delicate layer of areolar connective tissue, adheres to the inner surface of the fibrous pericardium.
- The **visceral layer of the serous pericardium** (also called the epicardium) is also composed of a simple squamous epithelium and an underlying delicate layer of areolar connective tissue. This serosal layer adheres directly to the heart. The two serosal layers are continuous with one another (near the great vessels of the heart) and separated by a potential space called the **pericardial cavity.**

The fibrous pericardium and the parietal layer of the serous pericardium together compose the more loosely fitting "bag," called the **pericardial sac,** that encloses the heart (see section 1.5e and figure 1.9).

The tough, fibrous pericardium of the pericardial sac both anchors the heart within the thoracic cavity and prevents the heart chambers from overfilling with blood. The two layers of the serous pericardium produce and release **serous fluid** into the pericardial cavity. This fluid has the consistency of an oily mixture, and it lubricates the serous membranes to minimize friction with every heartbeat (see section 1.5e).

WHAT DID YOU LEARN?

10 Describe the three layers that cover the heart. Where is the pericardial cavity relative to these layers?

INTEGRATE

LEARNING STRATEGY

Imagine your heart as a fist that is placed into a balloon (as shown). The two layers of the balloon are serous membranes. The portion adhered to your hand is the visceral layer, the outer portion is the parietal layer, and the space between them is the pericardial cavity. Now imagine a sandwich bag placed around your hand and balloon; this represents the fibrous pericardium. Together, the outer layer of the balloon and the sandwich bag form the pericardial sac.

INTEGRATE

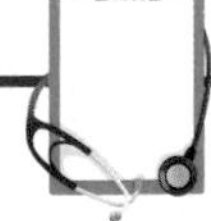

CLINICAL VIEW 19.2

Pericarditis

Pericarditis (per′i-kar-dī′tis; *itis* = inflammation) is an inflammation of the pericardium typically caused by viruses, bacteria, or fungi. The inflammation is associated with an increase in permeability of capillaries within the pericardium, which become more "leaky," resulting in excess fluid leaving the blood and accumulating within the pericardial cavity. At this point, the potential space of the pericardial cavity becomes an actual space as it fills with fluid. In severe cases, the excess fluid accumulation limits the heart's movement and keeps the heart chambers from filling with an adequate amount of blood. The heart is unable to pump blood, leading to a medical emergency called **cardiac tamponade** (tam′pō-nād′) and possibly resulting in heart failure and death.

A helpful physical finding in diagnosing pericarditis is **friction rub.** This is a crackling or scraping sound heard with a stethoscope that is caused by the movement of the inflamed pericardial layers against each other.

19.3 Heart Anatomy

When the pericardial sac is cut and reflected, the heart can be removed from the mediastinum and examined in greater detail. It is a relatively small, conical, muscular organ approximately the size of a person's clenched fist. In the average normal adult, it weighs about 300 grams (0.7 pound), but certain diseases may cause heart size to increase dramatically. Here we examine the heart's external and internal structures.

The superficial anatomic features of the heart are presented in **figure 19.7** from both an anterior and a posterior view in their normal position within the thoracic cavity. Both an illustration and a photo are included. The heart is rotated slightly within the thoracic cavity so that the right side of the heart is more visible from an anterior view (figure 19.7*a*) and the left side of the heart is more visible from a posterior view (figure 19.7*b*).

19.3a Superficial Features of the Heart

LEARNING OBJECTIVE

8. Compare the superficial features of the anterior and posterior aspects of the heart.

Anterior View: Chambers and Great Vessels

The right atrium and right ventricle are prominent when observing the heart from an anterior view (figure 19.7*a*). The portion of the right atrium that is most noticeable is its wrinkled, flaplike extension called the **right auricle** (aw′ri-kl; *auris* = ear). Portions of both the **left auricle** of the left atrium and the left ventricle are also visible. Also seen in this view are the positions of attachment for both the pulmonary trunk to the right ventricle and the aorta to the left ventricle. Note that the pulmonary trunk splits into the **right** and **left pulmonary arteries,** and that the aorta includes the **ascending aorta** (which extends superiorly from the heart), the curved **aortic arch,** and the **descending aorta** (which extends inferiorly through the trunk. (The three branches off the aortic arch are described in section 20.9a.)

Posterior View: Chambers and Great Vessels

The left atrium and left ventricle are prominent when observing the heart from a posterior view (figure 19.7*b*). The left atrium primarily forms the base on the posterosuperior surface of the heart. Also seen in this view are the point of attachment for the superior and inferior venae cavae to the right atrium and the pulmonary veins to the left atrium.

Sulci of the Heart

The atria are separated from the ventricles externally by a relatively deep groove called the **coronary** (kōr′o-nār-ē; *corona* = crown) **sulcus** (or *atrioventricular sulcus*), which extends around the circumference of the heart. It can be viewed on both the anterior and posterior view. An **interventricular** (in-ter-ven-trik′ū-lăr; *inter* = between) **sulcus** is a groove between the ventricles that extends inferiorly from the coronary sulcus toward the heart apex, and delineates the superficial boundary between the right and left ventricles. The **anterior interventricular sulcus** is located on the anterior side of the heart, and the **posterior interventricular sulcus** is located on the posterior side of the heart. Located within all of these sulci are coronary vessels associated with supplying blood to the heart wall. Coronary vessels are described in detail in section 19.4.

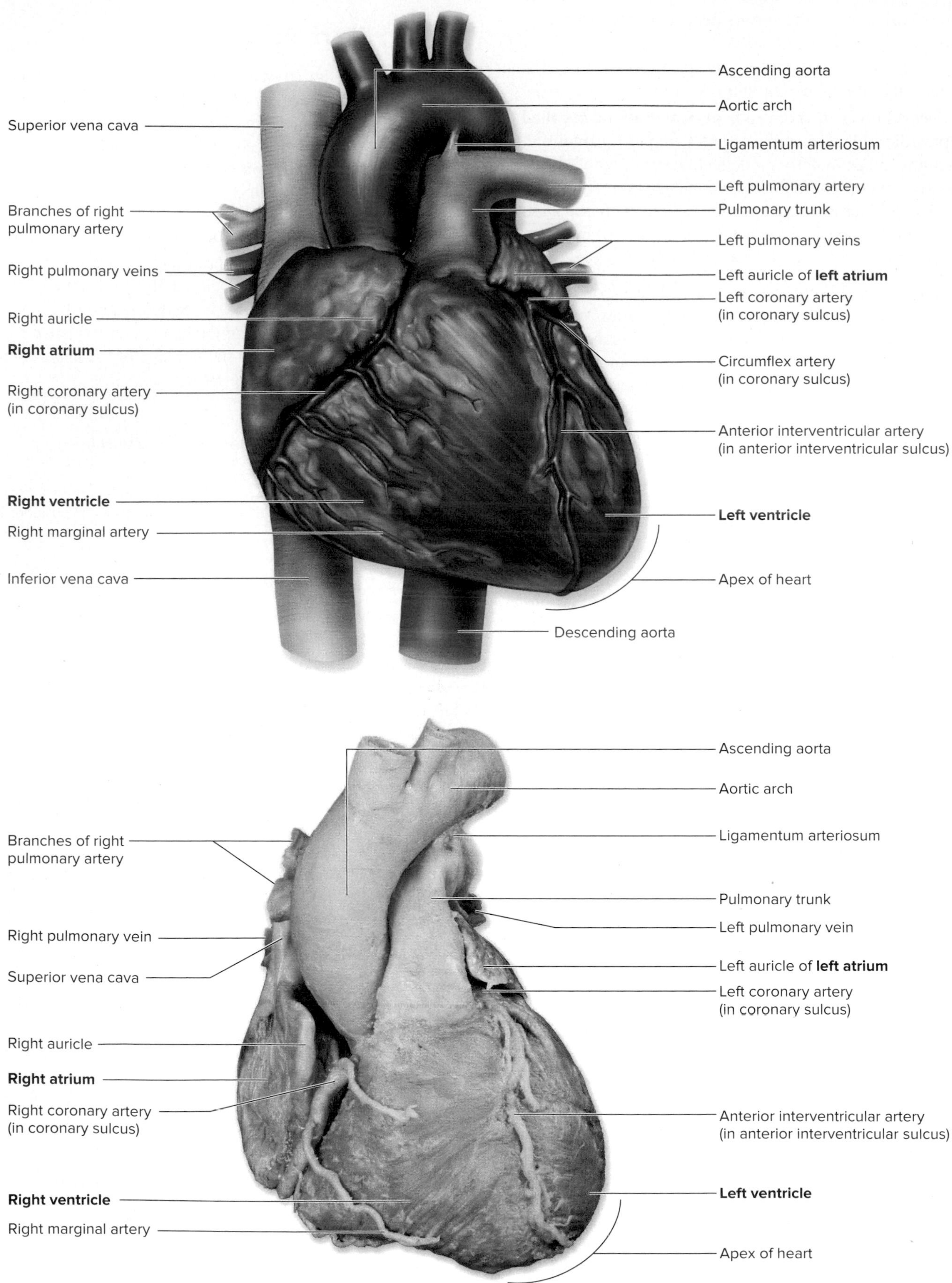

Figure 19.7 External Anatomy and Features of the Heart. A drawing and a cadaver photo show the superficial features of the heart. (*a*) Anterior view. Note: Adipose connective tissue, which supports the coronary vessels within the heart sulci, was removed prior to the taking of these photos. AP|R

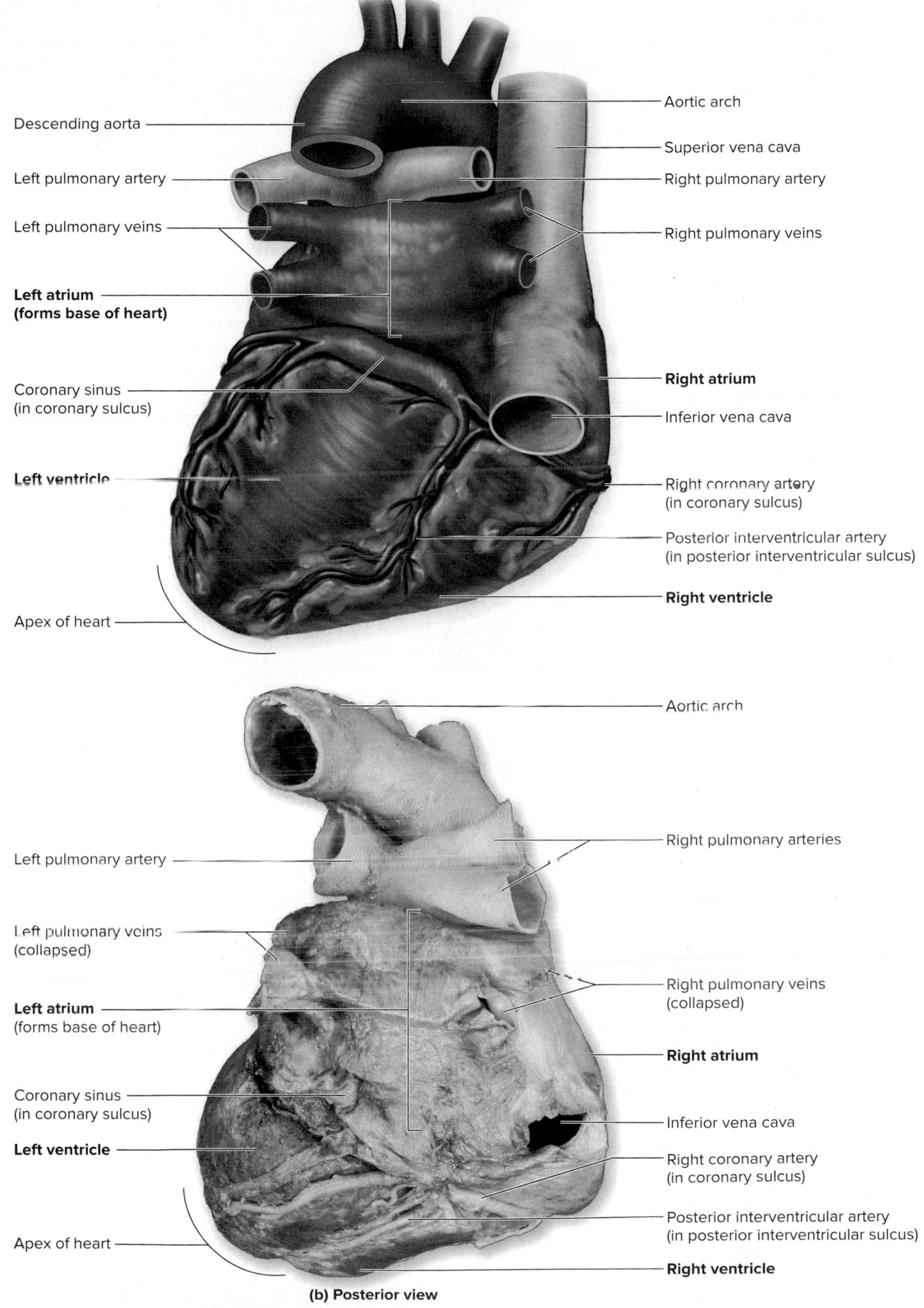

(b) Posterior view

Figure 19.7 External Anatomy and Features of the Heart *(continued).* (*b*) Posterior view. AP|R

WHAT DID YOU LEARN?

11 Which side of the heart is more visible on the anterior view of the heart? On the posterior view?

12 Which of the great vessels are more visible from the anterior view? From the posterior view?

13 What are the three primary sulci of the heart? What structures are within these superficial grooves of the heart?

19.3b Layers of the Heart Wall

LEARNING OBJECTIVE

9. Name the three layers of the heart wall and the tissue components of each.

A coronal section provides a view of the internal structures of the heart **(figure 19.8)**. First, observe the heart wall of the different heart chambers. Notice that the walls of the ventricles are thicker than the walls of the atria; this is because the ventricles are the "pumping chambers" and must exert force to move the blood from the heart through the cardiovascular circuits. Also note that the wall of the left ventricle is typically three times thicker than the right ventricular wall. This difference is because the left ventricle must generate enough pressure to force the blood through the entire systemic circulation, whereas the right ventricle merely has to pump blood to the nearby lungs as part of the pulmonary circulation. This difference in thickness can also be seen when a transverse section is made through the heart **(figure 19.9)**.

Three distinctive layers compose the wall of each chamber: an external epicardium, a thick middle myocardium, and an internal endocardium (figure 19.8*b*).

The **epicardium** (ep-i-kar′dē-ŭm; *epi* = upon) is the outermost heart layer and is also called the *visceral layer of serous pericardium.* This layer, as previously described in section 19.2b, is composed of simple squamous epithelium and an underlying layer of areolar connective tissue. As we age, the epicardium thickens as it becomes more invested with adipose connective tissue.

The **myocardium** (mī-ō-kar′dē-ŭm; *mys* = muscle) is the middle layer of the heart wall. It is composed of cardiac muscle tissue (see section 5.3) and is the thickest of the three heart wall layers. Contraction of cardiac muscle composing the myocardium generates the force necessary to pump blood. The ventricular myocardium may change in thickness as we age or if we participate in regular, rigorous exercise. For example, it hypertrophies in response to narrowing of systemic arteries because the heart must work harder to pump the blood. We consider the microscopic anatomy of cardiac muscle in section 19.3f.

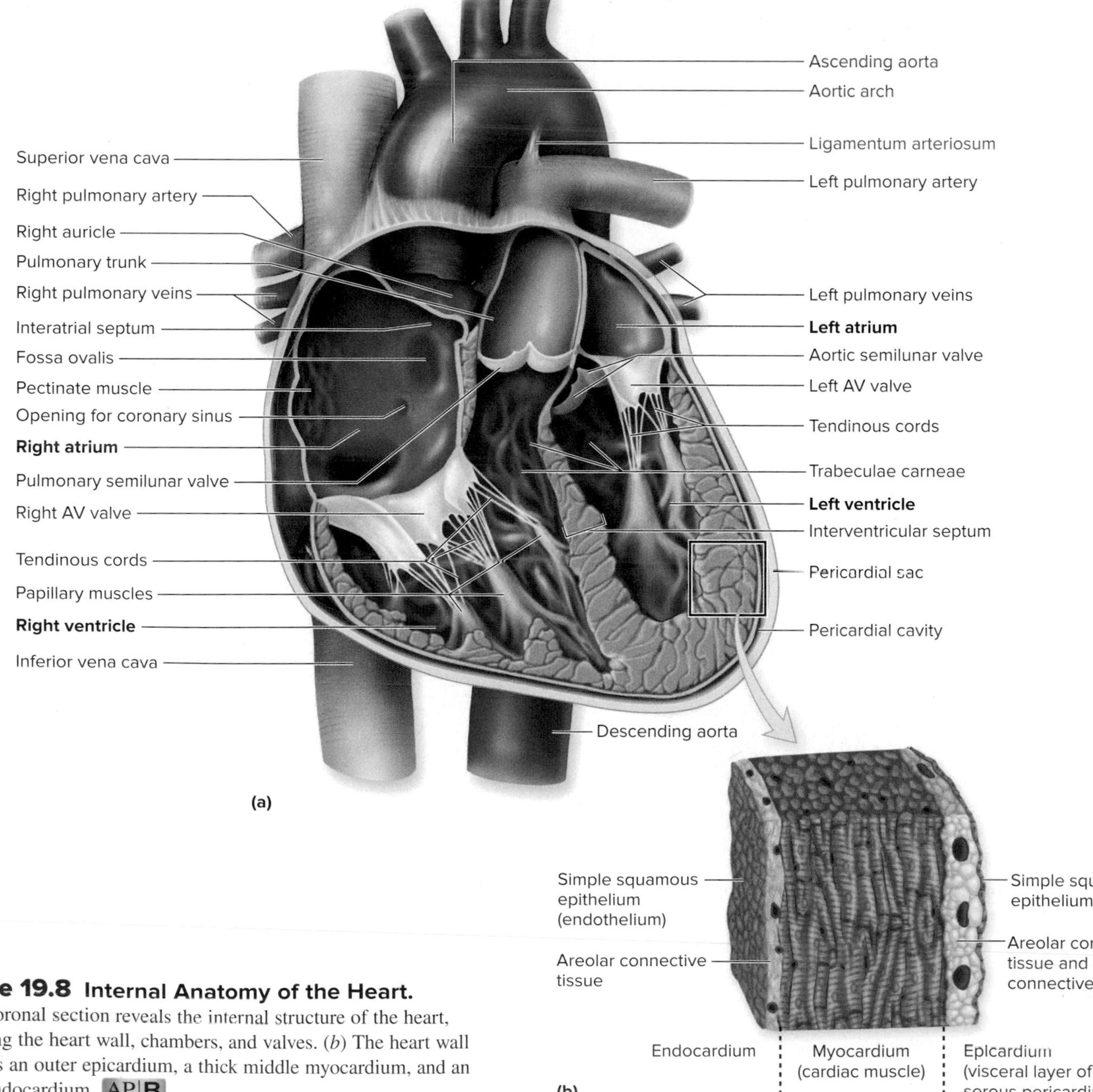

Figure 19.8 Internal Anatomy of the Heart. (*a*) A coronal section reveals the internal structure of the heart, including the heart wall, chambers, and valves. (*b*) The heart wall contains an outer epicardium, a thick middle myocardium, and an inner endocardium. AP|R

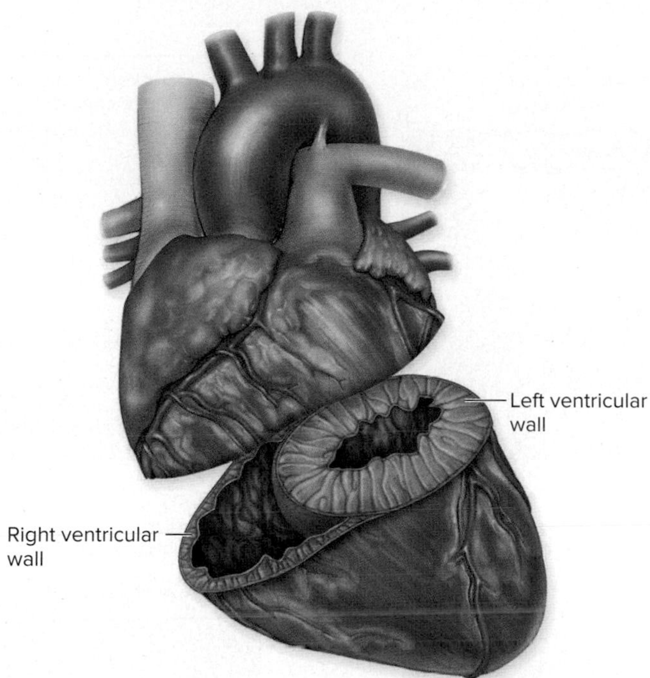

Figure 19.9 Comparison of Right and Left Ventricular Wall Thickness. The wall of the left ventricle is typically three times thicker than that of the right ventricle, because the left ventricle must generate a force sufficient to push blood through the systemic circulation.

The internal surface of the heart and the external surfaces of the heart valves are covered by **endocardium** (en-dō-kar′dē-ŭm; *endon* = within). The endocardium, like the epicardium, is composed of a simple squamous epithelium and an underlying layer of areolar connective tissue. The epithelial layer of the endocardium is continuous with the epithelial layer called the endothelium, which lines the blood vessels (see figure 20.1).

WHAT DID YOU LEARN?

14 What are the layers of the heart (in order) that a scalpel would pass through during dissection? What are the two names given to the outer layer of the heart wall?

19.3c Heart Chambers

LEARNING OBJECTIVE

10. Characterize the four chambers of the heart and their functions.

Figure 19.8 also depicts the internal anatomy and structural organization of the four heart chambers. The right and left atrial chambers are separated by a thin wall called the **interatrial** (in-ter-ā′trē-ăl) **septum,** and the right and left ventricles are separated by a thick wall called the **interventricular septum.** The position of the interventricular septum is delineated on the heart's superficial surface with the interventricular sulci (see figure 19.7*a, b*). (Thus, the interventricular groove on the surface is the landmark for the interventricular wall internal to it.)

Each of the four chambers plays a role in the continuous process of blood circulation and are described in the order that blood moves through them.

Right Atrium

The internal wall of the right atrium is smooth on its posterior surface, but it exhibits muscular ridges called **pectinate** (pek′ti-nāt; teeth of a comb) **muscles** on its anterior wall and within the auricle (figure 19.8*a*). Inspection of the interatrial septum reveals an oval depression called the **fossa ovalis** (fos′ă; trench; ō-va′lis). It occupies the former location of the fetal **foramen ovale** (ō-val′ē), which shunted blood from the right atrium to the left atrium, bypassing the lungs during fetal life (fetal circulation is described in section 20.12). Immediately inferior to the fossa ovalis is the opening of the *coronary sinus,* which drains deoxygenated blood from the heart wall. Openings of the superior and inferior venae cavae are also visible. Thus, three veins drain deoxygenated blood into the right atrium: the coronary sinus, superior vena cava, and inferior vena cava. Separating the right atrium from the right ventricle is the right atrioventricular opening that contains the right AV valve. Deoxygenated blood flows from the right atrium, through the right atrioventricular opening when the valve is open, into the right ventricle.

Right Ventricle

The internal wall surface of the right ventricle displays characteristic large, smooth, irregular muscular ridges, called the **trabeculae carneae** (tră-bek′ū-lē; *trabs* = beam; kar′nē-ē; *carne* = flesh). Extending from the internal wall of the right ventricle are typically three cone-shaped, muscular projections called **papillary** (pap′i-lăr-ē; *papilla* = nipple) **muscles.** (The number of papillary muscles in the right ventricle can range from two to nine.) Papillary muscles anchor thin strands of collagen fibers called **tendinous cords** or **chordae tendineae** (kōr′dē ten′di-nē-ē), which are attached to the free edge of the right atrioventricular valve. The superior portion of the right ventricle narrows into a smooth-walled region leading into the pulmonary trunk. The pulmonary semilunar valve is positioned between the right ventricle and pulmonary trunk. Deoxygenated blood is pumped from the right ventricle through the open pulmonary semilunar valve into the pulmonary trunk. This blood is delivered to the lungs for gas exchange and returned via pulmonary veins to the left atrium.

Left Atrium

The left atrium, like the right atrium, has pectinate muscles in its auricle. Openings of the pulmonary veins are visible. (Two are seen in figure 19.8.) Separating the left atrium from the left ventricle is the left atrioventricular opening, which contains the left AV valve. Oxygenated blood flows from the left atrium, through the left atrioventricular opening when the valve is open, into the left ventricle.

Left Ventricle

The internal surface of the left ventricle also displays characteristic trabeculae carneae. It has two papillary muscles that are anchored by tendinous cords. The entrance into the aorta is located at the superior aspect of the left ventricle. The aortic semilunar valve is positioned at the boundary of the left ventricle and ascending aorta. Oxygenated blood is pumped from the left ventricle through the open aortic semilunar valve into the pulmonary trunk.

WHAT DID YOU LEARN?

15 What is the structure that separates the two ventricles? What is the superficial landmark that identifies the location of this structure?

16 How are the papillary muscles, tendinous cords, and atrioventricular valve positioned relative to one another?

19.3d Heart Valves

LEARNING OBJECTIVE

11. Compare and contrast the structure and function of the two types of heart valves.

Effective blood flow requires valves to control the flow of blood and ensure it is "one-way." Recall that the two categories of heart valves are the atrioventricular (AV) valves and the semilunar valves. Each valve consists of endothelium-lined fibrous connective tissue flaps called cusps **(figure 19.10)**.

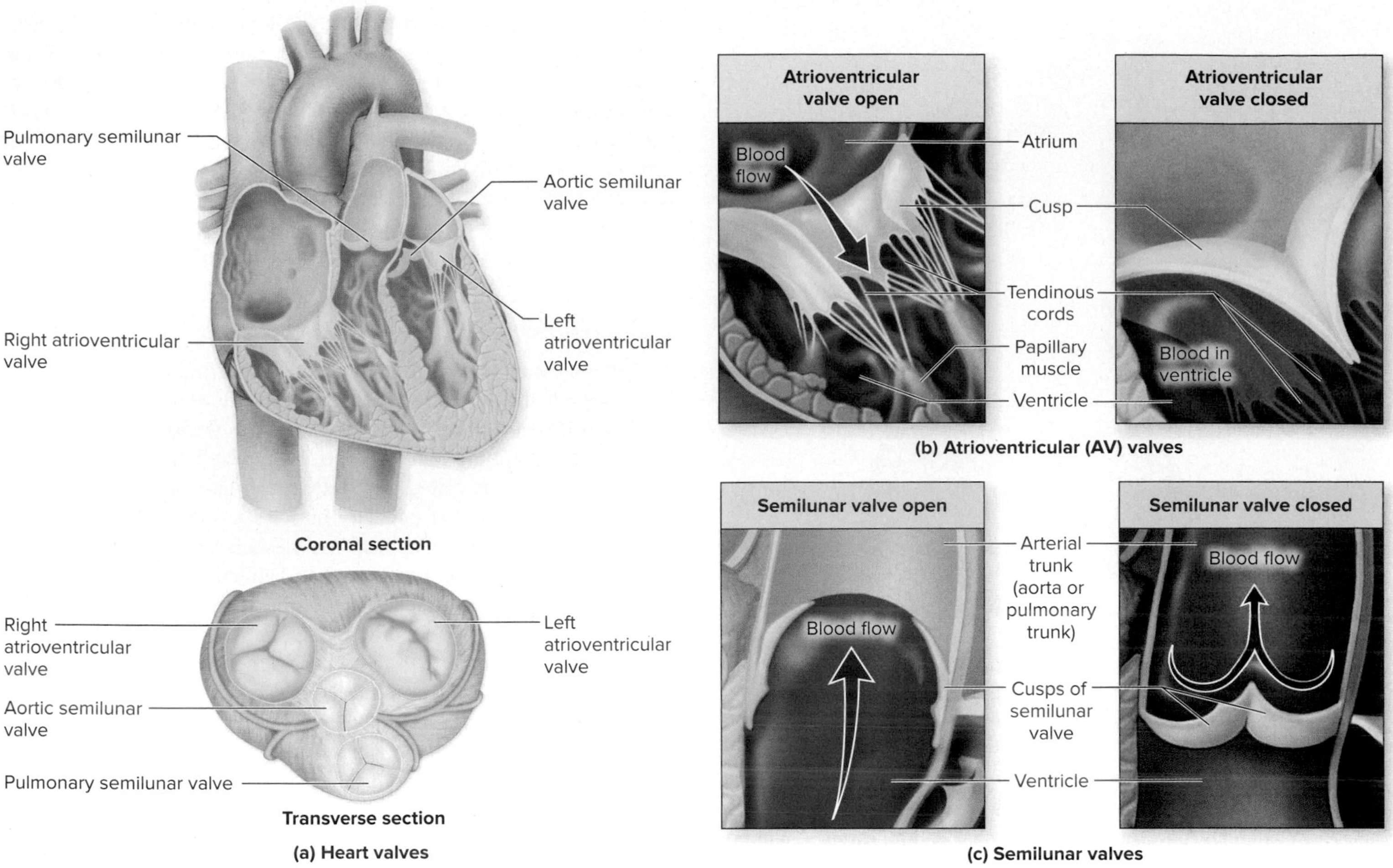

Figure 19.10 Heart Valves. (*a*) Location of heart valves as viewed in coronal section and transverse section. (*b*) Atrioventricular valves in open and closed positions. (*c*) Semilunar valves in open and closed positions. AP|R

INTEGRATE

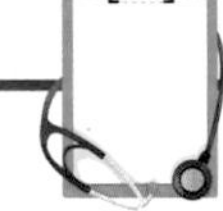

CLINICAL VIEW 19.3

Teenage Athletes and Sudden Cardiac Death

Sudden deaths of high school athletes have occurred during athletic events. These deaths were apparently caused by underlying, previously undetected cardiovascular disease. Previous studies indicated that no forewarning of a problem preceded these deaths. Autopsies revealed that most deaths were due to congenital heart defects and coronary artery anomalies. The result of these anomalies was cardiomegaly, which led to sudden death.

Cardiomegaly (kar′dē-ō-meg′ă-lē) is one term used to indicate an increase in the thickness of the heart muscle wall (hypertrophy) or an obvious increase or enlargement in heart size (dilation) due to stress applied to the heart. Many cardiovascular diseases can cause enlargement of the heart. However, the enlargement of the heart seen in young athletes is caused by a condition called **hypertrophic cardiomyopathy.** Other terms used are *hypertrophic heart, heart enlargement,* and *athlete's heart.* This condition involves both an enlargement of the heart walls (especially one or both ventricles) and a narrowing of the openings (outlets) for the blood to pass through, which result in a decrease in cardiac output. Strenuous exercise exacerbates the condition, and patients may develop symptoms of shortness of breath, chest pain, and fainting. In some cases, sudden death may occur, often while the individual is participating in a physically demanding activity.

A standard x-ray that may accompany a physical examination often reveals potential cardiomegaly. The confirmation of the cardiomegaly being associated with hypertrophic cardiomyopathy (athlete's heart) depends upon an echocardiogram (a specialized ultrasound of the heart) that is used to evaluate the heart's function and its structure.

As a result, many school districts require student athletes to undergo a physical exam with an accompanying x-ray or echocardiogram prior to being approved to participate on athletic teams.

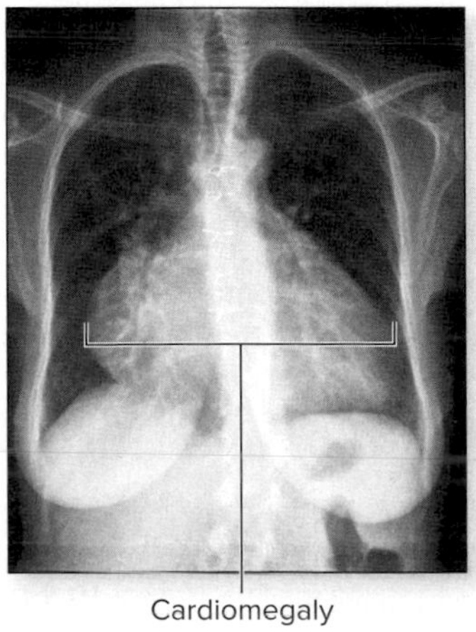

Cardiomegaly in an adult female. Note how the heart shadow encompasses most of the width of the thorax.
©ISM/Sovereign/Medical Images

INTEGRATE

CLINICAL VIEW 19.4

Heart Sounds and Heart Murmurs

There are two normal heart sounds associated with each heartbeat, which collectively form the lubb-dupp sounds. The *lubb* sound is also known as the **S1 sound** and represents the closing of the atrioventricular valves (see figure 19.21, step 2). The *dupp* sound is also known as the **S2 sound** and is the closing of the semilunar valves (see figure 19.21, step 4). These heart sounds provide clinically important information about heart activity and the action of heart valves.

The place where sounds from each AV valve and each semilunar valve may best be heard does not correspond with the location of the valve, because some overlap of valve sounds occurs near their anatomic locations:

- The aortic semilunar valve is best heard in the second intercostal space to the right of the patient's sternum.
- The pulmonary semilunar valve is best heard in the second intercostal space to the left of the patient's sternum.
- The right AV valve is best heard at the patient's inferior left sternal border in the fifth intercostal space.
- The left AV valve is best heard near the apex of the heart (at the level of the patient's left fifth intercostal space, about 9 centimeters from the midline of the sternum).

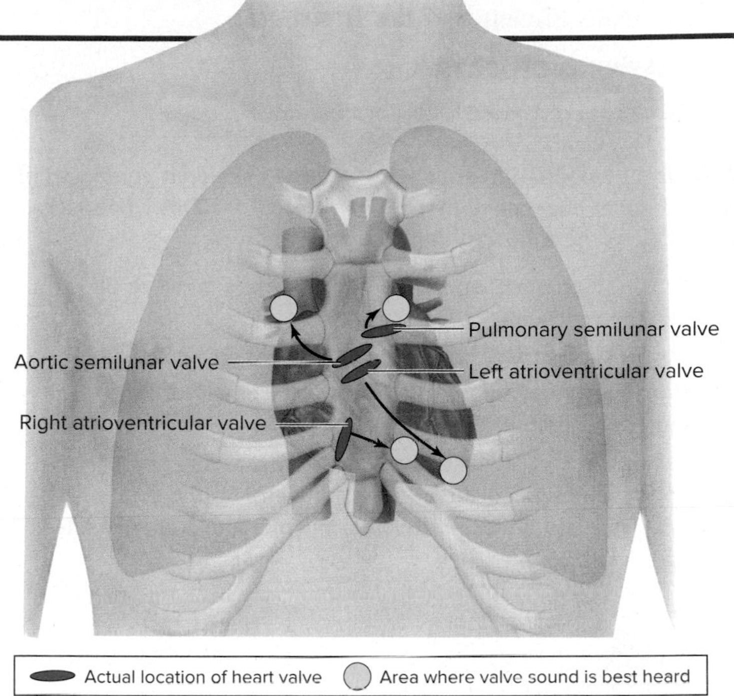

Locations of individual heart valves and the ideal listening sites for each valve.

An abnormal heart sound, generally called a **heart murmur,** is the first indication of heart valve problems. A heart murmur is usually the result of turbulence of the blood as it passes through the heart, and it may be caused by valvular leakage, decreased valve flexibility, or a misshapen valve. Sometimes heart murmurs are of little consequence, but all of them need to be evaluated to rule out a more serious heart problem. Two types of heart murmurs are valvular insufficiency and valvular stenosis.

Valvular insufficiency, also termed *valvular incompetence,* occurs when one or more of the cardiac valves leak because the valve cusps do not close tightly enough. Inflammation or disease may cause the free edges of the valve cusps to become scarred and constricted, allowing blood to regurgitate back through the valve and possibly causing further heart enlargement.

Valvular stenosis (ste-nō′sis; narrowing) is scarring of the valve cusps so that they become rigid or partially fused and cannot open completely. A stenotic valve is narrowed and presents resistance to the flow of blood, decreasing chamber output. Often the affected chamber undergoes hypertrophy and dilates—both conditions that may have dangerous consequences. A primary cause of valvular stenosis is **rheumatic** (rū-mat′ik) **heart disease,** which may follow a streptococcal infection of the throat.

Atrioventricular Valves

The right AV valve is located between the right atrium and the right ventricle, and it has three cusps (which is why it is also called the *tricuspid valve*). The left AV valve is located between the left atrium and left ventricle, but it has only two cusps (which is why it is also called the *bicuspid valve;* the name *mitral* is sometimes used for this valve because it resembles the headdress of bishops, which is called a mitre). The AV valves are shown in both the open and closed positions in figure 19.10*b*. When open, the cusps of the valve extend into the ventricles. This allows blood to move from an atrium into the ventricle. When the ventricles contract, blood is forced superiorly as ventricular pressure rises. This causes the AV valves to close. The papillary muscles secure the tendinous cords that attach to the lower surface of each AV valve cusp. This prevents the valve from inverting into the atrium when the valve is closed. By being properly held in place, the cusps of the AV valves prevent backflow of blood into the atrium.

Semilunar Valves

The pulmonary semilunar valve is located between the right ventricle and the pulmonary trunk, and the aortic semilunar valve is located between the left ventricle and the ascending aorta. Each valve is composed of three pocketlike cusps, which have the shape of a half-moon (figure 19.10*a, c*). Neither papillary muscles nor tendinous cords are associated with these valves. The semilunar valves are shown in both the open and closed positions in figure 19.10*b*.

The semilunar valves open when the ventricles contract and the force of the blood pushes the semilunar valves open and blood enters the arterial trunks. The valves close when the ventricles relax and the pressure in the ventricle becomes less than the pressure in an arterial trunk. Blood in the arteries begins to move backward toward the ventricle and is caught in the cusps of the semilunar valves, and they close. The closure of the semilunar valves prevents backflow of blood into the ventricle.

Both flexibility and elasticity of connective tissue composing heart valves decrease with aging (or disease). This may cause the heart valves to become inflexible. As a result, blood flow through the heart may be altered, and a heart murmur may be detected (see Clinical View 19.4: "Heart Sounds and Heart Murmurs").

WHAT DID YOU LEARN?

17 What are the functions of the tendinous cords and papillary muscles?

19.3e Fibrous Skeleton of the Heart

LEARNING OBJECTIVE

12. Describe the location and function of the fibrous skeleton.

The heart is supported internally by a **fibrous skeleton** composed of dense irregular connective tissue **(figure 19.11)**. This fibrous skeleton performs the following functions:

- Provides structural support at the boundary between the atria and the ventricles
- Forms supportive fibrous rings to anchor the heart valves
- Provides a rigid framework for the attachment of cardiac muscle tissue
- Acts as an electric insulator because it prevents propagation of action potentials directly from the atria to the ventricles, thus preventing the ventricles from contracting at the same time as the atria

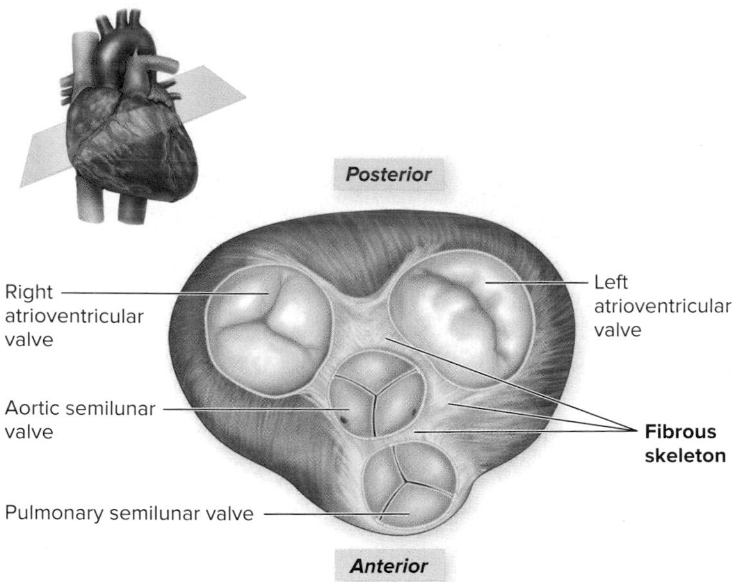

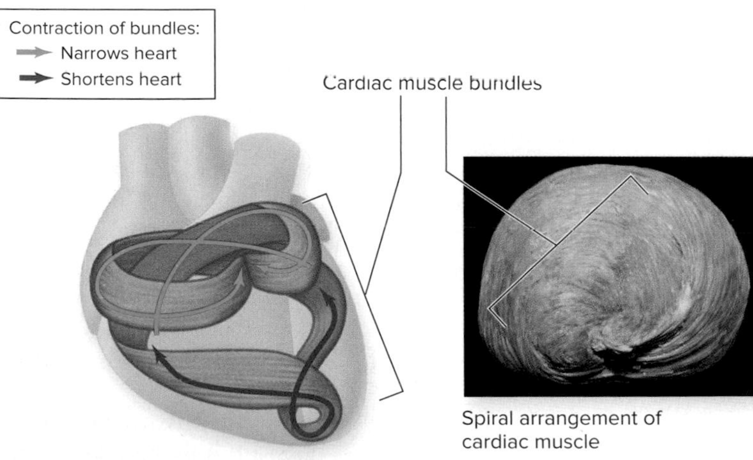

Figure 19.11 Fibrous Skeleton of the Heart.
(*a*) The fibrous skeleton is composed of dense irregular connective tissue that provides both mechanical support and electrical insulation within the heart. Shown in superior view, its structural support of heart valves is especially evident. (*b*) Diagram and photo of spiral pattern of cardiac muscle attached to the fibrous skeleton. (Note: The red dots on the aortic semilunar valve represent the openings for the right and left coronary arteries.)
(*b*) ©Dr. Carlos Baptista and Roy Schneider, University of Toledo HSC

Notice that cardiac muscle cells are arranged in spiral bundles around the heart chambers attached to the fibrous skeleton (figure 19.11*b*). When the atria contract, they compress the wall of the chambers inward, thus narrowing them and moving the blood inferiorly into the ventricles. When the ventricles contract, the action is similar to the wringing of a mop in that it begins at the apex of the heart and compresses superiorly, moving the blood into the great arteries.

WHAT DID YOU LEARN?

18 Which function of the fibrous skeleton allows the atria to contract separately from the ventricles?

19.3f Microscopic Structure of Cardiac Muscle

LEARNING OBJECTIVES

13. Describe the general structure of cardiac muscle.

14. Explain the intercellular structures of cardiac muscle.

15. Discuss how cardiac muscle meets its energy needs.

The myocardium is composed of cardiac muscle tissue (see section 5.3). This muscle tissue is made up of relatively short, branched cells that usually house one or two central nuclei **(figure 19.12)**. These cells are supported by areolar connective tissue, called an endomysium, that surrounds the cells. Other anatomic structures of cardiac muscle cells include the following:

- The sarcolemma (plasma membrane), which invaginates to form T-tubules that extend to the sarcoplasmic reticulum (SR). The T-tubules of cardiac muscle invaginate once per sarcomere and overlie Z discs.
- The SR surrounds bundles of myofilaments called myofibrils in cardiac muscle, but it is less extensive than the SR of skeletal muscle and lacks both terminal cisternae (the "end sacs" of sarcoplasmic reticulum) and a tight association with T-tubules (figure 19.12*c*).
- Myofilaments are arranged in sarcomeres—thus, cardiac muscle appears striated when viewed under a microscope (figure 19.12*d*). Interestingly, maximum overlap of thin and thick filaments within the sarcomeres does *not* occur when cardiac muscle is at rest (as is the case within skeletal muscle; see length-tension relationship in section 10.7c). Instead, maximum overlap of thin and thick filaments occurs when cardiac muscle is stretched as blood is added to a heart chamber. As we will see, this provides a means of forming additional crossbridges between the thin and thick filaments and cardiac muscle contracting with increasingly greater degrees of force as additional blood enters the chamber.

Intercellular Structures

Neighboring cardiac muscle cells have an extensively folded sarcolemma that permits adjoining membranes to interconnect, markedly increasing exposed surface areas between neighboring cells (figure 19.12*b*). (This can be demonstrated by interlocking your fingers.) This increases structural stability of the myocardium and facilitates communication between cardiac muscle cells.

Unique structures called **intercalated** (in-ter′kă-lā-ted) **discs** are found at these cell-to-cell junctions. They link cardiac muscle cells together both mechanically and electrically and contain two distinctive structural features:

- **Desmosomes.** These are protein filaments that anchor into a protein plaque located on the internal surface of the sarcolemma (see section 4.6d). They act as mechanical junctions to prevent cardiac muscle cells from pulling apart.

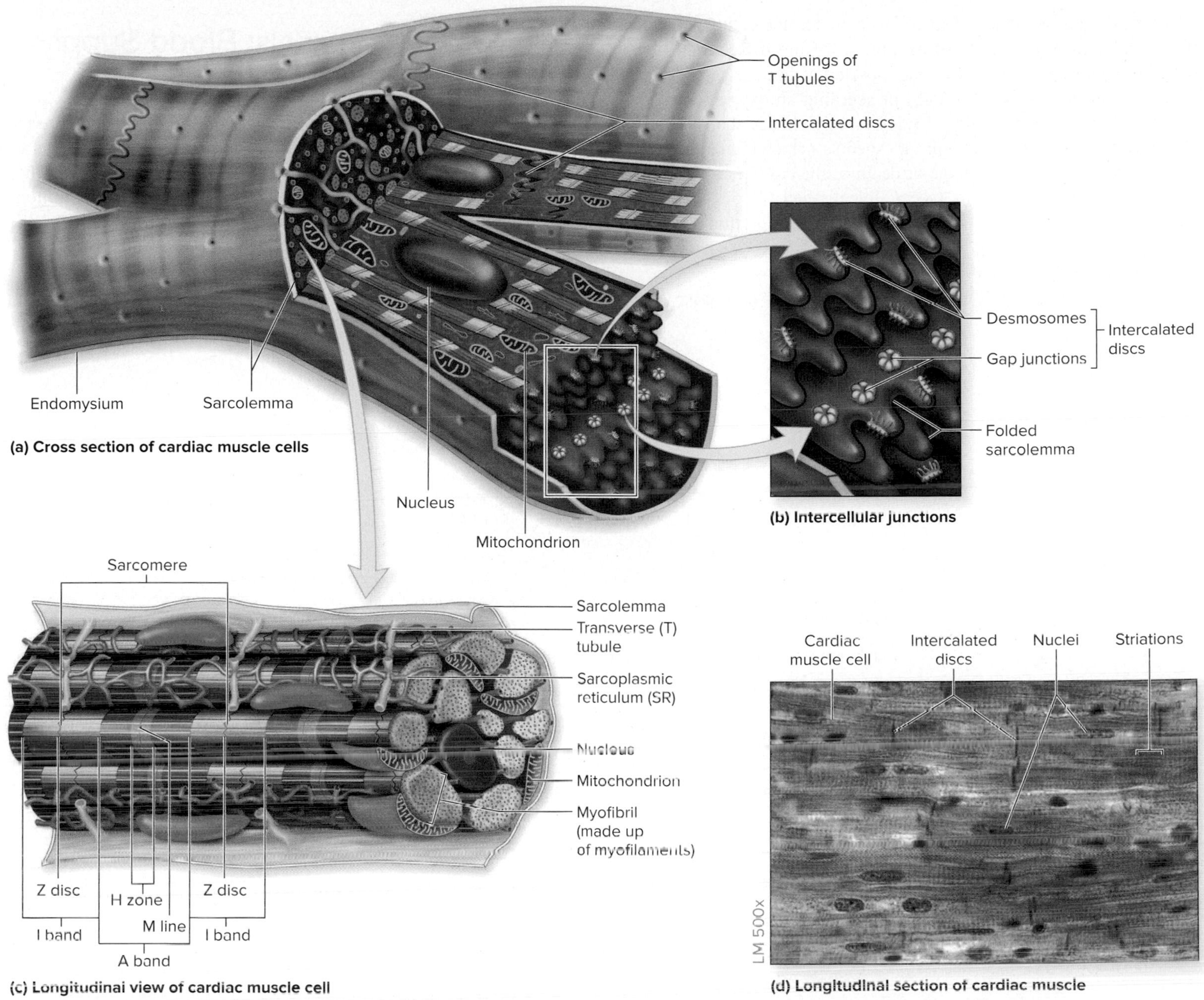

Figure 19.12 Histology of Cardiac Muscle. Cardiac muscle cells form the myocardium. (*a*) Individual cells are relatively short, branched, and striated. (*b*) They are connected to adjacent cells by intercalated discs composed of both desmosomes and gap junctions. (*c*) T-tubules are invaginations of the sarcolemma that extend internally to the sarcoplasmic reticulum. (*d*) Light micrograph of cardiac muscle in longitudinal section. Intercalated discs are visible as dark lines perpendicular to the cardiac muscle cells. AP|R

- **Gap junctions.** These are protein pores between the sarcolemma of adjacent cardiac muscle cells. They provide a low-resistance pathway for the flow of ions between the cardiac cells. Gap junctions allow an action potential to move continuously along the sarcolemma of cardiac muscle cells, resulting in essentially simultaneous stimulation of cardiac muscle cells within a chamber and, thus, synchronous contraction of that chamber. A chamber is called a **functional syncytium** (sin-si′shē-ŭm) because it functions as a single unit. (Note: A syncytium is defined as a multinucleated mass that is formed by the union of originally separate cells. Thus, to say that a heart chamber is a *functional syncytium* is to say that the chamber functions as if it were one cell.)

Metabolism of Cardiac Muscle

Cardiac muscle has features that support its great demand for energy, including an extensive blood supply, numerous mitochondria, and other structures such as **myoglobin** (which is a globular protein that binds oxygen when the muscle is at rest) and **creatine kinase** (which catalyzes the transfer of Pi from creatine phosphate to ADP, yielding ATP and creatine). Myoglobin was first described in section 10.2b, and creatine kinase in section 10.4a (see Clinical View 10.5: "Creatine Kinase Blood Levels as a Diagnostic Tool").

Cardiac muscle relies almost exclusively on aerobic cellular respiration (see section 3.4). Its cellular structures and metabolic processes support this. Cardiac muscle has a large number of mitochondria (comprising approximately 25% of its volume compared to about 2% of the volume in skeletal muscle). It is also versatile in being able to use different types of fuel molecules, including fatty acids, glucose, lactate, amino acids, and ketone bodies (see section 3.4h). The relative amounts of these molecules that cardiac muscle cells use fluctuate depending upon conditions. For example, during intense exercise when greater amounts of lactate are released into the blood by skeletal muscle, cardiac muscle will absorb the additional lactate from the blood and use this resource.

Yet, as a consequence of its reliance on aerobic cellular respiration, cardiac muscle is quite susceptible to failure if **ischemic** (low-oxygen) conditions prevail. Cardiac muscle has limited capability in using glycolysis (see section 3.4b) or accruing an oxygen debt (see section 10.4b) in its activities. Therefore, any change that interferes with blood flow to the heart muscle, such as narrowing of the coronary arteries, can cause damage or death of the cardiac muscle cells composing the myocardium (see Clinical View 19.5: "Coronary Heart Disease, Angina Pectoris, and Myocardial Infarction").

WHAT DID YOU LEARN?

19 Which features of cardiac muscle support aerobic cellular respiration?

19.4 Coronary Vessels: Blood Supply Within the Heart Wall

Although the heart continuously pumps blood through its chambers, it cannot absorb the oxygen and nutrients it requires from this blood. The diffusion of oxygen and nutrients at the needed rates through the thick wall of the heart is not possible; instead, an intricate distribution system called the **coronary circulation** handles the task **(figure 19.13)**. The vessels that transport oxygenated blood to the wall of the heart are called coronary arteries, whereas coronary veins transport deoxygenated blood away from the heart wall. Vessels on the posterior aspect of the heart are shown shaded in figure 19.13*a, b*.

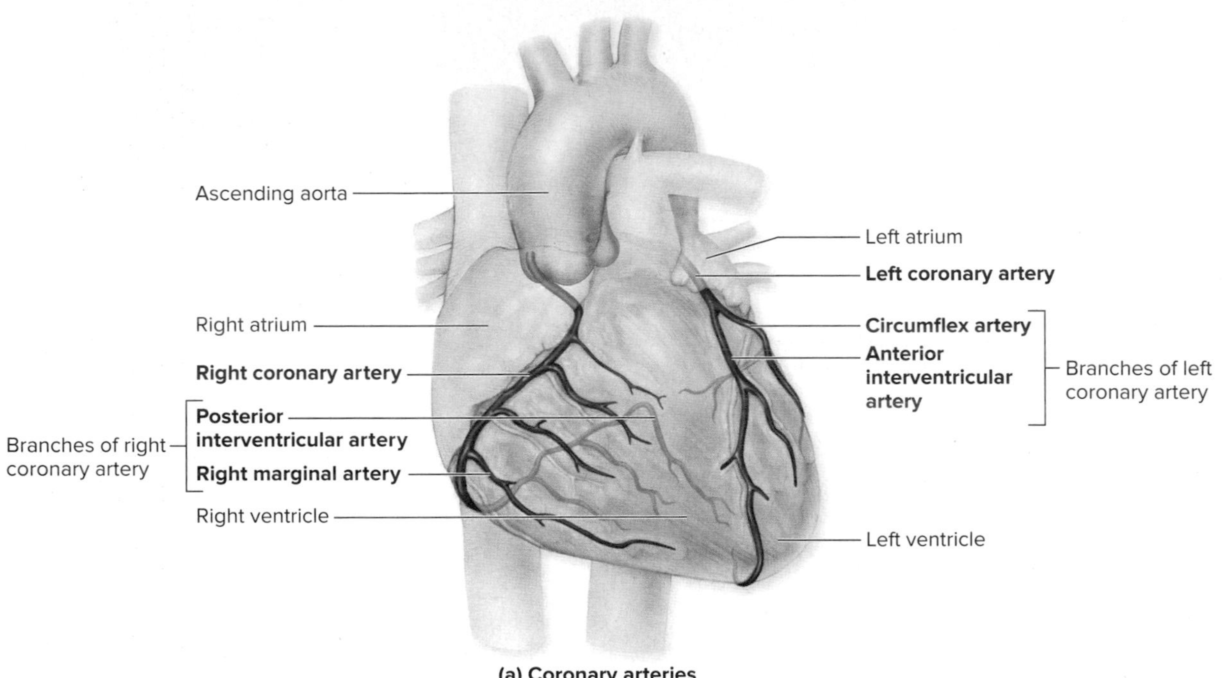

(a) Coronary arteries

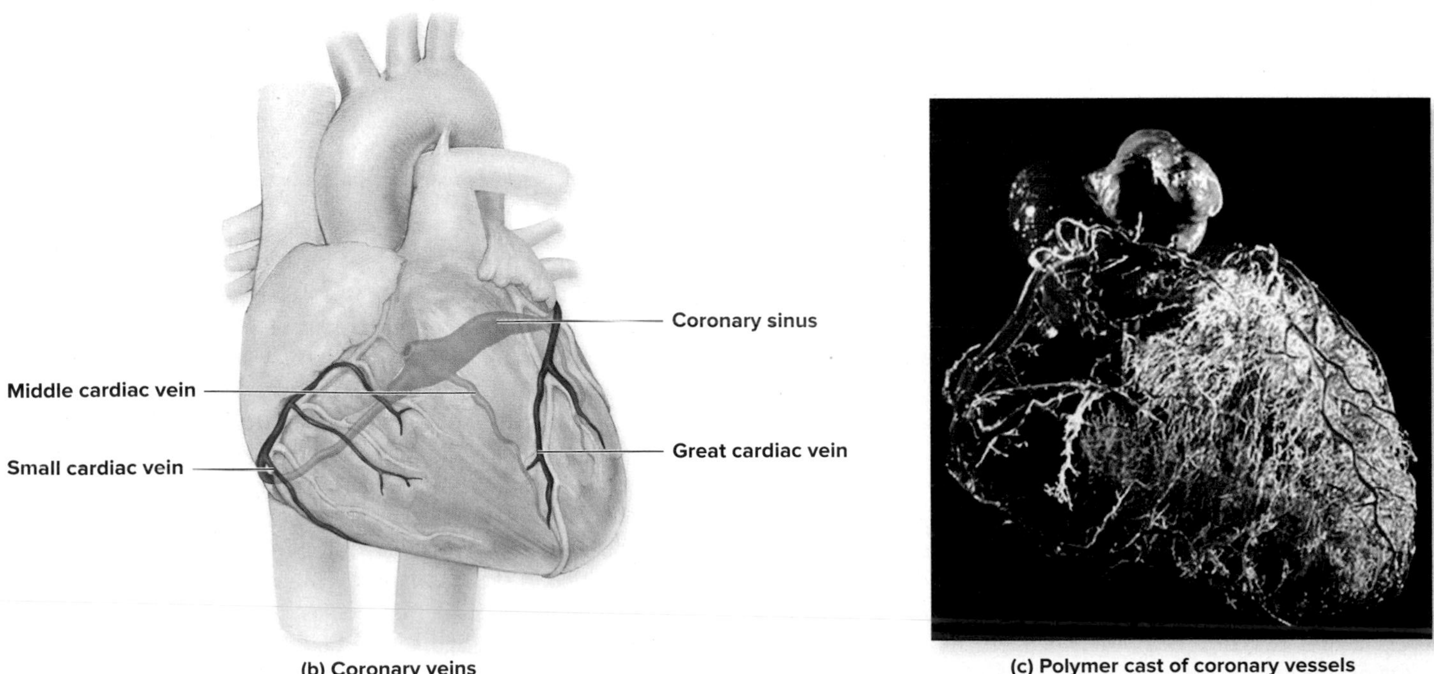

(b) Coronary veins

(c) Polymer cast of coronary vessels

Figure 19.13 Coronary Circulation. Diagram of the anterior view of (*a*) coronary arteries that transport blood to cardiac muscle tissue and (*b*) coronary veins that transport blood from cardiac muscle tissue. (Vessels on the posterior aspect of the heart are shown shaded.) (*c*) Photo of polymer cast of coronary vessels. APR

19.4a Coronary Arteries

LEARNING OBJECTIVES

16. Identify the coronary arteries, and describe the specific areas of the heart supplied by their major branches.
17. Explain the significance of coronary arteries as functional end arteries.
18. Describe blood flow through the coronary arteries.

Right and **left coronary arteries** are positioned within the coronary sulcus of the heart to supply the heart wall (figure 19.13*a*). These arteries are the first and only branches of the ascending aorta and originate immediately superior to the aortic semilunar valve.

The right coronary artery typically branches into the **right marginal artery** to supply the lateral wall of the right ventricle, and the **posterior interventricular artery** (or *posterior descending artery*) to supply the posterior wall of both the left and right ventricles. The left coronary artery typically branches into the **circumflex** (ser′kum-fleks; *circum* = around, *flexus* = to bend) **artery** to supply the lateral wall of the left ventricle, and the **anterior interventricular artery** (also called the *left anterior descending artery,* or *LAD*) to supply both the anterior wall of the left ventricle and most of the interventricular septum. However, this arterial pattern can vary greatly among individuals. Note that the anterior interventricular artery is nicknamed the *widowmaker*. This name reflects the fact that if this artery becomes occluded, there is a very high risk of a fatal heart attack, which makes a widow of the spouse. The coronary arteries and veins also are shown in figure 19.7*a, b*.

INTEGRATE

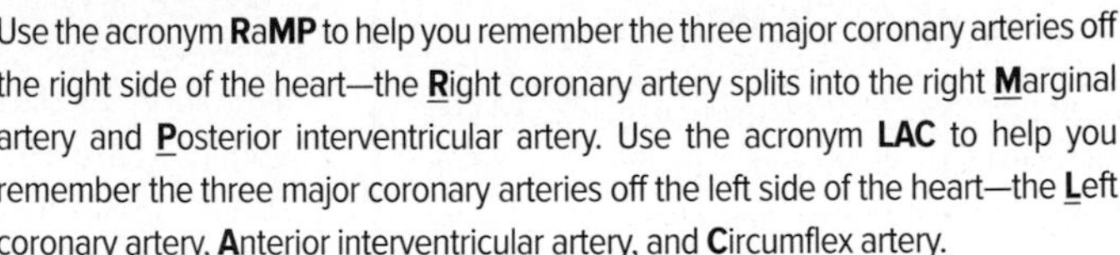

LEARNING STRATEGY

Use the acronym **RaMP** to help you remember the three major coronary arteries off the right side of the heart—the **R**ight coronary artery splits into the right **M**arginal artery and **P**osterior interventricular artery. Use the acronym **LAC** to help you remember the three major coronary arteries off the left side of the heart—the **L**eft coronary artery, **A**nterior interventricular artery, and **C**ircumflex artery.

Functional End Arteries

Body tissues are generally served by one artery; it is called an **end artery.** In comparison, some body tissues are served by two or more arteries; this is referred to as an **arterial anastomosis** (see section 20.1e). An arterial anastomosis provides two means of effectively delivering blood to a given region. The left and right coronary arteries provide blood to the myocardium and are described by some as an arterial anastomosis. However, if one of these arteries becomes blocked, this anastomosis is too small to shunt sufficient blood from one artery to the other, as may happen with coronary artery disease. As a result, the coronary arteries are more accurately called **functional end arteries.**

Blood Flow

Coronary arterial blood flow to the heart wall is intermittent. This occurs because coronary vessels are patent (open) when the heart is relaxed and blood flow is possible. However, coronary vessels are compressed when the heart contracts, temporarily interrupting blood flow. Thus, blood flow to the heart wall is not a steady stream; it is impeded and then flows, as the heart rhythmically contracts and relaxes.

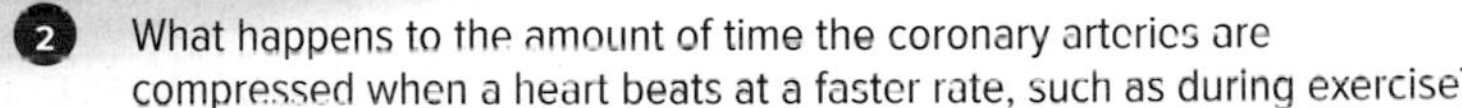

WHAT DO YOU THINK?

2 What happens to the amount of time the coronary arteries are compressed when a heart beats at a faster rate, such as during exercise?

WHAT DID YOU LEARN?

20 What areas of the heart are deprived of blood when there is a blockage in the posterior interventricular artery?

INTEGRATE

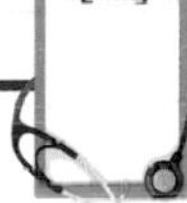

CLINICAL VIEW 19.5

Coronary Heart Disease, Angina Pectoris, and Myocardial Infarction

Coronary heart disease (which goes by many other names, including *coronary artery disease, atherosclerosis, hardening of the arteries,* and *heart disease*) is the buildup of a waxy substance, called plaque, within the coronary arteries. The lumen of coronary vessels becomes narrowed and occluded, which decreases blood flow through the coronary arteries that supply blood to the heart wall (see Clinical View 20.1: "Atherosclerosis," which discusses etiology, risk factors, and treatment options). In contrast, some individuals may experience coronary spasm, which is sudden narrowing of the vessels caused by smooth muscle contraction. Either atherosclerosis or coronary spasm can lead to angina pectoris or the more severe myocardial infarction.

Angina (an′ji-nă, an-jī′nă) **pectoris** is a poorly localized pain sensation in the left side of the chest, the left arm and shoulder, or sometimes the jaw and the back (although these symptoms may vary, especially in women). Generally, it results from strenuous activity, when workload demands on the heart exceed the ability of the narrowed coronary vessels to supply blood. The pain from angina is typically referred (referred pain; see section 16.2b) along the sympathetic pathways (T1–T5 spinal cord segments; see section 15.4), so an individual may experience pain in the chest region or down the left arm, where the T1 dermatome (see section 14.5a) is located. The pain diminishes shortly after the person stops the exertion, and normal blood flow to the heart is restored. Treatment may include medications that cause temporary vascular dilation, such as nitroglycerin. The prognosis and long-term therapy for angina depend upon the severity of the vascular narrowing or spasming.

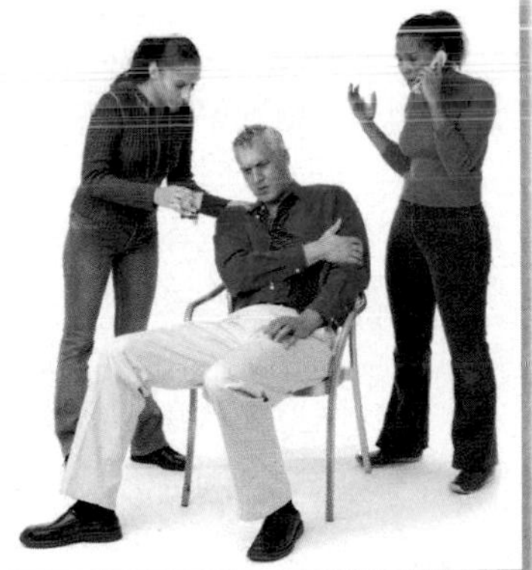

©McGraw-Hill Education/Ken Karp

The term **infarction** refers to the death of tissue due to lack of blood supply. **Myocardial infarction (MI),** commonly called a *heart attack,* is a potentially lethal condition resulting from sudden and complete occlusion of a coronary artery. A region of the myocardium is deprived of oxygen, and some of this tissue may die (undergo necrosis). Some people experiencing an MI report the sudden development of excruciating and crushing substernal chest pain that typically radiates into the left arm or left side of the neck. Women may not experience these symptoms, but instead be overcome with jaw pain, incredible fatigue, and flulike symptoms and thus may be misdiagnosed. Other immediate symptoms include weakness, shortness of breath, nausea, vomiting, anxiety, and marked sweating. Mature cardiac muscle cells have little or no capacity to regenerate (i.e., they cannot undergo cellular division), so when an MI results in cell death, scar tissue forms to fill the gap. If a large amount of tissue is lost, the person may die within a few hours or days because the heart has been profoundly and suddenly weakened.

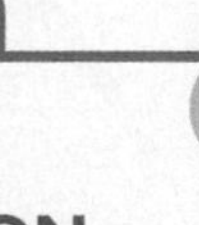

INTEGRATE

CONCEPT CONNECTION

A **sinus** is a modified vein (see section 20.1d) that has very thin walls. Its support is the surrounding tissue. The **coronary sinus** within the coronary sulcus of the heart is supported by adipose connective tissue (see figure 19.7*a*, *b* illustrations; in the photos, the adipose connective tissue has been removed). **Dural venous sinuses** are discussed in section 13.2. These sinuses are supported by the dura mater.

19.4b Coronary Veins

LEARNING OBJECTIVE

19. Identify the coronary veins, and describe the specific areas of the heart drained by their major branches.

Transport of deoxygenated blood from the myocardium occurs through one of several cardiac veins (figure 19.13*b*). These include the **great cardiac vein** within the anterior interventricular sulcus, positioned alongside the anterior interventricular artery; the **middle cardiac vein** within the posterior interventricular sulcus, positioned alongside the posterior interventricular artery; and a **small cardiac vein** positioned alongside the right marginal artery. These cardiac veins all drain into the **coronary sinus,** a large vein that lies within the posterior aspect of the coronary sulcus. The coronary sinus then returns this deoxygenated blood directly into the right atrium of the heart.

WHAT DID YOU LEARN?

21 What is the function of the coronary sinus?

19.5 Anatomic Structures Controlling Heart Activity

The heart pumps blood continuously and depends upon rhythmic stimulation of cardiac muscle cells. Precise electrical events are orchestrated by the heart's conduction system, beginning with stimulation by the sinoatrial node and then transmission of an action potential by specialized conduction fibers of the heart to ensure that the atria contract prior to the ventricles. These events are influenced by the activity of the autonomic nervous system.

19.5a The Heart's Conduction System

LEARNING OBJECTIVE

20. Identify and locate the components of the heart's conduction system.

Specialized cardiac muscle cells within the heart are located internal to the endocardium; they are collectively called the heart's **conduction system (figure 19.14*a*)**. These distinct cardiac cells do not contract, but rather they initiate and conduct electrical signals. The conduction system includes the following structures:

- The **sinoatrial** (sī′nō-ā′trē-ăl) **(SA) node** is located in the posterior wall of the right atrium, adjacent to the entrance of the superior vena cava. The cells here initiate the heartbeat and are commonly referred to as the *pacemaker* of the heart.
- The **atrioventricular (AV) node** is located in the floor of the right atrium between the right AV valve and the opening for the coronary sinus.
- The **atrioventricular (AV) bundle** (*bundle of His*) extends from the AV node into and through the interventricular septum. It divides into **left** and **right bundles.**
- The **Purkinje** (pŭr-kin′jē) **fibers** extend from the left and right bundles beginning at the apex of the heart and then continue through the walls of the ventricles.

WHAT DID YOU LEARN?

22 Why is the SA node referred to as the pacemaker?

19.5b Innervation of the Heart

LEARNING OBJECTIVE

21. Compare and contrast parasympathetic and sympathetic innervation of the heart.

While the heartbeat is initiated by the SA node, both the heart rate and the strength of contraction are regulated by the autonomic nervous system—specifically by the **cardiac center** of the **cardiovascular center** within the medulla oblongata (see section 13.5c). Sensory input is relayed from receptors (baroreceptors and chemoreceptors within specific blood vessels and the right atrium) to the cardiac center. Parasympathetic and sympathetic neurons extend from the cardiac center to the heart. The cardiac center houses both the **cardioinhibitory** and **cardioacceleratory centers** (figure 19.14*b*). The innervation from this center does not initiate the heartbeat; it merely modifies cardiac activity, including both the heart rate and its force of contraction.

Sinoatrial (SA) node
(pacemaker)
Right atrium
Atrioventricular (AV) node
Atrioventricular (AV)
bundle
Right and left bundles
Purkinje fibers
Purkinje fibers

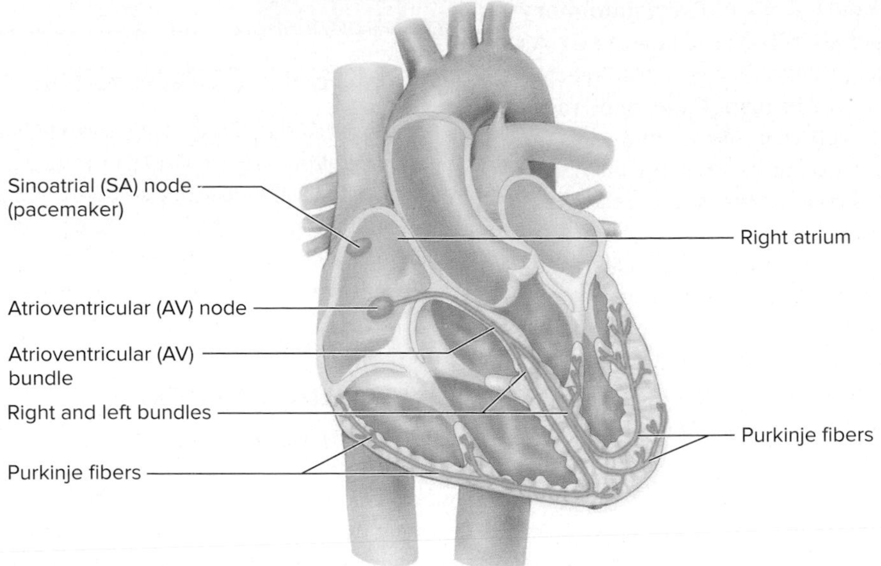

(a) Conduction system

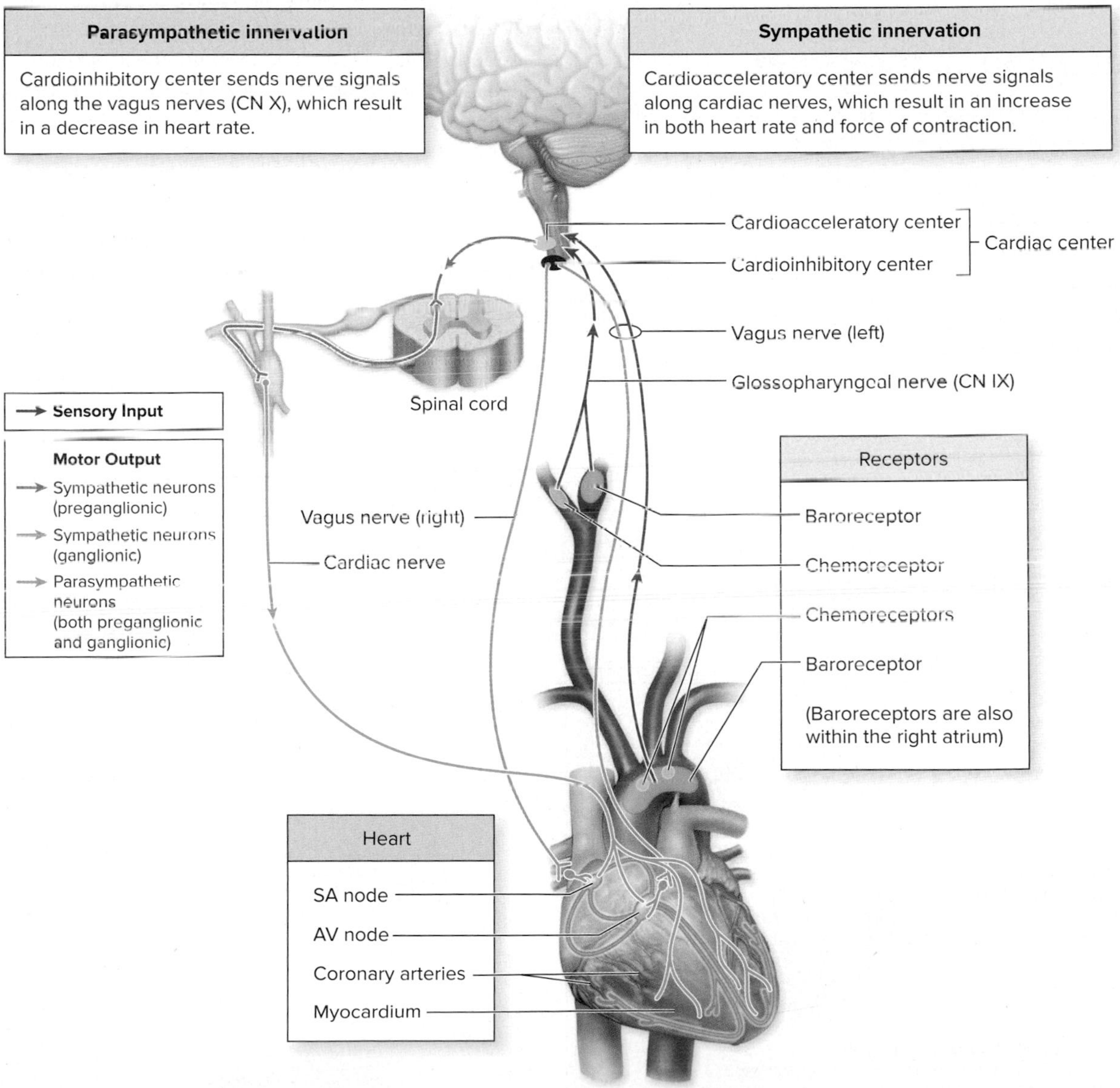

Parasympathetic innervation
Cardioinhibitory center sends nerve signals along the vagus nerves (CN X), which result in a decrease in heart rate.
Sympathetic innervation
Cardioacceleratory center sends nerve signals along cardiac nerves, which result in an increase in both heart rate and force of contraction.
Cardioacceleratory center
Cardioinhibitory center
Cardiac center
Vagus nerve (left)
Glossopharyngeal nerve (CN IX)
Spinal cord
Sensory Input
Motor Output
Sympathetic neurons (preganglionic)
Sympathetic neurons (ganglionic)
Parasympathetic neurons (both preganglionic and ganglionic)
Vagus nerve (right)
Cardiac nerve
Receptors
Baroreceptor
Chemoreceptor
Chemoreceptors
Baroreceptor
(Baroreceptors are also within the right atrium)
Heart
SA node
AV node
Coronary arteries
Myocardium

(b) Innervation of the heart

Figure 19.14 Anatomic Structures Controlling Heart Activity. (*a*) The heart's conduction system, composed of the sinoatrial (SA) node, atrioventricular (AV) node, right and left bundles, and Purkinje fibers, initiates and conducts the electrical activity to cause heart contraction. (*b*) Heart rate and force of contraction are modified by autonomic centers (cardioacceleratory and cardioinhibitory centers) in the medulla oblongata. Parasympathetic neurons extend from the cardioinhibitory center to both the SA node and AV node. Sympathetic neurons extend from the cardioacceleratory center to the SA node, AV node, myocardium, and coronary vessels.

Parasympathetic innervation comes from the **cardioinhibitory center** via the right and left vagus nerves (CN X; see table 13.5). As these nerves descend into the thoracic cavity, they give off branches that supply the heart (see section 15.3). Primarily, the right vagus nerve innervates the SA node, and the left vagus nerve innervates the AV node. Parasympathetic stimulation decreases heart rate but generally has no direct effect on the force of contraction (because the vagus nerve does not innervate the myocardium).

Sympathetic innervation arises from the **cardioacceleratory center.** Neurons within the T1–T5 segments of the spinal cord extend to the SA node, AV node, and the myocardium (see section 15.4). Stimulation by the sympathetic division increases both heart rate and the force of heart contraction. There is also sympathetic innervation to the coronary arteries, causing dilation of these vessels to support increased blood flow to the myocardium.

WHAT DID YOU LEARN?

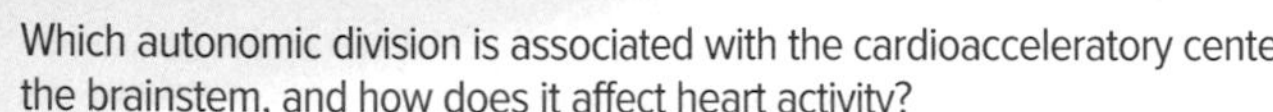

23 Which autonomic division is associated with the cardioacceleratory center in the brainstem, and how does it affect heart activity?

19.6 Stimulation of the Heart

The physiologic processes associated with heart contraction are organized into two major events (**figure 19.15**):

- **Conduction system.** Electrical activity is initiated at the SA node, and an action potential is then transmitted through the conduction system.
- **Cardiac muscle cells.** The action potential spreads across the sarcolemma of the cardiac muscle cells, causing sarcomeres within cardiac muscle cells to contract. These events occur *twice* in cardiac muscle cells during a heartbeat, first in the cells of the atria and then in the cells of the ventricles.

Here we discuss the events associated with the conduction system, including the physiologic conditions of nodal cells at rest and how they serve as the heart's pacemaker. The processes associated with cardiac muscle cells are described in section 19.7.

INTEGRATE

CONCEPT CONNECTION

A resting membrane potential (RMP) is the electrical charge difference at the plasma membrane when a cell is at rest (see section 4.4). The types of cells that exhibit an RMP include cardiac muscle cells, skeletal muscle fibers (see section 10.2d), neurons (see section 12.7b), and sensory receptor cells (see section 16.1a).

19.6a Nodal Cells at Rest

LEARNING OBJECTIVE

22. Describe a nodal cell at rest.

Nodal cells in the SA node are the pacemaker cells that initiate a heartbeat by spontaneously depolarizing to generate an action potential. These specialized cardiac cells exhibit several significant features. Refer to **figure 19.16*a*** as you read through this section.

One essential feature of nodal cells is the electrical charge difference across the plasma membrane; the intracellular fluid (cytosol) just inside the plasma membrane is relatively negative in comparison to the fluid outside the cell (interstitial fluid). This electrical charge difference when the nodal cell is at rest is called the **resting membrane potential (RMP).** Nodal cells have an RMP of about −60 millivolts (mV). An RMP, which is discussed in detail in section 4.4, is established and maintained by K^+ leak channels, Na^+ leak channels, and Na^+/K^+ pumps (not shown in figure 19.16*a*). The primary function of the Na^+/K^+ pumps is to maintain the concentration gradients for Na^+ (with more Na^+ outside the cell) and K^+ (with more K^+ inside the cell). Nodal cells also contain calcium ion (Ca^{2+}) pumps that establish a Ca^{2+} concentration gradient with more Ca^{2+} outside the cell than inside. It is important to note that nodal cells (unlike other cells) do not have a stable RMP, as described in section 19.6b.

Additionally, nodal cells contain specific voltage-gated channels, including slow voltage-gated Na^+ channels (which are open) and both fast voltage-gated Ca^{2+} channels and voltage-gated K^+ channels (which are closed).

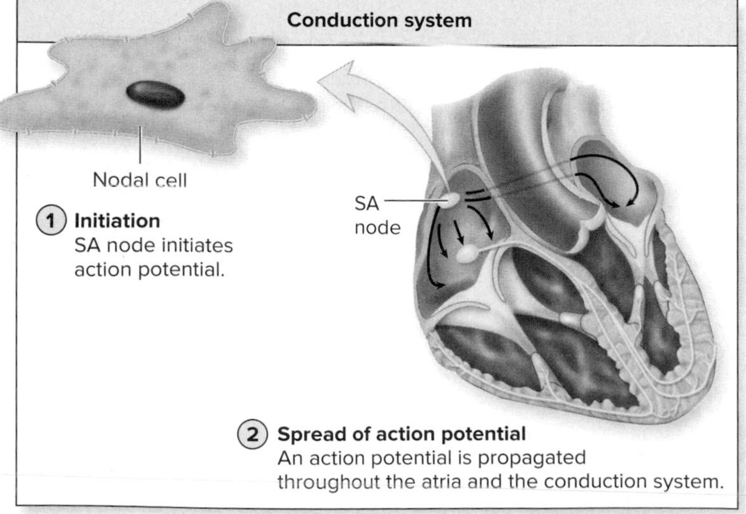

Conduction system
Nodal cell
SA node
1 Initiation
SA node initiates action potential.
2 Spread of action potential
An action potential is propagated throughout the atria and the conduction system.
(a)

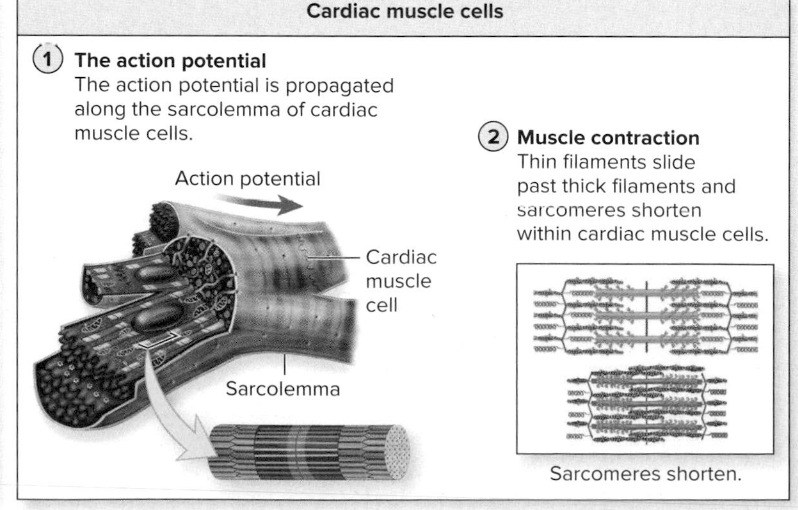

Cardiac muscle cells
1 The action potential
The action potential is propagated along the sarcolemma of cardiac muscle cells.
Action potential
Cardiac muscle cell
Sarcolemma
2 Muscle contraction
Thin filaments slide past thick filaments and sarcomeres shorten within cardiac muscle cells.
Sarcomeres shorten.
(b)

Figure 19.15 Physiologic Processes Associated with Heart Contraction. Both the conduction system and cardiac muscle cells have two significant physiologic processes that occur for heart contraction. (*a*) Initiation and spread of an action potential occurs in the conduction system. (*b*) Action potentials spread along the sarcolemma of cardiac muscle cells, triggering contraction of sarcomeres within these cardiac muscle cells. Events of cardiac muscle cells occur twice in one heartbeat: once in the cells of the atria and once in the cells of the ventricles.

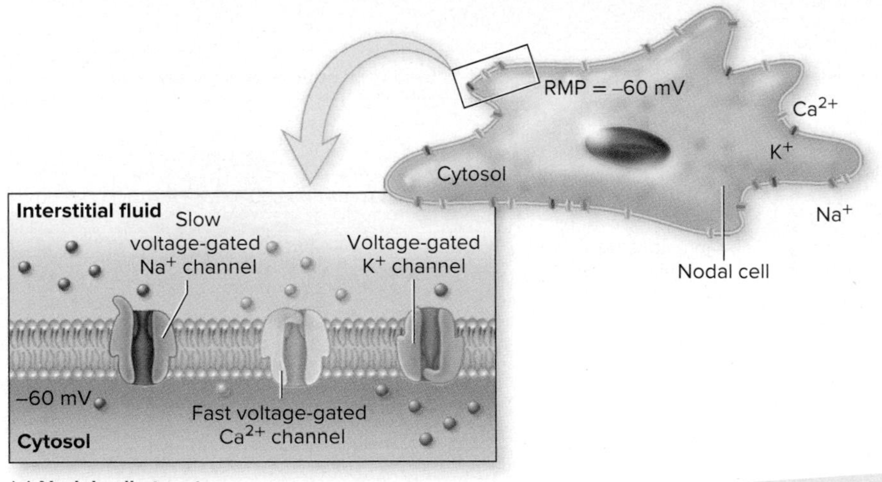

(a) Nodal cell at rest

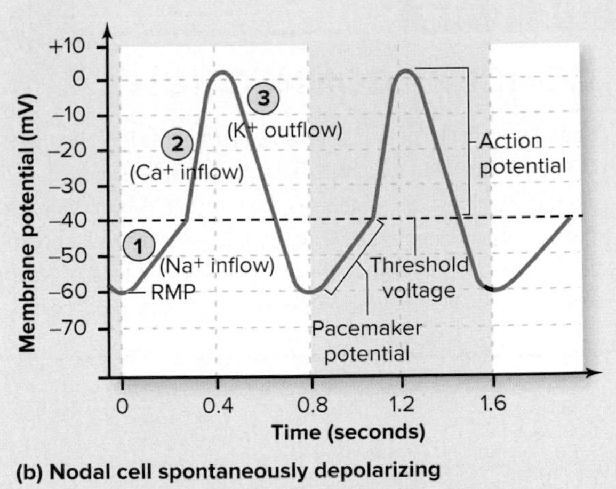

(b) Nodal cell spontaneously depolarizing

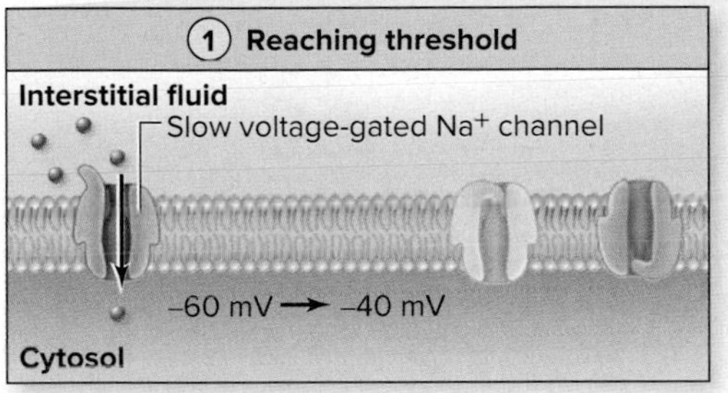

- Slow voltage-gated Na^+ channels open.
- Na^+ enters nodal cell.
- Threshold reached (–60 mV → –40 mV).
- Slow voltage-gated Na^+ channels close.

2 Depolarization

Fast voltage-gated Ca^{2+} channel

–40 mV → + mV

- Fast voltage-gated Ca^{2+} channels open.
- Ca^{2+} enters nodal cell.
- Depolarization occurs (–40 mV → just above 0 mV).
- Fast voltage-gated Ca^{2+} channels close.

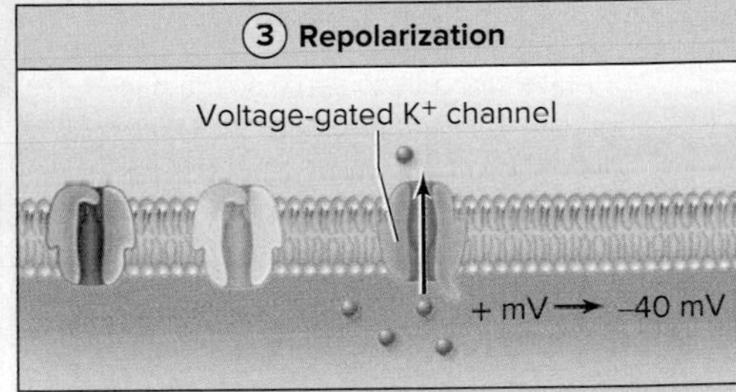

- Voltage-gated K^+ channels open.
- K^+ exits nodal cell.
- Repolarization occurs (+ mV → –60 mV).
- Voltage-gated K^+ channels close.

Figure 19.16 SA Node Cellular Activity. (*a*) Nodal cells have an unstable resting membrane potential of –60 mV. Voltage-gated Na^+ channels located in the plasma membrane of nodal cells open in response to changes in membrane potential. (*b*) Graph and accompanying diagrams of the sequential electrical changes occurring at the plasma membrane of SA nodal (pacemaker) cells to initiate stimulation of the heart. Note that ions not involved in a given step are not shown.

WHAT DID YOU LEARN?

24 What is the resting membrane potential (RMP) value of nodal cells?

19.6b Electrical Events at the SA Node: Initiation of the Action Potential

LEARNING OBJECTIVES

23. Define autorhythmicity.
24. Describe the steps for SA nodal cells to spontaneously depolarize and serve as the pacemaker cells.

SA nodal cells are unique in that they exhibit **autorhythmicity** (or *automaticity*), meaning that they are capable of depolarizing and initiating an action potential spontaneously without any external influence. The following series of events occur within SA nodal cells, as depicted in figure 19.16*b*:

(1) **Reaching threshold.** Slow voltage-gated Na^+ channels open (this is caused by repolarization from the previous cycle). The Na^+ flows into the nodal cells, changing the resting membrane potential from –60 mV to –40 mV, which is the **threshold** value. Notice that the threshold is reached without outside stimulation.

(2) **Depolarization.** Changing of the membrane potential to the threshold triggers the opening of fast voltage-gated Ca^{2+} channels, and Ca^{2+} entry into the nodal cell causes a change in the membrane potential from –40 mV to a slightly positive membrane potential (just above 0 mV). This reversal of polarity is called **depolarization.**

(3) **Repolarization.** Calcium channels close and voltage-gated K^+ channels open; K^+ flows out to change the membrane potential from a positive value to –60 mV, which is the RMP. The process of reestablishing the RMP is called **repolarization.** Repolarization triggers the reopening of slow voltage-gated Na^+ channels, and the process begins again.

This process typically takes approximately 0.8 second, at rest; this results in a resting heart rate of 75 beats per minute. Note that the inherent rhythm at which SA nodal cells spontaneously depolarize is at a much faster rate of approximately 100 times per minute. (This inherent rhythm is determined when excised cardiac muscle cells are placed into cell culture, without autonomic nerve innervation.) The normal resting heart rate of 75 beats per minute is due to continuous parasympathetic stimulation of the SA node by the vagus nerve. This slowing of the heart rate is called **vagal tone** (see section 15.7a).

WHAT DO YOU THINK?

3 What specific type of channel is unique to cardiac nodal cells? What is the significance of this channel?

Pacemaker Potential of Nodal Cells

Nodal cells do not have a stable resting membrane potential. Instead, the RMP gradually increases to threshold when slow Na^+ channels reopen. This ability to reach the threshold without stimulation is called a **pacemaker potential.** The pacemaker potential of nodal cells is responsible for initiating electrical signals, which will be propagated along the conduction system to stimulate the cardiac muscle cells of the heart to rhythmically contract. It is for this reason that initiation of heart contraction does not require stimulation by the autonomic nervous system.

INTEGRATE

CONCEPT CONNECTION

Nodal cells are like neurons because they have an RMP. However, a significant difference between them is that neurons require stimulation (in the form of neurotransmitters or a modality; see section 12.6a), whereas nodal cells spontaneously depolarize. Another difference between neurons and nodal cells is that depolarization results from the entrance of Na+ into neurons, whereas it results from the entrance of Ca2+ into nodal cells.

WHAT DID YOU LEARN?

25 What is autorhythmicity? Describe how nodal cells function as autorhythmic cells to serve as the pacemaker of the heart.

19.6c Conduction System of the Heart: Spread of the Action Potential

LEARNING OBJECTIVE

25. Describe the spread of the action potential through the heart's conduction system.

Cardiac muscle stimulation requires that the action potential initiated by the SA node be spread through the conduction system. The sequence of events occurs as follows (**figure 19.17**):

1. **Action potential is distributed throughout both atria.** The action potential initiated in the SA node is propagated first along the sarcolemma of cardiac muscle cells within the atria and spread between atrial cardiac muscle cells by gap junctions. This allows for almost instantaneous excitation of all cardiac muscle cells in the atrial walls and for both atria to contract at the same time.
2. **The action potential is relayed to the AV node and delayed.** The action potential arrives at the AV node. AV nodal cells have both smaller fiber diameters and fewer numbers of gap junctions—thus, they exhibit characteristics that *slow the conduction rate* of the action potential by serving as "a bottleneck." This is facilitated by the insulating characteristics of the fibrous skeleton, which only allow the action potential to move through the AV node. The delay in conduction (about 100 milliseconds, or 0.1 second) may seem very brief, but it is long enough to allow the atria to finish contracting and force blood into the ventricles to complete ventricular filling before the ventricles are stimulated to contract.
3. **The action potential travels from the AV node through the AV bundle to Purkinje fibers.** The action potential is propagated from the AV node along the AV bundle to the bundle branches to the Purkinje fibers.
4. **The action potential continues throughout both ventricles via gap junctions.** The action potential is then propagated along the sarcolemma of cardiac muscle cells within the ventricles and spread between ventricular cardiac muscle cells by gap junctions. This allows for the almost simultaneous stimulation of all the cardiac muscle cells in the ventricular walls and simultaneous contraction of both ventricles. Generally, these cells begin to contract within 120 to 200 milliseconds after the firing of the SA nodal cells.

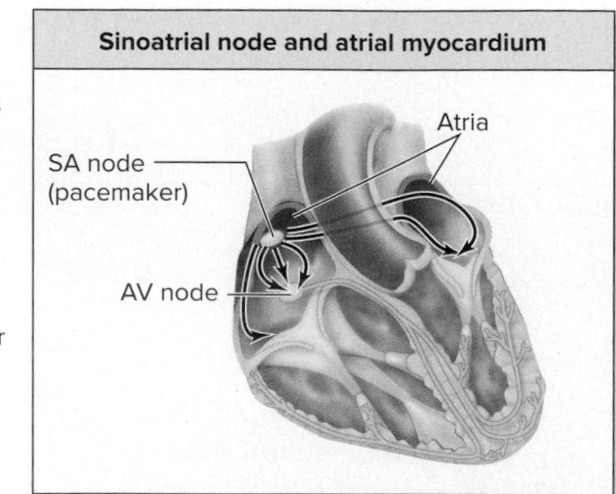

(1) An action potential is generated at the sinoatrial (SA) node. The action potential spreads via gap junctions between cardiac muscle cells throughout the atria to the atrioventricular (AV) node.
Sinoatrial node and atrial myocardium
Atria
SA node (pacemaker)
AV node

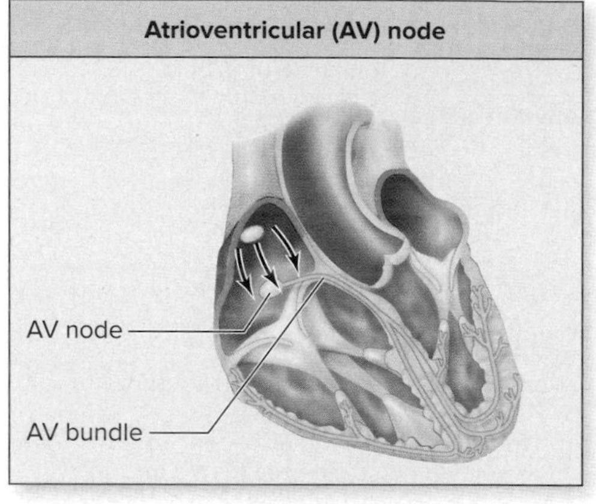

(2) The action potential is delayed at the AV node before it passes to the AV bundle within the interventricular septum.
Atrioventricular (AV) node
AV node
AV bundle

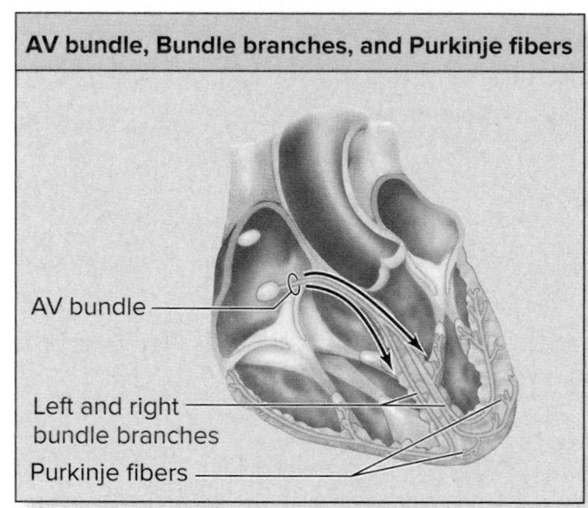

(3) The AV bundle conducts the action potential to the left and right bundle branches and then to the Purkinje fibers.
AV bundle, Bundle branches, and Purkinje fibers
AV bundle
Left and right bundle branches
Purkinje fibers

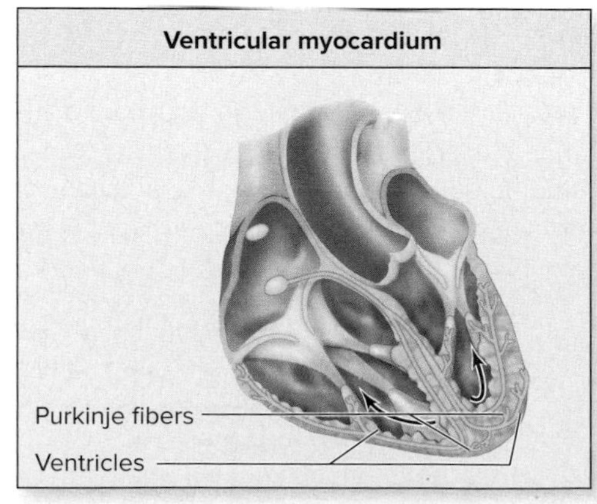

(4) The action potential is spread via gap junctions between cardiac muscle cells throughout the ventricles.
Ventricular myocardium
Purkinje fibers
Ventricles

Figure 19.17 Initiation and Spread of an Action Potential Through the Cardiac Conduction System. The initiation and spread of an action potential begins at the SA node and is propagated through the cardiac conduction system. The average rate at rest is one per 0.8 second. (The black arrows indicate the propagation of the action potential.) AP|R

INTEGRATE

CLINICAL VIEW 19.6

Ectopic Pacemaker

A pacemaker other than the SA node is called an **ectopic** (ek-top′ik; *ektos* = outside, *topos* = place) **pacemaker,** or *ectopic focus.* Cells of the conduction system other than the SA node and cardiac muscle cells also have the ability to spontaneously depolarize and serve as the pacemaker. However, they depolarize at slower rates than the SA node. The AV node has an inherent rhythm of 40 to 50 beats per minute, and cardiac muscle cells have a rate of 20 to 40 beats per minute. If the SA node is not functioning, the AV node becomes the "default" pacemaker, and the AV node then establishes the rhythm of the heart rate. Survival is possible because a rhythm of 40 to 50 beats per minute pumps sufficient blood to sustain life. However, if both the SA node and the AV node are not functioning, cardiac muscle cells will attempt to establish the rhythm. The heart rate will be only 20 to 40 beats per minute, which is almost always too slow to support life. A **mechanical pacemaker** is a small device that is surgically implanted in a patient in order to deliver repeated electrical impulses to heart muscle. It sustains a required heart rate and rhythm for the patient.

Specialized Features Associated with Ventricles

The efficient functioning of the ventricles requires a coordination of contraction that includes the following features:

- Purkinje fibers are relatively large in diameter, so action potential propagation is extremely rapid to the ventricular myocardium. Thus, the cardiac muscle cells of both ventricles contract at the same time.
- Papillary muscles within the ventricles are stimulated to contract immediately. These muscles anchor the tendinous cords to the AV valve cusps; they tighten the relaxed cords and cause them to start to pull on the AV valve cusps just prior to the increase in pressure within the ventricles. Thus, the valves are "braced" and better able to prevent backflow of blood into the atria.
- The stimulation of the ventricles begins at the apex of the heart. This feature ensures that blood is efficiently ejected superiorly toward the arterial trunks.

WHAT DID YOU LEARN?

26. What is the path of an action potential through the conduction system of the heart?
27. What anatomic features slow the conduction rate of the action potential as it passes through the AV node? What is the function of this delay?

19.7 Cardiac Muscle Cells

Two significant and interrelated events occur by cardiac muscle cells following their stimulation by the conduction system: propagation of an action potential at the sarcolemma and contraction of sarcomeres within the cardiac muscle cells. Remember that the events of cardiac muscle cells occur twice per heartbeat: first in the cardiac muscle cells of the atria (propagation of an action potential and contraction occur), and then in the cardiac muscle cells of the ventricles (propagation of an action potential and contraction occur).

19.7a Cardiac Muscle Cells at Rest

LEARNING OBJECTIVE

26. Describe the conditions at the sarcolemma of cardiac muscle cells at rest.

Cardiac muscle cells exhibit several significant features at their sarcolemma. Refer to **figure 19.18*a*** as you read through this section.

The sarcolemma of cardiac muscle cells, like nodal cells, has K^+ leak channels, Na^+ leak channels, and Na^+/K^+ pumps to establish and maintain a resting membrane potential with a greater concentration of Na^+ outside the cardiac muscle cells and a greater concentration of K^+ inside. The RMP value of cardiac muscle cells, however, is −90 millivolts (mV) (in comparison to −60 mV for nodal cells). Cardiac muscle cells also contain Ca^{2+} pumps that form a Ca^{2+} concentration gradient with more Ca^{2+} outside the cell than inside.

The sarcolemma of cardiac muscle cells has both fast voltage-gated Na^+ channels that participate in depolarization at the membrane and voltage-gated K^+ channels that participate in repolarization of the membrane. Additionally, cardiac muscle cells have slow voltage-gated Ca^{2+} channels within the sarcolemma. Slow voltage-gated Ca^{2+} channels, when open, allow Ca^{2+} into the cell. The movement of Ca^{2+} is significant in the normal function of cardiac muscle cells, as described in section 19.7b.

WHAT DID YOU LEARN?

28. In which direction does Ca^{2+} move in response to the opening of voltage-gated Ca^{2+} channels: into or out of the cardiac muscle cells?

19.7b Electrical and Mechanical Events of Cardiac Muscle Cells

LEARNING OBJECTIVES

27. List the electrical events of an action potential that occur at the sarcolemma.
28. Briefly summarize the mechanical events of muscle contraction.

We can distinguish certain electrical and mechanical events that occur as the heart contracts.

Electrical Events

Electrical events involving propagation of the action potential at the sarcolemma of cardiac muscle cells occur as follows (figure 19.18*b*):

1. **Depolarization.** An action potential transmitted through the conduction system (or via gap junctions) triggers the opening of fast voltage-gated Na^+ channels in the sarcolemma of cardiac muscle cells. Sufficient Na^+ enters the cardiac muscle cell to change the membrane potential from −90 mV to +30 mV. Voltage-gated Na^+ channels close to the inactivated state. This change from a relatively negative membrane potential to a relatively positive membrane potential is **depolarization** (see section 12.6a).
2. **Plateau.** Depolarization triggers the opening of voltage-gated K^+ channels, and K^+ leaves the cardiac muscle cells. There is a slight change in the membrane potential. Almost immediately, slow voltage-gated Ca^{2+} channels in the sarcolemma also open, and Ca^{2+} enters from the interstitial fluid into the cardiac muscle cells. The entering of Ca^{2+} stimulates the sarcoplasmic reticulum (SR) to release more Ca^{2+} (greater than 80% of the Ca^{2+} is released from the SR). The exit of positively charged K^+ from the sarcoplasm and the simultaneous entrance of positively charged Ca^{2+} into the sarcoplasm result in no electrical change at the sarcolemma. Thus, the sarcolemma of the cardiac muscle cell remains in a depolarized state. This "leveling off" is referred to as the **plateau.**
3. **Repolarization.** Voltage-gated Ca^{2+} channels then close, and K^+ channels remain open to complete repolarization. Sufficient K^+ exits the cardiac muscle cells to change the membrane

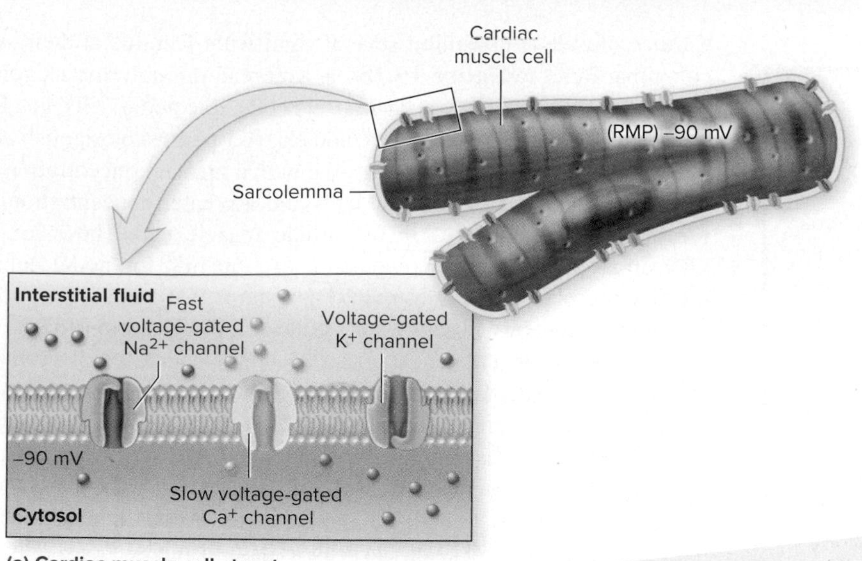

(a) Cardiac muscle cell at rest

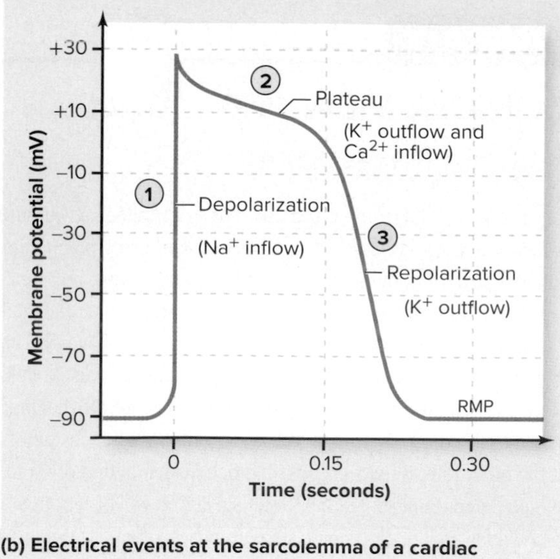

(b) Electrical events at the sarcolemma of a cardiac muscle cell

1 Depolarization

Fast voltage-gated Na^{2+} channel

–90 mV → +30 mV

- Fast voltage-gated Na^+ channels open.
- Na^+ rapidly enters cardiac cell.
- Depolarization occurs (–90 mV → +30 mV).
- Fast voltage-gated Na^+ channels close.

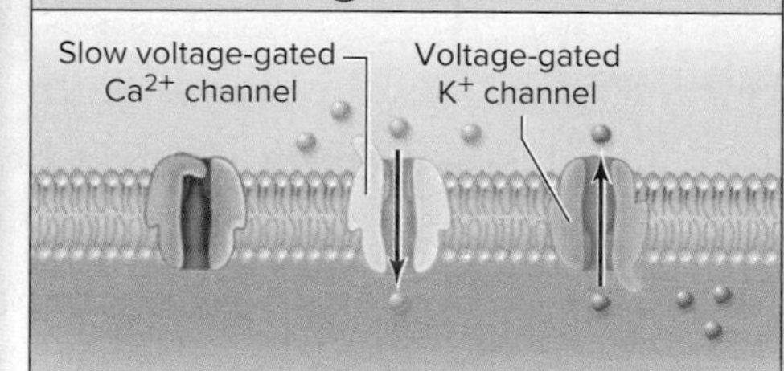

- Voltage-gated K^+ channels open.
- K^+ flows out of cardiac muscle cell *and*
- Slow voltage-gated Ca^{2+} channels open and
- Ca^{2+} enters the cardiac cell.
- No electrical change and the depolarized state is maintained

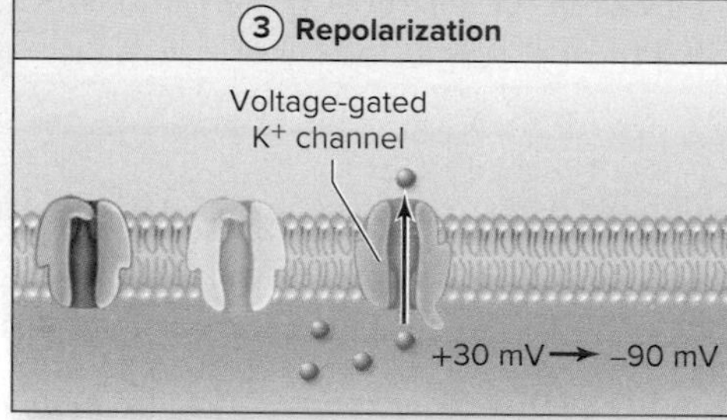

- Voltage-gated K^+ channels remain open.
- K^+ moves out of the cardiac muscle cell.
- Voltage-gated Ca^{2+} channels close.
- Repolarization occurs (+30 mV to –90 mV).

Figure 19.18 Electrical Events of Cardiac Muscle Cells. (*a*) Cardiac muscle cells have a resting membrane potential of –90 mV. Voltage-gated channels in the sarcolemma are closed when the cardiac muscle cell is at rest. (*b*) Graph and accompanying diagrams of the sequential electrical events of an action potential at the sarcolemma of cardiac muscle cells. (Cytosol is the fluid portion of the sarcoplasm within the cardiac muscle cell.)

INTEGRATE

CONCEPT CONNECTION

Conductivity involves an electrical change that is propagated along the plasma membrane as voltage-gated channels open sequentially during an action potential. Conductivity is exhibited along the sarcolemma of skeletal muscle fibers (see section 10.3b), the axolemma of neurons (see section 12.8c), and the sarcolemma of cardiac muscle cells.

potential from being relatively positive back to –90 mV (the RMP). This reversal of charge is called **repolarization.** It is repolarization that allows a muscle cell to propagate a new action potential when the cardiac muscle is stimulated again.

These three events of an action potential–depolarization, plateau, and repolarization–are repeated along the length of the sarcolemma as the voltage-gated channels are triggered to open sequentially. This propagation of the electrical signal along the plasma membrane is referred to as **conductivity.** Conductivity of the electrical signal serves to functionally connect the plasma membrane of the cardiac muscle cell (where stimulation occurs) to the interior of the cardiac muscle cell (where contraction occurs). These processes are similar to what occurs in skeletal muscle (see section 10.3b).

Mechanical Events (Crossbridge Cycling)

Cardiac muscle contraction is initiated with the entry of Ca^{2+} into the sarcoplasm from both the interstitial fluid and SR (as described in step 2). Calcium ions now bind to troponin to begin crossbridge cycling within a sarcomere, similar to the way in which skeletal muscle contracts (see section 10.3c).

A summary of the steps of muscle contraction involving sarcomeres is as follows (see figure 10.13):

Crossbridge formation: Myosin heads attached to actin form a crossbridge between the thick and thin filaments.

Powerstroke: The myosin head swivels (the powerstroke), which pulls the thin filament past the thick filament a short distance, which decreases the width of a sarcomere.

Release of myosin head: ATP binds to the myosin head to release the myosin head from actin.

Reset of myosin head: ATP is split by myosin ATPase, providing the energy to reset the myosin head.

Cardiac muscle relaxation is initiated with the closing of voltage-gated Ca^{2+} channels (as described in step 3). The continuous reuptake of Ca^{2+} from the sarcoplasm into the sarcoplasmic reticulum by Ca^{2+} pumps and the removal of Ca^{2+} from the cell by plasma membrane Ca^{2+} pumps decrease calcium levels within the sarcoplasm. Calcium is released from troponin with the subsequent decrease in crossbridges between the thin and thick filaments. Sarcomeres return to their resting length as the cardiac muscle cell now relaxes (see section 10.3d).

> **WHAT DO YOU THINK?**
>
> 4 Calcium channel–blocking drugs are sometimes given to patients with heart problems. If a patient is given a Ca^{2+} channel blocker, how will this affect both heart rate and force of contraction? Explain.

> **WHAT DID YOU LEARN?**
>
> 29 What three electrical events occur at the sarcolemma of cardiac muscle cells? Explain each event.

19.7c Repolarization and the Refractory Period

LEARNING OBJECTIVES

29. Define the refractory period.
30. Explain the significance of the plateau phase.

Cardiac muscle cells (unlike skeletal muscle fibers) *cannot* exhibit **tetany,** which is a sustained muscle contraction without relaxation (see section 10.6c). This distinction is critical for the heart to be able to function as a mechanical pump to move blood through the cardiovascular system. Here we compare the characteristics of skeletal muscle fibers to cardiac muscle cells to understand the reason for this difference. Refer to **figure 19.19** as you read through this section.

Examine the graphs in figure 19.19, and notice the following in the graphs of both skeletal muscle fibers and cardiac muscle cells:

- **Muscle tension,** which is depicted as the red line on each graph, includes both muscle contraction and muscle relaxation. (These mechanical events are due to sarcomeres within these muscle cells first shortening during contraction and then lengthening during relaxation.)

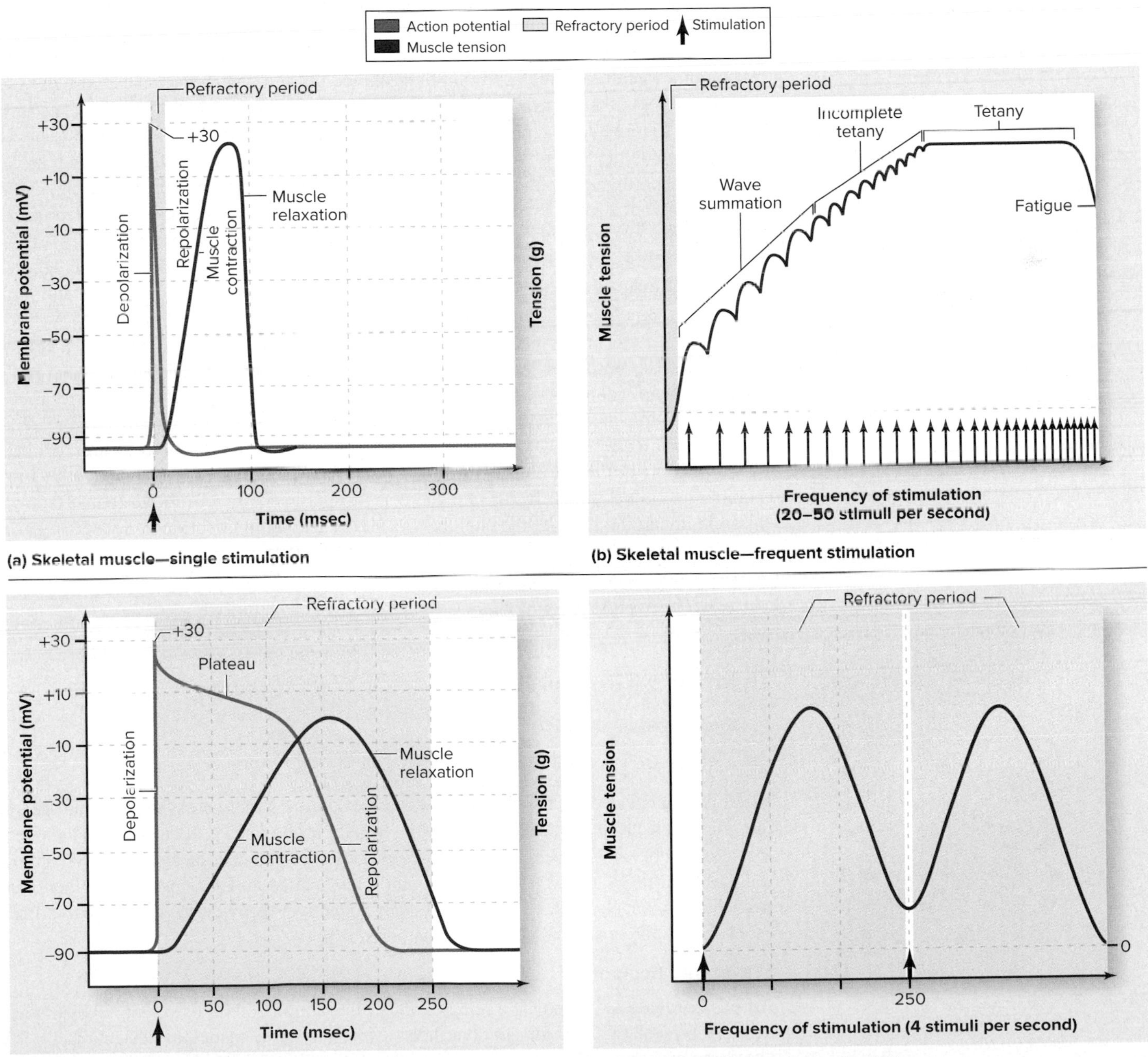

Figure 19.19 Comparison of Electrical and Mechanical Events in Skeletal Muscle Cells and Cardiac Muscle Cells. The electrical changes associated with an action potential propagated along the sarcolemma and the mechanical changes associated with generating muscle tension when a muscle contracts are shown for (*a*) skeletal muscle with a single stimulation, (*b*) skeletal muscle with frequent stimulation, (*c*) cardiac muscle with a single stimulation, and (*d*) cardiac muscle with frequent stimulation. The extended refractory period in cardiac muscle cells (due to the plateau) allows time for cardiac muscle to contract and relax before being stimulated again. Thus, a sustained contraction (tetany) is prevented.

- An **action potential,** which is shown as a blue line on the graph in figure 19.19*a, c,* includes both depolarization and repolarization. (These electrical events are occurring at the sarcolemma of these muscle cells.)
- The **refractory period,** which is the shaded green region on each graph, represents the time when the muscle cannot be re-stimulated to contract. (The refractory period is dependent upon how quickly depolarization and repolarization occur at the sarcolemma; the faster that repolarization occurs, the shorter the refractory period.)

Now observe just the two graphs that represent what is occurring in *skeletal muscle fibers* (figure 19.19*a, b*). Figure 19.19*a* represents the events of a single stimulation of skeletal muscle. Notice that muscle contraction and relaxation associated with sarcomeres within skeletal muscle fibers occur over approximately 100 milliseconds. Observe also that repolarization at the sarcolemma occurs immediately following depolarization, which results in a refractory period that is relatively short (approximately 1 to 2 milliseconds). Consequently, while sarcomeres are still contracting within skeletal muscle fibers, the sarcolemma has already been repolarized to allow for a new stimulation.

Figure 19.19*b* represents the events of frequent stimulation of skeletal muscle. Notice that skeletal muscle, because it has a relatively short refractory period, can be restimulated at a frequency that does not allow the skeletal muscle sufficient time to relax completely. In fact, skeletal muscle can be stimulated at a rate that the muscle remains contracted (without any relaxation). Recall that this condition of sustained muscle contraction is called *tetany*.

Now refer to the two graphs that represent what is occurring in *cardiac muscle cells* (figure 19.19*c, d*). Figure 19.19*c* represents the events of a single stimulation of cardiac muscle. Notice that muscle contraction and relaxation associated with sarcomeres within cardiac muscle cells occur over approximately 250 milliseconds (a longer period than in skeletal muscle). Note also that repolarization at the sarcolemma does not occur immediately following depolarization, which results in a refractory period that is relatively long (almost 250 milliseconds). This relatively long refractory period is due to the plateau event at the sarcolemma, which delays repolarization. Consequently, while sarcomeres are still contracting and relaxing within cardiac muscle, the sarcolemma has *not* been repolarized to allow for a new stimulation.

Figure 19.19*d* represents the events of frequent stimulation of cardiac muscle. Notice that cardiac muscle, because it has a relatively long refractory period, has an extended period of time in which it cannot be restimulated. This delay in restimulation allows time for the sarcomeres of cardiac muscle cells within the heart chamber wall to fully contract and relax before being stimulated again. Thus, cardiac muscle cells composing the myocardium of the heart chamber walls do *not* exhibit tetany, but continue to both contract and relax following each stimulation—an essential feature for the heart to pump the blood. (Note that if tetany were possible in cardiac muscle, the chambers of the heart would experience a sustained contraction—this would be similar to a pump ceasing or "locking up.")

WHAT DID YOU LEARN?

What is the significance of the extended refractory period in cardiac muscle?

19.7d The ECG Recording

LEARNING OBJECTIVE

31. Identify the components of an ECG recording.

Electrical changes within the heart can be detected during a routine physical examination using monitoring electrodes attached to the skin—usually at the wrists, the ankles, and six separate locations on the chest. The electrical signals are collected and charted as an **electrocardiogram** (ē-lek-trō-kar′dē-ō-gram; *gramma* = drawing), also called an **ECG** or **EKG.** When readings from the different electrodes are compared, they collectively provide an accurate, comprehensive assessment of the electrical changes of the heart.

Waves and Segments

An ECG provides a composite tracing of all cardiac action potentials generated by myocardial cells. A typical ECG tracing for one heart cycle has three principal deflections: a P wave above the baseline, a QRS complex that begins (Q) and ends (S) with small downward deflections from the baseline and has a large deflection (R) above the baseline, and a T wave above the baseline (**figure 19.20**). These waves indicate the electrical changes associated with depolarization and repolarization within specific heart regions:

1. The **P wave** reflects electrical changes of **atrial depolarization** that originates in the SA node. This event typically lasts 0.08 to 0.1 second.

Figure 19.20 The Electrocardiogram. (*a*) A tracing of an electrocardiogram with the location of three waves, including the P wave, QRS complex, and T wave. Two segments are located between the waves and include the P-Q segment and the S-T segment. The P-R interval and Q-T interval are also shown. (*b*) A graph of electrical changes within atria and ventricles is superimposed on an ECG to help you to see the relationship of these electrical changes and their relationship to the waves of an ECG.

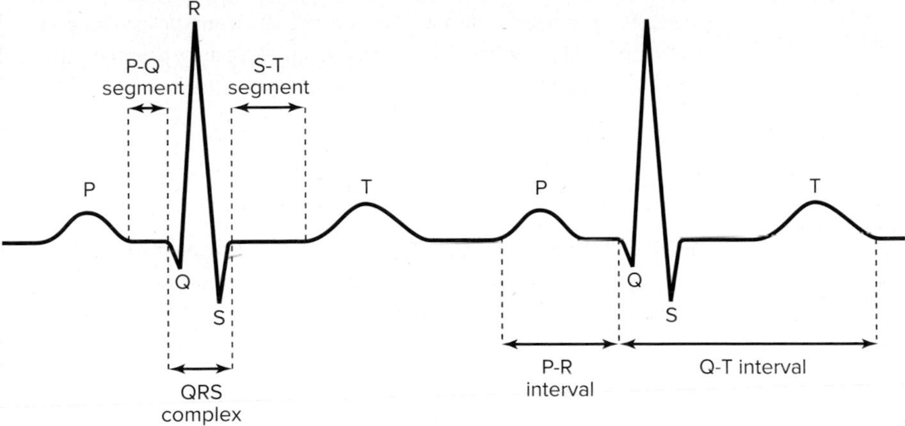

(a)

Electrical changes of atria

Electrical changes of ventricles

ECG

(b)

2. The **QRS complex,** which usually lasts between 0.06 and 0.1 second, represents the electrical changes associated with **ventricular depolarization.** Note that the atria are simultaneously repolarizing; however, this repolarization signal is masked by the greater electrical activity of the ventricles.
3. The **T wave** is the electrical change associated with **ventricular repolarization.**

The two segments between the waves correspond with a plateau (where there is essentially no electrical change). During these time periods, the sarcomeres are shortening within the cardiac muscle cells. The **P-Q segment** is associated with the atrial plateau at the sarcolemma when the cardiac muscle cells within the atria are contracting, and the **S-T segment** is the ventricular plateau when the cardiac muscle cells within the ventricles are contracting.

Figure 19.20*b* shows two of the graphs of the electrical changes associated with the atria and ventricles (see figure 19.18*b*) superimposed on the ECG. Notice how the waves of an ECG reflect electrical changes—either depolarization or repolarization—and how the flat or level portions in the segments correspond to a plateau when there is no electrical change. The flat line between cycles represents when the heart is resting between beats.

Intervals

Additional characteristics of an ECG include two intervals: the P-R interval and the Q-T interval (figure 19.20*a*). Changes in length of an interval may reflect abnormal changes to the heart.

The **P-R interval** represents the period of time from the beginning of the P wave (atrial depolarization) to the beginning of the QRS deflection (ventricular depolarization). This interval of time normally ranges from 0.12 to 0.20 second. It is the time required to transmit an action potential through the entire conduction system (i.e., from the SA nodal cells to the Purkinje fibers, which stimulate the ventricles). P-R intervals that extend longer than 0.20 second generally indicate an impaired conduction, often indicating a *heart block*. (See Clinical View 19.7: "Cardiac Arrhythmia.")

The **Q-T interval** represents the time from the beginning of the QRS (ventricular depolarization) to the end of the T wave (ventricular repolarization). This is the time required for the action potential to occur within the ventricles. This interval ranges from 0.2 to 0.4 second. A chronically longer Q-T interval may indicate a risk for a fast, irregular heart rate, which is called **tachyarrhythmia** (tak′ē′ă′ridh′mē-ă; *tachys* = quick).

INTEGRATE

CLINICAL VIEW 19.7

Cardiac Arrhythmia

Cardiac arrhythmia (ă-rith′mē-ă), also called *dysrhythmia,* is any abnormality in the electrical activity of the heart.

Heart blocks refer to an impairment within the heart's conducting system so the normal electrical activity is 'blocked' or slowed in its progression through the conduction system. Heart blocks may result in a feeling of light-headedness, fainting, irregular heartbeat, and chest palpitations. The three different types of heart blocks differ in the extent to which electrical signals are not transmitted through the conduction system.

First-degree AV block is also called *PR prolongation* because of a lengthened PR interval. The action potentials are *slowed* between the atria and ventricles. This type of block is generally asymptomatic.

Second-degree AV block is the failure of *some* atrial action potentials to be conducted to the ventricles. The PR interval may be either normal or prolonged.

Third-degree AV block is a complete heart block and is the failure of *all* action potentials to be conducted to the ventricles. This condition is life-threatening and requires medical intervention.

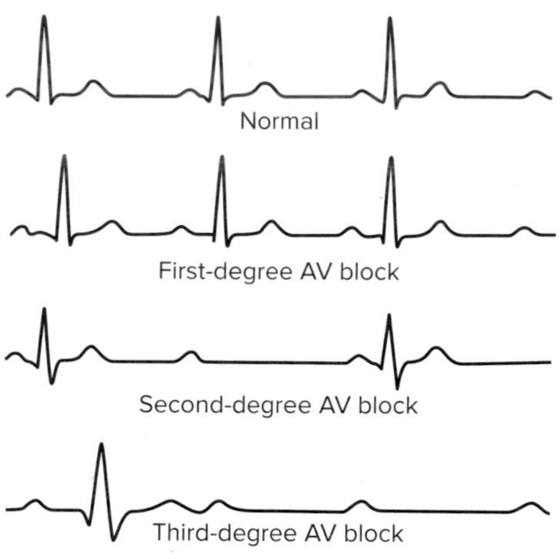

Premature ventricular contractions (PVCs) involve single or rapid bursts of abnormal action potentials initiated within the AV node or the ventricular conduction system instead of the SA node. PVCs often result from stress, stimulants such as caffeine, or sleep deprivation. They are not detrimental unless they occur in great numbers. Most PVCs go unnoticed, although occasionally one is perceived as the heart "skipping a beat" and then "jumping" in the chest.

P — VC

Atrial fibrillation (fī-bri-lā′shŭn) involves chaotic action potentials within the atria. The ventricles respond by increasing and decreasing contraction activities, which may lead to serious disturbances in cardiac rhythm.

Ventricular fibrillation is chaotic electrical activity within the ventricles. Muscle contraction of the cardiac muscle cells within the ventricles is uncoordinated, and the heart does not pump blood. Blood circulation stops (**cardiac arrest**), which leads to death of cardiac muscle (myocardial infarction), commonly called a "heart attack." To restore normal heart contractions, medical personnel apply a strong electrical shock to the chest using paddle electrodes in an attempt to synchronize the electrical events of cardiac muscle cells.

An **automated external defibrillator (AED)** can also be used to save an individual who is in sudden cardiac arrest (see Clinical View 19.4: "Coronary Heart Disease, Angina Pectoris, and Myocardial Infarction"). The most likely cause of a sudden cardiac arrest is ventricular fibrillation. An AED can be used to potentially restore a normal heart rhythm in ventricular fibrillation. An AED is typically available in public places where there are large numbers of people—for example, at airports, shopping malls, schools, hotels, and businesses.

An ECG is a common diagnostic tool that is used in physical exams and after surgery to detect many changes to the heart, including abnormal heart rhythms (fast, slow, arrhythmia), hypertrophy of the heart, low blood flow to the heart (ischemia), and damage to the heart from myocardial infarction or high blood pressure. Although it is a very effective tool in diagnosing many conditions, it does not detect all abnormalities. Thus, it is one of several diagnostic tools used by cardiologists and other physicians in determining a diagnosis and development of a treatment plan.

WHAT DID YOU LEARN?

What events in the heart are indicated by each of the following: P wave, QRS complex, and T wave? Identify the two segments of an ECG that reflect the plateau.

19.8 The Cardiac Cycle

A **cardiac cycle** is the inclusive changes within the heart from the initiation of one heartbeat to the start of the next. One heartbeat involves the contraction and relaxation of the heart chambers. Note that the term **systole** (sis′tō-lē) refers to contraction of a heart chamber, whereas **diastole** (dī-as′tō-lē; dilation) refers to relaxation of a heart chamber.

19.8a Overview of the Cardiac Cycle

LEARNING OBJECTIVES

32. Identify the two processes within the heart that occur due to pressure changes associated with the cardiac cycle.

33. List the five phases of the cardiac cycle.

The alternating contraction and relaxation of both the atria and ventricles cause pressure changes within their chambers. Pressure increases during contraction and decreases during relaxation. These alternating pressure changes are responsible for two significant physiologic processes:

- Unidirectional movement of blood through the heart chambers, as blood moves along a pressure gradient (i.e., from an area of greater pressure to an area of lesser pressure)
- Opening and closing of heart valves to ensure that blood continues to move in a "forward" direction without backflow

Keep in mind that contraction and relaxation of the *ventricles* are the most important driving force to continually move blood through the heart and to open and close heart valves during the cardiac cycle:

- **Ventricular contraction.** Ventricles contract and ventricular pressure rises. The one-way flow of blood occurs as the AV valves are closed, preventing backflow of blood into each atrium, and then the semilunar valves are opened, allowing ejection of blood from a ventricle into an arterial trunk (into the pulmonary trunk from the right ventricle and into the aorta from the left ventricle).
- **Ventricular relaxation.** Ventricles relax and ventricular pressure decreases. Now, the semilunar valves close, preventing backflow of blood into each ventricle, and then the AV valves open, allowing blood to again flow from each atrium into the right or left ventricle.

Here we organize the cardiac cycle into five phases: atrial contraction and ventricular filling, isovolumetric contraction, ventricular ejection, isovolumetric relaxation, and atrial relaxation and ventricular filling. These events are summarized in **figure 19.21**. Consider the following as you read through the description of each of the phases of the cardiac cycle: (1) whether the atria and ventricles are contracted or relaxed, (2) whether the pressure in the ventricles is higher or lower than the

> **INTEGRATE**
>
> **LEARNING STRATEGY**
>
> The action of heart valves is similar to the saying about doors (which originated from Alexander Graham Bell): "When one door closes, another opens." More specifically, as the ventricles contract, the AV valves close and the semilunar valves open. Then they reverse their position as the ventricles relax: The semilunar valves close and the AV valves open.

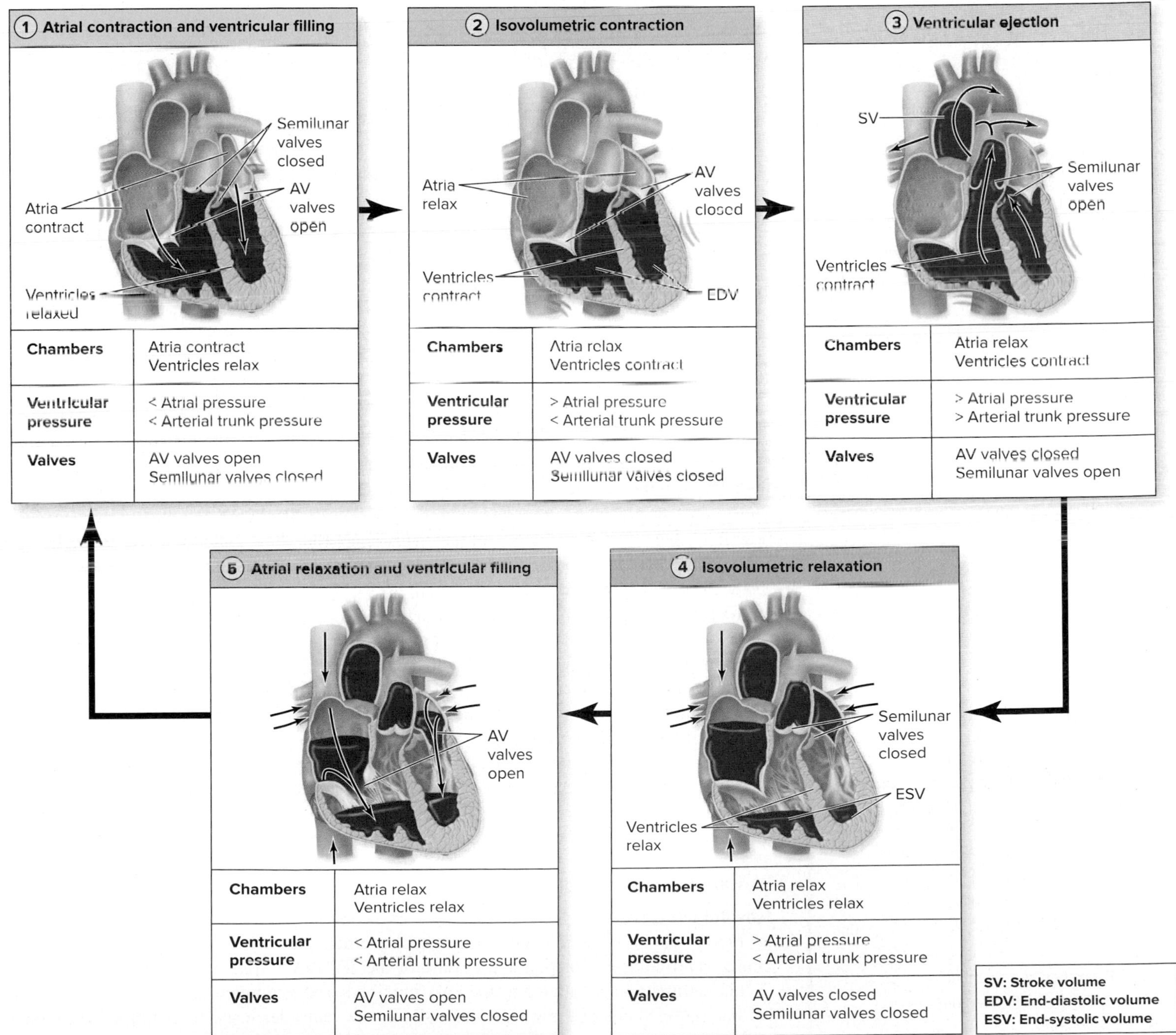

Figure 19.21 Phases of the Cardiac Cycle. Contraction and relaxation of heart chambers, changes in ventricular pressure, and opening and closing of heart valves occur during the phases of the cardiac cycle. AP|R

pressure in the atria and higher or lower than the pressure in the arterial trunks (aorta and pulmonary trunk), and (3) if the AV valves and semilunar valves are opened or closed.

WHAT DID YOU LEARN?

32 Pressure changes that occur during the cardiac cycle are responsible for what two physiologic processes within the heart?

19.8b Events of the Cardiac Cycle

LEARNING OBJECTIVES

34. List and describe what occurs during the five phases of the cardiac cycle.

35. Explain the significance of ventricular balance.

Because this is a cycle, we could start at any point and describe the events. However, for orientation and convenience we begin when all chambers are at rest and the atria are just initiating the contraction phase.

As the Cardiac Cycle Begins

We can note the following characteristics of the heart prior to atrial contraction as a new round of the cardiac cycle begins.

- All four chambers are at rest.
- Blood continues to return to the right atrium through the superior vena cava, inferior vena cava, and the coronary sinus, and to the left atrium through the pulmonary veins.
- Passive filling of the ventricles is under way. No contraction of the atria is needed at this time to assist with the continual ventricular filling.
- The AV valves are open because the pressure exerted by the blood filling the atria is greater than the pressure exerted by the blood in the resting ventricles.
- The semilunar valves are closed because the pressure exerted by the blood in the filling ventricles is lower than the pressure exerted by the blood in the arterial trunks.

Atrial Contraction and Ventricular Filling

This phase is distinguished from the resting conditions of the heart just described by two events: atrial contraction and completion of ventricular filling (phase 1 in figure 19.21). Atrial contraction is initiated by the SA node stimulating cardiac muscle cells of the atrial wall. Contraction of the atrial wall moves the remainder of blood that was within each atrium into the right or left ventricle.

Ventricular filling is complete upon termination of atrial contraction, and the ventricles now hold their maximum blood volume. This is appropriately called the **end-diastolic volume (EDV)** because it is the volume of blood within the ventricle at the end of diastole (or rest). EDV is labeled in phase 2 in figure 19.21. In a resting adult, the value of EDV is approximately 130 milliliters (mL) of blood.

Note that during atrial contraction, both the inflow of additional blood from the supplying veins into the atria (i.e., blood flow from the superior vena cava, inferior vena cava, and coronary sinus into the right atrium and blood flow from the pulmonary veins into the left atrium) and the backflow of blood from the atria into these veins are prevented because contraction of the atria compresses the openings for the great veins.

You will find it helpful to remember that at the end of this phase, the atria relax and will remain relaxed until the beginning of the next cardiac cycle, as shown in phases 2 through 5 in figure 19.21. These four phases (2 through 5) involve the changes associated with contraction and relaxation of the ventricles.

INTEGRATE

LEARNING STRATEGY

You may find this summary table of assistance in remembering what is occurring during each of the five phases of the cardiac cycle.

	1	2	3	4	5
Atria	C	R	R	R	R
Ventricles	R	C	C	R	R
AV valves	O	CL	CL	CL	O
Semilunar valves	CL	CL	O	CL	CL

R = Relaxation of a chamber
C = Contraction of a chamber
O = Valve open
CL = Valve closed

Isovolumetric Contraction

Isovolumetric (ī′sō-vol′yū-met′rik; *iso* = same) **contraction** involves no change in ventricular blood volume when the ventricles are contracting. Two distinctive changes occur in this phase: ventricular contraction and closure of the AV valves (phase 2 in figure 19.21). Ventricular contraction is initiated by the Purkinje fibers of the conduction system stimulating the cardiac muscle cells of the ventricular wall. As the ventricles begin to contract, the pressure within the ventricles increases and exceeds the pressure within the atria. Consequently, the AV valves are forced closed. Recall that these valves are braced by tendinous cords to the papillary muscles. The closure of these valves prevents backflow of blood from the ventricles into the atria. The semilunar valves remain closed because the pressure within the ventricle is still less than the pressure within the attached arterial trunk.

Note that all heart valves are closed at this time. Thus, although the ventricular cardiac muscles are contracting during this phase, no blood is entering or leaving the ventricle and the volume within the ventricular chambers remains the same.

Ventricular Ejection

Ventricular ejection, which is the movement of blood from the ventricles into the arterial trunks, occurs during this phase as ventricular contraction continues and the semilunar valves are forced open (phase 3 in figure 19.21). As the ventricles continue to contract, the pressure within the ventricles increases and exceeds the pressure within the arterial trunk. Consequently, the semilunar valves open, which allows blood to be ejected from the ventricles into their associated arterial trunk (i.e., blood is ejected from the left ventricle into the aorta and blood is ejected from the right ventricle into the pulmonary trunk). The AV valves remain closed because the pressure within the ventricles remains greater than the pressure within the atria. The amount of blood pumped out during ventricular contraction is termed the **stroke volume (SV).** Generally, this volume is approximately 70 mL of blood.

Not all of the blood in either ventricle is ejected. The blood remaining in a ventricle at the end of systole is appropriately called the **end-systolic volume (ESV)** because it is the volume of blood in the ventricle at the end of systole (or contraction). ESV is labeled in phase 4 in figure 19.21. ESV is determined by subtracting SV from EDV (EDV − SV = ESV): 130 mL − 70 mL = 60 mL.

Isovolumetric Relaxation

Isovolumetric relaxation involves no change in ventricular blood volume when the ventricles are relaxing. Two distinctive changes occur in this phase: ventricular relaxation and closure of the semilunar valves (phase 4 in figure 19.21). As the ventricles start to relax and the ventricular chamber expands back to its resting size, pressure within the ventricles decreases below the pressure within the attached arterial trunks. Blood initially flows backward slightly within the arterial trunks but is caught in the semilunar valves, causing them to close. This closure of the semilunar valves prevents blood backflow from the arterial trunks into the ventricles. The AV valves remain closed because the pressure within the ventricles is still greater than the pressure within the atria.

Note that all heart valves are again closed simultaneously. Thus, although the ventricular cardiac muscles are relaxing during this phase, no blood is entering or leaving the ventricle and the volume within the ventricular chambers remains the same.

Atrial Relaxation and Ventricular Filling

The final phase of the cardiac cycle is distinguished by the continued relaxation of the ventricles and the opening of the AV valves (phase 5 in figure 19.21). As the ventricles continue to relax, the pressure within the ventricles decreases below the pressure within the atria, and the AV valves open. The opening of these valves allows blood to once again move from the atria into the ventricles for ventricular filling. The semilunar valves remain closed because the pressure within the ventricles remains lower than the pressure with the arterial trunks. Amazingly, these events of the cardiac cycle, as organized and described in these five phases, are repeated with each heartbeat!

Figure 19.22 shows all of the principal events of one heartbeat, including the electrical events that control heart contraction, as depicted on an ECG; the contraction and relaxation of the heart chambers and position of the heart valves that occur during a cardiac cycle; the relative pressure within the left atrium, left ventricle, and aorta; and the changes in blood volume within the left ventricle.

Ventricular Balance

It is important to realize that *equal* amounts of blood are normally pumped by the two ventricles through the two circulations, a condition called **ventricular balance.** However, the right ventricle has to pump the blood only to the adjacent lungs (a relatively short circuit), whereas the left ventricle has to pump the blood through the systemic circulation throughout the body (a relatively long circuit). Thus, the left ventricle must be larger and stronger than the right ventricle to pump the blood farther—but it is the same amount of blood pumped from both sides of the heart. Sustained pumping of *unequal* amounts of blood may result in **edema** (e-dē′mă; *aoidema* = swelling), which is excess fluid in the interstitial space or within cells. (See Clinical View 19.1: "Congestive Heart Failure.")

INTEGRATE

LEARNING STRATEGY

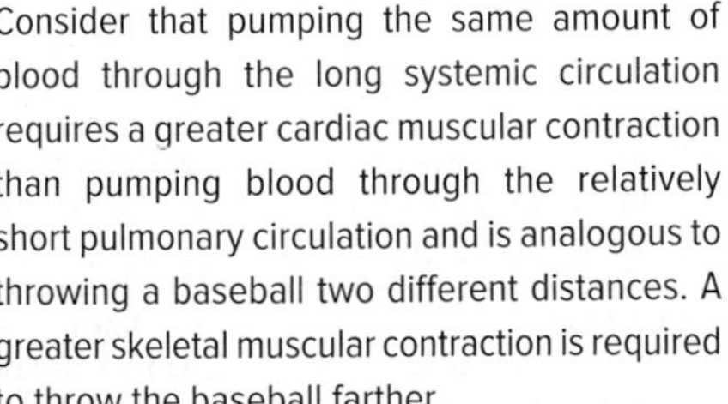

Consider that pumping the same amount of blood through the long systemic circulation requires a greater cardiac muscular contraction than pumping blood through the relatively short pulmonary circulation and is analogous to throwing a baseball two different distances. A greater skeletal muscular contraction is required to throw the baseball farther.

Cardiac Reserve

An increase in both heart rate and stroke volume results in an increase in cardiac output. During physical exertion, heart rate can be accelerated to more than 170 beats per minute. Likewise, stroke volume can be increased to more than 100 mL. **Cardiac reserve** is an increase in cardiac output above its level at rest. It can be determined by subtracting cardiac output at rest from the cardiac output during exercise.

Cardiac reserve is a measure of the level and duration of physical effort in which an individual can engage. Cardiac output may be increased by about four-fold in a healthy, nonathletic individual (to approximately 20 L/min), and up to seven-fold in a highly trained athlete (to approximately 35 L/min). In comparison, individuals with a weakened heart may have little cardiac reserve and thus experience limitations in exerting themselves.

WHAT DID YOU LEARN?

36 What are the two factors that determine cardiac output?

37 What is the cardiac output at rest and during exercise, and the cardiac reserve, if (a) the heart rate is 75 beats per minute and stroke volume is 70 mL at rest, and (b) if the heart rate is 150 beats per minute and stroke volume is 100 mL during exercise?

19.9b Variables That Influence Heart Rate

LEARNING OBJECTIVES

38. Define chronotropic agents, and describe how they affect heart rate.
39. Discuss how autonomic reflexes alter heart rate.

Cardiac output is directly influenced by both heart rate and stroke volume. Both heart rate and stroke volume, in turn, have variables that influence each of them. Here we describe the variables that influence heart rate and then in section 19.9c discuss the variables that influence stroke volume.

The heart rate can be altered by external factors that act on the SA node (the pacemaker) and the AV node. The primary external factors to increase and decrease heart rate come from autonomic nervous system innervation (both the sympathetic division and the parasympathetic division) and varying levels of some hormones. These factors that change heart rate are called **chronotropic** (kron′ō-trop′ik; *chrono* = time, *tropos* = change) **agents** and are classified as either positive chronotropic agents or negative chronotropic agents.

Positive chronotropic agents cause an increase in heart rate and include sympathetic nerve stimulation and certain types of hormonal stimulation. Review **figure 19.23** as you read through this section. Sympathetic axons release the neurotransmitter norepinephrine (NE) to act directly on the SA nodal cells (see section 15.6b). The sympathetic division also causes release of both epinephrine (EPI) and NE from the adrenal medulla into the blood (step 1). Both NE and EPI bind to β_1-adrenergic receptors of the heart (step 2). This binding initiates an intracellular pathway involving G protein that results in the activation of adenylate cyclase enzyme with the accompanying production of the second messenger, cAMP (see section 17.5b). Ultimately, protein kinase enzymes phosphorylate Ca^{2+} channels, causing them to open. Positively charged Ca^{2+} enters the nodal cells, and nodal cells reach threshold more quickly, increasing the firing rate of the SA node (step 3). (Recall that an influx of Ca^{2+} is responsible for depolarization of nodal cells; see section 19.6b). Sympathetic stimulation of the AV node also increases calcium influx into these cells (not shown in figure 19.23). The delay in the AV node is decreased, and the conduction rate is increased. Thus, heart rate increases.

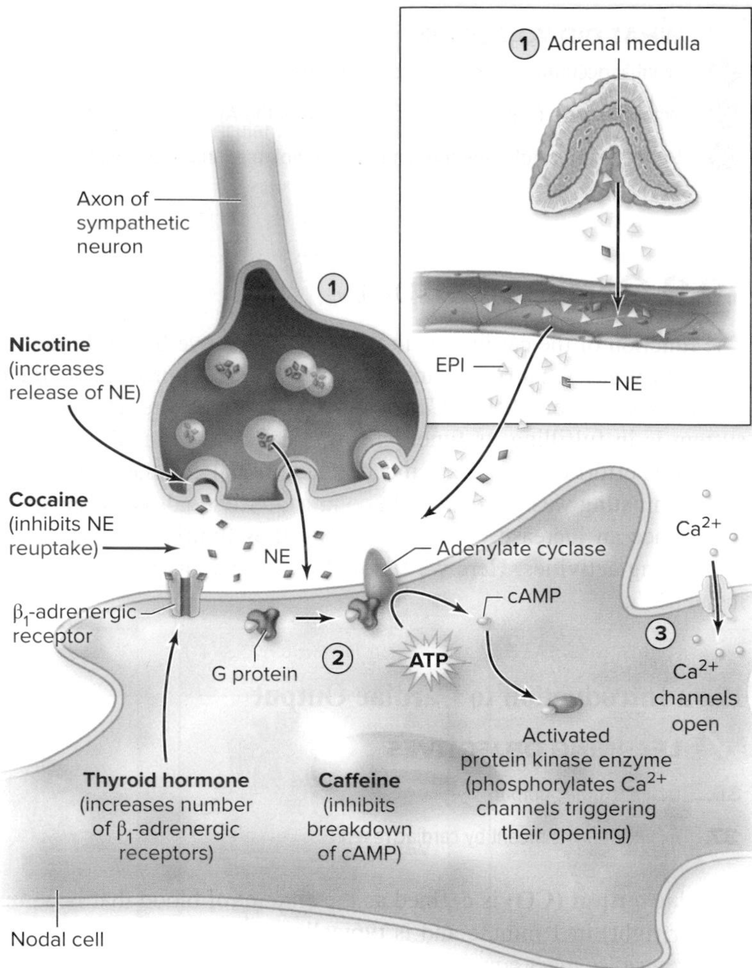

Figure 19.23 Sympathetic Innervation of Nodal Cells. Sympathetic axons release norepinephrine (NE) (and stimulate release of NE and epinephrine [EPI] from the adrenal medulla) (step 1), which acts as a positive chronotropic agent to increase heart rate. NE and EPI bind with receptors of nodal cells (step 2) to activate G protein and form a second messenger that results in activation of a protein kinase enzyme and opening of Ca^{2+} channels (step 3). Both thyroid hormone (TH) and certain chemical compounds (e.g., nicotine, cocaine, caffeine) increase heart rate at different steps within this pathway.

Thyroid hormone (see section 17.8b) is also a positive chronotropic agent because it makes nodal cells more responsive to NE and EPI by increasing the number of β_1-adrenergic receptors (at step 2). Additionally, several drugs act on this pathway to increase heart rate. Nicotine and cocaine both increase the amount of NE present in the synaptic cleft (at step 1). Nicotine functions by increasing the release of NE, and cocaine functions by inhibiting the reuptake of NE. Caffeine, in comparison, inhibits the breakdown of cAMP (at step 3). (Very high intake of caffeine in some "energy drinks" has resulted in several fatal heart attacks.)

In contrast, **negative chronotropic agents** decrease heart rate. Parasympathetic innervation is one of the most important of these agents. Parasympathetic axons release acetylcholine that binds to M2 muscarinic receptors (see section 15.6b), which are K^+ channels (see figure 19.16). These channels open, and K^+ moves down its concentration gradient to exit the cells. This loss of positive ion causes hyperpolarization (membrane potential becomes more negative) of nodal cells, and it takes a longer period of time for these cells to reach threshold—thus, the heart rate is slower. Beta-blocker drugs (which interfere with binding of norepinephrine and epinephrine to beta receptors) are another type of negative chronotropic agent and are used to treat high blood pressure.

Autonomic Reflexes

The ability of the autonomic nervous system to influence heart rate through the sympathetic and parasympathetic divisions is controlled through **autonomic reflexes** (see section 15.8). The cardiac center receives sensory input from baroreceptors (regarding stretch of blood vessels) and chemoreceptors (concerning blood carbon dioxide and hydrogen ion [H^+] levels). The cardiac center reflexively responds to this input by altering nerve signals relayed through sympathetic and parasympathetic neurons that innervate the heart to adjust heart rate and stroke volume as needed to maintain homeostasis (see figure 19.14).

One specific autonomic reflex is the **atrial reflex,** also called the *Bainbridge reflex,* which protects the heart from overfilling. It is initiated when baroreceptors in the atrial walls are stimulated by an increase in venous return. Nerve signals are increased along sensory neurons to the cardioacceleratory center, resulting in an increase in nerve signals relayed along sympathetic neurons to the heart. Heart rate increases so blood moves more quickly through the heart, thus decreasing atrial stretch. Arterial reflexes associated with regulating systemic blood pressure that are initiated by baroreceptors and chemoreceptors in the aorta and carotid are described in section 20.6a.

WHAT DID YOU LEARN?

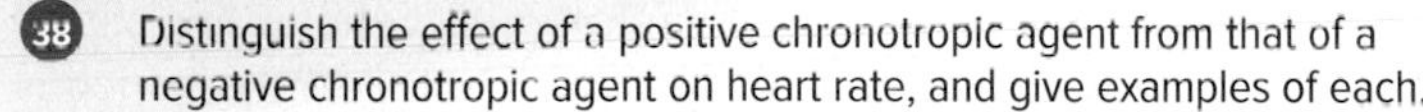

38 Distinguish the effect of a positive chronotropic agent from that of a negative chronotropic agent on heart rate, and give examples of each.

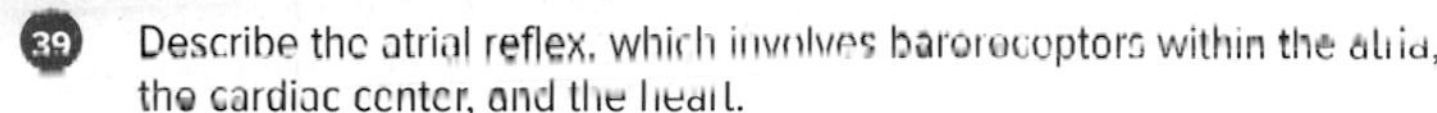

39 Describe the atrial reflex, which involves baroreceptors within the atria, the cardiac center, and the heart.

19.9c Variables That Influence Stroke Volume

LEARNING OBJECTIVES

40. List the three variables that may influence stroke volume.

41. Define each of the three variables, and describe the factors that influence each variable and how each variable affects stroke volume.

Stroke volume (SV) is the volume of blood pumped from a ventricle during a ventricular contraction (i.e., the amount of blood ejected per heartbeat) (**figure 19.24**). Recall from section 19.8b that stroke volume is dependent upon the volume of blood that enters the heart at the end of heart relaxation (called the *end-diastolic volume [EDV]*). The typical EDV in a resting adult is approximately 130 mL. However, not all blood in either ventricle is normally ejected from the heart during ventricular contraction. The blood remaining in a ventricle at the end of ventricular contraction is called the *end-systolic volume (ESV)*. The typical ESV in an adult is 60 mL. Thus, stroke volume is the difference between EDV and ESV:

$$\text{EDV} - \text{ESV} = \text{SV}$$

$$130\text{ mL} - 60\text{ mL} = 70\text{ mL}$$

These values for ESV, EDV, and SV are typical values for an adult. The specific volume of blood ejected as stroke volume varies and is influenced by several variables. These include (1) venous return, which is the amount of blood returned to the heart; (2) inotropic agents, which are external factors that alter the force of contraction of the myocardium; and (3) afterload, which is the resistance in the arteries to the ejection of blood from the heart.

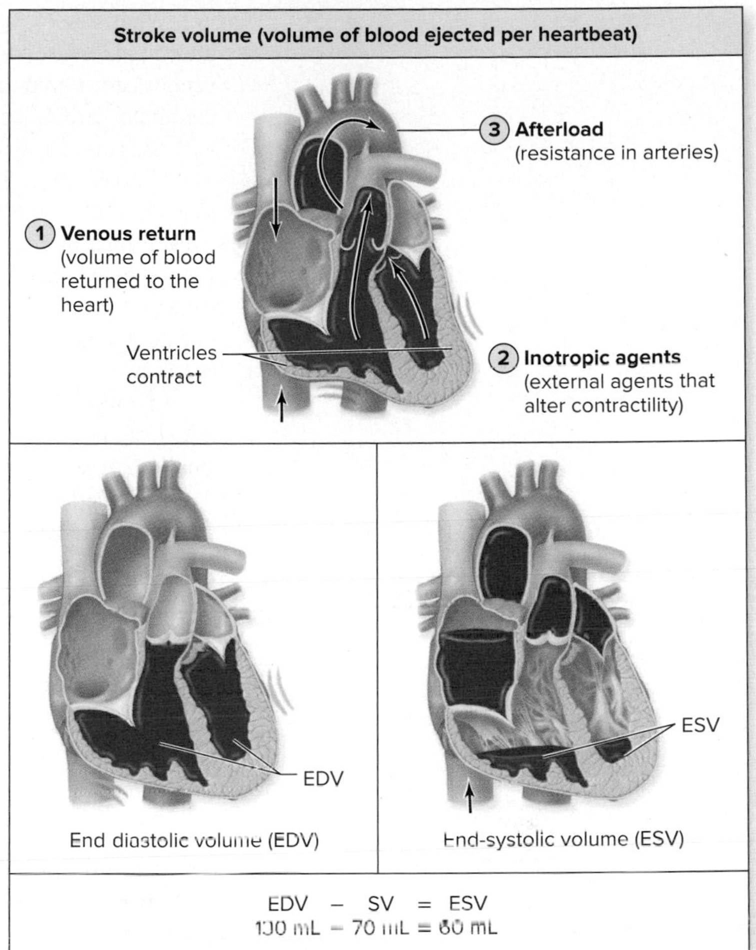

Figure 19.24 Relationship of EDV, ESV, and SV. Stroke volume is dependent upon venous return, presence of inotropic agents, and afterload. End diastolic volume (EDV) is the volume of blood in a ventricle at the end of rest. End-systolic volume (ESV) is the volume of blood in a ventricle directly following a contraction. Stroke volume (SV), which is the volume of blood pumped in one heartbeat, is equal to EDV – ESV.

Venous Return

Venous return is the volume of blood returned to the heart via the great veins and is directly related to stroke volume. Venous return determines the amount of blood in the ventricle at the end of rest immediately prior to contraction (i.e., end-diastolic volume, or EDV). This volume of blood, in turn, determines the preload on the heart. **Preload** is stretch of the heart wall due to the load to which a cardiac muscle is subjected before shortening. In other words, preload is the heart muscle wall stretch just prior to contraction.

The direct relationship of venous return and stroke volume can be explained by the **Frank-Starling law** of the heart (or simply *Starling's law*). This law essentially states that as the volume of blood entering the heart increases, there is greater stretch of the heart wall (or preload). This results in greater overlap of the thick and thin filaments in the sarcomeres of the cardiac muscle cells composing the myocardium, allowing formation of greater numbers of crossbridges. (Recall that cardiac muscle does not exhibit maximum overlap of thin and thick filaments at rest; see section 19.3f.) Consequently, a more forceful ventricular contraction is generated, and stroke volume increases.

As venous return decreases, there is less stretch of the heart wall (smaller preload), and this results in less overlap of the thick and thin filaments and fewer crossbridges form. Consequently, a less forceful ventricular contraction is generated, and stroke volume decreases.

What causes venous return, and thus preload, to either increase or decrease? Venous return is increased by either an increase in venous pressure or an increase in time to fill, and venous return may be decreased by a decrease in either of these two factors. You might find the analogy of filling a balloon helpful. The water pressure and the time to fill the balloon will determine how much water you can add to the balloon in a given period of time.

Venous return increases during exercise, for example, because of greater venous pressure. Veins are "squeezed" by skeletal muscles, which helps return blood to the heart (see section 20.5a). Greater muscular movement increases the action of the skeletal muscle pump. During exercise, venous return can approximately double in comparison to its rate at rest. Venous return also increases with a slower heart rate. The slower heart rate allows a greater amount of time for blood to enter the heart. This is most noticeable in highly trained athletes who have a very low resting heart rate.

In comparison, low blood volume (e.g., due to hemorrhage) or an abnormally high heart rate decreases venous return. The result is a smaller end-diastolic volume and preload, and a smaller stroke volume.

Balanced ventricular output is primarily a function of the inherent ability of the heart to contract more forcefully in response to increased venous return. Consider, for example, that when you first start to exercise, venous return from the body into the right side of the heart increases, causing it to contract more forcefully and increasing stroke volume. Increased amount of blood then moves through the pulmonary circulation and returns to the left ventricle. The left ventricle, in turn, will experience greater stretch of its chamber wall and will contract more forcefully.

Inotropic Agents

Stroke volume is similar to heart rate in that it is altered by external factors. The primary external factors to increase and decrease heart rate come from autonomic innervation and varying levels of some hormones. These factors that change stroke volume are called **inotropic** (in′ō-trop′ik; *ino* = fiber) **agents.** Inotropic agents alter **contractility,** which is the force of contraction at a given stretch of the cardiac muscle cells. An increase or a decrease in the force of contraction is generally due to a change in the available Ca^{2+} in the sarcoplasm. Changes in Ca^{2+} concentration alter the number of crossbridges formed and thus the force of contraction generated.

A positive inotropic agent increases Ca^{2+} concentration, which results in formation of additional crossbridges. Positive inotropic agents include norepinephrine that is released from sympathetic neurons and epinephrine and norepinephrine from the adrenal medulla. These ligands bind to β_1-adrenergic receptors to cause an increase in Ca^{2+} levels within cardiac muscle cells. Thyroid hormone is also a positive inotropic agent because it increases the number of β_1-adrenergic receptors in the cardiac muscle cells. Certain drugs (e.g., digitalis) are positive inotropic agents used to treat abnormally low cardiac output that accompanies some heart conditions (e.g., congestive heart failure).

In comparison, a negative inotropic agent decreases contractility by decreasing available Ca^{2+} and fewer numbers of crossbridges are formed. Electrolyte imbalances (see table 25.1 and sections 25.3b and 25.6), including an increase in either K^+ or H^+, act as negative inotropic agents. Certain drugs (e.g., nifedipine, a Ca^{2+} channel–blocking drug) are negative inotropic agents and are given to decrease cardiac output, typically in an effort to treat high blood pressure.

Afterload

Afterload is the resistance in arteries to the ejection of blood by the ventricles, and it represents the pressure that must be exceeded before blood is ejected from the chamber. Afterload generally becomes a consideration only in older individuals. Arteries typically develop atherosclerosis as we age. (However, because of the increased rate of obesity in teenagers and young adults, the prevalence of atherosclerosis is increasing in these age groups.) It is a condition in which plaque accumulates on the inner linings of a blood vessel. The smaller arterial lumen exerts greater resistance to the movement of blood into the arterial trunks, and stroke volume decreases. The relationship of these variables that influence stroke volume is integrated in **figure 19.25**.

 **WHAT DID YOU LEARN?**

 40 Which of the following increases stroke volume: (a) increased venous return, (b) increased Ca^{2+} in sarcoplasm, or (c) afterload? Explain.

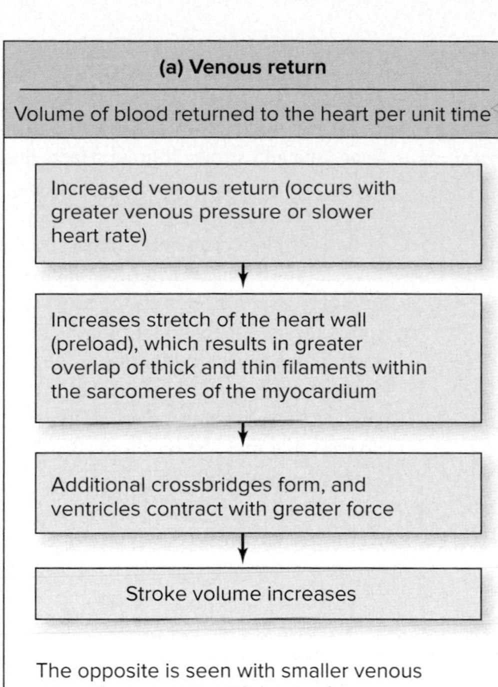
(a) Venous return
Volume of blood returned to the heart per unit time
Increased venous return (occurs with greater venous pressure or slower heart rate)
Increases stretch of the heart wall (preload), which results in greater overlap of thick and thin filaments within the sarcomeres of the myocardium
Additional crossbridges form, and ventricles contract with greater force
Stroke volume increases
The opposite is seen with smaller venous return (e.g., occurs with hemorrhage or extremely rapid heart rate)

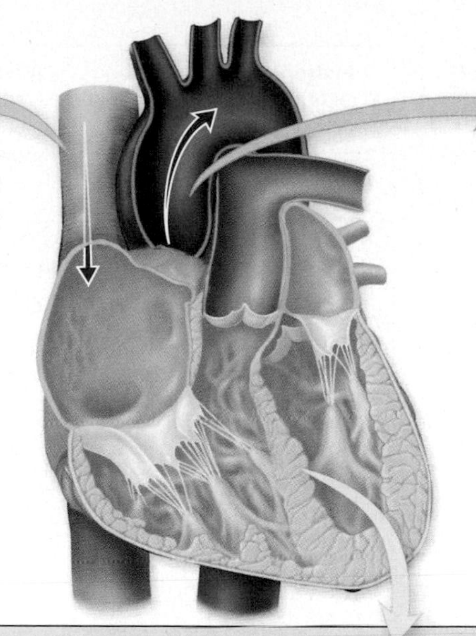

(c) Afterload

Resistance in arteries to ejection of blood

Atherosclerosis, which is deposition of plaque on the inner lining of arteries, is typically only a factor as we age

Arteries become more narrow in diameter

Increases the resistance to pump blood into the arteries

Stroke volume decreases

(b) Inotropic agents

Substances that act on the myocardium to alter contractility

Positive inotropic agents (e.g., stimulation by sympathetic nervous system)

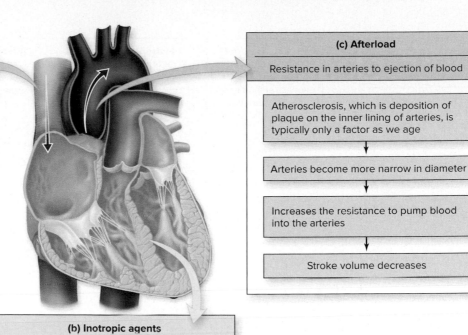

Increased Ca^{2+} levels in the sarcoplasm result in greater binding of Ca^{2+} to troponin of thin filaments within sarcomeres of the myocardium

Additional crossbridges form, and ventricles contract with greater force

Stroke volume increases

The opposite is seen with negative inotropic agents (e.g., calcium channel blockers)

Figure 19.25 Variables That Influence Stroke Volume. Three variables influence stroke volume, which is the amount of blood ejected from the heart per beat. These include (*a*) venous return, (*b*) inotropic agents, and (*c*) afterload.

19.9d Variables That Influence Cardiac Output

 **LEARNING OBJECTIVE**

42. Summarize the variables that influence cardiac output.

The factors that influence heart rate and stroke volume, and ultimately cardiac output, are integrated in a flowchart in **figure 19.26**. Note the following:

- **Heart rate.** An increase or a decrease in heart rate is dependent upon chronotropic agents that influence the *conduction system.* These agents stimulate the SA node to change its firing rate or the AV node to alter the amount of delay.
- **Stroke volume.** An increase or a decrease in stroke volume is generally due to changes in the *myocardium.* Venous return (which alters the stretch of the heart) and inotropic agents (which change the Ca^{2+} level in the sarcoplasm) influence the number of crossbridges, which alters the force of contraction. The only exception is afterload, which reflects increased resistance in arteries, making it more difficult for the heart to pump blood. Afterload is generally a factor only as we age.
- **Cardiac output.** Both heart rate and stroke volume are directly related to cardiac output. When both heart rate and stroke volume increase, cardiac output increases. In contrast, when both heart rate and stroke volume decrease, cardiac output decreases. It is not

INTEGRATE

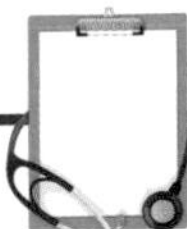

CLINICAL VIEW 19.8

Bradycardia and Tachycardia

A persistently low resting heart rate below 60 beats per minute in adults is called **bradycardia.** Bradycardia is considered a normal change in highly trained athletes who engage in a sustained aerobic exercise program. Abnormal conditions that cause bradycardia include hypothyroidism (see Clinical View 17.5: "Disorders of Thyroid Hormone Secretion"), electrolyte imbalances (see section 25.3b), and congestive heart failure (see Clinical View 19.1: "Congestive Heart Failure"). In comparison, a persistently high resting heart rate above 100 beats per minute in adults is called **tachycardia.** Tachycardia is caused by abnormal conditions such as heart disease, fever, or anxiety.

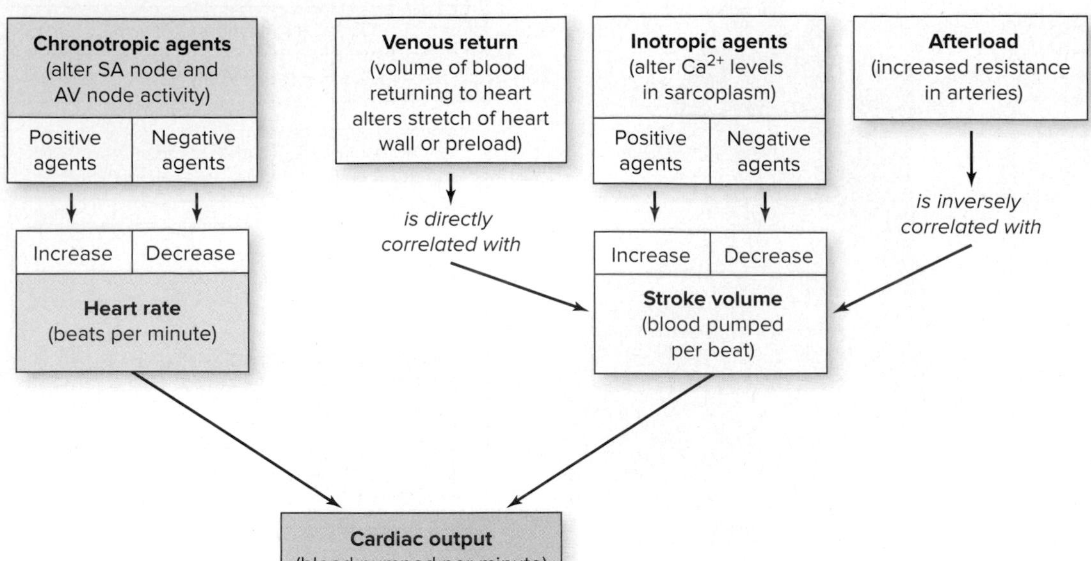

Figure 19.26 Factors Affecting Cardiac Output. Cardiac output is influenced by both heart rate and stroke volume. Heart rate is altered by chronotropic agents acting on the SA node and the AV node. Stroke volume is altered by both changes to the myocardium, which include an increase or a decrease in stretch of the heart wall (preload), and the amount of Ca^{2+} in the sarcoplasm caused by inotropic agents. Stroke volume is also influenced by changes in resistance (afterload) in the arteries that must be overcome for the heart to eject blood.

possible to predict the net effect on cardiac output (i.e., whether it will increase or decrease) if heart rate and stroke volume change in opposite directions (e.g., stroke volume decreases from blood loss, and heart rate increases in an attempt to maintain cardiac output).The net effect will be determined by the relative change to both heart rate and stroke volume.

WHAT DID YOU LEARN?

41 If both heart rate and stroke volume increase, does cardiac output (a) stay the same, (b) increase, or (c) decrease? Thus, is the relationship between these two variables and cardiac output direct or inverse?

19.10 Development of the Heart

LEARNING OBJECTIVES

43. Explain how postnatal heart structures develop from the primitive heart tube.

44. Describe septal defects that may occur during development.

Development of the heart commences in the third week, when the embryo becomes too large to receive its nutrients through diffusion alone. The steps involved in heart development are complex, because the heart must begin working before its development is complete (**figure 19.27**).

By day 19 (middle of week 3), two **heart tubes** (*endocardial tubes*) form from mesoderm in the embryo. By day 21, these paired tubes fuse, forming a single primitive heart tube (figure 19.27*a*). By day 22, the heart begins to beat, and later in the fourth week, this single heart tube bends and folds upon itself to begin to form the external heart shape (figure 19.27*b*). This tube develops the following named expansions that give rise to postnatal heart structures (listed from inferior to superior): **sinus venosus, primitive atrium, primitive ventricle,** and **bulbus cordis** (bŭl′bŭs kōr′dis). The sinus venosus and primitive atrium form parts of the left and right atria. The primitive ventricle forms most of the left ventricle. The bulbus cordis may be further subdivided into a **trabeculated part of the right ventricle,** which forms most of the right ventricle; the **conus cordis,** which forms the outflow tracts for the ventricles; and the **truncus arteriosus,** which forms the ascending aorta and pulmonary trunk (figure 19.27*c*).

The next major steps in heart development occur during weeks 5–8, when the single heart tube becomes partitioned into four chambers (two atria and two ventricles), and the great vessels form. This partitioning is complex, and errors in development lead to many of the more common congenital heart malformations.

The common atrium is subdivided into a left and right atrium by an interatrial septum, which consists of two parts (**septum primum** and **septum secundum**) that partially overlap (figure 19.27*d*). An opening in the septum secundum (which is covered by the septum primum) is called the **foramen ovale.** Because the embryonic lungs are not functional, much of the blood is shunted from the right atrium to the left atrium by moving through the foramen ovale and pushing the septum primum into the left atrium. When the baby is born and the lungs are fully functional, the blood from the left atrium pushes the septum primum and septum secundum together, closing the interatrial septum. The only remnant of the embryonic opening is an oval-shaped depression in the interatrial septum called the **fossa ovalis.**

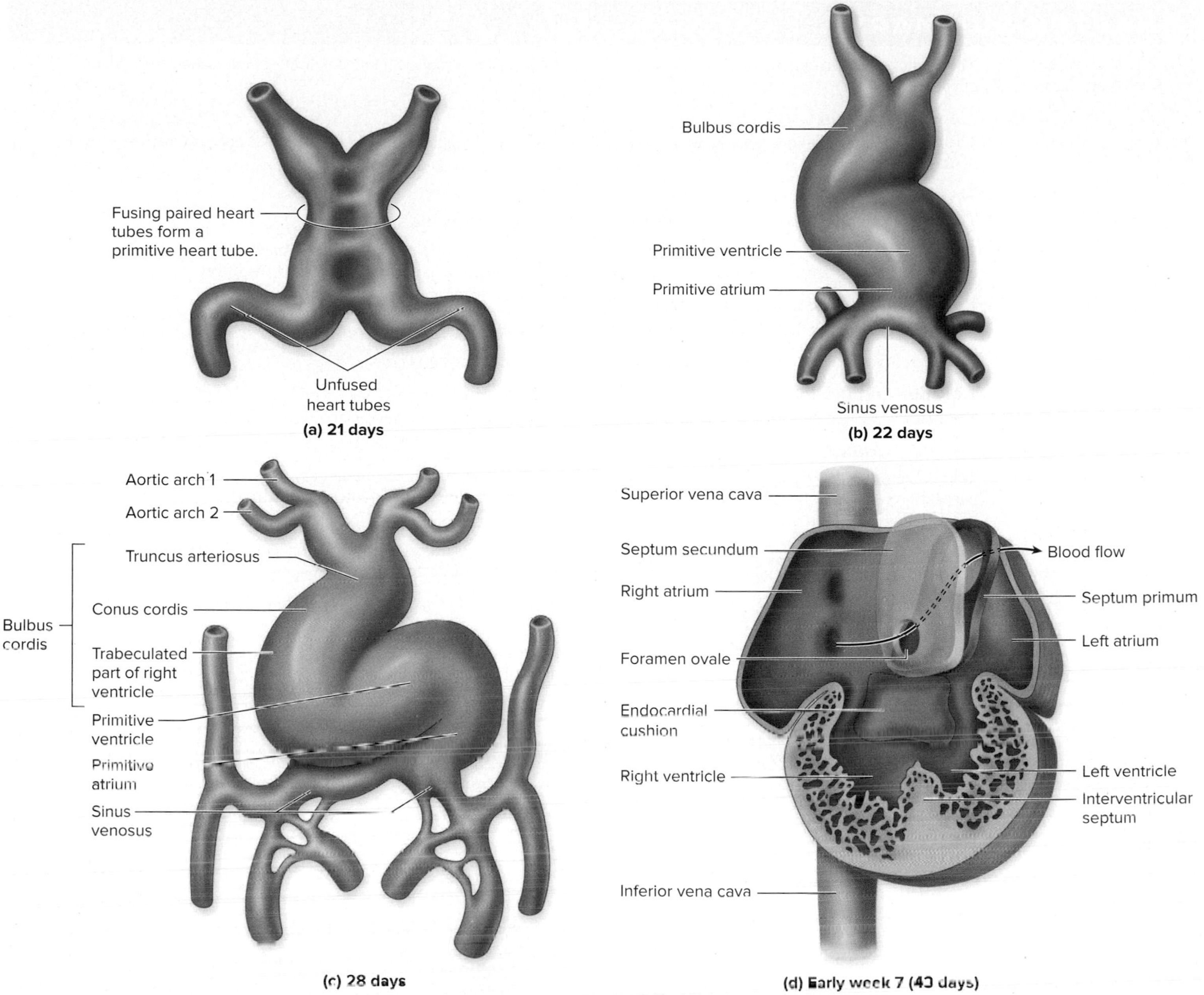

Figure 19.27 Development of the Heart. The heart develops from mesoderm. By day 19, paired heart tubes are present in the cardiogenic region of the embryo. (*a*) These paired tubes fuse by day 21. (*b*) The single heart tube bends and folds upon itself, beginning on day 22. (*c*) By day 28, the heart tube is S-shaped. (*d*) By early week 7, the interatrial septum is formed by two overlapping septa. The foramen ovale is a passageway that detours blood away from the pulmonary circulation into the systemic circulation prior to birth.

Left and right ventricles are partitioned by an **interventricular septum** that grows superiorly from the floor of the ventricles. The AV valves, papillary muscles, and tendinous cords all form from portions of the ventricular walls as well.

Many congenital heart malformations result from incomplete or faulty development during the early weeks of development. For example, in an **atrial septal defect,** the postnatal heart still has an opening between the left and right atria. Thus, blood from the left atrium (the higher-pressure system) is shunted to the right atrium (the lower-pressure system). This can lead to enlargement of the right side of the heart.

Ventricular septal defects can occur if the interventricular septum is incompletely formed. A common malformation called **tetralogy of Fallot** occurs when the aorticopulmonary septum divides the truncus arteriosus unevenly. As a result, the patient has a ventricular septal defect, a very narrow pulmonary trunk (pulmonary stenosis), an aorta that overlaps both the left and right ventricles, and an enlargement of the right ventricle (right ventricular hypertrophy).

WHAT DID YOU LEARN?

42 What would be the path of blood flow through the heart if the foramen ovale did not close shortly after birth?

CHAPTER SUMMARY

19.1 Introduction to the Cardiovascular System	• The cardiovascular system is composed of the heart and blood vessels organized into the pulmonary circulation and systemic circulation.
	19.1a General Function • The function of the cardiovascular system is to transport substances throughout the body and provide adequate perfusion to all tissues.
	19.1b Overview of Components • The primary types of blood vessels include arteries, capillaries, and veins. • The heart acts as two "side-by-side" pumps with the right side composed of a right atrium and right ventricle and the left side composed of a left atrium and left ventricle. Valves within the heart help ensure a one-way flow of blood.
	19.1c Pulmonary and Systemic Circulation • The circulation routes include the pulmonary circulation to the lungs and the systemic circulation to the body.
19.2 The Heart Within the Thoracic Cavity	• The heart is located in the thoracic cavity and enclosed within a fibroserous sac.
	19.2a Location and Position of the Heart • The heart is located left of the body midline, posterior to the sternum in the mediastinum with its apex projecting inferiorly.
	19.2b The Pericardium • The pericardium that encloses the heart includes the pericardial sac, which has an outer fibrous pericardium and an inner parietal layer of serous pericardium, and a visceral layer of serous pericardium (epicardium) that forms the outer layer of the heart wall. • The pericardial cavity is a potential space between the layers of the serous pericardium that contain serous fluid, which is produced by the serous membranes and lubricates the surfaces to reduce friction.
19.3 Heart Anatomy	• The heart is a relatively small, conical-shaped organ, which is approximately the size of a human fist.
	19.3a Superficial Features of the Heart • The right side of the heart is more visible from the anterior view, and the left side of the heart is more visible from the posterior view. • The coronary sulcus and interventricular sulci are visible from the external surface of the heart. These grooves house coronary vessels.
	19.3b Layers of the Heart Wall • The heart wall has an epicardium (visceral layer of serous pericardium), a myocardium, and an endocardium.
	19.3c Heart Chambers • Atria are separated by an interatrial septum, and ventricles are separated by an interventricular septum. • The four heart chambers include the right atrium, right ventricle, left atrium, and left ventricle.
	19.3d Heart Valves • The atrioventricular valves are located between an atrium and a ventricle, and the semilunar valves are between a ventricle and an arterial trunk (pulmonary trunk or aorta).
	19.3e Fibrous Skeleton of the Heart • The fibrous skeleton provides an attachment site for heart valves and cardiac muscle, and prevents action potentials from spreading between the atria and ventricles except through the AV node.
	19.3f Microscopic Structures of Cardiac Muscle • Cardiac muscle cells are small, have one or two centrally located nuclei, and are branched. • Intercalated discs, which are composed of desmosomes and gap junctions, tightly link the cardiac muscle cells together and permit the passage of action potentials, respectively. • Cardiac muscle cells rely almost exclusively on aerobic cellular respiration for supplying ATP, which makes them susceptible to failure if the oxygen supply is inadequate.
19.4 Coronary Vessels: Blood Supply Within the Heart Wall	• The coronary circulation is the circulation of blood to and from the heart wall.
	19.4a Coronary Arteries • Coronary arteries supply blood to the heart wall and include the left and right coronary arteries that branch off the ascending aorta.
	19.4b Coronary Veins • Venous return is through the cardiac veins into the coronary sinus, which drains into the right atrium of the heart.
19.5 Anatomic Structures Controlling Heart Activity	• Heart activity is initiated by the conduction system and altered by the autonomic nervous system.
	19.5a The Heart's Conduction System • Stimulation of the heart involves initiation of an action potential at the SA node and its transmission through the conduction system. • The conducting system includes sinoatrial (SA) node, atrioventricular (AV) node, AV bundle, bundle branches, and Purkinje fibers, which are composed of specialized cardiac cells that initiate and conduct action potentials, resulting in a heartbeat.
	19.5b Innervation of the Heart • Parasympathetic innervation comes from the cardioinhibitory center to decrease the heart rate. Sympathetic innervation comes from the cardioacceleratory center to increase both the heart rate and the force of contraction.

19.6 Stimulation of the Heart

- Heart contraction involves two events, which are the initiation and spread of an action potential through the conduction system and the spread of the action potential at the sarcolemma of cardiac muscle and cardiac muscle contraction.

19.6a Nodal Cells at Rest

- The pumps and channels associated with neurons are also in the plasma membrane of nodal cells. A type of channel unique to nodal cells is the slow voltage-gated Na^+ channel that allows the nodal cells to spontaneously depolarize.

19.6b Electrical Events at the SA Node: Initiation of the Action Potential

- The three events of SA nodal cells are (1) reaching the threshold as Na^+ enters the nodal cells through open voltage-gated Na^+ channels, (2) depolarization as Ca^{2+} enters the nodal cells through open voltage-gated Ca^{2+} channels, and (3) repolarization as K^+ exits the nodal cells through open voltage-gated K^+ channels.
- At rest, the parasympathetic nervous system decreases the inherent rhythm of nodal cells from a firing rate of 100 per minute to approximately 75 per minute.

19.6c Conduction System of the Heart: Spread of the Action Potential

- The action potential travels through the conduction system as follows: (1) SA node, (2) AV node, (3) AV bundle, (4) bundle branches, and (5) Purkinje fibers.

19.7 Cardiac Muscle Cells

- Following stimulation by the conduction system, there is a propagation of the action potential at the sarcolemma and contraction of sarcomeres within the cardiac muscle cells.

19.7a Cardiac Muscle Cells at Rest

- The pumps and channels associated with skeletal muscle fibers are also in the plasma membrane of cardiac muscle cells. A type of channel unique to the function of cardiac muscle cells in transmitting an action potential is the voltage-gated Ca^{2+} channel.

19.7b Electrical and Mechanical Events of Cardiac Muscle Cells

- The electrical events of cardiac muscle include depolarization, plateau, and repolarization at the sarcolemma.
- The mechanical events are similar to those of skeletal muscle fibers and involve crossbridge cycling and the shortening of sarcomeres within cardiac muscle cells.

19.7c Repolarization and the Refractory Period

- Cardiac muscle exhibits a longer refractory period than skeletal muscle fibers to allow time for contraction and relaxation of the muscle cells before they are stimulated again—a necessary requirement for the pumping action of the heart.

19.7d The ECG Recording

- An electrocardiogram (ECG) is a graphic recording of the electrical changes in the heart and is used in the diagnosis and treatment of abnormal heart function.

19.8 The Cardiac Cycle

- A cardiac cycle is the inclusive period of time from initiation of one heartbeat to the start of the next.

19.8a Overview of the Cardiac Cycle

- The cardiac cycle involves contraction and relaxation of heart chambers and associated pressure changes within the chambers, which result in the opening and closing of heart valves to allow for the one-way flow of blood through the heart.

19.8b Events of the Cardiac Cycle

- The cardiac cycle involves five phases: atrial contraction and ventricular filling, isovolumetric contraction, ventricular ejection, isovolumetric relaxation, and atrial relaxation and ventricular filling.

19.9 Cardiac Output

- Cardiac output is a measure of how effective the cardiovascular system is in transporting substances through the body

19.9a Introduction to Cardiac Output

- Cardiac output (CO) is defined as the amount of blood pumped by a single ventricle in 1 minute.
- Cardiac output is the heart rate multiplied by the stroke volume.
- Cardiac reserve is a measure of the ability of the heart to increase pumping capacity beyond the normal resting CO.

19.9b Variables That Influence Heart Rate

- Heart rate is altered through chronotropic agents, which change SA node and AV node activity. Positive chronotropic agents increase the heart rate, and negative chronotropic agents decrease the heart rate.

19.9c Variables That Influence Stroke Volume

- Stroke volume is influenced by venous return, inotropic agents, and afterload. Venous return is directly correlated with stroke volume. Positive inotropic agents increase stroke volume, negative inotropic agents decrease stroke volume, and afterload is inversely correlated with stroke volume.

19.9d Variables That Influence Cardiac Output

- Heart rate is altered by stimulating the conducting system, and stroke volume is altered by changes in the myocardium.
- Both heart rate and stroke volume are directly related to cardiac output.

19.10 Development of the Heart

- Development of the heart commences in the third week.
- Mesodermal cells in the cardiogenic region of the embryo form two heart tubes, which fuse by day 21, forming a single primitive heart tube.
- The foramen ovale connects the two atria, allowing most of the blood to bypass the pulmonary circulation. This foramen closes shortly after birth.

CHALLENGE YOURSELF

Create and Evaluate

Analyze and Apply

Understand and Remember

Do You Know the Basics?

1. Which of the following is the correct circulatory sequence for blood to pass through part of the heart?
   a. right atrium ⟶ left AV valve ⟶ right ventricle ⟶ pulmonary semilunar valve
   b. right atrium ⟶ right AV valve ⟶ right ventricle ⟶ pulmonary semilunar valve
   c. left atrium ⟶ right AV valve ⟶ left ventricle ⟶ aortic semilunar valve
   d. left atrium ⟶ left AV valve ⟶ left ventricle ⟶ pulmonary semilunar valve
2. The pericardial cavity is located between the
   a. fibrous pericardium and the parietal layer of the serous pericardium.
   b. parietal and visceral layers of the serous pericardium.
   c. visceral layer of the serous pericardium and the epicardium.
   d. myocardium and the visceral layer of the serous pericardium.
3. How is blood prevented from backflowing from the pulmonary trunk into the right ventricle?
   a. closing of the right AV valves
   b. opening of the pulmonary semilunar valve
   c. contraction of the right atrium
   d. closing of the pulmonary semilunar valve
4. Venous blood draining from the heart wall enters the right atrium through the
   a. coronary sinus.
   b. inferior vena cava.
   c. pulmonary veins.
   d. superior vena cava.
5. Calcium channels in the nodal cells function to
   a. cause depolarization and initiate the cardiac action potential.
   b. assure excess calcium can leave the cell.
   c. bring the cell quickly to its resting membrane potential.
   d. sustain contraction of the cell.
6. Action potentials are spread rapidly between cardiac muscle cells by
   a. sarcomeres.
   b. intercalated discs.
   c. chemical neurotransmitters.
   d. the fibrous skeleton.
7. Why is it necessary to stimulate papillary muscles in the ventricle slightly earlier than the rest of the ventricular wall myocardium?
   a. to assure rapid conduction speed of the action potential
   b. to pull on AV valve cusps to prevent backflow
   c. to assure blood will surge toward the semilunar valves
   d. to assure coordinated contraction of the ventricular myocardium
8. Preload is a measure of
   a. stretch of the heart chamber prior to contraction.
   b. contraction rate in cardiac muscle.
   c. reduced filling during exercise.
   d. autonomic nervous system stimulation of the heart.
9. All of the following occur when the ventricles contract *except*
   a. the AV valves close.
   b. blood is ejected into the aorta.
   c. the semilunar valves open.
   d. blood from the pulmonary trunk enters the atria.
10. What occurs during the atrial reflex?
    a. Atria slow their filling rate.
    b. There is a decrease in heart rate due to an increase in blood pressure.
    c. The SA node rhythm decreases.
    d. An increase in heart rate occurs in response to an increase in blood volume within the atria.
11. Describe and compare the differences between the pulmonary and systemic circulations.
12. Compare the structure, location, and function of the parietal and visceral layers of the serous pericardium.
13. Why are the tendinous cords required for the proper functioning of the AV valves?
14. Explain why the walls of the atria are thinner than those of the ventricles, and why the wall of the right ventricle is relatively thin when compared to the wall of the left ventricle.
15. Describe the structure and function of intercalated discs in cardiac muscle tissue.
16. Explain the general location and function of coronary vessels.
17. Describe the functional differences in the effects of the sympathetic and parasympathetic divisions of the autonomic nervous system on the activity of the heart.
18. Provide an overview of the two events for cardiac muscle contraction that include the conduction system and cardiac muscle cells.
19. List the five events of the cardiac cycle, and indicate for each if the atria are contracted or relaxed, if the ventricles are contracted or relaxed, if the AV valves are open or closed, and if the semilunar valves are open or closed.
20. Define cardiac output, and explain how it is influenced by both heart rate and stroke volume.

Can You Apply What You've Learned?

Use the following paragraph to answer questions 1 and 2.

A young man was doing some vigorous exercise when suddenly he felt chest pains just before he passed out. Upon being revived, he was taken to the hospital for examination. The medical personnel ran standard tests and ECG tracings, then said it would be fine for him to resume his normal activities and workouts.

1. Why might the increase in heart rate associated with a vigorous exercise program be the cause of his becoming unconscious?
   a. increase in blood pressure throughout the body
   b. failure of the conduction system
   c. increase in return volume of blood to the heart
   d. coronary blood flow reduction due to tachycardia
2. Which of the following tests would *not* be used to rule out that a myocardial infarction had occurred?
   a. doing an ECG
   b. performing constant blood pressure monitoring
   c. measuring blood levels of creatine kinase released from damaged heart muscle
   d. measuring blood levels of troponin released from damaged heart muscle
3. Calcium channel blockers are drugs that are given to cause
   a. an increase in the volume of blood being pumped.
   b. the afterload to be increased.
   c. contractility to decrease.
   d. preload to decrease.
4. A patient has been in a serious car accident and is hemorrhaging. Which of the following changes would be seen in this patient?
   a. a decrease in stroke volume
   b. an increase in heart rate
   c. a possible decrease in cardiac output
   d. All of these are correct.
5. During surgery, the right vagus nerve was accidently cut. Explain the effect on the heart rate.
   a. There is no change to the heart rate because the vagus nerve does not innervate the heart.
   b. The heart rate increases to the inherent rhythm of SA nodal cells.
   c. The heart stops beating, and the heart rate becomes zero.
   d. The heart rate decreases to the inherent rhythm of SA nodal cells.

Can You Synthesize What You've Learned?

1. A young couple that you are friends with gave birth to a new baby. They were told by their physician that the foramen ovale did not close between the right atrium and left atrium (i.e., that their baby has "a hole in her heart"). They know that you are a nurse and have come to you to help them to understand what is going on. Explain both the normal flow of blood through the heart and the flow with the foramen ovale still open. Include in your description the concept of oxygenated and deoxygenated blood.
2. Josephine is a 55-year-old overweight woman who has a poor diet and does not exercise. One day while walking briskly, she experienced pain in her chest and down her left arm. Her doctor told her that she was experiencing angina due to heart problems. Josephine asks you to explain what causes angina and why she was feeling pain in her arm even though the problem was with her heart. What do you tell her?
3. Your grandfather was told that his SA node (pacemaker) has stopped functioning. Explain how his heart is still beating at a rate of 40 to 50 times per minute. Are the atria stimulated to contract? Explain.

INTEGRATE

ONLINE STUDY TOOLS

connect | SMARTBOOK® | APR

The following study aids may be accessed through Connect.

Concept Overview Interactive: Figure 19.22: Changes Associated with a Cardiac Cycle

Clinical Case Study: A Middle-Aged Woman with Back Pain

Interactive Questions: This chapter's content is served up in a number of multimedia question formats for student study

SmartBook: Topics and terminology include introduction to the cardiovascular system; location of the heart and the pericardium; heart anatomy; coronary vessels: blood supply of the heart wall; anatomic structures controlling heart activity, stimulation of the heart, the cardiac cycle; cardiac output; development of the heart

Anatomy & Physiology Revealed: Topics include cardiovascular system overview; heart fly-through; blood flow through the heart; heart vasculature; internal features of the heart; cardiac muscle; cardiac cycle

Animations: Topics include action potentials in the sinoatrial node; cardiac cycle; mechanical events of the cardiac cycle; conduction system of the heart; cardiac excitation-contraction (EC) coupling

Design Elements: Learning Objective icon: ©McGraw-Hill Education; **What Do You Think?** icon: ©McGraw-Hill Education; **What Did You Learn?** icon: ©McGraw-Hill Education; **Learning Strategy** icon: ©Slavoljub Pantelic/Shutterstock RF; **Clinical View** icon: ©Laia Design Studio/Shutterstock RF; **Concept Connection** icon: ©McGraw-Hill Education; **Challenge Yourself** icon: ©McGraw-Hill Education; **Do You Know the Basics?** icon: ©McGraw-Hill Education; **Can You Apply What You've Learned?** icon: ©McGraw-Hill Education; **Can You Synthesize What You've Learned?** icon: ©McGraw-Hill Education; **Online Study Tools** icon: ©ayzek/Shutterstock RF

chapter

20 Cardiovascular System: Vessels and Circulation

Module 9: Cardiovascular System

INTEGRATE

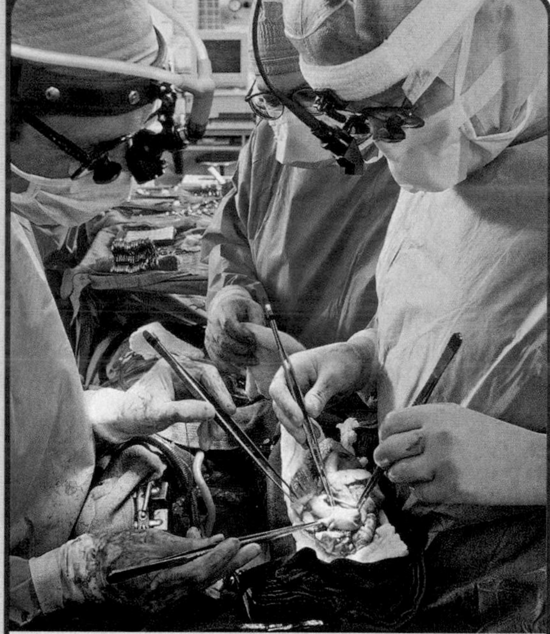

CAREER PATH
Cardiovascular Surgeon

A cardiovascular surgeon is a physician who specializes in diagnosing and treating disorders of the heart and blood vessels. One of the most common blood vessel disorders is atherosclerosis, where plaques develop in the vessels and cause lumen narrowing. If these vessels become completely blocked, the tissues they supply may die. Cardiovascular surgeons are trained to either remove blockages in vessels or add blood vessel bypasses to restore blood supply to an affected organ. They use medical imaging to help direct catheters through blood vessels and insert stents in blocked vessels.

Most adults are aware that high blood pressure (hypertension) is not healthy. Hypertension can damage blood vessels and lead to cardiovascular disease. But a minimal amount of blood pressure is needed to effectively pump blood (and the nutrients and respiratory gases it transports) throughout the body and deliver these materials to the body tissues. If blood pressure drops too low, the body will be deprived of nutrients and death may occur. Multiple body systems (including the endocrine, nervous, and urinary systems) participate in maintaining sufficient blood pressure to ensure that all tissues are adequately perfused (supplied with blood).

We begin this chapter by describing the general structure and function of blood vessels, blood flow velocity, processes of capillary exchange, and factors that influence both blood flow and blood pressure in vessels. Pulmonary circulation and the major arteries and veins of systemic circulation are then described for each body region, and we conclude with a comparison of fetal versus adult circulation.

20.1 Structure and Function of Blood Vessels

Blood vessels are classified into three primary types based on function. **Arteries** (ar′ter-ē) transport blood away from the heart to the capillaries. **Capillaries** (kap′i-lar-ē; *capillaris* = relating to hair) are microscopic, relatively porous blood vessels for the exchange of substances between blood and tissues. **Veins** (vān) drain blood from the capillaries, transporting it back to the heart.

20.1a General Structure of Vessels

LEARNING OBJECTIVES

1. Describe the three tunics common to most vessels.
2. Explain the distinguishing features of the tunics found in arteries, capillaries, and veins.

Vessel walls are composed of layers called **tunics** (tū′nik; *tunica* = coat). The tunics surround the **lumen** (lū′men), or inside space, of the vessel through which blood flows. The three tunics are the tunica intima, tunica media, and tunica externa **(figure 20.1)**. Arteries, capillaries, and veins differ in both the specific composition of their tunics and their functions.

Tunics

The innermost layer of a blood vessel wall is the **tunica intima** (tū′ni-kă in′tim-mă; *intimus* = inmost), or *tunica interna.* It is composed of an **endothelium** (a simple squamous epithelium; see section 5.1c) that is adjacent to the blood vessel lumen and a thin subendothelial layer of areolar connective tissue. The endothelium both provides a smooth surface as the blood moves through the lumen of the blood vessel and releases substances (e.g., nitric oxide, endothelin) to regulate contraction and relaxation of smooth muscle

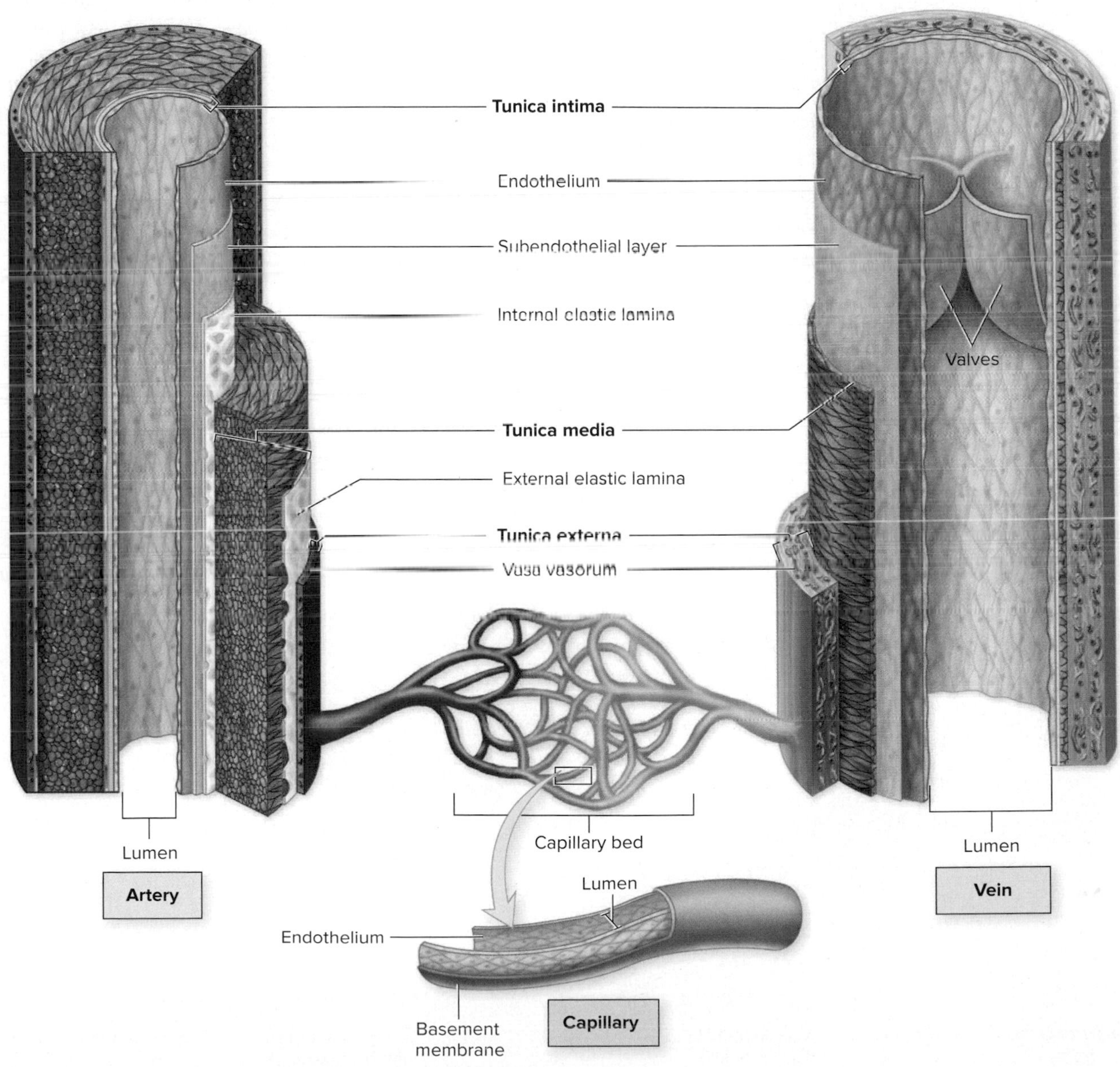

Figure 20.1 Walls of an Artery, a Capillary, and a Vein. Both arteries and veins have a tunica intima, tunica media, and tunica externa. However, an artery has a thicker tunica media and a relatively smaller lumen, whereas a vein's thickest layer is the tunica externa, and it has a larger lumen. Larger veins also have valves. Capillaries typically have only a tunica intima (basement membrane and endothelium), but they do not have a subendothelial layer. (Note: Different types of arteries vary in the distribution of elastic fibers within the tunica media. The array of elastic fibers depicted here, which are labeled elastic laminae, is the arrangement within a muscular artery.)

within the tunica media. Recall that the endothelium is continuous with the endocardium, which is the inner lining of the heart (see section 19.3b).

The **tunica media** (mē′dē-ă; *medius* = middle) is the middle layer of the vessel wall. It is composed predominantly of circularly arranged layers of smooth muscle cells that are supported by elastic fibers. Contraction of smooth muscle in the tunica media results in **vasoconstriction** (vā′sō-kon-strik′shŭn), or narrowing of the blood vessel lumen; relaxation of the smooth muscle causes **vasodilation** (vā′sō-dī-lā′shŭn), or widening of the blood vessel lumen.

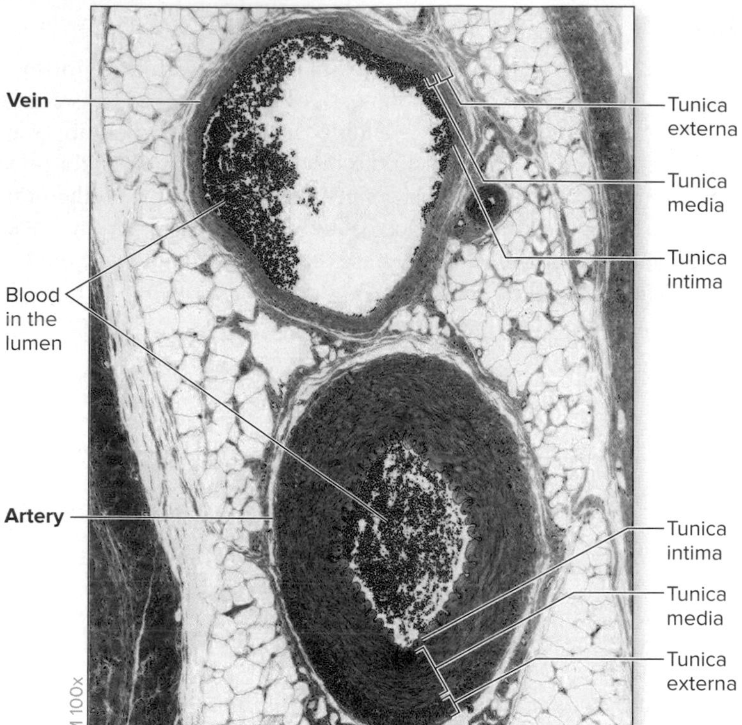

Figure 20.2 Microscopic Comparison of an Artery and a Vein. The artery generally maintains its shape in tissues as a result of its thicker wall. The wall of a companion vein often collapses when it is not filled with blood (although, in this photo, the vein is not collapsed). AP|R

The **tunica externa** (eks-ter′nă; *externe* = outside), or *tunica adventitia,* is the outermost layer of the blood vessel wall. It is composed of areolar connective tissue that contains elastic and collagen fibers. The tunica externa helps anchor the vessel to other structures. Very large blood vessels require their own blood supply to the tunica externa in the form of a network of small arteries called the **vasa vasorum** (vā′să vā-sŏr′ŭm; vessels of vessels). The vasa vasorum extends through the tunica externa.

Comparison of the Different Vessel Types

Arteries and veins that supply the same body region and tend to lie next to one another are called **companion vessels. Figure 20.2** is a histologic image of a companion artery and vein. Compared to their venous companions, arteries have a thicker tunica media, a narrower lumen, and more elastic and collagen fibers. These differences mean that arterial walls can spring back to shape and are more resilient and resistant to changes in blood pressure than are veins. In addition, an artery remains patent (open) even without blood in it.

In contrast, veins have a thicker tunica externa, a wider lumen, and less elastic and collagen fibers than a companion artery. The wall of a vein is typically collapsed if no blood is in the vessel. The characteristics of arteries and veins are summarized in **table 20.1**.

Capillaries are unique in that they contain only the tunica intima composed of an endothelium and its underlying basement membrane; there is no subendothelial layer. Having this thin barrier allows for rapid gas and nutrient exchange between the blood in capillaries and the tissues.

WHAT DID YOU LEARN?

1. What are three differences in anatomic structure between arteries and veins?

Table 20.1 Comparison of Companion Arteries and Veins

Characteristic	Artery	Vein
Lumen Diameter	Narrower than vein lumen	Wider than artery lumen
General Wall Thickness	Thicker than vein	Thinner than artery
Cross-Sectional Shape	Cross-sectional shape retained; even without blood in vessel	Cross-sectional shape tends to flatten out (collapse) without blood in vessel
Thickest Tunic	Tunica media	Tunica externa
Elastic and Collagen Fibers in Tunics	More than in vein	Less than in artery
Valves	None	Present in most veins
Blood Pressure Range	Higher than in veins (100 mm Hg in larger arteries to 40 mm Hg in smaller arterioles)	Lower than in arteries (20 mm Hg in venules to 0 mm Hg in the inferior vena cava)
Blood Flow	Transports blood away from heart to the body	Transports blood from the body toward the heart
Blood Oxygen Levels	Systemic arteries transport blood high in O_2 Pulmonary arteries transport blood low in O_2	Systemic veins transport blood low in O_2 Pulmonary veins transport blood high in O_2

20.1b Arteries

LEARNING OBJECTIVE

3. Distinguish among elastic arteries, muscular arteries, and arterioles.

Arteries progressively branch into smaller vessels as they extend from the heart to the capillaries. There is both a corresponding decrease in lumen diameter and a change in the composition of the tunic wall that includes both a decrease in the relative amount of elastic fibers and an increase in the relative amount of smooth muscle. Arteries may be classified into three basic types: elastic arteries, muscular arteries, and arterioles (**figures 20.3** and **20.4**).

Elastic Arteries

Elastic arteries are the largest arteries, with diameters ranging from 2.5 to 1 centimeters. They are also called *conducting arteries* because they conduct blood—from the heart to the smaller muscular arteries. As their name suggests, these arteries have a large proportion of

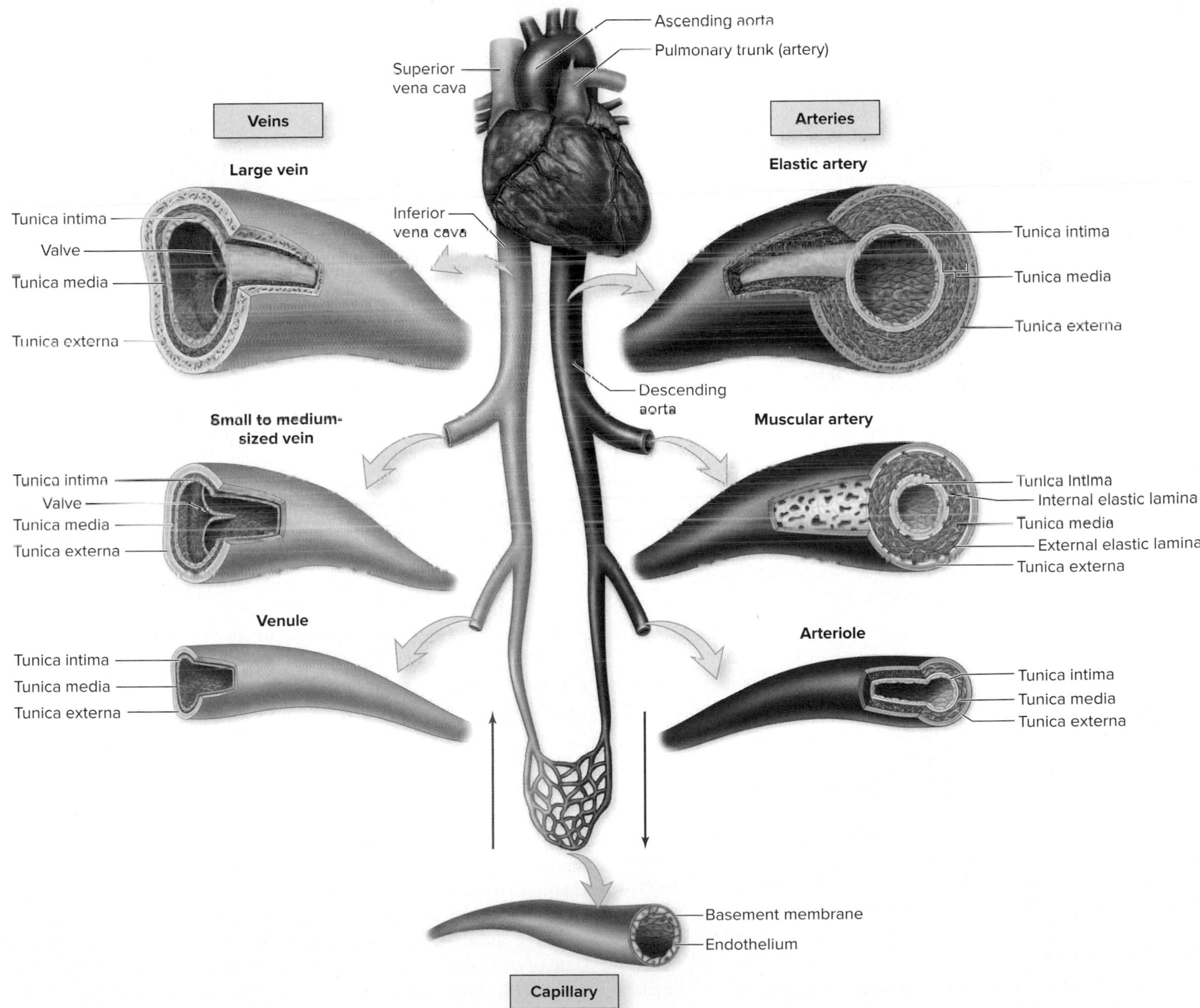

Figure 20.3 Comparison of Companion Vessels. The thickness of the tunics in companion arteries and veins differs, depending upon the size of the vessels. Capillaries are microscopic vessels and function as sites of exchange between the body tissues.

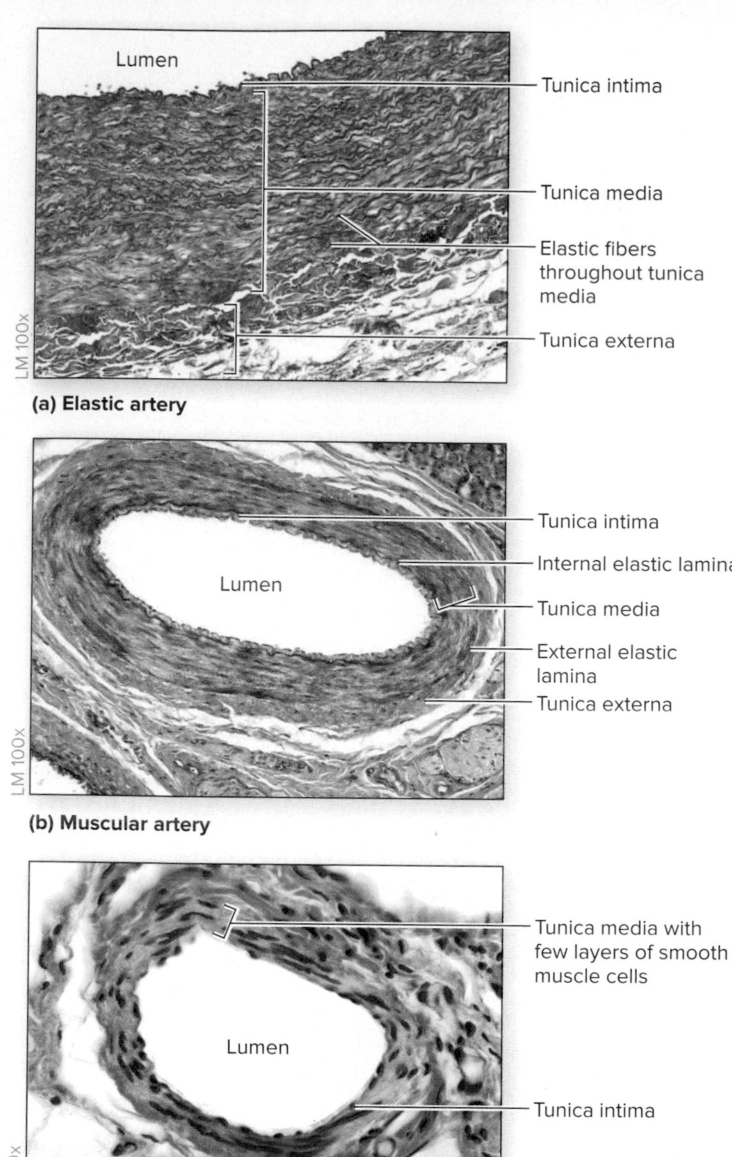

Figure 20.4 Types of Arteries. Light micrograph images of three types of arteries. (*a*) Elastic arteries have vast arrays of elastic fibers in their tunica media. (*b*) Muscular arteries have a tunica media with numerous layers of smooth muscle flanked by elastic laminae. (*c*) Arterioles typically have a tunica media composed of six or fewer layers of smooth muscle cells.
©McGraw-Hill Education/Al Telser

INTEGRATE

CLINICAL VIEW 20.1

Atherosclerosis

Atherosclerosis (ath′er-ō-skler-ō′sis; *athere* = gruel, sclerosis = hardening) is a progressive disease of the elastic and muscular arteries. It is characterized by the presence of an **atheroma** (ath′er-ō-ma; *oma* = tumor) (or *atheromatous plaque*), which leads to thickening of the tunica intima and narrowing of the arterial lumen.

Etiology

Although the etiology (cause) of atherosclerosis is not completely understood, the **response-to-injury hypothesis** is the most widely accepted. This proposal states that injury to the endothelium of an arterial wall, especially repeated injury caused by infection, trauma, or hypertension (high blood pressure), results in an inflammatory reaction, eventually leading to the development of an atheroma. The injured endothelium becomes more permeable, which encourages leukocytes and platelets to adhere to the lesion and initiate an

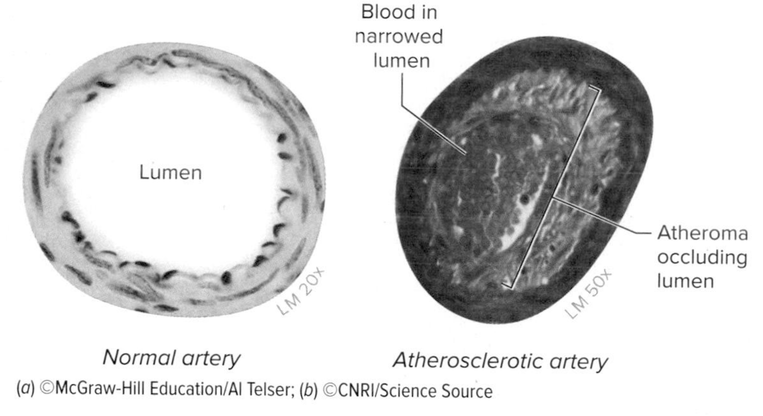

(*a*) ©McGraw-Hill Education/Al Telser; (*b*) ©CNRI/Science Source

elastic fibers; these are present throughout all three tunics, especially in the tunica media. The abundant elastic fibers allow the artery to stretch and accommodate the blood when a heart ventricle ejects blood into it during ventricular systole (contraction) and then recoil, which helps propel the blood through the arteries during ventricular diastole (relaxation).

The largest arteries close to the heart (e.g., aorta, pulmonary trunk, brachiocephalic, common carotid, subclavian) and the common iliac arteries are examples of elastic arteries (see figure 20.19*a*). Elastic arteries branch into muscular arteries.

Muscular Arteries

Muscular arteries typically have diameters ranging from 1 centimeter to 0.3 millimeter. These medium-sized arteries are also called *distributing arteries* because they distribute blood to specific body regions and organs.

Muscular arteries have a proportionately thicker tunica media, with multiple layers of smooth muscle cells. Unlike in elastic arteries, the elastic fibers in muscular arteries are confined to two circumscribed sheets: The **internal elastic lamina** (lam′i-nă) separates the tunica media from the tunica intima, and the **external elastic lamina** separates the tunica media from the tunica externa. The relatively greater amount of muscle and lesser amount of elastic tissue result in a better ability to vasoconstrict and vasodilate, although with a lessened ability to stretch in comparison to elastic arteries.

Most named arteries (e.g., the brachial, anterior tibial, coronary, and inferior mesenteric arteries) are examples of muscular arteries (see figure 20.19*a*). Muscular arteries branch into arterioles.

Arterioles

Arterioles are the smallest arteries, with diameters ranging from 0.3 millimeters to 10 micrometers; these vessels are not named. In general, arterioles have fewer than six layers of smooth muscle in their tunica media. Larger arterioles have all three tunics, whereas the smallest arterioles may have a tunica intima surrounded by a single layer of smooth muscle cells. Smooth muscle in the arterioles is slightly contracted (just as your skeletal muscles often are in a partial state of contraction; see section 10.7a). This contracted state is called **vasomotor tone** and is regulated by the vasomotor center in the brainstem (see section 13.5c). **Sympathetic motor tone** results in vasoconstriction, which allows for varying degrees of change from this slightly contracted state (see section 15.7a). Blood vessels can be

inflammatory response (see section 22.3d). Low-density lipoproteins and very-low-density lipoprotens (LDLs and VLDLs) (see section 27.6b) enter the tunica intima, combine with oxygen, and remain stuck to the vessel wall. This oxidation of these lipoproteins attracts monocytes, which adhere to the endothelium and migrate into the wall. As the monocytes migrate into the wall, they digest the lipids and develop into structures called **foam cells.** Eventually, smooth muscle cells from the tunica media migrate into the atheroma and proliferate, causing further enlargement of the atheroma and narrowing of the lumen of the blood vessel, thereby restricting blood flow to the regions the artery supplies.

Atherosclerosis is a progressive disease. The plaques typically begin to develop in early adulthood and grow and enlarge as we age. People are unaware of the plaques until they become large enough to restrict blood flow in an artery and cause vascular complications.

Risk Factors

Some individuals are genetically prone to atherosclerosis. **Hypercholesterolemia** (an increased amount of cholesterol in the blood; see Clinical View 27.4: "Blood Cholesterol Levels"), which also tends to run in families, has been positively associated with the rate of development and severity of atherosclerosis. Additionally, males tend to be affected more than females, and symptomatic atherosclerosis increases with age. Finally, smoking and hypertension cause vascular injury, which increases the risk.

Treatment Options

If an artery is occluded (blocked) in one or just a few areas, one form of treatment is an **angioplasty** (an′jē-ō-plas-tē; *angeion* = vessel, *plastos* = formed). A physician inserts a balloon-tip catheter into an artery and positions it at the site where the lumen is narrowed. Then the balloon is inflated, forcibly expanding the narrowed area, and a stent is placed in the vessel. A stent is a piece of wire-mesh that springs open to keep the vessel lumen open. For occluded coronary arteries, sometimes a much more invasive treatment known as coronary bypass surgery may be needed. A vein (e.g., the great saphenous vein) or artery (e.g., the internal thoracic artery) is detached from its original location and grafted from the aorta to the coronary artery system, thus bypassing the area(s) of atherosclerotic narrowing.

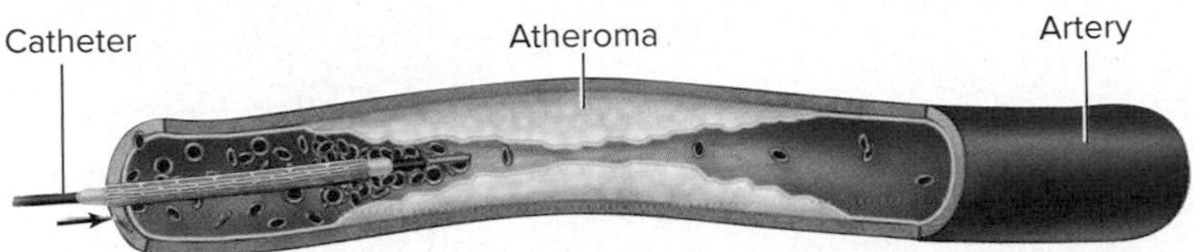

(1) An uninflated balloon and compressed stent are passed through a catheter to the area of the artery that is obstructed.

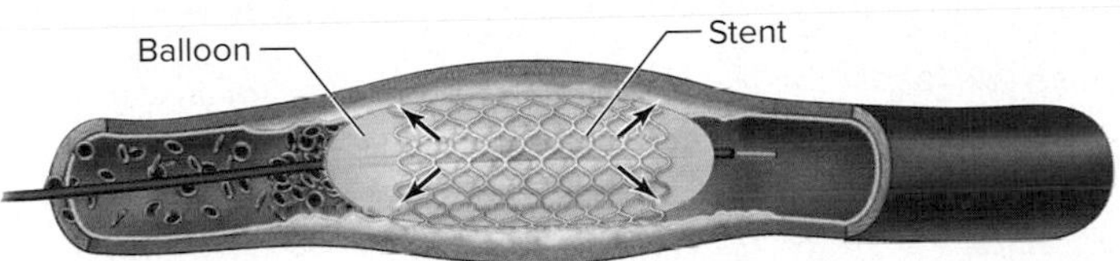

(2) The balloon inflates, which expands the stent, inserts it in place, and compresses the atheroma.

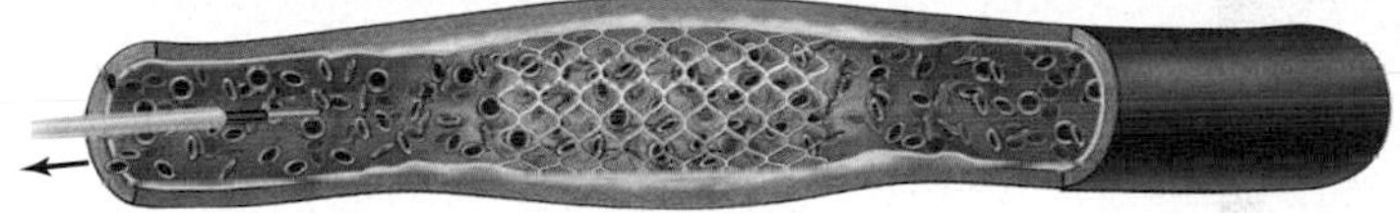

(3) The stent typically remains in the vessel as the balloon is deflated and the catheter is withdrawn.

Angioplasty is used to expand the narrowed region of an artery.

either vasoconstricted to a greater degree to decrease blood flow or vasodilated to allow more blood into an area. Arterioles have a significant role in regulating systemic blood pressure and blood flow to the different areas of the body.

WHAT DID YOU LEARN?

2 What changes are seen in the composition of the tunic wall of arteries as they branch into smaller and smaller vessels?

20.1e Capillaries

LEARNING OBJECTIVES

4. Describe the general anatomic structure and function of capillaries.
5. Compare the anatomic structure, function, and location of continuous capillaries, fenestrated capillaries, and sinusoids.
6. Trace the movement of blood through a capillary bed.

INTEGRATE

CLINICAL VIEW 20.2

Aneurysm

Elastic and muscular arteries are generally less able to withstand the forces from the pulsating blood as we age. Systolic blood pressure may increase with age, exacerbating this problem. As a result, older individuals are more apt to develop an **aneurysm** (an′ū-rizm), whereby part of the arterial wall thins and balloons out. This wall is more prone to rupture, which can cause massive bleeding and may lead to death. Aneurysms most commonly occur in the arteries at the base of the brain or in the aorta.

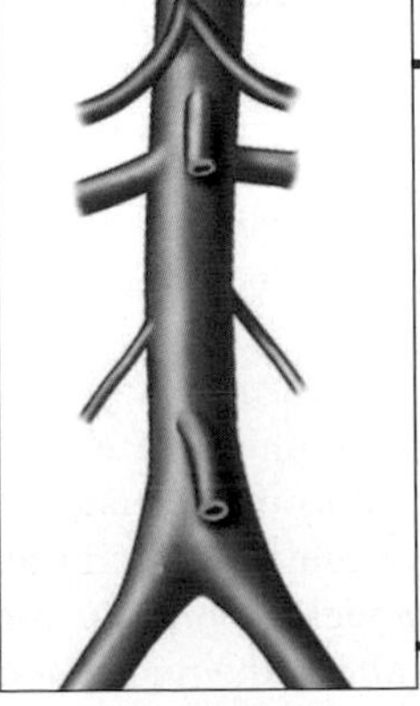

Normal aorta

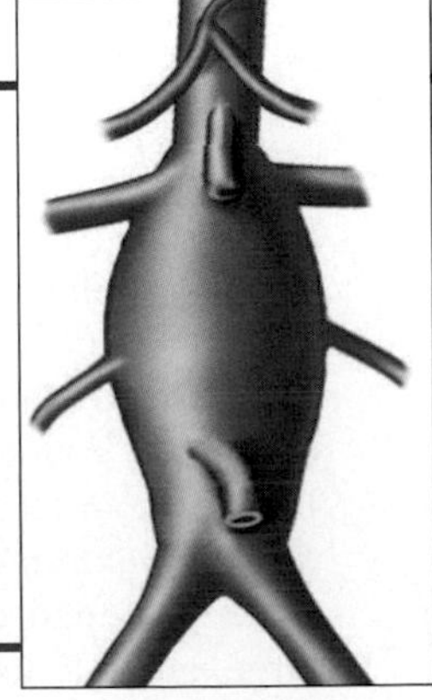

Large abdominal aortic aneurysm

Table 20.2 Types of Capillaries

Characteristics	(a) Continuous Capillary	(b) Fenestrated Capillary
Structure	Basement membrane; Pinocytotic vesicles; Erythrocyte; Intercellular cleft; Lumen; Nucleus of endothelial cell	Basement membrane; Fenestrations; Pinocytotic vesicles; Erythrocyte; Intercellular cleft; Lumen; Nucleus of endothelial cell
Description	Lining of endothelial cells is complete around lumen; basement membrane is complete; intercellular clefts between endothelial cells	Same as continuous capillary, except also contains fenestrations
Materials That Pass Through Vessel Wall	Plasma and its contents (except most proteins); some leukocytes	Large amounts of materials are filtered, released, or absorbed; some smaller proteins
Locations	Most capillaries (e.g., capillaries within muscles, skin, thymus, lungs, and central nervous system [CNS])	Small intestine; for absorbing nutrients Ciliary process; to produce aqueous humor in the eye Choroid plexus; to produce cerebrospinal fluid (CSF) in the brain Most endocrine glands; for release of hormones into the blood Kidneys; for filtering blood

Capillaries are the smallest blood vessels. They connect arterioles to venules (the smallest veins). The average capillary is approximately 1 mm in length with a diameter of 8 to 10 micrometers, just slightly larger than the diameter of a single erythrocyte. The narrow vessel diameter means erythrocytes must move in single file (termed **rouleau**) through each capillary (see section 18.3b). Capillaries consist solely of an endothelial layer (of simple squamous cells) resting on a basement membrane. The narrow vessel diameter and the thin wall are optimal for exchange of substances between blood and body tissues.

Types of Capillaries

Capillaries are differentiated based on their relative degree of permeability and include continuous capillaries, fenestrated capillaries, and sinusoids **(table 20.2)**.

Continuous capillaries are the most common type of capillary. The endothelial cells form a complete, *continuous* lining around the lumen that rests on a complete basement membrane. Tight junctions (see section 4.6d) secure endothelial cells to one another; however, they do not form a complete "seal." The gaps between the endothelial cells are called **intercellular clefts.** Materials can move into or out of the blood either through endothelial cells by membrane transport processes (e.g., diffusion, pinocytosis; see section 4.3) or between endothelial cells through intercellular clefts by diffusion and bulk flow (see section 20.3).

The size of intercellular clefts prevents the movement of large substances, including formed elements and plasma proteins, while allowing the movement of fluid containing small substances (smaller than about 5 nanometers), such as glucose, amino acids, and ions. Continuous capillaries are found, for example, in muscle, the skin, the thymus, the lungs, and the central nervous system.

Fenestrated (fen′es-trā′ted; *fenestra* = window) **capillaries** are also composed of a complete, continuous lining of endothelial cells and a complete basement membrane. However, small regions of the endothelial cells (typically 10 to 100 nanometers in diameter) are extremely thin; these thin areas are called **fenestrations** (or *pores*). Fenestrations are small enough to prevent formed elements from passing through the wall yet large enough to allow the movement of some smaller plasma proteins. Fenestrated

(c) Sinusoid
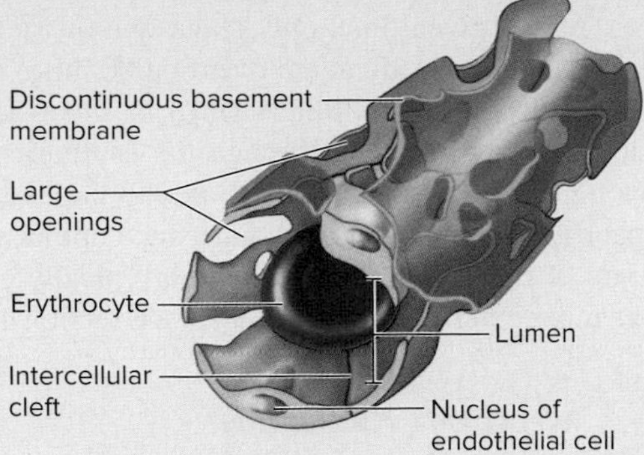
Lining of endothelial cells is incomplete around lumen; basement membrane is incomplete or absent
Large substances (formed elements, large plasma proteins) and plasma
Red bone marrow; for formed elements to enter the blood Liver and spleen; for removal of old erythrocytes from the cardiovascular circulation Some endocrine glands (anterior pituitary, adrenal, and parathyroid); for release of hormones into the blood

capillaries are seen where a great deal of fluid transport between the blood and interstitial tissue occurs. Examples of structures that contain fenestrated capillaries include the small intestine for the absorption of nutrients (see section 26.3b), the ciliary process of the eye in the production of aqueous humor (see section 16.4b), the choroid plexus of the brain in the production of cerebrospinal fluid (see section 13.2c), most of the endocrine glands to facilitate the absorption of hormones into the blood (see section 17.1a), and the kidneys for the filtering of blood (see section 25.5c).

Sinusoids (si′nŭ-soyd; *sinus* = cavity, *eidos* = appearance), or *discontinuous capillaries,* have an incomplete lining of the endothelial cells with large openings, or gaps, and the basement membrane is either discontinuous or absent. These openings allow for transport of large substances (formed elements, large plasma proteins), as well as plasma between the blood and tissues. Sinusoids are found in red bone marrow for the entrance of formed elements into the blood (see section 18.3a), the liver (see section 26.3c) and spleen (see section 21.4b) for the removal of aged erythrocytes from the cardiovascular circulation, and some endocrine glands (e.g., anterior pituitary, adrenal glands, parathyroid glands) for the movement of hormone molecules into the blood. A common feature of structures with sinusoids is their reddish color.

INTEGRATE

CONCEPT CONNECTION

The blood-brain barrier (BBB) (see section 13.2d) is formed by modified continuous capillaries that have thickened basement membranes and no intercellular clefts. Substances can pass through endothelial cells only by regulated membrane transport processes (e.g., facilitated diffusion, active transport). However, movement of nonpolar substances is not regulated by cells (see section 4.3a) and so these nonpolar substances (e.g., nicotine, alcohol) may pass through the cells by simple diffusion and enter the brain cells.

Capillary Beds

Capillaries do not function independently; rather, a group of capillaries (10 to 100) function together and form a **capillary bed** (**figure 20.5**). A capillary bed is fed by a **metarteriole** (met′ar-tēr′ē-ōl; *meta* = between), which is a branch of an arteriole. The proximal part of the metarteriole is encircled by scattered smooth muscle cells, whereas the distal part of the metarteriole (called the **thoroughfare channel**) has no smooth muscle cells. The thoroughfare channel connects to a **postcapillary venule** (ven′ūl, vē′nūl), which drains the capillary bed.

Vessels called **true capillaries** branch from the metarteriole and make up the bulk of the capillary bed. At the origin of each true capillary, a smooth muscle ring called the **precapillary sphincter** controls blood flow into the true capillaries. Sphincter relaxation permits blood to flow into the true capillaries, whereas sphincter contraction causes blood to flow directly from the metarteriole and thoroughfare channel into the postcapillary venule with blood bypassing the capillary bed. The precapillary sphincters go through cycles of contracting and relaxing at a rate of about 5 to 10 cycles per minute. This cyclical process is referred to as **vasomotion.**

At any given time, only about one-quarter of the capillary beds are open, because there are over 60,000 miles of capillaries and only about 250 to 300 milliliters (mL) of blood (about 5% of the total blood volume) moving through the capillaries. There is simply not enough blood available to fill all capillaries at the same time. The specific amount of blood entering capillaries per unit time per gram of tissue is called **perfusion,** typically expressed in milliliters per minute per gram (mL/min/g) (see section 19.1a).

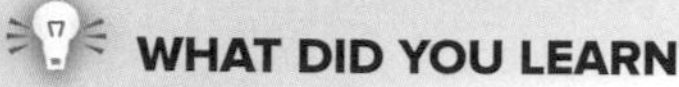

WHAT DID YOU LEARN?

3 What type of capillary is the most permeable, and where in the body are these capillaries found?

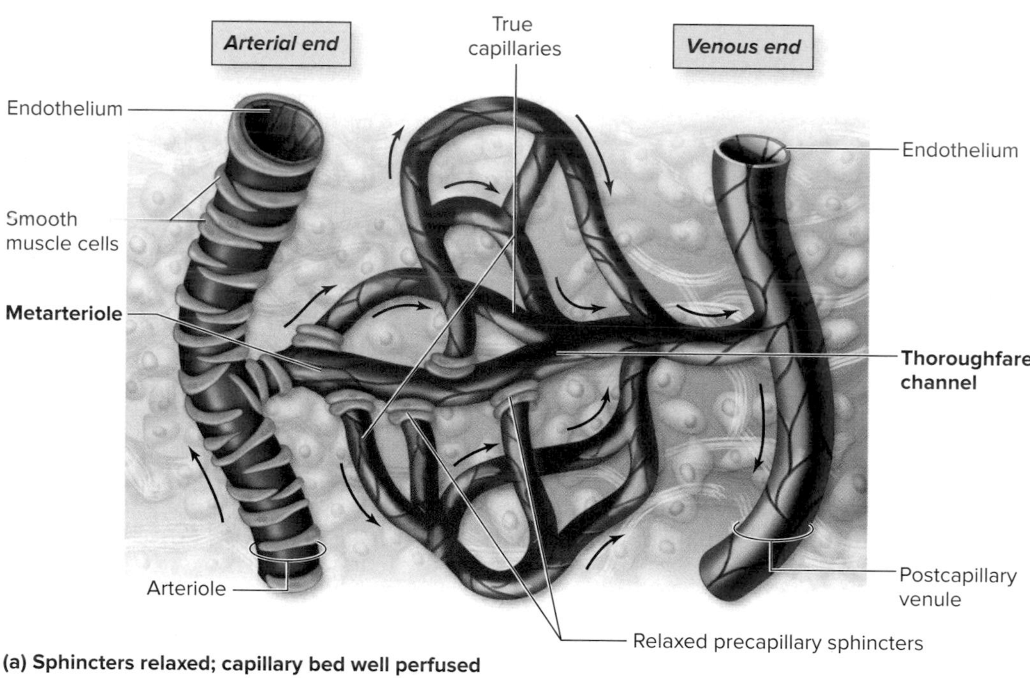

(a) Sphincters relaxed; capillary bed well perfused

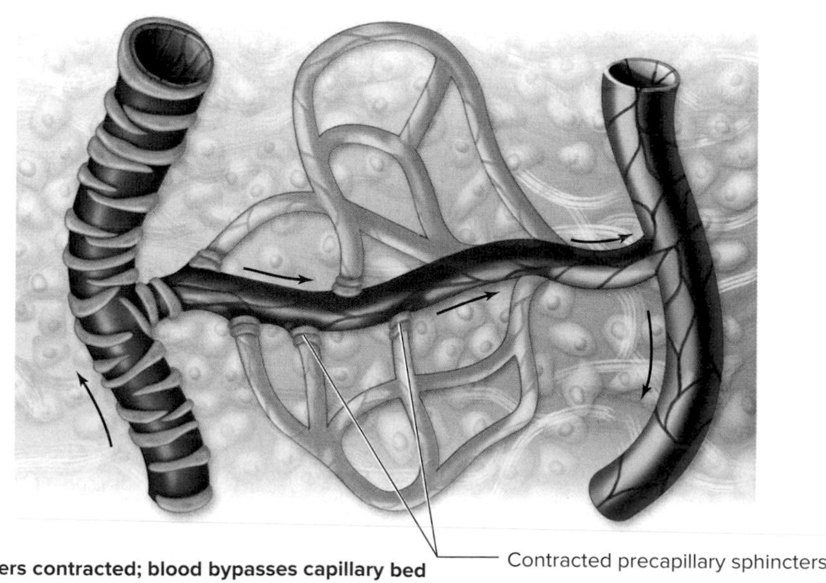

(b) Sphincters contracted; blood bypasses capillary bed

Figure 20.5 Capillary Bed Structure and Perfusion Through the Bed. A capillary bed originates from a metarteriole. The metarteriole continues as a thoroughfare channel that merges with the postcapillary venule. True capillaries branch from the metarteriole, and blood flow into these true capillaries is regulated by the precapillary sphincters. (*a*) A well-perfused capillary bed, with all of the precapillary sphincters relaxed, and (*b*) a capillary bed where most blood bypasses the capillary bed due to contracted precapillary sphincters.

20.1d Veins

LEARNING OBJECTIVES

7. Describe the structure and general function of veins.
8. Explain how veins serve as a blood reservoir for the cardiovascular system.

Veins merge into larger and larger vessels with a corresponding increase in lumen diameter as they extend from the capillaries to the heart (see figure 20.3).

Venules

Venules are the smallest veins, measuring from 8 to 100 micrometers in diameter. Venules are companion vessels with arterioles. The smallest venules, called postcapillary venules, drain capillaries (figure 20.5). Smaller venules merge to form larger venules. The largest venules have all three tunics. Venules merge to form veins (see figure 20.3).

Small, Medium-Sized, and Large Veins

A venule becomes a vein when its diameter is greater than 100 micrometers. Small and medium-sized veins are companion vessels with muscular arteries, whereas the largest veins are positioned alongside elastic arteries. Most veins contain numerous **valves,** so as to prevent blood from pooling in the limbs. The valves are formed primarily of tunica intima and strengthened by elastic and collagen fibers. Valves have an anatomic structure similar to semilunar valves of the heart (see section 19.3d).

Systemic Veins as Blood Reservoirs

The percentage of the total blood that is moving through each of the different components of the cardiovascular system while *at rest* is illustrated in **figure 20.6**. Relatively small amounts of blood are within the pulmonary circulation (about 18%) and the heart (about 12%). The largest percentage of blood is within the systemic circulation (about 70%), with the greatest amount (about 55%) within the body's systemic veins. The relatively large amount of blood within veins allows veins to function as **blood reservoirs.** Blood may be shifted from venous reservoirs into circulation through vasoconstriction of veins, when more blood is needed with increased physical exertion—and shifted back into venous reservoirs through vasodilation of veins, when less blood is needed at rest.

Figure 20.7 integrates how each blood vessel structure is adapted for the specific functions of that vessel.

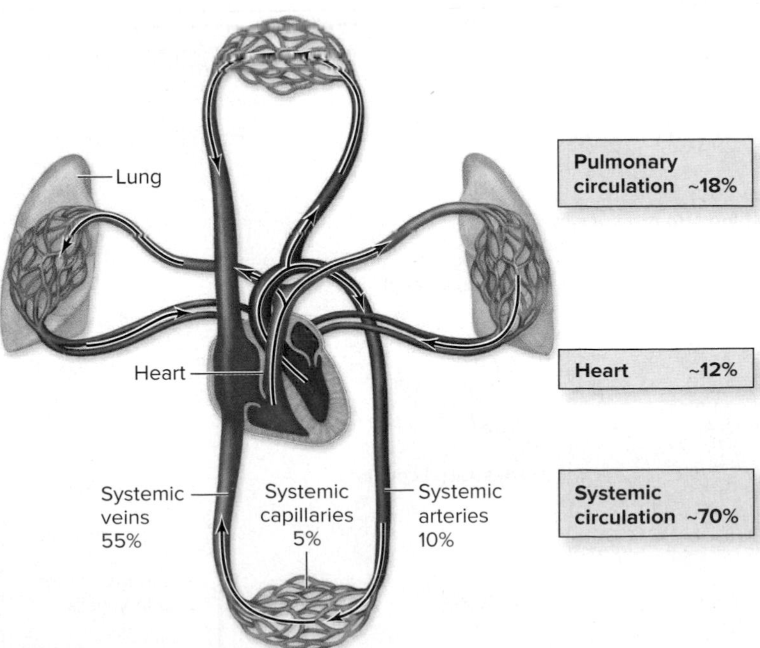

Figure 20.6 Blood Distribution at Rest. Different components of the cardiovascular system vary in the amounts of blood moving through them, with systemic veins containing the largest percentage.

INTEGRATE

CONCEPT CONNECTION

A modified vein that has very thin walls and no smooth muscle is referred to as a **sinus.** A sinus is supported by surrounding tissue. Two examples are the coronary sinus located in the coronary sulcus of the heart, supported by adipose connective tissue (see figure 19.7), and the dural venous sinuses of the brain, supported by the dense connective tissue of the dura mater (see figures 13.5 and 13.6).

Coronary sinus

WHAT DID YOU LEARN?

4. How does a vein serve as a blood reservoir?

20.1e Pathways of Blood Vessels

LEARNING OBJECTIVE

9. Compare and contrast the simple and alternative pathways of blood vessels.

Blood vessels are arranged in either simple or alternative pathways (**figure 20.8**). Each pathway is described here in detail.

Simple Pathway

In the simple pathway, one major artery delivers blood to the organ or body region and then branches into smaller and smaller arteries to become arterioles. Each arteriole feeds into a single capillary bed. A venule drains blood from the capillaries and merges with other venules to form one major vein that drains blood from the organ or body region. Thus, the simple pathway includes one artery, a capillary bed, and one vein associated with an organ or a body region.

Blood transported to and from the spleen is an example of a simple blood flow pathway. A single splenic artery delivers oxygenated blood to the spleen with the exchange made in a capillary bed of the spleen, and a single splenic vein drains deoxygenated blood from the spleen (see figure 21.7*a*). Arteries that provide only one pathway through which blood can reach an organ are referred to as **end arteries** (see section 19.4a).

Alternative Pathways

Several alternative circulatory pathways are possible, and these include various types of anastomoses and the portal systems. They differ from the simple pathway in the number of arteries, capillary beds, or veins that serve an organ or a body region.

INTEGRATE CONCEPT OVERVIEW

Figure 20.7 How Blood Vessel Form Influences Function. The structure and function of (*a*) arteries, (*b*) capillaries, and (*c*) veins are compared.

(a) Arteries
Systemic arteries transport oxygenated blood away from the heart. (Remember that pulmonary arteries transport deoxygenated blood to the lungs.) Larger arteries branch into progressively smaller arteries that lead into arterioles.
Elastic arteries stretch to accommodate the pulses of blood ejected from the heart and recoil to propel blood through the arteries.
Muscular arteries regulate distribution of blood through vasoconstriction and vasodilation.
Arterioles regulate distribution of blood through vasoconstriction and vasodilation.
Artery anatomy
Tunica intima
Endothelium
Subendothelial layer
Internal elastic lamina
Tunica media
External elastic lamina
Tunica externa
Vasa vasorum
Lumen
(c) Veins
Systemic veins transport deoxygenated blood toward the heart. (Remember that pulmonary veins transport oxygenated blood from the lungs.)
Large veins serve as a blood reservoir (at rest ~ 55% total blood).
Small/medium veins receive blood from venules; blood drains into small/medium veins and then into large veins.
Valves prevent backflow of blood.
Venules receive blood from capillaries.
Vein anatomy
Tunica intima
Endothelium
Subendothelial layer
Tunica media
Valves
Tunica externa
Vasa vasorum
Lumen
(b) Capillaries
Capillaries receive blood from arterioles and allow for exchange of substances between the blood and cells.
Precapillary sphincters regulate blood flow through capillary beds.
Precapillary sphincter contracted
Precapillary sphincter relaxed
Blood flow
Capillary bed
Continuous capillary
Least permeable
Fenestrated capillary
Varying degrees of permeability determine the size and amount of substances exchanged across capillary walls.
Sinusoid
Most permeable

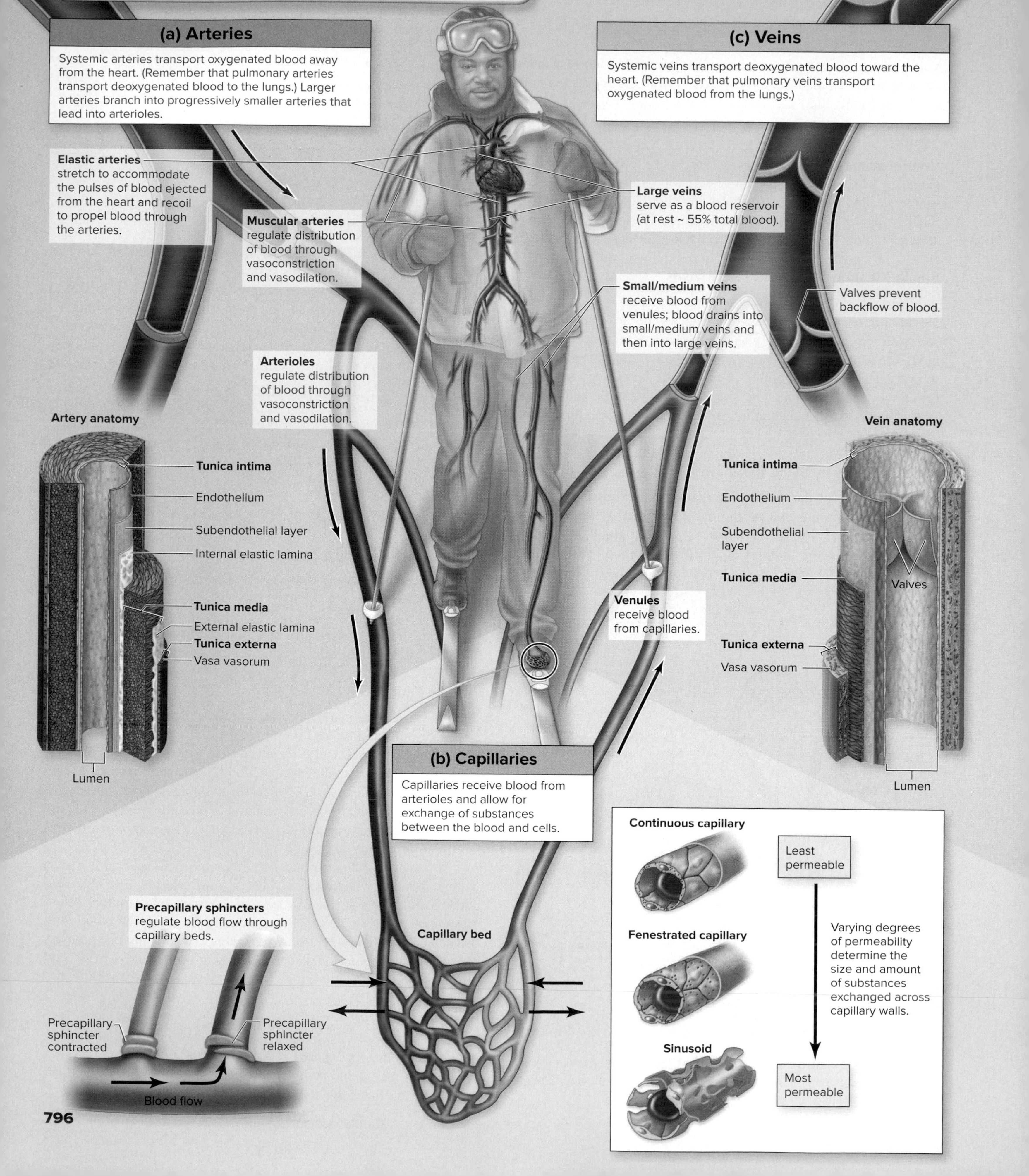

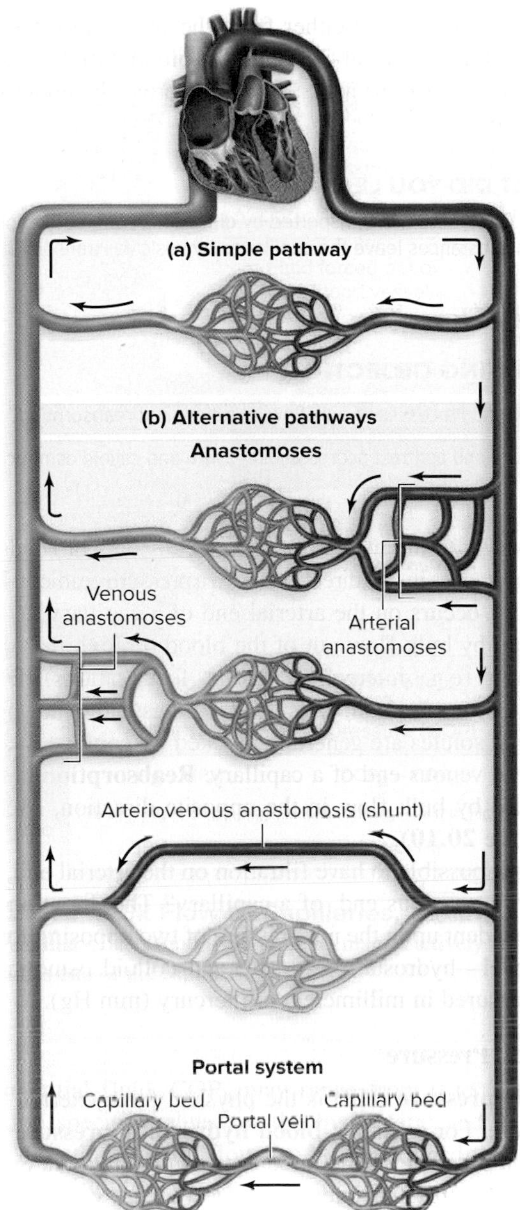

Figure 20.8 Comparison of Simple and Alternative Blood Flow Pathways. (*a*) Simple pathways use one major artery, one capillary bed, and one major vein to deliver blood to a body region. (*b*) Alternative pathways include anastomoses and portal systems. Anastomoses differ in the number of arteries, capillary beds, or veins that serve a body region. In a portal system, venous blood is directed through the capillary bed of another organ (like the liver) first before going back to the heart.

Three of the alternative pathways are designated as anastomoses. An **anastomosis** (a-nas′tō-mō′sis; pl., *anastomoses*) is the joining together of blood vessels. An **arterial anastomosis** includes two or more *arteries* converging to supply the same body region. (For example, figure 20.22*a* shows anastomoses among the superior and inferior epigastric arteries that serve the abdominal wall.) Other vessels, such as the coronary arteries, may have anastomoses that are so tiny that the function of the arteries may almost be considered end arteries; these arteries are called **functional end arteries** (see section 19.4a).

A **venous anastomosis** includes two or more *veins* draining the same body region. Veins tend to form many more anastomoses than do arteries. Veins that drain the upper limb including the basilic, brachial, and cephalic veins provide an example of a venous anastomosis (see figure 20.19*b*).

An **arteriovenous anastomosis,** or *shunt,* transports blood from an artery directly into a vein, bypassing the capillary bed. These shunts are present in the fingers, toes, palms, and ears, and they allow these areas to be bypassed if the body is becoming hypothermic (cold). It is the bypassing of these body structures, as blood is shunted through an arteriovenous anastomosis, that makes them particularly vulnerable to frostbite (see Clinical View 27.6: "Hyperthermia, Frostbite, and Dry Gangrene").

A different type of blood vessel arrangement is a **portal system.** Blood flows through *two* capillary beds, with the two capillary beds separated by a portal vein. A **portal vein** delivers blood to another organ first, before the blood is sent back to the heart. Thus, the sequence of blood vessels is as follows: artery, capillary bed, portal vein, capillary bed, and vein. One example is the hypothalamo-hypophyseal portal system that extends between the hypothalamus and the anterior pituitary (see section 17.7a). This portal system provides a more direct means of delivering hypothalamic regulatory hormones to the anterior pituitary. Another portal system is the hepatic portal system, which is described in section 20.10d.

WHAT DID YOU LEARN?

5 How does each of the various alternative blood vessel pathways differ from the simple pathway?

20.2 Total Cross-Sectional Area and Blood Flow Velocity

LEARNING OBJECTIVES

10. Describe the relationship of the total cross-sectional area and velocity of blood flow.
11. Predict the significance of slow blood flow in the capillaries.

The **cross-sectional area** of a vessel is the diameter of the vessel's lumen. The **total cross-sectional area** is estimated as the aggregate lumen diameter across the total number of a given type of vessel (artery, capillary, or vein) if they were all positioned side by side. You may be surprised to learn that, although the cross-sectional area of an individual artery is relatively large, the total cross-sectional area of arteries is relatively small. This observation is also accurate for veins, which have a relatively small total cross-sectional area. See the blue line on the graph in **figure 20.9**. In comparison, an individual

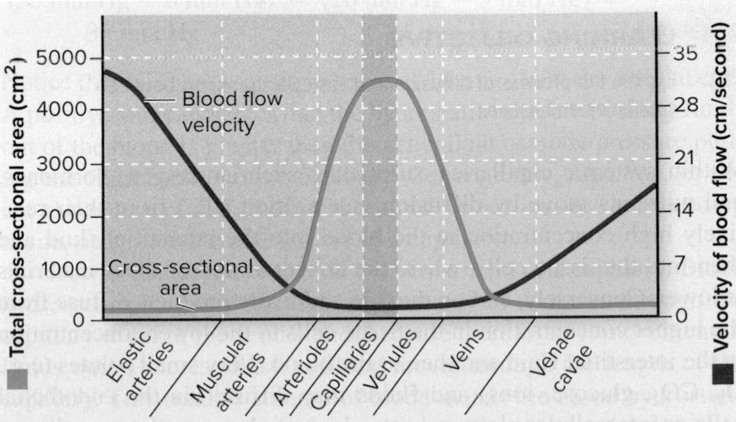

Figure 20.9 Relationship of Total Cross-Sectional Area and Velocity of Blood Flow. The greater the total cross-sectional area, the slower the blood flow. Notice that capillaries have the greatest cross-sectional area and the slowest blood flow—conditions that facilitate the exchange of substances between the blood and tissues.

Again, HP_{if} is assumed to be 0 mm Hg. Net filtration pressure (NFP) is calculated on the venous end as follows:

$$(16 \text{ mm Hg} - 0 \text{ mm Hg}) - (26 \text{ mm Hg} - 5 \text{ mm Hg}) = \text{NFP}$$
$$16 \text{ mm Hg} \quad - \quad 21 \text{ mm Hg} \quad = -5 \text{ mm Hg}$$

Notice that in this case, NFP has a negative value (–5 mm Hg). A negative value for NFP occurs because blood hydrostatic pressure is less than net osmotic pressure. Consequently, reabsorption occurs on the venous end of the capillary with a net movement of fluid back into the blood vessel from the surrounding tissue.

Keep in mind that the numbers for the pressures used to calculate net filtration pressure are examples only. The specific values depend upon the part of the body, the amount of blood entering a specific capillary bed, and the general health of the individual.

WHAT DO YOU THINK?

1 Is it possible for blood pressure to decrease to such a degree that capillary exchange ceases? Explain.

WHAT DID YOU LEARN?

9 How does the hydrostatic pressure change from the arterial end of a capillary to the venous end of a capillary? Do you see similar changes in colloid osmotic pressure?

10 Which two pressures have the largest values? Explain how each of these specifically influences filtration and reabsorption.

20.3d Role of the Lymphatic System

LEARNING OBJECTIVE

17. Explain the lymphatic system's role at the capillary bed.

Although net filtration occurs at the arterial end of a capillary and net reabsorption at its venous end, not all of the fluid is reabsorbed at the venous end of the capillary. The capillary typically reabsorbs only about 85% of the fluid that has passed into the interstitial fluid. What happens to the excess 15% of fluid that is *not* reabsorbed?

Another body system, the **lymphatic system,** is responsible for picking up this excess fluid and returning it to the blood. Lymph vessels reabsorb this excess fluid, filter it, and return it to the venous circulation (see section 21.1). The lymphatic system is described in detail in chapter 21.

WHAT DID YOU LEARN?

11 If these lymph vessels were nonfunctional, what would happen to the amount of interstitial fluid around the capillary bed?

20.4 Local Blood Flow

Recall that there is simply not enough blood in the body to fill all capillaries at the same time. Blood must therefore be directed to organs and tissues where it is most needed and away from areas where it is not. **Local blood flow** is the blood delivered locally to the capillaries of a specific tissue and is measured in milliliters per minute. Recall that the specific amount of blood entering capillaries per unit time per gram of tissue is called perfusion. The ultimate function of the cardiovascular system is *adequate perfusion of all tissues* (see section 19.1a).

The amount of blood delivered to a specific organ or tissue is dependent upon several factors, including (1) the degree of vascularization of the tissue, (2) the myogenic response, (3) local regulatory factors that alter blood flow, and (4) total blood flow.

20.4a Degree of Vascularization and Angiogenesis

LEARNING OBJECTIVES

18. Describe what is meant by degree of vascularization.

19. Explain the process of angiogenesis and how it aids perfusion.

The **degree of vascularization,** or the extent of blood vessel distribution within a tissue, determines the potential ability of blood delivery. Organs that are very active metabolically, such as the brain, skeletal muscle, the heart, and the liver, generally are highly vascularized. In comparison, some structures, such as tendons and ligaments, have little vascularization; blood delivery to these tissues is limited. Additionally, some structures contain no capillaries (are avascular); these include epithelial tissue, cartilage, and the cornea and lens of the eye.

The amount of vascularization in a given tissue may change over time through the process of angiogenesis. **Angiogenesis** (an′jē-ō-jen′ĕ-sis; *angio* = vessel, *genesis* = production) is the formation of new blood vessels in tissues that require them. This process helps provide adequate perfusion through long-term anatomic changes that occur over several weeks to months. For example, angiogenesis is stimulated in skeletal muscle in response to aerobic training; in adipose tissue, angiogenesis occurs in adipose connective tissue when an individual gains weight in the form of fat deposits. Angiogenesis also occurs in response to a slow, gradual occlusion (blockage) of coronary vessels, thus potentially providing alternative routes to deliver blood to the heart wall.

Regression (or return to previous state) of vessels is also possible. For example, some skeletal muscle blood vessels regress when an individual who was physically active becomes sedentary, or blood vessels in adipose tissue regress when the amount of adipose tissue is decreased through restriction of food and increased physical activity.

WHAT DID YOU LEARN?

12 In what ways is angiogenesis stimulated in skeletal muscle? In adipose tissue?

20.4b Myogenic Response

LEARNING OBJECTIVE

20. Describe the myogenic response that maintains normal blood flow through a tissue.

Systemic blood pressure, which is the driving force to move blood through blood vessels, changes under different conditions (e.g., when we get excited or become relaxed). However, blood flow into a tissue may remain relatively constant because of the **myogenic** (mī′-ō-jen′ik; *mys* = muscle) **response,** which is contraction and relaxation of smooth muscle within blood vessels in response to changes in stretch of the blood vessel wall (see section 10.10d).

An increase in systemic blood pressure causes an additional volume of blood to enter the blood vessel, which stretches the smooth muscle cells within the blood vessel wall. This stimulates the smooth muscle cells to contract, resulting in vasoconstriction. Thus, although systemic blood pressure is higher, which would drive additional blood into the blood vessel, the resulting vasoconstriction decreases the size of the blood vessel lumen offsetting the change, and blood flow into a tissue remains constant. In contrast, a decrease in systemic blood pressure causes a lower volume of blood to enter the blood vessel, with decreased stretch of the smooth muscle cells within the blood vessel wall. This stimulates the smooth muscle cells to relax, resulting in vasodilation. Thus, although systemic blood pressure is lower, which would drive less blood into the blood vessel, the resulting vasodilation increases the size of the blood vessel lumen offsetting the change, and blood flow into a tissue remains constant.

WHAT DID YOU LEARN?

13 Explain the myogenic response to an increase in system blood pressure.

20.4c Local, Short-Term Regulation

LEARNING OBJECTIVES

21. Compare and contrast a vasodilator and a vasoconstrictor.

22. Explain how a tissue autoregulates local blood flow based on metabolic needs.

23. Describe how local blood flow is altered by tissue damage and as part of the body's defense.

Local regulation of blood flow occurs on a continual basis in response to changes in metabolic activity of tissues. Local blood flow is also altered in response to tissue damage or as part of the body's defense systems. The stimulus is changing concentrations of certain chemicals, collectively called **vasoactive chemicals.** They are classified according to their action as either vasodilators or vasoconstrictors. **Vasodilators** are substances that cause smooth muscle relaxation, which results in both vasodilation of arterioles and opening of precapillary sphincters. Consequently, blood flow increases into a capillary bed. **Vasoconstrictors** are substances that cause smooth muscle contraction, which results in both arterioles vasoconstricting and precapillary sphincters closing. Thus, blood flow decreases into a capillary bed (see figure 20.5).

INTEGRATE

CLINICAL VIEW 20.3

Tumor Angiogenesis

Like normal cells of the body, abnormal cells of cancerous tumors require the delivery of oxygen and nutrients and the removal of their wastes. Thus, a critical event in development of a cancerous tumor is the formation of blood vessels within the tumor, a process called **tumor angiogenesis.** The process is initiated when cancerous cells release molecules that cause normal host cells to release growth factors that stimulate angiogenesis. One active area of cancer research is inhibiting tumor angiogenesis, with the intent of reducing or eliminating the tumor by depriving it of the oxygen and nutrients needed for growth.

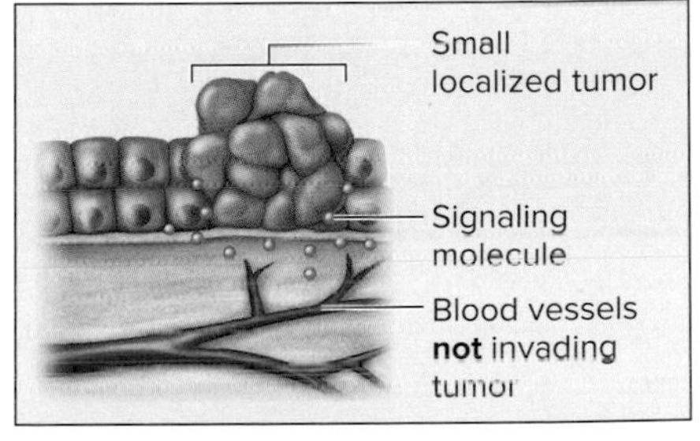

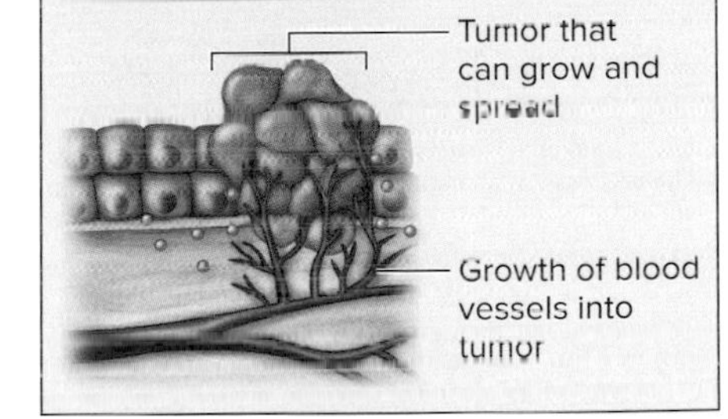

How is it possible to change the steepness of the pressure gradient? The blood pressure gradient is altered by changes in cardiac output. An increase in cardiac output will increase the pressure gradient. Conversely, a decrease in cardiac output will decrease the pressure gradient (see section 19.9).

WHAT DID YOU LEARN?

16 A 55-year-old female has an arterial blood pressure reading of 155/95 mm Hg. What are her pulse pressure and mean arterial pressure?

17 What is the physiologic significance of capillary blood pressure?

18 How is the small pressure gradient in veins overcome?

19 How is the pressure gradient to move blood through the systemic circulation calculated? What is the significance of this pressure gradient?

20.5b Resistance

LEARNING OBJECTIVE

29. Define resistance, and explain how it is influenced by blood viscosity, vessel length, and vessel radius.

Resistance also influences total blood flow. **Resistance** is defined as the amount of friction the blood experiences as it is transported through the blood vessels. Blood flow is always opposed by resistance. This friction is due to the contact between blood and the blood vessel wall. The term **peripheral resistance** is typically used when discussing the resistance of blood in the blood vessels (as opposed to the resistance of blood in the heart). Several factors affect peripheral resistance, including blood viscosity, blood vessel length, and the size of the lumen of blood vessels (as indicated by vessel radius).

Blood Viscosity

Viscosity refers to the resistance of a fluid to its flow. More generally, it is the "thickness" of a fluid. The thicker the fluid, the more viscous it is, and the greater its resistance to flow. The thickness is dependent upon the relative percentage of particles in the fluid and their interactions with one another. Because blood contains formed elements (erythrocytes, leukocytes, platelets) and plasma proteins, it is approximately 4.5 to 5.5 times more viscous than water. As a result, blood exhibits greater resistance to flow than water.

A change in blood's viscosity causes a change in resistance of blood flow through a vessel. For example, if someone is anemic (and has a lower than normal number of erythrocytes), then that person's blood viscosity is lower, and the blood has less resistance to flow. Conversely, if someone has a greater than normal concentration of erythrocytes in his or her blood (see Clinical View 18.1: "Blood Doping"), or the individual is dehydrated, then the blood is more viscous and has greater resistance to flow.

Vessel Length

Increasing vessel length increases resistance, because longer vessels result in greater friction, which the fluid experiences as it is transported through the vessel. In contrast, shorter vessels offer less resistance than longer vessels with comparable diameters. Normally, vessel length remains relatively constant. However, if a person gains a large amount of weight, the body must produce miles of additional vessel length by angiogenesis for blood to be transported through the extra fat. Thus, vessel resistance increases if one gains weight, and decreases if someone loses a lot of weight (because those vessels are no longer needed and regress).

Vessel Radius

Blood viscosity and vessel length remain relatively constant in a typical healthy individual. The major way resistance may be regulated is by altering vessel lumen radius (and thus changing the vessel diameter.)

How specifically does vessel radius influence resistance? Blood tends to flow fastest in the center of the vessel lumen, whereas blood near the sides of the vessels slows, because it encounters resistance from the nearby vessel wall. This difference in flow rate within a blood vessel (or in any conduit) is called **laminar flow.** You can see evidence of laminar flow by studying a river: The water flow near the banks, or edges, of the river is a bit slower or sluggish, whereas the water flow near the center of the river is quite fast in comparison. So, if vessel diameter increases, relatively less blood flows near the edges and overall blood flow increases. In contrast, if vessel radius decreases, then relatively more blood flows near the edges and overall blood flow decreases.

The relationship between blood flow and the radius of blood vessel lumen may be stated as follows (where the symbol $\propto$ means "is proportional to"):

$$F \propto r^4$$

where F = flow and r = radius of the lumen of a vessel (The radius is 1/2 the diameter of the lumen.). This mathematical expression reflects that flow is directly proportional to the fourth power of a radius. If a vessel vasodilates and its radius increases from 1 millimeter (mm) to 2 mm, the overall change in flow is 16 times greater: If $r =$ 1 mm, then $r^4 = 1$, and $F =$ 1 mm per second; and when $r =$ 2 mm, then $r^4 = 16$, and $F =$ 16 mm per second. In contrast, if a vessel vasoconstricts and its radius decreases from 2 mm to 1 mm, the overall change in flow is 16 times less.

Any vessel may vasoconstrict or vasodilate; however, resistance usually is regulated specifically by vasoconstriction and vasodilation of muscular arteries and arterioles, a change systemically controlled by the vasomotor center of the medulla oblongata (see sections 13.5c and 20.6a). Thus, because there are so many muscular arteries and arterioles (and the overall length of all these vessels is so great), even

INTEGRATE

LEARNING STRATEGY 

Drinking through a straw provides an easy way to demonstrate the variables that influence resistance:

- **Viscosity.** Water or soda moves through a straw more easily than does a thick vanilla milkshake.
- **Vessel length.** It is easier to drink through a relatively short straw compared with a very long, twisty straw.
- **Vessel radius.** Water can be sipped through a normal-sized straw more easily than through a hollow coffee stirrer with its smaller diameter.

©McGraw-Hill Education/Ken Cavanagh

Greater resistance is experienced when drinking thick liquids, using a longer straw, or using a more narrow straw. In the same way, an increase in resistance is exhibited with more viscous blood, increased total length of blood vessels, and vasoconstriction of blood vessels.

small changes in vessel radius significantly change resistance, and therefore can have dramatic effects on blood flow.

Note that atherosclerosis is a disease process where a plaque narrows the lumen of blood vessels. Thus, in individuals with atherosclerosis, the blood experiences greater resistance as it moves through the atherosclerotic blood vessels.

WHAT DID YOU LEARN?

20 How is resistance defined?

21 What are the three factors that alter resistance? How does each affect blood flow in vessels?

20.5c Relationship of Blood Flow to Blood Pressure Gradients and Resistance

LEARNING OBJECTIVES

30. Explain the relationship of both the blood pressure gradient and resistance to total blood flow.

31. Discuss why blood pressure increases with increased resistance in the systemic circulation.

Total blood flow is the amount of blood that moves through the cardiovascular system per unit time and is influenced by both blood pressure gradients and resistance, as previously described. This relationship is expressed mathematically as follows:

$$F \propto \frac{\Delta P}{R}$$

where F = blood flow, ΔP = pressure gradient ($P_1 - P_2$), and R = resistance.

Systemic Blood Pressure Gradient

The preceding formula demonstrates that blood flow is *directly* related to the pressure gradient. Thus, as the blood pressure gradient increases, total blood flow is greater, and as the blood pressure gradient decreases, total blood flow lessens (assuming resistance remains the same). Recall that an increase in cardiac output will increase the pressure gradient, whereas a decrease in cardiac output will decrease the pressure gradient. Consequently, increasing blood flow is possible with a steeper blood pressure gradient—however, it is a change exhibited only with greater effort exerted by the heart.

Resistance

The preceding mathematical expression also demonstrates that blood flow is inversely related to resistance. If resistance increases, blood flow lessens, whereas if resistance decreases, blood flow increases (assuming the pressure gradient remains the same).

Recall that resistance is increased with (1) an increase in blood viscosity (e.g., as occurs with greater concentration of erythrocytes); (2) an increase in blood vessel length (e.g., a change that accompanies weight gain); and (3) a decrease in vessel lumen diameter (e.g., a change that occurs with a net increase in vasoconstriction and in individuals with atherosclerosis).

Relationship of Systemic Blood Pressure and Resistance

Individuals with *sustained* increased resistance (e.g., that accompanies significant weight gain or atherosclerosis) generally exhibit elevated arterial blood pressure readings. This condition is clinically significant; it occurs because a greater pressure gradient must be produced to overcome the higher resistance and ensure normal blood flow and adequate perfusion of all tissues. The relationship of blood flow, blood pressure gradients, and resistance is summarized in **figure 20.13**.

WHAT DID YOU LEARN?

22 In general, would you predict a higher blood pressure, lower blood pressure, or normal blood pressure in individuals with sustained increased resistance? Explain.

Total blood flow $\propto$ **Pressure gradient** (established by the heart) / **Resistance** (experienced by blood as it moves through the vessels)

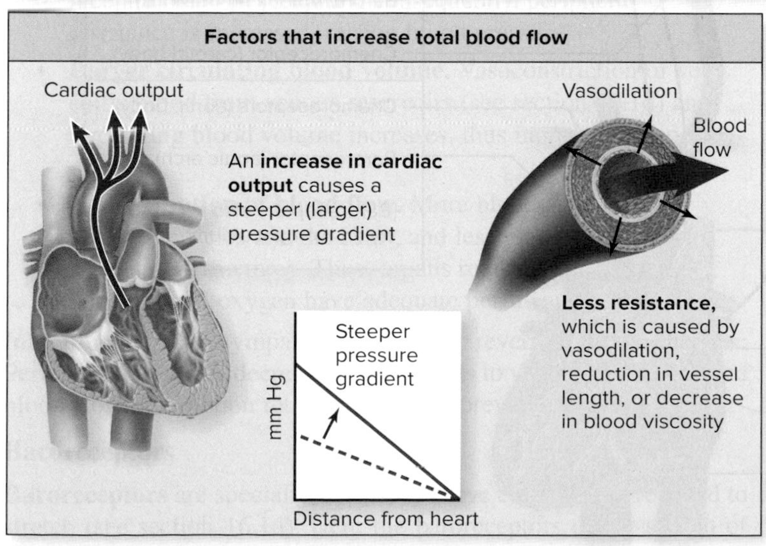

Factors that increase total blood flow
Cardiac output
Vasodilation
Blood flow
An increase in cardiac output causes a steeper (larger) pressure gradient
Less resistance, which is caused by vasodilation, reduction in vessel length, or decrease in blood viscosity
Steeper pressure gradient
mm Hg
Distance from heart
(a)

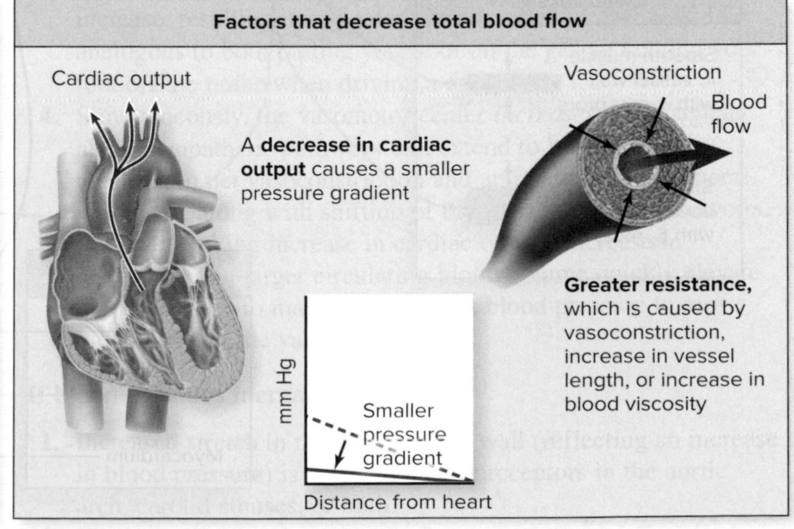

Factors that decrease total blood flow
Cardiac output
Vasoconstriction
Blood flow
A decrease in cardiac output causes a smaller pressure gradient
Greater resistance, which is caused by vasoconstriction, increase in vessel length, or increase in blood viscosity
Smaller pressure gradient
mm Hg
Distance from heart
(b)

Figure 20.13 Factors That Influence Total Blood Flow. Maintaining adequate perfusion of systemic tissues is dependent upon maintaining total blood flow. Total blood flow is a function of the pressure gradient and resistance and can be either (*a*) increased or (*b*) decreased.

20.10b Thoracic and Abdominal Walls

LEARNING OBJECTIVES

45. Describe the pairs of arteries that supply the thoracic wall.

46. Discuss the arteries that supply the abdominal wall.

47. List veins that drain the thoracic and abdominal walls and delineate their pathways.

The systemic circulation to the walls of the trunk region is primarily via paired vessels that have extensive anastomoses. The venous drainage is more complex than the arterial supply.

Arterial Supply

The thoracic and abdominal walls are both supplied by several pairs of arteries **(figure 20.22*a*)**. A left and right **internal thoracic artery** emerge from each subclavian artery to supply the anterior thoracic wall and mammary gland. Each internal thoracic artery is located lateral to the sternum and has the following branches: the first six **anterior intercostal arteries** and a **musculophrenic** (mŭs′kū-lō-fren′ik; *phren* = diaphragm) **artery** that divides into anterior intercostal arteries 7–9. The intercostal arteries supply the contents of the intercostal spaces. The internal thoracic artery then becomes the **superior epigastric** (ep-i-gas′trik; *epi* = upon, *gastric* = stomach) **artery,** which transports blood to the superior abdominal wall.

The **inferior epigastric artery,** a branch of the external iliac artery, supplies the inferior abdominal wall. This artery anastomoses extensively with the superior epigastric artery.

The **supreme intercostal artery** is a branch of the costocervical trunk; it branches into the first and second **posterior intercostal arteries.** Posterior intercostal arteries 3–11 are branches of the descending thoracic aorta. The posterior and anterior intercostal

Right subclavian artery
Brachiocephalic trunk
Internal thoracic artery
Anterior intercostal arteries (1–6) (anterior intercostal spaces)
Musculophrenic artery (anterior intercostal spaces)
Anterior intercostal arteries (7–9)
Superior epigastric artery (superior abdominal wall)
Inferior epigastric artery (inferior abdominal wall)
Aortic arch
Costocervical trunk
Left subclavian artery
Supreme intercostal artery
Posterior intercostal arteries (1 and 2) (posterior intercostal spaces)
Posterior intercostal arteries (3–11) (posterior intercostal spaces)
Descending thoracic aorta
Descending abdominal aorta
Lumbar arteries (5 pairs) (posterolateral abdominal wall)
Common iliac artery
Median sacral artery (sacrum and coccyx)

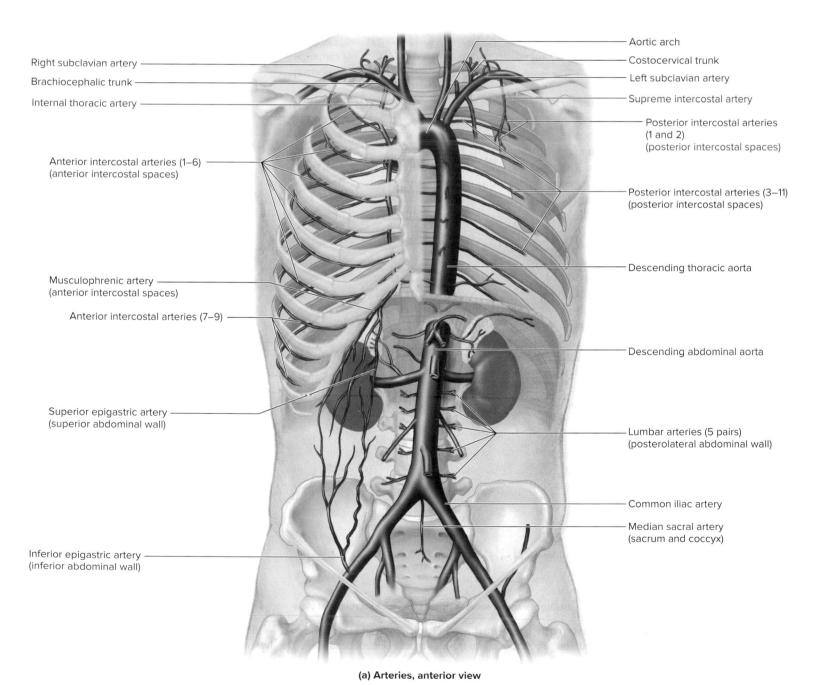

(a) Arteries, anterior view

Figure 20.22 Circulation to the Thoracic and Abdominal Body Walls. (*a*) The arteries that supply the thoracic and abdominal wall are shown. Notice the anastomoses of the superior and inferior epigastric arteries. AP|R

arteries anastomose, and each pair forms a horizontal vessel arc that spans a segment of the thoracic wall.

Finally, five pairs of **lumbar arteries** branch from the descending abdominal aorta to supply the posterolateral abdominal wall. In addition, a single **median sacral artery** arises at the bifurcation of the aorta in the pelvic region to supply the sacrum and coccyx.

Venous Drainage

Venous drainage of the thoracic and abdominal walls is a bit more complex than the arterial pathways (figure 20.22*b*). **Anterior intercostal veins,** a musculophrenic vein, and the **superior epigastric vein** all merge into the **internal thoracic vein,** which drains into the brachiocephalic vein.

The **inferior epigastric vein** merges with the **external iliac vein.** The first, second and third **posterior intercostal veins** merge with the **supreme intercostal vein** that drains into the brachiocephalic vein.

The **lumbar veins** and the remaining **posterior intercostal veins** drain into the azygos system of veins along the posterior thoracic wall. The **hemiazygos** (hem′ē-az′ī-gos) and **accessory hemiazygos veins** on the left side of the vertebrae drain the left-side veins. The **azygos vein** drains the right-side veins and receives blood from the hemiazygos veins. The azygos vein drains into the superior vena cava. A schematic for venous return can be viewed in figure 20.22*c*.

WHAT DID YOU LEARN?

34 Does the azygos system, which receives venous blood from the thoracic and abdominal walls, drain into the superior vena cava or the inferior vena cava?

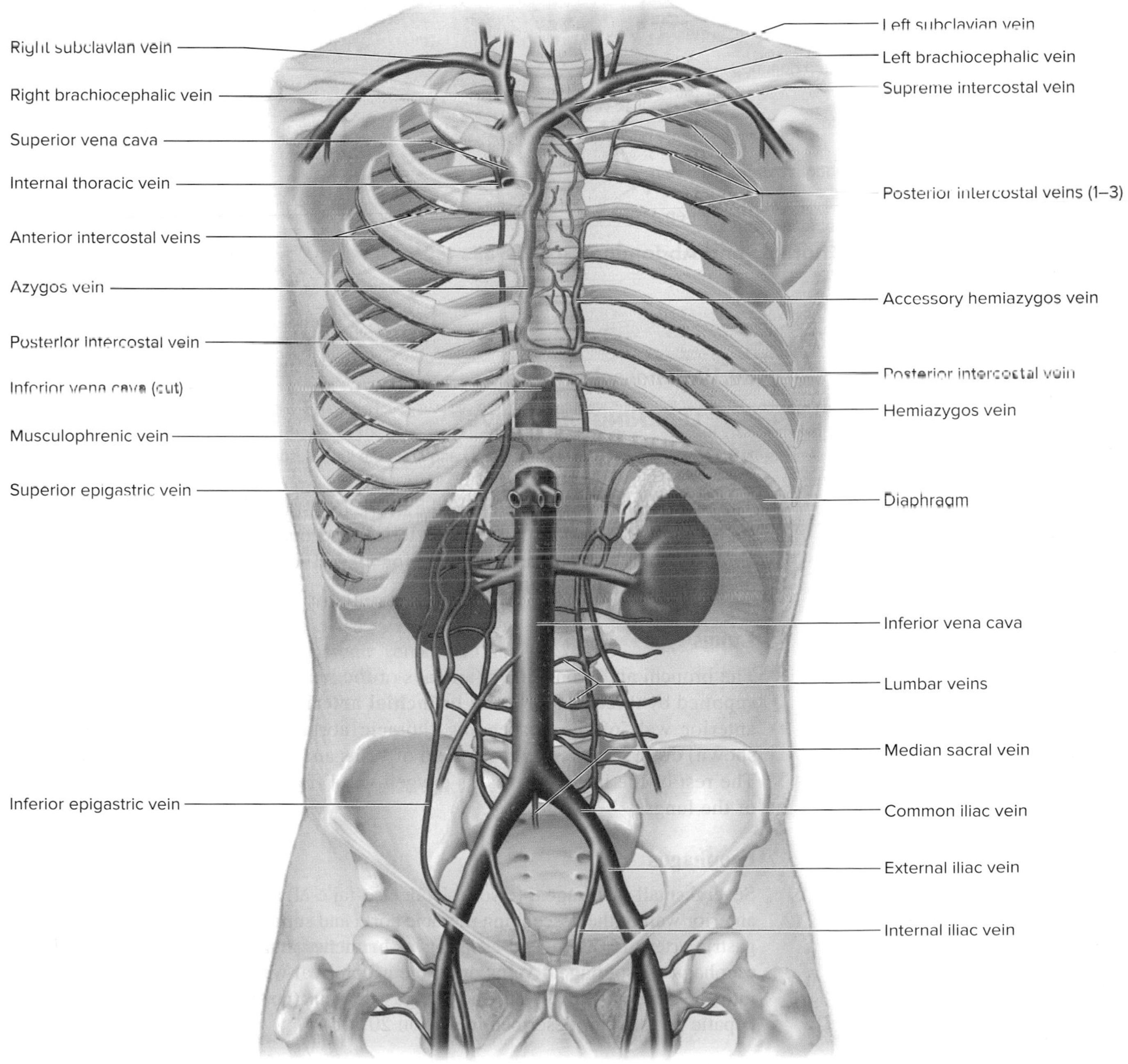

Figure 20.22 Circulation to the Thoracic and Abdominal Body Walls (*continued*). (*b*) The venous drainage of the thoracic and abdominal wall is more complex than its arterial supply. AP|R

20.11 Systemic Circulation: Upper and Lower Limbs

Blood flow through the upper limbs mirrors blood flow through the lower limbs. Both the upper and lower limbs (1) are supplied by a main arterial vessel: the subclavian artery for the upper limb, the external iliac artery for the lower limb; (2) have their main artery bifurcating at either the elbow or the knee; (3) have arterial and venous arches; and (4) have superficial and deep networks of veins.

20.11a Upper Limb

LEARNING OBJECTIVES

54. Trace the arteries of the upper limb from the subclavian artery to the fingers.

55. Compare and contrast the superficial venous drainage and the deep venous drainage of the upper limb.

Each upper limb is supplied by a subclavian artery and drained by a subclavian vein. The venous system includes both deep and superficial drainage.

Arterial Flow to the Upper Limb

A **subclavian artery** supplies blood to each upper limb. The left subclavian artery emerges directly from the aortic arch, while the right subclavian artery is a division of the brachiocephalic trunk (see figure 20.22*a*).

Before extending into the arm, the subclavian artery extends multiple branches to supply parts of the body's upper region: the vertebral artery, the thyrocervical trunks, the costocervical trunk, and the internal thoracic artery, as described in section 20.10a.

After the subclavian artery passes over the lateral border of the first rib, it is renamed the **axillary** (ak′sil-ār-ē) **artery** (**figure 20.27**). The axillary artery extends many branches to the shoulder and thoracic region. When the axillary artery passes the inferior border of the teres major muscle, it is renamed the **brachial** (brā′kē-ăl) **artery.** One of its branches is the **deep brachial artery,** which supplies blood to most brachial (arm) muscles. In the antecubital region, the brachial artery divides into a **radial artery** and an **ulnar artery.** Both arteries supply the forearm and wrist before they anastomose and form two arterial arches in the palm: the **deep palmar arch** (formed primarily from the radial artery) and the **superficial palmar arch** (formed primarily from the ulnar artery). **Digital arteries** emerge from the arches to supply the fingers.

WHAT DO YOU THINK?

If the left ulnar artery were cut, would any blood be able to reach the left hand and fingers? Why or why not?

Venous Drainage of the Upper Limb

Venous drainage of the upper limb, which converges on the axillary vein and into the subclavian vein, is accomplished through two groups of veins: superficial and deep.

Superficial Venous Drainage On the dorsum of the hand, a dorsal venous network (or arch) of veins drains into both the medially located **basilic** (ba-sil′ik) **vein** and the laterally located **cephalic** (se-fal′ik) **vein.** These veins drain into the axillary vein, and they also have perforating branches that allow them to connect to the deeper veins.

In the cubital region, an obliquely positioned **median cubital vein** connects the cephalic and basilic veins. The median cubital vein is a common site for venipuncture, in which a vein is punctured with a hollow needle to draw blood or inject fluids and medications. All of these superficial veins are highly variable among individuals and have multiple superficial tributaries draining into them.

Deep Venous Drainage The **digital veins** and **deep** and **superficial palmar venous arches** drain into *pairs* of **radial veins** and **ulnar veins** that run parallel to arteries of the same name. At the level of the antecubital region, the radial and ulnar veins merge to form a pair of **brachial veins** that are positioned alongside the brachial artery. Brachial veins and the basilic vein merge to form the **axillary vein.**

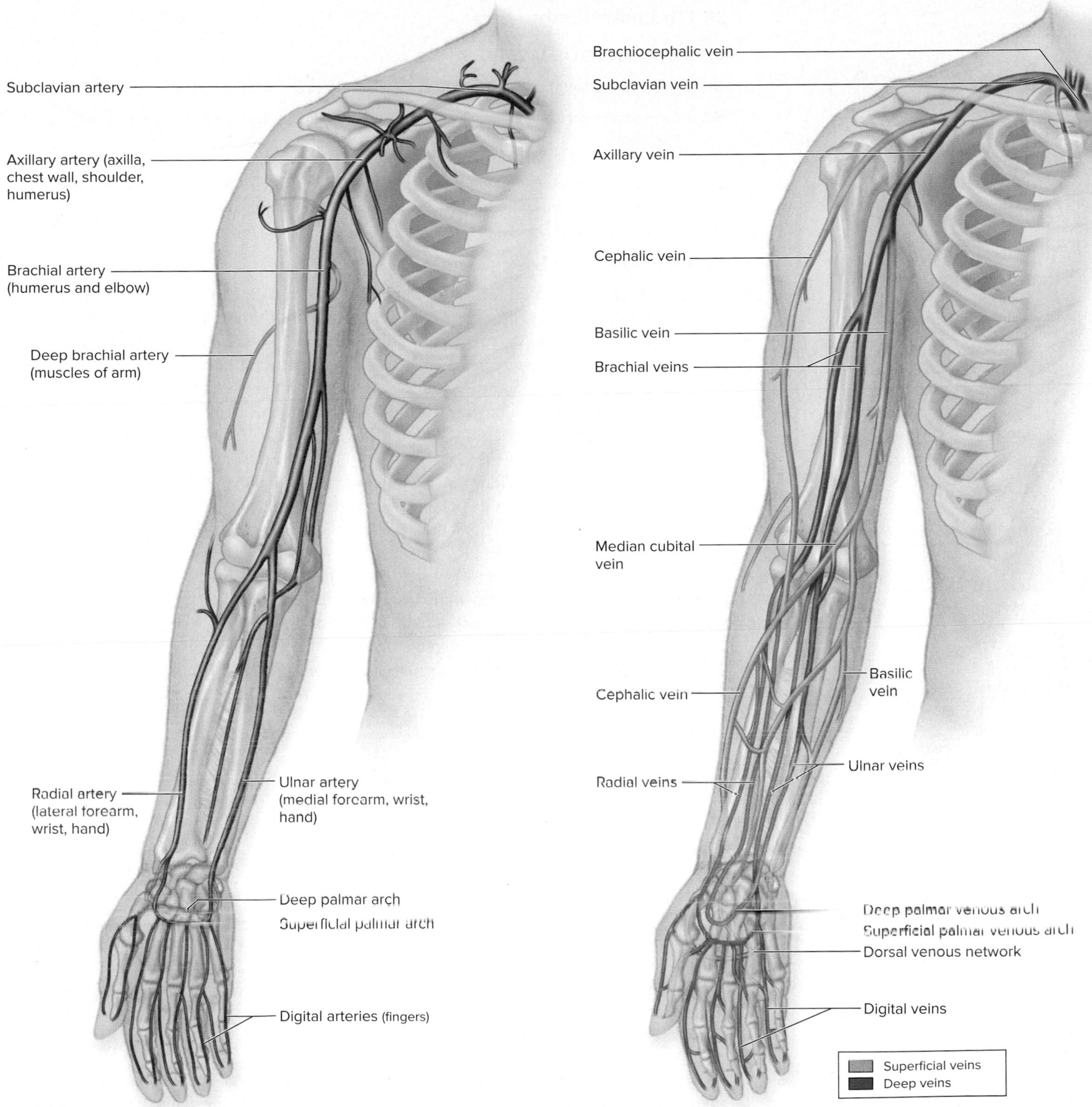

Subclavian artery

Axillary artery (axilla, chest wall, shoulder, humerus)

Brachial artery (humerus and elbow)

Deep brachial artery (muscles of arm)

Radial artery (lateral forearm, wrist, hand)

Ulnar artery (medial forearm, wrist, hand)

Deep palmar arch

Superficial palmar arch

Digital arteries (fingers)

(a) Arteries of right upper limb, anterior view

Brachiocephalic vein

Subclavian vein

Axillary vein

Cephalic vein

Basilic vein

Brachial veins

Median cubital vein

Cephalic vein

Basilic vein

Radial veins

Ulnar veins

Deep palmar venous arch

Superficial palmar venous arch

Dorsal venous network

Digital veins

Superficial veins

Deep veins

(b) Veins of right upper limb, anterior view

Figure 20.27 Vascular Supply to the Upper Limb. The subclavian artery transports oxygenated blood to the upper limb; veins merge to return deoxygenated blood to the heart. (*a*) Arteries that supply the upper limb. (*b*) Superficial and deep veins that return blood from the upper limb. **AP|R**

Superior to the lateral border of the first rib, the axillary vein is renamed the **subclavian vein**. When the subclavian vein and internal jugular veins of the neck merge, they form the brachiocephalic vein. As we have seen, the left and right brachiocephalic veins form the superior vena cava (see figure 20.19*b*).

WHAT DID YOU LEARN?

39 What is the sequence of vessels from the subclavian artery to the digital artery of the thumb?

40 What are the primary superficial veins of the upper limb?

20.11b Lower Limb

LEARNING OBJECTIVES

56. Trace the arteries of the lower limb from the external iliac artery to the toes.

57. Compare and contrast the superficial venous drainage and the deep venous drainage of the lower limb.

The arterial and venous blood flow of the lower limb is very similar to that of the upper limb. As we discuss lower limb blood flow, compare it with that of the upper limb.

Arterial Flow to the Lower Limb

The main arterial supply for the lower limb is the **external iliac artery,** which is a branch of the common iliac artery (**figure 20.28*a***). The external iliac artery is located inferior to the inguinal ligament, where it is renamed the **femoral** (fem′ŏ-răl) **artery.** The **deep femoral artery** (or *deep artery of the thigh*) emerges from the femoral artery to supply the hip joint (via medial and lateral femoral circumflex arteries) and many of the thigh muscles, before traversing posteromedially along the thigh. When the femoral artery enters the popliteal fossa, it is renamed the **popliteal** (pop-lit′ē-ăl, pop-li-tē′ăl) **artery.** The popliteal artery supplies the knee joint and muscles in this region.

The popliteal artery divides into an **anterior tibial artery,** which supplies the anterior compartment of the leg, and a **posterior tibial artery,** which supplies the posterior compartment of the leg. The posterior tibial artery extends a branch called the **fibular artery,** which supplies the lateral compartment leg muscles.

The posterior tibial artery continues to the plantar side of the foot, where it branches into **medial** and **lateral plantar arteries.** The anterior tibial artery crosses over the anterior surface of the ankle, where it is renamed the **dorsalis pedis artery** (*dorsal artery of the foot*). The dorsalis pedis artery and a branch of the lateral plantar artery unite to form the **plantar arterial arch** of the foot. **Digital arteries** extend from the plantar arch and supply the toes.

Venous Drainage of the Lower Limb

Venous drainage of the lower limb is similar to that of the upper limb in that there are both superficial and deep veins that drain the lower limb (figure 20.28*b*).

Superficial Venous Drainage On the dorsum of the foot, a **dorsal venous arch** drains into the **great saphenous** (săf-ĕ′nŭs, să-fē′nŭs) **vein** and the **small saphenous vein.** The great saphenous vein originates in the medial ankle and extends adjacent to the medial surface of the entire lower limb before it drains into the femoral vein. The small saphenous vein extends adjacent to the lateral ankle and then extends within the posterior calf, before draining into the popliteal vein. These superficial veins have perforating branches that connect to the deeper veins. If the valves in these veins (or the perforating branches) become incompetent, varicose veins develop. (See Clinical View 20.7: "Varicose Veins").

Deep Venous Drainage The **digital veins** and deep veins of the foot drain into pairs of **medial** and **lateral plantar veins.** These veins and a pair of **fibular veins** drain into a pair of **posterior tibial veins.** On the dorsum of the foot and ankle, deep veins drain into a pair of **anterior tibial veins,** which traverse alongside the anterior tibial artery. The **anterior** and **posterior tibial veins** merge to form a **popliteal vein** that curves to the anterior portion of the thigh and is renamed the **femoral vein.** Once this vein passes superiorly to the inguinal ligament, it is renamed the **external iliac vein.** The external and internal iliac veins merge in the pelvis, forming the common iliac vein. Left and right common iliac veins then merge to form the inferior vena cava.

WHAT DID YOU LEARN?

41. Can you trace one pathway that blood may be transported through from the external iliac artery to the digital arteries?

42. How do the great saphenous vein and small saphenous vein compare in terms of their anatomic position and length?

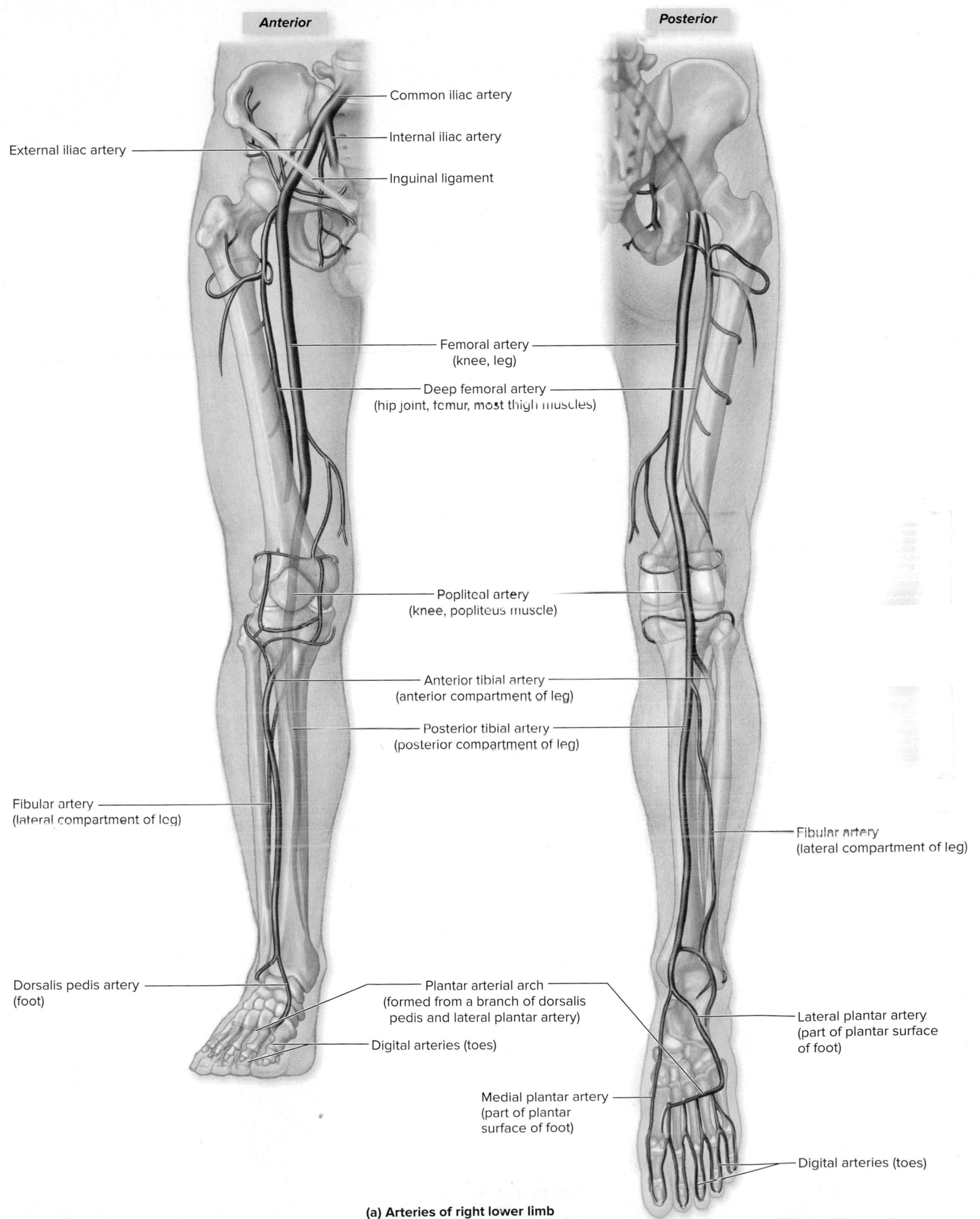

(a) Arteries of right lower limb

Figure 20.28 Vascular Supply to the Lower Limb. The external iliac artery transports oxygenated blood to the lower limb; veins merge to return deoxygenated blood to the heart. (*a*) Anterior and posterior views of arteries distributed throughout the lower limb. APR

(continued on next page)

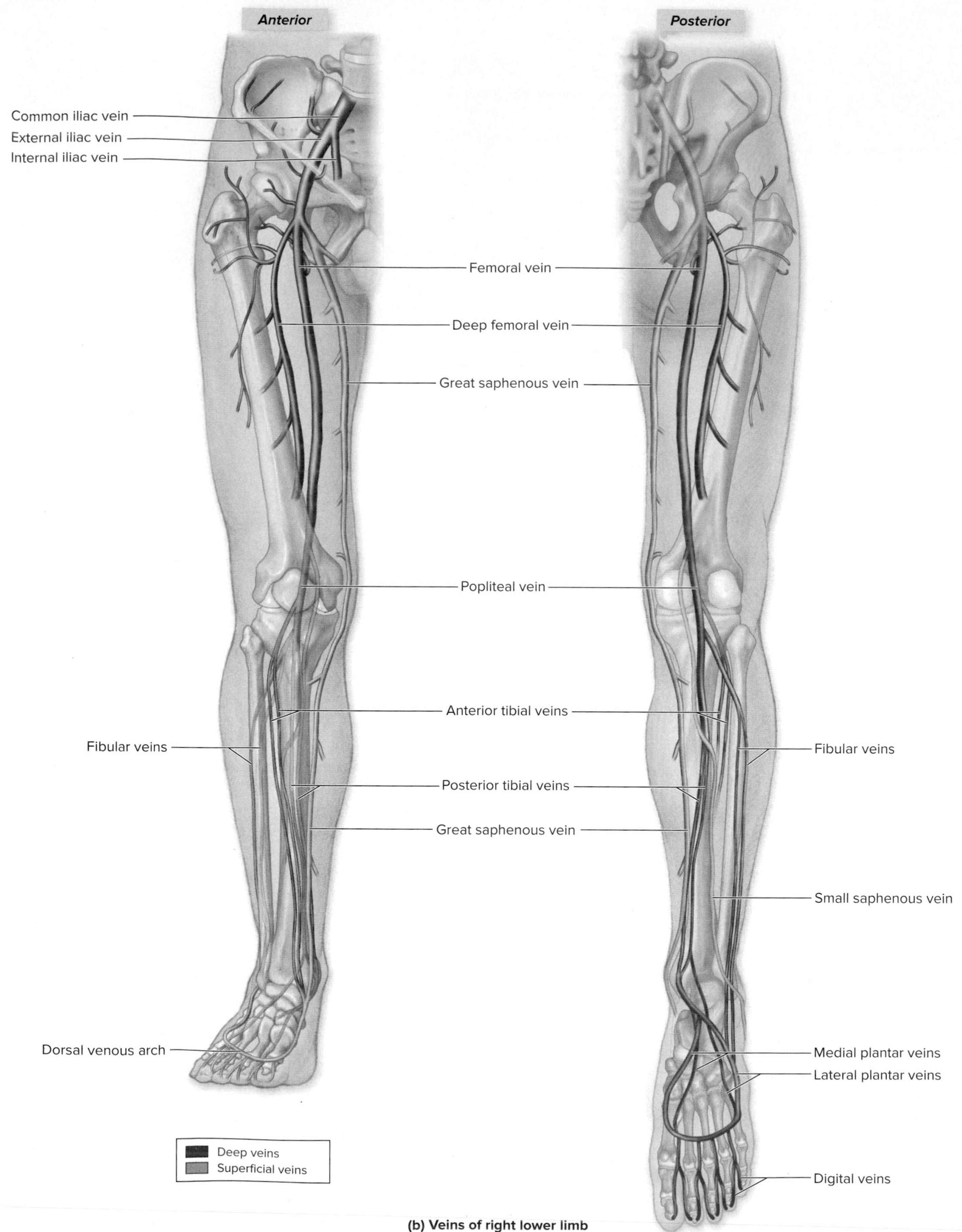

Figure 20.28 Vascular Supply to the Lower Limb (*continued*). (*b*) Anterior and posterior views of the superficial and deep veins that return blood from the lower limb.

20.12 Comparison of Fetal and Postnatal Circulation

The cardiovascular system of the fetus is structurally and functionally different from that of the newborn. Whereas the fetus receives oxygen and nutrients directly from the mother through the placenta, the newborn's postnatal cardiovascular system must be independent. Additionally, because the fetal lungs are not functional, the blood pressure in the pulmonary arteries and right side of the heart is greater than the pressure in the left side of the heart. Finally, several fetal vessels help shunt blood directly to the organs in need and away from the organs that are not yet functional. As a result, the fetal cardiovascular system has some structures that are modified or that cease to function once the baby is born.

20.12a Fetal Circulation

LEARNING OBJECTIVE

58. Trace the pathway of blood circulation in the fetus.

The fetal circulatory route is listed here and shown in **figure 20.29**:

1. Oxygenated blood from the placenta enters the body of the fetus through the **umbilical vein.**
2. The blood from the umbilical vein is shunted away from the liver and directly toward the inferior vena cava through the **ductus venosus** (dŭk′tŭs vē-nō′sŭs).
3. Oxygenated blood in the ductus venosus mixes with deoxygenated blood in the inferior vena cava.
4. Blood from the superior and inferior venae cavae empties into the right atrium.
5. Because pressure is greater on the right side of the heart than on the left side, most of the blood is shunted from the right atrium to the left atrium via the **foramen ovale.** This blood flows into the left ventricle and then is pumped out through the aorta.
6. A small amount of blood enters the right ventricle and then the pulmonary trunk, but much of this blood is shunted from the pulmonary trunk to the aorta through a vessel detour called the **ductus arteriosus** (ar-tēr′ē-ō′sŭs).
7. Blood is transported to the rest of the body, and the deoxygenated blood returns to the placenta through a pair of **umbilical arteries.**
8. Nutrient and gas exchange occurs at the **placenta** (see section 29.2e), and the cycle repeats.

At birth, the fetal circulation begins to change into the postnatal pattern. When the baby takes its first breath, pulmonary resistance drops, and the pulmonary arteries dilate. As a result, pressure on the right side of the heart decreases so that the pressure is greater on the left side of the heart, which handles the systemic circulation.

WHAT DID YOU LEARN?

43 List the five structures of fetal circulation, and identify the function of each.

20.12b Postnatal Changes

LEARNING OBJECTIVE

59. Describe the changes that occur after the baby is born and must utilize the pulmonary circulation.

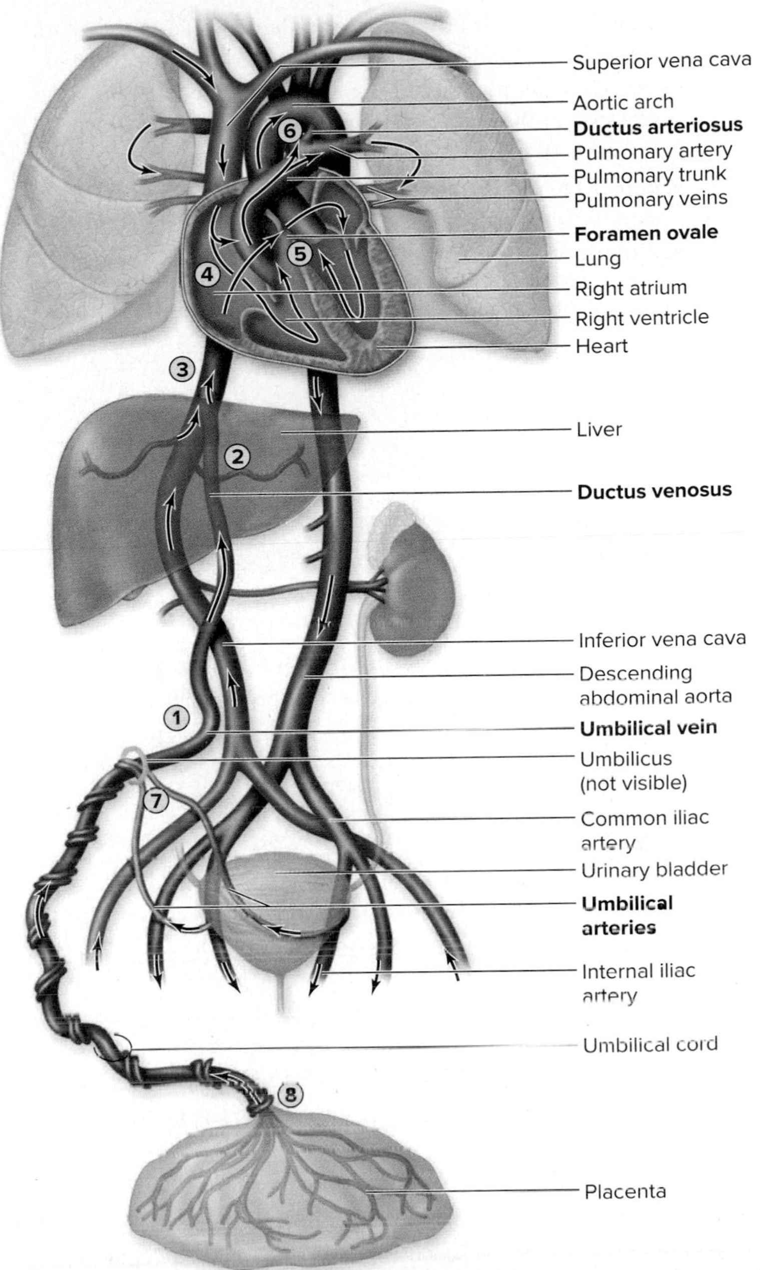

Fetal cardiovascular structure	Postnatal structure
Ductus arteriosus	Ligamentum arteriosum
Ductus venosus	Ligamentum venosum
Foramen ovale	Fossa ovalis
Umbilical arteries	Medial umbilical ligaments
Umbilical vein	Round ligament of liver (ligamentum teres)

Figure 20.29 Fetal Circulation. Structural changes in both the heart and blood vessels accommodate the different needs of the fetus and the newborn. The pathway of blood flow is indicated by black arrows. The chart at the bottom of the drawing summarizes the fate of each fetal cardiovascular structure following birth.

The postnatal changes occur as follows:

- The umbilical vein and umbilical arteries constrict and become nonfunctional. They turn into the **round ligament of the liver** (or *ligamentum teres*) (see figure 26.19*b*) and the **medial umbilical ligaments** (see figure 20.16), respectively.
- The ductus venosus ceases to be functional and constricts, becoming the **ligamentum venosum** (see figure 26.19*b*).

- Because pressure is now greater on the left side of the heart, the two flaps of the interatrial septum close off the foramen ovale. The only remnant of the foramen ovale is a thin, oval depression in the wall of the septum called the **fossa ovalis** (see figure 19.8).
- Within 10 to 15 hours after birth, the ductus arteriosus constricts and becomes a fibrous structure called the **ligamentum arteriosum** (see figure 19.8).

WHAT DID YOU LEARN?

44 Why is it important for the ductus arteriosus and foramen ovale to both close after birth?

INTEGRATE

CLINICAL VIEW 20.11

Patent Ductus Arteriosus

In some infants, especially those born prematurely, the ductus arteriosus fails to close after birth. A **patent** (open) **ductus arteriosus** serves as a conduit through which blood from the aorta can enter the pulmonary system; it can result in high blood pressure in the pulmonary circulation and in mixing of oxygenated blood from the aorta with deoxygenated blood from the pulmonary trunk. Because prostaglandins help keep the ductus arteriosus open during fetal life, a patent ductus arteriosus may be treated with prostaglandin-inhibiting medication. Surgery to remove the ductus arteriosus is a second treatment option.

CHAPTER SUMMARY

20.1 Structure and Function of Blood Vessels	• Arteries take blood away from the heart, capillaries are responsible for gas and nutrient exchange, and veins take blood toward the heart.
	20.1a General Structure of Vessels • Arteries and veins have a tunica intima (innermost layer), a tunica media (middle layer), and a tunica externa (outermost layer). • Capillaries have a tunica intima, composed of an endothelial layer and a basement membrane only.
	20.1b Arteries • Arteries progressively branch into elastic arteries, muscular arteries, and arterioles.
	20.1c Capillaries • Capillaries, the smallest blood vessels, connect arterioles with venules. Gas and nutrient exchange occurs in the capillaries. • The three types of capillaries are continuous capillaries, fenestrated capillaries, and sinusoids, which have different degrees of permeability and are arranged in capillary beds.
	20.1d Veins • Venules are small veins that merge into larger veins. One-way valves prevent blood backflow in veins. • Blood pressure is low in the veins, which act as reservoirs and hold about 55% of the body's blood at rest.
	20.1e Pathways of Blood Vessels • Blood vessels may be arranged in simple pathways (artery ⟶ capillary bed ⟶ veins ⟶ heart) or in alternative pathways, which include anastomoses and portal venous systems.
20.2 Total Cross-Sectional Area and Blood Flow Velocity	• The velocity of blood flow is inversely related to the total cross-sectional area of the blood vessels. • Blood flows slowest in the capillaries, increasing efficiency of nutrient and gas exchange.
20.3 Capillary Exchange	• Materials may pass through the capillary walls via diffusion, vesicular transport, and bulk flow.
	20.3a Diffusion and Vesicular Transport • Small solutes (e.g., oxygen, carbon dioxide, glucose, ions) move between blood and interstitial fluid via diffusion. • Certain solutes that are relatively large (e.g., insulin) are transported via vesicular transport.
	20.3b Bulk Flow • Bulk flow is the net filtration and reabsorption that occur at the capillary level due to hydrostatic pressure and osmotic pressure.
	20.3c Net Filtration Pressure • The net filtration pressure is the difference between the net hydrostatic pressure and the net colloid osmotic pressure. • The net filtration pressure is positive at the arterial end of the capillary, resulting in filtration, and is negative at the venous end of the capillary, resulting in reabsorption.
	20.3d Role of the Lymphatic System • Lymph vessels reabsorb interstitial fluid that is not picked up at the venous end of the blood capillaries. Lymph vessels eventually return this fluid back to the venous circulation.
20.4 Local Blood Flow	• Perfusion is the specific amount of blood per time per gram of tissue. • Local blood flow is dependent upon the degree of vascularization of the tissue, the myogenic response, local regulatory factors, and total blood flow.
	20.4a Degree of Vascularization and Angiogenesis • The degree of vascularization is tissue dependent, but the amount of vascularization may increase over a time through angiogenesis (the formation of new blood vessels) or decrease through regression.

	20.4b Myogenic Response • The myogenic response involves both contraction of smooth muscle in blood vessel walls in response to an increase in stretch and relaxation of smooth muscle in blood vessel walls in response to a decrease in stretch to maintain a constant blood flow.
	20.4c Local, Short-Term Regulation • Local, short-term regulation of blood flow is altered by autoregulation (which is the process of regulating local blood flow into a tissue by the tissue itself in response to vasoactive molecules or ions produced due to its changing metabolic needs) or in response to molecules released from either damaged tissue or as part of the body's defense system.
	20.4d Relationship of Local and Total Blood Flow • Total blood flow is relatively constant at rest, is the same as cardiac output (about 5.25 L/min), and may increase substantially during exercise to make additional blood available to tissues.
20.5 Blood Pressure, Resistance, and Total Blood Flow	• Blood flow is directly related to blood pressure and inversely related to resistance.
	20.5a Blood Pressure • Blood pressure is the force per unit area that blood places on the inside wall of a blood vessel, and it is measured in millimeters of mercury (mm Hg). • Systolic blood pressure is blood pressure in the vessel during ventricular systole (ventricular contraction); diastolic blood pressure is the pressure during ventricular diastole (ventricular relaxation). Pulse pressure is the difference between systolic pressure and diastolic pressure. • Mean arterial pressure (MAP) is the average pressure that pushes the blood from the heart to the tissues, and it is measured as the diastolic pressure plus one-third of the pulse pressure. • Blood pressure in the capillaries is approximately 40 mm Hg on the arterial end and about 20 mm Hg on the venule end. • Venous blood return is facilitated by valves, the skeletal muscle pump, and the respiratory pump.
	20.5b Resistance • Resistance is defined as the amount of friction the blood experiences as it is transported through the blood vessels. • Increased blood viscosity and increased vessel length both cause an increase in peripheral resistance. • Vasoconstriction causes an increase in peripheral resistance; vasodilation causes a decrease in peripheral resistance.
	20.5c Relationship of Blood Flow to Blood Pressure Gradients and Resistance • Blood flow (F) is proportional to the blood pressure gradient (ΔP) divided by the resistance (R) in the blood vessels: $F \propto \Delta\ P/R$. • A greater blood flow occurs with a steeper pressure gradient or lower peripheral resistance; a lesser blood flow occurs with a smaller pressure gradient or greater resistance.
20.6 Regulation of Blood Pressure and Blood Flow	• Blood pressure is dependent upon cardiac output, peripheral resistance, and blood volume, which are regulated by the short-term mechanisms of the nervous system, long-term mechanisms of the endocrine system, or both.
	20.6a Neural Regulation of Blood Pressure • Neural regulation of blood pressure involves baroreceptors located in the aorta and carotid arteries that detect changes in stretch of blood vessel walls. In response, nerve signals relayed along sensory neurons to the cardiovascular center are altered. Subsequently, motor output to both the heart and blood vessels is changed. • Chemoreceptor reflexes regulate blood pressure in response to changing blood chemistry levels.
	20.6b Hormonal Regulation of Blood Pressure • The renin-angiotensin system involves the activation of angiotensinogen to angiotensin II by the enzymes renin and angiotensin-activating enzyme (ACE). Angiotensin II increases resistance and blood volume both directly and through the release of aldosterone and antidiuretic hormone (ADH). • Both antidiuretic hormone and aldosterone decrease urine output to maintain blood volume and blood pressure. • Atrial natriuretic peptide (ANP) increases urine output, which decreases blood volume and causes vasodilation to lower blood pressure.
20.7 Blood Flow Distribution During Exercise	• Both total blood flow and distribution of blood change during exercise. • More blood is pumped to the heart wall, skeletal muscle, and skin during exercise, and relatively less blood is sent to the abdominal organs.
20.8 Pulmonary Circulation	**20.8a Blood Flow Through the Pulmonary Circulation** • Blood is pumped from the right ventricle to the lungs and back to the left side of the heart.
	20.8b Characteristics of the Pulmonary Circulation • The pulmonary circulation is a smaller circuit and has lower blood pressure than the systemic circulation.
20.9 Systemic Circulation: Vessels from and to the Heart	**20.9a General Arterial Flow Out of the Heart** • Oxygenated blood is pumped from the left ventricle through the ascending aorta; the coronary arteries branch from the ascending aorta. • Three branches of the aortic arch are the brachiocephalic trunk, the left common carotid, and the left subclavian. • The descending aorta (divided into the thoracic and abdominal aorta) bifurcates into the left and right common iliac arteries.
	20.9b General Venous Return to the Heart • Veins from the head, neck, upper limbs, and thoracic and abdominal walls merge to form left and right brachiocephalic veins, which merge to form the superior vena cava.

(continued on next page)

CHAPTER SUMMARY (continued)

20.10 Systemic Circulation: Head and Trunk	**20.10a Head and Neck** • The common carotid artery supplies most of the blood to the head and neck and divides into the external carotid artery and internal carotid artery. • The external carotid artery supplies superficial regions of the head and organs of the neck, whereas the internal carotid artery supplies the brain and orbits. • Blood from the cranium drains via the vertebral veins and the dural venous sinuses, which primarily drain into the internal jugular veins. Venous return is via the internal and external jugular veins.
	20.10b Thoracic and Abdominal Walls • Two internal thoracic arteries, as well as other arteries, supply the thoracic and abdominal walls. • Venous return from the thoracic and abdominal walls is complex and involves the azygos system of veins.
	20.10c Thoracic Organs • The lungs are supplied by bronchial arteries branching from the descending thoracic aorta; they are drained by bronchial veins. • The esophagus receives blood from the esophageal arteries that branch from the aorta and from the esophageal branches of the left gastric artery; it is drained via the esophageal veins. • The diaphragm is supplied by phrenic arteries and drained by the phrenic veins.
	20.10d Gastrointestinal Tract • The three major unpaired arteries from the descending abdominal aorta that supply the abdominal organs are the celiac trunk, the superior mesenteric artery, and the inferior mesenteric artery. • The hepatic portal vein receives oxygen-poor but nutrient-rich blood from the gastrointestinal (GI) organs and spleen; this vessel takes this blood to the liver for processing. Blood leaves the liver via the hepatic veins.
	20.10e Posterior Abdominal Organs, Pelvis, and Perineum • Paired branches of the descending abdominal aorta supply the posterior abdominal organs, pelvis, and perineum. Venous drainage is by veins of the same name as the arteries.
20.11 Systemic Circulation: Upper and Lower Limbs	• The upper and lower limbs are each supplied by a single artery that branches into multiple vessels. Both deep and superficial veins drain the limbs.
	20.11a Upper Limb • The upper limb is supplied by the subclavian artery and its branches and is drained into the subclavian vein.
	20.11b Lower Limb • The lower limb is supplied by the external iliac artery and is drained by the external iliac vein.
20.12 Comparison of Fetal and Postnatal Circulation	**20.12a Fetal Circulation** • In the fetus, oxygenated blood is supplied by the umbilical vein from the placenta, and deoxygenated blood leaves through a pair of umbilical arteries. • The fetal cardiovascular system bypasses the liver via the ductus venosus and bypasses the pulmonary circulation via the foramen ovale and ductus arteriosus.
	20.12b Postnatal Changes • In the newborn, the umbilical vein and arteries degenerate, and the ductus venosus, foramen ovale, and ductus arteriosus close.

CHALLENGE YOURSELF

Create and Evaluate

Analyze and Apply

Understand and Remember

Do You Know the Basics?

1. Which of the following is *not* a characteristic of a capillary?
   a. Fenestrated capillaries allow for larger amounts of materials to be exchanged.
   b. Sinusoid capillaries are the main type of capillary around the brain.
   c. Capillaries often are arranged in a capillary bed that is supplied by an arteriole.
   d. The capillary wall consists of an endothelium and a basement membrane only—there is no subendothelial layer.

2. Which statement is accurate about veins?
   a. Veins always transports deoxygenated blood.
   b. Veins drain into smaller vessels called venules.
   c. The largest tunic in a vein is the tunica externa.
   d. The lumen of a vein tends to be smaller than that of a comparably sized artery.

3. Vasa vasorum are found in the tunica __________ of a large blood vessel.
   a. intima
   b. media
   c. externa
   d. All of these are correct.

4. Which of the following decreases perfusion of a tissue?
   a. decreased total blood flow
   b. vasodilator
   c. angiogenesis
   d. increased carbon dioxide, decreased pH (increased H^+), and low O_2

5. A(n) __________ is a type of vessel with the smallest pressure gradient that must be overcome by contraction of skeletal muscles and breathing.
   a. artery
   b. vein
   c. capillary
   d. sinusoid

6. An increase in __________ will result in an increase in blood flow to an area.
   a. vessel length
   b. vessel diameter
   c. blood viscosity
   d. the number of formed elements

7. Which of the following is accurate?
   a. Total blood flow increases with a steeper pressure gradient (assuming resistance stays the same).
   b. Total blood flow decreases with increased resistance (assuming cardiac output stays the same).
   c. Total blood flow is significant in maintaining adequate perfusion of all tissues.
   d. All of these are correct.

8. Velocity of blood flow is the slowest in
   a. muscular arteries.
   b. capillaries.
   c. veins.
   d. elastic arteries.

9. Blood pressure is regulated by the
   a. cardiac center.
   b. vasomotor center.
   c. hormones.
   d. All of these are correct.

10. Choose the correct pathway that blood flows through in the upper limb arteries.
    a. subclavian ⟶ axillary ⟶ ulnar ⟶ radial ⟶ brachial
    b. subclavian ⟶ axillary ⟶ brachial ⟶ cephalic ⟶ basilic
    c. subclavian ⟶ ulnar ⟶ brachial ⟶ radial
    d. subclavian ⟶ axillary ⟶ brachial ⟶ radial and ulnar

11. List and describe the three tunics found in most blood vessels.

12. Compare and contrast arteries and veins with respect to function, tunic size, lumen size, and blood pressure.

13. Explain the difference between hydrostatic and osmotic pressures. How do these pressures change from the arteriole end of a capillary to the venule end of a capillary?

14. Write the formula for determining net filtration pressure in the capillaries.

15. Describe the relationship of vessel radius, vessel length, blood viscosity, and blood pressure to blood flow.

16. Explain how it is possible for the sympathetic division to cause vasoconstriction in most blood vessels but vasodilation in coronary and skeletal muscle blood vessels.

17. Briefly explain how changes in cardiac output, resistance, and blood volume influence blood pressure.

18. Compare how the cardiac center and vasomotor center regulate blood pressure and blood flow.

19. Compare the systemic circulation and pulmonary circulation. Discuss the function of arteries and veins in each system.

20. What postnatal changes occur in the heart and blood vessels? Why do these occur?

Can You Apply What You've Learned?

1. If a patient has cirrhosis of the liver and is unable to produce sufficient albumin and other plasma proteins, then what variable is changed in capillary exchange, and what is the effect?
   a. There is no change in the capillary exchange process.
   b. Hydrostatic pressure in the blood decreases, and fluid remains in the blood.
   c. Colloid osmotic pressure in the capillary increases, and blood volume increases in the blood vessels.
   d. Colloid osmotic pressure in the capillary decreases, and fluid remains in the interstitial fluid, potentially causing edema.

2. Arlene heads to the gym and initiates a vigorous exercise regimen. Which hormone will *not* be released?
   a. epinephrine
   b. atrial natriuretic peptide
   c. angiotensin II
   d. norepinephrine

3. After her workout, Arlene lies down on a mat to stretch out her leg muscles. She stands up suddenly after stretching and feels a bit light-headed. However, the light-headed feeling passes quickly. What physiologic process occurred that resulted in the passing of her light-headed feeling?
   a. The aortic bodies detected an increase in blood pressure in the head and initiated a chemoreceptor reflex that decreased blood pressure.
   b. The carotid bodies detected low CO_2 and high O_2 levels from exercising and stimulated the vasomotor center to vasoconstrict the vessels of the head and neck.
   c. Only α receptors in the head and neck initially were being stimulated, resulting in the light-headed feeling.
   d. The baroreceptors within the carotid detected decreased stretch in the carotid vessel wall and initiated an autonomic reflex to increase blood flow to the head and neck.

4. Over a period of 6 months, Harold loses 40 pounds. When Harold next visits his doctor, he discovers that his blood

pressure levels have fallen too. Which of the following best explains why Harold's blood pressure has dropped?

a. When fat is lost, the extra length in blood vessels supplying the adipose connective tissue is lost too. Decreased blood vessel length has resulted in decreased blood pressure.
b. Most of the fat that Harold lost was around his thoracic organs, which compressed these organs. By losing the weight, he removed this compressive force.
c. The resistance in the blood vessels increased due to the fat loss, resulting in decreased blood pressure.
d. The fat previously constricted the blood vessels. When Harold lost the fat, the blood vessels were allowed to dilate, and thus blood pressure was reduced.

5. Alejandro was hit by a car, which resulted in massive injury and bleeding to his right foot. You realize that you have to compress a lower limb artery so that Alejandro will not lose too much blood. Which vessel, if compressed, will completely stop the flow of blood to the foot?

a. anterior tibial artery
b. posterior tibial artery
c. deep femoral artery
d. popliteal artery

Can You Synthesize What You've Learned?

1. Your patient is Thomas, a 50-year-old man who rarely exercises, is overweight, and only occasionally eats healthy meals, which put him at risk for atherosclerosis. Explain to him the relationship between atherosclerosis and high blood pressure.
2. Arteries tend to have a lot of vascular anastomoses around body joints, such as the elbow and knee. Can you think of a reason this would be beneficial?
3. Explain why an overweight individual with high blood pressure is advised to lose weight. Include in your response the relationship of blood vessel length, resistance, and blood pressure.

INTEGRATE

ONLINE STUDY TOOLS

connect | SMARTBOOK® | AP|R

The following study aids may be accessed through Connect.

Clinical Case Study: A Man with Shortness of Breath Following an Airline Flight

Interactive Questions: This chapter's content is served up in a number of multimedia question formats for student study

SmartBook: Topics and terminology include structure and function of blood vessels; velocity of blood flow; capillary exchange; local blood flow; blood pressure, resistance, and total blood flow; regulation of blood pressure and blood flow; blood flow distribution during exercise; pulmonary circulation; systemic circulation: vessels from and to the heart; systemic circulation: head and trunk; systemic circulation: upper and lower limbs; comparison of fetal and postnatal circulation

Anatomy & Physiology Revealed: Topics include muscular artery and medium-sized vein; aneurysm; relationship between cross section and velocity of flow; capillary exchange; baroreceptor reflex; pulmonary and systemic circulation; cardiovascular system overview; brain arteries and veins; celiac trunk; inferior vena cava; abdominal aorta; hepatic portal system; shoulder and arm vasculature; hip and thigh vasculature

Animations: Topics include fluid exchange across walls of capillaries; arteriolar radius and blood flow; arteriolar resistance and blood pressure; baroreceptor reflex control of blood pressure; chemoreceptor reflex control of blood pressure

Design Elements: **Learning Objective** icon: ©McGraw-Hill Education; **What Do You Think?** icon: ©McGraw-Hill Education; **What Did You Learn?** icon: ©McGraw-Hill Education; **Learning Strategy** icon: ©Slavoljub Pantelic/Shutterstock RF; **Clinical View** icon: ©Laia Design Studio/Shutterstock RF; **Concept Connection** icon: ©McGraw-Hill Education; **Challenge Yourself** icon: ©McGraw-Hill Education; **Do You Know the Basics?** icon: ©McGraw-Hill Education; **Can You Apply What You've Learned?** icon: ©McGraw-Hill Education; **Can You Synthesize What You've Learned?** icon: ©McGraw-Hill Education; **Online Study Tools** icon: ©ayzek/Shutterstock RF

Lymphatic System

Module 10: Lymphatic System

INTEGRATE

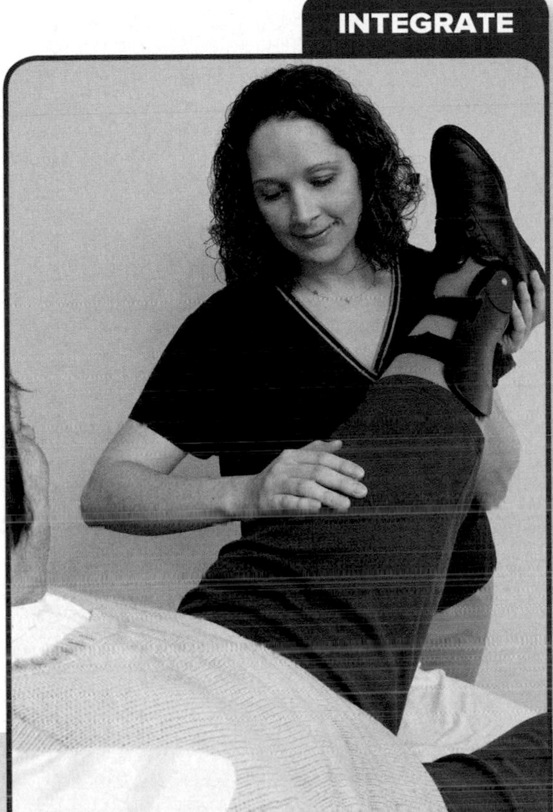

CAREER PATH

Lymphedema Therapist

A lymphedema therapist uses massage, compression wraps/garments, and exercise to help alleviate edema, a common condition if lymph vessels and lymph nodes are removed as a part of the treatment for cancer. Massage assists the movement of lymph in the lymph vessels and its return to the venous circulation. The certification to become a lymphedema therapist is typically earned by nurses, physical therapists, occupational therapists, and massage therapists.

It is not uncommon for young children to develop *tonsillitis*—a condition in which the tonsils become inflamed and enlarged. Often, the lymph nodes in the neck also are swollen, and the spleen may become enlarged. You might initially be curious about why the lymph nodes and the spleen are also affected when a child has tonsillitis. All of these structures—the tonsils, lymph nodes, and spleen—are components of the lymphatic system, and these structures function in helping the immune system defend the body against infectious agents. The inflammation and enlargement that occur in lymphatic structures are signs that these organs are actively engaged in defending the body against potentially harmful substances.

The lymphatic system has two significant functions, and both provide support to other body systems: (1) The lymphatic system transports and houses lymphocytes and other immune cells that help the immune system defend against potentially harmful substances, and (2) the lymphatic system aids the cardiovascular system by returning excess fluid to venous blood to maintain fluid balance, blood volume, and blood pressure. Interestingly, the lymphatic system has no unique function of its own.

Here, we describe lymph vessels and the absorption of excess fluid from the interstitial space and then discuss lymphatic organs, including lymph nodes and the spleen. Integrated within the chapter are brief descriptions of some of the common malfunctions of the lymphatic system. The details of the immune system are described in chapter 22.

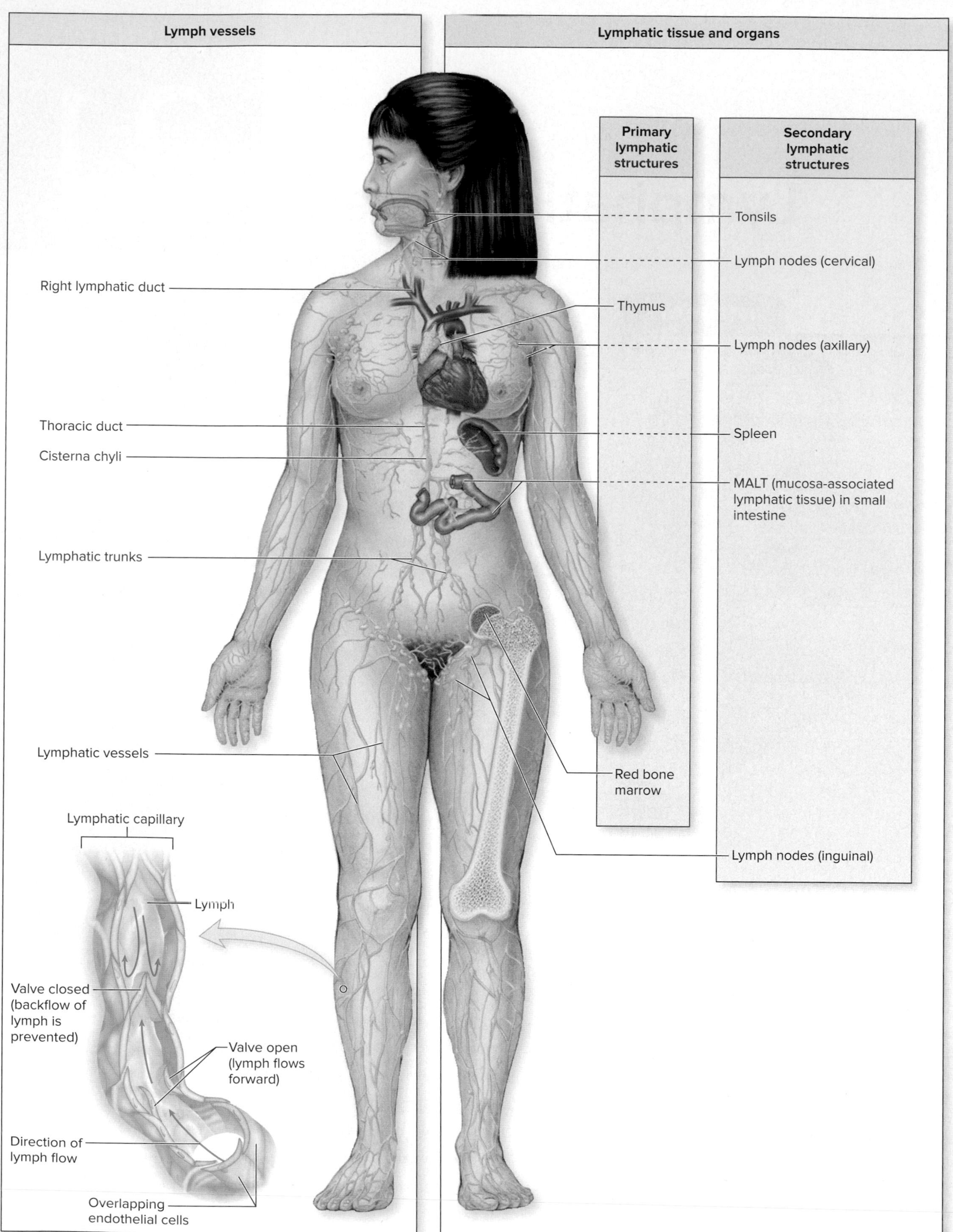

Figure 21.1 Lymphatic System. The lymphatic system is composed of both lymph vessels and lymphatic tissue and organs. Lymphatic tissue and organs are organized into primary and secondary lymphatic structures.

21.1 Lymph and Lymph Vessels

The **lymphatic** (lim-fat′ik) **system** is composed of (1) lymph vessels and (2) lymphatic tissues and organs (**figure 21.1**). Lymph is the fluid transported within lymph vessels. We begin by describing the formation and characteristics of lymph. We then discuss its transport through progressively larger lymph vessels until the lymph is returned to the cardiovascular system. Note that the term *lymph vessel* is used to indicate any type of vessel of the lymphatic system (e.g., lymphatic capillary, lymphatic vessel, lymphatic trunk), whereas the term *lymphatic vessels* is used when referring specifically to the vessels between the lymphatic capillaries and lymphatic trunks.

INTEGRATE

CLINICAL VIEW 21.1

Metastasis

Although the lymph vessels provide an essential function by rerouting excess interstitial fluid back into the blood, these vessels sometimes can participate in the spread of pathogens or disease. For example, cancerous cells can break free from a primary tumor and be transported in the lymph. These cancerous cells may establish secondary tumors that develop in other locations within the body, a process referred to as **metastasis** (mĕ-tas′tă-sis). For example, breast cancer may metastasize to the lungs, and this cancer in the lungs is referred to as metastasized breast cancer, not lung cancer. A biopsy of a lymph node that is positive for the presence of cancer cells from another body organ verifies that the cancer has metastasized.

21.1a Lymph and Lymphatic Capillaries

LEARNING OBJECTIVES

1. Describe lymph and its contents.
2. Discuss the location and anatomic structure of lymphatic capillaries.
3. Explain how fluid enters lymphatic capillaries.

Lymph originates as interstitial fluid surrounding tissue cells; it moves passively into the lymphatic capillaries due to a hydrostatic pressure gradient. Lymphatic capillaries merge to form larger lymph vessels.

Characteristics of Lymph

Approximately 15% of the fluid that enters the interstitial space surrounding the cells is not reabsorbed back into the blood capillaries during capillary exchange (see section 20.3d). This interstitial fluid amounts to about 3 liters daily and is normally absorbed into lymphatic capillaries.

Once inside the lymph vessels, the interstitial fluid is called **lymph** (limf; *lympha* = clear spring water). The components of lymph include water, dissolved solutes (e.g., ions), a small amount of protein (approximately 100 to 200 grams that leaked into the interstitial space during capillary exchange), sometimes foreign material that includes both cell debris and pathogens, and perhaps metastasized cancer cells (see Clinical View 21.1: "Metastasis").

Lymphatic Capillaries

The lymph vessel network begins with **lymphatic capillaries,** which are the smallest lymph vessels (**figure 21.2**). Lymphatic capillaries are microscopic, closed-ended vessels that absorb interstitial fluid. They are interspersed throughout areolar connective tissue among capillary networks, except those within the red bone marrow and avascular tissues (such as epithelia). In addition, recent research has found lymph vessels associated with the dural venous sinuses that drain blood away from the brain.

A lymphatic capillary resembles the anatomic structure of a blood capillary in that its wall is composed of an endothelium (figure 21.2*b*). However, lymphatic capillaries are typically larger in diameter than blood capillaries, lack a basement membrane, and have overlapping endothelial cells. These overlapping endothelial cells act as one-way flaps to allow fluid to enter the lymphatic capillary but

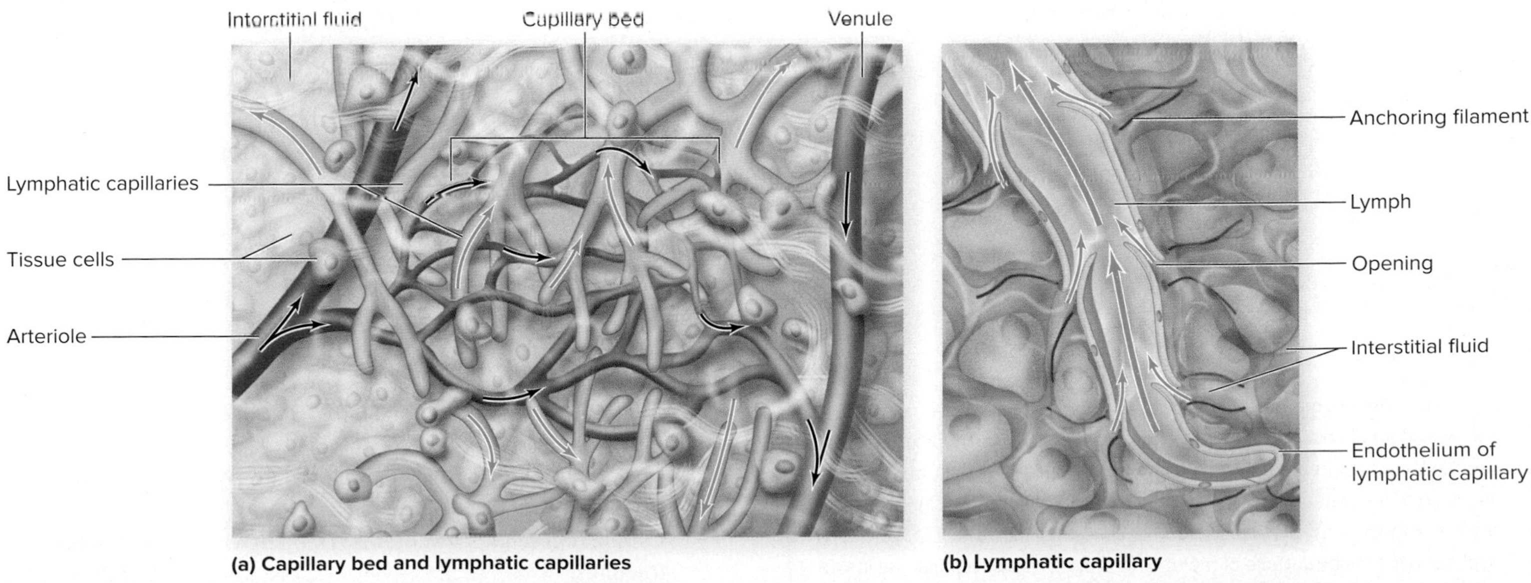

Figure 21.2 Lymphatic Capillaries. (*a*) Lymphatic capillaries begin as closed-end vessels within connective tissue among most blood capillary networks and absorb excess interstitial fluid left during capillary exchange. (*b*) A lymphatic capillary takes up excess interstitial fluid through overlapping endothelial cells. The fluid is then called lymph. Here, the black arrows show blood flow and the green arrows show lymph flow. AP|R

INTEGRATE

CONCEPT CONNECTION

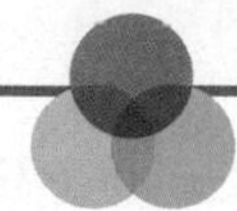

Lacteals are lymphatic capillaries within the small intestine that absorb dietary lipids and lipid-soluble vitamins that are unable to enter the blood directly from the gastrointestinal (GI) tract (see section 26.4c). The lipids are absorbed as part of chylomicrons (lipid droplets enveloped within protein).

prevent its loss. **Anchoring filaments** help hold these endothelial cells to the nearby structures. Lymphatic capillaries located within the gastrointestinal (GI) tract, called **lacteals** (lak′tē-ăl; *lactis* = milk), allow for the absorption of lipid-soluble substances from the GI tract, a concept detailed in section 26.4c.

Movement of Lymph into Lymphatic Capillaries

The driving force to move fluid into the lymphatic capillaries is an increase in hydrostatic pressure (see section 4.3b) within the interstitial space. Interstitial hydrostatic pressure rises as additional fluid is filtered from the blood capillaries (see section 20.3b). This pressure exerted by interstitial fluid at the margins of the lymphatic capillary endothelial cells "pushes" interstitial fluid into the lymphatic capillary lumen when the interstitial fluid hydrostatic pressure becomes greater than the lymph hydrostatic pressure. The higher the interstitial fluid pressure, the greater the amount of fluid that enters the lymphatic capillary. The **anchoring filaments** extending between lymphatic capillary cells and the surrounding tissue prevent the collapse of the lymphatic capillaries as pressure exerted by the interstitial fluid increases.

The pressure exerted by lymph after it enters the lymphatic capillary forces the endothelial cells of these vessels to close. Thus, lymph becomes "trapped" within the lymphatic capillary and cannot be released back into the interstitial space. Lymph is then transported through a network of increasingly larger vessels that include (in order) lymphatic capillaries, lymphatic vessels, lymphatic trunks, and lymphatic ducts.

WHAT DID YOU LEARN?

1. What substances typically are absorbed from the interstitial space into lymphatic capillaries?

2. How does fluid enter and become "trapped" in the lymphatic capillaries?

INTEGRATE

LEARNING STRATEGY

Fluid entry into a lymphatic capillary is analogous to the movement of the entryway door to your house or apartment. Imagine that the door is unlocked and the knob is turned. Putting pressure on the outside of the door (like the pressure of interstitial fluid on the outside of the lymphatic capillary wall) causes it to open to the inside so you can enter. Once inside, pressure applied to the inside surface of the door (or fluid pressure against the inside of the lymphatic capillary surface) causes it to close.

21.1b Lymphatic Vessels, Trunks, and Ducts

LEARNING OBJECTIVES

4. Explain the mechanisms that move lymph through lymphatic vessels, trunks, and ducts.
5. Name the five types of lymphatic trunks and the regions of the body from which they drain lymph.
6. Describe the regions that are drained by the right lymphatic duct and by the thoracic duct.

After lymph enters the lymphatic capillaries, it continues to flow into increasingly larger lymphatic vessels, trunks, and ducts. Ultimately, the lymph empties into the venous circulation.

Lymphatic Vessels

Lymphatic capillaries merge to form larger structures that are called **lymphatic vessels** (figure 21.1). Superficial lymphatic vessels are generally positioned adjacent to the superficial veins of the body; in contrast, deep lymphatic vessels are next to deep arteries and veins. Lymphatic vessels resemble small veins because both contain all three vessel tunics (intima, media, and externa) and have **valves** within their lumen. Valves are required to prevent lymph from pooling in these vessels and help prevent lymph backflow because the lymphatic vessel network is a low-pressure system. These valves are especially important in areas where lymph flow is against the direction of gravity, such as in the lower limbs.

The lymphatic system lacks a pump and, thus, relies on other mechanisms to move lymph through its vessels. These include (1) contraction of nearby skeletal muscles in the limbs (skeletal muscle pump) and the respiratory pump in the torso, which is similar to how blood movement is assisted through the venous circulation (see section 20.5a), (2) rhythmic contraction of smooth muscle within the walls of larger lymph vessels (trunks and ducts), which narrows the lumen and squeezes the lymph within the lymph vessel, and (3) pulsatile movement of blood in nearby arteries. All of these mechanisms are dependent upon valves within lymph vessels, which prevent the backflow of lymph, causing the lymph to move in one direction to be returned to venous blood circulation.

Some lymphatic vessels connect directly to lymphatic organs called lymph nodes. Foreign or pathogenic material is filtered as lymph passes through lymph nodes. They are arranged in a series along lymph vessels and are described in more detail in section 21.4a.

Lymphatic Trunks

Lymphatic vessels drain into **lymphatic trunks** on both the right and left sides of the body (**figure 21.3**). Each lymphatic trunk removes lymph from a specific major body region:

- *Jugular trunks* drain lymph from both the head and neck.
- *Subclavian trunks* remove lymph from the upper limbs, breasts, and superficial thoracic wall.
- *Bronchomediastinal trunks* drain lymph from deep thoracic structures.
- A single *intestinal trunk* drains lymph from most abdominal structures.
- *Lumbar trunks* drain lymph from the lower limbs, abdominopelvic wall, and pelvic organs.

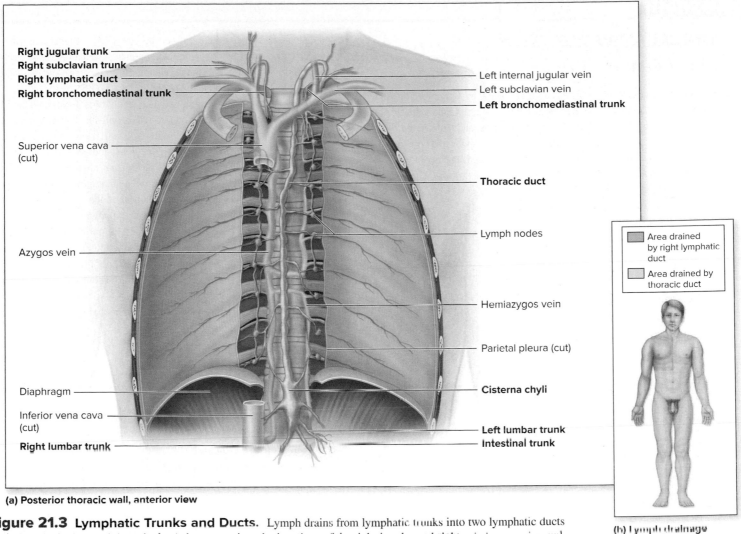

Right jugular trunk
Right subclavian trunk
Right lymphatic duct
Right bronchomediastinal trunk
Superior vena cava (cut)
Azygos vein
Diaphragm
Inferior vena cava (cut)
Right lumbar trunk
Left internal jugular vein
Left subclavian vein
Left bronchomediastinal trunk
Thoracic duct
Lymph nodes
Hemiazygos vein
Parietal pleura (cut)
Cisterna chyli
Left lumbar trunk
Intestinal trunk

(a) Posterior thoracic wall, anterior view

Area drained by right lymphatic duct
Area drained by thoracic duct

(b) Lymph drainage pattern

Figure 21.3 Lymphatic Trunks and Ducts. Lymph drains from lymphatic trunks into two lymphatic ducts (right lymphatic duct and thoracic duct) that empty into the junctions of the right jugular and right subclavian veins and left jugular and left subclavian veins, respectively. (*a*) An anterior view of the posterior thoracic wall illustrates the major lymphatic trunks and ducts and the site of lymph drainage into the venous circulation of the cardiovascular system. (*b*) Areas of lymph drainage into the right lymphatic duct and the thoracic duct. AP|R

WHAT DO YOU THINK?

Predict what may occur with respect to lymphatic drainage if a group of lymph nodes and their associated lymph vessels are surgically removed, as might occur when breast cancer has metastasized.

Lymphatic Ducts

Lymphatic trunks drain into the largest lymph vessels called **lymphatic ducts.** There are two lymphatic ducts: the right lymphatic duct and the thoracic duct. Both of these convey lymph back into the venous blood circulation.

Right Lymphatic Duct The **right lymphatic duct** is located near the right clavicle. It receives lymph from the lymphatic trunks that drain the following areas: (1) the right side of the head and neck, (2) the right upper limb, and (3) the right side of the thorax. It returns the lymph into the junction of the right subclavian vein and the right internal jugular vein. Thus, the right lymphatic duct drains lymph from the upper right quadrant of the body (figure 26.3*b*).

Thoracic Duct The larger of the two lymphatic ducts is the **thoracic duct.** It has a length of about 37.5 to 45 centimeters (15 to 18 inches) and extends from the diaphragm to the junction of the left subclavian and left jugular veins. The thoracic duct drains lymph from the remaining areas of the body (left side of the head and neck, left upper limb, left thorax, all of the abdomen, and both lower limbs).

At the base of the thoracic duct and anterior to the L_2 vertebra is a rounded, saclike structure called the **cisterna chyli** (sis-ter′nă kī′lī). The cisterna chyli gets its name from the milky, lipid-rich lymph called **chyle** (kīl; *chylos* = juice), which it receives from vessels that drain the small intestine of the gastrointestinal (GI) tract (see section 26.4c). Both the intestinal trunk and the left and right lumbar trunks drain into the cisterna chyli. The thoracic duct extends superiorly from the cisterna chyli and lies directly anterior to the vertebral bodies. It passes through the aortic opening of the diaphragm, then ascends to the left of the vertebral body midline.

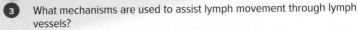

WHAT DID YOU LEARN?

3. What mechanisms are used to assist lymph movement through lymph vessels?
4. Which major body regions drain lymph to the right lymphatic duct?

INTEGRATE

CLINICAL VIEW 21.2

Lymphedema

Lymphedema (limf′e-dē′mă; *oidema* = a swelling) is recognized as an accumulation of interstitial fluid that occurs due to interference with lymph drainage in a part of the body. Interstitial fluid increases, and the affected area both swells and becomes painful. If lymphedema is left untreated, the protein-rich interstitial fluid may interfere with wound healing and can even contribute to an infection by acting as a growth medium for bacteria.

Most cases of lymphedema are *obstructive*, meaning they are caused by blockage of lymph vessels. There are several causes of obstructive lymphedema:

- Trauma or infection of the lymph vessels
- The spread of malignant tumors within the lymph nodes and lymph vessels
- Radiation therapy , which may cause scar formation in lymph vessels or nodes
- Any surgery that requires removal of a group of lymph nodes (e.g., breast cancer surgery when the axillary lymph nodes are removed)

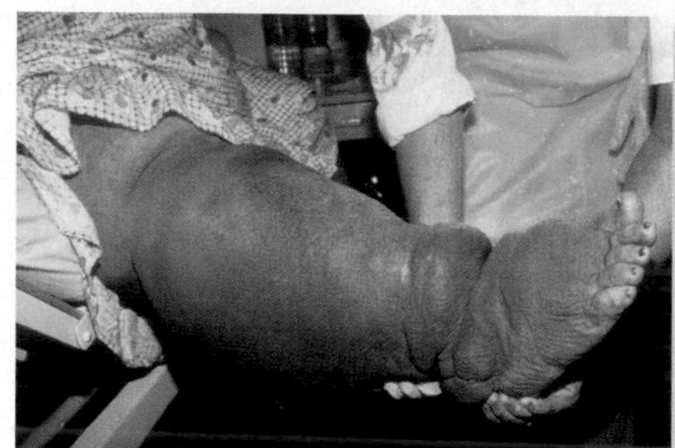

Elephantiasis caused by lymphatic filariasis.
©Andy Crumo, TDR, WHO/Science Source

Although lymphedema has no cure, it can be controlled. Patients may wear compression stockings or other compression garments, which reduce swelling and assist interstitial fluid return to the circulation. Certain exercise and massage regimens may improve lymph drainage as well.

Millions of individuals in Southeast Asia and Africa have developed lymphedema as a result of infection by threadlike, parasitic filarial worms. In **lymphatic filariasis** (fil-ă-rī′ă-sis; *filum* = thread), filarial worms lodge in the lymphatic system, live and reproduce there for years, and eventually obstruct lymph drainage. Mosquitoes are the most common vector for transmitting filariasis, although some filarial worms gain entrance through cracks in the skin of the foot. An affected body part can swell to many times its normal size. In these extreme cases, the condition is known as **elephantiasis** (el′ĕ-fan-tī′ă-sis; *elephas* = elephant). Patients are treated with antiparasitic medications to kill the filarial worms, although the damage to the lymphatic system may be irreversible.

21.2 Overview of Lymphatic Tissue and Organs

LEARNING OBJECTIVE

7. Name the two categories of lymphatic tissue and organs, and identify components of the body that belong to each category.

The lymphatic system also is made up of specialized lymphatic (lymphoid) tissue and organs. These tissues and organs include red bone marrow, the thymus, lymph nodes, the spleen, tonsils, lymphatic nodules, and MALT (mucosa-associated lymphatic tissue) (see figure 21.1; diffuse lymphatic nodules not shown).

Lymphatic system tissues and organs are categorized as either primary or secondary lymphatic structures:

- **Primary lymphatic structures** are involved in the formation and maturation of lymphocytes. Both the red bone marrow and thymus are considered primary lymphatic structures.
- **Secondary lymphatic structures** are not involved in lymphocyte formation; instead, they house both lymphocytes and other immune cells following their formation. Secondary lymphatic structures are the sites where an immune response is initiated (see section 22.6). The major secondary lymphatic structures include the lymph nodes spleen, tonsils, lymphatic nodules, and MALT.

Each component is described in more detail in the next several sections and is summarized in **table 21.1**.

WHAT DID YOU LEARN?

5 How are primary lymphatic structures and secondary lymphatic structures differentiated? What are examples of each?

Table 21.1 Lymphatic Structures

Component	Primary or Secondary Lymphatic Structures	Location	Function
Red bone marrow	Primary	Spongy bone of selected bones	Formation of all formed elements; site of B-lymphocyte maturation
Thymus	Primary	Superior mediastinum (in adults); anterior and superior mediastinum (in children)	Site of T-lymphocyte maturation and differentiation
Lymph nodes	Secondary	Along length of lymphatic vessels; clusters in axillary, inguinal, and cervical regions	Filter lymph; where immune response is initiated against a substance in the lymph
Spleen	Secondary	Left upper quadrant of abdomen, near 9th–11th ribs, "wraps" partially around stomach	Filters blood; where immune response is initiated against a substance in the blood; removes aged erythrocytes and platelets; serves as erythrocyte and platelet reservoir
Tonsils	Secondary	Oral cavity and pharynx (throat)	Protect against inhaled and ingested substances
Lymphatic nodules	Secondary	Every body organ and wall of appendix	Protect body organs
MALT (mucosa-associated lymphatic tissue)	Secondary	Clusters of lymphatic nodules within walls of gastrointestinal (GI), respiratory, urinary, and reproductive tracts	Protects mucosal membranes against foreign substances

21.3 Primary Lymphatic Structures

The general structure and function of the primary lymphatic structures are described in this section. These include the red bone marrow and the thymus.

21.3a Red Bone Marrow

LEARNING OBJECTIVES

8. Describe the location and general function of red bone marrow.
9. Identify the two major types of lymphocytes.

Red bone marrow is located within trabeculae in portions of spongy bone within the skeleton. In adults, these include the flat bones of the skull, the vertebrae, the ribs, the sternum, the ossa coxae, and proximal epiphyses of each humerus and femur (see section 7.2d).

Red bone marrow is responsible for **hemopoiesis** (or *hematopoiesis*), which is the production of formed elements. Hemopoiesis is described in detail in section 18.3a. Recall that the formed elements include erythrocytes, platelets, granulocytes (neutrophils, eosinophils, and basophils), and agranulocytes (monocytes and lymphocytes) (**figure 21.4**). Two of the major types of lymphocytes are **B-lymphocytes** (also called *B-cells*) and **T-lymphocytes** (also called *T-cells*). Most formed elements move directly from the red bone marrow into the blood following hemopoiesis. Unlike the other formed elements, T-lymphocytes must migrate to the thymus to complete their maturation. The functions of both B-lymphocytes and T-lymphocytes are described in detail in chapter 22. The "T" in the name *T-lymphocytes* originates from the requirement of these cells to complete their maturation in the thymus.

WHAT DID YOU LEARN?

6 Why is red bone marrow considered a primary lymphatic structure?

INTEGRATE

LEARNING STRATEGY

Lymphocytes can be identified according to the tissue or organ where they mature: **T**-lymphocytes mature in the **T**hymus. **B**-lymphocytes mature in the **B**one marrow. However, the "B" in the name *B-lymphocytes* did not originate because B-lymphocytes develop in the bone marrow. Rather, these cells were first discovered in chickens and were named for where they develop in chickens—in the *bursa of fabricius* (a lymphatic structure in birds). Humans lack the bursa of fabricius, and B-lymphocytes complete their maturation in the bone marrow.

21.3b Thymus

LEARNING OBJECTIVE

10. Describe the structure and general function of the thymus.

The **thymus** (thī′mŭs) is a bilobed organ that is located in the superior mediastinum and functions in T-lymphocyte maturation (**figure 21.5**). In infants and young children, the thymus is quite large and extends into the anterior mediastinum as well. The thymus continues to grow until puberty, when it reaches a maximum weight of 30 to 50 grams (approximately 1/10 of a pound). Cells within the thymus begin to regress after it reaches this size. Thereafter, much of the thymic tissue is replaced by adipose connective tissue (figure 21.5*c*).

INTEGRATE

CONCEPT CONNECTION

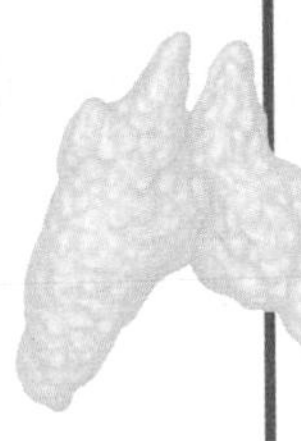

Recall from section 17.2a how some organs, in addition to their primary function, also house clusters of endocrine tissue that produce hormones. The thymus, for example, contains (1) lymphatic tissue and is considered part of the lymphatic system, and (2) epithelial tissue that produces and secretes hormones (e.g., thymosin, thymulin, thymopoietin) and is considered part of the endocrine system (see section 17.11c).

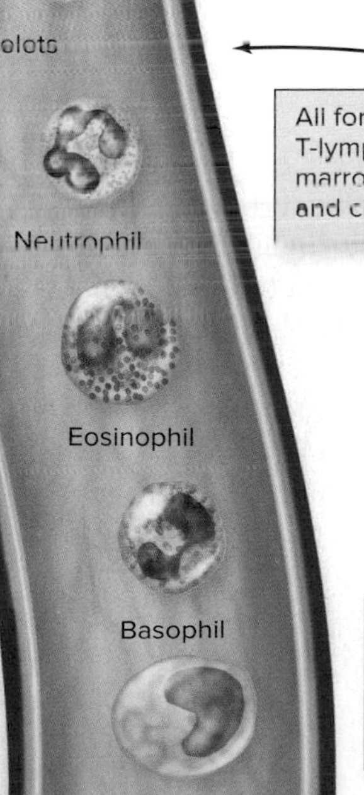

Erythrocyte
Platelets
Neutrophil
Eosinophil
Basophil
Monocyte
B-lymphocyte
T-lymphocyte
All formed elements (except T-lymphocytes) leave the bone marrow and directly enter and circulate in the blood.
Red bone marrow
Thymus
T-lymphocyte maturation

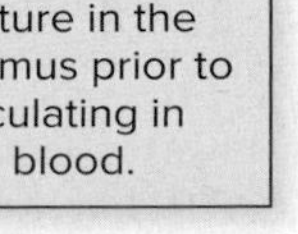

T-lymphocytes mature in the thymus prior to circulating in the blood.

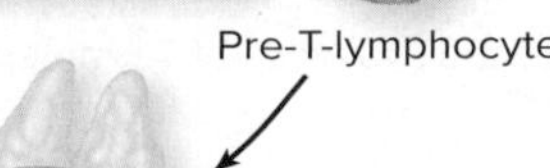

Pre-T-lymphocyte

Figure 21.4 Formed Elements of Hemopoiesis. Red bone marrow produces all formed elements through the process of hemopoiesis. Immature T-lymphocytes (pre-T-lymphocytes) migrate from the bone marrow to the thymus to complete their maturation. (Note that the figure does not accurately display the relative size of the different formed elements.)

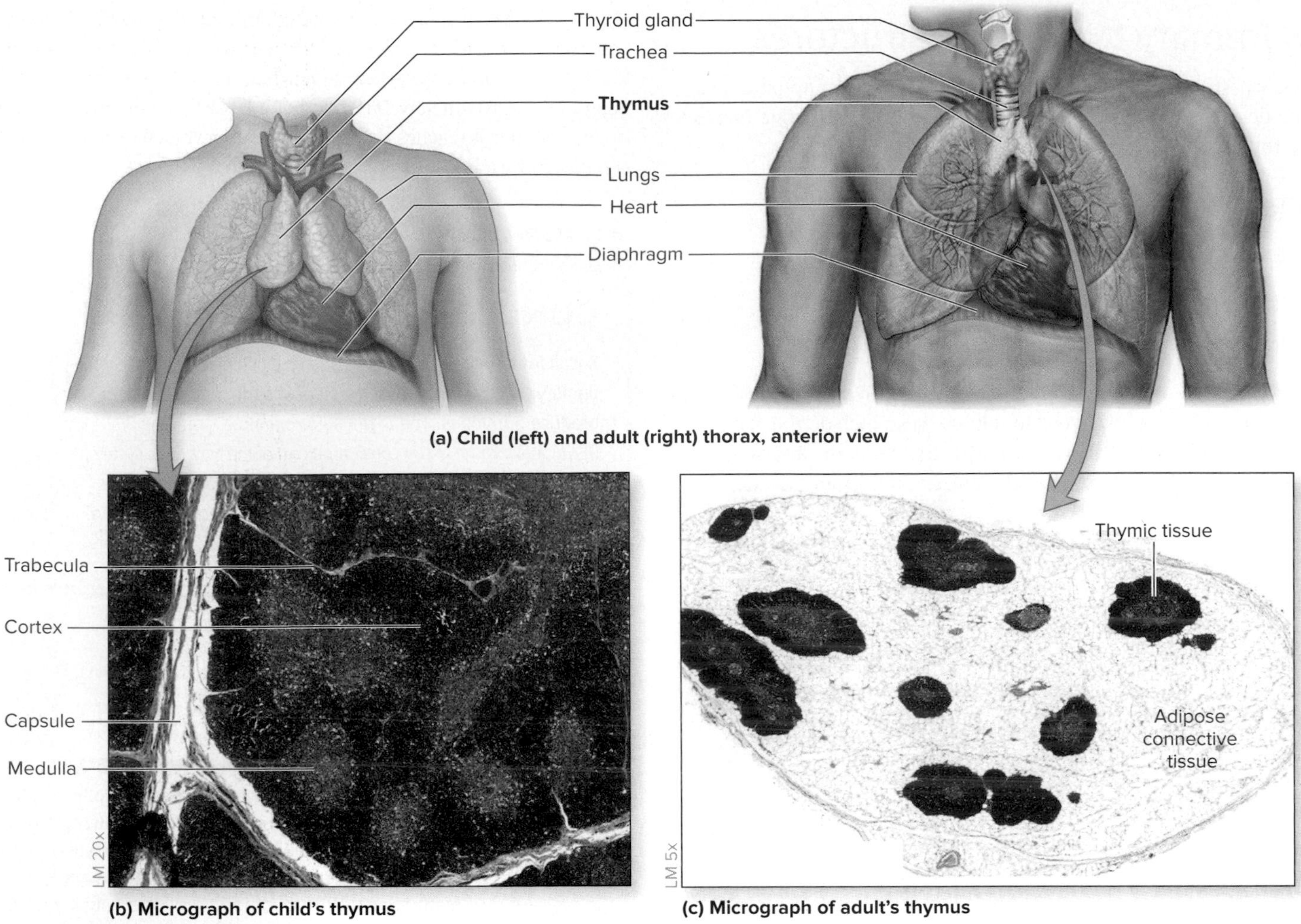

Thyroid gland
Trachea
Thymus
Lungs
Heart
Diaphragm

(a) Child (left) and adult (right) thorax, anterior view

Trabecula
Cortex
Capsule
Medulla
LM 20x

(b) Micrograph of child's thymus

Thymic tissue
Adipose connective tissue
LM 5x

(c) Micrograph of adult's thymus

Figure 21.5 Thymus. (*a*) The thymus is a bilobed lymphatic organ that is most prominent in children. (*b*) A micrograph of a child's thymus reveals the arrangement of the outer cortex and the central medulla within a lobule. (*c*) A micrograph of an adult's thymus showing loss of thymic tissue and gain of adipose connective tissue. **AP|R**

(*b*) ©McGraw-Hill Education/Al Telser: (*c*) ©Dr. Thomas Caceci, Virginia-Maryland Regional College of Veterinary Medicine

The thymus in a child consists of two fused **thymic lobes,** each surrounded by a connective tissue **capsule** (figure 21.5*b*). Fibrous extensions of the capsule, called **trabeculae** (tră-bek′ū-lē), or *septa,* subdivide the thymic lobes into **lobules.** Each lobule is arranged into an outer **cortex** and inner **medulla.** Both parts are composed primarily of epithelial tissue infiltrated with T-lymphocytes in varying stages of maturation. The cortex contains immature T-lymphocytes, and the medulla contains mature T-lymphocytes. The epithelial cells secrete thymic hormones (e.g., thymulin) that participate in the maturation of T-lymphocytes (see section 17.11c). Because the thymus contains both lymphatic cells and epithelial tissue, it is described as a *lymphoepithelial organ.* The details of T-lymphocyte maturation are described in section 22.5.

WHAT DID YOU LEARN?

7 How are the two types of T-lymphocytes arranged in the cortex and medulla?

21.4 Secondary Lymphatic Structures

Structures that house both lymphocytes and other immune cells are called *secondary lymphatic structures.* These structures are composed of lymphatic cells that are enmeshed within a reticular connective tissue matrix.

Secondary lymphatic structures are organized into both **lymphatic organs** and aggregates of *lymphatic nodules.* They are differentiated by the presence or absence of a capsule composed of dense irregular connective tissue that encloses the lymphatic structure. A *complete* capsule is present in lymphatic organs, which include lymph nodes and the spleen. A capsule is either *incomplete* or *absent* in other lymphatic structures, which include tonsils, diffuse lymphatic nodules, and MALT.

21.4a Lymph Nodes

LEARNING OBJECTIVES

11. Describe the structure of lymph nodes.
12. Explain the function of lymph nodes.

Lymph nodes are small, round or oval, encapsulated structures located along the pathways of lymphatic vessels, where they serve as the main lymphatic organ. Lymph nodes function in the filtering of lymph and removal of unwanted substances.

Lymph nodes vary in both their size (from 1 to 25 millimeters) and their number (estimated between 500 and 700 throughout the entire body). Lymph nodes are located both superficially and deep within the body and typically occur in clusters that receive lymph from selected body regions. Some examples of clustered lymph nodes include the *cervical lymph nodes* that receive lymph from the head and neck; the *axillary lymph nodes* in the armpit that receive lymph from the breast, axilla, and upper limb; and *inguinal lymph nodes* in the groin that receive lymph from the lower limb and pelvis (see figure 21.1). In addition to clusters, lymph nodes are individually distributed throughout the body.

Numerous **afferent lymphatic vessels** bring lymph into a lymph node (**figure 21.6**). There is typically only one **efferent lymphatic vessel,** which originates at the involuted portion of the lymph node called the **hilum** (hī′lŭm), or *hilus.* Lymph is drained via the efferent lymphatic vessel from this region of the lymph node.

The capsule of a lymph node is composed of dense irregular connective tissue that encapsulates the node. The capsule also has numerous internal extensions called **trabeculae,** which both subdivide the node into compartments and provide support for blood vessels and nerves that are within the lymph node.

The lymph node regions internal to the capsule are subdivided into an outer **cortex** and an inner **medulla.** The cortex is composed in part of multiple **lymphatic nodules.** Each lymphatic nodule within the lymph node is composed of reticular fibers, which support an inner **germinal center** that houses both proliferating B-lymphocytes and some macrophages. The germinal center is surrounded by an outer region called a **mantle zone,** which contains T-lymphocytes, macrophages, and dendritic cells (described in section 22.2a). The medulla differs because it has strands of connective tissue fibers that support B-lymphocytes, T-lymphocytes, and macrophages. These structures are called **medullary cords.**

Both the cortex and medulla of a lymph node also contain tiny, open channels called lymphatic sinuses (**cortical sinuses** and **medullary sinuses,** respectively). These spaces are lined with macrophages.

Lymph Flow Through Lymph Nodes

Lymph enters a lymph node through numerous afferent lymphatic vessels, then flows through the lymph node sinuses, before typically draining through one efferent lymphatic vessel. Numerous afferent lymphatic vessels and generally one efferent lymphatic vessel are significant in the normal function of a lymph node. Consider that in this anatomic arrangement the collective diameter of the "inflow pipes" (i.e., afferent lymphatic vessels) is larger than the diameter of the "outflow pipe" (i.e., efferent lymphatic vessel). This creates a higher fluid pressure in the intervening structures (i.e., the lymph node sinuses), which helps to force the lymph through the lymph node.

Lymph is continuously monitored for the presence of foreign or pathogenic material as it passes through nodes. Macrophages residing in the lymph node remove foreign debris from the lymph by phagocytosis (see section 22.3b). Lymph then exits the lymph node through an efferent lymphatic vessel. Recall that lymph nodes are often found in clusters, so after one lymph node receives and filters lymph, the lymph is then transported to another lymph node in the cluster, then to another, and so on. Thus, lymph is repeatedly screened for unwanted substances.

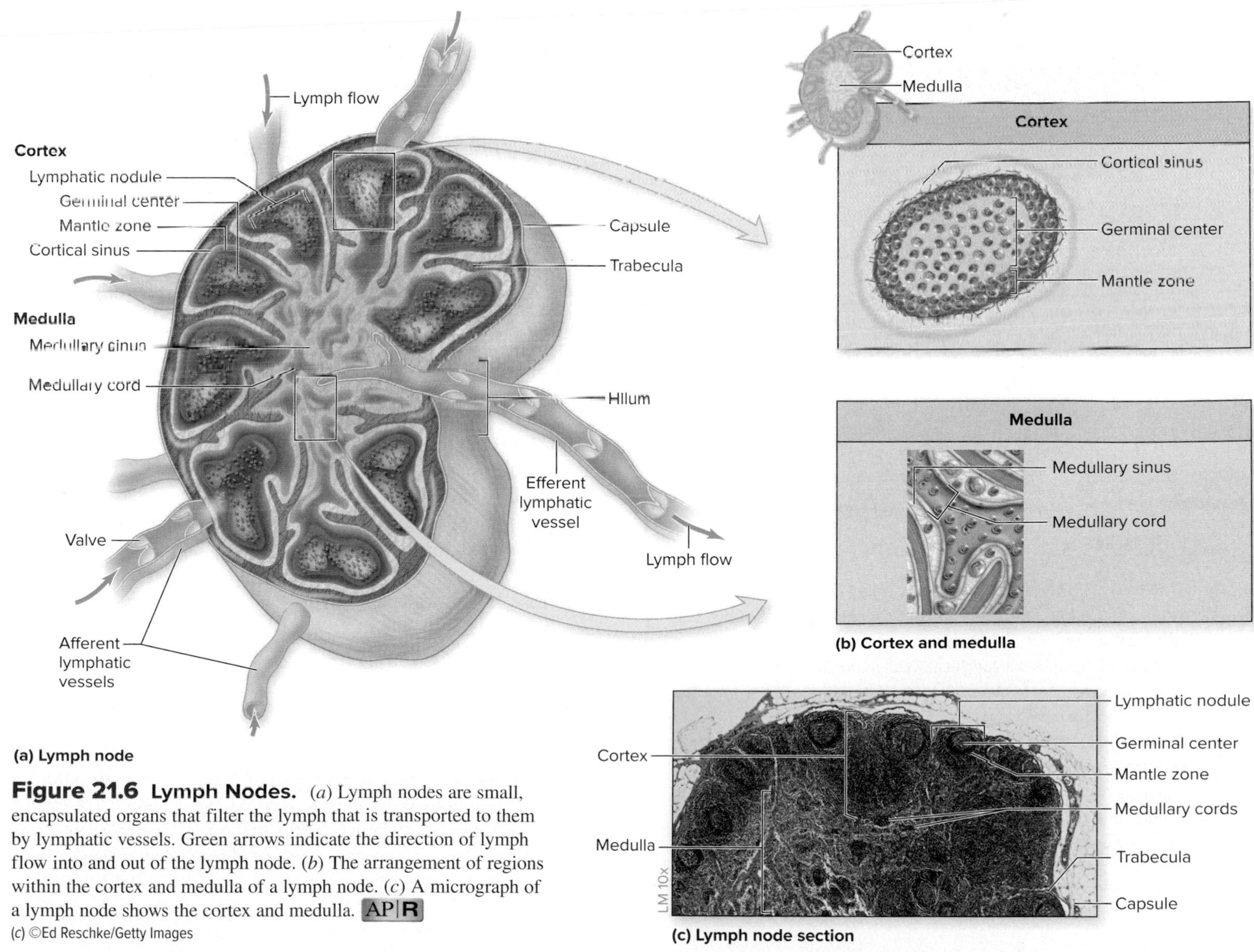

Figure 21.6 Lymph Nodes. (*a*) Lymph nodes are small, encapsulated organs that filter the lymph that is transported to them by lymphatic vessels. Green arrows indicate the direction of lymph flow into and out of the lymph node. (*b*) The arrangement of regions within the cortex and medulla of a lymph node. (*c*) A micrograph of a lymph node shows the cortex and medulla. AP|R

(c) ©Ed Reschke/Getty Images

INTEGRATE

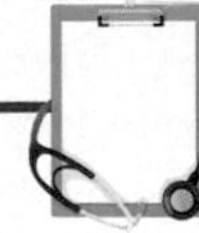

CLINICAL VIEW 21.3

Lymphoma

A **lymphoma** (lim-fō′mă; *oma* = tumor) is a malignant neoplasm (cancer) that develops within lymphatic structures. These tumors develop most often from abnormal B-lymphocytes, and less commonly from abnormal T-lymphocytes. Usually (but not always), a lymphoma presents as a nontender, enlarged lymph node, often in the neck or axillary region. Some patients have no further symptoms, whereas others may experience night sweats, fever, and unexplained weight loss in addition to the nodal enlargement. Lymphomas are grouped into two categories: Hodgkin lymphoma and non-Hodgkin lymphoma.

Hodgkin lymphoma (or *Hodgkin disease*) is characterized by the presence of the **Reed-Sternberg cell,** a large cell whose two nuclei resemble owl eyes, surrounded by lymphocytes within the affected lymph node. Hodgkin lymphoma affects young adults (ages 16–35) and people over 60. It arises in a lymph node and then spreads to other nearby lymph nodes. If caught early, Hodgkin lymphoma can be treated and cured by excision of the tumor, followed by radiation, chemotherapy, or both.

Non-Hodgkin lymphomas are much more common than Hodgkin lymphomas. Some kinds of non-Hodgkin lymphoma are aggressive and often fatal, whereas others are slow-growing and more responsive to treatment. Treatment depends upon the type of non-Hodgkin lymphoma, the extent of its spread at the time of discovery, and the rate of progression of the malignancy.

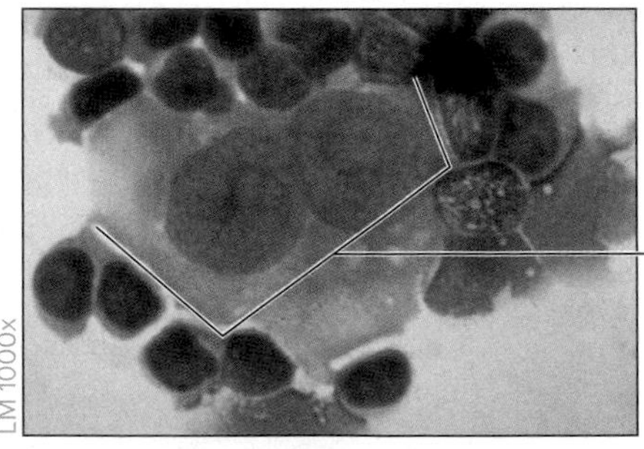

Reed-Sternberg cell, a characteristic of Hodgkin lymphoma.

INTEGRATE

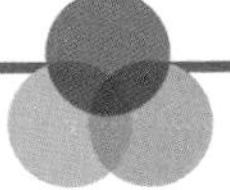

CONCEPT CONNECTION

Dendritic cells are specialized phagocytic cells that are formed in bone marrow and housed in both epithelial and connective tissue of the skin (see sections 6.1a and 6.1b) and mucosal membranes. Those cells within the epidermis of the skin are specifically called epidermal dendritic cells or Langerhans cells (see section 6.1a). Dendritic cells migrate from the skin and mucosal membranes to a lymph node following phagocytosis of foreign substances (see section 22.2a). **NK (natural killer) cells** are a type of lymphocyte formed in the lymphoid line (see section 18.3a). These innate immune cells (see section 22.3b) differentiate and mature within the thymus and secondary lymphatic structures before entering and circulating in the blood.

Lymphocytes housed within the lymph node also come into contact with foreign substances. An immune response may be initiated after this contact, during which lymphocytes undergo cellular division, especially in the germinal centers. Some of these new lymphocytes remain within the lymph node, whereas others are transported within the lymph and then enter the blood, to ultimately reach areas of infections (see section 22.6).

When a person has an infection, often some lymph nodes are swollen and tender to the touch. This is a condition erroneously termed *swollen glands*. These enlarged nodes are a sign that lymphocytes are proliferating and attempting to fight an infection. Swollen superficial lymph nodes, such as those in the neck and axilla, can generally be palpated (felt).

WHAT DID YOU LEARN?

8 How does lymph flow through a lymph node, and how is it monitored by macrophages and lymphocytes?

21.4b Spleen

LEARNING OBJECTIVES

13. Describe the spleen and its location.

14. Distinguish between white pulp and red pulp.

15. List the functions of the spleen.

The **spleen** (splēn) is the largest lymphatic organ in the human body. It is located in the left upper quadrant of the abdomen, inferior to the diaphragm and adjacent to ribs 9–11 (**figure 21.7*a***). This deep red organ lies lateral to the left kidney and posterolateral to the stomach. The spleen can vary considerably in size and weight, but typically is about 12 centimeters (5 inches) long and 7 centimeters (3 inches) wide.

The spleen's posterolateral aspect (called the *diaphragmatic surface*) is convex and rounded; the concave anteromedial border (the *visceral surface*) contains the hilum (or hilus), where blood vessels and nerves connect to the spleen. The **splenic** (splen′ik) **artery** delivers blood to the spleen, whereas blood is drained by the **splenic vein.**

The spleen is surrounded by a connective tissue capsule from which **trabeculae** extend into the organ. The spleen lacks a cortex and medulla. Rather, the trabeculae subdivide the spleen into regions of white pulp and red pulp. **White pulp** consists of spherical clusters of T-lymphocytes, B-lymphocytes, and macrophages, which surround a **central artery.**

The remaining splenic tissue, called **red pulp,** contains erythrocytes, platelets, macrophages, and B-lymphocytes. The cells in red pulp are housed in reticular connective tissue and form structures called **splenic cords** (*cords of Bilroth*). **Splenic sinusoids** are associated with red pulp. (Recall from section 20.1c that sinusoids are very permeable capillaries that have a discontinuous basal lamina [a component of the basement membrane], so blood cells can easily enter and exit across the vessel wall.) The sinusoids drain into small venules that ultimately lead into the splenic vein.

Red pulp of the spleen serves as a blood reservoir, including a storage site for both erythrocytes and platelets (about 30% of all platelets are stored in the spleen). In situations where more erythrocytes and platelets are needed, such as during hemorrhage, these stored formed elements reenter the blood (see section 18.3d).

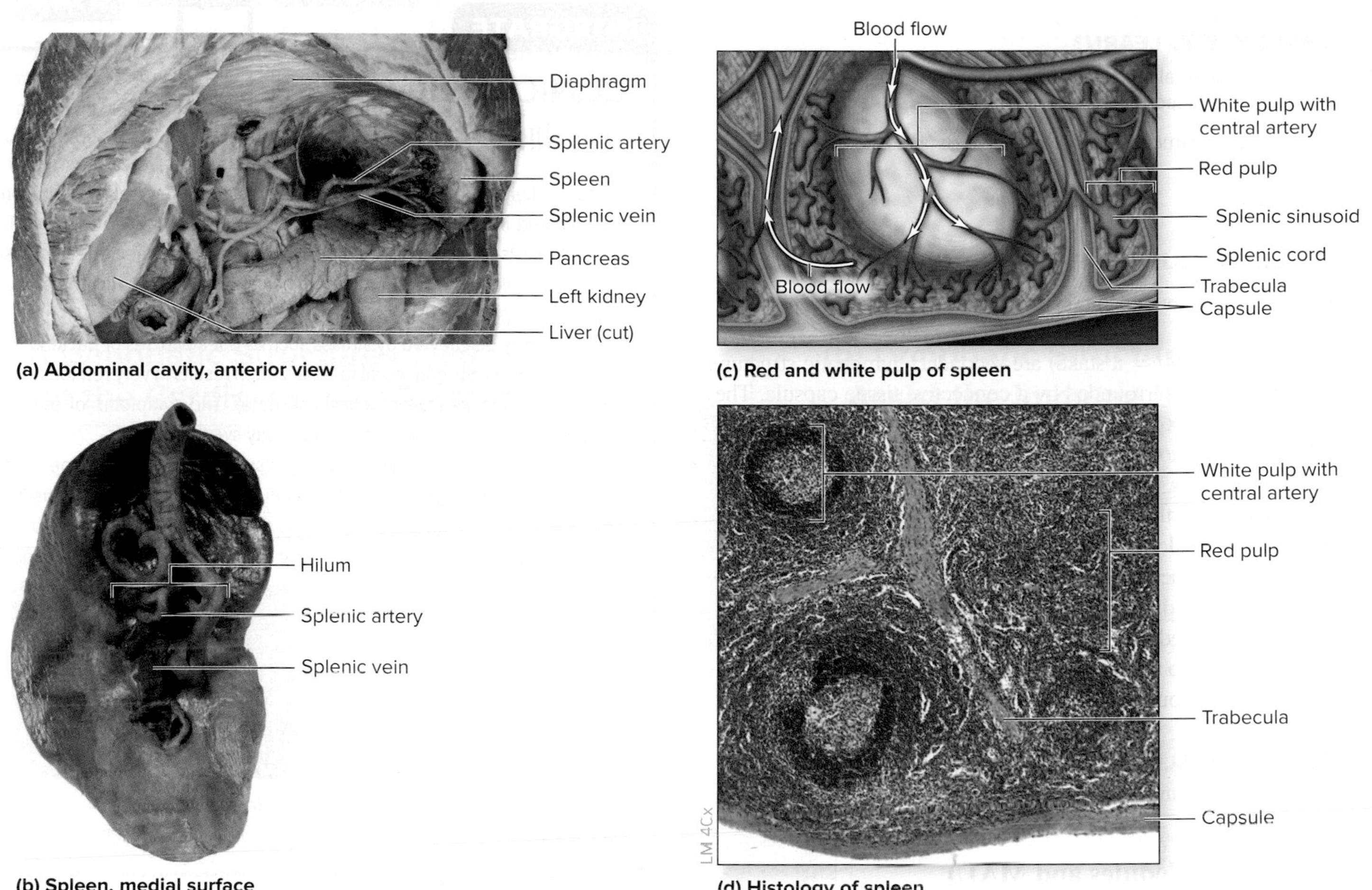

Figure 21.7 Spleen. (*a*) A cadaver photo shows the position of the spleen within the body. The pancreas has been moved inferiorly in this photo to show more clearly the splenic vessels. (*b*) A second photo shows the medial surface of the spleen, where the splenic artery and splenic vein extend from the hilum. A diagram (*c*) and micrograph (*d*) of spleen histology depict the microscopic arrangement of red pulp and white pulp. AP|R

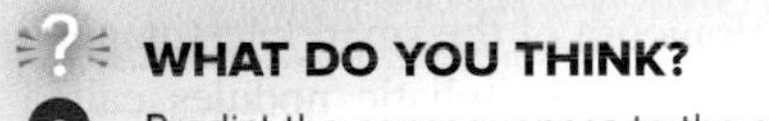

WHAT DO YOU THINK?

2 Predict the consequences to the spleen during an accident if an individual wears his or her seat belt more superiorly across the abdomen instead of more inferiorly across the pelvis.

Monitoring Blood as It Flows Through the Spleen

The spleen functions to filter *blood* (not lymph). As blood enters the spleen and flows through the central arteries, the white pulp monitors the blood for foreign materials, bacteria, and other potentially harmful substances.

After passing through a central artery, blood is transported through sinusoids of red pulp. As blood moves through the sinusoids, macrophages lining the sinusoids phagocytize bacteria and foreign debris from the blood, as well as both old and defective erythrocytes and platelets. Thus, the general flow of blood through the spleen is the splenic artery, the central artery (of white pulp), the splenic sinusoid (of red pulp), venules (that drain sinusoids), and ultimately the splenic vein (figure 21.7*c*).

In summary, the spleen serves several functions, including (1) phagocytosis of bacteria and other foreign materials in the blood as part of the body's defense (red and white pulp); (2) phagocytosis of old, defective erythrocytes and platelets from circulating blood (red pulp); and (3) a role as a blood reservoir and storage site for both erythrocytes and platelets (red pulp).

INTEGRATE

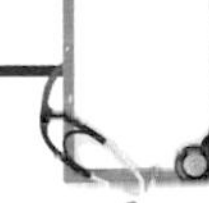

CLINICAL VIEW 21.4

Splenectomy

Splenectomy (splē-nek′tō-mē) is surgical removal of the spleen. This procedure may be performed for several reasons, including a severe splenic infection or cyst within the spleen, Hodgkin lymphoma or other types of cancers, and certain blood disorders (e.g., sickle-cell disease). The most common reason, however, is a ruptured spleen resulting from abdominal injury.

If the spleen is removed, other lymphatic organs may take over many of the spleen's functions, but the person may be more prone to contracting a serious or life-threatening infection. For this reason, individuals who have had their spleens removed are encouraged to receive vaccines against the flu and pneumonia and are placed on antibiotic therapy for an extended period of time, sometimes for their entire lives.

During fetal development through the fifth month, the spleen also engages in the formation of blood cells, a function performed after birth by the bone marrow. This function remains latent in the spleen and may be reactivated under certain conditions (e.g., some hematologic disorders). This process of reactivation is called **extramedullary hematopoiesis** (or *hemopoiesis*).

CHAPTER SUMMARY

	• The lymphatic system, which is composed of lymph vessels and lymphatic tissues and organs, supports the functions of the cardiovascular and immune system.
21.1 Lymph and Lymph Vessels	• Lymph is transported in lymph vessels.
	21.1a Lymph and Lymphatic Capillaries • Lymph is interstitial fluid containing solutes and sometimes foreign material that is absorbed into lymphatic capillaries, transported through lymph vessels, and returned to the blood. • Lymphatic capillaries are endothelium-lined vessels with overlapping endothelial cells where interstitial fluid enters lymph vessels to become lymph.
	21.1b Lymphatic Vessels, Trunks, and Ducts • Lymph is transported through a network of increasing larger vessels that include lymphatic capillaries, lymphatic vessels, lymphatic trunks, and lymphatic ducts. • Movement of lymph through lymph vessels is assisted by the presence of valves within the larger lymphatic vessels and trunks and several mechanisms that help propel the lymph.
21.2 Overview of Lymphatic Tissue and Organs	• Primary lymphatic structures are involved in the formation and maturation of lymphocytes. • Secondary lymphatic structures house lymphocytes and other immune cells, serving as sites to initiate the immune response.
21.3 Primary Lymphatic Structures	**21.3a Red Bone Marrow** • Red bone marrow produces all formed elements, including lymphocytes. **21.3b Thymus** • The thymus is a bilobed organ found in the mediastinum. It is most active in childhood until puberty, and then it declines in size and function. • The thymus is the site of T-lymphocyte maturation, which occurs in the presence of thymic hormones.
21.4 Secondary Lymphatic Structures	• Secondary lymphatic structures may be organized into lymphatic organs, which are enclosed within a complete capsule, and lymphatic nodules, which have an incomplete or absent capsule. • Both lymphatic organs and lymphatic nodules are composed of a reticular connective tissue matrix that houses lymphatic cells (e.g., B-lymphocytes).
	21.4a Lymph Nodes • Lymph nodes are numerous, small, encapsulated lymphatic organs that filter lymph. • Each lymph node is organized into an outer cortex and inner medulla that house lymphatic cells (e.g., B-lymphocytes).
	21.4b Spleen • The spleen is the largest lymphatic organ. It is partitioned into white pulp and red pulp that filters blood. • The spleen functions in the removal of bacteria and other foreign material from blood, the removal of old erythrocytes and platelets, and as a blood reservoir and storage site for both erythrocytes and platelets.
	21.4c Tonsils • Tonsils are large clusters of partially encapsulated lymphatic cells located in the pharynx and oral cavity to protect against potentially harmful substances that may be either inhaled or ingested. • Tonsils are named based on their specific location and include the pharyngeal tonsil, palatine tonsils, and lingual tonsils.
	21.4d Lymphatic Nodules and MALT • Lymphatic nodules can be found in every organ throughout the body. • MALT (mucosa-associated lymphatic tissue) is composed of large groups of lymphatic nodules housed in the mucosal-lined walls of the gastrointestinal tract, respiratory tract, urinary tract, and reproductive tract.

CHALLENGE YOURSELF

Create and Evaluate

Analyze and Apply

Understand and Remember

Do You Know the Basics?

1. What body systems are supported by the lymphatic system?
   a. cardiovascular and urinary
   b. cardiovascular and immune
   c. respiratory and cardiovascular
   d. respiratory and urinary

2. Lymph is drained into the thoracic duct from which of the following body regions?
   a. right lower limb
   b. right upper limb
   c. right side of the head
   d. right side of the thorax

3. The spleen is a secondary lymphatic structure located
   a. in the oral cavity.
   b. along lymphatic vessels.
   c. attached to the gastrointestinal tract.
   d. inferior to the diaphragm adjacent to the stomach.

4. What is the function of the thymus?
   a. It is the site of T-lymphocyte maturation.
   b. It filters lymph.
   c. It filters blood.
   d. It produces the formed elements of the blood.

5. Which type of lymph vessel consists solely of an endothelium and has one-way flaps that allow interstitial fluid to enter but not exit?
   a. lymphatic vessel
   b. lymphatic capillary
   c. lymphatic duct
   d. lymphatic trunk

6. Which statement is accurate about lymph nodes?
   a. Lymph nodes do not become swollen and tender.
   b. Lymph nodes filter blood.
   c. Lymph enters the lymph node through afferent lymphatic vessels.
   d. Lymphatic sinuses are located in the cortex of a lymph node only.

7. In a *Streptococcus* infection of the throat, all of the following structures may swell *except* the
   a. pharyngeal tonsil.
   b. spleen.
   c. cervical lymph node.
   d. palatine tonsil.

8. The lymphatic trunk that drains lymph from the upper limb, breasts, and superficial thoracic wall is the
   a. lumbar trunk.
   b. jugular trunk.
   c. subclavian trunk.
   d. bronchomediastinal trunk.

9. Aged erythrocytes are removed from circulation by the
   a. mucosa-associated lymphatic tissue.
   b. lymph nodes.
   c. thymus.
   d. spleen.

10. Interstitial fluid that is absorbed into lymph vessels will be monitored by the _______________ before the fluid is dumped into venous blood.
   a. mucosa-associated lymphatic tissue (MALT)
   b. spleen
   c. thymus
   d. lymph nodes

11. List the anatomic structures of the lymphatic system, including lymph vessels, primary lymphatic structures, and secondary lymphatic structures.

12. Explain what distinguishes a primary lymphatic structure from a secondary lymphatic structure.

13. Describe what lymph is, and draw a flowchart that illustrates what structures the lymph is transported through to return to the blood.

14. Which body regions have their lymph drained to the thoracic duct?

15. Describe how the thymus's anatomy changes as we age.

16. Describe the basic anatomy of a lymph node, how lymph enters and leaves the node, and the functions of this organ.

17. Compare and contrast the red and white pulp of the spleen with respect to the anatomy and functions of each.

18. Describe the specific locations of the tonsils.

19. Describe the location and function of diffuse lymphatic nodules and mucosa-associated lymphatic tissue (MALT).

20. Explain how the lymphatic system supports the functions of both the cardiovascular system and the immune system.

Can You Apply What You've Learned?

1. A tick has embedded itself in the scalp of a young boy. His mother is most likely to find which lymph nodes to be swollen?
   a. cervical
   b. inguinal
   c. axillary
   d. femoral

2. A child born without his thymus would not have mature
   a. macrophages.
   b. B-lymphocytes.
   c. dendritic cells.
   d. T-lymphocytes.

3. A young woman was in a car accident and had to have her spleen removed as a consequence of its rupture during the accident. She now has a greater risk of
   a. an overactive immune response.
   b. bacterial infections.
   c. low blood pressure.
   d. a pathogen surviving in her lymph.

4. One of the postoperative complications from the removal of lymph nodes during a mastectomy would be
   a. edema in the upper limb.
   b. regrowth of the lymph nodes.
   c. an overactive immune system.
   d. tender and swollen lymph nodes in other regions of the body.

5. All of the following can result in lymphedema (the accumulation of interstitial fluid due to interference with lymphatic drainage) *except*
   a. surgical removal of a group of lymph nodes.
   b. obstruction of lymph vessels that drain a lymph node, as might occur with a tumor or an infection.
   c. radiation therapy, which may cause scar formation of lymph vessels.
   d. exercise that increases the flow of lymph in the lymph vessels.

Can You Synthesize What You've Learned?

1. Arianna was diagnosed with mononucleosis, an infectious disease that targets B-lymphocytes. The doctor palpated her left side, just below the rib cage, and told Arianna she was checking to see if a certain organ was enlarged, a complication that can occur with mononucleosis. What lymphatic organ was the doctor checking, and why would it become enlarged? Include some explanation of the anatomy and histology of this organ in your answer.
2. Jordan has an enlarged lymph node along the side of his neck, and he is worried that the structure may be a lymphoma. Explain how malignant cells may have reached the lymph node.
3. Mark has come to the emergency care facility complaining of a sore throat. Upon examination, his tonsils appear swollen. When questioned further, he reports that although he has no history of his tonsils being swollen, his throat has been sore for almost a week. He is concerned that he will need to go to the hospital to have his tonsils removed. Explain the conditions that would generally require a tonsillectomy.

INTEGRATE

ONLINE STUDY TOOLS

connect | SMARTBOOK® | AP|R

The following study aids may be accessed through Connect.

Clinical Case Study: A Young Woman with a Neck Mass

Interactive Questions: This chapter's content is served up in a number of multimedia question formats for student study

SmartBook: Topics and terminology include lymph and lymph vessels; overview of lymphatic tissue and organs; primary lymphatic structures; secondary lymphatic structures

Anatomy & Physiology Revealed: Topics include lymphatic system overview; thoracic duct; thymus; lymph node; spleen; tonsils

Animations: Topics include lymphatic system overview

Immune System and the Body's Defense

INTEGRATE

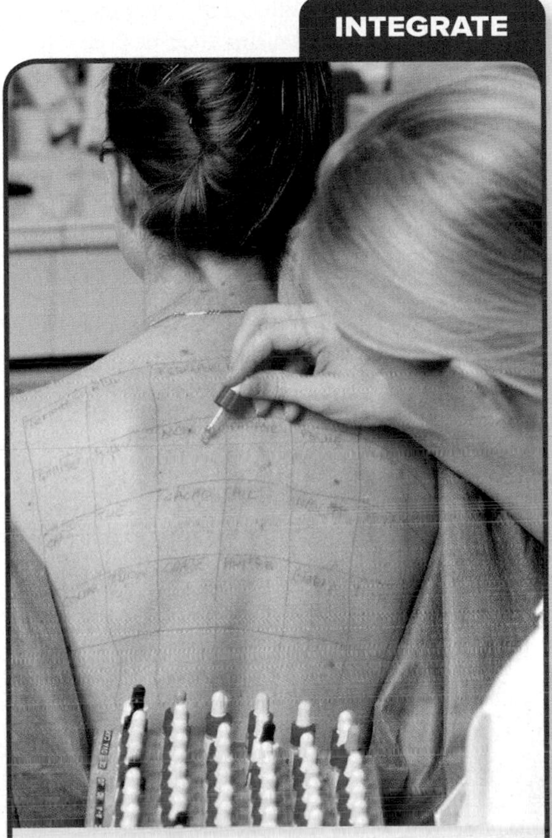

CAREER PATH
Allergist

An allergist is a physician who specializes in allergies, asthma, and other disorders of the immune system. Practicing specialists and research scientists who study the immune system analyze the normal activities, malfunctions, and best means of treating disorders involving this system. Hypersensitivity disorders, including allergies, are one common malfunction. Skin tests for allergies are performed on an area of the back by placing allergens into the epidermis and noting the reaction that occurs. The formation of a rash where the substance is tested indicates that the individual is allergic to that substance.

Anatomy & Physiology REVEALED® aprevealed.com

Module 10: Lymphatic System

There are children born without an **immune system**—a condition called *severe combined immune deficiency syndrome (SCIDS)*. The most famous of these children was a boy named David Vetter, born in 1971. He was dubbed by the media as "The Boy in the Bubble" because without an immune system, David literally had to live in a sterile plastic bubble. At the age of 12, David received a bone marrow transplant from his sister with the hope of his being able to live outside the bubble. This new bone marrow would have the ability to produce the immune cells that David's bone marrow was unable to form. Sadly, residing in his sister's bone marrow was a dormant (not active) virus called Epstein-Barr. This virus is generally kept under control by a healthy immune system. However, because David did not have a functioning immune system, the bone marrow with the Epstein-Barr virus induced formation of cancerous tumors throughout his body, and he died several months later.

Most of us may take for granted that we have a functioning immune system. While each of us may catch a cold or the flu on occasion, our immune system is—without our typically being aware of it—protecting us from infectious agents and other harmful substances. This system is unique because, unlike all other body systems, the immune system is not made up of organs. Rather, it is composed of numerous cellular and molecular structures located throughout the body that function together in the body's defense to provide us with *immunity*.

Our coverage of the immune system in this chapter is not comprehensive. Rather, our discussion is tailored to provide a general overview of how the immune system functions, as well as a description of some of the more common immune system malfunctions (e.g., hypersensitivities). We hope to help you to develop both an understanding of, and an appreciation for, this system that so diligently functions to protect us. We first provide a brief overview of the different infectious agents. This is a necessary preliminary step because immune system function is often dependent upon the specific type of infectious agent against which it must defend.

22.1 Overview of Diseases Caused by Infectious Agents

LEARNING OBJECTIVES

1. Compare and contrast the five major classes of infectious agents.
2. Describe prions, and name a disease they cause.

Infectious agents are organisms that cause damage, or possibly death, to the host organism that they invade. Infectious agents that cause harm to the host are said to be **pathogenic** (path′ō-jen′ik; *pathos* = disease, *genesis* = production). The five major categories of infectious agents that cause disease in humans are bacteria, viruses, fungi, protozoans, and multicellular parasites. These categories, including examples of each, are summarized and compared in **table 22.1** and described here:

- **Bacteria** are microscopic, single-celled organisms composed of **prokaryotic cells.** These cells are fundamentally different from the cells of humans and other living organisms. Prokaryotic cells are smaller in size, averaging between 1 and 2 micrometers (μm), which is approximately the size of a mitochondrion (see figure 4.2). Prokaryotic cells lack a nuclear envelope, and their cytoplasm and DNA are enclosed by both a plasma membrane and a *cell wall* (formed by complex carbohydrates cross-linked with peptides). Some bacteria also have an external, sticky polysaccharide *capsule,* which increases their **virulence** (vir′yū-lĕns)—the ability to cause serious illness. Pathogenic bacteria may also have **pili** (pī′lī), which are hairlike structures that act like Velcro for attaching to body structures (e.g., *E. coli* that cause urinary tract infections). Additionally, some pathogenic bacteria cause disease by releasing enzymes or toxins that interfere with the function of cells; an example is the bacterium *Clostridium tetani* (see Clinical View 10.3: "Muscular Paralysis and Neurotoxins"). (Note: The cells of humans and other living organisms are composed of *eukaryotic cells* (cells with a nucleus and organelles), and their structure is discussed in detail in chapter 4.)
- **Viruses** are *not* cells. Viruses are much smaller than a bacterial cell at about one-hundredth of a micrometer (see figure 4.2), and they are composed of DNA or RNA within a protein capsid, or shell. The protein capsid of some viruses may also be enclosed within a membrane. Viruses are **obligate intracellular parasites;** that is, they must enter a cell to replicate. The process of viral reproduction includes directing the infected cell to synthesize copies of both the viral DNA or RNA molecule and its capsid protein. New *viral particles* are then formed within the infected cells and released from them to enter surrounding cells. A virus, or the immune system's response to it, ultimately kills the cells it invades.

  Viruses cause different diseases depending upon the type of cell they infect. Examples of viral diseases include the common cold, chicken pox, ebola, and HIV.
- **Fungi** (fun′jī) are composed of eukaryotic cells that have a cell wall external to the plasma membrane. This group includes molds, yeasts, and multicellular fungi that produce spores. Proteolytic (protein-digesting) enzymes released from fungi induce inflammation (see section 22.3d) that causes redness and swelling of the infected area.

  Fungal diseases (*mycoses*) in healthy individuals in the United States are usually limited to superficial infections of the skin, scalp, and nails (e.g., ringworm, "athlete's foot"). Other fungal diseases involve infections of mucosal linings (e.g., vaginal yeast infections) or may cause internal fungal infections (e.g., histoplasmosis, which affects the respiratory system).
- **Protozoans** are microscopic, unicellular eukaryotic organisms that lack a cell wall. One example of a disease-causing protozoan is plasmodium, which causes malaria. This infectious agent is spread by mosquitoes and enters the blood, where it then infects erythrocytes, causing their destruction.
- **Multicellular parasites** are nonmicroscopic organisms (larger than a centimeter in size) that reside within a host from which they take nourishment. Parasitic worms such as tapeworms, for example, infect the intestinal tract of humans.

Prions (prī′on) are small fragments of infectious proteins that cause disease in nervous tissue. **Variant Creutzfeldt-Jakob disease** (also known as *bovine spongiform encephalopathy,* or "mad cow disease") is an example. This prion disease can be spread from cows to humans by consuming infected meat because nerves within the muscle are contaminated with prions. Prions are neither cells nor viruses, and research into how they cause disease is ongoing.

WHAT DID YOU LEARN?

1. Which pathogen must enter a cell to replicate? Which type of pathogen is composed of prokaryotic cells?

Table 22.1 Major Categories of Infectious Agents

Characteristic	Bacteria
Structure	
Cellular Characteristic	Prokaryotic
Significant Features	Intracellular and extracellular microbes—some have a sticky polysaccharide capsule and/or hairlike pili; some produce enzymes and/or toxins
Selected Diseases Caused by the Agent	Streptococcal infections (e.g., strep throat), staphylococcal infections, tuberculosis, syphilis, diphtheria, tetanus, Lyme disease, salmonella, and anthrax

INTEGRATE

LEARNING STRATEGY

You might find it helpful to consider that developing an understanding of the immune system is like putting together a puzzle. You will first need to become familiar with the numerous "pieces" of the immune system, many of which at first seem unrelated. It is only when these pieces are integrated into a whole that you are able to see the relationships between them.

22.2 Overview of the Immune System

Eleven organ systems are introduced in section 1.4c—and the immune system is not one them. This is because, unlike those 11 body systems, the immune system does not have organs (e.g., lungs of the respiratory system). Instead, numerous types of cells (including specialized cells called immune cells), secreted molecules (including cytokines), and plasma proteins (e.g., antibodies) are functionally integrated to form our immune system. The function of this system is to protect our body from infectious agents and other potentially harmful substances to provide us with **immunity** (i-myū′ni-tē). Here we provide a brief introduction to the location of the specialized immune cells and the secreted cytokine molecules that regulate them. We then give an overview of how the immune system is organized into two overlapping and complementary components: the innate immune system and the adaptive immune system.

22.2a Immune Cells and Their Locations

LEARNING OBJECTIVE

3. List the types of leukocytes of the immune system, and describe where they may be found.

Leukocytes (*white blood cells*) are the specialized cells of our immune system. They were first described in section 18.3c as the cells that help defend the body against pathogens. They are formed in the red bone marrow prior to circulating in the blood. Recall that leukocytes include (1) the three types of granulocytes (neutrophils, eosinophils, and basophils); (2) monocytes that become macrophages or dendritic cells when they exit blood vessels and take up residence in the tissues; and (3) the three types of lymphocytes, which include T-lymphocytes (or T-cells), B-lymphocytes (or B-cells), and NK (natural killer) cells.

Structures That House Immune System Cells

Most leukocytes are found in body tissues. The primary locations that house immune cells include secondary lymphatic structures, select organs, epithelial layers of the skin and mucosal membranes, and connective tissues of the body (**figure 22.1**). The housing of the specific types of immune cells varies by the location:

- **Secondary lymphatic structures.** *T-lymphocytes* and *B-lymphocytes* are housed in secondary lymphatic structures of lymph nodes, the spleen, tonsils, lymphatic nodules, and MALT (mucosa-associated lymphatic tissue) (see section 21.4). The presence of lymphocytes accounts for the collective name (lymphatic structures). Other immune cells, including macrophages, dendritic cells, and NK cells, are also present.

Viruses	Fungi	Protozoans	Multicellular Parasites
Not a cell; DNA or RNA within a capsid protein	Eukaryotic	Eukaryotic	Eukaryotic
Obligate intracellular parasites; must enter cell to replicate	Produce spores; release proteolytic enzymes	Intracellular and extracellular parasites that interfere with normal cellular functions	Live within a host; grow in size with nutrients provided by the host
Common cold, influenza, polio, mumps, measles, hepatitis, rubella, chicken pox, ebola, herpes, and HIV (which leads to AIDS)	Ringworm, diaper rash, jock itch, athlete's foot, yeast infections, and histoplasmosis	Malaria, toxoplasmosis, giardiasis, amoebiasis, leishmaniasis, trichomoniasis, and African sleeping sickness	Parasitic infection from tapeworms, lung flukes, liver flukes, blood flukes, hookworms, *Trichinella*, *Ascaris*, whipworms, and pinworms

(photos): (bacteria) Dr. William A. Clark/CDC; (viruses) Centers for Disease Control; (fungi) Centers for Disease Control; (protozoans) Janice Haney Carr/CDC; (multicellular parasites) CDC

22.3b Nonspecific Internal Defenses: Cells

LEARNING OBJECTIVE

8. Describe the cells that function as part of the nonspecific internal defenses in providing innate immunity.

Nonspecific internal defenses of the innate immune system include (1) activities of various types of cells; (2) chemicals such as interferon and complement; and (3) physiologic processes that include the inflammatory response and development of a fever (figure 22.2). The structure and function of the innate immune cells are organized and described here by their function. Refer to **figure 22.3** as you read through this section.

Phagocytic Cells

Phagocytic cells include **neutrophils, macrophages,** and **dendritic cells,** which engulf unwanted substances such as infectious agents and cellular debris through phagocytosis (figure 22.3*a*) (see sections 6.1a and 18.3c). The vesicle formed during phagocytosis that contains the unwanted substance (a *phagosome*) merges with a lysosome to form a *phagolysosome.* Within the phagolysosome, digestive enzymes contributed from the lysosome chemically digest the unwanted substances. Destruction of bacteria and viruses within neutrophils and macrophages is facilitated by the production of reactive oxygen-containing molecules (e.g., nitric oxide, hydrogen peroxide, superoxide); the release of these molecules is called a **respiratory burst** (or *oxidative burst*). Degraded residues of the engulfed substance are then released from the cell by exocytosis. Phagocytosis and the subsequent destruction of microbes are highly effective in protecting us. In fact, most microbes that enter the body are engulfed and destroyed by phagocytic cells. Participation in phagocytosis however, is fatal for neutrophils, which die soon after engulfing microbes. Dead neutrophils become the major component of the pus that is produced during some infections (see Clinical View 22.1: "Pus and Abscesses").

Macrophages and dendritic cells, in comparison, continue to function following phagocytosis. They serve an additional role, which is to present fragments of the microbe on their cell surface to T-lymphocytes. This process, called *antigen presentation,* is necessary for initiating adaptive immunity and is described in detail in section 22.4c.

Proinflammatory Chemical-Secreting Cells

Chemical-secreting cells that enhance inflammation (which is described in section 22.3d) include both **basophils** and **mast cells** (figure 22.3*b*). Recall from section 18.3c that basophils circulate in the blood and mast cells reside in connective tissue of the skin, mucosal linings, and various internal organs. Substances secreted by basophils and mast cells increase fluid movement from the blood to an injured tissue. They also serve as **chemotaxic chemicals,** which are molecules that attract immune cells as part of the inflammatory response.

Basophils and mast cells release granules during the inflammatory response. These granules contain various substances, including **histamine,** which increases both vasodilation and capillary permeability, and **heparin,** an anticoagulant. They also release **eicosanoids** (ī′kō-să-noydz) from their plasma membrane (see section 17.3b), which increase inflammation.

Apoptosis-Initiating Cells

NK (natural killer) cells, which are located within secondary lymphatic structures, destroy a wide variety of unhealthy or unwanted cells through apoptosis (see section 4.10). The types of cells eliminated by NK cells include virus-infected cells, bacteria-infected cells, tumor cells, and cells of transplanted tissue (figure 22.3*c*).

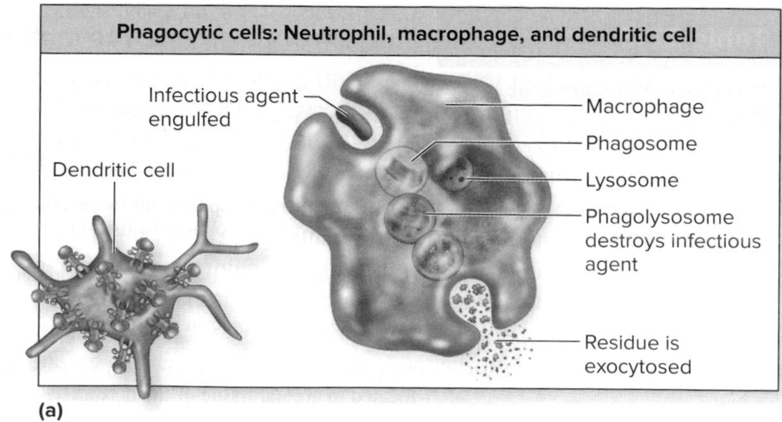

(a)

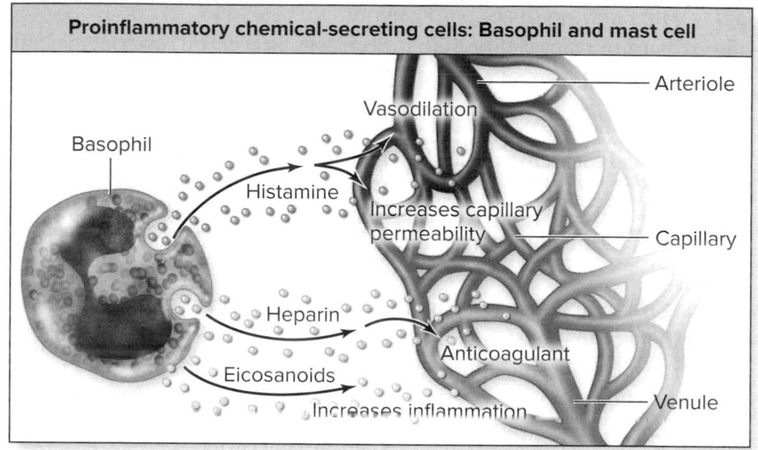

(b)

Apoptosis-initiating cells: NK cell

Perforin and granzyme

Perforin forms a transmembrane pore

Granzymes enter pore, causing apoptosis of cell

NK cell

Unhealthy or unwanted cell

Apoptosis

(c)

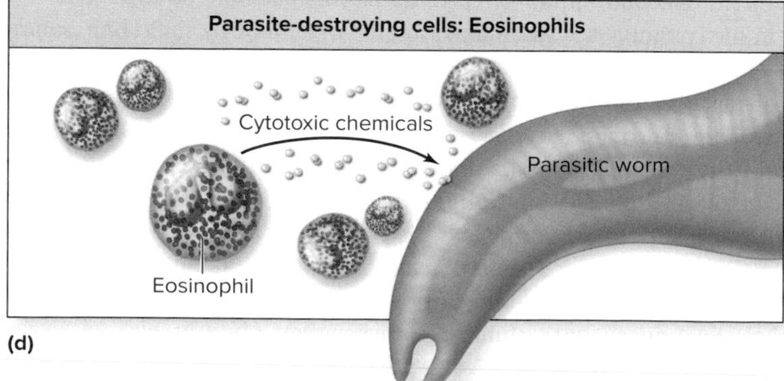

(d)

Figure 22.3 Cells of the Innate Immune System. The cells of the innate immune system use multiple tactics to combat pathogens, including (*a*) phagocytosis (example shown is a macrophage), (*b*) chemical secretion that increases inflammation (example shown is a basophil), (*c*) chemical secretion by NK cells that destroys unhealthy cells by inducing apoptosis, and (*d*) chemical secretion by eosinophils that helps eliminate parasites. AP|R

NK cells patrol the body in an effort to detect unhealthy cells, a process referred to as **immune surveillance.** NK cells make physical contact with unhealthy cells and destroy them by release of cytotoxic chemicals. These cytotoxic chemicals include *perforin,* which forms a transmembrane pore in the unwanted cells, and *granzymes,* which then enter the cell through the transmembrane pore initiating apoptosis. Apoptosis is a form of cellular death in which the cell does not lyse (i.e., rupture), but rather "shrivels"; this helps limit the spread of the infectious agent.

Eosinophils

Eosinophils (ē′ō-sin′ō-fil) target multicellular parasites, attacking the organisms' surfaces (figure 22.3*d*). Mechanisms of destruction include degranulation and release of enzymes and other substances (e.g., reactive oxygen-containing compounds, neurotoxins) from the eosinophils that are lethal to the parasite (see section 18.3c). Like NK cells, eosinophils can release proteins that form a transmembrane pore to destroy cells of the multicellular organism.

Eosinophils also participate in the immune response associated with allergy and asthma (see Clinical View 22.8: "Hypersensitivities") and engage in phagocytosis of antigen-antibody complexes (see section 22.8b).

Each of the innate immune cells is able to "see" many *different* types of microbes as foreign. How is this possible? Microbes (e.g., bacteria, viruses) possess molecular structures (or motifs) that they have in common. These molecular structures may be on the microbe's surface (e.g., bacterial cell wall) or within the microbe (e.g., viral RNA). These patterned molecular structures serve as "hazard signs" to the innate immune cells. Our innate immune cells possess *pattern recognition receptors* (e.g., toll-like receptors, or TLRs) on their cell surface, which allows them to make physical contact with these common molecular motifs, like two pieces of a puzzle fitting together. This binding of the immune cell to the microbe activates each type of the innate immune cells to function as just described.

INTEGRATE

CONCEPT CONNECTION

In section 18.3c, we discussed the differential count of white blood cells, a process that measures the amount of each type of leukocyte in your blood. Differential counts are useful in diagnosing different types of infections. For example:

- An increase in neutrophils is associated with an acute bacterial infection.
- Monocytes may increase with chronic inflammatory disorders or tuberculosis.
- An increase in eosinophils occurs in response to a parasitic infection.
- An elevated number of lymphocytes generally is associated with viral infections or with chronic bacterial infections.

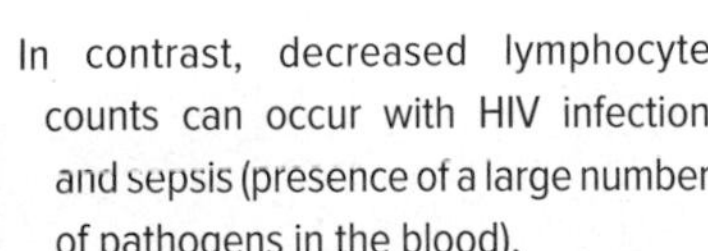

In contrast, decreased lymphocyte counts can occur with HIV infection and sepsis (presence of a large number of pathogens in the blood).

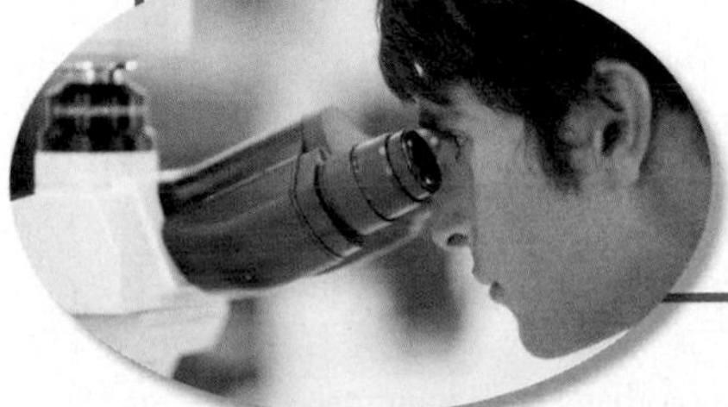

WHAT DID YOU LEARN?

6. What distinguishes neutrophils from dendritic cells? How do basophils differ from mast cells?
7. How do NK cells accomplish the task of eliminating unwanted cells?

22.3c Nonspecific Internal Defenses: Antimicrobial Proteins

LEARNING OBJECTIVES

9. Explain the general function of interferons.
10. Define the complement system, and describe how it is activated.
11. Describe the four major means by which complement participates in providing innate immunity.

Antimicrobial proteins are specific types of molecules of the innate immune system that function against microbes. Two of those, interferons and complement, are described in this section.

Interferons

Recall from section 22.2b that **interferons (IFNs)** are a category of cytokines (table 22.2). Here we describe the process of how IFNs serve as a nonspecific defense mechanism against the spread of any viral infection (**figure 22.4**).

Step 1: A body cell is first infected with a virus (see section 22.1).
Step 2: In response, the virus-infected cell releases both IFN-α and IFN-β.
Step 3a: IFN-α and IFN-β bind to receptors of neighboring body cells. This triggers these cells to synthesize enzymes that both destroy viral RNA or DNA and inhibit synthesis of viral proteins. Thus, these cells are protected from becoming infected with the virus.
Step 3b: IFN-α and IFN-β also stimulate NK cells to destroy the virus-infected cells by apoptosis (see section 22.3b).
Step 4: NK cells release IFN-γ to stimulate macrophages to aid in destroying the virus-infected cells through phagocytosis.

Complement System

Complement is a diverse array of proteins (at least 30) produced by our liver and released into the blood. Individual complement proteins

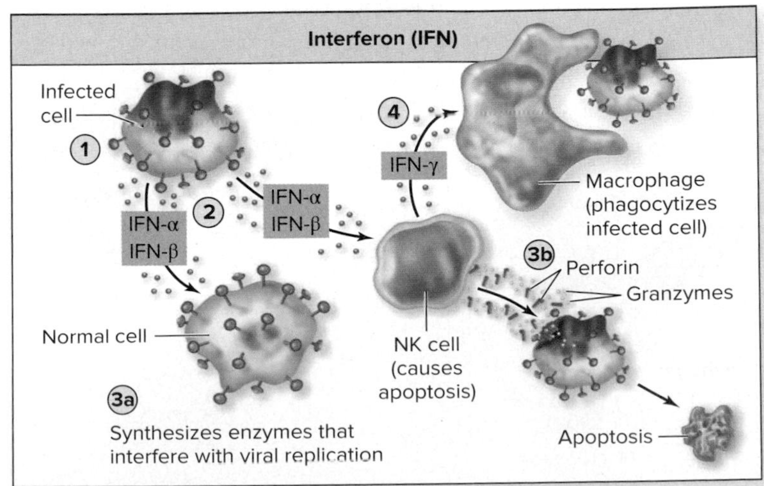

Figure 22.4 Effects of Interferon. Interferon (IFN) is one category of cytokines released from a variety of cells. A virus-infected cell releases IFN-α and IFN-β, which stimulate antiviral changes to neighboring cells to prevent their infection and induce NK cells to both destroy virus-infected cells and release IFN-γ to activate macrophages to destroy virus-infected cells.

are generally identified with the letter "C" followed by a number (e.g., C1, C2). These proteins are very abundant and make up approximately 10% of the blood serum proteins. The collective name for these proteins is derived from how they complement, or work along with, antibodies (proteins produced by differentiated B-lymphocytes, which are described in section 22.8).

The liver continuously synthesizes and releases inactive complement proteins into the blood. Once in the blood, inactive complement proteins are activated by an enzyme cascade. (Recall from section 18.4c that a similar process of an enzyme cascade of inactive proteins produced by the liver is also involved in blood clotting.)

Activation of complement occurs following entry of a pathogen into the body. Two of the major means of activation include the **classical pathway,** in which a complement protein binds to an antibody that has previously attached to a foreign substance (i.e., an antigen), and the **alternative pathway,** in which surface polysaccharides of certain bacterial and fungal cell walls (see section 22.1) bind directly with a complement protein. Note that antibody is required for activation by the classical pathway, but not for activation by the alternative pathway.

Following its activation, the complement system mediates several important defense mechanisms, and it is especially potent against bacterial infections (**figure 22.5**).

- **Opsonization** (op′sŏn-ī-zā′shŭn) is the binding of a protein (in this case, complement) to a portion of bacteria or other cell type that enhances phagocytosis. The binding protein is called an **opsonin** (op′sŏ-nin). The binding of complement makes it more likely that a substance is identified and engulfed by a phagocytic cell (e.g., macrophage). An opsonin functions as a red flag to indicate the tagged microbe.
- **Inflammation.** Complement increases the inflammatory response through the activation of mast cells and basophils and by attracting neutrophils and macrophages (see section 22.3d).
- **Cytolysis.** Various complement components (e.g., C5–C9) trigger direct killing of a target by forming a protein channel in the plasma membrane of a target cell called a **membrane attack complex (MAC).** The MAC protein channel compromises the cell's integrity, allowing an influx of fluid that causes lysis of the cell.
- **Elimination of immune complexes.** Complement links immune (antigen-antibody) complexes to erythrocytes so they may be transported to the liver and spleen. Erythrocytes are stripped of these complexes by macrophages within these organs, and the erythrocytes then continue circulating in the blood.

INTEGRATE

LEARNING STRATEGY 

You can remember the various actions of complement with the acronym **O-ICE:** **O**psonization, **I**nflammation, **C**ytolysis, and **E**limination of immune complexes.

WHAT DID YOU LEARN?

8. How is the complement system defined? What are the four major means by which complement participates in providing innate immunity?

22.3d Nonspecific Internal Defenses: Inflammation

LEARNING OBJECTIVES

12. Define inflammation, and discuss the basic steps involved, including the formation of exudate and its role in removing harmful substances.
13. Describe the benefits of inflammation.
14. List the cardinal signs of inflammation, and explain why each occurs.

Inflammation, or the **inflammatory response,** is an immediate, local, nonspecific event that occurs in vascularized tissue against a great variety of injury-causing stimuli. Inflammation occurs, for example, in response to a scratch of your skin, a bee sting, overuse of a body structure (e.g., pitching arm), or proteolytic enzymes released by fungi. This physiologic process is the *major* effector response of the innate immune system and is successful in helping to eliminate most infectious agents and other unwanted substances from the body!

Events of Inflammation

Inflammation involves several steps (**figure 22.6**). The first step is the *release of various chemicals.* Damaged cells of injured tissue, macrophages, dendritic cells, basophils, mast cells, and infectious organisms release numerous chemicals that promote inflammation (e.g., histamine, leukotrienes, prostaglandins, interleukins, TNFs, and chemotactic factors). **Table 22.4** lists various chemicals of inflammation, describes their function, and identifies their source.

The second step encompasses *vascular changes*. Released chemicals cause a variety of responses in local blood vessels, including vasodilation of arterioles, increase in capillary permeability, and

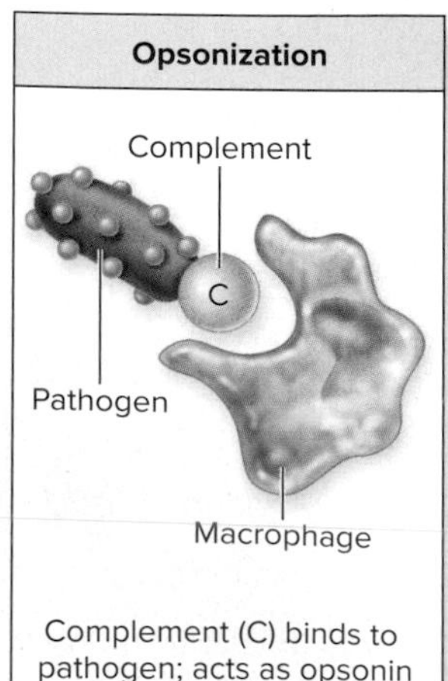

Opsonization
Complement
C
Pathogen
Macrophage
Complement (C) binds to pathogen; acts as opsonin

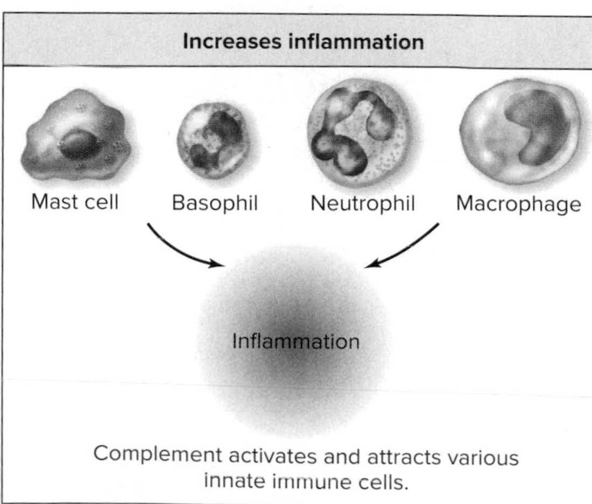

Increases inflammation
Mast cell
Basophil
Neutrophil
Macrophage
Inflammation
Complement activates and attracts various innate immune cells.

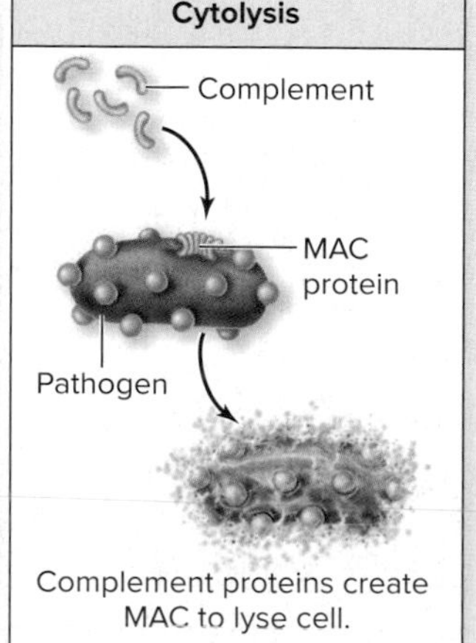

Cytolysis
Complement
MAC protein
Pathogen
Complement proteins create MAC to lyse cell.

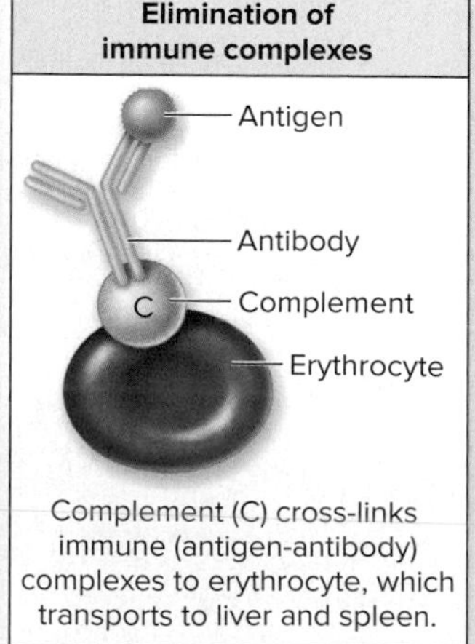

Elimination of immune complexes
Antigen
Antibody
C
Complement
Erythrocyte
Complement (C) cross-links immune (antigen-antibody) complexes to erythrocyte, which transports to liver and spleen.

Figure 22.5 Complement System. Upon activation, complement (C) proteins protect the body through various mechanisms, including opsonization, increasing inflammation, cytolysis of target cells, and elimination of immune complexes.

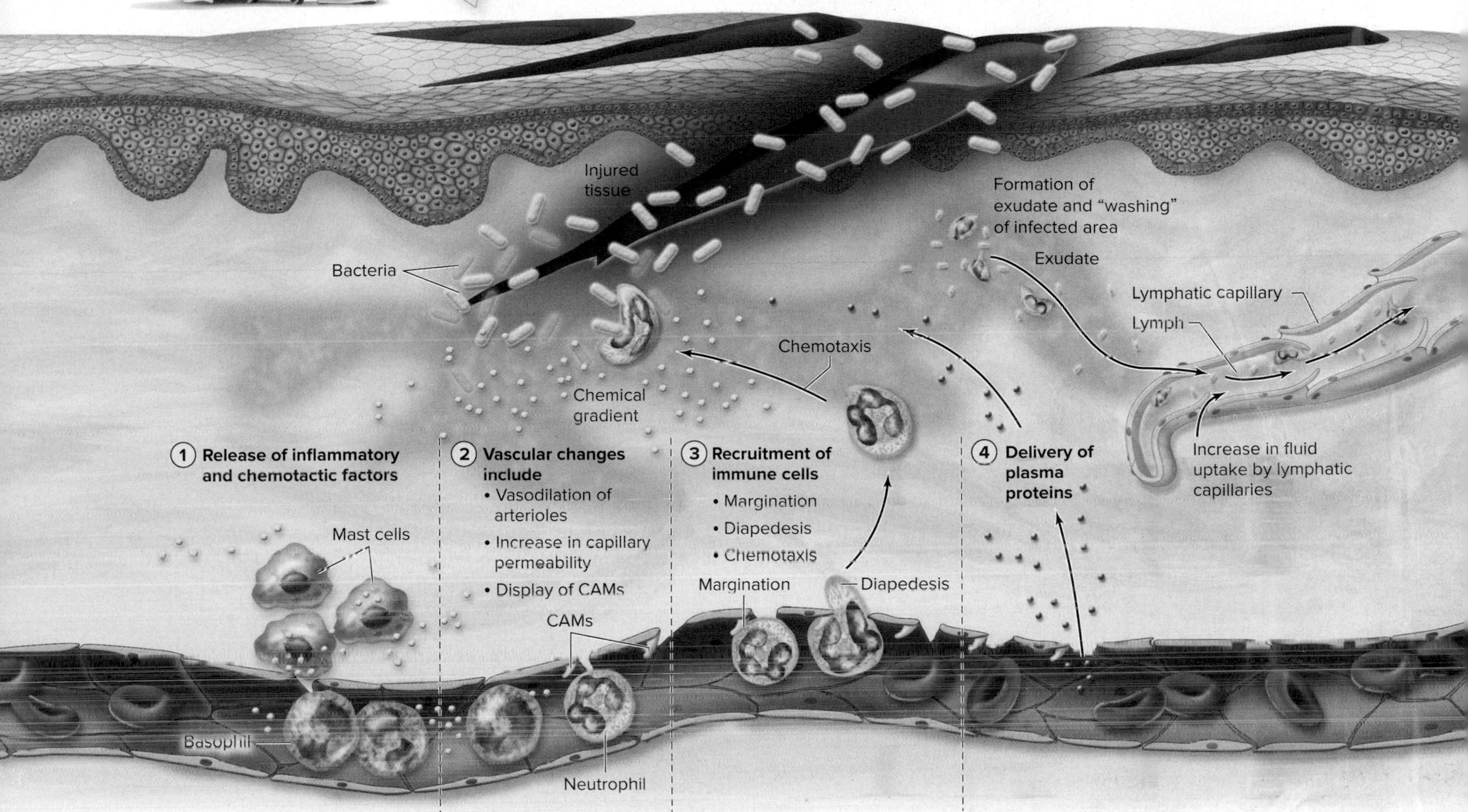

Figure 22.6 Inflammation. Inflammation involves several steps, including the release of chemicals, vascular changes, recruitment of cells, and delivery of plasma proteins. A critical event of inflammation is the formation of exudate that helps to "wash" the area of injurious or infectious agents.

stimulation of the capillary endothelium to provide molecules for leukocyte adhesion (**cell adhesion molecules,** or **CAMs**).

The third step involves the *recruitment of leukocytes*. Leukocytes make their way from the blood to the infected tissue through the following processes:

- **Margination** is the process by which CAMs on leukocytes adhere to CAMs on the endothelial cells of capillaries within the injured tissue. The result is similar to "cellular Velcro." Neutrophils are generally the first to arrive and are short-lived, followed later by the longer-lived macrophages.
- **Diapedesis** (dī′ă-pĕ-dē-sis) is the process by which cells exit the blood by "squeezing out" between vessel wall cells, usually in the postcapillary venules, and then migrate to the site of infection (see section 18.3c).
- **Chemotaxis** is migration of cells along a chemical gradient (see section 18.3c). Chemicals released from damaged cells, dead cells, or invading pathogens diffuse outward and form a chemical gradient that attracts immune cells. Recruited cells also participate in the inflammatory response through the release of specific cytokines, such as granulocyte-macrophage colony-stimulating factor (GM-CSF), that stimulate leukopoiesis within red bone marrow (see section 18.3a). This helps account for the increase in leukocyte count that occurs during an active infection. Macrophages may also release pyrogens, such as interleukin 1 (IL-1), that induce a fever (see section 22.3e).

Delivery of plasma proteins also occurs, as shown in the fourth step. Selective plasma proteins are brought into the injured or infected site, including immunoglobulins (described in section 22.8), complement (just described), clotting proteins, and kinins. **Clotting proteins** lead to formation of a clot that walls off microbes and prevents them from spreading into blood and other tissues (see section 18.4c). However, some bacterial species can dissolve clots. **Kinins** (kī′nin) are produced from kininogens, which are inactive plasma proteins produced by the liver (and released into and transported by the blood) and locally by numerous other cells. Kinins, including bradykinin, are activated by tissue injury and have similar effects to histamine; they increase capillary permeability and the production of CAMs by the capillary endothelium. Kinins also stimulate sensory pain receptors and are the most significant stimulus for causing the pain associated with inflammation.

Effects of Inflammation

One of the most important consequences produced by the inflammatory response is a net movement of additional fluid from the blood through the infected or injured area and then into the lymph. Increased fluid, immune cells, and protein leave the capillaries and then enter the

Table 22.4 Chemicals of Inflammation

Chemical	Function in Inflammation	Source of the Chemical
Histamine	Vasodilation; increased permeability of capillaries; conversion of an inactive plasma protein (kininogen) into active peptides called kinins; released early in inflammation	Mast cells, basophils, platelets
Kinins (e.g., bradykinin)	Vasodilation and increased permeability of capillaries; increase production of CAMs; stimulate sensory pain receptors	Plasma protein produced by the liver and other cells as kininogen; activated by tissue injury
Leukotrienes (slow-reacting substance of anaphylaxis [SRS-A])	Effects similar to histamine; released later in the inflammatory response than histamine and longer lasting	Eicosanoids produced from arachidonic acid molecules of mast cell and basophil plasma membranes
Prostaglandins	Vasodilation, fever, stimulate sensory pain receptors (categories include E, D, A, F, and B)	Eicosanoids produced from arachidonic acid molecules of mast cell and basophil plasma membranes
Chemotactic factor	Attracts immune cells; release of specific chemotactic factors attract a specific type of cell (e.g., neutrophil chemotactic factor attracts neutrophils early in the inflammatory response; with a parasitic infection, eosinophil chemotactic factor attracts eosinophils)	Mast cells and basophils
Serotonin	Effects similar to histamine	Platelets
Nitric oxide	Vasodilation; may inhibit mast cells and platelets	Endothelium of blood vessels
Alpha-1 antitrypsin	Inhibits damage to connective tissue by enzymes released from destroyed phagocytes	Plasma protein formed by the liver
C-reactive protein	Activates complement by binding to polysaccharides on bacteria surface	Liver
IL-1 and TNF-α	Increase cell-adhesion molecules (CAMs) to cause margination; cause endothelial cell contraction to facilitate diapedesis	Dendritic cells, macrophages

interstitial space of the tissue; this fluid and cellular/protein mix collectively is referred to as **exudate** (eks′ū-dāt). Exudate delivers immune cells and substances needed to eliminate the injurious agent and promote healing.

This increase in fluid movement is due to several factors, including the following:

- **Vasodilation,** which allows more blood into the infected area
- **Increased capillary permeability** as endothelial cells lining the blood vessel wall contract, which causes larger openings between the endothelial cells and allows more fluid to move from the blood into the interstitial fluid
- **Loss of plasma protein,** which decreases capillary osmotic pressure, resulting in less fluid being retained in the blood and reabsorbed back into the blood during capillary exchange (see section 20.3b)

INTEGRATE

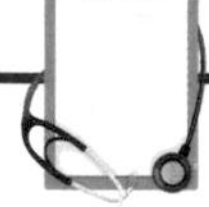

CLINICAL VIEW 22.1

Pus and Abscesses

Pus may form in severe infections, most typically from severe bacterial infections. Pus is exudate (i.e., excess fluid, protein, and immune cells that leave the capillaries and then enter the interstitial space of the tissue) that contains destroyed pathogens, dead neutrophils, macrophages, and cellular debris. Pus may be removed by the lymphatic system or through the skin (for surface injuries). If the pus is not completely cleared, an **abscess** may form in the area, whereby the pus is walled off with collagen fibers. If an abscess forms, it usually requires surgical intervention to remove.

INTEGRATE

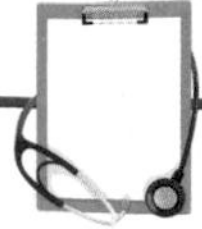

CLINICAL VIEW 22.2

Applying Ice for Acute Inflammation

The advice typically given for acute inflammation is to apply ice. Ice vasoconstricts blood vessels (decreasing the inflammatory response) and numbs the area so that it seems less painful.

Increased hydrostatic pressure exerted by the interstitial fluid causes additional fluid uptake by lymphatic capillaries (see section 21.1a). The newly formed lymph carries with it unwanted substances that include infectious agents, dead cells, and cellular debris. The contents of lymph can then be monitored as it passes through a series of lymph nodes. You may find it helpful to think of the inflammatory response as "washing" the infected or injured area.

The inflammatory response typically slows down and tissue healing begins within 3 days. Monocytes exit the blood and become macrophages to begin the cleanup of the affected area. Bacteria, damaged host cells, and dying neutrophils are engulfed and destroyed by macrophages. Tissue repair begins as fibroblasts multiply and synthesize collagen, forming new connective tissue. (Connective tissue formation may lead to the formation of scar tissue in the case of an extensive injury.)

Several benefits are associated with the inflammatory response, including helping to eliminate pathogens by limiting their spread; destroying infectious agents and removing cellular debris; and producing the conditions for tissue repair and healing.

WHAT DO YOU THINK?

1 How can you tell by looking at an injured area of the skin that inflammation has occurred?

Cardinal Signs of Inflammation

Inflammation is accompanied by certain **cardinal signs** (i.e., major representative characteristics) that may include the following:

- **Redness,** due to increased blood flow
- **Heat,** due to increased blood flow and increased metabolic activity within the area
- **Swelling,** resulting from increase in fluid loss from capillaries into the interstitial space

- **Pain,** which is caused by stimulation of pain receptors from compression due to accumulation of interstitial fluid, and chemical irritation by kinins, prostaglandins, and substances released by microbes
- **Loss of function** (which may occur in more severe cases of inflammation due to pain and swelling)

The inflammatory response typically lasts no longer than 8 to 10 days under normal conditions. The ending of the normal **acute inflammatory response** (the process just described) is necessary to prevent the unwanted detrimental effects of **chronic inflammation** (see Clinical View 22.3: "Chronic Inflammation").

WHAT DID YOU LEARN?

9 What is inflammation, and what are the basic steps involved in the inflammatory response?

10 In what ways does exudate assist in the body's defense?

22.3e Nonspecific Internal Defenses: Fever

LEARNING OBJECTIVES

15. Define a fever, and describe how it occurs.

16. List the benefits and risks of a fever.

A **fever** is defined as an abnormal elevation of body temperature **(pyrexia)** of at least 1°C (1.8°F) from the typically accepted core body temperature of 37°C (98.6°F). It results from release of fever-inducing molecules called **pyrogens** (e.g., IL-1, IL-6, TNF-α) that are released from either infectious agents (e.g., bacteria) or immune cells in response to infection, trauma, drug reactions, and brain tumors. A fever is a physiologic process of the innate immune system and may accompany the inflammatory response.

Events of Fever

Pyrogens are released from various types of cells and circulate in the blood; they target the hypothalamus (see section 13.4c) and cause release of prostaglandin E_2 (PGE_2) (a local hormone). It raises the temperature set point of the hypothalamus from the normal 37°C. The following stages of a fever occur in response: onset, stadium, and defervescence. (Keep in mind that these stages can be cyclical until the pathogen is eliminated or at least brought under control.)

During the **onset** of a fever, the hypothalamus stimulates blood vessels in the dermis of the skin to vasoconstrict to decrease heat loss through the skin, and a person shivers to increase heat production through muscle contraction (see section 1.6b). Consequently, body temperature rises. The person may experience chills during this stage, which leads to the shivering.

The period of time when the elevated temperature is maintained is referred to as **stadium.** The metabolic rate increases to promote physiologic processes of the innate and adaptive immune systems that are involved in eliminating the harmful substance. The liver and spleen bind zinc and iron (minerals needed by microbes) to slow microbial reproduction.

Defervescence (def′ĕr-ves′ents) occurs when the temperature returns to its normal set point. This happens when the hypothalamus is no longer stimulated by pyrogens, prostaglandin release decreases, and the temperature set point reverts to its normal value. The hypothalamus then stimulates the mechanisms to release heat from the body, including vasodilation of blood vessels in the skin and sweating. The person may appear flushed and the skin warm to the touch. An increase in fluid intake should occur during a fever to prevent dehydration caused by an increased loss of body fluid.

Benefits of Fever

Fever actually has numerous benefits. A fever inhibits replication of bacteria and viruses, promotes interferon activity, increases activity of lymphocytes, and accelerates tissue repair. Most recently, it has been demonstrated that a fever also increases CAMs on the endothelium of capillaries in the lymph nodes, resulting in additional immune cells migrating out of the blood and into the lymphatic tissue. Thus, it is not necessary (and may be detrimental) to treat a mild fever. Most physicians recommend letting a fever "run its course" and give fever-reducing medication only if the fever becomes very high or if the patient is in significant discomfort from the fever.

Risks of a High Fever

A fever is considered high (i.e., clinically significant) when it is above 100°F. High fevers (103°F in children, and slightly lower in an adult) are potentially dangerous because of the changes in metabolic pathways and denaturation of body proteins (see section 2.8b). Seizures may occur at sustained body temperature above 102°F (although generally they occur at much higher temperatures), irreversible brain damage may occur at body temperatures that are sustained at greater than 106°F, and death is likely when body temperature reaches 109°F.

Figure 22.7 is a visual summary of the means of providing innate immunity. Both the structures and processes that prevent entry (first line of defense) and the nonspecific internal defenses, including cells, antimicrobial proteins, and physiologic processes of inflammation and fever (second line of defense), are included. Remember that the processes of nonspecific internal defenses are generally effective in eliminating most pathogens.

WHAT DID YOU LEARN?

11 What is a fever, and what are the three stages of a fever?

12 What are the benefits and risks of a fever?

INTEGRATE

CLINICAL VIEW 22.3

Chronic Inflammation

Chronic inflammation is the condition in which inflammation continues for longer than 2 weeks. Chronic inflammation elicits all the discomfort of the inflammatory response without necessarily ridding the body of the foreign substance. Whereas the acute inflammatory response is generally characterized by neutrophils, chronic inflammation is generally characterized by the presence of cells that arrive later in the inflammatory response, such as macrophages and lymphocytes.

Rheumatoid arthritis

©Image Source/Getty Images RF

One primary cause of chronic inflammation is overuse injuries, which occur as a result of repetitive, minute trauma to a given area of the body. These include tennis elbow, swimmer's shoulder, and shin splints.

Some forms of chronic inflammation occur when the processes of acute inflammation do not eliminate the pathogen or injurious agent, as with tuberculosis, allergens, a splinter that is not removed, injury to a blood vessel, or an autoimmune disorder (e.g., rheumatoid arthritis, shown in the image). Unfortunately, chronic inflammation can lead to tissue destruction and formation of scar tissue (fibrosis).

INTEGRATE CONCEPT OVERVIEW

Figure 22.7 Innate Immunity.

Innate immunity is provided by the innate immune system. This system includes the (*a*) first line of defense (cellular and molecular structures that help prevent entry) and (*b*) second line of defense (all internal, nonspecific means of eliminating a foreign substance).

(a) First Line of Defense | Prevent Entry of Infectious Agents

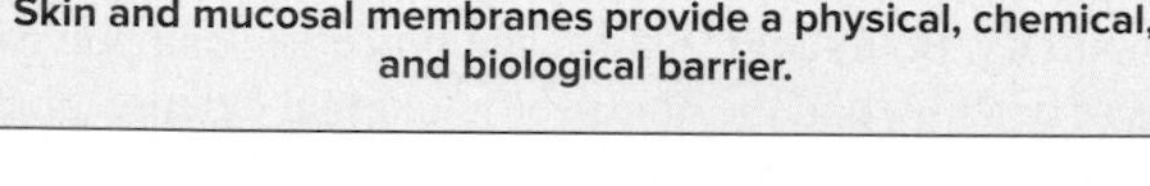

Skin and mucosal membranes provide a physical, chemical, and biological barrier.

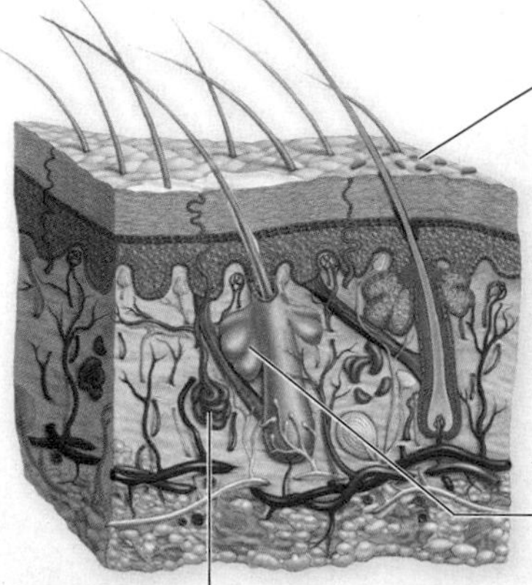

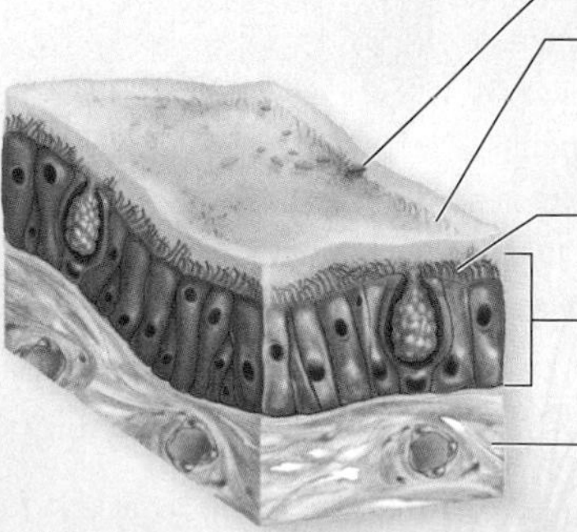

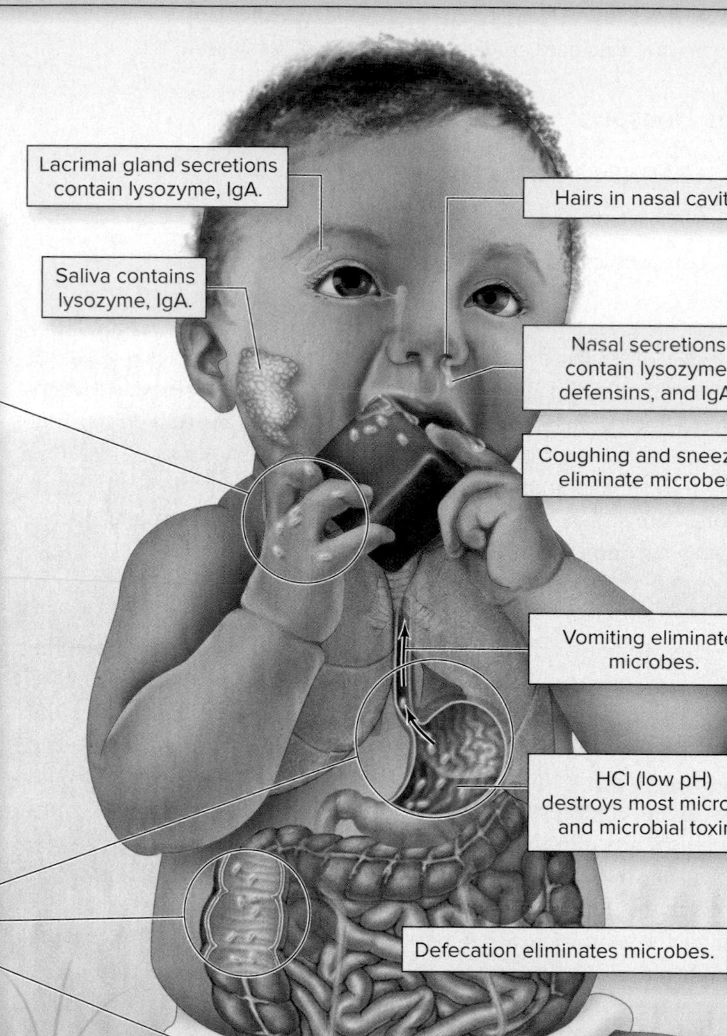

Skin: Covers body surface
Normal flora help prevent growth of pathogenic organisms.
Epidermis exfoliates, which removes potential pathogens.
Dermis contains gel-like hyaluronic acid; limits spread of microbes.
Sebaceous (oil) gland secretions called sebum are low in pH; interfere with microbial growth.
Sweat gland secretions help wash away microbes; contain lysozyme, defensins, and dermicidin, which inhibit microbial growth.

Mucosal membranes: Line organ system tracts
Normal flora
Mucus traps microbes, and contains lysozyme, defensins, and IgA to defend against potential pathogens.
Cilia sweep material along some tracts.
Epithelium provides a physical barrier.
Connective tissue contains hyaluronic acid; limits spread of infection.

Lacrimal gland secretions contain lysozyme, IgA.
Saliva contains lysozyme, IgA.
Hairs in nasal cavity.
Nasal secretions contain lysozyme, defensins, and IgA.
Coughing and sneezing eliminate microbes.
Vomiting eliminates microbes.
HCl (low pH) destroys most microbes and microbial toxins.
Defecation eliminates microbes.
Urine flushes potential pathogens from urinary tract.

(b) Second Line of Defense | Nonspecific Cellular and Molecular Defenses for Protecting the Body

FEVER

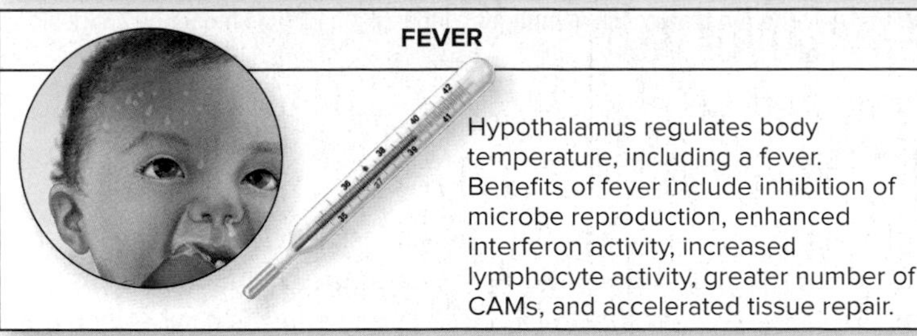

Hypothalamus regulates body temperature, including a fever. Benefits of fever include inhibition of microbe reproduction, enhanced interferon activity, increased lymphocyte activity, greater number of CAMs, and accelerated tissue repair.

INFLAMMATION

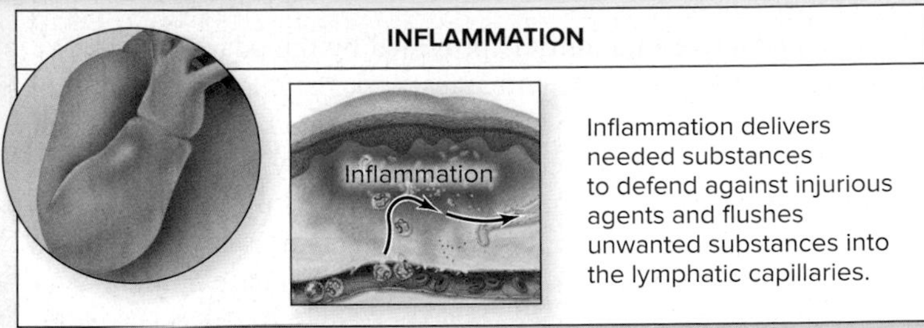
Inflammation

Inflammation delivers needed substances to defend against injurious agents and flushes unwanted substances into the lymphatic capillaries.

CELLULAR DEFENSES

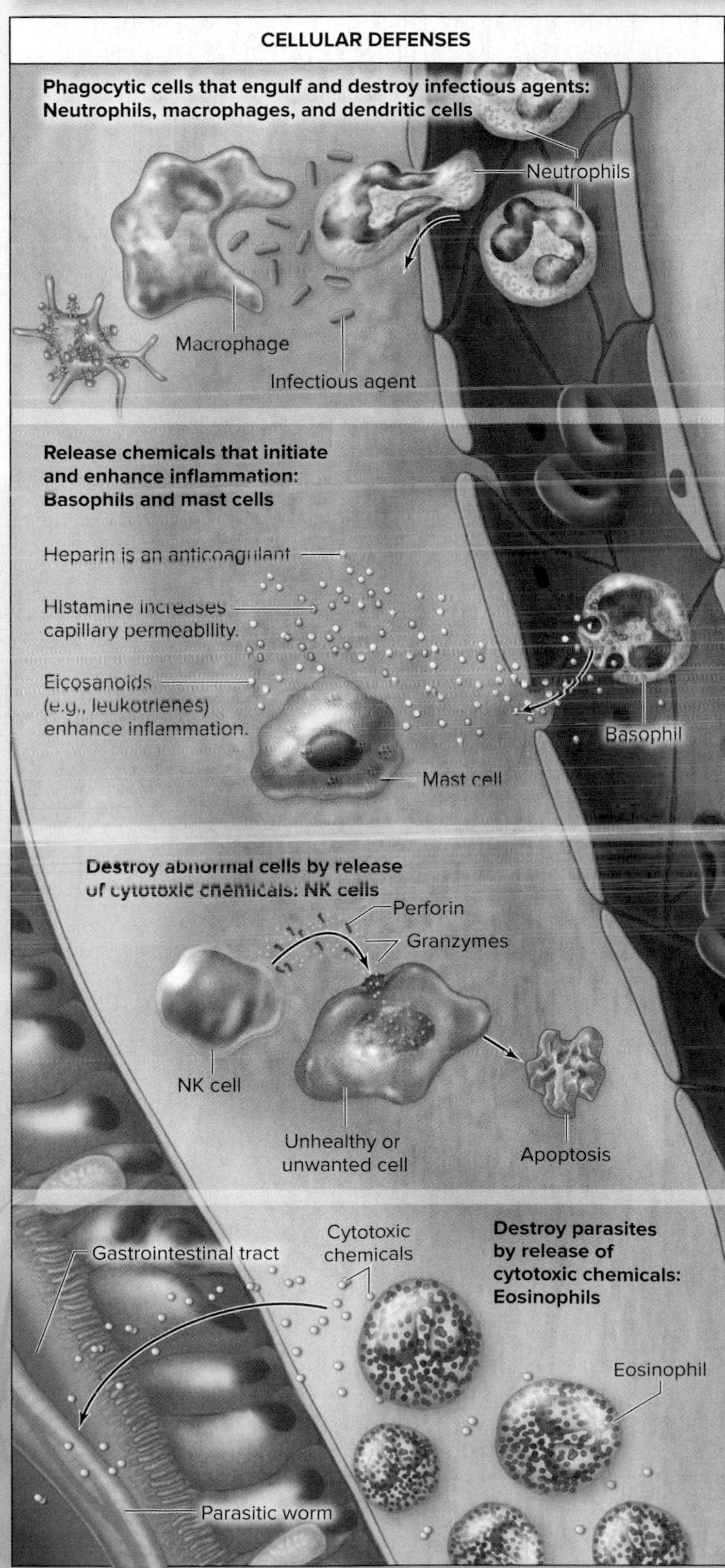
Phagocytic cells that engulf and destroy infectious agents: Neutrophils, macrophages, and dendritic cells
Neutrophils
Macrophage
Infectious agent
Release chemicals that initiate and enhance inflammation: Basophils and mast cells
Heparin is an anticoagulant
Histamine increases capillary permeability.
Eicosanoids (e.g., leukotrienes) enhance inflammation.
Basophil
Mast cell
Destroy abnormal cells by release of cytotoxic chemicals: NK cells
Perforin
Granzymes
NK cell
Unhealthy or unwanted cell
Apoptosis
Gastrointestinal tract
Cytotoxic chemicals
Destroy parasites by release of cytotoxic chemicals: Eosinophils
Eosinophil
Parasitic worm

ANTIMICROBIAL PROTEINS AND CHEMICALS

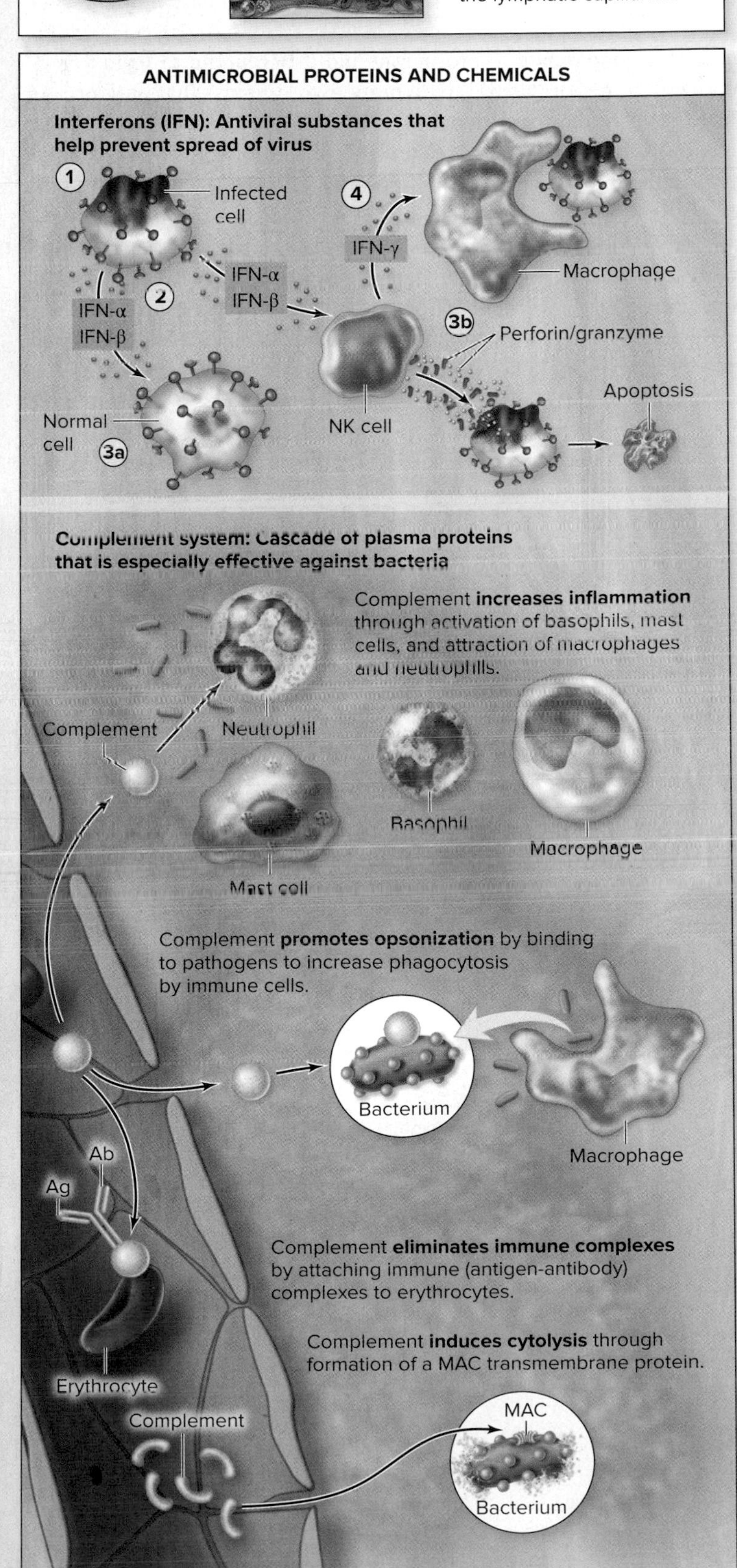
Interferons (IFN): Antiviral substances that help prevent spread of virus
1
Infected cell
2
IFN-α
IFN-β
IFN-α
IFN-β
Normal cell
3a
4
IFN-γ
Macrophage
3b
Perforin/granzyme
NK cell
Apoptosis
Complement system: Cascade of plasma proteins that is especially effective against bacteria
Complement increases inflammation through activation of basophils, mast cells, and attraction of macrophages and neutrophils.
Complement
Neutrophil
Basophil
Macrophage
Mast cell
Complement promotes opsonization by binding to pathogens to increase phagocytosis by immune cells.
Bacterium
Macrophage
Ab
Ag
Complement eliminates immune complexes by attaching immune (antigen-antibody) complexes to erythrocytes.
Erythrocyte
Complement induces cytolysis through formation of a MAC transmembrane protein.
Complement
MAC
Bacterium

22.4 Adaptive Immunity: An Introduction

Recall that adaptive immunity is provided by the adaptive immune system and involves T-lymphocytes and B-lymphocytes exclusively (see figure 22.2). The lymphocytes formed and the products they secrete in the body's defense are collectively referred to as the **immune response.** The adaptive immune response is initiated upon entry of a foreign substance (or antigen); however, it takes longer to effectively develop than the responses of the innate immune system. Lymphocytes must make contact with an antigen, which causes a lymphocyte to proliferate and differentiate to form a specialized clone, or "army," of lymphocytes against that antigen. At least several days are generally required to develop an immune response upon the first exposure to the antigen, and it is for this reason that providing adaptive immunity is considered the *third line of defense.*

Adaptive immunity may be more specifically divided into two parts: cell-mediated immunity and humoral immunity. **Cell-mediated immunity** (or *cellular immunity*) is the immune response specifically involving T-lymphocytes, which differentiate into helper T-lymphocytes and cytotoxic T-lymphocytes. In comparison, **humoral immunity** (or *antibody-mediated immunity*) is the immune response involving B-lymphocytes that develop into plasma cells to synthesize and release antibodies. A general overview of the two branches of adaptive immunity is shown in **figure 22.8**.

In this section, we first introduce several central concepts related to developing adaptive immunity, including a description of antigens, the general structure of lymphocytes, antigen-presenting cells and MHC molecules (structures that interact with T-lymphocytes for their activation), and an overview of the critical events in the life cycle of a lymphocyte.

22.4a Antigens

LEARNING OBJECTIVES

17. Describe the features of an antigen, and explain what is meant by antigenic determinant.
18. Explain immunogenicity, and list attributes that affect it.
19. Discuss how haptens stimulate immune responses.

Pathogenic organisms and other foreign substances are detected by T-lymphocytes and B-lymphocytes because they contain antigens. An **antigen** (an'ti-gen; *anti[body] + gen* = producing) is a substance that binds to a component of the adaptive immune system (T-lymphocyte or an antibody). Antigens are unique to each infectious agent and are usually proteins or large polysaccharide molecules (see section 2.7). Examples of antigens include parts of infectious agents such as the protein capsid of viruses, cell wall of bacteria or fungi, and bacterial toxins (see section 22.1). Tumor cells also contain antigens. In the case of cancerous cells, mutations occur that generally result in the production of unique (abnormal) proteins designated as **tumor antigens.**

Foreign antigens, or *nonself-antigens,* bind with the body's immune components because they are different enough in structure from the human body's molecules. In contrast, the body's molecules (structures called **self-antigens**) typically do not bind with the body's immune components. Lymphocytes are generally very effective in distinguishing a self-antigen from foreign antigen. One malfunction of the adaptive immune system, however, involves the lymphocytes reacting to self-antigens as if they were foreign. These conditions are collectively called **autoimmune disorders** (see Clinical View 22.4: "Autoimmune Disorders"). (Note that plastic and some metals are generally not antigens and are "ignored" by the immune system. This accounts for why plastics and metals, including titanium, stainless steel, and cobalt chrome, are used in artificial implants such as for a hip replacement.)

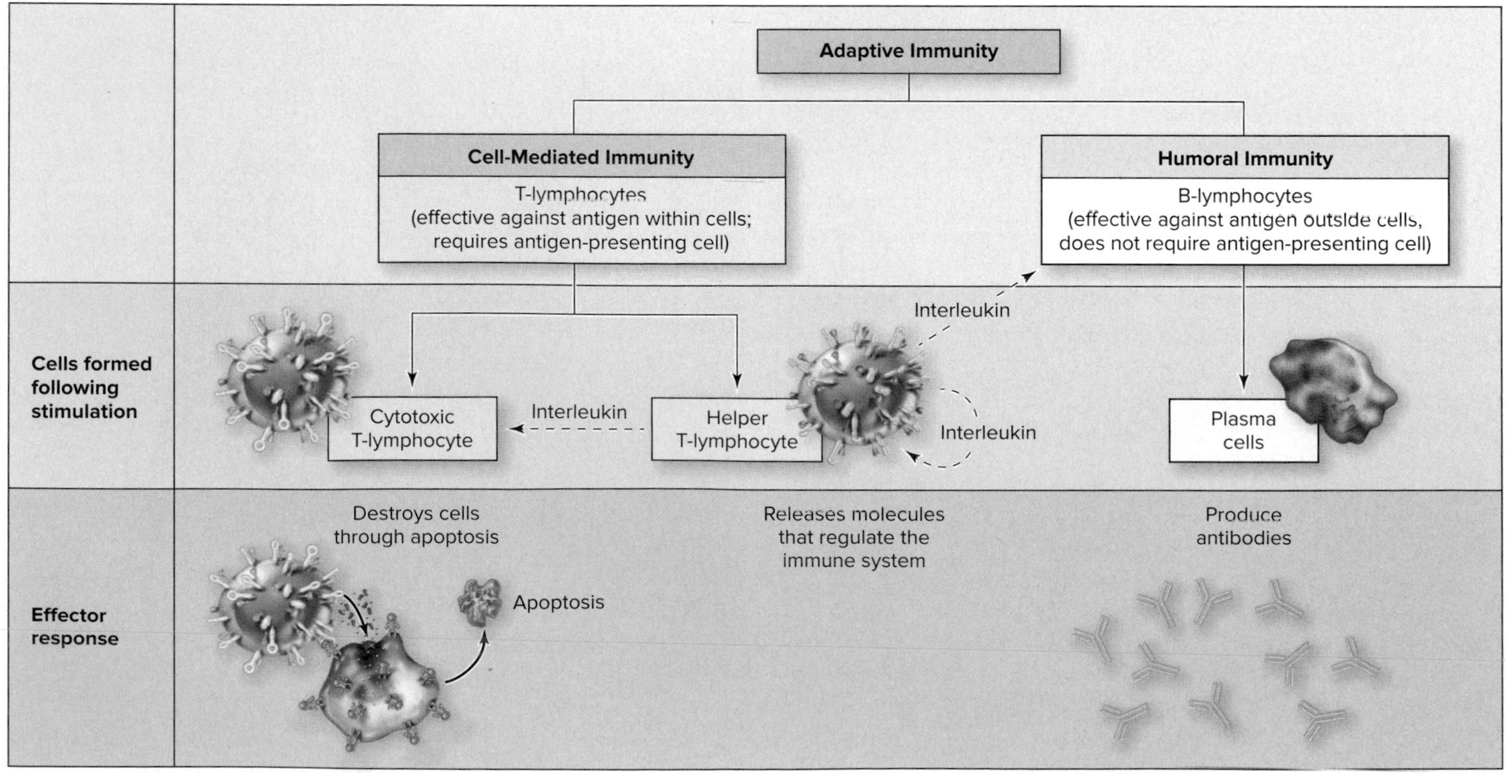

Figure 22.8 Two Branches of Adaptive Immunity. The two branches of adaptive immunity include cell-mediated immunity, which involves T-lymphocytes, and humoral immunity, which involves B-lymphocytes.

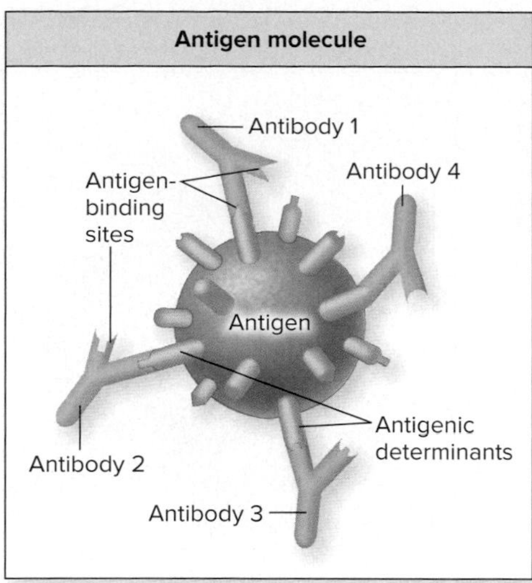

Figure 22.9 Antigens and Antigenic Determinants. An antigenic determinant is the specific portion of an antigen to which the components of the adaptive immune system bind (antibody binding shown here). Typically, each antigen has several antigenic determinants. AP|R

Lymphocytes normally have contact with only a portion of the antigen. The specific site on the antigen molecule that is recognized by lymphocytes (and antibodies) is referred to as the **antigenic determinant,** or *epitope.* Each type of antigenic determinant has a different shape, and a pathogenic organism can have numerous different antigenic determinants. **Figure 22.9** shows an antigen and several antigenic determinants.

An antigen that induces an immune response is more specifically called an **immunogen,** and its ability to cause an immune response is termed its **immunogenicity.** Important attributes that affect an antigen's immunogenicity include degree of foreignness, size, complexity, and quantity of the antigen. An increase in one or more of these attributes increases the antigen's ability to elicit an immune response, and thus its immunogenicity.

Some substances are too small to function as an antigen alone but, when attached to a carrier molecule in the host, become antigenic and trigger an immune response; these small molecules are called **haptens** (hap′ten; *hapto* = to grasp). An example is the lipid toxin in poison ivy, which penetrates the skin and combines with a body protein forming a molecular complex, which triggers an immune response. Hapten-stimulating immune responses account for hypersensitivity reactions to drugs, such as penicillin, and to chemicals in the environment, such as pollen, animal dander, mold, and snake or bee venom (see Clinical View 22.8: "Hypersensitivities").

WHAT DID YOU LEARN?

13 How is an antigenic determinant related to an antigen?

14 What distinguishes a hapten from an antigen?

22.4b General Structure of Lymphocytes

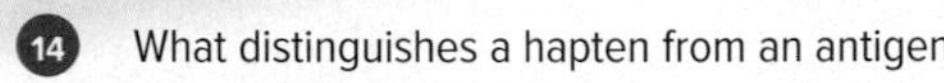

LEARNING OBJECTIVE

20. Describe receptors of both T-lymphocytes and B-lymphocytes.

T-lymphocytes and B-lymphocytes differ from other immune cells because each lymphocyte has a unique **receptor complex,** which are composed of several different and separate proteins. There are typically about 100,000 receptor complexes per cell. A receptor complex will bind one specific antigen. The antigen receptor (which is a portion of a receptor complex) of a T-lymphocyte is referred to as the **TCR** (or *T-cell receptor*), and the antigen receptor of a B-lymphocyte is called a **BCR** (or *B-cell receptor*) (**figure 22.10**).

WHAT DO YOU THINK?

2 If an antigen mutates, will the same lymphocytes recognize it?

The initial contact made between a TCR or BCR of a lymphocyte and the antigen it recognizes is different in T-lymphocytes and B-lymphocytes. T-lymphocytes must first have the antigen processed and presented in the plasma membrane of another type of cell. T-lymphocytes simply are not able to recognize the antigen without this preliminary step. In contrast, B-lymphocytes can make direct contact with an antigen.

INTEGRATE

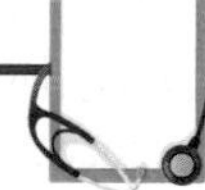

CLINICAL VIEW 22.4
Autoimmune Disorders

Autoimmune disorders occur when the immune system does not have tolerance for a specific self-antigen and subsequently initiates an immune response to these self-antigens as if they were foreign. Lack of tolerance and the development of autoimmune disorders can result from cross reactivity, from altered self-antigens, or when immune cells enter an area of immune privilege.

Cross-reactivity occurs when a foreign antigen is similar in structure to a self-antigen and the immune system is unable to distinguish between the two. For example, antigens of *Streptococcus* bacteria are similar to certain heart proteins, and immune cells damage the bicuspid (mitral) and aortic valves, resulting in *rheumatic heart disease*.

Altered self-antigens occur when a microbe induces changes in a specific protein in the body (self-antigen) and the immune cells now respond to it as if it were foreign. Examples of diseases include the following:

- *Type 1 (insulin-dependent) diabetes,* which is thought to result from a microbe's inducing changes in the proteins of the beta cells of the pancreatic islets in the pancreas; the immune system then destroys these cells (see Clinical View 17.8: "Conditions Resulting in Abnormal Blood Glucose Levels").
- *Multiple sclerosis,* which results from the destruction of the myelin sheath formed by oligodendrocytes; this destruction is caused by T-lymphocytes (see Clinical View 12.3: "Nervous System Disorders Affecting Myelin").

Areas of **immune privilege** are structures within the body that prevent or limit access of immune cells (e.g., the brain, eye, testes, ovaries, placenta). Previously, it was thought that these areas were protected passively by the blood-tissue barrier and lack of lymphatic drainage. However, recent studies have shown that these areas actively participate in maintaining immune privilege by producing various molecules, including specific immunosuppressive cytokines and special plasma proteins that actively destroy T-lymphocytes that infiltrate the area. If large numbers of immune cells enter an immune privileged area, they can destroy structures that are "perceived" as foreign. For example, if a male takes a forceful blow to the scrotum that destroys the blood-testis barrier, immune cells may destroy developing sperm cells and cause sterility.

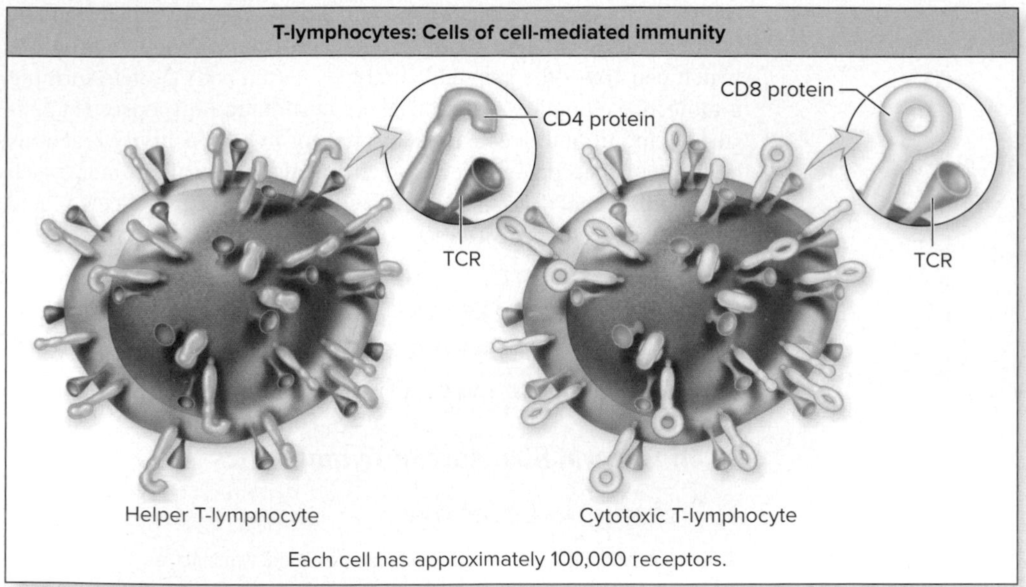

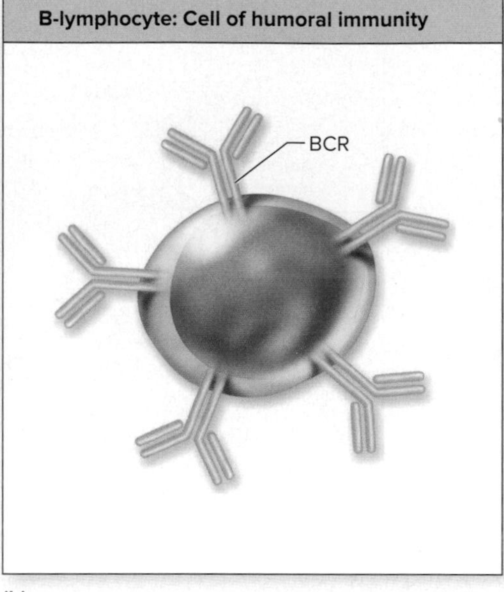

Figure 22.10 T-Lymphocytes and B-Lymphocytes. The receptors of T-lymphocytes and B-lymphocytes are plasma membrane molecules. (*a*) Helper T-lymphocytes contain TCRs (T-cell receptors) and CD4 proteins, whereas cytotoxic T-lymphocytes contain TCRs and CD8 protein. (*b*) B-lymphocytes contain BCRs (B-cell receptors). Note: There are many other receptors embedded within both T-lymphocytes and B-lymphocytes. This figure depicts only the TCR, CD4, or CD8 receptors of T-lymphocytes and the BCR of B-lymphocytes. AP|R

T-lymphocytes have additional receptor molecules (called coreceptors) that facilitate T-lymphocyte physical interaction with a cell presenting antigen. One significant category of coreceptors is the CD molecules. In fact, the two major types of T-lymphocytes—*helper T-lymphocytes* and *cytotoxic T-lymphocytes*—can be distinguished based on the specific CD protein associated with the TCR (figure 22.10*a*). The plasma membranes of helper T-lymphocytes contain the *CD4 protein,* and the plasma membranes of cytotoxic T-lymphocytes contain the *CD8 protein*.

Note that the terminology associated with T-lymphocytes can be confusing because they may have several designations. Keep in mind that their names reflect either the lymphocyte's function or the type of membrane protein receptor associated with the TCR:

- **Helper T-lymphocytes** (T_H) function to coordinate the immune response—helping to initiate both cell-mediated immunity and humoral immunity, as well as enhancing certain aspects of innate immunity (e.g., activate NK cells); it is for this reason they are called "helper" T-lymphocytes. Structurally, helper T-lymphocytes contain the CD4 plasma membrane protein and are classified as **CD4** (or *CD4+*) **cells.**
- **Cytotoxic T-lymphocytes** (T_C) release chemicals that are toxic to cells, resulting in their destruction. Because cytotoxic T-lymphocytes contain the CD8 plasma membrane protein, they are also classified as **CD8** (or *CD8+*) **cells.**

Various other types of T-lymphocytes are also formed, including (1) memory T-lymphocytes (both T_C and T_H), which cause a more rapid response to an antigen when future encounters of the same antigen occur, and (2) regulatory T-lymphocytes (Tregs), which function in suppressing the immune response. (The specific role of each is described later in the chapter.)

WHAT DID YOU LEARN?

15 What features distinguish the receptors of helper T-lymphocytes, cytotoxic T-lymphocytes, and B-lymphocytes?

22.4c Antigen-Presenting Cells and MHC Molecules

LEARNING OBJECTIVES

21. Define antigen presentation.
22. Describe antigen-presenting cells, and list cells that serve this function.
23. Explain the process of formation of MHC class I molecules in nucleated cells and MHC class II molecules in professional antigen-presenting cells.
24. Diagram the interaction of TCR and CD receptors of a T-lymphocyte with antigen associated with the MHC molecules of other cells.

Antigen presentation is the display of antigen on a cell's plasma membrane surface. This is a necessary process performed by other cells so that T-lymphocytes can recognize an antigen. Generally, two categories of cells present antigen to T-lymphocytes: all nucleated cells of the body (i.e., all cells except erythrocytes) and a category of cells called antigen-presenting cells. The term **antigen-presenting cell (APC)** is used to describe any immune cell that functions specifically to communicate the presence of antigen to *both* helper T-lymphocytes and cytotoxic T-lymphocytes. Dendritic cells and macrophages (see section 22.3b), as well as B-lymphocytes, function as APCs.

Antigen presentation requires the physical attachment of antigen to a specialized transmembrane protein called **MHC.** MHC is an abbreviation for **major histocompatibility** (his′tō-kom-pat′i-bil′i-tē; *histo* = tissue) **complex.** This name refers to the group of genes that code for MHC molecules embedded within plasma membranes. There are two primary categories of MHC molecules: MHC class I molecules and MHC class II molecules. All nucleated cells present antigen with MHC class I molecules, whereas APCs display antigen with both MHC class I molecules and MHC class II molecules (a molecule displayed only by APCs).

Synthesis and Display of MHC Class I Molecules on Nucleated Cells

MHC class I molecules are glycoproteins; they are genetically determined and are unique to each individual (see Clinical View 22.5: "Organ Transplants and MHC Molecules").

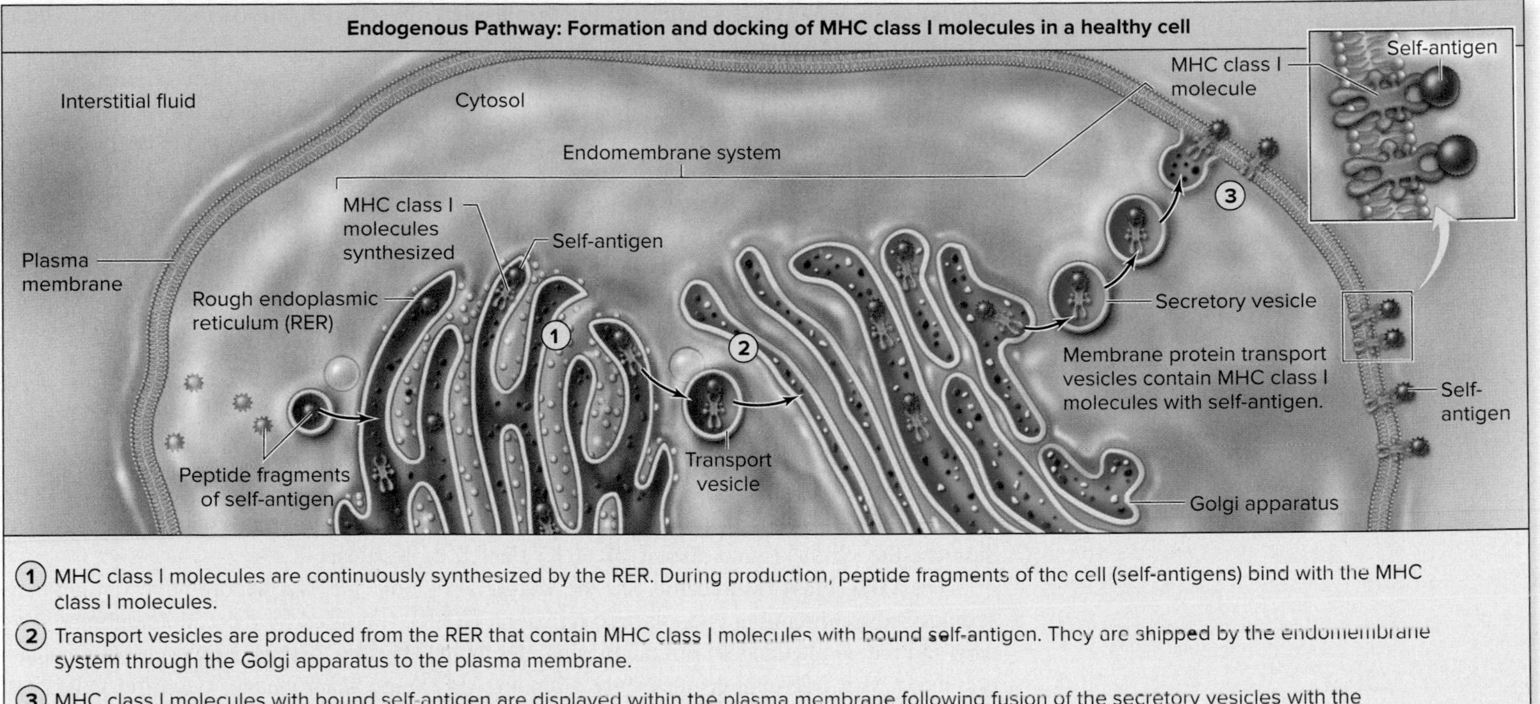

1. MHC class I molecules are continuously synthesized by the RER. During production, peptide fragments of the cell (self-antigens) bind with the MHC class I molecules.
2. Transport vesicles are produced from the RER that contain MHC class I molecules with bound self-antigen. They are shipped by the endomembrane system through the Golgi apparatus to the plasma membrane.
3. MHC class I molecules with bound self-antigen are displayed within the plasma membrane following fusion of the secretory vesicles with the plasma membrane.

(a)

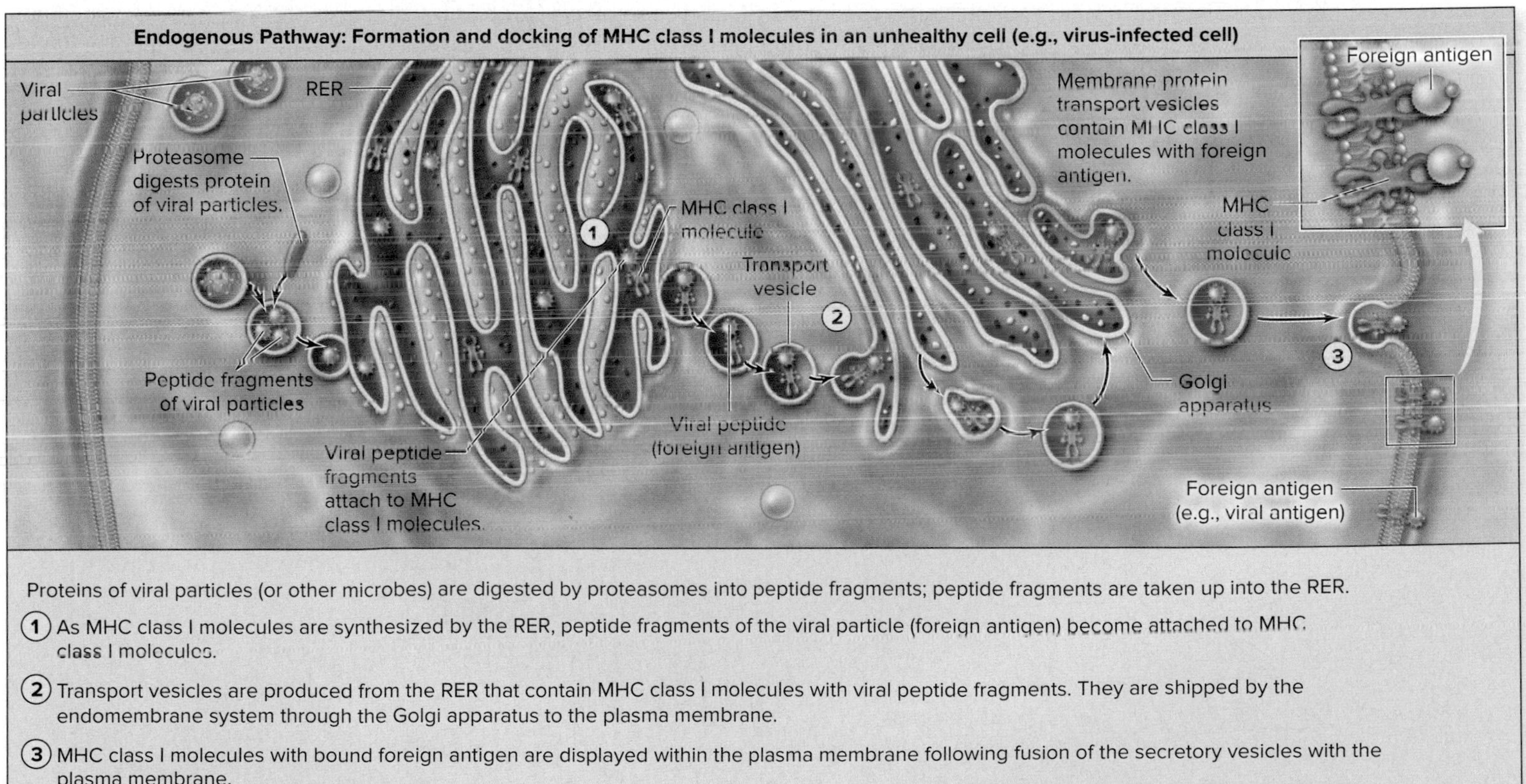

Proteins of viral particles (or other microbes) are digested by proteasomes into peptide fragments; peptide fragments are taken up into the RER.

1. As MHC class I molecules are synthesized by the RER, peptide fragments of the viral particle (foreign antigen) become attached to MHC class I molecules.
2. Transport vesicles are produced from the RER that contain MHC class I molecules with viral peptide fragments. They are shipped by the endomembrane system through the Golgi apparatus to the plasma membrane.
3. MHC class I molecules with bound foreign antigen are displayed within the plasma membrane following fusion of the secretory vesicles with the plasma membrane.

(b)

Figure 22.11 Formation and "Docking" of MHC Class I Molecules in Nucleated Cells. MHC class I molecules are produced and shipped to the plasma membrane of all nucleated cells. (*a*) A normal, healthy cell displays only self-antigens with the MHC class I molecules. (*b*) An infected or unhealthy cell displays foreign antigen with the MHC class I molecules. This "alerts" cytotoxic T-lymphocytes that this cell is infected or unhealthy and should be destroyed.

MHC class I molecules are continuously synthesized by the rough endoplasmic reticulum (RER), inserted into the ER, shipped within and modified by the endomembrane system (see section 4.6a), and then embedded within the plasma membrane for the purposes of displaying peptide fragments of **endogenous proteins** (proteins within the cell). This process is referred to as the **endogenous pathway** (**figure 22.11*a***).

A significant event occurs during the synthesis and transport of MHC class I molecules to the cell surface involving the endogenous pathway: Peptide fragments in the cell randomly bind with the MHC class I molecules. This occurs within the RER. These peptide fragments in uninfected, healthy cells are simply partially degraded proteins of the cell and are considered "self." Consequently, in *uninfected,*

INTEGRATE

LEARNING STRATEGY

You may find it helpful to consider the military analogy of an antigen-presenting cell (APC) as a "sentinel," a cell that keeps watch for potentially harmful substances and reports their presence to T-lymphocytes.

healthy cells, MHC class I molecules are displaying only self-antigens on their surface. These self-antigens are ignored or tolerated by the T-lymphocytes.

However, if the cell is infected, the antigens presented are foreign antigens (figure 22.11*b*). Proteins of an intracellular infectious agent (e.g., viral particle) are cleaved by a proteasome into peptide fragments of 3 to 15 amino acids (see section 4.6b); these degraded peptide fragments of the infectious agent are considered "nonself." The peptide fragments of the infectious agent that are in the cytosol are shipped into the RER, where the peptide fragments combine with MHC class I molecules within the RER. Through the endomembrane system, the MHC class I molecules with bound foreign antigen are shipped to the plasma membrane, where they are displayed at the cell surface. We will see that the display of foreign antigens with an MHC class I molecule provides the means of communicating specifically with *cytotoxic* T-lymphocytes and will result in the destruction of these infected cells.

Synthesis and Display of MHC Class II Molecules on Professional Antigen-Presenting Cells

Recall that APCs display both MHC class I and MHC class II molecules. MHC class I molecules are synthesized in an APC in a manner similar to other nucleated cells. Here we describe the synthesis and display of MHC class II molecules.

The **MHC class II molecule,** like the MHC class I molecule, is a glycoprotein continuously synthesized by the rough endoplasmic reticulum (RER), modified by the endomembrane system, and then embedded within the plasma membrane (**figure 22.12**). However, antigens are presented with MHC molecules only after an APC (e.g., macrophage, dendritic cell) first engulfs **exogenous antigens** (pathogens, cellular debris, or other potentially harmful substances located *outside* of cells). The process involving proteins that are engulfed from outside a cell is referred to as the **exogenous pathway.** (Recall from section 22.3b that innate immune cells, including dendritic cells and macrophages, recognize microbes through pattern recognition receptors, such as toll-like receptors, that are displayed on their cell surface, which bind with molecular motifs of a microbe.)

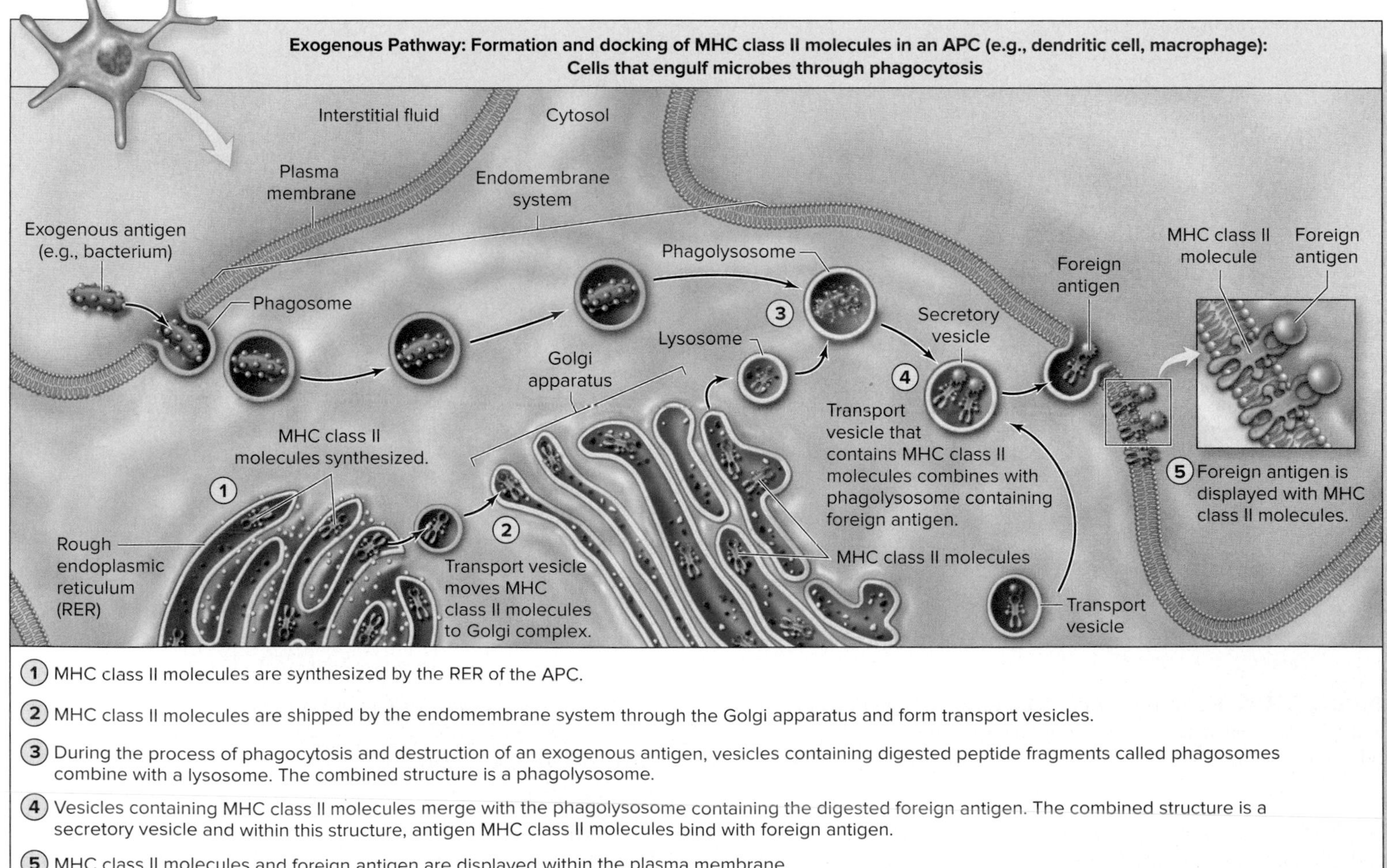
Exogenous Pathway: Formation and docking of MHC class II molecules in an APC (e.g., dendritic cell, macrophage): Cells that engulf microbes through phagocytosis
Interstitial fluid
Cytosol
Plasma membrane
Endomembrane system
Exogenous antigen (e.g., bacterium)
Phagosome
Phagolysosome
Lysosome
Golgi apparatus
Secretory vesicle
Foreign antigen
MHC class II molecule
Foreign antigen
MHC class II molecules synthesized.
Rough endoplasmic reticulum (RER)
Transport vesicle moves MHC class II molecules to Golgi complex.
Transport vesicle that contains MHC class II molecules combines with phagolysosome containing foreign antigen.
MHC class II molecules
Transport vesicle
Foreign antigen is displayed with MHC class II molecules.
1 MHC class II molecules are synthesized by the RER of the APC.
2 MHC class II molecules are shipped by the endomembrane system through the Golgi apparatus and form transport vesicles.
3 During the process of phagocytosis and destruction of an exogenous antigen, vesicles containing digested peptide fragments called phagosomes combine with a lysosome. The combined structure is a phagolysosome.
4 Vesicles containing MHC class II molecules merge with the phagolysosome containing the digested foreign antigen. The combined structure is a secretory vesicle and within this structure, antigen MHC class II molecules bind with foreign antigen.
5 MHC class II molecules and foreign antigen are displayed within the plasma membrane.

Figure 22.12 Formation and "Docking" of MHC Class II Molecules in Antigen-Presenting Cells. Antigen-presenting cells (APCs), which include dendritic cells, macrophages, and B-lymphocytes, display both MHC class I and MHC class II molecules. Here we see the engulfment of exogenous antigen (Ag) (antigen from outside a cell) and its presentation with MHC class II molecules in the plasma membrane.

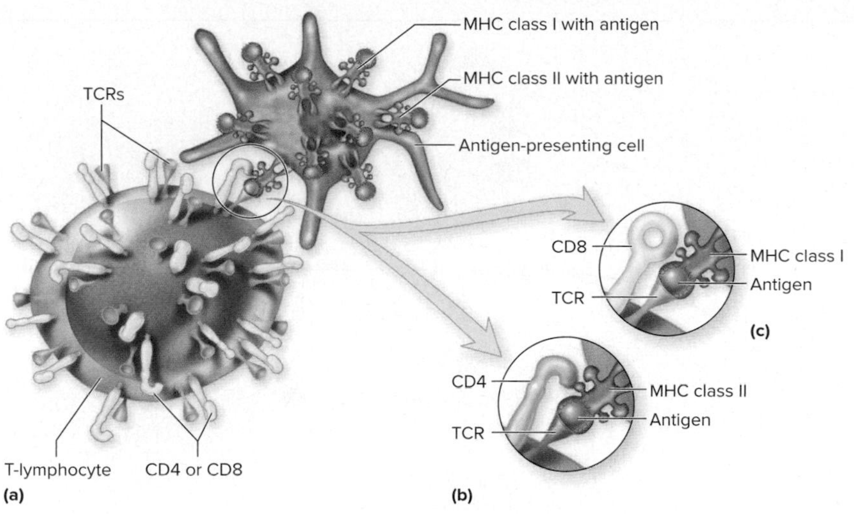

Figure 22.13 Interaction of Receptors of T-Lymphocytes with MHC Molecules of Other Cells. (*a*) A T-lymphocyte directly attaches to the MHC molecules of the antigen-presenting cell (APC). This helps keep the two cells together as the TCR of the T-lymphocyte examines the peptide fragment to determine if it is self-antigen or a foreign antigen. (*b*) CD4 interacts specifically with an MHC class II molecule. (*c*) CD8 interacts specifically with an MHC class I molecule.

A phagosome (vesicle) is formed as the APC engulfs the microbe through phagocytosis. The phagosome containing foreign antigen merges with a lysosome to form a phagolysosome, where the substance is digested into peptide fragments. The vesicle containing peptide fragments (antigens) then merges with vesicles containing newly synthesized MHC class II molecules. The peptide fragments are then "loaded" into the MHC class II molecules. These vesicles in turn then merge with the plasma membrane of the APC, with exogenous antigen now displayed bound to MHC class II molecules. This display of foreign antigen with an MHC class II molecule provides the means of communicating specifically with *helper* T-lymphocytes. A similar process occurs by APCs to display antigen with MHC class I molecules. However, this display of foreign antigen with an MHC class I molecule provides the means of communicating specifically with *cytotoxic* T-lymphocytes. In either case, the communication between the APCs and T-lymphocyte will trigger their activation (as described in detail in section 22.6).

Figure 22.13 illustrates how the receptors of a helper T-lymphocyte (TCR and CD4) interact with an MHC class II molecule that contains antigen on the surface of an APC (figure 22.13*b*), and how the receptors of a cytotoxic T-lymphocyte (TCR and CD8) interact with an MHC class I molecule that contains antigen on the surface of an APC (figure 22.13*c*). Details of these interactions are described in section 22.6.

INTEGRATE

LEARNING STRATEGY

Remember which CD of a T-lymphocyte interacts with which MHC molecule of an antigen-presenting cell by knowing that the two numbers multiplied always equals 8:

MHC class II (2) × CD 4 = 8

MHC class I (1) × CD 8 = 8

WHAT DID YOU LEARN?

16 Which type of MHC-class molecule is found on all nucleated cells and is used to communicate with cytotoxic T-lymphocytes? Which classes are displayed on APCs, and which class is used specifically to communicate with (a) helper T-lymphocytes and (b) cytotoxic T-lymphocytes?

INTEGRATE

CLINICAL VIEW 22.5

Organ Transplants and MHC Molecules

An **organ transplant** involves the transfer of an organ from one individual to another (see Clinical View 5.6: "Tissue Transplants"). Examples of transplanted organs include the kidney, liver, heart, and lungs. Prior to the transplant, the donor and recipient are tested for the major histocompatibility complex (MHC) molecules and the ABO blood group antigens. No two individuals (except identical twins) have exactly the same MHC molecules.

Organ transplants are risky because the MHC molecules of the cells of the transplanted tissue or organ may be detected by the immune system as foreign. Consequently, components of the innate and adaptive immune systems attempt to destroy it. Thus, the recipient's immune system is suppressed with drugs that make it less likely that it will detect the foreign antigens and cause rejection. However, these immunosuppressive drugs increase the risk that infectious disease and tumor cells will go undetected.

An important exception to these required precautions involves corneal transplants. The cornea of the eye is in an area of immune privilege (i.e., an area where immune cells are generally prevented from entering) (see Clinical View 22.4: "Autoimmune Disorders"). Thus, a cornea can be transplanted from one individual to another without the need for matching the tissue or administering immunosuppressive drugs.

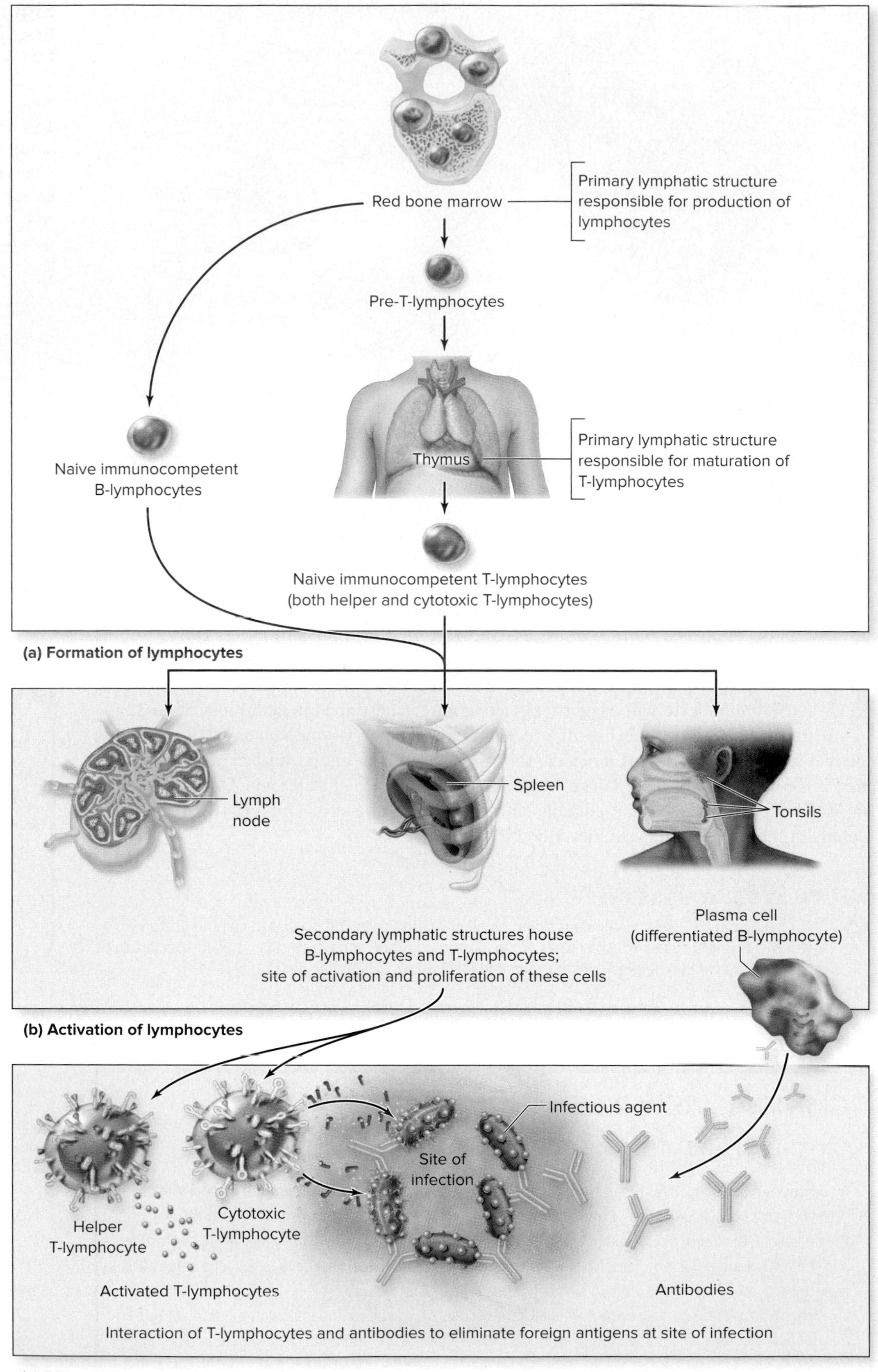

Figure 22.14 Overview of Life Events of Lymphocytes. (*a*) Lymphocyte formation occurs in primary lymphatic structures (bone marrow and thymus), and mature (immunocompetent) cells migrate to secondary lymphatic structures. (*b*) Activation occurs in secondary lymphatic structures. (*c*) The effector response occurs at the site of infection.

22.4d Overview of Life Events of Lymphocytes

LEARNING OBJECTIVE

25. Identify the three significant events that occur in the lifetime of a lymphocyte.

Participation of lymphocytes in the body's defense involves three significant events (**figure 22.14**):

- **Formation of lymphocytes.** The formation and maturation of lymphocytes occur within primary lymphatic structures (red bone marrow and the thymus; see section 21.3). Here T-lymphocytes and B-lymphocytes become able to recognize only one specific foreign antigen.
- **Activation of lymphocytes.** Following their formation, lymphocytes then migrate to secondary lymphatic structures (e.g., lymph nodes, the spleen, tonsils, MALT) where they are housed (see section 21.4). Typically, these locations are where lymphocytes have their first exposure to the antigen that they bind, and thus become activated. In response to activation, lymphocytes replicate to form many identical lymphocytes.
- **Effector response.** The effector response is the specific action of the T-lymphocytes and B-lymphocytes to help eliminate the antigen at the site of infection. T-lymphocytes leave the secondary lymphatic structures, migrating to the site of infection. B-lymphocytes (as differentiated plasma cells) primarily remain within the secondary lymphatic structures, synthesizing and releasing large quantities of antibodies against the antigen. The antibodies enter the blood and lymph and are transported to the site of infection.

Please note that these processes are sequential but generally differ in where they take place in the body: lymphocyte development in primary lymphatic structures, activation of lymphocytes in secondary lymphatic structures, and effector response at the site of infection.

WHAT DID YOU LEARN?

17 Where does a lymphocyte typically encounter an antigen for the first time: primary lymphatic structure, secondary lymphatic structure, or site of infection?

22.5 Formation and Selection of T-Lymphocytes in Primary Lymphatic Structures

Lymphocytes originate in red bone marrow (see section 18.3a). Following their formation, lymphocytes are "tested" to see whether they are able to bind antigen and respond to it—that is, whether they are *immunocompetent*. This process occurs primarily during development and shortly after birth in primary lymphatic structures (bone marrow and thymus). We describe how this occurs in T-lymphocytes.

22.5a Formation of T-Lymphocytes

LEARNING OBJECTIVE

26. Explain how T-lymphocytes mature.

T-lymphocytes originate in red bone marrow and then migrate to the thymus to complete their maturation. (The "T" of T-lymphocytes reflects the role of the thymus in its maturation.) Millions of pre-T-lymphocytes, which are called *thymocytes,* migrate from the red bone marrow to the thymus; they possess a unique TCR receptor *and* initially neither the CD4 nor the CD8 proteins (called "double negative"). Within the thymus, these cells will synthesize and display both CD4 and CD8 proteins (referred to as "double positive"). Thymocytes are immature T-lymphocytes with a TCR that was produced randomly through "gene shuffling," a concept beyond the scope of this text. Keep in mind that each lymphocyte has a unique TCR.

Each T-lymphocyte must have its TCR tested to determine not only whether it is able to bind to the MHC molecule with presented antigen but also whether it binds only to antigen that is foreign, or "nonself." This testing results in T-lymphocyte selection.

WHAT DID YOU LEARN?

18 Where does the maturation of T-lymphocytes take place?

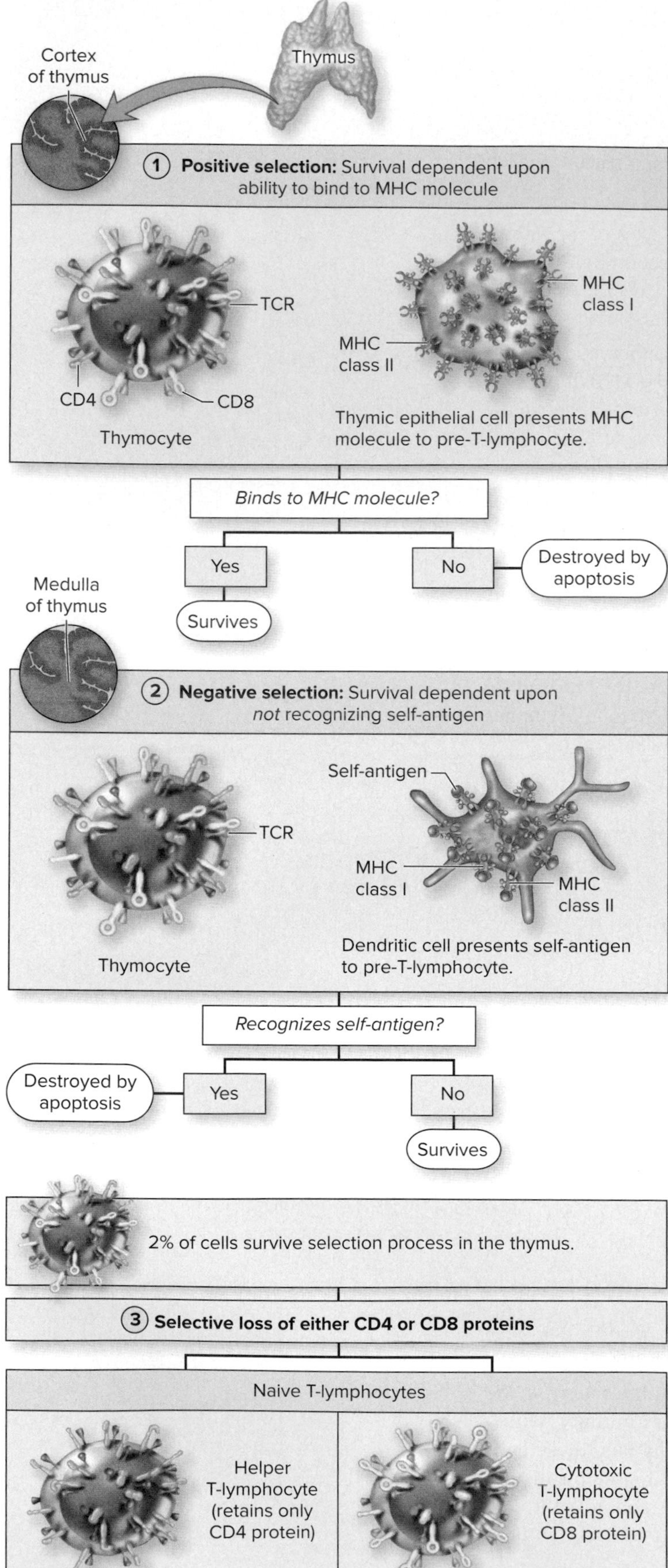

Figure 22.15 Thymic Selection. T-lymphocytes complete their maturation in the thymus to form naive T-lymphocytes that are immunocompetent. The process of thymic selection includes (1) positive selection, (2) negative selection, and (3) the selective loss of either CD4 or CD8 proteins.

22.5b Selection and Differentiation of T-Lymphocytes

LEARNING OBJECTIVE

27. Compare and contrast positive and negative selection of T-lymphocytes and what is meant by central tolerance.

Thymocytes go through two selection processes in the thymus to become T-lymphocytes, collectively called *thymic selection* (see **figure 22.15** as you read through this section):

1. **Positive selection.** Positive selection occurs within the outer cortex of the thymus. The TCR embedded in the plasma membrane of a T-lymphocyte must be able to recognize and bind an MHC molecule. This is tested by having thymocytes (pre-T-lymphocytes) bind with thymic epithelial cells that have MHC molecules. Those thymocytes that can bind an MHC molecule survive, and those that cannot are eliminated. Thus, those cells that perform this function are selected *for,* a process referred to as **positive selection** (figure 22.15, step 1). Thymocytes migrate into the medulla for negative selection.
2. **Negative selection.** These cells must also *not* bind to any self-antigens that are presented within an MHC molecule. This is tested by thymic dendritic cells presenting self-antigens with MHC class I and II molecules. If the thymocyte does bind to the self-antigen, then it is destroyed. Thus, those cells that perform this function are selected *against,* a process referred to as **negative selection** (figure 22.15, step 2). This is the process by which cells generally learn to "ignore" molecules of the body or self-antigens, a state referred to as **self-tolerance.** This process, which occurs in the primary lymphatic structures, is more specifically called **central tolerance.**

Consequently, thymocytes that survive both positive and negative selection can bind an MHC molecule and recognize foreign antigen. Only approximately 2% of the originally formed thymocytes survive both selection processes; the remaining 98% are eliminated in the thymus by apoptosis (see section 4.10).

3. The final step in T-lymphocyte selection is the *differentiation* of each thymocyte into either a helper T-lymphocyte (CD4 cell) by the selective loss of the CD8 protein, or a cytotoxic T-lymphocyte (CD8 cell) by the selective loss of CD4 protein (figure 22.15, step 3). Consequently, two primary types of T-lymphocytes leave the thymus: helper T-lymphocytes (that are CD4+) and cytotoxic T-lymphocytes (that are CD8+).

WHAT DID YOU LEARN?

19 What would happen if a thymocyte that failed the negative selection test was not destroyed and instead entered the blood to circulate as a T-lymphocyte?

22.5c Migration of T-Lymphocytes

LEARNING OBJECTIVES

28. Explain why T-lymphocyes leaving the thymus are called both *immunocompetent* and *naive*.
29. Describe the formation and function of T-lymphocytes (Tregs) in peripheral tolerance.

The T-lymphocytes that leave the thymus are **immunocompetent cells** (able to bind antigen and respond to it). However, each of these T-lymphocytes is also classified as a **naive T-lymphocyte.** (Here the term *naive* refers to T-lymphocytes that lack experience because they have not yet encountered the antigen that they recognize.) Naive immunocompetent helper T-lymphocytes and naive immunocompetent

INTEGRATE

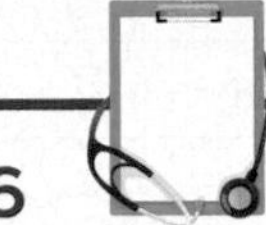

CLINICAL VIEW 22.6

Regulatory T-Lymphocytes and Tumors

Tumors have been shown to induce Tregs (specifically called tumor Tregs) to proliferate. The increased production of these cells results in greater than normal suppression of the immune system. Consequently, immune cells are less likely to detect the tumor cells, and the tumor continues to grow. Potential cancer treatments are in development that involve diminishing the inhibitory influences of tumor Tregs.

cytotoxic T-lymphocytes migrate from the thymus to secondary lymphatic structures, where they are housed (figure 22.14*b*).

It should be noted, however, that a subclass of CD4 cells, called **regulatory T-lymphocytes (Tregs),** is also formed. (These cells were previously called *suppressor T-lymphocytes*.) Tregs are formed from T-lymphocytes that bind self-antigens to a *moderate* extent compared to other CD4+ cells. Tregs migrate to the periphery (body structures outside the primary lymphatic structures), where they release inhibitory chemicals that turn off both the cell-mediated immune response and the humoral immune response. Tregs function in self-tolerance outside the primary lymphatic structures—a process that is more specifically called **peripheral tolerance.** Current studies have focused extensively on the role of Tregs in disease (e.g., microbial infections, autoimmune disorders, allergies, tumors; see Clinical View 22.6: "Regulatory T-Lymphocytes and Tumors").

A similar process to form and select B-lymphocytes occurs in the red bone marrow. (However, MHC is not involved in selection of B-lymphocytes.) Naive immunocompetent B-lymphocytes also then migrate to secondary lymphatic structures where they are housed and come in contact with foreign antigen that stimulates them to proliferate and differentiate. New T- and B-lymphocytes continue to form throughout one's lifetime, and these cells participate in the immune response when exposed to a particular foreign antigen.

 **WHAT DID YOU LEARN?**

 20 In general, how does central tolerance differ from peripheral tolerance?

22.6 Activation and Clonal Selection of Lymphocytes

Lymphocyte activation involves physical contact between a lymphocyte and the antigen it recognizes and the subsequent proliferation and differentiation of lymphocytes into a clone of identical cells that have the same TCR or BCR that matches that specific antigen. This process of forming a clone in response to a specific antigen is called **clonal selection.**

The first encounter between an antigen and a lymphocyte is called an **antigen challenge.** It typically occurs in secondary lymphatic structures. The specific secondary lymphatic structure in the body in which the antigen challenge occurs generally depends upon the point of entry of the antigen. Antigen in the blood is taken to the spleen; antigen that penetrates the skin is engulfed and transported by epidermal dendritic cells to a lymph node; and antigen that enters through the mucosal membrane of the respiratory, gastrointestinal, urinary, or reproductive tract comes into contact with the tonsils or MALT (mucosa-associated lymphatic tissue).

22.6a Activation of T-Lymphocytes

LEARNING OBJECTIVE

30. Describe how both helper T-lymphocytes and cytotoxic T-lymphocytes are activated, including the specific role of IL-2 in both activations.

Both types of T-lymphocytes must undergo activation before they can carry out immune system functions. Activation of both types of T-lymphocytes requires two signals; however, the specific process differs between the two types.

Activation of Helper T-Lymphocytes

The specifics of activation of helper T-lymphocytes is shown in **figure 22.16*b*.** The **first signal** is direct physical contact between the MHC class II molecule of an antigen-presenting cell (APC) and the TCR of a helper T-lymphocyte. Exogenous antigen previously engulfed by an APC is presented on its surface with MHC class II molecules (as described in section 22.4c). The APC either is housed in the secondary lymphatic structure (e.g., macrophage) or migrates there from the skin (e.g., dendritic cells) to make contact with the helper T-lymphocyte.

A helper T-lymphocyte binds to the APC to inspect the antigen: The specific TCR of a T-lymphocyte binds with the peptide fragment presented with an MHC *class II* molecule of the APC. This interaction is stabilized by the CD4 molecule of the helper T lymphocyte binding to other regions of the MHC class II molecule. If the TCR does not recognize the presented antigen, it disengages from the APC. If it does recognize the antigen, contact between the two cells lasts several hours.

The **second signal** takes place when other receptors of the APC (e.g., B7) interact with receptors of the helper T-lymphocyte (e.g., CD28). Ultimately, helper T-lymphocytes are induced to synthesize and release the cytokine interleukin 2 (IL-2), which occurs within about 24 hours. IL-2 acts as an autocrine hormone (see section 17.3b) to further stimulate the helper T-lymphocyte from which it was released.

T-lymphocytes are activated and proliferate to form clones of helper T-lymphocytes (T-lymphocytes that possess TCRs that bind that specific antigen). Some of the cells produced are *activated helper T-lymphocytes* that continue to produce IL-2, and some are *memory helper T-lymphocytes,* cells available for subsequent encounters with the specific antigen. (Note that lack of a second signal is thought to result in helper T-lymphocytes becoming Tregs. Recall that Tregs can also be formed in the thymus when CD4 cells bind self-antigen with moderate affinity.)

Activation of Cytotoxic T-Lymphocytes

The **first signal** for a cytotoxic T-lymphocyte is similar to the first stimulation for a naive helper T-lymphocyte (figure 22.16*a*). However, direct physical contact is made between the TCR of a cytotoxic T-lymphocyte and a peptide fragment presented with an MHC *class I* molecule of either an APC or an infected cell. This interaction is stabilized by the CD8 of the cytotoxic T-lymphocyte binding to other regions of the MHC class I molecule. The **second signal** is the binding of IL-2 that is released from *helper* T-lymphocytes. IL-2 acts as a paracrine hormone (see section 17.3b) to stimulate the cytotoxic T-lymphocyte. IL-2 is required for cytotoxic T-lymphocytes to become fully activated. (Note that only APCs [e.g., dendritic cells] are able to activate naive cytotoxic T-lymphocytes—that is, when cytotoxic T-lymphocytes are first exposed to the antigen they recognize.)

Upon activation, cytotoxic T-lymphocytes proliferate and differentiate into clones, some becoming *activated cytotoxic T-lymphocytes,* and others developing into *memory cytotoxic T-lymphocytes* that are activated upon reexposure to the same antigen.

Costimulation to activate T-lymphocytes for clonal selection
Cytotoxic T-lymphocyte
1 First signal: CD8 binds with MHC class I molecule of infected cell; TCR interacts with antigen within MHC class I molecule.
Antigen
MHC class I
APC
Naive cytotoxic T-lymphocyte
TCR
CD8
IL-2
2 Second signal: IL-2 released from activated helper T-lymphocyte activates the cytotoxic T-lymphocyte.
Activated cytotoxic T-lymphocyte proliferates and differentiates to form a clone of activated and memory cytotoxic T-lymphocytes.
(a)
Helper T-lymphocyte
1 First signal: CD4 binds with MHC class II molecule of APC; TCR interacts with antigen within MHC class II molecule.
Antigen
MHC class II
APC
Naive helper T-lymphocyte
TCR
CD4
IL-2
2 Second signal: Other receptors interact (not shown) and the helper T-lymphocyte releases IL-2, which binds with the helper T-lymphocyte.
Activated helper T-lymphocyte proliferates and differentiates to form a clone of activated and memory helper T-lymphocytes.
(b)

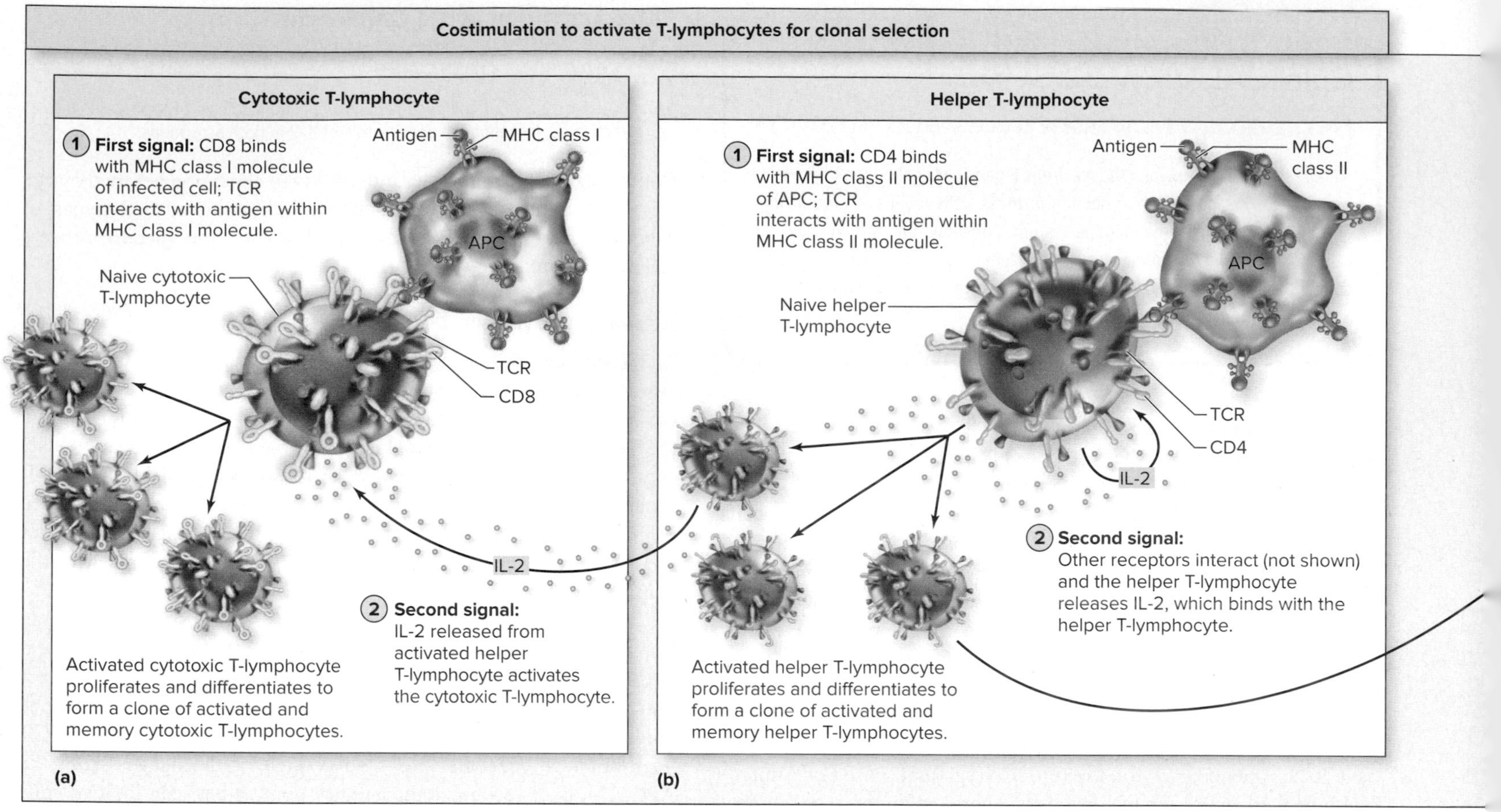

Figure 22.16 Activation of Lymphocytes. Activation of lymphocytes occurs in secondary lymphatic structures, usually the lymph nodes or spleen. Activation results in lymphocyte proliferation and differentiation to form a clone of identical cells that includes memory cells. Two signals (costimulation) are required to activate each type of lymphocyte: (*a*) cytotoxic T-lymphocyte, (*b*) helper T-lymphocyte, and (*c*) B-lymphocyte. AP|R

WHAT DID YOU LEARN?

21 What type of cell is required to activate both helper T-lymphocytes and naive cytotoxic T-lymphocytes?

22 How do cytokines released by helper T-lymphocytes participate in activation of both helper and cytotoxic T-lymphocytes?

22.6b Activation of B-Lymphocytes

LEARNING OBJECTIVE

31. Compare the activation of B-lymphocytes with that of T-lymphocytes.

Immunocompetent but naive B-lymphocytes are also activated by a specific antigen in secondary lymphatic structures. As with T-lymphocytes, two signals are required. However, B-lymphocytes do not require antigen to be presented by other nonlymphocyte cells. B-lymphocytes recognize and respond to antigens *outside of cells,* such as antigens of viral particles, bacterial toxins, or yeast spores.

The **first signal** occurs when antigen binds to the BCR, and the antigen cross-links BCRs (figure 22.16*c*). The stimulated B-lymphocyte engulfs, processes, and presents the antigen to the helper T-lymphocyte that recognizes that antigen. (This is similar to the action of other APCs; see section 22.4c.) The **second signal** occurs when an activated helper T-lymphocyte releases IL-4 to stimulate the B-lymphocyte.

Activation of B-lymphocytes causes the B-lymphocytes to proliferate and differentiate. Most of the *activated B-lymphocytes* differentiate into **plasma cells** that produce antibodies, and the remainder become *memory B-lymphocytes* that are activated upon reexposure of the same antigen. Memory B-lymphocytes differ from plasma cells in some respects: (1) The memory B-lymphocytes retain their BCRs, and (2) memory B-lymphocytes have a much longer life span (months to years) than plasma cells (typically 5 to 7 days). Note that B-lymphocytes can be stimulated by antigen without direct contact between a B-lymphocyte and helper T-lymphocyte under certain conditions. However, the production of memory B-lymphocytes and the various forms of antibodies (described in section 22.8) requires helper T-lymphocyte participation during B-lymphocyte activation. Observe figure 22.16 and notice the central role that helper T-lymphocytes play in activating both cytotoxic T-lymphocytes (cell-mediated branch of immunity) and B-lymphocytes (humoral branch of immunity).

WHAT DID YOU LEARN?

23 Is a separate APC required for B-lymphocyte activation, or is a B-lymphocyte able to serve the role of an APC?

24 Explain the role of cytokines that are released from helper T-lymphocytes in the activation of B-lymphocytes.

22.6c Lymphocyte Recirculation

LEARNING OBJECTIVE

32. Describe lymphocyte recirculation, and explain its general function.

One of the hurdles of developing adaptive immunity is the requirement of direct physical contact between antigen and the specific lymphocyte with the unique receptor that recognizes the antigen. It is estimated that only 1 in every 100,000 to 1 million T-lymphocytes or B-lymphocytes can bind with a specific antigen on the first exposure to that antigen—that is, during the antigen challenge. The "odds" for contact are increased because lymphocytes reside only temporarily in any given secondary lymphatic structure, and after a period of time they exit and then circulate through blood and lymph every several days. This process is referred to as

Costimulation to activate a B-lymphocyte for clonal selection

B-lymphocyte

Naive B-lymphocyte
Antigen
BCR
Antigen cross-links BCRs
Activated helper T-lymphocyte
IL-4
CD4
TCR
MHC II with antigen
Antigen is presented with MHC class II molecules.
Memory B-lymphocytes
Activated B-lymphocyte proliferates and differentiates to form a clone of plasma cells and memory B-lymphocytes.
Antibodies
Plasma cells produce antibodies.

(1) **First signal:** Free antigen binds to BCR; B-lymphocyte engulfs and presents antigen to activated helper T-lymphocyte.

(2) **Second signal:** IL-4 released from activated helper T-lymphocyte stimulates B-lymphocyte.

(c)

INTEGRATE

LEARNING STRATEGY

The "military" contribution of the different lymphocytes can be thought of as follows:

- **Helper T-lymphocytes** are generals of the army; they recruit and regulate other immune cells.
- **Cytotoxic T-lymphocytes** are highly trained soldiers; they engage in "cell-to-cell" combat against a specific foe.
- **B-lymphocytes** are the elite military forces that release "munitions" (antibodies) that are typically released from a distance.

lymphocyte recirculation. Lymphocyte recirculation provides a means of delivering different lymphocytes to secondary lymphatic structures, making it more likely that a lymphocyte will encounter its antigen, if present.

WHAT DID YOU LEARN?

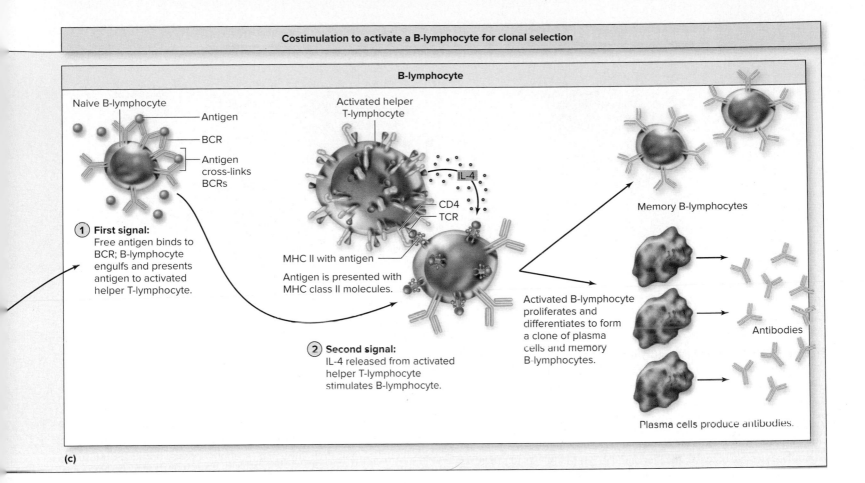

25 What advantage is provided by lymphocyte recirculation?

22.7 Effector Response at Infection Site

The **effector response** comprises the specific mechanisms that activated lymphocytes use to help eliminate the antigen. Each type of lymphocyte has a unique function. Helper T-lymphocytes release IL-2, IL-4, and other cytokines that regulate (or stimulate) cells of both the adaptive and innate immune systems. Cytotoxic T-lymphocytes destroy unhealthy cells by apoptosis. Plasma cells (differentiated B-lymphocytes) produce antibodies (see figure 22.8).

22.7a Effector Response of T-Lymphocytes

LEARNING OBJECTIVES

33. Explain the effector response of helper T-lymphocytes.
34. Explain how an unhealthy cell is destroyed by cytotoxic T-lymphocytes.
35. Explain why the processes of T-lymphocytes are collectively called the cell-mediated branch of adaptive immunity.

Just as with activation, the effector response of helper T-lymphocytes and cytotoxic T-lymphocytes differs, as is reflected in their names.

Effector Response of Helper T-Lymphocytes

Activated and memory helper T-lymphocytes leave the secondary lymphatic structure after several days of exposure to antigen. They migrate to the site of infection, where they continue to release the cytokines to regulate other immune cells **(figure 22.17*a*)**.

Although helper T-lymphocytes were named based on their function in helping activate B-lymphocytes, their contributions are much more encompassing. Helper T-lymphocytes activate cytotoxic T-lymphocytes, as described in section 22.6a, through the release of cytokines (e.g., IL-2); they also enhance formation and activity of cells of the innate immune system, including macrophages and NK cells. Thus, healthy helper T-lymphocytes play a central role in a normally functioning immune system (see Clinical View 22.9: "HIV and AIDS").

WHAT DO YOU THINK?

3 HIV (the virus that causes AIDS) specifically targets the helper T-lymphocytes, causing the destruction of these cells. Given the role of helper T-lymphocytes, why does this disease increase a person's susceptibility to infectious diseases?

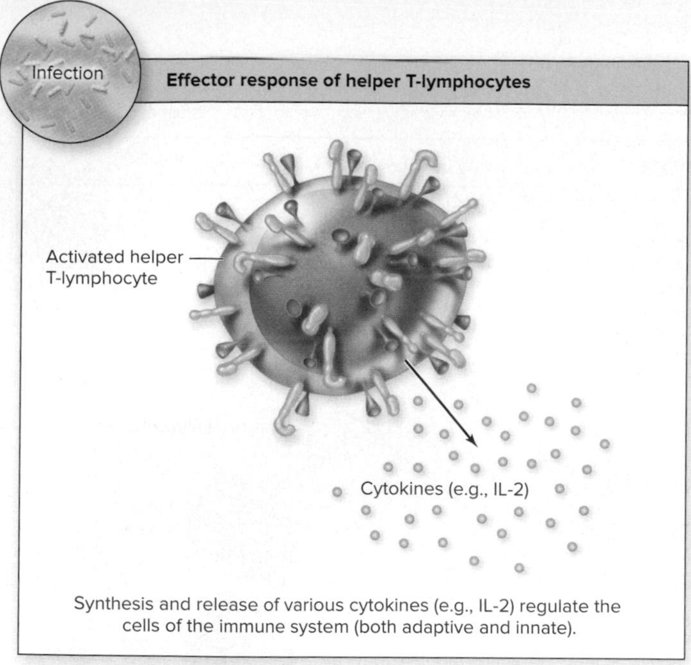

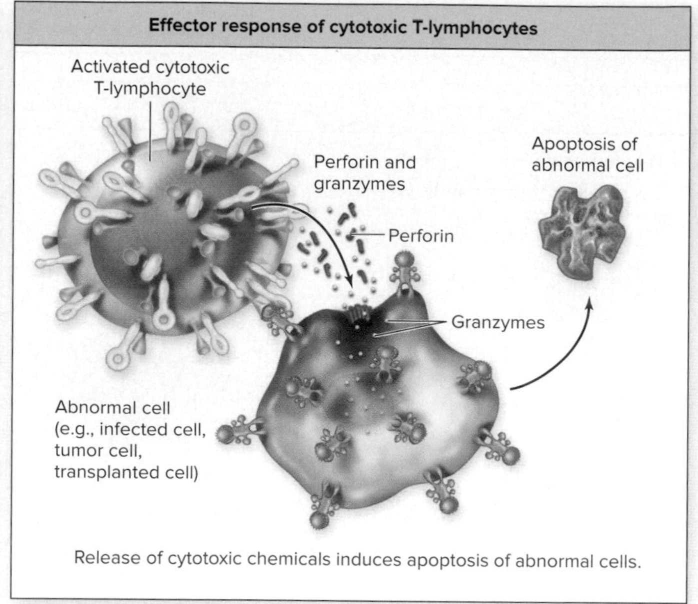

Figure 22.17 Effector Response of T-Lymphocytes. The effector response of helper T-lymphocytes and cytotoxic T-lymphocytes occurs at the site of infection. (*a*) Helper T-lymphocytes release various cytokines, and (*b*) cytotoxic T-lymphocytes destroy abnormal cells through the release of the cytotoxic chemicals perforin and granzymes, which induces apoptosis.

Effector Response of Cytotoxic T-Lymphocytes

Like helper T-lymphocytes, activated and memory cytotoxic T-lymphocytes also leave the secondary lymphatic structure after several days and migrate to the site of infection in the body's tissue. Cytotoxic T-lymphocytes destroy *unhealthy* or *infected cells* that display the antigen (e.g., a virus-infected cell, bacteria-infected cell, tumor cell, foreign transplanted cell; figure 22.17*b*). The effector response of cytotoxic T-lymphocytes is initiated when physical contact is made between a cytotoxic T-lymphocyte and the specific foreign antigen displayed on an unhealthy or a foreign cell.

If the cytotoxic T-lymphocyte recognizes the antigen presented by the infected cell (with MHC class I molecules), it destroys the cell by releasing granules containing the cytotoxic chemicals perforin and granzymes (the same substances released from NK cells described in section 22.3b). Perforin forms a channel in the target plasma membrane that increases the cell's permeability; granzymes enter the cell through the perforin channels. Granzymes induce cell death by apoptosis, which helps to limit spread of the infectious agent. It is because the immune response of T-lymphocytes is effective against antigens associated with cells that it is referred to as *cell-mediated immunity*.

WHAT DID YOU LEARN?

26 Are cells of both the innate and adaptive immune systems regulated by cytokines released by helper T-lymphocytes?

27 Cell-mediated immunity is specifically effective against what type of target? Provide examples.

22.7b Effector Response of B-Lymphocytes

LEARNING OBJECTIVES

36. Describe the function of plasma cells in the effector response of B-lymphocytes.
37. Define antibody titer.

Antibodies are the effectors that provide humoral immunity. Antibodies are formed primarily by plasma cells (although limited amounts are produced by B-lymphocytes). Plasma cells typically remain in the lymph nodes, continuing to synthesize and release antibodies. Antibodies circulate throughout the body in the lymph and blood, ultimately coming in contact with antigen at the site of infection. Plasma cells, over their life span of about 5 days, produce hundreds of millions of antibodies against the specific antigen.

The circulating blood concentration of antibody against a specific antigen is referred to as the **antibody titer.** This can be a measure of immune response. The details of antibody structure and function are described next.

WHAT DID YOU LEARN?

28 What is the specific role of plasma cells?

22.8 Immunoglobulins

An **antibody** is an **immunoglobulin** (im′ū-nō-glob′ū-lin) **(Ig)** protein produced against a particular antigen **(figure 22.18)**. The structure of antibodies ("bodies" against antigens) reflects their ability to target specific antigens that they may encounter. Antibodies do not destroy pathogens directly but facilitate the destruction by other immune cells.

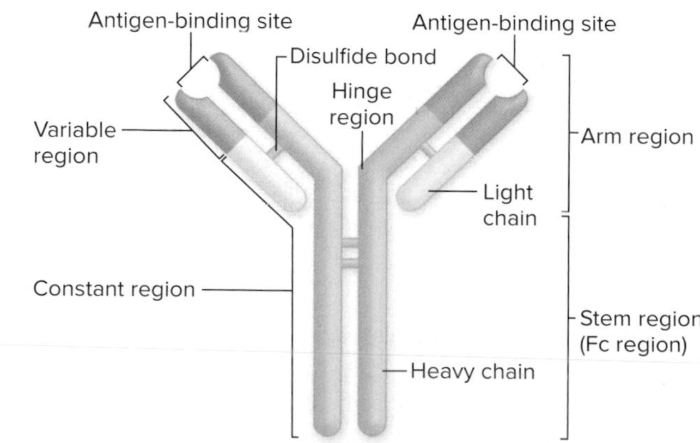

Figure 22.18 Antibody Structure. The structure of an antibody is a Y-shaped protein that includes two variable regions that bind antigen and one constant region that determines its biological activity.

You may find it helpful to think of the function of antibodies as "tagging" a specific antigen so that it can be eliminated. Here we consider both the structure of immunoglobulins and their actions.

22.8a Structure of Immunoglobulins

LEARNING OBJECTIVE

38. Describe the general structure of an immunoglobulin molecule, including its two functional regions.

An immunoglobulin molecule is a Y-shaped, soluble protein composed of four polypeptide chains: two identical heavy chains and two identical light chains, with flexibility at the hinge region of the two heavy chains. These four polypeptide chains are held together by covalent disulfide bonds (see section 2.8a) to form an **antibody monomer** (i.e., a single, Y-shaped protein). Two important functional regions of the antibody monomer are the variable regions and the constant region.

Variable Regions

The **variable regions** located at the ends of the "arms" of the antibody contain the antigen-binding site, which attach to a specific antigenic determinant of an antigen. The antigen-binding sites vary among antibodies that bind different antigens **(figure 22.19)**. Most antibodies have two antigen-binding sites, which allow each antibody to bind to two antigenic determinants. The variable region binds the antigen through weak intermolecular forces, including hydrogen bonds, ionic bonds, and hydrophobic interactions (see section 2.3d).

Constant Region

The **constant region** contains the **Fc region,** which is the portion of the antibody that determines the biological functions of the antibody. The constant region is the same or nearly the same in structure for antibody molecules of a given class; there are five major classes of immunoglobulins: IgG, IgM, IgA, IgD, and IgE. These classes are described in greater detail in section 22.8*c*.

WHAT DID YOU LEARN?

What is the significance of the variable regions of an immunoglobulin molecule?

22.8b Actions of Antibodies

LEARNING OBJECTIVE

39. List the functions of the antigen-binding site and Fc region of antibodies, and briefly describe how each occurs.

Antibodies are effective against antigens through the use of either (1) the antigen-binding region within the variable region (at the ends of the two arms of the Y-shaped antibody) or (2) the Fc constant region (stem of the Y-shaped antibody). The antibodies immobilize specific antigens and ultimately cause their elimination by other immune cells.

The binding of the antigen-binding site with the antigen can cause neutralization, agglutination, and precipitation **(Table 22.5)**. The three other functions first require the binding of antibody to an antigen; the Fc region of the antibody projects externally. The Fc region can then bind complement, bind to phagocytic cells to enhance phagocytosis, or bind to NK cells to trigger apoptosis.

Antibodies are especially effective in binding soluble antigens (e.g., viral particles, bacteria, toxins, yeast spores). It is because the immune response of B-lymphocytes is effective against soluble antigens (antigens dissolved in the body's "humors") that the action of antibodies produced by activated B-lymphocytes (i.e., plasma cells) is referred to as *humoral immunity*.

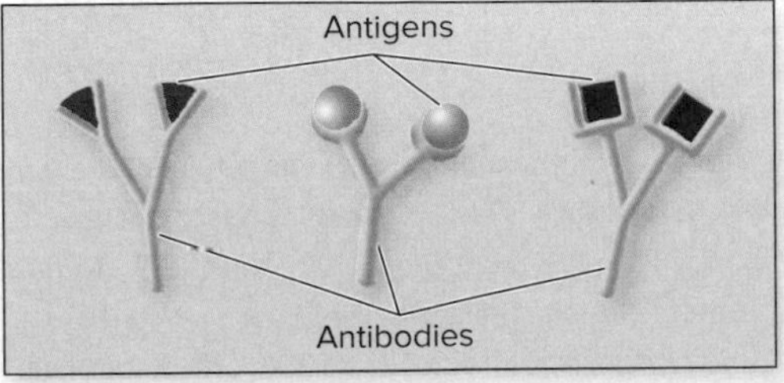

Figure 22.19 Variable Regions of an Antibody. The two variable regions of an antibody molecule are unique to the specific antigen to which they bind.

WHAT DID YOU LEARN?

What are the six major functions of antibodies? Which occur due to antigen-binding, and which depend on the Fc region?

22.8c Classes of Immunoglobulins

LEARNING OBJECTIVE

40. Describe the structure, location, and specific function of the five major classes of immunoglobulins.

The five classes of immunoglobulins are IgG, IgM, IgA, IgD, and IgE. These may be remembered with the acronym G-MADE. Each class of immunoglobulin is unique in a number of aspects. **Table 22.6** summarizes the characteristics of the five immunoglobulin classes.

IgG is a monomer (one Y-shaped protein) that is the major class of immunoglobulins. It makes up 75–85% of antibody in the blood and is the predominant antibody in the lymph, cerebrospinal fluid, serous fluid, and peritoneal fluid. IgG can participate in all of the functions previously listed for actions of antibodies, including the neutralization of toxins. Additionally, IgG antibodies cross the placenta and can be responsible for *hemolytic disease of the newborn* (see Clinical View 18.5: "Rh Incompatibility and Pregnancy").

IgM is normally a pentamer (composed of five monomers) found mostly in the blood. IgM is not as versatile in its biological functions as IgG. For example, IgM is not efficient at virus neutralization. However, IgM is the most effective at causing agglutination of cells and binding complement. Additionally, naturally occurring IgM antibodies are responsible for rejection of mismatched blood transfusions (see Clinical View 18.3: "Transfusions").

IgA is typically a dimer (composed of two monomers) and is found in areas exposed to the environment, such as mucosal membranes and tonsils, and it is produced in various secretions, including mucus, saliva, tears, and breast milk (see table 22.3). IgA plays a significant role in protecting the respiratory and gastrointestinal tract. It helps to prevent pathogens (e.g., viruses, bacteria) from adhering to epithelial tissue and penetrating underlying tissue through the process of neutralization. IgA is especially effective at agglutination.

Table 22.5 Actions of Antibodies Following Antigen Binding

BINDING OF ANTIGEN-BINDING SITE OF AN ANTIBODY WITH ANTIGEN CAUSES

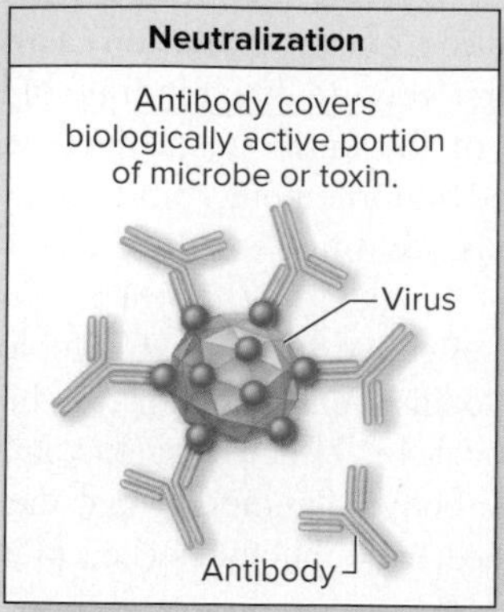	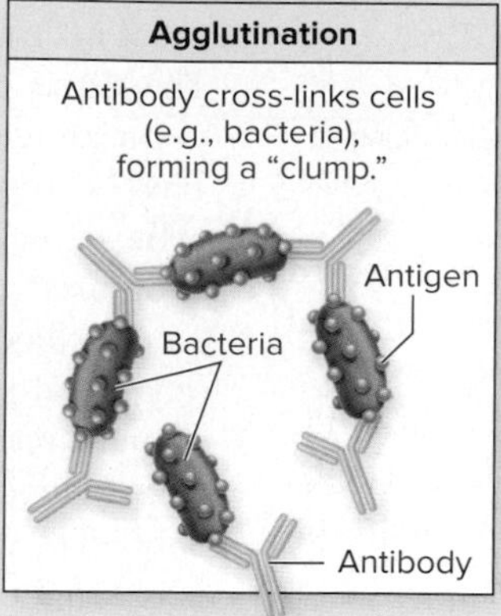	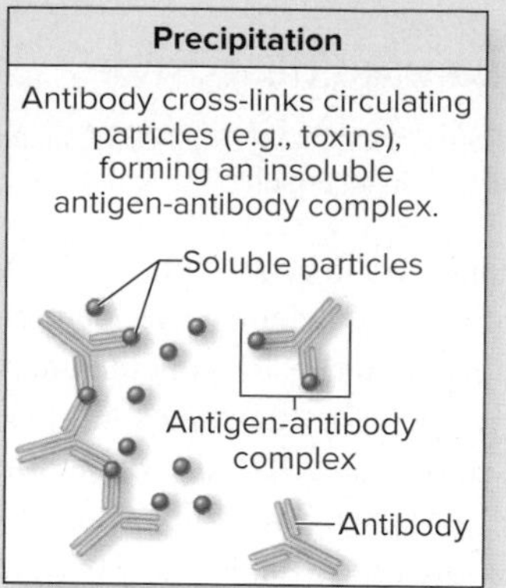
Neutralization. The antigen-binding site of an antibody physically covers an antigenic determinant of a pathogen to make it ineffective in establishing an infection or causing harm. For example, neutralization occurs when an antibody covers the region of a virus used to bind to a cell receptor, preventing entry of the virus into a cell (see section 22.1). A similar process neutralizes toxins (e.g., snake venom).	**Agglutination.** The antigen-binding site of an antibody cross-links antigens of foreign cells, causing them to agglutinate, or "clump." This is especially effective against bacterial cells and mismatched erythrocytes in a blood transfusion (see section 18.3b).	**Precipitation.** The antigen-binding site of an antibody can cross-link soluble, circulating antigens such as viral particles (not whole cells) to form an antigen-antibody complex. These complexes become insoluble and precipitate out of (drop out of) body fluids. The precipitated complexes are then engulfed and eliminated by phagocytic cells such as macrophages.

EXPOSED FC PORTION FOLLOWING ANTIGEN BINDING BY ANTIBODY PROMOTES

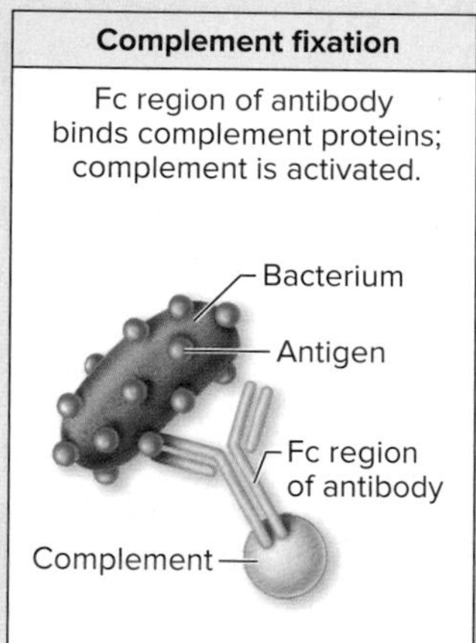	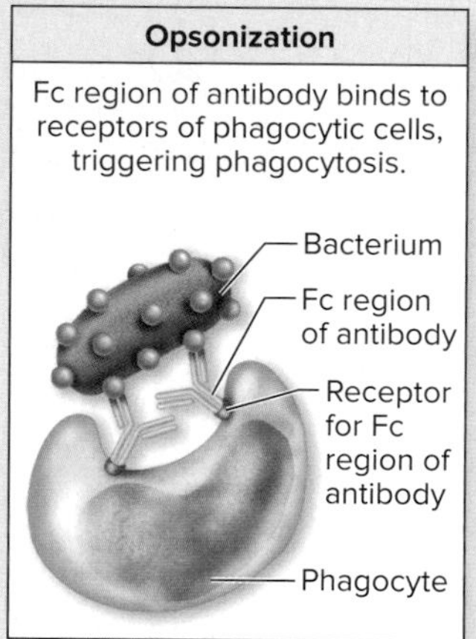	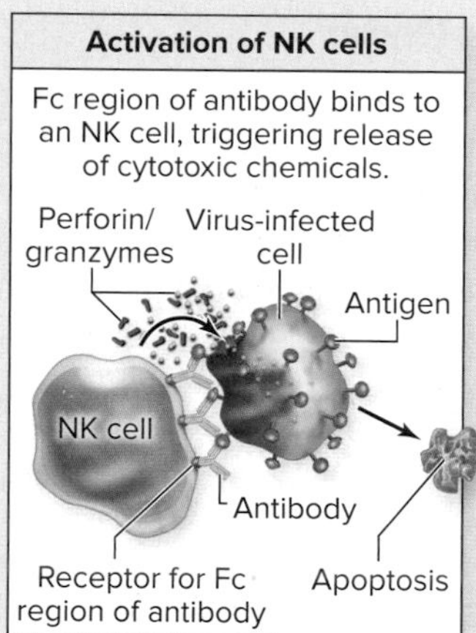
Complement fixation. The *Fc region* of certain classes of antibodies (IgG and IgM) can bind specific complement proteins to cause activation of complement by the classical pathway. The functions of complement are described in section 22.3c and include opsonization, increasing inflammation, inducing cytolysis, and elimination of immune complexes.	**Opsonization.** The *Fc region* of certain classes of antibodies (e.g., IgG) can also cause opsonization (making it more likely that a target is "seen" by phagocytic calls). Phagocytic cells such as neutrophils and macrophages have receptors for the Fc region of certain antibody classes. The phagocytic receptors bind in a zipperlike fashion to the Fc region of the antibodies to engulf both the antigen and antibody.	**Activation of NK cells.** The *Fc region* of certain classes of antibodies (IgG) binds to specific receptors on NK cells (much like phagocytes). This induces NK cells to destroy abnormal cells by the release of cytotoxic chemicals that cause apoptosis of the cell. This process is called **antibody-dependent cell-mediated cytotoxicity (ADCC).**

INTEGRATE

LEARNING STRATEGY

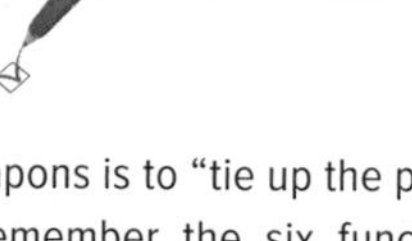

Generally, the role of antibodies as weapons is to "tie up the prisoner" until other help arrives. You can remember the six functions of an antibody with the acronyms NAP and CON: Neutralization, Agglutination, and Precipitation (NAP), as well as Complement, Opsonization, and NK cells (CON). Remember—a NAP can help you CONcentrate.

Table 22.6 Major Classes of Immunoglobulins

Characteristic	IgG (Monomer)	IgM (Pentamer)	IgA (Dimer)	IgD (Monomer)	IgE (Monomer)
Primary Locations	Body fluids, including blood, lymph, cerebrospinal fluid, serous fluid, peritoneal fluid, breast milk	Blood (monomer is B-lymphocyte receptor); breast milk	External secretions (mucus, saliva, tears, breast milk, and colostrum)	B-lymphocyte receptor	Blood
Actions	Neutralization (viruses, bacteria, toxins); agglutination; precipitation; complement activation; opsonization; natural killer cell activation (antibody-dependent cell-mediated cytotoxicity)	Neutralization, agglutination (great potential); complement binding (great potential)	Neutralization; agglutination (great potential)	BCR (B-cell receptor)	Produced during allergic reactions or as a result of a parasitic infection; activation of mast cells and basophils
Half-Life (Days) in Blood	23	5	5.5	2.8	2.0
Unique Characteristics	Used for passive immunity; crosses placenta	First produced antibody; only antibody produced in fetus	Associated with mucous membranes	Identifies when B-lymphocytes may be ready for activation	Causes release of products from basophils and mast cells; attracts eosinophils

IgD is a monomer that (along with a monomer form of IgM) functions as the antigen-specific B-lymphocyte receptor. It also identifies when B-lymphocytes may be ready for activation to participate in providing adaptive immunity.

IgE is a monomer that has a very low rate of synthesis. It is generally formed in response to allergic reactions and to parasitic infections. IgE causes release of histamine and other mediators of inflammation from basophils and mast cells, and it attracts eosinophils. The formation of IgE and its response to allergens is described in section 22.9c (see Clinical View 22.8: "Hypersensitivities").

Class Switching

Each plasma cell has the potential to produce different classes of antibodies. This process of changing the class of antibody produced from IgM to IgG, IgE, and IgA is called *class switching*. Direct contact between the helper T-lymphocytes (e.g., CD40 surface protein) and plasma calls (e.g., CD154) is required, along with the release of various cytokines from the helper T-lymphocytes. Specific cytokines determine the type of antibody class formed.

Figure 22.20 is a visual summary of adaptive immunity. Adaptive immunity is considered the third line of defense because, although its response is initiated immediately, on the first exposure to an antigen there is a lag time that occurs from the time of exposure to the development of the immune response. This process may take several days.

WHAT DID YOU LEARN?

31 Which subclass of antibodies is most prevalent? Which specific functions of antibodies can it engage in?

22.9 Immunologic Memory and Immunity

One of the central features of adaptive immunity is the development of "memory" that provides immunity against a specific antigen. Here we discuss immunologic memory and how immunity is obtained through both active and passive means.

22.9a Immunologic Memory

LEARNING OBJECTIVE

41. Define immunologic memory, and explain how it occurs.

Activation of lymphocytes requires direct physical contact between a lymphocyte and an antigen. On the first exposure to an antigen (the antigen challenge), limited numbers of helper T-lymphocytes, cytotoxic T-lymphocytes, and B-lymphocytes recognize the antigen (about a 1 in 100,000 to 1 million chance of being the cell that recognizes a specific antigen). Generally, a lag time occurs between the body's initial exposure to the antigen and the physical contact with lymphocytes required to develop an immune response.

The antigen challenge, however, causes the formation of memory cells in response to the activation of T-lymphocytes and B-lymphocytes, as described in sections 22.6a and 22.6b. These long-lived lymphocytes represent an "army" of thousands against specific antigens and are responsible for immunologic memory.

On subsequent exposures to an antigen, these vast number of memory cells make contact with the antigen more rapidly and produce a more powerful response, which is referred to as the

INTEGRATE CONCEPT OVERVIEW

Figure 22.20 Adaptive Immunity.

Adaptive immunity is considered the third line of defense. It is the specific means by which lymphocytes defend the body against specific antigens. Development of adaptive immunity requires three events in the life of lymphocytes: (*a*) formation of lymphocytes within primary lymphatic structures, (*b*) activation and clonal selection of lymphocytes within secondary lymphatic structures, and (*c*) effector response of T-lymphocytes and antibodies at the site of infection.

(a) Formation of Lymphocytes
Primary lymphatic structure
T-lymphocyte and B-lymphocyte formation and maturation into naive immunocompetent lymphocytes occur in primary lymphatic structures (red bone marrow and the thymus) primarily during development and shortly after birth, but they continue throughout one's lifetime. These cells migrate to secondary lymphatic structures.
T-lymphocytes (maturation is completed in thymus)
B-lymphocyte (maturation complete in red bone marrow)
CD4 receptor
TCR
CD8 receptor
TCR
BCR
Helper T-lymphocyte
Cytotoxic T-lymphocyte
Red bone marrow
Lymph node
Site of infection

(b) Activation and Clonal Selection of Lymphocytes

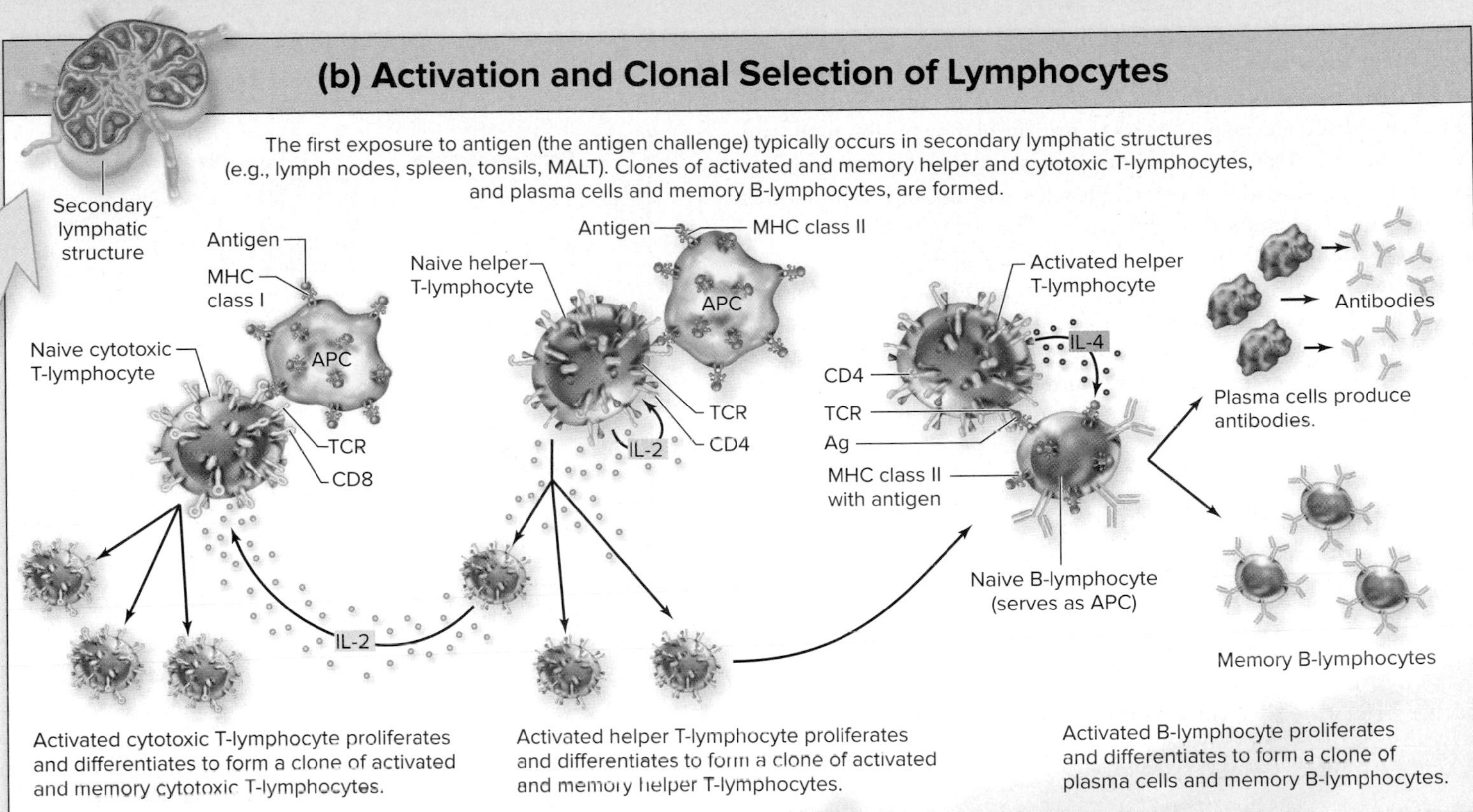

Secondary lymphatic structure
The first exposure to antigen (the antigen challenge) typically occurs in secondary lymphatic structures (e.g., lymph nodes, spleen, tonsils, MALT). Clones of activated and memory helper and cytotoxic T-lymphocytes, and plasma cells and memory B-lymphocytes, are formed.
Antigen
MHC class I
APC
Naive cytotoxic T-lymphocyte
TCR
CD8
Antigen
MHC class II
Naive helper T-lymphocyte
APC
TCR
CD4
IL-2
IL-2
Activated helper T-lymphocyte
IL-4
CD4
TCR
Ag
MHC class II with antigen
Naive B-lymphocyte (serves as APC)
Antibodies
Plasma cells produce antibodies.
Memory B-lymphocytes
Activated cytotoxic T-lymphocyte proliferates and differentiates to form a clone of activated and memory cytotoxic T-lymphocytes.
Activated helper T-lymphocyte proliferates and differentiates to form a clone of activated and memory helper T-lymphocytes.
Activated B-lymphocyte proliferates and differentiates to form a clone of plasma cells and memory B-lymphocytes.

(c) Effector Response

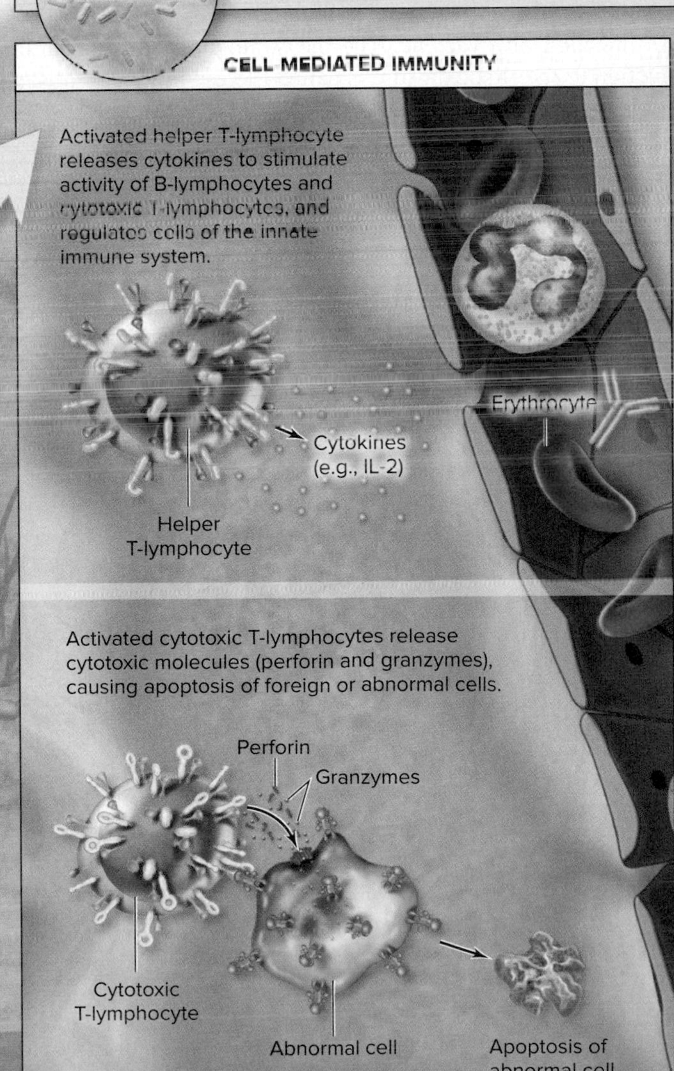

Site of infection
CELL MEDIATED IMMUNITY
Activated helper T-lymphocyte releases cytokines to stimulate activity of B-lymphocytes and cytotoxic T-lymphocytes, and regulates cells of the innate immune system.
Erythrocyte
Cytokines (e.g., IL-2)
Helper T-lymphocyte
Activated cytotoxic T-lymphocytes release cytotoxic molecules (perforin and granzymes), causing apoptosis of foreign or abnormal cells.
Perforin
Granzymes
Cytotoxic T-lymphocyte
Abnormal cell
Apoptosis of abnormal cell

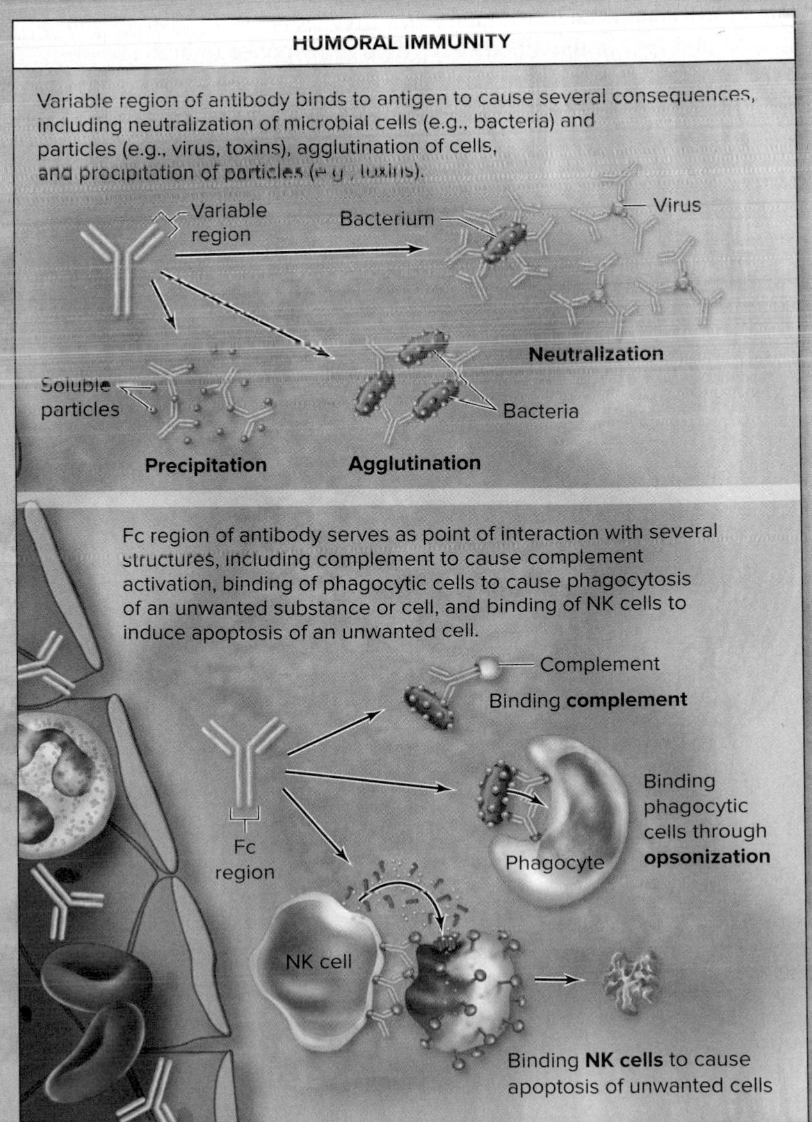

HUMORAL IMMUNITY
Variable region of antibody binds to antigen to cause several consequences, including neutralization of microbial cells (e.g., bacteria) and particles (e.g., virus, toxins), agglutination of cells, and precipitation of particles (e.g., toxins).
Variable region
Bacterium
Virus
Neutralization
Soluble particles
Bacteria
Precipitation
Agglutination
Fc region of antibody serves as point of interaction with several structures, including complement to cause complement activation, binding of phagocytic cells to cause phagocytosis of an unwanted substance or cell, and binding of NK cells to induce apoptosis of an unwanted cell.
Complement
Binding complement
Fc region
Phagocyte
Binding phagocytic cells through opsonization
NK cell
Binding NK cells to cause apoptosis of unwanted cells

secondary response, also known as the *memory response* or *anamnestic* (an′am-nes′tik; *an* = not, *amnesia* = forgetfulness) *response.* On each subsequent exposure to a specific pathogen, the pathogen is typically eliminated even before disease symptoms develop. For example, a person who develops measles will not develop measles again, even if reexposed to that virus. The virus is eliminated by activated memory T-lymphocytes, memory B-lymphocytes, and antibodies before it causes harm. This feature of immunologic memory makes adaptive immunity a highly potent means of protection. Vaccines are typically effective in developing memory, and a secondary response is seen on exposure to the substance vaccinated against (see Clinical View 22.7: "Vaccinations").

WHAT DID YOU LEARN?

32 Briefly describe immunologic memory, and explain its significance.

22.9b Measure of Immunologic Memory

LEARNING OBJECTIVE

42. Discuss the difference between the primary response and the secondary response to antigen exposure.

Antibody titer (concentration of antibody) in blood serum is one measure of immunologic memory. The graphs shown in **figure 22.21** reflect the changes in serum antibody titer (specifically, the amount of IgM and IgG in the blood) over time in response to both the initial exposure (the antigen challenge) and subsequent exposures to an antigen. The *degree* of protection is indicated by levels of circulating IgG.

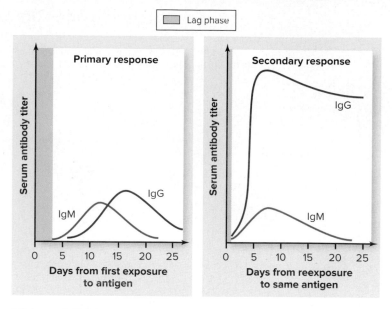

Figure 22.21 Primary and Secondary (Anamnestic) Response in Humoral Immunity. Graph of the primary response that shows the level of IgM and IgG antibody produced by plasma cells on the first exposure to a given antigen. Graph of the secondary response that depicts the level of these antibodies on all subsequent exposures to the same antigen.

Initial Exposure and the Primary Response

The initial exposure to a specific antigen can be in the form of an active infection or a vaccine. The measurable response of antibody production to the first exposure is called the **primary response:**

- **Lag or latent phase.** There is initially a period of no detectable antibody in the blood. This period may extend 3 to 6 days. Antigen detection, activation, proliferation, and

INTEGRATE

CLINICAL VIEW 22.7
Vaccinations

A **vaccine** is an attenuated (weakened) or dead microorganism, or some component of the microorganism (may be bioengineered), that is administered through one of several routes: oral, intradermal, intravenous (IV), intraperitoneal, or intranasal.

The function of a vaccine is to stimulate the immune system to develop memory B-lymphocytes (predominantly) while providing a relatively safe means for the initial exposure to a microorganism. The risk is relatively low because the microorganism (or its components) has no ability (or limited ability) to establish an infection. If an individual is later exposed to the same antigens, the secondary response is triggered and it will be swift and powerful; the individual is generally not even aware of contact with the microbe.

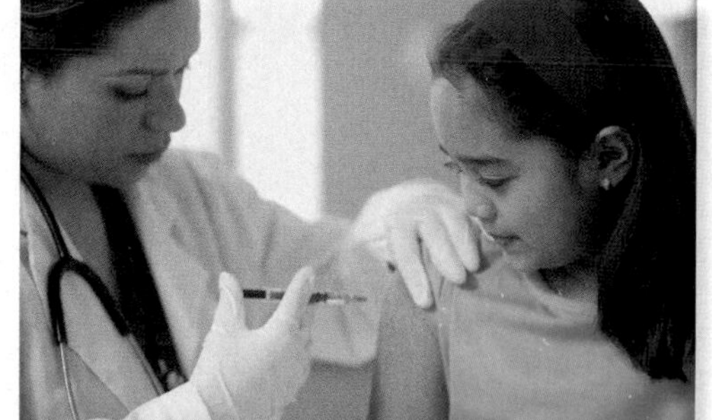

©Blend Images/Getty Images RF

However, a vaccine is different from having an active infection in these ways:

- The immune response to a vaccine is predominantly from the humoral-mediated branch. The reason is that B-lymphocytes bind unattached microbes to induce humoral-mediated immunity, but few, if any, of the weakened or dead microbes in the vaccine can infect cells to induce T-lymphocytes to establish cell-mediated immunity. (The B-lymphocytes present the antigen to T-lymphocytes in some instances.)
- Depending upon the life span of the particular memory B-lymphocytes, the vaccine may provide lifelong immunity, or periodic **booster shots** may be needed to ensure continued protection against the antigen. For example, booster shots are needed every 10 years for protection against tetanus.

differentiation of lymphocytes, including development of memory lymphocytes, occur during the lag phase.

- **Production of antibody.** Within 1 to 2 weeks, plasma cells produce IgM and then IgG. Antibody titer levels peak and then generally decrease over time.

Subsequent Exposures and the Secondary Response

Subsequent exposures to an antigen can occur after varying lengths of time following the initial exposure, and the measurable response to subsequent exposure is called the **secondary response:**

- **Lag or latent phase.** A much shorter lag phase occurs with subsequent exposures to the same antigen. This difference is due to the presence of memory lymphocytes.
- **Production of antibody.** Antibody levels rise more rapidly, with a greater proportion of the IgG class of antibodies. This higher level of IgG production may continue for longer periods, perhaps even years.

WHAT DID YOU LEARN?

33 How does the secondary response differ from the primary response? What is the advantage of the secondary response?

22.9c Active and Passive Immunity

LEARNING OBJECTIVES

43. Define active immunity and passive immunity.

44. Describe how both active and passive immunity can be acquired naturally and artificially.

Immunity can be active or passive **(figure 22.22)**. **Active immunity** results from a direct encounter with a pathogen or foreign substance that results in the production of memory cells and can be obtained either naturally or artificially (e.g., clinically). *Naturally* acquired active immunity occurs when an individual is directly exposed to the antigen of an infectious agent. *Artificially* acquired active immunity takes place when the exposure occurs through a vaccine. In both cases, memory cells against that specific antigen are formed.

In contrast to active immunity, **passive immunity** is obtained from another individual or an animal, and it can be obtained naturally or artificially. Naturally acquired passive immunity occurs from the transfer of antibodies from the mother to the fetus across the placenta (IgG) or to the baby in the mother's breast milk (IgA, IgM, and IgG). In contrast, when serum containing antibodies against a specific antigen is transferred from one individual to another, this process is referred to as artificially acquired passive immunity. For example, serum containing antibodies against the toxins associated with tetanus and botulism can be transferred to an individual who is at risk from one of these toxins. Antibodies to a poisonous snake venom (antivenom) can also be transferred to an individual who has been bitten by that species of snake. The antibodies neutralize the toxin or venom to prevent it from doing harm until the body is able to eliminate it.

In both types of passive immunity, the individual has not had an antigenic challenge and has not produced memory cells. Passive immunity lasts only as long as the antibody proteins are present in the individual. For example, the half-life of IgG in the blood is 23 days; IgM is 5 days (table 22.6).

WHAT DID YOU LEARN?

34 Which type of immunity—active or passive—results in the production of memory cells and generally provides lifelong protection from that antigen?

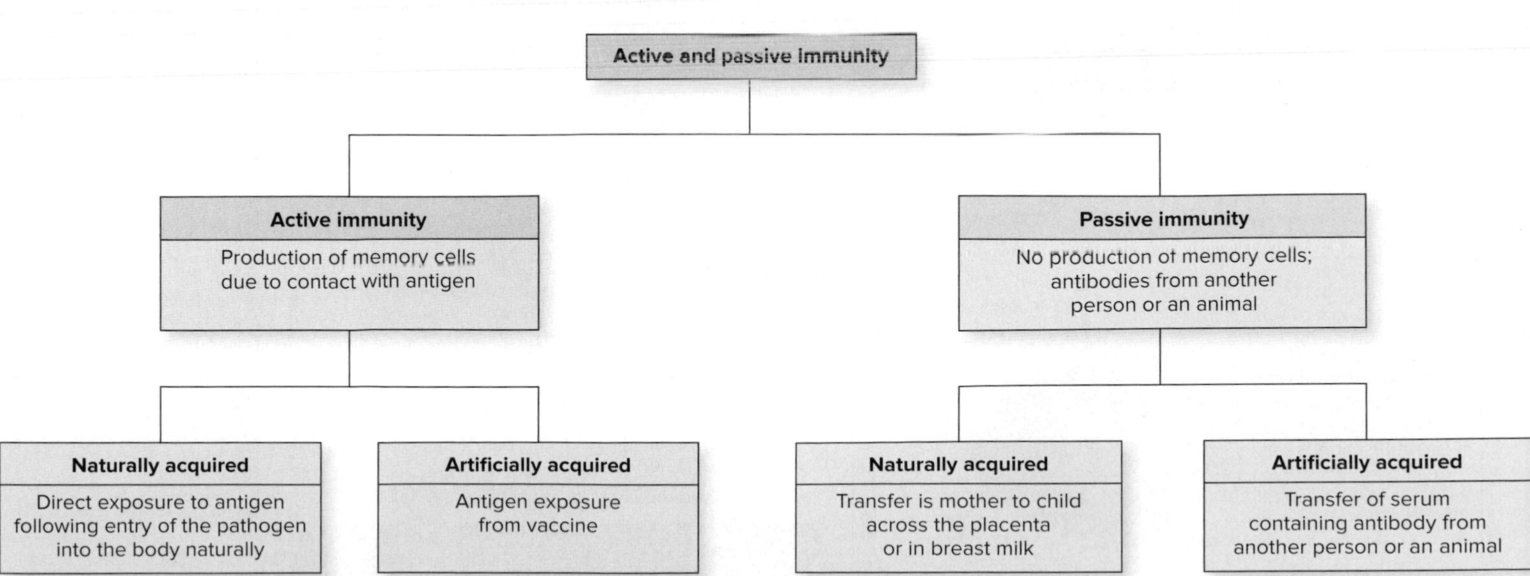

Figure 22.22 Active and Passive Immunity. Active immunity is distinguished from passive immunity by the production of memory cells. Memory cells are produced in active immunity, whereas in passive immunity they are not.

INTEGRATE

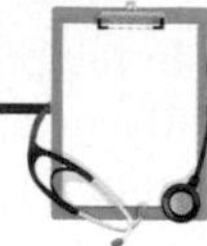

CLINICAL VIEW 22.8

Hypersensitivities

Hypersensitivity is an abnormal and exaggerated response of the immune system to an antigen. The various types of hypersensitivities are categorized based on the amount of time required for the immune response to occur following exposure to an antigen: **Acute hypersensitivities** occur within seconds; **subacute hypersensitivities** within 1 to 3 hours, and **delayed hypersensitivities** within 1 to 3 days. The difference in time of onset reflects the components of the immune system that are involved. Both acute and subacute hypersensitivity reactions involve humoral immunity. Immunoglobulin E (IgE) antibodies are involved in acute hypersensitivities, whereas subacute hypersensitivity reactions are triggered by immunoglobulin G (IgG) and immunoglobulin M (IgM). In contrast, delayed hypersensitivity reactions involve cell-mediated immunity. We limit our discussion to acute hypersensitivities.

Acute Hypersensitivity (Allergies)

An acute hypersensitivity is more commonly referred to as an **allergy** (or *type I hypersensitivity*), and it is an overreaction of the immune system to a noninfectious substance, or **allergen.** Examples of different allergens include pollen, latex, peanuts, and bee venom. Following exposure to the allergen, an allergic reaction occurs within seconds and continues for about a half hour.

There are three major phases associated with an allergic reaction:

1. **Sensitization phase.** An individual is exposed to an allergen. The allergen is engulfed by an antigen-presenting cell (APC) and presented to a helper T-lymphocyte (not shown in figure). The helper T-lymphocytes release cytokines that cause the B-lymphocytes to mature into plasma cells that produce IgE antibodies against the allergen. The IgE antibodies bind to basophils and mast cells (by the Fc region of the antibody) and may remain bound to these cells for several weeks or longer.
2. **Activation phase.** If the individual is reexposed to the same allergen, the allergen binds to the IgE antibodies that are bound to the basophils and mast cells, cross-linking the receptors.
3. **Effector phase.** The mast cells or basophils release chemicals (histamine, heparin, and eicosanoids) that cause an inflammatory response. The inflammatory response is responsible for the symptoms associated with allergies. The symptoms an individual experiences depend upon where the inflammatory response occurs in the body:

- Contact with the mucous membranes of the nasal passage and conjunctiva of the eye result in a runny nose and watery eyes (**allergic rhinitis,** or *hay fever*). This is the most common site, with approximately 20% of the general population experiencing allergic rhinitis.
- Exposure of the skin surface can result in red welts and itchy skin **(hives).**
- Entrance into the respiratory passageway causes bronchoconstriction and increases secretion of mucus, causing labored breathing and coughing **(allergic asthma).**
- Entrance into the gastrointestinal tract causes increased fluid secretions and peristalsis that may cause vomiting and diarrhea (not shown in figure).
- Circulation in the blood through a bee sting or injection by needle causes systemic vasodilation and inflammation. In extreme cases, extensive loss of fluid from the blood into the interstitial space results in a marked decrease in blood volume and blood pressure. Consequently, the individual may have insufficient blood pressure to maintain adequate perfusion **(anaphylactic shock).**

Allergies (or type I hypersensitivity)

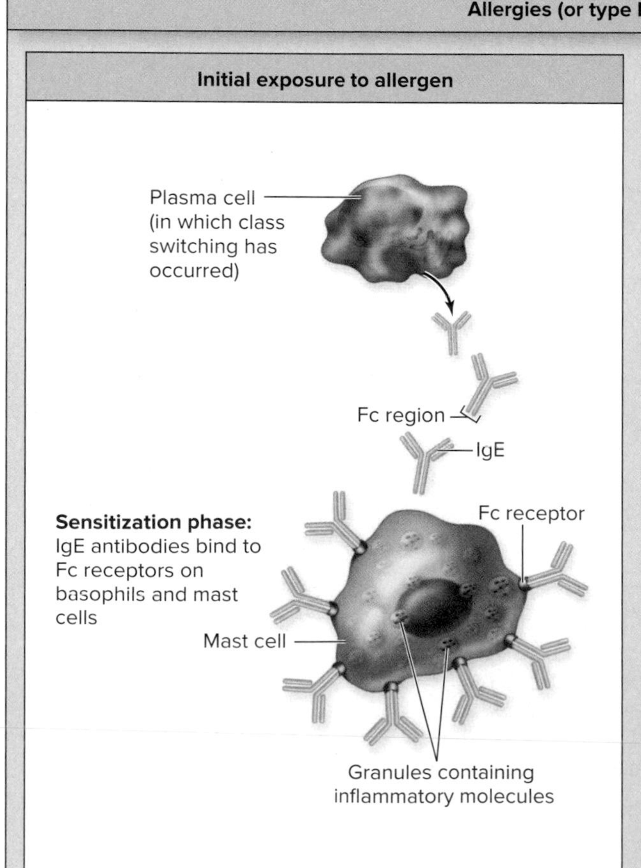

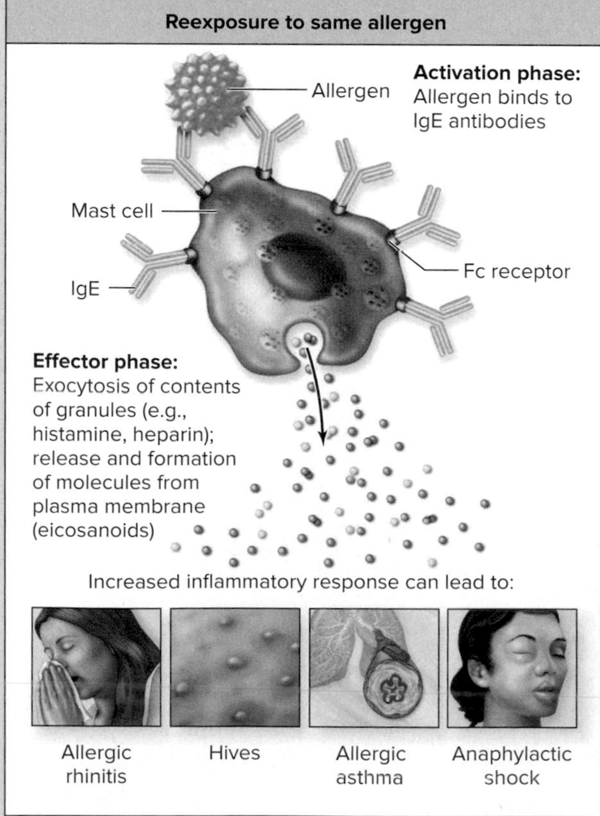

Acute hypersensitivity reactions involve the IgE antibody. The three stages of acute hypersensitivity are the sensitization phase, activation phase, and effector phase.

INTEGRATE

CLINICAL VIEW 22.9

HIV and AIDS

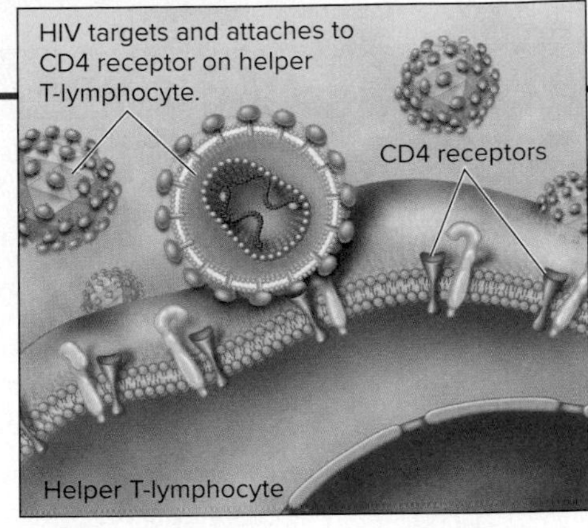

Process of the human immunodeficiency virus (HIV) infecting a cell.

AIDS (acquired immunodeficiency syndrome) is a life-threatening condition that is the result of the **human immunodeficiency virus (HIV).** An HIV infection targets the immune system—in particular, the helper T-lymphocyte (CD4 T-cell). HIV infects and destroys these helper T-lymphocytes over a period of time (months to years). Prolonged HIV infection leads to the devastating effects of AIDS.

Epidemiology

HIV resides in the body fluids of infected individuals, including their blood, semen, vaginal secretions, or even breast milk. HIV can be transmitted in the following ways: during unprotected sexual (vaginal or anal) intercourse, sharing hypodermic needles with other intravenous drug users, during development (because HIV crosses the placenta), or breastfeeding an infant. Current evidence indicates HIV is *not* spread by casual kissing, sharing eating utensils, using a public toilet, or casual physical contact. First seen in the United States among the homosexual male population, it is now a prominent disease among heterosexual populations. The United Nations Programme on AIDS (UNAIDS) estimates that 90% of all HIV infections are currently transmitted heterosexually. Prior to 1985, before HIV and AIDS were well known, HIV could be transmitted through the donated blood supply. Individuals who received blood transfusions sometimes received HIV-infected blood, thereby infecting them as well. This discovery has led to more stringent screening of blood donors.

Since the early 1980s, over 78 million people have become infected with HIV, and over 39 million people have died. The incidence of AIDS is increasing throughout the world, but the disease is particularly rampant on the continents of Africa and Asia. The AIDS epidemic in Africa has led to massive numbers of deaths, and children are frequently orphaned as both parents succumb to the disease. Asian countries are seeing a surge of new HIV and AIDS cases, especially among sex workers and their clients.

Prevention

The key to limiting the spread of HIV infection is to refrain from behaviors that allow the virus's transmission. Unprotected sexual intercourse (especially anal intercourse) can spread HIV, so individuals should either practice abstinence or protect themselves with the use of condoms. HIV can also be spread through oral sex. (Other contraceptives like birth control pills do *not* protect an individual from HIV infection.) Both partners in a monogamous relationship should be tested for presence of the HIV virus (done via a simple blood test) before engaging in sexual intercourse. Intravenous drug users should not share needles. Health-care workers should wear gloves and be careful around patient bodily fluids. HIV-infected pregnant women need special prenatal care to help prevent the transmission of the virus to their fetus. In addition, HIV-infected mothers are discouraged from breastfeeding their infants, as the virus is present in breast milk.

HIV cannot survive for long periods of time outside the human body. The virus can be eliminated from medical equipment or personal care items by cleaning them with a common disinfectant (such as bleach or hydrogen peroxide) or by heating them to temperatures above 135°F.

How HIV Causes Its Damage

Helper T-lymphocytes are destroyed by HIV infection. This destruction occurs in several ways. Some helper T-lymphocytes are programmed to produce HIV RNA at such a fast rate that the cell undergoes lysis, or bursts. Other helper T-lymphocytes are targeted and destroyed by other immune cells, such as macrophages or cytotoxic T-lymphocytes. Over a period of months to years, the helper T-lymphocyte population declines to a dangerously low level. Helper T-lymphocytes initiate and oversee the body's immune response; therefore, a decrease in helper T-lymphocytes results in a loss of normal immune function.

Early Symptoms

Approximately several weeks to several months after initial HIV infection, many individuals will experience flulike symptoms, including sore throat, fever, fatigue, headache, and swollen lymph nodes. Some people may also experience night sweats, whereas still others may be completely asymptomatic. Often, these symptoms disappear after a few weeks when the body's other immune cells target and destroy HIV infected cells. Healthy helper T-lymphocytes divide to replace those cells that were destroyed. However, HIV continues to replicate at a faster rate than the immune system can rid itself of infected cells; in addition, the virus mutates to avoid detection. Over a period of years, the helper T-lymphocyte population drops to very low numbers, setting the stage for AIDS.

What Do HIV Blood Tests Look For?

HIV blood tests look for the evidence of HIV antibodies in the blood. These antibodies are produced by plasma cells about a month after initial infection. These antibodies indicate the body is responding to HIV infection. It can take up to 6 months for antibody levels in the blood to rise to a point where they can be detected by the blood test. Thus, individuals who have been exposed to HIV but get tested within the first 6 months may receive false negative results and are still at risk for infecting others.

When Does HIV Become AIDS?

HIV infection is diagnosed as AIDS when either a person's helper T-lymphocyte count drops to below 200 cells per cubic milliliter (in comparison to 800 to 1200 cells per cubic millimeter for a healthy individual) or a person develops an opportunistic infection or illness.

Opportunistic infections are those that thrive due to the compromised immune system. Some examples of opportunistic illness include protozoan infections (e.g., toxoplasmosis and pneumonia caused by *Pneumocystis jiroveci*), fungal infections (e.g., candidiasis, histoplasmosis), some bacterial infections, and neoplasms (cancers such as Kaposi sarcoma, aggressive non-Hodgkin lymphoma, and cervical cancer). Opportunistic infections account for up to 80% of all AIDS-related deaths. Additionally, many AIDS patients have some form of central nervous system (CNS) complications, including meningitis, encephalitis, neurologic deficits, and neuropathies.

Treatment Options

There is no cure for HIV, so HIV infection is a lifelong illness. Current pharmaceutical treatments are "cocktails" of multiple drugs that alleviate symptoms or help prevent the spread of HIV in the body, but they cannot eradicate HIV from an infected individual. Most of these drugs also have unpleasant side effects.

Unfortunately, HIV drugs are expensive and not widely available in developing countries, where their need is greatest. One hopeful sign is that pharmaceutical companies are negotiating with these governments to make cheaper forms of the drugs available. In the meantime, education about preventing HIV infection continues throughout the world.

CHAPTER SUMMARY

- The immune system is a functional system composed of cells, plasma proteins, and other substances that protect the body from harmful agents.

22.1 Overview of Diseases Caused by Infectious Agents

- The five major classes of infectious agents are bacteria, viruses, fungi, protozoans, and multicellular parasites.
- Bacteria are composed of prokaryotic cells; viruses are composed of either DNA or RNA in a protein capsid; and fungi, protozoans, and parasites are composed of eukaryotic cells.

22.2 Overview of the Immune System

- The immune system is composed of immune cells and cytokines, and it is organized into the innate immune system and the adaptive immune system.

22.1a Immune Cells and Their Locations

- Immune cells circulate in the blood and are also located in body tissues, including lymphatic tissues, select organs, epithelial tissue of the skin and mucosal membranes, and connective tissue throughout the body.

22.2b Cytokines

- Cytokines are small, soluble proteins produced by immune cells that function similarly to hormones; cytokines include interleukins, tumor-necrosis factors, colony-stimulating factors, and interferons.

22.2c Comparison of Innate Immunity and Adaptive Immunity

- Innate immunity is provided by the innate immune system. The innate immune system encompasses defenses we are born with and includes barriers to prevent entry and nonspecific internal defenses.
- Adaptive immunity encompasses defenses developed by lymphocytes in response to exposure to specific antigens and includes both cell-mediated immunity (T-lymphocytes) and humoral immunity (B-lymphocytes).

22.3 Innate Immunity

- The advantages of innate immunity include an immediate response against a wide array of potentially harmful substances; however, the responses do not result in memory.

22.3a Preventing Entry

- The skin and mucous membranes provide a physical, chemical, and biological barrier that is usually successful in preventing entry of harmful substances. These are considered the body's first line of defense.

22.3b Nonspecific Internal Defenses: Cells

- Cells of the innate immune system include neutrophils, macrophages, dendritic cells, basophils and mast cells, NK (natural killer) cells, and eosinophils.

22.3c Nonspecific Internal Defenses: Antimicrobial Proteins

- Antimicrobial proteins include interferon and the complement system.

22.3d Nonspecific Internal Defenses: Inflammation

- Inflammation is an immediate, local, nonspecific response that occurs in vascularized tissue against a great variety of injury-causing stimuli. Inflammation is the major effector response of the innate immune system.

22.3e Nonspecific Internal Defenses: Fever

- A fever is an abnormal elevation of body temperature of at least 1°C (1.8°F), and three phases associated with a fever are onset, stadium, and defervescence. Mild fevers are beneficial, whereas high fevers may be detrimental.

22.4 Adaptive Immunity: An Introduction

- Adaptive immunity is initiated with stimulation of T-lymphocytes and B-lymphocytes by a specific antigen.

22.4a Antigens

- An antigen is a substance that binds to a component of the adaptive immune system (i.e., T-lymphocytes or antibodies).

22.4b General Structure of Lymphocytes

- Helper T-lymphocytes contain TCRs (T-cell receptors) and CD4 proteins, cytotoxic T-lymphocytes contain TCRs and CD8 proteins, and B-lymphocytes contain BCRs (B-cell receptors).
- TCRs bind with presented antigen, and BCRs bind with free antigen (e.g., viral particles).

22.4c Antigen-Presenting Cells and MHC Molecules

- Major histocompatibility complex (MHC) molecules, which are plasma membrane proteins, display antigen on a cell's surface so the antigen can be encountered by T-lymphocytes.
- All nucleated cells present antigen with MHC class I molecules, and antigen-presenting cells (APCs) present antigen with both MHC class I and MHC class II molecules.

22.4d Overview of Life Events of Lymphocytes

- Three significant events occur in the lifetime of a lymphocyte: (a) formation, which occurs in primary lymphatic structures; (b) activation and clonal selection, which occurs within secondary lymphatic structures; and (c) participation in an effector response, which occurs at the site of infection.

22.5 Formation and Selection of T-Lymphocytes in Primary Lymphatic Structures

22.5a Formation of T-Lymphocytes

- Formation of T-lymphocytes begins in the red bone marrow and is completed in the thymus to produce immunocompetent but naive lymphocytes, a process that contributes to central self-tolerance.

22.5 Formation and Selection of T-Lymphocytes in Primary Lymphatic Structures *(continued)*	**22.5b Selection and Differentiation of T-Lymphocytes** • T-lymphocytes undergo positive selection, in which their ability to recognize foreign antigen attached to an MHC molecule is determined, and negative selection, in which their tolerance for self-antigens attached to an MHC molecule is tested. Only those lymphocytes that successfully pass both types of selection are allowed to survive to become either helper T-lymphocytes or cytotoxic T-lymphocytes. (Tregs are formed from CD4 cells that moderately bind self-antigen.)
	22.5c Migration of T-Lymphocytes • Naive but immunocompetent T-lymphocytes migrate from the thymus to secondary lymphatic structures. • A similar process of selection takes place in red bone marrow to produce B-lymphocytes.
22.6 Activation and Clonal Selection of Lymphocytes	• Activation requires the binding of antigen by a specific lymphocyte. Once activated, large numbers of identical cells (clones) of the lymphocyte that binds the specific antigen are produced. • Antigen challenge is the first encounter between an antigen and a lymphocyte.
	22.6a Activation of T-Lymphocytes • Activation of both helper T-lymphocytes and cytotoxic T-lymphocytes requires presentation of antigen by APCs and stimulation by interleukin 2 (IL-2).
	22.6b Activation of B-Lymphocytes • Activation of B-lymphocytes requires binding and engulfing antigen that is presented to a helper T-lymphocyte that recognizes the same antigen, and stimulation by IL-4 released by the helper T-lymphocyte.
	22.6c Lymphocyte Recirculation • The probability of an encounter occurring between a lymphocyte and the antigen it responds to is increased by lymphocyte recirculation, the circulation of lymphocyte through the blood and lymph every several days.
22.7 Effector Response at Infection Site	**22.7a Effector Response of T-Lymphocytes** • Activated helper T lymphocytes play a central role in the immune response by releasing cytokines that regulate the activation of other lymphocytes and the activity of cells of the innate immune system. • Activated cytotoxic T-lymphocytes go to the site of infection and produce cytotoxic chemicals that kill the unwanted cells (infected cells, transplanted cells, or cancer cells) that contain antigen it recognizes.
	22.7b Effector Response of B-Lymphocytes • Plasma cells (differentiated B-lymphocytes) produce large amounts of antibody against a specific antigen.
22.8 Immunoglobulins	**22.8a Structure of Immunoglobulins** • An antibody is an immunoglobulin that is produced against a specific antigen. • An immunoglobulin is a Y-shaped protein. It has two identical variable regions in the arms that include the antigen-binding sites and an Fc region that determines the molecule's biological activity.
	22.8b Actions of Antibodies • An antibody binds antigen and cause neutralization, agglutination, and precipitation of pathogens. • Once antibody is bound with antigen, the Fc region of an antibody can bind complement, cause opsonization, and initiate destruction of a cell through NK cells.
	22.8c Classes of Immunoglobulins • There are five major classes of immunoglobulins: IgG, IgM, IgA, IgD, and IgE. Of these, IgG is the most prevalent in the blood and other body fluids.
22.9 Immunologic Memory and Immunity	• Memory is a significant feature of acquired immunity that protects us from subsequent exposures to a given antigen.
	22.9a Immunologic Memory • The level of immune response can be gauged by measuring the antibody titer of a specific antibody. The first exposure results in the primary response, and subsequent exposures to the same antigen result in the secondary response.
	22.9b Measure of Immunologic Memory • Blood concentration levels of IgM and IgG (i.e., antibody titers) can be used to measure the primary response and secondary response of the immune system.
	22.9c Active and Passive Immunity • Active immunity involves the production of memory cells and can be acquired through contact with an infectious agent (naturally acquired) or through a vaccine (artificially acquired). • Passive immunity does not involve the production of memory cells because protection comes from another individual—from the mother across the placenta or in breast milk (naturally acquired) or from the serum of another individual (artificially acquired).

CHALLENGE YOURSELF

Create and Evaluate
Analyze and Apply
Understand and Remember

Do You Know the Basics?

1. All of the following are phagocytic cells *except*
   a. neutrophils.
   b. T-lymphocytes.
   c. macrophages.
   d. dendritic cells.
2. This cell releases cytokines to activate B-lymphocytes, increases the activity of macrophages, and in general regulates the overall immune response.
   a. cytotoxic T-lymphocyte
   b. helper T-lymphocyte
   c. natural killer cell
   d. basophil
3. This cell is activated by binding antigen, and then engulfing and presenting the antigen with MHC class II molecules to helper T-lymphocytes. The helper T-lymphocytes release cytokines as the second form of stimulation.
   a. NK (natural killer) cell
   b. macrophage
   c. B-lymphocyte
   d. cytotoxic T-lymphocyte
4. These two cells destroy an infected cell by releasing chemicals that cause apoptosis.
   a. NK cell and cytotoxic T-lymphocyte
   b. macrophage and NK cell
   c. helper T-lymphocyte and cytotoxic T-lymphocyte
   d. B-lymphocyte and T-lymphocyte
5. All of the following are functions of antibodies *except*
   a. neutralization of pathogen.
   b. destruction of antigen.
   c. agglutination of antigen.
   d. opsonization.
6. The four characteristics of adaptive immunity include all of the following *except*
   a. activation by a specific antigen.
   b. memory.
   c. production of clones of cells that have the same TCR or BCR.
   d. effective against a wide array of pathogens.
7. During which process does additional fluid enter an injured or infected area from the blood and additional fluid is removed by the lymph vessels?
   a. fever
   b. clonal selection
   c. inflammation
   d. activation of helper T-lymphocytes
8. This chemical is released by virus-infected cells to decrease the spread of virus to nearby cells.
   a. interferon
   b. bradykinin
   c. perforin
   d. complement
9. The correct sequence of the major events in the life of a lymphocyte is
   a. effector response, formation, and activation.
   b. activation, formation, and effector response.
   c. activation, effector response, and formation.
   d. formation, activation, and effector response.
10. Two of the major actions of complement are
    a. increased inflammatory response and cytolysis.
    b. recognition and destruction of a specific antigen and cytolysis.
    c. production of antibody and increased inflammatory response.
    d. release of cytokines to increase the immune response and production of antibody.
11. Compare the general characteristics of innate immunity and adaptive immunity in terms of the cells involved, specificity, general mechanisms, and time required.
12. Define the inflammatory response, and explain its benefits.
13. Describe an antigen.
14. Describe class I and class II MHC molecules, and explain how they function in assisting T-lymphocytes in recognizing an antigen.
15. Explain positive and negative selection that occurs during the selection of T-lymphocytes.
16. Describe how helper T-lymphocytes play a pivotal role in a healthy, normally functioning immune system.
17. Explain the general function of cytotoxic T-lymphocytes.
18. Describe the functions of antibodies and complement in defending the body.
19. There are two branches of adaptive immunity: cell-mediated and humoral immunity. Distinguish the types of antigens they are each effective against.
20. Explain the difference between the primary and secondary immune response.

Can You Apply What You've Learned?

1. Maria, who is 3 years old, was stung by a bee. The area where the stinger entered the skin became red, warm, and swollen. This normal response to a foreign venom is called
   a. a fever.
   b. the complement cascade.
   c. an inflammatory response.
   d. an antigen challenge.
2. Jay, a young dad, takes his baby to the pediatrician several times in the first year of the child's life. These visits will stimulate the baby to make memory cells against specific antigens. Why are these visits necessary?
   a. The baby is being vaccinated.
   b. The visits verify that the baby has a normal inflammatory response.
   c. The baby must have an unusual immune deficit that must be monitored.
   d. The baby's blood is being filtered to remove foreign antigen.
3. A young woman has just been diagnosed with the human immunodeficiency virus (HIV). This virus is especially devastating because it infects the cells that regulate the immune response. The cells HIV infects are
   a. B-lymphocytes.
   b. helper T-lymphocytes.
   c. NK cells.
   d. cytotoxic T lymphocytes.
4. One-year-old Matthew always seems to be sick. When his blood is tested, there are no antibodies. The physician concludes that the child is lacking
   a. the ability to develop an inflammatory response.
   b. helper T-lymphocytes.
   c. cytotoxic T-lymphocytes.
   d. B-lymphocytes.
5. Upon further testing, it is found that Matthew in question 4 does have normal cellular immunity. However, without antibodies Matthew will be less able to
   a. destroy cancer cells.
   b. destroy virus-infected cells.
   c. bind viral particles.
   d. destroy intracellular pathogens.

Can You Synthesize What You've Learned?

1. Dianne is an avid tennis player but has recently been complaining of tendonitis in her elbow. She knows that you work in health care and asks you to explain what caused this flare-up.
2. Stephanie is in her first year of college and has recently come down with a cold. She is running a slight fever of 100°F. Explain to her why a fever is beneficial.
3. Describe the events that occur in an individual who has an allergy to ragweed.

INTEGRATE

ONLINE STUDY TOOLS

connect | SMARTBOOK® | AP|R

The following study aids may be accessed through Connect.

Concept Overview Interactive: Figure 22.20: Adaptive Immunity

Clinical Case Study: A Young Man with Memory Loss

Interactive Questions: This chapter's content is served up in a number of multimedia question formats for student study

SmartBook: Topics and terminology include overview of diseases caused by infectious agents; overview of immune system; innate immunity; adaptive immunity: an introduction; formation and selection of lymphocytes; activation and clonal selection of lymphocytes; effector response at infection site; immunoglobulins; immunologic memory and immunity

Anatomy & Physiology Revealed: Topics include immune response; antigen processing; cytotoxic T cells; helper T cells

Animations: Topics include antiviral activity of interferon; activation of complement; inflammatory response; antigenic (epitopes) determinants; antigen processing

chapter 23

Respiratory System

Module 11: Respiratory System

INTEGRATE

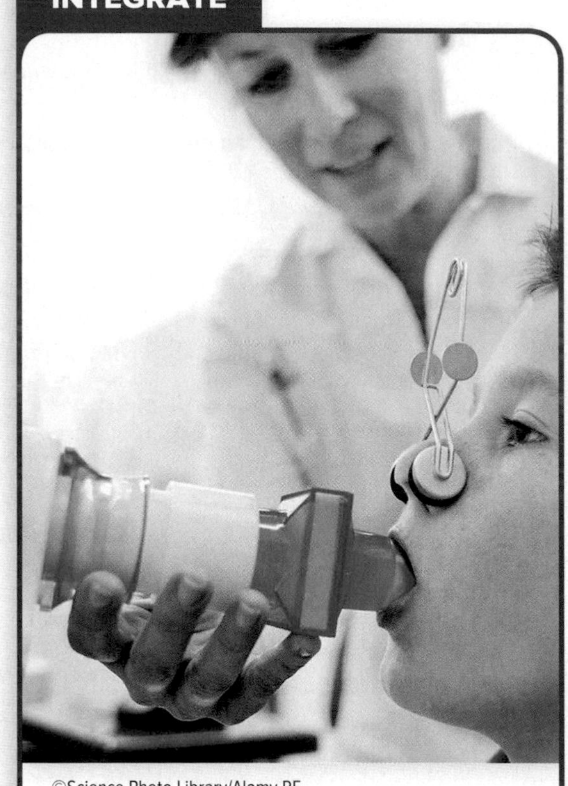

©Science Photo Library/Alamy RF

CAREER PATH
Respiratory Therapist

A respiratory therapist is an allied health-care professional who assesses, manages, and provides treatment to individuals with breathing irregularities or other cardiopulmonary disorders. These include both acute conditions that often accompany a stroke or a heart attack and chronic conditions such as bronchitis, asthma, and emphysema.

The **respiratory** (res′pi-ră-tōr′ē; *respire* = to breathe) **system** provides the means for gas exchange necessary for living cells. As cells engage in aerobic cellular respiration, they need both the uninterrupted supply of oxygen and the removal of carbon dioxide waste that is produced. This continuous exchange of oxygen and carbon dioxide between the atmosphere and body cells occurs through collective processes called **respiration.** Respiration requires both coordinated and integrated physiologic activities of a number of systems, including the respiratory, skeletal, muscular, nervous, and cardiovascular systems. The respiratory system is responsible for the exchange of gases between the atmosphere and the lungs. The skeletal and muscular systems alter the volume and pressure within the thoracic cavity to facilitate movement of air into and out of the lungs, whereas the nervous system stimulates and coordinates the contraction of skeletal muscles associated with breathing. The cardiovascular system transports oxygen and carbon dioxide between the lungs and the cells.

Our discussion begins by examining both the general functions and the anatomic structures of the respiratory system components. Next, we consider the processes involved in respiration, including (1) how the respiratory, skeletal, muscular, and nervous systems function together during breathing (pulmonary ventilation); (2) the exchange of respiratory gases between the lungs and blood (alveolar gas exchange) and between the blood and systemic cells (systemic gas exchange); and (3) gas transport by the cardiovascular system. We conclude by considering the influence of breathing rate on homeostasis.

23.1 Introduction to the Respiratory System

The respiratory system consists of the respiratory passageways extending through the head, neck, and trunk, as well as the lungs themselves. Here, we examine how the respiratory system serves numerous functions in the body, is organized both structurally and functionally, and is lined internally and protected with a mucous membrane.

23.1a General Functions of the Respiratory System

LEARNING OBJECTIVE

1. State the functions of the respiratory system.

The primary, and perhaps only, function most individuals would associate with the respiratory system is breathing. However, the respiratory system has several purposes that include the following:

- **Air passageway.** The respiratory tract is a passageway for air between the external environment and the alveoli (air sacs) of the lungs. Air is moved from the atmosphere to alveoli as we breathe in and then expelled into the atmosphere as we breathe out.
- **Site for the exchange of oxygen and carbon dioxide.** A thin barrier between the alveoli and the pulmonary capillaries provides the site for exchange of oxygen and carbon dioxide. Oxygen diffuses from the alveoli into the blood, and carbon dioxide diffuses from the blood into the alveoli.
- **Detection of odors.** Olfactory receptors located in the superior regions of the nasal cavity detect odors as air moves past them. Sensory input from these receptors is then relayed to various regions of the brain for interpretation (see section 16.3a).
- **Sound production.** The vocal cords of the larynx (voice box) vibrate as air moves across them to produce sounds; these sounds then resonate in upper respiratory structures.

WHAT DID YOU LEARN?

1. Which respiratory structure is associated with the exchange of respiratory gases?

23.1b General Organization of the Respiratory System

LEARNING OBJECTIVE

2. Distinguish between the structural organization and the functional organization of the respiratory system.

The respiratory system is organized *structurally* into two regions: an upper respiratory tract and a lower respiratory tract **(figure 23.1)**. The nose, nasal cavity, and pharynx form the **upper respiratory tract.** The larynx, trachea, bronchi, bronchioles (including terminal

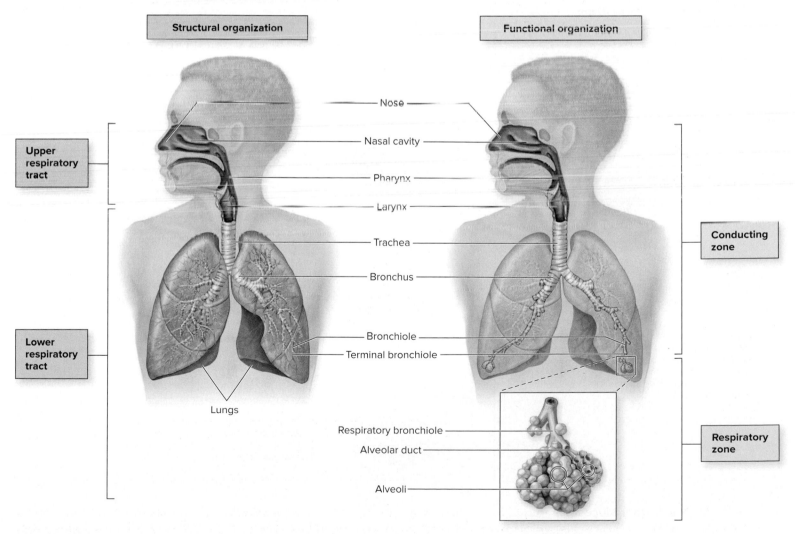

Figure 23.1 General Anatomy of the Respiratory System. Structurally, the respiratory system is organized into the upper respiratory tract and lower respiratory tract. Functionally, the respiratory system is divided into the conducting zone and respiratory zone.

INTEGRATE

LEARNING STRATEGY

You may find these hints helpful to remember the boundary structures in the general organization of the respiratory system:

Structural organization: The **L**ower respiratory tract begins at the **L**arynx and includes all structures to the alveoli.

Functional organization: The *terminal* bronchioles are the *end of* or *final portion* of the conducting zone.

and respiratory bronchioles), alveolar ducts, and alveoli are the components of the **lower respiratory tract.**

The structures of the respiratory system are also categorized based on *function.* Passageways that transport or conduct air are part of the **conducting zone;** these structures include the passageways from the nose to the end of the terminal bronchioles. Structures that participate in gas exchange with the blood—including the respiratory bronchioles, alveolar ducts, and alveoli—are part of the **respiratory zone.**

WHAT DID YOU LEARN?

2 What three structures compose the upper respiratory tract? What three structures compose the respiratory zone?

23.1c Respiratory Mucosa

LEARNING OBJECTIVES

3. Describe the structure of the mucosa that lines the respiratory tract and the structural changes observed along its length.
4. Explain the function of mucus produced by the mucosa.

The respiratory passageway is exposed to the external environment and is lined internally by a **mucosa,** also called a **mucous membrane** (see section 5.5b). In general, the mucosa is composed of an **epithelium** resting upon a basement membrane, and an underlying **lamina propria** composed of areolar connective tissue. The epithelium is ciliated (having **cilia**) in most portions of the respiratory tract conducting zone.

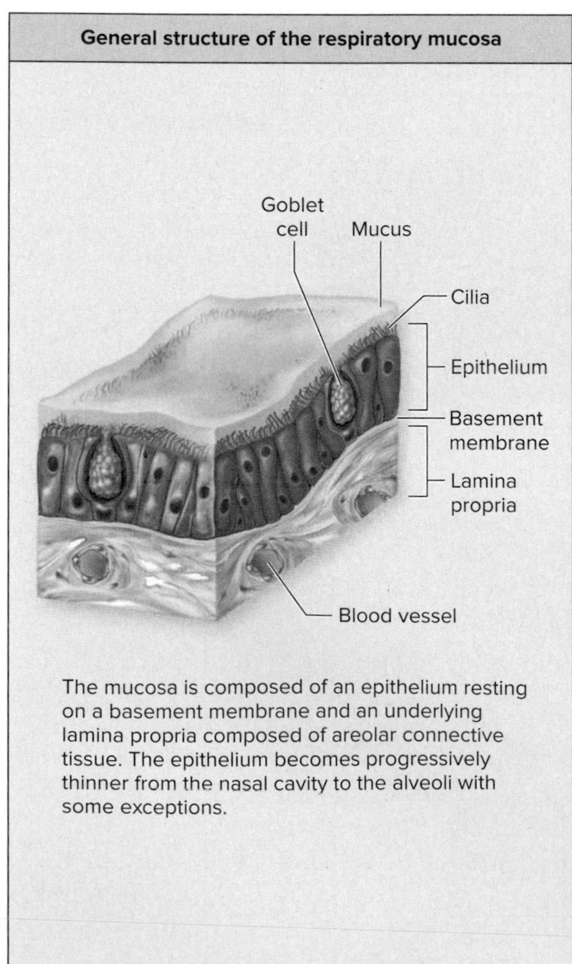
General structure of the respiratory mucosa
Goblet cell
Mucus
Cilia
Epithelium
Basement membrane
Lamina propria
Blood vessel
The mucosa is composed of an epithelium resting on a basement membrane and an underlying lamina propria composed of areolar connective tissue. The epithelium becomes progressively thinner from the nasal cavity to the alveoli with some exceptions.
(a)

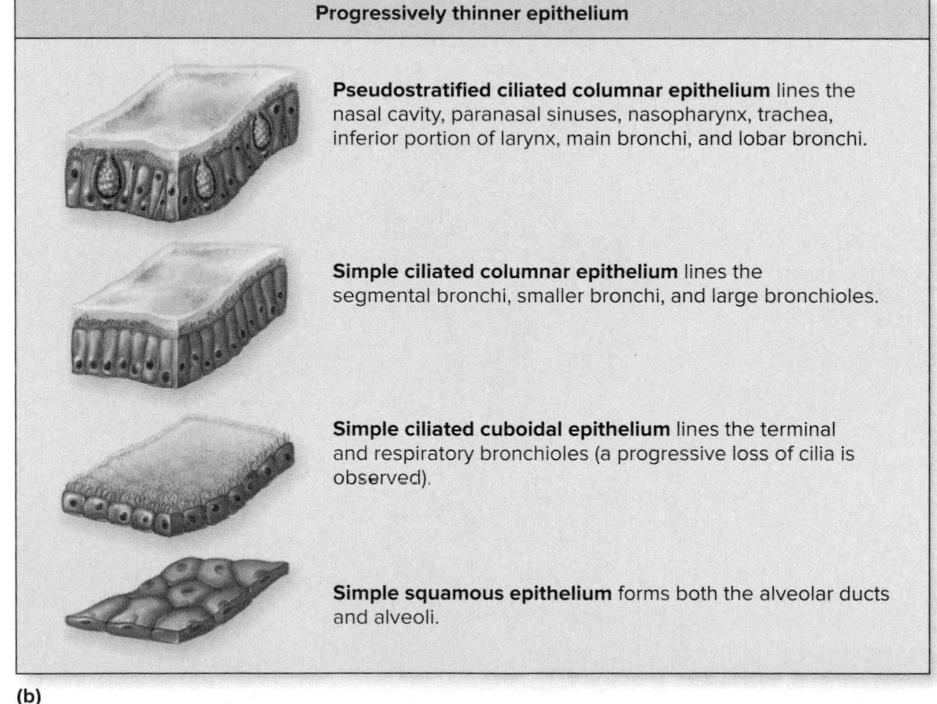
Progressively thinner epithelium
Pseudostratified ciliated columnar epithelium lines the nasal cavity, paranasal sinuses, nasopharynx, trachea, inferior portion of larynx, main bronchi, and lobar bronchi.
Simple ciliated columnar epithelium lines the segmental bronchi, smaller bronchi, and large bronchioles.
Simple ciliated cuboidal epithelium lines the terminal and respiratory bronchioles (a progressive loss of cilia is observed).
Simple squamous epithelium forms both the alveolar ducts and alveoli.
(b)

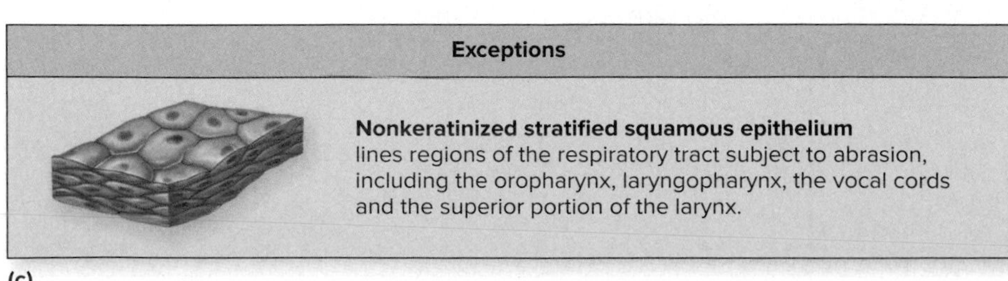
Exceptions
Nonkeratinized stratified squamous epithelium lines regions of the respiratory tract subject to abrasion, including the oropharynx, laryngopharynx, the vocal cords and the superior portion of the larynx.
(c)

Figure 23.2 Respiratory Mucosa. A mucosa forms the inner lining of the respiratory tract and is called the respiratory mucosa. (*a*) The general structure of the mucosa includes three major layers: an epithelium, a basement membrane, and an underlying lamina propria. (*b*) The types of epithelia in the mucosa become progressively thinner along the length of the respiratory tract. (*c*) Exceptions to the general thinning pattern occur in regions subject to abrasion. AP|R

A general pattern of structural change is observed in the epithelium along the length of the respiratory tract. The epithelium becomes progressively thinner from the nasal cavity to the alveoli; it changes from pseudostratified ciliated columnar to simple ciliated columnar to simple cuboidal to simple squamous (see section 5.1c). Exceptions to this general pattern occur in selected regions of the respiratory tract. These exceptions include (1) portions of the pharynx that serve as a passageway for both air and food (i.e., oropharynx and laryngopharynx) and (2) components of the larynx that include the vocal folds and the area immediately superior to them. These areas are lined by a nonkeratinized stratified squamous epithelium (instead of ciliated pseudostratified columnar epithelium) to withstand abrasion. **Figure 23.2** summarizes the types of epithelia found in the mucosa along the respiratory tract.

The epithelium lining most of the respiratory tract contains **goblet cells,** and the underlying lamina propria houses both mucous and serous glands. Mucus is produced from the combined secretions of these cells and glands. The amount of mucus produced daily is approximately 1 to 7 tablespoons, but that amount increases with exposure to irritants. Mucous secretions contain **mucin,** a protein that increases the viscosity of mucus to more effectively trap inhaled dust, dirt particles, microorganisms, and pollen. The secretions also contain specific substances to help defend the body against infectious agents (see section 22.1), including lysozyme (an antibacterial enzyme), defensins (antimicrobial proteins), and immunoglobulin A (antibodies) (see section 22.3a).

Both mucus and saliva entrap materials, which may be coughed up as a viscous substance called **sputum.** Physicians may request sputum samples from their patients to diagnose potential respiratory infections.

WHAT DID YOU LEARN?

3 In what ways does the epithelium of the upper respiratory tract differ from the epithelium in the alveoli?

23.2 Upper Respiratory Tract

The *upper respiratory tract,* as described previously, includes the nose, nasal cavity, and pharynx (**figure 23.3*a***).

23.2a Nose and Nasal Cavity

LEARNING OBJECTIVES

5. Describe the structure and function of the nose.
6. Provide a general description of the structure and function of the nasal cavity

The **nose** is the first structure of the conducting passageway for inhaled air (figure 23.3*b*); it is formed by bone, hyaline cartilage, and dense irregular connective tissue covered with skin externally. Paired nasal bones form the bridge of the nose and support it superiorly. Anteroinferiorly from the bridge, there is one pair of **lateral cartilages** and there are two pairs of **alar cartilages.** The flared components of the nose are composed of dense irregular connective tissue. The paired **nostrils,** or *nares* (nā′res; sing., *naris*) are the anterior openings that lead into the nasal cavity.

The **nasal cavity** is oblong-shaped, and it extends from the nostrils to paired openings called **choanae** (kō′an-ē; sing., *choana; choane* = a funnel) or *posterior nasal apertures* (figure 23.3*c, d*). The choanae are the "doorways" that lead from the nasal cavity into the pharynx. A single choana is labeled in figure 23.3*c*. The floor of the nasal cavity is formed by the hard and soft palates, and the roof is composed of the nasal, frontal, ethmoid, and sphenoid bones, and some cartilage of the nose. The **nasal septum** divides the nasal cavity into left and right portions (figure 23.3*d*). The septum is formed anteriorly by the **septal nasal cartilage** (figure 23.3*b*) and posteriorly by a thin, bony sheet composed of the perpendicular plate of the ethmoid superiorly and the vomer bone inferiorly (see section 8.2d).

WHAT DO YOU THINK?

1 What is a deviated nasal septum, and how would it affect breathing?

Figure 23.3 Upper Respiratory Tract. (*a*) Anatomic regions that compose the upper respiratory tract, (*b*) supporting structures of the nose, (*c*) midsagittal section of the nasal cavity, and (*d*) coronal view of the nasal cavity in a cadaver. AP|R
(*d*) ©McGraw-Hill Education/Christine Eckel

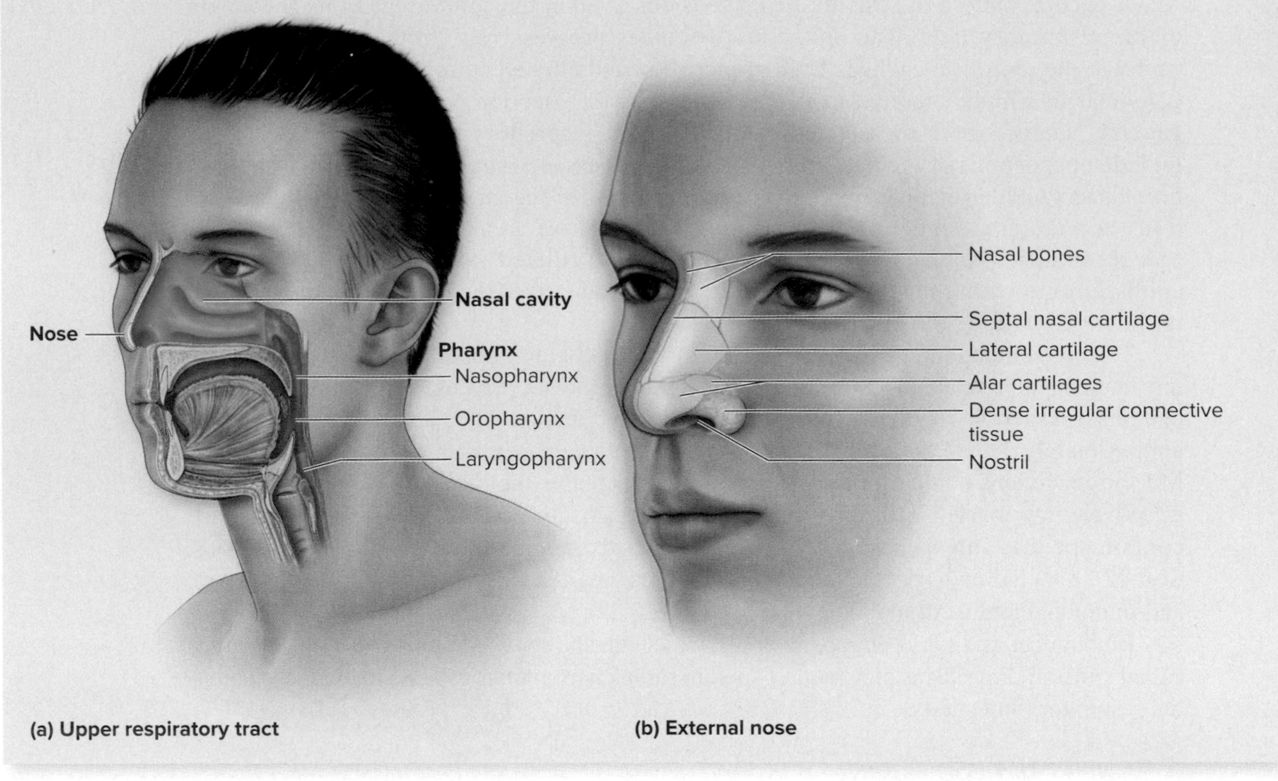

Three paired, bony projections are located along the lateral walls of the nasal cavity: the **superior, middle,** and **inferior nasal conchae** (kon′kē; sing., *concha;* a shell) (figure 23.3*c, d*). Because the conchae help produce turbulence in the inhaled air, they are sometimes called the *turbinate* bones. The conchae partition the nasal cavity into separate air passages (or "valleys"), each called a **nasal meatus** (mē-ā′tŭs). A meatus is located immediately inferior to its corresponding nasal concha. Be careful to distinguish the terms *choanae* and *conchae.* Choanae are the openings between the nasal cavity and pharynx; conchae are described here.

The nasal cavity is divided into three parts (figure 23.3*c*): the nasal vestibule, olfactory region, and respiratory region. The **nasal vestibule** is immediately internal to the nostrils and is lined by skin and coarse hairs called **vibrissae** (vi-bris′ē; sing., *vibrissa; vibro* = to quiver) to trap large particulates. This is the only normally visible portion of the nasal cavity.

The **olfactory region** is the superior portion of the nasal cavity. It contains the olfactory epithelium (which houses the olfactory receptors). Airborne molecules that dissolve in the mucus covering the olfactory epithelium stimulate olfactory receptors to detect different odors (see section 16.3a).

The **respiratory region** of the nasal cavity is lined by a mucosa composed of pseudostratified ciliated columnar epithelium. The lamina propria of this mucosal lining has an extensive vascular network. **Nosebleeds** (*epistaxis*) are especially common because of both the vast distribution of blood vessels and their superficial location (just deep to the epithelium), and they are more likely to occur during cold weather because the mucous membranes become dry and crack. Additionally, paired **nasolacrimal ducts** drain lacrimal secretions from the surface of each eye into the respiratory region of the nasal cavity (see figure 16.9 in section 16.4a).

A primary function of the nasal cavity is to *condition the air* (which means to warm, cleanse, and humidify the air) as it enters the respiratory tract. The air is warmed to body temperature by the extensive array of blood vessels within the nasal cavity lining. These vessels dilate in response to cold air, resulting in increased blood flow that helps to more effectively warm the inhaled air. The air is cleansed as inhaled microbes, dust, and other foreign material become trapped in the mucus covering the inner lining of the respiratory tract. Cilia then "sweep" the mucus and its trapped contents toward the pharynx to be swallowed (see section 22.3a). The air is also humidified as it passes through the moist environment of the nasal passageway. Conditioning of air is enhanced by conchae, which cause air turbulence that increases the amount of contact between the inhaled air and the mucosa (figure 23.3*c, d*).

WHAT DID YOU LEARN?

4 What changes occur to inhaled air as it passes through the nasal cavity?

5 What is the function of nasal conchae?

23.2b Paranasal Sinuses

LEARNING OBJECTIVE

7. Describe the structure and function of the four paired paranasal sinuses.

The **paranasal** (par-ă-nā′săl; *para* = alongside) **sinuses,** first described in section 8.2d, are associated with the nasal cavity **(figure 23.4)**. These sinuses, which are spaces within the skull bones, are named for the specific skull bones in which they are located. Thus, from a superior to inferior direction, they are the paired **frontal, ethmoidal,** and **maxillary sinuses;** the **sphenoidal**

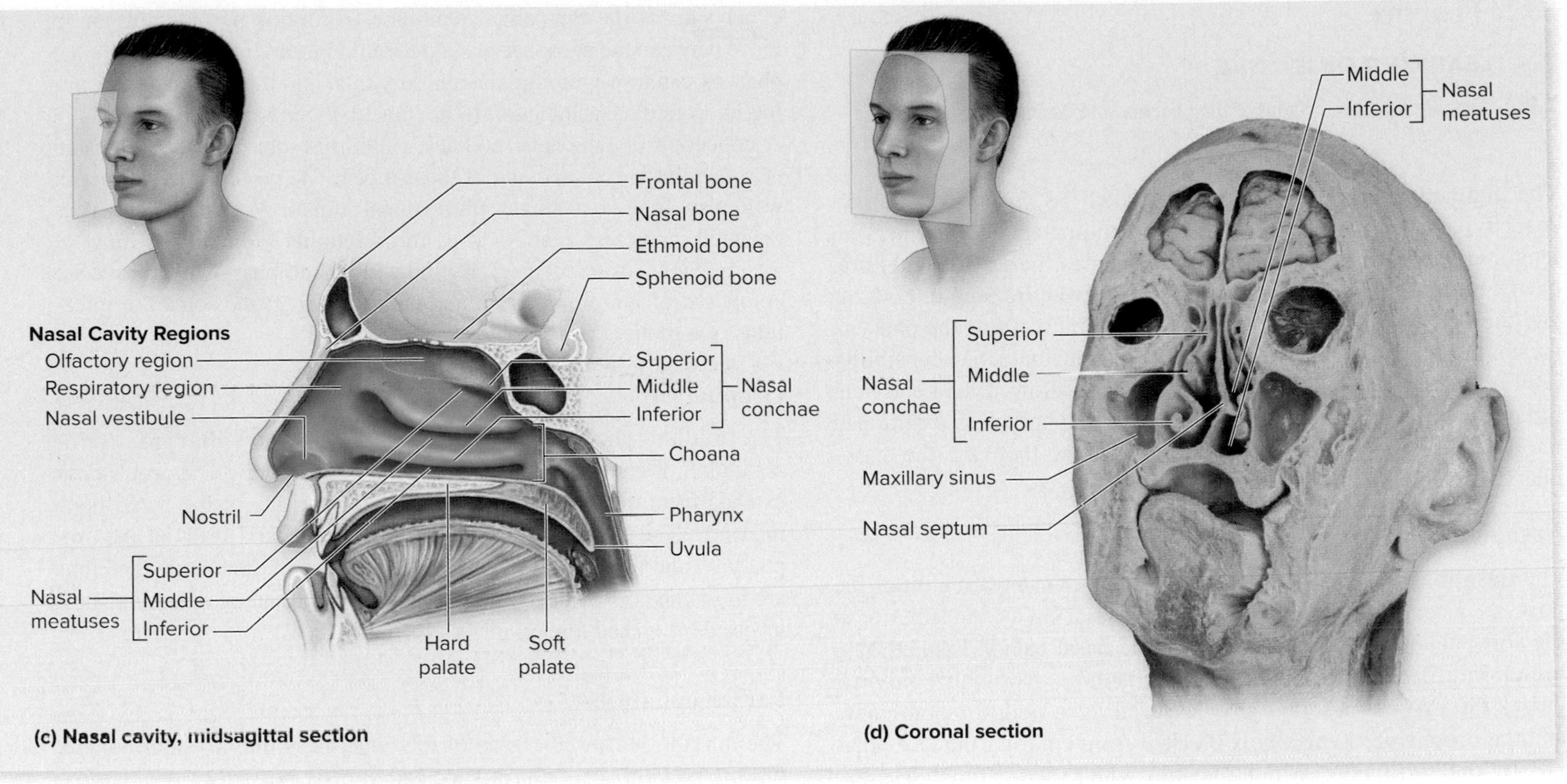

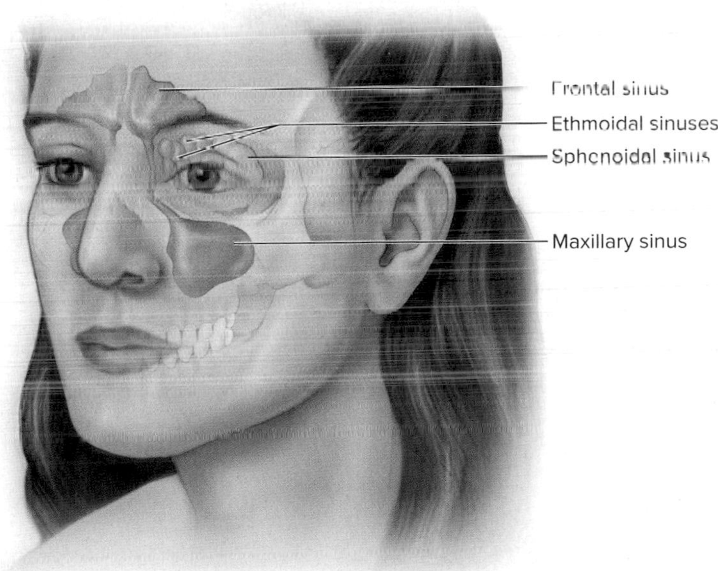

Figure 23.4 Paranasal Sinuses. The paranasal sinuses are air-filled cavities named for the bones in which they occur: frontal, ethmoidal, sphenoidal, and maxillary.

sinuses are located posterior to the ethmoidal sinuses. Ducts connect all paranasal sinuses to the nasal cavity. Both the paranasal sinuses and their ducts are lined by a pseudostratified ciliated columnar epithelium that is continuous with the mucosa of the nasal cavity. The mucus, with its trapped particulate matter, is swept by cilia from each paranasal sinus into the nasal cavity and then into the pharynx, where it is swallowed.

WHAT DID YOU LEARN?

6 How are the paranasal sinuses connected to the nasal cavity?

INTEGRATE

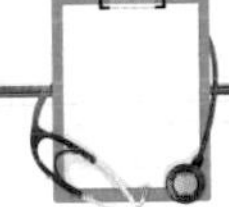

CLINICAL VIEW 23.1

Runny Nose

A **runny nose,** or **rhinorrhea** (rī′nō-rē-ă; *rhin* = nose, *rhoia* = flow), can occur as a result of (1) an increased production of mucus, such as occurs in response to a cold virus or allergy; (2) crying, due to increased secretions from the lacrimal glands that drain into the nasal cavity; or (3) exposure to cold air.

When cold air enters the nasal cavity and is exposed to the warm temperatures, water condensation occurs and the water mixes with mucus. In addition, the watery mucus is more likely to remain within the nasal cavity because the "chilled" cilia on epithelial cells are less able to sweep the mucus into the nasopharynx.

INTEGRATE

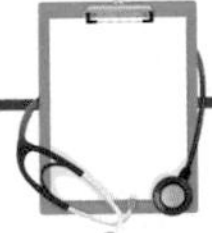

CLINICAL VIEW 23.2

Sinus Infections and Sinus Headaches

The mucosa of the ducts that drain from the paranasal sinuses into the nasal cavity can become inflamed in response to a respiratory infection or an allergy. Drainage of mucus is decreased and mucus accumulates in the paranasal sinuses as a result. A **sinus infection** can occur due to the lack of proper drainage of mucus. **Sinus headaches** result from increased pressure within the paranasal sinuses, due to the swelling of the mucosa. They also can result from pressure changes associated with swimming or high altitudes.

23.2c Pharynx

LEARNING OBJECTIVE

8. Compare the three regions of the pharynx, and describe their associated structures.

The **pharynx** (far′ingks), commonly called the *throat,* is a funnel-shaped passageway that averages 13 centimeters (5.1 inches) in length (**figure 23.5**). It is located posterior to the nasal cavity, oral cavity, and larynx. Air is conducted along its entire length, both air and food along its inferior portions. The lateral walls of the pharynx are composed of skeletal muscles that both contribute to distensibility (ability to stretch) needed to accommodate swallowed food and help force these materials into the esophagus. The pharynx is partitioned into three regions—from superior to inferior, they are the nasopharynx, oropharynx, and laryngopharynx.

Nasopharynx

The **nasopharynx** (nā′zō-far′inks) is the superiormost region of the pharynx. Located directly posterior to the nasal cavity and superior to the soft palate, the nasopharynx, like the nasal cavity, is lined by a pseudostratified ciliated columnar epithelium. Normally, only air passes through the nasopharynx. Material from both the oral cavity and the oropharynx typically is blocked from entering the nasopharynx by the soft palate, which elevates when we swallow (see section 26.2c). However, sometimes food or drink enters the nasopharynx and the nasal cavity, as when a person tries to swallow and then laughs at the same time. The soft palate cannot form a good seal for the nasopharynx, and the force from the laugh may propel some of the material into the nasal cavity. If the laugh is forceful enough, the material may come out the nostrils.

The nasopharynx lateral walls have paired openings into **auditory tubes** (*eustachian tubes,* or *pharyngotympanic tubes*) that connect the nasopharynx to the middle ear. These tubes equalize air pressure on either side of the tympanic membrane (eardrum) by allowing air to move between the nasopharynx and the middle ear. Infections within the pharynx can move through the auditory tube into the middle ear, resulting in a middle ear infection (see Clinical View 16.9: "Otitis Media"). A collection of lymphatic nodules, called the tubal tonsils, is located near the pharyngeal opening of these tubes. The posterior nasopharynx wall also houses a single **pharyngeal tonsil.** When this tonsil is enlarged, clinicians refer to it as the **adenoids** (ad′ĕ-noydz; *aden* = gland, *eidos* = resemblance). Both the tubal and pharyngeal tonsils are composed of lymphatic tissue and help to prevent the spread of infections (see section 21.4c).

Oropharynx

The middle pharyngeal region, called the **oropharynx** (ōr′ō-far′ingks), is immediately posterior to the oral cavity. The oropharynx extends from the level of the soft palate superiorly to the hyoid bone inferiorly. The **palatine tonsils** are located on lateral walls of the oropharynx, and the **lingual tonsils** are at the base of the tongue (and thus are in the anterior region of the oropharynx), providing defense against ingested or inhaled foreign materials (see figure 21.8).

Laryngopharynx

The inferior, narrowed region of the pharynx is the **laryngopharynx** (lă-ring′gō-far′ingks), which is located directly posterior to the larynx. It extends from the level of the hyoid bone and is continuous on its inferior end with both the larynx anteriorly and the esophagus posteriorly. Both the oropharynx and laryngopharynx serve as a common passageway for food and air. They are lined by a nonkeratinized stratified squamous epithelium to protect these regions of the pharynx from abrasion associated with swallowing food (see figure 23.2*c*).

WHAT DID YOU LEARN?

7 What two regions of the pharynx contain tonsils? What is their purpose?

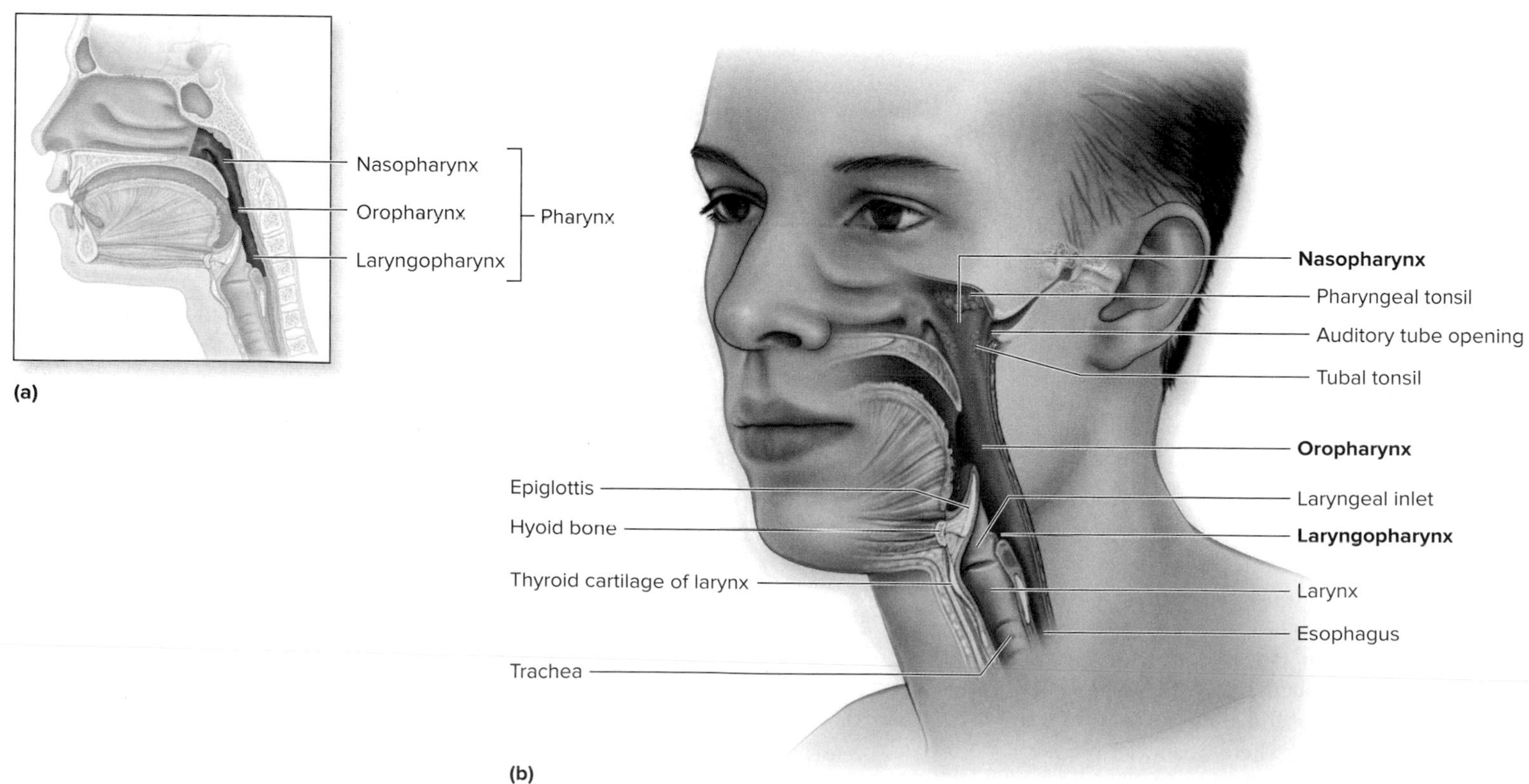

Figure 23.5 Pharynx. (*a*) The three specific regions of the pharynx (nasopharynx, oropharynx, and laryngopharynx) are highlighted in a sagittal section. (*b*) The pharynx is shown in relationship to the larynx, trachea, and esophagus in an anterolateral view. AP|R

23.3 Lower Respiratory Tract

The structures of the *lower respiratory tract* include both conducting pathways (the larynx, trachea, bronchi, and bronchioles) and those involved in gas exchange (respiratory bronchioles, alveolar ducts, and alveoli) (see figure 23.1).

23.3a Larynx

LEARNING OBJECTIVES

9. Describe the general functions and structure of the larynx.
10. Explain how the larynx functions in sound production.

The **larynx** (lar′ingks), also called the *voice box,* is a somewhat cylindrical structure that averages about 4 centimeters (1.6 inches) in length (figure 23.5). It is continuous superiorly with the laryngopharynx and inferiorly with the trachea. The superior opening from the laryngopharynx into the larynx is called the **laryngeal** (lă-rin′jē-ăl) **inlet,** *laryngeal aperture,* or *laryngeal aditus.*

Functions of the Larynx

The larynx has several major functions:

- **Serves as a passageway for air.** The larynx is normally open to allow the passage of air.
- **Prevents ingested materials from entering the respiratory tract.** During swallowing, the laryngeal inlet is covered by the epiglottis to prevent ingested materials from entering the lower respiratory passageway.
- **Produces sound for speech.** Ligaments within the larynx, called vocal cords, vibrate as air is passed over them during an expiration.
- **Assists in increasing pressure in the abdominal cavity.** The epiglottis of the larynx closes over the laryngeal inlet so air cannot escape, and simultaneously abdominal muscles contract to increase abdominal pressure. This action is referred to as the **Valsalva maneuver.** You can experience the increase in abdominal pressure associated with the Valsalva maneuver by holding your breath while forcefully contracting your abdominal muscles.
- **Participates in both a sneeze and cough reflex.** Both a sneeze and a cough result in an explosive blast of exhaled air. This occurs when the abdominal muscles contract forcefully and the vocal cords are initially closed and then open abruptly as the pressure increases in the thoracic cavity. Sneezing is a reflex (see sections 14.6a and 14.6b) initiated by irritants in the nasal cavity, whereas coughing is initiated by irritants in the trachea and bronchi. Both help remove irritants from the respiratory tract.

INTEGRATE

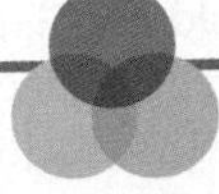

CONCEPT CONNECTION

The **Valsalva maneuver** facilitates several physiologic processes, including the elimination of both urine from the urinary bladder (see section 24.8c) and feces from the gastrointestinal tract (see section 26.3d), and the expulsion of a baby during childbirth (see section 29.6d).

Larynx Anatomy

The larynx is shown from three views relative to the hyoid bone (see section 8.3) and trachea in **figure 23.6**. Observe that the larynx is supported by a framework of nine pieces of cartilage that are supported and held in place by ligaments and muscles. The nine cartilages include the single thyroid, cricoid, and epiglottis cartilages and the paired arytenoid, corniculate, and cuneiform cartilages.

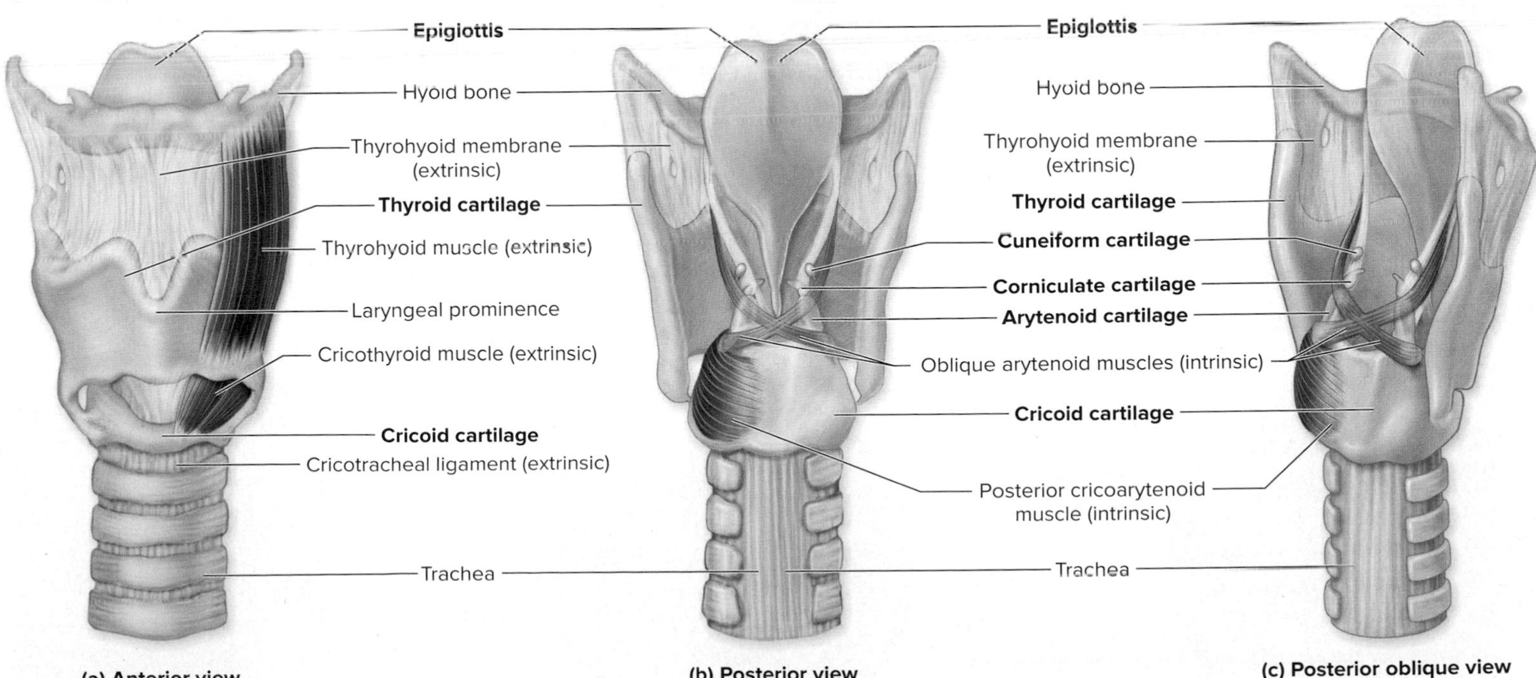

Figure 23.6 Larynx. The larynx shown in (*a*) an anterior view, (*b*) a posterior view, and (*c*) a posterior oblique view. It is composed of nine cartilages, various ligaments, and skeletal muscle. The nine cartilages, as well as extrinsic and intrinsic ligaments, form the flexible support of the larynx. Extrinsic muscles participate in the elevation of the larynx, and intrinsic muscles function in sound production. AP|R

The **thyroid cartilage** is the largest laryngeal cartilage. Shaped like a shield, it forms the anterior and lateral walls of the larynx. The almost V-shaped anterior projection of the thyroid cartilage is called the **laryngeal prominence** (commonly referred to as the *Adam's apple*). This protuberance is generally larger in males because (1) the laryngeal inlet is narrower in males (90 degrees) than in females (120 degrees), and (2) it enlarges at puberty due to testosterone-induced growth.

The thyroid cartilage is attached to the lateral surface of the ring-shaped **cricoid** (krī′koyd; *kridos* = a ring) **cartilage** located inferior to the thyroid cartilage. The large, spoon- or leaf-shaped **epiglottis** (ep-i-glot′is; *epi* = on, *glottis* = mouth of windpipe) is anchored to the inner aspect of the thyroid cartilage and projects posterosuperiorly into the pharynx. It closes over the laryngeal inlet during swallowing (figure 23.5*b*). The three smaller, paired cartilages, the **arytenoid** (ar-i-tē′noyd), **corniculate** (kōr′-ni-kū-lāt; *corniculatus* = horned), and **cuneiform** (kū′nē-i-fōrm; *cuneus* = wedge) **cartilages** are located internally (figure 23.6). All cartilages of the larynx, except the epiglottis, are composed of hyaline cartilage. The epiglottis, which opens and closes over the laryngeal inlet, is composed of the more flexible elastic cartilage (see section 5.2d).

Laryngeal ligaments are classified as either extrinsic ligaments or intrinsic ligaments. **Extrinsic ligaments** attach to the external surface of laryngeal cartilages and extend to other structures that include the superiorly located hyoid bone and inferiorly located trachea.

The **intrinsic ligaments** are located within the larynx and include both vocal ligaments and vestibular ligaments **(figure 23.7)**. The **vocal ligaments** are composed primarily of elastic connective tissue and extend anterior to posterior between the thyroid cartilage and the arytenoid cartilages. These ligaments are covered with a mucosa to form the **vocal folds.** Vocal folds also are called the *true vocal cords* because they produce sound when air passes between them. They are distinctive from the surrounding tissue because they are avascular and white. The opening between these folds is called the **rima glottidis** (rī′mă glo-tī′dis; *rima* = slit). Together the vocal folds and the rima glottidis form the **glottis.**

The **vestibular ligaments** form the other intrinsic ligaments. These extend between the thyroid cartilage to the arytenoid and corniculate cartilages. Together with the mucosa covering them, they form the **vestibular folds** located superior to the vocal folds. These folds also are called the *false vocal cords* because they have no function in sound production, but they protect the vocal folds. The opening between the vestibular folds is called the **rima vestibuli.**

Skeletal muscles compose part of the larynx wall and are classified as either extrinsic or intrinsic muscles. The **extrinsic muscles** attach to the thyroid cartilage on one end and extend either superiorly to attach to the hyoid bone (thyrohoid; figure 23.6*a*) or inferiorly to attach to the sternum (sternothyroid; see figure 11.11). Extrinsic muscles elevate the larynx during swallowing (see sections 11.3d and 26.2c).

The **intrinsic muscles** are located within the larynx and attach to both the arytenoid and corniculate cartilages (figure 23.6*b, c*). Contraction of the intrinsic muscles causes the arytenoid cartilages to pivot, resulting in a change in the dimension of the rima glottidis (figure 23.7). The opening becomes narrower as the vocal folds are adducted and becomes wider if the vocal folds are abducted. They function in voice production and help close off the larynx when swallowing.

Sound Production

Sound production (or *phonation*) originates as the vocal folds vibrate. This occurs when the intrinsic laryngeal muscles narrow the opening of the rima glottidis and air is forced past the vocal cords during an expiration.

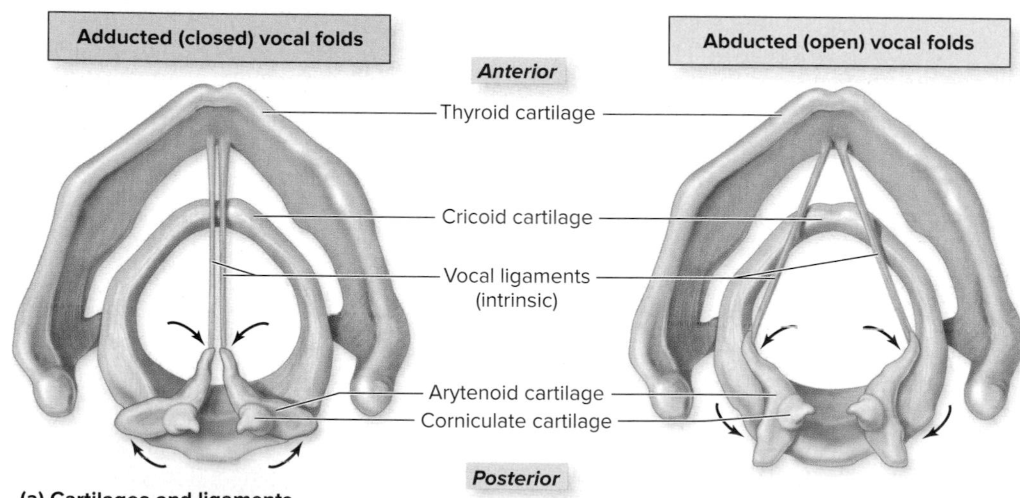

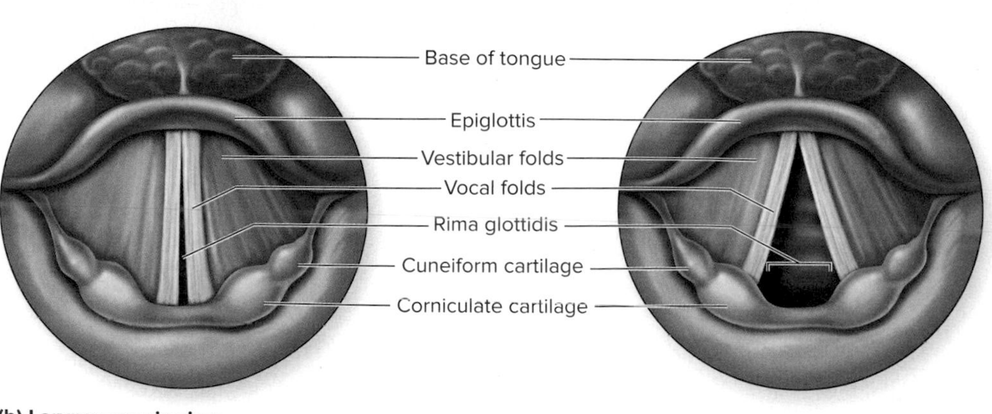

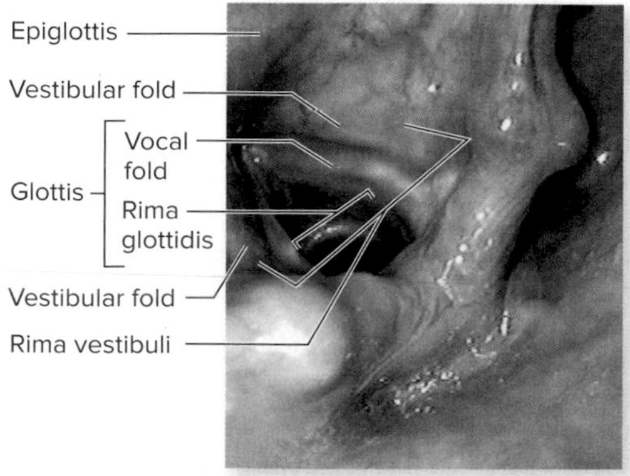

Figure 23.7 Vocal Folds. The vocal folds (true vocal cords) are elastic ligaments covered with a mucosa that extend between the thyroid and arytenoid cartilages. These folds surround the rima glottidis and are involved in sound production. Adducted (closed) and abducted (open) vocal folds are shown in (*a*) a superior view of the cartilages and ligaments only and (*b*) a diagrammatic laryngoscopic view of the coverings around these cartilages and ligaments. (*c*) A photo of a superolateral laryngoscopic view, showing the vocal folds and the rima glottidis, and the vestibular fold and rima vestibuli.

INTEGRATE

CLINICAL VIEW 23.3

Laryngitis

Laryngitis (lar-in-jī′tis) is an inflammation of the larynx that may extend to its surrounding structures. Viral or bacterial infection (see section 22.1) is the number one cause of laryngitis. Less frequently, laryngitis follows overuse of the voice, such as yelling for several hours at a football game. Symptoms include hoarse voice, sore throat, and sometimes fever. Severe cases can produce inflammation and swelling extending to the epiglottis. Children's airways are proportionately smaller, and a swollen and inflamed epiglottis (called *epiglottitis*) may lead to sudden airway obstruction and become a medical emergency.

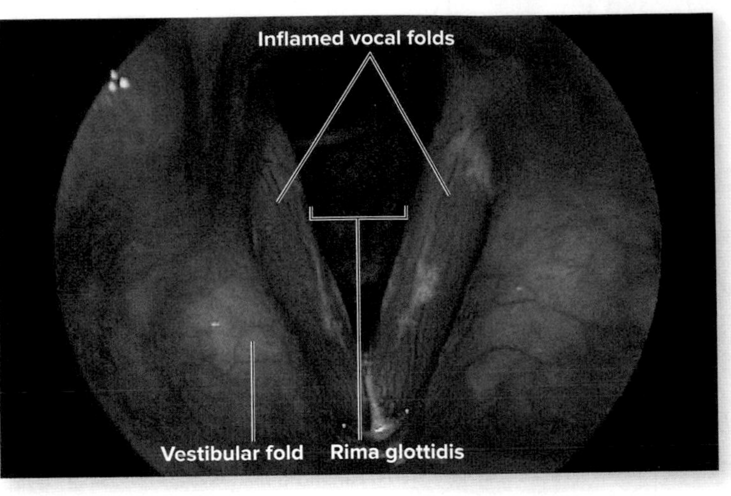

A laryngoscopic view shows the inflamed, reddened vocal folds characteristic of laryngitis.

©C. Richard Stasney, M.D.

The characteristics of the sound include range, pitch, and loudness. The **range** of a voice (be it soprano or bass) is determined by the length and thickness of the vocal folds. Males generally have longer and thicker folds than do females, and thus males produce sounds that are in a lower range. Our vocal folds increase in length as we grow, which is why our voices become lower or deeper as we mature to adulthood.

Pitch refers to the frequency of sound waves (see section 16.5b). Pitch is determined by the amount of tension or tautness on the vocal folds as regulated primarily by the intrinsic laryngeal muscles. Increasing the tension on the vocal folds causes the vocal folds to vibrate more when air passes by them and thus to produce a higher sound. Conversely, the less taut the vocal folds, the less they vibrate, and the lower the pitch of the sound.

Loudness depends on the force of the air passing across the vocal cords. A lot of air forced through the rima glottidis produces a loud sound, whereas a little air produces a soft sound. When you whisper, only the most posterior portion of the rima glottidis is open, and the vocal folds do not vibrate. Because the vocal folds are not vibrating, the whispered sounds are all of the same pitch.

Speech (or the *articulation* or *enunciation* of words) also requires the participation of the spaces in the pharynx, nasal and oral cavities, and paranasal sinuses, which serve as resonating chambers for sound, and the structures of the lips, teeth, and tongue that help form different sounds. Young children tend to have high, nasal-like voices because their sinuses are not yet well developed, so they lack large resonating chambers. When you hold your nose and speak, your voice sounds quite different because air doesn't pass through the nasal cavity.

WHAT DID YOU LEARN?

8. How does the larynx assist in increasing abdominal pressure?
9. What are the three unpaired cartilages in the larynx?
10. What are the structural and functional differences between the vocal folds and vestibular folds?

23.3b Trachea

LEARNING OBJECTIVES

11. Describe the structure of the trachea.
12. Explain the structure and function of the tracheal cartilages.

The **trachea** (trā′kē-ă; rough), which is commonly referred to as the *windpipe,* is a patent (open) tube that connects the larynx and the two main bronchi as it extends through the neck and into the mediastinum of the thoracic cavity, where it is partially protected by the sternum. (Its anatomic position relative to these structures is shown in figure 23.9.) The position of the trachea relative to the esophagus (the muscular tube of the gastrointestinal tract that leads from the mouth to the stomach) can be seen in **figure 23.8*a***. Notice that the trachea is anterior to the esophagus. Here, we discuss both the gross anatomy of the trachea and the histology of the tracheal wall.

Gross Anatomy of the Trachea

The trachea is a flexible, slightly rigid, tubular organ that averages approximately 13 centimeters (5.1 inches) in length and 2.5 centimeters (1 inch) in diameter. The anterior and lateral walls of the trachea are supported by 15 to 20 C-shaped rings of hyaline cartilage called **tracheal cartilages.** The tracheal cartilages are connected superiorly and inferiorly with one another by elastic connective tissue sheets called **anular** (an′ū-lăr; *anulus* = ring) **ligaments.**

Each C-shaped tracheal cartilage is ensheathed in a perichondrium (see section 7.2f) and a dense, fibrous membrane (not shown in figure 23.8). The open ends of the cartilage rings are positioned posteriorly (adjacent to the esophagus) and are connected to each other by both the **trachealis muscle** and an elastic ligamentous membrane (figure 23.8*b*). The C-shaped cartilage portion of each of these rings reinforces and provides structural support to the tracheal wall to ensure that the trachea remains open (patent) at all times. The more flexible trachealis muscle and ligamentous membrane on the posterior aspect of the trachea allow for distension during swallowing of food through the esophagus. The trachealis contracts during coughing to reduce the diameter of the trachea, thus facilitating the more rapid expulsion of air, helping to dislodge material (foreign objects or food) from the air passageway.

An internal ridge of mucosa-covered cartilage called the **carina** (kă-rī′nă) is located at the split of the trachea into the main bronchi (figure 23.8*c*). The carina has sensory receptors that are extremely sensitive and can induce a forceful cough when stimulated by irritants.

Histology of the Tracheal Wall

The innermost to outermost layers that form the wall of the trachea are (1) the **mucosa,** which is composed of a pseudostratified ciliated columnar epithelium with goblet cells and a lamina propria

INTEGRATE

CLINICAL VIEW 2

Bronchitis

Bronchitis (brong-kī'tis) is inflam bacterial infection (see section 22 chemicals, particulate matter, o divided into two categories: acut

Acute bronchitis develops r such as a cold. Symptoms include

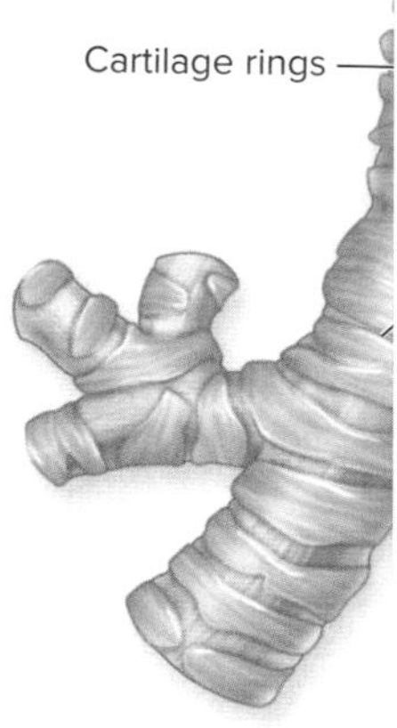

Figure 23.10 Structure of bronchioles do not contain cartilag constriction and bronchodilation th

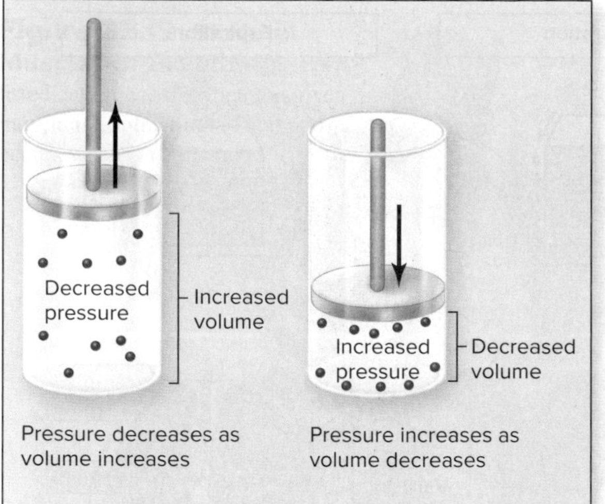

(a) Boyle's law

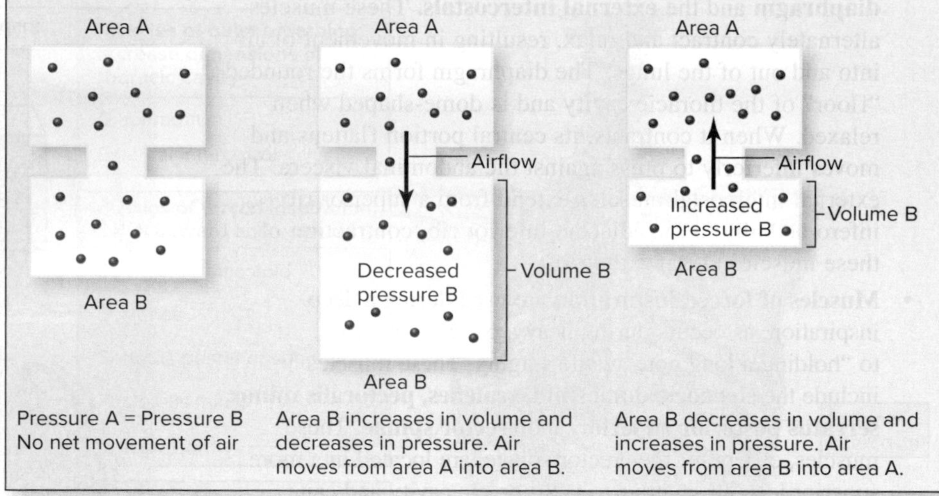

(b) Pressure gradients

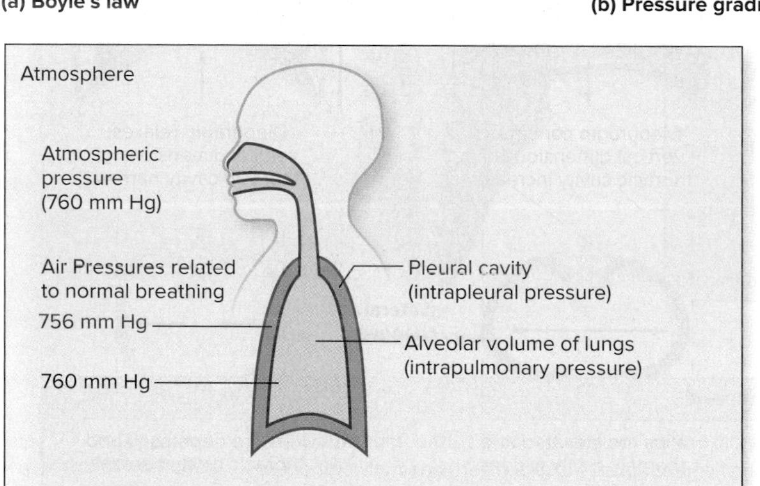

(c) Volumes and pressures with breathing (at the end of an expiration)

Figure 23.21 Boyle's Law and Pressure Gradients.
(*a*) Boyle's law states that volume and pressure are inversely related. (*b*) The figure shows that air does not move if the pressure is equal between two areas. It also shows how air moves from an area of high pressure to an area of low pressure when pressure gradients are established by changes in volume. (*c*) Three significant pressures are associated with breathing. Atmospheric pressure is 760 mm Hg at sea level. The other two are the intrapulmonary pressure and the intrapleural pressure, which both change during breathing as a result of changing volumes. AP|R

Pressure Gradients

An air pressure gradient occurs when the force per unit area is greater in one area than in another. If a pressure gradient exists between two regions, and they are interconnected, then air moves from the region of higher pressure to the region of lower pressure until the pressure in the two regions becomes equal. Figure 23.21*b* shows this relationship.

Volumes and Pressures Associated with Breathing

A similar relationship exists with respect to the atmosphere and the lungs, which are interconnected by the respiratory passageway (figure 23.21*c*).

The **atmosphere** is the air in the environment that surrounds us. **Atmospheric pressure** is the pressure (weight) gases in the air exert in the environment. Its value changes with altitude. The standard value is given at sea level because air becomes less dense, or "thins," with increased altitude. The "thinner air" exerts a correspondingly lower atmospheric pressure. The value for atmospheric pressure at sea level can be expressed in several ways: 14.7 pounds per square inch = 1 atmosphere (atm) = 760 mm Hg. The value used in this text is 760 mm Hg. (Note that 1 millimeter of mercury is expressed as 1 mm Hg, and it is the pressure exerted by a column of mercury that is 1 millimeter high in a glass tube.) Atmospheric pressure does not change in the context of breathing.

The thoracic cavity contains the lungs. The collective volume of the alveoli within the lungs is called the **alveolar volume,** and its associated pressure is the *intrapulmonary pressure* (described in section 23.4d). This pressure fluctuates with breathing and may be higher, lower, or the same as atmospheric pressure. Intrapulmonary pressure is equal to atmospheric pressure (at sea level = 760 mm Hg) at the end of both inspiration and expiration.

Recall from section 23.4c that the lungs are separated from the wall of the thoracic cavity by the pleural cavity. The pressure exerted within the pleural cavity is called the *intrapleural pressure*. Intrapleural pressure also fluctuates with breathing and is always lower than intrapulmonary pressure so that the lungs remain inflated. Prior to inspiration, it is generally about 4 mm Hg lower than intrapulmonary pressure (756 mm Hg).

Note that a volume change in the thoracic cavity occurring during inspiration and expiration establishes a pressure gradient between the atmosphere and the thoracic cavity that determines the direction of airflow. Thus, an increase in volume of the thoracic cavity, with an accompanying decrease in pressure, results in air moving into the lungs during inspiration. In contrast, a decrease in volume of the thoracic cavity, with an accompanying increase in pressure, results in air moving out of the lungs during expiration.

Integration of Concepts: Quiet Breathing

Quiet breathing was described as the normal breathing that occurs when you are relaxed (see section 23.5a). Refer to **figure 23.22** as you read through this description of the stepwise sequence of events that alter volume and pressure during quiet breathing:

Inspiration

1. Initially, both intrapulmonary pressure and atmospheric pressure are 760 mm Hg at sea level prior to inspiration. Intrapleural

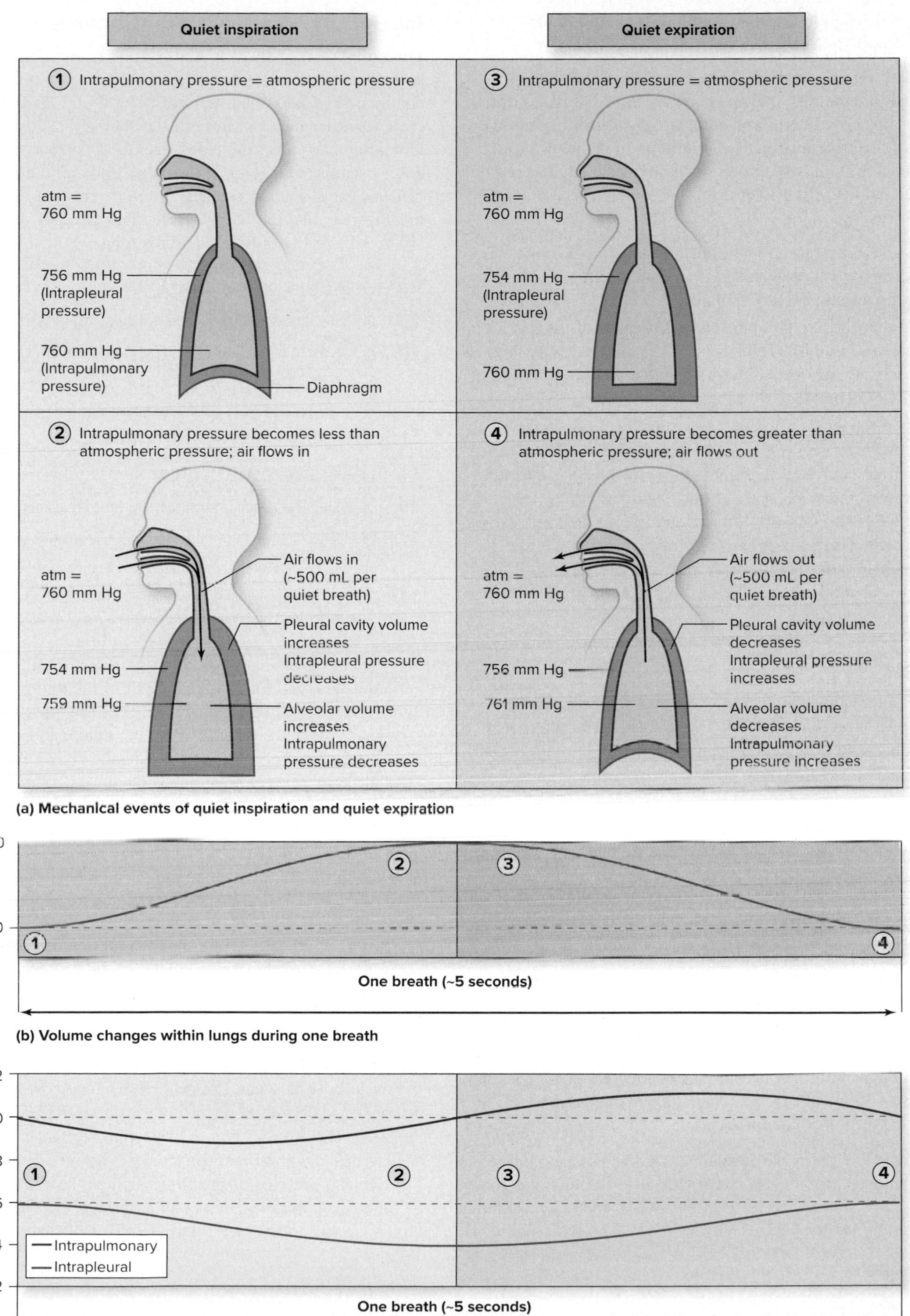

Figure 23.22 Volume and Pressure Changes Associated with the Mechanics of Quiet Breathing. Circled numbers correspond to the steps described in the text. (*a*) Schematic of changes in volume, pressure, and airflow that are associated with breathing. (*b*) Approximately 500 mL of air is inspired and then expired during quiet breathing; these volumes are associated with (*c*) relatively small changes in intrapulmonary and intrapleural pressures.

pressure is slightly lower at approximately 756 mm Hg, or about 4 mm Hg lower than intrapulmonary pressure.

2. The diaphragm contracts, increasing the thoracic cavity vertical dimensions, and the external intercostals contract, increasing both lateral and anterior-posterior dimensions. During quiet breathing, diaphragmatic movement may account for about two-thirds of the thoracic cavity volume change, and external intercostal movement for about one-third.
   - Muscle contraction results in pleural cavity volume increasing with an accompanying decrease in intrapleural pressure from approximately 756 to 754 mm Hg.
   - Simultaneously, the lungs expand because of the surface tension caused by serous fluid in the pleural cavity with an accompanying decrease in intrapulmonary pressure from 760 to 759 mm Hg.
   - When the intrapulmonary pressure decreases below atmospheric pressure, air moves down the pressure gradient from the environment into the alveoli, until the intrapulmonary pressure is once again equal to atmospheric pressure. The volume of air that moves from the atmosphere into the lungs during a single breath in quiet breathing is approximately 500 milliliters (mL), or 0.5 liter (L). This volume of air is called the *tidal volume*.

Expiration

3. Initially, the intrapulmonary pressure has equalized with atmospheric pressure (both at 760 mm Hg) prior to quiet expiration, and intrapleural pressure is about 754 mm Hg, still lower than intrapulmonary pressure.
4. The diaphragm and external intercostals relax, decreasing the dimensions of the thoracic cavity.
   - The pleural cavity volume decreases as the diaphragm relaxes and the thoracic wall recoils. Intrapleural pressure now increases from 754 mm Hg to return to 756 mm Hg.
   - Simultaneously, the alveolar volume decreases because the lungs are pulled inward by the recoil of elastic connective tissue in the lungs. Intrapulmonary pressure now increases from 760 to 761 mm Hg.
   - When intrapulmonary pressure exceeds atmospheric pressure, air is forced out of the alveoli into the atmosphere. This continues until the intrapulmonary pressure is once again equal to atmospheric pressure. Approximately 500 mL of air moves out of the lungs.

The changes in thoracic cavity volume and both intrapulmonary and intrapleural pressures that are associated with quiet breathing are shown in figure 23.22*b, c*. The events of quiet inspiration and quiet expiration are summarized in **table 23.3**.

Integration of Concepts: Forced Breathing

Forced breathing involves steps similar to quiet breathing. However, both forced inspiration and expiration are active processes, requiring contraction of additional muscles (see figure 23.19). Their activity causes greater changes in both the thoracic cavity volume and the intrapulmonary pressure. Consequently, a steeper air pressure gradient is established and more air moves into and out of the lungs. Significant chest volume changes are apparent that accompany forced breathing, unlike the barely perceptible changes in the chest cavity volume that occur during quiet breathing.

WHAT DID YOU LEARN?

22. Describe the sequence of events of quiet inspiration.
23. How are larger amounts of air moved between the lungs and atmosphere during forced inspiration and forced expiration? Is more energy expended during forced breathing? Why?

23.5c Nervous Control of Breathing

LEARNING OBJECTIVES

29. Describe the anatomic structures involved in regulating breathing.
30. Explain the physiologic events associated with controlling quiet breathing.
31. Explain the different reflexes that alter breathing rate and depth.
32. Distinguish between nervous system control of structures of the respiratory system and nervous system control of structures involved in breathing.

The *skeletal muscles* of breathing are coordinated by nuclei within the brainstem, and the rate and depth of breathing are either altered through reflexes or consciously controlled by the cerebrum (see section 13.3). Here we first describe the anatomic components of the (1) respiratory center, which regulates breathing (see section 13.5); (2) skeletal muscles of breathing and their innervation; and (3) receptors that detect stimuli and relay sensory input to the respiratory center to alter breathing rate and depth. We then integrate these structures to discuss how the nervous system controls breathing. Please refer to **figure 23.23** as you read through this section.

Anatomic Structures

The autonomic nuclei within the central nervous system (CNS) that control breathing are collectively called the **respiratory center** (figure 23.23*a*). One portion of this center, which is located within the medulla oblongata, is called the **medullary respiratory center.** Two groups of nuclei compose the medullary respiratory center. These are the **ventral respiratory group (VRG),** located within the anterior region of the medulla (which contains both inspiratory neurons and expiratory neurons), and the **dorsal respiratory group (DRG),** located posterior to the VRG. The other portion of the respiratory center is within the pons and is called the **pontine respiratory center** (or *pneumotaxic center*).

Skeletal muscles of breathing include the diaphragm, the external intercostal muscles, and other accessory muscles of breathing, as previously described (see figure 23.19). Upper motor neurons from

Table 23.3 Changes Associated with Quiet Breathing

Variable	Inspiration	Expiration
Diaphragm and external intercostals	Contracting (active)	Relaxing (passive)
Pleural cavity	Volume increases; pressure decreases	Volume decreases; pressure increases
Lungs	Volume increases; pressure decreases	Volume decreases; pressure increases
Air movement	Into lungs	Out of lungs

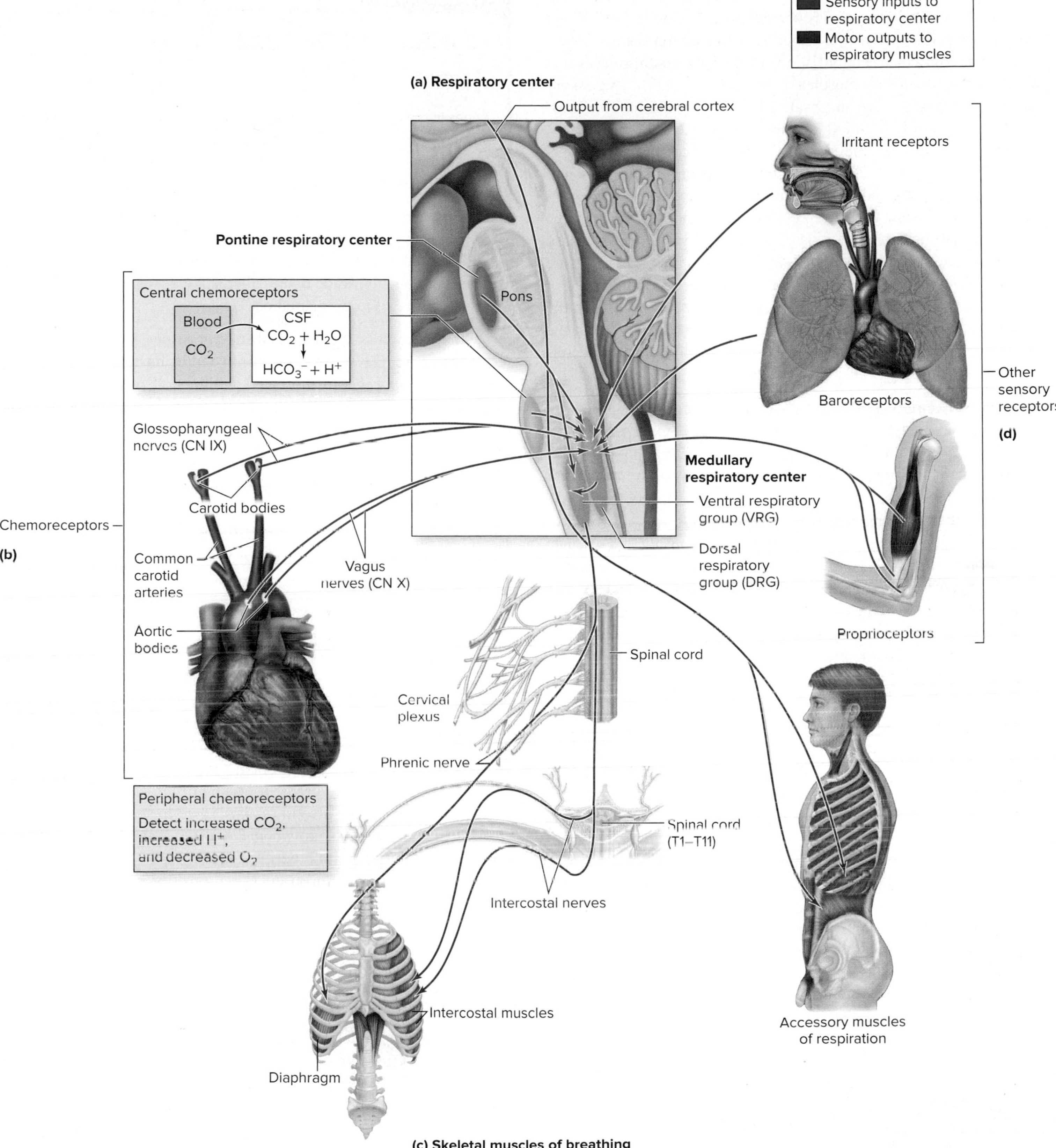

Figure 23.23 Respiratory Center. (*a*) The respiratory center rhythmically initiates nerve signals along motor neurons to the diaphragm and external intercostals to regulate quiet breathing. The respiratory center is stimulated to alter the breathing rate and depth primarily by (*b*) sensory input from central chemoreceptors in the brain and peripheral chemoreceptors within the carotid and aortic bodies. Based on this sensory input, nerve signals are relayed to (*c*) the skeletal muscles of breathing to alter the breathing rate or depth. Sensory input is also relayed to the respiratory center from (*d*) other sensory receptors, including irritant receptors of the mucosal lining of the respiratory tract, baroreceptors (stretch receptors) of the lungs and visceral pleura, and proprioceptors of muscles, tendons, and joints. Note that breathing can be consciously controlled by the cerebral cortex; its motor output bypasses the respiratory center to directly stimulate lower motor neurons innervating skeletal muscles of breathing. APR

the VRG synapse with lower motor neurons that extend from the spinal cord to the skeletal muscles of breathing (figure 23.23*c*). These lower motor neurons are within either the **phrenic nerves** that innervate the diaphragm (see section 14.5d) or the **intercostal nerves** that innervate the intercostal muscles (see section 14.5c). Accessory muscles of respiration are innervated by other individually named somatic nerves (not individually named in figure 23.23).

Chemoreceptors are the primary sensory receptors involved in altering breathing (figure 23.23*b*). **Chemoreceptors** monitor fluctuations in concentration of both hydrogen ions (H^+) and respiratory gases (P_{CO_2} and P_{O_2}) within both the cerebrospinal fluid (CSF) and the blood. Note that the concentrations of respiratory gases are expressed as the partial pressure of carbon dioxide (P_{CO_2}) and the partial pressure of oxygen (P_{O_2}). Partial pressures are discussed in section 23.6a. For now, just remember that the higher the partial pressure for a gas, the greater its concentration. Chemoreceptors are housed both within the brain (central chemoreceptors) and within specific blood vessels (peripheral chemoreceptors):

- **Central chemoreceptors** are within the medulla oblongata in close proximity to the medullary respiratory center. Central chemoreceptors monitor only H^+ changes of CSF induced by changes in blood P_{CO_2}. Carbon dioxide diffuses from the blood into the CSF. In the CSF, carbonic anhydrase catalyzes the formation of carbonic acid from carbon dioxide and water. The carbonic acid then dissociates into bicarbonate (HCO_3^-) and hydrogen ions (H^+). Thus, central chemoreceptors are monitoring H^+ concentration within the CSF, which is formed from CO_2 that diffuses there from the blood. It is important to note that unlike the blood, the CSF *lacks proteins* to buffer the gain or loss of H^+. (See the discussion of chemical buffers in section 25.5d.) Consequently, H^+ changes within the CSF most accurately reflect the changes in blood P_{CO_2}.
- **Peripheral chemoreceptors** are located both within the aortic arch (called the **aortic bodies**) and at the split of each common carotid artery into the external and internal carotid arteries (called **carotid bodies**). Peripheral chemoreceptors normally detect changes in the concentration of both H^+ and P_{CO_2} within arterial blood. The peripheral chemoreceptors differ from central chemoreceptors because they are stimulated by changes in H^+ produced independently of P_{CO_2}. This may, for example, occur due to kidney failure (kidneys normally eliminate H^+; see section 24.6d) or as result of uncontrolled diabetes mellitus (ketoacids are a by-product from metabolism of fatty acids; see section 25.6c). Peripheral chemoreceptors can also be stimulated by relatively large changes in blood P_{O_2}. When stimulated by changes in either blood H^+ or blood respiratory gases, the carotid bodies alter nerve signals relayed along the glossopharyngeal nerves, and the aortic bodies alter nerve signals relayed along the vagus nerves to the respiratory center.

INTEGRATE

LEARNING STRATEGY

This summary can help you to remember what is monitored by the two different types of chemoreceptors.

- **Central chemoreceptors** *monitor CSF:* H^+ levels in CSF (which is produced from blood CO_2) (lack proteins to buffer H^+).
- **Peripheral chemoreceptors** *monitor blood:* (1) CO_2 levels, (2) H^+ levels that are produced through metabolic processes (e.g., lactic acid, ketoacids), and (3) relatively large changes in O_2.

INTEGRATE

CLINICAL VIEW 23.12

Apnea

Apnea (ap′nē-ă; *a* = absence, *pnea* = breathing) is the absence of breathing. It can occur voluntarily (during swallowing or holding your breath), be drug-induced (e.g., anesthesia), or result from neurologic disease or trauma. **Sleep apnea** is the temporary cessation of breathing during sleep.

Other receptors include **irritant receptors** located within the respiratory passageways that are stimulated by dust and other particulate matter, **baroreceptors** located within both the visceral pleura and the bronchiole smooth muscle that are stretch receptors, and **proprioceptors** located within joints and muscles that are stimulated by body movement (figure 23.23*d*).

Physiology of Breathing

Although there is much still unknown about how breathing is completely controlled, the following description provides some insight into what is generally accepted.

Quiet Breathing *Quiet inspiration* is initiated when the inspiratory neurons of the VRG within the medullary respiratory center spontaneously depolarize, or are "turned on." Nerve signals initiated within the inspiratory center of the VRG are sent through the somatic nerve pathways to the skeletal muscles of quiet breathing for approximately 2 seconds. The intensity of the nerve signals increases over these 2 seconds. This stimulates both the diaphragm and external intercostal muscles to contract, resulting in an increase in thoracic cavity volume. Thus, a pressure gradient is established, and air moves from the atmosphere into the alveoli (as described in section 23.5b).

Quiet expiration occurs when the VRG is inhibited, or "turned off." This inhibition results because, during quiet inspiration, nerve signals from the VRG inspiratory neurons are also relayed to the VRG expiratory neurons. The VRG expiratory neurons in response will then send inhibitory signals back to the VRG inspiratory neurons (like an inhibitory feedback loop). This inhibition causes the VRG inspiratory neurons to "turn off." Consequently, nerve signals are no longer sent through the nerve pathways to the skeletal muscles of quiet breathing; this lasts typically for approximately 3 seconds. Lack of somatic nerve stimulation causes both the diaphragm and external intercostal muscles to relax, resulting in a decrease in thoracic cavity volume. Thus, a pressure gradient is established and air moves from the alveoli into the atmosphere (as described in section 23.5b).

A breathing rhythm that involves 2 seconds of inspiration followed by 3 seconds of expiration results in an average respiratory rate of 12 times per minute. The average range for the rate of quiet breathing is generally between 12 and 15 times per minute—a rate referred to as *eupnea*. (Note: Physiologists previously thought the roles of the VRG and DRG were reversed from what is explained here.)

The specific role of the pontine respiratory center in breathing is to relay nerve signals to the medullary respiratory center to facilitate a smooth transition between inspiration and expiration. Individuals who have experienced damage to this center are able to still breathe. However, the breathing is erratic and involves long, gasping inspirations followed by occasional expirations.

WHAT DO YOU THINK?

3 The phrenic nerves extend from the cervical plexus formed by the rami of spinal nerves C3–C5 (see section 14.5d), whereas the intercostal nerves are the anterior rami of spinal nerves T1–T11 (see section 14.5c). Predict the consequences to breathing for each condition if a spinal cord injury is (a) at C2 or above, (b) between C6 and T11, and (c) T12 or below.

Reflexes That Can Alter Breathing Rate and Depth

Altering Breathing Rate and Depth Through Reflexes Involving Chemoreceptors Breathing rate and depth are primarily altered by reflexes that respond to sensory input from chemoreceptors. Nerve signals from these receptors are sent along sensory neurons to the DRG. When the DRG is activated, nerve signals are subsequently relayed to the VRG, resulting in a change in the rate and depth of breathing. Change in the *rate* of breathing is accomplished by altering the amount of time spent in both inspiration and expiration, whereas altering the *depth* of breathing is accomplished through stimulation of accessory muscles, which results in greater thoracic volume changes.

Breathing rate and depth can be reflexively *increased* if either the central chemoreceptors detect an increase in H^+ concentration in the CSF or the peripheral chemoreceptors detect an increase in blood H^+ concentration, an increase in blood P_{CO_2}, or both. Central chemoreceptors relay additional nerve signals directly to the closely located DRG, and peripheral chemoreceptors relay additional nerve signals through both the glossopharyngeal nerves from the carotid bodies and the vagus nerves from the aortic bodies to the DRG. The DRG then relays this information to the VRG. Respiration rate and depth are increased and additional CO_2 is expelled, ultimately returning blood P_{CO_2} and CSF pH to normal levels. Conversely, a decrease in either H^+ or P_{CO_2} initiates fewer nerve signals relayed to the DRG, and respiration rate and depth are decreased.

The most important stimulus affecting breathing rate and depth is blood P_{CO_2}. The chemoreceptors are very sensitive to changes in blood P_{CO_2} levels; increases in P_{CO_2} levels as small as 5 mm Hg will double the breathing rate. The changes in blood P_{CO_2} can stimulate respiratory rate changes most powerfully when the carbon dioxide is joined to water, forming carbonic acid in the CSF. This is because, as described earlier in this section, changes in H^+ within the CSF most accurately reflect the changes in blood P_{CO_2} because the CSF (unlike the blood) lacks proteins to buffer H^+.

Generally, changes in blood P_{O_2} are *not* an independent means of regulating breathing. Note that the arterial oxygen level in the blood must decrease substantially from its normal P_{O_2} level of 95 mm Hg to an abnormally low level of 60 mm Hg before it can stimulate the chemoreceptors independently of P_{CO_2}. This relationship can have deadly consequences to swimmers who hyperventilate before swimming under water. This hyperventilation causes the blood P_{CO_2} levels to decrease to the point where chemoreceptors are not stimulated. The swimmer's blood P_{O_2} level is decreased at the same time by exertion, but not enough to stimulate chemoreceptors. A swimmer may lose consciousness and drown as a result.

P_{O_2} levels normally influence breathing rate by causing the chemoreceptors to be more sensitive to changes in blood P_{CO_2}. This relationship has a synergistic effect: The combination of decreased P_{O_2} and increased P_{CO_2}, along with the subsequent production of H^+, causes greater stimulation of the chemoreceptors.

Altering Breathing Patterns Through Reflexes Involving Other Receptors Receptors other than chemoreceptors alter breathing patterns. (1) Proprioceptors within joints and muscles, when stimulated by body movement, increase nerve signals to the respiratory center with a subsequent increase in breathing depth. (2) Baroreceptors within both the visceral pleura and bronchiole smooth muscle are stimulated by stretch. These sensory receptors initiate a reflex to prevent overstretching of the lungs by inhibiting inspiration activities. This reflex is referred to as the **inhalation reflex,** or *Hering-Breuer reflex.* It effectively protects the lungs from damage due to overinflation. When overstretched, these baroreceptors send nerve signals through the vagus nerves to the respiratory center to shut off inspiration activity, thus resulting in expiration. This may be a normal means of controlling respiration in infants, but it is thought to serve only as a protective reflex after infancy. (3) Irritant receptors, when stimulated, initiate either a sneezing or coughing reflex. A **sneeze reflex** is initiated by irritants within the nasal cavity and a **cough reflex** by irritants within the trachea and bronchi. Nerve signals are relayed by sensory neurons to the medulla oblongata (sneezing or coughing nuclei, respectively; see section 13.5c). These nuclei initiate nerve signals along motor neurons, which cause the vocal cords to close and the abdominal muscles to contract forcefully. The vocal cords, which are initially closed, then open abruptly as the pressure increases in the thoracic cavity. Both reflexes result in an explosive blast of exhaled air that potentially removes the irritant.

INTEGRATE

CLINICAL VIEW 23.13

Hypoxic Drive

Certain respiratory disorders (such as emphysema) result in decreased ability to expire carbon dioxide, and the blood P_{O_2} levels can become the stimulus for breathing. The blood P_{O_2} as the stimulus for breathing is called the **hypoxic** (hī-pok′sik) **drive.** It occurs as carbon dioxide levels in the blood become elevated and remain elevated over a long period of time. Chemoreceptors become less sensitive to P_{CO_2} and, by default, decreased P_{O_2} levels stimulate them. Since a low P_{O_2} is the stimulus for breathing, administering oxygen would elevate P_{O_2} and thus interfere with the person's ability to breathe on his or her own.

Action of Higher Brain Centers Higher brain centers, including the hypothalamus, limbic system, and cerebral cortex, can influence breathing rate. The hypothalamus increases the breathing rate if the body is warm and decreases it if the body is cold. The limbic system alters the breathing rate in response to emotions and emotional memories. The frontal lobe of the cerebral cortex controls voluntary changes in our breathing pattern for various activities, such as talking, singing, breath-holding, performing the Valsalva maneuver (see section 23.3a), and other actions. Unlike the other higher areas of the brain that relay impulses to the respiratory center, nerve signals from the cerebral cortex bypass the respiratory center to directly stimulate lower motor neurons in the spinal cord (figure 23.23).

Nervous Control of Anatomic Structures of the Respiratory System and Anatomic Structures of Breathing

We distinguish between the innervation to the anatomic structures of the respiratory system and the anatomic structures that function in breathing. Anatomic structures of the respiratory system, which are composed of both smooth muscle and glands, are innervated by the axons of lower motor neurons of the sympathetic and parasympathetic divisions of the autonomic nervous system and controlled by autonomic nuclei within the brainstem (see section 23.4b). In contrast, breathing muscles, which are skeletal muscles, are innervated by axons of the lower motor neurons of the somatic nervous system (see section 12.1b). However, the control of the breathing muscles comes from both autonomic nuclei in the brainstem and somatic nuclei in the cerebral cortex. The autonomic nuclei forming the respiratory center within the brainstem regulate normal breathing with their rhythmic output along the lower motor neurons of the phrenic and intercostal nerves, and this center alters breathing rate and depth in response to various sensory input, as described. The cerebral cortex consciously regulates breathing by directly stimulating lower motor neurons that extend to the skeletal muscles of breathing. This diverse motor output from the nervous system allows breathing to be controlled both reflexively and consciously.

WHAT DID YOU LEARN?

24 What are the functions of the VRG and DRG within the respiratory centers?

25 Which of the following stimuli will cause an increase in the respiratory rate: (a) increase in blood P_{CO_2}, (b) increase in blood H^+, (c) increase in H^+ within the CSF, and (d) increase in blood P_{O_2}?

26 Are the skeletal muscles of breathing innervated by somatic nerves or autonomic nerves? Explain.

23.5d Airflow, Pressure Gradients, and Resistance

LEARNING OBJECTIVES

33. Define airflow.
34. Explain how pressure gradients and resistance determine airflow.

Airflow is the amount of air that moves into and out of the respiratory tract with each breath. Sufficient exchange of air must be maintained for normal body functions; this exchange is in part determined by airflow and variables that affect it. Airflow is a function of two factors: (1) the *pressure gradient* established between atmospheric pressure and intrapulmonary pressure and (2) the *resistance* that occurs due to conditions within the respiratory tract, lungs, and chest wall. The formula for airflow is expressed here:

$$F = \frac{\Delta P}{R} \quad \text{or} \quad F = \frac{P_{atm} - P_{alv}}{R}$$

The parameters are F = flow, ΔP = difference in pressure between atmosphere (atm) and the intrapulmonary pressure within the alveoli (alv), and R = resistance.

This mathematical expression demonstrates that flow is *directly* related to the pressure gradient between atmosphere and lungs and *inversely* related to resistance. If the pressure gradient increases, airflow into the lungs increases, but if the pressure gradient decreases, airflow into the lungs lessens (assuming resistance remains the same). In contrast, if resistance increases, airflow lessens, whereas if resistance decreases, airflow increases (assuming the pressure gradient remains the same).

The **pressure gradient** (ΔP) is the difference between atmospheric pressure and intrapulmonary pressure ($P_{atm} - P_{alv}$). It can be changed by altering the volume of the thoracic cavity. The contraction of both the diaphragm and the external intercostals during quiet breathing causes small volume changes that allow approximately 500 mL of air to enter the lungs. The thoracic cavity volume is further increased if accessory muscles of forced inspiration are stimulated, causing a larger decrease in intrapulmonary pressure. Airflow into the lungs increases because a steeper pressure gradient is established between atmospheric pressure and intrapulmonary pressure.

INTEGRATE

CONCEPT CONNECTION

A similar mathematical relationship expresses blood flow as a function of a pressure gradient and resistance. The blood pressure gradient is established by the heart, and the resistance is experienced by the blood as it is transported through the blood vessels (see section 20.5c).

Airflow is always opposed by resistance. **Resistance** includes all the factors that make it more difficult to move air from the atmosphere through the respiratory passageway into the alveoli. Resistance may be altered in three ways: (1) a decrease in elasticity of the chest wall and lungs, (2) a change in the bronchiole diameter or the size of the passageway through which air moves, and (3) the collapse of alveoli.

A *decrease in elasticity of the chest wall and lungs* results in an increase in resistance. Young, healthy chest walls and lung tissue are naturally elastic. However, as we age, elastic connective tissue decreases in both the chest wall and the lungs. Elasticity of these structures is decreased when (1) an individual has vertebral column malformation, such as scoliosis; (2) arthritis develops within the thoracic cage; or (3) elastic connective tissue in the lungs is replaced with inflexible scar tissue, which occurs with pulmonary fibrosis.

Bronchiole diameter also influences resistance. Resistance increases with bronchoconstriction caused by parasympathetic stimulation, histamine release, or exposure to cold. In addition, decreasing lumen size through accumulation of mucus or inflammation in bronchioles also increases resistance. Resistance decreases with bronchodilation caused by sympathetic stimulation and the subsequent release of epinephrine from the adrenal medulla, or with external administration of epinephrine.

An increase in surface tension occurs if alveolar type II cells are not producing sufficient pulmonary surfactant because this condition increases resistance. This variable is generally important only with premature infants who are unable to produce sufficient pulmonary surfactant (healthy lungs continue to produce pulmonary surfactant beginning approximately 2 months prior to full-term birth). Without pulmonary surfactant, the alveoli in the lungs of these premature infants collapse with each expiration. With each inspiration, the high surface tension caused by the wet inner surface of the alveoli must be overcome for air to enter and reinflate the alveoli for gas exchange. These infants, therefore, experience greater resistance to airflow. This condition is referred to as **respiratory distress syndrome (RDS).**

Both surface tension and elasticity of the chest wall and lung determine compliance. **Compliance** is the ease with which the lungs and chest wall expand. Thus, the easier the lung expansion, the greater the compliance. In contrast, the more difficult it is for the lungs to expand, the lower the compliance.

WHAT DO YOU THINK?

4 Epinephrine is administered in the treatment of asthma. Does epinephrine increase or decrease airway resistance? Does epinephrine increase or decrease airflow?

Respiratory diseases and anatomic abnormalities that increase resistance to airflow are associated with either a decrease in the size of the lumen of bronchioles (e.g., asthma) or a decrease in compliance (e.g., pulmonary fibrosis), or both. These changes produce an increase in resistance. If adequate airflow is to be maintained, the increased resistance must be met with more forceful inspirations to establish a steeper pressure gradient.

The muscles of inspiration must work harder, and a greater amount of the body's metabolic energy must be spent on breathing, for more forceful inspirations to occur. Approximately 5% of the total energy expenditure of the body normally is spent for quiet breathing. This value increases as airway resistance increases, and it can reach

20–30% of energy expenditure. This four-fold to six-fold increase in energy is so demanding that individuals with these disorders can become exhausted simply from breathing.

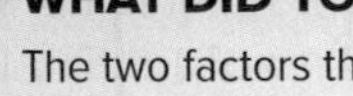

WHAT DID YOU LEARN?

27 The two factors that determine airflow are the pressure gradient and resistance. What are the three major factors that increase resistance to airflow? What changes to breathing must occur to maintain adequate airflow if resistance is increased?

23.5e Pulmonary and Alveolar Ventilation

LEARNING OBJECTIVES

35. Distinguish between pulmonary ventilation and alveolar ventilation, and discuss the significance of each.

36. Explain the relationship between anatomic dead space and physiologic dead space.

In section 3.5, we described the process of moving air into and out of the lungs and referred to it as pulmonary ventilation. The term **pulmonary ventilation** may also refer to the amount of air that is inhaled in 1 minute. The normal adult breathes approximately 500 mL per breath (tidal volume), and this occurs about 12 times per minute. The amount of air taken in during 1 minute (pulmonary ventilation) is calculated using the following formula:

$$\underset{\text{(amount of air per breath)}}{\text{Tidal volume}} \times \underset{\text{(number of breaths per minute)}}{\text{Respiration rate}} = \text{Pulmonary ventilation}$$

$$500\text{ mL} \times 12\text{ breaths/min} = 6000\text{ mL/min} = 6\text{ L/min}$$

WHAT DO YOU THINK?

5 Is *all* of the air taken in during pulmonary ventilation available for gas exchange? Why or why not?

Note that only the air reaching the alveoli is available for gas exchange with the blood. When air is moved from the atmosphere into the respiratory tract, a portion of it remains in the conducting zone. This collective space, where there is no exchange of respiratory gases, is referred to as the **anatomic dead space,** and it has an average volume of approximately 150 mL. The amount of air that reaches the alveoli and is available for gas exchange per minute is termed **alveolar ventilation.** This volume is less than pulmonary ventilation because the volume of air in the anatomic dead space must be subtracted from the volume of air inhaled with each breath. Thus, alveolar ventilation is calculated using the following mathematical formula:

$$(\text{Tidal volume} - \text{anatomic dead space}) \times \text{Respiration rate} = \text{Alveolar ventilation}$$

$$(500\text{ mL} - 150\text{ mL}) \times 12 = 350\text{ mL} \times 12 = 4200\text{ mL/min} = 4.2\text{ L/min}$$

Deeper breathing is more effective for maximizing alveolar ventilation than faster, shallower breathing. Assuming you take one deep breath, you have to overcome the dead air space only one time. All the additional air inhaled in that breath is available for gas exchange. If you take two quick breaths, you have to fill the dead air space twice.

Some respiratory disorders result in a decreased number of alveoli participating in gas exchange. This decrease can be due either to damage to the alveoli or to a change in the respiratory membrane, such as when fluid accumulates in the lungs with pneumonia (see Clinical View 23.7: "Pneumonia"). The difference in volume of air available for gas exchange is accounted for by the more inclusive term **physiologic dead space,** which is the normal anatomic dead space plus any loss of alveoli. The anatomic dead space is equivalent to the physiologic dead space in a healthy individual, because the loss of alveoli should be minimal.

WHAT DID YOU LEARN?

28 A person in yoga class is encouraged to take long, slow, deep breaths. Would this person have greater or lesser alveolar ventilation than an individual with more shallow breathing? Explain.

23.5f Volume and Capacity

LEARNING OBJECTIVES

37. Define the four different respiratory volume measurements.

38. Explain the four respiratory capacities that are calculated from the volume measurements.

39. Give the meaning of forced expiratory volume (FEV) and maximum voluntary ventilation (MVV).

The volume of air that enters and leaves the lungs can be measured with an instrument called a **spirometer.** Respiratory volumes vary throughout a 24-hour period and during different stages of your life. They also vary from individual to individual. The variation is significant enough to be used as a diagnostic tool for determining the health of an individual's respiratory system. Values for an individual are compared to standard values of a reference population. Respiratory measurements are often used to diagnose respiratory disease, monitor changes in respiratory impairment over time, and assess effectiveness of treatment.

Four major respiratory volumes are typically measured (**figure 23.24** and **table 23.4**). **Tidal volume (TV)** is the amount of air inhaled or exhaled per breath during quiet breathing. **Inspiratory reserve volume (IRV)** is the amount of air that can be forcibly inhaled beyond the tidal volume (after a normal inspiration). IRV is a measure of lung compliance. **Expiratory reserve volume (ERV)** is the amount that can be forcibly exhaled beyond the tidal volume (after a normal expiration). ERV is a measure of lung and chest wall elasticity. Finally, **residual volume (RV)** is the amount of air left in the lungs even after the most forceful expiration.

There are four major respiratory capacities that can be calculated from the summation of two or more of these respiratory volumes. **Inspiratory capacity (IC)** is the sum of the tidal volume plus the inspiratory reserve volume. **Functional residual capacity (FRC)** is the sum of the expiratory reserve volume plus the residual volume. It is the volume of air that is normally left in the lungs after a quiet expiration. **Vital capacity (VC)** is the sum of the tidal volume plus both

INTEGRATE

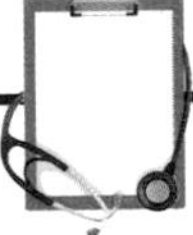

CLINICAL VIEW 23.14

Minimal Volume

Prior to birth, the fetus's respiratory system is nonfunctional because gas exchange occurs between fetal blood and maternal blood at the placenta. The lungs and pulmonary vessels of the fetus are collapsed; most of the blood is shunted to the systemic circulation.

When a newborn takes its first breath, the alveoli inflate. Thereafter, a small amount of air, termed **minimal volume,** remains in the lungs even if the lungs later collapse. In contrast, a stillborn infant never takes a breath. The test at autopsy of whether an infant has been stillborn is to remove the lungs and then to immerse the lungs in water. If the infant was stillborn, the lungs may sink because they contain no minimal volume of air.

6. Which areas of the brain contain the respiratory center?
   a. medulla oblongata and hypothalamus
   b. hypothalamus and pons
   c. medulla oblongata and pons
   d. medulla oblongata and cerebrum

7. The amount of which substance in the blood normally increases the respiratory rate?
   a. oxygen
   b. carbon dioxide
   c. hydrogen gas (H_2)
   d. bicarbonate

8. The movement of oxygen from the blood into the systemic cells is referred to as
   a. pulmonary ventilation.
   b. alveolar gas exchange.
   c. systemic gas exchange.
   d. gas transport.

9. Most carbon dioxide is transported in the blood
   a. dissolved within the plasma as carbon dioxide.
   b. in association with hemoglobin.
   c. as bicarbonate ion.
   d. in combination with oxygen.

10. All of the following are accurate statements about hemoglobin *except*
    a. hemoglobin carries oxygen on the Fe ion.
    b. hemoglobin carries carbon dioxide on the globin.
    c. hemoglobin carries only a small portion of the total carbon dioxide in the blood (less than 25%).
    d. hemoglobin releases oxygen at the level of the cell, making hemoglobin more saturated.

11. Explain how the respiratory tract is organized both structurally and functionally.

12. Describe the relationship of the visceral pleura, parietal pleura, pleural cavity, and serous fluid to keep the lungs inflated.

13. List the four processes of respiration, in order, for moving oxygen from the atmosphere to the body's tissues. List, in order, the processes for the movement of carbon dioxide.

14. Describe the muscles, volume changes, and pressure changes involved with quiet inspiration and expiration.

15. Explain how additional air is moved during a forced inspiration or expiration.

16. Describe how quiet breathing is controlled by the respiratory center.

17. Explain alveolar and systemic gas exchange.

18. List the two means by which oxygen is transported in the blood and the three means by which carbon dioxide is transported.

19. Describe the relationship of Po_2 and hemoglobin percent saturation.

20. List the variables that increase the release of oxygen (by decreasing the affinity of oxygen to hemoglobin) as blood passes through systemic capillaries.

Can You Apply What You've Learned?

1. Paramedics arrived at a car accident to find an elderly gentleman breathing erratically. He was not wearing his seat belt and had hit his head on the windshield. Given his erratic breathing, they were concerned that he had damage to the
   a. medulla oblongata.
   b. cerebral cortex.
   c. pons.
   d. spinal cord.

Use the following to answer questions 2–4.

Michelle, age 45, has been smoking for over 30 years. She now is beginning to have trouble breathing. Her physician sent her to have her respiratory function tested. The results showed that she has emphysema, which is characterized by a decreased surface area for alveolar gas exchange.

2. What component of the respiratory system is most affected in emphysema, causing her to struggle with breathing?
   a. Nasal passageways are inflamed, and less air can move into and out of the respiratory tract.
   b. Bronchi are inflamed.
   c. Bronchioles are dilated.
   d. Damage to the alveoli has occurred.

3. Because physiologic dead space is increased, she would be exhibiting
   a. increased resistance during inhaling.
   b. smaller amounts of air entering her lungs.
   c. harder and deeper breathing that leaves her feeling tired.
   d. decreased blood Pco_2.

4. Blood samples were also taken. Which results might be expected for her blood respiratory gases and blood pH?
   a. decrease in Po_2, increase in Pco_2, decrease in blood pH
   b. decrease in Po_2, decrease in Pco_2, decrease in blood pH
   c. decrease in Po_2, increase in Pco_2, increase in blood pH
   d. increase in Po_2, decrease in Pco_2, increase in blood pH

5. A 10-year-old boy named Michael is playing soccer and begins to have trouble breathing. His mother takes him to the pediatrician, who suspects that he has asthma. The pediatrician recommends that Michael see a respiratory specialist, and results from his pulmonary function tests indicate that Michael does have asthma. Michael's difficulty in breathing, as a result of the asthma, is caused by
   a. inflammation of his nasal passageways.
   b. inflammation and bronchoconstriction of his bronchioles.
   c. spastic closure of his epiglottis, temporarily closing off his trachea.
   d. damage to the alveoli, resulting in decreased surface area of the respiratory membrane.

Can You Synthesize What You've Learned?

1. Tiffany had returned to her college dorm and was having difficulty breathing. She knew she was having an asthma attack. What changes are occurring in her respiratory tract to cause her difficulty in breathing? What potential changes would you predict for her blood P_{O_2} and P_{CO_2}? What substance is generally given as a treatment, which will result in her bronchioles dilating?
2. The nerve to the sternocleidomastoid muscle was damaged during surgery. What type of breathing will be impaired—quiet inspiration, quiet expiration, forced inspiration, or forced expiration?
3. Mark is attempting his first serious mountain climb above 8000 feet. He begins to breathe harder during his ascent, feels light-headed, and has difficulty thinking clearly. What has stimulated him to breathe harder? What effect does this have on P_{CO_2}? What change in blood pH can occur? Why? Does more or less oxygen reach the brain? Explain.

INTEGRATE

ONLINE STUDY TOOLS

connect | SMARTBOOK® | AP|R

The following study aids may be accessed through Connect.

Concept Overview Interactive: Figure 23.31: The Movement of Oxygen and Carbon Dioxide

Clinical Case Study: How the Smallest Things Become Critically Important

Interactive Questions: This chapter's content is served up in a number of multimedia question formats for student study

SmartBook: Topics and terminology include upper and lower respiratory tracts, lungs, pulmonary ventilation, alveolar and systemic gas exchange, gas transport, and breathing rate and homeostasis

Anatomy & Physiology Revealed: Topics include pharynx and larynx, trachea and bronchi, alveolar pressure changes, partial pressure

INTEGRATE

CLINICAL VIEW 24.5

Renal Failure, Dialysis, and Kidney Transplant

Renal Failure

Renal failure refers to greatly diminished or absent renal functions caused by the destruction of about 90% of the kidney. Renal failure often results from a chronic disease that affects the glomerulus or the small blood vessels of the kidney, as a result of autoimmune conditions, high blood pressure, or diabetes. Once the kidney's structures have been destroyed, there is no chance they will regenerate or begin functioning again. Thus, the two main treatments are dialysis and kidney transplant.

Dialysis

The term **dialysis** (dī-al′i-sis) comes from a Greek word meaning "to separate agents or particles on the basis of their size." Two forms of dialysis are commonly used today: peritoneal dialysis and hemodialysis. In **peritoneal dialysis,** a catheter is permanently placed into the peritoneal cavity, to which a bag of dialysis fluid may be attached externally. A volume of a special dialysis fluid is introduced, and the harmful waste products in the blood are transferred, or dialyzed, into the fluid across the peritoneal membrane. After several hours, the fluid is drained from the peritoneal cavity and replaced with fresh fluid.

In **hemodialysis,** the patient's blood is cycled through a machine that filters the waste products across a specially designed membrane. The patient must remain stationary for the time it takes to cycle the blood through the dialysis unit while metabolic waste products are removed. Hemodialysis must be performed three or four times a week, and each treatment takes about 4 hours.

Kidney Transplant

A **kidney transplant** from a genetically similar person ("matched" for major histocompatability complex; see section 22.4c) may successfully restore renal function. The kidney is generally removed from the donor by a laparoscopic procedure. A recent development for obtaining the kidney is to remove the kidney through the umbilicus (navel, or belly button) after making a single small incision. This decreases the donor's recovery time from approximately 3 months to just under a month.

The replacement kidney is attached to an artery and a vein in the inferior abdominopelvic region, where it is relatively easy to establish a vascular connection. The new kidney rests either on the superior surface of or immediately lateral to the urinary bladder. Because a pelvic artery and vein connect the donor kidney to the patient's blood supply, having the kidney near the bladder means only a short segment of ureter is needed for the bladder connection. The diseased kidney is not removed. Because the transplanted kidney is a foreign tissue, immunosuppressant drugs are regularly administered to suppress the immune system's activity (see Clinical View 22.5: "Organ Transplants and MHC Molecules").

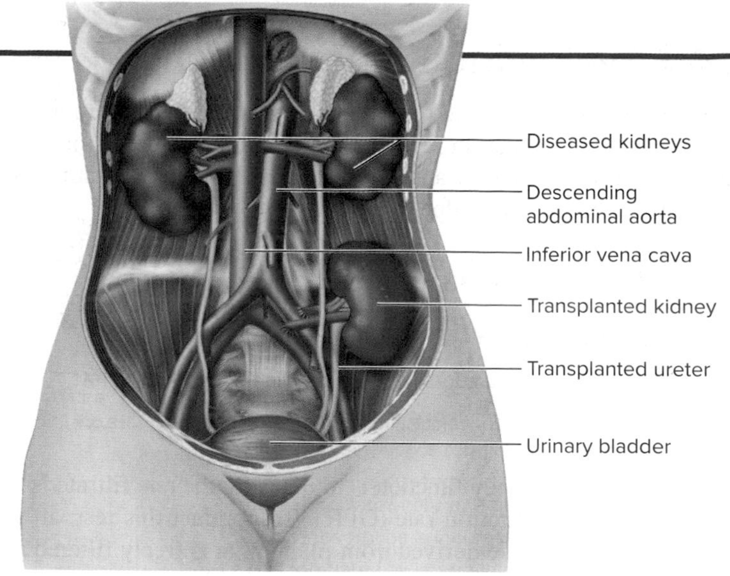

Location of a transplanted kidney in the abdominopelvic cavity.

tubular fluid and excreted in the urine. For example, creatinine has a renal plasma clearance of 140 mL/min, indicating that the substance is both filtered and secreted.

Many medications also have renal plasma clearances higher than the GFR because they are secreted. For this reason, it is important to determine renal plasma clearance to know the amount and timing of drug dosage. The higher the renal plasma clearance, the more often the medications must be given to maintain therapeutic levels.

Note that in clinical practice, renal plasma clearance of creatinine can be used to approximate glomerular filtration rate because its clearance is only slightly higher than GFR. Measuring creatinine clearance avoids the need to inject inulin into the patient's blood.

WHAT DID YOU LEARN?

35 What information is gained by measuring the renal plasma clearance for a specific substance (e.g., medication)?

24.8 Urine Characteristics, Transport, Storage, and Elimination

Urine is the fluid that exits the kidneys and passes into the ureters. Thereafter, it is transported through the ureters into the urinary bladder, where it is stored until it can be eliminated from the body through the urethra. The process of eliminating urine is referred to as micturition or urination. Here we describe the characteristics of urine, the components of the urinary tract, and the process of micturition.

24.8a Characteristics of Urine

LEARNING OBJECTIVES

51. Describe the composition of urine and its characteristics.
52. Explain what is meant by specific gravity.

Table 24.3 Abnormal Urinalysis

Abnormal Urine Constituent	Clinical Term	Possible Cause(s)
Glucose	Glucosuria or glycosuria	Diabetes mellitus
Ketones (acetone, acetoacetic acid, beta-hydroxybutyric acid)	Ketonuria	Diabetes mellitus, starvation (including anorexia), low-carbohydrate diet (e.g., Atkins, South Beach); ketones produced as by-products of fatty acid metabolism
Protein (Note that trace amounts of protein [5–10 mg/dL] are not clinically significant.)	Proteinuria	Increased glomerular permeability caused by kidney trauma, heavy metals, bacterial toxins, exertion (e.g., running a marathon), hypertension, cold exposure, glomerular nephritis
Bile pigment	Bilirubinuria	Liver pathology (e.g., hepatitis, cirrhosis), obstructed bile duct due to gallstones
Erythrocytes	Hematuria	Glomerular damage, kidney trauma, pathology along the urinary tract (as occurs with urinary tract stones), contamination from menstrual flow
Hemoglobin (Hb)	Hemoglobinuria	Accelerated rate of erythrocyte destruction (e.g., hemolytic anemia, a transfusion reaction), burns, renal damage, contamination from menstrual flow
Leukocytes	Pyuria	Urinary tract infection, acute glomerulonephritis, contamination from female reproductive tract
Nitrites	Nitrituria	Urinary tract infection because bacteria convert nitrates (NO_3^-) to nitrites (NO_2^-)
Myoglobin	Myoglobinuria	Rhabdomyolysis (a condition where skeletal muscle breaks down; caused by excessive exercise and is an uncommon side effect of statins)

Urine is the product of filtered and processed blood plasma. It typically is a sterile excretion unless contaminated with microbes in the kidney or urinary tract. Urine characteristics include its composition, volume, pH, specific gravity, color and turbidity, and smell.

Typical urine composition is approximately 95% water and 5% solutes. These solutes include salts (e.g., Na^+, Cl^-, K^+, Mg^{2+}, Ca^{2+}, SO_4^{2-}, $H_2PO_4^-$, NH_4^+), nitrogenous wastes (e.g., urea, uric acid, creatinine), some hormones, drugs, and small amounts of ketone bodies (a waste product of digesting fatty acids). **Table 24.3** lists some substances that should not normally be found within urine (i.e., they are abnormal constituents) and possible causes for their presence.

The average daily urine volume is normally 1–2 L. Variations occur due to many variables, including fluid intake, blood pressure, diet, body temperature, use of diuretics, diabetes, and the amount of fluid excreted by other means (e.g., heavy sweating, vomiting). A minimum of about 0.5 L of urine per day is required to eliminate the wastes from the body. If urine production drops below 0.40 L, wastes will accumulate in the blood. **Table 24.4** list factors that influence urine volume output.

The normal pH for urine ranges between 4.5 and 8.0; the average value is 6.0, which is slightly acidic. The pH of urine may be affected by diet. It generally decreases to become more acidic when we consume proteins and wheat, and it tends to increase and become more basic with a diet high in fruits and vegetables. Urine pH can also be influenced by other factors, such as metabolism and bacterial infections.

Specific gravity is the density (g/mL) of a substance compared to the density of water (1 g/mL). For example, if your urine were composed only of pure water, it would have a specific gravity of 1.000. The average specific gravity of urine is slightly higher, with levels ranging from 1.003 to 1.035 because solutes *are* normal components of urine. Levels vary due to time of day, amount of food and liquids consumed, and amount of exercise. Generally, if you are well hydrated, urine output increases and specific gravity values decrease. A specific gravity value below 1.010 indicates relative hydration, whereas a specific gravity value above 1.020 indicates relative dehydration.

The color of urine ranges from almost clear to dark yellow, depending upon the concentration of pigment from urobilin (a bilirubin breakdown product; see section 18.3b). An increased volume of urine generally is lighter in color, and a decreased volume of urine is darker in color. However, intake of some substances (e.g., beets, certain vitamins) can change the color of urine. Normal vaginal secretions, excessive substances in urine (cellular material, protein), crystallization or precipitation of salts if collected and left standing, and bacteria will increase the turbidity (cloudiness) of the urine.

Urinoid is the term used for the normal smell of fresh urine. Urine may develop an ammonia smell if allowed to stand because bacteria convert the nitrogen in urea into ammonia (NH_3). Asparagus and certain other foods can alter the smell of urine. One of the indications that a person has diabetes mellitus is that the urine smells fruity

Table 24.4 Factors That Influence Urine Volume

Decrease Urine Volume	Increase Urine Volume
Increase in ADH	Decrease in ADH
Increase in aldosterone	Decrease in aldosterone
Decrease in ANP	Increase in ANP
Decrease in fluid intake	Increase in fluid intake
Decrease in blood pressure	Increase in blood pressure
Increase in other fluid output (e.g., sweating, vomiting, diarrhea, hemorrhage)	Diabetes mellitus Diuretics (e.g., medications, alcohol)

(like acetone, which is found in fingernail polish) from the accelerated production of ketoacids from fatty acid metabolism (see section 3.4h).

WHAT DID YOU LEARN?

What characteristics are used to describe urine? What conditions cause variation in pH of urine?

INTEGRATE

CONCEPT CONNECTION

The inner epithelial lining of the urinary tract (except for distal portions of the male urethra) is composed of transitional epithelium (see section 5.1c). When stretched, the cells forming the transitional epithelium change from cuboidal or polyhedral to almost squamous in shape. In addition, the transitional epithelium contains tight junctions between the cells (see section 4.6d). This combination of the "stretchable" transitional epithelium and tight junctions function to prevent leakage of urine through the bladder wall.

24.8b Urinary Tract (Ureters, Urinary Bladder, Urethra)

LEARNING OBJECTIVES

53. Describe the structure and function of the ureters.

54. Explain the structure of the urinary bladder.

55. List distinguishing characteristics of the female urethra and male urethra.

The **urinary tract** consists of the ureters, urinary bladder, and urethra. The urinary tract is responsible for transporting the urine produced in the kidneys, storing the urine until the time and place are right for its elimination from the body, and transporting the urine from storage to the exterior of the body.

Ureters

The **ureters** (ū-rē′ter) are long, epithelial-lined, fibromuscular tubes that transport urine from the kidneys to the urinary bladder (**figure 24.25**). Each tube in an adult averages about 25 centimeters (about 10 inches) in length and is retroperitoneal. The ureters originate from the renal pelvis as it exits the hilum of the kidney and then extend inferiorly to enter the posterolateral wall of the base of the urinary bladder. The wall of the ureter is composed of three concentric tunics. From innermost to outermost, these tunics are the mucosa, muscularis, and adventitia. (Note that the ureter does not have a submucosa layer.)

The **mucosa** is formed by a transitional epithelium (see section 5.1c) that is both distensible and impermeable to the passage of urine. External to the transitional epithelium of the mucosa is the lamina propria, composed of a fairly thick layer of dense irregular connective tissue.

The middle **muscularis** consists of an inner longitudinal and an outer circular layer of smooth muscle cells. The presence of urine within the renal pelvis causes these muscle layers to contract rhythmically to propel the urine through the ureters into the urinary bladder. In areas where the ureter is not filled with urine, the mucosa folds to fill the lumen.

WHAT DO YOU THINK?

Why do the ureters use smooth muscle contraction to actively move urine to the urinary bladder? Why don't they rely only on gravity to move the urine to the inferiorly located bladder?

The external layer of the ureter wall is the **adventitia.** It is composed of a dispersed array of collagen and elastic fibers within areolar connective tissue. Some extensions of this areolar connective tissue layer also anchor the ureter to the posterior abdominal wall.

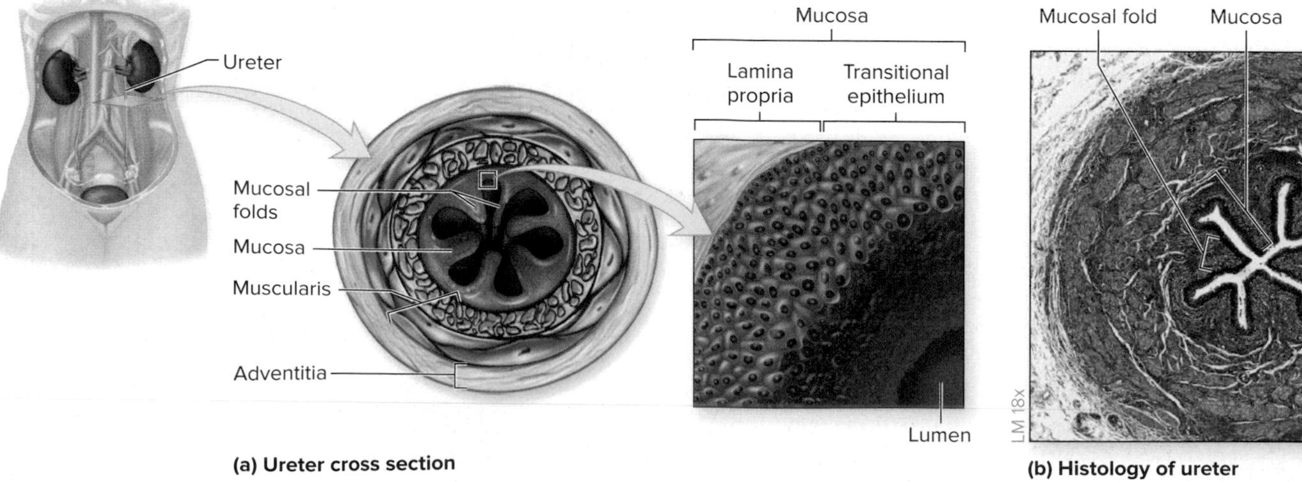

Figure 24.25 Ureters. The ureters transport urine from the kidneys to the urinary bladder for storage prior to elimination from the body. (*a*) Features of a ureter in cross-sectional view, including its transitional epithelium. (*b*) A photomicrograph of a ureter in cross section shows its mucosal folds and relatively thick muscularis. AP|R

(*b*) ©McGraw-Hill Education/Al Telser

INTEGRATE

CLINICAL VIEW 24.6

Intravenous Pyelogram

Sometimes a physician needs to visualize the kidneys, ureters, and urinary bladder, especially when the flow of urine from one or both of the kidneys into the bladder becomes blocked. A physician may ask for an x-ray study known as an **intravenous pyelogram** (pī′el-ō-gram; *pyelos* = tub, *gram* = recording), which is produced by injecting a small amount of radiopaque dye into a vein. As the dye passes through the kidneys and is cleared into the urine, a series of sequential abdominal x-rays provide a "time lapse" view of urinary system flow. If urine flow is normal, the entire pathway of the urinary tract should appear dark on the x-ray. However, if there is a blockage (as is the case with a renal calculus), there is a lack of dark coloration beginning at the blockage (see Clinical View 24.7: "Renal Calculi").

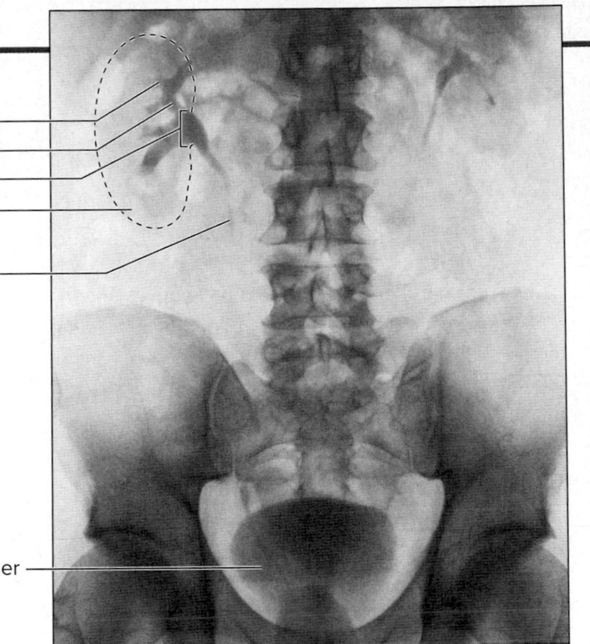

A pyelogram enables physicians to visualize the urinary system organs and identify the location of any blockages within the urinary system.
©SPL/Science Source

The ureters project through the bladder wall obliquely. Because of the oblique course of the ureters through the bladder wall, the ureter walls are compressed as the bladder distends, decreasing the likelihood of urine refluxing (backflowing) into the ureters from the bladder when it is emptying.

The ureters are innervated by the autonomic nervous system (see sections 15.3b and 15.4b). Sympathetic neurons extend from the T11–L2 segments of the spinal cord. Pain from the ureter (e.g., caused by a kidney stone lodged in the ureter) is referred to the T11–L2 dermatomes (see figure 16.4). These dermatomes are distributed along a "loin-to-groin" region, so "loin-to-groin" pain typically means ureter or kidney discomfort. The vagus nerve (CN X) innervates the superior region of the ureter, and the pelvic splanchnic nerves innervate the inferior region of the ureter. The effects of this parasympathetic stimulation are unknown.

Urinary Bladder

The **urinary bladder** is an expandable, muscular organ that serves as a reservoir for urine. The bladder is positioned immediately posterior to the pubic symphysis (see section 8.10). In females, the urinary bladder is anteroinferior to the uterus and directly anterior to the vagina (see figure 28.3); in males, the bladder is anterior to the rectum and superior to the prostate gland (see figure 28.14). The urinary bladder is a retroperitoneal organ (see section 24.2a), and only its superior surface is covered with the parietal peritoneum. When it is empty, the urinary bladder exhibits an inverted pyramidal shape (**figure 24.26**). As it fills with urine, the bladder distends superiorly until it assumes an oval shape.

A posteroinferior triangular area of the urinary bladder wall, called the **trigone** (trī′gōn; *trigonum* = triangle), is formed by imaginary lines connecting the two ureter openings and the urethral opening. The trigone remains immobile as the urinary bladder fills and evacuates. It functions as a funnel to direct urine into the urethra as the bladder wall contracts to eliminate the stored urine. Because the ureters and urethra form the three points of this triangular region, infections are more common within this area.

The four tunics that form the wall of the bladder are the mucosa, submucosa, muscularis, and adventitia. The innermost **mucosa** is adjacent to the bladder lumen; it is formed by a transitional epithelium, which accommodates the shape changes occurring with distension, and by a highly vascularized lamina propria that supports the mucosa. Additionally, **mucosal folds,** or *rugae* (rū′jē), allow for even greater distension. Within the trigone region, the mucosa is smooth, thick, and lacking mucosal folds. The **submucosa** lies immediately external to the mucosa and is formed by dense irregular connective tissue that supports the urinary bladder wall.

The **muscularis** is formed by three layers of smooth muscle, collectively called the **detrusor** (dē-trū′ser, -sōr; *detrudo* = to drive away) **muscle.** These smooth muscle bundles

INTEGRATE

CLINICAL VIEW 24.7

Renal Calculi

A **renal calculus** (pl., *calculi*), or *kidney stone,* is formed from crystalline minerals that build up in the kidney. Causes and risk factors for kidney stone formation include inadequate fluid intake and dehydration, reduced urinary flow and volume, frequent urinary tract infections, and certain abnormal chemical or mineral levels in the urine. In general, stones tend to develop more often in males than in females.

Very small renal calculi may be asymptomatic, and the person may excrete them without their being aware of the stones being eliminated in the urine. However, a larger stone can become obstructed in the kidney, renal pelvis, or ureter. Symptoms include severe pain along the "loin-to-groin" region and possibly nausea and vomiting. Following the intake of plenty of water (2 to 3 quarts per day) to assist movement, most stones smaller than about 4 millimeters in diameter eventually pass through the urinary tract on their own. Administration of intravenous fluids may also assist this process.

If the stone is too large to pass on its own, one common treatment is **lithotripsy** (lith'ō-trip-sē; *lithos* = stone, *tresis* = boring), in which sonic shock waves are directed toward the stones to pulverize them into smaller particles that can be expelled in the urine. Alternatively, using **ureteroscopy** (yu'ter-os'ko-pe), a scope is inserted from the urethra into the urinary bladder and ureter to break up and remove the stone. If these treatments aren't viable, traditional surgery may be required.

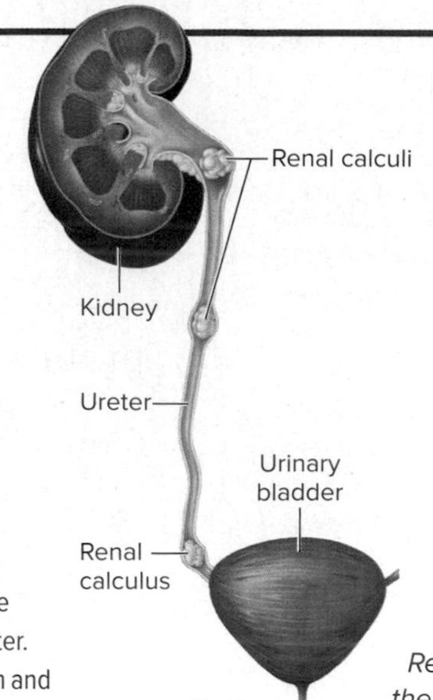

Renal calculi may become lodged at various sites along the urinary tract.

exhibit such complex orientations that it is difficult to delineate individual layers in histology sections. At the neck of the urinary bladder, an involuntary internal urethral sphincter is formed by the smooth muscle that encircles the urethral opening.

The **adventitia** is the outer layer of areolar connective tissue covering the urinary bladder. A peritoneal membrane covers only the superior surface of the urinary bladder.

Urethra

The **urethra** (ū-rē'thră) is an epithelial-lined, fibromuscular tube that extends from the anteroinferior surface of the urinary bladder to the urethral opening. The urethra transports urine to the exterior of the body **(figure 24.27)**.

Two urethral sphincters restrict the release of urine until the pressure within the urinary bladder is high enough and both involuntary and voluntary activities are activated to release the urine. The

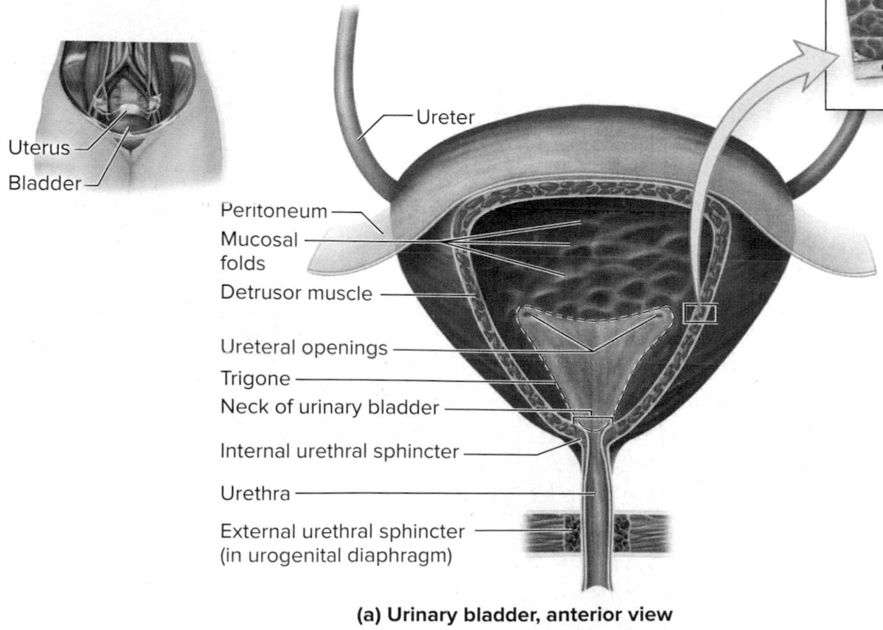

Uterus
Bladder
Ureter
Peritoneum
Mucosal folds
Detrusor muscle
Ureteral openings
Trigone
Neck of urinary bladder
Internal urethral sphincter
Urethra
External urethral sphincter (in urogenital diaphragm)

(a) Urinary bladder, anterior view

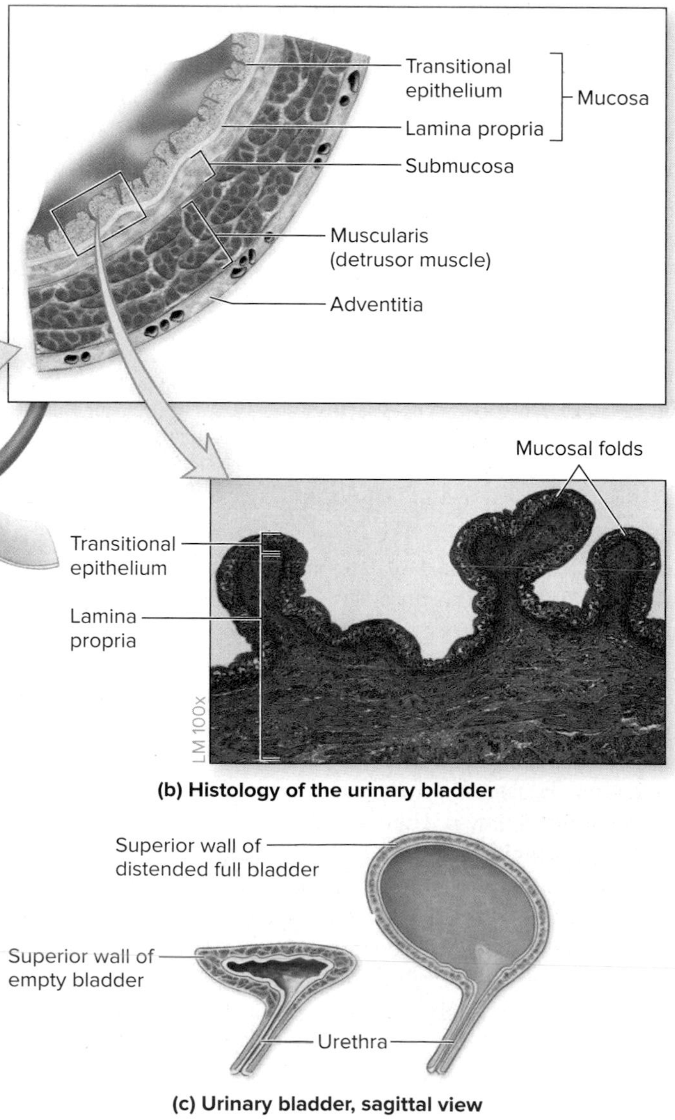

Transitional epithelium
Lamina propria
Mucosa
Submucosa
Muscularis (detrusor muscle)
Adventitia
Mucosal folds
Transitional epithelium
Lamina propria
LM 100x

(b) Histology of the urinary bladder

Superior wall of distended full bladder
Superior wall of empty bladder
Urethra

(c) Urinary bladder, sagittal view

Figure 24.26 Urinary Bladder. (*a*) The urinary bladder is an expandable, muscular sac. This view depicts a female bladder. (*b*) Illustration and photomicrograph of the urinary bladder wall show its tunics. (*c*) Sagittal views show that the urinary bladder expands superiorly and becomes more oval in shape as it fills with urine. AP|R

(*b*) ©Garry DeLong/Getty Images

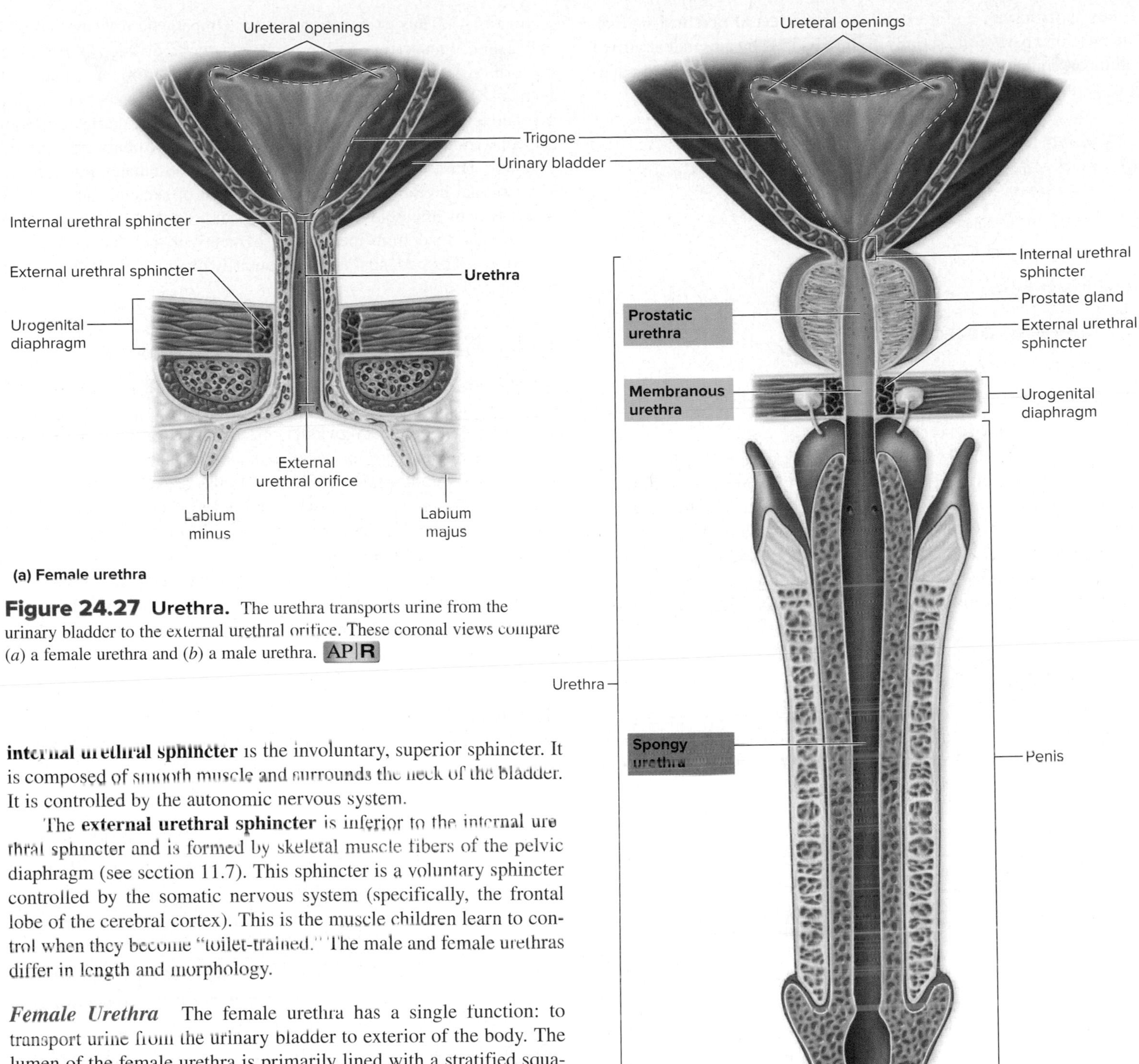

Figure 24.27 Urethra. The urethra transports urine from the urinary bladder to the external urethral orifice. These coronal views compare (*a*) a female urethra and (*b*) a male urethra. **AP|R**

internal urethral sphincter is the involuntary, superior sphincter. It is composed of smooth muscle and surrounds the neck of the bladder. It is controlled by the autonomic nervous system.

The **external urethral sphincter** is inferior to the internal urethral sphincter and is formed by skeletal muscle fibers of the pelvic diaphragm (see section 11.7). This sphincter is a voluntary sphincter controlled by the somatic nervous system (specifically, the frontal lobe of the cerebral cortex). This is the muscle children learn to control when they become "toilet-trained." The male and female urethras differ in length and morphology.

Female Urethra The female urethra has a single function: to transport urine from the urinary bladder to exterior of the body. The lumen of the female urethra is primarily lined with a stratified squamous epithelium. The urethra is approximately 4 centimeters (about 1.6 inches) long, and it opens to the outside of the body at the external urethral orifice located in the female perineum (see section 28.1c).

Male Urethra The male urethra has both urinary and reproductive functions because it serves as a passageway for both urine and semen (but not at the same time). It is approximately 19 centimeters long (7.5 inches) and is partitioned into three segments: the prostatic urethra, the membranous urethra, and the spongy urethra.

The **prostatic** (pros-tat′ik) **urethra** is approximately 3.5 centimeters (about 1.5 inches) in length and is the most dilatable portion of the urethra. It extends through the prostate gland immediately inferior to the male bladder, where multiple small prostatic ducts enter it. The urethra in this region is lined by a transitional epithelium. Two smooth muscle bundles surround the mucosa: an internal, longitudinal bundle and an external circular [illegible] thickened, circular muscular bundles are [illegible] cter at the bladder outlet. lar region of [illegible]

The **membranous** (mem′bră-nus) **urethra** is the shortest and least dilatable portion of the male urethra. It extends from the inferior surface of the prostate gland through the urogenital diaphragm. As a result, it is surrounded by skeletal muscle fibers that form the external urethral sphincter [illegible] urinary bladder. The epithelium in this region [illegible] either stratified columnar or pseudostratified columnar.

The **spongy urethra** (or *penile urethra*) is the longest part of the male urethra at approximately 15 centimeters long (about 6 inches). It is encased within a cylinder of erectile tissue in the penis called the corpus spongiosum, and it extends to the **external urethral orifice.** The proximal part of the spongy urethra is lined by a pseudostratified columnar epithelium, whereas the distal part has a stratified squamous epithelial lining.

WHAT DID YOU LEARN?

37 What are the major components of the urinary tract? What tunics are found in each of these?

38 How do the urethras of a male and female differ?

24.8c Micturition

LEARNING OBJECTIVES

56. Define micturition.

57. Compare and contrast the storage reflex and the micturition reflex.

58. Explain conscious control over micturition.

The expulsion of urine from the bladder is called **micturition** (mik-chū-rish′ŭn; *micturio* = to desire to make water), *urination,* or *voiding.* Two reflexes are associated with the process of micturition: the storage reflex and the micturition reflex. These reflexes are regulated by the sympathetic and parasympathetic divisions of the autonomic nervous system, respectively (see section 15.8).

Innervation of the Urinary Bladder and Urethral Sphincters

The urinary bladder and urethral sphincters are innervated by the nervous system. Sympathetic neurons extend from T11 to L2 segments of the spinal cord. Sympathetic stimulation causes relaxation of the detrusor muscle and contraction of the internal urethral sphincter (see section 15.4). Thus, stimulation by the sympathetic division inhibits micturition. Innervation by the parasympathetic division is from the micturition center located in the pons and extends from the spinal cord segments S2–S4 in the pelvic splanchnic nerves (see section 15.3b). The pelvic splanchnic nerves cause contraction of the detrusor muscle and relaxation of the internal urethral sphincter, so that urine can be expelled. Thus, the parasympathetic division stimulates micturition. The external urethral sphincter is composed of skeletal muscle. It is innervated by the somatic nervous system via the pudendal nerve (which is formed from the S2–S4 segments of the spinal cord; see table 14.6). The pudendal nerve voluntarily contracts this sphincter to prevent urination.

INTEGRATE

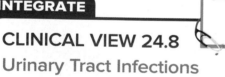

CLINICAL VIEW 24.8

Urinary Tract Infections

A **urinary tract infection (UTI)** occurs when either bacteria (most commonly, *E. coli*) or fungi (e.g., yeast) enter and multiply within the urinary tract. Women are more prone to UTIs because they have a short urethra that is close to the anus, allowing bacteria from the gastrointestinal (GI) tract to more readily enter the urethra. Sexual intercourse and medical use of a urinary catheter also increase the risk of UTIs.

A UTI often develops first in the urethra, an inflammation called **urethritis** (ū-rē-thrī′tis). If the infection spreads to the urinary bladder, **cystitis** (sis-tī′tis) results. Occasionally, bacteria from an untreated UTI can spread up the ureters to the kidneys, a condition termed **pyelonephritis** (pī′ĕ-lō-ne-frī′tis), which is inflammation of the kidney that is potentially fatal. Symptoms of a UTI include difficult and painful urination, called **dysuria** (dis-ū′rē-ă; *dys* = bad, difficult), frequent urination, and feeling of uncomfortable pressure in the pubic region. If the infection spreads to the kidneys, sharp back and flank pain, fever, and occasionally nausea and vomiting [illegible] occur. A UTI can be diagnosed through **urinalysis** (ū-ri-nal′i-sis), a test of the urine [illegible] reveal the presence of leukocytes, blood, and bacteria or fungi, as well as nitriturla [illegible] section 24.8a). Appropriate antibiotic therapy cures most UTIs caused by bac[illegible]

Storage Reflex

The storage of urine in the urinary bladder is controlled by both the autonomic and somatic nervous systems (**figure 24.28*a***). During the filling of the urinary bladder, urine moves through the ureters from the kidneys. Varying nerve signals conducted by sympathetic axons cause smooth muscle cells of (1) the detrusor muscle of the urinary bladder wall to relax, which allows the bladder to accommodate the urine, and (2) internal urethral sphincter to contract, so that urine is retained within the bladder.

This process is referred to as the **storage reflex.** The skeletal muscle of the external urethral sphincter is also continuously stimulated by nerve signals along the pudendal nerve so it remains contracted.

Micturition Reflex

The process of micturition is also controlled by both the autonomic and somatic nervous systems in a toilet-trained individual, and it usually proceeds as follows (figure 24.28*b*):

1. When the volume of urine retained within the bladder reaches approximately 200 to 300 mL, the bladder becomes distended, and baroceptors in the bladder wall are activated.
2. These baroceptors send nerve signals through visceral sensory neurons to stimulate the micturition center within the pons.
3. The micturition center alters nerve signals propagated down the spinal cord and through the pelvic splanchnic nerves (which are parasympathetic nerves).
4. Parasympathetic stimulation causes the smooth muscle cells composing the detrusor muscle to contract and the internal urethral sphincter to relax. In infants, urination occurs at this point because they lack voluntary control of the external urethral sphincter.

The sensation of having to urinate is relayed along sensory axons to the cerebral cortex.

Conscious Control of Urination

An individual's conscious decision to urinate is due to altering nerve signals relayed from the cerebral cortex through the spinal cord and along the pudendal nerve to cause relaxation of the external urethral sphincter. Expulsion of urine is facilitated by the voluntary contraction of muscles in the abdominal wall and expiratory muscles as part [illegible]Valsalva maneuver (see section 23.3a). Upon emptying of the [illegible] but about 10 mL), the detrusor muscle relaxes,

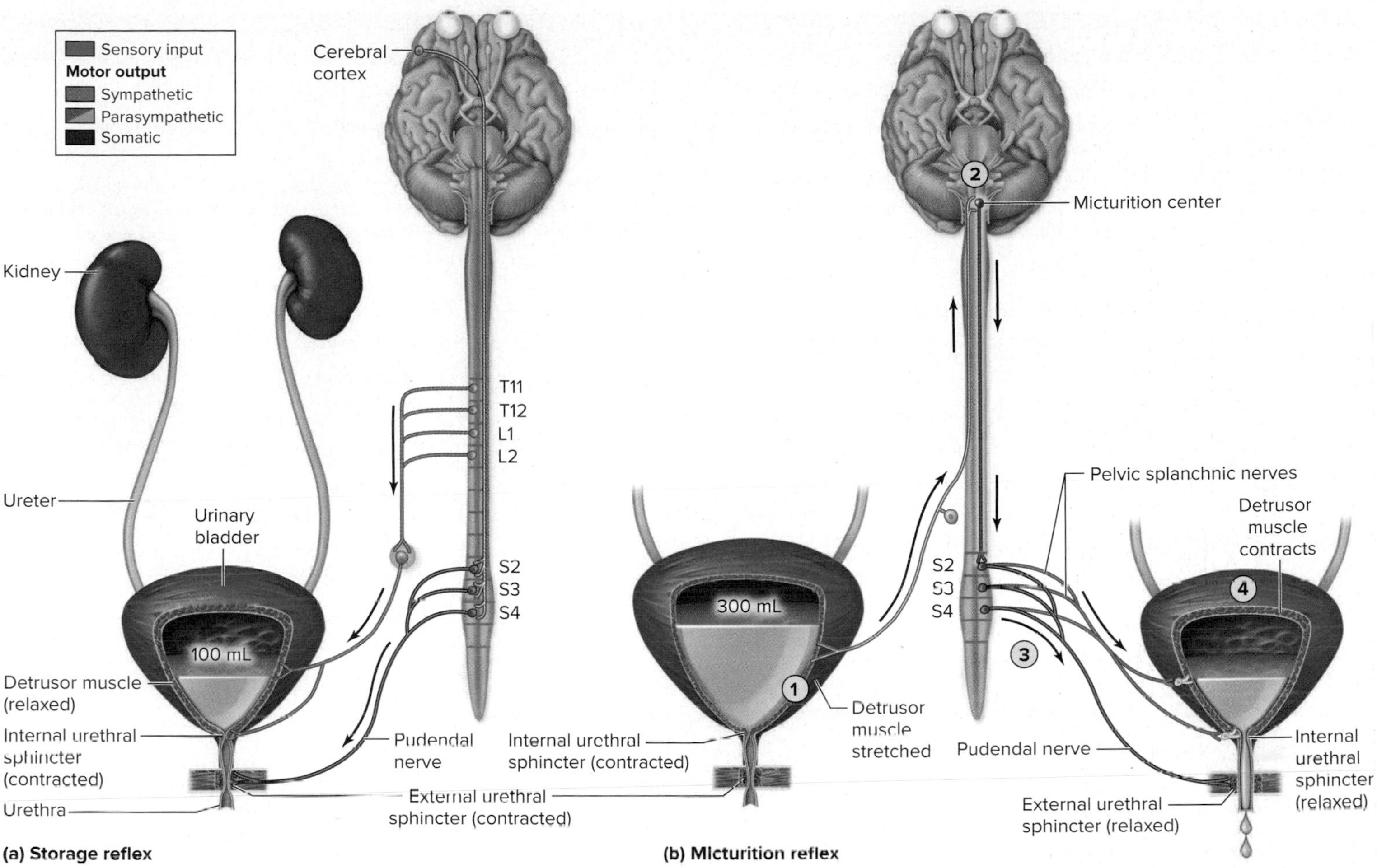

Figure 24.28 Micturition. (*a*) The storage reflex allows urine to be held in the urinary bladder until it can be eliminated from the body. (*b*) The micturition reflex is the involuntary portion of eliminating urine. The relaxation of the external urethral sphincter permits voluntary control of micturition. AP|R

and the neurons of the micturition reflex center are inactivated, while those of the storage reflex are activated.

If an individual does not urinate at the time of the first micturition reflex, the detrusor muscle relaxes as a consequence of the "stress-relaxation response" of smooth muscle (see section 10.10d). The bladder continues to fill with urine, and the micturition reflex is initiated again after another 200–300 mL of urine is added. This cycle continues until there is between 500 mL and 600 mL. At this point, urination occurs without conscious control.

Additionally, an individual may choose to empty the bladder prior to the initiation of the micturition reflex by contracting the abdominal muscles as part of the Valsalva maneuver (see section 23.3a) and compressing the bladder. This compression of the bladder stimulates baroreceptors within the bladder wall and the micturition reflex is initiated.

The urinary system, which is composed of the kidneys, ureters, urinary bladder, and urethra is responsible for filtering our blood and forming, storing, and eliminating urine. By adjusting the amount of urine that is formed, the kidneys play a critical role in regulating blood volume and blood pressure. These processes of the urinary system also function to maintain blood ion levels within normal homeostatic levels and eliminate both metabolic wastes and biologically active molecules—all of which function to keep our blood healthy.

INTEGRATE

CLINICAL VIEW 24.9

Impaired Urination

Two common terms are associated with impaired urination: incontinence and retention. **Incontinence** (in-kon'ti-nens; *in* = not, *contineo* = to hold together) is the inability to voluntarily control urination. Incontinence may occur as a result of childbirth (stress incontinence), stronger than normal detrusor muscle contractions (urge incontinence), or the secondary result of medications. Uncontrolled urination may also occur when an individual is frightened, as nerve signals initiated in the limbic system are sent to the micturition center, causing urination.

Retention (rē-ten'shŭn) involves failure to eliminate urine normally. Causes include side effects following general anesthesia and an enlarged prostate in a male that impedes urine flow through the prostatic urethra. If necessary, a small tube called a *catheter* may be placed into the urethra to allow urine to flow out of the bladder.

WHAT DID YOU LEARN?

39 What steps lead to micturition? At what point does the reflex overcome conscious control?

CHAPTER SUMMARY

	• The urinary system is composed of the kidneys and the urinary tract (ureters, urinary bladder, and the urethra).
24.1 Introduction to the Urinary System	• The kidneys filter blood to produce urine, which is subsequently eliminated from the body through the ureters, urinary bladder, and urethra. • The kidneys perform numerous functions, including eliminating metabolic wastes from the blood, regulating ion and acid-base balance, regulating blood pressure, eliminating biologically active molecules, producing the enzyme that forms calcitriol, producing and releasing erythropoietin, and having the potential to engage in gluconeogenesis to regulate blood glucose.
24.2 Gross Anatomy of the Kidney	• The kidneys are two bean-shaped, reddish organs.
	24.2a Location and Support • The kidneys are located along the posterior abdominal wall within the retroperitoneal space and are partially protected by the ribs. • The kidneys are surrounded by a fibrous capsule, perinephric fat, renal fascia, and paranephric fat.
	24.2b Sectional Anatomy of the Kidney • Each kidney is composed of an outer cortex and an inner medulla, and the medulla is subdivided into 8 to 15 renal pyramids (or 8 to 15 renal lobes). • Renal sinuses are spaces within the kidney and include the minor calyces, the major calyces, and a single renal pelvis.
	24.2c Innervation of the Kidney • Sympathetic axons innervate the afferent and efferent arterioles and the juxtaglomerular apparatus.
24.3 Functional Anatomy of the Kidney	• Nephrons, collecting tubules, collecting ducts, and associated structures form the functional anatomy of the kidney.
	24.3a Nephron • A nephron consists of a renal corpuscle, which is composed of a glomerulus plus glomerular capsule, and a renal tubule (proximal convoluted tubule, nephron loop, and distal convoluted tubule).
	24.3b Collecting Tubules and Collecting Ducts • Numerous nephrons drain into a collecting tubule, and numerous collecting tubules drain into a collecting duct.
	24.3c Juxtaglomerular Apparatus • The juxtaglomerular (JG) apparatus is composed of granular cells of the afferent arteriole and a macula densa of the distal convoluted tubule. It functions in the regulation of filtrate formation and systemic blood pressure.
24.4 Blood Flow and Filtered Fluid Flow	• Filtrate is formed from blood as it flows through the glomerulus.
	24.4a Blood Flow Through the Kidney • Blood is transported to the kidneys in renal arteries, which divide into segmental arteries and then interlobar arteries. Branching off the interlobar arteries are arcuate arteries (at the corticomedullary junction) and then interlobular arteries, which branch into afferent arterioles that supply a glomerulus. • Blood leaving the glomerulus follows this path: efferent arteriole, peritubular capillaries or vasa recta, interlobular veins, arcuate veins, interlobar veins, and a renal vein.
	24.4b Filtrate, Tubular Fluid, and Urine Flow • Filtrate is collected in the capsular space of the glomerular capsule. • Tubular fluid flows through the proximal convoluted tubule, nephron loop, distal convoluted tubule, and collecting tubules and ducts. • Urine flows from numerous collecting ducts into a papillary duct and then into the minor calyx, major calyx, renal pelvis, ureter, and urinary bladder (where it is stored), and it is excreted via the urethra.
24.5 Production of Filtrate Within the Renal Corpuscle	• The renal corpuscle is the site of filtrate production.
	24.5a Overview of Urine Formation • Filtration, reabsorption, and secretion are the three processes involved in urine formation.
	24.5b Filtration Membrane • The filtration membrane is composed of the fenestrated endothelium and basement membrane of the glomerulus and the visceral layer of the glomerular capsule.
	24.5c Formation of Filtrate and Its Composition • Filtrate is (essentially) protein-free, filtered plasma.
	24.5d Pressures Associated with Glomerular Filtration • The three pressures that determine net filtration pressure are glomerular (blood) hydrostatic pressure (HP_g), which promotes filtration, and both blood colloid osmotic pressure (OP_g) and capsular hydrostatic pressure (HP_c), which oppose filtration. • Net filtration pressure is determined by subtracting OP_g and HP_c from HP_g.
	24.5e Regulation of Glomerular Filtration Rate • Glomerular filtration rate (GFR) is maintained through renal autoregulation (myogenic and tubuloglomerular mechanisms), decreased by sympathetic stimulation, and increased by atrial natriuretic peptide hormone.

24.6 Reabsorption and Secretion in Tubules and Collecting Ducts	• The renal tubules and collecting ducts are responsible for modifying the filtrate into urine. • Processing of filtrate into urine may be categorized into three major categories: substances that are completely reabsorbed, substances that are regulated, and substances that are eliminated.
	24.6a Overview of Transport Processes • Reabsorption and secretion occur along the entire length of the nephron tubule, collecting tubule, and collecting ducts by paracellular and transcellular transport.
	24.6b Transport Maximum and Renal Threshold • Transport maximum is the maximum amount of a substance that can be reabsorbed (or secreted) in a given period of time (i.e., it is a rate of movement). • Renal threshold is the maximum plasma concentration of a substance that can be transported in the blood without exceeding the transport maximum of that substance and spilling over into the urine.
	24.6c Substances Reabsorbed Completely • Substances that typically are completely reabsorbed include nutrients (e.g., glucose) and proteins.
	24.6d Substances with Regulated Reabsorption • Substances that are regulated (not 100% reabsorbed) include sodium ions (Na^+), water (H_2O), potassium ions (K^+), calcium ions (Ca^{2+}), phosphate (PO_4^{3-}), hydrogen ions (H^+), and bicarbonate ions (HCO_3^-).
	24.6e Substances Eliminated as Waste Products • Substances eliminated as waste include nitrogenous wastes (urea, uric acid, and creatinine), certain drugs, other metabolic wastes, and some hormones.
	24.6f Establishing the Concentration Gradient • Establishing a concentration gradient in the interstitial fluid surrounding nephrons is dependent upon activity of the nephron loop, vasa recta, and urea recycling. • The concentration gradient establishes the osmotic "pull" to move water from the tubular fluid into the interstitial fluid (and then into capillaries) when antidiuretic hormone (ADH) is present.
24.7 Evaluating Kidney Function	• The effectiveness of kidney function can be determined by measuring glomerular filtration rate and measuring renal plasma clearance.
	24.7a Measuring Glomerular Filtration Rate • The GFR can be measured using the rate at which injected inulin, a freely filtered substance, appears in the urine.
	24.7b Measuring Renal Plasma Clearance • Renal plasma clearance is important in determining dosage of drugs. For some substances, such as creatinine, renal plasma clearance is close to being the same as GFR.
24.8 Urine Characteristics, Transport, Storage, and Elimination	• Urine is transported from the kidney to the outside of the body by the urinary tract.
	24.8a Characteristics of Urine • Urine characteristics include its chemical composition, volume, pH, specific gravity, color and turbidity, and smell.
	24.8b Urinary Tract (Ureters, Urinary Bladder, Urethra) • The ureters transport urine from the kidneys to the urinary bladder. • The urinary bladder stores urine and changes shape as the bladder fills. • The urethra transports urine from the bladder to the exterior of the body.
	24.8c Micturition • Micturition is the expulsion of urine. • The storage reflex is facilitated by the sympathetic division, and the micturition reflex is facilitated by the parasympathetic division of the autonomic nervous system. Voluntary control is regulated by the somatic nervous system.

CHALLENGE YOURSELF

Create and Evaluate

Analyze and Apply

Understand and Remember

Do You Know the Basics?

1. All of following are functions of the kidney *except*
   a. gluconeogenesis, the formation of glucose from noncarbohydrate sources.
   b. release of erythropoietin to control erythrocyte production.
   c. control of blood pressure through the release of renin.
   d. production of plasma proteins to control blood volume.
2. When the kidneys are described as being retroperitoneal, this refers to the fact that the kidneys
   a. are within the parietal peritoneal lining of the abdominal cavity.
   b. are posterior to the parietal peritoneal lining of the abdominal cavity.
   c. are superior to the peritoneal lining of the abdominal cavity.
   d. have no protective covering.
3. Which of the following is located within the renal medulla?
   a. collecting duct
   b. glomerulus
   c. renal corpuscle
   d. distal convoluted tubule
4. All of the following are capillaries *except*
   a. the glomerulus.
   b. the minor calyx.
   c. the vasa recta.
   d. peritubular capillaries.
5. Which of the following is a component of filtrate but not normally a component of urine?
   a. water
   b. erythrocytes
   c. nitrogenous waste
   d. glucose
6. If blood pressure in the glomerulus increases, then
   a. net filtration pressure decreases.
   b. the percentage of substances reabsorbed increases.
   c. urine production increases.
   d. renin is released.
7. Which hormone increases Na^+ and water reabsorption and K^+ secretion?
   a. antidiuretic hormone
   b. angiotensin II
   c. atrial natriuretic peptide
   d. aldosterone
8. If the tubular maximum is exceeded, then
   a. the carrier mechanism will work harder.
   b. the excess will be reabsorbed in the bladder.
   c. the excess will appear in the urine.
   d. its blood concentration will increase.
9. The function unique to the nephron loop is to
   a. regulate pH.
   b. excrete water.
   c. establish a concentration gradient in the medulla interstitial fluid.
   d. regulate the concentration of blood Ca^{2+}.
10. If antidiuretic hormone (ADH) concentration increases,
    a. urine volume increases.
    b. urine concentration decreases.
    c. urine volume decreases and urine concentration increases.
    d. urine volume increases and urine concentration decreases.
11. Trace blood flow into and out of the kidney. Identify the appropriate vessels.
12. Describe where filtrate, tubular fluid, and urine are found in the functioning kidney.
13. Describe the anatomic components of the juxtaglomerular apparatus.
14. Describe the filtration membrane, and explain the structures that do not normally pass through it.
15. Explain how glomerular filtration rate (GFR) is maintained by renal autoregulation, decreased by sympathetic stimulation, and increased by atrial natriuretic peptide.
16. Discuss the affect of aldosterone and antidiuretic hormone (ADH) on principal cells and their effect on urine production.
17. Explain how antidiuretic hormone (ADH) is dependent upon the concentration gradient in the medulla of the kidney.
18. Describe the significant differences among blood plasma, filtrate, and urine.
19. Identify all of the following that are functions of the kidney: (a) maintain blood pH; (b) regulate blood ion concentrations; (c) regulate blood volume and blood pressure; (d) eliminate wastes, some hormones, and certain drugs from the blood; (e) release renin; (f) release erythropoietin; and (g) stimulate the final step in calcitriol formation.
20. Explain the process of micturition.

Can You Apply What You've Learned?

Use the following paragraph to answer questions 1–3.

Maria is 8 months pregnant. When she goes to her doctor for her checkup, the nurse takes her blood pressure. It is unusually high: 200/100 mm Hg. The nurse also collects a urine sample and detects elevated levels of protein in her urine.

1. What can account for the protein in her urine?
   a. Protein is a normal component of urine and is no cause for concern.
   b. Excessive amounts of plasma proteins are filtered at the renal corpuscle.

c. Tubular cells produce additional protein in response to the elevated blood pressure.
d. The tubular cells have exocytosed protein into the tubular fluid.

2. What would you predict has happened to her plasma protein concentration?
a. An elevated level of protein in the urine has no effect.
b. Plasma protein levels have increased.
c. Plasma protein levels have decreased.
d. Plasma proteins levels have remained the same or increased.

3. What do you think has happened to her urine production? (Consider the pressures involved in determining net filtration pressure and osmotic force generated by solutes in the filtrate.)
a. Urine production has stayed the same.
b. Urine production has increased.
c. Urine production has decreased.
d. It is not possible to predict the effect on urine production.

4. Martin, a young man of 20, was in a car accident and is hemorrhaging. When he arrives by ambulance at the hospital, he is stabilized. However, they notice that over the next day he does not urinate. Which of the three variables for determining net filtration pressure has changed and can account for the lack of urine formation?
a. Glomerular hydrostatic (blood) pressure has decreased.
b. Capsular hydrostatic pressure has decreased.
c. Blood colloidal osmotic pressure has decreased.
d. Glomerular hydrostatic (blood) pressure has increased.

5. A 19-year-old male named Paul was in a diving accident and severed his spinal cord at the T1 segment. Which of the following describes one of the physiologic changes that he will experience?
a. His kidneys will no longer be able to filter his blood and produce urine.
b. His urinary bladder will no longer be able to contract to expel urine.
c. His filtration membrane will contract, and filtration will no longer occur.
d. He will no longer be able to consciously control urination.

Can You Synthesize What You've Learned?

1. A patient with cancer is treated with chemotherapy. This specific medication is eliminated by the kidney. Results from a renal plasma clearance test are higher than average. Will the medication need to be given more or less often? Explain.
2. Theoretically, which one of the following hormones could be administered to decrease blood volume: antidiuretic hormone, aldosterone, parathyroid hormone, or atrial natriuretic peptide? Explain.
3. Males who suffer from either benign prostatic hyperplasia (noncancerous prostate gland enlargement) or prostate cancer often have problems with urination. Based on your knowledge of the male urethra, hypothesize why these urination problems occur.

INTEGRATE

ONLINE STUDY TOOLS

connect | SMARTBOOK® | AP|R

The following study aids may be accessed through Connect.

Concept Overview Interactive: Figure 24.15: Glomerular Filtration

Clinical Case Study: A Young Child with a Fever and Foamy Urine

Interactive Questions: This chapter's content is served up in a number of multimedia question formats for student study

SmartBook: Topics and terminology include introduction to the urinary system; gross anatomy of the kidney; functional anatomy of the kidney; blood flow and filtered fluid flow; production of filtrate within the renal corpuscle; reabsorption and secretion in tubules and collecting ducts; evaluating kidney function; urine characteristics, transport, storage, and elimination

Anatomy & Physiology Revealed: Topics include urinary system overview; kidney gross anatomy; kidney microscopic anatomy; renal cortex; podocyte; urine formation; upper and lower urinary system; micturition reflex

Animations: Topics include urinary system overview; kidney gross anatomy; kidney microscopic anatomy; basic renal processes; urine formation; renal clearance; diffusion, facilitated diffusion; cotransport; sodium-potassium exchange pump

chapter

25 Fluid and Electrolytes

Our body's ability to maintain fluid balance is something most of us take for granted despite our sometimes erratic fluid intake. For example, we may drink several cups of coffee first thing in the morning, go the rest of the day without drinking anything, perhaps run several miles in the late afternoon, and then drink large quantities of fluid at the end of the day before retiring for the night. Fortunately, our body compensates for our inconsistent fluid intake by making physiologic adjustments throughout the day so that fluid levels are maintained within homeostatic limits. In a similar way, electrolyte levels, including those that determine pH, are being regulated to maintain their balance.

Fluid and electrolyte imbalances, however, can occasionally occur. Insufficient fluid intake, excessive sweating, vomiting, diarrhea, and abnormal changes in the function of an organ (e.g., kidney disease, congestive heart failure) all may stress the body to such an extent that fluid and electrolyte levels are driven out of balance. Here we describe the major concepts and mechanisms involved in maintaining fluid and electrolyte balance. These include the percentage, distribution, chemical composition, and movement of body fluid; maintenance of electrolyte levels; hormonal regulation of fluid balance and its influence on blood pressure; acid-base control and balance; and acid-base disturbances.

INTEGRATE

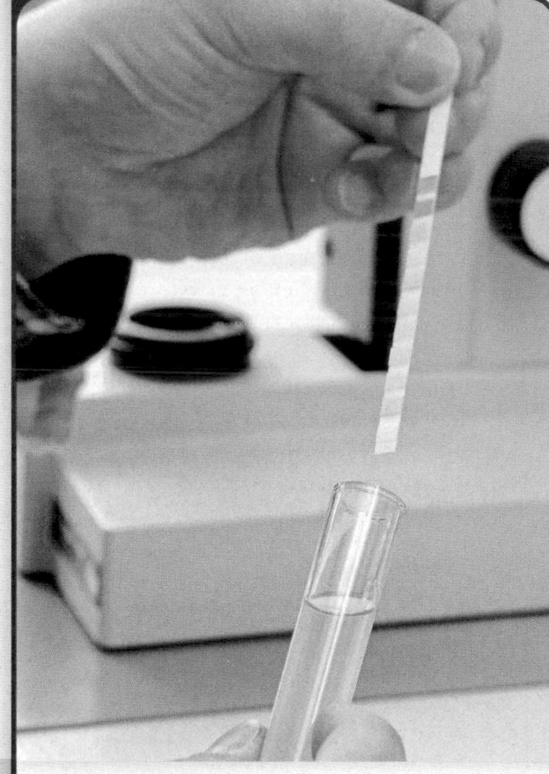

©CC Studio/Science Source

CAREER PATH
Medical Laboratory Technologist

A medical laboratory technologist is a health-care professional who performs biological, chemical, and hematological tests on body fluid samples or specimens, including urine, blood, stool, and pericardial fluid. Results are analyzed and reported to physicians. These medical care professionals are employed at hospitals, doctor's offices, and specialized medical laboratories.

Anatomy & Physiology REVEALED® apreveaIed.com

Module 13: Urinary System

25.1 Body Fluids

Optimal health requires both **fluid and electrolyte balance**—which is the state of maintaining homeostatic levels of body fluids and normal blood levels of ions (e.g., Na^+, K^+, H^+; see section 2.2a). Individuals who experience a fluid or electrolyte *imbalance* often seek medical assistance. Knowing the causes, effects, and symptoms associated with these imbalances, and understanding how to both prevent and treat them, is a critical part of patient care. Our discussion in this chapter is tailored to provide you with a general overview of the three central concepts related to this topic: fluid balance, electrolyte balance, and acid-base balance. We begin with fluid balance, which is examined in the first two sections. Body fluids and their distribution within fluid compartments are described in this section, and the specific variables that influence how fluid balance is maintained are discussed in section 25.2.

25.1a Percentage of Body Fluid

LEARNING OBJECTIVES

1. State the percentage of body fluid, and explain the significance of an individual's percentage relative to fluid balance.
2. List the factors that influence the percentage of body fluid.

The human body contains between 45% and 75% fluid by weight, with an average of about 65%. Consider, for example, an individual who weighs 120 pounds and has an average of 65% body fluid—78 of the 120 pounds are fluid! Understanding the percentage of the body's water content (and the variables that influence it) is important because the specific percentage of body fluid influences how susceptible a person is to a fluid imbalance. Individuals who have a lower percentage of body fluid are more likely to experience a fluid imbalance.

The specific percentage of body fluid depends upon two variables: the age of an individual and the relative amounts of adipose connective tissue and skeletal muscle tissue:

- **Age.** Infants have the highest percentage of fluid, at approximately 75% fluid by weight. In contrast, elderly individuals have the lowest percentage of fluid at 45%. Children, and both young and middle-aged adults, are usually somewhere in between these two extremes, with a general trend of decreasing percentage of body fluid seen with increasing age.
- **Relative amounts of adipose connective tissue to skeletal muscle tissue.** The percentage of fluid in the body at each age depends upon the ratio of adipose connective tissue to skeletal muscle tissue because of the difference in water content of these tissues. Adipose connective tissue is approximately 20% water, whereas skeletal muscle tissue is approximately 75% water. This accounts for the general differences in the percentage of body fluid that are observed between females and males of the same age after puberty.

  Lean adult females are, on average, typically composed of 55% body fluid, whereas lean adult males are, on average, typically composed of 60% body fluid. This difference reflects the relatively lower amounts of skeletal muscle and relatively higher amounts of adipose connective tissue in a lean adult female compared to a lean adult male.

WHAT DID YOU LEARN?

1. If you have an increase in muscle mass as a result of weight training, will your percentage of body fluid increase, decrease, or stay the same? Explain.

25.1b Fluid Compartments

LEARNING OBJECTIVES

3. Describe the two major body fluid compartments, and compare their chemical compositions.
4. Explain how fluid moves between the major body fluid compartments.

The fluid in our body is partitioned into two fluid compartments: intracellular fluid and extracellular fluid. **Intracellular fluid (ICF)** is the fluid within our cells. A majority, or approximately two-thirds, of the total fluid is within our cells (**figure 25.1**). The barrier

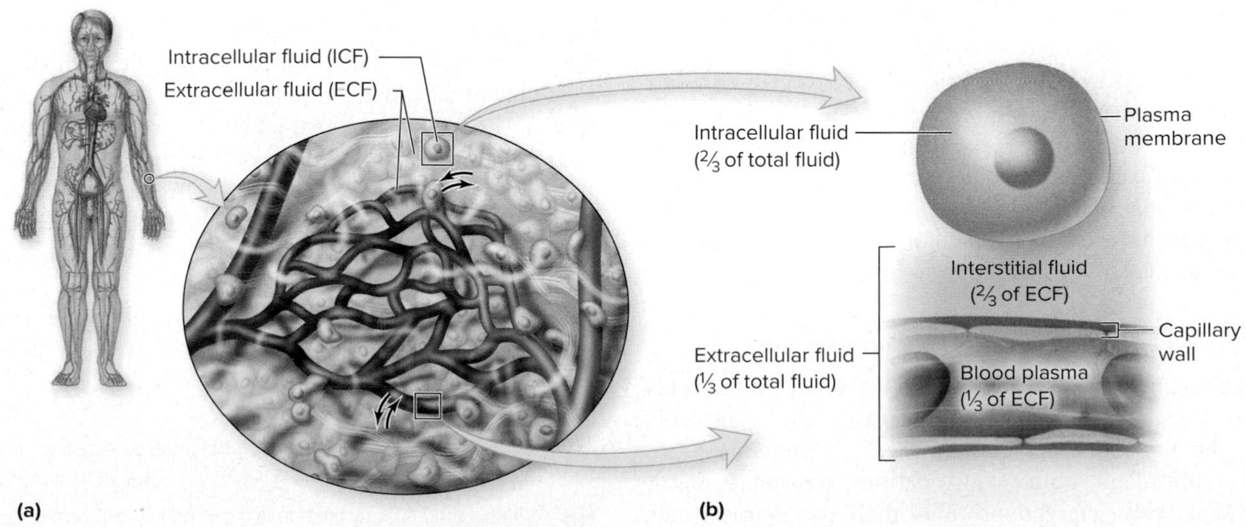

Figure 25.1 Fluid Compartments. (*a*) Fluid is contained within two major compartments: intracellular and extracellular fluid compartments. (*b*) Approximately two-thirds of the total body fluid is intracellular fluid (fluid within cells), and the other one-third is extracellular fluid (fluid outside of cells). Approximately two-thirds of the extracellular fluid is interstitial fluid (fluid around the cells), and one-third is plasma within blood vessels.

INTEGRATE

LEARNING STRATEGY

The "**rule of thirds**" can help you to remember the distribution of fluid: intracellular fluid (ICF) = two-thirds, and extracellular fluid (ECF) = one-third. When only the extracellular fluid is considered, the distribution within the ECF is two-thirds interstitial fluid and one-third blood plasma.

⅔ ICF

⅓ ECF

⅔ Interstitial fluid

⅓ Blood plasma

enclosing this fluid is the plasma membrane (see section 4.2). We have seen that this selectively permeable plasma membrane allows some, but not all, substances through it.

The fluid outside of cells is referred to collectively as **extracellular fluid (ECF).** Extracellular fluid includes both **interstitial fluid (IF),** the fluid that surrounds and "bathes" the cells, and **blood plasma,** which is the fluid within the blood vessels (see section 18.2).

Interstitial fluid composes approximately two-thirds of the extracellular fluid, and blood plasma about one-third. The barrier separating blood plasma from the interstitial fluid is the capillary vessel wall that consists of a single layer of simple squamous epithelium resting on a basement membrane (see section 20.1c). A capillary wall typically is more permeable than a plasma membrane. Consequently, the interstitial fluid and the blood plasma are more chemically similar to each other than they are to the intracellular fluid (as described shortly).

Examples of specific extracellular fluids include cerebrospinal fluid of the brain and spinal cord, synovial fluid of joints, aqueous and vitreous humor of the eye, fluids of the inner ear (perilymph and endolymph), and serous fluid within body cavities (pleural, pericardial, and peritoneal). These specific extracellular fluids are not typically subject to significant daily fluid gains and losses, and for this reason they are essentially ignored in this chapter. Only fluid in the two major compartments (i.e., the ICF and the ECF)—and the factors that influence them, such as lymph return to blood—is addressed.

Composition of Body Fluids in the Two Compartments

Each fluid compartment is chemically distinct based on the relative amounts of substances dissolved in it (**figure 25.2**). Intracellular fluid is the most distinct compartment; it contains more potassium (K^+) and magnesium (Mg^{2+}) cations, phosphate anion (PO_4^{3-}), and negatively charged proteins than the extracellular fluid (figure 25.2*a*).

These differences in chemical composition reflect both the processes within the cell and the regulatory activity of transport proteins within the plasma membrane that move substances into and out of the cell. For example, the relatively large amount of protein within intracellular fluid is a result of protein synthesis (see section 4.8), whereas the relatively high concentration of intracellular K^+ occurs as a result of sodium-potassium (Na^+/K^+) pumps embedded within the plasma membrane that transport K^+ into a cell as Na^+ is transported out (see section 4.3c).

The two fluids composing the extracellular fluid—interstitial fluid and blood plasma—are, in comparison, both distinct chemically from intracellular fluid and similar in chemical composition to one another (figure 25.2*b*). Both interstitial fluid and blood plasma have a high concentration of these ions: sodium (Na^+) cation, and chloride (Cl^-) and bicarbonate (HCO_3^-) anions.

Interstitial fluid and blood plasma exhibit one significant difference: protein is present in blood plasma, but there is very little protein within the interstitial fluid. The similarity in ionic composition and the difference in protein composition reflect the relative permeability of the capillary wall: Proteins are generally too large to move out of the blood through the openings in the capillary wall to enter the interstitial fluid, whereas fluids and ions move freely. Therefore, during capillary exchange, blood plasma and all of its dissolved substances—except for most proteins—are filtered to become part of the interstitial fluid (see section 20.3). Specific *percentages* of the different substances composing the intracellular fluid and extracellular fluid (both IF and blood plasma) are given in figure 25.2.

Fluid Movement Between Compartments

Fluid movement between compartments occurs continuously in response to changes in relative osmolarity (concentration). This happens when the fluid concentration in one fluid compartment becomes either hypotonic or hypertonic (see section 4.3b), with respect to another compartment; water immediately moves by osmosis between the two compartments until the water concentration is once again equal (**figure 25.3**). You may recall from section 4.3b that water always moves by osmosis from the hypotonic solution to the hypertonic solution. This movement of water between the compartments is possible because the plasma membranes and the capillary wall are both permeable to water.

When you drink water, it enters your blood from the gastrointestinal (GI) tract and becomes part of the blood plasma (see section 26.4e). Consequently, the plasma osmolarity decreases and blood plasma becomes hypotonic to both the interstitial fluid and the ICF. Thereafter, as blood moves through the capillaries, water first moves out of the blood plasma to become part of the interstitial fluid, then moves from the interstitial fluid into cells (figure 25.3*a*). This means there is a net movement of water from the blood plasma into the cells. In contrast, if water is lost from the body without being replaced in a timely manner, then dehydration occurs. Blood plasma osmolarity increases, and blood plasma becomes hypertonic to both the interstitial fluid and the cells. Consequently, a net movement of water occurs from the cells into the blood plasma. It moves first into the interstitial fluid and then into the blood plasma (figure 25.3*b*).

WHAT DID YOU LEARN?

2. Which ions are more prevalent in the intracellular fluid? Which are more prevalent in the extracellular fluid?
3. What is the major distinction in the chemical composition of blood plasma and interstitial fluid?
4. When you are dehydrated, is the net movement of fluid from the blood plasma into the cells or from the cells into the blood plasma?

Intracellular fluid

Cations: K^+ 75, Mg^{2+} 17, Other 8

Anions/Proteins: PO_4^{3-} 20, Proteins 27, Other 53

Percentage (%)

K^+, Mg^{2+}, PO_4^{3-}
Negatively charged protein

(a) Intracellular fluid

Interstitial fluid
Na^+, Cl^-, HCO_3^-
Little or no protein

Na^+, Cl^-, HCO_3^-
Protein

Blood plasma

(b) Extracellular fluid

Interstitial fluid

Cations: Na^+ 94, Other 6

Anions: Cl^- 77, HCO_3^- 18, Protein, Other 5

Blood plasma

Cations: Na^+ 94, Other 6

Anions/Proteins: Cl^- 69, HCO_3^- 16, Protein 10, Other 5

Percentage (%)

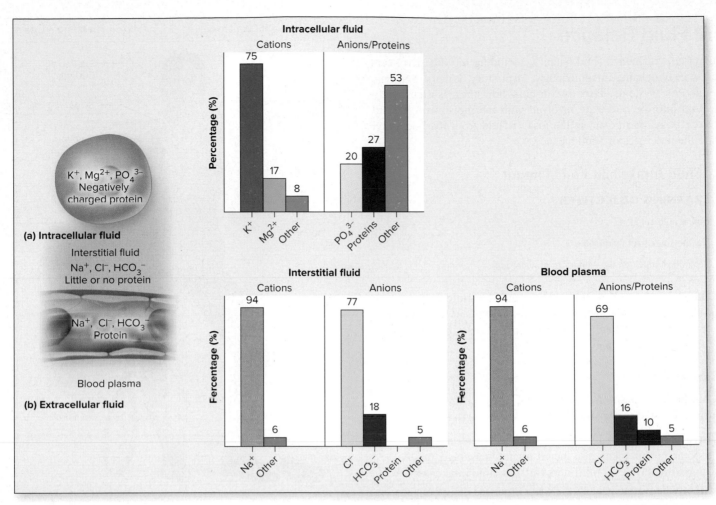

Figure 25.2 Percentages of Solutes in Body Fluids. (*a*) K^+ and Mg^{2+} cations, PO_4^{3-} anion, and proteins are more common in the intracellular fluid than in the extracellular fluid. (*b*) Na^+ cation and Cl^- and HCO_3^- anions are more prevalent in the extracellular fluid than in the intracellular fluid. The only significant difference between the interstitial fluid and blood plasma is the presence of protein within the blood plasma, with little or no protein in the interstitial fluid.

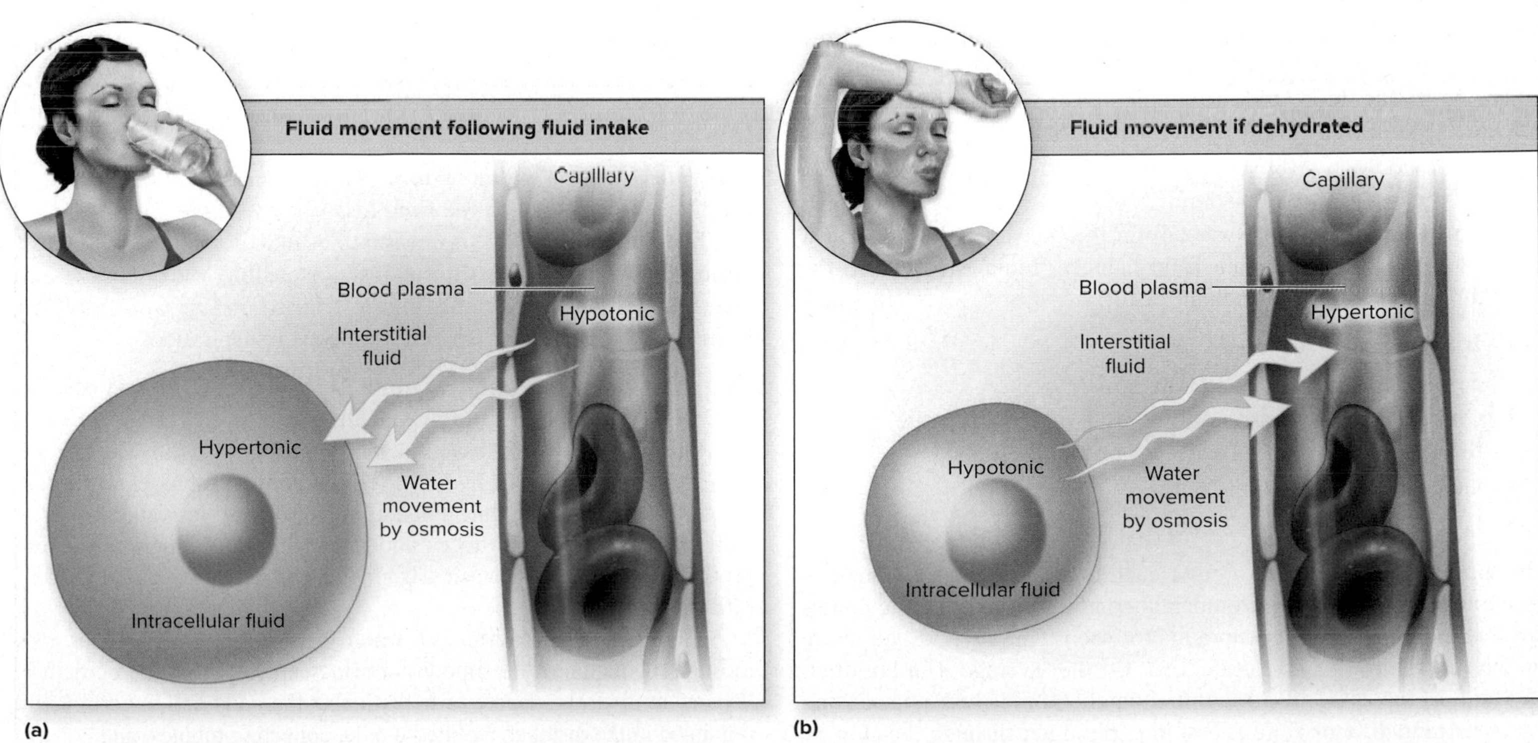

Figure 25.3 Fluid Movement Between Compartments. (*a*) A net movement of water from the blood plasma into the intracellular compartment occurs following fluid intake. (*b*) A net movement of water from the intracellular compartment into the blood plasma occurs when an individual is dehydrated.

25.2 Fluid Balance

Many systems influence fluid balance, including the digestive, cardiovascular, lymphatic, integumentary, respiratory, urinary, nervous, and endocrine systems. Here we discuss how these systems help bring fluid into the body, are involved with the movement of fluid within the body, participate in the loss of fluid from the body, and regulate the processes of fluid balance.

25.2a Fluid Intake and Fluid Output

LEARNING OBJECTIVES

5. Define fluid balance.
6. List the sources of fluid intake.
7. Distinguish between the categories of water loss.

Fluid balance exists when fluid intake is equal to fluid output, and a normal distribution of water and solutes is present in the two major fluid compartments. Our discussion begins by describing and comparing how fluid intake and fluid output occur.

Fluid Intake

Fluid intake is the addition of water to the body. It is divided into two categories, ingested water and metabolic water **(figure 25.4)**:

- **Ingested** (preformed) **water** includes the water absorbed from food and drink taken into the GI tract. On average, this is approximately 2300 milliliters (mL) of fluid intake per day.
- **Metabolic water** includes the water produced daily from aerobic cellular respiration (see section 3.4e) and dehydration synthesis (see section 2.7a). It is approximately 200 mL of fluid per day.

The sum of the average ingested water intake and metabolic water produced is 2500 mL. Given that fluid intake from both food and drink is approximately 2300 mL of the total 2500 mL, the fluid absorbed from the GI tract accounts for about 92% of daily fluid intake. A significant and perhaps obvious point is that normally the only way to significantly increase fluid in the body is through fluid intake from the food and drink that we consume (see Clinical View 25.1: "Intravenous [IV] Solution").

Fluid Output

Fluid output is the loss of water from the body. Fluid output must equal fluid intake to maintain fluid balance. Fluid is lost from the body through the normal mechanisms of

- Breathing (see section 23.5)
- Sweating (see sections 1.6b and 6.1d)
- Cutaneous transpiration (evaporation of water directly through the skin) (see sections 1.6b and 6.1d)
- Defecation (see section 26.3d)
- Urination (see section 24.8c)

The amount of water lost through each of these processes depends upon physical activity, environmental conditions, and internal conditions of the body. Average amounts for each type of fluid output are shown in figure 25.4. Notice that of the average fluid output, 1500 mL of the 2500 mL (or approximately 60%) is lost in urine, and the remaining 40% of fluid is lost in expired air, through the skin by sweat and cutaneous transpiration, and in feces.

Water loss can be described in two ways: as either sensible or insensible water loss and as either obligatory or facultative water loss.

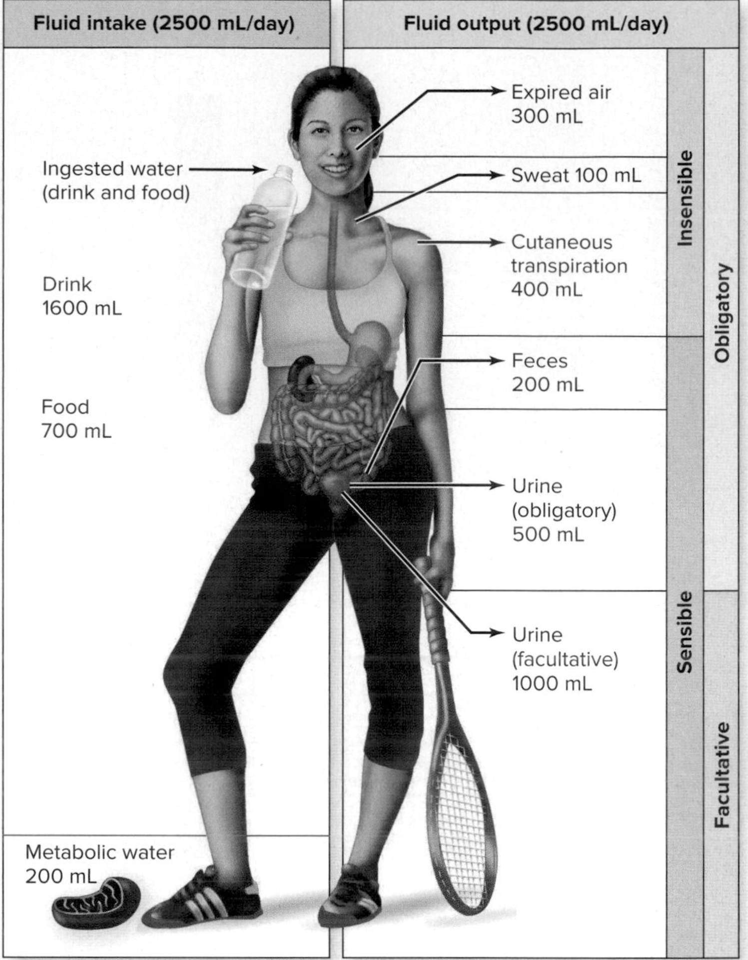

Figure 25.4 Fluid Intake and Fluid Output. Fluid balance is determined by relative amounts of fluid intake and fluid output. Fluid intake results from GI tract absorption of ingested food and drink, as well as water produced by metabolic processes. Fluid output includes fluid lost from the body as a component of expired air, through the skin by both sweat and cutaneous transpiration, and in the feces and urine. Fluid output is called sensible if it is measurable, insensible if it is not. It may also be categorized as obligatory if it is always lost, or facultative if the loss in the urine is regulated. (Values given are approximate averages.)

Sensible or Insensible Water Loss Sensible and insensible fluid loss reflect whether the fluid loss is *measurable.*

Sensible water loss is measurable, and it includes fluid lost through feces and urine. In contrast, **insensible water loss** is not measurable. It includes both fluid lost in expired air and fluid lost from the skin through sweat and cutaneous transpiration.

Obligatory or Facultative Water Loss Water loss may also be described as either obligatory or facultative.

Obligatory water loss is a loss of water that always occurs, regardless of the state of hydration of the body. It includes water lost through breathing and through the skin (insensible water loss), as well as fluid lost in the feces and in the minimal amount of urine produced to eliminate wastes from the body, approximately 0.5 L (500 mL) per day.

Facultative (fak′ŭl-tā-tiv) **water loss** is controlled water loss through regulation of the amount of urine expelled from the body. It is dependent upon the degree of hydration of the body and is hormonally regulated in the distal convoluted tubule, collecting tubules, and collecting ducts in the nephrons of the kidney (see sections 24.6d and 25.4).

Note that the only physiologic mechanism to control fluid output is the hormonal regulation of urine output. When the body is overhydrated,

INTEGRATE

CLINICAL VIEW 25.1

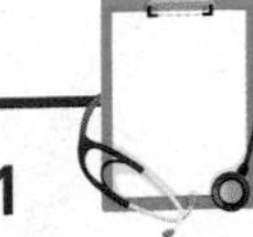

Intravenous (IV) Solution

A patient may be administered an **intravenous** (in'trā-vē'nŭs) **(IV) solution** as a clinical means of increasing fluid input directly into the blood. In this procedure, a needle is inserted into the individual's vein either in the hand or the antecubital region, with a tube extending from the needle to a bag. The two most common IV solutions are a 0.9% saline solution and a 5% dextrose, or D5W, solution.

A **0.9% saline solution** is a solution containing 0.9 gram of sodium chloride to every 100 mL of sterile water. This solution is the standard IV solution for fluid replacement because it has a similar osmolarity (see section 2.6b) to blood plasma. Thus, it will not cause a net shift of fluid between compartments following its addition.

A **5% dextrose,** or **D5W,** solution is composed of 5 grams of dextrose (glucose) sugar to every 100 mL of sterile water. However, this solution is a hypertonic solution (to blood plasma) when in the holding bag—but when infused into the body, the dextrose is metabolized, and the solution becomes hypotonic. This solution is used to supply the body with both water and an energy source (glucose).

hormonally controlled facultative water loss normally plays a significant role in eliminating excess fluid. In comparison, when the body is dehydrated, obligatory fluid output always occurs, regardless of the hydrated state of the body. Therefore, hormonal regulation of urine output can only decrease fluid loss when the body is dehydrated, but not inhibit it completely. Over time, the body continues to become more and more dehydrated unless fluids are replaced. Death may result in severe cases of dehydration.

WHAT DID YOU LEARN?

5 What are the two major sources of fluid intake? What are two ways that fluid output is categorized, and which one is based on the hydrated state of the body?

25.2b Fluid Imbalance

LEARNING OBJECTIVES

8. Name the different causes of fluid imbalance.
9. Compare and contrast the different types of fluid imbalances.
10. Explain what is meant by fluid sequestration.

A fluid imbalance occurs either when fluid output does not equal fluid intake or when fluid is distributed abnormally. Fluid imbalances can be organized into five categories: volume depletion, volume excess, dehydration, hypotonic hydration, and fluid sequestration. The first four of these categories (comprising imbalances that occur when fluid output does not equal fluid intake) can be differentiated using two criteria:

1. Does the fluid imbalance change the osmolarity (concentration) of body fluid?
2. Is the fluid imbalance caused by an excess or a deficiency of body fluid?

Fluid Imbalance with Constant Osmolarity

Fluid imbalances with constant osmolarity occur when *isotonic* fluid is lost or gained. **Volume depletion** occurs when isotonic fluid loss is greater than isotonic fluid gain. Examples of conditions that result in volume depletion include severe burns, chronic vomiting, diarrhea,

INTEGRATE

CLINICAL VIEW 25.2

Hemorrhaging

Hemorrhaging (or *haemorrhaging* or *bleeding*) is the loss of blood from the vessels of the cardiovascular system. The loss of blood from the inside of the body to the external environment can occur through the skin or an opening of the body (e.g., mouth, nose). **Internal hemorrhaging** is blood leaking from a damaged blood vessel and remaining within the body (e.g., gastrointestinal hemorrhage, cerebral hemorrhage). The physiologic processes involved in the stoppage of bleeding (**hemostasis**) normally occurs if damage is to relatively small vessels (see section 18.4). The risk of a hemorrhage is excessive loss of blood that results in **hypovolemia** (decrease in blood volume). Medical intervention (e.g., first aid, surgery) may be required to stop the bleeding and to replace the loss of fluid through an intravenous (IV) solution (see Clinical View 25.1: "Intravenous (IV) Solution") or a blood transfusion (see Clinical View 18.3: "Transfusions").

or the hyposecretion of aldosterone (a hormone that stimulates both Na^+ and water reabsorption in the kidney; see section 24.6d). Hemorrhaging (or loss of blood) also results in volume depletion (see Clinical View 25.2: "Hemorrhaging").

Volume excess occurs when isotonic fluid gain is greater than isotonic fluid loss. This typically results when fluid intake is normal, but there is decreased fluid loss through the kidneys (e.g., from either renal failure or aldosterone hypersecretion).

In both volume depletion and volume excess, there is no change in osmolarity. Consequently, there is no net movement of water between fluid compartments.

Fluid Imbalance with Changes in Osmolarity

Certain types of fluid imbalance involve fluid loss or gain that is not isotonic. **Dehydration** involves water *loss* that is greater than solute loss. Dehydration can result from profuse sweating, diabetes mellitus, intake of alcohol, hyposecretion of antidiuretic hormone (ADH—a hormone that stimulates water reabsorption in the kidney; see section 24.6d), insufficient water intake, or overexposure to cold weather. In each case, the greater loss of water than the loss of solutes results in the blood plasma becoming hypertonic (relatively more concentrated). Consequently, water moves between fluid compartments with a net movement of water from the cells into the interstitial fluid and then into blood plasma (figure 25.3*b*). Body cells may become dehydrated as a result (see Clinical View 25.3: "Dehydration in Infants and the Elderly").

Hypotonic hydration, also called *water intoxication* or *positive water balance,* involves water *gain* (or retention) that is greater than solute gain (or retention). Hypotonic hydration can result from ADH hypersecretion, but it is generally caused by drinking a large amount of plain water following excessive sweating. An example is an amateur athlete who runs a marathon and drinks excessive amounts of plain water. Both Na^+ and water are lost during sweating, and drinking water replaces only the water but not the solutes. The blood plasma becomes hypotonic to the other fluid compartments. Fluid moves from blood plasma into the interstitial fluid, and then into the cells (figure 25.3*a*). Cells may become swollen with fluid.

One of the consequences of extreme hypotonic hydration is cerebral edema. Brain cells become impaired as they swell with excess fluid. The person may experience headaches, nausea, or both. Convulsions, coma, or death may result in severe cases. In addition, some individuals have died after having been forced or enticed to drink excessive amounts of water (e.g., fraternity hazings, water-drinking contests).

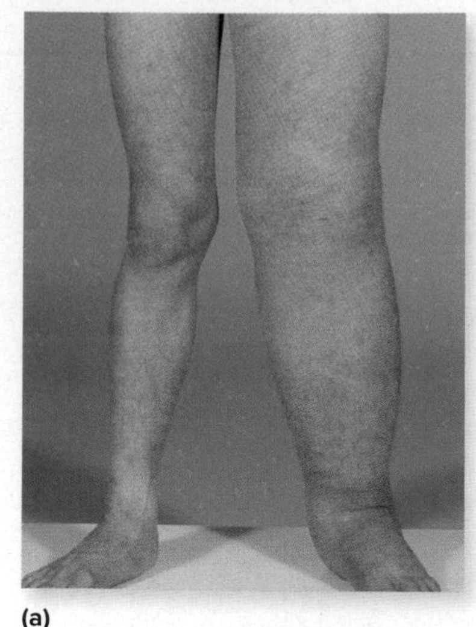
(a)

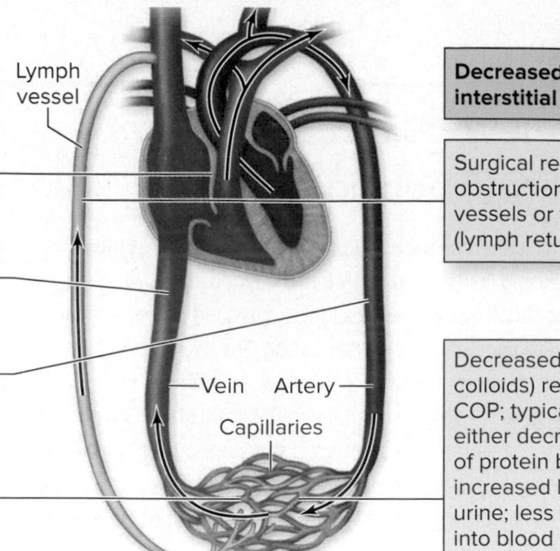

Increased formation of interstitial fluid
Congestive heart failure or incompetent heart valves (blood backs up in veins)
Faulty valves (blood return is impaired)
Increased blood volume (increased blood pressure forcing more fluid out)
Increased blood capillary permeability (e.g., caused by histamine release, bacterial toxins, increased age)
Lymph vessel
Vein
Artery
Capillaries
Decreased removal of interstitial fluid
Surgical removal or obstruction of lymph vessels or faulty valves (lymph return is impaired)
Decreased proteins (or colloids) result in decreased COP; typically caused by either decreased synthesis of protein by the liver or increased loss of protein in urine; less fluid is absorbed into blood capillaries

(b) (c)

Figure 25.5 Edema. (*a*) Edema is excess fluid within the interstitial space around cells that is characterized by swelling. The individual shown has edema in the left lower limb. Edema may result from either (*b*) increased formation of interstitial fluid or (*c*) decreased removal of interstitial fluid. (COP = Colloid osmotic pressure)

Fluid Sequestration

Fluid sequestration (sē′kwes-trā′shŭn; *sequestro* = to lay aside) differs from the other fluid imbalances because total body fluid may be normal, but it is distributed abnormally. Fluid accumulates in a particular location, and it is not available for use elsewhere.

Edema is an example of fluid sequestration in which fluid accumulates in the interstitial space around cells and is characterized by swelling. Anatomic or physiologic changes that can result in edema are depicted in **figure 25.5**. Notice as you review this figure that edema is due to either an increased formation of interstitial fluid or a decreased removal of interstitial fluid. These altered levels of interstitial fluid are generally a result of abnormal changes in the cardiovascular system (e.g., faulty valves within the heart or blood vessels), blood composition (e.g., decreased plasma proteins), or changes to lymph vessels (e.g., removal of lymph vessels). These changes alter the net filtration pressure (NFP) at systemic capillaries (see section 20.3c), causing additional fluid to either leave the capillaries or remain in the interstitial space. Edema can be associated with systemic cells of the body (*systemic edema*) or occur within the lungs (*pulmonary edema*). Both systemic edema and pulmonary edema are discussed in Clinical View 19.1: "Congestive Heart Failure."

Other examples of fluid sequestration include hemorrhage (specifically, internal hemorrhage), ascites, pericardial effusion, and pleural effusion. *Ascites* (ā-sī′tēz) is the accumulation of fluid within the peritoneal cavity (see Clinical View 26.9: "Cirrhosis of the Liver"), *pericardial effusion* is fluid within the pericardial cavity (see Clinical View 19.2: "Pericarditis"), and *pleural effusion* is the accumulation of fluid, sometimes up to several liters, in the pleural cavity as a consequence of lung infections (see Clinical View 23.10: "Pleurisy and Pleural Effusion").

WHAT DID YOU LEARN?

6. How would you distinguish fluid deficiency from dehydration in terms of changes in total body fluid, changes in blood osmolarity, and fluid movement between compartments?

INTEGRATE

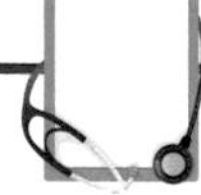

CLINICAL VIEW 25.3

Dehydration in Infants and the Elderly

Both infants and the elderly are more vulnerable to dehydration than are young and middle-aged adults.

Infants are especially vulnerable to dehydration for the following reasons:

- Infants have a greater ratio of skin surface area to volume in comparison to adults, and thus lose more fluid through sweating and transpiration relative to their body size.
- They have immature kidneys that cannot effectively concentrate urine; therefore, more water is required to eliminate wastes.
- They have a higher metabolic rate, which produces more metabolic waste to be eliminated.
- They do not have a completely developed homeostatic mechanism for temperature regulation (see section 1.6); thus, they develop fevers that are both higher and longer lasting. As a result, additional water is required to cool the body.

In comparison, the elderly are also susceptible to dehydration, but for different reasons:

- Due to the loss of skeletal muscle tissue, elderly individuals are on average composed of a smaller percentage of body fluid (average 45% by weight). Smaller losses in water thus may result in a fluid imbalance.
- The kidneys become less effective at concentrating and diluting urine as we age, so more fluid is lost in the urine.
- Cutaneous loss of fluid increases due to decreased thickness of the skin and loss of subcutaneous tissue.
- Fluid replacement may not occur with regularity because the thirst mechanism is less effective. Some elderly people, especially nonambulatory individuals, may also voluntarily restrict fluid intake due to concerns of incontinence.
- Elderly individuals are also at greater risk for hyperglycemia, with the elevated blood glucose causing the loss of additional water from the body (see Clinical View 24.3: "Glucosuria").

INTEGRATE

CONCEPT CONNECTION

You learned in section 20.3c that capillary dynamics are determined by the **net filtration pressure (NFP).** NFP is determined by subtracting the net colloid osmotic pressure from the net hydrostatic pressure. On the arterial end of the capillary, capillary hydrostatic pressure is sufficient to push fluid out of the capillary with an NFP of approximately 14 mm Hg. Thus, fluid exits the blood into the interstitial space, but most plasma proteins remain within the blood.

In contrast, at the venous end of capillaries, the NFP is approximately −5 mm Hg. This relatively negative pressure causes fluid to be pulled in the opposite direction from the interstitial space back into the blood. However, most but not all of the fluid returns to the blood on the venous end. The excess fluid remaining in the interstitial space, which is approximately 15% of what was originally filtered, is taken up by the lymphatic system (see section 21.1a). Changes to either NFP or lymph transport can result in edema (figure 25.5).

25.2c Regulation of Fluid Balance

LEARNING OBJECTIVES

11. Describe the stimuli that increase fluid intake.

12. Explain the conditions and stimuli that decrease fluid intake.

13. Identify the four hormones that are involved in regulating fluid output.

Maintaining fluid balance involves the regulation of both fluid intake and fluid output to prevent fluid imbalances. Because no receptors directly monitor either water volume or grams of dissolved solutes, such as Na^+ or K^+, the mechanisms for monitoring fluid are indirect. Fluid balance is typically regulated by monitoring blood volume, blood pressure, and blood plasma osmolarity (concentration of solutes in blood plasma). The relationships between these variables compared to the relative amount of fluid intake and fluid output can be described as follows:

- **Fluid intake is greater than fluid output.** When fluid intake is greater than fluid output, the following occur: an increase in blood volume and an increase in blood pressure—if water gain exceeds solute gain, then there is also a decrease in blood osmolarity.

  Fluid intake > Fluid output: ↑ blood volume ↑ blood pressure (perhaps ↓ blood osmolarity)

- **Fluid output is greater than fluid intake.** When fluid output is greater than fluid intake, the following occur: a decrease in blood volume and a decrease in blood pressure—if more water is lost than solutes, there is also an increase in blood osmolarity.

  Fluid output > Fluid intake: ↓ blood volume ↓ blood pressure (perhaps ↑ blood osmolarity)

It is helpful to keep these relationships in mind when reading through this section. Here we discuss how both fluid intake and fluid output regulation generally occur in response to the fluctuation of these variables.

Regulating Fluid Intake

Fluid intake is controlled by various stimuli that either activate or inhibit the **thirst center** located within the hypothalamus (see section 13.4c).

Stimuli to Turn on the Thirst Center Stimuli for activating the thirst center, which occurs when fluid intake is less than fluid output, include the following:

- **Decreased blood volume and blood pressure.** When fluid intake is less than fluid output, blood volume decreases, with an accompanying decrease in blood pressure. Renin is released from the kidney in response to a lower blood pressure (see section 20.6b). Renin (and the ACE enzyme) initiates the conversion of angiotensinogen to angiotensin II. An increase of 10–15% in the concentration of angiotensin II within the blood stimulates the thirst center. This mechanism is especially important when extreme volume depletion occurs—for example, when an individual is hemorrhaging (see Clinical View 25.2: "Hemorrhaging").
- **Increased blood osmolarity.** This occurs most commonly from insufficient water intake and dehydration. The increase in blood osmolarity directly stimulates sensory receptors in the thirst center within the hypothalamus and stimulates the hypothalamus to initiate nerve signals to the posterior pituitary to release antidiuretic hormone (ADH) (see section 17.7b). ADH activates the thirst center. This stimulation of the thirst center occurs with as little as a 2–3% increase in ADH.
- **Decreased salivary secretions.** A separate mechanism not related to blood volume, blood pressure, and blood osmolarity can also stimulate the thirst center. This additional stimulus is a relatively dry mouth. Mucous membranes are not as moist when less fluid is available and saliva production decreases. Sensory receptors in the mucous membranes of the mouth and throat relay sensory input to the thirst center.

When the thirst center is activated, nerve signals are relayed to the cerebral cortex, and we then become conscious of our thirst. *If* we take fluid into the body by drinking or eating, water is absorbed from the GI tract into the blood, and the water then moves into the interstitial space and ultimately into the cells (see figure 25.3*a*).

Stimuli to Turn off the Thirst Center Stimuli for inhibiting the thirst center are produced when fluid intake is greater than fluid output. All of these stimuli (except distension of the stomach, described here) oppose stimuli that activate the thirst center. These include the following:

- **Increased blood volume and blood pressure.** Blood volume and blood pressure increase with the addition of fluid. This rise in blood pressure inhibits the kidney from releasing renin, and the subsequent production of angiotensin II decreases. A decrease in angiotensin II results in a reduced stimulation of the thirst center.
- **Decreased blood osmolarity.** Blood osmolarity decreases when additional fluid enters the blood. In response, the thirst center is no longer stimulated directly, and the hypothalamus decreases stimulation of ADH release from the posterior pituitary.
- **Increased salivary secretions.** When body fluid level is high, salivary secretions increase, and the mucous membranes of the mouth and throat become moist. Sensory input to the thirst center decreases.
- **Distension of the stomach.** Fluid entering the stomach causes it to stretch, and nerve signals are relayed to the hypothalamus to inhibit the thirst center. (Note that an empty stomach does not stimulate the thirst center; rather, only a stretched stomach wall will inhibit the thirst center.)

The stimuli that inhibit the thirst center can be divided into two categories, depending upon both the time required to inhibit the thirst

INTEGRATE

LEARNING STRATEGY

The body can be thought of as a "leaky bucket" that we must continuously refill.

Fluid input. The only significant way to increase total water in the "bucket" is from food and drink, with fluid intake regulated through the thirst center. Fluid input can also be increased therapeutically through administration of an IV solution.

Fluid output (obligatory). Water continuously "leaks" from the body even when we are dehydrated. Obligatory water loss includes expired air, sweat, fluid loss through the skin (sweat and cutaneous transpiration), and defecation. Even when we are dehydrated, a minimum of urine production is also required to eliminate nitrogenous wastes and other substances. Abnormal fluid loss may also occur through vomiting or blood loss.

Fluid output (regulated). The primary means of regulating water loss is through hormonal stimulation of the kidney to control water loss. Three hormones decrease water loss (angiotensin II, antidiuretic hormone [ADH], and aldosterone), whereas atrial natriuretic peptide (ANP) increases water loss. This is like having a small tap in the "bucket" that can be opened or closed. However, even with this regulation, the bucket is still leaky; thus, fluids must be replaced daily.

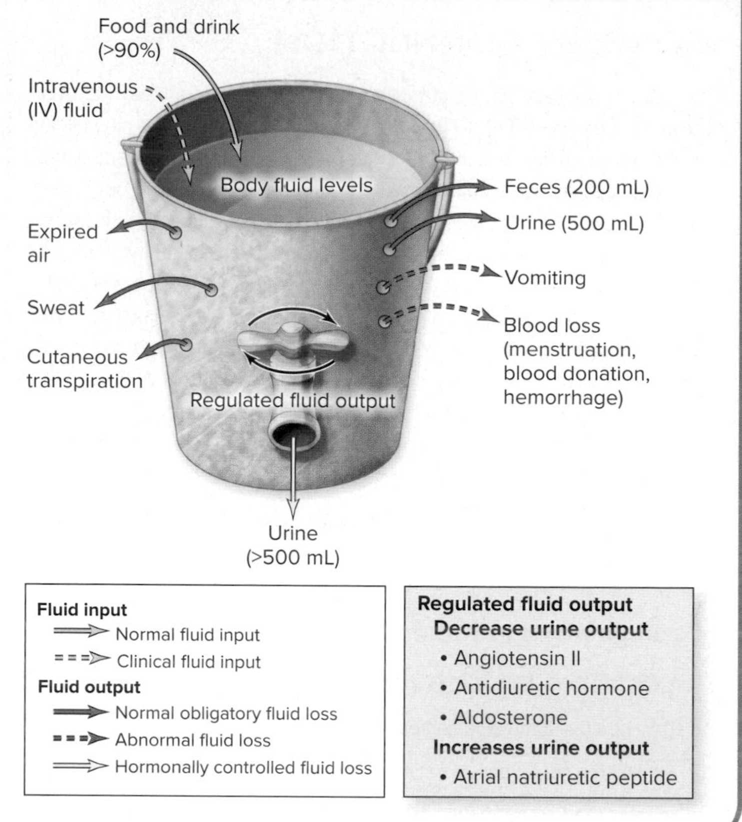

center and their level of accuracy in reflecting the hydrated state of the body. Stimuli that immediately inhibit the thirst center, but are less accurate concerning the hydrated state, are the moistening of the mucous membranes of the mouth and throat and the distension of the stomach. Signals from these stimuli will inhibit the thirst center for approximately 30 to 45 minutes, which is long enough for the absorption of fluids from the GI tract into the blood plasma.

Once fluids are absorbed, both blood volume and blood pressure increase (with an accompanying decrease in renin release and production of angiotensin II), and blood osmolarity decreases (with an accompanying decrease in ADH release). These are less immediate changes but more accurately reflect the body's state of hydration.

WHAT DO YOU THINK?

1. Which of the following would indicate that a person is thirsty: (a) increased blood volume, (b) increased saliva production, (c) increased blood osmolarity? Explain.

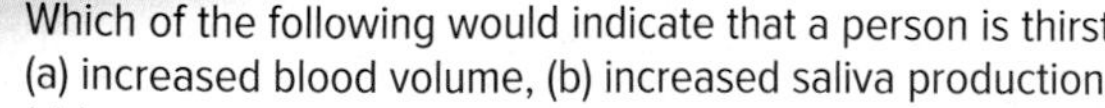

Regulating Fluid Output

Recall from section 25.2a that fluid output is regulated through the kidneys by controlling urine output through facultative water loss. Four major hormones are involved in regulating urine output: angiotensin II, antidiuretic hormone (ADH), aldosterone, and atrial natriuretic peptide (ANP).

Angiotensin II, ADH, and aldosterone help decrease urine output. These three hormones function to maintain both blood volume and blood pressure. In contrast, ANP increases urine output to decrease both blood volume and blood pressure. The specific mechanisms employed by each of these hormones in regulating fluid output in the kidneys also function in regulating some electrolytes (e.g., Na^+). Thus, it is more appropriate to describe the details of these hormones at the end of the next section, on electrolytes.

WHAT DID YOU LEARN?

7. What stimuli activate the thirst center?

8. Which of these four hormones—angiotensin II, antidiuretic hormone, aldosterone, and atrial natriuretic peptide—increases urine output?

25.3 Electrolyte Balance

Water movement between fluid compartments is caused by a relative difference in the concentration of solutes in those compartments. The greater the difference in the number of solutes, the greater the osmotic pressure (see section 4.3b). Because it is the relative number of solutes that determines the osmotic pressure, solutes are differentiated into nonelectrolytes and electrolytes. Nonelectrolytes and electrolytes were first introduced in section 2.4c.

25.3a Nonelectrolytes and Electrolytes

LEARNING OBJECTIVES

14. Describe the difference between a nonelectrolyte and an electrolyte.
15. Explain the general role of electrolytes in fluid balance.

Molecules that do not dissociate (or come apart) in solution are called **nonelectrolytes.** Most of these substances are covalently bonded organic molecules (e.g., glucose). In contrast, an **electrolyte** is any substance that dissociates in solution to form cations and anions. The term *electrolyte* refers directly to the ability of these substances, when dissolved and dissociated in solution, to conduct an electric current. Electrolytes include salts, acids, bases, and some negatively charged proteins. Examples of electrolytes and the dissociated ions they produce when placed into an aqueous solution (aq) include the following:

Salts: NaCl (aq) $\longrightarrow$ $Na^+ + Cl^-$ or $CaCl_2$ (aq) $\longrightarrow$ $Ca^{2+} + 2\,Cl^-$
Acid: HCl (aq) $\longrightarrow$ $H^+ + Cl^-$
Base: NaOH (aq) $\longrightarrow$ $Na^+ + OH^-$

NaCl dissociates into two ions, Na^+ and Cl^-. Because osmotic pressure is dependent upon the *number* of solutes, NaCl exerts twice the osmotic pressure (see section 4.3b) of the same concentration of a nonelectrolyte, such as glucose, which does not dissociate into ionic forms. Further, $CaCl_2$ in solution dissociates into three components, Ca^{2+} and two Cl^-, and exerts three times the osmotic pressure compared to that of glucose.

To account for this difference in exerting osmotic pressure, the concentration of electrolytes in solution is commonly expressed in milliequivalents per liter (mEq/L), which reflects the amount or equivalent number of electrical charges in 1 liter of solution. **Milliequivalents** is a measure of either the amount of H^+ an anion can bind or the amount of bicarbonate ion (HCO_3^-) a cation can bind in 1 liter of solution.

WHAT DID YOU LEARN?

9 Why do electrolytes exert a greater osmotic pressure than nonelectrolytes?

25.3b Major Electrolytes: Location, Functions, and Regulation

LEARNING OBJECTIVES

16. List the six major electrolytes found in body fluids, other than H^+ and HCO_3^-.
17. Explain why Na^+ is a critical electrolyte in the body.
18. Describe the variables that influence K^+ distribution.
19. Identify the main location, functions, and means of regulation for each of the common electrolytes.

The human body fluids contain common electrolytes. The common electrolytes include Na^+, K^+, Cl^-, Ca^{2+}, PO_4^{3-}, and Mg^{2+} (as well as H^+ and HCO_3^-). Each electrolyte has unique functions in the body in addition to its general function of contributing to the exertion of osmotic pressure. To carry out these functions effectively, each electrolyte must be maintained within a normal concentration range in the blood plasma (see table 18.3 for the normal ranges in mEq/L).

Regulating normal concentration of the different electrolytes in the blood plasma requires maintaining input equal to output, similar to what we discussed with respect to water balance. Here we present a description of the location, functions, and regulation of the major electrolytes, with the exception of H^+ and HCO_3^-, which are covered in section 25.5 on acid-base balance.

Sodium Ion (Na^+)

Sodium ions are located almost exclusively within the ECF—in both the interstitial fluid (IF) and blood plasma. Approximately 99% of Na^+ within the body is in the ECF and only 1% in the ICF, a gradient that is maintained by Na^+/K^+ pumps (see section 4.3c). Sodium is the principal cation within the ECF, composing 94% of the cations there (see figure 25.2). It typically dissociates from either sodium bicarbonate ($NaHCO_3$) or sodium chloride (NaCl).

Sodium functions in a number of physiologic processes, many of which have been presented in previous chapters (e.g., depolarization in muscles and neurons in chapters 10, 12, and 19 and cotransport in kidney tubules in chapter 24). We now describe Na^+ balance and how Na^+ functions in regulating fluid balance within the ECF.

Sodium Balance Our normal blood plasma Na^+ concentration is 135–145 mEq/L (**figure 25.6*a***). We obtain Na^+ from our diet. Although the daily dietary requirement for Na^+ is only 1.5 to 2.3 grams per day (g/day), our intake may vary from as little as 0.5 g on a low-sodium diet to as high as 20–25 g with high Na^+ intake. A typical American diet averages 3–7 g of Na^+ per day. Most of our Na^+

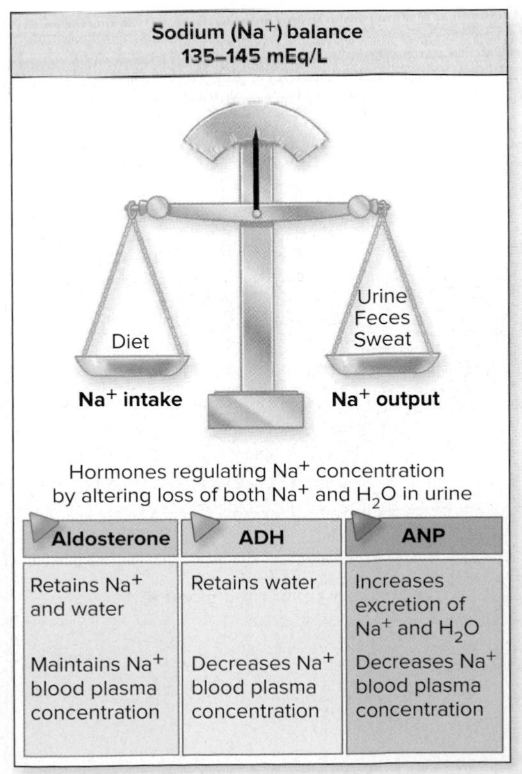

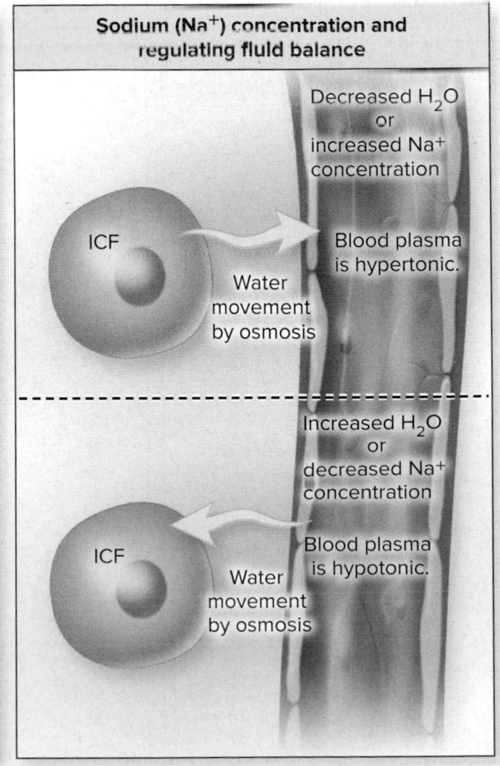

Figure 25.6 Sodium Balance. Sodium is the principal cation within the extracellular fluid (ECF). (*a*) Normal blood plasma Na^+ concentration is between 135 mEq/L and 145 mEq/L and is important in maintaining fluid balance within the ECF. Sodium level increases through the diet and decreases through urine, feces, and sweating. Sodium content and concentration are regulated by aldosterone, antidiuretic hormone (ADH), and atrial natriuretic peptide (ANP). (*b*) Changes in Na^+ concentration cause movement of water between fluid compartments.

comes from table salt and processed foods such as canned soups, lunch meats, and crackers. Sodium loss is through urine, feces, and sweat. The blood concentration of Na^+ is regulated by three hormones: aldosterone, ADH, and ANP. These hormones also indirectly help in regulating fluid balance and are described in detail in section 25.4.

WHAT DO YOU THINK?

2 Discuss why physicians might recommend a diet low in Na^+ for individuals with high blood pressure.

Sodium's Role in Blood Plasma Osmolarity As the most common cation within the ECF (see figure 25.2), Na^+ is the most important electrolyte in determining blood plasma osmolarity and in regulating fluid balance within the ECF. Consider that if Na^+ concentration increases from either elevated Na^+ intake or decreased water content, the ECF becomes temporarily more concentrated (hypertonic; figure 25.6*b*). Consequently, there is a net movement of water by osmosis from the cells into the blood plasma. This reestablishes the normal Na^+ concentration within the plasma (135–145 mEq/L).

In comparison, if Na^+ concentration decreases from either decreased Na^+ intake or increased water content, the ECF becomes temporarily less concentrated (hypotonic; figure 25.6*b*). The net movement of water by osmosis is then from the plasma into the cells until the normal Na^+ concentration is reestablished.

Note that the gain and loss of water relative to the plasma directly influence both blood volume and blood pressure. Thus, retention of Na^+ and water increases both blood volume and blood pressure, whereas the loss of Na^+ and water causes both a decrease in blood volume and blood pressure. For this reason, individuals with high blood pressure may be instructed by their physician to restrict their sodium intake. Although high blood pressure has many causes, and individuals retain Na^+ to varying degrees, studies have shown that decreasing dietary intake of Na^+ is effective in lowering blood pressure in some individuals.

Sodium imbalance is one of the most common types of electrolyte imbalances. A Na^+ imbalance occurs when the Na^+ concentration is either above the normal levels (**hypernatremia;** hī′per-nă-trē′mē-ă) or below normal levels (**hyponatremia;** hī′pō-nă-trē′mē-ă). Any condition that alters Na^+ intake or output, or any condition that alters water intake or output, may cause a change in Na^+ plasma concentration. Most occurrences result from changes in water content.

Potassium Ion (K^+)

Potassium ions are located almost exclusively within the ICF, with approximately 98% of potassium in the ICF and only 2% in the ECF. This gradient is also maintained by the Na^+/K^+ pump (see section 4.3c). Potassium is the principal cation within the ICF, composing 75% of the cations within the ICF (see figure 25.2).

Potassium is required for normal neuromuscular physiologic activities including, the reestablishment of the resting membrane potential through repolarization in skeletal muscle (see section 10.3), neurons (see section 12.8c), and cardiac muscle (see section 19.7b). Potassium also functions in establishing excitatory postsynaptic potentials in dendrites and cell bodies of neurons (see section 12.8a), and it has a significant role in controlling heart rhythm (see sections 19.6 and 19.7).

Potassium Balance Although the vast majority of the total body K^+ is located within the ICF, it is only the relatively small percentage of the K^+ within the ECF (about 2%) of the total K^+ that is continually regulated. The normal plasma value for K^+ within the ECF is between 3.5 mEq/L and 5.0 mEq/L (**figure 25.7*a***).

Both the total body K^+ and the distribution of K^+ must be regulated in order to maintain K^+ balance in our body fluids. **Total body potassium** is regulated by K^+ intake and output. The daily dietary requirement for K^+ is 40 mEq/L, although it may vary from 40 to 150 mEq/L. Potassium is generally obtained from fruits and vegetables, but input is increased with salt substitutes that contain K^+.

Total potassium (K^+) and its distribution

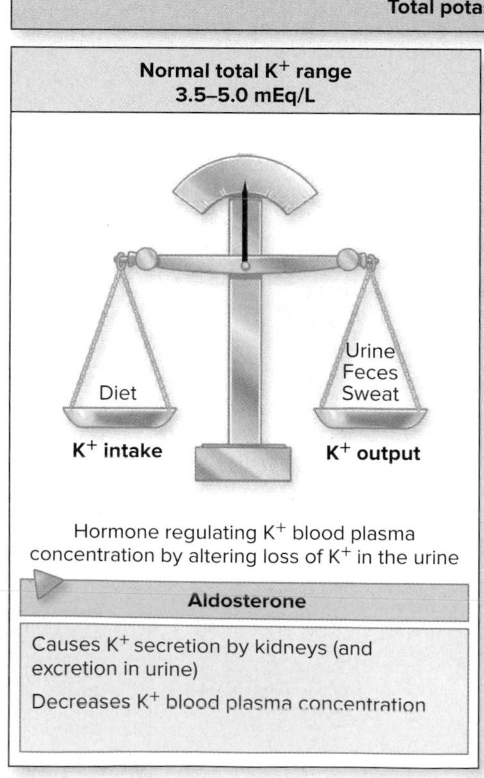

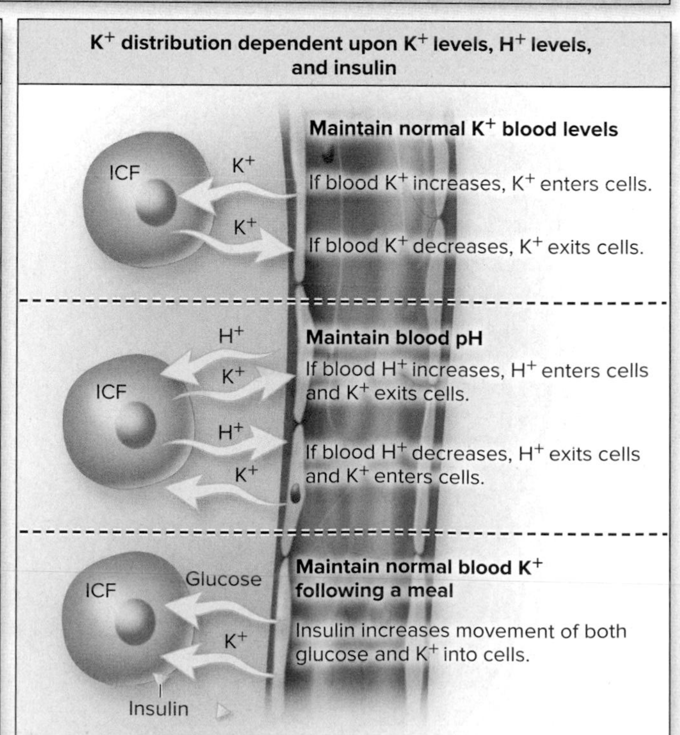

Figure 25.7 Potassium Balance. Potassium is the principal cation within the intracellular fluid (ICF). (*a*) Normal blood plasma K^+ concentration is 3.5–5.0 mEq/L. Potassium balance is dependent upon both total K^+ and its distribution in the fluid compartments. Total K^+ levels are a function of intake from the diet and loss through urine, feces, and sweat. (*b*) Potassium distribution changes in response to changing levels of K^+ within the extracellular fluid (ECF), changes in H^+ blood plasma concentration, and the presence of specific hormones (e.g., insulin).

Only small amounts of K^+ are lost from the body through sweat and in feces. Most K^+ (approximately 80–90%) is lost in urine. Some K^+ is always being lost because the body has no means to conserve all of its K^+. The amount of K^+ lost in the urine fluctuates (see section 24.6d), and greater amounts are lost during conditions of high blood plasma K^+, increased aldosterone secretion, and high blood pH.

Potassium distribution is the ratio of K^+ in the ICF to that in the ECF. This ratio is dependent upon both the level of activity of Na^+/K^+ pumps that actively pump K^+ into the cell and the K^+ leak channels that allow K^+ to flow back out of the cell (see section 4.4). Variables that influence K^+ distribution and how they may cause a potassium imbalance—*the most potentially lethal* of the electrolyte imbalances—are described next.

Potassium Shifts Potassium can redistribute between compartments under certain conditions that include changing blood plasma K^+ concentration, changing blood plasma H^+ concentration, and the presence of specific hormones in the blood (figure 25.7*b*).

Changes in blood plasma K^+ concentration cause shifts in K^+ between the ECF and ICF. An increase in blood plasma K^+ (referred to as **hyperkalemia;** hī'per kă lē'mē-ă) results in K^+ moving from the ECF into the ICF, whereas a decrease in blood plasma K^+ (referred to as **hypokalemia;** hī'pō-kă-lē'mē-ă) results in the movement of K^+ from the ICF into the ECF. Thus, the ICF serves as a reservoir for blood plasma K^+ levels, taking in excess K^+ in response to elevated blood plasma K^+ concentration and releasing K^+ when blood plasma K^+ concentration begins to decline.

Changes in blood plasma H^+ concentration also cause a shift in K^+ between the ECF and ICF. When H^+ increases (pH decreases) in the ECF (a condition called *acidosis;* see section 25.5), excess H^+ moves from the ECF into the ICF in an attempt to reestablish acid-base balance. At the same time, K^+ moves in the opposite direction, from the ICF into the ECF (potentially resulting in hyperkalemia). The reverse movement of both H^+ and K^+ occurs if H^+ decreases (pH increases). When H^+ decreases (pH increases) within the ECF (a condition called *alkalosis*), H^+ moves from the ICF into the ECF in an attempt to reestablish acid-base balance. At the same time, K^+ moves from the ECF into the ICF (potentially resulting in hypokalemia).

Notice these important relationships: (1) An increase in blood H^+ (acidosis) can result in increased blood K^+ (hyperkalemia), whereas (2) a decrease in blood H^+ (alkalosis) can result in decreased blood K^+ (hypokalemia). A primary reason an acid-base imbalance can be lethal is because of these shifts in K^+ that occur in response to pH changes.

Certain hormones also induce shifts in K^+ between the ECF and ICF; insulin is one important example. Insulin is typically released following a meal and decreases not only blood glucose (as described in section 17.10b) but also blood K^+. Insulin decreases blood plasma K^+ by stimulating the activity of Na^+/K^+ pumps in cells, thus increasing the transport of K^+ from the ECF into the ICF. Physiologists reason that this helps prevent the rapid development of hyperkalemia that can occur as K^+ is absorbed into the blood following a meal. Interestingly, individuals exhibiting hyperkalemia may be treated with the administration of insulin (along with glucose).

WHAT DO YOU THINK?

3 Bodybuilders have been known to inject insulin to increase muscle mass because it stimulates protein anabolism. What is one of the risks associated with this practice in terms of K^+ levels: elevated blood plasma K^+ (hyperkalemia) or decreased blood plasma K^+ (hypokalemia)? Could the practice of injecting insulin to increase muscle mass be fatal?

Chloride Ion (Cl^-)

Chloride is a common anion that is normally associated with Na^+ (as NaCl). It is the most abundant anion in the ECF (see figure 25.2). Chloride ions are found in the lumen of the stomach as hydrochloric acid (see section 26.2d), and they participate in the chloride shift within erythrocytes for transport of carbon dioxide in the form of HCO_3^- (see figure 23.28).

Normal blood plasma Cl^- concentration is 96–106 mEq/L. Chloride is obtained in the diet, primarily from table salt and processed foods. Chloride is normally lost in the sweat, stomach secretions, and urine. The amount lost in urine is primarily dependent upon blood plasma Na^+. Chloride levels are directly correlated with Na^+ levels, because Cl^- follows Na^+ by electrostatic interactions and is regulated by the same mechanisms as Na^+. Increased blood chloride levels are termed **hyperchloremia** (hī'per-klōr-ē'mē-ă), and decreased blood chloride levels are termed **hypochloremia** (hī'pō-klōr-ē'mē-ă).

Calcium Ion (Ca^{2+})

Calcium is the most abundant electrolyte in bone and teeth. Approximately 99% of all body Ca^{2+} (most commonly as calcium phosphate, $Ca_3[PO_4]_2$) is stored within the extracellular matrix of these structures, causing them to harden (see section 7.2e). To prevent hardening or calcification of other tissue, Ca^{2+} is normally moved by Ca^{2+} pumps either out of cells or into the sarcoplasmic reticulum within muscle cells. This prevents Ca^{2+} from binding to the abundant PO_4^{3-} within cells. In addition to hardening of both bones and teeth, Ca^{2+} is needed to initiate muscle contraction (see section 10.3c) and release neurotransmitters (see section 12.8d); it also serves as a second messenger (see section 17.5) and participates in blood clotting (see section 18.4c).

Normal blood plasma Ca^{2+} concentration is approximately 5 mEq/L. In blood plasma, Ca^{2+} can exist bound to protein such as albumin (about 35%), associated with anions such as PO_4^{3-} (about 15%), or in an ionized or unbound state (about 50%). Only the ionized form is physiologically active. Calcium is obtained in the diet from yogurt, milk, soy products, cheese, sardines, and green leafy vegetables (such as broccoli and collard greens). Calcium is lost from the body in urine, feces, and sweat. (The regulation of Ca^{2+} by parathyroid hormone, calcitriol, and calcitonin is discussed in sections 7.6 and 24.6d.) Increased blood calcium levels are called **hypercalcemia** (hī'per-kal-sē'mē-ă), and decreased blood calcium levels are called **hypocalcemia** (hī'pō-kal-sē'mē-ă). An important relationship to note regarding blood calcium levels is that an increase in blood calcium (hypercalcemia) results in *underexcitability* of muscles and nerves, whereas a decrease in blood calcium results in *overexcitability* of muscles and nerves, which can in severe cases cause laryngospasms and suffocation.

Phosphate Ion (PO_4^{3-})

Phosphate ions (PO_4^{3-}) occur as both hydrogen phosphate (HPO_4^{2-}) and dihydrogen phosphate ($H_2PO_4^-$). Phosphate is the most abundant anion in the ICF (see figure 25.2). Approximately 85% is stored in the extracellular matrix of bone and teeth as calcium phosphate, $Ca_3(PO_4)_2$. Phosphate is also a component of nucleotides of DNA and RNA (see section 2.7d) and of phospholipid molecules (see section 2.7b) within plasma membranes. It serves as an intracellular buffer against pH changes and is a common buffer in urine.

Normal blood plasma concentration of PO_4^{3-} ranges from 1.8 to 2.9 mEq/L. Most PO_4^{3-} is ionized (about 90%) in blood plasma and the rest is bound to plasma proteins such as albumin. Phosphorus is obtained in the diet from milk, meat, and fish. In addition, many food additives (used to prevent spoilage) and most soft drinks contain PO_4^{3-}. Phosphate is regulated by many of the same mechanisms as

Table 25.1 Blood Electrolyte Imbalances

Condition	Causes	Effects	Symptoms
Hypernatremia ($Na^+ > 145$ mEq/L)	Dehydration, diabetes insipidus (hyposecretion of ADH), administration of intravenous (IV) saline	Cell shrinkage, resulting in neurologic impairment; may result in excess fluid causing high blood pressure or edema	Confusion, coma, paralysis of breathing muscles, death
Hyponatremia ($Na^+ < 135$ mEq/L)	Excessive water intake, hypersecretion of antidiuretic hormone (ADH), diuretic abuse, severe diarrhea, burns, hyposecretion of aldosterone, excessive sweating	Cell swelling and decrease in blood volume and blood pressure	Nausea, lethargy, confusion, headache, muscle cramps, seizures, coma, death
Hyperkalemia ($K^+ > 5.0$ mEq/L)	Aldosterone hyposecretion, renal failure, acidosis, decreased insulin, extensive cellular trauma from a crushing injury or burn, transfusion of outdated blood, hemolytic anemia	A fast rise in K^+ results in K^+ diffusing into cells, altering the resting membrane potential; interferes with an action potential A slow rise in K^+ inactivates Na^+ channels; action potentials impaired, with muscles and nerves being less excitable	Nausea, vomiting, diarrhea, skeletal muscle weakness, tingling of skin, numbness of hands or feet, and irregular heartbeat that can lead to cardiac arrest[1] Can be asymptomatic, or an individual may have nausea, fatigue, muscle weakness
Hypokalemia ($K^+ < 3.5$ mEq/L)	Diuretic abuse, aldosterone hypersecretion, chronic vomiting, diarrhea, excessive laxative or enema use, heavy sweating, alkalosis, increased insulin, or formation of new tissue because K^+ is the predominant intracellular fluid (ICF) cation	Cells become hyperpolarized, interfering with neuron and muscle function	Nausea, vomiting; nervous and muscle tissue are less excitable, resulting in numbness, muscle weakness, decreased tone of smooth muscle, flaccid paralysis; decreased and irregular heart rate that may lead to cardiac arrest; muscle weakness can be severe enough to cause paralysis of diaphragm, leading to respiratory arrest
Hyperchloremia ($Cl^- > 107$ mEq/L)	Excess Cl^- in diet, intravenous (IV) solution containing NaCl, diarrhea, hyperaldosteronism, hyperparathyroidism, kidney disease; usually associated with metabolic acidosis	Decreased nerve excitability; causes changes in concentration of other ions	May be asymptomatic or the individual may have irregular breathing, weakness, and intense thirst; symptoms generally arise from other conditions associated with hyperchloremia
Hypochloremia ($Cl^- < 97$ mEq/L)	Heavy sweating, vomiting, gastrointestinal (GI) suction, kidney disease, metabolic alkalosis	Increased nerve excitability; causes changes in concentration of other ions	Many individuals are asymptomatic, but may experience tetany, hyperactive reflexes, muscle cramps; if severe—arrhythmias, seizures, coma, or respiratory arrest
Hypercalcemia ($Ca^{2+} > 5.2$ mEq/L)	Hyperparathyroidism, hypothyroidism (decreased calcitonin), bone loss from cancer, immobilization, excessive intake of vitamin D or calcium antacids, renal failure	Interferes with normal muscle and nerve function and causes underexcitability of muscles and nerves	May be asymptomatic; muscle weakness, fatigue, depression, confusion, loss of appetite, nausea, vomiting, constipation, and abnormal heart rhythms; severe calcification of soft tissue
Hypocalcemia ($Ca^{2+} < 4.5$ mEq/L)	Vitamin D deficiency, diarrhea, pregnancy, lactation; hypoparathyroidism or hyperthyroidism (increased calcitonin)	Overexcitability of muscles and nerves	Numbness, tingling, muscle spasms, increased heart rate; skeletal muscle also more excitable and can go into tetany, causing laryngospasms and suffocation
Hyperphosphatemia ($PO_4^{3-} > 2.9$ mEq/L)	Kidney impairment; increased dietary intake of supplements; laxatives or enemas; respiratory acidosis, cellular damage	Precipitation of Ca^{2+}, resulting in hypocalcemia	Generally asymptomatic by itself; however, may lead to hypocalcemia as PO_4^{3-} binds with Ca^{2+}; symptoms affect muscles, nerves, and bone calcification
Hypophosphatemia ($PO_4^{3-} < 1.8$ mEq/L)	Insufficient PO_4^{3-} intake; hyperparathyroidism; respiratory alkalosis	Many effects due to decreased production of ATP	Weakness and dysfunction of nervous system and both skeletal and cardiac muscle
Hypermagnesemia ($Mg^{2+} > 2.1$ mEq/L)	Kidney impairment, high intake of Mg^{2+}, such as in antacids or enemas	Changes in neuromuscular, cardiovascular, and central nervous system	Depressed neuromuscular activity, hypotension, cardiac arrest, weakness, lethargy, and respiratory depression
Hypomagnesemia ($Mg^{2+} < 1.4$ mEq/L)	Impaired absorption from the GI tract; increased loss in kidneys, or redistribution of Mg^{2+}; chronic alcoholism	Changes in neuromuscular, cardiovascular, and central nervous system	Muscle twitches, tremors, hyperactive reflexes, mental confusion and depression

1. Lethal injections for euthanizing animals and for capital punishment include high doses of K^+, causing cardiac arrest.

Ca^{2+}; this is because as just described 99% of body calcium is stored in bone, with the majority stored as calcium phosphate. Increased blood phosphate levels are called **hyperphosphatemia** (hī′per-fos-fă-tē′mē-ă), and decreased blood phosphate levels are called **hypophosphatemia** (hī′pō-fos-fă-tē′mē-ă).

Magnesium Ion (Mg^{2+})

Magnesium is primarily located within bone or within cells. After K^+, Mg^{2+} is the most abundant cation in the ICF (see figure 25.2). Magnesium participates in over 300 enzymatic reactions, including ATP synthesis, protein synthesis, and enzymatic reactions involving carbohydrate metabolism. It also assists in the movement of Na^+ and K^+ across the plasma membrane by the Na^+/K^+ pump, and it is important in muscle relaxation.

Normal blood plasma concentration of Mg^{2+} ranges from 1.4 to 2.1 mEq/L. Because most Mg^{2+} is located within cells, this represents only a small fraction of total body Mg^{2+}. In blood plasma, Mg^{2+} is either in a free, ionized form or bound to protein. The ionized form is physiologically active. Magnesium is obtained in the diet from beans and peas, leafy green vegetables, and other sources and is lost from the body in sweat and urine. Magnesium blood plasma levels are regulated through the kidney. Increased blood magnesium levels are called **hypermagnesemia** (hī′per-mag′nē-sē′mē-ă), and decreased blood magnesium levels are called **hypomagnesemia** (hī′pō-mag′nē-sē′mē-ă). **Table 25.1** summarizes the causes, effects, and symptoms of the major electrolyte imbalances.

 **WHAT DID YOU LEARN?**

 10 What is the net direction of K^+ movement in response to a decrease in pH? Explain.

25.4 Hormonal Regulation

Four hormones play a major role in homeostatic regulation of both fluid and electrolytes: angiotensin II, antidiuretic hormone, aldosterone, and atrial natriuretic peptide. We look closely at the details of regulatory mechanisms for each hormone, including the stimulus for its release and its subsequent effects on numerous variables, such as fluid intake, fluid output, blood volume, blood pressure, and blood osmolarity.

Before reading this section, recall from section 25.2c that when fluid intake is greater than fluid output, both blood volume and systemic blood pressure increase, and blood osmolarity may decrease if more water than solutes is taken in. The reverse occurs when fluid output is greater than fluid intake.

25.4a Angiotensin II

 **LEARNING OBJECTIVES**

20. Explain the means by which angiotensin II formation can be triggered.

21. List the four primary effects of angiotensin II.

Angiotensin II is a peptide hormone that is unique among the four hormones that affect fluid balance because (1) it has two significantly different means of stimulation for its formation, and (2) when produced, it stimulates the most diverse effects **(figure 25.8)**.

We previously described angiotensin II as a potent vasoconstrictor that helps regulate blood pressure (see sections 17.11c, 20.6b, and 24.5e). Recall that angiotensinogen is an inactive hormone synthesized and released continuously from the liver. Its activation, which occurs within the blood, is initiated by the enzyme renin. Renin is released from the juxtaglomerular (JG) apparatus of the kidneys in response to either (1) low blood pressure (as detected by decreased stretch of baroreceptors within granular cells or by decreased NaCl detected by chemoreceptors within macula densa cells; see section 24.3c), or (2) stimulation by the sympathetic division. The sequential action of renin and angiotensin-converting enzyme (ACE) (which is bound to the endothelial lining of blood vessels) causes the formation of angiotensin II (the active form of the hormone).

Angiotensin II has a number of effectors (target organs), and it initiates the following changes when it binds to these structures:

- **Blood vessels:** Stimulates vasoconstriction of systemic blood vessels to increase total peripheral resistance, which increases systemic blood pressure (see section 20.6b)
- **Kidneys:** Decreases urine output from the kidneys as a result of decreased glomerular filtration rate (GFR) in the nephrons by stimulating vasoconstriction of afferent arterioles and contraction of the mesangial cells within the glomerulus (see section 24.5e); this decreases urine output and helps to *maintain* systemic blood volume, and thus blood pressure
- **Thirst center:** Stimulates the thirst center in the hypothalamus; *if* fluid intake occurs, this increases blood volume, which increases systemic blood pressure
- **Hypothalamus and adrenal cortex:** Stimulates both the hypothalamus to activate the posterior pituitary to release ADH and the adrenal cortex to release aldosterone (both described shortly)

The formation and action of angiotensin II can be summarized as follows: It is synthesized either when blood pressure is low or the sympathetic division is activated. It causes an increase in resistance, a decrease in fluid output (which helps to maintain blood volume and blood pressure), and an increase in blood volume (if fluid intake occurs). Consequently, blood pressure increases. Increasing blood pressure is aided by the release of both ADH and aldosterone. As blood pressure returns to within normal homeostatic levels, both renin release and angiotensin II synthesis are decreased by negative feedback.

 **WHAT DID YOU LEARN?**

 11 How does angiotensin II alter fluid output and potentially alter fluid intake?

INTEGRATE

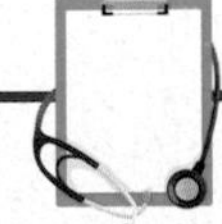

CLINICAL VIEW 25.4

Angiotensin-Converting Enzyme (ACE) Inhibitors

An **ACE inhibitor** is a medication used in the treatment of high blood pressure (typically prescribed in combination with other blood pressure medications). ACE inhibitors block the action of the ACE enzyme in the final step of angiotensin II formation. All actions of angiotensin II (see section 25.4a) are inhibited. Thus, blood vessels vasodilate, decreasing resistance, with an accompanying increased loss of sodium and water in the urine, which decreases blood volume. Both the decrease in resistance and the decrease in blood volume cause a decrease in blood pressure.

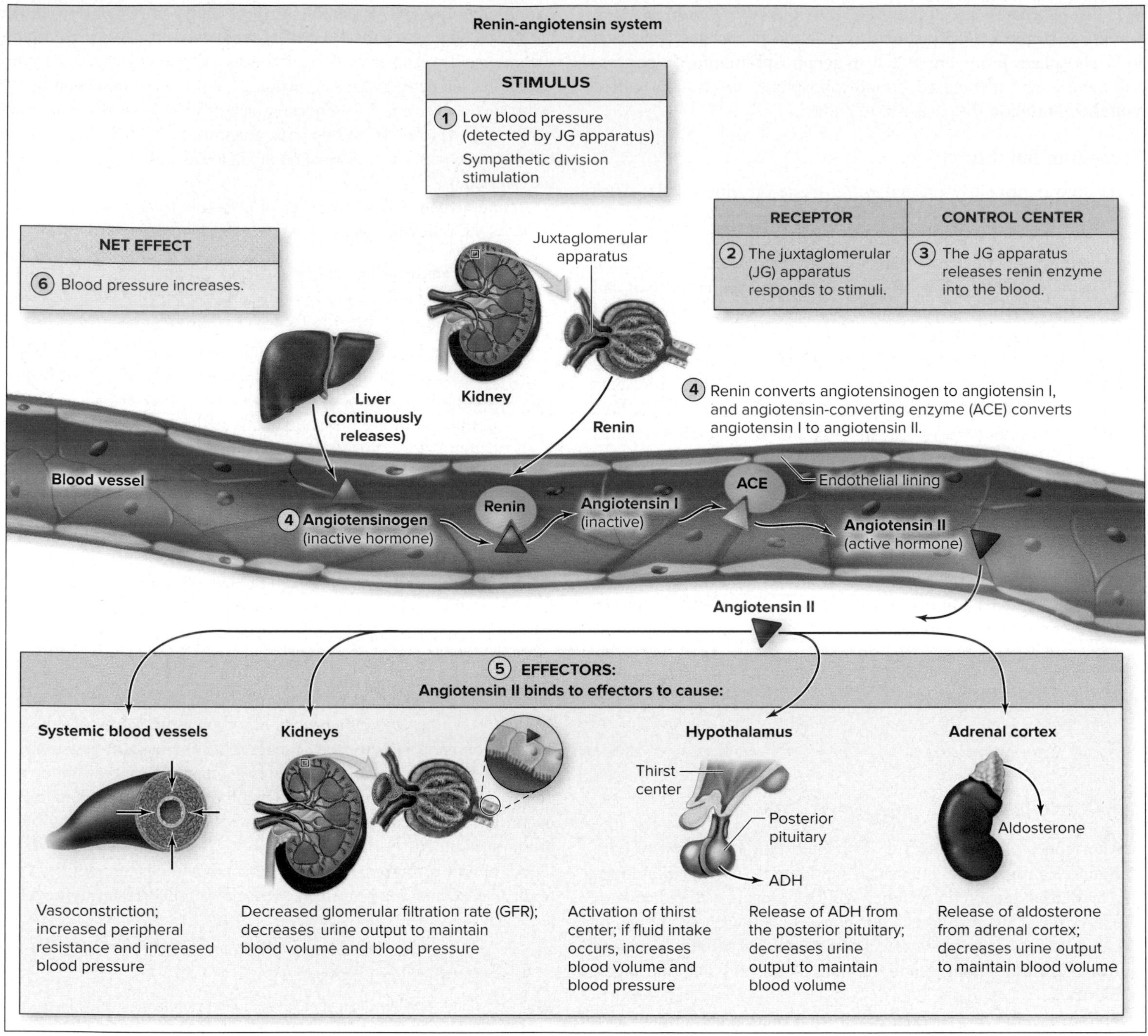
Renin-angiotensin system
STIMULUS
1 Low blood pressure (detected by JG apparatus)
Sympathetic division stimulation
RECEPTOR
2 The juxtaglomerular (JG) apparatus responds to stimuli.
CONTROL CENTER
3 The JG apparatus releases renin enzyme into the blood.
NET EFFECT
6 Blood pressure increases.
Juxtaglomerular apparatus
Liver (continuously releases)
Kidney
Renin
4 Renin converts angiotensinogen to angiotensin I, and angiotensin-converting enzyme (ACE) converts angiotensin I to angiotensin II.
Blood vessel
4 Angiotensinogen (inactive hormone)
Renin
Angiotensin I (inactive)
ACE
Endothelial lining
Angiotensin II (active hormone)
Angiotensin II
5 EFFECTORS:
Angiotensin II binds to effectors to cause:
Systemic blood vessels
Kidneys
Hypothalamus
Adrenal cortex
Thirst center
Posterior pituitary
ADH
Aldosterone
Vasoconstriction; increased peripheral resistance and increased blood pressure
Decreased glomerular filtration rate (GFR); decreases urine output to maintain blood volume and blood pressure
Activation of thirst center; if fluid intake occurs, increases blood volume and blood pressure
Release of ADH from the posterior pituitary; decreases urine output to maintain blood volume
Release of aldosterone from adrenal cortex; decreases urine output to maintain blood volume

Figure 25.8 Renin-Angiotensin System. Angiotensin II production is initiated when renin enzyme is released into the blood from the kidney in response to either decreased blood pressure or stimulation by the sympathetic division. The effects of angiotensin II are to increase peripheral resistance and to maintain and increase blood volume and blood pressure to within normal homeostatic limits. AP|R

25.4b Antidiuretic Hormone

LEARNING OBJECTIVES

22. Explain how release of antidiuretic hormone (ADH) occurs from the posterior pituitary.

23. Describe the three actions of antidiuretic hormone.

Antidiuretic hormone (ADH) (also called *vasopressin*) is a peptide hormone that is synthesized by the hypothalamus and then transported to the posterior pituitary, where it is stored (see sections 17.7b and 24.6d). Nerve signals from the hypothalamus stimulate the posterior pituitary to release ADH (**figure 25.9**).

Three primary types of stimuli—low blood pressure, low blood volume, or an increase in blood osmolarity—signal the hypothalamus regarding the need to retain fluid:

- *Angiotensin II* binds with receptors of cells of the hypothalamus, having been released in response to low blood pressure, as just described.

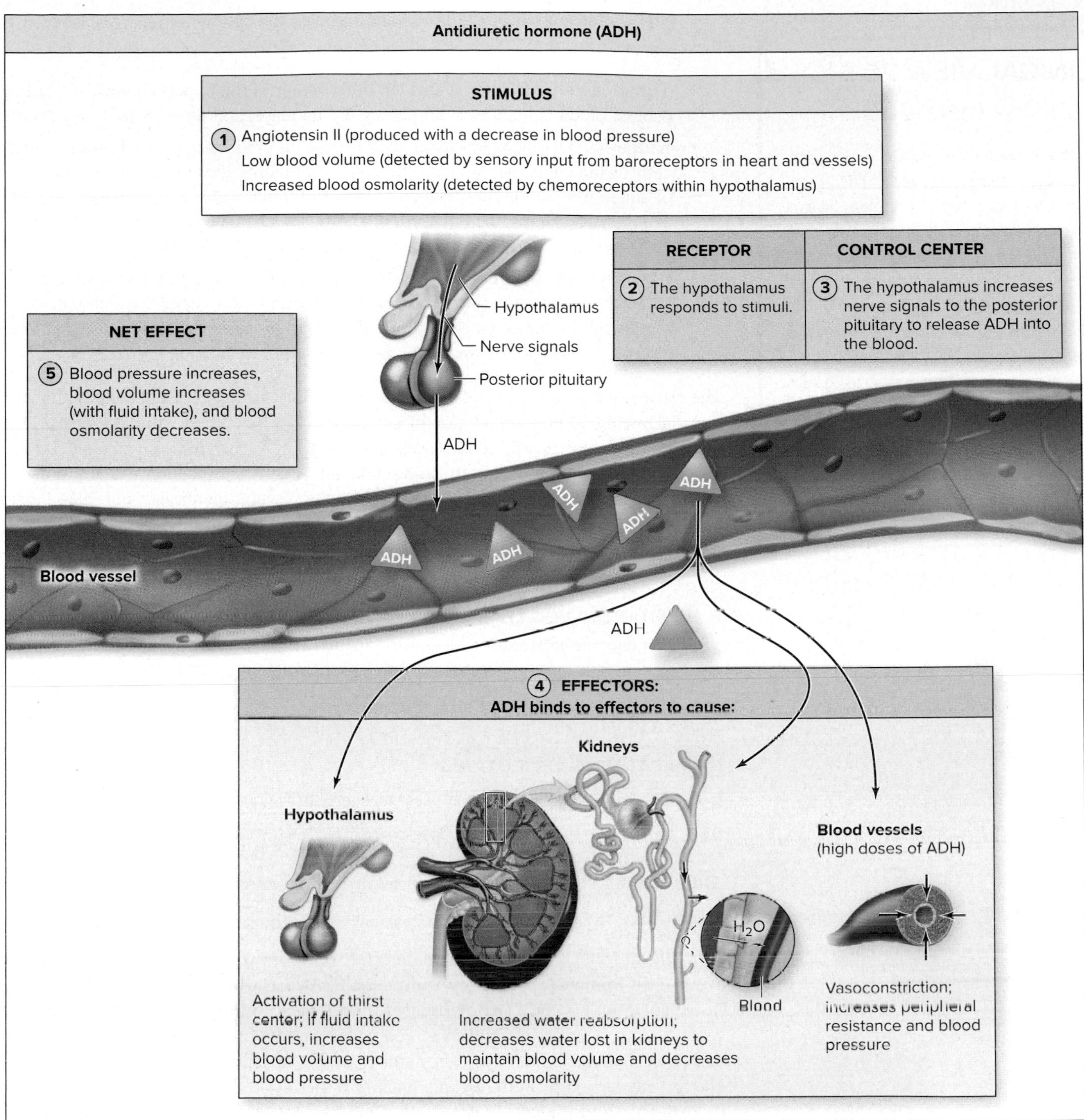

Figure 25.9 Actions and Effects of Antidiuretic Hormone. Antidiuretic hormone (ADH) is released in response to various stimuli, including angiotensin II, low blood volume, and high blood osmolarity. The effect of ADH on target organs is to maintain and possibly increase blood volume and blood pressure, and decrease blood osmolarity to within normal homeostatic limits.

- *Low blood volume* causes decreased stretch of baroreceptors within the atria of the heart and aorta and carotid blood vessels (see section 20.6a), which results in decreased nerve signals relayed along sensory neurons to the hypothalamus. Please note that this decrease in sensory nerve input from the heart and blood vessels to the hypothalamus is critical under conditions of severe blood loss, or hemorrhaging (see Clinical View 25.2: "Hemorrhaging").
- *Increased blood osmolarity* is detected directly by chemoreceptors within the hypothalamus. This is the primary stimulus for release of ADH.

In response to any of these stimuli (angiotensin II, low blood volume, or increased blood osmolarity), the hypothalamus increases nerve signals to the posterior pituitary to stimulate the release of ADH: The larger the change in these stimuli, the greater the amount of ADH released.

INTEGRATE

CLINICAL VIEW 25.5

Diabetes Insipidus

Diabetes insipidus (dī-ă-bē′tez = a siphon, *in-sip′i-dus* = tasteless), which affects approximately 3 per 100,000 in the general population, results from either hyposecretion of ADH from the posterior pituitary or the inability of the kidney to respond to ADH. The result is a decrease in fluid retention, with increased urine production. Individuals with diabetes insipidus typically lose approximately 3 liters of fluid/day, but in severe cases they can lose up to 20 liters per day if left untreated. (Be careful not to confuse diabetes insipidus with diabetes mellitus [me-lī′tŭs; sweetened with honey], which is discussed in Clinical View 17.8: "Conditions Resulting in Abnormal Blood Glucose Levels.")

ADH is transported in the blood and initiates the following changes when it binds to these various effectors:

- **Thirst center.** ADH stimulates the thirst center in the hypothalamus. *If* fluid intake occurs, blood volume and blood pressure increase, and blood osmolarity decreases.
- **Kidneys.** ADH increases water reabsorption in the kidneys. ADH binds to principal cells of the collecting tubules and ducts, stimulating these cells to increase the number of aquaporins in the tubular membrane. Additional water is reabsorbed through these aquaporins (as described in section 24.6d). This helps to both maintain blood volume by decreasing fluid loss in urine and decrease blood osmolarity.
- **Blood vessels.** High doses of ADH (which occur, for example, with severe hemorrhage) cause vasoconstriction of systemic blood vessels, which increases peripheral resistance (see section 20.5b). This action is the reason that ADH is also referred to as *vasopressin*. Systemic blood pressure increases as a result.

The release and action of ADH can be summarized as follows: ADH is released under conditions of low blood pressure (action of angiotensin II), low blood volume (detected by stretch receptors in the heart and blood vessels), and high blood osmolarity (detected directly by the hypothalamus) to stimulate both fluid intake and water reabsorption in the kidneys. If fluid intake occurs, blood volume increases, blood pressure increases, and blood osmolarity decreases. As the stimuli return to within normal homeostatic levels, ADH release is decreased through negative feedback.

WHAT DID YOU LEARN?

12 How does the homeostatic system involving ADH function? Include how it is released and its actions.

25.4c Aldosterone

LEARNING OBJECTIVES

24. List three conditions that lead to aldosterone release.

25. Describe the changes that occur in response to binding of aldosterone by kidney cells.

Aldosterone (ALDO) is normally released from the adrenal cortex in response to angiotensin II, decreased blood plasma Na^+ levels, or most importantly, increased blood plasma K^+ levels **(figure 25.10)**.

Aldosterone is a steroid hormone that is transported within the blood plasma and eventually binds to receptors within principal cells of the kidney, as described in sections 17.9a and 24.6d. The binding of aldosterone to these cells causes increased reabsorption and retention of both Na^+ and water, and increased secretion and then excretion of K^+. Aldosterone increases the number of Na^+/K^+ pumps and Na^+ channels, so more Na^+ is reabsorbed from the filtrate back into the blood. Water follows the Na^+ movement by osmosis. Fluid retention results in decreased urine output to help *maintain* blood volume and blood pressure. Because equal amounts of Na^+ and water are reabsorbed, blood osmolarity remains constant.

Note that K^+ excretion is normally increased except under conditions of low pH. In acidic conditions, as Na^+ and water are reabsorbed from the tubule into the blood, H^+ (instead of K^+) is secreted from the blood into the tubule. This loss of excess H^+ assists in returning blood pH to within normal homeostatic limits.

The release and action of aldosterone can be summarized as follows: Both low blood pressure and changes in Na^+ and K^+ blood plasma levels cause aldosterone release. Blood volume and blood pressure are maintained through the reabsorption of both Na^+ and water in the kidneys, with no change to osmolarity. K^+ secretion is increased (unless there is an increase in H^+). As blood volume, blood pressure, and both Na^+ and K^+ blood plasma levels return to normal ranges, aldosterone release is decreased. Thus, as with angiotensin II and ADH, aldosterone release and its effects are regulated by negative feedback.

WHAT DID YOU LEARN?

13 How does aldosterone influence the contents and volume of fluid output?

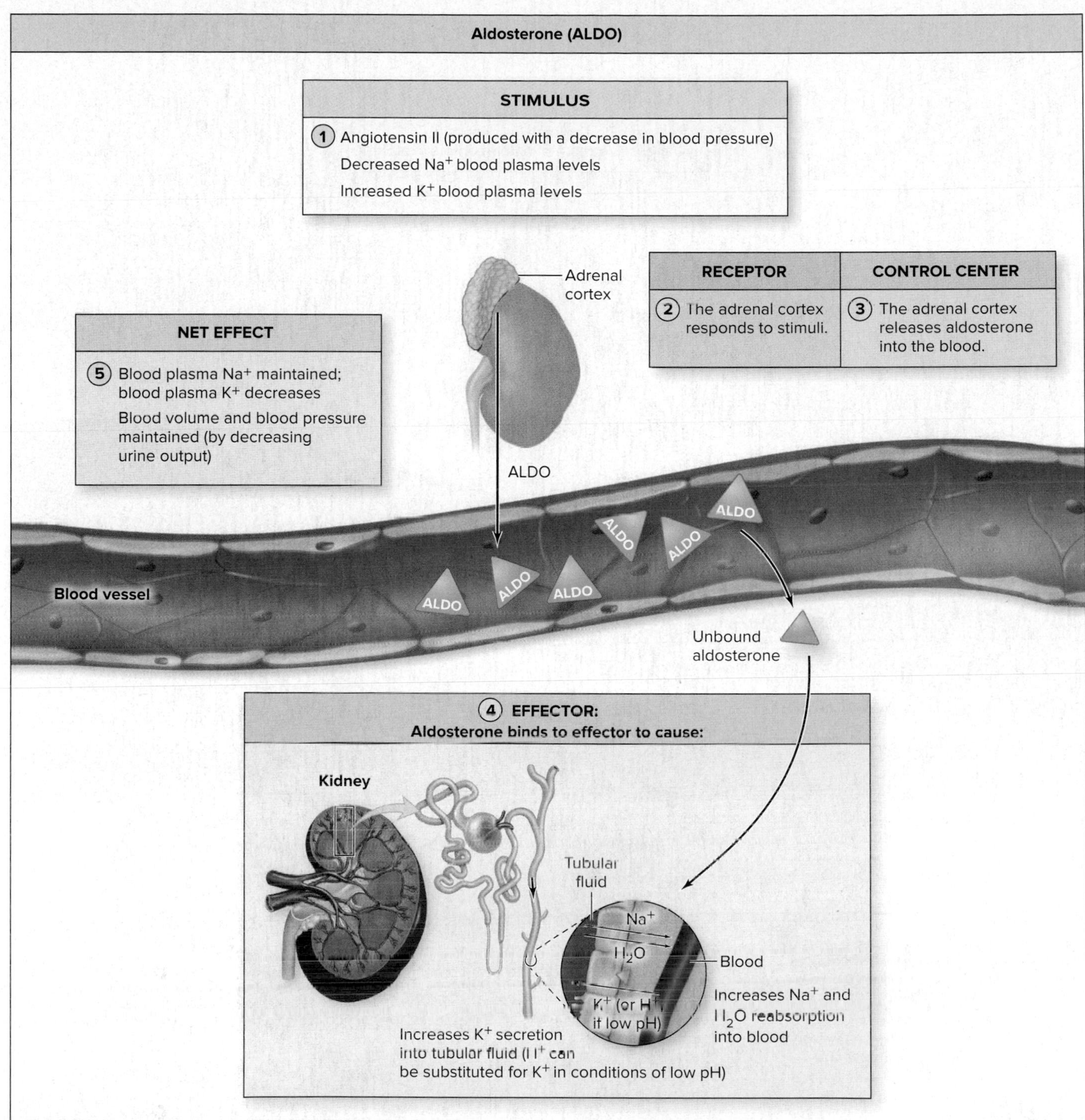

Figure 25.10 Actions and Effects of Aldosterone. Aldosterone is released from the adrenal cortex in response to various stimuli, including angiotensin II, decreased blood plasma levels of Na^+, or increased blood plasma levels of K^+. The effect of aldosterone is to decrease urine output to maintain blood volume and blood pressure, as well as to alter blood plasma concentration of Na^+ and K^+ to within normal homeostatic limits.

25.4d Atrial Natriuretic Peptide

LEARNING OBJECTIVES

26. Describe the stimulus for the release of atrial natriuretic peptide (ANP) and its three actions.
27. Explain the ways in which the effects of atrial natriuretic peptide differ from the effects of angiotensin II, ADH, and aldosterone.

Atrial natriuretic (nā′trē-yū-ret′ik; *natrium* = sodium, *uresis* = urination) **peptide (ANP)** is a peptide hormone that opposes the actions of the three hormones just discussed (**figure 25.11**). ANP is released into the blood from cells in the heart atria. The stimulus for its release is increased stretch of these chambers, which is an indication of both increased blood volume and increased blood pressure (see sections 17.11c, 20.6b, 24.5e, and 24.6d).

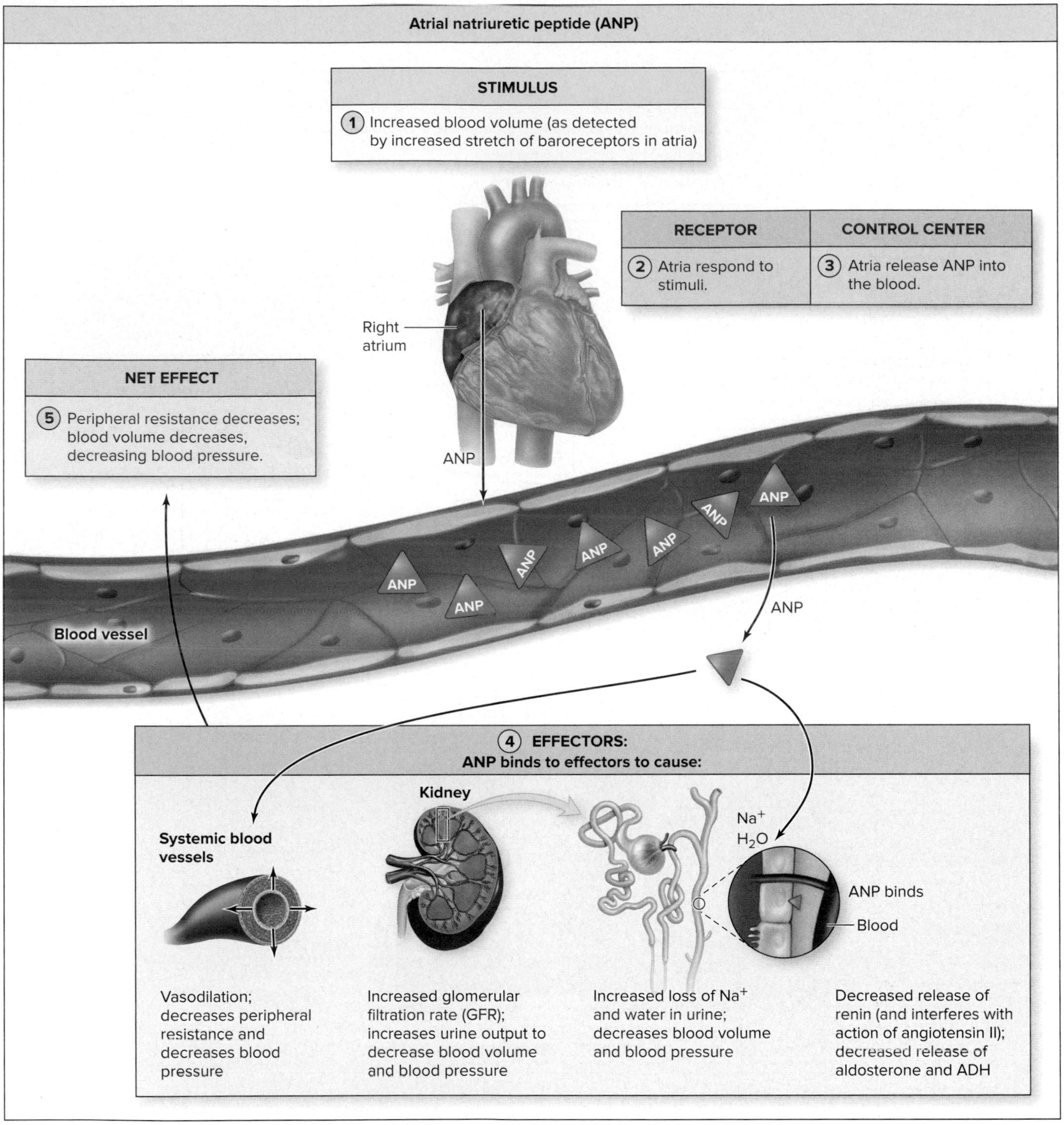

Figure 25.11 Actions and Effects of Atrial Natriuretic Peptide. Atrial natriuretic peptide (ANP) is released in response to increased blood volume and blood pressure. The effect of ANP is to decrease both peripheral resistance and blood volume with a resultant decrease in blood pressure to within normal homeostatic limits.

ANP decreases both blood volume and blood pressure by binding to these target organs and causing the following responses:

- **Blood vessels.** ANP dilates systemic blood vessels, resulting in decreased total peripheral resistance. Systemic blood pressure decreases as a result (see section 20.5b).
- **Kidneys.** ANP causes vasodilation of the afferent arterioles in the kidneys and relaxation of mesangial cells; both increase the glomerular filtration rate (see section 24.5e). Additionally, ANP inhibits Na^+ and water reabsorption by nephron tubules, resulting in additional loss of Na^+ and water (see section 24.6d). These changes increase urine output. Blood volume and systemic blood pressure decrease.

Atrial natriuretic peptide also inhibits the release of renin, the action of angiotensin II, and the release of ADH and aldosterone, thus preventing the actions of these hormones.

Note: The details for angiotensin II, ADH, aldosterone, and ANP are included in **table R.7**: "Regulating Fluid Balance, Blood Volume, and Blood Pressure," which directly follows chapter 17.

WHAT DID YOU LEARN?

14 How does ANP influence fluid output, blood volume, and systemic blood pressure?

INTEGRATE

CONCEPT CONNECTION

Blood pressure is one of the most fascinating and complex physiologic variables of the body. Blood pressure is similar to body temperature because it has a set point that is maintained through a variety of mechanisms. Three major variables influence blood pressure, as described in other chapters:

- Cardiac output (see section 19.9)
- Resistance within blood vessels (see section 20.5)
- Blood volume (see section 20.6b)

In this chapter, we have discussed and integrated how various hormones regulate both blood volume and resistance to alter blood pressure. Regulation by the nervous system of blood pressure, which controls cardiac output and resistance, is discussed in section 20.6.

INTEGRATE

CONCEPT CONNECTION

Recall from section 2.8b that normal three-dimensional folding of protein is pH dependent. If pH changes occur, the protein may permanently unfold or denature. Under acidic conditions, H^+ binds to the protein; under alkaline conditions, H^+ is released from the protein. In either case, the weak intermolecular bonds that hold the shape of a protein are broken, and it loses its three-dimensional shape and ability to function. Because most enzymes are proteins, one of the most serious consequences of an acid-base imbalance is decreased enzymatic activity in metabolic pathways due to denaturation. If severe enough, these disruptions can be fatal. **Buffers** are substances (or processes) that help prevent pH changes if either excess acid or excess base is added (see section 2.5c). Both the physiologic buffering systems and chemical buffers that maintain acid-base balance are discussed in section 25.5.

25.5 Acid-Base Balance

Acid-base balance is also called *pH balance*. It requires the regulation of hydrogen ion (H^+) concentration in body fluids in order to maintain a slightly alkaline arterial blood pH of 7.4, which is generally maintained between 7.35 and 7.45. The regulation of H^+ concentration is under the same constraints as other electrolytes; that is, the input must equal output. However, there are some unique considerations to acid-base regulation. Hydrogen ion concentration is altered by (1) the input of both acid and base, (2) the output of both acid and base (by the kidney), (3) the respiratory rate, and (4) chemical buffers. We describe here the categories of acids and the details of each of the factors that influence acid-base balance.

25.5a Categories of Acid

LEARNING OBJECTIVES

28. Distinguish between the two categories of acids in the body.

29. Name the two buffering systems that regulate each category.

You first encountered the concepts of acidity, alkalinity, and pH in section 2.5, and you've seen how maintaining a proper pH balance is critical to body functions. The pH of a solution was shown to be inversely related to H^+ concentration. Adding an acid increases the H^+ concentration, resulting in a lower pH; adding a base reduces the amount of H^+ in solution and results in a higher pH.

Two major categories of acid are present in the body: fixed acid and volatile acid. **Fixed acid** (also called *nonvolatile acid* or *metabolic acid*) is the wastes produced from metabolic processes (other than from carbon dioxide). Examples of fixed acids include lactic acid from glycolysis (see section 3.4b), phosphoric acid from nucleic acid metabolism, and keto-acids from metabolism of fat (see section 3.4h). Note that the term *fixed acids* was given to these types of acids because they are not produced from carbon dioxide, which is volatile and expired by the lungs. Rather, they are nonvolatile acids that are "fixed" in the body.

Volatile (vol'ă-til; to evaporate quickly) **acid** is *carbonic acid* produced when carbon dioxide combines with water, as shown:

$$CO_2 + H_2O \rightleftarrows H_2CO_3 \text{ (carbonic acid)}$$

This reaction occurs readily in the presence of the enzyme carbonic anhydrase; H_2CO_3 then dissociates into H^+ and HCO_3^-. The entire chemical reaction is as follows:

$$CO_2 + H_2O \rightleftarrows H_2CO_3 \rightleftarrows H^+ + HCO_3^-$$

The term *volatile acid* refers to the fact that carbonic acid is produced from a gas that is normally expired (or "evaporated"). Because CO_2 is readily converted to carbonic acid in the presence of carbonic anhydrase, CO_2 itself is often referred to as the volatile acid.

The level of each category of acid is altered by separate *physiologic* buffering systems:

- Fixed acid is regulated by the *kidney* (type A and type B intercalated cells) through the absorption and elimination of HCO_3^- and H^+.
- Volatile acid is eliminated by the *respiratory system*, and the amount expired is changed through the regulation of the respiratory rate and depth (see section 23.8).

We first consider the sources of fixed acids and their regulation by the kidney, and in section 25.5c we explore the regulation of volatile acids through the respiratory system.

WHAT DID YOU LEARN?

15 What is meant by acid-base balance?

16 How are fixed acids distinguished from volatile acids?

25.5b The Kidneys and Regulation of Fixed Acids

LEARNING OBJECTIVES

30. List the various sources of fixed acid.

31. Describe how the kidneys counteract increasing blood H^+.

32. Explain how the kidneys function in response to decreasing blood H^+.

Daily input of fixed acid and base varies, and the net effect can tend to either decrease or increase the body's pH. As a general rule, the daily processes that influence acid-base balance tend to increase blood H^+ concentration, thus decreasing blood pH. Maintaining acid-base balance requires that the normal excretion processes eliminate the excess H^+. We examine these typical conditions, looking first at input and then how pH is maintained.

Typical Conditions That Cause Blood H^+ Concentration to Increase

The input of acid into the blood, except that produced from CO_2, occurs from two major sources: nutrients absorbed by the gastrointestinal (GI) tract and body cells (**figure 25.12*a***). Typically, more acid is absorbed from the GI tract because most individuals, at

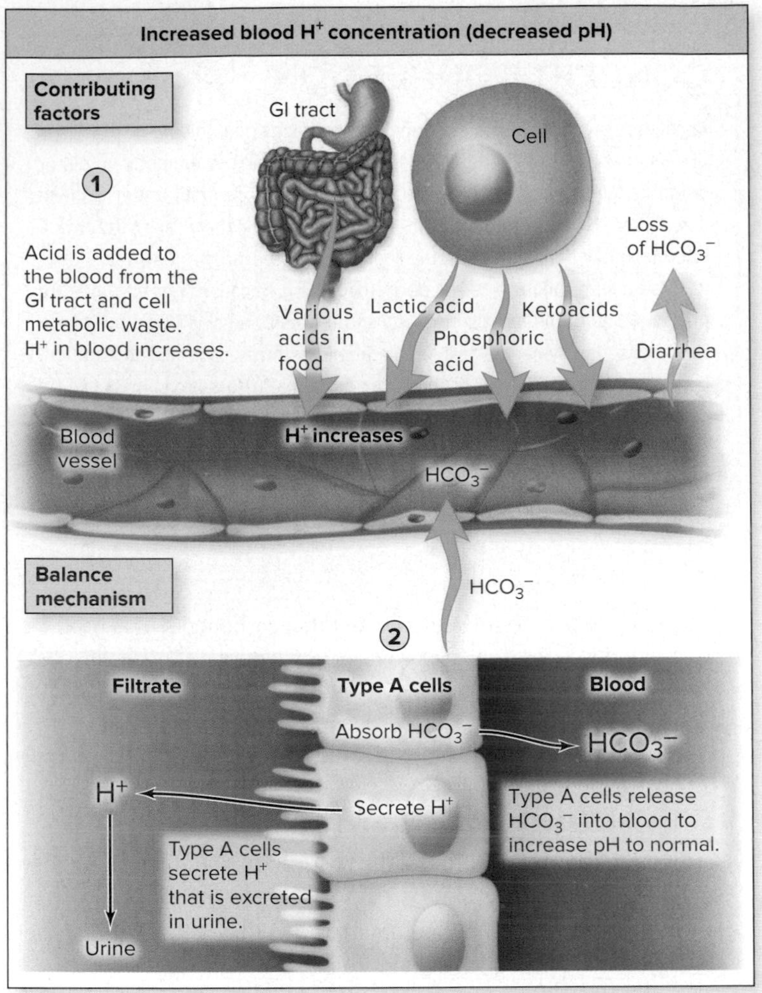

(a)

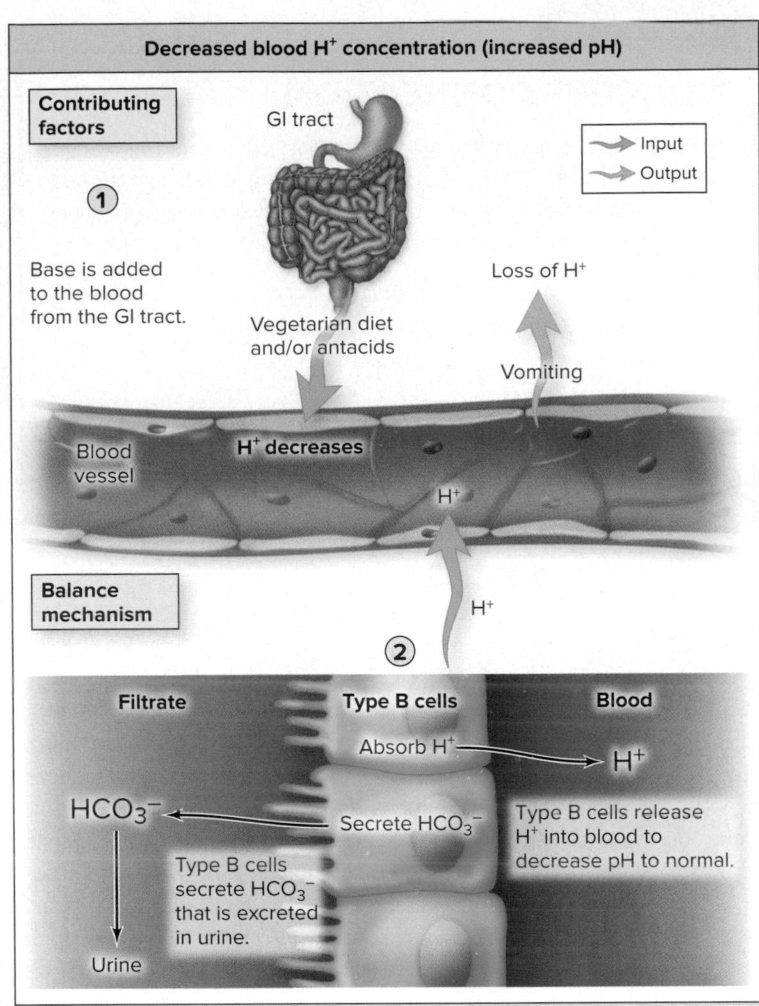

(b)

Figure 25.12 Altered Blood H^+ Concentration and Adjustments by the Kidneys. (*a*) Acid is normally added to the blood from the gastrointestinal (GI) tract as a result of diets rich in animal protein and wheat, as well as cellular metabolic wastes that include lactic acid from glycolysis, phosphoric acid from nucleic acid metabolism, and ketoacids from fat metabolism. Excessive loss of HCO_3^- as a consequence of diarrhea further increases blood H^+ concentration. The type A intercalated cells of the kidney adjust blood plasma H^+ concentration as they secrete H^+ (acid) and synthesize and absorb HCO_3^- (base) into the blood. (*b*) A vegetarian diet, high intake of antacids, or the loss of acid from vomiting decreases blood H^+ concentration. The type B intercalated cells of the kidney adjust blood H^+ concentration as they secrete HCO_3^- (base) and absorb H^+ (acid) into the blood. AP|R

least in the United States, consume a diet rich in animal protein and wheat. These various ingested items contribute H^+ to the blood. Cells also contribute acid as waste products from metabolic processes. These products include lactic acid, phosphoric acid, and ketoacids (as described in section 25.5a). Acidic conditions may also result from the excessive loss of HCO_3^- (a weak base) as a consequence of diarrhea. (Bicarbonate ion [HCO_3^-] is normally lost in the feces, but excess amounts can be lost when an individual has diarrhea; see Clinical View 25.6: "How Does Vomiting or Diarrhea Alter Blood H^+ Concentration?")

The additional H^+ must be eliminated (or the excessive loss of base replaced) to maintain acid-base balance. Recall from section 24.6d that intercalated *type A* cells of the distal convoluted tubule and collecting tubules respond to increased blood H^+ concentration, as follows (figure 25.12*a*):

- Secrete H^+ into the filtrate (which is then excreted in the urine)
- Synthesize new HCO_3^-, which is then absorbed into the blood

Thus, under conditions when blood H^+ concentration is increasing, kidneys help to maintain a normal blood pH by both eliminating excess H^+ (acid) and synthesizing and absorbing HCO_3^- (base). Both processes increase blood pH to normal.

Less Typical Conditions That Cause Blood H^+ Concentration to Decrease

Although the general daily input of acid tends to decrease pH, some conditions tend to increase pH as a result of increased base or loss of acid (figure 25.12*b*). Substances that increase blood pH also enter the blood from the GI tract. For example, this may occur if an individual regularly eats a vegetarian diet (that is rich in fruits and vegetables and low in animal protein) or regularly ingests antacids. These increase the absorption of alkaline substances into the blood, and blood H^+ concentration decreases. Alkaline conditions may also be caused by the abnormal loss of HCl as a result of vomiting (see Clinical View 25.6: "How Does Vomiting or Diarrhea Alter Blood H^+ Concentration?").

Intercalated *type B* cells of the distal convoluted tubule and collecting tubules respond to alkaline conditions as follows (figure 25.12*b*):

- Synthesize new HCO_3^- that is secreted into the filtrate (which is then excreted in the urine)
- Absorb H^+ into the blood

Thus, under conditions when blood H^+ concentration is decreasing, kidneys help to maintain a normal blood pH by both synthesizing and secreting HCO_3^- (base) and absorbing H^+ (acid). Both processes decrease blood pH to normal.

INTEGRATE

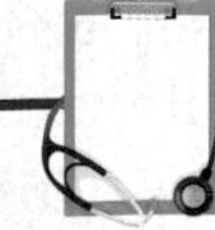

CLINICAL VIEW 25.6

How Does Vomiting or Diarrhea Alter Blood H^+ Concentration?

Although different regions of the gastrointestinal (GI) tract may contribute either acid or base to the blood, the overall effect of these secretions is generally neutral in pH. This balance changes, however, with vomiting and diarrhea.

Vomiting, or *emesis* (em'e'sis; *emeo* = to vomit), is the ejection of the contents of the stomach through the mouth that occurs by involuntary spasms of the stomach and forceful contraction of the abdominal muscles. The ejected material is referred to as *vomitus* and contains hydrochloric acid (HCl). The loss of HCl triggers cells of the epithelial lining of the stomach to replace the lost H^+. These cells are similar to the type A intercalated cells of the kidney; they excrete H^+ into the lumen of the stomach and simultaneously transport HCO_3^- into the blood. The addition of HCO_3^- to the blood increases blood pH. If severe enough, it can result in metabolic alkalosis.

Diarrhea is an excessive amount of watery stool. Severe cases of diarrhea are potentially life-threatening because of the excessive loss of HCO_3^- in the feces. The HCO_3^- loss triggers epithelial lining cells of the intestine to replace the lost HCO_3^-. These cells are similar to the type B intercalated cells of the kidney, and they both excrete HCO_3^- into the lumen of the intestine and simultaneously transport H^+ into the blood. The addition of H^+ into the blood decreases blood pH. Severe cases can result in metabolic acidosis.

Metabolic alkalosis and acidosis are described in section 25.6c.

The type A and type B intercalated cells of the kidneys therefore act as a physiologic buffering system to eliminate either excess acid or excess base from the body. The process is relatively slow, taking from *several hours to days;* however, this is the only way to eliminate fixed acid or base, and it provides the most powerful means of maintaining blood H^+ concentration and preventing pH changes.

WHAT DID YOU LEARN?

17 How do the kidneys regulate fixed acids to help maintain blood pH in response to increased blood H^+ concentration?

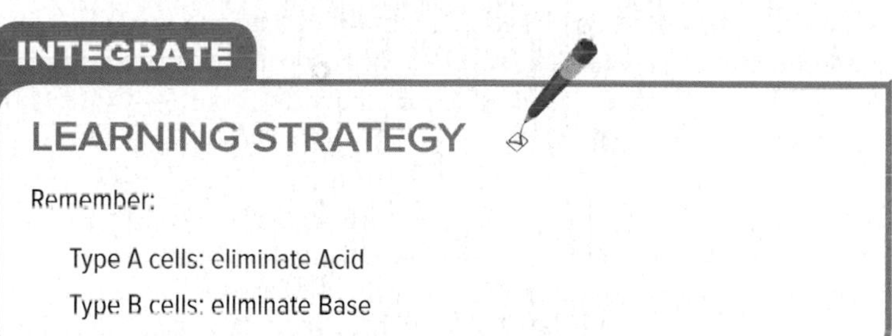

INTEGRATE

LEARNING STRATEGY

Remember:

Type A cells: eliminate Acid

Type B cells: eliminate Base

25.5c Respiration and Regulation of Volatile Acid

LEARNING OBJECTIVE

33. Explain the normal relationship between breathing rate and acid-base balance.

The respiratory system also serves as a physiologic buffering system to maintain acid-base balance, but it does so by regulating the level of the volatile carbonic acid (H_2CO_3), which is produced from CO_2 (see section 25.5a).

When the Body Is at Rest

Cells typically produce about 200 mL/min of CO_2 as a waste product during aerobic cellular respiration when the body is at rest (see section 23.7b). The resting breathing rate of 12-15 breaths per minute (eupnea) provides the means of eliminating this CO_2 (see section 23.5c). Thus, the resting breathing rate normally eliminates CO_2 from the lungs at rates that are equivalent to the production of CO_2 by body cells (**figure 25.13**).

During Exercise

During exercise, additional CO_2 is produced by body cells and enters the blood, causing a temporary increase in P_{CO_2} (see section 23.8b). Chemoreceptors are stimulated—central chemoreceptors in the brainstem (by an increase in H^+ produced from CO_2 that diffuses from the blood into the cerebrospinal fluid) and peripheral chemoreceptors in the aorta and carotid blood vessels (primarily by changes in blood H^+ and CO_2). Sensory input is relayed to the respiratory center, which responds by increasing breathing depth (referred to as **hyperpnea**), and additional CO_2 is expired. These adjustments occur relatively quickly, usually within several minutes. Because of these adjustments in breathing depth, blood levels of CO_2 during exercise (like at rest) are normally maintained within homeostatic limits. Consequently, whether the body is at rest or engaging in exercise, blood P_{CO_2} and H_2CO_3 do not usually have daily fluctuations and *do not normally* affect acid-base balance.

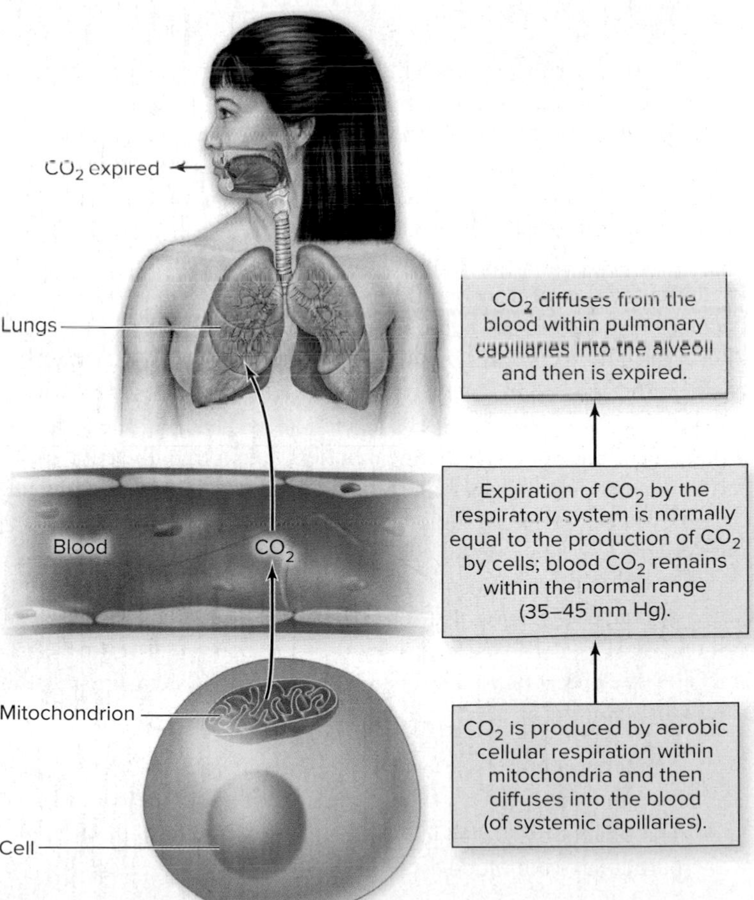

Figure 25.13 Respiration and Maintenance of Normal Blood CO_2 Levels. Normally, whether the body is at rest or engaging in exercise, the elimination of CO_2 from the lungs is equivalent to the CO_2 produced by body cells to maintain a normal blood CO_2 level.

 WHAT DID YOU LEARN?

 18 Explain why respiration does not normally influence acid-base balance.

INTEGRATE

CONCEPT CONNECTION

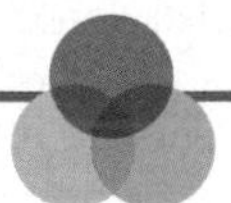

The pontine respiratory center of the brainstem receives sensory input from both central and peripheral chemoreceptors (see section 23.5c). These chemoreceptors are stimulated by increased H^+, and they initiate sensory input to the medullary respiratory center. Respiration rate and depth are increased to rid the body of more CO_2. Conversely, a decrease in H^+ reduces stimulation of chemoreceptors, causing a reduction in breathing rate and depth, resulting in more CO_2 being retained. Under normal conditions, adjustments in the respiratory rate maintain the partial pressure of carbon dioxide (P_{CO_2}) within a range of 35 to 45 mm Hg.

25.5d Chemical Buffers

 LEARNING OBJECTIVES

34. Describe the components of the protein buffering system and where and how they help prevent pH changes.
35. Explain the reactions of the phosphate buffering system within the ICF.
36. Discuss how the bicarbonate buffering system maintains acid-base balance in the ECF.

The physiologic buffering systems previously described are extremely effective in helping to maintain body fluid pH. Keep in mind that physiologic processes within the kidneys occur over several hours to several days in response to changes in fixed acids (and bases), and those of the respiratory system require a few minutes to respond to changing CO_2 levels to regulate volatile acid.

In contrast, **chemical buffering systems** act the most *rapidly* (within a fraction of a second) to help prevent pH changes that would occur in response to the addition of acid or base, as happens shortly after a meal, or as a result of abnormal loss of acid or base. However, chemical buffers are molecules that function to limit pH changes *temporarily* by absorbing excess H^+ when acid is added or releasing H^+ when base is added. Thus, chemical buffers help to rapidly, but temporarily, prevent pH changes—until the body can permanently eliminate the excess acid or base through the relatively slower physiologic buffering systems (i.e., the kidneys or the respiratory system).

The molecule or molecules of chemical buffering systems are composed of both a *weak base* that can bind excess H^+ and a *weak acid* that can release H^+.

The three most important chemical buffering systems include the following:

- The protein buffering system within *both* cells and the blood
- The phosphate (PO_4^{3-}) buffering system within cells
- The bicarbonate (HCO_3^-) buffering system within the ECF, particularly the blood

The three chemical buffering systems, although similar in the mechanism to buffer both acid and base, differ in their locations and in the molecules that compose them. Notice that the location of each chemical buffering system corresponds to where it is most prevalent in the fluid compartments, as described in section 25.1b and illustrated in figure 25.2.

INTEGRATE

LEARNING STRATEGY

You can think of chemical buffering systems as "sponges" for H^+. When pH decreases, the chemical buffering system soaks up H^+, and when pH increases, the chemical buffering system releases H^+. In this way, the H^+ concentration in solution may be kept within normal range, helping to prevent pH changes.

Protein Buffering System

The **protein buffering system** is a chemical buffering system composed of proteins within cells and blood plasma. It accounts for about three-quarters of the chemical buffering in body fluids. The **amine group (—NH_2)** of amino acids acts as a weak base to buffer acid, whereas the **carboxylic acid (—COOH)** of amino acids acts as a weak acid to buffer base (see section 2.8 for a review of amino acid and protein structures). The chemical change to each group is shown here:

$$\underset{\text{(weak base)}}{—NH_2} + \underset{\text{(strong acid)}}{H^+} \longrightarrow \underset{\text{(weak acid)}}{NH_3^+} \quad (1)$$

$$\underset{\text{(weak acid)}}{—COOH} + \underset{\text{(strong base)}}{OH^-} \longrightarrow \underset{\text{(weak base)}}{COO^-} + H_2O \quad (2)$$

With the addition of strong acid, shown in equation (1), the weak base (—NH_2) of the protein buffering system binds the H^+ that was added to the solution. This weak base becomes a weak acid (NH_3^+) as a result. The net effect is the elimination of a strong acid (H^+) and the production of a weak acid (NH_3^+). In comparison, the addition of strong base, as in equation (2), causes the weak acid (—COOH) of the protein buffering system to release H^+, and as it does so, it becomes a weak base (COO^-). The net effect is the removal of a strong base (OH^-) and the production of a weak base (COO^-).

Proteins are components within both cells (cellular proteins) and blood (both plasma proteins and hemoglobin within erythrocytes), and their buffering systems help minimize pH changes throughout the body. The most notable exception is in the cerebrospinal fluid (CSF), where there are no proteins (see section 13.2c).

Phosphate Buffering System

The **phosphate buffering system** is found within intracellular fluid (ICF). It is especially effective in buffering metabolic acid produced by cells because phosphate (PO_4^{3-}) is the most common anion within cells (see section 25.1b). The phosphate buffering system is also composed of both a weak base and a weak acid. Here **hydrogen phosphate (HPO_4^{2-})** is the weak base and **dihydrogen phosphate ($H_2PO_4^-$)** is the weak acid. The chemical change to HPO_4^{2-} and $H_2PO_4^-$ is shown here:

$$\underset{\text{(weak base)}}{HPO_4^{2-}} + \underset{\text{(strong acid)}}{H^+} \longrightarrow \underset{\text{(weak acid)}}{H_2PO_4^-} \quad (3)$$

$$\underset{\text{(weak acid)}}{H_2PO_4^-} + \underset{\text{(strong base)}}{OH^-} \longrightarrow \underset{\text{(weak base)}}{HPO_4^{2-}} + H_2O \quad (4)$$

The addition of acid, shown in equation (3), is buffered by the weak base (HPO_4^{2-}) that binds the H^+ to become a weak acid ($H_2PO_4^-$). In contrast, the addition of base, as in equation (4), causes the weak acid $H_2PO_4^-$ to release H^+. As it does so, it becomes the weak base HPO_4^{2-}, and water is formed. As with the protein buffering system, the net result is either a strong acid buffered to produce a weak acid or a strong base buffered to produce a weak base.

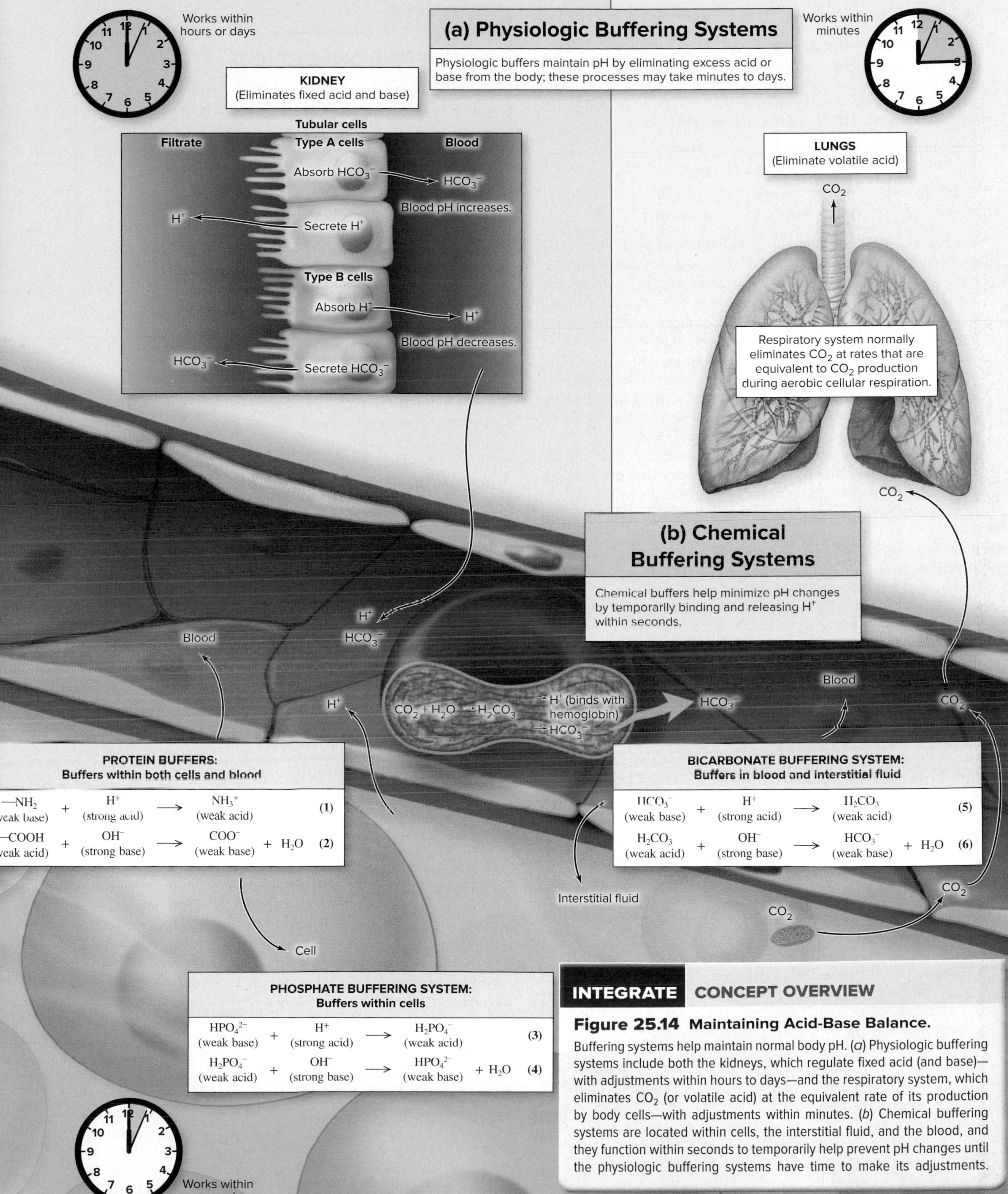

(a) Physiologic Buffering Systems
Physiologic buffers maintain pH by eliminating excess acid or base from the body; these processes may take minutes to days.
Works within hours or days
KIDNEY
(Eliminates fixed acid and base)
Tubular cells
Filtrate
Type A cells
Blood
Absorb HCO_3^-
HCO_3^-
Blood pH increases.
H^+
Secrete H^+
Type B cells
Absorb H^+
H^+
Blood pH decreases.
HCO_3^-
Secrete HCO_3^-
Works within minutes
LUNGS
(Eliminate volatile acid)
CO_2
Respiratory system normally eliminates CO_2 at rates that are equivalent to CO_2 production during aerobic cellular respiration.
CO_2
(b) Chemical Buffering Systems
Chemical buffers help minimize pH changes by temporarily binding and releasing H^+ within seconds.
H^+
HCO_3^-
Blood
H^+
$CO_2 + H_2O \rightarrow H_2CO_3$
H^+ (binds with hemoglobin)
HCO_3^-
HCO_3^-
Blood
CO_2
PROTEIN BUFFERS:
Buffers within both cells and blood
$—NH_2$ (weak base) + H^+ (strong acid) → NH_3^+ (weak acid) (1)
$—COOH$ (weak acid) + OH^- (strong base) → COO^- (weak base) + H_2O (2)
BICARBONATE BUFFERING SYSTEM:
Buffers in blood and interstitial fluid
HCO_3^- (weak base) + H^+ (strong acid) → H_2CO_3 (weak acid) (5)
H_2CO_3 (weak acid) + OH^- (strong base) → HCO_3^- (weak base) + H_2O (6)
Interstitial fluid
CO_2
CO_2
Cell
PHOSPHATE BUFFERING SYSTEM:
Buffers within cells
HPO_4^{2-} (weak base) + H^+ (strong acid) → $H_2PO_4^-$ (weak acid) (3)
$H_2PO_4^-$ (weak acid) + OH^- (strong base) → HPO_4^{2-} (weak base) + H_2O (4)
Works within seconds

INTEGRATE **CONCEPT OVERVIEW**

Figure 25.14 Maintaining Acid-Base Balance.
Buffering systems help maintain normal body pH. (*a*) Physiologic buffering systems include both the kidneys, which regulate fixed acid (and base)—with adjustments within hours to days—and the respiratory system, which eliminates CO_2 (or volatile acid) at the equivalent rate of its production by body cells—with adjustments within minutes. (*b*) Chemical buffering systems are located within cells, the interstitial fluid, and the blood, and they function within seconds to temporarily help prevent pH changes until the physiologic buffering systems have time to make its adjustments.

Bicarbonate Buffer System

The **bicarbonate buffering system** in the blood is the most important buffering system in the ECF. Bicarbonate ion and carbonic acid are the key components of this buffering system. **Bicarbonate (HCO_3^-)** serves as a weak base, whereas **carbonic acid (H_2CO_3)** acts as a weak acid. The chemical change to HCO_3^- and H_2CO_3 is shown here:

$$\underset{\text{(weak base)}}{HCO_3^-} + \underset{\text{(strong acid)}}{H^+} \longrightarrow \underset{\text{(weak acid)}}{H_2CO_3} \quad \textbf{(5)}$$

$$\underset{\text{(weak acid)}}{H_2CO_3} + \underset{\text{(strong base)}}{OH^-} \longrightarrow \underset{\text{(weak base)}}{HCO_3^-} + H_2O \quad \textbf{(6)}$$

The addition of acid is buffered in equation (5) by the weak base HCO_3^-. Bicarbonate binds the excess H^+ and becomes the weak acid $H_2CO_3^-$. In comparison, with the addition of base as in equation (6), the weak acid H_2CO_3 releases H^+. As it does so, it becomes a weak base (HCO_3^-), and water is formed. The net result again is that a strong acid is buffered to produce a weak acid, or a strong base is buffered to produce a weak base.

In summary, the phosphate buffering system is important in buffering against pH changes within cells, whereas the bicarbonate buffering system buffers against pH changes in the blood. Proteins buffer in both cells (intracellular proteins) and within the blood plasma (plasma proteins, hemoglobin) (figure 25.14*b*).

The moderation of pH changes by chemical buffers usually allows time for the kidneys to alter the excretion of H^+ or HCO_3^-, or for the respiratory system to adjust the expiration of CO_2. All the chemical buffering systems, however, are limited in the amount of acid or base that they can buffer. This is called the **buffering capacity.** If the buffering capacity is exceeded, then pH levels may decrease or increase beyond normal limits. These changes in pH are referred to as acid-base disturbances and are described in section 25.5.

The body's buffering systems—both physiologic buffering systems and chemical buffering systems—that are involved in maintaining acid-base balance are integrated in **figure 23.14**. Remember that the kidneys are responsible for regulating fixed acid (and base) and the respiratory system regulates volatile acid by eliminating CO_2 (to maintain carbonic acid [H_2CO_3] levels in the blood. The various chemical buffers, help prevent pH changes until kidneys or the respiratory systems can permanently eliminate the excess acid or base.

WHAT DID YOU LEARN?

19 What are the three chemical buffering systems, and where do they function?

20 What is the general amount of time required to maintain pH by (a) the chemical buffering systems, (b) the respiratory system, and (c) the kidneys?

INTEGRATE

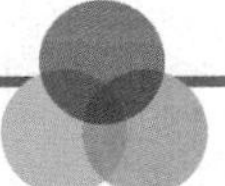

CONCEPT CONNECTION

The addition of either a strong acid or a strong base causes greater changes in pH than the addition of a weak acid or a weak base (see section 2.5b).

A **strong acid,** such as HCl, dissociates completely in solution, whereas a **weak acid,** such as H_2CO_3, only partially dissociates in solution. As a result, a strong acid releases more H^+ into solution, causing greater decreases in pH.

In a similar way, a **strong base,** such as NaOH, dissociates completely in solution, and a **weak base,** such as $NaHCO_3$, only partially dissociates. As a result, a strong base binds more H^+ from the solution, resulting in larger increases in pH.

By independently binding and releasing H^+, **buffers** help limit pH changes by converting strong acids to weak acids and strong bases to weak bases.

INTEGRATE

CONCEPT CONNECTION

Recall that following systemic gas exchange, CO_2 enters erythrocytes and binds to water, and then it dissociates from the water to produce H^+ and HCO_3^-. The HCO_3^- shifts out of the erythrocyte to be carried in the blood plasma, whereas the H^+ attaches to hemoglobin protein within the erythrocyte (see section 23.7b). This process then reverses when the blood reaches the lungs for alveolar gas exchange. During this process, hemoglobin acts as a protein buffer (first binding and then releasing H^+) to help maintain the concentration of free H^+ within both erythrocytes and blood plasma (see figure 23.28).

25.6 Disturbances to Acid-Base Balance

Severe vomiting, diarrhea, uncontrolled diabetes, emphysema, pneumonia, brainstem trauma, renal failure, congestive heart failure, climbing to high altitude, and diuretic abuse all have something in common: All of these conditions can cause an acid-base disturbance. **Acidosis,** or *acidemia,* is an arterial blood pH reading below 7.35, whereas **alkalosis,** or *alkalemia,* is an arterial blood pH reading above 7.45.

Before describing the different types of acid-base disturbances, we begin by describing some relevant terminology, including acid-base disturbance, compensation, and acid-base imbalance.

25.6a Overview of Acid-Base Imbalances

LEARNING OBJECTIVE

37. Explain acid-base disturbance, compensation, and acid-base imbalance.

An **acid-base disturbance** occurs when the buffering capacity of chemical buffering systems is exceeded. Consequently, there is a transient, or temporary, change in blood H^+ concentration, resulting in a change in blood pH beyond the normal range of 7.35 to 7.45. In response to the transient acid-base disturbance, the physiologic buffering system of the kidneys, the respiratory system, or both, attempts to offset the disturbance. The response of physiologic buffering systems to acid-base disturbances that results in the return of blood pH to *normal* is called **compensation.**

If these physiologic buffering systems are *not* effective in returning the pH to normal, then pH disturbance is referred to as **uncompensated** (or *partially compensated*). When an uncompensated, temporary pH disturbance results in a persistent pH change, it is referred to as an **acid-base imbalance.** An acid-base imbalance for any extended period of time is life-threatening. Without intervention, an acid-base imbalance continues to worsen, and if the pH reaches values below 7.0 or above 7.7, it is fatal within a few hours.

Four major types of acid-base disturbances are distinguished based on two criteria: whether the **primary disturbance** (which is the cause of the acid-base disturbance) is respiratory or metabolic in nature and whether the pH change is acidic or alkaline. The four categories are *respiratory acidosis, respiratory alkalosis, metabolic acidosis,* and *metabolic alkalosis.* We examine each of these in terms of the cause of the primary disturbance, and then we describe the compensation made by the physiologic buffering system (either the kidney or the respiratory system) that helps return the pH to normal to prevent an acid-base imbalance.

It is helpful to know the normal clinical values for arterial blood pH, blood P_{CO_2} (in the arteries), and blood HCO_3^-. These values represent the ranges for 95% of the normal, healthy individuals within the population.

Variable	Defined	Normal Range
pH	Relative amount of H^+ within arterial blood	7.4 (range 7.35–7.45)
$PaCO_2$	Partial pressure of carbon dioxide dissolved within arterial blood	35–45 mm Hg
$[HCO_3]$	Concentration of bicarbonate ion within arterial blood	22–26 mEq/L

WHAT DID YOU LEARN?

21 How does a compensated acid-base disturbance differ from an uncompensated acid-base imbalance?

25.6b Respiratory-Induced Acid-Base Disturbances

LEARNING OBJECTIVES

38. Define respiratory acidosis, identify some of the causes of this type of acid-base disturbance, and explain how it occurs.

39. Explain why infants are more susceptible to respiratory acidosis.

40. Define respiratory alkalosis, identify some of the causes of this type of acid-base disturbance, and explain how it occurs.

Respiratory acid-base disturbances occur when the expiration of CO_2 by the respiratory system is *not* equal to the production of CO_2 by cells as a consequence of changes in respiratory function. These acid-base disturbances result in either respiratory acidosis or respiratory alkalosis (**figure 25.15**).

Respiratory Acidosis

Respiratory acidosis is the most common acid-base disturbance and is clinically recognized as occurring when the P_{CO_2} in the arterial blood becomes elevated above 45 mm Hg.

Respiratory acidosis is a consequence of impaired respiratory function and can have many different causes, including

- **Hypoventilation,** which is breathing that is too slow or too shallow as a result of (1) disorders of the nerves or muscles involved with breathing (see section 23.5b) or (2) injury to the respiratory center (see section 23.5c) that is perhaps caused by trauma, drug overdose, or poliovirus infection
- **Decreased airflow** (see section 23.5d) caused by (1) an object lodged in the respiratory tract (choking), (2) severe bronchitis (see Clinical View 23.5: "Bronchitis"), or (3) severe asthma (see Clinical View 23.6: "Asthma")
- **Impaired alveolar gas exchange** (see section 23.6b) due to (1) reduced respiratory membrane surface area (e.g., emphysema; see Clinical View 23.16: "Emphysema") or (2) thickened width of the respiratory membrane (e.g., pneumonia; see Clinical View 23.17: "Respiratory Diseases and Efficiency of Alveolar Gas Exchange")

Let us consider how impaired respiratory function leads to respiratory acidosis, which is illustrated in figure 25.15*a*. An individual with impaired respiratory function expires less CO_2. Consequently, more CO_2 remains within the blood. The additional CO_2 drives the carbonic anhydrase reaction ($CO_2 + H_2O \rightleftarrows H_2CO_3 \rightleftarrows HCO_3^- + H^+$) to the *right,* with the most significant change being increased formation of H^+. If the buffering capacity of chemical buffers is exceeded (reached their limit to absorb H^+; see section 25.5d), and blood pH decreases below 7.35, acidosis results. Because the cause (primary disturbance) is impaired respiratory function, this acid-base disturbance is specifically called *respiratory acidosis.*

Infants are more susceptible to respiratory acidosis because their smaller lungs and lower residual volume (see section 23.5f) do not eliminate CO_2 as effectively those of as adults. CO_2 accumulates in the blood, with a subsequent increase in carbonic acid (H_2CO_3).

Respiratory Alkalosis

Respiratory alkalosis is clinically recognized as occurring when the P_{CO_2} decreases to levels below 35 mm Hg. Respiratory alkalosis occurs due to *hyperventilation,* which is when the breathing rate or depth is increased to eliminate greater amounts of CO_2 than are being produced by the body's cells (see section 23.8a). Causes include the following:

- **Severe anxiety**
- **Hypoxia,** which is insufficient oxygen delivery to body cells (e.g., as might occur when climbing to a high altitude where there is a decrease in the partial pressure of oxygen [P_{O_2}]; during congestive heart failure; as a result of severe anemia)
- **Aspirin overdose** (a condition that stimulates the respiratory center)

Let us consider how hyperventilation leads to respiratory alkalosis, which is illustrated in figure 25.15*b*. During hyperventilation, an individual expires more CO_2 than is being produced by body cells. Consequently, less CO_2 remains within the blood. The decrease in CO_2 drives the carbonic anhydrase reaction ($CO_2 + H_2O \rightleftarrows H_2CO_3 \rightleftarrows HCO_3^- + H^+$) to the *left,* with the most significant change being decreased formation of H^+. If the buffering capacity of chemical buffers is exceeded (reached their limit to release H^+; see section 25.5d) and blood pH increases above 7.45, alkalosis results. Because the cause (primary disturbance) is respiratory, this acid-base disturbance is specifically called *respiratory alkalosis.*

WHAT DID YOU LEARN?

22 Describe the change (increases, decreases, or stays the same) for each of the following variables if an individual hyperventilates: (a) blood CO_2, (b) blood H^+ concentration, and (c) blood pH.

25.6c Metabolic-Induced Acid-Base Disturbances

LEARNING OBJECTIVES

41. Explain how metabolic acid-base disturbances differ from respiratory acid-base disturbances.

42. Define both metabolic acidosis and metabolic alkalosis, identify some of the causes of each type of acid-base disturbance, and explain how each occurs.

Metabolic disturbances result from any acid-base disturbance that does *not* involve the respiratory system. These acid-base disturbances include metabolic acidosis and metabolic alkalosis.

Figure 25.15 Abnormal Respiration Rate and Blood pH. The respiratory system normally eliminates CO_2 at the same rate that cells produce it; thus, blood P_{CO_2} of 35–45 mm Hg is maintained under normal conditions. (*a*) Impaired respiratory function results in less CO_2 expired and additional CO_2 remaining within the blood. Consequently, blood P_{CO_2} increases, driving the carbonic anhydrase chemical reaction to the right. Subsequently, blood H^+ concentration increases, decreasing blood pH. (*b*) An abnormal increase in respiratory rate or depth results in more CO_2 expired and less CO_2 remaining within the blood. Consequently, blood P_{CO_2} decreases, driving the carbonic anhydrase chemical reaction to the left. Subsequently, blood H^+ concentration decreases, increasing blood pH.

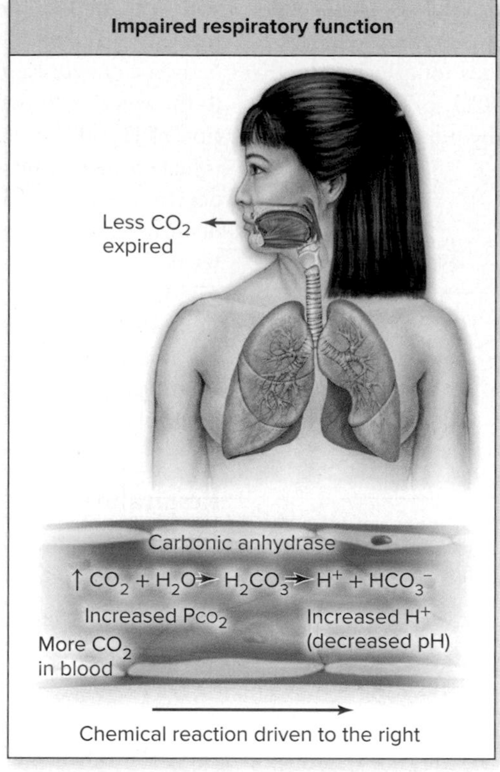

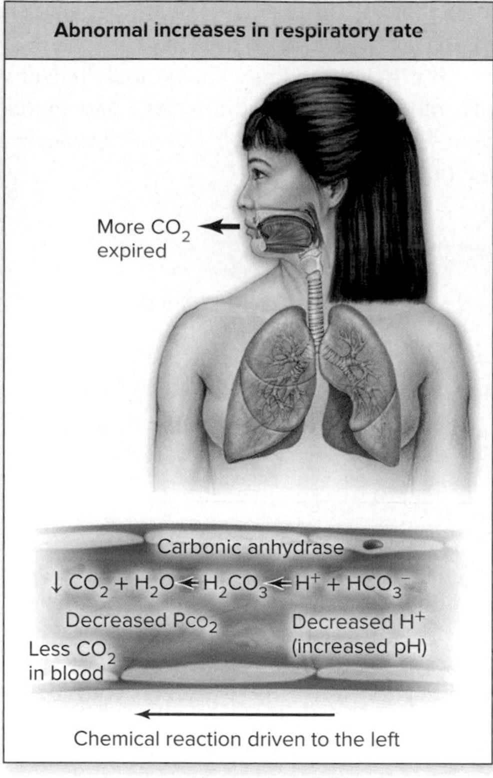

Metabolic Acidosis

Metabolic acidosis is more common than metabolic alkalosis and is clinically recognized when arterial blood levels of HCO_3^- (a weak base) fall below 22 mEq/L. This decrease (figure 25.12*a*)

- Generally occurs when there is an accumulation of fixed acid from (1) *increased production* of metabolic acids (e.g., ketoacidosis from diabetes mellitus, lactic acidosis, or acetic acidosis from excessive intake of alcohol; see Clinical View 25.7: "Lactic Acidosis and Ketoacidosis") or (2) decreased elimination of acid due to renal dysfunction
- May occur from an excessive loss of HCO_3^- (e.g., severe diarrhea because HCO_3^- is normally lost in the feces)

Let us consider how metabolic acidosis (HCO_3^- levels below 22 mEq/L) occurs with an accumulation of fixed acid. The abnormal increase in blood H^+ levels (from ingestion of fixed acid from the GI tract, or formation from cellular metabolism, or the kidneys failing to secrete H^+) binds with HCO_3^- and lowers blood HCO_3^- levels. If the buffering capacity of chemical buffers is exceeded (reached their limit to absorb H^+; see section 25.5d), blood pH decreases below 7.35, resulting in acidosis. Because the cause (primary disturbance) is *not* due to a respiratory function, this is acid-base disturbance is specifically called *metabolic acidosis.*

Metabolic Alkalosis

Metabolic alkalosis is clinically recognized when arterial blood levels of HCO_3^- exceed 26 mEq/L.

This increase (see figure 25.12*b*):

- Generally occurs due to loss of H^+ from vomiting or prolonged nasogastric suction (due to loss of stomach secretions)
- May occur from increased loss of acids by the kidneys with overuse of diuretics (medications that increase urine output)
- Occasionally results from increased base input from consuming large amounts of antacids

Let us consider how metabolic alkalosis (HCO_3^- levels above 26 mEq/L) occurs with a loss of fixed acid. The abnormal decrease in H^+ (from the stomach or by the kidney in response to certain diuretics) increases blood HCO_3^- levels. If the buffering capacity of chemical buffers is exceeded (reached their limit to release H^+; see section 25.5d), blood pH increases above 7.45, resulting in alkalosis. Because the cause (primary disturbance) is not due to a respiratory function, this acid-base disturbance is specifically called *metabolic alkalosis.*

Acid-Base Imbalances and Shifts in Potassium

One of the primary reasons that acid-base imbalances can be lethal is that they may cause a shift in potassium ion (as described in section 25.3b and illustrated in figure 25.7*b*). Recall that an increase in blood H^+ results in H^+ shifting into cells and K^+ shifting out of cells, which can result in hyperkalemia. In contrast, a decrease in blood H^+ results in H^+ shifting out of cells and K^+ shifting into cells, which can result in hypokalemia.

WHAT DID YOU LEARN?

23 What is the primary cause of metabolic alkalosis?

25.6d Compensation

LEARNING OBJECTIVE

43. Describe renal and respiratory compensation.

It is possible for both the kidneys and the respiratory system to make adjustments to compensate for changes in H^+ and HCO_3^- in an attempt to return blood pH to within its normal range. However, it is important to remember that both the initial primary disturbance *and* the system compensation are reflected in abnormal (low or high) concentrations of CO_2 and HCO_3^- in lab results that measure these variables. These changes in blood levels may be monitored with arterial blood gas (ABG) lab results (see Clinical View 25.8: "Arterial Blood Gas [ABG] and Diagnosing Different Types of Acid-Base Disturbances"). In general, respiratory disturbances are compensated for by the kidneys, whereas metabolic acid-base disturbances are compensated for by either the kidney or the respiratory system.

Renal Compensation

Renal compensation (physiologic adjustments of the kidney to changes in pH) occurs in response to elevated blood H^+ concentration due to a cause other than a renal dysfunction. We know that the normal physiologic activities of type A intercalated cells result in excretion of H^+ and reabsorption of HCO_3^- (see figure 25.12*a*). However, during renal compensation, this occurs to a greater degree than normal. Consequently, higher than normal levels of H^+ are excreted, and higher than normal amounts of HCO_3^- are synthesized and reabsorbed into the blood. Renal compensation results in elevated values for blood HCO_3^-. As expected, urine pH is lower than normal. (See the entry "Respiratory acidosis with renal compensation" in Clinical View 25.8: "Arterial Blood Gas [ABG] and Diagnosing Different Types of Acid-Base Disturbances.")

In comparison, renal compensation in response to a decrease in blood H^+ concentration involves the normal response of type B intercalated cells to excrete HCO_3^- and reabsorb H^+ (see figure 25.12*b*). During renal compensation, this activity occurs to a greater degree than normal; higher than normal levels of H^+ are reabsorbed into the blood, and higher amounts of HCO_3^- are excreted. Renal compensation results in lower than normal values for blood HCO_3^-. Urine pH in this case is higher than normal. (See the entry "Respiratory alkalosis with renal compensation" in Clinical View 25.8.)

Renal compensation is generally effective in neutralizing acid-base respiratory disturbances and those that are not due to renal dysfunction. Complete compensation occurs if the pH has been returned to *normal.* However, impaired renal function places an individual at higher risk for an acid-base imbalance. Renal function decreases as we age, and because of this, the elderly are at greater risk for pH imbalances.

Respiratory Compensation

Respiratory compensation occurs in response to metabolic acidosis that is caused by abnormally low levels of HCO_3^- (base). Breathing rate increases, and higher than normal amounts of CO_2 are expired and lower than normal levels of CO_2 (volatile acid) remain in the blood. Arterial blood gas readings will show lower than normal readings for both HCO_3^- and CO_2. (See the entry "Metabolic acidosis with respiratory compensation" in Clinical View 25.8.)

Respiratory compensation also occurs in response to metabolic alkalosis that is caused by increased levels of HCO_3^- (base). Breathing rate decreases, and lower than normal amounts of CO_2 are expired and higher than normal levels of CO_2 (volatile acid) remain in the blood. Arterial blood gas readings will show higher than normal readings for both HCO_3^- and CO_2. (See the entry "Metabolic alkalosis with respiratory compensation" in Clinical View 25.8.)

Generally, respiratory compensation is less effective in addressing metabolic acid-base disturbances than renal compensation. The ability to decrease respiratory rate in response to metabolic alkalosis is limited by the development of hypoxemia. As the respiratory rate decreases, blood Po_2 levels also decrease. When blood Po_2 levels decrease below critical values, the respiratory rate is stimulated to increase. This may prevent complete compensation for metabolic alkalosis.

Acid-base imbalances are summarized in **table 25.2**.

INTEGRATE

CLINICAL VIEW 25.7

Lactic Acidosis and Ketoacidosis

When insufficient oxygen is available (called *ischemia*), cells are forced to shift from aerobic cellular respiration to glycolysis (with increased production of lactic acid). This occurs during prolonged intense exercise or when the cardiovascular system is impaired, such as in congestive heart failure. Blood pH drops as a result. If substantial amounts of lactic acid are produced, acid-base balance is disturbed, resulting in the condition called **lactic acidosis.**

During conditions of insufficient cellular glucose uptake, cells are forced to shift from glucose metabolism to fat metabolism for their ATP production (ketoacids are a by-product). Individuals who have diabetes mellitus type 1 experience this condition because insufficient insulin is produced (see Clinical View 17.8: "Conditions Resulting in Abnormal Blood Glucose Levels"). Glucose is unable to enter most cells without insulin. This forces cells to use fat for ATP production. The greater the deficiency of insulin, the greater the degree of fat metabolism, and the larger the amount of ketoacids formed. If substantial amounts of ketoacids are produced, a disturbance in the acid-base balance may occur that is called **ketoacidosis.**

Ketoacidosis may also occur if blood glucose levels are too low, and the body cells must metabolize fat for ATP production. Ketoacidosis is one of the risks of very-low-carbohydrate diets that are not medically monitored.

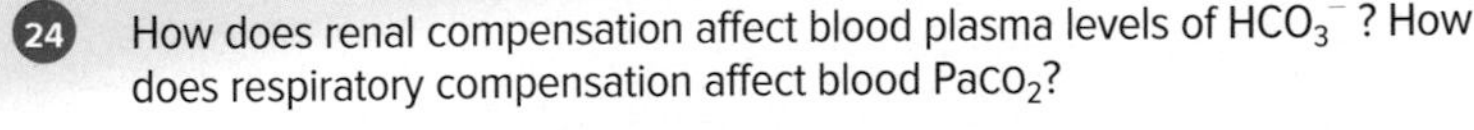

WHAT DID YOU LEARN?

24 How does renal compensation affect blood plasma levels of HCO_3^-? How does respiratory compensation affect blood $Paco_2$?

25 Using the normal arterial blood gas values (pH = 7.35–7.45; $Paco_2$ = 35–45 mm Hg; HCO_3^- = 22–26 mEq/L), identify both the primary disturbance and the degree of compensation for these three arterial blood gas readings:

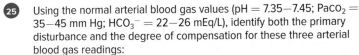

(a) pH = 7.20	(b) pH = 7.20	(c) pH = 7.36
$Paco_2$ = 35 mm Hg	$Paco_2$ = 25 mm Hg	$Paco_2$ = 20 mm Hg
$[HCO_3^-]$ = 17 mEq/L	$[HCO_3^-]$ = 17 mEq/L	$[HCO_3^-]$ = 17 mEq/L

26 Identify both the primary disturbance and degree of compensation for these three arterial blood gas readings:

(a) pH = 7.20	(b) pH = 7.20	(c) pH = 7.36
$Paco_2$ = 50 mm Hg	$Paco_2$ = 50 mm Hg	$Paco_2$ = 50 mm Hg
$[HCO_3^-]$ = 24 mEq/L	$[HCO_3^-]$ = 29 mEq/L	$[HCO_3^-]$ = 30 mEq/L

INTEGRATE

CLINICAL VIEW 25.8

Arterial Blood Gas (ABG) and Diagnosing Different Types of Acid-Base Disturbances

Measurements of **arterial blood gas (ABG)** are used to diagnose and monitor an acid-base disturbance and its compensation. These measurements include pH, $Paco_2$ (arterial Pco_2), and HCO_3^-. Changes in these values can help formulate a diagnosis, differentiate among the four types of acid-base disturbances, and determine whether compensation is occurring—and if so, whether it is complete or partial.

Normal Conditions

First, consider the ABG results if an individual is in acid-base balance (no disturbance or compensation):

- Blood pH is 7.4, within the normal range of 7.35 to 7.45.
- $Paco_2$ is 35 to 45 mm Hg, which reflects a normal respiratory rate and elimination of CO_2. By maintaining $Paco_2$, the H_2CO_3 concentration of approximately 3 μmol/dL is also regulated.
- Blood HCO_3^- concentration is 22–26 mEq/L, reflecting the effectiveness of the kidney in retaining or excreting HCO_3^-.

When an individual is in acid-base balance, a normal ratio between H_2CO_3 concentration and HCO_3^- concentration of 1:20 exists. A "balance" of acid and base with a normal ratio of 1:20 can be represented with a scale.

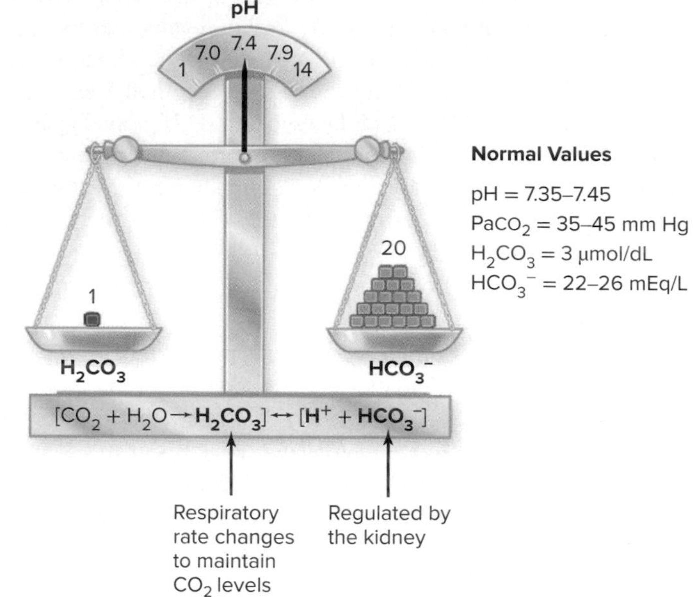

Acid-Base Disturbances and Compensation

Now consider what occurs to both ABG values and the ratio of H_2CO_3 to HCO_3^- in each of the four types of acid-base disturbances and the resulting compensation. When compensation is complete, pH returns to normal, but altered blood levels of HCO_3^- and CO_2 are still observed in the ABG values. These changes and their compensation are visually represented with scales. Each primary disturbance and the accompanying compensation are presented in a separate box. Each box contains two figures; the left figure represents the primary disturbance and the right figure represents compensation.

Notice that if compensation occurs, both respiratory acidosis and metabolic alkalosis have similar outcomes: an increase in H_2CO_3 and an increase in HCO_3^-. In a like manner, both respiratory alkalosis and metabolic acidosis have similar results if compensation occurs: a decrease in H_2CO_3 and a decrease in HCO_3^-. To differentiate between the pH imbalances that have similar lab results, you would need further information.

If compensation is not complete, you will be able to differentiate between the acid-base imbalances based on pH. For example, to distinguish respiratory acidosis from metabolic alkalosis, a low pH would occur if the primary disturbance were caused by respiratory acidosis, and a high pH would occur from metabolic alkalosis. It would also be possible to differentiate between these based on the diagnosis of the individual's condition. For example, you would most likely conclude that the pH imbalance was respiratory acidosis if the individual had emphysema, or that it was metabolic alkalosis if the person was taking diuretics to treat high blood pressure. Similar analysis could be used to differentiate respiratory alkalosis and metabolic acidosis.

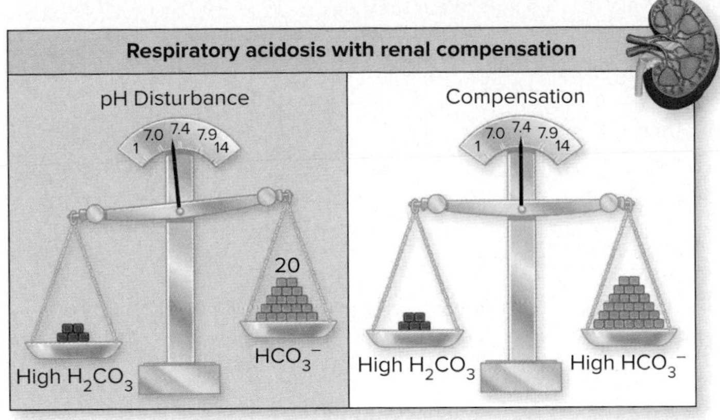

Respiratory acidosis with renal compensation.

In respiratory acidosis, respiratory function is impaired and blood CO_2 increases (see figure 24.15*a*). This drives the chemical equation to the right ($CO_2 + H_2O \longrightarrow H_2CO_3$), and blood H_2CO_3 increases. This increase in H_2CO_3 is represented with a "tipped" scale. The kidneys compensate with increased secretion of H^+ and increased reabsorption of HCO_3^-. The increase in blood HCO_3^- from renal compensation reestablishes acid-base balance. However, although blood pH is returned to normal and the ratio of H_2CO_3 to HCO_3^- is returned to 1:20, notice that blood levels of H_2CO_3 and HCO_3^- are both greater than normal.

Respiratory alkalosis with renal compensation
pH Disturbance
Compensation
1
7.0
7.4
7.9
14
20
Low H_2CO_3
HCO_3^-
Low H_2CO_3
Low HCO_3^-

Respiratory alkalosis with renal compensation.

In respiratory alkalosis, respiratory rate increases (hyperventilation) and blood CO_2 decreases (figure 25.15*b*). This drives the chemical equation to the left ($CO_2 + H_2O \longleftarrow H_2CO_3$), and H_2CO_3 decreases. Again, this decrease in H_2CO_3 is shown with a "tipped" scale. The kidneys compensate with increased secretion of HCO_3^- and increased reabsorption of H^+. The decrease in blood HCO_3^- from renal compensation reestablishes acid-base balance. However, although blood pH is returned to normal and the ratio of H_2CO_3 to HCO_3^- is returned to 1:20, both H_2CO_3 and HCO_3^- are lower than normal.

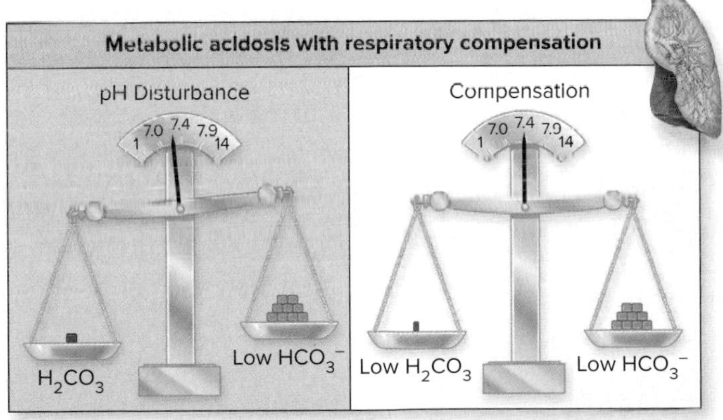

Metabolic acidosis with respiratory compensation.

In metabolic acidosis, blood H^+ increases, driving the chemical equation to the left ($CO_2 + H_2O \longleftarrow H_2CO_3 \longleftarrow H^+ + HCO_3^-$), and HCO_3^- decreases. The respiratory system compensates with an increased respiratory rate, and blood CO_2 (and H_2CO_3) decrease. The decrease in blood H_2CO_3 from respiratory compensation reestablishes acid-base balance. However, although blood pH is returned to normal and the ratio of H_2CO_3 to HCO_3^- is returned to 1:20, again notice that both H_2CO_3 and HCO_3^- are lower than normal.

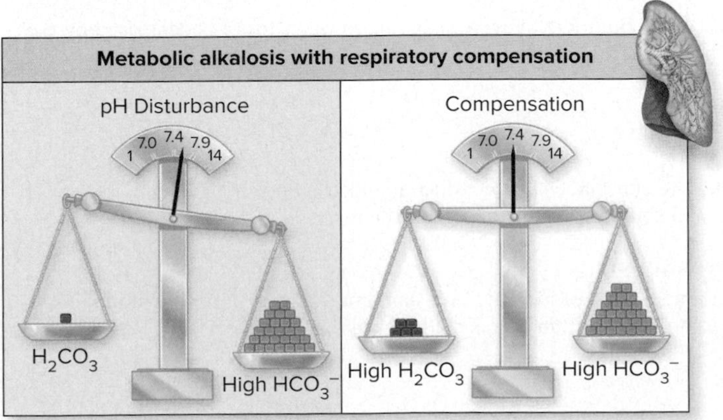

Metabolic alkalosis with respiratory compensation.

In metabolic alkalosis, blood H^+ has decreased, driving the chemical equation to the right ($CO_2 + H_2O \longrightarrow H_2CO_3 \longrightarrow H^+ + HCO_3^-$), and HCO_3^- increases. The respiratory system compensates with a decreased respiratory rate, and blood CO_2 (and H_2CO_3) increases. The increase in blood H_2CO_3 from respiratory compensation reestablishes acid-base balance. However, although blood pH is returned to normal and the ratio of H_2CO_3 to HCO_3^- is returned to 1:20, again notice that both H_2CO_3 and HCO_3^- are greater than normal.

Table 25.2 Acid-Base Imbalances

Variable	Normal Values	Respiratory Acidosis	Respiratory Alkalosis	Metabolic Acidosis	Metabolic Alkalosis
PRIMARY DISTURBANCE					
pH	7.35–7.45	< 7.35	> 7.45	< 7.35	> 7.45
$Paco_2$	35–45 mm Hg	**> 45 mm Hg (high)**	**< 35 mm Hg (low)**	Normal	Normal
HCO_3^-	22–26 mEq/L	Normal	Normal	**< 22 mEq/L (low)**	**> 26 mEq/L (high)**
COMPENSATION[1]					
pH	7.35–7.45	Normal	Normal	Normal	Normal
$Paco_2$	35–45 mm Hg	> 45 mm Hg (high)	< 35 mm Hg (low)	**< 35 mm Hg (low)**	**> 45 mm Hg (high)**
HCO_3^-	22–26 mEq/L	**> 26 mEq/L (high)**	**< 22 mEq/L (low)**	< 22 mEq/L (low)	> 26 mEq/L (high)

1. If completely compensated, pH returns to normal. The ratio of H_2CO_3 to HCO_3^- is once again 1:20.

CHAPTER SUMMARY

- Normal homeostatic levels of fluid and electrolytes are usually maintained despite sometimes erratic fluid intake.

25.1 Body Fluids

- The percentage of fluid that composes our body varies between individuals.

25.1a Percentage of Body Fluid

- Your total body water content is dependent upon your age and the relative amounts of adipose connective tissue to skeletal muscle tissue.

25.1b Fluid Compartments

- The two major fluid compartments are intracellular fluid (fluid inside a cell) and extracellular fluid (fluid outside a cell).
- Extracellular fluid is composed of interstitial fluid (fluid around cells) and fluid within the blood plasma.
- Fluid movement between compartments occurs continuously in response to change in the relative fluid concentrations between the compartments.

25.2 Fluid Balance

- Maintaining fluid balance involves the interaction of many systems of the body.

25.2a Fluid Intake and Fluid Output

- Fluid intake includes ingested (preformed) water in our food and drink and metabolic water that is produced within the body.
- Fluid output includes losses when we expire air and losses, through our skin, feces, and urine.
- Fluid output is categorized as sensible or insensible, depending upon if it is measurable, and as obligatory or facultative, depending upon if the loss is regulated based on state of hydration.

25.2b Fluid Imbalance

- A fluid imbalance occurs when fluid output does not equal fluid intake or when fluid is distributed abnormally.
- Fluid imbalances can be organized into five categories: volume depletion, volume excess, dehydration, hypotonic hydration, and fluid sequestration.

25.2c Regulation of Fluid Balance

- Fluid intake is regulated through stimuli that turn on and turn off the thirst center.
- Fluid output is hormonally regulated by altering the amount of fluid lost in urine (facultative water loss); is decreased by the renin-angiotensin system, aldosterone, and antidiuretic hormone; and is increased by atrial natriuretic peptide.

25.3 Electrolyte Balance

- Solutes are differentiated into nonelectrolytes and electrolytes.

25.3a Nonelectrolytes and Electrolytes

- Nonelectrolytes do not dissociate in solution. Electrolytes are substances that conduct an electric current following their dissociation in solution; they include salts, acids, bases, and some negatively charged proteins.

25.3b Major Electrolytes: Location, Functions, and Regulation

- Major electrolytes include sodium ion (Na^+), potassium ion (K^+), chloride ion (Cl^-), calcium ion (Ca^{2+}), phosphate ion (PO_4^{3-}), and magnesium ion (Mg^{2+}). Each is normally maintained within homeostatic levels.

25.4 Hormonal Regulation	• Hormonal regulation of fluid and electrolyte balance occurs primarily through: angiotensin II, antidiuretic hormone, aldosterone, and atrial natriuretic peptide.
	25.4a Angiotensin II • Angiotensin II stimulates vasoconstriction to increase resistance, decreases fluid output by the kidney to maintain blood volume, activates the thirst center, and stimulates the release of aldosterone and antidiuretic hormone; the net effect is an increase in blood pressure.
	25.4b Antidiuretic Hormone • Antidiuretic hormone (ADH) stimulates the thirst center for us to increase fluid intake, decreases fluid output from the kidney to maintain blood volume, and decreases blood osmolarity. In high doses, ADH is a vasoconstrictor.
	25.4c Aldosterone • Aldosterone (ALDO) stimulates reabsorption of both Na^+ and water to maintain blood volume and blood pressure. However, blood osmolarity is unchanged. Potassium ions are normally secreted except under conditions of low pH; then excess H^+ is excreted.
	25.4d Atrial Natriuretic Peptide • Atrial natriuretic peptide (ANP) decreases blood volume and blood pressure by causing vasodilation and increasing fluid output—both processes that occur from stimulation of ANP directly—and through the inhibition of the renin-angiotensin system, antidiuretic hormone, and aldosterone.
25.5 Acid-Base Balance	• Acid-base balance requires regulating H^+ concentration in body fluids to maintain arterial blood pH between 7.35 and 7.45.
	25.5a Categories of Acid • Acid can be classified as fixed or volatile. Fixed acid is acid from a source other than CO_2 and is regulated by the kidney, and volatile acid is produced from CO_2 and is regulated by changes in breathing rate and depth by the respiratory system.
	25.5b The Kidneys and Regulation of Fixed Acids • Acidic and alkaline input occurs through the diet with absorption of foods from the gastrointestinal (GI) tract. Acidic input also occurs through by-products of metabolism. Adjustments are made to maintain acid-base balance by the output of the kidney as it alters its secretion and reabsorption of H^+ and HCO_3^-.
	25.5c Respiration and Regulation of Volatile Acid • Normally, the amount of CO_2 eliminated by the respiratory system is equal to the amount of CO_2 produced by the body cell during aerobic cellular respiration.
	25.5d Chemical Buffers • Chemical buffers help prevent pH changes within fractions of a second and include proteins, the phosphate system, and the bicarbonate system until the kidney or respiratory system can alter the blood pH through physiologic mechanisms.
25.6 Disturbances to Acid-Base Balance	• A transient acid-base disturbance, in which the buffering capacity of chemical buffers is temporarily exceeded, is usually compensated for by the kidneys or respiratory system to prevent permanent acid-base imbalances.
	25.6a Overview of Acid-Base Imbalances • There are four types of acid-base imbalances, including respiratory acidosis, respiratory alkalosis, metabolic acidosis, and metabolic alkalosis.
	25.6b Respiratory-Induced Acid-Base Disturbances • Respiratory acidosis is the most common acid-base disturbance and occurs due to impaired respiratory function. P_{CO_2} is above 45 mm Hg, causing a decrease in pH. • Respiratory alkalosis occurs due to hyperventilation. P_{CO_2} is below 35 mm Hg, resulting in an increase in pH.
	25.6c Metabolic-Induced Acid-Base Disturbances • Metabolic acidosis is the most common metabolic acid-base disturbance and occurs when HCO_3^- is below 22 mEq/L, resulting in a decrease in pH. • Metabolic alkalosis occurs when HCO_3^- is above 26 mEq/L, resulting in an increase in pH.
	25.6d Compensation • Renal compensation eliminates excess H^+ and reabsorbs HCO_3^- in response to low pH, or it eliminates excess HCO_3^- and reabsorbs H^+ in response to high pH. Blood HCO_3^- levels will be higher or lower than normal, respectively. • Respiratory compensation eliminates additional CO_2 in response to metabolic acidosis, or it decreases the elimination of CO_2 in response to metabolic alkalosis. Blood CO_2 (and H_2CO_3) will be lower or higher than normal, respectively.

CHALLENGE YOURSELF

Create and Evaluate

Analyze and Apply

Understand and Remember

Do You Know the Basics?

1. Which of the following individuals is most likely to contain the largest percentage of water?
   a. a frail 76-year-old woman
   b. a chunky 52-year-old male athlete
   c. a healthy 88-year-old man
   d. a lean 35-year-old male athlete
2. The fluid compartment with the largest percentage of fluid is
   a. intracellular.
   b. interstitial.
   c. plasma.
   d. extracellular.
3. Which of the following would result in fluid moving from the blood into the interstitial space?
   a. dehydration
   b. increased blood pressure
   c. burns
   d. vomiting
4. If an individual has decreased saliva production, increased blood osmotic pressure, decreased blood volume, and decreased blood pressure, he or she is experiencing
   a. edema.
   b. dehydration.
   c. hypotonic hydration.
   d. acidosis.
5. Which hormone decreases total body fluid, blood volume, and blood pressure?
   a. antidiuretic hormone (ADH)
   b. aldosterone
   c. atrial natriuretic peptide (ANP)
   d. angiotensin II
6. Which of the following describes an electrolyte? An electrolyte
   a. does not conduct an electric current.
   b. is covalently bonded and usually contains carbon.
   c. does not dissociate in solution.
   d. plays a significant role in regulating fluid balance.
7. The major role of this electrolyte is to maintain osmotic pressure in the extracellular fluid (ECF).
   a. HCO_3^-
   b. Ca^{2+}
   c. Na^+
   d. K^+
8. An increase in blood CO_2 levels is followed by a(n) ______ in blood H^+ levels and a(n) _____ in blood pH.
   a. decrease, decrease
   b. increase, increase
   c. decrease, increase
   d. increase, decrease
9. Which of the following is *not* a chemical buffer in the body?
   a. carbohydrate
   b. protein
   c. phosphate
   d. hemoglobin
10. The kidney can act to buffer the blood by
    a. secreting H^+.
    b. reabsorbing H^+.
    c. producing HCO_3^-.
    d. All of these are correct.
11. List the three variables that determine the percentage of total water in the body.
12. Describe the movement of water between the compartments after taking a drink of water.
13. List the stimuli that turn on, and turn off, the thirst center.
14. Explain the homeostatic system involving the renin-angiotensin system, ADH, and aldosterone.
15. Describe how ANP is regulated and how it opposes the action of the other three hormones (angiotensin II, ADH, and aldosterone).
16. Describe the functions of Na^+ and how it is regulated.
17. Describe what occurs in the kidney to maintain acid-base balance when there is an increase in blood plasma $[H^+]$.
18. Explain how increasing breathing rate increases blood pH.
19. List the three chemical buffers, and describe how and where they buffer pH.
20. Describe respiratory acidosis and its compensation.

Can You Apply What You've Learned?

1. Maria brings her baby to the emergency room. She reports that her daughter Sophia has had a fever for 2 days. She has also been vomiting and had diarrhea, and she refuses to drink any fluid. Sophia is at risk for
   a. volume excess.
   b. volume depletion.
   c. dehydration.
   d. hypotonic hydration.
2. A young man has burns over 25% of his body, destroying his cells. Which electrolyte is released from the damaged cells, resulting in elevated blood plasma levels? Results from the lab would indicate elevated levels of this common intracellular electrolyte.
   a. Ca^{2+}
   b. K^+
   c. Na^+
   d. Cl^-

3. Harriet, a young, poor, single mother, brings her baby to the emergency room because the baby is crying constantly. She reports that she is unable to buy formula and has been unable to breastfeed. Instead, she has been giving the baby a bottle with only water. What is the potential fluid imbalance this infant might be experiencing?

   a. volume excess
   b. dehydration
   c. volume depletion
   d. hypotonic hydration

4. Harold has been suffering from diabetes mellitus for approximately 15 years. As a result of having recently lost his job and being under a great deal of stress, he has not been adequately managing his diabetes. His breath smells sweet, a sign of producing ketoacids. Harold is most likely suffering from which type of pH imbalance?

   a. respiratory acidosis
   b. metabolic acidosis
   c. respiratory alkalosis
   d. metabolic alkalosis

5. Harold's (from question 4) blood lab results would most likely show ________ caused by the primary disturbance and ________ caused by compensation.

   a. decreased HCO_3^-, decreased P_{CO_2}
   b. decreased HCO_3^-, increased P_{CO_2}
   c. increased HCO_3^-, decreased P_{CO_2}
   d. increased HCO_3^-, increased P_{CO_2}

Can You Synthesize What You've Learned?

1. Morgan is a nurse at the local hospital. She received the lab results back from a patient that said the patient has hyperaldosteronism, a high level of aldosterone. Based on this diagnosis, explain what you would expect in regard to blood levels of Na^+ and K^+ as well as blood pressure. Explain why.

2. Ms. Taylor, 68 years old, has been vomiting for 2 days. On the day of admission to the hospital, she feels agitated and anxious. Her lab results, and normal lab values, are given in the following table. What is the primary disturbance: respiratory acidosis, respiratory alkalosis, metabolic acidosis, or metabolic alkalosis? How is her body attempting to compensate, and is it successful? Explain.

Variable	Patient's Lab Results	Normal Values
Blood pH	7.38	7.35–7.45
HCO_3^-	30 mEq/L	22–26 mEq/L
Pa_{CO_2}	60 mm Hg	35–45 mm Hg

INTEGRATE

ONLINE STUDY TOOLS

connect | SMARTBOOK® | AP|R

The following study aids may be accessed through Connect.

Clinical Case Study: An Elderly Woman Who Suddenly Develops Severe Shortness of Breath

Interactive Questions: This chapter's content is served up in a number of multimedia question formats for student study

SmartBook: Topics and terminology include body fluids; fluid balance; electrolyte balance; hormonal regulation; acid-base balance; disturbances to acid-base balance

Anatomy & Physiology Revealed: Topics include renal corpuscle: macula densa; renal corpuscle: distal convoluted tubule

Design Elements: **Learning Objective** icon: ©McGraw-Hill Education; **What Do You Think?** icon: ©McGraw-Hill Education; **What Did You Learn?** icon: ©McGraw-Hill Education; **Learning Strategy** icon: ©Slavoljub Pantelic/Shutterstock RF; **Clinical View** icon: ©Laia Design Studio/Shutterstock RF; **Concept Connection** icon: ©McGraw-Hill Education; **Challenge Yourself** icon: ©McGraw-Hill Education; **Do You Know the Basics?** icon: ©McGraw-Hill Education; **Can You Apply What You've Learned?** icon: ©McGraw-Hill Education; **Can You Synthesize What You've Learned?** icon: ©McGraw-Hill Education; **Online Study Tools** icon: ©ayzek/Shutterstock RF

chapter 26

Digestive System

INTEGRATE

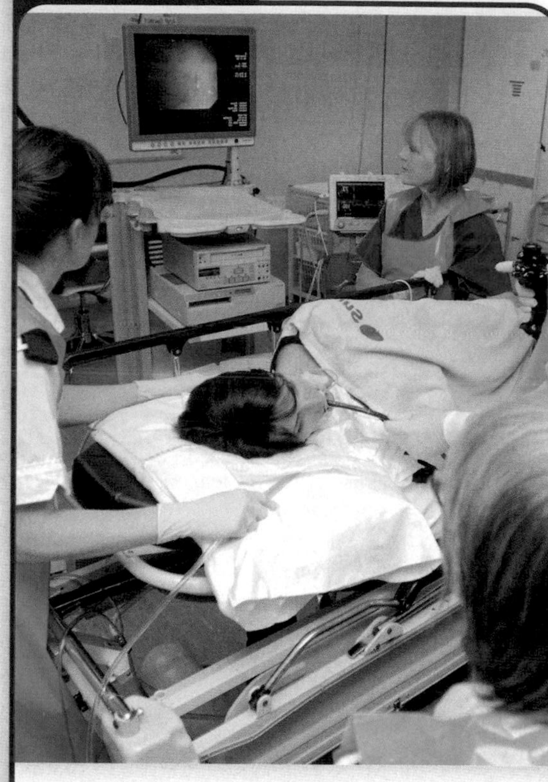

©By Ian Miles-Flashpoint Pictures/Alamy

CAREER PATH

Gastroenterologist

A gastroenterologist specializes in the diagnosis and treatment of gastrointestinal (GI) tract disorders, as well as disorders of the pancreas, liver, and gallbladder. These medical specialists treat the various malfunctions of the digestive system, including gastroesophageal reflux (heartburn), peptic ulcers, irritable bowel syndrome, hepatitis, colitis, colon polyps, and cancer. This photo shows a gastroenterologist viewing a video monitor during an endoscopy procedure, which shows the inner lining of the upper GI tract.

Anatomy & Physiology | REVEALED®
aprevealed.com

Module 12: Digestive System

Each time we eat a meal and drink fluids, our body takes in the nutrients required for our survival. However, these nutrients are typically unusable by us in their original forms. The **digestive system** provides the means to break down ingested nutrients and then absorb them.

Here we present the many organs and processes associated with the digestive system. We begin with an introduction to the system and its organs, and then explore each component in more detail: the upper gastrointestinal (GI) tract, lower GI tract, and accessory digestive organs associated with them. The chapter concludes with a review of the major nutrients and the specific details of their digestion and absorption, including the structures of the digestive system in which these functions occur. This will provide you with an opportunity to view the "big picture" of enzyme pathways that occur in the different organs for each type of molecule—an approach that integrates form and function.

26.1 Introduction to the Digestive System

The **digestive system** includes the organs that ingest the food; mix and move the ingested materials; add secretions to facilitate digestion of these materials into smaller, usable components; absorb the necessary nutrients into the blood or lymph; and expel the waste products from the body.

26.1a Organization of the Digestive System

LEARNING OBJECTIVES

1. Identify the six organs that make up the gastrointestinal (GI) tract.
2. List the accessory digestive organs and structures involved in the digestive process.

The digestive system has two separate categories of organs: those composing the gastrointestinal tract, and the accessory digestive organs (**figure 26.1**). The **gastrointestinal** (gas′trō-in-tes′tin-ăl; *gastro* = the stomach) **(GI) tract** is also called either the *digestive tract* or the *alimentary canal.* The GI tract organs essentially form a continuous tube that includes the oral cavity (mouth), pharynx (throat), esophagus, stomach, small intestine, and large intestine. It ends at the anus. It is typically about 30 feet in length in an adult cadaver—although, due to smooth muscle tone (see section 10.10c), it is significantly shorter in a living individual. Within the **lumen** (inner opening) of the GI tract, ingested food is broken down into smaller components that can then be absorbed along its length. Thus, the disassembly of molecules for absorption occurs within the lumen of the GI tract, and optimal digestion and absorption are dependent upon regulating and maintaining the environmental conditions within this space. Note that materials within the lumen of the GI tract are not considered part of the body until they are absorbed.

Accessory digestive organs assist in the breakdown of food. Accessory digestive glands produce secretions that empty into the lumen of the GI tract and include the salivary glands, liver, and pancreas. Other accessory digestive organs are not glands. They include the teeth and tongue, which participate in the chewing and swallowing of food, and the gallbladder, which concentrates and stores the secretions of the liver.

WHAT DID YOU LEARN?

1. How is the gastrointestinal (GI) tract distinguished from accessory digestive organs? List the structures that compose each category.

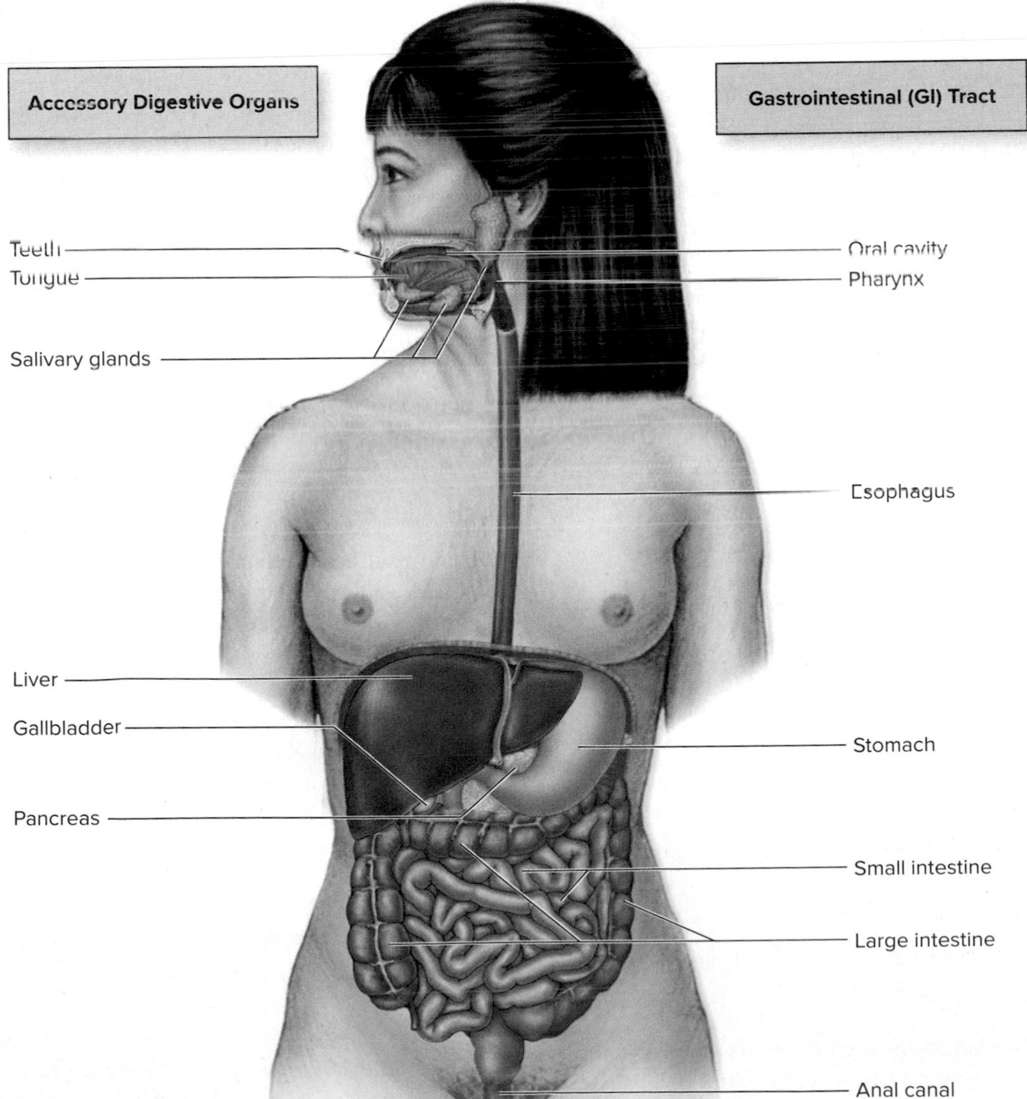

Figure 26.1 Digestive System. The digestive system is composed of the gastrointestinal (GI) tract and accessory digestive organs that assist the GI tract in the process of digestion. AP|R

26.1b General Functions of the Digestive System

LEARNING OBJECTIVE

3. List and describe the six general functions of the digestive system.

The digestive system performs six main functions: ingestion, motility, secretion, digestion, absorption, and elimination of wastes.

- **Ingestion** (in-jes′chŭn; *ingero* = to carry in) is the introduction of solid and liquid nutrients into the oral cavity (*mouth*). It is the first step in the process of digesting and absorbing nutrients.
- **Motility** (mō-til′i-tē) is a general term describing both voluntary muscular contractions (by skeletal muscle) and involuntary muscular contractions (by smooth muscle) for mixing and moving materials through the GI tract.
- **Secretion** (se-krē′shŭn) is the process of producing and releasing substances that facilitate both digestion and the movement of contents within the GI tract. Secretions are produced by both the accessory digestive glands (salivary glands, liver, pancreas) and the wall of the GI tract.
- **Digestion** is the breakdown of ingested food into smaller components that may be absorbed from the GI tract. Digestion is categorized as either mechanical digestion or chemical digestion. **Mechanical digestion** is the breaking of ingested material into smaller pieces without changing its chemical structure (i.e., no enzymes are involved). This is similar to an ice cube being crushed into ice chips. **Chemical digestion** involves the activity of specific enzymes (see section 3.3) to break down complex molecules into smaller molecules so that they can be absorbed. (Note that chemical digestion is also performed by bacteria within the large intestine.)
- **Absorption** (ab-sōrp′shŭn) involves membrane transport of digested molecules, electrolytes, vitamins, and water across the epithelial lining of the GI tract into the blood or lymph. Absorption occurs primarily within the small intestine.
- **Elimination** is the expulsion of indigestible components through the anal canal.

WHAT DID YOU LEARN?

What is the primary difference between mechanical digestion and chemical digestion?

26.1c Gastrointestinal Tract Wall

LEARNING OBJECTIVES

4. List and describe the four tunics (layers) that make up the gastrointestinal wall.
5. Briefly describe the general process of absorption.
6. Distinguish the action of the muscularis mucosae from that of the muscularis tunic.

The GI tract from the esophagus through the large intestine is a hollow tube composed of four general layers, called **tunics.** From innermost (adjacent to the lumen) to outermost, the four general tunics are the mucosa, submucosa, muscularis, and adventitia or serosa **(figure 26.2)**.

Mucosa

The **mucosa** (mū-kō′sa) is the inner-lining mucous membrane. It typically consists of an epithelium, an underlying layer called the lamina propria, and a thin layer of muscularis mucosae.

The **epithelium** is in contact with the contents within the lumen. It is a simple columnar epithelium for most of the GI tract (stomach, small intestine, large intestine). Recall from section 5.1c that this type of epithelium allows for secretion and absorption. The portions of the GI tract that must withstand abrasion (such as the esophagus) are lined by a nonkeratinized, stratified squamous epithelium.

The underlying **lamina propria** consists of areolar connective tissue that contains small blood and lymph vessels and fine, small branches of nerves. Absorption occurs when substances are moved through the simple columnar epithelial cells that line the GI tract wall and are absorbed into blood or lymphatic capillaries located within the lamina propria (figure 26.2*b*).

The **muscularis** (mŭs-kū-lā′ris) **mucosae** is the deepest layer of the mucosa and is composed of a thin layer of smooth muscle. Contractions of this smooth muscle layer cause slight movements in the mucosa to gently “shake things up,” which (1) facilitates the release of secretions from the mucosa into the lumen and (2) increases contact of materials in the lumen with the epithelial layer of the mucosa for more efficient absorption.

Submucosa

The **submucosa** is composed of areolar and dense irregular connective tissue, the relative amounts of each vary depending upon the specific region of the GI tract. Many large blood vessels, lymph vessels, nerves, and glands are within the submucosa. Fine branches of the nerves extend into the mucosa and, along with their associated autonomic ganglia, are collectively referred to as the **submucosal nerve plexus,** or *Meissner plexus.* These nerves innervate both the smooth muscle and glands of the mucosa and submucosa.

The areolar connective tissue of both the lamina propria of the mucosa and the submucosa house mucosa-associated lymphatic tissue (MALT). In the small intestine (and especially in its last portion, the ileum), larger aggregates of lymphatic nodules in the submucosa are called **Peyer patches.** The presence of MALT protects us from potentially harmful agents by preventing ingested microbes from crossing the GI tract wall and entering the body (see section 21.4d).

Muscularis

The **muscularis** is located deep to the submucosa and is composed of smooth muscle. The smooth muscle cells are arranged in an **inner circular layer,** which contains muscle cells oriented circumferentially within the GI tract wall, and an **outer longitudinal layer,** which is composed of muscle cells oriented lengthwise within the GI tract wall. Fine branches of nerves and their associated autonomic ganglia are located between these two layers of smooth muscle; these nerve branches control contractions of the muscularis and are collectively referred to as the **myenteric** (mī-en-ter′ik; *mys* = muscle, *enteron* = intestine) **nerve plexus,** or *Auerbach plexus.*

INTEGRATE

LEARNING STRATEGY

The relative orientation of the two layers of smooth muscle within the muscularis tunic can be mimicked by forming an “O” with your left hand (as the inner circular layer) and then placing your right hand with the fingers oriented upward along the outer edge of your left hand (as the outer longitudinal layer).

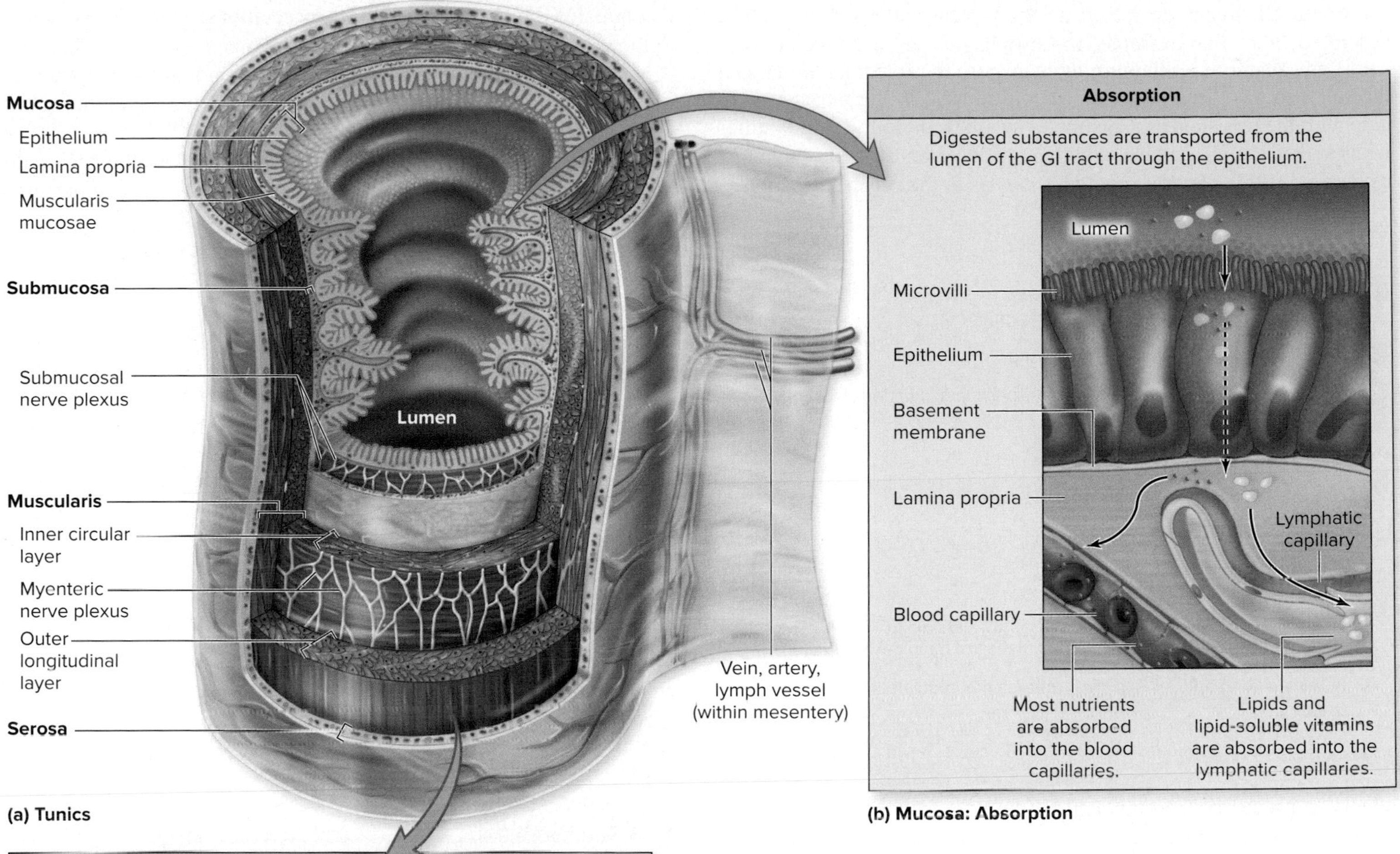

(a) Tunics

(b) Mucosa: Absorption

(c) Muscularis: Motility

Figure 26.2 Tunics of the Abdominal GI Tract. (*a*) The wall of the abdominal GI tract has four tunics: the mucosa, submucosa, muscularis, and adventitia or serosa. (*b*) Substances need only to cross the epithelium of the mucosa through membrane transport processes to be absorbed into the blood capillaries or lymphatic capillaries. (*c*) Motility includes both mixing, which is a type of muscular contraction that facilitates the blending of materials within the GI tract, and propulsion, which moves material through the lumen of the GI tract. **AP|R**

The function of the muscularis is motility. If you think of the GI tract as a hollow tube, then contractions of the circular layer constrict the lumen of the tube, whereas contractions of the longitudinal layer shorten the tube. The collective contractions of these smooth muscle layers are associated with two primary types of motility: mixing and propulsion (figure 26.2*c*):

- **Mixing** is a "backward-and-forward" motion that blends secretions with ingested material within the GI tract, but does not result in directional movement of the lumen contents. Mixing includes *mixing waves* (by the stomach) and *segmentation* (by the small intestine).
- **Propulsion**, in comparison, is the directional movement of materials through the GI tract, and it occurs by the muscularis of the GI tract by peristalsis. **Peristalsis** (per′i-stal′sis; *stalsis* = constriction) is the sequential contraction of the muscularis within the GI tract wall that moves like a wave within the different regions of the GI tract (the esophagus, stomach, small intestine, and large intestine). Persistalsis results in one-way movement of the lumen contents from the esophagus to the anus.)

Additionally, the inner circular muscle layer is greatly thickened at several locations along the GI tract to form a **sphincter.** A sphincter is typically positioned between regions of the GI tract. These rings of smooth muscle relax (open) and contract (close) to (1) control the movement of materials into the next section of the GI tract and (2) prevent its backflow. The pyloric sphincter, for example, regulates the movement of material from the stomach into the small intestine.

Adventitia or Serosa

The outermost tunic may be either an adventitia or a serosa. An **adventitia** (ad-ven-tish′ă) is composed of areolar connective tissue with dispersed collagen and elastic fibers. A **serosa** (se-rō′să) has the same composition as the adventitia, but it is completely covered by a serous membrane called the *visceral peritoneum*. Only GI structures that are *intraperitoneal organs* (as described in section 26.1e) have a serosa as their outermost tunic.

Some GI organs deviate from the typical pattern of the tunics described here. For example, the esophagus has a nonkeratinized stratified squamous epithelium in its mucosa to protect its lining, and the stomach has three layers of smooth muscle in the muscularis to help with the mechanical digestion of swallowed food. Being familiar with the basic tunic pattern, and then discovering how an organ may deviate from this pattern, provides clues as to the organ's function.

WHAT DID YOU LEARN?

3. What specific layer(s) must substances cross to enter the blood or lymphatic capillaries during their absorption?
4. How does peristalsis differ from mixing?
5. What purpose is served by muscular sphincters at various locations along the length of the GI tract?

26.1d Overview of the Regulation of the Digestive System

LEARNING OBJECTIVES

7. Describe the general function of the enteric nervous system and autonomic nervous system in the regulation of the digestive system.
8. Compare long reflexes and short reflexes that regulate the digestive system.
9. List the major hormones that regulate the processes of digestion.

Digestive processes are regulated by the nervous and endocrine systems. Here we (1) provide a general description of the enteric nervous system and the autonomic nervous system in regulating digestion through nerve reflexes and (2) identify the primary hormones that control digestion through endocrine reflexes.

Enteric Nervous System

Recall from section 15.5b that the **enteric nervous system (ENS)** is an array of both sensory neurons and motor neurons, which extends from the esophagus to the anus. This network of neurons is located within the submucosal plexus and the myenteric plexus of the gastrointestinal (GI) tract wall (figure 26.2*a*). It innervates the smooth muscle and glands of the GI tract and mediates the complex coordinated reflexes for the mixing and propulsion of materials through the GI tract. (Some researchers consider the ENS a completely separate nervous system both because of the number of neurons that compose it and because it can function independently of the central nervous system [CNS]).

Autonomic Nervous System (ANS)

The GI tract wall is also innervated by both the parasympathetic and sympathetic divisions of the autonomic nervous system (ANS). The parasympathetic and sympathetic axons synapse with smooth muscle and glands of the GI tract wall (to control these structures directly) and with neurons within the ENS (to regulate these structures indirectly). In general, parasympathetic innervation *promotes* GI tract activity: It stimulates GI motility and relaxes GI tract sphincters (see table 15.6). In contrast, sympathetic innervation *opposes* GI tract activity: It inhibits GI tract motility, contracts GI tract sphincters, and vasoconstricts blood vessels within the GI tract wall. Thus, any conditions that activate the sympathetic division (e.g., exercise, anger, stress) may slow or interfere with digestion.

Nerve Reflexes

Both the ENS and ANS control the GI tract wall through nerve reflexes (which were first described in section 14.6). The receptors that initiate GI reflexes are either **baroreceptors,** which detect stretch of the GI tract wall, or **chemoreceptors,** which monitor the chemical content of the material within the lumen, including the presence of protein and acid (see section 16.1d). In response to stimulation, either a short reflex or a long reflex is initiated.

A **short reflex** is a local reflex that only involves the ENS (and does not involve the central nervous system). Sensory input detected by either baroreceptors or chemoreceptors is relayed to neurons within the ENS to alter smooth muscle contraction and gland secretion. These reflexes function in coordinating *small segments* of the GI tract to changes in stimuli.

A **long reflex** involves sensory input relayed to the central nervous system (CNS), which serves as the integration center. Autonomic motor output is then relayed to alter smooth muscle contraction and gland secretion of the GI tract wall. Note that autonomic motor output is often relayed to other structures, including the accessory digestive organs (e.g., salivary glands, pancreas, liver). The results are coordinated smooth muscle contractions and secretory activity of potentially many different components of the digestive system.

Hormonal Control

Several primary hormones participate in the regulation of the processes of digestion: gastrin released from the stomach and secretin, cholecystokinin, and motilin released from the small intestine. The specific functions of these hormones are described in sections 26.2d and 26.3b, c.

INTEGRATE

CONCEPT CONNECTION

Three of the four cranial nerves (see sections 13.9 and 15.3a) containing parasympathetic axons are involved in regulating digestive activities:

The **facial nerve (CN VII)** extends to the sublingual and submandibular glands to stimulate salivary secretions.

The **glossopharyngeal nerve (CN IX)** extends to the parotid gland to stimulate salivary secretions.

The **vagus nerve (CN X)** innervates most of the organs of the digestive system (e.g., stomach, small intestine, most of the large intestine, pancreas, liver) and stimulates their digestive activities.

WHAT DID YOU LEARN?

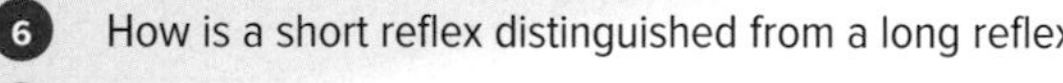

6. How is a short reflex distinguished from a long reflex?

7. What primary hormones regulate digestive processes?

26.1e Serous Membranes of the Abdominal Cavity

LEARNING OBJECTIVES

10. Describe the structure of the serous membranes associated with the GI tract.
11. Distinguish between intraperitoneal and retroperitoneal organs.
12. Explain the function of the mesentery, and describe the five individual mesenteries of the abdominopelvic cavity.

The serous membrane associated with the abdominopelvic cavity is termed **peritoneum** (per′i-tō-nē′ŭm; *periteino* = to stretch over) and was first introduced in section 1.5e. The peritoneum consists of parietal and visceral components that are continuous with one another along the posterior abdominal wall. The **parietal peritoneum** lines the inner surface of the abdominal wall, whereas the **visceral peritoneum** covers the surface of internal organs within the abdominopelvic cavity. Between the parietal and visceral peritoneum is a potential space called the **peritoneal cavity,** which contains serous fluid (that is produced by the peritoneum). This fluid lubricates both the internal abdominal wall and the external organ surfaces. It allows the abdominal organs to move freely and reduces friction resulting from this movement.

INTEGRATE

CONCEPT CONNECTION

The peritoneum is very similar to the pericardium (see section 19.2b) and the pleura (see section 23.4c). These membranes have an outer parietal layer lining the inside of the cavity and an inner visceral layer that covers the external surface of the organ. The space between the parietal and visceral layers is where serous fluid is secreted, and this fluid acts as a lubricant to prevent friction between the layers as the organs move.

Intraperitoneal and Retroperitoneal Organs

Organs within the abdomen that are completely surrounded by visceral peritoneum are called **intraperitoneal** (in′tră-per′i-tō-nē′ăl) **organs.** They include the stomach, most of the small intestine, and parts of the large intestine (transverse colon and sigmoid colon). **Retroperitoneal** (re-trō-per′i-tō-nē′ăl) **organs** lie outside the parietal peritoneum directly against the posterior abdominal wall, so only their anterolateral portions are covered with the parietal peritoneum. Retroperitoneal digestive organs include the pancreas, esophagus (abdominal portion), most of the duodenum (the first part of the small intestine), parts of the large intestine (ascending and descending colon), and the rectum (see figure 1.9).

 **WHAT DO YOU THINK?**

 Are there any organs within the peritoneal cavity?

INTEGRATE

LEARNING STRATEGY

Organs that are retroperitoneal are positioned against the posterior wall outside of the parietal peritoneum. These organs are associated with the digestive system, urinary system, and cardiovascular system. A common medical mnemonic used to remember these organs is SAD PUCKER:

S = **S**uprarenal glands (adrenal glands)
A = **A**orta and inferior vena cava
D = **D**uodenum (most of)

P = **P**ancreas
U = **U**rinary bladder and ureters
C = **C**olon (ascending and descending)
K = **K**idneys
E = **E**sophagus (abdominal portion)
R = **R**ectum

Mesentery

The general term **mesentery** (mes′en-ter-ē) refers to the double layer of peritoneum that supports, suspends, and stabilizes the intraperitoneal GI tract organs. Blood vessels, lymph vessels, and nerves that supply the GI tract are sandwiched between the two folds. Several terms are used to identify the individual mesenteries associated with specific organs (**figure 26.3**; see also figure 1.9):

- The **greater omentum** (ō-men′tŭm) extends inferiorly like an apron from the inferolateral surface of the stomach (greater curvature) and covers most of the abdominal organs. It often accumulates large amounts of adipose connective tissue; thus, it is referred to as the "fatty apron" and insulates the abdominal organs and stores fat.
- The **lesser omentum** connects the superomedial surface of the stomach (lesser curvature) and the proximal end of the duodenum to the liver.
- The **falciform** (fal′si-fōrm; *falx* = sickle) **ligament** is a flat, thin, crescent-shaped peritoneal fold that attaches the liver to the internal surface of the anterior abdominal wall.
- The **mesentery proper,** which is sometimes referred to as the *mesentery,* is a fan-shaped fold of peritoneum that suspends most of the small intestine (the jejunum and the ileum) from the internal surface of the posterior abdominal wall.
- The **mesocolon** (mez′ō-kō′lon) is a fold of the peritoneum that attaches the large intestine to the posterior abdominal wall. The mesocolon has several distinct sections, each named for the portion of the colon it suspends. For example, transverse mesocolon is associated with the transverse colon, whereas sigmoid mesocolon is associated with the sigmoid colon.

Now that an overview of the main functions and features of the GI tract has been presented, we are ready to proceed with a more detailed exploration through the system, beginning with the upper GI tract.

 **WHAT DID YOU LEARN?**

 What is the difference between intraperitoneal and retroperitoneal organs? List the digestive organs that are intraperitoneal organs.

 Where is the greater omentum located?

INTEGRATE

CLINICAL VIEW 26.1

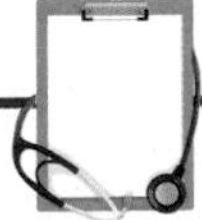

Peritonitis

Peritonitis is an inflammation of the peritoneum and is associated with abdominal pain. The most common cause is perforation of the GI tract, which allows the contents of the GI tract to leak out and come in contact with the peritoneum. Perforations may result from within—for example, ulcers of the stomach or duodenum that "eat through" the tunics of the GI tract wall, a sharp object that has been ingested and pierces through the GI tract wall, or a ruptured appendix (which is the most common cause). Perforations may also result from without—for example, from a gunshot wound or during abdominal surgery. If the inflammatory response (see section 22.3d) is unsuccessful in eliminating infectious agents, surgery and extensive antibiotic therapy are required. Without medical intervention, peritonitis that spreads can be fatal.

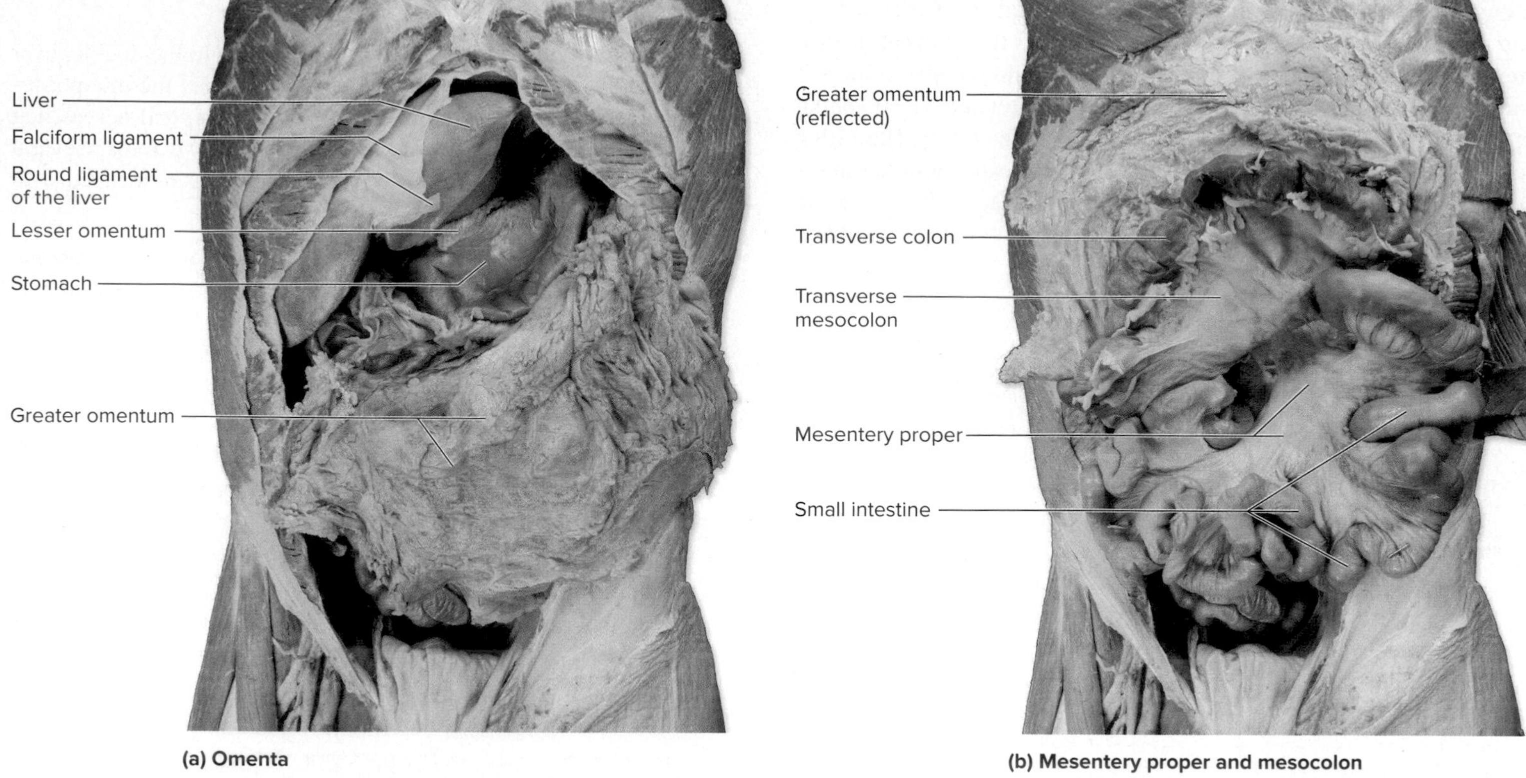

Figure 26.3 Serous Membranes. Many abdominal organs are held in place or are insulated by double-layered serous membranes called mesenteries, which include (*a*) the falciform ligament, the greater and lesser omenta, and (*b*) the mesentery proper and the mesocolon.

26.2 Upper Gastrointestinal Tract

The **upper gastrointestinal tract** consists of the oral cavity and salivary glands, pharynx, esophagus, stomach, and duodenum. It is where the initial mechanical and chemical processing of ingested material takes place.

26.2a Overview of the Upper Gastrointestinal Tract Organs

LEARNING OBJECTIVE

13. Describe the components of the upper gastrointestinal tract.

A superficial view of the upper GI tract organs and accessory structures helps integrate their general structures with their digestive activities and functions (see figure 26.1):

- **Oral cavity and salivary glands.** Mechanical digestion (mastication) begins in the oral cavity. Saliva is secreted from the salivary glands in response to food being present within the oral cavity. It is mixed with the ingested materials to form a globular, wet mass called a *bolus*. One component of saliva is salivary amylase, an enzyme that initiates the chemical digestion of starch (amylose).
- **Pharynx** (far′ingks). The bolus is moved into the pharynx during swallowing. Mucus secreted in saliva and in the superior part of the pharynx provides lubrication to facilitate swallowing.
- **Esophagus.** The bolus is transported from the pharynx through the esophagus into the stomach. Mucus secretion by the esophagus lubricates the passage of the bolus.
- **Stomach.** The bolus is mixed with gastric secretions as the muscularis in the stomach wall contracts. These secretions are released into the stomach lumen by epithelial cells of the stomach mucosa and include acid (hydrochloric acid [HCl]), digestive enzymes, and mucin. The mixing continues as an acidic "purée" called *chyme* is formed.

Note that the first part of the small intestine is the duodenum. It is also included in the upper GI tract; however, it will be described with the rest of the small intestine in section 26.3b.

WHAT DID YOU LEARN?

10 What structures are considered part of the upper GI tract? How is the ingested material referred to as it moves through each of the structures of the upper GI tract?

26.2b Oral Cavity and Salivary Glands

LEARNING OBJECTIVES

14. Identify the anatomic structures of the oral cavity.

15. Describe the structure and function of salivary glands and how the release of saliva is regulated.

16. Explain the process of mastication.

17. Discuss the structure and development of the teeth.

The Oral Cavity

The **oral cavity,** or *mouth,* is the entrance to the GI tract **(figure 26.4)**. Food is ingested into the oral cavity, where it undergoes the initial processes of mechanical and chemical digestion.

Gross Anatomy The oral cavity has two distinct spatial regions: (1) the vestibule (or buccal cavity), which is the space between the gums, lips, and cheeks; and (2) the oral cavity proper, which lies central to the teeth. The oral cavity is bounded laterally by the cheeks and anteriorly by the teeth and lips, and it leads posteriorly into the oropharynx.

The cheeks are covered externally by the integument and contain the buccinator muscles (see figure 11.5). These muscles compress the

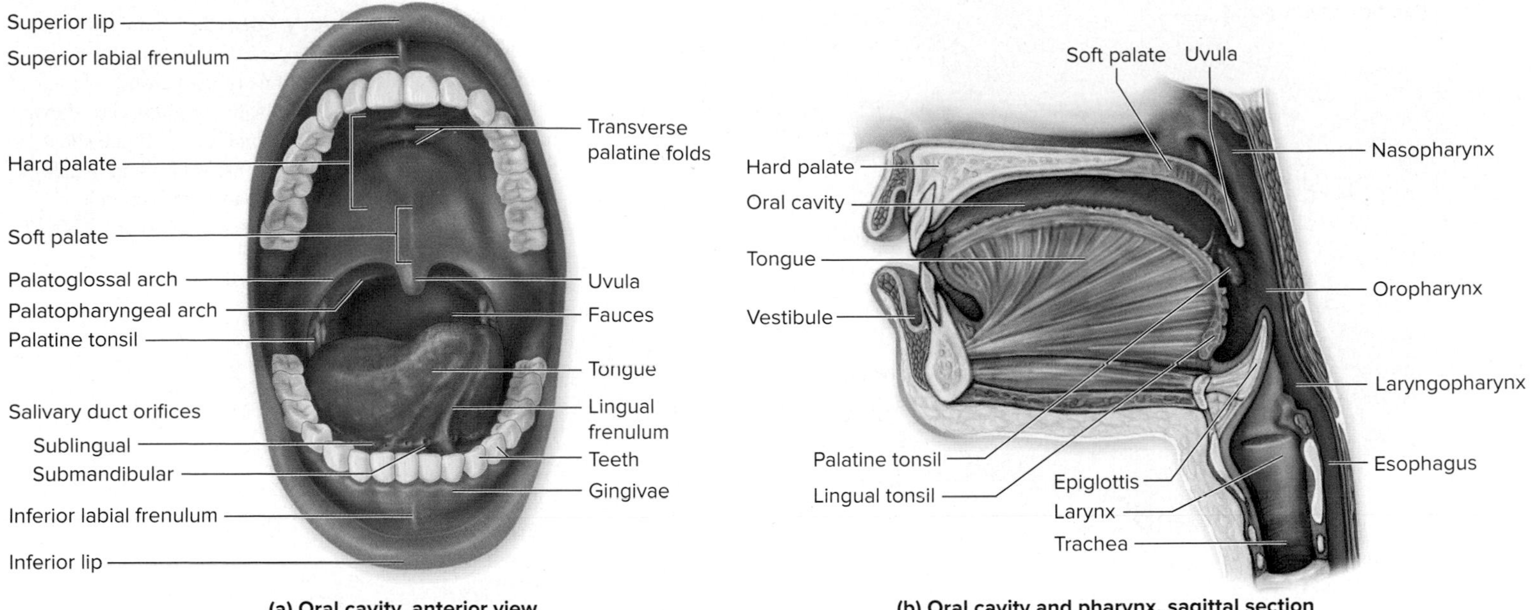

Figure 26.4 Oral Cavity. Ingested food and drink enter the GI tract through the oral cavity and move into the pharynx. (*a*) An anterior view shows the structures of the oral cavity. (*b*) A sagittal section shows the structures of both the oral cavity and the pharynx. AP|R

cheeks against the teeth to hold solid materials in place during mastication (chewing). The cheeks terminate at the fleshy **lips** (or *labia*) that are formed primarily by the orbicularis oris muscle (see section 11.3a). Lips have a reddish hue because of their abundant supply of superficial blood vessels and the reduced amount of keratin within their outer skin. The internal surfaces of both the superior and inferior lips each are attached to the gingivae (gums) by a thin mucosa fold in the midline, called the **labial frenulum** (lā'bē-ăl; *labium* = lip, fren'ū-lŭm; *frenum* = bridle). You can feel the labial frenulum associated with the upper lips by firmly pressing your tongue into the midline of the vestibule.

The **palate** (pal'ăt) forms the superior boundary, or "roof," of the oral cavity and acts as a barrier to separate it from the nasal cavity. The anterior two-thirds of the palate is hard and bony (called the hard palate), whereas the posterior one-third is soft and muscular (called the soft palate). You can feel the distinction between the hard palate and soft palate by moving your tongue along the roof of your mouth. The **hard palate** is formed by fusion of the two palatine processes of the maxillae and the horizontal plates of the two palatine bones (see section 8.2b). Failure of these bones to fuse results in a cleft palate, which is associated with swallowing issues. (See Clinical View 8.1: "Cleft Lip and Palate.") Prominent **transverse palatine folds,** or *friction ridges,* on the anterior hard palate assist the tongue in manipulating ingested materials prior to swallowing. Extending inferiorly from the posterior part of the soft palate is a cone-shaped medial projection called the **uvula** (ū'vū-lă). When you swallow, the soft palate and the uvula elevate to close off the posterior entrance into the nasopharynx and prevent ingested materials from entering the nasal region (see section 26.2c).

The **fauces** (faw'sēz) represent the opening (or "doorway") between the oral cavity and the oropharynx. The fauces are bounded by paired muscular folds or arches: the anterior **palatoglossal** (păl'a-tō-glos'ăl; *glossa* = tongue) **arch** and the posterior **palatopharyngeal** (păl'a-tō-fa-rin'jē-ăl; *pharynx* = throat) **arch.** The name of each of these folds reflects its proximity to the tongue and pharynx, respectively. The palatine tonsils are housed between the arches. These tonsils are clusters of lymphatic tissue (see section 21.4c) that protect us by monitoring ingested food and drink for potentially harmful agents (see section 22.4).

The inferior surface, or floor, of the oral cavity houses the tongue. The **tongue** is formed primarily from skeletal muscle. Both extrinsic and intrinsic muscles move the tongue (see section 11.3c). Numerous small projections called **papillae** (pă-pil'ē; sing., *papilla; papula* = pimple) cover the superior (dorsal) surface of the tongue and are involved in the sense of taste (see section 16.3b). The posteroinferior region of the tongue also contains clusters of lymphatic tissue called the lingual tonsils (see section 21.4c). The inferior surface of the tongue attaches to the floor of the oral cavity by a thin vertical mucous membrane, the **lingual frenulum.** The tongue manipulates and mixes ingested materials during chewing and helps compress the partially digested materials against the palate to assist in mechanical digestion. The tongue also performs important functions in both swallowing and speech production (as well as when infants are breastfeeding). The teeth and gums are described in detail at the end of this section.

Histology The epithelial lining of the oral cavity is a stratified squamous epithelium that protects against the abrasive activities associated with chewing. The nonkeratinized type of epithelium lines most of the oral cavity; the keratinized type lines the lips, portions of the tongue, and a small region of the hard palate.

Salivary Glands

Salivary glands, which produce saliva, are located both within the oral cavity (intrinsic salivary glands) and outside the oral cavity (extrinsic salivary glands). **Intrinsic salivary glands** are unicellular glands that continuously release relatively small amounts of secretions independent of the presence of food. Only the secretions from the intrinsic salivary glands contain **lingual lipase,** an enzyme that begins the digestion of triglycerides. Most saliva, however, is produced from multicellular exocrine glands outside the oral cavity called **extrinsic salivary glands.**

Gross Anatomy Three pairs of multicellular salivary glands are located external to the oral cavity: the parotid, submandibular, and sublingual glands (**figure 26.5*a***).

The **parotid** (pă-rot'id; *para* = beside, *ot* = ear) **salivary glands** are the largest salivary glands. Each parotid gland is located anterior and inferior to the ear, partially overlying the masseter muscle. The parotid salivary glands produce a portion (about 25–30%) of the saliva, which is transported through the **parotid duct** to the oral cavity. The parotid duct extends from the gland, across the external surface of the masseter muscle, before penetrating the buccinator muscle and opening into the vestibule of the oral cavity near the second upper molar. Infection of the parotid glands by a virus (see section 22.1) called

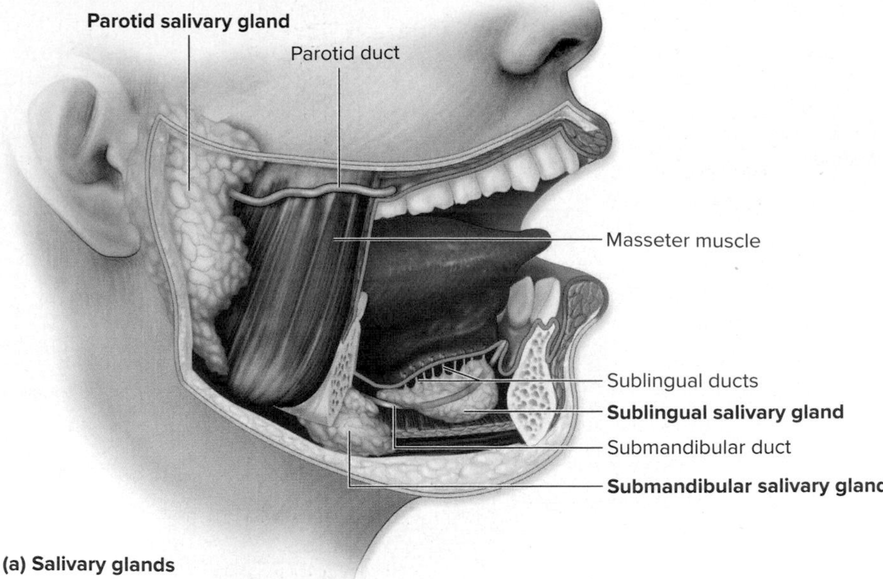

(a) Salivary glands

Figure 26.5 Salivary Glands. Saliva is produced primarily by three paired extrinsic salivary glands. (*a*) The relative locations of the parotid, submandibular, and sublingual salivary glands are shown in a side view. (*b*) Serous and mucous alveoli are shown in a diagrammatic representation of salivary gland histology. (*c*) Both mucous and serous cells may be seen in a micrograph of the submandibular salivary gland. AP|R

(c) ©McGraw-Hill Education/Al Telser

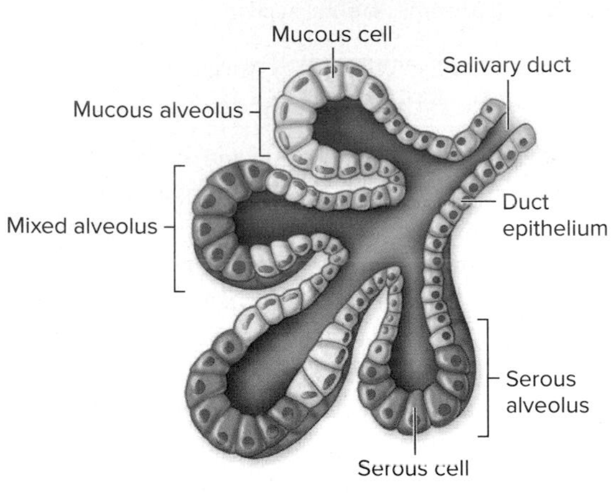

(b) Salivary gland histology

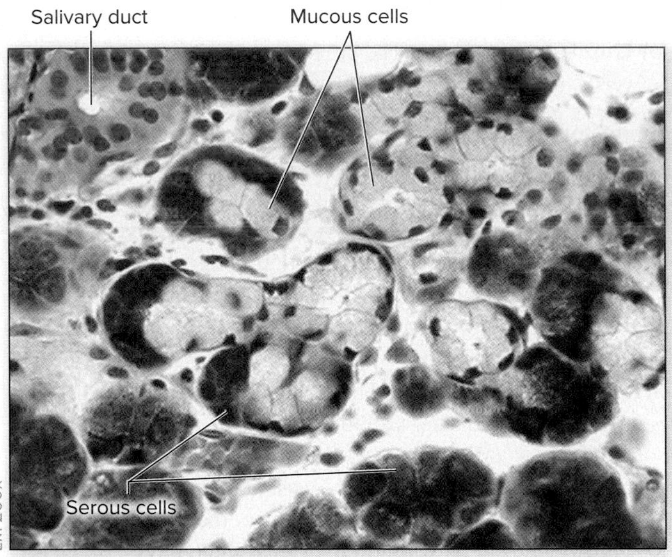

(c) Submandibular salivary gland

myxovirus causes **mumps.** Children are protected against mumps when immunized with the MMR (measles, mumps, and rubella) vaccine.

The **submandibular salivary glands** are both inferior to the floor of the oral cavity and medial to the body of the mandible, as their name suggests. The submandibular salivary glands produce most of the saliva (about 60–70%). A **submandibular duct** opens from each gland through a papilla in the floor of the oral cavity on either side of the lingual frenulum.

The **sublingual salivary glands** are inferior to the tongue, and medial and anterior to the submandibular salivary glands. Each sublingual salivary gland extends multiple tiny sublingual ducts that open onto the inferior surface of the oral cavity, posterior to the submandibular duct papilla. These tiny glands contribute only a small amount (about 3–5%) of the total saliva.

Histology Two types of secretory cells are housed within the large, paired salivary glands and collectively produce the components of saliva: mucous cells and serous cells (figure 26.5*b, c*). Mucous cells secrete mucin, which forms mucus upon hydration, whereas serous cells secrete a watery fluid containing electrolytes and salivary amylase. The proportion of mucous cells to serous cells varies among the three types of salivary glands. The parotid glands produce only serous secretions, whereas the submandibular and sublingual glands produce both mucus and serous secretions.

Saliva The volume of **saliva** (sa-li′va) secreted daily ranges between 1 and 1.5 liters. (Imagine that volume of saliva in a liter drink container!) Most saliva is produced during mealtime, but smaller amounts are produced continuously to ensure that the oral cavity mucous membrane remains moist. Saliva is composed of 99.5% water and a mixture of solutes. Saliva is formed as water and electrolytes are filtered from plasma within blood capillaries, then through cells (acini) of a salivary gland. Other components are added by cells of the salivary glands, including salivary amylase, mucin, and lysozyme. These components permit saliva to participate in various functions:

- Moistens ingested food as it is formed into a **bolus** (bō′lŭs; *bolos* = lump), a globular, wet mass of partially digested material that is more easily swallowed
- Initiates the chemical breakdown of starch in the oral cavity because of the **salivary amylase** it contains
- Acts as a watery medium into which food molecules are dissolved so taste receptors may be stimulated (see section 16.3b)
- Cleanses the oral cavity structures
- Helps inhibit bacterial growth in the oral cavity because it contains antibacterial substances, including **lysozyme** and IgA antibodies (IgA is formed by plasma cells in the lamina propria and transported across the epithelial cells [see section 22.8c])

Regulation of Salivary Secretions The salivary nuclei within the brainstem (see section 13.5c) regulate salivation. A basal level of salivation in response to parasympathetic stimulation ensures that the oral cavity remains moist (see section 15.3a). Input to the salivary nuclei is received from chemoreceptors or baroreceptors in the upper GI tract. These receptors detect various types of stimuli, including the introduction of substances into the oral cavity, especially those that are acidic, such as a lemon; and arrival of foods into the stomach lumen, especially foods that are spicy or acidic. If one eats spoiled food, bacterial toxins within the stomach stimulate receptors that initiate sensory nerve signals to the salivary nuclei. Input is also received by the salivary nuclei from the higher brain centers in response to the thought, smell, or sight of food. Stimulation of the salivary nuclei by either sensory receptors or higher brain centers results in increased nerve signals relayed along parasympathetic neurons within both the facial nerve (CN VII), which innervates the submandibular and sublingual salivary glands, and the glossopharyngeal nerve (CN IX), which innervates the parotid salivary glands, and additional saliva is released.

Sympathetic stimulation, which occurs during exercise or when an individual is excited or anxious (see section 15.4), results in a more viscous saliva by decreasing the water content of saliva. (This occurs because sympathetic stimulation constricts the capillaries of the salivary gland and decreases the fluid added to saliva.)

WHAT DO YOU THINK?

2 Research suggests that a dry mouth (inadequate production of saliva) is correlated with an increase in both bad breath (halitosis) and dental problems, such as cavities. What are the possible reasons for this correlation?

Mechanical Digestion: Mastication

Mechanical digestion in the oral cavity is called **mastication** (mas′ti-kā′shŭn; *mastico* = to chew) or *chewing*. It requires the coordinated activities of teeth, skeletal muscles in lips, tongue, cheeks, and jaws that are controlled by nuclei within the medulla oblongata and pons, collectively called the **mastication center.**

The primary function in chewing the food is to mechanically reduce its bulk into smaller particles to facilitate swallowing. Chemical digestion and absorption are affected very little by chewing, except that the surface area of the food is increased, which facilitates exposure to and action by digestive enzymes. Mastication also promotes salivation to help soften and moisten the food to form a bolus.

Note that medications composed of small, nonpolar molecules (e.g., nitroglycerin, to treat angina pectoris) may be absorbed directly into the blood from the mouth. When these medications are placed under the tongue, they pass through the oral cavity epithelium by simple diffusion (see section 4.3a) and are absorbed into the blood.

Teeth

The **teeth** are collectively known as the **dentition** (den-tish′ŭn; *dentition* = teething). A tooth has an exposed **crown,** a constricted **neck,** and one or more **roots** that anchor it to the jaw **(figure 26.6*a*)**. The roots of the teeth fit tightly into dental alveoli, which are sockets within the alveolar processes of both the maxillae and the mandible. Collectively, the roots, the dental alveoli, and the periodontal ligament that binds the roots to the alveolar processes form a gomphosis joint (see section 9.2a).

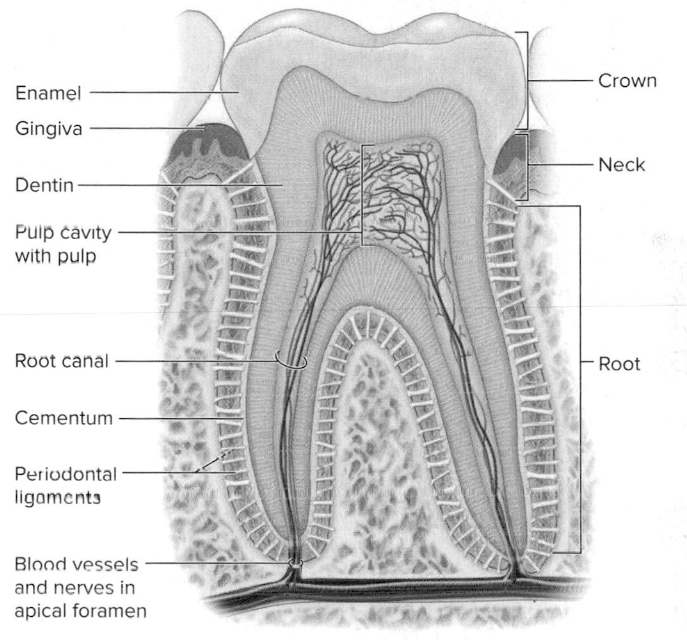

(a) Anatomy of a molar

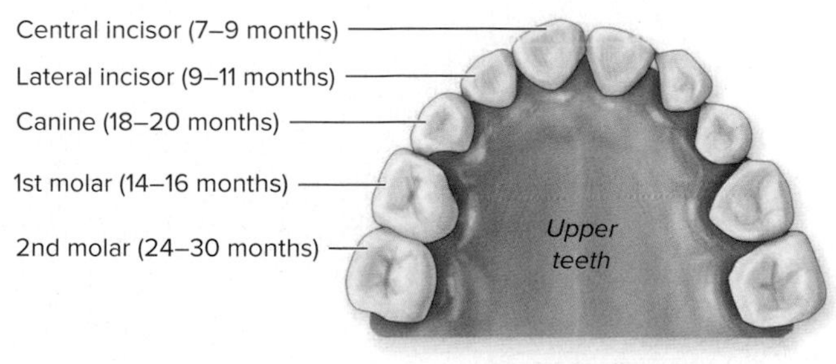

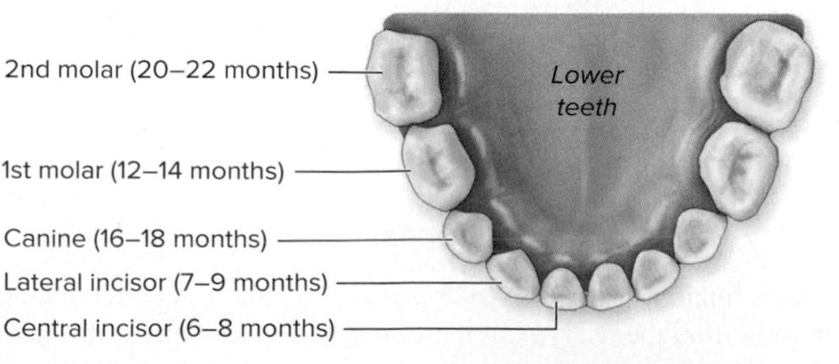

(b) Deciduous teeth

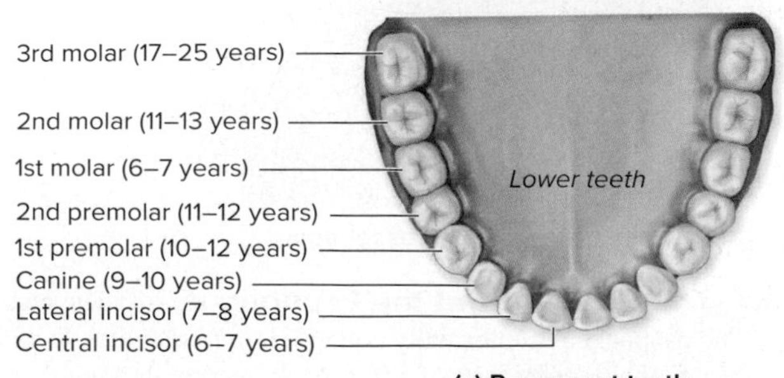

(c) Permanent teeth

Figure 26.6 Teeth. Ingested food is chewed by the teeth in the oral cavity. (*a*) Anatomy of a molar. (*b, c*) Comparison of the average dentition of deciduous and permanent teeth, including the approximate age at eruption for each tooth. AP|R

Dentin (den′tin; *dens* = tooth) forms the primary mass of a tooth. Dentin is comparable to bone but harder. On the external surface of the dentin, a tough, durable layer of **enamel** forms the crown of the tooth. Enamel is the hardest substance in the body and is primarily composed of calcium phosphate crystals. The center of the tooth is a **pulp cavity** that contains a connective tissue called **pulp.**

A **root canal** is continuous with the pulp cavity and opens into the connective tissue surrounding the root through an opening called the apical foramen. Blood vessels and nerves housed in the pulp pass through the apical foramen. Each root of a tooth is ensheathed within hardened material called **cementum** (se-men′tŭm). **Dental caries** (*tooth decay* or *tooth cavities*) is damage to the dentin, tooth enamel, or cementum. Dental caries are promoted by bacteria within the mouth, which metabolize (breakdown) ingested carbohydrate to form acidic products that damage the teeth.

Deciduous and Permanent Teeth

Two sets of teeth develop and erupt during a normal lifetime (figure 26.6*b, c*). In an infant, 20 **deciduous** (dē-sid′ū-ŭs; *deciduus* = falling off) **teeth,** also called *milk teeth,* erupt between 6 months and 30 months after birth. These teeth are eventually lost and replaced by 32 **permanent teeth.** As figure 26.6*b* shows, the more anteriorly placed permanent teeth tend to appear first, followed by the posteriorly placed teeth. (The major exception to this rule is the first molars, which appear at about age 6 and sometimes are referred to as the *6-year molars.*) The last teeth to erupt are the third molars, often called *wisdom teeth,* in the late teens or early 20s. The jaw often lacks space to accommodate these final molars, and they may either emerge only partially or grow at an angle and become impacted (wedged against another structure). Impacted teeth cannot erupt properly because of the angle of their growth. Pressure or pain caused by impacted teeth may require removal of the impacted tooth (or teeth) by surgical extraction.

The most anteriorly placed permanent teeth are called **incisors** (in-sī′zōr; *incido* = to cut into). They are shaped like a chisel and have a single root. They are designed for slicing or cutting into food. Immediately posterolateral to the incisors are the **canines** (kā′nīn; *canis* = dog) that have a pointed tip for puncturing and tearing food. **Premolars** are located posterolateral to the canines and anterior to the molars. They have flat crowns with prominent ridges called **cusps** that are used to crush and grind ingested materials. Premolars may have one or two roots. The **molars** are the thickest and most posteriorly placed teeth. They have large, broad, flat crowns with distinctive cusps and three or more roots. Molars are also adapted for grinding and crushing ingested materials. If the oral cavity is divided into quadrants, each quadrant contains the following number of permanent teeth: two incisors, one canine, two premolars, and three molars.

The **gingivae** (jin′ji-vă, -vē) are the *gums.* They are composed of dense irregular connective tissue, with an overlying nonkeratinized stratified squamous epithelium that covers the alveolar processes of the upper and lower jaws and surrounds the neck of the teeth. **Gingivitis** is inflammation of the gingivae—the gums appear red and swollen, and they may bleed. It is important to have gingivitis treated because it can lead to dental disease and tooth loss.

WHAT DID YOU LEARN?

11 What are the roles of the tongue, teeth, and salivary glands in forming a bolus?

26.2c Pharynx and Esophagus

LEARNING OBJECTIVE

18. Discuss the anatomy of the pharynx and esophagus and their complementary activities in the process of swallowing.

The pharynx and the esophagus connect the oral cavity to the stomach (**figure 26.7*a***).

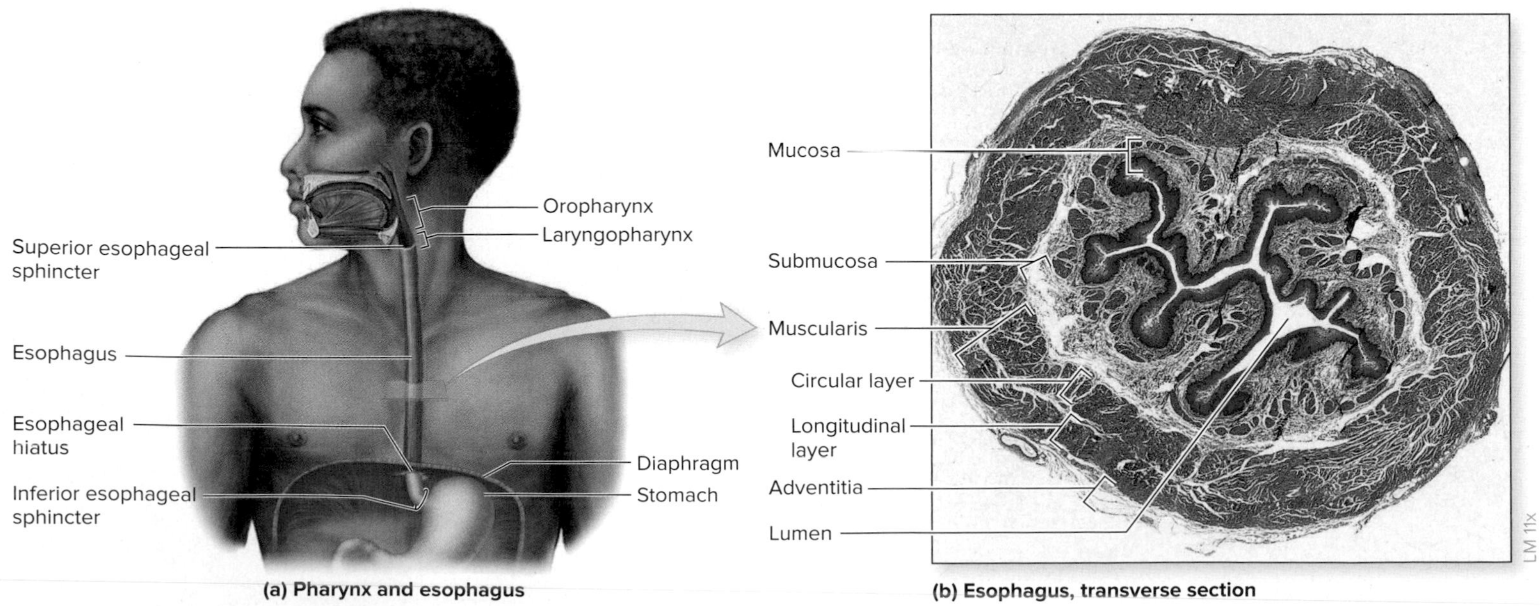

Figure 26.7 The Pharynx and the Esophagus. (*a*) The pharynx is an open, funnel-shaped, muscular passageway that connects the mouth to the esophagus. The esophagus is a normally collapsed, muscular tube that extends inferiorly from the pharynx to the stomach and functions in the passage of food and drink. (*b*) A photomicrograph of a transverse section through the esophagus identifies the tunics in its wall. The esophagus is shown here in its normal collapsed position. AP|R

(*b*) ©Alfred Pasieka/Science Source

INTEGRATE

CLINICAL VIEW 26.2

Reflux Esophagitis and Gastroesophageal Reflux Disease (GERD)

Reflux esophagitis is an inflammation of the esophagus caused by backflow (or reflux) of acidic stomach contents into the esophagus. Because the pain is felt posterior to the sternum and may be so intense that it is mistaken for a heart attack, this condition is commonly known as *heartburn.*

Unlike the stomach epithelium, the esophageal epithelium is poorly protected against acidic contents and easily becomes inflamed and irritated. Reflux esophagitis is seen most frequently in overweight individuals, smokers, those who have eaten a very large meal (especially just before bedtime), and people with **hiatal hernias** (hī-ā'tăl her'nē-ă; rupture), in which a portion of the stomach protrudes through the diaphragm into the thoracic cavity. Eating spicy foods, or ingesting too much caffeine, may exacerbate the symptoms in people affected by reflux esophagitis. Preventive treatment includes lifestyle changes such as losing weight, quitting smoking, limiting meal size, and not lying down until 2 hours after eating. Sleeping with the head of the bed elevated 4 to 6 inches, so that the body lies at an angle rather than flat, also may alleviate symptoms.

Chronic reflux esophagitis may lead to **gastroesophageal reflux disease (GERD).** Frequent gastric reflux erodes the esophageal tissue in this condition, so over a period of time, scar tissue builds up in the esophagus, leading to narrowing of the esophageal lumen. In more advanced cases, the esophageal epithelium may change from stratified squamous to columnar secretory epithelium, a condition known as **Barrett esophagus.** The secretions of secretory columnar epithelium may provide protection from the erosive gastric secretions. Unfortunately, this metaplasia increases the risk of cancerous growths.

GERD may be treated with a series of medications. Proton pump inhibitors (e.g., omeprazole [Prilosec], esomeprazole [Nexium]) limit acid secretion in the stomach by acting on the proton (hydrogen ion [H^+]) pumps that help produce acid. Histamine (H_2) blockers (e.g., famotidine [Pepcid], nizatidine [Axid], ranitidine [Zantac]) also help limit acid secretion in the stomach. Antacids help neutralize stomach acid.

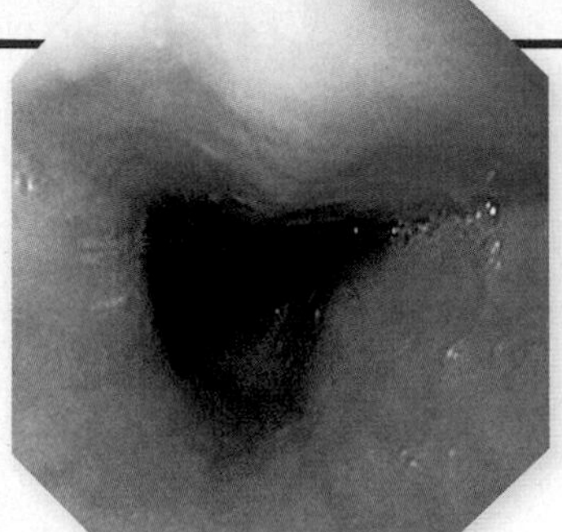

Endoscopic view of a normal esophagus.

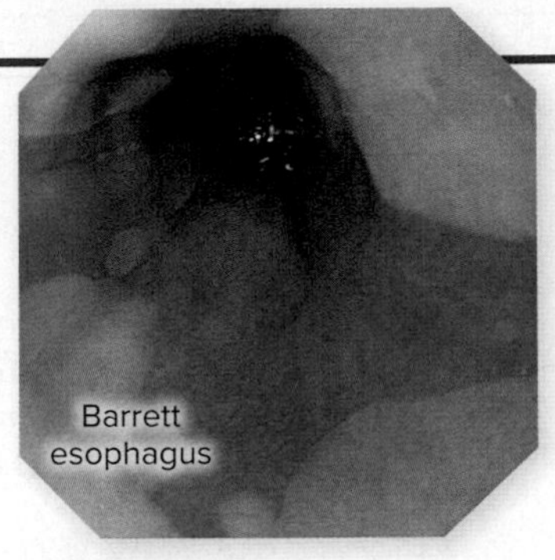

Endoscopic view of the esophagus shows the signs of Barrett esophagus.

Gross Anatomy of the Pharynx

The pharynx was described in detail in section 23.2c. It is a funnel-shaped, muscular passageway with distensible (stretchable) lateral walls that serves as the passageway for both air and food. Three *skeletal muscle* pairs called the superior, middle, and inferior **pharyngeal constrictors** form the wall of the pharynx (see section 11.3c and figure 11.10). The oropharynx and laryngopharynx are lined with nonkeratinized stratified squamous epithelium that provides protection against abrasion associated with swallowing ingested materials.

Gross Anatomy of the Esophagus

The **esophagus** (ĕ-sof'ă-gŭs) is a normally collapsed, tubular passageway. It is about 25 centimeters (10 inches) long in an adult and begins at approximately the level of the cricoid cartilage of the larynx (see figure 23.5), with most of its length within the thoracic cavity. This tube is directly anterior to the vertebral bodies and posterior to the trachea (see figure 23.8*a*, *b*), until it passes through the diaphragm. The inferior region of the esophagus connects to the stomach, where it passes through an opening in the diaphragm called the **esophageal hiatus** (hī-ā'tŭs; to yawn). Only the last 1.5 centimeters (slightly more than 1/2 inch) of the esophagus is located within the abdominal cavity.

The **superior esophageal sphincter** (or *pharyngoesophageal sphincter*) is a contracted ring of circular skeletal muscle at the superior end of the esophagus. It is the area where the esophagus and the pharynx meet. This sphincter is closed during inhalation of air, so air does not enter the esophagus and instead enters the larynx and trachea.

The **inferior esophageal sphincter** (*gastroesophageal,* or *cardiac, sphincter*) is a contracted ring of circular smooth muscle at the inferior end of the esophagus. This sphincter is not strong enough alone to prevent materials from refluxing back into the esophagus; instead, the muscles of the diaphragm at the esophageal opening contract to help prevent materials from regurgitating from the stomach into the esophagus (see Clinical View 26.4: "Reflux Esophagitis and Gastroesophageal Reflux Disease [GERD]").

Histology

The mucosa of the esophagus is lined with a nonkeratinized stratified squamous epithelium (**figure 26.7*b***). It protects this region from abrasion as food is swallowed.

The esophagus submucosa is thick and composed of abundant elastic fibers that permit distension during swallowing (and recoil after the bolus has passed through). It also houses numerous mucous glands that provide thick, lubricating mucus for the epithelium. The ducts of these glands project through the mucosa and open into the lumen.

The muscularis of the esophagus is unique in that it contains a blend of both skeletal and smooth muscle. The two layers of muscle in the superior one-third of the esophageal muscularis are skeletal, rather than smooth, to ensure that the swallowed material moves rapidly out of the pharynx and into the esophagus before the next respiratory cycle begins. (Remember that smooth muscle contracts more slowly than does skeletal muscle; see section 10.10c.) Skeletal muscle and smooth muscle cells intermingle in the middle one-third of the esophageal muscularis, and only smooth muscle is found within the wall of the inferior one-third of this muscularis. This transition marks the beginning of a continuous smooth muscle muscularis that extends throughout the stomach and the small and large intestines to the anus. The outermost layer of the esophagus is an adventitia.

Motility: The Swallowing Process

Swallowing is called *deglutition* (dē-glū-tish'ŭn). It is the process of moving ingested materials from the oral cavity to the stomach.

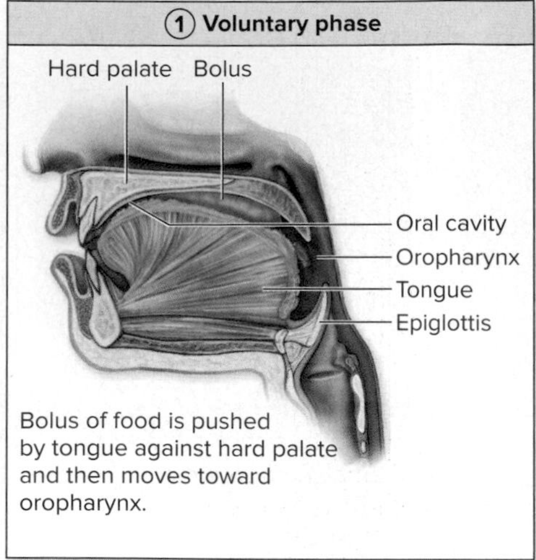

① Voluntary phase
Hard palate
Bolus
Oral cavity
Oropharynx
Tongue
Epiglottis
Bolus of food is pushed by tongue against hard palate and then moves toward oropharynx.

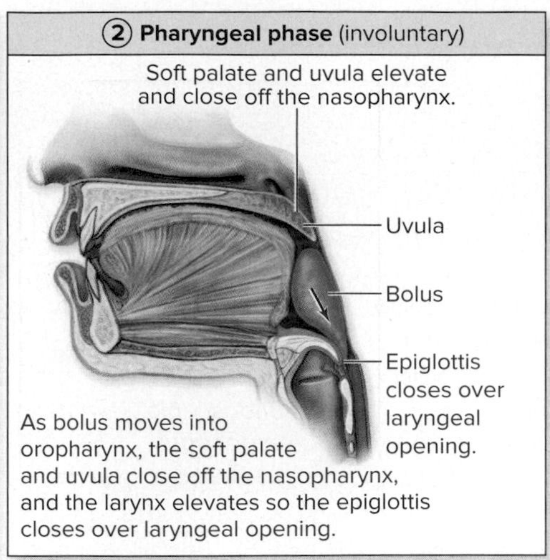

② Pharyngeal phase (involuntary)
Soft palate and uvula elevate and close off the nasopharynx.
Uvula
Bolus
Epiglottis closes over laryngeal opening.
As bolus moves into oropharynx, the soft palate and uvula close off the nasopharynx, and the larynx elevates so the epiglottis closes over laryngeal opening.

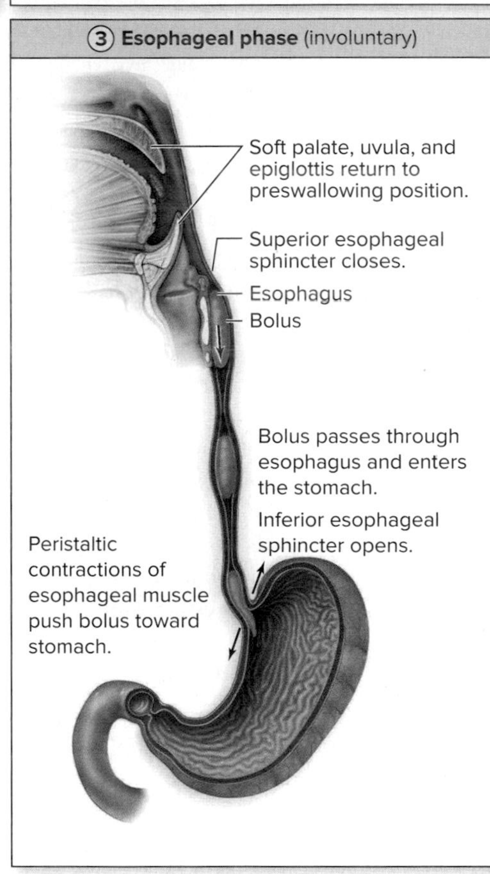

③ Esophageal phase (involuntary)
Soft palate, uvula, and epiglottis return to preswallowing position.
Superior esophageal sphincter closes.
Esophagus
Bolus
Bolus passes through esophagus and enters the stomach.
Inferior esophageal sphincter opens.
Peristaltic contractions of esophageal muscle push bolus toward stomach.

Figure 26.8 Phases of Swallowing. Swallowing occurs as a result of coordinated muscular activities that force the bolus from the oral cavity into the stomach. The process is organized into three phases: (1) voluntary phase, (2) pharyngeal phase, and (3) esophageal phase.

Swallowing has three phases: the voluntary phase, the pharyngeal phase, and the esophageal phase (**figure 26.8**).

The **voluntary phase** occurs after ingestion. It is controlled by the cerebral cortex (primarily the temporal lobes and motor cortex of the frontal lobe). Ingested materials and saliva mix in the oral cavity. Chewing forms a bolus that is mixed and manipulated by the tongue and then pushed superiorly against the hard palate. Transverse palatine folds in the hard palate help direct the bolus posteriorly toward the oropharynx.

The arrival of the bolus at the entryway to the oropharynx (fauces) initiates the swallowing reflex of the **pharyngeal phase.** The pharyngeal phase is involuntary. Tactile sensory receptors around the fauces are stimulated by the bolus and initiate nerve signals along sensory neurons to the **swallowing center** (or *deglutition center*) in the medulla oblongata (see section 13.5c). Nerve signals are then relayed along motor neurons to effectors to cause the following response:

- Entry of the bolus into the oropharynx
- Elevation of the soft palate and uvula to block the passageway between the oropharynx and nasopharynx
- Elevation of the larynx by the extrinsic muscles (see section 23.3a) move the larynx anteriorly and superiorly, resulting in the epiglottis covering the laryngeal inlet; this prevents ingested material from entering the trachea

In addition, nerve signals are relayed to the respiratory center within the medulla oblongata to assure that a breath is not taken during swallowing. During this time, the bolus passes quickly and involuntarily through the pharynx to the esophagus—about 1 second elapses in this phase. Sequential contraction of the pharyngeal constrictors decreases the diameter of the pharynx, beginning at its superior end and moving toward its inferior end. This creates a pressure difference, forcing swallowed material from the pharynx into the esophagus.

The **esophageal phase** is also involuntary. It is the time during which the bolus passes through the esophagus and into the stomach—about 5 to 8 seconds. The presence of the bolus within the lumen of the esophagus stimulates sequential waves of muscular contraction that assist in propelling the bolus toward the stomach. Higher pressure occurs in the superior region of the esophagus relative to the inferior region.

The superior and inferior esophageal sphincters are normally closed at rest. When the bolus is swallowed, these sphincters relax to allow it to pass through the esophagus. The inferior esophageal sphincter contracts after passage of the bolus, helping to prevent reflux of materials and fluids from the stomach into the esophagus.

INTEGRATE

CLINICAL VIEW 26.3

Achalasia

Achalasia (ak-ă-lā′zē′ă) is failure of the smooth muscle of a sphincter to relax (preventing passage of material). The term typically refers to esophageal achalasia when not specified. **Esophageal achalasia** (also called *achalasia cardiae, cardiospasm,* or *esophageal aperistalsis*) is failure of the lower esophageal sphincter to relax and smooth muscle within the wall of the esophagus to contract (resulting in decreased peristalsis). Impaired ability to swallow (dysphagia), which becomes progressively worse, and regurgitation characterize esophageal achalasia. Treatments include calcium channel blockers, botulinum toxin (Botox), and surgery (Heller myotomy, which is a lengthwise cut along the esophagus).

WHAT DID YOU LEARN?

12. How do the tunics of the esophagus differ from the "default" tunic pattern in both the mucosa and muscularis?

13. How is the bolus moved from the oral cavity into the stomach, as described in the three phases of swallowing?

26.2d Stomach

LEARNING OBJECTIVES

19. Describe the gross anatomy and histology of the stomach.
20. Explain the two general functional activities of the stomach.
21. Describe the phases that regulate motility and secretion in the stomach.

The **stomach** (stŭm′ŭk) is a holding sac in the superior left quadrant of the abdomen immediately inferior to the diaphragm (see figure 26.1). Under normal conditions, between 3 and 4 liters of food, drink, and saliva enter the stomach daily and generally spend between 2 and 6 hours there, depending upon the amount and composition of the ingested material. It mixes the ingested food with secretions released from the stomach wall and mechanically digests the contents into a semifluid mass called *chyme*. Chemical digestion of both protein and fat begins in the stomach, but absorption from it is limited to small, nonpolar substances that are in contact with the mucosa of the stomach. Both alcohol and aspirin are examples of substances that are absorbed in the stomach. One significant function of the stomach is to serve as a "holding bag" for controlled release of partially digested materials into the small intestine, where most chemical digestion and absorption occur. One of the most vital functions performed by the stomach is the release of **intrinsic factor** (a substance required for the absorption of vitamin B_{12}, which occurs within the small intestine).

Gross Anatomy of the Stomach

The stomach is a muscular, J-shaped organ (**figure 26.9**). It has both a larger, convex inferolateral surface called the **greater curvature** and a smaller, concave superomedial surface called the **lesser curvature.** This organ is composed of four regions:

- The **cardia** (kar′dē-ă) is a small, narrow, superior entryway into the stomach lumen from the esophagus. The internal opening where the cardia meets the esophagus is called the **cardiac orifice**, which is the location of the inferior esophageal sphincter (also known as the *cardiac sphincter*) (see section 26.2c).
- The **fundus** (fŭn′dŭs; bottom) is the dome-shaped region lateral and superior to the esophageal connection with the stomach. Its superior surface contacts the inferior surface of the thoracic diaphragm. The fundus has both weaker muscular contractions and a higher pH in its lumen area than other regions of the stomach.
- The **body** is the largest region of the stomach; it is inferior to the cardiac orifice and the fundus and extends to the pylorus.
- The **pylorus** (pī-lōr′ŭs; *pylorus* = gatekeeper) is the narrow, funnel-shaped terminal region of the stomach. Its opening into the duodenum of the small intestine is called the **pyloric orifice.** Surrounding this pyloric orifice is a thick ring of circular smooth muscle called the **pyloric sphincter.** The pyloric sphincter regulates the movement of material from the stomach into the small intestine.

INTEGRATE

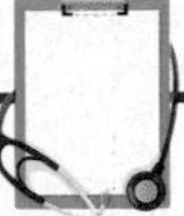

CLINICAL VIEW 26.4

Gastric Bypass

Gastric bypass is a treatment used in severely obese individuals to assist in weight loss. It is a surgical procedure that involves sectioning off a small part of the stomach (so the individual eats smaller portions) and attaching it to a lower part of the small intestine (so less nutrients are absorbed). Several changes are noted following surgery, including a decrease in appetite and altered response to hormones, including insulin. One of the most surprising changes is that the surgery can induce type 2 diabetes into remission—often within a few days of the surgery. The International Diabetes Foundation now endorses gastric bypass surgery for treatment of type 2 diabetes (see Clinical View 17.8: "Conditions Resulting in Abnormal Blood Glucose Levels").

The internal stomach lining is composed of numerous **gastric folds,** or *rugae* (rū′jē; *ruga* = wrinkle). These gastric folds are seen only when it is empty. They allow the stomach to expand greatly when it fills with food and drink and then return to its normal J-shape when it empties. In addition, the stomach is able to accommodate varying quantities of food due to the stress-relax response exhibited by the smooth muscle within the stomach wall (see section 10.10d).

Two serous membranes structures are associated with the stomach: the *greater omentum* and the *lesser omentum,* which were described in section 26.1e. The greater omentum extends inferiorly from the greater curvature of the stomach, forming the fatty apron that covers the anterior surface of abdominal organs. The lesser omentum extends superiorly from the lesser curvature of the stomach and duodenum to attach these structures to the liver.

Histology

The mucosa of the stomach is only 1.5 millimeters at its thickest region (about the thickness of a nickel). This inner lining has three significant features (**figure 26.10**):

- It is lined by a simple columnar epithelium supported by lamina propria. The transition from stratified squamous in the esophagus to simple columnar epithelium in the stomach is abrupt (figure 26.9*b*). The simple columnar epithelial cells are replaced often (usually within a week) because of the harsh acidic environment of the stomach contents.
- The lining is indented by numerous depressions called **gastric pits.**
- Several **gastric glands** extend deep into the mucosa from the base of each gastric pit. The muscularis mucosae partially surrounds the gastric glands and helps expel gastric gland secretions when it contracts.

The muscularis of the stomach varies from the general GI tract pattern in that it is composed of *three* smooth muscle layers instead of two: an inner oblique layer, a middle circular layer, and an outer longitudinal layer. The presence of a third (oblique) layer of smooth muscle assists the continued churning and blending of the swallowed bolus to help mechanically digest the food. The muscularis becomes increasingly thicker (and stronger) as it progresses from the body to the pylorus.

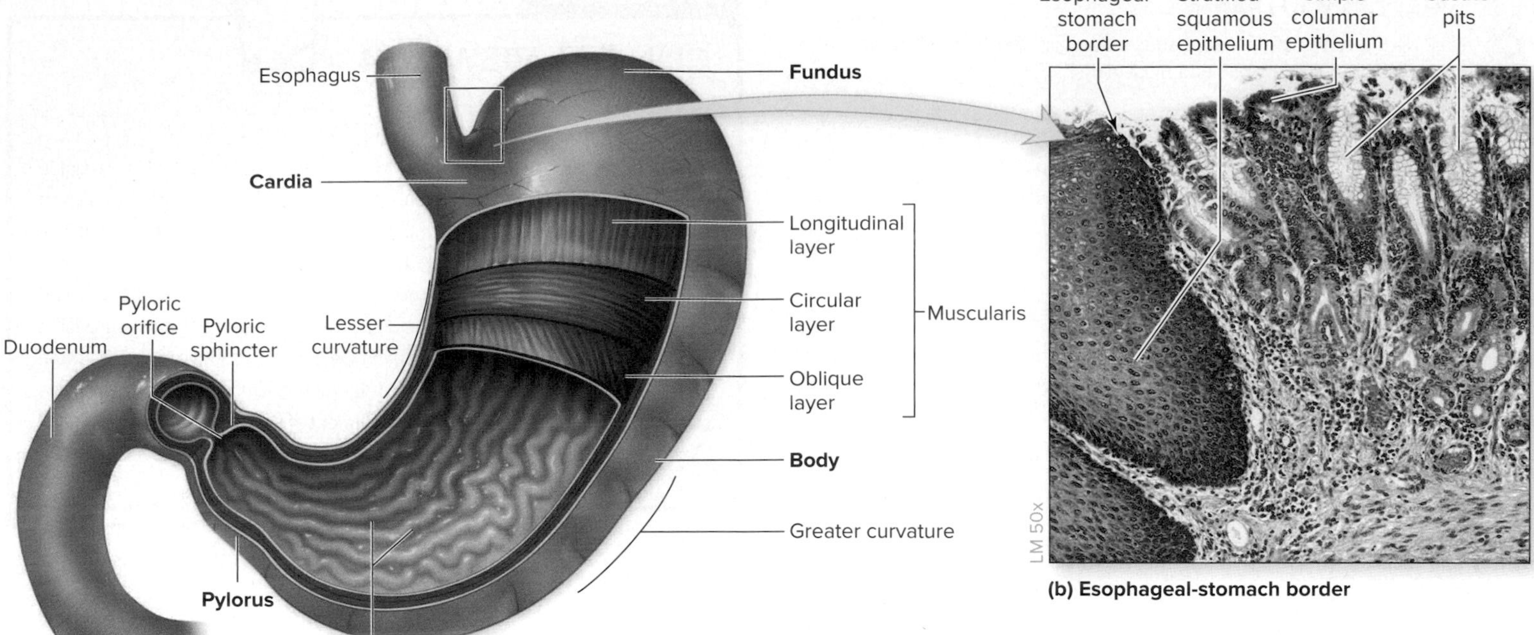

(a) Stomach regions, anterior view

(b) Esophageal-stomach border

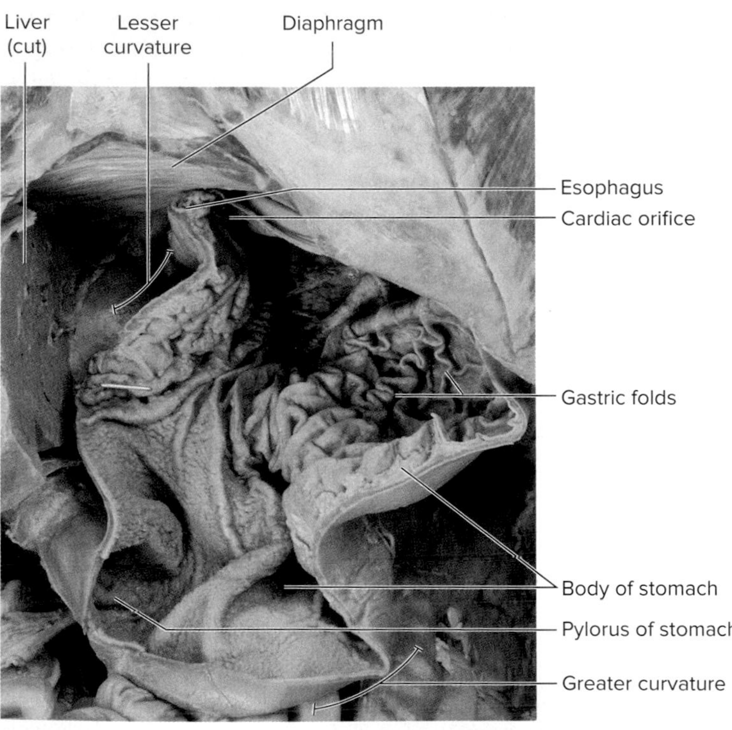

(c) Gross anatomy of stomach (cut open)

Figure 26.9 Gross Anatomy of the Stomach. The stomach is a muscular sac where mechanical and chemical digestion of the bolus occur. (*a*) The major regions of the stomach are the cardia, the fundus, the body, and the pylorus. Three layers of smooth muscle make up the muscularis tunic. (*b*) A photomicrograph of the abrupt transition from stratified squamous epithelium in the esophagus to simple columnar in the stomach. (*c*) A cadaver photo shows an anterior, open section of the stomach, revealing the gastric folds and the cardiac orifice (opening of stomach connected to esophagus). AP|R

(*b*) ©Victor P. Eroschenko; (*c*) ©McGraw-Hill Education/Christine Eckel

The outermost layer of the stomach is a serosa called the visceral peritoneum because the stomach is intraperitoneal. It produces serous fluid that lubricates the external surface of the stomach to decrease friction associated with stomach motility.

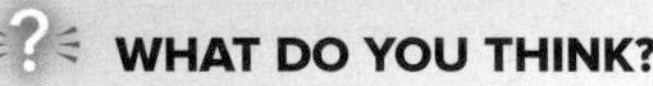

WHAT DO YOU THINK?

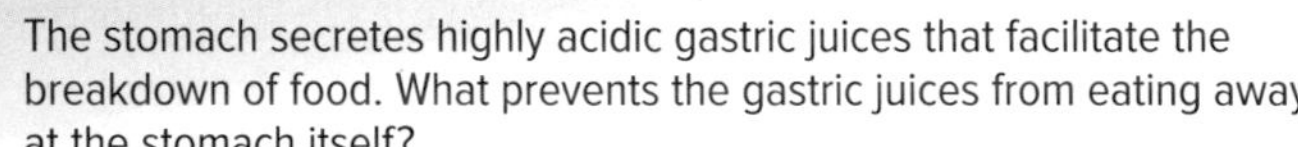

3 The stomach secretes highly acidic gastric juices that facilitate the breakdown of food. What prevents the gastric juices from eating away at the stomach itself?

Gastric Secretions

Five types of secretory cells of the gastric epithelium are integral contributors to the process of digestion (figure 26.10*c*). Four of these cell types produce the approximately 3 liters per day of **gastric juice** that are released into the stomach lumen. The fifth type of cell (G-cell) secretes a hormone into the blood.

Surface Mucous Cells **Surface mucous cells** line the stomach lumen and extend into the gastric pits. They continuously secrete an alkaline product containing mucin onto the gastric surface. Mucin becomes hydrated, producing a 1- to 3-millimeter mucus layer that coats the epithelial lining. This mucus layer, along with a high rate of cell turnover in the mucosa, helps to prevent ulceration of the stomach lining upon exposure to both the high acidity of the gastric fluid and gastric enzymes.

Mucous Neck Cells **Mucous neck cells** are located immediately deep to the base of the gastric pit and are interspersed among the parietal cells (discussed next). Mucous neck cells produce an acidic mucin that differs structurally and functionally from the alkaline mucin secreted by the surface mucous cells. The acidic mucin helps maintain the acidic conditions resulting from the secretion of hydrochloric acid by parietal cells. The mucus produced from mucin secretion by both types of mucous cells has lubricating properties to protect the stomach lining from abrasion or mechanical injury.

Parietal Cells **Parietal cells** (also called *oxyntic cells*) are responsible for the addition of two substances into the lumen of the stomach:

- **Intrinsic factor.** The production of this glycoprotein is the *only* essential function performed by the stomach. Intrinsic

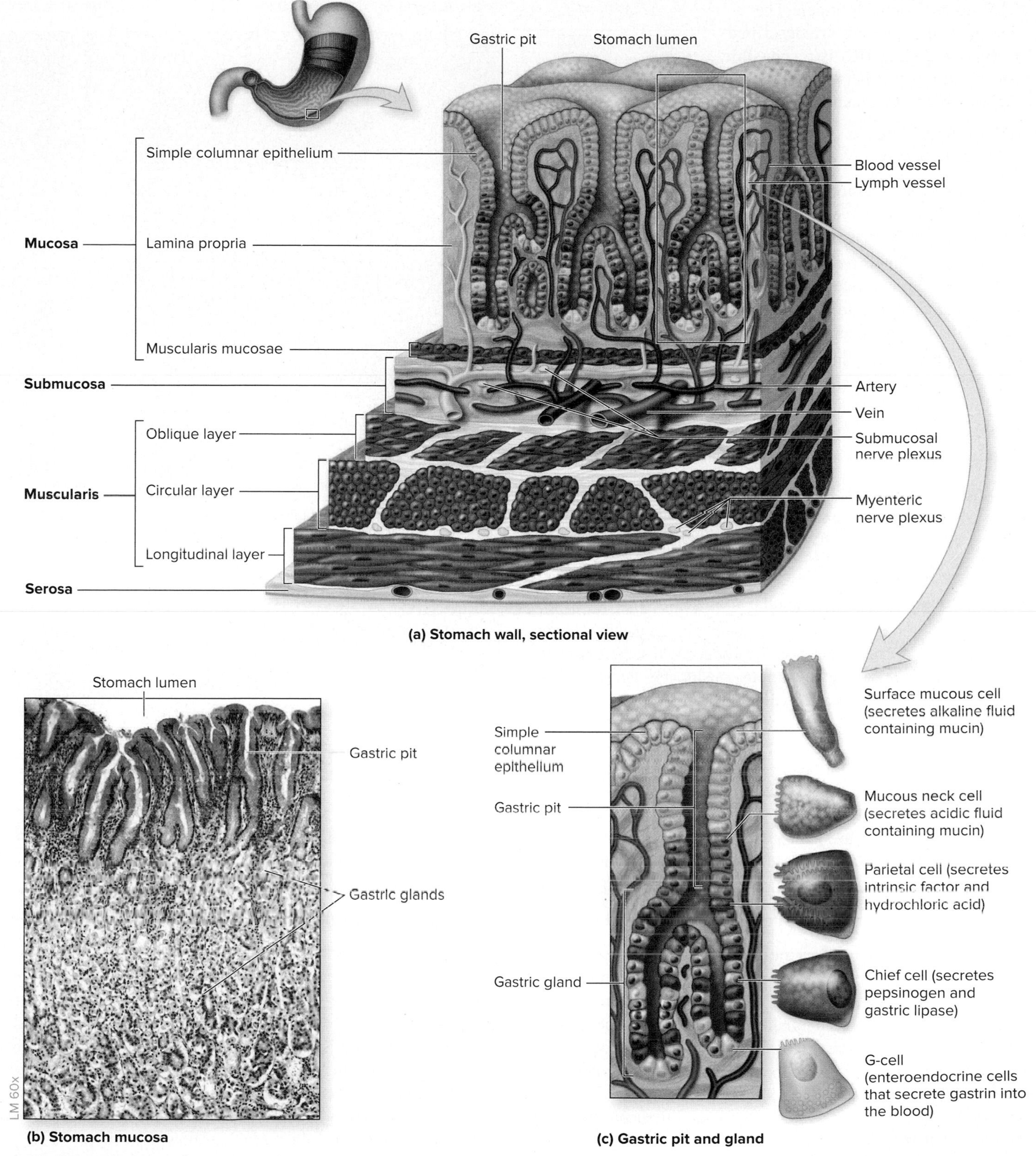

Figure 26.10 Histology of the Stomach Wall. (*a*) The stomach wall contains invaginations within the mucosa called gastric pits that lead into gastric glands. (*b*) A photomicrograph shows the cells lining the gastric pit and gastric glands. (*c*) A diagrammatic section of a gastric gland shows its structure and the distribution of different secretory cells. AP|R

factor is required for absorption of vitamin B_{12} in the ileum (the final portion of the small intestine). B_{12} is necessary for production of normal erythrocytes. A critical decrease or absence of B_{12} results in pernicious anemia (see Clinical View 18.2: "Anemia").

- **Hydrochloric acid (HCl).** HCl is not formed within the parietal cell; it would destroy the cell. Instead, the parietal cell forms H^+ and releases both H^+ and

Cl^- into the stomach lumen. The details of this process are described in **figure 26.11**. HCl is responsible for the low pH of between 1.5 and 2.5 within the stomach.

Hydrochloric acid facilitates the digestive processes of the stomach by several means:

- It contributes to the breakdown of relatively tough plant cell walls and animal connective tissue.
- It denatures proteins (see section 2.8b) by causing them to unfold, thus facilitating chemical digestion by the proteolytic enzyme pepsin.
- The low pH converts the inactive enzyme pepsinogen (released from chief cells) into active pepsin.
- HCl creates the optimal pH environment for enzymatic activity of both pepsin and acidic lipases.

The very low pH (pH ~2) also protects us from infectious agents (see section 22.1). Most bacteria, bacterial toxins, and other microbes that enter the stomach cannot survive in the harsh acidic stomach environment.

Chief Cells **Chief cells** (also called *zymogenic cells,* or *peptic cells*) are the most numerous secretory cells within the gastric glands, hence the name "chief" cells. These cells produce and secrete packets of zymogen granules primarily containing pepsinogen. Pepsinogen is the inactive precursor of the proteolytic enzyme pepsin. Pepsin must be produced in this inactive form to prevent the destruction of chief cell proteins.

Pepsinogen is activated following its release into the stomach. It is activated by both the low pH and active pepsin molecules already present within the stomach. The pepsin chemically digests denatured proteins in the stomach into smaller peptide fragments (oligopeptides) (see section 26.4b). Chief cells also produce **gastric lipase,** an enzyme that has a limited role in fat digestion (digests about 10–15% of the ingested fat) (see section 26.4c).

G-Cells G-cells are **enteroendocrine** (en′ter-ō-en′dō-krin; *enteron* = gut, intestine) **cells** that are hormone-producing cells in the gastric glands of the stomach. G-cells secrete the hormone **gastrin** into the *blood.* Gastrin stimulates stomach motility and secretions. Other enteroendocrine cells (there are at least eight types) produce other hormones, such as somatostatin, a peptide hormone that modulates the function of nearby enteroendocrine and exocrine cells.

The five primary types of secretory cells and their products are summarized in **figure 26.12**.

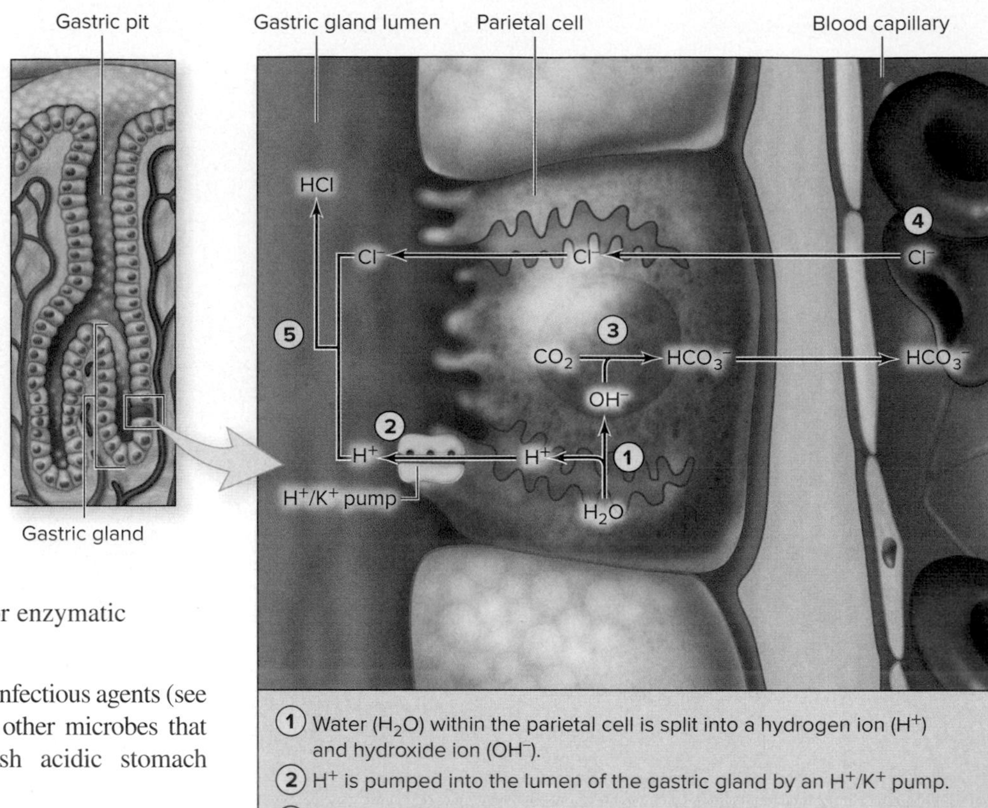

Gastric pit
Gastric gland
Gastric gland lumen
Parietal cell
Blood capillary
HCl
Cl⁻
Cl⁻
Cl⁻
CO₂
HCO₃⁻
HCO₃⁻
OH⁻
H⁺
H⁺
H⁺/K⁺ pump
H₂O
1 Water (H₂O) within the parietal cell is split into a hydrogen ion (H⁺) and hydroxide ion (OH⁻).
2 H⁺ is pumped into the lumen of the gastric gland by an H⁺/K⁺ pump.
3 OH⁻ bonds with carbon dioxide (CO₂) to form bicarbonate ion (HCO₃⁻).
4 An exchange occurs as HCO₃⁻ is transported out of the parietal cell (HCO₃⁻ then enters the blood), while chloride ion (Cl⁻) is transported into the parietal cell; Cl⁻ then enters the lumen of the gastric gland.
5 Within the lumen of the gastric gland, Cl⁻ combines with H⁺ to form hydrochloric acid (HCl).

Figure 26.11 Formation of HCl from Parietal Cells. H^+ and HCO_3^- are produced within parietal cells of the stomach. The net movement is H^+ into the lumen of the gastric gland and HCO_3^- into the blood. Additionally, Cl^- is transferred from the blood into the gastric gland lumen, where it combines with H^+ to form hydrochloric acid (HCl). AP|R

Motility in the Stomach

Smooth muscle activity in the stomach wall has two primary functions: (1) mixing the bolus with gastric juice to form chyme and (2) emptying chyme from the stomach into the small intestine (**figure 26.13**).

Gastric mixing is a form of mechanical digestion that changes the semidigested bolus into **chyme** (kīm; *chymos* = juice). Chyme has the consistency of a pastelike soup. Contractions of the stomach's thick muscularis layer churn and mix the bolus with the gastric secretions, leading to a reduction in the size of swallowed particles.

INTEGRATE

LEARNING STRATEGY

To distinguish the products formed from parietal cells and chief cells, remember:

Parietal cells are like "parental" cells because they are *protective.* These cells release (1) *intrinsic factor* to protect from pernicious anemia and (2) *H^+ and Cl^-* that create a low pH environment within the stomach that protects us from most ingested pathogens (which cannot survive in the hostile acidic environment of the stomach).

Chief cells are both "**pep**py" (produces **pep**sinogen) and "**lip**py" (produce gastric **lip**ase).

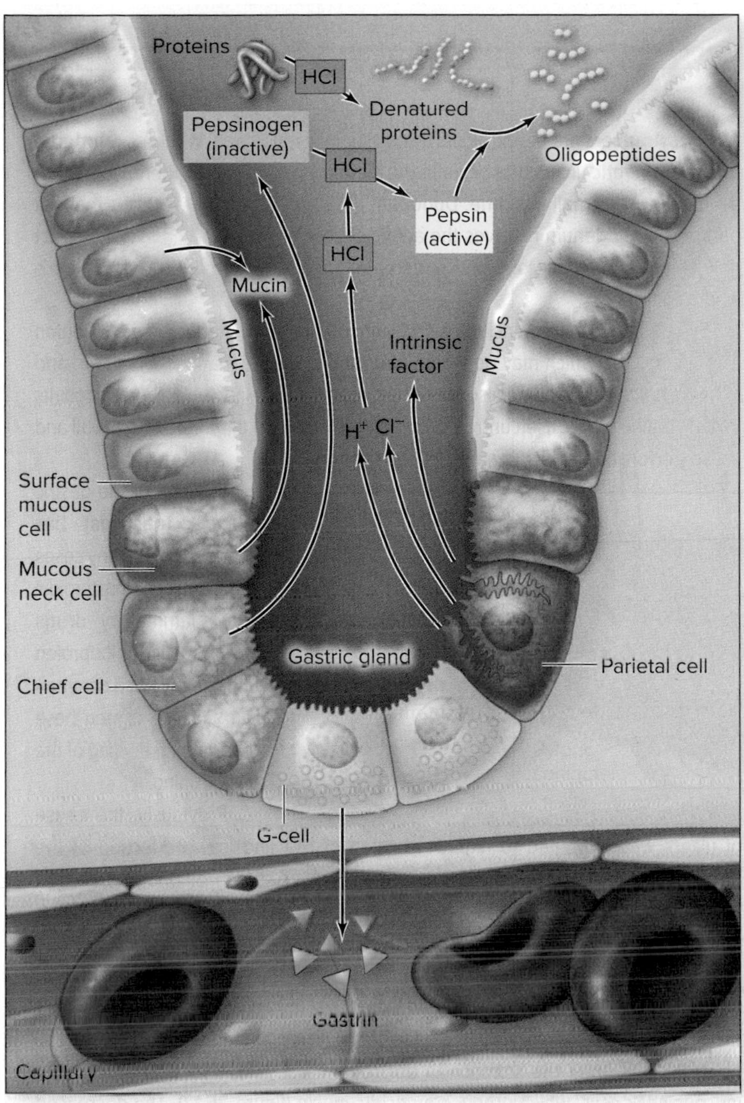

Figure 26.12 Gastric Secretions. Secretory cells and their products are identified. Secretions enter the lumen of the stomach, except the hormone gastrin, which enters the blood.

Gastric emptying is the movement of acidic chyme from the stomach through the pyloric sphincter into the duodenum, which is facilitated by the progressive thickening of the muscularis layer in the pyloric region. As a wave of peristaltic muscular contraction moves through the pylorus toward the pyloric sphincter, a pressure gradient is established that drives the stomach contents toward the small intestine.

The interaction here is unique: The peristaltic wave establishes a greater pressure on the contents in the pylorus than the pressure exerted by the pyloric sphincter to stay closed and prevent movement. Consequently, a few milliliters (about 3 milliliters) of chyme are emptied into the small intestine. After the peristaltic wave has moved past the pyloric sphincter, the pressure of the sphincter is once again greater than the pressure on the contents, and the pyloric sphincter closes. As this sphincter closes, stomach contents are squeezed back toward the stomach body. This reverse flow event is called **retropulsion.** Retropulsion not only results in the prevention of further chyme moving into the small intestine but also contributes to additional mixing of the stomach contents to further reduce the size of food particles.

Regulation of the Digestive Processes in the Stomach

The stomach is essentially a holding bag for partially digested food until the food is moved into the small intestine, where its digestion will be completed. Both the stomach's motility and the release of its secretions are highly regulated so that ingested material is "pulverized" into chyme, which can then be effectively processed within the small intestine.

Pacemaker cells *(interstitial cells of Cajal)* in the stomach wall spontaneously depolarize less than four times per minute and establish its basic muscular contraction rhythm. Electrical signals spread via gap junctions (see section 4.6d) between the smooth muscle cells in the muscularis layer of the stomach. These muscular contractions by the stomach wall are regulated by both nervous reflexes and hormones, which alter the force but not rate of contraction, which is constant. Secretory activity of gastric glands is also altered. How these occur is organized into three phases: cephalic phase, gastric phase, and intestinal phase. The cephalic and gastric phases involve the events before and during a meal, whereas the intestinal phase involves the events that occur after a meal, as the

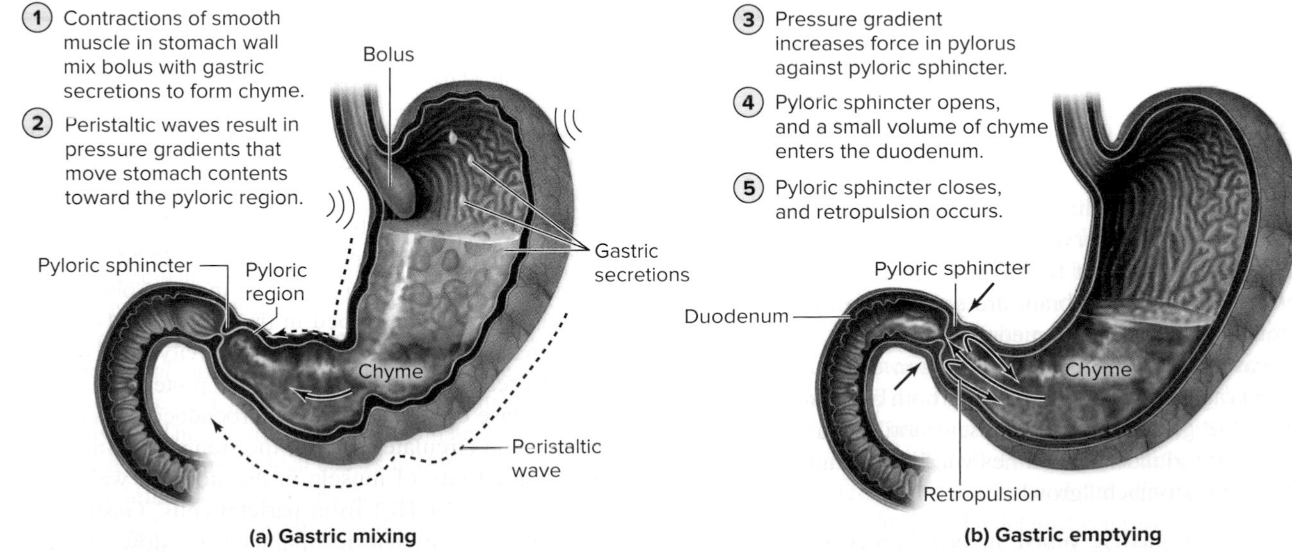

Figure 26.13 Motility in the Stomach: Gastric Mixing and Emptying. (*a*) Chyme is formed in the stomach as the bolus is mixed with gastric secretions. (*b*) A small volume is then forced into the duodenum through the partially open pyloric sphincter, and then the sphincter closes and retropulsion occurs.

INTEGRATE

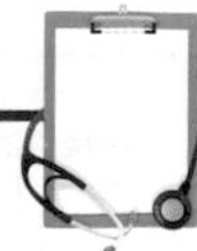

CLINICAL VIEW 26.6

Vomiting

Vomiting is the rapid expulsion of gastric contents through the oral cavity. Prior to vomiting, heart rate and sweating increase, nausea is felt, and a noticeable increase in saliva production occurs. The vomiting reflex is a complicated act that is controlled by the vomiting center in the medulla oblongata (see section 13.5c). This brainstem region responds to head injury, motion sickness, infection, toxicity (e.g., alcohol, drugs, bacterial toxins), or food irritation in the stomach and intestines.

Vomiting is initiated following a deep inspiration and the closure of both nasal cavities by the soft palate and the larynx (laryngeal inlet) by the epiglottis. Skeletal muscle contraction (abdominal muscles and diaphragm) increase pressure within the stomach and thus supply the primary force for expulsion of digestive tract contents. As the pressure increases in the stomach, the acidic gastric contents are forced into and through the esophagus and out of the oral cavity.

Care must be taken that vomit is not aspirated into the respiratory tract, a risk for a semiconscious or unconscious individual. Thus, it is critical that individuals undergoing surgical procedures have an empty stomach and small intestine because general anesthesia has an associated risk of inducing nausea and vomiting. Additionally, vomiting causes increased formation of HCl; this results in increased HCO_3^- in the blood, raising blood pH. Extensive vomiting can lead to metabolic alkalosis (see Clinical View 25.6: "How Does Vomiting or Diarrhea Alter Blood H^+ Concentration?").

regulated stomach activity. However, this hormone is now thought to primarily regulate the release of insulin in response to increased glucose concentration in the contents of the small intestine. To better reflect its role, it has been renamed **glucose-dependent insulinotropic peptide (GIP).**

WHAT DID YOU LEARN?

14 List the secretory cell types in the stomach, their products, and the function of the products.

15 Which neural reflex is initiated by food in the stomach, and what does it control?

26.3 Lower Gastrointestinal Tract

The **lower gastrointestinal (GI) tract** continues the processes of digestion and importantly functions in the absorption of nutrients. Material that cannot be digested and absorbed is then eliminated.

26.3a Overview of the Lower Gastrointestinal Tract Organs

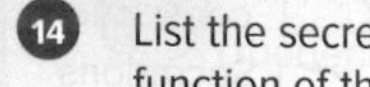

LEARNING OBJECTIVE

22. Describe the three components of the lower gastrointestinal tract.

First, we present a superficial view of the lower GI tract organs and accessory digestive organs to help integrate their structures with their digestive activities and functions (**figure 26.15**):

- **Small intestine.** The small intestine is divided into three continuous regions (duodenum, jejunum, and ileum). Recall from section 26.2a that the duodenum is considered part of the upper GI tract, but it is described in this section on the lower GI tract organs. The small intestine receives acidic chyme from the stomach that is then mixed with accessory digestive organ secretions. Most chemical digestion of macromolecules and absorption of nutrients, water, and electrolytes occur within the small intestine.
- **Accessory digestive organs.** Accessory digestive organ secretions include bile and pancreatic juice. Bile is produced by the liver and then stored, concentrated, and released by the gallbladder. Pancreatic juice contains numerous digestive enzymes and is produced and released by the pancreas. Accessory digestive organ secretions—both bile and pancreatic juice—contain HCO_3^- (a weak base), which neutralizes acidic chyme entering the duodenum.
- **Large intestine.** The large intestine primarily absorbs water, electrolytes, and vitamins (including vitamins B and K produced by bacteria within the large intestine). The digestive process is completed as the semifluid mass of partly digested food is converted to feces and then eliminated through the anus.

WHAT DID YOU LEARN?

16 What organs are considered part of the lower GI tract?

26.3b Small Intestine

LEARNING OBJECTIVES

23. Describe the anatomy of the small intestine.
24. List the glands found in the small intestine and their secretions.
25. Explain motility within the small intestine.

The **small intestine,** also called the *small bowel,* is a long tube that is inferior to the stomach and located medially within the abdominal cavity. Generally, about *9 to 10 liters* of ingested food, water, and digestive system secretions enter the small intestine daily. Ingested nutrients typically spend at least 12 hours in the small intestine. The small intestine finishes chemical digestion and is responsible for absorbing almost all of the nutrients and a large percentage of the water, electrolytes, and vitamins.

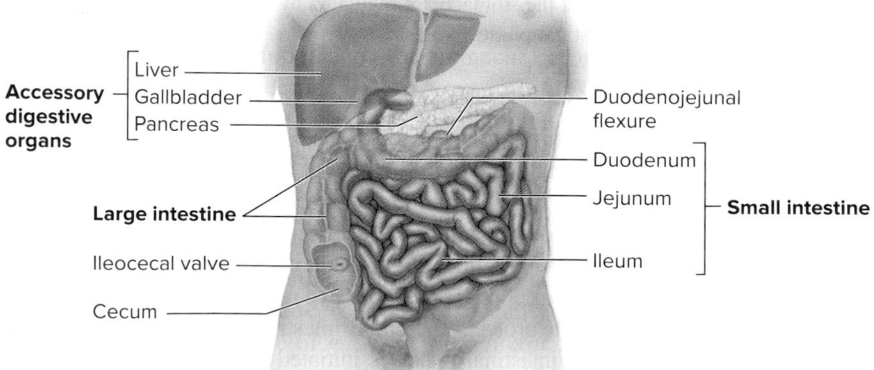

Figure 26.15 Gross Anatomy of the Lower GI Tract Organs and Accessory Digestive Organs. The three regions of the small intestine—duodenum, jejunum, and ileum—are continuous and framed within the abdominal cavity by the large intestine. Accessory digestive organs within the abdominal cavity release secretions into the duodenum.

INTEGRATE

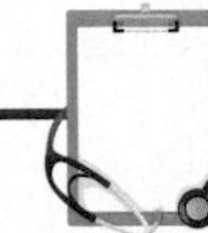

CLINICAL VIEW 26.7

Inflammatory Bowel Disease and Irritable Bowel Syndrome

The term **inflammatory bowel disease (IBD)** applies to two autoimmune disorders, Crohn disease and ulcerative colitis. In both of these disorders, selective regions of the intestine become inflamed.

Crohn disease is a condition of young adults characterized by inflammation of the inner lining of the GI tract. Although any region of the GI tract may be involved, the distal ileum is the most frequently and severely affected site. Inflammation involves the entire thickness of the intestinal wall, extending from the mucosa to the serosa. For reasons that are not clear, lengthy regions of the intestine having no trace of injury or inflammation may be followed abruptly by several inches of markedly diseased intestine. Symptoms include intermittent and relapsing episodes of abdominal cramping and pain, severe diarrhea, fatigue, weight loss, and malnutrition.

The age distribution and symptoms of **ulcerative colitis** are similar to those of Crohn disease, but ulcerative colitis involves only the large intestine. The rectum and descending colon are the first to show signs of inflammation and are generally the most severely affected. Also, in ulcerative colitis the inflammation is confined to the mucosa, instead of the full thickness of the intestinal wall. Finally, unlike Crohn disease, ulcerative colitis is associated with a profoundly increased risk of colon cancer.

Crohn disease and ulcerative colitis are distinctly different from a much more common disorder called **irritable bowel syndrome (IBS).** IBS is characterized by abnormal function of the colon with symptoms of crampy abdominal pain, bloating, constipation, and diarrhea. It occurs in about one in every five people in the United States, and is more common in women than men. Irritable bowel syndrome may be diagnosed if a medical evaluation has ruled out Crohn disease and ulcerative colitis. Although neither a cause nor a cure for IBS is known, most people can control their symptoms by reducing stress, changing their diet, and using certain medications. Recently, fecal transplants are being studied as a means to treat Crohn disease, ulcerative colitis, and IBS (see Clinical View 26.13: "Fecal Transplant").

Gross Anatomy of the Small Intestine

The smal intestine is a coiled, thin-walled tube about 1 inch in diameter and approximately 6 meters (20 feet) in length in the unembalmed cadaver. (It is much shorter in a living individual due to smooth muscle tone; see section 10.10c.) It extends from the pylorus of the stomach to the cecum of the large intestine; thus, it occupies a significant portion of the abdominal cavity. The small intestine consists of three specific segments: the duodenum, jejunum, and ileum.

The **duodenum** (dū-ō-dē′nŭm, dū-od′ĕ-nŭm; breadth of 12 fingers) forms the first segment of the small intestine. It is approximately 25 centimeters (10 inches) long and originates at the pyloric sphincter, which regulates movement of chyme from the stomach into the small intestine (see figure 26.9). The duodenum is arched into a C shape around the head of the pancreas and becomes continuous with the jejunum at the **duodenojejunal flexure** (flek′sher; *fleksura* = bend) (figure 26.15). Most of the duodenum is retroperitoneal, although the very initial portion is intraperitoneal.

The most significant function of the duodenum is to serve as an "anatomic blender" that allows for efficient chemical digestion. The duodenum receives (1) *acidic chyme* from the stomach and (2) the secretions from the accessory digestive organs. These secretions include *bile* from the liver and gallbladder and *pancreatic juice* from the pancreas. Thus, it is within the lumen of the duodenum where all of these substances are mixed, and where digestive enzymes (with many released from the pancreas) have contact with ingested molecules and chemical digestion primarily occurs.

The **jejunum** (jĕ-jū′nŭm; *jejunus* = empty) is the middle region of the small intestine (figure 26.15). Extending approximately 2.5 meters (7.5 feet), it makes up about two-fifths of the small intestine's total length. The jejunum is the primary region within the small intestine for nutrient absorption (i.e., where the chemically digested contents that are received from the duodenum are moved from the small intestine lumen into either the blood or the lymph; see figure 26.2*b*).

The **ileum** (il′ē-ŭm; *eiles* = twisted) is the last region of the small intestine. At about 3.6 meters (10.8 feet) in length, the ileum forms approximately three-fifths of the small intestine. Its distal end terminates at the **ileocecal** (il′ē-ō-sē′kăl) **valve,** a sphincter that controls the entry of materials from the small intestine into the large intestine. Absorption of digested materials continues in the ileum along with the absorption of bile salts (see section 26.4c) and vitamin B_{12} (see section 26.4e). Both the jejunum and the ileum are intraperitoneal organs and are suspended within the abdomen by the mesentery proper (see figure 26.3).

Small Intestine Structures That Increase Surface Area

Absorption of substances from within the small intestine into the blood or the lymph requires vast amounts of surface area of the epithelial lining (see figure 26.2*b*). Three structures increase this surface area: circular folds, villi, and microvilli. **Circular folds** (also called *plicae circulares*) are macroscopic structures that are easily seen by the naked eye: They are projections formed by both the mucosal and submucosal tunics of the small intestine (**figure 26.16**). Circular folds also act as "speed bumps" to slow down the movement of chyme and ensure that it remains within the small intestine for maximal nutrient absorption. Circular folds are more numerous in the duodenum and jejunum and least numerous in the ileum. Villi and microvilli are both microscopic structures described in the histology section.

WHAT DO YOU THINK?

4 Why are the circular folds much more numerous in the duodenum and least numerous in the ileum? How does the abundance of circular folds relate to the main functions of the duodenum?

Histology

Villi are microscopic structures of the small intestine mucosa. A **villus** (vil′ŭs; pl., vil′ī) is a small, fingerlike projection of the simple columnar epithelium and lamina propria of the mucosa. They, like circular folds, increase the surface area of the epithelial lining through which nutrients are absorbed. Villi are larger and most numerous in the jejunum, where much of the absorption takes place. The epithelium and lamina propria of each villus appears analogous to a glove (epithelium) covering a finger (lamina propria). Each villus contains an arteriole, a rich blood capillary network, and a venule. Most nutrients are absorbed into these blood capillaries. A **lacteal** is a type of lymphatic capillary also within the villus (described in section 21.1a). A lacteal is responsible for absorbing lipids and lipid-soluble vitamins that are too large to be absorbed by the blood capillaries (see sections 26.4c, e).

Microvilli (mī-krō-vil′ī; *micros* = small) are microscopic extensions of the plasma membrane of the simple columnar epithelial cells lining the small intestine (and average about 1 μm high) (see section 4.6c). Microvilli further increase the surface area of the small intestine. Individual microvilli are not clearly visible in light micrographs of the small intestine; instead, they appear as a fuzzy edge of the simple columnar cells called the **brush border.** Embedded within this brush border are various enzymes that complete the chemical digestion of most nutrients immediately before absorption (see section 26.4). Collectively, these are

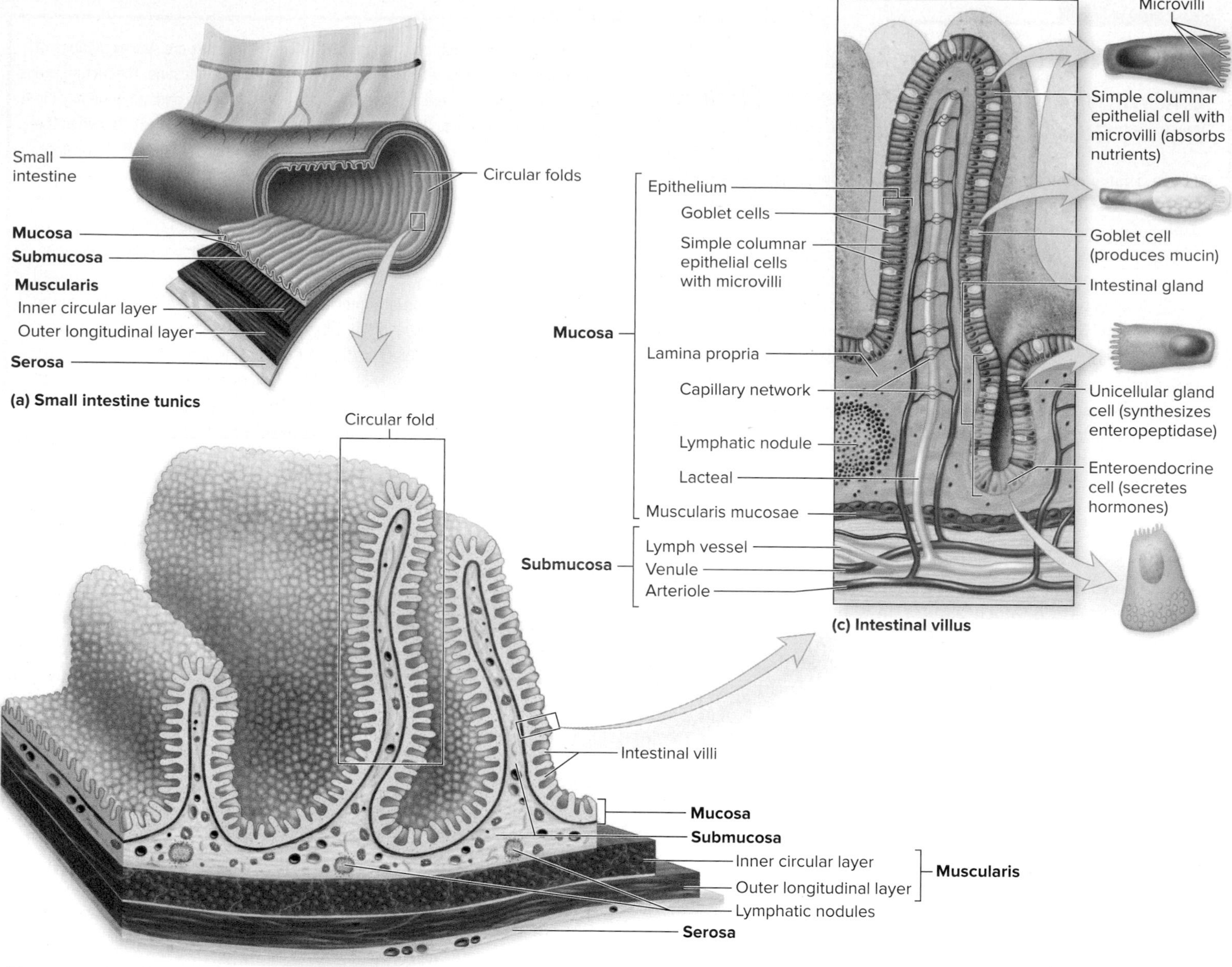

Figure 26.16 Histology of the Small Intestine. (*a*) The wall of the small intestine is formed by four tunics: mucosa, submucosa, muscularis, and serosa. (*b*) Circular folds are inward projections of both the mucosa and submucosa, whereas villi are fingerlike projections of only the epithelium and lamina propria of the mucosa. Both circular folds and villi increase the surface area of the small intestine. (*c*) The epithelial cells covering the surface of each villus have microvilli to further increase the surface area. The lamina propria within each villus houses both blood capillaries and lymphatic capillaries (lacteals) where substances are absorbed. AP|R

called **brush border enzymes.** Located in close proximity and also embedded within the plasma membrane are the required proteins for membrane transport of digested molecules. Histologic images of both villi and microvilli are shown in **figure 26.17**. Between the intestinal villi are invaginations of the mucosa called **intestinal glands** (also known as *intestinal crypts* or *crypts of Lieberkühn*), which secrete intestinal juice. These glands extend to the base of the mucosa and slightly resemble the anatomy of the gastric glands of the stomach (figure 26.16*c*).

Small Intestine Secretions

Four types of secretory cells of the intestinal epithelium contribute to the process of digestion (figure 26.16*c*). Three of these cell types produce **intestinal juice.** The fourth type of cell secretes hormones into the blood.

- **Goblet cells** produce mucin that when hydrated form mucus, which lubricates and protects the intestinal lining. These cells increase in number from the duodenum to the ileum, because more lubrication is needed as digested materials (and water) are absorbed and undigested materials (and less water) remain in the lumen.
- **Unicellular gland cells** synthesize **enteropeptidase,** an enzyme described in section 26.4b.
- The **enteroendocrine cells** release hormones such as CCK and secretin into the blood. We have already discussed the stomach-associated functions of these hormones (see section 26.2d). Several other functions are described in section 26.3c.

Another type of gland housed within the submucosal layer and found only in the proximal duodenum is called a **duodenal submucosal gland** (or *Brunner gland*) (not shown in figure 26.16). This gland produces a viscous, alkaline mucus secretion that protects the duodenum from the acidic chyme entering the duodenum from the stomach.

Motility of the Small Intestine

Smooth muscle activity of the muscularis within the small intestine wall has three primary functions: (1) mixing chyme with accessory gland

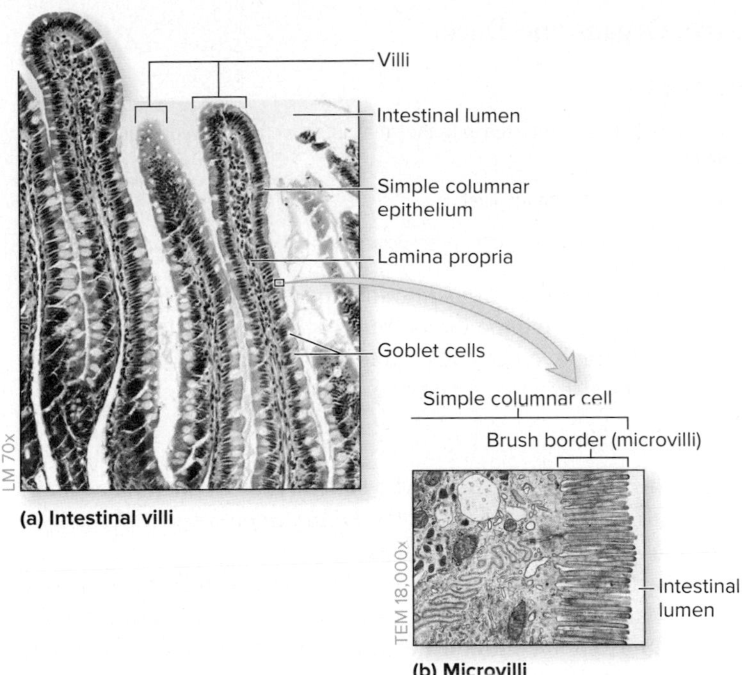

Figure 26.17 Intestinal Villi and Microvilli. Photomicrographs show (*a*) the internal structure of villi projecting into the intestinal lumen and (*b*) microvilli epithelium's apical surface.
(*a*) ©McGraw-Hill Education/Al Telser; (*b*) ©Dr. Lee Peachey

secretions, (2) moving the chyme continually against the brush border, and (3) propelling the contents through the small intestine toward the large intestine.

All these functions facilitate chemical digestion and absorption, employing the processes of segmentation and peristalsis. When chyme first enters the small intestine, segmentation is more prevalent than peristalsis. **Segmentation** mixes chyme with accessory gland secretions through a "backward-and-forward" motion (see figure 26.2*c*).

Peristalsis then propels material within the GI lumen by alternating contraction of the circular and longitudinal muscle layers in small regions. The rhythm of muscular contractions is more frequent in the duodenum than in the ileum; thus, the net movement of intestinal contents is toward the large intestine.

Regulation of Small Intestine Motility During the Intestinal Phase

The small intestine is the primary portion of the GI tract for both (1) the disassembly of complex molecules into a smaller, simpler form (chemical digestion) and (2) absorption of these smaller, simpler molecules (and the absorption of most of the water, electrolytes, and vitamins). Thus, for optimal chemical digestion and absorption, it is essential that small intestine motility is highly regulated.

Early Intestinal Phase and Segmentation *Segmentation* is more prevalent *early in the intestinal phase* (when chyme is entering from the stomach into the small intestine; see figure 26.17*c*). The function of segmentation is to thoroughly mix the chyme, accessory gland secretions, and intestinal juice. Recall that segmentation is the "back-and-forth" motion of a few centimeters for mixing contents. Imagine these events as placing the ingredients for a cake mix into an elongated balloon, then using your hands on the outside of the balloon to squeeze the contents back and forth until they are mixed.

The muscular contractions are initiated by **pacemaker cells** (*interstitial cells of Cajal*), which are located between the smooth muscle layers of the muscularis within the small intestine wall (as they are in the stomach wall). Electrical signals spread through the smooth muscle cells in the muscularis layer of the stomach via gap junctions to allow the single-unit smooth muscle (see section 10.10e) to contract. This establishes the basic rhythm of muscular contraction of the small intestine to cause segmentation. The specific rate the pacemaker cells spontaneously depolarize is more frequent than the rate within the stomach (which is less than four times per minute), but it differs depending upon the segment of the small intestine. The rhythm is the most frequent in the duodenum (about 12–14 times per minute) and declines along the length of the small intestine to the ileum (about 8–9 times per minute). The significance of the difference is that because contractions occur more frequently in the proximal portion of the small intestine, there is a slow net movement of chyme from the duodenum to the ileum—allowing time for optimal digestion and absorption. Note that while the frequency is typically unchanging, the *intensity* (force) of the contractions can be altered by both short reflexes involving the enteric nervous system and long reflexes involving the autonomic nervous system (see section 26.1d).

Late Intestinal Phase and Peristalsis *Peristalsis* is more prevalent (and segmentation decreases) *late in the intestinal phase* after most substances have been digested and absorbed. The function of peristalsis is to move the remaining undigested material, sloughed off epithelial cells, and bacteria toward the ileum for its movement into the large intestine. Recall that peristalsis is a directional movement. Imagine these events like placing your hands on an elongated balloon and squeezing from one end to the other—to empty the contents from the balloon.

The peristaltic muscular contractions are initiated by **motilin** hormone, which is released from the duodenum in progressively greater amounts late in the intestinal phase. Peristalsis is first initiated in the proximal portion of the duodenum. A peristaltic wave of muscle contraction squeezes on the contents and moves approximately 2 feet before the wave of contraction wanes and ceases. Another peristaltic wave is then initiated a small distance (distally) from the original point of initiation. This peristaltic wave travels about 2 feet before waning and ceasing. This pattern repeats itself down the length of the small intestine, moving the contents within the lumen toward the ileum. These successive waves of contractions are called a **migrating motility complex,** which requires about 2 hours from duodenum to ileum. The migrating motility complex repeats until all of the content is moved into the large intestine. Note that these events are also regulated by the enteric nervous system and long reflexes involving the autonomic nervous system (see section 26.1d).

Moving Chyme from the Small Intestine into the Large Intestine

The ileocecal valve, which is located between the ileum and the cecum (the first part of the large intestine) is typically contracted and closed, preventing movement of chyme from the small intestine into the large intestine. Regulatory processes including the gastroileal reflex, which is initiated during the gastric phase, function to open this valve. The **gastroileal** (gas′trō-il′ē-ăl) **reflex** (which is thought to have both short reflexes and long reflexes that involve the medulla oblongata) is initiated by food entering the stomach. As part of this reflex, the ileum contracts, ileocecal valve relaxes, and the cecum (the first part of the large intestine) relaxes. Thus, contents within the GI tract are moved from the ileum through the open ileocecal valve into the cecum. Then, the ileocecal valve contracts to prevent backflow from the cecum into the ileum. During the intestinal phase, the release of CCK causes relaxation of the ileocecal valve.

WHAT DID YOU LEARN?

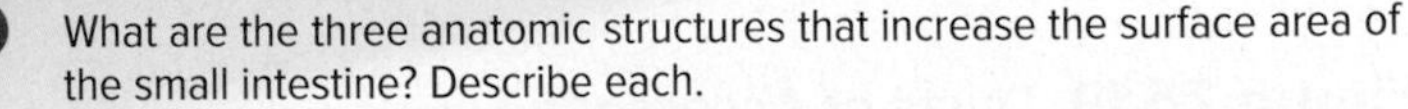
17 What are the three anatomic structures that increase the surface area of the small intestine? Describe each.

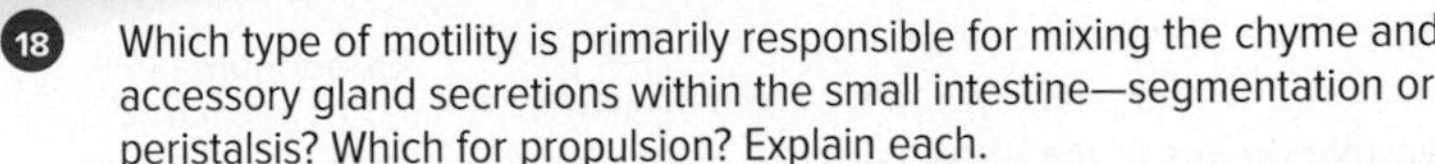
18 Which type of motility is primarily responsible for mixing the chyme and accessory gland secretions within the small intestine—segmentation or peristalsis? Which for propulsion? Explain each.

26.3c Accessory Digestive Organs and Ducts

LEARNING OBJECTIVES

26. Describe the accessory digestive organs associated with the small intestine and the contributions of each to digestive processes.

27. Explain how both blood and bile flow through the liver.

28. Discuss the regulation of the accessory digestive glands associated with the small intestine.

Three accessory digestive organs release secretions into the duodenum: the liver, the gallbladder, and the pancreas. Here we examine the ducts from each organ to the duodenum, the anatomy and histology of each organ, and their products conveyed to the small intestine that contribute to the digestion of the chyme arriving from the stomach.

Accessory Digestive Organ Ducts

A series of ducts deliver secretions from the accessory digestive organs to the duodenum of the small intestine **(figure 26.18)**. These ducts include the *biliary apparatus* from the liver and gallbladder, and the *pancreatic ducts* from the pancreas.

The **biliary** (bil′ē-ăr-ē) **apparatus** is a network of thin ducts that include the right and left hepatic ducts, which drain the right and left lobes of the liver, respectively. The right and left **hepatic ducts** merge to form a single **common hepatic duct.** The union of the **cystic duct** from the gallbladder and the common hepatic duct forms the **common bile duct**, which joins with the hepatopancreatic ampulla (described shortly).

The **pancreatic ducts** include both the main pancreatic duct and the accessory pancreatic duct. The **main pancreatic duct** transports the majority of the pancreatic juice. It, along with the common bile duct, joins with the hepatopancreatic ampulla. The **accessory pancreatic duct** is an an alternative way that a small amount of pancreatic juice may also enter the duodenum. This duct penetrates the duodenal wall, forming the **minor duodenal papilla.**

The **hepatopancreatic ampulla** (or *ampulla of vater*) is a swelling either adjacent to or within the posterior duodenal wall, which penetrates through the duodenal wall forming a small projection called the **major duodenal papilla.** The hepatopancreatic ampulla receives bile from the common bile duct (coming from the liver and gallbadder) and pancreatic juice from the main pancreatic duct. The release of these accessory gland secretions (bile and pancreatic juice) is regulated by the **hepatopancreatic sphincter** that is located within the ampulla. This sphincter is normally closed, preventing accessory digestive gland secretions from entering the duodenum. Relaxation and opening of this sphincter permits the flow of these secretions into the duodenum (a process stimulated by the cholecystokinin (CCK) hormone).

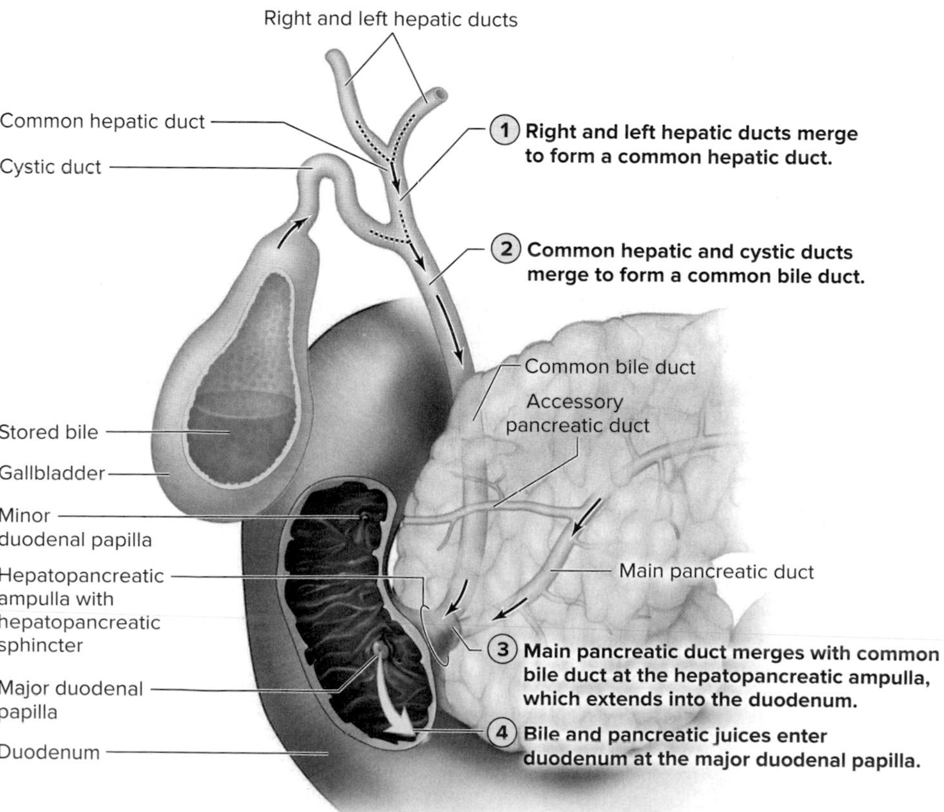

Right and left hepatic ducts
Common hepatic duct
Cystic duct
1 Right and left hepatic ducts merge to form a common hepatic duct.
2 Common hepatic and cystic ducts merge to form a common bile duct.
Common bile duct
Accessory pancreatic duct
Stored bile
Gallbladder
Minor duodenal papilla
Hepatopancreatic ampulla with hepatopancreatic sphincter
Main pancreatic duct
3 Main pancreatic duct merges with common bile duct at the hepatopancreatic ampulla, which extends into the duodenum.
Major duodenal papilla
Duodenum
4 Bile and pancreatic juices enter duodenum at the major duodenal papilla.

Figure 26.18 Ducts of Accessory Digestive Organs. Various ducts merge to transport bile and pancreatic juice from accessory digestive organs to the duodenum.

Liver

The **liver** is an accessory digestive organ located in the right upper quadrant of the abdomen, immediately inferior to the diaphragm (see figure 26.1). It has numerous functions (described in section 27.6), but its main function in digestion is the production of bile.

Gross Anatomy of the Liver The liver is the largest internal organ, weighing 1 to 2 kilograms (2 to 4 pounds), and it constitutes approximately 2% of an adult's body weight. The liver is covered by a connective tissue capsule except at the porta hepatis. Covering the connective tissue capsule is a layer of visceral peritoneum, except for a small region on its diaphragmatic surface called the bare area **(figure 26.19)**.

The liver is composed of four partially separated lobes and is supported by two ligaments. The major lobes are the **right lobe** and the **left lobe.** The right lobe is separated from the smaller left lobe by the falciform ligament, a peritoneal fold that secures the liver to the internal surface of the anterior abdominal wall, as discussed in section 26.1e. In the inferior free edge of the falciform ligament lies the **round ligament of the liver** (*ligamentum teres*), which represents the remnant of the fetal umbilical vein (see section 20.12a). Within the right lobe are the **caudate** (kaw′dāt; *cauda* = tail) **lobe** and the **quadrate** (kwah′drāt; *quadrates* = square) **lobe.** The caudate lobe is adjacent to the inferior vena cava, and the quadrate lobe is adjacent to the gallbladder.

Along the inferior surface of the liver are several structures that collectively resemble the letter H. The gallbladder and the round ligament of the liver form the vertical superior parts of the H; the **inferior vena cava** and the **ligamentum venosum** form the vertical inferior parts. (Recall from section 20.12 that the ligamentum venosum is a remnant of the ductus venosus in the embryo. This vessel shunted blood from the umbilical vein to the inferior vena cava.) Finally, the **porta** (pōr′tă; gate); **hepatis** (hep′ă-tis; *hepatikos* = liver), the horizontal crossbar of the H, is the site at which blood and lymph vessels, bile ducts, and nerves (not shown) extend from the liver. In particular, the hepatic portal vein and branches of the hepatic artery proper enter at the porta hepatis.

Histology The liver's connective tissue capsule branches throughout the organ and forms septa that partition the liver into thousands of microscopic polyhedral **hepatic lobules,** which are the structural and functional units of the liver **(figure 26.20)**. Within hepatic lobules are liver cells called **hepatocytes** (hep′ă-tō-sīt). At the periphery of each lobule are several **portal triads,** composed of a bile ductule, and *microscopic branches* of both the hepatic portal vein and the hepatic artery. At the center of each lobule is a **central vein** that drains the blood flow from the lobule. Central veins collect the blood and merge throughout the liver to form numerous **hepatic veins** that eventually empty into the inferior vena cava.

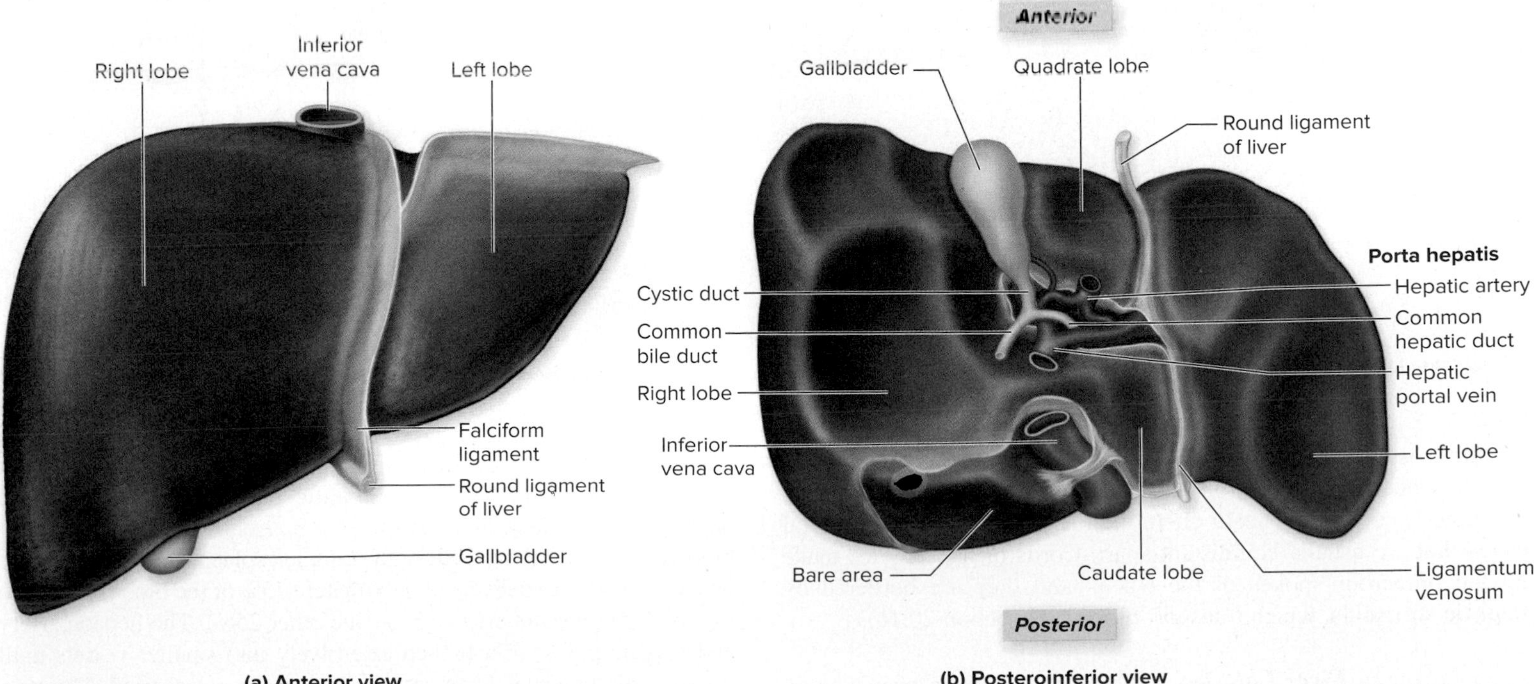

Figure 26.19 Gross Anatomy of the Liver. The liver is located in the upper right quadrant of the abdomen. (*a*) Anterior and (*b*) posteroinferior views show the four lobes of the liver. AP|R

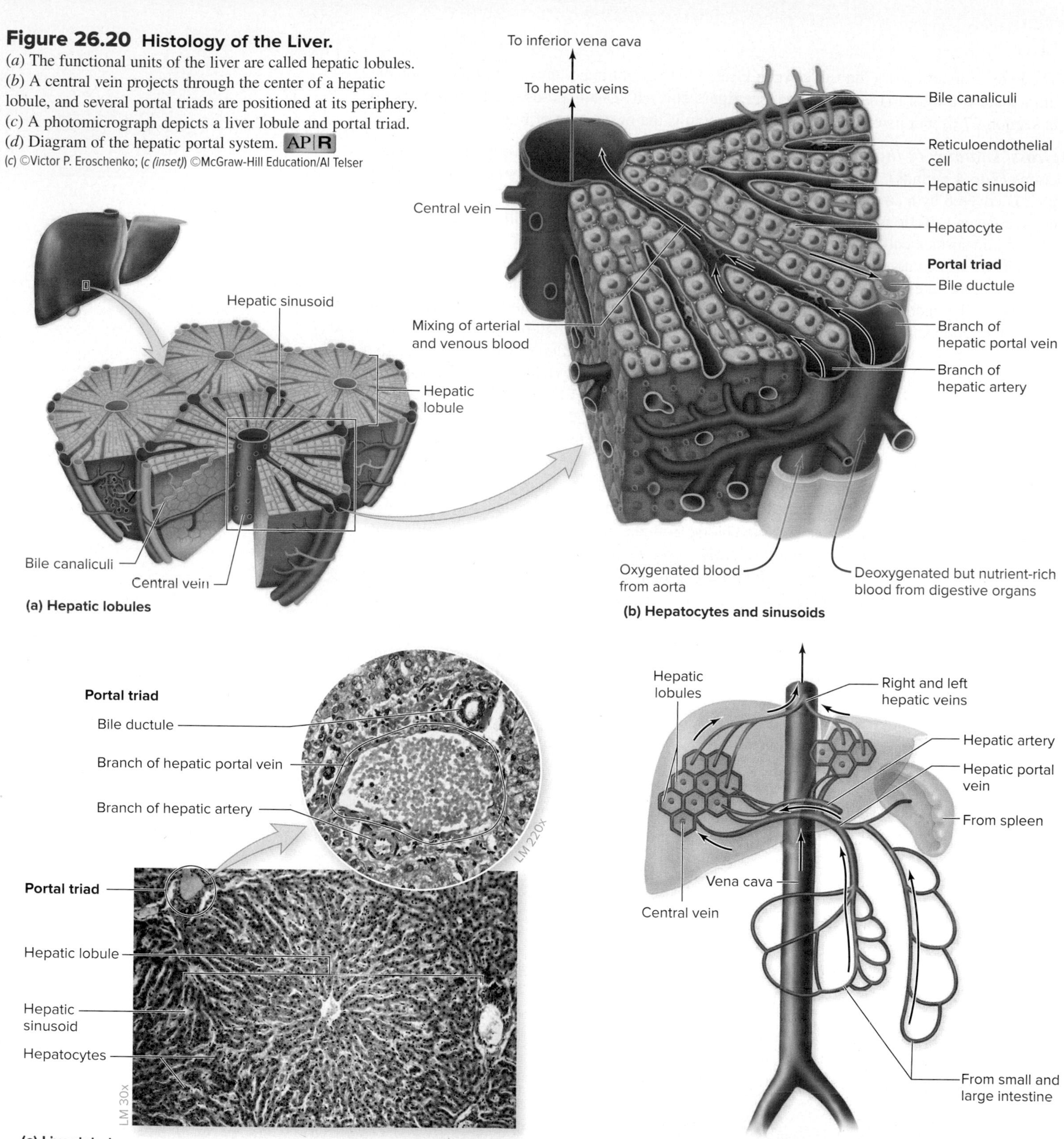

Figure 26.20 Histology of the Liver. (*a*) The functional units of the liver are called hepatic lobules. (*b*) A central vein projects through the center of a hepatic lobule, and several portal triads are positioned at its periphery. (*c*) A photomicrograph depicts a liver lobule and portal triad. (*d*) Diagram of the hepatic portal system. AP|R

(c) ©Victor P. Eroschenko; (c *(inset)*) ©McGraw-Hill Education/Al Telser

In cross section, a hepatic lobule looks like a side view of a bicycle wheel. The central vein is like the hub of the wheel. At the circumference of the wheel (where the tire would be) are the portal triads that are usually equidistant apart. Cords of hepatocytes make up the numerous spokes of the wheel, and they are bordered by **hepatic sinusoids,** which transport blood (see section 20.1c).

Blood Flow in Liver Lobules The cells of the liver receive blood from two sources; one is oxygenated and the other is deoxygenated (see section 20.10d). The **hepatic artery** is a branch of the celiac trunk that extends off of the descending abdominal aorta and transports oxygenated blood to the liver. The **hepatic portal vein** is part of the hepatic portal system and transports deoxygenated and nutrient-rich blood from the capillary beds of the GI tract, spleen, and pancreas. The hepatic portal vein delivers approximately 75% of the blood volume to the liver (the hepatic artery brings the other 25%). The hepatic artery and hepatic portal vein branch extensively into smaller vessels until microscopic branches form components of the portal triad. The oxygenated blood from the hepatic artery branch (within the portal triad) and the deoxygenated blood of the hepatic portal vein branch (within

INTEGRATE

CONCEPT CONNECTION

The hepatic portal system is a venous network that drains the digestive organs and shunts the blood to the liver (see section 20.10d). The blood exits the liver through hepatic veins that merge with the inferior vena cava.

The GI tract absorbs digested nutrients. These nutrients must be transported to and processed within the liver. The liver also detoxifies any potentially harmful agents that have been absorbed into the GI blood vessels. It is much more efficient to transport blood containing these substances first to the liver via the hepatic portal system for processing, before it is distributed throughout the body.

the portal triad) both enter a sinusoid where the blood is "processed." (Recall from section 20.1c that sinusoids are thin-walled capillaries with large gaps between these cells, which make the sinusoids more permeable than other capillaries.) Blood then drains into the central vein of the lobule. Central veins collect the blood from each lobule and merge throughout the liver to ultimately form several **hepatic veins** that empty into the inferior vena cava.

Several significant events occur as blood is transported through hepatic sinusoids:

- Nutrients are absorbed from the sinusoids and enter the hepatocytes.
- Oxygen is delivered to hepatocytes for aerobic cellular respiration (see section 3.4).
- **Stellate cells** (or *Kupffer cells*), which are macrophages that line the liver sinusoid, and engage in phagocytosis of potentially harmful substances (e.g., microbes). (Stellate cells are fixed macrophages of the immune system (see section 22.2a.)

Bile: Its Formation and Flow in Liver Lobules **Bile** is a yellowish-green, alkaline fluid containing mostly water, bicarbonate ions (HCO_3^-), bile salts, bile pigments (e.g., bilirubin), cholesterol, lecithin (a phospholipid), and mucin. Hepatocytes of the liver produce bile at a rate of 0.5 to 1 liter per day.

Bile is released from hepatocytes into bile canaliculi (figure 26.20*b*). These small channels transport bile to bile ductules of portal triads. (Observe that bile flow, which is away from the central vein to the portal triad, is in the opposite direction of blood flow, which moves from the portal triad to the central vein.) Bile within the bile ductules flows into progressively larger bile ducts until reaching either the right or left hepatic duct (see figure 26.18). The accessory digestive organ ducts transport the bile to the duodenum.

The release of bile into the small intestine has several functions, including

- Neutralizing acidic chyme within the small intestine through bicarbonate ions
- Mechanical digestion of lipids by bile salts and lecithin (see section 26.4c)
- Elimination of bilirubin, a waste product of erythrocyte destruction (see section 18.3b)

Gallbladder

Attached to the inferior surface of the liver (figure 26.19), the **gallbladder** (or *cholecyst*) is a saclike organ that stores, concentrates, and releases

INTEGRATE

CLINICAL VIEW 26.8

Cirrhosis of the Liver

Liver cirrhosis (sir-rō′sis; *kirrhos* = yellow) results when hepatocytes have been destroyed and are replaced by fibrous scar tissue. This scar tissue, which results from collagen synthesis by stellate cells in the liver, often surrounds isolated nodules of regenerating hepatocytes. The fibrous scar tissue also compresses (1) the blood vessels, resulting in **hepatic portal hypertension** (high blood pressure in the hepatic portal venous system), and (2) bile ducts in the liver, which impedes bile flow.

Liver cirrhosis is caused by injury to the hepatocytes, as may result from chronic alcoholism, liver diseases, or certain drugs or toxins. Most frequently, viral infections from either **hepatitis B** or **hepatitis C** produce chronic hepatitis. Other disorders that result in liver cirrhosis include some inherited diseases, chronic biliary obstruction, and biliary cirrhosis.

Early stages of liver cirrhosis may be asymptomatic. However, once liver function begins to falter, the individual complains of fatigue, weight loss, and nausea and may have pain in the right upper quadrant. During an exam, a doctor may palpate an abnormally small and hard liver. To confirm the diagnosis, a liver biopsy is done to obtain a small portion of liver tissue through a needle passed into the liver; the cells are then examined microscopically. The fibrosis and scarring of liver cirrhosis are irreversible. However, further scarring may be slowed or prevented by treating the cause of the cirrhosis (e.g., alcoholism, hepatitis).

Advanced liver cirrhosis may have a variety of complications :

- **Jaundice** (yellowing of the skin and sclerae of the eyes) occurs when the liver's ability to eliminate bilirubin (a component of bile) is impaired.
- **Edema** (e-dē′mă), the accumulation of fluid in body tissues, is evident due to reduced formation and release of albumin (which results in a decrease in colloid osmotic pressure; see section 20.3).
- **Ascites** (ă-sī′tēz; fluid accumulation in the abdomen) may also develop because of decreased albumin production.
- Intense itching occurs when bile products are deposited in the skin.
- Toxins in the blood and brain accumulate because the liver cannot effectively process them.
- Hepatic portal hypertension may lead to dilated veins of the inferior esophagus (esophageal varices).

End-stage liver cirrhosis may be treated only with a liver transplant. Otherwise, death results either from progressive liver failure or from the complications.

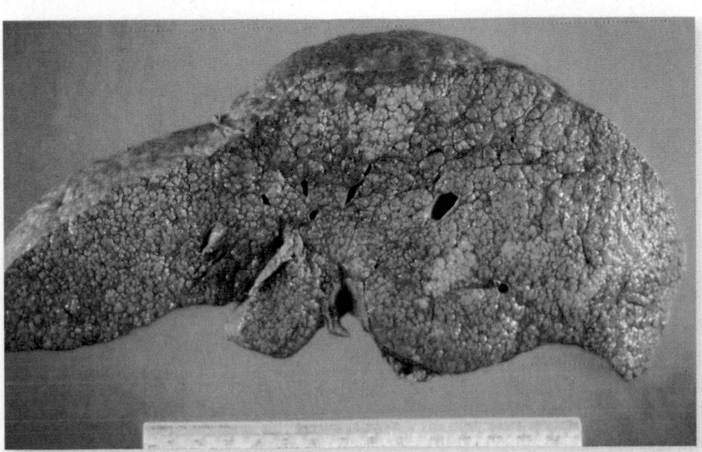

This gross specimen depicts a type of nodular cirrhosis of the liver.
©Yoav Levy/Medical Images

INTEGRATE

CLINICAL VIEW 26.9

Gallstones (Cholelithiasis)

Concentration of bile within the gallbladder may lead to the eventual formation of **gallstones.** Gallstones occur twice as frequently in women as in men and are more prevalent in developed countries. Obesity, increasing age, female sex hormones, Caucasian ethnicity, and lack of physical activity are all risk factors for developing gallstones.

The term **cholelithiasis** (kō′lē-li-thī′ă-sis; *chol* = bile, *lithos* = stone, *iasis* = condition) refers to the presence of gallstones in either the gallbladder or the biliary apparatus. Gallstones are typically formed from condensations of either cholesterol or calcium and bile salts. These stones may vary from the tiniest grains to almost golf-ball-sized. The majority of gallstones are asymptomatic until a gallstone becomes lodged in the neck of the cystic duct, causing the gallbladder to become inflamed **(cholecystitis)** and dilated. The most common symptom is severe pain (called biliary colic) perceived in the right hypochondriac region (see figure 1.10), between the scapulae, or in the area of the right posterior deltoid region. Nausea and vomiting may occur, along with indigestion and bloating. Symptoms are typically worse after eating a fatty meal. Treatment consists of surgical removal of the gallbladder, called **cholecystectomy** (kō′lē-sis-tek′tō-mē; *kystis* = bladder, *ektome* = excision). Following surgery, the liver continues to produce bile, even in the absence of the gallbladder, but concentrating bile (which is a function of the gallbladder) no longer occurs.

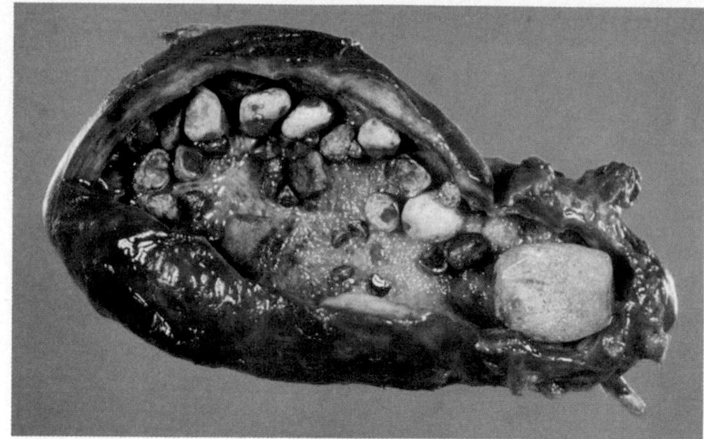

A photograph of gallstones in a gallbladder.

bile that the liver produces. The gallbladder has three tunics: an inner mucosa, a middle muscularis, and an external serosa. The mucosa is thrown into folds that permit distension of the wall as the gallbladder fills with bile. The gallbladder drains bile into the **cystic** (sis′tik; *cysto* = bladder) **duct,** which connects to the common bile duct (see figure 26.18).

At the neck of the gallbladder, a sphincter valve controls the flow of bile into and out of the gallbladder. Bile enters the gallbladder when the hepatopancreatic sphincter associated with the hepatopancreatic ampulla is closed. It backs up through both the common bile duct and the cystic duct into the gallbladder. The gallbladder can hold approximately 40 to 60 milliliters of concentrated bile. Concentrated bile is transported from the gallbladder through the cystic duct and then the common bile duct through the hepatopancreatic ampulla into the duodenum.

WHAT DO YOU THINK?

5 If your gallbladder were surgically removed, how would this affect your digestion of fatty meals? What diet alterations might you have to make after this surgery?

Pancreas

The **pancreas** has both endocrine and exocrine functions. Endocrine cells produce and secrete hormones such as insulin and glucagon (see section 17.10b). Exocrine cells (called *acinar cells*) produce pancreatic juice to assist with digestive activities. The pancreas is the "workhorse" for providing digestive enzymes into the small intestine for chemical digestion. Disorders that affect either (1) the pancreatic ducts that lead from the pancreas into the duodenum (e.g., Cystic fibrosis; see Clinical View 26.17: "Cystic Fibrosis and the Pancreas") or (2) the pancreas (e.g., pancreatic cancer; see Clinical View 26.10: "Pancreatic Cancer") have serious and potentially fatal effects on the ability to digest and absorb nutrients.

Gross Anatomy of the Pancreas The pancreas is approximately 5 to 6 inches in length and about 1 inch thick. It is a retroperitoneal organ that extends horizontally from the duodenum toward the left side of the abdominal cavity, where it has contact with the spleen. The pancreas exhibits a wide **head** adjacent to the curvature of the duodenum; a central, elongated **body** projecting toward the left lateral abdominal wall; and a **tail** that tapers as it approaches the spleen **(figure 26.21)**.

Histology The pancreas contains modified simple cuboidal epithelial cells called **acinar** cells that are arranged in saclike **acini** (sing., *acinus* = grape) **(figure 26.22)**. These cells are organized into large clusters termed **lobules.** Acinar cells produce and release digestive enzymes. Small ducts lead from each acinus into larger ducts that empty into either the main pancreatic duct or the accessory pancreatic duct, which lead to the duodenum (as described). The simple cuboidal epithelial cells lining the pancreatic ducts have the important function of secreting alkaline fluid containing bicarbonate ion (HCO_3^-). This weak base functions to neutralize acidic chyme entering the small intestine.

Pancreatic Secretions Together, secretions of acinar cells and cells that line the pancreatic ducts form **pancreatic juice.** Pancreatic juice (approximately 1 to 1.5 liters per day) is an alkaline fluid containing mostly water, HCO_3^-, and a versatile mixture of digestive enzymes, which are described in detail in section 26.4. These enzymes include the following:

- Pancreatic amylase to digest starch
- Pancreatic lipase for the digestion of triglycerides
- Inactive proteases (trypsinogen, chymotrypsinogen, and procarboxypeptidase) that, when activated, digest protein
- Nucleases for the digestion of nucleic acids (DNA and RNA)

Regulation of Accessory Structures Several of the same processes that regulate stomach motility and secretions described in sections 26.2d and 26.3b also control the release of secretions from the accessory digestive organs. Recall that regulation of the stomach is organized into three phases: cephalic phase, gastric phase, and intestinal phase. The increase in vagal stimulation in the *cephalic phase* and *gastric phase,* in addition to stimulating stomach motility and secretion, also activates the pancreas to release pancreatic juice. Recall that in the intestinal phase, both cholecystokinin (CCK) and secretin are released. Cholecystokinin is a hormone released from the small intestine primarily in response to free fatty acids in chyme. The functions of CCK include:

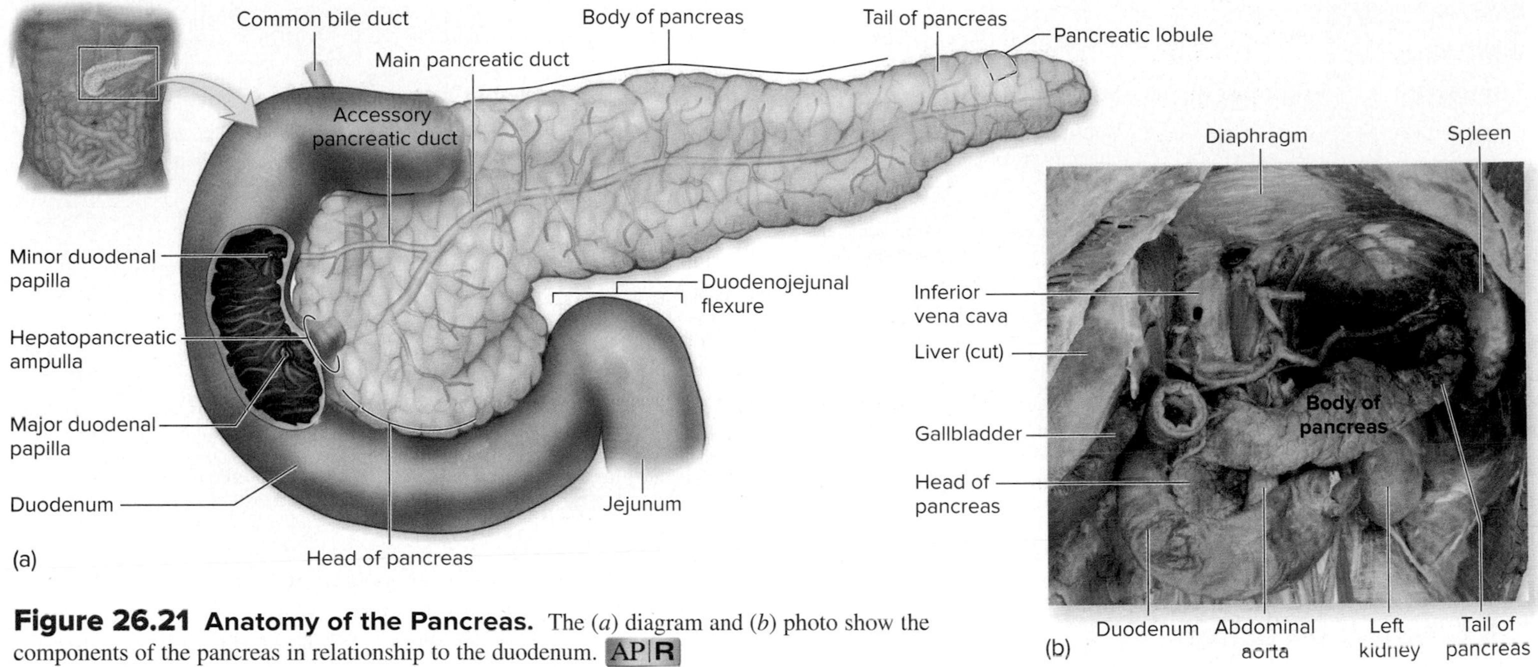

Figure 26.21 Anatomy of the Pancreas. The (*a*) diagram and (*b*) photo show the components of the pancreas in relationship to the duodenum. AP|R

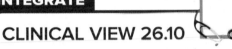

(*b*) ©McGraw Hill Education/Christine Eckel

- Initiating smooth muscle within the gallbladder wall to strongly contract, causing the release of concentrated bile (this primary function of stimulating the gallbladder, also called the *cholecyst*, is how the name cholecystokinin is derived)
- Stimulating the pancreas to release enzyme-rich pancreatic juice
- Relaxing the smooth muscle within the hepatopancreatic ampulla, allowing entry of bile and pancreatic juice into the small intestine

Secretin is released from the small intestine primarily in response to an increase in chyme acidity. Secretin primarily causes the release of an alkaline solution that contains HCO_3^- from both the liver and ducts of the pancreas. Upon entering the small intestine, this alkaline fluid helps neutralize the acidic chyme. (Recall from section 26.2d that CCK and secretin also inhibit stomach motility and release of gastric secretions.) Hormones that regulate digestion are summarized in **table 26.1**.

WHAT DID YOU LEARN?

19 Where do deoxygenated, nutrient-rich blood and oxygenated blood first come together within a liver lobule?

20 Does the liver produce digestive enzymes? If not, what substance does it produce that assists in digestion?

21 What are the primary functions of pancreatic juice?

INTEGRATE

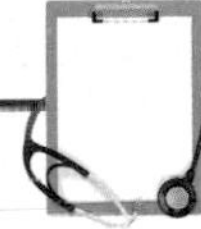

CLINICAL VIEW 26.10

Pancreatic Cancer

The pancreas is both an exocrine gland (producing digestive enzymes that are released into the duodenum) and an endocrine gland (producing the hormones insulin and glucagon). **Pancreatic cancer** is cancer of pancreatic cells, which typically originates in exocrine cells of the pancreas (in approximately 95% of cases). The prognosis for pancreatic cancer, like most other types of cancers, is more favorable with early detection—before the cancer has metastazied to the lymph nodes. Unfortunately, early detection of pancreatic cancer is difficult due to the (1) lack of a screening test for pancreatic cancer and (2) the absence of signs and symptoms associated with pancreatic cancer in the early stages. It is not until advanced stages of the disease that an individual might experience symptoms or signs (e.g., abdominal pain, jaundice from blockage of the accessory organ ducts, loss of appetite, weight loss). Consequently, pancreatic cancer is often detected in advanced stages of the disease when complete surgical removal of the tumor is no longer possible. It is for this reason (its late detection) that pancreatic cancer is often fatal.

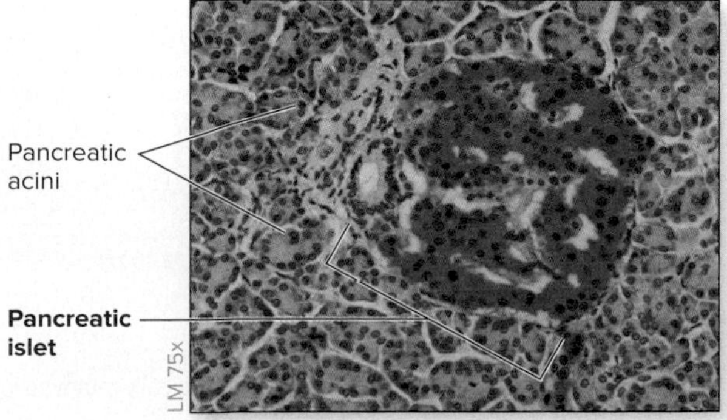

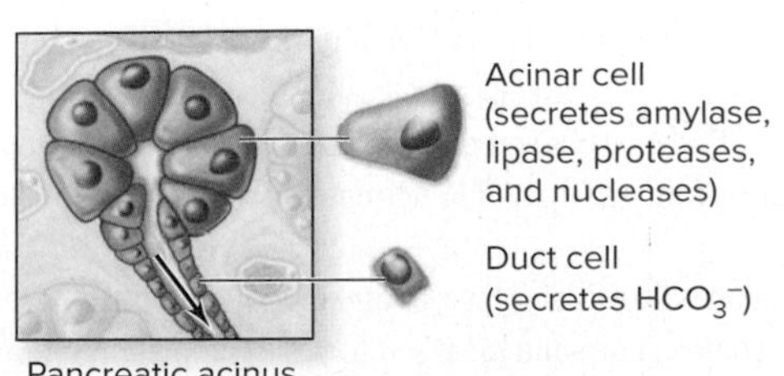

Figure 26.22 Histology of the Pancreas. Photomicrograph of the histology of pancreatic acini and pancreatic islets, with diagram of pancreatic acinus.

©Carolina Biological Supply Company/Medical Images

Table 26.1	Hormones That Control Digestion[1]		
Hormone	**Secreted by**	**Stimulus for Release**	**Primary Target(s) and Effects**
Gastrin	G-cells in stomach	Bolus in stomach—especially if bolus contains proteins	Stomach: Stimulates stomach motility and release of gastric secretions from parietal cells (hydrochloric acid [HCl]) and chief cells (pepsinogen) Pyloric sphincter: Stimulates contraction
Cholecystokinin (CCK)	Enteroendocrine cells of small intestine	Chyme containing amino acids and fatty acids entering small intestine	Stomach: Inhibits stomach motility and gastric secretion Gallbladder: Stimulates release of bile Pancreas: Stimulates release of enzyme-rich pancreatic juice Hepatopancreatic sphincter: Causes relaxation Ileocecal valve: Causes relaxation
Secretin	Enteroendocrine cells of small intestine	Primarily with increase in acidity of chyme entering small intestine	Stomach: Inhibits gastric secretion and stomach motility Liver: Stimulates secretion of alkaline solution Pancreas: Stimulates secretion of alkaline solution from pancreatic ducts

1. Motilin is not included (see section 26.3b).

26.3d Large Intestine

LEARNING OBJECTIVES

29. Name the three major regions of the large intestine and four segments of the colon of the large intestine.

30. Explain the distinguishing histologic features of the large intestine.

31. Describe the bacterial action that takes place in the large intestine.

The **large intestine**, also called the *large bowel,* is a relatively wide tube that is significantly shorter than the small intestine. It is called the "large" intestine because its diameter is greater than that of the small intestine. Approximately 1 liter of digested material passes from the small intestine to the large intestine daily. Most nutrients (including water, electrolytes, and vitamins) have been absorbed within the small intestine. The large intestine absorbs water (about 0.8 liter per day, for an adult). The large intestine also absorbs some electrolytes (primarily sodium [Na^+] and chloride [Cl^-] ions) from the remaining digested material that enters it from the small intestine (as well as vitamins B and K, which are synthesized by bacteria within the large intestine). It is estimated that only 200 milliliters of the water entering the colon daily is lost in feces (see figure 25.4). Thus, the watery chyme that first enters the large intestine, including all of the undigested materials as well as the waste products secreted by the accessory digestive organs (e.g., bilirubin by the liver), solidifies and is compacted into feces, or fecal material. The large intestine then stores this fecal material until it is eliminated through defecation.

Gross Anatomy of the Large Intestine

The large intestine has a diameter of 6.5 centimeters (2.5 inches) and an approximate length of 1.5 meters (5 feet) from its origin at the ileocecal junction to its termination at the anus (**figure 26.23**). Three major regions constitute the large intestine: the cecum, the colon, and the rectum.

Cecum The **cecum** (sē′kŭm; *caecus* = blind) is a blind sac. It is the first portion of the large intestine and located in the right lower abdominal quadrant. This pouch extends inferiorly from the ileocecal valve. Chyme enters the cecum from the ileum. Projecting inferiorly from the posteromedial region of the cecum is the **vermiform** (ver′mi-fōrm; *vermis* = worm) **appendix** (ă-pen′diks; appendage), a thin, fingerlike sac lined by lymphocyte-filled lymphatic nodules (see section 21.4d). Both the cecum and the vermiform appendix are intraperitoneal organs. Research studies suggest that the appendix may harbor bacteria that are beneficial to the function of the colon.

Colon At the level of the ileocecal valve, the second region of the large intestine, the **colon,** begins and forms an inverted U-shaped

INTEGRATE

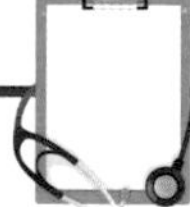

CLINICAL VIEW 26.11

Appendicitis

Inflammation of the appendix is called **appendicitis** (ă-pen-di-sī′tis). Most cases of appendicitis occur because fecal matter obstructs the appendix, although sometimes an appendix becomes inflamed without any obstruction. As the tissue becomes inflamed, the appendix swells, the blood supply is compromised, and bacteria may proliferate in the wall. Untreated, the appendix may burst and release its contents into the peritoneum, causing a massive and potentially deadly infection called peritonitis (see Clinical View 26.1: "Peritonitis").

During the early stages of acute appendicitis, the smooth muscle wall contracts and goes into spasms. Because this smooth muscle is innervated by the autonomic nervous system, pain is referred to the T10 dermatome around the umbilicus (see section 16.2b).

As the inflammation worsens and the parietal peritoneum becomes inflamed as well, the pain becomes sharp and localized to the right lower quadrant of the abdomen. Individuals with appendicitis typically experience nausea or vomiting, abdominal tenderness in the inferior right quadrant, a low fever, and an elevated leukocyte count. An inflamed appendix is surgically removed in a procedure called an **appendectomy.**

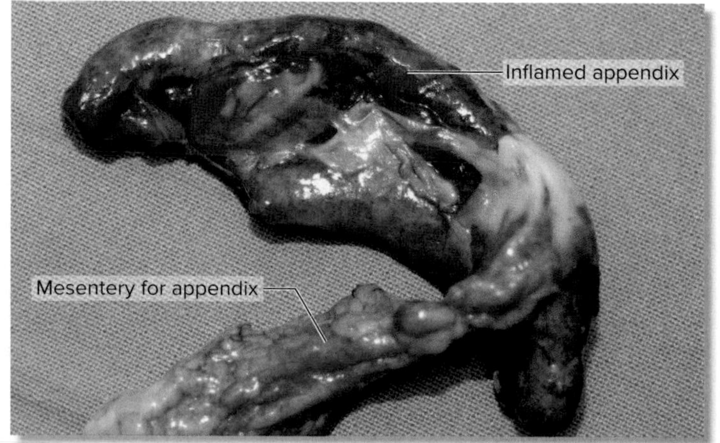

Inflamed appendix.
©Medicimage/Medical Images

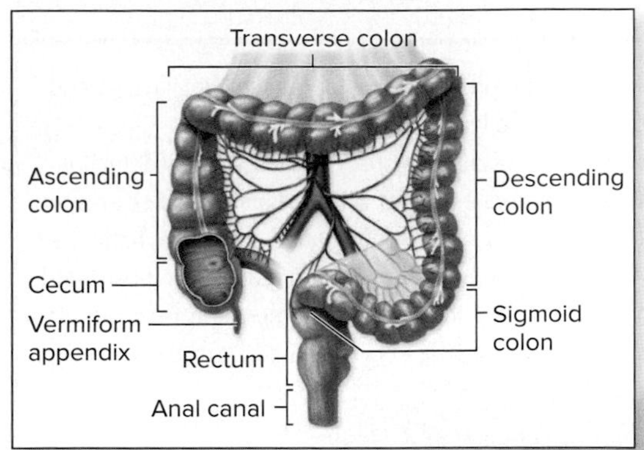

Figure 26.23 Gross Anatomy of the Large Intestine. (*a*) Anterior view of the large intestine that forms the distal end of the GI tract. (*b*) Details of the anal canal. APR

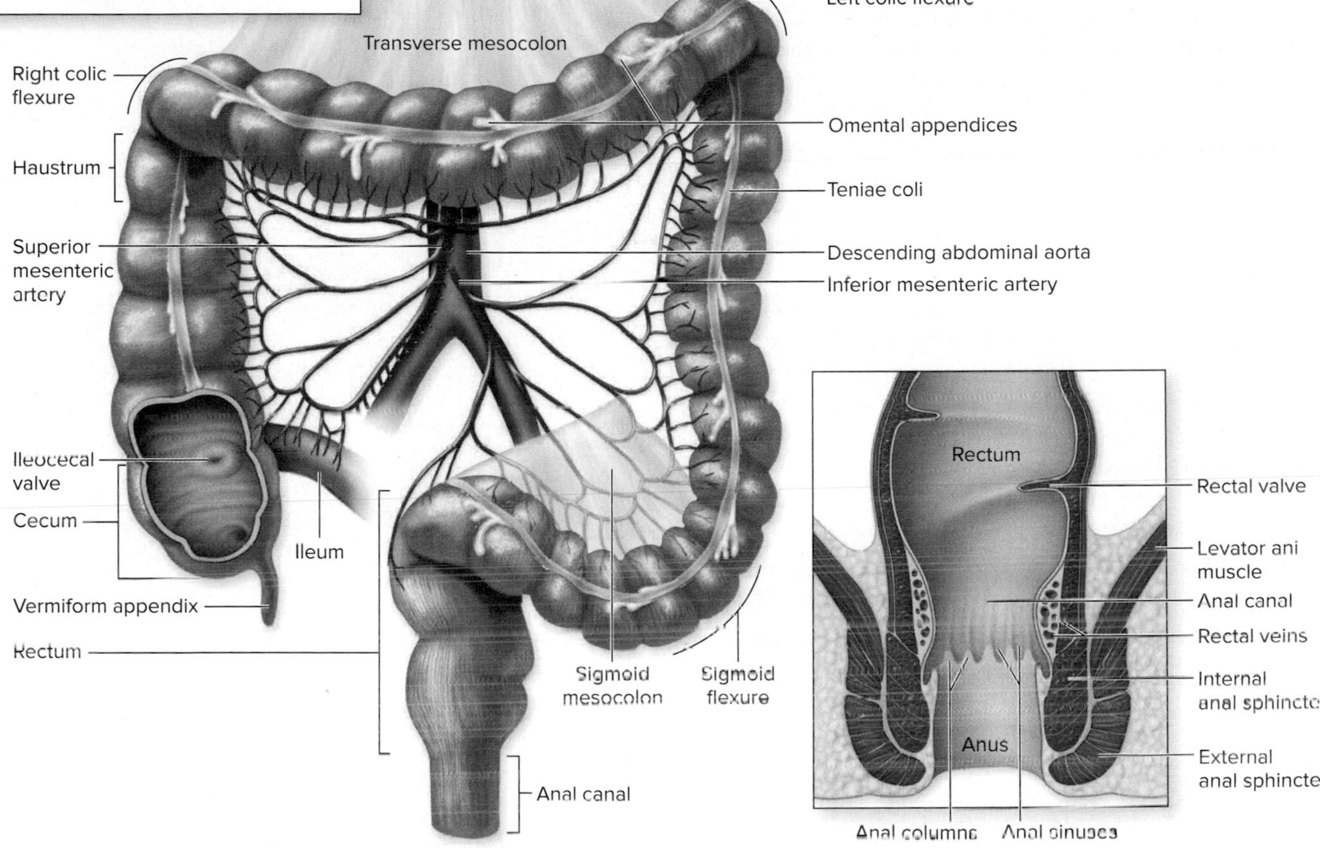

arch. The colon is partitioned into four segments: the ascending colon, transverse colon, descending colon, and sigmoid colon.

The **ascending colon** originates at the ileocecal valve and extends superiorly from the superior edge of the cecum along the right lateral border of the abdominal cavity. The ascending colon is retroperitoneal, since its posterior wall directly adheres to the posterior abdominal wall, and only its anterior surface is covered with peritoneum. As it approaches the inferior surface of the liver, the ascending colon makes a 90-degree turn toward the left side and anterior region of the abdominal cavity. This bend in the colon is called the **right colic** (kol′ik) **flexure,** or the *hepatic flexure.*

The **transverse colon** originates at the right colic flexure and curves slightly anteriorly as it projects horizontally to the left across the anterior region of the abdominal cavity. The transverse colon is intraperitoneal. As the transverse colon approaches the spleen in the left upper quadrant of the abdomen, it makes a 90-degree turn inferiorly and posteriorly. The resulting bend in the colon is called the **left colic flexure,** or the *splenic flexure.*

The **descending colon** is retroperitoneal and located along the left side of the abdominal cavity and slightly posterior. It originates at the left colic flexure and descends vertically to the sigmoid colon.

The **sigmoid** (sig′moyd; resembling letter S) **colon** originates at the **sigmoid flexure** and turns inferomedially into the pelvic cavity. The sigmoid colon, like the transverse colon, is intraperitoneal. The sigmoid colon terminates at the rectum. Recall from section 26.1e that a type of mesentery, called the *mesocolon,* attaches each section of the colon to the posterior abdominal wall, with the mesocolon of each region specifically named (e.g., ascending mesocolon, transverse mesocolon).

Rectum The **rectum** (rek′tŭm; *rectus* = straight) is the third major region of the large intestine. It is a retroperitoneal structure that extends from the sigmoid colon. The rectum is a muscular tube that readily expands to store accumulated fecal material prior to defecation. Three thick transverse folds of the rectum, called **rectal valves,** ensure that fecal material is retained during the passing of gas.

The **anal canal** makes up the terminal few centimeters of the large intestine. The anal canal is lined by a stratified squamous epithelium, and it passes through an opening in the levator ani muscles of the pelvic floor (see section 11.7) and terminates at the anus. The internal lining of the anal canal contains relatively thin longitudinal ridges, called **anal columns,** between which are small depressions termed **anal sinuses.** As fecal material passes through the anal canal during defecation, pressure exerted on the anal sinuses causes their cells to release mucin to form mucus. The extra mucus lubricates the anal canal during defecation. At the base of the anal canal are the

INTEGRATE

CLINICAL VIEW 26.12

Colorectal Cancer

Colorectal cancer is the second most common type of cancer in the United States, with over 140,000 cases occurring annually and 60,000 of those resulting in death. The term **colorectal cancer** refers to a malignant growth anywhere along the large intestine (colon or rectum). The majority of colorectal cancers appear in the distal descending colon, sigmoid colon, and rectum, which are the segments of the large intestine that have the longest contact with fecal matter before it is expelled from the body. Most colorectal cancers arise from **polyps** (pol′ip), which are outgrowths from the colon mucosa. Note, however, that colon polyps are very common, and most of them never become cancerous. Low-fiber diets have also been implicated in increasing the risk of colon cancer, because decreased dietary fiber leads to decreased stool bulk and longer time for stools to remain in the large intestine. Theoretically, this condition exposes the large intestine mucosa to toxins in the stools for longer periods of time. Other traditional risk factors include a family history of colorectal cancer, personal history of ulcerative colitis, and increased age (most patients are over the age of 40).

A recent American Cancer Society study showed an increase in the number of individuals in their 20s and 30s who are being diagnosed with the disease. The reasons for this rise are unclear but may be an increase in obesity rates or a change in the large intestine bacterial flora (due to dietary trend changes).

Initially, most patients are asymptomatic. Later, they may notice rectal bleeding (often evidenced as blood in the stool or on the toilet paper) and a persistent change in bowel habits (typically constipation). Eventually, the person may experience abdominal pain, fatigue, unexplained weight loss, and anemia.

The cancerous growth must be removed surgically, and sometimes radiation or chemotherapy is used as well. Colorectal cancers that are limited to the mucosa have a 5-year survival rate, but the prognosis is poor for cancers that have spread into deeper colon wall tunics or to the lymph nodes.

The key to an increased survival rate for colorectal cancer is early detection. If caught early, colorectal cancer is very treatable. An individual should see his or her doctor if rectal bleeding or any persistent change in bowel habits is experienced. Screening tests recommended by age 50 (or earlier, for those who have symptoms or a family history of colorectal cancer) include a **fecal occult** (ō-kŭlt′, ok′ŭlt) **blood test, sigmoidoscopy** (sig′moy-dos′kō-pē; *skopeo* = to view), and **colonoscopy** (kō-lon-os′kō-pē).

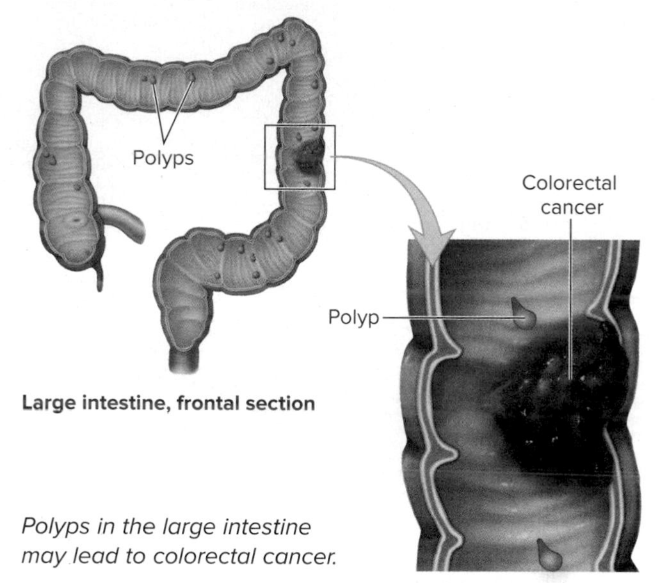

Polyps in the large intestine may lead to colorectal cancer.

involuntary smooth muscle **internal anal sphincter** and voluntary skeletal muscle **external anal sphincter,** which generally close off the opening to the anal canal. The muscles composing these sphincters relax and allow the sphincter to open during defecation.

Specialized Structures of the Large Intestine

Several features are unique to the large intestine, including teniae coli, haustra, and epiploic appendages. **Teniae** (tē′nē-ē; ribbons, band) **coli** (kō′lī) are thin, distinct, longitudinal bundles of smooth muscle. The teniae coli act like elastic in a waistband—they help bunch up the large intestine into many sacs, collectively called **haustra** (haw′stră; sing., *haustrum; haustus* = to drink up). Hanging off the external surface of the haustra are lobules of fat called **omental appendices,** or *epiploic* (e-pi-plō′ik; membrane-covered) *appendages.*

Histology

The mucosa of the large intestine is lined by a simple columnar epithelium with numerous goblet cells **(figure 26.24)**. Unlike the small intestine, the large intestine mucosa is smooth and lacks intestinal villi, yet it resembles the small intestine in that it has epithelial cells and numerous **intestinal glands** (or crypts) that extend inward toward the muscularis mucosae. The glands' mucous cells secrete mucin into the lumen to lubricate the undigested material and facilitate its passage through the large intestine. Many lymphatic nodules and lymphatic cells occupy the lamina propria of the large intestine.

The muscularis of the cecum and colon has two layers of smooth muscle, but the outer longitudinal layer is discontinuous and does not completely surround the colon and cecum. Instead, these longitudinal smooth muscle fibers form the teniae coli, just described.

Bacterial Action in the Large Intestine

Numerous normal bacterial flora inhabit the large intestine; they are termed the *indigenous microbiota.* These bacteria are responsible for the chemical breakdown of complex carbohydrates, proteins, and lipids that remain in the chyme after it has passed through the small intestine. Bacterial actions produce gases (carbon dioxide, hydrogen, hydrogen sulfide, and methane) from digestion of carbohydrates, and indoles and skatoles from digestion of protein. Some of these substances account for the odor of feces. Additionally, B vitamins and vitamin K are produced by the bacterial flora, which are then absorbed from the large intestine into the blood. (Note that these vitamins are also absorbed in the small intestine from the foods that we eat.) Feces is the final product formed and then eliminated from the GI tract. **Feces** (fē′cēz; *faex* = dregs) are composed of water, salts, epithelial cells (sloughed from GI lining), bacteria, and undigested material (see Clinical View 26.13: "Fecal Transplant").

INTEGRATE

CONCEPT CONNECTION

Recall that heme is broken down into bilirubin and absorbed from the blood in the liver (see section 18.3b). Bilirubin is then released as a component of bile into the small intestine. It is either reabsorbed back into the blood and eliminated as urobilin in the urine, or it continues through the GI tract into the large intestine, where bacteria convert it to stercobilin. Stercobilin is a brown pigment that contribute to the normal color of feces.

Figure 26.24 Histology of the Large Intestine. (*a*) The luminal wall of the large intestine is composed of four tunics: mucosa, submucosa, muscularis, and serosa. (*b*) A photomicrograph shows the histology of the mucosa and submucosa of the large intestine wall.

(*b*) ©McGraw-Hill Education/Al Telser

Opening to intestinal gland
Goblet cells
Lumen
Simple columnar epithelium
Intestinal gland
Lamina propria
Lymphatic nodule
Muscularis mucosae
Mucosa
Submucosa
Muscularis
Circular layer
Longitudinal layer (tenia coli)
Nerves
Arteriole
Venule
Serosa

(a) Large intestine tunics

Lumen
Opening to intestinal gland
Goblet cells
Simple columnar epithelium
Intestinal gland
Muscularis mucosae
LM 80x

(b) Large intestine mucosa and submucosa

INTEGRATE

CLINICAL VIEW 26.13

Fecal Transplant

A **fecal transplant** (or *fecal microbiota transplant* or *bacteriotherapy*) involves a relatively low-cost, low-risk procedure of collecting the fecal matter of a stool from a healthy donor (stool contains normal microbiota), combining it with a solution (e.g., saline solution), straining it, and then placing it in the colon of a patient (e.g., through an enema or a colonoscopy).

Fecal transplants are done to treat recurrent *Clostridium difficile* (*C. difficile*) colitis, which is an often debilitating pathogenic bacterial infection that sometimes results in fatal diarrhea (in an estimated 4–14% of cases). *C. difficile* colitis is usually the result of a previous antibiotic treatment that decreased or eliminated the normal microbiota within the colon.

Only qualified physicians with an Investigational New Drug (IND) application are permitted by the Federal Drug Administration (FDA) to perform fecal transplants, as of 2013—and only for treating recurrent *C. difficile* infections. Research studies are investigating the use of fecal transplants in treating intestinal disorders (e.g., Crohn disease, ulcerative colitis, irritable bowel syndrome; see Clinical View 26.7: "Inflammatory Bowel Disease and Irritable Bowel Syndrome"). Results have shown that fecal transplant may be an effective treatment, and research continues into the types of conditions that might benefit from this treatment.

INTEGRATE

CLINICAL VIEW 26.14

Diverticulosis and Diverticulitis

Diverticulosis (dī′ver-tik′yū-lō′sis) is the presence of **diverticula** (small "bulges") in the intestinal lining. These are formed typically when the colon tightens and narrows in response to low amounts of fiber or bulk in the colon. **Diverticulitis** is inflammation of diverticula. Diverticulitis may be life-threatening if the diverticula rupture and the intestinal contents leak into the abdominal cavity.

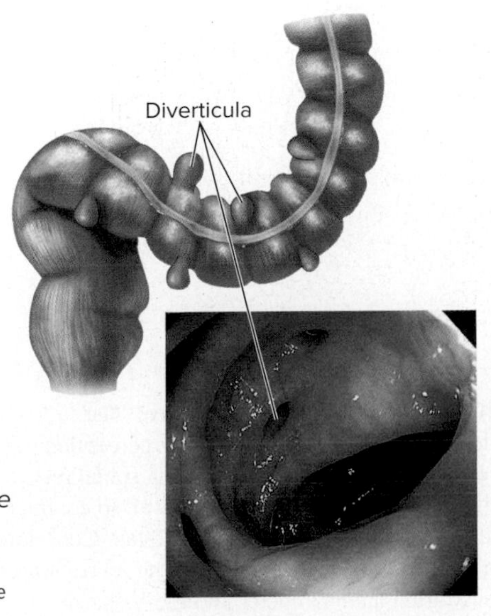

An external view and endoscopic view of the sigmoid colon show diverticula.

©Gastrolab/SPL/Science Source

Motility and Regulation of the Large Intestine

Several types of movements are noted in the large intestine:

- *Peristalsis* of the large intestine is usually weak and sluggish, but otherwise it resembles the peristalsis that occurs in the wall of the small intestine.
- **Haustral churning** occurs after a relaxed haustrum fills with digested or fecal material until its distension stimulates reflex contractions in the muscularis. These contractions increase churning and move the material to more distal haustra.
- **Mass movements** are powerful, peristaltic-like contractions involving the teniae coli, which propel fecal material toward the rectum. A wave of contraction begins in the middle of the transverse colon, forcing a large amount of fecal matter into the descending colon, sigmoid colon, and rectum. Generally, mass movements occur two or three times a day, often during or immediately after a meal.

Two major reflexes are associated with motility in the large intestine:

- The **gastrocolic** (gas′trō-kol′ik) **reflex** is initiated by stomach distension to cause a mass movement (motility in the large intestine, just described).
- The **defecation reflex** is the elimination of feces from the GI tract by the process of **defecation** (def-ĕ-kā′shŭn; *defaeco* = to remove the dregs) **(figure 26.25)**. Filling of the rectum initiates the urge to defecate. This stimulus results in transmission of nerve signals from the receptors to the spinal cord. In response, increased nerve signals are relayed along parasympathetic motor neurons, which causes both the sigmoid colon and rectum to contract and the internal (involuntary) anal sphincter to relax. (The defecation reflex is an example of a monosynaptic reflex; see section 14.6c.)

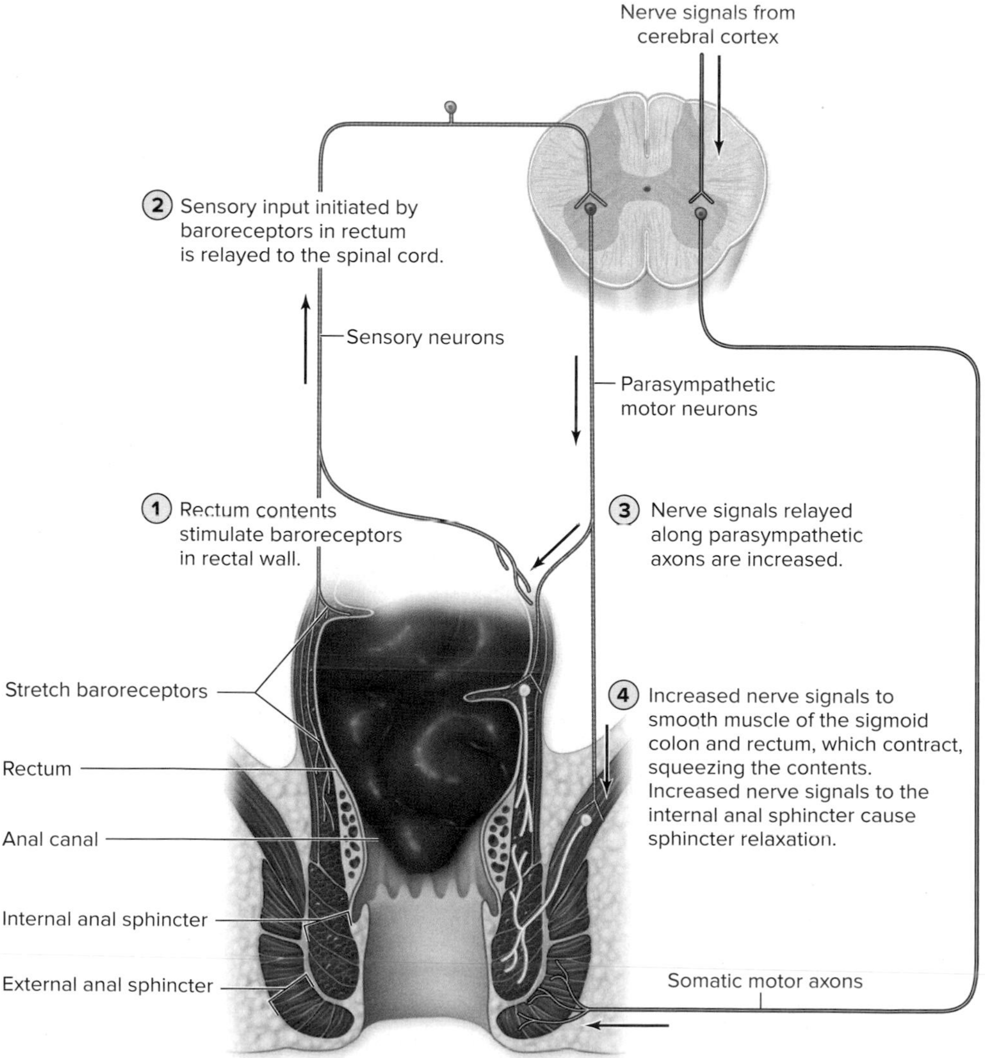
Nerve signals from cerebral cortex
2 Sensory input initiated by baroreceptors in rectum is relayed to the spinal cord.
Sensory neurons
Parasympathetic motor neurons
1 Rectum contents stimulate baroreceptors in rectal wall.
3 Nerve signals relayed along parasympathetic axons are increased.
Stretch baroreceptors
Rectum
Anal canal
Internal anal sphincter
External anal sphincter
4 Increased nerve signals to smooth muscle of the sigmoid colon and rectum, which contract, squeezing the contents. Increased nerve signals to the internal anal sphincter cause sphincter relaxation.
Somatic motor axons
5 The conscious decision to defecate is controlled by the cerebral cortex. External anal sphincter relaxes and Valsalva maneuver is initiated, eliminating the feces.

Figure 26.25 Defecation. The defecation reflex involves sensory perception of stretch in the rectum that initiates a spinal reflex, stimulating muscles in the rectal wall to contract and the internal anal sphincter to relax. Conscious regulation of defecation requires the relaxation of the external anal sphincter and the Valsalva maneuver.

INTEGRATE

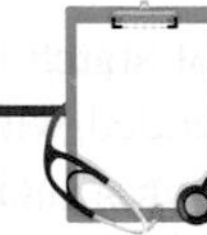

CLINICAL VIEW 26.15
Constipation and Diarrhea

Constipation is typically a temporary, impaired ability to defecate and generally results in compacted feces that are difficult to eliminate. Constipation may result from a combination of factors, including a diet low in fiber (insufficient bulk), dehydration, lack of exercise, and improper bowel habits (not defecating when the urge arises, allowing additional water to be absorbed from the feces). Constipation is also a common side effect of general anesthesia. Part of postoperative care is to verify that GI motility has returned to normal by determining that the patient has had a bowel movement. Additionally, long-term use of opiate medications (e.g., oxycontin) causes constipation. This is because the large intestine contains opioid receptors and when opiates bind to them, GI motility decreases and water absorption from the feces is greater. Constipation increases the risk of both **hemorrhoids** (hem′o′roydz; *rhoia* = flow), which are dilated and inflamed veins around the rectum and anus, and **anal fissures,** which are oval-shaped tears of the anus that often bleed.

In contrast, **diarrhea** may result as a consequence of disruption in normal mechanisms to absorb intestinal water, or excessive amounts of osmotically active solutes that keep water in the intestinal lumen and prevent its reabsorption. One example of such an osmotically active solute is sorbitol, a sugar substitute often found in sugar-free ice cream, diet sodas, cough syrups, and sugar-free gum. Sorbitol is considered a nonstimulant laxative because it draws water into the large intestine and stimulates bowel movements. Immodium AD essentially works by binding to the opiate receptors and thereby decreasing GI motility and limiting diarrhea symptoms.

Voluntary elimination of feces from the body typically is learned sometime after age 3 years. The conscious decision (initiated by the cerebral cortex) to defecate involves both the Valsalva maneuver (see section 23.3a) and relaxation of the external anal sphincter.

WHAT DID YOU LEARN?

22. What is the pathway of chyme from its entry into the large intestine until feces is eliminated?
23. What are the general functions of bacteria within the large intestine?
24. Which substances are typically absorbed by the large intestine?

26.4 Nutrients and Their Digestion

The term **essential nutrients** indicates substances that must constitute part of the diet for survival. The six essential nutrients are carbohydrates, proteins, lipids, minerals, vitamins, and water. These nutrients and their functions in the body are described in depth in sections 27.1 through 27.3.

This section discusses the chemical digestion of the foods that we eat. In addition to the mechanism of their breakdown, we describe their absorption and general use by the body. We describe the breakdown of carbohydrates, lipids, and proteins, as well as nucleic acids, the last of which is not an essential nutrient. All are digested by the process of *hydrolysis,* the breaking of chemical bonds with an accompanying addition of a water molecule (see section 2.7a).

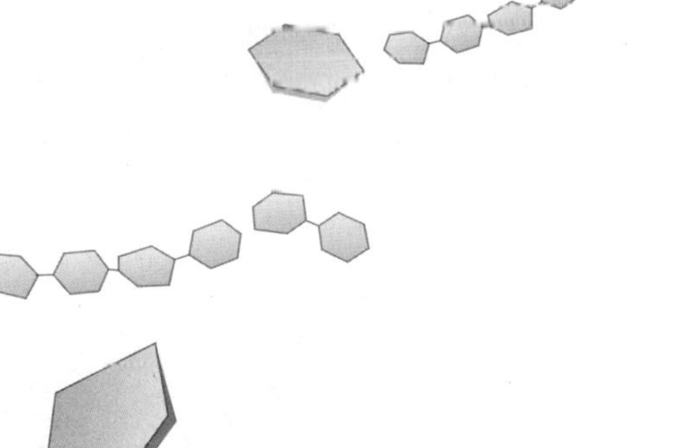

26.4a Carbohydrate Digestion

LEARNING OBJECTIVES

32. Name the three classes of carbohydrates.
33. Explain the processing in the oral cavity that initiates carbohydrate digestion.
34. Describe the chemical digestion of carbohydrates that occurs in the small intestine.

Carbohydrates are organized based upon the number of repeating units of simple sugars. You learned in section 2.7c that carbohydrates may be classified as **monosaccharides** (e.g., glucose, fructose, and galactose), **disaccharides** (e.g., sucrose, maltose, and lactose), and **polysaccharides** (e.g., starch and cellulose). Chemical digestion of carbohydrates consists of (1) the breakdown of starch into individual glucose molecules and (2) the breakdown of disaccharides into the individual monosaccharides that compose them. The oral cavity and small intestine are the main sites of carbohydrate digestion.

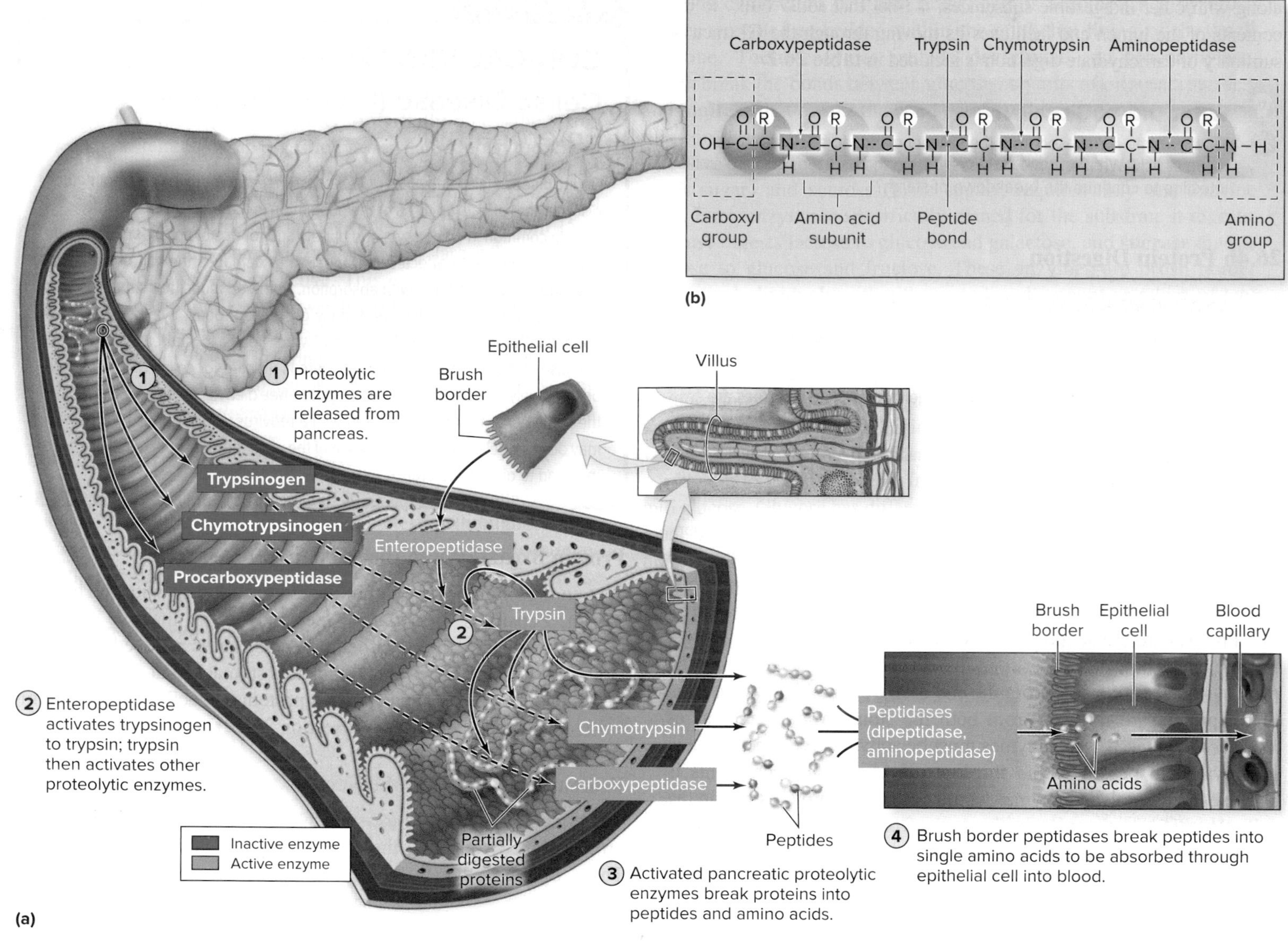

Figure 26.27 Protein Digestion in the Small Intestine. (*a*) Pancreatic enzymes for protein digestion are produced and secreted into the small intestine in an inactive form. After reaching the small intestine, these enzymes are activated. (*b*) The locations where different protein-digesting enzymes break peptide bonds are highlighted in purple.

synthesized and released from the pancreas into the small intestine in inactive forms—trypsinogen, chymotrypsinogen, and procarboxypeptidase (**figure 26.27**). Once these inactive forms of the enzymes reach the small intestine, trypsinogen is activated by the enzyme **enteropeptidase** (en′tĕr-ō-pep′ti-dās; previously called *enterokinase*), an enzyme previously synthesized by the small intestine and released into the lumen of the small intestine (see section 26.3b). Enteropeptidase activates trypsinogen to **trypsin.** Trypsin in turn activates additional molecules of trypsinogen to trypsin, as well as chymotrypsinogen to **chymotrypsin** (kī-mō-trip′sin) and procarboxypeptidase to **carboxypeptidase** (kar-bok′sē-pep′ti-dās).

Trypsin and chymotrypsin break the bonds between specific amino acids within the protein to produce smaller strands of amino acids called peptides. (Trypsin cleaves a protein specifically at positively charged amino acids of arginine and lysine, whereas chymotrypsin cleaves specifically at hydrophobic amino acids, including phenylalanine, tryptophan, and tyrosine; see figure 2.25, which shows the 20 different amino acids.) Carboxypeptidase is restricted to breaking the bond only between an amino acid on the carboxyl end and the remaining protein (it releases one amino acid at a time). Dipeptides and free amino acids are the breakdown products of carboxypeptidase.

The brush border enzyme **dipeptidase** breaks the final bond between the two amino acids of a dipeptide so that both may be absorbed. The brush border enzyme **aminopeptidase** generates free amino acids from the amino end of peptides.

Free amino acids are absorbed across the small intestine epithelial lining and enter into the blood. Amino acids can be used as building blocks of new proteins by cells, or if excess amino acids are absorbed, they are either converted into glucose (by gluconeogenesis in the liver primarily or kidney; see figure 27.5*b*) or deaminated (amine [—NH_2] group is removed in the liver; see figure 27.5*c*) and used as fuel for cellular respiration (see figure 27.6). A summary of protein digestion is included in table 26.2.

 **WHAT DID YOU LEARN?**

 How are proteolytic enzymes activated in the stomach and in the small intestine? Explain why this is necessary.

26.4c Lipid Digestion

 **LEARNING OBJECTIVES**

38. Explain the role of bile salts in mechanical digestion of lipids and the role of pancreatic lipase in the chemical digestion of triglycerides.

39. Discuss the process by which lipids are absorbed.

Lipids are highly variable structures that contain different arrangements of their building blocks (see section 2.7b). Lipids have one unifying property, which is that they are not water-soluble. Two major ingested lipids are triglycerides (or neutral fats) and cholesterol. Triglycerides are composed of a glycerol molecule and three fatty acids bonded to the glycerol, and enzymes are required to break the bonds between glycerol and fatty acids. Chemical digestion of cholesterol is not required for its absorption.

Lipid Breakdown in the Stomach

Lingual lipase (produced by intrinsic salivary glands in the mouth; see section 26.2b) is a component of saliva in the oral cavity. The optimal pH of this enzyme (4.5 to 5.4), however, means it is not activated until it reaches the stomach. Once in the stomach, triglycerides undergo limited digestion by both lingual lipase and gastric lipase (an enzyme produced by chief cells of the stomach). These "acidic lipases" digest approximately 30% of the triglycerides to diglyceride and a fatty acid. Neither of these lipase enzymes requires the participation of bile salts.

Lipid Breakdown in the Small Intestine

The majority of triglyceride digestion occurs in the small intestine and is facilitated by **pancreatic lipase,** an enzyme produced by the pancreas and released into the duodenum. Pancreatic lipase digests each triglyceride into a monoglyceride and two free fatty acids. However, because triglycerides are lipids and do not dissolve in the luminal fluids of the digestive system, triglycerides form relatively large lipid masses. For example, when butter is added to water, the butter does not mix with the water but remains separate. Thus, the large lipid droplets must first be mechanically separated into smaller droplets before pancreatic lipase effectively digests the fat. This process is called **emulsification** (ē-mŭl-si-fĭ-kā′shun; **figure 26.28**). (This is similar to breaking an ice cube into ice chips.) Emulsification occurs by the action of **bile salts,** which are part of bile. Bile is produced by the liver and stored, concentrated, and released from the gallbladder. Bile salts are amphipathic molecules composed of a polar head and a nonpolar tail. The nonpolar tails position themselves around the fat with the polar heads next to the aqueous fluid in the lumen (like an inverted spiked ball with the fat in the center; see section 2.4c). This structure is called a **micelle** (mi-sel′, mī-sel; *micella* = small morsel). Thus, the function of bile salts is to emulsify fats so that pancreatic lipase has greater "access" to the triglyceride molecules and may more effectively chemically digest the fat molecules. (Note that the process of emulsification is facilitated by *lecithin,* which is a type of phospholipid molecule within bile.) Cholesterol is also within the micelle, but it is not chemically digested. No brush border enzymes are required in the breakdown of triglycerides.

Eventually, in the last portion of the ileum, bile salts are recovered from the GI tract back into the blood by active transport and ultimately recycled to the liver for reuse (see section 26.4e). New bile salts are synthesized by hepatocytes to replace those inevitably lost during elimination of feces. Table 26.2 summarizes lipid digestion.

Lipid Absorption

Digested triglycerides (monoglycerides and free fatty acids), cholesterol, other lipids (e.g., lecithin), and fat-soluble vitamins are contained within micelles. Micelles transport lipids to the simple columnar epithelial lining of the small intestine. Here, the lipids enter the epithelial cells, whereas the bile salts remain in the small intestine lumen to be recycled and reused.

chapter

27 Nutrition and Metabolism

INTEGRATE

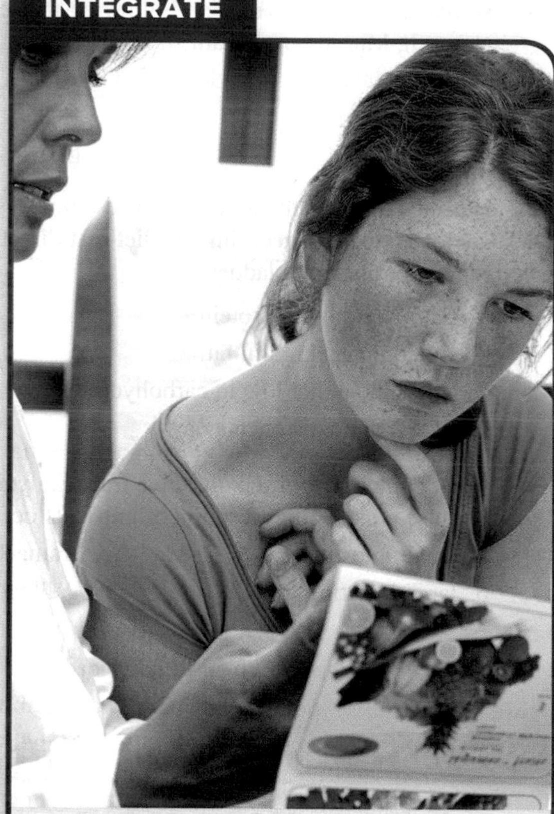

©BURGER/PHANIE/Science Source

CAREER PATH
Dietitians and Nutritionists

Dietitians and nutritionists are health-care professionals who are concerned with proper diet to maintain good health. They evaluate and assess the needs of individual patients and help develop appropriate nutritional programs. Often, they recommend modifications of the diet to help prevent or treat illnesses. Dietitians manage and provide dietary services in hospitals, nursing care facilities, and schools.

Module 12: Digestive System

Many individuals express the desire to "eat healthy." Given the amount of media coverage regarding dietary intake, most of us are aware that eating healthy requires a diet with plenty of fruits, vegetables, and whole grains. This diet must also include small portions of protein, and it should be low in saturated fat, sugar, and sodium. By eating healthy, we provide our bodies with the nutrients required for both the synthesis of new molecules and the energy for maintenance, growth, and repair, and we avoid foods that may be unhealthy for us.

Here we provide an overview to nutrition. Our discussion includes a description of nutrients, various guidelines for obtaining adequate nutrition, and how the body regulates nutrient levels in the blood during and following a meal. We also examine the functions of the liver, including its essential role in regulating nutrient blood levels, the central role of cellular respiration in nutrient metabolism, the production of heat from metabolic processes, and the regulation of body temperature.

27.1 Introduction to Nutrition

LEARNING OBJECTIVES

1. Define both nutrition and nutrients.
2. Distinguish macronutrients from micronutrients and essential from nonessential nutrients.
3. Explain the meaning of recommended daily allowance (RDA).

Nutrition is the study of the means by which living organisms obtain and utilize the nutrients they need to grow and sustain life. **Nutrients** include most biological macromolecules (e.g., carbohydrates, lipids, and proteins; see section 2.7), vitamins, and minerals that the body needs for development and growth, maintenance of anatomic structures and physiologic processes, and repair of damaged tissues. Water is also considered to be a nutrient because the body requires 2 to 3 liters of it daily. Water and its properties are described in section 2.4, and water balance is discussed in section 25.2.

Nutrients can be described in two ways: as (1) either macronutrients or micronutrients and as (2) either essential nutrients or nonessential nutrients. Macronutrients and micronutrients reflect the daily amounts that are required. **Macronutrients** must be consumed in relatively large quantities. All macronutrients are the biological macromolecules described in the previous paragraph. **Micronutrients** must be consumed in relatively small quantities and include both vitamins and minerals.

Nutrients are also organized into two categories, depending upon whether it is vital that you obtain them in your diet. **Essential nutrients** (see section 26.4) must be obtained and absorbed by the processes of the digestive system, and thus it is required (essential) that these nutrients be part of your dietary intake. Essential nutrients include some macronutrients and some micronutrients. **Nonessential nutrients** can be adequately provided by biochemical processes within the body, and for this reason they are not required to be part of your dietary intake.

Recommended Daily Allowance

Federal government agencies have established values for the amount of each nutrient that must be obtained every day called the **recommended daily allowances (RDAs)**. These were originally established by the Food and Nutrition Board (FNB) as directed by the National Academy of Sciences (NAS), and they are reviewed and updated periodically based on additional data from nutrition studies. RDA information may be accessed online at the USDA's food and nutrition information center (http://www.health.gov/dietaryguidelines/). These government-established values are used for food planning, food labeling, clinical dietetics, food programs, and educational programs on nutrition. Although RDA values are currently based on population studies, in the future these RDA levels could be established for each individual based on one's specific genetic makeup.

WHAT DID YOU LEARN?

1. List the six nutrients required by the body.
2. Distinguish between macronutrients and micronutrients.
3. What is meant by the RDA of nutrients?

27.2 Macronutrients

All macronutrients (i.e., carbohydrates, lipids, proteins) have a common function; they provide fuel for the processes of cellular respiration to form ATP (see section 3.4); thus, these molecules provide energy to the body. The amount of energy is measured in calories. A **calorie** is defined as the amount of heat required to raise the temperature of 1 gram of water by 1 degree Celsius. A **kilocalorie** is equal to 1000 calories, or 1 Calorie (with a capital "C"). Body weight is maintained when there is balance between calories consumed in the diet and those expended through metabolic processes and physical activity.

27.2a Carbohydrates

LEARNING OBJECTIVE

4. Identify the categories that are dietary sources of carbohydrates, and give examples of each category.

We described carbohydrates in section 2.7c. Recall that **carbohydrates** are structurally classified as **monosaccharides** (the monomer); **disaccharides,** which have two monosaccharides; and **polysaccharides,** formed from chains of monosaccharides. However, when describing dietary sources of carbohydrates, three different categories are recognized, as follows:

- **Sugars.** These carbohydrates include both the monosaccharides glucose, fructose, and galactose and the disaccharides sucrose (table sugar, maple syrup, and fruits), lactose (milk sugar), and maltose (found in cereals). Other sugars (or sweeteners) include dextrose, brown sugar, honey, malt syrup, corn syrup, corn sweetener, high fructose corn syrup, invert sugar, molasses, raw sugar, turbinado sugar, and trehalose.
- **Starch.** This carbohydrate is a polysaccharide polymer of glucose molecules found within certain types of foods, including tubers (e.g., potatoes, carrots, bananas), grains (e.g., wheat, barley, rice, corn), and beans and peas (kidney beans, garbanzo beans, lentils). Breads and pasta are also primarily composed of starch. Refined starches are sometimes added as thickeners and stabilizers. Cornstarch is an example of a refined starch.
- **Fiber.** This type of carbohydrate includes the fibrous molecules (e.g., cellulose) of both plants and animals that cannot be chemically digested and absorbed by the gastrointestinal (GI) tract. Sources of fiber include vegetables, lentils, peas, beans, whole grains, oatmeal, berries, and nuts.

Sugars and starch are usually converted to glucose, which is one of the primary nutrients supplying energy to cells. We have seen that each glucose molecule is oxidized during the process of cellular respiration to net approximately 30 ATP molecules (see section 3.4f). Stated in terms of calories, this is the equivalent of approximately 4 kilocalories of energy per gram of glucose. Interestingly, glucose is not considered an essential nutrient because it can be synthesized within hepatocytes from other monosaccharides (e.g., fructose, galactose) and noncarbohydrate molecules (e.g., amino acids, fatty acids) by the process of gluconeogenesis (see section 2.7c).

The fiber in our diet serves a different purpose. Fiber is neither chemically digested nor absorbed. Rather, it remains within the lumen of the GI tract and adds bulk. This bulk stimulates peristalsis of the large intestine (see section 26.3d), which facilitates the movement of the GI tract contents and their ultimate elimination as a component of feces. In other words, fiber helps to "keep you regular." In addition, some types of fiber (water-soluble fiber, including beta-glucan, psyllium, pectin, guar gum) decrease low-density lipoprotein (LDL) cholesterol levels (see section 27.6c). It is thought that water-soluble fiber interferes with the reabsorption of bile acids (which are formed from cholesterol) and the bile acids are then eliminated in the feces (see section 27.6a). Consequently, the liver must absorb additional cholesterol from the blood to replace the bile acids that have been lost.

WHAT DID YOU LEARN?

4. What is the general function of both sugars and starch in our diet?
5. How does dietary fiber differ from other dietary carbohydrates?

27.2b Lipids

LEARNING OBJECTIVES

5. Identify the types and dietary sources of triglycerides, and describe their functions.
6. Describe the sources and functions of cholesterol.

Lipids are a diverse group of biologically active, hydrophobic molecules that include triglycerides, phospholipids, steroids (including cholesterol), and eicosanoids (see section 2.7b). Here we describe both triglycerides and cholesterol.

Triglycerides (or fats) are composed of glycerol and fatty acids. Fatty acids are organized into three categories, which depend upon their degree of saturation (see figure 2.18):

- **Saturated fatty acids** have no double bonds (each carbon in the fatty acid chain is completely saturated with hydrogen atoms). Sources of saturated fats are generally solid at room temperature, and dietary sources include the fat in meat, milk, cheese, coconut oil, and palm oil.
- **Unsaturated** (also called *monounsaturated*) **fatty acids** have one double bond. Unsaturated fats are typically liquid at room temperature. Dietary unsaturated fats include nuts and certain vegetable oils (e.g., canola oil, olive oil, sunflower oil).
- **Polyunsaturated fatty acids** have two or more double bonds. Sources of polyunsaturated fats are also liquid at room temperature, and dietary sources include some vegetable oils (e.g., soybean oil, corn oil, safflower oil).

Triglycerides are also a primary nutrient supplying energy to cells. Oxidation of triglyceride molecules yields approximately 9 kilocalories of energy per gram of fat—more than twice that of glucose. Fats are also necessary for the absorption of fat-soluble vitamins (vitamins A, D, E, and K) (see section 26.4e).

In general, fat molecules and the fatty acids that compose them can be synthesized by the body. The only two fatty acids that the body needs, but cannot normally synthesize, are alpha-linoleic acid and linoleic acid. However, other fatty acids (e.g., docosahexaenoic acid) may be required in the diet in certain disease conditions.

Cholesterol is required as a component of the plasma membrane of our cells (see section 4.2a). It is also the precursor molecule for the formation of steroid hormones (see section 17.3a), bile salts (see section 27.6b), and vitamin D (see section 7.6a). Cholesterol either is made available through our diet (a component of animal-based products, including meat, eggs, and milk) or is synthesized by metabolic pathways within the liver. Both cholesterol synthesis and transport of lipids are described in section 27.6.

WHAT DID YOU LEARN?

6. What are two primary functions of fats?

27.2c Proteins

LEARNING OBJECTIVES

7. Describe why protein is required in our diet and the general amount that is needed.
8. Explain the difference between a complete protein and an incomplete protein.
9. Discuss nitrogen balance, and include the difference between a positive and negative nitrogen balance.

Proteins are the most structurally and functionally diverse molecules. (There are an estimated 50,000 different proteins in the body; see table 2.6.) Proteins are synthesized from 20 different amino acids (see figure 2.25). Adequate dietary intake of protein provides the amino acids to synthesize new proteins to replace body protein structures. The specific amount of protein required in the diet is generally dependent upon one's age and sex, but ranges between approximately 45 and 60 grams (about 1/10 pound) daily. Required protein is increased when fighting an infection, following an injury, under conditions of stress, and during pregnancy. It is also higher in infants and children to synthesize protein molecules needed for growth.

Eight of the 20 amino acids (isoleucine, leucine, lysine, methionine, phenylalanine, threonine, tryptophan, and valine) are essential

INTEGRATE

CLINICAL VIEW 27.1
High Fructose Corn Syrup

©Fuse/Getty Images RF

The manufacture of **high fructose corn syrup** begins with cornstarch that is liquefied and broken down into glucose, and some of the glucose is converted to fructose. High fructose corn syrups vary in the relative amounts of glucose and fructose they contain.

Individuals in the United States have a much higher intake of high fructose corn syrup today as compared to 30 years ago. The reason is that many manufacturers of food products now sweeten their products with high fructose corn syrup, which is sweeter and typically less expensive than sucrose. Many processed foods contain some high fructose corn syrup, including soft drinks, cereals, breads, soups, breakfast bars, yogurt, lunch meats, cookies, and condiments (e.g., ketchup, mustard).

The intake of large amounts of high fructose corn syrup, and its potential effects on an individual's health, remains controversial. On the one hand, some claim that high fructose corn syrup leads to higher blood levels of triglycerides, is partially responsible for the obesity epidemic in the United States (see Clinical View 27.3: "Obesity"), and exacerbates the condition of diabetes mellitus. On the other hand, the manufacturers of high fructose corn syrup claim high fructose corn syrup is natural and is fine in moderation. Research continues to investigate whether health issues are associated with a large intake of this sweetener.

in adults and therefore must be obtained from the diet. The other 12 can be synthesized within the body. (Histidine is also essential in infants because they are not able to synthesize it.) Dietary protein sources vary in whether they provide all of the essential amino acids, and they are identified as either complete proteins or incomplete proteins. **Complete proteins** contain all of the essential amino acids, whereas **incomplete proteins** do not. Generally, animal proteins (meats, poultry, fish, eggs, milk, cheese, yogurt) are complete proteins, and plant proteins (legumes, vegetables, grains) tend to be lacking in one or more of the essential amino acids and thus they are incomplete proteins. For example, the protein in corn has low amounts of both lysine and tryptophan. Combinations of dishes containing plant proteins (such as legumes and grains) can provide all of the essential amino acids (e.g., beans and corn, soybeans and rice, red beans and rice).

Note that there is no storage of excess amino acids. Consequently, all of the amino acids required for protein synthesis must be supplied regularly either through animal proteins or through plant protein combinations. A **pescatarian** is an individual who does not eat meat or poultry but does consume fish as the only form of animal protein. A **vegetarian** is an individual who does not eat meat, fish, or poultry. **Lacto-ovo vegetarians** do not eat animal flesh but do eat milk, eggs, and cheese. **Vegans** do not eat any animal products. A concern for vegetarians is that plant-based protein sources are often individually incomplete. These individuals must be sure to obtain the essential amino acids from a variety of complementary protein sources. These balanced food combinations do not need to be eaten in the same meal, but when eaten regularly they may supply the necessary essential amino acids.

Nitrogen Balance

Proteins are also a source of **nitrogen** (recall that each amino acid contains a $-NH_2$ functional group; see section 2.7e). Nitrogen is a chemical element that is needed for synthesizing other nitrogen containing molecules, such as nitrogenous bases of nucleic acids (DNA, RNA) (see section 2.7d), and porphyrin (a component of heme in hemoglobin) (see section 18.3b). Adequate availability of nitrogen is expressed as nitrogen balance. **Nitrogen balance** occurs when equilibrium exists between its dietary intake and its loss in urine and feces. Input of nitrogen must equal output of nitrogen to maintain a nitrogen balance. A **positive nitrogen balance** occurs when an individual absorbs more nitrogen than is excreted, as occurs in individuals who are growing, pregnant, or recovering from injury. A **negative nitrogen balance** occurs when more nitrogen is excreted than is absorbed; this is a condition that might result from malnutrition or blood loss, for example. An individual in sustained negative nitrogen balance has insufficient nitrogen to synthesize nitrogen-containing molecules, and a decrease of as little as a few hundred grams of total nitrogen may be fatal.

 **WHAT DID YOU LEARN?**

 Explain the general difference in animal proteins and plant proteins in terms of obtaining essential amino acids.

 How may a vegetarian obtain all of the essential amino acids?

27.3 Micronutrients

Micronutrients, which include both vitamins and minerals, are primarily obtained in the foods that we eat. Plants tend to be very good dietary sources of micronutrients. Each type of food may contain a variety of vitamins and minerals, but no one food contains all micronutrients that are required in our RDA. Consequently, it is necessary to have variety in the diet to meet the body's needs. Some individuals may also obtain vitamins and minerals through supplements.

27.3a Vitamins

LEARNING OBJECTIVES

10. Distinguish between water-soluble and fat-soluble vitamins.
11. List examples of how both water-soluble vitamins and fat-soluble vitamins function in the body.
12. Describe the difference between essential and nonessential vitamins.

Vitamins are organic molecules required for normal metabolism, yet they are present in only small amounts in food. Vitamins are categorized in two ways: (1) water-soluble vitamins or fat-soluble vitamins and (2) essential vitamins or nonessential vitamins. **Table 27.1** lists the water-soluble and fat-soluble vitamins and their sources, the recommended daily allowance (RDA), and deficiency symptoms. Good sources of vitamins include meat, eggs, beans, legumes, nuts, vegetables, and fruits.

 **WHAT DO YOU THINK?**

 Predict what happens to the level of fat-soluble vitamins stored in adipose connective tissue as a result of taking megadoses of vitamins.

Water-Soluble and Fat-Soluble Vitamins

Water-soluble vitamins dissolve in water: They include both the B vitamins and vitamin C. These vitamins are easily absorbed into the blood from the digestive tract. If dietary intake of water-soluble vitamins exceeds what is needed by the body, the excess is excreted into the urine. There are several different types of B vitamins, each of which is designated with a number and with a name (e.g., B_1 is thiamine; B_2 is riboflavin). B vitamins serve as coenzymes in various enzymatic chemical reactions. For example, vitamin B_3, also called niacin, is a necessary hydrogen carrier in mitochondria during adenosine triphosphate (ATP) synthesis (see section 3.4).

Vitamin C (or ascorbic acid) is required for the synthesis of collagen, which is an important protein in connective tissue (see section 5.2). This vitamin, along with vitamins A and K, functions as an antioxidant by removing free radicals (damaging chemical structures that contain unpaired electrons).

Fat-soluble vitamins dissolve in fat (not in water) and include vitamins A, D, E, and K (you can remember the fat-soluble vitamins with the acronym D.A.K.E.). They are absorbed from the gastrointestinal tract within the lipid of micelles and ultimately enter the lymphatic capillaries (lacteals) (see section 26.4c). If dietary intake of fat-soluble vitamins exceeds body requirements (generally caused by excessive intake through supplements), the excess is stored within the body fat and may reach toxic levels (a condition termed **hypervitaminosis**).

The functions of fat-soluble vitamins vary.

- Vitamin A (retinol) is a precursor molecule for the formation of the visual pigment retinal (see section 16.4d).
- Vitamin D (calciferol) is modified to form calcitriol: This is a hormone that increases calcium absorption from the gastrointestinal tract (see section 7.6).
- Vitamin E (tocopherol) helps stabilize and prevent damage to cell membranes (see section 4.2).
- Vitamin K is required for synthesis of specific blood clotting proteins (see section 18.4c).

Essential and Nonessential Vitamins

Vitamins are also categorized based on their requirement in the diet. Essential vitamins must be provided in the diet. If an essential

Table 27.1 Vitamins Required by Adults

Vitamin	Source	RDA (mg)	Deficiency Symptoms
WATER-SOLUBLE VITAMINS			
B-complex vitamins			
Vitamin B_1 (thiamine)	Meat, nuts, whole grains, eggs, seeds, legumes	1.1–1.2	Weakened heart, beriberi, edema (see section 25.2b)
Vitamin B_2 (riboflavin)	Meat, whole grains, eggs, greens	1.1–1.3	Ariboflavinosis—bright pink tongue, cracked lips, throat swelling, bloodshot eyes, anemia
Vitamin B_3 (niacin)	Meat, liver, whole grains	14–16	Nerve inflammation, pellagra
Vitamin B_5 (pantothenic acid)	Meat, eggs, whole grains, legumes		Fatigue, decreased coordination; paraesthesia—prickling and burning sensations in hands and feet
Vitamin B_6 (pyridoxine)	Meat, whole grains, legumes	1.3–1.7	Anemia (see Clinical View 18.2), convulsions
Vitamin B_{12} (cyanocobalamin)	Dairy products, meat, eggs	2.4	Pernicious anemia (see Clinical View 18.2 and section 26.2d); hypoalbuminemia; damage to nervous tissue
Biotin (vitamin B_7)	Meat, eggs, cheese, legumes, nuts	0.03–0.10	Skin rash, hair loss, anemia, mental conditions (including hallucinations)
Folate (folic acid; B_9)	Many green vegetables, liver, eggs, whole grains, seeds, legumes	400	Risk of neural tube defects to unborn child if mother is deficient during pregnancy (see Clinical View 13.2)
Vitamin C	Leafy green vegetables, fruit, citrus, broccoli, tomatoes, cabbage	75–90	Scurvy (see Clinical View 5.1)—skin spots, bleeding gums, loss of teeth, fever, lethargy, and death
FAT-SOLUBLE VITAMINS			
Vitamin A (retinol)	Liver, milk products, green vegetables, fish oil, eggs, margarine	700–900	Night blindness; dry, flaky skin; child blindness
Vitamin D (calciferol)	Cod liver oil, dairy products, UV rays (when synthesized by keratinocytes of epidermis)	15	Rickets (see Clinical View 7.4), osteomalacia, bone abnormalities
Vitamin E (tocopherol)	Leafy green vegetables, seeds, margarine, butter, fish oil, wheat germ	15	Rare
Vitamin K	Leafy green vegetables, liver (synthesized by bacteria in colon)	90–120	Abnormal bleeding (see Clinical View 18.7)

vitamin is lacking in our intake, or its absorption is impaired, then a vitamin deficiency disease results (**avitaminosis**). The vitamins in table 27.1 are all essential vitamins.

Nonessential vitamins are cofactors that the body is able to produce and recycle as needed. Examples are the NADH and $FADH_2$ that occur in the citric acid cycle (see section 3.4d).

WHAT DID YOU LEARN?

9 Which vitamins are water-soluble? Which are fat-soluble? Which of these groups may be dangerous in excess?

27.3b Minerals

LEARNING OBJECTIVES

13. Define minerals, and list examples of how minerals absorbed in the small intestine function in the body.
14. Distinguish between major minerals and trace minerals.

Minerals are inorganic ions (see section 2.2a) such as iron, calcium, sodium, potassium, iodine, zinc, magnesium, and phosphorus. Many foods that are a good source of vitamins are also a good source of minerals. Minerals have diverse functions in the body—for example,

- **Iron** is present both in hemoglobin within erythrocytes, where it binds oxygen (see section 18.3b), and within the mitochondria in the electron transport system to bind electrons (see section 3.4e).
- **Calcium** is required for the formation and maintenance of the skeleton (see section 7.2e), muscle contraction (see section 10.3c), exocytosis of neurotransmitters (see section 12.8d), and blood clotting (see section 18.4c).
- **Sodium** and **potassium** function to maintain a resting membrane potential in excitable cells and are required in the propagation of an action potential (see sections 10.3b and 12.8c).
- **Iodine** is needed to produce thyroid hormone (see section 17.8b).
- **Zinc** has roles in both protein synthesis (see section 4.8) and wound healing (see section 6.3).

All minerals are essential and must be obtained from the diet. We have a daily requirement for all minerals, and the amount required places minerals into one of two categories: **major minerals,** which are needed at levels greater than 100 milligrams (mg) per day, and **trace minerals,** which are required at less than 100 mg per day. Major minerals include calcium, chloride, magnesium, phosphorus, potassium, sodium, and sulfur; trace minerals include chromium, cobalt, copper, fluoride, iodine, iron, manganese, molybdenum, selenium, and zinc.

Minerals absorbed from the GI tract are stored to varying degrees within the body. These reserves are "hedges" against unexpected decreases in availability; however, if stored minerals are depleted, clinical problems may occur. **Table 27.2** lists the minerals and their sources, recommended daily allowance, and deficiency symptoms.

Table 27.2 Minerals Required by Adults

Mineral	Dietary Sources	RDA (mg)	Deficiency Symptoms
MAJOR MINERALS			
Calcium	Green vegetables, dairy products	1000–1300	Loss of bone mass, muscle weakness, depressed nerve activity
Chloride	Cured meat, table salt (NaCl)	2399	Muscle cramps
Magnesium	Green leafy vegetables, grains (whole)	310–420	Muscle weakness, nervous system disturbances
Phosphorus	Meat, nuts, dairy products, grains	700	Loss of appetite; anxiety, fatigue
Potassium	Meat, fruits, grains, vegetables, dairy products, salt substitutes	4700	Muscle weakness, abnormal cardiac function
Sodium	Table salt (NaCl), canned foods	< 2300	Neuromuscular disturbances, hypertension
Sulfur	Kale, cauliflower, cabbage	Unknown	Symptoms associated with protein deficiency
TRACE MINERALS			
Chromium	Whole grains, meat, cereals	0.05–0.25	Impaired glucose tolerance associated with insulin resistance
Cobalt	Fish, green vegetables, nuts	Unknown	Anemia
Copper	Seafood, legumes, liver, nuts	900	Anemia
Fluoride	Fluorinated water, fish, tea	1.5–4.0	Increased risk of dental caries (cavities)
Iodine	Iodized table salt	0.15	Goiter
Iron	Red meat, poultry, fish, dark green leafy vegetables, nuts, whole grains	8–18	Anemia, weakness, fatigue, pale skin, headaches, sensitivity to cold
Manganese	Vegetables, nuts, tea, fruits	2.5–5.0	Behavioral changes, abnormal bone formation
Molybdenum	Whole grains, nuts, beans	0.005–0.25	Rare
Selenium	Cereals, nuts, meat	55	Rare
Zinc	Meat, seeds	8–11	Changes in skin, eyes; hair loss; diarrhea

CLINICAL VIEW 27.2

Iron Deficiency

Iron (Fe) has several critical functions within the body: (1) It is required in the structure of hemoglobin in erythrocytes and myoglobin in muscle cells. (2) Iron is a component of specific electron transport proteins (cytochromes) within mitochondria. (3) It is required for some enzymes in the synthesis of certain hormones, neurotransmitters, and amino acids.

Iron is obtained in the diet from diverse sources, including red meat, poultry, fish, dark green leafy vegetables, nuts, and whole grains. Although it is an essential mineral, iron is toxic in its free or unattached form. The toxic effects of iron are avoided by having it bound to specific proteins. **Mucosal ferritin** binds iron within epithelial cells in the small intestine lining following its absorption from the gastrointestinal (GI) tract. When iron is needed in the body, mucosal ferritin transfers iron to **mucosal transferrin** (of epithelial cells of the small intestine), which then transfers it to blood transferrin. **Blood transferrin** (a plasma protein) transports iron to tissues throughout the body, including red bone marrow for hemoglobin synthesis, or to the liver for storage. Within the liver, iron is bound to **ferritin.**

Iron deficiency is one of the most common nutritional deficiencies. Symptoms of iron deficiency typically include fatigue, weakness, pale skin, headaches, and sensitivity to cold temperatures. It results from insufficient intake of iron based on loss of iron from the body. Iron is obtained in the diet. How much is obtained is dependent upon the quantities present in the consumed food and the amount absorbed from the GI tract. Heme iron (iron obtained in eating animal flesh of meats, poultry, and fish) is absorbed much more readily than nonheme iron (iron not associated with heme, which is obtained by eating iron-rich plants such as spinach and beans). Absorption may be altered by other food substances that are also present in the GI tract. For example, vitamin C increases the absorption of iron, whereas tea, which contains tannic acid, interferes with absorption of iron. In fact, high intake of tea may lead to iron deficiency. Changes to the intestinal lining that accompany certain diseases, such as celiac sprue (an autoimmune disorder) or Crohn disease (an inflammatory disease) (see Clinical View 26.7: "Inflammatory Bowel Disease and Irritable Bowel Syndrome") also interfere with iron absorption.

Iron is lost in feces (from the cells of the intestinal lining), urine, and sweat. Excessive loss of iron from bleeding (e.g., heavy or frequent menses, internal bleeding, trauma, blood donation) also may lead to an iron deficiency. Women are more prone to iron deficiencies because of the loss of blood with each menses. Pregnant women and women who are breastfeeding have an even higher risk because of the extra demands for iron placed on them.

Note that some foods are **fortified** by adding one or more essential nutrients, which may or may not be normally present in the food, to increase its nutritional value. **Fortification** may include, for example, vitamins (e.g., vitamin D in milk to increase calcium absorption, folic acid [B_9] in flour to prevent neural tube defects) or minerals (e.g., iodine in salt to prevent goiter).

WHAT DID YOU LEARN?

10 What are the major minerals? The trace minerals? How are these two groups distinguished?

27.4 Guidelines for Adequate Nutrition

LEARNING OBJECTIVES

15. Describe MyPlate, which was developed by the USDA to help people eat healthy.

16. Identify the items that are included on a food label.

The amount of each type of nutrient biomolecule, vitamin, and mineral required by the body is not the same for all individuals; rather, it is dependent upon a number of factors, including an individual's age, sex, body mass, level of physical activity, and health, as well as whether an individual is pregnant. A number of guidelines (recommended daily allowance, the USDA MyPlate, and required nutritional labels on packaged foods) have been provided or are required by scientific governmental agencies in the United States.

MyPlate

Not only may malnutrition be avoided, but an individual's state of health may be maximized when all of the nutrients are obtained at appropriate levels in the diet. Previously, food groups were organized by the United States Department of Agriculture (USDA) into a food guide pyramid that consisted of breads/grains, vegetables, fruits, dairy, proteins, and fats/oils/sweets. This organization provided a general approach for the number of servings that should be consumed daily from each category.

In 2011, the USDA revised the pyramid to **MyPlate (figure 27.1)**. This visual of a divided plate helps us to see at a glance the relative proportions of the types of foods we need to consume in order to stay healthy. One half of the plate is vegetables and fruits, and the other half is protein and grains, with dairy off to the side. Examples of each category of food include the following:

- **Vegetables**—broccoli, lettuce, peppers, and carrots
- **Fruits**—apples, blueberries, bananas, and oranges
- **Protein**—poultry, fish, beans, eggs, and nuts
- **Grains**—bread, cereals, rice, and pasta
- **Dairy**—skim milk, yogurt, and cheese

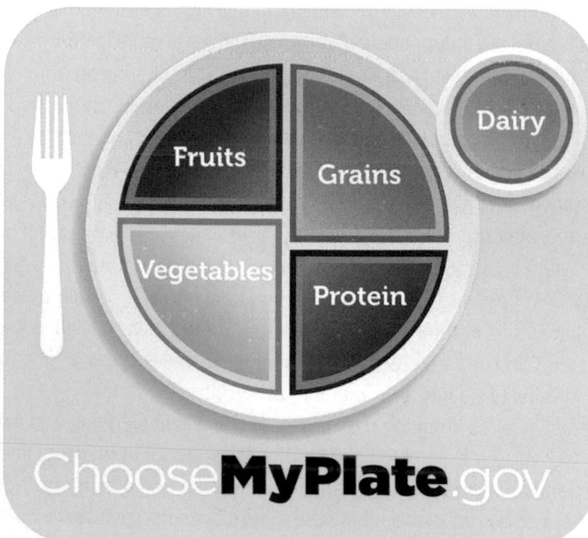

Figure 27.1 MyPlate. MyPlate shows the relative proportions of fruits, vegetables, grains, protein, and dairy for eating healthy.
Source: USDA

You will notice several additional recommendations listed if you go to the USDA website (www.choosemyplate.gov). These include the suggestion to balance calories by eating less, switch from whole milk to low-fat or no-fat milk, decrease consumption of saturated fats (see section 27.2b), reduce the intake of foods that are high in salt (e.g., canned soup), and cut back on sugary drinks by drinking water instead.

Nutritional Food Labels

Specific details on the composition of prepackaged items are made available on food labels **(figure 27.2)**. This information is helpful for individuals who are (1) interested in eating a healthy diet, (2) meal planning for weight-loss programs, and (3) restricting intake of nutrients such as sugar or sodium. These labels are required by law for most (but not all) products, and it is mandatory that the following information be included:

- Servings per container and calories per serving—this enables you to determine if there is more than one serving per container and how many calories are being consumed

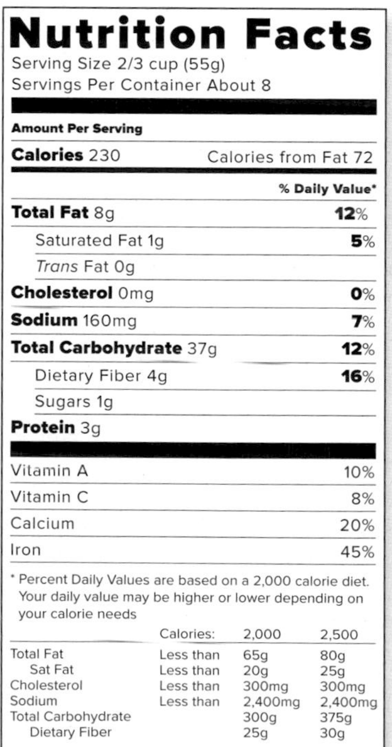

Nutrition Facts
Serving Size 2/3 cup (55g)
Servings Per Container About 8

Amount Per Serving	
Calories 230	Calories from Fat 72
	% Daily Value*
Total Fat 8g	**12%**
Saturated Fat 1g	**5%**
Trans Fat 0g	
Cholesterol 0mg	**0%**
Sodium 160mg	**7%**
Total Carbohydrate 37g	**12%**
Dietary Fiber 4g	**16%**
Sugars 1g	
Protein 3g	
Vitamin A	10%
Vitamin C	8%
Calcium	20%
Iron	45%

* Percent Daily Values are based on a 2,000 calorie diet. Your daily value may be higher or lower depending on your calorie needs

	Calories:	2,000	2,500
Total Fat	Less than	65g	80g
Sat Fat	Less than	20g	25g
Cholesterol	Less than	300mg	300mg
Sodium	Less than	2,400mg	2,400mg
Total Carbohydrate		300g	375g
Dietary Fiber		25g	30g

Nutrition Facts
8 servings per container
Serving size 2/3 cup (55g)

Amount per 2/3 cup
Calories 230

% DV*	
12%	**Total Fat** 8g
5%	**Saturated Fat** 1g
	***Trans* Fat** 0g
0%	**Cholesterol** 0mg
7%	**Sodium** 160mg
12%	**Total Carbs** 37g
14%	**Dietary Fiber** 4g
	Sugars 1g
	Added Sugars 0g
	Protein 3g
10%	**Vitamin D** 2mcg
20%	**Calcium** 260mcg
45%	**Iron** 8mg
5%	**Potassium** 235mg

* Footnote on Daily Values (DV) and calories reference to be inserted here.

(a) Current nutrition label **(b) Proposed nutrition label**

Figure 27.2 Food Label. A food label is included on prepackaged foods and provides serving size, calories, and nutrient information. (*a*) A nutrition label introduced over 20 years ago and (*b*) a recently proposed nutrition label are shown. (The abbreviation *g* is grams and *mcg* is micrograms [which is one-millionth of a gram]).

INTEGRATE

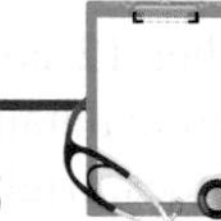

CLINICAL VIEW 27.3

Obesity

Obesity is a complex medical disorder whereby too much body fat may have a negative effect on health. It is often determined by measuring the **body mass index (BMI),** which is calculated by dividing an individual's weight in kilograms by his or her height in meters squared. You can determine your own BMI using pounds and inches as follows: *pounds/height (inches²) × 703*. For example, an individual who weighs 150 pounds and is 5′7″ (67 inches) tall has a BMI of 24 (150 pounds/4489 inches^2 × 703 = 23.5). Obesity is defined as having a BMI of 30 or greater. However, it is important to note that the BMI does not measure the relative amount of body fat in an individual. One of the issues with the BMI is that some muscular individuals may be classified as being overweight due to the BMI, and some people with excess body fat may still have normal BMI. An accurate measurement of percentage body fat requires expensive equipment to perform either underwater weighing (or hydrostatic weighing) and dual-energy X-ray absorptiometry (DEXA). A less precise measurement uses simpler and less expensive tools, including both skin calipers and bioelectric impedance analysis (BIA).

Obesity is a major public health problem that has many causes, including (1) an individual's genetics (e.g., leptin deficiency), (2) overeating (especially fat and simple sugar), (3) lack of exercise or physical activity, (4) specific medications (e.g., antidepressants), (5) insufficient sleep, (6) excess eating when bored, angry, stressed, or sad, (7) certain diseases (e.g., hypothyroidism), (8) being poor and unable to afford high-quality, nutritious food, and (9) being overweight as a child or teen. It is estimated that the prevalence of obesity in the United States is one out of every three individuals.

Numerous health risks are associated with obesity. In addition to an increased likelihood of a shortened life span, there is an increased risk of

- Diabetes (see Clinical View 17.8)
- High blood pressure (see Clinical View 20.10)
- High levels of cholesterol in the blood (see Clinical View 27.4)
- Myocardial infarction or heart attack (see Clinical View 19.5)
- Cerebrovascular accident or stroke (see Clinical View 13.9)
- Congestive heart failure (see Clinical View 19.1)
- Gallstones (see Clinical View 26.9)
- Osteoarthritis and gout (see Clinical View 9.9)
- Sleep apnea (see Clinical View 23.12)
- Colon cancer (see Clinical View 26.12)
- Breast cancer (see Clinical View 28.6)

Treating obesity generally involves multiple approaches, including lifelong changes in both an individual's eating habits and regular exercise, which result in a healthier, if not necessarily ideal, weight. Lifelong changes are more likely if an individual works with health-care and fitness professionals (e.g., physician, dietician, exercise specialist). Medications are available to facilitate weight loss, but they should be taken only after consulting with a physician. Removing or reducing the size of the stomach is a surgical treatment for obesity that can also induce type 2 diabetes into remission (see Clinical View 26.4).

- Total fat and the different types of fat (e.g., unsaturated fat, saturated fat, trans fat) and cholesterol
- Carbohydrates, including grams of dietary fiber
- Protein
- Vitamins
- Some minerals (e.g., sodium)

The label also provides both the Percent Daily Value, which is based on a diet of 2000 or 2500 calories, and the product ingredients. Product ingredients are listed in order of product weight—those having the greatest weight listed first. The list of product ingredients helps us to determine if the product contains healthy ingredients (e.g., 100% whole grains) or unhealthy ingredients. Unhealthy ingredients include both relatively large amounts of sugar (which, recall from section 27.2a, comes in many forms) and trans fats, which are produced during the process of hydrogenation of unsaturated fats to saturated fats. Product ingredients are also helpful to individuals with allergies (e.g., peanut allergies) to make sure that they avoid these products.

It has been proposed that nutrition labels should include information to reflect the actual serving size that an individual would eat—not what he or she should eat—how much sugar is added, and the amount of potassium and vitamin D that the food contains. Both the original food label that was introduced over 20 years ago and the recently revised nutrition label are shown in figure 27.2. Observe how much easier it is to determine both the calories per serving and the number of servings in a food product with the new food label.

Foods that do not typically require food labels include (1) those prepared on site (e.g., bakery, deli, candy store) and (2) those with limited nutrients (e.g., coffee, tea, some spices). In addition, some businesses are exempt from having to provide food labels because of either their small size or low annual sales income. In addition, dietary supplements (e.g., multivitamins) are required to have a supplemental label instead of a food label.

WHAT DID YOU LEARN?

11 What categories of food are shown on the USDA MyPlate? Which food category takes up the largest portion of the plate?

12 What is the purpose of the requirement for nutritional labeling on packaged foods?

27.5 Regulating Blood Levels of Nutrients

Recall from section 18.1c that blood is composed of plasma and formed elements. Plasma contains water, proteins, electrolytes, nutrients, respiratory gases, and waste products (see section 18.2). The percentage of these substances is regulated by several processes, many of which have been described in the preceding chapters. Here we limit our discussion to regulation of levels of nutrients as they vary during the day relative to the intake of meals. Generally, regulation of blood nutrients is dependent upon the length of time since you have eaten. These conditions are referred to as the absorptive state and the postabsorptive state.

27.5a Absorptive State

LEARNING OBJECTIVE

17. Explain when the absorptive state occurs and how nutrient levels are regulated during this time.

The **absorptive state** includes the time you are eating, digesting, and absorbing nutrients. It usually lasts approximately 4 hours after a

which are described in the next section, or (2) synthesized into bile salts (bile acids) and released as a component of bile into the small intestine (see section 26.4c). A majority of the bile salts are reabsorbed back into the blood primarily while moving through the ileum (and to a limited extent while moving through the large intestine). A small amount of bile salts (about 10%) continue into the large intestine and are removed from the body as a component of feces. This provides a means of eliminating excess cholesterol from the body and lowering blood cholesterol levels.

 **WHAT DID YOU LEARN?**

 Is cholesterol synthesis increased or decreased in response to high dietary intake of cholesterol?

 What are the two fates of cholesterol following its synthesis?

27.6b Transport of Lipids

 **LEARNING OBJECTIVES**

20. Define a lipoprotein, and provide a general overview of their function in the body.

21. Describe the transport of lipids within the blood.

Lipids are hydrophobic molecules and are insoluble in blood. Their transportation within the blood requires that they are first wrapped in a water-soluble protein. The lipid and the protein "wrap" are collectively called a **lipoprotein:** These are a general category of structures that contain triglycerides, cholesterol, and phospholipids within the "confines" of a protein. Thus, lipoproteins provide the means to transport lipids within the body.

We previously discussed one group of lipoproteins—chylomicrons (see section 26.4c). Recall that the absorption of lipids from the small intestine requires the formation of chylomicrons within epithelial cells lining the small intestine. A chylomicron is mostly composed of triglycerides and some cholesterol enveloped in protein. After its formation, it is absorbed into a lacteal and transported within the lymph until it enters venous blood at the junction of the jugular and subclavian vein (see section 21.1b). Chylomicrons deliver lipids to the liver and other tissues (e.g., adipose connective tissue for storage and skeletal muscle and cardiac muscle for energy use). Chylomicron remnants are then taken up by the liver.

Various other lipoproteins are formed within the liver. The relative density of these structures is used to classify these lipoproteins. The three broad categories of lipoproteins are (1) very-low-density lipoproteins (VLDLs), which contain the most lipid; (2) low-density lipoproteins (LDLs), with somewhat less lipid; and (3) high-density lipoproteins (HDLs), with the least amount of lipid. These function in the transport of lipids between the liver and peripheral tissues.

Transport from the Liver to Peripheral Tissues

Both very-low-density lipoproteins and low-density lipoproteins are associated with the transport of lipids from the liver to the peripheral tissues (**figure 27.4**):

- **Very-low-density lipoproteins (VLDLs)** contain various types of lipids (e.g., triglycerides, cholesterol) packaged with protein. VLDLs are assembled within the liver and then released into the blood. These "lipid delivery vehicles" circulate in the blood to release triglycerides to all cells of peripheral tissues, but primarily to adipose connective tissue. A change in density accompanies the release of triglycerides from these structures, and the lipoprotein is then called a low-density lipoprotein.
- **Low-density lipoproteins (LDLs)** contain relatively high amounts of cholesterol. LDLs deliver cholesterol to cells. LDLs bind to LDL receptors displayed within the plasma membrane of

INTEGRATE

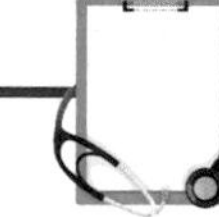

CLINICAL VIEW 27.4

Blood Cholesterol Levels

Blood cholesterol levels are tested to determine blood levels for total cholesterol, low-density lipoproteins (LDLs), and high-density lipoproteins (HDLs). An important comparison also includes the relative levels of HDLs and LDLs. The results are clinically significant because high blood levels for either total cholesterol or LDLs, as well as low blood levels for HDLs, are considered risk factors for cardiovascular disease.

A total cholesterol value above 200 mg/dL is considered "high." In addition, individual LDL and HDL levels provide a more specific measure of cholesterol levels. A normal value for LDL blood levels for a healthy adult is less than 100 mg/dL, and for those at risk for cardiovascular disease an acceptable value is below 70 mg/dL. In comparison, a normal value for HDL blood levels is between 40 and 50 mg/dL for adult males and between 50 and 60 mg/dL for adult females. A healthy ratio of HDL to LDL is considered to be above 0.3, or even better above 0.4.

A healthy ratio of HDL to LDL is a more accurate reading than total cholesterol because of the difference in the function of LDLs and HDLs. LDLs are considered to be the "bad cholesterol" because excess cholesterol not needed by peripheral tissue is deposited on the inner arterial walls, an event that may lead to development of plaque and arteriosclerosis (a risk factor for heart attacks and strokes). HDLs are considered the "good cholesterol" because they transport lipid (e.g., cholesterol, triglycerides) from the arterial wall to the liver for eventual elimination from the body through the GI tract.

Dietary intake of various types of fats influences blood cholesterol levels. The omega-3 fatty acids found in some cold-water fish, for example, lower total blood cholesterol levels. In comparison, high intake of saturated fats (especially the trans fats) has been shown to decrease HDL blood levels and increase LDL blood levels. Other risks factors correlated with elevated blood cholesterol include cigarette smoking, caffeine intake, and stress.

Statin drugs (e.g., Zocor, Lipitor, Crestor) were developed for the purpose of lowering blood cholesterol. They are typically prescribed for individuals with LDL blood levels higher than 130 mg/dL. Statin drugs act as a competitive inhibitor for HMG-CoA (3-hydroxy-3-methyglutaryl CoA) reductase, a specific enzyme within the metabolic pathway for cholesterol synthesis (figure 27.3). Thus, these drugs help decrease cholesterol synthesis. Subsequently, liver cells increase their uptake of cholesterol, which lowers blood cholesterol levels. Statin drugs do have side effects (e.g., muscle weakness, nausea, difficulty sleeping), so they should be taken only by individuals with high blood cholesterol or those who are at risk for cardiovascular disease.

Note that elevated blood cholesterol is only one of many risk factors for developing cardiovascular disease (see Clinical View 20.1: "Atherosclerosis"). Given that cardiovascular disease is the cause of death in approximately 50% of disease-related deaths in the United States, research continues in determining the most accurate risk factors and how to help individuals to lower their risk.

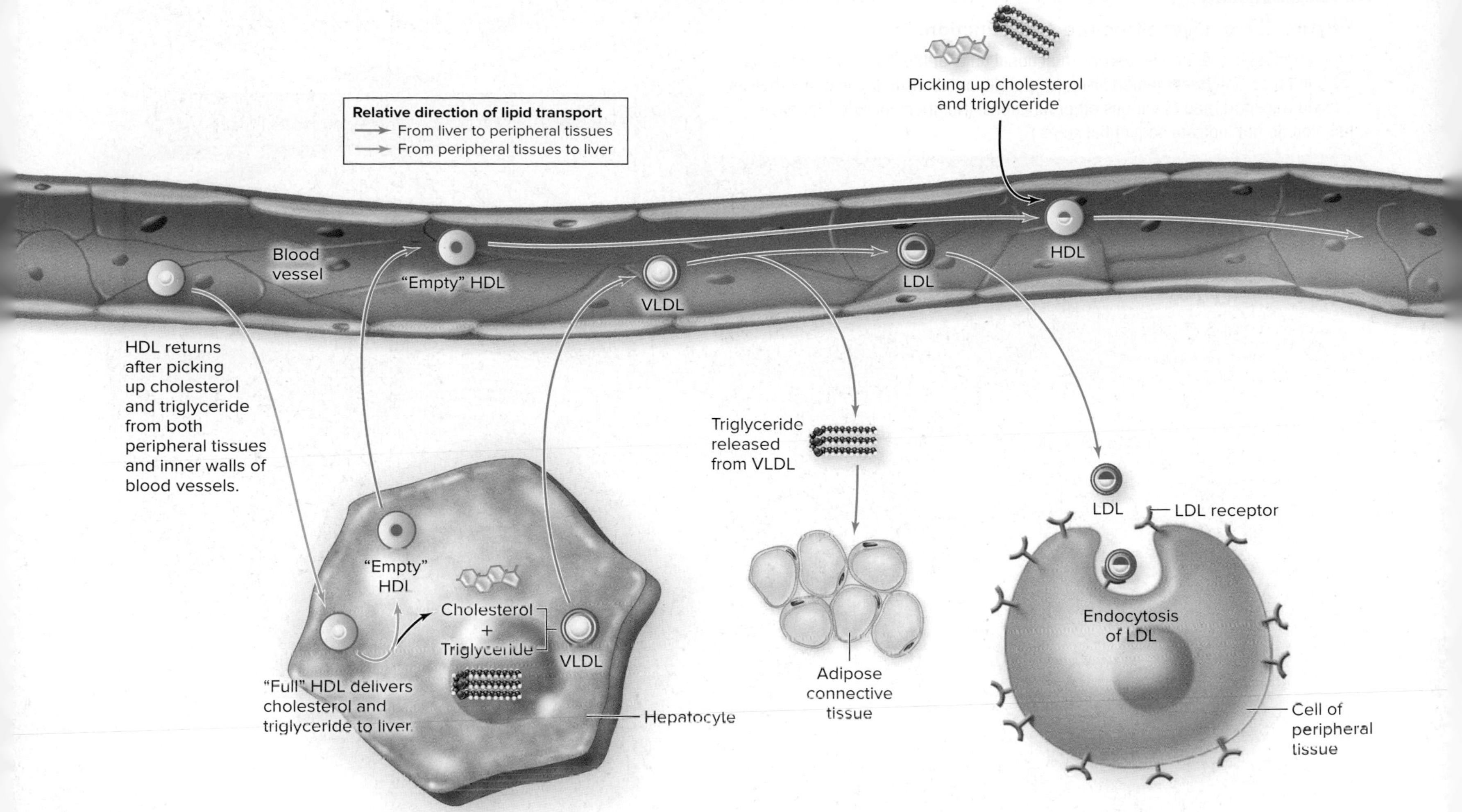

Figure 27.4 **Lipid Transport.** Very-low-density lipoproteins (VLDLs), which become low-density lipoproteins (LDLs), participate in transporting lipid from the liver to the peripheral tissues. In contrast, high-density lipoproteins (HDLs) transport lipid from peripheral tissues to the liver.

a cell and are subsequently engulfed into the cell by receptor-mediated endocytosis (see section 4.3c). Cholesterol is incorporated into the plasma membrane of all cells or is used by certain tissues (e.g., testes, ovaries, the adrenal cortex) to produce steroid hormones (see section 17.5a).

Transport from Peripheral Tissues to the Liver

High-density lipoproteins (HDLs) are associated with the transport of lipid from the peripheral tissues to the liver. These structures are produced in stages and have a function that opposes both VLDLs and LDLs. The proteins are formed in the liver and are released into the blood without the addition of lipid (i.e., these structures are "empty" HDLs). HDL molecules circulate throughout the blood and "fill" with lipids (e.g., cholesterol and triglycerides) from peripheral tissue and the inner lining of arterial walls. HDLs then transport these lipids to the liver. Excess cholesterol is converted to bile salts within the liver and released into the small intestine as a component of bile (recall that this is the means of eliminating cholesterol from the body). Note that HDLs, like LDLs, make cholesterol available to steroid-producing tissues. However, in contrast to LDLs, HDLs have cholesterol removed without being engulfed by the cell (not shown in figure).

WHAT DO YOU THINK?

2 Which lab test result is more informative: (a) one that gives total cholesterol or (b) one that provides the levels of both LDL and HDL? Explain.

WHAT DID YOU LEARN?

17 Which molecular structure (VLDL, LDL, or HDL) is responsible for transporting cholesterol from peripheral tissues to the liver?

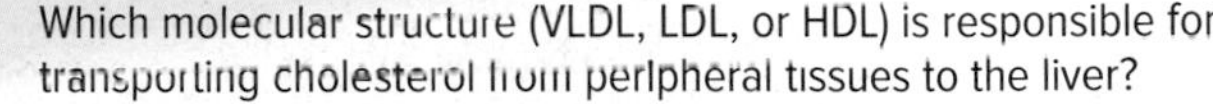

27.6c Integration of Liver Structure and Function

LEARNING OBJECTIVE

22. Identify and briefly describe the numerous roles of the liver in metabolism.

The liver is responsible for many different metabolic processes. Its structure and some of its many functions are integrated, visually represented, and described in **figure 27.5**. A summary of liver functional categories include the following:

- Carbohydrate metabolism
- Protein metabolism
- Lipid metabolism
- Transport of lipids
- Other functions (e.g., storage, drug detoxification)

WHAT DID YOU LEARN?

18 Provide an example for each of the five categories of liver functions integrated in figure 27.5.

INTEGRATE

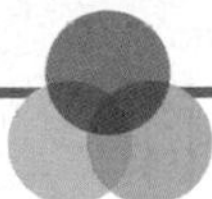

CONCEPT CONNECTION

We introduced the details of cellular respiration early in the text in section 3.4. This information provided the necessary background to understand later sections, including muscle metabolism in section 10.4a, muscle fiber types in section 10.5a, and metabolism of cardiac muscle in section 19.3f. We also discussed how cellular respiration is supported by the general processes of respiration, and how an individual's ability to produce sufficient energy (ATP) is dependent upon healthy respiratory and cardiovascular systems.

27.7 Central Role of Cellular Respiration

The biochemical pathways of cellular respiration were discussed in detail in section 3.4 for the oxidation of glucose to generate ATP. Here we first review ATP generation through the metabolic pathways of cellular respiration. We then discuss how these pathways are also used to oxidize other nutrient molecules (e.g., fatty acids, amino acids) both to generate ATP and to interconvert between various biomolecules and their building blocks.

27.7a ATP Generation

LEARNING OBJECTIVES

23. Describe where the following nutrient molecules enter the metabolic pathway of cellular respiration: glucose, the breakdown products of triglycerides, and amino acids.

24. Explain deamination of proteins.

Recall from our discussion of cellular respiration that it involves four stages **(figure 27.6)**:

(1) **Glycolysis** is a metabolic pathway that occurs in the cytosol of a cell and does not require oxygen. Glucose is oxidized to form two pyruvate molecules. Two ATP molecules are produced (by substrate-level phosphorylation), and hydrogen is transferred to 2 NAD^+ molecules to form 2 NADH molecules during this process (see section 3.4b). If insufficient oxygen is available to the cell, then the pyruvate is converted to lactate (see section 3.4g). The other steps in cell respiration occur within the mitochondria and require oxygen.

(2) The **intermediate stage** is a multienzyme step during which pyruvate is converted to acetyl CoA, and a molecule of carbon dioxide (CO_2) is formed by decarboxylation. One molecule of NADH is produced per pyruvate (see section 3.4c).

(3) The **citric acid cycle** is the metabolic pathway in which acetyl CoA binds to oxaloacetic acid (OAA) to form citric acid. Thereafter, two molecules of carbon dioxide (CO_2) are formed by decarboxylation. One ATP molecule, 3 NADH molecules, and 1 $FADH_2$ molecule are produced per "turn" of the cycle. Remember that both the intermediate stage and the citric acid cycle occur twice in the breakdown of a glucose molecule because there are two pyruvate molecules formed from the oxidation of each glucose molecule entering glycolysis (see section 3.4d).

(4) The **electron transport system** involves the transfer of hydrogen ions and an electron from the coenzymes NADH and $FADH_2$. ATP formation subsequently occurs through the process of oxidative phosphorylation (see section 3.4e).

Glycerol and Fatty Acids

The building blocks of triglycerides are glycerol and fatty acids. They may enter the cellular respiration pathway at certain stages and release their chemical energy to generate ATP. Glycerol specifically enters the pathway of glycolysis. Glycerol is converted to glucose through gluconeogenesis within the liver. The carbons of fatty acids are removed two at a time to form acetyl CoA (through beta-oxidation). Acetyl CoA molecules then enter the citric acid cycle.

Amino Acids

The amino acids that compose a protein may be used at various stages of cellular respiration to generate ATP. However, recall that amino acids contain nitrogen in an amine group ($—NH_2$). The amine group is first removed by a process called **deamination** (dē-am′i-nā′shŭn; *de* = remove) within hepatocytes of the liver. You might find it helpful to think of the amine group as a "contaminant" that must be taken off in order for amino acids to be used in cellular respiration to generate ATP.

The remaining portion of the amino acid following deamination enters the metabolic pathway of cellular respiration at one of several different steps, depending upon the specific amino acid. The modified amino acid may enter (1) into the pathway of glycolysis, (2) at the intermediate stage, or (3) at specific points within the citric acid cycle. The amine group (—NH2) is converted to urea (urea cycle) and is eliminated through the kidney as a component of urine (see section 24.8a).

WHAT DID YOU LEARN?

19 Where in the biochemical pathway of cellular respiration (glycolysis, intermediate stage, or citric acid cycle) does each of the following enter: (a) glycerol, (b) fatty acids, and (c) deaminated amino acids?

27.7b Interconversion of Nutrient Biomolecules and Their Building Blocks

LEARNING OBJECTIVE

25. Describe the physiologic advantages of the ability to interconvert nutrient biomolecules.

Individuals who eat excess amounts of carbohydrates, including sweets, can deposit additional triglycerides into their adipose connective tissue and gain weight. How is consumed sugar turned into fat? Some weight-loss plans advocate following a low-carbohydrate diet (e.g., the Atkins diet). Considering that nervous tissue primarily uses glucose, how is it possible that individuals on low-carbohydrate diets are able to provide the cells of the nervous tissue with the needed glucose?

Interconversion of nutrient biomolecules, which is the changing of one nutrient biomolecule to another, is possible because of the biochemical pathways that are associated with cellular respiration. Notice in figure 27.6 how the three nutrient biomolecules can be converted to each other through pathways that involve cellular respiration. This is indicated with arrows extending in both directions between the different pathways. For example, if energy is not needed, glucose can be broken down to acetyl CoA, which is then synthesized into triglycerides and stored, instead of entering the citric acid cycle.

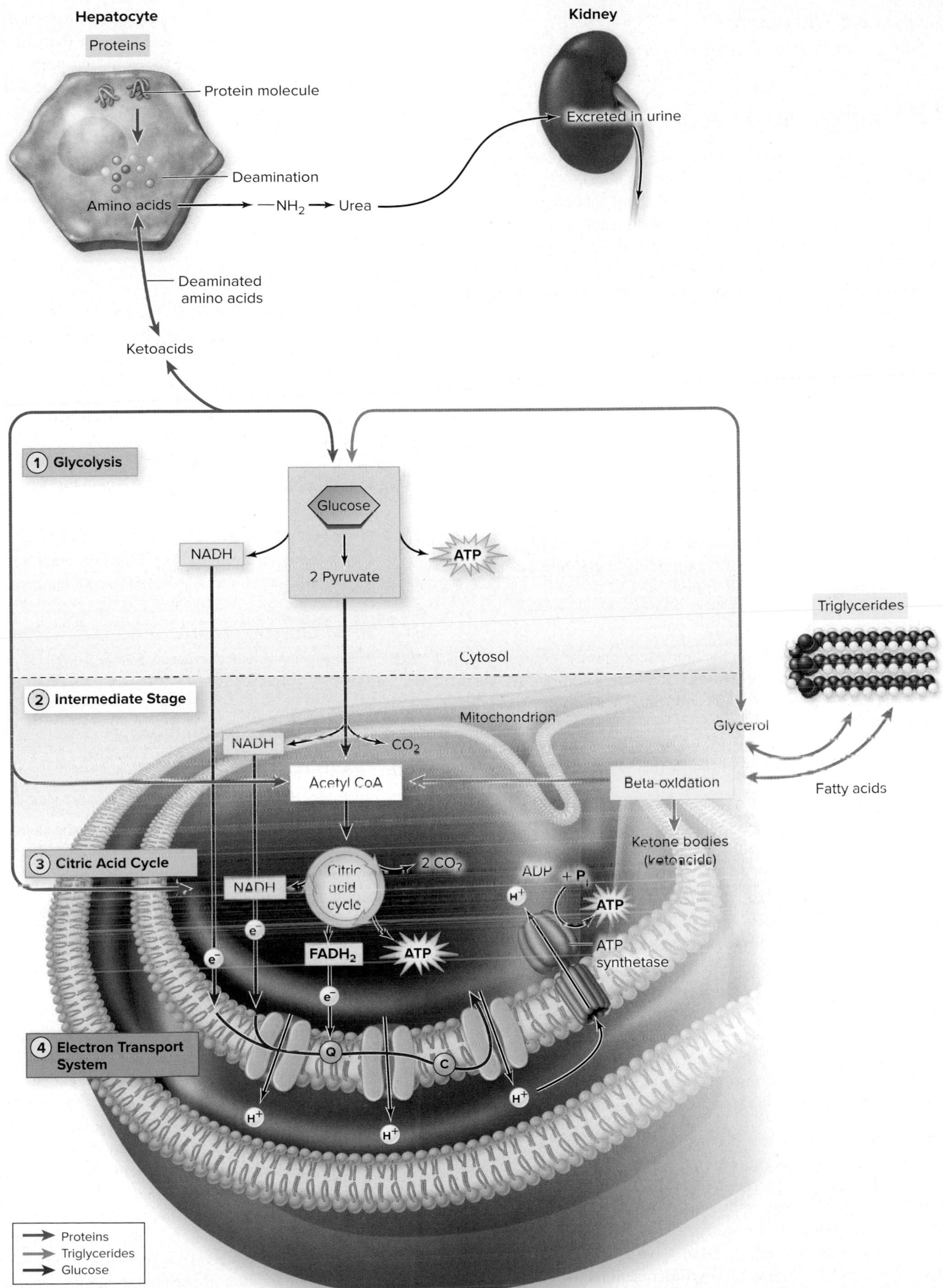

Figure 27.6 Cellular Respiration: Generation of ATP Molecules and Interconversion of Nutrient Molecules. The metabolic pathways of cellular respiration both generate ATP molecules through the oxidation of glucose, triglycerides, and proteins and provide a means of converting one type of nutrient biomolecule to another.

WHAT DID YOU LEARN?

20 How is excess sugar (glucose) converted to fat (triglycerides)?

27.8 Energy and Heat

How energy may be converted from one form to another was first discussed in section 3.1. For example, the chemical energy in food is first converted into ATP (another form of chemical energy), which is used to generate the mechanical energy of muscle contraction that occurs as we walk. We have previously stated that energy conversions result in the production of heat (according to the second law of thermodynamics). Here we describe how heat is related to metabolic rate and how we regulate body temperature, which has normal variations among individuals.

27.8a Metabolic Rate

LEARNING OBJECTIVES

26. Define metabolic rate.

27. Explain how both basal metabolic rate and total metabolic rate are measured and the variables that influence each.

The **metabolic rate** is the measure of energy used in a given period of time. There are two means of expressing metabolic rate: basal metabolic rate and total metabolic rate.

Basal Metabolic Rate

Basal metabolic rate (BMR) is the amount of energy required when an individual is at rest (and not eating). Resting conditions are determined as follows: The individual has not eaten for 12 hours, is reclining and relaxed, and is exposed to specific environmental conditions, including a room temperature between 20°C and 25°C (68°F to 77°F).

BMR may be measured by either of two methods:

- A **calorimeter,** which is a water-filled chamber into which an individual is placed. Heat released from the person's body alters the temperature of the water, and the change in temperature is measured. This is considered a direct method because heat is directly measured.
- A **respirometer,** which is an instrument to measure oxygen consumption. It is used to indirectly measure BMR because a relationship exists between oxygen consumption and heat production. Oxygen is used to produce ATP in aerobic cellular respiration, and ATP is utilized in metabolic processes that produce heat. Typically, for every liter of oxygen consumed, 4.8 kilocalories of heat are produced. This is considered an indirect method because the amount of heat produced is inferred from oxygen consumption.

The BMR of individuals varies because of their age, lean body mass, sex, and levels of various hormones in the blood. The BMR decreases as we age, with about a 3% decrease each decade beginning at age 30. Individuals with greater lean body mass (greater amounts of skeletal muscle relative to adipose connective tissue) tend to have a higher BMR. This is why males tend to have a higher BMR than females (average about 5–10% higher). Thyroid hormone increases BMR with an accompanying increase in lipolysis occurring within adipose connective tissue (see section 17.8b and **table R.4:** "Regulating Metabolism with Thyroid Hormone"). Individuals with hypothyroidism have a lower than normal BMR and tend to gain weight, whereas those with hyperthyroidism have a higher than normal BMR and tend to lose weight.

Another important variable in BMR, however, is body surface area. The reason is that heat is lost through the surface of the skin. The greater the surface area of the skin, the more heat that is lost. Heat lost through the surface is offset by the metabolic rate of cells, within limits. The more heat that is being lost, the more metabolically active body cells must be to maintain body temperature.

WHAT DO YOU THINK?

3 Which individual would you predict to have a higher basal metabolic rate, a man who is 6 feet tall weighing 200 pounds or a man who is 5 feet, 9 inches tall weighing 160 pounds? Explain your answer.

Total Metabolic Rate

Total metabolic rate (TMR) is the amount of energy used by the body, including energy needed for physical activity. Thus, TMR is the BMR plus metabolism associated with physical activity. The TMR varies widely, depending upon several factors:

- **The amount of skeletal muscle and its activity.** For example, a rapid elevation in TMR occurs during vigorous exercise and stays elevated for several hours after exercise.
- **Food intake.** Metabolic rate increases following ingestion of a meal but decreases after the absorption of nutrients has been completed.
- **Changing environmental conditions.** Metabolic rate increases, for example, when one is exposed to cold temperatures.

WHAT DID YOU LEARN?

21 Would your total metabolic rate increase when you exited your home on a cold winter day? Explain.

27.8b Temperature Regulation

LEARNING OBJECTIVES

28. Define core body temperature, and explain why it must be maintained.

29. Explain the neural and hormonal controls of temperature regulation.

We have just seen that a direct relationship exists between the metabolic rate of an individual and heat production—as the metabolic rate increases, more heat is produced. Because metabolic rate fluctuations are influenced by level of activity, food intake, and environmental temperature, the amount of heat produced also varies. Consequently, it is imperative that body temperature be maintained within certain physiologic limits in response to these changes. Homeostatic mechanisms usually keep our body temperature near the normal value of 98.6°F (37°C). This temperature regulation is accomplished through a balance of heat production and heat loss, and it occurs through both neural and hormonal controls. Mechanisms of regulating body temperature were discussed in numerous chapters, including sections 1.6b, 2.4b, 3.1c, 6.1c, 13.4c, and 18.1.

One of the critical aspects of regulating body temperature is maintaining **core body temperature,** which is the temperature of the vital portions of the body, or *core,* which consists of the head and torso. The temperature of these regions is kept relatively constant, or stable, to assure that life is maintained. This generally occurs by allowing fluctuations in the temperature of peripheral regions, such as the limbs.

Nervous System Control

Nervous system control of body temperature is mediated primarily through the hypothalamus (see section 13.4c). Motor pathways extend from the hypothalamus to the sweat glands in the skin, skeletal muscles, and peripheral blood vessels (see figure 1.13). The hypothalamus detects changes in body temperature either by monitoring the temperature of blood as it passes through the hypothalamus or by monitoring nerve signals received from the skin.

An increase in metabolic rate causes a subsequent increase in body temperature, and heat must be released. The hypothalamus responds by stimulating sweat glands to release sweat onto the surface of the body to draw heat away by both evaporation and transpiration (see section 25.2a) and stimulating vasodilation of peripheral blood vessels to bring heat to the skin surface (see Clinical View 27.5: "Heat-Related Illnesses").

In contrast, when metabolic rate decreases, it causes a subsequent decrease in body temperature, and additional heat must be generated. Now the hypothalamus inhibits sweat gland activity; stimulates constriction of peripheral blood vessels, thereby reducing blood circulation and heat loss at the periphery; and induces both smooth muscle contraction of arrector pili (to cause "goosebumps") and skeletal muscle contraction through shivering to generate heat. An individual can develop hypothermia either in a cold environment or when in a cool environment for an extended period. With extreme cold, the result of vasoconstriction to the peripheral regions may be frostbite, leading to dry gangrene (see Clinical View 27.6: "Hypothermia, Frostbite, and Dry Gangrene").

Conscious changes in behavior that are initiated by the cerebral cortex can help regulate body temperature. When we are hot, heat release is expedited by removing clothing, stopping work to rest, going inside where it is air-conditioned, or jumping into a pool; but when cold we attempt to conserve heat by putting on additional clothing or curling up under a blanket, and we generate additional heat by becoming more active.

INTEGRATE

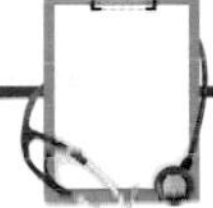

CLINICAL VIEW 27.5

Heat-Related Illnesses

The inability to release excess body heat can result in **heat-related illnesses,** which include (from mildest to most severe) heat cramps, heat exhaustion, and heatstroke. Children under 4 years of age and adults over 65 are at greatest risk of heat-related illness. Heat-related illnesses are described next in order of their severity.

Heat cramps are involuntary muscle spasms (usually in the thigh and calf muscles) that result from excessive loss of fluid and electrolytes (from sweating) while engaging in intense physical activity in hot weather, especially with accompanying high humidity. Heat cramps can occur during or following the exercise. Treating heat cramps includes resting in a cool environment, drinking fluids, and stretching the cramping muscle. Heat cramps are the first sign of heat-related illness and may lead to the more severe forms.

An individual experiencing **heat exhaustion** may have a core body temperature from 98.6°F to 104°F and experiences symptoms the include lightheadedness, headache, weakness, a rapid heart rate, nausea, and vomiting. Cooling body temperature to normal can be facilitated through a cool shower, bath, sponge bath, or ice towels.

Heat exhaustion, if left, untreated can lead to heatstroke, which is the most severe of the heat-related illness. A **heatstroke** is a medical emergency characterized by the inability to control body temperature (body temperature of 104°F or higher), fever, lack of sweating, confusion, and rapid breathing. An individual experiencing heatstroke requires medical assistance. Cooling body temperature as much as possible and transporting the individual to a hospital are essential. If left untreated, heatstroke can damage internal organs and can be fatal.

Hormonal Control

Temperature regulation is also mediated by hormone secretion, including thyroid hormone, epinephrine and norepinephrine, growth hormone, and testosterone. The most significant is thyroid hormone. Recall from section 17.8b that thyroid hormone establishes the metabolic rate, causing us to produce a certain amount of heat to keep the body warm. As your body temperature begins to drop, the hypothalamus releases thyrotropin-releasing hormone (TRH); TRH stimulates the anterior pituitary to release thyroid-stimulating hormone (TSH); and TSH stimulates the thyroid gland to release the thyroid hormones (T_3 and T_4) (see figure 17.17). Thyroid hormone is able to help maintain body temperature by increasing the metabolic rate of almost all cells, especially neurons. Neurons are specifically stimulated to increase their number of sodium-potassium (Na^+/K^+) pumps. Because there are more Na^+/K^+ pumps, more energy is utilized as the pumps use ATP to move the ions, then more heat is produced, and body temperature is maintained.

WHAT DID YOU LEARN?

22 What is the core body temperature, and why must it be maintained?

23 Explain the role of the hypothalamus in regulating the response to an increase in body temperature.

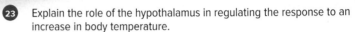

INTEGRATE

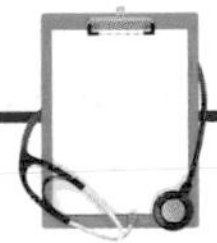

CLINICAL VIEW 27.6

Hypothermia, Frostbite, and Dry Gangrene

Prolonged exposure to cool or cold temperature can result in **hypothermia,** which is a body temperature below 95°F. Hypothermia is more likely for an individual submerged in cold water (heat is lost at a much greater rate in water than in air). Symptoms and signs include a decrease in heart rate, slower breathing, confusion, drowsiness, and fatigue. Notably, shivering, which happens when we are cold, ceases in hypothermia to conserve energy.

The risk of hypothermia increases with a decrease in temperature, but it also depends upon other factors. The following all increase the risk for hypothermia: a person's age (both infants and older adults are at greater risk), health (those with diabetes, thyroid malfunctions), mental illness, and drug and alcohol use.

Frostbite is damage to superficial cells due to exposure to extreme cold and the skin freezes. The skin appears white and may be blistered. Loss of sensation may also occur. It may lead to dry gangrene in severe cases. **Dry gangrene** (necrosis or pathologic death) is a form of gangrene (see Clinical View 5.5: "Gangrene") in which the involved body part is desiccated (loss of water), sharply demarcated (distinct change in coloration), and shriveled. These changes are usually due to extensive and prolonged vasoconstriction of blood vessels as a result of exposure to extreme cold. This results in oxygen deprivation and death of the tissue.

Photo of frostbite and dry gangrene.
©Images & Stories/Alamy

9. All of the following are functions of the liver *except* the
   a. release of insulin.
   b. formation of plasma proteins (e.g., albumin).
   c. formation of LDL particles.
   d. elimination of bilirubin.
10. Thyroid hormone is the primary hormone that regulates body temperature. It does so by
   a. stimulating muscle contraction.
   b. increasing the number of Na^+/K^+ pumps in neurons.
   c. vasoconstriction of peripheral blood vessels.
   d. sweating.
11. Define nutrition.
12. Describe the general functions of vitamin C.
13. Distinguish between essential and nonessential amino acids.
14. Explain the risk of excessive intake of fat-soluble vitamins.
15. Define minerals, and give examples of their functions in the body.
16. Define nitrogen balance, and compare positive nitrogen balance and negative nitrogen balance.
17. Define the postabsorptive state. What is the major hormone that regulates blood nutrients during this time, and what is its role?
18. Explain the function of the liver in the transport of lipid in the blood.
19. Briefly describe the role of the liver in carbohydrate metabolism, protein metabolism, and lipid metabolism.
20. Explain the neural- and hormonal-induced changes that occur when you are cold.

Can You Apply What You've Learned?

1. An individual has recently adopted a vegetarian lifestyle. She has been on this type of diet for approximately 6 months and admits that she has not been conscientious about making healthy food choices. She is feeling weak, appears gaunt, and is losing body mass. What type of nutrient biomolecule is most likely missing from her diet?
   a. nonessential amino acids
   b. essential amino acids
   c. fats (triglycerides)
   d. carbohydrates
2. A young man explains to his doctor that he has been having headaches and fluctuations in body temperature, going from shivering to sweating, that are not induced by changes in his environment. After a great deal of diagnostic work, the physician orders a magnetic resonance imaging (MRI) scan. A mass is most likely detected in which area of his brain?
   a. hypothalamus
   b. pons
   c. thalamus
   d. medulla oblongata
3. Derrick is a young man who lifts weights for several hours a day. He explains that remaining physically fit is very important to him and has been taking megadoses of vitamins as part of maintaining his health. You are concerned about
   a. reduced carbohydrate intake that may result in ketoacidosis.
   b. toxic buildup of water-soluble vitamins.
   c. loss of protein mass.
   d. toxic buildup of fat-soluble vitamins.
4. An elderly gentleman has recently been diagnosed with cirrhosis of the liver. He admits to being a heavy drinker. This man is your patient. Given the damage to his liver, you would have concerns about all of the following *except*
   a. regulating his ability to maintain body temperature, because of decreased thyroid hormone levels.
   b. internal bleeding, because he has impaired ability to synthesize clotting proteins.
   c. edema, because of impaired synthesis of plasma proteins (e.g., albumin).
   d. jaundice, because his liver is less efficient at removing bilirubin from the blood.
5. Howard is a middle-aged man and has been slowly gaining weight after reaching the age of 30. He has decided to lose weight by following a low-carbohydrate diet. What physiologic process allows his brain to still have sufficient glucose to function normally?
   a. maintenance of nitrogen balance
   b. hypothalamus regulation
   c. gluconeogenesis
   d. transamination

Can You Synthesize What You've Learned?

1. As a dietician in the hospital, explain MyPlate to a patient.
2. Explain the challenge in meeting dietary requirements to an individual who has adopted a vegetarian lifestyle.
3. Explain to a patient who is being admitted to the hospital with frostbite what has occurred.

INTEGRATE

ONLINE STUDY TOOLS

connect | SMARTBOOK® | AP|R

The following study aids may be accessed through Connect.

Clinical Case Study: A Young Man with Lung Disease and Bruising

Interactive Questions: This chapter's content is served up in a number of multimedia question formats for student study

SmartBook: Topics and terminology include nutrients; obtaining nutrients from food; regulating blood levels of nutrients; functions of the liver; central role of cellular respiration; energy and heat

Anatomy & Physiology Revealed: Topics include liver

Animations: Topics include B vitamins; complementary proteins; anatomy of a food label

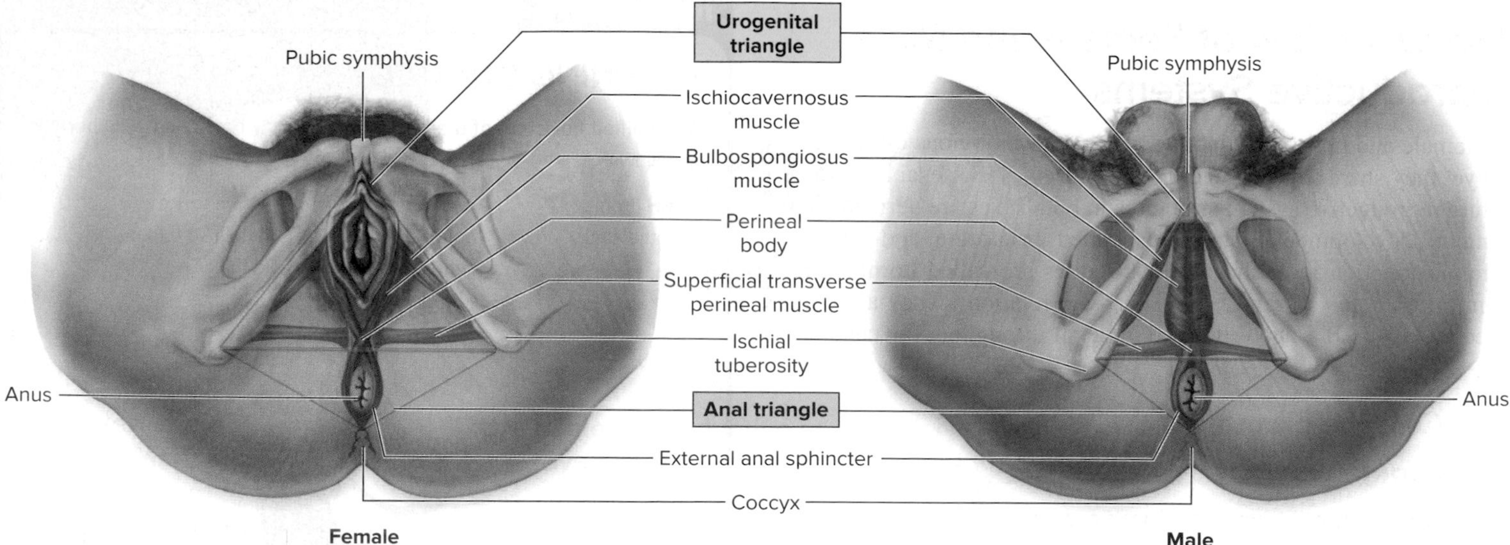

Figure 28.1 Perineum. In both females and males, the perineum is the diamond-shaped area between the thighs, extending from the pubis anteriorly to the coccyx posteriorly, and bordered laterally by the ischial tuberosities. An imaginary horizontal line extending from the ischial tuberosities subdivides the perineum into a urogenital triangle anteriorly and an anal triangle posteriorly. AP|R

symphysis, laterally by the ischial tuberosities, and posteriorly by the coccyx **(figure 28.1)**. It is an area that may be partially torn during childbirth, due to the extensive stretching needed to expel the baby. (Sometimes, this region may have to be surgically cut during birth if there is difficulty expelling the baby; see section 29.6d.) Two distinct triangle bases are formed by an imaginary horizontal line extending between the ischial tuberosities of the ossa coxae. Both triangles house specific structures in the floor of the pelvis:

- The anterior triangle, called the **urogenital triangle,** contains the urethral and vaginal orifices in females and the base of the penis and the scrotum in males. Within the urogenital triangle are the muscles that surround the external genitalia, called the ischiocavernosus, bulbospongiosus, and superficial transverse perineal muscles.
- The posterior triangle, called the **anal triangle,** is the location of the anus in both sexes. Surrounding the anus is the external anal sphincter.

The external anal sphincter, bulbospongiosus, and superficial transverse perineal muscles are partly anchored by a dense connective tissue structure called the **perineal body.** Review table 11.12 and figure 11.17 in section 11.7 when learning these structures.

WHAT DID YOU LEARN?

3 What are the components of the urogenital triangles in females and males?

28.2 Gametogenesis

Gametogenesis (gam′ĕ-tō-jen′ĕ-sis; *gameto* = gamete, *genesis* = beginning) is the process of forming human sex cells, called gametes. Female gametes are called *secondary oocytes* (commonly referred to as "eggs"), whereas male gametes are called *sperm.* The process of gametogenesis begins with a specific type of cell division called *meiosis.* The events of meiosis generally are similar in both females and males, with a few differences. Here we describe the basics of heredity and meiosis; sex-specific differences are described throughout this chapter.

28.2a A Brief Review of Heredity

LEARNING OBJECTIVES

4. Distinguish between autosomes and sex chromosomes.
5. Explain why somatic cells contain $2n$ chromosomes but gametes must contain n chromosomes.

Humans pass on their traits to a new individual when they reproduce. This hereditary information is carried on chromosomes. Human somatic (body) cells contain 23 pairs of chromosomes—22 pairs of autosomes and 1 pair of sex chromosomes, for a total of 46 chromosomes (see figure 29.8):

- **Autosomes** contain genes that code for cellular functions. These genes also help determine most human characteristics, such as eye color, hair color, height, and skin pigmentation. A pair of matching autosomes is called **homologous chromosomes** (hō-mol′ō-gŭs; *homos* = same, *logos* = relation).
- The pair of **sex chromosomes** consists of either two X chromosomes or an X and a Y chromosome. These chromosomes primarily determine whether an individual is female (two X chromosomes) or male (one X chromosome and one Y chromosome), although they also contain genes that code for cellular functions.

One member of each pair of chromosomes is inherited from each parent. In other words, if you examined one of your body cells, you would discover that 23 of the chromosomes came from your mother, and the other 23 chromosomes in this same cell came from your father. A cell that contains 23 pairs of chromosomes is said to be **diploid** (dip′loyd; *diploos* = double) and to have $2n$ chromosomes, where n is the unpaired chromosome number.

If the gametes were diploid, then each new individual would receive two sets of paired chromosomes, or $4n$. This situation is not what happens in nature, and it is not compatible in life. The gametes from either sex are **haploid** (hap′loyd; *haplos* = simple, *eidos* = appearance) because they contain 23 chromosomes only, not 23 pairs, and thus their chromosome number is designated as $1n$, or just n. (Heredity is discussed in more detail in section 29.9.)

INTEGRATE

LEARNING STRATEGY

Meiosis I (the first meiotic division) randomly separates maternal and paternal *pairs* of replicated (double-stranded) chromosomes, whereas meiosis II (the second meiotic division) separates replicated chromosomes into single chromosomes. Also, meiosis II is very similar to mitosis. Thus, if you remember the steps of mitosis, you can figure out the steps of meiosis II.

WHAT DID YOU LEARN?

4 How do sex chromosomes differ from autosomes?

5 Why do gametes have to be haploid instead of diploid?

28.2b An Overview of Meiosis

LEARNING OBJECTIVES

6. Compare and contrast meiosis and mitosis.
7. Describe events during interphase, before cell division begins, and the specific steps of meiosis.
8. Explain the difference between homologous chromosome pairs and sister chromatids

Meiosis (mī-ō′sis; *meiosis* = lessening) is sex cell division that starts off with a diploid parent cell and produces haploid daughter cells called *gametes.* Mitosis (somatic cell division, described in section 4.9) and meiosis (sex cell division) differ in the following ways:

- Mitosis produces *two* daughter cells that are genetically *identical* to the parent cell. In contrast, meiosis produces *four* daughter cells that are genetically *different* from the parent cell.
- Mitosis produces daughter cells that are *diploid,* whereas meiosis produces daughter cells that are *haploid.*
- Meiosis includes a process called **crossing over,** whereby genetic material is exchanged between homologous chromosomes.

The process of *crossing over* helps to "shuffle the genetic deck of cards," so to speak. Thus, crossing over is a means of combining different genes from both parents on one of the homologous chromosomes. Crossing over does not occur in mitosis.

Meiosis begins with a diploid parent cell located in a gonad (ovary or testis). In this cell, 23 chromosomes came from the organism's mother (23 maternal chromosomes), and 23 from the father (23 paternal chromosomes). To produce haploid gametes, this parent cell must undergo meiosis.

Interphase

The cell cycle phase called **interphase** occurs prior to meiosis (see section 4.9b). During interphase, the DNA in each chromosome is replicated, or duplicated exactly, in the parent cell. These **replicated chromosomes** (also known as *double-stranded chromosomes*) are composed of two identical structures called **sister chromatids** (krō′mă-tid; *chromo* = color, *id* = resembling). Each sister chromatid contains an identical copy of DNA at this point. The sister chromatids are attached at a specialized region termed the **centromere.**

Note that sister chromatids are *not* the same as a *pair* of chromosomes. A chromosome composed of sister chromatids resembles a written letter X, whereas a homologous *pair* of chromosomes is composed of a matching maternal chromosome and paternal chromosome,

Before Birth

The process of oogenesis begins in a female fetus before birth. At this time, the ovary cortex contains primordial germ cells called **oogonia** (ō-ō-gō′nē-ă; sing., *oogonium*), which are diploid cells (they have 23 pairs of chromosomes). (The oogonia are within the primordial follicles.) During the fetal period, the oogonia divide by mitosis to produce primary oocytes. Primary oocytes are the structures that start the process of meiosis, but they are arrested in prophase I. At this point, the cells are the primary oocytes described in the preceding section. At birth, the ovaries of a female child are estimated to contain a total of approximately 1.5 million primordial follicles within its cortex. The primary oocytes in the primordial follicles remain arrested in prophase I until after puberty.[1]

Childhood

During childhood, a female's ovaries are relatively inactive. In fact, the continuing event that occurs during childhood is **atresia** (ă-trē′zē-ă; *a* = not, *tresis* = a perforation), in which some primordial follicles regress. By the time a female child reaches puberty, only about 200,000–400,000 primordial follicles remain in the ovaries.

From Puberty to Menopause

When a female reaches puberty, the hypothalamus increases its release of gonadotropin-releasing hormone (GnRH). GnRH stimulates the anterior pituitary to release follicle-stimulating hormone (FSH) and luteinizing hormone (LH). The levels of FSH and LH vary in a cyclical pattern and help produce a monthly sequence of events in follicle development called the **ovarian cycle.**[2] The three phases of the ovarian cycle are the follicular phase, ovulation, and the luteal phase **(figure 28.7)**.

Follicular Phase The **follicular phase** occurs during days 1–13 of an approximate 28-day ovarian cycle. Prior to (up to one year prior) until the beginning of the follicular phase, molecular signals from the oocyte and follicle cells stimulate up to about 20 primordial follicles to develop into primary follicles. It is unclear why some of the primordial follicles are stimulated to develop while the rest remain unaffected by the FSH and LH secretion. As the ovarian follicles develop, their granulosa cells release the hormone **inhibin.** Inhibin (and the initial low levels of estrogen secreted by the primary follicles) helps inhibit FSH production, thus preventing excessive ovarian follicle development and allowing the current primary follicles to mature. Shortly thereafter, a few of these primary follicles mature and become secondary follicles. The primary follicles that do not mature undergo atresia. While the earlier stages of follicular development (primordial to primary to secondary) primarily are the result of molecular signals between the oocyte and follicle cells, the later stages of follicular development (secondary to antral to mature follicle) are more directly dependent on increased FSH and LH levels. Thus, FSH and LH now directly act on the secondary follicles. Typically, only one secondary follicle in an ovary develops into an antral follicle. Late in the follicular phase, this antral follicle develops into a mature follicle. The volume of fluid increases within the antrum of this follicle, and the oocyte is forced toward one side of the follicle, where it is surrounded by the cumulus oophorus. The innermost layer of the cumulus oophorus cells is the corona radiata.

As the antral follicle develops into a mature follicle, its primary oocyte finishes meiosis I (where pairs of replicated chromosomes are separated), and two cells form (figure 28.6). One of these cells receives a minimal amount of cytoplasm and forms a **polar body,** which is a nonfunctional cell that later regresses. The other cell receives the bulk of the cytoplasm and becomes the secondary oocyte, which continues to develop and reaches metaphase II of meiosis before it is arrested again. This secondary oocyte does not complete meiosis II (where sister chromatids are separated) unless it is fertilized by a sperm. If the oocyte is never fertilized, it breaks down and regresses about 24 hours later.

Ovulation **Ovulation** (ov′yū-lā′shŭn) occurs on day 14 of a 28-day ovarian cycle and is defined as the release of the secondary oocyte from a mature follicle (figure 28.7). Typically, only one ovary ovulates each month. Ovulation is induced only when there is a peak in LH secretion. As the time of ovulation approaches, the granulosa cells of the mature follicle increase their rate of fluid secretion into the antrum, forming a larger antrum and causing further swelling

[1]Research has suggested some primordial follicles develop into primary follicles prior to puberty, but these follicles do not play a role in the ovarian cycle and the development of viable gametes.

[2]Although we simplistically discuss the ovarian cycle as happening within a single 28-day time frame, individual follicle maturation takes almost a year. It takes up to 290 days for a primary follicle to develop into a late secondary follicle, and then another 60 days of continued development before ovulation of the secondary oocyte from the mature follicle.

INTEGRATE

CLINICAL VIEW 28.2

Ovarian Cancer

Ovarian cancer refers to a primary malignancy in the ovaries. About 90% of all ovarian cancer are *carcinomas,* meaning the cancer derived from epithelium (in this case, the germinal epithelium of the ovary). Not all tumors that develop in the ovaries are malignant (or cancerous), but when the tumors are malignant, a diagnosis of ovarian cancer is made. Ovarian cancer is the eighth most common cancer and the fifth most common cancer cause of death in women.

Women who have the inherited mutation in the *BRCA1* and *BRCA2* genes (which normally produce tumor suppressor proteins) have a lifetime risk of about 10–40% of developing ovarian cancer (compared to a 1–2% risk for those without the mutation in these genes). (About .2–.3% of the female population have one or both gene mutations.) Other risk factors for the disease include increasing age, obesity, infertility, and a family history of ovarian cancer, breast cancer, or colorectal cancer. Birth control medication (in premenopausal women) reduces the risk of ovarian cancer, whereas postmenopausal estrogen therapy increases the risk.

Ovarian cancer is difficult to diagnose early because initial symptoms may be nonspecific and because there is no reliable test for the disease. Initial symptoms may include constipation, increased need and urgency to urinate, abdominal swelling and bloating, pelvic pain, indigestion, nausea, unexplained back pain, and changes in weight. As all of these symptoms may be nonspecific, the causes of the symptoms may not be correctly attributed to ovarian cancer. A blood test for CA-125 (a cancer antigen protein found on ovarian cancer cells) may indicate the likelihood of ovarian cancer, but because CA-125 levels also may be elevated in numerous other noncancerous conditions (such as endometriosis; see Clinical View 28.4), a high CA-125 test alone is not considered reliable for a diagnosis of cancer.

Once a diagnosis of cancer is made, treatment typically involves a combination of surgery and chemotherapy. If it is diagnosed early, 5-year-survival-rate prognoses are very good. Unfortunately, many ovarian cancers are not detected until after the cancer has metastasized to other organs, which dramatically reduces the chances for good long-term survival rates. Therefore, women who experience the initial symptoms described above should seek medical consultation promptly.

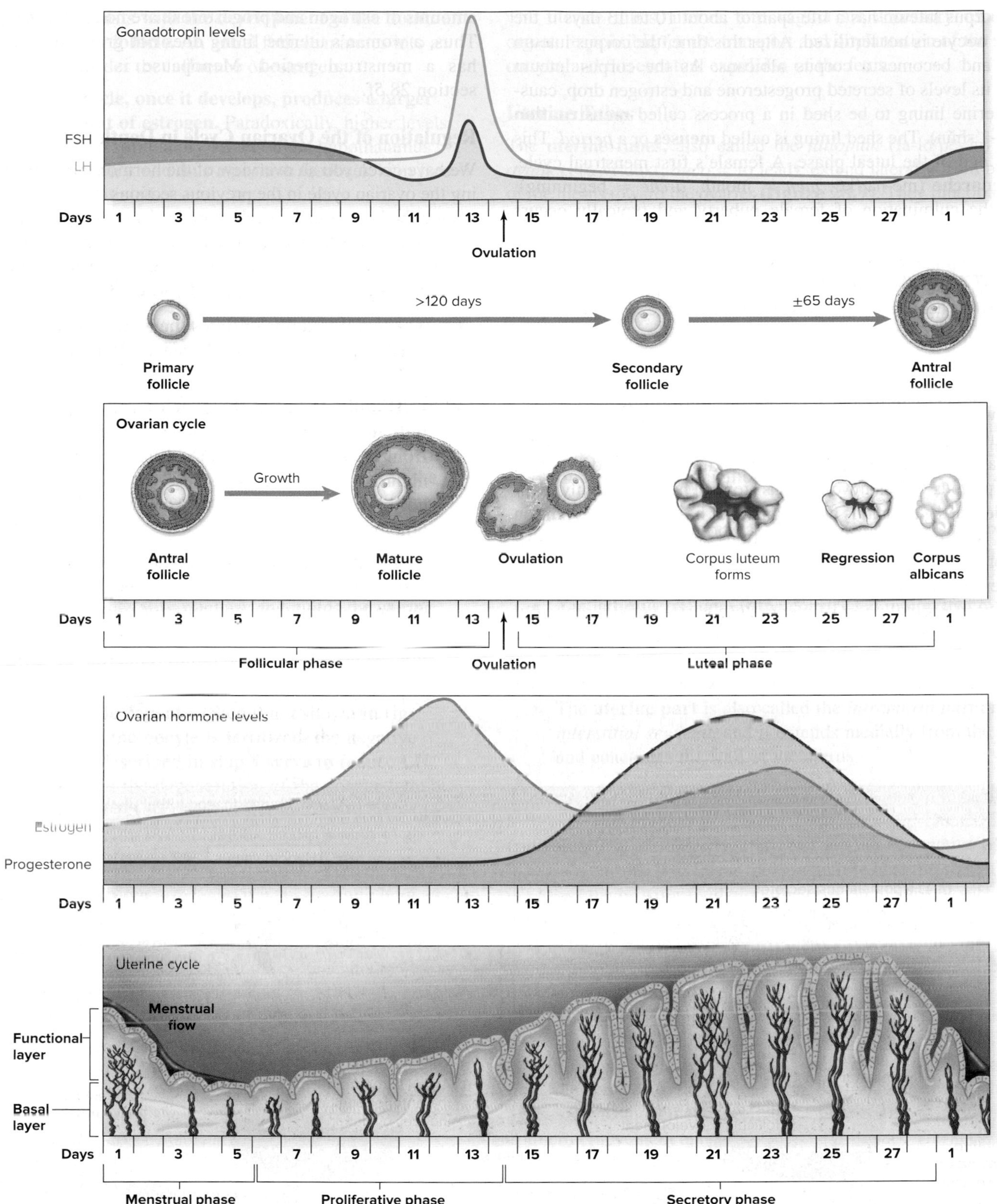

Figure 28.7 Hormonal Changes in the Female Reproductive System. Follicle-stimulating hormone (FSH) stimulates development of ovarian follicles during the follicular phase of the ovarian cycle. Estrogen (secreted from ovarian follicles) stimulates the proliferative phase in the uterine cycle, whereas a peak in luteinizing hormone (LH) promotes ovulation and the development of the corpus luteum, which produces both progesterone and estrogen to promote uterine lining growth. AP|R

within the follicle. The edge of the follicle that continues to expand at the ovarian surface becomes quite thin and eventually ruptures, expelling the secondary oocyte.

Luteal Phase The **luteal phase** occurs during days 15–28 of the ovarian cycle, when the remaining follicle cells in the ruptured mature follicle become the corpus luteum.

The corpus luteum is essentially a temporary endocrine gland. It secretes progesterone and estrogen that stabilize and build up the uterine lining, and prepare for possible implantation of a fertilized oocyte. It also secretes inhibin, which along with the high levels of estrogen and progesterone, inhibits the hypothalamus and anterior pituitary from secreting their reproductive hormones.

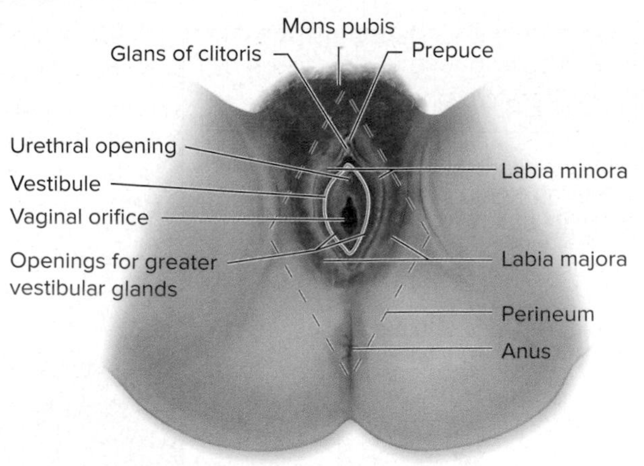

(a) Superficial structures of external genitalia

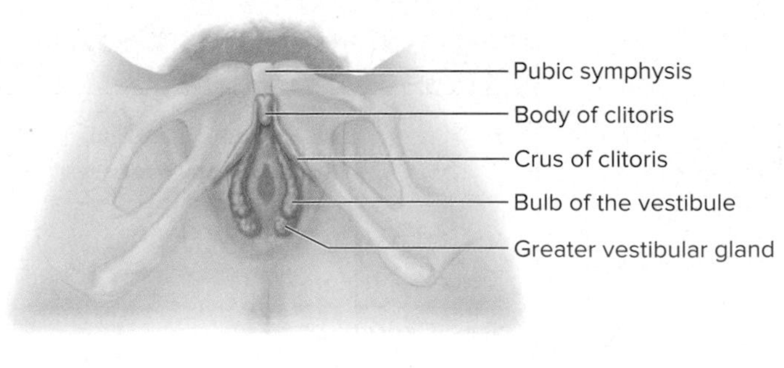

(b) Deep structures of external genitalia

Figure 28.12 Female External Genitalia. (*a*) Inferior view of the external genitalia, illustrating the urethral opening and the vaginal opening, which are within the vestibule and bounded by the labia minora. (*b*) Deep structures associated with the external genitalia.

The space between the labia minora is called the **vestibule.** Within the vestibule are the **urethral opening** and the vaginal orifice. On either side of the vaginal orifice is an erectile body called the **bulb of the vestibule,** which engorges with blood and increases in sensitivity during sexual intercourse. A pair of **greater vestibular glands** (previously called *glands of Bartholin*) are housed within the posterolateral walls of the vestibule and secrete mucin, which forms mucus to act as a lubricant for the vagina. Secretion increases during sexual intercourse, when additional lubrication is needed. These secretory structures are homologous to the male bulbourethral glands (which are discussed with the male reproductive system).

The **clitoris** (klit′ō-ris) is a small erectile body, usually less than 2 centimeters in length, located at the anterior regions of the labia minora. It is homologous to the penis of the male. Two small erectile bodies called **corpora cavernosa** form the **body** of the clitoris. The corpora cavernosa become engorged with blood and increase in sensitivity during sexual intercourse. Extending from each of these bodies posteriorly are elongated masses, each called the **crus** (krūs) of the clitoris, which attach to the pubic arch. Capping the body of the clitoris is the **glans** (glanz; acorn). The many specialized sensory nerve receptors housed in the clitoris provide pleasure to the female during sexual intercourse. The **prepuce** (prē′pūs; foreskin) is an external fold of the labia minora that forms a hoodlike covering over the clitoris.

WHAT DID YOU LEARN?

21 What are the individual components of the female external genitalia and their functions?

28.3f Mammary Glands

LEARNING OBJECTIVES

23. Explain the gross anatomy of the mammary glands.

24. Compare the hormones responsible for milk production and milk ejection.

Each **mammary gland,** or *breast,* is located within the anterior thoracic wall and is composed of a compound tubuloalveolar exocrine

INTEGRATE

CLINICAL VIEW 28.6
Breast Cancer

Breast cancer affects approximately one in every eight women in the United States, and it also occurs infrequently in males. Some well-documented risk factors for breast cancer include maternal relatives with breast cancer, longer reproductive span (early menarche coupled with delayed menopause), obesity, nulliparity (i.e., never having been pregnant), late age at first pregnancy, and having mutations in specific genes (*BRCA1* and *BRCA2*) that normally produce tumor suppressor proteins. Most of these risk factors (except for the *BRCA1* and *BRCA2* mutations) are related to an increased exposure to estrogen over a long period of time. Almost all breast cancers are *carcinomas,* meaning the cancer derived from epithelium (in this case, the epithelium that lines the ducts and lobules of the breast).

Monthly self-examination is an important means of early detection of breast malignancies. Mammography, which is an x-ray of the breast that can detect small areas of increased tissue density, can identify many small malignancies that are not yet palpable in a self-examination. Recommendations vary, but most physicians agree that women over the age of 40 should have a mammogram done every 1 to 2 years. Women with a family history of breast cancer should consider regular mammography before the age of 40.

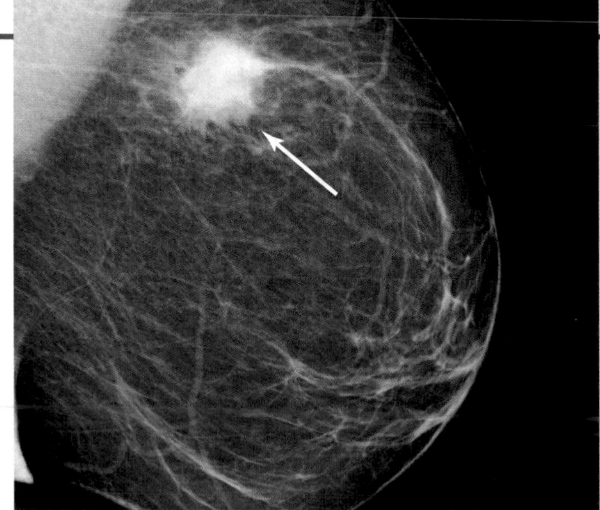

Mammogram showing a tumor (white arrow).

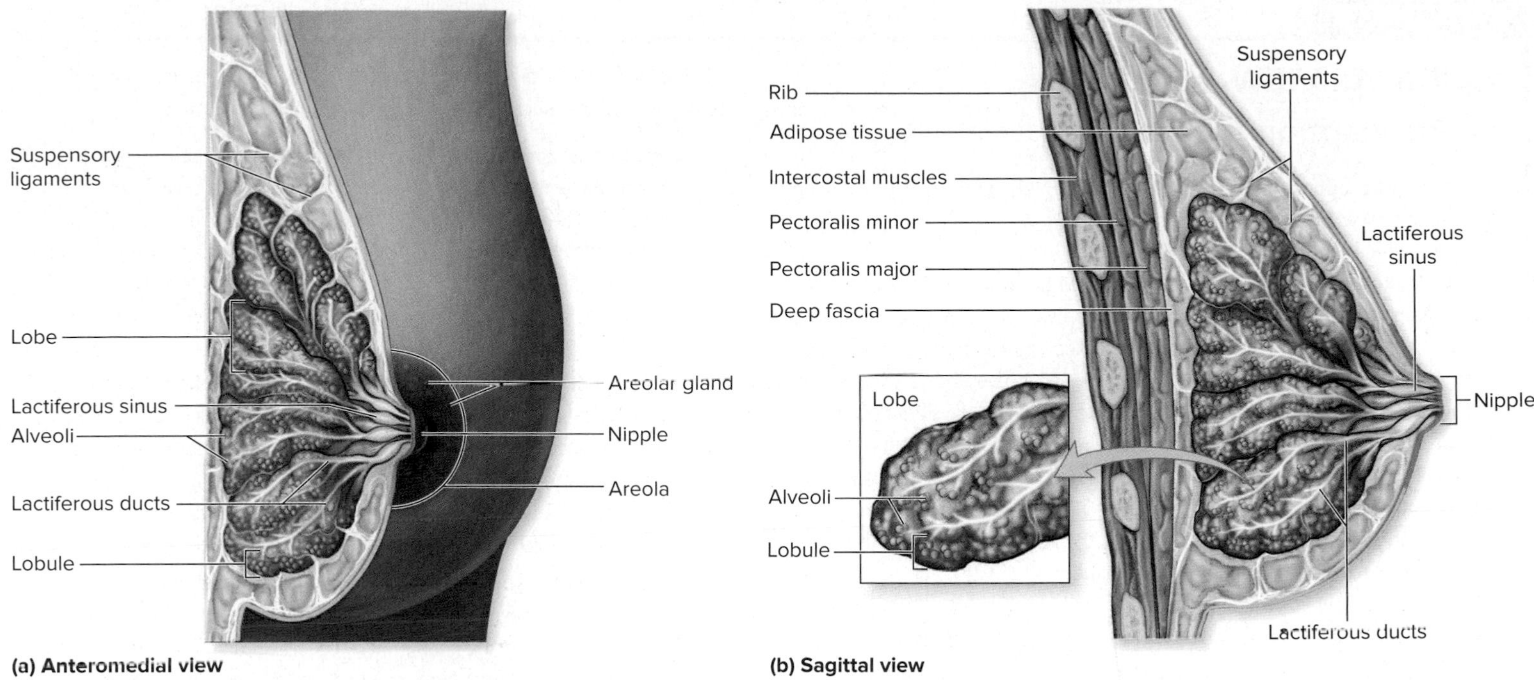

Figure 28.13 Mammary Glands. The mammary glands are composed of glandular tissue and a variable amount of fat. (*a*) An anterior view is partially cut away to reveal internal structures. (*b*) A diagrammatic sagittal section of a mammary gland shows the distribution of alveoli within lobules and the extension of ducts to the nipple. AP|R

gland (**figure 28.13**). The gland's secretory product is called breast milk and it contains proteins, fats, and lactose sugar to provide nutrition to infants.

The **nipple** is a cylindrical projection on the center of the mammary gland. It contains multiple, tiny openings of the excretory ducts that transport breast milk. The **areola** (ă-rē′ō-lă; small area) is the pigmented, rosy or brownish ring of skin around the nipple. Its surface often appears uneven and grainy due to the numerous sebaceous glands, called **areolar glands,** immediately internal to the surface. The color of the areola may vary, depending upon whether or not a woman has given birth. In a **nulliparous** (nŭl-ip′ă-rŭs; *nullus* = none, *pario* = to bear) woman, one who has never given birth, the areola is rosy or light brown in color. In a **parous** (par′ŭs) woman, one who has given birth, the areola may change to a darker color.

Internally, the mammary glands are supported by fibrous connective bands called **suspensory ligaments.** These thin bands extend from the skin and attach to the deep fascia overlying the pectoralis major muscle. Thus, the mammary gland and the pectoralis major muscle are structurally linked.

The mammary glands are subdivided into **lobes,** which are further subdivided into smaller compartments called **lobules.** Lobules contain secretory units termed **alveoli** that produce milk in the lactating female. Alveoli become more numerous and larger during pregnancy. Tiny ducts drain milk from the alveoli and lobules. The tiny ducts of the lobules merge and form 10 to 20 larger channels called **lactiferous** (lak-tif′er-ŭs; *lact* = milk, *fero* = to bear) **ducts.** A lactiferous duct drains breast milk from a single lobe. As each lactiferous duct approaches the nipple, its lumen expands to form a **lactiferous sinus,** a space where milk is stored prior to release from the nipple.

Breast milk is released by a process called **lactation** (lak-tā′shŭn; *lactatio* = to suckle), which occurs in response to a complex sequence of internal and external stimuli. Normally, a woman starts to produce breast milk when she has recently given birth. Recall from section 17.7c that the hormone **prolactin** is produced in the anterior pituitary and is responsible for milk production. Thus, when the amount of prolactin increases, the mammary gland grows and forms more expanded and numerous alveoli. The hormone **oxytocin,** produced by the hypothalamus and released from the posterior pituitary, is responsible for milk ejection.

 **WHAT DID YOU LEARN?**

 What is the relationship between lobes, suspensory ligaments, and lactiferous ducts in the mammary gland?

 What are the effects of prolactin and oxytocin on the female mammary gland?

28.3g Female Sexual Response

 **LEARNING OBJECTIVE**

25. Explain how the female sexual response and orgasm is elicited.

The female sexual response refers to a series of physiologic events that occur during stimulation of the female reproductive organs. The response begins with the **excitement phase,** where parasympathetic innervation (via pelvic splanchnic nerves; see section 15.3b) of reproductive structures such as the mammary glands, clitoris, vaginal wall, bulbs of the vestibule, and labia become engorged with blood. The nipples become erect as a result of the blood engorgement in the mammary glands. The vestibular glands and the glands within the vaginal wall both produce mucin for lubrication. The uterus shifts from an anteverted position to a more erect position within the pelvic cavity.

As the excitement phase continues, the erectile tissue of the clitoris swells as it engorges with blood and becomes very sensitive to tactile stimuli. The inferior part of the vaginal wall constricts slightly. The woman's heart rate, blood pressure, and respiratory rate increase during this time as orgasm nears. Both divisions of the autonomic

INTEGRATE

CLINICAL VIEW 28.7

Contraception Methods

The term **contraception** (kon-tră-sep′shun) refers to birth control, or the prevention of pregnancy. A wide range of birth control methods are available, and they have varying degrees of effectiveness.

Abstinence (ab′sti-nens; *abstineo* = to hold back) means refraining from sexual intercourse. Abstinence is the only 100% proven way to prevent pregnancy.

The **rhythm method** requires avoiding sexual intercourse during the time when a woman is ovulating. Because sperm can live for several days in the female reproductive tract, it is best to avoid intercourse both a few days prior and a few days after ovulating. The rhythm method requires that a woman know when she is ovulating, which may be difficult to determine because there is great variation in ovarian cycles. As a result, this method has a high failure rate (~25%).

The **withdrawal method,** also known as the *pull-out method,* is where the male removes his penis from the vagina before he ejaculates. This method has a high failure rate, in part because it may be difficult to predict the point prior to ejaculation. Additionally, it may be possible for pre-ejaculate (emissions secreted prior to ejaculation) to collect enough sperm in the urethra from a prior ejaculation to impregnate a female. The failure rate for the withdrawal method is about 19%.

Lactation is the production of milk for nursing a baby, and can prevent ovulation and menstruation for many months after childbirth *if* a woman nurses her child constantly (i.e., much more than five times a day). Frequent lactation sends signals to the hypothalamus to prevent follicle-stimulating hormone (FSH) and luteinizing hormone (LH) from being secreted, thus preventing ovulation. If a woman is lactating, she should always use another form of birth control as well, because she will not know when her ovulation cycle begins again until many weeks later, when she begins menstruating again.

Barrier methods of birth control use a physical barrier to prevent sperm from reaching the uterine tubes. Barrier methods include the following:

- **Condoms,** when used properly, collect the sperm and prevent them from entering the female reproductive tract. They are also the only birth control method that helps protect against sexually transmitted infections (STIs), such as human papillomavirus (HPV), herpes, and HIV. Condoms for males fit snugly on the erect penis, whereas **vaginal condoms** are placed in the vagina prior to sexual intercourse. The typical failure rate for condoms is about 15%.
- **Spermicidal foams and gels** are chemical barrier methods that kill sperm before they travel to the uterine tubes. They are inserted into the vagina or placed on the penis prior to sexual intercourse. Foams and gels are not the most effective method of birth control (when used alone, the failure rate is about 25%); rather, they should be used in conjunction with a physical barrier method.
- **Diaphragms** and **cervical caps** are structures made of rubber or silicone that are inserted into the vagina and placed over the cervix prior to sexual intercourse. Spermicidal gel is used around the edges to help prevent sperm from entering the cervix. Some women find it difficult to correctly place the diaphragm or cervical cap, and incorrect placement can result in sperm entering the uterus. The failure rate may reach as high as 20% if the structure is not inserted properly.

Intrauterine devices (IUDs) are T-shaped, flexible plastic structures inserted into the uterus by a health-care provider. Once in place, the IUD prevents fertilization from occurring, although researchers aren't sure specifically how. The IUD may contain copper or levonorgestrel (a synthetic progesterone). IUDs containing copper are effective for up to 10 years, and those containing

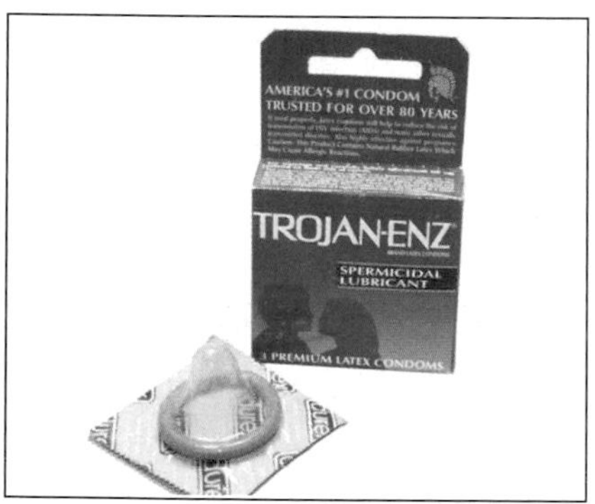

(a) Condoms

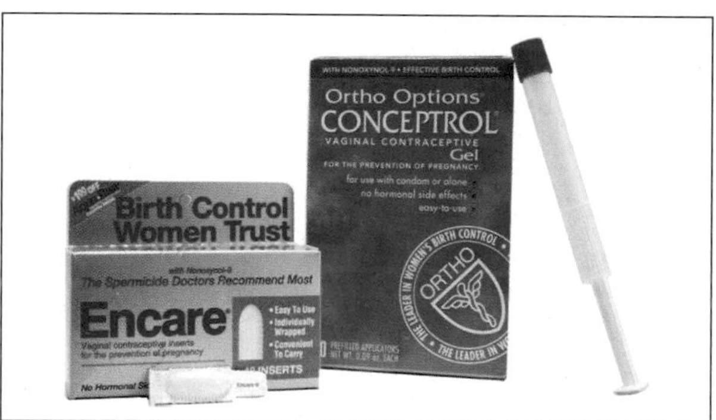

(b) Spermicidal foams

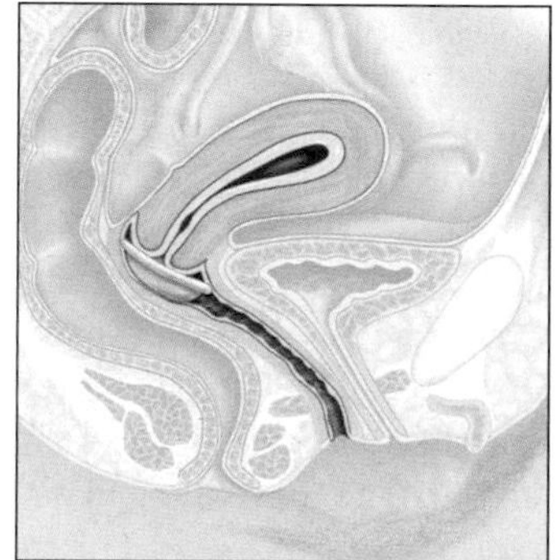

(c) Diaphragm

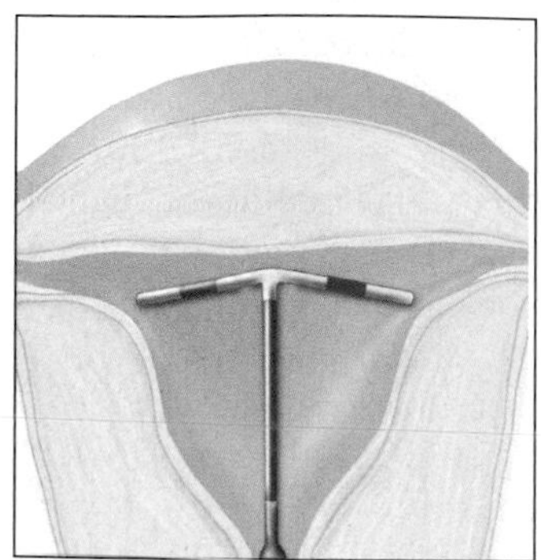

(d) Intrauterine device (IUD)

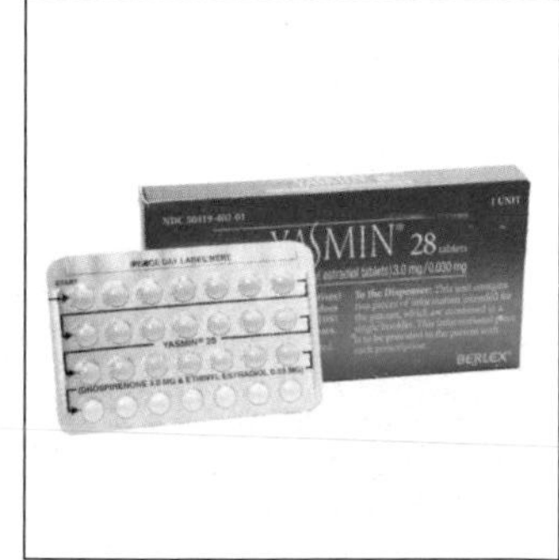

(e) Oral contraceptive

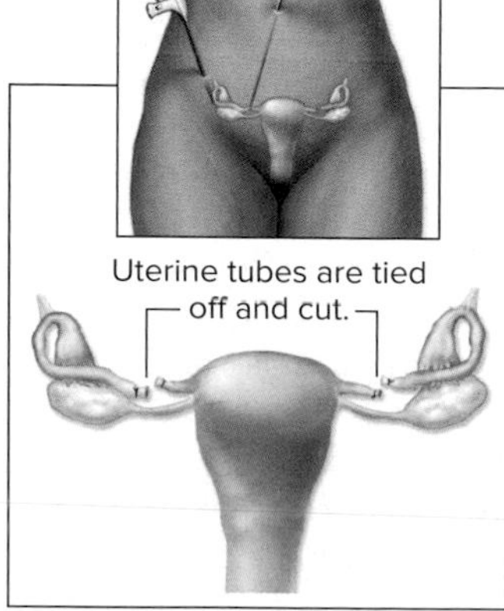

(f) Tubal ligation

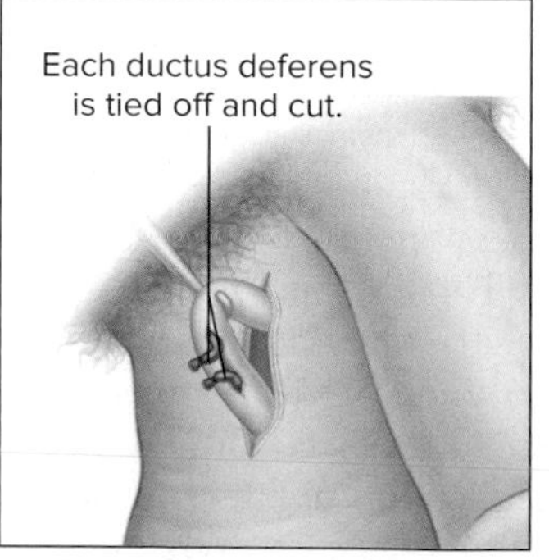

(g) Vasectomy

Contraception includes barrier, chemical, and surgical methods.

(*a, b, e*) ©McGraw-Hill Education/Jill Braaten, photographer levonorgestrel (e.g., Skyla or Mirena) are effective for 3 to 5 years. Their failure rate typically is low (1–2%).

Chemical methods of birth control are very effective if used properly. They include the following:

- **Oral contraceptives** are commonly called *birth control pills* and typically come in 28-day packets. The first 21 days of pills contain low levels of estrogen and progestins, and the last 7 days are sugar pills. (Progesterone is one type of progestin.) The low levels of estrogen and progestins prevent the LH "spike" needed for ovulation. Thus, oral contraceptives prevent ovulation. During the 7 days of sugar pills, the circulating levels of estrogen and progestins drop, and menstruation occurs. Menstrual flow typically is much lighter when a woman takes oral contraceptives because the circulating levels of hormones are lower than normal during her cycle, so the uterine lining is not as thick. Oral contraceptives require a woman to take a pill a day, at about the same time each day. If she misses one or more days of pills, ovulation may occur. In addition, some medications (such as antibiotics) may interfere with their effectiveness, so a backup method of contraception should be used when taking these medications. When used properly, the failure rate is very low (0.1%) but may increase to 7% if the timing of taking the pills varies or if the woman is on medication that may interfere with the pill's effectiveness.
- **Estrogen/progestin** (prō-jes′tin; *pro* = before; gestation) **patches** are alternatives to the daily oral contraceptive. A patch placed on the body delivers a regular amount of estrogen and progestin through the skin (transdermally). The patch is replaced each week and failure rate is 1%.
- **Injected/implanted progestins** help prevent pregnancy by preventing ovulation and thickening the mucus around the cervix (thus creating a slight physical barrier to the sperm). Medroxyprogesterone (Depo-Provera) is an injectable contraceptive given once every 3 months, whereas etonogestrel (Implanon) is an implantable contraceptive that lasts for up to 3 years. The drawback is that ovulation may not occur for many months after stopping the injections. Typical use failure rate is 0.3%.
- **Morning-after pills** containing levonorgestrel (e.g., Preven or Plan B One-Step) can be taken within 72 hours after having unprotected intercourse. These pills work by inhibiting ovulation, altering the menstrual cycle to delay ovulation, or irritating the uterine lining to prevent implantation. Many of these brands are available over the counter (without a prescription).
- **Mifepristone** (Mifeprex in the United States; RU-486 in Europe) may be used during the first 7 weeks of pregnancy. Mifepristone blocks progesterone receptors, so progesterone cannot attach to these receptors and thereby maintain a pregnancy. When taken with a prostaglandin drug, mifepristone induces a miscarriage. Mifepristone's availability and use is very politically charged, with each side of the abortion debate arguing for or against it.

Surgical methods of contraception are both **tubal ligation** for women and **vasectomy** (va-sek′tō-mē) for men. A tubal ligation requires that uterine tubes are cut and the ends are tied off or cauterized shut to prevent both sperm from reaching the oocyte and the oocyte from reaching the uterus. The uterine tubes are ligated either laparoscopically (shown in the figure) or via entry through the uterus (which is a much more difficult procedure for the surgeon.) A vasectomy is an outpatient procedure whereby each ductus deferens is cut, a short segment is removed, and then the ends are tied off. Sperm cannot leave the testis and thus are broken down and resorbed. Both surgeries are effective birth control methods, but they are usually permanent and irreversible, so they are not options for people who wish to have more children.

INTEGRATE

CONCEPT CONNECTION

The male and female sexual response each requires an intricate interplay among various body systems beyond the reproductive system. The nervous system (both autonomic and somatic components) is involved with coordinating the entire event and controlling the numerous physiologic processes. The cardiovascular system is responsible for the engorgement of erectile tissue with blood, and heart rate and blood pressure are increased just prior to orgasm. In addition, the respiratory system is involved as breathing rate increases prior to orgasm. For the male, there is an additional step as the internal urethral sphincter (part of the urinary system) is contracted so no urine enters the urethra during ejaculation.

nervous system control many of these physiologic responses, and the excitement phase also is facilitated by somatosensory signals, such as the sensation of the penis in the vagina or the caressing of parts of the body.

Orgasm in the female refers to the time period where there are intense feelings of pleasure, a feeling of a release of tension, perhaps a feeling of warmth, and some pelvic throbbing. The vagina and uterus contract rhythmically for a period of many seconds.

The **resolution phase** follows orgasm. The uterus returns to its original position and the vaginal wall relaxes at this time. The excess blood leaves the other reproductive organs. The cycle may begin again. Unlike men, women do not have a refractory period, which is a period of time where the body cannot have another orgasm; thus, women have the potential to have multiple orgasms during a single sexual experience.

WHAT DID YOU LEARN?

24 What are the three phases of female sexual response?

28.4 Male Reproductive System

The primary reproductive organs in the male are the **testes** (tes′tēz; sing., *testis*). The accessory reproductive organs include a complex set of ducts and tubules leading from the testes to the penis, a group of male accessory glands, and the penis, which is the organ of copulation (**figure 28.14**).

28.4a Scrotum

LEARNING OBJECTIVE

26. Describe the gross anatomy and function of the scrotum.

The ideal temperature for producing and storing sperm is about 2–3°C lower than internal body temperature. The **scrotum** (skrō′tŭm), a skin-covered sac between the thighs, provides the cooler environment needed for normal sperm development and maturation (**figure 28.15**). Embryologically, the scrotum is homologous to the labia majora in the female.

Externally, the scrotum contains a distinct, ridgelike seam at its midline, called the **raphe** (rā′fē; *rhaphe* = seam). The raphe extends in an anterior direction along the inferior surface of the penis and in a posterior direction to the anus. The wall of the scrotum is composed of an external layer of skin, a thin layer of superficial fascia immediately

diameter. The testes produce sperm and androgens (male sex hormones), the most common of which is testosterone. The details for testosterone are included in the **table R.10**: "Regulating the Male Reproductive System."

 **WHAT DO YOU THINK?**

3 If a man's testes were removed, would he still be able to produce androgens?

Each testis is covered anteriorly and laterally by a serous membrane, the **tunica vaginalis** (văj-in-ăl′ĭs; ensheathing) **(figure 28.16*a*)**. This membrane is derived from the peritoneum of the abdominal cavity. The tunica vaginalis has an outer **parietal layer** and an inner **visceral layer** that are separated by a cavity filled with serous fluid. A thick, whitish, fibrous capsule called the **tunica albuginea** covers the testis and lies immediately deep to the visceral layer of the tunica vaginalis. At the posterior margin of the testis, the tunica albuginea thickens and projects into the interior of the organ as the **mediastinum testis.** Blood vessels, a system of ducts, within lymph vessels, and some nerves enter or leave each testis within the mediastinum testis.

The tunica albuginea projects internally into the testis and forms delicate connective tissue **septa,** which subdivide the internal space into about 250 separate **lobules.** Each lobule contains up to four extremely convoluted, thin, and elongated **seminiferous** (sem′in-if′er-ŭs; *semen* = seed, *fero* = to carry) **tubules.** The seminiferous tubules contain two types of cells: (1) a group of nondividing support cells called the **sustentacular** (sŭs-ten-tak′ū-lăr; *sustento* = to hold upright) **cells,** which also are termed *Sertoli cells,* or *nurse cells;* and (2) a population of dividing germ cells (i.e., spermatogonia) that continuously replicate and develop into sperm beginning at puberty.

The sustentacular cells provide a protective environment for the developing sperm, and their cytoplasm helps nourish the developing sperm (figure 28.16*b*). In addition, sustentacular cells release the hormone **inhibin** when sperm count is high. Inhibin inhibits FSH secretion and thus regulates sperm production. Conversely, when sperm count declines, inhibin secretion decreases.

The sustentacular cells are bound together by tight junctions, which form a **blood-testis barrier** that is similar to the blood-brain barrier (see section 13.2d). The blood-testis barrier helps protect developing sperm from materials in the blood. It also protects the sperm from the body's leukocytes, which may perceive the sperm as foreign because they have different chromosome numbers and proteins. (In contrast, the female oocyte likely is protected from materials in the blood by the ovarian follicle structures that surround the oocyte.)

The spaces surrounding the seminiferous tubules are called **interstitial spaces.** Within these spaces reside the **interstitial cells** (or *Leydig cells*). Luteinizing hormone stimulates the interstitial cells to produce hormones called **androgens** (an′drō-jen; *andros* = male human). There are several types of androgens; the most common one is testosterone. Although the adrenal cortex secretes a small amount of androgens in both sexes (see section 17.9a), the vast majority of androgen release is via the interstitial cells in the testis in males, beginning at puberty.

Hormonal Regulation of Androgen Production and Sperm Development

Like the hormone interactions involved in the ovarian cycle, the interplay among the hormones involved in regulating the spermatogenesis and androgen production is intricate and involves several negative feedback mechanisms. A stepwise description of these hormonal effects is listed here and shown in **figure 28.17**:

1. **The hypothalamus initiates spermatogenesis by secreting GnRH,** which stimulates the anterior pituitary to secrete FSH and LH.

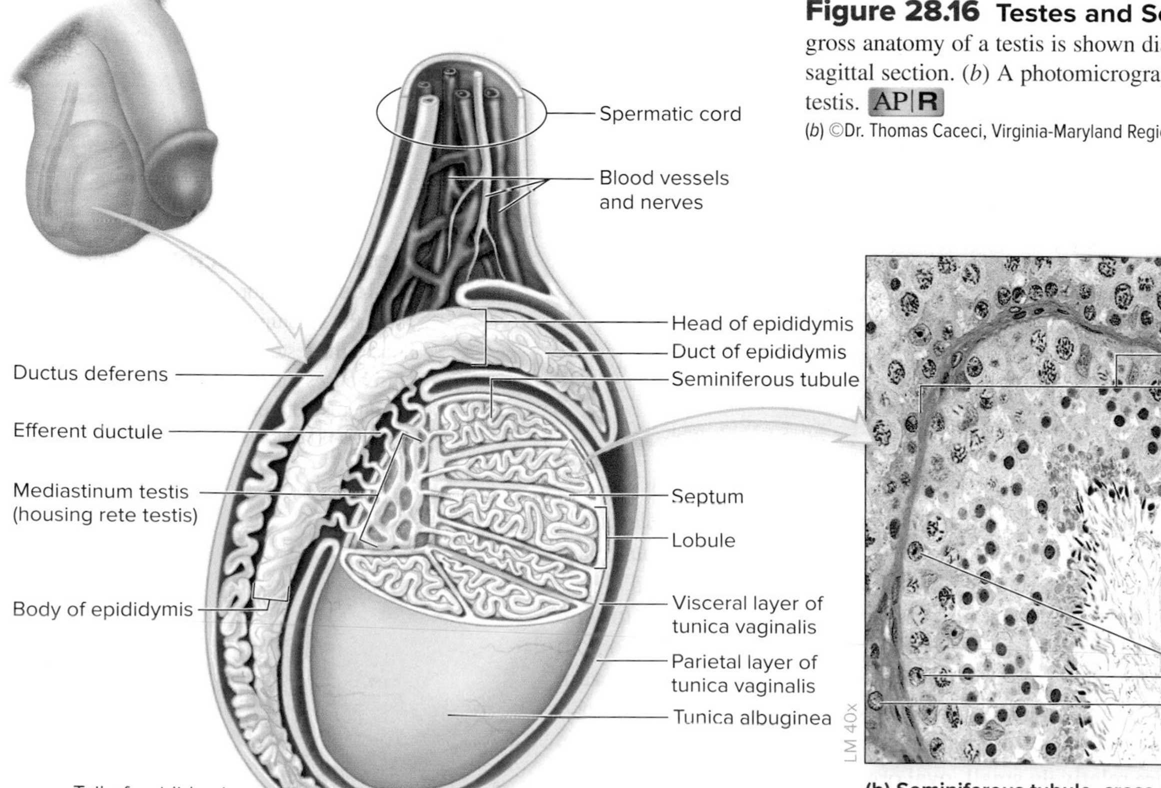

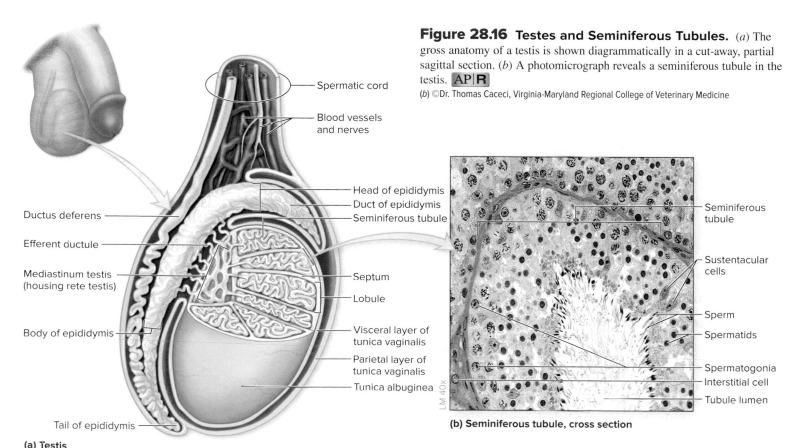

Figure 28.16 Testes and Seminiferous Tubules. (*a*) The gross anatomy of a testis is shown diagrammatically in a cut-away, partial sagittal section. (*b*) A photomicrograph reveals a seminiferous tubule in the testis. AP|R

(*b*) ©Dr. Thomas Caceci, Virginia-Maryland Regional College of Veterinary Medicine

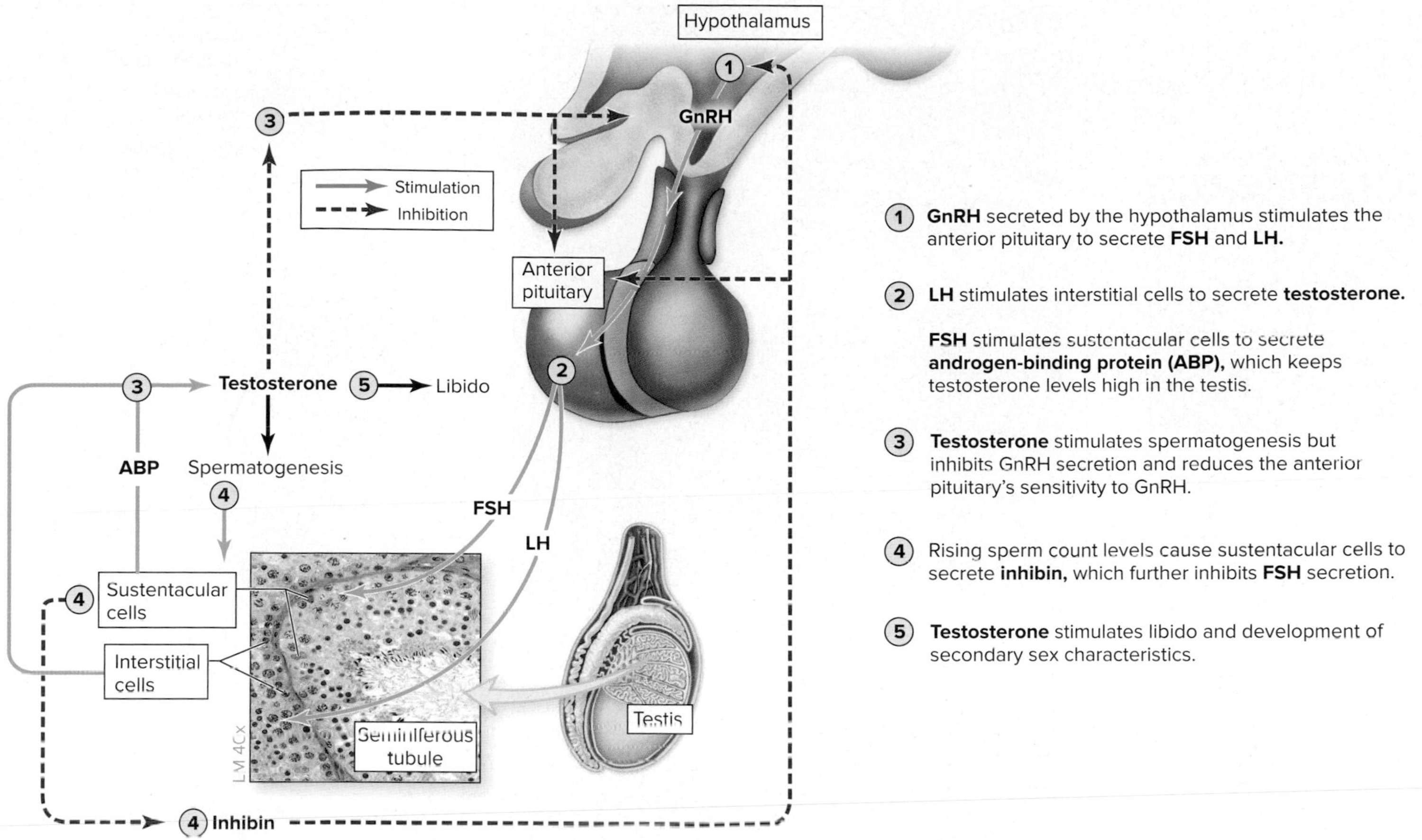

Hypothalamus
GnRH
Stimulation
Inhibition
Anterior pituitary
Testosterone
Libido
ABP
Spermatogenesis
FSH
LH
Sustentacular cells
Interstitial cells
Seminiferous tubule
Testis
LM 4Cx
Inhibin
1 GnRH secreted by the hypothalamus stimulates the anterior pituitary to secrete FSH and LH.
2 LH stimulates interstitial cells to secrete testosterone.
FSH stimulates sustentacular cells to secrete androgen-binding protein (ABP), which keeps testosterone levels high in the testis.
3 Testosterone stimulates spermatogenesis but inhibits GnRH secretion and reduces the anterior pituitary's sensitivity to GnRH.
4 Rising sperm count levels cause sustentacular cells to secrete inhibin, which further inhibits FSH secretion.
5 Testosterone stimulates libido and development of secondary sex characteristics.

Figure 28.17 Hormonal Regulation of Spermatogenesis and Androgen Production. The hypothalamus secretes gonadotropin-releasing hormone (GnRH), which stimulates the anterior pituitary to release follicle-stimulating hormone (FSH) and luteinizing hormone (LH). FSH and LH stimulate spermatogenesis and testosterone (and other androgen) production, as shown in this image.

② **FSH and LH target the testes and stimulate both spermatogenesis and androgen production.** Specifically, FSH stimulates the sustentacular cells to secrete androgen-binding protein (ABP) and LH stimulates interstitial cells in the testis to secrete testosterone. ABP binds to testosterone (and other androgens) to ensure that testosterone levels remain high in the testes. Therefore, the testis and its cells are the effectors, and they release stimuli (testosterone and ABP) in this cycle.

③ **The increased levels of testosterone have various immediate effects on the body.** The high testosterone levels facilitate spermatogenesis. However, this same circulating testosterone inhibits GnRH secretion and reduces anterior pituitary sensitivity to GnRH. Thus, rising levels of testosterone have a negative feedback effect on the entire cycle.

④ **Sustentacular cells respond to rising sperm count levels and secrete inhibin.** Inhibin primarily causes inhibition of FSH secretion from the anterior pituitary, and it serves as an additional negative feedback mechanism on the cycle.

⑤ **Circulating testosterone stimulates libido (sex drive) and development of secondary sex characteristics.** Testosterone acts on the brain so there is an increased desire for and sensitivity to sexual stimulation. Secondary sex characteristics include growth and development of hair in the pubic and axillary regions, a deeper voice (because testosterone affects larynx development), and the growth of facial hair.

Note that sperm development and androgen production are controlled by negative feedback only, whereas the ovarian cycle uses both positive and negative feedback mechanisms.

Development of Sperm: Spermatogenesis and Spermiogenesis

Spermatogenesis (sper′mă-tō-jen′ĕ sis) is the process of sperm development that occurs within the seminiferous tubule of the testis. Spermatogenesis does not begin until puberty, when significant levels of FSH and LH stimulate the testis to begin gamete development.

The process of spermatogenesis is shown in **figure 28.18*a***. All sperm develop from primordial germ (stem) cells called **spermatogonia** (sper′mă-tō-gō′nē-ă; sing., *spermatogonium; sperma* = seed, *gone* = generation). Spermatogonia are diploid cells (meaning they have 23 pairs of chromosomes for a total of 46). These cells lie near the base of the seminiferous tubule, surrounded by the cytoplasm of a sustentacular cell. To produce sperm, spermatogonia first divide by mitosis. One of the cells produced is a new spermatogonium (a new germ cell), to ensure that the numbers of spermatogonia never become depleted, and the other cell is a committed cell that becomes a **primary spermatocyte.** Primary spermatocytes are diploid and are the cells that undergo meiosis I.

When a primary spermatocyte undergoes meiosis I, the two cells produced are called **secondary spermatocytes.** Secondary spermatocytes are haploid cells, meaning they each have 23 chromosomes only. These cells are still surrounded by the sustentacular cell cytoplasm, but they are relatively closer to the lumen of the seminiferous tubule as opposed to the base of the seminiferous tubule.

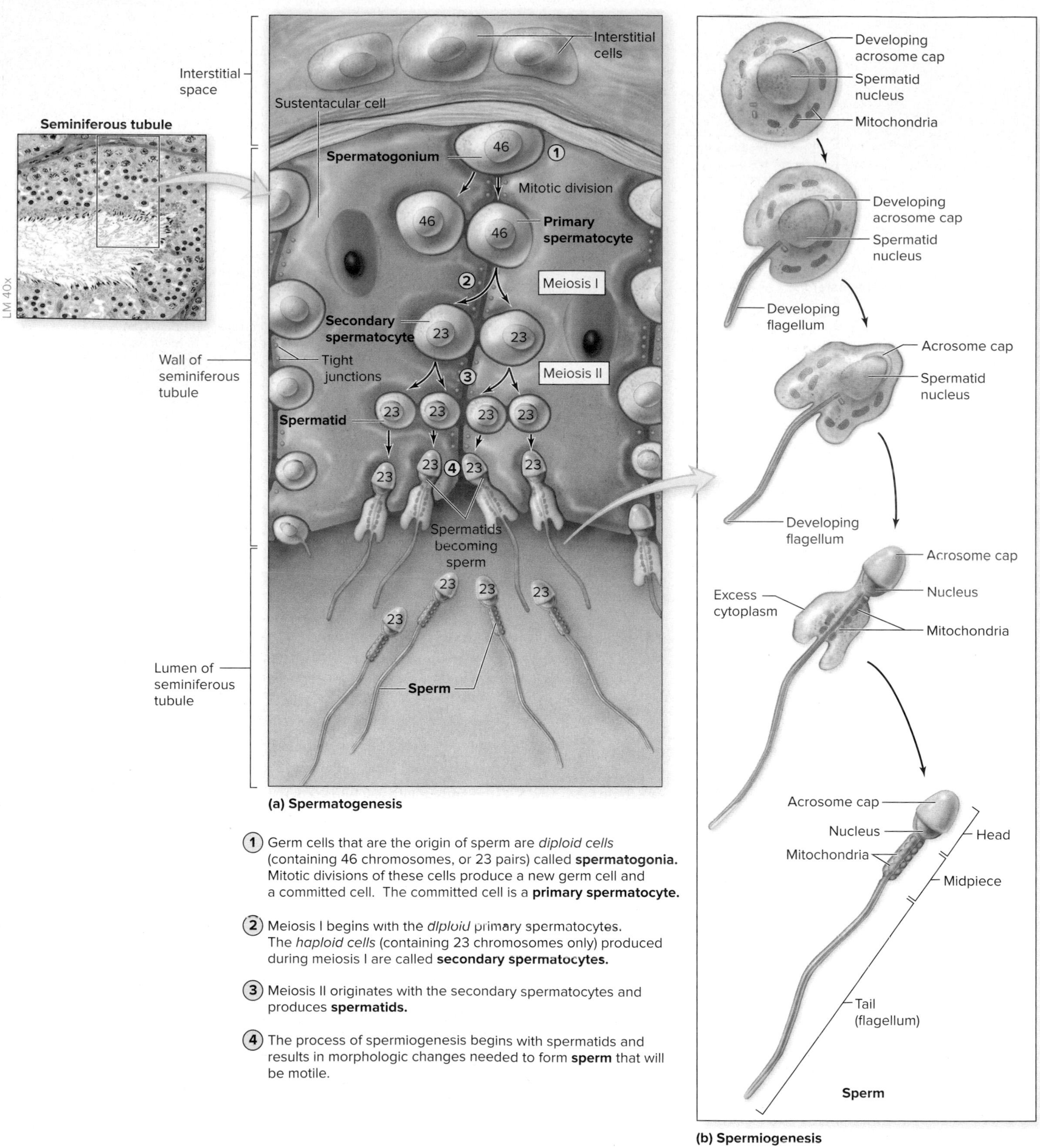
Seminiferous tubule
LM 40x
Interstitial space
Wall of seminiferous tubule
Lumen of seminiferous tubule
Interstitial cells
Sustentacular cell
Spermatogonium
46
1
Mitotic division
46
46
Primary spermatocyte
2
Meiosis I
Secondary spermatocyte
23
23
Tight junctions
3
Meiosis II
Spermatid
23
23
23
23
23
23
4
23
23
Spermatids becoming sperm
23
23
23
23
Sperm
(a) Spermatogenesis
1 Germ cells that are the origin of sperm are *diploid cells* (containing 46 chromosomes, or 23 pairs) called **spermatogonia.** Mitotic divisions of these cells produce a new germ cell and a committed cell. The committed cell is a **primary spermatocyte.**
2 Meiosis I begins with the *diploid* primary spermatocytes. The *haploid cells* (containing 23 chromosomes only) produced during meiosis I are called **secondary spermatocytes.**
3 Meiosis II originates with the secondary spermatocytes and produces **spermatids.**
4 The process of spermiogenesis begins with spermatids and results in morphologic changes needed to form **sperm** that will be motile.
Developing acrosome cap
Spermatid nucleus
Mitochondria
Developing acrosome cap
Spermatid nucleus
Developing flagellum
Acrosome cap
Spermatid nucleus
Developing flagellum
Acrosome cap
Nucleus
Excess cytoplasm
Mitochondria
Acrosome cap
Nucleus
Mitochondria
Head
Midpiece
Tail (flagellum)
Sperm
(b) Spermiogenesis

Figure 28.18 Spermatogenesis and Spermiogenesis. (*a*) The processes of spermatogenesis and spermiogenesis take place within the wall of the seminiferous tubule. (*b*) Structural changes occur during spermiogenesis as a spermatid becomes a sperm. AP|R

(*inset*) ©Dr. Thomas Caceci, Virginia-Maryland Regional College of Veterinary Medicine

Secondary spermatocytes complete meiosis II to form **spermatids** (sper′mă-tid). A spermatid is a haploid cell and is surrounded by the sustentacular cell cytoplasm, very near to the lumen of the seminiferous tubule. The spermatids still have a circular appearance, rather than the sleek shape of mature sperm.

In the final stage of spermatogenesis, a process called **spermiogenesis** (sper′mē-ō-jen′ĕ-sis), the newly formed spermatids differentiate to become anatomically mature **spermatozoa**

INTEGRATE

CLINICAL VIEW 28.8

Sexually Transmitted Infections

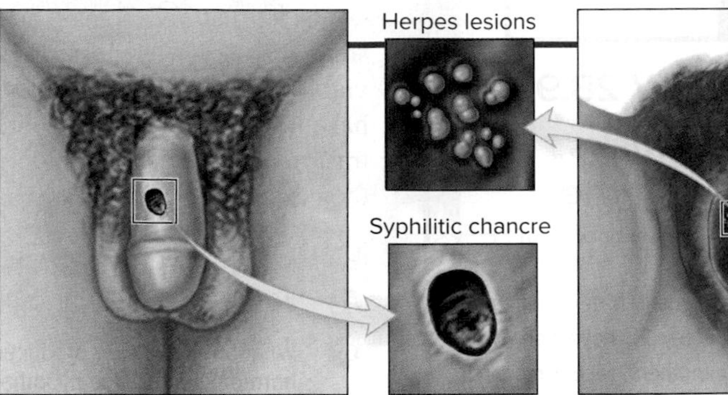

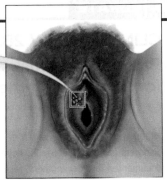

Syphilitic chancre on the penis of a male and herpes lesions on the vulva of a female.

Sexually transmitted infections (STIs), also known as *sexually transmitted diseases* or *venereal diseases,* are a group of infectious diseases that are usually transmitted via sexual contact. Often, the symptoms of STIs are not immediately noticeable, so infected individuals may infect someone else without realizing it. Mothers also may transmit STIs to their newborns, either directly across the placenta or at the time of delivery. Condoms have been shown to help prevent the spread of STIs, but they are not 100% effective.

STIs are a leading cause of **pelvic inflammatory disease** in women, in which the reproductive organs become infected. Should bacteria from an STI infect the uterus and uterine tubes, scarring is likely to follow, leading to blockage of the tubes and infertility. We've already discussed two types of STIs: human papillomavirus infections (see Clinical View 28.5: "Cervical Cancer") and AIDS (see Clinical View 22.9: "HIV and AIDS"). Here we describe other common STIs.

Chlamydia (kla-mid'ē-ă) is the most frequently reported bacterial STI in the United States. The responsible agent is the bacterium *Chlamydia trachomatis.* Most infected people are asymptomatic; the rest develop symptoms within 1 to 3 weeks after exposure. These symptoms include abnormal vaginal discharge, painful urination (in both men and women), and low back pain. Chlamydia is treated with antibiotics.

Genital herpes (her'pēz; *herpo* = to creep) is caused by herpes simplex virus type 1 (HSV-1) or type 2 (HSV-2). Infected individuals undergo cyclic outbreaks of blister formation in the genital and anal regions; the blisters are filled with fluid containing millions of infectious viral particles. The blisters then break and turn into tender sores that remain for 2 to 4 weeks. Typically, future cycles of blistering are less severe and shorter in duration than the initial episode. There is no cure for herpes, but antiviral medications can lessen the severity and length of an outbreak.

Gonorrhea (gon-ō-rē'ă) is caused by the bacterium *Neisseria gonorrhoeae,* and it is spread either by sexual contact or from mother to newborn at the time of delivery. Symptoms include painful urination and a yellowish discharge from the penis or vagina. Gonorrhea is treated with antibiotics, although in recent years many gonorrhea strains have become resistant to some antibiotics. If untreated, women may develop pelvic inflammatory disease, and men may develop epididymitis, a painful condition of the epididymis that can lead to infertility. If a newborn acquires the disease, then blindness, joint problems, or a life-threatening blood infection may result.

Syphilis (sif'i-lis) is caused by the corkscrew-shaped bacterium *Treponema pallidum.* The bacterium is spread sexually via contact with a syphilitic sore called a **chancre** (shan'ker), or a newborn may acquire it *in utero.* Babies can acquire congenital syphilis from their mothers and are often stillborn, but if they live, they have a high incidence of skeletal malformations and neurologic problems. Syphilis can be treated with antibiotics.

(sing., *spermatozoon;* sper'mă-to-zo'on) or **sperm** (figure 28.18*b*). During spermiogenesis, the spermatid sheds excess cytoplasm and its nucleus elongates. A structure called the **acrosome cap** (ak'rō-sōm; *akros* = tip, *soma* = body) forms over the nucleus. This structure contains digestive enzymes that help penetrate the secondary oocyte for fertilization. As the spermatid elongates, a **tail,** also called the *flagellum,* forms from the organized microtubules within the cell. The tail is attached to a **midpiece** or *neck* region containing mitochondria and a centriole. These mitochondria provide the energy to move the tail.

Although the sperm look mature, they do not yet have all of the characteristics needed to successfully travel through the female reproductive tract and fertilize an oocyte. The sperm must leave the seminiferous tubule through a network of ducts (described next) and reside in the epididymis for a period of time in order to become fully motile. **Table 28.6** summarizes the stages of spermatogenesis.

Table 28.6 Stages of Spermatogenesis

Cell Type	Number of Chromosomes	Haploid or Diploid	Action
Spermatogonium	23 pairs (46)	Diploid	Divides by mitosis to produce a new spermatogonium and a primary spermatocyte
Primary spermatocyte	23 pairs (46) of replicated chromosomes	Diploid	Completes meiosis I to produce secondary spermatocytes
Secondary spermatocyte	23 replicated chromosomes only	Haploid	Completes meiosis II to produce spermatids
Spermatid	23 single-stranded chromosomes only	Haploid	Undergoes spermiogenesis, where most of its cytoplasm is shed and a midpiece, tail, and head form
Spermatozoon (sperm)	23 single-stranded chromosomes only	Haploid	Leaves seminiferous tubule and matures in epididymis

the distal end of the penis, the skin is attached to the raised edge of the glans and forms a circular fold called the **prepuce** (prē′pūs), or *foreskin.*

Within the shaft of the penis are three cylindrical erectile bodies. The paired **corpora cavernosa** (kav′er-nō-sa′; sing., *corpus cavernosum; caverna* = grotto) are located dorsolaterally. Ventral to them in the midline is the single **corpus spongiosum** (spŭn′jē-ō-sŭm), which contains the spongy urethra. Each corpus cavernosum terminates in the shaft of the penis, whereas the corpus spongiosum continues within the glans. The erectile bodies are ensheathed by a thin layer of dense connective tissue called the **tunica albuginea,** which also provides an attachment to the skin over the shaft of the penis.

WHAT DID YOU LEARN?

33 What are the similarities and differences between the corpora cavernosa and the corpus spongiosum of the penis?

28.4f Male Sexual Response

LEARNING OBJECTIVES

36. Compare and contrast the processes of erection and ejaculation.

37. Explain how the male sexual response (and ejaculation) is elicited.

The male sexual response begins with an excitement phase. The erectile bodies of the penis are composed of a complex network of **venous spaces** surrounding a central artery. During sexual excitement, blood enters the erectile bodies and fills the venous spaces. As these venous spaces become engorged with blood, the erectile bodies become rigid, a process called **erection** (ē-rek′shŭn; *erecto* = to set up). The rigid erectile bodies compress the veins that drain blood away from the venous spaces. Thus, the spaces fill with blood, but the blood cannot leave the erectile bodies until the sexual excitement ceases. Parasympathetic innervation (through the pelvic splanchnic nerves; see section 15.3b) is responsible for increased blood flow and thus the erection of the penis. Specifically, the parasympathetic innervation facilitates local release of nitric oxide (NO) into the tissues, which assists with erection. (Recall from section 20.4c that NO is a vasodilator.) Near the end of the excitement phase but before orgasm, increases occur in heart rate, blood pressure, and respiratory rate.

Orgasm in the male refers to the time period during which there are intense feelings of pleasure, a feeling of a release of tension, and expulsion of semen. In the beginning phases of orgasm, the ductus deferens undergoes peristalsis and moves the sperm toward the urethra. This peristalsis is facilitated by a burst of oxytocin released from the posterior pituitary, which induces smooth muscle contraction in the male reproductive ducts. Later in this phase, the accessory glands secrete their components of seminal fluid, which combine with the sperm to form semen. The internal urethral sphincter of the urinary bladder contracts to ensure that no urine enters the urethra at this time. **Ejaculation** (ē-jak-ū-lā′shŭn) typically occurs at the ending stage of an orgasm and is the process by which semen is expelled from the penis with the help of rhythmic contractions of the smooth muscle in the wall of the urethra. Sympathetic innervation (from the lumbar splanchnic nerves; see section 15.4a) is responsible for ejaculation.

INTEGRATE

LEARNING STRATEGY

One way to remember the autonomic innervation for the penis is the phrase "Point and Shoot." The **p** in **p**oint (erection) also stands for **p**arasympathetic innervation, while the **s** in **s**hoot (ejaculation) stands for **s**ympathetic innervation.

Although in most body systems sympathetic and parasympathetic innervation tend to perform opposite functions, the male reproductive system is an exception. Here, parasympathetic innervation is necessary to achieve an erection, while sympathetic innervation promotes ejaculation. Reduction of autonomic activity after sexual excitement reduces blood flow to the erectile bodies and shunts most of the blood to other veins, thereby returning the penis to its flaccid condition.

Following an orgasm, there is a resolution phase, which is marked by feelings of intense relaxation. The sympathetic division is stimulated to contract the central artery of the penis and contract small muscles around the erectile tissue, which expels the engorged blood. Gradually, the penis becomes soft and flaccid again. Resolution in men is followed by a **refractory period,** during which the man cannot attain another erection. This period may last for many minutes or for hours. This refractory period becomes longer as men age. (Women, in contrast, have no refractory period and thus have the potential to experience multiple orgasms in a single sexual experience.)

WHAT DID YOU LEARN?

34 How do erection and ejaculation differ?

35 How do both parasympathetic and sympathetic stimulation contribute to penile function during sexual arousal?

28.5 Development and Aging of the Female and Male Reproductive Systems

The female and male reproductive structures originate from the same basic embryonic primordia, but they differentiate into either female or male structures, depending upon the molecular signals the primordia receive. To better explain this process, we must first distinguish between the genetic and phenotypic sex of an individual.

28.5a Genetic Versus Phenotypic Sex

LEARNING OBJECTIVES

38. Compare and contrast genetic versus phenotypic sex.

39. List the gene(s) responsible for producing a phenotypic male.

Genetic sex is also called *genotypic sex,* and it refers to the sex of an individual based on the sex chromosomes inherited. An individual with two X chromosomes is a genetic female, whereas a person with one X and one Y chromosome is a genetic male. Genetic sex is determined at fertilization.

In contrast, **phenotypic** (fē′nō-tip-ik, fen′ō-) **sex** refers to the appearance of an individual's internal and external genitalia. A person with ovaries and female external genitalia (labia) is a phenotypic female, whereas a person with testes and male external genitalia (penis, scrotum) is a phenotypic male. Phenotypic sex starts to become apparent after the seventh week of development.

What determines the development of the primordial tissue into female reproductive organs or male reproductive organs? In males, the **sex-determining region Y (SRY) gene** is located within the larger **testis-determining factor (TDF) region** on the Y chromosome. If the Y chromosome is present and the SRY gene is appropriately expressed, this gene produces proteins to stimulate the production of androgens that initiate male phenotypic development. If a Y chromosome is absent, or if the Y chromosome is either lacking or has an abnormal SRY gene, a female phenotypic sex results.

WHAT DID YOU LEARN?

36 What gene normally present on the Y chromosome is responsible for a male phenotypic sex?

28.5b Formation of Indifferent Gonads and Genital Ducts

LEARNING OBJECTIVE

40. Describe what anatomic structures are formed from the mesonephric and paramesonephric ducts.

Early in the fifth week of embryonic development, paired **genital ridges,** also called *gonadal ridges,* form from **intermediate mesoderm** (see section 5.6a and figure 29.12). The genital ridges will form the gonads. These longitudinal ridges are medial to the developing kidneys at about the level of the tenth thoracic vertebra **(figure 28.21**, *top*). Between weeks 5 and 6, primordial **germ cells** migrate from the yolk sac to the genital ridges. These germ cells will form the future gametes (either oocytes or sperm). Shortly thereafter, two sets of duct systems are formed.

- The **mesonephric** (mez-ō-nef′rik) **ducts,** also known as *Wolffian ducts,* form most of the male duct system. The mesonephric ducts also connect the mesonephros to the developing urinary bladder.

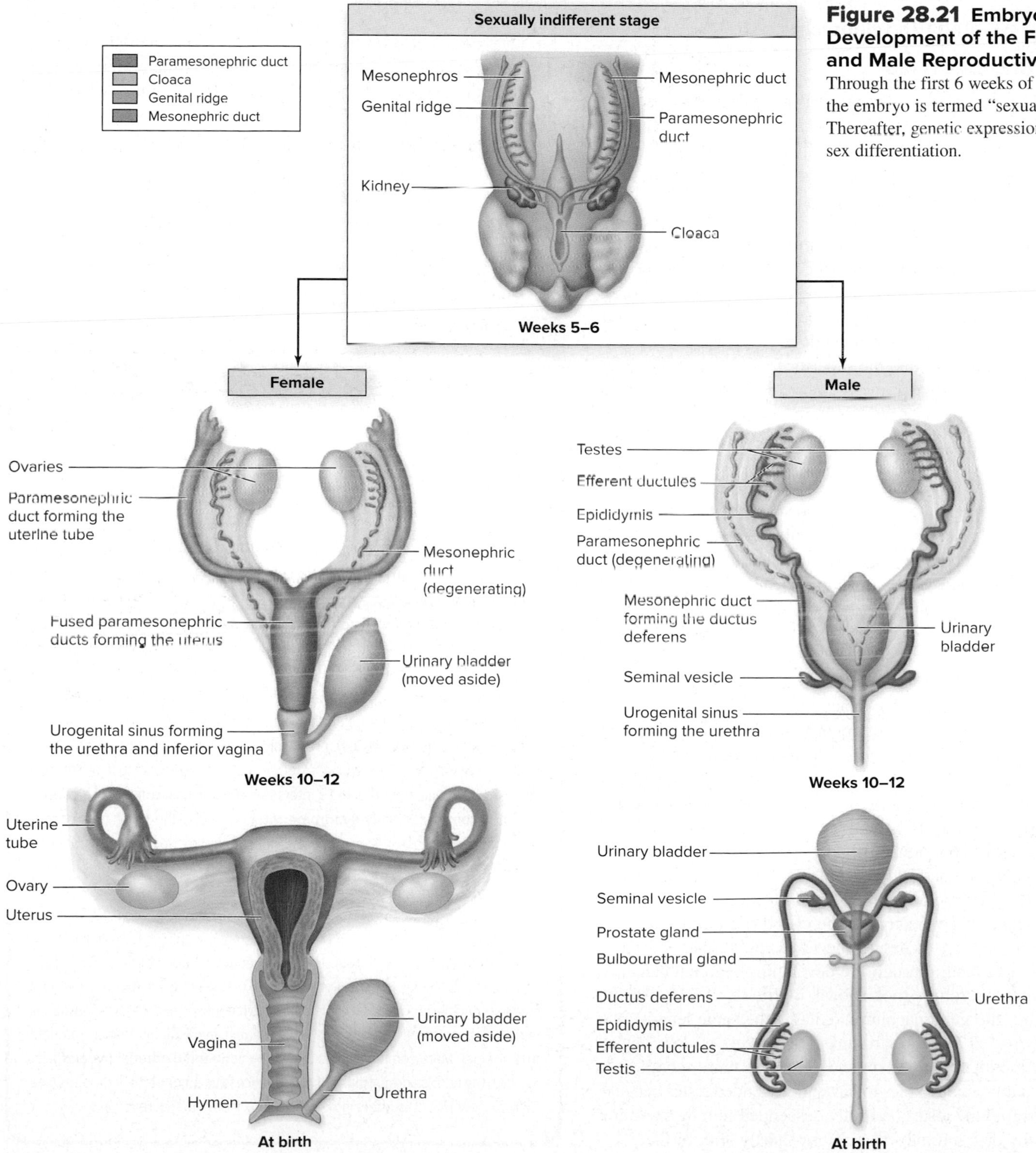

Figure 28.21 Embryonic Development of the Female and Male Reproductive Tracts. Through the first 6 weeks of development, the embryo is termed "sexually indifferent." Thereafter, genetic expression determines sex differentiation.

- The **paramesonephric ducts,** or *Müllerian ducts,* form most of the female duct system, including the uterine tubes, uterus, and superior part of the vagina. These ducts appear lateral to the mesonephric ducts.

All human embryos develop both duct systems, but only one of the duct systems remains in the fetus. If the embryo is female, the paramesonephric ducts develop, and the mesonephric ducts degenerate. If the embryo is male, the mesonephric ducts grow and differentiate into male reproductive structures, while the paramesonephric ducts degenerate.

WHAT DID YOU LEARN?

 37 Which duct system persists in female embryos? In male embryos?

28.5c Internal Genitalia Development

LEARNING OBJECTIVES

41. Describe the events that cause the female internal reproductive organs to develop.

42. Identify the hormone responsible for inducing paramesonephric duct degeneration, and identify the cells responsible for secreting this hormone.

The development of the female internal reproductive structures is traced in figure 28.21, *left.* Because no SRY proteins are produced in the developing female, the mesonephric ducts degenerate. Between weeks 8 and 20 of development, the paramesonephric ducts develop and differentiate. The caudal (inferior) ends of the paramesonephric ducts fuse, forming the uterus and the superior part of the vagina. The cranial (superior) parts of the paramesonephric ducts remain separate and form two uterine tubes. The remaining inferior part of the vagina is formed from the urogenital sinus, which also forms the urinary bladder and urethra.

By about week 7 of development in the male, the SRY gene on the Y chromosome begins influencing the indifferent gonad to become a testis, which then forms sustentacular cells and interstitial cells. Once the sustentacular cells form, they begin secreting **anti-Müllerian hormone (AMH)** (also known as *Müllerian inhibiting substance*), which inhibits the development of the paramesonephric ducts (figure 28.21, *right*). These paramesonephric ducts degenerate, and between weeks 8 and 12, the mesonephric ducts begin to form the male duct system—efferent ductules, epididymides, vasa deferens, seminal vesicles, and ejaculatory ducts.

The prostate and bulbourethral glands do not form from the mesonephric ducts. Instead, they begin to form as endodermal "buds" or outgrowths of the developing urethra between weeks 10 and 13. As the prostate gland and bulbourethral glands develop, they incorporate mesoderm into their structures as well.

Note that the indifferent gonad originates near the level of the T_{10} vertebra. Throughout prenatal development, the developing testis descends inferiorly from the abdominal region toward the developing scrotum. A thin band of connective tissue called the **gubernaculum** (gū′ber-nak′ū-lŭm; helm) attaches to the testis and assists the testis descent from the abdomen, through the developing inguinal canal, to its placement in the scrotum (not shown in figure 28.21). As the embryo grows (but the gubernaculums remains the same length), the testis is passively pulled into the scrotum. This process is slow, beginning in the third month and not completed until the ninth month. It is common for premature male babies to have undescended testes because they were born before the testes had fully descended into the scrotum. Their testes usually descend into the scrotum shortly after birth.

WHAT DID YOU LEARN?

 38 What is the fate of the paramesonephric ducts in a female embryo?

 39 What hormone do the sustentacular cells secrete in the male embryo, and how does this affect internal genitalia development?

28.5d External Genitalia Development

LEARNING OBJECTIVE

43. List the common primordial external genitalia structures, and compare their development in females and males.

As with the internal genitalia, both female and male external genitalia develop from the same primordial structures **(figure 28.22)**. By the sixth week of development, the following external structures are seen:

- The **urogenital folds** (or *urethral folds*) are paired, elevated structures on either side of the urogenital membrane, a thin partition separating the urogenital sinus from the outside of the body.
- The **genital tubercle** is a rounded structure anterior to the urogenital folds.
- The **labioscrotal swellings** (or *genital swellings*) are paired elevated structures lateral to the urethral folds.

INTEGRATE

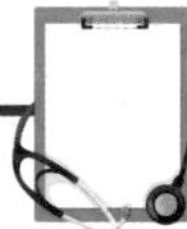

CLINICAL VIEW 28.12

Intersex Conditions (Disorders of Sex Development)

Intersex conditions, or *disorders of sex development,* are a series of disorders where there is a discrepancy between a person's genotype (and gonad development) and their external genitalia. (The previous term to describe this condition is *hermaphrodite* (her-maf′rō-dīt), which is derived from the Greek name Hermaphroditus, the mythological son of the Greek god Hermes and the goddess Aphrodite.) **True gonadal intersex,** previously known as *true hermaphroditism,* refers to an individual with both ovarian and testicular structures (which typically do not have the potential for fertility) and ambiguous (or female) external genitalia. The person may be a genetic male (XY) or a genetic female (XX). True gonadal intersex is rare.

A **46 XY intersex** individual, previously known as a *male pseudohermaphrodite,* is a genetic male (XY) whose external genitalia resemble those of a female (female phenotypic sex). The 46 XY intersex condition most commonly is caused by **androgen insensitivity syndrome (AIS),** whereby the body cells cannot respond to androgens, because the androgen receptors do not form or function properly. Another cause may be either a reduction or lack of male hormones (e.g., testosterone) during development.

A **46 XX intersex** individual, previously known as a *female pseudohermaphrodite,* is a genetic female (XX) with external genitalia that resemble those of a male (male phenotypic sex). The clitoris enlarges to look like a small penis, and the two labia may become partially fused to resemble a scrotum. The 46 XX intersex condition may result if the female fetus is exposed to excessive androgens (e.g., if the pregnant mother was given certain medications to help prevent miscarriage). More commonly, it is caused by **congenital adrenal hyperplasia,** in which the fetus's adrenal glands produce excessive amounts of androgens (see Clinical View 17.6: "Disorders in Adrenal Cortex Hormone Secretion").

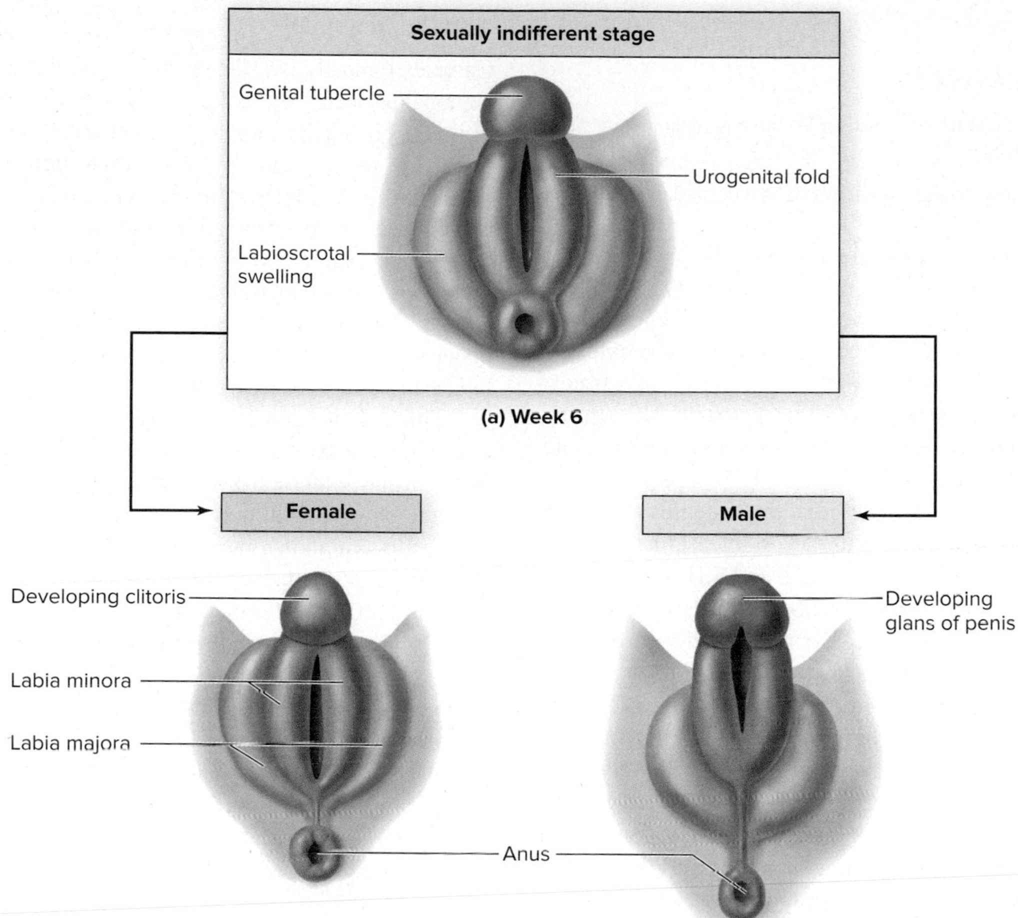

(a) Week 6

(b) Week 12: Urogenital folds begin to fuse in the male

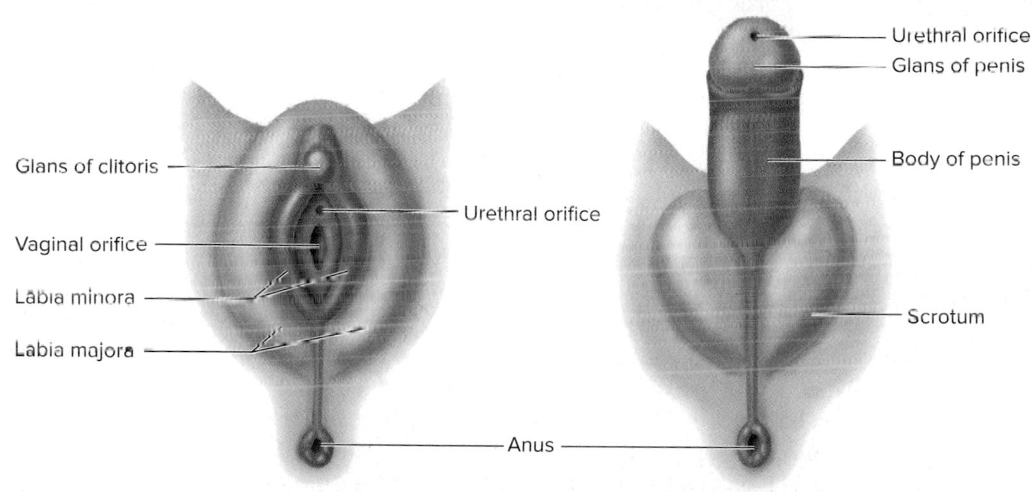

(c) Week 20: External genitalia well differentiated

Figure 28.22 Development of External Genitalia. (*a*) At 6 weeks of development, the external genitalia are undifferentiated. (*b*) By 12 weeks, the urogenital folds begin to fuse in the male and remain open in the female. (*c*) By 20 weeks, external genitalia are well differentiated.

The external genitalia appear very similar in both females and males until about week 12 of development: They do not become clearly differentiated until about week 20. In the absence of testosterone, female external genitalia develop. The genital tubercle becomes the clitoris. The urogenital folds do not fuse, but become the labia minora. Finally, the labioscrotal folds also remain unfused and become the labia majora.

Production and circulation of testosterone in the male cause the primitive external structures to differentiate. The genital tubercle enlarges and elongates, forming the glans of the penis and part of the dorsal side of the penis. The urogenital folds grow and fuse around the developing urethra and form most of the body of the penis. Finally, the labioscrotal swellings fuse at the midline, forming the scrotum.

WHAT DID YOU LEARN?

40 What causes the differentiation of tissues into male reproductive structures?

28.5e Puberty

LEARNING OBJECTIVES

44. Give the common definition for puberty, and list the age range during which it commonly occurs.
45. List some of the common developmental events that occur during puberty.

Recall that both the female and male reproductive systems are primarily nonfunctional or dormant until puberty. **Puberty** (pū′ber-tē; *puber* = grown up) is a period in adolescence where the reproductive organs become fully functional and the external sex characteristics become more prominent, such as breast enlargement in females and pubic hair in both sexes. The timing of puberty may be affected by genetics, environmental factors, and overall health of the individual.

Puberty is initiated when the hypothalamus begins secreting GnRH, which stimulates the anterior pituitary to release the gonadotropins FSH and LH. Prior to puberty, FSH and LH are virtually nonexistent in girls and boys. As levels of FSH and LH increase, the gonads produce significant levels of sex hormones and start the processes of gamete maturation and sexual maturation.

The earliest signs of puberty are the development of breast buds in girls and the appearance of pubic and axillary hair in both boys and girls. Menarche is one of the later events of puberty and tends to occur about 2 years after the first signs appear. Boys experience growth in the testicles and penis, and they may begin to get erections at this time and experience ejaculations during the night. Male voices start to change and become lower in pitch, as the increase in testosterone causes rapid growth of the laryngeal structures.

The timing of puberty differs between females and males and varies among different populations. Girls generally reach puberty about 2 years prior to boys. African-American girls tend to experience puberty about 1 year earlier than their Caucasian counterparts. Additionally, children in better economic conditions experience puberty earlier than children who are underprivileged, presumably because the former group experience better nutrition and heath care. Over the past 100 years, the age of puberty has dropped, so current-day children are starting puberty about 2 years earlier than children in the past. Currently, puberty typically begins between ages 8 and 12 for girls and between ages 9 and 14 for boys (although puberty may occur much later in either population).

Precocious puberty is apparent when the signs of puberty develop much earlier than normal—such as before ages of 7 to 8 for girls, or before age 9 for boys. Precocious puberty may be idiopathic (no known cause) or due to brain injury or infection, or a tumor in the pituitary or gonads. If a child experiences signs of precocious puberty, he or she should be taken to a pediatrician for an evaluation to see if any treatment is needed.

WHAT DID YOU LEARN?

41 What factors affect the age that menarche first occurs?

28.5f Menopause and Male Climacteric

LEARNING OBJECTIVES

46. Give the common definition and symptoms for menopause.
47. Describe events of the male climacteric.

After reaching sexual maturity, the female and male reproductive systems exhibit marked differences in their response to aging.

Female Menopause

Gametes typically stop maturing in females by their 40s or 50s, and menopause occurs. In section 28.3b we discussed that when a woman has stopped having monthly menstrual cycles for 1 year and is not pregnant, she is said to be in **menopause** (men′ō-pawz; *pauses* = cessation). A reduction in sex hormone production that accompanies menopause causes some atrophy of the reproductive organs and the breasts. The vaginal wall thins and there is a reduction in glandular secretions for maintaining a moist, lubricated lining. The uterus shrinks and atrophies, becoming much smaller than it was before puberty. The woman's endometrial lining does not grow, and she no longer has a menstrual period.

The lack of significant amounts of estrogen and progesterone in a menopausal woman also affects other organs and body systems. Women may experience **hot flashes,** in which their bodies perceive periodic elevations in body temperature, and they may develop thinning scalp hair and an increase in facial hair. Menopausal women are at greater risk for osteoporosis and heart disease due to the drop in estrogen and progesterone levels.

Formerly, **hormone replacement therapy (HRT)** in the form of estrogen and progesterone supplements was routinely offered to menopausal women to help diminish these symptoms and risks. However, studies indicate that the risks associated with HRT (e.g., increased risk of breast cancer and lack of protection against heart disease) may outweigh the benefits in older women. As a result, many physicians have stopped prescribing HRT for menopausal symptoms, although some women still choose to receive HRT if their symptoms are severe.

Male Climacteric

In contrast to females, males do not experience the relatively abrupt change in reproductive system function that females do. A slight decrease in the size of the testes parallels a reduction in the size of the seminiferous tubules and the number of interstitial cells. As a consequence of the reduced number of interstitial cells, decreased testosterone levels in men in their 50s signal a change called the **male climacteric** (klī-mak′ter-ik, klī-mak-ter′ik). Most men experience few symptoms, but some may experience mood swings, decreased sex drive, and hot flashes and sweating episodes. However, men generally do not stop producing gametes as women do following menopause. Additionally, while men experience a reduction in testosterone levels, this reduction is gradual and not as steep or sudden as the estrogen and progesterone drop seen in menopausal women.

Most men experience prostate enlargement (either benign or cancerous) as they age. This prostate enlargement can interfere with sexual and urinary functions. Also associated with aging are **erectile dysfunction** and **impotence,** which refer to the inability to achieve or maintain an erection. Besides aging, other risk factors for this condition include heart disease, diabetes, smoking, and prior prostate surgery. Many drugs (e.g., sildenafil [Viagra]) have entered the market that treat erectile dysfunction by prolonging vasodilation of the penile arteries and thus inhibit relaxation of the erectile bodies. (Patients who have heart disease and take nitrates for chest pain are discouraged from taking sildenafil, because the combination of drugs can cause a dangerous drop in blood pressure.)

WHAT DID YOU LEARN?

42 How do the female and male reproductive systems differ with respect to aging?

CHAPTER SUMMARY

28.1 Overview of Female and Male Reproductive Systems	• The female and male reproductive systems function to propagate the next generation. **28.1a Common Elements of the Two Systems** • Both reproductive systems have gonads that produce gametes (sex cells) and sex hormones and accessory reproductive organs (that transport or sustain the gametes). **28.1b Sexual Maturation in Females and Males** • Sexual maturation begins at puberty when the hypothalamus begins secreting gonadotropin-releasing hormone (GnRH). This hormone stimulates the anterior pituitary to release follicle-stimulating hormone (FSH) and luteinizing hormone (LH). • FSH and LH stimulate the gonads to release sex hormones and produce gametes. **28.1c Anatomy of the Perineum** • The perineum is a diamond-shaped area between the thighs that houses the urogenital and anal triangles.
28.2 Gametogenesis	• Gametogenesis is the process of forming sex cells. **28.2a A Brief Review of Heredity** • Humans have 22 pairs of autosomes and 1 pair of sex chromosomes in the body's diploid cells. The sex chromosomes, termed X and Y, determine genetic sex: XX is female, and XY is male. • The gametes, or sex cells, are haploid cells; they contain 22 chromosomes and 1 sex chromosome, not pairs of chromosomes. When gametes unite in fertilization, the diploid condition is restored. **28.2b An Overview of Meiosis** • Meiosis is sex cell division that produces haploid gametes from diploid parent cells. • Meiosis involves two rounds of division: meiosis I and meiosis II. As a result of these two rounds of division, one diploid cell gives rise to four haploid cells. **28.2c Meiosis I: Reduction Division** • Chromosomes become replicated prior to meiosis I so that each replicated chromosome consists of two sister chromatids held together by a single centromere. • The pairs of replicated homologous chromosomes separate during meiosis I. The stages are prophase I, metaphase I, anaphase I, and telophase I and cytokinesis. • Crossing over and exchange between homologous chromosomes occurs in prophase I and helps produce gametes that are genetically different from the parent cell. **28.2d Meiosis II: Separation of Sister Chromatids** • The sister chromatids (composing a replicated chromosome) separate in meiosis II. The stages are prophase II, metaphase II, anaphase II, and telophase II and cytokinesis. • The result of the two sequences of meiosis is that four haploid cells are produced that may go on to form gametes.
28.3 Female Reproductive System	• Female reproductive organs include paired ovaries and uterine tubes, a uterus, a vagina, external genitalia, and the mammary glands. **28.3a Ovaries** • The cortex of the ovary houses ovarian follicles that consist of an oocyte surrounded by follicle (or granulosa) cells. **28.3b Oogenesis and the Ovarian Cycle** • GnRH stimulates release of FSH and LH, which bring about follicle maturation. As the follicles mature, they release estrogen, and when these levels reach a threshold, the estrogen positively stimulates the hypothalamus and anterior pituitary to release more of their hormones. • Changing levels of FSH and LH cause a primordial follicle to mature into a primary follicle, secondary follicle, antral follicle, and then into a mature follicle. • A peak in LH causes the secondary oocyte to be released from the mature follicle at ovulation; remaining follicular cells become the hormone-producing corpus luteum. • The ovarian cycle consists of the follicular phase, ovulation, and the luteal phase. **28.3c Uterine Tubes, Uterus, and Vagina** • The uterine tubes extend from the uterus to each ovary; the uterine tube is the site of fertilization. • The uterus is a thick-walled muscular organ that functions as the site of pre-embryo implantation, supports and nourishes the embryo/fetus, and is the site of menstruation. • The uterine wall consists of an inner mucosa, the endometrium; a thick-walled middle muscular layer, the myometrium; and an outer serosa, the perimetrium. • The vagina is an epithelial-lined fibromuscular tube that serves as the birth canal for the fetus, the organ of copulation during intercourse, and the passageway for menstrual discharge. **28.3d Uterine (Menstrual) Cycle and Menstruation** • The endometrium has a functional layer that is sloughed off as menses and a deeper basal layer that regenerates a new functional layer during the next uterine cycle. • The menstrual and proliferative phases of the menstrual cycle occur in conjunction with the follicular and ovulation phases of the ovarian cycle. The secretory phase of the menstrual cycle occurs at the same time as the luteal phase of the ovarian cycle.

(continued on next page)

- Transmission of maternal antibodies to the developing embryo or fetus (see section 22.8c)
- Production of estrogen and progesterone to maintain and build the uterine lining

The placenta begins to form during the second week of development. The fetal portion of the placenta develops from the chorion, whereas the maternal portion of the placenta forms from the functional layer of the uterus. The early organism is connected to the placenta via a structure called the **connecting stalk** (figure 29.7*a*, *b*). This connecting stalk eventually contains the umbilical arteries and umbilical vein that distribute blood to and from the embryo or fetus. The connecting stalk is the precursor to the future **umbilical cord** (figure 29.7*c*).

Figure 29.7 illustrates how the components of the placenta become better defined during the embryonic period. Fingerlike structures called **chorionic villi** (sing., *villus*) form from the chorion. The chorionic villi contain branches of the umbilical vessels. Adjacent to the chorionic villi is the functional layer of the endometrium, which contains maternal blood. Note that fetal blood and maternal blood do not mix; however, the two bloodstreams are so close to one another that exchange of gases and nutrients can occur. The concentration of O_2 and nutrients is higher in the maternal blood, and therefore these diffuse into the fetal blood. Conversely, the concentration of CO_2 and waste products is higher in the fetal blood, so these materials diffuse from the fetal blood into the maternal circulation.

Although the placenta first forms during the pre-embryonic period, most of its growth and development occur during the fetal period (the last 30 weeks of development). When the placenta matures, it is disc shaped and adheres firmly to the wall of the uterus. Immediately after the baby is born, the placenta is also expelled from the uterus. The expelled placenta is often called the *afterbirth.*

The placenta may be thought of as a selectively permeable structure. Certain materials enter freely through the placenta into the fetal blood, whereas other substances are effectively blocked. For example, respiratory gases and nutrients may freely cross the placental barrier, but many microorganisms (e.g., viruses, bacteria) and high levels of maternal hormones are prevented from crossing this barrier into the developing fetus. Unfortunately, a number of undesirable substances, such as some viruses (e.g., HIV and rubella; see section 22.1) and bacteria (e.g., *Treponema,* the bacterium that causes syphilis) *can* cross the placental barrier, infecting the fetus and sometimes causing birth defects or death. Most drugs, alcohol, and the toxins from smoking can pass through the placental barrier as well.

Some fetuses may be more susceptible to materials that cross the placental barrier than other fetuses. Additionally, the *dose* of the material crossing the placental barrier and the timing of this crossing both affect fetus susceptibility. These facts help explain why some newborns are strongly affected by materials that cross the placental barrier, whereas other newborns are relatively unaffected.

Prior to implantation, the blastocyst is not harmed by undesirable substances because it does not yet have a connection with the mother's uterine lining. However, once implantation begins and the placenta starts to form, the developing organism is exposed to most of the substances to which the mother is exposed. For these reasons, pregnant women are strongly urged to quit smoking and to refrain from taking drugs and drinking alcohol during their pregnancies.

INTEGRATE

LEARNING STRATEGY

The *third week* of development produces an embryo with *three* primary germ layers: ectoderm, mesoderm, and endoderm.

WHAT DID YOU LEARN?

10 What are the main functions of the placenta?

29.3 Embryonic Period

The embryonic period begins in week 3 with the establishment of the three primary germ layers (see section 5.6a) through the process of *gastrulation.* Subsequent interactions and rearrangements among the cells of the three layers prepare for the formation of all body tissues and organs, a process called organogenesis. By week 4, the embryo has a beating heart, and by the end of the embryonic period (week 8), the main organ systems have been established, and the major features of the external body form are recognizable. **Table 29.2** summarizes the events that occur during the embryonic period.

Table 29.2 Events in Embryonic Development

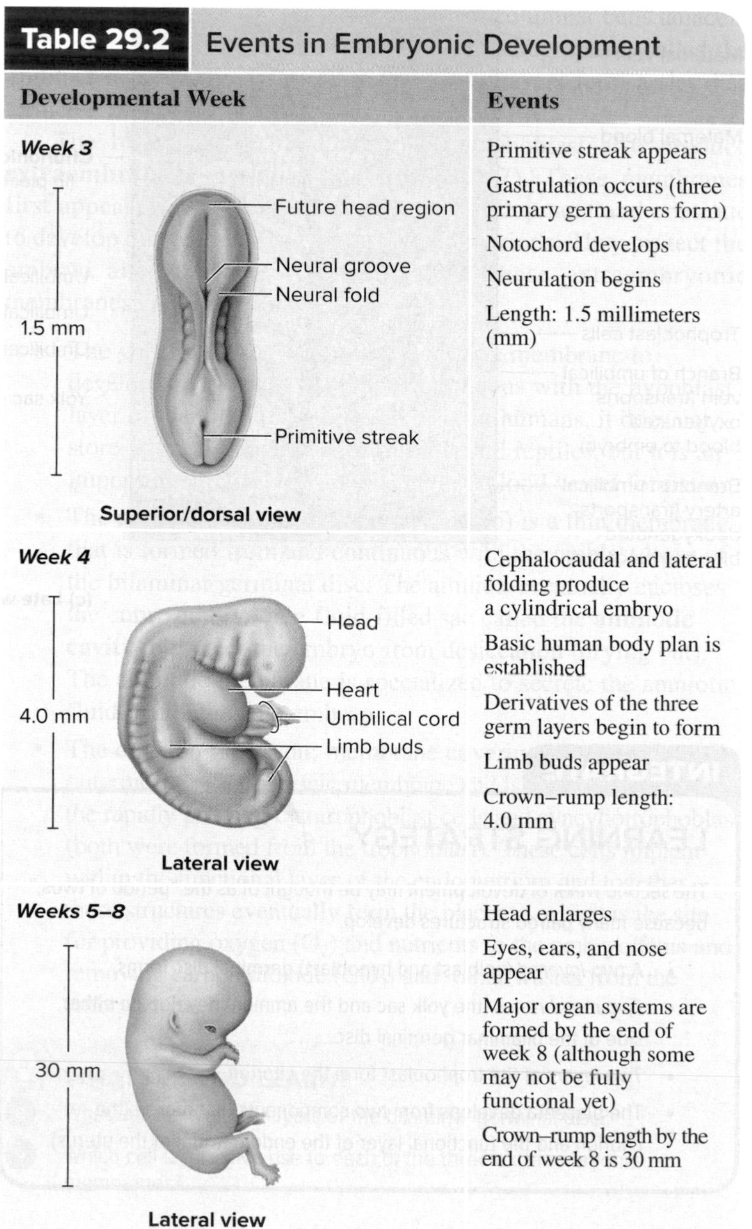

Developmental Week	Events
Week 3 (Superior/dorsal view)	Primitive streak appears Gastrulation occurs (three primary germ layers form) Notochord develops Neurulation begins Length: 1.5 millimeters (mm)
Week 4 (Lateral view)	Cephalocaudal and lateral folding produce a cylindrical embryo Basic human body plan is established Derivatives of the three germ layers begin to form Limb buds appear Crown–rump length: 4.0 mm
Weeks 5–8 (Lateral view)	Head enlarges Eyes, ears, and nose appear Major organ systems are formed by the end of week 8 (although some may not be fully functional yet) Crown–rump length by the end of week 8 is 30 mm

29.3a Gastrulation and Formation of the Primary Germ Layers

LEARNING OBJECTIVES

12. Describe the process of gastrulation.

13. List the three primary germ layers that compose the embryo.

Gastrulation (gas-trū-lā′shŭn; *gaster* = belly) occurs during the third week of development immediately after implantation, and it is one of the most critical periods in the development of the embryo. Gastrulation is a process by which the cells of the epiblast migrate and form the three **primary germ layers,** which are the cells from which all body tissues develop (see section 5.6a). The three primary germ layers are called ectoderm, mesoderm, and endoderm. Once these three layers have formed, the developing trilaminar (three-layered) structure may be called an **embryo.**

Gastrulation begins with formation of the **primitive streak,** a thin depression on the surface of the epiblast (**figure 29.8*a, b***). The cephalic (head) end of the streak, known as the **primitive node,** consists of a slightly elevated area. Within the center of the primitive node is a depression called the **primitive pit.**

Cells detach from the epiblast layer and migrate through the primitive streak between the epiblast and hypoblast layers. This inward movement of cells is known as **invagination.** Migrating cells first displace the hypoblast and form the **endoderm** (en′dō-derm; *endo* = inner, *derma* = skin). Next, more epiblast cells invaginate and form a new primary germ layer known as **mesoderm** (mez′ō-derm; *meso* = middle). Cells remaining in the epiblast then form the **ectoderm** (ek′tō-derm; *ektos* = outside). Thus, the epiblast, through the process of gastrulation, is the source of the three primary germ layers from which all body tissues and organs eventually derive (figure 29.8*c*).

WHAT DID YOU LEARN?

11 What events occur during gastrulation? What are the locations of the three germ layers?

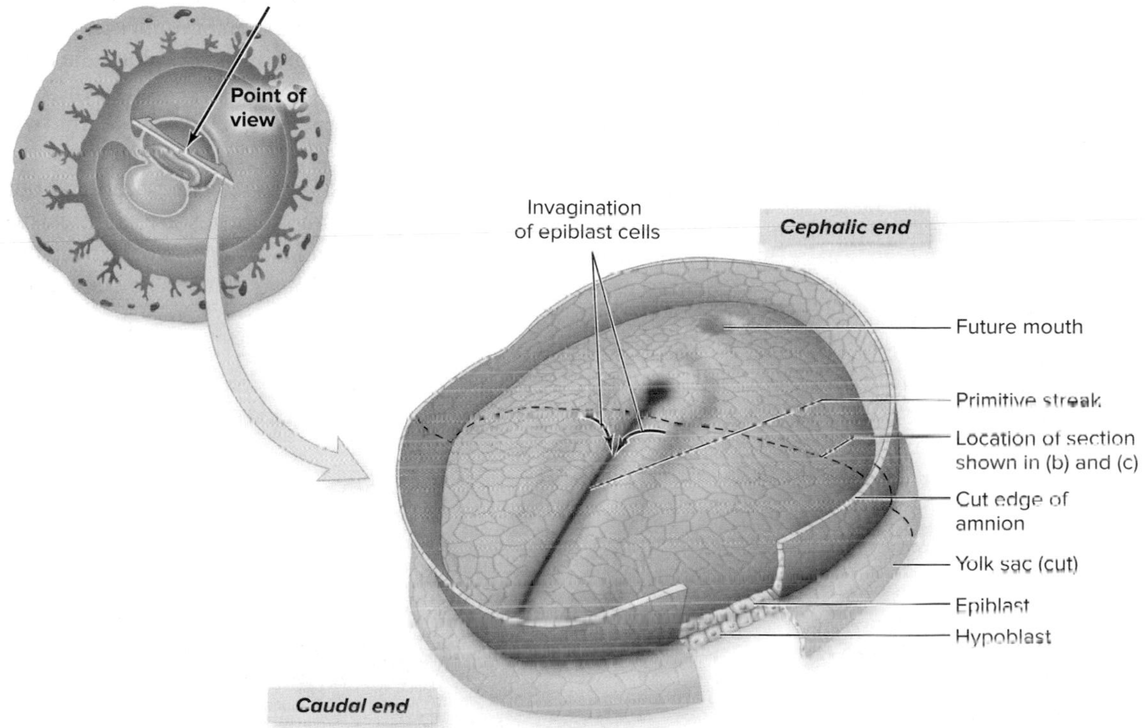

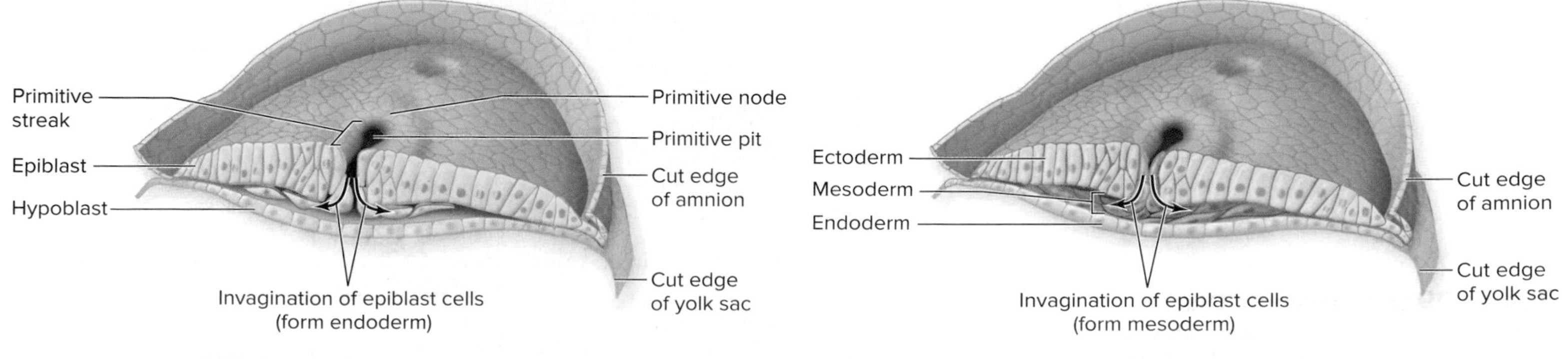

Figure 29.8 The Role of the Primitive Streak in Gastrulation. (*a*) The primitive streak is a raised groove on the epiblast surface of the bilaminar germinal disc that appears early in the third week. (*b, c*) During gastrulation, epiblast cells migrate toward the primitive streak, where some become embryonic endoderm, and others form mesoderm between the epiblast and the new endoderm. Cells remaining in the epiblast become the ectoderm. Thus, the epiblast forms all three primary germ layers.

exam by 4 weeks after fertilization, and by 12 weeks (the end of the first trimester) the uterus is just superior to the level of the pubic symphysis and is about the size and shape of a large grapefruit. As the uterus expands, it impinges on the space occupied by the urinary bladder, so more frequent urination during this trimester is a common complaint. Recall that the developing fetus has a crown-to-rump length of only 9 cm at this time, so the bulk of the enlargement is due to myometrium muscle hypertrophy and hyperplasia, placental growth, and amniotic fluid production.

By 16 weeks (4 months) (figure 29.13*b*), the uterus has expanded into the abdominal cavity and its fundus typically is at a midpoint level between the pubic symphysis and umbilicus. The uterus continues to expand and reaches the level of the umbilicus by about week 22. During this second trimester, the superior growth of the uterus temporarily decreases pressure on the urinary bladder, so the frequent need to urinate may lessen during this trimester.

The expansion continues during the third trimester, as now the fetus is preparing for its own rapid growth sequence. By about 28 weeks (7 months), the fundus of the uterus is superior to the umbilicus, and by the ninth month the fundus of the uterus is at the level of the xiphoid process of the sternum (figure 29.13*d*). The enlarged uterus pushes against the diaphragm and compresses many of the abdominopelvic organs, resulting in certain gastrointestinal (GI) ailments (described in the next section). In addition, now the uterus is so large that it compresses the bladder, so frequent urination becomes common again.

WHAT DO YOU THINK?

2 Some pregnant women experience difficulty with deep breathing during the last trimester. Can you think of any reasons deep breathing (versus shallow breathing) is more difficult during this time?

The mammary glands typically are tender and sore during the first trimester due to the increasing levels of estrogen and progesterone. The placenta secretes **melanocyte-stimulating hormone (MSH),** which is in part responsible for the darkening of the areolae and nipples during this time (figure 29.13*d*). MSH also is responsible for a darkening of the linea alba, the ligamentous connection between left and right rectus sheaths, turning it into a temporary vertical dark line called the **linea nigra** (nī′gră; *niger* = black). Mammary glandular tissue grows and additional acini develop, especially under the influence of prolactin. It is typical for a pregnant woman to increase one bra cup size during this time due to the mammary gland growth. Further discussion of lactation occurs in section 29.8c.

WHAT DID YOU LEARN?

19 What impact does the growth of the uterus during pregnancy have on other organs of the abdomen?

20 What hormones affect the mammary glands of a pregnant woman?

29.5d Digestive System, Nutrient, and Metabolic Changes

LEARNING OBJECTIVES

23. Describe the effects of HPL and other hormones on the pregnant woman's ability to utilize glucose.
24. List some common GI changes that occur during pregnancy and their causes.

Human placental lactogen (HPL) affects how the mother metabolizes certain nutrients, so that she metabolizes more fatty acids instead of glucose, leaving the glucose for the developing fetus, as noted in section 29.5b. The elevated levels of HPL, as well as higher levels of corticosteroids (e.g., cortisol), estrogen, and progesterone, also result in increased insulin resistance in the pregnant mother. This increased insulin resistance in some cases can lead to gestational diabetes in the mother (see Clinical View 29.3: "Gestational Diabetes").

INTEGRATE

CLINICAL VIEW 29.4

Hyperemesis Gravidarum

Hyperemesis (hī-per-em-ē′sis; *emesis* = vomiting) **gravidarum (HG)** (grav′ĭ-dăr-ŭm; *gravid* = a pregnant woman) is severe, continuous nausea and vomiting during pregnancy that results in dehydration, electrolyte imbalance, and weight loss in the mother. In severe cases, the woman may have to be hospitalized and given intravenous fluids or be fed with a nasogastric tube until either symptoms subside or she is ready to give birth. Fortunately, hyperemesis gravidarum is not common, compared to the milder forms of morning sickness.

Many expectant mothers may experience **morning sickness** during the first trimester of their pregnancy. Contrary to its name, morning sickness does not occur just in the morning, although typically symptoms tend to be most severe then. Symptoms may vary dramatically among women—some may experience a little nausea for a few months only, whereas others may have dehydrating vomiting that lasts throughout their pregnancy and may be debilitating. In the most severe cases, a woman may need to be hospitalized and replenished with IV fluids until the pregnancy comes to term.

There are several hypotheses about the cause of morning sickness, but none have been definitively proven. One suggests that the high levels of circulating hormones, especially hCG, are the cause. However, several research studies have not been able to demonstrate this. Others have suggested that morning sickness is an evolutionary adaptation to protect the developing fetus from harmful toxins in the food. Women with morning sickness tend to prefer bland carbohydrates (which may be less likely to be spoiled or have toxins) and may have aversions to meat and eggs (which may be more likely to have toxins or be spoiled with pathogens).

The higher circulating levels of progesterone result in relaxed smooth muscle and thus slowed intestinal motility, so digested materials remain in the GI tract for longer periods of time. In the later stages of pregnancy, the expanding uterus compresses the abdominal organs and may impinge on part of the intestines. All of these factors may result in heartburn and indigestion; mothers also may experience constipation (see Clinical View 26.17: "Constipation and Diarrhea"). Chronic constipation, as well as problems with venous circulation (caused by compression of the lower veins by the fetus) also can lead to hemorrhoids (see Clinical View 20.7: "Varicose Veins").

A pregnant woman typically needs about 300 extra calories daily to supply both herself and her fetus. Adequate levels of folic acid, calcium, protein, and iron are especially important during pregnancy. A woman experiences weight gain during pregnancy, but only about 20 pounds of this weight gain is due to the fetus, placenta, breast and uterine enlargement, and fluid retention. Any additional weight gain typically is additional adipose tissue, fluid retention, or both.

WHAT DID YOU LEARN?

21 Why is it important to check a woman's blood glucose levels during pregnancy?

22 What conditions other than overeating might explain weight gain during pregnancy?

INTEGRATE

CLINICAL VIEW 29.5

Preeclampsia

©PunchStock/BananaStock RF

Preeclampsia (prē-ē-klamp′sē-ă), also known as *toxemia* or *pregnancy-induced hypertension,* is high blood pressure that occurs by the second half of pregnancy and is marked by protein in the urine (proteinuria). It occurs in about 2–6% of all pregnancies, and risk factors include maternal obesity, diabetes, advanced age, and having a previous preeclampsic birth. The cause is unknown, but researchers suspect that endothelial cell injury, due to a poorly perfused placenta, initiates the disorder. The high blood pressure carries with it the risks of general hypertension for the mother, and poor perfusion of the placenta means the fetus may not get an adequate supply of nutrients. The only cure for preeclampsia is giving birth, so depending upon when symptoms occur, a woman may be put on bedrest and medications (to keep blood pressure levels lower), or labor may be induced if the baby is close to term.

Eclampsia is high blood pressure that causes seizures or convulsions in the mother and is a medical emergency. The concern for every preeclampsia patient is to make sure her symptoms do not develop into eclampsia.

29.5e Cardiovascular and Respiratory System Changes

LEARNING OBJECTIVES

25. List the cardiovascular changes a woman typically exhibits during pregnancy.
26. Explain the changes to the respiratory system during pregnancy.

Because of the needs of both the mother and the embryo/fetus, the cardiovascular system of the mother undergoes dramatic changes throughout pregnancy. The mother's respiratory system function also becomes altered to meet increased requirements for gas exchange.

Cardiovascular System

The mother's cardiovascular system must distribute respiratory gases and nutrients to both the mother and the growing fetus. More blood is needed, so plasma volume increases by about 50% throughout pregnancy. The heart must work harder to circulate the increased blood volume. Cardiac output (amount of blood pumped per minute) increases 30–50% beginning at week 6 of pregnancy and peaks about weeks 24–28 of pregnancy, before dropping slightly. To increase cardiac output, the body increases both heart rate (on average an increase of 10 to 20 beats per minute) and stroke volume.

The increase in blood volume may initially cause an increase in blood pressure during the first trimester. However, by the second trimester the blood pressure drops because of a decrease in peripheral vascular resistance that results from both a decrease in blood viscosity (a pregnant woman has a lower hematocrit than a nonpregnant female due to relatively greater proportion of plasma to erythrocytes; see section 18.1c) and a decreased sensitivity to the hormone angiotensin (see sections 20.6b and 25.4a).

By the third trimester, the uterus and fetus compress the abdominal blood vessels, so venous return from the lower part of the body may be impaired. As a result, some pregnant women may develop varicose veins, hemorrhoids, and edema in the lower limbs.

Respiratory System

Earlier we mentioned that in the later part of pregnancy, the expanding uterus prevents the diaphragm from fully descending and the lungs from fully expanding with air. **Dyspnea** (disp′nē-a) is shortness of breath and may occur during periods of exertion. Increased estrogen levels may cause increased blood circulation and fluid retention in the nasal cavity mucosa, resulting in nasal congestion. The expectant mother also may experience **epistaxis** (nosebleeds) due to the increased blood circulation.

Progesterone increases the sensitivity of central chemoreceptors to blood carbon dioxide (CO_2) levels, ultimately functioning to lower the blood CO_2 levels. These lower blood CO_2 levels facilitate the diffusion of gases across the placenta from the fetal blood to the mother's blood. To lower the blood CO_2 levels, the tidal volume increases by 30–40% as breathing depth increases and pulmonary ventilation (number of breaths per minute) increases (see section 23.5). Additionally, the mother's oxygen consumption increases about 20–30% to meet the oxygen demands of both mother and fetus. These alterations provide enough oxygen to mother and fetus, as well as facilitate gas exchange between the placenta and the maternal blood.

WHAT DID YOU LEARN?

 23 How is a woman's cardiovascular system altered during pregnancy?

 24 Why is it beneficial for CO_2 chemoreceptor sensitivity to be increased when a woman is pregnant?

29.5f Urinary System Changes

LEARNING OBJECTIVE

27. Describe the effects of pregnancy on the mother's urinary system.

The pregnant woman's urinary system is responsible for eliminating not only her own metabolic waste products but also the waste products of her fetus. In addition, up to 50% more plasma volume must be filtered by the kidneys. Glomerular filtration rate (GFR; see section 24.5d) thus increases about 30–50% during pregnancy and urine output increases slightly.

Recall that compression by the expanding uterus on the urinary bladder can lead to frequent urination in both the first and third trimesters. In contrast, the uterus places relatively less pressure on the bladder during the second trimester.

Progesterone causes smooth muscle relaxation in the ureters, which may cause expansion of the ureters and renal pelvis of the kidneys. This dilation and the increased urine volume may result in urine stasis (slowing or stopping) from the kidneys to the urinary bladder. In addition, compression of the ureter or kidney by the uterus can result in urine drainage issues. All of these factors put pregnant women at much greater risk for urinary tract infections (UTIs), which are perhaps the most common type of bacterial infection during pregnancy.

WHAT DID YOU LEARN?

 25 What factors make pregnant women more likely to get urinary tract infections?

29.6 Labor (Parturition) and Delivery

Labor is also known as *parturition* (par-tūr-ĭsh′ŭn; *parturitio* = to be in labor) and is the physical expulsion of the fetus and placenta from the uterus. True labor typically occurs at 38 weeks for a full-term pregnancy, but not all uterine contractions lead to true labor. We first consider factors that initiate labor, and then compare the contractions of false labor with those of true labor; we conclude with a description of the three stages of labor.

INTEGRATE

CONCEPT CONNECTION

Pregnancy has an effect on every body system in the pregnant woman, including the reproductive, endocrine, digestive, cardiovascular, respiratory, urinary, and even integumentary systems, which respond and adapt to the changes brought about during pregnancy. The skeletal and muscular systems adjust to her body's increasing demand for calcium, several skeletomuscular changes develop as well (e.g., **lordosis,** or accentuated lumbar curve seen in late-term pregnancies, see Clinical View 8.3: "Spinal Curvature Abnormalities"). The immune system typically is suppressed slightly so the woman's body will not mistakenly attack the fetus as something foreign. In fact, pregnancy is a wonderful example of how all body systems adapt and develop their own forms of homeostasis in response to such a physiologic challenge.

29.6a Factors That Lead to Labor

LEARNING OBJECTIVE

28. Explain the physiologic processes that initiate labor.

Throughout the pregnancy, as the uterus enlarges and stretches, the uterine myometrium prepares itself for uterine contractions. In the later stages of pregnancy, the increasing levels of estrogen counteract the calming influence of progesterone on the uterine myometrium and increase the uterine myometrium sensitivity. In addition, the rising levels of estrogen stimulate the production of oxytocin receptors on the smooth muscle cells of the uterine myometrium, so as the levels of oxytocin also rise, more receptors are available on the uterus for binding this hormone.

All of these factors result in the uterine myometrium becoming more sensitive and "irritable" in the later stages of pregnancy, and contractions begin to occur. These contractions typically are weak and irregular, but as levels of estrogen and oxytocin continue to rise in the later stages, they become more intense and frequent. Thus, weak contractions may occur and be noticed as soon as the second trimester of pregnancy.

Premature labor refers to labor that occurs prior to 38 weeks. Premature labor (and giving birth to a premature infant) is not desirable because the infant's body systems, especially the lungs, may not be fully developed. Very premature infants are at greater risk for morbidity and mortality due to their underdeveloped organ systems. Thus, the ideal outcome is for labor to begin as close as possible to full term.

WHAT DID YOU LEARN?

26 How do progesterone, estrogen, and oxytocin interact to eventually bring about labor?

29.6b False Labor

LEARNING OBJECTIVE

29. List the signs and characteristics of false labor.

False labor is defined as uterine contractions that do *not* result in the three stages of labor and the expulsion of the fetus. The contractions of false labor are known as **Braxton-Hicks contractions,** named for a nineteenth-century gynecologist. It may be difficult for a woman to know whether she is experiencing Braxton-Hicks contractions or true labor contractions, and it is not uncommon for a woman to mistakenly think she is about to give birth.

In general, Braxton-Hicks contractions have the following characteristics:

- They tend to be irregularly spaced and do not become more frequent as time passes.
- They tend to be relatively weak, do not increase in intensity, and may stop entirely if the woman changes position or activity (e.g., goes for a walk).
- The pain from these contractions is usually limited to the lower abdomen and pelvic region, instead of radiating through the entire abdominal region and back (as with true labor contractions).
- The pain from the contractions may stop or change in response to movement.
- They do not lead to the cervical changes seen in the three stages of labor (described in section 29.6d).

INTEGRATE

CLINICAL VIEW 29.6

Inducing Labor

If a woman has not undergone labor within 2 weeks after her due date, her physician may recommend inducing labor. The woman typically is admitted to the hospital the night before labor is to progress, and a prostaglandin gel is smoothed onto her cervix. The prostaglandin gel assists with cervical dilation (thinning and widening of the cervix) and may even "kick start" the true labor process. If true labor has not begun the following morning, the physician may administer an intravenous (IV) drip of synthetic oxytocin (called Pitocin) to initiate true labor.

Recall from section 28.4d that semen contains high levels of prostaglandins. Thus, initiating sexual intercourse close to the fetus's due date may help induce labor as well.

WHAT DID YOU LEARN?

27 What are five signs of false labor?

29.6c Initiation of True Labor

LEARNING OBJECTIVES

30. Explain the signs and characteristics of true labor.

31. Describe the positive feedback mechanism of true labor.

True labor is defined as uterine contractions that increase in intensity and regularity, and that result in changes to the cervix. The mother and the fetus both have an active role in initiating true labor.

As the pregnancy nears term, the mother's hypothalamus triggers the posterior pituitary to release increasing levels of oxytocin. (This increase in oxytocin levels is in response to a cascade of changes in both fetal and maternal hormones responsible for maintaining pregnancy.) Near the beginning of true labor, the fetus's hypothalamus is also triggering release of oxytocin from the fetus's posterior pituitary. Oxytocin from both the mother and fetus stimulate the placenta to secrete **prostaglandins** (pros-tă-glan′dĭn). Prostaglandins are eicosanoids that act as local hormones to stimulate smooth muscle contraction, most notably uterine muscle contraction (see sections 17.3b and 20.4c). Prostaglandins are also responsible for the softening and dilating of the cervix. The combined actions of maternal oxytocin, fetal oxytocin, and the rising levels of prostaglandins initiate the rhythmic contractions of true labor.

In comparison to Braxton-Hicks contractions, true labor contractions have the following characteristics:

- They tend to be regularly spaced and increase in frequency over time. So, a woman who starts with contractions that occur roughly every 15 minutes eventually will have contractions that occur about every 5 minutes.
- Contractions increase in intensity as labor progresses.
- The pain from the contractions tends to radiate from the upper abdomen inferiorly to the lower back (or vice versa), instead of being localized in the lower abdomen or groin (as with Braxton-Hicks contractions).
- The pain from the contractions does not go away or change in response to movement.
- The contractions facilitate cervical dilation and expulsion of the fetus and placenta.

True labor also initiates a positive feedback mechanism (**figure 29.14**). The more intense uterine contractions result in the fetus's head pushing against the cervix, stimulating the

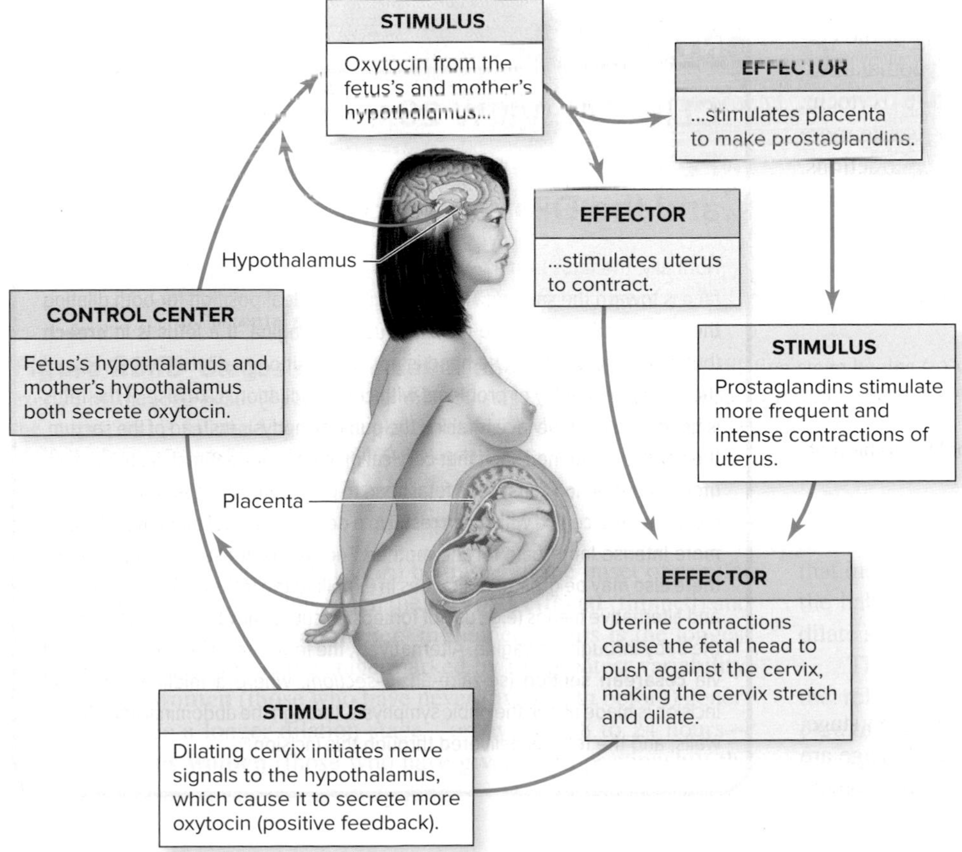
STIMULUS
Oxytocin from the fetus's and mother's hypothalamus...
EFFECTOR
...stimulates placenta to make prostaglandins.
EFFECTOR
...stimulates uterus to contract.
Hypothalamus
CONTROL CENTER
Fetus's hypothalamus and mother's hypothalamus both secrete oxytocin.
STIMULUS
Prostaglandins stimulate more frequent and intense contractions of uterus.
Placenta
EFFECTOR
Uterine contractions cause the fetal head to push against the cervix, making the cervix stretch and dilate.
STIMULUS
Dilating cervix initiates nerve signals to the hypothalamus, which cause it to secrete more oxytocin (positive feedback).

Figure 29.14 Positive Feedback Mechanism of True Labor. Rising levels of maternal and fetal oxytocin together initiate the positive feedback mechanism of true labor. Oxytocin stimulates uterine contractions as well as the release of prostaglandins, which promote cervical stretching and facilitate uterine contractions. Sensory input relayed from the uterus to the hypothalmus stimulates it to release more oxytocin from the posterior pituitary. The process escalates until the fetus is expelled from the uterus.

Expulsion Stage

The **expulsion stage** (figure 29.15*c*) begins with the complete dilation of the cervix and ends with the expulsion of the fetus from the mother's body. This stage may last as little as several minutes but typically takes 30 minutes to several hours. As with the dilation stage, nulliparous women typically have a longer expulsion stage than parous women. The uterine contractions help push the fetus through the vagina and may be facilitated if the woman "bears down" (i.e., uses the Valsalva maneuver to increase abdominal pressure as she pushes) with each contraction.

When the first part of the baby's calvarium (see section 8.2a) distends the vagina, this is referred to as **crowning.** The head is then followed by the rest of the body. If there is difficulty in expelling the baby from the vagina, then an **episiotomy** (e-piz-ē-ot′ō-mē; *episeion* = pundenda, *tome* = incision) may be performed, which is where the perineal muscles are surgically incised to create a wider opening for the baby to pass through. (This cut is sutured after the birth.) When the baby's body is fully expelled, the umbilical cord is clamped and tied off.

Placental Stage

The **placental stage** (figure 29.15*d*) occurs after the baby is expelled. The uterus continues to contract, and these contractions help compress uterine blood vessels and help displace the placenta from the uterine wall. The placenta and remaining fetal membranes (e.g., the amnion) collectively are referred to as the **afterbirth.** The expulsion of the afterbirth typically is completed within 30 minutes. The obstetrician or other birth practitioner carefully examines the afterbirth to make sure all portions of the placenta have been expelled from the uterus, because fragments of placenta left in the uterus can lead to extensive bleeding or other postpartum complications.

WHAT DID YOU LEARN?

30 What are the three stages of labor?

31 Why is the placental stage a critical part of labor?

29.7 Postnatal Changes for the Newborn

LEARNING OBJECTIVES

33. Describe the respiratory events that occur as the newborn adjusts to life outside of the uterus.
34. Compare and contrast the fetal circulatory pattern with the newborn circulatory pattern.

Once the fetus is expelled from the uterus, it is now known as a newborn, or **neonate** (nē′ō-nāt; *neo* = new, *natalis* = relating to birth). A variety of respiratory and cardiovascular changes must occur quickly after birth in order for the neonate to adjust to life outside of the uterus.

Prior to birth, respiratory gases were exchanged between maternal and fetal circulation at the placenta. The fetal lungs are not fully inflated because they are not yet fully functional. However, within about 10 seconds after being born, the neonate typically takes its first breath. This first breath is thought to be caused by the central nervous system responding to the change in environment and temperature. This process may be facilitated by a general respiratory acidosis (caused by clamping of the umbilical vessels and constriction of the umbilical vessels prior to birth), but note that the first breath typically occurs regardless of whether the umbilical vessels have been clamped or not.

Once this first breath is taken, the lungs become inflated and the surfactant that is present in the alveoli keeps the alveoli patent (open) (see section 23.3d). Thus, every breath after the first is easier now that the alveoli remain patent. Premature infants born earlier than 28 weeks are not producing sufficient levels of surfactant to keep their alveoli patent, so these infants may need to placed on a ventilator until their lungs mature.

WHAT DO YOU THINK?

3 In the past, obstetricians would lift a neonate by its feet and slap it on the back, allegedly to start it breathing. The baby's crying was supposed to be an indication that the procedure had worked. Was this procedure necessary? Could it actually be harmful?

The fetal circulatory pattern was described in detail in section 20.12a. Given that the fetal lungs are not functional, other pathways (i.e., ductus arteriosus, foramen ovale) shunt blood away from the nonfunctional lungs and directly to the fetal circulation. As a result, the fetal cardiovascular system has some structures that are modified or that cease to function once the fetus is born.

At birth, the fetal circulation begins to change into the postnatal pattern. When the neonate takes its first breath, pulmonary resistance drops, and the pulmonary arteries dilate. As a result, pressure on the right side of the heart decreases and the pressure is then greater on the left side of the heart, which handles the systemic circulation. The specific changes are described in section 20.12b.

WHAT DID YOU LEARN?

32 What event must occur when the neonate takes its first breath?

33 How does the fetal circulation change once the neonate begins breathing?

29.8 Changes in the Mother After Delivery

Postpartum refers to the time period after giving birth. It is a time when a woman's body must undergo further transformative changes to both feed the neonate and return to pre-pregnancy form and function. Here we describe most of the postpartum changes that typically occur over the first 6 weeks after giving birth.

29.8a Hormonal Changes

LEARNING OBJECTIVE

35. Compare and contrast the hormonal levels of a woman prior to birth and after birth.

Within a few days after giving birth, estrogen and progesterone levels plummet, because the uterine lining no longer needs to be maintained for pregnancy. Many experts believe this precipitous drop in sex hormone levels may account for the **baby blues,** feelings of varying degrees of sadness and depression some women experience immediately after giving birth.

As estrogen and progesterone levels plummet, the integumentary system is affected. Recall from section 6.2b that high levels of these hormones prevented the normal cyclical hair loss. After giving birth, the hair reverts back to its normal cyclical growth and hair loss cycle. And, in fact, some of the hair that was prevented from falling out during pregnancy may fall out rather abruptly for some women after giving birth. A peak in hair loss may be experienced by some women about 3 to 4 months after delivery, and it may take up to 12 months for the normal cyclical hair growth and loss cycle to resume. Thus, many new mothers may experience temporary hair thinning during this time.

The decrease in progesterone also affects the respiratory system. Without the high levels of progesterone, the chemoreceptors are less sensitive to CO_2 levels. As a result, respiratory rate, tidal volume, and pulmonary ventilation return to pre-pregnancy levels.

Additionally, the levels of corticotropin-releasing hormone (CRH) drop dramatically, now that there is no longer a placenta producing copious amounts of this hormone. Recent research has suggested that high levels of CRH during pregnancy are associated with an increased risk of **postpartum depression,** a serious disorder in which the mother experiences severe depression and possibly suicidal thoughts; unlike baby blues, this condition should be treated as soon as possible.

Prolactin levels and oxytocin levels drop after birth as well. However, because both of these hormones are involved in lactation, periodic surges occur in these hormone levels each time a baby nurses. These surges are described in greater detail in section 29.8c.

WHAT DID YOU LEARN?

34 Which hormone levels drop in a postpartum woman, and what are the effects of this reduction in hormone levels?

29.8b Blood Volume and Fluid Changes

LEARNING OBJECTIVE

36. List the various ways that the mother loses the excess fluids gained during pregnancy.

A pregnant woman retains a great deal of fluid throughout the 9 months of pregnancy. In addition to the fluid retained in the amniotic sac and some excess fluid found in the interstitial spaces, most additional fluid is due to the increased blood volume acquired during pregnancy. After a woman gives birth, she no longer has need for this additional fluid and must expel it in a relatively quick and efficient manner. The amniotic fluid is quickly expelled by the end of the first stage of labor. But what about the rest?

A portion of the blood volume, as well as mucus and hypertrophied endometrial tissue, is released from the uterus as **lochia** (lō′kē-ă; *lochos* = childbirth). Lochia is similar to a menstrual period, in that blood and some endometrial tissue are expelled from the uterus via the vagina. However, lochia results in much heavier bleeding than a typical menstrual period, because the uterine lining buildup occurred over a 9-month period, instead of a typical 28-day cycle. Thus, the first 5 days of lochia typically result in very heavy bleeding, after which it lightens but progresses for at least 2 to 3 weeks. For some women, it may take 4 to 6 weeks before the flow finally stops. As the blood volume decreases, the woman's cardiac output returns to pre-pregnancy levels.

A woman also may expel excess fluids via increased urination. The decline in CRH after birth results in a decline in aldosterone (see section 25.4c), which precipitates the overall drop in blood volume and interstitial fluid levels. The lymphatic system cycles some of the excess interstitial fluid into the blood circulation, where it may be filtered by the kidneys and secreted as urine. Within about 24 hours after giving birth, most women experience copious, frequent urination, the result of the kidneys "working overtime," in filtering out this excess fluid. Urination levels typically return to normal by the end of the first week after birth.

Another common symptom new mothers experience is profuse sweating for the first 2 weeks after giving birth. This abundant sweating is yet another way the body eliminates the excess fluid gained during pregnancy.

WHAT DID YOU LEARN?

35 What are four ways by which excess fluid is eliminated from a postpartum woman's body?

29.8c Lactation

LEARNING OBJECTIVE

37. Describe the process by which lactation occurs.

Lactation (lak-tā′shŭn; *lactatio* = suckle) refers to the production and release of breast milk from the mammary glands. Prolactin is produced by the anterior pituitary and is responsible for milk production (see section 17.7c). In nonpregnant women and in men, the secretion of significant amounts of prolactin is inhibited by prolactin-inhibiting hormone (also known as dopamine), secreted by cells in the hypothalamus.

High levels of estrogen positively influence the secretion of prolactin, so as estrogen levels rise during pregnancy, so do prolactin levels. Both estrogen and prolactin cause mammary gland acini proliferation and branching of the lactiferous ducts. Paradoxically, the high levels of estrogen and progesterone are also responsible for preventing breast milk secretion until after birth. It isn't until levels of estrogen and progesterone drop that prolactin works unopposed to stimulate breast milk production.

During late pregnancy and for the first few days after birth, the substance produced by the mammary glands isn't breast milk per se. It is a watery, yellowish, milklike substance called **colostrum** (kō-los′trŭm; *foremilk*), and it has lower concentrations of fat than true breast milk but is rich in immunoglobulins, especially immunoglobulin A (IgA; see section 22.8). By drinking colostrum, the infant acquires passive immunity from the mother (see section 22.9c). IgA resists breakdown in the infant's stomach and is believed to protect the infant against ingested pathogens. Colostrum also has a laxative effect and facilitates the infant's first bowel movement shortly after birth.

A few days postpartum, the true breast milk starts to be produced. It has a higher fat content than colostrum, and it contains several growth factors, essential fatty acids (needed for optimal brain growth and development), specific enzymes to aid in digestion of the milk, and an array of immunoglobulins. Breast milk typically is more easily digestible for an

INTEGRATE

CONCEPT CONNECTION

Many organ systems work in tandem to help bring the mother's body back to a pre-pregnancy homeostasis. Excess fluids are removed from the cardiovascular system and excreted by both the urinary system and the integumentary system. The endocrine system secretes oxytocin to promote uterine contractions and shrinkage to a near pre-pregnancy state. The immune system produces immunoglobulins that are found in abundance in the newly produced breast milk. Many of the pregnancy digestive system challenges, such as heartburn, constipation, and morning sickness, resolve once the baby is born. Finally, if the mother is breastfeeding, the bones of the skeletal system serve as the reservoir for calcium for the breast milk.

©Blend Images/Getty Images RF

infant than other types of breast milk substitutes (e.g., cow's milk, soy milk) and it remains the optimal source of nutrition for an infant, if the mother is able to breastfeed. One vitamin that is *not* abundant in breast milk is vitamin D, and recently physicians have recommended that infants of breastfeeding mothers receive a vitamin D supplement.

WHAT DO YOU THINK?

4 The recommendation that breastfeeding infants receive vitamin D supplementation is a recent one. Can you think of any change in behaviors that may have made this recommendation necessary? (Hint: How does the body produce vitamin D?)

Lactation requires both prolactin for breast milk production and oxytocin for breast milk secretion (see section 28.3f). Although prolactin levels drop after birth by about 50%, surges in prolactin continue as long as the baby continues to breastfeed. Thus, the continual production and release of breast milk is a positive feedback mechanism that is maintained by regular breastfeeding **(figure 29.16*a*)**.

INTEGRATE

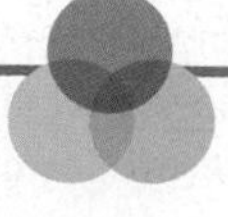

CONCEPT CONNECTION

Recall from section 22.8c that there are five classes of immunoglobulins (or antibodies). The fetus produces only IgM antibodies, and during pregnancy the mother supplies the fetus with IgG, the only antibody that crosses the placenta. After birth, the nursing mother supplies the newborn with IgG, IgA, and IgM antibodies in her breast milk. It isn't until the baby is about 2 months old that he or she is able to produce enough IgA and IgM on his or her own.

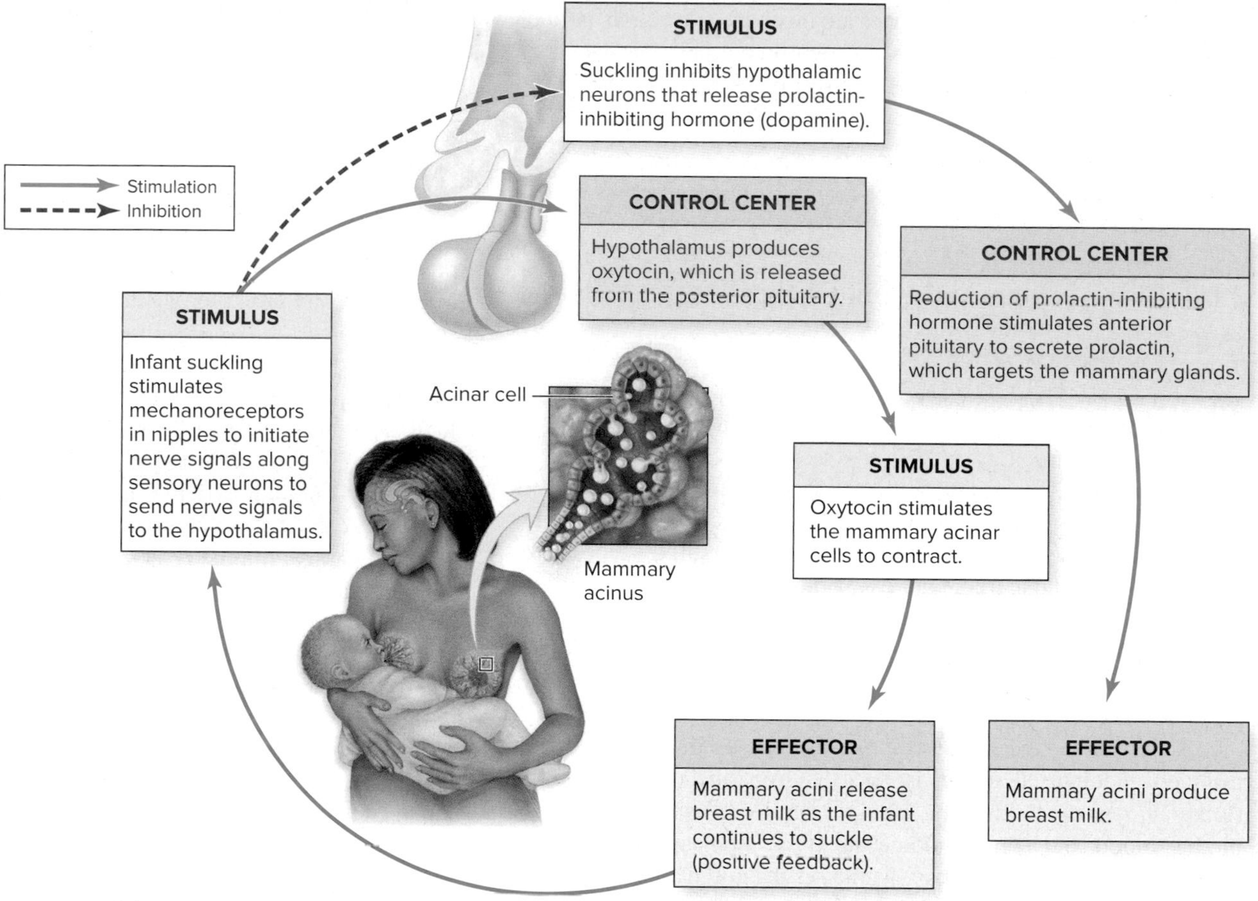

Figure 29.16 Lactation. (*a*) Positive feedback mechanism of lactation. (*b*) Prolactin levels during lactation. After labor and delivery, prolactin levels drop by about 50%. However, each time the infant nurses, prolactin levels spike, thus ensuring that milk production will continue for the next feeding.

The release of breast milk is referred to as **milk letdown,** or the *letdown reflex,* because it involves this positive feedback mechanism. When the infant suckles at the breast, mechanoreceptors in the nipple and areola are stimulated, and they send nerve signals along sensory neurons to the hypothalamus. The hypothalamus is stimulated to produce oxytocin, and the posterior pituitary is stimulated to release oxytocin into the blood (see section 17.7b). The oxytocin targets special cells in the mammary glands called *myoepithelial cells,* which surround the mammary acini. Specifically, oxytocin stimulates the myoepithelial cells to contract, thereby releasing breast milk from the mammary acini. As milk is released from the breast, the infant may continue to nurse. Milk will continue to be released from the breast as long as the infant continues to nurse. Once the infant stops suckling (and the nipple and areola are no longer being mechanically stimulated), oxytocin levels drop and milk letdown stops.

The milk letdown reflex may be initiated in some mothers simply in response to hearing a baby cry. Note that since a baby usually cries when it is hungry, the mother's milk letdown response may have adapted to prepare for the eventual feeding of this baby.

Note in figure 29.16*a* that as the infant breastfeeds, prolactin-inhibiting hormone (dopamine) release is inhibited by the hypothalamus. This inhibition of prolactin-inhibiting hormone (dopamine) secretion stimulates the anterior pituitary to secrete large amounts of prolactin. Figure 29.16*b* illustrates that spikes, or peaks, in prolactin production occur each time a baby breastfeeds. This prolactin promotes new breast milk production, so a new supply of breast milk will be available to the baby at the next feeding.

Most women who regularly breastfeed (i.e., more than four to five feedings a day) often do not ovulate at this time. Researchers believe that either the hormones or the sensory stimuli that initiate nerve signals to the hypothalamus involved with breastfeeding inhibit release of gonadotropin-releasing hormone (GnRH) from the hypothalamus. If GnRH is not released, then FSH and LH will not be released by the anterior pituitary, and so ovulation is prevented. This mechanism may have evolved as a way of spacing births, so that a nursing mother's body will not be additionally burdened by a new pregnancy. However, breastfeeding is not a reliable form of birth control because ovulation can and does occur in breastfeeding mothers, especially those who may have reduced numbers of breastfeedings, and it is difficult to determine that ovulation has occurred. Thus, most lactating women are encouraged to use some form of birth control if they do not want to become pregnant again at that time.

 **WHAT DID YOU LEARN?**

 How does the positive feedback mechanism in breastfeeding ensure that the infant receives breast milk?

29.8d Uterine Changes

 **LEARNING OBJECTIVE**

38. Explain the mechanisms by which the uterus returns close to its pre-pregnancy size.

The uterus increases in size due to both hyperplasia and hypertrophy during the 9 months of pregnancy. It takes 6 weeks following birth for the uterus to shrink to close to its pre-pregnancy size. Oxytocin facilitates this shrinkage by stimulating uterine contractions. These contractions tend to be most severe the first week after giving birth and are referred to as **afterpains.** Afterpains may be most severe and noticeable when a woman breastfeeds, since oxytocin is involved in milk ejection. After the first week, these contractions will continue but typically will be less noticeable.

The spike in oxytocin that occurs each time a woman breastfeeds not only expels the milk but also stimulates uterine contractions. Thus, regular breastfeeding facilitates rapid, efficient shrinkage of the uterus. Although these contractions occur in women who don't breastfeed, they may not be as frequent or as efficient, and the uterus may not return to its pre-pregnancy size as quickly.

The changes that occur in the mother to both accommodate a pregnancy and return the body back to normal are quite remarkable. Refer to **figure 29.17** to review these processes.

 **WHAT DID YOU LEARN?**

 How are prolactin and oxytocin related to both the uterine changes and lactation?

29.9 Heredity

Heredity is the transmission of genetic characteristics from parent to child. **Genetics** is the field of biology that studies heredity and its transmission patterns. Here we provide a basic overview of human genetics and introduce you to some of the concepts and vocabulary necessary to understand patterns of inheritance.

29.9a Overview of Human Genetics

39. Become familiar with common genetic terminology.

We previously discussed the mechanisms of somatic and sex cell division in sections 4.9 and 28.2, respectively. A human somatic cell has 23 pairs of chromosomes. The display of the chromosome pairs, which are ordered and arranged by size and similar features (e.g., centromere location, banding patterns), is called the **karyotype** (kar′ē ō tīp; *karyon* = nucleus) (**figure 29.18**). The paired chromosomes are called **homologous chromosomes** (or *homologues;* see section 28.2a), as they contain genes for equivalent biological characteristics, such as eye color. Twenty-two of these pairs of chromosomes are called **autosomes;** they have no genes that determine the sex of the individual. The last two chromosomes are called the **sex chromosomes,** which contain genes that specify the sex of an individual. (In mammals, two X chromosomes occur in cells of females, and an X and a Y are found in cells of males.)

Genes were described in section 4.7b as discrete units of DNA that provide the instructions for the production of specific proteins. Genes are located on chromosomes; the specific place where each gene is located on a chromosome is called its **locus** (lō′kūs; pl. = *loci*). Variants of one gene found at the same locus on homologous chromosomes are called **alleles** (ă-lēl′; *allelon* = reciprocally). For simple patterns of inheritance, you inherit one allele from each of your parents for a combination of two alleles that determines the expression of a particular trait. For example, the two alleles at a locus may determine whether you have the A or O blood type, or whether your hair is straight or wavy.

How do two alleles code for one trait? In the most straightforward situation, one allele is **dominant,** and the other allele is **recessive.** The dominant allele **expresses,** or physically shows, the trait (e.g., blood type), whereas the recessive trait is masked. The recessive allele is expressed only if it is present on both homologous chromosomes.

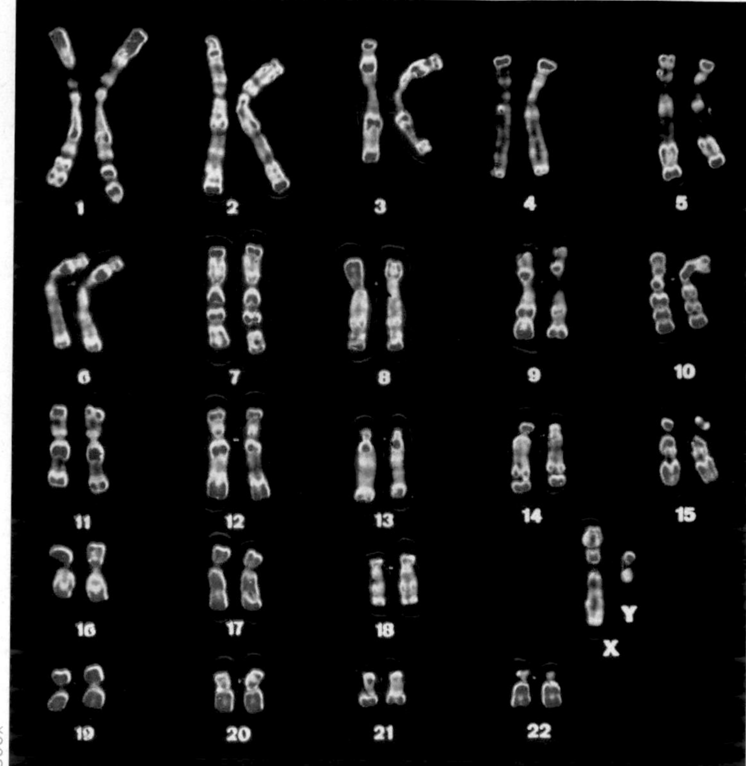

Figure 29.18 Karyotype. A karyotype is a conventional representation of pairs of homologous chromosomes as they appear in metaphase of the cell cycle and arranged by size and other structural features.
©CNRI/Science Source

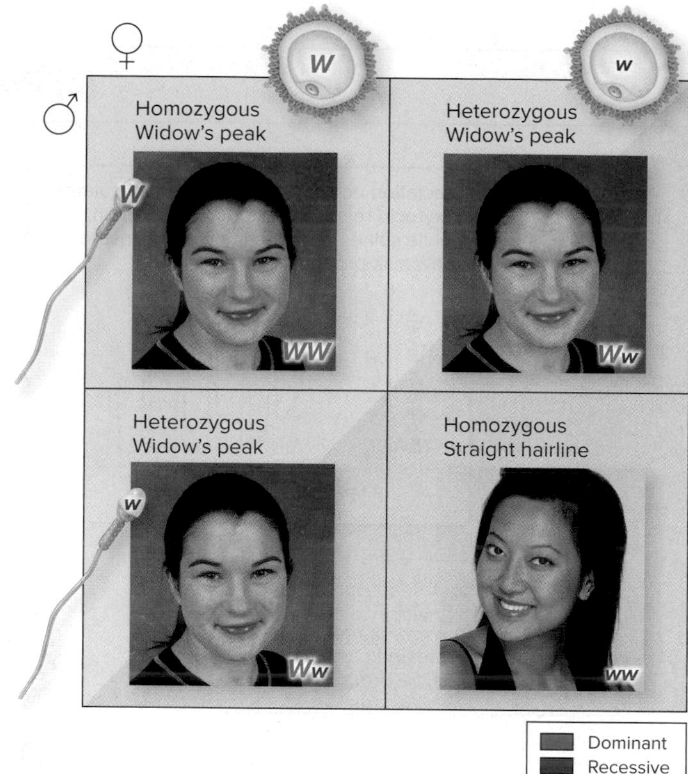

Figure 29.19 Dominant Versus Recessive Alleles. A widow's peak is a dominant trait, and the allele for this trait is designated as *W*. A straight hairline allele is designated as *w*. The Punnett square shown here illustrates the probabilities that two parents who are heterozygous for the widow's peak trait will have offspring that also have a widow's peak. There is a 1 in 4 chance that a child born to these parents will have a straight hairline, and a 3 in 4 chance that a child will have a widow's peak.
(*Widow's peak*) ©Aaron Haupt/Science Source; (*Straight hairline*) ©Brownstock/Alamy RF

Conventionally, a dominant allele is expressed by a capital letter, and the recessive allele is represented by a lowercase letter. For example, having a *widow's peak,* in which the hairline on the forehead comes to a point, or V, in the center, is a dominant trait. So the allele for a widow's peak would be shown as *W*. The recessive allele for a strait forehead hairline would be represented by a *w*. **Figure 29.19** illustrates how these two traits would be expressed. The box used to sort out the inheritance pattern of alleles is called a **Punnett square,** and it shows the specific gene combinations resulting from two parents. The Punnett square can also give the probability that a particular gene combination can occur.

If identical alleles are present, the individual is said to be **homozygous** (hō-mō-zī′gŭs; *homos* = same, *zygotes* = yoke) for the trait. So in this example, *WW* (widow's peak) and *ww* (straight hairline) are both homozygous, because the alleles are identical. Individuals who are **heterozygous** for a trait have both a dominant allele and a recessive allele (e.g., *Ww*), but only the dominant allele is expressed. Expression of the recessive allele may appear to skip generations, because its phenotype is masked by the dominant allele.

In a mating between heterozygous individuals, or heterozygotes, there is a 1 in 4 chance (or 25%) the child would be homozygous for a widow's peak, a 2 in 4 chance (50%) of being heterozygous for a widow's peak, and a 1 in 4 chance (25%) the child would have a straight hairline. If we simply look at how the children may appear, then there is a 3 in 4 chance (75%) that a child would have a widow's peak (includes both *WW* and *Ww*), and a 1 in 4 chance (25%) that a child would have a straight hairline (*ww*).

The genetic makeup of an individual is called the **genotype** (jen′ō-tīp; *genos* = birth). So in this example, individuals may have the genotype *WW, Ww,* or *ww*. The physical expression of the genotype, however, is called the **phenotype** (fē′nō-tīp; *phaino* = to appear, show forth). The phenotype for a *WW* and a *Ww* genotype is to exhibit a widow's peak, whereas the phenotype for a *ww* genotype is a straight hairline. Thus, genotype and phenotype describe the difference between the genetic constitution and the actual observed traits of an individual.

WHAT DID YOU LEARN?

38 What would be the results in terms of the genotype and phenotype if one parent were heterozygous for the widow's peak and the other parent were homozygous for a straight hairline?

29.9b Patterns of Inheritance

LEARNING OBJECTIVE

40. Compare and contrast the types of inheritance patterns.

Not all types of inheritance follow the relatively simple pattern just described. In fact, inheritance of most traits involves the interaction of multiple genes. Here we describe some of the typical patterns of inheritance.

Strict dominant-recessive inheritance has been illustrated in figure 29.19. This type of inheritance is also referred to as *Mendelian inheritance* after Gregor Mendel, who first described this pattern. The dominant allele is always expressed in the phenotype, regardless of whether the individual is homozygous or heterozygous for that trait. Examples of traits that follow this strict dominant-recessive inheritance are found in **table 29.4**. Relatively few human traits follow this pattern of inheritance, but instead may involve the interaction of

Table 29.4	Traits That Follow a Strict Dominant-Recessive Inheritance
Dominant Trait	**Recessive Trait**
Dimples in cheeks	Absence of dimples
Cleft chin	Uncleft chin
Widow's peak	Straight hairline
Normal skin and hair pigmentation	Albinism (absence of skin and hair pigmentation)
Huntington disease (leads to progressive neurodegeneration and death)	Absence of Huntington disease
Astigmatism	Nondistorted vision
Achondroplasia (a type of dwarfism)	Normal endochondral ossification of bones
Rh^+ blood type (e.g., O^+)	Rh^- blood type (e.g., O^-)
Free earlobes	Attached earlobes
Polydactyly (extra digits)	Normal number of digits
Extensive flexibility in thumb joints ("hitchhiker's thumb")	Inability to extend distal phalanx of thumb past 180 degrees

multiple genes (and thus multiple alleles). In addition, the expression of many traits may be affected by environmental factors.

For some traits, the phenotype of two heterozygous alleles is intermediate between the phenotypes of homozygous dominant or recessive alleles. This trait is said to exhibit **incomplete dominance.** Thus, three phenotypes are possible: (1) the phenotype produced by homozygous dominant alleles, (2) the phenotype produced by homozygous recessive alleles, and (3) the phenotype produced by heterozygous alleles.

One example of incomplete dominance is the sickle-cell trait, which affects the shape of erythrocytes. Most individuals have two identical alleles (each designated *A*) that code for the normal hemoglobin A in erythrocytes. The sickling allele (designated *s*) produces an abnormal form of hemoglobin, hemoglobin S. This hemoglobin crystallizes when oxygen levels are low and results in the erythrocyte becoming brittle and forming a sickle shape. The sickling allele is recessive, and individuals who are homozygous recessive exhibit **sickle-cell disease,** also known as *sickle-cell anemia* (see Clinical View 18.2: "Anemia"). Erythrocytes in these individuals are inefficient; they may fragment in blood vessels and obstruct them, and they don't have as long a life span as normal erythrocytes. Individuals with sickle-cell disease may experience greater morbidity and mortality than individuals who code for normal hemoglobin A.

Heterozygous individuals are said to carry the **sickle-cell trait,** whereby under very low oxygen conditions, some (but not all) of their erythrocytes may develop this sickle shape. Researchers have found that individuals with the sickle-cell trait experience less mortality from malaria than do individuals without the trait, so the sickle-cell trait is somewhat adaptive for individuals who live in malaria-prone environments.

In some cases, two or more alleles appear to be equally dominant, and the trait is said to exhibit **codominant inheritance.** Both alleles are expressed in the phenotype. An example of codominance is the ABO blood group. Blood types A and B are codominant, so if an individual receives an *A* allele from one parent and a *B* allele from the other parent, the individual will have an AB blood type. A third allele, *i,* is recessive; the combination *ii* results in the O blood type (see section 18.3b).

Polygenic inheritance, also known as *multiple gene inheritance,* occurs when multiple genes interact to produce a phenotypic trait. These multiple genes may be on the same chromosome or on different chromosomes. Most human traits are the result of polygenic inheritance. For example, eye color was originally believed to be a strict dominant-recessive inheritance involving a single gene, but we now know that eye color is the result of interactions of many genes. Other traits such as height, skin color, and predisposition to certain diseases, as well as many others, are traits under polygenic influence.

INTEGRATE

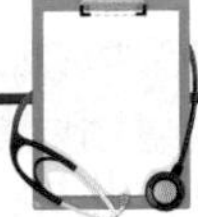

CLINICAL VIEW 29.9

Genetics of Familial Hypercholesteremia

Another example of incomplete dominance in humans is the condition **familial hypercholesterolemia (FH).** In this clinical disorder, a person who has two abnormal alleles lacks receptors for low-density lipoprotein (LDL) and cannot transport LDL from the blood; thus, the individual has very high serum cholesterol and often develops atherosclerosis and heart disease early in life (see Clinical View 20.1: "Atherosclerosis"). These individuals are at greater risk of heart attack at a young age. In contrast, the heterozygote for this condition has about half the number of receptors for LDL. This individual exhibits some LDL transport-related problems related to slightly elevated cholesterol blood levels, but these levels may be managed with diet, exercise, and statin medications.

WHAT DID YOU LEARN?

39 How does codominant inheritance differ from incomplete dominance?

29.9c Sex-Linked Inheritance

LEARNING OBJECTIVE

41. Describe sex-linked inheritance, and give a clinical example of this type of inheritance.

Sex-linked traits are traits expressed by genes on the X or Y chromosomes. You can see from figure 29.18 that the X and Y chromosomes are physically different—and these differences are also represented in the number of genes each carries. The National Institutes of Health estimates there are between 900 and 1400 genes on the X chromosome, whereas the Y chromosome contains somewhere between 70 and 200 genes. Further, since only males have a Y chromosome, most of the Y chromosome genes code for male-specific development. In contrast, most of the genes on the X chromosome are not involved in sex determination. Thus, although either sex chromosome may be responsible for a sex-linked trait, it is far more common for the X chromosome to be involved.

X-linked traits may be recessive or dominant. An **X-linked recessive trait** is always expressed in a male, because he has only one X chromosome. In contrast, an X-linked recessive trait is expressed in a female *only* if she has two recessive alleles for that trait; typically, the probability of this occurring is very low. A woman is said to be a **carrier** if she has one X-linked recessive allele only and does not

9. Skin color is a trait that is determined by
   a. strict dominant-recessive inheritance.
   b. incomplete dominant inheritance.
   c. codominant inheritance.
   d. polygenic inheritance.

10. A woman is a carrier for the color-blindness gene, found on the X chromosome. She does not have color blindness, however. What are the chances that her female child will also inherit this gene?
   a. no chance (0%)
   b. 1 in 4 chance (25%)
   c. 2 in 4 chance (50%)
   d. 100% chance

11. Briefly describe the process of fertilization, mentioning a hallmark event that occurs in each phase.

12. List the five regions of the mesoderm, and identify some major body parts derived from each region.

13. Explain why teratogens are especially harmful to the developing organism during the embryonic period. What events occur during this period?

14. Describe the differences between the embryonic period and the fetal period.

15. List the major hormone levels that change during pregnancy and their major functions during pregnancy.

16. List the three stages of labor, and describe a major event that occurs in each stage.

17. Explain the positive feedback mechanism involved with lactation and milk letdown, and describe the function of the hormones prolactin and oxytocin during this process.

18. Describe the various ways by which the mother's body expels the excess fluids gained during pregnancy.

19. Compare and contrast strict dominant-recessive inheritance, incomplete dominance, codominance, and polygenic inheritance, and give an example of each.

20. Explain the difference between X-linked recessive traits and X-linked dominant traits. What are the possibilities that a female child will be affected and express each of the traits?

Can You Apply What You've Learned?

Use the following paragraph to answer questions 1–5.

Ashley is a 29-year-old pregnant woman who is in the second trimester of her pregnancy. At her regular OB/GYN checkup, the doctor records the following about Ashley:

Heart rate: 92

Urinalysis: Negative for bacteria and protein in urine

Weight gain since beginning of pregnancy: 15 pounds

Blood pressure: 115/79 mm Hg

Uterine fundus height: At a midpoint between the umbilicus and the xiphoid process of the sternum

Blood glucose level: 95

1. Which of the following medical assessments is abnormal and would be of concern for the physician?
   a. Blood pressure is above normal.
   b. Urinalysis results are abnormal.
   c. Ashley's blood glucose levels are above normal.
   d. All of the medical findings are within normal ranges for a pregnant woman.

2. The physician determines that Ashley is in her twentieth week of pregnancy. Which of the following best describes the development of the fetus at this time (refer to table 29.3)?
   a. The crown–rump length is 19 centimeters.
   b. The heart and lungs have fully developed, but the GI tract has yet to form.
   c. The testes have descended into the scrotum.
   d. The average weight of the fetus at this time is about 2.5 kilograms.

3. Months pass as Ashley has an uneventful pregnancy. Eventually, Ashley goes into labor and arrives at the hospital to give birth. As Ashley progresses through the first stage of labor, all of the following occur *except*
   a. the fetus's head pushes against the cervix, promoting the cervical dilation.
   b. oxytocin levels decrease as labor progresses.
   c. the amniotic sac may rupture.
   d. contractions become more frequent and more intense.

4. Ashley gives birth to a 7.5-pound baby boy. In the first few weeks after giving birth, Ashley notices she develops cramping in the pelvic region whenever she breastfeeds her child. What is the likely cause of the cramping?
   a. The suckling stimulates oxytocin production, which both releases the milk and causes uterine contractions.
   b. Breastfeeding results in a rise in prolactin, which also causes cramping of the abdominal muscles.
   c. The drop in estrogen and progesterone levels causes the pelvic cramping.
   d. The tugging on the breast (caused by the suckling) pulls on the abdominal muscles, causing the cramping.

5. As Ashley cares for her son, she notices some problems in his development. During one of the baby's checkups, Ashley voices her concerns to the pediatrician. The pediatrician performs a series of assessments on the baby and orders a genetic workup. The genetic test results demonstrate that the baby has Duchenne muscular dystrophy, which is an X-linked recessive trait.

   Based on your knowledge of heredity, if the baby boy has Duchenne muscular dystrophy, then what else must be true?
   a. The baby's father is a carrier of the Duchenne muscular dystrophy allele and passed along this trait to the baby.
   b. Duchenne muscular dystrophy is an autosomal-dominant disorder.
   c. Ashley is a carrier of the Duchenne muscular dystrophy allele but does not display symptoms because she has two X chromosomes, one of which is normal.
   d. If Ashley had a baby girl (instead of a baby boy), the baby girl would be at equal risk for having Duchenne muscular dystrophy.

Can You Synthesize What You've Learned?

1. In the late 1960s, a number of pregnant women in Europe and Canada were prescribed a drug called thalidomide. Many of these women gave birth to children with amelia (no limbs) or meromelia (malformed upper and/or lower limbs). It was later discovered that thalidomide is a teratogen that can cause limb defects in an unborn baby. Based on this information, during what period of their pregnancy (pre-embryonic, embryonic, or fetal) do you think these women took thalidomide? During which of these periods would thalidomide cause the most harm to limb development?

2. A 22-year-old woman consumes large quantities of alcohol at a party and loses consciousness. Three weeks later, she misses her second consecutive period, and a pregnancy test is positive. Should she be concerned about the effects of her binge-drinking episode on her baby?

3. Simon and Carla are siblings who recently discovered that their father was diagnosed with Huntington disease. This disease is autosomal dominant and results in gradual neurologic degeneration and death, usually within 5 years of diagnosis. There is no cure for the disease. Their mother does not have the disease. Both Simon and Carla discuss the possibility of getting tested to see if they will develop Huntington disease as well. Based on your knowledge of heredity and genetics, what is the likelihood that Simon or Carla will develop Huntington disease?

INTEGRATE

ONLINE STUDY TOOLS

connect | SMARTBOOK® | AP|R

The following study aids may be accessed through Connect.

Clinical Case Study: Queen Victoria's Family Tree

Interactive Questions: This chapter's content is served up in a number of multimedia question formats for student study

SmartBook: Topics and terminology include overview of prenatal development; pre-embryonic period; embryonic period; fetal period; effects of pregnancy on the mother; labor (parturition) and delivery; postnatal changes for the newborn; changes to the mother after delivery; heredity

Anatomy & Physiology Revealed: Ovulation through implantation

Animations: Fetal development and risk

Design Elements: **Learning Objective** icon: ©McGraw-Hill Education; **What Do You Think?** icon: ©McGraw-Hill Education; **What Did You Learn?** icon: ©McGraw-Hill Education; **Learning Strategy** icon: ©Slavoljub Pantelic/Shutterstock RF; **Clinical View** icon: ©Laia Design Studio/Shutterstock RF; **Concept Connection** icon: ©McGraw-Hill Education; **Challenge Yourself** icon: ©McGraw-Hill Education; **Do You Know the Basics?** icon: ©McGraw-Hill Education; **Can You Apply What You've Learned?** icon: ©McGraw-Hill Education; **Can You Synthesize What You've Learned?** icon: ©McGraw-Hill Education; **Online Study Tools** icon: ©ayzek/Shutterstock RF

CHAPTER 11

1. Since the omohyoid attaches to the scapula (which is part of the shoulder), the prefix *omo-* means "shoulder."
2. When you breathe deeply, the diaphragm contracts and pushes inferiorly on the GI tract (abdominal viscera). If these viscera are bulging with food, the diaphragm has less room to contract fully, making it hard to take deep breaths.
3. The brachialis is an anterior arm muscle. Because anterior arm muscles tend to flex the elbow joint, we can surmise that the brachialis flexes the elbow joint.
4. Remember that there are leg muscles that also move the toes. In this case, the extensor digitorum longus (a leg muscle) attaches to toes 2–5 and helps move them all.

CHAPTER 12

1. The generation of IPSPs, which are produced in receptive segments either by the loss of K+ or the gain of Cl–, make it less likely that a nerve signal will be sent because the resting membrane potential is made relatively more negative inside the neuron, which changes the membrane potential so that it is further away from the threshold (the value that must be reached for a nerve signal to be initiated).
2. A drug that blocks acetylcholinesterase (an enzyme that breaks down acetylcholine molecules in the synaptic cleft) increases the amount of time acetylcholine molecules remain in the synaptic cleft. This allows an increase in the duration of stimulation of neurons of the brain. Given that acetylcholine increases attention (or mental alertness), a drug that increases the half-life of acetylcholine in the synaptic cleft would be predicted to increase attention.

CHAPTER 13

1. Although all the meningeal layers give some support and protection to the brain, the dura mater provides the most support. This layer is the thickest and most durable, and it also forms the cranial dural septa that support the brain components.
2. Cutting the corpus callosum dramatically reduces communication between the right and left cerebral hemispheres, but some communication is maintained between the hemispheres via the much smaller anterior and posterior commissures.
3. If the primary somatosensory cortex was damaged, general somatic sensory information would not be perceived. An individual likely would experience anesthesia (numbness) in the region that transmits sensations to the damaged cortex. In contrast, if the somatosensory association area was damaged, a person could still detect sensations, but would be unable to tell the *difference* in those sensations. A person may perceive that an object is in one's hand, but would be unable to tell if that object was smooth or rough, cold or hot, round or square.
4. If there were no thalamus, the cerebrum would still receive sensory stimuli, but the information would not be decoded first. So, the cerebrum would not be able to distinguish taste information from touch or vision information. In addition, there would be no filtering of sensory information so the person would experience sensory overload.
5. Severe injury to the medulla oblongata would most likely cause death, because the medulla oblongata is responsible for basic reflex and life functions, including breathing and heartbeat.

CHAPTER 14

1. The two layers of the cranial dura mater split to form the dural venous sinuses, which are large veins that drain blood away from the brain. The spinal cord lacks dural venous sinuses, so it does not need two layers of dura.
2. An anterior ramus is larger than a posterior ramus because the posterior rami only innervate deep back muscles and the skin of the back, whereas the anterior rami innervate almost all other body structures (e.g., the limbs and the anterior and lateral trunk).
3. A nerve plexus houses axons from several different spinal nerves. Thus, damage to a single segment of the spinal cord or damage to a single spinal nerve generally does not result in complete loss of innervation to a particular muscle or region of skin.

CHAPTER 15

1. The pterygopalatine ganglion is nicknamed the "hay fever ganglion" because when it is overstimulated, it causes some of the classic allergic reactions, including watery eyes, runny and itchy nose, sneezing, and scratchy throat.
2. Sympathetic innervation causes vasoconstriction of most blood vessels. When blood vessels are constricted, it takes more force and pressure to pump blood through the vessels, so blood pressure rises.
3. Nicotine binds to nictotinic receptors. Many neurons in the brain and peripheral nervous system are affected, but the overall effect usually is an increase in blood pressure and heart rate, and vasoconstriction (in other words, a general stimulation of the sympathetic division). Nicotine also works on the reward centers of the brain, stimulating a release of the neurotransmitter dopamine, which is responsible for the pleasurable, relaxed sensations most smokers get while lighting up. These effects are short lived, so a smoker must smoke fairly continuously to continue to get the increased dopamine in his or her system.

CHAPTER 16

1. Olfaction plays a major role in detecting tastes. Most taste is due to our perception of the odor, rather than the taste, of the food. If your nose is stuffed up and you can't smell the food, your taste perception is impaired as well.
2. A deer sacrifices depth perception for a wider field of vision, thus it has a greater visual range.
3. Air pressure increases when the airplane descends, which is why you may feel greater pressure outside your ears. The "popping" noise results from your auditory tubes opening and equalizing the pressure on either side of the tympanic membrane.

CHAPTER 17

1. These examples of prostaglandin activity are given in the text: (1) stimulating the hypothalamus to raise body temperature to induce a fever, (2) inhibiting stomach acid secretion, (3) acting on mast cells to release molecules that increase inflammation, and (4) stimulating pain receptors.

   An individual might take aspirin to decrease a fever, to decrease inflammation, or to decrease painful sensations. Increased stomach acid secretions would be a side effect.

2. Hormones typically are eliminated from the blood by either the liver or the kidney. Thus, impaired liver or kidney function could potentially result in increased blood levels of hormones.

3. If a drug binds to a specific hormone receptor of a cell to initiate its affect, the cell will down regulate receptors. Over time, more of the drug will be required for the same response.

4. Drinking alcohol inhibits the release of ADH (a hormone that decreases urine formation). One side effect of drinking alcohol is an increase in urine formation.

5. Increased lipolysis is caused by growth hormone. This results in loss of stored adipose connective tissue and the "thinning out" experienced by many during puberty.

6. Iodine is absorbed into the blood and ultimately used for thyroid hormone synthesis.

7. The affected individual with hyperthyroidism would have a high temperature, elevated pulse, elevated breathing rate, and a thin body.

8. Elevated insulin levels typically results in decreased levels of all nutrients in the blood and an increase in the storage form of these molecules within the cells of the body. Thus, elevated serum insulin leads to protein anabolism in the muscle cells, resulting in increased muscle bulk. An insulin overdose may lead to dangerously low serum glucose levels and ATP levels. If ATP cannot be generated at the rate needed, death could result. (It also can result in a potassium imbalance, as described in chapter 25.)

CHAPTER 18

1. A woman who has 5 L of blood would donate approximately 10% of her blood. Low-weight individuals are not allowed to donate blood because these individuals have relatively less blood volume, and by donating 1 pint of blood they would be donating much more than 10% of their total blood volume.

2. By losing its nucleus and other organelles the erythrocyte has more space for hemoglobin, enabling the erythrocyte to bind to and transport greater amounts of O_2 and CO_2 more efficiently. Because the erythrocyte has lost its nucleus it will no longer be able to synthesize proteins or divide via mitosis. The oxygen being transported in an erythrocyte will not be used up by aerobic cellular respiration (since erythrocytes don't have mitochondria).

3. A person with type O^- blood is considered a "universal donor" because his or her erythrocytes have no surface antigens. Without surface antigens, the type O^- erythrocytes will not be destroyed through agglutination and hemolysis by antibodies in the recipient's plasma. Likewise, a person with type AB^+ blood is considered a "universal recipient" because his or her blood plasma has no antibodies to the ABO or Rh blood types. Thus, the AB^+ recipient may receive any type of blood and not worry about the donor's erythrocytes being destroyed.

4. The formation of the platelet plug is an example of positive feedback. The platelets release chemicals that stimulate more platelets to arrive at the site of injury and release their chemicals.

5. An individual who is confined to a wheelchair may have impaired blood flow, as the skeletal muscle pump is less effective. If blood is not moving through the blood vessels properly and begins to pool in the vessels, the clotting cascade may be initiated.

CHAPTER 19

1. The aorta and pulmonary veins are vessels attached to the heart that contain oxygenated blood. The aorta is an artery and the pulmonary veins are veins.

2. Coronary arteries are compressed more often as heart rate increases.

3. Slow Na^+ voltage-gated channels are unique to nodal cells. These channels open and Na^+ enters the nodal cells. When threshold is reached, Ca^{2+} voltage-gated channels open and Ca^{2+} enters the cells and causes depolarization. Thus, nodal cells are able to depolarize without external stimulation.

4. Ca^{2+} channel blockers decrease the intracellular calcium in a cardiac muscle cell, thus decreasing the contractility of the heart. Ca^{2+} channel blockers may also block calcium movement into conduction cells, decreasing heart rate. (Blood pressure is lowered as a result.)

CHAPTER 20

1. A decrease in blood pressure (blood hydrostatic pressure) is dependent on the pumping action of the heart, resistance that occurs as blood moves through the blood vessels, and the total blood volume. A significant decrease in any of these (e.g., congestive heart failure, severe hemorrhage) can result in lowering blood pressure to such an extent that blood hydrostatic pressure is not sufficient to overcome blood colloid osmotic pressure. When this occurs, filtration (and capillary exchange) ceases.

2. A smoker likely would have elevated blood pressure, since nicotine increases firing rate of the SA node and increased contractility of myocardial cells, which increases cardiac output. Nicotine also causes vasoconstriction of arterioles, which increases resistance. Both an increase in cardiac output and an increase in resistance result in increasing blood pressure.

3. ACE inhibitors are a medication given to treat high blood pressure (hypertension). These medications interfere with the production of angiotensin II (a potent vasoconstrictor), thus helping to prevent an increase in resistance. Angiotensin II also regulates blood volume by influencing urine production. A decrease in angiotensin II would increase fluid loss from the blood and blood volume decreases. By preventing an increase in resistance and decreasing blood volume, ACE inhibitors help to lower blood pressure.

4. If the left ulnar artery were cut, the left hand and fingers could still receive blood via the left radial artery, since both vessels contribute to the superficial and deep palmar arches.

CHAPTER 21

1. The removal of lymph nodes interferes with the drainage of fluid from specific regions of the body and will result in lymphedema.

2. The spleen, located in the left upper quadrant, is a soft organ that is not protected by bone. It could easily be ruptured by the pressure of the seatbelt pushing against the abdomen in an automobile accident.

CHAPTER 22

1. Inflammation can be observed if the area is red and swollen. It also may feel warm to the touch.
2. If an antigen mutates, it is likely that it will not be recognized by the same lymphocytes. Each lymphocyte recognizes a specific molecular shape of an antigen. (An example of a virus that mutates often is HIV, the virus that causes AIDS. It is for this reason that vaccines have not been produced against HIV.)
3. Helper T-lymphocytes play an essential role in both the innate and adaptive immune responses. These cells function to activate B cells, cytotoxic T-lymphocytes, as well as macrophages and NK cells. Without the proper functioning of all of these cells, an individual is highly susceptible to infection.

CHAPTER 23

1. A deviated nasal septum occurs when the nasal septum—the bone and cartilage that divide the nasal cavity of the nose in half—is off-center so one side of the nasal cavity is larger than the other. This alters the normal flow of air through the nose, and if the narrower side becomes blocked, nasal congestion and sinus problems may result.
2. The epithelium changes because a stratified squamous epithelium is more sturdy and protective against smoke than a pseudostratified, ciliated columnar epithelium. Unfortunately, since a stratified squamous epithelium lacks both goblet cells and cilia, there is less mucus produced, and no cilia are present to propel particles away from the bronchi towards the pharynx. Thus, the main way to eliminate the particles is by coughing, leading to the chronic "smoker's cough."
3. The phrenic nerves extend from each cervical plexus to the diaphragm. Since the cervical plexus includes axons of C3–C5, spinal cord injuries at or above C2 prevent nerve stimulation of the diaphragm, and breathing stops. Spinal cord injuries between C6 and T12 result in nerve signals being sent along the phrenic nerves but not along all intercostal nerves. The degree of loss will depend on where the level of injury has occurred. The further down the spinal cord, the closer the breathing will be to normal (since additional intercostal nerves will be functioning). Spinal cord injuries below T12 should not normally interfere with normal quiet breathing, because these represent injuries that do not interfere with nerve signals being relayed along the phrenic or intercostal nerves.
4. Epinephrine causes bronchodilation; consequently resistance is decreased, with a subsequent increase in airflow.
5. No, not all of the air that we breathe in is available for gas exchange. Only the air that reaches the alveoli is available for gas exchange. There is no gas exchange with air in the anatomic dead space (air that remains in the conducting zone).
6. Blood PO_2 is lower in systemic capillaries than when it left pulmonary capillaries, because bronchial veins dump small amounts of deoxygenated blood into the pulmonary veins prior to the blood being returned to the heart and subsequently pumped by the left ventricle through the systemic circuit.
7. Since hemoglobin is 98% saturated with oxygen if $Po_2 = 104$ mm Hg (normal value at sea level), additional administered oxygen can only increase the % saturation of hemoglobin 2%. This can occur only with an increase from 1 atm to 3 atm, pressures that are only generally reached in hyperbaric oxygen chambers. (In addition, the limiting factor is usually not the ability to load oxygen in the lungs but the ability to deliver the oxygen to the body's tissues.) Therefore, it would not be expected that the athletic performance would be improved.
8. During exercise, the Po_2 decreases as cells engage in elevated levels of cellular respiration. The decrease in Po_2 results in more oxygen delivered to the cells. As a result the oxygen reserve (the amount remaining attached to hemoglobin) is lower.
9. Exercise increases venous return of blood and lymph by the action of both the muscular pump and respiratory pump. Subsequently, more blood enters the atria triggering the atrial (Bainbridge) reflex that causes increased sympathetic output from the cardiac center to the SA node, thus increasing heart rate.

CHAPTER 24

1. A loss of kidney function could: (a) impair the ability of the kidneys to filter and eliminate wastes; (b) induce anemia as erythropoietin production may be altered; (c) decrease urine production with an accompanying increase in blood volume and blood pressure; (d) lead to a disruption of the normal pH balance in the blood as the kidneys play a role in maintaining acid-base (pH) balance.
2. A substance that is not filtered remains in the blood and exits the glomerulus by the efferent arteriole.
3. Cirrhosis of the liver may lead to a drop of blood colloid osmotic pressure (OP_g). OP_g opposes the filtration pressure and draws fluid back into the glomerulus. A decrease in OP_g will induce less reabsorption of fluid and therefore more filtrate will be formed.
4. Glycosuria is the excretion of glucose in the urine and it is a classical sign of diabetes mellitus. Glucose molecules are not reabsorbed and they act as an osmotic diuretic, pulling water into the tubular fluid and causing increased loss of fluid in the urine.
5. An individual with renal disease that results in damage to the filtration membrane will have a lower concentration of plasma proteins because tubule cellular structures involved in protein uptake will be saturated and unable to reclaim all proteins that have been filtered. If renal filtration is lower than normal, there is a decrease in filtration of small plasma proteins resulting in an increase in plasma proteins.
6. If a substance in the blood is filtered and not all of it is completely reabsorbed back into the blood, the level of the substance in the blood decreases. One of the critical functions of the kidney is to regulate blood levels of specific substances.
7. When we are lying down, gravity is unable to passively transport urine to the urinary bladder. Thus, peristalsis is needed so that urine can be moved actively from the ureters to the urinary bladder, no matter the position of the body.

CHAPTER 25

1. A person would experience thirst with (c), an increased blood osmolarity, which indicates a loss of fluid and a relative increase in blood concentration.
2. If individuals have high blood pressure, their physicians would want to recommend the following to reduce their blood pressure. Reduction of sodium intake would decrease Na^+ concentration in blood plasma in order to reestablish normal Na^+ concentration in the blood. The normal Na^+ concentration is reestablished and plasma blood volume would be reduced, thus decreasing blood pressure.
3. Since insulin increases K^+ uptake by cells, insulin injection could result in low blood K^+ (hypokalemia). Severe hypokalemia could be life threatening (fatal), as potassium is required for cells, especially nerve and muscle cells, to function normally.

CHAPTER 26

1. No. The peritoneal cavity is the thin potential space between the parietal peritoneum and visceral peritoneum that contains serous fluid. Abdominal organs are located in either the intraperitoneal or the retroperitoneal spaces.
2. Decreased saliva production can cause bad breath and dental problems, such as cavities, because saliva helps to both cleanse the oral cavity structures and inhibit bacterial growth in the oral cavity (saliva contains antibacterial substances, including lysozyme and IgA antibodies).
3. The stomach is lined with specialized surface mucous cells that continuously secrete an alkaline product containing mucin onto the gastric surface. The mucin hydrates to form a 1–3 millimeter mucus layer that prevents acid from coming in contact with the stomach wall. In addition, the stomach epithelial lining is constantly regenerating to replace any cells that are damaged.
4. The duodenum has many circular folds so that movement of digested materials can be slowed down and may be adequately mixed with pancreatic juice and bile. By the time chyme reaches the ileum, most nutrient absorption has occurred and the circular folds are not needed to slow digested materials' movement.
5. When the gallbladder is removed, bile is still produced by the liver. However, if the gallbladder has been removed, the more concentrated bile is no longer available for facilitating the mechanical digestion of a fatty meal. Therefore, patients who have their gallbladder removed are instructed to limit their fat intake.

CHAPTER 27

1. Fat-soluble vitamins include vitamin A, vitamin D, vitamin E, and vitamin K. Fat-soluble vitamins are stored in body fat. Excessive consumption of fat-soluble vitamins may induce toxicity (hypervitaminosis).
2. A lipid profile that includes both HDL and LDL levels is more informative than a total cholesterol count. HDLs are considered the "good cholesterol" because they function to transport lipids from the peripheral tissue to the liver. LDLs function to transport cholesterol to the cells for storage or use within the cell. LDL is considered "bad" cholesterol as some of this lipid tends to be deposited on the inner walls of arteries, increasing risk for heart disease and high blood pressure. The ratio provides the information regarding the relative amounts of HDLs and LDLs and is a better measure of cardiovascular risk.
3. A man who is 6 feet tall and 200 pounds would have a higher BMR than a man who is 5 feet 9 inches tall weighing 160 pounds. BMR is the amount of energy that an individual uses when their body is at rest. One of the most important variables in determining BMR is body surface area. A larger individual has a greater surface area of skin and thus loses more energy as heat. When more heat is lost, the body must increase its metabolic rate to maintain body temperature.

CHAPTER 28

1. A woman can become pregnant as long as she has one remaining functioning ovary. However, she may not ovulate every month, since the ovaries typically "take turns" ovulating each month.
2. Stress, age, medications, and body weight all can affect a woman's monthly uterine (menstrual) cycle. Stress and excessively lean body mass can lead to amenorrhea (absence of periods).
3. If a man's testes were removed, the adrenal glands could still produce a small amount of androgens. However, because the testes produce the overwhelming majority of androgens, the small amount produced by the adrenal glands would have little effect.
4. If a man has a vasectomy, sperm still form in the seminiferous tubule and then mature in the epididymis. However, since the sperm are not ejaculated, they die, and their components are broken down and resorbed in the epididymis. An individual who has had a vasectomy ejaculates seminal fluid only, not semen (which contains sperm).

CHAPTER 29

1. If two sperm penetrate the secondary oocyte, the condition called polyspermy occurs. In this case, the fertilized cell contains 23 triplets of chromosomes, instead of the normal 23 pairs of chromosomes. In humans, a cell with 69 chromosomes (3 sets of 23 chromosomes) will not survive.
2. In the last trimester, the uterus continues to expand and compress abdominal organs. The uterus also prevents the diaphragm from moving too far inferiorly, such as is needed when one takes a deep breath. The uterus and fetus are physical barriers that prevent full lung expansion, and thus deep breathing, for some women.
3. It is unnecessary to slap newborns to "make" them breathe, because the breathing reflex generally occurs in about 10 seconds in response to the change in environment and temperature. It can be harmful because crying may actually make the baby's breathing more difficult, to say nothing of the shock and fear response that could occur.
4. As parents have become more diligent about protecting infants' skin from sun exposure, one potential negative effect is that these infants cannot synthesize vitamin D from the UV rays. Because breast milk is low in vitamin D, and these infants aren't synthesizing enough vitamin D from sunlight, supplementation of this vitamin is recommended.

-oma	tumor	carcinoma (cancerous tumor derived from epithelial cells)
-ose	full of	adipose (fatty: consisting of, resembling, or relating to fat)
-osis	condition, state	osteoporosis (atrophy of bone characterized by decreased bone mass)
osmo-	push	osmosis (movement of water through a semipermeable membrane into a region of higher solute concentration)
osse-, osteo-	bone	osteoblast (bone-forming cell)
-ous	much, full of	nitrogenous (containing nitrogen)
oxy-	oxygen	hypoxia (condition of decreased normal levels of oxygen in inspired gases)
para-	near, beside	paranasal (adjacent to the nose)
-pathy	disease	neuropathy (nerve disease)
pelv-	basin	renal pelvis (funnel-shaped expansion of the upper end of the ureter)
-penia	deficiency	leukopenia (deficiency in the number of leukocytes in the blood)
peri-	around	periosteum (fibrous membrane covering surface of the bone)
phag-	eat	phagocytosis (ingestion and digestion of solid substances by cells)
phil-	have an affinity for	lipophilic (associates with fat)
phobo-	fearing, repelled by	hydrophobic (tending not to dissolve in water)
physio-	nature, natural cause	physiology (the science of the function of living systems)
-physis	growth	diaphysis (the shaft of a long bone)
plasm-	formed	cytoplasm (gel-like substance between cell membrane and the nucleus)
-plegia	paralyze, stroke	paraplegia (paralysis of lower extremities)
pneumo-	air, gas, lungs	pneumothorax (air in the pleural cavity)
-poie, -poiesis	produce, formation of	erythropoietin (hormone that stimulates erythrocyte production)
poly-	many	polycythemia (relative increase in erythrocytes)
pro-	before in time, place	proacrosomal (early stage in the development of the acrosome)
pseudo-	false	pseudostratified (appears to be layered but it is not)
quad-	fourfold	quadriceps femoris (four-headed muscle of anterior side of the thigh)
rami-	branch	ramus (primary division of a nerve or blood vessel)
rect-	straight	rectus abdominis (straight muscle of abdomen)
reno-	kidney	renal (of the kidney)
reti-	network	reticular (fibers that crosslink to form a fine meshwork in soft tissues)
retro-	backward, behind	retroperitoneal (external or posterior to the peritoneum)
sclero-	hard	arteriosclerosis (hardening of the arteries)
semi-	half	semilunar (half-moon shaped)
serrate	saw-edged, toothed	serratus anterior (muscle of thorax)
somato-	body	somatotropin (protein hormone from anterior pituitary: growth hormone)
splanchno-	viscera	splanchnic (a term used to describe visceral organs)
spleno-	spleen	splenius capitis (a broad, straplike muscle in the back of the neck)
stasi, stati-	put, remain	homeostasis (regulate and maintain stable, constant environment)
steno-	narrow	stenosis (narrowing of opening or a canal)
stria-	stripe, band	striated (marked with stripes)
sub-	under	subcutaneous (beneath the skin)
super-, supra-	above, upper	suprarenal (superior to the kidney)
sym-, syn-	together, with	symphysis (growing together)
tachy-	rapid, fast	tachycardia (rapid beating of the heart)
therm-	heat	thermometer (tool to measure temperature)
thorac-	chest	thoracic cavity (body cavity housing the heart and lungs)
-tomy	cut, incise	appendectomy (removal of appendix)
tono-	tension, pressure	isotonic (solutions possessing the same osmotic pressure)
topo-	place, topical	ectopic (an organ that is not in the proper position or place)
trans-	across	transdermal (entering through the skin)
tri-	three	triceps brachii (three-headed muscle)
troph-	food, nourishment	trophoblast (cells providing nutrients to the embryo)
-tropic	having an affinity for	gonadotropic (effecting the gonads)
tunica-	layer, coat	tunica intima (inner part of blood vessel wall)
-ul	small	tubule (very small tube or tubular structure)
uni-	one	unicellular (single cell)
vas-	vessel	vasodilation (widening of lumen of blood vessel)
villo-	hair	microvilli (small hair-like projections of some epithelial cell membranes)
viscer-	internal organ	visceral (of the internal organs)
vivi-	life, alive	in vivo ("within the living body")
zyg-	yoked, paired, union	azygos (unpaired anatomical structure that occurs singly)

glossary

Pronunciation Key

Pronouncing a word correctly is as important as knowing its spelling and its contextual meaning. The mastery of all three allows a student to take ownership of the word. The system employed in this text is basic and consists of the following conventions:

1. Vowels marked with a line above the letter are pronounced as follows:
   ā day, base
   ē be, feet
   ī pie, ivy
   ō so, pole
   ū unit, cute
2. Vowels marked with the breve (˘) are pronounced as follows:
   ă above, about
   ĕ genesis, bet
   ĭ it, sip
   ŏ collide
   ŭ cut, bud
3. Vowels not marked are pronounced as follows:
   a mat
   o not, ought
   e term
4. Other phonetic symbols used include the following:
   ah father
   aw fall
   oo food
   ow cow
   oy void
5. For consonants, the following key was employed.
   b bad
   ch child
   d dog
   dh this
   f fit
   g got
   h hit
   j jive
   k keep
   ks tax
   kw quit
   l learn
   m mice
   n no
   ng ring
   p put
   r right
   s so
   sh shoe
   t tight
   th thin
   v very
   w wet
   y yes
   z zero
   zh measure
6. The principal stressed syllables are followed by a prime (′); single-syllable words do not have a stress mark. Nonstressed syllables are separated by a hyphen.
7. Acceptable alternate pronunciations are given as needed.

A

abduction (ab-dŭk′shŭn) Movement of a body part away from the median plane of the body.

absolute refractory (rē-frak′tōr-ē) **period** Time period when an excitable cell cannot be restimulated to respond.

absorption (ab-sōrp-shŭn) Process of moving substances, such as products of digestion, into the blood or lymph.

accommodation (a-kom′ŏ-dā′shŭn) Changing the lens shape in order to focus on a nearby object.

acetylcholine (a-sē′til-kō′lēn) **(ACh)** Neurotransmitter produced by the central and peripheral nervous systems.

acid Substance that releases a hydrogen ion when added to a solution.

acid-base balance Maintaining hydrogen ion concentration of body fluids within normal limits; also called pH balance.

acidosis (as-i-dō′sis) Condition in which the pH of arterial blood is below 7.35.

acrosome (ak′rō-sōm) A cap-like structure on the anterior two-thirds of the sperm nucleus that contains digestive enzymes for penetrating an oocyte.

actin (ak′-tin) Contractile protein forming the major part of the thin filaments in a sarcomere.

action potential Self-propagating change in membrane potential occurring in excitable cells (e.g., neurons, muscle cells).

active immunity (i-myū′ni-tē) Activation of the immune system by a vaccine or by exposure to the naturally occurring infectious agent. Offers long-term protection because memory cells are formed. *Compare to passive immunity.*

active site Region of an enzyme where substrate binds.

active transport Method of a protein pump transporting a substance across the membrane, against its concentration gradient.

acute (ă-kyūt′) Takes place over a short period of time. *Compare to chronic.*

adaptation (ad-ap-tā′shŭn) Advantageous change of an organ or tissue to meet new conditions.

adduction (ăd-dŭk′shŭn) Medial movement of a body part toward the midline.

adenosine triphosphate (ă-den′ō-sēn trī-fos′fāt) **(ATP)** Stores and releases chemical energy in a cell; composed of adenine, ribose, and three phosphate groups.

adipocyte (ad′i-pō-sīt) Fat storage cell.

adrenergic (ad-rĕ-ner′jik) Relating to nerve cells that release norepinephrine neurotransmitter.

adventitia (ad-ven-tish′ă) Outermost covering of some organs.

aerobic (ār-o′-bik) **respiration** Breakdown of glucose or other nutrients to produce ATP, water, and carbon dioxide. The process requires oxygen.

afferent (af′er-ent) Inflowing or going toward a center.

agglutination (ă-glū-ti-nā′shŭn) Process by which cells clump due to cross-linking by antibodies.

agonist (ag′on-ist) Muscle that contracts to produce a particular movement; also called prime mover.

albumin (al-bū′min) Plasma protein important in regulating fluid balance.

alimentary (al-i-men′ter-ē) Relating to food or nutrition.

alkalosis (al-kă-lō′-sis) Condition in which the pH of arterial blood is above 7.45.

allele (ă-lēl′) Variation of a gene found on the same locus of homologous chromosomes.

allergen (al′er-gen) Noninfectious substance that elicits an excessive response by the immune system (allergic reaction).

alveolus (al-vē′ō-lŭs; pl., alveoli, -ō-lī) Small cavity. Air sac in the lungs; also, a milk-secreting portion of a mammary gland.

amino (ă-mē′nō) **acid** Organic molecule used to build proteins; contains both an amine group and carboxyl group.

amnion (am′nē-on) Extra-embryonic membrane that envelops the embryo.

amphiarthrosis (am′fē-ar-thrō′sis) A slightly movable joint.

amphipathic (am-fē-path′ik) Molecule that contains a hydrophobic region and a hydrophilic region.

ampulla (am-pul′lă; pl., ampullae, -lē) Saccular dilation of a canal or duct, such as the ductus deferens in the male reproductive system.

anabolism (ă-nab′ō-lizm) Formation of large, complex molecules from simple molecules.

anaerobic (an-ār-ō′bik) Process that does not require oxygen.

anastomosis (ă-nas′tō-mō′sis; pl., anastomoses, -sēz) Union of two structures, such as blood vessels, to supply the same region.

anatomic (ă-na-tom′ik) **position** Standing upright, arms at the sides with palms forward and feet flat on the floor; reference position for naming body regions.

anatomy (ă-nat′ŏ-mē) Study of structures in the human body.

androgen (an′drō-jen) Generic term for a hormone that stimulates the activity of accessory male sex organs or the development of male sex characteristics.

anemia (ă-nē′mē-ă) Any condition in which the number of erythrocytes is below normal.

aneurysm (an′ū-rizm) Ballooning of an artery due to a weakened vessel wall; it is susceptible to rupture, leading to severe bleeding.

angiogenesis (an′jē-ō-jen′ĕ-sis) Process of forming new blood vessels.

angioplasty (an′jē-ō-plas′tē) The reopening of a blood vessel through a variety of means.

angiotensin II (an′jē-ō-ten′sin) A peptide hormone, derived from angiotensin I, that increases blood pressure.

anion (an′ī-on) Negatively charged ion; e.g., Cl^-.

antagonist (an-tag′ō-nist) Muscle (or hormone) that opposes or resists the action of another.

antebrachium (an-te-brā′kē-ŭm) Forearm.

anterior Toward the front of the body.

antibody (an′tē-bod-ē) Immunoglobulin that binds to a specific antigen; released by plasma cells (activated B lymphocytes).

anticodon (antē-kō′don) Group of three nucleotide bases on a transfer RNA molecule; base pairs with a complementary codon on messenger RNA.

antidiuretic (an′tē-dī-ū-ret′-ik) **hormone (ADH)** Hormone released by the posterior pituitary gland; increases water reabsorption in the kidney and reduces urine production.

antigen (an′ti-jen) Substance that causes a state of sensitivity or responsiveness and reacts with antibodies or immune cells of the affected subject.

antigen-presenting cell (APC) Immune cell that presents (displays) an antigen to T-lymphocytes; e.g., a macrophage.

aorta (ā-ōr′tă) Main trunk of the systemic arterial system, beginning at the left ventricle and ending when it forks at its inferior end to form the common iliac arteries.

aortic (ā-ōr′tik) **body** Structure composed of neurons sensitive to changing levels of blood pH, CO_2, and O_2; located in the aortic arch.

aperture (ap′er-chūr) Open gap or hole.

apex (ā′peks) Extremity of a conical or pyramidal structure; e.g., the inferior, conical end of the heart.

apical (ap′i-kăl) Related to the tip or extremity of a conical or pyramidal structure; opposite of basal.

apnea (ap′nē ă) Cessation of breathing while sleeping.

apocrine (ap′ō-krin) **gland** Gland that releases a substance by pinching off the apical membrane of a cell and a portion of its cytoplasm; e.g., apocrine sweat glands.

aponeurosis (ap′ō-nu-rō′sis; pl., aponeuroses, -sēz) Fibrous sheet or flat, expanded tendon.

apoptosis (ap′op-tō′sis) Programmed cell death.

appendicular (ap′en-dik′-ū-lăr) Relating to an appendage or limb; e.g., the appendicular skeleton.

appositional (ap-ō-zish′ŭn-ăl) Being placed or fitted together; e.g., appositional growth of bone.

aquaporins (ak′wă-pōr-in) Protein channels in the plasma membrane allowing the passage of water.

arachnoid mater (ă-rak′noyd mah′ter) Spider weblike meningeal layer; located between the dura mater and the pia mater.

arcuate (ar′kū-āt) Having a shape that is arched or bowed.

arteriole (ar-tēr′ē-ōl) The smallest type of artery.

artery (ar′ter-ē) Blood vessel that transports blood away from the heart.

articular (ar-tik′ū-lăr) Relating to a joint.

articulation (ar-tik-ū-lā′shŭn) Joint or connection between bones.

astrocyte (as′trō-sīt) Largest and most abundant glial cell of the nervous system.

atherosclerosis (ath′er-ō-skler-ō′sis) Disease in which an artery wall thickens, leaving a smaller lumen for blood flow.

atlas (at′las) The first cervical vertebra.

atom The smallest particle that displays properties of an element; composed of electrons, protons, and neutrons (except in hydrogen).

atomic mass unit (AMU) Mass of a specific atom.

atomic number Indicates the number of protons in one atom of a specific element. Value appears above each atomic symbol in the periodic table.

ATP (adenosine triphosphate) Chemical that transfers energy within a cell.

atrioventricular node (ā′trē-ō-ven-trik′ū-lar nōd) **(AV)** Group of specialized heart cells, located in the inferior right atrium, that delays action potentials to the AV bundle.

atrium (ā′trē-ŭm; pl., atria, ā′trē-ă) Chamber or cavity to which are connected other chambers or passageways; e.g., the heart has both a right atrium and a left atrium that are thin-walled, superior chambers that receive blood returning to the heart.

atrophy (at′rō-fē) Wasting of tissues, organs, or the entire body.

auricle (aw′ri-kl) The external ear; also flaplike extensions on the anterior part of each atrium of the heart.

autocrine (aw′tō-krin) Hormonal secretions that stimulate the same cell that released it; also called a local hormone (does not enter the blood). *Compare to endocrine.*

autoimmune (aw-tō-i-mūn′) **disease** Disease in which the immune system attacks self-antigens as if they were foreign.

autolysis (aw-tol′i-sis) Digestion of cells by enzymes present within the cell itself.

autonomic (aw-tō-nom′ik) **nervous system** The part of the nervous system that regulates processes that occur below the conscious level; e.g., the activity of cardiac muscle, smooth muscle, and glands.

autophagy (aw-tof′ā-jē) Segregation and disposal of damaged organelles within a cell.

autoregulation Intrinsic ability of an organ to regulate its activity.

axial (ak′sē-ăl) Relating to or situated in the central part of the body—the head, neck, and trunk; e.g., the axial skeleton.

axis (ak′sis) The second cervical vertebra.

axolemma (ak′sō-lem′ă) Plasma membrane of an axon.

axon (ak′son) Process of a nerve cell that propagates nerve signals away from the cell body.

axoplasm (ak′sō-plazm) Cytoplasm within the axon.

B

B-lymphocyte (lim′fō-sīt) Adaptive immune system cell that functions in humoral immunity; matures into a plasma cell.

baroreceptor (bar′ō-rē-sep′ter, -tōr) Sensory receptor that detects stretch.

basal nuclei (nū′klē′ī) *See cerebral nuclei.*

base Substance that accepts a hydrogen ion.

basement membrane Selective molecular layer that attaches epithelial tissue to underlying connective tissue.

basophil (bā′sō-fil) The least common of the white blood cells. Basophils release proinflammatory agents; e.g., histamine, heparin.

benign (bē-nīn′) Term that denotes the mild character of an illness or the nonmalignant character of a neoplasm.

bile (bīl) Fluid secreted by the liver, stored, concentrated, and released from the gallbladder into the duodenum.

bilirubin (bil-i-rū′bin) Waste product derived from the breakdown of hemoglobin and excreted in bile.

blood-brain barrier Structure formed by capillary endothelial cells and astrocytes that regulates what can enter the interstitial fluid in the brain: helps prevent transport of harmful substances from the blood into the brain.

blood pressure Measure of the force of blood pushing against the blood vessel wall; commonly measured in the brachial artery.

bolus (bō′lŭs) A single quantity of something, such as a mass of food swallowed.

brachial (brā′kē-ăl) Relating to the region between the shoulder and the elbow.

bradycardia (brad-ē-kar′dē-ă) A resting heart rate below 60 beats per minute.

brainstem Brain region composed of the midbrain, pons, and medulla oblongata.

bronchiole (brong′kē-ōl) Small tubules that branch from bronchi in the lungs.

bronchus (brong′kŭs) Airways that deliver air from the trachea to the bronchioles.

buffer Substance that minimizes a change in pH after an acid or base is added.

bursa (ber′să; pl., bursae, ber′sē) Closed, fluid-filled sac lined with a synovial membrane; usually found in areas subject to friction.

C

calcification (kal′si-fi-kā′shŭn) Process in which structures in the body become hardened as a result of deposited calcium salts; normally occurs only in the formation of bone and teeth.

callus (kal′ŭs) Composite mass of cells and extracellular matrix that forms at a fracture site to establish continuity between the bone ends.

calorie The amount of energy required to raise the temperature of 1 g of water by 1°C.

calyx (kā′liks; pl., calyces or calices, kal′i-sēz) Cup-shaped cavity or structure.

canaliculus (kan-ă-lik′ū-lŭs; pl., canaliculi, -lī) Small canal or channel.

cancer Disease involving a malignant neoplasm that can metastasize (spread to other organs).

capacitation (kă-pas′i-tā′shŭn) A period of conditioning whereby the sperm cell membrane is modified while in the female reproductive tract prior to being able to fertilize the secondary oocyte.

capillary (kap′i-lār-ē) The smallest blood vessel, its thin walls allow movement of substances between blood and interstitial fluid.

carbohydrate An organic molecule composed of carbon, hydrogen, and oxygen.

cardiac cycle Events taking place in the heart during one heart beat.

cardiac output (CO) Volume of blood ejected by the ventricle in 1 minute; calculated by multiplying heart rate times stroke volume.

carotene (kar′ō-tēn) Class of yellow-red pigments widely distributed in plants and animals.

carotid (ka-ro′tid) **body** Contains neurons sensitive to changing levels of blood pH, CO_2, and O_2; located at the bifurcation of the common carotid artery.

carotid sinus (sī′nŭs) Structure composed of neurons sensitive to changes in blood pressure; located in the internal carotid arteries.

carpal (kar′păl) Relating to the wrist.

carrier Protein that binds a molecule to transport the molecule across the plasma membrane or within the blood; also, a person with one recessive mutant allele (when in a homozygous form causes a phenotypical change).

cartilaginous (kar-ti′laj′i-nŭs) Relating to or consisting of cartilage.

catabolism (kă-tab′ō-lizm) Breakdown of complex molecules into simple molecules.

catalyst (kat′ă-list) Substance that speeds up a chemical reaction.

cataract (kat′ă-rakt) Complete or partial opacity of the lens.

catecholamines (kat-ĕ-kōl′ă-mēnz) Class of neurotransmitters (includes epinephrine, norepinephrine, and dopamine).

cation (kat′-ī-on) Ion with a positive charge; e.g., Na^+.

cauda equina (kaw′dă ē-kwī′nă) Spinal nerve roots within the vertebral canal inferior to the tapered inferior end of the spinal cord.

cecum (sē′kŭm) Blind pouch forming the first part of the large intestine.

cell Basic structural and functional unit of a living organism.

cell-mediated immunity Immune response involving T-lymphocytes.

cellular respiration Multistep metabolic pathway in which organic molecules (e.g., glucose) are disassembled in a controlled manner by a series of enzymes to eventually form ATP.

central nervous system (CNS) Composed of the brain and spinal cord.

centriole (sen′trē-ōl) Organelle that participates in the separation of chromosome pairs during cell division.

centromere (sen′trō-mēr) The nonstaining constriction of a chromosome that is the point of attachment of the spindle fiber.

cephalic (se-fal′ik) Relating to the head area.

cerebellum (ser-e-bel′ŭm) The second largest part of the brain; develops posteriorly to the pons in the metencephalon.

cerebral cortex (se-rē′bral kor′teks) Superficial layer of gray matter in the cerebrum.

cerebral nuclei Paired irregular masses of gray matter buried deep within central white matter in the basal region of the cerebral hemispheres inferior to the floor of the lateral ventricle.

cerebrospinal (sĕ-rē′brō-spī-năl) **fluid (CSF)** A clear, colorless fluid that circulates in the ventricles, subarachnoid space, and central canal to protect and support the brain and spinal cord.

cerebrovascular (sĕ-rē′brō-vas′kū-lăr) **accident (CVA)** Altered brain activity caused by reduced blood supply to the brain. It may result in permanent damage.

cerebrum (ser′ē-brŭm, ser′ĕ-brŭm) The largest, most superior part of the brain; composed of the left and right cerebral hemispheres; location of conscious thought processes and origin of all complex intellectual functions.

cerumen (sĕ-rū′men) Soft, waxy secretion of the ceruminous gland; found in the external auditory meatus.

cervical (ser′vĭ-kal) Relating to the neck.

chemical energy Energy stored in the chemical bonds of a molecule.

chemical equilibrium (ē-kwi-lib′rē-ŭm) State of a chemical reaction in which there is no net change in formation of products or reactants.

chemical reaction Process during which chemical bonds of a molecule are broken and new ones are formed.

chemical synapse (sin′aps) Junction between two communicating cells (neurons, and neurons and effectors); neurotransmitter released at this type of synapse.

chemoreceptor (kē′mō-rē-sep′tor) Peripheral sensory receptor or specialized cells within the brain that detect specific chemicals in a fluid; e.g., taste receptors.

chemotaxis (kē-mo-tak′sis) Movement in response to chemicals.

cholecalciferol (kō′lē-kal-sif′er-ol) Vitamin D.

cholecystokinin (kō′lē-sis-to-kī′nin) **(CCK)** Hormone released from the duodenum in response to lipid-rich chyme; stimulates secretion from the pancreas and gallbladder.

cholesterol Type of steroid found in the plasma membrane.

cholinergic (kol-in-er′jik) Relating to neurons that use acetylcholine as their neurotransmitter.

chondroblast (kon′drō-blast) Actively mitotic form of a matrix-forming cell found in developing and growing cartilage.

chondrocyte (kon′drō-sīt) Mature, nondividing cartilage cell.

chorion (kō′rē-on) Multilayered, outermost extraembryonic membrane; together with the functional layer of the endometrium it forms the placenta, the site through which nourishment and waste are exchanged between mother and developing fetus; attachment to the uterus.

chromatid (krō′mă-tid) One of the two strands of a chromosome joined by a centromere.

chromatin (krō′ma-tin) Genetic material of the nucleus in a nondividing cell.

chromosome (krō′mō-sōm) The most compact form of genetic material; a single long molecule of DNA and associated proteins; becomes visible only when the cell is dividing.

chronic (krŏn′ik) Takes place over a long period of time. *Compare to acute.*

chronic obstructive pulmonary disease (COPD) Group of respiratory diseases involving obstruction of the airways.

chyle (kīl) Chylomicron-containing lymph drained from gastrointestinal tract.

chyme (kīm) Mixture of partially digested food and gastric juice within the gastrointestinal tract.

cilia (sil′ē-ă) Plasma membrane extensions containing cytoplasm and microtubules that move materials past the cell.

circumduction (ser-kŭm-dŭk′shŭn) Movement of a body part in a circular direction.

cisterna (sis-ter′nă; pl., cisternae, -ter′nē) Enclosed microscopic space; e.g., between the parallel membranes of the Golgi apparatus.

citric acid cycle A cyclic metabolic pathway that occurs in the matrix of mitochondria during which energy in the bonds of acetyl CoA is transferred to form ATP, NADH, and FADH.

cleavage (klēv′ij) Series of mitotic cell divisions occurring in the zygote immediately following its fertilization.

climacteric (klī-mak′ter-ik, klī-mak-ter′ik) Changes occurring in the male and female reproductive systems, beginning at around 50 years, resulting in altered levels of reproductive hormones.

coagulation (kō-ag-yū-lā′-shŭn) Formation of a blood clot.

codon (kō′don) Group of three nucleotide bases in a messenger RNA molecule.

coenzyme (kō-en′zīm) Organic molecules assisting in enzyme function; e.g., NAD^+.

cofactor Inorganic chemical structure attached to an enzyme that aids in enzyme function; e.g., Zn^{2+}.

cognition (kog-ni′shŭn) Mental activities associated with thinking, learning, and memory.

coitus (kō-i′tŭs; koy′tus) Sexual union between a male and a female.

collagen (kol′lă-jen) **fibers** Strong, flexible protein located within many connective tissues.

colloid (kol′oyd) Opaque mixture composed of water and solute (usually protein); substance within thyroid follicles.

columnar (kol-ŭm′năr) Relating to epithelial cells that are taller than they are wide.

commissure (kom′i-shūr) Bundle of axons passing from one side to the other in the brain or spinal cord.

complement (kom′plĕ-ment) Group of plasma proteins working together during an innate immune response; may lead to increased inflammation and cytolysis of the invading cell.

concentration gradient (grā′dē-ent) Difference in the concentration of a substance between two areas.

conducting zone Respiratory system passageways transporting air from the nose to the terminal bronchioles. *Compare to respiratory zone.*

conductivity (kon-dŭk-tiv′i-tē) Propagating an electrical signal along the plasma membrane.

cone Cone-shaped cell in the retina that provides color vision; functions best in bright light.

congenital (kon-jen′i-tăl) Present at birth.

conjunctiva (kon-jŭnk-tī′vă) Epithelial layer covering the anterior surface of the eyeball and the posterior surface of the eyelids.

contralateral (kon-tră-lat′er-ăl) Relating to the opposite side.

control center Structure in a feedback cycle interpreting information from a receptor and sending information to an effector to produce a response.

cornea (kōr′nē-ă) Transparent structure that forms the anterior surface of the eye.

coronal (kōr′ō-năl) A vertical plane that divides the body into anterior and posterior parts; also called frontal plane.

coronary (kōr′o-nār-ē) Denoting the blood vessels or other structures and activities related to the heart.

cortex (kor′teks) Outer region of an organ; e.g., cerebral cortex, adrenal cortex.

corticosteroid (kōr-ti-kō-stēr′-oyd) Hormones secreted by the adrenal cortex; e.g., cortisol.

costal (kos′tăl) Relating to a rib.

covalent (kō-văl′ent) **bond** Chemical bond formed when atoms share electrons.

cranial (krā′nē-ăl) Relating to the skull.

cranium (krā′nē-ŭm) Region of the skull composed of the frontal, parietal, temporal, occipital, ethmoid, and sphenoid bones.

creatine (krē′-ă-tēn; -tin) Chemical providing muscle cells with energy; present as creatine phosphate.

creatinine (krē-at′i-nēn) Nitrogenous waste product resulting from the breakdown of creatine and excreted in urine. Used to estimate glomerular filtration rate by the kidney.

cross-section Transverse section.

cubital (kū′bi-tal) Relating to the elbow.

cuboidal (kū-boy′dăl) Relating to cells that are cube-shaped.

current Movement of charged particles, such as ions.

cutaneous (kū-tā′nē-ŭs) Relating to the skin.

cyclic adenosine monophosphate (sī′klik ă-den′ō-sēn mon-ō-fos′-fāt) **(cAMP)** A second messenger used when some hormones or neurotransmitters act on a target cell. Formed when adenylate cyclase reacts with ATP.

cytokine (sī′tō-kīn) Protein that regulates and facilitates immunity system activity.

cytokinesis (sī′tō-ki-nē′sis) Division of the cytoplasm during cell division.

cytology (sī-tol′ō-jē) Study of cells.

cytoplasm (sī′tō-plazm) All cellular contents contained between the plasma membrane and the nucleus; includes cytosol, organelles, and inclusions.

cytoskeleton (sī-tō-skel′ĕ-ton) Organized network of protein filaments and hollow tubules that provide organization, support, and movement throughout the cell.

cytosol (sī′tō-sol) The viscous, syruplike fluid medium with dissolved solutes in the cytoplasm.

cytotoxic T (T_C) lymphocyte (sī′tō-tok′-sik lĭm′fō′sīt) A class of T-lymphocytes that release chemicals to destroy unhealthy cells; also called CD8 cells.

D

deamination (dē-am-i-nā′shŭn) Process of removing the amine group of an amino acid.

deciduous (dē-sid′ū-ŭs) Not permanent; e.g., deciduous teeth.

decussation (dē-kŭ-sā′shŭn) Any crossing over or intersection of parts.

defecation (def-ĕ-kā′shŭn) Discharge of feces from the rectum.

deglutition (dē-glū-tish′ŭn) Swallowing.

dehydration synthesis Chemical reaction in which water is formed during formation of a complex molecule.

denaturation (dē-nā-tū-ra′shŭn) A change in a protein's complex three-dimensional shape that causes its biological activity to be impaired or to cease; may occur with changes in pH or increased temperature.

dendrite (den′drīt) Process on a neuron where graded potentials are initiated.

dendritic (den-dri′tik) **cell** Phagocytic cells of the skin and mucous membranes.

dentition (den-tish′ŭn) Natural teeth in the dental arch, considered collectively.

deoxyribonucleic (dē-oks′ē-rībō-nū-klē̆ic′) **acid (DNA)** A double-stranded nucleic acid, composed of deoxyribonucleotide monomers; directs protein synthesis.

depolarization (dē-pō′lăr-i-zā-shŭn) Change in membrane potential or voltage to a more positive value.

depression (dē-presh′ŭn) Downward or inward displacement of a body part.

dermatome (der′mă-tōm) Specific segment of skin supplied by a spinal nerve. Also, during embryonic development, the cells that form the connective tissue of the skin.

dermis (der′mis) Connective tissue layer of skin internal to the epidermis; contains blood and lymph vessels, nerves and nerve endings, glands, and usually hair follicles.

desmosome (dez′mō-sōm) One type of adhesion between two epithelial cells; a type of intercellular junction that holds cells together at a single point (like a button).

detrusor (dē-trū′ser, -sōr) Smooth muscle within urinary bladder wall that expels urine.

diabetes (dī-ă-bē′tēz) **insipidus** Disease involving reduced ADH release or reduced kidney response to ADH; leads to excessive urine production.

diabetes mellitus (me-lī′tŭs) Disease involving reduced insulin release or reduced tissue response to insulin; leads to elevated blood and urine glucose levels.

diapedesis (dī′ă-pĕ-dē′sis) Passage of blood or its formed elements through the intact blood vessel wall.

diaphragm Muscle that separates the thoracic cavity and abdominopelvic cavity; aids in breathing.

diaphysis (dī-af′i-sis; pl., diaphyses, -sēz) Elongated, usually cylindrical part of a long bone between its two ends; the shaft of a long bone.

diarthrosis (dī-ar-thrō′sis; pl., diarthroses, -sēz) A freely movable (synovial) joint.

diastole (dī-as′tō-lē) The relaxation phase of a heart chamber.

diastolic pressure (dī-ă-stol′ik) Blood pressure measured in an artery during diastole.

diencephalon (dī-en-sef′ă-lon) Brain region deep to the cerebrum; contains the thalamus, hypothalamus, and epithalamus.

diffusion (di-fū′zhŭn) Random movement of molecules or particles down their concentration gradient.

diploë (dip′lō-ē) Central layer of spongy bone between the two layers of compact bone (outer and inner plates) of the flat cranial bones.

diploid (dip′loyd) State of a cell containing pairs of homologous chromosomes. In humans, the diploid number of chromosomes is 46 (23 pairs).

disaccharide (dī-sak′ă-rīd) Carbohydrate composed of two monosaccharides; e.g., sucrose.

dislocation (dis-lō-kā′shŭn) Complete displacement of an organ (e.g., a bone) from its normal position.

distal (dis′tăl) Far from the point of attachment at the trunk.

distal attachment: the (typically more movable) attachment site of an appendicular skeletal muscle (*previously called the insertion*).

diuretic (dī-ū-ret′ik) Agent that increases the excretion of urine.

diverticulum (dī′ver-tik′yū-lŭm; pl., diverticula) A pouch or sac opening from a tubular or saccular organ, such as the gut or bladder; small bulge in the intestinal wall.

dorsal (dōr′săl) Toward the back.

dorsiflexion (dōr-si-flek′shŭn) Upward movement of the foot or toes, or of the hand or fingers.

duodenum (dū-ō-dē′nŭm, dū-od′ĕ-nŭm) First section of small intestine.

dura mater (dū′ră mah′ter) Tough, fibrous membrane forming the outer meningeal layer of the central nervous system.

E

ectoderm (ek′tō-derm) Outermost of the three primary germ layers of the embryo.

ectopic (ek-top′ik) Out of place; e.g., in an ectopic pregnancy, the pre-embryo implants in the uterine tube rather than in the uterus.

edema (e-dē′mă) Localized swelling of a tissue.

effector (ē-fek′tŏr, -tōr) Peripheral tissue or organ that responds to nervous or hormonal stimulation.

efferent (ef′er-ent) Outgoing or moving away from a center.

eicosanoids (ī′kō′-să-noydz) Local hormones derived from fatty acids.

ejaculation (ē-jak-ū-lā′shŭn) Expulsion of semen from the penis.

elastic fiber Protein that allows tissues to stretch and return to their original shape; located within connective tissues.

electrical energy Movement of charged particles (such as ions).

electrical synapse (sin′aps) Junction between two communicating cells. Electrical signal moves across this type of synapse through gap junctions.

electrochemical gradient Electrical charge difference across a membrane.

electrolyte (ē-lek′trō-līt) Chemical that dissociates when added to water and can conduct an electrical current; includes salts, bases, and acids.

electrolyte balance Maintenance of the appropriate levels of electrolytes (such as sodium ions) in body fluids.

electron Subatomic particle with a negative charge; found orbiting the nucleus of an atom.

element Substance composed of only one type of atom.

elevation (el-ĕ-vā′shŭn) Superior movement of a body part.

embolus (em-bō′lŭs) Dislodged blood clot or air bubble traveling through the blood.

embryo (em′brē-ō) Organism in the early stages of development; in humans, the embryonic stage extends from the third to the eighth week of development.

embryology (em-brē-ol′ō-jē) Study of the origin and development of the organism, from fertilization of the secondary oocyte until birth.

end-diastolic (dī-ă-stol′ik) **volume (EDV)** Volume of blood in the ventricle at the end of diastole (relaxation).

end-systolic (sys-tol′ik) **volume (ESV)** Volume of blood in the ventricle at the end of systole (contraction).

endergonic (en′der-gon′ik) **reaction** Chemical reaction that requires the input of energy.

endocardium (en-dō-kar′dē-ŭm) Covering of the internal surface of the heart wall and external surface of heart valves.

endochondral ossification (en-dō-kon′drāl os′i-fi-kā′shŭn) Bone formation that takes place within hyaline cartilage; used to form most bones of the body.

endocrine (en′dō-krin) Hormonal secretions that are transported by the blood. *Compare to exocrine.*

endocytosis (en′dō-sī-tō′sis) Movement of substances from the extracellular environment into the cell through the formation of a vesicle.

endoderm (en′dō-derm) Innermost of the three primary germ layers of the embryo.

endolymph (en′dō-limf) Fluid within the membranous labyrinth of the inner ear.

endometrium (en′do-me′trē-ŭm) Mucous membrane forming the inner layer of the uterine wall.

endomysium (en′dō-miz′ē-ŭm, -mis′ē-ŭm) Areolar connective tissue layer surrounding a muscle fiber.

endoneurium (en-dō-nū′re-ŭm) Areolar connective tissue that surrounds each axon of a peripheral nerve.

endoplasmic reticulum (en′-dō-plas′mik re-tik′ū-lum) **(ER)** Organelle composed of an extensive network of connected membranes; involved in synthesis, transport, and storage of macromolecules, and detoxification of drugs: present as smooth ER or rough ER.

endosteum (en-dos′tē-ŭm) Layer of cells lining the inner surface of bone in the medullary cavity.

endothelium (en-dō-thē′lē-ŭm) The simple squamous epithelium that lines the lumen of blood vessels, lymphatic vessels, and heart chambers and valves.

energy Capacity to do work.

enzyme (en′zīm) Protein that catalyzes a chemical reaction by lowering the activation energy.

eosinophil (ē-ō-sin′ō-fil) White blood cell that destroys parasitic worms and phagocytizes antibody-antigen complexes.

ependymal (ep-en′di-mal) Relating to the cellular lining of the brain ventricles and central canal of the spinal cord; assists in production and circulation of CSF.

epicardium (ep-i-kar′dē-ŭm) The visceral (outermost) layer of the heart; also called the serous pericardium.

epidermis (ep-i-derm′is) The epithelium (keratinized, stratified squamous) of the integument.

epimysium (ep-i-mis′ē-ŭm) A layer of dense, irregular connective tissue surrounding a skeletal muscle.

epinephrine (ep′i-nef′rin) Hormone released by the adrenal medulla during activation of the sympathetic nervous system.

epineurium (ep-i-nū′rē-ŭm) Outermost supporting connective tissue layer of peripheral nerves.

epiphyseal (ep-i-fiz′ē-ăl) **line** The remnant of the epiphyseal plate that remains when long bone growth ceases; thin, defined area of compact bone.

epiphyseal plate Layer of hyaline cartilage located between a long bone diaphysis and epiphysis; allows longitudinal growth of the bone.

epiphysis (e-pif′i sis; pl., epiphyses, -sēz) Expanded, knobby region at the end of a long bone.

erection (ē-rek′shŭn) Erectile tissues in the penis fill with blood and cause the penis to enlarge and become firm.

erythrocyte (ĕ-rith′rō-sīt) Mature red blood cell.

erythropoiesis (ĕ-rith′rō-poy-ē′sis) Formation of erythrocytes.

erythropoietin (ĕ-rith-rō-poy′ĕ-tin) Hormone that stimulates erythropoiesis.

esophagus (ē-sof′ă-gŭs) Portion of the gastrointestinal tract between the pharynx and the stomach.

excitability The ability of a cell to respond to a stimulus.

excretion (eks-krē′-shŭn) Process of removing waste products from the body.

exergonic (ek′ser-gon′ik) **reaction** Chemical reaction in which chemical energy is released.

exocrine (ek′sō-krin) Glandular secretions delivered to an apical or luminal surface through a duct.

exocytosis (ek′sō-sī-to′sis) Process whereby contents within a vesicle are released from a cell.

exon (ek′-son) A portion of a DNA molecule that codes for a section of the future messenger RNA molecule; these "coding" regions in pre-messenger RNA are joined together to form mature messenger RNA.

expiration (eks-pi rā′shŭn) To breathe out; exhalation.

extension (eks-ten′shŭn) Movement that increases the articulating angle.

exteroceptor (eks′ter-ō-sep′ter, -tōr) Terminal end of sensory neuron in the skin or mucous membrane that responds to external stimuli.

extracellular (eks-tră-sel′ū-lăr) **fluid (ECF)** Fluid located outside of cells; plasma and interstitial fluid.

extracellular matrix (mā′triks) Protein fibers and ground substance in the extracellular space of connective tissue.

extrinsic (eks trin′sik) Originates outside of an organ; e.g., extrinsic eye muscles.

F

facet (fas′et, fă-set′) Small, flat, shallow articulating surface; smooth area on a bone.

facilitated diffusion Passive transport process using carrier proteins or channel proteins to move a chemical across the plasma membrane.

falciform (fal′si-fōrm) Having a crescent or sickle shape; e.g., falciform ligament.

falx cerebri (falks se-rē′brē) Portion of the dura mater septa that projects into the longitudinal fissure between the right and left cerebral hemispheres of the brain.

fascia (fash′ē-ă; pl., fasciae, -ē-ē) Sheath of fibrous connective tissue that envelops the body internal to the skin; encloses muscles, and separates their various layers or groups.

fascicle (fas′i-kl) Bundle of muscle fibers or axons.

fauces (faw′sēz) Space between the oral cavity and the pharynx.

feces (fē′sēz) Material discharged from the GI tract during defecation.

fertilization (fer′til-i-zā′shŭn) Process of sperm penetration of the secondary oocyte.

fetal (fē′tăl) Relating to the fetus; in humans, the fetal period extends from the eighth week after conception until birth.

fibrin (fī′brin) The supporting meshwork of insoluble fibrous protein within a blood clot.

fibroblast (fī′brō-blast) Large, flat, connective tissue cells with tapered ends that produce the fibers and ground substance components of the extracellular matrix.

fibrosis (fī-brō′sis) Formation of fibrous connective tissue as a repair or reactive process.

filtrate (fil′trāt) The materials that pass through a filter.

fimbria (fim′brē-ă; pl., fimbriae, -brē-ē) Any fringelike structure; e.g., the fimbriae of the infundibulum enclose the ovary at the time of ovulation.

fissure (fish′ur) Deep furrow, cleft, or slit.

flagellum (flă-jel′-ŭm; pl., flagella, -ă) Whiplike locomotory extension of the plasma membrane; enables a sperm cell to move.

flexion (flek′shŭn) Movement that decreases the angle between the articulating bones.

flexure (flek′sher) Bend in an organ or structure.

follicle Roughly spherical group of cells enclosing a cavity.

fontanelle (fon′tă-nel′) One of several membranous intervals at the margins of the cranial bones in an infant.

foramen (fō-rā′men) Hole in a bone; e.g., foramen magnum, obturator foramen.

formed elements Erythrocytes, leukocytes, and platelets within blood.

fornix (fōr′niks; pl., fornices, -ni-sēz) Arch-shaped structure.

fossa (fos′ă; pl., fossae, -fos′ē) Depression, often more or less longitudinal in shape, below the level of the surface of a part.

G

G protein Specific protein that acquires its energy from guanosine triphosphate; when activated by a membrane receptor, it relays the signal to another membrane protein and alters the activity of that protein.

gamete (gam′ēt) A sex cell with the haploid number of chromosomes.

gametogenesis (gam′ĕ-tō-jen′ĕ-sis) Formation and development of gametes (sex cells).

ganglion (gang′glē-on; pl., ganglia, -glē-ă) Group of neuron cell bodies in the peripheral nervous system.

gastric (gas′trik) Relating to the stomach.

gastrulation (gas-trū-lā′shŭn) Formation of the three primary germ layers.

gene Segment of DNA containing information to direct synthesis of a specific protein; functional unit of DNA.

genotype (jen′ō-tīp) The genetic constitution of an individual that, along with environmental influences, contributes to the phenotype.

gestation (jes-tā′shŭn) Pregnancy.

gland Organ or individual cells that secrete a substance.

glomerulus (glō-mer′yū-lŭs) A capillary network within the renal corpuscle in a kidney.

glucocorticoid (glū-kō-kōr′ti-kōyd) Group of hormones released from the adrenal cortex; help regulate glucose levels; e.g., cortisol, corticosteroid.

gluconeogenesis (glū′kō-nē′ō-jen′ĕsis) Formation of glucose from a noncarbohydrate source.

glucose (glū′kōs) A monosaccharide; primary nutrient source for cellular respiration.

glycerol (glis′er-ol) Chemical component of a triglyceride molecule.

glycocalyx (glī-kō-kā′liks) Filamentous coating on the apical surface of certain types of cells.

glycogen (glī′kō-jen) Polysaccharide formed from glucose monomers.

glycogenesis (glī′kō-jen′ĕ-sis) Formation of glycogen from glucose.

glycogenolysis (glī′kō-je-nol′i-sis) Breakdown of glycogen into glucose.

glycolipid (glī-kō-lip′id) Lipid with an attached carbohydrate group found on the extracellular side of the plasma membrane.

glycolysis (glī-kol′i-sis) Stage of cellular respiration in which glucose is partially catabolized to form pyruvic acid and to form ATP molecules.

glycosuria (glī-kō-sū′rē-ă) Presence of glucose in urine; symptom of diabetes mellitus (also called glucosuria).

goblet cells Unicellular glands that secrete mucin.

Golgi (gol′jē) **apparatus** Series of saclike membranes that act as a center to package, sort, and modify molecules arriving from the endoplasmic reticulum in a transport vesicle.

gonad (gō′nad) Organ that produces sex cells and sex hormones; the ovaries in a female and the testes in a male.

gonadocorticoids (gō-na′dō-kōr′ti-kōyd) Group of sex hormones secreted from the adrenal cortex, including estrogen and androgen.

graded potential Small deflection in the resting membrane potential in excitable cells due to the movement of small amounts of ions across the plasma membrane; it may be due to either a depolarization or hyperpolarization.

gradient A difference from one area to another; e.g., concentration gradient, pressure gradient.

gray matter Brain or spinal cord tissue composed of neuron cell bodies, dendrites, and unmyelinated axons.

gustation (gŭs-tā′shŭn) Sense of taste.

gyrus (jī′rŭs; pl., gyri, -rī) Prominent rounded surface elevation of the cerebral hemispheres.

H

hair cell Specialized sensory cell in the inner ear.

half-life Time required to reduce substance by one-half of its original quantity.

haploid (hap′loyd) The number of chromosomes in a sperm or secondary oocyte. In humans, the haploid number is 23.

hapten (hap′ten) Small substance that attaches to another molecule to initiate an immune reaction.

helper T (T_H) lymphocyte (lim′fō-sīt) A T-lymphocyte that releases cytokines to regulate humoral immunity and cellular immunity as well as cells of innate immunity; also called CD4 cells.

hematocrit (hē′mă-tō-krit, hem′ă-) Percentage of whole blood attributed to erythrocytes.

hematoma (hē-mă-tō′mă) Mass of blood outside of the blood vessels; e.g., subdural hematoma.

heme (hēm) Portion of a hemoglobin molecule that binds iron, which transports oxygen.

hemocytoblast (hē′mō-sī′tō-blast) Immature cells that produce all types of formed elements in blood.

hemoglobin (hē-mō-glō′bin) A red-pigmented protein that transports oxygen and carbon dioxide; responsible for characteristic red color of blood.

hemolysis (hē-mol′i-sis) The process of rupture and destruction of erythrocytes.

hemopoiesis (hē′mō poy-ē′sis) Formation and development of blood cells.

hemorrhage (hem′ŏ-rij) Abnormal loss of blood from a blood vessel.

hemostasis (hē′mō-stā′sĭs) Process of stopping bleeding; steps include vascular spasm, platelet plug formation, and coagulation.

heparin (hep′a-rĭn) Chemical that prevents blood clot formation; released by basophils and mast cells.

hepatic (he-pat′ik) Relating to the liver.

hernia (her′nē-ă) Protrusion of a part through the structures normally surrounding or containing it.

heterozygous (het′er-ō-zī′gŭs) Having both dominant and recessives alleles for a trait.

hiatus (hī-ā′tŭs) Opening.

hillock (hil′lok) Any small elevation or prominence; e.g., axon hillock.

hilum (hī′lŭm) The part of an organ where structures such as blood vessels, nerves, and lymph vessels enter or leave.

histamine (his′tă-mēn) Chemical that increases capillary permeability and causes vasodilation; released by basophils and mast cells.

histology (his-tol′ō-jē) Study of tissues formed by cells and cell products.

homeostasis (hō′mē-ō-stā′sis, -os′tă-sis) A consistent internal environment in the body with respect to various functions and the chemical composition of fluids and tissues.

homologous (hŏ-mol′ō-gŭs) Alike in certain critical attributes; e.g., homologous chromosomes.

homozygous (hō-mō-zī′gŭs) Having two identical alleles for a trait.

hormone (hōr′mōn) Chemical formed by an endocrine gland and transported by the blood to its target organ to alter cellular activity.

humoral (hyū′mōr-ăl) **immunity** Immune response involving B-lymphocytes; involves release of antibodies from plasma cells.

hyaline (hī′a-lin, -lēn) Clear, homogeneous substance; a type of cartilage.

hydrogen bond A weak attraction formed when a hydrogen atom of one molecule is attracted to a slightly negative atom within either the same molecule or a different molecule.

hydrolysis (hī-drol′i-sis) Chemical reaction in which water is used during the breakdown of a complex molecule.

hydrophilic (hī-drō-fil′ik) **molecule** Substance that does dissolve in water.

hydrophobic (hī-drō-fōb′ik) **molecule** Substance that does not dissolve in water.

hydroxyapatite (hī-drok′sē-ap-ă-tīt) Natural mineral structure that the crystal lattice of bone and teeth closely resembles.

hyperextension (hī′per-eks-ten′shŭn) Extension of a body part beyond 180 degrees.

hyperplasia (hī-per-plā′zhē-ă) Increase in the number of cells in a tissue.

hyperpolarization (hī′pĕr-pō′lăr-i-zā-shŭn) Change in the membrane potential to a value more negative than the resting potential.

hypertension (hī′pĕr-ten′shŭn). Persistently elevated blood pressure.

hypertonic (hī′pĕr-ton′ik) **solution** Solution with a lower water concentration and higher solute concentration than that of the cytosol.

hypertrophy (hī-pĕr′trō-fē) Increase in the size of cells in a tissue.

hyperventilation (hī-pĕr-ven′ti-lā′shŭn) Increase in the breathing rate or depth.

hypodermis (hī-pō-der′mis) Subcutaneous layer of tissue internal to and not part of the integument.

hypogastric (hī-pō-gas′trik) Relating to the lower abdomen.

hypothalamus (hī′pō-thal′ă-mŭs) Region of the brain in the diencephalon; regulates body temperature, autonomic nervous system, and endocrine system.

hypotonic (hī′pō-ton′ik) **solution** Solution with a lower solute concentration and a higher water concentration than that of the cytosol.

hypoventilation Decrease in the breathing rate or depth.

hypoxia (hī-pok′sē-ă) Decreased oxygen level.

I

idiopathic (id′ē-ō-path′ik) Describing a disease of unknown cause.

immunity (i-myū′ni-tē) Ability of the body to protect itself against potentially harmful substances; includes humoral immunity and innate immunity.

immunocompetence (im′ū-nō-kom′pĕ-tens) Ability of immune cells to recognize and bind to an antigen.

immunoglobulin (im′ū-nō-glob′ū-lin) **(Ig)** *See antibody.*

implantation Embedding of the pre-embryo in the uterine wall.

inclusion A temporary store of molecules in the cytosol.

infarction (in-fark′shŭn) Tissue death resulting from reduced or absent blood supply; e.g., myocardial infarction.

inferior Toward the feet.

inferior attachment: the (typically less movable) attachment site of an axial skeletal muscle (*previously called the origin*)

inflammation A nonspecific, localized response of the innate immune system to tissue injury.

infundibulum (in-fŭn-dib′ū-lŭm) Funnel-shaped structure or passage.

ingestion (in-jes′chŭn) Introduction of food and drink into the GI tract.

inguinal (ing′gwi-năl) Relating to the groin area.

innervation (in-er-vā′shŭn) Functional connection of a neuron with a structure of the body.

inorganic Chemical substance that is not organic; e.g., water, acids, bases.

insertion (in-ser′shŭn) The usually distal and more movable attachment of a muscle. (This term typically replaced with phrase *inferior attachment* or *distal attachment* in this text.)

inspiration (in-spi-rā′shŭn) To breathe in; inhalation.

integument (in-teg′ū-ment) The membrane that covers the body; includes the epidermis, dermis, and all derivatives of the epidermis; also called skin or the cutaneous membrane.

intercalated disc (in-ter′kă-lā-ted disk) Intercellular junction between cardiac muscle cells composed of desmosomes and gap junctions that mechanically and electrically link these cells.

interneuron (in′ter-nū′ron) Type of neuron that resides completely within the CNS and coordinates activity between sensory and motor neurons.

interoceptor (in′ter-ō-sep′ter) Sensory receptors, within the walls of the viscera and blood vessels, detect stimuli of the body's internal environmnent.

interphase First phase of the cell cycle during which the cell carries out normal activities and prepares for cell division.

interstitial (in-ter-stish′ăl) Relating to spaces within a tissue or organ, but not a body cavity; e.g., interstitial fluid occupies the extracellular environment of the cells.

intervertebral (in-ter-ver′te-brăl) Between vertebrae.

intracellular fluid (ICF) *See cytosol.*

intramembranous (in′tră-mem′brā-nŭs) **ossification** Bone formation that takes place within a membrane; the formation of flat bones.

intraperitoneal (in′tră-per′i-tō-nē′ăl) Location of a structure or organ that is completely covered with visceral peritoneum.

intrinsic (in-trin′sik) Originates inside of an organ; e.g., intrinsic muscles of the hand.

intrinsic factor Chemical, produced by stomach cells, required for absorption of vitamin B_{12}.

intron (in′tron) A portion of DNA that lies between two exons; noncoding regions in pre-messenger RNA. These are removed during pre-messenger RNA processing.

invaginate (in-vaj′i-nāt) To infold or insert a structure within itself.

ion (ī′on) An atom with a net positive or negative charge.

ionic bond Chemical bond formed when a cation is electrostatically attracted to an anion.

ipsilateral (ip-si-lat′er-ăl) Relating to the same side.

ischemia (is-kē′mē-ă) Reduced blood flow; may lead to hypoxia.

isomer (ī′sō-mer) Molecules composed of the same number and types of atoms but with a different arrangement; e.g., glucose and galactose.

isometric (ī-sō-met′rik) **contraction** Muscle contraction during which its length does not change because tension does not exceed resistance (load).

isotonic (ī-sō-ton′ik) **contraction** Muscle contraction during which tension exceeds the resistance, resulting in movement.

isotonic solution Contains the same relative solute and water concentration as the cytosol.

isotope (ī′sō-tōp) Atoms of an element that have a different number of neutrons.

K

keratin (ker′ă-tin) Protein that strengthens the epidermis, hair, and nails; produced by keratinocytes.

keratinization (ker′ă-tin-i- zā′shŭn) Process during which keratinocytes fill with the protein keratin.

keratinocyte (ke-rat′i-nō-sīt) The most abundant cell type in the epidermis and found throughout the epidermis; cells that produce keratin.

ketone (kē′tōn) **bodies** Waste product of fatty acid breakdown; also called ketoacids.

ketoacidosis (kē′tō-as-i-dō′sis) Accumulation of ketone bodies in the blood; a symptom of diabetes mellitus.

kinase (kī′nās) Type of enzyme that transfers a phosphate group from one molecule to another.

kinetic (ki-net′ik) **energy** Energy of movement. *Compare to potential energy.*

L

labia (lā′bē-ă; sing. labium) Tissue fold having the appearance of a lip.

lacrimal (lak′ri-măl) Relating to tears.

lactate Chemical produced during glycolysis.

lactation (lak-tā′shŭn) Production of milk.

lacteal (lak′tē-al) Lymphatic capillary where lipids and lipid-soluble vitamins are absorbed from the small intestine.

lacuna (lă-kū′nă; pl., lacunae, -kū′nē) Small space, cavity, or depression.

lamella (lă-mel′ă; pl., lamellae, -mel′ē) Layer of bone connective tissue; forms concentric rings.

lamina (lam′i-nă) Thin layer (i.e., epithelial tissue basal lamina).

lanugo (lă-nū′gō) Fine, soft, unpigmented fetal hair.

larynx (lar′ingks) Organ of voice production that lies between the pharynx and the trachea.

lateral (lat′er-ăl) Away from the midline.

lateralization (lat′er-al-ī-zā′shŭn) Process whereby certain asymmetries of structure and function occur.

lemniscus (lem-nis′kŭs; pl., lemnisci, -nis′ī) Bundle of axons ascending from sensory relay nuclei to the thalamus.

lesion (lē′zhŭn) Pathologic change in a tissue.

lethargy (leth′ar-jē) Mild impairment of consciousness characterized by reduced awareness and alertness.

leukocyte (lū′kō-sīt) Any one of several types of white blood cells.

leukopoiesis (lū′kō-poy-ē′sis) Formation and development of leukocytes.

ligament (lig′ă-ment) Band or sheet of dense regular fibrous tissue that connects structures; usually bones.

ligand (lig′and, lī′gand) Chemical released from one cell that binds to receptors on another cell; e.g., hormone, neurotransmitter.

lingual (lin′gwăl) Relating to the tongue.

lipase (lip′ās) Enzyme that digests triglycerides.

lipid Group of biological macromolecules including triglycerides, phospholipids, steroids, and ecosanoids.

lipogenesis (līp′o-jĕn′e-sis) Formation of triglycerides from glycerol and fatty acids.

lipolysis (li-pol′i-sis) Breakdown of triglycerides to glycerol and fatty acids.

lobe (lōb) Subdivision of an organ, bounded by some structural demarcation.

lumbar (lŭm′bar) Relating to the lower back.

lumen (lū′men) The space inside a structure, such as where blood is transported within a blood vessel.

lymph (limf) Usually transparent fluid found in lymph vessels; derived from interstitial fluid.

lymphocyte (lim′fō-sīt) White blood cell involved in immune responses (i.e. T-lymphocyte, B-lymphocyte).

lysosome (lī′sō-sōm) Organelle containing digestive enzymes.

lysozyme (lī′-sō-zīm) Enzyme that attacks the bacterial cell wall. Present in saliva, tears, sweat, and nasal secretions.

M

macromolecules Large organic molecules; e.g., carbohydrates, nucleic acids, proteins.

macrophage (mak′rō-fāj) Phagocytic cell derived from monocytes that serves as an antigen-presenting cell for T-lymphocytes.

major histocompatibility (his′to-kom-pat′i-bil′i-tē) **complex (MHC)** Plasma membrane proteins used to mark cells as self and to present antigens.

malignant (mă-lig′nant) In reference to a neoplasm, having the property of invasiveness and spread.

mass number Total number of neutrons and protons in the nucleus of an atom.

mast cells Resident cells that secrete heparin and histamine during inflammation; often found near blood vessels.

mastication (mas-ti-kā′shŭn) Chewing ingested food in preparation for swallowing.

matrix (mā′triks; pl., matrices, mā′tri-sēz) Surrounding substance within which cells or structures are contained or embedded.

meatus (mē-ā′tŭs) Passage or opening in the body; e.g., external acoustic meatus.

mechanoreceptor (mek′ă-nō-rē-sep′tŏr) Sensory receptor that responds to distortion of the plasma membrane caused by touch, vibration, pressure, or stretch.

medial (mē′dē-ăl) Toward the midline.

mediastinum (me′dē-as-tī′nŭm) Median space of the thoracic cavity between the lungs.

medulla (me-dūl′ă) Inner region of an organ; e.g., adrenal medulla.

medulla oblongata (ob-long-gah′tă) Brain region that transmits information between the spinal cord and higher brain centers; controls heart rate, blood pressure, breathing.

meiosis (mī-ō′sis) Process of sex cell division that results in haploid number of chromosomes.

melanin (mel′ă-nin) Any of the dark brown-black pigments or yellow-red pigments that occur in the skin, hair, and retina.

melanocyte (mel′ă-nō-sīt) Pigment-producing cell in the basal layer of the epidermis.

membrane potential Difference in charges across the plasma membrane.

memory cells Lymphocytes that produce a strong, rapid response during the second exposure to an invader.

menarche (me-nar′kē) Time of the first menstrual period.

meninx (mē′ninks; pl., meninges, -jēz) Any one of the membranes covering the brain and spinal cord.

meniscus (mĕ-nis′kŭs; pl., menisci, mĕ-nis′sī) Crescent- shaped fibrocartilage found in certain joints.

menopause (men′ō-pawz) Permanent cessation of menses.

menses (men′sēz) Periodic physiologic hemorrhage from the uterine mucous membrane; commonly referred to as the menstrual period.

menstruation (men-strū-ā′shŭn) Cyclic endometrial shedding and discharge of bloody fluid.

mesenchyme (mez′en-kīm) An embryonic connective tissue.

mesentery (mes′en-ter-ē) Fan-shaped fold of peritoneum attaching the small intestine to the posterior abdominal wall; may refer to other membranes associated with organs of the abdominal cavity.

mesoderm (mez′ō-derm) The middle of the three primary germ layers of the embryo.

mesothelium (mez-ō-thē′lē-ŭm) The simple squamous epithelium that lines serous cavities.

metabolic acid Acid produced as a byproduct of metabolism; e.g., ketoacid.

metabolic rate Amount of energy used in a given time period.

metabolic water Water formed during dehydration reactions and during aerobic cellular respiration; contributes approximately 8% of total fluid intake.

metabolism (mĕ-tab′ō-li-zem) All chemical reactions taking place in the body, including both anabolic and catabolic reactions.

metaphysis (mĕ-taf′i-sis; pl., metaphyses, -sēz) Section of a long bone between the epiphysis and the diaphysis.

metaplasia (met-ă-plā′zē-ă) Abnormal change in a tissue.

metastasis (mĕ-tas′tă-sis) The movement or spread of malignant cells from one part of the body to another.

microfilament (mī-krō-fil′ă- ment) Smallest structural protein of the cytoskeleton; composed of actin.

microglia (mī-krog′lē-ă) Category of small glial cells in the central nervous system; wander and exhibit phagocytic activity.

microscopy (mī-kros′kŏ-pē) Investigation of very small objects by means of a microscope.

microtubule (mī-krō-too′būl) Hollow cylinders of tubulin protein that are part of the cytoskeleton; able to lengthen and shorten.

microvilli (mī-krō-vil′ī; sing., microvillus -vil′ŭs) Microscopic extensions of the plasma membrane that increase the surface area for secretion or absorption.

micturition (mik-chū-rish′ŭn) Urination.

midsagittal (mid-saj′-i-tăl) Vertical plane cutting the body or body part into an even left and right half.

mineralocorticoid (min′er-al-ō-kōr′ti-kōyd) Group of hormones released from the adrenal cortex that help regulate electrolyte levels in the body.

mitochondrion (mī-tō-kon′drē-on; pl., mitochondria, -ă) Organelle associated with the production of ATP during aerobic cellular respiration.

mitosis (mī-tō′sis) Process of somatic cell division.

mixed nerve Nerve composed of both sensory and motor neurons.

molarity (mō-lar′i-tē) Number of moles in 1 liter of solution.

mole Unit composed of 6.023×10^{23} particles; its mass in grams is equal to the element's atomic mass or the compound's molecular mass.

molecule Composed of atoms or ions held together by an attraction or chemical bond.

monocyte (mon′ō-sīt) White blood cell that develops into a macrophage and phagocytizes bacteria and viruses.

monomers Identical or similar molecules that repeat within a polymer.

monosaccharide (mon-ō-sak′ă-rīd) The simplest carbohydrate molecules; e.g., glucose, ribose.

mons (monz) Anatomic prominence or slight elevation above the general level of the surface; e.g., mons pubis.

motor (efferent) nerves Composed of neurons that transmit motor output from the central nervous system.

motor unit One motor neuron and all the muscle cells it innervates.

mucin (mū′sin) A carbohydrate-rich glycoprotein that forms mucus when hydrated.

mucosa (mū-kō′să) Mucous membrane composed of an epithelium and lamina propria that lines the body tracts (e.g., respiratory) that open to the external environment.

multipotent (mŭl-tip′ŏ-tent) **stem cells** Adult stem cell that can differentiate into a few specific cell types.

muscle fiber Muscle cell.

muscularis (mŭs-kū-lā′ris) Muscular layer in the wall of a hollow organ or tubular structure.

mutation An alteration in a segment of DNA; may lead to production of an abnormal protein.

myelin (mī′ĕ-lin) Lipoproteinaceous material of the myelin sheath.

myoblast (mī′ō-blast) Undifferentiated muscle cell with the potential of becoming a muscle fiber.

myocardium (mī-ō-kar′dē-ŭm) Middle layer of the heart wall, consisting of cardiac muscle.

myofibril (mī-ō-fī′bril) Bundles of myofilaments within skeletal and cardiac muscle cells.

myofilament (mī-ō-fil′ă-ment) A protein filament that makes up the myofibrils in both skeletal and cardiac muscle cells.

myoglobin (mī-ō-glō′bin) Oxygen-carrying and -storing molecule in muscle.

myometrium (mī-ō-mē′trē-ŭm) Middle tunic (muscular wall) of the uterus.

N

naris (nā′ris; pl., nares, -res) Anterior opening to the nasal cavity.

necrosis (nĕ-krō′sis) Pathologic death of cells or a portion of a tissue or organ.

negative feedback Control mechanism that keeps a variable within normal levels. *Compare to positive feedback*

neoplasia (nē-ō-plā′zē-ă) Process that results in the formation of a neoplasm or abnormal growth.

neoplasm (nē′ō-plazm) An abnormal tissue that grows by cell proliferation more rapidly than normal.

nephron (nef′ron) Functional filtration unit in the kidney; composed of a renal corpuscle, a proximal and distal convoluted tubule, and nephron loop.

nerve A bundle of axons within the peripheral nervous system.

nerve plexus (plek′sŭs) Group of interconnecting spinal nerves; e.g., brachial plexus, lumbar plexus.

nerve signal Propagation of an action potential along a neuron.

net filtration pressure The aggregate pressure that causes fluid to move out across a blood capillary wall; calculated by subtracting net osmotic pressure from net hydrostatic pressure.

neurilemma (nūr-i-lem′a) The delicate outer membrane sheath around an axon.

neurofibril (nūr-o-fi′bril) Filamentous protein structure in a neuron composed of neurofilaments and microtubules.

neurofilament (nūr-o-fil′a- ment) Group of intermediate-sized filaments in a neuron.

neuroglia (nū-rog′lē-a) Nervous system cells that support neurons.

neuromuscular (nūr-ō-mus′ku-lar) Relationship between a nerve and a muscle; e.g., a neuromuscular junction where a neuron and a muscle cell interface.

neuron (nūr′on) Functional unit of the nervous system.

neurotoxin (nūr-ō-tok′sin) Chemical that alters normal activity of a neuron.

neurotransmitter (nū′rō-trans-mit′er) Chemical released from a neuron that binds to receptors of another cell to initiate a cellular change.

neurulation (nū′rū-lā′shun) Formation of the neural plate and its closure to form the neural tube.

neutron (nū′tron) Subatomic particle with a neutral charge; found in the nucleus of an atom.

neutrophil (nū′trō-fil) The most common of the white blood cells. Cells that phagocytize bacteria.

nociceptor (nō-si-sep′ter, -tōr) Peripheral sensory receptor for the detection of painful stimuli.

nonpolar molecules Molecules containing nonpolar covalent chemical bonds.

norepinephrine (nōr-ep-i-nef′rin) Neurotransmitter released by the sympathetic nervous system; hormones released from adrenal medulla.

nucleic acid (nū-klē′ik) A biological macromolecule composed of nucleotide monomers; in DNA and RNA. These store genetic information in the cell.

nucleolus (nū′klē′ō-lŭs; pl., nucleoli, -lī) Spherical, dark body within the nucleus where subunits of ribosomes are made.

nucleotide (nū′klē-ō-tīd) Building blocks of DNA and RNA; composed of a nitrogenous base, a phosphate group, and a sugar.

nucleus (nū′klē-ŭs) Cellular structure housing DNA or a group of cell bodies in the central nervous system.

nutrient Chemicals obtained from food that are necessary for normal function in cells.

O

oblique (ob-lek′) Slanted, at an angle.

occult (ŏ-kŭlt′, ok′ŭlt) Hidden or concealed.

octet rule Tendency of atoms to maintain an outer shell with eight electrons.

olfaction (ol-fak′shŭn) Sense of smell.

oligodendrocyte (ol′i-gō-den′drō-sīt) Category of large glial cells in the central nervous system that wrap around and insulate axons.

omentum (ō-men′tŭm) Fold of peritoneum from the stomach to either cover the region between the liver and stomach, or extends to cover most abdominal organs on their anterior surface.

oocyte (ō′-ō-sīt) Female gamete released from the ovary during ovulation.

oogenesis (ō-ō-jen′ĕ-sis) Formation and development of oocytes.

oogonium (ō-ō-gō′nē-ŭm; pl., oogonia) Primitive germ cell of the oocyte.

opposition (op′pō-si′shŭn) Movement of the thumb across the palm to touch the palmar side of the fingertips.

optic Relating to the eye.

orbit Bony socket housing the eye.

organ Composed of two or more tissue types that perform a specific function for the body.

organ system A group of organs working together to coordinate and perform specific function(s).

organelle (or′gă-nel) Complex, organized structures in the cytoplasm of a cell with unique characteristic shapes; called "little organs."

organic molecule Molecule containing carbon atoms; e.g., carbohydrates, proteins, lipids.

organism A living being.

organogenesis (ōr′gă-nō-jen′ĕ-sis) Formation of organs during development.

orifice (or′i-fis) Aperture or opening.

origin (ōr′i-jin) The less movable of the two points of attachment of a muscle. (This term typically replaced with phrase *superior attachment* or *proximal attachment* in this text.)

os (os; pl., ossa, os′ă) An opening into a hollow organ or canal.

osmosis (os-mō′sis) Process by which water moves through a semipermeable membrane from a hypotonic solution to a hypertonic one.

osmotic pressure Pressure exerted by water movement across a membrane as it moves toward an area of lower water concentration.

osseous (os′ē-ŭs) Bony.

ossicle (os′i-kl) Small bone within the ear.

ossification (os′i-fi-kā′shŭn) Bone formation; also called osteogenesis.

osteoblast (os′tē-ō-blast) Bone-forming cell.

osteoclast (os′tē-ō-klast) Large cell type that functions in the absorption and removal of bone connective tissue.

osteocyte (os′tē-ō-sīt) Bone cell in a lacuna within bone matrix.

osteogenesis (os′tē-ō-jen′ĕ-sis) Production of new bone.

osteon (os′tē-on) Functional unit of compact bone tissue; also called a Haversian system.

osteopenia (os′tē-ō-pē′nē-ă) Decreased calcification or density of bone.

osteoporosis (os′tē-ō-pō-rō′sis) Medical condition characterized by decreased bone mass and increased susceptibility to fracture.

osteoprogenitor (os′tē-ō-prō-jen′i-ter) Precursor to bone cells.

ovulation (ov′ū-lā′shŭn) Release of a secondary oocyte from the ovarian follicle.

ovum (ō′vŭm; pl., ova, -vă) Female sex cell that has been fertilized and has completed meiosis II.

oxidation (ok-si-dā′shŭn) Occurs when a chemical loses an electron.

oxidation-reduction reaction Exchange reaction involving transfer of electrons from one chemical to another; also called a redox reaction.

oxidative phosphorylation (ok-si-dā-tiv fos′fōr-i-lā′shŭn) Process of forming ATP by using coenzymes to transfer energy.

P

palate (pal′ăt) Partition between the oral and nasal cavities; the roof of the mouth.

palpation (pal-pā′shŭn) Using the sense of touch to identify or examine internal body structures.

papilla (pă-pil′ă; pl., papillae, -pil′ē) Small, nipplelike process.

paracrine (par′a-krīn) Hormonal secretions that stimulate neighboring cells; also called a local hormone (does not enter the blood). *Compare to endocrine.*

parietal (pă-rī′ĕ-tăl) Relating to the wall of any ventral body cavity.

partial pressure Pressure exerted by a gas in a mixture of gases.

passive immunity Activation of the immune system by administration of antibodies produced by another individual; offers short-term protection since memory cells are not formed. *Compare to active immunity.*

patent (pa′tent, pā′tent) Open or unblocked.

pathogen (path′ō-jen) Disease-causing substance or organism.

pathologic (path-ō-loj′-ik) Pertaining to disease.

pectoral (pek′tō-răl) Relating to the chest area.

pedal (ped′ăl) Relating to the foot.

peptide (pep′tīd) **bond** Chemical bond joining amino acid monomers.

perfusion (per-fyū′zhŭn) Blood flow through a tissue; measured in milliliters per minute per gram (mL/min/g).

pericardium (per-i-kar′dē-ŭm) Fibroserous membrane covering the heart.

perichondrium (per′i-kon′drē-ŭm) Layer of dense irregular connective tissue around the surface of cartilage.

perilymph (per′i-limf) Fluid within the osseous labyrinth that surrounds and protects the membranous labyrinth.

perimetrium (per-i-mē′trē-ŭm) Serous membrane of the uterus.

perimysium (per-i-mis′ē-ŭm, -miz′ē-ŭm) Fibrous sheath enveloping each of the fascicles of skeletal muscle fibers.

perineurium (per-i-nū′rē-ŭm) Fibrous sheath enveloping each of the fascicles within nerves.

periosteum (per-ē-os′tē-ŭm) Thick, fibrous membrane covering the entire external surface of a bone, except for the articular cartilage on the epiphyses.

peripheral (pĕ-rif′-ĕ-răl) **nervous system (PNS)** Composed of nerves and ganglia that are outside of the central nervous system.

peristalsis (per-i-stal′sis) Rhythmic contractions of the muscularis that propel material through tubes in the gastrointestinal tract.

peritoneum (per′i-tō-nē′ŭm) Serous sac that lines the abdominopelvic cavity and covers most of the viscera within.

permeability (per′mē-ă-bil′i-tē) Capacity of a membrane to allow passage of a substance.

peroxisome (per-ok′si-sōm) Membrane-bound organelle containing oxidative enzymes.

pH Value indicating the relative hydrogen ion (H^+) concentration of a solution; expressed as a number between 0 and 14.

phagocytosis (fāg′ō-sī-tō′sis) A form of endocytosis by which cells ingest and digest solid substances; "cell eating."

phalanx (fā′langks; pl., phalanges, -jēz) Long bone of a digit.

pharynx (far′ingks) Funnel-shaped muscular tube extending from the posterior nasal cavity to the esophagus and larynx; composed of the nasopharynx, oropharynx, and laryngopharynx.

phenotype (fē′nō-tīp) Observable characteristics of an individual.

phospholipid (fos-fō-lip′id) Lipid that forms bilayers of the plasma membrane.

phosphorylation (fos′fōr-i-lā′shŭn) Process of adding a phosphate group to a molecule.

photoreceptor Specialized receptor cells in the eye that detect light. *See rod; cone.*

physiology (fiz-ē-ol′ō-jē) Study of how body parts function together.

pinocytosis (pin′ō-sī-tō′sis) A form of endocytosis by which cells ingest liquid; "cell drinking."

placenta (plă-sen′tă) Organ of exchange between the embryo or fetus and the mother.

plasma (plaz′mă) The liquid portion of blood.

plasma cell Cell that produces and releases specific antibodies; derived from B-lymphocytes.

plasma membrane Barrier separating the cytoplasm from the interstitial fluid.

platelet (plāt′let) Irregularly shaped cell fragment that participates in hemostasis.

pleura (plūr′ă; pl., pleurae, ploor′ē) Serous membranes enveloping the lungs and lining the internal walls of the thoracic cavity.

pleuripotent (plūr′i′pŏ-tent) **cells** Embryonic stem cells that can differentiate into any cell type (except cells of the placenta).

plexus (plek′sŭs) Network of nerves, blood vessels, or lymph vessels.

polar (po′lăr) **molecule** Molecules containing polar covalent chemical bonds.

polycythemia (pol′ē-sī-thē′mē-ă) More than the normal number of erythrocytes.

polymer (pol′-i-mer) Molecule composed of many repeating subunits (or monomers).

polypeptide (pol-ē-pep′tīd) Chain of 21–199 amino acids linked by peptide bonds.

polysaccharide (pol-ē-sak′ă-rīd) Class of carbohydrates composed of three or more monosaccharide monomers; e.g., glycogen.

portal (pōr′tăl) **system** Unique organization of blood vessels sending blood from a capillary bed through a vein to another capillary bed.

positive feedback Control mechanism that increases the original change in a variable. *Compare to negative feedback.*

posterior Toward the back of the body.

postsynaptic (pōst-si-nap′tik) After a synaptic cleft.

potential energy Stored energy.

pressure gradient Difference in pressure between two adjacent areas.

presynaptic (prē-si-nap′tik) Before or prior to a synaptic cleft.

prion (prī′on) Small disease-causing protein.

pronation (pro-nā′shŭn) Rotational movement of the forearm such that the palm of the hand is posterior or inferior.

proprioceptor (prō′prē-ō-sep′ter) Sensory receptor within joints, muscles, and tendons that detect body and limb movements.

prostaglandin (pros-tă-glan′din) Local hormone derived from a fatty acid of the plasma membrane.

protein Biological macromolecule composed of one or more chains of amino acid monomers.

proton (prō′ton) Subatomic particle with a positive charge; found in the nucleus of an atom.

protraction (prō-trak′shun) Movement of a body part anteriorly in a horizontal plane; e.g., thrusting out of the mandible (chin).

protuberance (prō-tū′ber-ans) Swelling or knoblike outgrowth.

proximal (prok′si-măl) Near the point of attachment to the trunk.

proximal attachment: the (typically less movable) attachment site of an appendicular skeletal muscle (*previously called the origin*).

puberty Time period when reproductive organs become fully functional and secondary sex characteristics become more prominent.

pulmonary (pŭl′mō-nār-ē) Relating to the lung.

pulse (pŭls) Rhythmic dilation of an artery resulting from passage of increased blood volume during heart contraction.

pupil (pū′pl) Circular hole in the center of the iris of the eye.

pus (pŭs) Substance composed of dead pathogens, white blood cells, and cellular debris.

pyrogen (pī′rō-jen) Fever-producing substance.

R

radioactivity Emission of high energy radiation from a radioisotope.

radioisotope (rā′dē-ō-ī′sō-tōp) Unstable isotope that emits high-energy radiation (e.g., gamma rays) until it decays into a stable isotope.

ramus (rā′mŭs; pl., rami, rā′mī) One of the primary divisions of a nerve or blood vessel. Also, part of an irregularly shaped bone that forms an angle.

raphe (rā′fē) Line of union between two contiguous, bilaterally symmetrical structures; e.g., the raphe of the scrotum.

receptor Cellular protein that binds a ligand (e.g., hormone, neurotransmitter); also structure that detects a stimulus.

reduction Occurs when a chemical gains an electron.

referred pain Pain sensed, not from the organ, but from an unrelated region of the body.

reflex Fast, predetermined response to a stimulus.

refraction (rē-frak′shŭn) Bending of light waves as they pass through a material.

refractory (rē-frak′tōr-ē) **period** Time period when a cell is either unresponsive or less responsive to a stimulus.

renal (rē′năl) Relating to the kidney.

repolarization (rē′pō-lăr-i-zā′shŭn) Change in membrane potential from a depolarized value back to the resting value.

respiration Exchange of O_2 and CO_2 between air and systemic body cells.

respiratory zone Respiratory system passageways transporting air from respiratory bronchioles to the alveoli; airways that participate in gas exchange. *Compare to conducting zone.*

resting membrane potential (RMP) Voltage measured across the plasma membrane of an excitable cell at rest.

reticular (re-tik′ū-lăr) **fibers** Proteins that form a flexible network in tissues; found in some connective tissues.

reticulocyte (re-tik′ū-lō-sīt) Immature erythrocyte.

retroperitoneal (re′trō-per′i-tō nē′ăl) External or posterior to the peritoneum.

ribonucleic (rī′bō-nū-klē′ik) **acid (RNA)** Nucleic acid composed of ribonucleotide monomers; used to direct protein synthesis based on instructions in DNA; types include messenger RNA, ribosomal RNA, and transfer RNA.

ribosome (rī′bō-sōm) Organelle composed of protein and rRNA that is the site of protein synthesis.

rod Rod-shaped cell in retina that detects light and functions best in dim light.

rotation (rō-tā′shŭn) Movement of a part around its axis.

rouleau (rū-lō′; pl., rouleaux) Aggregation of erythrocytes in single file.

ruga (rū′gă; pl., rugae, rū′gē) Fold, ridge, or crease.

S

sacroiliac (sā-krō-il′ē-ak) Pertaining to the sacrum and the ilium; e.g., sacroiliac joint.

sacrum (sā′krŭm) Next-to-last group of bones in the vertebral column; formed of five fused vertebrae.

sagittal (saj′i-tăl) Relating to a plane cutting the body into a left and a right part.

salt Substance formed by an ionic bond.

saltatory (sal′tă-tōr-ē) **conduction** Transmission of an action potential along the plasma membrane of a myelinated axon.

sarcolemma (sar′kō-lem′ă) Plasma membrane of a muscle cell.

sarcomere (sar′kō-mēr) Functional unit of skeletal muscle.

sarcoplasm (sar′kō-plazm) Cytoplasm of a muscle cell.

sarcoplasmic reticulum (sar-ko′-plaz′mik re-tik′ū-lŭm) **(SR)** Endoplasmic reticulum found in muscle cells; stores calcium ions required for contraction.

sclera (skler′ă) Portion of the fibrous layer forming the outer covering of the eyeball, except for the cornea; "white" of the eye.

scoliosis (skō-lē-ō′sis) Abnormal lateral and rotational curvature of the vertebral column.

sebum (sē′bŭm) Secretion of a sebaceous gland.

second messenger Intracellular chemical modifying activity within a cell after a first messenger (hormone or neurotransmitter) binds to the plasma membrane.

selective permeability Ability of a cell to regulate what can cross a membrane.

semen (sē′men) Secretion composed of sperm and seminal fluid.

semipermeable (sem′ē-per′mē-ă-bl) Freely permeable to some molecules, but relatively non-permeable to other molecules.

sensation (sen-sā′shun) Conscious perception of a stimulus.

sensory (afferent) nerve Nerves sending information to the central nervous system.

sensory receptor Structure that detects a stimulus.

septum (sep′tŭm; pl., septa) Wall of tissue separating nearby chambers; e.g., nasal septum.

serosa (se-rō′să) Outermost coat of a visceral structure that lies in a closed body cavity.

serous (sēr′ŭs) Producing a substance that has a watery consistency.

serum (sēr′ŭm) Clear, watery fluid that remains after clotting proteins are removed from plasma.

set point Normal value of a variable.

simple diffusion Passive transport process used when small nonpolar molecules cross the plasma membrane unassisted.

sinoatrial (sī′nō-ā′trē-ăl) **(SA) node** Group of cells in right atrium that set the heart rate.

sinus (sī′nŭs) Cavity or hollow space; also, a channel for the passage of blood or lymph.

sinusoid (sī′nŭ-soyd) Resembling a sinus; extremely permeable capillary in certain organs, such as the liver or the spleen.

skeleton (skel′ĕ-tŏn) Bony framework of the body; collectively, all the bones of the body.

solute (sol′ūt) Substance dissolving in a solvent.

solution A homogeneous mixture composed of a solvent and a solute.

solvent (sol′vent) Substance (e.g., water) holding a solute in solution.

somite (sō′mīt) One of the paired segments consisting of cell masses formed in the early embryonic mesoderm on the sides of the neural tube.

sonography (sŏ-nog′ră-fē) Radiographic technique using ultrasound waves.

sperm (spermatozoon) Male gamete produced in the testes.

spermatogenesis (sper′mă-tō-jen′ĕ-sis) Process by which stem cells called spermatogonia become sperm.

sphincter (sfingk′ter) Muscle that encircles a duct, tube, or orifice such that its contraction constricts the lumen or orifice and restricts movement through the tube.

sphygmomanometer (sfig′mō-mă-nom′ĕ-ter) Instrument for measuring arterial blood pressure.

splanchnic (splangk′nik) Relating to the viscera.

sprain Tearing or overstretching ligaments without fracturing the nearby bone; results in localized pain and swelling.

squamous (skwā′mŭs) Referring to cell or area that is flat.

stem cells Unspecialized, immature cells.

stenosis (ste-nō′sis) Inability of a tube or valve to open completely; e.g., valvular stenosis.

steroid (stēr′oyd, ster′oyd) Group of lipids including bile salts, cholesterol, and some hormones; molecule is composed of four attached hydrocarbon rings.

stimulus Change in a regulated variable; event that provokes a cellular response.

stria (strī′ă; pl., striae, strī′ē) Stripes, bands, streaks, or lines distinguished by a difference in color, texture, or elevation from surrounding tissue.

stupor (stū′per) State of impaired consciousness from which the individual can be aroused only by continual stimulation.

subcutaneous (sŭb-kū-tā′nē-ŭs) Immediately internal to the integument.

subluxation (sŭb-lŭk-sā′shŭn) Incomplete dislocation of a body part from its normal position.

substrate (sŭb′strāt) Substance binding to the active site of an enzyme; reactant in a chemical reaction.

sulcus (sŭl′kŭs; pl., sulci, sŭl′sī) Groove or furrow; e.g., on the surface of the brain.

summation (sŭm-ā′shŭn) Accumulation of stimuli occurring in neurons or muscle cells; e.g., wave summation.

superficial (sū-per-fish′ăl) Toward the surface.

superior attachment: the (typically more movable) attachment site of an axial skeletal muscle (*previously called the insertion*).

supination (sū′pi-nā′shŭn) Rotation of the forearm such that the palm of the hand is anterior or superior.

suspension Mixture of a solvent with large materials that do not dissolve, e.g., formed elements in blood.

suture (sū′chūr) Synarthrosis in which cranial bones are united by a dense regular connective tissue membrane.

symphysis (sim′fi-sis; pl., symphyses, -sēz) Cartilaginous joint in which the two bones are separated by a pad of fibrocartilage.

synapse (sin′aps) Functional contact of a neuron with another neuron or effector (muscle or gland).

synaptic vesicle (si-nap′tik ves′i-kl) Package of plasma membrane enclosing neurotransmitter molecules in the synaptic knob.

syndesmosis (sin′dez-mō′sis; pl., syndesmoses, -sēz) Fibrous joint in which the opposing surfaces of articulating bones are united by ligaments.

synergist (sin′er-jist) Structure, muscle, agent, or process that aids the action of another.

synostosis (sin-os-tō′sis; pl., synostoses, -sēz) Osseous union between two bones that were initially separate.

system Group of interacting structures working together; e.g., urinary system.

systemic (sis-tem′ik) Relating to the entire organism as opposed to any of its individual parts.

systole (sis′tō-lē) Contraction of the heart.

systolic (sis-tol′ik) **pressure** Blood pressure measured in an artery during systole.

T

T-lymphocyte (lim′fō-sīt) Adaptive immune system cell that functions in cell-mediated immunity.

tachycardia (tak′i-kar′dē-ă) Resting heart rate above 100 beats per minute.

tactile (tak′til) Relating to touch.

tendon (ten′dŏn) Cord of dense regular connective tissue that connects muscle to bone.

teratogen (ter′ă-tō-jen, te-ră′tō-jēn) Substance that may cause death or malformation of an embryo.

testis (tes′tis; pl., testes, -tēz) Male organ for producing gametes and androgens.

tetanus (tet′ă-nŭs) Sustained contraction that may lead to spastic paralysis.

thalamus (thal′ă-mŭs) Region of the brain in the diencephalon involved in motor control and relaying sensory information to higher brain centers.

thermoreceptor (ther′mo-rē-sep′ter, -tōr) Sensory receptor that is sensitive to heat.

thoracic (thō-ras′ik) Relating to the thorax, the area between the neck and the abdomen.

thrombopoiesis (throm′bō-poy-ē′sis) Formation of blood platelets.

thrombus (throm′bŭs) Blood clot formed within a blood vessel during hypercoagulation; may become an embolus if dislodged.

tissue Groups of similar types of cells performing a common function for the body.

tomography (tō-mog′ră-fē) A radiographic image of a plane constructed by means of reciprocal linear or curved motion of the x-ray tube and film cassette; used in producing a CT scan.

totipotent (tō-tip′ŏ-tent) **cells** Embryonic stem cells that can differentiate into any cell of the body or placenta.

trabecula (tră-bek′ū-lă; pl., trabeculae, -lē) Meshwork; e.g., trabeculae within spongy bone.

tract (trakt) Bundle of axons in the central nervous system.

transcription (tran-skrip′shŭn) Copying information from DNA to form an RNA molecule.

transducer (tranz-dū′sĕr) Device or organ that converts energy from one form to another.

translation Process involving RNA and ribosomes to produce a new protein.

transverse (trans-vers′) Relating to a plane cutting a structure into a superior and inferior portion.

transverse (T) tubule (tū′būl) Invaginations of the sarcolemma that allow action potentials to approach and stimulate the sarcoplasmic reticulum.

triglyceride (trī-glis′er-īd) Lipid providing long-term energy storage in adipose connective tissue; composed of glycerol and three fatty acids.

trophoblast (trof′ō-blast, trō′fō- blast) Cell layer covering the blastocyst that will allow the embryo to receive nourishment from the mother.

tubercle (tū′ber-kl) Nodule or slight elevation from the surface of a bone that allows attachment of a tendon or ligament.

tuberosity (tū′ber-os′i-tē) Large tubercle or rounded elevation, as on the surface of a bone.

tumor Mass of new tissue cells resulting from abnormal cell division.

tunic (tū′nik) One of the layers of a body part, such as within the blood vessel wall.

U

ultrastructure (ŭl-tră-strŭk′chūr) Cell structure viewable via the electron microscope.

umbilical (ŭm-bil′i-kăl) Relating to the umbilicus (navel).

unipotent (ū-ni′pŏ-tent) **stem cells** Adult stem cell that can differentiate into only one specific cell type.

unmyelinated (ŭn-mī′ĕ-li-nā-ted) Not covered by a myelin sheath.

urea (ū-rē′ă) A nitrogenous waste product produced by amino acid metabolism and eliminated in urine.

ureter (ū-rē′ter, ū′rē-ter) Tubes that connect the kidney to the urinary bladder.

urethra (ū-rē′thră) Tube that extends from the bladder for transport of urine to the exterior of the body.

uric (ūr′ik) **acid** Waste product of nucleic acid metabolism eliminated in urine.

urinalysis (ū-ri-nal′i-sis) Analysis of the urine to help assess the state of a person's health.

urine (ūr′in) Fluid and dissolved substances excreted by the kidney.

B

C

D

E

F

G

H

I

J

K

L

M

N

O

P

Q

R

S

T

U

V

W

Z